Essential Formulas

1. $\int x^\alpha \, dx = \dfrac{x^{\alpha+1}}{\alpha+1} + C, \quad \alpha \neq -1$
2. $\int \dfrac{1}{x} \, dx = \ln |x| + C$
3. $\int e^x \, dx = e^x + C$
4. $\int \sin x \, dx = -\cos x + C$
5. $\int \cos x \, dx = \sin x + C$
6. $\int \tan x \, dx = \ln|\sec x| + C$
7. $\int \sec x \, dx = \ln|\sec x + \tan x| + C$
8. $\int \sec^2 x \, dx = \tan x + C$
9. $\int \sec x \tan x \, dx = \sec x + C$
10. $\int \dfrac{dx}{\sqrt{1-x^2}} = \sin^{-1} x + C$
11. $\int \dfrac{dx}{1+x^2} = \tan^{-1} x + C$
12. $\int \dfrac{dx}{x\sqrt{x^2-1}} = \sec^{-1} x + C$

Useful Formulas

13. $\int \cot x \, dx = \ln|\sin x| + C$
14. $\int \csc x \, dx = \ln|\csc x - \cot x| + C$
15. $\int \csc^2 x \, dx = -\cot x + C$
16. $\int \csc x \cot x \, dx = -\csc x + C$
17. $\int \dfrac{dx}{\sqrt{a^2-x^2}} = \sin^{-1} \dfrac{x}{a} + C, \quad a > 0$
18. $\int \dfrac{dx}{a^2+x^2} = \dfrac{1}{a} \tan^{-1} \dfrac{x}{a} + C, \quad a > 0$
19. $\int \dfrac{dx}{x\sqrt{x^2-a^2}} = \dfrac{1}{a} \sec^{-1} \dfrac{x}{a} + C, \quad a > 0$
20. $\int a^x \, dx = \dfrac{a^x}{\ln a} + C, \quad a > 0, \quad a \neq 1$
21. $\int \sinh x \, dx = \cosh x + C$
22. $\int \cosh x \, dx = \sinh x + C$
23. $\int \operatorname{sech}^2 x \, dx = \tanh x + C$
24. $\int \operatorname{csch}^2 x \, dx = -\coth x + C$
25. $\int \operatorname{sech} x \tanh x \, dx = -\operatorname{sech} x + C$
26. $\int \operatorname{csch} x \coth x \, dx = -\operatorname{csch} x + C$
27. $\int \dfrac{dx}{a+bx} = \dfrac{1}{b} \ln|a+bx| + C$
28. $\int \dfrac{x\,dx}{a+bx} = \dfrac{1}{b^2}(a + bx - a \ln|a+bx|) + C$
29. $\int \dfrac{x\,dx}{(a+bx)^2} = \dfrac{a}{b^2(a+bx)} + \dfrac{1}{b^2} \ln|a+bx| + C$
30. $\int \dfrac{x^2\,dx}{(a+bx)^2} = \dfrac{1}{b^3}\left(a + bx - \dfrac{a^2}{a+bx} - 2a \ln|a+bx|\right) + C$
31. $\int \dfrac{dx}{x(a+bx)^2} = \dfrac{1}{a(a+bx)} - \dfrac{1}{a^2} \ln\left|\dfrac{a+bx}{x}\right| + C$
32. $\int \dfrac{dx}{x^2(a+bx)} = -\dfrac{1}{ax} + \dfrac{b}{a^2} \ln\left|\dfrac{a+bx}{x}\right| + C$
33. $\int x(a+bx)^n \, dx = \dfrac{(a+bx)^{n+1}}{b^2}\left(\dfrac{a+bx}{n+2} - \dfrac{a}{n+1}\right) + C, \quad n \neq -1, -2$
34. $\int \dfrac{x\,dx}{(a+bx)(c+dx)} = \dfrac{1}{bc-ad}\left(-\dfrac{a}{b} \ln|a+bx| + \dfrac{c}{d} \ln|c+dx|\right) + C, \quad bc - ad \neq 0$
35. $\int \dfrac{x\,dx}{(a+bx)^2(c+dx)} = \dfrac{1}{bc-ad}\left[\dfrac{a}{b(a+bx)} + \dfrac{c}{bc-ad} \ln\left|\dfrac{a+bx}{c+dx}\right|\right] + C, \quad bc - ad \neq 0$
36. $\int \dfrac{dx}{a^2-x^2} = \dfrac{1}{2a} \ln\left|\dfrac{x+a}{x-a}\right| + C$
37. $\int \dfrac{dx}{x^2-a^2} = \dfrac{1}{2a} \ln\left|\dfrac{x-a}{x+a}\right| + C$
38. $\int \dfrac{dx}{(a^2 \pm x^2)^n} = \dfrac{1}{2(n-1)a^2}\left[\dfrac{x}{(a^2 \pm x^2)^{n-1}} + (2n-3) \int \dfrac{dx}{(a^2 \pm x^2)^{n-1}}\right], \quad n \neq 1$
39. $\int \dfrac{dx}{(x^2-a^2)^n} = \dfrac{1}{2(n-1)a^2}\left[-\dfrac{x}{(x^2-a^2)^{n-1}} - (2n-3) \int \dfrac{dx}{(x^2-a^2)^{n-1}}\right], \quad n \neq 1$

Forms Containing $\sqrt{a+bx}$

40. $\int x\sqrt{a+bx}\,dx = \frac{2}{15b^2}(3bx-2a)(a+bx)^{3/2}+C$

41. $\int x^n\sqrt{a+bx}\,dx = \frac{2}{b(2n+3)}\left[x^n(a+bx)^{3/2} - na\int x^{n-1}\sqrt{a+bx}\,dx\right]$

42. $\int \frac{x\,dx}{\sqrt{a+bx}} = \frac{2}{3b^2}(bx-2a)\sqrt{a+bx}+C$

43. $\int \frac{x^2\,dx}{\sqrt{a+bx}} = \frac{2}{15b^3}(8a^2+3b^2x^2-4abx)\sqrt{a+bx}+C$

44. $\int \frac{x^n\,dx}{\sqrt{a+bx}} = \frac{2x^n\sqrt{a+bx}}{b(2n+1)} - \frac{2na}{b(2n+1)}\int \frac{x^{n-1}\,dx}{\sqrt{a+bx}}$

45. $\int \frac{dx}{x\sqrt{a+bx}} = \begin{cases} \frac{1}{\sqrt{a}}\ln\left|\frac{\sqrt{a+bx}-\sqrt{a}}{\sqrt{a+bx}+\sqrt{a}}\right| + C, & a>0 \\ \frac{2}{\sqrt{-a}}\tan^{-1}\sqrt{\frac{a+bx}{-a}} + C, & a<0 \end{cases}$

46. $\int \frac{dx}{x^n\sqrt{a+bx}} = -\frac{\sqrt{a+bx}}{a(n-1)x^{n-1}} - \frac{b(2n-3)}{2a(n-1)}\int \frac{dx}{x^{n-1}\sqrt{a+bx}}$

47. $\int \frac{\sqrt{a+bx}}{x}\,dx = 2\sqrt{a+bx} + a\int \frac{dx}{x\sqrt{a+bx}}$

48. $\int \frac{\sqrt{a+bx}}{x^2}\,dx = -\frac{\sqrt{a+bx}}{x} + \frac{b}{2}\int \frac{dx}{x\sqrt{a+bx}}$

Forms Containing $\sqrt{x^2 \pm a^2}$

49. $\int \sqrt{x^2 \pm a^2}\,dx = \frac{x}{2}\sqrt{x^2 \pm a^2} \pm \frac{a^2}{2}\ln\left|x+\sqrt{x^2 \pm a^2}\right| + C$

50. $\int x\sqrt{x^2 \pm a^2}\,dx = \frac{1}{3}(x^2 \pm a^2)^{3/2} + C$

51. $\int x^2\sqrt{x^2 \pm a^2}\,dx = \frac{x}{8}(2x^2 \pm a^2)\sqrt{x^2 \pm a^2} - \frac{a^4}{8}\ln\left|x+\sqrt{x^2 \pm a^2}\right| + C$

52. $\int \frac{\sqrt{x^2+a^2}}{x}\,dx = \sqrt{x^2+a^2} - a\ln\left|\frac{a+\sqrt{x^2+a^2}}{x}\right| + C$

53. $\int \frac{\sqrt{x^2-a^2}}{x}\,dx = \sqrt{x^2-a^2} - a\sec^{-1}\left|\frac{x}{a}\right| + C$

54. $\int \frac{\sqrt{x^2 \pm a^2}}{x^2}\,dx = -\frac{\sqrt{x^2 \pm a^2}}{x} + \ln\left|x+\sqrt{x^2 \pm a^2}\right| + C$

55. $\int \frac{dx}{\sqrt{x^2 \pm a^2}} = \ln\left|x+\sqrt{x^2 \pm a^2}\right| + C$

56. $\int \frac{x^2\,dx}{\sqrt{x^2 \pm a^2}} = \frac{x}{2}\sqrt{x^2 \pm a^2} \mp \frac{a^2}{2}\ln\left|x+\sqrt{x^2 \pm a^2}\right| + C$

57. $\int \frac{dx}{x\sqrt{x^2+a^2}} = -\frac{1}{a}\ln\left|\frac{a+\sqrt{x^2+a^2}}{x}\right| + C$

58. $\int \frac{dx}{x\sqrt{x^2-a^2}} = \frac{1}{a}\sec^{-1}\left|\frac{x}{a}\right| + C$

(continued inside back cover)

Calculus  Analytic Geometry

THIRD EDITION

Abe Mizrahi

INDIANA UNIVERSITY NORTHWEST

Michael Sullivan

CHICAGO STATE UNIVERSITY

Wadsworth Publishing Company

BELMONT, CALIFORNIA

A DIVISION OF WADSWORTH, INC.

To our wives, Caryl and Mary

Mathematics Editor: Anne Scanlan-Rohrer
Assistant Editor: Tamiko Verkler
Editorial Assistant: Leslie With
Production: Cece Munson, The Cooper Company
Print Buyer: Randy Hurst
Cover and Interior Designer: Al Burkhardt
Copy Editor: Carol Reitz
Technical Illustrator: Scientific Illustrators
Compositor: Polyglot Compositors Pte., Ltd.
Product Manager: Robin Levy O'Neill
Signing Representative: Tom Braden
Cover: Max Bill. *Unlimited and Limited.* 1947. Oil on canvas, $43\frac{3}{4} \times 40\frac{1}{2}''$.

Credits: Page 529, Gateway Arch, St. Louis, Missouri, photo © UPI/Bettmann Newsphotos; page 712, Golden Gate Bridge, San Francisco, California, photo © Elizabeth Hamlin, provided by Stock, Boston, Inc.; page 714, photo courtesy of TS; page 721, Colosseum, Rome, photo by Bruce Kokernot; page 931, contour map of Menan Buttes, Idaho, courtesy of U.S. Geological Survey; page 931, Pacific hurricane Ava, courtesy of U.S. Department of Commerce, National Oceanic and Atmospheric Administration.

Printed in the United States of America

2 3 4 5 6 7 8 9 10—94 93 92 91 90

Library of Congress Cataloging-in-Publication Data

Mizrahi, Abe.
Calculus and analytic geometry/Abe Mizrahi & Michael Sullivan.—3rd ed.
p. cm.
Includes index.
ISBN 0-534-11646-9
1. Calculus. 2. Geometry, Analytic. I. Sullivan, Michael, 1942– II. Title.
QA303.M6878 1990
515'.15—dc20 89-14691
CIP

Contents

1 Introduction

2 The Limit of a Function

3 The Derivative

4 Applications of the Derivative

5 The Definite Integral

6 Applications of the Integral

7 Exponential and Logarithmic Functions

8 Inverse Trigonometric Functions; Hyperbolic Functions

9

Techniques of Integration

10

Indeterminate Forms; Improper Integrals; Taylor Polynomials

11

Infinite Series

12 Conics

13 Polar Coordinates; Parametric Equations

14 Vectors; Analytic Geometry in Space

15 Vector Functions

16 Functions of Several Variables

17 Directional Derivative, Gradient, and Extrema

18 Multiple Integrals

19 Topics in Vector Calculus

20 Differential Equations

APPENDICES

About the Authors

Abe Mizrahi and Michael Sullivan both received Ph.D.'s in mathematics from the Illinois Institute of Technology. Dr. Mizrahi is a professor of mathematics at Indiana University, Northwest Campus, and Dr. Sullivan is a professor of mathematics at Chicago State University. They are also the authors of *Mathematics for Business and Social Sciences, An Applied Approach* (Wiley) and *Finite Mathematics with Applications for Business and Social Sciences* (Wiley), and Sullivan is the author of *Precalculus*, *College Algebra*, *College Algebra and Trigonometry*, *Trigonometry*, and *College Algebra with Review* (Dellen).

Preface to the Instructor

A New Era in Teaching Calculus

The third edition of this text is being published at the start of an exciting new era in teaching calculus. Calculus instructors widely agree that this decade will see changes in the way calculus is taught. But leaders in calculus reform are moving in different directions, and it is not clear what the changes will finally be. While experiments are conducted at a number of colleges, the teaching of calculus for the most part adheres to traditional syllabi, with forays in some new directions. While ours is a "traditional" calculus text, we felt it important in this edition to take initial steps into the realm of the "new calculus" by including explorations in the use of graphing calculators and computers.

Continuing Features

In our first edition we set out to write a straightforward, mathematically sound book that would nevertheless capture the interest of students. We felt—and still feel—that there is a place for a mature, mainstream text with the readability and practicality usually associated with lower-level books. The text can readily adapt to either semester or quarter systems. There is flexibility in the order and depth in which the material can be presented.

As before, we present concepts in both theorem–proof style and in the vernacular. We have often found that when students are presented with an alternative explanation, a mental block is unlocked, and better appreciation of the material results. Explanations in plain English also serve to provide the qualitative perspective from which analysis must at times be viewed.

We have retained and augmented our store of interesting applications. These applications are located at the most strategic places—where students are most likely to be fatigued by the presentation of difficult material or where an application can bring to life a particular aspect of mathematics.

We have also kept the historical sections in order to provide perspective on how calculus actually evolved. We have avoided the capsule biography approach to historical notes by including exercises in these optional sections. We want teachers to be able to give their students a taste of what it was like to invent the calculus. This Socratic style of instruction is, in our experience, more instructive than the simple reading of footnotes.

We continue to provide several exercises taken directly from the best-selling texts in physics as well as many exercises that call for the use of a programmable calculator or computer.

Changes Made Throughout the Text

Each chapter concludes with Review Exercises (a set of approximately 40 to 50 exercises representing typical test questions for each chapter) and Challenge Exercises (10 to 20 very difficult, thought-provoking exercises).

More "B" level exercises have been added to the section exercise sets throughout the text. The text now contains over 7,400 exercises and over 700 examples.

Graphing calculator essays with exercises have been added to select chapters in the text.

Computer essays and exercises have been added to select chapters in the text.

Full color has been added in Chapter 6 and certain other chapters to enhance the three-dimensional effect of the illustrations.

Theorems and definitions are identified with new numbering sequences at the left. Formulas are given a single number on the right. Wherever possible, theorems are identified by names.

A new 8 × 10 inch trim size allows for a more open design.

Chapter-by-Chapter Changes

Chapter 1: The material on trigonometry, formerly in the Appendix, has been rewritten and now appears in Chapter 1.

Chapter 2: A section on the limits of trigonometric functions has been added, parallel to the additional trigonometry material in Chapter 1. Accordingly, more examples and exercises involving trigonometry appear throughout the first seven chapters. We have added a calculator vignette, "Functions and Limits Explored Using Graphing Calculators."

Chapter 3: Four sections from the second edition (3.1, 3.2, 3.3, and 3.12) have been condensed into two sections to improve the flow of material and save time teaching the derivative. The proof for the derivative of x to a rational exponent has been revised. The sections on the Chain Rule and implicit differentiation were rewritten with additional explanation, examples, and problems. We have added a calculator vignette, "Finding Derivatives on the HP-28."

Chapter 4: The material on sketching graphs has been expanded into an entire section to make the difficult subject of curve sketching easier. We have added a calculator vignette, "Exploring Asymptotes and Limits at Infinity Using Graphing Calculators," and a computer vignette, "Using a Microcomputer to Explore Graphs of Functions."

Chapter 5: The material on summation notation was rewritten and expanded. The section on the average value of a function was moved here from Chapter 6. A theorem on the integration of even and odd functions now appears in the text. We have added a computer vignette, "Using Microcomputer Graphics to Investigate Approximations to Definite Integrals."

Chapter 6: Applications involving the center of mass and the centroid of an object have been included. Examples and exercises dealing with the revolution of a region about a line other than the x-axis or the y-axis have been added. We have added a calculator vignette, "Finding Arc Length on the HP-28."

Chapter 7: Some of the differential equations material has been postponed and now appears in the new Chapter 20. An alternative derivation of the derivative of $y = x^\alpha$ was added.

Chapter 8: Sections 8.1 and 8.2 have been combined to reflect the emphasis on the inverse trigonometric functions and the interrelatedness of integration and differentiation. New examples on the derivative and integrals of inverse trigonometric hyperbolic functions appear.

Chapter 9: Sections 9.8 and 9.9 involving miscellaneous substitutions have been removed from the text and now appear as exercises.

Chapter 10: A three-step procedure for the use of L'Hospital's rule has been added.

Chapter 11: The discussion of series of positive terms was reorganized and expanded to improve clarity. The Root Test is now covered in the text rather than in the exercises.

Chapter 12: Several new exercises have been added.

Chapter 13: The discussion of curvature has been postponed to Chapter 15. The section on graphing polar equations was entirely rewritten to include more examples. The section on the angle ψ has been removed. We have added a calculator vignette, "Graphing Polar and Parametric Curves on a Graphing Calculator."

Chapter 14: Vectors in the plane and vectors in space are discussed concurrently. The exercises have been upgraded and new applications have been added.

Chapter 15: This chapter was reorganized to improve readability. The topics of curvature in the plane and in space are now discussed concurrently. A discussion of osculating circles appears in the text and a discussion of the binormal vector $\mathbf{B} = \mathbf{T} \times \mathbf{N}$ is given in the exercises. Additional exercises involving radicals and rational functions now appear.

Chapter 16: More epsilon–delta exercises and examples were added. All definitions and theorems have been rewritten to be consistent with earlier material. Additional exercises involving radicals and rational functions now appear. We have added a calculator vignette, "Finding Partial Derivatives on the HP-28."

Chapter 17: A discussion of maximum and minimum problems occurring on the boundary has been added. We have also added a computer vignette, "Graphical Solutions to a Constrained Optimization Problem."

Chapter 18: A new section on Jacobians was added.

Chapter 19: A new section on vector fields appears. The similarities of Green's Theorem with the Divergence Theorem and Green's Theorem with Stokes' Theorem are discussed. The section on surface integrals was rewritten to improve clarity.

Chapter 20: This new chapter provides an introduction to differential equations.

In addition to these substantive changes, all three-dimensional graphics have been redrawn. There are two types of three-dimensional computer-generated graphs: surfaces defined by equations of the form $z = f(x, y)$ and quadric surfaces. The graphs of the first type show lines of constant x and lines of constant y on the surface. This is done by drawing the portion of the graph nearest the viewer first, while maintaining a record of the outline of what is presently visible; further portions that are visible are then drawn and added to the outline, whereas new "back" portions that fall within the outline would be hidden and are not drawn. This technique is computationally more efficient than hidden-surface removal algorithms for arbitrary polygonal surfaces. Some of the quadric surfaces are drawn this way also; the rest are generated by "clipping" against the plane that separates the visible and invisible parts of a quadric when viewed from infinity. These graphics have been provided by Douglas Dunham of the University of Minnesota, Duluth.

The three-dimensional airbrushed pieces were done by Ron Kempke of Scientific Illustrators. Each piece was first described mathematically by the appropriate equation, a three-dimensional graph to be used as an accurate template was generated according to the methods described above, and then the airbrush paintings were done. Thus, all three-dimensional artwork is mathematically and dimensionally accurate.

Much of the two-dimensional line art has been redone by Brian Morris, director of computer operations at Scientific Illustrators. All graphs represented by an equation $y = f(x)$ were plotted on a Hewlett-Packard 7550 $x - y$ plotter using the software package PLOTR developed by Scientific Illustrators. In addition, all answer art has been competely replotted to ensure accuracy using the latest technology.

Ancillaries

1. A *Student Solutions Manual*, prepared by Richard Fritz and Richard Tucker, accompanies this text. We have found the manual an invaluable aid for students who have weak backgrounds in algebra. The manual includes worked-out solutions to every other odd-numbered problem in the text. (The answers to odd-numbered problems appear in the back of the book.)
2. A complete *Solutions Manual* is also available to instructors.
3. A *Student Study Guide*, prepared by Richard St.André, accompanies the text. This study guide tests students on both key concepts and possible algebraic difficulties. It is a strategic aid and is not as comprehensive in solving remedial difficulties as the *Student Solutions Manual*. Our experience has been that the study guide is of great help at urban campuses where students may miss a few classes.
4. A test items booklet, prepared by George Feissner of SUNY Cortland, contains one multiple-choice and one fill-in test per chapter.
5. EXP-Test, a computerized test bank for IBM PCs and compatible hardware, contains all of the test questions in the test items booklet and is available to adopters of the text.
6. Calculus software from *The Math Lab* by Avery, Barker, and Soler is being provided by the publisher to adopters of the text. This software follows an exploratory model that allows students to go far beyond algebraic processes. Concepts such as limits, differentiation, and integration are numerically and graphically performed at the touch of a key. The student focuses on the correct

approach to a problem, and the underlying principles of a problem take precedence over algebraic procedure. This software is available in Apple II and IBM PC ($3\frac{1}{2}$ inch and $5\frac{1}{4}$ inch) versions.

Acknowledgments

Books that are effective teaching tools (for us) and learning tools (for our students) are not so much written as they are developed out of classroom experience. The more experiences of colleagues an author is exposed to, the more the text is thoroughly developed into a reliable teaching tool. This work received an invaluable contribution through the suggestions, criticisms, and encouragement of the following colleagues:

Reviewers of the Third Edition

Anthony Barcellos
American River College

Richard Barshinger
Pennsylvania State University, Scranton

David Baughman
College of DuPage

Steven Blasberg
West Valley College

Franklin D. Cheek
University of Wisconsin, Platteville

Barbara Chudilowsky
De Anza College

Jack Crowell
Delta College

Joe Davis
East Carolina University

Garret Etgen
University of Houston

Russell Euler
Northwest Missouri State University

Laurene Fausett
Florida Institute of Technology

Kim Hughes
California State University, San Bernardino

Howard Jones
Lansing Community College

Maryann E. Justinger
Erie Community College, South

Lois G. Leonard
Erie Community College, South

Robert Lindahl
Morehead State University

Patti Frazer Lock
St. Lawrence University

Marie Loftus
Marywood College

Ben Manvel
Colorado State University

Reza Mirdamadi
Shepherd College

William Richardson
Wichita State University

Robert Schmoyer
State University College at Buffalo

Raymond Southworth
College of William and Mary

Wesley W. Tom
Chaffey College

John R. Unbehaun
University of Wisconsin, La Crosse

Lee Welch
Cuesta College

August Zarcone
College of DuPage

Reviewers of the First and Second Editions

Daniel D. Anderson
University of Iowa, Iowa City

Robert Anderson
University of Quebec at Montreal

William C. Belvel
Sierra College

William B. Bickford
Arizona State University

Robert Blatz
Marywood College

David Boyd
University of South Carolina, Sumter

Blaine Butler
Purdue University

Glenn Calkins
Western State College

Joseph Cannon
Elizabethtown College

Peter G. Casazza
University of Alabama

Benjamin R. Cato, Jr.
College of William and Mary

Charles Chapman
Cedar Crest College

Charles K. Chui
Texas A & M University

Philip S. Clarke
Los Angeles Valley College

Douglas B. Crawford
College of San Mateo

John Dinkins
Wallace Community College

Russell Euler
Northwest Missouri State University, Maryville
Henri Feiner
Northrop Institute of Technology
Russell Floyd
Texas Wesleyan College
Richard A. Fritz
Moraine Valley Community College
Anton Glaser
Pennsylvania State University, Abington
Stuart Goldenberg
California Polytechnic State University, San Luis Obispo
Michael B. Gregory
University of North Dakota, Grand Forks
Daniel R. Gustafson
Wayne State University
Alfred W. Hales
University of California, Los Angeles
Donald Hall
California State University, Sacramento
Ray Hamlett
East Central Oklahoma University
Vern Heeren
American River College
William Higgins
Wittenberg University
Kim Hughes
California State University, San Bernardino
Glenn E. Johnston
Morehead State University
Ken Kalmanson
Montclair State College
Arthur Kaufman
College of Staten Island
William G. Koellner
Montclair State College
Steven Krantz
Pennsylvania State University
V. V. Krishnan
San Francisco State University
Thomas J. Kyrouz
Salem State College
Jeuel LaTorre
Clemson University
Stanley Lukawecki
Clemson University
George Luna
California Polytechnic State University, San Luis Obispo
Thomas R. Lupton
University of North Carolina, Wilmington
Thomas R. McCabe
Harper College
Mary McCammon
Pennsylvania State University, University Park
Otis McCowan
Belmont College
Kendall McDonald
Northwest Missouri State University, Marysville
Fred Martens
University of Alabama, Birmingham
John Mathews
California State College, Fullerton
Glenda Merhoff
Vanderbilt University
Eldon Miller
University of Mississippi
John Muth
Honolulu Community College
Thomas O'Neil
California Polytechnic State University
Thomas J. O'Reilly
St. Joseph's University
LeRoy Peterson
Indiana University Northwest
Perry G. Phillips
Pinebrook Junior College
Margaret S. Piedem
Somerset County College
Charles S. Rees
University of New Orleans
Dan Rinne
California State University, San Bernardino
Richard St.André
Central Michigan University
Richard Savage, Jr.
Tennessee Technological University
John T. Scheick
Ohio State University
L. Schiefelbusch
Indiana University Northwest
Zeev Schuss
Tel Aviv University
Thomas S. Shores
University of Nebraska
Chanchal Singh
St. Lawrence University
Arthur G. Sparks
Georgia Southern College
David R. Stone
Georgia Southern College
Keith D. Stroyan
University of Iowa, Iowa City
Donna M. Szott
Community College of Allegheny
Willie E. Taylor, Jr.
Texas Southern University, Houston
Mel Tuscher
West Valley College

Paul Vicknair
California State University, San Bernardino
Bob Allen Wake
University of California, Santa Cruz
J. Norman Wells
Georgia Southern College
W. Thurmon Whitley
University of New Haven

The authors wish to especially acknowledge the efforts of Tom Braden, who was instrumental in bringing us to Wadsworth, and of Rich Jones, who orchestrated the development of the first edition. In addition, this text owes a considerable debt to Robert Brown, Richard Johnsonbaugh, Marta Kongsle, Gloria Langer, Phyllis Marmont, Eldon Miller, John Muth, Jane Scott, David Wend, and Thurmon Whitley for its first edition. We would also like to acknowledge the special efforts of Thomas O'Neil, Richard Fritz, and Stanley Lukawecki, whose advice was invaluable.

The Historical Exercises in this text were contributed by Ken Abernethy, from an unpublished paper on the history of the calculus. This outstanding work is unique in that it consistently "reinvents" the calculus, rather than merely talking about the calculus. We hope that some of its contribution is evident in this textbook.

Many professors and students contributed their time and expertise to making this edition as error-free as possible. In particular, we would like to thank the following: Bobbi J. Barry, David L. Baughman, Steve Blasberg, Thomas Brown, Frank Cheek, George Christopher, Jennifer L. Dydo, Michael Ecker, William Hosch, Kim Hughes, Howard Jones, Maryann Justinger, Eleanor Killam, Patti Fraser Lock, Marie Loftus, Richard P. Savage, Jr., Raymond Southworth, Sarah Wada, Lee Welch, and August Zarcone.

We thank the following people who worked on the solutions manuals, also contributing to the accuracy of the text: Miriam Byers, Northwestern University; Bonny Ernst, Wichita State University; Richard A. Fritz, Moraine Valley Community College; David Popp and Frank Svara, students at Moraine Valley Community College; Raymond Southworth, College of William and Mary; and Richard Tucker, Shepherds College.

We would like to thank Eugene Schlereth for his reading of the theorems, proofs, and definitions; Ken Seydel and Kim Hughes for their extensive work on the review and challenge problems; and Steve Blasberg, Anthony Barcellos, David Baughman, and August Zarcone for their special reviews of the text. We would also like to thank the following professors for their special reviews of the differential equations chapter: Garret Etgen, Marie Loftus, and Wesley Tom.

We are grateful to Gregory P. Foley of The Ohio State University for contributing the graphing calculator vignettes and Carl Leinbach of Gettysburg College for contributing the computer vignettes.

We were fortunate to have Anne Scanlan-Rohrer as editor of this project. Her extensive expertise in mathematical publishing was evident throughout and was particularly helpful in providing meaningful reviews of the manuscript. In addition, we would like to express our appreciation to Cece Munson, whose expertise was invaluable in the production of this book.

Finally, we would like to thank the many teachers and students who wrote us unsolicited letters about our book. We have used many of their criticisms, but we would be more than human if we didn't admit we were most gratified by some of the comments of students who told us that their good experiences with the calculus made them eager to take additional mathematics courses. We don't pretend to take much credit for that, but it is a pleasure to imagine that we've had something to do with introducing a new generation to the power and beauty of the calculus.

Using Hand-Held Computers in Calculus and Analytic Geometry*

Computer technology has developed to the point where it can help students learn calculus. Graphing calculators, such as the Casio fx-7000G, fx-7500G, and fx-8000G, Hewlett-Packard 28C and 28S, and Sharp EL-5200, are really hand-held computers. They combine the capabilities of a scientific calculator, a programmable computer, an interactive graphics computer system, and in the case of the Hewlett-Packards, a computer algebra system. These devices can perform many of the computations of calculus and analytic geometry. They can readily produce the graphs of functions and, with some programming, the graphs of polar equations, parametric equations, and conics. These hand-held machines permit the user to explore a wide variety of examples in a relatively short time. They permit interactive experimentation for learning mathematical concepts, solving problems, and generating conjectures. They can make mathematics more oriented toward concept development and problem solving and less oriented toward paper-and-pencil computation. The use of hand-held computers can also be applied to many other areas of undergraduate mathematics: statistics, linear algebra, differential equations, probability, number theory, and mathematical modeling.

In many chapters of this book there are hand-held computer vignettes—brief asides that explain how a pocket computer can be used to provide a different approach to a topic just covered in the text. Only selected applications of supercalculators to calculus are explained, however. Many other uses are possible. For additional ideas, refer to the documentation provided with your machine or read the sources listed in the bibliography at the back of the book. Hewlett-Packard also publishes some applications booklets for the HP-28, which you may find helpful.

The Casios

The Casio fx-7000G, fx-7500G, and fx-8000G are all quite similar. In each case, the screen displays eight lines of numbers, words, and symbols in the text window. The graphics display is 95 × 63 pixels. The 8000 can be linked to a printer or tape recorder (to save programs on tape) using special interface equipment. The 7500 has the fastest graphics and the most memory. They all have the following features that are useful in calculus and analytic geometry:

1. *Big Screen Computation.* You can see both problems and answers on the screen at the same time. The screen is large enough to display up to four computations, both inputs and outputs, at once. If an answer doesn't make sense or the machine displays an error message, you can edit your input without having to reenter the entire command and then reexecute the problem.
2. *Interactive Graphics.* You can create virtually any mathematical graph: functions, relations, geometric figures, even three-dimensional graphs. The viewing rectangle and the scales are set using the **Range** feature.

* Gregory D. Foley
The Ohio State University

Trace allows pixel-to-pixel movement along the most recently drawn graph, with the computer displaying the x- or y-coordinate associated with each pixel along the way. It is easy to stop, reverse direction, and switch from x-readout to y or vice versa. The automatic zoom feature or the **Factor** command can be used to zoom in or zoom out about a plotted or traced-to point, or as a default, about the center of the current viewing rectangle. Early versions of the 7000 did not have automatic zoom, but now all models have this important feature.

3. *On-Screen Programming.* The Casio programming language is simple and easy to learn. On-screen editing makes programming on this little machine like programming on a microcomputer rather than on past generations of programmable calculators. A few fundamentals will carry you a long way. Programming is useful for repeated calculations and for polar, parametric, conic, and three-dimensional graphing. As an example of combining programming and graphics, an epitrochoid as it would appear on the Casio is shown here.

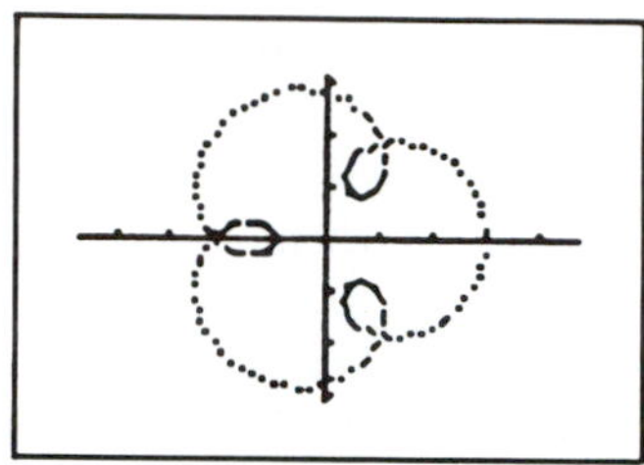

The Sharp

The Sharp EL-5200 has four lines of textual display and 96 × 32 pixels of graphics display. The graphics are slower than on the Casios, and the user does *not* see the graph being drawn. The Sharp, however, has a scrolling screen, a **SOLVE** key, and built-in matrix capabilities. The SOLVE key is especially helpful for many problems in calculus. The programming is awkward compared to the Casio, but the memory capacity is larger even than the Casio fx-7500G.

The Hewlett-Packard

The HP-28C, which is very similar to the 28S but with a smaller memory, is no longer manufactured. Like the Sharp, the HP-28S has only four lines of text. The graphics screen is 137 × 32 pixels, and graphs are shown as they are being drawn. The 28S can solve equations and operate on matrices and, in addition, can find derivatives and definite integrals, generate Taylor series, and determine antiderivatives of polynomials. The 28S can also handle complex numbers. The programming is nice for experienced programmers, permitting the commands of BASIC, FORTRAN, and Pascal. It is the most powerful, and the most expensive, of the machines described here.

Bibliography

Demana, F., & Waits, B. K. (1988). The Ohio State University Calculator and Computer Precalculus Project: The mathematics of tomorrow today! *The AMATYC Review*, *10*(1), 46–55.

Demana, F., & Waits, B. K. (1988). Pitfalls in graphical computation, or why a single graph isn't enough. *College Mathematics Journal, 19*, 236–241.

Foley, G. D. (1987). Future shock: Hand-held computers. *The AMATYC Review, 9*(1), 53–57.

Foley, G. D. (1987). Reader reflections: Zoom revisited. *Mathematics Teacher, 80*, 606.

Nievergelt, Y. (1987). The chip with the college education: The HP-28C. *American Mathematical Monthly, 94*, 895–902.

Nievergelt, Y. (1988). The HP-28S brings computations and theory back together in the classroom. *Notices of the American Mathematical Society, 35*, 799–804.

Potter, D. (1987, August). Seven ounces of pocket graphics: Casio's new fx-8000G and friends. *Pico: The Magazine of Portable Computing*, pp. 22–23.

Potter, D. (1987, August). Casio's fx-8000G goes to high school. *Pico: The Magazine of Portable Computing*, pp. 23–24.

Shumway, R. J. (1988). Graphics calculators: Skills versus concepts. In J. de Lang & M. Doorman (Eds.), *Senior Secondary Mathematics Education* (pp. 136–140). Utrecht, The Netherlands: University Press.

Small, D., Hosack, J., & Lane, K. (1986). Computer algebra systems in undergraduate instruction. *College Mathematics Journal, 17*, 423–433.

Smith, D. A., Porter, G. J., Leinbach, L. C., & Wenger, R. H. (Eds.). (1988). *Computers and mathematics: The use of computers in undergraduate instruction* [MAA Notes No. 9]. Washington, DC: Mathematical Association of America.

Steen, L. A. (1987, October 14). Who still does math with paper and pencil? *The Chronicle of Higher Education*, p. A48.

Steen, L. A. (Ed.). (1988). *Calculus for a new century: A pump, not a filter* [MAA Notes No. 8]. Washington, DC: Mathematical Association of America.

Tucker, T. (1987, January–February). Calculators with a college education? *Focus: The Newsletter of the Mathematical Association of America*, pp. 1, 5.

Waits, B. K., & Demana, F. (1988). Manipulative algebra—the culprit or the scapegoat? *Mathematics Teacher, 81*, 322–334.

Using Computing in Teaching Calculus*

Why Do It?

A colleague and I were having a conversation. My colleague said, "I don't use the computer at all in my calculus course. I want my students to think." My reply was that I use computing extensively for exactly the same reason. This brief conversation contains the extreme positions held within the mathematical community. As the warden said to Luke in the movie *Cool Hand Luke*, "What we have here is a failure to communicate." Both my colleague and I have the same goal—to have our students understand and appreciate one of the premier achievements of human intellectual activity. Both of us are sincere in our desire to achieve this goal. The communications failure is a result of our concept of the role that technology can play in achieving the

* L. Carl Leinbach
Gettysburg College

goal. Obviously, there are many, quite possibly the majority of, calculus teachers who hold positions somewhere between that of my colleague and me. The computing sections of this text are directed to this group.

The fundamental objects for study of the calculus are functions. Students of the calculus spend their time examining functions and the effects of certain operations on these functions. A student who successfully completes a calculus course should be able to understand at a deep level the information conveyed by any function, not just those presented in a closed algebraic form. The medical practitioner should know what it means for the concentration of a drug to decay at an exponential rate. The economist should understand what is meant when one is talking about economic "forces." The social scientist should understand trends and the limitations of linear projections. The physicist and chemist should be able to use a knowledge of present states to make meaningful predictions about future states of a system. The mathematician should have an intimacy with a very important class of mathematical objects. This knowledge is gained not by the inclusion of specific examples in a book (no matter how carefully chosen) but by a thorough understanding of functions independent of the manner in which they are presented.

Unfortunately, students come to us with very little intuition about functions. They have concentrated on algebraic formulas and see, for example, the replacement of the variable x by the expression $(x + a)$ in an equation as an algebraic operation and not a translation of the solution set. The exercises that are assigned in most calculus courses perpetuate this perspective. This is especially unfortunate, since it is from the exercises that students create their impressions about what is really important in the subject. The differentiation and integration operators become means of transforming, sometimes rather painfully, one algebraic expression into another. Seldom does a student spend time studying the graph of a function and deciphering the information that is contained in that object. Even more seldom does the student come to the realization that this information is valuable in understanding the given relationships.

By no means am I against algebra! What is needed is a balance between the algebraic knowledge and the knowledge that is gained by visualization. The computing packages provide us with an opportunity to provide that balance. It is easy to generate data in the form of graphs of functions. The student can study functions presented in this form, make conjectures, and test those conjectures with more data. In this mode, the student acts like a laboratory scientist. Herein lies the beauty of using computing in this way; the student then submits conjectures, supported by laboratory observation, for mathematical verification. If we have done our job correctly, some of these conjectures will fail, but for the most part they will succeed. The result is that the student, in addition to developing some intuition about functions, has developed a vested interest in the class presentation. This fact alone makes computer use important.

How Can It Be Done?

This text contains four sections specifically devoted to using the computer to explore concepts presented in the more traditional manner. The first on Newton's method is rather conventional in its approach. The method is used to find a numerical solution to an equation that would be very difficult to solve by algebraic methods. The second section illustrates the problems that can be encountered by a strict reliance on the evidence of a computer-generated graph. The subsequent investigation of that graph is designed to enforce the idea that the analysis done in calculus can give us insights that are not necessarily obvious to the observer of a computer-generated graph. In the third section, the student will use the graphical and numerical capabilities of the computing package to anticipate the fundamental theorem of calculus. The approximating sum for the definite integral is used to investigate the very nature of what is being approximated, not just to make a numerical approximation. The student is encouraged to make a hypothesis that is verified by the proof of the fundamental theorem. The fourth section deals with the explanation of the LaGrange multiplier method for constrained optimization of a function of two variables. An estimate made using graphical evidence is enhanced with numerical techniques.

These four sections are designed to guide students through the investigation and lead them to an understanding of the analysis done in class. As such, the sections need to be read interactively. The student should be in front of a computer generating the data that are presented in the text. The exercises following the section should also be done in front of a computer. Students should be encouraged to generate their own exercises. It is vital that the student become involved in the process of generating data and giving an initial test to hypotheses.

Software Packages

The original intent was to write an essay on packages that are available for use in a calculus course. Actually there are too many to mention. The philosophy of use is much more important than the particular product that is used. My personal choice and the one that was used to generate the four sections in this book is *MicroCalc* by Harley Flanders of the University of Michigan. This package contains a rich assortment of routines that are easily accessed by the user. The package contains both the graphical and numerical routines that are required for the types of investigations done in this book. It even contains a rudimentary symbolic differentiation routine that proves useful in Newton's method and Lagrange multiplier investigations. Another option is to use a computer algebra system such as DERIVE from the Soft Warehouse in Honolulu, Hawaii, Maple from the Symbolic Computation Group at Waterloo University, or Mathematica from Wolfram Enterprises. These packages contain good numerical and graphics routines as well as very good routines for symbolically solving equations, symbolic differentiation, and symbolic integration among other mathematical operations. The possession of one of these packages can expand one's horizon far beyond the traditional calculus with computing course, but that is another story.

Preface to the Student

I hear . . . . . and I forget
I see . . . . . and I remember
I do . . . . . and I understand

CHINESE SAYING

Introduction

1

This chapter contains material needed for the study of calculus. Most of this material is review, but some of it may be new to you.

1.1 Real Numbers

Sets

We begin with the idea of a *set*. A *set* is a collection of objects considered as a whole. The objects of a set S are called *elements* of S, or *members* of S. The set that has no elements, called the *empty set* or *null set*, is denoted by the symbol $\varnothing$.

If a is an element of the set S, we write $a \in S$, which is read "a is an element of S" or "a is in S." To indicate that a is not an element of S, we write $a \notin S$, which is read "a is not an element of S" or "a is not in S."

Ordinarily, a set S can be written in either of two ways. These two methods are illustrated by the following example: Consider the set D that has the elements

$$0, 1, 2, 3, 4, 5, 6, 7, 8, 9$$

In this case, we write

$$D = \{0, 1, 2, 3, 4, 5, 6, 7, 8, 9\}$$

This expression is read "D is the set consisting of the elements 0, 1, 2, 3, 4, 5, 6, 7, 8, 9." Here, we list or display the elements of the set D.

Another way of writing this same set D is

$$D = \{x \mid x \text{ is a nonnegative integer less than } 10\}$$

This is read "D is the set of all x such that x is a nonnegative integer less than 10." Here, we have described the set D by giving a property that every element of D has and that no element not in D can have.

Now consider the two sets

$$A = \{1, 2\} \qquad B = \{1, 2, 3, 4\}$$

where each element of A is also found in B. When the elements of A are also elements of B, we say *A is a subset of B* and use the notation $A \subseteq B$ to denote this fact. In addition, it is customary to consider $\varnothing \subseteq A$ for any set A. For example,

$$\{1, 3, 5\} \subseteq \{1, 3, 5, 7, 9\} \qquad \{1, 2, 3\} \subseteq \{1, 2, 3\} \qquad \varnothing \subseteq \{1, 2, 3\}$$

We now describe some important sets of numbers.

The set of *integers* is $\{0, 1, -1, 2, -2, \ldots\}$. The set of *nonnegative integers*, $\{0, 1, 2, \ldots\}$, forms a subset of the set of integers, as does the set of *negative integers*, $\{-1, -2, -3, \ldots\}$.

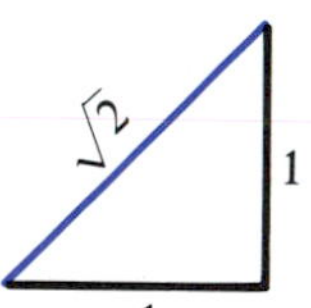

Figure 1

Rational numbers are ratios of integers. For a rational number a/b, the integer a is called the *numerator*, and the integer b, which cannot be 0, is called the *denominator*.

Although rational numbers occur frequently in nature, numbers that are not rational also occur. For example, the number $\sqrt{2}$ is not rational, but $\sqrt{2}$ is equal to the length of the hypotenuse of an isosceles right triangle with legs that are each of length 1 (see Fig. 1). As another example, the number π, which may be described as the ratio of the circumference of a circle to its diameter, is not a rational number (see Fig. 2). These numbers, $\sqrt{2}$ and π, are *irrational numbers*—numbers that are not rational.

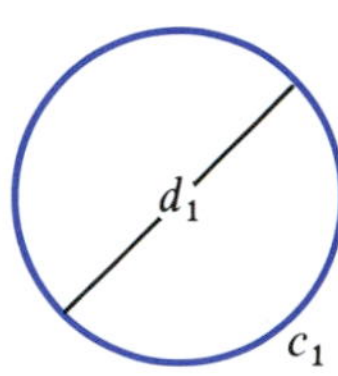

$$\frac{c_1}{d_1} = \frac{c_2}{d_2} = \pi$$

Figure 2

The irrational numbers and rational numbers together form the *set $\mathbb{R}$ of real numbers*. To represent each real number, we use what is commonly referred to as a *decimal representation*, or simply a *decimal*. For example, the decimal representations of the rational numbers $\frac{3}{4}$, $\frac{5}{2}$, $\frac{2}{3}$, and $\frac{7}{66}$ are $\frac{3}{4} = 0.75$, $\frac{5}{2} = 2.5$, $\frac{2}{3} = 0.666\ldots$, and $\frac{7}{66} = 0.10606\ldots$. We observe (and can prove) that the decimal representation of a rational number is always one of two types: (*1*) *terminating*, or ending ($\frac{3}{4}$, $\frac{5}{2}$, etc.); or (*2*) eventually *repeating* (in $\frac{2}{3}$, the 6's repeat; in $\frac{7}{66}$, the block 06 repeats).

At first, it may appear that these two types represent all possible decimals. However, it is relatively easy to construct a decimal that neither terminates nor eventually repeats. For example, the decimal 0.123456789101112 . . . , where we write down the positive integers one after the other, will neither terminate nor eventually repeat. In fact, there are an infinite number of such decimals, and they represent the irrational numbers. For example, the real numbers

$$\sqrt{2} = 1.414213 \ldots \qquad \pi = 3.14159 \ldots$$

are irrational, having decimals that neither terminate nor eventually repeat.

Thus, the set of real numbers may be thought of as the set of all possible decimals. This feature of real numbers gives them their practicality.* In the physical world, many changing magnitudes, like the length of a heated rod or the velocity of a particle, are assumed to pass through every possible magnitude from the initial one to the final one. Since the precise measurement of a magnitude is naturally given by a decimal, the equivalent of all possible magnitudes is all possible decimals (real numbers).

In practice, it is usually necessary to represent real numbers by approximations. For example, using the symbol $\approx$, read "approximately equal to ," we can write

$$\sqrt{2} \approx 1.4142 \qquad \pi \approx 3.1416$$

Coordinates

Real numbers can be represented geometrically on a horizontal line. We begin by selecting an arbitrary point O, called the *origin*, and associate it with the real number 0. We then establish a scale by marking off line segments of equal length (units) on each side of 0. By agreeing that the positive direction is to the right of 0 and the negative direction is to the left of 0, we can successively associate the integers 1, 2, 3, . . . with each mark to the right of 0 and the integers $-1, -2, -3, \ldots$ with each mark to the left of 0 (see Fig. 3).

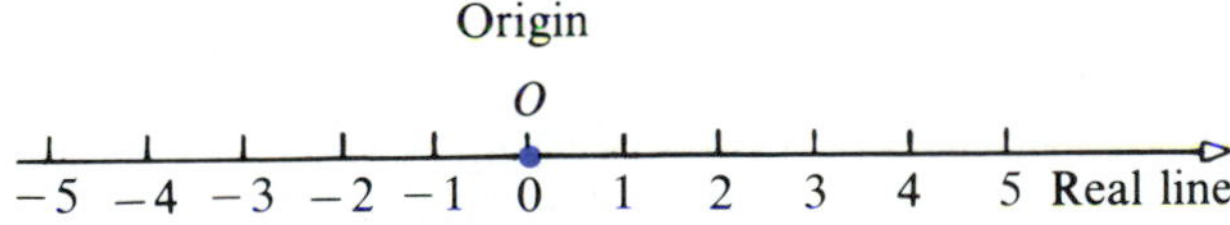

Figure 3

By subdividing these segments, we can locate rational numbers such as $\frac{1}{2}$ and $-\frac{3}{2}$. The irrational numbers are located by geometric construction (as in the case of $\sqrt{2}$) or by other means. In this way, every point P on the line is associated with a unique real number x, called the *coordinate of P* (see Fig. 4).

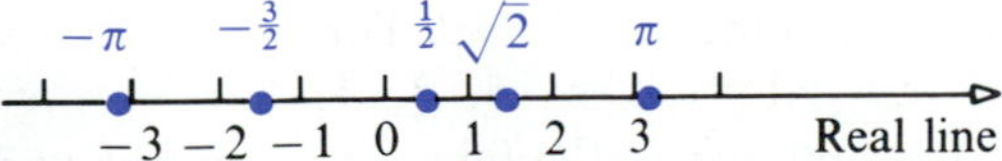

Figure 4

---

* Other number systems exist that have many applications (such as the complex numbers). In this book, however, we limit our discussion to problems in which only real numbers are used.

Inequalities

There is a natural ordering of the real numbers; as we move along the real line from left to right, the numbers increase in size. To describe the relative size of two real numbers, we use the ordering symbols $<$ (less than) and $\leq$ (less than or equal to), which are called *inequalities*. These symbols are defined as follows:

[1.1.1] DEFINITION / *Inequalities.*

If a and b are real numbers, then

$$a < b \quad \text{means} \quad b - a \quad \text{is positive}$$

$$a \leq b \quad \text{means} \quad a < b \quad \text{or} \quad a = b$$

Of course, the inequality $a < b$, which is read "a is less than b," is the same as "b is greater than a," and we may write $b > a$. Similarly, the inequality $a \leq b$, which is read "a is less than or equal to b," is the same as "b is greater than or equal to a," written as $b \geq a$.

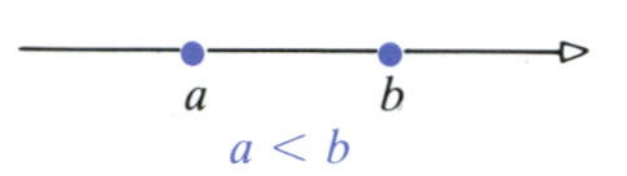

Figure 5

Geometrically, coordinates on the number line establish an ordering for the real numbers; that is, if a and b are coordinates of two points P and Q, respectively, then $a < b$ means that P lies to the left of Q on the line (see Fig. 5).

Inequalities obey the following properties, which are stated without a proof.

1. ***Addition Property.***
   If $a \leq b$, then $a + c \leq b + c$ for any choice of c. That is, the addition of a number to each side of an inequality will not affect the sense or direction of the inequality.
2. ***Multiplication Property.***
   (a) If $a \leq b$ and $c > 0$, then $a \cdot c \leq b \cdot c$.
   (b) If $a \leq b$ and $c < 0$, then $a \cdot c \geq b \cdot c$.
   When each side of an inequality is multiplied by a number, the sense or direction of the inequality remains the same if we multiply by a positive number; it is reversed if we multiply by a negative number.
3. ***Division Property.***
   (a) If $a > 0$, then $\frac{1}{a} > 0$. That is, the reciprocal of a positive number is positive.
   (b) If a and b are both positive or both negative and $a < b$, then $\frac{1}{a} > \frac{1}{b}$.
4. ***Trichotomy Property.*** For any two real numbers a and b, one and only one of the following is true: $a < b$, $a = b$, $b < a$.
5. ***Transitive Property.*** If $a < b$ and $b < c$, then $a < c$.

Table 1 illustrates some of these properties.

Table 1

Given Inequality	Operation	Property Used	Resulting Inequality
$-3 < 5$	Add 6 to both sides	Addition	$3 < 11$
$-3 < 5$	Add -6 to both sides	Addition	$-9 < -1$
$-3 < 5$	Multiply both sides by 2	Multiplication	$-6 < 10$
$-3 < 5$	Multiply both sides by -2	Multiplication	$6 > -10$
$3 < 5$	Take reciprocal of both sides	Division	$\frac{1}{3} > \frac{1}{5}$
$-5 < -3$	Take reciprocal of both sides	Division	$-\frac{1}{5} > -\frac{1}{3}$
$3 < 5$ and $5 < 7$		Transitive	$3 < 7$

If $a < b$ and $b < c$, we write $a < b < c$. This says that b is between a and c. Similarly, if $a \le b$ and $b \le c$, we write $a \le b \le c$. The inequalities $a \le b < c$ and $a < b \le c$ are given similar interpretations.

Geometrically, $a < b < c$ states that b is to the right of a and c is to the right of b on the line (see Fig. 6).

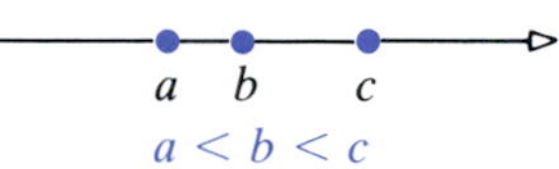

Figure 6

Variables

A *variable* is a symbol (usually a letter, x, y, etc.) used to represent any real number. An *inequality* is a statement involving one or more variables and one of the inequality symbols ($<, \le, >, \ge$). To *solve* an inequality means to find all possible numbers that the variables can assume to make the statement true. The set of all such numbers is called the *solution* (or *solution set*). Two inequalities with the same solution are called *equivalent.*

To find the solution of an inequality, we apply the properties of inequalities. For example, the statement $x + 2 \le 3x - 5$ is an inequality with one variable; it can be solved by obtaining a series of equivalent inequalities until we get to an inequality with an obvious solution. Thus, for the inequality $x + 2 \le 3x - 5$, we have

$$x + 2 \le 3x - 5$$

$$-2x \le -7 \qquad \text{Subtract 2 and then } 3x \text{ from both sides.}$$

$$x \ge \frac{7}{2} \qquad \text{Multiply by } -\tfrac{1}{2} \text{ and remember to reverse the inequality because we multiplied by a negative number.}$$

The solution is the set of all real numbers to the right of $\frac{7}{2}$, including $\frac{7}{2}$ (see Fig. 7).

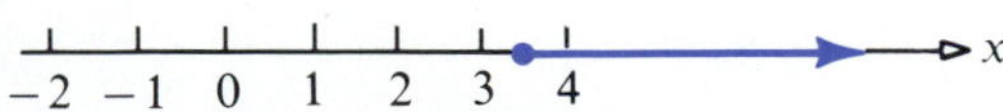

Figure 7

Solving other types of inequalities may be more difficult. Some examples are given here.

EXAMPLE 1

Solve the inequality: $x^2 + x - 12 > 0$

Solution

We factor the left side, obtaining

$$(x-3)(x+4) > 0$$

The product of two real numbers is positive either when both factors are positive or when both factors are negative.

Both Positive			***or***	***Both Negative***		
$x-3>0$	and	$x+4>0$		$x-3<0$	and	$x+4<0$
$x>3$	and	$x>-4$		$x<3$	and	$x<-4$

The numbers x that are greater than 3 and -4 are simply

$$x > 3$$

or

The numbers x that are less than 3 and -4 are simply

$$x < -4$$

The solution is $\{x \mid x > 3 \quad \text{or} \quad x < -4\}$ (see Fig. 8).

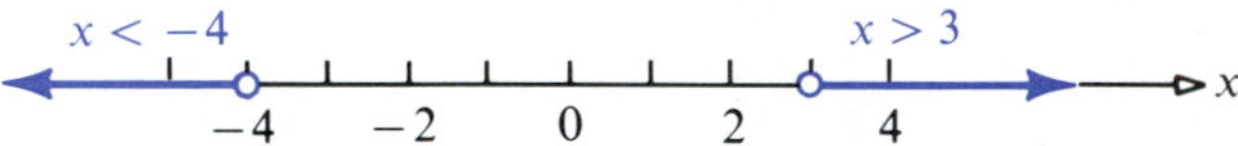

Figure 8

We also can obtain the solution of the inequality of Example 1 by another method. The left-hand side of the inequality is factored so that it becomes $(x-3)(x+4) > 0$, as before. We then construct a graph that uses the numbers $x = 3$ and $x = -4$ as cutoff points (see Fig. 9). These numbers are the solutions of the equation

$$x^2 + x - 12 = (x-3)(x+4) = 0$$

and they separate the line into three parts: $x < -4$, $-4 < x < 3$, and $x > 3$. In the part of the line where $x < -4$, we deduce that both the quantities $(x-3)$ and $(x+4)$ are always negative, so their product must always be positive. Therefore, $x < -4$ is a solution of the inequality. In the part of the line where $-4 < x < 3$, we deduce that $(x-3)$ is always negative and $(x+4)$ is always positive, so their product is always negative. We conclude that the numbers between -4 and 3 are not solutions of the

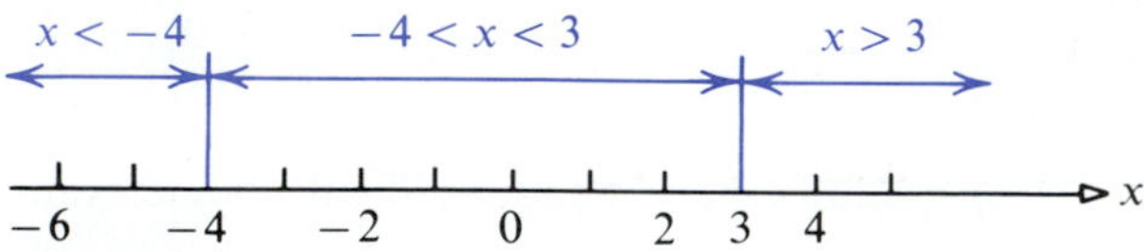

Figure 9

inequality. In the part of the line where $x > 3$, we deduce that both the quantities $(x - 3)$ and $(x + 4)$ are always positive, so their product is always positive. Hence, numbers greater than 3 are solutions of the inequality. Table 2 summarizes these results.

Table 2

	Sign of $x - 3$	Sign of $x + 4$	Sign of $(x - 3)(x + 4)$	Conclusion
$x < -4$	$-$	$-$	$+$	$x < -4$ is a solution
$-4 < x < 3$	$-$	$+$	$-$	$-4 < x < 3$ is not a solution
$x > 3$	$+$	$+$	$+$	$x > 3$ is a solution

This method of solving inequalities is particularly appealing when the inequality contains more than two factors. It can also be used to solve inequalities that involve ratios.

EXAMPLE 2

Solve: $\dfrac{x}{x-4} < 3$

Solution

We begin by transforming the inequality to get 0 on the right side:

$$\frac{x}{x-4} - 3 < 0$$

$$\frac{x - 3(x-4)}{x-4} < 0$$

$$\frac{-2(x-6)}{x-4} < 0$$

$$\frac{x-6}{x-4} > 0$$

The quantities $(x - 4)$ and $(x - 6)$ supply two cutoff points, 4 and 6, as shown in Figure 10.

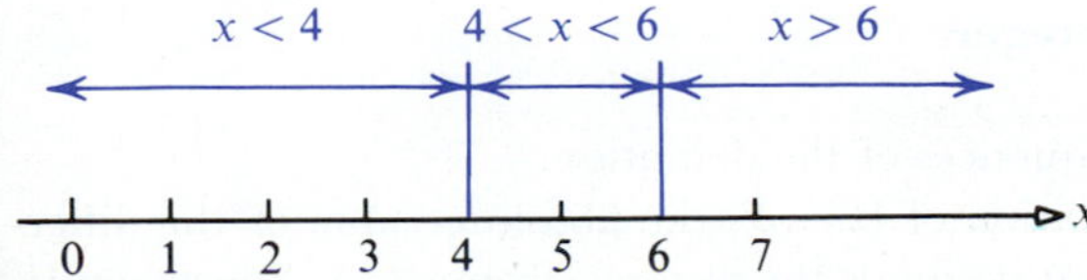

Figure 10

Now we construct a table to find out where $(x-6)/(x-4)$ is positive (see Table 3). The solution is $\{x \mid x < 4 \quad \text{or} \quad x > 6\}$ (see Fig. 11).

Table 3

	Sign of $x-4$	Sign of $x-6$	Sign of $\dfrac{x-6}{x-4}$	Conclusion
$x < 4$	$-$	$-$	$+$	$x < 4$ is a solution
$4 < x < 6$	$+$	$-$	$-$	$4 < x < 6$ is not a solution
$x > 6$	$+$	$+$	$+$	$x > 6$ is a solution

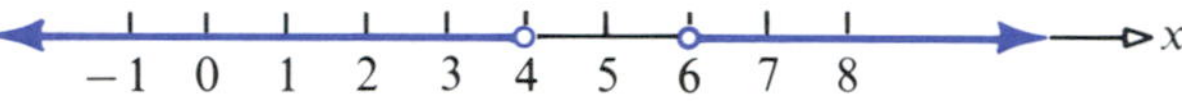

Figure 11

Absolute Value

The absolute value of a number is the magnitude, or distance from 0, of that number. Thus, the absolute value of 5 is 5, and the absolute value of -6 is 6.

[1.1.2] DEFINITION / ***Absolute Value.***

The *absolute value* of a real number x, denoted by $|x|$, is defined as

$$|x| = \begin{cases} x & \text{if} \quad x \geq 0 \\ -x & \text{if} \quad x < 0 \end{cases}$$

For example,

$$|8| = 8 \qquad |0| = 0 \qquad |-4| = -(-4) = 4 \qquad |1-\sqrt{2}| = -(1-\sqrt{2}) = \sqrt{2}-1$$

Since $|x|^2 = x^2$ by the definition of absolute value, there is a relationship between square root and absolute value:

$$|x| = \sqrt{x^2}$$

Because the absolute value of a negative number is positive, it follows that the absolute value of any number is either positive or 0. Some other properties of absolute value are listed here.

If a and x are real numbers, then:

(a) $|x-a| = |a-x|$ **(b)** $|ax| = |a||x|$

(c) $-|x| \leq x \leq |x|$ **(d)** $\left|\dfrac{x}{a}\right| = \dfrac{|x|}{|a|} \quad a \neq 0$ **(1)**

(e) $|x|^n = |x^n|$ ***n* an integer**

These properties are consequences of the definition.

A geometric interpretation of $|x-a|$, the absolute value of the difference between x and a, is that it equals the distance from a to x. For example, if $|x-3| = 6$, then x lies either 6 units to the right of 3 or 6 units to the left of 3 (see Fig. 12). That is, the solution of $|x-3| = 6$ is $\{-3, 9\}$.

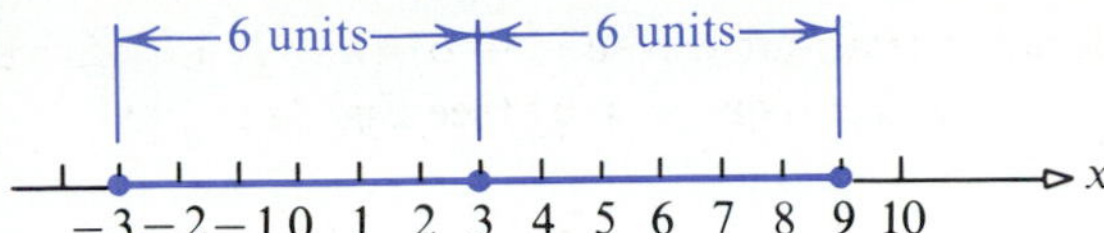

Figure 12

The general rule follows:

If the absolute value of a mathematical expression equals some positive number p, then the expression itself equals either p or $-p$.

For example,

$$|x| = 3 \quad \text{means} \quad x = 3 \quad \text{or} \quad x = -3$$

Similarly,

$$|x - 8| = 6 \quad \text{means} \quad x - 8 = 6 \qquad \text{or} \quad x - 8 = -6$$

$$x = 6 + 8 \qquad\qquad x = -6 + 8$$

$$= 14 \qquad\qquad = 2$$

Let's turn now to some examples of inequalities that involve absolute value.

EXAMPLE 3

Find all numbers x for which $|x| < 7$.

Solution

Here, we are asked to find all numbers x for which the distance from 0 is less than 7. Any x between -7 and 7 satisfies this condition. Consequently, the solution is $\{x \mid -7 < x < 7\}$.

In general, for any number $a > 0$:

$$|x| < a \quad \text{is equivalent to} \quad -a < x < a$$
$$|x| \leq a \quad \text{is equivalent to} \quad -a \leq x \leq a \qquad \textbf{(2)}$$

Similarly,

$$|x| > a \quad \text{is equivalent to} \quad x < -a \quad \text{or} \quad x > a$$
$$|x| \geq a \quad \text{is equivalent to} \quad x \leq -a \quad \text{or} \quad x \geq a \qquad \textbf{(3)}$$

See Figure 13 for a geometric interpretation.

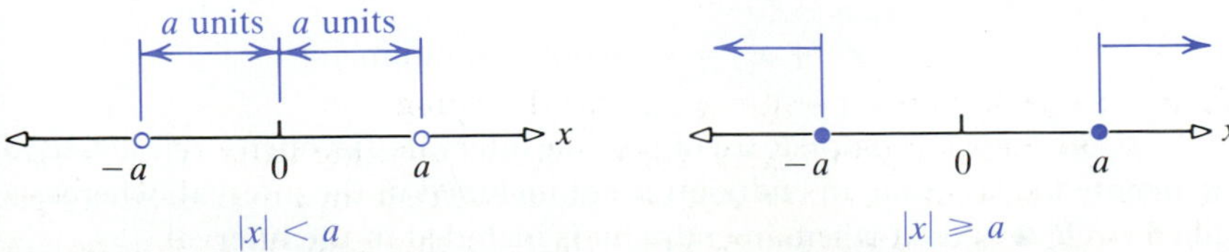

Figure 13

EXAMPLE 4

Find all x for which $|x-3| \leq 6$.

Solution

Here, think of $(x-3)$ as a single unknown quantity so that by applying expression (2), we find

$$-6 \leq x-3 \leq 6$$

If we add 3 to each term in this inequality, we get

$$-6+(3) \leq (x-3)+(3) \leq 6+(3)$$

which simplifies to $-3 \leq x \leq 9$. The solution is $\{x \mid -3 \leq x \leq 9\}$.

Intervals

Let a and b be two real numbers with $a < b$. A *closed interval* $[a, b]$ is the set of all real numbers x from a to b, inclusive; that is,

$$[a, b] = \{x \mid a \leq x \leq b\}$$

An *open interval* (a, b) consists of all real numbers x between a and b, exclusive of both a and b; that is,

$$(a, b) = \{x \mid a < x < b\}$$

Finally, the *half-open* (*semi-open*) or *half-closed* (*semi-closed*) intervals are defined by

$$[a, b) = \{x \mid a \leq x < b\} \qquad (a, b] = \{x \mid a < x \leq b\}$$

In these definitions, a is the *left endpoint* and b is the *right endpoint* of each interval.

For a real number a, the notation $[a, +\infty)$ denotes the set of all real numbers greater than or equal to a; that is,

$$[a, +\infty) = \{x \mid x \geq a\}$$

The symbol $+\infty$, read "plus infinity," is not a real number but is merely a notational device. We define

$$(a, +\infty) = \{x \mid x > a\}$$

$$(-\infty, a] = \{x \mid x \leq a\}$$

$$(-\infty, a) = \{x \mid x < a\}$$

$$(-\infty, +\infty) = \mathbb{R} \text{ (set of real numbers)}$$

Table 4 gives a complete list of possible intervals. An open circle ○ is used to denote the fact that an endpoint is not included in the interval, whereas a filled circle ● is used when an endpoint is included in the interval.

Table 4

Interval Notation	Set Notation	Geometric Picture
$[a, b]$	$\{x \mid a \le x \le b\}$	$[a, b]$
(a, b)	$\{x \mid a < x < b\}$	(a, b)
$[a, b)$	$\{x \mid a \le x < b\}$	$[a, b)$
$(a, b]$	$\{x \mid a < x \le b\}$	$(a, b]$
$[a, +\infty)$	$\{x \mid x \ge a\}$	$[a, +\infty)$
$(a, +\infty)$	$\{x \mid x > a\}$	$(a, +\infty)$
$(-\infty, a]$	$\{x \mid x \le a\}$	$(-\infty, a]$
$(-\infty, a)$	$\{x \mid x < a\}$	$(-\infty, a)$
$(-\infty, +\infty)$	$\{x \mid x \in \mathbb{R}\}$	$(-\infty, +\infty)$

EXAMPLE 5

Find all x for which $|x - 2| \ge 3$.

Solution

Here, we ask for all real numbers for which the distance from 2 is greater than or equal to 3. Using Figure 14 as an aid, we conclude that

$$x \le -1 \qquad \text{or} \qquad x \ge 5$$

Thus, the solution is $(-\infty, -1]$ together with $[5, +\infty)$.

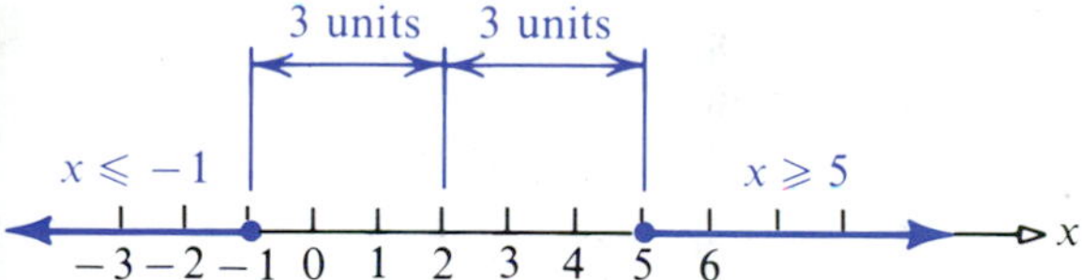

Figure 14

We are now ready for an inequality involving the absolute value of a sum. This inequality will be useful in later chapters.

Triangle Inequality.* **If x and y are real numbers, then**

$$|x + y| \le |x| + |y| \tag{4}$$

* This inequality is an algebraic expression of the fact that the length of any side of a triangle does not exceed the sum of the lengths of the other two sides.

Proof By property (c) in (1), we have $-|x| \le x \le |x|$ and $-|y| \le y \le |y|$. Adding these, we find

$$-(|x| + |y|) \le x + y \le |x| + |y|$$

Hence, from (2) we have $|x + y| \le |x| + |y|$.

Another useful property involving the absolute value of a difference is stated below. Its proof is left as an exception.

If x and y are real numbers, then

$$|x - y| \ge |x| - |y| \quad (5)$$

EXERCISE 1.1

In Problems 1–10 label each real number as rational or irrational.

1. $\frac{21}{8}$ **2.** $\frac{8}{21}$ **3.** $2.151515\ldots$ **4.** $\sqrt{2} + 5$ **5.** $\pi - 2$

6. $0.999\ldots$ **7.** $-5.613613613\ldots$ **8.** $3.01252525\ldots$ **9.** $\sqrt{371}$ **10.** $\sqrt{283}$

In Problems 11–14 replace the * by <, >, or =, whichever gives a true statement.

11. $\frac{1}{3} * 0.33$ **12.** $\frac{1}{4} * 0.25$ **13.** $3 * \sqrt{9}$ **14.** $\pi * \frac{22}{7}$

In Problems 15–54 find the solution.

15. $3x + 5 \le 2$ **16.** $-3x + 5 \le 2$ **17.** $3x + 5 \ge 2$ **18.** $4 - 5x \ge 3$

19. $6x - 3 \ge 8x + 5$ **20.** $8 - 2x \le 5x - 6$ **21.** $14x - 21x + 16 \le 3x - 2$

22. $10x - 3x \le 2x + 5 - 15$ **23.** $x^2 - 5x + 6 \ge 0$ **24.** $x^2 + 2x \ge 0$

25. $x^2 + 7x < -12$ **26.** $x^2 - x < 12$ **27.** $|x| = 5$

28. $|x| = 6$ **29.** $|x| \le 3$ **30.** $|x| \le 4$

31. $|x - 3| < 4$ **32.** $|x + 2| \le 6$ **33.** $|2x - 4| + 5 \le 9$

34. $6 + |3x - 7| \le 10$ **35.** $|x - 3| > 4$ **36.** $|x + 2| > 3$

37. $|2x - 4| \ge -8$ **38.** $|3x + 4| \ge -5$ **39.** $|x - 3| < 0.01$

40. $|2x - 1| < 0.02$ **41.** $|\frac{1}{2} - 2x| \le 4$ **42.** $|4 - \frac{1}{2}x| < 5$

43. $\frac{1}{x} < 3$ **44.** $\frac{2}{x} \ge -5$ **45.** $\frac{2}{x - 2} \le -5$ **46.** $\frac{2}{x + 3} > 6$

47. $\frac{2x + 1}{x - 3} < 1$ **48.** $\frac{\frac{3}{2}x - 2}{x + 5} > 1$ **49.** $\frac{2}{3 - x} < 1$ **50.** $\frac{2}{3 - x} \le 4$

51. $\frac{3}{x} < \frac{2}{x - 1}$ **52.** $\frac{3}{x + 1} < \frac{2}{x - 1}$ **53.** $\left|\frac{1}{x}\right| < 2$ **54.** $\left|\frac{2}{x}\right| < 3$

55. When does $|x| = x$ hold? **56.** When does $|x| = -x$ hold?

In Problems 57 and 58 find all numbers x for which the given expression has meaning.

57. $\sqrt[4]{x^2 - 3x + 2}$

58. $\sqrt{x^2 - 4}$

The following steps show how to recover the fractional form of the repeating decimal 3.151515 . . . . Set

$$x = 3.1515\ldots$$

$$100x = 315.15\ldots \qquad \text{Multiply } x \text{ by } 100.$$

$$100x - x = 99x = 312 \qquad \text{Subtract.}$$

$$x = \frac{312}{99} = \frac{104}{33} \qquad \text{Solve for } x.$$

In Problems 59–64 use a similar sequence of steps to recover the rational number of each repeating decimal.

59. 2.151515 . . .

60. 5.363363363 . . .

61. 2.105232323 . . .

62. 3.2173232 . . .

63. 0.9999 . . .

64. 0.252525 . . .

65. Solve: $|x| + |x - 3| < 4$.

66. Solve: $2x + |x - 5| < 6$.

67. If $|x - 2| < \frac{1}{5}$ and $|y - 3| < \frac{1}{10}$, show that $|xy - 6| < \frac{41}{50}$.

68. If $|x - 2| < \frac{1}{100}$, show that $|x^2 - 4| < \frac{1}{10}$ and $|x^3 - 8| < \frac{13}{100}$.

69. Show that if $|x - a| < \frac{1}{3}$ and $|a - y| < \frac{1}{3}$, then $|x - y| < \frac{2}{3}$. [*Hint:* $|x - y| = |(x - a) + (a - y)|$.]

70. Verify the triangle inequality (4) for:

(a) $x = 2, \quad y = 3$
(b) $x = -2, \quad y = 3$
(c) $x = 2, \quad y = -3$
(d) $x = -2, \quad y = -3$

71. If $a < b$, prove that $a < \dfrac{a + b}{2} < b$. The number $\dfrac{a + b}{2}$ is called the *arithmetic mean of a and b*. Show that the arithmetic mean is equidistant from a and b.

72. If $0 < a < b$, prove that $a < \sqrt{ab} < b$. The number $\sqrt{ab}$ is called the *geometric mean of a and b*. Show that the geometric mean is less than the arithmetic mean. [*Hint:* Use the fact that $(\sqrt{b/2} - \sqrt{a/2})^2 > 0$.]

73. If $0 < a < b$, prove that $a < h < b$, where h is defined by

$$\frac{1}{h} = \frac{1}{2}\left(\frac{1}{a} + \frac{1}{b}\right)$$

The number h is called the *harmonic mean of a and b*. Show that h equals the ratio of the geometric mean squared to the arithmetic mean.

74. Show that, for a fixed perimeter, a square encloses more area than any rectangle.

75. Show that, for a fixed perimeter, a circle encloses more area than a square.

76. Prove that $|x - y| \geq |x| - |y|$. [*Hint:* Show that $|x| = |x - y + y| \leq |x - y| + |y|$.]

77. If $a \leq b$ and $c < 0$, prove that $ac \geq bc$.

78. If a and b are positive and if $a < b$, prove that $1/a > 1/b$.

79. If $0 < a \leq b$, prove that $a^2 \leq b^2$.

80. If a and b are positive and if $a^2 \leq b^2$, prove that $a \leq b$.

Graphing 1.2

Rectangular Coordinates

Consider two lines, one horizontal and the other vertical. Call the horizontal line the *x-axis* and the vertical line the *y-axis*. Assign coordinates to points on these lines, as described previously, by using their point of intersection as the

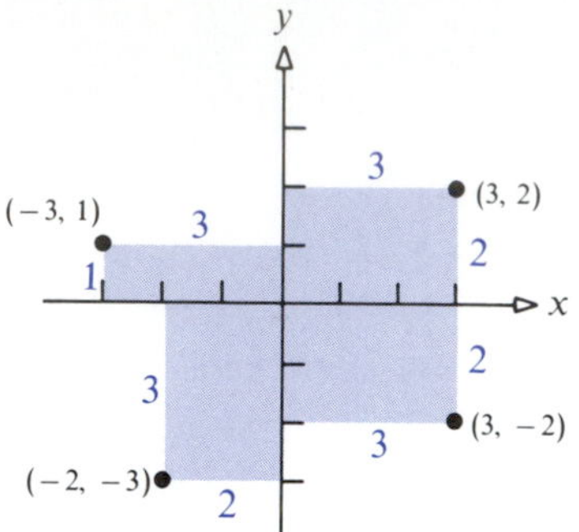

Figure 15

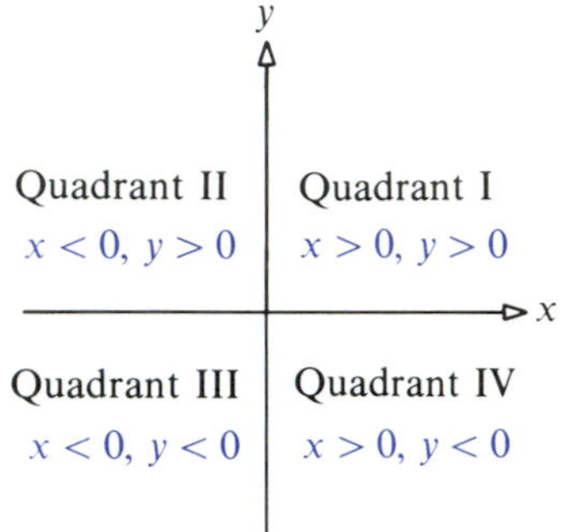

Figure 16

origin O and using a convenient scale on each. We follow the usual convention that points on the x-axis to the right of O are associated with positive real numbers, those to the left of O with negative numbers, those on the y-axis above O are associated with positive real numbers, and those below O with negative real numbers. This gives the origin a value of 0 on both the x-axis and the y-axis.

Any point P in the plane formed by the x-axis and y-axis can then be located by using an *ordered pair* of real numbers. Let x denote the signed distance of P from the y-axis (signed in the sense that if P is to the right of the y-axis, then $x > 0$ and if P is to the left of the x-axis, then $x < 0$); and let y denote the signed distance of P from the x-axis. The ordered pair (x, y), the *coordinates of* P, then gives us enough information to locate the point P. We can assign ordered pairs of real numbers to every point P, as shown in Figure 15.

If (x, y) are the coordinates of a point P, then x is called the *abscissa of* P and y is the *ordinate of* P. For example, the coordinates of the origin O are $(0, 0)$. The abscissa of any point on the y-axis is 0; the ordinate of any point on the x-axis is 0.

The coordinate system described here is a *rectangular* or *cartesian coordinate system* and divides the plane into four sections called *quadrants* (see Fig. 16). In quadrant I, for example, both the abscissa x and the ordinate y of all points are positive.

Graph of Equations

If an equation involves the variables x and y, then the *graph* of the equation consists of the set of points (x, y) in the plane with coordinates that satisfy the equation.

EXAMPLE 1

Graph the equation: $y = 2x + 5$

Solution

We want to find all points (x, y) for which the ordinate y equals twice the abscissa x plus 5. To locate some of these points (and thus get an idea of the pattern of the graph), let us assign some numbers to x and find corresponding values for y:

$$\begin{aligned}
&\text{If } x = 0, && \text{then } y = 2(0) + 5 = 5 \\
&\text{If } x = 1, && \text{then } y = 2(1) + 5 = 7 \\
&\text{If } x = -5, && \text{then } y = 2(-5) + 5 = -5 \\
&\text{If } x = 10, && \text{then } y = 2(10) + 5 = 25
\end{aligned}$$

By connecting these points,* we obtain the graph of the equation (a straight line), as shown in Figure 17 on page 15.

* Merely connecting points will not give the complete picture for most problems. An important application of calculus is as an aid in graphing by identifying various characteristics of the graph (see Chap. 4).

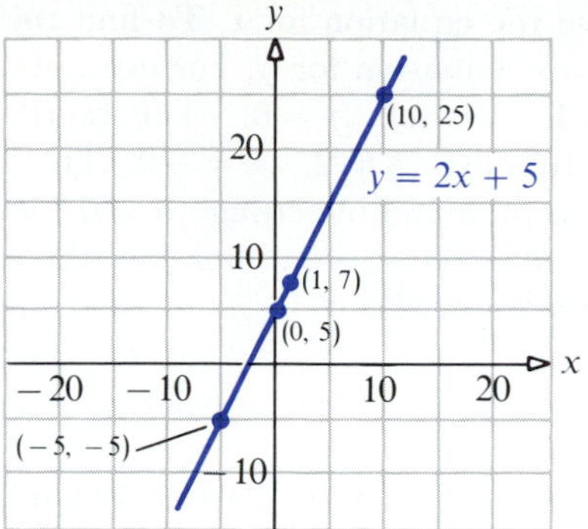

Figure 17

EXAMPLE 2

Graph the equation: $y = x^2$

Solution

A table provides several points on the graph:

x	0	1	2	3	4	−1	−2	−3
y	0	1	4	9	16	1	4	9

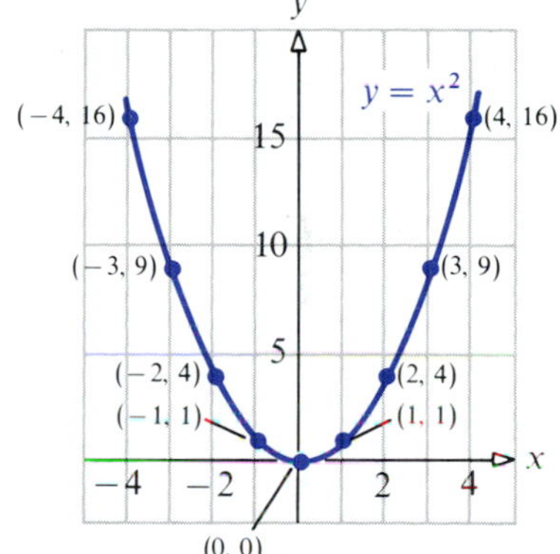

Figure 18

In Figure 18 we list these points and, by connecting them with a smooth curve, we obtain the graph (a *parabola*).

Two useful tools for obtaining the graph of an equation are *intercepts* and *symmetry*.

Intercepts

The points at which a graph intersects the coordinate axes are called the *intercepts*. The abscissa of a point at which the graph crosses the x-axis is an x-intercept, and the ordinate of a point at which the graph crosses the y-axis is a *y-intercept* (see Fig. 19). To find the x-intercept(s) of an equation,

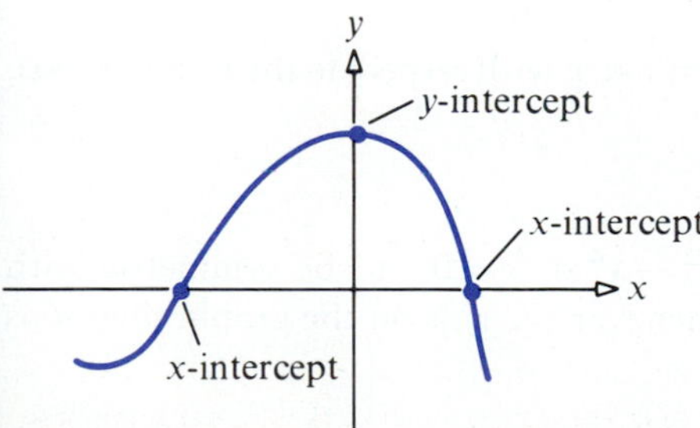

Figure 19

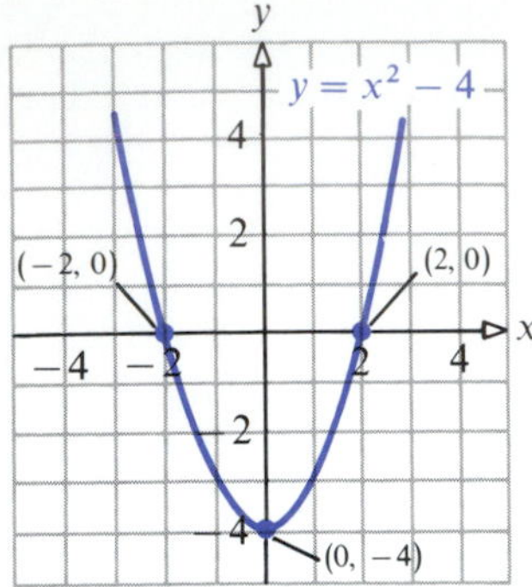

Figure 20

we set $y = 0$ in the equation and solve the equation for x. To find the y-intercept(s), we set $x = 0$ and solve the equation for y. For example, to find the x-intercept(s) of $y = x^2 - 4$, we set $y = 0$. The resulting equation, $x^2 - 4 = 0$, has two solutions: $x = 2$, $x = -2$. Thus, the x-intercepts are 2 and -2. The y-intercept, found by setting $x = 0$ in the equation, is $y = -4$. The graph of $y = x^2 - 4$ thus has three intercepts: $(2, 0)$, $(-2, 0)$, and $(0, -4)$. See Figure 20.

Symmetry

Another useful tool for graphing equations is *symmetry*, particularly symmetry with respect to the x-axis, y-axis, and origin.

1. ***Symmetry with Respect to the x-Axis.*** For every point (x, y) on a graph, the point $(x, -y)$ is also on the graph.
2. ***Symmetry with Respect to the y-Axis.*** For every point (x, y) on a graph, the point $(-x, y)$ is also on the graph.
3. ***Symmetry with Respect to the Origin.*** For every point (x, y) on a graph, the point $(-x, -y)$ is also on the graph.

Figure 21 illustrates some of the possibilities that can occur.

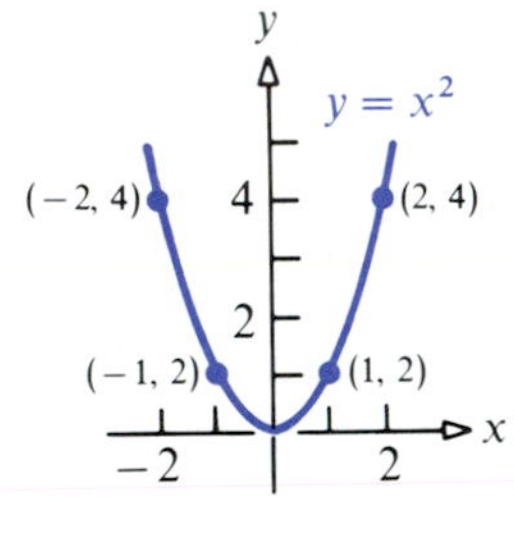

(a) Symmetry with respect to the y-axis

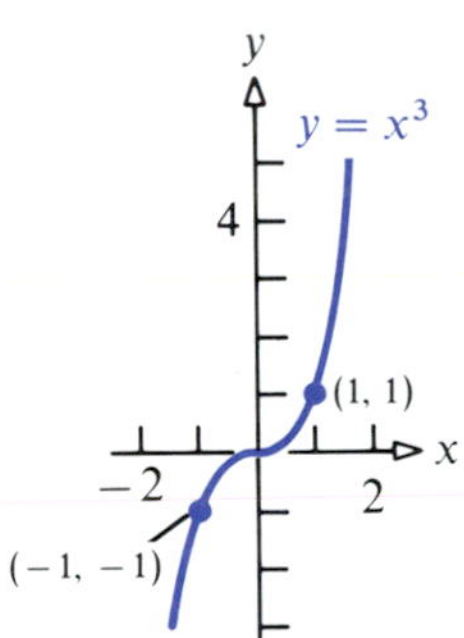

(b) Symmetry with respect to the origin

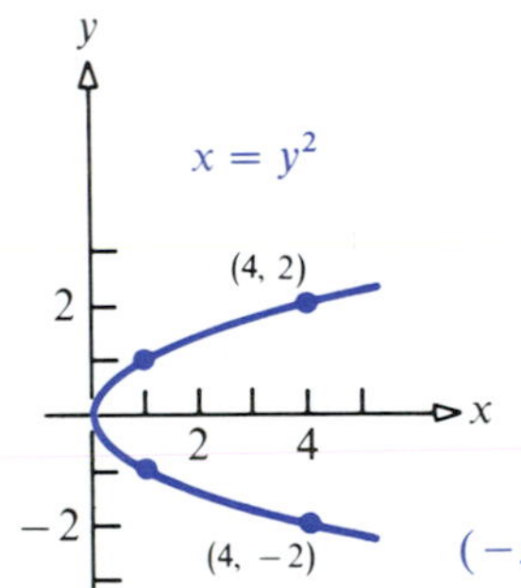

(c) Symmetry with respect to the x-axis

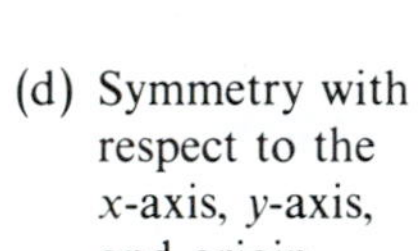

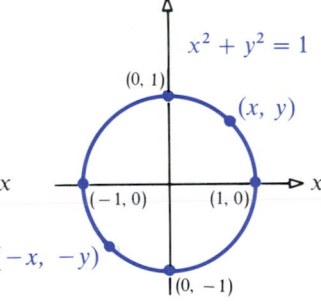

(d) Symmetry with respect to the x-axis, y-axis, and origin

Figure 21

EXAMPLE 3

Examine $xy^2 - x^4 + 5 = 0$ for symmetry with respect to the x-axis, y-axis, and origin.

Solution

For the graph of the equation $xy^2 - x^4 + 5 = 0$ to be symmetric with respect to the x-axis requires that whenever (x, y) is on the graph, then so is $(x, -y)$. This requires that

$$\text{if} \quad xy^2 - x^4 + 5 = 0 \quad \text{then} \quad x(-y)^2 - x^4 + 5 = 0$$

This is the case, since $(-y)^2 = y^2$. The graph of $xy^2 - x^4 + 5 = 0$ is symmetric with respect to the x-axis.

For symmetry with respect to the y-axis, we require that

$$\text{if} \quad xy^2 - x^4 + 5 = 0 \quad \text{then} \quad (-x)y^2 - (-x)^4 + 5 = 0$$

This implication is not true, so the graph of $xy^2 - x^4 + 5 = 0$ is not symmetric with respect to the y-axis.

Similarly, we can show that the graph of $xy^2 - x^4 + 5 = 0$ is not symmetric with respect to the origin.

Knowing that the graph of $xy^2 - x^4 + 5 = 0$ is symmetric with respect to the x-axis means that once we know its graph above the x-axis, we automatically know its graph below the x-axis.

Sometimes it is possible to obtain the graph of an equation by a simple *translation*. For example, the graph of the equation $y = x^2 + 1$ may be easily obtained by "lifting" the graph of $y = x^2$ one unit. Some examples are given in Figure 22, where we show translations of some of the graphs given in Figure 21.

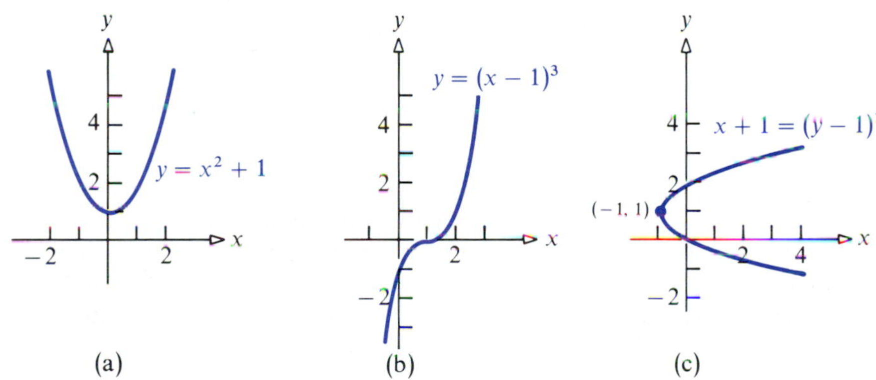

Figure 22

Another graphing technique is that of adding ordinates. To graph the equation $y = x^2 + x$, we can graph the two equations $y = x^2$ and $y = x$ and then add the heights. Figure 23 illustrates this procedure.

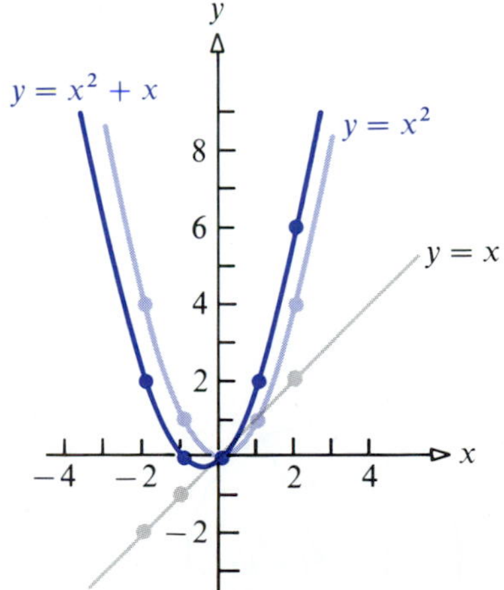

Figure 23

Distance Between Points

Let (x_1, y_1) be the coordinates of point P_1 and let (x_2, y_2) be the coordinates of point P_2. In moving from P_1 to P_2, the abscissa changes from x_1 to x_2. In calculus we denote this *change in x* by the symbol Δx (read "delta x"); that is, $\Delta x = x_2 - x_1$. Similarly, in moving from P_1 to P_2, the ordinate changes from y_1 to y_2, and this *change in y* is denoted by $\Delta y = y_2 - y_1$.* For example, if $P_1 = (5, -2)$ and $P_2 = (4, 7)$, then the change in

* The symbols Δx and Δy measure the change in x and the change in y, respectively, and do not mean the product of Δ and x or the product of Δ and y.

x from P_1 to P_2 is $\Delta x = 4 - 5 = -1$. The change in y from P_1 to P_2 is $\Delta y = 7 - (-2) = 9$.

If the same scale is used on both the x-axis and the y-axis, then all distances in the plane can be measured by using this same scale. In fact, by using the theorem of Pythagoras, we find:

[1.2.1] THEOREM / *Distance Between Two Points.*

The distance between the two points $P_1 = (x_1, y_1)$ and $P_2 = (x_2, y_2)$, denoted by $|P_1P_2|$, is

$$|P_1P_2| = \sqrt{(x_2 - x_1)^2 + (y_2 - y_1)^2} = \sqrt{(\Delta x)^2 + (\Delta y)^2} \qquad (1)$$

See Figure 24.

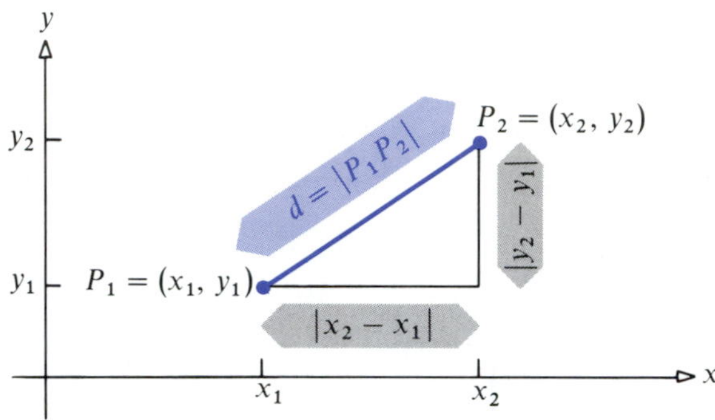

Figure 24

Thus, to compute the distance between two points, find the change in their abscissas (Δx), square it, and add this to the square of the change in their ordinates (Δy). The square root of this sum is the distance. For example, to find the distance d between the points $(-2, 5)$ and $(3, 2)$, we compute

$$\Delta x = 3 - (-2) = 5 \qquad \text{and} \qquad \Delta y = 2 - 5 = -3$$

Then

$$d = \sqrt{(\Delta x)^2 + (\Delta y)^2} = \sqrt{(5)^2 + (-3)^2} = \sqrt{34}$$

The distance between two points is never a negative number. Furthermore, the only time the distance between two points is 0 is when the two points are identical. Finally, it makes no difference whether the distance is computed from P_1 to P_2 or from P_2 to P_1; that is, $|P_1P_2| = |P_2P_1|$.

Circles

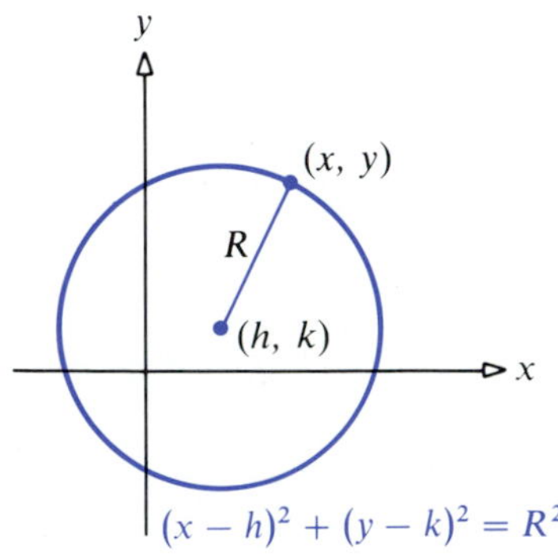

Figure 25

Figure 25 is the graph of all points (x, y) that are a fixed distance R from a fixed point (h, k). We recognize this as the graph of a *circle* with its center at (h, k) and with radius R.

We can find the equation of this circle by using the distance formula (1). If (x, y) is any point on the circle, then

$$\sqrt{(x - h)^2 + (y - k)^2} = R$$

or, equivalently,

$$(x - h)^2 + (y - k)^2 = R^2 \tag{2}$$

This equation is referred to as the *standard equation* of a circle with radius R and center (h, k).

For example, the set of points defined by

$$(x-2)^2 + (y+3)^2 = 9$$

is a circle with center $(2, -3)$ and radius 3.

If the center is at the origin (that is, $h = 0$, $k = 0$) and the radius R is of unit length $(R = 1)$, then we obtain

$$x^2 + y^2 = 1$$

which is the equation of the *unit circle*.

EXAMPLE 4

Discuss the graph of $x^2 + y^2 + 4x - 6y + 12 = 0$.

Solution

First, we group the terms as follows:

$$(x^2 + 4x) + (y^2 - 6y) = -12$$

We proceed to complete the square of each parenthetical term:

$$(x^2 + 4x + 4) + (y^2 - 6y + 9) = -12 + 4 + 9$$
$$(x+2)^2 + (y-3)^2 = 1$$

This is a circle with center $(-2, 3)$ and radius 1 (see Fig. 26).

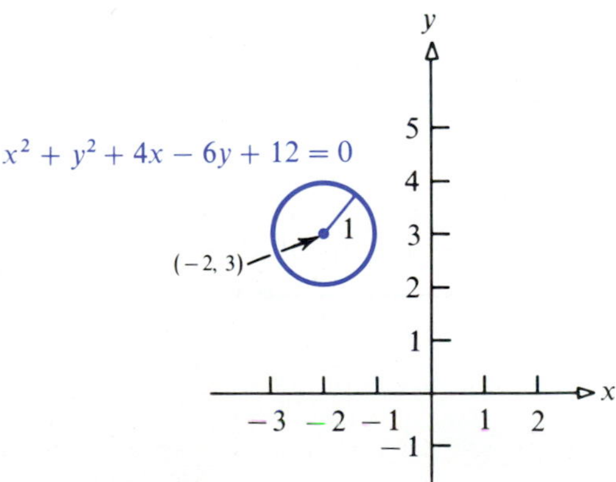

Figure 26

It can be shown that the graph of any equation of the form

$$x^2 + y^2 + ax + by + c = 0$$

is a circle, a point, or no graph at all. For example, the graph of the equation $x^2 + y^2 = 0$ is the single point (0, 0). As another example, the equation $x^2 + y^2 + 5 = 0$, or $x^2 + y^2 = -5$, has no graph, since sums of squares are never negative. When the graph is a circle, the equation

$$x^2 + y^2 + ax + by + c = 0$$

is referred to as the *general equation of a circle.*

Midpoint Formula

We close this section with the formula for the coordinates of the *midpoint of a line segment.* If $P_1 = (x_1, y_1)$ and $P_2 = (x_2, y_2)$ are the endpoints of a line segment, the point M that is equidistant from P_1 and P_2 has coordinates

$$\left(\frac{x_1 + x_2}{2}, \frac{y_1 + y_2}{2}\right)$$

To see how we obtained this result, look at Figure 27. There, triangles P_1AM and MBP_2 are congruent, since M is the midpoint and the three angles are congruent. Hence, corresponding sides are equal in length. That is,

$$x - x_1 = x_2 - x \qquad \text{and} \qquad y - y_1 = y_2 - y$$

By solving for x and y, we obtain

$$x = \frac{x_1 + x_2}{2} \qquad \text{and} \qquad y = \frac{y_1 + y_2}{2} \tag{3}$$

Thus, to find the midpoint of a line segment joining two points, we average the abscissas and ordinates of the points.

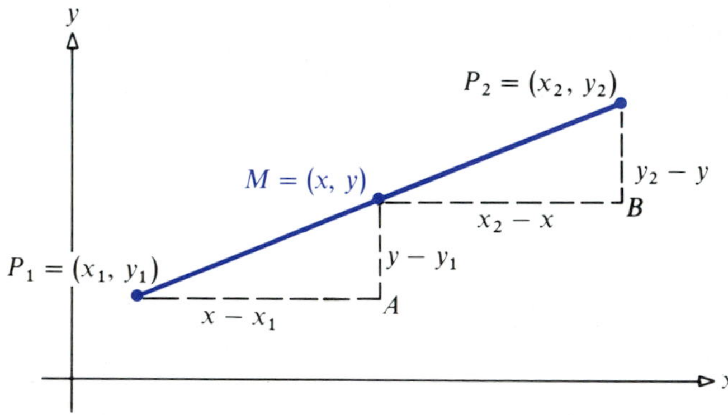

Figure 27

EXERCISE 1.2

In Problems 1–4 find the distance $|P_1P_2|$ between the points P_1 and P_2.

1. $P_1 = (3, -4)$; $P_2 = (3, 1)$

2. $P_1 = (-1, 0)$; $P_2 = (2, 1)$

3. $P_1 = (-0.6, 2)$; $P_2 = (-0.4, -0.2)$

4. $P_1 = (-5, 1.2)$; $P_2 = (0.6, -0.5)$

In Problems 5–12 graph each equation.

5. $y = 3x$ **6.** $y = -2x$ **7.** $y = 2x - 3$ **8.** $y = 2x - 4$

9. $3y + 2x + 1 = 0$ **10.** $4y - 2x - 5 = 0$ **11.** $y = x^2 + 4$ **12.** $y = 2x^2 - 4$

In Problems 13–16 find the general equation of the circle determined by the indicated center (h, k) and radius R.

13. $(h, k) = (2, -3)$; $R = 4$

14. $(h, k) = (-1, 2)$; $R = 2$

15. $(h, k) = (1, -2)$; $R = 1$

16. $(h, k) = (0, 3)$; $R = 3$

In Problems 17–20 find the center (h, k) and radius R of each circle.

17. $(x - 1)^2 + y^2 = 4$

18. $(x - 3)^2 + (y + 4)^2 = 9$

19. $x^2 + y^2 + 4x - 6y - 3 = 0$

20. $x^2 + y^2 + 8y + 15 = 0$

In Problems 21–24 find the midpoint of the line segment joining the points P_1 and P_2.

21. $P_1 = (2, 3)$; $P_2 = (6, 5)$

22. $P_1 = (1, -3)$; $P_2 = (3, 5)$

23. $P_1 = (0, 1)$; $P_2 = (1, 0)$

24. $P_1 = (3, 0)$; $P_2 = (6, 2)$

In Problems 25–28 use the accompanying graph.

25. Add to the graph to make it symmetric with respect to the x-axis.

26. Add to the graph to make it symmetric with respect to the y-axis.

27. Add to the graph to make it symmetric with respect to the origin.

28. Add to the graph to make it symmetric with respect to the origin, x-axis, and y-axis.

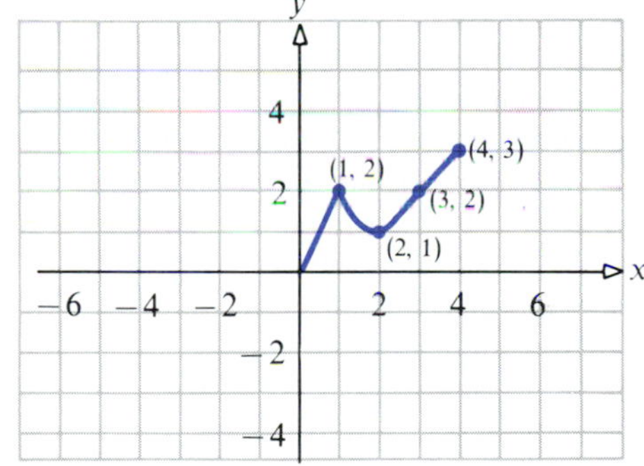

In Problems 29–36 follow the steps of the solution for Example 3 to examine each equation for symmetry with respect to the x-axis, y-axis, and origin.

29. $x^2y + y^2x = 5$ **30.** $x^2y^2 = 4$ **31.** $x^2 - 2xy = y^4$ **32.** $x^3 + y^3 = 1$

33. $3x^2 + 6y = 2$ **34.** $4y^2 - 6x^2 = x$ **35.** $y = -x^5 + 3x$ **36.** $y = (x^2 + 1)^2 - 1$

In Problems 37–42 find the general equation of each circle.

37. Center at $(1, -2)$ and passing through the point $(2, -1)$

38. Center at $(4, 1)$ and passing through the point $(5, 2)$

39. Tangent to the y-axis with center at $(2, 3)$

40. Center at $(3, 1)$ and passing through $(0, 0)$

41. Center at $(-3, -2)$ and tangent to the line $y = 5$

42. Passing through the points $(2, 4)$, $(4, 8)$, and $(2, 8)$

43. (a) Find the center and radius of the circle with equation

$$x^2 + ax + y^2 + by = 0$$

(b) Find a condition on a, b, and c so that

$$x^2 + ax + y^2 + by + c = 0$$

is the equation of a circle of positive radius.

44. The line with equation $y = \frac{1}{2}x + 2$ is tangent to a circle at $(0, 2)$. The line $y = 2x - 7$ is tangent to the same circle at $(3, -1)$. Find the center of this circle.

45. Find the general equation of a line containing the centers of the circles

$$x^2 + y^2 + 4x - 8y + 4 = 0 \quad \text{and}$$

$$x^2 + y^2 - 2x - 2y = 0$$

46. Find the general equation of a line containing the centers of the circles

$$x^2 + y^2 + 2x - y + 1 = 0 \quad \text{and}$$

$$x^2 + y^2 - 4x - 4y = 0$$

47. Find the lengths of the medians* of the triangle with vertices at $(0, 0)$, $(0, 6)$, and $(8, 0)$.

48. If two vertices of an equilateral triangle* are $(4, -3)$ and $(0, 0)$, find the third vertex. How many of these triangles are possible?

In Problems 49–52 find the length of each side of the triangle determined by the three points P_1, P_2, and P_3; and state whether the triangle is an isosceles triangle,* a right angle triangle,* neither of these, or both.

49. $P_1 = (2, 1)$; $P_2 = (-4, 1)$; $P_3 = (-4, -3)$

50. $P_1 = (-1, 4)$; $P_2 = (6, 2)$; $P_3 = (4, -5)$

51. $P_1 = (-2, -1)$; $P_2 = (0, 7)$; $P_3 = (3, 2)$

52. $P_1 = (7, 2)$; $P_2 = (-4, 0)$; $P_3 = (4, 6)$

53. In a study of conduction of heat through a wall, the ordered pair (x, y) represents the temperature x at each side of the wall and the corresponding thermal conductivity y. The formula for thermal resistance depends on the midpoint of the line segment determined by the temperatures and thermal conductivities on the two sides of the wall. If on one side of the wall the temperature is 300°F and the thermal conductivity is 0.22, while on the other side the temperature is 180°F and the thermal conductivity is 0.10, find the coordinates of the midpoint.

54. The earth is represented on a map of a portion of the solar system so that its surface is a circle with equation $x^2 + y^2 - 2x + 4y - 4091 = 0$. A satellite circles 0.6 unit above the earth in a circular orbit with its center the center of the earth. Find the equation for the orbit of the satellite on this map.

55. If r is a real number, prove that the coordinates of the point $P = (x, y)$ that divides the line segment from $P_1 = (x_1, y_1)$ to $P_2 = (x_2, y_2)$ in the ratio r (that is, $|P_1P|/|P_1P_2| = r$) are

$$x = (1 - r)x_1 + rx_2 \qquad y = (1 - r)y_1 + ry_2$$

[*Hint:* Use similar triangles.]

In Problems 56–60 use the result of Problem 55.

56. Verify that the midpoint divides the line segment from $P_1 = (x_1, y_1)$ to $P_2 = (x_2, y_2)$ in the ratio $r = \frac{1}{2}$.

57. What point P divides the line segment from P_1 to P_2 in the ratio $r = 1$?

58. What point P divides the line segment from P_1 to P_2 in the ratio $r = 0$?

59. Find the point P on the line joining $P_1 = (1, 4)$ and $P_2 = (5, 6)$ that is twice as far from P_1 as P_2 is from P_1 and lies on the same side of P_1 as P_2 does.

60. Find the point(s) P on the line joining $P_1 = (0, 4)$ and $P_2 = (-1, 1)$ that is three times as far from P_1 as P_2 is from P_1.

---

* The *medians* of a triangle are the line segments from each vertex to the midpoint of the opposite side. An *equilateral triangle* is one in which all three sides are of equal length. An *isosceles triangle* is one in which two of the sides are of equal length. A *right angle triangle* is one in which one of the angles is 90°. The Pythagorean theorem holds for all right triangles.

The Straight Line

1.3

Slopes of Lines

We begin with the result from plane geometry that there is one and only one line L containing two distinct points P_1 and P_2. If P_1 and P_2 are each represented by ordered pairs of real numbers, the following definition can be given:

[1.3.1] DEFINITION / ***Slope.***

Let P_1 and P_2 be two distinct points with coordinates (x_1, y_1) and (x_2, y_2), respectively. The *slope m* of the line L containing P_1 and P_2 is defined by the formula*

$$m = \frac{y_2 - y_1}{x_2 - x_1} \quad \text{if} \quad x_1 \neq x_2 \tag{1}$$

If $x_1 = x_2$, the slope m of L is *undefined* (since this results in division by 0) and L is a *vertical line.*

Using the facts that $\Delta x = x_2 - x_1$ and $\Delta y = y_2 - y_1$, we can write the slope m of a nonvertical line as

$$m = \frac{\Delta y}{\Delta x}$$

See Figure 28.

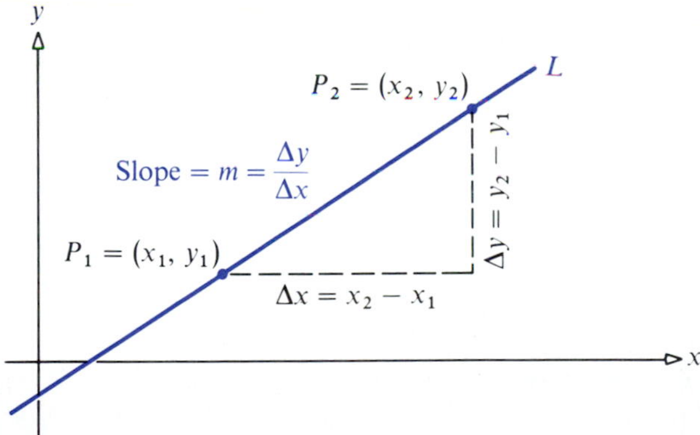

Figure 28

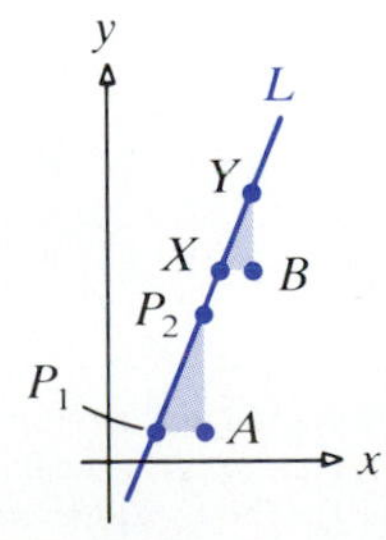

* The following argument, involving similar triangles, shows that the slope of a line L is the same no matter what two distinct points are used: Let L be a nonvertical line joining P_1 and P_2, and let X and Y be any other two distinct points on L. Construct the triangles depicted in the figure. Since triangle P_1P_2A is similar to triangle XYB (why?), it follows that the lengths of the corresponding sides are in proportion. That is, $|AP_2|/|BY| = |AP_1|/|BX|$ or $|AP_2|/|AP_1| = |BY|/|BX|$. But the slope m of L is $|AP_2|/|AP_1|$, and by the foregoing equality, we see that $m = |BY|/|BX|$. In other words, since X and Y are *any* two points, the slope m of a line L is the same no matter what points on L are used to compute m.

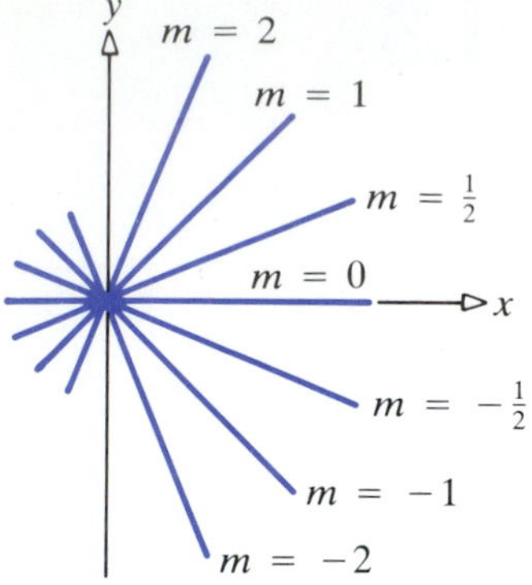

Figure 29

That is, the slope m of a nonvertical line L is the ratio of the change in the ordinates from P_1 to P_2 to the change in the abscissas from P_1 to P_2. Since

$$\frac{y_2 - y_1}{x_2 - x_1} = \frac{y_1 - y_2}{x_1 - x_2}$$

the result is the same whether the changes are computed from P_1 to P_2 or from P_2 to P_1. For example, the slope m of the line joining the points $(1, 2)$ and $(5, -3)$ may be computed as

$$m = \frac{-3 - 2}{5 - 1} = \frac{-5}{4} \qquad \text{or as} \qquad m = \frac{2 - (-3)}{1 - 5} = \frac{5}{-4} = \frac{-5}{4}$$

Figure 29 illustrates the slopes of several lines.

EXAMPLE 1

Compute the slopes of the lines L_1, L_2, L_3, and L_4 containing the given pairs of points. Graph each line.

$$L_1\text{:} \quad P = (2, 3); \quad Q_1 = (-1, -2)$$

$$L_2\text{:} \quad P = (2, 3); \quad Q_2 = (3, -1)$$

$$L_3\text{:} \quad P = (2, 3); \quad Q_3 = (5, 3)$$

$$L_4\text{:} \quad P = (2, 3); \quad Q_4 = (2, -2)$$

Solution

Let m_1 be the slope of L_1, m_2 the slope of L_2, and so on. Then,

$$m_1 = \frac{-2 - 3}{-1 - 2} = \frac{-5}{-3} \approx 1.66$$

$$m_2 = \frac{-1 - 3}{3 - 2} = \frac{-4}{1} = -4$$

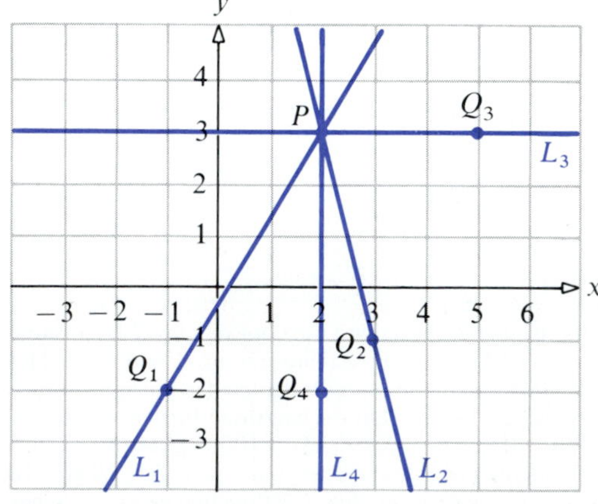

Figure 30

$$m_3 = \frac{3-3}{5-2} = \frac{0}{3} = 0$$

m_4 is undefined, since $x_1 = x_2 = 2$

These lines are graphed in Figure 30. Note that when the slope m is positive, the line slants upward from left to right (L_1); when m is negative, the line slants downward from left to right (L_2); when $m = 0$, the line is horizontal (L_3); and when m is undefined, the line is vertical (L_4).

Equations of Lines

A vertical line is given by the equation

$$x = a \tag{2}$$

where a is a given real number.

For example, the graph of the equation $x = 3$ is a vertical line (see Fig. 31).

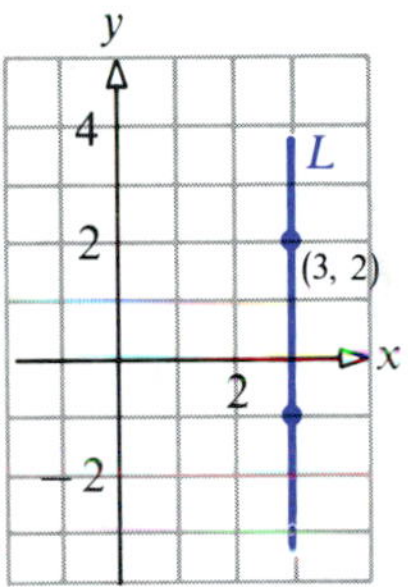

Figure 31

Now let L be a nonvertical line with slope m and containing (x_1, y_1). For (x, y) any other point on L, we have

$$m = \frac{y - y_1}{x - x_1} \qquad \text{or} \qquad y - y_1 = m(x - x_1)$$

Point–Slope Form. **An equation of a nonvertical line of slope m that passes through the point (x_1, y_1) is**

$$y - y_1 = m(x - x_1) \tag{3}$$

For example, an equation of the line with slope 4 and passing through the point $(1, 2)$ is

$$y - 2 = 4(x - 1)$$
$$y = 4x - 2$$

EXAMPLE 2

Find an equation of the line passing through $(2, 3)$ and $(-4, 5)$.

Solution

Here, since two points are given, we first compute the slope of the line:

$$m = \frac{5-3}{-4-2} = \frac{2}{-6} = \frac{-1}{3}$$

Using the point $(2, 3)$ (we could just as well use $(-4, 5)$), we find the equation of the line to be

$$y - 3 = \left(\frac{-1}{3}\right)(x - 2)$$
$$y = \left(\frac{-1}{3}\right)x + \frac{11}{3}$$

Another useful equation of a line is obtained when the slope m and y-intercept b are known. Since in this event we know both the slope m of the line and a point $(0, b)$ on the line, we may use the point–slope form (3) to obtain the following equation:

$$y - b = m(x - 0) \qquad \text{or} \qquad y = mx + b$$

Slope–Intercept Form. **An equation of a nonvertical line L with slope m and y-intercept b is**

$$y = mx + b \tag{4}$$

When an equation of a line is given in the slope–intercept form, the slope and the y-intercept are read directly from the equation. The slope is the coefficient of x and the y-intercept is the constant:

$$y = \underset{\substack{\uparrow \\ \text{Slope}}}{m}x + \underset{\substack{\uparrow \\ y\text{-intercept}}}{b}$$

EXAMPLE 3

Table 5 shows how to find the slope and y-intercept of a line when the equation is written in slope–intercept form.

Table 5

Equation	Slope	y-intercept
$y = 2x + 5$	$m = 2$	$b = 5$
$y = -x + \frac{3}{2}$	$m = -1$	$b = \frac{3}{2}$
$y = 0.1x - 2$	$m = 0.1$	$b = -2$
$y = x$	$m = 1$	$b = 0$
$y = 5$	$m = 0$	$b = 5$

Sometimes we write the equation of a line L in *general form*—namely,

$$Ax + By + C = 0$$

where A, B, and C are three real numbers with either $A \neq 0$ or $B \neq 0$. This is referred to as the *general form* because every line has an equation that can be written this way.

EXAMPLE 4

Find the slope m and y-intercept b of the line L given by $2x + 4y - 8 = 0$. Graph the line.

Solution

To obtain the slope and y-intercept, we transform the equation to its slope–intercept form. Thus, we need to solve for y:

$$2x + 4y - 8 = 0$$

$$4y = -2x + 8$$

$$y = \left(\frac{-1}{2}\right)x + 2$$

The coefficient of x, $-\frac{1}{2}$, is the slope, and the y-intercept is 2. To graph this line, we need two points. Normally, the easiest points to locate are the intercepts. Since the y-intercept is 2, we know one point is $(0, 2)$. To obtain the x-intercept, we set $y = 0$ and solve for x:

$$2x - 8 = 0$$

$$x = 4$$

Thus, the intercepts are $(4, 0)$ and $(0, 2)$, as shown in Figure 32.

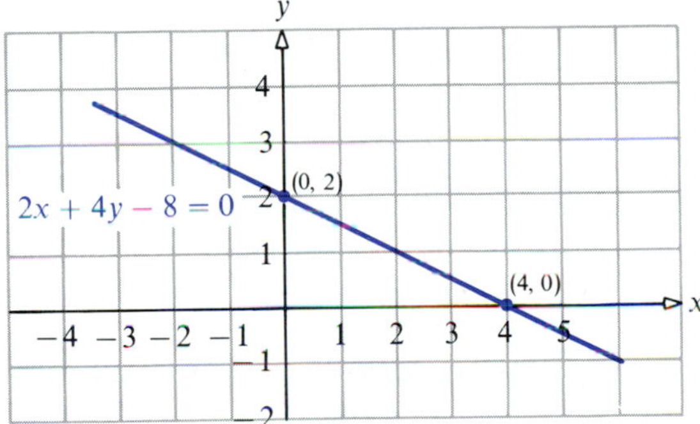

Figure 32

Intersecting Lines; Parallel and Perpendicular Lines

Let L_1 and L_2 be two (distinct) lines. If L_1 and L_2 have exactly one point P in common, then L_1 and L_2 are said to *intersect* and the common point P is called the *point of intersection.*

EXAMPLE 5

Find the point of intersection of the lines

$$L_1:\quad x + y - 5 = 0 \qquad \text{and} \qquad L_2:\quad 2x + y - 6 = 0$$

Solution

Let the coordinates of the point P of intersection of L_1 and L_2 be (x_0, y_0). Since (x_0, y_0) is on both L_1 and L_2, then

$$x_0 + y_0 - 5 = 0 \qquad\qquad 2x_0 + y_0 - 6 = 0$$

so that

$$y_0 = 5 - x_0 \qquad\qquad y_0 = 6 - 2x_0$$

Setting these equal, we obtain

$$5 - x_0 = 6 - 2x_0$$

$$x_0 = 1$$

If $x_0 = 1$, then $y_0 = 5 - 1 = 4$. Thus, the point P of intersection of L_1 and L_2 is $(1, 4)$ (see Fig. 33).

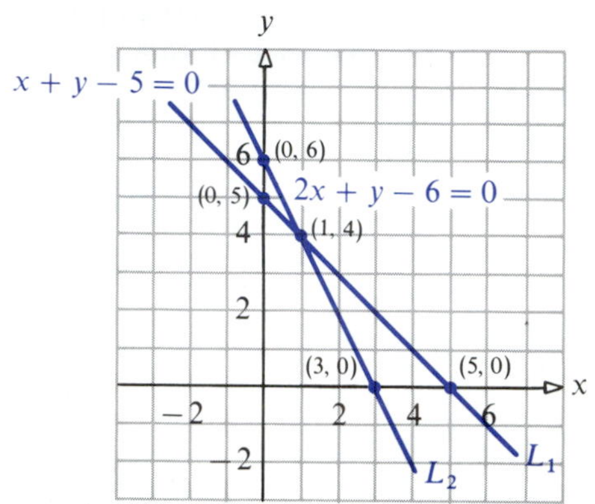

Figure 33

The technique given in Example 5 for finding the point P of intersection of two lines is sometimes called the *substitution technique*, since in reality the value for y in one equation is substituted for the value of y in the other equation.

If two distinct lines (in a plane) do not intersect, they are said to be *parallel*.

[1.3.2] THEOREM / *Parallel Lines.*

Two distinct nonvertical lines are parallel if and only if their slopes are equal.

Proof Let $y = mx + b$ and $y = mx + b'$ $(b \neq b')$ be the equations of two lines with the same slope. These equations have no common point; hence, the lines do not intersect. Thus, two different lines with the same slope are parallel. On the other hand, if two lines have equations $y = m_1x + b_1$ and $y = m_2x + b_2$, and $m_1 \neq m_2$ (here b_1 may equal b_2), then it can be shown that the equations have the common point whose coordinates are

$$x = \frac{b_2 - b_1}{m_1 - m_2} \quad \text{and} \quad y = \frac{m_1b_2 - m_2b_1}{m_1 - m_2}$$

Hence, the lines intersect at the point $\left(\dfrac{b_2 - b_1}{m_1 - m_2}, \dfrac{m_1b_2 - m_2b_1}{m_1 - m_2}\right)$ and they are not parallel.

For example, the two lines

$$2x + 3y - 6 = 0 \quad \text{and} \quad 4x + 6y = 0$$

are parallel, since each has slope $-\frac{2}{3}$ (see Fig. 34).

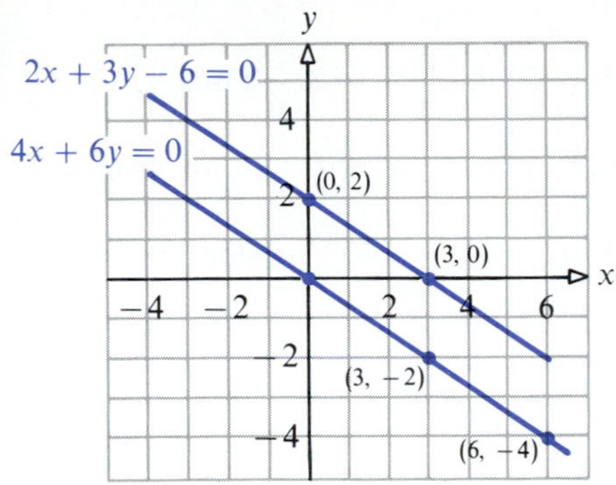

Figure 34

Slopes can also be used to determine whether two lines are *perpendicular*—that is, meet at a right angle.

[1.3.3] THEOREM / *Perpendicular Lines.*

Two nonvertical lines are perpendicular if and only if the product of their slopes is −1. Thus, two lines with slopes m_1 and m_2 are perpendicular if and only if

$$m_1 = -\frac{1}{m_2}$$

The proof can readily be obtained using trigonometry; therefore, we postpone it until Section 1.8.

EXAMPLE 6

Show that the line L_1 through (0, 0) and (2, 4) is perpendicular to the line L_2 joining (3, 1) and (9, −2).

Solution

The slopes of these two lines are

$$L_1: \quad m_1 = \frac{4-0}{2-0} = 2 \qquad L_2: \quad m_2 = \frac{-2-1}{9-3} = \frac{-3}{6} = \frac{-1}{2}$$

Since $m_1 m_2 = 2(-\frac{1}{2}) = -1$, the lines are perpendicular.

EXAMPLE 7

The characteristics of four lines are given below. Find the slope–intercept form of the equation of each line.

(a) Slope = −5 passing through (0, 3)
(b) Slope = 1 passing through the origin
(c) Passing through (2, 5) and parallel to $y = 2x - 5$
(d) Passing through (2, 5) and perpendicular to $y = -\frac{1}{3}x + 2$

Solution

(a) From the given condition, $m = -5$ and $b = 3$. Thus, from (4), $y = -5x + 3$.

(b) From the given condition, $m = 1$ and $b = 0$. Thus, from (4), $y = x + 0$ or $y = x$.

(c) Since the desired line is parallel to $y = 2x - 5$, its slope is $m = 2$. Using (3) and simplifying, we get

$$y - 5 = 2(x - 2)$$

$$y = 2x + 1$$

(d) Since the desired line is perpendicular to $y = -\frac{1}{3}x + 2$, its slope is $m = 3$. Using (3) and simplifying, we get

$$y - 5 = 3(x - 2)$$

$$y = 3x - 1$$

EXAMPLE 8

Use slopes to show that the points $A = (3, -1)$, $B = (10, 4)$, and $C = (5, 11)$ are the vertices of a right triangle.

Solution

The line through A and B has slope

$$m_1 = \frac{4 - (-1)}{10 - 3} = \frac{5}{7}$$

and the line through B and C has slope

$$m_2 = \frac{11 - 4}{5 - 10} = -\frac{7}{5}$$

Since $m_1 m_2 = -1$, the line through A and B is perpendicular to the line through B and C. Thus, ABC is a right triangle.

EXERCISE 1.3

In Problems 1–12 find the general equation of the line having the given properties.

1. Slope = 2; passing through $(-2, 3)$
2. Slope = 3; passing through $(4, -3)$
3. Slope = $-\frac{2}{3}$; passing through $(1, -1)$
4. Slope = $\frac{1}{2}$; passing through $(3, 1)$
5. Passing through $(1, 3)$ and $(-1, 2)$
6. Passing through $(-3, 4)$ and $(2, 5)$
7. Slope = -3; y-intercept = 3
8. Slope = -2; y-intercept = -2
9. x-intercept = 2; y-intercept = -1
10. x-intercept = -4; y-intercept = 4
11. Slope undefined; passing through $(1, 4)$
12. Slope undefined; passing through $(2, 1)$

In Problems 13–18 find the slope and y-intercept of the given line if they exist.

13. $3x - 2y = 6$

14. $4x + y = 2$

15. $x + 2y = 4$

16. $-x - y = 4$

17. $x = 4$

18. $y = 3$

19. Find the general equation of the line passing through $(1, 2)$ and parallel to $2x - y = 6$.

20. Find the general equation of the line passing through $(-1, 3)$ and parallel to $x + y = 4$.

In Problems 21–26 find the slope of a line perpendicular to the given line.

21. $3x - 2y = 6$

22. $3x + y = 4$

23. $x + 2y = -4$

24. $x - y = 1$

25. $x = 4$

26. $y = 2$

In Problems 27–32 determine whether the lines are parallel or intersecting. If they intersect, find the point of intersection.

27. L_1: $2x - 3y + 6 = 0$
L_2: $4x - 6y + 7 = 0$

28. L_1: $4x - y + 2 = 0$
L_2: $3x + 2y = 0$

29. L_1: $-x + 3y + 6 = 0$
L_2: $x - 6y - 12 = 0$

30. L_1: $2x + 3y - 5 = 0$
L_2: $5x - 6y + 1 = 0$

31. L_1: $3x - 3y + 10 = 0$
L_2: $x + y - 2 = 0$

32. L_1: $2x - 5y - 1 = 0$
L_2: $x - 2y - 1 = 0$

In Problems 33 and 34 use slopes to determine whether the given points lie on the same line.

33. $(2, 0)$, $(7, -4)$, $(22, -16)$

34. $(1, 5)$, $(9, 8)$, $(-15, 1)$

35. Use slopes to show that $(-3, -4)$, $(0, 2)$, and $(6, -1)$ are vertices of a right triangle.

36. Use slopes to show that $(4, -6)$, $(8, -5)$, $(5, 2)$, and $(1, 1)$ are vertices of a parallelogram.

37. If $(3, h)$ is a point on the line containing the points $(2, 4)$ and $(6, 5)$, find h.

38. If $(k, 4)$ is a point on the line with slope $m = 3$ and passing through $(2, 5)$, determine k.

39. Find h if the line through $(3, h)$ and $(4, -5)$ is parallel to the line through $(8, 2)$ and $(-2, 6)$.

40. Find k if the line through $(k, 4)$ and $(-2, -9)$ is perpendicular to the line through $(4, 1)$ and $(0, 3)$.

41. Show that the lines $x + 2y + 1 = 0$, $6x - 3y = 5$, $y = 2x - 1$, and $4x + 8y + 7 = 0$ form a rectangle.

42. Show that the lines $2x - 3y + 2 = 0$, $4x - 2y = 3$, $4x - 6y - 1 = 0$, and $y - 2x - 2 = 0$ form a parallelogram.

43. Show in two ways that the triangle with vertices at $A = (1, -6)$, $B = (8, 8)$, and $C = (-7, -2)$ is a right triangle.

44. Determine whether the points $(1, 8)$, $(2, 16)$, and $(-1, 2)$ are *collinear* (that is, lie on the same line) by:

(a) Calculating slopes
(b) Using the distance formula

45. A point (a, b) lies at a distance 4 units from the line $15x + 8y - 34 = 0$. Find an equation involving a and b (two solutions).

46. Find the coordinates of the point on the y-axis that is equidistant from the points $(-3, 5)$ and $(2, 4)$.

47. Express by an equation the fact that the point (x, y) is always at a distance 4 units from the point $(1, -2)$.

48. The point $(2, -5)$ is at a distance $\sqrt{65}$ from the midpoint of the segment joining $(4, 2)$ and $(x, 4)$. Find x.

49. If the line through $(x, 4)$ and $(-2, 1)$ is perpendicular to the line through $(2, 3)$ and $(-1, y)$, find an equation relating x and y.

50. Determine the coordinates of the point on the line $3x + 2y = 0$ that is equidistant from $(0, 0)$ and $(-2, 3)$.

51. Find an equation for the tangent line to the circle $x^2 + y^2 + 2x - 6y - 3 = 0$ at the point $(2, 1)$.

52. Find an equation of each of the two lines having slope $-\frac{2}{3}$ that are tangent to the circle $x^2 + y^2 + 2x - 4y - 5 = 0$.

53. Find an equation for the tangent line to the circle with center at $(2, -1)$ at the point $(3, 2)$.

54. Find an equation for the tangent line to the circle with center at (4, 5) at the point (6, −3).

55. Mr. Nicholson has just retired and needs $12,000 per year in income to live on. He has $100,000 to invest and can invest in AAA bonds at 14% interest annually or in savings and loan certificates at 10% interest a year. How much money should be invested in each so that he realizes exactly $12,000 in income per year?

56. One solution is 15% acid and another is 5% acid. How many cubic centimeters of each should be mixed to obtain 100 cubic centimeters of an 8% solution?

57. The relationship between the Celsius (°C) and Fahrenheit (F°) temperature scales is a straight line. Find the equation relating °C and °F if 0°C corresponds to 32°F and 100°C corresponds to 212°F. Use the equation to find the Celsius measure of 70°F.

58. The annual sales of Motors, Inc., for the past 5 years are given in the table.

Years	Units Sold (Thousands)
1985	2,200
1986	2,800
1987	3,100
1988	3,200
1989	3,400

(a) Graph this information using the x-axis for years and the y-axis for units sold. (For convenience, use a different scale on each axis.)
(b) Draw a line L that passes through two of the points and comes close to passing through the remaining points.
(c) Find the equation of the line L.
(d) Using the equation of the line, what is your estimate for units sold in 1990?

1.4 Functions and Their Graphs

In many applications, a correspondence exists between two sets of numbers. For example, the volume V of a sphere of radius R is given by the formula (correspondence) $V = \frac{4}{3}\pi R^3$.

As another example, suppose a man standing on the moon throws a rock 20 meters (almost 22 yards) up and starts a stopwatch just as the rock begins to fall back down. Let x represent the number of seconds shown on the stopwatch, and let y represent the height (in meters) of the rock above the surface of the moon. Then there is a correspondence between the time and the height—that is, between the numbers x and the numbers y. When the time is 0, the rock is at its highest point of 20 meters; therefore, $x = 0$ corresponds to $y = 20$. But to what heights do the numbers $x = 1$, $x = 2.5$, and $x = 5$ correspond? To find approximate answers to these questions without actually sending someone to the moon, we may use the following formula:

$$y = 20 - 0.8x^2$$

The height corresponding to $x = 1$ is found when we replace x in the formula by the number 1, as follows:

$$y = 20 - 0.8(1)^2 = 19.2$$

Thus, when the stopwatch shows 1 second, the rock is still 19.2 meters above the surface of the moon. Similarly, when $x = 2.5$, the height is

$$y = 20 - 0.8(2.5)^2 = 15$$

When $x = 5$, the height is

$$y = 20 - 0.8(5)^2 = 0$$

and the rock has again reached the surface of the moon. (If you think that the rock falls to the moon more slowly than it would fall to the earth, you are right. See Problem 61.)

An important point made by this example is that if X is the set of times from 0 to 5 seconds and Y is the set of heights from 0 to 20 meters, then each element of X corresponds to one and only one element of Y. The correspondence $y = 20 - 0.8x^2$ is called a *function from X into Y*.

[1.4.1] DEFINITION / *Function; Domain; Range.*

Let X and Y be two sets of numbers. A *function from X into Y* is a correspondence that associates with each element of X a unique element of Y. The set X is called the *domain* of the function. For each element x in X, the corresponding element y in Y is called the *value* of the function at x, or the *image* of x. The set of all images of the elements of the domain is called the *range* of the function.

Since there may be elements in Y that are the image of no x in X, it follows that the range of a function is a subset of Y.

Functions are often denoted by letters such as f, F, g, G, and so on. If f is a function from X into Y, then for each number x in X the corresponding image in the set Y is designated by the symbol $f(x)$, read "f of x." We refer to $f(x)$ as the *value of f at the number x*. For example, in the case of the falling rock, we may designate the function by the letter H (to remind us of the word *height*). Then for each x in X, $H(x)$ designates the value of H at x; that is, $H(x)$ designates the height of the rock at time x. In symbols, we write

$$H(x) = 20 - 0.8x^2$$

How do we designate the value of H at the times $x = 1$, $x = \frac{5}{4}$, $x = \sqrt{2}$? These are the heights $H(1)$, $H(\frac{5}{4})$, $H(\sqrt{2})$, and they may be computed by using the formula as follows:

$$H(1) = 20 - 0.8(1)^2 = 19.2$$

$$H(\tfrac{5}{4}) = 20 - 0.8(\tfrac{5}{4})^2 = 18.75$$

$$H(\sqrt{2}) = 20 - 0.8(\sqrt{2})^2 = 18.4$$

The expression $H(1) = 19.2$ is read "the value of H at 1 is 19.2" or "1 second corresponds to 19.2 meters." Other ways to write this fact are by using *arrow notation*,

$$1 \rightarrow 19.2$$

or *ordered-pair notation*,

$$(1, 19.2)$$

Each of these can be read "1 corresponds to 19.2."

It is convenient to use ordered-pair notation to show the difference between correspondences that *are* functions and a correspondence that is *not* a function. For example, let $X = \{1, 2\}$ and $Y = \{4, 5\}$. The correspondence given by the pairs

$$(1, 5) \qquad \text{and} \qquad (2, 4)$$

is a function from X into Y because each element of X has a unique value in Y. Similarly, the correspondence given by

$$(1, 5) \qquad \text{and} \qquad (2, 5)$$

is a function. (Notice that 4 is *not* in the range of this function because the value of the function at 1 and 2 is the same—namely, 5. Nevertheless, the condition that each element of X has one and only one corresponding value is still satisfied.) The correspondence given by the pairs

$$(1, 4) \qquad (1, 5) \qquad (2, 5)$$

is *not* a function because the element 1 in X *corresponds* to *more than one value* in Y.

By using ordered-pair notation, we can also consider a function as a set of *ordered pairs* (x, y) in which no different pairs have the same first element. The set of all first elements is the *domain* and the set of all second elements is the *range* of the function. Thus, there is associated with each element x in the domain a unique element y in the range. An example is the set of all ordered pairs (x, y) such that $y = x^2$. Some of the pairs in this set are

$$(2, 2^2) = (2, 4) \quad (0, 0^2) = (0, 0) \quad (-2, (-2)^2) = (-2, 4) \quad (\tfrac{1}{2}, (\tfrac{1}{2})^2) = (\tfrac{1}{2}, \tfrac{1}{4})$$

In this set no two different pairs have the same *first* element (even though there are different pairs that have the same *second* element). This set is the squaring function, which associates with each real number x the value x^2.

The ordered pairs (x, y) for which $y^2 = x$ do not represent a function because there are ordered pairs with the same first number but different second numbers. For example, $(1, 1)$ and $(1, -1)$ are ordered pairs satisfying the relationship $y^2 = x$ with the same first number, but different second numbers.

The element x that appears in the first position of the ordered pair (x, y) is often called the *independent variable*, since it can be assigned any of the permissible numbers from the domain; the second member of the pair is called the *dependent* variable, since the value of y depends on the number x.

Another advantage of expressing a function (or any correspondence) as a set of ordered pairs is that we can then graph the set of pairs to make a "picture" of the function. For example, the graph of $y = x^2$ is shown in Figure 21(a) on page 16.

To summarize, we have determined that a function f associates with real numbers x other real numbers y, and we now agree to use the notation

$$y = f(x)$$

to denote the rule that associates x and y. The set of all ordered pairs (x, y), where $y = f(x)$ is the ordinate and x is the abscissa, is called the *graph* of the function f.

Vertical Line Test

Regardless of whether a function is described by a formula, by some rule, or by other means, it will always have a graph. However, not every collection of points is the graph of a function. In fact, a graph provides a visual technique for determining whether a collection of ordered pairs is a function. ***If any***

vertical line intersects the graph in more than one point, the graph is not that of a function.

The reason for this is simple: A vertical line has the equation $x = a$, and the graph of a function f has at most one point $(a, f(a))$ whose x-coordinate is a. Compare Figures 35(a) and 35(b). The graph in Figure 35(a) is the graph of a function; the one in Figure 35(b) is not the graph of a function, since some vertical line $(x = 2)$ intersects the graph twice.

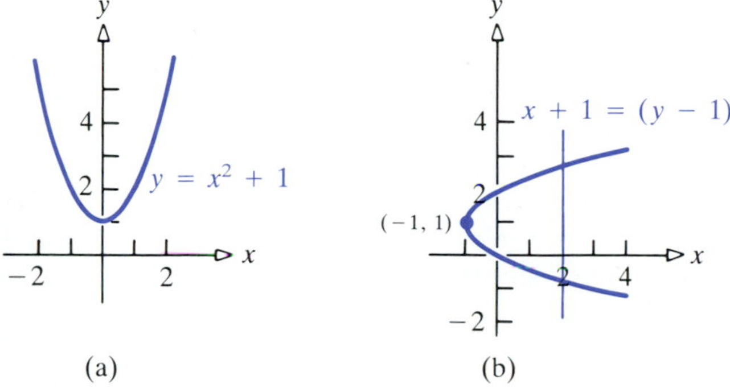

Figure 35

Up to this point we have discussed functions that could be described by a *single* formula or rule. For example, during the time $0 \le x \le 5$, the height of the falling rock is given by $H(x) = 20 - 0.8x^2$. But what is the height after the time $x = 5$ when the rock strikes the ground? Assuming that the rock does not bounce, we have a second rule that the height is 0 if $x > 5$. Therefore, to specify the height for *all* $x \ge 0$, we need a new function, say K, that incorporates *both* rules. The function K may be defined as follows:

$$K(x) = \begin{cases} 20 - 0.8x^2 & \text{if } 0 \le x \le 5 \\ 0 & \text{if } x > 5 \end{cases}$$

What is the difference between the functions H and K? The answer lies in the fact that they have different domains. The domain of H contains only those times from 0 to 5, whereas the domain of K contains all nonnegative times.

Here is another example of a function that is given by more than one rule.

EXAMPLE 1

Graph the function f given by the three rules:

$$f(x) = \begin{cases} \frac{x}{2} & \text{if } -1 \le x < 1 \\ 2 & \text{if } x = 1 \\ x + \frac{1}{2} & \text{if } x > 1 \end{cases}$$

Solution

Here, the domain of f is all real numbers $x \ge -1$. Its graph is given in Figure 36. We use a filled circle ● to indicate that at $x = 1$, the value of f is $f(1) = 2$; we use an open circle ○ to illustrate that the function does not assume either of the values $\frac{1}{2}$ or $\frac{3}{2}$ at $x = 1$.

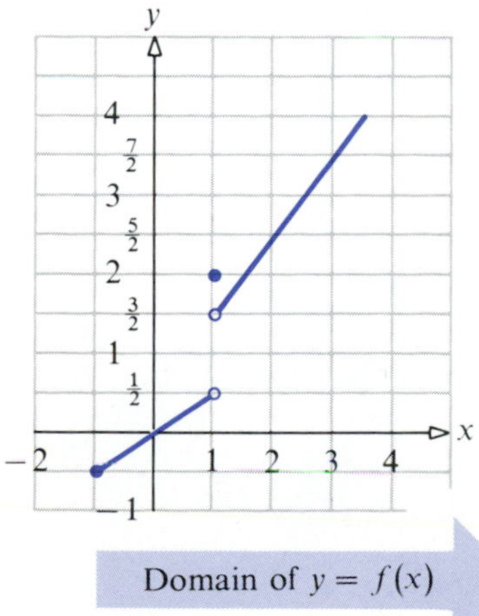

Figure 36

Domain of a Function

In this book we will sometimes give directions such as "Graph the function $f(x) = x^2$." We actually mean "Graph the equation $y = x^2$, in which the numbers x are restricted to the numbers in the domain of f." But how can we graph a function if its domain is not specified? The answer is simply this: When the domain of a function is *not* specified but a rule of association is known, we automatically assume that the domain is the largest set of real numbers for which the rule *makes sense* (or more precisely, for which we can compute $f(x)$ as a unique real number). For example, the operation of squaring can be performed on *any* real number x. Therefore, to associate x with x^2 makes sense for *every* real number x, and thus the domain of $y = f(x) = x^2$ is the set $\mathbb{R}$ of *all* real numbers.

What is the domain of $f(x) = 1/x$? We can divide any nonzero real number into 1. Hence, it makes sense to associate x with $1/x$ as long as $x \neq 0$. The domain of $y = f(x) = 1/x$ is therefore $\{x \mid x \in \mathbb{R}$ and $x \neq 0\}$—that is, all real numbers x except $x = 0$.

EXAMPLE 2

Find the domain and range, and graph the *square root function*:

$$y = f(x) = \sqrt{x}$$

Solution

To find the domain D of f, we ask the question: What are the numbers x for which we can compute $\sqrt{x}$? Now, we know it is impossible (in the universe of real numbers) to find the square root of a negative number. Thus, we can compute $\sqrt{x}$ only if $x \geq 0$, and the domain is $\{x \mid x \geq 0\}$. Now we look for the range. To each number x in the domain, there is associated exactly one nonnegative number y. (This number y is nonnegative because of the definition of square root.) In fact, as x runs through the domain, $\sqrt{x}$ runs through *all* nonnegative numbers. Hence, the range is the set of nonnegative real numbers. The graph is shown in Figure 37.

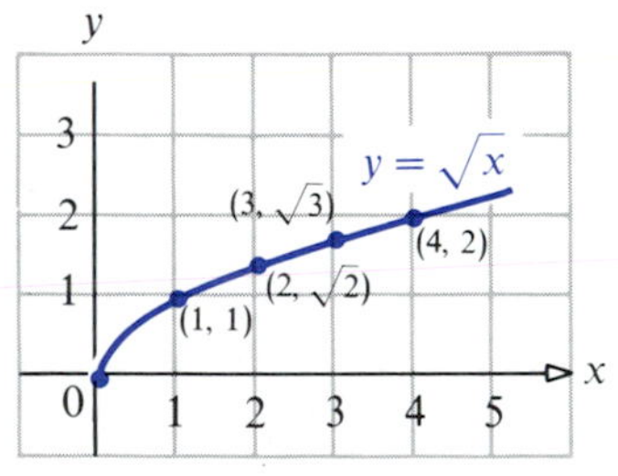

Figure 37

EXAMPLE 3

Find the domain and graph the function: $f(x) = \sqrt{x^2 + 2x - 3}$

Solution

For $\sqrt{x^2 + 2x - 3}$ to be a real number, the domain of f has to consist of all numbers x for which $x^2 + 2x - 3 = (x + 3)(x - 1)$ is 0 or positive. The expression is 0 for $x = -3$ and $x = 1$. What remains is to solve the inequality $(x + 3)(x - 1) > 0$, which we do by setting up a table (Table 6) and using -3 and 1 as cutoff points, since $(x + 3)$ and $(x - 1)$ are the factors of $x^2 + 2x - 3$ (see Fig. 38). The preceding discussion shows that the domain of f consists of all real numbers in the intervals $(-\infty, -3]$ and $[1, +\infty)$. Figure 39 is the graph.

Table 6

	Sign of $x + 3$	Sign of $x - 1$	Sign of $x^2 + 2x - 3$	Conclusion
$x < -3$	$-$	$-$	$+$	$x < -3$ is a solution
$-3 < x < 1$	$+$	$-$	$-$	$-3 < x < 1$ is not a solution
$x > 1$	$+$	$+$	$+$	$x > 1$ is a solution

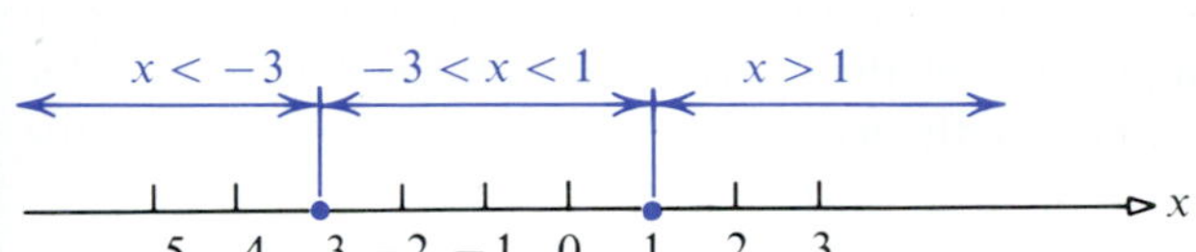

Figure 38

Figure 39

Observe the similarity between the graph of $y = \sqrt{x^2 + 2x - 3}$ in Figure 39 and that of the square root function in Figure 37. Note also the symmetry about the line $x = -1$ in Figure 39.

EXAMPLE 4

Graph the function:

$$f(x) = \begin{cases} x & \text{if } x \geq 0 \\ -x & \text{if } x < 0 \end{cases} \quad \textbf{(1)}$$

Solution

If $x \geq 0$, then f is represented by the line $y = x$ (slope 1); if $x < 0$, then f is represented by the line $y = -x$ (slope -1). The graph of f is given in Figure 40. This function is called the *absolute value function* and is written as

$$f(x) = |x|$$

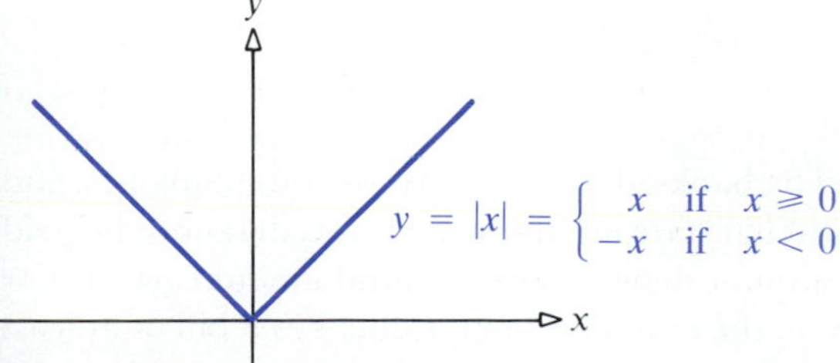

Figure 40

The next function, which has the graph shown in Figure 41, occurs frequently enough in mathematics and in applications that it merits a special name, the *greatest integer function.*

[1.4.2] DEFINITION / *Greatest Integer Function.*

The *greatest integer function*, denoted by $[\![x]\!]$ and read "bracket x," is defined as the greatest integer less than or equal to x.

For example,

$$[\![5]\!] = 5 \qquad [\![4.9]\!] = 4 \qquad [\![-2.1]\!] = -3$$

The domain of this function is the set of real numbers, while its range is the set of integers (see Fig. 41). From the graph of the greatest integer function, we can see why it is sometimes referred to as a *step function.* The greatest integer function exhibits what we shall refer to as *discontinuities* at $x = 0, \pm 1, \pm 2$, and so on—that is, points where the function suddenly jumps from one value to another without taking on any of the intermediate values. This occurs, for example, at $x = 3$, where to the left of 3, the y values are at 2 and to the right of 3, the y values are at 3.

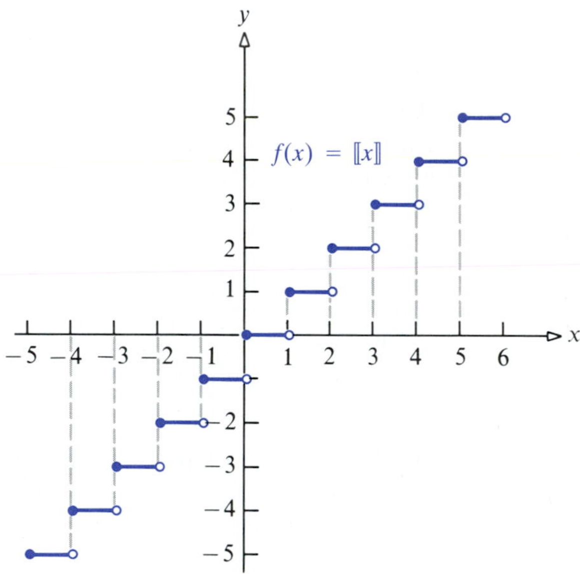

Figure 41

EXAMPLE 5

Holders of credit cards issued by banks, department stores, oil companies, and so on, receive bills each month that state minimum amounts that must be paid by their due dates. The minimum depends on the total amount owed. For instance, for a bill of up to \$10, the entire amount is due. For a bill of at least \$10 but less than \$500, the minimum is \$10. There is a minimum of \$15 on a bill of at least \$500 but less than \$1000; a minimum of \$20 for \$1000 up to \$1500; and a minimum of \$25 on bills of \$1500 or more. The function f that describes the minimum payment on a bill of \$$x$ is

$$f(x) = \begin{cases} x & \text{if } 0 \le x < 10 \\ 10 & \text{if } 10 \le x < 500 \\ 15 & \text{if } 500 \le x < 1000 \\ 20 & \text{if } 1000 \le x < 1500 \\ 25 & \text{if } 1500 \le x \end{cases}$$

The graph is given in Figure 42.

The card holder may pay any amount between the minimum and the total owed. The organization issuing the card charges the card holder interest of $1\frac{1}{2}\%$ a month for the first \$1000 owed and 1% a month on any unpaid balance above \$1000. Thus, if $g(x)$ is the amount of interest charged for a month on a balance of x, then $g(x) = 0.015x$ for $0 \le x \le 1000$. The amount of the unpaid balance above \$1000 is $x - 1000$. If the balance due is $x > 1000$, then the interest is $0.015(1000) + 0.01(x - 1000) = 5 + 0.01x$, so

$$g(x) = \begin{cases} 0.015x & \text{if } 0 \le x \le 1000 \\ 5 + 0.01x & \text{if } x > 1000 \end{cases}$$

See Figure 43.

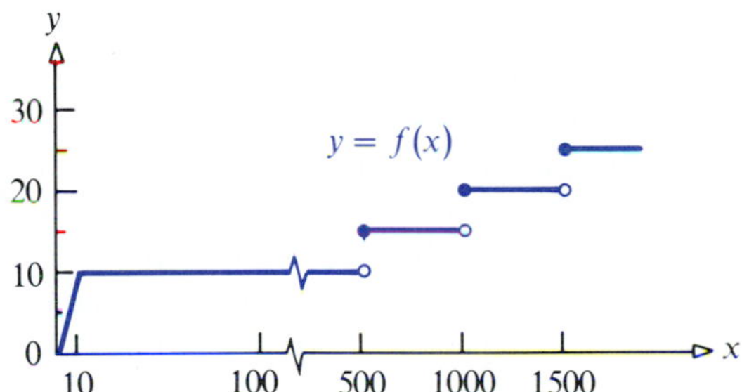

Figure 42

Figure 43

Function Notation

One important use of function notation is illustrated in the next two examples.

EXAMPLE 6

For the function $f(x) = 3x + 1$, find:

(a) $f(x + \Delta x)$ (b) $f(x + \Delta x) - f(x)$ (c) $\dfrac{f(x + \Delta x) - f(x)}{\Delta x}$ $\Delta x \ne 0$

Solution

(a) The function $f(x) = 3x + 1$ tells us to multiply x by 3 and then add 1. To find $f(x + \Delta x)$, we multiply $(x + \Delta x)$ by 3 and then add 1. Thus,

$$f(x + \Delta x) = 3(x + \Delta x) + 1 = 3x + 3\Delta x + 1$$

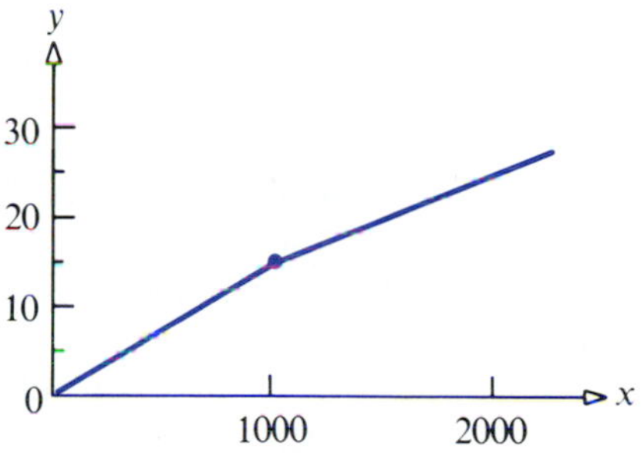

Notice that x has been replaced by the quantity $(x + \Delta x)$.

(b) $f(x+\Delta x) - f(x) = (3x + 3\Delta x + 1) - (3x + 1) = 3\Delta x$

(c) $\dfrac{f(x+\Delta x) - f(x)}{\Delta x} = \dfrac{3\Delta x}{\Delta x} = 3 \qquad \Delta x \neq 0$

EXAMPLE 7

For the function $f(x) = 1/x$, find:

(a) $f(x+\Delta x)$ (b) $f(x+\Delta x) - f(x)$ (c) $\dfrac{f(x+\Delta x) - f(x)}{\Delta x}$, $\Delta x \neq 0$

Solution

(a) The function $f(x) = 1/x$ tells us to find the reciprocal of x. Thus, for $f(x+\Delta x)$ we should find the reciprocal of $(x+\Delta x)$. That is,

$$f(x+\Delta x) = \frac{1}{x+\Delta x}$$

(b) $f(x+\Delta x) - f(x) = \dfrac{1}{x+\Delta x} - \dfrac{1}{x} = \dfrac{x - (x+\Delta x)}{x(x+\Delta x)} = \dfrac{-\Delta x}{x(x+\Delta x)}$

(c) $\dfrac{f(x+\Delta x) - f(x)}{\Delta x} = \dfrac{-\Delta x}{x(x+\Delta x)}\left(\dfrac{1}{\Delta x}\right) = \dfrac{-1}{x(x+\Delta x)}$

Many formulas that occur in mathematics and the sciences determine functions. For example, $A = \pi R^2$ is a formula that gives the area A of a circle in terms of its radius R. Similarly, if we know both the height h and the radius R of a right circular cylinder (such as a soup can), then we can find its volume V by the formula $V = \pi R^2 h$. In the formula $A = \pi R^2$, π is a constant (approximately 3.14159) and A is the dependent variable. Since its value depends only on the single independent variable R, A is called a *function of one variable*. In the formula $V = \pi R^2 h$, V is the dependent variable. Its value depends on the *two* independent variables R and h; therefore, V is a *function of two variables*.

We shall discuss functions of one variable in Chapters 2–13. In later chapters, we deal with functions of two or more variables.

EXERCISE 1.4

1. For the function $f(x) = 3x - 2$, find:

(a) $f(3)$ (b) $f(-2)$ (c) $f(0)$

(d) $f(x+2)$ (e) $f(x+\Delta x)$ (f) $f\left(\dfrac{1}{x}\right)$

2. For the function $f(x) = 3x^2 + 1$, find:

(a) $f(1)$ (b) $f(-2)$ (c) $f(0)$

(d) $f(x+4)$ (e) $f(x+\Delta x)$ (f) $f\left(\dfrac{1}{x}\right)$

3. For the function $f(x) = \begin{cases} 3x & \text{if } x \leq 2 \\ x^2 & \text{if } x > 2 \end{cases}$ find:

(a) $f(-5)$ (b) $f(1.9)$ (c) $f(2.1)$

(d) $f(5)$ (e) $f(2+\Delta x)$ $\Delta x > 0$

4. For the function $f(x) = \begin{cases} x & \text{if } x \leq 0 \\ 1 - x & \text{if } 0 < x < 1 \\ x & \text{if } x \geq 1 \end{cases}$

find:

(a) $f(0.1)$ (b) $f(0)$ (c) $f(\frac{1}{2})$

(d) $f(1)$ (e) $f(2)$

In Problems 5–14 use the graph of the function f shown here.

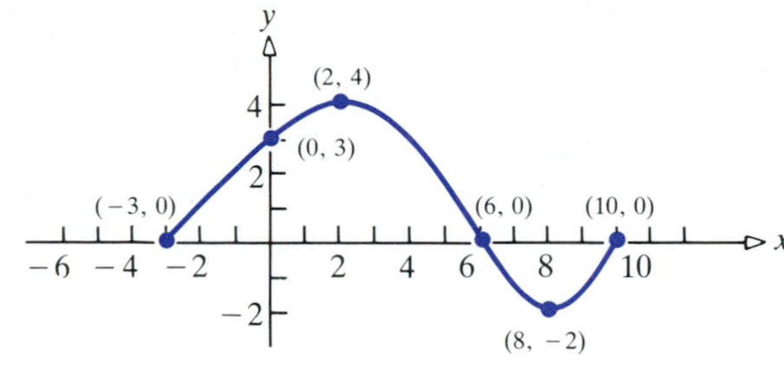

5. Find $f(0)$ and $f(2)$.

6. Find $f(8)$ and $f(-3)$.

7. Is $f(-2)$ positive or negative?

8. Is $f(7)$ positive or negative?

9. What is the y-intercept, if any, of f?

10. What are the x-intercepts, if any, of f?

11. Where is the graph above the x-axis?

12. Where is the graph below the x-axis?

13. What is the domain of f?

14. What is the range of f?

In Problems 15–28 determine whether the given correspondence determines a function $y = f(x)$.

15. $y = x^2 + 2x + 1$ **16.** $y = x^3 - 3x$ **17.** $y = \frac{2}{x}$ **18.** $y = \frac{3}{x} - 4$

19. $y^2 = 1 - x^2$ **20.** $y = \pm\sqrt{1 - 2x}$ **21.** $x^2 + y = 1$ **22.** $x + y^2 = 1$

23. $x^2y^2 = 5$ **24.** $x^2y = 4$ **25.** $y = |x - 2|$ **26.** $y = \sqrt{x^2}$

27. $\{(1, 5), (2, 5), (5, 1)\}$ **28.** $\{(2, 2), (2, 3), (3, 4), (4, 5)\}$

In Problems 29–38 find the domain of the function f.

29. $f(x) = 3x + 5$ **30.** $f(x) = x^2 + 1$ **31.** $f(x) = \sqrt{x - 1}$ **32.** $f(x) = \sqrt{2x + 5}$ **33.** $f(x) = \sqrt{x^2 + 4}$

34. $f(x) = \sqrt{x^2 - 4}$ **35.** $f(x) = \frac{2x}{x - 2}$ **36.** $f(x) = \frac{x^2}{x^2 - 4}$ **37.** $f(x) = \sqrt{\frac{3}{x}}$ **38.** $f(x) = \frac{3x^2}{x^4 + 1}$

For the functions in Problems 39–52 find the domain, and graph each function.

39. $f(x) = \begin{cases} 2x - 3 & \text{if } x < 0 \\ x - 3 & \text{if } 0 \le x < 5 \end{cases}$

40. $f(x) = \begin{cases} 1 & \text{if } x \ge 0 \\ -1 & \text{if } x < 0 \end{cases}$

41. $f(x) = \begin{cases} 4x + 5 & \text{if } -2 \le x < 0 \\ 4 & \text{if } x = 0 \\ 2x & \text{if } x > 0 \end{cases}$

42. $f(x) = \begin{cases} 4 - x & \text{if } x \le 2 \\ x - 2 & \text{if } 2 < x \end{cases}$

43. $f(x) = \begin{cases} x^2 & \text{if } x \le 0 \\ \sqrt{x + 1} & \text{if } x > 0 \end{cases}$

44. $f(x) = \begin{cases} x^2 + 2 & \text{if } x \le 0 \\ \sqrt{x + 4} & \text{if } x > 0 \end{cases}$

45. $f(x) = x - [\![x]\!]$ **46.** $f(x) = x + [\![x]\!]$ **47.** $f(x) = |x + 4|$ **48.** $f(x) = |x - 2|$

49. $f(x) = \frac{|x + 4|}{x + 4}$ **50.** $f(x) = \frac{|x - 5|}{x - 5}$ **51.** $f(x) = |x^2 - 1|$ **52.** $f(x) = x + |x - 1|$

For the functions in Problems 53–56 find:

(a) $f(x + \Delta x)$ (b) $f(x + \Delta x) - f(x)$ (c) $\frac{f(x + \Delta x) - f(x)}{\Delta x}$ $\Delta x \ne 0$

53. $f(x) = 2x + 5$ **54.** $f(x) = x^2 + 3$ **55.** $f(x) = x^2 + 3x + 4$ **56.** $f(x) = x + \frac{1}{x}$

57. For the function $f(x) = \sqrt{x}$, show that

$$\frac{f(x + \Delta x) - f(x)}{\Delta x} = \frac{1}{\sqrt{x + \Delta x} + \sqrt{x}} \qquad \Delta x \neq 0$$

58. For the function $f(x) = \sqrt{x + 3}$, show that

$$\frac{f(x + \Delta x) - f(x)}{\Delta x} = \frac{1}{\sqrt{x + 3 + \Delta x} + \sqrt{x + 3}} \qquad \Delta x \neq 0$$

59. If $f(x) = 2x^3 + Ax^2 + Bx - 5$ and if $f(2) = 3$ and $f(-2) = -37$, what is the value of $A + B$?

60. If a rock falls from a height of 20 meters on the planet Jupiter, its height after x seconds is approximately

$$H(x) = 20 - 13x^2$$

(a) What is the height of the rock when $x = 1$ second? $x = 1.1$ seconds? $x = 1.2$ seconds? $x = 1.3$ seconds?
(b) When does the rock strike the ground?
(c) Compare these results with the results obtained at the beginning of this section for the rock falling on the moon.
(d) Write a function that gives the height of the rock for all times $x \geq 0$.

61. If a rock falls from a height of 20 meters here on the earth, the height H after x seconds is approximately

$$H(x) = 20 - 4.9x^2$$

Use this function to answer parts (a)–(d) of Problem 60.

62. Express the perimeter P and area A of a semicircle as a function of the diameter x.

63. A rectangular field requires 3000 feet of fence to enclose it. If the length of the field is x feet, express the area A as a function of x. What is the domain of A?

64. A trucking company transports goods between Chicago and New York, a distance of 960 miles. The company's policy is to charge, for each pound, \$0.50 per mile for the first 100 miles, \$0.40 per mile for the next 300 miles, \$0.25 per mile for the next 400 miles, and no charge for the remaining 160 miles. Graph the relationship between the cost of transportation and mileage over the entire 960 mile route. Find the cost as a function of mileage for hauls between 100 and 400 miles from Chicago. Find the cost as a function of mileage for hauls between 400 and 800 miles from Chicago.

65. A page with dimensions 11 inches by 7 inches has a border of uniform width x surrounding the printed matter of the page. Write a formula for the area A of the printed part as a function of the width x of the border. Give the domain and range of A.

66. Sketch the graph of
$f(x) = \text{Minimum}(x - [\![x]\!], 1 - x + [\![x]\!])$.

67. A strip of nickel 200 centimeters long and 16 centimeters wide is to be made into a rain gutter by turning up all four edges to form a trough with a rectangular cross section. If the height of the bent-up edge is x centimeters, express the volume of the trough as a function of x.

68. A gardener wishes to fence a rectangular garden along a straight river. No fence is required along the river. The gardener has enough wire to build a fence 200 meters long. If the length of the side bordering on the river is represented by the variable x, express the area of the garden as a function of x.

69. If a Norman window (a rectangle surmounted by a semicircle) has a perimeter of 100 centimeters and if the length of the side that is not the diameter of the semicircle is represented by x, express the area of the window as a function of x.

70. A wire 10 meters long is cut in two parts: one part is bent into the shape of a square and the other into the shape of the circumference of a circle. If the perimeter of the square is represented by p, express the sum of the areas of both the square and the circle as a function of p.

71. The modulus of elasticity of structural steel is 29×10^6 pounds per square inch at a temperature of 70°F. At 900°F the modulus is 25×10^6 pounds per square inch. Between these temperatures, the change in modulus with temperature is known to be practically linear. Based on the above, give a quantitative formula for the modulus of elasticity of structural steel between the temperatures of 70°F and 900°F.

72. A trapezoid is inscribed in a circle of radius 4 centimeters with one base coinciding with a diameter of the circle. Express the area of the trapezoid as a function of its altitude.

73. For the function

$$f(x) = \begin{cases} \dfrac{|x|}{x} & \text{if } x \neq 0 \\ 1 & \text{if } x = 0 \end{cases}$$

find:

(a) $f(1)$ (b) $f(-1)$ (c) $f(3)$ (d) $f(-3)$
(e) $f(5)$ (f) $f(-5)$ (g) $f(0 + \Delta x)$, $\Delta x > 0$
(h) $f(0 + \Delta x)$, $\Delta x < 0$

In Problems 74–81 decide which graphs are graphs of functions.

74.

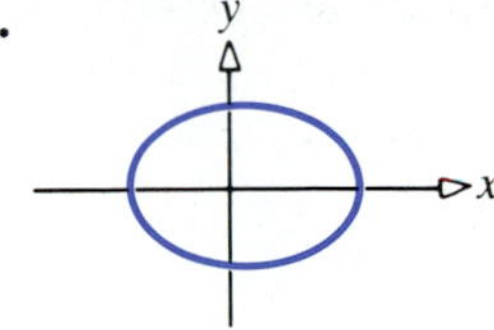

75.

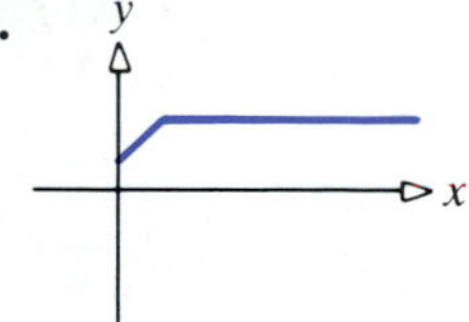

76.

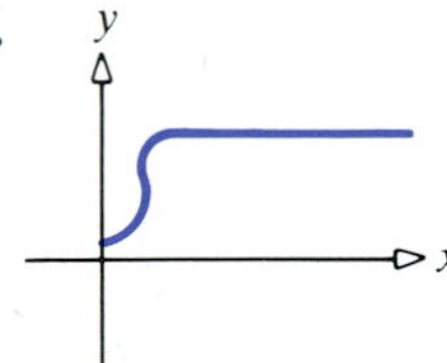

77.

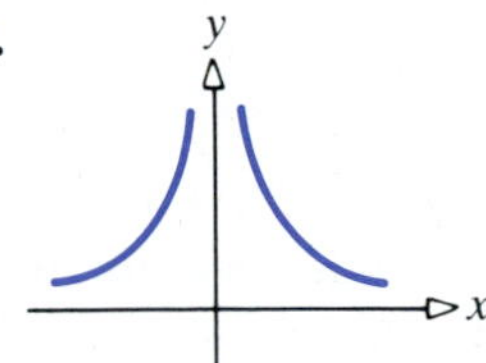

78.

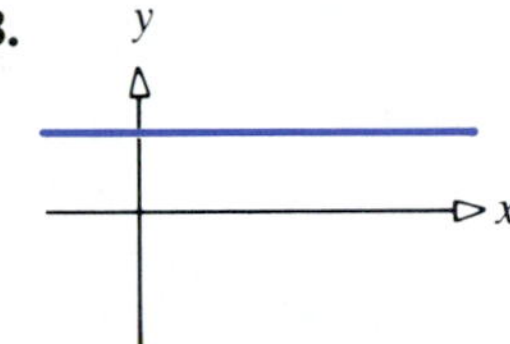

79.

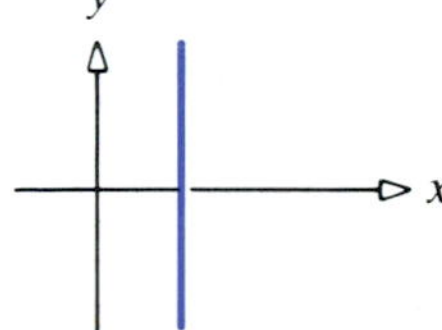

80.

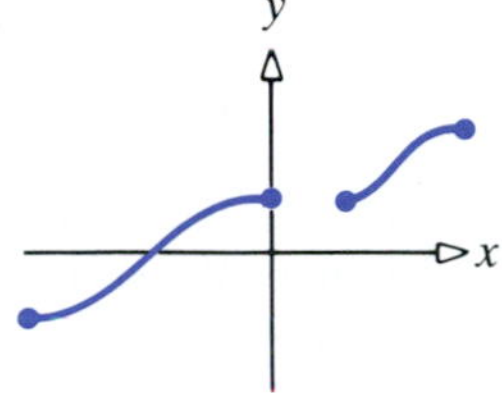

81.

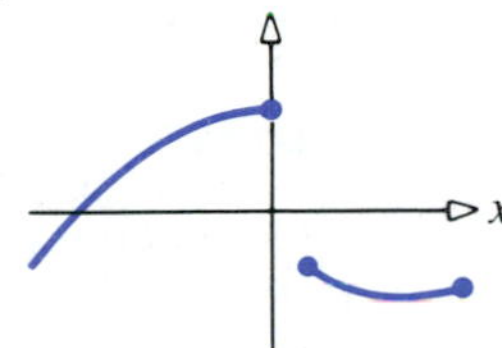

Operations on Functions; Types of Functions 1.5

Operations on Functions

In this section we introduce some operations on functions. We shall see that functions, like numbers, can be added, subtracted, multiplied, and divided.

[1.5.1] DEFINITION / *Operations on Functions.*

If f and g are functions, their *sum*, $f + g$; their *difference*, $f - g$; their *product*, $f \cdot g$; and their *quotient*, f/g, are defined by

$$(f + g)(x) = f(x) + g(x) \qquad (f - g)(x) = f(x) - g(x)$$

$$(f \cdot g)(x) = f(x) \cdot g(x) \qquad (f/g)(x) = f(x)/g(x)$$

In each case, the *domain* of the resulting function consists of the numbers x that are common to the domains of f and g, but the numbers x for which $g(x) = 0$ must be excluded from the domain of the quotient f/g.

EXAMPLE 1

Let f and g be two functions defined as

$$f(x) = \sqrt{x + 2} \qquad \text{and} \qquad g(x) = \sqrt{x - 3}$$

Find the following, and in each case determine the domain:

(a) $(f+g)(x)$ (b) $(f-g)(x)$ (c) $(f\cdot g)(x)$ (d) $(f/g)(x)$

Solution

(a) $(f+g)(x) = \sqrt{x+2} + \sqrt{x-3}$

(b) $(f-g)(x) = \sqrt{x+2} - \sqrt{x-3}$

(c) $(f\cdot g)(x) = (\sqrt{x+2})(\sqrt{x-3}) = \sqrt{(x+2)(x-3)}$

(d) $(f/g)(x) = \dfrac{\sqrt{x+2}}{\sqrt{x-3}} = \sqrt{\dfrac{x+2}{x-3}}$

The domain of f is the interval $[-2, +\infty)$ and that of g is $[3, +\infty)$. The x common to both these domains is the interval $[3, +\infty)$; and, as a result, this is the domain of the sum $f+g$, the difference $f-g$, and the product $f\cdot g$. For part (d), the domain is the interval $(3, +\infty)$, since for $x=3$ the denominator function g has the value 0.

Types of Functions

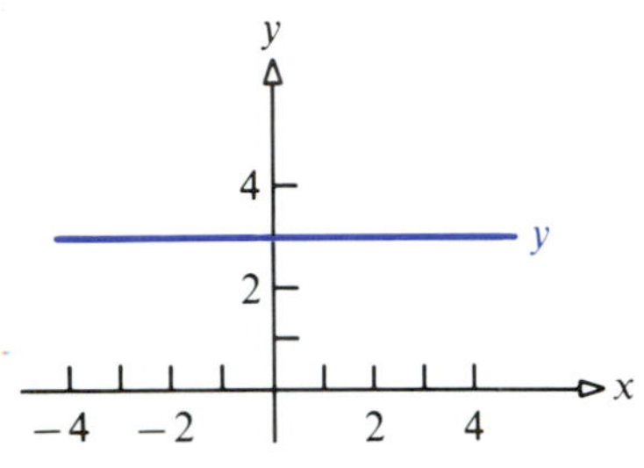

Figure 44

Many situations lead to functions that can be easily classified. We start with the *constant function*. By a *constant function* we mean a function $f(x) = A$ with domain the set of real numbers and range consisting of only one number, A. The graph of a constant function is a straight line parallel to the x-axis. For example, the function $f(x) = 3$ is a constant function; its graph is given in Figure 44.

A function P is a *polynomial function* if

$$P(x) = a_n x^n + a_{n-1} x^{n-1} + \cdots + a_1 x + a_0 \tag{1}$$

for all x, where the coefficients $a_0, a_1, \ldots, a_n$ are real numbers and the exponents are nonnegative integers. The domain of a polynomial function is the set of real numbers. If $a_n \neq 0$, then a_n is called the *leading coefficient of* f, and we say that the polynomial has *degree n*. For example, the function P defined by

$$P(x) = 2x^7 - 3x^2 + \tfrac{1}{2}x - 2$$

is a polynomial of degree 7 with leading coefficient 2. The coefficients of $P(x)$ are $a_7 = 2$, $a_6 = a_5 = a_4 = a_3 = 0$, $a_2 = -3$, $a_1 = \frac{1}{2}$, and $a_0 = -2$.

The constant function $f(x) = A$, $A \neq 0$, is a polynomial function of degree 0. The constant function $f(x) = 0$ is the *zero polynomial function* and has no degree.

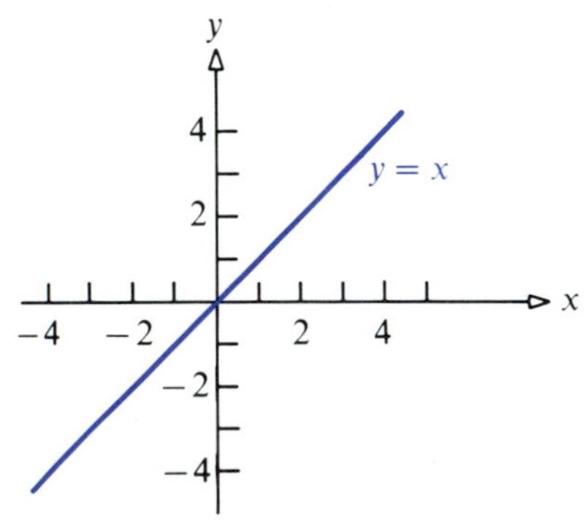

Figure 45

If the degree of a polynomial function is 1, then the function is called a *linear function* and is of the form $P(x) = ax + b$, where $a \neq 0$. From the discussion in previous sections, we know that the graph of this function is a straight line with slope $m = a$ and y-intercept $= b$. If $a = 1$, $b = 0$, we get the linear function $P(x) = x$, which is known as the *identity function*. Its graph is given in Figure 45.

Any polynomial function P of degree 2 may be written as

$$P(x) = ax^2 + bx + c$$

where a, b, and c are constants and $a \neq 0$. Such polynomials are also called *quadratic functions*. The graph of a quadratic function is known as a *parabola*. Figure 18 on page 15 illustrates a typical parabola.

Obtaining the graphs of most polynomials of degree 3 or higher is generally not easy with the tools we now have available. We would have to locate several well-chosen points on the graphs (this is easier said than done) and then hope that, by connecting them with a smooth curve, we would obtain an accurate picture. This tedious method is imprecise, and we will soon show how the power of calculus can be used to get accurate graphs without requiring a random selection of many points.

Rational Functions

A *rational function* is a function of the form

$$R(x) = \frac{P(x)}{Q(x)} = \frac{a_n x^n + \cdots + a_1 x + a_0}{b_m x^m + \cdots + b_1 x + b_0} \tag{2}$$

where P is a polynomial function of degree n and Q is a nonzero polynomial function of degree m.

To find the domain of a rational function, remember that the only time we will not be able to compute a value for $R(x)$ is when the x chosen gives a 0 in the denominator. Thus, the domain of R is $\{x \mid x \in \mathbb{R} \quad \text{and} \quad Q(x) \neq 0\}$.

The graphs of most rational functions, like those of most polynomial functions, require the use of calculus. We postpone a discussion of their graphs to Chapter 4. As we continue in this book, other types of functions will be encountered, classified, and discussed. For most of them, calculus will be not only useful but also necessary to obtain a complete description.

EXERCISE 1.5

For the functions f and g in Problems 1–6 find: (a) $f + g$ (b) $f - g$ (c) $f \cdot g$ (d) f/g
Determine the domains of f, g, $f + g$, $f \cdot g$, and f/g.

1. $f(x) = x - 1$; $g(x) = 2x^2$

2. $f(x) = \sqrt{x+1}$; $g(x) = \sqrt{x^2 - 1}$

3. $f(x) = \sqrt{x+1}$; $g(x) = x + 1$

4. $f(x) = |x|$; $g(x) = |x - 1|$

5. $f(x) = \dfrac{1}{x}$; $g(x) = \dfrac{1}{x} + 1$

6. $f(x) = (x^2 - 3x + 1)^5$; $g(x) = \sqrt{x^4 + 1}$

In Problems 7–12 indicate which are polynomial functions.

7. $f(x) = 2x^5 - 3x + 4$

8. $f(x) = \dfrac{1}{x^2}$

9. $f(x) = 2x^2 - \sqrt{x}$

10. $f(x) = 2x + \dfrac{3}{x} - 2$

11. $f(x) = \sqrt{x} - 2$

12. $f(x) = 3x^2 + 5x$

In Problems 13–16 find the domain of each function.

13. $f(x) = \dfrac{3x}{x+2}$

14. $f(x) = \dfrac{2x+1}{3x^2 - 5x - 2}$

15. $f(x) = \dfrac{2}{x^2 - 4}$

16. $f(x) = \dfrac{x^4}{x^3 - 8}$

In Figure 18 (p. 15) the graph of $f(x) = x^2$ is given. In Problems 17–22 graph each function using Figure 18 as a guide.

17. $f(x) = x^2 + 4$

18. $f(x) = x^2 - 4$

19. $f(x) = x^2 + x$

20. $f(x) = x^2 - x$

21. $f(x) = |x^2 - 4|$

22. $f(x) = [\![x^2 - 4]\!]$

A function f is said to be *even* if $f(-x) = f(x)$ for every number x in the domain of f. A function f is said to be *odd* if $f(-x) = -f(x)$ for every x in the domain. (In both cases, it is understood that for each x in the domain, $-x$ must also be in the domain.)

In Problems 23–28 determine whether the given function f is even, odd, or neither.

23. $f(x) = 4x^3$

24. $f(x) = 3x^2 - 2x + 1$

25. $f(x) = |x|$

26. $f(x) = \dfrac{\sqrt{x^2 - 1}}{\sqrt{x^2 + 1}}$

27. $f(x) = \dfrac{x - 1}{x + 1}$

28. $f(x) = (x + 1)^2$

In Problems 29–34 give examples to illustrate each fact.

29. The sum of two odd functions is odd.

30. The difference of two odd functions is odd.

31. The product of two even functions is even.

32. The product of two odd functions is even.

33. The function f/g is even if f is odd and g is odd.

34. The function f/g is even if f is even and g is even.

35. Given $f(x) = ax^2 + bx + c$, find numbers a, b, and c such that

$$f(x + y) = f(x) + f(y)$$

36. Given $f(x) = 3x^2 + 2x - 1$ and $g(x) = (A + B)x^2 + Cx + D$, with A, B, C, D real numbers, under what conditions does $f(x) = g(x)$?

37. Given $f(x) = 3x + 1$ and $(f + g)(x) = 6 - \frac{1}{2}x$, find $g(x)$.

38. For what numbers a, b, c is the function $f(x) = ax^2 + bx + c$ even? Odd?

39. Graph any function f having the following properties: the domain of f is $(-1, 1)$; the range of f consists of exactly four numbers; $f(x) > 0$ when $x > 0$; and $f(x) < 0$ when $x < 0$.

40. Given $f(x) = 4x + 1$ and $g(x) = 2x - 5$, write each of the following in interval notation:

(a) $f(x) > g(x)$ (b) $f(x) \leq g(x)$
(c) $f(x) \neq g(x)$ (d) $|g(x) - f(x)| < 1$

41. A function f is defined on the closed interval from -3 to 3 and has the graph shown.

(a) On the axes provided, sketch the entire graph of $y = |f(x)|$.
(b) On the axes provided, sketch the entire graph of $y = f(|x|)$.
(c) On the axes provided, sketch the entire graph of $y = f(-x)$.
(d) On the axes provided, sketch the entire graph of $y = f(x - 1)$.

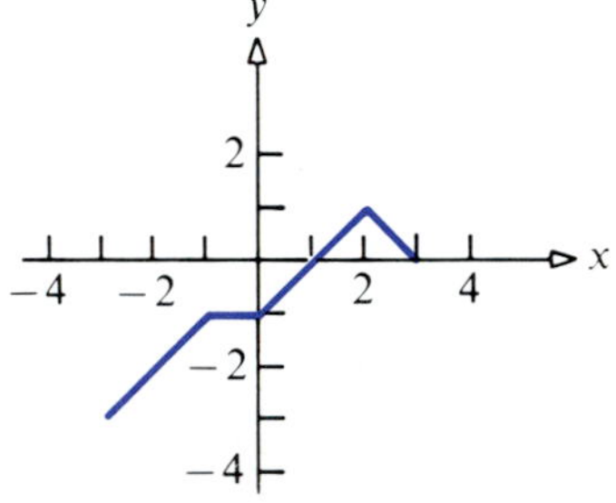

1.6 Composite Functions

Consider the function $y = (2x + 3)^2$. If we write $y = f(u) = u^2$ and $u = g(x) = 2x + 3$, then by a substitution process we can obtain the original function—namely, $y = f(u) = f(g(x)) = (2x + 3)^2$. This process is called *composition*. In general, suppose that f and g are two functions, and suppose that x is a number in the domain of g. By applying g to x, we get $g(x)$. If $g(x)$ is in the domain of f, then we may apply f to $g(x)$ and thereby obtain the value $f(g(x))$. If we do this for all such x's in the domain of g, the resulting correspondence is called a *composite function.*

[1.6.1] DEFINITION / *Composite Function.*

Given the two functions f and g, the *composite function*, denoted by $f \circ g$ (read "f circle g"), is defined by

$$(f \circ g)(x) = f(g(x))$$

where the domain of $f \circ g$ is the set of all numbers x in the domain of g such that $g(x)$ is in the domain of f.

Figures 46 and 47 illustrate the definition. Some examples will give you the idea.

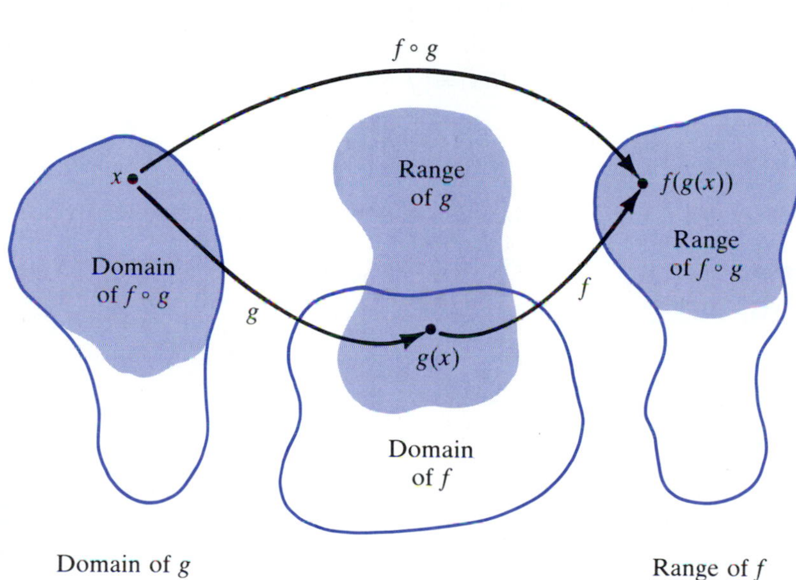

Figure 46

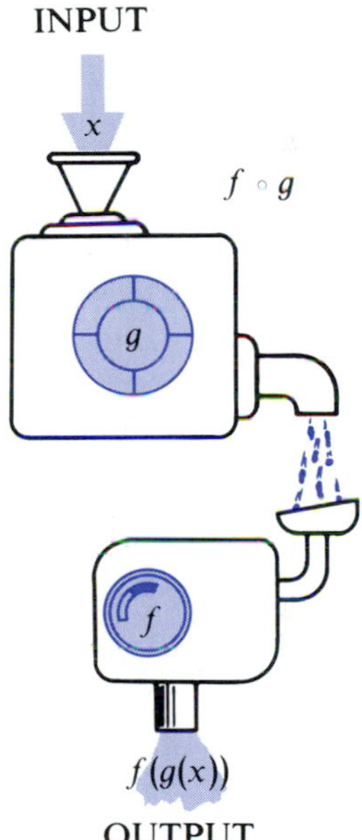

Figure 47

EXAMPLE 1

Suppose $f(x) = \sqrt{x}$ and $g(x) = x^3 - 1$. Find the following composite functions, and then find the domain of each composite function:

(a) $f \circ g$ (b) $g \circ f$ (c) $f \circ f$ (d) $g \circ g$

Solution

(a) The function f is the square root function, so the composite function $f \circ g = f(g(x))$ means to take the square root of $g(x)$; thus,

$$(f \circ g)(x) = f(g(x)) = \sqrt{g(x)} = \sqrt{x^3 - 1}$$

The domain of $f \circ g$ is the interval $[1, +\infty)$ and is found by determining those x in the domain of g for which $x^3 - 1 \geq 0$.

(b) The function g tells us to cube x and then subtract 1. The composite function $(g \circ f)(x) = g(f(x))$ tells us to cube $f(x)$ and then subtract 1. Thus,

$$(g \circ f)(x) = g(f(x)) = (f(x))^3 - 1 = (\sqrt{x})^3 - 1 = x^{3/2} - 1$$

The domain of $g \circ f$ is $[0, +\infty)$.

(c) $$(f \circ f)(x) = f(f(x)) = \sqrt{f(x)} = \sqrt{\sqrt{x}} = \sqrt[4]{x}$$

The domain of $f \circ f$ is $[0, +\infty)$.

(d) $$(g \circ g)(x) = g(g(x)) = (g(x))^3 - 1 = (x^3 - 1)^3 - 1$$

The domain of $g \circ g$, evidently, is the set of all real numbers.

Some techniques in calculus require that we be able to determine the components of a composite function. For example, the function $H(x) = \sqrt{x + 1}$ is the composition of the functions f and g, where $f(x) = \sqrt{x}$ and $g(x) = x + 1$, because $H(x) = (f \circ g)(x) = f(g(x)) = \sqrt{g(x)} = \sqrt{x + 1}$.

EXAMPLE 2

Find functions f and g such that $f \circ g = H$ if $H(x) = (x^2 + 1)^{50}$.

Solution

Set $f(x) = x^{50}$ and $g(x) = x^2 + 1$. Then

$$(f \circ g)(x) = f(g(x)) = (g(x))^{50} = (x^2 + 1)^{50} = H(x)$$

EXAMPLE 3

Find functions f and g so that $f \circ g = H$ and $H(x) = 1/(x + 1)$.

Solution

If we set $f(x) = 1/x$ and $g(x) = x + 1$, we find that

$$(f \circ g)(x) = \frac{1}{g(x)} = \frac{1}{x + 1} = H(x)$$

Other functions f and g also have the above property. For example, if $f(x) = 1/(x - 2)$ and $g(x) = x + 3$, then

$$(f \circ g)(x) = \frac{1}{g(x) - 2} = \frac{1}{(x + 3) - 2} = \frac{1}{x + 1}$$

Although the answer to Example 3 and other problems involving composite functions is not unique, there is usually a "natural" selection for f and g—one that comes to mind first. More will be said later about this aspect of composition. In the meantime, it is sufficient to be able to write some functions f and g whose composite is a given function H. You will most likely find that your selection is the "natural" one.

We end this section by describing two broad classes of functions—*algebraic* and *transcendental*. A function f is called *algebraic* if it can be expressed in terms of sums, differences, products, quotients, powers, or roots of polynomial functions. For example, the function f defined by

$$f(x) = \frac{3x^3 - x^2(x+1)^{4/3}}{\sqrt{x^4+2}}$$

is an algebraic function. Functions that are not algebraic are termed *transcendental* functions. Examples of transcendental functions are trigonometric functions, studied at the end of this chapter, and logarithmic functions and exponential functions, studied in Chapter 7.

EXERCISE 1.6

In Problems 1–8 functions f and g are given. In each problem find:

(a) $f \circ g$ (b) $g \circ f$ (c) $g \circ g$ (d) $f \circ f$

1. $f(x) = 3x + 1$; $g(x) = x^2$

2. $f(x) = \sqrt{x+1}$; $g(x) = \dfrac{1}{x^2}$

3. $f(x) = \sqrt{x}$; $g(x) = x^2 - 1$

4. $f(x) = \dfrac{1}{\sqrt{x-1}}$; $g(x) = (x^2+1)^3$

5. $f(x) = \dfrac{x-1}{x+1}$; $g(x) = \dfrac{1}{x}$

6. $f(x) = \sqrt{x}$; $g(x) = \dfrac{1}{x}$

7. $f(x) = 3x^4 - 2x^2$; $g(x) = \dfrac{2}{\sqrt{x}}$

8. $f(x) = \dfrac{1}{3x+2}$; $g(x) = \dfrac{3}{2x-5}$

In Problems 9–14 find f and g such that $f \circ g = H$.

9. $H(x) = \sqrt{x^2 + x - 1}$

10. $H(x) = (1 + x^2)^{-3}$

11. $H(x) = (x^2 - 1)^7$

12. $H(x) = \left(1 - \dfrac{1}{x^2}\right)^2$

13. $H(x) = \dfrac{1}{(3x-5)^2}$

14. $H(x) = \sqrt[3]{2 - 3x}$

15. If $f(x) = x^3$, find a function g such that $f(g(x)) = x$ for every x in the domain of g.

16. If $f(x) = \sqrt{x}$, find a function g such that $f(g(x)) = x$ for every x in the domain of g.

17. Let $f(x) = 3 - 2x$. Find:

(a) $f \circ f$ (b) $f^2 = f \cdot f$

18. Give an example of two functions f and g for which $f \circ g = g \circ f$. Does $f \circ g = g \circ f$ for every choice of f and g?

19. If $f(x) = 2x^3 + 3x^2 + 4x + 5$ and $g(x) = 2$, find $g(f(x))$ and $f(g(x))$.

20. If $f(x) = x/(x+1)$, find $f(1/x)$, $f(f(x))$, and $f(1/f(x))$.

21. Determine p so that $(f \circ g)(x) = (g \circ f)(x)$, where $f(x) = 3x + 2$ and $g(x) = 2x - p$.

22. If $f(x) = mx + b$ and $g(x) = sx + c$, where m, b, s, and c are real numbers, show that $f \circ g$ and $g \circ f$ are linear functions.

23. If f denotes an even function and g denotes an odd function, determine whether the following functions are even, odd, or neither:

(a) $f + g$ (b) $f \cdot g$ (c) $f \circ f$
(d) $g \circ g$ (e) $f \circ g$ (f) $g \circ f$

1.7 Inverse Functions

Recall that a function f from X into Y is a correspondence that associates with each element of X a unique element of Y. The set X is called the *domain of f*. For each element x in X, the corresponding element y in Y is called the *value of f at x*. The set of all values is called the *range of f*. If a function f has the additional property that corresponding to each element in the range there is exactly one element in the domain, then f is called a *one-to-one function*.

[1.7.1] DEFINITION / ***One-to-One.***

A function f is said to be *one-to-one* if for any two numbers x_1 and x_2 in the domain of f:

$$\textbf{if} \qquad x_1 \neq x_2 \qquad \textbf{then} \qquad f(x_1) \neq f(x_2)$$

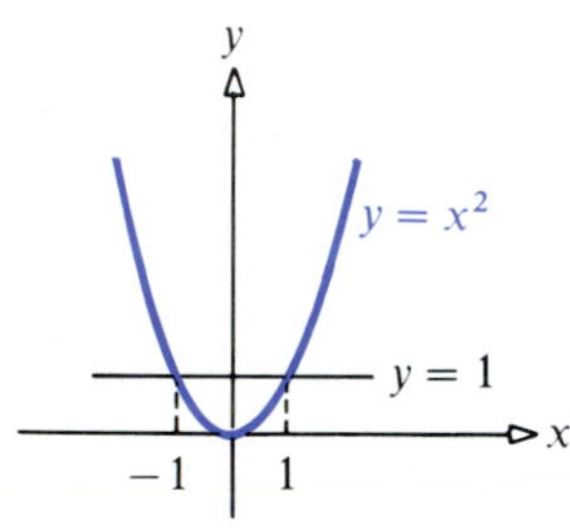

(a) Horizontal line strikes graph twice

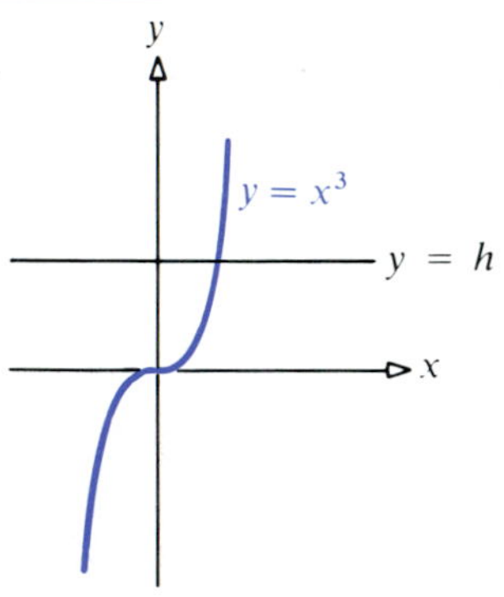

(b) Every horizontal line strikes graph once

Figure 48

If the graph of a function f is known, there is a simple test to determine whether f is one-to-one. If any horizontal line of height h strikes the graph of f more than once, then the value of $y = h$ corresponds to more than one x and f is *not* one-to-one. Figure 48 illustrates this test for the functions $y = x^2$ and $y = x^3$. It is easy to see that $y = x^2$ is not one-to-one, whereas $y = x^3$ is one-to-one.

Analytically, this geometric interpretation means that when the equation of a one-to-one function $y = f(x)$ is solved for x in terms of y, then x is also a *function* of y. For example, the function $f(x) = 3x - 4$ is one-to-one, since when we solve the equation $y = 3x - 4$ for x, x is expressed as a function of y. In this case,

$$x = \tfrac{1}{3}(y + 4) \tag{1}$$

The function $f(x) = x^2 - 4$ is not one-to-one because when we solve the equation $y = x^2 - 4$ for x, we get

$$x = \pm\sqrt{y + 4}$$

This equation does not define x as a function of y; there are two numbers x for values of $y > -4$.

For the one-to-one function $f(x) = 3x - 4$, the function g defined by (1)—namely, $g(y) = \frac{1}{3}(y + 4)$—is called the *inverse of f*. As the preceding examples illustrate, a function must be one-to-one to have an inverse. In fact, the inverse of a one-to-one function is unique. It can be proven that the inverse function $x = g(y)$ of a one-to-one function $y = f(x)$ is unique. If y is in the range of f, then f must take on the value y at some number x. Since f is one-to-one, this number x is unique. We have called this number $g(y)$ and hence g is unique.

[1.7.2] DEFINITION / ***Inverse Function.***

Let f be a one-to-one function. The *inverse of f*, denoted by f^{-1}, is the unique function defined on the range of f for which

$$x = f^{-1}(y) \qquad \textbf{if and only if} \qquad y = f(x)$$

In $x = f^{-1}(y)$ substitute $y = f(x)$. Then

$$x = f^{-1}(f(x)) \qquad \text{for all } x \text{ in the domain of } f \tag{2}$$

Similarly,

$$y = f(f^{-1}(y)) \qquad \text{for all } y \text{ in the range of } f$$

As a result of definition (1.7.2), it follows that

$$\text{Domain of } f = \text{Range of } f^{-1} \qquad \text{Range of } f = \text{Domain of } f^{-1}$$

See Figure 49 for an illustration.

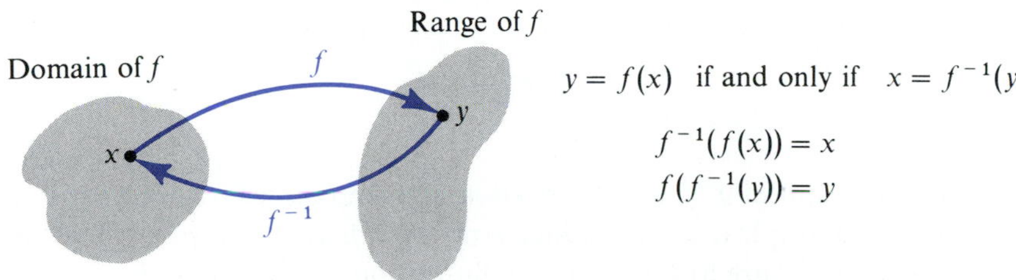

Figure 49

Example 1 demonstrates how to find the inverse of a given function f.

EXAMPLE 1

Find the inverse of $y = f(x) = 2x + 3$.

Solution

Since the graph of the line $y = 2x + 3$ is not horizontal, any horizontal line intersects it exactly once. Thus, the function is one-to-one. To find the inverse, we solve for x. Since $y = 2x + 3$, we find

$$2x = y - 3$$

$$x = \tfrac{1}{2}(y - 3)$$

The inverse of $f(x) = 2x + 3$ is therefore $x = f^{-1}(y) = \frac{1}{2}(y - 3)$. Since the symbol traditionally used to represent the independent variable of a function is x, it is convenient to replace y by x in f^{-1}. That is, $f^{-1}(x) = \frac{1}{2}(x - 3)$.

We can verify that the function f^{-1} is the inverse of the function f by checking to see that $f^{-1}(f(x)) = x$. For the function in Example 1,

$$f^{-1}(f(x)) = \tfrac{1}{2}[f(x) - 3] = \tfrac{1}{2}[(2x + 3) - 3] = \tfrac{1}{2}(2x) = x$$

Geometric Interpretation

If we sketch the graphs of f and f^{-1} of Example 1 (illustrated in Fig. 50), we notice an interesting fact:

The graphs of f and f^{-1} are symmetric with respect to the line $y = x$.

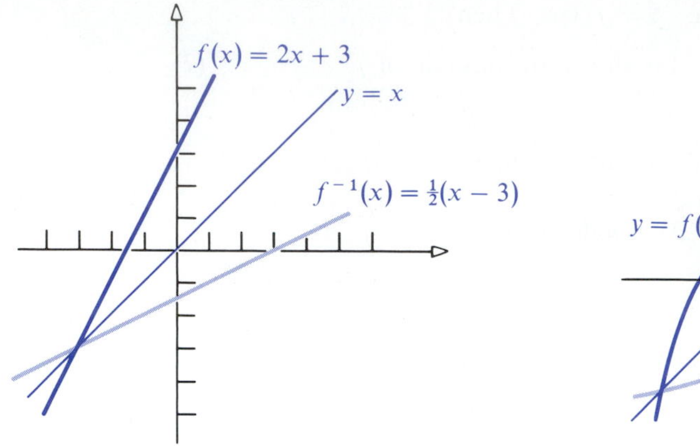

Figure 50

Figure 51

That is, the graph of f^{-1} is the reflection of the graph of f about the line $y = x$. (Each graph is a mirror image of the other, the mirror being the line $y = x$.) See Figure 51 for a general illustration of this situation.

EXAMPLE 2

Find the inverse of $y = f(x) = x^2$ if $x \geq 0$.

Solution

The function $f(x) = x^2$ is not one-to-one. However, if we restrict f to only that part of its domain for which $x \geq 0$, we have a one-to-one function. If we solve for x, obtaining $x = \sqrt{y}$ (the minus sign is excluded, since $x \geq 0$), and replace y by x, we find the inverse of the new function to be

$$f^{-1}(x) = \sqrt{x}$$

Figure 52 shows the graphs of $f(x) = x^2$ and $f^{-1}(x) = \sqrt{x}$.

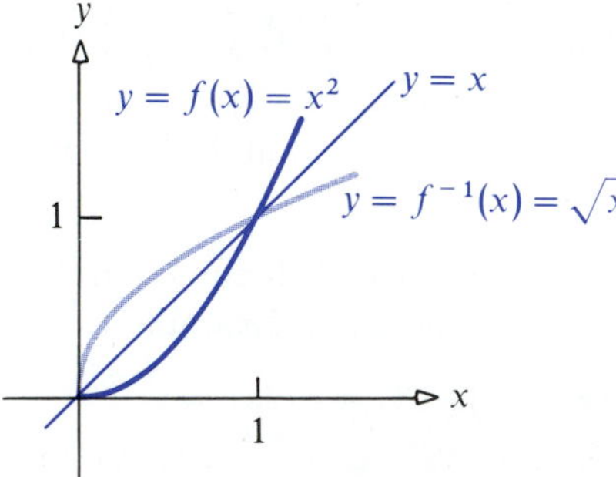

Figure 52

The idea expressed in Example 2 of restricting the domain of a function that is not one-to-one is a common practice. We use this concept in Chapter 8 when inverse trigonometric functions are discussed.

EXERCISE 1.7

In Problems 1–8 use a graph of each function to determine whether it is one-to-one.

1. $y = 3x - 1$ **2.** $y = 5x + 3$ **3.** $y = x^2 + 3$ **4.** $y = x^2 - 4$

5. $y = x^3$ **6.** $y = x^3 + 1$ **7.** $y = x^n$, n even **8.** $y = x^n$, n odd

In Problems 9–16 the function f is one-to-one. Find f^{-1}. For the functions in Problems 9–14 graph f and use it to get the graph of f^{-1}.

9. $f(x) = 3x - 1$ **10.** $f(x) = 5x + 3$ **11.** $f(x) = x^3$ **12.** $f(x) = \sqrt[3]{x}$

13. $f(x) = \dfrac{1}{x}$, $x > 0$ **14.** $f(x) = \dfrac{1}{1-x}$, $x > 1$ **15.** $f(x) = \dfrac{x}{x+1}$, $x > -1$ **16.** $f(x) = \dfrac{x}{x-1}$, $x > 1$

In Problems 17–22 show that f is one-to-one. Then find the inverse and state its domain.

17. $f(x) = 3x + 5$ **18.** $f(x) = ax + b$ **19.** $f(x) = \dfrac{2x+3}{5x-6}$

20. $f(x) = \dfrac{2x-1}{3x+2}$ **21.** $f(x) = \dfrac{1}{x}$ **22.** $f(x) = \sqrt[3]{x}$

23. A function f has an inverse. If the graph of f lies in the first quadrant, in what quadrant does the graph of f^{-1} lie?

24. A function f has an inverse. If the graph of f lies in the second quadrant, in what quadrant does the graph of f^{-1} lie?

25. To convert from x degrees Celsius to y degrees Fahrenheit, we use the formula $y = f(x) = \frac{9}{5}x + 32$. To convert from u degrees Fahrenheit to v degrees Celsius, we use the formula $v = g(u) = \frac{5}{9}(u - 32)$. Show that f and g are inverse functions.

26. Prove that if f is a *periodic function*—that is, if there is a positive number a so that $f(x + a) = f(x)$, for all x in the domain of f—then f does not have an inverse.

27. If the function f is defined by $f(x) = x^5 - 1$, find f^{-1}.

28. If $f(x) = 3x/(x - 2)$, find f^{-1} and the domains of f and f^{-1}.

29. Let h be the function defined by $h(x) = x^n$, where n is an integer. What condition, if any, must be placed on n for h to have an inverse?

30. Find f^{-1} if $f(x) = (ax + b)/(cx + d)$. What happens if $ad - bc = 0$?

31. Show that $(f \circ g)^{-1} = g^{-1} \circ f^{-1}$ if the three inverses exist.

Trigonometric Functions 1.8

It will be presumed here that you have a certain familiarity with the nature and use of trigonometric functions. Nevertheless, we will review some of this material, including the fundamental identities, which are important in the treatment of the calculus of trigonometric functions. Recall that the sine and cosine functions are introduced in trigonometry to relate the angles in right triangles to ratios of the lengths of their sides. In calculus these functions are used to relate angles to the coordinates of points in planes. For this purpose it is convenient to give angles counterclockwise or clockwise orientations and to consider angles that are larger than those in triangles. It is also convenient in calculus to measure angles in *radians* rather than in degrees.

Angles

An *angle* is formed when two *rays* (*half-lines*) have the same endpoint or *vertex*. One ray is called the *initial side* and the other is called the *terminal side*. The angle may be measured by the amount of rotation needed for the initial side to coincide with the terminal side. We agree that when this rotation is counterclockwise, the angle is measured positively; when the rotation is clockwise, the angle is measured negatively (see Fig. 53).

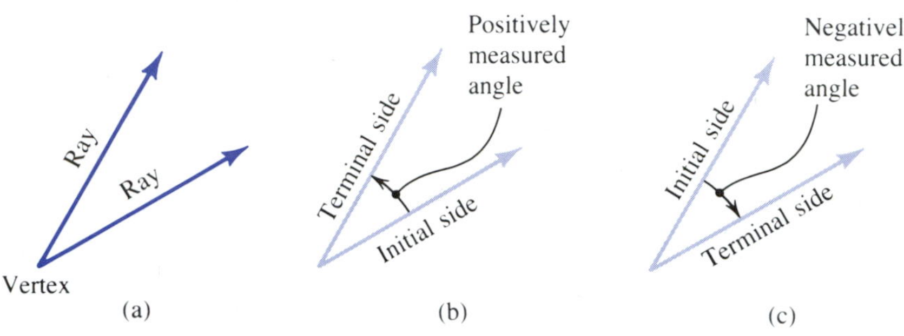

Figure 53

An angle is in *standard position* when its vertex is placed at the origin and its initial side coincides with the positive x-axis of a rectangular coordinate system (see Fig. 54).

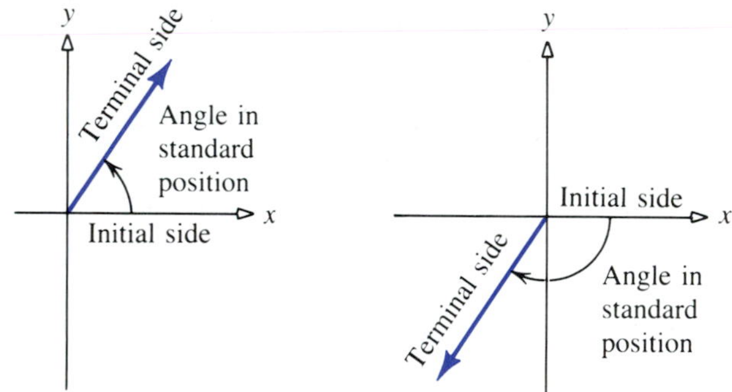

Figure 54

The *radian measure* of an angle is then defined in the following manner:

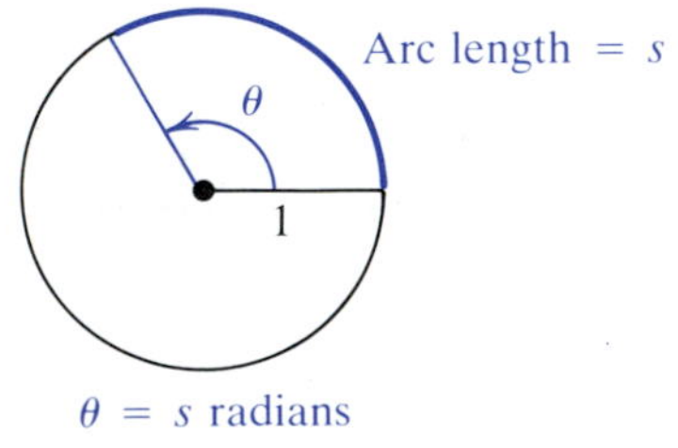

$\theta = s$ radians

Figure 55

[1.8.1] DEFINITION / *Radian.*

Consider the unit circle. If an angle θ has a counterclockwise orientation, its *radian measure* is the length of the arc it subtends. See Figure 55. If the angle has a clockwise orientation, its radian measure is the negative of the length of the arc it subtends.

Since the circumference of a circle of radius 1 is 2π, a full counterclockwise revolution is 2π radians, half of a counterclockwise revolution is π

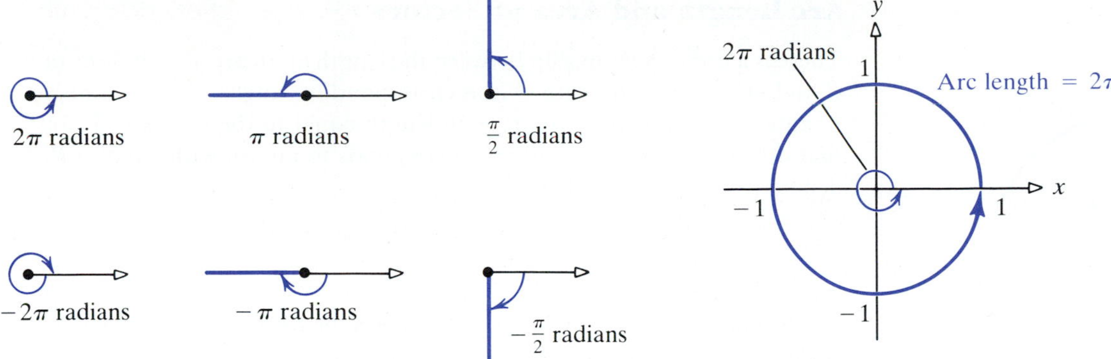

Figure 56

Figure 57

radians, and a counterclockwise right angle is $\pi/2$ radians. The measurements of clockwise-oriented angles are the negatives of these numbers (see Fig. 56).

To convert from the degree system for measuring angles to the radian system, we proceed as follows: Observe that one complete revolution around a unit circle corresponds to an angle of 360° in the degree system and to 2π radians in the radian system (see Fig. 57). Thus,

$$360° = 2\pi \text{ radians}$$

It follows that

$$1° = \frac{\pi}{180} \text{ radians}$$

$$1 \text{ radian} = \frac{180}{\pi} \text{ degrees}$$

If we use a calculator to approximate $\pi/180$ and $180/\pi$, we obtain

$$1° \approx 0.017453 \text{ radian} \qquad \text{and} \qquad 1 \text{ radian} \approx 57.2958°$$

In general,

1. To change radian measure to degrees, multiply by $180/\pi$.
2. To change degree measure to radians, multiply by $\pi/180$.

Table 7 lists the degree and radian measures of some frequently encountered angles.

Angles greater than 2π radians can be measured in these systems. For example, an angle of $480° = 2\pi + (2\pi/3) = 8\pi/3$ radians corresponds to one and one-third complete revolutions (see Fig. 58).

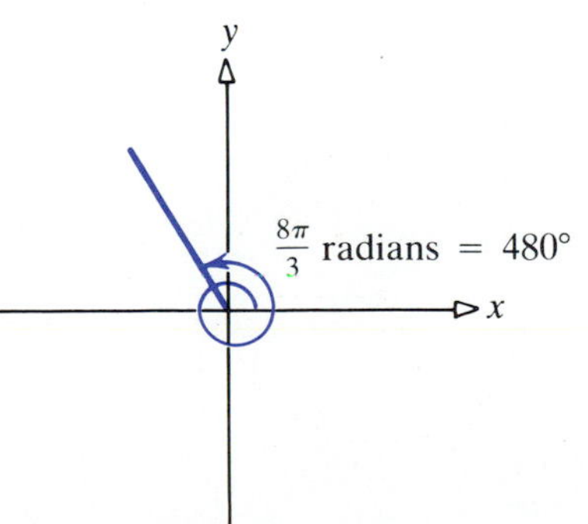

Figure 58

Table 7

Degrees	0°	30°	45°	60°	90°	180°	270°	360°
Radians	0	$\pi/6$	$\pi/4$	$\pi/3$	$\pi/2$	π	$3\pi/2$	2π

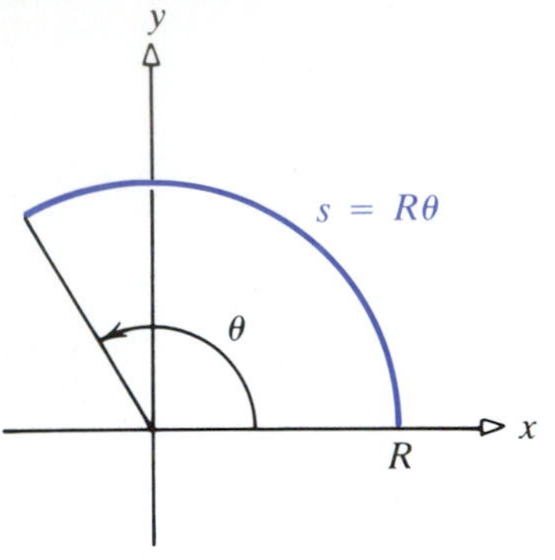

Figure 59

Arc Length and Area of Sectors

There is a direct relationship between the length of an arc of a circle of radius R and the radian measure of the corresponding angle. A central angle of 1 radian corresponds to an arc with length equal to the radius R. It follows that a central angle of θ radians corresponds to an arc with length θR. If s denotes the arc length, then

$$s = \theta R \tag{1}$$

See Figure 59.

Similarly, we can calculate the area of a sector. Recall that the area of a circle of radius R is πR^2. Since this is an area with a central angle of 2π radians, we have

$$\text{Area of a sector of } 2\pi \text{ radians} = \pi R^2$$

If we take proportional parts, we find that

$$\text{Area of a sector of 1 radian} = \frac{\pi R^2}{2\pi} = \frac{R^2}{2}$$

Similarly, if the central angle measures θ radians, then

$$\text{Area of a sector of } \theta \text{ radians} = \frac{\theta R^2}{2} \tag{2}$$

See Figure 60.

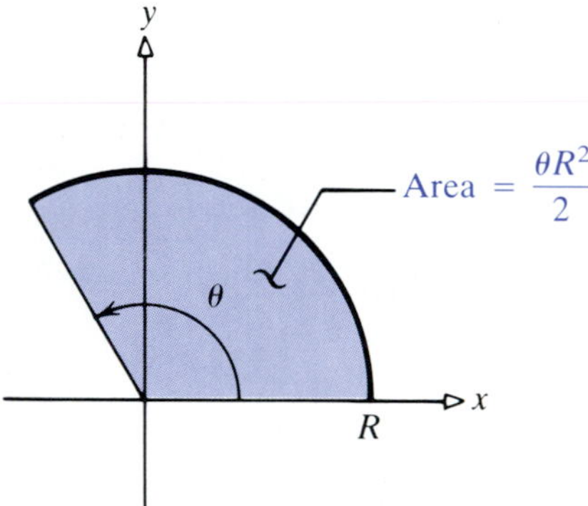

Figure 60

The Trigonometric Functions

We now define the sine and the cosine functions for any real number.

[1.8.2] DEFINITION / ***Sine; Cosine.***

Let θ be any real number. Place an angle having radian measure θ in standard position, and let the point P be at the intersection of the terminal side of the angle with the unit circle having its center at the origin. If (x, y) are the coordinates of the point P, then

the *sine function* is defined by

$$\sin \theta = y \tag{3}$$

and the *cosine function* is defined by

$$\cos \theta = x \tag{4}$$

See Figure 61. The domain of both the sine and cosine functions is the set of real numbers because $\sin \theta = y$ and $\cos \theta = x$ exist for every real number θ.

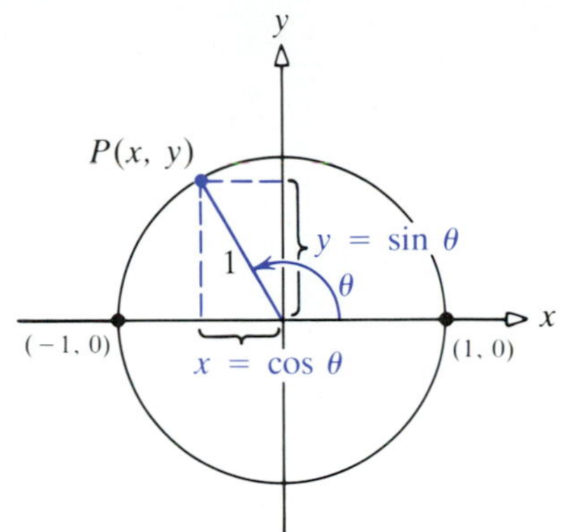

Figure 61

Note that if $P = (x, y)$ is a point on the unit circle, then $|x| \leq 1$ and $|y| \leq 1$. This implies that

$$|\sin \theta| \leq 1 \qquad |\cos \theta| \leq 1 \tag{5}$$

for every θ in the domains of these functions. Also, $\sin \theta$ and $\cos \theta$ take on every value between -1 and 1.

In general, values of the sine and cosine are difficult to compute. Fortunately, most calculators are designed to give values of $\sin \theta$ and $\cos \theta$ with a high degree of precision. However, the sine and cosine of $\pi/6$, $\pi/3$, and $\pi/4$ are readily read off from 30°–60°–90° triangles and isosceles right triangles. This is because the ratios of the sides of such triangles can be found from elementary geometry. See Figure 62.

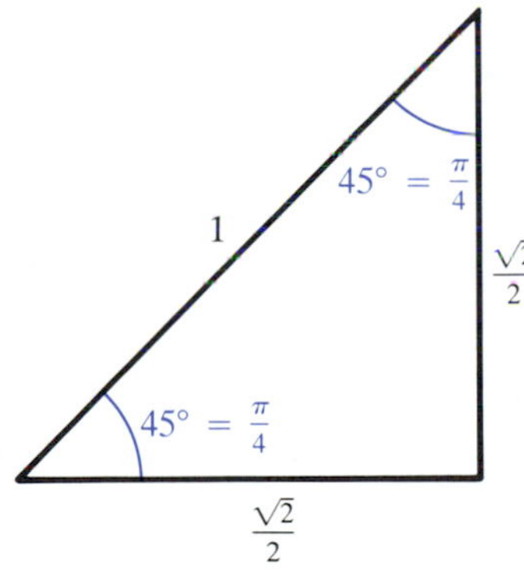

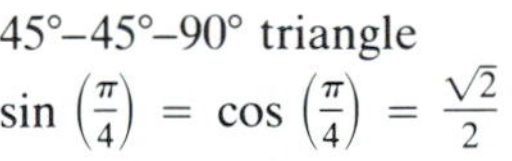
45°–45°–90° triangle

$\sin\left(\frac{\pi}{4}\right) = \cos\left(\frac{\pi}{4}\right) = \frac{\sqrt{2}}{2}$

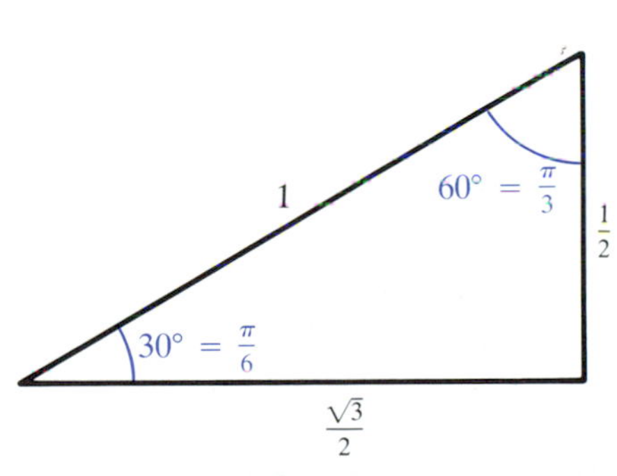

30°–60°–90° triangle

$\cos\left(\frac{\pi}{3}\right) = \sin\left(\frac{\pi}{6}\right) = \frac{1}{2}$

$\sin\left(\frac{\pi}{3}\right) = \cos\left(\frac{\pi}{6}\right) = \frac{\sqrt{3}}{2}$

Figure 62

EXAMPLE 1

Find the values of the sine and cosine functions at:

(a) $\theta = \dfrac{\pi}{2}$ (b) $\theta = \dfrac{3\pi}{4}$ (c) $\theta = \dfrac{7\pi}{6}$ (d) $\theta = \dfrac{3\pi}{2}$

Solution

See Figure 63.

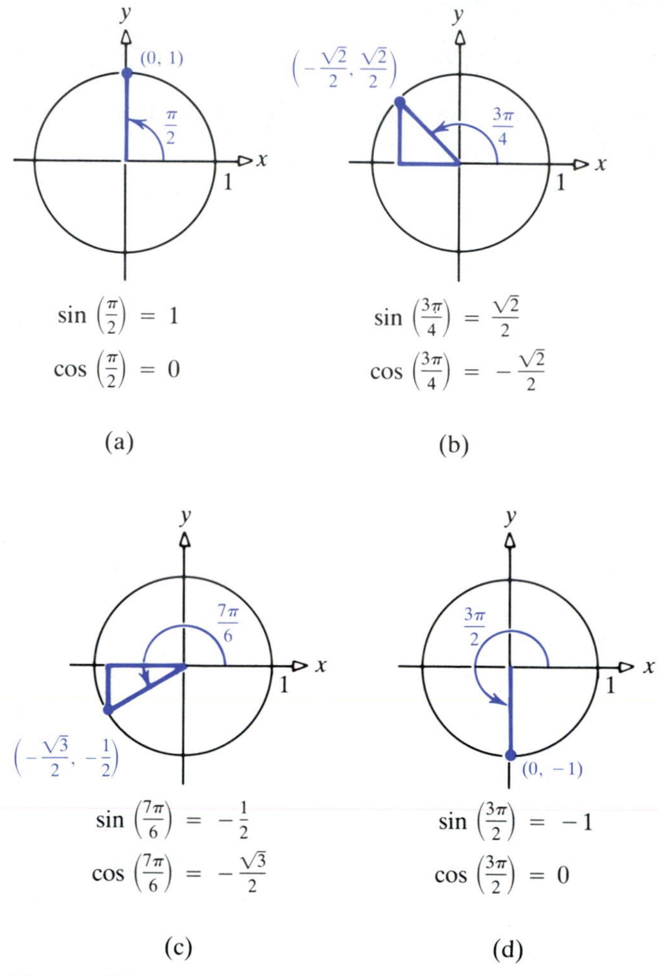

Figure 63

Periodic Functions

Since the circumference of the unit circle is 2π, the same point $P = (x, y)$ is obtained for $\theta + 2n\pi$ for every integer n. Hence the values of the sine and cosine repeat in successive intervals of length 2π. The following is true for the sine and cosine functions:

$$\sin(\theta + 2\pi) = \sin\theta \qquad \cos(\theta + 2\pi) = \cos\theta \tag{6}$$

Property (6) is described by saying that $\sin\theta$ and $\cos\theta$ are *periodic* with period 2π. A function f is said to be *periodic* if there exists a positive number p such that

$$f(x + p) = f(x) \tag{7}$$

whenever $f(x)$ is defined. The smallest such p is called the *period* of f.

Graphs of the Sine and Cosine Functions

Since we normally use x to represent points in the domain of the function, we will usually follow that convention for sine and cosine functions and replace θ by x. Thus, we consider the sine and cosine as functions of a real variable x. For each value x, we define $\sin x$ and $\cos x$ to be the functions of the corresponding angle measured in radians. (For example, $\sin(\pi/6) = \sin(\pi/6 \text{ radians}) = 1/2$.)

By plotting a few points on the graphs of the sine and cosine functions and using (6), we obtain the complete graphs of these functions shown in Figure 64.

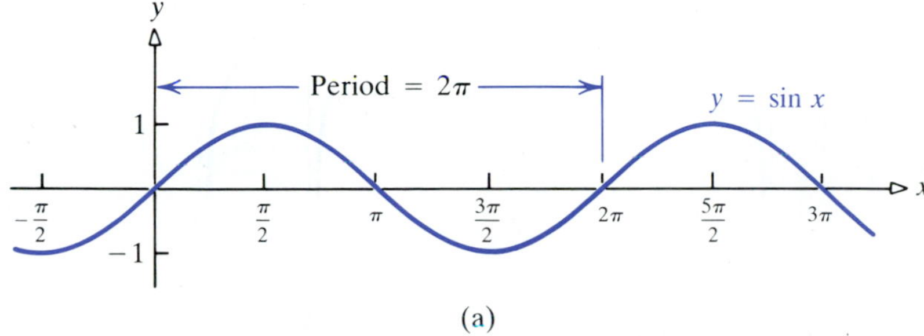

(a)

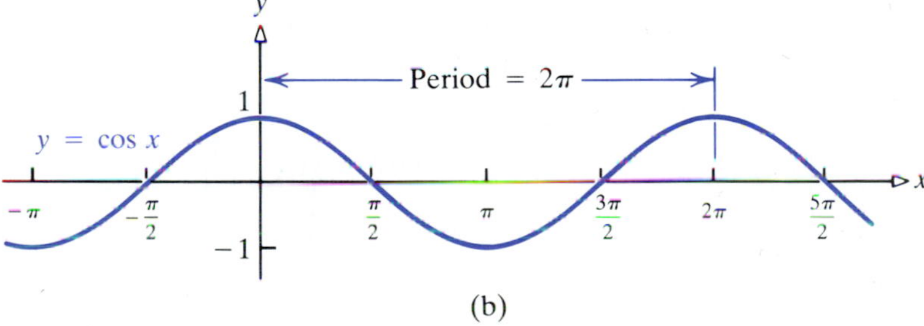

(b)

Figure 64

Other Trigonometric Functions

There are four other basic trigonometric functions, which are defined in terms of the sine and the cosine. The *tangent* and *secant* functions are defined by

$$\tan x = \frac{\sin x}{\cos x} \qquad \sec x = \frac{1}{\cos x} \tag{8}$$

for all real numbers x for which $\cos x \neq 0$. The *cotangent* and *cosecant* functions are defined by

$$\cot x = \frac{\cos x}{\sin x} \qquad \csc x = \frac{1}{\sin x} \tag{9}$$

for all real numbers x for which $\sin x \neq 0$. The condition $\cos x \neq 0$ implies that the values $\pm\pi/2, \pm 3\pi/2, \ldots$ are not in the domain of either the tangent function or the secant function. Similarly, $\sin x \neq 0$ implies that $0, \pm\pi, \pm 2\pi, \ldots$ are not in the domain of either the cotangent function or the cosecant function. The graphs of these four functions are shown in Figure 65.

Table 8 summarizes some values of the trigonometric functions, while Table 9 summarizes the important properties of the trigonometric functions.

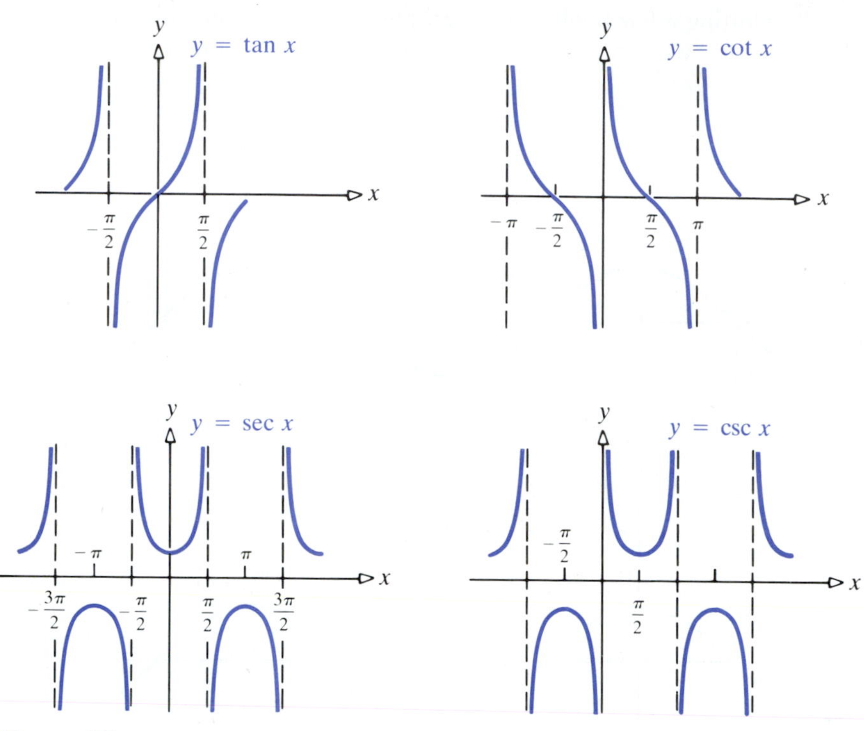

Figure 65

Table 8

	θ							
	0	$\pi/6$	$\pi/4$	$\pi/3$	$\pi/2$	π	$3\pi/2$	2π
$\sin\theta$	0	$1/2$	$\sqrt{2}/2$	$\sqrt{3}/2$	1	0	-1	0
$\cos\theta$	1	$\sqrt{3}/2$	$\sqrt{2}/2$	$1/2$	0	-1	0	1
$\tan\theta$	0	$1/\sqrt{3}$	1	$\sqrt{3}$	Not defined	0	Not defined	0
$\sec\theta$	1	$2/\sqrt{3}$	$\sqrt{2}$	2	Not defined	-1	Not defined	1
$\csc\theta$	Not defined	2	$\sqrt{2}$	$2/\sqrt{3}$	1	Not defined	-1	Not defined
$\cot\theta$	Not defined	$\sqrt{3}$	1	$1/\sqrt{3}$	0	Not defined	0	Not defined

Table 9

Function	Domain	Range	Period	Symmetry
$y = \sin x$	All reals	$-1 \le y \le 1$	2π	Origin
$y = \cos x$	All reals	$-1 \le y \le 1$	2π	y-axis
$y = \tan x$	$x \ne \ldots, -\pi/2, \pi/2, 3\pi/2, \ldots$	All reals	π	Origin
$y = \sec x$	$x \ne \ldots, -\pi/2, \pi/2, 3\pi/2, \ldots$	$y \le -1, \quad y \ge 1$	2π	y-axis
$y = \csc x$	$x \ne \ldots, -\pi, 0, \pi, 2\pi, \ldots$	$y \le -1, \quad y \ge 1$	2π	Origin
$y = \cot x$	$x \ne \ldots, -\pi, 0, \pi, 2\pi, \ldots$	All reals	π	Origin

Trigonometric Identities

There are many equations, called *trigonometric* identities, that describe relationships among the various trigonometric functions. If in the equation of the unit circle $x^2 + y^2 = 1$ we make the substitution $x = \cos\theta$ and $y = \sin\theta$, we obtain the Pythagorean identity

$$\sin^2\theta + \cos^2\theta = 1 \tag{10}$$

Note that $\cos^2\theta$ and $\sin^2\theta$ stand for $(\cos\theta)^2$ and $(\sin\theta)^2$, respectively. Other forms of the Pythagorean identity are obtained by dividing (10) by $\sin^2\theta$ and $\cos^2\theta$, respectively:

$$1 + \tan^2\theta = \sec^2\theta \qquad \cot^2\theta + 1 = \csc^2\theta \tag{11}$$

Because the terminal sides of angles θ and $-\theta$ are symmetric about the x-axis, we have, for all θ,

$$\cos(-\theta) = \cos\theta \qquad \sin(-\theta) = -\sin\theta \tag{12}$$

See Figure 66.

We list other trigonometric identities that will be useful later in this text. Proofs may be found in books on trigonometry.

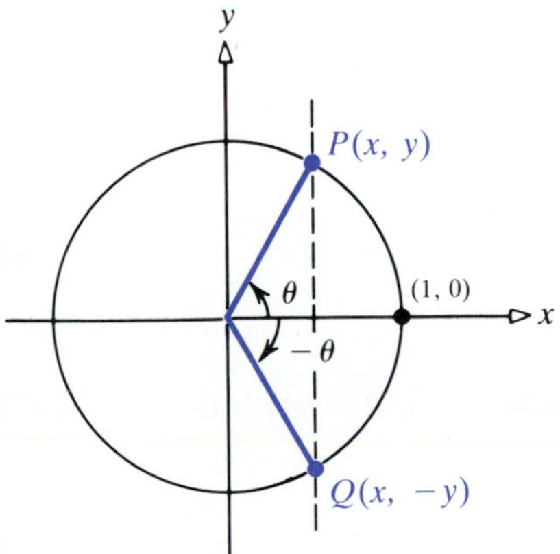

Figure 66

Sum and Difference Formulas

$$\cos(\theta + \phi) = \cos\theta\cos\phi - \sin\theta\sin\phi \tag{13}$$

$$\cos(\theta - \phi) = \cos\theta\cos\phi + \sin\theta\sin\phi \tag{14}$$

$$\sin(\theta + \phi) = \sin\theta\cos\phi + \cos\theta\sin\phi \tag{15}$$

$$\sin(\theta - \phi) = \sin\theta\cos\phi - \cos\theta\sin\phi \tag{16}$$

Double-Angle Formulas

Setting $\phi = \theta$ in (13) and (15), we obtain

$$\cos 2\theta = \cos^2\theta - \sin^2\theta \tag{17}$$

$$\sin 2\theta = 2\sin\theta\cos\theta \tag{18}$$

Using the Pythagorean identity, we can express $\cos 2\theta$ in the form

$$\cos 2\theta = 1 - 2\sin^2\theta \tag{19}$$

or

$$\cos 2\theta = 2\cos^2\theta - 1 \tag{20}$$

Half-Angle Identities

If we let $\theta = \phi/2$ in (19) and (20) and then solve for $\sin\phi/2$ and $\cos\phi/2$, respectively, we obtain

$$\sin\frac{\phi}{2} = \pm\sqrt{\frac{1-\cos\phi}{2}} \tag{21}$$

$$\cos\frac{\phi}{2} = \pm\sqrt{\frac{1+\cos\phi}{2}} \tag{22}$$

The sign before the radical is determined by the quadrant of $\phi/2$.

Product Formulas

Adding (13) and (14), we obtain

$$\cos\theta\cos\phi = \tfrac{1}{2}[\cos(\theta+\phi) + \cos(\theta-\phi)] \tag{23}$$

Subtracting (13) from (14), we obtain

$$\sin\theta\sin\phi = \tfrac{1}{2}[\cos(\theta-\phi) - \cos(\theta+\phi)] \tag{24}$$

Finally, adding (15) and (16), we get

$$\sin\theta\cos\phi = \tfrac{1}{2}[\sin(\theta+\phi) + \sin(\theta-\phi)] \tag{25}$$

Angle of Inclination

We now establish a relationship between the slope of a line and the angle the line makes with the positive x-axis. To establish this relationship we need the following definition:

[1.8.3] DEFINITION / *Angle of Inclination.*

For a line L not parallel to the x-axis, the *angle of inclination* ϕ is the smallest angle measured counterclockwise from the positive direction of the x-axis to L (see Fig. 67). If the line L is parallel to the x-axis, we take $\phi = 0$.

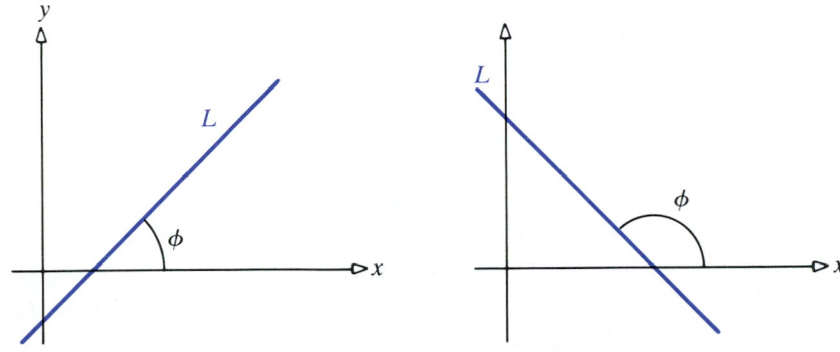

Figure 67

We emphasize that if ϕ is the angle of inclination of a line, then $0° \leq \phi < 180°$ or, in radians, $0 \leq \phi < \pi$.

We now tie together the notions of slope and angle of inclination.

[1.8.4] THEOREM

If m is the slope of a nonvertical line L and ϕ is the angle of inclination, then

$$m = \tan \phi \qquad (26)$$

Proof If L is parallel to the x-axis, then $m = 0$ and $\phi = 0°$. Thus,

$$\tan \phi = \tan 0 = 0 = m$$

so (26) holds in this case. If the line is not parallel to the x-axis, refer to Figure 68, which shows the given line L whose angle of inclination is ϕ and

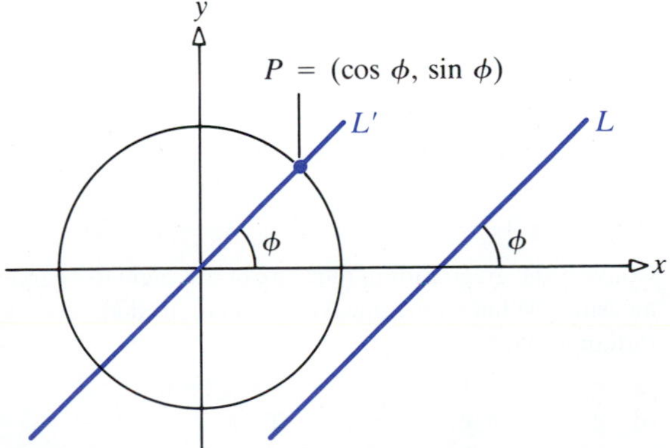

Figure 68

whose slope is m. (In the figure we drew a line L' passing through the origin and parallel to L, with an angle of inclination ϕ and a slope m.) The line L' intersects the unit circle at the point $P = (\cos\phi, \sin\phi)$. Since the points $(0, 0)$ and $(\cos\phi, \sin\phi)$ lie on L', we can use them to compute the slope m. This gives

$$m = \frac{\sin\phi - 0}{\cos\phi - 0} = \frac{\sin\phi}{\cos\phi} = \tan\phi$$

which proves theorem (1.8.4).

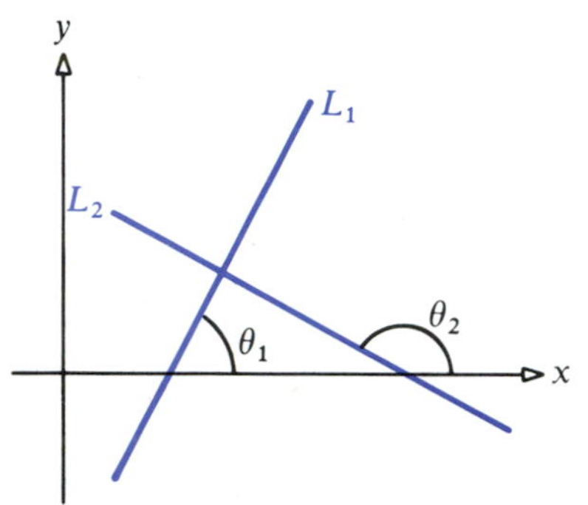

Figure 69

If the line L is parallel to the y-axis, then the angle of inclination is $\phi = \pi/2$, and $\tan(\pi/2)$ is not defined. This is consistent with the fact that the slope of a vertical line is undefined.

Using the angle of inclination, we now prove that two lines with slopes m_1 and m_2 are perpendicular if and only if $m_1 = -1/m_2$.

Proof Let θ_1 and θ_2 denote the inclinations of two nonvertical lines L_1 and L_2, respectively. We shall take θ_1 to be the smaller inclination. Then, as Figure 69 illustrates, if L_1 and L_2 are perpendicular, we have

$$\theta_2 = \theta_1 + \frac{\pi}{2}$$

Thus,

$$\begin{aligned} m_2 = \tan\theta_2 = \tan\left(\theta_1 + \frac{\pi}{2}\right) &= \frac{\sin[\theta_1 + (\pi/2)]}{\cos[\theta_1 + (\pi/2)]} \\ &= \frac{\sin\theta_1\cos(\pi/2) + \cos\theta_1\sin(\pi/2)}{\cos\theta_1\cos(\pi/2) - \sin\theta_1\sin(\pi/2)} \\ &= -\frac{\cos\theta_1}{\sin\theta_1} = -\cot\theta_1 \\ &= \frac{-1}{\tan\theta_1} = \frac{-1}{m_1} \end{aligned}$$

The proof that if $m_1 = -1/m_2$ holds, then L_1 and L_2 are perpendicular is left as an exercise (see Problem 61).

EXERCISE 1.8

1. Convert the given degree measure of an angle to radian measure. Express the answer as a multiple of π and then obtain an approximation by using 3.14 for π.

(a) 45° (b) −90° (c) 135°
(d) 180° (e) −210° (f) 300°
(g) −720° (h) 450° (i) 10°
(j) −80° (k) 2° (l) 120°

2. Convert the given radian measure of an angle to degree measure. Where an approximation is needed, take 1 radian as 57.3°.

(a) $\pi/3$ (b) $3\pi/4$ (c) $-5\pi/6$
(d) 2 (e) $-3\pi/2$ (f) π
(g) $\pi/6$ (h) $-\pi/2$ (i) 0.3
(j) $-\sqrt{2}$ (k) $11\pi/3$ (l) $5\pi/4$

3. Draw, in standard position, the angle whose measure is:

(a) $30°$ (b) $-225°$ (c) $5\pi/4$
(d) $-\pi$ (e) $3\pi/2$ (f) $480°$
(g) $-300°$ (h) $\frac{1}{2}$ (i) -2

4. Determine the exact numerical value of:

(a) $\sin\theta$, $\cos\theta$, and $\tan\theta$; $\theta = \pi/6$
(b) $\sin x$, $\cos x$, and $\sec x$; $x = 2\pi/3$
(c) $\cos\phi$, $\tan\phi$, and $\csc\phi$; $\phi = -\pi/3$
(d) $\sin t$, $\cos t$, and $\cot t$; $t = 3\pi/4$
(e) $\sin\alpha$, $\tan\alpha$, and $\sec\alpha$; $\alpha = 7\pi/6$
(f) $\cos\beta$, $\cot\beta$, and $\csc\beta$; $\beta = -3\pi/4$
(g) $\sin y$, $\tan y$, and $\cot y$; $y = 5\pi/6$
(h) $\sin w$, $\cos w$, and $\tan w$; $w = 11\pi/6$

5. Determine $\sin x$, $\cos x$, $\tan x$, $\cot x$, $\sec x$, and $\csc x$ for the given angle.

(a) $x = 0$ (b) $x = \pi/2$
(c) $x = \pi$ (d) $x = 3\pi/2$

6. Determine the "sign" of each of $\sin\theta$, $\cos\theta$, and $\tan\theta$, given that:

(a) $0 < \theta < \pi/2$
(c) $\pi < \theta < 3\pi/2$
(b) $\pi/2 < \theta < \pi$
(d) $3\pi/2 < \theta < 2\pi$

7. Determine the value (or values) of x in the interval $[0, 2\pi)$ for which:

(a) $\sin x = \frac{1}{2}$ (b) $\cos x = \sqrt{2}/2$
(c) $\tan x = 0$ (d) $\sin x = -\sqrt{3}/2$
(e) $\cos x = -\frac{1}{2}$ (f) $\tan x = -1$
(g) $\cot x = -\sqrt{3}$ (h) $\sec x = \frac{2}{3}\sqrt{3}$
(i) $\csc x = \sqrt{2}$ (j) $\cot x = 1$
(k) $\sec x = 1$ (l) $\csc x = -1$

8. Determine the value (or values) of x in the interval $(-\pi, \pi]$ for which:

(a) $\sin x = -\frac{1}{2}$ (b) $\cos x = 0$
(c) $\tan x = -1$ (d) $\sin x = -1$
(e) $\cos x = -\sqrt{3}/2$ (f) $\cot x = 0$
(g) $\sec x = -2$ (h) $\csc x = 2$
(i) $\sec x = -1$

9. Determine the value of the other five trigonometric functions of θ, given that:

(a) $\sin\theta = \frac{3}{5}$; $\pi/2 < \theta < \pi$
(b) $\cos\theta = -\frac{5}{13}$; $-\pi < \theta < -\pi/2$
(c) $\tan\theta = \frac{12}{5}$; $0 < \theta < \pi/2$
(d) $\sin\theta = -\frac{1}{3}$; $-\pi/2 < \theta < 0$
(e) $\cos\theta = \frac{2}{5}$; $3\pi/2 < \theta < 2\pi$
(f) $\tan\theta = -\frac{1}{10}$; $\pi/2 < \theta < \pi$

10. Determine:

(a) $\cos x$, given that $\sin x = 1/n$, where $n > 1$ and $0 < x < \pi/2$
(b) $\cos\theta$, given that $\sin\theta = \sqrt{1 - x^2}$, where $0 \le x \le 1$ and $\pi/2 \le \theta \le \pi$
(c) $\sin\alpha$, given that $\cos\alpha = 2x$, where $-\frac{1}{2} \le x \le 0$ and $\pi/2 \le \alpha \le \pi$
(d) $\tan\theta$, given that $\sec\theta = 4x$, where $x \ge \frac{1}{4}$ and $0 \le \theta < \pi/2$
(e) $\sin\beta$, given that $\tan\beta = 3x$, where $x \le 0$ and $-\pi/2 < \beta \le 0$
(f) $\tan\theta$, given that $\cos\theta = x - 1$, where $1 \le x \le 2$ and $3\pi/2 < \theta \le 2\pi$

11. Find the length of the arc intercepted on a circle of radius 2 by each central angle.

(a) $\pi/3$ (b) $2\pi/3$ (c) $60°$ (d) $120°$

12. Find the area of the sector determined by each central angle in Problem 11.

In Problems 13–20 reduce each expression to a trigonometric function of θ.

13. $\sin(\theta + \pi)$

14. $\tan\left(\frac{3\pi}{2} + \theta\right)$

15. $\cos\left(\frac{3\pi}{2} - \theta\right)$

16. $\sin(2\pi - \theta)$

17. $\cos(\theta + \pi)$

18. $\cos(\theta - \pi)$

19. $\sin\left(\theta + \frac{3\pi}{2}\right)$

20. $\tan\left(\theta + \frac{\pi}{2}\right)$

In Problems 21–32 prove that the given statement is true by transforming the expression on the left side of the equality into the expression on the right side.

21. $\sec\alpha - \cos\alpha = \tan\alpha\sin\alpha$

22. $(1 - \sin^2\alpha)\csc^2\alpha = \cot^2\alpha$

23. $\cos^4(2x) - \sin^4(2x) = \cos(4x)$

24. $\dfrac{2\tan A}{1 + \tan^2 A} = \sin(2A)$

25. $\tan x + \tan y = \dfrac{\sin(x + y)}{\cos x\cos y}$

26. $\cos\left(x - \frac{\pi}{4}\right) = \frac{\sqrt{2}}{2}(\sin x + \cos x)$

27. $\cot\theta - \tan\theta = 2\cot(2\theta)$

28. $\dfrac{1+\sec\beta}{\sec\beta} = 2\cos^2(\tfrac{1}{2}\beta)$

29. $\cos(7\alpha) + \cos(5\alpha) = 2\cos(6\alpha)\cos\alpha$

30. $\cos(2\theta) - \cos\theta = -2\sin(3\theta/2)\sin(\theta/2)$

31. $\sin^2 x - \sin^2 y = \sin(x+y)\sin(x-y)$

32. $\cos(3\theta) = 4\cos^3\theta - 3\cos\theta$

In Problems 33–48 sketch the graph of the given equation for the indicated values of x. Give the period of the function.

33. $y = 2\sin x;\quad x$ in $[-\pi, 2\pi]$

34. $y = -\frac{1}{2}\cos x;\quad x$ in $[-2\pi, 2\pi]$

35. $y = 3\sin(2x);\quad x$ in $[-\pi, \pi]$

36. $y = 2\cos(2x);\quad x$ in $[0, 2\pi]$

37. $y = 4\cos(x/2);\quad x$ in $[-\pi, 4\pi]$

38. $y = \sin(3x);\quad x$ in $[0, 3\pi]$

39. $y = \tan(2x);\quad x$ in $[-\pi, 2\pi]$

40. $y = \sec(2x);\quad x$ in $[0, 2\pi]$

41. $y = 4\sin(x/2);\quad x$ in $[0, 8\pi]$

42. $y = 2\csc x;\quad x$ in $[-2\pi, 2\pi]$

43. $y = 2\cot(x/2);\quad x$ in $[-\pi, 4\pi]$

44. $y = \sin^2 x;\quad x$ in $[0, 2\pi]$

45. $y = \sin(x/4);\quad x$ in $[-\pi, 2\pi]$

46. $y = |\sin x|;\quad x$ in $[0, 3\pi]$

47. $y = 2\sin(\pi x);\quad x$ in $[-2, 4]$

48. $y = \frac{1}{2}\cos(2\pi x);\quad x$ in $[-1, 2]$

In Problems 49–54 solve the given trigonometric equation. For example, to solve $\sin(2x) = \frac{1}{2}$, either

$$2x = \frac{\pi}{6} + 2\pi n \quad \text{or} \quad 2x = \frac{5\pi}{6} + 2\pi n \quad n \text{ an integer}$$

Therefore,

$$x = \frac{\pi}{12} + \pi n \quad \text{or} \quad x = \frac{5\pi}{12} + \pi n \quad n \text{ an integer}$$

49. $\sin(2x) = 1$

50. $\sin x - \cos x = 0$

51. $\cos(3x) = \sqrt{3}/2$

52. $\tan(2x) = -1$

53. $\sin^2 x = \frac{1}{4}$

54. $\cos(2x+1) = \frac{1}{2}$

In Problems 55–58 use the following facts to help solve the given equations.

If $\sin\alpha = \sin\beta$, then either $\alpha = \beta + 2\pi n$ or $\alpha = (\pi - \beta) + 2\pi n$, n an integer.

If $\cos\alpha = \cos\beta$, then either $\alpha = \beta + 2\pi n$ or $\alpha = -\beta + 2\pi n$, n an integer.

If $\tan\alpha = \tan\beta$, then $\alpha = \beta + \pi n$, n an integer.

55. $\sin x = \sin(2x)$

56. $\cos x = \cos(3x)$

57. $\tan x = \tan\left(3x + \dfrac{\pi}{3}\right)$

58. $\sin(2x) = \sin(3x - 1)$

59. Let f and g be functions that are periodic with period p. Prove that:

(a) $(f+g)(x+p) = (f+g)(x)$
(b) $(f\cdot g)(x+p) = (f\cdot g)(x)$
(c) $(f/g)(x+p) = (f/g)(x)$

60. Let $f(x) = \dfrac{1}{k}\cos(kx)$. For what number k does f have period 3?

61. Prove that if the product of the slopes of two nonvertical lines is -1, then the lines are perpendicular. [*Hint:* One slope—say, m_1—is positive and the other, m_2, is negative. Thus, if $m_1 = \tan\theta_1$ and $m_2 = \tan\theta_2$, then $0 < \theta_1 < \pi/2$ and $\pi/2 < \theta_2 < \pi$. Now use the identity $\tan[\theta_2 - (\pi/2)] = -1/\tan\theta_2$.]

62. Let L_1 and L_2 be two nonvertical, intersecting lines. If θ is the acute angle between L_1 and L_2, show that

$$\tan\theta = \frac{m_2 - m_1}{1 + m_1 m_2}$$

where m_1 and m_2 are the slopes of L_1 and L_2, respectively. See the figure.

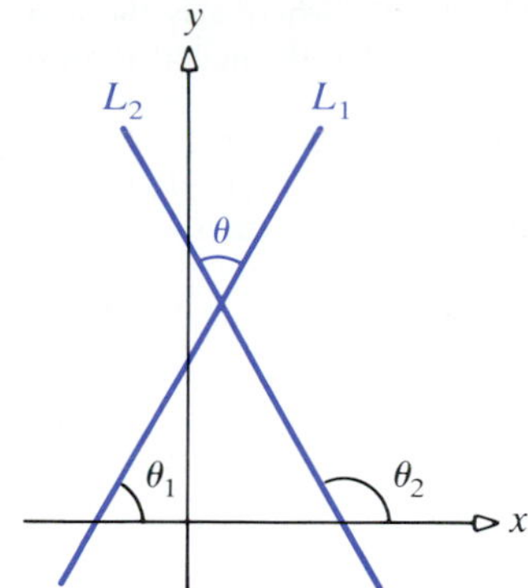

In Problems 63–70 use the result of Problem 62 to find the tangent of the angle between the lines whose slopes are:

63. 1 and 2 **64.** -1 and 2 **65.** $-\frac{1}{2}$ and $\frac{2}{3}$ **66.** $\frac{1}{4}$ and $-\frac{2}{5}$

67. $\frac{1}{2}$ and -1 **68.** $\frac{3}{2}$ and $\frac{1}{4}$ **69.** $\frac{1}{4}$ and 3 **70.** $\frac{1}{5}$ and $-\frac{1}{2}$

In Problems 71–74 use the result of Problem 62 to find the tangent of the acute angle from the line L_1 to the line L_2.

71. L_1: $2x + y = 2$, L_2: $x - y = 4$

72. L_1: $x + y = 2$, L_2: $2x - y = 0$

73. L_1: $2x + 3y = 4$, L_2: $x - y = 6$

74. L_1: $x + 3y = 4$, L_2: $2x - y = 5$

75. Use the result of Problem 62 and a calculator to find, to the nearest degree, the measurement of the angle between the lines whose slopes are given in Problems 63 and 64.

76. Repeat Problem 75 for the slopes given in Problems 69 and 70.

77. Use the result of Problem 62 to find, to the nearest degree, the measurements of the interior angles of the triangle having vertices $(-3, -4)$, $(0, 2)$, and $(3, -7)$.

78. Find the angle between the lines given by the equations $2x + 3y - 7 = 0$ and $3x - 5y - 2 = 0$.

79. Plot the points $A = (1, 1)$, $B = (5, 3)$, $C = (3, 7)$, and $D = (-1, 5)$.

(a) Find the slopes of the lines containing A and B, containing B and C, containing C and D, and containing A and D.
(b) Which of these lines are parallel? Which are perpendicular?
(c) Find the tangent of the angle between the line containing AB and the line containing AD.
(d) Find the point of intersection of the line containing AC and the line containing BD.
(e) Find the distance between the line containing AB and the line containing CD.
(f) Find the midpoint of the line segment AB.

80. Consider a general triangle with angles A, B, C and corresponding opposite sides a, b, c. Show that the following two important properties hold:

Law of sines: $$\frac{\sin A}{a} = \frac{\sin B}{b} = \frac{\sin C}{c}$$

Law of cosines: $$a^2 + 2bc\cos A = b^2 + c^2$$

[*Hint:* To prove the law of sines, drop a perpendicular from one vertex to the opposite side (or possibly its extension) and consider the two right triangles so determined. To prove the law of cosines, compute the square of the distance from $(b, 0)$ to $(c\cos A, c\sin A)$ (see the figure).]

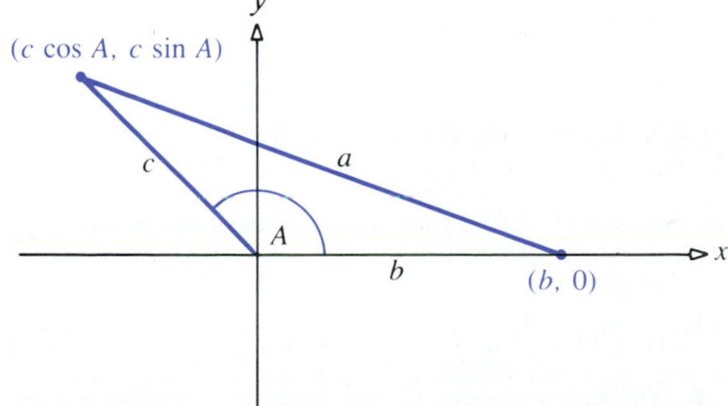

81. Show that the area A of an isosceles triangle is given by the formula $A = \frac{1}{2}a^2 \sin\theta$, where a is the length of each of the equal sides and θ is the included angle.

82. A pendulum of length L is displaced through an angle θ with the vertical (see the figure). Show that the vertical displacement s of the end of the pendulum is given by the formula $s = 2L\sin^2(\theta/2)$.

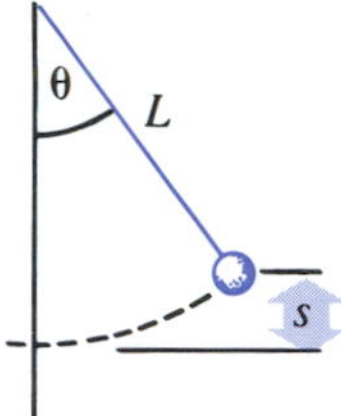

83. A plank leans against a wall and forms an angle θ with the wall (see the figure). Show that the distance h that the top of the plank is below its highest point is given by $h = s\tan(\theta/2)$, where s is the distance of the bottom of the plank from the wall. Determine a formula for h if the angle α that the plank forms with the ground is used instead of θ.

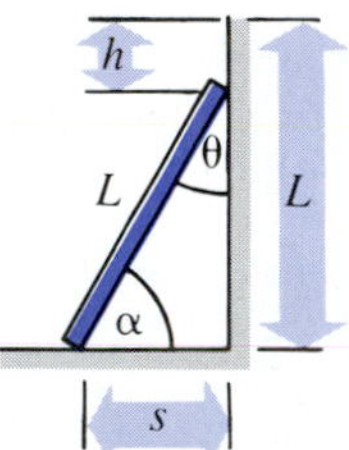

84. The minute and hour hands of a clock have lengths a and b, respectively $(a > b)$. Show that the distance s between the tips of the minute and hour hands is given by $s = \sqrt{a^2 + b^2 - 2ab\cos(\frac{11}{6}\pi t)}$, where t is time in hours. Assume that $t = 0$ when both hands point to 12. Determine the times when both hands point in the same direction.

85. Rod OA (see the figure) rotates about the fixed point O so that point A travels on a circle of radius R. Connected to point A is another rod AB of length $L \geq R$. Point B is connected to a piston that slides in a cylinder. Show that the distance x between point O and point B is given by $x = R\cos\theta + \sqrt{R^2\cos^2\theta + L^2 - R^2}$, where θ is the angle of rotation of rod OA.

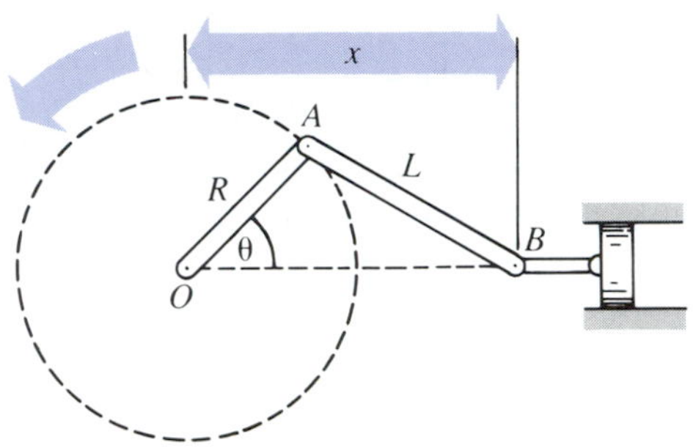

86. *Calculator Problem.* Use "sin x" to stand for the "sine of an angle of x radians," and complete the following table:

x	1.5	1	0.1	0.01	0.001	0.0001
$\sin x$						
$(\sin x)/x$						

Can you conjecture what happens to $(\sin x)/x$ as x approaches 0 from the right?

REVIEW EXERCISES

In Problems 1–5 find the solution of each inequality.

1. $\left|\dfrac{2}{x+4}\right| > 3$ **2.** $\left|\dfrac{3}{x-2}\right| \leq 5$ **3.** $\dfrac{1}{x^2-4} < 0$ **4.** $\dfrac{1}{x(x^2-2x-3)} > 0$ **5.** $|x^3 - 36| < 28$

6. Develop a formula for the distance between the parallel lines $y = mx + b_1$ and $y = mx + b_2$, where $b_1 \neq b_2$.

For the pairs of points in Problems 7–10, determine (a) the distance between them, (b) the midpoint of the line segment joining them, (c) the equation of the straight line through them, and (d) the equation of the perpendicular bisector of the line segment joining them.

7. $(0, 0)$ and $(-1, -2)$

8. $(1, 1)$ and $(5, 3)$

9. $(2, 3)$ and $(7, 3)$

10. $(\pi, 1)$ and $(1, \pi)$

In Problems 11–14 determine (a) the circle having the indicated line segment as a diameter and (b) the circle C having the same radius as the circle in part (a) and with the indicated line as a tangent line in the indicated location.

11. The line segment from Problem 7; the line from Problem 7, part (c), lying below C

12. The line segment from Problem 8; the line from Problem 8, part (c), lying above C

13. The line segment from Problem 9; the line from Problem 9, part (d), lying to the left of C

14. The line segment from Problem 10; the line from Problem 10, part (c), lying above C

15. The point $(2, 3)$ is a distance 4 units from the midpoint of the segment with endpoints $(-1, 4)$ and $(1, y)$. Find y.

16. Let P and Q be points in the plane with $|PQ| = d$. Let d_1 and d_2 be positive and real with $d_1 + d_2 = d$. Find the point R on the line PQ lying d_1 units away from P and d_2 units away from Q.

17. Evaluate: $\sin(\pi/8)$, $\cos(\pi/8)$, $\cos(\pi/16)$

18. Find the equation of the circle with center at (h, k) and passing through (x_0, y_0).

19. Find the set of all points equidistant from the intersection of the line L with equation $3x + 2y = 1$ and the circle C with equation $x^2 + 2x + y^2 - 2y = 4$.

20. Find the equation of a line containing the points of intersection of the circles

$$x^2 + y^2 + 4x - 8y + 4 = 0$$

and

$$x^2 + y^2 - 2x - 2y = 0$$

21. Find the center of the circle circumscribing the triangle with sides $x + y = 4$, $y = x$, and $x + 2y = 9$.

22. Develop a formula for the general equation of the tangent line to the circle $x^2 + y^2 = R^2$ at a point (x_0, y_0) on the circle.

23. Let $f(x) = \sqrt{4 - x}$. Simplify $\dfrac{f(x + \Delta x) - f(x)}{\Delta x}$ for $\Delta x \neq 0$, dividing out Δx in some manner.

24. Let $f(x) = \sin x$. Show that $\dfrac{f(x + \Delta x) - f(x)}{\Delta x}$ can be written as $\cos\left(x + \dfrac{\Delta x}{2}\right) \cdot \dfrac{\sin\left(\dfrac{\Delta x}{2}\right)}{\dfrac{\Delta x}{2}}$. What happens to $\dfrac{\sin\left(\dfrac{\Delta x}{2}\right)}{\dfrac{\Delta x}{2}}$ as Δx is taken closer and closer to 0?

25. Suppose

$$f(x) = \begin{cases} -x & \text{if } x < 0 \\ -2 & \text{if } x = 0 \\ \dfrac{-1}{x + 1} & \text{if } x > 0 \end{cases}$$

(a) Show graphically that f has an inverse.
(b) Find the domains of f and f^{-1}.
(c) Graph f^{-1} and find equations for f^{-1} over various parts of its domain.

26. Let $f(x) = (x^2 - 1)/(x - 1)$ and $g(x) = x + 1$. Are these the same functions? Why or why not?

27. Find the domain of the following functions.

(a) $f(x) = \sqrt{\dfrac{x^4 - 1}{x^2 - 1}}$ (b) $g(x) = \sqrt{\dfrac{1 - x^6}{1 + x^3}}$

28. Find the domain of each function and draw its graph.

(a) $f(x) = \sqrt{|x|}$ (b) $f(x) = \sqrt{|x^2 - 4|}$
(c) $f(x) = \sqrt{\sin x}$

29. Give three functions satisfying the relation $x^2 + (f(x))^2 = 1$. How many $f(x)$ satisfying this equation are there in all?

30. If f^{-1} and g^{-1} exist, need $(f \circ g)^{-1}$ exist?

31. A *fixed point* of f is a number x such that $f(x) = x$.

(a) If x_0 is a fixed point of f, show that $(f \circ f)(x_0) = x_0$.

(b) Find the fixed points of $f(x) = (2x + 3)/(x + 4)$.

(c) Show that, in general, $f(x) = (ax + b)/(cx + d)$, where a, b, c, and d are constants, has at most two fixed points. Discuss.

32. (a) Show that for angles θ and ϕ, we have

$$\tan(\theta + \phi) = \frac{\tan\theta + \tan\phi}{1 - \tan\theta\tan\phi}.$$

(b) Use part (a) to evaluate $\tan 105°$ as a radical expression.

33. A right circular cylinder of radius x and height y just fits inside a right circular cone of radius b and height h. Draw a plane section through the common axis of the cone and cylinder, and from the geometry of the figure read off the relation that expresses y as a function of x (b and h are regarded as constants).

34. If a^2 is an even integer, prove that a must also be even. [*Hint:* Assume the contrary.]

35. Prove that $\sqrt{2}$ is irrational. [*Hint:* Assume the contrary and use the result of Problem 34.]

36. Solve for x: $\cot x + \tan x = 2$.

37. Prove properties (a)–(d) for absolute values on page 8.

38. Let $P(x)$ be a polynomial. Write $P(x)$ as a sum of an even polynomial and an odd polynomial. See exercise set 1.5.

39. Let a, b, and c be real.

(a) Show that c being in an interval $a - b < c < a + b$ is equivalent to $|c - a| < b$.

(b) Show that c being in an interval $a - b < c < a + b$ where $c \neq a$ is equivalent to $0 < |c - a| < b$.

40. Let $P(x)$ and $Q(x)$ be polynomials. Show that each of the following is a polynomial.

(a) $P(x) \pm Q(x)$

(b) $P(x) \cdot Q(x)$

(c) Is $P(Q(x))$ a polynomial?

CHALLENGE EXERCISES

1. Let the line L have equation $Ax + By = C$ and let the point $P = (x_1, y_1)$ be given. Show that the distance from P to L is given by

$$d = \frac{|Ax_1 + By_1 - C|}{\sqrt{A^2 + B^2}}$$

[*Hint:* This formula can be derived using techniques from calculus and also techniques from vector arithmetic, but it can also be done without either!]

2. Consider a point $P = (x_1, y_1)$ and the line L given by the equation $Ax + By = C$. The distance d between P and L is the shortest distance between P and any point Q on L. Show that the nearest point Q_0 on L lies on a line M that is perpendicular to L.

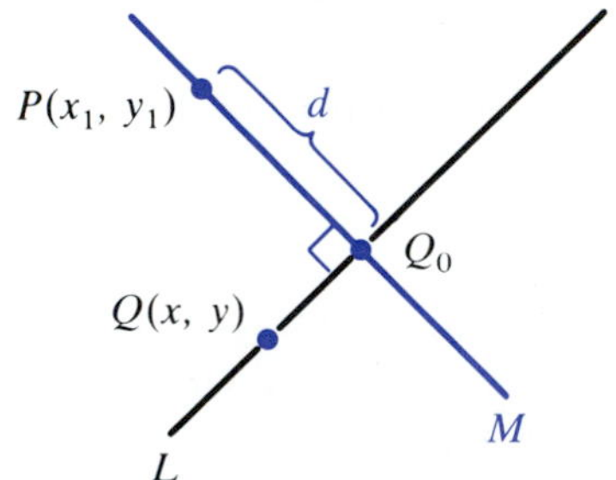

3. Let C be a circle with center P, and let Q be on C. Show that the line L determined by P and Q is perpendicular to the tangent line T to C at Q. Use the result of Problem 2.

4. Show that if the point P lies outside of the circle C with center Q and if two lines tangent to C meet at P, then these tangent lines make the same angle with the line PQ. Use the result of Problem 3.

5. Find the equations of all circles tangent to both $y = mx$ and $y = x/m$ for $m \neq \pm 1$ or 0. What can you say when $m = 0$? Use Problems 1 and 4.

6. Find the equation of the tangent line to the graph of $xy = 1/2$ at $x = \sqrt{2}/2$ using Problem 3. Here we take a tangent line to mean a line touching the graph of $xy = 1/2$ at $(\sqrt{2}/2; \sqrt{2}/2)$ and lying otherwise entirely above or below the graph of $xy = 1/2$ for some interval of x values, $a < x < b$, that includes $x = \sqrt{2}/2$.

7. Let $P = (x_1, y_1)$ and $Q = (x_2, y_2)$ be points in the plane not at the origin and where $x_1y_2 - x_2y_1 \neq 0$. Find the area of the parallelogram determined by O, P, and Q. Use the result of Problem 1.

8. Let points $P=(x_1, y_1)$ and $Q=(x_2, y_2)$ be given, with neither at the origin.

(a) For $x_1y_2 - x_2y_1 \neq 0$, find the center of the circle passing through P, Q, and the origin O.
(b) Show that if $x_1y_2 - x_2y_1 = 0$, then the points P and Q do not lie on a circle including the origin; rather they lie on a straight line through the origin.

9. Let $P=(x_1, y_1)$ and $Q=(x_2, y_2)$, where neither P nor O is at the origin O. Show that O, P, and Q are collinear if and only if $x_1y_2 - x_2y_1 = 0$. See Problem 7.

10. Consider the line L with equation $Ax + By = C$. Show that for (x, y) not on L, we have $Ax + By > C$ in one half-plane determined by L and $Ax + By < C$ in the other.

11. Find the domain of

$$h(x) = 1 + \cfrac{1}{1 + \cfrac{1}{1 + \cfrac{1}{1 + x}}}$$

12. Let $f(x)$ and $g(x)$ be functions. Define $\min(f, g)$ as

$$\min(f, g)(x) = \begin{cases} f(x) & \text{if } f(x) \leq g(x) \\ g(x) & \text{if } g(x) \leq f(x) \end{cases}$$

Show that

$$\min(f, g)(x) = \frac{f(x) + g(x)}{2} - \frac{|f(x) - g(x)|}{2}.$$

Find a formula for $\max(f, g)$, where

$$\max(f, g)(x) = \begin{cases} f(x) & \text{if } f(x) \geq g(x) \\ g(x) & \text{if } f(x) \leq g(x) \end{cases}$$

13. Let $f(x)$ be any function with domain symmetric about $x = 0$. Show that $f(x)$ can be written in exactly one way as a sum of an even and an odd function with the same domain. (See the exercise set for Section 1.5.)

14. Consider the setup from Problem 62, p. 42. Find the rectangle of maximum area.

15. Consider the setup for Problem 68, p. 42. Find the garden of maximum area.

16. We say that y is an algebraic function of x if it is a function that satisfies an irreducible equation

$$P_0(x)y^n + P_1(x)y^{n+1} + \cdots + P_{n-1}(x)y + P_n(x) = 0$$

where the $P_i(x)$ are polynomials for $0 \leq i \leq n$.

(a) Show that $y = \sqrt{x}$ is algebraic.

(b) Show that $y = \dfrac{-x + \sqrt{x^2 - 4}}{2}$ is algebraic.

(c) Is $y = \sqrt{x} + \sqrt[3]{x}$ algebraic?

2 The Limit of a Function

The concept of the limit of a function is what bridges the gap between the mathematics of algebra and geometry and the mathematics of calculus. Although the idea of a limit is difficult to understand at first, it can be mastered. In fact, the *evaluation* of limits is fairly easy (after practice). For this reason, we will develop limits intuitively at first, then develop a technique for finding limits, and finally look at a precise formulation of limits and limit theorems.

2.1 Limits from an Intuitive Point of View

Calculus was developed in response to the difficulty of answering certain questions in geometry. Two of these questions are:

1. Given a function f and a point P on its graph, what is the slope of the line tangent to the graph of the function at the point P? See Figure 1.

2. Given a nonnegative function f whose domain is the closed interval $[a, b]$, what is the area enclosed by the graph of f, the x-axis, and the vertical lines $x = a$ and $x = b$? See Figure 2.

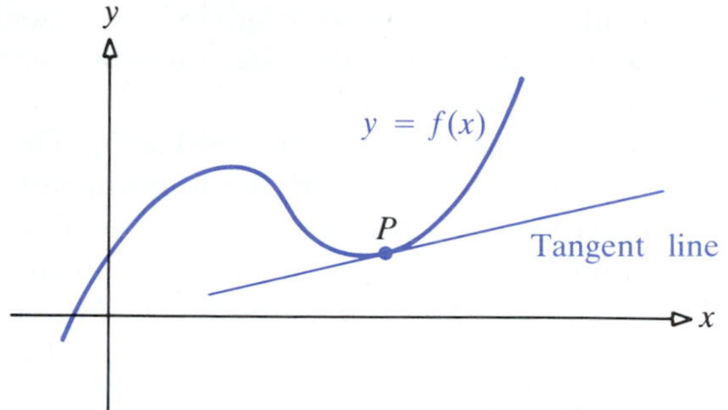

Figure 1

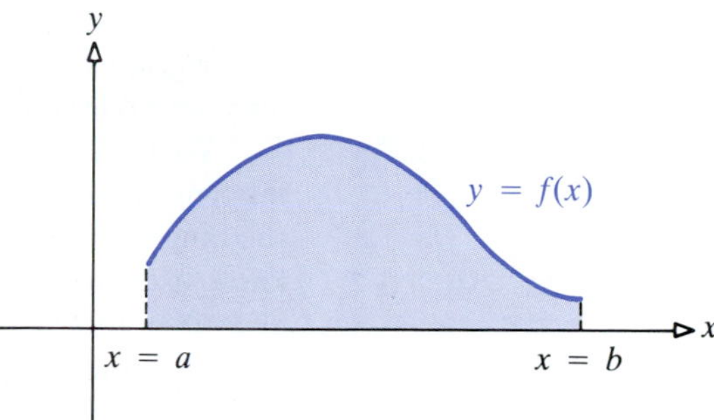

Figure 2

These problems, traditionally called the *tangent problem* and the *area problem*, were solved by Gottfried Wilhelm von Leibniz (1646–1716) and Sir Isaac Newton (1642–1727).* Their solutions were shown to be intimately related not only to each other but also to many other problems concerning the behavior of functions. Here, we will discuss the tangent problem; the area problem is postponed until Chapter 5.

The Tangent Problem

Consider the function whose graph appears in Figure 1. Through the point P on the graph we have drawn a line that just touches the graph of f. We call this unique line the *tangent*† *line* to the graph of f at P. Our first problem is to give a satisfactory definition of a tangent line.

Let's start with some basics. In plane geometry, a tangent line to a circle is defined as a line having exactly one point in common with the circle (see Fig. 3). However, this definition is not satisfactory for graphs in general. For example, consider the graph shown in Figure 4. There are many

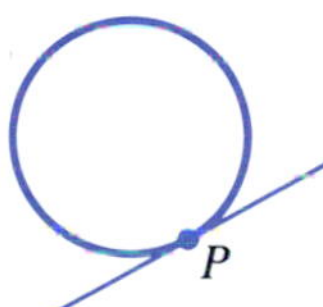

Tangent line to circle at P

Figure 3

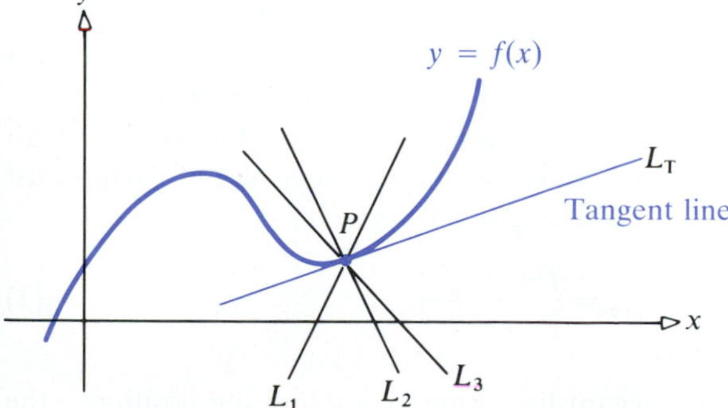

Figure 4

* The Historical Perspectives section at the end of this chapter traces the various solutions to the tangent problem from the Greeks to Fermat, who preceded Leibniz and Newton.

† *Tangere* is a Latin word meaning "to touch."

lines—for instance, L_1, L_2, L_3—that pass through the point P and have exactly one point in common with the graph, but they do not meet the requirement of just touching the graph at P. Furthermore, the line L_T, which does just touch the graph of f at P, also intersects the graph at another point.

It is evident that the definition from plane geometry for circles will not do for graphs in general. We need a different definition that will work not only for circles but also for the graph of any function.

We begin with a function f and a point P on its graph (see Fig. 5). The tangent line L_T to f at P will necessarily pass through P. But to distinguish the tangent line L_T from all the other lines that pass through P, we need to know its slope $m_{\tan}$. If we use the number c to represent the x-coordinate of P, then the coordinates of P are $(c, f(c))$. To get the slope $m_{\tan}$ of L_T, we need to know two points on L_T. However, we cannot find two points on L_T, so we look for another way to find $m_{\tan}$.

Suppose $Q = (x, f(x))$ is any point different from P on the graph of f. Then the line through P and Q is called a *secant line* of f (see Fig. 6).

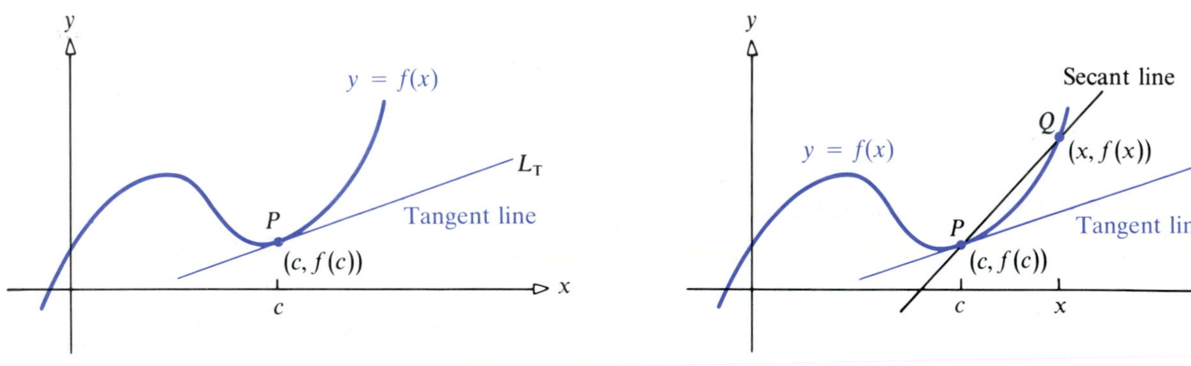

Figure 5

Figure 6

Now, in Figure 7, we show three different points and the secant lines L_1, L_2, and L_3. As x gets closer to c, the secant lines $L_1, L_2, L_3, \ldots$ tend to a limiting position. The line L_T, the limiting position of these secant lines, is the *tangent line to the graph of f at c*. As Figure 8 illustrates, this new idea of a tangent line coincides with the traditional one when applied to circles.

But what is the slope of the tangent line L_T? We know L_T passes through the point $(c, f(c))$. Furthermore, we know that the slope $m_{\sec}$ of each secant line joining the points $(c, f(c))$ and $(x, f(x))$ is given by

$$m_{\sec} = \frac{f(x) - f(c)}{x - c} \tag{1}$$

As x gets closer to c, the secant lines approach a limiting position—the tangent line L_T. So we may expect the limiting value of the slopes of the secant lines to equal the slope of the tangent line. That is, (1) suggests that

$$m_{\tan} = \begin{bmatrix} \text{Slope of tangent line} \\ \text{to } f \text{ at } c \end{bmatrix} = \begin{bmatrix} \text{Limiting value of } \dfrac{f(x) - f(c)}{x - c} \\ \text{as } x \text{ gets close to } c \end{bmatrix}$$

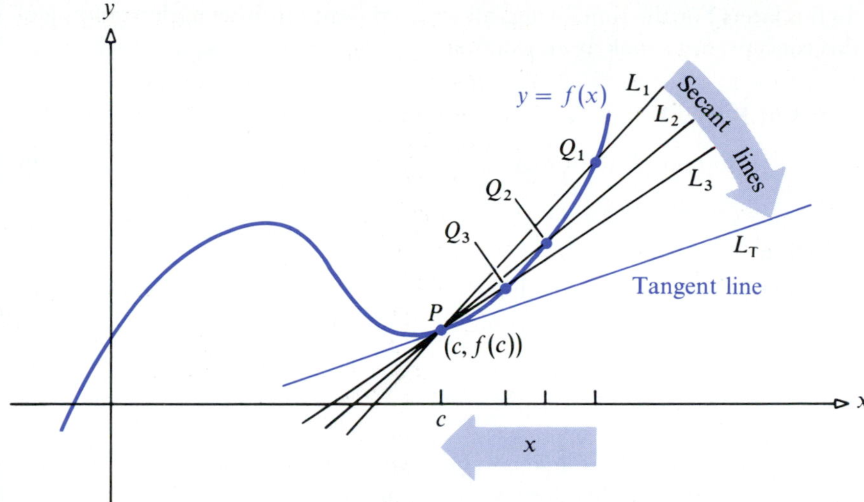

Figure 7

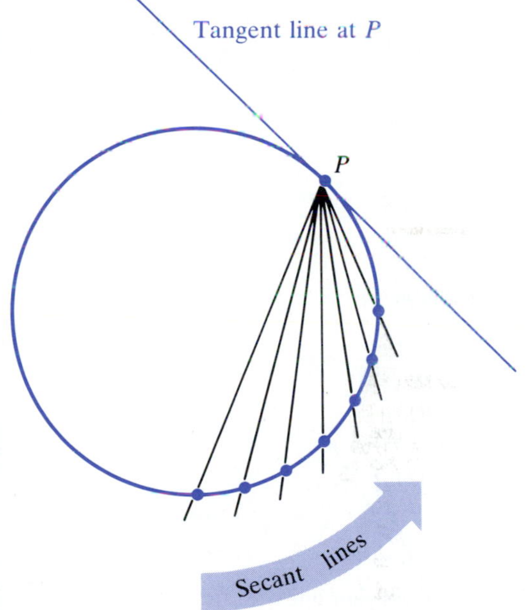

Figure 8

In symbols, we write

$$m_{\tan} = \lim_{x \to c} \frac{f(x) - f(c)}{x - c} \tag{2}$$

The notation "lim" with "$x \to c$" beneath it (or set to the side as "$\lim_{x \to c}$") is read "the limit as x approaches c of . . . ."

In order to study the tangent problem in more detail, the notion of limiting value or limit will have to be made mathematically precise. We will do this later. For the time being, all we need is an intuitive understanding of this concept. Let's look at an example.

EXAMPLE 1

Find an equation of the tangent line to the graph of $f(x) = x^2$ at the point (2, 4).

Solution

The tangent line passes through the point (2, 4). The slopes of the secant lines joining (2, 4) to points $(x, f(x))$ on the graph of $y = x^2$ are

$$m_{\text{sec}} = \frac{f(x) - f(2)}{x - 2} = \frac{f(x) - 4}{x - 2} = \frac{x^2 - 4}{x - 2}$$

Since the point $(x, f(x))$ is taken to be different from the point (2, 4), we conclude that $x \neq 2$. Therefore, we may simplify to get

$$m_{\text{sec}} = \frac{(x-2)(x+2)}{x-2} = x + 2$$

Now comes the important step in this procedure! As x gets closer to 2, the values of $m_{\text{sec}} = x + 2$ get close to 4. That is, the slope $m_{\tan}$ of the tangent line at (2, 4) is

$$m_{\tan} = \lim_{x \to 2} \frac{x^2 - 4}{x - 2} = \lim_{x \to 2} (x + 2) = 4$$

Therefore, an equation of this tangent line is

$$y - 4 = 4(x - 2)$$
$$y = 4x - 4$$

See Figure 9.

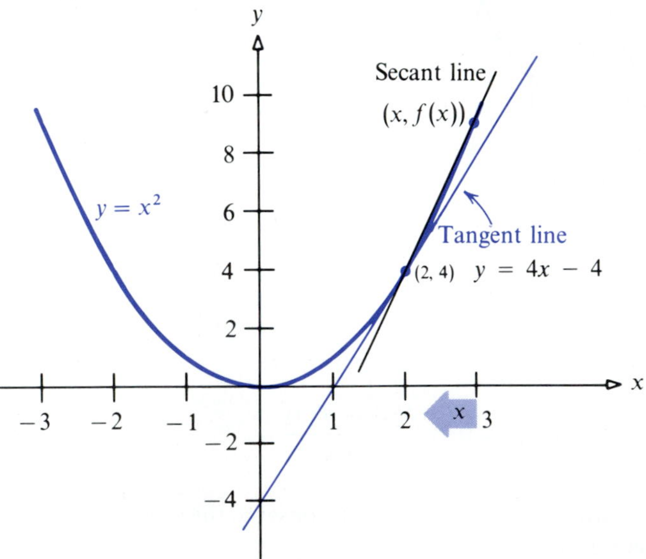

Figure 9

Now we can turn our attention to a more detailed description of the concept of limit. Our approach will proceed in stages that eventually lead to a precise definition of limit.

The Idea of a Limit

We begin by asking a question: What does it mean for a function f to have a limit L as x approaches some fixed number c? To find an answer, we need to be more precise about f, L, and c. The function f must be defined in an open interval near the number c, but it does not have to be defined at c itself. The limit L is some number. With these restrictions in mind, we introduce the symbolism

$$\lim_{x \to c} f(x) = L \tag{3}$$

which is read "the limit of $f(x)$ as x approaches c equals the number L." This indicates that f has a limit L as x approaches c. We may describe (3) in two ways:

For all x approximately equal to c, but $x \neq c$, the value $f(x)$ is approximately equal to L.

For all x sufficiently close to c, but unequal to c, the value $f(x)$ can be made as close as we please to L.

In Figure 10(a) we show the graph of the function f and observe that as x gets closer to c, the value of f, as measured along the y-axis, gets closer to the number L. This is the key idea behind the notion of a limit. Note that the value of f at c does not matter. As Figure 10(b) illustrates, even though f is not defined at c, it is still true that as x gets closer to c, the value of f gets closer to L.

Numerical Approach

Based on our interpretation of (3), it is natural to try a numerical approach as an aid in calculating limits.

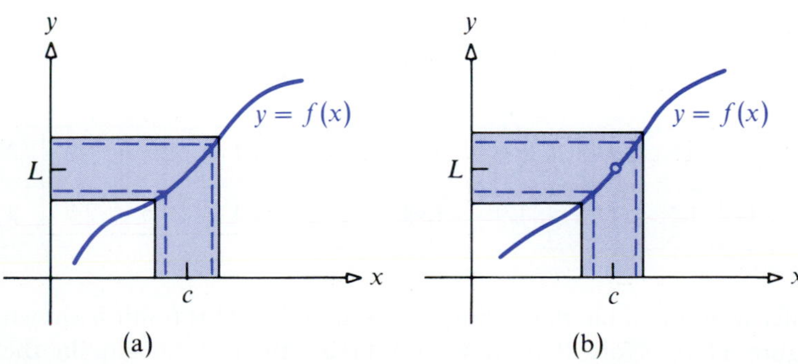

Figure 10

EXAMPLE 2

Let $f(x) = x^3$ and $c = 2$. The numerical approach uses a table and a calculator to guess $\lim_{x\to 2} x^3$.

x	1	1.5	1.6	1.75	1.8	1.9	1.99	1.999	1.9999
$f(x) = x^3$	1	3.375	4.096	5.359	5.832	6.859	7.8806	7.988	**7.9988**

x	3	2.5	2.4	2.25	2.2	2.1	2.01	2.001	2.0001
$f(x) = x^3$	27	15.625	13.824	11.3906	10.648	9.261	8.1206	8.012	**8.0012**

We infer that for x "sufficiently close" to 2, the value of $f(x) = x^3$ can be made "as close as we please" to 8; that is, $\lim_{x\to 2} x^3 = 8$.

EXAMPLE 3

Use a numerical approach to find

(a) $\lim_{x\to 1} f(x)$, where $f(x) = \dfrac{x^2 - 1}{x - 1}$

(b) $\lim_{x\to 1} g(x)$, where $g(x) = x + 1$

Solution

(a) As in Example 2, we construct a table. The choices for x, though arbitrary, are selected so that they are close to 1; some are chosen less than 1 and some greater than 1.

x	0	0.5	0.75	0.9	0.99	1.01	1.1	1.25	1.5	2
$\dfrac{x^2 - 1}{x - 1}$	1	1.5	1.75	1.9	**1.99**	**2.01**	2.1	2.25	2.5	3

We infer from the table that $\lim_{x\to 1}[(x^2 - 1)/(x - 1)] = 2$. This result is shown graphically in Figure 11(a).

(b) Similarly, we construct a table for $\lim_{x\to 1}(x + 1)$.

x	0	0.5	0.75	0.9	0.99	1.01	1.1	1.25	1.5	2
$x + 1$	1	1.5	1.75	1.9	**1.99**	**2.01**	2.1	2.25	2.5	3

We infer from the table that $\lim_{x\to 1}(x + 1) = 2$. This result is shown in Figure 11(b). Note that in Figure 11(b), there is no gap in the graph of the function $g(x) = x + 1$.

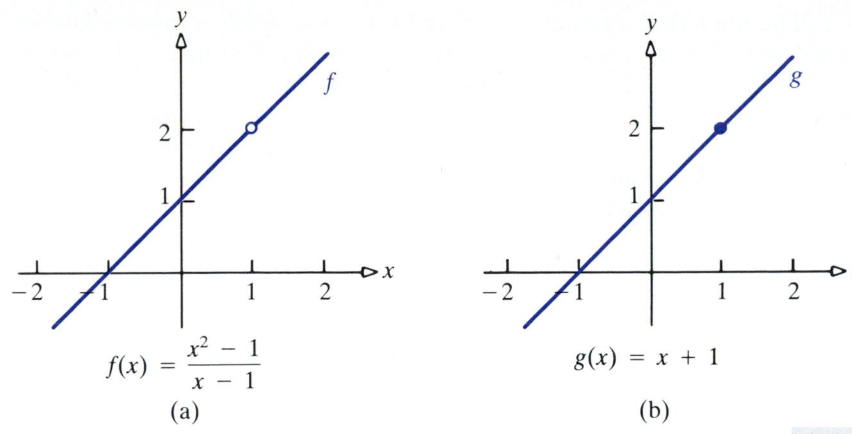

$f(x) = \dfrac{x^2 - 1}{x - 1}$
(a)

$g(x) = x + 1$
(b)

Figure 11

You are right if you make the observation that as long as x is close to 1 but not equal to 1, f can be simplified as

$$f(x) = \frac{x^2 - 1}{x - 1} = \frac{(x - 1)(x + 1)}{(x - 1)} = x + 1 = g(x) \qquad x \neq 1$$

Now it is easy to see that for x close to 1, $f(x)$ will be close to $1 + 1 = 2$.

There are limits, though, for which this type of simplification does not work.

EXAMPLE 4

Use a numerical approach to find $\lim_{x \to 0} \dfrac{\sin x}{x}$.

Solution

The functions $\sin x$ and x are familiar, and their quotient $(\sin x)/x$ can be calculated easily at every point except 0. At 0, the value of both functions is 0, and the meaningless ratio (0/0) presents us with a problem. However, a careful calculation (using a calculator) for numbers x close to 0 yields the following table:

x (radians)	1	0.5	0.1	0.01	−0.01	−0.1	−0.5	−1
$\sin x$	0.8415	0.4794	0.0998	0.0100	−0.0100	−0.0998	−0.4794	−0.8415
$\dfrac{\sin x}{x}$	0.8415	0.9589	0.9983	**1.0000**	**1.0000**	0.9983	0.9589	0.8415

We infer that the ratio $(\sin x)/x$ approaches the value 1 as x tends to 0, and we write $\lim_{x \to 0} \dfrac{\sin x}{x} = 1$.

The numerical approach we have been using involves making guesses based on a table of numerical values. It is possible that if we had used numbers x even closer to 0 in the table of Example 4, the value of $(\sin x)/x$ might have been shown to be some number that is not close to 1. What we really need is a proof that $\lim_{x\to 0}(\sin x)/x = 1$. This proof is given in Section 2.5. For now, though, let's continue our intuitive development of limits. This time we use a graphic approach.

Graphic Approach

Let's look at some more examples. In these examples, we will graph each function and use the graph to find the limit.

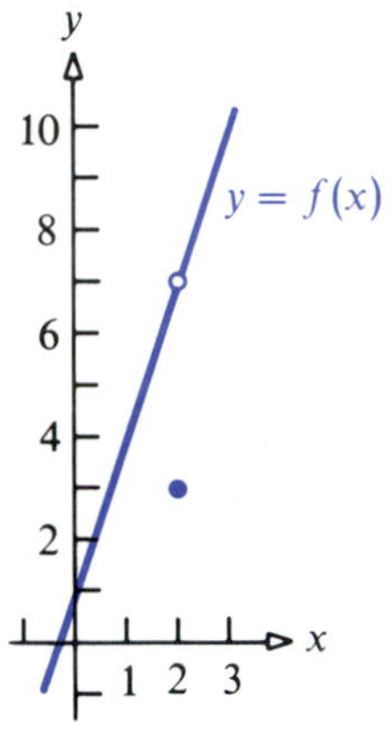

EXAMPLE 5

Use the graph of the function below to find $\lim_{x\to 2} f(x)$:

$$f(x) = \begin{cases} 3x + 1 & \text{if} \quad x \neq 2 \\ 3 & \text{if} \quad x = 2 \end{cases}$$

Solution

First, we graph f in Figure 12. We observe that for x near 2, the value of the function f is near 7. In fact, by choosing x close enough to 2, we can force the value of f to get as close as we please to 7. We conclude that $\lim_{x\to 2} f(x) = 7$. Note that the value of f at 2 is $f(2) = 3$.

Figure 12

Example 5 illustrates the following important principle about limits:

The limit L of a function $y = f(x)$ as x approaches the number c *does not* depend on the value of f at c.

If there is *no single* number that the value of f approaches, we say that f *has no limit as x approaches c*, or, more simply, that the *limit does not exist at c*. The next example illustrates this for $c = 0$.

EXAMPLE 6

Use the graph of the function below to find $\lim_{x\to 0} f(x)$, if it exists:

$$f(x) = \begin{cases} -1 & \text{if} \quad x < 0 \\ 1 & \text{if} \quad x > 0 \end{cases}$$

Solution

Figure 13 shows the graph of f. We observe that if x is close to 0 and negative, the value of f equals -1. On the other hand, if x is close to 0 and positive, the value of f equals 1. Since there is no single number that the values of f are close to when x is close to 0, we conclude that $\lim_{x\to 0} f(x)$ does not exist at 0.

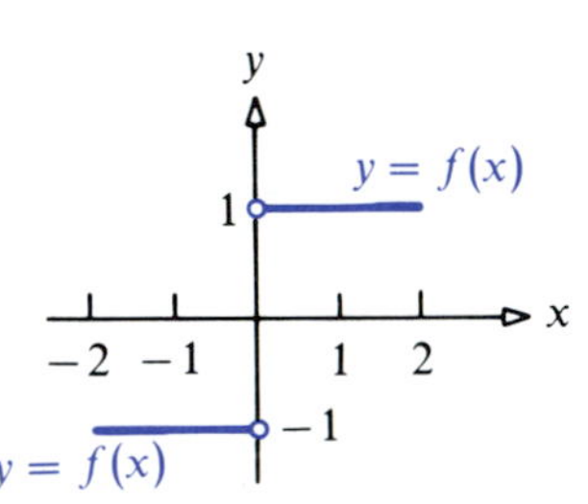

Figure 13

In the next two examples we use the graphic approach to obtain two limits that we will rely on quite heavily in the next section.

EXAMPLE 7

Use a graphic approach to show that for a constant function $f(x) = A$ and any real number c,

$$\lim_{x \to c} f(x) = A \tag{4}$$

Solution

The function f is the constant function $f(x) = A$, whose graph is a horizontal line (see Fig. 14). For any choice of c, it is clear that if x is close to c, the value of f remains at A. Therefore, we conclude that $\lim_{x \to c} A = A$.

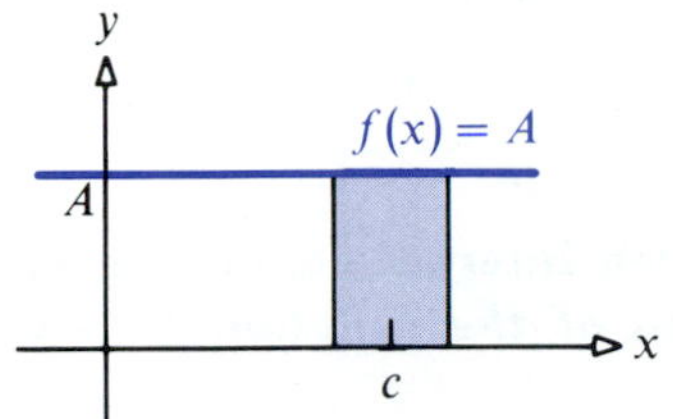

For x close to c, the value of f remains at A
$\lim_{x \to c} A = A$

Figure 14

Using the result (4) in Example 7 we see that

$$\lim_{x \to 5} 2 = 2 \qquad \lim_{x \to \sqrt{2}} \frac{1}{3} = \frac{1}{3} \qquad \lim_{x \to 5} (-\pi) = -\pi$$

EXAMPLE 8

Use a graphic approach to show that

$$\lim_{x \to c} x = c \qquad c \text{ a real number} \tag{5}$$

Solution

The function f is the identity function $f(x) = x$, whose graph is the straight line illustrated in Figure 15. For any choice of c, we see that if x is close to c, the value of f is just as close to c. We conclude that $\lim_{x \to c} x = c$.

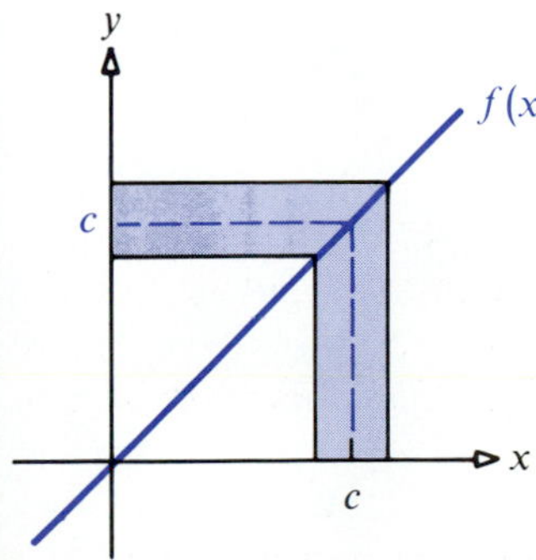

For x close to c, the value of f is just as close to c
$\lim_{x \to c} x = c$

Figure 15

Using the result (5) in Example 8 we see that

$$\lim_{x \to -5} x = -5 \qquad \lim_{x \to \sqrt{3}} x = \sqrt{3} \qquad \lim_{x \to 5} x = 5$$

In each of these examples, we draw conclusions about $\lim_{x \to c} f(x)$ by asking whether the values of f can be forced to be "as close as we please" to a single number L when x is "sufficiently close to" c. But the expressions "sufficiently close to" and "as close as we please" are too vague for practical purposes. To treat the concept precisely, we need a *formal* definition of *limit*. For the sake of completeness and accuracy, we now give the definition, but we will postpone a detailed discussion of it until Section 2.6. As is customary, the definition uses the Greek letters ε (epsilon) and δ (delta) and is sometimes referred to as the ε, δ *definition of a limit*.

[2.1.1] DEFINITION / ***Limit of a Function.***

Let f be a function defined on an open interval containing c, except possibly for c itself. Then the *limit of the function f as x approaches c* is the number L, written

$$\lim_{x \to c} f(x) = L$$

if, for any given number $\varepsilon > 0$, a number $\delta > 0$ exists such that

$$|f(x) - L| < \varepsilon \qquad \text{whenever} \qquad 0 < |x - c| < \delta$$

EXERCISE 2.1

In Problems 1–10 complete each table and use it to evaluate the indicated limit. A calculator will be helpful.

1.

x	0.9	0.99	0.999	1.001	1.01	1.1
$f(x) = 2x$						

$\lim_{x \to 1} 2x = ?$

2.

x	1.9	1.99	1.999	2.001	2.01	2.1
$f(x) = x + 3$						

$\lim_{x \to 2} (x + 3) = ?$

3.

x	0.1	0.01	0.001	−0.001	−0.01	−0.1
$f(x) = x^2 + 2$						

$\lim_{x \to 0} (x^2 + 2) = ?$

4.

x	-1.1	-1.01	-1.001	-0.999	-0.99	-0.9
$f(x)=x^2-2$						

$\lim_{x\to -1}(x^2-2)=?$

5.

x	-2.5	-2.9	-2.99	-3.01	-3.1	-3.5
$f(x)=\dfrac{x^2-9}{x+3}$						

$\lim_{x\to -3}\dfrac{x^2-9}{x+3}=?$

6.

x	-0.2	-0.1	-0.01	0.01	0.1	0.2
$f(x)=\dfrac{x^2+x}{x}$						

$\lim_{x\to 0}\dfrac{x^2+x}{x}=?$

7.

x (radians)	-0.2	-0.1	-0.01	0.01	0.1	0.2
$f(x)=\dfrac{\tan x}{x}$						

$\lim_{x\to 0}\dfrac{\tan x}{x}=?$

8.

x (radians)	-0.2	-0.1	-0.01	0.01	0.1	0.2
$f(x)=\dfrac{1-\cos x}{x}$						

$\lim_{x\to 0}\dfrac{1-\cos x}{x}=?$

9.

x (radians)	-0.2	-0.1	-0.01	0.01	0.1	0.2
$f(x)=\dfrac{\sin x}{1+\tan x}$						

$\lim_{x\to 0}\dfrac{\sin x}{1+\tan x}=?$

10.

x (radians)	-0.2	-0.1	-0.01	0.01	0.1	0.2
$f(x) = \dfrac{\tan x}{\cos x}$						

$$\lim_{x\to 0} \frac{\tan x}{\cos x} = ?$$

In Problems 11–18 use each graph to determine whether $\lim_{x\to c} f(x)$ exists at the number c.

11.

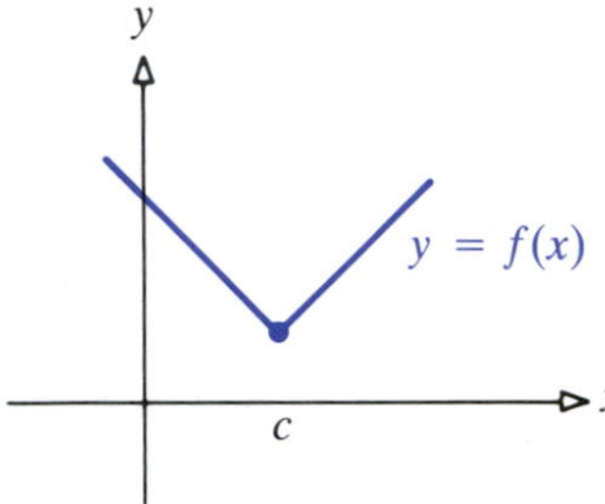

12.

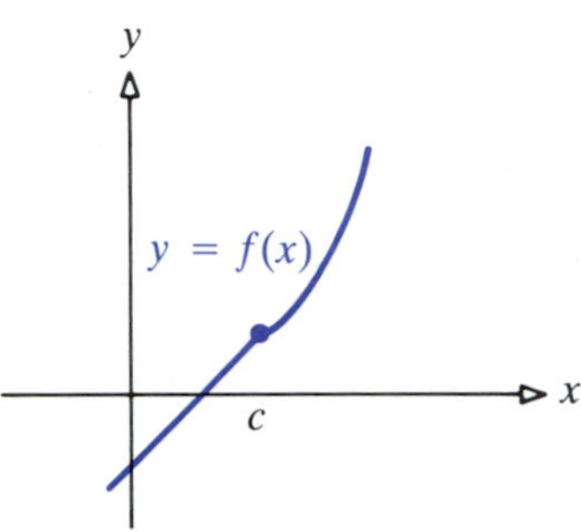

13.

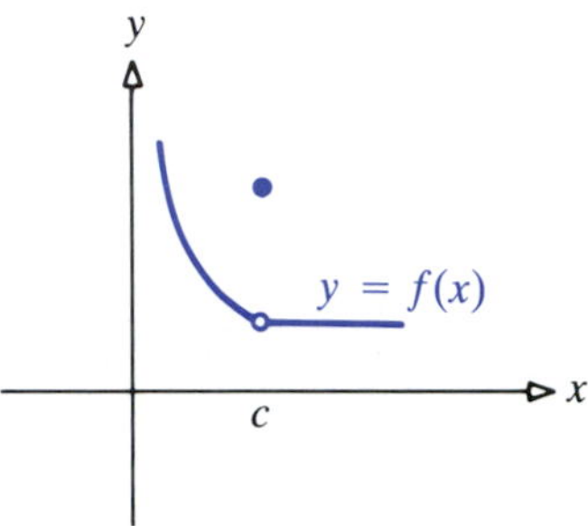

14.

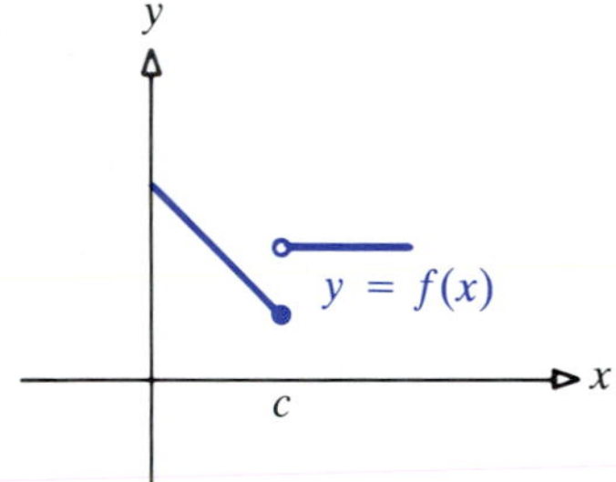

15.

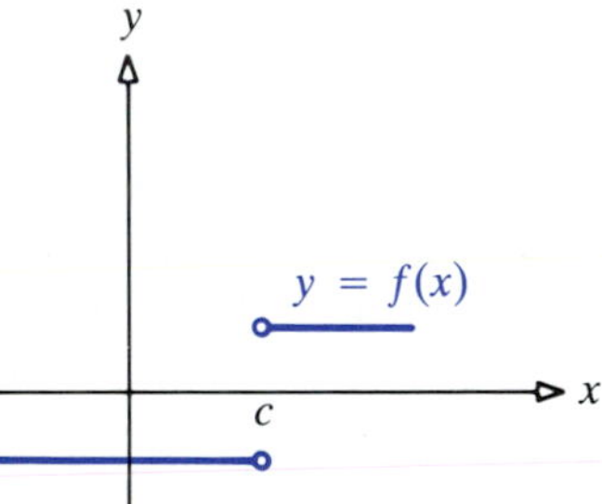

16.

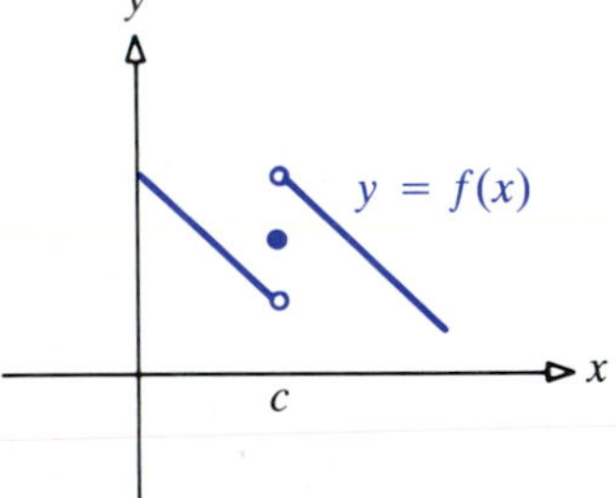

17.

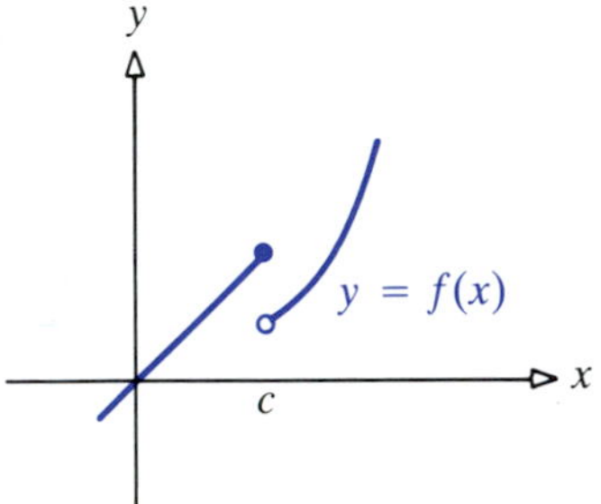

18.

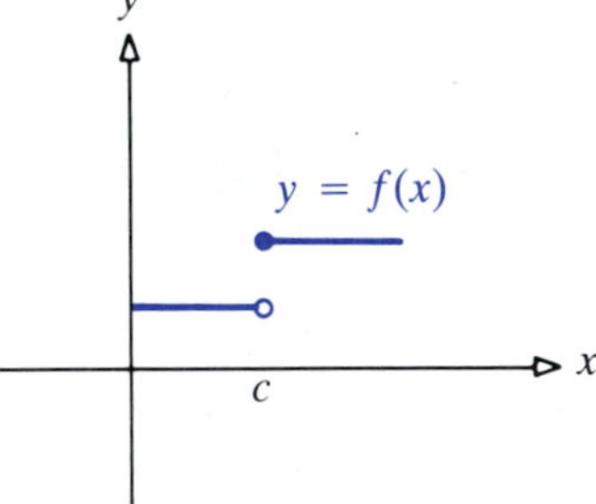

In Problems 19–26 use a graph to determine whether $\lim_{x\to c} f(x)$ exists at the given number c.

19. $f(x) = \begin{cases} 2x + 5 & \text{if } x \le 2 \\ 4x + 1 & \text{if } x > 2 \end{cases}$ $\quad c = 2$

20. $f(x) = \begin{cases} 2x + 1 & \text{if } x \le 0 \\ 2x & \text{if } x > 0 \end{cases}$ $\quad c = 0$

21. $f(x) = \begin{cases} 3x - 1 & \text{if } x < 1 \\ 4 & \text{if } x = 1 \\ 2x & \text{if } x > 1 \end{cases}$ $\quad c = 1$

22. $f(x) = \begin{cases} 3x - 1 & \text{if } x < 1 \\ 2 & \text{if } x = 1 \\ 2x & \text{if } x > 1 \end{cases}$ $\quad c = 1$

23. $f(x) = \begin{cases} 3x - 1 & \text{if } x < 1 \\ \text{Not defined} & \text{if } x = 1 \\ 2x & \text{if } x > 1 \end{cases} \quad c = 1$

24. $f(x) = \begin{cases} 3x - 1 & \text{if } x < 1 \\ 2 & \text{if } x = 1 \\ 3x & \text{if } x > 1 \end{cases} \quad c = 1$

25. $f(x) = \begin{cases} x^2 & \text{if } x \le 0 \\ 2x + 1 & \text{if } x > 0 \end{cases} \quad c = 0$

26. $f(x) = \begin{cases} x^2 & \text{if } x < -1 \\ 2 & \text{if } x = -1 \\ -3x + 2 & \text{if } x > -1 \end{cases} \quad c = -1$

In Problems 27–30 use the results of Examples 7 and 8 to find each limit.

27. $\lim_{x \to 0} (-3)$

28. $\lim_{x \to -2} (-6)$

29. $\lim_{x \to -1} x$

30. $\lim_{x \to 0} x$

31. For $f(x) = 3x^2$:

(a) Find the slope of the secant line joining the points (2, 12) and (3, 27).

(b) Use the method illustrated in Example 1 to find an equation of the tangent line to the graph of f at 2.

(c) Graph f, show the tangent line at 2, and show the secant line from part (a).

32. For $f(x) = x^3$:

(a) Find the slope of the secant line joining the points (2, 8) and (3, 27).

(b) Use the method illustrated in Example 1 to find an equation of the tangent line to the graph of f at 2.

(c) Graph f, show the tangent line at 2, and show the secant line from part (a).

33. For $f(x) = \frac{1}{2}x^2 - 1$:

(a) Determine the slope m_{sec} of the secant line joining the points $P = (2, f(2))$ and $Q = (2 + h, f(2 + h))$.

(Problem 33 continues in the next column)

33. *(continued)*

(b) Use the result found in part (a) to determine the slopes of the secant lines for $h = -0.5$ and $h = 0.5$.

(c) Use the results in parts (a) and (b) to complete the following table:

h	-0.5	-0.1	-0.001	0.001	0.1	0.5
m_{sec}						

(d) Determine the limiting value of the slope of the secant line found in part (a) as $h \to 0$.

(e) Find an equation for the tangent line to f at the point $P = (2, f(2))$.

(f) Draw the graph of the function f and the tangent line to f at $P = (2, f(2))$.

34. What is $\lim_{x \to \sqrt{2}} (\sin^2 x + \cos^2 x)$?

Algebraic Techniques for Finding Limits 2.2

The results from Examples 7 and 8 in Section 2.1 and the algebraic properties of limits stated in the next theorem enable us to compute the limits of many functions encountered in calculus. Some of these properties are proved in Appendix I. In Theorems 2.2.1 through 2.2.7 we assume that f and g are two functions for which $\lim_{x \to c} f(x) = L$ and $\lim_{x \to c} g(x) = M$, with L and M being two real numbers.

[2.2.1] THEOREM / ***Limit of a Sum.***

$$\lim_{x \to c}[f(x) + g(x)] = \lim_{x \to c} f(x) + \lim_{x \to c} g(x) = L + M \qquad (1)$$

That is, the limit of the sum of two functions equals the sum of their limits.

EXAMPLE 1

$$\lim_{x\to 1}(x+5) = \lim_{x\to 1} x + \lim_{x\to 1} 5 = 1+5=6$$

By (1) — By (4) and (5) in Section 2.1

[2.2.2] THEOREM / *Limit of a Difference.*

$$\lim_{x\to c}[f(x)-g(x)] = \lim_{x\to c} f(x) - \lim_{x\to c} g(x) = L - M \tag{2}$$

The limit of the difference of two functions equals the difference of their limits.

EXAMPLE 2

$$\lim_{x\to 3}(4-x) = \lim_{x\to 3} 4 - \lim_{x\to 3} x = 4-3=1$$

By (2) — By (4) and (5) in Section 2.1

[2.2.3] THEOREM / *Limit of a Constant Times a Function.*

If k is any real number,

$$\lim_{x\to c}[kf(x)] = k\lim_{x\to c} f(x) = kL \tag{3}$$

The limit of a constant times a function equals the constant times the limit of the function.

EXAMPLE 3

$$\lim_{x\to -2}(3x+5) = \lim_{x\to -2}(3x) + \lim_{x\to -2} 5 = 3\lim_{x\to -2} x + \lim_{x\to -2} 5 = 3(-2)+5 = -1$$

By (1) — By (3) — By (4) and (5) in Section 2.1

[2.2.4] THEOREM / *Limit of a Product.*

$$\lim_{x\to c}[f(x)g(x)] = \left[\lim_{x\to c} f(x)\right]\left[\lim_{x\to c} g(x)\right] = LM \tag{4}$$

The limit of the product of two functions equals the product of their limits.

EXAMPLE 4

$$\lim_{x\to 1}[(2x)(x+4)] = \left[\lim_{x\to 1}(2x)\right]\left[\lim_{x\to 1}(x+4)\right] = (2)(5) = 10$$

By (4) — By (3) and (1)

[2.2.5] THEOREM / *Limit of a Quotient.*

If $\lim_{x\to c} g(x) = M \neq 0$,

$$\lim_{x\to c} \frac{f(x)}{g(x)} = \frac{\lim_{x\to c} f(x)}{\lim_{x\to c} g(x)} = \frac{L}{M} \tag{5}$$

If the limit of the denominator is not 0, the limit of the quotient of two functions equals the quotient of their limits.

EXAMPLE 5

$$\lim_{x\to 2} \frac{3x+4}{x+3} \underset{\text{By (5)}}{=} \frac{\lim_{x\to 2}(3x+4)}{\lim_{x\to 2}(x+3)} = \frac{10}{5} = 2$$

[2.2.6] THEOREM / *Limit of $[f(x)]^n$.*

If n is a positive integer,

$$\lim_{x\to c}[f(x)]^n = \left[\lim_{x\to c} f(x)\right]^n = L^n \tag{6}$$

EXAMPLE 6

$$\lim_{x\to 1}(2x-3)^3 \underset{\text{By (6)}}{=} \left[\lim_{x\to 1}(2x-3)\right]^3 = (-1)^3 = -1$$

[2.2.7] THEOREM / *Limit of $\sqrt[n]{f(x)}$.*

If $n \geq 2$ is an integer,

$$\lim_{x\to c} \sqrt[n]{f(x)} = \sqrt[n]{\lim_{x\to c} f(x)} = \sqrt[n]{L} \tag{7}$$

(Here we require $f(x) \geq 0$ and $L \geq 0$ if n is even.)

EXAMPLE 7

$$\lim_{x\to 4} \sqrt[3]{x^2+11} \underset{\text{By (7)}}{=} \sqrt[3]{\lim_{x\to 4}(x^2+11)} = \sqrt[3]{16+11} = 3$$

EXAMPLE 8

Show that $\lim_{x\to 0}|x| = 0$.

Solution

Since $|x| = \sqrt{x^2}$, we have

$$\lim_{x\to 0}|x| = \lim_{x\to 0}\sqrt{x^2} \underset{\text{By (7)}}{=} \sqrt{\lim_{x\to 0} x^2} = \sqrt{0} = 0$$

It is useful to observe in (6) the special case for $f(x) = x$, namely,

$$\lim_{x\to c} x^n = c^n \qquad n \text{ a positive integer} \tag{8}$$

By using this result together with (3), we conclude

$$\lim_{x\to c} kx^n = kc^n \qquad k \text{ a real number, } n \text{ a positive integer} \tag{9}$$

These results demonstrate that *some* limits may be evaluated by merely substituting c for x. This is true for polynomial functions:

[2.2.8] THEOREM / *Limit of a Polynomial.*

For every polynomial $P(x) = a_nx^n + a_{n-1}x^{n-1} + \cdots + a_1x + a_0$ of degree n, we have

$$\lim_{x\to c} P(x) = P(c) \tag{10}$$

We can see that this is true because

$$\begin{aligned}\lim_{x\to c} P(x) &= \lim_{x\to c}(a_nx^n + a_{n-1}x^{n-1} + \cdots + a_1x + a_0)\\ &\underset{\text{By repeated use of (1)}}{=} \lim_{x\to c}(a_nx^n) + \lim_{x\to c}(a_{n-1}x^{n-1}) + \cdots + \lim_{x\to c}(a_1x) + \lim_{x\to c} a_0\\ &\underset{\text{By repeated use of (9)}}{=} a_nc^n + a_{n-1}c^{n-1} + \cdots + a_1c + a_0 = P(c)\end{aligned}$$

EXAMPLE 9

(a) $\lim_{x\to 3}(4x^2 - x + 2) = 4(3)^2 - 3 + 2 = 35$

(b) $\lim_{x\to -1}(7x^5 + 4x^3 - 2x^2) = 7(-1)^5 + 4(-1)^3 - 2(-1)^2 = -13$

(c) $\lim_{x\to 0}(10x^6 - 4x^5 - 8x + 5) = 10(0)^6 - 4(0)^5 - 8(0) + 5 = 5$

[2.2.9] THEOREM

If c is in the domain of a rational function P/Q, so that $Q(c) \neq 0$, then

$$\lim_{x\to c} \frac{P(x)}{Q(x)} = \frac{P(c)}{Q(c)} \tag{11}$$

EXAMPLE 10

(a) $\lim_{x\to 1} \dfrac{3x^3 - 2x + 1}{4x^2 + 5} = \dfrac{3 - 2 + 1}{4 + 5} = \dfrac{2}{9}$

(b) $\lim_{x\to -2} \dfrac{2x + 4}{3x^2 - 1} = \dfrac{-4 + 4}{12 - 1} = 0$

These examples might lead one to conclude that the evaluation of limits is simply a question of substituting the number that x approaches into the function. Although this is often the case, the next few examples are a reminder that substitution cannot always be used.

EXAMPLE 11

Find: $\lim_{x \to -2} \dfrac{x^2 + 5x + 6}{x^2 - 4}$

Solution

$$\lim_{x \to -2} (x^2 - 4) = \lim_{x \to -2} x^2 - \lim_{x \to -2} 4 = 4 - 4 = 0$$

Since the limit of the denominator function is 0, we cannot use (11). However, this does not mean that the limit does not exist! We can factor the numerator and denominator and find

$$\frac{x^2 + 5x + 6}{x^2 - 4} = \frac{(x + 3)(x + 2)}{(x - 2)(x + 2)}$$

Recall that we are interested only in the limit as x approaches -2, and *not* in the value when x equals -2. Thus, the factor $(x + 2)$ is not 0 and we can cancel. Then

$$\lim_{x \to -2} \frac{x^2 + 5x + 6}{x^2 - 4} = \lim_{x \to -2} \frac{x + 3}{x - 2} \underset{\substack{\uparrow \\ \text{By (5) or (11)}}}{=} \frac{-2 + 3}{-2 - 2} = -\frac{1}{4}$$

EXAMPLE 12

Find: $\lim_{x \to 1} \dfrac{x^2 - 1}{x^2 - x}$

Solution

$$\lim_{x \to 1} \frac{x^2 - 1}{x^2 - x} \underset{\substack{\uparrow \\ \text{Factor}}}{=} \lim_{x \to 1} \frac{(x - 1)(x + 1)}{x(x - 1)} \underset{\substack{\uparrow \\ \text{Cancel}}}{=} \lim_{x \to 1} \frac{x + 1}{x} \underset{\substack{\uparrow \\ \text{By (11)}}}{=} \frac{1 + 1}{1} = 2$$

The next example illustrates how another algebraic manipulation may be used to find a limit.

EXAMPLE 13

Find: $\lim_{x \to 5} \dfrac{\sqrt{x} - \sqrt{5}}{x - 5}$

Solution

Note that

$$\frac{\sqrt{x} - \sqrt{5}}{x - 5} = \frac{(\sqrt{x} - \sqrt{5})}{x - 5} \frac{(\sqrt{x} + \sqrt{5})}{(\sqrt{x} + \sqrt{5})} = \frac{x - 5}{x - 5} \frac{1}{(\sqrt{x} + \sqrt{5})} = \frac{1}{\sqrt{x} + \sqrt{5}}$$

Therefore,

$$\lim_{x\to 5}\frac{\sqrt{x}-\sqrt{5}}{x-5}=\lim_{x\to 5}\frac{1}{\sqrt{x}+\sqrt{5}}\underset{\text{By (7)}}{=}\frac{1}{2\sqrt{5}}$$

EXAMPLE 14

For $f(x) = x^2 + 2x$, find: $\displaystyle\lim_{h\to 0}\frac{f(x+h)-f(x)}{h}$

Solution

$$f(x+h) = (x+h)^2 + 2(x+h) = x^2 + 2xh + h^2 + 2x + 2h$$

$$f(x+h) - f(x) = (x^2 + 2xh + h^2 + 2x + 2h) - (x^2 + 2x) = 2xh + h^2 + 2h$$

$$\frac{f(x+h)-f(x)}{h} = \frac{2xh + h^2 + 2h}{h} = \frac{h(2x+h+2)}{h} = 2x + h + 2$$

Therefore,

$$\lim_{h\to 0}\frac{f(x+h)-f(x)}{h} = \lim_{h\to 0}(2x+h+2) = 2x+2$$

We close this section with two results we will need later in Chapter 4. Their proof, which uses the ε, δ definition of a limit, is given in Section 2.6.

[2.2.10] THEOREM

If $\lim_{x\to c} f(x) > 0$, then there is an open interval about c, possibly excluding c itself, on which $f(x) > 0$.

[2.2.11] THEOREM

If $\lim_{x\to c} f(x) < 0$, then there is an open interval about c, possibly excluding c itself, on which $f(x) < 0$.

These results may be summarized by saying that if the limit of a function is nonzero at c, then, for x sufficiently close to c, the value of the function will be nonzero and have the same sign as the limit of $f(x)$ at c.

EXERCISE 2.2

In Problems 1–46 evaluate each limit by using algebraic techniques.

1. $\lim_{x\to 3} 2(x+4)$
2. $\lim_{x\to 4} 5x$
3. $\lim_{x\to 1}(3x^2 - 2x + 4)$
4. $\lim_{x\to 0}(-3x^4 + 2x + 1)$
5. $\lim_{x\to -2}(x^4 + 2x)$
6. $\lim_{x\to -1}(5x^6 - 2x^2 + x)$
7. $\lim_{x\to 1/2}(2x^4 - 8x^3 + 4x - 5)$
8. $\lim_{x\to -1/3}(27x^3 + 9x + 1)$
9. $\lim_{x\to 4} 3\sqrt{x}$
10. $\lim_{x\to 8} \frac{1}{4}\sqrt[3]{x}$
11. $\lim_{x\to 3}\sqrt{x^2 + x + 4}$
12. $\lim_{t\to -2}\sqrt{3t^2 + 4}$

13. $\lim_{t\to 2} t\sqrt{t^3-4}$

14. $\lim_{t\to -1} t^2\sqrt[3]{t}$

15. $\lim_{x\to 2}(\sqrt{x^3+1}-\sqrt{x^2+5})$

16. $\lim_{x\to 1}(\sqrt[3]{2-x}-\sqrt[3]{1-x})$

17. $\lim_{x\to 2}\dfrac{x^2+4}{x}$

18. $\lim_{x\to 5}\dfrac{x^2+5}{3x}$

19. $\lim_{x\to -2}\dfrac{2x^3+5x}{3x-2}$

20. $\lim_{x\to 1}\dfrac{2x^4-1}{3x^3+2}$

21. $\lim_{x\to 2}\dfrac{x^2-4}{x-2}$

22. $\lim_{x\to -1}\dfrac{x^3-x}{x+1}$

23. $\lim_{x\to 2}\dfrac{x^3-8}{x^2+x-6}$

24. $\lim_{x\to 1}\dfrac{x^3-3x^2+3x-1}{x^3-x}$

25. $\lim_{x\to 0}\dfrac{3x^3-4x}{x^2+x}$

26. $\lim_{x\to -1}\dfrac{x^3+x^2}{x^2-1}$

27. $\lim_{x\to -3}\dfrac{x+3}{x^2-x-12}$

28. $\lim_{x\to 4}\dfrac{2x^2-32}{x^3-4x^2}$

29. $\lim_{x\to -1}\dfrac{x^2+4x+3}{x^2+5x+4}$

30. $\lim_{x\to 2}\dfrac{x^3-2x^2+x-2}{x-2}$

31. $\lim_{x\to -8}\left(\dfrac{2x}{x+8}+\dfrac{16}{x+8}\right)$

32. $\lim_{x\to 2}\left(\dfrac{3x}{x-2}-\dfrac{6}{x-2}\right)$

33. $\lim_{x\to -2}\dfrac{x^3+8}{x+2}$

34. $\lim_{x\to 1}\dfrac{x^4-1}{x-1}$

35. $\lim_{x\to 2}\dfrac{\sqrt{x}-\sqrt{2}}{x-2}$

36. $\lim_{x\to -2}\dfrac{x+2}{x^2-4}$

37. $\lim_{h\to 0}\dfrac{\dfrac{1}{x+h}-\dfrac{1}{x}}{h}$

38. $\lim_{h\to 0}\dfrac{(x+h)^2-x^2}{h}$

39. $\lim_{h\to 0}\dfrac{\sqrt{x+h}-\sqrt{x}}{h}$

40. $\lim_{h\to 0}\dfrac{\dfrac{1}{(x+h)^3}-\dfrac{1}{x^3}}{h}$

41. $\lim_{x\to 0}\dfrac{1}{x}\left(\dfrac{1}{4+x}-\dfrac{1}{4}\right)$

42. $\lim_{x\to 1}\dfrac{x^8-1}{x-1}$

43. $\lim_{x\to 4}\dfrac{\sqrt{x+5}-3}{x-4}$

44. $\lim_{x\to 3}\dfrac{\sqrt{x+1}-2}{x-3}$

45. $\lim_{x\to 7}\dfrac{x-7}{\sqrt{x+2}-3}$

46. $\lim_{x\to 2}\dfrac{x-2}{\sqrt{x+2}-2}$

In Problems 47–50 use the facts that $\lim_{x\to c} f(x)=5$ and $\lim_{x\to c} g(x)=2$ to find each limit.

47. $\lim_{x\to c}[2f(x)]$

48. $\lim_{x\to c}[f(x)-g(x)]$

49. $\lim_{x\to c}[g(x)]^3$

50. $\lim_{x\to c}\dfrac{f(x)}{g(x)-f(x)}$

In Problems 51–56 find $\lim_{h\to 0}\dfrac{f(x+h)-f(x)}{h}$ for the indicated function f.

51. $f(x)=4x-3$

52. $f(x)=3x+5$

53. $f(x)=3x^2+4x+1$

54. $f(x)=2x^2+x$

55. $f(x)=\dfrac{2}{x}$

56. $f(x)=\dfrac{3}{x^2}$

57. Find: $\lim_{x\to 2}\dfrac{x^2-4}{3-\sqrt{x^2+5}}$

58. Find: $\lim_{x\to -1}\dfrac{2-\sqrt{x^2+3}}{1-x^2}$

59. Find: $\lim_{x\to a}\dfrac{x^n-a^n}{x-a}$

60. Find: $\lim_{x\to -a}\dfrac{x^n+a^n}{x+a}$

61. Find: $\lim_{x\to 1}\dfrac{x^m-1}{x^n-1}$

62. Find: $\lim_{x\to 0}\dfrac{\sqrt[3]{1+x}-1}{x}$

63. Find: $\lim_{x\to 0}\dfrac{\sqrt{(1+ax)(1+bx)}-1}{x}$

64. Find: $\lim_{x\to 0}\dfrac{\sqrt{(1+a_1x)(1+a_2x)\cdots(1+a_nx)}-1}{x}$

65. Let $f(x) = \dfrac{\sqrt{x} - 2}{x - 4}$. Use a calculator to complete the following table:

x	3.9	3.99	3.999	4.001	4.01	4.1
$f(x)$						

What is a reasonable guess for the value of $\lim_{x\to 4} f(x)$? Defend it by evaluating the limit.

66. The graph of $y = \dfrac{x-3}{3-x}$ is a straight line with a point punched out. What straight line and what point?

67. Find: $\displaystyle\lim_{h\to 0} \frac{f(h) - f(0)}{h}$ if $f(x) = x|x|$

2.3 One-Sided Limits

In defining $\lim_{x\to c} f(x)$ in (4) in Section 2.1 we were careful to restrict x to an open interval containing c; that is, we studied the behavior of f on both sides of c. However, in some cases it is necessary to investigate *one-sided limits*: the *left-hand* limit, $\lim_{x\to c^-} f(x)$, and the *right-hand* limit, $\lim_{x\to c^+} f(x)$.

The idea behind the left-hand limit $\lim_{x\to c^-} f(x) = L$, read "the limit of $f(x)$, as x approaches c from the left, equals L," is that for all x sufficiently close to c, but *less than* c, the value $f(x)$ can be made as close as we please to L (see Fig. 16).

The idea behind the right-hand limit $\lim_{x\to c^+} f(x) = R$, read "the limit of $f(x)$, as x approaches c from the right, equals R," is that for all x sufficiently close to c, but *greater than* c, the value of $f(x)$ can be made as close as we please to R (see Fig. 17).

The rules for calculating left-hand and right-hand limits are the same as those used to calculate limits in the preceding section.

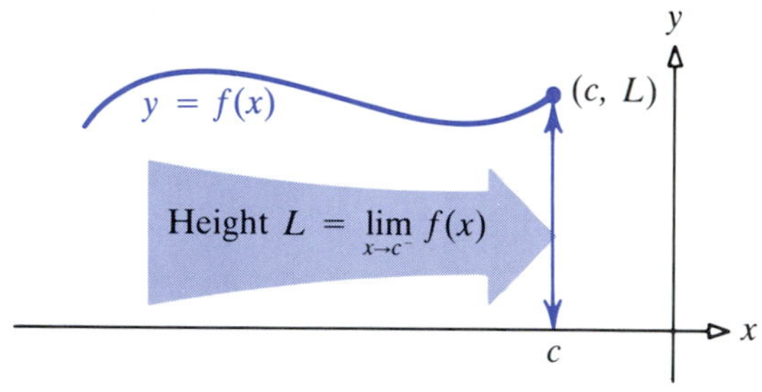

Figure 16

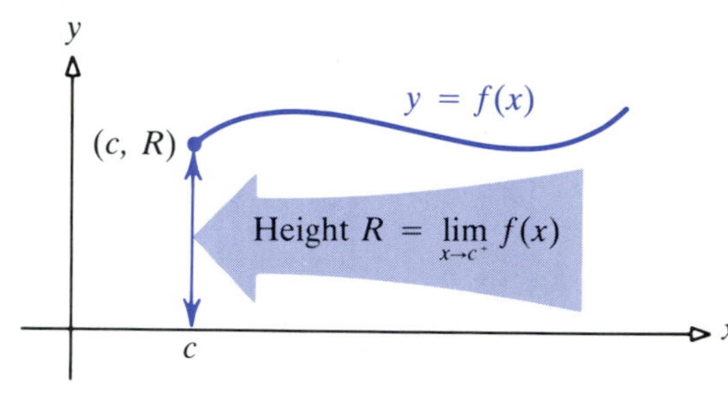

Figure 17

EXAMPLE 1

For the function below, calculate $\lim_{x\to 0^-} f(x)$ and $\lim_{x\to 0^+} f(x)$:

$$f(x) = \begin{cases} -1 & \text{if } x < 0 \\ 1 & \text{if } x > 0 \end{cases}$$

Solution

For $\lim_{x\to 0^-} f(x)$, x must remain negative. But, if $x < 0$, then $f(x)$ is constantly equal to -1 so that $\lim_{x\to 0^-} f(x) = \lim_{x\to 0^-}(-1) = -1$. Similar reasoning gives $\lim_{x\to 0^+} f(x) = \lim_{x\to 0^+} 1 = 1$.

For the function in Example 1, the fact that $\lim_{x\to 0^-} f(x) = -1$ and $\lim_{x\to 0^+} f(x) = 1$ means that there can be no single number L that $f(x)$ is close to when x is close to 0. Consequently, $\lim_{x\to 0} f(x)$ does not exist (see Fig. 18).

As Figure 19 illustrates, one-sided limits provide a way to determine whether the limit of a function exists. We state this criterion now:

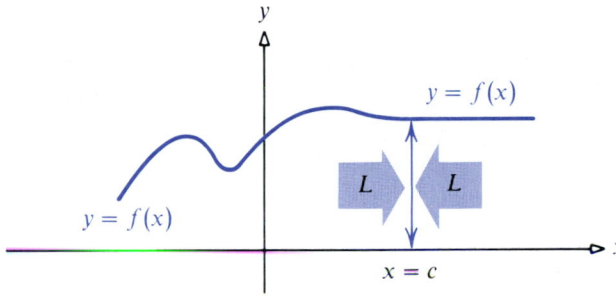

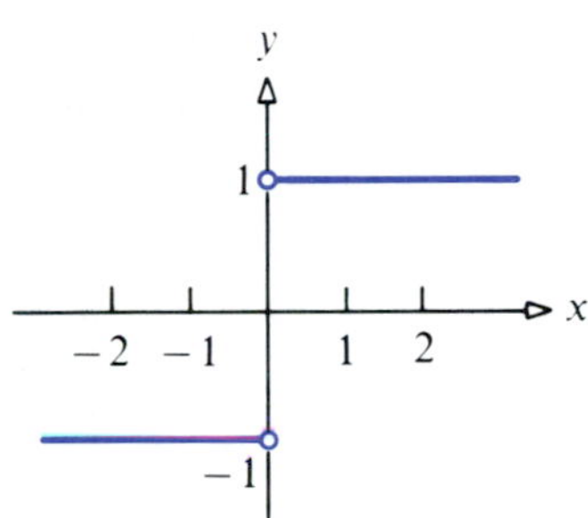

Figure 18

Figure 19

[2.3.1] THEOREM

$$\lim_{x\to c} f(x) = L \quad \text{if and only if} \quad \lim_{x\to c^-} f(x) = L \quad \text{and} \quad \lim_{x\to c^+} f(x) = L$$

The theorem can be proved by using the definition of limit and is given in Section 2.6.

EXAMPLE 2

Determine whether $\lim_{x\to 2} f(x)$ and $\lim_{x\to 4} f(x)$ exist if

$$f(x) = \begin{cases} 2x + 1 & \text{if} \quad 0 \le x \le 2 \\ 7 - x & \text{if} \quad 2 < x < 4 \\ x & \text{if} \quad 4 \le x \le 6 \end{cases}$$

Solution

$\lim_{x\to 2} f(x)$: Since the rule for $f(x)$ depends on whether $x < 2$ or $x > 2$, we need to look at the one-sided limits $\lim_{x\to 2^-} f(x)$ and $\lim_{x\to 2^+} f(x)$ to obtain information about $\lim_{x\to 2} f(x)$.

$$\lim_{x\to 2^-} f(x) = \lim_{x\to 2^-} (2x + 1) = 5 \qquad \lim_{x\to 2^+} f(x) = \lim_{x\to 2^+} (7 - x) = 5$$

Thus, by theorem (2.3.1), we conclude that $\lim_{x\to 2} f(x)$ exists and equals 5.

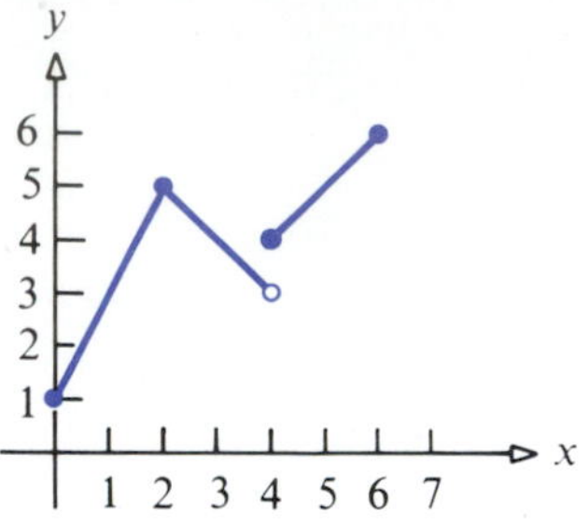

Figure 20

$\lim_{x\to 4} f(x)$: The rule for $f(x)$ changes at $x = 4$. Hence, to calculate $\lim_{x\to 4} f(x)$, we look at the one-sided limits:

$$\lim_{x\to 4^-} f(x) = \lim_{x\to 4^-} (7 - x) = 3$$

$$\lim_{x\to 4^+} f(x) = \lim_{x\to 4^+} x = 4$$

Since the limits are not equal, we conclude that $\lim_{x\to 4} f(x)$ does not exist. Figure 20 illustrates these conclusions.

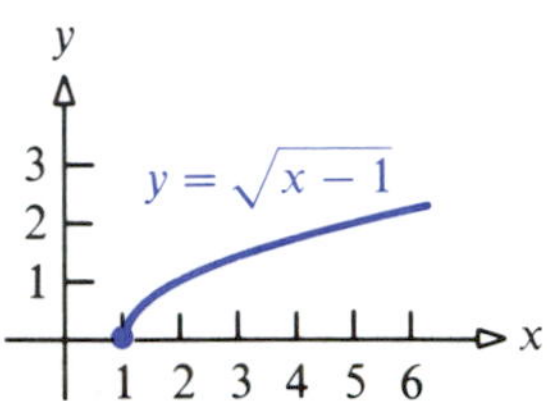

Figure 21

One-sided limits must be used to describe the behavior of functions like $f(x) = \sqrt{x-1}$ near $x = 1$, since f is defined for only $x \geq 1$. For this function, $\lim_{x\to 1^-} \sqrt{x-1}$ makes no sense. However, the right-hand limit does make sense, and $\lim_{x\to 1^+} \sqrt{x-1} = 0$ describes how $\sqrt{x-1}$ behaves near and to the right of 1 (see Fig. 21).

The ε, δ definitions of left-hand and right-hand limits are given in Appendix I.

EXERCISE 2.3

In Problems 1–20 evaluate each one-sided limit, if it exists.

1. $\lim_{x\to 3^-} (x^2 - 4)$ **2.** $\lim_{x\to 2^+} (3x^2 + x)$ **3.** $\lim_{x\to -2^+} (2x^3 + x - 1)$ **4.** $\lim_{x\to -4^-} (x^2 - 1)$

5. $\lim_{x\to -4^-} \dfrac{3x}{x-4}$ **6.** $\lim_{x\to -4^+} \dfrac{3x}{x-4}$ **7.** $\lim_{x\to 3^-} \dfrac{x^2-9}{x-3}$ **8.** $\lim_{x\to 3^+} \dfrac{x^2-9}{x-3}$

9. $\lim_{x\to 3^-} (\sqrt{9-x^2} + x)$ **10.** $\lim_{x\to 2^+} (2\sqrt{x^2-4} + 3x)$ **11.** $\lim_{x\to 5^+} \dfrac{|x-5|}{x-5}$ **12.** $\lim_{x\to 5^-} \dfrac{|x-5|}{x-5}$

13. $\lim_{x\to 1/2^-} [\![2x]\!]$ **14.** $\lim_{x\to 1/2^+} [\![2x]\!]$ **15.** $\lim_{x\to 2/3^-} [\![2x]\!]$ **16.** $\lim_{x\to 2/3^+} [\![2x]\!]$

17. $\lim_{x\to 2^+} \sqrt{|x| - x}$ **18.** $\lim_{x\to 2^-} \sqrt{|x| - x}$ **19.** $\lim_{x\to 2^+} \sqrt[3]{[\![x]\!] - x}$ **20.** $\lim_{x\to 2^-} \sqrt[3]{[\![x]\!] - x}$

In Problems 21–32 evaluate $\lim_{x\to c^-} f(x)$ and $\lim_{x\to c^+} f(x)$ for the given number c. Based on your answer, determine whether $\lim_{x\to c} f(x)$ exists.

21. $f(x) = \begin{cases} 2x & \text{if } x \neq 0 \\ 1 & \text{if } x = 0 \end{cases}$ $c = 0$

22. $f(x) = \begin{cases} 2x & \text{if } x \neq 0 \\ 0 & \text{if } x = 0 \end{cases}$ $c = 0$

23. $f(x) = \begin{cases} \dfrac{x^2-9}{x-3} & \text{if } x \neq 3 \\ 6 & \text{if } x = 3 \end{cases}$ $c = 3$

24. $f(x) = \begin{cases} \dfrac{x-2}{x^2-4} & \text{if } x \neq 2 \\ 1 & \text{if } x = 2 \end{cases}$ $c = 2$

25. $f(x) = \begin{cases} 2x - 3 & \text{if } x \leq 1 \\ 3 - x & \text{if } x > 1 \end{cases}$ $c = 1$

26. $f(x) = \begin{cases} 5x + 2 & \text{if } x < -2 \\ 1 + 3x & \text{if } x \geq -2 \end{cases}$ $c = -2$

27. $f(x) = \begin{cases} 3x - 1 & \text{if } x < 1 \\ 4 & \text{if } x = 1 \\ 2x & \text{if } x > 1 \end{cases} \quad c = 1$

28. $f(x) = \begin{cases} 3x - 1 & \text{if } x < 1 \\ 2 & \text{if } x = 1 \\ 2x & \text{if } x > 1 \end{cases} \quad c = 1$

29. $f(x) = \begin{cases} 3x - 1 & \text{if } x < 1 \\ \text{Not defined} & \text{if } x = 1 \\ 2x & \text{if } x > 1 \end{cases} \quad c = 1$

30. $f(x) = \begin{cases} 3x - 1 & \text{if } x < 1 \\ 2 & \text{if } x = 1 \\ 3x & \text{if } x > 1 \end{cases} \quad c = 1$

31. $f(x) = \begin{cases} \sqrt{x^2 - 9} & \text{if } x \geq 3 \\ \sqrt{9 - x^2} & \text{if } x < 3 \end{cases} \quad c = 3$

32. $f(x) = \begin{cases} \dfrac{|x - 1|}{x - 1} & \text{if } x \neq 1 \\ 0 & \text{if } x = 1 \end{cases} \quad c = 1$

In Problems 33–38 use the function below to evaluate each limit, if it exists.

$$f(x) = \begin{cases} \sqrt{15 - 5x} & \text{if } x < 2 \\ \sqrt{5} & \text{if } x = 2 \\ \sqrt{9 - x^2} & \text{if } 2 < x < 3 \\ x - 2 & \text{if } 3 \leq x \end{cases}$$

33. $\lim_{x \to 2^-} f(x)$

34. $\lim_{x \to 2^+} f(x)$

35. $\lim_{x \to 3^-} f(x)$

36. $\lim_{x \to 3^+} f(x)$

37. $\lim_{x \to 2} f(x)$

38. $\lim_{x \to 3} f(x)$

For Problems 39 and 40 use the function below to evaluate each limit, if it exists.

$$f(x) = \begin{cases} 3x + 5 & \text{if } x \leq 2 \\ 13 - x & \text{if } x > 2 \end{cases}$$

39. $\lim_{h \to 0^-} \dfrac{f(2 + h) - f(2)}{h}$

40. $\lim_{h \to 0^+} \dfrac{f(2 + h) - f(2)}{h}$

For Problems 41–44 use the function f defined by

$$f(x) = \begin{cases} 1 & \text{if } x \text{ is an integer} \\ 0 & \text{if } x \text{ is not an integer} \end{cases}$$

41. Does $\lim_{x \to 2} f(x)$ exist?

42. Does $\lim_{x \to 1/2} f(x)$ exist?

43. Does $\lim_{x \to 3} f(x)$ exist?

44. Does $\lim_{x \to 0} f(x)$ exist?

45. Show by example that $\lim_{x \to c}[f(x) + g(x)]$ may exist even though $\lim_{x \to c} f(x)$ and $\lim_{x \to c} g(x)$ do not exist.

46. Show by example that $\lim_{x \to c}[f(x)g(x)]$ may exist even though $\lim_{x \to c} f(x)$ and $\lim_{x \to c} g(x)$ do not exist.

47. Show by example that $\lim_{x \to c}|f(x)|$ may exist even though $\lim_{x \to c} f(x)$ does not exist.

Continuous Functions 2.4

We have seen that sometimes $\lim_{x \to c} f(x)$ equals $f(c)$ and sometimes it does not. In fact, sometimes $f(c)$ is not even defined and $\lim_{x \to c} f(x)$ exists. Then, what is the relationship between $\lim_{x \to c} f(x)$ and $f(c)$? For the answer, we look at the possibilities (see Fig. 22):

(a) $\lim_{x\to c} f(x)$ exists and equals $f(c)$

(b) $\lim_{x\to c} f(x)$ exists and does not equal $f(c)$

(c) $\lim_{x\to c} f(x)$ exists and $f(c)$ is not defined

(d) $\lim_{x\to c} f(x)$ does not exist and $f(c)$ is defined

(e) $\lim_{x\to c} f(x)$ does not exist and $f(c)$ is not defined

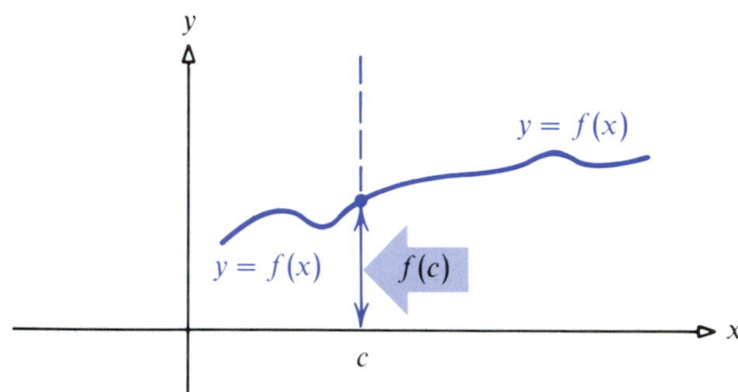

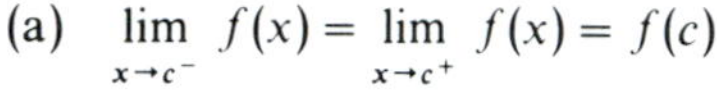
(a) $\lim_{x\to c^-} f(x) = \lim_{x\to c^+} f(x) = f(c)$

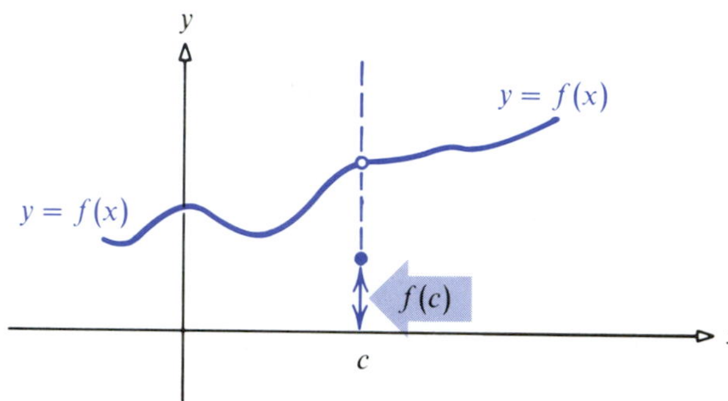

(b) $\lim_{x\to c^-} f(x) = \lim_{x\to c^+} f(x) \neq f(c)$

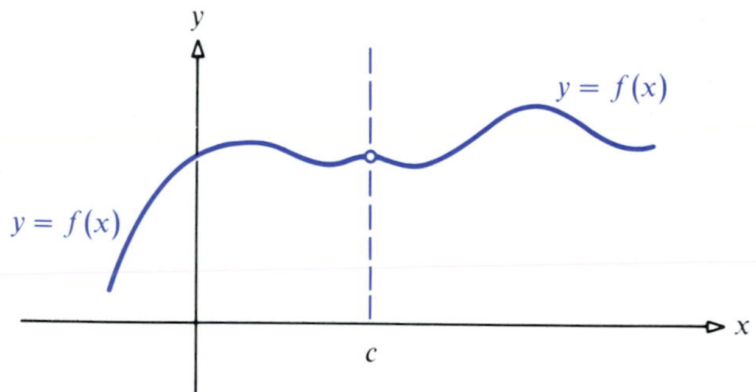

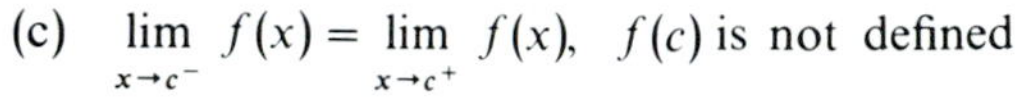
(c) $\lim_{x\to c^-} f(x) = \lim_{x\to c^+} f(x)$, $f(c)$ is not defined

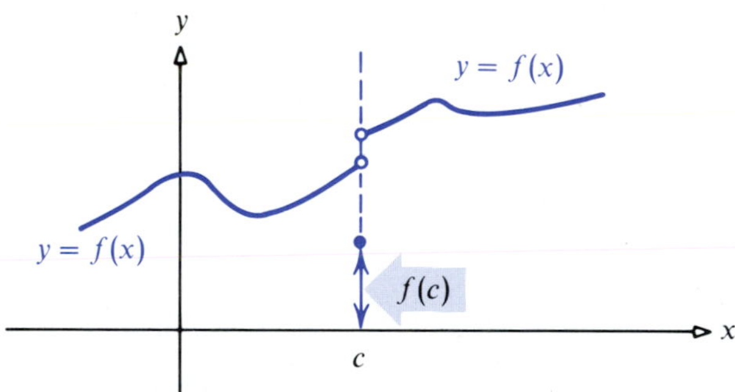

(d) $\lim_{x\to c^-} f(x) \neq \lim_{x\to c^+} f(x)$, $f(c)$ is defined

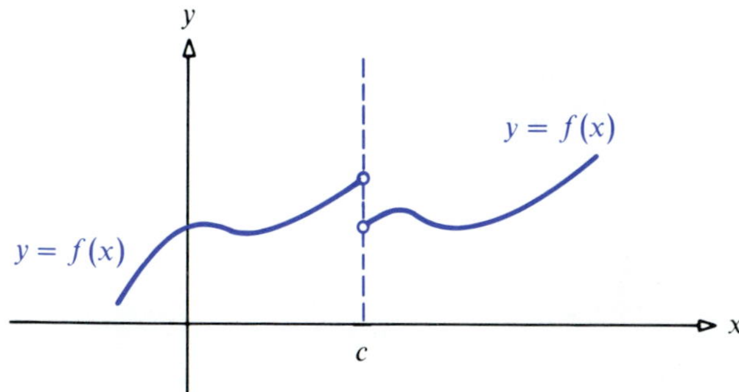

(e) $\lim_{x\to c^-} f(x) \neq \lim_{x\to c^+} f(x)$, $f(c)$ is not defined

Figure 22

Of these five situations, the "nicest" one is (a). There, $\lim_{x\to c} f(x)$ both exists and is equal to $f(c)$. Functions that have these two properties are said to be *continuous at c*. This appears to agree with the intuitive notion that a function is continuous if its graph can be drawn without lifting the pencil. The functions in (b), (c), (d), and (e) are not continuous at c, since each has a break in the graph at c. This leads us to definition (2.4.1).

[2.4.1] DEFINITION / *Continuous Function.*

Let $y = f(x)$ be a function defined on an open interval. If

1. $f(c)$ is defined and 2. $\lim_{x\to c} f(x)$ exists and 3. $\lim_{x\to c} f(x) = f(c)$

then the function is said to be *continuous at c*.

If *any one* of these three conditions is not satisfied, then the function is said to be *discontinuous at c*.

EXAMPLE 1

Show that the function $f(x) = 3x^3 - 5x + 4$ is continuous at 1.

Solution

We can show that f is continuous at 1, since

$$\lim_{x\to 1} f(x) = \lim_{x\to 1}(3x^3 - 5x + 4) = 2 \quad \text{and} \quad f(1) = 2$$

EXAMPLE 2

Let $f(x) = \dfrac{x^2 + 2}{x^2 - 4}$. Determine the numbers at which f is continuous.

Solution

Observe that f is a rational function. Since the denominator of f is 0 for $x = 2$ and $x = -2$, f is defined for all numbers except 2 and -2. Consequently, f is continuous at every number except 2 and -2.

In fact, because of theorem (2.2.8), polynomial functions are everywhere continuous. And because of theorem (2.2.9) it follows that rational functions are continuous everywhere on their domains.

Let's look at some more examples.

EXAMPLE 3

Discuss the continuity of the function below at 3:

$$f(x) = \begin{cases} \dfrac{x^2 - 9}{x - 3} & \text{if } x \neq 3 \\ 6 & \text{if } x = 3 \end{cases}$$

Solution

The function f is defined at 3, since $f(3) = 6$. Also,

$$\lim_{x\to 3^+} \frac{x^2-9}{x-3} = \lim_{x\to 3^+} \frac{(x+3)(x-3)}{x-3} = \lim_{x\to 3^+} (x+3) = 6$$

$$\lim_{x\to 3^-} \frac{x^2-9}{x-3} = \lim_{x\to 3^-} \frac{(x+3)(x-3)}{x-3} = \lim_{x\to 3^-} (x+3) = 6$$

Therefore $\lim_{x\to 3} f(x) = f(3) = 6$. Thus, the three conditions in definition (2.4.1) are satisfied, and hence the function is continuous at 3 (see Fig. 23(a)).

Note that the function

$$g(x) = \frac{x^2-9}{x-3} \qquad x \neq 3$$

is not continuous at 3, since condition 1 of definition (2.4.1) is not satisfied. We say, therefore, that g is discontinuous at 3 (see Fig. 23(b)).

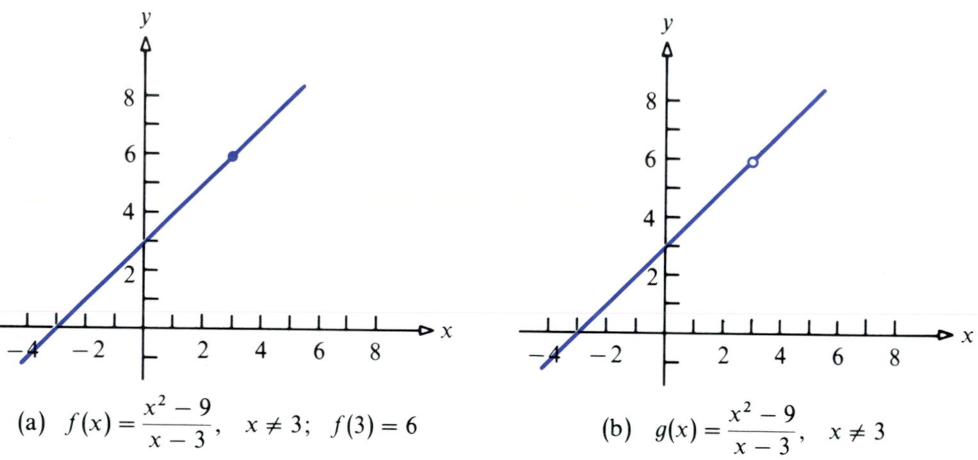

(a) $f(x) = \dfrac{x^2-9}{x-3}$, $x \neq 3$; $f(3) = 6$

(b) $g(x) = \dfrac{x^2-9}{x-3}$, $x \neq 3$

Figure 23

EXAMPLE 4

Show that $f(x) = |x|$ is continuous at every real number c.

Solution

Since $f(x) = |x| = \sqrt{x^2}$, we have

$$\lim_{x\to c} f(x) = \lim_{x\to c} |x| = \lim_{x\to c} \sqrt{x^2}$$

$$= \sqrt{\lim_{x\to c} x^2} = \sqrt{c^2} = |c| = f(c)$$

By (7) in Section 2.2

Hence, by definition (2.4.1), f is continuous at c.

EXAMPLE 5

Discuss the continuity of the function below at 0:

$$f(x) = \begin{cases} x^2 + 1 & \text{if} \quad x \neq 0 \\ 2 & \text{if} \quad x = 0 \end{cases}$$

Solution

The value of the function at 0 is $f(0) = 2$. Also,

$$\lim_{x\to 0^-} f(x) = \lim_{x\to 0^-} (x^2 + 1) = 1$$

$$\lim_{x\to 0^+} f(x) = \lim_{x\to 0^+} (x^2 + 1) = 1$$

Thus, $\lim_{x\to 0} f(x) = 1$ and $f(0) = 2$. Since condition 3 of definition (2.4.1) does not hold, the function is discontinuous at 0 (see Fig. 24).

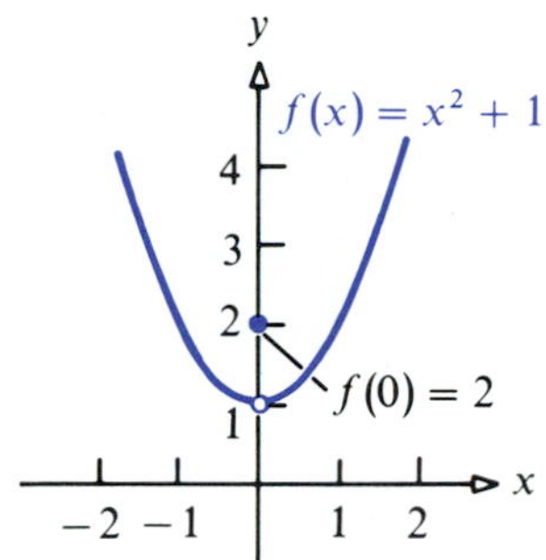

Figure 24

EXAMPLE 6

Discuss the continuity of $f(x) = [\![x]\!]$ at 1.

Solution

The function f is defined at 1 and $f(1) = 1$. But

$$\lim_{x\to 1^-} f(x) = 0 \qquad \text{and} \qquad \lim_{x\to 1^+} f(x) = 1$$

Therefore $\lim_{x\to 1} f(x)$ does not exist. Since condition 2 of definition (2.4.1) is not met, we say that f is discontinuous at 1. In fact, as Figure 25 indicates, the function is discontinuous at each integer.

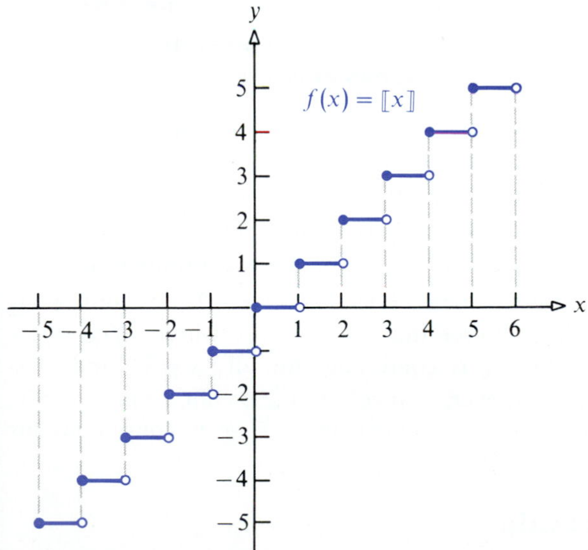

Figure 25

The discontinuities encountered in Examples 5 and 6 illustrate a more general result. If a function f is defined on an open interval that contains c, then f may be discontinuous at c for only one of two reasons: (*1*) $\lim_{x\to c} f(x)$ does not exist (Example 6) or (*2*) $\lim_{x\to c} f(x)$ exists but does not equal $f(c)$ (Example 5). The latter type of discontinuity is called a *removable discontinuity*, since we can redefine f at c to equal $\lim_{x\to c} f(x)$ and thus make f continuous at c. For instance, in Example 5, if we redefine $f(0)$ to be 1, then f will be continuous at 0.

There are more subtle types of discontinuity. For instance, the function below is nowhere continuous:

$$f(x) = \begin{cases} 1 & \text{if} \quad x \text{ is rational} \\ 0 & \text{if} \quad x \text{ is irrational} \end{cases}$$

Let's concentrate, however, on characteristics of functions that *are* continuous.

Properties of Continuous Functions

Because the definition of a continuous function is based on a limit, theorems that apply to limits also apply to continuous functions. We state these results now.

[2.4.2] THEOREM / ***Continuity of a Sum, Product, and Difference.***

If f and g are continuous at c, then so are their sum $f + g$, difference $f - g$, and product $f \cdot g$.

[2.4.3] THEOREM / ***Continuity of a Quotient.***

If f and g are continuous at c and $g(c) \neq 0$, then the quotient f/g is also continuous at c.

[2.4.4] THEOREM / ***Continuity of a Composite Function.***

If f is continuous at $g(c)$ and g is continuous at c, then the composite function $f(g(x))$ is continuous at c.

For example, the function $h(x) = \sqrt{3x^2 + 5}$ is continuous for all x. To see this, observe that $h = f \circ g$, where $f(x) = \sqrt{x}$ and $g(x) = 3x^2 + 5$. Now, g is continuous for all x and f is continuous for all $x \geq 0$. Since $g(x) > 0$ for all x, $f \circ g$ is defined for all x. It follows from theorem (2.4.4) that $(f \circ g)(x) = \sqrt{3x^2 + 5}$ is continuous for all x.

As another example, the function $h(x) = \sqrt{x^2/(x-1)}$ is continuous for all $x > 1$. To see this, observe that $h = f \circ g$, where $f(x) = \sqrt{x}$ and $g(x) = x^2/(x-1)$. Now g is continuous for all $x \neq 1$ and f is continuous for $x \geq 0$. However, since $g(x) \geq 0$ for only $x > 1$, $f \circ g$ is defined for only $x > 1$. Consequently, $f \circ g$ is continuous for $x > 1$.

Continuity on Intervals

In definition (2.4.1) we defined what is meant by a function f being continuous *at a number c*. Here, we define what is meant by a function f being continuous *on an interval*. As you will see, the definition requires that special attention be given to the endpoints, if they are contained in the interval. Figure 26 illustrates definition (2.4.5).

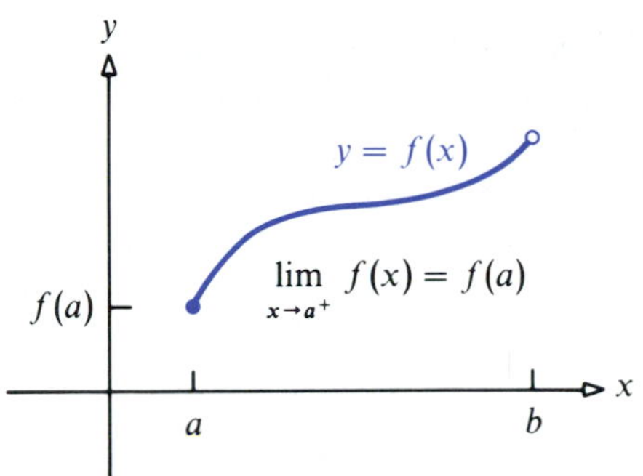

(a) f is continuous on $[a, b)$

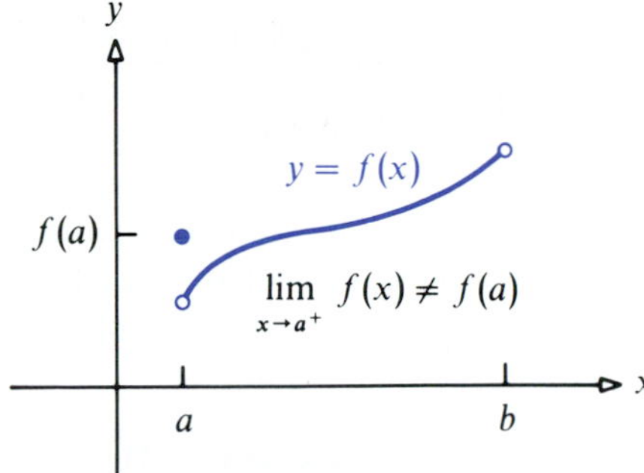

(b) f is not continuous on $[a, b)$

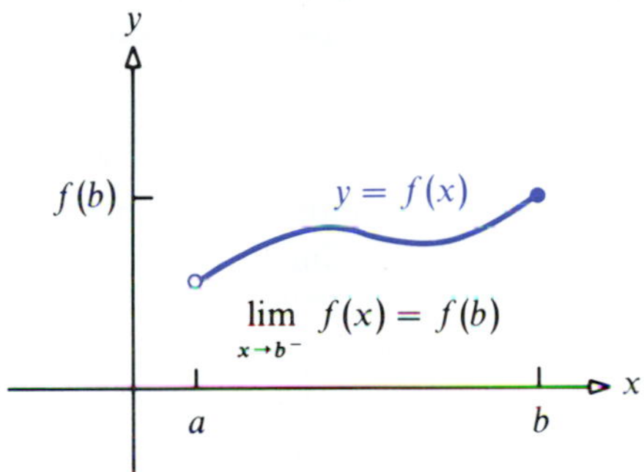

(c) f is continuous on $(a, b]$

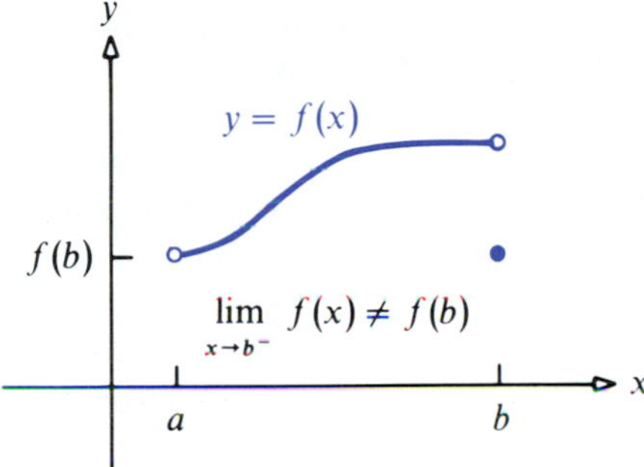

(d) f is not continuous on $(a, b]$

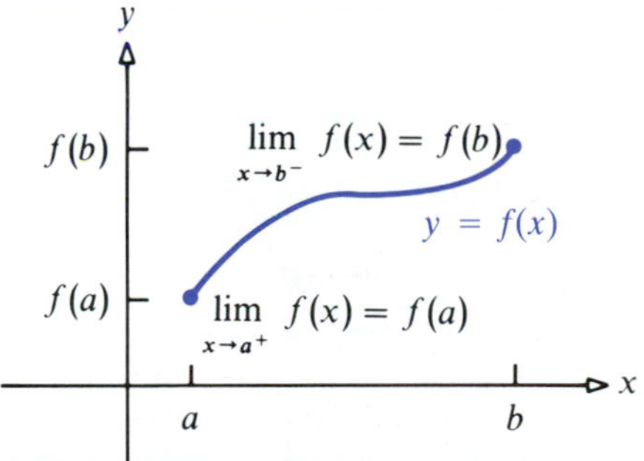

(e) f is continuous on $[a, b]$

Figure 26

[2.4.5] DEFINITION / *Continuous on an Interval.*

A function f *is continuous on an open interval* (a, b) if f is continuous at every number c between a and b.

A function f *is continuous on an interval* $[a, b)$ if f is continuous on (a, b) and $\lim_{x\to a^+} f(x) = f(a)$.

A function f *is continuous on an interval* $(a, b]$ if f is continuous on (a, b) and $\lim_{x \to b^-} f(x) = f(b)$.

A function f *is continuous on a closed interval* $[a, b]$ if f is continuous on (a, b), $\lim_{x \to a^+} f(x) = f(a)$, and $\lim_{x \to b^-} f(x) = f(b)$.

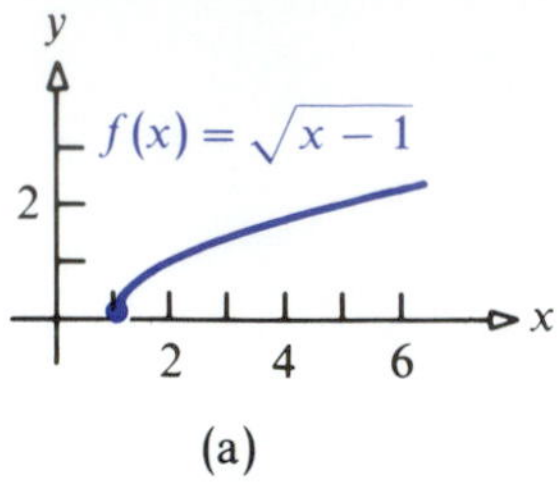

(a)

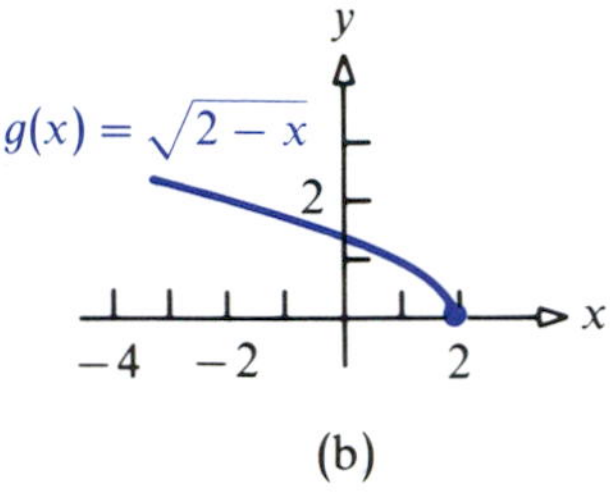

(b)

Figure 27

For example, the function $f(x) = \sqrt{x-1}$ is continuous on $[1, +\infty)$, whereas $g(x) = \sqrt{2-x}$ is continuous on $(-\infty, 2]$ (see Fig. 27). As Example 6 illustrates, the greatest integer function $f(x) = [\![x]\!]$ is continuous on every open interval $(n, n+1)$ where n is an integer.

Functions that are continuous on a closed interval have many important properties. One of them is the *intermediate value theorem*, which we shall use in later chapters. (The proof of this result may be found in most books on advanced calculus.)

[2.4.6] THEOREM / *Intermediate Value Theorem.*

Let f denote a function that is continuous on the closed interval $[a, b]$ and suppose $f(a) \neq f(b)$. If N is any number between $f(a)$ and $f(b)$, then there is at least one number c between a and b so that $f(c) = N$.

For an interpretation of this result, consider the following scenario: If you climbed a mountain, starting at 2000 meters and ending at 5000 meters, then no matter how many ups and downs you took in getting from the bottom to the top, there was some time when your altitude was 3765.6 meters—or any other number between 2000 and 5000 you may choose. In other words, a continuous function whose domain is a closed interval $[a, b]$ must take on all values between $f(a)$ and $f(b)$. Figure 28 illustrates this situation and why the continuity of the function is crucial for the validity of the result.

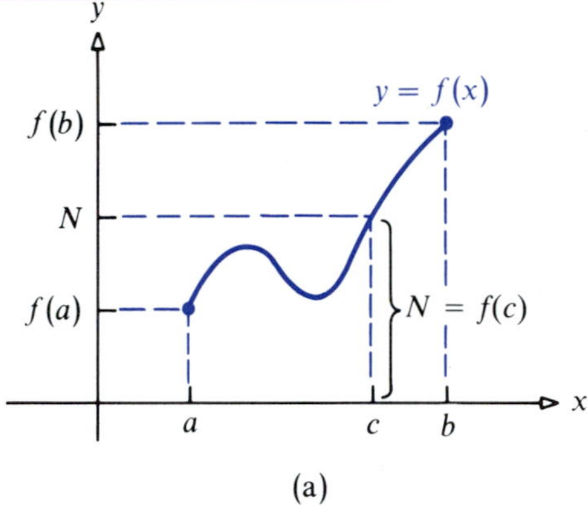

(a)

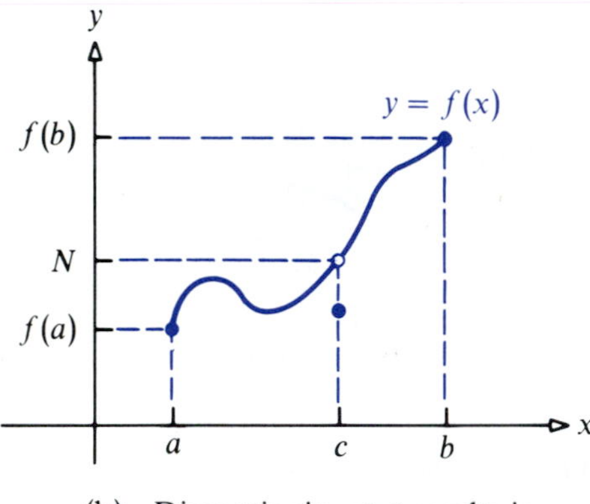

(b) Discontinuity at c results in no number x in $[a, b]$ so that $f(x) = N$

Figure 28

One immediate application of the intermediate value theorem involves the location of the zeros* of a function. If a function f is continuous on the

---

* r is a *zero* of f if $f(r) = 0$.

closed interval $[a, b]$ and if $f(a)$ and $f(b)$ are of opposite sign, there is at least one c between a and b so that $f(c) = 0$; that is, f has a zero between a and b. By "squeezing" the interval, better and better approximations of this zero may be obtained. For example, the function $f(x) = x^3 + x^2 - x - 2$ is continuous on the closed interval $[0, 2]$. Since $f(0) = -2$ and $f(2) = 8$ are of opposite sign, the intermediate value theorem implies $f(c) = 0$ for at least one number c in the interval $[0, 2]$. Therefore our first approximation to the zero is that it lies in the interval $[0, 2]$. We continue by testing the midpoint 1 of the interval $[0, 2]$. Since $f(1) = -1$, the function f must have a zero between 1 and 2. Our second approximation to the zero is that it lies in the interval $[1, 2]$. We could continue in this way (testing $\frac{3}{2}$ next) to obtain increasingly smaller intervals that contain the zero (see Fig. 29).

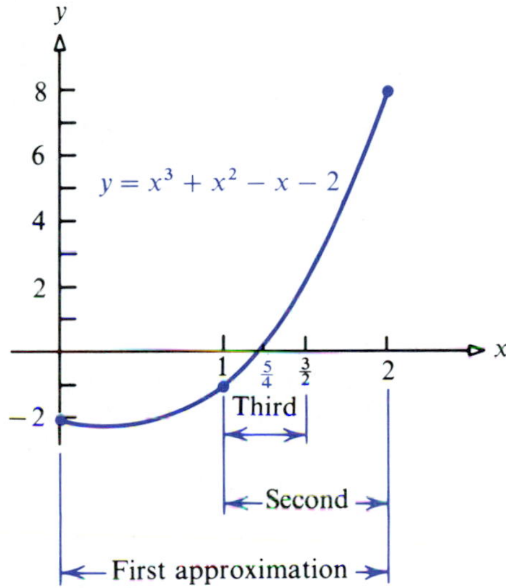

Figure 29

EXERCISE 2.4

In Problems 1–18 determine whether the function f is continuous at c.

1. $f(x) = 2x^2 + x - 5 \quad c = 1$

2. $f(x) = 3 - 2x^3 \quad c = -2$

3. $f(x) = \dfrac{x^2 - 9}{x - 3} \quad c = -3$

4. $f(x) = 1 + \dfrac{1}{x} \quad c = 1$

5. $f(x) = |x - 5| \quad c = 5$

6. $f(x) = \sqrt{x^2 + 9} \quad c = 0$

7. $f(x) = \begin{cases} 2x + 5 & \text{if } x \le 2 \\ 4x + 1 & \text{if } x > 2 \end{cases} \quad c = 2$

8. $f(x) = \begin{cases} 2x + 1 & \text{if } x \le 0 \\ 2x & \text{if } x > 0 \end{cases} \quad c = 0$

9. $f(x) = \begin{cases} 3x - 1 & \text{if } x < 1 \\ 4 & \text{if } x = 1 \\ 2x & \text{if } x > 1 \end{cases} \quad c = 1$

10. $f(x) = \begin{cases} 3x - 1 & \text{if } x < 1 \\ 2 & \text{if } x = 1 \\ 2x & \text{if } x > 1 \end{cases} \quad c = 1$

11. $f(x) = \begin{cases} 3x - 1 & \text{if } x < 1 \\ \text{Not defined} & \text{if } x = 1 \\ 2x & \text{if } x > 1 \end{cases}$ $c = 1$

12. $f(x) = \begin{cases} 3x - 1 & \text{if } x < 1 \\ 2 & \text{if } x = 1 \\ 3x & \text{if } x > 1 \end{cases}$ $c = 1$

13. $f(x) = \begin{cases} x^2 & \text{if } x \le 0 \\ 2x & \text{if } x > 0 \end{cases}$ $c = 0$

14. $f(x) = \begin{cases} x^2 & \text{if } x < -1 \\ 2 & \text{if } x = -1 \\ -3x + 2 & \text{if } x > -1 \end{cases}$ $c = -1$

15. $f(x) = \begin{cases} 4 - 3x^2 & \text{if } x < 0 \\ 4 & \text{if } x = 0 \\ \sqrt{16 - x^2} & \text{if } 0 < x < 4 \end{cases}$ $c = 0$

16. $f(x) = \begin{cases} \sqrt{4 + x} & \text{if } x \le 4 \\ \sqrt{\dfrac{x^2 - 16}{x - 4}} & \text{if } x > 4 \end{cases}$ $c = 4$

17. $f(x) = [\![x]\!]$ $c = 1$

18. $f(x) = [\![2x]\!]$ $c = \frac{1}{2}$

In Problems 19–22 the function f is not defined at c. For each function, decide how to define $f(c)$ so that f is continuous at c.

19. $f(x) = \dfrac{x^2 - 4}{x - 2}$ $c = 2$

20. $f(x) = \dfrac{x^2 + x - 12}{x - 3}$ $c = 3$

21. $f(x) = \begin{cases} 2x & \text{if } x > 1 \\ 1 + x & \text{if } x < 1 \end{cases}$ $c = 1$

22. $f(x) = \begin{cases} x^2 + 5x & \text{if } x < -1 \\ x - 3 & \text{if } x > -1 \end{cases}$ $c = -1$

In Problems 23–40 find all numbers x for which f is continuous.

23. $f(x) = 3x^5 - 2x^3 + x - 2$

24. $f(x) = -6x^3 + 4x$

25. $f(x) = \dfrac{x}{x^2 + 4}$

26. $f(x) = \dfrac{2 - x}{x^2 + x + 1}$

27. $f(x) = \dfrac{x}{x - 2}$

28. $f(x) = \dfrac{x^2 - 4}{x^2 - 1}$

29. $f(x) = \dfrac{x^2}{x^2 - 4}$

30. $f(x) = \dfrac{x}{x^3 - 8}$

31. $f(x) = |x|$

32. $f(x) = \dfrac{1}{x}$

33. $f(x) = \sqrt{x^2 + 1}$

34. $f(x) = \sqrt[3]{1 + x}$

35. $f(x) = \sqrt{\dfrac{x^2 + 1}{2 - x}}$

36. $f(x) = \sqrt{\dfrac{4}{x^2 - 1}}$

37. $f(x) = \dfrac{x - 9}{\sqrt{x} - 3}$

38. $f(x) = \dfrac{x - 4}{\sqrt{x} - 2}$

39. $f(x) = \dfrac{9x^2 - 4}{3x - 2}$

40. $f(x) = \dfrac{x^2 - 4x + 3}{x - 3}$

In Problems 41–44 use the function

$$f(x) = \begin{cases} \sqrt{15 - 3x} & \text{if } x < 2 \\ \sqrt{5} & \text{if } x = 2 \\ \sqrt{9 - x^2} & \text{if } 2 < x < 3 \\ [\![x - 2]\!] & \text{if } 3 \le x \end{cases}$$

41. Is f continuous at 0? Why or why not?

42. Is f continuous at 4? Why or why not?

43. Is f continuous at 2? Why or why not?

44. Is f continuous at 3? Why or why not?

45. Sketch the graph of the function f below and determine where f is continuous:

$$f(x) = \begin{cases} 1 - x^2 & \text{if } |x| \le 1 \\ x^2 - 1 & \text{if } |x| > 1 \end{cases}$$

46. Suppose: $f(x) = \dfrac{x^2 - 6x - 16}{(x^2 - 7x - 8)\sqrt{x^2 - 4}}$

(a) For what numbers x is f defined?
(b) For what numbers x is f discontinuous?
(c) Which of those numbers x found in part (b) are removable?

47. Find constants A and B so that the function below is continuous for all x. Sketch the graph of the resulting function.

$$f(x) = \begin{cases} (x-1)^2 & \text{if } -\infty < x < 0 \\ A - x^2 & \text{if } 0 \le x < 1 \\ x + B & \text{if } 1 \le x < +\infty \end{cases}$$

48. Suppose a function f is defined and continuous on the closed interval $[a, b]$. Is the domain of $h(x) = 1/f(x)$ also the closed interval $[a, b]$? What do you conclude about the continuity of h?

In Problems 49–54 use the intermediate value theorem (2.4.6) to determine which of the functions f must have zeros in the given intervals. Indicate those for which the theorem gives no information. Do not attempt to locate the zeros.

49. $f(x) = x^3 - 3x$ on $[-2, 2]$

50. $f(x) = x^4 - 1$ on $[-2, 2]$

51. $f(x) = \dfrac{x}{(x+1)^2} - 1$ on $[10, 20]$

52. $f(x) = x^3 - 2x^2 - x + 2$ on $[3, 4]$

53. $f(x) = \sqrt{x^3+3} - \sqrt{x^3-1} - 1$ on $[1, 10]$

54. $f(x) = \sqrt{x^2 - 3x} - 2$ on $[3, 5]$

55. Let $f(x) = \dfrac{1}{x-1} + \dfrac{1}{x-2}$. Use the intermediate value theorem to prove that there is a real number c between 1 and 2 for which $f(c) = 0$.

56. Prove that there exists a real number c between 2.64 and 2.65 such that $c^2 = 7$.

57. Graph a function that is continuous on the interval $[-1, 2]$, that is positive at both endpoints, and that has exactly two zeros in this interval.

58. Suppose that f and g are continuous in $[a, b]$, $f(a) < g(a)$, and $f(b) > g(b)$. Prove that the graphs of $y = f(x)$ and $y = g(x)$ intersect somewhere between $x = a$ and $x = b$.

59. For the function below, find k so that f is continuous at 2.

$$f(x) = \begin{cases} \dfrac{\sqrt{2x+5} - \sqrt{x+7}}{x-2} & \text{if } x \neq 2 \text{ and } x \ge -\dfrac{5}{2} \\ k & \text{if } x = 2 \end{cases}$$

60. Given the two functions f and h such that

$$f(x) = x^3 - 3x^2 - 4x + 12 \qquad h(x) = \begin{cases} \dfrac{f(x)}{x-3} & \text{if } x \neq 3 \\ p & \text{if } x = 3 \end{cases}$$

(a) Find all zeros of the function f.
(b) Find the number p so that the function h is continuous at $x = 3$. Justify your answer.
(c) Determine whether h, with the number found in (b) is even, odd, or neither. Justify your answer.

61. The function $f(x) = |x|/x$ is not defined at 0. Tell why it is impossible to define $f(0)$ so that f is continuous at 0.

62. If f and g are each continuous at c, prove that $f + g$ is continuous at c.

63. Discover two functions f and g that are each continuous at c, and yet f/g is not continuous at c.

Limits and Continuity of Trigonometric Functions 2.5

Squeezing Theorem

Until now, in many of the examples involving limits, we were able to evaluate the limit of the function either directly or by using algebra. However, many limit problems arise that cannot be directly evaluated by algebraic techniques. These sometimes require geometric arguments (such as to prove*

* Remember that the numerical argument given in Section 1.1 was not a proof.

that $\lim_{x\to 0}(\sin x)/x = 1)$ and sometimes entirely new theories (such as for $\lim_{x\to 0^+} x^x$).

In many instances, the following result, called the *squeezing theorem*, may be helpful for evaluating limits. Its proof is given in Appendix I.

[2.5.1] THEOREM / *Squeezing Theorem.*

Let f, g, and h be functions such that $f(x) \le g(x) \le h(x)$ for all numbers x in some open interval containing c, except possibly at c. If $\lim_{x\to c} f(x) = L$ and $\lim_{x\to c} h(x) = L$, then $\lim_{x\to c} g(x) = L$.

The idea behind this theorem is that it provides a technique for evaluating $\lim_{x\to c} g(x)$ if we know, or can find, simpler approximating functions f and h with g "sandwiched" between f and h for all x close to c. In this event, if f and h have the same limit L as x approaches c, then g is "squeezed" to this same limit L as x approaches c (see Fig. 30).

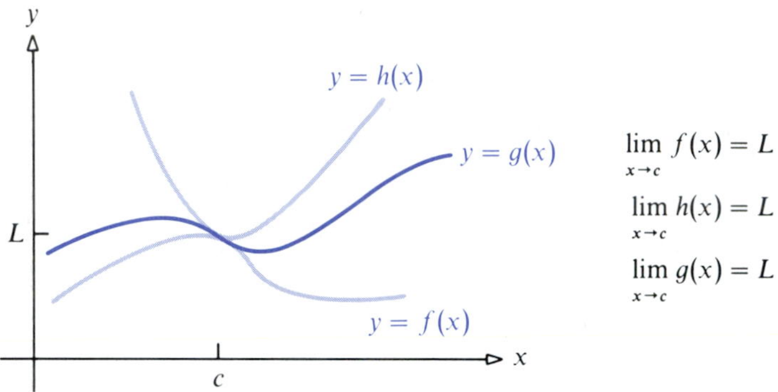

Figure 30

For example, suppose we wish to find $\lim_{x\to 0} f(x)$ and it can be shown that $-x^2 \le f(x) \le x^2$ for all $x \ne 0$. Since $\lim_{x\to 0}(-x^2) = 0$ and $\lim_{x\to 0} x^2 = 0$, it follows from the squeezing theorem that $\lim_{x\to 0} f(x) = 0$ (see Fig. 31).

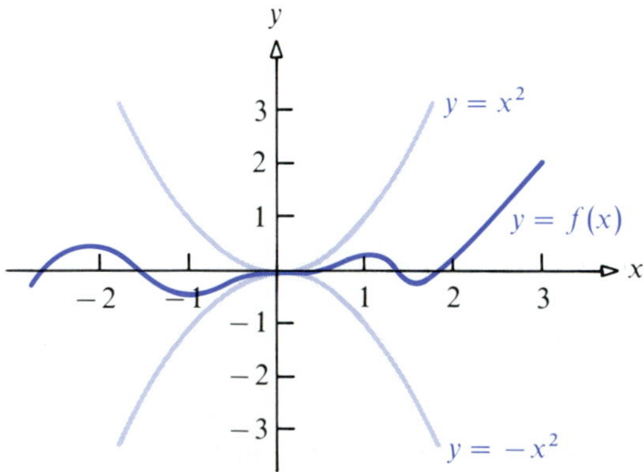

Figure 31

EXAMPLE 1

Show that: $\lim_{x\to 0}\left(x \sin \frac{1}{x}\right) = 0$

Solution

If $x \neq 0$, we know that $|\sin(1/x)| \leq 1$. Thus, for all $x \neq 0$, we have

$$\left|x \sin \frac{1}{x}\right| = |x|\left|\sin \frac{1}{x}\right| \leq |x| \qquad \text{so that} \qquad \left|x \sin \frac{1}{x}\right| \leq |x|$$

Consequently,

$$-|x| \leq x \sin \frac{1}{x} \leq |x|$$

Since $\lim_{x\to 0}(-|x|) = 0$ and $\lim_{x\to 0}|x| = 0$, it follows from theorem (2.5.1) that

$$\lim_{x\to 0}\left(x \sin \frac{1}{x}\right) = 0$$

Limits of Trigonometric Functions

Next we state and prove two trigonometric limits that we will need to establish the continuity of the trigonometric functions. We assume that the angle θ in the forthcoming formula is measured in radians.

[2.5.2] THEOREM

$$\lim_{\theta\to 0} \sin \theta = 0 \tag{1}$$

Proof Suppose $0 < \theta < \pi/2$. Refer to Figure 32, which shows the unit circle $x^2 + y^2 = 1$ and the points $A = (1, 0)$ and $P = (x, y)$ on the unit circle. If the radian measure of the angle AOP is θ, then the length of the arc from A to P is also θ.

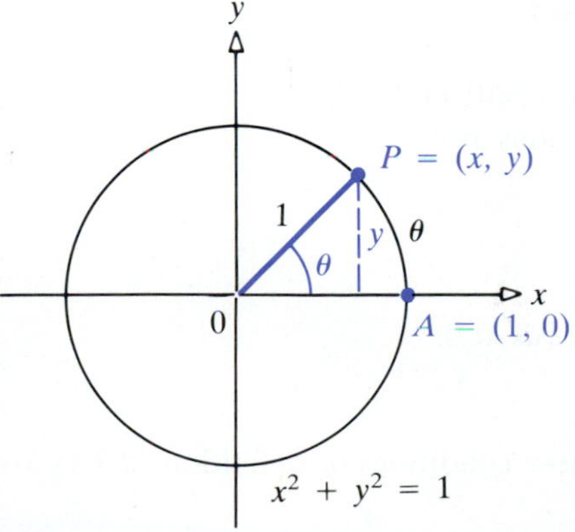

Figure 32

We see from the figure that

$$0 < y < \theta$$

or, since $y = \sin\theta$,

$$0 < \sin\theta < \theta$$

Since we know that $\lim_{\theta\to 0^+} \theta = 0$ and $\lim_{\theta\to 0^+} 0 = 0$, it follows from the squeezing theorem (2.5.1) that $\lim_{\theta\to 0^+} \sin\theta = 0$.

To show that $\lim_{\theta\to 0^-} \sin\theta = 0$, we proceed as follows: If $-\pi/2 < \theta < 0$, then $0 < -\theta < \pi/2$ and hence

$$0 < \sin(-\theta) < -\theta$$

If we multiply the last inequality by -1, and use the fact that $\sin(-\theta) = -\sin\theta$, we get

$$\theta < \sin\theta < 0$$

This inequality, together with the fact that $\lim_{\theta\to 0^-} \theta = 0$ and the squeezing theorem, shows that $\lim_{\theta\to 0^-} \sin\theta = 0$. Thus,

$$\lim_{\theta\to 0} \sin\theta = 0$$

EXAMPLE 2

Show that

$$\lim_{\theta\to 0} \cos\theta = 1 \qquad \textbf{(2)}$$

Solution

From the identity $\sin^2\theta + \cos^2\theta = 1$, it follows that $\cos\theta = \pm\sqrt{1-\sin^2\theta}$. For $-\pi/2 \le \theta \le \pi/2$, $\cos\theta$ is nonnegative, and hence, $\cos\theta = \sqrt{1-\sin^2\theta}$. Thus,

$$\lim_{\theta\to 0} \cos\theta = \lim_{\theta\to 0} \sqrt{1-\sin^2\theta} = \sqrt{\lim_{\theta\to 0}(1-\sin^2\theta)}$$
$$= \sqrt{1-0} = \sqrt{1} = 1$$

We use theorem (2.5.2) and the result of Example 2 to establish the continuity of the sine and cosine functions at 0.

EXAMPLE 3

Prove that the sine function is continuous at 0.

Solution

If we let $f(\theta) = \sin\theta$, then the three conditions of definition (2.4.1) are satisfied:

1. $f(0) = \sin 0 = 0$

2. $\lim_{\theta \to 0} f(\theta) = \lim_{\theta \to 0} \sin \theta = 0$

Thus, the limit exists.

3. $\lim_{\theta \to 0} \sin \theta = \sin 0 = 0$

Thus, $f(\theta) = \sin \theta$ is continuous at 0.

In a similar way we can show that $f(\theta) = \cos \theta$ is continuous at 0. That is,

$$\lim_{\theta \to 0} \cos \theta = 1 = \cos 0$$

In Problem 33 you are asked to show that the sine and cosine functions are continuous everywhere.

Next we establish the following result, which we conjectured earlier.

[2.5.3] THEOREM

If θ is measured in radians, then

$$\lim_{\theta \to 0} \frac{\sin \theta}{\theta} = 1 \qquad (3)$$

Proof To evaluate this limit, we start by taking θ to be a positive acute central angle of a circle with radius $R = 1$. As shown in Figure 33, $\widehat{OAB}$ represents a sector of the circle.

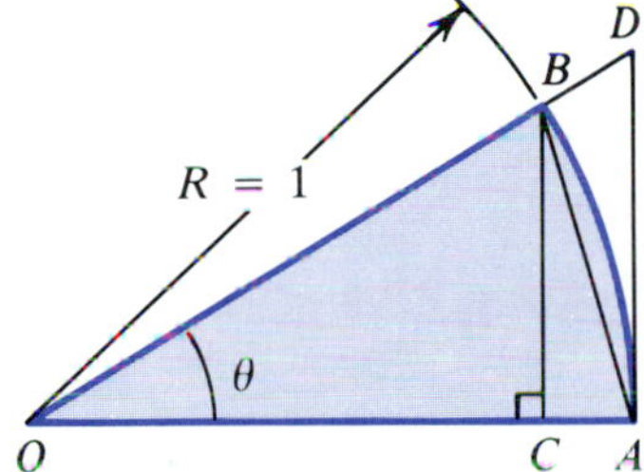

Figure 33

$$\text{Area of } \triangle OAB < \text{Area of } \widehat{OAB} < \text{Area } \triangle OAD \qquad (4)$$

If we use vertical bars | | to denote the length of a line segment, then, since the radius $R = 1$, we find that $|OB| = |OA| = 1$. Furthermore,

$$\sin \theta = \frac{|CB|}{|OB|} = \frac{|CB|}{1} = |CB| \qquad \tan \theta = \frac{|AD|}{|OA|} = \frac{|AD|}{1} = |AD|$$

Thus, in terms of θ, the areas in (4) may be expressed as

$$\text{Area of } \triangle OAB = \frac{1}{2}|OA||CB| = \frac{1}{2}(1) \sin \theta = \frac{1}{2} \sin \theta$$

$$\text{Area of } \widehat{OAB} = \frac{1}{2} R^2\theta = \frac{1}{2}(1)\theta = \frac{1}{2}\theta \qquad (5)$$

$$\text{Area of } \triangle OAD = \frac{1}{2}|OA||AD| = \frac{1}{2}(1) \tan \theta = \left(\frac{1}{2}\right)\frac{\sin \theta}{\cos \theta}$$

By substituting (5) in (4), we obtain

$$\frac{1}{2} \sin \theta < \frac{\theta}{2} < \left(\frac{1}{2}\right)\frac{\sin \theta}{\cos \theta}$$

By dividing by $\frac{1}{2} \sin \theta$, we obtain

$$1 < \frac{\theta}{\sin \theta} < \frac{1}{\cos \theta} \qquad 0 < \theta < \frac{\pi}{2} \qquad (6)$$

Since $0 < \theta < \pi/2$, the quantities $\theta/(\sin\theta)$ and $1/(\cos\theta)$ are positive. As a result, if we take the reciprocal of each expression in (6), the inequalities will be reversed. That is,

$$1 > \frac{\sin\theta}{\theta} > \cos\theta \qquad \text{or} \qquad \cos\theta < \frac{\sin\theta}{\theta} < 1$$

We know that $\lim_{\theta\to 0^+} \cos\theta = 1$. Since the quantity $(\sin\theta)/\theta$ is between a quantity approaching 1 and 1 itself, by the squeezing theorem (2.5.1) it also must approach 1. Thus,

$$\lim_{\theta\to 0^+} \frac{\sin\theta}{\theta} = 1 \tag{7}$$

Since $\sin(-\theta) = -\sin\theta$, we have

$$\frac{\sin(-\theta)}{-\theta} = \frac{-\sin\theta}{-\theta} = \frac{\sin\theta}{\theta}$$

Hence, as $\theta \to 0^-$, we must have

$$\lim_{\theta\to 0^-} \frac{\sin\theta}{\theta} = \lim_{\theta\to 0^+} \frac{\sin(-\theta)}{-\theta} = \lim_{\theta\to 0^+} \frac{\sin\theta}{\theta} = 1 \tag{8}$$

Together, (7) and (8) yield

$$\lim_{\theta\to 0} \frac{\sin\theta}{\theta} = 1 \tag{9}$$

The result just obtained states that for values of θ close to 0, $(\sin\theta)/\theta$ is close to 1. This means that for values of θ close to 0, $\sin\theta \approx \theta$. Table 1 shows this.

Table 1

θ	$\sin\theta$	$(\sin\theta)/\theta$
0.1	0.0998334	0.998334
0.05	0.0499792	0.999583
0.04	0.0399893	0.999733
0.03	0.0299955	0.999850
0.02	0.0199987	0.999933
0.01	0.0099998	0.999983

EXAMPLE 4

Evaluate: $\displaystyle\lim_{\theta\to 0} \frac{\sin 3\theta}{\theta}$

Solution

Since $(\sin 3\theta)/\theta$ is not exactly in the form required by (3), we cannot use it immediately. However, by using the substitution $t = 3\theta$, we find that

$$\lim_{\theta\to 0} \frac{\sin 3\theta}{\theta} \underset{\uparrow}{=} \lim_{t\to 0} \frac{\sin t}{t/3} = \lim_{t\to 0}\left(3\,\frac{\sin t}{t}\right) = 3 \lim_{t\to 0} \frac{\sin t}{t} \underset{\uparrow}{=} (3)(1) = 3$$

Set $t = 3\theta$
As $\theta \to 0$, then $t \to 0$

Apply (3)

Note that $\sin 3\theta \neq 3\sin\theta$.

EXAMPLE 5

Evaluate: $\lim_{\theta \to 0} \dfrac{\tan 5\theta}{\tan \theta}$

Solution

Recall that

$$\frac{\tan 5\theta}{\tan \theta} = \frac{\sin 5\theta}{\cos 5\theta} \cdot \frac{\cos \theta}{\sin \theta} = \frac{\sin 5\theta}{\sin \theta} \cdot \frac{\cos \theta}{\cos 5\theta}$$

We know that

$$\lim_{\theta \to 0} \frac{\sin \theta}{\theta} = 1 \qquad \text{and} \qquad \lim_{\theta \to 0} \cos \theta = 1$$

Now

$$\frac{\sin 5\theta}{\sin \theta} = \frac{5 \cdot \dfrac{\sin 5\theta}{5\theta}}{\dfrac{\sin \theta}{\theta}}$$

If we then substitute $5\theta = t$, we see that $t \to 0$ when $\theta \to 0$ so that

$$\lim_{\theta \to 0} 5 \cdot \frac{\sin 5\theta}{5\theta} = \lim_{t \to 0} 5 \cdot \frac{\sin t}{t} = 5 \cdot 1 = 5$$

and

$$\lim_{\theta \to 0} \cos 5\theta = \lim_{t \to 0} \cos t = 1$$

Therefore,

$$\lim_{\theta \to 0} \frac{\tan 5\theta}{\tan \theta} = 5$$

EXAMPLE 6

Establish the formula:

$$\lim_{\theta \to 0} \frac{\cos \theta - 1}{\theta} = 0 \tag{10}$$

Solution

To obtain this result, we proceed as follows:

$$\frac{\cos \theta - 1}{\theta} = \left(\frac{\cos \theta - 1}{\theta}\right)\left(\frac{\cos \theta + 1}{\cos \theta + 1}\right) = \frac{\cos^2\theta - 1}{\theta(\cos \theta + 1)} = \frac{-\sin^2\theta}{\theta(\cos \theta + 1)}$$

$$= (-\sin \theta)\left(\frac{\sin \theta}{\theta}\right)\left(\frac{1}{\cos \theta + 1}\right)$$

Thus,

$$\lim_{\theta\to 0}\frac{\cos\theta-1}{\theta}=\lim_{\theta\to 0}\left[(-\sin\theta)\left(\frac{\sin\theta}{\theta}\right)\left(\frac{1}{\cos\theta+1}\right)\right]$$

$$=\left[\lim_{\theta\to 0}(-\sin\theta)\right]\left[\lim_{\theta\to 0}\frac{\sin\theta}{\theta}\right]\left[\lim_{\theta\to 0}\frac{1}{\cos\theta+1}\right]$$

$$=(0)(1)(\tfrac{1}{2})=0$$

EXERCISE 2.5

In Problems 1–18 evaluate each limit.

1. $\lim_{x\to 0}(x^3+\sin x)$

2. $\lim_{x\to 0}(x^2-\cos x)$

3. $\lim_{x\to 0}\sin(\cos x)$

4. $\lim_{x\to 0}\cos(\sin x)$

5. $\lim_{x\to\pi/3}(\cos x+\sin x)$

6. $\lim_{x\to\pi/3}(\sin x-\cos x)$

7. $\lim_{x\to\pi/6}\csc x\cot^2 x$

8. $\lim_{\theta\to -3\pi}\theta^3\sin^4\theta$

9. $\lim_{x\to 0}\frac{\sin 7x}{x}$

10. $\lim_{x\to 0}\frac{\sin(x/3)}{x}$

11. $\lim_{x\to 0}\frac{\cos x}{1+\sin x}$

12. $\lim_{x\to 0}\frac{\sin x}{1+\cos x}$

13. $\lim_{x\to 0}\frac{\sin x^2}{x}$

14. $\lim_{x\to 0}\frac{\sin^2 2x}{x^2}$

15. $\lim_{x\to\pi}\frac{\sin x}{\pi-x}$

16. $\lim_{x\to 0}\frac{2x-5\sin 3x}{x}$

17. $\lim_{\theta\to 0}\frac{\sin\theta}{\theta+\tan\theta}$

18. $\lim_{x\to 0}\frac{x}{\sin(x/2)}$

In Problems 19–22 show that each equation is true.

19. $\lim_{x\to 0}\frac{\sin ax}{\sin bx}=\frac{a}{b}$

20. $\lim_{x\to 0}\frac{\cos ax}{\cos bx}=1$

21. $\lim_{x\to 0}\frac{\sin ax}{bx}=\frac{a}{b}$

22. $\lim_{x\to 0}\frac{1-\cos ax}{bx}=0$

23. The function $f(x)=(\sin\pi x)/x$ is not defined at 0. Decide how to define $f(0)$ so that f is continuous at 0.

24. How must we define $f(0)$ and $f(1)$ so that the function below is continuous on $0\le x\le 1$?

$$f(x)=\frac{\sin\pi x}{x(1-x)}$$

25. If $1-x^2\le f(x)\le\cos x$ for all x in the interval $-\pi/2<x<\pi/2$, show that $\lim_{x\to 0}f(x)=1$.

26. If $f(x)=\begin{cases}1 & \text{if } x \text{ is rational}\\ 0 & \text{if } x \text{ is irrational}\end{cases}$

show that $\lim_{x\to 0}[xf(x)]=0$.

27. If $0\le f(x)\le 1$ for every x, show that $\lim_{x\to 0}[x^2f(x)]=0$.

28. If $0\le f(x)\le M$ for every x, show that $\lim_{x\to 0}[x^2f(x)]=0$.

29. Show that $\lim_{x\to 0}[x^n\sin(1/x)]=0$, n a positive integer. [*Hint:* Look first at Problem 27.]

30. Let $f(x)=\begin{cases}\dfrac{\sin x}{x} & \text{if } x\ne 0\\ 1 & \text{if } x=0\end{cases}$

Is f continuous at 0?

31. Let $f(x)=\begin{cases}\dfrac{1-\cos x}{x} & \text{if } x\ne 0\\ 0 & \text{if } x=0\end{cases}$

Is f continuous at 0?

32. Suppose we are given points A and B with coordinates (0, 0) and (1, 0), respectively. Let n be a fixed number greater than 0, and let θ be an angle such that $0<\theta<\pi/(1+n)$. We can construct a triangle ABC such that AC and AB form the angle θ, and CB and

(*Problem 32 continues on page 113*)

32. (*continued*)

AB form the angle $n\theta$ (see the figure). Let D be the point of intersection of AB with the perpendicular from C to AB. What is the limiting position of D as θ approaches 0?

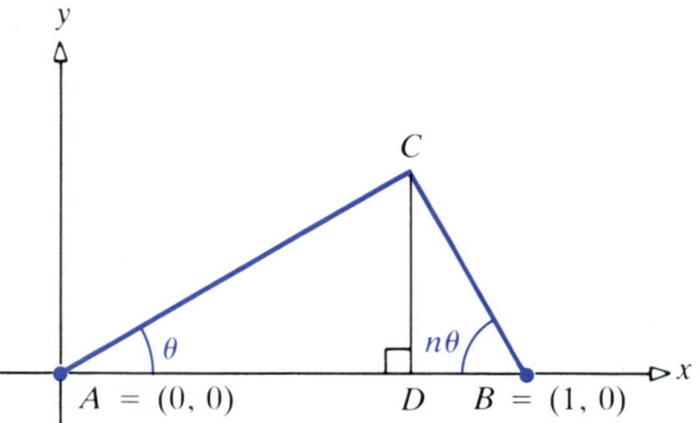

33. Use identities (15) and (13) from Section 1.8 to show that the sine and cosine functions are continuous everywhere.

34. Show that $\lim_{\theta\to 0} \frac{\sin\theta}{\theta^2}$ does not exist; show that $\lim_{\theta\to 0} \frac{\cos\theta - 1}{\theta^2} = -\frac{1}{2}$.

Formal Definition of Limit

2.6

In this section we take a closer look at the definition of a limit. Let's begin with an example. Consider the function f defined by

$$f(x) = \begin{cases} 3x + 1 & \text{if } x \neq 2 \\ 3 & \text{if } x = 2 \end{cases}$$

Then, $\lim_{x\to 2} f(x) = 7$, as can be inferred from the graph (see Fig. 12, p. 80) or from the following table:

x	1.75	1.9	1.95	1.99	1.995	1.999	1.9999	2.0001	2.001	2.005	2.01	2.1	2.25	2.5
$f(x)$	6.25	6.7	6.85	6.97	6.985	6.997	**6.9997**	**7.0003**	7.003	7.015	7.03	7.3	7.75	8.5

Notice from the table that as x gets closer and closer to 2, $f(x)$ gets closer and closer to 7. In fact, we can make $f(x)$ as close to 7 as we please by taking x close enough to 2. For example, suppose we want $f(x)$ to differ from 7 by less than 0.3; that is,

$$7 - 0.3 < f(x) < 7 + 0.3$$

$$6.7 < f(x) < 7.3$$

First, we must require $x \neq 2$ because when $x = 2$, then $f(x) = f(2) = 3$ (by definition) and we obtain $6.7 < 3 < 7.3$, which is impossible.

From the table, we can see that 6.7 corresponds to 1.9, and 7.3 corresponds to 2.1. Thus,

$$6.7 < f(x) < 7.3 \qquad \text{whenever} \qquad x \neq 2 \quad \text{and} \quad 1.9 < x < 2.1$$

In other words, $f(x)$ differs from 7 by less than 0.3 whenever $x \neq 2$ and x differs from 2 by less than 0.1. Similarly, the table indicates that $f(x)$ differs from 7 by less than 0.003 whenever $x \neq 2$ and x differs from 2 by less than 0.001.

Now, let us ask a question that cannot be answered from the table. For $x \neq 2$, how close to 2 must x be in order to guarantee that $f(x)$ differs from 7 by less than some *arbitrary* positive number ε (ε might be extremely small)? Well, the words $f(x)$ *differs from 7 by less than* ε may be written as

$$7 - \varepsilon < f(x) < 7 + \varepsilon$$

If $x \neq 2$, $f(x) = 3x + 1$. Hence,

$$7 - \varepsilon < 3x + 1 < 7 + \varepsilon$$

By subtracting 1 from all parts, we obtain $6 - \varepsilon < 3x < 6 + \varepsilon$. And, finally, we multiply by $\frac{1}{3}$ to obtain

$$2 - \tfrac{1}{3}\varepsilon < x < 2 + \tfrac{1}{3}\varepsilon$$

Thus, the answer to our question is $\frac{1}{3}\varepsilon$. That is, $f(x)$ differs from 7 by less than ε whenever x differs from 2 by less than $\frac{1}{3}\varepsilon$, provided that $x \neq 2$.

For example, $f(x)$ differs from 7 by less than $\varepsilon = \frac{1}{10}$ whenever $x \neq 2$ and x differs from 2 by less than $\frac{1}{3}\varepsilon = (\frac{1}{3})(\frac{1}{10}) = \frac{1}{30}$; $f(x)$ differs from 7 by less than $\varepsilon = 0.3$ whenever $x \neq 2$ and x differs from 2 by less than $\frac{1}{3}\varepsilon = \frac{1}{3}(0.3) = 0.1$; $f(x)$ differs from 7 by less than $\varepsilon = 0.003$ whenever $x \neq 2$ and x differs from 2 by less than $\frac{1}{3}\varepsilon = 0.001$. (These last two statements are verified by the table.)

We now conjecture the following: If ε is any given positive number, then there is a positive number—say, δ—such that

$$\begin{bmatrix} f(x) \text{ differs from 7} \\ \text{by less than } \varepsilon \end{bmatrix} \qquad \text{whenever} \qquad \begin{bmatrix} x \neq 2 \quad \text{and } x \text{ differs from 2} \\ \text{by less than } \delta \end{bmatrix}$$

(For the example just discussed, $\delta = \varepsilon/3$.)

This statement can be shortened by using the appropriate mathematical notation. We shorten the phrase "ε is any given positive number" by writing simply $\varepsilon > 0$. The difference between $f(x)$ and 7 is $(f(x) - 7)$. Thus, the statement "$f(x)$ differs from 7 by less than ε" may be written $|f(x) - 7| < \varepsilon$. Similarly, the difference between x and 2 is $(x - 2)$, and the statement "x differs from 2 by less than δ" may be written $|x - 2| < \delta$. To say that $x \neq 2$ is simply to say that the difference between them is positive; that is, $0 < |x - 2|$. Therefore, for the given function, we have the following: If $\varepsilon > 0$ is given, then there is a $\delta > 0$ such that

$$|f(x) - 7| < \varepsilon \qquad \text{whenever} \qquad 0 < |x - 2| < \delta$$

In fact, $\delta = \frac{1}{3}\varepsilon$ works. Since this is possible for any $\varepsilon > 0$, we conclude that the limit of $f(x)$ as x approaches 2 is equal to 7, and we write $\lim_{x\to 2} f(x) = 7$.

The above discussion leads to the ε, δ definition of the limit of a function, which we now restate.

[2.6.1] DEFINITION / ***Limit of a Function.***

Let f be a function defined on an open interval containing c, except possibly for c itself. Then $\lim_{x\to c} f(x) = L$ if, for any given number $\varepsilon > 0$, a number $\delta > 0$ exists such that

$$|f(x) - L| < \varepsilon \qquad \text{whenever} \qquad 0 < |x - c| < \delta$$

Figure 34 illustrates the definition for two choices of ε. Notice that in part (b), the smaller ε requires a smaller δ than in part (a).

Figure 35 illustrates what happens if the δ is too large for the choice of ε; there are values of $f(x)$—for example, at x_1 and x_2—for which $|f(x) - L| \not< \varepsilon$.

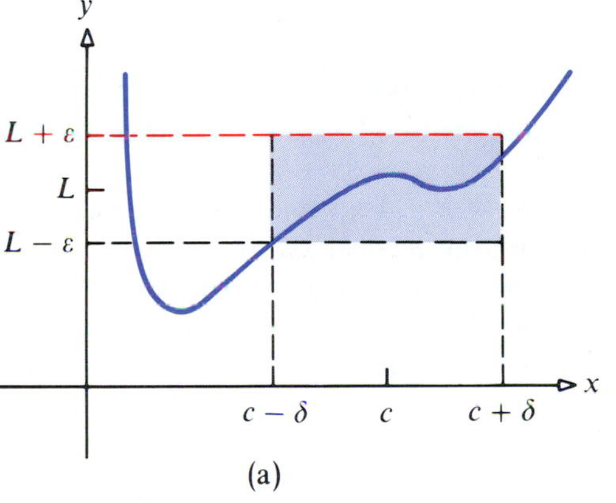

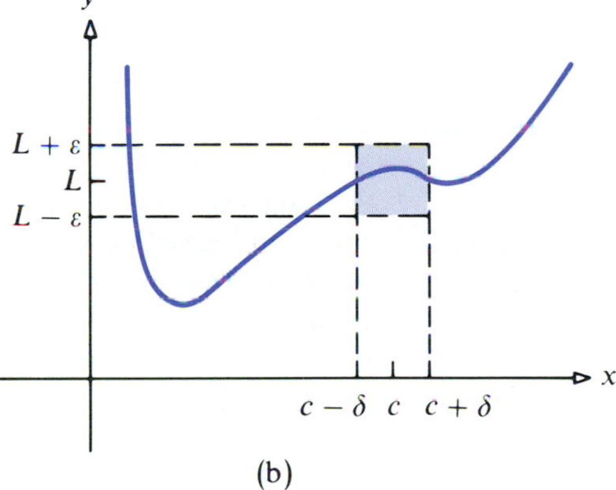

Figure 34

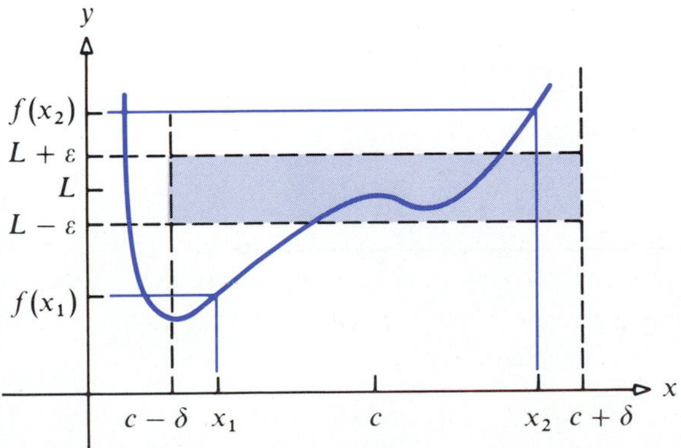

Figure 35

Let's look at our original example again to see how this definition of limit is applied.

EXAMPLE 1

Using the definition of limit, prove that $\lim_{x\to 2} f(x) = 7$ for the function

$$f(x) = \begin{cases} 3x + 1 & \text{if} \quad x \neq 2 \\ 3 & \text{if} \quad x = 2 \end{cases}$$

Solution

We seek a number $\delta > 0$ such that

$$|(3x + 1) - 7| < \varepsilon \qquad \text{whenever} \qquad 0 < |x - 2| < \delta$$

To find a connection between

$$|x - 2| \qquad \text{and} \qquad |(3x + 1) - 7|$$

we simplify the last expression to get

$$|(3x + 1) - 7| = |3x - 6| = |3(x - 2)| = 3|x - 2|$$

We can make this expression less than ε by making $|x - 2|$ less than $\frac{1}{3}\varepsilon$. This suggests that we set $\delta = \frac{1}{3}\varepsilon$. Thus, we see that for any ε, we can find a δ so that whenever $0 < |x - 2| < \delta = \frac{1}{3}\varepsilon$, then $|(3x + 1) - 7| < \varepsilon$. Hence, we have established that $\lim_{x\to 2} f(x) = 7$. A geometric interpretation is shown in Figure 36. We see that $f(x)$ on the vertical axis will lie between the horizontal lines $y = 7 + \varepsilon$ and $y = 7 - \varepsilon$, whenever x on the horizontal axis lies between $2 - \delta$ and $2 + \delta$. Thus, $\lim_{x\to 2} f(x) = 7$ describes the behavior of f near 2—namely, that the value of f is close to 7 when x is close to 2. Notice that $\lim_{x\to 2} f(x) \neq f(2)$.

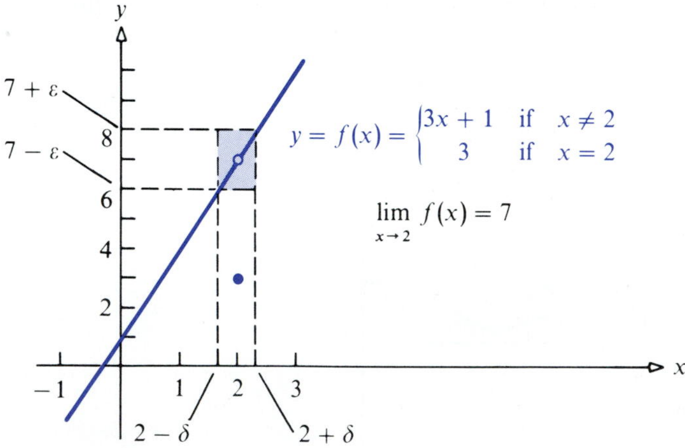

Figure 36

The next example proves the important results given in (4) and (5) in Section 2.1.

EXAMPLE 2

Use the definition of limit to prove that:

(a) $\lim_{x \to c} A = A$, A and c real numbers
(b) $\lim_{x \to c} x = c$, c a real number

Solution

(a) The function f is the constant function $f(x) = A$ whose graph is a horizontal line. Let $\varepsilon > 0$ be given. We must find $\delta > 0$ so that whenever $0 < |x - c| < \delta$, then $|f(x) - A| = |A - A| < \varepsilon$. Since $|A - A| = 0$, no matter what δ is used, it always happens that $|f(x) - A| < \varepsilon$. That is, any choice of δ guarantees that whenever $0 < |x - c| < \delta$, then $|f(x) - A| < \varepsilon$.

(b) The function f is the identity function $f(x) = x$. Let $\varepsilon > 0$ be given. We must find δ so that whenever $0 < |x - c| < \delta$, then $|f(x) - c| = |x - c| < \varepsilon$. The obvious choice for δ is ε itself. That is, whenever $0 < |x - c| < \delta = \varepsilon$, it follows that $|f(x) - c| = |x - c| < \varepsilon$.

The following observations about the definition of a limit are important:

The limit of the function in no way depends on the value of the function at c.

Recall from Example 1 that $\lim_{x \to 2} f(x) = 7$ and yet $f(2) \neq 7$.

In general, the size of δ depends on the size of ε.

For example, if $\delta = \varepsilon/3$ (as in Example 1), then δ grows smaller as ε does.

For a given ε, if a suitable δ has been found, any *smaller positive number* will also work. That is, δ is not uniquely determined when ε is given.

Refer to our earlier discussion of the function in Example 1, where $L = 7$ and $c = 2$. For $\varepsilon = \frac{1}{10}$, we found $\delta = \frac{1}{30}$, which is actually the *maximum* permissible δ for this ε. For $\varepsilon = 0.003$, we found $\delta = 0.001$, which again is the maximum permissible δ for this ε.

Let's look at some more examples.

EXAMPLE 3

Prove that $\lim_{x \to -1} (1 - 2x) = 3$.

Solution

To begin, we assume that a number $\varepsilon > 0$ is given. We must show that there exists a number $\delta > 0$ such that if $0 < |x - (-1)| < \delta$, then

$$|(1 - 2x) - 3| < \varepsilon$$

The idea is to find a connection between

$$|x - (-1)| \qquad \text{and} \qquad |(1 - 2x) - 3|$$

We begin by noting that

$$|x - (-1)| = |x + 1|$$

and

$$|(1 - 2x) - 3| = |-2(x + 1)| = |-2||x + 1| = 2|x + 1|$$

so that

$$|(1 - 2x) - 3| = 2|x - (-1)|$$

In other words,

$$|(1 - 2x) - 3| < \varepsilon \qquad \text{whenever} \qquad |x - (-1)| < \frac{\varepsilon}{2}$$

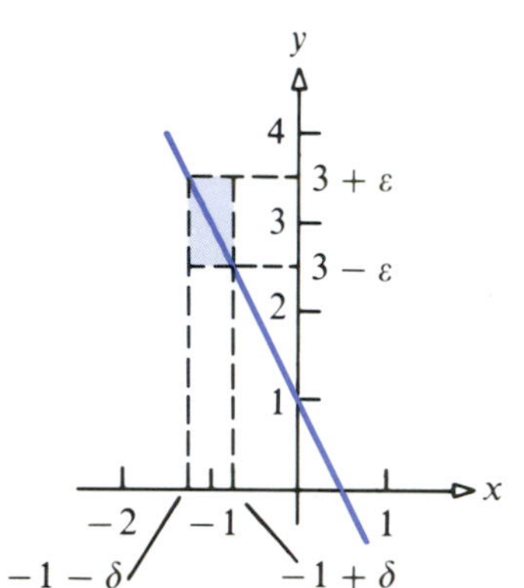

Figure 37

Thus, we may choose $\delta = \varepsilon/2$ (see Fig. 37). This completes the argument that $\lim_{x \to -1}(1 - 2x) = 3$.

EXAMPLE 4

Prove that $\lim_{x \to 2} x^2 = 4$.

Solution

To begin, we assume that a number $\varepsilon > 0$ is given. We must show that there exists a number $\delta > 0$ such that

$$|x^2 - 4| < \varepsilon \qquad \text{whenever} \qquad 0 < |x - 2| < \delta$$

We need to establish a connection between $|x^2 - 4|$ and $|x - 2|$. We note that

$$|x^2 - 4| = |(x + 2)(x - 2)| = |x + 2||x - 2|$$

If we can find a number K such that $|x + 2| < K$, the choice of δ is clear—namely, $\delta < \varepsilon/K$. If x is confined to some interval centered about 2, then K can be found. For example, suppose $|x - 2| < 1$—that is, $1 < x < 3$. Then we add 2 to each part to get $1 + 2 < x + 2 < 3 + 2$, and, in particular, $|x + 2| = x + 2 < 5$. It then follows that whenever $|x - 2| < 1$,

$$|x^2 - 4| = |(x + 2)(x - 2)| < 5|x - 2|$$

If $|x - 2| < \varepsilon/5$ also, then

$$|x^2 - 4| < 5|x - 2| < 5\left(\frac{\varepsilon}{5}\right) = \varepsilon$$

as desired. But the quantity $|x - 2|$ now has two restrictions—namely,

$$|x - 2| < 1 \qquad \text{and} \qquad |x - 2| < \frac{\varepsilon}{5}$$

To ensure that both inequalities are obeyed, we select δ to be the smaller of 1 and $\varepsilon/5$, abbreviated $\delta = \min(1, \varepsilon/5)$. With this choice of δ, whenever

$$|x - 2| < \delta = \min\left(1, \frac{\varepsilon}{5}\right)$$

we have $|x^2 - 4| < \varepsilon$.

In this example, the choice of restricting x so that $|x - 2| < 1$ is completely arbitrary. The reader should verify that if we had restricted x so that $|x - 2| < \frac{1}{3}$, then the choice for δ would be less than or equal to the smaller of $\frac{1}{3}$ and $3\varepsilon/13$; that is, $\delta \leq \min(\frac{1}{3}, 3\varepsilon/13)$.

EXAMPLE 5

Prove that $\lim\limits_{x \to c} \dfrac{1}{x} = \dfrac{1}{c}$, $c > 0$.

Solution

For a given $\varepsilon > 0$, we wish to find a δ such that $|(1/x) - (1/c)| < \varepsilon$ whenever $0 < |x - c| < \delta$. Now, for $x \neq 0$ we have

$$|f(x) - L| = \left|\frac{1}{x} - \frac{1}{c}\right| = \left|\frac{c - x}{xc}\right| = \frac{|x - c|}{|x||c|} = \frac{|x - c|}{c|x|}$$

The idea here is to find a connection between

$$|x - c| \qquad \text{and} \qquad \frac{|x - c|}{c|x|}$$

We proceed as in Example 4. Since we are interested only in x near c, we restrict x so that $|x - c| < c/2$. It follows that $-c/2 < x - c < c/2$, or

$$\frac{c}{2} < x < \frac{3c}{2}$$

Therefore, if $|x - c| < c/2$, then $|x| > c/2$, so that $1/|x| < 2/c$ and

$$\left|\frac{1}{x} - \frac{1}{c}\right| = \frac{|x - c|}{c|x|} < \frac{2}{c^2}|x - c|$$

We can make $|(1/x) - (1/c)| < \varepsilon$ by choosing

$$|x - c| < \frac{\varepsilon}{2/c^2} = \frac{c^2\varepsilon}{2}$$

since then,

$$\left|\frac{1}{x} - \frac{1}{c}\right| < \frac{2}{c^2}\left(\frac{c^2\varepsilon}{2}\right) = \varepsilon$$

But $|x - c|$ now has two restrictions: $|x - c| < c/2$ and $|x - c| < c^2\varepsilon/2$.

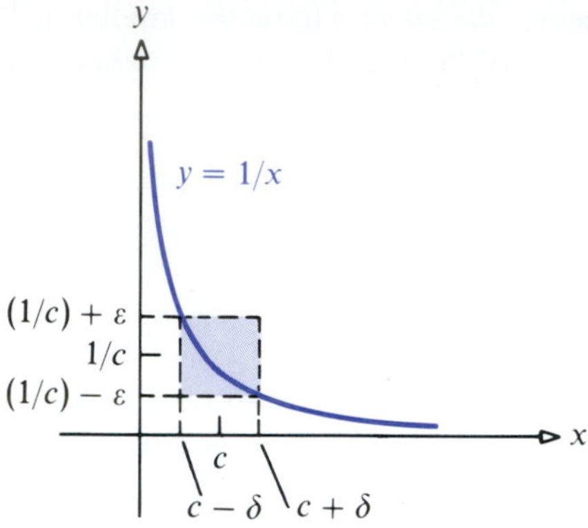

Figure 38

Hence, given $\varepsilon > 0$, we choose $\delta = \min(c/2, c^2\varepsilon/2)$. Then, whenever $|x - c| < \delta$, we have $|(1/x) - (1/c)| < \varepsilon$ (see Fig. 38).

The next example illustrates how a mistake in computing the limit can be discovered.

EXAMPLE 6

Prove that the statement $\lim_{x \to 3}(4x - 5) = 10$ is false.

Solution

This is a proof by contradiction. Thus, if we can show that for one ε, no number δ can be found that satisfies the definition of the limit, the proof will be complete. Suppose $\varepsilon = 1$. If $\lim_{x \to 3}(4x - 5) = 10$, then there is a $\delta > 0$ such that

$$|(4x - 5) - 10| < 1 \qquad \text{whenever} \qquad 0 < |x - 3| < \delta$$

The first inequality leads to

$$3.5 < x < 4 \qquad \text{whenever} \qquad x \neq 3 \qquad \text{and} \qquad |x - 3| < \delta$$

But the second inequality must be obeyed and, regardless of the value of δ, the inequality $0 < |x - 3| < \delta$ contains a number x that is less than 3, in contradiction of the fact that $3.5 < x < 4$. This impossibility shows us that $\lim_{x \to 3}(4x - 5) \neq 10$.

The next example uses an ε, δ proof to show that a limit does not exist.

EXAMPLE 7

Prove that for the function given below, $\lim_{x \to c} f(x)$ does not exist.

$$f(x) = \begin{cases} 1 & \text{if} \quad x \text{ is rational} \\ 0 & \text{if} \quad x \text{ is irrational} \end{cases}$$

Solution

Here, we use the technique of proof by contradiction. Suppose a limit does exist, that is, $\lim_{x\to c} f(x) = L$ for some number c. Let $\varepsilon = \frac{1}{2}$; then there must exist a $\delta > 0$ such that

$$|f(x) - L| < \tfrac{1}{2} \qquad \text{whenever} \qquad 0 < |x - c| < \delta$$

Now, let x_1 be a rational number satisfying $0 < |x_1 - c| < \delta$ and let x_2 be an irrational number satisfying $0 < |x_2 - c| < \delta$. From the definition of the function, we know that

$$f(x_1) = 1 \qquad \text{and} \qquad f(x_2) = 0$$

Thus, by combining, we must have

$$|f(x_1) - L| = |1 - L| < \tfrac{1}{2} \qquad \text{and} \qquad |f(x_2) - L| = |0 - L| < \tfrac{1}{2}$$

From the first inequality, we have $L > \frac{1}{2}$. From the second, we have $L < \frac{1}{2}$. But it is clearly impossible for both of these inequalities to be satisfied.

The next example proves theorem (2.2.10), which was stated on page 90. In Problem 24, you are asked to prove theorem (2.2.11).

EXAMPLE 8

Prove that if $\lim_{x\to c} f(x) > 0$, then there is an open interval about c, possibly excluding c itself, on which $f(x) > 0$.

Solution

Suppose $\lim_{x\to c} f(x) = L > 0$. Then, given any $\varepsilon > 0$, there is a $\delta > 0$ such that

$$|f(x) - L| < \varepsilon \qquad \text{whenever} \qquad 0 < |x - c| < \delta$$

Choose $\varepsilon = L/2$. Then there is a $\delta > 0$ such that

$$|f(x) - L| < \frac{L}{2} \qquad \text{whenever} \qquad 0 < |x - c| < \delta$$

$$-\frac{L}{2} < f(x) - L < \frac{L}{2} \qquad \text{whenever} \qquad 0 < |x - c| < \delta$$

$$\frac{L}{2} < f(x) < \frac{3L}{2} \qquad \text{whenever} \qquad 0 < |x - c| < \delta$$

Since $L/2 > 0$, the last statement proves our assertion that $f(x) > 0$ for all x satisfying $0 < |x - c| < \delta$.

EXERCISE 2.6

1. For the function $f(x) = 4x - 1$, $\lim_{x\to 3} f(x) = 11$. For each $\varepsilon > 0$, find a $\delta > 0$ such that

$$|(4x - 1) - 11| < \varepsilon \quad \text{whenever} \quad 0 < |x - 3| < \delta$$

(a) $\varepsilon = 0.1$ (b) $\varepsilon = 0.01$ (c) $\varepsilon = 0.001$
(d) $\varepsilon > 0$ is arbitrary

2. For the function $f(x) = 2 - 5x$, $\lim_{x\to -2} f(x) = 12$. For each $\varepsilon > 0$, find a $\delta > 0$ such that

$$|(2 - 5x) - 12| < \varepsilon \quad \text{whenever} \quad 0 < |x + 2| < \delta$$

(a) $\varepsilon = 0.2$ (b) $\varepsilon = 0.02$ (c) $\varepsilon = 0.002$
(d) $\varepsilon > 0$ is arbitrary

3. For the function $f(x) = \dfrac{x^2 - 9}{x + 3}$, $\lim_{x\to -3} f(x) = -6$. For each $\varepsilon > 0$, find a $\delta > 0$ such that

$$\left|\frac{x^2 - 9}{x + 3} - (-6)\right| < \varepsilon \quad \text{whenever} \quad 0 < |x + 3| < \delta$$

(a) $\varepsilon = 0.1$ (b) $\varepsilon = 0.01$
(c) $\varepsilon > 0$ is arbitrary

4. For the function $f(x) = \dfrac{x^2 - 4}{x - 2}$, $\lim_{x\to 2} f(x) = 4$. For each $\varepsilon > 0$, find a $\delta > 0$ such that

$$\left|\frac{x^2 - 4}{x - 2} - 4\right| < \varepsilon \quad \text{whenever} \quad 0 < |x - 2| < \delta$$

(a) $\varepsilon = 0.1$ (b) $\varepsilon = 0.01$
(c) $\varepsilon > 0$ is arbitrary

For the limits in Problems 5–8 find a δ that is less than the largest δ that "works" for the given ε.

5. $\lim_{x\to 1}(2x) = 2$, $\varepsilon = 0.01$

6. $\lim_{x\to 2}(-3x) = -6$, $\varepsilon = 0.01$

7. $\lim_{x\to 2}(6x - 1) = 11$, $\varepsilon = \frac{1}{2}$

8. $\lim_{x\to -3}(2 - 3x) = 11$, $\varepsilon = \frac{1}{3}$

In Problems 9–20 give an ε, δ proof for each limit.

9. $\lim_{x\to 2}(3x) = 6$

10. $\lim_{x\to 3}(4x) = 12$

11. $\lim_{x\to 0}(2x + 5) = 5$

12. $\lim_{x\to -1}(2 - 3x) = 5$

13. $\lim_{x\to -3}(-5x + 2) = 17$

14. $\lim_{x\to 2}(2x - 3) = 1$

15. $\lim_{x\to 2}(x^2 - 2x) = 0$

16. $\lim_{x\to 0}(x^2 + 3x) = 0$

17. $\lim_{x\to 1} \dfrac{1 + 2x}{3 - x} = \dfrac{3}{2}$

18. $\lim_{x\to 2} \dfrac{2x}{4 + x} = \dfrac{2}{3}$

19. $\lim_{x\to 0} \sqrt[3]{x} = 0$

20. $\lim_{x\to 1} \sqrt{2 - x} = 1$

21. Show that the statement $\lim_{x\to 3}(3x - 1) = 12$ is false.

22. Show that the statement $\lim_{x\to -2}(4x) = -7$ is false.

23. Show that $\left|\dfrac{1}{x^2 + 9} - \dfrac{1}{18}\right| < \dfrac{7}{234}|x - 3|$ if $2 < x < 4$.

Use this to show that $\lim_{x\to 3} \dfrac{1}{x^2 + 9} = \dfrac{1}{18}$.

24. Prove that if $\lim_{x\to c} f(x) < 0$, then there is an open interval about c, possibly excluding c itself, on which $f(x) < 0$. See Example 8 in Section 2.6.

25. Use the definition of limit to show that $\lim_{x\to 1} x^2 \neq 1.31$. [*Hint:* Use $\varepsilon = 0.1$.]

26. If m and b are any constants, prove that

$$\lim_{x\to c}(mx + b) = mc + b$$

27. Use the definition of limit to prove that $\lim_{x\to 0}(4 - x^2) = 4$.

Historical Perspectives

Historically, one of the important problems for the development of the calculus is the *tangent problem*—the problem of finding a line tangent to the graph of a function at a given point on it. The tangent problem was solved for some special cases by the Greeks more than 2000 years ago. However, their methods were almost entirely geometric, as opposed to the calculus method discussed in this chapter. After the decline of the Greek civilization, very little progress was made on the tangent problem until the invention of analytic geometry by two Frenchmen, René Descartes* (1596–1650) and Pierre de Fermat† (1601–1665).

One can hardly overestimate the importance of the new mathematical tool, analytic geometry, for the development of calculus. In the early seventeenth century, one great hindrance to solving some of the basic mathematical problems was the need to translate these problems into geometric terms. The Greek methods, with their total dependence on geometry (though modified considerably), still formed the basis for the mathematical analysis done prior to Descartes and Fermat.

Both Descartes and Fermat provided solutions to the tangent problem, working independently and using quite dissimilar methods. Descartes' method was purely algebraic, and while it was a definite improvement over the Greek methods, its application was limited to a relatively small number of graphs. Fermat, on the other hand, developed a method that could be applied to a wide variety of graphs. In fact, his method was close in spirit to the method used in this chapter. Fermat's method was improved later by one of the two principal founders of calculus, Sir Isaac Newton (1642–1727) of England. And another Frenchman, Blaise Pascal (1623–1662), a contemporary of Descartes and Fermat, did work on the tangent problem that later inspired Gottfried Wilhelm von Leibniz (1646–1716) of Germany, who shares with Newton the honor of founding the calculus.

You are now invited to explore the history of the tangent problem in the following set of exercises.

---

* René Descartes, who came from a moderately wealthy family, was trained from an early age in a Jesuit school. Although Descartes showed academic promise early on, he did not settle down to scholarly work until he was past 30. Descartes moved to Holland at the age of 32, and for the next 20 years, he lived and worked there in relative obscurity, devoting most of his energy to the development of a deductive philosophical system. In 1649, Descartes was persuaded to go to Sweden as tutor to Queen Christine. His duties there required him to begin his day at 5:00 AM in an unheated library. This schedule, so contrary to his lifelong habit of sleeping most of the morning, combined with a most severe winter in Sweden, was more than the frail Descartes could take, and he died of a lung inflammation a few months after his arrival in Sweden.

† Pierre Fermat ranks as the greatest "amateur" mathematician ever. We say "amateur" because Fermat was a lawyer by training and trade, and his considerable contributions to mathematics were made in his spare time! This fact becomes even more impressive when we consider those contributions. In addition to his work in analytic geometry, he was cofounder (with Pascal) of the theory of probability, an area of mathematics that has become increasingly important during the present century; he established many fundamental results in the field of number theory; and finally, since his work on calculus was the best done before Newton and Leibniz, he must be considered one of the principal founders of the subject. Besides his interest in mathematics he was a classical scholar, linguist, and poet.

Greek Methods for Tangents

1. The Greek method for finding the tangent line to a circle depended on the fact that at any point on a circle, the radius and the tangent lines are perpendicular. Use this method to find an equation of the tangent line to the circle $x^2 + y^2 = 4$ at the point $(1, \sqrt{3})$. See Figure 39. [*Hint:* First find the slope of L_2; then write the slope of L_1 and use the point–slope formula for the equation of L_1.]

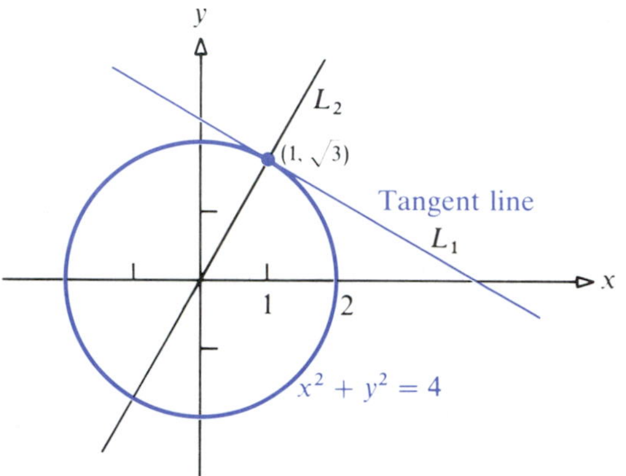

Figure 39

2. Use the method of Problem 1 to find an equation of the tangent line to the circle $(x - 1)^2 + (y - 2)^2 = 4$ at the point $(0, 2 + \sqrt{3})$.

3. The Greek method for finding a tangent line to a parabola is a bit more involved than that for a circle. If we seek the tangent line at P (see Fig. 40), we can show that it also passes through the point B on the axis

(*Problem 3 continues in the next column*)

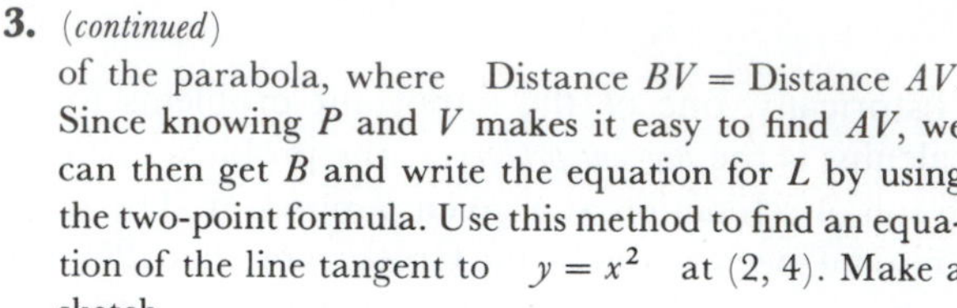
3. (*continued*)
of the parabola, where Distance BV = Distance AV. Since knowing P and V makes it easy to find AV, we can then get B and write the equation for L by using the two-point formula. Use this method to find an equation of the line tangent to $y = x^2$ at (2, 4). Make a sketch.

4. Use the method of Problem 3 to find an equation of the tangent line to the graph of $x = y^2 + 1$ at the point (2, 1).

5. Use the method of Problem 3 to find an equation of the tangent line to the graph of $y = x^2$ at *any* point (x_0, y_0).

Note: Similar methods for finding tangents to ellipses and hyperbolas are outlined in Figures 41 and 42. The Greeks also had some success with curves other than conics. For example, Archimedes (287–212 BC) developed a method for finding tangents to the Archimedean spiral.

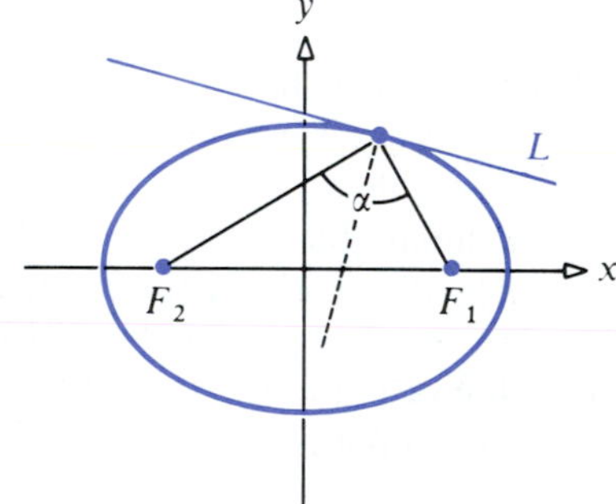

Figure 41

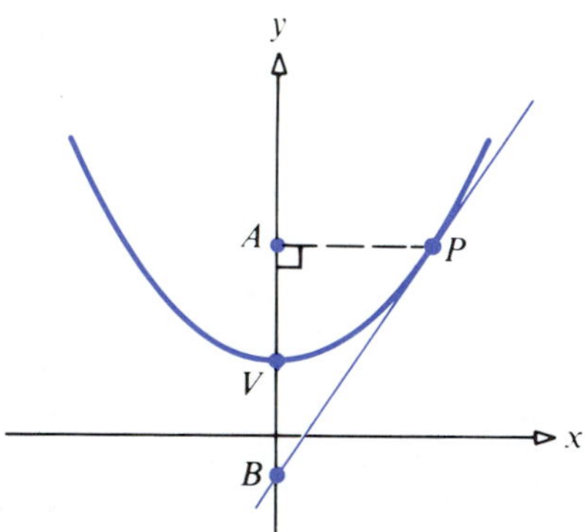

Figure 40

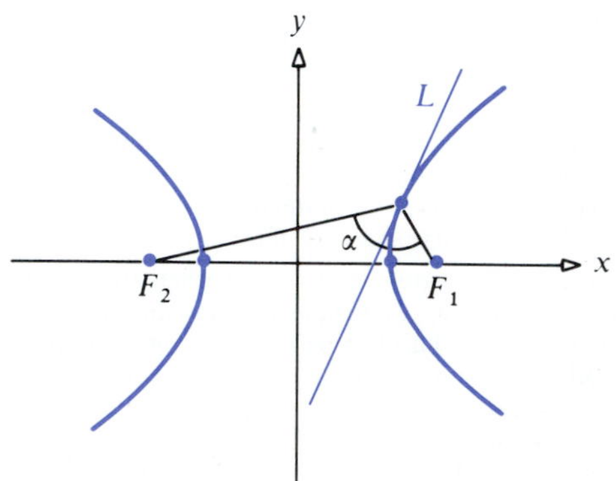

Figure 42

Descartes' Method of Equal Roots

6. Descartes' method for finding tangents depends on the idea that for many graphs, the tangent line at a given point is the *unique* line that intersects the graph at that point only. We will apply his method to find an equation of the tangent line to the parabola $y = x^2$ at the point (2, 4). First, we know the equation of the tangent line must be in the form $y = mx + b$. Using the fact that the point (2, 4) is on the line, we can solve for b in terms of m and get the equation $y = mx + (4 - 2m)$. Now we want (2, 4) to be the *unique* solution to the system

$$\begin{cases} y = x^2 \\ y = mx + 4 - 2m \end{cases}$$

From this system, we get $x^2 = mx + 4 - 2m$ or $x^2 - mx + (2m - 4) = 0$. By using the quadratic formula, we get

$$x = \frac{m \pm \sqrt{m^2 - 4(2m - 4)}}{2}$$

In order to obtain a *unique* solution for x, the *two roots must be equal*; in other words, the expression $m^2 - 4(2m - 4)$ must be 0. Complete the work to get m and write an equation of the tangent line. Compare this answer with the result of Problem 3.

7. Repeat Problem 4 using Descartes' method of equal roots.

8. Descartes' method of equal roots will not work for $y = x^3$. Prove this by finding the tangent line at the point (2, 8) using the methods of this chapter and showing that it intersects the curve at *two* points, rather than at a single point.

9. Use the method of equal roots to find an equation of the tangent line to the graph of $x^2 + y^2 = 25$ at the point (3, 4).

Fermat's Method for Tangents

10. We will illustrate Fermat's method by using it to find an equation of the tangent line to the graph of $y = x^2$ at (2, 4). See Figure 43. Choose Q on the tangent line

(*Problem 10 continues in the next column*)

10. (*continued*)

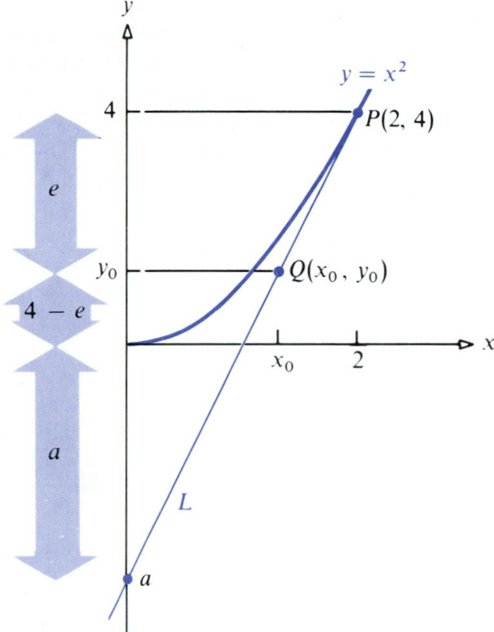

Figure 43

below P. Now, Q is outside the parabola, $y_0 < x_0^2$. (Why?) So we get $4/y_0 > 4/x_0^2$. But $y_0 = 4 - e$, so

$$\frac{4}{4-e} > \frac{4}{x_0^2} = \left(\frac{2}{x_0}\right)^2$$

By using similar triangles, $2/x_0 = (4 + a)/(a + 4 - e)$, and combining, we get

$$\frac{4}{4-e} > \left(\frac{4+a}{a+4-e}\right)^2$$

$$\frac{4}{4-e} > \frac{16 + 8a + a^2}{a^2 + 8a + 16 - 2ae - 8e + e^2}$$

Now, when Q is near P on the tangent line, it is not very far outside the parabola; thus, the inequality very nearly becomes an equality. In this case, we write

$$\frac{4}{4-e} \approx \frac{16 + 8a + a^2}{a^2 + 8a + 16 - 2ae - 8e + e^2} \qquad \textbf{(1)}$$

To complete the problem, we must calculate a (the y-intercept of the tangent line) and then write the equation of the tangent line. To do this, we need three steps: (*1*) cross-multiply in (*1*) to clear the fractions and then cancel like terms; (*2*) divide through by e; (*3*) set $e = 0$ and solve for a. Complete this procedure for this problem. Compare this answer with the results obtained in Problems 3 and 6.

11. Use Fermat's method to find an equation of the tangent line to the graph of $y = x^2$ at $(3, 9)$.

12. Fermat's method was criticized by many (especially by Descartes) because in one step he divides by e, and in the next step he sets $e = 0$. You should be able to defend the method by pointing out that one does not really set $e = 0$. Elaborate. [*Hint:* What major concept of today's calculus is missing in Fermat's argument?]

13. Fermat's method is quite general. Apply it to find an equation of the tangent line to the graph of $y = x^3$ at the point $(2, 8)$. [*Hint:* Begin with $8/y_0 > 8/x_0^3 = (2/x_0)^3$. (Why is this true?)] Compare your answer with the result of Problem 8.

Note: Though Fermat's method is quite close in spirit to the limiting secant line concept developed in this chapter, it falls short in one fundamental way: it is basically *geometric* in nature. What was needed for calculus to develop fully was the analytic approach provided by limits. This approach was not perfected until 200 years after Fermat, with the work of Cauchy, Bolzano, and others.

REVIEW EXERCISES

1. Make a table of values near 0 and guess $\lim_{x\to 0} \dfrac{1-\cos x}{x^2}$.

In Problems 2 and 3 use a graph to determine whether $\lim_{x\to c} f(x)$ exists.

2. $f(x) = \begin{cases} 2x - 5 & \text{if } x < 1 \\ 6 - 9x & \text{if } x \geq 1 \end{cases}$ $\quad c = 1$

3. $f(x) = \begin{cases} x^2 + 2 & \text{if } x < 2 \\ 2x + 1 & \text{if } x \geq 2 \end{cases}$ $\quad c = 2$

In Problems 4 and 5 find the limits.

4. $\lim_{x\to -2} \pi$

5. $\lim_{x\to -1} x$

6. For $f(x) = x^2 - 3$:

(a) Find the slope of the secant line joining $(1, -2)$ and $(2, 1)$.
(b) Use a limit process to find the equation of the tangent line at $(1, -2)$.

In Problems 7 and 8 find $\lim_{h\to 0} \dfrac{f(x+h) - f(x)}{h}$ for each of the following:

7. $f(x) = 3/x$

8. $f(x) = 3x^2 + 2x$

9. Find $\lim_{x\to 0} f(x)$ if, for $-\pi/2 < x < \pi/2$, $1 + \sin x \leq f(x) \leq |x| + 1$.

In Problems 10–20 find the limits.

10. $\lim_{x\to 0} \dfrac{1}{x}\left[\dfrac{1}{(x+3)^2} - \dfrac{1}{9}\right]$

11. $\lim_{x\to 3} \dfrac{x^3 - 27}{x - 3}$

12. $\lim_{x\to 3} \left(\dfrac{x^2}{x-3} - \dfrac{3x}{x-3}\right)$

13. $\lim_{x\to 2} \dfrac{x^2 - 4}{x - 2}$

14. $\lim_{x\to -1} \dfrac{x^2 + 3x + 2}{x^2 + 4x + 3}$

15. $\lim_{x\to 4} \dfrac{x^3 - 2x^2 - 7x - 4}{x^2 - 6x + 8}$

16. $\lim_{x\to -2} \dfrac{x^3 + 5x^2 + 6x}{x^2 + x - 2}$

17. $\lim_{x\to 1} \dfrac{x^3 - 3x^2 + 3x - 1}{x^2 - 2x + 1}$

18. $\lim_{x\to 2} \dfrac{3 - \sqrt{x^2 + 5}}{x^2 - 4}$

19. $\lim_{x\to 0} \dfrac{1}{x}\left[\dfrac{1}{(2+x)^2} - \dfrac{1}{4}\right]$

20. $\lim_{x\to 0} \dfrac{(x+3)^2 - 9}{x}$

In Problems 21–27 find the one-sided limits if they exist.

21. $\lim_{x\to 3^+} \frac{2x}{x-3}$

22. $\lim_{x\to 3^-} \frac{2x}{x-3}$

23. $\lim_{x\to -2^+} \frac{x^2+5x+6}{x+2}$

24. $\lim_{x\to 5^+} \frac{|x-5|}{x-5}$

25. $\lim_{x\to 1^-} \frac{|x-1|}{x-1}$

26. $\lim_{x\to 3/2^-} [\![2x]\!]$

27. $\lim_{x\to 4^-} \frac{x^2-16}{x-4}$

In Problems 28–30 find $\lim_{x\to c^-} f(x)$ and $\lim_{c\to c^+} f(x)$ for the given c. Determine whether $\lim_{x\to c} f(x)$ exists.

28. $f(x) = \begin{cases} 2x+3 & \text{if } x<2 \\ 9-x & \text{if } x \geq 2 \end{cases} \quad c=2$

29. $f(x) = \begin{cases} 3x+1 & \text{if } x<3 \\ 10 & \text{if } x=3 \\ 4x-2 & \text{if } x>3 \end{cases} \quad c=3$

30. $f(x) = \begin{cases} 5x-2 & \text{if } x<1 \\ 5 & \text{if } x=1 \\ 2x+1 & \text{if } x>1 \end{cases} \quad c=1$

In Problems 31–37 determine whether f is continuous at c.

31. $f(x) = \begin{cases} 5x-2 & \text{if } x<1 \\ 5 & \text{if } x=1 \\ 2x+1 & \text{if } x>1 \end{cases} \quad c=1$

32. $f(x) = \begin{cases} 3x+1 & \text{if } x<3 \\ 10 & \text{if } x=3 \\ 4x-2 & \text{if } x>3 \end{cases} \quad c=3$

33. $f(x) = \begin{cases} x^2 & \text{if } x<-1 \\ 2 & \text{if } x=-1 \\ -3x-2 & \text{if } x>-1 \end{cases} \quad c=-1$

34. $f(x) = \begin{cases} 4-3x^2 & \text{if } x<0 \\ 4 & \text{if } x=0 \\ \sqrt{16-x^2} & \text{if } 4 \geq x > 0 \end{cases} \quad c=0$

35. $f(x) = \begin{cases} \sqrt{4+x} & \text{if } -4 \leq x \leq 4 \\ \sqrt{\dfrac{x^2-16}{x-4}} & \text{if } x>4 \end{cases} \quad c=4$

36. $f(x) = [\![2x]\!]; \quad c = \frac{1}{2}$

37. $f(x) = |x-5|; \quad c=5$

38. Graph $f(x) = x - [\![x]\!]$ on $[-3, 3]$ and discuss its discontinuities.

39. Define a function on the interval $[-1, 1]$, continuous on $[-1, 1]$ except at 0, negative at -1, positive at 1, but with no zeros. Does this contradict the intermediate value theorem (2.4.6)?

In Problems 40–42 find all values x for which $f(x)$ is continuous.

40. $f(x) = \frac{x}{x^3-27}$

41. $f(x) = \frac{x^2-3}{x^2+5x+6}$

42. $f(x) = \frac{2x+1}{x^3+4x^2+4x}$

43. Use the intermediate value theorem to determine whether $2x^3+3x^2-23x-42=0$ has a root in $[3, 4]$.

44. Find $\lim_{x\to 0^+} \frac{|x|}{x}(1-x)$ and $\lim_{x\to 0^-} \frac{|x|}{x}(1-x)$. What can you say about $\lim_{x\to 0} \frac{|x|}{x}(1-x)$?

45. Find $\lim_{x\to 2}\left(\frac{x^2}{x-2} - \frac{2x}{x-2}\right)$. Then comment on the statement that this limit is given by

$$\lim_{x\to 2} \frac{x^2}{x-2} - \lim_{x\to 2} \frac{2x}{x-2}$$

46. Determine where f is continuous if

$$f(x) = \begin{cases} \sqrt{4-x^2} & \text{if } |x| \leq 2 \\ |x|-2 & \text{if } |x|>2 \end{cases}$$

47. Prove that $y=x^3$ and $y=1-x^2$ intersect somewhere between $x=0$ and $x=1$.

48. Find $\lim_{h\to 0} \dfrac{f(x+h)-f(x)}{h}$ for $f(x)=\sqrt{x}$.

49. For $\lim_{x\to 3}(2x+1)=7$, find the largest possible δ that "works" for $\varepsilon = 0.01$.

50. Show that $|(2+x)^2-4| \le 5|x|$ if $-1<x<1$. Use this to prove $\lim_{x\to 0}(2+x)^2=4$.

51. Show that $\left|\dfrac{1}{x^2+9}-\dfrac{1}{13}\right| \le \dfrac{1}{26}|x-2|$ if $1<x<3$. Use this to prove $\lim_{x\to 2} \dfrac{1}{x^2+9}=\dfrac{1}{13}$.

In Problems 52–54 give an ε, δ proof for each limit.

52. $\lim_{x\to 1} \dfrac{2x-1}{x+3}=\dfrac{1}{4}$

53. $\lim_{x\to 3} \dfrac{2x}{x+3}=1$

54. $\lim_{x\to 2} \sqrt{x+2}=2$

In Problems 55–59 evaluate the limits.

55. $\lim_{x\to 0} \cos(\tan x)$

56. $\lim_{x\to \pi/6} (\sin x+\cos x)$

57. $\lim_{x\to 0} \dfrac{\sin \frac{x}{4}}{x}$

58. $\lim_{x\to 0} \dfrac{\tan 3x}{\tan 4x}$

59. $\lim_{x\to 0} \dfrac{\cos \frac{x}{3}-1}{x}$

CHALLENGE EXERCISES

1. Give an ε, δ, proof that $\lim_{x\to 1}(4x^3+3x^2-24x+22)=5$.

2. If $\lim_{x\to c} f(x)=L$ and $\lim_{x\to c} g(x)=M$, prove that $\lim_{x\to c}[f(x)+g(x)]=L+M$.

3. Show that the existence of $\lim_{h\to 0} \dfrac{f(a+h)-f(a)}{h}$ implies that $f(x)$ is continuous at $x=a$.

4. Find constants A, B, C, and D so that the function below is continuous for all x. Sketch the graph of the resulting function.

$$f(x)=\begin{cases} \dfrac{x^2+x-2}{x-1} & \text{if } -\infty<x<1 \\ A & \text{if } x=1 \\ B(x-C)^2 & \text{if } 1<x<4 \\ D & \text{if } x=4 \\ 2x-8 & \text{if } 4<x<+\infty \end{cases}$$

5. Let f be a function for which $0\le f(x)\le 1$ for all x in $[0, 1]$. If f is continuous on $[0, 1]$, show that there exists at least one number c in $[0, 1]$ such that $f(c)=c$. [*Hint:* Let $g(x)=x-f(x)$.]

6. Suppose f is defined on an interval (a, b) and there is a number K such that $|f(x)-f(c)|\le K|x-c|$ for all c in (a, b) and x in (a, b). Such a constant K is called a *Lipschitz constant*. Find a Lipschitz constant for $f(x)=x^3$ on $(0, 2)$.

7. Use the definition of limit to prove that no number L exists such that $\lim_{x\to 0}(1/x)=L$.

8. Use the definition of limit to prove that the linear function $f(x)=ax+b$ is continuous everywhere.

9. Let

$$f(x)=\begin{cases} x & \text{rational} \\ -x & \text{irrational} \end{cases}$$

Discuss the continuity of this function.

Functions and Limits Explored Using Graphing Calculators*

* Gregory D. Foley
The Ohio State University

In this vignette, we use graphing calculators to extend the graphical approach to limits introduced in Section 2.1 and to gain another perspective on the squeezing theorem and the intermediate value theorem. In place of graphing calculators, Calculus Blackboard or other computer graphing packages can be used. Feel free to try the ideas and methods presented here on any interactive computer graphics system.

Let's begin by exploring the graph of $y = (\sin x)/x$ using the Casio graphing calculator. The goal of the exploration is to find $\lim_{x\to 0}[(\sin x)/x]$, which was the goal of Example 4 in Section 2.1.

EXAMPLE 1

Find: $\lim\limits_{x\to 0} \dfrac{\sin x}{x}$

Solution

Turn on the Casio and adjust the contrast so that you can easily read the display. Then put the machine in radian mode by pressing

[MODE] [5] [EXE]

Press [Range]. The Range settings window should appear; it shows the minimum and maximum x- and y-values and the scaling units for each axis. Press [SHIFT] [Mcl]. (The delete key, [DEL], performs the Memory Clear function, [Mcl], when preceded by [SHIFT]. Look above each key to see its SHIFT function.) You have now set the viewing rectangle to the Casio default of $[-4.7, 4.7] \times [-3.1, 3.1]$ with a scaling unit of 1 on each axis. In general, a viewing rectangle is a rectangular portion of the cartesian plane; it gives us one of many possible views of the graph we are interested in. In this case, we want to obtain one view of the graph of $y = (\sin x)/x$ so that we can begin our exploration of its behavior in the neighborhood of $x = 0$. Now press [Range] to toggle back out of the Range settings window.

Press [Graph] [sin] [ALPHA] [X] [÷] [ALPHA] [X] [EXE]. (The alphabetic characters, accessed by [ALPHA], appear just below the basic function of each key. The letter X is associated with the addition key, [+].) A graph of $y = (\sin x)/x$ should appear on the Casio screen. Notice that it is drawn from left to right, in the direction of increasing values of x. To see the coordinates of the points just plotted, press [SHIFT] [Trace] (or just [Trace] on the fx – 7500 G). "X = –4.7" should appear at the bottom of the screen. Also, the leftmost point (pixel) of the graph should be blinking. To move closer to $x = 0$, press [⇒] repeatedly ([▷] on the fx – 7500 G). This should cause the blinking pixel to move to the right and the x-coordinate readout to increase in value (become less negative). How much does x change with each press of the button? Keep pressing [⇒] (or [▷]) until you reach $x = -0.1$. Now press the button once more. Notice that the blinking pixel "jumped over" the y-axis and the x-readout skipped $x = 0$. Why do you think this happened?

To obtain a readout of y-values, press [SHIFT] [X↔Y]. Compare the $y =$ readouts for the pixels just to the left and just to the right of the y-axis. The function $y = (\sin x)/x$ appears to be an even function. (See Problem 33 in Exercise 1.5 of Chapter 1.) Now compare the values obtained from the Trace readout with those in the table associated with Example 4 in Section 2.1. With the blinking pixel at (0.1, 0.9983), press

SHIFT × (SHIFT followed by the multiplication key) to zoom in on the point. Press Range to see the new viewing rectangle parameters. Then press Range again to return to the graphics screen. You may zoom in farther by pressing SHIFT × again (or zoom back out by pressing SHIFT ÷). Experiment with zooming and the Trace feature until you convince yourself that $\lim_{x\to 0}[(\sin x)/x] = 1$ is a reasonable conclusion. Is this a proof?

In the next example, we revisit Example 1 from Section 2.5, which illustrated the use of the squeezing theorem (2.5.1).

EXAMPLE 2

Show graphically that $-|x| \leq x \sin \dfrac{1}{x} \leq |x|$ and hence

$$\lim_{x\to 0}\left(x \sin \frac{1}{x}\right) = 0$$

Solution

Set the Casio to its default viewing rectangle by pressing Range SHIFT Mcl Range. Then press Graph SHIFT Abs ALPHA X : Graph (−) SHIFT Abs ALPHA X : Graph ALPHA X sin (1 ÷ ALPHA X) EXE. (The key (−) is used for negatives, and − is used for subtraction.) Now zoom in by pressing SHIFT ×; watch as the three graphs are drawn. Repeat this several times, observing the behavior of the graphs each time. As we zoom in on the point (0, 0), the shape of the two absolute value graphs remains the same, but the graph of $y = x \sin(1/x)$ oscillates wildly near the origin. Nevertheless, its limit approaches 0 as x approaches 0 because it is squeezed between the two absolute value graphs.

The final example extends the discussion that follows the intermediate value theorem (2.4.6). It shows how to speed up the approximation of zeros guaranteed by the intermediate value theorem.

EXAMPLE 3

Approximate the zero of $f(x) = x^3 + x^2 - x - 2$ that lies between $x = 0$ and $x = 2$, to at least three decimal places of accuracy.

Solution

Let's begin by trying to replicate Figure 29 on the Casio. Go back and study the graph and the portion of the plane shown in Figure 29. Then set the Range parameters as follows:

```
X  min:  0
   max:  2.5
   scl:  1
```

```
Y  min:  −3
   max:  8
   scl:  1
```

To do this press [Range] 0 [EXE] 2.5 [EXE] 1 [EXE] [(−)] 3 [EXE] 8 [EXE] 1 [EXE]. To graph the function key in [Graph] [ALPHA] [X] [x^y] 3 [+] [ALPHA] [X] [x^2] [−] [ALPHA] [X] [−] 2. The screen should look like this:

```
Graph  Y=Xxʸ3+X²−
X−2
```

Once it does, press [EXE]. Now Trace to the two adjacent points on the graph that *appear* to be on the x-axis as the graph crosses the axis. Their coordinates should be $(1.20, -0.05)$ and $(1.22, 0.10)$ rounded to two decimal places. Should the points with these coordinates be *on* the x-axis? Although they appear to be on the axis because of the limited resolution of the screen, we know mathematically that $(1.20, -0.05)$ is below the x-axis and $(1.22, 0.10)$ is above. Why?

We can use the x- and y-coordinates of these points to set new Range parameters as follows:

```
X  min:  1.20
   max:  1.22
   scl:  0.01
Y  min:  −0.05
   max:  0.10
   scl:  0.01
```

After keying in these new Range parameters in a manner similar to that used above, the screen should look this:

```
Graph  Y=Xxʸ3+X²−
X−2
                    done
```

Now press [EXE] to redraw the graph in the new, smaller viewing rectangle. What is the shape of the curve? Check the y-values associated with the two adjacent pixels that are on the screen's x-axis. The x-value of the point with the smaller absolute y-value should be $x = 1.205531915$. The solution to the problem is $x = 1.2055 \pm 0.0002$.

Many equations cannot be solved exactly, but virtually all equations can be solved approximately using this sort of graphical approach. What is required to use this graphical approach? Put the equation in the form $f(x) = 0$. Find two numbers a and b such that $f(a)$ and $f(b)$ are oppositely signed (like $f(0) = -2$ and $f(2) = 8$ in Example 3). If the function f is continuous on $[a, b]$, then the intermediate value theorem guarantees that there is a solution between $x = a$ and $x = b$. If we graph $y = f(x)$ on the interval $[a, b]$, there must be at least one x-intercept. By zooming in on that intercept, we obtain successively better approximations to the solution of $f(x) = 0$.

On the Sharp EL-5200, there is a SOLVE key that automates this process. In Example 3, after graphing the function in any viewing rectangle that contains the x-intercept, simply press the SOLVE key. After a brief wait the solution $X = 1.205569431$ will appear on the screen.

It is interesting to note that the Sharp will not find solutions that do not correspond to sign changes in the associated function. Said another way, the Sharp finds x-intercepts only if the graph of the function crosses the x-axis at the intercept point. For example, it cannot solve $x^2 - 2x + 1 = 0$ because $x^2 - 2x + 1 = (x - 1)^2$ is positive except at $x = 1$; that is, the graph of $y = x^2 - 2x + 1$ lies entirely above the x-axis except for the intercept point $(1, 0)$. Much of the equation solving done on the computer is based on the intermediate value theorem.

EXERCISES

Solve using a graphing calculator or computer graphing software.

1. Find $\lim_{x \to 2}[(\sqrt{x} - \sqrt{2})/(x - 2)]$ by graphing the function involved and then zooming in in the neighborhood of $x = 2$. Compare your result with the one you obtain algebraically by solving Problem 35 in Exercise 2.2.

2. Graph the function given by $f(x) = |x|/x$. Use this to solve Problem 61 in Exercise 2.4.

3. Solve Problem 44 from the Review Exercises in Chapter 2.

4. Approximate the solution of $\cos x = \tan x$ that lies between $x = 0$ and $x = 1$, to at least three decimal places of accuracy. What is the exact solution?

The Derivative

3

Prelude to the Derivative

3.1

One of the most prominent features of nature is change. In order to describe natural processes mathematically, the idea of rate of change is necessary. In this section, we will see how limits can be used to describe rates of change. We will discuss three, seemingly unrelated, interpretations of rate of change and see how all lead to the same idea.

Average and Instantaneous Velocity (First Interpretation)

Average velocity is a familiar example of average rate of change. For example, consider the rectilinear motion of a particle along a straight line. It is convenient to think of rectilinear motion along a horizontal line with the positive direction to the right. The position of the particle at time $t = 0$ is called the *initial position* of the particle. Usually the initial position is selected to be the origin O on the line. If we assume the distance of the particle from O is known for all times t (in seconds), then we can represent its motion by a function $s = f(t)$, where s is the signed or directed distance of the particle from O at t seconds (see Fig. 1).

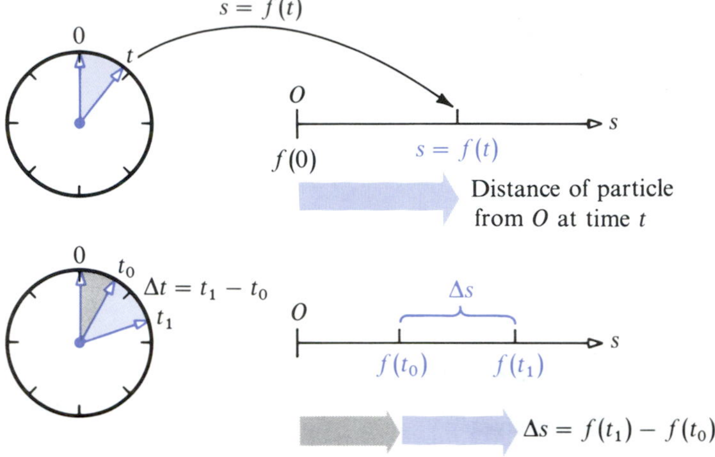

Figure 1

If at time t_0 the particle is at $s_0 = f(t_0)$ and at time t_1 the particle is at the position $s_1 = f(t_1)$, then the change in time is $\Delta t = t_1 - t_0$ and the change in position is $\Delta s = s_1 - s_0 = f(t_1) - f(t_0)$. **The rate of change of position with respect to time, $\dfrac{\Delta s}{\Delta t} = \dfrac{f(t_1) - f(t_0)}{t_1 - t_0}$, is called the *average velocity** of the particle.**

For example, suppose the function $s = f(t) = 16t^2$ describes the distance s (in feet) an object dropped from a tall building has traveled after a time t (in seconds) for $0 \le t \le 8$. The average velocity of the object over the entire time interval (8 seconds) is

$$\frac{\Delta s}{\Delta t} = \frac{f(8) - f(0)}{8 - 0} = \frac{16 \cdot 8^2 - 0}{8} = 128 \text{ feet per second}$$

This average accurately describes the velocity of the object over the 8 second interval, but it gives no information about the velocity at any particular instant of time. We now wish to define the velocity of the object at a particular instant.

To see how this might be done, we will seek the exact velocity of the object at the instant when $t = 3$. This is called the *instantaneous velocity at 3.*

So far we have no *mathematical* method for finding instantaneous velocities. However, we can *estimate* the instantaneous velocity at $t = 3$ seconds by computing the average velocities for some intervals of time beginning at $t = 3$. For example, let's compute the average velocity for the

---

* *Speed* and *velocity* are not the same thing. Speed is defined as the *absolute value of velocity* and therefore is always nonnegative. Velocity, on the other hand, may be a positive or a negative number, indicating direction to the right or left. Thus, speed is a measure of how fast a particle is moving, whereas velocity is a measure of both its speed and its direction.

1 second interval beginning at $t = 3$ and ending at $t = 4$. Here, $\Delta t = 1$. At $t = 3$, the distance of the object from the starting position is

$$s = f(3) = 16(9) = 144 \text{ feet}$$

At $t = 4$,

$$s = f(4) = 16(16) = 256 \text{ feet}$$

Thus, over the 1 second interval from $t = 3$ to $t = 4$,

$$\text{Average velocity} = \frac{\Delta s}{\Delta t} = \frac{f(4) - f(3)}{1} = \frac{256 - 144}{1} = 112 \text{ feet per second}$$

The average velocities for the smaller intervals of time $\Delta t = 0.5, 0.1, 0.01, 0.0001$ may be found similarly, and Table 1 gives the five estimates obtained for the instantaneous velocity. We can see from the table that for this example, the larger the time interval Δt, the larger the average velocity $\Delta s/\Delta t$. The most accurate estimates for instantaneous velocity will correspond to very small time intervals Δt. For example, over the interval $\Delta t = 0.0001$ second, we would not expect the velocity of the object to change very much. Thus, the average velocity of 96.0016 feet per second during the very short time interval $\Delta t = 0.0001$ should be very close to the instantaneous velocity at $t = 3$.

Table 1

Start, t_0	End, t	Δt	Average Velocity, $\frac{\Delta s}{\Delta t} = \frac{f(t) - f(t_0)}{t - t_0}$
3	4	1	$\frac{\Delta s}{\Delta t} = \frac{f(4) - f(3)}{4 - 3} = \frac{16(16)^2 - 16(9)}{1} = 112$
3	3.5	0.5	$\frac{\Delta s}{\Delta t} = \frac{f(3.5) - f(3)}{3.5 - 3} = \frac{16(3.5)^2 - 16(9)}{0.5} = 104$
3	3.1	0.1	$\frac{\Delta s}{\Delta t} = \frac{f(3.1) - f(3)}{3.1 - 3} = \frac{16(3.1)^2 - 16(9)}{0.1} = 97.6$
3	3.01	0.01	$\frac{\Delta s}{\Delta t} = \frac{f(3.01) - f(3)}{3.01 - 3} = \frac{16(3.01)^2 - 16(9)}{0.01} = 96.16$
3	3.0001	0.0001	$\frac{\Delta s}{\Delta t} = \frac{f(3.0001) - f(3)}{3.0001 - 3} = \frac{16(3.0001)^2 - 16(9)}{0.0001} = 96.0016$

But what is the exact or instantaneous velocity at $t = 3$? It must be close to 96.0016, but is it 96.0016? Or is it 96.0001? Or what?

To obtain the precise answer, we first use some algebra. Specifically, we find the average velocity for the object over the time interval that begins at $t_0 = 3$ and ends at t, where $\Delta t = t - 3 \neq 0$ represents any small interval of time:

$$f(t) = 16t^2$$

$$f(3) = 16(9) = 144$$

Thus,

$$\text{Average velocity} = \frac{\Delta s}{\Delta t} = \frac{f(t) - f(3)}{t - 3} = \frac{16t^2 - 144}{t - 3} = \frac{16(t + 3)(t - 3)}{(t - 3)}$$

We can cancel $t - 3$ because the increment $t - 3$ does not equal 0. As a result,

$$\text{Average velocity} = \frac{\Delta s}{\Delta t} = 16(t + 3)$$

Now comes the important step in this procedure! As t gets closer and closer to 3, the values of $\Delta s/\Delta t = 16(t + 3)$ get closer and closer to 96. The average velocity will never equal 96 because we must have a nonzero time interval in order to compute an average velocity. Nevertheless, we can make $\Delta s/\Delta t = 16(t + 3)$ as close as we please to 96 by taking t sufficiently close to 3.

In mathematical terminology (as introduced in Chap. 2), the number 96 obtained in this way is called the *limit* of the average velocity $(\Delta s/\Delta t)$ *as* Δt *approaches* 0. In symbols, we write

$$\lim_{\Delta t \to 0} \frac{\Delta s}{\Delta t} = \lim_{t \to 3} \frac{f(t) - f(3)}{t - 3} = \lim_{t \to 3} 16(t + 3) = 96$$

Intuition tells us that the limit 96 feet per second is what we mean by the (instantaneous) velocity of the object at time $t = 3$ seconds. Thus, we are led to the following definition:

[3.1.1] DEFINITION / *Instantaneous Velocity.*

If $s = f(t)$ is a function that describes the distance s a particle travels in time t, the (*instantaneous*) *velocity* v of the particle at time t_0 is defined as the limit of the average velocity $\Delta s/\Delta t$ as Δt approaches 0. Specifically, the velocity v at time t_0 is

$$v = \lim_{\Delta t \to 0} \frac{\Delta s}{\Delta t} = \lim_{t \to t_0} \frac{f(t) - f(t_0)}{t - t_0} \quad \textbf{(1)}$$

provided this limit exists.

EXAMPLE 1

A ball thrown into the air has height $s = f(t) = 38t - 16t^2$ feet above the ground after t seconds. Find the velocity v at

(a) $t_0 = 0$ (b) $t_0 = 1$ (c) $t_0 = 2$

Solution

Using (1), we have

(a) $$v = \lim_{\Delta t \to 0} \frac{\Delta s}{\Delta t} = \lim_{t \to 0} \frac{f(t) - f(0)}{t - 0} = \lim_{t \to 0} \frac{38t - 16t^2 - 0}{t - 0}$$

$$= \lim_{t \to 0} (38 - 16t) = 38 \text{ feet per second}$$

(b) $v = \lim_{\Delta t \to 0} \frac{\Delta s}{\Delta t} = \lim_{t \to 1} \frac{f(t) - f(1)}{t - 1} = \lim_{t \to 1} \frac{38t - 16t^2 - 22}{t - 1}$

$= \lim_{t \to 1} \frac{-2(8t^2 - 19t + 11)}{t - 1} = \lim_{t \to 1} \frac{-2(8t - 11)(t - 1)}{t - 1}$

$= \lim_{t \to 1} -2(8t - 11) = 6$ feet per second

(c) $v = \lim_{t \to 2} \frac{f(t) - f(2)}{t - 2} = \lim_{t \to 2} \frac{38t - 16t^2 - 12}{t - 2}$

$= \lim_{t \to 2} \frac{-2(8t - 3)(t - 2)}{t - 2}$

$= \lim_{t \to 2} - 2(8t - 3) = -26$ feet per second

The positive velocities indicate that the ball is going up, and the negative velocity indicates that it is coming down.

Tangent Lines (Second Interpretation)

Let's review the concept of tangent lines. Consider the graph of the function $y = f(x)$ as shown in Figure 2. The line connecting two points $(c, f(c))$ and $(d, f(d))$ on the graph of the function f is called a *secant line* and its slope is

$$m_{\text{sec}} = \frac{f(d) - f(c)}{d - c} \quad \textbf{(2)}$$

There may be many secant lines passing through the point $(c, f(c))$ (see Fig. 3). Specifically, suppose $(x, f(x))$ is any point on the graph of $y = f(x)$ with $x \neq c$. Next, suppose the variable x assumes values

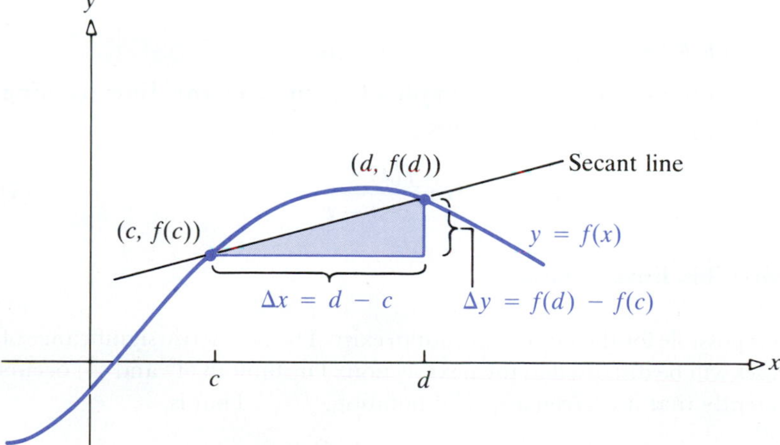

Figure 2

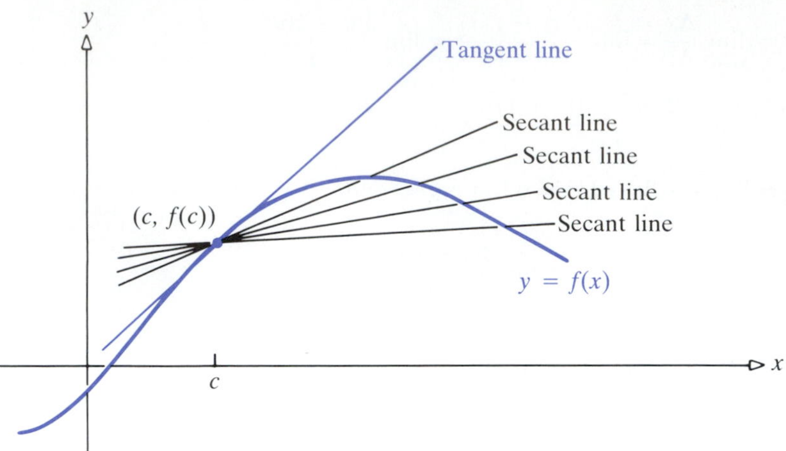

Figure 3

closer and closer to c, so that the point $(x, f(x))$ moves along the graph of $y = f(x)$ and approaches the point $(c, f(c))$. As $(x, f(x))$ approaches $(c, f(c))$, the secant line connecting these two points also moves and approaches a limiting position called the *tangent line* to the graph of $y = f(x)$ at $x = c$ (again, see Fig. 3).

Similarly, as x approaches c, the slope m_{sec} of the secant line approaches the slope $m_{\tan}$ of the tangent line. Since the slope of the secant line connecting $(c, f(c))$ and $(x, f(x))$ is

$$m_{\text{sec}} = \frac{\Delta y}{\Delta x} = \frac{f(x) - f(c)}{x - c} \tag{3}$$

we write

$$m_{\tan} = \lim_{x \to c} m_{\text{sec}} = \lim_{x \to c} \frac{f(x) - f(c)}{x - c}$$

[3.1.2] DEFINITION / ***Tangent Line.***

The tangent line to the graph of f at c is the line passing through $(c, f(c))$ and having slope

$$m_{\tan} = \lim_{x \to c} \frac{f(x) - f(c)}{x - c} \tag{4}$$

provided this limit exists.

It is possible for the limit in (4) not to exist. The geometric significance of such cases will be discussed in the next section. The limit in (1) and (4) occurs so frequently that it is given a special notation, $f'(c)$. That is,

$$f'(c) = \lim_{x \to c} \frac{f(x) - f(c)}{x - c}$$

where $f'(c)$ is read "f prime of c." Figure 4 illustrates this definition.

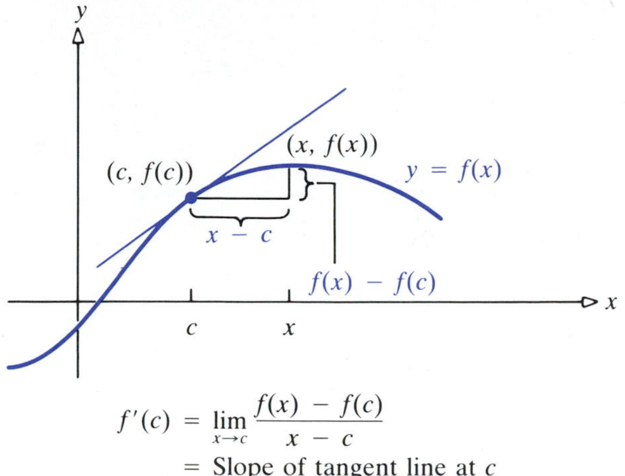

$$f'(c) = \lim_{x \to c} \frac{f(x) - f(c)}{x - c}$$
$$= \text{Slope of tangent line at } c$$

Figure 4

By using the point–slope form of the equation of a line, we find that an equation of the tangent line to the graph of f at $(c, f(c))$ is

$$y - f(c) = f'(c)(x - c) \tag{5}$$

EXAMPLE 2

Let $f(x) = x^2 + 1$.

(a) Find the slope of the tangent line at $c = 1$, 0, and -2.
(b) Use the results in part (a) to find the equation of the tangent line passing through each point.
(c) Graph the function and show the tangent lines at the given points.

Solution

(a) At $c = 1$,

$$\begin{aligned}\text{Slope of tangent line at 1} &= f'(1) = \lim_{x \to 1} \frac{f(x) - f(1)}{x - 1} \\ &= \lim_{x \to 1} \frac{x^2 + 1 - 2}{x - 1} = \lim_{x \to 1} \frac{x^2 - 1}{x - 1} \\ &= \lim_{x \to 1} \frac{(x + 1)(x - 1)}{x - 1} = \lim_{x \to 1} (x + 1) = 2\end{aligned}$$

At $c = 0$,

$$\begin{aligned}\text{Slope of tangent line at 0} &= f'(0) = \lim_{x \to 0} \frac{f(x) - f(0)}{x - 0} \\ &= \lim_{x \to 0} \frac{x^2 + 1 - 1}{x - 0} = \lim_{x \to 0} \frac{x^2}{x} = \lim_{x \to 0} x = 0\end{aligned}$$

At $c = -2$,

$$\text{Slope of tangent line at } -2 = f'(-2) = \lim_{x\to -2} \frac{f(x) - f(-2)}{x - (-2)}$$

$$= \lim_{x\to -2} \frac{x^2 + 1 - 5}{x + 2} = \lim_{x\to -2} \frac{(x+2)(x-2)}{x+2}$$

$$= \lim_{x\to -2} (x - 2) = -4$$

(b) By using (5) we find an equation of the tangent line passing through $(1, f(1)) = (1, 2)$ to be

$$y - 2 = 2(x - 1)$$

$$y = 2x$$

The equation of the tangent line passing through $(0, f(0)) = (0, 1)$ is

$$y - 1 = 0(x - 0)$$

$$y = 1$$

The equation of the tangent line passing through $(-2, f(-2)) = (-2, 5)$ is

$$y - 5 = -4(x + 2)$$

$$y = -4x - 3$$

(c) See Figure 5.

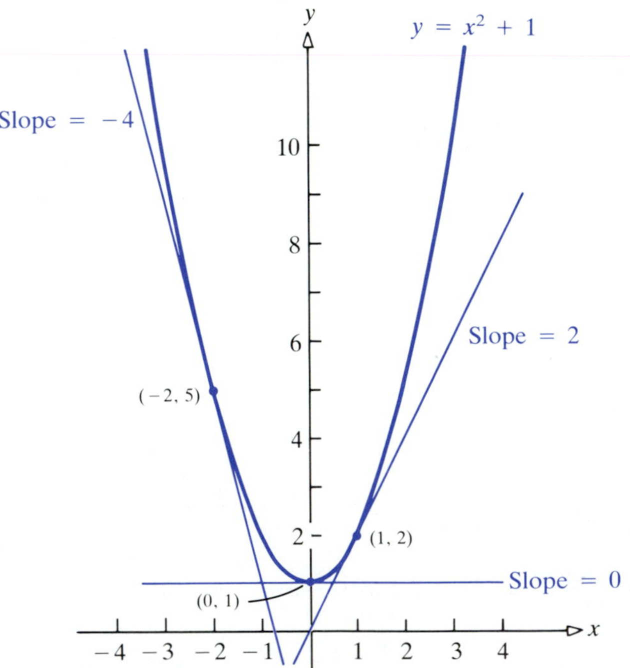

Figure 5

Rates of Change (Third Interpretation)

Although it is natural to think of rates of change in terms of motion and time, there is no need to be so restrictive. We can define the average rate of change for any function $y = f(x)$ over any interval in its domain and define the instantaneous rate of change as the limit of the average rate of change whenever the limit exists.

[3.1.3] DEFINITION / ***Average Rate of Change.***

The *average rate of change* of a function $y = f(x)$ over the interval from c to x is

$$\text{Average rate of change} = \frac{f(x) - f(c)}{x - c} \tag{6}$$

The *instantaneous rate of change* of f at c is

$$f'(c) = \lim_{x \to c}(\text{Average rate of change}) = \lim_{x \to c} \frac{f(x) - f(c)}{x - c} \tag{7}$$

provided the limit exists.

EXAMPLE 3

Compute the instantaneous rate of change for the function $f(x) = x^2 + 5x$ at:

(a) $c = 0$ (b) $c = 2$ (c) c any real number

Solution

(a) We first compute the average rate of change by using (6). For $c = 0$, we have

$$f(x) = x^2 + 5x$$
$$f(c) = f(0) = 0$$

so that

$$\begin{array}{c}\text{Average rate of change}\\ \text{over the interval } [0, x]\end{array} = \frac{f(x) - f(0)}{x - 0} = \frac{x^2 + 5x - 0}{x - 0} = x + 5$$

To compute the instantaneous rate of change, we use (7) with $c = 0$. Thus,

$$\begin{array}{c}\text{Instantaneous rate of}\\ \text{change at } c = 0\end{array} = f'(0) = \lim_{x \to 0}(x + 5) = 5$$

(b) As in part (a), we first compute the average rate of change.

$$f(x) = x^2 + 5x$$
$$f(c) = f(2) = 2^2 + 5(2) = 14$$

so that

$$\begin{array}{l}\text{Average rate of change}\\ \text{over the interval } [2, x]\end{array} = \frac{f(x) - f(c)}{x - c} = \frac{x^2 + 5x - 14}{x - 2}$$

$$= \frac{(x + 7)(x - 2)}{(x - 2)} = x + 7$$

To compute the instantaneous rate of change, we use (7) with $c = 2$.

$$\begin{array}{l}\text{Instantaneous rate of}\\ \text{change at } c = 2\end{array} = f'(2) = \lim_{x \to 2}(x + 7) = 9$$

(c) Similarly, for c any real number, the average rate of change is

$$f(x) = x^2 + 5x$$

$$f(c) = c^2 + 5c$$

so that

$$\begin{array}{l}\text{Average rate of change}\\ \text{over the interval } [c, x]\end{array} = \frac{f(x) - f(c)}{x - c} = \frac{x^2 + 5x - (c^2 + 5c)}{x - c}$$

$$= \frac{x^2 - c^2 + 5(x - c)}{x - c}$$

$$= \frac{(x - c)(x + c) + 5(x - c)}{x - c} = (x + c) + 5$$

To compute the instantaneous rate of change, we use (7).

$$\begin{array}{l}\text{Instantaneous rate of}\\ \text{change at } c\end{array} = f'(c) = \lim_{x \to c}(x + c) + 5 = 2c + 5$$

Observe that we can use the answer to part (c) to check the answers found in parts (a) and (b).

Instantaneous rates of change are usually called merely *rates of change* and sometimes just *rates*.

EXAMPLE 4

In a metabolic experiment, the mass M of glucose decreases according to the formula $M(t) = 4.5 - 0.03t^2$, where M is measured in grams and t is time (in hours). Find the reaction rate at 1 hour.

Solution

The reaction rate at $t = 1$ is $M'(1)$. Thus,

$$M'(1) = \lim_{t \to 1} \frac{M(t) - M(1)}{t - 1} = \lim_{t \to 1} \frac{(4.5 - 0.03t^2) - (4.5 - 0.03)}{t - 1}$$

$$= \lim_{t \to 1} \frac{(-0.03)(t^2 - 1)}{t - 1} = \lim_{t \to 1} \frac{(-0.03)(t + 1)(t - 1)}{t - 1}$$

$$= (-0.03)(2) = -0.06$$

The reaction rate at $t = 1$ is -0.06; that is, the mass M at $t = 1$ is decreasing at the rate of 0.06 gram per hour.

EXERCISE 3.1

1. The distance s (in meters) that a particle moves in time t (in seconds) is given by $s = f(t) = 3t^2 + 4t$ Find the velocity at $t = 0$. At $t = 2$. At any time t.

2. The distance s (in meters) that a particle moves in time t (in seconds) is $s = f(t) = t^2 - 4t$. Find the velocity at $t = 0$. At $t = 3$. At any time t.

3.*The motion of a certain body along the x-axis is described by the equation $s = 10t^2$ (s in centimeters). Compute the instantaneous velocity of the body at time $t = 3$ seconds by letting Δt first equal 0.1 second, then 0.01 second, and finally 0.001 second. What limiting value do the results seem to be approaching?

4. At a certain instant the speedometer of an automobile reads V miles per hour. During the next $\frac{1}{4}$ second the automobile travels 20 feet. Estimate V from this information.

5. A ball is thrown upward. Let the height in feet of the ball be given by $s(t) = 100t - 16t^2$, where t is the time elapsed in seconds. What is the velocity when $t = 0$, $t = 1$, and $t = 4$? At what time does the ball strike the ground? At what time does the ball reach its highest point? (At this time the ball should be "stationary" so that its velocity is 0.)

6.*A body moves along a straight line, its distance from the origin at any instant being given by the equation $s = 8t - 3t^2$, where s is in centimeters and t is in seconds. Find the velocity of the body at $t = 1$ second and $t = 4$ seconds.

7. Suppose the distance s a person can walk in time t can be represented by $s = 4\sqrt{t}$, where s is measured in kilometers and t is measured in hours. What is the person's velocity at $t = 16$ hours? From $t = 1$ to $t = 4$ hours? From $t = 1$ to $t = 2$ hours?

8. A rock is dropped from a height of 88.2 meters and falls toward earth in a straight line. In t seconds the rock falls $9.8t^2$ meters.

(a) How long does it take for the rock to hit the ground?
(b) What is the average velocity of the rock during the time it is falling?
(c) What is the average velocity of the rock for the first 2 seconds?
(d) What is the velocity of the rock when it hits the ground?

9. If a ball is dropped from the top of the Empire State Building, 1002 feet above the ground, then its height in feet after t seconds is $s(t) = 1002 - 16t^2$.

(a) How long does it take for the ball to hit the ground?
(b) What is the average velocity of the ball during the time it is falling?
(c) What is the average velocity of the ball for the first 2 seconds?
(d) What is the velocity of the ball when it hits the ground?

10. What is the average velocity of an automobile traveling from New York City to Miami if half the distance is traversed at 45 miles per hour and the other half at 55 miles per hour? [*Hint:* The answer is not 50 miles per hour.]

11. Let $f(x) = x^2 - 1$.

(a) Find the slope of the tangent line at 0, 1, and -1.
(b) Use the result in part (a) to find the equation of the tangent line at each point.
(c) Graph the function and show the tangent lines at the given points.

12. Let $f(x) = 2x^2 + 1$.

(a) Find the slope of the tangent line at 0, $\frac{1}{2}$, and 1.
(b) Use the result in part (a) to find the equation of the tangent line at each value.
(c) Graph the function and show the tangent lines at the given points.

13. Let $f(x) = \dfrac{1}{x}$.

(a) Find the slope of the tangent line at 1, 2, and 3.
(b) Use the result in part (a) to find the equation of the tangent line at each point.
(c) Graph the function and show the tangent lines at the given points.

14. Let $f(x) = \dfrac{1}{x - 1}$.

(a) Find the slope of the tangent line at 2, 3, and 4.
(b) Use the result in part (a) to find the equation of the tangent line at each point.
(c) Graph the function and show the tangent lines at the given points.

---

* Adapted from F. W. Sears, M. W. Zemansky, and H. D. Young, *University Physics* (Reading, Mass.: Addison-Wesley Publishing Co., 1976), p. 63. Reprinted by permission of the publisher.

15. Let $f(x) = x^3$.

(a) Find the slope of the tangent line at 0, 1, and -1.
(b) Use the result in part (a) to find the equation of the tangent line at each point.
(c) Graph the function and show the tangent lines at the given points.

16. Let $f(x) = x^2 + x$.

(a) Find the slope of the tangent line at 0, 1, and 2.
(b) Use the result in part (a) to find the equation of the tangent line at each point.
(c) Graph the function and show the tangent lines at the given points.

In Problems 17–22 find an equation for the tangent line to the graph of each function at the indicated point. Graph each function and show this tangent line.

17. $f(x) = x^2$, at $(3, 9)$

18. $f(x) = x^2$, at $(-1, 1)$

19. $f(x) = x^2 + 2x + 1$, at $(1, 4)$

20. $f(x) = x^3 + 1$, at $(1, 2)$

21. $f(x) = \dfrac{1}{x}$, at $(1, 1)$

22. $f(x) = \sqrt{x}$, at $(4, 2)$

In Problems 23–26 find an equation of the tangent line to the graph of each function at the indicated point.

23. $f(x) = \dfrac{1}{x+5}$, at $\left(1, \dfrac{1}{6}\right)$

24. $f(x) = \dfrac{2}{x+4}$, at $\left(1, \dfrac{2}{5}\right)$

25. $f(x) = \dfrac{1}{\sqrt{x}}$, at $(1, 1)$

26. $f(x) = \dfrac{2}{\sqrt{x}}$, at $(1, 2)$

27. Does the tangent line to the graph of $y = x^2$ at $(1, 1)$ pass through the point $(2, 5)$?

28. Does the tangent line to the graph of $y = x^3$ at $(1, 1)$ pass through the point $(2, 5)$?

29. The equation of the tangent line to the graph of a function f at $(2, 6)$ is $y = -3x + 12$. What is $f'(2)$?

30. The equation of the tangent line of a function f at $(3, 2)$ is $y = \frac{1}{3}x + 1$. What is $f'(3)$?

31. Compute the instantaneous rate of change for the function $f(x) = 5x - 2$ at:

(a) $c = 0$ (b) $c = 2$ (c) c any real number

32. Compute the instantaneous rate of change for the function $f(x) = x^2 - 1$ at:

(a) $c = -1$ (b) $c = 1$ (c) c any real number

33. Compute the instantaneous rate of change for the function $f(x) = \dfrac{x^2}{x+3}$ at:

(a) $c = 0$ (b) $c = 1$

(c) c any real number, $c \neq -3$

34. Compute the instantaneous rate of change for the function $f(x) = \dfrac{x}{x^2 - 1}$ at:

(Problem 34 continues in the next column)

34. *(continued)*

(a) $c = 0$ (b) $c = 2$

(c) c any real number, $c \neq \pm 1$

35. A human being's respiration rate R (in breaths per minute) is given by $R = -10.35 + 0.59p$, where p is the partial pressure of carbon dioxide in the lungs. Find the rate of change in respiration when $p = 50$.

36. A metal cube with an edge length x is expanding uniformly as a consequence of being heated.

(a) Find the average rate of change of volume of the cube with respect to an edge length as x increases from 2.00 to 2.01 centimeters.
(b) Find the instantaneous rate of change of volume of the cube with respect to an edge length at the instant when $x = 2$ centimeters.

37. A protein disintegrates into amino acids according to the formula $M = 28/(t + 2)$, where M is the mass of the protein (in grams) and t is time (in hours). Find the average reaction rate, $\Delta M/\Delta t$, from $t = 0$ to $t = 2$ hours. Interpret your answer.

38. In a metabolic experiment, the mass M of glucose decreases according to the formula $M = 4.5 - 0.03t^2$, where M is measured in grams and t is time (in hours). Find the average reaction rate, $\Delta M/\Delta t$, from $t = 0$ to $t = 2$ hours. Interpret your answer.

The Derivative 3.2

Instantaneous velocity, slope of a tangent line, and instantaneous rate of change lead to the same limit—namely,

$$\lim_{x \to c} \frac{f(x) - f(c)}{x - c}$$

The common theme of the three ideas is the mathematical concept of *the derivative* of a function.

[3.2.1] DEFINITION / *Derivative of a Function.*

Let $y = f(x)$ be a function and let c be in the domain of f. The derivative of f at c, denoted by $f'(c)$ and read "f prime of c," is the number

$$f'(c) = \lim_{x \to c} \frac{f(x) - f(c)}{x - c} \tag{1}$$

provided this limit exists.

The derivative of a function can be interpreted in three ways:

1. *Physical Interpretation* When the position of a body at time t is given by $s = f(t)$, $f'(t)$ is the instantaneous velocity of the body at t.
2. *Geometric Interpretation* $f'(c)$ is the slope of the tangent line to the graph of $y = f(x)$ at the point $(c, f(c))$.
3. *Rate of Change Interpretation* $f'(c)$ is the instantaneous rate of change of a function $f(x)$ at $x = c$.

EXAMPLE 1

Find the derivative of the function $f(x) = x^2 + 1$ at 2.

Solution

From (1), we have

$$f'(2) = \lim_{x \to 2} \frac{f(x) - f(2)}{x - 2} = \lim_{x \to 2} \frac{x^2 + 1 - 5}{x - 2} = \lim_{x \to 2} \frac{(x+2)(x-2)}{x - 2} = \lim_{x \to 2} (x+2) = 4$$

EXAMPLE 2

Find $f'(2)$ if $f(x) = x^2 + 2x$.

Solution

$f(2) = 4 + 4 = 8$

$$f'(2) = \lim_{x \to 2} \frac{f(x) - f(2)}{x - 2} = \lim_{x \to 2} \frac{x^2 + 2x - 8}{x - 2} = \lim_{x \to 2} \frac{(x-2)(x+4)}{x - 2} = \lim_{x \to 2} (x+4) = 6$$

The Derivative As a Function

An equivalent way of stating the definition of the derivative of a function is obtained if we write $x = c + \Delta x$. Then $f(x) = f(c + \Delta x)$, $x - c = \Delta x$, and

$$f'(c) = \lim_{x \to c} \frac{f(x) - f(c)}{x - c} = \lim_{x \to c} \frac{f(c + \Delta x) - f(c)}{\Delta x}$$

However, x approaches c if and only if Δx approaches 0. Hence,

$$f'(c) = \lim_{\Delta x \to 0} \frac{f(c + \Delta x) - f(c)}{\Delta x} \tag{2}$$

Figure 6 illustrates the definition.

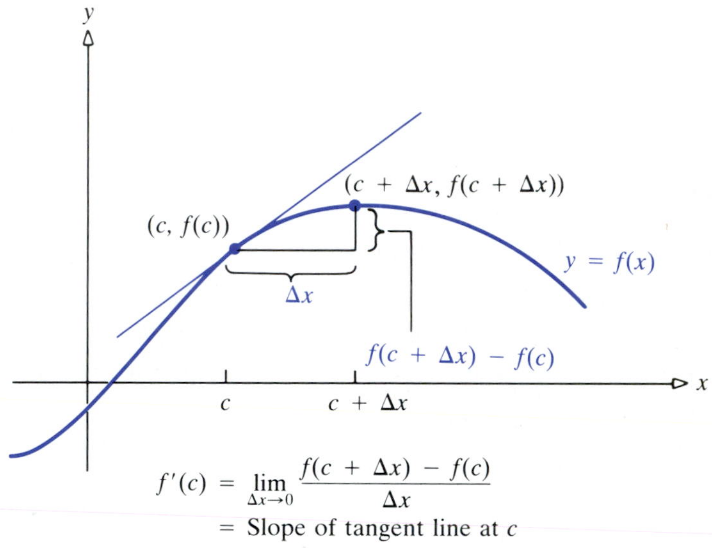

$$f'(c) = \lim_{\Delta x \to 0} \frac{f(c + \Delta x) - f(c)}{\Delta x}$$
$$= \text{Slope of tangent line at } c$$

Figure 6

In Example 1 we calculated the derivative of $f(x) = x^2 + 1$ at 2. It is often just as easy to find the derivative at an arbitrary number c, as the next example illustrates.

EXAMPLE 3

Find the derivative of $f(x) = x^2 + 1$ at c.

Solution

From (2), we have

$$f'(c) = \lim_{\Delta x \to 0} \frac{f(c + \Delta x) - f(c)}{\Delta x} = \lim_{\Delta x \to 0} \frac{(c + \Delta x)^2 + 1 - (c^2 + 1)}{\Delta x}$$

$$= \lim_{\Delta x \to 0} \frac{c^2 + 2c\Delta x + (\Delta x)^2 + 1 - c^2 - 1}{\Delta x} = \lim_{\Delta x \to 0} \frac{2c\Delta x + (\Delta x)^2}{\Delta x}$$

$$= \lim_{\Delta x \to 0} (2c + \Delta x) = 2c$$

Notice that for $c = 2$, $f'(c) = 2 \cdot c = 2 \cdot 2 = 4$, which agrees with the answer found in Example 1.

Since the limit

$$f'(c) = \lim_{\Delta x \to 0} \frac{f(c + \Delta x) - f(c)}{\Delta x} = 2c$$

exists for any choice of c, it is convenient to replace c by x, so that we can write $f'(x) = 2x$. In general, the derivative $f'(x)$ of a function f at x is itself a function, since it gives a rule for associating a number x with a number $f'(x)$. This function f', called the *derived function* of f, or the *derivative* of f, is defined by

$$f'(x) = \lim_{\Delta x \to 0} \frac{f(x + \Delta x) - f(x)}{\Delta x} \tag{3}$$

The domain of the derived function, which is always contained in the domain of the original function, consists of all numbers x for which the limit (3) exists. At such numbers, the function f is said to be *differentiable.*

In general, to "differentiate f" means to "find the derivative of f."

EXAMPLE 4

Differentiate: $f(x) = x^2 + 2x$

Solution

To differentiate f, we compute

$$f'(x) = \lim_{\Delta x \to 0} \frac{f(x + \Delta x) - f(x)}{\Delta x}$$

$$= \lim_{\Delta x \to 0} \frac{[(x + \Delta x)^2 + 2(x + \Delta x)] - (x^2 + 2x)}{\Delta x}$$

$$\underset{\text{Simplify.}}{=} \lim_{\Delta x \to 0} \frac{[x^2 + 2x\Delta x + (\Delta x)^2 + 2x + 2\Delta x] - x^2 - 2x}{\Delta x}$$

$$\underset{\text{Simplify.}}{=} \lim_{\Delta x \to 0} \frac{2x\Delta x + (\Delta x)^2 + 2\Delta x}{\Delta x}$$

By factoring Δx from the numerator and canceling it with the Δx in the denominator, we obtain

$$f'(x) = \lim_{\Delta x \to 0} \frac{\Delta x(2x + \Delta x + 2)}{\Delta x} = \lim_{\Delta x \to 0} (2x + \Delta x + 2) = 2x + 2$$

Thus, $f'(x) = 2x + 2$.

EXAMPLE 5

Differentiate: $f(x) = \dfrac{1}{x^2}$

Solution

$$f'(x) = \lim_{\Delta x \to 0} \frac{f(x + \Delta x) - f(x)}{\Delta x} = \lim_{\Delta x \to 0} \frac{\dfrac{1}{(x + \Delta x)^2} - \dfrac{1}{x^2}}{\Delta x}$$

$$\underset{\substack{\uparrow \\ \text{Simplify.}}}{=} \lim_{\Delta x \to 0} \frac{\dfrac{x^2 - (x + \Delta x)^2}{x^2(x + \Delta x)^2}}{\Delta x} \underset{\substack{\uparrow \\ \text{Simplify.}}}{=} \lim_{\Delta x \to 0} \frac{-2x\Delta x - (\Delta x)^2}{x^2(x + \Delta x)^2 \Delta x}$$

$$\underset{\substack{\uparrow \\ \text{Factor } \Delta x \text{ in numerator.}}}{=} \lim_{\Delta x \to 0} \frac{(-2x - \Delta x)\Delta x}{x^2(x + \Delta x)^2 \Delta x} \underset{\substack{\uparrow \\ \Delta x/\Delta x = 1}}{=} \lim_{\Delta x \to 0} \frac{-2x - \Delta x}{x^2(x + \Delta x)^2} = \frac{-2x}{x^4} = \frac{-2}{x^3}$$

Thus, $f'(x) = -2/x^3$.

EXAMPLE 6

Find the derivative of $f(x) = \sqrt{x}$ and determine the domain of f'.

Solution

We first compute $\Delta y/\Delta x$:

$$\frac{\Delta y}{\Delta x} = \frac{f(x + \Delta x) - f(x)}{\Delta x} = \frac{\sqrt{x + \Delta x} - \sqrt{x}}{\Delta x}$$

In its present form, we cannot cancel the Δx's. Let's see what happens if we rationalize the numerator by multiplying the numerator and denominator by $\sqrt{x + \Delta x} + \sqrt{x}$:

$$\frac{\Delta y}{\Delta x} = \left(\frac{\sqrt{x + \Delta x} - \sqrt{x}}{\Delta x}\right)\left(\frac{\sqrt{x + \Delta x} + \sqrt{x}}{\sqrt{x + \Delta x} + \sqrt{x}}\right)$$

$$= \frac{(x + \Delta x) - x}{\Delta x(\sqrt{x + \Delta x} + \sqrt{x})} = \frac{\Delta x}{\Delta x(\sqrt{x + \Delta x} + \sqrt{x})}$$

Since $\Delta x \neq 0$, we can cancel the Δx's:

$$\frac{\Delta y}{\Delta x} = \frac{1}{\sqrt{x + \Delta x} + \sqrt{x}}$$

Now, when we let Δx approach 0, the quantity $\sqrt{x + \Delta x}$ approaches $\sqrt{x}$, so that

$$f'(x) = \lim_{\Delta x \to 0} \frac{\Delta y}{\Delta x} = \lim_{\Delta x \to 0} \frac{1}{\sqrt{x + \Delta x} + \sqrt{x}} = \frac{1}{2\sqrt{x}}$$

This limit does not exist when $x = 0$. But for all other x in the domain of f $(x \geq 0)$, the limit does exist. Hence, the domain of the function f' is all $x > 0$ and the function f has a derivative for all positive x.

Example 6 demonstrates a general result: the domain of the derived function is a subset of the domain of the function itself.

EXAMPLE 7

Show that the rate of change of the area of a circle with respect to its radius is equal to its circumference.

Solution

The area A of a circle of radius R is $A = \pi R^2$. Thus, the rate of change of area with respect to radius is

$$\begin{aligned} A'(R) &= \lim_{\Delta R \to 0} \frac{\pi(R + \Delta R)^2 - \pi R^2}{\Delta R} \\ &= \lim_{\Delta R \to 0} \frac{\pi[R^2 + 2R(\Delta R) + (\Delta R)^2] - \pi R^2}{\Delta R} \\ &= \lim_{\Delta R \to 0} \frac{2\pi R(\Delta R) + \pi(\Delta R)^2}{\Delta R} = \lim_{\Delta R \to 0} \frac{\pi \Delta R(2R + \Delta R)}{\Delta R} \\ &= \lim_{\Delta R \to 0} \pi(2R + \Delta R) = 2\pi R = \text{Circumference} \end{aligned}$$

Functions That Are Not Differentiable at *c*

So far we have only been concerned with functions f that are differentiable at a number c. We look now at some functions that are not differentiable at c—that is, functions f for which

$$\lim_{x \to c} \frac{f(x) - f(c)}{x - c}$$

does not exist. Two of the most common ways that the derivative may fail to exist are:*

1. When f *has a corner at* c (see Fig. 7)
2. When f *is not continuous at* c (see Fig. 8)

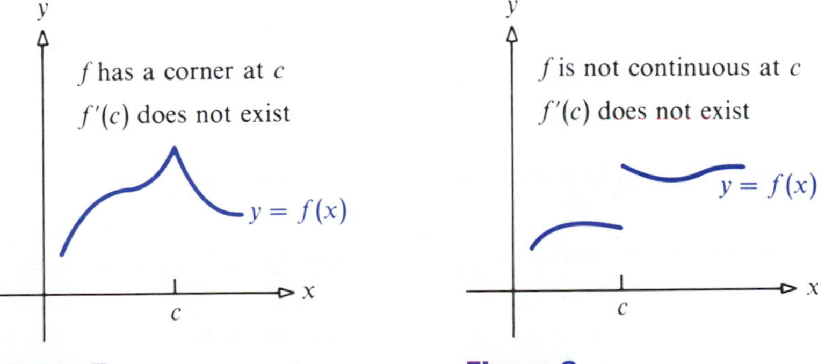

Figure 7 **Figure 8**

---

* There are other ways, but they will be discussed later.

The first of these situations is demonstrated in the next two examples. The function $f(x) = |x|$ is continuous at 0 (see Example 4 in Section 2.4), but it is not differentiable at 0. Figure 9 suggests that this is because of the corner point at $(0, f(0)) = (0, 0)$.

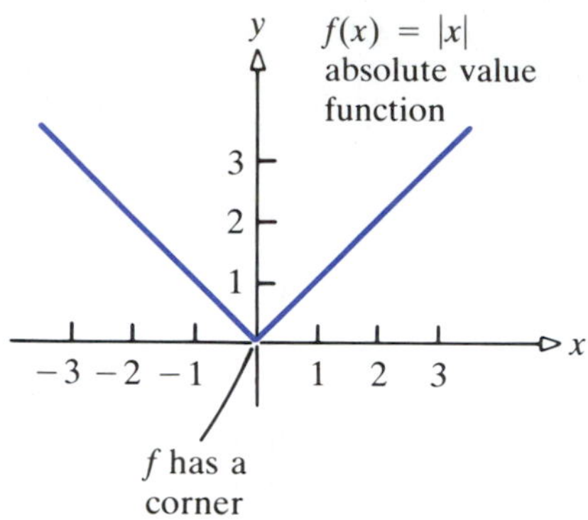

Figure 9

EXAMPLE 8

Show that $f(x) = |x|$ is not differentiable at 0.

Solution

To show that f is not differentiable at 0, we must show that

$$\lim_{x\to 0} \frac{f(x) - f(0)}{x - 0} = \lim_{x\to 0} \frac{|x| - |0|}{x - 0} = \lim_{x\to 0} \frac{|x|}{x}$$

does not exist. We note that

$$|x| = \begin{cases} -x & \text{if} \quad x < 0 \\ x & \text{if} \quad x > 0 \end{cases}$$

Therefore,

$$\lim_{x\to 0^-} \frac{|x|}{x} = \lim_{x\to 0^-} \frac{-x}{x} = \lim_{x\to 0^-} (-1) = -1$$

and

$$\lim_{x\to 0^+} \frac{|x|}{x} = \lim_{x\to 0^+} \frac{x}{x} = \lim_{x\to 0^+} (1) = 1$$

Since the left-hand limit at 0 does not equal the right-hand limit at 0,

$$\lim_{x\to 0} \frac{|x|}{x}$$

does not exist. Thus, $f(x) = |x|$ is not differentiable at 0.

We can give an intuitive explanation of why the graph in Figure 9 has no tangent line at 0 by using a simple mechanical example. Consider a line attached to a spool of thread. If the line is pulled taut, the point Q where the line is held taut and the point P where the line leaves the spool determine the tangent to the spool at P. Therefore, as we change the location of the point Q, the point P will also change, since P and Q must lie on the tangent to the spool (see Fig. 10). On the other hand, if a line is attached to a triangular device as in Figure 11, the line can be pulled taut from any number of points Q and still pivot about the vertex P.

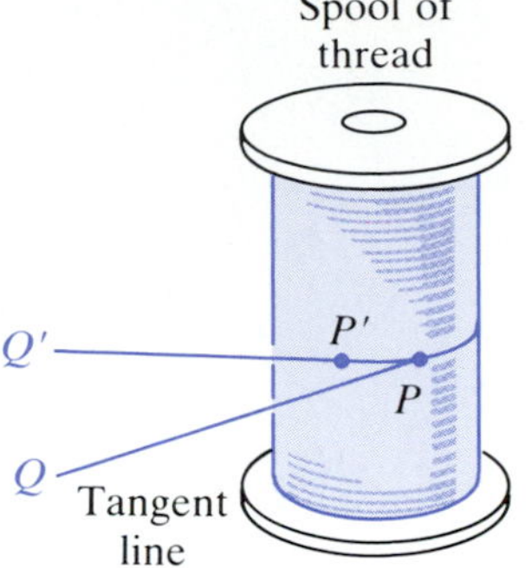

Figure 10

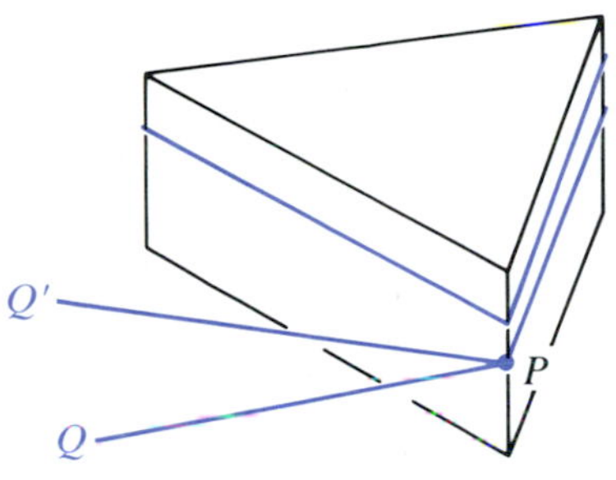

Figure 11

Here is another example of a function whose graph has a corner point.

EXAMPLE 9

For the function below, determine whether $f'(1)$ exists.

$$f(x) = \begin{cases} -2x^2 + 4 & \text{if } x < 1 \\ x^2 + 1 & \text{if } x \geq 1 \end{cases}$$

Solution

We need to find

$$\lim_{x \to 1} \frac{f(x) - f(1)}{x - 1} = \lim_{x \to 1} \frac{f(x) - 2}{x - 1}$$

But we face a difficulty. If $x < 1$, then $f(x) = -2x^2 + 4$; if $x \geq 1$, then $f(x) = x^2 + 1$. Consequently, we calculate the following limits of f at 1:

$$\lim_{x \to 1^-} \frac{f(x) - f(1)}{x - 1} = \lim_{x \to 1^-} \frac{(-2x^2 + 4) - 2}{x - 1} = \lim_{x \to 1^-} \frac{-2(x^2 - 1)}{x - 1} = -4 \quad \textbf{(4)}$$

$$\lim_{x \to 1^+} \frac{f(x) - f(1)}{x - 1} = \lim_{x \to 1^+} \frac{(x^2 + 1) - 2}{x - 1} = \lim_{x \to 1^+} \frac{(x - 1)(x + 1)}{x - 1} = 2 \quad \textbf{(5)}$$

Since the limit in (4) is not equal to the limit in (5), $f'(1)$ does not exist.

Figure 12 illustrates the graph of the function f in Example 9. At 1, where the derivative does not exist (and hence there is no tangent line), the graph of f has a corner. We usually say that the graph of f is not *smooth* at a corner.

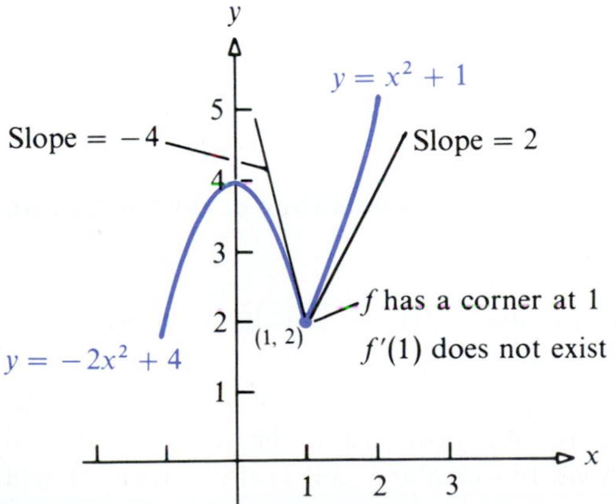

Figure 12

Derivatives and Continuity

The preceding two examples illustrate that if a function is continuous at a point but has a corner at that point, then it has no derivative there. It is also true that if a function is *not* continuous at some point, then it has no derivative there.

[3.2.2] THEOREM / *Relationship Between Differentiability and Continuity.*

If a function f has a derivative at c, then it is continuous at c.

Proof To show that f is continuous at c, we need to verify that $\lim_{x \to c} f(x) = f(c)$. We begin by observing that $x \neq c$, so

$$f(x) - f(c) = \left[\frac{f(x) - f(c)}{x - c}\right](x - c)$$

Now we take the limit of both sides as $x \to c$ and use the fact that the limit of a product equals the product of the limits:

$$\begin{aligned}\lim_{x \to c}[f(x) - f(c)] &= \lim_{x \to c}\left\{\left[\frac{f(x) - f(c)}{x - c}\right](x - c)\right\} \\ &= \left[\lim_{x \to c}\frac{f(x) - f(c)}{x - c}\right]\left[\lim_{x \to c}(x - c)\right]\end{aligned}$$

Since f has a derivative at c, we know that

$$\lim_{x \to c}\frac{f(x) - f(c)}{x - c} = f'(c)$$

is a number. Furthermore, since $\lim_{x \to c}(x - c) = 0$, we find

$$\lim_{x \to c}[f(x) - f(c)] = [f'(c)](0) = 0$$

That is, $\lim_{x \to c} f(x) = f(c)$, so f is continuous at c. ■

Thus, every differentiable function is continuous. Or, to put it another way:

[3.2.3] COROLLARY

If a function f is not continuous at a number c, then it has no derivative at c.

For example, the greatest integer function $f(x) = [\![x]\!]$ has no derivative at c if c is an integer, since f is not continuous at c.

Corollary (3.2.3) may save some time if you are asked to find the derivative of a function you suspect is not continuous. If the function is indeed not continuous at a number c, then by corollary (3.2.3) the function f will have no derivative at c.

Mathematicians for a long time were concerned with relationships between the continuity and differentiability of functions. While they were able to come up with a continuous function that has no derivative at a point, or even at a finite number of points (such as the function illustrated in Figure 13), they found it remarkable that there exist functions that are continuous at every point in an open interval but have no derivative at any point in the interval. The first such function was constructed by the Bohemian priest and mathematician Bernard Bolzano (1781–1848), followed by another example from the famous mathematician Karl Weierstrass (1815–1897). An example of such a function is beyond the scope of this book but can be found in advanced calculus books. Even to draw such a function is a formidable task. Imagine the graph of a function that has to be drawn as a continuous curve (connected?) and yet nowhere does it have a well-defined tangent line.

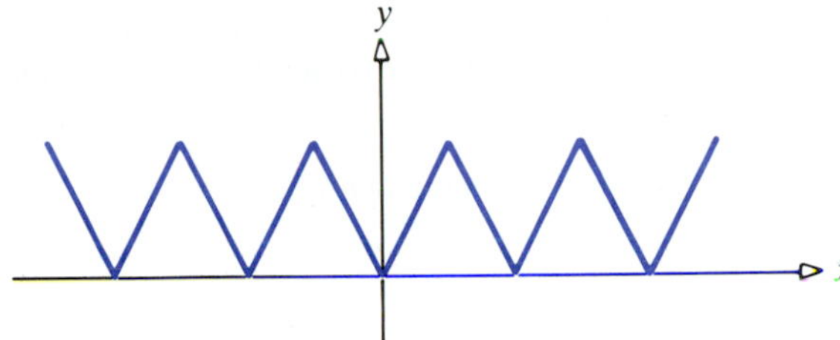

Figure 13

EXERCISE 3.2

In Problems 1–10 find the derivative of each function at the given number by using (1).

1. $f(x) = 2x + 3$, at 1

2. $f(x) = 3x - 5$, at 2

3. $f(x) = x^2 - 2$, at 0

4. $f(x) = 2x^2 + 4$, at 1

5. $f(x) = 3x^2 + x + 5$, at -1

6. $f(x) = 2x^2 - x - 7$, at -1

7. $f(x) = \sqrt{x}$, at 4

8. $f(x) = \dfrac{1}{x^2}$, at 2

9. $f(x) = \dfrac{2 - 5x}{1 + x}$, at 0

10. $f(x) = \dfrac{2 + 3x}{2 + x}$, at 1

In Problems 11–26 find the derivative of each function by using (3).

11. $f(x) = 2x + 3$

12. $f(x) = 3x - 5$

13. $f(x) = x^2 - 2$

14. $f(x) = 2x^2 + 4$

15. $f(x) = 3x^2 + x + 5$

16. $f(x) = 2x^2 - x - 7$

17. $f(x) = 5$

18. $f(x) = -2$

19. $f(x) = 5\sqrt{x}$

20. $f(x) = \dfrac{4}{x^3}$

21. $f(x) = mx + b$

22. $f(x) = ax^2 + bx + c$

23. $f(x) = \dfrac{x - 1}{x + 1}$

24. $f(x) = x - \dfrac{2}{x}$

25. $f(x) = \dfrac{5}{1 + x^2}$

26. $f(x) = \dfrac{1}{\sqrt{x - 1}}$

In Problems 27–34 each limit represents the derivative of some function f at some number c. State f and c in each case.

27. $\lim_{\Delta x \to 0} \dfrac{(2 + \Delta x)^2 - 4}{\Delta x}$

28. $\lim_{\Delta x \to 0} \dfrac{(2 + \Delta x)^3 - 8}{\Delta x}$

29. $\lim_{x \to 1} \dfrac{x^2 - 1}{x - 1}$

30. $\lim_{x \to 1} \dfrac{x^{14} - 1}{x - 1}$

31. $\lim_{x \to \pi/6} \dfrac{\sin x - \frac{1}{2}}{x - \frac{\pi}{6}}$

32. $\lim_{x \to \pi/4} \dfrac{\cos x - \frac{\sqrt{2}}{2}}{x - \frac{\pi}{4}}$

33. $\lim_{x \to 0} \dfrac{2(x + 2)^2 - (x + 2) - 6}{x}$

34. $\lim_{x \to 0} \dfrac{3x^3 - 2x}{x}$

35. Let $f(x) = x^2 + 2$. Find all points on the graph of f for which the tangent line passes through the origin.

36. Let $f(x) = x^2 - 2x + 1$. Find all points on the graph of f for which the tangent line passes through the point $(1, 2)$.

37. Use the graph of a function f given in the figure below to answer the following questions.

(a) For which numbers c does $\lim_{x \to c} f(x)$ exist but f is not continuous at c?

(b) For which number c is f continuous at c but not differentiable at c?

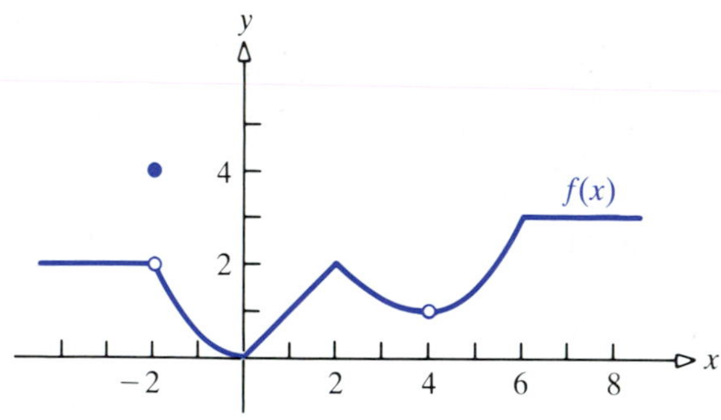

38. Repeat Problem 37 for the graph below.

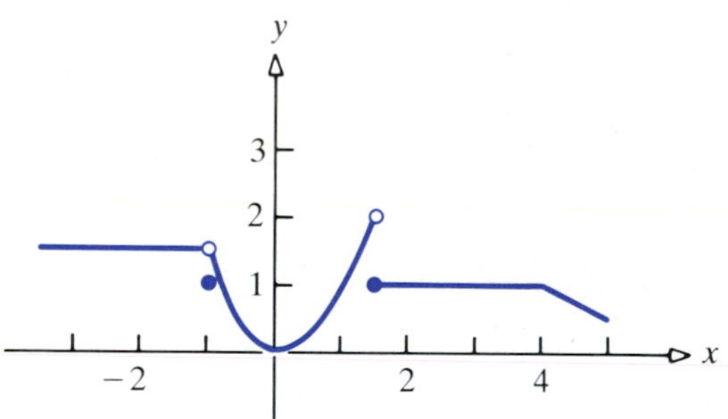

In Problems 39–46 determine whether the given function f has a derivative at c. Graph each function.

39. $f(x) = \begin{cases} 2x + 3 & \text{if } x < 1 \\ x^2 + 4 & \text{if } x \geq 1 \end{cases} \quad c = 1$

40. $f(x) = \begin{cases} 3 - 4x & \text{if } x < -1 \\ 2x + 9 & \text{if } x \geq -1 \end{cases} \quad c = -1$

41. $f(x) = \begin{cases} -4 + 2x & \text{if } x \leq \frac{1}{2} \\ 4x^2 - 4 & \text{if } x > \frac{1}{2} \end{cases} \quad c = \frac{1}{2}$

42. $f(x) = \begin{cases} 2x^2 + 1 & \text{if } x < -1 \\ -1 - 4x & \text{if } x \geq -1 \end{cases} \quad c = -1$

43. $f(x) = |x^2 - 4| \quad c = 2$

44. $f(x) = |x^2 - 4| \quad c = -2$

45. $f(x) = \begin{cases} 2x^2 + 1 & \text{if } x < -1 \\ 2 + 2x & \text{if } x \geq -1 \end{cases} \quad c = -1$

46. $f(x) = \begin{cases} 5 - 2x & \text{if } x < 2 \\ x^2 & \text{if } x \geq 2 \end{cases} \quad c = 2$

47. For the function

$$f(x) = \begin{cases} x^3 & \text{if } x \le 0 \\ x^2 & \text{if } x > 0 \end{cases}$$

determine whether:

(a) f is continuous at 0; (b) $f'(0)$ exists.
(c) Graph the function and give a geometric interpretation to the answers found in parts (a) and (b).

48. Repeat Problem 47 for the function

$$f(x) = \begin{cases} 2x & \text{if } x \le 0 \\ x^2 & \text{if } x > 0 \end{cases}$$

49. Calculate analytically the velocity (in feet per second) of an automobile whose position is given by

$$s = \begin{cases} t^3 & \text{if } 0 \le t < 5 \\ 125 & \text{if } 5 \le t \end{cases}$$

This could represent a crash test in which a vehicle is accelerated into a brick wall. Find the velocity just before and just after impact. Are the formulas quoted accurate during impact?

50. If the line $3x - 4y = 0$ is tangent to the graph of $y = x^3 + k$ in the first quadrant, find k.

51. For what nonnegative number b is the line given by $y = -\frac{1}{3}x + b$ normal to the graph of $y = x^3$? See the comment about normal lines after Problem 61.

52. A simple model for population growth states that the rate of change of population size with respect to time is proportional to the population size. Express this statement as an equation involving a derivative.

53. If f is an even function that is differentiable at c, show that $f'(-c) = -f'(c)$.

54. If f is an odd function that is differentiable at c, show that $f'(-c) = f'(c)$.

55. Let f be a function defined for all x. Suppose f has the following properties: *(1)* $f(u + v) = f(u) \cdot f(v)$, *(2)* $f(0) = 1$, *(3)* $f'(0)$ exists. Show that $f'(x)$ exists for all x. Also show that $f'(x) = f'(0)f(x)$.

56. A dive bomber is flying from right to left along the graph of $y = x^2$. When a rocket bomb is released, it follows a path that is approximately along the tangent line. Where should the pilot release the bomb if the target is at (1, 0)?

57. Answer the question in Problem 56 if the plane is flying from right to left along the graph of $y = x^3$.

58. Atmospheric pressure decreases as the distance from the surface of the earth increases, and the rate of change of

(Problem 58 continues in the next column)

58. *(continued)*
pressure with respect to height is proportional to the pressure. Express this law as an equation involving a derivative.

59. Under certain conditions, an electric current will die out at a rate (with respect to time) that is proportional to the current remaining. Express this law as an equation involving a derivative.

60. A circle of radius R has area $A = \pi R^2$ and circumference $C = 2\pi R$. If the radius changes from R to $R + \Delta R$, find the:

(a) Change in area
(b) Change in circumference
(c) Average rate of change of area with respect to radius
(d) Average rate of change of circumference with respect to radius
(e) Rate of change of circumference with respect to radius

61. The volume V of a sphere of radius R is $V = 4\pi R^3/3$. If the radius changes from R to $R + \Delta R$, find the:

(a) Change in volume
(b) Average rate of change of volume with respect to radius
(c) Rate of change of volume with respect to radius

Normal Lines The *normal line* to the graph of a function f at a point $(c, f(c))$ is defined as the line through $(c, f(c))$ and perpendicular to the tangent line to the graph of f at $(c, f(c))$. See the figure. Thus, if f is a function whose derivative at c is $f'(c) \ne 0$, the slope of the normal line to the graph of f at $(c, f(c))$ is the negative reciprocal of $f'(c)$—namely, $-1/f'(c)$—and an equation of the normal line to the graph of f at $(c, f(c))$ is

$$y - f(c) = \frac{-1}{f'(c)}(x - c)$$

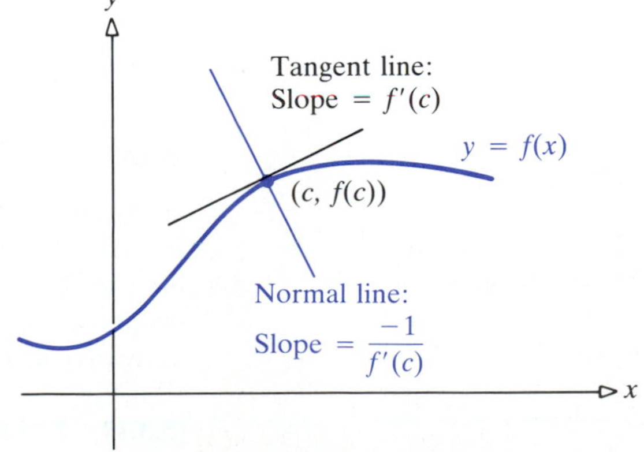

In Problems 62–67 find the slope of the normal line to the graph of each function at the indicated point. Graph each function and show this normal line.

62. $f(x) = x^2 + 1$, at $(1, 2)$

63. $f(x) = x^2 - 1$, at $(-1, 0)$

64. $f(x) = x^2 - 2x$, at $(-1, 3)$

65. $f(x) = 2x^2 + x$, at $(1, 3)$

66. $f(x) = \dfrac{1}{x}$, at $(1, 1)$

67. $f(x) = \sqrt{x}$, at $(4, 2)$

3.3 Formulas for Finding Derivatives

The derivative of a function f is the limit of a difference quotient—namely,

$$f'(x) = \lim_{\Delta x \to 0} \frac{\Delta y}{\Delta x} = \lim_{\Delta x \to 0} \frac{f(x + \Delta x) - f(x)}{\Delta x} \tag{1}$$

provided this limit exists. But calculating the derivative of a function by this definition can become a tedious chore, particularly if the function f is complicated. In this section and the next, we develop some formulas that make this calculation a relatively straightforward procedure for any algebraic function.

In the development of many of the formulas for derivatives, the required computations are easier to perform if Δx is replaced by h. Hence, we rewrite formula (1) as

$$f'(x) = \lim_{h \to 0} \frac{f(x + h) - f(x)}{h} \tag{2}$$

Steps for Calculating a Derivative

We calculate the derivative of a function f in four steps:

Step 1 **Find $f(x + h)$.**

Step 2 **Subtract $f(x)$ from $f(x + h)$ to get $f(x + h) - f(x)$.**

Step 3 **Divide this result by $h \neq 0$ to get $\dfrac{f(x + h) - f(x)}{h}$ and simplify, if possible.**

Step 4 **Find the limit (if it exists) of $\dfrac{f(x + h) - f(x)}{h}$ as h approaches 0.**

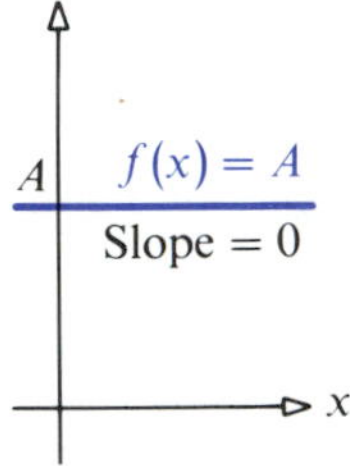

Figure 14

Constant Function

We begin with the constant function $f(x) = A$. A little geometry will tell us what to expect. Since the graph of the constant function f is a horizontal line (see Fig. 14), the tangent line to f at any point is also a horizontal line (slope equals 0). Since the derivative is the slope of the tangent line, the derivative of f should be 0.

[3.3.1] THEOREM / *Derivative of a Constant.*

If f is a constant function $f(x) = A$, then $f'(x) = 0$.

This result is sometimes restated to say "the derivative of a constant is 0." To prove this, we follow the four-step method using $f(x) = A$.

Step 1 $f(x+h) = A$

Step 2 $f(x+h) - f(x) = A - A = 0$

Step 3 $\dfrac{f(x+h) - f(x)}{h} = \dfrac{0}{h} = 0$

Step 4 $f'(x) = \lim\limits_{h \to 0} \dfrac{f(x+h) - f(x)}{h} = \lim\limits_{h \to 0} 0 = 0$

Other Forms of Notation

Besides the *prime notation* f', there are several other ways to denote the derivative of a function $y = f(x)$. The most common ones are

$$y' \qquad \frac{dy}{dx} \qquad Df(x)$$

The *Leibniz notation* dy/dx may be interpreted as

$$\frac{dy}{dx} = \frac{d}{dx}(y) = \frac{d}{dx} f(x)$$

where d/dx is an instruction to compute the derivative (with respect to the independent variable x) of the function $y = f(x)$.

In the *operator notation* $Df(x)$, D is said to *operate* on the function, and the result is the derivative of f. To emphasize that the operation is being performed with respect to the independent variable x, we sometimes write $Df(x) = D_x f(x)$.

In these new forms of notation, theorem (3.3.1) may be stated as

$$\frac{d}{dx} A = 0 \qquad \text{or} \qquad DA = 0 \tag{3}$$

where A is a constant function. For example, if $f(x) = 6$, then $f'(x) = 0$. That is,

$$\frac{d}{dx} 6 = 0$$

In subsequent work with derivatives we shall use the prime notation or the Leibniz notation, or sometimes a mixture of the two, depending on which is most convenient. We shall not use the operator notation in this book.

The Simple Power Rule

The formula for the derivative of x^n, $n > 0$ an integer, is as follows:

[3.3.2] THEOREM / *Derivative of x^n.*

If n is a positive integer, then

$$\frac{d}{dx} x^n = nx^{n-1} \tag{4}$$

To prove this, we follow the four-step method using $f(x) = x^n$ for $n = 1$, $n = 2$, $n = 3$, and $n > 3$.

Step 1

$n=1$ $\quad f(x)=x \quad f(x+h)=x+h$

$n=2$ $\quad f(x)=x^2 \quad f(x+h)=(x+h)^2=x^2+2xh+h^2$

$n=3$ $\quad f(x)=x^3 \quad f(x+h)=(x+h)^3=x^3+3x^2h+3xh^2+h^3$

$n>3$ $\quad f(x)=x^n \quad f(x+h)=(x+h)^n$

By applying the binomial theorem, we find the expansion of $(x + h)^n$ to be

$$(x+h)^n = x^n + nx^{n-1}h + \frac{n(n-1)}{2}x^{n-2}h^2 + \cdots + nxh^{n-1} + h^n$$

Step 2

$n = 1$ $\quad f(x+h) - f(x) = (x+h) - x = h$

$n = 2$ $\quad f(x+h) - f(x) = (x^2 + 2xh + h^2) - x^2 = 2xh + h^2$

$n = 3$
$$f(x+h) - f(x) = (x^3 + 3x^2h + 3xh^2 + h^3) - x^3 = 3x^2h + 3xh^2 + h^3$$

$n > 3$
$$\begin{aligned} f(x+h) - f(x) &= (x+h)^n - x^n \\ &= \left[x^n + nx^{n-1}h + \frac{n(n-1)}{2}x^{n-2}h^2 + \cdots + nxh^{n-1} + h^n\right] - x^n \\ &= nx^{n-1}h + (\text{Assorted terms})(h^2) \end{aligned}$$

After subtracting the x^n terms, we factor h^2 out of all terms except the first.

Step 3

$n = 1$
$$\frac{f(x+h) - f(x)}{h} = \frac{h}{h} = 1$$

$n = 2$
$$\frac{f(x+h) - f(x)}{h} = \frac{2xh + h^2}{h} = \frac{h(2x+h)}{h} = 2x + h$$

$n = 3$
$$\begin{aligned} \frac{f(x+h) - f(x)}{h} &= \frac{3x^2h + 3xh^2 + h^3}{h} \\ &= \frac{h(3x^2 + 3xh + h^2)}{h} \\ &= 3x^2 + 3xh + h^2 \end{aligned}$$

$n > 3$
$$\begin{aligned} \frac{f(x+h) - f(x)}{h} &= \frac{nx^{n-1}h + (\text{Assorted terms})h^2}{h} \\ &= nx^{n-1} + (\text{Assorted terms})(h) \end{aligned}$$

Step 4 $n = 1$ $\quad f'(x) = \lim_{h \to 0} \frac{f(x+h) - f(x)}{h} = \lim_{h \to 0} 1 = 1$

$n = 2$ $\quad f'(x) = \lim_{h \to 0} \frac{f(x+h) - f(x)}{h} = \lim_{h \to 0}(2x + h) = 2x$

$n = 3$ $\quad f'(x) = \lim_{h \to 0} \frac{f(x+h) - f(x)}{h}$

$$= \lim_{h \to 0}(3x^2 + 3xh + h^2) = 3x^2$$

$n > 3$ $\quad f'(x) = \lim_{h \to 0} \frac{f(x+h) - f(x)}{h}$

$$= \lim_{h \to 0}[nx^{n-1} + (\text{Assorted terms})(h)]$$

$$= \lim_{h \to 0} nx^{n-1} + \lim_{h \to 0}[(\text{Assorted terms})(h)] = nx^{n-1}$$

This last formula for $n > 3$ agrees with the preceding three formulas, thus establishing the general formula (4) for n a positive integer; that is,

$$\frac{d}{dx} x^n = nx^{n-1}$$

This result tells us that **the derivative of x raised to the power n is n times x raised to the power $n - 1$.**

EXAMPLE 1

(a) $\frac{d}{dx} x^5 = 5x^4$ (b) $\frac{d}{dx} x^{13} = 13x^{12}$

(c) $\frac{d}{dx} x^{65} = 65x^{64}$ (d) $\frac{d}{dx} x = 1 \cdot x^0 = 1$

Some General Differentiation Formulas

The following theorem, which has a simple proof, is used often:

[3.3.3] THEOREM / ***Derivative of a Constant Times a Function.***

If f is a differentiable function and if $F(x) = kf(x)$, where k is a constant, then F is differentiable and

$$F'(x) = kf'(x) \tag{5}$$

Proof We follow the four-step method using the function $y = F(x) = kf(x)$.

Step 1 $\quad F(x + h) = kf(x + h)$

Step 2 $\quad F(x + h) - F(x) = kf(x + h) - kf(x) = k[f(x + h) - f(x)]$

Step 3 $$\frac{F(x+h)-F(x)}{h} = k\frac{[f(x+h)-f(x)]}{h}$$

Step 4 Since f is differentiable, the limit of $\dfrac{f(x+h)-f(x)}{h}$ as $h \to 0$ equals the derivative $f'(x)$. Taking the limit as $h \to 0$ and using the fact that the limit of a constant times a function equals the constant times the limit of a function (see (3) in Section 2.2), we have

$$\lim_{h\to 0}\frac{F(x+h)-F(x)}{h} = \lim_{h\to 0}\left[k\frac{f(x+h)-f(x)}{h}\right]$$
$$= k\left[\lim_{h\to 0}\frac{f(x+h)-f(x)}{h}\right] = kf'(x)$$

Theorem (3.3.3) may be restated as follows:

The derivative of a constant times a differentiable function f equals the constant times the derivative of f. That is, if k is a constant, then

$$\frac{d}{dx}[kf(x)] = k\left[\frac{d}{dx}f(x)\right] \tag{6}$$

EXAMPLE 2

(a) $\dfrac{d}{dx}(3x^2) = 3\dfrac{d}{dx}x^2 = (3)(2x) = 6x$

(b) $\dfrac{d}{dx}\left(\dfrac{1}{3}\right)x^3 = \left(\dfrac{1}{3}\right)(3x^2) = x^2$

(c) $\dfrac{d}{dx}(-4x^3) = (-4)(3x^2) = -12x^2$

(d) $\dfrac{d}{dx}\left(\dfrac{5}{2}x^8\right) = \left(\dfrac{5}{2}\right)(8x^7) = 20x^7$

The next theorem provides a formula for finding the derivative of a function expressed as the sum of two functions whose derivatives are known.

[3.3.4] THEOREM / ***Derivative of a Sum.***

If f and g are differentiable functions and if $F(x) = f(x) + g(x)$, then F is differentiable and

$$F'(x) = f'(x) + g'(x)$$

Proof We follow the four-step method using the function $F(x) = f(x) + g(x)$.

Step 1 $$F(x+h) = f(x+h) + g(x+h)$$

Step 2 $$F(x+h) - F(x) = f(x+h) + g(x+h) - [f(x) + g(x)]$$
$$= [f(x+h) - f(x)] + [g(x+h) - g(x)]$$

Step 3 $$\frac{F(x+h)-F(h)}{h} = \frac{[f(x+h)-f(x)] + [g(x+h)-g(x)]}{h}$$
$$= \frac{f(x+h)-f(x)}{h} + \frac{g(x+h)-g(x)}{h}$$

Step 4 Taking the limit as $h \to 0$ and using the fact that the limit of a sum equals the sum of the limits (see (1) in Section 2.2), we have

$$F'(x) = \lim_{h\to 0} \frac{F(x+h) - F(x)}{h}$$

$$= \lim_{h\to 0} \frac{f(x+h) - f(x)}{h} + \lim_{h\to 0} \frac{g(x+h) - g(x)}{h}$$

The limits on the right are $f'(x)$ and $g'(x)$, respectively. Hence,

$$F'(x) = f'(x) + g'(x)$$

Theorem (3.3.4) may be restated as follows:

The derivative of the sum of two differentiable functions equals the sum of their derivatives. In the Leibniz notation,

$$\frac{d}{dx}[f(x) + g(x)] = \frac{d}{dx} f(x) + \frac{d}{dx} g(x) \tag{7}$$

EXAMPLE 3

Find the derivative of $f(x) = 3x^2 + 8$.

Solution

Here, f is the sum of $3x^2$ and 8. Hence,

$$f'(x) = \frac{d}{dx}(3x^2 + 8) \underset{\text{By (7)}}{=} \frac{d}{dx}(3x^2) + \frac{d}{dx} 8 \underset{\text{By (6) and (3)}}{=} 3 \frac{d}{dx} x^2 + 0 = (3)(2x) = 6x$$

If f and g are two differentiable functions and if $F(x) = f(x) - g(x)$, then $F'(x) = f'(x) - g'(x)$. That is,

$$\frac{d}{dx}[f(x) - g(x)] = \frac{d}{dx} f(x) - \frac{d}{dx} g(x) \tag{8}$$

The proof is left as an exercise (Problem 45 in Exercise 3.3).

The results given above extend to sums (or differences) of more than two differentiable functions. Thus, if $f_1, f_2, \ldots, f_n$ are n differentiable functions, we have the formula

$$\frac{d}{dx}[f_1(x) + f_2(x) + \cdots + f_n(x)] = \frac{d}{dx} f_1(x) + \frac{d}{dx} f_2(x) + \cdots + \frac{d}{dx} f_n(x)$$

Our final example for this section illustrates how these results may be combined.

EXAMPLE 4

Find the derivative of $f(x) = 6x^4 - 3x^2 + 2x - 5$.

Solution

$$f'(x) = \frac{d}{dx}(6x^4 - 3x^2 + 2x - 5)$$

$$= \frac{d}{dx}(6x^4) - \frac{d}{dx}(3x^2) + \frac{d}{dx}(2x) - \frac{d}{dx}5 = 24x^3 - 6x + 2$$

EXERCISE 3.3

In Problems 1–24 find the derivative of the function f by using the formulas of this section.

1. $f(x) = 3$

2. $f(x) = \sqrt{2}$

3. $f(x) = \sqrt{2}x + \sqrt{3}$

4. $f(x) = -\sqrt{3}x + 1$

5. $f(x) = 3x^{15}$

6. $f(x) = \frac{1}{3}x^{12}$

7. $f(x) = 3x + 2$

8. $f(x) = 5x - \frac{1}{2}$

9. $f(x) = x^2 + 3x - 4$

10. $f(x) = 4x^4 + 2x^2 - 2$

11. $f(x) = 8x^5 - 5x + 1$

12. $f(x) = 9x^3 - 2x^2 + 4x + 4$

13. $f(x) = \frac{1}{3}x^4 - 3x + \frac{3}{2}$

14. $f(x) = -3x^4 - 2x^3$

15. $f(x) = \pi x^3 + \frac{3}{2}x^2$

16. $f(x) = 4 - \pi x^2$

17. $f(x) = \frac{1}{3}(x^5 - 8)$

18. $f(x) = \dfrac{x^3 + 2}{5}$

19. $f(x) = \frac{1}{5}(x^7 - 3x^2 + 2)$

20. $f(x) = \dfrac{x^3 + 2x + 1}{7}$

21. $f(x) = \dfrac{x^7 - 5x}{9}$

22. $f(x) = \dfrac{1}{a}(ax^2 + bx + c)$

23. $f(x) = ax^2 + bx + c$

24. $f(x) = ax^3 + bx^2 + cx + d$

In Problems 25–30 find the indicated derivative.

25. $\dfrac{d}{dx}(\sqrt{3}x + \frac{1}{2})$

26. $\dfrac{d}{dx}\left(\dfrac{2x^4 - 5}{8}\right)$

27. $\dfrac{dA}{dR}$ if $A = \pi R^2$

28. $\dfrac{dC}{dR}$ if $C = 2\pi R$

29. $\dfrac{dV}{dR}$ if $V = \frac{4}{3}\pi R^3$

30. $\dfrac{dP}{dT}$ if $P = 0.2T$

In Problems 31 and 32 find the slope of the tangent line to the graph of the function f at the indicated point. What is an equation of the tangent line?

31. $f(x) = x^3 + 3x - 1$, at $(0, -1)$

32. $f(x) = x^4 + 2x - 1$, at $(1, 2)$

In Problems 33–36 find those x, if any, at which the graph of the function f has a horizontal tangent line—that is, at which $f'(x) = 0$.

33. $f(x) = 3x^2 - 12x + 4$

34. $f(x) = x^2 + 4x - 3$

35. $f(x) = x^3 - 3x + 2$

36. $f(x) = x^4 - 4x^3$

In Problems 37–40 find an equation for the tangent line(s) to the graph of the function f that is (are) parallel to the line L.

37. $f(x) = 3x^2 - x$, L: $y = 5x$

38. $f(x) = 2x^3 + 1$, L: $y = 6x - 1$

39. $f(x) = \frac{1}{3}x^3 - x^2$, L: $y - 3x + 2 = 0$

40. $f(x) = x^3 - x$, L: $x + y = 0$

41. In t seconds, the position of an object is a distance of s meters from the origin, where $s = t^3 - t + 1$. Find the velocity at $t = 0$. At $t = 5$.

42. In t seconds, the position of an object is a distance of s meters from the origin, where $s = t^4 - t^3 + t$. Find the velocity at $t = 0$. At $t = 1$.

43. What is $\lim_{h \to 0} \dfrac{5(\frac{1}{2} + h)^8 - 5(\frac{1}{2})^8}{h}$?

44. What is $\lim_{h \to 0} \dfrac{6(2 + h)^5 - 6(2)^5}{h}$?

45. Prove that if f and g are differentiable functions and if $F(x) = f(x) - g(x)$, then $F'(x) = f'(x) - g'(x)$.

46. The velocity v of a liquid flowing through a cylindrical tube is given by the formula $v = k(R^2 - r^2)$, where R is the radius of the tube, k is a constant that depends on the length of the tube and the velocity of the liquid at its ends, and r is the variable distance of the liquid from the center of the tube. Find the rate of change of v with respect to r at the center of the tube. What is the rate halfway from the center to the wall of the tube? What is it at the wall of the tube?

47. Let $f(x) = x^n$, where n is a positive integer. Use (4) and a factoring principle to show that $f'(c) = nc^{n-1}$.

48. Find the constants a, b, and c so that the graph of $y = ax^2 + bx + c$ passes through the point $(-1, 1)$ and is tangent to the line $y = 2x$ at $(0, 0)$.

Formulas for Finding Derivatives (Continued) 3.4

In this section we will develop general formulas for differentiating functions that are products and quotients.

[3.4.1] THEOREM / ***Derivative of a Product.***

If f and g are differentiable functions and if $F(x) = f(x)g(x)$, then F is differentiable and

$$F'(x) = f(x)g'(x) + f'(x)g(x)$$

Proof We proceed directly to Step 3 of the four-step method:

$$\frac{F(x + h) - F(x)}{h} = \frac{f(x + h)g(x + h) - f(x)g(x)}{h}$$

Now, we subtract and add the term $f(x + h)g(x)$ in the numerator and factor (the reason for this will become apparent shortly):

$$\frac{F(x + h) - F(x)}{h} = \frac{f(x + h)g(x + h) - f(x + h)g(x) + f(x + h)g(x) - f(x)g(x)}{h}$$

$$= f(x + h)\frac{[g(x + h) - g(x)]}{h} + \frac{[f(x + h) - f(x)]}{h}g(x)$$

Then we take the limit as h approaches 0 and apply properties of limits:

$$F'(x) = \lim_{h \to 0} \frac{F(x + h) - F(x)}{h} = \lim_{h \to 0}\left\{f(x + h)\frac{[g(x + h) - g(x)]}{h}\right\} + \lim_{h \to 0}\left\{\frac{[f(x + h) - f(x)]}{h}g(x)\right\}$$

$$= \left\{\overset{①}{\lim_{h \to 0}} f(x + h)\right\}\left\{\overset{②}{\lim_{h \to 0}} \frac{g(x + h) - g(x)}{h}\right\} + \left\{\overset{③}{\lim_{h \to 0}} \frac{f(x + h) - f(x)}{h}\right\}\left\{\overset{④}{\lim_{h \to 0}} g(x)\right\}$$

We have numbered the limits for convenience. The limit ② is $g'(x)$, the derivative of g; the limit ③ is $f'(x)$, the derivative of f; limit ④ is the limit of $g(x)$, which equals $g(x)$ since x is fixed while $h \to 0$. The first limit requires a more careful look since, as we saw in Chapter 2, it may not be possible to merely replace h by 0.

①

$$\lim_{h\to 0} f(x+h) = \lim_{h\to 0}[f(x+h) - f(x) + f(x)] = \lim_{h\to 0}[f(x+h) - f(x)] + \lim_{h\to 0} f(x)$$

$$= \lim_{h\to 0}\left[h\,\frac{f(x+h) - f(x)}{h}\right] + f(x)$$

$$= \left[\lim_{h\to 0} h\right]\left[\lim_{h\to 0}\frac{f(x+h) - f(x)}{h}\right] + f(x)$$

$$= (0)[f'(x)] + f(x) = f(x)$$

Hence, $\lim_{h\to 0} f(x+h) = f(x)$, and the four limits reduce to

$$F'(x) = f(x)g'(x) + f'(x)g(x)$$

Theorem (3.4.1) may be restated as follows:

The derivative of the product of two differentiable functions equals the first function times the derivative of the second plus the derivative of the first function times the second function. In the Leibniz notation,

$$\frac{d}{dx}[f(x)g(x)] = f(x)\left[\frac{d}{dx}g(x)\right] + \left[\frac{d}{dx}f(x)\right]g(x) \qquad \textbf{(1)}$$

Observe that, unlike the situation with limits, the derivative of a product does not equal the product of the derivatives.

EXAMPLE 1

Find the derivative of $F(x) = (x^2 + 2x - 5)(x^3 - 1)$.

Solution

The function F is the product of the two polynomial functions $f(x) = x^2 + 2x - 5$ and $g(x) = x^3 - 1$ so that, by (1), we have

$$F'(x) = (x^2 + 2x - 5)\left[\frac{d}{dx}(x^3 - 1)\right] + \left[\frac{d}{dx}(x^2 + 2x - 5)\right](x^3 - 1)$$

$$= (x^2 + 2x - 5)(3x^2) + (2x + 2)(x^3 - 1)$$

$$= 5x^4 + 8x^3 - 15x^2 - 2x - 2$$

Now that you know the rule for the derivative of a product, be careful not to use it unnecessarily. When one of the factors is a constant, you should use theorem (3.3.3). For example, it is easier to work

$$\frac{d}{dx}[5(x^2 + 1)] = 5\,\frac{d}{dx}(x^2 + 1) = (5)(2x) = 10x$$

than it is to work

$$\frac{d}{dx}[5(x^2+1)] = 5\left[\frac{d}{dx}(x^2+1)\right] + \left[\frac{d}{dx}5\right](x^2+1) = (5)(2x) + (0)(x^2+1) = 10x$$

The next theorem provides a formula for finding the derivative of the quotient of two functions.

[3.4.2] THEOREM / ***Derivative of a Quotient.***

If f and $g \neq 0$ are differentiable functions and if $F(x) = f(x)/g(x)$, then F is differentiable and

$$F'(x) = \frac{g(x)f'(x) - f(x)g'(x)}{[g(x)]^2}$$

Proof We proceed directly to Step 3 of the four-step method:

$$\frac{F(x+h) - F(x)}{h} = \frac{\dfrac{f(x+h)}{g(x+h)} - \dfrac{f(x)}{g(x)}}{h}$$

$$= \frac{g(x)f(x+h) - f(x)g(x+h)}{g(x)g(x+h)h}$$

We subtract and add $g(x)f(x)$ in the numerator and factor (the reason for doing this will become apparent shortly):

$$\frac{g(x)f(x+h) - g(x)f(x) + g(x)f(x) - f(x)g(x+h)}{g(x)g(x+h)h}$$

$$= \frac{g(x)[f(x+h) - f(x)] - f(x)[g(x+h) - g(x)]}{g(x)g(x+h)h}$$

$$= \left[\frac{g(x)}{g(x)g(x+h)}\right]\left[\frac{f(x+h) - f(x)}{h}\right] - \left[\frac{f(x)}{g(x)g(x+h)}\right]\left[\frac{g(x+h) - g(x)}{h}\right]$$

Taking the limit as h approaches 0 and applying properties of limits give

$$F'(x) = \frac{g(x)f'(x)}{[g(x)]^2} - \frac{f(x)g'(x)}{[g(x)]^2} = \frac{g(x)f'(x) - f(x)g'(x)}{[g'x)]^2}$$

Justification of this statement is left to you. (Refer to the proof of theorem (3.4.1) for a hint.)

Theorem (3.4.2) may be restated as follows:

The derivative of a quotient of two functions is the denominator times the derivative of the numerator minus the numerator times the derivative of the denominator, all divided by the denominator squared. In the Leibniz notation,

$$\frac{d}{dx}\left[\frac{f(x)}{g(x)}\right] = \frac{g(x)\left[\dfrac{d}{dx}f(x)\right] - f(x)\left[\dfrac{d}{dx}g(x)\right]}{[g(x)]^2} \qquad (2)$$

Observe that unlike the situation with limits the derivative of the quotient of two functions is not the quotient of the derivatives of the two functions.

EXAMPLE 2

Find the derivative of $F(x) = \dfrac{x^2+1}{x-3}$.

Solution

Here, F is the quotient of $f(x) = x^2 + 1$ over $g(x) = x - 3$. Thus,

$$F'(x) = \frac{d}{dx}\left(\frac{x^2+1}{x-3}\right) = \frac{(x-3)\left[\dfrac{d}{dx}(x^2+1)\right] - (x^2+1)\left[\dfrac{d}{dx}(x-3)\right]}{(x-3)^2}$$

$$= \frac{(x-3)(2x) - (x^2+1)(1)}{(x-3)^2}$$

$$= \frac{2x^2 - 6x - x^2 - 1}{(x-3)^2} = \frac{x^2 - 6x - 1}{(x-3)^2}$$

We now state a corollary of the theorem on the derivative of a quotient.

[3.4.3] COROLLARY

If $g \neq 0$ is a differentiable function, then

$$\frac{d}{dx}\left[\frac{1}{g(x)}\right] = \frac{-g'(x)}{[g(x)]^2} \tag{3}$$

For a proof, merely set $f(x) = 1$ in (2).

We may state (3) in words:

The derivative of the reciprocal of a function is the negative of the derivative of the function divided by the square of the function.

EXAMPLE 3

(a) $\dfrac{d}{dx}\left(\dfrac{1}{3x+5}\right) = \dfrac{-3}{(3x+5)^2}$ (b) $\dfrac{d}{dx}\left(\dfrac{1}{x^2+1}\right) = \dfrac{-2x}{(x^2+1)^2}$

We may use (3) to show that the formula $(d/dx)(x^n) = nx^{n-1}$ holds when n is a negative integer. Since in this event $-n$ is a positive integer, we see that

$$\frac{d}{dx}(x^n) \underset{\substack{\uparrow \\ \text{Definition}}}{=} \frac{d}{dx}\left(\frac{1}{x^{-n}}\right) \underset{\substack{\uparrow \\ \text{By (3)}}}{=} \frac{-\dfrac{d}{dx}(x^{-n})}{(x^{-n})^2} \underset{\substack{\uparrow \\ \text{By theorem (3.3.2)}}}{=} \frac{-(-n)x^{-n-1}}{x^{-2n}} \underset{\substack{\uparrow \\ \text{Simplify.}}}{=} nx^{n-1}$$

Now, we have the more general formula

$$\frac{d}{dx}(x^n) = nx^{n-1} \qquad \textbf{for } n \textbf{ any integer} \tag{4}$$

EXAMPLE 4

(a) $\frac{d}{dx}(x^{-1}) = -x^{-2} = -\frac{1}{x^2}$ (b) $\frac{d}{dx}\left(\frac{1}{x^2}\right) = \frac{d}{dx}(x^{-2}) = -2x^{-3} = -\frac{2}{x^3}$

(c) $\frac{d}{dx}\left(\frac{1}{x^3}\right) = \frac{d}{dx}(x^{-3}) = -3x^{-4} = -\frac{3}{x^4}$

(d) $\frac{d}{dx}\left(\frac{4}{x^5}\right) = \frac{d}{dx}(4x^{-5}) = 4(-5)x^{-6} = -\frac{20}{x^6}$

Note that, in general, the derivative of the reciprocal of a function f is *not* the reciprocal of the derivative. That is,

$$\frac{d}{dx}\frac{1}{f(x)} \neq \frac{1}{f'(x)}$$

One last remark: A change in the symbol used for the independent variable does not affect the formula. For example,

$$\frac{d}{dt}(t^{-2}) = -2t^{-3} = -\frac{2}{t^3} \qquad \frac{d}{ds}\left(\frac{3}{s^4}\right) = 3(-4)s^{-5} = -\frac{12}{s^5}$$

$$\frac{d}{du}(6u^3 - 5u^{-2}) = 18u^2 + 10u^{-3} = 18u^2 + \frac{10}{u^3}$$

As a matter of fact, each of the derivative formulas given so far can be written without reference to the independent variable of the function by using the prime notation. That is, for differentiable functions f and g, we have

$$(f+g)' = f' + g' \qquad (fg)' = fg' + f'g \qquad \left(\frac{f}{g}\right)' = \frac{gf' - fg'}{g^2}$$

EXERCISE 3.4

In Problems 1–24 find the derivative of the function f by using the formulas of this section.

1. $f(x) = (x^2 + 1)(x^3 - 1)$

2. $f(x) = (x^4 - 2)(x + 5)$

3. $f(x) = (3x^2 - 5)(2x + 1)$

4. $f(x) = (3x - 2)(4x + 5)$

5. $f(t) = (2t^5 - t)(t^3 + 1)$

6. $f(u) = (u^4 - 3u^2 + 1)(u^2 - u + 2)$

7. $f(t) = t^{-3}$

8. $f(u) = u^{-4}$

9. $f(x) = \frac{10}{x^4} + \frac{3}{x^2}$

10. $f(x) = \frac{2}{x^5} - \frac{3}{x^3}$

11. $f(s) = \frac{2s}{s+1}$

12. $f(z) = \frac{z+1}{2z}$

13. $f(x) = \frac{4x^2 - 2}{3x + 4}$

14. $f(x) = \frac{-3x^3 - 1}{2x^2 + 1}$

15. $f(t) = 3t + \frac{1}{3t}$

16. $f(u) = 4u - \frac{1}{4u}$

17. $f(u) = \frac{1 - 2u}{1 + 2u}$

18. $f(w) = \frac{1 - w^2}{1 + w^2}$

19. $f(x) = 3x^3 - \frac{1}{3x^2}$

20. $f(x) = x^5 - \frac{5}{x^5}$

21. $f(t) = \frac{1}{t} - \frac{1}{t^2} + \frac{1}{t^3}$

22. $f(v) = \left(\frac{1 - v}{v}\right)(1 - v^2)$

23. $f(w) = \frac{1}{w^3 - 1}$

24. $f(v) = \frac{1}{v^2 + 5}$

In Problems 25 and 26 find the slope of the tangent line to the function f at the point indicated. What is an equation of the tangent line?

25. $f(x) = \dfrac{x^3}{x+1}$, at $(1, \frac{1}{2})$

26. $f(x) = \dfrac{x^2}{x-1}$, at $(-1, -\frac{1}{2})$

In Problems 27 and 28 find those x, if any, at which the function f has a horizontal tangent line—that is, at which $f'(x) = 0$.

27. $f(x) = \dfrac{x^2}{x+1}$

28. $f(x) = \dfrac{x^2+1}{x}$

29. If $y = x^2(3x-2)$, find y' by:

(a) Using the derivative of a product formula
(b) Multiplying the two factors first and then differentiating
(c) Compare the answers from parts (a) and (b).

30. If $y = (x^2+2)(x-1)$, find y' by:

(a) Using the derivative of a product formula
(b) Multiplying the two factors first and then differentiating
(c) Compare the answers from parts (a) and (b).

31. The intensity of illumination I on a surface is inversely proportional to the square of the distance r from the surface to the source of light. If the intensity is 1000 units when the distance is 1 meter, find the rate of change of the intensity with respect to the distance when the distance is 10 meters.

32. Prove that if $g \neq 0$ is a differentiable function, then

$$\frac{d}{dx}\left[\frac{1}{g(x)}\right] = \frac{-g'(x)}{[g(x)]^2}$$

33. Prove that if f, g, and h are differentiable functions, then

$$\frac{d}{dx}[f(x)g(x)h(x)]$$
$$= f(x)g(x)h'(x) + f(x)g'(x)h(x) + f'(x)g(x)h(x)$$

From this, deduce that

$$\frac{d}{dx}[f(x)]^3 = 3[f(x)]^2 f'(x)$$

In Problems 34–39 use the result of Problem 33 to find dy/dx.

34. $y = (x^2+1)(x-1)(x+5)$

35. $y = (x-1)(x^2+5)(x^3-1)$

36. $y = (x^4+1)^3$

37. $y = (x^3+1)^3$

38. $y = (3x+1)\left(1+\dfrac{1}{x}\right)(x^{-5}+1)$

39. $y = \left(1-\dfrac{1}{x}\right)\left(1-\dfrac{1}{x^2}\right)\left(1-\dfrac{1}{x^3}\right)$

40. Write a formula for the derivative of the product of four differentiable functions. That is, find a formula for $(d/dx)[f_1(x)f_2(x)f_3(x)f_4(x)]$.

41. If f and g are differentiable functions, prove that if $k(x) = 1/[f(x)g(x)]$, then

$$k'(x) = -k(x)\left[\frac{f'(x)}{f(x)} + \frac{g'(x)}{g(x)}\right]$$

3.5 Higher-Order Derivatives

Earlier, we noted that the derivative of a differentiable function $y = f(x)$ is also a function, called the *derivative function* f'.

The derivative (if there is one) of the function f' is called the *second derivative* of f and is denoted by f''. For example, if

$$f(x) = 6x^3 - 3x^2 + 2x - 5$$

then

$$f'(x) = 18x^2 - 6x + 2$$
$$f''(x) = \frac{d}{dx}f'(x) = \frac{d}{dx}(18x^2 - 6x + 2) = 36x - 6$$

By continuing in this fashion, we can find the *third derivative* f''', the *fourth derivative* $f^{(4)}$, and so on, provided that these derivatives exist. These are collectively called *higher-order derivatives*. For example, the first, second, third, and fourth derivatives of

$$f(x) = x^4 + 3x^3 - 2x^2 + 5x - 6$$

are

$$f'(x) = 4x^3 + 9x^2 - 4x + 5$$

$$f''(x) = \frac{d}{dx}f'(x) = 12x^2 + 18x - 4$$

$$f'''(x) = \frac{d}{dx}f''(x) = 24x + 18$$

$$f^{(4)}(x) = 24$$

All derivatives for this function of order 5 or more equal 0.

The result obtained in this example can be generalized: For a polynomial function f of degree n, we have

$$f(x) = a_nx^n + a_{n-1}x^{n-1} + \cdots + a_1x + a_0$$
$$f'(x) = na_nx^{n-1} + (n-1)a_{n-1}x^{n-2} + \cdots + a_1$$

Thus, the first derivative of a polynomial function of degree n is a polynomial function of degree $(n - 1)$. By continuing the differentiation process, it follows that the nth-order derivative of f is

$$f^{(n)}(x) = n(n-1)(n-2)\cdots(3)(2)(1)a_n$$

a polynomial of degree 0—a constant function. Therefore, all derivatives of order greater than n will equal 0.

In some applications it is important to find both the first and second derivatives of a function and to solve for those numbers x that make these derivatives equal 0.

EXAMPLE 1

For $f(x) = 4x^3 - 12x^2 + 2$, find those x, if any, at which $f'(x) = 0$. For what numbers x will $f''(x) = 0$?

Solution

$$f'(x) = 12x^2 - 24x = 12x(x - 2) = 0 \quad \text{when} \quad x = 0 \quad \text{or} \quad x = 2$$

$$f''(x) = 24x - 24 = 24(x - 1) = 0 \quad \text{when} \quad x = 1$$

Other Forms of Notation

The symbols f', f'', and so on, for higher-order derivatives of $y = f(x)$ have parallels in the Leibniz notation:

$$y' = f'(x) = \frac{dy}{dx} = \frac{d}{dx} f(x)$$

$$y'' = f''(x) = \frac{d^2y}{dx^2} = \frac{d^2}{dx^2} f(x)$$

$$y''' = f'''(x) = \frac{d^3y}{dx^3} = \frac{d^3}{dx^3} f(x)$$

$$y^{(n)} = f^{(n)}(x) = \frac{d^ny}{dx^n} = \frac{d^n}{dx^n} f(x)$$

Acceleration in Rectilinear Motion

Suppose the position of an object at time t is a distance s from the origin, where s is given as a function of t—say, as $s = f(t)$. Then from definition (3.1.1), the first derivative ds/dt is the velocity v of the particle.

[3.5.1] DEFINITION / *Acceleration.*

The *acceleration* a of this particle is defined as the rate of change of velocity with respect to time. That is,

$$a = \frac{dv}{dt} = \frac{d}{dt} v = \frac{d}{dt}\left(\frac{ds}{dt}\right) = \frac{d^2s}{dt^2}$$

In other words, acceleration is the second derivative of the function $s = f(t)$ with respect to time.

EXAMPLE 2

A ball is thrown vertically upward with an initial velocity of 19.6 meters per second. The distance s (in meters) of the ball above the ground is $s = -4.9t^2 + 19.6t$, where t is the number of seconds elapsed from the moment that the ball is thrown.

(a) What is the velocity of the ball at the end of 1 second?
(b) When will the ball reach its highest point?
(c) What is the maximum height the ball reaches?
(d) What is the acceleration of the ball at any time t?
(e) How long is the ball in the air?
(f) What is the velocity of the ball upon impact?
(g) What is the total distance traveled by the ball?

Solution

(a) $v = \frac{ds}{dt} = -9.8t + 19.6$

At $t = 1$, $v = 9.8$ meters per second.

(b) The ball will reach its highest point when $v = 0$.

$$v = -9.8t + 19.6 = 0 \qquad \text{when} \qquad t = 2 \text{ seconds}$$

(c) At $t = 2$, $s = -4.9(4) + 19.6(2) = 19.6$ meters.

(d) $a = \dfrac{d^2s}{dt^2} = -9.8$ meters per second per second

(e) We can answer this question in two ways. First, since it takes 2 seconds for the ball to reach its maximum height, it follows that it will take another 2 seconds to reach the ground, for a total time of 4 seconds in the air. The second way is to set $s = 0$ and solve for t:

$$-4.9t^2 + 19.6t = 0$$

$$t = 0 \qquad \text{or} \qquad t = \frac{19.6}{4.9} = 4$$

The ball is at ground level when $t = 0$ and when $t = 4$.

(f) Upon impact, $t = 4$. Hence, when $t = 4$,

$$v = (-9.8)(4) + 19.6 = -19.6 \text{ meters per second}$$

The minus sign here indicates that the direction of the velocity is downward.

(g) The total distance traveled is

$$\text{Distance up} + \text{Distance down} = 19.6 + 19.6 = 39.2 \text{ meters}$$

See Figure 15 for an illustration.

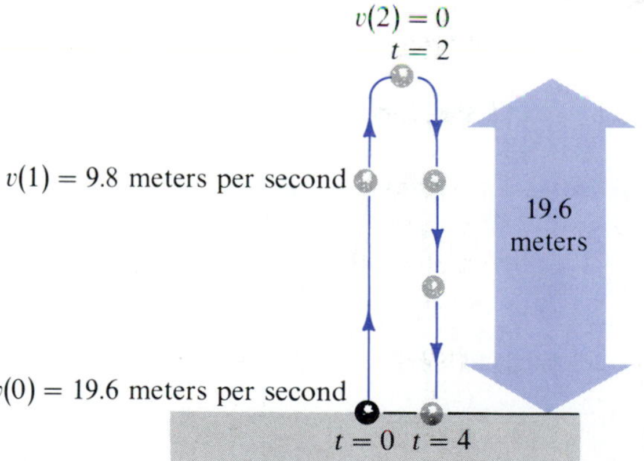

Figure 15

For Example 2, the acceleration of the ball is constant. This is approximately true for all falling bodies provided air resistance is ignored. In fact, the constant is even the same for all falling bodies, as Galileo (1564–1642) discovered in the sixteenth century. We can use calculus to see this. Galileo found by experimentation that all falling bodies obey the law that the distance they fall when dropped is proportional to the square of the time t

it takes to fall that distance. Of great importance is the fact that the constant of proportionality c is the same for all bodies. Thus, Galileo's law states that the distance s a body falls in time t is given by

$$s = -ct^2$$

The reason for the minus sign is that the body is falling and we have chosen our coordinate system so that the positive s direction is up, along the y-axis.

The velocity v of this freely falling body is

$$v = \frac{ds}{dt} = -2ct$$

and its acceleration a is

$$a = \frac{dv}{dt} = \frac{d^2s}{dt^2} = -2c$$

Thus, the acceleration of a freely falling body is a constant. Usually, we denote this constant by $-g$ so that

$$a = -g$$

The number g is called the *acceleration of gravity*. For our planet, g may be approximated by 32 feet per second per second, or 980 centimeters per second per second.* On the planet Jupiter, $g \approx 2600$ centimeters per second per second, and on our moon, $g \approx 160$ centimeters per second per second.

EXERCISE 3.5

In Problems 1–12 find f' and f'' for each function.

1. $f(x) = 2x + 5$

2. $f(x) = 3x + 2$

3. $f(x) = 3x^2 + x - 2$

4. $f(x) = -5x^2 - 3x$

5. $f(x) = x + \dfrac{1}{x}$

6. $f(x) = x - \dfrac{1}{x}$

7. $f(t) = \dfrac{t}{t + 1}$

8. $f(u) = \dfrac{u + 1}{u}$

9. $f(x) = \dfrac{x^2}{x + 1}$

10. $f(x) = \dfrac{x^3}{x - 1}$

11. $f(x) = \dfrac{4}{3x + 5}$

12. $f(x) = \dfrac{10}{-2x + 1}$

13. Find y' and y'' for the following:

(a) $y = 5x^4 - 2x^3 + x$ (b) $y = \dfrac{1}{x}$

(c) $y = (2x + 1)(x^3 + 5)$ (d) $y = \dfrac{2x - 5}{x}$

14. Find dy/dx and d^2y/dx^2 for the following:

(a) $y = -5x^3 + x^2 - 6$ (b) $y = \dfrac{5}{x^2}$

(c) $y = (3x - 5)(x^2 - 2)$ (d) $y = \dfrac{2 - 3x}{x}$

15. Find y''' for the following:

(a) $y = 4x^3 - 3x^2 + x$

(b) $y = ax^3 + bx^2 + cx + d$

16. Find d^3y/dx^3 for the following:

(a) $y = -10x^4 + x^3 - 4$

(b) $y = ax^4 + bx^3 + cx^2 + dx + e$

* The earth, as you know, is not perfectly round; it bulges slightly at the equator. But neither is it perfectly oval, and its mass is not distributed uniformly. As a result, the acceleration of any freely falling body varies slightly from these constants.

In Problems 17–22 find the indicated derivative.

17. $f^{(4)}(x)$ if $f(x) = x^3 - 3x^2 + 2x - 5$

18. $f^{(5)}(x)$ if $f(x) = 4x^3 + x^2 - 1$

19. $\dfrac{d^{20}}{dx^{20}}(8x^{19} - 2x^{14} + 2x^5)$

20. $\dfrac{d^{14}}{dx^{14}}(x^{13} - 2x^{10} + 5x^3 - 1)$

21. $\dfrac{d^8}{dx^8}(\frac{1}{8}x^8 - \frac{1}{7}x^7 + x^5 - x^3)$

22. $\dfrac{d^6}{dx^6}(x^6 + 5x^5 - 2x + 4)$

In Problems 23–26 find the velocity v and acceleration a of an object whose position s at time t is given.

23. $s = 16t^2 + 20t$

24. $s = 16t^2 + 10t + 1$

25. $s = 4.9t^2 + 4t + 4$

26. $s = 4.9t^2 + 5t$

27. Find the second derivative of $f(x) = x^2g(x)$, where g' and g'' exist.

28. Find the second derivative of $f(x) = g(x)/x$, where g' and g'' exist.

In Problems 29–32 the given function describes the position of a particle moving on a straight line.

(a) Find the velocity of the particle at any time t.
(b) When is the particle at rest?

29. $s(t) = 2 - 5t + t^2$

30. $s(t) = t^3 - \frac{9}{2}t^2 + 15t + 4$

31. $s(t) = \dfrac{1}{t^2 - 1}$

32. $s(t) = \dfrac{t}{t^2 + 1}$

33. An object is propelled vertically upward with an initial velocity of 39.2 meters per second. The distance s (in meters) of the object from the ground after t seconds is $s = -4.9t^2 + 39.2t$.

(a) What is the velocity of the object at any time t?
(b) When will the object reach its highest point?
(c) What is the maximum height?
(d) What is the acceleration of the object at any time t?
(e) How long is the object in the air?
(f) What is the velocity of the object upon impact?
(g) What is the total distance traveled by the object?

34. A ball is thrown vertically upward with an initial velocity of 80 feet per second. The distance s (in feet) of the ball from the ground after t seconds is $s = 6 + 80t - 16t^2$.

(a) What is the velocity of the ball after 2 seconds?
(b) When will the ball reach its highest point?
(c) What is the maximum height the ball reaches?
(d) What is the acceleration of the ball at any time t?
(e) How long is the ball in the air?
(f) What is the velocity of the ball upon impact?
(g) What is the total distance traveled by the ball?

35. Develop a formula for F'' if $F(x) = f(x)g(x)$. Assume that f', g', f'', and g'' exist.

36. Develop a formula for F'' if $F(x) = f(x)/g(x)$. Assume that f', g', f'', and g'' exist.

37. Show that if $f(x) = 1/x$, then

$$f^{(n)}(x) = \frac{(-1)^n(n)(n-1)(n-2)\cdots(3)(2)(1)}{x^{n+1}}$$

where $(-1)^n = 1$ if n is even, and $(-1)^n = -1$ if n is odd.

38. What is the nth derivative of $f(x) = 1/(2 + x)$ at 0?

39. A particle moves in a straight line according to the law $s = \dfrac{t}{t^2 - 1}$. Find its velocity and acceleration.

40. Find y'' in terms of y alone if $b^2x^2 + a^2y^2 = a^2b^2$.

Derivatives of Trigonometric Functions 3.6

In order to find the derivatives of the functions $y = \sin x$ and $y = \cos x$, we need the limits

$$\lim_{x \to 0} \frac{\sin x}{x} = 1 \quad \text{and} \quad \lim_{x \to 0} \frac{\cos x - 1}{x} = 0 \qquad \textbf{(1)}$$

which were established in Section 2.5.

Derivative of $y = \sin x$ and $y = \cos x$

[3.6.1] THEOREM / *Derivative of* $y = \sin x$.

The derivative of $y = \sin x$ is $y' = \cos x$. That is,

$$\frac{d}{dx} \sin x = \cos x \tag{2}$$

Proof We follow the procedure established in (2) of Section 3.3 and calculate the difference quotient for $f(x) = \sin x$:

$$\frac{f(x+h) - f(x)}{h} = \frac{\sin(x+h) - \sin x}{h}$$

Using the fact that $\sin(x + h) = \sin x \cos h + \cos x \sin h$, we have

$$\begin{aligned}\frac{f(x+h) - f(x)}{h} &= \frac{\sin x \cos h + \cos x \sin h - \sin x}{h} \\ &= \frac{\sin x \cos h - \sin x}{h} + \frac{\cos x \sin h}{h} \\ &= (\sin x)\left(\frac{\cos h - 1}{h}\right) + (\cos x)\left(\frac{\sin h}{h}\right)\end{aligned}$$

Now, using limit theorems for sums and products with the formulas in (1), we have

$$\begin{aligned}\frac{d}{dx}(\sin x) &= \lim_{h \to 0}\left[\sin x\left(\frac{\cos h - 1}{h}\right) + \cos x\left(\frac{\sin h}{h}\right)\right] \\ &= \sin x \lim_{h \to 0}\left(\frac{\cos h - 1}{h}\right) + \cos x \lim_{h \to 0} \frac{\sin h}{h} \\ &= (\sin x)(0) + (\cos x)(1) = \cos x\end{aligned}$$

EXAMPLE 1

Find y' if:

(a) $y = x + 4 \sin x$ (b) $y = x^2 \sin x$ (c) $y = \dfrac{\sin x}{x}$

Solution

(a) We use the rule for differentiating a sum to get

$$y' = 1 + 4 \cos x$$

(b) We use the rule for differentiating a product to get

$$y' = 2x \sin x + x^2 \cos x$$

(c) We use the rule for differentiating a quotient to get

$$y' = \frac{x \cos x - \sin x}{x^2}$$

We obtain the derivative of $y = \cos x$ the same way we obtained the derivative of $y = \sin x$. Thus, if $f(x) = \cos x$, then

$$\frac{f(x+h) - f(x)}{h} = \frac{\cos(x+h) - \cos x}{h}$$

$$= \frac{\cos x \cos h - \sin h \sin x - \cos x}{h}$$

$$= \cos x\left(\frac{\cos h - 1}{h}\right) - \sin x\left(\frac{\sin h}{h}\right)$$

Now,

$$\frac{d}{dx}\cos x = \lim_{h \to 0} \frac{f(x+h) - f(x)}{h} = (\cos x)(0) - (\sin x)(1) = -\sin x$$

We have established the following result:

[3.6.2] THEOREM / *Derivative of* $y = \cos x$.
The derivative of $y = \cos x$ is $y' = -\sin x$. That is,

$$\frac{d}{dx}\cos x = -\sin x \tag{3}$$

EXAMPLE 2

Find y' if:

(a) $y = x^2 \cos x$ (b) $y = \dfrac{\cos x}{1 - \sin x}$

Solution

(a) $y' = 2x \cos x + x^2(-\sin x) = 2x \cos x - x^2 \sin x$

(b) $y' = \dfrac{(1 - \sin x)(-\sin x) - (\cos x)(-\cos x)}{(1 - \sin x)^2} = \dfrac{-\sin x + \sin^2 x + \cos^2 x}{(1 - \sin x)^2}$

$$= \frac{1 - \sin x}{(1 - \sin x)^2} = \frac{1}{1 - \sin x}$$

Derivatives of Other Trigonometric Functions

EXAMPLE 3

Establish the formulas:

(a) $\dfrac{d}{dx}\tan x = \sec^2 x$ **(b)** $\dfrac{d}{dx}\sec x = \sec x \tan x$ **(4)**

(c) $\dfrac{d}{dx}\cot x = -\csc^2 x$ **(d)** $\dfrac{d}{dx}\csc x = -\csc x \cot x$

Solution

(a) $$\frac{d}{dx}\tan x = \frac{d}{dx}\frac{\sin x}{\cos x} = \frac{\cos x \cos x - \sin x(-\sin x)}{\cos^2 x}$$

$$= \frac{\cos^2 x + \sin^2 x}{\cos^2 x} = \frac{1}{\cos^2 x} = \sec^2 x$$

(b) $$\frac{d}{dx}\sec x = \frac{d}{dx}\frac{1}{\cos x} \underset{\uparrow}{=} \frac{-\frac{d}{dx}\cos x}{(\cos x)^2}$$

By corollary (3.4.3)

$$= \frac{\sin x}{\cos^2 x} = \left(\frac{1}{\cos x}\right)\left(\frac{\sin x}{\cos x}\right) = \sec x \tan x$$

(c) $$\frac{d}{dx}\cot x = \frac{d}{dx}\frac{\cos x}{\sin x} = \frac{(\sin x)[(d/dx)\cos x] - (\cos x)[(d/dx)\sin x]}{(\sin x)^2}$$

$$= \frac{(\sin x)(-\sin x) - (\cos x)(\cos x)}{(\sin x)^2}$$

$$= \frac{-(\sin^2 x + \cos^2 x)}{(\sin x)^2}$$

$$= \frac{-1}{\sin^2 x} = -\csc^2 x$$

(d) In Problem 56 in Exercise 3.6 you are asked to establish this formula.

EXAMPLE 4

Find y' if $y = x^3 \tan x$.

Solution

$$y' = 3x^2 \tan x + x^3 \sec^2 x$$

EXAMPLE 5

Find $f''\left(\frac{\pi}{4}\right)$ if $f(x) = \sec x$.

Solution

$$f'(x) = \sec x \tan x$$

$$f''(x) = \sec x \tan x \tan x + \sec x \sec^2 x = \sec x \tan^2 x + \sec^3 x$$

Thus,

$$f''\left(\frac{\pi}{4}\right) = (\sqrt{2})(1)^2 + (\sqrt{2})^3 = 3\sqrt{2}$$

EXERCISE 3.6

In Problems 1–24 find y'.

1. $y = 3 \sin x - 2 \cos x$

2. $y = 4 \tan x + \sin x$

3. $y = \sin x \cos x$

4. $y = x \cos x$

5. $y = \sec x \tan x$

6. $y = x \tan x$

7. $y = \dfrac{\cot x}{x}$

8. $y = \dfrac{\csc x}{x}$

9. $y = x^2 \tan x$

10. $y = x^2 \sin x$

11. $y = x \tan x - 3 \sec x$

12. $y = x \sec x + 2 \cot x$

13. $y = \dfrac{\sin x}{1 - \cos x}$

14. $y = \dfrac{x}{\cos x}$

15. $y = \dfrac{\sin x}{1 + x}$

16. $y = \dfrac{\tan x}{1 + x}$

17. $y = \dfrac{\sin x + \cos x}{\sin x - \cos x}$

18. $y = \dfrac{\sin x - \cos x}{\sin x + \cos x}$

19. $y = \dfrac{\sec x}{1 + x \sin x}$

20. $y = \dfrac{\csc x}{1 + x \cos x}$

21. $y = \csc x \cot x$

22. $y = \tan x \cos x$

23. $y = \dfrac{1 + \tan x}{1 - \tan x}$

24. $y = \dfrac{\csc x - \cot x}{\csc x + \cot x}$

In Problems 25–36 find y''.

25. $y = \sin x$

26. $y = \cos x$

27. $y = \sec x$

28. $y = \csc x$

29. $y = x \sin x$

30. $y = x \cos x$

31. $y = 2 \sin x - 3 \cos x$

32. $y = 3 \sin x + 4 \cos x$

33. $y = x^2 \sin x$

34. $y = x^2 \cos x$

35. $y = a \sin x + b \cos x$

36. $y = x \sec x$

In Problems 37–40 find the derivative of f at c.

37. $f(x) = 2 \sin x + \cos x, \quad c = \dfrac{\pi}{2}$

38. $f(x) = x - \sin x, \quad c = \pi$

39. $f(x) = \dfrac{\cos x}{1 + \sin x}, \quad c = \dfrac{\pi}{3}$

40. $f(x) = \dfrac{\sin x}{1 + \cos x}, \quad c = \dfrac{5\pi}{6}$

In Problems 41–48 find the equation of the tangent line to the graph of f at the indicated point.

41. $f(x) = \sin x$, at $(0, 0)$

42. $f(x) = \sin x$, at $\left(\dfrac{\pi}{6}, \dfrac{1}{2}\right)$

43. $f(x) = \cos x$, at $\left(\dfrac{\pi}{3}, \dfrac{1}{2}\right)$

44. $f(x) = \cos x$, at $(0, 1)$

45. $f(x) = \tan x$, at $(0, 0)$

46. $f(x) = \tan x$, at $\left(\dfrac{\pi}{4}, 1\right)$

47. $f(x) = \sin x + \cos x$, at $\left(\dfrac{\pi}{4}, \sqrt{2}\right)$

48. $f(x) = \sin x - \cos x$, at $\left(\dfrac{\pi}{4}, 0\right)$

In Problems 49 and 50 find the nth derivative of each function.

49. $f(x) = \sin x$

50. $f(x) = \cos x$

51. What is $\displaystyle\lim_{h \to 0} \frac{\cos\left(\dfrac{\pi}{2} + h\right) - \cos \dfrac{\pi}{2}}{h}$?

52. What is $\displaystyle\lim_{h \to 0} \frac{\sin(\pi + h) - \sin \pi}{h}$?

53. If $y = A \sin t + B \cos t$, where A and B are constants, show that $y'' + y = 0$.

54. If $y = \sin x$ and $y^{(n)}$ is the nth derivative of y with respect to x, find the smallest positive integer n for which $y^{(n)} = y$.

55. Use the identity

$$\sin A - \sin B = 2 \cos \frac{A+B}{2} \sin \frac{A-B}{2}$$

with $A = x + h$, $B = x$, to prove that

$$\frac{d}{dx} \sin x = \lim_{h \to 0} \frac{\sin(x+h) - \sin x}{h} = \cos x$$

56. Establish formula (d) in Example 3:

$$\frac{d}{dx} \csc x = -\csc x \cot x$$

57. Let $f(x) = \cos x$. Show that finding $f'(0)$ is the same as finding $\lim_{x \to 0} \frac{\cos x - 1}{x}$.

58. Let $f(x) = \sin x$. Show that finding $f'(0)$ is the same as finding $\lim_{x \to 0} \frac{\sin x}{x}$.

59. A particle moves in a straight line according to the law $s = \frac{1}{8} \cos 4\pi t$. Find its velocity and acceleration.

3.7 The Chain Rule

We would have a difficult time using the differentiation formulas developed so far to compute the derivative of the function

$$y = (x^3 - 2x + 1)^{100}$$

In this section we derive a result that will enable us to compute rather easily the derivative of this function as well as a large number of other functions. The idea behind the result is illustrated below.

Suppose that $y = f(u)$ is a function of u and $u = g(x)$ is a function of x. That is, y is a composite function $y = f(g(x))$ or $y = (f \circ g)(x)$. What then is the derivative of $f(g(x))$? It turns out that the derivative of the composite function $f \circ g$ can be written as the product of $f'(g(x))$ and $g'(x)$. This fact is one of the most important of the differentiation rules and is called the chain rule. The following discussion is another way to motivate the chain rule when we interpret derivatives as rates of change.

View $du/dx = g'(x)$ as the rate of change of u with respect to x, and $dy/du = f'(u)$ as the rate of change of y with respect to u. We now may ask: What is the rate of change of y with respect to x? That is, what is dy/dx? In this case the chain rule has a simple interpretation. If u is moving du/dx times as fast as x, and if y is moving dy/du times as fast as y, then y is moving $(du/dx)(dy/du)$ times as fast as x; thus,

$$\frac{dy}{dx} = \frac{dy}{du} \cdot \frac{du}{dx}$$

[3.7.1] THEOREM / ***Chain Rule.***

If f and g are differentiable functions, then the composite function $f \circ g$ is differentiable, and

$$\frac{d}{dx} f(g(x)) = f'(g(x)) \cdot g'(x) \tag{1}$$

The statement of this theorem in the Leibniz notation is more elegant. If $y = f(u)$ and $u = g(x)$ are both differentiable functions, then

$$\frac{dy}{dx} = \frac{dy}{du} \cdot \frac{du}{dx} \tag{2}$$

Partial Proof of the Chain Rule If x is changed by a small amount Δx and the corresponding change in $u = g(x)$ is Δu, we know that

$$g'(x) = \frac{du}{dx} = \lim_{\Delta x \to 0} \frac{\Delta u}{\Delta x}$$

Corresponding to the small change Δu is a change Δy in $y = f(u)$. If $\Delta u \neq 0$, then

$$f'(u) = \frac{dy}{du} = \lim_{\Delta u \to 0} \frac{\Delta y}{\Delta u}$$

To calculate dy/dx, we write

$$\frac{dy}{dx} = \lim_{\Delta x \to 0} \frac{\Delta y}{\Delta x} = \lim_{\Delta x \to 0} \frac{\Delta y}{\Delta u} \cdot \frac{\Delta u}{\Delta x} = \left(\lim_{\Delta x \to 0} \frac{\Delta y}{\Delta u}\right)\left(\lim_{\Delta x \to 0} \frac{\Delta u}{\Delta x}\right)$$

In the first factor we replace $\Delta x \to 0$ by $\Delta u \to 0$, since the differentiable function g is continuous, and so $\Delta u \to 0$ as $\Delta x \to 0$. Thus,

$$\frac{dy}{dx} = \left(\lim_{\Delta u \to 0} \frac{\Delta y}{\Delta u}\right)\left(\lim_{\Delta x \to 0} \frac{\Delta u}{\Delta x}\right) = \frac{dy}{du} \cdot \frac{du}{dx}$$

The proof is not complete. The Δu determined by Δx could well be 0, and division by 0 is not allowed. This part of the proof is given in Appendix I.

EXAMPLE 1

Find the derivative of the composite function $f \circ g$ for $y = f(u) = u^2 - 4$ and $u = g(x) = \cos x$.

Solution

$$\frac{dy}{du} = \frac{d}{du} f(u) = \frac{d}{du}(u^2 - 4) = 2u = 2 \cos x \qquad \frac{du}{dx} = \frac{d}{dx} g(x) = \frac{d}{dx} \cos x = -\sin x$$

↑ Remember, $u = \cos x$

Thus, by (2),

$$\frac{d}{dx}(f \circ g)(x) = \frac{dy}{dx} = \frac{dy}{du} \cdot \frac{du}{dx} = 2 \cos x(-\sin x) = -2 \cos x \sin x$$

EXAMPLE 2

Find the derivative of $y = (x^3 - 2x + 1)^{100}$:

(a) Using (1)
(b) Using (2)

Solution

(a) $y = (x^3 - 2x + 1)^{100}$ is a composite function where

$$f(x) = x^{100} \quad \text{and} \quad g(x) = x^3 - 2x + 1$$

Then

$$f'(x) = 100x^{99} \quad \text{and} \quad g'(x) = 3x^2 - 2$$

So from (1),

$$\begin{aligned} y' &= f'(g(x)) \cdot g'(x) \\ &= 100(x^3 - 2x + 1)^{99} \cdot (3x^2 - 2) \end{aligned}$$

(b) Set $y = u^{100}$ and $u = x^3 - 2x + 1$. Then

$$\frac{dy}{du} = 100u^{99} \quad \text{and} \quad \frac{du}{dx} = 3x^2 - 2$$

So from (2),

$$\frac{dy}{dx} = \frac{dy}{du} \cdot \frac{du}{dx} = 100u^{99}(3x^2 - 2) = 100(x^3 - 2x + 1)^{99}(3x^2 - 2)$$

When using (1) to find the derivative of a composite function, we need to decide how to choose f and g. In Example 2, we chose $g(x)$ so that it became a simple part of the function to be differentiated, and this determined what $f(x)$ had to be. Furthermore, to compute $(d/dx)f(g(x))$, we first computed $f'(x)$ and then replaced x by $g(x)$ to obtain $f'(g(x))$. With practice, the choice for f and g can be made mentally and not written down.

It may be helpful to express the formula $(d/dx)f(g(x)) = f'(g(x))g'(x)$ in words. If we refer to f as the outside function and g as the inside function, then (1) states: **The derivative of $f(g(x))$ is the derivative of the outside function evaluated at the inside function times the derivative of the inside function.**

In Example 2, we note that the computations of the derivative of the composite function $y = (x^3 - 2x + 1)^{100}$ are the same using either (1) or (2). The choice of which formula to use is a personal one. Some people prefer to use (1), since it does not require the introduction of a new variable, whereas others find it easier to use (2).

EXAMPLE 3

Find the derivative of $y = \dfrac{1}{(x^3 - 1)^{50}}$:

(a) Using (1)
(b) Using (2)

Solution

(a) $y = (x^3 - 1)^{-50}$ is a composite function where

$$\underbrace{f(x) = x^{-50}}_{\text{Outside function}} \quad \text{and} \quad \underbrace{g(x) = x^3 - 1}_{\text{Inside function}}$$

Then

$$\frac{d}{dx}f(g(x)) = \underbrace{-50(x^3-1)^{-51}}_{\substack{\text{Derivative}\\ \text{of the}\\ \text{outside}\\ \text{evaluated at}\\ \text{the inside}}} \cdot \underbrace{3x^2}_{\substack{\text{Derivative}\\ \text{of the}\\ \text{inside}}} = -150x^2(x^3-1)^{-51} = \frac{-150x^2}{(x^3-1)^{51}}$$

(b) Set $y = u^{-50}$ and $u = x^3 - 1$. Then

$$\frac{dy}{du} = -50u^{-51} \qquad \text{and} \qquad \frac{du}{dx} = 3x^2$$

So from (2),

$$\frac{dy}{dx} = \frac{dy}{du} \cdot \frac{du}{dx} = -50u^{-51} \cdot 3x^2 = -150x^2(x^3-1)^{-51} = \frac{-150x^2}{(x^3-1)^{51}}$$

Power Rule for Functions

The chain rule is often used to establish other differentiation formulas. We start by showing how to obtain the formula for the derivative of a power of a function.

[3.7.2] THEOREM / *Power Rule for Functions.*

If g is a differentiable function and n is any integer, then

$$\frac{d}{dx}[g(x)]^n = n[g(x)]^{n-1}g'(x) \tag{3}$$

Proof To prove theorem (3.7.2), we use the chain rule. If

$$y = [g(x)]^n$$

we set

$$y = u^n \qquad \text{and} \qquad u = g(x)$$

Then

$$\frac{dy}{du} = nu^{n-1} = n[g(x)]^{n-1} \qquad \text{and} \qquad \frac{du}{dx} = g'(x)$$

By the chain rule, we have

$$\frac{d}{dx}[g(x)]^n = \frac{dy}{dx} = \frac{dy}{du} \cdot \frac{du}{dx} = n[g(x)]^{n-1}g'(x)$$

The alternative version of the power rule for functions is

$$\frac{d}{dx}(u^n) = nu^{n-1} \cdot \frac{du}{dx} \tag{4}$$

Notice the similarity between the power rule for functions (3.7.2) and theorem (3.3.2), $(d/dx)(x^n) = nx^{n-1}$; the main difference is the third factor, $g'(x)$.

EXAMPLE 4

Find the derivative of $f(x) = (x^2 + x + 1)^3$.

Solution

The function $f(x) = (x^2 + x + 1)^3$ is the function $g(x) = x^2 + x + 1$ raised to the power $n = 3$. By the power rule for functions,

$$\frac{d}{dx}f(x) = \frac{d}{dx}(x^2 + x + 1)^3 = (3)(x^2 + x + 1)^2\left[\frac{d}{dx}(x^2 + x + 1)\right]$$
$$= 3(x^2 + x + 1)^2(2x + 1)$$

Additional examples of the derivatives of composite functions using (3) are:

1. If $f(x) = (3 - x^3)^{-5}$, then
$$f'(x) = (-5)(3 - x^3)^{-6}(-3x^2) = 15x^2(3 - x^3)^{-6}$$

2. If $f(x) = (2x + 3)^2$, then
$$f'(x) = (2)(2x + 3)^1(2) = 4(2x + 3)$$

3. If $f(x) = (x^3 - 3x^2 + 1)^5$, then
$$f'(x) = (5)(x^3 - 3x^2 + 1)^4(3x^2 - 6x) = 15x(x - 2)(x^3 - 3x^2 + 1)^4$$

We often use other differentiation rules along with the power rule for functions to differentiate a function. Here is an example.

EXAMPLE 5

Find the derivative of $f(x) = x(x^2 + 1)^3$.

Solution

Here, the function is the product of x and $(x^2 + 1)^3$. Thus, we begin by using the rule for differentiating a product—that is,

$$f'(x) = (x)\left[\frac{d}{dx}(x^2 + 1)^3\right] + (x^2 + 1)^3$$

By applying the power rule for functions, we have

$$f'(x) = (x)[3(x^2 + 1)^2(2x)] + (x^2 + 1)^3$$
$$= 6x^2(x^2 + 1)^2 + (x^2 + 1)^3 = (x^2 + 1)^2(6x^2 + x^2 + 1)$$
$$= (x^2 + 1)^2(7x^2 + 1)$$

EXAMPLE 6

Find the derivative of $f(x) = \left(\dfrac{3x + 2}{4x^2 - 5}\right)^5$.

Solution

Here, f is the quotient $(3x + 2)/(4x^2 - 5)$ raised to the power 5. Thus, we begin by using the power rule for functions and then use the rule for differentiating a quotient.

$$f'(x) = (5)\left(\frac{3x+2}{4x^2-5}\right)^4\left[\frac{d}{dx}\left(\frac{3x+2}{4x^2-5}\right)\right]$$

Apply power rule for functions.

$$= (5)\left(\frac{3x+2}{4x^2-5}\right)^4\left[\frac{(4x^2-5)(3)-(3x+2)(8x)}{(4x^2-5)^2}\right]$$

Apply quotient rule.

$$= \frac{5(3x+2)^4(-12x^2-16x-15)}{(4x^2-5)^6}$$

Whenever a problem involves finding the derivative of a function raised to a power, either the power rule for functions or the chain rule may be used. However, the power rule for functions, as a special case of the chain rule, will not always work for composite functions involving nonalgebraic functions.

EXAMPLE 7

Find y' if: (a) $y = \sin x^2$ (b) $y = \sin^2 x$

Solution

(a) We use the chain rule with $y = \sin u$ and $u = x^2$. Then

$$y' = \frac{dy}{dx} = \frac{dy}{du} \cdot \frac{du}{dx} = (\cos u)(2x) = 2x \cos x^2$$

(b) Since $y = \sin^2 x = (\sin x)^2$, we may use either the power rule for functions or the chain rule. First, we try the power rule for functions:

$$y' = 2 \sin x\left(\frac{d}{dx} \sin x\right) = 2 \sin x \cos x$$

Using the chain rule, we set $y = u^2$ and $u = \sin x$:

$$y' = \frac{dy}{du} \cdot \frac{du}{dx} = 2u(\cos x) = 2 \sin x \cos x$$

Example 7 is an important one. Although the solution in part (b) may be obtained by either the power rule for functions or the chain rule, the solution in part (a) can be obtained only by using the chain rule. To summarize, whenever the power rule for functions can be used, so can the chain rule, but there are many times when only the chain rule will do the job!

In Example 7, part (a), we used the combination of the chain rule and the fact that $(d/dx) \sin x = \cos x$ to find the derivative of $y = \sin x^2$. In general, if $y = \sin u$, where u is a differentiable function, then by (2),

$$\frac{d}{dx}(\sin u) = \cos u \frac{du}{dx}$$

In a similar way we can derive the following general derivative formulas for the remaining trigonometric functions:

$$\frac{d}{dx}(\cos u) = -\sin u \frac{du}{dx}$$

$$\frac{d}{dx}(\tan u) = \sec^2 u \frac{du}{dx} \qquad \frac{d}{dx}(\cot u) = -\csc^2 u \frac{du}{dx}$$

$$\frac{d}{dx}(\sec u) = \sec u \tan u \frac{du}{dx} \qquad \frac{d}{dx}(\csc u) = -\csc u \cot u \frac{du}{dx}$$

In terms of prime notation, the chain rule formula is

$$(f \circ g)'(x) = f'(g(x))g'(x) \tag{5}$$

The next example illustrates a use for (5).

EXAMPLE 8

Suppose $h = f \circ g$. If

$$f(1) = 2 \qquad f'(1) = 3 \qquad f'(2) = -4$$
$$g(1) = 2 \qquad g'(1) = -3 \qquad g'(2) = 5$$

calculate $h'(1)$.

Solution

Based on (5), we have

$$h'(x) = [f'(g(x))][g'(x)]$$

When $x = 1$,

$$h'(1) = [f'(g(1))][g'(1)] = f'(2)g'(1) = (-4)(-3) = 12$$

Note that only a portion of the given information is required.

The formula

$$\frac{dy}{dx} = \frac{dy}{du} \cdot \frac{du}{dx} \qquad u = g(x)$$

can be extended. For example, if

$$y = f(u) \qquad u = g(v) \qquad v = h(x)$$

then the composite $y = (f \circ g \circ h)(x)$ is a function of x and

$$\frac{dy}{dx} = \frac{dy}{du} \cdot \frac{du}{dv} \cdot \frac{dv}{dx} \tag{6}$$

This "chain" of factors is the basis for the name *chain rule*.

EXAMPLE 9

Find dy/dx if $y = u^4$, $u = 4v^3 - 2$, and $v = 2/x^2$.

Solution

$$\frac{dy}{dx} = \frac{dy}{du} \cdot \frac{du}{dv} \cdot \frac{dv}{dx} = (4u^3)(12v^2)\left(-\frac{4}{x^3}\right) \underset{\uparrow\ u = 4v^3 - 2}{=} 4(4v^3 - 2)^3(12v^2)\left(-\frac{4}{x^3}\right)$$

$$\underset{\uparrow\ v = 2/x^2}{=} 4\left(\frac{32}{x^6} - 2\right)^3(12)\left(\frac{4}{x^4}\right)\left(-\frac{4}{x^3}\right) = \frac{-768}{x^7}\left(\frac{32}{x^6} - 2\right)^3$$

EXAMPLE 10

Find y' if: (a) $y = 5\cos^2(3x + 2)$ (b) $y = \sin^3 5x$

Solution

(a) We use the chain rule, where $y = 5u^2$, $u = \cos v$, and $v = (3x + 2)$:

$$y' = \frac{dy}{dx} = \frac{dy}{du} \cdot \frac{du}{dv} \cdot \frac{dv}{dx} = 10u(-\sin v)(3)$$

$$= -30 \cos v \sin v = -30 \cos(3x + 2)\sin(3x + 2)$$

(b) We use the chain rule where $y = u^3$, $u = \sin v$, and $v = 5x$:

$$y' = \frac{dy}{dx} = \frac{dy}{du} \cdot \frac{du}{dv} \cdot \frac{dv}{dx} = (3u^2)(\cos v)(5) = 15 \sin^2 v \cos v$$

$$= 15 \sin^2 5x \cos 5x$$

Application to Rectilinear Motion

An important formula involving acceleration is derived by using the chain rule. We have seen that the acceleration of a body at time t is given by $a = dv/dt$, where v is the velocity of the object. Sometimes it is useful to express the velocity v as a function of the distance s. In this case, $v = v(s)$ and the acceleration a may be expressed as

$$a = \frac{dv}{dt} \underset{\uparrow}{=} \frac{dv}{ds} \cdot \frac{ds}{dt} \underset{\uparrow}{=} v\frac{dv}{ds}$$

Apply chain rule to $v = v(s)$. $\qquad v = ds/dt$

Thus, we have the following alternate formula for acceleration:

$$a = v\frac{dv}{ds}$$

We shall have occasion to refer to this formula later.

EXERCISE 3.7

In Problems 1–8 for the functions f and g, find the derivative of $y = (f \circ g)(x)$ by using the chain rule.

1. $y = f(u) = u^5, \quad u = g(x) = x^3 + 1$

2. $y = f(u) = u^3, \quad u = g(x) = 2x + 5$

3. $y = f(u) = \dfrac{u}{u+1}, \quad u = g(x) = x^2 + 1$

4. $y = f(u) = \dfrac{u-1}{u}, \quad u = g(x) = x^2 - 1$

5. $y = f(u) = (u+1)^2, \quad u = g(x) = \dfrac{1}{x}$

6. $y = f(u) = (u^2 - 1)^3, \quad u = g(x) = \dfrac{1}{x+2}$

7. $y = f(u) = (u^3 - 1)^5, \quad u = g(x) = x^{-2}$

8. $y = f(u) = (u^2 + 4)^4, \quad u = g(x) = x^{-2}$

In Problems 9–56 find the derivative of the function.

9. $f(x) = (3x+5)^2$

10. $f(x) = (2x-5)^3$

11. $f(x) = (6x-5)^{-3}$

12. $f(x) = (4x+1)^{-2}$

13. $f(x) = (x^2+5)^4$

14. $f(x) = (x^3-2)^5$

15. $f(t) = (t^5 - t^2 + t)^7$

16. $f(u) = (u^4 - u^2 + u - 1)^6$

17. $f(x) = \left(x - \dfrac{1}{x}\right)^3$

18. $f(x) = \left(x + \dfrac{1}{x}\right)^3$

19. $f(z) = \left(\dfrac{z}{z+1}\right)^3$

20. $f(w) = \dfrac{(3w+1)^4}{w}$

21. $f(x) = \tan^2 x$

22. $f(x) = \sec^3 x$

23. $f(t) = \sin^2 t - \cos^2 t$

24. $f(z) = (\sin z + \cos z)^2$

25. $f(x) = \left(\dfrac{3x^2+1}{x^3-1}\right)^2$

26. $f(x) = \left(\dfrac{x^3+4}{x^2-1}\right)^3$

27. $f(x) = (x^2+4)^2(2x^3-1)^3$

28. $f(x) = (x^2-2)^3(3x^4+1)^2$

29. $f(x) = [5x + (3x + 6x^2)^3]^4$

30. $f(x) = [2x - (3x^2 + 4x^3)^4]^2$

31. $f(x) = \dfrac{1}{x^4 - 2x + 1}$

32. $f(x) = \dfrac{3}{x^5 + 2x^2 - 3}$

33. $f(x) = \dfrac{1}{(x^2 - 2x + 1)^5}$

34. $f(x) = \dfrac{1}{(x^3 - 4x + 1)^{10}}$

35. $y = \sin 4x$

36. $y = \cos 5x$

37. $y = \tan 5x$

38. $y = \csc(x^3 + 1)$

39. $y = \sin(3x^2 + 4)$

40. $y = 5\cos(x^2 - 4)$

41. $y = 4\sin^2 3x$

42. $y = 2\cos^2(x^2)$

43. $y = 2\sin(x^2 + 2x - 1)$

44. $y = \frac{1}{2}\cos(x^3 - 2x + 5)$

45. $y = \csc^2(1 + 3x)$

46. $y = \cot^2(1 + 3x^2)$

47. $y = x^2 \sin 4x$

48. $y = x^2 \cos 4x$

49. $y = \sin \dfrac{1}{x}$

50. $y = \sin \dfrac{3}{x}$

51. $y = x^2 \sin^2 x$

52. $y = x \sin \dfrac{1}{x}$

53. $f(x) = \dfrac{(1 - x^2)^{11}}{1 + x^2}$

54. $f(x) = \dfrac{(3x^3 - 2x)^{-3}}{(x^2 + 5)^{-2}}$

55. $f(x) = (1 + \cos^3 x^2)^2$

56. $f(x) = [x \sin x - x^2 \cos 3x]^{12}$

57. Find the derivative y' of $y = (x^3 + 1)^2$ by:
(a) Using the chain rule
(b) Using the power rule for functions
(c) Expanding and then differentiating
(d) Compare the answers from parts (a)–(c).

58. Follow the directions in Problem 57 for the function $y = (x^2 - 2)^3$.

In Problems 59–64 find the equation of the tangent line at the given point.

59. $y = (x^2 - 2x + 1)^5$, at $(1, 0)$

60. $y = (x^3 - x^2 + x - 1)^{10}$, at $(0, 1)$

61. $y = \dfrac{x}{(x^2 - 1)^3}$, at $(2, \frac{2}{27})$

62. $y = \dfrac{x^2}{(x^2 - 1)^2}$, at $(2, \frac{4}{9})$

63. $y = \sin 2x + \cos x$, at $(0, 1)$

64. $y = \cot^2 x$, at $\left(\dfrac{\pi}{4}, 1\right)$

In Problems 65–70 find dy/dx by using (6).

65. $y = u^3$, $u = 3v^2 + 1$, $v = \dfrac{4}{x^2}$

66. $y = 3u$, $u = 3v^2 - 4$, $v = \dfrac{1}{x}$

67. $y = u^2 + 1$, $u = \dfrac{4}{v}$, $v = x^2$

68. $y = u^3 - 1$, $u = -\dfrac{2}{v}$, $v = x^3$

69. $y = \tan^3 2x$

70. $y = \sec^2(x^3)$

In Problems 71 and 72 compute the indicated derivatives.

71. $\dfrac{d^2}{dx^2} \cos(x^5)$

72. $\dfrac{d^3}{dx^3} \sin^3 x$

73. Find the nth-order derivative of $f(x) = (2x + 3)^n$.

74. Find the nth-order derivative of $f(x) = 1/(3x - 4)$.

75. Prove that if a differentiable function f is odd, then f' is even. [*Hint:* $f(x) = -f(-x)$.]

76. Prove that if a differentiable function f is even, then f' is odd.

In Problems 77–84 find the indicated derivative.

77. $\dfrac{d}{dx} f(x^2 + 1)$ [*Hint:* Let $u = x^2 + 1$.]

78. $\dfrac{d}{dx} f(1 - x^2)$

79. $\dfrac{d}{dx} f\left(\dfrac{x + 1}{x - 1}\right)$

80. $\dfrac{d}{dx} f\left(\dfrac{1 - x}{1 + x}\right)$

81. $\dfrac{d}{dx} f(\sin x)$

82. $\dfrac{d}{dx} f(\tan x)$

83. $\dfrac{d^2}{dx^2} f(\cos x)$

84. $\dfrac{d^2}{dx^2} f(\sec x)$

85. Suppose $h = f \circ g$. If $f'(2) = 6$, $f(1) = 4$, $g(1) = 2$, and $g'(1) = -2$, calculate $h'(1)$.

86. Suppose $h = f \circ g$. If $f'(3) = 4$, $f(1) = 1$, $g(1) = 3$, $g'(1) = 3$, calculate $h'(1)$.

87. If $y = u^5 + u$ and $u = 4x^3 + x - 4$, find dy/dx at $x = 1$.

88. The resistance R (measured in ohms) of an 80 meter long electric wire of radius x (in centimeters) is given by the formula $R = 0.0048/x^2$. In turn, the radius x varies with the absolute temperature T according to the rule $x = 0.1991 + 0.000003T$. How fast does R change with respect to T when $T = 320°$? (Do not eliminate x in obtaining your answer.)

89. A bullet is fired horizontally into a bale of paper. The distance s (in meters) the bullet travels into the bale of paper in t seconds is given by $s = 8 - (2 - t)^3$ for $0 \le t \le 2$. Find the velocity of the bullet after 1 second. Find the acceleration of the bullet at any time t.

90. If $y = A \sin \omega t + B \cos \omega t$, where A, B, and ω are constants, show that $y'' + \omega^2 y = 0$.

91. Find the acceleration of a car if its position s on a highway at time t is given by

$$s = \frac{80}{3}\left[t + \left(\frac{3}{\pi}\right) \sin \frac{\pi}{6} t\right]$$

92. Find the nth derivative of: (a) $f(x) = \sin ax$ (b) $f(x) = \cos ax$

93. Use the chain rule and the fact that

$\cos x = \sin\left(\frac{\pi}{2} - x\right)$ to show that $\frac{d}{dx} \cos x = -\sin x$.

94. The equation of motion of a particle is given by $s = A\cos(\omega t + \phi)$.

(a) Find the velocity of the particle at time t.
(b) When is the velocity 0?

95. If a function f has the properties that (1) $f(u + v) = f(u)f(v)$ for all choices of u and v, and (2) $f(x) = 1 + xg(x)$, where $\lim_{x\to 0} g(x) = 1$, show that $f' = f$.

96. If

$$\frac{d}{dx} f(x) = g(x) \qquad \text{and} \qquad h(x) = x^2$$

find $(d/dx)f(h(x))$.

97. If $f(x) = \sin x$ and $F(t) = f(t^2 - 1)$, find $F'(1)$.

3.8 Implicit Differentiation

So far we have considered only functions whose law of correspondence is expressed in the form $y = f(x)$. This expression of the relationship between x and y is said to be in *explicit form* because we have solved for the dependent variable y. If the functional relationship between the independent variable x and the dependent variable y is not of this form, we say that x and y are related *implicitly*. For example, x and y are related implicitly in the equations

$$3x + 4y - 5 = 0 \qquad 3y^2 + x^2 - 1 = 0 \qquad xy - 4 = 0$$

If x and y are related implicitly, and if it is possible to solve for y—say, as $y = f(x)$—then the replacement of y by $f(x)$ in the implicit equation results in an identity. For example, in the implicit equation $3x + 4y - 5 = 0$, the explicit form is $y = \frac{1}{4}(5 - 3x)$. This may be verified by showing that the replacement of y by $f(x)$ in $3x + 4y - 5 = 0$ results in an identity:

$$3x + 4[\tfrac{1}{4}(5 - 3x)] - 5 = 3x + 5 - 3x - 5 = 0$$

We will now develop a procedure for finding the derivative of y with respect to x when the functional relationship between x and y is given implicitly. This procedure, called *implicit differentiation*, is illustrated next.

EXAMPLE 1

Find dy/dx if $3x + 4y - 5 = 0$.

Solution

We begin by assuming that there is a differentiable function $y = f(x)$ implied by the above relationship. That is, the expression

$$3x + 4f(x) - 5 = 0$$

is an identity. We differentiate both sides of this identity with respect to x:

$$\frac{d}{dx}[3x + 4f(x) - 5] = \frac{d}{dx}(0)$$

$$\frac{d}{dx}(3x) + \frac{d}{dx}[4f(x)] - \frac{d}{dx}(5) = \frac{d}{dx}(0)$$

$$3 + (4)\left[\frac{d}{dx}f(x)\right] = 0$$

$$\frac{d}{dx}f(x) = -\frac{3}{4}$$

So, we have $dy/dx = -\frac{3}{4}$.

In Example 1 the function $y = f(x)$, whose existence was assumed, can be found by solving for y. In this case,

$$y = \frac{1}{4}(5 - 3x) = \frac{5}{4} - \frac{3x}{4}$$

so that $dy/dx = -\frac{3}{4}$, which agrees with the result obtained using implicit differentiation.

Often, though, it is very difficult or even impossible to actually solve for y in terms of x. Forthcoming examples illustrate this difficulty (try to solve for y and you will see). However, we still make the assumption that there is a differentiable function $y = f(x)$ that obeys the implicit equation.

In these examples we will use the power rule for functions,

$$\frac{d}{dx}[f(x)]^n = n[f(x)]^{n-1}f'(x) \qquad n \text{ an integer}$$

which, if $y = f(x)$, looks like this:

$$\frac{d}{dx}y^n = ny^{n-1}\frac{dy}{dx} \tag{1}$$

For example, if $y = f(x)$, then

$$\frac{d}{dx}y^2 = 2y\frac{dy}{dx} \qquad \text{and} \qquad \frac{d}{dx}y^3 = 3y^2\frac{dy}{dx}$$

EXAMPLE 2

Find dy/dx if $3x^2 + 4y^2 = 2x$.

Solution

We again assume that there is a differentiable function $y = f(x)$ so that when y is replaced by $f(x)$ in the expression $3x^2 + 4y^2 = 2x$, we obtain an identity in x. We proceed to differentiate both sides of this identity with respect to x:

$$\frac{d}{dx}(3x^2 + 4y^2) = \frac{d}{dx}(2x)$$

$$\frac{d}{dx}(3x^2) + \frac{d}{dx}(4y^2) = 2$$

$$6x + 4\left[\frac{d}{dx}(y^2)\right] = 2$$

By using (1), we obtain

$$6x + 4\left(2y\frac{dy}{dx}\right) = 2$$

$$8y\frac{dy}{dx} = 2 - 6x$$

This is a linear equation in dy/dx. Solving for dy/dx, we have

$$\frac{dy}{dx} = \frac{1 - 3x}{4y} \qquad \text{provided} \quad y \neq 0$$

In Examples 1 and 2 we assumed that the given relationships determine y as a differentiable function of x. This very plausible statement is in fact false, unless the proper interpretation is given to the statement.

To clarify the matter, we consider the following equation:

$$x^2 + y^2 = 1$$

If we solve this equation for y, we get

$$y = \pm\sqrt{1 - x^2}$$

which is not a function of x. See Figure 16(a). The equation does, however,

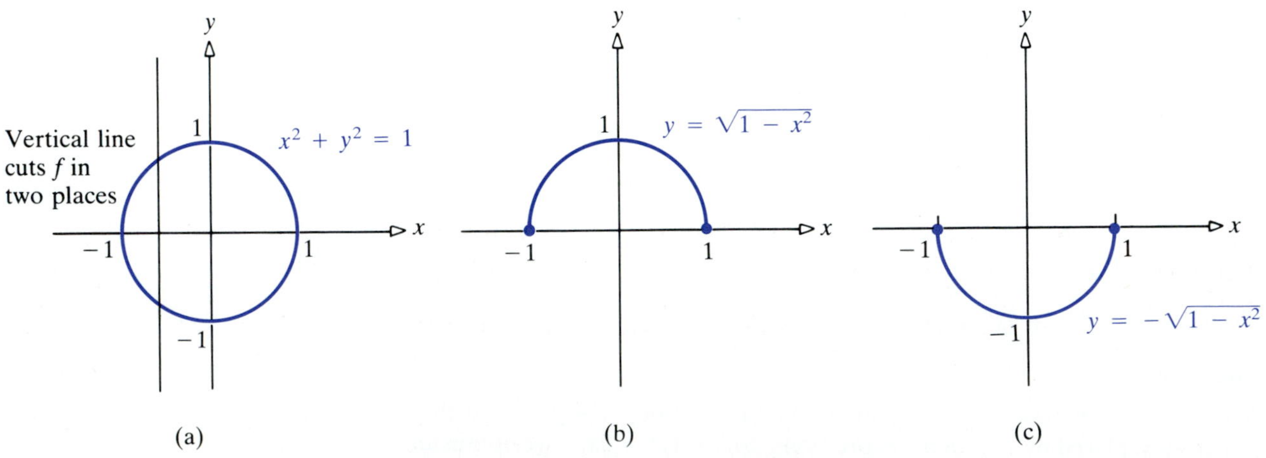

Figure 16

define two functions of x:*

$$y = f_1(x) = \sqrt{1 - x^2} \qquad y = f_2(x) = -\sqrt{1 - x^2}$$

The graph of the first function is the upper semicircle (see Fig. 16(b)), and the graph of the second is the lower semicircle (see Fig. 16(c)).

Generally speaking, an equation such as $x^2 + y^2 = 1$ or the one given in Example 2 gives rise to more than one function that will satisfy the relationship. Among these there will be at least one that is differentiable. In what follows we will always assume there is at least one such differentiable function. When we say that the equation implicitly determines y as a function of x, we mean that we have selected one of these differentiable functions.

EXAMPLE 3

Find dy/dx if $x^2 + y^2 = 1$.

Solution

We differentiate both sides of $x^2 + y^2 = 1$ with respect to x and equate the results:

$$2x + 2y\frac{dy}{dx} = 0$$

Then we solve for dy/dx:

$$\frac{dy}{dx} = -\frac{x}{y} \tag{2}$$

The formula for dy/dx, even though containing both x and y, is just as useful as one containing only x. For example, (2) tells us that the slope of the tangent line to the circle $x^2 + y^2 = 1$ at the point $(1/2, \sqrt{3}/2)$ is

$$\frac{dy}{dx} = -\frac{\frac{1}{2}}{\frac{\sqrt{3}}{2}} = -\frac{1}{\sqrt{3}}$$

Note that if $y = \pm\sqrt{1 - x^2}$, then

$$\frac{dy}{dx} = \frac{-x}{\pm\sqrt{1 - x^2}} = -\frac{x}{y}$$

in agreement with (2). Thus, (2) simultaneously gives us the derivatives of both the functions $y = +\sqrt{1 - x^2}$ and $y = -\sqrt{1 - x^2}$ defined implicitly by the equation $x^2 + y^2 = 1$.

The assumption that the selected function is differentiable is important, as the next example illustrates.

---

* Actually, the equation $x^2 + y^2 = 1$ defines an infinite number of functions of x. The two that we consider are the only ones that are continuous for $-1 \leq x \leq 1$.

EXAMPLE 4

Consider the equation

$$x^2 + y^2 + 1 = 0 \tag{3}$$

If we differentiate implicitly with respect to x, we obtain

$$2x + 2y \cdot \frac{dy}{dx} = 0 \qquad \text{or} \qquad \frac{dy}{dx} = -\frac{x}{y}$$

But this answer is meaningless because there are no real numbers x and y satisfying (3). (x^2 is nonnegative, y^2 is nonnegative, and 1 is positive so their sum cannot equal 0.) Thus, implicit differentiation makes sense only when y is a differentiable function of x.

The theoretical justification for implicit differentiation more properly belongs in an advanced calculus course and is omitted here.

Steps to Differentiate a Function Implicitly

We summarize the procedure of implicit differentiation as follows:

Step 1 **To find dy/dx when x and y are related implicitly, assume that y is a differentiable function of x.**

Step 2 **Differentiate both sides of the relationship with respect to x by using the power rule for functions or the chain rule and other properties of differentiation.**

Step 3 **Solve the resulting equation, which is linear in dy/dx, for dy/dx.**

EXAMPLE 5

Find dy/dx if:

(a) $xy^2 - x + y^3x + 5y = 10$ (b) $(x^3 + y)^4 = x$ (c) $\cos(xy) = x$

Solution

(a)
$$\left[\frac{d}{dx}(xy^2)\right] - \frac{d}{dx}x + \left[\frac{d}{dx}(y^3x)\right] + \frac{d}{dx}(5y) = \frac{d}{dx}(10)$$

$$\left[x\frac{d}{dx}y^2 + y^2\frac{d}{dx}x\right] - 1 + \left[y^3\frac{d}{dx}x + x\frac{d}{dx}y^3\right] + 5\frac{dy}{dx} = 0$$

$$\left[x\left(2y\frac{dy}{dx}\right) + y^2\right] - 1 + \left[y^3 + x\left(3y^2\frac{dy}{dx}\right)\right] + 5\frac{dy}{dx} = 0$$

$$(2xy + 3xy^2 + 5)\frac{dy}{dx} = 1 - y^2 - y^3$$

Hence, if $2xy + 3xy^2 + 5 \neq 0$, we find

$$\frac{dy}{dx} = \frac{1 - y^2 - y^3}{2xy + 3xy^2 + 5}$$

(b)
$$\frac{d}{dx}(x^3 + y)^4 = 4(x^3 + y)^3\left[\frac{d}{dx}(x^3 + y)\right] = \frac{d}{dx}x$$

$$4(x^3 + y)^3\left(3x^2 + \frac{dy}{dx}\right) = 1$$

Then, if $(x^3 + y)^3 \neq 0$, we have

$$3x^2 + \frac{dy}{dx} = \frac{1}{4(x^3 + y)^3}$$

$$\frac{dy}{dx} = \frac{1}{4(x^3 + y)^3} - 3x^2$$

(c) We use implicit differentiation and the chain rule to get

$$[-\sin(xy)]\left(y + x\frac{dy}{dx}\right) = 1$$

$$[-x\sin(xy)]\frac{dy}{dx} - y\sin(xy) = 1$$

$$\frac{dy}{dx} = \frac{1 + y\sin(xy)}{-x\sin(xy)} \qquad \text{if} \quad x\sin(xy) \neq 0$$

The next example shows how implicit differentiation can be used to find the slope of the tangent line to the graph of a function that is defined implicitly.

EXAMPLE 6

Find the slope of the tangent line to the graph of $x^3 + xy + y^3 = 5$ at the point $(-1, 2)$.

Solution

We differentiate with respect to x, obtaining

$$3x^2 + x\frac{dy}{dx} + y + 3y^2\frac{dy}{dx} = 0$$

By solving for dy/dx, we get

$$\frac{dy}{dx} = \frac{-(3x^2 + y)}{x + 3y^2} \qquad \text{provided} \quad x + 3y^2 \neq 0$$

The derivative dy/dx equals the slope of the tangent line to the graph at any point (x, y) for which $x + 3y^2 \neq 0$. In particular, for $x = -1$ and $y = 2$, we find the slope of the tangent line to the graph at $(-1, 2)$ to be

$$\frac{dy}{dx} = \frac{-(3+2)}{-1+12} = -\frac{5}{11}$$

The prime notation y', y'', and so on, is usually used in finding higher-order derivatives for implicitly defined functions.

EXAMPLE 7

Using implicit differentiation, find y' and y'' in terms of x and y if $xy + y^2 - x^2 = 5$.

Solution

$$y + xy' + 2yy' - 2x = 0 \tag{4}$$

$$y'(x + 2y) = 2x - y$$

$$y' = \frac{2x - y}{x + 2y} \qquad \text{provided} \quad x + 2y \neq 0 \tag{5}$$

It is easier to find y'' by differentiating (4) than using (5):

$$y' + y' + xy'' + 2y'(y') + 2yy'' - 2 = 0$$

$$y''(x + 2y) = 2 - 2y' - 2(y')^2$$

$$y'' = \frac{2 - 2y' - 2(y')^2}{x + 2y} \qquad \text{provided} \quad x + 2y \neq 0$$

To express y'' in terms of x and y, use (5). Then

$$y'' = \frac{2 - 2\left(\dfrac{2x - y}{x + 2y}\right) - 2\left(\dfrac{2x - y}{x + 2y}\right)^2}{x + 2y} = \frac{-10(x^2 - xy - y^2)}{(x + 2y)^3} \underset{\substack{\uparrow \\ x^2 - xy - y^2 = -5}}{=} \frac{50}{(x + 2y)^3}$$

Derivative of an Inverse Function

There is a simple relationship between the derivative of a function and the derivative of the inverse function.* Let f be the function and let f^{-1} be its inverse function. Set $g = f^{-1}$ to simplify notation. Then, as a result of (2) in Section 1.7, we have

$$f^{-1}(f(x)) = x \qquad \text{or} \qquad g(f(x)) = x$$

* You may wish to review Section 1.7 of Chapter 1 before proceeding further.

If both g and f are differentiable, we may apply the chain rule (3.7.1) and conclude that

$$[g'(f(x))][f'(x)] = 1$$

Since the product is not 0, this shows that each function has a nonzero derivative.

It turns out that, conversely, if one of the functions has a nonzero derivative, the other also has a nonzero derivative. (See Appendix I for a proof.) More precisely, we have the following theorem:

[3.8.1] THEOREM

Let $y = f(x)$ and $x = g(y)$ be inverse functions. Assume that f is differentiable on an open interval containing x_0 and that $y_0 = f(x_0)$. If $f'(x_0) \neq 0$, then g is differentiable at y_0 and

$$g'(y_0) = \frac{1}{f'(x_0)} \tag{6}$$

Let's look at another use of (6). In the Leibniz notation, this formula assumes the simple form

$$\frac{dx}{dy} = \frac{1}{dy/dx} \tag{7}$$

An advantage of (6) is that it enables us to find the derivative of the inverse function at y_0, where $y_0 = f(x_0)$, without explicitly knowing a formula for $g = f^{-1}$, provided we can determine x_0 and thus calculate $f'(x_0)$. Let's look at an example.

EXAMPLE 8

The function $f(x) = x^5 + x$ has an inverse function g. Find $g'(2)$.

Solution

To find $g'(2)$, we use (6). Then

$$g'(2) = \frac{1}{f'(x_0)} \qquad \text{where} \quad 2 = f(x_0)$$

By inspection, we find that a solution of the equation

$$f(x_0) = x_0^5 + x_0 = 2$$

is $x_0 = 1$. Since $f'(x) = 5x^4 + 1$, it follows that

$$g'(2) = \frac{1}{f'(1)} = \frac{1}{(5)(1^4) + 1} = \frac{1}{6}$$

Observe in Example 8 that we calculated the derivative of the inverse g without actually knowing a formula for g.

EXERCISE 3.8

In Problems 1–34 find dy/dx by using implicit differentiation.

1. $x^2 + y^2 = 4$
2. $3x^2 + 2y^2 = 6$
3. $x^2y = 5$
4. $x^3y = 8$
5. $x^2 - y^2 - xy = 2$
6. $x^2y + xy^2 = x + 1$
7. $x^2 - 4xy + y^2 = y$
8. $x^2 + 2xy + y^2 = x$
9. $3x^2 + y^3 = 4$
10. $y^4 - 4x^2 = 4$
11. $4x^3 + 2y^3 = x$
12. $5x^2 + xy - y^2 = 0$
13. $\dfrac{1}{x^2} - \dfrac{1}{y^2} = 1$
14. $\dfrac{1}{x^2} + \dfrac{1}{y^2} = 1$
15. $\dfrac{1}{x} + \dfrac{1}{y} = 1$
16. $\dfrac{1}{x} - \dfrac{1}{y} = 4$
17. $(x^2 + y)^3 = y$
18. $(x + y^2)^3 = 3x$
19. $\dfrac{x}{y} + \dfrac{y}{x} = 4$
20. $x^2 + y^2 = \dfrac{2y}{x}$
21. $x^2 = \dfrac{y^2}{y^2 - 1}$
22. $x^2 + y^2 = \dfrac{2y^2}{x^2}$
23. $y = x \sin y$
24. $y = x \cos y$
25. $x + xy + \sin(2x + 3y) = 0$
26. $x + y = \cos(x - y)$
27. $y = \tan(x - y)$
28. $y = \cos(x + y)$
29. $y = \sin(x + y) + \cos(x - y)$
30. $\cos(x + y) = y \sin x$
31. $\tan^2(xy^3 + y) = x$
32. $\dfrac{xy^2}{1 + \sec y} = 1 + y^3$
33. $\sin(x + y) = y^2 \cos x$
34. $\cos(x + y) = y^2 \sin x$

In Problems 35–38 find y' and y'' in terms of x and y.

35. $x^2 + y^2 = 4$
36. $x^2 - y^2 = 1$
37. $xy + yx^2 = 2$
38. $4xy = x^2 + y^2$

In Problems 39 and 40 find the slope of the tangent line at the indicated point. Write an equation for this tangent line.

39. $x^2 + y^2 = 5$, at $(1, 2)$
40. $x^2 - y^2 = 8$, at $(3, 1)$

41. Use implicit differentiation to show that the tangent line to a circle $x^2 + y^2 = R^2$ at any point P on the circle is perpendicular to OP, where O is the center of the circle.

42. For ideal gases, *Boyle's law* states that pressure is inversely proportional to volume. A more realistic relationship between pressure P and volume V is given by *van der Waals equation*:

$$P + \frac{a}{V^2} = \frac{C}{V - b}$$

where C is the constant of proportionality, a is a constant that depends on molecular attraction, and b is a constant that depends on the size of the molecules. Find the compressibility of the gas, which is measured by dV/dP.

43. Given the equation $x + xy + 2y^2 = 6$:
(a) Find an expression for the slope of the tangent line at any point (x, y) on the graph.
(b) Write an equation for the line tangent to the graph at the point $(2, 1)$.
(c) Find the coordinates of all other points on this graph with slope equal to the slope at $(2, 1)$.

44. The graph of the function $(x^2 + y^2)^2 = x^2 - y^2$ contains exactly four points at which the tangent line is horizontal. Find them.

In Problems 45–48 the functions f and g are inverse functions. Calculate the indicated derivative.

45. If $f(0) = 4$, $f'(0) = -2$, find $g'(4)$.
46. If $f(1) = -2$, $f'(1) = 4$, find $g'(-2)$.
47. If $g(3) = -2$, $g'(3) = \frac{1}{2}$, find $f'(-2)$.
48. If $g(-1) = 0$, $g'(-1) = -\frac{1}{3}$, find $f'(0)$.

49. The function $f(x) = x^3 + 2x$ has an inverse function g. Find g. Find $g'(0)$ and $g'(3)$.

50. The function $f(x) = 2x^3 + x - 3$ has an inverse function g. Find $g'(-3)$ and $g'(0)$.

51. For a freely falling object that is falling from rest (in a vacuum), the distance s the object falls in time t is given by $s = \frac{1}{2}gt^2$. Its velocity v after time t is $v = gt$. Find dv/ds without eliminating t from the given equations.

52. Find y' and y'' at the point $(-1, 1)$ if $3x^2y + 2y^3 = 5x^2$.

53. A police car approaching an intersection at 80 feet per second spots a suspicious vehicle on the cross street.

(*Problem 53 continues in the next column*)

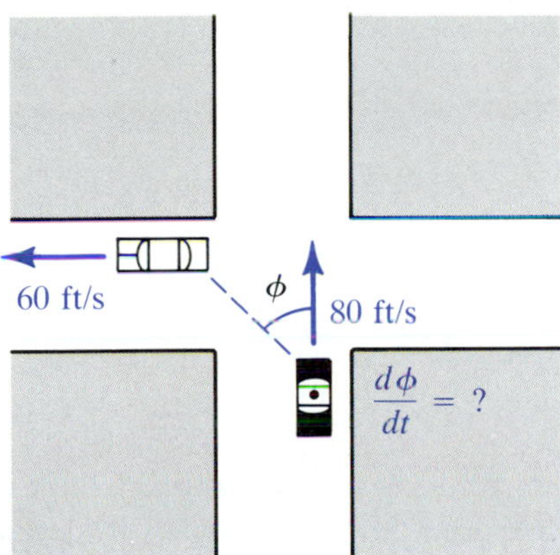

53. (*continued*)
When the police car is 210 feet from the intersection, the policeman turns his spotlight on the vehicle, which is at that time just crossing the intersection at a constant rate of 60 feet per second. See the figure. How fast must the light beam be turning 2 seconds later in order to follow the suspicious vehicle?

54. As a particle of mass m moves along the x-axis, its position s and velocity $v = ds/dt$ obey the equation

$$m(v^2 - v_0^2) = k(s_0^2 - s^2)$$

where k is a positive constant and v_0 and s_0 are the initial velocity and position, respectively. Show that if $v > 0$, then

$$ma = -ks$$

where $a = d^2x/dt^2$ is the acceleration of the particle. [*Hint:* Differentiate the expression $m(v^2 - v_0^2) = k(s_0^2 - s^2)$ with respect to t.]

55. Prove that if y is a differentiable function of x, such that x and y are related by the equation $x^ny^m + x^my^n = k$ (k a constant), then

$$\frac{dy}{dx} = -\frac{y(nx^r + my^r)}{x(mx^r + ny^r)} \qquad \text{where} \quad r = n - m$$

Derivative of Functions with Rational Exponents; One-Sided Derivatives 3.9

In Section 3.4 we showed that the formula

$$\frac{d}{dx}x^n = nx^{n-1} \tag{1}$$

is true if n is any integer. We can use the technique of implicit differentiation to extend (1) where n is any rational number. More precisely, we will show that if p/q is a rational number, then

$$\frac{d}{dx}x^{p/q} = \frac{p}{q}x^{(p/q)-1} \tag{2}$$

provided that $x^{p/q}$ and $x^{(p/q)-1}$ are real functions.

We start with the function

$$y = x^{p/q}$$

If we raise both sides of this equation to the power q, we get

$$y^q = x^p$$

We now assume that y is a differentiable function of x and differentiate implicitly, obtaining

$$\frac{d}{dx}(y^q) = \frac{d}{dx}(x^p)$$

or

$$qy^{q-1}\frac{dy}{dx} = px^{p-1}$$

Thus,

$$\frac{dy}{dx} = \frac{p}{q}\,\frac{x^{p-1}}{y^{q-1}} \underset{\uparrow}{=} \frac{p}{q}\,\frac{x^{p-1}}{(x^{p/q})^{q-1}} = \frac{p}{q}\,\frac{x^{p-1}}{x^{p-(p/q)}} = \frac{p}{q}x^{p-1-p+(p/q)} = \frac{p}{q}x^{(p/q)-1}$$

$$y^{q-1} = (x^{p/q})^{q-1}$$

Since $y = x^{p/q}$, we have

$$\frac{d}{dx}(x^{p/q}) = \frac{p}{q}x^{(p/q)-1}$$

Therefore, the formula*

$$\frac{d}{dx}(x^n) = nx^{n-1} \tag{3}$$

holds when $n = p/q$.

Our proof of this formula is incomplete because we assumed, but did not prove, that y' exists. To give a complete proof, the definition of the derivative can be used.

EXAMPLE 1

(a) $\dfrac{d}{dx}\sqrt{x} = \dfrac{d}{dx}x^{1/2} = \dfrac{1}{2}x^{(1/2)-1} = \dfrac{1}{2}x^{-1/2} = \dfrac{1}{2x^{1/2}} = \dfrac{1}{2\sqrt{x}}$

(b) $\dfrac{d}{dx}x^{1/5} = \dfrac{1}{5}x^{-4/5}$

(c) $\dfrac{d}{dx}\sqrt[3]{x} = \dfrac{d}{dx}x^{1/3} = \dfrac{1}{3}x^{-2/3} = \dfrac{1}{3\sqrt[3]{x^2}}$

(d) $\dfrac{d}{dx}x^{5/2} = \dfrac{5}{2}x^{3/2}$

(e) $\dfrac{d}{dx}x^{2/3} = \dfrac{2}{3}x^{-1/3}$

(f) $\dfrac{d}{dx}x^{-3/2} = -\dfrac{3}{2}x^{-5/2}$

(g) $\dfrac{d}{dx}x^{-4/3} = -\dfrac{4}{3}x^{-7/3}$

---

* Equation (3) is valid for any real exponent a (the proof is given in Chap. 7); that is,

$$\frac{d}{dx}x^a = ax^{a-1} \qquad a \text{ any real number}$$

Thus,

$$\frac{d}{dx}x^{\sqrt{2}} = \sqrt{2}x^{\sqrt{2}-1} \qquad \text{and} \qquad \frac{d}{dx}x^{\pi} = \pi x^{\pi-1}$$

Having established that $(d/dx)(x^n) = nx^{n-1}$ where n is a rational number, we can use the chain rule to obtain a more general formula.

[3.9.1] THEOREM

If u is a differentiable function of x, then

$$\frac{d}{dx} u^n = nu^{n-1} \frac{du}{dx} \tag{4}$$

provided u^n and u^{n-1} are defined.

EXAMPLE 2

(a) $$\frac{d}{dx}(x^3 - 2x + 1)^{5/3} = \frac{5}{3}(x^3 - 2x + 1)^{2/3} \frac{d}{dx}(x^3 - 2x + 1)$$

$$= \frac{5}{3}(x^3 - 2x + 1)^{2/3}(3x^2 - 2)$$

(b) $$\frac{d}{dx}(\sqrt[3]{x^4 - 2x + 5}) = \frac{d}{dx}(x^4 - 2x + 5)^{1/3}$$

$$= \frac{1}{3}(x^4 - 2x + 5)^{-2/3} \frac{d}{dx}(x^4 - 2x + 5)$$

$$= \frac{4x^3 - 2}{3(x^4 - 2x + 5)^{2/3}}$$

(c) $$\frac{d}{dx}(\tan 3x)^{-3/4} = -\frac{3}{4}(\tan 3x)^{-7/4} \cdot (\sec^2 3x) \cdot 3$$

↓ By chain rule

$$= -\frac{9}{4}(\tan 3x)^{-7/4}(\sec^2 3x)$$

EXAMPLE 3

For the function $f(x) = \sqrt{x^2 + 4}$ find those x, if any, where $f'(x) = 0$ and at which $f''(x) = 0$.

Solution

From (4),

$$f'(x) = \frac{1}{2}(x^2 + 4)^{-1/2}(2x) = \frac{1}{2\sqrt{x^2 + 4}}(2x) = \frac{x}{\sqrt{x^2 + 4}}$$

If $f'(x) = 0$, then $x/\sqrt{x^2 + 4} = 0$, or $x = 0$. Next, by using the rule for differentiating a quotient, we get

$$f''(x) = \frac{(1)\sqrt{x^2 + 4} - x(x/\sqrt{x^2 + 4})}{x^2 + 4} = \frac{x^2 + 4 - x^2}{(x^2 + 4)^{3/2}} = \frac{4}{(x^2 + 4)^{3/2}}$$

Notice that $f''(x)$ is never 0; in fact, $f''(x) > 0$ for all x.

Because of (4), we can now find the derivative of any algebraic function—that is, any function composed of a finite number of sums, products, quotients, powers, roots of x, and constants.

EXAMPLE 4

Differentiate: $f(x) = \sqrt{3x^8 - 3 + \dfrac{1}{\sqrt{x^2+4}}}$

Solution

The basic form of the function is $f(x) = u^{1/2}$, where

$$u = 3x^8 - 3 + \frac{1}{\sqrt{x^2+4}}$$

Thus, the first step is to use (4), obtaining

$$f'(x) = \frac{1}{2}u^{-1/2}\frac{du}{dx}$$

where

$$\frac{du}{dx} = \frac{d}{dx}\left(3x^8 - 3 + \frac{1}{\sqrt{x^2+4}}\right) = 24x^7 + \frac{d}{dx}\left(\frac{1}{\sqrt{x^2+4}}\right)$$

The final derivative can be calculated by the power rule for functions (3) from Section 3.7:

$$\frac{d}{dx}\left(\frac{1}{\sqrt{x^2+4}}\right) = \frac{d}{dx}(x^2+4)^{-1/2} = -\frac{1}{2}(x^2+4)^{-3/2}(2x) = \frac{-x}{(x^2+4)^{3/2}}$$

The derivative of the original function is

$$\frac{1}{2}u^{-1/2}\left[24x^7 - \frac{x}{(x^2+4)^{3/2}}\right] = \frac{1}{2}\left(3x^8 - 3 + \frac{1}{\sqrt{x^2+4}}\right)^{-1/2}\left[24x^7 - \frac{x}{(x^2+4)^{3/2}}\right]$$

EXAMPLE 5

Find dy/dx for:

(a) $y = |x|$ (b) $y = |g(x)|$

Solution

(a) Using the fact that

$$y = |x| = \sqrt{x^2} = (x^2)^{1/2}$$

we get

$$\frac{dy}{dx} = \frac{1}{2}(x^2)^{-1/2}(2x) = \frac{x}{(x^2)^{1/2}} = \frac{x}{|x|} \qquad \text{for } x \neq 0$$

(b) Using part (a) and the chain rule, we get

$$\frac{dy}{dx} = \frac{g(x)g'(x)}{|g(x)|} \qquad \text{for } g(x) \neq 0$$

One-Sided Derivatives

The derivative of a function f at c is given by

$$f'(c) = \lim_{x \to c} \frac{f(x) - f(c)}{x - c}$$

A criterion for this limit to exist is that the one-sided limits

$$\lim_{x \to c-} \frac{f(x) - f(c)}{x - c} \quad \text{and} \quad \lim_{x \to c+} \frac{f(x) - f(c)}{x - c}$$

exist and are equal. These limits, when they exist, are referred to as the *left derivative of f at c* and the *right derivative of f at c*, respectively. Collectively, these are called *one-sided derivatives of f at c*.

Consider the function $f(x) = x^{3/2}$. Can we find the slope of the tangent line to the graph of f at $(0, 0)$? By (3),

$$f'(x) = \tfrac{3}{2}x^{1/2} = \tfrac{3}{2}\sqrt{x}$$

If we attempt to evaluate $f'(x)$ at 0, we get

$$f'(0) = 0$$

However, looking at Figure 17, we see that the graph of $f(x) = x^{3/2}$ stops abruptly at $(0, 0)$, which implies that we cannot compute $f'(x)$ at 0, since in order for the derivative to exist, the two one-sided limits have to exist and be equal. We may, however, define the slope at $(0, 0)$ as the limit of the slopes of the secant lines containing the points $(0, 0)$ and $Q(x, y)$ on the graph. That is,

$$m_{\text{sec}} = \frac{y - 0}{x - 0} = \frac{x^{3/2}}{x} = x^{1/2}$$

As $Q(x, y) \to (0, 0)$, we have

$$\lim_{x \to 0^+} m_{\text{sec}} = \lim_{x \to 0^+} x^{1/2} = 0 \tag{5}$$

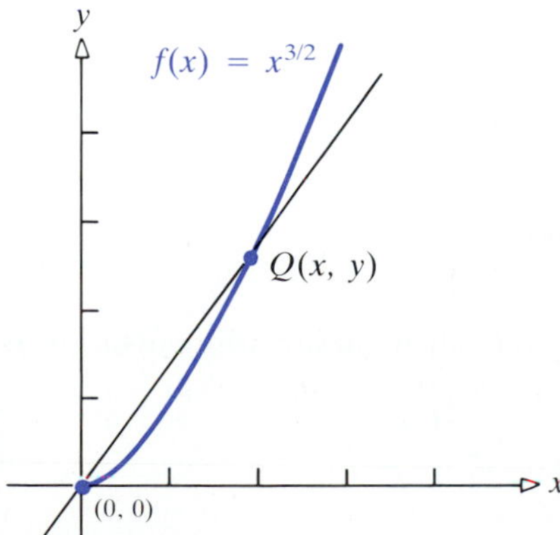

Figure 17

which agrees with the fact that $f'(0) = 0$. Thus, at $x = 0$, the graph has a horizontal tangent line.

One-sided derivatives allow us to extend the notion of differentiability to any point of the domain of a function. We say that a function f defined on a closed interval $[a, b]$ is differentiable at a if its right derivative exists at a. Similarly, we say that f is differentiable at b if its left derivative exists at b.

Application: Calculating the Rate of Spreading of an Oil Spill

Figure 18

The situation we wish to analyze is that of an oil spill from a tanker at sea. Our question is: At what rate will the oil spread? We make some assumptions to simplify and clarify the situation. First, we assume that a fixed volume V of oil is spilled and that the spill occurs in a short period of time. Second, we assume that the sea is calm and that the oil spreads in the shape of a circle with uniform thickness that may vary over time. Figure 18 illustrates the situation.

Let

$V =$ Fixed volume of oil spilled

$R = R(t)$, the radius of the circle at time t

$h = h(t)$, the thickness of the oil at time t

Since the thickness of the oil is uniform, the spill takes on the shape of a cylinder. Thus,

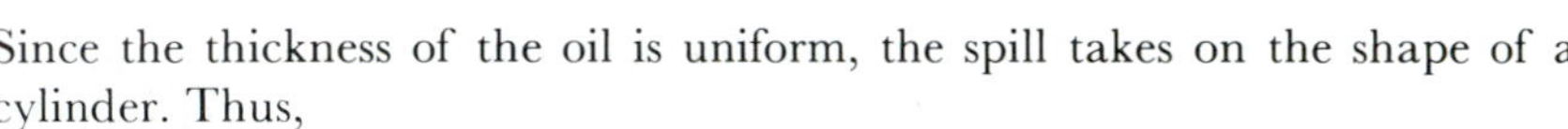

$$V = \pi R^2(t)h(t) \tag{6}$$

In the early stages of an oil spill, the thickness h of the oil may be calculated from the formula

$$h(t) = \frac{k}{t^{1/2}} \qquad k \text{ a constant}$$

Replacing $h(t)$ by $k/t^{1/2}$ in (6) gives

$$\begin{aligned} V &= \pi R^2(t)\frac{k}{t^{1/2}} \\ Vt^{1/2} &= \pi k R^2(t) \end{aligned} \tag{7}$$

We wish to find the rate of spreading of the oil. Since this is measured by the time rate of change of the radius R, we differentiate (7) with respect to t to get dR/dt:

$$\frac{V}{2}t^{-1/2} = 2\pi k R(t)\frac{dR}{dt}$$

By solving for dR/dt and using (7), we find the rate of oil spreading to be given by the formula

$$\frac{dR}{dt} = \frac{1}{4}\left(\frac{V}{\pi k R}\right)t^{-1/2} \underset{\uparrow}{=} \frac{1}{4}\frac{R}{t}$$

$$\frac{V}{\pi k R} = Rt^{-1/2}$$

EXERCISE 3.9

In Problems 1–50 find dy/dx.

1. $y = x^{2/3} + 4$
2. $y = x^{1/3} - 1$
3. $y = \sqrt[3]{x^2}$
4. $y = \sqrt[4]{x^5}$
5. $y = \dfrac{2}{x^{1/2}} + \dfrac{3}{x^{1/3}} - \dfrac{4}{x^{3/2}} + \dfrac{8}{x^{3/4}}$
6. $y = 2x^{1/2} + 3x^{1/3} - 4x^{3/2}$
7. $y = \sqrt[3]{x} - \dfrac{1}{\sqrt[3]{x}}$
8. $y = 3\sqrt{x^3} + 4\sqrt[3]{x}$
9. $y = \sqrt{\dfrac{1}{3x}}$
10. $y = \sqrt{x} + \dfrac{1}{\sqrt{x}}$
11. $y = (x^3 - 1)^{1/2}$
12. $y = (x^2 - 1)^{1/3}$
13. $y = \sqrt{\sin x}$
14. $y = \sqrt[3]{\tan x}$
15. $y = \sec\sqrt{x}$
16. $y = \cos\sqrt{1 + x}$
17. $y = x\sqrt{x^2 - 1}$
18. $y = (\sqrt{x^2 + 4})^{2/3}$
19. $y = x\sqrt{x^3 + 1}$
20. $y = x^2\sqrt{x + 1}$
21. $y = \sqrt{3 - x^2} + \sqrt{4 - x^2}$
22. $y = \sqrt{x^2 + 2x - 1}$
23. $y = \sqrt{x^2 + 1}$
24. $y = \dfrac{\sqrt{x^2 + 1}}{x}$
25. $y = \sqrt{\dfrac{x + 1}{x - 1}}$
26. $y = \sqrt[3]{\dfrac{x}{x + 1}}$
27. $y = \sqrt{x^3(8x + 1)}$
28. $y = \dfrac{\sqrt[3]{3x + 1}}{\sqrt{3x + 2}}$
29. $y = \sqrt{4 + \sqrt[3]{x}}$
30. $y = \sqrt{4 + \sqrt{x}}$
31. $y = |3x|$
32. $y = |x^5|$
33. $y = |2x - 1|$
34. $y = |5 - x^2|$
35. $y = |\cos x|$
36. $y = |\sin x|$
37. $y = \sin|x|$
38. $|x| + |y| = 1$
39. $x + \sin|y| = 1$
40. $x + \tan|y| = 1$
41. $y = (x^2 \cos x)^{3/2}$
42. $y = (x^2 \sin x)^{3/2}$
43. $y = \sin(\cos\sqrt{x^2 + 1})$
44. $y = \cos(\sin\sqrt{x^2 - 1})$
45. $y = (x^2 - 3)^{1/2}(6x + 1)^{1/3}$
46. $y = (3x + 4)^{3/2}(x^3 - 4)^{2/3}$
47. $y = \dfrac{(2x^3 - 1)^{2/3}}{(3x + 4)^{1/2}}$
48. $y = \dfrac{(4x^2 - 1)^{1/4}}{(3x + 5)^{3/2}}$
49. $y = \sqrt[3]{2x^3 + x\sqrt[4]{x^2 + 1}}$
50. $y = \sqrt[3]{5x - 4\sqrt[4]{2x^2 + x}}$

In Problems 51 and 52 find y' and y'' for the given function.

51. $y = \sqrt{x^2 + 1}$
52. $y = \sqrt{4 - x^2}$

In Problems 53 and 54 find an equation of the tangent line at the given point.

53. $x^{1/3} + y^{1/3} = 1$, at $(8, -1)$
54. $x^{2/3} + y^{2/3} = \frac{1}{2}$, at $(\frac{1}{8}, \frac{1}{8})$
55. Show that the slope of the tangent line to

$$x^{2/3} + y^{2/3} = a^{2/3} \qquad a > 0$$

at any point is $-y^{1/3}/x^{1/3}$.

56. At what point does the graph of $y = 1/\sqrt{x}$ have a tangent line parallel to the line $x + 16y = 5$?
57. Let $f(x) = x^3 + x$. If h is the inverse function of f, find $h'(2)$.

58. Another way of computing the derivative of $\sqrt[n]{x}$ is to use inverse functions. The function $y=f(x)=x^n$, n a positive integer, has the derivative nx^{n-1}. Hence, if $x \neq 0$, then $f'(x) \neq 0$. The inverse function of f—namely, $x=g(y)=\sqrt[n]{y}$—is defined for all y if n is odd, and for all $y \geq 0$ if n is even. But this inverse function is differentiable for all $y \neq 0$ and

$$g'(y)=\frac{d}{dy}\sqrt[n]{y}=\frac{1}{f'(x)}=\frac{1}{nx^{n-1}}$$

(*Problem 58 continues in the next column*)

58. (*continued*)
But $nx^{n-1}=n(\sqrt[n]{y})^{n-1}=ny^{(n-1)/n}=ny^{1-(1/n)}$. Therefore,

$$\frac{d}{dy}\sqrt[n]{y}=\frac{d}{dy}y^{1/n}=\frac{1}{ny^{1-(1/n)}}=\frac{1}{n}y^{(1/n)-1}$$

Use the result from above and the chain rule to prove the formula:

$$\frac{d}{dx}x^{p/q}=\frac{p}{q}x^{(p/q)-1}$$

3.10 Newton's Method of Solving Equations

Newton's method will enable us to find, to any desired degree of accuracy, the real roots of many equations.

Suppose we let $y=f(x)$ denote a function whose derivative f' is continuous, and we wish to find the real roots of the equation $f(x)=0$. Graphically, this means that we are to find the x-intercepts of the graph. Now suppose that, from a graph or from tables or by trial calculations, we have found that the equation $f(x)=0$ has a real root in some open interval containing the number $x=c_1$. We draw the tangent line to the graph of f at the point $P_1=(c_1, f(c_1))$ and let P_2 be the point where this tangent line intersects the x-axis (see Fig. 19). Then, in general, if $x=c_1$ is a fair *first approximation* to the required root of the given equation $f(x)=0$, the x-intercept $c_2=\overline{OP_2}$ of the tangent line will give a better, or *second, approximation* to the root. Graphically, this is the idea behind *Newton's method* of solving equations.

We may derive a formula for calculating the approximation c_2 as follows: The coordinates of P_1 are $(c_1, f(c_1))$, and the slope of the tangent line to the graph of f at P_1 is $f'(c_1)$. Therefore, the equation of this tangent

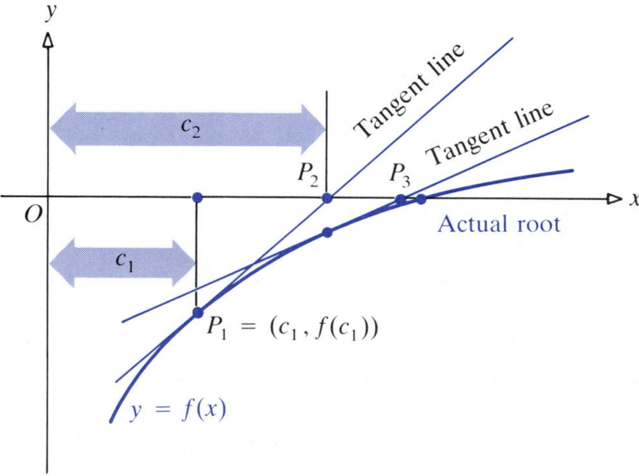

Figure 19

line at P_1 is

$$y - f(c_1) = f'(c_1)(x - c_1)$$

By putting $y = 0$ in this equation and solving for x in order to find the x-intercept, we obtain

$$x = c_2 = c_1 - \frac{f(c_1)}{f'(c_1)}$$

Hence,

[3.10.1] THEOREM

If $x = c_1$ is a sufficiently close first approximation to a real root of the equation $f(x) = 0$, then the formula

$$c_2 = c_1 - \frac{f(c_1)}{f'(c_1)} \tag{1}$$

gives a second approximation to the root.

We may now use c_2 in (1) in place of c_1 and get a third, and possibly closer, approximation to the required root of the equation—namely,

$$c_3 = c_2 - \frac{f(c_2)}{f'(c_2)}$$

The process may be repeated as often as required to give the desired degree of accuracy.*

EXAMPLE 1

Use Newton's method to find a third approximation to the real positive root r of the equation $x^3 - 2x - 5 = 0$, where $2 \leq r \leq 3$.

Solution

Let $f(x) = x^3 - 2x - 5$. We find $f(2) = -1$ and $f(3) = 16$, which indicates that the given equation has a root between 2 and 3, probably nearer 2 than 3. Therefore, we take $c_1 = 2$ and apply Newton's formula. Then, $f(c_1) = f(2) = -1$, and since $f'(x) = 3x^2 - 2$, we have $f'(c_1) = f'(2) = 10$. By (1),

$$c_2 = 2 - \frac{(-1)}{10} = 2.1 \tag{2}$$

Now we apply (1) again, with $c_1 = 2.1$. We find $f(2.1) = 0.061$ and $f'(2.1) = 11.23$. Hence, a third approximation to the root is

$$c_3 = 2.1 - \frac{0.061}{11.23} = 2.1 - 0.0054 = 2.0946$$

To see how good this approximation is, we use a calculator and find that $f(2.0946) = 0.0005416$.

---

* Successive approximations may not get successively closer to the actual solution, even though they do approach the solution (as in a limit process). In addition, there are instances in which successive approximations do not approach the solution. Consult a numerical analysis book for details.

EXAMPLE 2

Use Newton's method to find a third approximation to the real root of the equation $\sin x + x - 1 = 0$.

Solution

To get the first approximation c_1, we note that the real root of

$$\sin x + x - 1 = 0$$

occurs at the intersection of $y = \sin x$ and $y = 1 - x$, which is some number between 0 and 1 (see Fig. 20). We begin with $c_1 = 0$. To use (1), we set $f(x) = \sin x + x - 1$. Then $f'(x) = \cos x + 1$ and

$$c_2 = c_1 - \frac{f(c_1)}{f'(c_1)} = 0 - \frac{f(0)}{f'(0)} = -\frac{-1}{2} = 0.5$$

Since

$$f(0.5) = \sin \frac{1}{2} + \frac{1}{2} - 1 \approx -0.0206$$

Use a calculator.

and

$$f'(0.5) = \cos \frac{1}{2} + 1 \approx 1.8776$$

we find the third approximation to be

$$c_3 = 0.5 - \frac{-0.0206}{1.8776} \approx 0.5 + 0.0110 = 0.5110$$

Using a calculator, we find that $f(0.5110) = 0.0000497$.

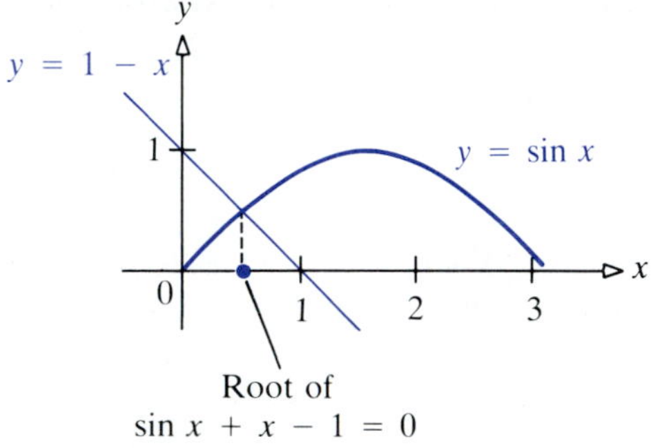

Figure 20

Estimating Roots on a Computer

To approximate a real root for the equation $f(x) = 0$, Newton's method requires an initial estimate (c_1) for the root and repeated (frequently time-consuming) applications of the formula

$$c_{n+1} = c_n - \frac{f(c_n)}{f'(c_n)}$$

The successive applications of this formula produce the sequence of values $c_1, c_2, c_3, \ldots,$ which usually will approach the desired root for $f(x) = 0$. That is, $|f(c_n)|$ will approach 0 as n increases.

A number of problems are associated with Newton's method:

1. The initial estimate c_1 may not be "good enough." A poor first value can invalidate the whole process (and produce a really strange sequence of values).
2. A poor first value may also cause the sequence of values to approach the root so slowly that we have to repeat the formula calculation hundreds of times before we get close enough to the root to be satisfied.
3. One of the values in the sequence may fall at a place where the derivative is 0 (i.e., a horizontal tangent line). We must stop the process at that point, since the formula is not defined for $f'(c_n) = 0$.

To see a few examples of these problems, use a hand calculator and Newton's formula to calculate a few steps in the sequence of estimates for each of the following functions:

1. $f(x) = -x^3 + 6x^2 - 9x + 6$ with an initial estimate of $c_1 = 2.9$ and then with $c_1 = 3.0$ (The root is near $x = 4.2$.)
2. $f(x) = x^8 - 1$ with an initial estimate of $c_1 = 0.1$
3. $f(x) = (x - 1)^{1/3}$ with an initial estimate of $c_1 = 2.0$

Although the problems listed above are not the only problems that can occur, they are sufficient to discourage the use of Newton's method even when the calculations are done on a hand calculator. However, now that microcomputers are readily available, we can have a computer do all the tedious calculations, while we just sit back and observe how well Newton's method works.

Programming a computer to repeat a sequence of calculations is easily done in languages such as BASIC and Pascal. Each step (calculating the next estimate) is just another application of Newton's formula. However, we need a set of conditions that tell the computer when to stop doing the calculations. There are many different stopping conditions that we could choose. The set of conditions we will use is one that seems at first to be the most intuitively acceptable (and one that we might use if we had to do the calculations by hand). We will instruct the computer to stop if:

1. $|f(c_n)| <$ eps, where eps(ilon) is a measure of how close $f(c_n)$ is to 0*

---

* The primary stopping condition $|f(c_n)| <$ eps does not give us much information about the size of the actual error ($|c_n -$ Root$|$). Thus, in most implementations of Newton's method, the primary condition for terminating the loop is when $|c_{n+1} - c_n| < \delta$. Here, δ is a user-supplied value that specifies how close we want our estimates to be to the root, instead of how close we want the $f(c_n)$ value to 0.

2. The number of estimates (steps in the sequence) calculated so far reaches the maximum allowed (max)
3. The derivative is 0; that is, $f'(c_n) = 0$

The algorithm (procedure, or list of instructions) we will use when we write the program that tells the computer what to do consists of a very simple loop:

```
while no stopping condition is satisfied
   calculate the next estimate
   print the values of c_n and f(c_n)
   check the stopping conditions
```

Sample computer programs in BASIC and Pascal are shown here, which implement the strategy outlined above. Both were run on a standard microcomputer using the MS DOS operating system. Find a microcomputer and try to get one of these programs running. Then do some of the problems in Section 3.10. See how fast (or how slowly) Newton's method approaches each root.

Another alternative is to see whether your school already owns a software package (i.e., computer program) that will do Newton's method for functions defined by the user. Often these programs will offer a variety of different methods using other techniques (bisection, secants, etc.) for finding roots. Also, many of these programs are designed to show you graphically what each method does as it approximates a root.

Programs

The programs given here were not written for speed or efficiency. We simply want to display the basic algorithm as clearly as possible. We anticipate that the version of BASIC you use will be an interpreted form of the language and that Pascal will probably have to be a compiled language. Therefore, the BASIC version is a stripped-down, but perfectly functional, version of the Pascal program.

Both programs are currently set up to apply Newton's method to the function $f(x) = x^3 - 2x - 5$. In order to apply these programs to other functions, the function definitions in the programs for $f(x)$ and $f'(x)$ must be changed.

The identifiers used in the programs are:

c1	= Initial estimate for the desired root
cn	= Current estimate for the root
eps	= Tolerance limit for how close we want $\lvert f(c_n)\rvert$ to be to 0
max	= Maximum number of estimation steps that will be done
n	= Current step in the sequence of estimates
abs( )	= Absolute value function

In the BASIC program, the values for c_1, eps, and max are assigned in lines 120–140. In the Pascal program, the user must enter these values at the computer keyboard after the program has begun.

```
100 DEF FNF(X)=X^3-2*X-5
110 DEF FNP(X)=3*X^2-2
120 CN=.1
130 EPS=.0001
140 MAX=20
150 PRINT "STEP","X","F(X)"
160 N=1
170 PRINT N,CN,FNF(CN)
180 IF (ABS(FNF(CN)) < EPS) OR (FNP(CN) = 0) OR (N >= MAX) THEN GOTO 500
190     N=N+1
200     CN=CN-FNF(CN)/FNP(CN)
210     PRINT N,CN,FNF(CN)
220 GOTO 180
500 IF ABS(FNF(CN)) < EPS THEN PRINT "F(X) WITHIN TOLERANCE" : END
510 IF FNP(CN) = 0  THEN PRINT "DERIVATIVE = 0 - LOOP TERMINATED" : END
520 IF N >= MAX THEN PRINT "MAXIMUM NUMBER OF STEPS"
```

BASIC Program

```
program newton_method;

function f(x: real) : real;
begin (* f *)
     f := x * x * x  - 2.0 * x - 5.0
end; (* f *)

function fp(x: real) : real;
begin (* fprime *)
     fp := 3.0 * x * x  - 2.0
end; (* fp = f prime *)

var
     cn, eps : real;
     n, max : integer;

begin (* main *)
     write('Enter the initial estimate for the root --> ');
     readln(cn);
     write('Enter the maximum number of approximation steps --> ');
     readln(max);
     write('Enter the tolerance limit for  : f(cn) :  --> ');
     readln(eps);
     writeln('step','x':10,'f(x)':20);
     n := 1;
     writeln(n:3, cn:18:7, f(cn):18:7);
     while (abs(f(cn)) >= eps) and (fp(cn) <> 0.0) and (n < max) do
     begin
          cn := cn - f(cn) / fp(cn);
          n := n + 1;
          writeln(n:3, cn:18:7, f(cn):18:7);
     end; (* while *)
     if abs(f(cn)) < eps then
          writeln('f(x) value within tolerance of 0')
     else if fp(cn) = 0.0 then
          writeln('derivative = 0 -- cannot continue')
     else
          writeln('maximum number of steps allowed')
end. (* main *)
```

Pascal Program

EXERCISE 3.10

Solve Problems 1–8 by using Newton's method. In each equation find a third approximation to the root indicated.

1. $x^3 + 3x - 5 = 0$; root between 1 and 2

2. $x^3 - 4x + 2 = 0$; root between 1 and 2

3. $2x^3 + 3x^2 + 4x - 1 = 0$; root between 0 and 1

4. $x^3 - x^2 - 2x + 1 = 0$; root between 0 and 1

5. $x^3 - 6x - 12 = 0$; root between 3 and 4

6. $3x^3 + 5x - 40 = 0$; root between 2 and 3

7. $x^4 - 2x^3 + 21x - 23 = 0$; root between 1 and 2

8. $x^4 - x^3 + x - 2 = 0$; root between 1 and 2

9. Use Newton's method to find a second approximation to the positive root of the equation $2x - 3\sin x = 0$. Use $c_1 = 1.5$ as your first approximation.

10. Follow the directions of Problem 9 for the equation $x^2 = 2\cos x$, using $c_1 = \pi/4$ as your first approximation.

Newton's method can also be used to approximate the value of radicals. In Problems 11–14 use the method to find the value of the given radical to four decimal places.

11. $\sqrt{11}$ by solving the equation $x^2 - 11 = 0$

12. $\sqrt{3}$ by solving the equation $x^2 - 3 = 0$

13. $\sqrt[3]{7}$ by solving the equation $x^3 - 7 = 0$

14. $\sqrt[3]{5}$ by solving the equation $x^3 - 5 = 0$

15. The volume of a spherical segment is given by

$$V = \tfrac{1}{3}\pi h^2(3R - h)$$

where R is the radius of the sphere and h is the height of the segment. If $R = 4$ feet and $V = 12$ cubic feet, find a second approximation to h. Use $c_1 = 1$ as your first approximation.

16. A solid wooden sphere of diameter d and specific gravity S sinks in water to a depth h, which is determined by the equation $2x^3 - 3x^2 + S = 0$, where $x = h/d$. Find a second approximation to h for a maple ball of diameter 6 inches for which $S = 0.786$.

17. The equation $x - p\sin x = M$, called *Kepler's equation*, occurs in astronomy. Find a second approximation to x when $p = 0.2$ and $M = 0.85$. Use $c_1 = 1$ as your first approximation.

REVIEW EXERCISES

In Problems 1–45 find the derivative. When a or b appears, it is a constant.

1. $y = (ax + b)^n$

2. $y = \sqrt{2ax}$

3. $y = x\sqrt{1 - x}$

4. $y = \dfrac{1}{\sqrt{x^2 + 1}}$

5. $f(x) = \dfrac{x^2}{x + 1}$

6. $F(z) = \dfrac{2}{\sqrt{z}}$

7. $u = \dfrac{r}{a^2\sqrt{a^2 - r^2}}$

8. $s = \dfrac{t^3}{t - 2}$

9. $y = (x^2 + 4)^{3/2}$

10. $y = x(a^2 + x^2)\sqrt{a^2 - x^2}$

11. $y = 3x^{-2} + 2x^{-1} + 1$

12. $w = (z^2 + 1)^{5/2}$

13. $u = \dfrac{1}{z^2 + 1}$

14. $s = (t - 1)^{5/2}$

15. $f(x) = \dfrac{x^2}{\sqrt{x^2 - 1}}$

16. $\phi(x) = \dfrac{\sqrt{x + 1}}{x}$

17. $q = \dfrac{\sqrt{p^2 - 1}}{a^2 p}$

18. $r = (x + \sqrt{x^2 - 1})^n$

19. $y = x^2(a^2 + x^2)^{3/2}$

20. $z = \dfrac{\sqrt{2ax - x^2}}{x}$

21. $y = \sqrt{x + 2}$

22. $y = \sqrt{x} + \sqrt[3]{x}$

23. $u = \dfrac{y}{\sqrt{y^2 + 9}}$

24. $w = \dfrac{v - 1}{v^2 + 1}$

25. $f(x) = \sqrt{1 - x^2}$

26. $F(x) = \dfrac{2}{(a^2 - x^2)^{3/2}}$

27. $w = \dfrac{1}{1 - z + z^2}$

28. $r = (t - 1)\sqrt{t^2 + 1}$

29. $y = (1 - x^3)^{1/2}$

30. $\phi(x) = \dfrac{(x^2 - a^2)^{3/2}}{x^3}$

31. $v = (1 + u)^{3/2}$

32. $y = x\sqrt{3 - x}$

33. $f(x) = \dfrac{x^2}{(x - 1)^2}$

34. $g(x) = x^2\sqrt{2ax - x^2}$

35. $u = (a^{1/2} - x^{1/2})^2$

36. $v = (2r - 1)^{5/2}$

37. $y = x \sin 2x$

38. $u = \cos^3 x$

39. $v = \tan u + \sec u$

40. $y = \sqrt{a^2 \sin(x/a)}$

41. $\phi(z) = \sqrt{1 + \sin z}$

42. $u = \sin v - \frac{1}{3}\sin^3 v$

43. $y = \sqrt{1 + \sqrt{1 + \sqrt{1 + x}}}$

44. $y = \sin\sqrt{\pi/x}$

45. $y = (1 + \sqrt{1 + \cos^3 x})^{2001}$

46. Let $f(x) = (x^6 - x^4 + x^2)/(x^4 + 1)$. Perform a long division and then find $f'(x)$ without the quotient rule.

47. Find y' if $x = y^5 + y$.

48. Find y' if $x = \cos^5 y + \cos y$.

49. If $f(x) = (x - 1)/(x + 1)$ for all $x \neq -1$, find $f'(1)$.

50. If $f(x) = \sqrt{1 - \sin^2 x}$, find the domain of $f'(x)$.

51. If $f(x) = x^{1/2}(x - 2)^{3/2}$ for all $x \geq 2$, find the domain of f'.

52. If $f(x) = 2 + |x - 3|$ for all x, determine whether the derivative $f'(x)$ exists at $x = 3$.

53. If $\tan(xy) = x$, find dy/dx.

54. If $y = x + \sin(xy)$, find dy/dx.

55. Let $f(x) = 4x^3 - 3x - 1$.
   (a) Find the x-intercepts of the graph of f.
   (b) Write an equation for the tangent line to the graph of f at $x = 2$.
   (c) Write an equation of the graph that is the reflection across the y-axis of the graph of f.

56. Let f be the function defined by $f(x) = \sqrt{1 + 6x}$.
   (a) Give the domain and range of f.
   (b) Determine the slope of the line tangent to the graph of f at $x = 4$.
   (c) Determine the y-intercept of the line tangent to the graph of f at $x = 4$.
   (d) Give the coordinates of the point on the graph of f where the tangent line is parallel to $y = x + 12$.

57. If $f(x) = 1/(1 - x)$, find a formula for the nth derivative—that is, $f^{(n)}(x)$.

58. Let f and g be two differentiable functions at c. State the relationship between their tangent lines at c if:
   (a) $f'(c) = g'(c)$
   (b) $f'(c) = -\dfrac{1}{g'(c)}$

59. A particle moves in a straight line according to the equation $s = 2t^3 - 15t^2 + 24t + 3$, where t is measured in minutes and s in meters. Determine:
   (a) When the particle is at rest
   (b) Acceleration when $t = 3$

60. Find the value of the limit below and specify the function f for which this is the derivative.

$$\lim_{\Delta x \to 0} \frac{[4 - 2(x + \Delta x)]^2 - (4 - 2x)^2}{\Delta x}$$

61. Find equations of the tangent and normal lines to the graph of $y = x\sqrt{x + (x - 1)^2}$ at the point $(2, 2\sqrt{3})$.

CHALLENGE EXERCISES

1. If f and g are differentiable functions, find the derivative of $fg/(f+g)$.

2. Let f and g have derivatives up to the fourth order. Compute the first four derivatives of fg and simplify your answers. In particular, show that the fourth derivative is

$$f^{(4)}g + 4f^{(3)}g^{(1)} + 6f^{(2)}g^{(2)} + 4f^{(1)}g^{(3)} + fg^{(4)}$$

Identify a pattern.

3. Let $f_1(x), \ldots, f_n(x)$ be differentiable.

(a) Find $(f_1(x) \cdots f_n(x))'$.

(b) Find $\left(\dfrac{1}{f_1(x) \cdots f_n(x)}\right)'$.

4. Let $f_1(x), \ldots, f_n(x)$ be differentiable. Find the derivative of $f_1(f_2(f_3(\ldots(f_n(x)\ldots))))$.

5. If n is an odd positive integer, show that the tangent lines to the graph of $y = x^n$ at $(1, 1)$ and at $(-1, -1)$ are parallel.

6. If n is an even positive integer, show that the tangent line to the graph of $y = \sqrt[n]{x}$ at $(1, 1)$ is perpendicular to the tangent line to the graph of $y = x^n$ at $(-1, 1)$.

7. Show that the tangent line to the graph of $y = x^n$ at $(1, 1)$ has y-intercept $1 - n$.

8. Find a, b, c, d so that the tangent line to the graph of the cubic $y = ax^3 + bx^2 + cx + d$ at the point $(1, 0)$ is $y = 3x - 3$ and at the point $(2, 9)$ is $y = 18x - 27$.

9. The graphs of two functions are said to be *orthogonal* (perpendicular to each other) if their tangent lines are perpendicular at each point of intersection.

(a) Show that the graphs of $xy = c_1$ and $-x^2 + y^2 = c_2$ are orthogonal, where c_1 and c_2 are constants.

(b) Plot the graphs on one coordinate system for $c_1 = 1, 2, 3$ and $c_2 = 1, 9, 25$.

10. Show that the parabolas $y^2 = 2ax + a^2$ and $y^2 = a^2 - 2ax$ intersect at right angles.

11. If $f(x) = Ax^2 + B$, then:

(a) Find c in terms of A such that the tangent lines to the graph of f at $(c, f(c))$ and $(-c, f(-c))$ are perpendicular.

(b) Find the slopes of the tangent lines in part (a).

(c) Find the coordinates, in terms of A and B, of the point of intersection of the tangent lines in part (a).

12. A function f is defined for all real numbers and has the following three properties: (*1*) $f(1) = 5$, (*2*) $f(3) = 21$, and (*3*) for all real a and b,

$$f(a + b) - f(a) = kab + 2b^2$$

where k is a fixed real number independent of a and b.

(a) Use $a = 1$ and $b = 2$ to find k.

(b) Find $f'(3)$.

(c) Find $f'(x)$ and $f(x)$ for all real x.

13. For a differentiable function f, let f^* be the function defined by

$$f^*(x) = \lim_{h \to 0} \frac{f(x + h) - f(x - h)}{h}$$

(a) Determine $f^*(x)$ for $f(x) = x^2 + x$.

(b) Determine $f^*(x)$ for $f(x) = \cos x$.

(c) Write an equation that expresses the relationship between the functions f^* and f', where f' denotes the derivative of f. Justify your answer.

14. The line $x = c$, where $c > 0$, intersects the cubic $y = 2x^3 + 3x^2 - 9$ at point P and the parabola $y = 4x^2 + 4x + 5$ at point Q.

(a) If a line tangent to the cubic at point P is parallel to the line tangent to the parabola at point Q, find the number c.

(b) Write the equations of the two tangent lines in part (a).

15. A function f is periodic if there is a positive number p so that $f(x + p) = f(x)$ for all x. Show that if f is a differentiable function of x, then f' is also of period p.

16. Let T be the line tangent to the graph of $y = x^3$ at the point $(\frac{1}{2}, \frac{1}{8})$. At what point Q does the line T intersect the graph? What is the angle between the tangent line at Q and the line T?

17. Let N be the line normal to the graph of $y = x^2$ at the point $(-2, 4)$. At what point Q does N meet the graph?

18. At what point(s), if any, is the line $y = x - 1$ parallel to the tangent to the graph of $y = \sqrt{25 - x^2}$?

19. If $f(x) = \sqrt{x - 1}$, let A and B be points on the graph of f with $x = 2$ and $x = 5$, respectively. A line is moved upward on the graph such that it remains parallel to the secant line AB. Find the coordinates of the last point on the graph of f before the secant line loses contact with the graph.

20. Let $f(x) = 1 + \cfrac{1}{1 + \cfrac{1}{1 + \cfrac{1}{1+x}}}$. Find $f'(x)$ and state its domain.

21. Let

$$f(x) = \begin{cases} x^2 \sin \dfrac{1}{x} & \text{if } x \neq 0 \\ 0 & \text{if } x = 0 \end{cases}$$

Show that $f'(0)$ exists but that $f'(x)$ is not continuous at 0.

22. Let $f(x) = \sin x$.
(a) Find $f^{(245)}(x)$.
(b) Find $f^{(n)}(x)$ for n a positive integer.

23. Let $f(t) = mt$, $m > 0$. Let $F(x)$ be defined for $x > 0$ as the area of the shaded region in the figure. Find $F'(x)$.

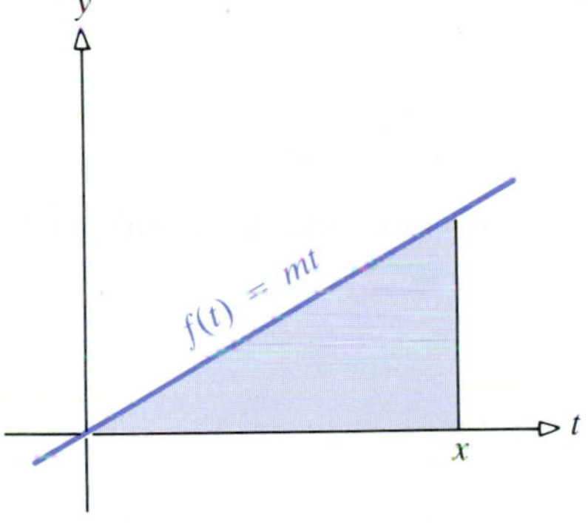

24. We say that y is an algebraic function of x if it is a function that satisfies an irreducible equation.[†]

$$P_0(x)y^n + P_1(x)y^{n-1} + \cdots + P_{n-1}(x)y + P_n(x) = 0$$

where the $P(x)$'s are polynomials. For example, $y = \sqrt{x}$ satisfies

$$y^2 - x = 0$$

Use implicit differentiation to obtain a formula for the derivative of an algebraic function.

25. Let a, b, c, and d be real numbers. Define

$$\begin{vmatrix} a & b \\ c & d \end{vmatrix} = ad - bc$$

This is called a *determinant* and it arises in the study of linear equations. Let $f_1(x)$, $f_2(x)$, $f_3(x)$, and $f_4(x)$ be differentiable and set

$$D(x) = \begin{vmatrix} f_1(x) & f_2(x) \\ f_3(x) & f_4(x) \end{vmatrix}$$

Show that

$$D'(x) = \begin{vmatrix} f'_1(x) & f'_2(x) \\ f_3(x) & f_4(x) \end{vmatrix} + \begin{vmatrix} f_1(x) & f_2(x) \\ f'_3(x) & f'_4(x) \end{vmatrix}$$

[†] An equation is said to be irreducible if it cannot be factored.

Finding Derivatives on the HP-28*

One of the simplest operations on the HP-28 is computing derivatives. Derivatives can be computed in two ways using the HP-28: one step at a time or all at once. The initial temptation is to use the full power of the machine and seek the completed derivative immediately. One reason to use the step-by-step differentiation, however, is to gain an understanding of multistep chain rule problems, like the following example.

* Gregory D. Foley
The Ohio State University

EXAMPLE 1

Find dy/dx for $y = \cos^2(x^3)$.

Solution

Let's first review how to "walk through" the steps of the derivation using the chain rule. Since $\cos^2(x^3) = [\cos(x^3)]^2$, we have

$$\begin{aligned}\frac{dy}{dx} &= \frac{d}{dx}\{[\cos(x^3)]^2\}\\ &= 2[\cos(x^3)]^{2-1}\frac{d}{dx}[\cos(x^3)]\\ &= 2[\cos(x^3)][-\sin(x^3)]\frac{d}{dx}(x^3)\\ &= 2\cos(x^3)[-\sin(x^3)](3x^2)\\ &= -6x^2\cos(x^3)\sin(x^3).\end{aligned}$$

The HP can emulate this step-by-step procedure. First, clear the stack by pressing [ ] [CLEAR], and check to see whether the radian mode annunciator (2π) is displayed at the top of the screen. If not, press [ ] [MODE] [RAD]. Then, key in the expression for the derivative:

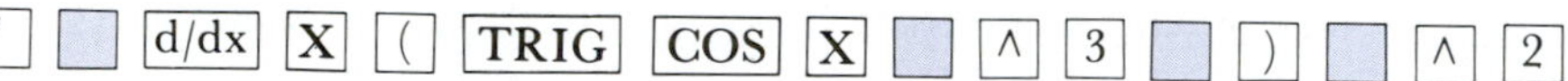

Press [ENTER] to move the expression from the command line to the stack. The stack should look like this:

```
3:
2:
1:  '∂X(COS(X^3)^2)'
```

Notice that the HP automatically closed off the delimiters; that is, it inserted the)′ at the end of the expression. The X after the HP derivative symbol ∂ and before the function indicates that we are differentiating with respect to X.

Now to compute the derivative step-by-step, press [EVAL]. Compare the expression that appears on the screen with the first step in the standard derivation given above. Repeat this EVAL step three times, pausing after each step to make a comparison with the standard derivation. The following four expressions should appear on the HP, in order:

```
'∂X(COS(X^3))*2*COS(X^3)^(2-1)'
'-SIN(X^3)*∂X(X^3)*2*COS(X^3)'
'-(SIN(X^3)*(∂X(X)*3*X^(3-1))*2*COS(X^3))'
'-(SIN(X^3)*(3*X^2)*2*COS(X^3))'
```

If the expression ($\pi/180$) appears within the expressions on your screen, your calculator is not in radian mode.

In the next chapter the goal is to apply derivatives to a variety of problem situations. The focus is not on how to find derivatives, but rather on how to use them. The HP-28 can be used to compute derivatives of functions in one quick step so that our attention can remain focused on the application at hand. This is particularly relevant when the function involved is rather complicated. Let's try Problem 41 from the Review Exercises in Chapter 3 using the one-step HP method.

EXAMPLE 2

Find the derivative of $\phi(z) = \sqrt{1 + \sin z}$.

Solution

As before, you should clear the stack and check to see that the radian mode annunciator (2π) is displayed. Then,

(a) **Key in the function:**

(b) **Key in the variable of differentiation:**

['] [Z] [ENTER]

The stack should look like this:

```
3:
2:  '√(1+SIN(Z))'
1:              'Z'
```

Now compute the derivative by pressing [ ] [d/dx]. The HP should quickly respond with the following:

```
2:
1:  'COS(Z)/(2*√(1+SIN(Z
    )))'
```

So, the derivative is $\phi'(z) = \dfrac{\cos z}{2\sqrt{1 + \sin z}}$.

EXERCISES

Solve using the HP-28.

1. Find dy/dx for $y = \cos^2(u^3)$ using the step-by-step process. Compare the steps with those shown in Example 1. How do the steps differ due to the variable of differentiation, x, being different from the independent variable, u, of the function?

2. Solve both parts of Example 10 from Section 3.7 using the method of Example 2 above.

3–20. Solve Problems 23–40 from the Review Exercises in Chapter 3.

4 Applications of the Derivative

In this chapter we present some important applications of the techniques developed in Chapters 2 and 3. The first sections give applications of the derivative to related rates and differentials. Then we consider maxima and minima of functions and the related subject of graphing functions. Finally, we introduce the concept of antiderivatives, which is the basis for our later study of integral calculus.

4.1 Related Rates

In all of the natural sciences and many of the social and behavioral sciences, we encounter quantities that are related but vary with time. For example, the pressure of an ideal gas of fixed volume is proportional to temperature, yet each of these quantities may change over a period of time.

Problems involving rates of related variables are called *related rate problems.* In such problems we normally want to find the rate at which one of the variables is changing at a certain time, while the rates at which the other variables are changing are known.

The usual procedure in such problems is to write an equation that relates all the time-dependent variables involved. Such a relationship is often obtained by investigating the geometric and/or physical conditions imposed by the problem. When this relationship is differentiated with respect to the time t, a new equation that involves the variables and their rates of change with respect to time is obtained.

For example, suppose x and y are two differentiable functions of time t—that is, $x = x(t)$ and $y = y(t)$. And suppose they obey the equation

$$x^3 - y^3 + 2y - x - 99 = 0$$

If we differentiate with respect to time t (remembering that x and y are functions of t), we obtain

$$3x^2 \frac{dx}{dt} - 3y^2 \frac{dy}{dt} + 2\frac{dy}{dt} - \frac{dx}{dt} = 0$$

This equation is valid for all times t under consideration and involves the derivatives of x and y with respect to t, as well as the variables themselves. Because the derivatives are related by this equation, we call them *related rates.* We can solve for one of these rates once the value of the other rate and the values of the variables are known. For example, if in the above equation at a specific time t, we know that $x = 5$, $y = 3$, and $dx/dt = 2$, then by direct substitution we find that $dy/dt = \frac{148}{25}$.

EXAMPLE 1

A child throws a stone into a still millpond, causing a circular ripple to spread (see Fig. 1). If the radius of the circle increases at the constant rate of 0.5 meter per second, how fast is the area of the ripple increasing when the radius of the ripple is 20 meters?

Figure 1

Solution

The variables involved are:

t = Time (in seconds) elapsed from the time the stone hits the water

R = Radius of the ripple (in meters) after t seconds

A = Area of the ripple (in square meters) after t seconds

The rates involved are:

$\frac{dR}{dt}$ = Rate (in meters per second) at which the radius is increasing

$\frac{dA}{dt}$ = Rate (in square meters per second) at which the area is increasing

We wish to find dA/dt when $R = 20$—that is, the rate at which the area of the ripple is increasing at the instant when $R = 20$. The relationship between A and R is given by the formula for the area of a circle:

$$A = \pi R^2 \tag{1}$$

Since A and R are functions of t, we differentiate both sides of (1) with respect to t to obtain

$$\frac{dA}{dt} = 2\pi R\frac{dR}{dt} \tag{2}$$

Since the radius increases at the rate of 0.5 meter per second, we know that

$$\frac{dR}{dt} = 0.5 \tag{3}$$

By substituting (3) into (2), we get

$$\frac{dA}{dt} = 2\pi R(0.5) = \pi R$$

Thus, when $R = 20$, the area of the ripple is increasing at the rate

$$\frac{dA}{dt} = \pi(20) = 20\pi \approx 62.8 \text{ square meters per second}$$

Example 1 illustrates some general guidelines that will prove helpful for solving related rate problems:

Guidelines for Solving Related Rate Problems

Step 1 **If possible, draw a picture to illustrate the problem.**

Step 2 **Identify the variables and assign symbols to them.**

Step 3 **Identify and interpret rates of change as derivatives.**

Step 4 **Express all relationships among the variables by equations.**

Step 5 **Obtain additional relationships among the variables and their derivatives by differentiating.**

Step 6 Substitute numerical values for the variables and the derivatives. Solve for the unknown rate.

Note: It is important to remember that the substitution of numerical values (Step 6) must occur *after* the differentiation process (Step 5).

EXAMPLE 2

A balloon in the form of a sphere is being inflated at the rate of 10 cubic meters per minute. Find the rate at which the surface area of the sphere is increasing at the instant when the radius of the sphere is 3 meters.

Solution

The variables of the problem are:

t = Time (in minutes) measured from the moment inflation of the balloon begins

R = Length (in meters) of the radius of the balloon at time t

V = Volume (in cubic meters) of the balloon at time t

S = Surface area (in square meters) of the balloon at time t

The rates of change are:

$$\frac{dR}{dt} = \text{Rate of change of radius with respect to time}$$

$$\frac{dV}{dt} = \text{Rate of change of volume with respect to time}$$

$$\frac{dS}{dt} = \text{Rate of change of surface area with respect to time}$$

We are given that $dV/dt = 10$ cubic meters per minute, and we seek dS/dt when $R = 3$ meters. At any time t the volume V of the balloon (a sphere) is $V = \frac{4}{3}\pi R^3$ and the surface area S of the balloon is $S = 4\pi R^2$. By differentiating each of these equations with respect to the time t, we have

$$\frac{dV}{dt} = 4\pi R^2 \frac{dR}{dt} \qquad \text{and} \qquad \frac{dS}{dt} = 8\pi R \frac{dR}{dt}$$

In the equation for dV/dt, we solve for dR/dt and substitute this quantity into the equation for dS/dt. Then

$$\frac{dS}{dt} = 8\pi R \frac{dV/dt}{4\pi R^2} = \frac{2}{R}\frac{dV}{dt}$$

At $R = 3$ and $dV/dt = 10$, we have

$$\frac{dS}{dt} = \left(\frac{2}{3}\right)(10) \approx 6.67 \text{ square meters per minute}$$

Thus, the surface area is increasing at the rate of 6.67 square meters per minute when the radius is 3 meters.

EXAMPLE 3

A rectangular swimming pool 10 meters long and 5 meters wide is 3 meters deep at one end and 1 meter deep at the other. (A cross-sectional view of the pool is illustrated in Fig. 2.) If water is pumped into the pool at the rate of 300 liters per minute, at what rate is the water level rising when it is 1.5 meters deep at the deep end? (*Note:* 1 liter of water = 10^{-3} cubic meter.)

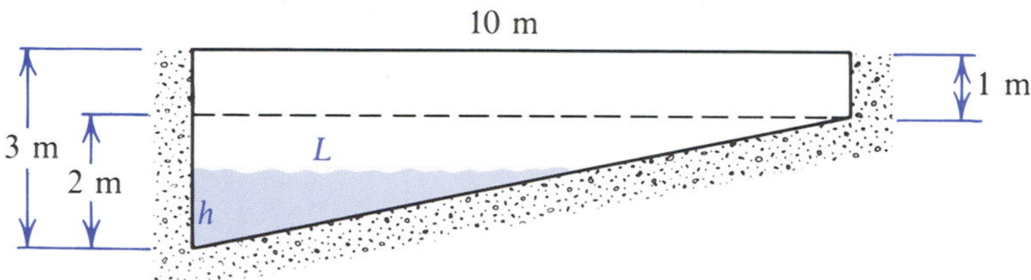

Figure 2

Solution

First, the variables involved are:

t = Time (in minutes) measured from the moment water begins to flow into the pool

h = Water level (in meters) measured at the deep end

L = Distance (in meters) from the deep end to the short end measured at water level

V = Volume (in cubic meters) of water in the pool

The rates of change are:

$\frac{dV}{dt}$ = Rate of increase in volume at a given instant

$\frac{dh}{dt}$ = Rate of increase in height at a given instant

The volume V is related to L and h by the formula

$$V = (\text{Cross-sectional triangular area})(\text{Width}) = \left(\tfrac{1}{2}Lh\right)(5) \text{ cubic meters} \qquad \textbf{(4)}$$

Using similar triangles, we see from Figure 2 that L and h are related by the equation

$$\frac{L}{h} = \frac{10}{2}$$

$$L = 5h$$

By replacing L by $5h$ in (4), we have

$$V = \tfrac{1}{2}(5h)(h)(5) = \tfrac{25}{2}h^2 \text{ cubic meters} \qquad \textbf{(5)}$$

Here, V and h are each functions of time t. By differentiating (5) with respect to t, we obtain

$$\frac{dV}{dt} = 25h\frac{dh}{dt} \text{ cubic meters per minute}$$

We seek the rate at which the water level is rising, dh/dt, when $h = 1.5$ meters and the rate of water pumped into the pool is $dV/dt = 300$ liters per minute $= 300(10^{-3})$ cubic meter per minute. Thus, the water level is rising at a rate of

$$\frac{dh}{dt} = \frac{(300)(10^{-3})}{25(1.5)} = 0.008 \text{ meter per minute}$$

EXAMPLE 4

A person is standing on a pier and pulling a boat inward by pulling a rope at the rate of 2 meters per second. The end of the rope is 3 meters above water level (see Fig. 3). How fast is the boat approaching the base of the pier when 5 meters of rope are left to pull in? Disregard sagging of the rope and assume the rope is attached to the boat at water level.

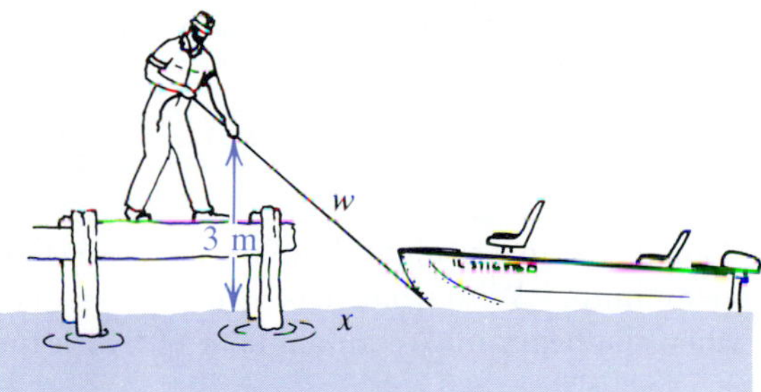

Figure 3

Solution

The variables of the problem are:

t = Time (in seconds)

x = Distance (in meters) from the boat to the base of the pier

w = Distance (in meters) from the boat to the person (that is, the length of rope)

The rates of change are:

$$\frac{dx}{dt} = \text{Rate at which the boat is approaching the pier}$$

$$\frac{dw}{dt} = \text{Rate at which the rope is being pulled}$$

From Figure 3 we see that

$$w^2 = 9 + x^2 \tag{6}$$

The variables w and x are functions of time t. By differentiating (6) with respect to time t, we obtain

$$2w\frac{dw}{dt} = 2x\frac{dx}{dt}$$

$$w\frac{dw}{dt} = x\frac{dx}{dt} \tag{7}$$

We seek to find dx/dt, the rate at which the boat is approaching the pier, when $w = 5$ meters and $dw/dt = -2$ meters per second. (The negative sign is used to indicate that the length of the rope is *decreasing* at the rate of 2 meters per second.) The value of x when $w = 5$ is found from equation (6) to be $x = \sqrt{25 - 9} = 4$ meters. By substituting these values into (7), we find

$$4\frac{dx}{dt} = 5(-2)$$

$$\frac{dx}{dt} = -2.5 \text{ meters per second}$$

Thus, the boat is approaching the pier at the rate of 2.5 meters per second.

EXAMPLE 5

A revolving light located 5 kilometers from a straight shoreline has a constant angular velocity of 3 radians per minute.* With what velocity does the spot of light move along the shore when the beam makes an angle of 60° with the shoreline?

Solution

Using Figure 4, we find that the variables are:

t = Time (in minutes)

x = Distance (in kilometers) of the beam of light from B

θ = Angle (in radians) the beam of light makes with AB

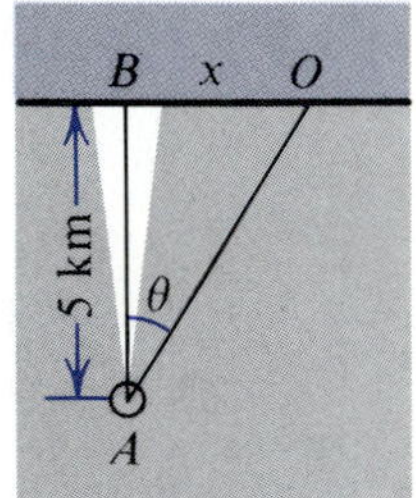

Figure 4

The rates of change are:

$\frac{dx}{dt}$ = Velocity of the spot of light along the shore

$\frac{d\theta}{dt}$ = Angular velocity of the beam of light

---

* *Angular velocity* ω is defined as the rate of change of angle with respect to time—that is, $\omega = \lim_{\Delta t \to 0}(\Delta\theta/\Delta t) = d\theta/dt$, with θ in radians.

From Figure 4, we see that

$$\tan\theta = \frac{x}{5}$$

$$x = 5\tan\theta$$

Hence,

$$\frac{dx}{dt} = 5\sec^2\theta\,\frac{d\theta}{dt} \tag{8}$$

We are given that $d\theta/dt = 3$ radians per minute. We seek dx/dt when angle $AOB = 60°$ $(\theta = 30° = \pi/6$ radian). From (8), we have

$$\frac{dx}{dt} = 5\sec^2\theta\,\frac{d\theta}{dt} = \frac{5}{\cos^2\theta}\,\frac{d\theta}{dt} = \frac{5(3)}{(\cos\pi/6)^2} = \frac{15}{\frac{3}{4}} = 20 \text{ kilometers per minute}$$

Thus, when $\theta = 30°$, the velocity of the light along the shore is 20 kilometers per minute.

EXERCISE 4.1

In Problems 1–4 assume x and y are differentiable functions of t. Find dx/dt when $x = 3$, $y = 4$, and $dy/dt = 2$.

1. $x^2 + y^2 = 25$

2. $x^2 - y^2 = -7$

3. $x^3y^2 = 432$

4. $x^2y^3 = 576$

5. Suppose h is a differentiable function of t and suppose that when $h = 3$, $dh/dt = \frac{1}{12}$. Find dV/dt if $V = 80h^2$.

6. Suppose x is a differentiable function of t and suppose that when $x = 15$, $dx/dt = 3$. Find dy/dt if $y^2 = 625 - x^2$.

7. Suppose h is a differentiable function of t and suppose that $dh/dt = \frac{5}{16}\pi$ when $h = 8$. Find dV/dt if $V = \frac{1}{12}\pi h^3$.

8. Suppose x and y are differentiable functions of t and suppose that when $t = 20$, $dx/dt = 5$, $dy/dt = 4$, $x = 150$, and $y = 80$. Find ds/dt if $s^2 = x^2 + y^2$.

9. If each edge of a cube is increasing at the constant rate of 3 centimeters per second, how fast is the volume increasing when x, the length of an edge, is 10 centimeters long?

10. If the radius of a sphere is increasing at 1 centimeter per second, find the rate of change of its volume when the radius is 6 centimeters.

11. Consider a right triangle with hypotenuse of (fixed) length 45 centimeters and variable legs of lengths x and y, respectively. If the leg of length x increases at the rate of 2 centimeters per minute, how fast is y changing when x is 4 centimeters long?

12. Air is pumped into a balloon with a spherical shape at the rate of 80 cubic centimeters per second. How fast is the surface area of the balloon increasing when the radius is 10 centimeters?

13. A spherical balloon filled with gas has a leak that permits the gas to escape at a rate of 1.5 cubic meters per minute. How fast is the surface area of the balloon shrinking when the radius is 4 meters?

14. When a metal plate is heated, it expands. If the shape of the metal is circular and if its radius, as a result of expansion, increases at the rate of 0.02 centimeter per second, at what rate is the area of the top surface increasing when the radius is 3 centimeters?

15. A public swimming pool has a rectangular shape with the following dimensions: length 30 meters, width 15 meters, depth 3 meters at the adult side and 1 meter at the children's side. If water is pumped into the pool at the rate of 15 cubic meters per minute, how fast is the water level rising when it is 2 meters deep at the adult side?

16. A radar antenna, making one revolution every 5 seconds, is located on a ship that is 6 kilometers from a straight shoreline. How fast is the radar beam moving along the shoreline when the beam makes an angle of 45° with the shore?

17. A ball is hit along the third-base line with a speed of 100 feet per second. At what rate is the ball's distance from first base changing when it crosses third base? (See the figure.)

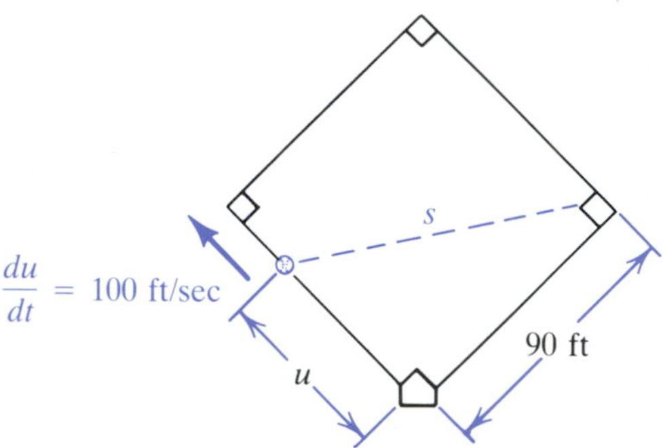

18. A falcon flies up from its trainer at an angle of 60° until it has flown 200 feet. It then levels off and continues to fly away. If the speed of the bird is 132 feet per second, how fast is the falcon moving away from the falconer after 6 seconds? (See the figure.)

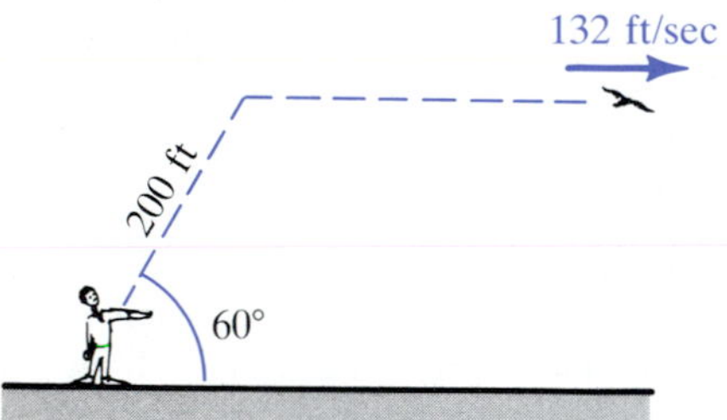

19. An object moves along the curve $y = x^3$. At what points on the curve are the x- and y-coordinates of the object changing at the same rate?

20. A ferris wheel is 50 feet in diameter and has its center located 30 feet above the ground. If the wheel revolves once every 2 minutes, how fast is a passenger rising when he is 42.5 feet above the ground? How fast is he moving horizontally?

21. The dome of an observatory is 60 feet in diameter. A boy is playing near the observatory at sunset. He throws a ball upward so that its shadow climbs to the highest point on the dome. How fast is the shadow moving along the dome $\frac{1}{2}$ second after the ball begins to fall? How did you use the fact that it was sunset in solving the problem? *Note:* A ball falling from rest covers a distance $s = 16t^2$ feet after t seconds.

22. An elevated train on a track 20 feet above the ground crosses a street at a rate of 25 feet per second and at an angle of 30°. Five seconds later a car crosses under the tracks going 40 feet per second. How fast are the train and the car separating 3 seconds later? (See the figure.)

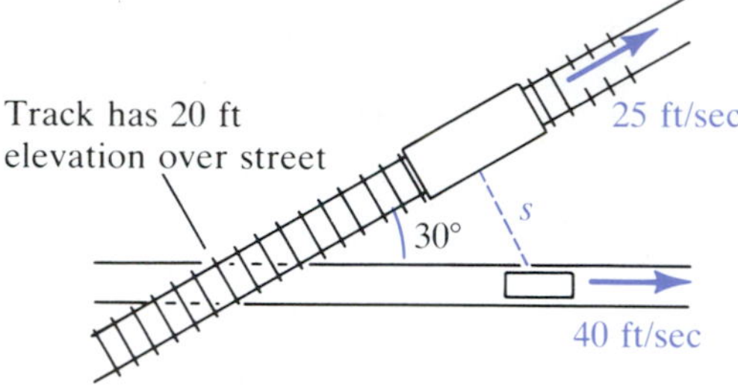

23. A light in a lighthouse 2000 meters from a straight shoreline is rotating at 2 revolutions per minute. How fast is the beam moving along the shore when it passes a point 500 meters from the point on shore opposite the lighthouse? [*Hint:* 1 revolution $= 2\pi$ radians.]

24. Water is flowing into a vertical cylindrical tank of diameter 6 meters at the rate of 5 cubic meters per minute. Find the rate at which the depth of the water is rising.

25. Consider a container in the form of a right circular cone (vertex down) with radius $R = 4$ meters and height $h = 16$ meters (see the figure). If water is poured into the container at the constant rate of 16 cubic meters per minute, how fast is the water level rising when the water is 8 meters deep? [*Hint:* The volume V of a cone of radius R and height h is $V = \frac{1}{3}\pi R^2 h$.]

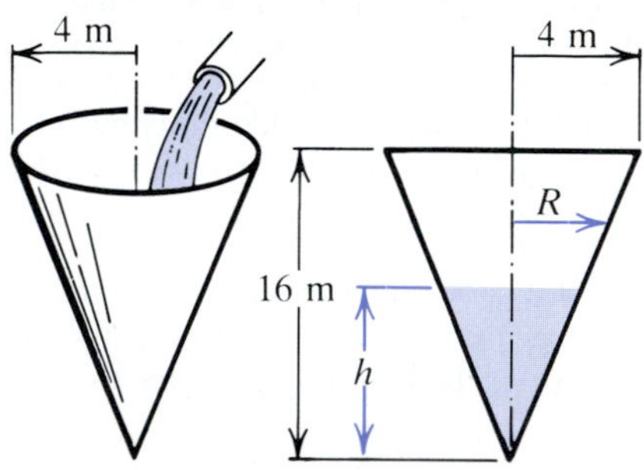

26. Sand is being poured on the ground, forming a conical pile with height equal to one-fourth of the diameter of the base. If the sand is falling at the rate of 20 cubic centimeters per second, how fast is the height increasing when it is 3 centimeters?

27. A cistern in the shape of a cone 4 meters deep and 2 meters in diameter at the top is being filled with water at the rate of 3 cubic meters per minute. If at the time the water is 3 meters deep, it is observed to be rising 0.5 meter per minute, at what rate is the water leaking away?

28. A conical funnel 15 centimeters in diameter and 15 centimeters deep is filled with a liquid that runs out at the rate of 5 cubic centimeters per minute. How fast is the surface falling when the depth of the liquid is 8 centimeters?

29. An 8 meter ladder is leaning against a vertical wall (see the figure). If a person pulls the base of the ladder away from the wall at the rate of 0.5 meter per second, how fast is the top going down the wall when the base of the ladder is:

(a) 3 meters from the wall?
(b) 4 meters from the wall?
(c) 6 meters from the wall?

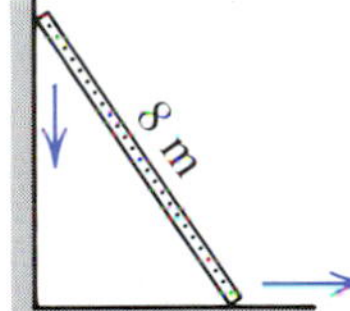

30. A boy is walking toward the base of a pole 20 meters high at the rate of 4 kilometers per hour. At what rate (in meters per second) is the distance from his feet to the top of the pole changing when he is 5 meters from the pole?

31. A girl flies a kite at a height of 30 meters above her hand. If the kite flies horizontally away from the girl at the rate of 2 meters per second, at what rate is the string being let out when the length of the string released is 70 meters? Assume that the string remains taut.

32. A street light hangs 6 meters high. A child (1 meter tall) is walking directly under it. If the child walks away from the light at the rate of 40 meters per minute, how fast is the child's shadow lengthening?

33. An isosceles triangle has equal sides 4 centimeters long and the included angle is θ. If θ increases at the rate of $2°$ per minute, how fast is the area of the triangle changing when θ is $30°$?

34. A particle P is moving along the parabola $y^2 = 4(3 - x)$. When P passes the point $(-1, 4)$, its y-coordinate is increasing at the rate of 3 units per second. How fast is the distance from P to the origin changing at that instant?

35. A gas is said to be compressed adiabatically if there is no gain or loss of heat. When such a gas is diatomic (has two atoms per molecule), it satisfies the equation $PV^{1.4} = k$, where k is a constant, P is the pressure, and V is the volume. At a given instant, the pressure is 20 kilograms per square centimeter, the volume is 32 cubic centimeters, and the volume is decreasing at the rate of 2 cubic centimeters per minute. At what rate is the pressure changing?

36. Two cars approach an intersection, one heading east at the rate of 30 kilometers per hour and the other heading south at the rate of 40 kilometers per hour. At what rate are the two cars approaching each other at the instant when the first car is 100 meters from the intersection and the second car is 75 meters from the intersection? Assume the cars maintain their respective speeds.

37. In order to lift a container to the third floor, which is 10 meters above the top of the container, a rope is attached to the container and, with the help of a pulley, hoists the container up (see the figure). If a person holds the end of the rope and walks away from beneath the pulley at the rate of 2 meters per second, how fast is the container rising when the person is 5 meters away? We assume that the end of the rope in the person's hand was originally at the same height as the top of the container.

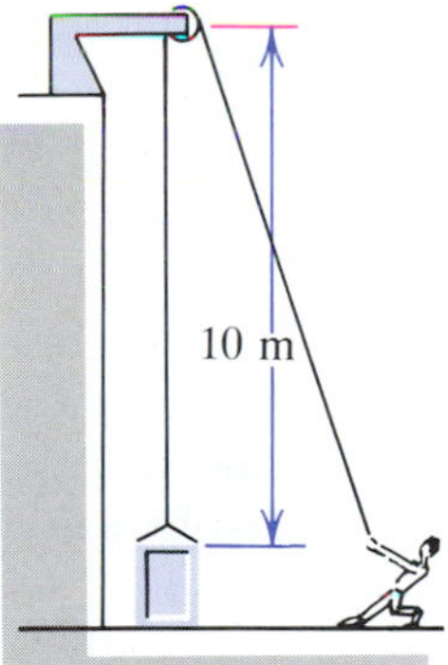

38. In a rectangle with a diagonal 15 centimeters long, one side is increasing at the rate of $2\sqrt{5}$ centimeters per second. Find the rate of change of the area when that side is 10 centimeters long.

39. Consider the following situation: An elevator in a building is located on the fifth floor, which is 25 meters above the ground. A delivery truck is positioned directly beneath the elevator at street level. If, simultaneously, the elevator goes up at a speed of 5 meters per second and the truck pulls away at a speed of 8 meters per second, how fast will the elevator and the truck be separating 1 second later? Assume the speeds remain constant at all times.

40. A soldier at an antiaircraft battery observes an airplane flying toward him at an altitude of 4500 feet (see the figure). When the angle of elevation of the battery is 30°, the soldier must increase the angle of elevation by 1° per second to keep the plane in sight. What is the ground speed of the plane?

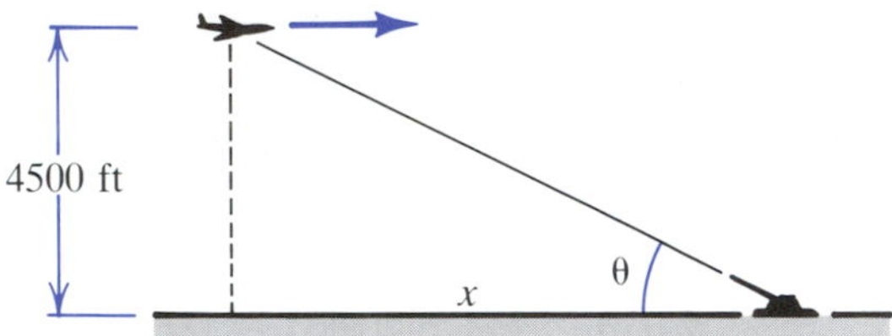

41. An airplane 2000 meters high is flying horizontally with a speed of 300 meters per second. How fast is the angle of elevation of the plane changing when this angle is 45°?

42. In a certain piston engine the distance x in meters between the center of the driving shaft and the head of the piston is given by $x = \cos\theta + \sqrt{16 - \sin^2\theta}$, where θ is the angle between the crank and the path of the piston head. If θ increases at the constant rate of 45 radians per second, what is the speed of the piston head when θ is $\pi/6$?

43. An object that weighs K pounds on the surface of the earth weighs approximately

$$W(R) = K\left(\frac{4000}{4000 + R}\right)^2$$

pounds when it is a distance of R miles from the earth's surface. Find the rate at which the weight of an object weighing 1000 pounds on the earth's surface is changing when it is 50 miles above the earth's surface and is being lifted at the rate of 10 miles per second.

4.2 Differentials

In studying the derivative of a function $y = f(x)$, we use the notation dy/dx to represent the derivative. The symbols dy and dx, called *differentials*, which appear in this notation, may also be given their own meanings.

To pursue this, recall that for a differentiable function f, the derivative is defined as

$$\frac{dy}{dx} = f'(x) = \lim_{\Delta x \to 0} \frac{\Delta y}{\Delta x} = \lim_{\Delta x \to 0} \frac{f(x + \Delta x) - f(x)}{\Delta x}$$

That is, the derivative f' is the limit of the ratio of the change in y to the change in x as Δx tends to 0, but $\Delta x \neq 0$. In other words, for Δx sufficiently close to 0, we can make $\Delta y/\Delta x$ as close as we please to $f'(x)$. We express this fact by writing

$$\frac{\Delta y}{\Delta x} \approx f'(x) \qquad \text{when} \qquad \Delta x \approx 0 \quad (\Delta x \neq 0)$$

Another way of writing this is

$$\Delta y \approx f'(x)\Delta x \qquad \text{when} \quad \Delta x \approx 0 \quad (\Delta x \neq 0)$$

The quantity $f'(x)\Delta x$ is called the *differential of* y.

[4.2.1] DEFINITION / *Differential.*

Let f denote a differentiable function and let Δx denote a change in x.

(a) The *differential of y*, denoted by dy, is defined as $dy = f'(x)\Delta x$.
(b) The *differential of x*, denoted by dx, is defined as $dx = \Delta x \neq 0$.

Thus, using the notation of differentials, we have

$$\boldsymbol{dy = f'(x)\,dx} \tag{1}$$

Since $dx \neq 0$, this can be written as

$$\frac{dy}{dx} = f'(x) \tag{2}$$

The expression in (2) should look very familiar. Interestingly enough, we have given an independent meaning to the symbols dy and dx in such a way that, when dy is divided by dx, their *quotient* will be equal to the derivative. That is, the differential of y divided by the differential of x is equal to the derivative $f'(x)$. For this reason, **we may formally regard the derivative as a quotient of differentials.**

Note that the differential dy is a function of both x and dx. For example, the differential dy of the function $y = x^3$ is

$$dy = 3x^2\,dx$$

so that

if	$x = 1$	and	$dx = 2,$	then	$dy = 3(1)^2(2) = 6$
if	$x = 0.5$	and	$dx = 0.1,$	then	$dy = 3(0.5)^2(0.1) = 0.075$
if	$x = 2$	and	$dx = -5,$	then	$dy = 3(2)^2(-5) = -60$

EXAMPLE 1

(a) If $y = x^2 + 3x - 5$, then $dy = (2x + 3)\,dx$.
(b) If $y = x \sin x$, then $dy = (\sin x + x \cos x)\,dx$.
(c) If $y = \sqrt{x^2 + 4}$, then $dy = \dfrac{x}{\sqrt{x^2 + 4}}\,dx$.

Geometric Interpretation

We use Figure 5 to arrive at a geometric interpretation of the differentials dx and dy and their relationship to Δx and Δy. From the definition, the differential dx and the change Δx are equal. Therefore, we concentrate on the relationship between dy and Δy.

In Figure 5(a), $P = (x, y)$ is a point on the graph of $y = f(x)$ and $Q = (x + \Delta x, y + \Delta y)$ is a nearby point that is also on the graph of f. The slope of the tangent line to the graph of f at P is $f'(x)$. From the figure, it follows that

$$f'(x) = \frac{dy}{\Delta x} = \frac{dy}{dx} \qquad \text{or} \qquad dy = f'(x)\,dx$$

Figure 5(a) illustrates the case for which $dy < \Delta y$ (concave up) and $\Delta x > 0$. The case for which $dy > \Delta y$ and $\Delta x > 0$ is illustrated in Figure 5(b). The remaining cases, in which $\Delta x = dx < 0$, have similar graphical representations.

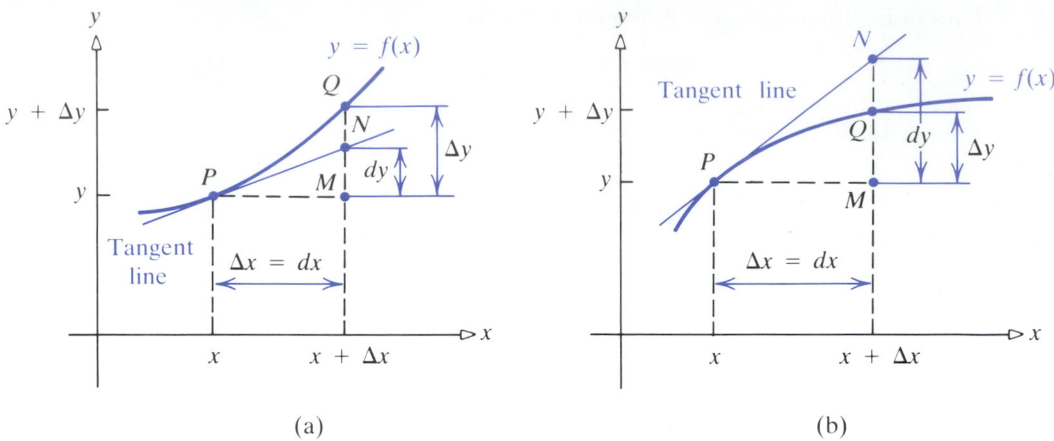

Figure 5

Linear Approximations

Let us now establish a relationship between Δy and dy. In Figure 5(a), $\Delta x = dx$ and the increment Δy is represented by the length of the line segment $|MQ|$. Thus, $\Delta y - dy$ is just the length of the line segment $|NQ|$. The size of $|NQ|$ equals the amount by which the graph departs from its tangent line. In fact, for $dx = \Delta x$ sufficiently small, the graph does not depart very much from its tangent line. Thus the linear function whose graph is this tangent line is referred to as the *linear approximation to f near P*.

For $dx = \Delta x$ sufficiently small, the differential dy is a good approximation to Δy, in the sense that dy differs from Δy by a small percentage of dx. That is,

$$\Delta y \approx dy \qquad \text{if} \quad \Delta x \approx 0 \tag{3}$$

We can use (3) to obtain the linear approximation to a function f near a point $P = (x_0, y_0)$ on f. Since

$$dy = f'(x_0)\, dx = f'(x_0)\Delta x = f'(x_0)(x - x_0)$$

we find from (3) that

$$\begin{aligned} \Delta y &\approx dy \\ f(x) - f(x_0) &\approx f'(x_0)(x - x_0) \\ f(x) &\approx f(x_0) + f'(x_0)(x - x_0) \end{aligned}$$

[4.2.2] THEOREM / *Linear Approximation.*

The linear approximation to $f(x)$ near $x = x_0$ is given by

$$y = f(x_0) + f'(x_0)(x - x_0) \tag{4}$$

for x sufficiently close to x_0.

EXAMPLE 2

Find the linear approximation to $f(x) = x^2 + 2x$ near $x = 1$. Graph f and the linear approximation.

Solution

First, $f(1) = 3$ and $f'(1) = 2(1) + 2 = 4$. By (4), the linear approximation to f near $x = 1$ is $f(1) + f'(1)(x - 1)$. Therefore,

$$f(x) \approx 3 + 4(x - 1) = 4x - 1$$

See Figure 6.

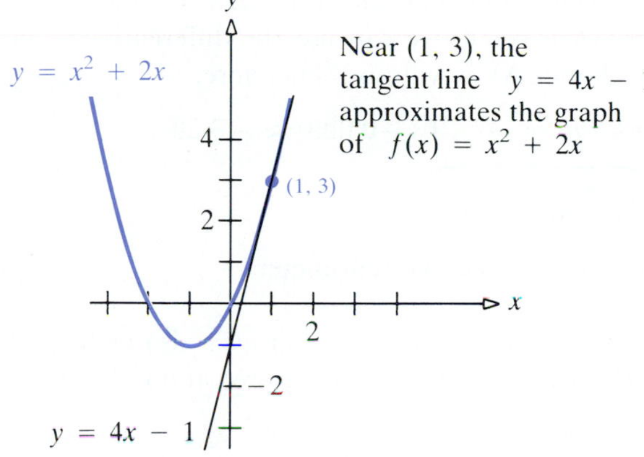

Figure 6

Whenever $f(x)$ is difficult to compute and $f(x_0)$ and $f'(x_0)$ are easy to compute, (4) may be used to obtain a numerical approximation.

EXAMPLE 3

Use differentials to approximate $\sqrt{123}$.

Solution

The closest perfect square to 123 is 121, and $\sqrt{121} = 11$. What we wish to know is the value of $y = f(x) = \sqrt{x}$ when $x = 123$. The change in x from a square root we know $(x_0 = 121)$ to a number whose square root we seek $(x = 123)$ is $x - x_0 = 2$. At $x_0 = 121$,

$$f(x_0) = \sqrt{121} = 11 \qquad \text{and} \qquad f'(x_0) = \frac{1}{2\sqrt{121}} = \frac{1}{22}$$

From (4), we have

$$f(123) = \sqrt{123} \approx f(x_0) + f'(x_0)(x - x_0) = 11 + \frac{1}{22}(2)$$

$$= 11 + 0.0909 = 11.0909$$

(On a calculator, $\sqrt{123} = 11.090537$.)

Since $\sqrt{123}$ is so easily approximated on a calculator, Example 3 may seem to be of little practical use. The next two examples use (3) in a way that is useful even when a calculator is available.

EXAMPLE 4

A bearing with a spherical shape has a radius of 3 centimeters when it is new. Find the approximate volume of the metal lost after it wears down to a radius of 2.971 centimeters.

Solution

The exact volume of metal lost equals the change, ΔV, in volume V of the sphere, where $V = \frac{4}{3}\pi R^3$. The change in radius R is $\Delta R = -0.029$ centimeter. Since the change ΔR is small, we can use the differential dV of volume to approximate the change ΔV in volume. Therefore,

$$\Delta V \approx dV = 4\pi R^2\, dR \underset{\substack{R=3 \\ dR = \Delta R = -0.029}}{=} (4\pi)(9)(-0.029) \approx -3.28$$

The approximate loss in volume is 3.28 cubic centimeters.

The use of dy to approximate Δy when dx is small may also be helpful in approximating *errors*. If Q is the quantity to be measured and if ΔQ is the change in Q, we define

Relative error in $Q = \dfrac{|\Delta Q|}{Q}$ **Percentage error in $Q = \dfrac{|\Delta Q|}{Q}(100\%)$**

For example, if $Q = 50$ units and the change ΔQ in Q is measured to be 5 units, then

Relative error in $Q = \frac{5}{50} = 0.10$ Percentage error in $Q = 10\%$

EXAMPLE 5

Suppose a company manufactures spherical ball bearings with radius 3 centimeters, and the percentage error in the radius must be no more than 1%. What is the approximate percentage error for the surface area of the ball bearing?

Solution

If S is the surface area of a sphere of radius R, then $S = 4\pi R^2$. The actual error $\Delta S/S$ we seek may be approximated by the use of differentials. That is,

$$\frac{\Delta S}{S} \approx \frac{dS}{S} = \frac{8\pi R\, dR}{4\pi R^2} = \frac{2\, dR}{R} = \frac{2\Delta R}{R} = 2(0.01) = 0.02$$

The percentage error in the surface area is 2%.

In Example 5 the percentage error of 1% in the radius of the sphere means the radius will lie somewhere between 2.97 and 3.03 centimeters. But the percentage error of 2% in the surface area means the surface area lies within a factor of $\pm(0.02)$ of $S = 4\pi R^2 = 36\pi$; that is, it lies between $35.28\pi = 110.84$ and $36.72\pi = 115.36$ square centimeters. A rather small error in the radius results in a more significant range of possibilities for the surface area!

Differential Formulas

All the formulas derived earlier for finding derivatives carry over to differentials. We use the symbol df to indicate the differential of the function f. Thus, $d(x^2 + 5) = 2x\,dx$. The following list gives formulas for differentials next to the corresponding derivative formulas.

Derivative	***Differential***
(1) $\frac{d}{dx} c = 0$	(1′) $dc = 0$, if c is constant
(2) $\frac{d}{dx}(kx) = k$	(2′) $d(kx) = k\,dx$, if k is constant
(3) $\frac{d}{dx}(u + v) = \frac{du}{dx} + \frac{dv}{dx}$	(3′) $d(u + v) = du + dv$
(4) $\frac{d}{dx}(uv) = u\frac{dv}{dx} + v\frac{du}{dx}$	(4′) $d(uv) = u\,dv + v\,du$
(5) $\frac{d}{dx}\left(\frac{u}{v}\right) = \frac{v\frac{du}{dx} - u\frac{dv}{dx}}{v^2}$	(5′) $d\left(\frac{u}{v}\right) = \frac{v\,du - u\,dv}{v^2}$
(6) $\frac{d}{dx} x^r = rx^{r-1}$	(6′) $d(x^r) = rx^{r-1}\,dx$, r a rational number
(7) $\frac{d}{dx} \sin x = \cos x$	(7′) $d(\sin x) = (\cos x)\,dx$
(8) $\frac{d}{dx} \cos x = -\sin x$	(8′) $d(\cos x) = (-\sin x)\,dx$
(9) $\frac{d}{dx} \tan x = \sec^2 x$	(9′) $d(\tan x) = (\sec^2 x)\,dx$
(10) $\frac{d}{dx} \sec x = \sec x \tan x$	(10′) $d(\sec x) = (\sec x \tan x)\,dx$
(11) $\frac{d}{dx} \csc x = -\csc x \cot x$	(11′) $d(\csc x) = (-\csc x \cot x)\,dx$
(12) $\frac{d}{dx} \cot x = -\csc^2 x$	(12′) $d(\cot x) = (-\csc^2 x)\,dx$

From now on, to find the differential of a function $y = f(x)$, either find the derivative dy/dx and then multiply by dx or use (1′)–(12′). For example, if $y = x^3 + 2x + 1$, then

$$\frac{dy}{dx} = 3x^2 + 2 \qquad \text{so that} \qquad dy = (3x^2 + 2)\,dx$$

By using the differential formulas, we get

$$\begin{aligned} dy &= d(x^3 + 2x + 1) = d(x^3) + d(2x) + d(1) \\ &= 3x^2\,dx + 2\,dx + 0 = (3x^2 + 2)\,dx \end{aligned}$$

But be careful! The use of dy on the left side of an equation requires dx on the right side. Thus, $dy = 3x^2 + 2$ is incorrect.

The symbol d is an instruction to take the differential.

EXAMPLE 6

(a) $d(x^2 - 3x) = (2x - 3)\,dx$ (b) $d(3y^4 - 2y + 4) = (12y^3 - 2)\,dy$

(c) $d(\sqrt{z^2 + 1}) = \dfrac{z}{\sqrt{z^2 + 1}}\,dz$

The differential can be used as an alternative to implicit differentiation to find the derivative of a function that is defined implicitly.

EXAMPLE 7

Find dy/dx and dx/dy if $x^2 + y^2 = 2xy^2$.

Solution

We take the differential of each side:

$$\begin{aligned} d(x^2 + y^2) &= d(2xy^2) \\ 2x\,dx + 2y\,dy &= 2(y^2\,dx + 2xy\,dy) \\ (y - 2xy)\,dy &= (y^2 - x)\,dx \\ \frac{dy}{dx} &= \frac{y^2 - x}{y - 2xy} \qquad \text{provided} \quad y - 2xy \neq 0 \\ \frac{dx}{dy} &= \frac{y - 2xy}{y^2 - x} \qquad \text{provided} \quad y^2 - x \neq 0 \end{aligned}$$

EXERCISE 4.2

In Problems 1–6 find the differential dy.

1. $y = x^3 - 2x + 1$

2. $y = 4(x^2 + 1)^{3/2}$

3. $y = \dfrac{x - 1}{x^2 + 2x - 8}$

4. $y = \sqrt{x^2 - 1}$

5. $y = 3 \sin 2x + x$

6. $y = \cos^2 3x - x$

In Problems 7–14 find dy/dx and dx/dy by means of differentials.

7. $xy = 7$

8. $3x^2y + 2x - 9 = 0$

9. $x^2 + y^2 = 4$

10. $4xy^2 + yx^2 + 2 = 0$

11. $x^3 + y^3 = 3x^2y$

12. $2x^2 + y^3 = xy^2$

13. $\sin 3y = 2x$

14. $y \sin 2x + x \cos 2y = 1$

In Problems 15–18 find the indicated differential.

15. $d(\sqrt{x} + 2)$

16. $d\left(\frac{1+x}{1-x}\right)$

17. $d(x^3 + x - 2)$

18. $d(x^2 + 2)^{2/3}$

In Problems 19–24 find the linear approximation of f near x_0. Graph f and the linear approximation.

19. $f(x) = x^2 - 2x + 1, \quad x_0 = 2$

20. $f(x) = x^3 - 1, \quad x_0 = 0$

21. $f(x) = \sqrt{x}, \quad x_0 = 4$

22. $f(x) = x^{2/3}, \quad x_0 = 1$

23. $f(x) = \sin x, \quad x_0 = \frac{\pi}{6}$

24. $f(x) = \cos x, \quad x_0 = \frac{\pi}{3}$

25. Use theorem (4.2.2) to approximate:

(a) $\sqrt{35}$ (b) $\sqrt{26.2}$ (c) $1/\sqrt{1.2}$

(d) $\sin 29°$ (Use radians.)

26. Use theorem (4.2.2) to approximate:

(a) $\sqrt[3]{126}$ (b) $\sqrt[3]{123}$ (c) $\sqrt[4]{15}$ (d) $\cos 31°$

27. Use theorem (4.2.2) to find the approximate change in:

(a) $y = f(x) = x^2$ as x changes from 3 to 3.001
(b) $y = f(x) = 1/(x + 2)$ as x changes from 2 to 1.98

28. Use theorem (4.2.2) to find the approximate change in:

(a) $y = x^3$ as x changes from 3 to 3.01
(b) $y = 1/(x - 1)$ as x changes from 2 to 1.98

29. A circular plate is heated and expands. If the radius of the plate increases from $R = 10$ centimeters to $R = 10.1$ centimeters, find the approximate increase in area of the top surface.

30. In a wooden block 3 centimeters thick, an existing circular hole with a radius of 2 centimeters is enlarged to a hole with a radius of 2.2 centimeters. Approximately what volume of wood is removed?

31. Find the approximate change in volume of a spherical balloon of radius 3 meters as the balloon swells to a radius of 3.1 meters.

32. A bee flies around the circumference of a circle traced on a ball with a radius of 7 centimeters at a constant distance of 2 centimeters from the ball. An ant travels along the circumference of the same circle on the ball. Approximately how many more centimeters does the bee travel in one trip around than does the ant?

33. If the percentage error in measuring the edge of a cube is 2%, what is the percentage error in computing its volume?

34. The radius of a spherical ball is computed by measuring the volume of the sphere (by finding how much water it displaces). The volume is found to be 40 cubic centimeters, with a percentage error of 1%. Compute the corresponding percentage error in the radius (due to the error in measuring the volume).

35. A manufacturer produces paper cups in the shape of a right circular cone with radius equal to one-fourth its height. Specifications call for the cups to have a diameter of 4 centimeters. After production, it is discovered that the diameters measure only 3 centimeters. Assuming that the radius is still one-fourth of the height, what is the approximate loss in the capacity of the cup?

36. The oil pan of a car is shaped in the form of a hemisphere with a radius of 8 centimeters. The depth h of the oil is found to be 3 centimeters, with a percentage error of 10%. Approximate the percentage error in the volume. [*Hint:* The volume V for a spherical segment is $V = \frac{1}{3}\pi h^2(3R - h)$, where R is the radius.]

37. To find the height of a building, the length of the shadow of a 3 meter pole placed 9 meters from the building is measured (see the figure). This measurement is found to be 1 meter, with a percentage error of 1%. What is the estimated height of the building? What is the percentage error in the estimate?

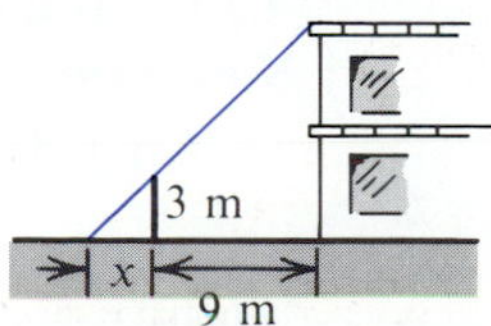

38. The period of the pendulum of a grandfather clock is $T = 2\pi\sqrt{l/g}$, where l is the length (in meters) of the pendulum, T is the period (in seconds), and g is the acceleration due to gravity (9.8 meters per second per second). Suppose the length of the pendulum, a thin wire, increases by 1% due to an increase in temperature. What is the corresponding percentage error in the period? How much time will the clock lose each day?

39. Refer to Problem 38. If the pendulum of a grandfather clock is normally 1 meter long and the length is increased by 10 centimeters, how many minutes will the clock lose each day?

40. What is the approximate volume enclosed by a hollow sphere if its inner radius is 2 meters and its outer radius is 2.1 meters?

41. (a) If $y = f(x)$, $y' = f'(x)$, and so on, why does $dy' = y''\,dx$?
(b) What are dy'' and $d(y'^2)$ when expressed with dx as a factor?

4.3 Maxima and Minima*

Preliminary Definitions

Consider the function f defined on the closed interval $[a, b]$, whose graph appears in Figure 7. Note the behavior of the graph of the function f at the numbers x_1, x_2, x_3, and x_4. In a sufficiently small open interval surrounding x_1, the value of the function is greatest at x_1; the same remark holds true for x_3 and x_4. We say that at x_1, x_3, and x_4, f has *local maxima*—local in the sense that the value of f is greatest (maximum) in some open interval about x_1, x_3, and x_4. Similarly, f has a *local minimum* at x_2.

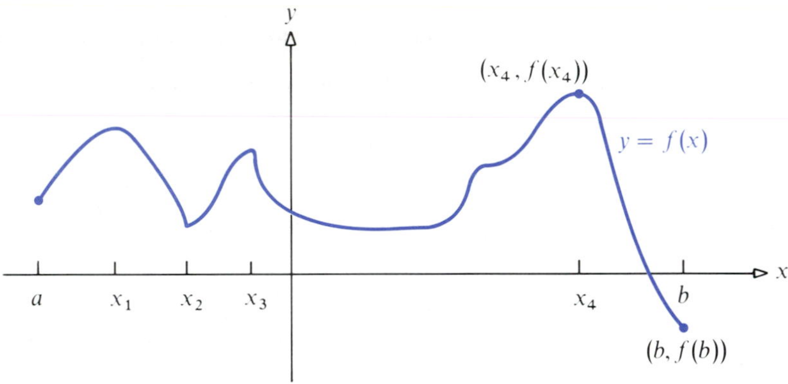

Figure 7

On the closed interval $[a, b]$, the largest value of f is $f(x_4)$, while the smallest value of f is $f(b)$. These are called, respectively, the *absolute maximum* and *absolute minimum* of f on $[a, b]$.

[4.3.1] DEFINITION / *Absolute Maximum and Absolute Minimum.*

Let f denote a function defined on some interval I.

* *Maxima* is the plural of *maximum*; *minima* is the plural of *minimum*.

If there is a number u in I for which $f(u) \geq f(x)$ for all x in I, then $f(u)$ is the *absolute maximum* of f on I and we say that the absolute maximum of f occurs at u.

If there is a number v in I for which $f(v) \leq f(x)$ for all x in I, then $f(v)$ is the *absolute minimum* of f on I and we say that the absolute minimum of f occurs at v.

The values $f(u)$ and $f(v)$ are sometimes referred to as the *absolute extrema** or the *extreme values* of f on I (see Fig. 8). The *absolute* maximum and *absolute* minimum, if they exist, are the largest and smallest values, respectively, that a function f assumes *on the entire interval* over which f is defined. Contrast this idea with that of a *local* maximum and a *local* minimum. These are the largest and smallest values a function f assumes in *some open interval contained in the interval* over which f is defined. The next definition makes this precise.

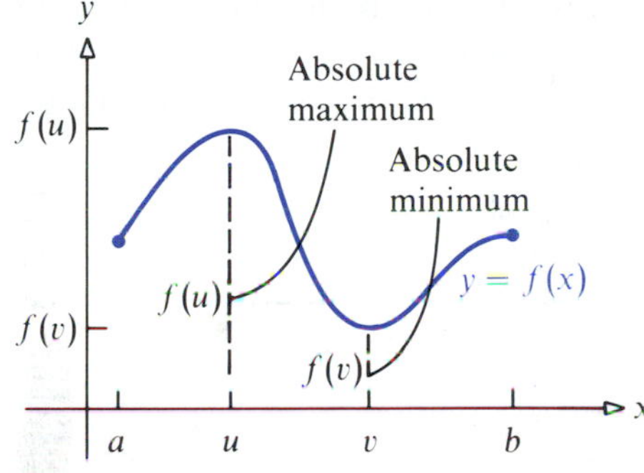

f is defined on $[a, b]$
For all x in $[a, b]$, $f(u) \geq f(x)$.
For all x in $[a, b]$, $f(v) \leq f(x)$;
$f(u)$ is the absolute maximum of f;
$f(v)$ is the absolute minimum of f.

Figure 8

[4.3.2] DEFINITION / *Local Maximum; Local Minimum.*

Let f be a function defined on some interval I and let u and v be numbers in I.

If there is some open interval in I containing u so that $f(u) \geq f(x)$ for all x in this open interval, then f has a *local maximum* at u.

If there is some open interval in I containing v so that $f(v) \leq f(x)$ for all x in this open interval, then f has a *local minimum* at v.

Figure 9 illustrates definition (4.3.2) for a function f defined on $[a, b]$. There is an open interval containing u_1 so that $f(u_1) \geq f(x)$ for all x in this open interval. There is also an open interval containing u_2 so that $f(u_2) \geq f(x)$ for all x in this open interval. Thus, f has a local maximum at u_1 and a local maximum at u_2. There is an open interval containing v so that $f(v) \leq f(x)$ for all x in this open interval; thus, f has a local minimum at v.

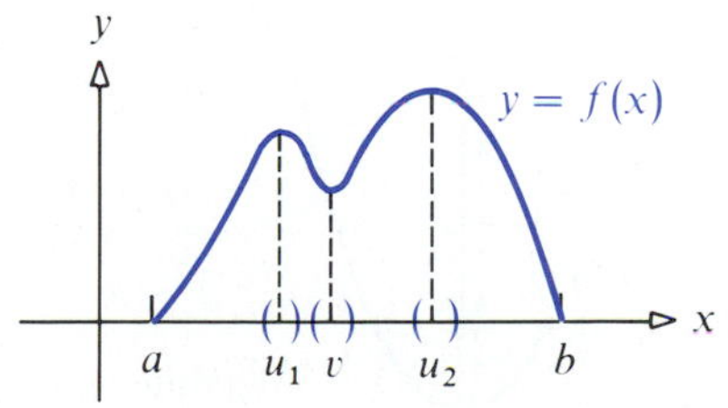

Figure 9

---

* *Extrema* is the plural of *extremum*.

Note that the endpoints a and b are not considered, since f is not defined on any open interval containing a and b.

We call attention to the fact that the interval that contains u in the definition is required to be *open* and that the value $f(u)$ must be larger than or equal to *all* the other values in this open interval. The word *local* is used to emphasize that $f(u)$ is larger than other values of $f(x)$ "around u" or in some (possibly small) open interval containing u. Of course, similar remarks hold for a local minimum.

We shall use the term *local extremum* to describe either a local maximum of f or a local minimum of f.

Examples

EXAMPLE 1

The graph of the function $f(x) = 3x^4 - 4x^3 - 12x^2 + 8$, $-2 \le x \le 3$, is shown in Figure 10. The graph shows that f has both a local minimum and an absolute minimum at 2, a local minimum at -1, and an absolute maximum at -2. The absolute minimum is -24; the absolute maximum is 40.

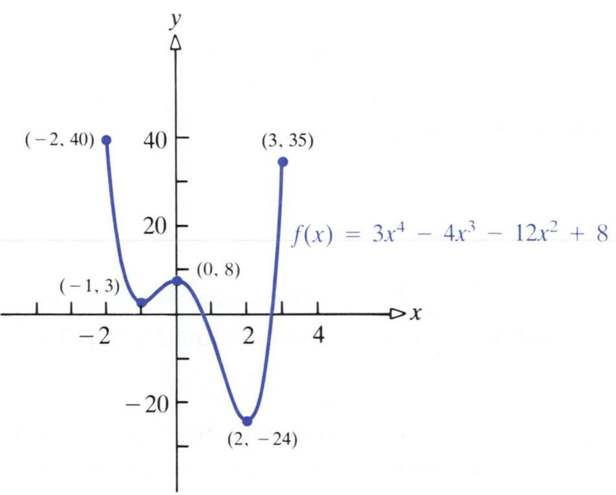

Figure 10

EXAMPLE 2

The function $f(x) = x^2 + 1$, $-1 \le x \le 2$, has an absolute minimum and a local minimum at 0 and an absolute maximum at 2. The absolute minimum is 1; the absolute maximum is 5. See Figure 11.

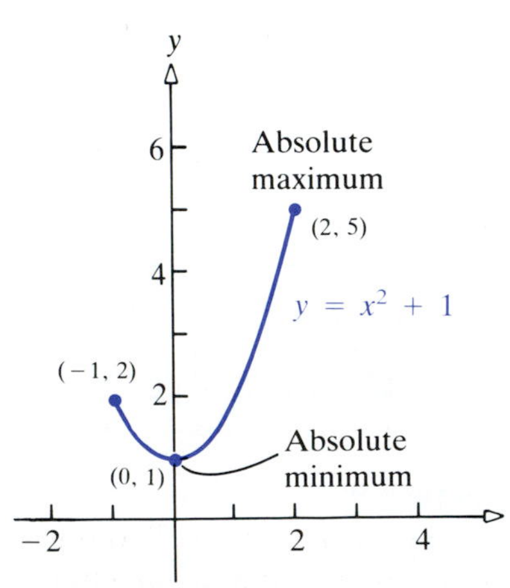

Figure 11

EXAMPLE 3

The function $f(x) = x^2$, $0 < x < 2$, has neither local extrema nor absolute extrema (see Fig. 12 on page 237).

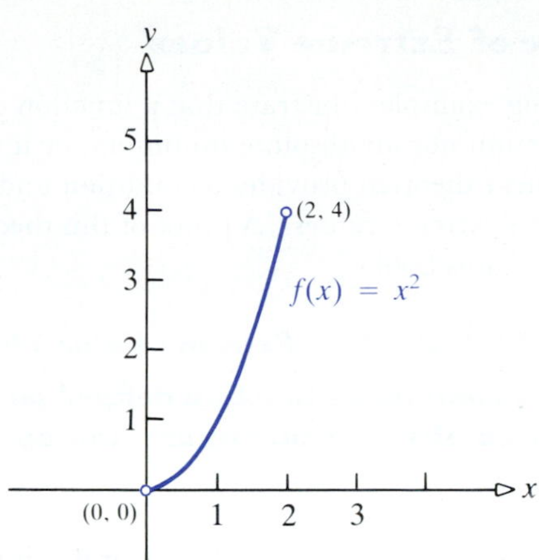

Figure 12

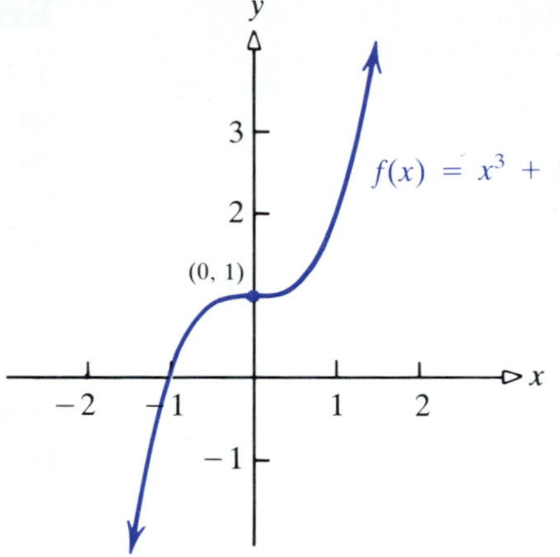

Figure 13

EXAMPLE 4

The function $f(x) = x^3 + 1$, shown in Figure 13 above, has no absolute maximum, no absolute minimum over any open interval, and no local extrema.

EXAMPLE 5

The function $f(x) = 2x^3 - 3x^2 + 1$, $-1 < x < 2$, shown in Figure 14, has local extrema, but no absolute extrema.

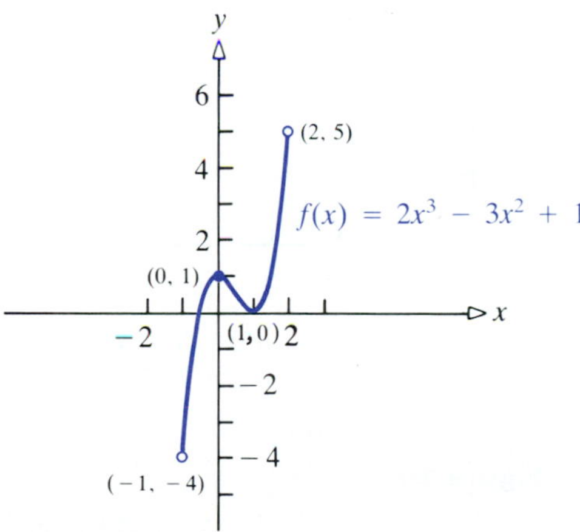

Figure 14

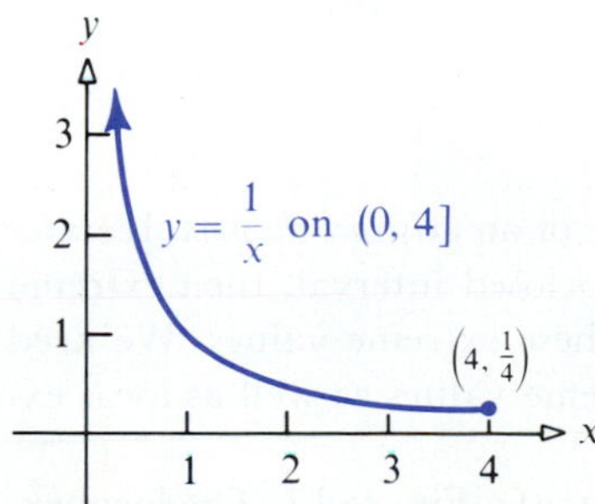

Figure 15

EXAMPLE 6

The function $f(x) = 1/x$, $0 < x \leq 4$, is shown in Figure 15. This function has an absolute minimum of $\frac{1}{4}$ at 4, but no absolute maximum and no local extrema.

Existence of Extreme Values

The foregoing examples illustrate that a function f may have neither an absolute maximum nor an absolute minimum, or it may have one without the other. The next theorem provides a condition under which a function f will always possess extreme values. (A proof of this theorem may be found in most advanced calculus books.)

[4.3.3] THEOREM / ***Extreme Value Theorem.***

If f is a continuous function defined on a closed interval $[a, b]$, then f has an absolute maximum and an absolute minimum on $[a, b]$.

If f fails to be continuous on $[a, b]$, or if f is not defined on a closed interval, then the theorem does not apply, as demonstrated by the two functions in Figure 16. The function f in Figure 16(a) is not continuous at 1. It has an absolute minimum of 1 at 2 over $[0, 2]$, but there is no absolute maximum over $[0, 2]$. This does not contradict the extreme value theorem because f is not continuous on $[0, 2]$.

In Figure 16(b) the function $g(x) = x$, $0 < x < 2$, is continuous for $0 < x < 2$, but it has neither an absolute maximum nor an absolute minimum on $(0, 2)$. This does not contradict the extreme value theorem, since the interval $(0, 2)$ is not closed.

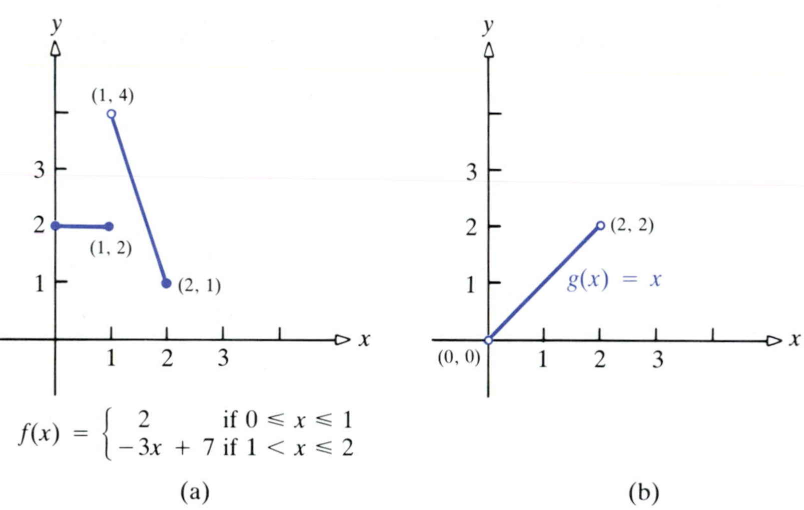

Figure 16

The extreme value theorem is an example of an *existence theorem*. It states that, if a continuous function is defined on a closed interval, then extreme values exist. It does not tell us how to find these extreme values. We need a tool that will enable us to locate these extreme values as well as local extrema. We start by looking at local extrema.

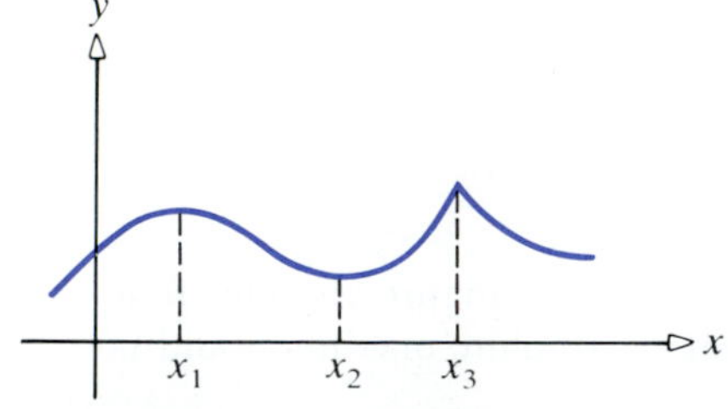

Figure 17

Consider the function whose graph is shown in Figure 17. The local extrema occur at x_1 and x_2, at which the derivative is 0, and at x_3, at which the derivative fails to exist. This is what the next theorem says in general.

[4.3.4] THEOREM / *A Necessary Condition for Local Extrema.*

If a function f has a local maximum or a local minimum at the number c, then either $f'(c) = 0$ or $f'(c)$ does not exist.

Proof Suppose f has a local maximum at c. Then, by definition, $f(x) \leq f(c)$ for all x in some open interval containing c. As a result, $f(x) - f(c) \leq 0$. The derivative of f at c may be written as

$$\lim_{x \to c} \frac{f(x) - f(c)}{x - c}$$

If this limit does not exist, then $f'(c)$ does not exist and there is nothing further to prove. If this limit does exist, then

$$\lim_{x \to c^-} \frac{f(x) - f(c)}{x - c} = \lim_{x \to c^+} \frac{f(x) - f(c)}{x - c}$$

In the limit on the left side, $x < c$ and $f(x) - f(c) \leq 0$. Hence, the quantity

$$\frac{f(x) - f(c)}{x - c} \geq 0$$

and so

$$\lim_{x \to c^-} \frac{f(x) - f(c)}{x - c} \geq 0^* \qquad (1)$$

In the limit on the right side, $x > c$ and $f(x) - f(c) \leq 0$. Hence, the quantity

$$\frac{f(x) - f(c)}{x - c} \leq 0$$

and so

$$\lim_{x \to c^+} \frac{f(x) - f(c)}{x - c} \leq 0 \qquad (2)$$

Since the limits (1) and (2) are required to be equal, we must have

$$\lim_{x \to c} \frac{f(x) - f(c)}{x - c} = f'(c) = 0$$

The proof when f has a local minimum at c is similar and is left as an exercise (Problem 63).

For differentiable functions, theorem (4.3.4) has the following form:

[4.3.5] THEOREM

If a differentiable function f has a local maximum or a local minimum at c, then $f'(c) = 0$.

---

* If $\lim_{x \to c^-} \frac{f(x) - f(c)}{x - c} < 0$ there would be some left-sided interval about c on which $\frac{f(x) - f(c)}{x - c} < 0$, which is not possible. See Theorem (2.2.10).

In other words, for differentiable functions, a local maximum or a local minimum occurs at a point where the tangent line to the graph of f is horizontal. We shall make use of this fact a little later.

As theorem (4.3.4) shows, the numbers at which a function f has a 0 derivative, or at which f' does not exist, provide a clue for locating where f has local extrema. Unfortunately, the fact that the derivative is 0 at a number will not guarantee a local extremum at this number. Nor will the nonexistence of the derivative guarantee a local extremum. For example, in Figure 18 $f'(x_3) = 0$, but f has neither a local maximum nor a local minimum at x_3. Similarly, $f'(x_4)$ does not exist, but f has neither a local maximum nor a local minimum at x_4.

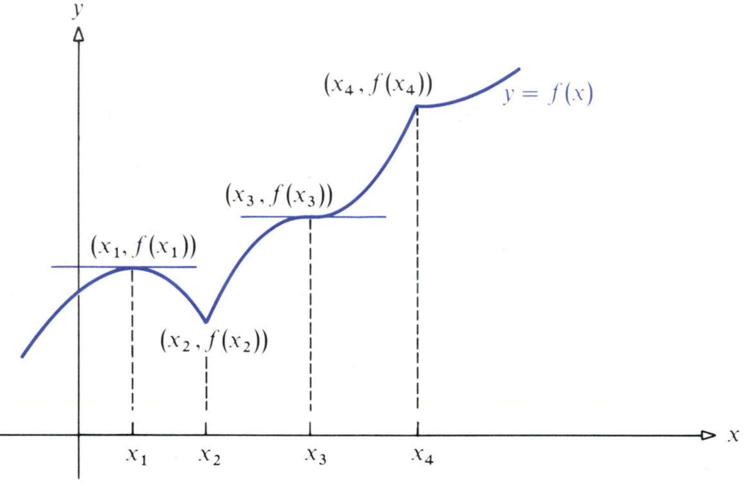

$f'(x_1) = 0$; f has a local maximum at x_1
$f'(x_2)$ does not exist; f has a local minimum at x_2
$f'(x_3) = 0$; f has neither a local maximum nor a local minimum
$f'(x_4)$ does not exist; f has neither a local maximum nor a local minimum

Figure 18

Critical Numbers

Even though local extrema may not be found at all numbers x for which $f'(x) = 0$ or $f'(x)$ does not exist, these numbers do provide *all* the *candidates* at which f *might* have local extrema. For this reason, we give a special name to the numbers x at which $f'(x) = 0$ or $f'(x)$ does not exist; they are called *critical numbers.*

[4.3.6] DEFINITION / ***Critical Number.***

A number c in the domain of a function f for which either $f'(c) = 0$ or $f'(c)$ does not exist is called a *critical number* of the function f.

EXAMPLE 7

Find the critical numbers of the following functions:

(a) $f(x) = 6 - x^2$ (b) $f(x) = x^3 - 6x^2 + 9x + 2$

(c) $f(x) = \dfrac{1}{x-2}$ (d) $f(x) = \dfrac{(x-2)^{2/3}}{x}$ (e) $f(x) = \sin x$

Solution

(a) For $f(x) = 6 - x^2$, the derivative $f'(x) = -2x$ exists everywhere. Since $f'(x) = 0$ only at $x = 0$, 0 is the only critical number.

(b) For $f(x) = x^3 - 6x^2 + 9x + 2$, the derivative

$$f'(x) = 3x^2 - 12x + 9 = 3(x-1)(x-3)$$

exists everywhere. Since $f'(x) = 0$ at $x = 1$ and $x = 3$, 1 and 3 are the only critical numbers.

(c) For $f(x) = 1/(x-2)$, the derivative $f'(x) = -1/(x-2)^2$ exists throughout the domain of f and is never 0. (Note that 2 is not in the domain of f and hence is excluded.) Therefore, there are no critical numbers.

(d) For $f(x) = (x-2)^{2/3}/x$, the derivative is

$$f'(x) = \frac{(x)(\frac{2}{3})(x-2)^{-1/3} - (x-2)^{2/3}}{x^2}$$

$$= \frac{2x - 3(x-2)}{3x^2(x-2)^{1/3}} = \frac{6-x}{3x^2(x-2)^{1/3}}$$

Clearly, 6 is a critical number, since $f'(6) = 0$. The derivative f' does not exist for those numbers x that give a 0 denominator. There are two such numbers, 0 and 2. However, of these, we must exclude 0, since 0 is not in the domain of f. Thus, f has only two critical numbers, 6 and 2.

(e) For $f(x) = \sin x$, the derivative $f'(x) = \cos x$ exists everywhere. Since $f'(x) = 0$ at $x = \pm\pi/2, \pm 3\pi/2, \pm 5\pi/2, \ldots$, f has infinitely many critical numbers.

We are not yet ready to give a procedure for determining whether a function f has a local maximum or a local minimum or neither at a critical number. This requires the mean value theorem, which is the subject of the next section. However, the critical numbers do help us find the extreme values of a function f.

Finding the Absolute Maximum and the Absolute Minimum

The absolute maximum (and absolute minimum) of a continuous function f defined on a closed interval $[a, b]$ will occur either at an endpoint of the interval $[a, b]$ or at a critical number in the open interval (a, b). Hence, we have the following test:

[4.3.7] THEOREM | ***Test for Absolute Maximum and Absolute Minimum.***

If a continuous function f is defined on a closed interval $[a, b]$, the absolute maximum and the absolute minimum of f are, respectively, the largest and the smallest values found among the following:

(a) The values of f at the critical numbers in the open interval (a, b)
(b) $f(a)$ and $f(b)$, the values of f at the endpoints a and b

Let's use theorem (4.3.7) to find the extreme values of some specific functions.

EXAMPLE 8

Find the absolute maximum and absolute minimum of the functions

(a) $f(x) = x^3 - 6x^2 + 9x + 2, \quad \text{on } [0, 2]$

(b) $f(x) = \dfrac{(x-2)^{2/3}}{x}, \quad \text{on } [1, 10]$

Solution

(a) The function f is continuous on $[0, 2]$. Thus, by theorem (4.3.3), f has an absolute maximum and an absolute minimum. From Example 7, part (b), the critical numbers of f are 1 and 3. However, we exclude 3, since we are interested in the extrema of f only on $[0, 2]$. We calculate

$$f(1) = 6$$

At the endpoints 0 and 2, we have

$$f(0) = 2 \qquad \text{and} \qquad f(2) = 4$$

Thus, the absolute maximum of f, which occurs at 1, is 6; the absolute minimum of f, which occurs at 0, is 2.

(b) The function f is continuous on $[1, 10]$. Thus, by theorem (4.3.3), f has an absolute maximum and an absolute minimum. From Example 7, part (d), the critical numbers of f are 6 and 2. We calculate

$$f(6) = \frac{4^{2/3}}{6} \approx 0.42 \qquad f(2) = 0$$

At the endpoints 1 and 10, we have

$$f(1) = 1 \qquad f(10) = \tfrac{4}{10} = 0.4$$

Thus, the absolute maximum of f, which occurs at 1, is 1; the absolute minimum of f, which occurs at 2, is 0.

For functions f given by more than one rule, we need to be careful at the number where the split occurs. Let's look at an example.

EXAMPLE 9

Find the absolute maximum and absolute minimum of the function

$$f(x) = \begin{cases} 2x - 1 & \text{if } 0 \le x \le 2 \\ x^2 - 5x + 9 & \text{if } 2 < x \le 3 \end{cases}$$

Solution

The function f is continuous on $[0, 3]$. (You should check this.) To find the critical numbers of f on $[0, 3]$, we observe that for $0 < x < 2$, $f'(x) = 2$, and for $2 < x < 3$, $f'(x) = 2x - 5$. Thus, $f'(x)$ is never 0 on $(0, 2)$, whereas $f'(x) = 0$ at $\frac{5}{2}$ on $(2, 3)$. At 2, the derivative does not exist (check this), so 2 is a critical number. Hence, the critical numbers of f are $\frac{5}{2}$ and 2. We list the values of $f(x)$ at critical numbers and endpoints of the interval:

$$f(2) = (2)(2) - 1 = 3$$

$$f(\tfrac{5}{2}) = (\tfrac{5}{2})^2 - 5(\tfrac{5}{2}) + 9 = \tfrac{11}{4}$$

$$f(0) = (2)(0) - 1 = -1$$

$$f(3) = 9 - 15 + 9 = 3$$

The absolute maximum is 3 and occurs at 2 and at 3; the absolute minimum is -1 and occurs at 0. The graph of f is given in Figure 19.

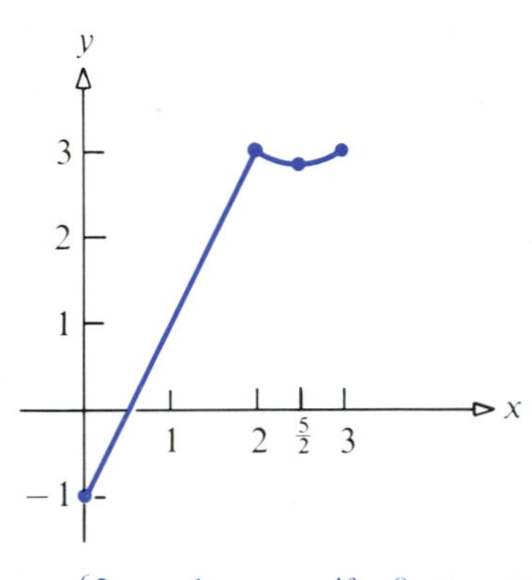

$f(x) = \begin{cases} 2x - 1 & \text{if } 0 \le x \le 2 \\ x^2 - 5x + 9 & \text{if } 2 < x \le 3 \end{cases}$

Figure 19

EXERCISE 4.3

In Problems 1 and 2 use the functions whose graphs are shown to determine whether the function has absolute or local extrema at the indicated numbers.

1.

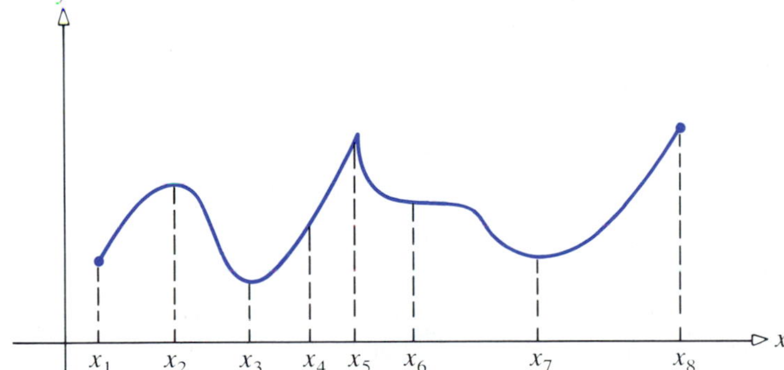

2.

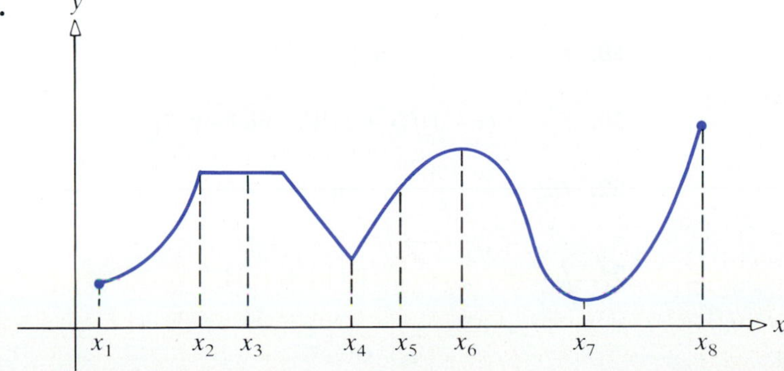

In Problems 3–32 find all the critical numbers for the given functions.

3. $f(x) = x^2 + 2x$

4. $f(x) = x^2 - 8x$

5. $f(x) = 1 - 6x + x^2$

6. $f(x) = 4 - 2x - x^2$

7. $f(x) = x^3 - 3x^2$

8. $f(x) = x^3 - 6x$

9. $f(x) = x^4 - 2x^2 + 1$

10. $f(x) = 3x^4 - 4x^3$

11. $f(x) = x^{2/3}$

12. $f(x) = x^{1/3}$

13. $f(x) = 2\sqrt{x}$

14. $f(x) = 4 - \sqrt{x}$

15. $f(x) = x\sqrt{1 - x^2}$

16. $f(x) = x^2\sqrt{2 - x}$

17. $f(x) = \dfrac{x^2}{x - 1}$

18. $f(x) = \dfrac{x}{x^2 - 1}$

19. $f(x) = (x + 3)^2(x - 1)^{2/3}$

20. $f(x) = (x - 1)^2(x + 1)^{1/3}$

21. $f(x) = \dfrac{(x - 3)^{1/3}}{x - 1}$

22. $f(x) = \dfrac{(x + 3)^{2/3}}{x + 1}$

23. $f(x) = \dfrac{\sqrt[3]{x^2 - 9}}{x}$

24. $f(x) = \dfrac{\sqrt[3]{4 - x^2}}{x}$

25. $f(x) = \cos x, \quad 0 \le x \le 2\pi$

26. $f(x) = \tan x, \quad -\dfrac{\pi}{4} \le x \le \dfrac{\pi}{4}$

27. $f(x) = \sin 4x, \quad -\dfrac{\pi}{2} \le x \le \dfrac{\pi}{2}$

28. $f(x) = \cos\dfrac{x}{2}, \quad -\pi \le x \le \pi$

29. $f(x) = |x - 1|$

30. $f(x) = |x + 1|$

31. $f(x) = \begin{cases} 3x & \text{if } 0 \le x < 1 \\ 4 - x & \text{if } 1 \le x \le 2 \end{cases}$

32. $f(x) = \begin{cases} x^2 & \text{if } 0 \le x < 1 \\ 1 - x^2 & \text{if } 1 \le x \le 2 \end{cases}$

In Problems 33–58 find the absolute maximum and absolute minimum of each function f on the indicated interval. Note that the functions in Problems 33–54 are the same as those in Problems 3–24.

33. $f(x) = x^2 + 2x, \quad \text{on } [-3, 3]$

34. $f(x) = x^2 - 8x, \quad \text{on } [-1, 10]$

35. $f(x) = 1 - 6x + x^2, \quad \text{on } [0, 4]$

36. $f(x) = 4 - 2x - x^2, \quad \text{on } [-2, 2]$

37. $f(x) = x^3 - 3x^2, \quad \text{on } [1, 4]$

38. $f(x) = x^3 - 6x, \quad \text{on } [-1, 1]$

39. $f(x) = x^4 - 2x^2 + 1, \quad \text{on } [0, 2]$

40. $f(x) = 3x^4 - 4x^3, \quad \text{on } [-2, 0]$

41. $f(x) = x^{2/3}, \quad \text{on } [-1, 1]$

42. $f(x) = x^{1/3}, \quad \text{on } [-1, 1]$

43. $f(x) = 2\sqrt{x}, \quad \text{on } [1, 4]$

44. $f(x) = 4 - \sqrt{x}, \quad \text{on } [0, 4]$

45. $f(x) = x\sqrt{1 - x^2}, \quad \text{on } [-1, 1]$

46. $f(x) = x^2\sqrt{2 - x}, \quad \text{on } [0, 2]$

47. $f(x) = \dfrac{x^2}{x - 1}, \quad \text{on } [-1, \tfrac{1}{2}]$

48. $f(x) = \dfrac{x}{x^2 - 1}, \quad \text{on } [-\tfrac{1}{2}, \tfrac{1}{2}]$

49. $f(x) = (x + 3)^2(x - 1)^{2/3}, \quad \text{on } [-4, 5]$

50. $f(x) = (x - 1)^2(x + 1)^{1/3}, \quad \text{on } [-2, 7]$

51. $f(x) = \dfrac{(x - 3)^{1/3}}{x - 1}, \quad \text{on } [2, 11]$

52. $f(x) = \dfrac{(x + 3)^{2/3}}{x + 1}, \quad \text{on } [-4, -2]$

53. $f(x) = \dfrac{\sqrt[3]{x^2 - 9}}{x}, \quad \text{on } [3, 6]$

54. $f(x) = \dfrac{\sqrt[3]{4 - x^2}}{x}, \quad \text{on } [-4, -1]$

55. $f(x) = \begin{cases} 2x+1 & \text{if} \quad 0 \le x < 1 \\ 3x & \text{if} \quad 1 \le x \le 3 \end{cases}$

56. $f(x) = \begin{cases} x+3 & \text{if} \quad -1 \le x \le 2 \\ 2x+1 & \text{if} \quad 2 < x \le 4 \end{cases}$

57. $f(x) = \begin{cases} x^2 & \text{if} \quad -2 \le x < 1 \\ x^3 & \text{if} \quad 1 \le x \le 2 \end{cases}$

58. $f(x) = \begin{cases} x+2 & \text{if} \quad -1 \le x < 0 \\ 2-x & \text{if} \quad 0 \le x \le 1 \end{cases}$

59. A truck has a top speed of 75 miles per hour and, when traveling at the rate of x miles per hour, it consumes gasoline at the rate of $\frac{1}{200}(1600/x + x)$ gallon per mile. If the length of the trip is 200 miles and the price of gasoline is \$1.60 per gallon, the cost $C(x)$ (in dollars) is

$$C(x) = (1.60)\left(\frac{1600}{x} + x\right)$$

What is the most economical speed for the truck? Use the interval $[10, 75]$.

60. If the driver of the truck in Problem 59 is paid \$8.00 per hour and his salary is added to the cost, what is the most economical speed for the truck?

61. The function $f(x) = Ax^2 + Bx + C$ has a local minimum at 0, and its graph passes through the points $(0, 2)$ and $(1, 8)$. Find A, B, and C.

62. Let $f(x) = \sqrt{1 + x^2} + |x - 2|$. Find the absolute maximum and absolute minimum of f on $[0, 3]$, and determine where each occurs.

63. Prove that if f has a local minimum at c, then either $f'(c) = 0$ or $f'(c)$ does not exist.

64. Discuss the domain of the function

$$f(x) = [(16 - x^2)(x^2 - 9)]^{1/2}$$

and find the absolute maximum of f in its domain.

65. Show that if f has a local minimum at c, then $g(x) = -f(x)$ has a local maximum at c.

Rolle's Theorem; Mean Value Theorem 4.4

Rolle's Theorem

The next theorem, which is due to the French mathematician Michel Rolle (1652–1719), is important because of its theoretical value. We shall use it to prove the mean value theorem, which then will be used to derive tests for locating local extrema.

[4.4.1] THEOREM / *Rolle's Theorem.*

Let f be a function defined on a closed interval $[a, b]$. If

(a) f is continuous on $[a, b]$,
(b) f is differentiable on (a, b), and
(c) $f(a) = f(b)$,

then there is at least one number c in the open interval (a, b) for which $f'(c) = 0$.

Before we give a proof of Rolle's theorem, let's look at Figure 20. The graphs in parts (a) and (b) meet conditions (a), (b), and (c) of theorem (4.4.1). The existence of at least one number c at which $f'(c) = 0$ is apparent. The graphs in parts (c), (d), and (e) demonstrate that the conclusion of Rolle's theorem may not hold when one or more of the conditions (a), (b), and (c) are not met.

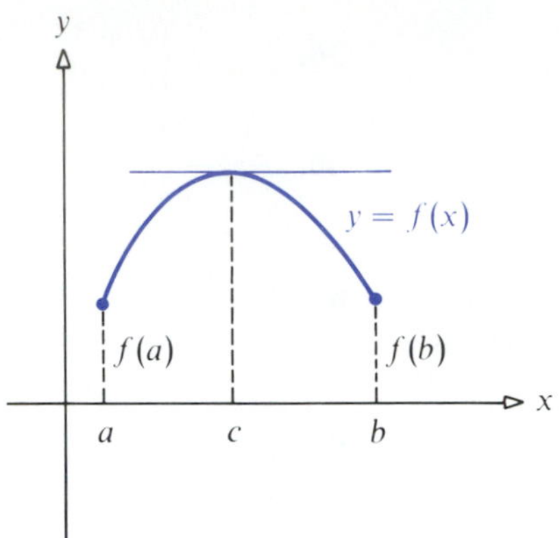

(a) $f'(c) = 0$

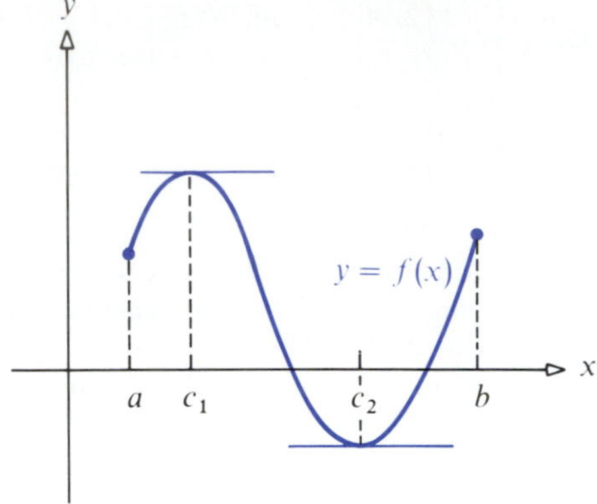

(b) $f'(c_1) = 0$; $f'(c_2) = 0$

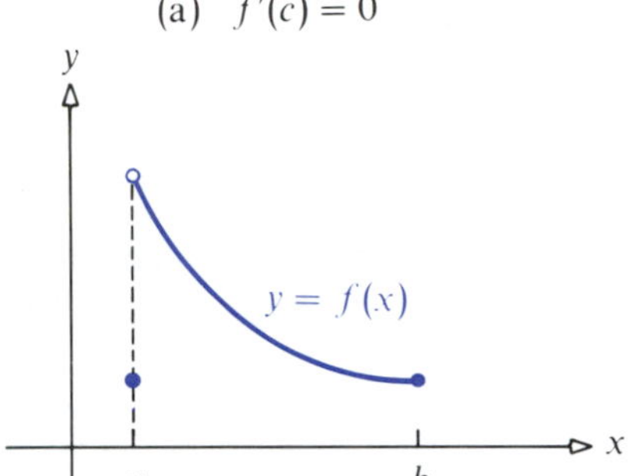

(c) f is not continuous at a
f is differentiable on (a, b)
$f(a) = f(b)$
No c in (a, b) at which $f'(c) = 0$

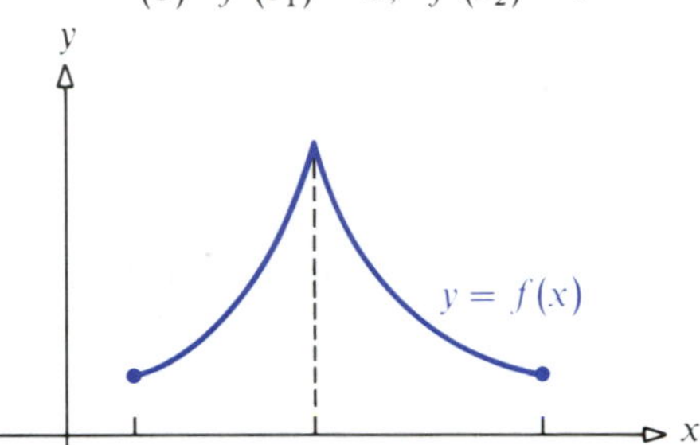

(d) f is continuous on $[a, b]$
f is not differentiable on (a, b), no derivative at c
$f(a) = f(b)$
No c in (a, b) at which $f'(c) = 0$

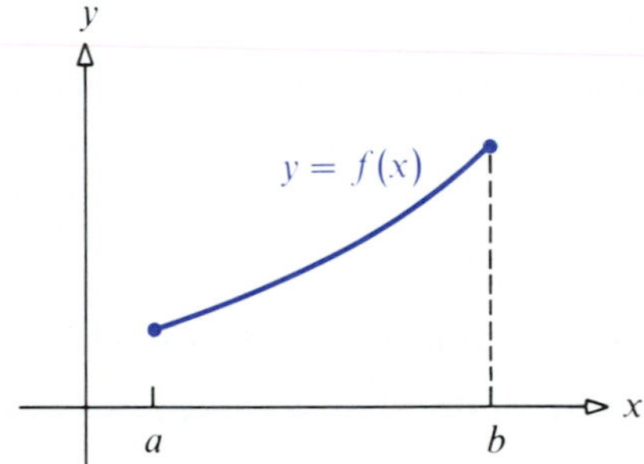

(e) f is continuous on $[a, b]$
f is differentiable on (a, b)
$f(a) \neq f(b)$
No c in (a, b) at which $f'(c) = 0$

Figure 20

Proof of Rolle's Theorem Because f is continuous on a closed interval $[a, b]$, we know that f has an absolute maximum and an absolute minimum (the extreme value theorem, (4.3.3)). We look at the possibilities:

1. If f is a constant function on $[a, b]$, then $f(x) = f(a)$ for all x in $[a, b]$ and, hence, $f'(x) = 0$ for all x in (a, b).

2. If f is not constant on $[a, b]$, then, because $f(a) = f(b)$, either the absolute maximum or the absolute minimum must be attained in the open interval (a, b)—say, at c. But at c, f has a local maximum (or a local minimum). From theorem (4.3.4), we know that a necessary condition for a local extremum is that $f'(c) = 0$ or $f'(c)$ does not exist. Since f is differentiable on (a, b) and c is in (a, b), then we must have $f'(c) = 0$.

Let's look at a specific function that satisfies the conditions of Rolle's theorem (4.4.1).

EXAMPLE 1

For the function $f(x) = x^2 - 5x + 6$, find the two x-intercepts and show that $f'(x) = 0$ for some number between these two intercepts.

Solution

To find the x-intercepts, we set $f(x) = 0$ to get

$$f(x) = x^2 - 5x + 6 = (x - 2)(x - 3) = 0$$

Thus, $f(2) = 0$ and $f(3) = 0$. The function f is continuous on $[2, 3]$ and differentiable on $(2, 3)$ because it is a polynomial. Since f satisfies the three conditions of Rolle's theorem, the theorem guarantees the existence of a number c in the interval $(2, 3)$ such that $f'(c) = 0$. We can find c as follows:

$$f'(x) = 2x - 5 = 0$$

$$x = \tfrac{5}{2}$$

Since $\frac{5}{2}$ is in the interval $(2, 3)$ and since $f'(\frac{5}{2}) = 0$, the required number is $c = \frac{5}{2}$ (see Fig. 21).

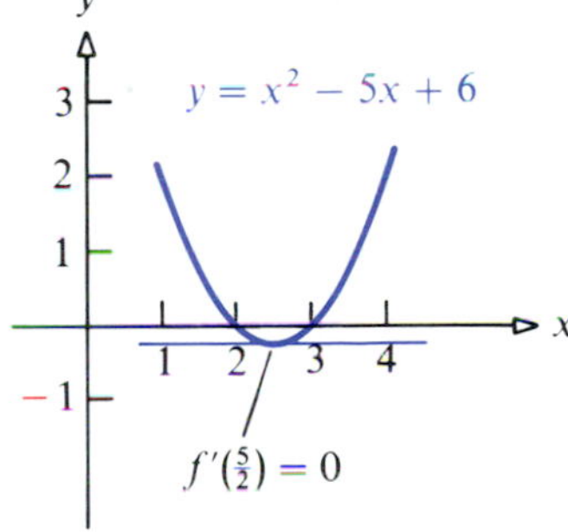

Figure 21

Physical Interpretation of Rolle's Theorem

Suppose $s = f(t)$ denotes the distance a particle has traveled at time t. If the particle is in the same place at two different times $t = a$ and $t = b$, then $f(a) = f(b)$. Rolle's theorem says that there is some time $t = c$ between a and b when $v = f'(c) = 0$—that is, when the velocity is 0. For example, if a ball is thrown directly upward, then at some instant it changes direction—namely, the instant when $f'(c) = 0$. This is also the time when the height of the object is greatest.

Mean Value Theorem

The significance of Rolle's theorem is its theoretical value in obtaining other results, many of which have wide-ranging application. Perhaps the most important of these is the *mean value theorem*, or the *theorem of the mean for derivatives*. This theorem asserts the geometric condition that if f is continuous on $[a, b]$ and differentiable on (a, b), then there is at least one point $(c, f(c))$ between $(a, f(a))$ and $(b, f(b))$ on the graph of f at which the slope $f'(c)$ of

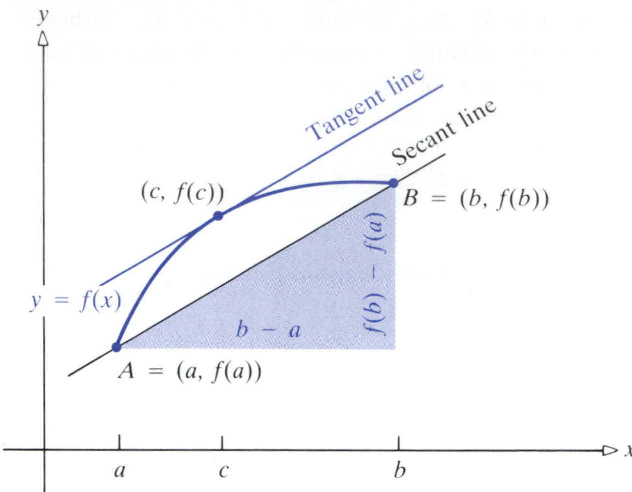

Figure 22

the tangent line equals the slope of the secant line joining $A = (a, f(a))$ and $B = (b, f(b))$. That is, as Figure 22 illustrates,

$$f'(c) = \text{Slope of } AB = \frac{f(b) - f(a)}{b - a}$$

[4.4.2] THEOREM / *Mean Value Theorem.*

Let f be a function defined on a closed interval $[a, b]$. If

(a) f is continuous on $[a, b]$ and
(b) f is differentiable on (a, b),

then there is at least one number c in the open interval (a, b) at which

$$f'(c) = \frac{f(b) - f(a)}{b - a}$$

Proof of Mean Value Theorem We begin by constructing a function g that satisfies the conditions of Rolle's theorem. We define g as

$$g(x) = f(x) - \left[f(a) + \frac{f(b) - f(a)}{b - a}(x - a) \right]$$

for all x on $[a, b]$. This function g has geometric significance: Its value equals the directed vertical distance from the graph of $y = f(x)$ to the secant line joining $(a, f(a))$ to $(b, f(b))$. (Refer back to Fig. 22.)

Since f is continuous on $[a, b]$ and is differentiable on (a, b), it follows that g is also continuous on $[a, b]$ and is differentiable on (a, b). Furthermore,

$$g(a) = f(a) - \left[f(a) + \frac{f(b) - f(a)}{b - a}(a - a) \right] = 0$$

$$g(b) = f(b) - \left[f(a) + \frac{f(b) - f(a)}{b - a}(b - a) \right] = 0$$

Since g satisfies conditions (a), (b), and (c) of Rolle's theorem, there is a number c in (a, b) at which $g'(c) = 0$. Also, since $f(a)$ and $[f(b) - f(a)]/(b - a)$ are numbers, we calculate $g'(x)$ to be

$$g'(x) = f'(x) - \frac{f(b) - f(a)}{b - a}$$

Hence,

$$g'(c) = f'(c) - \left[\frac{f(b) - f(a)}{b - a}\right] = 0$$

so that

$$f'(c) = \frac{f(b) - f(a)}{b - a}$$

The mean value theorem does not tell how to find the number c for which $f'(c) = 0$. However, the number c can often be found by algebra.

EXAMPLE 2

Verify that the function $f(x) = x^3 - 3x + 5$ on the interval $[-1, 1]$ satisfies the conditions of the mean value theorem, and find the number(s) in the interval $(-1, 1)$ at which

$$f'(c) = \frac{f(1) - f(-1)}{1 - (-1)}$$

Solution

Since f is a polynomial function, we know that f is continuous on $[-1, 1]$ and differentiable on $(-1, 1)$. Thus,

$$f'(x) = 3x^2 - 3 \qquad f'(c) = 3c^2 - 3$$

and

$$f(1) = 3 \qquad f(-1) = 7$$

By the mean value theorem, there is a number c in $(-1, 1)$ so that

$$f'(c) = \frac{f(1) - f(-1)}{1 - (-1)} \qquad \text{or} \qquad 3c^2 - 3 = \frac{3 - 7}{2}$$

$$3c^2 - 3 = -2$$

$$3c^2 = 1$$

There are two numbers in the interval $(-1, 1)$ that satisfy the requirement—namely, $-1/\sqrt{3} = -\sqrt{3}/3$ and $1/\sqrt{3} = \sqrt{3}/3$ (see Fig. 23 on page 250).

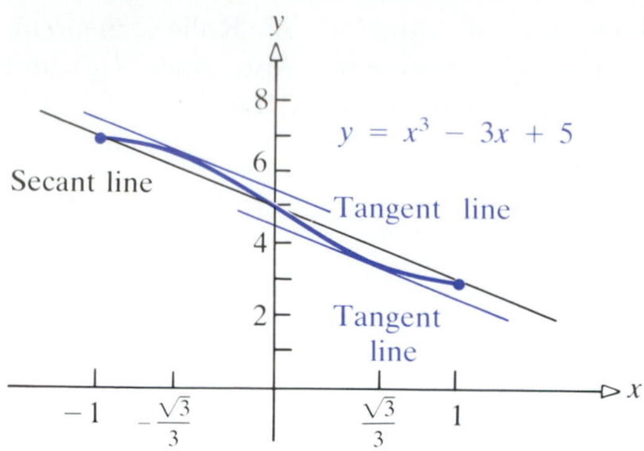

Figure 23

Physical Interpretation of Mean Value Theorem

Suppose $s = f(t)$ denotes the distance a particle has traveled at time t. If f is continuous on $[a, b]$ and differentiable on (a, b), the mean value theorem tells us that **a moving particle must attain its average velocity over an interval of time at some instant during the interval**. To see this, notice that

$$\text{Average velocity} = \frac{f(t_2) - f(t_1)}{t_2 - t_1}$$

$$\text{Instantaneous velocity } v(t) = \frac{ds}{dt} = f'(t)$$

By the mean value theorem, there exists a time t_0 such that $t_1 < t_0 < t_2$ and

$$f'(t_0) = \frac{f(t_2) - f(t_1)}{t_2 - t_1}$$

Therefore $v(t_0) = f'(t_0)$ is the average velocity over the interval.

For example, if a car travels 110 miles in 2 hours and $f(t)$ represents the distance traveled after t hours, the *average velocity* during this time period is:

$$\text{Average velocity} = \frac{f(2) - f(0)}{2 - 0} = \frac{110 - 0}{2 - 0} = 55 \text{ miles per hour}$$

We may therefore conclude, by the mean value theorem, that the car has velocity $v(t_0) = 55$ miles per hour at least once during the 2 hour period.

EXERCISE 4.4

In Problems 1–12 the function f defined on the indicated interval $[a, b]$ satisfies the conditions of Rolle's theorem. Find all numbers c in (a, b) at which $f'(c) = 0$.

1. $f(x) = x^2 - 3x$, on $[0, 3]$

2. $f(x) = x^2 + 2x$, on $[-2, 0]$

3. $f(x) = x^2 - 2x - 2$, on $[0, 2]$

4. $f(x) = x^2 + 1$, on $[-1, 1]$

5. $f(x) = x^3 - x$, on $[-1, 0]$

6. $f(x) = x^3 - 4x$, on $[-2, 2]$

7. $f(x) = x^3 - x + 2$, on $[-1, 1]$

8. $f(x) = x^4 - 3$, on $[-2, 2]$

9. $f(x) = x^4 - 2x^2 + 1$, on $[-2, 2]$

10. $f(x) = x^4 + x^2$, on $[-2, 2]$

11. $f(x) = \sin 2x$, on $[0, \pi]$

12. $f(x) = \sin x + \cos x$, on $[0, 2\pi]$

State why Rolle's theorem cannot be applied to the functions in Problems 13–16.

13. $f(x) = x^2 - 2x + 1$, on $[-2, 1]$

14. $f(x) = x^3 - 3x$, on $[2, 4]$

15. $f(x) = x^{1/3} - x$, on $[-1, 1]$

16. $f(x) = x^{2/5}$, on $[-1, 1]$

In Problems 17–26 the function f defined on the indicated interval $[a, b]$ satisfies the conditions of the mean value theorem. Find all numbers c in (a, b) for which $f'(c) = \dfrac{f(b) - f(a)}{b - a}$.

17. $f(x) = x^2$, on $[-1, 2]$

18. $f(x) = x^2 + 1$, on $[0, 2]$

19. $f(x) = \sqrt{x}$, on $[0, 4]$

20. $f(x) = x^4 + 5$, on $[0, 1]$

21. $f(x) = x^3 - 5x^2 + 4x - 2$, on $[1, 3]$

22. $f(x) = x^3 - 7x^2 + 5x$, on $[-2, 2]$

23. $f(x) = \dfrac{x + 1}{x}$, on $[1, 3]$

24. $f(x) = \dfrac{x^2}{x + 1}$, on $[0, 1]$

25. $f(x) = \sqrt[3]{x^2}$, on $[1, 8]$

26. $f(x) = \sqrt{x - 2}$, on $[2, 4]$

27. Consider $f(x) = |x|$ on the interval $[-1, 1]$. Here, $f(1) = f(-1) = 1$, but there is no c in $(-1, 1)$ at which $f'(c) = 0$. Explain why this does not contradict Rolle's theorem.

28. Consider $f(x) = x^{2/3}$ on the interval $[-1, 1]$. Verify that there is no c in $(-1, 1)$ for which

$$f'(c) = \frac{f(1) - f(-1)}{1 - (-1)}$$

Explain why this does not contradict the mean value theorem.

29. To demonstrate that the conclusion of the mean value theorem may not hold, draw the graph of a function f that is continuous on $[a, b]$ and not differentiable on (a, b).

30. Repeat Problem 29 for a function f that is differentiable on (a, b) but not continuous on $[a, b]$.

31. Use the mean value theorem to verify that

$$\tfrac{1}{9} < \sqrt{66} - 8 < \tfrac{1}{8}$$

[*Hint:* Consider $f(x) = \sqrt{x}$ on the interval $[64, 66]$.]

32. Prove that there is no k for which the function

$$f(x) = x^3 - 3x + k$$

has two distinct zeros in the interval $[0, 1]$.

33. Apply Rolle's theorem to the function $f(x) = (x - 1)\sin x$ on $[0, 1]$. Thereby conclude that the equation $\tan x + x = 1$ has a solution in (0, 1).

34. Explain why the equation $ax^4 + bx^3 + cx^2 + dx + e = 0$ must have a root between 0 and 1 if

$$\frac{a}{5} + \frac{b}{4} + \frac{c}{3} + \frac{d}{2} + e = 0$$

35. Explain why the equation $x^n + ax + b = 0$, n a positive even integer, has at most two distinct real roots.

36. Explain why the equation $x^n + ax + b = 0$, n a positive odd integer, has at most three distinct real roots.

37. Explain why the equation $x^n + ax^2 + b = 0$, n a positive odd integer, has at most three distinct real roots.

38. Explain why the equation $x^n + ax^2 + b = 0$, n a positive even integer, has at most four distinct real roots.

39. Let f be a function that is continuous on $[a, b]$ and differentiable on (a, b). If $f(x) = 0$ for three different numbers x in (a, b), show that there must be at least two numbers in (a, b) at which f has a 0 derivative.

40. An automobile travels 20 miles down a straight road at an average speed of 40 miles per hour. Show that the automobile must have a velocity of exactly 40 miles per hour at some time during the trip. (Assume that the distance function is differentiable.)

41. At 4:00 PM a car's speedometer reads 40 miles per hour. At 4:20 PM it reads 60 miles per hour. Show that at some time between 4:00 and 4:20 PM the acceleration was exactly 60 miles per hour per hour.

42. Two race cars start a race at the same time and finish in a tie. If $f_1(t)$ is the position of one car at time t and $f_2(t)$ is the position of the second car, prove that at some time during the race they had the same velocity. [*Hint:* Set $f(t) = f_2(t) - f_1(t)$.]

4.5 Increasing and Decreasing Functions; First Derivative Test

In Section 4.3 we learned that all local extrema of a function f occur at critical numbers. Each of these critical numbers is a candidate for locating a local extremum for f. But how do we sift out those that locate local extrema from those that do not? And then how do we determine whether each local extremum thus located is a local maximum or a local minimum? A study of Figure 24 will provide a clue. If you look from left to right along the graph of the continuous function f, you will notice that parts of the graph are increasing and parts are decreasing. The function f is increasing to the left of x_1, where a local maximum occurs, and is decreasing to its right; the function f is decreasing to the left of x_2, where a local minimum occurs, and is increasing to its right. Apparently, knowing when a function f is increasing and when it is decreasing will enable us to identify the local maxima and the local minima.

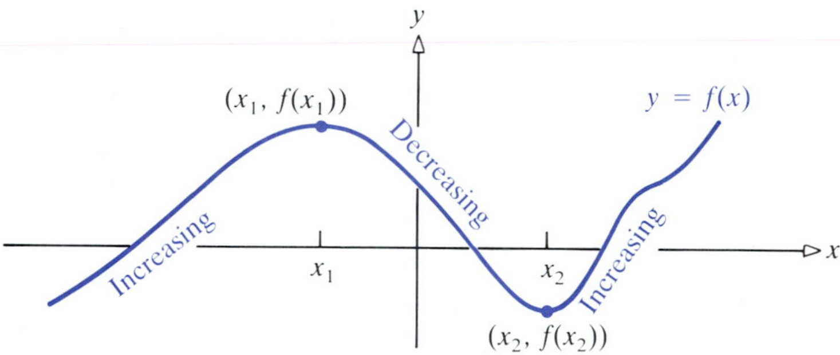

Figure 24

[4.5.1] DEFINITION | ***Increasing Function; Decreasing Function.***

(a) The function f *is increasing on an interval I* if, for any choice of x_1, x_2 in I with $x_1 < x_2$, we have $f(x_1) < f(x_2)$.
(b) The function f *is decreasing on an interval I* if, for any choice of x_1, x_2 in I with $x_1 < x_2$, we have $f(x_1) > f(x_2)$.

The terms *increasing* and *decreasing* describe the possible behavior of the graph of a function as we examine it from left to right (see Fig. 25(a), (b)). If there is an interval on which f remains constant, we say that the graph of f is *horizontal on the interval* (see Fig. 25(c)).

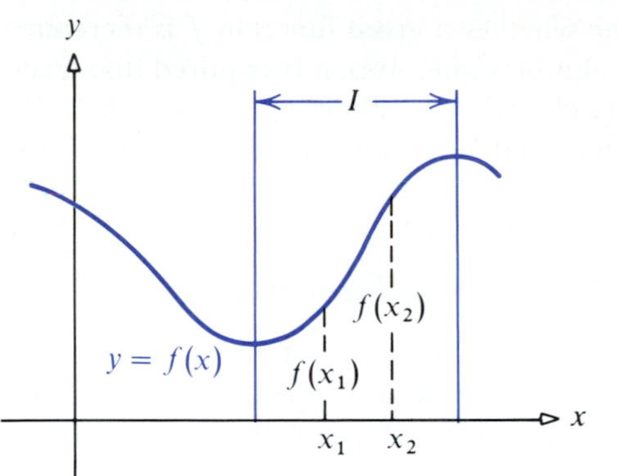

(a) Graph of f is increasing on I

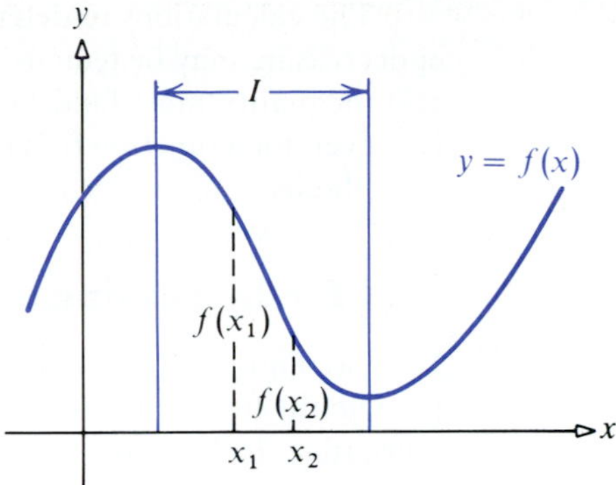

(b) Graph of f is decreasing on I

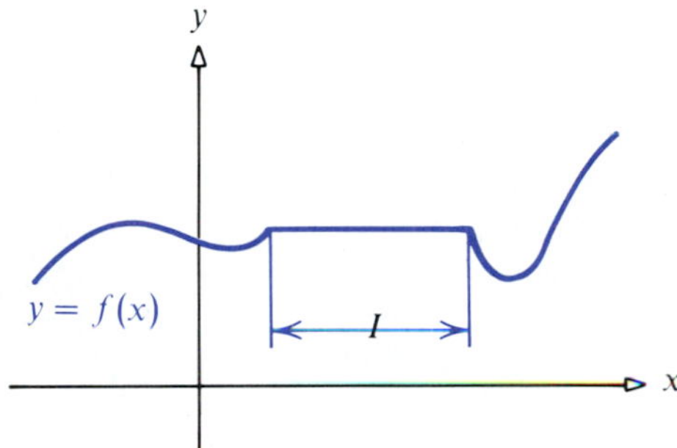

(c) Graph of f is horizontal on I

Figure 25

The function $f(x) = x^2$ is increasing on the interval $[0, +\infty)$, since whenever $0 \leq x_1 < x_2$, we have $x_1^2 < x_2^2$ (see Fig. 26(a)). The function $g(x) = 2 - x$ is decreasing on the interval $(-\infty, +\infty)$, since whenever $x_1 < x_2$, we have $2 - x_1 > 2 - x_2$ (see Fig. 26(b)).

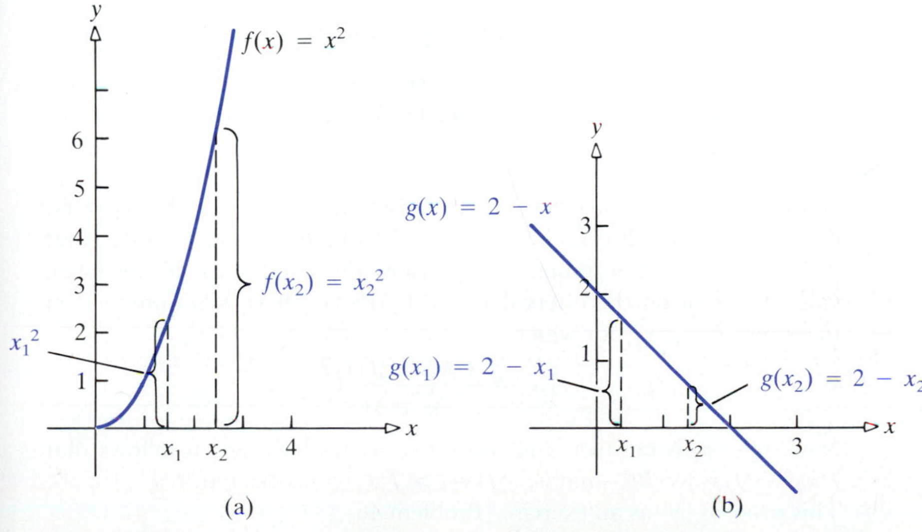

(a) (b)

Figure 26

The calculations to determine whether a given function f is increasing or decreasing may be tedious and cumbersome, since it is required that a certain inequality must hold for *every* choice of numbers x_1, x_2 in an interval. However, for a function that is differentiable, a relatively straightforward test is available.

Test for Increasing and Decreasing Function

Look at Figure 27. It appears that where the slope of the tangent line is positive the function f is increasing, and where the slope of the tangent line is negative the function f is decreasing. But the slope of the tangent line at x is just $f'(x)$. So it appears that whenever $f'(x) > 0$ the function increases, and whenever $f'(x) < 0$ the function decreases. We now state and prove a theorem that tells us this is true in general.

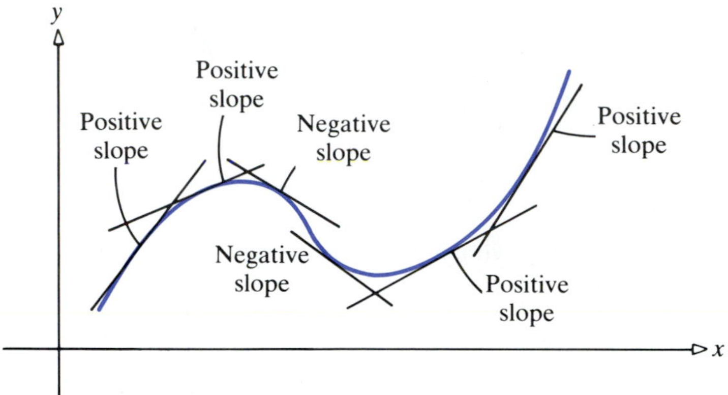

Figure 27

[4.5.2] THEOREM / ***Test for Increasing and Decreasing Function.***

Suppose f is continuous on $[a, b]$ and is differentiable on (a, b).

(a) If $f'(x) > 0$ throughout (a, b), then f is increasing on $[a, b]$.
(b) If $f'(x) < 0$ throughout (a, b), then f is decreasing on $[a, b]$.

Proof

(a) Here $f'(x) > 0$ throughout (a, b). Let x_1 and x_2, with $x_1 < x_2$, denote any two distinct numbers in $[a, b]$. We want to show that $f(x_1) < f(x_2)$. The function f satisfies the conditions of the mean value theorem on the interval $[x_1, x_2]$. Hence, there is a number c in the interval (x_1, x_2) for which

$$f'(c) = \frac{f(x_2) - f(x_1)}{x_2 - x_1}$$

Now, $x_2 - x_1$ is positive, and since $f'(c)$ is also positive, it follows that $f(x_2) - f(x_1) > 0$; that is, $f(x_2) > f(x_1)$, as desired.

(b) This proof is left as an exercise (Problem 46).

Let's look at a specific function and determine where it is increasing and where it is decreasing.

EXAMPLE 1

Determine where the function $f(x) = 2x^3 - 9x^2 + 12x - 5$ is increasing and where it is decreasing. Use this information to graph f.

Solution

$$f'(x) = 6x^2 - 18x + 12 = 6(x-2)(x-1)$$

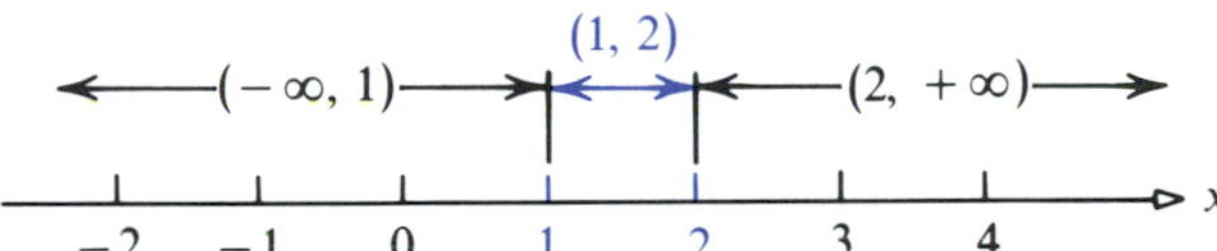

Figure 28

Table 1

Interval	Sign of $x-1$	Sign of $x-2$	Sign of $f'(x) = 6(x-2)(x-1)$	Conclusion
$(-\infty, 1)$	Negative $(-)$	Negative $(-)$	Positive $(+)$	f is increasing
$(1, 2)$	Positive $(+)$	Negative $(-)$	Negative $(-)$	f is decreasing
$(2, +\infty)$	Positive $(+)$	Positive $(+)$	Positive $(+)$	f is increasing

The critical numbers of f are 1 and 2. We use these as cutoffs, which in turn partition the real line into three intervals, as illustrated in Figure 28 and Table 1. On each of these intervals we calculate the sign of the factors of $f'(x)$, which by theorem (4.5.2) tells us whether f is increasing or decreasing. Thus, f is increasing on $(-\infty, 1]$ and on $[2, +\infty)$, and f is decreasing on $[1, 2]$. Since f is increasing to the left of 1 and decreasing to the right of 1, we conclude that at 1, f has a local maximum. Similarly, at 2, f has a local minimum. See Figure 29 for the graph of f.

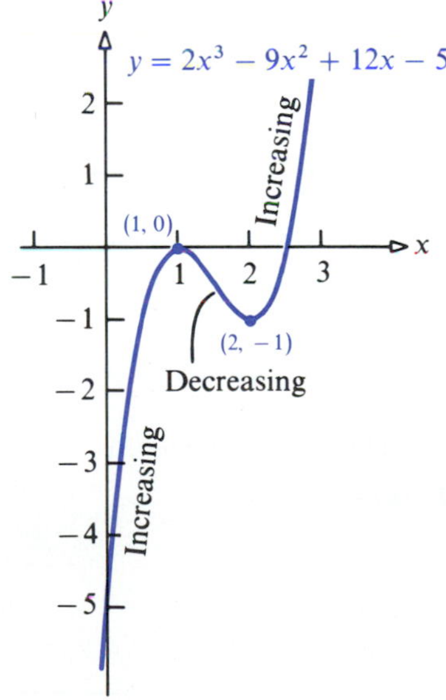

Figure 29

EXAMPLE 2

Determine where the function $f(x) = (x^2 - 1)^{2/3}$ is increasing and where it is decreasing. Graph f.

Solution

$$f'(x) = \frac{2}{3}(x^2 - 1)^{-1/3}(2x) = \frac{4x}{3(x^2 - 1)^{1/3}}$$

Since $f'(x) = 0$ at $x = 0$ and f' does not exist at $x = \pm 1$, the critical numbers of f are -1, 0, and 1. Table 2 analyzes the behavior of the graph of f, which is given in Figure 30.

Table 2

Interval	Sign of $4x$	Sign of $(x^2-1)^{1/3}$	Sign of $f'(x) = \dfrac{4x}{3(x^2-1)^{1/3}}$	Conclusion
$(-\infty, -1)$	Negative $(-)$	Positive $(+)$	Negative $(-)$	f is decreasing
$(-1, 0)$	Negative $(-)$	Negative $(-)$	Positive $(+)$	f is increasing
$(0, 1)$	Positive $(+)$	Negative $(-)$	Negative $(-)$	f is decreasing
$(1, +\infty)$	Positive $(+)$	Positive $(+)$	Positive $(+)$	f is increasing

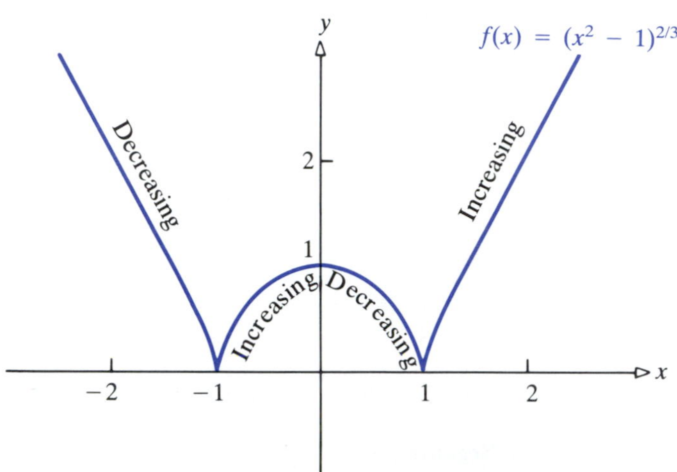

Figure 30

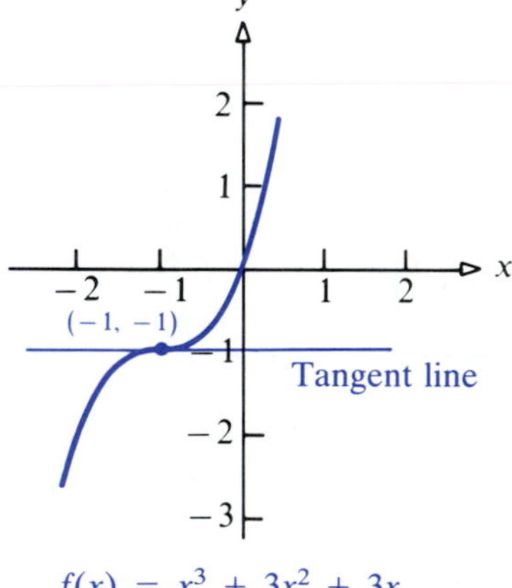

Figure 31

EXAMPLE 3

Show that $f(x) = x^3 + 3x^2 + 3x$ is increasing for all x.

Solution

$$f'(x) = 3x^2 + 6x + 3 = 3(x^2 + 2x + 1) = 3(x+1)^2$$

This function f has only one critical number, and it is -1. However, since

$$f'(x) > 0 \quad \text{for} \quad x < -1$$

and

$$f'(x) > 0 \quad \text{for} \quad x > -1$$

we conclude that f is increasing for all x. Figure 31 illustrates the graph of $y = x^3 + 3x^2 + 3x$. Note the horizontal tangent line at the point $(-1, -1)$.

First Derivative Test

In Example 1 we noticed that at the critical number 1 the function f has a local maximum, and at 2 the function has a local minimum. However, in Example 3 the critical number -1 did not give rise to a local extremum. We therefore need a test that will tell us when a critical number will give rise to local extremum.

The following result is usually referred to as the *first derivative test*, since it relies on information obtained from the first derivative of a function.

[4.5.3] THEOREM / *First Derivative Test.*

Let c be a critical number of a function f and let (a, b) denote an open interval containing c. The function f is assumed to be continuous on the closed interval $[a, b]$ and differentiable on the open interval (a, b), except possibly at c.

(a) If $f'(x)$ is positive for $a < x < c$ and is negative for $c < x < b$, then f has a local maximum at c.
(b) If $f'(x)$ is negative for $a < x < c$ and is positive for $c < x < b$, then f has a local minimum at c.
(c) If $f'(x)$ is positive on both sides of c or is negative on both sides of c, then f has neither a local maximum nor a local minimum at c.

Proof

(a) The function f is increasing on $a \le x \le c$, since $f'(x) > 0$ for $a<x<c$; in addition, f is decreasing on $c \le x \le b$, since $f'(x) < 0$ on $c < x < b$. Hence, for all x in $[a, b]$, we have $f(x) \le f(c)$; that is, f has a local maximum at c. See Figure 32(a).
(b) This is left as an exercise (Problem 47). See Figure 32(b).
(c) This is also left as an exercise (Problem 48). See Figures 32(c) and 32(d).

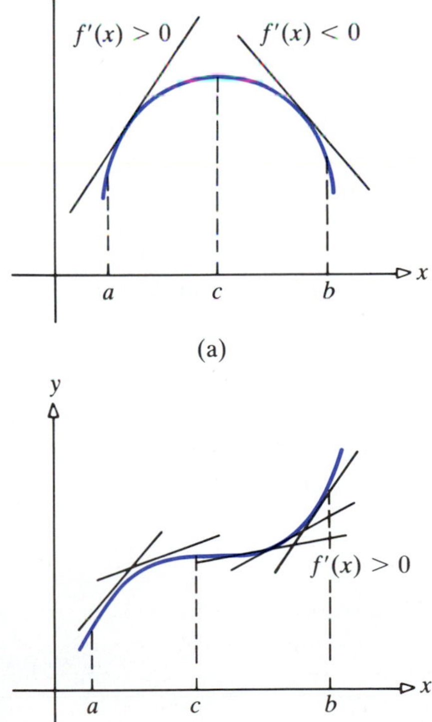

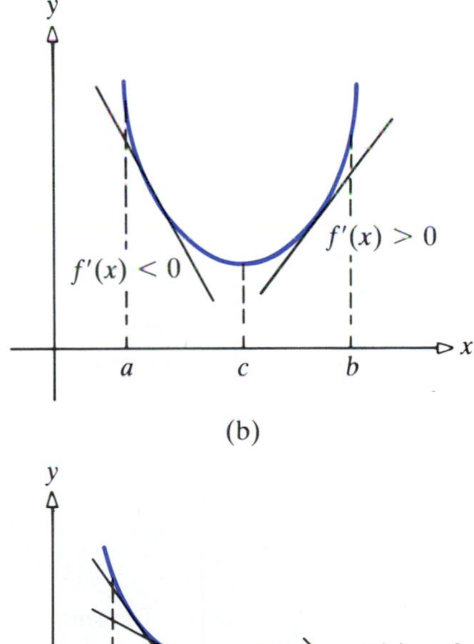

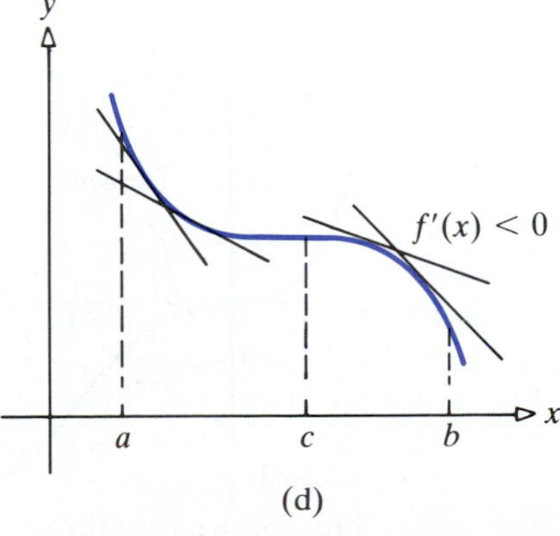

Figure 32

EXAMPLE 4

Find the local extrema of $f(x) = x^4 - 4x^3$ and sketch a graph of f.

Solution

$$f'(x) = 4x^3 - 12x^2 = 4x^2(x - 3)$$

The critical numbers are 0 and 3. As in Example 1, we use Figure 33 and Table 3 to help determine where f is increasing and where it is decreasing.

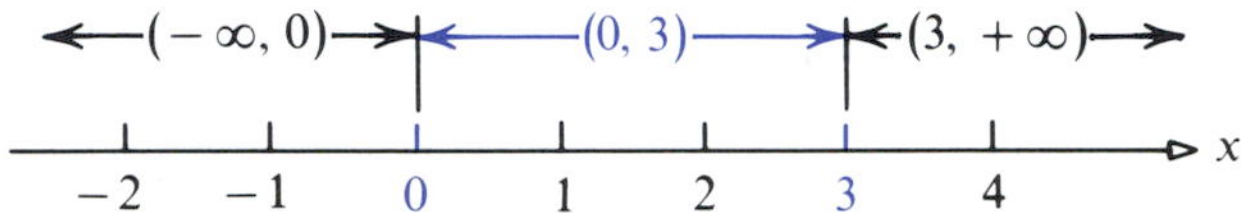

Figure 33

Table 3

Interval	Sign of x^2	Sign of $x - 3$	Sign of $f'(x) = 4x^2(x - 3)$	Conclusion
$(-\infty, 0)$	+	−	−	f is decreasing
$(0, 3)$	+	−	−	f is decreasing
$(3, +\infty)$	+	+	+	f is increasing

Thus, by the first derivative test, f has neither a local maximum nor a local minimum at 0, and f has a local minimum at 3. It is advisable to locate the intercepts and test for possible symmetries before sketching the graph of f. By inspection, the y-intercept is $y = 0$. The x-intercepts, which obey the equation $x^4 - 4x^3 = 0$, are $x = 0$ and $x = 4$. The graph is not symmetric with respect to the x-axis, the y-axis, or the origin. See Figure 34 for a sketch of the graph. Note the horizontal tangent line at $(0, 0)$.

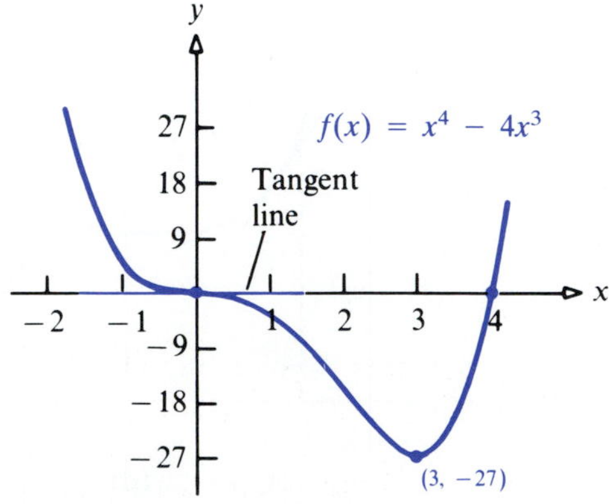

Figure 34

EXAMPLE 5

Find the local extrema of $f(x) = x^{2/3}(x - 5)$ and sketch the graph of f.

Solution

$$f'(x) = x^{2/3} + \left(\frac{2}{3}\right)x^{-1/3}(x - 5) = \frac{3x + 2(x - 5)}{3x^{1/3}} = \frac{5}{3}\left(\frac{x - 2}{x^{1/3}}\right)$$

The critical numbers are 0 and 2. Table 4 summarizes the behavior of f. By the first derivative test, f has a local maximum at 0 and a local minimum at 2. At the point $(2, -3\sqrt[3]{4})$, the tangent line is horizontal; at 0, the derivative does not exist. The x-intercepts are

$$x = 0 \qquad \text{and} \qquad x = 5$$

The y-intercept is 0. There is no symmetry. See Figure 35 for a sketch of the graph.

Table 4

Interval	Sign of $x - 2$	Sign of $x^{1/3}$	Sign of $f'(x) = \frac{5}{3}\left(\frac{x-2}{x^{1/3}}\right)$	Conclusion
$(-\infty, 0)$	$-$	$-$	$+$	f is increasing
$(0, 2)$	$-$	$+$	$-$	f is decreasing
$(2, +\infty)$	$+$	$+$	$+$	f is increasing

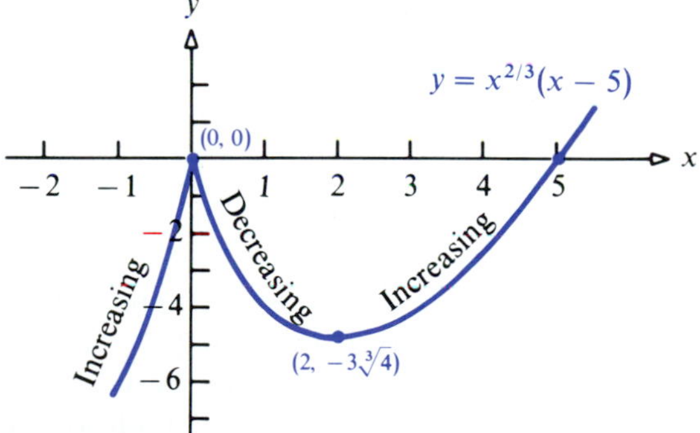

Figure 35

Rectilinear Motion

It is advantageous to use the notion of increasing and decreasing functions in discussing the motion of a particle. Suppose the directed distance s of a particle from the origin at time t is given by $s = f(t)$. As is customary, we think of the motion as along a horizontal line with the positive direction to the right. The velocity v of this particle is $v = ds/dt$. If $v = ds/dt > 0$, the directed distance s is increasing as time t increases, and hence the particle will move to the right. If $v = ds/dt < 0$, the directed distance s is decreasing as time t increases, and the particle will move to the left. This information may be used with the first derivative test to find the local extrema of f and determine at what times the direction of motion of the particle changes.

The acceleration of the particle is $a = dv/dt$. If $a = dv/dt > 0$, the velocity is increasing; if $a = dv/dt < 0$, the velocity is decreasing. This information, along with the use of the first derivative test, can be used to find the local extrema of the velocity.

If both $a < 0$ and $v < 0$, the velocity is decreasing (becoming more negative) and the particle is moving to the left. However, the speed $|v|$ actually may be increasing. The next example illustrates this phenomenon.

EXAMPLE 6

Suppose the directed distance s of a particle from the origin at time t is given by

$$s = t^3 + 3t^2 - 9t + 3$$

Determine the time interval during which the particle is moving to the right. When is it moving to the left? When does it reverse direction? When is its velocity increasing? When is it decreasing? Draw a figure to illustrate the motion of the particle.

Solution

The velocity v of the particle is

$$v = \frac{ds}{dt} = 3t^2 + 6t - 9 = 3(t^2 + 2t - 3) = 3(t + 3)(t - 1)$$

Table 5 summarizes the motion of the particle. The particle reverses direction when $t = -3$ and when $t = 1$. The acceleration a of the particle is

$$a = \frac{dv}{dt} = 6t + 6 = 6(t + 1)$$

When $-\infty < t < -1$, then $a < 0$, and the velocity is decreasing. When $-1 < t < +\infty$, then $a > 0$, and the velocity is increasing. Note that for $-3 < t < -1$, we have $a < 0$ and $v < 0$. Yet the speed $|v| = 3|(t + 3)(t - 1)|$ over this time interval is increasing. Figure 36 illustrates the motion of the particle.

Table 5

Time Interval	Velocity, v	Motion of Particle
$(-\infty, -3)$	+	To the right
$(-3, 1)$	−	To the left
$(1, +\infty)$	+	To the right

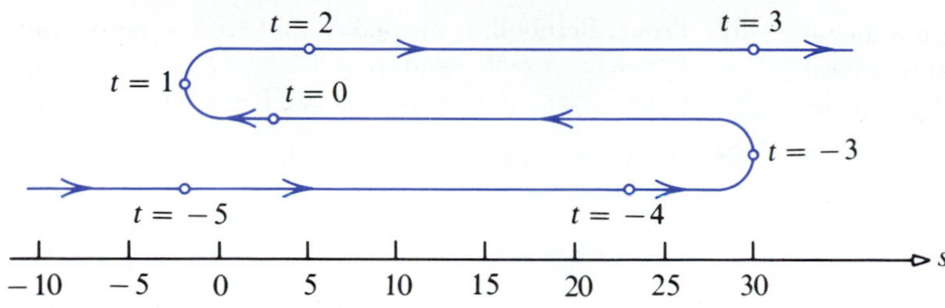

Figure 36

EXERCISE 4.5

1. Show that the function $f(x)=2x^3-6x^2+6x-5$ is increasing for all x.

2. Show that the function $f(x)=x^3-3x^2+3x$ is increasing for all x.

3. Show that the function $f(x)=x/(x+1)$ is increasing on any interval not containing $x \neq -1$.

4. Show that the function $f(x)=(x+1)/x$ is decreasing on any interval not containing $x \neq 0$.

In Problems 5–34 find (a) the critical numbers of f. (b) Determine the intervals on which f is increasing or decreasing, and find the local extrema of f. (c) Sketch the graph of f.

5. $f(x)=x^2-x-2$
6. $f(x)=-x^2+4x-3$
7. $f(x)=-2x^2+4x-5$
8. $f(x)=-3x^2-12x+2$
9. $f(x)=2x^3+3x^2+4$
10. $f(x)=x^3-6x^2+2$
11. $f(x)=x^3+6x^2+12x+1$
12. $f(x)=-x^3-3x^2+4$
13. $f(x)=3x^4-12x^3+5$
14. $f(x)=x^4-4x+2$
15. $f(x)=3x^4-4x^3$
16. $f(x)=x^4+2x^3-3$
17. $f(x)=x^3+3x^2-9x+1$
18. $f(x)=2x^3+3x^2-36x+4$
19. $f(x)=x^2(x+1)$
20. $f(x)=x(x^2-1)$
21. $f(x)=x^4+2x^2$
22. $f(x)=x^4-2x^2$
23. $f(x)=x^{2/3}+x^{1/3}$
24. $f(x)=\frac{1}{2}x^{2/3}-x^{1/3}$
25. $f(x)=x^{2/3}(x-10)$
26. $f(x)=x^{1/3}(x-8)$
27. $f(x)=x^{2/3}(x^2-4)$
28. $f(x)=x^{1/3}(x^2-7)$
29. $f(x)=|x^2-1|$
30. $f(x)=|x^2-4|$
31. $f(x)=3\sin x$
32. $f(x)=\cos 2x$
33. $f(x)=\sin x-2\cos x$
34. $f(x)=x+2\sin x$

In Problems 35–42 the distance s of a particle from the origin at time t is given. Determine the time interval during which the particle is moving to the right. When is it moving to the left? When does it reverse direction? When is its velocity increasing? When is it decreasing? Draw a figure to illustrate the motion of the particle.

35. $s=t^2-2t+3$
36. $s=2t^2+8t-7$
37. $s=2t^3+6t^2-18t+1$
38. $s=3t^4-16t^3+24t^2$
39. $s=2t-\dfrac{6}{t}, \quad t>0$
40. $s=3\sqrt{t}-\dfrac{1}{\sqrt{t}}, \quad t>0$
41. $s=2\sin 3t, \quad 0 \le t \le \dfrac{2\pi}{3}$
42. $s=3\cos \pi t, \quad 0 \le t \le 2$

43. If $f(x)=ax^2+bx+c$, $a \neq 0$, prove that f has a local maximum at $-b/(2a)$ if $a<0$ and has a local minimum at $-b/(2a)$ if $a>0$.

44. If $f(x)=ax^3+bx^2+cx+d$, $a \neq 0$, how does the quantity b^2-3ac determine the number of potential local extrema?

45. If $f(x) = ax^3 + bx^2 + cx + d$, $a \neq 0$, determine a, b, c, and d so that f has a local minimum at 0, a local maximum at 4, and a graph that passes through the points (0, 5) and (4, 33).

46. Prove part (b) of theorem (4.5.2).

47. Prove part (b) of theorem (4.5.3).

48. Prove part (c) of theorem (4.5.3).

49. Show that 0 is the only critical number of $f(x) = \sqrt[3]{x}$ and that f has no local extrema.

50. Show that 0 is the only critical number of $f(x) = \sqrt[3]{x^2}$ and that f has a local minimum at 0.

51. Find a number x, $0 < x < 1$, so that the function below is as small as possible.

$$f(x) = \frac{2}{x} + \frac{8}{1 - x}$$

52. Prove Bernoulli's inequality $(1 + x)^n > 1 + nx$ for $x > -1$, $x \neq 0$, and $n > 1$.
[*Hint:* Set $f(x) = (1 + x)^n - (1 + nx)$.]

53. If $x > 0$ and $n > 1$, can the expression

$$x^n - n(x - 1) - 1$$

ever be negative?

54. Show that $\sin x \leq x$, $0 \leq x \leq 2\pi$.
[*Hint:* Let $f(x) = x - \sin x$.]

55. Show that $1 - (x^2/2) \leq \cos x$, $0 \leq x \leq 2\pi$.
[*Hint:* Use the result of Problem 54.]

56. Show that $2\sqrt{x} > 3 - (1/x)$, for $x > 1$.

4.6 Concavity; Second Derivative Test

Concavity

We begin by looking at the graphs of two familiar functions: $y = x^2$ $(x \geq 0)$ and $y = \sqrt{x}$ (see Fig. 37). Each graph starts at the origin, passes through the point (1, 1), and is increasing. But there is a noticeable difference. The graph of $y = x^2$ opens up—is *concave up*—whereas the graph of $y = \sqrt{x}$ opens down—is *concave down*.

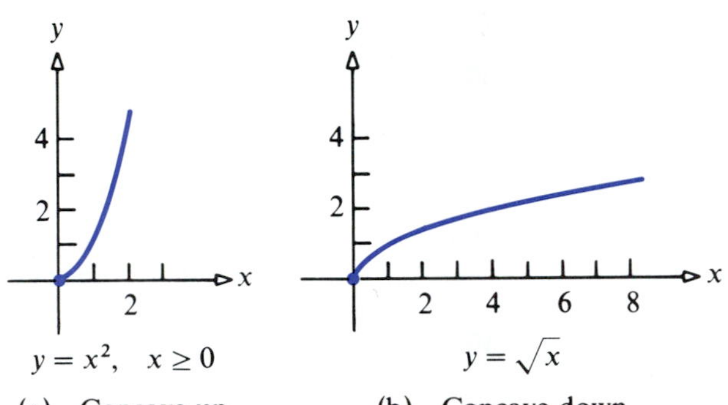

(a) Concave up (b) Concave down

Figure 37

In this section we discuss a test to determine where a function is concave up and where it is concave down. First, though, we need a definition. The definition is based on the illustration in Figure 38. We observe in this figure that when tangent lines are drawn to the graph where it opens up (is concave up), the graph lies above the tangent lines (except at the point of tangency). Similarly, where the graph opens down (is concave down), the graph lies

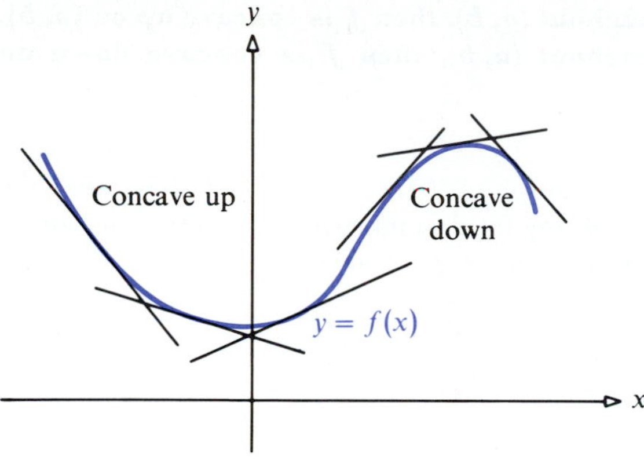

Figure 38

below the tangent lines (except at the point of tangency). This leads us to formulate definition (4.6.1).

[4.6.1] DEFINITION / ***Concave Up; Concave Down.***

Let f be a function that is continuous on a closed interval $[a, b]$ and is differentiable on the open interval (a, b).

(a) f is *concave up* on $[a, b]$ if, throughout (a, b), the graph of f lies above the tangent lines to f.
(b) f is *concave down* on $[a, b]$ if, throughout (a, b), the graph of f lies below the tangent lines to f.

We may formulate a useful test for determining where a function f is concave up or concave down, providing f'' exists. Since f'' equals the rate of change of f', it follows that if in some open interval, we have f'' positive, then f' will be increasing on that interval. But f' equals the slope of the tangent line to the graph of f. Hence, if the slope is increasing, the tangent line will turn in a counterclockwise sense, as indicated in Figure 39. The graph of f will therefore lie above the tangent line, and f will be concave up. These observations lead to the test for concavity.

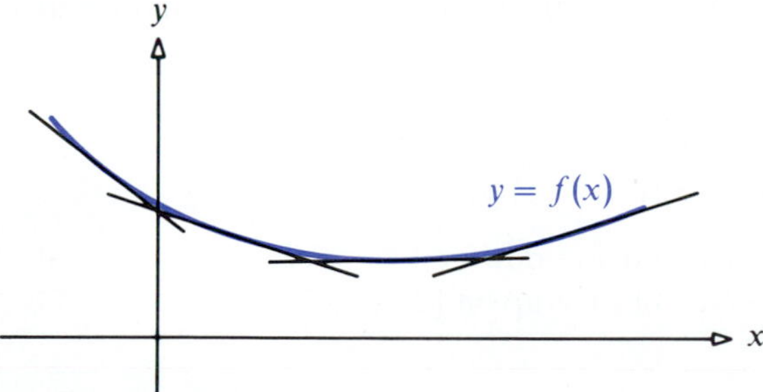

Figure 39

[4.6.2] THEOREM / ***Test for Concavity.***

Let f denote a function that is continuous on a closed interval $[a, b]$ and is twice differentiable on the open interval (a, b).

(a) If $f''(x) > 0$ throughout (a, b), then f is concave up on $[a, b]$.
(b) If $f''(x) < 0$ throughout (a, b), then f is concave down on $[a, b]$.

Proof

(a) We need to show that the graph of f throughout (a, b) lies above the tangent lines to f. Let c be any fixed number in (a, b). The equation of the tangent line to f at the point $(c, f(c))$ is

$$y = f(c) + f'(c)(x - c)$$

We must establish that

$$f(x) \geq f(c) + f'(c)(x - c) \qquad \text{for all } x \text{ in } (a, b)$$

If $x = c$, this is obviously true. If x is different from c, then, by applying the mean value theorem to the function f, there is a number x_1 between c and x so that

$$f'(x_1) = \frac{f(x) - f(c)}{x - c} \qquad \text{where } x_1 \text{ is between } c \text{ and } x$$

$$f(x) = f(c) + f'(x_1)(x - c) \tag{1}$$

There are two possibilities: either $c < x_1 < x$ or $x < x_1 < c$. Suppose $c < x_1 < x$. Since f'' is positive throughout (a, b), it follows from theorem (4.5.2) that f' is increasing on (a, b). For $x_1 > c$, this means that $f'(x_1) > f'(c)$. Therefore, we may write (1) as

$$f(x) > f(c) + f'(c)(x - c)$$

That is, the graph of f lies above its tangent line to the right of c in (a, b). The case where $x < x_1 < c$ is left as an exercise (Problem 64).

(b) We also leave this part of the proof as an exercise (Problem 65).

Let's look at a specific function and determine where it is concave up and where it is concave down.

EXAMPLE 1

Determine where $f(x) = x^3 - 6x^2 + 9x + 30$ is concave up and where it is concave down.

Solution

$$f'(x) = 3x^2 - 12x + 9 \qquad f''(x) = 6x - 12$$

$f''(x) < 0$ if $x < 2$, so f is concave down on $(-\infty, 2]$.
$f''(x) > 0$ if $x > 2$, so f is concave up on $[2, +\infty)$.

Inflection Points

[4.6.3] DEFINITION | ***Inflection Points.***

An *inflection point* of a function f is a point on the graph of f at which the concavity of f changes.

If f has an inflection point at c* and if $f'(c)$ exists, then the derivative f' must have a local maximum or minimum at c. In either case, it follows that $f''(c) = 0$ or $f''(c)$ does not exist. This test is stated more formally next.

[4.6.4] THEOREM / *Test for Inflection Point.*

Let f denote a function that is continuous on a closed interval $[a, b]$ and is differentiable on the open interval (a, b). If f has an inflection point at the number c in (a, b), then either $f''(c) = 0$ or f'' does not exist at c.

Note the wording. If you *know* there is an inflection point at c, the second derivative at c is 0 or does not exist. The converse is not necessarily true. In other words, numbers at which $f''(x)$ is 0 or does not exist will not always determine points of inflection. Thus, to find the points of inflection of a function f:

1. Find all numbers in the domain of f at which $f''(x) = 0$ or $f''(x)$ does not exist.
2. Use the test for concavity to determine the concavity on either side of each of these numbers.
3. If the concavity changes, there is an inflection point; otherwise, there is not.

For the function given in Example 1, $f''(x) = 6x - 12 = 0$ at 2. The function f has an inflection point at (2, 32), since f is concave down to the left of 2 and concave up to the right of 2. You may verify that the sketch of the graph of f is as given in Figure 40.

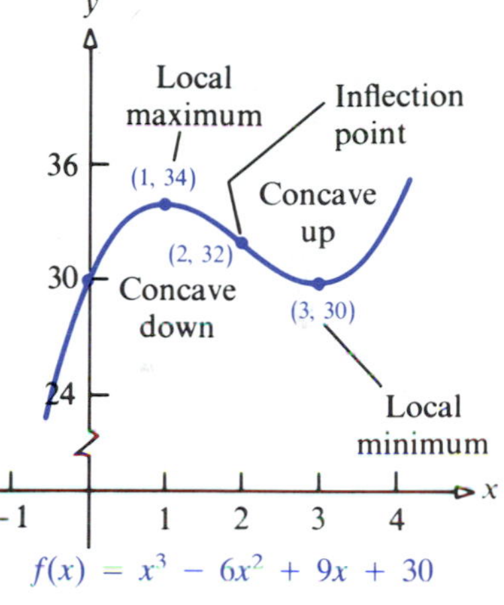

Figure 40

EXAMPLE 2

Find all inflection points of $f(x) = x^{5/3}$. Sketch the graph.

Solution

$$f'(x) = \frac{5}{3}x^{2/3} \qquad f''(x) = \frac{10}{9}x^{-1/3} = \frac{10}{9x^{1/3}}$$

The second derivative of f is never 0 and does not exist when $x = 0$. Thus, the only candidate for an inflection point is 0. Now,

if $x < 0$, then $f''(x) < 0$ so f is concave down on $(-\infty, 0]$

if $x > 0$, then $f''(x) > 0$ so f is concave up on $[0, +\infty)$

Hence, f has an inflection point at 0. The sketch of the graph of f is given in Figure 41. Note that $f'(x) > 0$ for all x, so f is increasing on $(-\infty, +\infty)$.

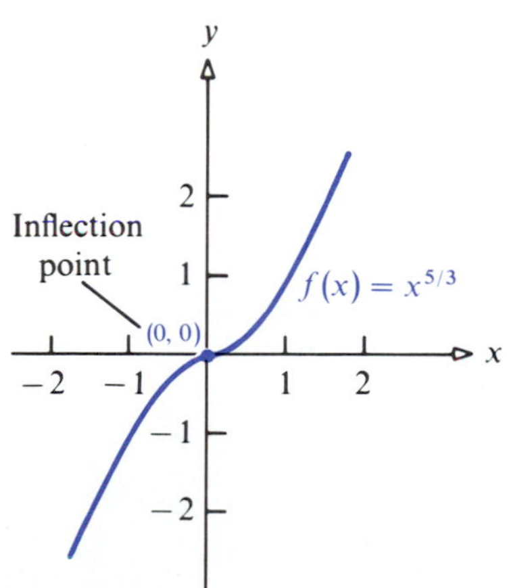

Figure 41

---

* By this we mean "at the point $(c, f(c))$ on the graph of f."

The fact that a change in concavity occurs is not, of itself, a guarantee that there is an inflection point. For example, the function $f(x) = 1/x$ is concave down on $(-\infty, 0)$ and concave up on $(0, +\infty)$, since $f''(x) = 2/x^3 < 0$ if $x < 0$ and $f''(x) = 2/x^3 > 0$ if $x > 0$. Yet f has no inflection points. This is because f is not defined at 0.

Second Derivative Test

Suppose c is a critical number of f; that is, $f'(c) = 0$. Geometrically, this means that the graph of f has a horizontal tangent line at the point $(c, f(c))$. If f is also twice differentiable on an open interval containing c and $f''(c) > 0$, then the graph of f is concave up at $(c, f(c))$. See Figure 42(a). Intuitively it would seem that $f(c)$ is a local minimum of f. If, on the other hand, $f''(c) < 0$, the graph of f is concave down at $(c, f(c))$, and $f(c)$ would appear to be a local maximum of f. See Figure 42(b).

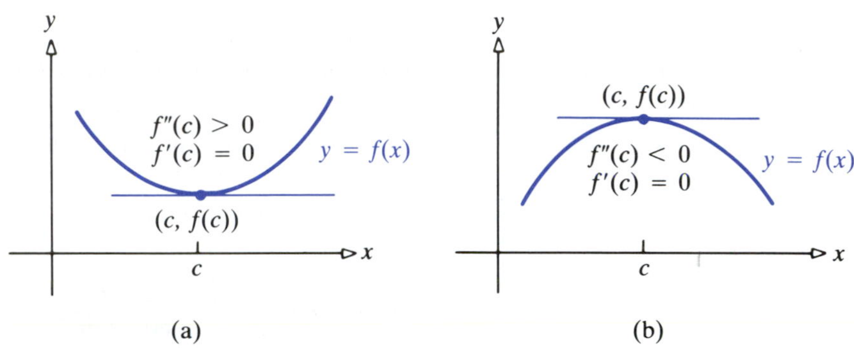

Figure 42

[4.6.5] THEOREM / ***Second Derivative Test.***

Let f be a function that is twice differentiable on an open interval (a, b) containing a critical number c. Suppose further that $f''(c)$ exists.

(a) If $f''(c) < 0$, then f has a local maximum at c.
(b) If $f''(c) > 0$, then f has a local minimum at c.

Proof

(a) Since $f''(c)$ exists and is negative, we have

$$f''(c) = \lim_{x \to c} \frac{f'(x) - f'(c)}{x - c} < 0$$

Referring back to (2.2.12), on some interval about c we must have

$$\frac{f'(x) - f'(c)}{x - c} < 0$$

But c is a critical number, so $f'(c) = 0$. Hence,

$$\frac{f'(x)}{x - c} < 0$$

Now for $x < c$ on this interval, $f'(x) > 0$, and for $x > c$ on this interval, $f'(x) < 0$. By the first derivative test, f has a local maximum at c.

(b) This proof is left as an exercise (Problem 66).

EXAMPLE 3

Find all local extrema of $f(x) = x^3 - 6x^2 + 9x + 1$ by using the second derivative test. Sketch the graph.

Solution

$$f'(x) = 3x^2 - 12x + 9 = 3(x^2 - 4x + 3) = 3(x - 1)(x - 3)$$

The critical numbers are 1 and 3.

$$f''(x) = 6x - 12$$

At the critical numbers, the second derivatives are $f''(1) < 0$ and $f''(3) > 0$. Thus, f has a local maximum at 1 and a local minimum at 3. Figure 43 illustrates the graph of f.

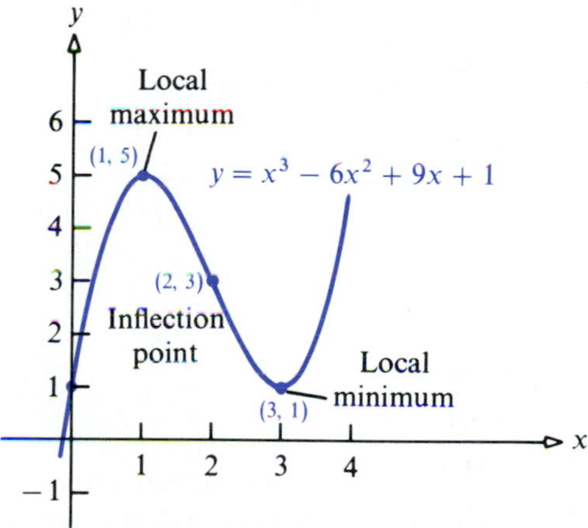

Figure 43

EXAMPLE 4

Determine where $f(x) = x - 2\cos x$, $0 \le x \le 2\pi$, is concave up and where it is concave down. Use the second derivative test to determine all local extrema.

Solution

$$f'(x) = 1 + 2\sin x \qquad f''(x) = 2\cos x$$

If $\sin x = -\frac{1}{2}$, then $f'(x) = 0$. Since $0 \le x \le 2\pi$, we have

$$x = \frac{7\pi}{6} \quad \text{or} \quad x = \frac{11\pi}{6}$$

If $\cos x = 0$, then $f''(x) = 0$. Since $0 \le x \le 2\pi$, we have

$$x = \frac{\pi}{2} \qquad \text{or} \qquad x = \frac{3\pi}{2}$$

If $0 < x < \pi/2$, then $f''(x) > 0$, so f is concave up on $[0, \pi/2]$. If $\pi/2 < x < 3\pi/2$, then $f''(x) < 0$, so f is concave down on $[\pi/2, 3\pi/2]$. If $3\pi/2 < x < 2\pi$, then $f''(x) > 0$, so f is concave up on $[3\pi/2, 2\pi]$. The critical numbers are $7\pi/6$ and $11\pi/6$.

$$f''\left(\frac{7\pi}{6}\right) < 0 \qquad \text{and} \qquad f''\left(\frac{11\pi}{6}\right) > 0$$

Hence, f has a local maximum at $7\pi/6$ and a local minimum at $11\pi/6$. See Figure 44.

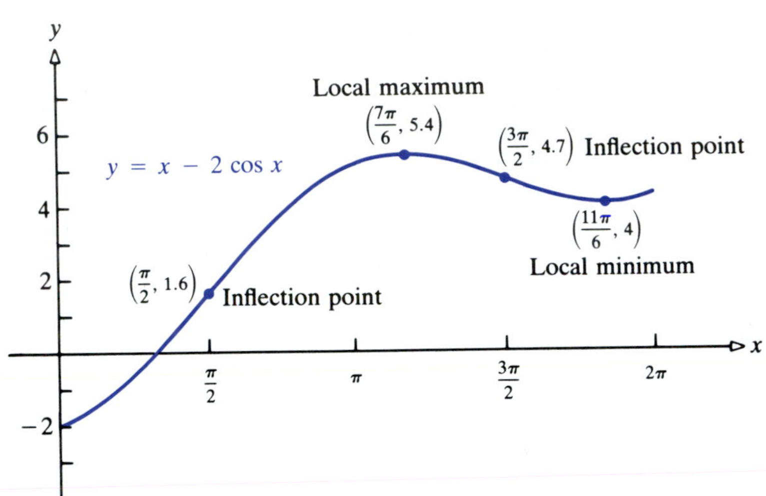

Figure 44

The second derivative test should be used when the second derivative of f is easily calculated. If the second derivative equals 0 or does not exist, then the test gives no information. In such cases, the first derivative test must be used.

EXAMPLE 5

Find all local extrema of $f(x) = x(3 - x)^{5/3}$. Sketch the graph of f.

Solution

The reader should verify that

$$f'(x) = \tfrac{1}{3}(3 - x)^{2/3}(9 - 8x) \qquad \text{and} \qquad f''(x) = -\tfrac{10}{9}(3 - x)^{-1/3}(9 - 4x)$$

The critical numbers are 3 and $\frac{9}{8}$. At $x = \frac{9}{8}$, $f''(\frac{9}{8}) = -5(\frac{15}{8})^{-1/3} < 0$ and hence by theorem (4.6.5), f has a local maximum at $\frac{9}{8}$. Since $f''(3)$ does not exist, the second derivative test gives no information about whether there is a local extremum at 3. However, by the first derivative test, there is not a local extremum at 3. It can be shown that $(3, 0)$ and $(\frac{9}{4}, \frac{9}{4}(\frac{3}{4})^{5/3})$ are inflection points of the graph of f. See Figure 45.

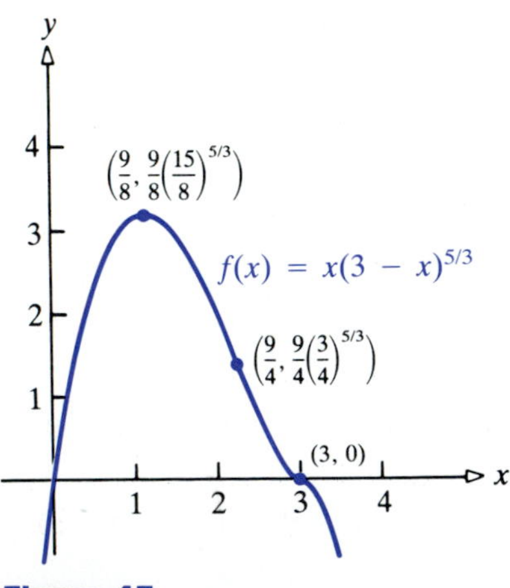

Figure 45

EXERCISE 4.6

In Problems 1–12 determine the intervals on which the function f is concave up and on which f is concave down. Find all points of inflection.

1. $f(x) = x^2 - 2x + 5$ **2.** $f(x) = x^2 + 4x - 2$ **3.** $f(x) = x^3 - 9x^2 + 2$

4. $f(x) = x^3 - 6x^2 + 9x + 1$ **5.** $f(x) = x^4 - 4x^3 + 10$ **6.** $f(x) = 3x^4 - 8x^3 + 6x + 1$

7. $f(x) = x + \dfrac{1}{x}$ **8.** $f(x) = 2x^2 - \dfrac{1}{x}$ **9.** $f(x) = 3x^{1/3} + 2x$

10. $f(x) = (x - 1)^{3/2}$ **11.** $f(x) = 3 - \dfrac{4}{x} + \dfrac{4}{x^2}$ **12.** $f(x) = x^{4/3} - 8x^{1/3}$

In Problems 13–40 use the second derivative test, if possible, to find the local extrema of f. Determine the intervals on which f is concave up and on which f is concave down. Find all points of inflection. Graph each function.

13. $f(x) = 2x^3 - 6x^2 + 6x - 3$ **14.** $f(x) = 2x^3 + 9x^2 + 12x - 4$ **15.** $f(x) = -2x^3 + 15x^2 - 36x + 7$

16. $f(x) = x^3 + 10x^2 + 25x - 25$ **17.** $f(x) = x^4 - 4x$ **18.** $f(x) = x^4 + 4x$

19. $f(x) = 5x^4 - x^5$ **20.** $f(x) = 4x^6 + 6x^4$ **21.** $f(x) = 3x^5 - 20x^3$ **22.** $f(x) = 3x^5 + 5x^3$

23. $f(x) = 6x^{4/3} - 3x^{1/3}$ **24.** $f(x) = x^{2/3} - x^{1/3}$ **25.** $f(x) = x^{2/3}(x - 10)$ **26.** $f(x) = x^{1/3}(x - 3)$

27. $f(x) = x^{2/3}(x^2 - 8)$ **28.** $f(x) = x^{1/3}(x^2 - 2)$ **29.** $f(x) = \dfrac{x^2}{1 + x^2}$ **30.** $f(x) = \dfrac{x}{(1 + x^2)^{5/2}}$

31. $f(x) = \dfrac{\sqrt{x}}{1 + x}$ **32.** $f(x) = x^2\sqrt{1 - x^2}$ **33.** $f(x) = \sqrt{x - x^2}$ **34.** $f(x) = x\sqrt{1 - x}$

35. $f(x) = \dfrac{x^2}{3x + 1}$ **36.** $f(x) = \dfrac{2x}{x^2 + 2}$ **37.** $f(x) = \sin^2 x$ **38.** $f(x) = \cos^2 x$

39. $f(x) = x - 2\sin x,\quad 0 \le x \le 2\pi$ **40.** $f(x) = 2\cos^2 x - \sin^2 x,\quad 0 \le x \le 2\pi$

In Problems 41–52 sketch the graph of a continuous function f that has the given properties. More than one correct sketch is possible.

41. The graph of f is concave up and increasing on $(-\infty, 0]$ and concave up and decreasing on $[0, +\infty)$; $f(0) = 1$.

42. The graph of f is concave up and decreasing on $(-\infty, 0]$ and concave down and increasing on $[0, +\infty)$; $f(0) = 1$.

43. The graph of f is concave down and decreasing on $(-\infty, 0]$, concave down and increasing on $[0, 1]$, and concave up and increasing on $[1, +\infty)$; $f(0) = 1$; $f(1) = 2$.

44. The graph of f is concave down and increasing on $(-\infty, 0]$, concave up and increasing on $[0, 1]$, and concave up and increasing on $[1, +\infty)$; $f(0) = 1$; $f(1) = 2$.

45. $f'(x) > 0$ if $x < 0$; $f'(x) < 0$ if $x > 0$; $f''(x) > 0$ if $x < 0$; $f''(x) > 0$ if $x > 0$; $f(0) = 1$

46. $f'(x) > 0$ if $x < 0$; $f'(x) > 0$ if $x > 0$; $f''(x) > 0$ if $x < 0$; $f''(x) < 0$ if $x > 0$; $f(0) = 1$

47. $f''(0) = 0$; $f'(0) = 0$; $f''(x) > 0$ if $x < 0$; $f''(x) > 0$ if $x > 0$; $f(0) = 1$

48. $f''(0) = 0$; $f'(x) > 0$ if $x \neq 0$; $f''(x) < 0$ if $x < 0$; $f''(x) > 0$ if $x > 0$; $f(0) = 1$

49. $f'(0) = 0$; $f'(x) < 0$ if $x \neq 0$; $f''(x) > 0$; if $x < 0$; $f''(x) < 0$ if $x > 0$; $f(0) = 1$

50. $f''(0) = 0$; $f'(0) = \frac{1}{2}$; $f''(x) > 0$ if $x < 0$; $f''(x) < 0$ if $x > 0$; $f(0) = 1$

51. $f'(0)$ does not exist; $f''(x)>0$ if $x<0$; $f''(x)>0$ if $x>0$; $f(0)=1$

52. $f'(0)$ does not exist; $f''(0)$ does not exist; $f''(x)<0$ if $x<0$; $f''(x)>0$ if $x>0$; $f(0)=1$

53. For the function $f(x)=ax^3+bx^2$ determine a and b so that the point $(1, 6)$ is a point of inflection of f.

54. For the cubic polynomial function $f(x)=ax^3+bx^2+cx+d$, determine a, b, c, and d so that the point $(0, 4)$ is a critical point and the point $(1, -2)$ is a point of inflection.

55. Use calculus to show that $x^2-8x+21>0$ for all x.

56. Use calculus to show that $3x^4-4x^3-12x^2+40>0$ for all x.

57. Show that the function $f(x)=ax^2+bx+c$, $a\neq 0$, has no inflection points. For what values of a is f concave up? For what values of a is f concave down?

58. Show that the function $f(x)=(ax+b)/(cx+d)$ has no critical points and no inflection points.

59. Find the local extrema and the points of inflection of $y=\sqrt{3}\sin x+\cos x$ on the interval $(0, 2\pi)$.

60. Show that every polynomial of degree 3,

$$f(x)=ax^3+bx^2+cx+d, \quad a\neq 0$$

has exactly one inflection point.

61. Prove that a polynomial of degree $n\geq 3$ has at most $(n-1)$ critical numbers and at most $(n-2)$ inflection points.

62. Show that the function $f(x)=(x-a)^n$, a a constant, has exactly one point of inflection if n is odd, $n\geq 3$.

63. Show that the function $f(x)=(x-a)^n$, a a constant, has no points of inflection if n is even.

64. Complete the proof of part (a) of the test for concavity (4.6.2).

65. Prove part (b) of the test for concavity (4.6.2).

66. Prove part (b) of the second derivative test (4.6.5).

4.7 Limits at Infinity; Infinite Limits; Asymptotes

In Chapter 2 we described $\lim_{x\to c} f(x)=L$ by saying that the value of $f(x)$ can be made as close as we please to L by choosing numbers x sufficiently close to c. It is understood that L and c are numbers. In this section we extend the language of limits to allow c to be $+\infty$ or $-\infty$ (*limits at infinity*) and to allow L to be $+\infty$ or $-\infty$ (*infinite limits*).* These limits, it turns out, are useful for locating *asymptotes* and hence aid in obtaining the graph of a function.

We begin with limits at infinity.

Limits at Infinity

Let's look at a familiar function, $f(x)=1/x$, whose domain is $x\neq 0$ (see Fig. 46). This function has the property that the value $f(x)$ can be made as close as we please to 0 when the number x is sufficiently positive. The following table illustrates this fact for selected numbers x:

x	1	10	100	1,000	10,000	100,000
$f(x)=1/x$	1	0.1	0.01	0.001	0.0001	0.00001

* You are reminded that the symbols $+\infty$ (plus infinity) and $-\infty$ (minus infinity) are *not numbers*. Plus infinity expresses the idea of unboundedness in the positive direction; minus infinity expresses the idea of unboundedness in the negative direction.

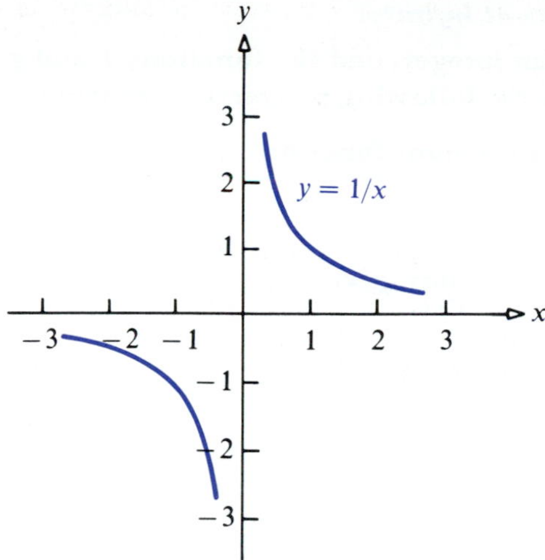

Figure 46

This phenomenon is expressed by saying that $f(x) = 1/x$ has the limit 0 as x approaches $+\infty$ and is symbolized by writing

$$\lim_{x \to +\infty} \frac{1}{x} = 0 \tag{1}$$

Informally, this statement means that $1/x$ can be made as close to 0 as we please by taking x sufficiently positive. For example, to make $1/x < 0.0001$, we take $x > 10{,}000$; to make $1/x < 0.00001$, we take $x > 100{,}000$. In an analogous way, we write

$$\lim_{x \to -\infty} \frac{1}{x} = 0 \tag{2}$$

to indicate that $1/x$ can be made as close as we please to 0 by selecting numbers x sufficiently negative. We summarize (1) and (2) by saying that $f(x) = 1/x$ has *limits at infinity*.

We now state an informal working definition of limits at infinity. The precise definition can be found at the end of this section.

[4.7.1] DEFINITION / ***Limits at Infinity.***

The statement $\lim_{x \to +\infty} f(x) = L$ means that values of $f(x)$ approach the number L as x increases without bound. The statement $\lim_{x \to -\infty} f(x) = M$ means that the values of $f(x)$ approach the number M as x decreases without bound.

The following theorem states that limits at infinity have the same algebraic properties stated in Chapter 2 if "$x \to c$" is replaced by "$x \to +\infty$" or "$x \to -\infty$." Although the properties are stated for limits as $x \to +\infty$, they are also valid for limits as $x \to -\infty$.

[4.7.2] THEOREM / *Limits at Infinity.*

If k is a real number, n is an integer, and the functions f and g have limits as $x \to +\infty$, then the following properties are true:

(a) $\lim_{x \to +\infty} A = A$ **where A is a constant function $f(x) = A$**

(b) $\lim_{x \to +\infty} kf(x) = k \lim_{x \to +\infty} f(x)$

(c) $\lim_{x \to +\infty} [f(x) \pm g(x)] = \lim_{x \to +\infty} f(x) \pm \lim_{x \to +\infty} g(x)$

(d) $\lim_{x \to +\infty} [f(x)g(x)] = \left[\lim_{x \to +\infty} f(x)\right]\left[\lim_{x \to +\infty} g(x)\right]$

(e) $\lim_{x \to +\infty} \dfrac{f(x)}{g(x)} = \dfrac{\lim_{x \to +\infty} f(x)}{\lim_{x \to +\infty} g(x)}$ **if** $\lim_{x \to +\infty} g(x) \neq 0$

(f) $\lim_{x \to +\infty} [f(x)]^n = \left[\lim_{x \to +\infty} f(x)\right]^n$

(g) $\lim_{x \to +\infty} \sqrt[n]{f(x)} = \sqrt[n]{\lim_{x \to +\infty} f(x)}$ **($f(x) \geq 0$ if n is even)**

Using these properties of limits at infinity, we can extend the statement $\lim_{x \to +\infty}(1/x) = 0$ to $\lim_{x \to +\infty}(1/x^p) = 0$ for $p > 0$ a real number. In fact,

$$\lim_{x \to +\infty} \frac{k}{x^p} = 0 \qquad p > 0, \quad k \text{ any real number} \tag{3}$$

As examples,

$$\lim_{x \to +\infty} \frac{4}{x^2} = 0 \qquad \lim_{x \to +\infty} \frac{-10}{\sqrt{x}} = 0$$

Limits as $x \to -\infty$ are handled in the same way as limits as $x \to +\infty$. For example, if p is a positive number for which x^p, $x < 0$, is defined and k is any real number, then

$$\lim_{x \to -\infty} \frac{k}{x} = 0 \qquad \lim_{x \to -\infty} \frac{k}{x^p} = 0 \tag{4}$$

We shall use (3) and (4) together with theorem (4.7.2) in the examples that follow.

EXAMPLE 1

Find: $\lim_{x \to +\infty} \dfrac{3x - 2}{4x - 1}$

Solution

We evaluate this limit by first dividing each term of both the numerator and the denominator by the highest power of x that appears in the denominator (in this case x). Then

$$\lim_{x\to+\infty}\frac{3x-2}{4x-1}=\lim_{x\to+\infty}\frac{3-(2/x)}{4-(1/x)}=\frac{\lim_{x\to+\infty}[3-(2/x)]}{\lim_{x\to+\infty}[4-(1/x)]}$$

$$=\frac{\lim_{x\to+\infty}3-\lim_{x\to+\infty}(2/x)}{\lim_{x\to+\infty}4-\lim_{x\to+\infty}(1/x)}=\frac{3-0}{4-0}=\frac{3}{4}$$

EXAMPLE 2

Find: $\displaystyle\lim_{x\to-\infty}\frac{4x^2-5x}{3x^3-2x+9}$

Solution

We divide the numerator and the denominator by x^3:

$$\lim_{x\to-\infty}\frac{4x^2-5x}{3x^3-2x+9}=\lim_{x\to-\infty}\frac{(4x^2/x^3)-(5x/x^3)}{(3x^3/x^3)-(2x/x^3)+(9/x^3)}$$

$$=\frac{\lim_{x\to-\infty}[(4/x)-(5/x^2)]}{\lim_{x\to-\infty}[3-(2/x^2)+(9/x^3)]}=\frac{0-0}{3-0+0}=0$$

From the preceding two examples, we see that the idea of dividing the numerator and the denominator by the highest power of x that appears in the denominator reduces the problem to looking at just the term in the numerator with the highest exponent and the term in the denominator with the highest exponent. These terms are said to *dominate* the other terms *near* $-\infty$ and $+\infty$.* For example, the limit in Example 1 can be found as

$$\lim_{x\to+\infty}\frac{3x-2}{4x-1}=\lim_{x\to+\infty}\frac{3x}{4x}=\frac{3}{4}$$

since

$$\frac{3x-2}{4x-1}\approx\frac{3x}{4x}\qquad\text{for } x \text{ sufficiently positive}$$

The limit in Example 2 can be found as

$$\lim_{x\to-\infty}\frac{4x^2-5x}{3x^3-2x+9}=\lim_{x\to-\infty}\frac{4x^2}{3x^3}=\lim_{x\to-\infty}\frac{4}{3x}=0$$

since

$$\frac{4x^2-5x}{3x^3-2x+9}\approx\frac{4x^2}{3x^3}\qquad\text{for } x \text{ sufficiently negative}$$

* The expression "near $+\infty$" means "for all numbers x greater than some positive number," and "near $-\infty$" means "for all numbers x less than some negative number."

Let's look at some other examples:

$$\lim_{x\to+\infty} \frac{2x^3-5x+4}{3x^3+2x^2-10} = \lim_{x\to+\infty} \frac{2x^3}{3x^3} = \frac{2}{3}$$

$$\lim_{x\to-\infty} \frac{5x^4-10x^3+5}{2x^3+10x-3} = \lim_{x\to-\infty} \frac{5x^4}{2x^3} = \lim_{x\to-\infty} \frac{5x}{2} = -\infty$$

$$\lim_{x\to+\infty} \frac{-10x^2+5x+2}{5x^3+2x-1} = \lim_{x\to+\infty} \frac{-10x^2}{5x^3} = \lim_{x\to+\infty} \frac{-2}{x} = 0$$

$$\lim_{x\to-\infty} \frac{5x^4-10x^2+1}{-3x^3+10x^2+50} = \lim_{x\to-\infty} \frac{5x^4}{-3x^3} = \lim_{x\to-\infty} \frac{5x}{-3} = +\infty$$

$$\lim_{x\to+\infty} \frac{10x^{3/2}+50x-2}{2x^2+1} = \lim_{x\to+\infty} \frac{10x^{3/2}}{2x^2} = \lim_{x\to+\infty} \frac{5}{x^{1/2}} = 0$$

$$\lim_{x\to-\infty} \frac{2-3x}{\sqrt{3+4x^2}} = \lim_{x\to-\infty} \frac{-3x}{\sqrt{4x^2}} \underset{\sqrt{x^2}=|x|}{=} \lim_{x\to-\infty} \frac{-3x}{2|x|} \underset{|x|=-x \text{ if } x<0}{=} \lim_{x\to-\infty} \frac{-3x}{2(-x)} = \frac{3}{2}$$

Horizontal Asymptotes

Limits at infinity have an interesting geometric interpretation. When $\lim_{x\to+\infty} f(x) = L$, it means that as x becomes sufficiently positive, the value of $f(x)$ can be made as close as we please to L; that is, the graph of $y = f(x)$ for x sufficiently positive is as close as we please to the horizontal line $y = L$. Similarly, $\lim_{x\to-\infty} f(x) = M$ means that the values of $f(x)$ can be made as close as we please to M for x sufficiently negative.

[4.7.3] DEFINITION / *Horizontal Asymptote.*

The line $y = L$ is a *horizontal asymptote* for the graph of the function f if either

$$\lim_{x\to+\infty} f(x) = L \qquad \text{or} \qquad \lim_{x\to-\infty} f(x) = M$$

In Figure 47 $y=L$ is a horizontal asymptote because

$$\lim_{x\to+\infty} f(x) = L$$

The line $y = M$ is also a horizontal asymptote because

$$\lim_{x\to-\infty} f(x) = M$$

We find horizontal asymptotes by calculating the limits at infinity.

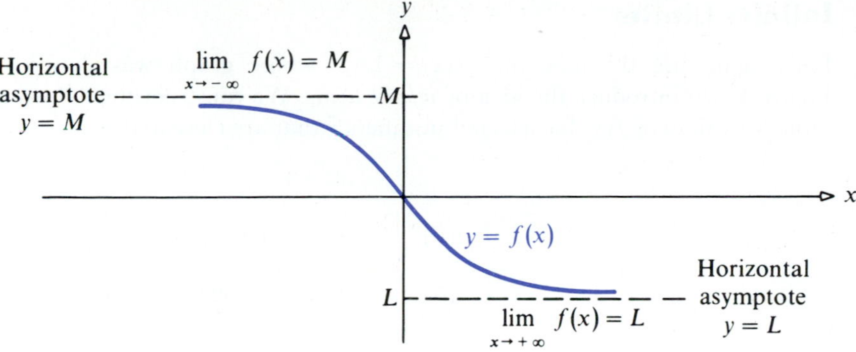

Figure 47

EXAMPLE 3

Find the horizontal asymptotes of

$$y = f(x) = \frac{x}{\sqrt{x^2 + 4}}$$

Solution

For x sufficiently positive, $\sqrt{x^2 + 4} \approx \sqrt{x^2} = x$. Thus,

$$\lim_{x \to +\infty} \frac{x}{\sqrt{x^2 + 4}} = \lim_{x \to +\infty} \frac{x}{\sqrt{x^2}} = \lim_{x \to +\infty} \frac{x}{x} = 1$$

Thus, $y = 1$ is a horizontal asymptote for x sufficiently positive. Similarly, for x sufficiently negative, $\sqrt{x^2 + 4} \approx \sqrt{x^2} = -x$. Thus,

$$\lim_{x \to -\infty} \frac{x}{\sqrt{x^2 + 4}} = \lim_{x \to -\infty} \frac{x}{\sqrt{x^2}} = \lim_{x \to -\infty} \frac{x}{-x} = -1$$

Thus, $y = -1$ is a horizontal asymptote for x sufficiently negative (see Fig. 48).

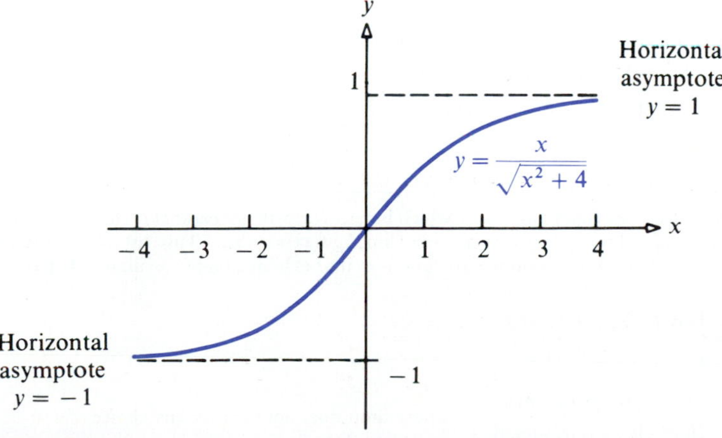

Figure 48

Infinite Limits

Let's again use the function $f(x) = 1/x$, whose graph was shown in Figure 46, to introduce the idea of *infinite limits*. We construct the following table for values of $f(x)$ for selected numbers x that are close to 0:

x	1	0.1	0.01	0.001	0.0001	0.00001
$f(x) = 1/x$	1	10	100	1,000	10,000	100,000

We see that as x gets closer to 0 from the right, the value of $f(x) = 1/x$ can be made as positive as we please; that is, $1/x$ becomes unbounded in the positive direction. We express this fact by writing

$$\lim_{x\to 0^+} \frac{1}{x} = +\infty \tag{5}$$

Similarly, we use the notation

$$\lim_{x\to 0^-} \frac{1}{x} = -\infty \tag{6}$$

to indicate that $1/x$ becomes unbounded in the negative direction by selecting numbers x sufficiently close to 0 but less than 0. We summarize (5) and (6) by saying that $f(x) = 1/x$ has *one-sided infinite limits* at 0.

As another example, consider $f(x) = 1/x^2$ as x approaches 0. When x approaches 0 from the left or when x approaches 0 from the right, the value of $1/x^2$ becomes positively infinite so that

$$\lim_{x\to 0^-} \frac{1}{x^2} = +\infty$$

$$\lim_{x\to 0^+} \frac{1}{x^2} = +\infty$$

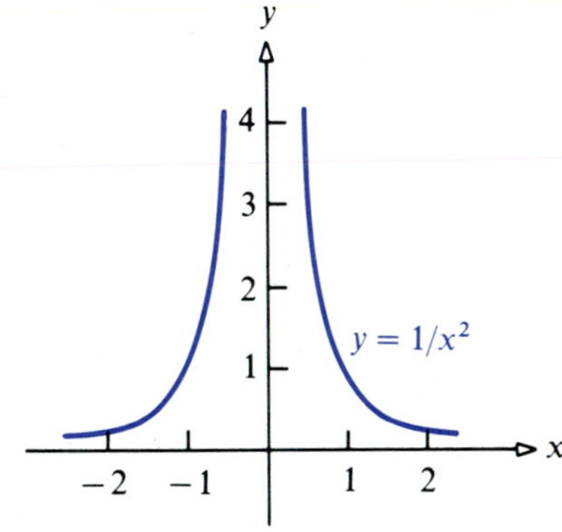

Figure 49

Even though we shall write $\lim_{x\to 0}(1/x^2) = +\infty$, we still say that the *limit as* $x \to 0$ *of* $1/x^2$ *does not exist* because* $+\infty$ *is not a real number.* Figure 49 illustrates the graph of $f(x) = 1/x^2$.

---

* So far, we have seen two conditions under which $\lim_{x\to c} f(x)$ will not exist: *(1)* $\lim_{x\to c^-} f(x) \neq \lim_{x\to c^+} f(x)$ or *(2)* $\lim_{x\to c} f(x) = +\infty$ or $\lim_{x\to c} f(x) = -\infty$. This list of conditions is by no means complete. Other examples are functions that exhibit a highly oscillatory behavior, such as

$$f(x) = \begin{cases} \sin(1/x) & \text{if } x \neq 0 \\ 0 & \text{if } x = 0 \end{cases} \quad \text{at } x = 0$$

and $f(x) = \begin{cases} 1 & \text{if } x \text{ is rational} \\ 0 & \text{if } x \text{ is irrational} \end{cases}$ whose limit does not exist for any choice of x

We now state an informal working definition of infinite limits. The precise definition (4.7.8) can be found at the end of this section.

[4.7.4] DEFINITION / ***Infinite Limits.***

Let f be a function defined on an open interval containing c, except possibly at c itself. Then

$$\lim_{x \to c} f(x) = +\infty$$

means that the values of $f(x)$ can be made arbitrarily large in the positive sense by taking x sufficiently close to c $(x \neq c)$. Similarly,

$$\lim_{x \to c} f(x) = -\infty$$

means that the value of $f(x)$ can be made arbitrarily large in the negative sense by taking x sufficiently close to c $(x \neq c)$.

The one-sided infinite limits

$$\lim_{x \to c^+} f(x) = +\infty \qquad \lim_{x \to c^+} f(x) = -\infty$$

$$\lim_{x \to c^-} f(x) = +\infty \qquad \lim_{x \to c^-} f(x) = -\infty$$

are defined similarly.

Vertical Asymptotes

Infinite limits are used to find vertical asymptotes.

[4.7.5] DEFINITION / ***Vertical Asymptote.***

The line $x = c$ is a *vertical asymptote* for the graph of the function f if either

$$\lim_{x \to c^-} f(x) = \pm\infty \qquad \text{or} \qquad \lim_{x \to c^+} f(x) = \pm\infty$$

Figure 50 on page 278 illustrates the possibilities that can occur when a function has infinite limits.

We state three useful results for locating vertical asymptotes:

1. If n is an even positive integer, then $\displaystyle\lim_{x \to c} \frac{1}{(x-c)^n} = +\infty$.

2. If n is an odd positive integer, then $\displaystyle\lim_{x \to c^-} \frac{1}{(x-c)^n} = -\infty$.

3. If n is an odd positive integer, then $\displaystyle\lim_{x \to c^+} \frac{1}{(x-c)^n} = +\infty$.

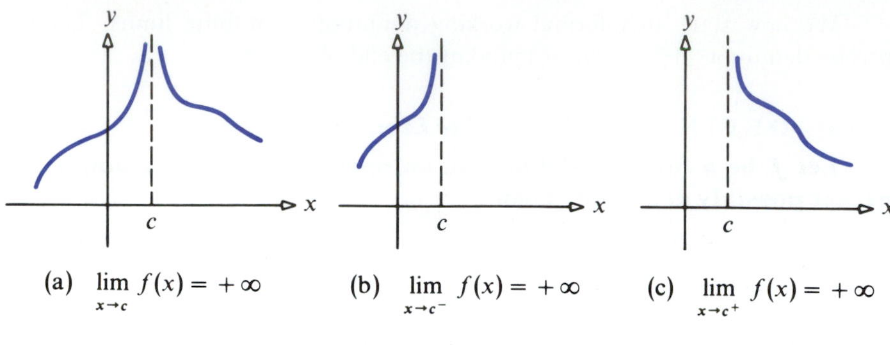

(a) $\lim_{x \to c} f(x) = +\infty$ (b) $\lim_{x \to c^-} f(x) = +\infty$ (c) $\lim_{x \to c^+} f(x) = +\infty$

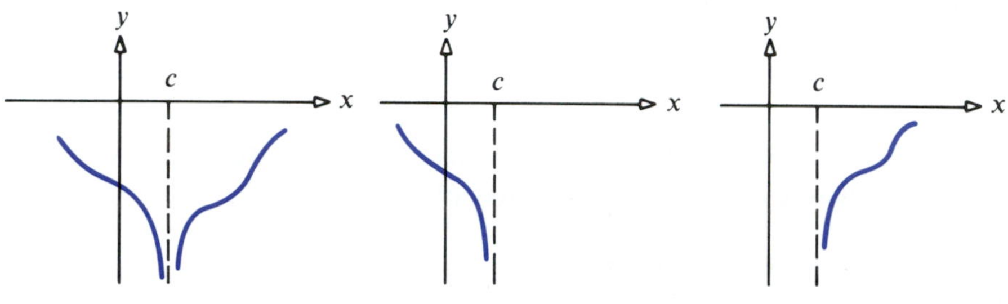

(d) $\lim_{x \to c} f(x) = -\infty$ (e) $\lim_{x \to c^-} f(x) = -\infty$ (f) $\lim_{x \to c^+} f(x) = -\infty$

Figure 50

EXAMPLE 4

Locate all vertical and horizontal asymptotes of $f(x) = x/(x-3)$ and graph f.

Solution

Since $x = 3$ is the only number for which the denominator of f equals 0, we evaluate the one-sided limits of f as $x \to 3$ to determine whether $x = 3$ is a vertical asymptote:

$$\lim_{x \to 3^-} \frac{x}{x-3} = -\infty \qquad \text{and} \qquad \lim_{x \to 3^+} \frac{x}{x-3} = +\infty$$

Hence, the line $x = 3$ is a vertical asymptote for the graph. To locate the horizontal asymptotes, if any, we look at the limits of f at infinity:

$$\lim_{x \to +\infty} \frac{x}{x-3} = \lim_{x \to +\infty} \frac{x}{x} = 1 \qquad \text{and} \qquad \lim_{x \to -\infty} \frac{x}{x-3} = \lim_{x \to -\infty} \frac{x}{x} = 1$$

Thus, the line $y = 1$ is a horizontal asymptote for x sufficiently positive and for x sufficiently negative. At $x = 0$, we have $f(0) = 0$, and this is the only x-intercept. Putting all this information together, we obtain the graph of f depicted in Figure 51.

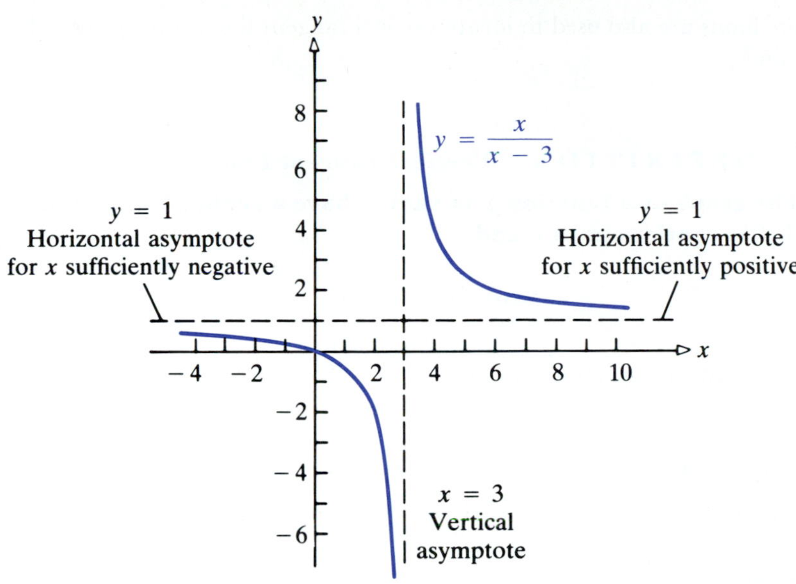

Figure 51

Vertical Tangent Lines

Vertical tangent lines can occur in a variety of ways, as Figure 52 illustrates.

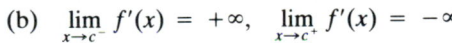

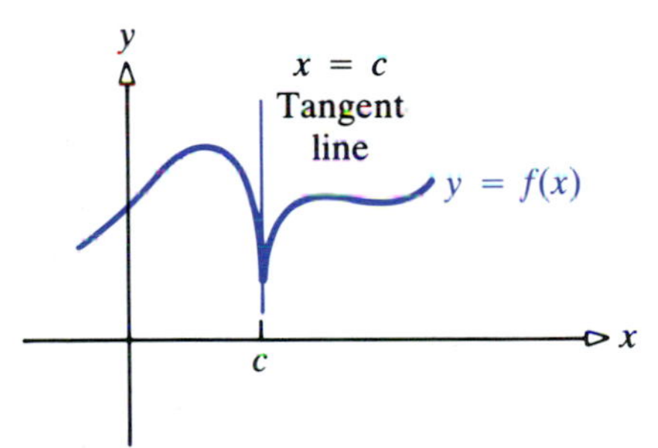

(a) $\lim_{x\to c^-} f'(x) = -\infty$, $\lim_{x\to c^+} f'(x) = +\infty$

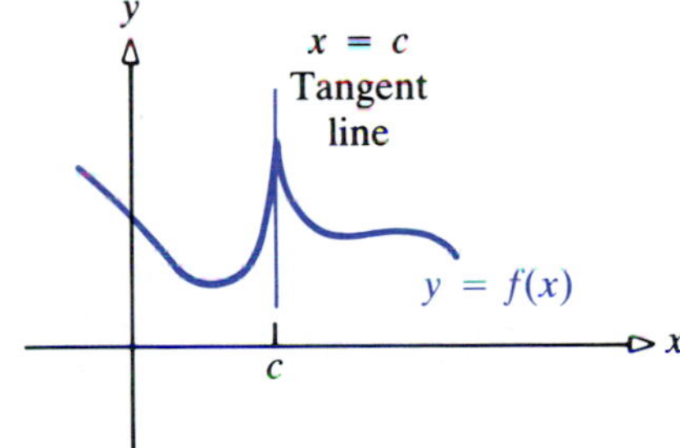

(b) $\lim_{x\to c^-} f'(x) = +\infty$, $\lim_{x\to c^+} f'(x) = -\infty$

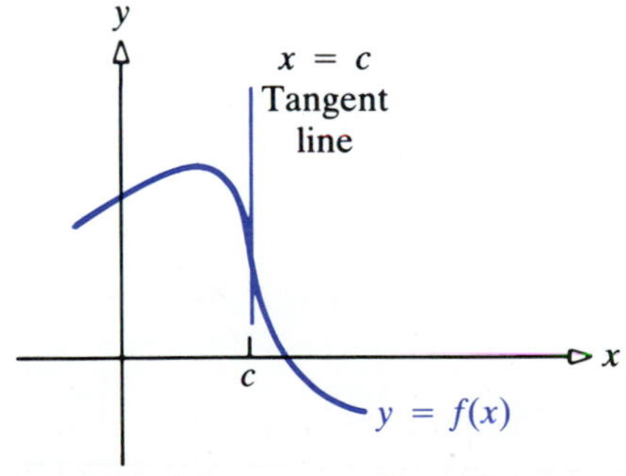

(c) $\lim_{x\to c^-} f'(x) = -\infty$, $\lim_{x\to c^+} f'(x) = -\infty$

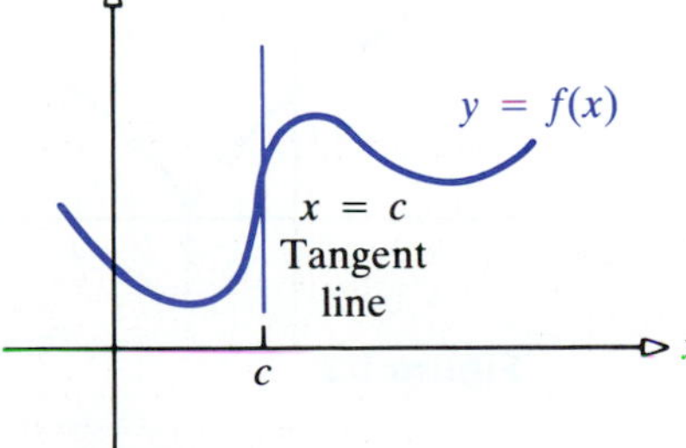

(d) $\lim_{x\to c^-} f'(x) = +\infty$, $\lim_{x\to c^+} f'(x) = +\infty$

Figure 52

Infinite limits are also used to locate vertical tangent lines to the graph of a function.

[4.7.6] DEFINITION / *Vertical Tangent Line.*

The graph of a function f is said to have a *vertical tangent line* at c if f is continuous at c and

$$\lim_{x \to c} |f'(x)| = +\infty$$

Let's look at a specific example.

EXAMPLE 5

Show that the graph of $f(x) = (x - 2)^{2/3}$ has a vertical tangent line at 2.

Solution

The function f is continuous at 2, and

$$f'(x) = \left(\frac{2}{3}\right)(x - 2)^{-1/3} = \frac{2}{3(x - 2)^{1/3}}$$

At 2, we have

$$\lim_{x \to 2} |f'(x)| = \lim_{x \to 2} \left| \frac{2}{3(x - 2)^{1/3}} \right| = +\infty$$

Hence, f has a vertical tangent line at 2. See Figure 53 for a sketch of the graph of $f(x) = (x - 2)^{2/3}$.

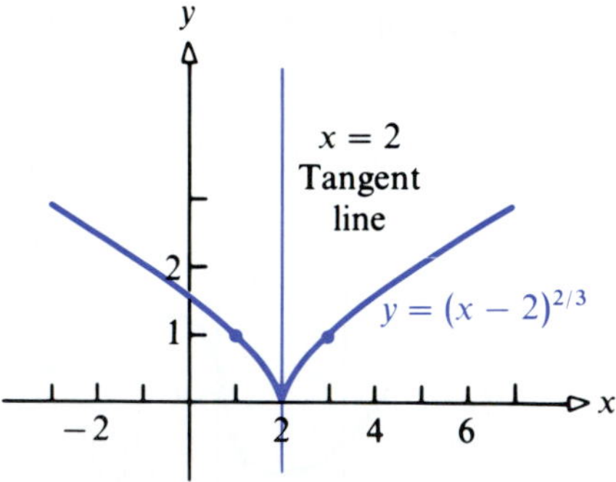

Figure 53

EXAMPLE 6

Show that the graph of $f(x) = (x - 2)^{1/3}$ has a vertical tangent line at 2.

Solution

The function f is continuous at 2, and

$$f'(x) = \frac{1}{3(x-2)^{2/3}}$$

At 2, we have

$$\lim_{x \to 2} |f'(x)| = \lim_{x \to 2} \frac{1}{3(x-2)^{2/3}} = +\infty$$

Thus, f has a vertical tangent at 2. See Figure 54 for a sketch of $f(x) = (x-2)^{1/3}$

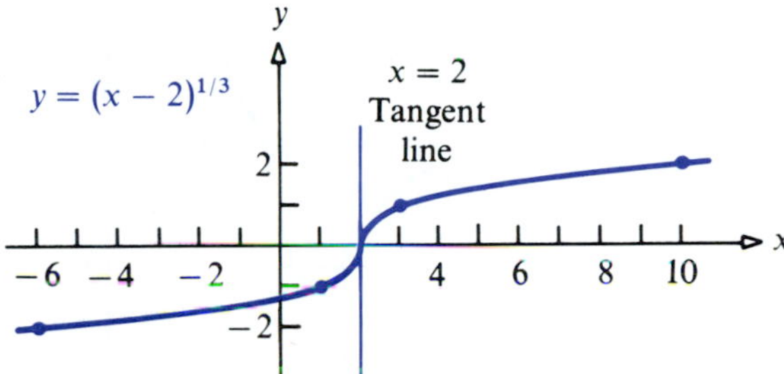

Figure 54

We close this section with the precise definitions of limits at infinity and infinite limits.

[4.7.7] DEFINITION / *Limits at Infinity.*

Let f be a function defined on the open interval $(b, +\infty)$. Then

$$\lim_{x \to +\infty} f(x) = L$$

if, for any given $\varepsilon > 0$, there is a positive number M so that

$|f(x) - L| < \varepsilon$ **whenever** $x > M$ **and x is in the domain of f**

See Figure 55.

If f is a function defined on the open interval $(-\infty, a)$, then

$$\lim_{x \to -\infty} f(x) = L$$

if, for any given $\varepsilon > 0$, there is a negative number N so that

$|f(x) - L| < \varepsilon$ **whenever** $x < N$ **and x is in the domain of f**

See Figures 55 and 56 on page 282.

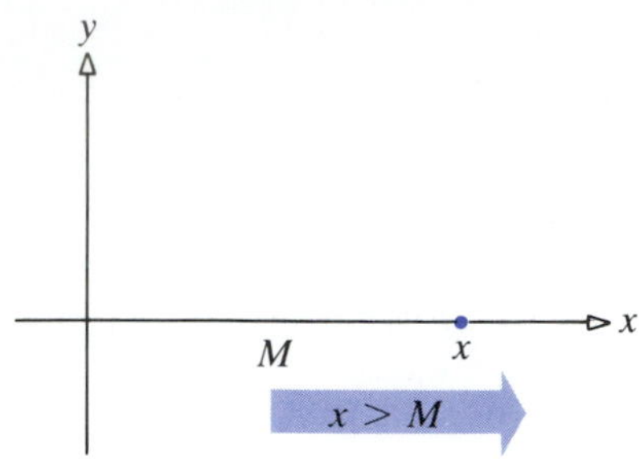

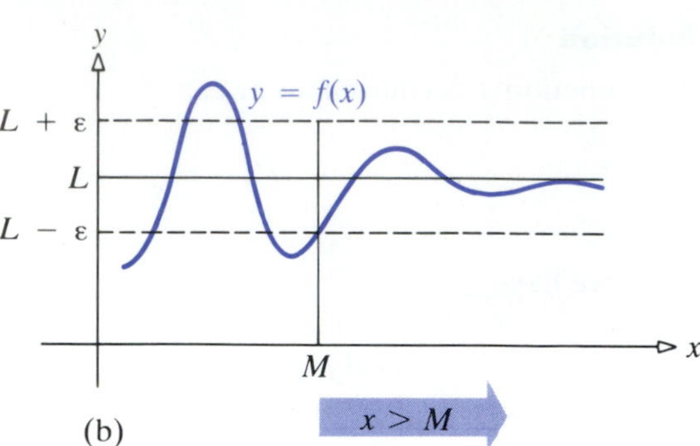

(a) For any $\varepsilon > 0$ there is a positive number M so that whenever $x > M$ and x is in the domain of f, then $|f(x) - L| < \varepsilon$

(b)

Figure 55

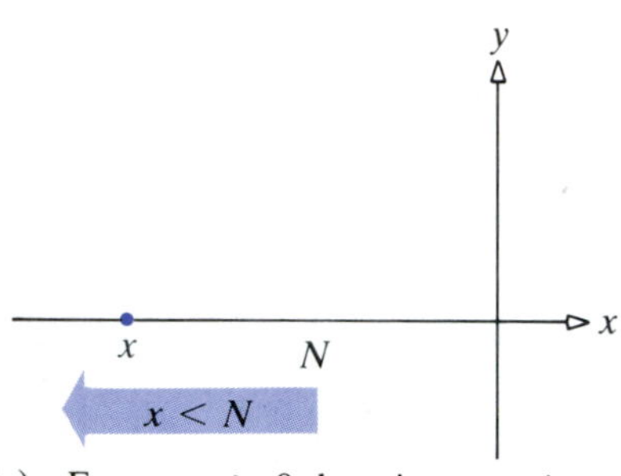

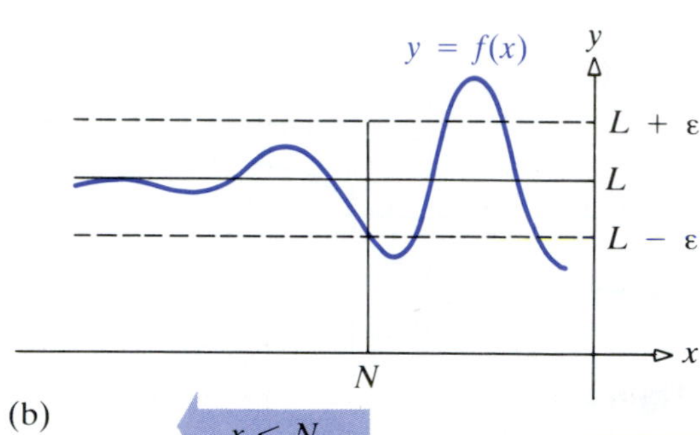

(a) For any $\varepsilon > 0$ there is a negative number N so that whenever $x < N$ and x is in the domain of f, then $|f(x) - L| < \varepsilon$

(b)

Figure 56

EXAMPLE 7

Use definition (4.7.7) to prove that

$$\lim_{x \to +\infty} \frac{1}{x} = 0 \tag{7}$$

Solution

To prove (7), we must show that definition (4.7.7) holds for $f(x) = 1/x$ and $L = 0$. That is, we must show that for every $\varepsilon > 0$, there exists a number $M > 0$, such that

$$\text{if} \quad x > M \quad \text{then} \quad \left|\frac{1}{x} - 0\right| < \varepsilon$$

Given any $\varepsilon > 0$, choose $M = 1/\varepsilon$. Then whenever $x > M = 1/\varepsilon$, we have

$$\left|\frac{1}{x} - 0\right| = \frac{1}{|x|} < \varepsilon$$

Thus, we have proved that $\lim_{x \to +\infty} 1/x = 0$.

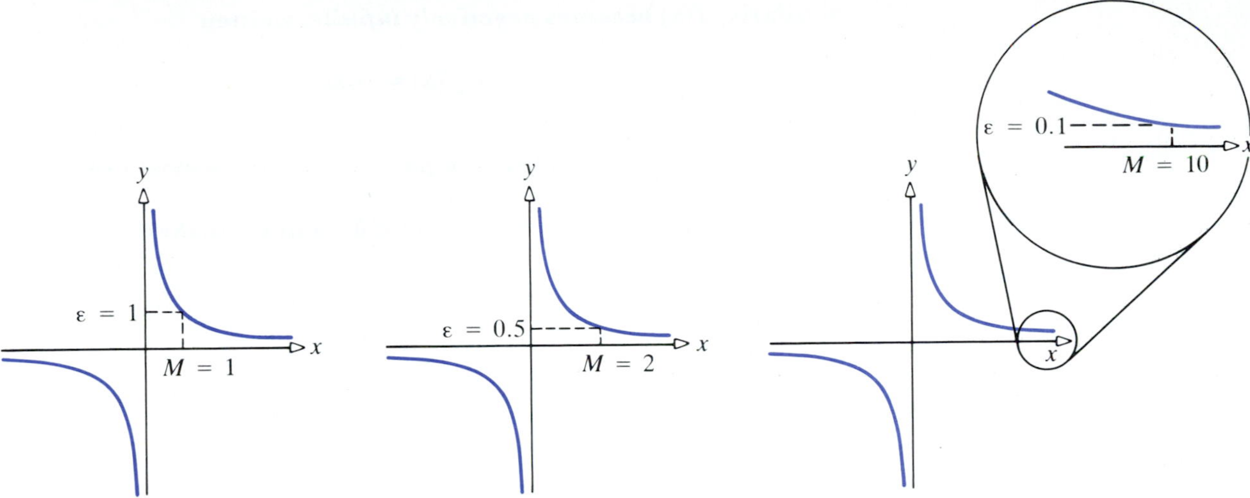

Figure 57

Figure 57 illustrates the proof by showing some values of ε and the corresponding values of M.

[4.7.8] DEFINITION / ***Infinite Limits.***

Let f be a function defined on an open interval containing c. Then $f(x)$ becomes *positively infinite as x approaches c*, written

$$\lim_{x \to c} f(x) = +\infty$$

if, for every positive number M, a positive number δ exists such that

$$f(x) > M \quad \text{whenever} \quad 0 < |x - c| < \delta \quad \text{and } x \text{ is in the domain of } f$$

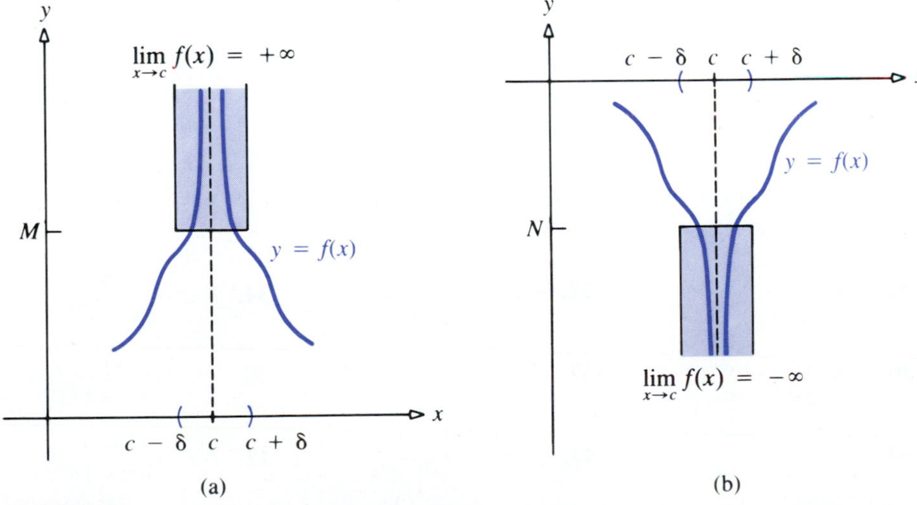

Figure 58

Similarly, $f(x)$ becomes *negatively infinite*, written

$$\lim_{x \to c} f(x) = -\infty$$

if, for any negative number N, a positive number δ exists such that

$$f(x) < N \quad \text{whenever} \quad 0 < |x - c| < \delta \quad \text{and } x \text{ is in the domain of } f$$

See Figure 58 for an illustration.

EXERCISE 4.7

In Problems 1–30 find the indicated limit.

1. $\lim_{x \to +\infty} \dfrac{x^3 + x^2 + 2x - 1}{x^3 + x + 1}$
2. $\lim_{x \to +\infty} \dfrac{2x^2 - 5x + 2}{5x^2 + 7x - 1}$
3. $\lim_{x \to +\infty} \dfrac{2x + 4}{x - 1}$
4. $\lim_{x \to +\infty} \dfrac{x + 1}{x}$
5. $\lim_{x \to +\infty} \dfrac{3x^2 - 1}{x^2 + 4}$
6. $\lim_{x \to -\infty} \dfrac{x^2 - 2x + 1}{x^3 + 5x + 4}$
7. $\lim_{x \to -\infty} \dfrac{5x^3 - 1}{x^2 + 1}$
8. $\lim_{x \to -\infty} \dfrac{x^2 + 1}{x^3 - 1}$
9. $\lim_{x \to +\infty} \dfrac{x^{3/2} + 1}{x - 2}$
10. $\lim_{x \to +\infty} \dfrac{x^2 - 1}{x^{1/2} + 4}$
11. $\lim_{x \to +\infty} \dfrac{\sqrt{x^2 + 4}}{3x - 1}$
12. $\lim_{x \to +\infty} \dfrac{\sqrt[3]{x^3 - 4}}{2x + 5}$
13. $\lim_{x \to 2^-} \dfrac{3x}{x - 2}$
14. $\lim_{x \to 2^+} \dfrac{x}{x^2 - 4}$
15. $\lim_{x \to -1^+} \dfrac{5x + 3}{x(x + 1)}$
16. $\lim_{x \to -3^-} \dfrac{1}{x^2 - 9}$
17. $\lim_{x \to 2^+} \dfrac{x + 4}{\sqrt{x - 2}}$
18. $\lim_{x \to 3^+} \dfrac{1 - x}{(3 - x)^2}$
19. $\lim_{t \to -\infty} \dfrac{3 - t}{\sqrt{4 + 5t^2}}$
20. $\lim_{t \to -\infty} \dfrac{3 + t}{\sqrt{4 + 5t^2}}$
21. $\lim_{x \to -\infty} \dfrac{1 + \sqrt[7]{x}}{1 - \sqrt[7]{x}}$
22. $\lim_{x \to +\infty} \dfrac{\sqrt[5]{x}}{x^4 + 4}$
23. $\lim_{x \to +\infty} (x - \sqrt{x^2 + 4})$ [*Hint:* Rationalize.]
24. $\lim_{x \to +\infty} (x - \sqrt{x^2 - x})$
25. $\lim_{x \to 0^+} \left(\dfrac{1}{x + 1} - \dfrac{1}{x}\right)$
26. $\lim_{x \to 0^-} \left(\dfrac{1}{x + 1} - \dfrac{1}{x}\right)$
27. $\lim_{x \to +\infty} \dfrac{\sin x}{x}$
28. $\lim_{x \to +\infty} \dfrac{\cos x}{x}$
29. $\lim_{x \to 0^+} \cot x$
30. $\lim_{x \to \pi/2^-} \tan x$

In Problems 31–46 locate all horizontal and vertical asymptotes, if any, of the function f.

31. $f(x) = 3 + \dfrac{1}{x}$
32. $f(x) = 2 - \dfrac{1}{x^2}$
33. $f(x) = \dfrac{2}{(x - 1)^2}$
34. $f(x) = \dfrac{5}{(x + 2)^2}$
35. $f(x) = \dfrac{3x - 1}{x + 1}$
36. $f(x) = \dfrac{x^2}{x^2 - 4}$
37. $f(x) = \dfrac{x}{x^2 - 1}$
38. $f(x) = \dfrac{x^2}{x^2 + 1}$
39. $f(x) = \dfrac{x^2 + 4}{x^2 + 1}$
40. $f(x) = \dfrac{x^4}{x^3 - 1}$
41. $f(x) = \dfrac{3x^4 + 1}{x^3}$
42. $f(x) = \dfrac{x^3 - 1}{x^4 + 1}$
43. $f(x) = \dfrac{x^5}{x^2 + 1}$
44. $f(x) = \dfrac{2x^2 - 1}{x^2 - 1}$
45. $f(x) = \dfrac{ax + b}{cx + d}$
46. $f(x) = \dfrac{x^n}{(x - a)^n}$

In Problems 47–52 verify that the graph of each function has a vertical tangent line at the indicated number.

47. $f(x) = x^{1/3}$, at 0

48. $f(x) = x^{1/4} + 2$, at 0

49. $f(x) = \sqrt{x+4}$, at -4

50. $f(x) = x + x^{1/3}$, at 0

51. $f(x) = (x-3)^{2/3} + 2$, at 3

52. $f(x) = \sqrt[5]{x} - 7$, at 0

53. Explain why a rational function, where numerator and denominator have no common zeros, will have vertical asymptotes at each point of discontinuity.

54. Explain why a nonconstant polynomial function cannot have any asymptotes.

55. Show that the tangent line to the circle $x^2 + y^2 = 1$ is vertical at $x = 1$ and at $x = -1$.

56. If P and Q are polynomials of degrees m and n, respectively, discuss $\lim_{x\to+\infty}[P(x)/Q(x)]$ when: (a) $m > n$ (b) $m = n$ (c) $m < n$

Summary: Sketching Graphs 4.8

We can make quick and accurate sketches of many graphs by applying the information developed thus far.

Checklist of Information for Sketching the Graph of $y = f(x)$

1. Determine the domain of f.
2. Locate the x-intercepts and y-intercept.
3. Check for symmetry.
4. Find all horizontal and vertical asymptotes.
5. Locate all critical numbers.
6. Find all intervals on which f is increasing $(f'(x) > 0)$ and all intervals on which f is decreasing $(f'(x) < 0)$.
7. Find all local maxima and local minima.
8. Find all intervals on which the graph is concave up $(f''(x) > 0)$ and all intervals on which the graph is concave down $(f''(x) < 0)$.
9. Find all points of inflection.
10. Sketch the graph using items 1–9. If additional points are needed, evaluate the function at those points. It may also be helpful to compute the value of the derivative at such points.

EXAMPLE 1

Discuss the graph of the function $f(x) = \dfrac{x^2}{x^2 - 1}$, using items 1–9.

Solution

1. The domain of f consists of all x except 1 and -1.
2. The only intercept is $(0, 0)$.
3. Since $f(-x) = f(x)$, the graph is symmetric with respect to the y-axis.
4. Since

$$\lim_{x\to+\infty} \frac{x^2}{x^2-1} = 1 \quad \text{and} \quad \lim_{x\to-\infty} \frac{x^2}{x^2-1} = 1$$

the line $y = 1$ is a horizontal asymptote for x positive and for x negative. The lines $x = -1$ and $x = 1$ are vertical asymptotes, since

$$\lim_{x \to -1^-} \frac{x^2}{x^2 - 1} = +\infty \qquad \lim_{x \to -1^+} \frac{x^2}{x^2 - 1} = -\infty,$$

$$\lim_{x \to 1^-} \frac{x^2}{x^2 - 1} = -\infty \qquad \lim_{x \to 1^+} \frac{x^2}{x^2 - 1} = +\infty$$

5. $$f'(x) = \frac{(x^2 - 1)(2x) - x^2(2x)}{(x^2 - 1)^2} = \frac{-2x}{(x^2 - 1)^2}$$

There is a critical number at $x = 0$ ($x = -1$ and $x = 1$ are not in the domain of f).

6. We find that $f'(x) > 0$ for $x < 0$ and $f'(x) < 0$ for $x > 0$. Hence, f is increasing for $x < -1$ and $-1 < x < 0$, and decreasing for $0 < x < 1$ and $x > 1$.

7. By the first derivative test, f has a local maximum at $x = 0$.

8. $$f''(x) = (-2)\left[\frac{(x^2 - 1)^2(1) - (x)(2)(x^2 - 1)(2x)}{(x^2 - 1)^4}\right] = \frac{2(3x^2 + 1)}{(x^2 - 1)^3}$$

The following table provides an analysis of the concavity of f:

Interval	Sign of f''	Concavity
$(-\infty, -1)$	+	Up
$(-1, 1)$	−	Down
$(1, +\infty)$	+	Up

9. Since $f''(x)$ exists for all $x \neq \pm 1$ and is never 0, we conclude that f has no inflection points.

10. Figure 59 illustrates the graph of f.

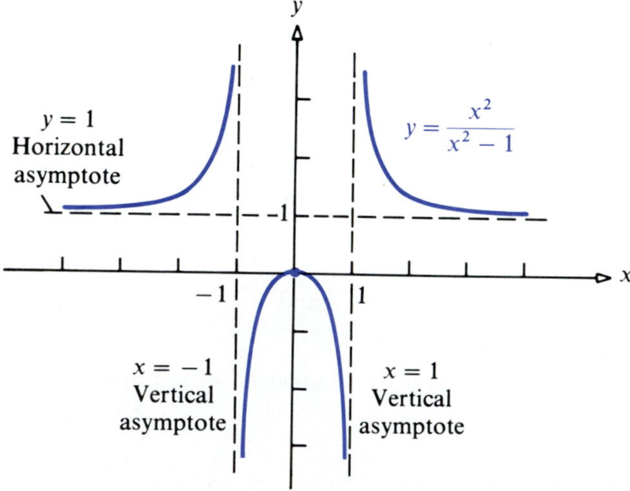

Figure 59

EXAMPLE 2

Discuss the graph of $f(x) = \dfrac{6x^2 - 6}{x^3}$.

Solution

1. The domain of f consists of all x, except 0.
2. The x-intercepts are $(-1, 0)$ and $(1, 0)$. There is no y-intercept.
3. Since $f(-x) = -f(x)$, f is an odd function and the graph is symmetric with respect to the origin.
4. Since

$$\lim_{x\to+\infty}\left(\frac{6x^2-6}{x^3}\right) = 0 \quad \text{and} \quad \lim_{x\to-\infty}\left(\frac{6x^2-6}{x^3}\right) = 0$$

the line $y = 0$ is a horizontal asymptote. Since

$$\lim_{x\to0^-}\left(\frac{6x^2-6}{x^3}\right) = +\infty \quad \text{and} \quad \lim_{x\to0^+}\left(\frac{6x^2-6}{x^3}\right) = -\infty$$

the line $x = 0$ is a vertical asymptote.

5. $$f'(x) = \frac{6(3 - x^2)}{x^4}$$

The critical numbers are $x = -\sqrt{3}$, $x = \sqrt{3}$. (*Note:* $x = 0$ is not in the domain of f.)

6. Since $f'(x) > 0$ for $-\sqrt{3} < x < 0$ and $0 < x < \sqrt{3}$, f is increasing on the intervals $[-\sqrt{3}, 0)$ and $(0, \sqrt{3}]$; and since $f'(x) < 0$ for $x < -\sqrt{3}$ and $x > \sqrt{3}$, f is decreasing on the intervals $(-\infty, -\sqrt{3}]$ and $[\sqrt{3}, +\infty)$.
7. By the first derivative test, f has a local maximum at $\sqrt{3}$ and a local minimum at $-\sqrt{3}$.
8. $$f''(x) = \frac{12(x^2 - 6)}{x^5}$$

The following table provides an analysis of the concavity of f:

Interval	Sign of f''	Concavity
$x < -\sqrt{6}$	$-$	Down
$-\sqrt{6} < x < 0$	$+$	Up
$0 < x < \sqrt{6}$	$-$	Down
$x > \sqrt{6}$	$+$	Up

9. Since $f''(x) = 0$ at $x = -\sqrt{6}$ and $x = \sqrt{6}$, the graph has an inflection point and a change in concavity at $-\sqrt{6}$ and at $\sqrt{6}$.
10. See Figure 60.

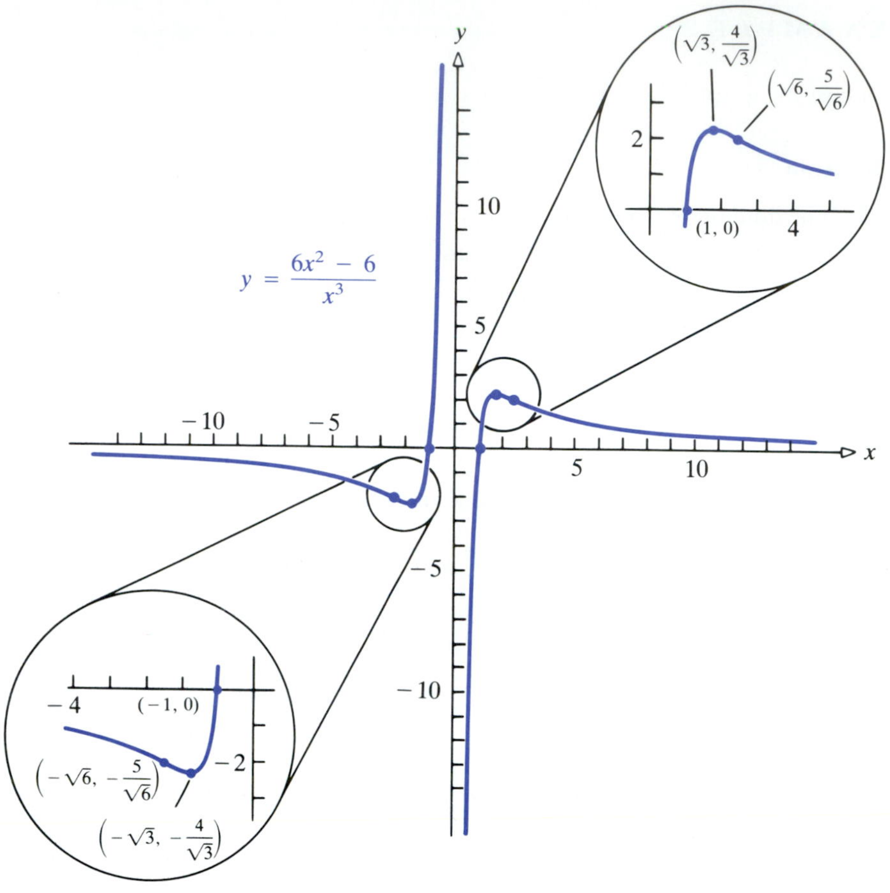

Figure 60

EXAMPLE 3

Discuss the graph of $f(x) = \dfrac{x^2 - 2x + 6}{x - 3}$.

Solution

1. The domain of f consists of all x except 3.
2. There are no x-intercepts, since $x^2 - 2x + 6 > 0$ for all x. Since $f(0) = 6/-3 = -2$, the y-intercept is $(0, -2)$.
3. There is no symmetry.
4. The line $x = 3$ is a vertical asymptote; there are no horizontal asymptotes.
5. $$f'(x) = \frac{x(x-6)}{(x-3)^2}$$

   The critical numbers are $x = 0$; $x = 6$. Note that $x = 3$ is not in the domain of f.

6.

Interval	Sign of x	Sign of $x - 6$	Sign of $f'(x) = \dfrac{x(x-6)}{(x-3)^2}$	Conclusion
$x < 0$	$-$	$-$	$+$	f is increasing
$0 < x < 6$	$+$	$-$	$-$	f is decreasing
$x > 6$	$+$	$+$	$+$	f is increasing

7. By the first derivative test, f has a local maximum at $x = 0$ and a local minimum at $x = 6$.

8.

$$f''(x) = \frac{18}{(x-3)^3}$$

Since $f''(x)$ is never 0, there are no points of inflection.

9. Since $f''(x) > 0$ for $x > 3$, f is concave up on the interval $x > 3$. Since $f''(x) < 0$ for $x < 3$, f is concave down on the interval $x < 3$.

10. See Figure 61.

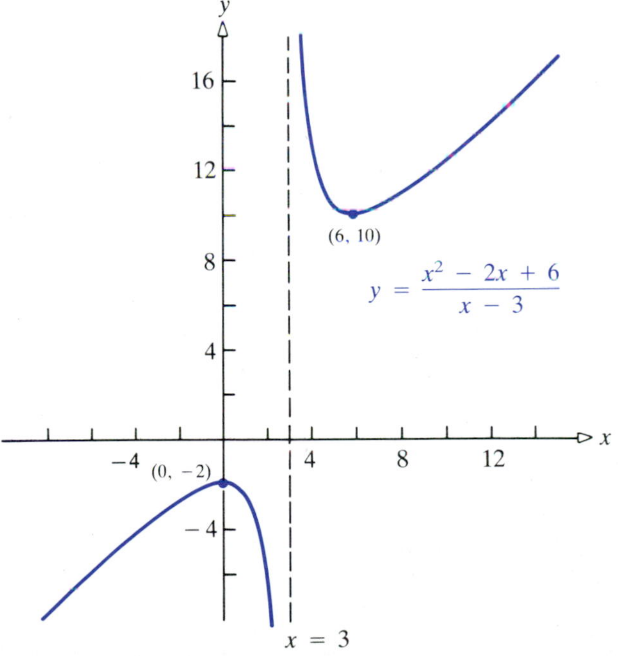

Figure 61

A rational function in which the degree of the numerator exceeds the degree of the denominator is called *improper*. An improper rational function may have an *oblique asymptote*. To find the oblique asymptote, we use long division and express the rational function as the sum of a polynomial and a proper rational function. If the polynomial is a first-degree polynomial, its graph, a straight line, is the oblique asymptote of the graph. In Example 3,

$$\frac{x^2 - 2x + 6}{x - 3} = \underbrace{x + 1}_{\substack{\text{Equation of}\\ \text{oblique}\\ \text{asymptote}}} + \underbrace{\frac{9}{x - 3}}_{\substack{\text{Goes to 0}\\ \text{as}\quad x \to +\infty}}$$

In Figure 62 we draw the oblique asymptote $y = x + 1$ using Figure 61.

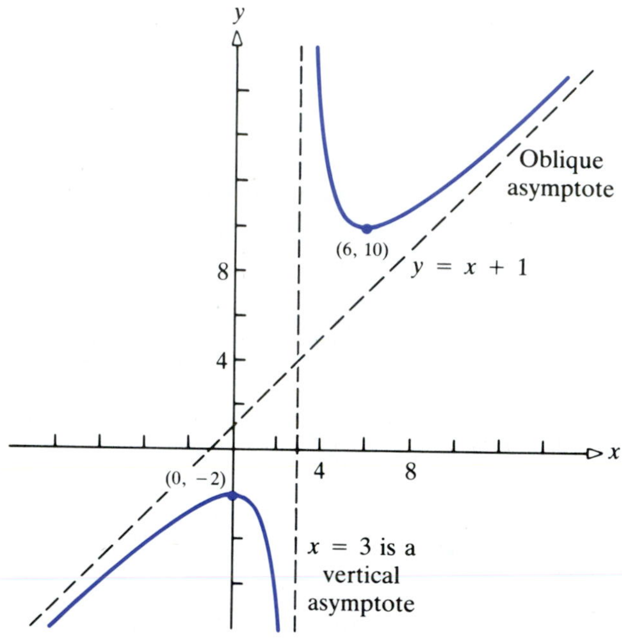

Figure 62

The next example is of another rational function with an oblique asymptote.

EXAMPLE 4

Discuss the graph of $f(x) = \dfrac{x^2 + 1}{x}$.

Solution

1. The domain of f is $x \neq 0$.
2. There are no x- or y-intercepts.
3. The graph is symmetric with respect to the origin.
4. f has a vertical asymptote at $x = 0$. There is no horizontal asymptote.

Since

$$\frac{x^2+1}{x} = \underbrace{x}_{\text{Oblique asymptote}} + \frac{1}{x}$$

the oblique asymptote is $y = x$.

5. $$f'(x) = \frac{x^2-1}{x^2} = \frac{(x-1)(x+1)}{x^2}$$

The critical numbers are $x = -1$, $x = 1$.

6.

Interval	Sign of $x-1$	Sign of $x+1$	Sign of $f'(x) = \frac{(x-1)(x+1)}{x^2}$	Conclusion
$x < -1$	$-$	$-$	$+$	f is increasing
$-1 < x < 0$	$-$	$+$	$-$	f is decreasing
$0 < x < 1$	$-$	$+$	$-$	f is decreasing
$1 < x$	$+$	$+$	$+$	f is increasing

7. By the first derivative test, f has a local maximum at $x = -1$ and a local minimum at $x = 1$.

8. $f''(x) = 2/x^3$. Since $f''(x) > 0$ for $x > 0$, f is concave up for all $x > 0$; and since $f''(x) < 0$ for $x < 0$, f is concave down for all $x < 0$.

9. $f''(x) \neq 0$. There are no inflection points.

10. See Figure 63.

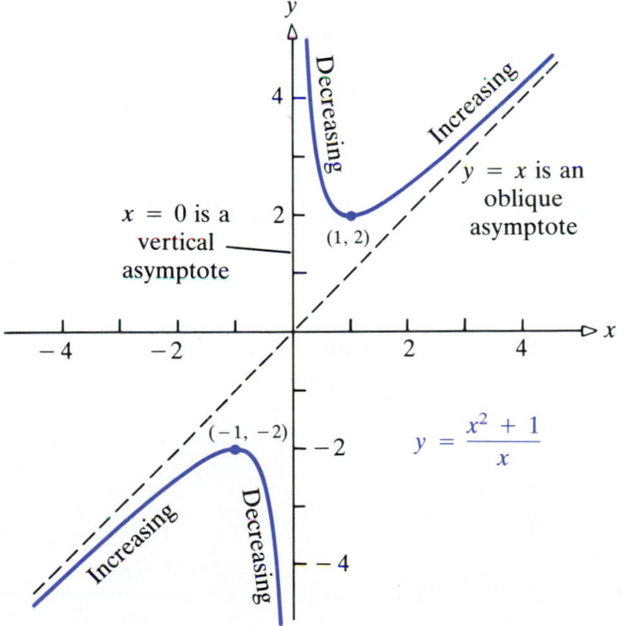

Figure 63

EXAMPLE 5

Discuss the graph of $f(x) = \dfrac{x}{\sqrt{x^2+4}}$.

Solution

1. The domain of f is all real numbers.
2. The only intercept is $(0, 0)$.
3. Since $f(-x) = -f(x)$, the graph is symmetric with respect to the origin.
4. The horizontal asymptotes are $y = -1$ and $y = 1$, since $\lim_{x\to-\infty} f(x) = -1$ and $\lim_{x\to+\infty} f(x) = 1$.
5. $f'(x) = \dfrac{4}{(x^2+4)^{3/2}}$. Since $f'(x) \neq 0$, there are no critical numbers.
6. Since $f'(x) > 0$ for all x, f is increasing on $(-\infty, +\infty)$.
7. There are no local extrema.
8. $f''(x) = \dfrac{-12x}{(x^2+4)^{5/2}}$. Since $f''(x) > 0$ for $x < 0$, f is concave up for all $x < 0$; and since $f''(x) < 0$ for $x > 0$, f is concave down for all $x > 0$.
9. At $x = 0$, f has an inflection point.
10. See Figure 64.

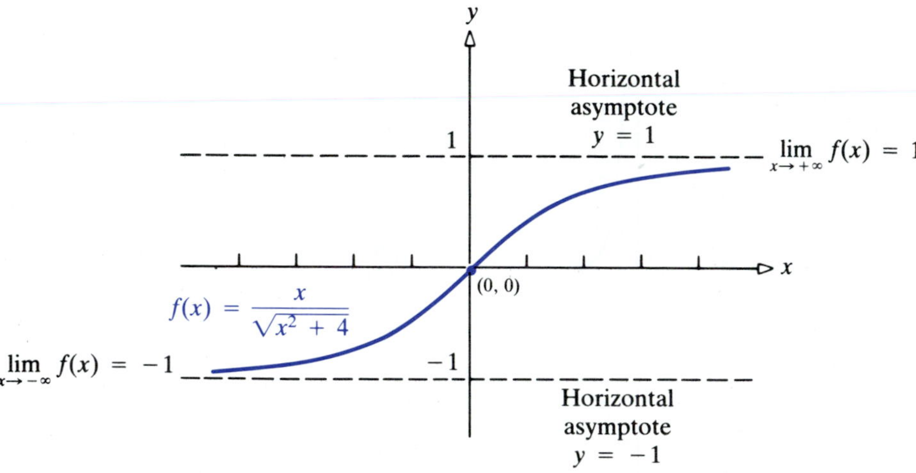

Figure 64

EXAMPLE 6

Discuss the graph of $f(x) = 4x^{1/3} - x^{4/3}$.

Solution

1. The domain of f is all real numbers.

2. The x-intercepts are $(0, 0)$ and $(4, 0)$; the y-intercept is $(0, 0)$.
3. There is no symmetry.
4. There are no asymptotes.
5. $f'(x) = \frac{4}{3}x^{-2/3} - \frac{4}{3}x^{1/3} = \frac{4}{3}x^{-2/3}(1 - x)$. The critical numbers are $x = 0$ and $x = 1$.
6. Since $f'(x) > 0$ for $x < 0$ and $0 < x < 1$, f is increasing on the interval $(-\infty, 1]$; since $f'(x) < 0$ for $x > 1$, f is decreasing on the interval $[1, +\infty)$. There is a vertical tangent line at $x = 0$.
7. f has a local maximum at $x = 1$.
8. $$f''(x) = -\tfrac{4}{9}x^{-5/3}(2 + x)$$

Interval	Sign of $-\frac{4}{9}x^{-5/3}$	Sign of $2 + x$	Sign of $f''(x) = -\frac{4}{9}x^{-5/3}(2 + x)$	Conclusion
$x < -2$	+	−	−	f is concave down
$-2 < x < 0$	+	+	+	f is concave up
$0 < x$	−	+	−	f is concave down

9. The inflection points are at $x = -2$ and $x = 0$.
10. See Figure 65.

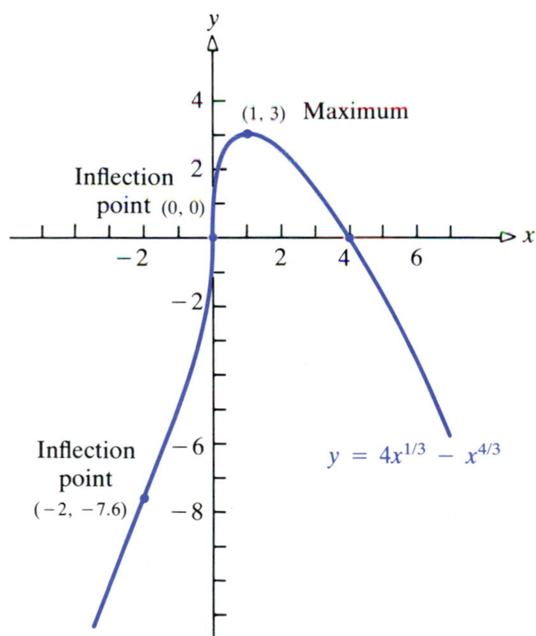

Figure 65

EXAMPLE 7

Discuss the graph of $f(x) = \cos^2 x - \sin x$.

Solution

1. The domain of f is the set of all real numbers.
2. The y-intercept is $(0, 1)$; the x-intercepts are obtained when

$$\cos^2 x - \sin x = 1 - \sin^2 x - \sin x = 0$$

or when

$$\sin x = -\frac{1-\sqrt{5}}{2}$$

$\left(\sin x \text{ can never be } -\dfrac{1+\sqrt{5}}{2}\right)$.

3. No symmetry. However, since $\sin x$ and $\cos x$ are periodic functions with period 2π, so is f; that is, $f(x + 2\pi) = f(x)$. As a result, we need only sketch the graph of f on an interval of length 2π—say, the interval $[0, 2\pi]$—and the rest of the graph will be a repetition of this part of the graph.
4. There are no asymptotes.
5. $$f'(x) = -2\cos x \sin x - \cos x = -\cos x(2\sin x + 1)$$

The critical numbers obey

$$\cos x = 0 \qquad \text{or} \qquad \sin x = -\tfrac{1}{2}$$

Now,

$$\cos x = 0 \qquad \text{on } [0, 2\pi] \text{ only if } \quad x = \frac{\pi}{2} \quad \text{or} \quad x = \frac{3\pi}{2}$$

$$\sin x = -\tfrac{1}{2} \qquad \text{on } [0, 2\pi] \text{ only if } \quad x = \frac{7\pi}{6} \quad \text{or} \quad x = \frac{11\pi}{6}$$

Hence, the critical numbers of f on $[0, 2\pi]$ are

$$\frac{\pi}{2}, \quad \frac{3\pi}{2}, \quad \frac{7\pi}{6}, \quad \frac{11\pi}{6}$$

6. We choose to use the second derivative test on these critical numbers. The second derivative of f is given by

$$\begin{aligned} f''(x) &= 2(\sin^2 x - \cos^2 x) + \sin x \\ &= 2(\sin^2 x - 1 + \sin^2 x) + \sin x \\ &= 4\sin^2 x + \sin x - 2 \end{aligned}$$

7. By the second derivative test, we verify that

$$f''\left(\frac{\pi}{2}\right) = 4 + 1 - 2 = 3 > 0 \qquad \text{so that } f \text{ has a local minimum at } \frac{\pi}{2}$$

$$f''\left(\frac{3\pi}{2}\right) = 4 - 1 - 2 = 1 > 0 \qquad \text{so that } f \text{ has a local minimum at } \frac{3\pi}{2}$$

$$f''\left(\frac{7\pi}{6}\right) = 1 - \frac{1}{2} - 2 = -\frac{3}{2} < 0 \qquad \text{so that } f \text{ has a local maximum at } \frac{7\pi}{6}$$

$$f''\left(\frac{11\pi}{6}\right) = 1 - \frac{1}{2} - 2 = -\frac{3}{2} < 0 \qquad \text{so that } f \text{ has a local maximum at } \frac{11\pi}{6}$$

8–10. By using the above information and the computations from the following table, we are able to sketch the graph, as shown in Figure 66.

x	0	$\pi/2$	π	$7\pi/6$	$3\pi/2$	$11\pi/6$	2π
$\sin x$	0	1	0	$-\frac{1}{2}$	-1	$-\frac{1}{2}$	0
$\cos^2 x$	1	0	1	$\frac{3}{4}$	0	$\frac{3}{4}$	1
$\cos^2 x - \sin x$	1	-1	1	$\frac{5}{4}$	1	$\frac{5}{4}$	1

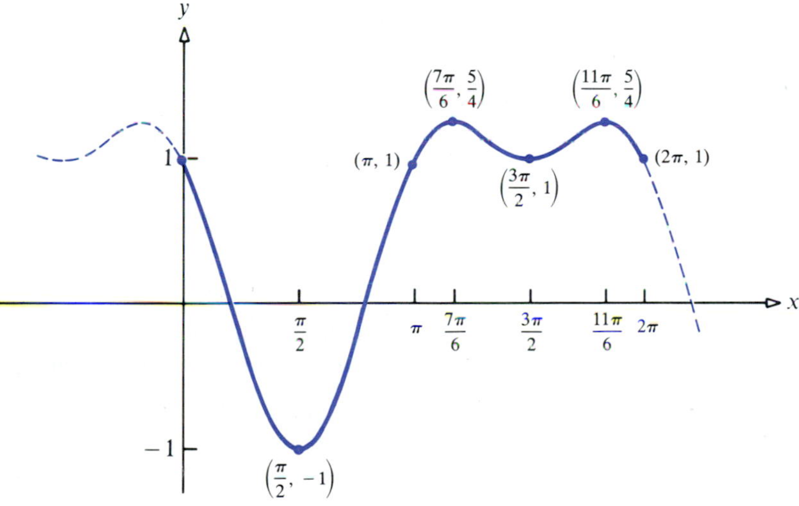

Figure 66

EXERCISE 4.8

In Problems 1–56 discuss the graph of the given function. Follow the list of items 1–10 given in this section.

1. $f(x) = \dfrac{1}{x-2}$

2. $f(x) = \dfrac{2}{x+2}$

3. $f(x) = \dfrac{x}{x-1}$

4. $f(x) = \dfrac{1}{x-2} - 3$

5. $f(x) = \dfrac{2}{x^2-4}$

6. $f(x) = \dfrac{1}{x^2-1}$

7. $f(x) = \dfrac{2x-1}{x+1}$

8. $f(x) = \dfrac{x-2}{x}$

9. $f(x) = \dfrac{x}{x^2+1}$

10. $f(x) = \dfrac{2x}{x^2-4}$

11. $f(x) = \dfrac{8}{x^2-16}$

12. $f(x) = \dfrac{x^2}{4-x^2}$

13. $f(x) = \dfrac{x^2+1}{2x}$

14. $f(x) = \dfrac{x^2-1}{2x}$

15. $f(x) = \dfrac{x^4+1}{x^2}$

16. $f(x) = \dfrac{x^2+1}{x+1}$

17. $f(x) = \dfrac{x^2-3}{2x-4}$

18. $f(x) = \dfrac{x^2+x-2}{x-2}$

19. $f(x) = \dfrac{x^2-2x+4}{x-2}$

20. $f(x) = \dfrac{-x^3+x^2+4}{x^2}$

21. $xy = x^2 + 2$

22. $xy = x^2 + x - 1$

23. $f(x) = \dfrac{x^2+1}{x}$

24. $f(x) = \dfrac{x^2+4}{x^2-1}$

25. $f(x) = \dfrac{x^2}{x+3}$

26. $f(x) = \dfrac{3x^2-1}{x-1}$

27. $f(x) = \dfrac{x^{2/3}}{x-1}$

28. $f(x) = \dfrac{x^{1/3}}{x-1}$

29. $f(x) = 1 + \dfrac{1}{x} + \dfrac{1}{x^2}$

30. $f(x) = \dfrac{2}{x} + \dfrac{1}{x^2}$

31. $f(x) = \sqrt{3-x}$

32. $f(x) = x\sqrt{x+2}$

33. $f(x) = x + \sqrt{x}$

34. $f(x) = \sqrt{x} - \sqrt{x+1}$

35. $f(x) = x\sqrt{2-x}$

36. $f(x) = x\sqrt{2-x^2}$

37. $f(x) = \dfrac{x^2}{\sqrt{x+1}}$

38. $f(x) = \dfrac{x}{\sqrt{x^2+2}}$

39. $f(x) = \dfrac{1}{(x+1)(x-2)}$

40. $f(x) = \dfrac{1}{(x-1)(x+3)}$

41. $f(x) = \dfrac{1}{x^2(x+4)}$

42. $f(x) = \dfrac{1}{x^3-9x}$

43. $f(x) = |x^2-1|$

44. $f(x) = |x^2-4|$

45. $f(x) = x^{2/3} + 3x^{1/3} + 2$

46. $f(x) = x^{5/3} - 5x^{2/3}$

47. $f(x) = \sin x - \cos x$

48. $f(x) = \sin x + \tan x$

49. $f(x) = \sin^2 x - \cos x$

50. $f(x) = \cos^2 x + \sin x$

51. $f(x) = \sin x - \tan x$

52. $f(x) = \sec x - \tan x$

53. $f(x) = x + \sin 2x$

54. $f(x) = x - \sin x$

55. $f(x) = x - 2\cos x$

56. $f(x) = \sec x + \tan x$

57. Sketch the graph of $y = x\sqrt{6-x}$ after answering the questions below. Also, show the graph of $y^2 = x^2(6-x)$ on the same drawing.

(a) What is the domain?

(b) Where does the graph intersect the x-axis?

(c) Find all extreme values of y; identify all horizontal and vertical tangents.

(Problem 57 continues in the next column)

57. *(continued)*

(d) Show that $y'' = 3(x-8)/4(6-x)^{3/2}$. Comment on the statement that an inflection point occurs at $x = 8$ and discuss the concavity of the graph.

58. Discuss the graph of $y^2 = x^2(4-x^2)$.

In Problems 59–62 sketch a graph of a function f defined and continuous for $2 \le x \le 5$ that satisfies the given conditions.

59. $f'(2)$ does not exist; $f'(3) = -1$; $f''(3) = 0$; $f'(5) = 0$; $f''(x) < 0$ if $2 < x < 3$; $f''(x) > 0$ if $x > 3$.

60. $f'(2) = 0$; $f''(2) = 0$; $f'(3)$ does not exist; $f'(5) = 0$; $f''(x) > 0$ if $2 < x < 3$; $f''(x) > 0$ if $x > 3$.

61. $f'(2) = 0$; $\lim_{x\to 3^-} f'(x) = +\infty$; $\lim_{x\to 3^+} f'(x) = +\infty$; $f'(5) = 0$; $f''(x) > 0$ if $x < 3$; $f''(x) < 0$ if $x > 3$.

62. $f'(2) = 0$; $\lim_{x\to 3^-} f'(x) = -\infty$; $\lim_{x\to 3^+} f'(x) = -\infty$; $f'(5) = 0$; $f''(x) < 0$ if $x < 3$; $f''(x) > 0$ if $x > 3$.

63. Sketch the graph of a function f defined and continuous for $-1 \le x \le 2$ that satisfies the following conditions: $f(-1) = 1$, $f(1) = 2$, $f(2) = 3$, $f(0) = 0$, $f(\frac{1}{2}) = 3$; $\lim_{x\to -1^+} f'(x) = -\infty$, $\lim_{x\to 1^-} f'(x) = -1$, $\lim_{x\to 1^+} f'(x) = +\infty$; f has a local minimum at 0, f has a local maximum at $\frac{1}{2}$.

Applied Extrema Problems

4.9

We begin with an example.

EXAMPLE 1

A farmer with 4000 meters of available fencing wishes to enclose a rectangular plot that borders on a straight river (see Fig. 67). If the farmer does not fence the side along the river, what is the largest rectangular area that can be enclosed?

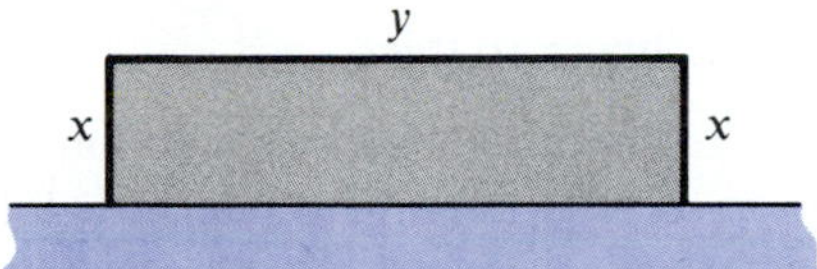

Figure 67

Solution

The quantity to be maximized is the area. We denote it by A, and we denote the dimensions of the rectangle by x and y, with y the length of the side parallel to the river. The area A is therefore

$$A = xy$$

But we need to express A in terms of a single variable. Since the length of available fence is 4000 meters, the variables x and y are related by the equation

$$x + y + x = 4000$$
$$y = 4000 - 2x$$

Thus, the area A is

$$A = x(4000 - 2x) = 4000x - 2x^2$$

The restrictions on x are $x \geq 0$ and $x \leq 2000$ (if $x > 2000$, then $y < 0$). The problem therefore is to maximize $A = 4000x - 2x^2$ on the closed interval $[0, 2000]$. The critical numbers obey

$$A'(x) = 4000 - 4x = 0$$

which gives the critical number $x = 1000$. The maximum value of A must occur either at the critical number or at an endpoint of the interval $[0, 2000]$.

$$A(1000) = 2{,}000{,}000 \qquad A(0) = 0 \qquad A(2000) = 0$$

So by definition (4.3.1) the maximum value is $A(1000) = 2{,}000{,}000$.

We could have reached the same conclusion by noting that $A''(x) = -4 < 0$ for all x, so A is always concave down, and the local maximum at $x = 1000$ must be the absolute maximum. The maximum area that can be enclosed is 2,000,000 square meters.

In general, each problem we discuss in this section requires that some quantity be minimized or maximized. We assume that this quantity can be represented by a function. Once this function is determined, the problem can be reduced to determining at what number the function assumes its absolute maximum or absolute minimum. Even though each applied problem has its unique features, it is possible to outline in a rough way a procedure for obtaining a solution. This five-step procedure is given here:

Step 1 Identify the quantity for which a maximum or a minimum value is to be found and assign a symbol to represent it.

Step 2 Assign symbols to represent other variables in the problem. If possible, use a picture to assist you.

Step 3 Determine the relationships among these variables.

Step 4 Express the quantity to be maximized or minimized as a function of one of the variables, and determine the domain of meaningful numbers for this variable.

Step 5 Apply the techniques of the previous sections to this function to determine the absolute maximum or absolute minimum relative to the domain found in Step 4.

The following examples illustrate this procedure.

EXAMPLE 2

From each corner of a square piece of sheet metal 18 centimeters on a side, remove a small square and turn up the edges to form an open box. What should be the dimensions of the box so as to maximize its volume?

Solution

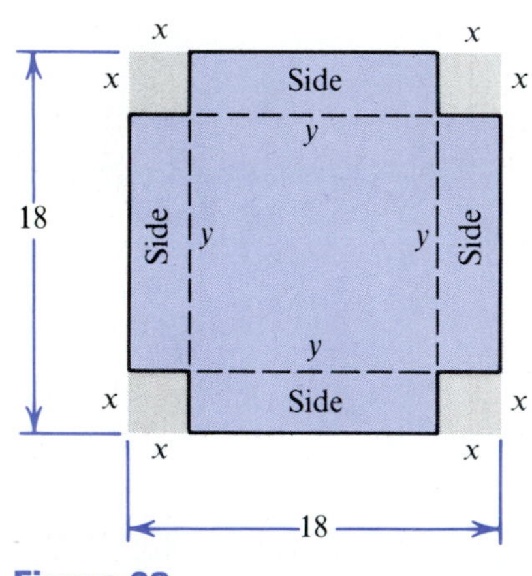

Figure 68

The quantity to be maximized is the volume of the box. Denote the volume by V, and denote the dimension of each side of the small square by x, as shown in Figure 68. Although the total area of the piece of sheet metal is fixed, the length of the sides of the squares can be changed, and this is treated as a variable. If y denotes the length left after cutting out the squares, we have

$$y = 18 - 2x$$

The height of the box is x, and the area of the base of the box is y^2. The volume V (height times area of base) is therefore

$$V = x(y^2)$$

This function of two variables can be expressed in terms of one variable by substituting for y in the formula for the volume. This gives

$$V = x(18 - 2x)^2$$

which is the function to be maximized. The only real numbers x that make sense in this case are those between 0 and 9. Thus, we want to find the absolute maximum of

$$V = x(18 - 2x)^2 \qquad 0 \le x \le 9$$

To find the x that maximizes V, we differentiate and find the critical numbers:

$$V'(x) = (18 - 2x)^2 + 2x(18 - 2x)(-2) = (18 - 2x)(18 - 6x)$$
$$= 12(9 - x)(3 - x)$$

Now, set $V'(x) = 0$ and solve for x. The solutions are

$$x = 9 \qquad \text{or} \qquad x = 3$$

The only critical number in the open interval $(0, 9)$ is $x = 3$. Thus, we calculate

$$V(0) = 0 \qquad V(3) = 3(18 - 6)^2 = 432 \qquad V(9) = 0$$

The maximum volume is 432 cubic centimeters, and the dimensions of the box that yield the maximum volume are

$$\text{Height} = x = 3 \text{ centimeters}$$
$$\text{Base} = y^2 = 12^2 = 144 \text{ square centimeters}$$

EXAMPLE 3*

A certain manufacturer makes a flexible square playpen that can be opened at a corner and attached at right angles to a wall (the side of a house, for example). Each side is 1 unit long, so that when the playpen is used in its square shape, its area is 1 square unit. When the playpen is placed as in Figure 69, the area enclosed is 2 square units, which doubles the child's play area. Is there a configuration that will do better than double the child's play area?

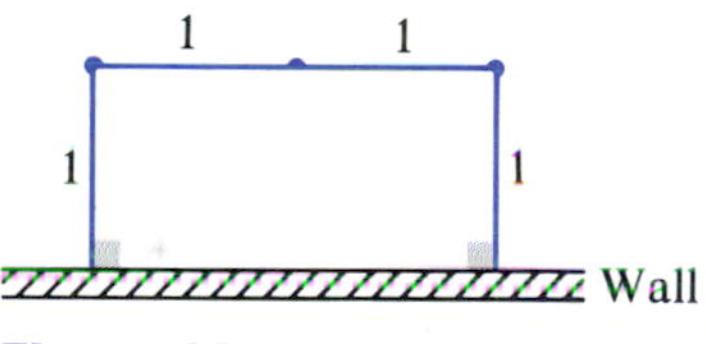

Figure 69

Solution

Since the playpen must be attached at right angles to the wall, the possible configurations depend on the amount of wall used as a fifth side for the playpen (see Fig. 70). Let x represent half the length of wall used as a fifth side. The area A is a function of x and is the sum of two rectangles (with sides 1 and x) and two right triangles (with hypotenuse 1 and base x). Thus, the quantity to be maximized is

$$A(x) = 2x + x\sqrt{1 - x^2} \qquad 0 \le x \le 1$$

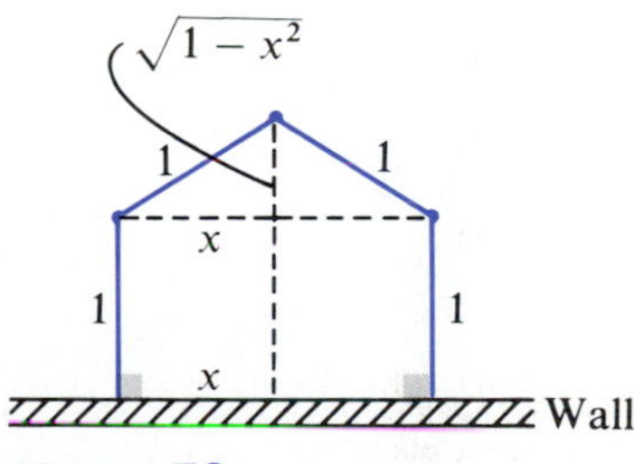

Figure 70

To compute the maximum area, we note that

$$A'(x) = 2 + \sqrt{1 - x^2} - \frac{x^2}{\sqrt{1 - x^2}} = \frac{2\sqrt{1 - x^2} + 1 - 2x^2}{\sqrt{1 - x^2}}$$

* Adapted from *Proceedings, Summer Conference for College Teachers on Applied Mathematics* (University of Missouri—Rolla, 1971).

The critical numbers obey

$$\begin{aligned} 2\sqrt{1-x^2}+1-2x^2&=0\\ 2\sqrt{1-x^2}&=2x^2-1\\ 4(1-x^2)&=4x^4-4x^2+1\\ 4x^4&=3\\ x&=\sqrt[4]{\tfrac{3}{4}}\approx 0.931 \end{aligned}$$

Thus, the only critical number in the open interval $(0, 1)$ is $\sqrt[4]{\frac{3}{4}} \approx 0.931$. Now, compute $A(x)$ at the endpoints $x = 0$ and $x = 1$ and at the critical number $x \approx 0.931$. The results are

$$A(0) = 0 \qquad A(1) = 2 \qquad A(0.931) \approx 2.20$$

Thus, a wall of length approximately $2x = 1.862$ will maximize the area, and a configuration like the one in Figure 70 increases the play area by about 10% (from 2 to 2.20 square units).

EXAMPLE 4

A can company wishes to produce a cylindrical container with a capacity of 1000 cubic centimeters. The top and bottom of the container must be made of material that costs \$0.05 per square centimeter, while the sides of the container can be made of material costing \$0.03 per square centimeter. Find the dimensions that will minimize the total cost of the container.

Solution

Figure 71 shows a cylindrical container and the area of its top, bottom, and lateral surfaces. As indicated in the figure, if we let h stand for the height of the can and R for the radius, then the total area of the bottom and top is $2\pi R^2$, and the area of the lateral surface of the can is $2\pi Rh$. The total cost

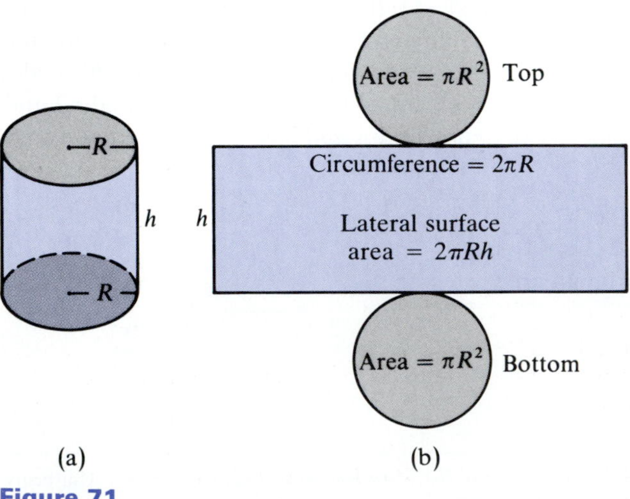

Figure 71

C (in cents) of manufacturing the can is therefore

$$C = (5)(2\pi R^2) + (3)(2\pi Rh) = 10\pi R^2 + 6\pi Rh$$

This is the function we wish to minimize. The cost function is a function of two variables, h and R, but there is a relationship between h and R, since the volume of the cylinder is fixed at 1000 cubic centimeters. That is,

$$V = 1000 = \pi R^2 h$$

$$h = \frac{1000}{\pi R^2}$$

By substituting for h in the cost function C, we obtain

$$C = 10\pi R^2 + (6\pi R)\left(\frac{1000}{\pi R^2}\right) = 10\pi R^2 + \frac{6000}{R}$$

The only restriction on R is that $R > 0$. To find the R that gives the minimum cost, we differentiate C with respect to R. Thus,

$$C'(R) = 20\pi R - \frac{6000}{R^2} = \frac{20\pi R^3 - 6000}{R^2}$$

The critical numbers obey $C'(R) = 0$, or

$$20\pi R^3 - 6000 = 0$$

$$R^3 = \frac{300}{\pi}$$

$$R = \sqrt[3]{\frac{300}{\pi}} \approx 4.57 \text{ centimeters}$$

By using the second derivative test (4.6.5), we have

$$C''(R) = 20\pi + \frac{12{,}000}{R^3}$$

$$C''\left(\sqrt[3]{\frac{300}{\pi}}\right) = 20\pi + \frac{12{,}000\pi}{300} > 0$$

Thus, for $R = \sqrt[3]{300/\pi} \approx 4.57$ centimeters, the cost has a local minimum. Since $C''(R)$ is clearly positive for all positive R, it follows that the graph of C is everywhere concave up, and hence the local minimum is the absolute minimum. The corresponding height of this can is

$$h = \frac{1000}{\pi R^2} \approx \frac{1000}{20.89\pi} \approx 15.24 \text{ centimeters}$$

These are the dimensions that will minimize the cost of the material.

If the costs of the materials for the top, bottom, and lateral surfaces of a cylindrical container are all the same, then the minimum total cost occurs when the surface area is minimum. It can be shown (see Problem 14 in Exercise 4.9) that for any fixed volume, the minimum surface area is obtained when the height equals twice the radius.

In Example 4, we can also use the first derivative test to conclude that $R = \sqrt[3]{300/\pi}$ yields an absolute minimum. Observe that $C'(R) < 0$ for $R < \sqrt[3]{300/\pi}$ and $C'(R) > 0$ for $R > \sqrt[3]{300/\pi}$, so C is decreasing for all R to the left of the critical number and increasing for all R to the right. Hence, $C(\sqrt[3]{300/\pi})$ is an absolute minimum.

Thus far we have used the first and second derivative tests to locate local extrema. In Example 4, we also used these tests to locate an absolute minimum. We now state a variant of the first derivative test that applies to absolute extrema.

[4.9.1] THEOREM

Suppose that c is a critical number of a continuous function f defined on an open interval I containing c.

(a) If $f'(x) > 0$ for all $x < c$ and $f'(x) < 0$ for all $x > c$, then $f(c)$ is the absolute maximum of f on I.
(b) If $f'(x) < 0$ for all $x < c$ and $f'(x) > 0$ for all $x > c$, then $f(c)$ is the absolute minimum value of f on I.

EXAMPLE 5

Find the coordinates of the points on the graph of $y = x^2 - 1$ that are nearest the origin.

Solution

See Figure 72. Let d be the distance from the point (x, y) to $(0, 0)$. By the distance formula, we have

$$d = \sqrt{(x-0)^2 + (y-0)^2} \tag{1}$$

But $y = x^2 - 1$, so we can rewrite (1) as

$$d = \sqrt{x^2 + (x^2-1)^2} = \sqrt{x^4 - x^2 + 1}$$

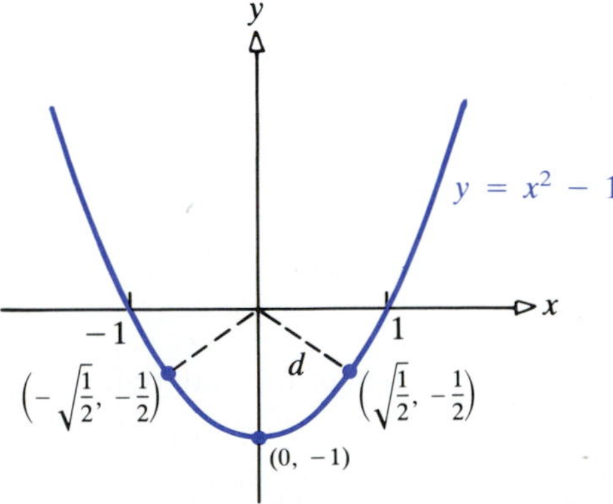

Figure 72

Since d is smallest when the expression inside the radical is smallest, we need only to find the critical numbers of

$$f(x) = x^4 - x^2 + 1 \tag{2}$$

Note that the domain of f is all real numbers. Differentiating f yields

$$f'(x) = 4x^3 - 2x = 2x(2x^2 - 1)$$

The critical numbers are $x = 0$, $x = -\sqrt{\frac{1}{2}}$, and $x = \sqrt{\frac{1}{2}}$. By the first derivative test, $x = 0$ yields a local maximum, while both $x = -\sqrt{\frac{1}{2}}$ and $x = \sqrt{\frac{1}{2}}$ yield a local minimum. Hence, the closest points are $(-\sqrt{\frac{1}{2}}, -\frac{1}{2})$ and $(\sqrt{\frac{1}{2}}, -\frac{1}{2})$, as shown in Figure 72.

EXAMPLE 6

A company charges \$200 for each set of tools on orders of 150 or fewer sets. The cost to the buyer on every set is reduced by \$1 for each set in excess of 150. For what size order is revenue maximum?

Solution

For an order of exactly 150 sets, the company's revenue is

$$\$200(150) = \$30{,}000$$

For an order of 160 sets (which is 10 in excess of 150), the per set charge is $\$200 - 10(\$1) = \$190$ and the revenue is

$$\$190(160) = \$30{,}400$$

To solve the problem, let x denote the number of sets sold. The revenue R is

$$R = (\text{Number of sets})(\text{Cost per set}) = x(\text{Cost per set})$$

The cost per set is

$$\$200 - \$1(\text{Number of sets in excess of } 150) = 200 - 1(x - 150) = 350 - x$$

Hence, the revenue R is

$$R = x(350 - x) = 350x - x^2$$

The only meaningful values for x are $150 \le x \le 350$. To find the number of sets leading to maximum revenue, we find the critical numbers of R:

$$R'(x) = 350 - 2x = 0 \qquad \text{when} \quad x = 175$$

We evaluate the revenue R at the critical number and at the endpoints:

$$R(175) = 175(175) = \$30{,}625 \qquad R(150) = 150(200) = \$30{,}000$$

$$R(350) = 350(0) = 0$$

The company's revenue is a maximum when 175 sets are sold. (Of course, the company would set this figure as the most it would allow anyone to purchase on this plan, since revenue to the company starts to decrease for orders in excess of 175.)

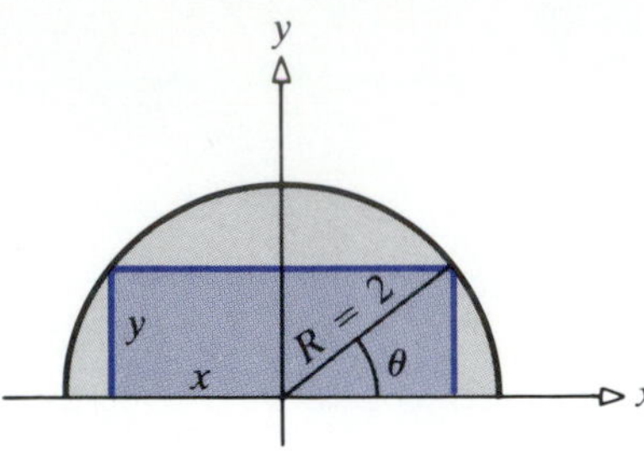

Figure 73

EXAMPLE 7

If a rectangle is inscribed in a semicircle of radius 2, find the dimensions of the rectangle that will have the maximum area.

Solution

Locate the semicircle so that its center is at the origin and its diameter is along the x-axis. Then the length of the inscribed rectangle is $2x$ and its height is y (see Fig. 73). We shall give two methods of solution; the first one is algebraic, and the second uses trigonometry.

Method I From Figure 73 it follows that

$$A = 2xy \qquad \text{and} \qquad x^2 + y^2 = 4 \qquad y \geq 0$$

We solve for y in $x^2 + y^2 = 4$ to get $y = \sqrt{4 - x^2}$, so we can use this in $A = 2xy$. As a result, the area A to be maximized can be expressed in terms of x alone as

$$A = 2x\sqrt{4 - x^2} \qquad 0 \leq x \leq 2$$

The critical numbers of A obey

$$A'(x) = 2\left[\sqrt{4 - x^2} + \frac{x(\frac{1}{2})(-2x)}{\sqrt{4 - x^2}}\right] = 2\,\frac{(4 - x^2) - x^2}{\sqrt{4 - x^2}}$$

$$= \frac{-4(x^2 - 2)}{\sqrt{4 - x^2}} = 0$$

The only critical number in the open interval $(0, 2)$ is $\sqrt{2}$, so we list the values of $A(x)$ at the endpoints and at the critical number:

$$A(2) = 0 \qquad A(0) = 0 \qquad A(\sqrt{2}) = 4$$

The maximum area is 4, and the corresponding dimensions of the rectangle are

$$\text{Length} = 2x = 2\sqrt{2} \qquad \text{Height} = y = \sqrt{2}$$

Method II From Figure 73 we see that $x = 2\cos\theta$ and $y = 2\sin\theta$ with $0 \leq \theta \leq \pi/2$. The area A of the rectangle is

$$A = 2xy = 2(2\cos\theta)(2\sin\theta) = 8\cos\theta\sin\theta = 4\sin 2\theta$$

$$\uparrow$$

$$\sin 2\theta = 2\sin\theta\cos\theta$$

To obtain the critical numbers, we differentiate $A = A(\theta)$ with respect to θ and set the result equal to 0:

$$A'(\theta) = 8\cos 2\theta = 0$$

$$\cos 2\theta = 0$$

$$\theta = \frac{\pi}{4}$$

By using the second derivative test (4.6.5), we get

$$A''(\theta) = -16 \sin 2\theta$$

$$A''\left(\frac{\pi}{4}\right) = -16 \sin \frac{\pi}{2} = -16$$

Thus, at $\theta = \pi/4$ the area is maximum, and

$$A = 4 \sin 2 \frac{\pi}{4} = 4$$

The dimensions of the rectangle are:

$$\text{Length} = 2\sqrt{2} \quad \text{Height} = \sqrt{2}$$

In the final example of this section, we prove *Snell's law of refraction.*

EXAMPLE 8

Light travels at different speeds in different media (air, water, glass, and so on). Suppose that light travels from a point A in one medium, where its speed is c_1, to a point B in another medium, where its speed is c_2 (see Fig. 74). Use Fermat's principle that light always travels along the path requiring least time to prove Snell's law of refraction—namely, that

$$\frac{\sin \theta_1}{c_1} = \frac{\sin \theta_2}{c_2}$$

Solution

Let the light pass from one medium to the other at the point P. Then the path taken by the light is made up of two line segments—from A to P and from P to B—since the shortest distance between two points is a line. We position our coordinate system as illustrated in Figure 75. By using the formula

$$\text{Time} = \frac{\text{Distance}}{\text{Speed}}$$

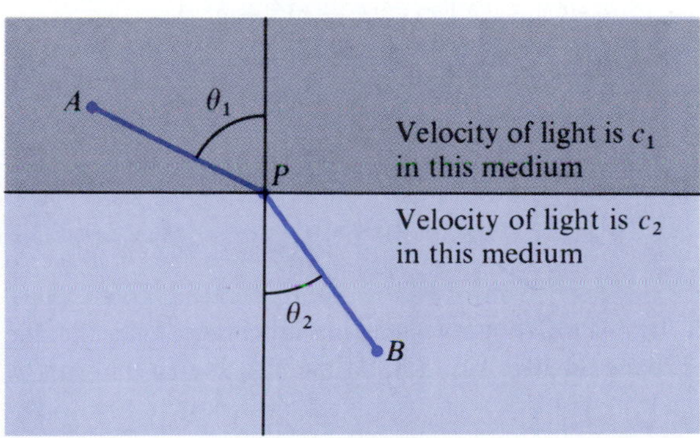

Figure 74

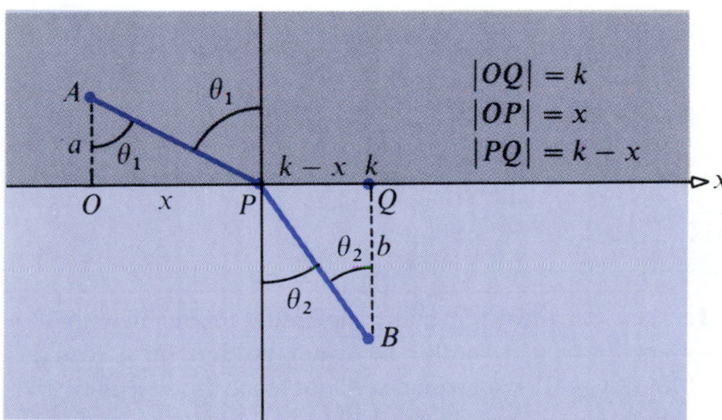

Figure 75

we see from Figure 75 that the travel time from A to P is

$$t_1 = \frac{\sqrt{x^2 + a^2}}{c_1}$$

and the travel time from P to B is

$$t_2 = \frac{\sqrt{(k - x)^2 + b^2}}{c_2}$$

The total time is therefore given by the formula

$$t = t_1 + t_2 = \frac{\sqrt{x^2 + a^2}}{c_1} + \frac{\sqrt{(k - x)^2 + b^2}}{c_2}$$

To find the least time, we compute dt/dx and set it equal to 0:

$$\frac{dt}{dx} = \frac{x}{c_1\sqrt{x^2 + a^2}} - \frac{k - x}{c_2\sqrt{(k - x)^2 + b^2}} = 0 \tag{3}$$

From Figure 75, we see that

$$\frac{x}{\sqrt{x^2 + a^2}} = \sin\theta_1 \quad \text{and} \quad \frac{k - x}{\sqrt{(k - x)^2 + b^2}} = \sin\theta_2$$

Hence,

$$\frac{dt}{dx} = \frac{\sin\theta_1}{c_1} - \frac{\sin\theta_2}{c_2} = 0$$

$$\frac{\sin\theta_1}{c_1} = \frac{\sin\theta_2}{c_2}$$

To ensure that the minimum t occurs when $dt/dx = 0$, we need to show that $d^2t/dx^2 > 0$. By using (3), we find that

$$\frac{d^2t}{dx^2} = \frac{d}{dx}\left[\frac{x}{c_1\sqrt{x^2 + a^2}}\right] - \frac{d}{dx}\left[\frac{k - x}{c_2\sqrt{(k - x)^2 + b^2}}\right]$$

$$= \frac{a^2}{c_1(x^2 + a^2)^{3/2}} + \frac{b^2}{c_2[(k - x)^2 + b^2]^{3/2}} > 0$$

EXERCISE 4.9

1. A farmer with 3000 meters of available fencing wishes to enclose a rectangular plot that borders on a straight highway. If the farmer does not fence the side along the highway, what is the largest area that can be enclosed?

2. If the farmer in Problem 1 decides also to fence the side along the highway, what is the largest area that can be enclosed?

3. Find the dimensions of the rectangle with the largest area that can be enclosed on all sides by L meters of fencing.

4. A builder wishes to fence in 60,000 square meters of land in a rectangular shape. For security reasons, the fence along the front part of the land will cost \$20.00 per meter, while the fence for the other three sides will cost \$10.00 per meter. How much of each type of fence will the builder have to buy in order to minimize the cost of the fence? What is the minimum cost?

5. A gardener with 20,000 meters of available fencing wishes to enclose a rectangular field and then divide it into two plots with a fence parallel to one of the sides, as shown in the figure. What is the largest area that can be enclosed?

6. A realtor wishes to enclose 600 square meters of land in a rectangular plot and then divide it into two plots with a fence parallel to one of the sides. What are the dimensions of the rectangular plot that require the least amount of fence?

7. An isosceles triangle has a perimeter of fixed length L. What should the dimensions of the triangle be if its area is to be a maximum?

8. An open box with a square base is to be made from a square piece of cardboard that measures 12 centimeters on each side. A square will be cut out from each corner of the cardboard and the sides will be turned up to form the box. Find the dimensions that yield the maximum volume of the box.

9. An open box with a square base is to be made from a square piece of cardboard 24 centimeters on a side by cutting out a square from each corner and turning up the sides. Find the dimensions that yield the maximum volume.

10. An open box with a square base is to have a volume of 2000 cubic centimeters. What should the dimensions of the box be if the amount of material used is to be a minimum?

11. If the box in Problem 10 is to be closed on top, what should the dimensions of the box be if the amount of material used is to be a minimum?

12. A cylindrical container that has a capacity of 10 cubic meters is to be produced. The top and bottom of the container are to be made of a material that costs \$20.00 per square meter, while the side of the container is to be made of material costing \$15.00 per square meter. Find the dimensions that will minimize the total cost of the container.

13. A cylindrical container that has a capacity of 4000 cubic centimeters is to be produced. The top and bottom of the container are to be made of material that costs \$0.50 per square centimeter, while the side of the container is to be made of material costing \$0.40 per square centimeter. Find the dimensions that will minimize the total cost of the container.

14. Prove that a cylindrical container of fixed volume V requires the least material (minimum surface area) when its height is twice its radius.

15. A car rental agency has 24 cars (identical model). The owner of the agency finds that at a price of \$18 per day, all the cars can be rented; however, for each \$1 increase in rental cost, one of the cars is not rented. What should the agency charge to maximize income?

16. A charter flight club charges its members \$200 per year. But for each new member in excess of 60, the charge for every member is reduced by \$2. What number of members leads to a maximum revenue?

17. Find the coordinates of the points on the graph of the parabola $y = x^2$ that are closest to the point $(2, \frac{1}{2})$.

18. Find the coordinates of the points on the graph of the parabola $y = 2x^2$ that are closest to the point $(1, 4)$.

19. Find the coordinates of the points on the graph of the parabola $y = 4 - x^2$ that are closest to the point $(6, 2)$.

20. Find the coordinates of the points on the graph of $y = \sqrt{x}$ that are closest to the point $(4, 0)$.

21. A heavy object of mass m is to be dragged along a horizontal surface by a rope making an angle θ with the horizontal. The force F required to move the object is given by the formula

$$F = \frac{mc}{c \sin \theta + \cos \theta}$$

where c is the *coefficient of friction* of the surface. Show that the force is least when $\tan \theta = c$.

22. A self-catalytic chemical reaction results in the formation of a product that causes its formation rate to increase. The reaction rate V of many self-catalytic chemicals obeys the relationship

$$V = kx(a - x) \qquad 0 \leq x \leq a$$

where k is a positive constant, a is the initial amount of the chemical, and x is the variable amount of the chemical. For what value of x is the reaction rate a maximum?

23. A truck has a top speed of 75 miles per hour and, when traveling at the rate of x miles per hour, consumes gasoline at the rate of $\frac{1}{200}[(1600/x) + x]$ gallon per mile. This truck is to be taken on a 200 mile trip by a driver who is to be paid at the rate of $\$b$ per hour plus a commission of $\$c$. Since the time required for this trip at x miles per hour is $200/x$, the total cost, if gasoline costs $\$a$ per gallon, is

$$C(x) = \left(\frac{1600}{x} + x\right)a + \frac{200}{x}b + c$$

Find the most economical possible speed under each of the following sets of conditions:

(a) $a = \$1.50, \quad b = 0, \quad c = 0$
(b) $a = \$1.50, \quad b = \$8.00, \quad c = \$500$
(c) $a = \$1.60, \quad b = \$10.00, \quad c = 0$

24. Find the largest area of a rectangle with one vertex on the parabola $y = 9 - x^2$, another at the origin, and the remaining two on the positive x-axis and positive y-axis, respectively.

25. A telephone company is asked to provide telephone service to a customer whose house is located 2 kilometers away from the road along which the telephone lines run. The nearest telephone box is located 5 kilometers down the road. As shown in the figure, let $5 - x$ denote the distance from the box to the connection so that x is the distance from this point to the point on the road closest to the house. If the cost to connect the telephone line is

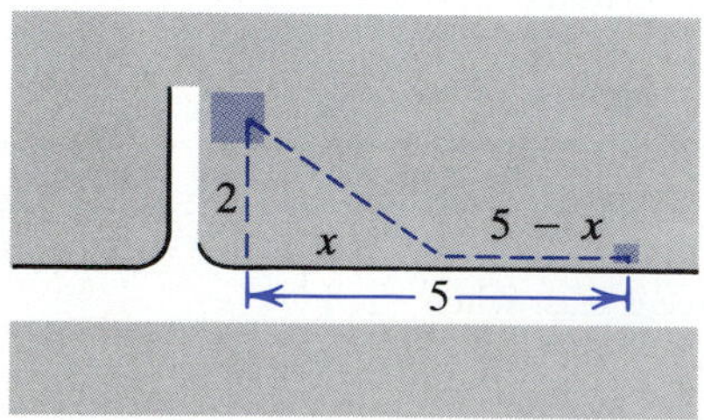

(*Problem 25 continues in the next column*)

25. (*continued*)
\$50 per kilometer along the road and \$60 per kilometer away from the road, where along the road from the box should the company connect the telephone line so as to minimize construction cost?

26. A small island is 3 kilometers from the nearest point P on the straight shoreline of a large lake. A town is 12 kilometers down the shore from P. If a person on the island can row a boat 2.5 kilometers per hour and can walk 4 kilometers per hour, where should the boat be landed so that the person arrives in town in the shortest time?

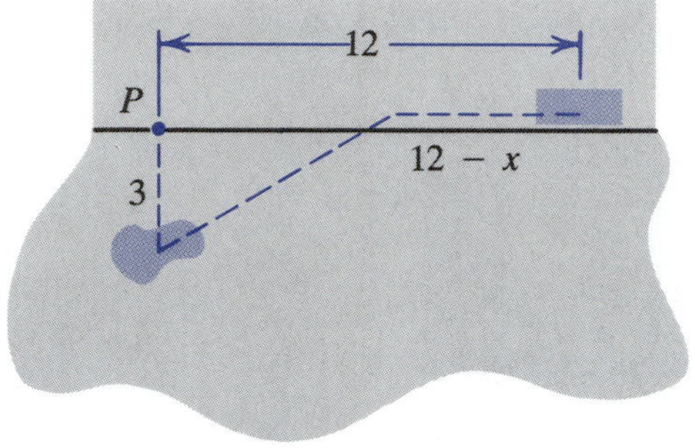

27. Two houses A and B on the same side of a road are a distance p apart, with distances q and r, respectively, from the road. Find the length of the shortest path that goes from A to the road and then on to the other house B.

(a) Use the derivative techniques introduced in this section.
(b) Use only elementary geometry.
[*Hint:* Introduce an imaginary house C on the other side of the road such that the midpoint between B and C is on the road and the segment BC is perpendicular to the road; that is, "reflect" B across the road to become C.]

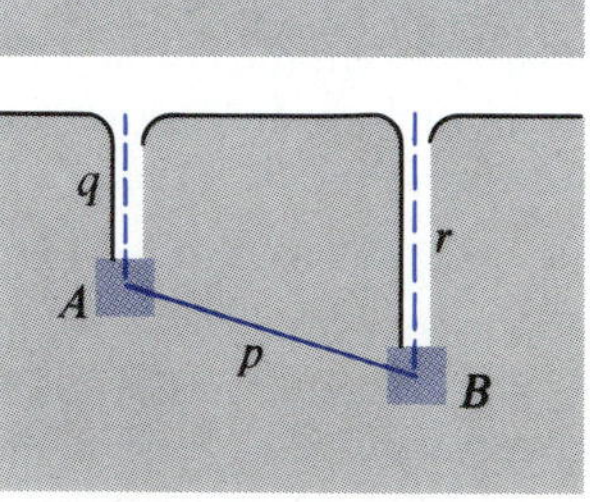

28. The strength of a rectangular beam is proportional to the product of the width and the cube of its depth. Find the dimensions of the strongest beam that can be cut

(*Problem 28 continues on page 309*)

28. *(continued)* from a log whose cross section has the form of a circle of fixed radius R.

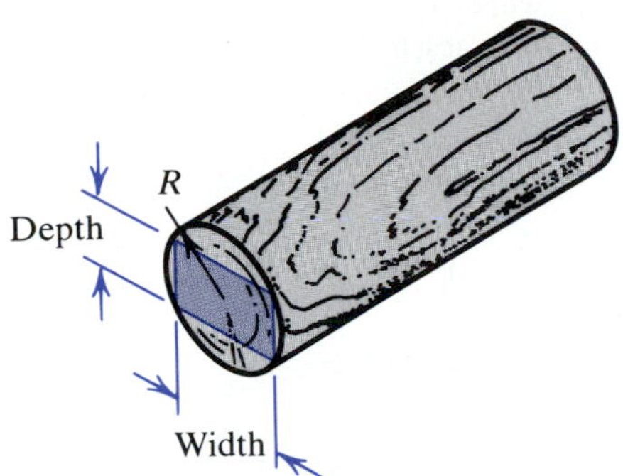

29. If the strength of a rectangular beam is proportional to the product of its width and the square of its depth, find the dimensions of the strongest beam that can be cut from a log whose cross section has the form of the ellipse $10x^2+9y^2=90$. [*Hint:* Choose Width $=2x$ and Depth $= 2y$.]

30. The strength of a beam made from a certain wood is proportional to the product of its width and the cube of its depth. Find the dimensions of the rectangular cross section of the beam with maximum strength that can be cut from a log whose original cross section is in the form of the ellipse $b^2x^2 + a^2y^2 = a^2b^2$, $a \geq b$.

31. The figure shows two corridors meeting at a right angle. One has width 1 meter, and the other, width 8 meters. Find the length of the longest pipe that can be carried horizontally from one hall, around the corner, and into the other hall.

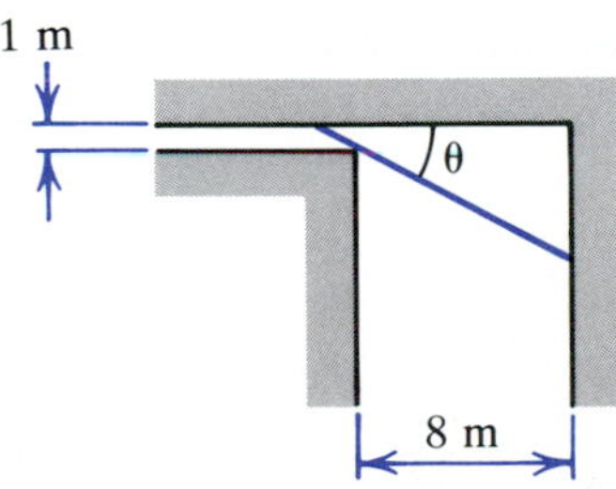

32. Find the length of the shortest beam that can be used to brace a wall if the beam is to pass over a second wall 2 meters high and 5 meters from the first wall.

33. The sides of a **V**-shaped trough are 28 centimeters wide. Find the angle between the sides of the trough that results in maximum capacity.

34. A metal rain gutter is to have 10 centimeter sides and a 10 centimeter horizontal bottom, with the sides making equal angles with the bottom (see the figure). How wide should the opening across the top be for maximum carrying capacity?

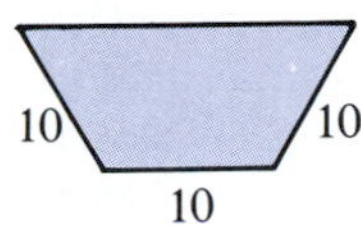

35. An open irrigation ditch of a given (fixed) cross-sectional area is to be lined with concrete to prevent seepage. If the two equal sides are perpendicular to the flat bottom, find the relative dimensions that require the least amount of concrete.

36. The intensity of illumination at a point varies inversely as the square of the distance between the point and the light source. Two lights, one having an intensity 8 times that of the other, are 6 meters apart. At what point between the two lights is the total illumination least?

37. A proposed tunnel of a given (fixed) cross-sectional area is to have a horizontal floor, vertical walls of equal height, and a ceiling that is a semicircular cylinder. If the ceiling costs 3 times as much per square meter to build as the vertical walls and the floor, find the most economical ratio of the diameter of the semicircular cylinder to the height of the vertical walls.

38. An observatory is to be in the form of a right circular cylinder surmounted by a hemispherical dome. If the hemispherical dome costs 3 times as much per square meter as the cylindrical wall, what are the most economical dimensions for a given volume? Neglect the floor.

39. A wire is to be cut into two pieces. One piece will be bent into a square, and the other piece will be bent into a circle. If the total area enclosed by the two pieces is to be 64 square centimeters, what is the minimum length of wire that can be used? What is the maximum length of wire that can be used?

40. A wire is to be cut into two pieces. One piece will be bent into an equilateral triangle, and the other piece will be bent into a circle. If the total area enclosed by the two pieces is to be 64 square centimeters, what is the minimum length of wire that can be used? What is the maximum length of wire that can be used?

41. A wire 35 centimeters long is cut into two pieces. One piece is bent into the shape of a square, and the other piece is bent into the shape of a circle. How should the wire be cut so that the area enclosed is minimum? How should it be cut to maximize the area?

42. A wire 35 centimeters long is cut into two pieces. One piece is bent into the shape of an equilateral triangle, and the other piece is bent into the shape of a circle. How should the wire be cut so that the area enclosed is maximum? How should it be cut to minimize the area?

43. A Norman window has the shape of a rectangle surmounted by a semicircle of diameter equal to the width of the rectangle (see the figure). If the perimeter of the window is 10 meters, what dimensions will admit the most light?

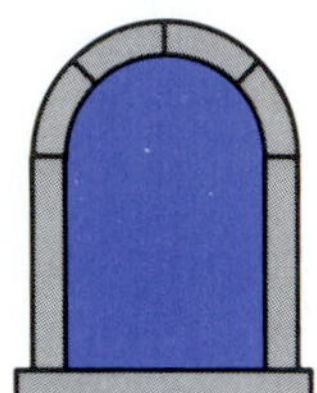

44. In the figure a circular area of radius 20 feet is surrounded by a walk. A light is placed above the center of the area. What height most strongly illuminates the walk? The intensity of illumination is given by $I = \sin\theta/s$, where s is the distance from the source and θ is the angle at which the light strikes the surface.

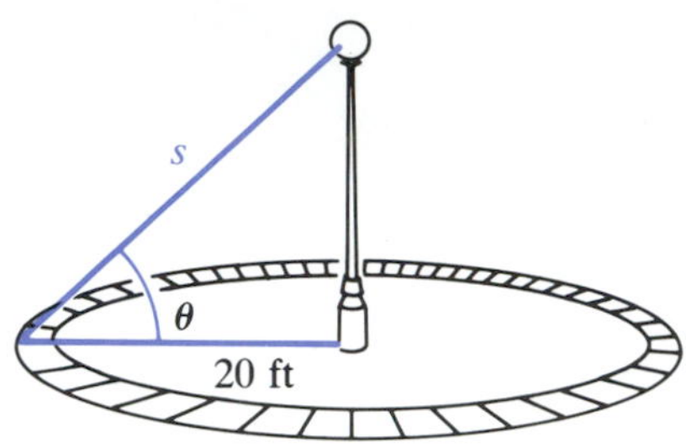

45. Let a and b be two positive real numbers. Find the line through (a, b) connecting points $(0, y_0)$ and $(x_0, 0)$ and having minimum length. (In general, x_0 and y_0 will depend on the line.)

4.10 Antiderivatives

We have already learned that for each differentiable function f there is a corresponding derivative function f'. We now ask the following question: If a function f is given, can we find a function F whose derivative is f? That is, is it possible to find a function F so that $F' = dF/dx = f$? If such a function F can be found, it is called an *antiderivative of f*.

[4.10.1] DEFINITION / *Antiderivative.*

A function F is called an *antiderivative* of the function f *if* $F' = f$.

For example, an antiderivative of $2x$ is x^2, since

$$\frac{d}{dx}x^2 = 2x$$

Another function whose derivative is $2x$ is $x^2 + 3$, since

$$\frac{d}{dx}(x^2 + 3) = 2x$$

This leads us to suspect that the function $f(x) = 2x$ has an unlimited number of antiderivatives. Indeed, any of the functions x^2, $x^2 + \frac{1}{2}$, $x^2 + 2$, $x^2 + \sqrt{5}$, $x^2 - \pi$, $x^2 - 1, \ldots, x^2 + C$, where C is any constant, has the property that its derivative is $2x$. We conjecture that all the antiderivatives of $2x$ are given by $x^2 + C$, where C is any constant. The following theorem assures us that this conjecture is correct.

[4.10.2] THEOREM

If f and g are differentiable functions, and

$$f'(x) = g'(x) \qquad \textbf{for all } x \textbf{ in } [a, b]$$

then there is a constant C so that

$$f(x) = g(x) + C$$

for all x in $[a, b]$.

Before we prove the theorem, let's interpret it geometrically. The hypothesis of the theorem states that at corresponding points in $[a, b]$ the slopes of the tangent lines to the graphs of f and g must be the same (see Fig. 76). If we plot the graphs of f and g from parts (a) and (b) of the figure, along with their tangent lines, on the same set of coordinate axes, we obtain Figure 76(c). We can see that the values $f(x)$ and $g(x)$ differ by a constant amount as x varies. This constant amount is simply the vertical difference between the two graphs.

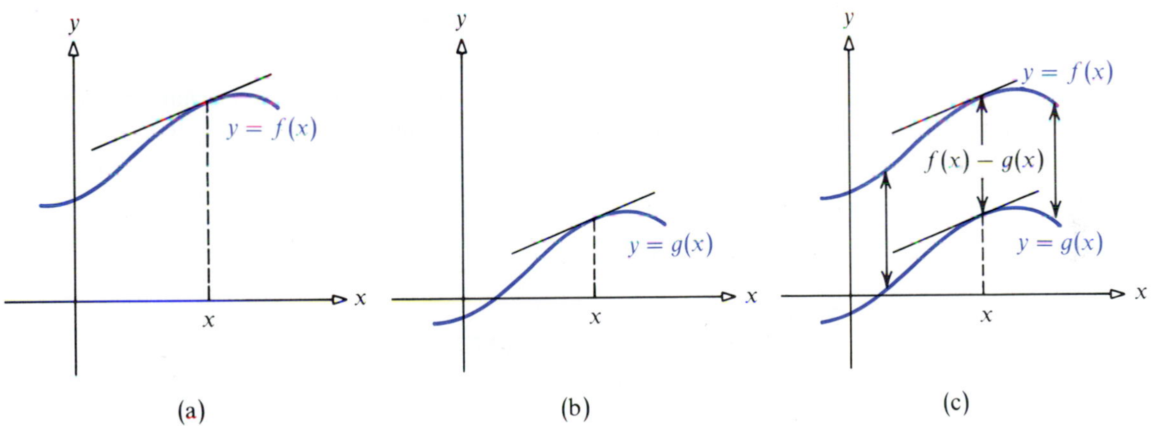

Figure 76

Proof Define the function h by

$$h(x) = f(x) - g(x)$$

By differentiating both sides with respect to x and using the fact that $f'(x) = g'(x)$ for all x, we find

$$h'(x) = f'(x) - g'(x) = 0 \qquad \text{for all } x \text{ in } [a, b]$$

and hence

$$h'(x) = 0 \qquad \text{for all } x \text{ in } [a, b]$$

Now, we let x be any number for which $a < x \leq b$. By the mean value theorem (4.4.2) there exists a number c between a and x such that

$$h(x) - h(a) = (x - a)h'(c) = (x - a)0 = 0$$

Thus, $h(x) = h(a)$, a constant, for all x in $[a, b]$. By setting $C = h(a)$, we get

$$C = f(x) - g(x)$$

$$f(x) = g(x) + C \qquad \text{for all } x \text{ in } [a, b]$$

As a result of theorem (4.10.2), we have the following theorem:

[4.10.3] THEOREM

If F is an antiderivative of f in an interval I, then any other antiderivative has the form $F(x) + C$, where C is an (arbitrary) constant.

All the antiderivatives of f may be obtained from the expression $F(x) + C$ by letting C range over all real numbers. For example, all the antiderivatives of x^5 are of the form $(x^6/6) + C$, where C is a constant.

In the course of proving theorem (4.10.2), the following corollary of the mean value theorem was also proved:

[4.10.4] COROLLARY

If $h'(x) = 0$ for every x in an interval I, then h is a constant function on I.

We shall be using corollary (4.10.4) in this and later chapters. For now, we give some examples of antidifferentiation.

EXAMPLE 1

Find all the antiderivatives of $f(x) = x^{1/2}$.

Solution

Recall that the derivative of $\frac{2}{3}x^{3/2}$ is

$$\left(\tfrac{2}{3}\right)\left(\tfrac{3}{2}x^{1/2}\right) = x^{1/2}$$

Hence, all the antiderivatives of $f(x) = x^{1/2}$ are of the form

$$\tfrac{2}{3}x^{3/2} + C \qquad \text{where } C \text{ is a constant}$$

In Example 1, you may ask how we knew to choose $\frac{2}{3}x^{3/2}$. We arrived at the answer in two stages. First, we know that

$$\frac{d}{dx}x^r = rx^{r-1}$$

That is, differentiation reduces the exponent by 1. Antidifferentiation is the inverse process, so it should increase the exponent by 1. This is how we obtain the $x^{3/2}$ part of $\frac{2}{3}x^{3/2}$. Second, the $\frac{2}{3}$ factor is needed so that when we differentiate $x^{3/2}$, we get $x^{1/2}$ and not $\frac{3}{2}x^{1/2}$.

Now, let $f(x) = x^r$, where r is a rational number, $r \neq -1$. (The case for which $r = -1$ requires special attention and is taken up in Chapter 7.) Then the function $F(x)$ defined by F:

$$F(x) = \frac{x^{r+1}}{r+1}$$

is an antiderivative of f, since $F'(x) = f(x)$. By virtue of theorem (4.10.2), *every* antiderivative of f on an interval is of the form

$$\frac{x^{r+1}}{r+1} + C$$

where C is a constant.

Table 6 includes this result along with the antiderivatives of some other functions.

Table 6

Function f	Antiderivative F of f
$f(x) = x^r, \quad r \neq -1$	$F(x) = \dfrac{x^{r+1}}{r+1} + C$
$f(x) = \cos x$	$F(x) = \sin x + C$
$f(x) = \sin x$	$F(x) = -\cos x + C$
$f(x) = \sec^2 x$	$F(x) = \tan x + C$
$f(x) = \sec x \tan x$	$F(x) = \sec x + C$
$f(x) = \csc x \cot x$	$F(x) = -\csc x + C$
$f(x) = \csc^2 x$	$F(x) = -\cot x + C$

Because of the relationship between the processes of differentiation and finding antiderivatives, we are able to state the following results:

[4.10.5] THEOREM

If functions F_1 and F_2 are antiderivatives of f_1 and f_2, respectively, then $F_1 + F_2$ is an antiderivative of $f_1 + f_2$.

[4.10.6] THEOREM

If k is a constant and if F is an antiderivative of f, then kF is an antiderivative of kf.

These statements are a consequence of the properties of derivatives. They may be more easily remembered in the following forms:

An antiderivative of a sum of functions equals the sum of the antiderivatives of the functions.

An antiderivative of a constant times a function equals the constant times an antiderivative of the function.

Theorem (4.10.5) can be extended to any finite sum of functions. A similar result is also true for differences. Let's see how these are used by looking at an example.

EXAMPLE 2

If $f(x) = x^5 - x^{1/2} + (6/x^2) - \sin x$, find all its antiderivatives.

Solution

We have already found that $x^6/6$ and $\frac{2}{3}x^{3/2}$ are antiderivatives of x^5 and $x^{1/2}$, respectively. For the term $6/x^2$, we write $6/x^2 = 6x^{-2}$. By using theorem (4.10.6) and Table 6, we find that an antiderivative of $6x^{-2}$ is

$$\frac{6x^{-1}}{-1} = \frac{-6}{x}$$

Finally, an antiderivative of $\sin x$ is $-\cos x$. Hence, all the antiderivatives of the function f are given by

$$\frac{x^6}{6} - \frac{2}{3}x^{3/2} - \frac{6}{x} + \cos x + C \qquad C \text{ a constant} \tag{1}$$

Note that the constant C in Example 2 is the sum of the constants resulting from each term, since the antiderivatives of x^5 are $(x^6/6) + C_1$, the antiderivatives of $x^{1/2}$ are $\frac{2}{3}x^{3/2} + C_2$, the antiderivatives of $6/x^2$ are $(-6/x) + C_3$, the antiderivatives of $\sin x$ are $-\cos x + C_4$. Combining these expressions, we get (1) by letting

$$C_1 - C_2 + C_3 - C_4 = C$$

Differential Equations

In scientific studies of physical, chemical, biological, and other phenomena, attempts are made, on the basis of long observation, to deduce mathematical laws that describe and predict natural behavior. Such laws often involve the derivatives of some unknown function F, which is to be found. For example, it may be required to find all functions $y = F(x)$ so that $dy/dx = f(x)$. This equation is an example of what is called a *differential equation*, and a function $y = F(x)$ for which $dy/dx = f(x)$ is a *solution* of the differential equation. The *general solution* of $dy/dx = f(x)$ consists of all the antiderivatives of f.

For example, the general solution of the differential equation

$$\frac{dy}{dx} = 5x^2 + 2 \tag{2}$$

is

$$y = F(x) = \frac{5}{3}x^3 + 2x + C$$

A *particular solution* of $dy/dx = f(x)$ occurs when C is assigned a particular value. When a particular solution is required, we use a *boundary condi-*

tion on F. For example, in the differential equation (2) we might require the general solution to satisfy the condition that $y = 5$ when $x = 3$. Then

$$5 = (\tfrac{5}{3})(27) + (2)(3) + C$$

$$C = -46$$

The particular solution of (2) with the boundary condition that $y = 5$ when $x = 3$ is therefore

$$y = \tfrac{5}{3}x^3 + 2x - 46$$

EXAMPLE 3

Solve the differential equation below with the boundary condition that $y = -1$ when $x = 3$.

$$\frac{dy}{dx} = x^2 + 2x + 1$$

Solution

The general solution of the differential equation is

$$y = \frac{x^3}{3} + x^2 + x + C$$

To determine the number C, we use the boundary condition. Then

$$-1 = \frac{3^3}{3} + 3^2 + 3 + C$$

$$C = -22$$

The particular solution of the differential equation with the boundary condition that $y = -1$ when $x = 3$ is

$$y = \frac{x^3}{3} + x^2 + x - 22$$

Rectilinear Motion

Suppose the functions $s = s(t)$, $v = v(t)$, and $a = a(t)$ represent the position, velocity, and acceleration, respectively, of a particle at time t. The three quantities s, v, and a are related by the differential equations

$$\frac{ds}{dt} = v(t) \quad \textbf{and} \quad \frac{dv}{dt} = a(t)$$

Thus, if the acceleration $a = a(t)$ of a particle is a known function of the time t, its velocity may be found by solving the differential equation $dv/dt = a(t)$. Similarly, if the velocity $v = v(t)$ of a particle is a known function of t, its position $s = s(t)$ at time t is the solution of the differential equation $ds/dt = v(t)$.

In the next example we illustrate how to determine the position $s = s(t)$ of a particle when the velocity or the acceleration function and some boundary conditions are given. In a physical problem, boundary conditions are often the values of the velocity v and position s at time $t = 0$. In such cases, $v(0)$ and $s(0)$ are referred to as *initial conditions*. They can also be written as v_0 and s_0.

EXAMPLE 4

Find the position of a particle at any time t if its acceleration a is known to be

$$a(t) = 8t - 3 \text{ meters per second per second}$$

and the initial conditions are given as

$$v_0 = v(0) = 4 \text{ meters per second} \qquad \text{and} \qquad s_0 = s(0) = 1 \text{ meter}$$

Solution

Since $dv/dt = a(t) = 8t - 3$, we have $v(t) = 4t^2 - 3t + C_1$ for some constant C_1. The initial conditions state that when $t = 0$, then $v_0 = v(0) = 4$. Thus,

$$v_0 = v(0) = 4(0)^2 - 3(0) + C_1 = 4$$
$$C_1 = 4$$

The velocity at any time t is therefore

$$v(t) = 4t^2 - 3t + 4$$

The position of the particle at time t obeys the differential equation

$$\frac{ds}{dt} = v(t) = 4t^2 - 3t + 4$$

Hence,

$$s(t) = \tfrac{4}{3}t^3 - \tfrac{3}{2}t^2 + 4t + C_2 \qquad \text{for some constant } C_2$$

Applying the initial condition that $s_0 = s(0) = 1$, we find

$$s_0 = s(0) = 0 - 0 + 0 + C_2 = 1$$
$$C_2 = 1$$

Hence,

$$s = \tfrac{4}{3}t^3 - \tfrac{3}{2}t^2 + 4t + 1$$

is the position of the particle at any time t.

EXAMPLE 5

When the brakes of a car are applied, they produce a deceleration at the constant rate of 10 meters per second per second. If the car is to stop within 20 meters after the brakes are applied, what is the maximum allowable speed?

Solution

Let $s(t)$ represent the distance in meters the car has traveled t seconds after the brakes are applied. Let v_0 be the speed of the car at the time the brakes are applied $(t = 0)$. Since the car decelerates at the rate of 10 meters per second per second, its acceleration a is

$$a = -10 \qquad \text{or} \qquad \frac{dv}{dt} = -10$$

By solving the differential equation for v, we find

$$v(t) = -10t + C_1$$

At $t = 0$, $v(0) = v_0$, the speed of the car when the brakes are applied. Thus,

$$v(t) = -10t + v_0 \qquad \text{or} \qquad \frac{ds}{dt} = -10t + v_0$$

By solving the differential equation for s, we get

$$s(t) = -5t^2 + v_0 t + C_2 \qquad C_2 \text{ a constant}$$

At $t = 0$, $s(0) = 0$, since we start measuring distance at the point at which the brakes are applied. Hence,

$$s(t) = -5t^2 + v_0 t \tag{3}$$

The car stops completely when the speed is equal to 0—that is, when

$$\frac{ds}{dt} = -10t + v_0 = 0$$

$$t = \frac{v_0}{10}$$

This is the time that must elapse for the car to come to rest. If we substitute $v_0/10$ for t in (3), we get the total distance the car has traveled:

$$s\left(\frac{v_0}{10}\right) = -5\left(\frac{v_0}{10}\right)^2 + v_0\left(\frac{v_0}{10}\right) = \frac{v_0^2}{20}$$

But, according to the original conditions, this distance cannot exceed 20 meters; that is, $v_0^2/20 \leq 20$. Thus, the maximum allowable speed v_0 for the car is

$$v_0^2 = 400$$

$$v_0 = 20 \text{ meters per second} = \left(\frac{20 \text{ meters}}{\text{second}}\right)\left(\frac{1 \text{ kilometer}}{1000 \text{ meters}}\right)\left(\frac{3600 \text{ seconds}}{1 \text{ hour}}\right)$$

$$= 72 \text{ kilometers per hour} \approx \left(\frac{72 \text{ kilometers}}{\text{hour}}\right)\left(\frac{1 \text{ mile}}{1.6 \text{ kilometer}}\right)$$

$$= 45 \text{ miles per hour}$$

Freely Falling Bodies

A common example of motion with (nearly) constant acceleration is a body falling toward the earth. In the absence of air resistance we find that all bodies, regardless of their size, weight, or composition, fall with the same acceleration at the same point of the earth's surface, and if the distance covered is not too great, the acceleration remains constant throughout the fall. This ideal motion, in which air resistance and the small change in acceleration with altitude are neglected, is called *free fall*. The acceleration of a freely falling body is called the *acceleration due to gravity* and is denoted by the symbol g. Near the earth's surface its magnitude is approximately 32 feet per second per second, or 9.8 meters per second per second, and it is directed down toward the center of the earth.

If F is the weight of a body of mass m, then according to Galileo, assuming air resistance is negligible, a freely falling body obeys the relationship

$$F = -mg \tag{4}$$

The minus sign is chosen because the body is falling, and we choose upward movement to be positive. Also, according to Newton, $F = ma$ (this is called *Newton's law of motion*), so (4) can be written as

$$ma = -mg$$

$$a = -g$$

where a is acceleration.

We wish to obtain formulas for the velocity v and position s of a falling body at time t. To fix our ideas, suppose an object at $t = 0$ has speed v_0 and is at a height s_0 above the ground. By solving the differential equation $dv/dt = a = -g$ for v, we get

$$v = -gt + v_0$$

Since $v = ds/dt$, we find the distance s above ground level of the object to be

$$s = (-\tfrac{1}{2})gt^2 + v_0 t + s_0$$

The next example illustrates how problems involving falling bodies are solved in practice.

EXAMPLE 6

A rock is thrown straight up with an initial velocity of 9.8 meters per second from the roof of a building 14.7 meters above ground level. How long does it take the rock to reach its maximum height? What is the maximum height of the rock? If the rock misses the edge of the building on the way down and eventually strikes the ground, what is the total time the rock is in the air?

Solutions

In Figure 77 we start measuring time at the moment the rock is released. If s is the distance in meters of the rock from the ground, then, since the rock is

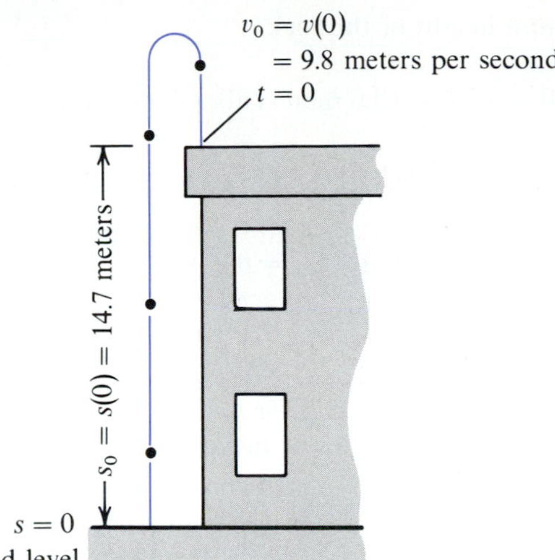

Figure 77

released at a height of 14.7 meters, we have

$$s_0 = s(0) = 14.7 \tag{5}$$

The initial velocity of the rock is

$$v_0 = v(0) = 9.8 \tag{6}$$

If we ignore air resistance, the only force acting on the rock is gravity. Since gravity ($g = 9.8$ meters per second per second) is acting in a direction opposite to that of the motion of the rock, the acceleration a of the rock is

$$a = -9.8$$

But $a = dv/dt$, so

$$\frac{dv}{dt} = -9.8 \qquad v(t) = -9.8t + v_0$$

By using the initial condition (6), we find the velocity of the rock at any time t:

$$v(t) = -9.8t + 9.8 \tag{7}$$

But $v = ds/dt$, so

$$\frac{ds}{dt} = -9.8t + 9.8 \qquad s(t) = -4.9t^2 + 9.8t + s_0$$

By using the initial condition (5), we find the distance of the rock from the ground at any time t:

$$s(t) = -4.9t^2 + 9.8t + 14.7 \tag{8}$$

The rock reaches its maximum height when its velocity is 0. From (7), this happens when $t = 1$ second. To obtain the maximum height, we use (8) with $t = 1$. The maximum height of the rock is

$$s(1) = -4.9 + 9.8 + 14.7 = 19.6 \text{ meters above the ground}$$

The total time the rock is in the air is found by setting $s = 0$. From (8), we find

$$\begin{aligned} -4.9t^2 + 9.8t + 14.7 &= 0 \\ t^2 - 2t - 3 &= 0 \\ (t-3)(t+1) &= 0 \end{aligned}$$

The only meaningful solution (since t cannot be negative) is $t = 3$. Thus, the rock is in the air for 3 seconds.

Law of Inertia

We close this section with a special case of the *law of inertia*, which was originally stated by Galileo.

[4.10.7] THEOREM

A body acted upon by no force remains at rest or in a state of uniform motion along a straight line.

The force F acting upon a body of mass m is given by Newton's law of motion $F = ma$, where a is the acceleration of the body. If there is no force acting upon the body, then $F = 0$. In this case, the acceleration a must be 0. But $a = dv/dt$, where v is the speed of the body. Hence,

$$\frac{dv}{dt} = 0 \qquad \text{or} \qquad v = \text{Constant}$$

That is, the body is at rest $(v = 0)$ or else in a state of uniform motion (v is a nonzero constant).

EXERCISE 4.10

In Problems 1–18 find all the antiderivatives of each function.

1. $f(x) = 4x^5$
2. $f(x) = x^{4/3}$
3. $f(x) = 5x^{3/2}$
4. $f(x) = x^{5/2}$
5. $f(x) = 2x^{-2}$
6. $f(x) = 3x^{-3}$
7. $f(x) = \sqrt{x}$
8. $f(x) = \dfrac{1}{\sqrt{x}}$
9. $f(x) = 4x^3 - 3x^2 + 1$
10. $f(x) = x^2 - x$
11. $f(x) = (2 - 3x)^2$
12. $f(x) = (3x - 1)^2$
13. $f(x) = \dfrac{3x - 2}{\sqrt{x}}$
14. $f(x) = \dfrac{4x^{3/2} - 1}{x^2}$
15. $f(x) = \dfrac{x^2 + 10x + 21}{3x + 9}$
16. $f(x) = \dfrac{x^3 - 5x + 8}{x^5}$
17. $f(x) = 2x - 3 \cos x$
18. $f(x) = \sin x - \cos 2x$

In Problems 19–26 find the solution of each differential equation having the given boundary conditions.

19. $\dfrac{dy}{dx} = 3x^2 - 2x + 1$, if $y = 1$ when $x = 0$

20. $\dfrac{dy}{dx} = x^{1/3} + x\sqrt{x} - 2$, if $y = 2$ when $x = 0$

21. $\dfrac{dv}{dt} = 3t^2 - 2t + 1$, if $v = 5$ when $t = 1$

22. $\dfrac{ds}{dt} = t^4 + 4t^3 - 5$, if $s = 5$ when $t = 2$

23. $\dfrac{ds}{dt} = t^3 + \dfrac{1}{t^2}$, if $s = 2$ when $t = 1$

24. $\dfrac{dy}{dx} = \sqrt{x} - x\sqrt{x} + 1$, if $y = 0$ when $x = 1$

25. $\dfrac{dy}{dx} = x - 2 \sin x$, if $y = 0$ when $x = 0$

26. $\dfrac{dy}{dx} = x^2 - \sin 2x$, if $y = 0$ when $x = 0$

In Problems 27–30 find the distance $s = s(t)$ under the stated conditions.

27. $a = -32$ feet per second per second, $s(0) = 0$ feet, $v(0) = 128$ feet per second

28. $a = -980$ centimeters per second per second, $s(0) = 5$ centimeters, $v(0) = 1980$ centimeters per second

29. $a = 3t$ meters per second per second, $s(0) = 2$ meters, $v(0) = 18$ meters per second

30. $a = 5t - 2$ feet per second per second, $s(0) = 0$ feet, $v(0) = 8$ feet per second

31. Using the fact that

$$\frac{d}{dx}(x \cos x + \sin x) = -x \sin x + 2 \cos x$$

find F if

$$\frac{dF}{dx} = -x \sin x + 2 \cos x \quad \text{and} \quad F(0) = 1$$

32. Using the fact that

$$\frac{d}{dx} \sin x^2 = 2x \cos x^2$$

find h if

$$\frac{dh}{dx} = x \cos x^2 \quad \text{and} \quad h(0) = 2$$

33. A car decelerates at a constant rate of 10 meters per second per second when its brakes are applied. If the car must stop within 15 meters after applying the brakes, what is the maximum allowable speed for the car?

34. A car can accelerate from 0 to 60 kilometers per hour in 60 seconds. If the acceleration is constant, how far does the car travel during this time?

35. The 2 meter high jump is rather commonplace today. If this event were held on the moon, where the acceleration due to gravity is 1.6 meters per second per second, what height would be attained? Assume the athlete can propel herself with the same force on the moon as on the earth.

36. A ball is thrown straight up from ground level with an initial velocity of 19.6 meters per second. How high is the ball thrown? How long will it take the ball to return to ground level?

37. A child throws a ball straight up. If the ball is to reach a height of 9.8 meters, what is the minimum velocity that must be imparted to the ball? Assume the initial height of the ball is 1 meter.

38. A ball thrown directly down from a roof 49 meters high reaches the ground in 3 seconds. What is the initial velocity?

39. A constant force is applied to a particle that is initially at rest. If the mass of the particle is 4 grams and if its velocity after 6 seconds is 12 centimeters per second, determine the force applied to it.

40. Starting from rest, with what constant acceleration must a car proceed to go 2 kilometers in 2 minutes? (Give your answer in centimeters per second per second.)

41. A child on top of a building 24 meters high drops a rock and then 1 second later throws another rock straight down. What initial velocity must be second rock be given so that the dropped rock and the thrown rock hit the ground at the same time?

REVIEW EXERCISES

1. A spherical snowball is melting at the rate of 2 cubic centimeters per minute. How fast is the surface area changing when the radius is 5 centimeters?

2. A lighthouse is 3 kilometers from a straight shoreline. Its light makes one revolution every 8 seconds. How fast is the light moving along the shore when it makes an angle of 30° with the shoreline?

3. Two planes are approaching an airport, one from the north and one from the west. The plane from the north is flying at 250 miles per hour and is 30 miles from the airport. The plane from the west is flying at 200 miles per hour and is 20 miles from the airport. How fast are the planes approaching each other at that instant?

4. Suppose that the volume of a spherical ball of ice decreases (by melting) at a rate proportional to its surface area. Show that its radius decreases at a constant rate.

5. If p is the period of a pendulum of length L, the acceleration due to gravity may be computed by the formula $g = (4\pi^2 L)/p^2$. If L is measured with negligible error, but a 2% error may occur in the measurement of p, what is the approximate percentage error in the computation of g?

6. A motorcycle accelerates at a constant rate from 0 to 72 kilometers per hour in 10 seconds. How far has it traveled in that time?

7. Suppose that in a town with a population P, the number of people with an infectious disease increases at a rate proportional to both the number who have the disease and the number who do not.

(a) Write the above information as a differential equation.

(Problem 7 continues in the next column)

7. *(continued)*

(b) If N is the number of people who have the disease at time t, show that the graph of N as a function of t has an inflection point when half the town is infected. Sketch such a graph, showing its asymptote.

8. Find the differential dy for $x^3 + 2y^2 = x^2y$.

9. Use differentials to approximate $\sqrt[3]{129}$.

10. If the percentage error in measuring a cube is 5%, what is the percentage error in computing its volume?

11. Find all the critical numbers for $f(x) = \dfrac{x^2}{2x - 1}$.

12. Find all the critical numbers for $f(x) = \cos 2x$ on $[0, \pi]$.

13. Verify the hypothesis for Rolle's theorem and find the coordinates of the point at which there is a horizontal tangent to $f(x) = x^3 - 4x^2 + 4x$ on $[0, 2]$.

14. Verify the hypothesis for the mean value theorem and find a point on the function that has a tangent with a slope the same as that of the secant on the interval:

$$f(x) = \frac{2x - 1}{x} \qquad \text{on } [1, 4]$$

15. Does the mean value theorem (4.4.2) apply to the function $f(x) = \sqrt{x}$ in the interval $[0, 9]$? If not, why not? If so, find the number c referred to in the theorem.

16. Show that when the mean value theorem is applied to the function $f(x) = Ax^2 + Bx + C$ in the interval $[a, b]$, the number c referred to in the theorem is the midpoint of the interval.

In Problems 17 and 18 sketch the graph of each function.

17. $f(x) = -x^3 - x^2 + 2x$

18. $f(x) = x^{1/3}(x^2 - 9)$

19. The distance s of a particle from the origin at time t is given by

$$s = t^4 + 2t^3 - 36t^2$$

Draw a figure to illustrate the motion of the particle.

20. Without finding them, explain why the function $f(x) = \sqrt{x(2 - x)}$ must have an absolute maximum and an absolute minimum. Then find them in two ways (with and without the help of calculus).

21. Without finding the derivative, prove that if $f(x) = (x^2 - 4x + 3)(x^2 + x + 1)$, then $f'(x) = 0$ for at least one number between 1 and 3. Check by finding the derivative and applying the intermediate value theorem (2.4.6).

22. Suppose that the domain of f is an open interval and $f'(x) > 0$ for all x in the interval. Prove that f cannot have an extreme value.

In Problems 23 and 24 determine the intervals on which f is concave up and on which f is concave down.

23. $f(x) = x^4 + 12x^2 + 36x - 11$

24. $f(x) = 3x^4 - 2x^3 - 24x^2 - 7x + 2$

25. Let g be a continuous function on the closed interval $[0, 1]$. Let $g(0) = 1$ and $g(1) = 0$. Which of the following is *not* necessarily true?

(a) There exists a number u in $[0, 1]$ such that $g(u) \geq g(x)$ for all x in $[0, 1]$.
(b) For all a and b in $[0, 1]$, if $a = b$, then $g(a) = g(b)$.
(c) There exists a number c in $[0, 1]$ such that $g(c) = \frac{1}{2}$.
(d) There exists a number c in $[0, 1]$ such that $g(c) = \frac{3}{2}$.
(e) For all c in the open interval $(0, 1)$, $\lim_{x\to c} g(x) = g(c)$.

26. If f is a continuous function on the closed interval $[a, b]$, which of the following is necessarily true?

(a) f' exists on (a, b).
(b) If $f(u)$ is a maximum of f, then $f'(u) = 0$.
(c) $\lim_{x\to c} f(x) = f(\lim_{x\to c} x)$, for $a < c < b$.
(d) $f'(x) = 0$, for some x, $a \leq x \leq b$.
(e) The graph of f' is a straight line.

27. If y is a function of x such that $y' > 0$ for all x and $y'' < 0$ for all x, which of the following could be part of the graph of $y = f(x)$? See illustrations (a) to (e).

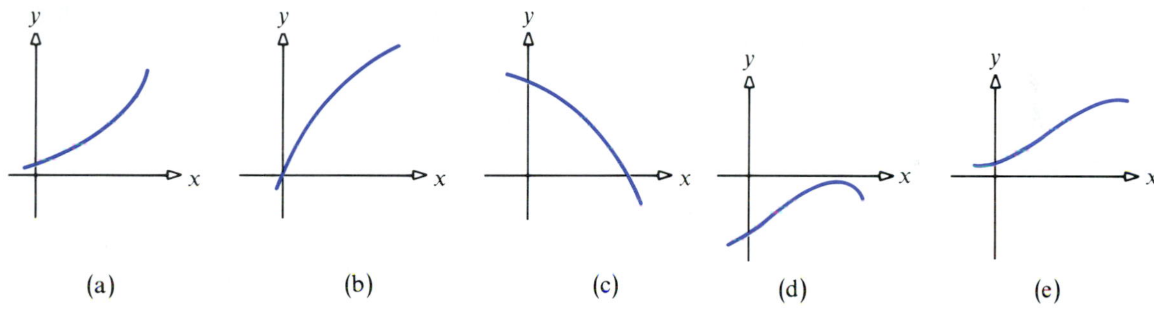

(a) (b) (c) (d) (e)

28. If a function f is continuous for all x and if f has a local maximum at $(-1, 4)$ and a local minimum at $(3, -2)$, which of the following statements must be true?

(a) The graph of f has a point of inflection somewhere between $x = -1$ and $x = 3$.
(b) $f'(-1) = 0$.
(c) The graph of f has a horizontal asymptote.
(d) The graph of f has a horizontal tangent line at $x = 3$.
(e) The graph of f intersects both axes.

29. Sketch the graph of a function f that has the following properties:

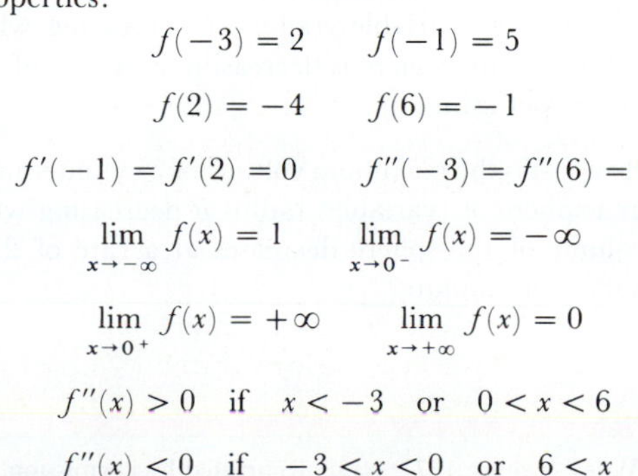

$$f(-3) = 2 \qquad f(-1) = 5$$
$$f(2) = -4 \qquad f(6) = -1$$
$$f'(-1) = f'(2) = 0 \qquad f''(-3) = f''(6) = 0$$
$$\lim_{x\to-\infty} f(x) = 1 \qquad \lim_{x\to 0^-} f(x) = -\infty$$
$$\lim_{x\to 0^+} f(x) = +\infty \qquad \lim_{x\to+\infty} f(x) = 0$$
$$f''(x) > 0 \text{ if } x < -3 \text{ or } 0 < x < 6$$
$$f''(x) < 0 \text{ if } -3 < x < 0 \text{ or } 6 < x$$

30. Sketch the graph of a function f that has the following properties:

$$f(-2) = 2 \qquad f(5) = 1 \qquad f(0) = 0$$
$$f'(x) > 0 \text{ if } x < -2 \text{ or } 5 < x$$
$$f'(x) < 0 \text{ if } -2 < x < 2 \text{ or } 2 < x < 5$$
$$f'(5) = 0 \qquad f''(x) > 0 \text{ if } x < 0 \text{ or } 2 < x$$
$$f''(x) < 0 \text{ if } 0 < x < 2$$
$$\lim_{x\to 2^-} f(x) = -\infty \qquad \lim_{x\to 2^+} f(x) = +\infty$$
$$\lim_{x\to-\infty} f(x) = 0$$

31. For the function $f(x) = x\sqrt{x+1}$, $0 \leq x \leq b$, the number c satisfying the mean value theorem is $c = 3$. Find b.

In Problems 32 and 33 find all vertical and horizontal asymptotes of f.

32. $f(x) = \dfrac{4x - 2}{x + 3}$

33. $f(x) = \dfrac{2x}{x^2 - 4}$

In Problems 34 and 35 use the appropriate methods discussed in the chapter to sketch the graph.

34. $f(x) = \dfrac{x^2 - 4}{x - 3}$

35. $f(x) = x\sqrt{x - 3}$

36. An open box is to be made from a piece of cardboard by cutting squares out of each corner and folding up the sides. If the size of the cardboard is 2 feet by 3 feet, what size squares (in inches) should be cut out to maximize the volume of the box?

37. Find the point on the graph of $2y = x^2$ nearest to (4, 1).

38. At what number x does the *derivative* of $f(x) = (x^4/3) - (x^5/5)$ attain its maximum value?

39. Find the antiderivative of $f(x) = 4x^3 - 9x^2 + 10x - 3$.

40. Find the antiderivative of $f(x) = (5 - 2x)^3$.

41. A box moves down an inclined plane with an acceleration of $t^2(t - 3)$ centimeters per second per second. It covers a distance of 10 centimeters in 2 seconds. What was the original velocity?

42.*Two bodies begin a free fall from rest at the same height 1.0 second apart. How long after the first body begins to fall will the two bodies be 10 meters apart?

43. A manufacturer has determined that the cost of producing x items is given by $C = 200 + 35x + 0.02x^2$ dollars. Each item can be sold for \$78. How many items should she produce to maximize profit?

44. If $f'(x)$ and $g'(x)$ exist and $f'(x) > g'(x)$ for all real x, then which of the following statements must be true about the graph of $y = f(x)$ and the graph of $y = g(x)$?

(a) They intersect exactly once.
(b) They intersect no more than once.
(c) They do not intersect.
(d) They could intersect more than once.
(e) They have a common tangent at each point of intersection.

45. Given

$$f(x) = \frac{ax^n + b}{cx^n + d} \quad \text{with} \quad ad - bc \neq 0$$

determine the critical numbers and where the function is increasing and where it is decreasing.

46. Let P_n be the perimeter of a regular n-sided polygon inscribed in a circle of radius R. Show that:

(a) $P_n = 2nR \sin\left(\dfrac{\pi}{n}\right)$

(b) $\lim_{n \to +\infty} P_n = 2\pi R$ (the perimeter of a circle of radius R)

CHALLENGE EXERCISES

1. A lamp is on a post 10 meters high. A ball is thrown straight up at a distance of 5 meters from the lamp and 20 meters from a wall with initial velocity 19.6 meters per second. The acceleration due to gravity is $a = -9.8$ meters per second per second. If the ball is thrown up from an initial height of 1 meter above ground, how fast is the shadow of the ball moving on the wall 3 seconds after the ball is released? Is the ball moving up or down? How far is the ball above ground at $t = 3$ seconds?

2. How fast is the maximum volume of a cone inscribed in a sphere of (variable) radius R decreasing when the volume of the sphere is decreasing at a rate of 9 cubic meters per minute?

3. How fast is the maximum volume of a cylinder inscribed in a sphere of (variable) radius R decreasing when the volume of the sphere decreases at a rate of 27 cubic meters per minute?

* Adapted from D. Halliday and R. Resnick, *Physics*, part I (New York: John Wiley & Sons, 1977) p. 52. Reprinted by permission.

4. Which point of the semicircle $y = \sqrt{25 - x^2}$ is farthest from the chord AB if $A = (0, 5)$ and $B = (3, 4)$?

5. A railroad train is running 15 miles an hour past a station 800 feet long, with the track having the form of the parabola

$$y^2 = 600x$$

and situated as shown in the figure. If the sun is just rising in the east, find how fast the shadow S of the locomotive L is moving along the wall of the station at the instant it reaches the end of the wall.

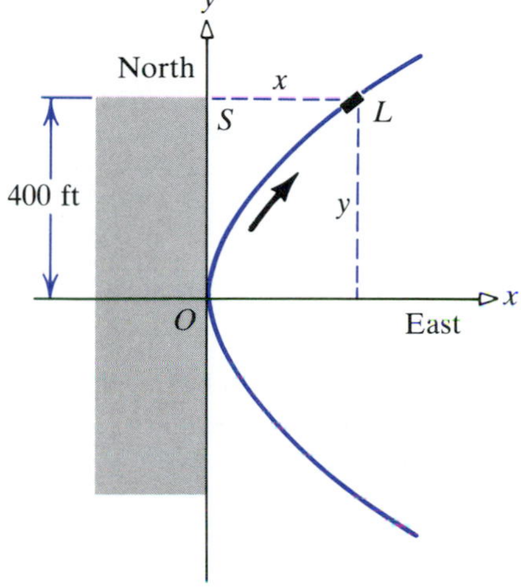

6. Find where the general cubic $f(x) = ax^3 + bx^2 + cx + d$ is increasing and where it is decreasing by considering cases depending on the value of $b^2 - 3ac$.

7. (a) Prove that a rational function of the form $f(x) = (ax^{2n} + b)/(cx^n + d)$ has at most five critical numbers.

(b) Give an example of such a rational function with exactly five critical numbers.

8.[†] The engineer of a train moving at a speed v_1 sights a freight train a distance d ahead of him on the same track moving in the same direction with a slower speed v_2. He puts on the brakes and gives his train a constant deceleration a. Show that:

If $d > \dfrac{(v_1 - v_2)^2}{2a}$, there will be no collision.

If $d < \dfrac{(v_1 - v_2)^2}{2a}$, there will be a collision.

9.[‡] Two cars, A and B, travel in a straight line. The distance of A from the starting point is given as a function of time by $s_A = 4t + t^2$, and the distance of B from the starting point is $s_B = 2t^2 + 2t^3$.

(a) Which car is ahead just after they leave the starting point?

(b) At what times are the cars at the same point?

(c) At what times is the velocity of B relative to A 0?

(d) At what times is the distance from A to B neither increasing nor decreasing?

10. Let A_n be the area bounded by a regular n-sided polygon inscribed in a circle of radius R. Show that:

(a) $A_n = \left(\dfrac{n}{2}\right) R^2 \sin\left(\dfrac{2\pi}{n}\right)$

(b) $\lim\limits_{n \to +\infty} A_n = \lim\limits_{n \to +\infty} \left(\dfrac{n}{2}\right) R^2 \sin\left(\dfrac{2\pi}{n}\right) = \pi R^2$

(the area of a circle of radius R)

Exploring Asymptotes and Limits at Infinity Using Graphing Calculators*

In this vignette, we extend the methods introduced in the graphing calculator vignette in Chapter 2. We examine the asymptotic behavior of functions by graphing functions in appropriate viewing rectangles. Graphs and the patterns within their coordinates can suggest solutions to infinite limits and limits at infinity.

To begin our graphical exploration of asymptotes and the associated limits involving infinity we revisit Example 1 from Section 4.8.

[†] Adapted from D. Halliday and R. Resnick, *Physics*, part I (New York: John Wiley & Sons, 1977), p. 51. Reprinted by permission.

[‡] Adapted from F. W. Sears, M. W. Zemansky, and H. D. Young, *University Physics* (Reading, Mass.: Addison-Wesley Publishing Co., 1976), p. 67. Reprinted by permission.

* Gregory D. Foley
The Ohio State University

EXAMPLE 1

Investigate fully the graph of the function $f(x) = \dfrac{x^2}{x^2 - 1}$.

Solution

Use a graphing calculator to graph the function as shown in Figure 59 of Chapter 4. Figure 59 suggests a viewing rectangle of $[-4, 4] \times [-3, 4]$ with a scaling unit of 1 on each axis. You may wish to refresh your memory of how to enter a function and how to set a viewing rectangle by reviewing the graphing calculator vignette in Chapter 2. After obtaining a facsimile of Figure 59, try the viewing rectangles and scale settings listed in the following table:

Xmin	Xmax	Xscale	Ymin	Ymax	Yscale
−50	50	10	0	2	1
−1000	1000	100	0	2	1
−2	0	1	−10	10	1
−1.1	−0.9	0.01	−200	200	50
−1.01	−0.99	0.001	−2000	2000	500
0	2	1	−10	10	1
0.9	1.1	0.01	−200	200	50
0.99	1.01	0.001	−2000	2000	500

Explain the resulting graphs in terms of the asymptotic behavior of f. Use the Trace feature of your calculator to get a y-coordinate readout as you move from point to point along the obtained graphs.

In the next example we use the graph obtained on a graphing calculator to determine the value of a limit at infinity. The goal of the exploration is to find $\lim_{x\to+\infty} x \sin(1/x)$.

EXAMPLE 2

Find $\lim_{x\to+\infty} x \sin(1/x)$.

Solution

Use a graphing calculator to graph $y = x \sin(1/x)$. Study the graph shown

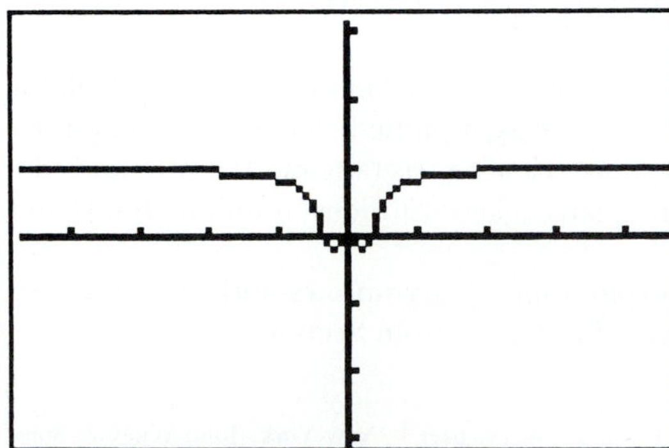

The graph of $y = x \sin(1/x)$ as it appears in the Casio default viewing rectangle of $[-4.7, 4.7] \times [-3.1, 3.1]$

in the figure. As x increases in the first quadrant portion of the graph, the associated y-values increase toward a value of 1. This suggests that $\lim_{x\to+\infty} x\sin(1/x) = 1$. This conjecture can be verified by letting $u = (1/x)$ in the limit expression. This substitution yields

$$\lim_{x\to+\infty} x \sin\frac{1}{x} = \lim_{u\to 0^+} \frac{\sin u}{u} = 1$$

EXERCISES

1–108. Solve Problems 1–52 from Exercise 4.7 and Problems 1–56 from Exercise 4.8 with the aid of a graphing calculator or computer graphing software. [*Hint:* Use the root key when rational exponents are involved. For example, in Problem 51 from Exercise 4.7, enter $f(x)$ as $\sqrt[3]{(X-3)^2} + 2$.]

Using a Microcomputer to Explore Graphs of Functions*

Most software packages for graphing functions make it easy for a user to enter a function, a set of domain values, and the desired range for the graph. After entering these values, the user will see the graph of the function in a fairly short time. Usually the display on the computer screen is a good representation of the graph of the function. However, there are times when the actual graph of the function and the screen display are not in such close agreement. It is at these times that the user must be more discriminating in the analysis of the function. A graph produced by a microcomputer is simply not sufficient evidence for the analysis of a function.

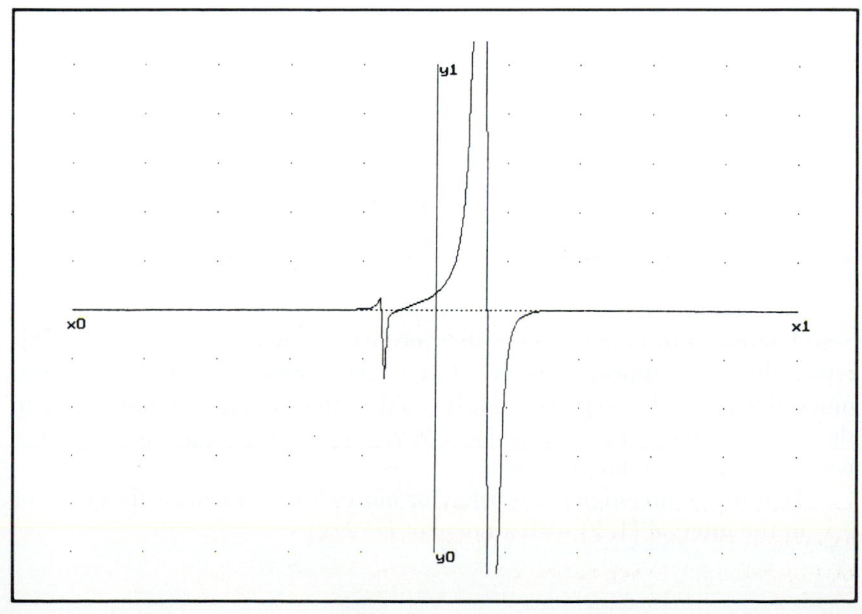

* L. Carl Leinbach
Gettysburg College

To illustrate this point, consider the following function:

$$F(x) = \frac{x^3 + x^2 - 12x - 12}{2x^6 - 3x^5 - 4x^4 + 9x^2 + 12x - 18}$$

The software display of this graph over the interval $[-10, 10]$ is displayed in the figure on page 327. It appears that the graph crosses the x-axis at $x = -1$ and that there are two vertical asymptotes: one somewhere in the vicinity of -1.5 and one at 1.5. Closer examination of the numerator of $F(x)$ reveals that it can be factored as $(x + 1)(x^2 - 12)$. This means that $F(x)$ should also cross the axis at $x = -1$, $x = 2\sqrt{3}$, and $x = -2\sqrt{3}$. In Problem 1 you will be asked to investigate the graph of $F(x)$ at these points.

Our project is to locate the vertical asymptote that appears to lie between $x = 1$ and $x = 2$. If we draw the graph of the denominator of $F(x)$,

$$g(x) = 2x^6 - 3x^5 - 4x^4 + 9x^2 + 12x - 18$$

for x in $[-2, 2]$ and y in $[-25, 5]$, we see that it crosses the x-axis between -2 and -1 as expected.

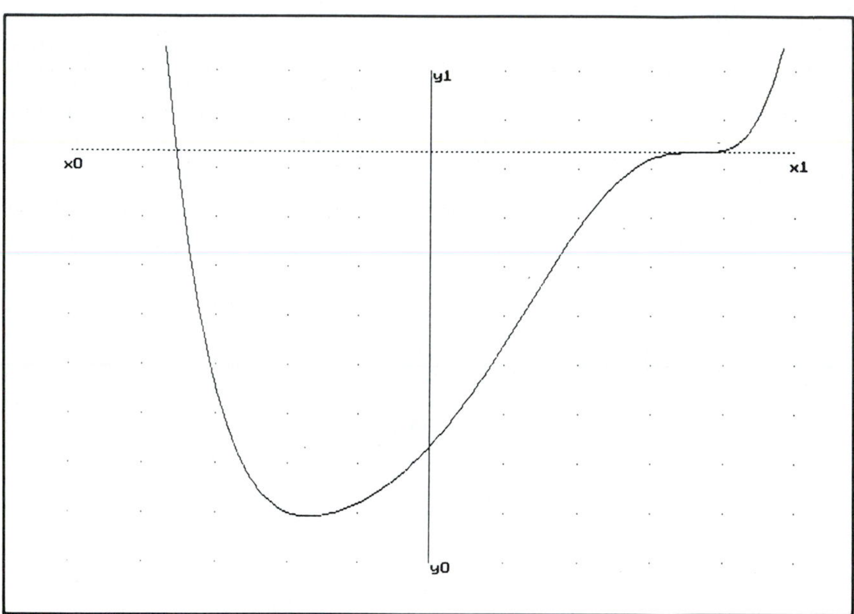

There is, however, an interesting occurrence between 1 and 2. In this region the graph appears to be very flat and coincides with the x-axis. This is unusual behavior for a polynomial! It could be that the range is just too big to show the approach of $g(x)$ to the axis. *Note:* x^n for large odd n may exhibit a behavior similar to this at $x = 0$.

In order to investigate this behavior more closely, we draw the graph of $g(x)$ in the interval $[1, 2]$ with a range of $[-2, 2]$.

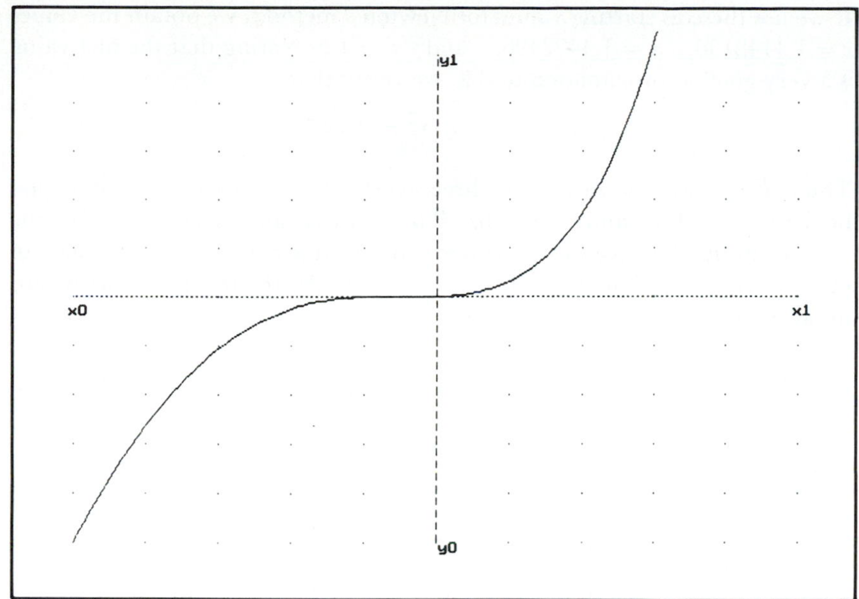

The function still seems to exhibit this unusual behavior. We can make a more accurate assessment that the part of the graph that seemingly coincides with the x-axis lies in the interval $[1.4, 1.5]$. We draw the graph for x in this region and magnify the vertical dimension. We define the range of y-values to be $[-0.002, 0.001]$.

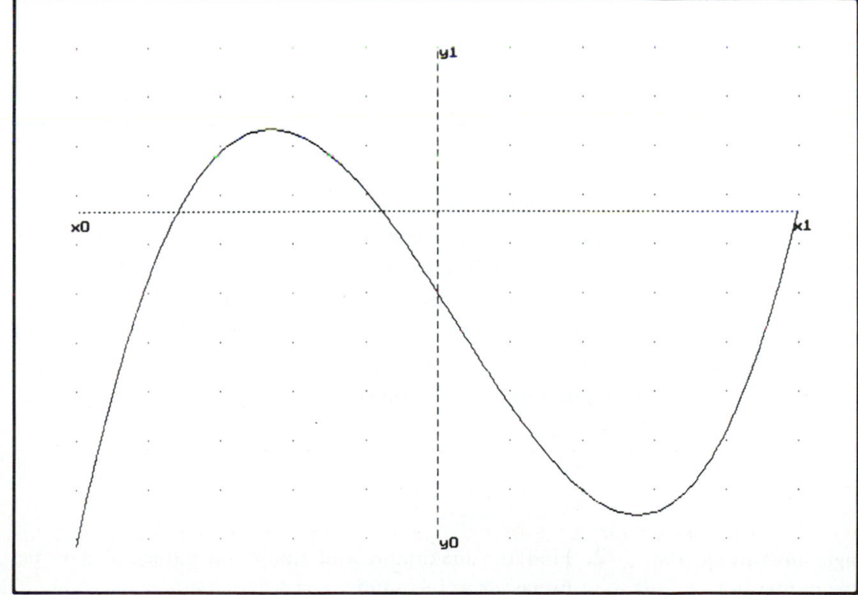

In this graph we see that $g(x)$ is neither constant nor monotonic. It has three zeros in this interval as well as a local maximum and a local minimum. The zeros of $g(x)$ appear to be located in the vicinity of 1.41, 1.44, and 1.5.

If we use these as starting values for Newton's method, we obtain the values $x = 1.4142136$, $x = 1.4422496$, and $x = 1.5$. Noting that the first value is a very good approximation to $\sqrt{2}$, we verify that

$$g(x) = (2x - 3)(x^2 - 2)(x^3 - 3)$$

Thus, $F(x)$ has not one, but three, vertical asymptotes in the region between $x = 1.4$ and $x = 1.6$. This fact was not at all evident from the first graph. Any attempt to draw the graph in this region by restricting x to $[1.4, 1.55]$ and greatly expanding y to $[-10{,}000, 10{,}000]$ is met with very unsatisfying results.

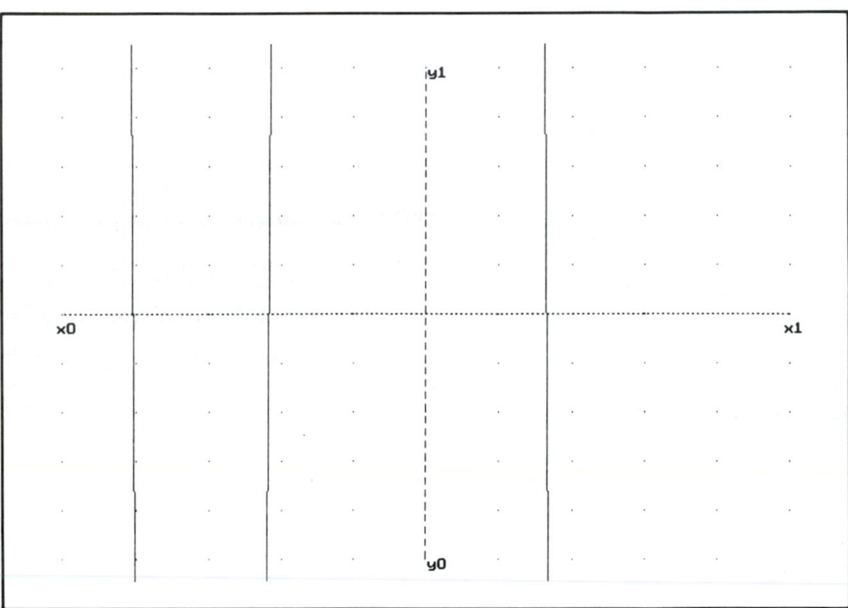

The message of this section is that although computer graphics displays can be of great value in investigating the behavior of functions, the screen display can be limited. Absolute accuracy is impossible on a device that is capable of displaying only a finite number of function values. This is true no matter how large a number of values can be displayed. The methods of analyzing continuous functions developed in this course can provide accurate information about the behavior of the function.

EXERCISES

1. By restricting the domain and range, investigate the behavior of $F(x)$ about the roots of the numerator.

2. Find the maximum and minimum values of $g(x)$ between $x = 1.4$ and $x = 1.5$.

3. Most calculus-based computer graphics packages have basic symbolic differentiation routines. Use this routine to differentiate $F(x)$, and use the graphics routines to help you locate the maximum and minimum values of $F(x)$.

4. Construct a polynomial, such as $g(x)$, with integer coefficients that has roots located very close together and whose graph does not show these distinct roots immediately.

The Definite Integral

5

The development of the integral was motivated to a large extent by attempts to solve a basic problem in geometry—namely, the *area problem*. As we stated in Chapter 2, the question is: Given a nonnegative function f whose domain is the closed interval $[a, b]$, what is the area enclosed by the graph of f, the

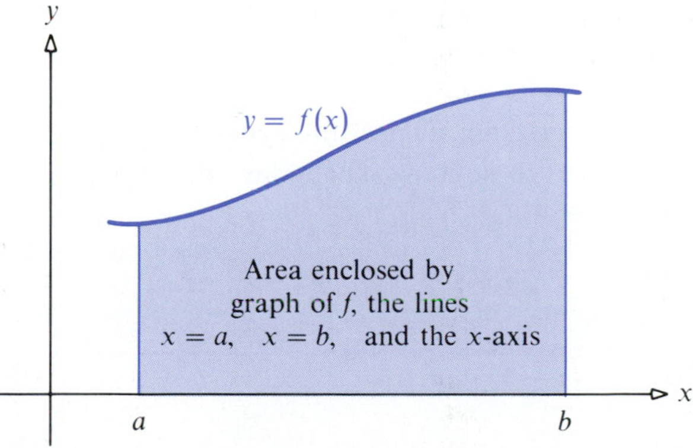

Figure 1

x-axis, and the vertical lines $x = a$ and $x = b$? Figure 1 illustrates the area to be found.

In Sections 5.2–5.4 we show how the concept of the integral develops from the area problem. At first glance, the area problem and the tangent problem, which helped motivate the invention of the derivative (see Chap. 2), may look quite dissimilar. However, much of calculus is built upon a surprising relationship between the two problems and their associated concepts. This relationship is the basis for the *fundamental theorems of calculus*.

5.1 Summation Notation

To facilitate some of the work in the forthcoming sections, it is convenient to introduce *summation notation*. The use of this notation permits us to express sums of many terms in a concise form. This notation is sometimes called *sigma notation* because it uses the Greek capital letter sigma, $\sum$, as a *summation sign*.

[5.1.1] DEFINITION / *Summation Notation.*

The sum of n terms $a_1, a_2, a_3, \ldots, a_n$ is written as

$$\sum_{i=1}^{n} a_i = a_1 + a_2 + a_3 + \cdots + a_n \qquad (1)$$

With function notation, definition (5.1.1) can be written as

$$\sum_{i=1}^{n} f(i) = f(1) + f(2) + f(3) + \cdots + f(n-1) + f(n) \qquad (2)$$

Thus, the symbol $\sum$ is an instruction to add; the variable i is called the *index* of the sum. The expression on the left in (2) should be read "the sum of $f(i)$ from $i = 1$ through $i = n$." For example, we can use summation notation to represent the sum $1^2 + 2^2 + \cdots + 20^2$ concisely as

$$\sum_{i=1}^{20} i^2 = 1^2 + 2^2 + \cdots + 20^2$$

Letters other than i can be used as the index. For example,

$$\sum_{j=1}^{20} j^2 \quad \text{and} \quad \sum_{k=1}^{20} k^2$$

represent the same sum. Also, for convenience, the symbols $\sum_{j=1}^{20} j^2$ and $\sum_{k=1}^{20} k^2$ may be used to represent these sums. The index i does not have to start with 1; any integer is allowed.

EXAMPLE 1

(a) $\sum_{i=1}^{n} i = 1 + 2 + 3 + \cdots + (n-1) + n$

(b) $\sum_{j=0}^{6} 3^j = 3^0 + 3^1 + 3^2 + 3^3 + 3^4 + 3^5 + 3^6$

(c) $\sum_{k=0}^{10} 4 = 4+4+4+4+4+4+4+4+4+4+4 = 44$

(d) $\sum_{j=1}^{n} \frac{1}{j} = 1 + \frac{1}{2} + \frac{1}{3} + \cdots + \frac{1}{n-1} + \frac{1}{n}$

(e) $\sum_{i=1}^{5} \frac{i+1}{i^2-2} = \frac{2}{-1} + \frac{3}{2} + \frac{4}{7} + \frac{5}{14} + \frac{6}{23}$

(f) $\sum_{k=0}^{10} (-1)^k \frac{1}{2k+1} = 1 - \frac{1}{3} + \frac{1}{5} - \frac{1}{7} + \frac{1}{9} - \frac{1}{11} + \frac{1}{13} - \frac{1}{15} + \frac{1}{17} - \frac{1}{19} + \frac{1}{21}$

(g) $\sum_{i=1}^{n} f(x_i)\Delta x = f(x_1)\Delta x + f(x_2)\Delta x + \cdots + f(x_n)\Delta x$

Now for the rules that sums obey.

[5.1.2] THEOREM / ***Summation Properties.***

If f and g are functions, k is a real number, and n is an integer, we have the following rules for sums:

$$\sum_{i=1}^{n} kf(i) = k\sum_{i=1}^{n} f(i) \tag{3}$$

$$\sum_{i=1}^{n} [f(i) + g(i)] = \sum_{i=1}^{n} f(i) + \sum_{i=1}^{n} g(i) \tag{4}$$

$$\sum_{i=1}^{n} [f(i) - g(i)] = \sum_{i=1}^{n} f(i) - \sum_{i=1}^{n} g(i) \tag{5}$$

$$\sum_{i=1}^{n} f(i) = \sum_{i=1}^{j} f(i) + \sum_{i=j+1}^{n} f(i) \qquad \text{when} \quad 1 < j < n \tag{6}$$

EXAMPLE 2

Find: $\sum_{i=1}^{n} 1$

Solution

$$\sum_{i=1}^{n} 1 = \underbrace{1 + 1 + \cdots + 1}_{n \text{ times}} = n \tag{7}$$

In addition to the summation properties given in theorem (5.1.2), we need to know some summation formulas.

[5.1.3] THEOREM / ***Summation Formulas.***

$$\sum_{i=1}^{n} k = kn \tag{8}$$

$$\sum_{i=1}^{n} i = 1 + 2 + 3 + \cdots + n = \frac{n(n+1)}{2} \tag{9}$$

$$\sum_{i=1}^{n} i^2 = 1^2 + 2^2 + 3^2 + \cdots + n^2 = \frac{n(n+1)(2n+1)}{6} \tag{10}$$

$$\sum_{i=1}^{n} i^3 = 1^3 + 2^3 + \cdots + n^3 = \frac{n^2(n+1)^2}{4} \tag{11}$$

Proof We give a proof of (9). The others may be verified for any positive integer n by mathematical induction. We write the sums in increasing and decreasing order and add corresponding terms vertically.

$$\begin{aligned}
\sum_{i=1}^{n} i &= 1 + 2 + 3 + \cdots + (n-1) + n \\
\sum_{i=1}^{n} i &= n + (n-1) + (n-2) + \cdots + 2 + 1 \\
2\sum_{i=1}^{n} i &= \underbrace{(n+1) + (n+1) + (n+1) + \cdots + (n+1) + (n+1)}_{n \text{ terms}}
\end{aligned}$$

Therefore,

$$\sum_{i=1}^{n} i = \frac{n(n+1)}{2}$$

Formula (9) provides us with a straightforward method for evaluating certain sums. For example,

$$\sum_{i=1}^{75} i = 1 + 2 + \cdots + 75 = \frac{75(75+1)}{2} = 2850$$

Let's look at two examples of how the properties and formulas are used in practice.

EXAMPLE 3

Express the sum $\sum_{i=1}^{n}(i-1)$ as a function of n.

Solution

$$\sum_{i=1}^{n}(i-1) \underset{\text{By (5)}}{=} \sum_{i=1}^{n} i - \sum_{i=1}^{n} 1 \underset{\text{By (9), (7)}}{=} \frac{n(n+1)}{2} - n = \frac{1}{2}(n^2 - n)$$

EXAMPLE 4

Express the sum $\sum_{i=1}^{n}(i-1)^2\left(\frac{4}{n}\right)^3$ as a function of n.

Solution

$$\sum_{i=1}^{n}(i-1)^2\left(\frac{4}{n}\right)^3 \underset{\text{By (3)}}{=} \frac{64}{n^3}\sum_{i=1}^{n}(i-1)^2 = \frac{64}{n^3}\sum_{i=1}^{n}(i^2 - 2i + 1) \underset{\text{By (3), (4), (5)}}{=} \frac{64}{n^3}\left(\sum_{i=1}^{n} i^2 - 2\sum_{i=1}^{n} i + \sum_{i=1}^{n} 1\right)$$

Now, by using (10), (9), and (8), we get

$$\sum_{i=1}^{n} (i-1)^2\left(\frac{4}{n}\right)^3 = \frac{64}{n^3}\left[\frac{n(n+1)(2n+1)}{6} - \frac{2n(n+1)}{2} + n\right]$$
$$= \frac{64}{n^3}(n)\left[\frac{(n+1)(2n+1) - 6(n+1) + 6}{6}\right]$$
$$= \frac{64}{n^2}\left(\frac{2n^2 - 3n + 1}{6}\right) = \frac{32}{3}\left(\frac{2n^2 - 3n + 1}{n^2}\right)$$

EXERCISE 5.1

In Problems 1–10 express the sum in expanded form.

1. $\sum_{i=1}^{7} \sqrt[3]{i}$ **2.** $\sum_{i=1}^{6} \sqrt{i}$ **3.** $\sum_{i=1}^{5} 2^i$ **4.** $\sum_{i=1}^{8} (i-1)^2$ **5.** $\sum_{i=1}^{6} \frac{1}{2i-1}$

6. $\sum_{i=1}^{10} i^8$ **7.** $\sum_{j=1}^{n} (j^2 - 1)$ **8.** $\sum_{j=1}^{9} (-1)^{j+1}$ **9.** $\sum_{j=1}^{n} (-1)^j \frac{1}{j+1}$ **10.** $\sum_{i=1}^{n} \frac{(i-1)}{2}\Delta x$

In Problems 11–20 express each sum by using summation notation in two ways in which the index begins at (a) $i = 0$ and (b) $i = 1$.

11. $1^2 + 2^2 + 3^2 + \cdots + n^2$

12. $1^3 + 2^3 + 3^3 + 4^3 + \cdots + n^3$

13. $1 + 2 + 4 + 8 + 16 + \cdots + 2^n$

14. $1 + 3 + 9 + 27 + 81 + \cdots + 3^n$

15. $1 + \frac{1}{2} + \frac{1}{4} + \frac{1}{8} + \frac{1}{16} + \cdots + \frac{1}{2^n}$

16. $1 + 3 + 5 + 7 + 9 + \cdots + (2n - 1)$

17. $1 \cdot 2 + 2 \cdot 3 + 3 \cdot 4 + \cdots + n(n+1)$

18. $\frac{1}{2} + \frac{2}{3} + \frac{3}{4} + \cdots + \frac{n}{n+1}$

19. $1 + x^2 + x^4 + \cdots + x^{2n}$

20. $x - x^3 + x^5 - x^7 + x^9 - \cdots + (-1)^{n+1}x^{2n-1}$

In Problems 21–28 express each sum as a function of n.

21. $\sum_{i=1}^{n} 3i(i+1)$ **22.** $\sum_{i=1}^{n} 3i(i-1)$ **23.** $\sum_{i=1}^{n} (i^3 - 2i + 1)$ **24.** $\sum_{i=1}^{n} (i^3 + i + 1)$

25. $\sum_{i=1}^{n} \left(\frac{i}{n}\right)^3\left(\frac{1}{n}\right)$ **26.** $\sum_{i=1}^{n} \left(\frac{i}{n}\right)^2\left(\frac{1}{n}\right)$ **27.** $\sum_{i=1}^{n} \left[\left(\frac{i}{n}\right)^2 - \frac{i}{n}\right]\left(\frac{1}{n}\right)$ **28.** $\sum_{i=1}^{n} \left(\frac{i+1}{2}\right)^2\left(\frac{1}{n}\right)$

29. Show that $\sum_{i=1}^{n} [(i+1)^2 - i^2] = (n+1)^2 - 1$.
[*Hint:* Write out the terms of the sum.]

30. Show that $\sum_{i=1}^{n} [(i+1)^4 - i^4] = n^4 + 4n^3 + 6n^2 + 4n$.
[*Hint:* Write out the terms of the sum.]

31. Show that $\sum_{i=1}^{n} \left(\frac{1}{i} - \frac{1}{i+1}\right) = 1 - \frac{1}{n+1}$.
[*Hint:* Write out the terms of the sum.]

32. Show that $\sum_{i=1}^{n} (2^{i+1} - 2^i) = 2^{n+1} - 2$.
[*Hint:* Write out the terms of the sum.]

In Problems 33–36 find the limit.

33. $\lim_{n \to +\infty} \sum_{i=1}^{n} \left(\frac{i}{n}\right)^3\left(\frac{1}{n}\right)$

34. $\lim_{n \to +\infty} \sum_{i=1}^{n} \left(\frac{i}{n}\right)^2\left(\frac{1}{n}\right)$

35. $\lim_{n\to+\infty}\sum_{i=1}^{n}\left(\frac{1}{i}-\frac{1}{i+1}\right)$

36. $\lim_{n\to+\infty}\sum_{i=1}^{n}\left[\left(\frac{i+1}{n}\right)^2-\left(\frac{i}{n}\right)\right]\frac{1}{n}$

37. Use the fact that $i^3-(i-1)^3=3i^2-3i+1$ to prove that

$$\sum_{i=1}^{n} i^2=\frac{n(n+1)(2n+1)}{6}$$

38. Prove that $\sum_{i=1}^{n}(C_i-C_{i-1})=C_n-C_0$.

39. Use the trigonometric identity

$$2\sin\tfrac{1}{2}x\cos ix=\sin(i+\tfrac{1}{2})x-\sin(i-\tfrac{1}{2})x$$

and the result of Problem 38 to show that

$$\sum_{i=1}^{n}\cos ix=\frac{\sin(n+\frac{1}{2})x-\sin\frac{1}{2}x}{2\sin\frac{1}{2}x}$$

where x is not an integer multiple of 2π. Also show that

$$\sum_{i=1}^{n}\cos ix=\frac{\sin\frac{1}{2}nx\cos\frac{1}{2}(n+1)x}{\sin\frac{1}{2}x}$$

40. Use the idea behind the solution of Problem 39 to show that

$$\sum_{i=1}^{n}\sin ix=\frac{\sin\frac{1}{2}nx\sin\frac{1}{2}(n+1)x}{\sin\frac{1}{2}x}$$

where x is not an integer multiple of 2π.

41. Derive a formula for $\sum_{i=1}^{n} r^i$ by using the fact that

$$\sum_{i=1}^{n} r^i=r+r^2+r^3+\cdots+r^n$$

$$r\sum_{i=1}^{n} r^i=r^2+r^3+r^4+\cdots+r^{n+1}$$

5.2 Area

In this section we discuss the problem of finding area. To be more precise, we wish to compute the area of the region enclosed by the graph of a non-negative function $y=f(x)$, the lines $x=a$ and $x=b$, and the x-axis.

If the graph of $y=f(x)$ is a horizontal line—say, $f(x)=h$, with h positive—the region is a rectangle and its area (A) is the product of the height (h) and the width $(b-a)$, as shown in Figure 2. If the graph of $y=f(x)$ consists of three horizontal lines, each of positive height (see Fig. 3), the area A of the region enclosed by the graph of $y=f(x)$, the lines $x=a$ and $x=b$, and the x-axis may be computed by adding up

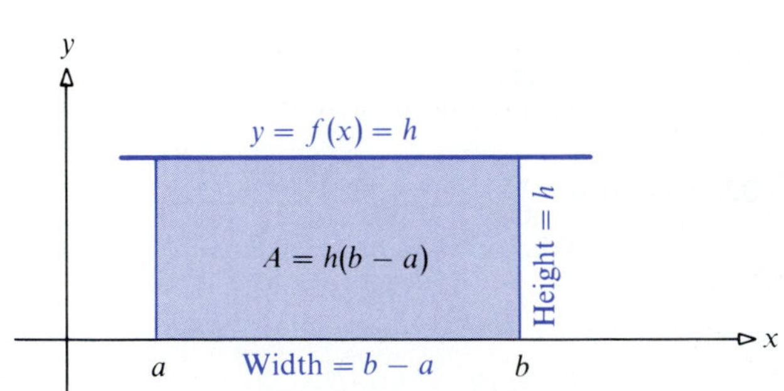

Figure 2

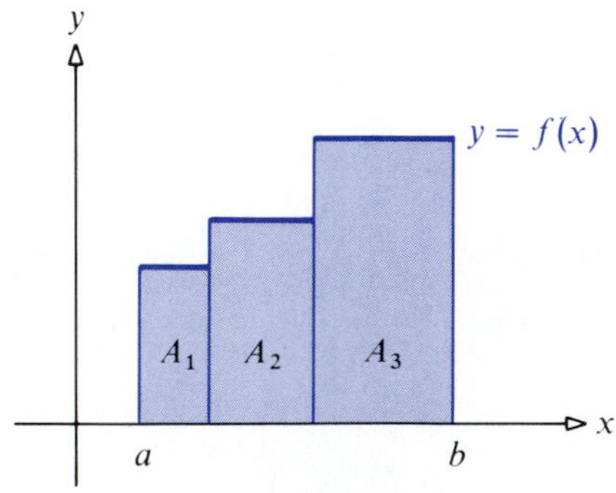

Figure 3

the rectangular areas A_1, A_2, and A_3. In general, the procedure for computing the area of any region enclosed by the graph of a nonnegative function $y = f(x)$, the lines $x = a$ and $x = b$, and the x-axis is based on this idea of adding up rectangular areas. For convenience, we shall refer to this region as the *area under the graph of f from a to b*.

To fix our ideas, we make two assumptions:

1. The function f is continuous on the closed interval $[a, b]$.
2. The function f is nonnegative on the closed interval $[a, b]$.

Since we know how to compute the area of a rectangle, we approximate the area under the graph of f from a to b by using rectangles. We do this in the following way: We pick $(n - 1)$ numbers between a and b and label them $x_1, x_2, \ldots, x_{n-1}$, with $a < x_1 < x_2 < \ldots < x_{n-1} < b$ (see Fig. 4).

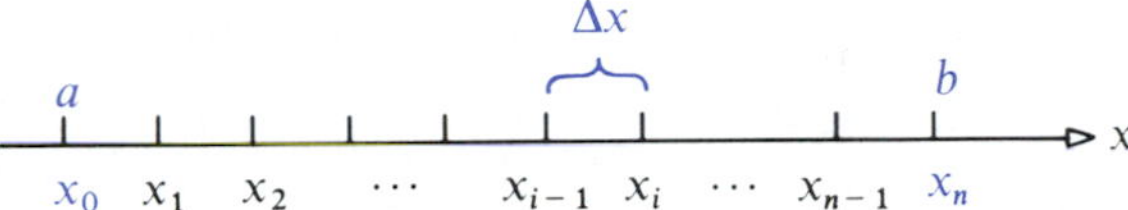

Figure 4

For convenience, we set $a = x_0$ and $b = x_n$. The numbers thus selected divide, or *partition*, the interval $[a, b]$ into n subintervals:

$$[x_0, x_1], \quad [x_1, x_2], \quad \ldots, \quad [x_{n-1}, x_n]$$

The selection of numbers between a and b is arbitrary, but for simplicity we select them so that the length of each subinterval is the same. If we denote this common length by Δx, so that $\Delta x = x_1 - x_0 = x_2 - x_1 = \cdots = x_n - x_{n-1}$, it follows that

$$\Delta x = \frac{b - a}{n}$$

Since f is continuous on the closed interval $[a, b]$, it is continuous on every subinterval $[x_{i-1}, x_i]$ of $[a, b]$. Because of the extreme value theorem (4.3.3), there is a number in each of these subintervals at which f attains its absolute minimum. We denote these numbers by $c_1, c_2, c_3, \ldots, c_n$, so that $f(c_i)$ is the absolute minimum of f in the subinterval $[x_{i-1}, x_i]$. We now construct n rectangles, each having a subinterval as a base and $f(c_i)$ as an altitude (see Fig. 5). In doing this, we obtain n thin strips of uniform width

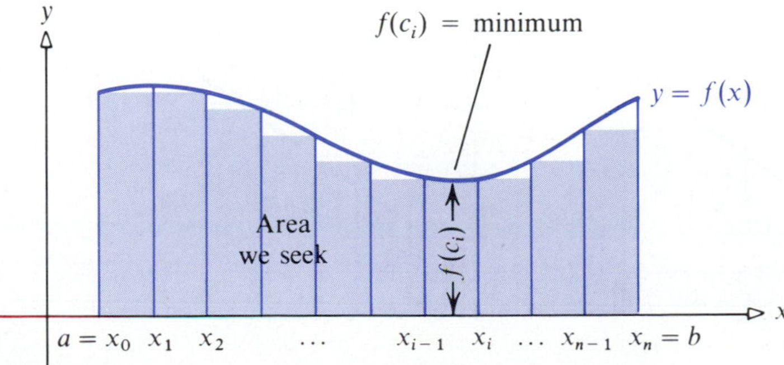

Figure 5

$\Delta x = (b - a)/n$ and altitudes $f(c_1), f(c_2), \ldots, f(c_n)$, respectively, so that their areas may be calculated as follows:

$$\text{Area of first rectangle} = f(c_1)(x_1 - x_0) = f(c_1)\Delta x$$

$$\text{Area of second rectangle} = f(c_2)(x_2 - x_1) = f(c_2)\Delta x$$

$$\vdots$$

$$\text{Area of } n\text{th (and last) rectangle} = f(c_n)(x_n - x_{n-1}) = f(c_n)\Delta x$$

The sum s_n of the areas of these n inscribed rectangles gives an approximation of the area A we seek. That is, the area A is approximately equal to

$$s_n = f(c_1)\Delta x + f(c_2)\Delta x + \cdots + f(c_i)\Delta x + \cdots + f(c_n)\Delta x = \sum_{i=1}^{n} f(c_i)\Delta x$$

Since the rectangles we used to approximate the area A are *inscribed rectangles*, the sum s_n *underestimates* the area A of the region. Thus, we conclude that

$$s_n \leq A$$

Let's look at an example before continuing the discussion.

EXAMPLE 1

Approximate the area under the graph of $f(x) = 3x$ from 0 to 10 by computing s_n for $n = 2$, $n = 5$, $n = 10$.

Solution

For $n = 2$, we partition the closed interval [0, 10] into two subintervals of equal length, [0, 5] and [5, 10], as shown in Figure 6(a). The length of each of these subintervals is $\Delta x = (10 - 0)/2 = 5$. To compute s_2, we need to know where f attains its minimum in each subinterval. Since f is increasing, the minimum is attained at the left endpoint of each subinterval. Thus, for $n = 2$, the minimum of f on [0, 5] occurs at 0 and the minimum of f on [5, 10] occurs at 5. The value of s_2 is therefore

$$s_2 = \sum_{i=1}^{2} f(c_i)\Delta x = f(c_1)\Delta x + f(c_2)\Delta x = f(0)(5) + f(5)(5)$$
$$= (0)(5) + (15)(5) = 75$$

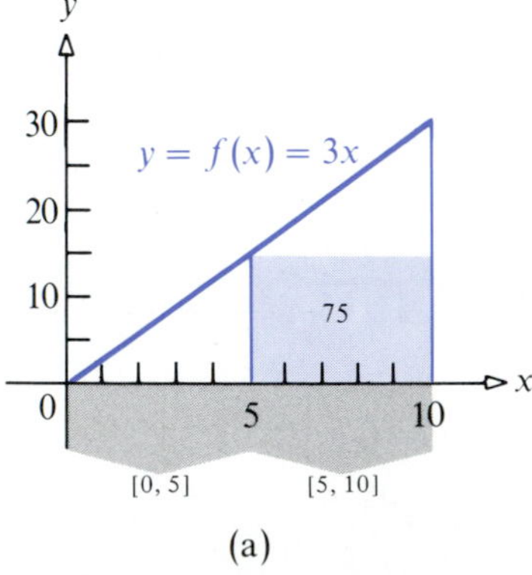

(a)

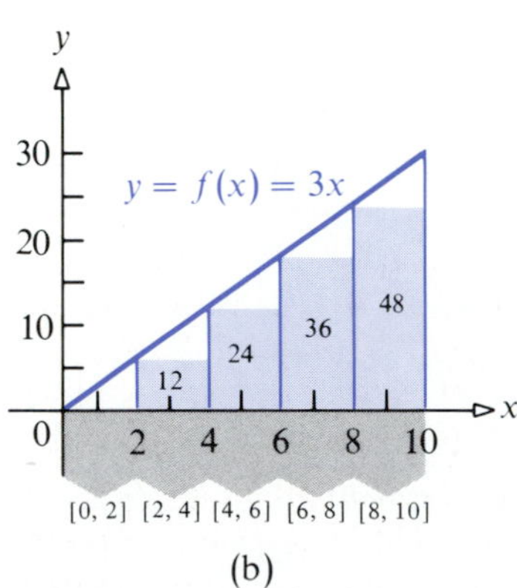

(b)

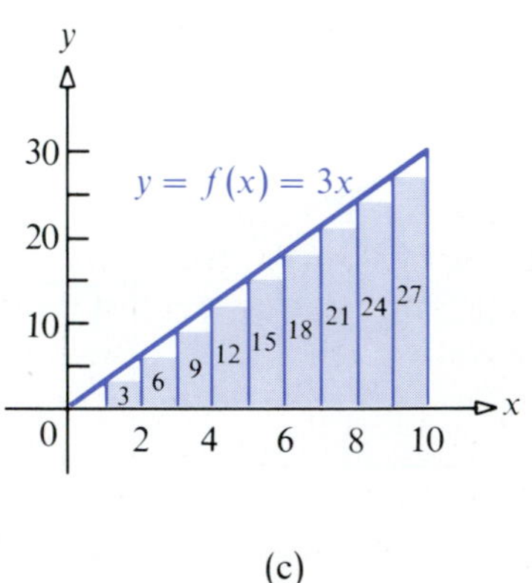

(c)

Figure 6

For $n = 5$, we partition [0, 10] into five subintervals of equal length—namely, [0, 2], [2, 4], [4, 6], [6, 8], and [8, 10], as shown in Figure 6(b). The length of each of these is $\Delta x = (10 - 0)/5 = 2$. The sum s_5 is

$$s_5 = \sum_{i=1}^{5} f(c_i)\Delta x = f(c_1)\Delta x + f(c_2)\Delta x + f(c_3)\Delta x + f(c_4)\Delta x + f(c_5)\Delta x$$

$$= f(0)(2) + f(2)(2) + f(4)(2) + f(6)(2) + f(8)(2)$$

$$= (0)(2) + (6)(2) + (12)(2) + (18)(2) + (24)(2) = 120$$

For $n = 10$, we partition [0, 10] into 10 subintervals of equal length—namely, [0, 1], [1, 2], [2, 3], . . . , [9, 10], as shown in Figure 6(c). The length of each subinterval is $\Delta x = (10 - 0)/10 = 1$. The sum s_{10} is

$$s_{10} = \sum_{i=1}^{10} f(c_i)\Delta x = f(0)(1) + f(1)(1) + f(2)(1) + \cdots + f(9)(1)$$

$$= 0 + 3 + 6 + 9 + 12 + 15 + 18 + 21 + 24 + 27 = 135$$

The region under the graph of $f(x) = 3x$ from 0 to 10 is, of course, a triangle with base 10 and height 30. Thus, the *actual area* is

$$A = (\tfrac{1}{2})(10)(30) = 150$$

The following table summarizes the results of Example 1:

n	2	5	10
s_n	75	120	135

Observe that as the number n of subintervals increases, the estimates s_n of the area get closer to the actual area. For $n = 2$, the error in estimating A is 75 square units, whereas for $n = 10$, the error has been reduced to 15 square units.

In general, the error due to using inscribed rectangles occurs when a portion of the region lies outside the inscribed rectangles (see Fig. 7). It is this error that makes the sum s_n less than the actual area A. To get a better approximation of the area A, we must decrease such errors, and we can usually do this by increasing the number of subintervals. For example, suppose the number n of subintervals is doubled; that is, each interval of the first subdivision is itself subdivided to give a finer subdivision. In doing so we double the number of inscribed rectangles; for example, compare Figures 6(b) and 6(c). The result is that a greater portion of the region is covered, and the error is smaller than it was with the first subdivision. By further subdivision, we can reduce the error even more. Thus, by taking a finer and finer subdivision of the interval $[a, b]$ (we do this by increasing n without bound), we can make the sum of the areas of the inscribed rectangles as close as we please to the area A. (You can find the proof of this statement in books on advanced calculus.) With this analysis in mind, we now give the definition of the area under the graph of a function f from a to b.

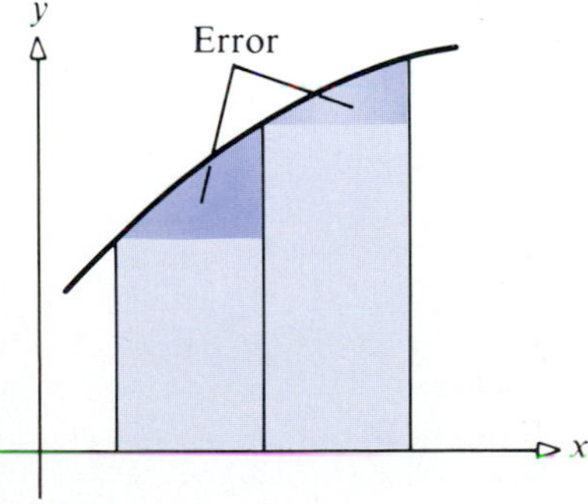

Figure 7

[5.2.1] DEFINITION / *Area under a Graph.*

Let f denote a function that is both continuous and nonnegative on a closed interval $[a, b]$. Divide the interval $[a, b]$ into n subintervals $[x_0, x_1], [x_1, x_2], [x_2, x_3], \ldots, [x_{n-1}, x_n]$, each of length

$$\Delta x = \frac{b-a}{n}$$

In each subinterval $[x_{i-1}, x_i]$, let $f(c_i)$ denote the absolute minimum of f on this subinterval. Form the sum

$$s_n = \sum_{i=1}^{n} f(c_i)\Delta x = f(c_1)\Delta x + \cdots + f(c_n)\Delta x$$

The area A under the graph of $y = f(x)$ from a to b is the number

$$A = \lim_{n \to +\infty} s_n \qquad (1)$$

provided this limit exists.

In the above discussion we calculate the area A of a region by the use of inscribed rectangles. By a parallel argument, we can use circumscribed rectangles to compute the area A, as shown in Figure 8. In this case, the values $C_1, C_2, \ldots, C_n$ are chosen so that the altitude $f(C_i)$ of the ith rectangle is the absolute maximum of f on the ith subinterval. The corresponding sum S_n of the areas of the *circumscribed rectangles* is an *overestimate* of the area A. In this case, $S_n \geq A$. It can be shown that as n increases without bound, the limit of the sum S_n approaches the same limit as the one obtained from the inscribed rectangles; that is,

$$\lim_{n \to +\infty} s_n = \lim_{n \to +\infty} S_n = A$$

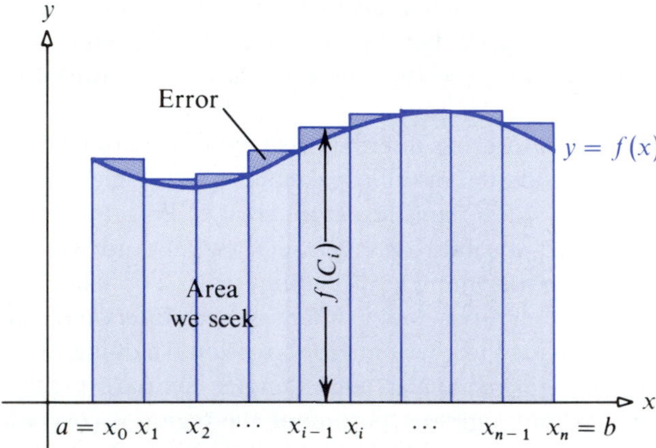

Figure 8

Expressions such as $\lim_{n\to+\infty} s_n$ (and $\lim_{n\to+\infty} S_n$) require further comment. In limits like this, where s_n is defined for only positive integers n, $\lim_{n\to+\infty} s_n = A$ means that for any given $\varepsilon > 0$, there is a positive number N so that $|s_n - A| < \varepsilon$ whenever $n > N$, n being a positive integer. Based on the similarity between this definition and the preceding discussion (in Chap. 4) on limits at infinity, the procedures adopted for evaluating limits at infinity will be used to evaluate limits such as (1).

To evaluate the area A under the graph of a continuous nonnegative function $y = f(x)$ defined on a closed interval $[a, b]$, we need to find the limit of a certain sum—namely,

$$A = \lim_{n\to+\infty} s_n = \lim_{n\to+\infty} \sum_{i=1}^{n} f(c_i)\Delta x$$

$$= \lim_{n\to+\infty} [f(c_1)\Delta x + f(c_2)\Delta x + \cdots + f(c_n)\Delta x] \qquad \textbf{(2)}$$

where Δx is the length of each subinterval $[x_{i-1}, x_i]$ of $[a, b]$ and $f(c_i)$ is the minimum of f on $[x_{i-1}, x_i]$, $i = 1, 2, \ldots, n$.

To evaluate the limit in (2), we require that the sum $\sum_{i=1}^{n} f(c_i)\Delta x$ be expressed as a function of n.

EXAMPLE 2

Find the area under the graph of $f(x) = 3x$ from 0 to 10 by using the definition of area, $A = \lim_{n\to+\infty} s_n$ (inscribed rectangles).

Solution

The region whose area is to be computed is illustrated in Figure 9. We divide the closed interval [0, 10] into n equal subintervals

$$[x_0, x_1], \quad [x_1, x_2], \quad \ldots, \quad [x_{n-1}, x_n]$$

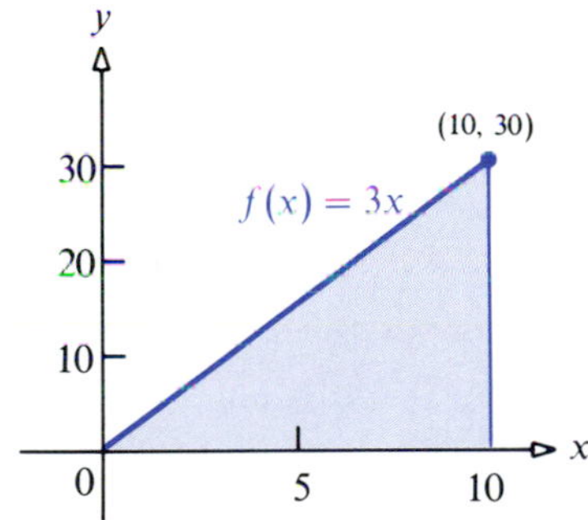

Figure 9

where

$$0 = x_0 < x_1 < x_2 < \cdots < x_{n-1} < x_n = 10$$

and

$$x_1 - x_0 = x_2 - x_1 = \cdots = x_n - x_{n-1} = \Delta x$$

Since the length of each subinterval is

$$\Delta x = \frac{10 - 0}{n} = \frac{10}{n}$$

the coordinates of the endpoints (as Fig. 10 illustrates) can be expressed in terms of n as follows:

$$x_0 = 0, \quad x_1 = \frac{10}{n}, \quad x_2 = 2\left(\frac{10}{n}\right), \quad x_3 = 3\left(\frac{10}{n}\right), \quad \ldots, \quad x_{i-1} = (i-1)\left(\frac{10}{n}\right),$$

$$x_i = i\left(\frac{10}{n}\right), \quad \ldots, \quad x_{n-1} = (n-1)\left(\frac{10}{n}\right), \quad x_n = n\left(\frac{10}{n}\right) = 10$$

To find area A by using the limit of s_n, we need to compute the absolute minimum of f on each subinterval $[x_{i-1}, x_i]$, $i = 1, 2, \ldots, n$. In this

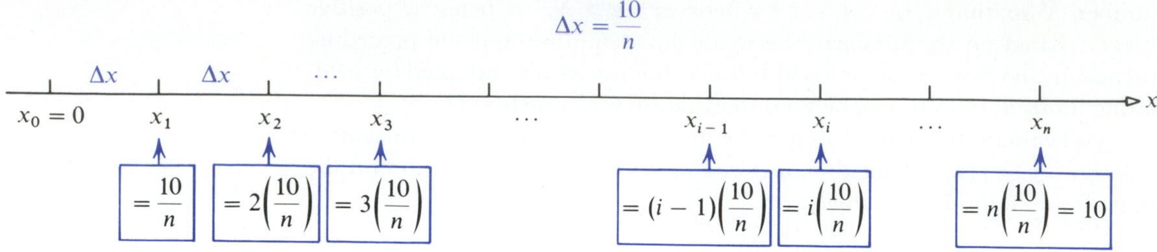

Figure 10

example, since the function $f(x) = 3x$ is an increasing function, it will assume its absolute minimum at the left endpoint x_{i-1} of each subinterval. Since $x_{i-1} = (i-1)(10/n)$, the absolute minimum of f on $[x_{i-1}, x_i]$ is

$$f(x_{i-1}) = f\left[(i-1)\left(\frac{10}{n}\right)\right] = 3(i-1)\left(\frac{10}{n}\right)$$

Thus,

$$s_n = \sum_{i=1}^{n} f(x_{i-1})\Delta x = \sum_{i=1}^{n} 3(i-1)\left(\frac{10}{n}\right)\left(\frac{10}{n}\right) = \frac{300}{n^2}\sum_{i=1}^{n}(i-1)$$

$$\uparrow \; \Delta x = \frac{10}{n}$$

Using the result obtained in Example 3 of Section 5.1, we find

$$s_n = \frac{300}{n^2}\left(\frac{n^2-n}{2}\right) = 150\left(\frac{n^2-n}{n^2}\right) = 150\left(1-\frac{1}{n}\right)$$

By letting $n \to +\infty$, the area A of the region under the graph of $f(x) = 3x$ from 0 to 10 is

$$A = \lim_{n\to+\infty} s_n = \lim_{n\to+\infty} 150\left(1-\frac{1}{n}\right) = 150 \text{ square units}$$

This answer is in agreement with the answer obtained earlier by using the formula for the area of a triangle—namely, $A = (\frac{1}{2})bh = (\frac{1}{2})(10)(30) = 150$.

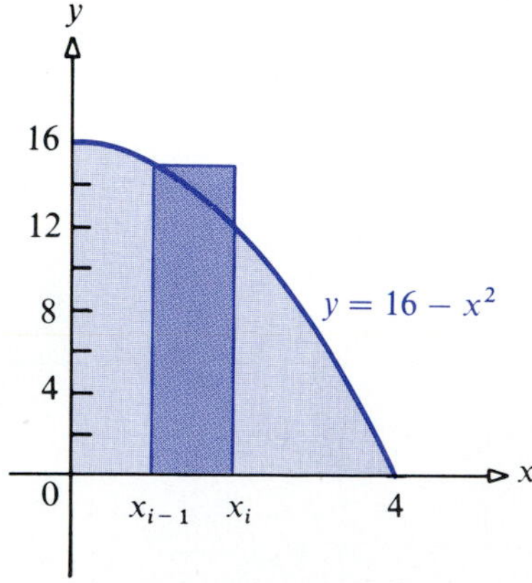

Figure 11

EXAMPLE 3

Find the area A of the region under the graph of $y = f(x) = 16 - x^2$ from 0 to 4 by using $A = \lim_{n\to+\infty} S_n$ (circumscribed rectangles).

Solution

Figure 11 illustrates the region and a typical circumscribed rectangle with base of length $\Delta x = x_i - x_{i-1}$. We divide the closed interval [0, 4] into n subintervals of equal length,

$$[x_0, x_1], \quad [x_1, x_2], \quad \ldots, \quad [x_{n-1}, x_n]$$

where

$$0 = x_0 < x_1 < \cdots < x_{n-1} < x_n = 4$$

and

$$x_1 - x_0 = x_2 - x_1 = \cdots = x_n - x_{n-1} = \Delta x$$

The length of each subinterval is $\Delta x = (4 - 0)/n = 4/n$. The coordinates of the endpoints (as Fig. 12 illustrates) can be expressed in terms of n as follows:

$$x_0 = 0, \quad x_1 = \frac{4}{n} \quad x_2 = 2\left(\frac{4}{n}\right), \quad \ldots, \quad x_{i-1} = (i-1)\left(\frac{4}{n}\right),$$

$$x_i = i\left(\frac{4}{n}\right), \quad \ldots, \quad x_{n-1} = (n-1)\left(\frac{4}{n}\right), \quad x_n = n\left(\frac{4}{n}\right) = 4$$

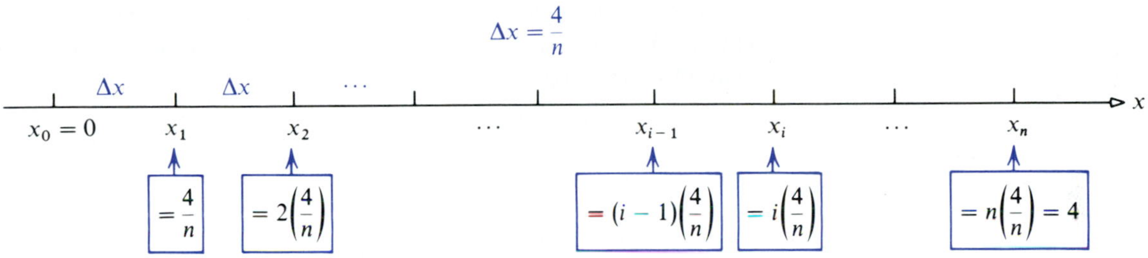

$\Delta x = \frac{4}{n}$

Δx Δx ...

$x_0 = 0$, $x_1 = \frac{4}{n}$, $x_2 = 2\left(\frac{4}{n}\right)$, ..., $x_{i-1} = (i-1)\left(\frac{4}{n}\right)$, $x_i = i\left(\frac{4}{n}\right)$, ..., $x_n = n\left(\frac{4}{n}\right) = 4$, x

Figure 12

To use circumscribed rectangles, we need to compute the absolute maximum of f on each subinterval $[x_{i-1}, x_i]$, $i = 1, 2, \ldots, n$. In this example, since the function f is a decreasing function, it will assume its absolute maximum at the left endpoint x_{i-1} of each subinterval. Thus,

$$S_n = \sum_{i=1}^{n} f(x_{i-1})\Delta x$$

Since $x_{i-1} = (i-1)(4/n)$, $f(x) = 16 - x^2$, and $\Delta x = 4/n$, we calculate S_n to be

$$S_n = \sum_{i=1}^{n} f(x_{i-1})\,\Delta x = \sum_{i=1}^{n} \left\{16 - \left[(i-1)\left(\frac{4}{n}\right)\right]^2\right\}\left(\frac{4}{n}\right)$$

$$= \sum_{i=1}^{n} \frac{64}{n} - \sum_{i=1}^{n} (i-1)^2\left(\frac{64}{n^3}\right) = \frac{64}{n}\sum_{i=1}^{n} 1 - \frac{64}{n^3}\sum_{i=1}^{n} (i-1)^2$$

By using the result obtained in Section 5.1, formula (8) and Example 4, we find

$$S_n = \frac{64}{n}(n) - \frac{64}{n^3}(n)\left(\frac{2n^2 - 3n + 1}{6}\right) = 64 - \frac{32}{3}\left(2 - \frac{3}{n} + \frac{1}{n^2}\right) = \frac{128}{3} + \frac{32}{n} - \frac{32}{3n^2}$$

Taking the limit as $n \to +\infty$, we get

$$A = \lim_{n\to+\infty} S_n = \lim_{n\to+\infty}\left(\frac{128}{3} + \frac{32}{n} - \frac{32}{3n^2}\right) = \frac{128}{3}$$

The area of the region is therefore $\frac{128}{3}$ square units.

If we had used $A = \lim_{n\to+\infty} s_n$ (inscribed rectangles) in Example 3, we would have obtained the same answer. Of course, in using $A = \lim_{n\to+\infty} s_n$, we must find the number c_i in $[x_{i-1}, x_i]$ at which the function f attains its absolute minimum. In this example, since $f(x) = 16 - x^2$ is decreasing on $[0, 4]$, we find $c_i = x_i = i(4/n)$.

The next example illustrates how to obtain the coordinates of the partition when the initial left endpoint is other than 0.

EXAMPLE 4

Find the area A of the region under the graph of $y = f(x) = 4x$ from 1 to 3 by using $A = \lim_{n\to+\infty} S_n$ (circumscribed rectangles).

Solution

Figure 13 illustrates the region and a typical circumscribed rectangle with base of length $\Delta x = x_i - x_{i-1}$. We divide the closed interval $[1, 3]$ into n subintervals of equal length

$$[x_0, x_1], \quad [x_1, x_2], \quad \ldots, \quad [x_{n-1}, x_n]$$

where

$$1 = x_0 < x_1 < \cdots < x_{n-1} < x_n = 3$$

and

$$x_1 - x_0 = x_2 - x_1 = \cdots = x_n - x_{n-1} = \Delta x$$

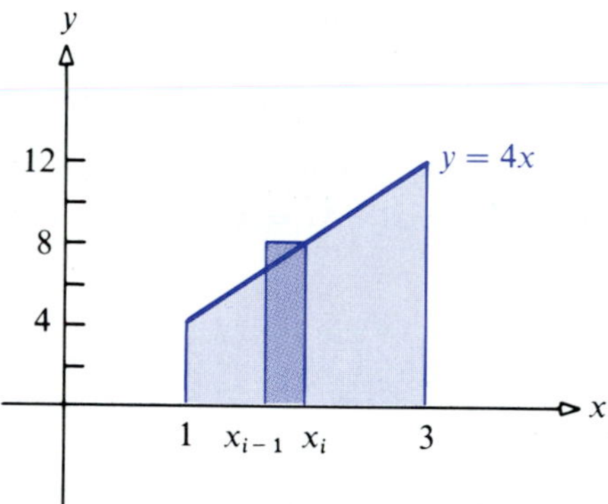

Figure 13

The length of each subinterval is $\Delta x = (3 - 1)/n = 2/n$. The coordinates of the endpoints (as Fig. 14 illustrates) can be expressed in terms of n as follows:

$$x_0 = 1, \quad x_1 = 1 + \frac{2}{n}, \quad x_2 = 1 + 2\left(\frac{2}{n}\right), \quad \ldots,$$

$$x_i = 1 + i\left(\frac{2}{n}\right), \quad \ldots, \quad x_n = 1 + n\left(\frac{2}{n}\right) = 1 + 2 = 3$$

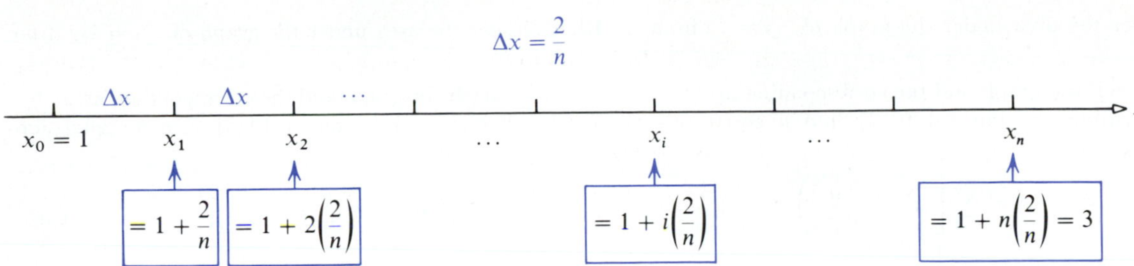

Figure 14

To use circumscribed rectangles, we need to compute the absolute maximum of f on each subinterval $[x_{i-1}, x_i]$, $i = 1, 2, \ldots, n$. In this example, since f is increasing, it assumes its absolute maximum at the right endpoint of each subinterval. Thus,

$$S_n = \sum_{i=1}^{n} f(x_i)\Delta x$$

Since $x_i = 1 + (i)(2/n)$, $f(x) = 4x$, and $\Delta x = 2/n$, we find S_n to be

$$\begin{aligned} S_n &= \sum_{i=1}^{n} f(x_i)\Delta x = \sum_{i=1}^{n} 4\left[1 + (i)\left(\frac{2}{n}\right)\right]\left(\frac{2}{n}\right) \\ &= \left(\frac{8}{n}\right)\left[\sum_{i=1}^{n} 1 + \left(\frac{2}{n}\right)\sum_{i=1}^{n} i\right] \\ &= \left(\frac{8}{n}\right)\left[n + \left(\frac{2}{n}\right)\frac{n(n+1)}{2}\right] \\ &= \left(\frac{8}{n}\right)[n + (n+1)] = 16 + \frac{8}{n} \end{aligned}$$

The area A is therefore

$$A = \lim_{n\to+\infty} S_n = 16 \text{ square units}$$

EXERCISE 5.2

In Problems 1–4 partition the given interval into n subintervals of equal length.

1. $[1, 4]$; $n = 3$ **2.** $[0, 9]$; $n = 9$ **3.** $[-1, 4]$; $n = 10$ **4.** $[-4, 4]$; $n = 16$

In Problems 5–10 approximate the area under the graph of each function from a to b by computing s_n and S_n for $n = 4$ and $n = 8$.

5. $f(x) = 2x + 5$, on $[2, 6]$ **6.** $f(x) = -x + 10$, on $[0, 8]$ **7.** $f(x) = x^2$, on $[0, 2]$

8. $f(x) = x^3$, on $[0, 8]$ **9.** $f(x) = \frac{1}{x}$, on $[1, 5]$ **10.** $f(x) = \frac{1}{x^2}$, on $[2, 6]$

11. Consider the area under the graph of $y = x$ from 0 to 3.

(a) Sketch the graph and the corresponding area.
(b) Partition the interval [0, 3] into n equal subintervals.

(c) Show that $s_n = \sum_{i=1}^{n} \left[(i-1)\frac{3}{n}\right]\left(\frac{3}{n}\right)$.

(d) Show that $S_n = \sum_{i=1}^{n} \left[i\left(\frac{3}{n}\right)\right]\left(\frac{3}{n}\right)$.

(e) Complete the following table:

n	5	10	50	100
s_n S_n				

(f) Show that $\lim_{n\to+\infty} s_n = \lim_{n\to+\infty} S_n = \frac{9}{2}$.

12. Consider the area under the graph of $y = 4x$ from 0 to 5.

(a) Sketch the graph and the corresponding area.
(b) Partition the interval [0, 5] into n equal subintervals.

(c) Show that $s_n = \sum_{i=1}^{n} (i-1)\frac{100}{n^2}$.

(d) Show that $S_n = \sum_{i=1}^{n} i\frac{100}{n^2}$.

(e) Complete the following table:

n	5	10	50	100
s_n S_n				

(f) Show that $\lim_{n\to+\infty} s_n = \lim_{n\to+\infty} S_n = 50$.

13. Use circumscribed rectangles to find the area under the graph of $f(x) = 3x$ from 0 to 10.

14. Use inscribed rectangles to find the area under the graph of $f(x) = 16 - x^2$ from 0 to 4.

In Problems 15–24 use inscribed rectangles to find the area under the graph of $y = f(x)$ from a to b. A graph will be helpful.

15. $f(x) = 2x + 1, \quad a = 0, \quad b = 4$

16. $f(x) = 3x + 1, \quad a = 0, \quad b = 4$

17. $f(x) = x^2, \quad a = 0, \quad b = 2$

18. $f(x) = x^2 + 2, \quad a = 0, \quad b = 3$

19. $f(x) = 4 - x^2, \quad a = 0, \quad b = 2$

20. $f(x) = 12 - x^2, \quad a = 0, \quad b = 3$

21. $f(x) = x^3, \quad a = 0, \quad b = 2$

22. $f(x) = x^3 + 3, \quad a = 0, \quad b = 1$

23. $f(x) = 2x + 1, \quad a = 1, \quad b = 3$

24. $f(x) = 4 + 3x, \quad a = 1, \quad b = 2$

In Problems 25–34 use circumscribed rectangles to find the area under the graph of $y = f(x)$ from a to b.

25. $f(x) = 2x + 1, \quad a = 0, \quad b = 4$

26. $f(x) = 3x + 1, \quad a = 0, \quad b = 4$

27. $f(x) = x^2, \quad a = 0, \quad b = 2$

28. $f(x) = x^2 + 2, \quad a = 0, \quad b = 3$

29. $f(x) = 4 - x^2, \quad a = 0, \quad b = 2$

30. $f(x) = 12 - x^2, \quad a = 0, \quad b = 3$

31. $f(x) = x^3, \quad a = 0, \quad b = 2$

32. $f(x) = x^3 + 3, \quad a = 0, \quad b = 1$

33. $f(x) = 2x + 1, \quad a = 1, \quad b = 3$

34. $f(x) = 4 + 3x, \quad a = 1, \quad b = 2$

35. Use a method of this section to find the area of a right triangle of height H and base B.

36. Use a method of this section to find the area of a trapezoid of height H and bases B_1 and B_2.

37. Find the area under the graph of $y = f(x) = x^2$ from $a = 0$ to $b = 3$ by evaluating

$$\lim_{n\to+\infty} \sum_{i=1}^{n} f(m_i)\Delta x$$

where $\Delta x = 3/n$ and m_i is the midpoint of the ith subinterval.

38. Follow the instructions of Problem 37 to find the area under the graph of $f(x) = 9 - x^2$ from $a = 0$ to $b = 2$.

39. Approximate the area under the graph of $f(x) = x$ from a to b by computing s_n and S_n for a partition of $[a, b]$ into n subintervals of equal length. Show that

$$s_n < \frac{b^2 - a^2}{2} < S_n$$

40. Approximate the area under the graph of $f(x) = x^2$ from $a \geq 0$ to b by computing s_n and S_n for a partition of $[a, b]$ into n subintervals of equal length. Show that

$$s_n < \frac{b^3 - a^3}{3} < S_n$$

The Definite Integral 5.3

We have defined the area A under the graph of $y = f(x)$ from a to b as the limit

$$A = \lim_{n \to +\infty} \sum_{i=1}^{n} f(c_i)\Delta x \tag{1}$$

where the following assumptions were made:

I. The function f is continuous on $[a, b]$.

II. The function f is nonnegative on $[a, b]$.

III. The subintervals of $[a, b]$ are all of equal length, $\Delta x = (b - a)/n$.

IV. In each subinterval $[x_{i-1}, x_i]$ the number c_i in $[x_{i-1}, x_i]$ is the one at which f has its absolute minimum.

We also noted that

$$A = \lim_{n \to +\infty} \sum_{i=1}^{n} f(C_i)\Delta x$$

where in each subinterval $[x_{i-1}, x_i]$ the number C_i in $[x_{i-1}, x_i]$ is the one at which f has its absolute maximum.

Riemann Sums

This procedure for calculating area involves partitioning an interval into n subintervals, finding the area of each inscribed (or circumscribed) rectangle, adding them up, and taking the limit as $n \to +\infty$. Interestingly enough, similar processes enable us to find the volume of a solid of revolution, the length of a graph, the work done by a variable force, and other quantities. For this reason, we now study sums of the form

$$\sum_{i=1}^{n} f(u_i)\Delta x_i \tag{2}$$

using the following more general assumptions:

I. The function f is not necessarily continuous on $[a, b]$.

II. The function f does not have to be nonnegative on $[a, b]$.

III. The lengths Δx_i of the subintervals $[x_{i-1}, x_i]$ of $[a, b]$ do not have to be equal.

IV. The number u_i may be any number in $[x_{i-1}, x_i]$; it is not necessarily the number at which f has its absolute minimum (or maximum).

Observe that the sum in (1) is a special case of the sums in (2), which we will study in this section. The new sums are called *Riemann sums* for f on $[a, b]$; they are named after the German mathematician Georg Friedrich Bernhard Riemann (1826–1866).

Partition; Norm

We begin with a function f defined on a closed interval $[a, b]$, and we divide the interval $[a, b]$ into n subintervals

$$[x_0, x_1], \quad [x_1, x_2], \quad [x_2, x_3], \quad \ldots, \quad [x_{n-1}, x_n]$$

where

$$a = x_0 < x_1 < x_2 < \cdots < x_{n-1} < x_n = b$$

These subintervals are not necessarily of the same length. As a result, we denote the length of the first interval by $\Delta x_1 = x_1 - x_0$, the length of the second interval by $\Delta x_2 = x_2 - x_1$, and so on. In general, the length of the ith subinterval is

$$\Delta x_i = x_i - x_{i-1} \qquad i = 1, 2, \ldots, n$$

The set of all such subintervals of the interval $[a, b]$ is called a *partition P* of $[a, b]$. Three examples of possible partitions of the interval $[0, 2]$ are:

(a) $[0, 1], [1, 2]$

(b) $[0, \frac{1}{2}], [\frac{1}{2}, 1], [1, \frac{3}{2}], [\frac{3}{2}, 2]$

(c) $[0, \frac{1}{4}], [\frac{1}{4}, \frac{1}{3}], [\frac{1}{3}, \frac{1}{2}], [\frac{1}{2}, \frac{7}{8}], [\frac{7}{8}, 1], [1, \frac{5}{4}], [\frac{5}{4}, \frac{3}{2}], [\frac{3}{2}, \frac{7}{4}], [\frac{7}{4}, 2]$

The length of the largest subinterval in a partition P is called the *norm* of the partition and is denoted by $\|P\|$. For example, $\|P\| = 1$ in (a), $\|P\| = \frac{1}{2}$ in (b), $\|P\| = \frac{3}{8}$ in (c).

Suppose in each subinterval $[x_{i-1}, x_i]$ of a partition P we choose a number u_i, $i = 1, 2, \ldots, n$, and form the Riemann sums

$$\sum_{i=1}^{n} f(u_i)\Delta x_i = f(u_1)\Delta x_1 + f(u_2)\Delta x_2 + \cdots + f(u_n)\Delta x_n \tag{3}$$

Notice that once the interval $[a, b]$ and the function f have been chosen, the choices for the partition P are unlimited. Furthermore, once the partition has been chosen, the choices for the u_i's are unlimited. The value of the Riemann sums (3) depends on *all* of these choices.

Now, suppose P is a partition for which the norm $\|P\|$ is close to 0. Then, since $\|P\|$ is the length of the largest subinterval, the lengths of *all* subintervals in P will be close to 0. In particular, when the norm $\|P\|$ approaches 0, the effect is similar to that produced by repeatedly refining the original partition.

Definite Integral

For many functions, when $\|P\|$ approaches 0, the Riemann sums approach a limit—say, I. In this event, we write

$$\lim_{\|P\|\to 0} \sum_{i=1}^{n} f(u_i)\Delta x_i = I \tag{4}$$

In words, this means that values of the Riemann sums can be made as close to I as we please by choosing a partition P whose norm $\|P\|$ is sufficiently close to 0. This choice of P is made independently of the choice of u_i.

We now state the meaning of (4) more precisely.

[5.3.1] DEFINITION

Let f be a function defined on a closed interval $[a, b]$ and I be a number. The statement

$$\lim_{\|P\|\to 0} \sum_{i=1}^{n} f(u_i)\Delta x_i = I \tag{5}$$

means that for any given $\varepsilon > 0$, there is a positive number δ so that if P is a partition of $[a, b]$ for which $\|P\| < \delta$, then

$$\left|\sum_{i=1}^{n} f(u_i)\Delta x_i - I\right| < \varepsilon$$

for any choice of numbers u_i in the subintervals $[x_{i-1}, x_i]$ of P.

The number I in (5) plays such a major role in mathematics that a special name and symbol are given to it.

[5.3.2] DEFINITION / ***Definite Integral.***

Let f be a function defined on the closed interval $[a, b]$. If the limit in (5) exists, then the number I is called the *definite integral of f* from a to b and is denoted by $\int_a^b f(x)\,dx$. That is,

$$\int_a^b f(x)\,dx = \lim_{\|P\|\to 0} \sum_{i=1}^{n} f(u_i)\Delta x_i \tag{6}$$

For the definite integral $\int_a^b f(x)\,dx$, the number a is called the *lower limit of integration*, the number b is called the *upper limit of integration*,* the symbol $\int$ (an elongated S to remind you of summation) is called the *integral sign*, and $f(x)$ is called the *integrand*. The variable used in the definite integral is an *artificial*, or *dummy*, variable because it may be replaced by any other letter. Thus, for example,

$$\int_a^b f(x)\,dx, \qquad \int_a^b f(t)\,dt, \qquad \int_a^b f(s)\,ds$$

---

* The terms *upper limit* and *lower limit* of integration simply refer to the endpoints of the closed interval $[a, b]$ and have nothing to do with the highly specialized concept of limit as introduced in Chapter 2.

all denote the definite integral of f from a to b, and if any of them exist, they are all equal to the same number.

In defining the definite integral $\int_a^b f(x)\,dx$, we have assumed that $a < b$. To remove this restriction, we give the following definition.

[5.3.3] DEFINITION

If $f(a)$ is defined, then

$$\int_a^a f(x)\,dx = 0$$

If $a > b$ and if $\int_b^a f(x)\,dx$ exists, then

$$\int_a^b f(x)\,dx = -\int_b^a f(x)\,dx$$

Thus, interchanging the limits of integration will reverse the sign of the integral. Here are some specific examples to illustrate the definition:

$$\int_1^1 x^2\,dx = 0 \qquad \int_3^2 x^2\,dx = -\int_2^3 x^2\,dx$$

We can specify a condition on the function f that will guarantee that the limit in (6) exists. (The proof of this result may be found in advanced calculus texts.)

[5.3.4] THEOREM / *Existence of the Definite Integral.*

If a function f is continuous on a closed interval $[a, b]$, then the definite integral

$$\int_a^b f(x)\,dx = \lim_{\|P\|\to 0} \sum_{i=1}^{n} f(u_i)\Delta x_i \tag{7}$$

exists.

Two items deserve special mention here. First, f is defined on a *closed* interval, and second, f is *continuous* on that interval. There are some functions that are continuous on an open interval (or even a half-open interval) for which the limit in (7) does not exist. For example, the definite integral $\int_0^1 (1/x^2)\,dx$ does not exist, yet $f(x) = 1/x^2$ is continuous on $(0, 1)$ (and on $(0, 1]$). Also, there are examples of functions that are discontinuous at some numbers in the closed interval $[a, b]$ and yet the limit in (7) does exist (see Problems 17 and 18 in Exercise 5.3).* To summarize, theorem (5.3.4) states that if f is continuous on $[a, b]$, then we are guaranteed that $\int_a^b f(x)\,dx$ exists.

In the event that $\int_a^b f(x)\,dx$ exists, the limit in (6) will exist for any choice of u_i in the ith subinterval. This means that we are free to choose the u_i in any manner we please—such as the left endpoint of each subinterval, or the right

---

* Additional discussion of the definition and existence of $\int_a^b f(x)\,dx$, where f is not continuous on $[a, b]$, is given in Chapter 10.

endpoint, or the midpoint. Furthermore, if the limit in (6) exists, it is independent of the partitions P of the closed interval $[a, b]$, provided $\|P\|$ is close to 0. It is this feature that enables the definite integral to play such an important role in applications to engineering, physics, chemistry, geometry, and economics.

In evaluating the limit in (6), we will usually use a partition that divides the interval $[a, b]$ into n subintervals of the same length. We refer to such a partition as a *regular partition.* For a regular partition, the norm is

$$\|P\| = \frac{b-a}{n}$$

From this relationship, it follows that for a regular partition, the two statements

$$\|P\| \to 0 \qquad \text{and} \qquad n \to +\infty$$

are interchangeable. As a result, we may write

$$\int_a^b f(x)\,dx = \lim_{\|P\|\to 0} \sum_{i=1}^{n} f(u_i)\,\Delta x_i = \lim_{n\to+\infty} \sum_{i=1}^{n} f(u_i)\,\Delta x$$

where $\Delta x = \Delta x_i = (b-a)/n$.

EXAMPLE 1

Evaluate: $\int_0^2 (3x-8)\,dx$

Solution

Since the integrand $f(x) = 3x - 8$ is continuous on the closed interval $[0, 2]$, we know from theorem (5.3.4) that the definite integral exists. To evaluate it, we may use any partition of $[0, 2]$ whose norm can be made as close to 0 as we please, and we may choose any u_i in each subinterval. We elect to use a regular partition of $[0, 2]$, and we will choose the u_i as the right endpoint of each subinterval. As a result, we partition $[0, 2]$ into n subintervals,

$$[0, x_1], \quad [x_1, x_2], \quad \ldots, \quad [x_{i-1}, x_i], \quad \ldots, \quad [x_{n-1}, 2]$$

each of length $\Delta x = 2/n$. The coordinates of the partition, in terms of n, are

$$x_0 = 0, \quad x_1 = \frac{2}{n}, \quad x_2 = 2\left(\frac{2}{n}\right), \quad \ldots, \quad x_i = i\left(\frac{2}{n}\right), \quad \ldots, \quad x_n = n\left(\frac{2}{n}\right) = 2$$

The Riemann sum of $f(x) = 3x - 8$ from 0 to 2, using $u_i = x_i = i(2/n)$, is

$$\begin{aligned}\sum_{i=1}^{n} f(u_i)\,\Delta x_i &= \sum_{i=1}^{n} f(x_i)\,\Delta x_i = \sum_{i=1}^{n} (3x_i - 8)\,\Delta x = \sum_{i=1}^{n} \left[3i\left(\frac{2}{n}\right) - 8\right]\left(\frac{2}{n}\right) \\ &= \left(\frac{12}{n^2}\right)\sum_{i=1}^{n} i - \left(\frac{16}{n}\right)\sum_{i=1}^{n} 1 = \left(\frac{12}{n^2}\right)\frac{n(n+1)}{2} - \left(\frac{16}{n}\right)(n) = -10 + \frac{6}{n}\end{aligned}$$

Therefore,

$$\int_0^2 (3x - 8)\, dx = \lim_{||P|| \to 0} \sum_{i=1}^{n} f(u_i)\Delta x_i = \lim_{n \to +\infty} \left(-10 + \frac{6}{n}\right) = -10$$

Observe that the integrand $f(x) = 3x - 8$ in Example 1 is negative on the interval [0, 2]. As a result, we may *not* interpret $\int_0^2 (3x - 8)\, dx$ as representing an area. The fact that our answer (-10) is not positive is further evidence that we do not have an area problem here. Thus, when looking at a definite integral, do not presume it represents area. As you will see in Section 5.5 and in Chapter 6, the definite integral may have many interpretations.

EXERCISE 5.3

In Problems 1–4 use (3) to calculate each Riemann sum of f for the partition P and the numbers u_i listed.

1. $f(x) = x$, $[0, 2]$; for P: $x_0 = 0$, $x_1 = \frac{1}{4}$, $x_2 = \frac{1}{2}$, $x_3 = \frac{3}{4}$, $x_4 = 1$, $x_5 = \frac{5}{4}$, $x_6 = \frac{3}{2}$, $x_7 = \frac{7}{4}$, $x_8 = 2$; $u_1 = \frac{1}{8}$, $u_2 = \frac{3}{8}$, $u_3 = \frac{5}{8}$, $u_4 = \frac{7}{8}$, $u_5 = \frac{9}{8}$, $u_6 = \frac{11}{8}$, $u_7 = \frac{13}{8}$, $u_8 = \frac{15}{8}$

2. $f(x) = x$, $[0, 2]$; for P: $x_0 = 0$, $x_1 = \frac{1}{2}$, $x_2 = 1$, $x_3 = \frac{3}{2}$, $x_4 = 2$; $u_1 = \frac{1}{2}$, $u_2 = 1$, $u_3 = \frac{3}{2}$, $u_4 = 2$

3. $f(x) = x^2$, $[-2, 1]$; for P: $x_0 = -2$, $x_1 = -1$, $x_2 = 0$, $x_3 = 1$; $u_1 = -\frac{3}{2}$, $u_2 = -\frac{1}{2}$, $u_3 = \frac{1}{2}$

4. $f(x) = x^2$, $[1, 2]$; for P: $x_0 = 1$, $x_1 = \frac{5}{4}$, $x_2 = \frac{3}{2}$, $x_3 = \frac{7}{4}$, $x_4 = 2$; $u_1 = \frac{5}{4}$, $u_2 = \frac{3}{2}$, $u_3 = \frac{7}{4}$, $u_4 = 2$

In Problems 5–10 evaluate each definite integral. Use a regular partition and choose u_i any way you wish.

5. $\int_0^1 (x - 4)\, dx$

6. $\int_0^3 (3x - 1)\, dx$

7. $\int_0^{-4} (2x^2)\, dx$

8. $\int_0^{-1} (x^2 + 1)\, dx$

9. $\int_{-2}^{1} (3x^2 - x)\, dx$

10. $\int_{-1}^{1} (2x^2 + x)\, dx$

In Problems 11–16 find the Riemann sum for the given function and interval. Use a regular partition to divide the interval $[a, b]$ into n subintervals, and always choose u_i as the right endpoint of the ith subinterval $[x_{i-1}, x_i]$. Leave your answer in summation notation.

11. $f(x) = \sqrt{x}$, $[0, 1]$

12. $f(x) = \sqrt{x - 1}$, $[1, 2]$

13. $f(x) = \dfrac{1}{x}$, $[2, 3]$

14. $f(x) = \dfrac{1}{x + 3}$, $[-2, 1]$

15. $f(x) = \dfrac{2}{x^2}$, $[1, 4]$

16. $f(x) = x^{1/3}$, $[0, 8]$

17. The function $f(x) = [\![x]\!]$ is not continuous on [0, 4]. Show that $\int_0^4 f(x)\, dx$ exists.

18. Consider the function f, where

$$f(x) = \begin{cases} 0 & \text{if } x \text{ is rational} \\ 1 & \text{if } x \text{ is irrational} \end{cases}$$

Show that $\int_0^1 f(x)\, dx$ does not exist. [*Hint:* Evaluate the Riemann sum in two different ways: first by using rational numbers for u_i and then by using irrational numbers for u_i.]

19. Use the definition of a definite integral in terms of Riemann sums to prove that $\int_a^b k\, dx = k(b - a)$, where k is a constant.

20. Find an approximate value of $\int_1^2 dx/x$ by computing the Riemann sums corresponding to a partition of [1, 2] into four subintervals of equal length and evaluating the integrand at the midpoint of each subinterval. Compare with the true value, which is 0.6931 . . . (found by methods not yet discussed).

Using Microcomputer Graphics to Investigate Approximations to Definite Integrals*

A Closer Look at the Approximating Sums

In the previous sections we have seen that the definite integral $\int_a^b f(x)\,dx$ may be approximated by the sum

$$\sum_{i=1}^{n} f(u_i)(x_i - x_{i-1})$$

where u_i is some point in the interval $[x_{i-1}, x_i]$. In particular, when an approximate value for the definite integral was desired, specific values were assigned to n and the x_i. Then u_i was chosen to be some distinguished point in the interval $[x_{i-1}, x_i]$, such as the left endpoint, the right endpoint, or the midpoint. From these data, the approximation to the definite integral was made.

In this section we are going to look at the approximating sum from a different perspective. Instead of trying to compute an extremely accurate approximation to a particular definite integral $\int_a^b f(x)\,dx$, we will fix n and investigate the behavior of the approximating sum for the definite integral of f over the interval $[0, x]$ as a function of x. In particular, given a function $f(x)$, we will define a new function,

$$F_N(x) = \frac{x}{N}\left\{f\left[\frac{x}{2N}\right] + f\left[\frac{3x}{2N}\right] + \cdots + f\left[\frac{(2N-1)x}{2N}\right]\right\}$$

that is the approximating sum for the definite integral of f over the interval $[0, x]$ using N subintervals of equal width and choosing the u_i to be the midpoints of the intervals.

For computational convenience, we will choose $N = 4$ for our investigation. Admittedly, the approximations will not be very accurate; however, our goal is to observe the general behavior of the approximation. Thus, given a function $f(x)$, we will consider the function

$$F_4(x) = \frac{x}{4}\left\{f\left[\frac{x}{8}\right] + f\left[\frac{3x}{8}\right] + f\left[\frac{5x}{8}\right] + f\left[\frac{7x}{8}\right]\right\}$$

For the purpose of illustrating the nature of the investigation, consider the function $f(x) = x^3 - x$ whose graph is given on the next page.

Constructing the associated function, $F_4(x)$, we have

$$\begin{aligned} F_4(x) &= \frac{x}{4}\left\{\left[\frac{x^3}{512} - \frac{x}{8}\right] + \left[\frac{27x^3}{512} - \frac{3x}{8}\right] + \left[\frac{125x^3}{512} - \frac{5x}{8}\right] + \left[\frac{343x^3}{512} - \frac{7x}{8}\right]\right\} \\ &= \frac{31x^4 - 64x^2}{128} \end{aligned}$$

* L. Carl Leinbach
Gettysburg College

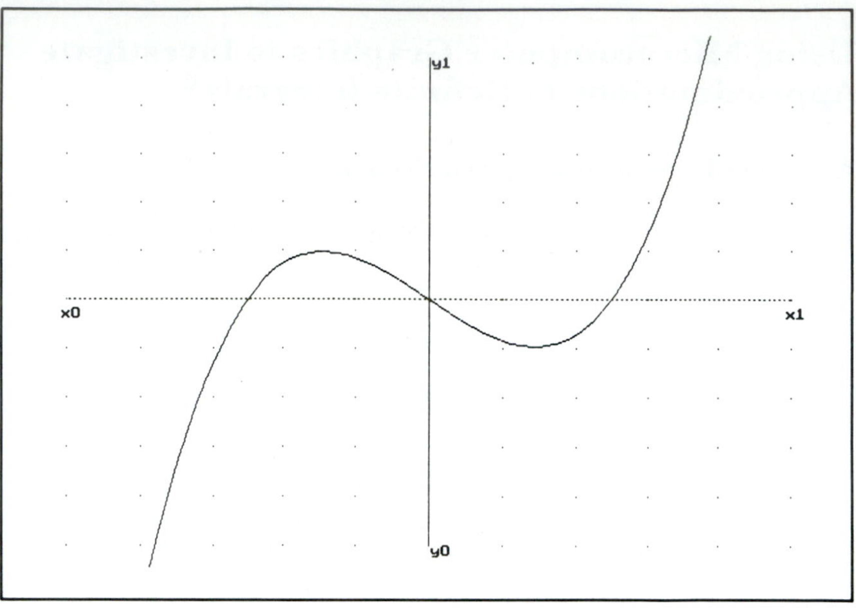

Looking at the graph of $F_4(x)$ in the figure below, we see that it has, as expected, a shape that is fairly typical of a fourth-degree polynomial. In fact, we can easily show that F_N is a fourth-degree polynomial for *any* value of N.

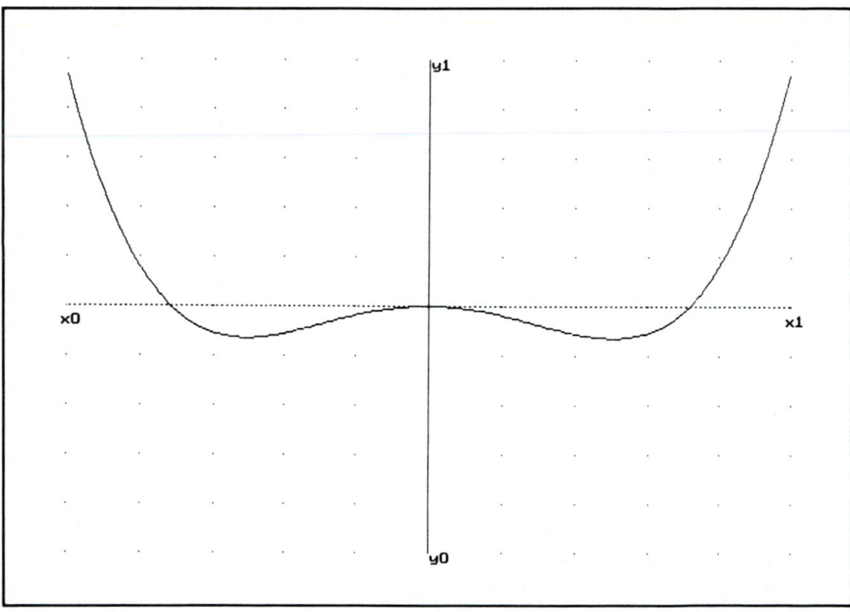

The observation about the degree of $F_4(x)$ is of only mild interest. A much more interesting relationship between $F_N(x)$ and $f(x)$ can be hypothesized when we superimpose the graph of $f(x)$ on the graph of $F_4(x)$, as shown on the next page.

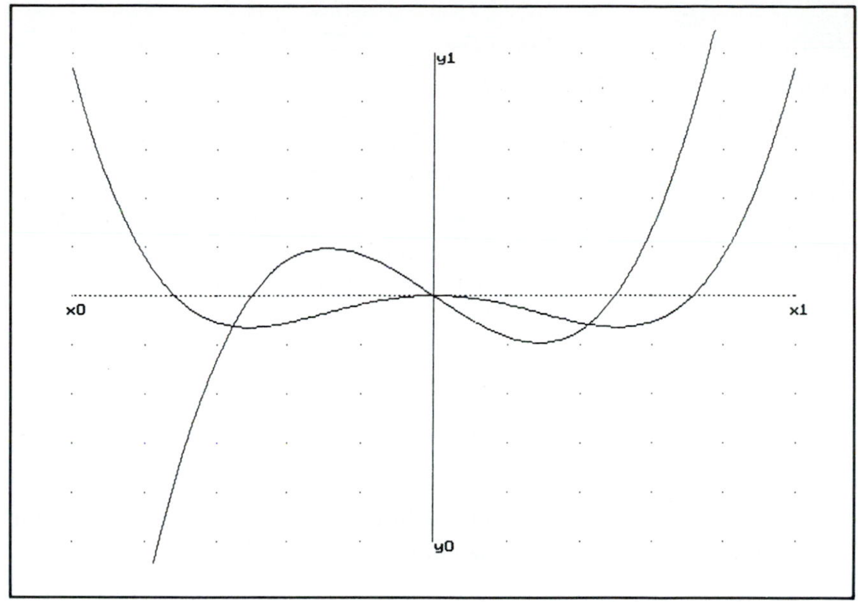

Note that $f(x) = x^3 - x$ has three zeros: $x = -1$, $x = 0$, and $x = 1$. The graph of $F_4(x)$ has minima in the vicinity of $x = -1$ and $x = 1$ (actually at $x = -1.016$ and $x = 1.016$). At $x = 0$ the graph of $F_4(x)$ has a local maximum. If we used larger values for N, we would see that the x-coordinates of the minima of $F_N(x)$ converge to -1 and 1, respectively. In particular, the graphical evidence leads us to make the following observations:

x	Graph of $x^3 - x$	Graph of $F_N(x)$
$x < -1$	Below the x-axis	Decreasing
$x = -1$	Crossing the x-axis	Minimum
$-1 < x < 0$	Above the x-axis	Increasing
$x = 0$	Crossing the x-axis	Local maximum
$0 < x < 1$	Below the x-axis	Decreasing
$x = 1$	Crossing the x-axis	Minimum
$x > 1$	Above the x-axis	Increasing
$x = \frac{1}{\sqrt{3}}$	Local minimum	Inflection point
$x = -\frac{1}{\sqrt{3}}$	Local maximum	Inflection point

In the table the correspondence was made based on the observations from the preceding graph. These observations made without any further theoretical justification do not prove that any particular relationship exists between the function $f(x)$ and the function $F(x) = \int_0^x f(t)\, dt$. They do, however, suggest a very interesting hypothesis for investigation: $f(x)$ is the derived function for $F(x) = \int_0^x f(t)\, dt$—that is, $F'(x) = f(x)$.

This hypothesis is the basis for the investigation in the next section. Its proof is the primary result of this text. A corollary of this result provides a convenient and efficient way to evaluate definite integrals.

EXERCISES

1. Repeat the investigation of this section for $f(x)$ and $F_5(x)$. Superimpose the two graphs and construct a table similar to the one in this section.

2. For the following functions defined on the given interval $[x_0, x_1]$, draw the graphs of $f(x)$ and $F_4(x)$. In each case construct a table. Does the hypothesis still seem to be true?

(Problem 2 continues in the next column)

2. *(continued)*
   (a) $f(x) = \cos(x)$, $[-\pi, \pi]$
   (b) $f(x) = x^4 - x^2$, $[-2, 2]$
   (c) $f(x) = \sin(x^2)$, $[-\pi, \pi]$

3. Consider the graph of $F_4(x)$ when $f(x) = \cos(x)$. What function is suggested by this graph? How does your observation agree with the hypothesis at the bottom of page 354?

5.4 The Fundamental Theorems of Calculus

The following properties of the definite integral are needed for the discussion of the fundamental theorems of calculus:

[5.4.1] THEOREM

If a function f is continuous on an interval containing the numbers a, b, and c, then

$$\int_a^b f(x)\,dx = \int_a^c f(x)\,dx + \int_c^b f(x)\,dx$$

A proof of theorem (5.4.1) is given in Appendix I. In particular, if f is continuous and nonnegative on $[a, b]$ and if c is between a and b, then theorem (5.4.1) has a simple geometric interpretation, as seen in Figure 15.

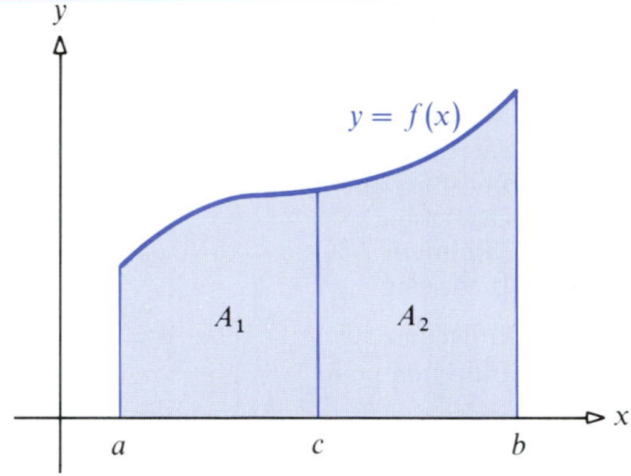

Figure 15

A = Area under f from a to b = $A_1 + A_2$

$$\int_a^b f(x)\,dx = \int_a^c f(x)\,dx + \int_c^b f(x)\,dx$$

[5.4.2] THEOREM

If a function f is continuous on a closed interval $[a, b]$ and if m and M denote the absolute minimum and absolute maximum of f on $[a, b]$, respectively, then

$$m(b-a) \le \int_a^b f(x)\,dx \le M(b-a) \tag{1}$$

A proof of theorem (5.4.2) is given in Appendix I. If f is nonnegative on $[a, b]$, then theorem (5.4.2) may be illustrated geometrically. In Figure 16 the area of the shaded region is $\int_a^b f(x)\,dx$. The area of the smaller rectangle, which has $\text{Width} = (b - a)$ and $\text{Height} = m$, is $m(b - a)$; the area of the larger rectangle, which has $\text{Width} = (b - a)$ and $\text{Height} = M$, is $M(b - a)$. These three areas are numerically related by the inequalities in (1).

Interestingly enough, the area under the graph of f from a to b—namely, $\int_a^b f(x)\,dx$—*is equal to* the area of a certain rectangle of width $(b - a)$ and height $f(u)$ for a special choice (or choices) of u in $[a, b]$ (see Fig. 17). In terms of definite integrals, this result is known as the *mean value theorem for integrals*.

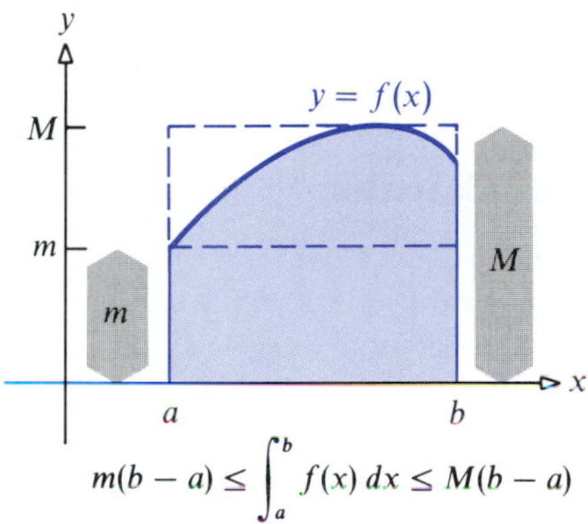

Figure 16

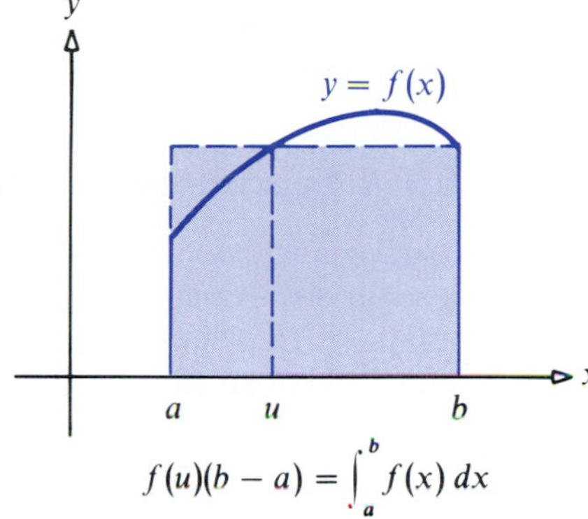

Figure 17

[5.4.3] THEOREM / ***Mean Value Theorem for Integrals.***

If f is continuous on a closed interval $[a, b]$, there exists a number u, $a \leq u \leq b$, so that

$$\int_a^b f(x)\,dx = f(u)(b - a)$$

Note the similarity between theorems (5.4.3) and (4.4.2), the mean value theorem for derivatives. The proof of theorem (5.4.3), which follows, is independent of the geometric interpretation.

Proof If f is a constant function on $[a, b]$, the result is true for any choice of u in $[a, b]$. Suppose f is not identically constant on $[a, b]$. Since f is continuous on the closed interval $[a, b]$, by the extreme value theorem (4.3.3), f attains its extreme values on $[a, b]$. Suppose f assumes its absolute minimum m at the number c so that $f(c) = m$; and suppose f assumes its absolute maximum M at the number C so that $f(C) = M$. Then for all x in $[a, b]$, by theorem (5.4.2) we have

$$m(b - a) \leq \int_a^b f(x)\,dx \leq M(b - a) \qquad \textbf{(2)}$$

If we divide (2) by $(b - a)$ and replace m by $f(c)$ and M by $f(C)$, then

$$f(c) \leq \frac{1}{b - a}\int_a^b f(x)\,dx \leq f(C)$$

Since $[1/(b-a)]\int_a^b f(x)\,dx$ is a number between $f(c)$ and $f(C)$, it follows from the intermediate value theorem (2.4.6) that there is a number u between c and C, and hence in $[a, b]$, so that

$$f(u) = \frac{1}{b-a}\int_a^b f(x)\,dx$$

That is, there is a u, $a \le u \le b$, such that

$$\int_a^b f(x)\,dx = f(u)(b-a)$$

The number u is not necessarily unique. However, the value $f(u)$ in theorem (5.4.3), which is unique, is called the *average value* or *mean value* of f on $[a, b]$. The reasons for this name will be made clear in Section 5.5.

First Fundamental Theorem of Calculus

We begin with a function f that is continuous on a closed interval $[a, b]$. As a result of theorem (5.3.4), the definite integral $\int_a^b f(t)\,dt$ exists and is a real number. If x denotes any number in $[a, b]$, the definite integral $\int_a^x f(t)\,dt$ will also exist, and it will depend on x. That is, $\int_a^x f(t)\,dt$ is a function—say, I—of x; namely,

$$I(x) = \int_a^x f(t)\,dt$$

whose domain is the closed interval $[a, b]$. Note that the integral above has a *variable upper limit* x; the t that appears is a dummy variable.

The relationship between the functions I and f on $[a, b]$ is stated in the next theorem.

[5.4.4] THEOREM / *First Fundamental Theorem of Calculus.*

Let f be a continuous function defined on the closed interval $[a, b]$. The function I defined by

$$I(x) = \int_a^x f(t)\,dt$$

has the property that

$$I'(x) = \frac{d}{dx}\left[\int_a^x f(t)\,dt\right] = f(x) \qquad \text{for all } x \text{ in } [a, b]$$

Proof If x and $(x+h)$, $h \ne 0$, are in $[a, b]$, form the difference

$$I(x+h) - I(x) = \int_a^{x+h} f(t)\,dt - \int_a^x f(t)\,dt \underset{\substack{\uparrow \\ \text{By definition (5.3.3)}}}{=} \int_a^{x+h} f(t)\,dt + \int_x^a f(t)\,dt \underset{\substack{\uparrow \\ \text{By theorem (5.4.1)}}}{=} \int_x^{x+h} f(t)\,dt$$

To form the quotient $\dfrac{I(x+h) - I(x)}{h}$, we divide the last equality by $h \ne 0$.

Then

$$\frac{I(x+h) - I(x)}{h} = \frac{1}{h} \int_x^{x+h} f(t)\, dt \qquad \textbf{(3)}$$

We apply theorem (5.4.3), the mean value theorem for integrals, to the integral on the right. There are two possibilities: $h > 0$ or $h < 0$. If $h > 0$, there exists a u, $x \le u \le x + h$, so that

$$\int_x^{x+h} f(t)\, dt = f(u)h \qquad \text{or} \qquad \frac{1}{h}\int_x^{x+h} f(t)\, dt = f(u)$$

Therefore, from (3),

$$\frac{I(x+h) - I(x)}{h} = f(u) \qquad \textbf{(4)}$$

Suppose we let $h \to 0^+$ in (4). Since $x \le u \le x + h$, as $h \to 0^+$, u must tend to x^+. Thus, $\lim_{h\to 0^+} f(u) = \lim_{u\to x^+} f(u)$. But f is continuous. Therefore $\lim_{u\to x^+} f(u) = f(x)$. Hence,

$$\lim_{h\to 0^+} \frac{I(x+h) - I(x)}{h} = f(x) \qquad \textbf{(5)}$$

Similarly, if $h < 0$, then

$$\lim_{h\to 0^-} \frac{I(x+h) - I(x)}{h} = f(x) \qquad \textbf{(6)}$$

Since the two one-sided limits (5) and (6) are equal, we conclude that

$$\lim_{h\to 0} \frac{I(x+h) - I(x)}{h} = f(x)$$

We recognize this limit as the derivative of the function I. Thus, $I'(x) = f(x)$ for all x in $[a, b]$.

If f is nonnegative, a geometric justification of theorem (5.4.4) may be given, as shown in Figure 18.

We can summarize theorem (5.4.4) as follows: Let f be continuous on $[a, b]$. If $I(x) = \int_a^x f(t)\, dt$, then $I'(x) = f(x)$, $a \le x \le b$. Stated another way,

$$\frac{d}{dx}[I(x)] = \frac{d}{dx}\left[\int_a^x f(t)\, dt\right] = f(x) \qquad \textbf{(7)}$$

EXAMPLE 1

(a) $\displaystyle \frac{d}{dx}\int_0^x \sqrt{t^2+1}\, dt = \sqrt{x^2+1}$ (b) $\displaystyle \frac{d}{dx}\int_2^x \frac{s^3-1}{2s^2+s+1}\, ds = \frac{x^3-1}{2x^2+x+1}$

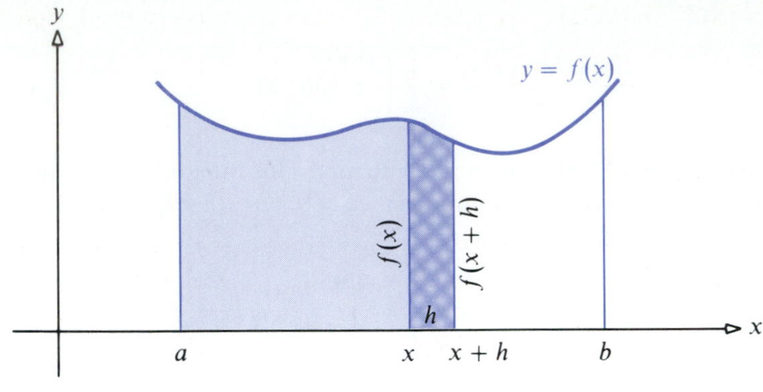

$$\begin{aligned} I(x+h) &= \text{Area from } a \text{ to } x+h \\ I(x) &= \text{Area from } a \text{ to } x \\ I(x+h) - I(x) &= \text{Area from } x \text{ to } x+h \\ \frac{I(x+h) - I(x)}{h} &= \frac{\text{Area from } x \text{ to } x+h}{h} \approx f(x) \quad \text{if } h \text{ is small} \end{aligned}$$

Figure 18

EXAMPLE 2

Find $\displaystyle \frac{d}{dx}\int_4^{3x^2+1} \sqrt{\sqrt{t}+2}\, dt.$

Solution

Here we have to be careful to use the chain rule in conjunction with (7). Let $u = 3x^2 + 1$. Then

$$\frac{d}{dx}\int_4^{3x^2+1} \sqrt{\sqrt{t}+2}\, dt = \frac{d}{dx}\int_4^{u} \sqrt{\sqrt{t}+2}\, dt \underset{\substack{\uparrow \\ \text{Chain rule}}}{=} \left[\frac{d}{du}\int_4^{u} \sqrt{\sqrt{t}+2}\, dt\right]\frac{du}{dx}$$

$$\underset{\substack{\uparrow \\ \text{By (7)}}}{=} \sqrt{\sqrt{u}+2}\cdot\frac{du}{dx} = \sqrt{\sqrt{3x^2+1}+2}\cdot 6x$$

EXAMPLE 3

Find $\displaystyle \frac{d}{dx}\int_{x^3}^{5} (t^4+1)^{1/3}\, dt.$

Solution

To use (7) we first reverse the order of integration and then use the chain rule.

$$\frac{d}{dx}\int_{x^3}^{5} (t^4+1)^{1/3}\, dt = \frac{d}{dx}\left[-\int_5^{x^3} (t^4+1)^{1/3}\, dt\right] = -\frac{d}{dx}\int_5^{x^3} (t^4+1)^{1/3}\, dt$$

$$= -[(x^3)^4+1]^{1/3}\frac{d}{dx}(x^3) = -3x^2(x^{12}+1)^{1/3}$$

Note that in both Examples 2 and 3 the differentiation is with respect to the variable upper limit. Furthermore, observe that the dummy variable plays no role—the answer is a function of the variable upper limit.

Second Fundamental Theorem of Calculus

Since the function I defined in theorem (5.4.4) is differentiable, it follows by theorem (3.2.2) on page 152 that I is also continuous. But theorem (5.4.4) does more than relate the derivative and the definite integral; it also provides a method for calculating definite integrals without using definition (5.3.2). This method is so valuable it is referred to as the *second fundamental theorem of calculus*.

[5.4.5] THEOREM / *Second Fundamental Theorem of Calculus.*

Let f be a continuous function defined on the closed interval $[a, b]$. If F is any antiderivative of f on $[a, b]$, then

$$\int_a^b f(x)\,dx = F(b) - F(a)$$

Proof By theorem (5.4.4), we know that $\int_a^x f(t)\,dt$ is an antiderivative of f. Since F is any antiderivative of f, we may write

$$F(x) = \int_a^x f(t)\,dt + C \tag{8}$$

where C is some constant. (Here we use the fact that if two functions have the same derivative, they differ by a constant.) Now we replace x by a and then replace x by b in (8). The results are

$$F(a) = \int_a^a f(t)\,dt + C \qquad F(b) = \int_a^b f(t)\,dt + C$$

But by definition (5.3.3), $\int_a^a f(t)\,dt = 0$. Hence, by subtracting $F(a)$ from $F(b)$ we obtain

$$F(b) - F(a) = \int_a^b f(t)\,dt$$

Since t is an artificial variable, we may replace t by x, and the result follows.

Thus, if F is an antiderivative of f, we have the formula

$$\int_a^b f(x) = F(b) - F(a)$$

As an aid in computation, we introduce the notation

$$\int_a^b f(x)\,dx = F(x)\Big|_a^b = F(b) - F(a) \tag{9}$$

This notation suggests that we first find an antiderivative $F(x)$ of $f(x)$. Then we write $F(x)|_a^b$ as an aid in computing $F(b) - F(a)$. Specifically, we first replace x by the upper limit b, obtaining $F(b)$, and from this subtract the value $F(a)$ obtained by setting $x = a$.

EXAMPLE 4

(a) $x^2\Big|_{-2}^{1} = 1^2 - (-2)^2 = 1 - 4 = -3$

(b) $(1 - 3x^2)\Big|_{-1}^{5} = [1 - 3(5)^2] - [1 - 3(-1)^2] = -74 - (-2) = -72$

(c) $\sin^2 x\Big|_{0}^{\pi/6} = \left(\sin\frac{\pi}{6}\right)^2 - (\sin 0)^2 = \left(\frac{1}{2}\right)^2 - 0 = \frac{1}{4}$

We observe that for any constant,

$$[F(x) + C]\Big|_{a}^{b} = [F(b) + C] - [F(a) + C] = F(b) - F(a)$$

In other words, it does not matter which of the antiderivatives of f is chosen in applying the second fundamental theorem of calculus, since the same answer is obtained for every antiderivative. Let's look at some more examples.

EXAMPLE 5

Use the second fundamental theorem of calculus to evaluate:

(a) $\int_{-2}^{1} x^2\, dx$ (b) $\int_{0}^{\pi/6} \cos x\, dx$

(c) $\int_{-2}^{1} (x^2 - 2x + 1)\, dx$ (d) $\int_{1}^{2} \frac{x^3 - 4}{x^2}\, dx$ (e) $\int_{-3}^{4} |x|\, dx$

Solution

(a) An antiderivative of x^2 is $x^3/3$. By applying (9) with $F(x) = x^3/3$, we get

$$\int_{-2}^{1} x^2\, dx = \frac{x^3}{3}\Big|_{-2}^{1} = \frac{1}{3} - \frac{(-2)^3}{3} = \frac{1}{3} + \frac{8}{3} = \frac{9}{3} = 3$$

(b) An antiderivative of $\cos x$ is $\sin x$. By applying (9), we get

$$\int_{0}^{\pi/6} \cos x\, dx = \sin x\Big|_{0}^{\pi/6} = \sin\frac{\pi}{6} - \sin 0 = \frac{1}{2}$$

(c) An antiderivative of $x^2 - 2x + 1$ is $(x^3/3) - x^2 + x$. By applying (9), we get

$$\begin{aligned}\int_{-2}^{1} (x^2 - 2x + 1)\, dx &= \left[\frac{x^3}{3} - x^2 + x\right]\Bigg|_{-2}^{1} \\ &= \left[\frac{(1)^3}{3} - (1)^2 + 1\right] - \left[\frac{(-2)^3}{3} - (-2)^2 - 2\right] \\ &= \frac{1}{3} + \frac{26}{3} = \frac{27}{3} = 9\end{aligned}$$

(d) $\displaystyle\int_1^2 \frac{x^3-4}{x^2}\,dx = \int_1^2 \left(x - \frac{4}{x^2}\right) dx = \int_1^2 (x - 4x^{-2})\,dx = \left(\frac{x^2}{2} - \frac{4x^{-1}}{-1}\right)\Bigg|_1^2$

$$= \left(\frac{x^2}{2} + \frac{4}{x}\right)\Bigg|_1^2 = \left[\frac{(2)^2}{2} + \frac{4}{2}\right] - \left(\frac{1}{2} + \frac{4}{1}\right) = 4 - \frac{9}{2} = -\frac{1}{2}$$

(e)
$$|x| = \begin{cases} x & \text{if } x \geq 0 \\ -x & \text{if } x < 0 \end{cases}$$

We use theorem (5.4.1) to split the integral with $c = 0$:

$$\int_{-3}^4 |x|\,dx = \int_{-3}^0 |x|\,dx + \int_0^4 |x|\,dx = \int_{-3}^0 -x\,dx + \int_0^4 x\,dx$$
$$= -\int_{-3}^0 x\,dx + \int_0^4 x\,dx$$
$$= -\left[\frac{x^2}{2}\right]\Bigg|_{-3}^0 + \left[\frac{x^2}{2}\right]\Bigg|_0^4$$
$$= 0 - \left(-\frac{9}{2}\right) + \left(\frac{16}{2} - 0\right) = \frac{25}{2}$$

We have just seen that the second fundamental theorem of calculus provides a short way of evaluating the limit of some Riemann sums—namely,

$$\lim_{\|P\|\to 0} \sum_{i=1}^{n} f(u_i)\Delta x_i = \int_a^b f(x)\,dx = F(b) - F(a)$$

provided the antiderivative F of f has a closed form. This is what is so astonishing about the second fundamental theorem and the reason for its name: Evaluating the limit of a sum and the inverse of the process of differentiation (antidifferentiation) are intimately related.

EXERCISE 5.4

In Problems 1–36 apply the second fundamental theorem of calculus to evaluate each definite integral (a and b are constants).

1. $\displaystyle\int_{-2}^3 dx$
2. $\displaystyle\int_{-2}^3 2\,dx$
3. $\displaystyle\int_{-1}^2 x^7\,dx$
4. $\displaystyle\int_1^3 \frac{dx}{x^7}$

5. $\displaystyle\int_1^2 (3x-1)\,dx$
6. $\displaystyle\int_1^2 (2x+1)\,dx$
7. $\displaystyle\int_0^1 3x^2\,dx$
8. $\displaystyle\int_{-2}^0 (x+x^2)\,dx$

9. $\displaystyle\int_0^1 \sqrt{u}\,du$
10. $\displaystyle\int_1^8 \sqrt[3]{y}\,dy$
11. $\displaystyle\int_0^{\pi/3} \sin x\,dx$
12. $\displaystyle\int_0^{\pi/2} \cos x\,dx$

13. $\displaystyle\int_0^1 (t^2 - t^{3/2})\,dt$
14. $\displaystyle\int_1^4 (\sqrt{x} - a^2x)\,dx$
15. $\displaystyle\int_{-2}^3 (x-1)(x+3)\,dx$
16. $\displaystyle\int_0^1 (z^2+1)^2\,dz$

17. $\int_1^2 \frac{x^2-12}{x^4}\,dx$ **18.** $\int_1^3 \frac{2-x^2}{x^4}\,dx$ **19.** $\int_0^1 (\sqrt[5]{t^2}+1)\,dt$ **20.** $\int_1^4 (\sqrt{u}+a)\,du$

21. $\int_1^4 \frac{x+1}{\sqrt{x}}\,dx$ **22.** $\int_1^9 \frac{\sqrt{x}+1}{x^2}\,dx$ **23.** $\int_0^1 (ax^4+b)\,dx$ **24.** $\int_{-1}^1 (x+1)^3\,dx$

25. $\int_a^b (x+2)^2\,dx$ **26.** $\int_{-1}^0 (x^2+4a)\,dx$ **27.** $\int_{-2}^3 (x+|x|)\,dx$ **28.** $\int_0^3 |x-1|\,dx$

29. $\int_0^2 |3x-1|\,dx$ **30.** $\int_0^2 |2-x|\,dx$

31. $\int_{-1}^1 f(x)\,dx$ where $f(x)=\begin{cases}1 & \text{if } x<0\\ x^2+1 & \text{if } x\ge 0\end{cases}$ **32.** $\int_{-1}^1 f(x)\,dx$ where $f(x)=\begin{cases}x+1 & \text{if } x<0\\ x^2+1 & \text{if } x\ge 0\end{cases}$

33. $\int_{-2}^2 f(x)\,dx$ where $f(x)=\begin{cases}3x & \text{if } -2\le x<0\\ 2x^2 & \text{if } 0\le x\le 2\end{cases}$ **34.** $\int_0^4 h(x)\,dx$ where $h(x)=\begin{cases}x-2 & \text{if } 0\le x\le 2\\ 2-x & \text{if } 2<x\le 4\end{cases}$

35. $\int_{-2}^1 H(x)\,dx$ where $H(x)=\begin{cases}1+x^2 & \text{if } -2\le x<0\\ 1+3x & \text{if } 0\le x\le 1\end{cases}$

36. $\int_{-\pi/2}^{\pi/2} f(x)\,dx$ where $f(x)=\begin{cases}x^2+x & \text{if } -\frac{\pi}{2}\le x\le 0\\ \sin x & \text{if } 0<x<\frac{\pi}{4}\\ \frac{\sqrt{2}}{2} & \text{if } \frac{\pi}{4}\le x<\frac{\pi}{2}\end{cases}$

In Problems 37 and 38 find the area under the graph of $y=f(x)$ from a to b. Draw a graph first.

37. $f(x)=x^2+9$ from $a=-2$ to $b=2$ **38.** $f(x)=x^3+4$ from $a=-1$ to $b=3$

In Problems 39–50 find the indicated derivative by using the first fundamental theorem of calculus.

39. $\frac{d}{dx}\int_1^x \sqrt{t^2+1}\,dt$ **40.** $\frac{d}{dx}\int_3^x \frac{t+1}{t}\,dt$ **41.** $\frac{d}{dt}\left[\int_0^t (3+x^2)^{3/2}\,dx\right]$

42. $\frac{d}{dx}\left[\int_{-4}^x (t^3+8)^{1/3}\,dt\right]$ **43.** $\frac{d}{dx}\left[\int_1^x f(u)\,du\right]$ **44.** $\frac{d}{dt}\left[\int_4^t g(x)\,dx\right]$

45. $\frac{d}{dx}\left[\int_1^{2x^3} \sqrt{t^2+1}\,dt\right]$ **46.** $\frac{d}{dx}\left[\int_1^{\sqrt{x}} \sqrt{t^4+5}\,dt\right]$ **47.** $\frac{d}{dx}\left[\int_2^{x^5} \sec t\,dt\right]$

48. $\frac{d}{dx}\left[\int_3^{1/x} \sin^5 t\,dt\right]$ **49.** $\frac{d}{dx}\left[\int_{x^3}^3 (t^2-5)^{10}\,dt\right]$ **50.** $\frac{d}{dx}\left[\int_{\tan x}^5 \sin(t^6)\,dt\right]$

51. Given that

$f(x)=(2x^3-3)^2$ and $f'(x)=12x^2(2x^3-3)$
use the second fundamental theorem of calculus to find $\int_0^2 12x^2(2x^3-3)\,dx$.

52. Given that

$f(x)=(x^2+5)^3$ and $f'(x)=6x(x^2+5)^2$
use the second fundamental theorem of calculus to find $\int_{-1}^2 6x(x^2+5)^2\,dx$.

In Problems 53–56 find the number(s) u referred to in the mean value theorem for integrals (5.4.3).

53. $\int_0^3 6x^2\,dx$ **54.** $\int_0^2 4x^3\,dx$ **55.** $\int_0^2 (x^2-4)\,dx$ **56.** $\int_0^4 (4-x)\,dx$

In Problems 57–62 use theorem (5.4.2) to obtain a lower estimate and an upper estimate for each of the given integrals.

57. $\int_1^3 (5x + 1)\, dx$

58. $\int_0^1 (1 - x)\, dx$

59. $\int_{\pi/4}^{\pi/2} \sin x\, dx$

60. $\int_{\pi/6}^{\pi/3} \cos x\, dx$

61. $\int_0^1 \sqrt{1 + x^2}\, dx$

62. $\int_{-1}^1 \sqrt{1 + x^4}\, dx$

63. If the interval $[1, 5]$ is divided into eight subintervals of equal length, what is the largest Riemann sum of $f(x) = x^2$ that can be computed using this partition? The smallest? Compute the average of these sums. What integral has been approximated and what is its exact value?

64. For all real b find $\int_0^b |2x|\, dx$.

65. Use integration to find the area enclosed by the graph of $y = 3 - |x|$ and the x-axis. Check by elementary geometry.

66. If $f(x) = \int_0^x 1/\sqrt{t^3 + 2}\, dt$, which of the following is *false*?

(a) $f(0) = 0$
(b) f is continuous at x for all $x \geq 0$
(c) $f(1) > 0$
(d) $f'(1) = 1/\sqrt{3}$
(e) $f(-1) > 0$

67. Find $(d/dx) \int_x^1 (t - 1)^2\, dt$ without integrating. Then check by integrating before differentiating.

68. If u and v are differentiable functions and f is a continuous function, develop a formula for

$$\frac{d}{dx}\left[\int_{u(x)}^{v(x)} f(x)\, dt\right]$$

69. If f is continuous on $[a, b]$, show that the functions defined by

$$F(x) = \int_c^x f(t)\, dt \qquad G(x) = \int_d^x f(t)\, dt$$

for any choice of c and d in (a, b), always differ by a constant, and also show that

$$F(x) - G(x) = \int_c^d f(t)\, dt$$

70. If f'' is continuous on $[a, b]$, show that

$$\int_a^b xf''(x)\, dx = bf'(b) - af'(a) - f(b) + f(a)$$

[*Hint:* Look at the derivative of the function $F(x) = xf'(x) - f(x)$.]

71. If f' is continuous in $[a, b]$, show that

$$\int_a^b f(x)f'(x)\, dx = \tfrac{1}{2}\{[f(b)]^2 - [f(a)]^2\}$$

Properties of the Definite Integral; Average Value of a Function 5.5

Properties of the Definite Integral

In this section we list several properties of the definite integral that are useful. Although you already know some of them, we list them again for completeness.

[5.5.1] THEOREM

If a function f is continuous on an interval containing a, b, and c, then

$$\int_a^b f(x)\, dx = \int_a^c f(x)\, dx + \int_c^b f(x)\, dx \qquad (1)$$

This, of course, is a duplication of theorem (5.4.1).

EXAMPLE 1

If f is continuous on the closed interval [2, 7], then

$$\int_2^7 f(x)\,dx = \int_2^4 f(x)\,dx + \int_4^7 f(x)\,dx$$

The choice of c in (1) does not have to lie between a and b.

EXAMPLE 2

If g is continuous on the closed interval [3, 25], then

$$\int_3^{10} g(x)\,dx = \int_3^{25} g(x)\,dx + \int_{25}^{10} g(x)\,dx$$

[5.5.2] THEOREM

If the functions f and g are continuous on the closed interval $[a, b]$, then

$$\int_a^b [f(x) + g(x)]\,dx = \int_a^b f(x)\,dx + \int_a^b g(x)\,dx \qquad (2)$$

EXAMPLE 3

$$\int_0^1 (x^2 + x)\,dx = \int_0^1 x^2\,dx + \int_0^1 x\,dx$$

[5.5.3] THEOREM

If a function f is continuous on a closed interval $[a, b]$ and if k is a constant, then

$$\int_a^b kf(x)\,dx = k\int_a^b f(x)\,dx \qquad (3)$$

EXAMPLE 4

$$\int_0^1 3x^2\,dx = 3\int_0^1 x^2\,dx$$

By repeated use of (2) and (3), we have the extended result:

[5.5.4] THEOREM

If the functions $f_1, f_2, \ldots, f_n$ are continuous on a closed interval $[a, b]$ and if $k_1, k_2, \ldots, k_n$ are constants, then

$$\int_a^b [k_1 f_1(x) + k_2 f_2(x) + \cdots + k_n f_n(x)]\,dx$$

$$= k_1 \int_a^b f_1(x)\,dx + k_2 \int_a^b f_2(x)\,dx + \cdots + k_n \int_a^b f_n(x)\,dx \qquad (4)$$

[5.5.5] THEOREM

If a function f is continuous on a closed interval $[a, b]$ and if $f(x) \geq 0$ on $[a, b]$, then

$$\int_a^b f(x)\,dx \geq 0 \tag{5}$$

This property asserts the geometric fact that the area under the graph of a nonnegative function is never negative.

[5.5.6] THEOREM

If the functions f and g are continuous on a closed interval $[a, b]$ and if $f(x) \leq g(x)$ on $[a, b]$, then

$$\int_a^b f(x)\,dx \leq \int_a^b g(x)\,dx \tag{6}$$

A proof of (1)–(6) will require the use of definition (5.3.2). (See Appendix I for such a proof of (1).) However, if we assume that each of the integrands has an antiderivative, then we can use theorem (5.3.4) and the second fundamental theorem of calculus (5.4.5). We shall prove (2) and (5) here with this added assumption.

Proof of (2) Since f and g have antiderivatives F and G respectively on $[a, b]$, then $F' = f$ and $G' = g$ on $[a, b]$. Further, $F + G$ is an antiderivative of $f + g$ on $[a, b]$, since $(F + G)' = F' + G' = f + g$. Consequently,

$$\begin{aligned}\int_a^b [f(x) + g(x)]\,dx &= [F(x) + G(x)]\Big|_a^b = [F(b) + G(b)] - [F(a) + G(a)] \\ &= [F(b) - F(a)] + [G(b) - G(a)] \\ &= \int_a^b f(x)\,dx + \int_a^b g(x)\,dx\end{aligned}$$

Proof of (5) Since $f(x) \geq 0$ on $[a, b]$ and f has an antiderivative F, then

$$F'(x) = f(x) \geq 0 \qquad \text{on } [a, b]$$

Thus, F is a non-decreasing function on $[a, b]$, which implies that $F(b) \geq F(a)$. Consequently,

$$\int_a^b f(x)\,dx = F(b) - F(a) \geq 0$$

Average Value of a Function

We now examine how the definite integral can be applied to calculate averages. For example, at the U.S. Weather Bureau, continuous readings of the temperature over a 24 hour period are taken daily. To obtain the average daily temperature, 12 readings may be taken at 2 hour intervals beginning at midnight: $f(0), f(2), f(4), \ldots, f(20), f(22)$. The average temperature is then calculated as

$$\frac{f(0)+f(2)+f(4)+\cdots+f(20)+f(22)}{12}$$

This number represents a good approximation to the true average as long as there are no drastic temperature changes over short periods of time. To improve the approximation, readings may be taken every hour. The average in this case would be

$$\frac{f(0)+f(1)+\cdots+f(22)+f(23)}{24}$$

An even better approximation would be obtained if readings were recorded every half hour.

In general, if $y=f(x)$ is a continuous function defined on the interval $[a, b]$, we can obtain the *average of f on* $[a, b]$ as follows: We partition the closed interval $[a, b]$ into n subintervals

$$[a, x_1], \quad [x_1, x_2], \quad \ldots, \quad [x_{i-1}, x_i], \quad \ldots, \quad [x_{n-1}, b]$$

each of length $\Delta x=(b-a)/n$. We pick a number u_i in the ith subinterval $[x_{i-1}, x_i]$. An approximation of the average value of f over the interval $[a, b]$ is then the sum

$$\frac{f(u_1)+f(u_2)+\cdots+f(u_n)}{n} \tag{7}$$

If we multiply and divide the expression in (7) by $(b-a)$, we get

$$\begin{aligned}
\frac{f(u_1)+f(u_2)+\cdots+f(u_n)}{n} \\
&= \frac{1}{b-a}\left[f(u_1)\frac{b-a}{n}+f(u_2)\frac{b-a}{n}+\cdots+f(u_n)\frac{b-a}{n}\right] \\
&= \frac{1}{b-a}[f(u_1)\Delta x+f(u_2)\Delta x+\cdots+f(u_n)\Delta x] \\
&= \frac{1}{b-a}\sum_{i=1}^{n} f(u_i)\Delta x
\end{aligned}$$

The sum obtained gives an approximation to the average value. As the length of each subinterval gets smaller and smaller, this sum becomes a better and better approximation to the average value of f on $[a, b]$. However, this sum is a Riemann sum, so that its limit is a definite integral. This suggests the following definition:

[5.5.7] DEFINITION / ***Average Value of a Function over an Interval.***

The average value $\bar{y}$ of a continuous function f over $[a, b]$ is

$$\bar{y}=\frac{1}{b-a}\int_a^b f(x)\,dx \tag{8}$$

EXAMPLE 5

The average value of $f(x) = x^3$ over the interval $[0, 2]$ is

$$\bar{y} = \frac{\int_0^2 x^3\,dx}{2-0} = \frac{4}{2} = 2$$

The average value $\bar{y}$ of a function f, as defined in (8), equals the value $f(u)$ referred to in theorem (5.4.3), the mean value theorem for integrals. Let's review the geometric interpretation.

We begin by rearranging the formula for $\bar{y}$ as

$$\bar{y}(b-a) = \int_a^b f(x)\,dx \tag{9}$$

If $f(x) \geq 0$ on $[a, b]$, the right side of (9) represents the area enclosed by the graph of $y = f(x)$, the x-axis, the line $x = a$, and the line $x = b$. The left side of the equation can be interpreted as the area of a rectangle of height $\bar{y}$ and base $(b - a)$. Hence, theorem (5.4.3) asserts that $\bar{y}$, the average value of the function, is the height of a rectangle whose base is $(b - a)$ and whose area is equal to the area under the graph of f (see Fig. 19).

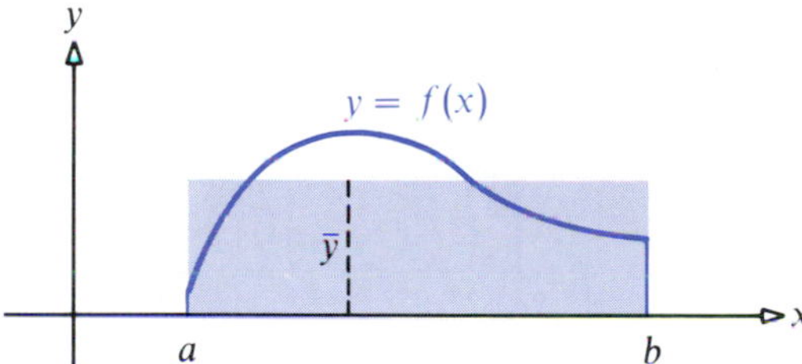

Figure 19

The next example illustrates that in computing averages it is important to indicate the variable with respect to which the average is to be computed.

EXAMPLE 6

For a freely falling body starting from rest, calculate:

(a) The average velocity $\bar{v}_t$ with respect to time over $[0, t_1]$
(b) The average velocity $\bar{v}_s$ with respect to distance over $[0, s_1]$, where s_1 is the distance the body falls in time t_1

Solution

(a) Since we wish to compute the average velocity $\bar{v}_t$ relative to time t, we need to express the velocity v as a function of t. The distance is $s = \frac{1}{2}gt^2$, so it follows that

$$v(t) = \frac{ds}{dt} = gt$$

Then

$$\bar{v}_t = \frac{1}{t_1 - 0}\int_0^{t_1} v(t)\,dt = \frac{1}{t_1}\int_0^{t_1} gt\,dt = \left(\frac{1}{t_1}\right)\frac{1}{2}gt^2\Big|_0^{t_1} = \frac{1}{2}gt_1 = \frac{1}{2}v(t_1)$$

(b) To find the average velocity $\bar{v}_s$ relative to distance s, we note that $s = \frac{1}{2}gt^2$, so that $t = \sqrt{2s/g}$ and $v(s) = g\sqrt{2s/g} = \sqrt{2gs}$. Then,

$$\begin{aligned}\bar{v}_s &= \frac{1}{s_1 - 0}\int_0^{s_1} v(s)\,ds = \frac{1}{s_1}\int_0^{s_1} \sqrt{2g}\sqrt{s}\,ds \\ &= \frac{\sqrt{2g}}{s_1}\left(\frac{s^{3/2}}{\frac{3}{2}}\right)\Big|_0^{s_1} = \frac{2\sqrt{2g}}{3}s_1^{1/2} = \frac{2}{3}\sqrt{2gs_1} = \frac{2}{3}v(s_1)\end{aligned}$$

EXERCISE 5.5

In Problems 1–4 use properties of the definite integral to verify each statement. Assume that all integrals involved exist.

1. $\int_3^{11} f(x)\,dx - \int_7^{11} f(x)\,dx = \int_3^{7} f(x)\,dx$

2. $\int_{-2}^{6} f(x)\,dx - \int_3^{6} f(x)\,dx = \int_{-2}^{3} f(x)\,dx$

3. $\int_0^{4} f(x)\,dx - \int_6^{4} f(x)\,dx = \int_0^{6} f(x)\,dx$

4. $\int_{-1}^{3} f(x)\,dx - \int_5^{3} f(x)\,dx = \int_{-1}^{5} f(x)\,dx$

In Problems 5–10 evaluate each definite integral if it is known that $\int_1^3 f(x)\,dx = 5$, $\int_1^3 g(x)\,dx = -2$, $\int_3^5 f(x)\,dx = 2$, $\int_3^5 g(x)\,dx = 1$.

5. $\int_1^3 [f(x) + g(x)]\,dx$

6. $\int_1^3 [f(x) - g(x)]\,dx$

7. $\int_1^3 [5f(x) - 3g(x)]\,dx$

8. $\int_1^3 [3f(x) + 4g(x)]\,dx$

9. $\int_1^5 [2f(x) - 3g(x)]\,dx$

10. $\int_1^5 [f(x) - g(x)]\,dx$

11. Without evaluating the definite integral, show that $\int_0^1 x\,dx \ge \int_0^1 x^3\,dx$.

12. Without evaluating the definite integral, show that $\int_1^2 x^3\,dx \ge \int_1^2 x\,dx$.

13. Prove (3). Assume f has an antiderivative on $[a, b]$.

14. Prove (4). Assume $f_1, f_2, \ldots, f_n$ have antiderivatives on $[a, b]$.

15. Prove (6). Assume f and g have an antiderivative on $[a, b]$.

16. Prove that if f is continuous on $[a, b]$, then

$$\left|\int_a^b f(x)\,dx\right| \le \int_a^b |f(x)|\,dx$$

What is the geometric interpretation?

17. If $f(x) = \begin{cases} x + 1 & \text{if } x < 0 \\ x^2 + 1 & \text{if } x \ge 0 \end{cases}$, find $\int_{-1}^{1} f(x)\,dx$.

18. If $f(x) = \begin{cases} 1 & \text{if } x < 0 \\ x^2 + 1 & \text{if } x \ge 0 \end{cases}$, find $\int_{-1}^{1} f(x)\,dx$.

In Problems 19–28 find the average value of each function f over the given interval.

19. $f(x) = x^2$, over $[0, 1]$

20. $f(x) = 2x^2$, over $[-4, 2]$

21. $f(x) = 1 - x^2$, over $[-1, 1]$

22. $f(x) = 16 - x^2$, over $[-4, 4]$

23. $f(x) = 3x$, over $[1, 5]$

24. $f(x) = 4x$, over $[-5, 5]$

25. $f(x) = -5x^4 + 4x - 10$, over $[-2, 2]$

26. $f(x) = 10x^4 - 2x + 7$, over $[-1, 2]$

27. $f(x) = \sin x$, over $[0, \pi/2]$

28. $f(x) = \cos x$, over $[0, \pi/2]$

29. Find the average velocity with respect to time of a freely falling object that falls from rest for 5 seconds. What is its average velocity with respect to distance? (Use the International System of Metric Units.)

30. If the object in Problem 29 falls from rest for 3 seconds, what is its average velocity with respect to time? What is its average velocity with respect to distance? (Use the International System of Metric Units.)

31. A rod 3 meters long is heated to $25x$ degrees Celsius, where x is the distance (in meters) from one end of the rod. Calculate the average temperature of the rod.

32. The rainfall per day, x days after the beginning of the year, is $0.00002(6511 + 366x - x^2)$, measured in centimeters. By integration, estimate the average daily rainfall for the first 180 days of the year.

33. A car starting from rest accelerates at the rate of 3 meters per second per second. Find its average speed over the first 8 seconds.

34. What is the average area of all circles whose radii are between 1 and 3 meters?

35. The mass density of a metal bar of length 3 meters is given by $\rho(x) = 1000 + x - \sqrt{x}$ kilograms per cubic meter, where x is the distance in meters from one end of the bar. What is the average mass density over the length of the entire bar?

36. The acceleration at time t of a particle moving on the x-axis is $4\pi \cos t$. If the velocity is 0 at $t = 0$, what is the average velocity of the particle over the interval $0 \le t \le \pi$?

37. Use the definition of average value of a function to find the *average slope* of the graph of $y = f(x)$, $a \le x \le b$. (Assume that f' is continuous.) What is the geometric interpretation?

38. What theorem guarantees that the average slope found in Problem 37 is equal to $f'(u)$ for some u in $[a, b]$? What *different* theorem guarantees the same thing? (The connection between these theorems should now be apparent!)

The Indefinite Integral; Method of Substitution 5.6

Through the second fundamental theorem of calculus, we discovered an intimate relationship between definite integrals and antiderivatives; namely, the definite integral $\int_a^b f(x)\, dx$ can be evaluated in many cases by finding an antiderivative of f. Because of this, it has become customary to use the integral symbol $\int$ as a direction to find all antiderivatives of a function. That is:

The notation $\int f(x)\, dx$ is used to denote all antiderivatives of f, and we refer to $\int f(x)\, dx$ as the *indefinite integral of f*.

For example,

$$\int x^2\, dx = \frac{x^3}{3} + C \qquad \int (x^2 + 1)\, dx = \frac{x^3}{3} + x + C$$

where C is a constant called the *constant of integration*.

The process of evaluating either the indefinite integral $\int f(x)\, dx$ or the definite integral $\int_a^b f(x)\, dx$ is called *integration* and the function f is called the *integrand*. It is important to distinguish between the definite integral $\int_a^b f(x)\, dx$ and the indefinite integral $\int f(x)\, dx$. The definite integral is a number that depends on the limits of integration a and b. On the other hand, the indefinite integral of f is defined as a collection of functions $F(x) + C$, C a constant, such that $F'(x) = f(x)$. For example,

$$\int_0^2 x^2\, dx = \frac{x^3}{3}\bigg|_0^2 = \frac{8}{3} \qquad \int x^2\, dx = \frac{x^3}{3} + C$$

Again, we remind you that an antiderivative F of f and the definite integral $\int_a^b f(x)\,dx$ of f are related by the second fundamental theorem of calculus—namely,

$$\int_a^b f(x)\,dx = F(x)\Big|_a^b = F(b) - F(a)$$

Properties of the Indefinite Integral

We now state several rules concerning indefinite integrals. The first is a consequence of the definition of $\int f(x)\,dx$ as all antiderivatives of f:

$$\frac{d}{dx}\int f(x)\,dx = f(x) \qquad (1)$$

For example,

$$\frac{d}{dx}\int \sqrt{x^2+1}\,dx = \sqrt{x^2+1}$$

Also, we have the following property:

$$\int kf(x)\,dx = k\int f(x)\,dx \qquad \textbf{where } k \textbf{ is a real number} \qquad (2)$$

This formula states that to find the indefinite integral of a constant k times a function f, we first find the indefinite integral of f and then multiply by k. To prove this statement, we differentiate the right side of (2):

$$\frac{d}{dx}\left[k\int f(x)\,dx\right] = k\left[\frac{d}{dx}\int f(x)\,dx\right] = kf(x)$$

Another property is

$$\int [f(x) + g(x)]\,dx = \int f(x)\,dx + \int g(x)\,dx \qquad (3)$$

This formula states that the indefinite integral of a sum equals the sum of the indefinite integrals.

The next formula is a consequence of (3) in Section 3.9:

$$\int x^r\,dx = \frac{x^{r+1}}{r+1} + C \qquad r \neq -1 \quad \textbf{a rational number} \qquad (4)$$

This is easily verified, since

$$\frac{d}{dx}\left(\frac{x^{r+1}}{r+1} + C\right) = x^r$$

EXAMPLE 1

$$\int (2x^{1/3} + 5x^{2/3})\,dx = \int 2x^{1/3}\,dx + \int 5x^{2/3}\,dx = 2\int x^{1/3}\,dx + 5\int x^{2/3}\,dx$$

$$= \frac{2x^{4/3}}{\frac{4}{3}} + \frac{5x^{5/3}}{\frac{5}{3}} + C = \frac{3x^{4/3}}{2} + 3x^{5/3} + C$$

Sometimes an appropriate algebraic manipulation is required before (4) can be used.

EXAMPLE 2

$$\int\left(\frac{12}{x^5}+\frac{1}{\sqrt{x}}\right)dx = 12\int\frac{1}{x^5}\,dx+\int\frac{1}{\sqrt{x}}\,dx = 12\int x^{-5}\,dx+\int x^{-1/2}\,dx$$

$$= 12\left(\frac{x^{-4}}{-4}\right)+\frac{x^{1/2}}{\frac{1}{2}}+C = \frac{-3}{x^4}+2x^{1/2}+C$$

Now, since the differentiation formulas for $\sin x$ and $\cos x$ are

$$\frac{d}{dx}\sin x = \cos x \qquad \text{and} \qquad \frac{d}{dx}\cos x = -\sin x$$

we have the following integral formulas:

$$\int \cos x\,dx = \sin x + C \qquad \int \sin x\,dx = -\cos x + C \tag{5}$$

Method of Substitution; Change of Variable

Indefinite integrals that cannot be evaluated directly by using (4) and (5) may sometimes be evaluated by the *substitution technique.* This technique involves the introduction of a function that changes the integrand into one to which the formulas apply. For example, to evaluate $\int(x^2+5)^3 2x\,dx$, we use the substitution $u = x^2+5$ so that $du = 2x\,dx$:

$$\int(x^2+5)^3 2x\,dx = \int u^3\,du = \frac{u^4}{4}+C = \frac{(x^2+5)^4}{4}+C$$

Note that to make the substitution method work, the differentials must also be substituted for. Of course, once a substitution is chosen, its differential is determined. We may verify the correctness of the answer as follows:

$$\frac{d}{dx}\left[\frac{(x^2+5)^4}{4}+C\right] = \frac{4(x^2+5)^3(2x)}{4} = (x^2+5)^3 2x$$

A Justification of the Substitution Method

A justification for the substitution method is based on the chain rule, which states that

$$f'(g(x))g'(x) = \frac{d}{dx}f(g(x))$$

This formula provides us with a rule for integration—namely,

$$\int f'(g(x))g'(x)\,dx = f(g(x)) + C \qquad C \text{ a constant}$$

If we let $u = g(x)$ in the above and note that $du = g'(x)\,dx$, we have

$$\int f'(g(x))g'(x)\,dx = \int f'(u)\,du = f(u) + C = f(g(x)) + C$$

In other words, if we replace $g(x)$ by u and $g'(x)\,dx$ by du in $\int f'(g(x))g'(x)\,dx$, we obtain an easily evaluated integral, $\int f'(u)\,du$. This replacement of $g(x)$ by u is called the *substitution method*, since we substitute u for $g(x)$. Sometimes this technique is referred to as the *change of variable method*, since the variable u has replaced the variable x.

EXAMPLE 3

Evaluate:

(a) $\int (2x+1)^2\,dx$ (b) $\int \dfrac{dx}{\sqrt{5-2x}}$ (c) $\int \sin(3x+2)\,dx$

Solution

(a) Let $u = 2x + 1$. Then $du = 2\,dx$, so that $dx = du/2$. Hence,

$$\int (2x+1)^2\,dx = \int u^2\left(\frac{du}{2}\right) = \frac{1}{2}\int u^2\,du = \frac{1}{2}\left(\frac{u^3}{3}\right) + C = \frac{(2x+1)^3}{6} + C$$

(b) Let $u = 5 - 2x$. Then $du = -2\,dx$, so that $dx = du/(-2)$. Hence,

$$\int \frac{dx}{\sqrt{5-2x}} = \int \frac{du/(-2)}{\sqrt{u}} = -\frac{1}{2}\int \frac{du}{\sqrt{u}}$$

$$= -\frac{1}{2}\int u^{-1/2}\,du = -\frac{1}{2}\left(\frac{u^{1/2}}{\frac{1}{2}}\right) + C = -\sqrt{5-2x} + C$$

(c) Let $u = 3x + 2$. Then $du = 3\,dx$, so that $dx = du/3$. Hence,

$$\int \sin(3x+2)\,dx = \int (\sin u)\frac{du}{3} = \frac{1}{3}\int \sin u\,du$$

$$= \left(\frac{1}{3}\right)(-\cos u) + C = \left(-\frac{1}{3}\right)\cos(3x+2) + C$$

EXAMPLE 4

Evaluate: $\int x\sqrt{x^2+1}\,dx$

Solution

We use the substitution $u = x^2 + 1$. Then $du = 2x\,dx$, so $x\,dx = du/2$. Hence,

$$\int x\sqrt{x^2+1}\,dx = \int \sqrt{x^2+1}\;x\,dx = \int \sqrt{u}\,\frac{du}{2} = \frac{1}{2}\int u^{1/2}\,du$$

$$= \frac{1}{2}\left(\frac{u^{3/2}}{\frac{3}{2}}\right) + C = \frac{(x^2+1)^{3/2}}{3} + C$$

Note that in using the substitution $u = x^2 + 1$ in Example 4 we must substitute not only for the integrand $x\sqrt{x^2 + 1}$, but also for the differential dx. In addition, the existence of x as part of the integrand makes the substitution $u = x^2 + 1$ work. For example, if we used this same substitution to evaluate $\int \sqrt{x^2 + 1}\, dx$, we would obtain $u = x^2 + 1$, $du = 2x\, dx$. Since $x = \sqrt{u - 1}$, we have

$$\int \sqrt{x^2 + 1}\, dx = \int \sqrt{u} \frac{du}{2\sqrt{u - 1}} = \int \frac{\sqrt{u}}{2\sqrt{u - 1}}\, du$$

In this case, the substitution we used results in an integrand that does not conform to (4). In fact, it results in an integrand that is *more* complicated than the original one.

Thus, the idea behind the substitution method is to obtain an integral $\int h(u)\, du$ that is simpler than the original integral $\int f(x)\, dx$. When a substitution does not simplify the integral, other substitutions should be tried. If these do not work, other integration methods should be applied (we will discuss some of these other techniques in Chap. 9). Thus, since integration—unlike differentiation—has no prescribed method, some ingenuity and a lot of practice are required.

To illustrate that the integral has no prescribed method for evaluation, we use two different substitutions in the next example.

EXAMPLE 5

Evaluate: $\displaystyle\int x\sqrt{4 + x}\, dx$

Solution

Substitution I Let $u = 4 + x$. Then $du = dx$ and $x = u - 4$, so that

$$\int x\sqrt{4 + x}\, dx = \int (u - 4)\sqrt{u}\, du = \int (u^{3/2} - 4u^{1/2})\, du$$

$$= \frac{u^{5/2}}{\frac{5}{2}} - \frac{4u^{3/2}}{\frac{3}{2}} + C = \frac{2(4 + x)^{5/2}}{5} - \frac{8(4 + x)^{3/2}}{3} + C$$

Substitution II Let $u = \sqrt{4 + x}$, so that $u^2 = 4 + x$. Then we have $2u\, du = dx$, $x = u^2 - 4$, and

$$\int x\sqrt{4 + x}\, dx = \int (u^2 - 4)(u)(2u\, du) = 2\int (u^4 - 4u^2)\, du$$

$$= \frac{2u^5}{5} - \frac{8u^3}{3} + C = \frac{2(4 + x)^{5/2}}{5} - \frac{8(4 + x)^{3/2}}{3} + C$$

EXAMPLE 6

Evaluate: $\displaystyle\int_0^2 x\sqrt{4 - x^2}\, dx$

Solution

First we evaluate the indefinite integral $\int x\sqrt{4-x^2}\,dx$ using the substitution $u = 4 - x^2$. Then $du = -2x\,dx$ and

$$\int x\sqrt{4-x^2}\,dx = \int \sqrt{4-x^2}\,x\,dx = \int \sqrt{u}\left(-\frac{du}{2}\right)$$
$$= -\frac{1}{2}\left(\frac{u^{3/2}}{\frac{3}{2}}\right) + C = \frac{-(4-x^2)^{3/2}}{3} + C$$

By the fundamental theorem, we find

$$\int_0^2 x\sqrt{4-x^2}\,dx = \left.\frac{-(4-x^2)^{3/2}}{3}\right|_0^2 = 0 + \frac{4^{3/2}}{3} = \frac{8}{3}$$

Changing Limits of Integration

When the substitution method is used to evaluate definite integrals, it is sometimes easier to change the limits of integration. For example, in order to evaluate $\int_0^2 x\sqrt{4-x^2}\,dx$ in Example 6, we let $u = 4 - x^2$ and $du = -2x\,dx$. But now we use the equation $u = 4 - x^2$ to change the limits of integration. Thus, when $x = 0$ (the *old* lower limit of integration), we have $u = 4 - 0^2 = 4$ (the *new* lower limit of integration). Similarly, when $x = 2$ (the old upper limit of integration), we have $u = 0$ (the new upper limit of integration). By using this technique,

$u = 4 - x^2 = 4 - 2^2 = 0$

$$\int_0^2 x\sqrt{4-x^2}\,dx = \int_4^0 \sqrt{u}\left(-\frac{du}{2}\right) = -\frac{1}{2}\left.\left(\frac{u^{3/2}}{\frac{3}{2}}\right)\right|_4^0 = -\frac{1}{3}(0 - 4^{3/2}) = \frac{8}{3}$$

$u = 4 - x^2 = 4 - 0 = 4$

A word to the wise! This second way of evaluating a definite integral may be faster, but it also requires more care. Be certain not to forget to change the limits of integration when using this method.

EXAMPLE 7

Evaluate: $\displaystyle\int_0^{\pi/2} \frac{1-\cos 2\theta}{2}\,d\theta$

Solution

$$\int_0^{\pi/2} \frac{1-\cos 2\theta}{2}\,d\theta = \frac{1}{2}\int_0^{\pi/2} (1-\cos 2\theta)\,d\theta = \frac{1}{2}\int_0^{\pi/2} d\theta - \frac{1}{2}\int_0^{\pi/2} \cos 2\theta\,d\theta$$

In the second integral on the right, make the substitution $u = 2\theta$. Then $du = 2\,d\theta$ and

$$\int_0^{\pi/2} \frac{1-\cos 2\theta}{2}\,d\theta = \frac{1}{2}\theta\Big|_0^{\pi/2} - \frac{1}{2}\int_0^{\pi} (\cos u)\,\frac{du}{2}$$
$$= \left(\frac{1}{2}\right)\left(\frac{\pi}{2}\right) - \left(\frac{1}{4}\right)\sin u\Big|_0^{\pi} = \frac{\pi}{4}$$

The next theorem describes a use for the substitution method for definite integrals of functions that are even or odd.

[5.6.1] THEOREM / *Integration of Even and Odd Functions.*

Let f be continuous on $[-a, a]$, where $a > 0$.

(a) If f is an even function, then

$$\int_{-a}^{a} f(x)\,dx = 2\int_{0}^{a} f(x)\,dx$$

(b) If f is an odd function, then

$$\int_{-a}^{a} f(x)\,dx = 0$$

Proof We prove property (a) and leave the proof of property (b) as an exercise (Problem 97). Now using theorem (5.5.1), we have

$$\int_{-a}^{a} f(x)\,dx = \int_{-a}^{0} f(x)\,dx + \int_{0}^{a} f(x)\,dx = -\int_{0}^{-a} f(x)\,dx + \int_{0}^{a} f(x)\,dx \quad \textbf{(6)}$$

In $-\int_0^{-a} f(x)\,dx$, we make the substitution $u = -x$; then $du = -dx$. Also when $x = -a$, $u = a$. Therefore,

$$-\int_{0}^{-a} f(x)\,dx = \int_{0}^{a} f(-u)\,du$$

Since f is even, we know that $f(u) = f(-u)$. Thus,

$$-\int_{0}^{-a} f(x)\,dx = \int_{0}^{a} f(u)\,du = \int_{0}^{a} f(x)\,dx \quad \textbf{(7)}$$

Combining (6) and (7), we have

$$\int_{-a}^{a} f(x)\,dx = \int_{0}^{a} f(x)\,dx + \int_{0}^{a} f(x)\,dx = 2\int_{0}^{a} f(x)\,dx$$

In using theorem (5.6.1), two conditions must be met:

1. f must be even or odd.
2. The integration must take place from $-a$ to a, $a > 0$.

EXAMPLE 8

Evaluate: $\displaystyle\int_{-3}^{3} (x^7 - 4x^3 + x)\,dx$

Solution

Since

$$f(-x) = (-x)^7 - 4(-x)^3 + (-x) = -(x^7 - 4x^3 + x) = -f(x)$$

f is an odd function. By theorem (5.6.1), we have

$$\int_{-3}^{3} (x^7 - 4x^3 + x)\,dx = 0$$

EXERCISE 5.6

In Problems 1–56 evaluate each indefinite integral.

1. $\int 6\,dx$
2. $\int 3\,dx$
3. $\int 3x\,dx$
4. $\int 6x^2\,dx$
5. $\int t^{-4}\,dt$
6. $\int u^{-1/2}\,du$
7. $\int (x^2+2)\,dx$
8. $\int (3x^3-1)\,dx$
9. $\int (x^7+1)\,dx$
10. $\int (4\sqrt{x}+1)\,dx$
11. $\int (4x^3-3x^2+5x-2)\,dx$
12. $\int (3x^5-2x^4-x^2-1)\,dx$
13. $\int (3\sqrt{z}+z)\,dz$
14. $\int (4t^{3/2}+t^{1/2})\,dt$
15. $\int \left(x-\frac{1}{x^2}\right)dx$
16. $\int \left(3x^2-\frac{1}{\sqrt{x}}\right)dx$
17. $\int u(u-1)\,du$
18. $\int t^2(t+1)\,dt$
19. $\int \frac{3x^5+1}{x^2}\,dx$
20. $\int \frac{x^2+2x+1}{x^4}\,dx$
21. $\int \frac{t^2-4}{t-2}\,dt$
22. $\int \frac{z^3-8}{z-2}\,dz$
23. $\int (2x+1)^5\,dx$
24. $\int (1-3x)^3\,dx$
25. $\int x\sqrt{x^2-9}\,dx$
26. $\int (1-t^2)^6 t\,dt$
27. $\int \frac{x}{\sqrt{1+x^2}}\,dx$
28. $\int \frac{x}{\sqrt[5]{1-3x^2}}\,dx$
29. $\int x\sqrt{x+3}\,dx$
30. $\int x\sqrt{4-x}\,dx$
31. $\int \sin 3x\,dx$
32. $\int (1-\cos 4x)\,dx$
33. $\int x\sin x^2\,dx$
34. $\int \sin x\cos^2 x\,dx$
35. $\int x^2\sqrt{x+1}\,dx$
36. $\int x^2\sqrt{3x-1}\,dx$
37. $\int \frac{x}{\sqrt{1+x}}\,dx$
38. $\int \frac{x}{\sqrt{x+2}}\,dx$
39. $\int (s-5)^{1/2}\,ds$
40. $\int \frac{2x+1}{(x^2+x-5)^2}\,dx$
41. $\int \frac{1}{\sqrt{x}(1+\sqrt{x})^4}\,dx$
42. $\int \frac{x+4x^3}{\sqrt{x}}\,dx$
43. $\int (x-5)\sqrt{x+1}\,dx$
44. $\int \frac{x^3}{\sqrt{2x+1}}\,dx$
45. $\int \sqrt{t}\sqrt{4+t\sqrt{t}}\,dt$
46. $\int \frac{z\,dz}{z+\sqrt{z^2+4}}$
47. $\int [\sqrt{(z^2+1)^4-3}][(z^2+1)^3]z\,dz$
48. $\int \frac{x\sin\sqrt{2x^2-7}}{\sqrt{2x^2-7}}\,dx$
49. $\int \sin^5 x\cos x\,dx$
50. $\int \frac{\sin\sqrt{x}}{\sqrt{x}}\,dx$
51. $\int \cos^4 x\sin x\,dx$
52. $\int \tan^2 x\sec^2 x\,dx$
53. $\int x^2\sin(1-x^3)\,dx$
54. $\int \cos(3-x)\,dx$
55. $\int \sin(2x+5)\,dx$
56. $\int \sin x\sin(\cos x)\,dx$

In Problems 57–76 evaluate each definite integral.

57. $\int_{-2}^{0} \frac{x}{(x^2+3)^2}\,dx$
58. $\int_{-1}^{1} (s^2-1)^5 s\,ds$
59. $\int_{-1/3}^{0} x\sqrt[3]{3x+1}\,dx$
60. $\int_{0}^{26} x\sqrt[3]{x+1}\,dx$

61. $\int_6^1 x\sqrt{x+3}\,dx$

62. $\int_6^2 x^2\sqrt{x-2}\,dx$

63. $\int_2^6 \frac{4}{\sqrt{3x-2}}\,dx$

64. $\int_0^4 \frac{x}{\sqrt{x^2+9}}\,dx$

65. $\int_{-1}^7 x(1+x)^{1/3}\,dx$

66. $\int_1^3 \frac{1}{x^2}\sqrt{1-\frac{1}{x}}\,dx$

67. $\int_0^{\pi/3} \sin\frac{x}{2}\,dx$

68. $\int_0^{\pi/6} \cos 2x\,dx$

69. $\int_0^{\sqrt{\pi}} 6x\cos x^2\,dx$

70. $\int_{-\pi/4}^{\pi} \sin\theta\cos\theta\,d\theta$

71. $\int_{\pi/12}^{\pi/9} \sec^2 3\theta\,d\theta$

72. $\int_0^{\pi/4} \frac{\sin 2x}{\sqrt{5-2\cos 2x}}\,dx$

73. $\int_{-\pi/2}^{\pi/2} \cos(2x+\pi)\,dx$

74. $\int_{-\pi/4}^{\pi/4} \sin 7\theta\,d\theta$

75. $\int_0^1 \frac{(z^2+5)(z^3+15z-3)\,dz}{\sqrt{194-(z^3+15z-3)^2}}$

76. $\int_2^{17} \frac{dx}{\sqrt{\sqrt{x-1}+(x-1)^{5/4}}}$

77. If $\int_1^b t^2(5t^3-1)^{1/2}\,dt = \frac{38}{45}$, find b.

78. If $\int_a^3 t\sqrt{9-t^2}\,dt = 6$, find a.

79. Find the area under the graph of $f(x) = \sin x$ from 0 to $\pi/2$.

80. Find the area under the graph of $f(x) = \cos x$ from $-\pi/2$ to $\pi/2$.

81. Find the area under the graph of $f(x) = \sqrt{x+4}$ from 0 to 5.

82. Find the area under the graph of $f(x) = x^2/\sqrt{2x+1}$ from 0 to 4.

83. Find the area under the graph of $f(x) = \sqrt{x-1}$ from 1 to 2.

84. Find the area under the graph of $f(x) = x/(x^2+1)^2$ from 0 to 2.

In Problems 85 and 86 evaluate each indefinite integral by both (a) using the substitution method and (b) expanding the integrand. Compare the two results. How do the constants of integration differ?

85. $\int (x+1)^2\,dx$

86. $\int (x^2+1)^2 x\,dx$

87. Verify that $\int x\sqrt{x}\,dx \neq (\int x\,dx)(\int \sqrt{x}\,dx)$.

88. Verify that $\int x(x^2+1)\,dx \neq x\int (x^2+1)\,dx$.

89. Verify that $\int \frac{x^2-1}{x-1}\,dx \neq \frac{\int (x^2-1)\,dx}{\int (x-1)\,dx}$.

90. The electric field strength a distance z out along the axis of a ring of radius R carrying a charge Q is given by the formula

$$E(z) = \frac{Qz}{(R^2+z^2)^{3/2}}$$

If the electric potential V is related to E by $E = -dV/dz$, what is $V(z)$?

91. For the function given below find $\int_{-1}^1 f(x)\,dx$.

$$f(x) = \begin{cases} x+1 & \text{if } x < 0 \\ \cos\pi x & \text{if } x \geq 0 \end{cases}$$

92. If f is continuous on $[a, b]$, show that $\int_a^b f(x)\,dx = \int_a^b f(a+b-x)\,dx$.

93. Use an appropriate substitution to show that $\int_0^{\pi/2} \sin^n\theta\,d\theta = \int_0^{\pi/2} \cos^n\phi\,d\phi$.

94. Use an appropriate substitution to show that $\int_0^1 x^m(1-x)^n\,dx = \int_0^1 x^n(1-x)^m\,dx$.

95. If $\int_0^2 f(x-3)\,dx = 8$, find $\int_{-3}^{-1} f(x)\,dx$.

96. If f is a continuous function defined on the interval $[0, 1]$, show that

$$\int_0^{\pi} xf(\sin x)\,dx = \frac{\pi}{2}\int_0^{\pi} f(\sin x)\,dx$$

97. If f is an odd function, show that $\int_{-a}^a f(x)\,dx = 0$.

Historical Perspectives

The fundamental idea of integral calculus can be traced back more than 2000 years to the Greek method of "exhaustion," which is described in Euclid's *Tenth Book*. The method is illustrated in Problems 6–8 at the end of this section. The great Greek mathematician Archimedes (287–212 BC) made extensive use of this method to establish many theorems on areas and volumes, but the method could not be fully developed until the discovery of algebra and its relationship to geometry, which was not to come until several hundred years later in the late sixteenth and early seventeenth centuries.

One of the first to rediscover and exploit the method of exhaustion was the Italian mathematician Bonaventura Cavalieri (1598–1647), who developed an *infinitesimal* method (characterized by the use of infinitesimal, or infinitely small, length, area, and volume units) for finding areas and volumes (see Problems 4 and 5). At about the same time, many others, including the German astronomer Johannes Kepler (1571–1630), the French mathematicians Fermat and Roberval, and the Italian mathematician and astronomer Evangelista Torricelli (1608–1647), were developing their own infinitesimal methods for finding areas and volumes. All of these methods built upon the Greek geometric method of exhaustion by adding analytic or algebraic features to the analysis.

However, old prejudices die hard and not every mathematician of the seventeenth century was willing to abandon the classical Greek methods and embrace the new algebraic methods. An important example is provided by Newton's teacher, Sir Isaac Barrow (1630–1677). Barrow vigorously opposed the new approach and pushed the old geometric methods further than anyone before him. In fact, he discovered the second fundamental theorem of calculus in geometric form, and if he had been more receptive to the new methods, he might well have gained recognition as the founder of calculus. But, in 1669, he gave up his notes and his position as professor at Cambridge to Newton in order to devote his time to the study of theology, leaving that distinction to his successor.

HISTORICAL EXERCISES

The problem of finding the area of a plane region was suggested to Cavalieri by his teacher, Galileo. In his work on falling objects, Galileo had recognized that the distance an object fell could be interpreted as the area under the graph of its velocity function. In Problems 1–3 we explore in some simple cases the relationship between velocity and area. (Keep in mind the intimate relationship between velocity and the derivative.) Note that Problems 2 and 3 provide a kind of "proof" of the second fundamental theorem for the special case $\int_0^{t_0} kt\,dt$.

1. Suppose an object moves in a straight line with a constant velocity (speed) of 4 feet per second.

(a) How far does the object travel in 5 seconds? Graph the velocity function v over the interval $0 \le t \le 5$ and interpret the distance traveled as an area.

(b) Shade the area that represents the distance traveled during the third second.

2. Suppose an object that was originally at rest is moving in a straight line and is accelerating uniformly (a falling object, for example), so that its velocity is being increased by a constant amount each second; that is, the velocity function is of the form $v(t) = kt$.

(*Problem 2 continues on page 381*)

2. *(continued)*

(a) Graph the particular velocity function $v(t) = 3t$ over the interval $0 \le t \le 4$. What is the velocity at the start of the interval? At the end? At the midpoint? What is the acceleration; that is, how much does the velocity increase each second?

(b) It can be shown that under the assumption of a uniform rate of acceleration, the average velocity over any interval equals the velocity at the midpoint of the interval. That is, the object will end up at the same point it would have if it had had a constant velocity over the entire interval equal to its velocity at the midpoint. Verify this for the velocity function in part (a).

(c) Graph the velocity function $v_1 = 6t$ over $0 \le t \le 4$. On the same axes graph the constant velocity function v_2, which represents the average velocity over this interval. Calculate the areas under the two velocity graphs. How far did the object travel during the interval?

3. Consider the accompanying figure.

(a) Verify that (no matter what k and t_0 are) the area of triangle ABE equals the area of $ABCD$, by directly calculating each area.

(b) Verify the equality in part (a) by integrating to get the area of the triangle.

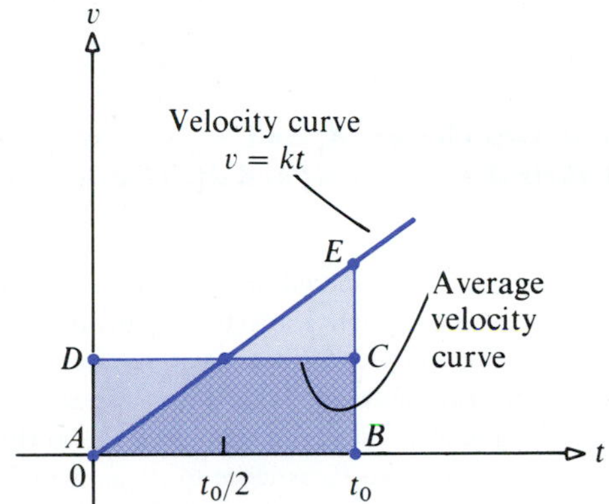

Cavalieri studied under Galileo and became a professor at Bologna in 1629, a position he held until his death. Credit for the first general theorem of calculus goes to Cavalieri. He derived (in our notation) the theorem

$$\int_0^a x^n\,dx = \frac{a^{n+1}}{n+1}$$

for positive integers n, a result that had been demonstrated for some special cases as far back as Archimedes. Although his theories were not complete or rigorously defensible, in 1635 Cavalieri published the first textbook in what we would now call *integral calculus*.

4. In this exercise we give a simple application of Cavalieri's method of *indivisibles* to "prove" a theorem from elementary geometry. Consider the parallelogram shown here. The theorem from geometry is: Area of $\triangle ABC$ = Area of $\triangle ADC$. Cavalieri would have considered $\triangle ABC$ to be generated by the infinite set of lines (like OP) parallel to AD that we get by letting point O move from A to B. He would have called these lines the *indivisibles* that make up the triangle. State an argument to "prove" the theorem after considering the diagram and thinking about what can be said about OP and $O'P'$ when $AO = CO'$. (Prove your conclusion concerning OP and $O'P'$.)

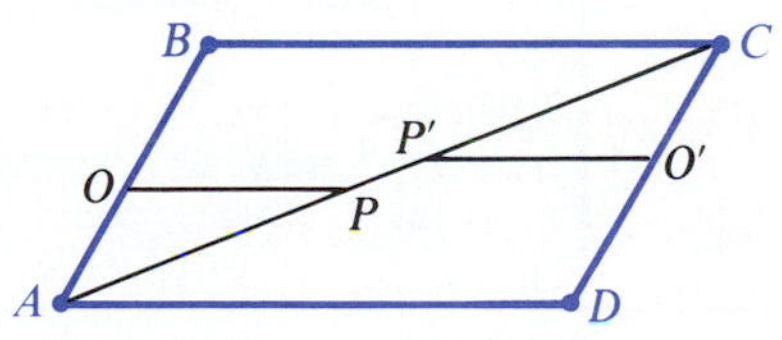

5. A well-known theorem of elementary geometry is due to Cavalieri and is often called *Cavalieri's principle*. Referring to the figure, the theorem asserts that if $f(x) - g(x)$ and $h(x) - k(x)$ are in a constant proportion—say, $f(x) - g(x) = M[h(x) - k(x)]$—for all x between a and b, then Area $A = M$(Area B). Cavalieri's proof of this theorem occupies several pages in his textbook on his method of indivisibles. Give a one-line proof using integration.

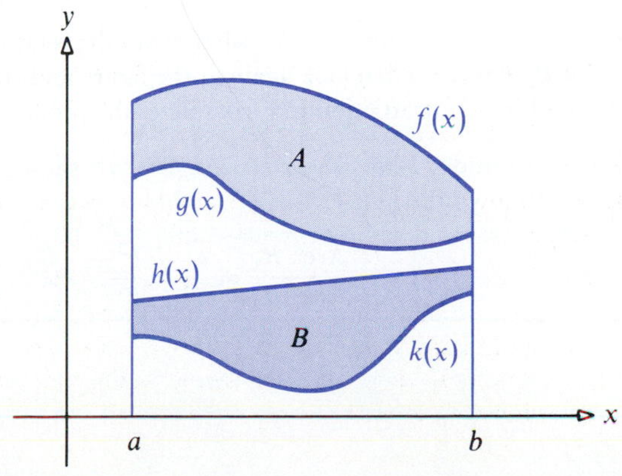

To illustrate the Greek method of exhaustion, we sketch the proof of a theorem in geometry by using that method. Problems 6–8 complete the proof and provide a proof of a corollary to the theorem. The method of exhaustion was usually used in giving indirect proofs, which managed to avoid any need for an argument involving infinite processes. This kind of indirect proof, in which we argue to a contradiction from an assumption and hence conclude that the assumption is false, is called *reductio ad absurdum.*

THEOREM

The areas of two circles, A_1 and A_2, are in the same proportion as the squares of their diameters, d_1^2 and d_2^2. That is, $A_1/A_2 = d_1^2/d_2^2$.

Proof Consider the figures shown here. Let us assume that $A_1/A_2 \neq d_1^2/d_2^2$. We will show that this assumption leads to a contradiction. Now, consider a circle, different from the first circle, with area A so that $A/A_2 = d_1^2/d_2^2$, and suppose $A < A_1$ (a similar argument will work if we suppose $A_1 < A$). Then we inscribe in the circle with area A_1, a polygon with area P_1, so that $A < P_1 < A_1$. We can do this because we can make P_1 as close to A_1 as we wish by doubling the number of sides of the polygon as many times as needed (this is the core idea of the method of exhaustion—notice the lack of the need for an infinite process here). Now, if we inscribe a similar polygon with area P_2 inside the circle with area A_2, we can show that $P_1/P_2 = d_1^2/d_2^2$ (see Problem 7). But $d_1^2/d_2^2 = A/A_2$ and so $P_1/P_2 = A/A_2$. However, $P_1 > A$, so what can we conclude about the relative sizes of P_2 and A_2 (see Problem 6)?

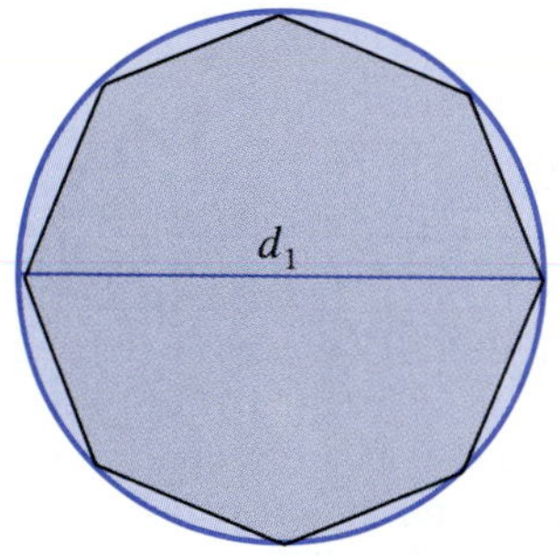

Circle area $= A_1$
Diameter $= d_1$
Polygon area $= P_1$

Circle area $= A$

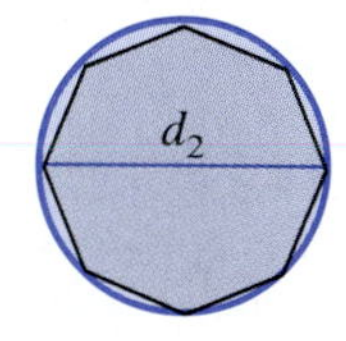

Circle area $= A_2$
Diameter $= d_2$
Polygon area $= P_2$

6. Draw the appropriate conclusion about the relative sizes of P_2 and A_2. Then look again at the figure and state the resulting contradiction that completes the proof.

7. (a) Consider two *similar rectangles*, as shown here. Show that

$$\frac{\text{Area } R_1}{\text{Area } R_2} = \frac{a^2}{A^2} = \frac{b^2}{B^2}$$

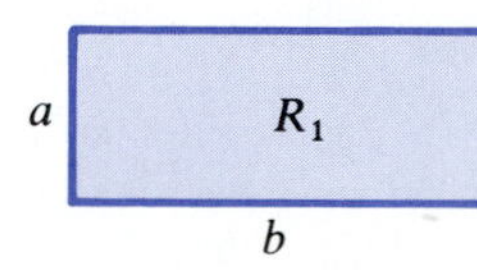

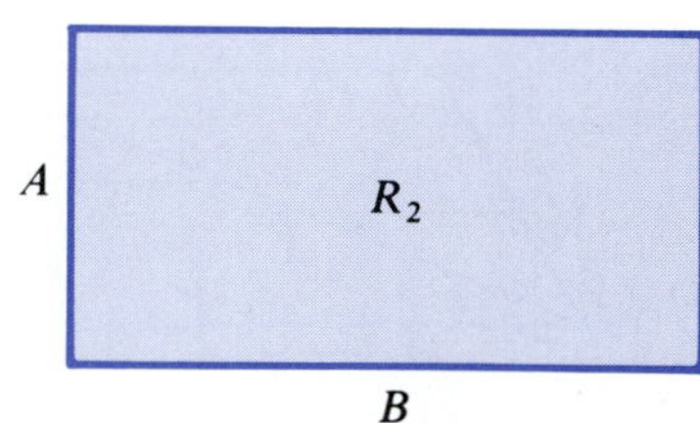

(*Problem 7 continues on page 383*)

7. *(continued)*

(b) Consider two *similar triangles*, as shown here. Show that

$$\frac{\text{Area } T_1}{\text{Area } T_2} = \frac{b^2}{B^2} = \frac{h^2}{H^2}$$

(c) Use part (b) to argue that $P_1/P_2 = d_1^2/d_2^2$ in the above proof.

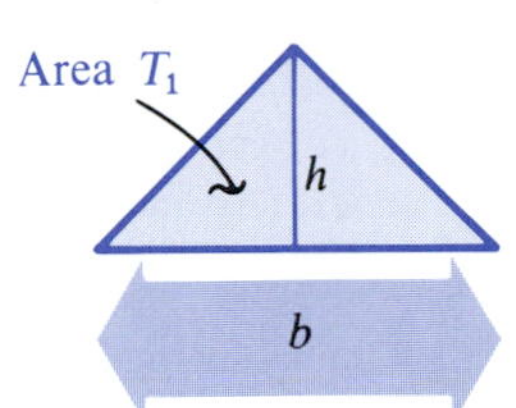

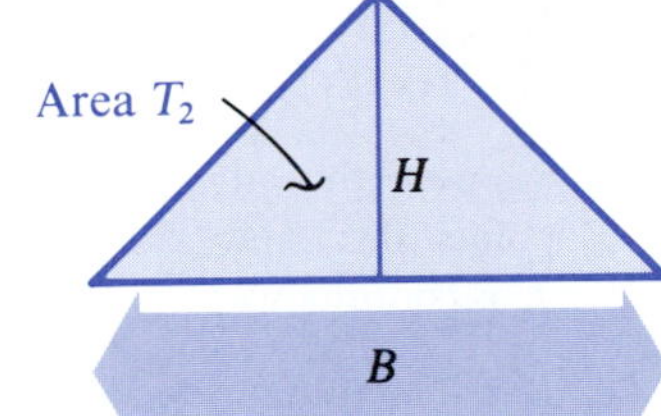

8. Use the theorem proved in Problem 7 to prove that for any circle, the area equals a constant C times the radius squared, and C is the same constant for all circles. (Of course, we usually call C by the name π.)

REVIEW EXERCISES

In Problems 1–11, evaluate each integral.

1. $\int \sqrt[n]{a + bx}\, dx$; a, b real, $b \neq 0$

2. $\int \sqrt[3]{x^3 + 3\cos x}\,(x^2 - \sin x)\, dx$

3. $\int \left(2\sqrt{x^2+3} - \frac{4}{x} + 9\right)^6 \left(\frac{x}{\sqrt{x^2+3}} + \frac{2}{x^2}\right) dx$

4. $\int \frac{x^3\, dx}{\sqrt{x^2 - 9}}$

5. $\int \frac{x^6 + 3x^4 + 3x^2 + x + 1}{(x^2+1)^2}\, dx$

6. $\int \frac{y\, dy}{(y-2)^3}$

7. $\int \frac{x}{(2-3x)^3}\, dx$

8. $\int \sqrt{\frac{1+x}{x^5}}\, dx$

9. $\int_{\frac{\pi^2}{4}}^{4\pi^2} \frac{1}{\sqrt{x}} \sin\sqrt{x}\, dx$

10. $\int_0^1 \frac{1}{\sqrt{x}} \cos \frac{\pi\sqrt{x}}{2}\, dx$

11. $\int_1^{\sqrt[3]{1/2}} \frac{[1 - (1/t^3)]^3}{t^4}\, dt$

12. Find the approximate value of $\int_0^2 \sqrt{4 - x^2}\, dx$ by computing the Riemann sums corresponding to a partition of $[0, 2]$ into 16 subintervals of equal length and evaluating the integrand at the left endpoint of each subinterval. Round to the nearest thousandth.

13. Compute: $\lim_{n \to +\infty} \sum_{i=1}^{n} \frac{2n}{(n+2i)^2}$

14. Suppose that P is a partition of $[0, \pi/2]$ into n subintervals and u_i is an arbitrary point of the typical subinterval $[x_{i-1}, x_i]$, $i = 1, 2, \ldots, n$. Explain why

$$\lim_{\|P\| \to 0} \sum_{i=1}^{n} (\cos u_i)\Delta x_i = 1$$

15. The interval $[0, 4]$ is divided into n subintervals of equal length Δx and a point u_i is chosen in the typical subinterval $[x_{i-1}, x_i]$, $i = 1, 2, \ldots, n$. What is the limit of $\sum_{i=1}^{n} \sqrt{u_i}\, \Delta x$ as $n \to +\infty$?

16. Find k, $0 \le k \le 3$, so that

$$\int_0^3 \frac{x}{\sqrt{x^2+16}}\, dx = \frac{3k}{\sqrt{k^2+16}}$$

17. It can be shown (with a certain amount of work) that if $f(x)$ is integrable on $[a, b]$, then so is $|f(x)|$. Is the converse true?

18. Let $f(x) = x^4 - 2x^3 - x^2 + 2x$ on $[-1, 2]$. Find the area enclosed by the graph of $f(x)$ and the x-axis.

19. If n is a nonnegative integer, for what n does $\int_0^1 x^n\,dx = \int_0^1 (1-x)^n\,dx$?

20. Given the function f defined for all real numbers x by $f(x) = 2|x-1|x^2$:

(a) What is the range of the function?
(b) For what x is the function continuous?
(c) For what x is the derivative of f continuous?
(d) Determine the value of $\int_0^1 f(x)\,dx$.

21. Let $f(x) = x^3 - 6x^2 + 11x - 6$. Find $\int_1^3 |f(x)|\,dx$.

22. Describe a method for computing $\int_a^b |f(x)|\,dx$ in terms of $F(x) = \int f(x)\,dx$ when $f(x)$ has finitely many zeros.

23. If F is a function whose derivative is continuous for all real x, find

$$\lim_{h\to 0} \frac{1}{h}\int_c^{c+h} F'(x)\,dx$$

24. Let $f(x) = k\sin(kx)$, where k is a positive constant.

(a) Find the area of the region enclosed by one arch of the graph of f and the x-axis.
(b) Find the area of the triangle formed by the x-axis and the tangents to one arch of f at the points where the graph of f crosses the x-axis.

25. Let $f(x) = \displaystyle\int_0^x \frac{dt}{\sqrt{1-t^2}}$. Which is false?

(a) f is odd.
(b) f is continuous.
(c) f has domain $[-1, 1]$.
(d) f is increasing.

26. Let $f(x) = \displaystyle\int_0^x \frac{dt}{\sqrt{1-t^2}}$.

(a) Find $\dfrac{d}{dx} f(\sin x)$.
(b) Is f one-to-one? (See Problem 25.)
(c) Does f have an inverse?

27. Find $f''(x)$ if $f(x) = \int_0^x \sqrt{1-t^2}\,dt$.

28. Suppose that $F(x) = \int_0^x \sqrt{t}\,dt$ and $G(x) = \int_1^x \sqrt{t}\,dt$. Explain why $F(x) - G(x)$ is constant and find the constant.

29. Given $y = \sqrt{x^2-1}(4-x)$, $1 \le x \le a$, for what number a will $\int_1^a y\,dx$ have a maximum value?

30. If $\int_1^2 f(x-c)\,dx = 5$, where c is a constant, find $\int_{1-c}^{2-c} f(x)\,dx$.

31. Suppose that the graph of $y = f(x)$ contains the points $(0, 1)$ and $(2, 5)$. Find $\int_0^2 f'(x)\,dx$. (Assume that f' is continuous.)

32. If f is continuous for all x, which of the following integrals necessarily have the same value?

I. $\displaystyle\int_a^b f(x)\,dx$ **II.** $\displaystyle\int_0^{b-a} f(x+a)\,dx$

III. $\displaystyle\int_{a+c}^{b+c} f(x+c)\,dx$

33. Let $a < c < b$ and let f be differentiable on $[a, b]$. Which of the following is *not* necessarily true?

(a) $\displaystyle\int_a^b f(x)\,dx = \int_a^c f(x)\,dx + \int_c^b f(x)\,dx$
(b) There exists d in $[a, b]$ such that

$$f'(d) = \frac{f(b) - f(a)}{b - a}.$$

(c) $\displaystyle\int_a^b f(x)\,dx \ge 0$
(d) $\lim_{x\to c} f(x) = f(c)$
(e) If k is a real number, then $\int_a^b kf(x)\,dx = k\int_a^b f(x)\,dx$.

34. Find the area under the graph of $y = 1/\sqrt{x}$ from $x = 1$ to $x = r$ (where $r > 1$). Then examine the behavior of this area as $r \to +\infty$.

35. Find the area under the graph of $y = 1/x^2$ from $x = 1$ to $x = r$ (where $r > 1$). Then examine the behavior of this area as $r \to +\infty$.

In Problems 36 and 37 use theorem (5.4.2) to find lower and upper estimates for each integral.

36. $\displaystyle\int_{-4}^{-1} (2x^3 + 9x^2 + 12x + 32)\,dx$

37. $\displaystyle\int_0^3 x^2(x^2-1)^{1/3}\,dx$

38. If n is a known positive integer, for what number c is $\int_1^c x^{n-1}\,dx = 1/n$?

39. Approximate $\int_{-1}^3 (x^3 - 3x^2 + 3)\,dx$ using a regular partition with four subintervals. Choose u_i as the number at which $(x^3 - 3x^2 + 3)$ assumes its minimum value in the ith subinterval.

40. The formula $(d/dx)\int f(x)\,dx = f(x)$ says that if a function is integrated and the result is differentiated, the original function is returned. What about the other way around? Is the formula $\int f'(x)\,dx = f(x)$ correct?

41. Prove that if f is continuous on $[a, b]$ and $\int_a^b f(x)\,dx = 0$, there is at least one number u in $[a, b]$ such that $f(u) = 0$.

42. Give a counterexample to the statement in Problem 41 if f is not required to be continuous.

43. If in our work we allowed only regular partitions, then we could not always partition an interval $[a, b]$ in such a manner as to also automatically partition subintervals $[a, c]$ and $[c, b]$ for $a < c < b$. Why not? On account of this fact, what theorem on the definite integral would be affected?

CHALLENGE EXERCISES

1. Let $b > 0$. Suppose $L_b(x)$ is a function defined for $x > 0$ that obeys the following rules:

(a) $L_b(b) = 1;\quad L_b(1) = 0$

(b) $L_b(xy) = L_b(x) + L_b(y)$

(c) $L_b(x^r) = r\,L_b(x),\quad r$ rational

(d) $L_b\left(\dfrac{1}{x}\right) = -L_b(x)$

(a) Using these properties, deduce that for rational x and Δx,

$$\lim_{\substack{\Delta x \to 0 \\ \Delta x \text{ rational}}} \frac{L_b(x + \Delta x) - L_b x}{\Delta x}(1) = \frac{1}{x} \lim_{\substack{\Delta x \to 0 \\ \Delta x \text{ rational}}} L_b\left(1 + \frac{\Delta x}{x}\right)^{x/\Delta x}$$

provided either limit exists.

(b) Assume that $L_b(x)$ is a continuous function of x and that $\lim_{\Delta x \to 0}\left(1 + \dfrac{\Delta x}{x}\right)^{x/\Delta x}$ exists. Denote the value of this limit by e. Find $L_b'(x)$, provided it exists and is continuous. What choice of b gives $L_b'(x) = 1/x$?

2. Let $L(x) = \int_1^x (dt/t)$.

(a) Find $L'(x)$.
(b) Show that $L(x)$ is increasing.
(c) Give the domain of $L(x)$.
(d) Show that $L(x)$ is continuous.

3. Let $L(x) = \int_1^x (dt/t)$.

(a) Find $L(1)$.
(b) Show that if $F(x) = \int_1^{ax} (dt/t)$, then $F'(x) = L'(x)$. Deduce that $L(ax) = L(a) + L(x)$ and hence that for $a, b > 0$, we have $L(ab) = L(a) + L(b)$.
(c) Show by a similar technique to that in part (b) that for $a>0$ and r rational, we have $L(a^r) = rL(a)$.
(d) Show that $L(1/a) = -L(a)$ for $a > 0$.
(e) Give the range of L.

4. Let $L(x) = \int_1^x (dt/t)$.

(a) L is increasing (see Problem 2, part (b)). Thus, L has an inverse. Set $E = L^{-1}$. Find E'.
(b) Using Problem 2, part (c), and Problem 3, part (e), give the domain and range of E.
(c) Show that $E(a + b) = E(a)E(b)$.

5. For this exercise, you need the results of Problems 1 through 4.

(a) Show that

$$\lim_{\substack{\Delta x \to 0 \\ \Delta x \text{ rational}}} \left(1 + \frac{\Delta x}{x}\right)^{x/\Delta x}$$

exists and evaluate it in terms of E.

(b) Obtain an estimate of the limit in part (a) by taking $\Delta x/x = 10^{-9}$.
(c) Refer to the limit of parts (a) and (b) as e. What is $L(e)$? For what choice of b does the function $L(x) = \int_1^x (dt/t)$ satisfy all the properties (a)–(d) listed in Problem 1 for $L_b(x)$?

6. Let $A(x) = \displaystyle\int_0^x \frac{dt}{1 + t^2}$.

(a) Show that A is increasing.
(b) Since A is increasing, it has an inverse T. Find T'.
(c) Show that T satisfies the differential equation $y' = 1 + y^2$. There is another familiar function that also satisfies this differential equation; find it.

7. Use the substitution $\sqrt{x} = \sin y$ to calculate $\int_0^{1/2} (\sqrt{x}/\sqrt{1 - x})\,dx$. [*Hint:* $\sin^2 y = (1 - \cos 2y)/2$.]

8. (a) The evaluation of $\int_0^2 \sqrt{4 - x^2}\,dx$ by Riemann sums or by the fundamental theorem is difficult. However, this integral may be interpreted as a certain area and evaluated by elementary geometry. What is the result?

(b) Of course, in regard to part (a), it is only fair to point out that the elementary formula referred to is

(*Problem 8 continues on page 386*)

8. *(continued)*
derived using calculus in the first place, although that fact is not usually stated in elementary geometry. There is a substitution that will help to evaluate this integral—namely, $x = 2 \sin u$. Work out the integral using this substitution; see the hint in Problem 7.

9. What conditions on f and f' guarantee that

$$f(x) = \int_0^x f'(t)\,dt$$

10. Suppose that F is an antiderivative of f in $[a, b]$ and partition $[a, b]$ into n subintervals.

(a) Apply the mean value theorem for derivatives to F in each subinterval $[x_{i-1}, x_i]$ to show that there is a point u_i in the subinterval such that $F(x_i) - F(x_{i-1}) = f(u_i)\Delta x_i$.
(b) Show that $\sum_{i=1}^{n}[F(x_i) - F(x_{i-1})] = F(b) - F(a)$.
(c) Use parts (a) and (b) to explain why $\int_a^b f(x)\,dx = F(b) - F(a)$.

11. (a) Give reasons for the steps in the following argument, which yields an estimate for the error in approximating an integral by a Riemann sum: Let f' be continuous on the closed interval $[a, b]$ and let M be the maximum of $|f'(x)|$ on $[a, b]$. Also, let P be a regular partition of $[a, b]$ and let $R = \sum_{i=1}^{n} f(u_i)\Delta x_i$ be a Riemann sum. Then,

$$R - \int_a^b f(x)\,dx = R - \sum_{i=1}^{n} \int_{x_{i-1}}^{x_i} f(x)\,dx \qquad \text{(a) Why?}$$

$$= R - \sum_{i=1}^{n} f(t_i)\Delta x_i \qquad \text{(b) Why?}$$

$$= \sum_{i=1}^{n} f'(c_i)(u_i - t_i)\Delta x_i \qquad \text{(c) Why?}$$

(Problem 11 continues in the next column)

11. *(continued)*
Therefore,

$$\left|R - \int_a^b f(x)\,dx\right| \le \sum_{i=1}^{n} |f'(c_i)||u_i - t_i|\Delta x_i \qquad \text{(d) Why?}$$

$$\le M \sum_{i=1}^{n} (\Delta x_i)^2 \qquad \text{(e) Why?}$$

$$= Mn\left(\frac{b-a}{n}\right)^2 \qquad \text{(f) Why?}$$

$$= \frac{M(b-a)^2}{n}$$

(b) Let $f(x) = 2x$ on the closed interval $[1, 3]$. Let P be the regular partition consisting of ten subintervals. Let u_i be the left endpoint of the ith interval. Compare the actual error,

$$\left|\sum_{i=1}^{n} f(u_i)\Delta x_i - \int_a^b f(x)\,dx\right|,$$

with the estimate given in part (a).
(c) Repeat part (b) with $f(x) = x^2$ on the closed interval $[0, 1]$.

12. Let

$$f(x) = \begin{cases} 0 & \text{if } x \text{ irrational} \\ \dfrac{1}{q} & \text{if } x \text{ rational with } x = p/q \text{ in lowest terms} \end{cases} \qquad \text{for } x \text{ in } [0, 1]$$

Show that $\int_0^1 f(x)\,dx$ exists and that $\int_0^1 f(x)\,dx = 0$.

Applications of the Integral

6

In the applications that follow, we rely on two basic facts from Chapter 5. The first is that for a continuous function f on a closed interval $[a, b]$, the limit of the Riemann sum—the definite integral—exists. That is,

$$\lim_{\|P\| \to 0} \sum_{i=1}^{n} f(u_i)\Delta x_i = \int_a^b f(x)\, dx = \text{Some number} \tag{1}$$

The second basic fact is that if F is an antiderivative of a continuous function f defined on $[a, b]$, then, by the second fundamental theorem of calculus,

$$\int_a^b f(x)\, dx = F(b) - F(a)$$

When $f(x) \geq 0$ on $[a, b]$, we can use this result to calculate the area under the graph of $y = f(x)$ from a to b. In this chapter we extend this result to calculate the area enclosed by the graphs of two or more functions. Furthermore, we show how (1) can be applied to find the volume of a solid, to compute work and fluid pressure, to find the length of a graph, and to calculate center of mass and centroid. For each of these applications, we can approximate the required quantity by a Riemann sum and then use (1).

6.1 Area

Suppose we want to calculate the area enclosed by the graphs of $y = f(x)$ and $y = g(x)$ and the lines $x = a$ and $x = b$. As indicated in Figure 1, we assume that f and g are continuous on $[a, b]$ and that $f(x) \geq g(x)$ for all x in $[a, b]$. We begin by partitioning the interval $[a, b]$ into n subintervals,

$$[a, x_1], \quad [x_1, x_2], \quad \ldots, \quad [x_{i-1}, x_i], \quad \ldots, \quad [x_{n-1}, b]$$

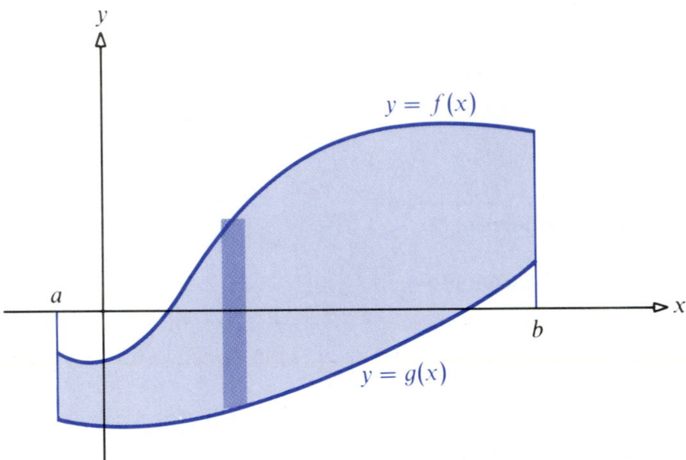

Figure 1

We denote the length of the ith subinterval by $\Delta x_i = x_i - x_{i-1}$. For each i, $1 \leq i \leq n$, we select a number u_i in the ith subinterval $[x_{i-1}, x_i]$ and construct n rectangles, each of width Δx_i and height $f(u_i) - g(u_i)$. The area of each rectangle is then $[f(u_i) - g(u_i)]\Delta x_i$ (see Fig. 2).

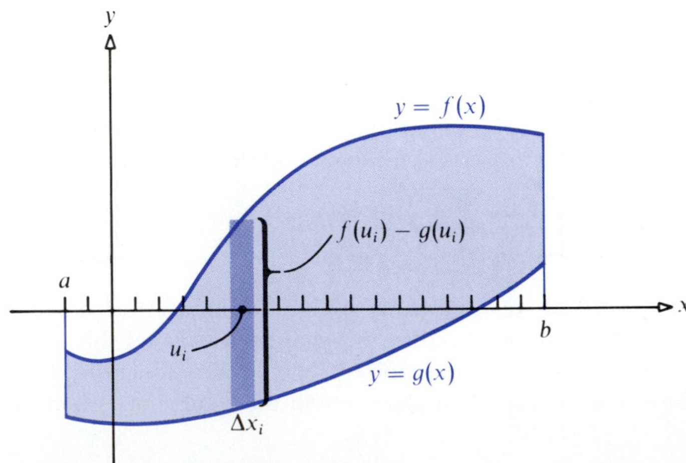

Figure 2

The sum of the areas of these n rectangles, $\sum_{i=1}^{n}[f(u_i) - g(u_i)]\Delta x_i$, gives an approximation to the area we seek. As the length of each subinterval gets smaller and smaller—that is, as the norm $\|P\|$ of the partition approaches 0—the sum $\sum_{i=1}^{n}[f(u_i) - g(u_i)]\Delta x_i$ is a better and better approximation to the area we seek. However, this sum is a Riemann sum, so, as $\|P\| \to 0$, its limit is a definite integral. This suggests the following definition:

[6.1.1] DEFINITION / *Area.*

The area A enclosed by the graphs of $y = f(x)$ and $y = g(x)$ and the lines $x = a$ and $x = b$, where f and g are continuous on $[a, b]$ and $f(x) \geq g(x)$ on $[a, b]$, is

$$A = \lim_{\|P\| \to 0} \sum_{i=1}^{n} [f(u_i) - g(u_i)]\Delta x_i = \int_a^b [f(x) - g(x)]\, dx$$

An interesting aspect of this result is that it works whether the graphs lie above the x-axis, below the x-axis, or partially above and partially below the x-axis. The next example illustrates this fact.

EXAMPLE 1

Find the area enclosed by the graphs of $y = f(x) = 10x - x^2$ and $y = g(x) = 3x - 8$.

Solution

First we graph the two functions (see Fig. 3). Before we can compute the area, we need to locate the points of intersection of the two graphs. Thus, we need to solve

$$f(x) = g(x)$$

$$10x - x^2 = 3x - 8$$

$$x^2 - 7x - 8 = 0 \qquad \text{so that} \quad x = -1, \quad x = 8$$

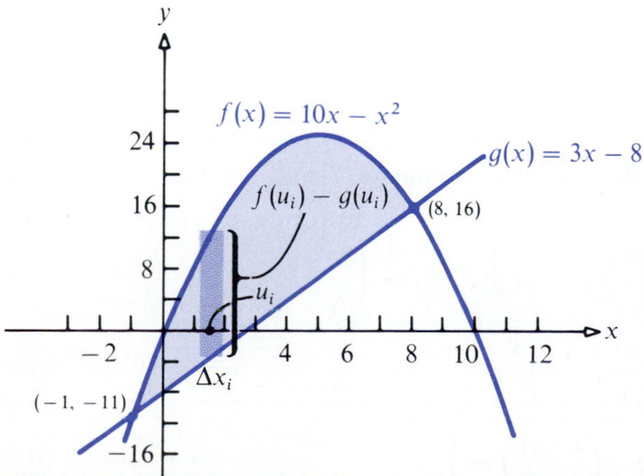

Figure 3

From Figure 3, we note that $f(x) \geq g(x)$ on $[-1, 8]$ so that the area we seek is given by

$$\begin{aligned}
\lim_{\|P\| \to 0} \sum_{i=1}^{n} [f(u_i) - g(u_i)]\Delta x_i &= \int_{-1}^{8} [f(x) - g(x)]\, dx \\
&= \int_{-1}^{8} [(10x - x^2) - (3x - 8)]\, dx \\
&= \int_{-1}^{8} (-x^2 + 7x + 8)\, dx \\
&= \left(\frac{-x^3}{3} + \frac{7x^2}{2} + 8x\right)\Bigg|_{-1}^{8} \\
&= \left(\frac{-512}{3} + 224 + 64\right) - \left(\frac{1}{3} + \frac{7}{2} - 8\right) \\
&= -171 + 296 - \frac{7}{2} = \frac{243}{2}
\end{aligned}$$

The application of definition (6.1.1) requires that one graph lies above the other graph on $[a, b]$. The next example illustrates how to proceed when this is not the case.

EXAMPLE 2

Find the area enclosed by the graphs of $f(x) = x^3$ and $g(x) = x$.

Solution

First we graph the two functions (see Fig. 4). The points of intersection obey

$$\begin{aligned}
x^3 &= x \\
x^3 - x &= 0 \\
x(x^2 - 1) &= 0
\end{aligned}$$

so they occur at

$$x = -1 \qquad x = 0 \qquad x = 1$$

On $[-1, 0]$, $f(x) \geq g(x)$, whereas on $[0, 1]$, we have $g(x) \geq f(x)$. Thus, the area A_1 enclosed by the graphs of f and g from $x = -1$ to $x = 0$ is

$$\begin{aligned}
A_1 &= \int_{-1}^{0} [f(x) - g(x)]\, dx = \int_{-1}^{0} (x^3 - x)\, dx \\
&= \left(\frac{x^4}{4} - \frac{x^2}{2}\right)\Bigg|_{-1}^{0} \\
&= 0 - \left(\frac{1}{4} - \frac{1}{2}\right) = \frac{1}{4}
\end{aligned}$$

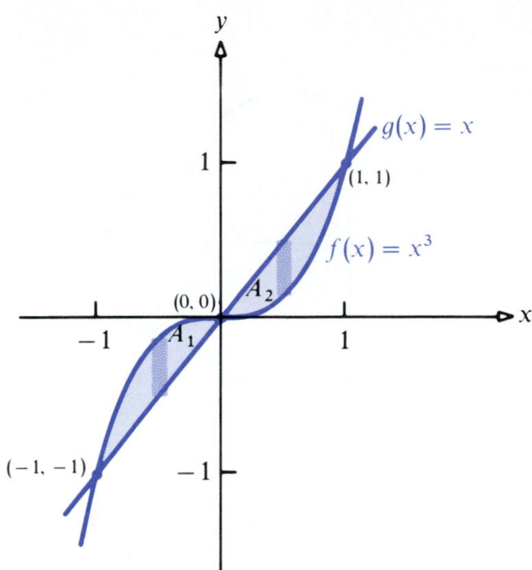

Figure 4

The area A_2 enclosed by the graphs of f and g from $x = 0$ to $x = 1$ is

$$A_2 = \int_0^1 [g(x) - f(x)]\, dx = \int_0^1 (x - x^3)\, dx = \left(\frac{x^2}{2} - \frac{x^4}{4}\right)\Bigg|_0^1$$
$$= \left(\frac{1}{2} - \frac{1}{4}\right) - 0 = \frac{1}{4}$$

The area we seek is therefore $A_1 + A_2 = \frac{1}{2}$.

The result of Example 2 can be more easily obtained by using symmetry. By examining Figure 4, we can see that $A_1 = A_2$. Thus,

$$A = A_1 + A_2 = 2A_2 = 2\int_0^1 (x - x^3)\, dx = 2\left(\frac{1}{4}\right) = \frac{1}{2}$$

The importance of graphing when doing an area problem cannot be overemphasized. For the functions in Example 2, a thoughtless application of definition (6.1.1) would give

$$\int_{-1}^1 (x^3 - x)\, dx = \left(\frac{x^4}{4} - \frac{x^2}{2}\right)\Bigg|_{-1}^1$$
$$= \left(\frac{1}{4} - \frac{1}{2}\right) - \left(\frac{1}{4} - \frac{1}{2}\right) = 0$$

which is a ridiculous answer. A graph will prevent this kind of mistake.

Finding the area A enclosed by the graph of $y = g(x)$, the x-axis, and the lines $x = a$ and $x = b$, when $g(x) \leq 0$ on $[a, b]$, is the same as finding the area enclosed by the graphs of $y = f(x) = 0$ (the x-axis) and

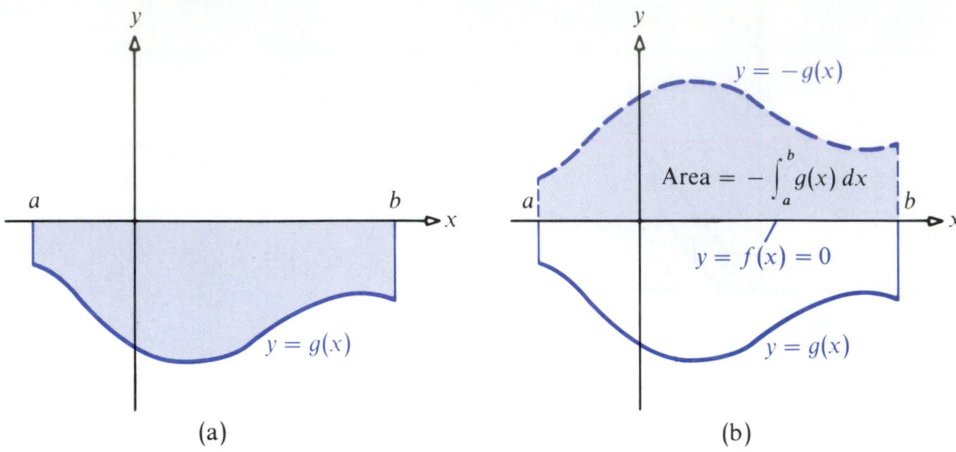

Figure 5

$y = -g(x)$ and the lines $x = a$ and $x = b$ (see Fig. 5). Then

$$A = \int_a^b [0 - g(x)]\,dx = \int_a^b -g(x)\,dx = -\int_a^b g(x)\,dx$$

We use this fact in the next example.

EXAMPLE 3

Find the area enclosed by the graph of $f(x) = x^2 - 4$, the x-axis, and the lines $x = 0$ and $x = 4$.

Solution

On the interval [0, 4], the graph crosses the x-axis at $x = 2$. As we can see from Figure 6, $f(x) \leq 0$ from $x = 0$ to $x = 2$, and $f(x) \geq 0$ from $x = 2$ to $x = 4$. Thus, the area A_1 enclosed by the graph of

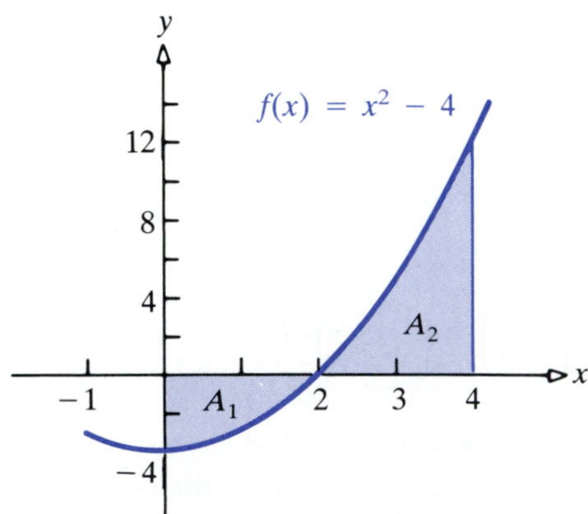

Figure 6

$f(x) = x^2 - 4$ and the x-axis from $x = 0$ to $x = 2$ is

$$A_1 = -\int_0^2 (x^2 - 4)\,dx = -\left(\frac{x^3}{3} - 4x\right)\Bigg|_0^2 = -\left(\frac{8}{3} - 8\right) = \frac{16}{3}$$

The area A_2 enclosed by the graph of $f(x) = x^2 - 4$ and the x-axis from $x = 2$ to $x = 4$ is

$$A_2 = \int_2^4 (x^2 - 4)\,dx = \left(\frac{x^3}{3} - 4x\right)\Bigg|_2^4 = \left(\frac{64}{3} - 16\right) - \left(\frac{8}{3} - 8\right) = \frac{56}{3} - 8 = \frac{32}{3}$$

The total area is therefore

$$A = A_1 + A_2 = \frac{16}{3} + \frac{32}{3} = \frac{48}{3} = 16$$

EXAMPLE 4

Find the area in the first quadrant enclosed by the graphs of $y = f(x) = \sin x$, $y = g(x) = \cos x$, and the y-axis, $0 \le x \le \pi$.

Solution

First we graph the two functions (see Fig. 7). Next we locate the point of intersection of the two graphs:

$$f(x) = g(x)$$
$$\sin x = \cos x$$
$$\tan x = 1$$
$$x = \frac{\pi}{4}$$

From Figure 7, we note that $g(x) \ge f(x)$ on $[0, \pi/4]$, so that the area we seek is given by

$$\int_0^{\pi/4} (\cos x - \sin x)\,dx = (\sin x + \cos x)\Bigg|_0^{\pi/4} = \left(\frac{\sqrt{2}}{2} + \frac{\sqrt{2}}{2}\right) - (0 + 1) = \sqrt{2} - 1$$

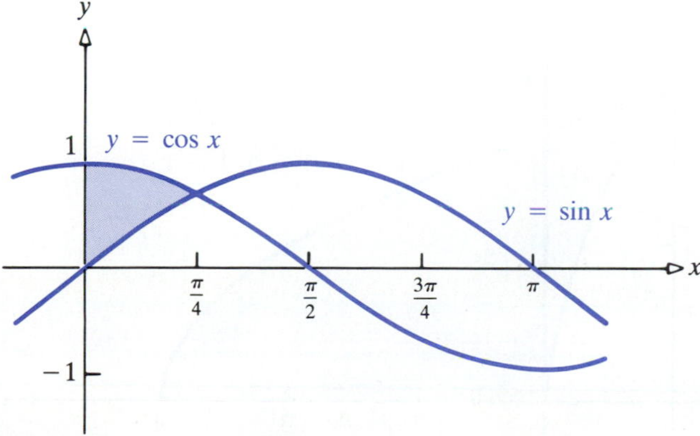

Figure 7

Sometimes it is better to partition the y-axis instead of the x-axis. The next example illustrates this.

EXAMPLE 5

Find the area enclosed by the graphs of $y = f(x) = \sqrt{4 - 4x}$, $y = g(x) = \sqrt{4 - x}$, and the x-axis.

Solution

First, we will solve the problem by partitioning the x-axis, and then we will solve it by partitioning the y-axis.

Partition of x-axis: Refer to Figure 8. There, we see that the area we seek is the sum of two areas:

A_1 = Area enclosed by $y = \sqrt{4 - x}$, $y = \sqrt{4 - 4x}$, and $x = 1$

A_2 = Area enclosed by $y = \sqrt{4 - x}$, the x-axis, and $x = 1$

The area A we seek is therefore,

$$\begin{aligned} A = A_1 + A_2 &= \int_0^1 (\sqrt{4 - x} - \sqrt{4 - 4x})\, dx + \int_1^4 \sqrt{4 - x}\, dx \\ &= \int_0^1 \sqrt{4 - x}\, dx - \int_0^1 \sqrt{4 - 4x}\, dx + \int_1^4 \sqrt{4 - x}\, dx \\ &= \int_0^4 \sqrt{4 - x}\, dx - \int_0^1 \sqrt{4 - 4x}\, dx \end{aligned}$$

Now,

$$\int_0^4 \sqrt{4 - x}\, dx \underset{\uparrow}{=} -\int_4^0 u^{1/2}\, du = \frac{-2}{3} u^{3/2} \Big|_4^0 = \frac{-2}{3}(0 - 8) = \frac{16}{3}$$

Set $u = 4 - x$.

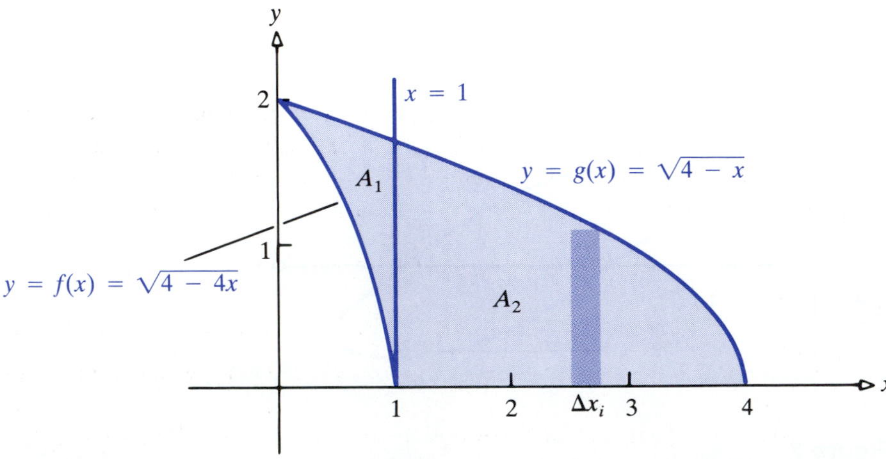

Figure 8

$$\int_0^1 \sqrt{4-4x}\,dx \underset{\text{Set } u=4-4x.}{=} -\frac{1}{4}\int_4^0 u^{1/2}\,du = -\frac{2u^{3/2}}{4(3)}\bigg|_4^0 = \frac{-1}{6}(0-8) = \frac{4}{3}$$

The required area A is

$$A = \frac{16}{3} - \frac{4}{3} = 4$$

Partition of y-axis: When we partition the y-axis, we obtain horizontal rectangles of width Δy and length $x_2 - x_1$, where x_2 is the distance from the y-axis to $y = g(x)$ and x_1 is the distance from the y-axis to $y = f(x)$ (refer to Fig. 9). The area A we seek is

$$\begin{aligned} A = \lim_{\|P\|\to 0} \sum_{i=1}^{n} (x_2 - x_1)\Delta y_i &= \int_0^2 \left[(4-y^2) - \frac{1}{4}(4-y^2)\right] dy \\ &= \int_0^2 \frac{3}{4}(4-y^2)\,dy = \frac{3}{4}\left(4y - \frac{y^3}{3}\right)\bigg|_0^2 \\ &= \frac{3}{4}\left(8 - \frac{8}{3}\right) = 4 \end{aligned}$$

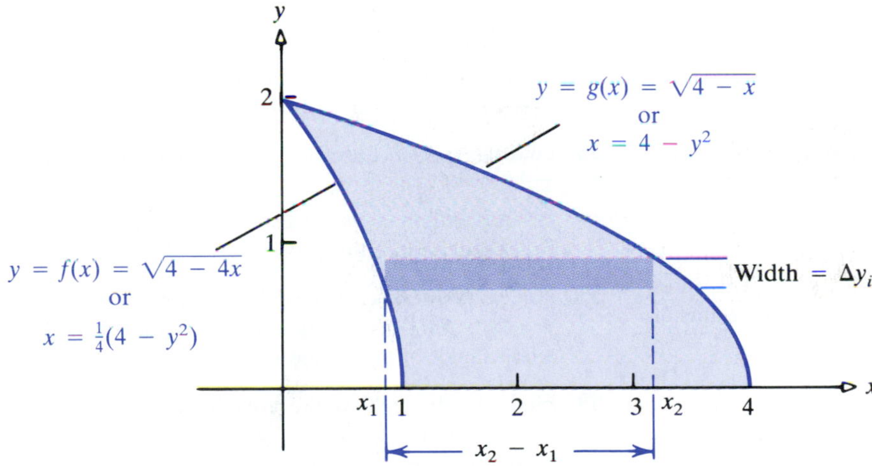

Figure 9

As Example 5 illustrates, a partition of the y-axis is simpler. But how would you have known this in advance? Usually the configuration of the graph will make it clear whether to partition the x-axis or the y-axis. However, as is so often the case, there is no substitute for experience.

EXERCISE 6.1

In Problems 1–26 find the area enclosed by the graphs of the given functions and lines. Draw a sketch first.

1. $f(x) = x, \quad g(x) = 2x, \quad x = 0, \quad x = 1$

2. $f(x) = x, \quad g(x) = 3x, \quad x = 0, \quad x = 3$

3. $f(x) = x^2, \quad g(x) = x$

4. $f(x) = x^2, \quad g(x) = 4x$

5. $f(x) = x^2 + 1, \quad g(x) = x + 1$

6. $f(x) = x^2 + 1, \quad g(x) = 4x + 1$

7. $f(x) = \sqrt{x}, \quad g(x) = x^3$

8. $f(x) = x^2, \quad g(x) = x^3$

9. $f(x) = x^2, \quad g(x) = x^4$

10. $f(x) = \sqrt{x}, \quad g(x) = x^2$

11. $f(x) = x^2 - 4x, \quad g(x) = -x^2$

12. $f(x) = x^2 - 8x, \quad g(x) = -x^2$

13. $f(x) = 4 - x^2, \quad g(x) = x + 2$

14. $f(x) = 2 + x - x^2, \quad g(x) = -x - 1$

15. $f(x) = x^3, \quad g(x) = 4x$

16. $f(x) = x^3, \quad g(x) = 16x$

17. $y = x^2, \quad y = x, \quad y = -x$

18. $y = x^2 - 1, \quad y = x - 1, \quad y = -x - 1$

19. $y = \sqrt{9 - x}, \quad y = \sqrt{9 - 3x}, \quad x\text{-axis}$

20. $y = \sqrt{16 - 2x}, \quad y = \sqrt{16 - 4x}, \quad x\text{-axis}$

21. $y^2 = x, \quad y = x - 6$

22. $y^2 = x + 16, \quad y = -x - 4$

23. $y = \cos x, \quad y = 1, \quad x = \pi/6$

24. $y = \sin x, \quad y = 1, \quad x = 0$

25. $f(x) = \sin(x/2), \quad g(x) = \cos(x/2), \quad x = 0, \quad x = \pi$

26. $f(x) = \sin x, \quad g(x) = \sin 2x, \quad x = 0, \quad x = \pi/2$

In Problems 27–32 find the area enclosed by the graphs of the given functions by using a partition of the y-axis.

27. $y = \sqrt{2x - 6}, \quad y = \sqrt{x - 2}, \quad x\text{-axis}$

28. $y = \sqrt{2x - 5}, \quad y = \sqrt{4x - 17}, \quad x\text{-axis}$

29. $y^2 = x, \quad x + y = 2$

30. $y^2 = x, \quad x + y = 6$

31. $y^2 = 4x, \quad 4x - 3y - 4 = 0$

32. $y^2 = 4x + 1, \quad x = y + 1$

33. Find the area enclosed by $x^{1/2} + y^{1/2} = 1$, the x-axis, and the y-axis.

35. Find c, $0 < c < 1$, so that the area under the graph of $y = x^2$ from 0 to c equals the area under the same graph from c to 1.

37. Show that the shaded area in the figure is two-thirds of the area of the parallelogram $ABCD$. (This illustrates a result due to Archimedes concerning sectors of parabolas.)

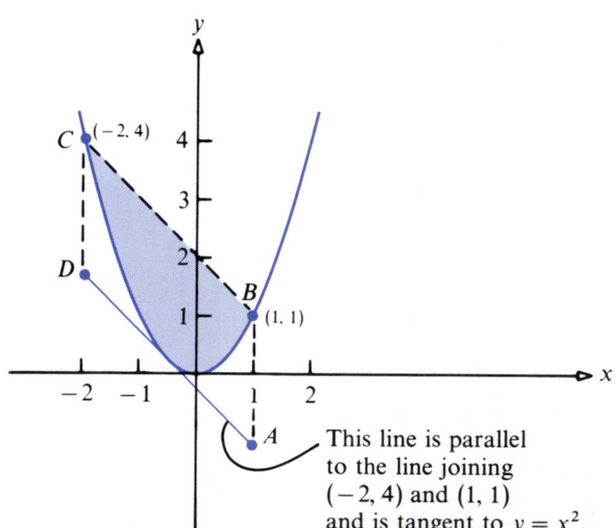

34. Find the area of the "triangle" in the first quadrant enclosed by $y = \sin 2x$, $y = \cos 2x$, and $x = \pi/4$.

36. Find the area enclosed by $x = y^2$, the y-axis, $y = 1$, and $y = 2$ by using:

(a) A partition of the x-axis
(b) A partition of the y-axis

38. Find h if the area enclosed by the graphs of $y = x$, $y = 8x$, and $y = 1/x^2$ is equal to that of an isosceles triangle of base 1 and height h.

39. Find $b > 0$ so that the area in the first quadrant enclosed by the graph of $y = 1 + b - bx^2$ and the coordinate axes is a minimum.

Volume of a Solid of Revolution: Disk Method

6.2

A solid of revolution can be generated by revolving the area under the graph of a continuous, nonnegative function $y = f(x)$ from a to b about the x-axis (see Fig. 10). A common example of a solid of revolution is a *cone*, which may be generated by revolving the area under a line that passes through the origin from 0 to h about the x-axis (see Fig. 11). Another common example of a solid of revolution is a *cylinder*, which may be generated by revolving the area under a horizontal line of positive height from a to b about the x-axis (see Fig. 12).

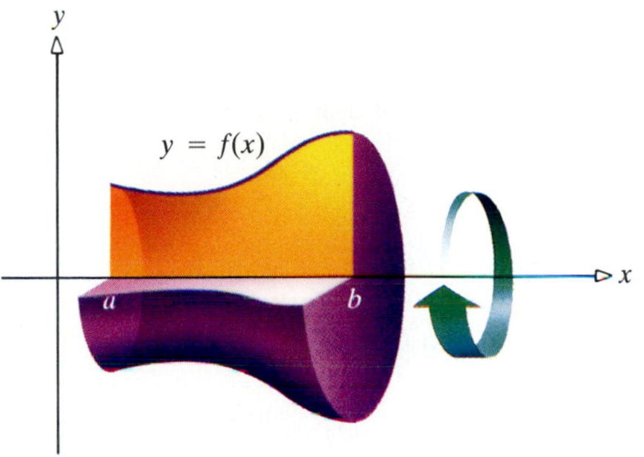

Figure 10

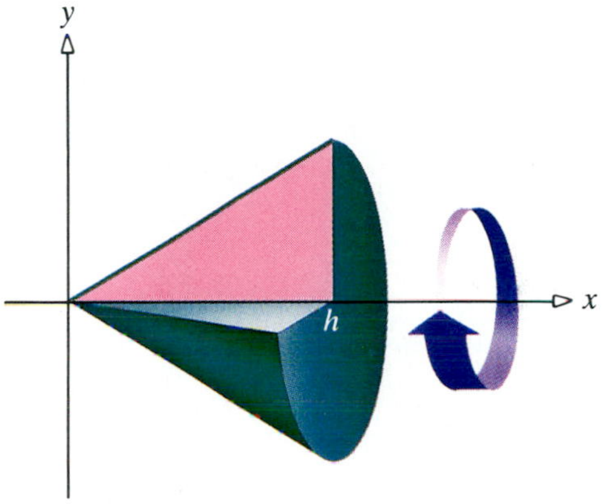

Figure 11

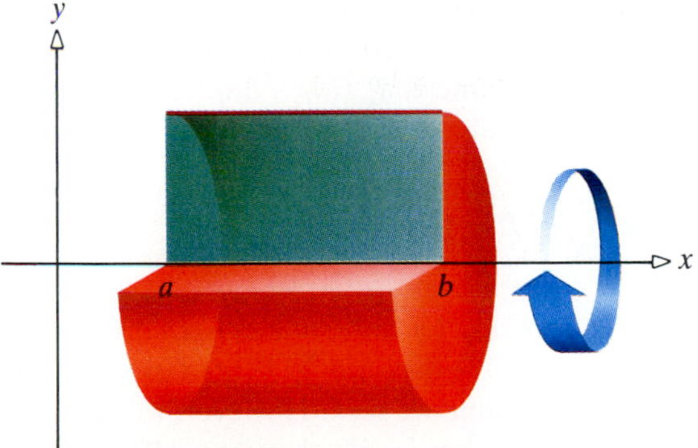

Figure 12

Revolving about the x-Axis

We seek a formula for finding the volume V of a solid of revolution. To fix our ideas, suppose the area under the graph of $y = f(x)$, $f(x) \geq 0$, from a to b is revolved about the x-axis, thus generating a solid of revolution. If we cut this solid with planes perpendicular to it and the x-axis, we find that a typical cross section is a circle and a typical slice is a *disk*. For this reason, the method developed in this section is referred to as the *disk method*.

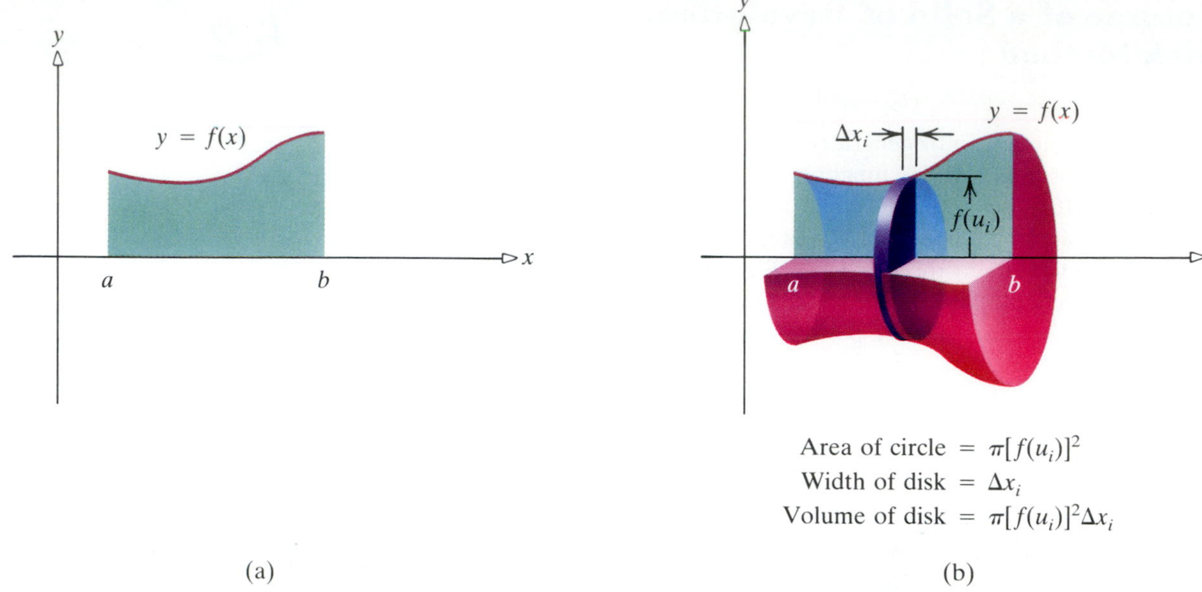

Figure 13

We begin by partitioning the interval $[a, b]$ into n subintervals,

$$[a, x_1], \quad [x_1, x_2], \quad \ldots, \quad [x_{n-1}, b]$$

As before, we denote the length of the ith subinterval by $\Delta x_i = x_i - x_{i-1}$, and for each i, $1 \le i \le n$, we select a number u_i in the ith subinterval $[x_{i-1}, x_i]$ (see Fig. 13). The circle obtained by cutting the solid at $x = u_i$ has radius $f(u_i)$, and its area is $\pi[f(u_i)]^2$. The solid of revolution obtained from the ith subinterval is a disk; its volume is $\pi[f(u_i)]^2\Delta x_i$. Hence, an approximation to the desired volume V of the solid of revolution is

$$V \approx \sum_{i=1}^{n} \pi[f(u_i)]^2\Delta x_i$$

As the length of each subinterval gets smaller and smaller—that is, as the norm $\|P\|$ of the partition approaches 0—the sum $\sum_{i=1}^{n} \pi[f(u_i)]^2\Delta x_i$ becomes a better approximation to the volume we seek. However, this sum is a Riemann sum. Hence, if f is continuous on $[a, b]$, then as $\|P\| \to 0$, its limit is a definite integral. This suggests the following definition:

[6.2.1] DEFINITION / *Volume: Disk Method.*

The volume V of the solid of revolution obtained by revolving the area under the graph of a continuous nonnegative function $y = f(x)$ from a to b about the x-axis is

$$V = \lim_{\|P\| \to 0} \sum_{i=1}^{n} \pi[f(u_i)]^2\Delta x_i = \pi \int_a^b [f(x)]^2\, dx \qquad (1)$$

EXAMPLE 1

Find the volume of the solid of revolution generated by revolving the area under the graph of $f(x) = \sqrt{x}$ from $x = 0$ to $x = 5$ about the x-axis.

Solution

First, we graph the area to be revolved and the resultant solid of revolution, as shown in Figure 14. The desired volume V is

$$V = \int_0^5 \pi[f(x)]^2\,dx$$

$$= \int_0^5 \pi(\sqrt{x})^2\,dx$$

$$= \int_0^5 \pi x\,dx = \frac{\pi x^2}{2}\bigg|_0^5 = \frac{25\pi}{2}$$

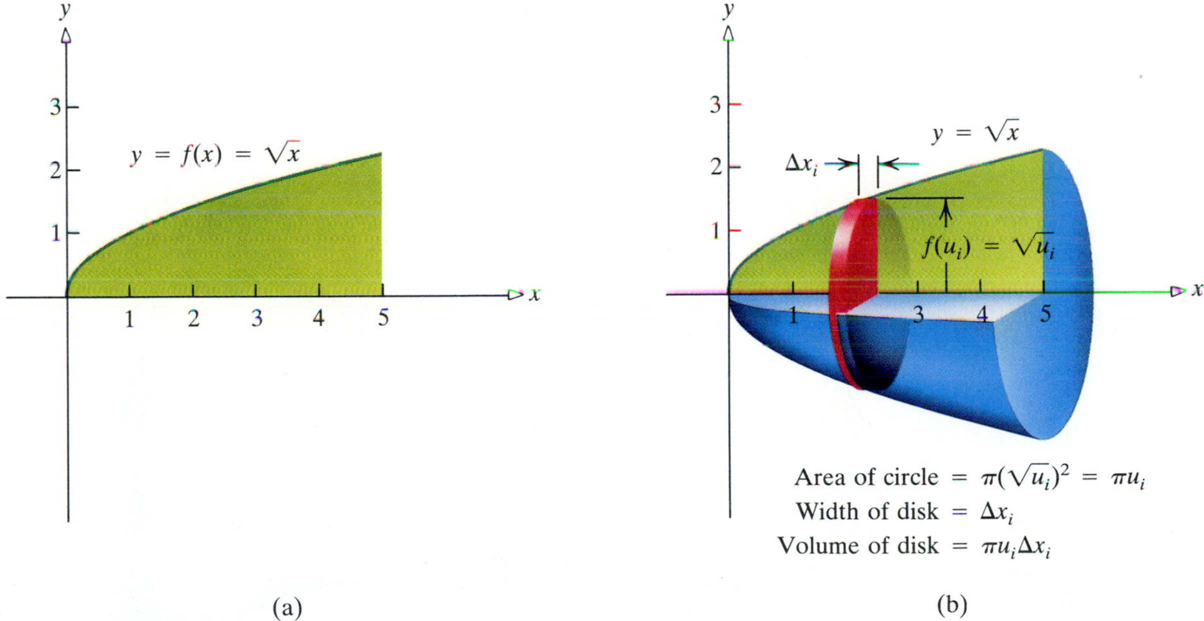

Figure 14

The function f in definition (6.2.1) does not have to be nonnegative. As Figure 15 illustrates, the volume V of the solid of revolution obtained by revolving about the x-axis the area enclosed by the graph of a continuous function f, the x-axis, $x = a$, and $x = b$ (Fig. 15(a)) will equal the volume of the solid of revolution obtained by revolving about the x-axis the area under the graph of $y = |f(x)|$ from a to b (Fig. 15(b)). Since $|f(x)|^2 = [f(x)]^2$, the formula for V is the same as the one given by (1).

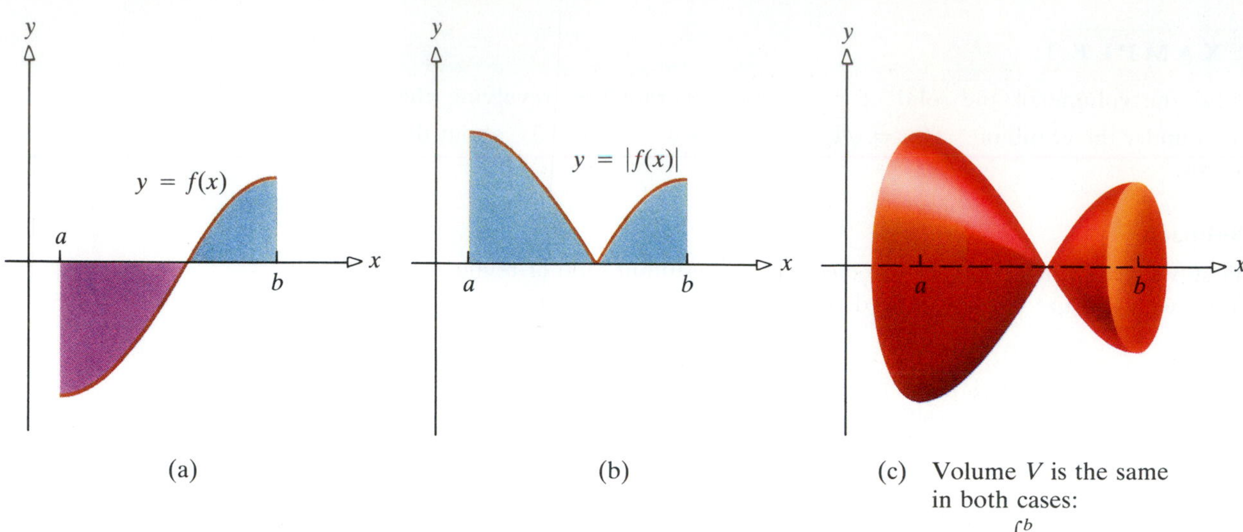

(a)

(b)

(c) Volume V is the same in both cases:

$$V = \pi\int_a^b [f(x)]^2\, dx$$

Figure 15

EXAMPLE 2

Find the volume of the solid of revolution generated by revolving the area enclosed by $y = x^3$, the x-axis, $x = -1$, and $x = 2$ about the x-axis.

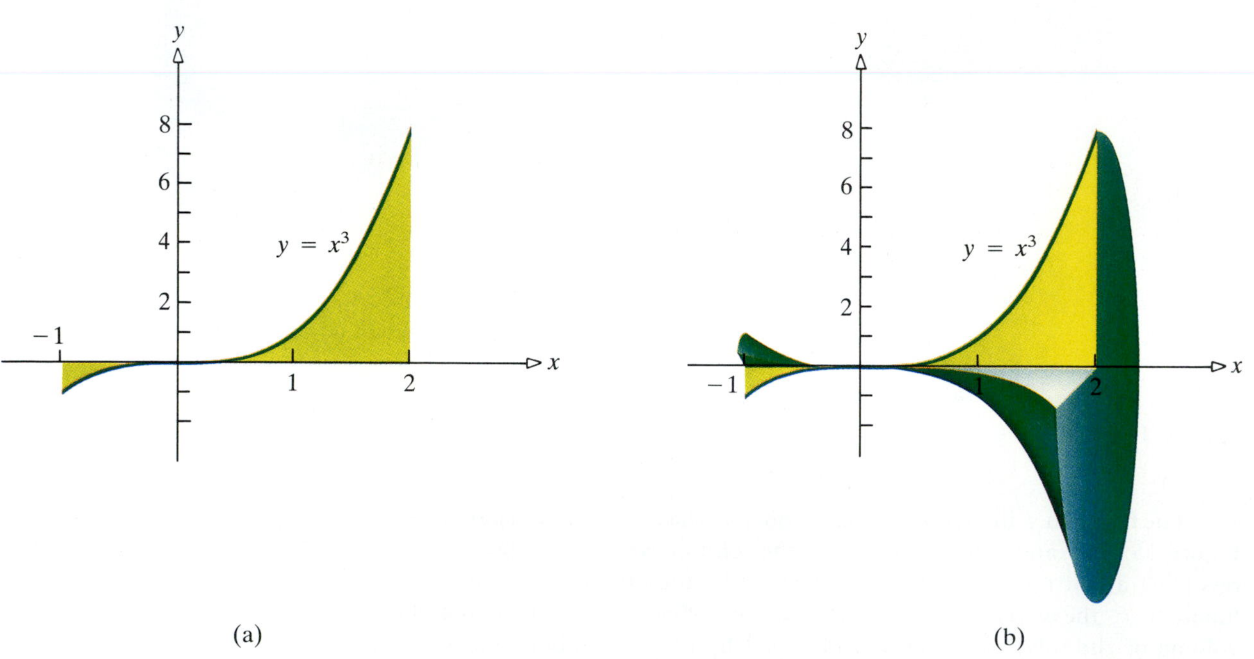

(a)

(b)

Figure 16

Solution

Figure 16 illustrates the area to be revolved and the resultant solid of revolution. The desired volume V is

$$V = \pi \int_{-1}^{2} (x^3)^2\, dx = \pi \frac{x^7}{7}\bigg|_{-1}^{2} = \frac{\pi}{7}(128 + 1) = \frac{129}{7}\pi$$

Revolving about the y-Axis

A solid of revolution can be generated by revolving an area about any line. In particular, the volume V of the solid of revolution generated by revolving the area enclosed by the graph of $x = g(y)$, $g(y) \geq 0$, the y-axis, $y = c$, and $y = d$ about the y-axis may be obtained by partitioning the y-axis and taking slices of width Δy_i perpendicular to the y-axis (see Fig. 17). In this case, the area of a typical slice is $\pi x_i^2 = \pi[g(v_i)]^2$, and the volume of a typical disk is $\pi[g(v_i)]^2 \Delta y_i$. By summing all the volumes and taking the limit, we find the required volume V to be

$$V = \pi \int_c^d [g(y)]^2\, dy \tag{2}$$

We use (2) in the next example.

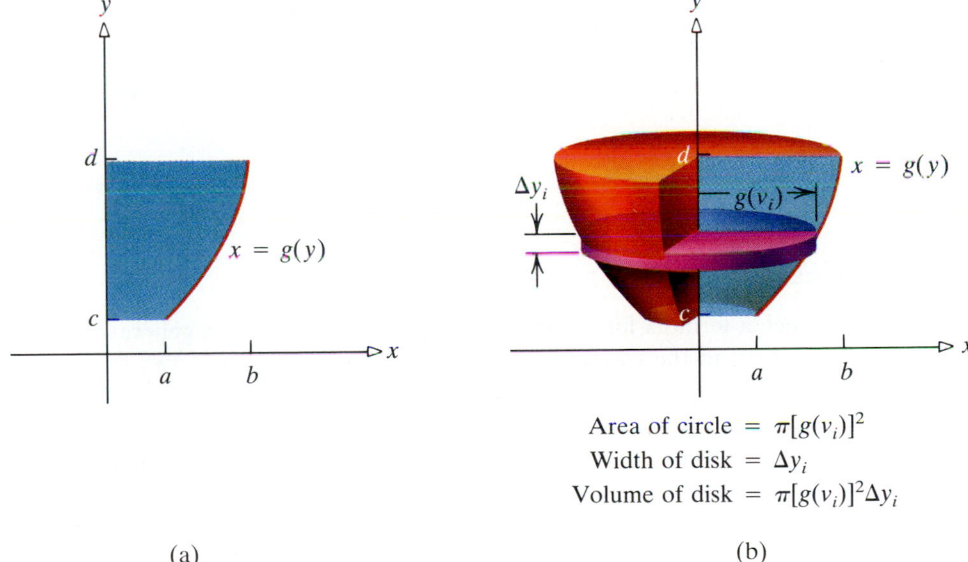

Figure 17

EXAMPLE 3

Find the volume of the solid of revolution generated by revolving the area enclosed by the graphs of $y = x^3$, the y-axis, $y = 1$, and $y = 8$ about the y-axis.

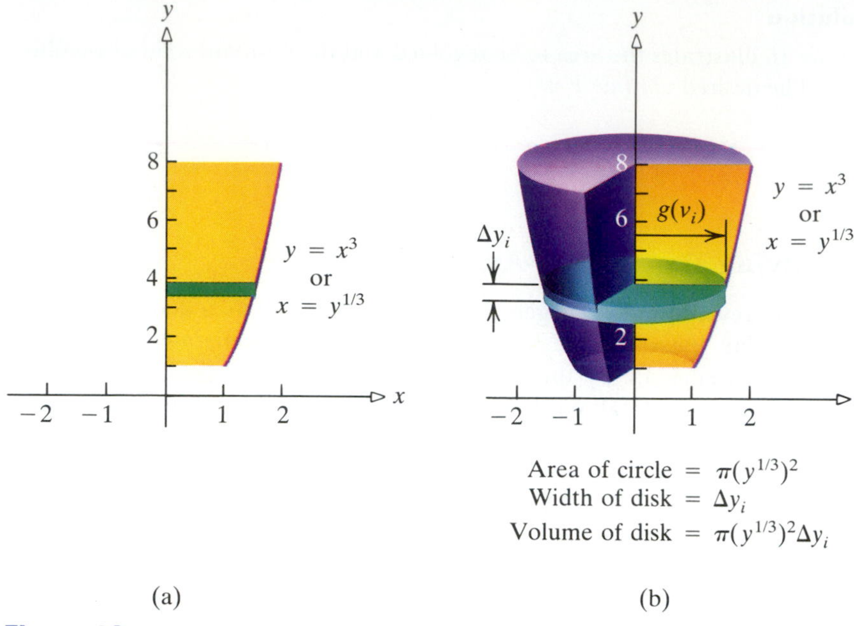

Figure 18

Solution

Figure 18 illustrates the situation. By using $x = g(y) = y^{1/3}$ in (2), we find the volume V to be

$$V = \pi \int_c^d [g(y)]^2\, dy = \pi \int_1^8 y^{2/3}\, dy = \left.\frac{\pi y^{5/3}}{\frac{5}{3}}\right|_1^8$$

$$= \frac{3\pi}{5}(32 - 1) = \frac{93}{5}\pi$$

Washers

We seek a formula for the volume of a solid of revolution generated by revolving about the x-axis the area enclosed by the graphs of two continuous functions $y = f(x)$ and $y = g(x)$, $f(x) \geq g(x) \geq 0$, and the lines $x = a$ and $x = b$. Figure 19(a) illustrates the situation.

As has been our practice, we begin by partitioning the interval $[a, b]$ into n subintervals,

$$[a, x_1], \quad [x_1, x_2], \quad \ldots, \quad [x_{n-1}, b]$$

The length of the ith subinterval is $\Delta x_i = x_i - x_{i-1}$. For each i, $1 \leq i \leq n$, select a number u_i in the ith subinterval $[x_{i-1}, x_i]$ (see Fig. 19(b)). The area obtained by cutting the solid at $x = u_i$ is the difference of the area of the outer circle, $\pi[f(u_i)]^2$, and the area of the inner circle, $\pi[g(u_i)]^2$—namely, $\pi[f(u_i)]^2 - \pi[g(u_i)]^2$. The solid of revolution obtained from the ith subinterval is called a *washer*, and its volume is $\{\pi[f(u_i)]^2 - \pi[g(u_i)]^2\}\Delta x_i$. By adding up the volumes of all the washers and taking the limit as the norm of the partition approaches 0, we obtain the following definition:

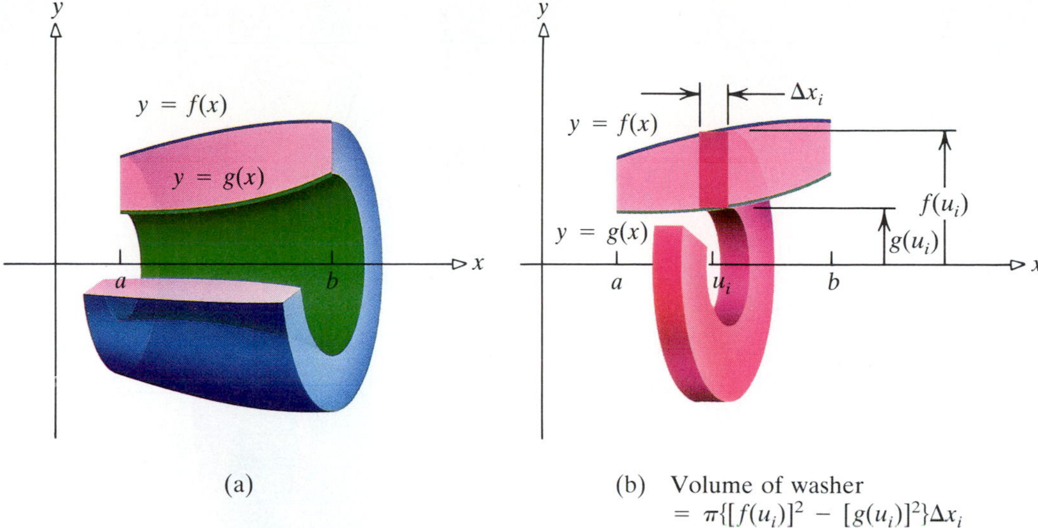

(a)

(b) Volume of washer $= \pi\{[f(u_i)]^2 - [g(u_i)]^2\}\Delta x_i$

Figure 19

[6.2.2] DEFINITION / *Volume: Washer Method.*

The volume V of the solid of revolution obtained by revolving about the x-axis the area enclosed by the graphs of two continuous functions $y = f(x)$ and $y = g(x)$, $f(x) \geq g(x) \geq 0$, and the lines $x = a$ and $x = b$ is

$$V = \lim_{\|P\| \to 0} \sum_{i=1}^{n} \{\pi[f(u_i)]^2 - \pi[g(u_i)]^2\} \, \Delta x_i$$

$$= \int_a^b \pi\{[f(x)]^2 - [g(x)]^2\} \, dx \qquad (3)$$

Because of the way we obtained definition (6.2.2), it is sometimes referred to as the *washer method*. In using (3), it may be helpful to remember the general formula:

Volume of a washer

$$= \pi[(\text{Outer radius})^2 - (\text{Inner radius})^2](\text{Thickness}) \qquad (4)$$

EXAMPLE 4

Find the volume of the solid of revolution generated by revolving the area enclosed by the graphs of $y = 2/x$ and $y = 3 - x$ about the x-axis.

Solution

Figure 20 illustrates the situation. The points of intersection of the graphs

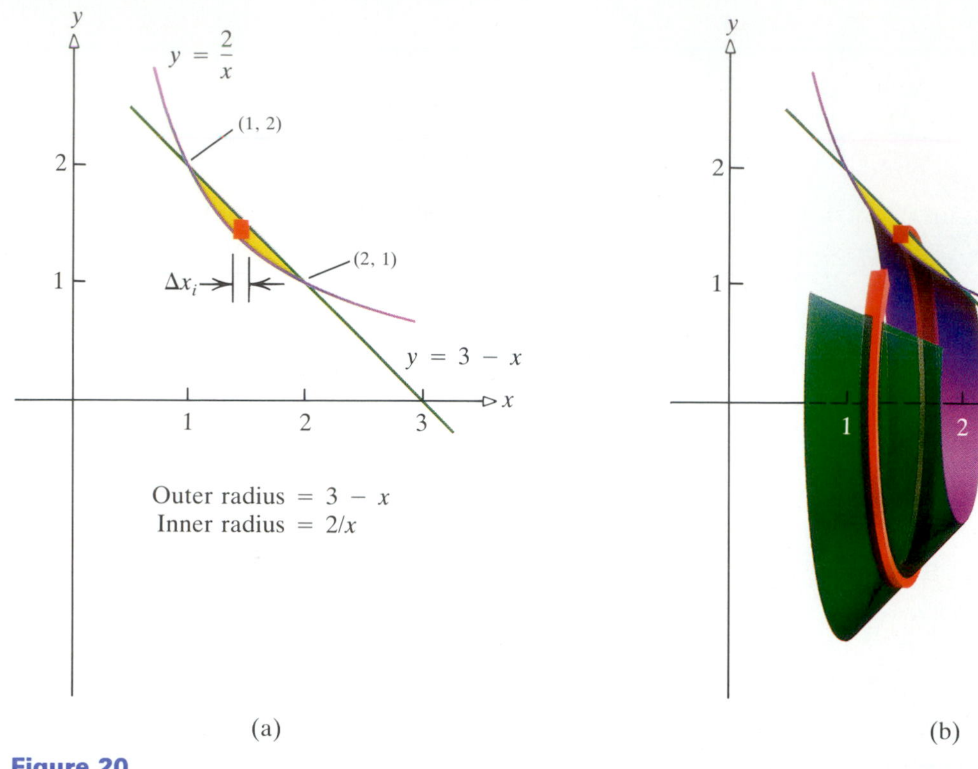

Figure 20

obey

$$\frac{2}{x} = 3 - x$$

$$x^2 - 3x + 2 = 0$$

$$(x-2)(x-1) = 0$$

So, the area to be revolved lies between $x = 1$ and $x = 2$. Partitioning the x-axis gives the volume of a typical washer:

$$\pi\{[f(u_i)]^2 - [g(u_i)]^2\}\Delta x_i = \pi\left[(3-u_i)^2 - \left(\frac{2}{u_i}\right)^2\right]\Delta x_i$$

The desired volume V is

$$V = \int_1^2 \pi\left[(3-x)^2 - \left(\frac{2}{x}\right)^2\right]dx = \pi\int_1^2\left(9 - 6x + x^2 - \frac{4}{x^2}\right)dx$$

$$= \pi\left(9x - 3x^2 + \frac{x^3}{3} + \frac{4}{x}\right)\bigg|_1^2 = \frac{\pi}{3}$$

If the solid of revolution is obtained by revolving about the y-axis, the partition will be along the y-axis and the thickness of the washer in (4) will be Δy_i. Let's look at an example.

EXAMPLE 5

Find the volume of the solid of revolution generated by revolving about the y-axis the area enclosed by the graphs of $y = 2x$ and $y = x^2$.

Solution

Figure 21 illustrates the situation. The points of intersection of the graphs obey

$$x^2 = 2x$$
$$x^2 - 2x = 0$$
$$x(x-2) = 0$$

So, the area to be revolved lies between $x = 0$ and $x = 2$, or between $y = 0$ and $y = 4$. Partitioning the y-axis gives the volume of a typical washer as

$$\pi[(\text{Outer radius})^2 - (\text{Inner radius})^2]\Delta y_i = \pi\left[(\sqrt{v_i})^2 - \left(\frac{v_i}{2}\right)^2\right]\Delta y_i$$

The desired volume V is

$$V = \int_0^4 \pi\left[(\sqrt{y})^2 - \left(\frac{y}{2}\right)^2\right]dy = \pi\int_0^4\left(y - \frac{y^2}{4}\right)dy = \pi\left(\frac{y^2}{2} - \frac{y^3}{12}\right)\Big|_0^4 = \frac{8\pi}{3}$$

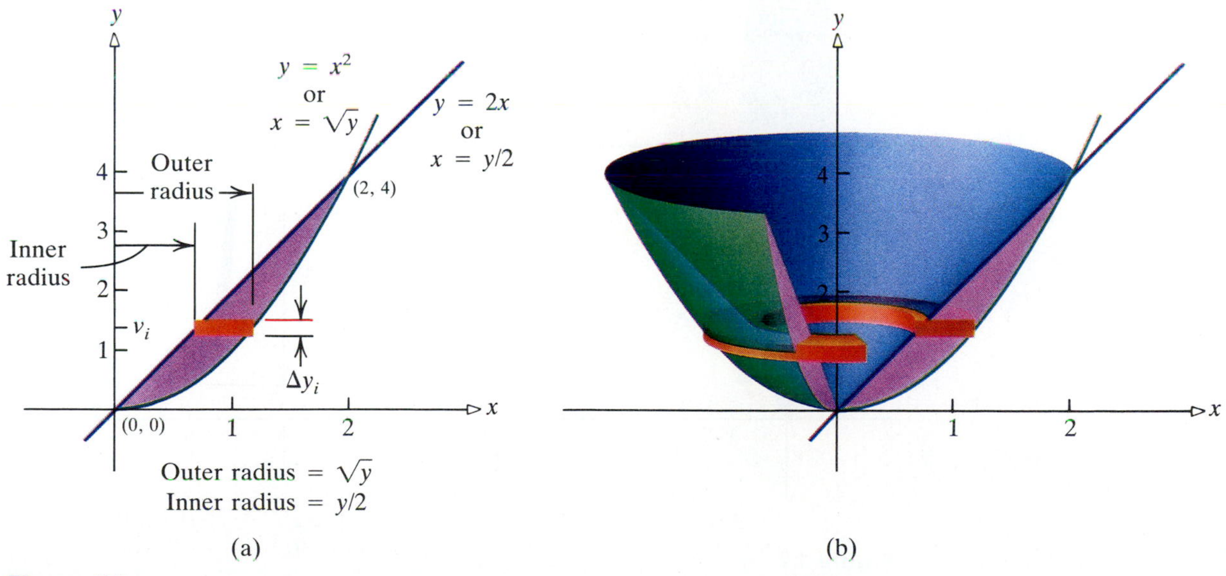

Figure 21

The next two examples illustrate how to obtain the solid of revolution by revolving about a line other than the x- or y-axis.

EXAMPLE 6

Find the volume of the solid of revolution generated by revolving about the line $y = 5$ the area enclosed by the graphs of $y = 2x$ and $y = x^2$.

Solution

Figure 22 illustrates the solid of revolution and a typical washer. Note that in this example the inner radius is $5 - 2x$ and the outer radius is $5 - x^2$. Partitioning the x-axis, the volume of a typical washer is

$$\pi[(5 - u_i^2)^2 - (5 - 2u_i)^2]\Delta x_i$$

and so the volume V is

$$V = \pi \int_0^2 [(5 - x^2)^2 - (5 - 2x)^2]\, dx = \pi \int_0^2 (x^4 - 14x^2 + 20x)\, dx$$

$$= \pi\left[\frac{x^5}{5} - \frac{14x^3}{3} + 10x^2\right]\Bigg|_0^2 = \frac{136}{15}\pi$$

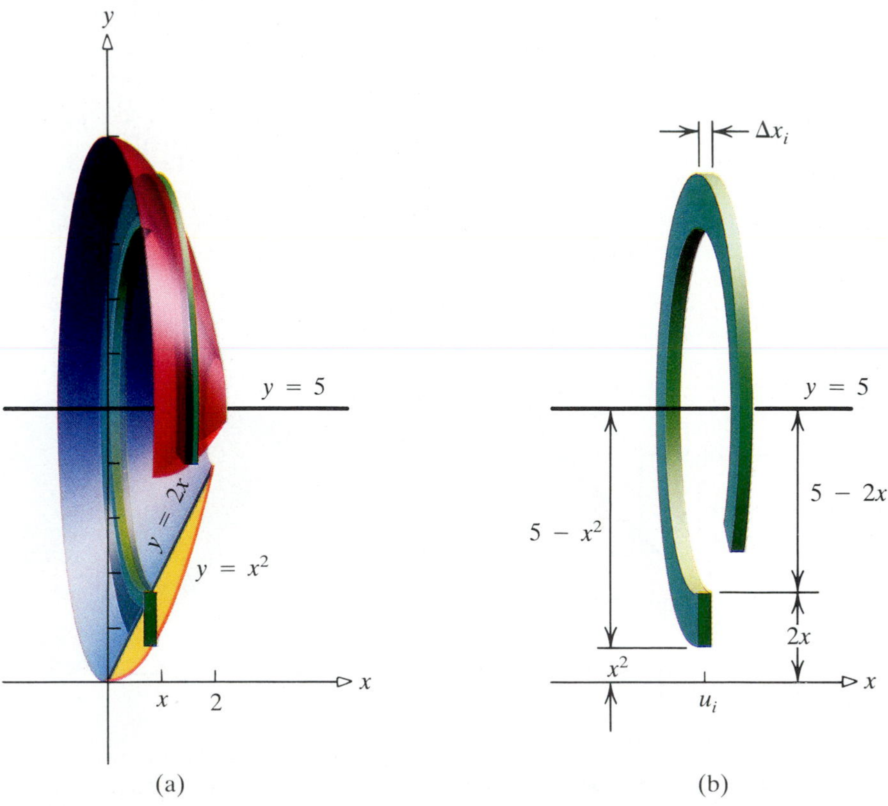

Figure 22

EXAMPLE 7

Find the volume of the solid of revolution generated by revolving about the line $x = 2$ the area enclosed by the graphs of $y = 2x$ and $y = x^2$.

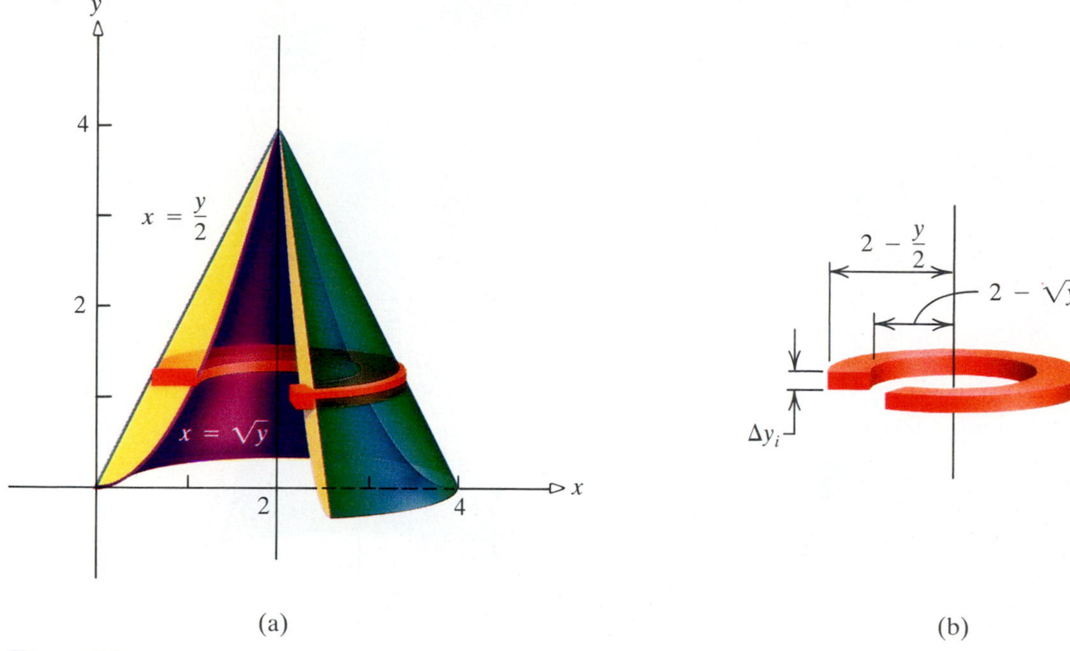

Figure 23

Solution

The situation is similar to Example 6, except that the roles of x and y are interchanged. We see from Figure 23 that the volume V is

$$\begin{aligned}
V &= \int_0^4 \left[\pi\left(2 - \frac{y}{2}\right)^2 - \pi(2 - \sqrt{y})^2\right] dy \\
&= \pi \int_0^4 \left[\left(4 - 2y + \frac{y^2}{4}\right) - (4 - 4\sqrt{y} + y)\right] dy \\
&= \pi \int_0^4 \left(\frac{y^2}{4} - 3y + 4\sqrt{y}\right) dy = \pi\left[\frac{y^3}{12} - \frac{3}{2}y^2 + \frac{8}{3}y^{3/2}\right]\Bigg|_0^4 \\
&= \pi\left[\frac{16}{3} - 24 + \frac{64}{3}\right] = \frac{8}{3}\pi
\end{aligned}$$

Application: Cooling Towers*

A proposed design for the cooling towers at a nuclear power plant is a branch of a hyperbola† rotated about an axis (see Fig. 24). The equation of the branch of the hyperbola—assuming it is rotated about the y-axis, its vertex is on the x-axis, and its center is at the origin—is $x = \sqrt{a^2 + by^2}$, where $a = 147$ and $b = 0.16$. The base of the tower is 442 feet below the vertex,

* Data obtained from U.S. Nuclear Regulatory Commission.

† Refer to Section 12.4 of Chapter 12, where a detailed discussion of the hyperbola is given.

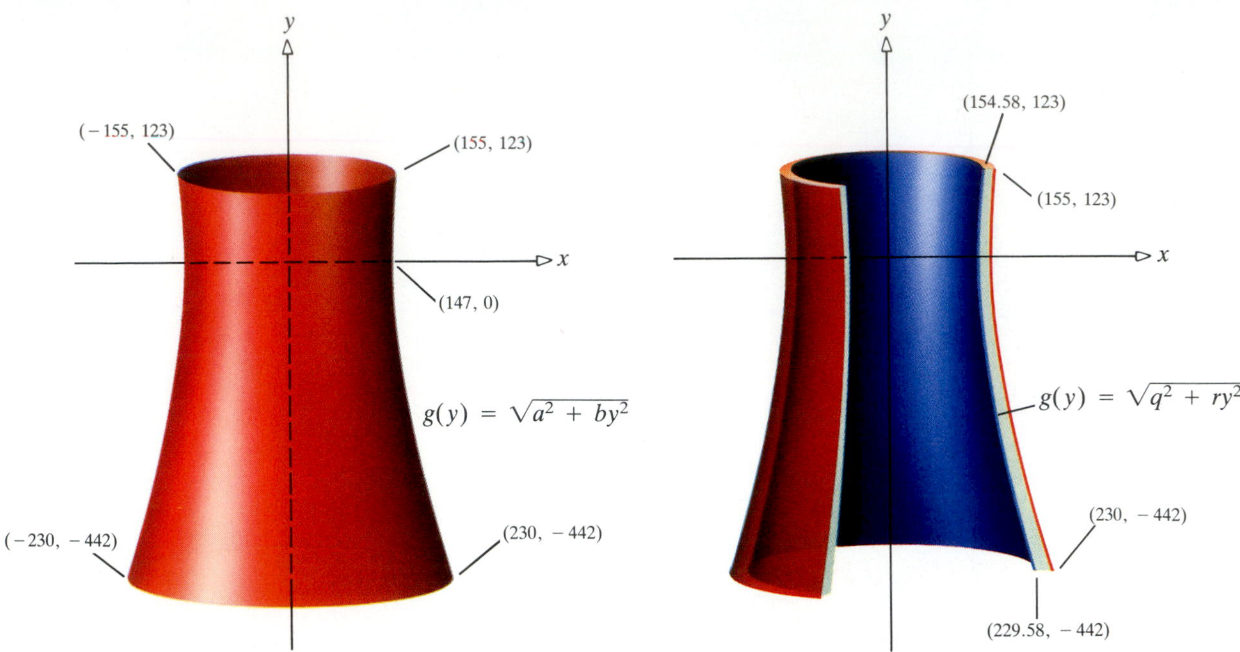

Figure 24

Figure 25

and the top of the tower is 123 feet above the vertex. Then, using (2), $V = \pi \int_c^d [g(y)]^2\, dy$, we find the volume of the cooling tower to be

$$V = \pi \int_{-442}^{123} (\sqrt{a^2 + by^2})^2\, dy = \pi \int_{-442}^{123} (a^2 + by^2)\, dy$$

$$= \pi\left(a^2y + \frac{by^3}{3}\right)\Bigg|_{-442}^{123} = \pi(16{,}913{,}712) \approx 53{,}135{,}993 \text{ cubic feet}$$

Now, let's assume that the walls of this cooling tower are a constant 5 inches (≈ 0.42 foot) thick. Then the interior hyperbola (see Fig. 25) has as its equation $x = \sqrt{q^2 + ry^2}$, where $q = 146.58$ and $r = 0.16$. Again using $V = \pi \int_c^d [g(y)]^2\, dy$, we find that the volume of the space between the interior walls—that is, the space inside the cooling tower—is

$$V = \pi \int_{-442}^{123} (\sqrt{q^2 + ry^2})^2\, dy = \pi \int_{-442}^{123} (q^2 + ry^2)\, dy$$

$$= \pi\left(q^2y + \frac{ry^3}{3}\right)\Bigg|_{-442}^{123} = \pi(16{,}844{,}045) \approx 52{,}917{,}129 \text{ cubic feet}$$

By combining this result with the previous result, we find that the volume of the walls themselves is approximately 218,864 cubic feet, so 218,864 cubic feet of reinforced concrete are needed to construct the walls of the tower.

EXERCISE 6.2

In Problems 1–14 find the volume of the solid of revolution generated by revolving the area enclosed by the graph of each function about the indicated axis. Draw a sketch first.

1. $y = x^2$, the x-axis, $x = 0$, and $x = 2$; about the x-axis

2. $y = 2x^2$, the x-axis, $x = 0$, and $x = 1$; about the x-axis

3. $y = x^3$, the x-axis, $x = 0$, and $x = 2$; about the x-axis

4. $y = 2x^4$, the x-axis, $x = 0$, and $x = 1$; about the x-axis

5. $y = 2\sqrt{x}$, the x-axis, $x = 1$, and $x = 4$; about the x-axis

6. $y = \sqrt{x}$, the x-axis, $x = 4$, and $x = 9$; about the x-axis

7. $y = 1/x$, the x-axis, $x = 1$, and $x = 2$; about the x-axis

8. $y = x^{2/3}$, the x-axis, $x = 0$, and $x = 8$; about the x-axis

9. $y = x^2$, the y-axis, $y = 1$, and $y = 4$; about the y-axis

10. $y = 2\sqrt{x}$, the y-axis, $y = 1$, and $y = 4$; about the y-axis

11. $y = 1/x$, the y-axis, $y = 1$, and $y = 4$; about the y-axis

12. $y = x^{2/3}$, the y-axis, $y = 1$, and $y = 4$; about the y-axis

13. $y = \sec x$, $y = 1$, $x = -1$, and $x = 1$; about the x-axis

14. $y = \tan x$, $y = 1$, and $x = 0$; about the x-axis

In Problems 15–18 find the volume of the solid of revolution generated by revolving the area enclosed by the graphs $y = f(x)$ and $y = g(x)$ and the two vertical lines about the x-axis.

15. $f(x) = 3x$, $g(x) = x^3$, $x = 0$, $x = 1$

16. $f(x) = 2x + 1$, $g(x) = x$, $x = 0$, $x = 3$

17. $f(x) = -x$, $g(x) = x^2$, $x = -1$, $x = 0$

18. $f(x) = \cos x$, $g(x) = \sin x$, $x = 0$, $x = \pi/4$

19. Find the volume of the solid of revolution generated by revolving the area enclosed by the graphs of $y = x$ and $y = x^2$ about the x-axis.

20. Find the volume of the solid of revolution generated by revolving the area enclosed in the first quadrant by the graphs of $y = x$ and $y = x^3$ about the x-axis.

In Problems 21–26 find the volume of the solid of revolution generated by revolving the area enclosed by the graph of each function about the indicated axis.

21. $y = x^2$, the y-axis, and $y = 4$; about the x-axis

22. $y = 2x^2$, the y-axis, and $y = 2$; about the x-axis

23. $y = x^3$, the x-axis, and $x = 2$; about the y-axis

24. $y = 2x^4$, the x-axis, and $x = 1$; about the y-axis

25. $y = 2\sqrt{x}$, the y-axis, and $y = 4$; about the x-axis

26. $y = \sqrt{x}$, the y-axis, and $y = 9$; about the x-axis

In Problems 27–35 find the volume of the solid of revolution generated by revolving the area enclosed by the graph of each function about the line indicated.

27. $y = x^2$, the x-axis, $x = 0$, and $x = 1$; about $x = 1$

28. $y = x^3$, the x-axis, $x = 0$, and $x = 1$; about $x = 1$

29. $y = \sqrt{x}$, the x-axis, $x = 0$, and $x = 4$; about $x = 4$

30. $y = 1/\sqrt{x}$, the x-axis, $x = 1$, and $x = 4$; about $x = 4$

31. $y = 1/x^2$, $y = 0$, $x = 1$, and $x = 4$; about $y = 4$

32. $y = \sqrt{x}$, $y = 0$, and $0 \le x \le 4$; about $y = 4$

33. $y = \cos x$, $y = 0$, $x = 0$, and $x = \pi/2$; about $y = 1$ $\left[\textit{Hint:}\ \cos^2 x = \dfrac{1 + \cos 2x}{2}\right]$

34. $y = \cos x$, $y = 0$, $x = 0$, and $x = \pi/2$; about $y = -1$ [See hint in Problem 33.]

35. $y = x$, $y = x^2$, the x-axis, $x = 0$, and $x = 1$; about $x = 1$

In Problems 36–39 find the volume of the solid of revolution generated by revolving the area enclosed by the graph of each function.

36. $y = \sqrt{x}$, $x = 4$, and $y = 0$ about:

(a) The x-axis
(b) The y-axis
(c) The line $x = 4$
(d) The line $y = 2$

37. $y = x^2$, $y = 4$, and $x = 0$ about:

(a) The x-axis
(b) The y-axis
(c) The line $y = 4$
(d) The line $x = 2$

38. $y = x^3$, $y = 8$, and $x = 0$ about:

(a) The x-axis
(b) The y-axis
(c) The line $x = 2$
(d) The line $y = 8$

39. $y = x^2$ and $y = x$ for $0 \le x \le 1$ about:

(a) The x-axis
(b) The y-axis
(c) The line $x = 3$
(d) The line $y = 1$

40. In the figure, the shaded area A is enclosed by the graphs of $xy = 1$, $x = 1$, $x = 2$, and $y = 0$. Find the volume of the solid generated by revolving the area A about the x-axis.

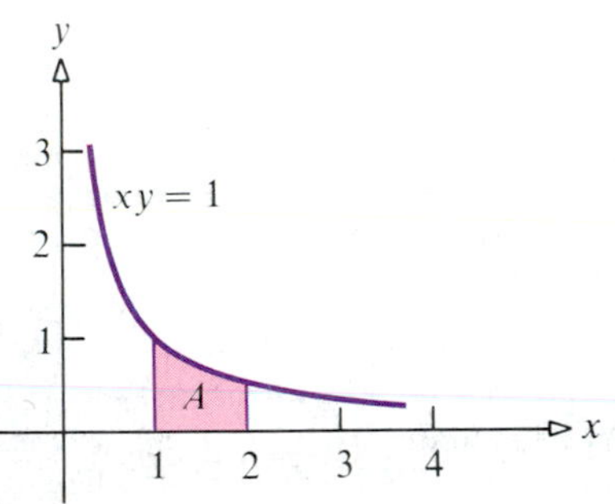

41. Let A be the area in the first quadrant enclosed by the x-axis and the graph of $y = kx - x^2$, where $k > 0$.

(a) In terms of k, find the volume produced when A is revolved around the x-axis.
(b) In terms of k, find the volume produced when A is revolved around the y-axis.
(c) Find the number k for which the volumes found in parts (a) and (b) are equal.

42. (a) Find all numbers b for which the graphs of $y = 2x + b$ and $y^2 = 4x$ intersect in two distinct points.
(b) If $b = -4$, find the area enclosed by the graphs of $y = 2x - 4$ and $y^2 = 4x$.
(c) If $b = 0$, find the volume of the solid generated by revolving about the x-axis the area enclosed by the graphs of $y = 2x$ and $y^2 = 4x$.

6.3 Volume of a Solid of Revolution: Shell Method

In this section we approach the problem of finding volumes of solids of revolution by using *cylindrical shells*. A cylindrical shell is the solid region between two concentric cylinders. For example, the water pipes in a house are cylindrical shells (see Fig. 26). If the inner radius is r_1 and the outer radius is r_2, the volume V of a cylindrical shell of height h is

$$V = \pi r_2^2 h - \pi r_1^2 h$$

We shall find it convenient to write this formula as

$$V = \pi h(r_2^2 - r_1^2) = \pi h(r_2 + r_1)(r_2 - r_1) = 2\pi h\left(\frac{r_2 + r_1}{2}\right)(r_2 - r_1)$$

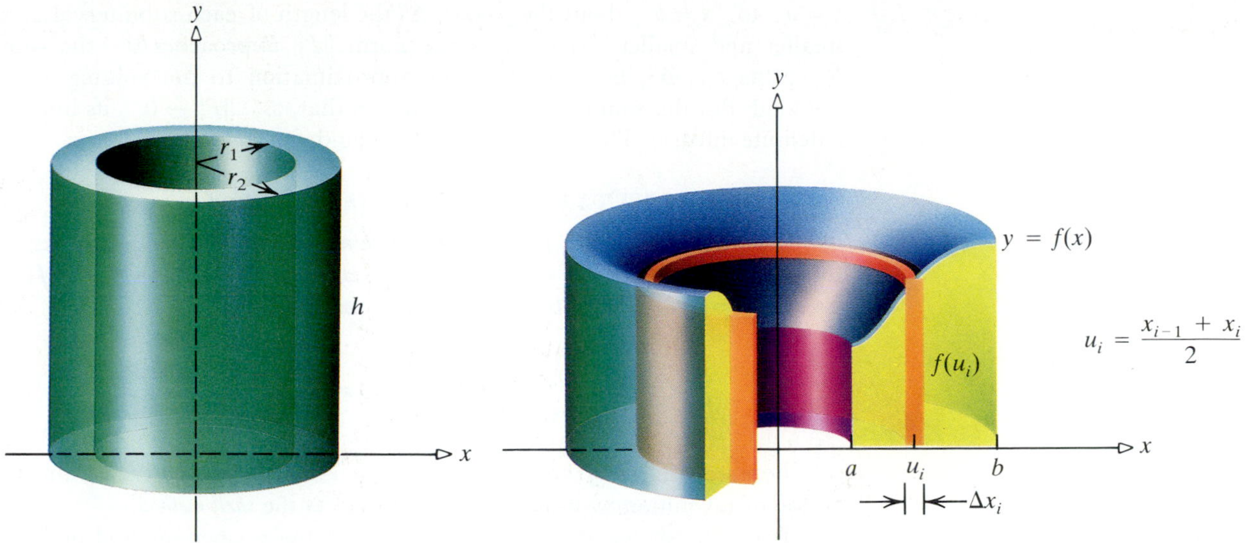

Figure 26 **Figure 27**

As an aid to remembering this formula, it helps to state it as

$$V = 2\pi(\text{Height})(\text{Average radius})(\text{Thickness}) \tag{1}$$

Revolving about the y-Axis

We use (1) to find the volume of the solid generated by revolving the area under the graph of a function $y = f(x)$ from $x = a$ to $x = b$ about the y-axis. We assume f is nonnegative and continuous on $[a, b]$ and $a \geq 0$ (see Fig. 27). To find the volume of such a solid, we partition the interval $[a, b]$ into n subintervals

$$[a, x_1], \quad [x_1, x_2], \quad \ldots, \quad [x_{i-1}, x_i], \quad \ldots, \quad [x_{n-1}, b]$$

The length of the ith subinterval is $\Delta x_i = x_i - x_{i-1}$, $i = 1, 2, \ldots, n$. Now, we concentrate on the rectangle whose base is the subinterval $[x_{i-1}, x_i]$ and whose height is $f(u_i)$, where $u_i = (x_{i-1} + x_i)/2$ is the midpoint of the subinterval $[x_{i-1}, x_i]$. When this rectangle is revolved about the y-axis, it generates a cylindrical shell of inner radius x_{i-1}, outer radius x_i, and height $f(u_i)$. From (1), the volume of this cylindrical shell is

$$V = 2\pi(\text{Height})(\text{Average radius})(\text{Thickness}) = 2\pi f(u_i)\frac{(x_{i-1}+x_i)}{2}(x_i - x_{i-1})$$

$$= 2\pi f(u_i)u_i\Delta x_i = 2\pi u_i f(u_i)\Delta x_i$$

The sum of all the volumes due to each subinterval is

$$\sum_{i=1}^{n} 2\pi u_i f(u_i)\Delta x_i$$

This sum represents an approximation to the volume V of the solid generated by revolving the area under the graph of $y = f(x)$ from

$x = a$ to $x = b$ about the y-axis. As the length of each subinterval gets smaller and smaller—that is, as the norm $\|P\|$ approaches 0—the sum $\sum_{i=1}^{n} 2\pi u_i f(u_i)\Delta x_i$ becomes a better approximation to the volume V of the solid. But this sum is a Riemann sum, so that, as $\|P\| \to 0$, its limit is a definite integral. This suggests the following definition:

[6.3.1] DEFINITION / ***Volume: Shell Method.***

The volume V of the solid generated by revolving the area under the graph of a continuous nonnegative function $y = f(x)$ from $x = a$ to $x = b$ about the y-axis is

$$V = \lim_{\|P\| \to 0} \sum_{i=1}^{n} 2\pi u_i f(u_i)\Delta x_i = \int_a^b 2\pi x f(x)\, dx \qquad (2)$$

Because of the way we obtained definition (6.3.1), finding the volume of a solid of revolution by using (2) is referred to as the *shell method.*

It can be shown that this definition and the washer method of Section 6.2 both lead to the same answer.* The advantage of having two equivalent, yet different, formulas is that on occasion one might be easier to use. The next example illustrates just such a situation.

EXAMPLE 1

Find the volume of the solid generated by revolving the area under the graph of $y = x^2 + 2x$ from $x = 0$ to $x = 1$ about the y-axis.

Solution

By the shell method: Figure 28 illustrates the situation. Note that for the shell method, a revolution about the y-axis requires a partition of the x-axis. The height of the rectangle is $f(u_i)$, where $f(x) = x^2 + 2x$. The required volume V is

$$V = 2\pi \int_0^1 x(x^2 + 2x)\, dx = 2\pi \int_0^1 (x^3 + 2x^2)\, dx$$

$$= 2\pi\left(\frac{x^4}{4} + \frac{2x^3}{3}\right)\Bigg|_0^1 = \frac{11\pi}{6}$$

By the washer method: Figure 29 illustrates the situation. Note that for the washer method, a revolution about the y-axis requires a partition of the y-axis. To use the washer method, we need to solve for x in the expression $y = x^2 + 2x$. We treat this as a quadratic equation in the variable x and use the quadratic formula† with $a = 1$, $b = 2$, and $c = -y$ to obtain $x = -1 \pm \sqrt{1 + y}$. Since we require $x > 0$, we use only the + sign.

---

* This topic is discussed in detail in an article by Charles A. Cable, "The Disk and Shell Method," *American Mathematical Monthly* 91, no. 2 (Feb. 1984): 139.

† If $ax^2 + bx + c = 0$, then $x = \dfrac{-b \pm \sqrt{b^2 - 4ac}}{2a}$.

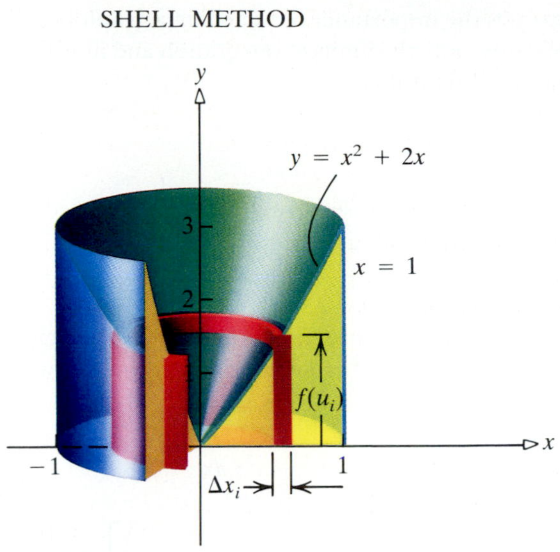

SHELL METHOD

Figure 28

WASHER METHOD

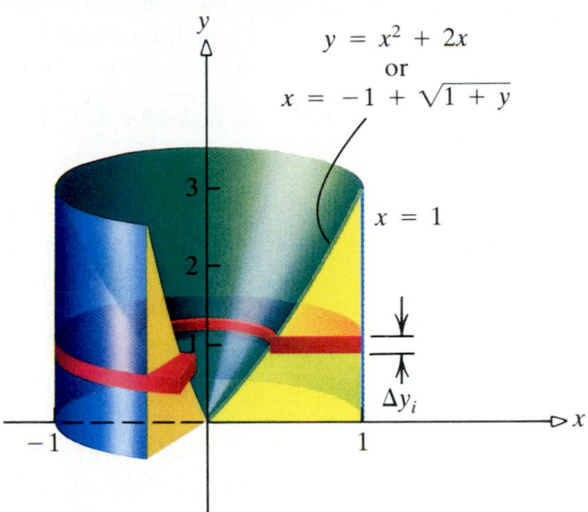

Figure 29

Then

$$V = \pi \int_0^3 [(1)^2 - (-1 + \sqrt{1+y})^2]\, dy$$

$$= \pi \int_0^3 [1 - (1 - 2\sqrt{1+y} + 1 + y)]\, dy$$

$$= \pi \int_0^3 [2\sqrt{1+y} - 1 - y]\, dy$$

$$= 2\pi \int_0^3 \sqrt{1+y}\, dy - \pi \int_0^3 (1+y)\, dy$$

Now

$$2\pi \int_0^3 \sqrt{1+y}\, dy \underset{\uparrow}{=} 2\pi \int_1^2 u(2u\, du) = 4\pi \left.\frac{u^3}{3}\right|_1^2 = \frac{28\pi}{3}$$

Set $u^2 = 1 + y$.

$$\pi \int_0^3 (1+y)\, dy = \pi\left(y + \frac{y^2}{2}\right)\Bigg|_0^3 = \frac{15\pi}{2}$$

so the required volume V is

$$V = \frac{28\pi}{3} - \frac{15\pi}{2} = \frac{11\pi}{6}$$

This example gives a clue as to when the shell method is preferable to the washer method—namely, when it may be difficult to solve $y = f(x)$ for x in terms of y. For example, if the function given in Example 1 had been $y = x^5 + x^2 + 1$, the only practical choice would have been the shell method.

The next example illustrates the importance of sketching a graph before blindly using a formula. Note especially the limits of integration and how they come about when the washer method is used.

EXAMPLE 2

Find the volume of the solid generated by revolving the area enclosed by $y = x^2$ and $y = 12 - x$ to the right of $x = 1$ about the y-axis.

Solution

By the shell method: Figure 30 illustrates the situation. The height of a typical rectangle is $(12 - x) - (x^2) = 12 - x - x^2$, and the integration is with respect to x from 1 to 3. The desired volume V is

$$V = 2\pi \int_1^3 x(12 - x - x^2)\,dx = 2\pi \int_1^3 (12x - x^2 - x^3)\,dx$$

$$= 2\pi\left(6x^2 - \frac{x^3}{3} - \frac{x^4}{4}\right)\Bigg|_1^3 = 2\pi\left[\left(54 - 9 - \frac{81}{4}\right) - \left(6 - \frac{1}{3} - \frac{1}{4}\right)\right] = \frac{116\pi}{3}$$

By the washer method: Figure 31 illustrates the situation. The integration is with respect to y from 1 to 11, with a change occurring at $y = 9$. The required volume V is obtained by subtracting the inner volume from the outer volume. Thus,

$$V = \pi \int_1^9 (\sqrt{y})^2\,dy + \pi \int_9^{11} (12 - y)^2\,dy - \pi \int_1^{11} (1)^2\,dy$$

$$= \pi \frac{y^2}{2}\Bigg|_1^9 + \pi\left[144y - 24\left(\frac{y^2}{2}\right) + \frac{y^3}{3}\right]\Bigg|_9^{11} - \pi y\Bigg|_1^{11}$$

$$= 40\pi + \pi\left[288 - (12)(40) + \frac{602}{3}\right] - \pi(10) = -162\pi + \frac{602\pi}{3} = \frac{116\pi}{3}$$

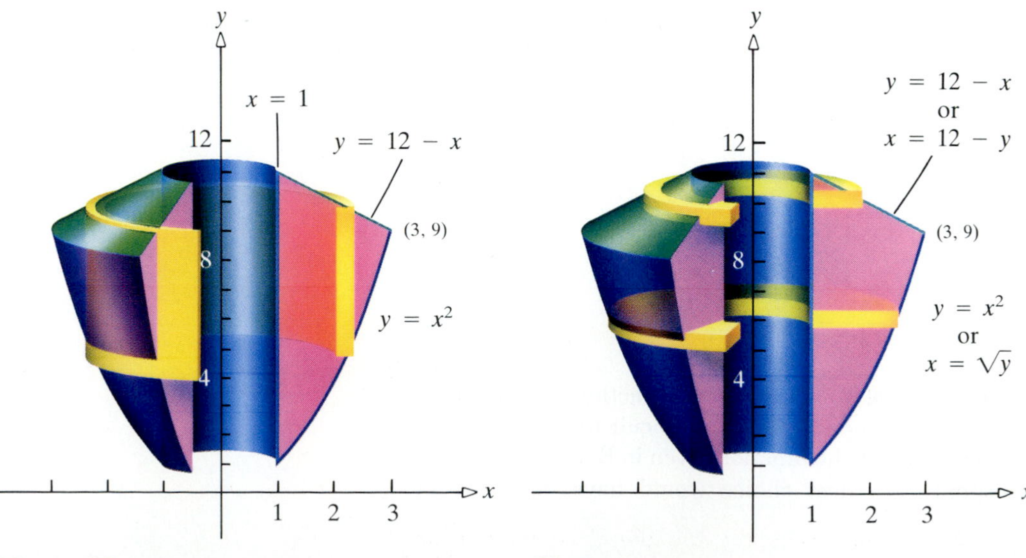

Figure 30

Figure 31

Revolving about the *x*-Axis

The shell method can also be used when the solid is generated by a revolution about the x-axis, as shown in Figure 32.

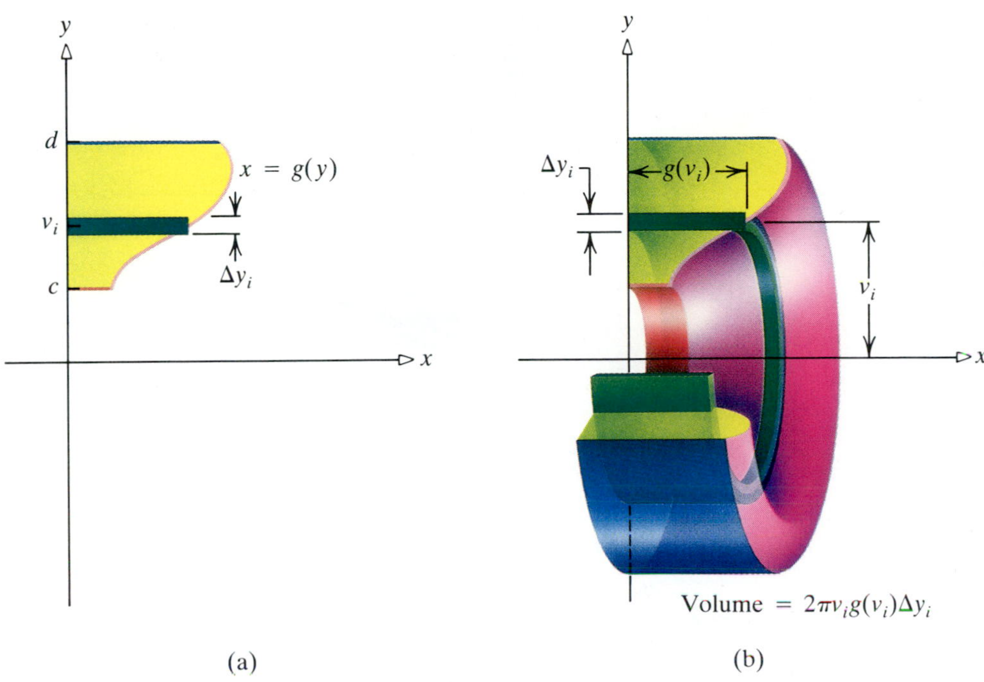

Figure 32

[6.3.2] DEFINITION / *Volume.*

The volume V of the solid generated by revolving the area enclosed by the graph of $x = g(y)$, $g(y) \geq 0$, the y-axis, $y = c$, and $y = d$ about the x-axis is

$$V = 2\pi \int_c^d yg(y)\,dy \tag{3}$$

Figure 32 illustrates how (3) is obtained.

EXAMPLE 3

Find the volume generated by revolving the area enclosed in the first quadrant by $x^2/a^2 + y^2/b^2 = 1$, about the x-axis. Here, a and b are positive constants.

Solution

By the shell method: Figure 33 illustrates the situation. The integration is with respect to y from 0 to b and the function g is $g(y) = (a/b)\sqrt{b^2 - y^2}$.

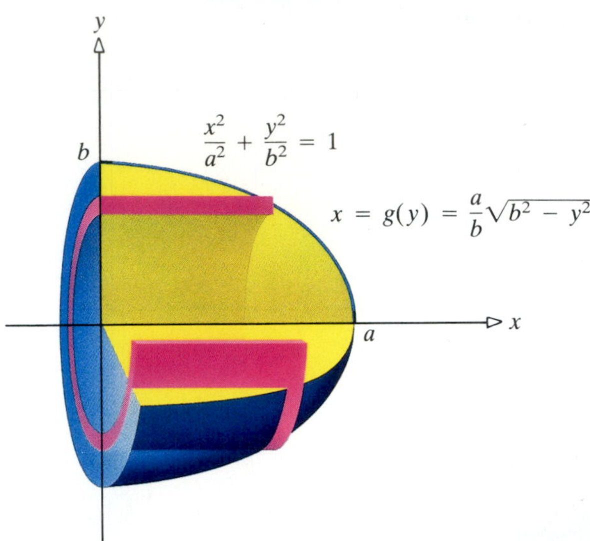

Figure 33

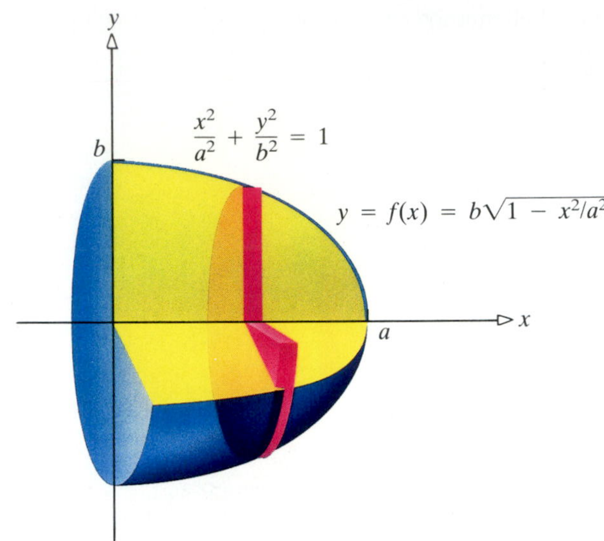

Figure 34

The required volume V is

$$V = 2\pi \int_0^b y g(y)\, dy = 2\pi \int_0^b y\left(\frac{a}{b}\sqrt{b^2 - y^2}\right) dy$$

$$\underset{\substack{\uparrow \\ \text{Set } u^2 = b^2 - y^2}}{=} \frac{2\pi a}{b}\int_b^0 u(-u\, du) = \frac{2\pi a}{b}\left(\frac{-u^3}{3}\right)\Bigg|_b^0 = \frac{2\pi a}{b}\left(\frac{b^3}{3}\right) = \frac{2\pi ab^2}{3}$$

By the washer method: Figure 34 illustrates the situation. The integration is with respect to x from 0 to a. The required volume V is

$$V = \pi \int_0^a b^2\left(1 - \frac{x^2}{a^2}\right) dx = \pi b^2\left(x - \frac{x^3}{3a^2}\right)\Bigg|_0^a = \pi b^2\left(a - \frac{a}{3}\right) = \frac{2\pi ab^2}{3}$$

If $a = b$ in Example 3, the solid generated is a hemisphere and its volume is $2\pi a^3/3$. The volume of a sphere of radius R is therefore $4\pi R^3/3$.

Table 1 summarizes the washer and shell methods.

Table 1

	Washer Method	Shell Method
Revolution about x-axis	Partition the x-axis; use vertical strips	Partition the y-axis; use horizontal strips
Revolution about y-axis	Partition the y-axis; use horizontal strips	Partition the x-axis; use vertical strips

EXAMPLE 4

Find the volume of the solid of revolution generated by revolving the area under the graph of $y = \cos x$ from $x = 0$ to $x = \pi/2$ about the x-axis.

Solution

Figure 35 illustrates the situation. We choose to use the disk method here because the shell method requires that we solve for x in the equation $y = \cos x$, the solution of which we will not study until Chapter 8. Thus, we partition along the x-axis, obtaining disks of radius $\cos u_i$ and width Δx_i. The required volume V is

$$V = \pi \int_0^{\pi/2} \cos^2 x\, dx \underset{\uparrow}{=} \pi\left(\frac{x}{2} + \frac{\sin 2x}{4}\right)\Bigg|_0^{\pi/2} = \frac{\pi^2}{4} \approx 2.467$$

$$\cos^2 x = \tfrac{1}{2}(1 + \cos 2x)$$

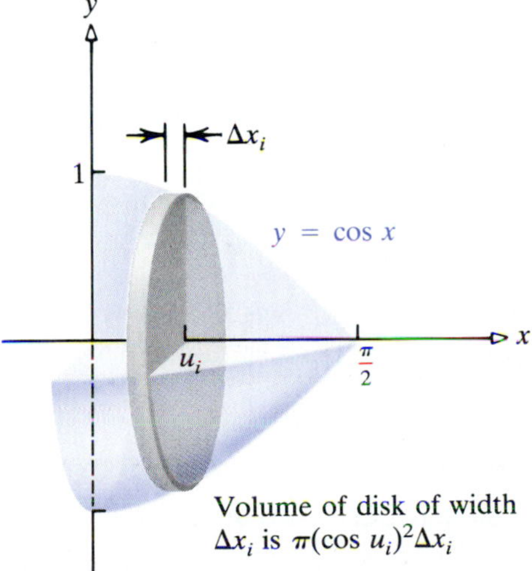

Figure 35

EXAMPLE 5

Find the volume of the solid of revolution generated by revolving the area enclosed by $y = f(x) = 2x - 2x^2$ and $y = 0$ about the line $x = 2$.

Solution

Figure 36 illustrates a rectangle with a base $[x_{i-1}, x_i]$ revolved about $x = 2$ to form a cylindrical shell with average radius $2 - x_i$, height $2x_i - 2x_i^2$, and width Δx_i. The desired volume V is

$$V = 2\pi \int_0^1 (2 - x)(2x - 2x^2)\, dx$$

$$= 4\pi \int_0^1 (x^3 - 3x^2 + 2x)\, dx = 4\pi\left[\frac{x^4}{4} - x^3 + x^2\right]\Bigg|_0^1 = \pi$$

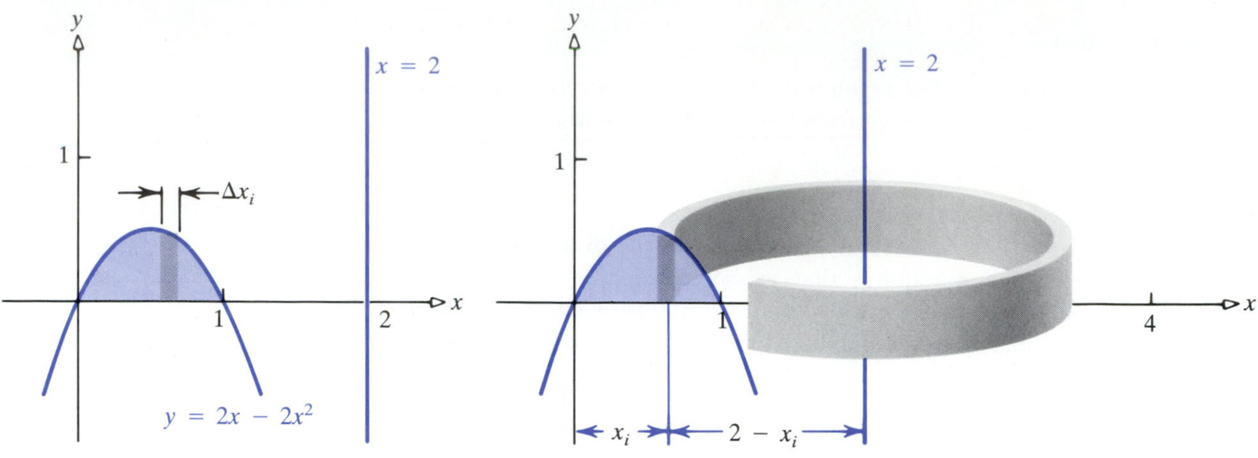

Figure 36

EXERCISE 6.3

In Problems 1–12 use the shell method to find the volume of the solid of revolution generated by revolving the area enclosed by the graph of each function about the indicated axis.

1. $y = x^2 - 1$, the x-axis, $x = 1$, and $x = 3$; about the y-axis

2. $y = x^3 + x$, the x-axis, $x = 0$, and $x = 1$; about the y-axis

3. $y = \sqrt{x} + x$, the x-axis, $x = 1$, and $x = 4$; about the y-axis

4. $y = x^{2/3} + x^{1/3}$, the x-axis, $x = 1$, and $x = 8$; about the y-axis

5. $y = x^3$ and $y = x^2$; about the y-axis

6. $y = \sqrt{x}$ and $y = x^2$; about the y-axis

7. $y = x^3$, the y-axis, and $y = 8$; about the x-axis

8. $y = \sqrt{x}$, the y-axis, and $y = 2$; about the x-axis

9. $x = \sqrt{y}$, the y-axis, and $y = 1$; about the x-axis

10. $x = 4\sqrt{y}$, the y-axis, and $y = 4$; about the x-axis

11. $y = x$ and $y = x^2$; about the x-axis

12. $y = x$ and $y = x^3$; in the first quadrant about the x-axis

In Problems 13–22 use either the shell method or the washer method to find the volume of the solid of revolution generated by revolving the area enclosed by the graph of each function about the indicated axis.

13. $y = \sqrt{x}$, the y-axis, and $y = 4$; about the x-axis

14. $y = 1/x$, the x-axis, $x = 1$, and $x = 4$; about the y-axis

15. $y = x^3$ and $y = x$ to the right of $x = 0$; about the y-axis

16. $y = x^3$ and $y = x^2$ to the right of $x = 0$; about the x-axis

17. $y = 3x^2$ and $y = 30 - x$ to the right of $x = 1$; about the y-axis

18. $y = 3x^2$ and $y = 30 - x$ to the right of $x = 1$; about the x-axis

19. $y = x^2$ and $y = 8 - x^2$ to the right of $x = 1$; about the y-axis

20. $y = x^2$ and $y = 8 - x^2$ to the right of $x = 1$; about the x-axis

21. $y = \sqrt{x}$ and $y = 18 - x^2$ to the right of $x = 1$; about the y-axis

22. $y = \sqrt{x}$ and $y = 18 - x^2$ to the right of $x = 1$; about the x-axis

In Problems 23–28 use the shell method to find the volume of the solid of revolution generated by revolving the area enclosed by the graph of each function about the line indicated.

23. $y = x^2$ and $y = 4x - x^2$; about $x = 2$

24. $y = x^2$ and $y = 4x - x^2$; about $x = 3$

25. $y = x^2$, $y = 0$, $x = 1$, and $x = 2$; about $x = 1$

26. $y = x^2$, $y = 0$, $x = 1$, and $x = 2$; about $x = 2$

27. $x = y - y^2$ and the y-axis; about $y = 1$

28. $x = y - y^2$ and the y-axis; about $y = -1$

29. Use the shell method to find the outside volume of the cooling tower described in the application at the end of Section 6.2. [*Hint:* The equation $x = \sqrt{a^2 + by^2}$ can be rewritten as $y = \pm\sqrt{(x^2 - a^2)/b}$.]

30. Let A be the area of the first quadrant enclosed by the x-axis and the graph of $y = 2x - x^2$.

(a) Find the volume produced when A is revolved about the x-axis.

(b) Find the volume produced when A is revolved about the y-axis.

In Problems 31–34 find the volume of the solid of revolution generated by revolving the area enclosed by the graph of each function about the indicated axis by first using the shell method and then the disk method.

31. $y = 2x - x^2$ and the x-axis about:

(a) The x-axis (b) The y-axis

32. $y = x^2$ and $y = 2$ about the line $y = 4$

33. $y = x^2$ and $y = 8 - x^2$ about:

(a) The y-axis (b) The line $x = 4$

Use first quadrant only.

34. $y = x^2$ and $y = x$ about:

(a) The y-axis (b) The line $x = 3$

Use first quadrant only.

35. Suppose $f(x) \geq 0$ for $x \geq 0$, and the area below f from $x = 0$ to $x = k$ is revolved about the x-axis. If, for each $k > 0$, the volume of the resulting solid is $\frac{1}{5}k^5 + k^4 + \frac{4}{3}k^3$, find f.

36. Consult the accompanying figure. The area enclosed by the graphs of $y_1 = g(x)$, $y_2 = f(x)$, and the y-axis is to be revolved about the y-axis. Show that the

(*Problem 36 continues in the next column*)

36. (*continued*)
resulting volume V is given by:

(a) $V = \pi a^2[f(a) - g(a)] + 2\pi \int_a^b x[f(x) - g(x)]\, dx$
if the shell method is used

(b) $V = \pi \int_{g(a)}^{g(b)} [g^{-1}(y)]^2\, dy + \pi \int_{g(b)}^{f(a)} [f^{-1}(y)]^2\, dy$
if the disk method is used

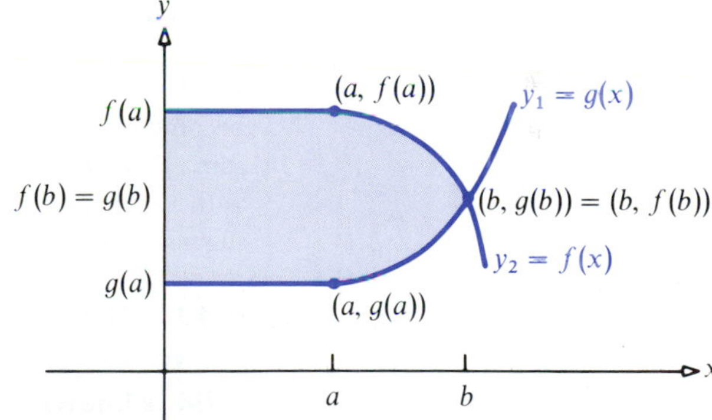

Volume by Slicing 6.4

In this section we consider the problem of computing the volume of a solid by the *method of slicing*. The idea is to cut the solid into thin slices using planes perpendicular to the x-axis and then add up the volumes of these slices to obtain the total volume (see Fig. 37).

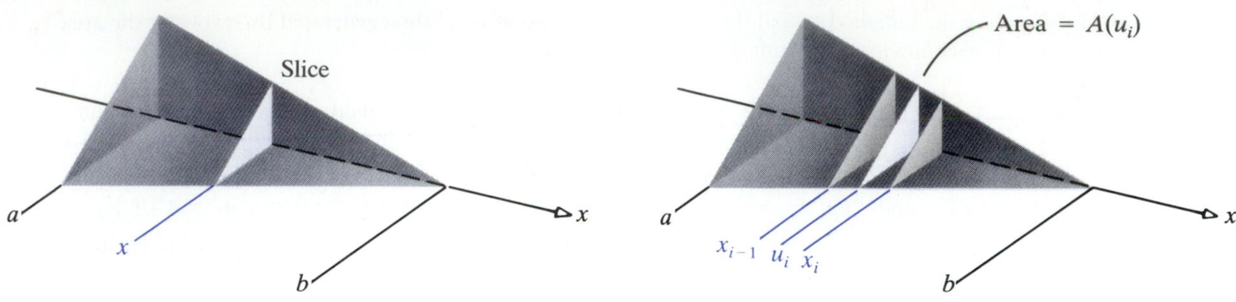

Figure 37

Figure 38

We begin by partitioning the interval $[a, b]$ into n subintervals

$$[a, x_1], \quad [x_1, x_2], \quad \ldots, \quad [x_{i-1}, x_i], \quad \ldots, \quad [x_{n-1}, b]$$

The length of each subinterval is

$$\Delta x_i = x_i - x_{i-1} \qquad i = 1, \ldots, n$$

In the ith subinterval, we pick a number u_i and consider the slice (cross section) cut from the solid at $x = u_i$ by a plane perpendicular to the x-axis. We denote the area of this cross section by $A(u_i)$, as indicated in Figure 38. The volume of the thin slice from x_{i-1} to x_i may then be approximated by $A(u_i)\Delta x_i$, and the sum of all the volumes due to each subinterval is

$$\sum_{i=1}^{n} A(u_i)\Delta x_i \tag{1}$$

This sum represents an approximation to the volume V of the solid from $x = a$ to $x = b$. As the length of each subinterval gets smaller and smaller—that is, as the norm $\|P\|$ approaches 0—the sum $\sum_{i=1}^{n} A(u_i)\Delta x_i$ becomes a better approximation to the volume V of the solid. But the sum (1) is a Riemann sum. Hence, if the cross sections $A(x)$ vary continuously with x, then, as $\|P\| \to 0$, the limit of this sum is a definite integral. This suggests the following definition:

[6.4.1] DEFINITION / *Volume: Slicing Method.*

If for each x in $[a, b]$, the area $A(x)$ of the cross section of a solid is known and is continuous on $[a, b]$, then the volume V of the solid is

$$V = \lim_{\|P\| \to 0} \sum_{i=1}^{n} A(u_i)\,\Delta x_i = \int_a^b A(x)\,dx \tag{2}$$

Formula (2) may be used if the solid whose volume we seek is not a solid of revolution. Although the volume of such a solid usually requires the use of a double integral or a triple integral (the subjects of Chap. 18), in certain situations a single definite integral will suffice. For example, this is the case whenever parallel cross sections of the solid all have the same simple geometric configuration (all are semicircles, or triangles, or squares, etc.). The result is that the area $A(x)$ of the cross section is easy to calculate, and can then be used to obtain the volume of the solid.

We use (2) when the area of the cross section varies (that is, when the areas of the slices are not all equal). For solids of constant cross-sectional areas, such as the one in Figure 39, the volume is simply the area of the cross section times its thickness—that is,

$$V = (\text{Area of cross section})(b - a)$$

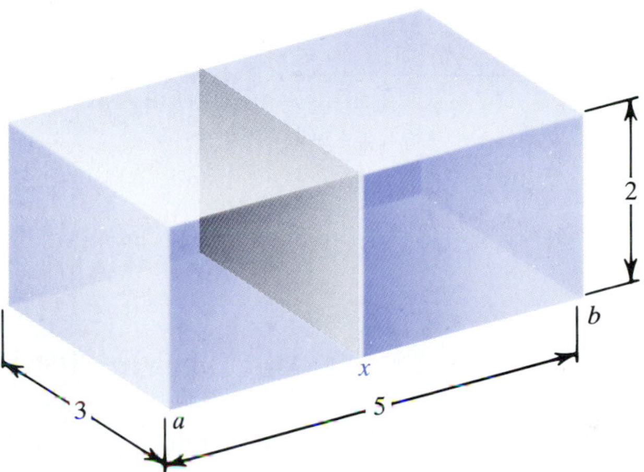

Figure 39

For example, the volume of the solid in Figure 39 is $(6)(5) = 30$, since the cross section $A(x)$ is always equal to $(2)(3) = 6$ and the length $b - a$ is 5.

The use of definition (6.4.1) and (2) to calculate the volume of a solid is usually referred to as the *slicing method*. We illustrate its use in the next example.

EXAMPLE 1

Find the volume of a right circular cone having radius R and height h.

Solution

We position the cone in such a way that its vertex is at the origin and its axis coincides with the x-axis (see Fig. 40). Therefore, the cone extends from $x = 0$ to $x = h$. The cross section at any number x is a circle. To obtain its area $A(x)$, we must find its radius $r(x)$, which depends on x. Because we have similar triangles, as shown in Figure 40(b), we have

$$\frac{r(x)}{x} = \frac{R}{h}$$

$$r(x) = \frac{xR}{h}$$

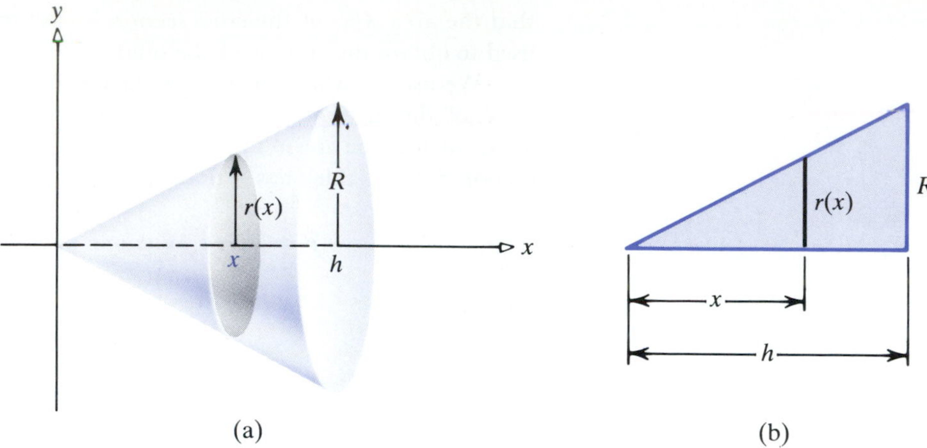

Figure 40

so that

$$A(x) = \pi[r(x)]^2 = \pi\left(\frac{xR}{h}\right)^2$$

Using (2), we find that the volume of the cone is

$$V = \int_a^b A(x)\,dx = \int_0^h \frac{\pi x^2 R^2}{h^2}\,dx = \frac{\pi R^2}{h^2}\left(\frac{x^3}{3}\right)\bigg|_0^h = \frac{\pi R^2 h}{3}$$

Of course, a right circular cone is a solid of revolution, and its volume can also be found by the disk method (see Problem 11).

Example 1 illustrates that the way in which the solid is positioned relative to the x-axis is important if the slice $A(x)$ is to be easily found and integrated. The next two examples further illustrate the importance of good positioning.

EXAMPLE 2

A solid has a circular base of radius 3 units. Find the volume of the solid if every plane cross section that is perpendicular to a fixed diameter is an equilateral triangle.

Solution

Position the circular base as shown in Figure 41(a), with the x-axis as the fixed diameter. The equation of the circle is then $x^2 + y^2 = 9$. The cross section of the solid is an equilateral triangle of side $2y$ and area $A(x) = \sqrt{3}y^2$, as indicated in Figure 41(b). Since $y^2 = 9 - x^2$, we have $A(x) = \sqrt{3}(9 - x^2)$, so that the volume is

$$V = \int_a^b A(x)\,dx \underset{\uparrow}{=} 2\int_0^3 \sqrt{3}(9 - x^2)\,dx = 2\sqrt{3}\left(9x - \frac{x^3}{3}\right)\bigg|_0^3 = 36\sqrt{3}$$

Use symmetry.

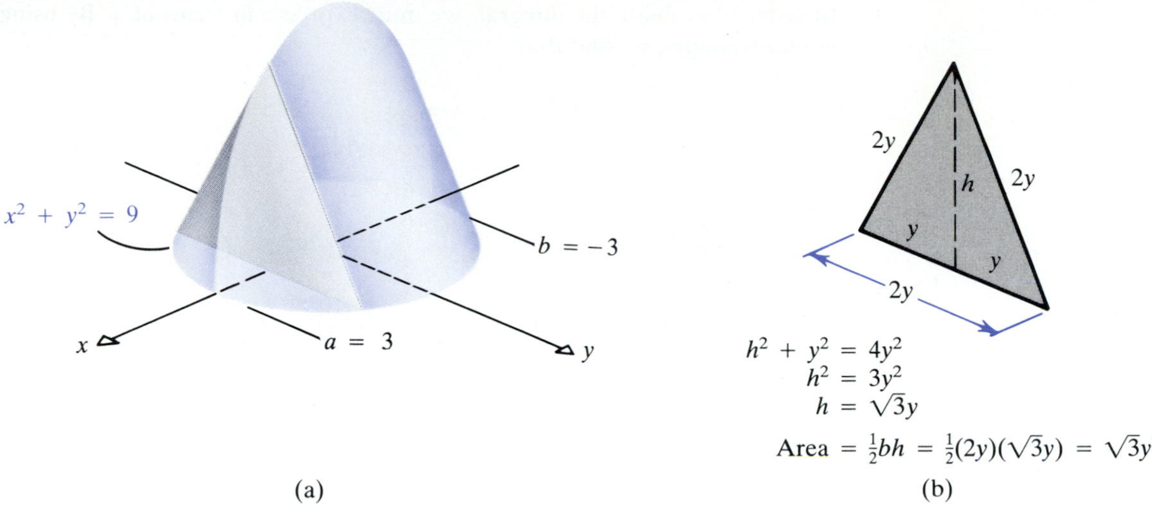

Figure 41

EXAMPLE 3

Find the volume of a pyramid having a height of length h and a square base, with each side of length b.

Solution

Position the pyramid so that its vertex is at the origin and its height is along the positive x-axis. Then a typical cross section $A(x)$ is a square with side s, where s is a function of x (see Fig. 42). The volume of the pyramid is

$$V = \int_0^h s^2\,dx$$

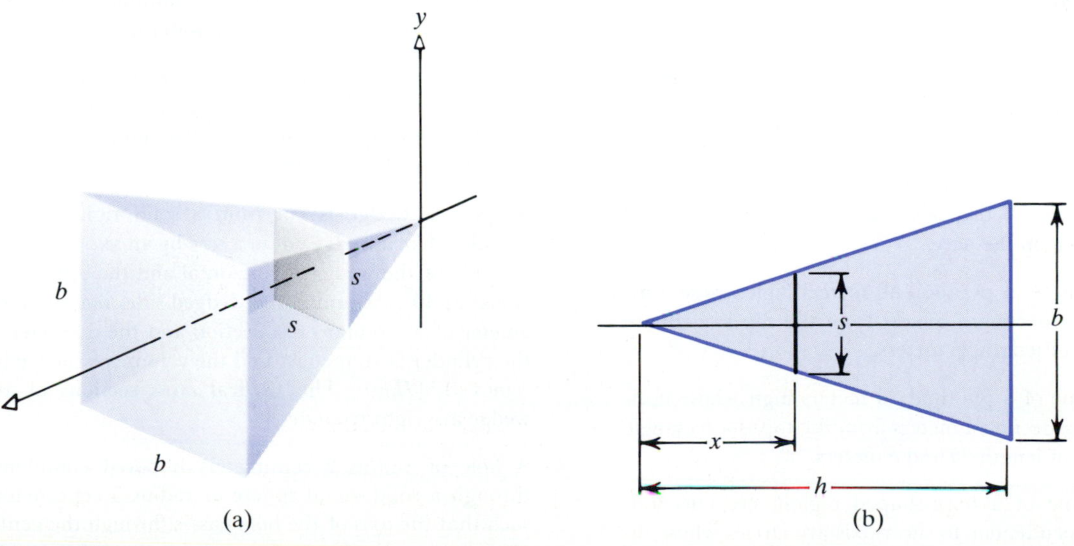

Figure 42

In order to evaluate the integral, we must express s in terms of x. By using similar triangles, we find that

$$\frac{s}{x} = \frac{b}{h}$$

$$s = \frac{b}{h}x$$

As a result,

$$V = \int_0^h \frac{b^2}{h^2}x^2\,dx = \frac{b^2}{h^2}\left(\frac{x^3}{3}\right)\Big|_0^h = \frac{1}{3}b^2h$$

EXERCISE 6.4

In Problems 1–10 compute the volume of each solid by the method of slicing.

1. The base is a square with each side of length 2; cross sections taken perpendicular to the base are rectangles of height 1.

2. The base is a square with each side of length 1; cross sections taken perpendicular to the base are rectangles of height 2.

3. The volume of a solid whose base is the area enclosed by the circle $x^2 + y^2 = 1$; cross sections taken perpendicular to the base and parallel to the y-axis are squares.

4. Rework Problem 3 if the cross sections are isosceles triangles of constant altitude h. [*Hint:* $\int_{-1}^{1} \sqrt{1 - x^2}\,dx$ is the area of the upper half of the circle $x^2 + y^2 = 1$.]

5. The volume of a solid whose base is the area enclosed by $y = x^2$, $x = 1$, and $y = 0$; cross sections taken perpendicular to the base and parallel to the y-axis are semicircles.

6. The base is a circle of radius 2; cross sections taken perpendicular to the base are isosceles right triangles with one leg on the base.

7. The volume of a pyramid 40 meters high whose horizontal cross section h meters from the top is a square with sides of length $2h$ meters.

8. The volume of a pyramid 20 meters high whose horizontal cross section h meters from the top is a rectangle with sides of length $2h$ and h meters.

9. The volume of a horn-shaped region; cross sections taken perpendicular to the x-axis are circles whose diameters extend from the graph of $y = x^{1/2}$ to the graph of $y = \frac{4}{3}x^{1/3}$ from $x = 0$ to $x = 1$.

10. The volume of a horn-shaped region; cross sections taken perpendicular to the x-axis are circles whose diameters extend from the graph of $y = x^{1/3}$ to $y = \frac{3}{2}x^{1/3}$ from $x = 0$ to $x = 1$.

11. Use the disk method to verify that the volume V of a right circular cone having radius R and height h is $V = \pi R^2 h/3$ (see Example 1).

12. Find the volume of a parallelepiped with edge lengths a, b, and c such that the edges having lengths a and b make an acute angle θ with each other, and the edge of length c makes an acute angle of ϕ with the diagonal of the parallelogram formed by a and b.

13. A hemispherical bowl of radius R contains water to the depth h. Find the volume of the water in the bowl.

14. Suppose a cylindrical glass full of water is tipped until the water level bisects the base and touches the rim. What is the volume of the water remaining? Set up the integral; do not evaluate.

15. Suppose a wedge is cut from a solid right circular cylinder (like a wedge cut in a tree by an axe) such that one side of the wedge is horizontal and the other is inclined at 30°. Assuming the wedged sides meet in a diameter of a circular cross section and the diameter of the cylinder is 10 meters, find the volume of the wedge removed. [*Hint:* The vertical cross sections of the wedge are right triangles.]

16. A hole of radius 2 centimeters is bored completely through a solid metal sphere of radius 5 centimeters, such that the axis of the hole passes through the center of the sphere. Find the volume of the metal removed by the drilling.

Arc Length

6.5

In this section we find a formula for measuring the length of the graph of a function (*arc length*). As the Greeks discovered, the formula for the circumference of a circle is $C = \pi d$, where d is the diameter. The Greeks arrived at this by the use of inscribed polygons; that is, they picked appropriate points on a circle and connected them with straight lines of equal length (see Fig. 43). By adding up the lengths of these line segments, they obtained an approximation to the length (circumference) of the circle. And by choosing more and more points on the circle, they obtained a better and better approximation to the actual length, so they eventually obtained the formula $C = \pi d$.

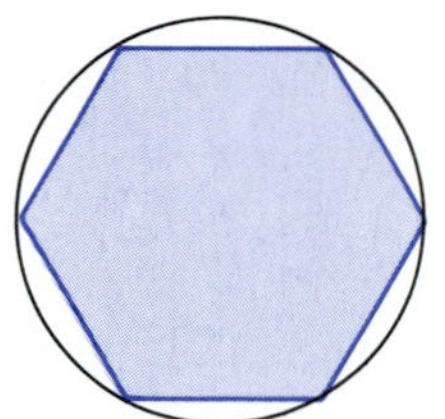

(a) 6 points

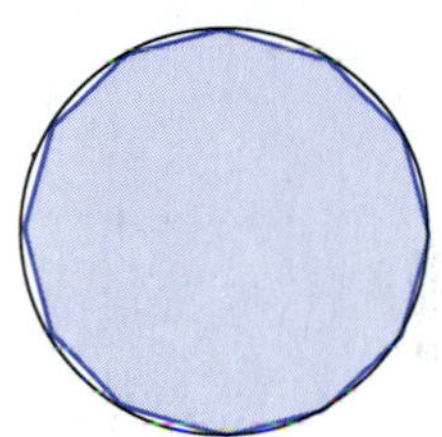

(b) 12 points

Figure 43

To find a formula for the length of the graph of $y = f(x)$ from $x = a$ to $x = b$, where the function f has a continuous derivative on $[a, b]$,* we use the same approach. First, we partition the closed interval $[a, b]$ into n subintervals,

$$[a, x_1], \quad [x_1, x_2], \quad \ldots, \quad [x_{i-1}, x_i], \quad \ldots, \quad [x_{n-1}, b]$$

We denote the length of the ith subinterval by $\Delta x_i = x_i - x_{i-1}$. Corresponding to each number $a, x_1, x_2, \ldots, x_{n-1}, b$ in the partition, there is a succession of points $P_0, P_1, P_2, \ldots, P_{n-1}, P_n$ on the graph (see Fig. 44). When we join each point to its successor by a line segment, the sum L of the lengths of these line segments provides an approximation to the length of the graph of $y = f(x)$ from $x = a$ to $x = b$. This sum may be written as

$$L = d(P_0, P_1) + d(P_1, P_2) + \cdots + d(P_{n-1}, P_n) = \sum_{i=1}^{n} d(P_{i-1}, P_i) \qquad \textbf{(1)}$$

where $d(P_{i-1}, P_i)$ is the length of the line segment joining P_{i-1} to P_i.

By the formula for the distance between two points, it follows that the length of the ith line segment is

$$d(P_{i-1}, P_i) = \sqrt{(x_i - x_{i-1})^2 + (y_i - y_{i-1})^2} = \sqrt{(\Delta x_i)^2 + (\Delta y_i)^2} = \sqrt{1 + \left(\frac{\Delta y_i}{\Delta x_i}\right)^2}\,\Delta x_i \qquad \textbf{(2)}$$

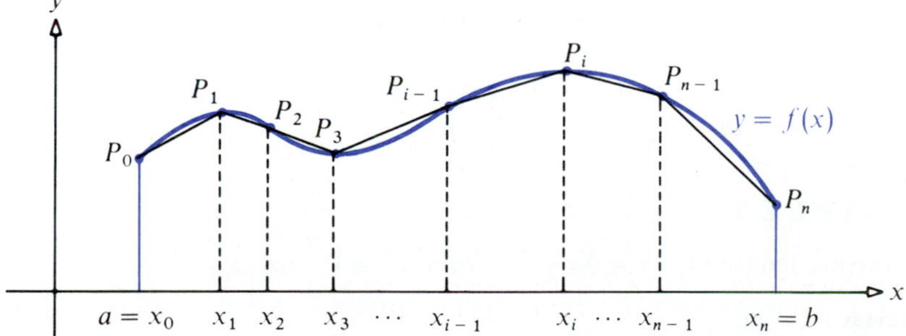

Figure 44

* We restrict our discussion of arc length to functions for which f' is continuous on $[a, b]$ in order to eliminate some complications that would otherwise occur.

where $\Delta x_i = x_i - x_{i-1}$ and $\Delta y_i = y_i - y_{i-1} = f(x_i) - f(x_{i-1})$. By combining (1) and (2), we find

$$L = \sum_{i=1}^{n} \sqrt{1 + \left(\frac{\Delta y_i}{\Delta x_i}\right)^2}\,\Delta x_i \tag{3}$$

Ordinarily, we would now let the norm $\|P\|$ of the partition approach 0. However, this will not be helpful at this stage because the sum (3) is not in the form of (2) in Section 5.3—it is not a Riemann sum; that is, taking the limit will not result in a definite integral. However, the sum (3) can be put in a proper form as follows: Since we assume the function f has a derivative on $[a, b]$, it follows that f has a derivative on each subinterval $[x_{i-1}, x_i]$. As a result, we can apply the mean value theorem (4.4.2). For each subinterval $[x_{i-1}, x_i]$, there is a number u_i between x_{i-1} and x_i such that

$$f(x_i) - f(x_{i-1}) = f'(u_i)(x_i - x_{i-1})$$

$$\Delta y_i = f'(u_i)\Delta x_i$$

$$\frac{\Delta y_i}{\Delta x_i} = f'(u_i)$$

Thus, the sum L of the lengths of the line segments in (3) becomes

$$L = \sum_{i=1}^{n} \sqrt{1 + [f'(u_i)]^2}\,\Delta x_i \tag{4}$$

where u_i is some number in the subinterval $[x_{i-1}, x_i]$.

The sum (4) gives an approximation to the length of the graph of $y = f(x)$ from $x = a$ to $x = b$. As the length of each subinterval gets smaller and smaller—that is, as the norm $\|P\|$ of the partition approaches 0—the sum (4) becomes a better approximation to the length of the graph of $y = f(x)$ from $x = a$ to $x = b$. However, the sum (4) is now a Riemann sum, so that, as $\|P\| \to 0$, its limit is a definite integral. This suggests the following definition:

[6.5.1] DEFINITION / *Length of a Graph.*

Consider a function f, which has a continuous derivative on $[a, b]$. The length s of the graph of $y = f(x)$ from $x = a$ to $x = b$ is

$$s = \int_a^b \sqrt{1 + [f'(x)]^2}\,dx \tag{5}$$

We refer to the length s of the graph of $y = f(x)$ from $x = a$ to $x = b$ as *the arc length s of $y = f(x)$ from $x = a$ to $x = b$.*

EXAMPLE 1

Find the arc length of $y = 4x + 5$ from $x = 0$ to $x = 1$.

Solution

Here $f'(x) = 4$ and the arc length s is

$$s = \int_0^1 \sqrt{1 + 16}\,dx = \sqrt{17}x\Big|_0^1 = \sqrt{17}$$

For this function, whose graph is a straight line, we can verify our answer by using the distance formula. We find that the distance from (0, 5) to (1, 9) is $d = \sqrt{4^2 + 1^2} = \sqrt{17}$.

EXAMPLE 2

Find the arc length of $y = x^{2/3}$ from $x = 1$ to $x = 8$.

Solution

The derivative of $f(x) = x^{2/3}$ is $f'(x) = \frac{2}{3}x^{-1/3} = 2/(3x^{1/3})$. The arc length from $x = 1$ to $x = 8$ is

$$\int_1^8 \sqrt{1 + \frac{4}{9x^{2/3}}}\, dx$$

Since

$$\sqrt{1 + \frac{4}{9x^{2/3}}} = \sqrt{\frac{9x^{2/3} + 4}{9x^{2/3}}} = \frac{1}{3}\sqrt{9x^{2/3} + 4}\ x^{-1/3}$$

we use the substitution $u = 9x^{2/3} + 4$, $du = 6x^{-1/3}\, dx$. Also, $u = 13$ when $x = 1$, and $u = 40$ when $x = 8$. Thus,

$$\int_1^8 \sqrt{1 + \frac{4}{9x^{2/3}}}\, dx = \int_1^8 \frac{1}{3}\sqrt{9x^{2/3} + 4}\, x^{-1/3}\, dx = \int_{13}^{40} \frac{1}{3}\sqrt{u}\, \frac{du}{6}$$

$$= \frac{1}{18}\left(\frac{u^{3/2}}{\frac{3}{2}}\right)\Bigg|_{13}^{40} = \frac{u^{3/2}}{27}\Bigg|_{13}^{40} = \frac{1}{27}(80\sqrt{10} - 13\sqrt{13})$$

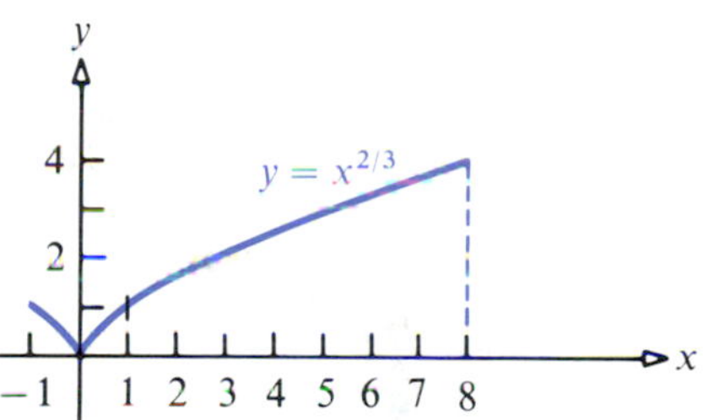

Figure 45

Figure 45 illustrates the graph of $y = x^{2/3}$. Observe that if we had been asked to find the arc length of $y = x^{2/3}$ from, say, -1 to 8, we would not have been able to use definition (6.5.1), since the derivative does not exist when $x = 0$. However, in such cases we may be able to do the problem by using a partition of the y-axis. We give a precise statement of the formula now, and then we use it in Example 3 to solve this problem.

[6.5.2] DEFINITION / *Length of a Graph.*

If x is a function of y, $x = g(y)$, $c \le y \le d$, and g' is continuous on $[c, d]$, the arc length s of the graph of $x = g(y)$ from c to d is given by the formula

$$s = \int_c^d \sqrt{1 + [g'(y)]^2}\, dy \qquad \textbf{(6)}$$

EXAMPLE 3

Compute the arc length of $y = x^{2/3}$ from $x = -1$ to $x = 8$.

Solution

The derivative of $y = x^{2/3}$ is $y' = 2/(3x^{1/3})$. Since the derivative does not exist at $x = 0$, we cannot use (5) to calculate the arc length (see

Fig. 45). In this situation we apply (6) twice: first, to calculate the arc length of

$$x_1 = g_1(y) = -y^{3/2}$$

from $y = 0$ to $y = 1$; and second, to calculate the arc length of

$$x_2 = g_2(y) = y^{3/2}$$

from $y = 0$ to $y = 4$. The sum of these lengths is the arc length of $y = x^{2/3}$ from $x = -1$ to $x = 8$. By (6), the arc length s_1 of $x_1 = g_1(y) = -y^{3/2}$ from $y = 0$ to $y = 1$ is

$$s_1 = \int_0^1 \sqrt{1 + [g_1'(y)]^2}\, dy = \int_0^1 \sqrt{1 + \frac{9y}{4}}\, dy = \frac{1}{2}\int_0^1 \sqrt{4 + 9y}\, dy$$

Use the substitution $u = 4 + 9y$ so that $du = 9\, dy$. Then, since $u = 4$ when $y = 0$, and $u = 13$ when $y = 1$, we have

$$s_1 = \frac{1}{2}\int_4^{13} \sqrt{u}\,\frac{du}{9} = \frac{1}{18}\left(\frac{u^{3/2}}{\frac{3}{2}}\right)\Bigg|_4^{13} = \frac{1}{27}(13\sqrt{13} - 8)$$

Similarly, the arc length s_2 of $x_2 = g_2(y) = y^{3/2}$ from $y = 0$ to $y = 4$ is

$$s_2 = \int_0^4 \sqrt{1 + [g'(y)]^2}\, dy = \int_0^4 \sqrt{1 + \frac{9y}{4}}\, dy = \frac{1}{2}\int_0^4 \sqrt{4 + 9y}\, dy$$

$$\underset{\substack{\uparrow \\ u = 4 + 9y}}{=} \frac{1}{2}\int_4^{40} \sqrt{u}\,\frac{du}{9} = \frac{1}{18}\left(\frac{u^{3/2}}{\frac{3}{2}}\right)\Bigg|_4^{40} = \frac{1}{27}(80\sqrt{10} - 8)$$

Thus, the total length s of $y = x^{2/3}$ from $x = -1$ to $x = 8$ is the sum

$$s = s_1 + s_2 = \tfrac{1}{27}(80\sqrt{10} + 13\sqrt{13} - 16)$$

As Example 3 illustrates, when one or more vertical tangents occur between the ends of a portion of a graph whose length is to be found, we use (6) to calculate the lengths of the portions of the graph between the vertical tangents and add them.

EXERCISE 6.5

In Problems 1–4 find the arc length of each line between the points indicated. Verify your answer by using the distance formula.

1. $y = 3x - 1$, from $(1, 2)$ to $(3, 8)$

2. $y = -4x + 1$, from $(-1, 5)$ to $(1, -3)$

3. $2x - 3y + 4 = 0$, from $(1, 2)$ to $(4, 4)$

4. $3x + 4y - 12 = 0$, from $(0, 3)$ to $(4, 0)$

In Problems 5–20 find the indicated arc length.

5. $y = x^{2/3} + 1$, from $x = 1$ to $x = 8$

6. $y = x^{2/3} + 6$, from $x = 1$ to $x = 8$

7. $y = x^{3/2}$, from $x = 0$ to $x = 4$

8. $y = x^{3/2} + 4$, from $x = 1$ to $x = 4$

9. $9y^2 = 4x^3$, from $x = 0$ to $x = 1$; $y \geq 0$

10. $y = \dfrac{x^3}{6} + \dfrac{1}{2x}$, from $x = 1$ to $x = 3$

11. $y = \frac{2}{3}(x^2 + 1)^{3/2}$, from $x = 1$ to $x = 4$

12. $y = \frac{1}{3}(x^2 + 2)^{3/2}$, from $x = 2$ to $x = 4$

13. $y = \frac{2}{9}\sqrt{3}(3x^2 + 1)^{3/2}$, from $x = -1$ to $x = 2$

14. $y = (1 - x^{2/3})^{3/2}$, from $x = \frac{1}{8}$ to $x = 1$

15. $8y = x^4 + \dfrac{2}{x^2}$, from $x = 1$ to $x = 2$

16. $9y^2 = 4(1 + x^2)^3$, from $x = 0$ to $x = 2\sqrt{2}$; $y \geq 0$

17. $y = x^{2/3}$, from $x = 0$ to $x = 1$

18. $y = x^{2/3}$, from $x = -1$ to $x = 0$

19. $(x + 1)^2 = 4y^3$, from $y = 0$ to $y = 1$; $x \geq -1$

20. $x = \frac{2}{3}(y - 5)^{3/2}$, from $y = 5$ to $y = 6$

21. Find the total length of the *hypocycloid* $x^{2/3} + y^{2/3} = a^{2/3}$, $a > 0$.

22. Find the distance between $(0, 1)$ and $(3, 7)$ along $2x - y + 1 = 0$:

(a) By the distance formula (b) By (5)

23. Find the distance between $(1, 1)$ and $(3, 3\sqrt{3})$ along $y^2 = x^3$.

24. Find the perimeter of the area enclosed by $y^3 = x^2$ and $x + 3y - 4 = 0$.

25. Find the perimeter of the area enclosed by $y = 3(x - 1)^{3/2}$ and $y = 6(x - 1)$.

26. Find the length of the loop of $9y^2 = x(x - 3)^2$, $0 \leq x \leq 3$.

27. Find the length of $6xy = y^4 + 3$ from $y = 1$ to $y = 2$.

28. Find the length of the loop of $9y^2 = x^2(2x + 3)$, $-\frac{3}{2} \leq x \leq 0$.

In Problems 29–32 use (5) to set up the integral for the arc length. Do not attempt to integrate. (Techniques for evaluating these integrals are given in Chap. 9.)

29. $y = x^2$, from $x = 0$ to $x = 2$

30. $x = y^2$, from $y = 1$ to $y = 3$

31. $y = \sqrt{25 - x^2}$, from $x = 0$ to $x = 4$

32. $x = \sqrt{4 - y^2}$, from $y = 0$ to $y = 1$

33. The graph whose equation is given by $(x^2/a^2) + (y^2/b^2) = 1$ is called an *ellipse*. Use (5) to set up the arc length of this ellipse from $x = 0$ to $x = a/2$. This integral, which is approximated by numerical techniques (see Chap. 9), is called an *elliptic integral of the second kind.*

34. The arc length formula does not apply to the function f given below. Why not?

$$f(x) = \begin{cases} \sin 1/x & \text{if } 0 < x \leq 1 \\ 0 & \text{if } x = 0 \end{cases}$$

35. In each case below, $P_1 = (x_1, y_1)$ and $P_2 = (x_2, y_2)$ are points on the circle $x^2 + y^2 = 1$, with neither coordinate 0. Express the length of the counterclockwise arc P_1P_2 in terms of integrals of the form

$$\int_u^v \frac{1}{\sqrt{1 - t^2}}\, dt \qquad -1 < u < v < 1$$

(a) When P_1 is in quadrant I and P_2 is in quadrant II
(b) When P_1 and P_2 are both in quadrant III and $y_1 < y_2$
(c) When P_1 is in quadrant II and P_2 is in quadrant IV

Work 6.6

The work W done by a *constant force* F in moving an object a distance x along a straight line in the direction of F is defined to be

$$W = Fx \qquad (1)$$

One unit of work is the work done by a unit force in moving an object a unit distance in the direction of the force. In the International System of Metric Units (abbreviated SI, for Système International d'Unités), the unit of work is 1 newton-meter, which is generally called 1 joule. In terms of customary U.S. units, the unit of work is the foot-pound.* These terms are summarized in Table 2.

Table 2

	Work	= Force	× Distance
SI	joule (J)	newton (N)	meter (m)
U.S.	foot-pound (ft-lb)	pound (lb)	foot (ft)

For example, the work required to lift an object weighing 80 pounds a distance of 5 feet would be $(80)(5) = 400$ foot-pounds.

When the force F acts in the same direction as the motion, the work done is positive; if the force F acts in a direction opposite to the motion, the work is negative. In some cases, a force F acts along the line of motion of an object, but the magnitude of the force varies depending on the position of the object. For example, the force required to stretch a spring depends on how far the spring has already been stretched from its normal length. This is an example of a *variable* force. Similarly, suppose a cylindrical tank full of water is to be emptied by pumping the water over the top. The work required to do this depends on the weight of the column of water, a variable, and the distance it is to be lifted.

In general, suppose the force F is given by a continuous function $F = F(x)$, $a \le x \le b$, of the position x of the object it acts upon. We seek a procedure for calculating the work done by a variable force F in moving an object from one position to another. We begin by partitioning $[a, b]$ into n subintervals

$$[a, x_1], \quad [x_1, x_2], \quad \ldots, \quad [x_{i-1}, x_i], \quad \ldots, \quad [x_{n-1}, b]$$

The length of the ith subinterval is $\Delta x_i = x_i - x_{i-1}$, $i = 1, 2, \ldots, n$.

We now consider the ith subinterval $[x_{i-1}, x_i]$ and choose a number u_i in $[x_{i-1}, x_i]$. If the length of this subinterval is small enough, the force $F = F(x)$ will not change too much, so that the work done by F as the object moves from x_{i-1} to x_i is, by (1), approximately $F(u_i)(x_i - x_{i-1}) = F(u_i)\Delta x_i$ (see Fig. 46). Thus, a good approximation to the total work W done by F as the object moves from a to b may be found by adding up the work done by F on each subinterval. That is,

$$W \approx F(u_1)\Delta x_1 + F(u_2)\Delta x_2 + \cdots + F(u_n)\Delta x_n = \sum_{i=1}^{n} F(u_i)\Delta x_i \qquad \textbf{(2)}$$

As the length of each subinterval gets smaller and smaller—that is, as the norm $\|P\|$ approaches 0—the sum $\sum_{i=1}^{n} F(u_i)\Delta x_i$ becomes a better

---

* 1 newton is the force required to accelerate a 1 kilogram mass at 1 meter per second per second; 1 joule = 0.7376 foot-pound; 1 foot-pound = 1.356 joules.

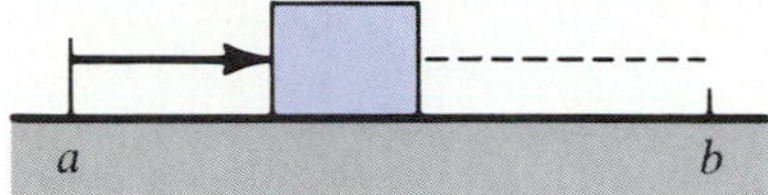

(a) Constant force
$W = F(b - a)$

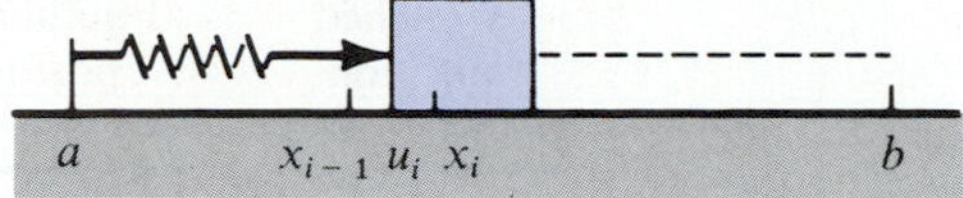

(b) Variable force
$W = \int_a^b F(x)\,dx$

Figure 46

approximation to the work done by the force $F = F(x)$ in moving an object along a line from $x = a$ to $x = b$ in the direction of F. However, this sum is a Riemann sum, so that, as $\|P\| \to 0$, its limit is a definite integral. This suggests the following definition:

[6.6.1] DEFINITION / *Work.*

The work W done by a continuously varying force $F = F(x)$ acting upon an object, as that object moves along a straight line in the direction of F from $x = a$ to $x = b$, is

$$W = \int_a^b F(x)\,dx \tag{3}$$

Hooke's Law

A common example of the work done by a variable force is found in the extension or compression of an elastic spring. In the case of an elastic spring of *stiffness** k and negligible mass, the force F supported by the spring at any deformation x—either compression or extension—is given by *Hooke's law*:

$$F(x) = kx \tag{4}$$

where the constant k, measured in newtons per meter (N/m, in SI units) or pounds per foot (lb/ft, in U.S. units), depends on the type of spring. As Figure 47 illustrates, the distance x in (4) is measured from the equilibrium, or unstretched, position of the spring.

As an example, if a certain spring without any force applied to it is 0.8 meter long, and a force of 2 newtons stretches the spring to a length of 1.2 meters, the constant k for this spring is

$$2 = k(1.2 - 0.8)$$

$$\frac{2}{0.4} = k$$

$$k = 5 \text{ newtons per meter}$$

Thus, the force F required to hold the spring stretched to a length of 3 meters is

$$F = kx = (5)(3 - 0.8) = (5)(2.2) = 11 \text{ newtons}$$

* Sometimes called the *spring constant k*.

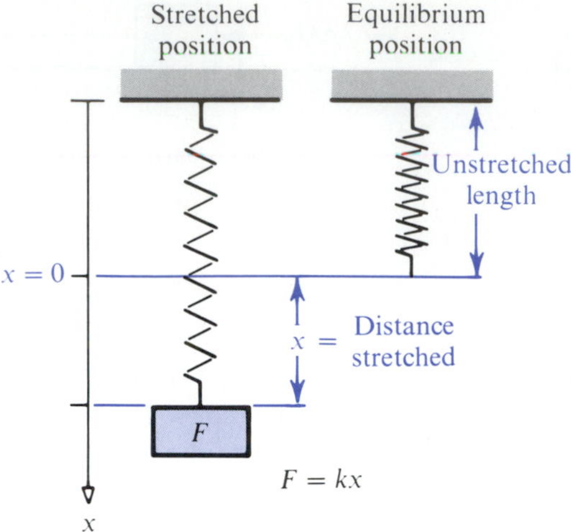

Figure 47

EXAMPLE 1

Consider a spring of stiffness $k = 5$ newtons per meter, whose unstretched length is 0.8 meter.

(a) How much work is required to stretch the spring to a length of 1.4 meters?
(b) How much work is required to stretch it from 2 to 3 meters?

Solution

(a) By using (3) and (4) and remembering that the coordinate system sets $x = 0$ at the unstretched length, we find that the work W required to stretch the spring from $x = 0$ (equilibrium) to $x = 1.4 - 0.8 = 0.6$ is

$$W = \int_0^{0.6} 5x\,dx = \tfrac{5}{2}x^2 \Big|_0^{0.6} = \tfrac{5}{2}(0.36) = 0.9 \text{ joule}$$

(b) The work required to stretch the spring from 2 to 3 meters equals the work required to stretch it to 3 meters from equilibrium less the work required to stretch it to 2 meters from equilibrium. The work W required is therefore,

$$W = \int_0^{3-0.8} 5x\,dx - \int_0^{2-0.8} 5x\,dx = \int_{1.2}^{2.2} 5x\,dx = \tfrac{5}{2}x^2 \Big|_{1.2}^{2.2}$$
$$= \tfrac{5}{2}(4.84 - 1.44) = 8.5 \text{ joules}$$

Pumping Liquids

As a second example of a situation in which force is variable, we take up the problem of the work needed to pump a liquid out of a tank. We rely on the basic idea that the work needed to lift an object a given distance is the product of the weight (force) of the object times the distance it is lifted. That is,

$$\textbf{Work} = \textbf{(Weight of object)(Distance lifted)} \qquad \textbf{(5)}$$

Let's see how this formula works with a liquid in a cylindrical container.

EXAMPLE 2

An oil tank in the shape of a right circular cylinder, with height 40 meters and radius 5 meters, is half full of oil. How much work is required to pump all the oil over the top of the tank?

Solution

Position the cylinder as illustrated in Figure 48. Note that we position the top of the tank at $x = 0$ so that the bottom of the tank is at $x = 40$. The work required to pump a certain volume of the oil over the top depends on the weight of this amount of oil and its distance from the top. The oil fills the tank from $x = 20$ to $x = 40$. Therefore, we partition the interval $[20, 40]$ into n subintervals and let Δx_i denote the length of the ith subinterval. In the ith subinterval, we view the oil as consisting of a thin layer of height Δx_i. Now we choose a number u_i in the ith subinterval. Then,

$$\text{Area of } i\text{th layer} = \pi(\text{Radius})^2 = 25\pi$$

$$\text{Volume of } i\text{th layer} = (\text{Area of layer})(\text{Height}) = 25\pi\Delta x_i$$

$$\text{Weight* of } i\text{th layer} = \rho g(\text{Volume}) = \rho g(25\pi\Delta x_i)$$

$$\text{Distance } i\text{th layer is lifted} = u_i$$

$$\text{Work done in lifting } i\text{th layer} = 25\pi\rho g u_i \Delta x_i$$

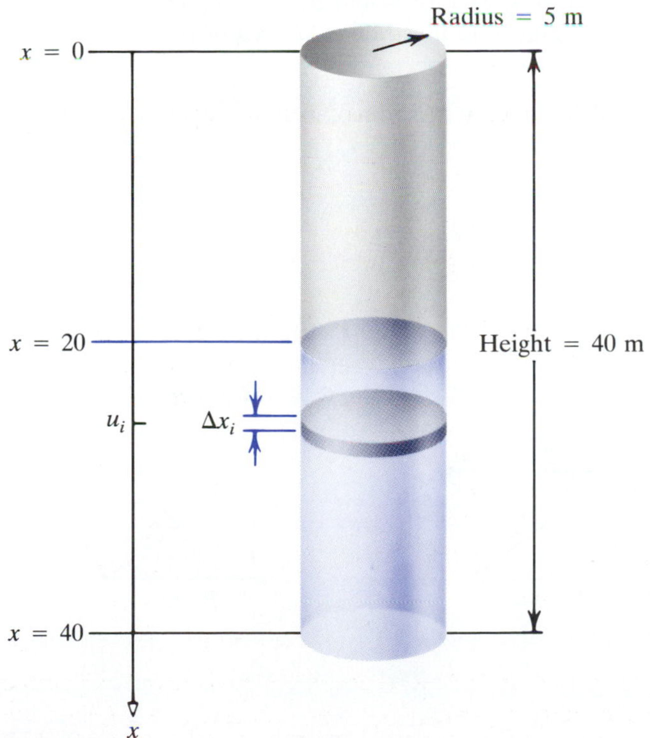

Figure 48

* Weight $= \rho g(\text{Volume})$, where ρ is the mass density (mass per unit volume, a constant that depends on the type of liquid involved) and $g \approx 9.8 \text{ m/sec}^2$ (the acceleration of gravity).

The layers of oil occur from 20 to 40. Therefore, the work W required to pump all the oil over the top is

$$W = \int_{20}^{40} 25\pi\rho g x\, dx = 25\pi\rho g \frac{x^2}{2}\bigg|_{20}^{40} = 25\pi\rho g(800 - 200) = 15{,}000\pi\rho g \text{ joules}$$

Let's look at another example.

EXAMPLE 3

A water tank in the shape of a hemisphere of radius 2 meters is full of water. How much work is required to pump all the water to a level 3 meters above the tank?

Solution

We choose our coordinate system so that the top of the tank is positioned at $x = 0$ and the bottom is at $x = 2$ (see Fig. 49). The work required to pump a certain volume of the water to a level 3 meters above the top of the tank will depend on the weight of this amount of water and its distance from the level 3 meters above the tank. The water fills the container from $x = 0$ to $x = 2$. We partition the interval [0, 2] into n subintervals, and let Δx_i be the length of the ith interval. We may view the water in the tank as composed of circular layers, each of height Δx_i. We choose a number u_i in the ith subinterval. As Figure 49 illustrates, the area of the circular layer u_i meters from the top of the tank is

$$\text{Area of } i\text{th layer} = \pi(\text{Radius})^2 = \pi(4 - u_i^2)$$

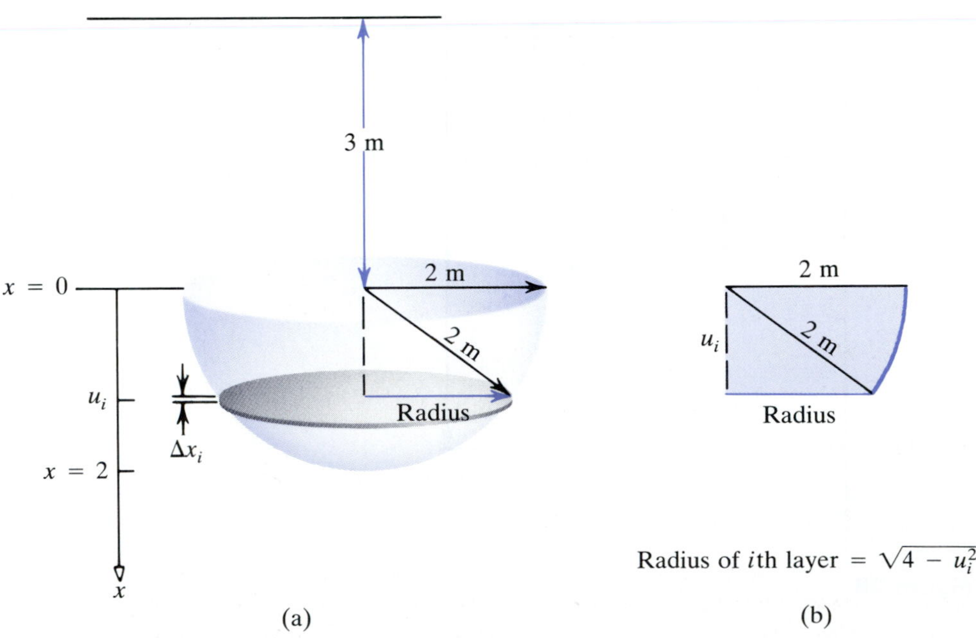

Figure 49

Then,

$$\text{Volume of } i\text{th layer} = (\text{Area of layer})\,\Delta x_i = \pi(4 - u_i^2)\,\Delta x_i$$

The mass density of water is $\rho = 1000$ kilograms per cubic meter; hence,

$$\text{Weight of } i\text{th layer} = \rho g(\text{Volume}) = 1000g\pi(4 - u_i^2)\,\Delta x_i$$

$$\text{Distance } i\text{th layer is lifted} = u_i + 3$$

$$\text{Work done in lifting } i\text{th layer} = 1000g\pi(4 - u_i^2)(u_i + 3)\,\Delta x_i$$

The work W required to lift all the water is thus

$$W = \int_0^2 1000g\pi(4 - x^2)(x + 3)\,dx = 1000g\pi \int_0^2 (-x^3 - 3x^2 + 4x + 12)\,dx$$

$$= 1000g\pi(20) \approx 615{,}752 \text{ joules}$$

$\uparrow$ $g \approx 9.8$

In general, to compute the work required to pump liquid from a container, think of the liquid as composed of thin layers of height Δx_i and area $A(x)$. If the liquid is to be lifted a height h above the tank, the work required is

$$\mathbf{W = \int_a^b \rho g A(x)(x + h)\,dx} \qquad \textbf{(6)}$$

where ρ is the mass density of the liquid and the liquid fills the container from $x = a$ to $x = b$ (see Fig. 50).

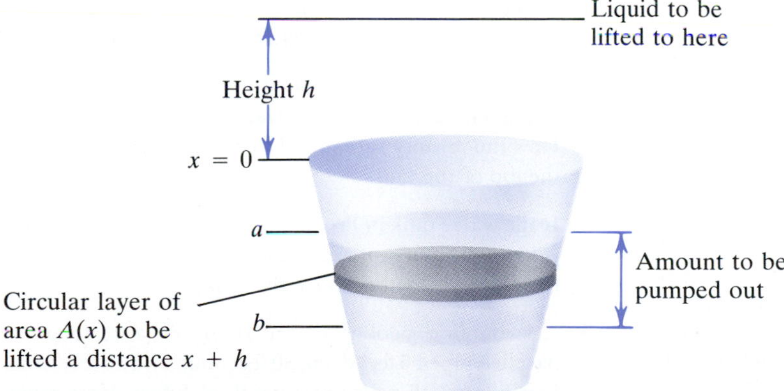

Figure 50

Gravitational Forces

One of the conclusions of Newton's law of universal gravitation is that the force required to move an object—say, a rocket—of mass m (in kilograms) at a point x meters above the center of the earth is $F(x) = GMm/x^2$, where G is the gravitational constant of the earth and M is the mass (in kilograms) of the earth. Hence, using (3), we see that the work required to move an object of mass m from the surface of the earth to a distance r meters from the center of the earth (where R is the radius of the earth in meters) is

$$W = \int_R^r \frac{GMm}{x^2}\,dx = -\frac{GMm}{x}\bigg|_R^r = \frac{GMm}{R} - \frac{GMm}{r} = GMm\left(\frac{1}{R} - \frac{1}{r}\right) \qquad (7)$$

Physicists know that $GM = gR^2$, where $g \approx 9.8$ meters per second per second is the acceleration of gravity of the earth and $R \approx 6.37 \times 10^6$ meters. Further, if d is the distance the object is to be moved above the surface of the earth, then $r = R + d$. Consequently, from (7), we find that the work required to move a body of mass m to a distance d meters above the surface of the earth is

$$W = gRm\left(1 - \frac{R}{R+d}\right)\text{joules}$$

Observe that even though d may be extremely large, W will never be bigger than $gRm \approx 62.4 \times 10^6 m$ joules.

EXERCISE 6.6

1. A spring, whose unstretched length is 1 meter, stretches to a length of 3 meters when a force of 3 newtons is applied. Find the work needed to stretch the spring to a length of 2 meters from its natural length.

2. How much work is required to stretch the spring in Problem 1 to a length of 4 meters?

3. A spring, whose unstretched length is 2 meters, is compressed to a length of $\frac{1}{2}$ meter when a force of 10 newtons is applied. Find the work required to compress the spring to a length of 1 meter.

4. How much work is required to compress the spring in Problem 3 to a length of 1.5 meters?

5. A spring, whose unstretched length is 4 feet, stretches to a length of 8 feet when a force of 2 pounds is applied. If 9 foot-pounds of work is expended to stretch this spring from an unstretched position, what is its total length?

6. If 8 foot-pounds of work is expended on the spring in Problem 5, how far is it stretched?

7. How much work is required to pump all the water over the top of a full cylindrical tank that is 4 meters in diameter and 6 meters high?

8. How much work is required to pump half the water over the top of the tank in Problem 7?

9. A full water tank in the shape of an inverted right circular cone is 8 meters across the top and 4 meters high. How much work is required to pump all the water over the top of the tank?

10. If the surface of the water in the tank of Problem 9 is 2 meters below the top of the tank, how much work is required to pump all the water over the top of the tank?

11. A water tank in the shape of a hemispherical bowl of radius 4 meters is filled with water to a depth of 2 meters. How much work is required to pump all the water over the top of the tank?

12. If the water tank in Problem 11 is completely filled with water, how much work is required to pump all the water to a height 2 meters above the tank?

13. A swimming pool is in the shape of a rectangular parallelepiped 6 feet deep, 30 feet long, and 20 feet wide. It is filled with water to a depth of 5 feet. How much work is required to pump all the water over the top? [*Hint:* $\rho g = 62.5$ pounds per cubic foot.]

14. A 1 horsepower motor can do 550 foot-pounds of work per second. Using this motor, how long does it take to pump all the water out of a swimming pool 5 feet deep, 25 feet long, and 15 feet wide if the pool is filled to a depth of 4 feet? The pool is in the shape of a rectangular parallelepiped.

Use (7) to do Problems 15 and 16.

15. The minimum energy required to move an object of mass 30 kilograms a distance 500 kilometers above the surface of the earth is equal to the work required to accomplish this. Find the work required. (The radius of the earth is approximately 6370 kilometers.)

16. The minimum energy required to move a rocket of mass 1000 kilograms a distance of 800 kilometers above the surface of the earth is equal to the work required to do this. Find the work required.

17. By *Coulomb's law*, a positive charge m of electricity repels a unit of positive charge at a distance x with the force m/x^2. What is the work done when the unit charge is carried from $x = 2a$ to $x = a$?

18. In raising a leaky bucket from the bottom of a well 25 feet deep, one-fourth of the water is lost. If the bucket weighs 1.5 pounds, the water in the bucket at the start weighs 20 pounds, and the amount that has leaked out is assumed to be proportional to the displacement, find the work done in raising the bucket.

19. The stiffness of a spring on a bumping post in a freight yard is 300,000 newtons per meter. Find the work done in compressing the spring 0.1 meter.

20. A cable of uniform linear mass $\rho = 9$ kilograms per meter is being unwound from a cylindrical drum. If 15 meters are already unwound, what is the work done by gravity in unwinding 60 meters more?

21. A uniform chain 30 feet long and weighing 30 pounds is hanging from the top of a building 30 feet high. If a bucket filled with cement weighing 150 pounds is attached to the end of the chain, how much work is required to pull the bucket and chain to the top of the building?

22. In Problem 21, if a uniform chain 30 feet long and weighing 20 pounds is used instead, how much work is required to pull the bucket and chain to the top of the building?

The pressure p (in pounds per square inch, lb/in.^{2}) and the volume v (in cubic inches) of an adiabatic expansion of a gas are related by $pv^k = c$, where k and c are constants that depend on the gas. If the gas expands from $v = a$ to $v = b$, the work done (in inch-pounds, in.-lb) is $W = \int_a^b p\,dv$. Use this fact in Problems 23 and 24.

23. The pressure of 1 pound of a gas is 100 pounds per square inch and the volume is 2 cubic feet. Find the work done by the gas in expanding to double its volume according to the law $pv^{1.4} = c$ (in inch-pounds).

24. The pressure and volume of a certain gas obey the law $pv^{1.2} = 120$ (in inch-pounds). Find the work done when the gas expands from $v = 2.4$ to $v = 4.6$ cubic inches.

25. A container is formed by revolving the area under the graph of $y = x^2$, $0 \leq x \leq 2$, about the y-axis.

(*Problem 25 continues in the next column*)

25. (*continued*)

How much work is required to fill the container with water from a source 2 units below the x-axis by pumping through a hole in the bottom of the container?

26.* The force in newtons required to stretch a certain spring a distance of x meters beyond its unstretched length is given by $F = 100x$.

(a) What force will stretch the spring 0.1 meter? 0.2 meter? 0.4 meter?

(b) How much work is required to stretch the spring 0.1 meter? 0.2 meter? 0.4 meter?

Liquid Pressure and Force 6.7

When containers are built to hold liquids in place, it is important to know the force F caused by liquid pressure on the sides of the containers. The pressure P exerted by a liquid of mass density ρ at a depth h below the surface of the liquid is defined as

$$P = \rho g h \tag{1}$$

* Adapted from F. W. Sears, M. W. Zemansky, and H. D. Young, *University Physics* (Reading, Mass.: Addison-Wesley Publishing Co., 1976), p. 132. Reprinted by permission.

where $g \approx 9.8 \text{ m/sec}^2 \approx 32.2 \text{ ft/sec}^2$ is the acceleration of gravity. Table 3 summarizes the units of measure needed for these calculations.

Table 3

	Pressure	= Mass Density	× g	× Depth
SI	newton/meter^2 (N/m^2)	kilogram/meter^3 (kg/m^3)	9.8 m/sec^2	meter (m)
U.S.	pound/foot^2 (lb/ft^2)	slug/foot^3 (slug/ft^3)	32.2 ft/sec^2	foot (ft)

If a flat plate of area A is suspended horizontally in a liquid at a depth h, the force F (weight) caused by the liquid on one face of the plate is

$$F = PA = \rho ghA \tag{2}$$

For example, the mass density of water is about $\rho = 1.94$ slugs per cubic foot, so the pressure due to water on a plate suspended horizontally at a depth of 4 feet is $(4)(1.94)g = 4(62.5) = 250$ pounds per square foot. If the plate has an area of 2 square feet, the force (weight) exerted on one side of the plate is $(2)(250) = 500$ pounds.

On the other hand, if a plate is suspended vertically in a liquid, the force caused by the liquid on one face of the plate is more difficult to find because the force at the bottom of the plate is greater than that at the top. To solve this problem, we proceed as follows: Suppose a plate is suspended vertically in a liquid of mass density ρ. Suppose further that the plate is bounded by $y = c$, $y = d$, $x = g(y)$, and $x = f(y)$, where f and g are continuous functions on $c \le y \le d$ and $f(y) \ge g(y)$ on $c \le y \le d$ (see Fig. 51). The surface of the liquid is the line $y = H$, where $H \ge d$, so that the top of the plate is at a depth of $(H - d)$ and the bottom of the plate is at a depth of $(H - c)$.

We partition the interval $[c, d]$ into n subintervals

$$[c, y_1], \quad [y_1, y_2], \quad \ldots, \quad [y_{i-1}, y_i], \quad \ldots, \quad [y_{n-1}, d]$$

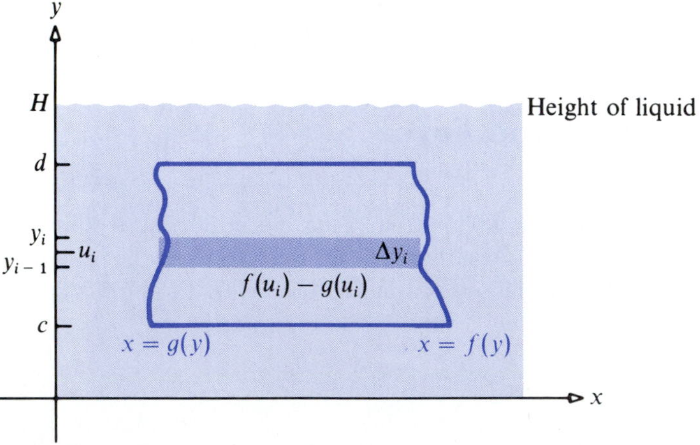

Figure 51

The length of the ith subinterval is $\Delta y_i = y_i - y_{i-1}$. And, as usual, we select a number u_i in the ith subinterval $[y_{i-1}, y_i]$. If the length Δy_i of the ith subinterval is small, then all points in the ith portion of the plate are roughly the same distance $(H - u_i)$ from the surface, so that the pressure P of the liquid on this portion of the plate is approximately $P = \rho g(H - u_i)$. By (2), the force due to liquid pressure on this portion of the plate is approximately

$$\rho g(H - u_i)[f(u_i) - g(u_i)]\Delta y_i$$

A good approximation to the force F on the plate can be found by adding up the forces on each subinterval. That is,

$$F \approx \sum_{i=1}^{n} \rho g(H - u_i)[f(u_i) - g(u_i)]\Delta y_i \tag{3}$$

As the length of each subinterval gets smaller and smaller—that is, as the norm $\|P\|$ approaches 0—the sum (3) becomes a better and better approximation of the force due to liquid pressure on the plate. However, the sum (3) is a Riemann sum, so that, as $\|P\| \to 0$, its limit is a definite integral. This suggests the following definition:

[6.7.1] DEFINITION / ***Liquid Pressure.***

The force F due to a liquid of mass density ρ on a plate of the type illustrated in Figure 51, where f and g are continuous on $[c, d]$, is

$$F = \int_c^d \rho g(H - y)[f(y) - g(y)]\, dy \tag{4}$$

As an aid in using (4), it is useful to remember that

Force on ith rectangle = Mass density $\times g \times$ Depth $\times$ Area of ith rectangle

EXAMPLE 1

A trough, whose cross section is a trapezoid, is 2 meters across at the bottom, 4 meters across at the top, and 2 meters deep. If the trough is filled with a liquid of mass density ρ, what is the force due to liquid pressure on one end of the trough?

Solution

It is convenient to position the trough as shown in Figure 52. The sides of the end of the trough are lines that pass through the points (0, 2), (1, 0) and (3, 0), (4, 2), respectively. Their equations are

$$y - 0 = \frac{-2}{1}(x - 1) \qquad \text{and} \qquad y - 0 = \frac{2}{1}(x - 3)$$

$$x = g(y) = \frac{-1}{2}y + 1 \qquad\qquad x = f(y) = \frac{1}{2}y + 3$$

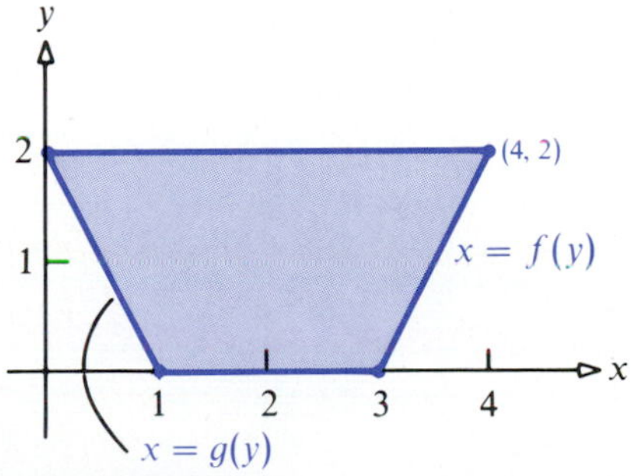

Figure 52

The height of the liquid is $H = 2$, so by (4), the force F due to the liquid on an end of the trough is

$$F = \int_0^2 \rho g(2 - y)[f(y) - g(y)]\, dy = \int_0^2 \rho g(2 - y)(y + 2)\, dy$$

$$= \rho g \int_0^2 (-y^2 + 4)\, dy = \rho g\left(\frac{-y^3}{3} + 4y\right)\Big|_0^2$$

$$= \rho g\left(\frac{-8}{3} + 8\right) = \frac{16}{3}\rho g \text{ newtons}$$

In the development of the definition of liquid pressure, we positioned the coordinate system so that the submerged plate was located in the first quadrant. As the next example illustrates, the coordinates may be placed in any convenient position. Keep in mind that the essential idea behind the formula for force due to liquid pressure is: Force $= \rho g \times$ Depth $\times$ Area.

EXAMPLE 2

A cylindrical sewer pipe of radius 2 meters is half full of water. Find the force exerted on one side of a gate that is used to seal off the sewer.

Solution

Figure 53 illustrates the situation. The equation of the circle with center at (0, 0) and radius 2 is $x^2 + y^2 = 4$. Thus, we have

$$x = g(y) = -\sqrt{4 - y^2} \qquad \text{and} \qquad x = f(y) = \sqrt{4 - y^2}$$

The force F exerted on one side of the gate by the water is

$$F = \int_{-2}^{0} \rho g(0 - y)[2\sqrt{4 - y^2}]\, dy$$

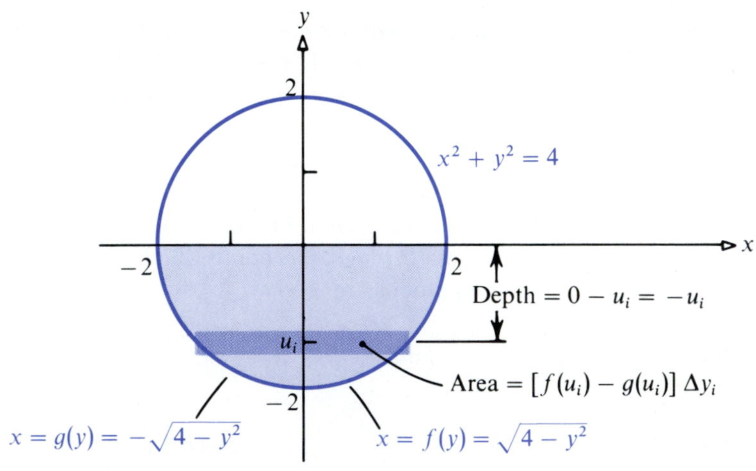

Figure 53

Now we use the substitution $u = 4 - y^2$ so that $du = -2y\,dy$. Since $u = 0$ when $y = -2$, and $u = 4$ when $y = 0$, we find that

$$F = \rho g \int_0^4 2\sqrt{u}\,\frac{du}{2} = \rho g \left.\frac{u^{3/2}}{\frac{3}{2}}\right|_0^4$$

$$= \rho g \frac{16}{3} \approx 52{,}266.7 \text{ newtons}$$

$$\uparrow \quad \rho = 1000 \text{ kg/m}^3, \quad g = 9.8 \text{ m/sec}^2$$

EXERCISE 6.7

1. A rectangular plate of width 2 meters and depth 6 meters is suspended vertically in a pool of water so that the top of the plate is even with the surface of the water. What is the force due to water pressure on one side of the plate?

2. If the plate in Problem 1 is suspended in the water so that the top of the plate is 1 meter below the water surface, what is the force due to water pressure on one side of the plate?

3. A swimming pool is in the shape of a rectangular parallelepiped 6 feet deep, 30 feet long, and 20 feet wide. If the pool is full of water, what is the force due to water pressure on one short side of the pool? [*Hint:* $\rho \approx 1.94$ slugs per cubic foot.]

4. For the pool in Problem 3, what is the force due to water pressure on one long side of the pool?

5. A trough, whose cross section is a trapezoid, is 1 meter across at the bottom, 5 meters across at the top, and 2 meters deep. If the trough is filled with water, what is the force due to water pressure on one end of the trough?

6. A trough, whose cross section is an equilateral triangle with side 2 meters long, is filled with water. What is the force due to water pressure on one end of the trough?

7. A trough, whose cross section is a semicircle of radius 2 meters, is filled with water. What is the force due to water pressure on one end of the trough?

8. A viewing plate in a submarine is a circle of radius 1 foot. If the depth of the center of the viewing plate is 5 feet below the surface of the water, what is the force due to water pressure on one side of the plate? Assume the viewing plate is vertical. (See the hint for Problem 4 in Exercise 6.4, p. 424.)

9. Find the force on the face of a vertical floodgate in the shape of an isosceles triangle whose base is 1.5 meters and whose altitude is 1 meter, if its base is on the surface of the water. (See the figure.)

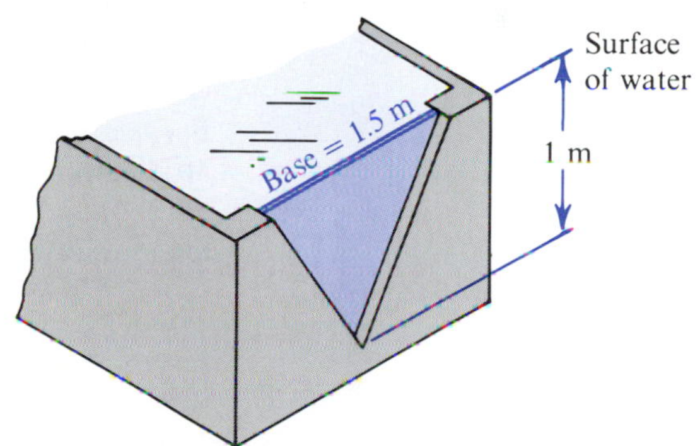

10. A vertical masonry dam in the form of an isosceles trapezoid is 200 meters long at the surface of the water, 150 meters long at the bottom, and 60 meters high. What force must it withstand? Express the result in newtons. [*Hint:* Use similar triangles.]

11. A tanker truck carrying a half load of oil is in the shape of a right circular cylinder. If the radius of the cylinder is 4 feet, what is the force due to the pressure of the oil on one end of the cylinder? Assume that the mass density of the oil is $\rho = 1.86$ slugs per cubic foot.

12. The gas tank in a sports car is a cylinder lying on its side. If the diameter of the tank is 1 meter and if the tank is filled with gasoline to within 0.5 meter of the top, find the force on one end of the tank. (The mass density of gasoline is 600 kilograms per cubic meter.)

6.8 Center of Mass and Centroid

Center of Mass

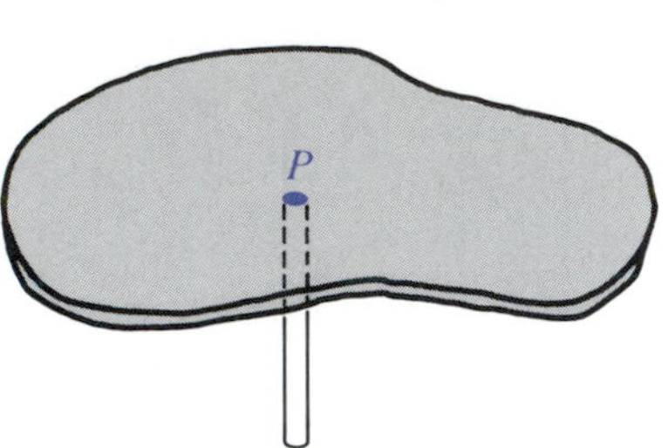

Figure 54

The central problem we pose and solve in this section is the following: Given a regularly shaped flat sheet of material (often called a lamina), what is its center of mass or gravity?

Intuitively, we know that the *center of mass* or *gravity* of a lamina is a point P on the flat surface on which the lamina will be in balance, as shown in Figure 54.

To reduce the problem to one of integration, we must first consider a finite system of point masses. Consider the situation illustrated in Figure 55, where three weights are placed on a weightless seesaw board. The center of mass will be some point $(\bar{x}, 0)$ at which a fulcrum balances the three weights.

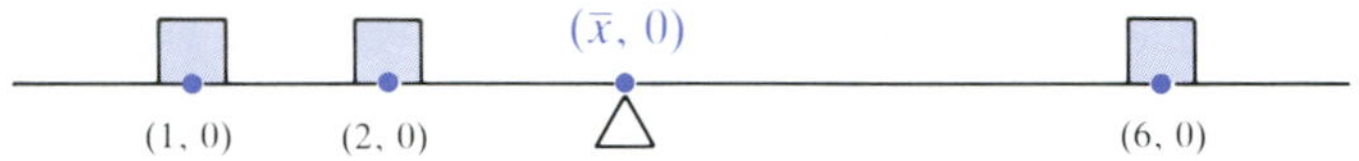

Figure 55

For example, if a mass of 90 kilograms is placed at (6, 0) and masses of 40 kilograms and 110 kilograms are placed on the opposite side of the fulcrum at (2, 0) and (1, 0), respectively, then it follows from elementary physics that for equilibrium we must have

$$90(6 - \bar{x}) = 40(\bar{x} - 2) + 110(\bar{x} - 1)$$

or

$$\bar{x}(90 + 40 + 110) = 540 + 80 + 110$$

$$\bar{x} = \tfrac{730}{240} \simeq 3.04$$

If the three weights are replaced by three general masses m_1, m_2, and m_3 placed at the points $(x_1, 0)$, $(x_2, 0)$, and $(x_3, 0)$, respectively, then for equilibrium, we must have

$$\bar{x}(m_1 + m_2 + m_3) = x_1 m_1 + x_2 m_2 + x_3 m_3 \tag{1}$$

or

$$\bar{x} = \frac{x_1 m_1 + x_2 m_2 + x_3 m_3}{m_1 + m_2 + m_3} \tag{2}$$

The numbers $m_1 x_1$, $m_2 x_2$, and $m_3 x_3$ are called the *moments* of the masses m_1, m_2, and m_3 (with respect to the origin), and (2) says that the center of mass $\bar{x}$ is obtained by adding the moments of the masses and dividing by the total mass $m_1 + m_2 + m_3$. In general, if we have a system of n masses $m_1, m_2, \ldots, m_n$ with coordinates $(x_1, 0), (x_2, 0), \ldots, (x_n, 0)$, it can be shown that the center of mass of this system is located at

$$\bar{x} = \frac{\sum_{i=1}^{n} m_i x_i}{\sum_{i=1}^{n} m_i} = \frac{\sum_{i=1}^{n} m_i x_i}{M} \qquad (3)$$

where $M = \sum_{i=1}^{n} m_i$ is the total mass of the system.

Now consider a system of n point masses $m_1, m_2, \ldots, m_n$ in the plane located at the points $(x_1, y_1), (x_2, y_2), \ldots, (x_n, y_n)$ as shown in Figure 56.

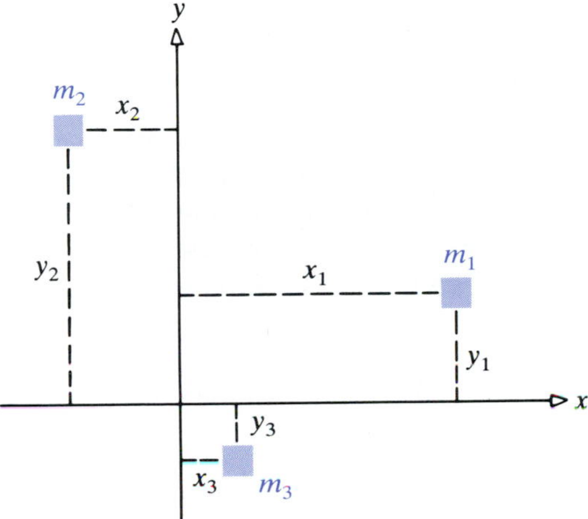

Figure 56

The center of mass $(\bar{x}, \bar{y})$ of this system of n masses is given by

$$\bar{x} = \frac{\sum_{i=1}^{n} x_i m_i}{M} \qquad \bar{y} = \frac{\sum_{i=1}^{n} y_i m_i}{M} \qquad (4)$$

where $M = \sum_{i=1}^{n} m_i$ is the total mass of the system. If we denote by M_y the *moment* about the y-axis and by M_x the moment about the x-axis, then (4) can be written as

$$\bar{x} = \frac{M_y}{M} \qquad \bar{y} = \frac{M_x}{M} \qquad (5)$$

where $M_y = \sum_{i=1}^{n} x_i m_i$ and $M_x = \sum_{i=1}^{n} y_i m_i$. Physically, M_y measures the tendency of the system to rotate about the y-axis, and M_x measures the tendency of the system to rotate about the x-axis.

EXAMPLE 1

Find the moments and center of mass of the system of particles having masses 4, 6, and 9 kilograms, located at the points $(-2, 1)$, $(3, -2)$, and $(4, 3)$.

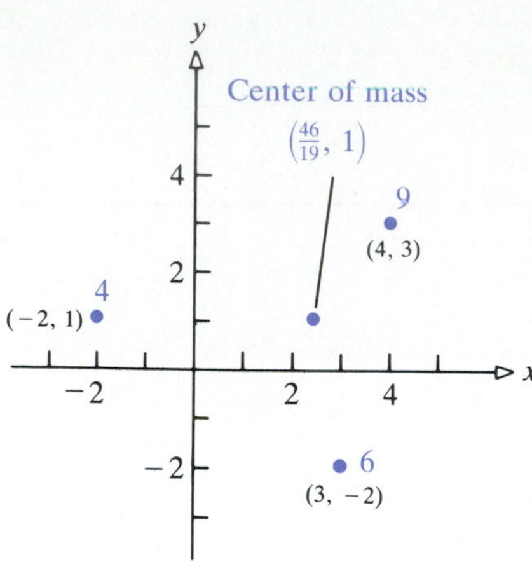

Figure 57

Solution

$$M_y = \sum_{i=1}^{3} x_i m_i = 4(-2) + 6(3) + 9(4) = 46$$

$$M_x = \sum_{i=1}^{3} y_i m_i = 4(1) + 6(-2) + 9(3) = 19$$

By (4), we have

$$\bar{x} = \frac{46}{4+6+9} = \frac{46}{19} \qquad \bar{y} = \frac{19}{19} = 1$$

Thus, the center of mass is $(\frac{46}{19}, 1)$. See Figure 57.

With the aid of these formulas we can approximate the center of mass of a thin sheet of material and express its center of mass in terms of certain integrals. We will assume that the matter is continuously distributed throughout the lamina; that is, the density* function is continuous. If the material is of constant density, the lamina is called *homogeneous.* The mass of a homogeneous lamina is ρA, where A is the area of the lamina and ρ is its constant mass density.

In general, however, substances are not homogeneous and so the mass density is variable. To develop the general case, we will be guided by two principles:

1. The moment of the union of two or more nonoverlapping regions is the sum of the moments of the individual regions.
2. *The Symmetry Principle.* If a region R is symmetric about a line L, then the center of mass of R lies on L (provided it is homogeneous).

Suppose a lamina is determined by a certain region R enclosed by a continuous function f, the x-axis, $x = a$, and $x = b$ (see Fig. 58(a)).

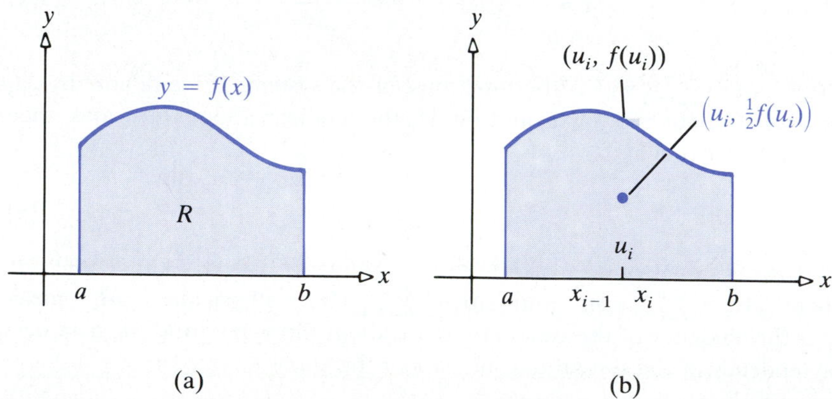

Figure 58

* The mass density of a two-dimensional material is defined as the mass per unit area of the material. In SI units, mass density is measured in kilograms per square meter (kg/m^2); in U.S. customary units, it is measured in slugs per square foot (slug/ft^2).

Suppose the density is variable and depends on only the x-coordinate; that is, the density $\rho(x)$ is continuous and varies horizontally but not vertically. (The case for which the density varies also with the y-coordinates requires double integration and is treated in Chap. 18.)

We begin by partitioning the interval $[a, b]$ into n subintervals $[a_1, x_1]$, $[x_1, x_2], \ldots, [x_{i-1}, x_i], \ldots, [x_{n-1}, b]$, $1 \le i \le n$, and select the number u_i to be the midpoint of the ith subinterval—namely $u_i = (x_{i-1} + x_i)/2$. The center of mass of the ith rectangular lamina of material will be of the form $(u_i, \frac{1}{2}f(u_i))$ because the center of mass must be located halfway up the rectangular slab (see Fig. 58(b)). The mass of the ith rectangular slab will be

$$\rho(x)f(u_i)\Delta x_i$$

The moment of R_i (the ith rectangle) about the y-axis, $M_y(R_i)$, is the product of the mass of the ith rectangle and the distance from $(u_i, \frac{1}{2}f(u_i))$ to the y-axis, which is u_i. Then,

$$M_y(R_i) = \rho(x)f(u_i)\Delta x_i u_i = \rho(x)u_i f(u_i)\Delta x_i$$

Adding these moments and taking the limit as $\|P\| \to 0$, we obtain the moment M_y about the y-axis:

$$M_y = \lim_{\|P\| \to 0} \sum_{i=1}^{n} \rho(u_i)u_i f(u_i)\Delta x_i = \int_a^b f(x)\rho(x)x\,dx$$

In a similar way, we compute the moment of R_i about the x-axis, $M_x(R_i)$, as the product of its mass and the distance from $(u_i, \frac{1}{2}f(u_i))$ to the x-axis, which is $\frac{1}{2}f(u_i)$:

$$M_x(R_i) = \rho(x)f(u_i)\Delta x_i \tfrac{1}{2}f(u_i) = \rho(x)\tfrac{1}{2}[f(u_i)]^2\Delta x_i$$

Again, if we add these moments and take the limit as $\|P\| \to 0$, we obtain the moment of R about the x-axis:

$$M_x = \lim_{\|P\| \to 0} \sum_{i=1}^{n} \rho(x)\tfrac{1}{2}[f(u_i)]^2\Delta x_i = \int_a^b \tfrac{1}{2}[f(x)]^2\rho(x)\,dx$$

Recall from (5) that the center of mass $(\bar{x}, \bar{y})$ for a finite system is

$$\bar{x} = \frac{M_y}{M} \qquad \bar{y} = \frac{M_x}{M}$$

For the region R the total mass of the lamina is

$$M = \int_a^b f(x)\rho(x)\,dx \tag{6}$$

Therefore in summary, we have the following result:

[6.8.1] THEOREM / *Center of Mass of Lamina.*

Let R be a region enclosed by a continuous function f, the x-axis, $x = a$, and $x = b$. Suppose further that the density at (x, y) is $\rho(x)$, where ρ is a continuous function on $[a, b]$. Then the

center of mass $(\bar{x}, \bar{y})$ of R is given by

$$\bar{x} = \frac{\int_a^b xf(x)\rho(x)\,dx}{\int_a^b f(x)\rho(x)\,dx} \tag{7}$$

$$\bar{y} = \frac{\int_a^b \frac{1}{2}[f(x)]^2\rho(x)\,dx}{\int_a^b f(x)\rho(x)\,dx} \tag{8}$$

Centroid

In the special case in which the density function ρ is constant—that is, the sheet is homogeneous—the center of mass of R is called the *centroid* of R. In this case, the density of the lamina does not play a role and the centroid becomes a purely geometric concept. Specifically, we have:

[6.8.2] DEFINITION / ***Centroid of a Region.***

Let R be a region enclosed by a continuous function f, the x-axis, $x = a$, and $x = b$. Then the centroid of R, $(\bar{x}, \bar{y})$ is defined by

$$\bar{x} = \frac{1}{A}\int_a^b xf(x)\,dx \tag{9}$$

$$\bar{y} = \frac{1}{A}\int_a^b \frac{1}{2}[f(x)]^2\,dx \tag{10}$$

where $A = \int_a^b f(x)\,dx$.

EXAMPLE 2

Find the centroid of a region enclosed by the graph of $f(x) = x^2$, the x-axis, and $x = 1$.

Solution

The area of the region is

$$A = \text{area} = \int_0^1 x^2\,dx = \frac{1}{3}$$

Using (9) and (10), we get

$$\bar{x} = \frac{1}{A}\int_0^1 xf(x)\,dx = 3\int_0^1 x^3\,dx = \frac{3}{4}$$

$$\bar{y} = \frac{1}{A}\int_0^1 \frac{1}{2}[f(x)]^2\,dx = \frac{3}{2}\int_0^1 x^4\,dx = \frac{3}{10}$$

EXAMPLE 3

Find the centroid of a quarter circular plate,of radius R.

Solution

In order to use (9) and (10), we place the quarter circle as shown in Figure 59 with $f(x) = \sqrt{R^2 - x^2}$, $0 \le x \le R$, $a = 0$, and $b = R$. Because the quarter of a circular plate is symmetric with respect to the line $y = x$, the centroid must lie on this line so that $\bar{x} = \bar{y}$. We choose to compute $\bar{y}$, since it is easier to integrate. The area of a quarter circular region is $\pi R^2/4$, so

$$\bar{y} = \frac{4}{\pi R^2}\int_0^R \frac{1}{2}[f(x)]^2\,dx = \frac{2}{\pi R^2}\int_0^R (R^2 - x^2)\,dx = \frac{2}{\pi R^2}\left[R^2 x - \frac{x^3}{3}\right]\Bigg|_0^R$$

$$= \frac{2}{\pi R^2}\cdot\frac{2}{3}\cdot R^3 = \frac{4}{3\pi}R$$

Hence, $\bar{y} = \bar{x} = 4R/(3\pi)$ (see Fig. 59).

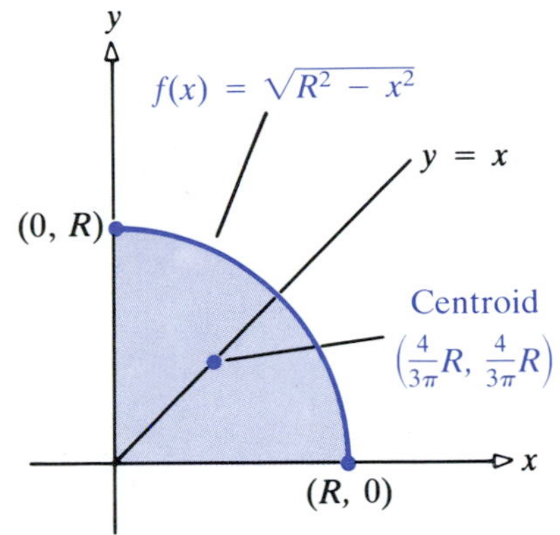

Figure 59

EXAMPLE 4

Find the centroid of a right triangle.

Solution

We place the triangle as shown in Figure 60. The equation of the line connecting $(0, h)$ and $(k, 0)$, is $f(x) = -(h/k)x + h$, $0 \le x \le k$. The area of the triangle is $\frac{1}{2}hk$, so

$$\bar{x} = \frac{\int_0^k xf(x)\,dx}{\frac{1}{2}hk} = \frac{\int_0^k x\left(-\frac{h}{k}x + h\right)dx}{\frac{1}{2}hk} = \frac{2}{hk}\int_0^k \left(-\frac{h}{k}x^2 + hx\right)dx$$

$$= \frac{2}{k}\int_0^k \left(x - \frac{x^2}{k}\right)dx = \frac{2}{k}\left[\frac{x^2}{2} - \frac{x^3}{3k}\right]\Bigg|_0^k = \frac{k}{3}$$

It is left to the reader to show that $\bar{y} = h/3$. Hence, the centroid is $(k/3, h/3)$. Note that it is located at a point that is one-third the distance from each vertex to the opposite leg of the triangle.

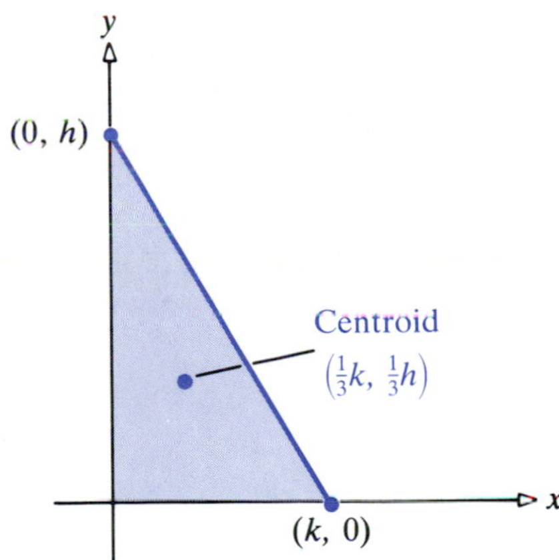

Figure 60

If the region R is enclosed by the graphs of $f(x)$ and $g(x)$, as shown in Figure 61, then (9) and (10) are modified accordingly, and the centroid of R is $(\bar{x}, \bar{y})$, where

$$\bar{x} = \frac{1}{A}\int_a^b x[f(x) - g(x)]\,dx \tag{11}$$

$$\bar{y} = \frac{1}{A}\int_a^b \frac{1}{2}[(f(x))^2 - (g(x))^2]\,dx \tag{12}$$

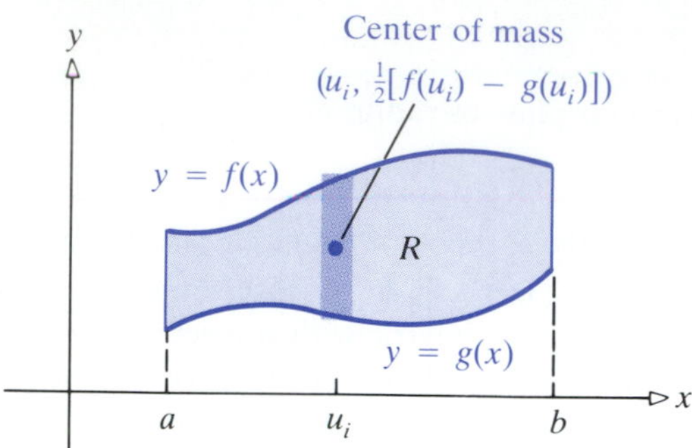

Figure 61

EXAMPLE 5

Find the centroid of the lamina enclosed by the graphs $y = \sqrt{x}$ and $y = x$.

Solution

The lamina is illustrated in Figure 62. We use (11) and (12) with $f(x) = \sqrt{x}$, $g(x) = x$, $a = 0$, and $b = 1$. The area A is

$$A = \int_0^1 (\sqrt{x} - x)\, dx = \left[\frac{2x^{3/2}}{3} - \frac{x^2}{2}\right]\Bigg|_0^1 = \frac{1}{6}$$

Then,

$$\bar{x} = \frac{1}{A}\int_0^1 x[f(x) - g(x)]\, dx = \frac{1}{\frac{1}{6}}\int_0^1 (x^{3/2} - x^2)\, dx = 6\left[\frac{2x^{5/2}}{5} - \frac{x^3}{3}\right]\Bigg|_0^1 = \frac{6}{15}$$

$$\bar{y} = \frac{1}{A}\int_0^1 \frac{1}{2}[(f(x))^2 - (g(x))^2]\, dx = \frac{1}{\frac{1}{6}}\int_0^1 \frac{1}{2}(x - x^2)\, dx = 3\left[\frac{x^2}{2} - \frac{x^3}{3}\right]\Bigg|_0^1 = \frac{1}{2}$$

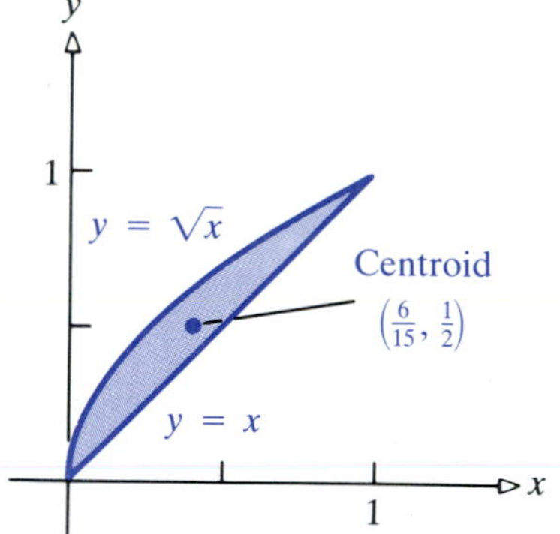

Figure 62

The centroid is $(\frac{6}{15}, \frac{1}{2})$.

Centroids are useful not only in determining centers of masses of homogeneous laminas but also in computing volumes. The next theorem was named after the Greek mathematician Pappus of Alexandria (c. 300 AD), whose mathematical collection contains a record of much of classical Greek mathematics. It shows a relationship between volume and centroids.

[6.8.3] THEOREM / ***Pappus Theorem about Volume.***

Let R be a plane region and let V be the volume of the solid of revolution formed by revolving R about a line L that does not intersect R. Then, the volume V is given by

$$V = 2\pi\, lA \tag{13}$$

where A is the area of R and l is the distance between the centroid of R and the line L of revolution.

Proof We give a proof for the special case where R is the area enclosed by the graph of f, the x-axis, $x = a$, and $x = b$ and where L is the y-axis (see Fig. 63). Using the method of cylindrical shells to find volumes, we have

$$V = \int_a^b 2\pi x f(x)\, dx = 2\pi \int_a^b x f(x)\, dx$$

$$= 2\pi(\bar{x}A)$$

↑

by (9)

where $\bar{x} = l$ is the distance of the centroid $(\bar{x}, \bar{y})$ of R from the y-axis.

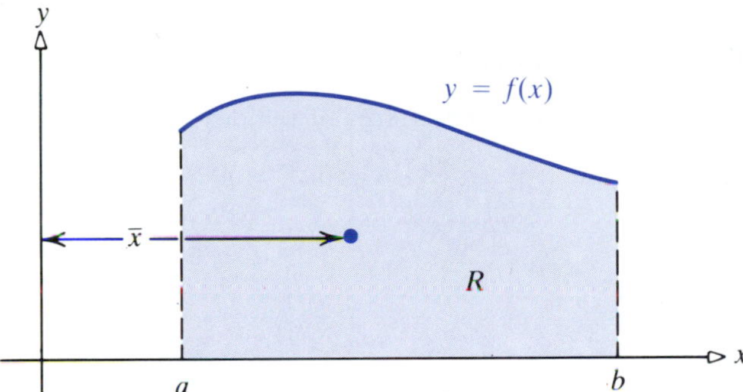

Figure 63

EXAMPLE 6

A torus (doughnut-shaped solid) is formed by revolving the region enclosed by the circle

$$(x - 3)^2 + y^2 = 1$$

about the y-axis. Use the Pappus theorem (6.8.3) to find the volume of the torus.

Solution

The centroid of the circular region is the center of the circle, the point $(3, 0)$. See Figure 64. The distance from the centroid to the axis of revolution, which is the y-axis, is $d = 3$.

The area of the circle is

$$A = \pi \cdot R^2 = \pi 1^2 = \pi$$

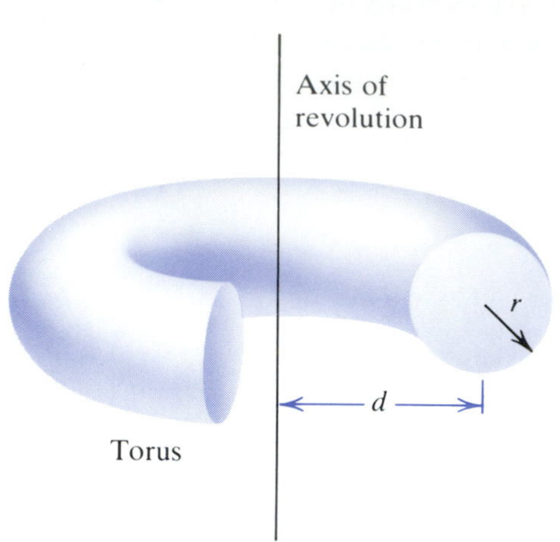

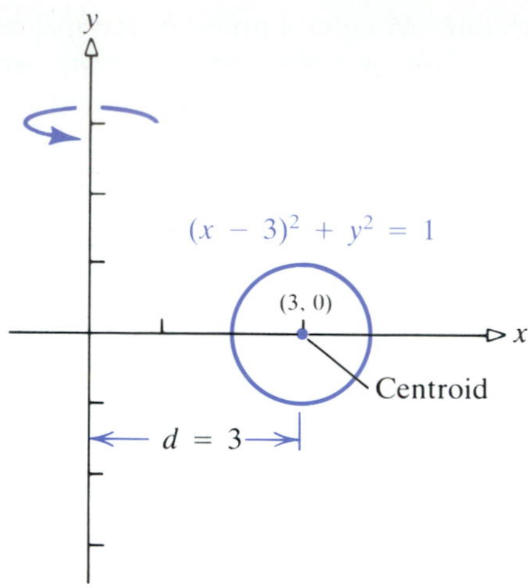

Figure 64

It follows from the Pappus theorem that

$$V = \text{Volume of torus} = 2\pi d \cdot (\text{Area of region})$$
$$= 2\pi \cdot 3 \cdot \pi = 6\pi^2 \approx 59.2$$

EXERCISE 6.8

In Problems 1–4 find the moments M_x and M_y and the center of mass of each system.

1. $m_1 = 4$, $m_2 = 3$, $m_3 = 3$, $m_4 = 5$ located respectively at $(-1, 0)$, $(2, 0)$, $(4, 0)$, $(3, 0)$

2. $m_1 = 7$, $m_2 = 3$, $m_3 = 2$, $m_4 = 4$ located respectively at $(6, 0)$, $(-2, 0)$, $(-4, 0)$, $(-1, 0)$

3. $m_1 = 4$, $m_2 = 3$, $m_3 = 3$, $m_4 = 5$ located respectively at $(-1, 2)$, $(2, 3)$, $(4, 5)$, $(3, 6)$

4. $m_1 = 8$, $m_2 = 6$, $m_3 = 3$, $m_4 = 5$ located respectively at $(-4, 4)$, $(0, 5)$, $(6, 4)$, $(-3, -5)$

In Problems 5–10 find the centroid of the region enclosed by the graph of each function.

5. $y = x^2$, $y = 0$, $x = 3$

6. $y = \sqrt{x}$, $x = 4$, $y = 0$

7. $y = 2x + 3$, $y = 0$, $x = -1$, $x = 2$

8. $y = 4x - x^2$ and the x-axis

9. $y = x^3$, $x = 2$, $y = 0$

10. $y = x^2 + x + 1$, $x = 0$, $x = 4$, the x-axis

In Problems 11–18 compute the center of mass of the region with density ρ enclosed by the graphs of f and g on the given interval.

11. $f(x) = x$, $g(x) = -1$; $[0, 3]$; $\rho(x) = x$

12. $f(x) = 3x - 1$, $g(x) = x - 3$; $[1, 4]$; $\rho(x) = x$

13. $f(x) = 2x$, $g(x) = x^2$; $[0, 1]$; $\rho(x) = x + 1$

14. $f(x) = x$, $g(x) = \sqrt{x}$; $[1, 4]$; $\rho(x) = x + 1$

15. $f(x) = 4 - x^2$, $g(x) = x + 2$; $[-2, 1]$; $\rho(x) = 2$

16. $f(x) = 2 - x^2$, $g(x) = |x|$; $[-1, 1]$; $\rho(x) = 2$

17. $f(x) = \cos x$, $g(x) = 1/2$; $[-\frac{\pi}{3}, \frac{\pi}{3}]$; $\rho(x) = 1$

18. $f(x) = \sin x$, $g(x) = 0$; $[0, \pi]$; $\rho(x) = 1$

In Problems 19–24 compute the moments M_x and M_y and the center of mass of each homogeneous lamina illustrated.

19.

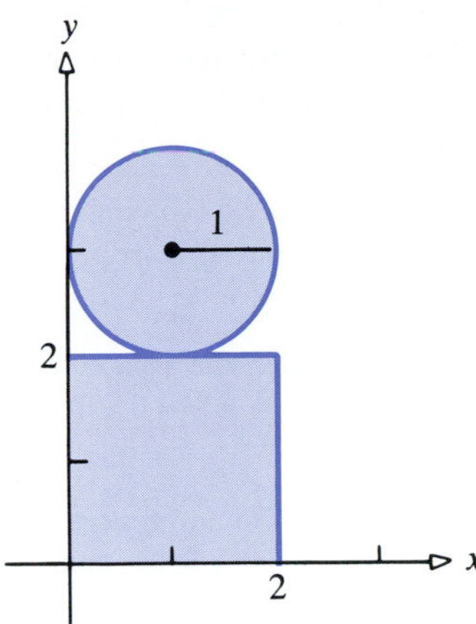

20.

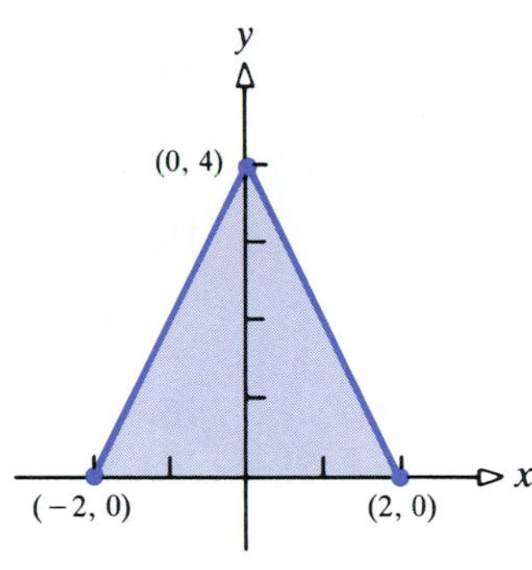

21.

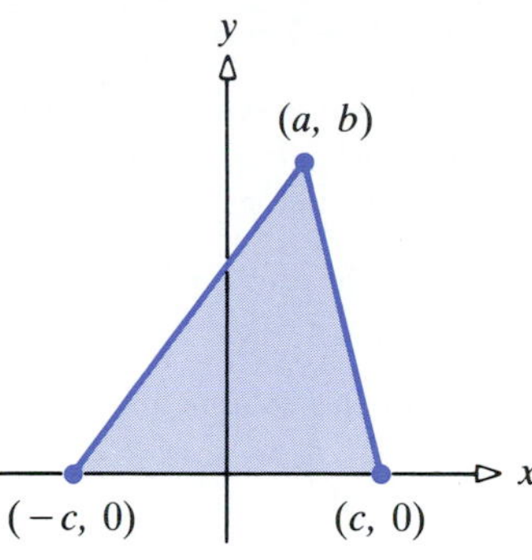

22.

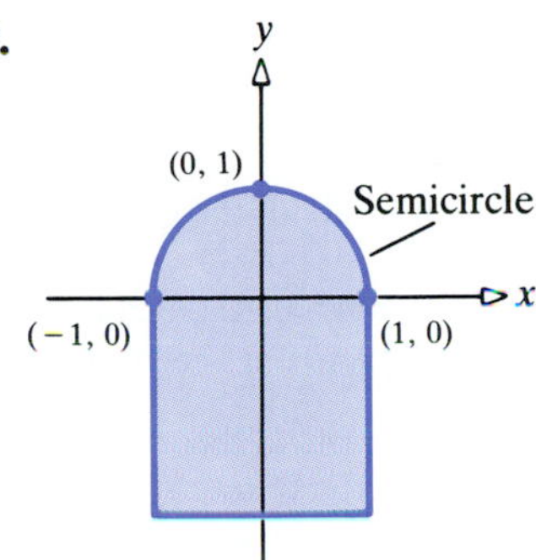

23.

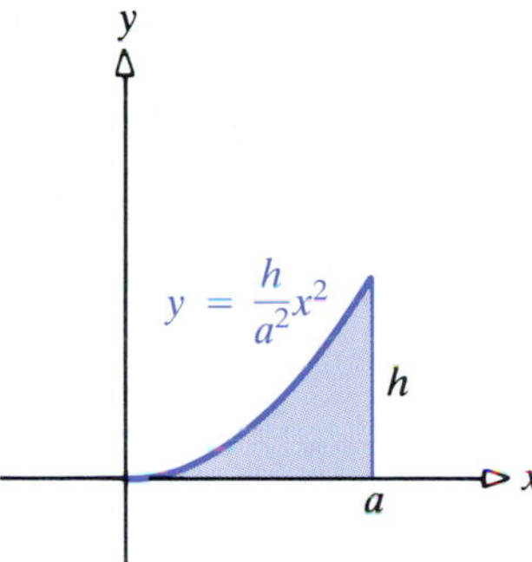

24.

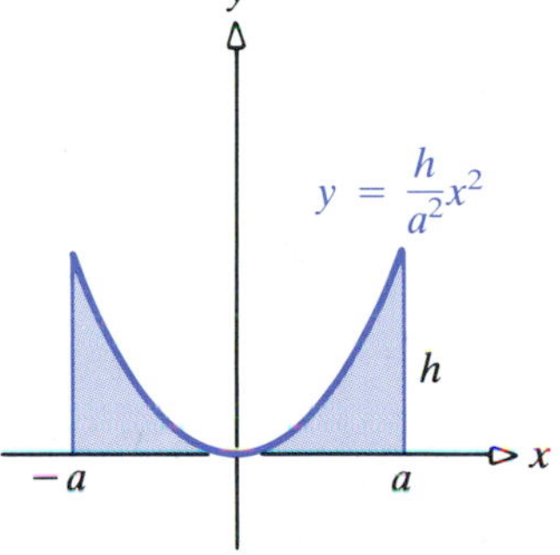

In Problems 25–30 use the Pappus theorem to find the volume of the solid of revolution.

25. The torus formed by revolving the circle $(x-4)^2 + y^2 = 9$ about the y-axis

26. The torus formed by revolving the circle $x^2 + (y-2)^2 = 1$ about the x-axis

27. The solid formed by revolving the region enclosed by the graph $y=\sqrt{x-1}$, $y=0$, and $x=2$ about the y-axis

28. The solid formed by revolving the triangle enclosed by the lines $x = 3$, $y = 2$, and $x + 2y = 9$ about the y-axis

29. A right circular cone of radius R and height H

30. The solid formed by revolving the triangle whose vertices are $(1, 1)$, $(5, 3)$, and $(3, 3)$ about the x-axis

REVIEW EXERCISES

In Problems 1–4 find the areas enclosed by the graphs.

1. $y = \sin x, \quad y = 1, \quad x = \pi/6$

2. $y = x\sqrt{4 - x^2}, \quad y = -x, \quad 0 \le x \le 2$

3. $y = x^2, \quad y = 18 - x^2$

4. $x = 2(y-1)^2, \quad (y-1)^2 = x - 1$

5. Find b such that the area under the graph of $y = (x+1)\sqrt{x^2 + 2x + 4}$ is $\frac{56}{3}$ for $0 \le x \le b$.

6. Find h such that the area under the graph of $y^2 = x^3$, $0 \le x \le 64$, $y \ge 0$, is equal to the area of a rectangle of base 4 and height h.

7. Use integration to find the area of the triangle formed by $x - y + 1 = 0$, $3x + y - 13 = 0$, and $x + 3y - 7 = 0$.

8. If P is a polynomial that is positive for $x > 0$, and for each $k > 0$ the area under the graph of P from $x = 0$ to $x = k$ is $k^3 + 3k^2 + 6k$, find P.

9. Find the area enclosed by the axes and $\sqrt{x} + \sqrt{y} = \sqrt{a}$.

10. Find the volume generated by revolving the area enclosed by $xy = 1$, $x = 1$, $x = 2$, and $y = 0$ about the x-axis.

11. Find the volume generated by revolving the area enclosed by $y = x^2 - 4$ and the x-axis about the line $y = -4$.

12. Find the volume of the cone obtained by revolving the area enclosed by $y = x/2$, $y = 0$, and $x = 4$ about the x-axis.

13. Find the volume generated by revolving one arch of $y = \sin x$ about the x-axis.

14. Find the volume generated by revolving the area enclosed by $y = 4x - x^2$ and the x-axis about the x-axis.

15. Find the volume generated by revolving the area enclosed by $y^2 = 8x$, $y = 0$, and $x = 2$ about the line $x = 2$.

16. Find the volume generated when the area enclosed by $y = -x^2 - 3x + 6$ and $x + y - 3 = 0$ is revolved about the x-axis.

17. Use horizontal slices to find the volume generated when the area enclosed by $y = x^3/2$, $y = 0$, and $x = 2$ is revolved about the y-axis.

In Problems 18–20 use cylindrical shells to find the volume generated when each region is revolved about the given line.

18. $y = x^2$, $y = 4x - x^2$; about $x = 5$

19. $y = x^2 - 5x + 6$, $y = 0$; about the y-axis.

20. $y = -x^2 - 3x + 6$, $x + y - 3 = 0$; about $x = 3$

21. Use vertical slices to find the volume generated when the area enclosed by $x = 0$, $y = 8$, and $y^2 = x^3$ is revolved about $x = 4$.

22. A solid has a circular base of radius 4 units. Every cross section perpendicular to a fixed diameter is an equilateral triangle. Find its volume.

23. A hole of radius 3 is bored through the center of a sphere with radius 4. How much volume has been removed?

24. Find the volume of an elliptical cone with base $x^2/4 + y^2/1 = 1$ and height 5. [*Hint:* The area of an ellipse with semiaxes a and b is πab.]

25. The base of a solid is enclosed by $4x + 5y = 20$, $x = 0$, $y = 0$. Every cross section perpendicular to the x-axis is a semicircle. Find its volume.

In Problems 26–28 find the arc length.

26. $y = x^{3/2} + 4$ from $x = 2$ to $x = 5$

27. $y = \dfrac{x^3}{6} + \dfrac{1}{2x}$ from $x = 2$ to $x = 6$

28. $y^3 = 8x^2$ from $x = 1$ to $x = 8$

29. Find the point P on $y = \frac{2}{3}x^{3/2}$ to the right of the y-axis such that the length of the curve from 0 to P is $\frac{52}{3}$.

30. Find the work done in raising a 800 pound anchor 150 feet with a chain weighing 20 pounds per foot.

31. Find the work done in raising 1000 pounds of gold ore from a mine 1200 feet deep with a cable weighing 3 pounds per foot.

32. A hemispherical water tank has a diameter of 12 feet. It is filled to a depth of 4 feet. How much work is done in pumping all the water over the edge? (Use $\rho = 62.5$ pounds per cubic foot.)

33. A spring with an unstretched length of 0.6 meter requires a force of 4 newtons to stretch it to 0.8 meter. How much work is done in stretching it to 1.4 meters?

34. Find the unstretched length of a spring if the work required to stretch the spring from 1.0 meter to 1.4 meters is half the work required to stretch it from 1.2 meters to 1.8 meters.

35. A trough of trapezoidal cross section is 2 feet wide at the bottom, 4 feet wide at the top, and 3 feet deep. What is the force due to liquid pressure on the end, if it is full of water?

36. A cylindrical tank is on its side. It has a diameter of 10 feet and is full to a depth of 4 feet with oil that has a density of 56 pounds per cubic foot. What is the force due to liquid pressure on the end?

37. A dam is built in the shape of a trapezoid 1000 ft long at the top, 700 ft long at the bottom, and 80 ft deep. Determine the force due to liquid on the dam if

(a) The reservoir behind the dam is full.
(b) The reservoir behind the dam has a depth of 60 ft.

38. Find the centroid of the region enclosed by $y = \sqrt{x}$, $y = 0$, and $x = 9$.

39. Find the center of mass of the region enclosed by $y=\sqrt{x+1}$, $x=0$, and $y=3$, where $\rho(x)=x$.

40. Find the volume of the torus with an outer diameter of 5 inches and an inner diameter of 2 inches.

41. Find the volume generated when the triangle formed by $x = 3$, $y = 0$, and $2x + y - 12 = 0$ is revolved about the y-axis.

CHALLENGE EXERCISES

1. Suppose a plane region of area A to the right of the y-axis is revolved about the y-axis, generating a solid of volume V. If this same area is revolved about the line $x = -k$, $k > 0$, show that the solid thus generated has volume $V + 2\pi kA$.

2. Find the volume generated by revolving the area enclosed by the parabola $y^2 = 8x$ and its latus rectum about the latus rectum. (A latus rectum is the chord through the focus perpendicular to the axis; see Section 12.2.)

(a) Use the disk method.
(b) Use cylindrical shells.

3. The axes of two pipes of equal radii r intersect at right angles. Find their common volume.

4. Find the volume of a cone with an elliptical base with a major axis of $2a$ and a minor axis of $2b$. (See Section 12.3.)

5. Find the area enclosed by the curve $y^2 = x^2 - x^4$.

6. *Calculator Problem.* Inscribe a regular polygon of 2^n sides in a circle of radius 1. The figure illustrates the situation for $n = 3$. Show that the indicated angle $\alpha = 45/2^{n-2}$ degrees. Show that the formula $2^{n+1}\sin(45/2^{n-2})$ gives a good approximation to the arc length (circumference) of the circle. Evaluate this approximation for several values of n. Compare your answers with the exact value.

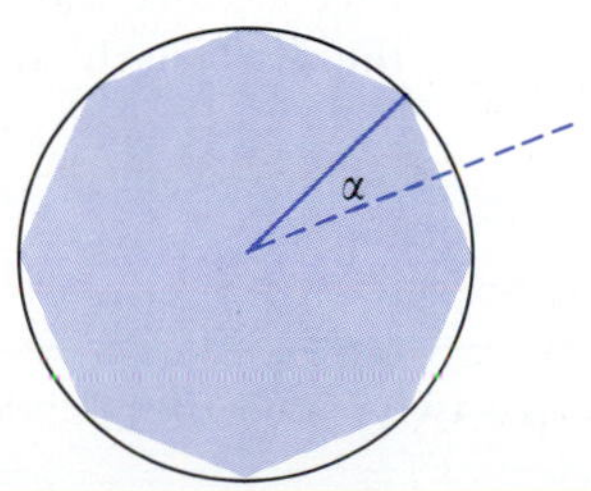

7. *Calculator Problem.* By considering Problem 6 and the half-angle formula, show that $2^{n+2}a_n$ gives an approximation to π, where

$$a_0 = \frac{1}{\sqrt{2}}$$

and

$$a_{n+1} = \sqrt{\frac{1 - \sqrt{1 - a_n^2}}{2}} \qquad n = 0, 1, 2, \ldots$$

Evaluate $2^{n+2}a_n$ for several values of n and compare your answer with the exact value of π.

8. A continuous function f on $[a, b]$ satisfying (i) $f(x) \geq 0$ for x in $[a, b]$ and (ii) $\int_b^a f(x)\,dx = 1$ is called a *probability density function* on $[a, b]$. If $a \leq c < d \leq b$, the probability of obtaining a value between c and d is defined as $\int_c^d f(x)\,dx$.

(a) Find a constant c so that $f(x) = cx$ is a probability density function on $[0, 2]$.
(b) Find the probability of obtaining a value between 1 and 1.5.

9. Refer to Problem 8. If f is a probability density function on $[a, b]$, the *distribution function* F for f is defined as

$$F(x) = \int_a^x f(t)\,dt \qquad a \leq x \leq b$$

Find the distribution function F for the probability density function f of Problem 8, part (a).

10. For the distribution function $F(x) = x - 1$, on $[1, 2]$, find:

(a) The probability density function f corresponding to F
(b) The probability of obtaining a value between 0.5 and 0.7

Finding Arc Length on the HP-28*

It is often difficult to evaluate the definite integrals arising in arc length problems using the fundamental theorem of calculus because the antiderivative is either difficult to find or impossible to express in terms of elementary functions. Problems 29–33 in Exercise 6.5 illustrate this point. Such difficulties are dealt with in the last two sections of Chapter 9. This vignette shows how the HP-28 can be used to approximate definite integrals. This approach is especially useful when the fundamental theorem fails to provide a value. Let's begin by revisiting Problem 29 from Exercise 6.5.

EXAMPLE 1

Find the arc length of $y = x^2$ from $x = 0$ to $x = 2$.

Solution

The derivative of $y = x^2$ is $y' = 2x$. So

$$s = \int_0^2 \sqrt{1 + (2x)^2}\, dx = \int_0^2 \sqrt{1 + 4x^2}\, dx$$

In order to evaluate this, or any, definite integral, the HP needs to be given: (a) the integrand, (b) the variable of integration along with the limits of integration, and (c) the desired error-to-answer ratio. Before beginning, you may wish to clear the stack by pressing [ ] [CLEAR]. Then:

(a) Key in the integrand:

['] [ ] [√] [(] [1] [+] [4] [×] [X] [ ] [∧] [2] [ENTER]

(b) Key in the variable and limits of integration:

[{] [X] [SPACE] [0] [SPACE] [2] [ENTER]

(c) Key in the desired error-to-answer ratio in decimal form:

For example, 0.00001 [ENTER]

The stack should look like this:

```
3:  '√(1+4*X^2)'
2:       { X 0 2 }
1:          .00001
```

Now evaluate the integral by pressing [ ] [∫]. After a short wait, the screen should reveal the approximate value of the arc length sought together with a bound on the error:

```
2:      4.64678413555
1:  4.64631703544E-5
```

The arc length is about 4.65 units or, more precisely, 4.64678 ± 0.00005.

* Gregory D. Foley
The Ohio State University

Example 1 can actually be solved exactly using the fundamental theorem of calculus. To find the antiderivative, you can use the method of substitution in combination with formula 49 in the Table of Integrals (inside the covers of this book) or integration by trigonometric substitution together with (6) in Section 9.2. In either case, after some effort, you should find the exact value of the arc length to be $\sqrt{17} + \frac{1}{4}\ln(4 + \sqrt{17})$, which is approximately 4.646783762. This reveals the seven-digit accuracy of the approximate value found on the HP; the actual error of 3.73×10^{-7} is substantially less than the error bound of 4.65×10^{-5} given by the HP. Actual errors are always less than or equal to their associated error bounds.

Now try finding the arc length of $y = x^3$ from $x = 0$ to $x = 2$. Although this is quite similar to the example problem, in this case the antiderivative cannot be expressed in terms of elementary functions. So an exact answer cannot be found. But the answer can be approximated to a great degree of accuracy using the HP or other numerical techniques.

You may wish to try using the HP on a few problems for which you already know, or can easily find, the exact answers. For example, try finding the arc length of $y = 3x$ from $x = -1$ to $x = 4$. In this case, you can find the arc length using the distance formula, the fundamental theorem of calculus, or the HP-28. Try all three methods. Compare the ease and the exactness of the solutions.

To be sure, virtually any definite integral, not just arc length integrals, can be approximated using the HP in the way illustrated here. To see this, go back to Chapter 5 and try a few definite integral problems that you may have found difficult—for instance, $\int_0^3 |x - 1|\,dx$. (The absolute value function, ABS, is on the REAL menu.) This problem lends itself to a variety of solutions. See how many different approaches you can find. Does $|x - 1|$ have an antiderivative?

Your instructor may suggest other activities for using the definite integration capabilities of the HP.

7 Exponential and Logarithmic Functions

7.1 Introduction

In this section we review the laws for exponents and logarithms, as presented in precalculus mathematics, in preparation for the definitions of the logarithmic and exponential functions given in Sections 7.2 and 7.3.

Let a be a positive number. The following definitions should be familiar:

$$a^n = \underbrace{a \cdot a \cdot \cdots \cdot a}_{n \text{ factors}} \qquad \text{for } n \text{ a positive integer}$$

$$a^0 = 1$$

$$a^{n/m} = (\sqrt[m]{a})^n = \sqrt[m]{a^n} \qquad \text{for } m \text{ and } n \text{ positive integers}$$

$$a^{-r} = \frac{1}{a^r} \qquad \text{for } r \text{ a rational number}$$

From these definitions we can obtain the following laws, which hold for any positive real numbers a and b and any rational numbers r and s:

$$\boldsymbol{a^r a^s = a^{r+s} \qquad (a^r)^s = a^{rs} \qquad (ab)^r = a^r b^r} \tag{1}$$

At this point in most precalculus courses, the student is asked to *assume* that if $a > 0$, then a^x can be defined for *any* real number x. However, making this assumption causes difficulty for two reasons: First, the development is not rigorous; and second, we have stepped beyond our intuition. For example, what is meant by $2^{\sqrt{5}}$ or 3^π?

But if we do not object to this "definition" of a^x, $a > 0$, then it can be used to "define" the logarithm as follows: Consider the exponential relationship $x = a^y$, $a > 0$, $a \neq 1$. The name we give the solution for y when x is given is the *logarithm of* x, *to the base* a, written $y = \log_a x$. That is,

$$\boldsymbol{y = \log_a x \qquad \text{if and only if} \qquad x = a^y}$$

For example,

$$
\begin{array}{llllll}
y = \log_2 1 & \text{means} & 2^y = 1, & \text{that is,} & y = 0 \\
y = \log_2 4 & \text{means} & 2^y = 4, & \text{that is,} & y = 2 \\
y = \log_3 \frac{1}{3} & \text{means} & 3^y = \frac{1}{3}, & \text{that is,} & y = -1 \\
y = \log_a 1 & \text{means} & a^y = 1, & \text{that is,} & y = 0 \\
y = \log_a a & \text{means} & a^y = a, & \text{that is,} & y = 1
\end{array}
$$

The last two results may be generalized as follows:

$$\boldsymbol{\log_a 1 = 0 \qquad \log_a a = 1 \qquad \text{for} \quad a > 0, \quad a \neq 1} \tag{2}$$

For positive real numbers a, M, and N, with $a \neq 1$, the following laws for logarithms may be derived from the laws for exponents:

$$\boldsymbol{\log_a(MN) = \log_a M + \log_a N}$$

$$\boldsymbol{\log_a \frac{M}{N} = \log_a M - \log_a N}$$

$$\boldsymbol{\log_a M^r = r \log_a M \qquad \text{for } r \text{ any real number}} \tag{3}$$

Here are some examples of (3):

$$\log_a[x(x+1)] = \log_a x + \log_a(x+1)$$

$$\log_a \frac{x^2+4}{x+1} = \log_a(x^2+4) - \log_a(x+1)$$

$$\log_a x^{3/2} = \tfrac{3}{2} \log_a x$$

Until recently, *common* logarithms—that is, logarithms to the base 10—were used to facilitate arithmetic computations. However, the development of hand calculators has made this particular use of logarithms less important than it once was. On the other hand, *natural* logarithms—that is, logarithms to the base $e = 2.718\ldots$—remain important for two major reasons: First, they arise naturally in the study of many real-world phenomena; and, second, their use simplifies many mathematical calculations.

Since we now have calculus as a tool, the approach we shall take to logarithms will be based on the definite integral. Although this approach is different from the approach taken in precalculus courses, it has the advantage that it avoids the lack of rigor of the precalculus approach.

7.2 The Natural Logarithm as an Integral

We begin with a definition of the natural logarithm function, which we choose to symbolize as $\ln x$ (instead of $\log_e x$, $e = 2.718\ldots$).* The basis for the definition is the integration problem

$$\int \frac{1}{t}\,dt$$

Recall that the integration formula $\int t^n\,dt = t^{n+1}/(n+1) + C$ does not hold when $n = -1$, in which event $t^n = t^{-1} = 1/t$. However, since the function $f(t) = 1/t$ is continuous when $t > 0$, it follows from definition (5.3.4) that the definite integral $\int_1^x (1/t)\,dt$ exists for all positive x. This integral therefore defines a function whose domain is $x > 0$.

[7.2.1] DEFINITION / ***Natural Logarithm.***

For $x > 0$, the *natural logarithm of* x is

$$\ln x = \int_1^x \frac{1}{t}\,dt \qquad (1)$$

As it turns out, this function $\int_1^x (1/t)\,dt$ has all the properties of a logarithm function, as stated in (2) and (3) on page 457, along with those familiar from precalculus mathematics.

[7.2.2] THEOREM / ***Properties of Logarithm.***

For positive real numbers a and b and any rational number r:

(a) $\ln 1 = 0$ **(b)** $\ln(ab) = \ln a + \ln b$

(c) $\ln a^r = r \ln a$ **(d)** $\ln \frac{a}{b} = \ln a - \ln b$

Proof

(a) Replace x by 1 in (1) to get $\ln 1 = \int_1^1 (1/t)\,dt = 0$.

(b) Replace x by ab in (1). The result is

$$\ln(ab) = \int_1^{ab} \frac{1}{t}\,dt = \int_1^{a} \frac{1}{t}\,dt + \int_a^{ab} \frac{1}{t}\,dt = \ln a + \int_a^{ab} \frac{1}{t}\,dt \qquad (2)$$

* The letters ln, read "ell-en," in $\ln x$ are an abbreviation for the Latin *logarithmus naturalis.*

In $\int_a^{ab}(1/t)\,dt$, we use the substitution $t = au$ so that $dt = a\,du$. When $t = a$, $u = 1$, and when $t = ab$, $u = b$. Thus,

$$\int_a^{ab} \frac{1}{t}\,dt = \int_1^b \frac{1}{au}\,a\,du = \int_1^b \frac{1}{u}\,du \underset{\text{By (1)}}{=} \ln b$$

By using this result in (2), we have

$$\ln(ab) = \ln a + \ln b$$

(c) We replace x by a^r in (1). The result is

$$\ln a^r = \int_1^{a^r} \frac{1}{t}\,dt$$

If $r=0$, the result is true. If $r\neq 0$, we use the substitution $t=u^r$, so that $dt = ru^{r-1}\,du$. When $t = 1$, $u = 1$, and when $t = a^r$, $u = a$. Hence,

$$\ln a^r = \int_1^{a^r} \frac{1}{t}\,dt = \int_1^a \frac{1}{u^r}\,ru^{r-1}\,du = r\int_1^a \frac{du}{u} = r\ln a$$

(d) The proof of part (d) is left as an exercise (Problem 70).

The Derivative of ln *x*

A straightforward application of (7) in Section 5.4 yields

$$\frac{d}{dx}\ln x = \frac{d}{dx}\int_1^x \frac{1}{t}\,dt = \frac{1}{x}$$

[7.2.3] THEOREM / *Derivative of* $y = \ln x$.

The derivative of the logarithm function $y = \ln x$ is $dy/dx = 1/x$. That is,

$$\frac{d}{dx}\ln x = \frac{1}{x} \tag{3}$$

EXAMPLE 1

Find y' if: (a) $y = x\ln x$ (b) $\ln x + \ln y = 2x$

Solution

(a) We use the rule for differentiating a product to get

$$y' = (x)\left(\frac{1}{x}\right) + (1)(\ln x) = 1 + \ln x$$

(b) We use implicit differentiation to get

$$\frac{1}{x} + \left(\frac{1}{y}\right)y' = 2$$

$$y' = y\left(2 - \frac{1}{x}\right) = \frac{y(2x-1)}{x}$$

If we apply the chain rule to the composite function $y = \ln u$, where u is a differentiable function, we find

$$\frac{dy}{dx} = \frac{dy}{du}\frac{du}{dx} = \frac{1}{u}\frac{du}{dx} \tag{4}$$

In particular, if $u = g(x)$, $g > 0$ is a differentiable function, we have

$$\frac{d}{dx}\ln g(x) = \frac{g'(x)}{g(x)} \tag{5}$$

EXAMPLE 2

Find y' if:

(a) $y = \ln(x^2 + 1)$ (b) $y = \ln\sqrt{x^2 + 1}$

(c) $y = (\ln x)^2$ (d) $y = \sin(\ln x)$

Solution

(a) We may use either (5) or the chain rule. By (5), $g(x) = x^2 + 1$ and $g'(x) = 2x$, so that

$$y' = \frac{2x}{x^2 + 1}$$

By the chain rule, we set $y = \ln u$ and $u = x^2 + 1$. Then

$$y' = \frac{dy}{du}\frac{du}{dx} = \left(\frac{1}{u}\right)(2x) = \frac{2x}{x^2 + 1}$$

(b) Always simplify first if possible! Here we may use part (c) of theorem (7.2.2) to write $y = \ln\sqrt{x^2 + 1}$ as $y = \frac{1}{2}\ln(x^2 + 1)$. Then we may use the solution to part (a) above to get $y' = x/(x^2 + 1)$.

(c) We may use either the power rule or the chain rule.

Power rule: $y = [g(x)]^2$ and $g(x) = \ln x$

$$y' = 2\ln x\left(\frac{d}{dx}\ln x\right) = \frac{2\ln x}{x}$$

Chain rule: Set $y = u^2$ and $u = \ln x$; then

$$y' = \frac{dy}{du}\frac{du}{dx} = (2u)\left(\frac{1}{x}\right) = \frac{2\ln x}{x}$$

(d) We use the chain rule with $y = \sin u$ and $u = \ln x$. Then

$$y' = \frac{dy}{du}\frac{du}{dx} = (\cos u)\left(\frac{1}{x}\right) = \frac{\cos(\ln x)}{x}$$

Sometimes the chain rule must be used repeatedly to obtain a derivative.

EXAMPLE 3

Find y' if $y = \cos[\ln(4x - 3)]$, $x > \frac{3}{4}$.*

Solution

We use the chain rule with $y = \cos u$, $u = \ln v$, and $v = 4x - 3$. Then

$$y' = \frac{dy}{du} \cdot \frac{du}{dv} \cdot \frac{dv}{dx} = (-\sin u)\left(\frac{1}{v}\right)(4) = -4[\sin(\ln v)]\frac{1}{v} = \frac{-4\sin[\ln(4x - 3)]}{4x - 3}$$

Logarithmic Differentiation

The properties of logarithms (7.2.2) can sometimes be used to simplify the work needed to find the derivative of certain algebraic functions.

EXAMPLE 4

Find the derivative of $y = \ln[(2x - 1)^3(2x + 1)^5]$.

Solution

Rather than attempt to use (5) or the chain rule, we first use the fact that a logarithm transforms products into sums. That is, we may write

$$y = \ln(2x - 1)^3 + \ln(2x + 1)^5 = 3\ln(2x - 1) + 5\ln(2x + 1)$$

By theorem (7.2.2(b)) By theorem (7.2.2(c))

Now we differentiate by using (5):

$$y' = (3)\left(\frac{2}{2x - 1}\right) + (5)\left(\frac{2}{2x + 1}\right) = \frac{6}{2x - 1} + \frac{10}{2x + 1} = \frac{4(8x - 1)}{4x^2 - 1}$$

As Example 4 illustrates, some thought should be given to the possibility of simplification before differentiating. The next example illustrates a somewhat more subtle procedure.

EXAMPLE 5

Find the derivative of $y = x^2\sqrt{5x + 1}/(3x - 2)^3$.

Solution

As you will see, it is easier to take the natural logarithm of both sides before differentiating; that is, look instead at

$$\ln y = \ln \frac{x^2\sqrt{5x + 1}}{(3x - 2)^3}$$

---

* In subsequent examples, we do not explicitly state the domain of a function containing a natural logarithm. Instead, we assume that the variable is restricted so that the given expression makes sense.

By using some of the properties in theorem (7.2.2), we may write the above expression as

$$\ln y = \ln x^2 + \ln\sqrt{5x+1} - \ln(3x-2)^3 = 2\ln x + \tfrac{1}{2}\ln(5x+1) - 3\ln(3x-2)$$

We now use implicit differentiation and (5) to find y':

$$\frac{y'}{y} = \frac{2}{x} + \frac{5}{2(5x+1)} - \frac{(3)(3)}{3x-2}$$

$$y' = \frac{x^2\sqrt{5x+1}}{(3x-2)^3}\left[\frac{2}{x} + \frac{5}{2(5x+1)} - \frac{9}{3x-2}\right]$$

We refer to the procedure used in Example 5 as *logarithmic differentiation.* This procedure was first used in 1697 by Johann Bernoulli (1667–1748).*

Here are the basic steps used in logarithmic differentiation:

		Comment
Step 1	$y = f(x)$	The function f consists of products and quotients.
Step 2	$\ln y = \ln f(x)$	Take the natural logarithm of both sides and simplify by using theorem (7.2.2).
Step 3	$\dfrac{y'}{y} = \dfrac{d}{dx}\ln f(x)$	Differentiate by using (5).
Step 4	$y' = f(x)\dfrac{d}{dx}\ln f(x)$	Multiply both sides by $y = f(x)$.

Let's do another example to illustrate the procedure.

EXAMPLE 6

Find the derivative of $y = \sqrt{4x+3}/(2x-5)^3$.

Solution

$$\ln y = \ln\left[\frac{(4x+3)^{1/2}}{(2x-5)^3}\right] = \ln(4x+3)^{1/2} - \ln(2x-5)^3$$

$$= \frac{1}{2}\ln(4x+3) - 3\ln(2x-5)$$

$$\frac{y'}{y} = \frac{1}{2}\left(\frac{4}{4x+3}\right) - 3\left(\frac{2}{2x-5}\right)$$

$$y' = \frac{\sqrt{4x+3}}{(2x-5)^3}\left(\frac{2}{4x+3} - \frac{6}{2x-5}\right)$$

---

* Johann was only one of the famous Bernoulli brothers. See the Historical Perspectives section in Chapter 11 for a discussion of their work.

The Graph of $f(x) = \ln x$

We now proceed to obtain the graph of the natural logarithm function. From definition (7.2.1), we know that the domain of $f(x) = \ln x$ is the set of *positive real numbers*, and because of theorem (5.4.4) (p. 358), $f(x) = \ln x$ is both continuous and differentiable. Since $(d/dx) \ln x = 1/x > 0$, we conclude that $f(x) = \ln x$ is an increasing function. Moreover, since $f''(x) = -1/x^2 < 0$, it follows that $f(x) = \ln x$ is concave down on its domain.

Since $f(x) = \ln x$ is increasing on its domain and $\ln 1 = 0$, we conclude that $\ln x > 0$ for $x > 1$, and $\ln x < 0$ for $0 < x < 1$. This conclusion can also be obtained by using a geometric (area) argument, which we now give.

If $x > 1$, then

$$\ln x = \int_1^x \frac{1}{t}\,dt = \text{Area under graph of } y = \frac{1}{x} \text{ from 1 to } x$$

See Figure 1(a). Hence, for $x > 1$, $\ln x > 0$ and the graph of $y = \ln x$ lies above the x-axis.

If $0 < x < 1$, then

$$\ln x = \int_1^x \frac{1}{t}\,dt = -\int_x^1 \frac{1}{t}\,dt = -\left(\text{Area under graph of } y = \frac{1}{x} \text{ from } x \text{ to } 1\right)$$

See Figure 1(b). Hence, for $0 < x < 1$, $\ln x < 0$ and the graph of $y = \ln x$ lies below the x-axis.

Let's summarize what we know about the graph of $y = \ln x$:

1. Domain: positive real numbers
2. $y = \ln x$ is increasing and continuous
3. $y = \ln x$ is concave down
4. $y = \ln x < 0$ if $0 < x < 1$
5. $\ln x = 0$ if $x = 1$
6. $y = \ln x > 0$ if $x > 1$

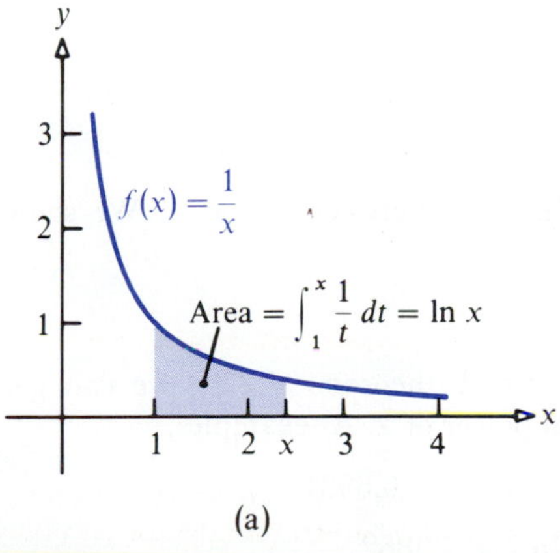

(a)

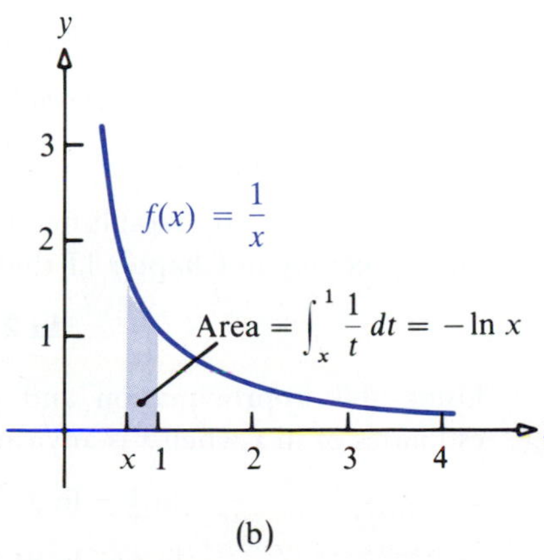

(b)

Figure 1

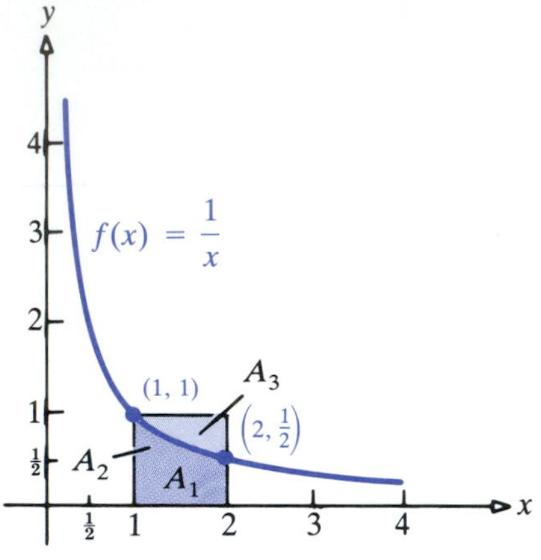

Figure 2

We now proceed to get additional information about $y = \ln x$. First, let's find out how to locate other points on the graph. One way is by approximating ln 2. As Figure 2 illustrates, the areas A_1, A_2, and A_3 obey the inequality

$$A_1 < A_2 < A_3$$

These areas are computed to be

$$(1)\left(\frac{1}{2}\right) < \int_1^2 \frac{1}{t}\,dt < (1)(1)$$

$$\frac{1}{2} < \ln 2 < 1 \tag{6}$$

[7.2.4] THEOREM

$$\textbf{(a)} \quad \lim_{x \to +\infty} \ln x = +\infty \qquad \textbf{(b)} \quad \lim_{x \to 0+} \ln x = -\infty \tag{7}$$

Proof We use (6)

(a) Since $f(x) = \ln x$ is an increasing function, it follows that

$$\ln x > \ln 4^n \qquad \text{whenever} \qquad x > 4^n$$

Since $\ln 4^n = n \ln 4$, this becomes

$$\ln x > n \ln 4 \qquad \text{whenever} \qquad x > 4^n$$

But $\ln 4 > 1$, since $\ln 4 = 2 \ln 2$ and $\ln 2 > \frac{1}{2}$. Thus,

$$\ln x > n \qquad \text{whenever} \qquad x > 4^n$$

By taking $N = 4^n$, we find that for any $n > 0$,

$$\ln x > n \qquad \text{whenever} \qquad x > N = 4^n$$

In other words, $\lim_{x \to +\infty} \ln x = +\infty$.

(b) To determine $\lim_{x \to 0^+} \ln x$, we set $x = 1/n$. Then, as $x \to 0^+$, n approaches $+\infty$ and

$$\lim_{x \to 0^+} \ln x = \lim_{n \to +\infty} \ln \frac{1}{n} \underset{\substack{\uparrow \\ \text{By theorem (7.2.2(d))}}}{=} \lim_{n \to +\infty} (\ln 1 - \ln n) \underset{\substack{\uparrow \\ \ln 1 = 0}}{=} -\lim_{n \to +\infty} \ln n \underset{\substack{\uparrow \\ \text{By (7(a))}}}{=} -\infty$$

Inequality (6) establishes that ln 2 lies between $\frac{1}{2}$ and 1. We show more precisely in Chapter 11 that

$$\mathbf{\ln 2 = 0.6931 \ldots}$$

Using this approximation and part (c) of theorem (7.2.2), we can get estimates of $\ln x$ when x is any rational power of 2. As examples,

$$\ln \tfrac{1}{2} = \ln 2^{-1} = -\ln 2 \approx -0.6931$$

$$\ln 4 = \ln 2^2 = 2 \ln 2 \approx 1.3862$$

$$\ln 8 = \ln 2^3 = 3 \ln 2 \approx 2.0793$$

We conclude from part (b) of (7) that the line $x = 0$ is a vertical asymptote to the graph of $y = \ln x$. From (7) and the intermediate value theorem (2.4.6), we conclude that the range of the natural logarithm is $(-\infty, +\infty)$. Putting together all of the foregoing facts, we obtain the graph of $y = \ln x$ as shown in Figure 3.

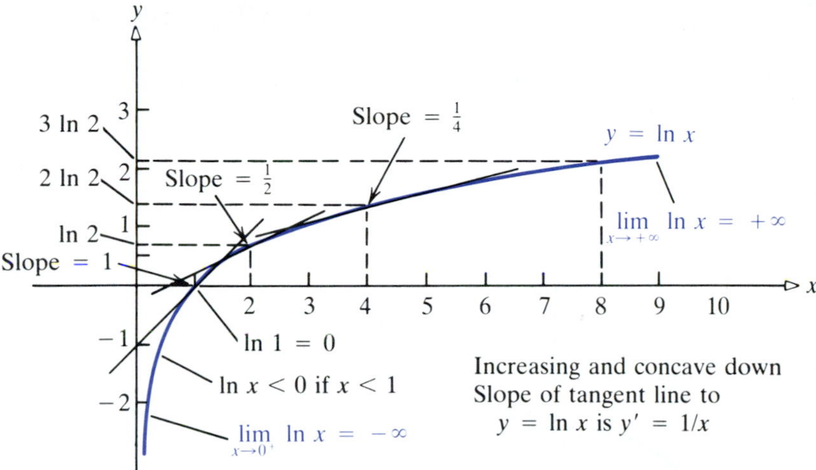

Figure 3

EXAMPLE 7

Sketch the graph of $y = \ln(9 - x^2)$ using the steps outlined in Section 4.8.

Solution

1. The domain is all x satisfying $9 - x^2 > 0$—namely, $-3 < x < 3$.
2. The y-intercept is $f(0) = \ln 9$. To find the x-intercept, we let $y=0$:

$$\ln(9 - x^2) = 0$$

Since $\ln 1 = 0$, we write

$$\ln(9 - x^2) = \ln 1$$

$$9 - x^2 = 1$$

$$x^2 = 8$$

Thus, the x-intercepts are $\pm 2\sqrt{2}$.

3. Since $f(-x) = f(x)$, f is an even function and therefore its graph is symmetric with respect to the y-axis.
4. There are no horizontal asymptotes. Since

$$\lim_{x\to 3^-} \ln(9 - x^2) = -\infty \qquad \text{and} \qquad \lim_{x\to -3^+} \ln(9 - x^2) = -\infty$$

we have two vertical asymptotes at $x = -3$ and $x = 3$.

5. $$f'(x) = \frac{-2x}{9 - x^2}$$

Thus, $x = 0$ is the only critical number.

6. $f'(x) > 0$ when $-3 < x < 0$ and $f'(x) < 0$ when $0 < x < 3$, so f is increasing on $(-3, 0)$ and decreasing on $(0, 3)$.
7. Since $f'(x) > 0$ for $x < 0$ and $f'(x) < 0$ for $x > 0$, by the first derivative test, $(0, \ln 9)$ is a local maximum.
8. $$f''(x) = \frac{(9 - x^2)(-2) + 2x(-2x)}{(9 - x^2)^2} = \frac{-2x^2 - 18}{(9 - x^2)^2} = \frac{-(2x^2 + 18)}{(9 - x^2)^2}$$

   Thus, $f''(x) < 0$ for all x, so the graph is concave down on $(-3, 3)$ and there is no inflection point.
9. See Figure 4.

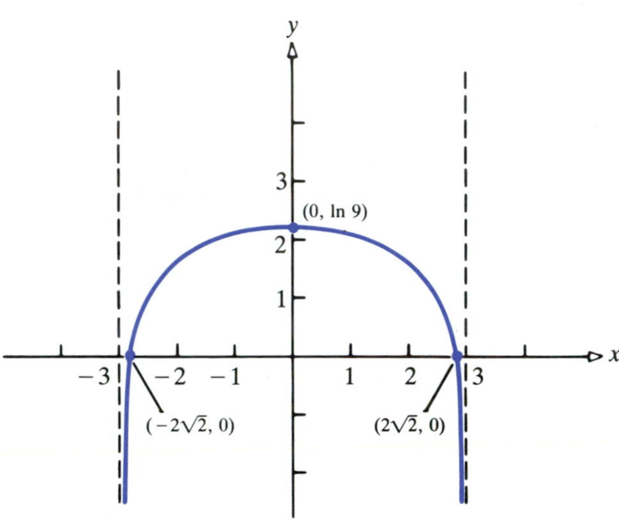

Figure 4

The Number e

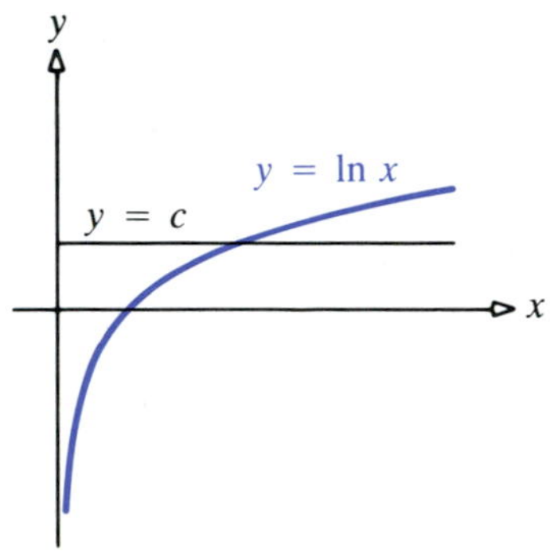

Figure 5

Since the natural logarithm function $f(x) = \ln x$ is continuous on its domain $(x > 0)$ and its range includes all real numbers, it follows that the graph must cross each horizontal line $y = c$ at a point $(c, \ln c)$ to the right of the y-axis, as in Figure 5. Moreover, since $f(x) = \ln x$ is an increasing function, its graph cannot cross any such horizontal line more than once. Thus, we have the following theorem:

[7.2.5] THEOREM

For every real number c, there is a unique positive real number x such that

$$\ln x = c$$

[7.2.6] DEFINITION / *The Number e.*

In particular, there is a unique number x for which $\ln x = 1$. This number is designated by e (see Fig. 6). Thus,

$$e \text{ is that unique number for which } \ln e = 1$$

We have already discovered in (6) that $\ln 2 < 1 < \ln 4$, so that $\ln 2 < \ln e < \ln 4$. Hence, we conclude that

$$2 < e < 4 \tag{8}$$

In Section 7.6 we show that e may be expressed as the limit

$$e = \lim_{n \to +\infty} \left(1 + \frac{1}{n}\right)^n \tag{9}$$

and we use this formula to approximate e as

$$e = 2.71828\ldots$$

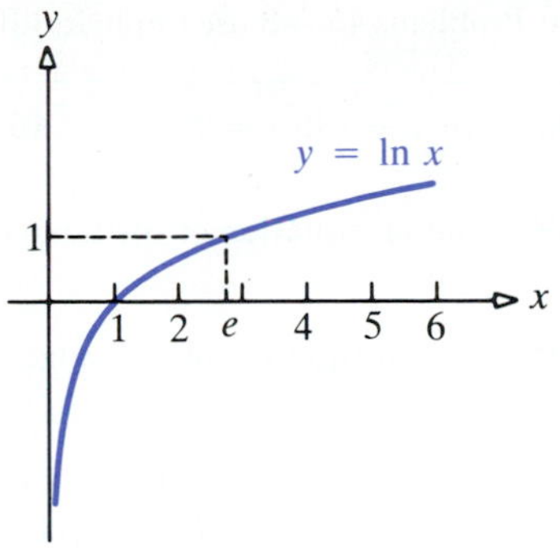

Figure 6

EXERCISE 7.2

In Problems 1–8 let $a = \ln 2$ and $b = \ln 5$. Express each logarithm in terms of a and b only.

1. $\ln 10$ **2.** $\ln 25$ **3.** $\ln \sqrt[2]{5}$ **4.** $\ln \sqrt[3]{10}$

5. $\ln 0.05$ **6.** $\ln(0.05)^3$ **7.** $\ln \frac{5}{2}$ **8.** $\ln \frac{1}{25}$

In Problems 9–12 sketch the graph.

9. $y = -\ln x$ **10.** $y = \ln x - x$

11. $y = \ln 3x$ **12.** $y = \ln(x + 2)$

In Problems 13–36 find the derivative of each function.

13. $y = \ln 3x$ **14.** $y = 5 \ln x$ **15.** $y = x \ln(x^2 + 4)$ **16.** $y = x \ln(x^2 + 5x + 1)$

17. $y = x \ln\sqrt{x^2 + 1}$ **18.** $y = x \ln(\sqrt[3]{3x + 1})$ **19.** $y = \frac{1}{2} \ln\left(\frac{1 + x}{1 - x}\right)$ **20.** $y = \frac{1}{2} \ln\left(\frac{1 + x^2}{1 - x^2}\right)$

21. $y = \ln(\ln x)$ **22.** $y = \ln\left(\ln \frac{1}{x}\right)$ **23.** $y = \ln\left(\frac{x}{\sqrt{x^2 + 1}}\right)$ **24.** $y = \ln\left(\frac{4x^3}{\sqrt{x^2 + 4}}\right)$

25. $y = \ln\left[\frac{(x^2 + 1)^2}{x\sqrt{x^2 - 1}}\right]$ **26.** $y = \ln\left[\frac{x\sqrt{3x - 1}}{(x^2 + 1)^3}\right]$ **27.** $y = \ln(\sin x)$ **28.** $y = \ln(\cos x)$

29. $y = (\ln x)^{1/2}$ **30.** $y = (\cos x)(\ln x)$ **31.** $y = x \ln\sqrt{\cos 2x}$ **32.** $y = x^2 \ln\sqrt{\sin 2x}$

33. $y = \ln(x + \sqrt{x^2 + 4})$ **34.** $y = \ln(\sqrt{x + 1} + \sqrt{x})$ **35.** $y = \ln(x + \sqrt{x^2 + a^2})$ **36.** $y = \ln(\sqrt{x + a} + \sqrt{x})$

In Problems 37–44 use logarithmic differentiation to find y'.

37. $y = (x^2 + 1)^2(2x^3 - 1)^4$ **38.** $y = (3x^2 + 4)^3(x^2 + 1)^4$ **39.** $y = (x^3 + 1)(x - 1)(x^4 + 5)$

40. $y = \sqrt{x^2 + 1}\,(x^3 - 5)(3x + 4)$ **41.** $y = \frac{x^2(x^3 + 1)}{\sqrt{x^2 + 1}}$ **42.** $y = \frac{\sqrt{x}(x^3 + 2)^2}{\sqrt[3]{3x + 4}}$

43. $y = \frac{x \cos x}{(x^2 + 1)^3 \sin x}$ **44.** $y = (x \sin x)(\cos x)(\ln x)$

In Problems 45–48 use implicit differentiation to find dy/dx.

45. $x \ln y + y \ln x = 2$

46. $\ln(x^2 + y^2) = x + y$

47. $\ln\left(\frac{y}{x}\right) - \ln\left(\frac{x}{y}\right) = 1$

48. $\ln(x^2 - y^2) = x - y$

49. Find an equation of the tangent line to $y = \ln 5x$ at $(\frac{1}{5}, 0)$.

50. Find an equation of the tangent line to $y = x \ln x$ at $(1, 0)$.

51. Show that for $x > 1$, $\ln x < 2(\sqrt{x} - 1)$. [*Hint:* Compare $\int_1^x (1/t)\, dt$ with $\int_1^x (1/\sqrt{t})\, dt$ for $x > 1$.]

52. Show that $0 < \ln x < x$ for $x > 1$.

53. Use Problem 51 to evaluate $\lim_{x \to +\infty} (\ln x/x)$.

54. Use Problem 53 to evaluate $\lim_{x \to 0^+} x(\ln x)$. [*Hint:* Let $x = 1/u$.]

55. Prove that $x \le \ln(1 + x)^{1+x}$ for all $x > -1$. [*Hint:* Consider $f(x) = -x + \ln(1 + x)^{1+x}$.]

56. Show that $1 + \ln x < x$ whenever $x > 1$.

57. Find $\frac{d^{10}}{dx^{10}}(x^9 \ln x)$.

58. If $f(x) = \ln(x - 1)$, find $f^{(n)}(x)$.

In Problems 59–64 discuss the graph of each function.

59. $y = \ln(4 - x^2)$

60. $y = \ln(\sin x)$

61. $y = \ln(\cos x)$

62. $y = \ln(x\sqrt{x - 1})$

63. $y = x \ln x$

64. $y = (1/x) \ln x$

65. If $y = \ln(x^2 + y^2)$, find the value of dy/dx at the point $(1, 0)$.

66. If $y = \tan u$, $u = v - (1/v)$, and $v = \ln x$, what is the value of dy/dx at $x = e$?

67. A point moves on the x-axis in such a way that its velocity at time t $(t > 0)$ is given by $v = (\ln t)/t$. At what time t does v attain its maximum?

68. Find a formula for the nth derivative of $y = \ln(ax)$, where a is a positive constant.

69. A telephone cable is made up of a core of copper wires covered by an insulating material. If x is the ratio of the radius of the core to the thickness of the insulating material, the speed v of signaling is

$$v = kx^2 \ln \frac{1}{x}$$

where k is a constant. Determine the ratio x that results in maximum speed.

70. Use theorem (7.2.2), properties (b) and (c), to show that

$$\ln \frac{a}{b} = \ln a - \ln b \qquad \text{where } a \text{ and } b \text{ are positive real numbers}$$

7.3 The Exponential Function

The Graph of $y = e^x$

Let r be any rational number. Using the facts that $\ln e^r = r \ln e$ and $\ln e = 1$, we have

$$\ln e^r = r$$

Consistent with this equation and the fact that the range of $\ln x$ is all real numbers, we state the following definition:

[7.3.1] DEFINITION

If α is a real number, we define e^α to be that (unique) number for which

$$\ln e^\alpha = \alpha \tag{1}$$

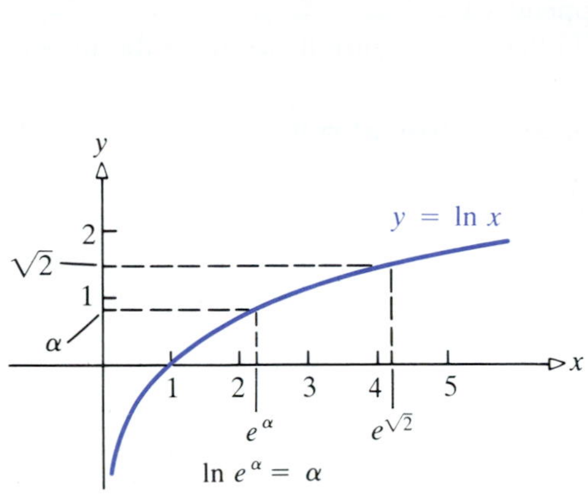

Figure 7

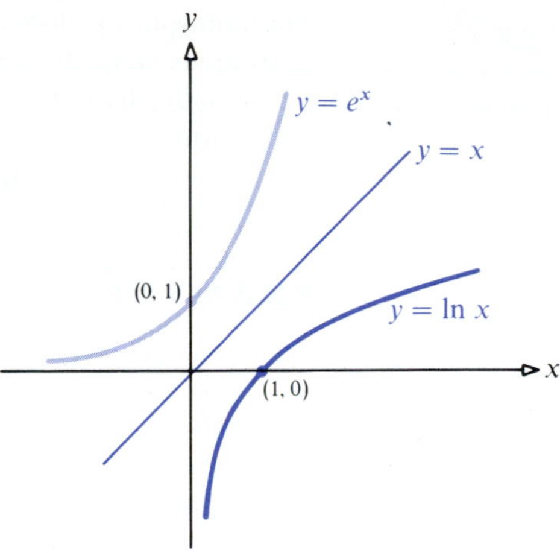

Figure 8

With this definition, we have given meaning to expressions such as $e^{\sqrt{2}}$, which is the unique number for which $\ln e^{\sqrt{2}} = \sqrt{2}$ (see Fig. 7).

The function $y = e^x$ is called the *exponential function with base e*.* Since, by definition,

$$\ln e^x = x \qquad \text{for all real } x \tag{2}$$

it follows that:

[7.3.2] THEOREM

The exponential function $y = e^x$ and the natural logarithm function $y = \ln x$ are inverses.

We may also write this fact as

$$e^{\ln x} = x \qquad \text{for all} \quad x > 0 \tag{3}$$

We use the fact that $y = \ln x$ and $y = e^x$ are inverses to get information about the graph of $y = e^x$. Indeed, we can obtain the graph of $y = e^x$ by simply reflecting the graph of $y = \ln x$ about the line $y = x$ (see Fig. 8). Because $y = \ln x$ and $y = e^x$ are inverses, we conclude that:

Domain of $y = e^x$ equals the range of $y = \ln x$ equals all real numbers.

Range of $y = e^x$ equals the domain of $y = \ln x$ equals the positive real numbers.

Furthermore, since $y = \ln x$ crosses the x-axis at $(1, 0)$, $y = e^x$ crosses the y-axis at $(0, 1)$. Also, $y = \ln x$ is increasing, concave down, and

---

* Some books use the notation $y = \exp x$ to represent the exponential function.

continuous on its domain. Hence, $y = e^x$ is increasing, concave up, and continuous on its domain (see Appendix I).

Finally, based on theorem (7.2.4) and Figure 8, we have the important limits

$$\lim_{x\to+\infty} e^x = +\infty \qquad \lim_{x\to-\infty} e^x = 0 \tag{4}$$

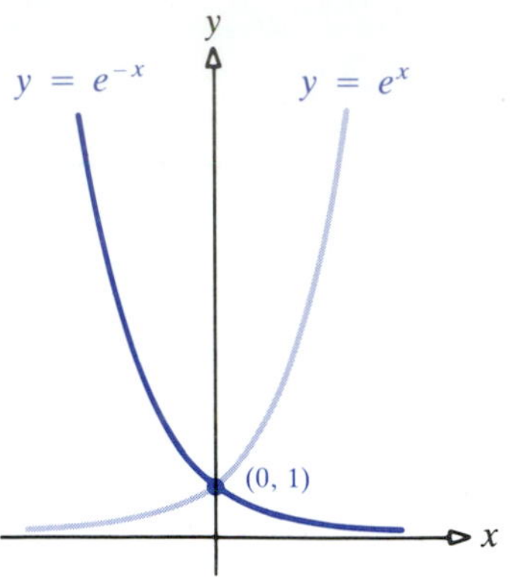

Figure 9

EXAMPLE 1

Sketch the graph of $y = e^{-x}$.

Solution

The replacement of x by $-x$ in the equation $y = e^x$ results in the reflection of the graph of this equation about the y-axis. Thus, the graph of this equation is as shown in Figure 9.

Properties of the Exponential Functions $y = e^x$

The following limits, which are consequences of (4), are evident from the graph of e^{-x}:

$$\lim_{x\to+\infty} e^{-x} = \lim_{x\to+\infty} \frac{1}{e^x} = 0 \qquad \lim_{x\to-\infty} e^{-x} = \lim_{x\to-\infty} \frac{1}{e^x} = +\infty \tag{5}$$

The familiar laws of exponents hold for e^x.

[7.3.3] THEOREM / ***Properties of*** $y = e^x$.

For any real numbers α and β:

$$\text{(a)}\quad e^\alpha e^\beta = e^{\alpha+\beta} \qquad \text{(b)}\quad \frac{e^\alpha}{e^\beta} = e^{\alpha-\beta} \qquad \text{(c)}\quad (e^\alpha)^\beta = e^{\alpha\beta}$$

Proof To prove part (a), we use the properties of the natural logarithm function. Since

$$\ln(e^\alpha e^\beta) = \ln e^\alpha + \ln e^\beta = \alpha + \beta = \ln e^{\alpha+\beta}$$

and since $\ln x$ is a one-to-one function, it follows that

$$e^\alpha e^\beta = e^{\alpha+\beta}$$

The proofs of parts (b) and (c) are left as exercises (Problems 95 and 96).

Derivative of $y = e^x$

We are now ready to find the derivative of $y = e^x$. Although the formula is quite simple, it is one of the most important results in calculus.

[7.3.4] THEOREM / ***Derivative of*** $y = e^x$.

The derivative of the exponential function $y = e^x$ is $dy/dx = e^x$. That is,

$$\frac{d}{dx} e^x = e^x \tag{6}$$

Proof Since the natural logarithm function is differentiable and its derivative is never 0, it follows by theorem (3.8.1) (p. 195) that its inverse, the exponential function, is also differentiable. By (2), we know that

$$\ln e^x = x \qquad \text{for all real } x$$

We differentiate both sides with respect to x, applying (5) in Section 7.2 to the left side, to get

$$\frac{\frac{d}{dx} e^x}{e^x} = 1$$

Therefore,

$$\frac{d}{dx} e^x = e^x$$

Combining theorem (7.3.4) with the chain rule gives

$$\frac{d}{dx} e^u = e^u \frac{du}{dx} \tag{7}$$

where $u = u(x)$ is a differentiable function.

Let's now look at some derivative problems.

EXAMPLE 2

Find y' if:

(a) $y = x^3 e^x$ (b) $y = e^{x^2}$ (c) $e^y + e^x = 4x$ (d) $y = \cos e^x$

(e) $y = \sin e^{\sqrt{x^2+4}}$

Solution

(a) We use the rule for differentiating a product to get

$$y' = x^3\left(\frac{d}{dx} e^x\right) + \left(\frac{d}{dx} x^3\right)e^x = (x^3)(e^x) + (3x^2)(e^x) = x^2e^x(x + 3)$$

(b) We use the chain rule with $y = e^u$ and $u = x^2$ to get

$$y' = \frac{dy}{du} \cdot \frac{du}{dx} = (e^u)(2x) = 2xe^{x^2}$$

(c) We use implicit differentiation to get

$$e^y y' + e^x = 4$$

$$y' = \frac{4 - e^x}{e^y}$$

(d) We use the chain rule with $y = \cos u$ and $u = e^x$ to get

$$y' = \frac{dy}{du} \cdot \frac{du}{dx} = (-\sin u)(e^x) = -e^x \sin e^x$$

(e) We use the chain rule with $y = \sin u$, $u = e^v$, and $v = \sqrt{x^2+4}$ to get

$$y' = \frac{dy}{du} \cdot \frac{du}{dv} \cdot \frac{dv}{dx} = (\cos u)(e^v)\left(\frac{x}{\sqrt{x^2+4}}\right) = \frac{xe^{\sqrt{x^2+4}} \cos e^{\sqrt{x^2+4}}}{\sqrt{x^2+4}}$$

Curve Sketching

EXAMPLE 3

Graph the function $f(x) = xe^{-x}$ using the steps outlined in Section 4.8.

Solution

1. The domain is all real numbers.
2. The origin is the only intercept.
3. There is no symmetry.
4. Since $\lim_{x\to+\infty}(x/e^x) = 0$, then $y = 0$ is a horizontal asymptote (see Problem 77). There are no vertical and no oblique asymptotes.
5. Since $f'(x) = x(-e^{-x}) + (1)e^{-x} = (1-x)e^{-x}$, then $x = 1$ is the only critical number.
6. Since e^{-x} is always positive, $f'(x)$ has the same sign as $(1-x)$; therefore, $f'(x) > 0$ for $x < 1$ and $f'(x) < 0$ for $x > 1$. Thus, f is increasing for $x < 1$ and decreasing for $x > 1$.
7. By the first derivative test, f has a local maximum at $x = 1$. The value of the function at $x = 1$ is $f(1) = e^{-1} \approx 0.368$.
8. Since $f''(x) = (1-x)(-e^{-x}) + (-1)e^{-x} = (x-2)e^{-x}$, f is concave down for $x < 2$ and concave up for $x > 2$. The graph has an inflection point at $x = 2$.
9. If we put all this information together, we obtain the graph in Figure 10.

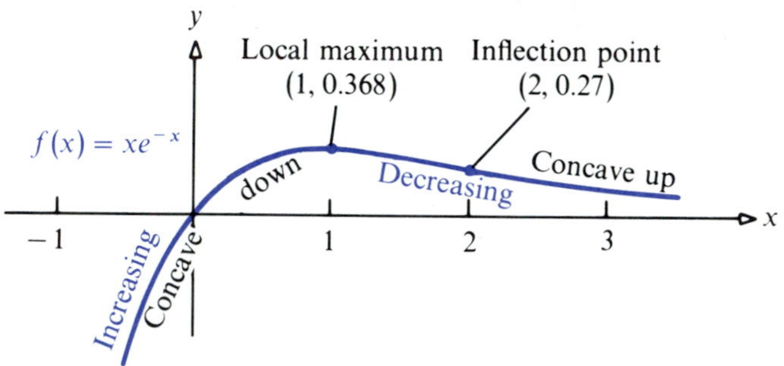

Figure 10

EXAMPLE 4

The damped vibration curve $y = e^{-ax}\sin bx$, a, b positive constants, is of importance in many applications, such as the motion of musical strings, the vibration of a pendulum, and the current in a radio circuit. Graph this function.

Solution

We seek the intercepts first. If $x = 0$, then $y = 0$. If $y = 0$, then $\sin bx = 0$, since e^{-ax} can never be 0. From $\sin bx = 0$, we find $bx = n\pi$, or $x = n\pi/b$, where n is any integer. Thus, the x-intercepts are a distance π/b apart. As x increases, $\sin bx$ oscillates regularly between -1 and 1, but e^{-ax} decreases and approaches 0 as $x \to +\infty$. That is, $\lim_{x\to+\infty} e^{-ax}\sin bx = 0$. Hence, the curve oscillates regularly, but with a constantly decreasing amplitude.

When $\sin bx = 1$ or $x = \pi/2b, 5\pi/2b, 9\pi/2b$, and so on, we have $y = e^{-ax}$, so that the damped vibration curve is in contact with the exponential curve $y = e^{-ax}$ at these x's. Similarly, at $x = 3\pi/2b$, $7\pi/2b$, $11\pi/2b$, and so on, the damped vibration curve is in contact with the exponential curve $y = -e^{-ax}$. The damped vibration curve therefore oscillates between the two exponential guiding curves.

We find $y' = e^{-ax}(b \cos bx - a \sin bx)$. If $y' = 0$, then, since e^{-ax} cannot be 0, we must have $b \cos bx - a \sin bx = 0$, or $\sin bx/\cos bx = b/a$, or $\tan bx = b/a$; these numbers x, at which alternate maximum and minimum points of the damped vibration curve occur, are to the left of the points of contact with the guiding curves, $y = \pm e^{-ax}$.

Making use of all these facts, we obtain the graph shown in Figure 11.

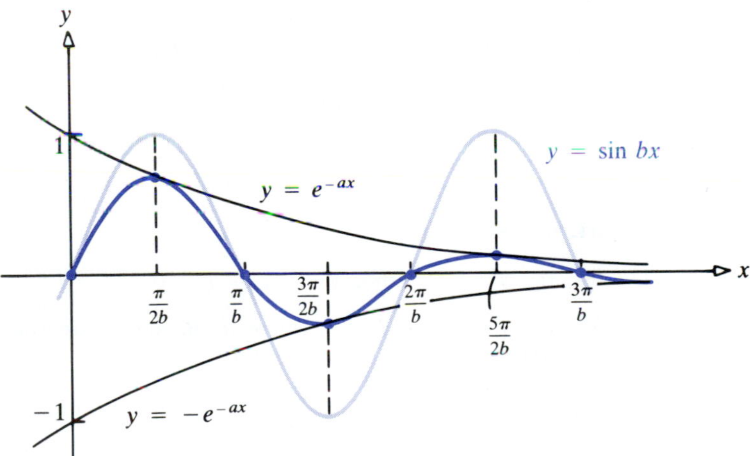

Figure 11

EXERCISE 7.3

In Problems 1–10 use (3) and definition (7.3.1) to simplify each expression as far as possible.

1. $\ln e^{x-x^2}$ **2.** $e^{-\ln(1/x)}$ **3.** $e^{\ln(1/x)}$ **4.** $e^{\ln 5 - \ln 2}$ **5.** $\ln \dfrac{1}{e^{x^2}}$

6. $e^{\ln x^2}$ **7.** $e^{x+\ln x}$ **8.** $e^{\ln x - 10 \ln y}$ **9.** $\ln e^{\ln 2}$ **10.** $e^{-\ln(xy^{-2})}$

In Problems 11–14 solve for x.

11. $\ln e^{-\ln x^2} = 5$ **12.** $\ln x^{5/2} + \ln\sqrt{x} = 3$ **13.** $e^{2x} - 3e^x + 2 = 0$ **14.** $\ln(1 + x) = 5$

In Problems 15–22 use (4) to find the indicated limit.

15. $\lim_{x\to+\infty} 3e^{3x}$ **16.** $\lim_{x\to-\infty} 3e^{3x}$ **17.** $\lim_{x\to+\infty} 3e^{-3x}$ **18.** $\lim_{x\to-\infty} 3e^{-3x}$

19. $\lim_{x\to+\infty} (1 - e^{-x^3})$ **20.** $\lim_{x\to-\infty} (1 - e^{-x^3})$ **21.** $\lim_{x\to+\infty} e^{1/x}$ **22.** $\lim_{x\to-\infty} e^{1/x}$

In Problems 23–52 find y'. (a and b are constants when they appear.)

23. $y = 5e^{3x}$ **24.** $y = xe^{2x}$ **25.** $y = \dfrac{e^x + e^{-x}}{2}$ **26.** $y = \dfrac{e^x - e^{-x}}{2}$

27. $y = e^{-3x}\ln(2x)$ **28.** $y = \frac{1}{2}\ln(4x) - 5e^{2x}$ **29.** $y = e^x \sin x$ **30.** $y = e^{2x}\cos(3x)$

31. $y = e^{ax}\sin bx$ **32.** $y = \dfrac{e^x + e^{-x}}{e^x - e^{-x}}$ **33.** $y = e^{\sqrt{x^2-9}}$ **34.** $y = e^{\cos(4x)}$

35. $y = e^{1/x}$ **36.** $y = \sqrt{e^x}$ **37.** $y = \dfrac{e^{ax} - 1}{e^{ax} + 1}$ **38.** $y = \ln(e^{ax} + e^{-ax})$

39. $e^{x+y} = y$ **40.** $x^2y = e^{xy}$ **41.** $ye^x = y - x$ **42.** $e^{x+y} = x^2$

43. $y = \dfrac{100}{1 + 99e^{-x}}$ **44.** $y = \dfrac{1}{1 + 2e^{-x}}$ **45.** $y = \cos e^{x^2}$ **46.** $y = \dfrac{\cos(e^{\sqrt{x}})}{\sqrt{x}}$

47. $y = \ln(\sin e^x)$ **48.** $y = \sin e^{\ln x}$ **49.** $\ln(\ln y) = e^x$ **50.** $e^y = \sin x$

51. $e^x \sin y + e^y \cos x = 4$ **52.** $e^y \cos x + e^{-x}\sin y = 10$

In Problems 53 and 54 find a general formula for the nth derivative of y.

53. $y = e^{ax}$

54. $y = e^{-ax}$

55. If $y = e^{2x}$, show that $y'' - 4y = 0$.

56. If $y = e^{-2x}$, show that $y'' - 4y = 0$.

57. If $y = Ae^{2x} + Be^{-2x}$, where A and B are constants, show that $y'' - 4y = 0$.

58. If $y = Ae^{ax} + Be^{-ax}$, where A, B, and a are constants, show that $y'' - a^2y = 0$.

59. If $y = Ae^{2x} + Be^{3x}$, where A and B are constants, show that $y'' - 5y' + 6y = 0$.

60. If $y = Ae^{-2x} + Be^{-x}$, where A and B are constants, show that $y'' + 3y' + 2y = 0$.

61. If $y = e^{-at}(A \sin \omega t + B \cos \omega t)$, where A, B, a, and ω are constants, find y'.

62. If $\ln T = kt$, where k is a constant, show that $dT/dt = kT$.

In Problems 63–68 graph the given function, and label local extrema, concavity, and inflection points.

63. $y = 3e^{3x}$ **64.** $y = 3e^{-3x}$ **65.** $y = e^{-x^2}$

66. $y = e^{-x}\cos x, \quad 0 \le x \le 2\pi$ **67.** $y = e^{1/x}$ **68.** $y = e^{|x|}$

69. Find the absolute maximum and absolute minimum of $f(x) = e^x - 3x$ on $[0, 1]$.

70. Find the local maxima, local minima, absolute maximum, and absolute minimum of $y = e^{\cos x}$ on $[-\pi, 2\pi]$.

71. Let $y = 2e^{\cos x}$.

(a) Calculate dy/dx and d^2y/dx^2.

(b) If x and y both vary with time in such a way that y increases at a steady rate of 5 units per second, at what rate is x changing when $x = \pi/2$?

72. A particle moves along the x-axis in such a way that at time $t > 0$ its position is $x = \sin e^t$.

(a) Find the velocity and acceleration of the particle at time t.

(b) At what time does the particle first have 0 velocity?

(c) What is the acceleration of the particle at the time determined in part (b)?

73. The formula $i = I_0 e^{(-R/L)t}$ measures the current i in amperes after t seconds in an electric circuit containing no capacitors, a resistance of R ohms, and an inductance of L henrys. The current in amperes at time $t = 0$ (when the electromotive force is cut off) is represented by I_0. Show that the rate of change of the current is proportional to the current.

74. Use Newton's method to solve $e^{-x} = \ln x$ correct to one decimal place.

75. Show that $e^x > 1 + x$ for all $x > 0$. [*Hint:* Show that $f(x) = e^x - 1 - x$ is an increasing function for $x > 0$.]

76. Show that $e^x > x^2$ for all $x > 0$.

77. Show that $\lim_{x\to+\infty} \dfrac{x}{e^x} = 0$.

$$\left[\textit{Hint:}\ \text{From Problem 76, we know that}\quad 0 \le \frac{x}{e^x} < \frac{x}{x^2} = \frac{1}{x}.\right]$$

78. Find the triangle of largest area that has two sides along the positive coordinate axes and its hypotenuse tangent to the function $f(x) = 3e^{-x}$.

79. Show that $2x - \ln(3 + 6e^x + 3e^{2x}) = c - 2\ln(1 + e^{-x})$ for some constant c. [*Hint:* Show that both sides have the same derivative.]

80. Use Newton's method to estimate the value of e by finding the root of the equation $\ln x - 1 = 0$. Start with $c_1 = 3$, and calculate c_4. Compare your results with the approximation 2.71828 for e.

81. Use the intermediate value theorem (2.4.6) to show that $f(x) = x + e^x$ has a root.

82. Use Newton's method to find the root in Problem 81 correct to five decimal places.

83. Use Newton's method to find the root of $f(x) = x - e^{-x}$ correct to five decimal places.

84. Find the point on the graph of $y = e^{-x}$ where the normal line to the graph passes through the origin.

85. Show that the line perpendicular to the x-axis passing through the point (x, y) on the graph of $y = e^x$ and the tangent to $y = e^x$ at the point (x, y) intersect the x-axis 1 unit apart. See the figure.

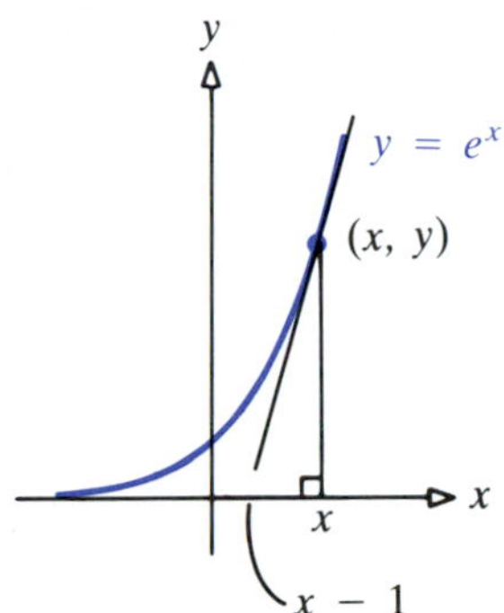

86. A pipe in the shape of an hourglass is formed by revolving the function given by $y = e^{3x}$, $-10 \le x \le 10$, about the line whose equation is $y = x$. What is the radius of the largest ball bearing that can pass through this hourglass pipe? [*Hint:* The radius will be equal to the minimum distance from the function to the line $y = x$.]

87. Define f by

$$f(x) = \begin{cases} e^{-1/x^2} & \text{if } x \ne 0 \\ 0 & \text{if } x = 0 \end{cases}$$

Show that f is differentiable on $(-\infty, +\infty)$ and find $f'(x)$ for each value of x. [*Hint:* for $f'(0)$ you must use the definition of the derivative. Show that $1 < x^2 e^{1/x^2}$ for $x \ne 0$.]

88. The function $f(x) = (1/\sqrt{2\pi})e^{-x^2/2}$, encountered in probability theory, is called the *standard normal density function.* Determine where this function is increasing and decreasing, find all local maxima and local minima, find all inflection points, and determine intervals of concavity. Then graph the function.

89. The vertices of a rectangle are at $(0, 0)$, $(0, e^x)$, $(x, 0)$, and (x, e^x). If x increases at the rate of 1 unit per second, at what rate is the area increasing when $x = 10$?

90. The atmospheric pressure at a height of x meters above sea level is $P(x) = 10^4 e^{-0.00012x}$ kilograms per square meter. What is the rate of change of the pressure with respect to the height at $x = 500$ meters? At $x = 750$ meters?

91. Show that the largest rectangle that can be inscribed under the graph of $y = e^{-x^2}$ has two of its vertices at the points of inflection of the graph.

92. The sales of a new car model over a period of time are expected to follow the so-called logistic curve,

$$f(x) = \frac{20{,}000}{1 + 50e^{-x}} \qquad x \geq 0$$

where x is measured in months. Determine the month in which the sales rate is a maximum. Graph the function.

93. The sales of a new stereo system over a period of time are expected to follow the so-called logistic curve,

$$f(x) = \frac{5000}{1 + 5e^{-x}} \qquad x \geq 0$$

where x is measured in years. Determine the year in which the sales rate is a maximum. Graph the function.

94. In a town of 50,000 people, the number of people at time t who have influenza is

$$N(t) = \frac{10{,}000}{1 + 9999e^{-t}}$$

where t is measured in days. Note that the flu is spread by the one person who has it at $t = 0$. At what time t is the rate of spreading of flu the greatest? Graph the function.

95. Prove part (b) of theorem (7.3.3).

96. Prove part (c) of theorem (7.3.3).
[*Hint:* $\ln(e^{\alpha})^{\beta} = \beta \ln e^{\alpha}$]

97. What is wrong with the following? If $x + y = e^{x+y}$, then $1 + y' = e^{x+y}(1 + y')$ or $y'[1 - e^{x+y}] = e^{x+y} - 1$ or $y' = \dfrac{e^{x+y} - 1}{1 - e^{x+y}} = -1$, and therefore $x + y = e^{x+y}$ must be a line of slope -1.

98. The function $f(x) = e^x$ has the property $f'(x) = f(x)$. Give an example of another function $g(x)$ such that $g(x)$ is defined for all real x, $g'(x) = g(x)$, and $g(x) \neq f(x)$.

99. (a) Use the areas illustrated in the figure to show that

$$\frac{1}{x+1} < \int_x^{x+1} \frac{1}{t}\,dt < \frac{1}{x} \qquad x > 0$$

(b) Use the result in part (a) to show that

$$\frac{1}{x+1} < \ln\left(1 + \frac{1}{x}\right) < \frac{1}{x} \qquad x > 0$$

(c) Use the result in part (b) to show that

$$e^{x/(x+1)} < \left(1 + \frac{1}{x}\right)^x < e \qquad x > 0$$

(d) Use the result of part (c) and the squeezing theorem to show that

$$\lim_{x \to +\infty} \left(1 + \frac{1}{x}\right)^x = e$$

(e) Use the result in part (c) to show that

$$\left(1 + \frac{1}{x}\right)^x < e < \left(1 + \frac{1}{x}\right)^{x+1} \qquad x > 0$$

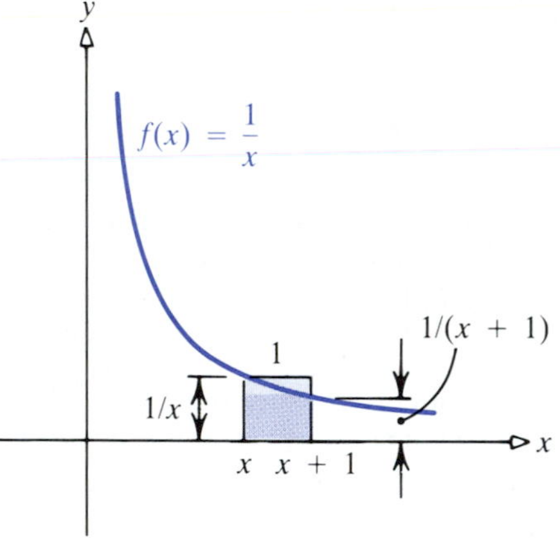

7.4 Related Integrals

Corresponding to every differentiation formula there is a related integration formula. Thus, as a consequence of theorem (7.3.4), $(d/dx)e^x = e^x$, we have the formula

$$\int e^x\,dx = e^x + C \qquad \text{where } C \text{ is a constant} \tag{1}$$

Let's look at the differentiation formula (7.2.3)—namely,

$$\frac{d}{dx}\ln x = \frac{1}{x} \tag{2}$$

which holds when $x > 0$. On the other hand, if $x < 0$, we may use (5) in Section 7.2 to obtain the formula

$$\frac{d}{dx}\ln(-x) = \left(\frac{1}{-x}\right)(-1) = \frac{1}{x} \tag{3}$$

By combining (2) and (3) into one formula, we get

$$\frac{d}{dx}\ln|x| = \frac{1}{x} \qquad x \neq 0 \tag{4}$$

The differentiation formula (4) yields the related integration formula

$$\int \frac{1}{x}\,dx = \ln|x| + C \qquad \text{where } C \text{ is a constant} \tag{5}$$

Let's look at some examples that require the use of (1) or (5).

EXAMPLE 1

Evaluate each integral:

(a) $\int e^{3x+4}\,dx$ (b) $\int \frac{5x^2\,dx}{4x^3 - 1}$ (c) $\int_0^1 \frac{e^x\,dx}{e^x + 4}$

(d) $\int_1^4 \frac{e^{\sqrt{x}}\,dx}{\sqrt{x}}$ (e) $\int e^{3x}\cos e^{3x}\,dx$

Solution

(a) We use the substitution $u = 3x + 4$. Then $du = 3\,dx$, and

$$\int e^{3x+4}\,dx = \int e^u\left(\frac{du}{3}\right) = \frac{1}{3}e^u + C = \frac{1}{3}e^{3x+4} + C$$

(b) We note that the numerator, except for a constant factor, is the derivative of the denominator. Hence, we use the substitution $u = 4x^3 - 1$, so that $du = 12x^2\,dx$ or $5x^2\,dx = \frac{5}{12}\,du$. Then,

$$\int \frac{5x^2\,dx}{4x^3 - 1} = \int \frac{5}{u}\frac{du}{12} = \frac{5}{12}\ln|u| + C = \frac{5}{12}\ln|4x^3 - 1| + C$$

(c) $$\int_0^1 \frac{e^x\,dx}{e^x + 4} = \int_5^{e+4} \frac{du}{u} = \ln|u|\Big|_5^{e+4}$$

Set $u = e^x + 4$.

$$= \ln(e + 4) - \ln 5 \approx 1.9048 - 1.6094 = 0.2954$$

Use a calculator.

(d) $$\int_1^4 \frac{e^{\sqrt{x}}\,dx}{\sqrt{x}} \underset{\uparrow}{=} \int_1^2 e^u(2\,du) = 2e^u\Big|_1^2 = 2(e^2 - e) \underset{\uparrow}{\approx} 2(7.389 - 2.718) = 9.342$$

Set $u = \sqrt{x}$. Use a calculator.

(e) $$\int e^{3x}\cos e^{3x}\,dx \underset{\uparrow}{=} \int (\cos u)\frac{du}{3} = \frac{1}{3}\sin u + C = \frac{1}{3}\sin e^{3x} + C$$

Set $u = e^{3x}$.

In parts (b) and (c) of Example 1, you may have noticed a general integration rule that evolves from (5) in Section 7.2—namely,

$$\int \frac{g'(x)}{g(x)}\,dx = \ln|g(x)| + C \tag{6}$$

In other words, whenever the numerator is the derivative of the denominator, the integral results in a logarithm. Note in both (5) and (6) that we must use absolute value bars, since the domain of the logarithm function is the positive real numbers. Of course, if in (6) the function $g(x)$ is known to be positive, we need not use the absolute value bars.

In Example 2 we use (6) to get two basic integration formulas.

EXAMPLE 2

Use (6) to show that:

(a) $$\int \tan x\,dx = -\ln|\cos x| + C = \ln|\sec x| + C$$

(b) $$\int \cot x\,dx = \ln|\sin x| + C \tag{7}$$

Solution

(a) $$\int \tan x\,dx = \int \frac{\sin x}{\cos x}\,dx \underset{\uparrow}{=} \int -\frac{du}{u}$$

Set $u = \cos x$.

$$= -\ln|u| + C = -\ln|\cos x| + C = \ln|\cos x|^{-1} + C$$

$$= \ln|\sec x| + C$$

(b) $$\int \cot x\,dx = \int \frac{\cos x}{\sin x}\,dx \underset{\uparrow}{=} \int \frac{du}{u} = \ln|u| + C = \ln|\sin x| + C$$

Set $u = \sin x$.

EXAMPLE 3

Find the volume of the solid of revolution generated by revolving the area under the graph of $y = e^x$ from $x = 0$ to $x = 1$ about the x-axis.

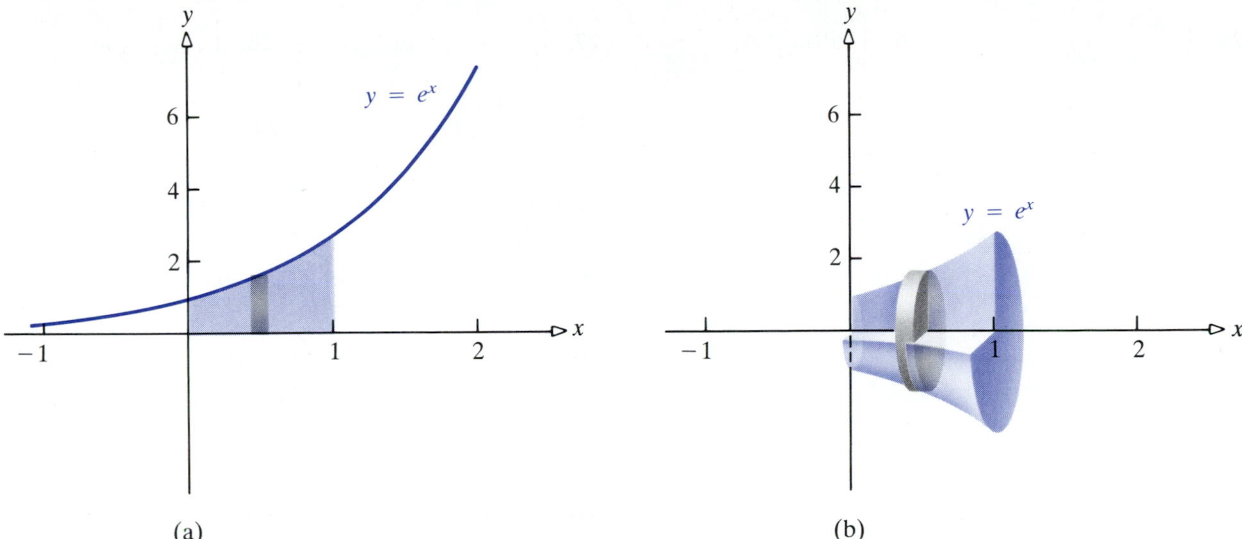

Figure 12

Solution

Figure 12 illustrates the situation. We choose to use the disk method because the shell method leads to an integral we cannot yet evaluate. (Check this out yourself.*) Thus, by the disk method,

$$V = \pi \int_0^1 (e^x)^2\,dx = \pi \int_0^1 e^{2x}\,dx = \pi \left.\frac{e^{2x}}{2}\right|_0^1 = \frac{\pi}{2}(e^2 - 1) \underset{\uparrow}{\approx} \frac{\pi}{2}(6.389) = 10.036$$

Use a calculator.

EXERCISE 7.4

In Problems 1–34 evaluate each integral.

1. $\int e^{3x}\,dx$
2. $\int e^{-2x}\,dx$
3. $\int e^{3x+1}\,dx$
4. $\int e^{bx+1}\,dx$
5. $\int (x + e^{7x})\,dx$
6. $\int (x^2 + e^{-7x})\,dx$
7. $\int_1^2 \frac{x+2}{3x^2}\,dx$
8. $\int \frac{dx}{3x-1}$
9. $\int \frac{dx}{1-5x}$
10. $\int \frac{x\,dx}{x^2-1}$
11. $\int \frac{5x\,dx}{1-x^2}$
12. $\int_0^2 \frac{e^{2x}}{e^{2x}+1}\,dx$
13. $\int_1^3 \frac{e^{3x}}{e^{3x}-1}\,dx$
14. $\int_0^1 x^2 e^{x^3+1}\,dx$
15. $\int_0^1 xe^{x^2-2}\,dx$
16. $\int \frac{e^{1/x}}{x^2}\,dx$
17. $\int \frac{e^{\sqrt[3]{x}}}{\sqrt[3]{x^2}}\,dx$
18. $\int \frac{e^x + e^{-x}}{e^x - e^{-x}}\,dx$
19. $\int \frac{e^{-x}}{6+e^{-x}}\,dx$
20. $\int \frac{3e^x}{\sqrt[4]{e^x - 1}}\,dx$
21. $\int \frac{e^x}{\sqrt{1+e^x}}\,dx$
22. $\int_2^3 \frac{dx}{x \ln x}$
23. $\int_2^3 \frac{dx}{x(\ln x)^2}$
24. $\int \frac{dx}{x(\ln x)^n}$, $n > 0$, n an integer

* The integral $\int \ln x\,dx$ is discussed in Chapter 9.

25. $\int \frac{(\ln 5x)^3}{x}\,dx$

26. $\int_0^{\pi} e^x \cos e^x\,dx$

27. $\int_0^{\pi} e^{-x}\cos e^{-x}\,dx$

28. $\int \frac{1}{2+3e^{-x}}\,dx$

29. $\int_1^2 \frac{x-4}{x^2}\,dx$

30. $\int \frac{dx}{\sqrt{x}(1+\sqrt{x})}$

31. $\int \frac{4}{1+e^x}\,dx$

32. $\int \frac{dx}{(x-2)^{1/2}+(x-2)}$

33. $\int \frac{\cos x\,dx}{2\sin x - 1}$

34. $\int \frac{1-\cos 3x}{3x-\sin 3x}\,dx$

35. Find the area enclosed by the graphs of $y = e^x$, $y = e^{-x}$, and $x = \ln 2$.

36. What is the area enclosed by the graph of $y = e^{2x}$ and the lines $x = 1$ and $y = 1$?

37. Find the area under the graph of $y = (\sin x)/(2 - \cos x)$ from $x = 0$ to $x = \pi/2$.

38. Find the area under the graph of $y = x/(2 - x^2)$ from $x = 0$ to $x = 1$.

39. Find the area enclosed by the graphs of $y = e^x$, $y = e^{3x}$, and $x = 2$.

40. Find the area under the graph of $y = 1/(4 - x)$ from $x = 0$ to $x = 3$.

41. Find the volume of a solid generated by revolving the area under the graph of $y = e^{-x}$ from $x = 0$ to $x = 2$ about the x-axis.

42. Find the volume of a solid generated by revolving the area under the graph of $y = e^{-x^2}$ from $x = 0$ to $x = 2$ about the y-axis.

43. Find the average value of $f(x) = e^x$ on $[0, 1]$.

44. Find the average value of $f(x) = 1/x$ on $[1, e]$.

7.5 Exponentials and Logarithms to Other Bases

The Function $f(x) = x^\alpha$, α Real

In our development thus far, we have given meaning to expressions such as x^2, x^{-4}, $x^{1/2}$, and $x^{8/3}$, but we have not given meaning to expressions such as $x^{\sqrt{2}}$, x^π, or, in general, x^α, where α is irrational. Just as we used the logarithm function to give meaning to e^x, we use it to give meaning to x^α, where α is irrational. This definition of x^α, α irrational, avoids the difficulty mentioned in the opening of this chapter and is motivated by the property

$$x^r = (e^{\ln x})^r = e^{r \ln x} \qquad r \text{ rational} \tag{1}$$

Consistent with this equation and the fact that the domain of $\ln x$ is $x > 0$, we present the following:

[7.5.1] DEFINITION

$$x^\alpha = e^{\alpha \ln x} \qquad \alpha \text{ real}, \quad x > 0 \tag{2}$$

For example,

$$x^{\sqrt{2}} = e^{\sqrt{2}\ln x} \qquad x^\pi = e^{\pi \ln x} \qquad 3^\pi = e^{\pi \ln 3}$$

The familiar properties of exponents hold for x^α, α real. For example, if α and β are real, then

$$x^\alpha x^\beta = x^{\alpha+\beta}$$

because

$$x^\alpha x^\beta = e^{\alpha \ln x} e^{\beta \ln x} \underset{\uparrow}{=} e^{\alpha \ln x + \beta \ln x} = e^{(\alpha+\beta)\ln x} \underset{\uparrow}{=} x^{\alpha+\beta}$$

By theorem (7.3.3(a)) By (2)

The proofs of the other properties of exponents are left as exercises.

Derivative of x^α

To obtain the derivative of x^α, α a real number,* we use (2). Then,

$$\frac{d}{dx} x^\alpha \underset{\uparrow}{=} \frac{d}{dx} e^{\alpha \ln x} \underset{\uparrow}{=} e^{\alpha \ln x} \frac{d}{dx}(\alpha \ln x) = x^\alpha \frac{\alpha}{x} = \alpha x^{\alpha-1}$$

By (2) Chain rule

Thus,

$$\boldsymbol{\frac{d}{dx} x^\alpha = \alpha x^{\alpha-1} \qquad \alpha \text{ a real number}} \tag{3}$$

Another way of deriving this result is to use logarithmic differentiation on $y = x^\alpha = e^{\alpha \ln x}$. Then

$$\ln y = \ln e^{\alpha \ln x} = \alpha \ln x$$

$$\frac{y'}{y} = \frac{\alpha}{x}$$

$$y' = \frac{\alpha y}{x} = \frac{\alpha x^\alpha}{x} = \alpha x^{\alpha-1}$$

Let's look at an example of how to find the derivative of $y = x^x$, $x > 0$.

EXAMPLE 1

Find the derivative of $y = x^x$, $x > 0$.

Solution

To find y' we use logarithmic differentiation:

$$\ln y = \ln x^x = \ln e^{x \ln x} = x \ln x$$

We differentiate with respect to x; then

$$\frac{1}{y}\left(\frac{dy}{dx}\right) = x\left(\frac{1}{x}\right) + \ln x$$

$$\frac{dy}{dx} = y(1 + \ln x) = x^x(1 + \ln x)$$

---

* Recall from (2) in Section 3.9 that thus far, we have found the derivative of x^r only for r a *rational number.*

Another way of solving Example 1 is to use (2). Since $y = x^x = e^{x \ln x}$, we have

$$y' = e^{x \ln x} \frac{d}{dx}(x \ln x) = e^{x \ln x}\left[x\left(\frac{1}{x}\right) + \ln x\right] = x^x\left(1 + \ln x\right)$$

The Function $f(x) = a^x$, $a > 0$, $a \neq 1$

Since $e^{x \ln a} = (e^{\ln a})^x = a^x$, we define

$$a^x = e^{x \ln a} \qquad a > 0, \quad a \neq 1 \tag{4}$$

We use the chain rule on (4) to get

$$\frac{d}{dx} a^x = \frac{d}{dx} e^{x \ln a} = e^{x \ln a} \cdot (\ln a) = a^x(\ln a) \qquad a > 0, \quad a \neq 1 \tag{5}$$

Another way of deriving this result is to use logarithmic differentiation on $y = a^x$. Taking the logarithm of both sides, we have

$$\ln y = \ln a^x = x \ln a$$

$$\frac{y'}{y} = \ln a$$

$$y' = y \ln a = a^x \ln a$$

EXAMPLE 2

Find y' if: (a) $y = 2^x$ (b) $y = 3^{x^2}$

Solution

(a) By (5), we have $y' = 2^x \ln 2$.
(b) We use the chain rule with $y = 3^u$ and $u = x^2$ to get

$$y' = \frac{dy}{du} \cdot \frac{du}{dx} = (3^u \ln 3)(2x) = (2 \ln 3)x\, 3^{x^2}$$

In view of (5), we have the related integration formula

$$\int a^x\, dx = \frac{a^x}{\ln a} + C \qquad a > 0, \quad a \neq 1 \tag{6}$$

EXAMPLE 3

$$\int 3^{2x+1}\, dx = \int 3^u \frac{du}{2} = \frac{1}{2}\left(\frac{3^u}{\ln 3}\right) + C = \frac{3^{2x+1}}{2 \ln 3} + C$$

↑ Set $u = 2x + 1$.

The Graph of $y = a^x$

From (5), it follows that when $a > 1$,

$$\frac{d}{dx} a^x > 0$$

and when $0 < a < 1$,

$$\frac{d}{dx}a^x < 0$$

Hence, the function $y = a^x$, $a > 1$, is increasing, whereas $y = a^x$, $0 < a < 1$, is decreasing. Furthermore, $a^x > 0$ for all x, so that its graph lies above the x-axis. The y-intercept is $(0, 1)$. Figure 13 illustrates the graph of the function $y = a^x$ for selected values of a.

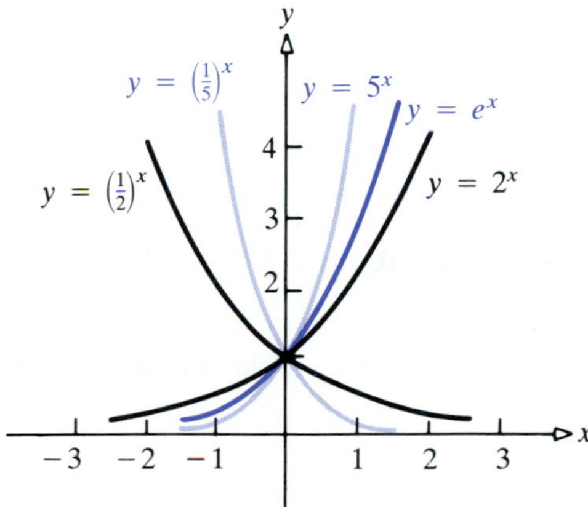

Graph of $y = a^x$ for $0 < a < 1$ is decreasing
Graph of $y = a^x$ for $a > 1$ is increasing

Figure 13

The Function $f(x) = \log_a x$, $a > 0$, $a \neq 1$

We define $f(x) = \log_a x$ as the inverse of the function $y = a^x$. See Figure 14 for the graphs for selected values of a.

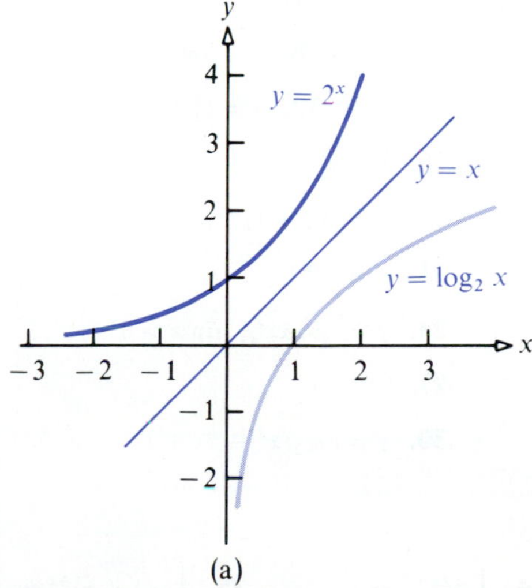

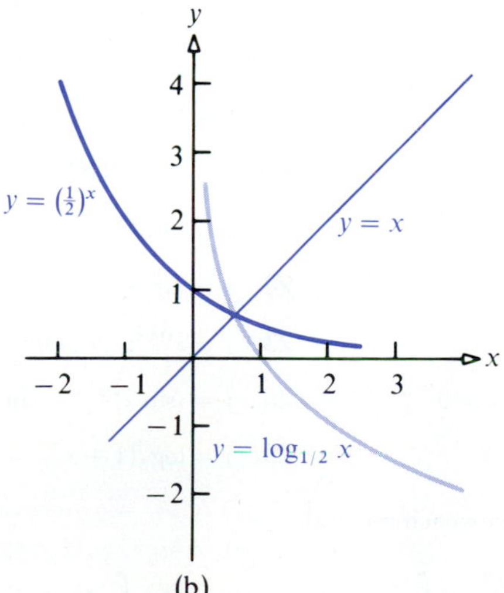

Figure 14

The function $f(x) = \log_a x$ can always be calculated from the natural logarithms of a and x by the following formula:

$$\log_a x = \frac{\ln x}{\ln a} \qquad a > 0, \quad a \neq 1 \tag{7}$$

This formula can be derived in the following way. If $y = \log_a x$, then $a^y = x$. Therefore,

$$\ln a^y = \ln x$$

$$y \ln a = \ln x$$

$$y = \frac{\ln x}{\ln a}$$

which is (7). It follows that

$$\frac{d}{dx} \log_a x = \frac{1}{x \ln a} \qquad a > 0, \quad a \neq 1 \tag{8}$$

Notice that when $a = e$, $\log_e x = \ln x$, and we have the simpler formula

$$\frac{d}{dx} \ln x = \frac{1}{x}$$

From (7), it also follows that

$$\log_a a^x = \frac{\ln a^x}{\ln a} = \frac{x \ln a}{\ln a} = x$$

EXERCISE 7.5

In Problems 1–30 find y'.

1. $y = x^{\sqrt{2}}$
2. $y = x^{\pi}$
3. $y = (1 + x^2)^{\sqrt{2}}$
4. $y = (x^2 - 1)^{\sqrt{3}}$
5. $y = (\frac{1}{2})^x$
6. $y = 2^{-x}$
7. $y = \log_2 x$
8. $y = \log_3 x$
9. $y = \log_2(1 + x^2)$
10. $y = \log_2(x^2 - 1)$
11. $y = 2^{-x}\sin x$
12. $y = (\frac{1}{2})^x \cos x$
13. $y = (3x)^x, \quad x > 0$
14. $y = (x^2 + 1)^{2x}, \quad x > 0$
15. $y = x^{\ln x}$
16. $y = x^{x^2}$
17. $y = (3x)^{\sqrt{x}}, \quad x > 0$
18. $y = (x^2 + 1)^{x/2}, \quad x > 0$
19. $y = x^{e^x}$
20. $y = (x^2 + 1)^{e^x}$
21. $y = 2^{\sin x}$
22. $y = 2^{\cos x}$
23. $y = x^{\sin x}, \quad x > 0$
24. $y = (\sin x)^x, \quad \sin x > 0$
25. $y = (\sin x)^{\cos x}, \quad \sin x > 0$
26. $y = (\sin x)^{\tan x}, \quad \sin x > 0$
27. $2^{xy} = x$
28. $x^y = 4$
29. $y = \log_3(1 + x^2)$
30. $y = \log_2(x^2 - 2x + 1)$

In Problems 31–40 evaluate each integral.

31. $\int 2^x \, dx$
32. $\int 3^x \, dx$
33. $\int 2^{3x+5} \, dx$
34. $\int 3^{2x+7} \, dx$
35. $\int_0^1 (1 + x)^{3\pi} \, dx$

36. $\int_0^1 x\,10^{-x^2}\,dx$ **37.** $\int_3^9 \frac{dx}{x \log_3 x}$ **38.** $\int_{10}^{100} \frac{dx}{x \log_{10} x}$ **39.** $\int_{10}^{100} \frac{dx}{x \ln x}$ **40.** $\int_{10}^{100} \frac{dx}{x \log_5 x}$

41. Prove that $\log_b a = 1/\log_a b$, $a > 0$, $a \neq 1$, $b > 0$, $b \neq 1$.

42. Prove that $\log_a x^\alpha = \alpha \log_a x$, α a real number.

43. Estimate $\ln a$ in terms of x_1 and x_2 if $e^{x_1} < a < e^{x_2}$.

44. Estimate e^a in terms of x_1 and x_2 if $\ln x_1 < a < \ln x_2$.

45. If f and g are differentiable functions, and if $f(x) > 0$, show that

$$\frac{d}{dx} f(x)^{g(x)} = g(x) f(x)^{g(x)-1} f'(x) + f(x)^{g(x)} \ln f(x) g'(x)$$

46. Use (2) to show that $(x^\alpha)^\beta = x^{\alpha\beta}$, α and β real.

47. Use (2) to show that $(xy)^\alpha = x^\alpha y^\alpha$, α real.

48. If α and β are real, prove that $x^\alpha / x^\beta = x^{\alpha - \beta}$.

The Number e as a Limit; Continuously Compounded Interest

7.6

The Number e as a Limit

In this section, we show that the number e defined in (7.2.6) can also be expressed as a limit. In fact, there are two equivalent ways of writing e as a limit, each of which we shall have occasion to refer to later.

[7.6.1] THEOREM

$$\text{(a)}\quad \lim_{h \to 0} (1 + h)^{1/h} = e \qquad \text{(b)}\quad \lim_{n \to +\infty} \left(1 + \frac{1}{n}\right)^n = e$$

Proof

(a) We use the definition of a derivative to calculate the derivative of $f(x) = \ln x$ at $x = 1$. Of course, we already know that the answer is $f'(1) = \frac{1}{1} = 1$.

$$1 = f'(1) = \lim_{h \to 0} \frac{\ln(1 + h) - \ln 1}{h} = \lim_{h \to 0} \left[\left(\frac{1}{h}\right) \ln(1 + h)\right] = \lim_{h \to 0} [\ln(1 + h)^{1/h}]$$

By theorem (7.2.2.(c))

Now we use the fact that $y = \ln x$ is continuous, so that

$$\lim_{x \to c} [\ln f(x)] = \ln[\lim_{x \to c} f(x)]$$

Also, since $\ln e = 1$, we have

$$1 = \ln e = \ln\left[\lim_{h \to 0} (1 + h)^{1/h}\right]$$

Since $y = \ln x$ is a one-to-one function, we conclude that

$$e = \lim_{h \to 0} (1 + h)^{1/h}$$

(b) The limit derived in part (a) is valid when $h \to 0^+$. Hence, if we set $n = 1/h$, then $h \to 0^+$ implies $n \to +\infty$, and

$$e = \lim_{h \to 0^+} (1 + h)^{1/h} = \lim_{n \to +\infty} \left(1 + \frac{1}{n}\right)^n$$

Table 1

n	$1 + \frac{1}{n}$	$\left(1 + \frac{1}{n}\right)^n$
1	2	2.00000
100	1.01	2.70481
10,000	1.0001	2.71815
100,000	1.00001	2.71827
1,000,000	1.000001	2.71828

Table 1 lists some approximate values for e that were calculated by using theorem (7.6.1), part (b). The last value in the table is correct to five decimal places. Correct to nine decimal places, it turns out that

$$e = 2.718281828 \ldots$$

We hasten to point out that this pattern for e does not repeat. In fact, it can be shown that e is an irrational number.

Theorem (7.6.1) can be used to evaluate certain limits in terms of the number e.

EXAMPLE 1

Express each limit in terms of the number e.

(a) $\lim_{h \to 0}(1 + 2h)^{1/h}$ (b) $\lim_{n \to +\infty} \left(1 + \frac{3}{n}\right)^{2n}$

Solution

(a) Observe the resemblance of the limit in question to that of theorem (7.6.1), part (a). In fact, note that

$$(1 + 2h)^{1/h} = [(1 + 2h)^{1/2h}]^2$$

Since $h \to 0$ is equivalent to $2h = k \to 0$, we discover that

$$\lim_{h \to 0}(1 + 2h)^{1/h} = \lim_{h \to 0}[(1 + 2h)^{1/2h}]^2 \underset{\uparrow}{=} \lim_{k \to 0}[(1 + k)^{1/k}]^2$$

Set $2h = k$.

$$= [\lim_{k \to 0}(1 + k)^{1/k}]^2 = e^2 \approx 7.389$$

(b) Observe the resemblance of the limit in question to that of theorem (7.6.1), part (b). Furthermore,

$$\left(1 + \frac{3}{n}\right)^{2n} = \left[\left(1 + \frac{3}{n}\right)^{n/3}\right]^6 = \left[\left(1 + \frac{1}{n/3}\right)^{n/3}\right]^6$$

Since $n \to +\infty$ is equivalent to $n/3 = m \to +\infty$, we find that

$$\lim_{n \to +\infty} \left(1 + \frac{3}{n}\right)^{2n} = \lim_{n \to +\infty} \left[\left(1 + \frac{1}{n/3}\right)^{n/3}\right]^6$$

$$\underset{\uparrow}{=} \lim_{m \to +\infty} \left[\left(1 + \frac{1}{m}\right)^m\right]^6 = \left[\lim_{m \to +\infty} \left(1 + \frac{1}{m}\right)^m\right]^6 = e^6 \approx 403.429$$

Set $m = n/3$.

Continuously Compounded Interest

One use of the fact that $e = \lim_{h\to 0^+}(1+h)^{1/h}$ is found in finance. Suppose a principal P is to be invested at an annual rate of interest r, which is compounded n times per year for t years. The interest earned on the principal P at each compounding period is then $P(\frac{r}{n})$. The amount A after one compounding period is

$$A = P + P\left(\frac{r}{n}\right) = P\left(1 + \frac{r}{n}\right)$$

After 2 compoundings,* $$A = P\left(1+\frac{r}{n}\right) + P\left(1+\frac{r}{n}\right)\left(\frac{r}{n}\right) = P\left(1+\frac{r}{n}\right)^2$$

After 3 compoundings, $$A = P\left(1+\frac{r}{n}\right)^2 + P\left(1+\frac{r}{n}\right)^2\left(\frac{r}{n}\right) = P\left(1+\frac{r}{n}\right)^3$$

After k compoundings, $$A = P\left(1+\frac{r}{n}\right)^k$$

In t years there are nt compounding periods.

[7.6.2] THEOREM / *Compound Interest Formula.*

The amount A after t years accrued on a principal P when it is invested at an annual rate of interest r and is compounded n times per year is

$$A = P\left(1+\frac{r}{n}\right)^{nt} \tag{1}$$

Formula (1) is referred to as the *compound interest formula.*

Table 2 on page 488 lists the results of investing \$1000 at an annual rate of 10% for 1 year for various compounding periods.

Now we ask what happens to the amount after t years as the number of times n that the interest is compounded per year gets larger and larger. In other words, we seek to calculate $\lim_{n\to+\infty}[P(1+r/n)^{nt}]$.

$$\lim_{n\to+\infty}\left[P\left(1+\frac{r}{n}\right)^{nt}\right] = P\lim_{n\to+\infty}\left[\left(1+\frac{r}{n}\right)^{n}\right]^t = P\lim_{n\to+\infty}\left[\left(1+\frac{r}{n}\right)^{n/r}\right]^{rt}$$

$$= P\left[\lim_{n\to+\infty}\left(1+\frac{r}{n}\right)^{n/r}\right]^{rt} \underset{h = r/n}{=} P\left[\lim_{h\to 0^+}(1+h)^{1/h}\right]^{rt} = Pe^{rt}$$

Therefore,

$$A = Pe^{rt} \tag{2}$$

* Remember, the new principal is now $P(1 + r/n)$.

Table 2

P = Principal = \$1,000; r = Annual Rate of Interest = 10% = 0.10; t = 1 Year = 1

n = Number of Times Compounded per Year		A = Amount after 1 Year
1	Annual compounding	$A = P(1 + r) = 1{,}000(1 + 0.1) = \$1{,}100.00$
2	Semiannual compounding	$A = P\left(1 + \frac{r}{2}\right)^2 = 1{,}000(1 + 0.05)^2 = \$1{,}102.50$
4	Quarterly compounding	$A = P\left(1 + \frac{r}{4}\right)^4 = 1{,}000(1 + 0.025)^4 = \$1{,}103.81$
12	Monthly compounding	$A = P\left(1 + \frac{r}{12}\right)^{12} = 1{,}000(1 + 0.00833)^{12} = \$1{,}104.71$
365	Daily compounding*	$A = P\left(1 + \frac{r}{365}\right)^{365} = 1{,}000(1 + 0.000274)^{365} = \$1{,}105.16$

* Often banks use 360 days for daily compounding.

Thus, no matter how often the interest is compounded during the year, the amount after t years has a definite ceiling, Pe^{rt}. When interest is compounded so that the amount after t years is Pe^{rt}, we say that the interest is *compounded continuously*.

For example, the amount A due to investing \$1000 for 1 year at an anual rate of 10% compounded continuously is

$$A = 1000e^{0.1} = \$1105.17 \tag{3}$$

The phrase *effective rate of interest* is often used. This is the equivalent annual rate of interest with compounding. When interest is compounded annually, there is no difference between the annual rate and the effective rate; however, when interest is compounded more than once a year, the effective rate always exceeds the annual rate. For example, by using the results in Table 2 and (3), we find that when the annual rate is 10%, the effective rates are those listed in Table 3. It is worth noting that although the difference between a bank's paying interest yearly (almost none do now) versus compounding quarterly or monthly is fairly substantial, the difference between daily and continuous compounding is practically negligible.

Table 3

	Annual Rate (%)	Effective Rate (%)
Annual compounding	10	10
Semiannual compounding	10	10.25
Quarterly compounding	10	10.381
Monthly compounding	10	10.471
Daily compounding	10	10.516
Continuous compounding	10	10.517

EXERCISE 7.6

In Problems 1–4 express each limit in terms of e.

1. $\lim_{n\to+\infty}\left(1+\frac{1}{n}\right)^{2n}$

2. $\lim_{n\to+\infty}\left(1+\frac{1}{n}\right)^{n/2}$

3. $\lim_{n\to+\infty}\left(1+\frac{1}{3n}\right)^{n}$

4. $\lim_{n\to+\infty}\left(1+\frac{4}{n}\right)^{n}$

5. Find the amount after 1 year if \$500 is invested at 6% compounded continuously for 1 year. Use a calculator to obtain the answer if the rate is $6\frac{1}{4}$% compounded quarterly. Which is better?

6. Find the amount after 2 years if \$1000 is invested at 8% compounded continuously. Use a calculator to obtain the amount if the rate is $8\frac{1}{2}$% compounded quarterly. Which is better?

7. What principal P should be invested at 6% compounded continuously in order to have \$1000 after 1 year? What principal is required if the interest is compounded quarterly?

8. What principal P should be invested at 10% compounded continuously in order to have \$2000 after 3 years? What principal is required if the interest is compounded quarterly?

9. How long (in months) does it take for a principal P to double if it is invested at 10% compounded continuously? How long does it take if the compounding is quarterly?

10. Rework Problem 9 if the rate of interest is 8%.

11. Graph $y = [1 + (1/x)]^x$ using a graphing calculator. Then overlay the graph of $y = e$. Compare your result with theorem 7.6.1, part (b).

7.7 Differential Equations for Growth and Decay

In studies of physical, chemical, biological, and other phenomena, scientists attempt, on the basis of long observation or by other means, to deduce mathematical laws that describe or predict nature's behavior. In many situations the amount A of a substance varies with time t in such a way that the time rate of change of A is proportional to A itself. We may state this in the form of the differential equation

$$\frac{dA}{dt} = kA \tag{1}$$

where $k \neq 0$ is a real number. If $k > 0$, then (1) asserts that the time rate of change of A is positive, so that the amount A of the substance is increasing; if $k < 0$, then (1) asserts that the time rate of change of A is negative, so that the amount A of the substance is decreasing.

We assume that the initial amount A_0 is known, giving us the boundary condition $A = A_0$ when $t = 0$. By separating the variables and integrating, we get

$$\frac{dA}{A} = k\,dt$$

$$\int \frac{dA}{A} = \int k\,dt + C$$

$$\ln A = kt + C$$

The boundary condition requires that $A = A_0$ when $t = 0$. Thus, $C = \ln A_0$, so that

$$\ln A = kt + \ln A_0 = \ln A_0 + kt = \ln A_0 + \ln e^{kt} = \ln A_0 e^{kt}$$

or

$$A = A_0 e^{kt} \tag{2}$$

Therefore, this is the solution to the differential equation (1), where A_0 is the original amount.

When a function $A = A(t)$ varies according to the law (1), or its equivalent (2), it is said to follow the *exponential law*, or the *law of uninhibited growth or decay*, or the *law of continuously compounded interest*. Figure 15 illustrates the graphs of (2) for both $k > 0$ and $k < 0$.

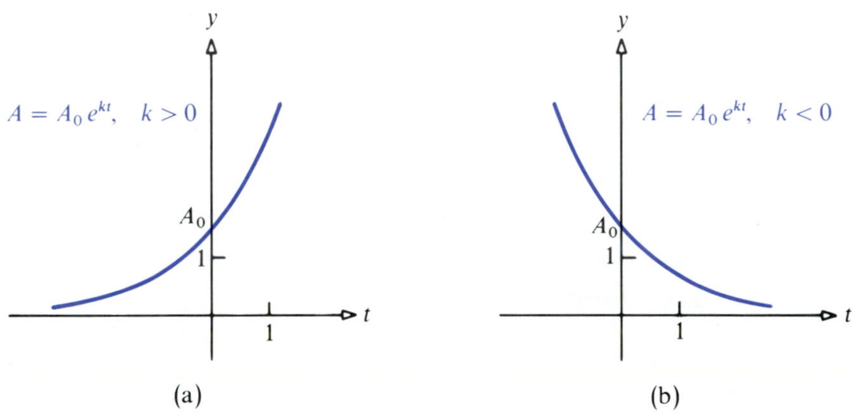

Figure 15

Bacterial Growth

EXAMPLE 1

Assume that a colony of bacteria grows at a rate proportional to the number present.* If the number of bacteria doubles in 5 hours, how long will it take for the number of bacteria to triple?

Solution

Let $N(t)$ be the number of bacteria present at time t. Then the assumption that this colony of bacteria grows at a rate proportional to the number present can be mathematically written as the differential equation

$$\frac{dN}{dt} = kN \tag{3}$$

* This is a model of uninhibited growth. However, after enough time has passed, growth will not continue at a rate proportional to the number present. Other factors, such as lack of living space, dwindling food supply, and so on, will start to affect the rate of growth. The model presented accurately reflects the way growth occurs in the early stages.

where k is a positive constant of proportionality. The solution of (3) is

$$N(t) = N_0 e^{kt}$$

where N_0 is the initial number of bacteria in this colony. Since the number of bacteria doubles in 5 hours, we have

$$\begin{aligned} N(5) &= 2N_0 \\ N_0 e^{5k} &= 2N_0 \\ e^{5k} &= 2 \\ k &= (\tfrac{1}{5})\ln 2 \end{aligned}$$

The time t required for this colony to triple must satisfy the equation

$$\begin{aligned} N(t) &= 3N_0 \\ N_0 e^{kt} &= 3N_0 \\ e^{kt} &= 3 \\ t &= \left(\frac{1}{k}\right)\ln 3 = 5\frac{\ln 3}{\ln 2} = 5\frac{1.0986}{0.6931} = 7.925 \text{ hours} \end{aligned}$$

Radioactive Decay

Our next application is related to the problem of dating archaeological specimens and organic fossils by measuring radioactive decay. For a radioactive substance, the *rate of decay is proportional to the amount present at a given time t.* That is, if A represents the amount of a radioactive substance at time t, we have

$$\frac{dA}{dt} = kA \tag{4}$$

where the constant k is negative and depends on the atomic structure of the substance. The *half-life* of a radioactive substance is the time required for half of the substance to decay.

In carbon dating, we use the fact that all living organisms contain two kinds of carbon, carbon-12 (a stable carbon) and carbon-14 (a radioactive carbon). When an organism dies, the amount of carbon-12 present remains unchanged, while the amount of carbon-14 begins to decrease. This change in the amount of carbon-14 present relative to the amount of carbon-12 present makes it possible to calculate the time at which the organism lived.

EXAMPLE 2

In the skull of an animal found in an archaeological dig, it was determined that about 20% of the original amount of carbon-14 was still present. If the half-life of carbon-14 is 5730 years, find the approximate age of the animal.

Solution

Let A be the amount of carbon-14 present in the skull at time t. Then A

satisfies the differential equation $dA/dt = kA$, whose solution is

$$A = A_0 e^{kt}$$

where A_0 is the amount of carbon-14 present at time $t = 0$. To determine the constant k, we use the fact that when $t = 5730$, half of the original amount A_0 will remain. Thus,

$$\begin{aligned} \tfrac{1}{2}A_0 &= A_0 e^{5730k} \\ \tfrac{1}{2} &= e^{5730k} \\ 5730k &= \ln \tfrac{1}{2} \\ k &= -0.000121 \end{aligned}$$

The relationship between the amount A of carbon-14 and the time t is therefore

$$A = A_0 e^{-(0.000121)t}$$

If the amount A of carbon-14 is 20% of the original amount A_0, we have

$$\begin{aligned} 0.2A_0 &= A_0 e^{-(0.000121)t} \\ 0.2 &= e^{-(0.000121)t} \end{aligned}$$

By taking the natural logarithm of both sides, we have

$$\begin{aligned} -(0.000121)t &= \ln 0.2 \\ -(0.000121)t &= -1.6094 \\ t &= \frac{1.6094}{0.000121} \approx 13{,}300 \text{ years} \end{aligned}$$

Thus, the animal lived approximately 13,000 years ago.

Newton's Law of Cooling

Newton's *law of cooling* states that the rate of decrease of the temperature of an object is continuous and proportional to the difference between the temperature of the object and that of the surrounding medium. That is, if u is the temperature of the body at time t and if T $(u > T)$ is the (constant) temperature of the surrounding medium, then

$$\frac{du}{dt} = k(u - T) \qquad \textbf{(5)}$$

where k is a negative constant* that depends on the object.

---

* If u is the temperature of a body at time t and if T $(u < T)$ is the constant temperature of the surrounding medium, the *law of heating* states that $du/dt = k(T - u)$, where k is a positive constant.

We seek u as a function of t; that is, we seek a rule for finding the temperature of the object at any time t. Using an argument similar to the one used to solve (1), we separate the variables in order to rewrite (5) as

$$\frac{du}{u - T} = k\,dt$$

Then

$$\int \frac{du}{u - T} = \int k\,dt$$

from which

$$\ln(u - T) = kt + C \tag{6}$$

We use the boundary condition that when $t = 0$, the initial temperature of the body is u_0. As a result, $C = \ln(u_0 - T)$. By substituting this into (6) and simplifying, we obtain

$$\ln(u - T) = kt + \ln(u_0 - T)$$

$$\ln(u - T) = \ln e^{kt} + \ln(u_0 - T)$$

$$\ln(u - T) = \ln(u_0 - T)e^{kt}$$

$$\mathbf{u - T = (u_0 - T)e^{kt}} \tag{7}$$

Let's look at a specific example.

EXAMPLE 3

Suppose an object is heated to 90°C and allowed to cool in a room with air temperature at 20°C. If after 10 minutes the temperature of the object is 60°C, what will be its temperature after 20 minutes?

Solution

When $t = 0$, $u_0 = 90°C$, and when $t = 10$, $u = 60°C$. The temperature T of the medium is 20°C. By using this information in (7), we get

$$(60 - 20) = (90 - 20)e^{10k}$$

$$\tfrac{40}{70} = e^{10k}$$

$$10k = \ln \tfrac{4}{7} = -0.5596$$

$$k = -0.05596$$

The relationship between the temperature u and time t is therefore

$$u - 20 = 70e^{-0.05596t}$$

When $t = 20$, the temperature u of the object is

$$u = 70e^{-0.05596(20)} + 20 = 70(0.3265) + 20 = 22.86 + 20 = 42.86°C$$

EXERCISE 7.7

1. The half-life of radium is 1690 years. If 8 grams of radium is present now, how much will be present in 100 years?

2. If 25% of a radioactive substance disappears in 10 years, what is the half-life of the substance?

3. A piece of charcoal is found to contain 30% of the carbon-14 it originally had. When did the tree from which the charcoal came die? Use 5730 years as the half-life of carbon-14.

4. A fossilized leaf contains 70% of a normal amount of carbon-14. How old is the fossil?

5. The population growth of a colony of mosquitoes obeys the uninhibited growth equation (3). If there are 1500 mosquitoes initially, and there are 2500 mosquitoes after 24 hours, what is the size of the mosquito population after 3 days?

6. The population of a suburb doubled in size in an 18 month period. If this growth continues and the current population is 8000, what will the population be in 4 years?

7. The number of bacteria in a culture is growing at a rate of $3000e^{2t/5}$ per unit of time t. At $t = 0$, the number of bacteria present is 7500. Find the number present at $t = 5$.

8. At any time t, the rate of increase in the area of a culture of bacteria is twice the area of the culture. If the initial area of the culture is 10, then what is the area at time t?

9. The rate of change in the number of bacteria in a culture is proportional to the number present. In a certain laboratory experiment, a culture has 10,000 bacteria initially, 20,000 bacteria at time t_1 minutes, and 100,000 bacteria at $(t_1 + 10)$ minutes.
   (a) In terms of t only, find the number of bacteria in the culture at any time t minutes $(t \geq 0)$.
   (b) How many bacteria are there after 20 minutes?
   (c) At what time are 20,000 bacteria observed? That is, find the value of t_1.

10. Salt (NaCl) dissociates in water into sodium (Na^+) and chloride (Cl^-) ions at a rate proportional to its mass. The initial amount of salt is 25 kilograms, and after 10 hours, 15 kilograms are left.
    (a) How much salt will be left after 1 day?
    (b) After how many hours will there be less than $\frac{1}{2}$ kilogram of salt left?

11. Radioactive beryllium is sometimes used to date fossils found in deep-sea sediment. The decay of beryllium satisfies the equation $dA/dt = -\alpha A$, where $\alpha = 1.5 \times 10^{-7}$ and t is measured in years. What is the half-life of beryllium?

12. The voltage of a certain condenser decreases at a rate proportional to the voltage. If the initial voltage is 20, and 2 seconds later it is 10, what is the voltage at time t? When will the voltage be 5?

13. A thermometer reading 4°C is brought into a room where the temperature reading is 30°C. If the thermometer reads 10°C after 2 minutes, determine the temperature reading 5 minutes after the thermometer is first brought into the room.

14. A thermometer reading 70°F is taken outside where the temperature is 22°F. Four minutes later the reading is 32°F.
    (a) Find the thermometer reading 7 minutes after the thermometer was brought outside.
    (b) Find the time for the reading to change from 70°F to within $\frac{1}{2}$°F of the air temperature.

15. Atmospheric pressure is a function of altitude above sea level and is given by the equation $dP/da = \beta P$, where β is a constant. The pressure is measured in millibars (mb). At sea level $(a = 0)$, $P(0)$ is 1013.25 mb, which means that the atmosphere at sea level will support a column of mercury 1013.25 millimeters high at a standard temperature of 15°C. At an altitude of $a = 1500$ meters, the pressure is 845.6 millibars.
    (a) What is the pressure at $a = 4000$ meters?
    (b) What is the pressure at 10 kilometers?
    (c) In California, the highest and lowest points are Mount Whitney (4418 meters) and Death Valley (86 meters below sea level). What is the difference in their atmospheric pressures?
    (d) What is the atmospheric pressure at Mount Everest (elevation 8848 meters)?
    (e) At what elevation is the atmospheric pressure equal to 1 millibar?

16. The number y of bacteria in a culture at time t is given approximately by

$$y = 1000(25 + te^{-t/20}) \qquad 0 \leq t \leq 100$$

    (a) Find the largest number and the smallest number of bacteria in the culture during the interval.
    (b) At what time during the interval is the rate of change in the number of bacteria a minimum?

REVIEW EXERCISES

In Problems 1–15 differentiate and simplify each expression (a is a positive constant when it appears).

1. $v = \ln(y^2 + 1)$

2. $z = \ln(u - \sqrt{u^2 + a^2})$

3. $y = x^{1/a} + a^{1/x}$

4. $y = x^a + a^x$

5. $y = \ln(\sin 2x)$

6. $f(y) = e^{-y}\sin y$

7. $g(x) = \ln(x^2 - 2x)$

8. $y = x \ln e^x$

9. $y = \ln \dfrac{x^2 + 1}{x^2 - 1}$

10. $y = e^{-x}\ln x$

11. $w = \ln(\sqrt{x + a} - \sqrt{x})$

12. $y = \dfrac{1}{a} \ln \dfrac{x}{x + \sqrt{a^2 - x^2}}$

13. $f(x) = \dfrac{1}{a} \log_b\left(\dfrac{x}{\sqrt[3]{x^2 + a^2}}\right)$

14. $f(x) = 10^{\ln x}$

15. $f(x) = \log_{10} e^x$

16. Find y' if $\ln x + \ln y = x \cos y$.

17. If $f(x) = (x^2 + 1)^{(2-3x)}$, find $f'(1)$.

18. Find: $\lim\limits_{h \to 0}\left[\dfrac{1}{h} \ln\left(\dfrac{2 + h}{2}\right)\right]$

In Problems 19–26 evaluate each integral.

19. $\displaystyle\int \frac{e^x + 1}{e^x - 1}\, dx$

20. $\displaystyle\int \frac{dx}{\sqrt{x}(1 - 2\sqrt{x})}$

21. $\displaystyle\int_{1/5}^{3} \frac{\log_3(5x)}{x}\, dx$

22. $\displaystyle\int_{-1}^{1} \frac{5^{-x}}{2^x}\, dx$

23. $\displaystyle\int \frac{dx}{x \cdot \ln x \cdot \ln \ln x}$

24. $\displaystyle\int \frac{dx}{x \log_{10} x}$

25. $\displaystyle\int \frac{dx}{x \log_3 \sqrt[5]{x}}$

26. $\displaystyle\int e^{x + e^x}\, dx$

27. Give a different proof of the formula $\ln(ab) = \ln a + \ln b$ (where a and b are positive) as follows:

(a) Confirm that $\ln ax$ and $\ln x$ have the same derivative for all $x > 0$.

(b) Why does it follow that $\ln ax - \ln x = \ln a$ for all $x > 0$? Let $x = b$ to finish the proof.

28. Give a different proof of the formula $\ln a^r = r \ln a$ (where $a > 0$ and r is rational) as follows:

(a) Confirm that $\ln x^r$ and $r \ln x$ have the same derivative for all $x > 0$.

(b) Why does it follow that $\ln x^r - r \ln x = 0$ for all $x > 0$? Let $x = a$ to finish the proof.

29. Sketch the graph of $y = x^3 - 3 \ln x$ after discussing domain, asymptotes, extrema, and concavity.

30. Suppose that a, b, and c are positive constants. Show that the minimum value of $ae^{cx} + be^{-cx}$ is $2\sqrt{ab}$.

31. If $f'(x) = -f(x)$ and $f(1) = 1$, find $f(x)$.

32. Find $y = f(x)$ if the arc length s of $y = f(x)$ from 0 to x satisfies $s = e^x - y + 4$ and $f(0) = 1$.

33. A function $P(x) = y = kx^2$ is symmetric with respect to the y-axis and passes through $(0, 0)$ and (b, e^{-b^2}), where $b > 0$.

(Problem 33 continues in the next column)

33. *(continued)*

(a) Write an equation for P.

(b) The area enclosed by P and the line $y = e^{-b^2}$ is revolved about the y-axis to form a solid. Compute its volume.

(c) For what number b is the volume of the solid in part (b) a maximum? Justify your answer.

34. At $x = 0$, which of the following is true of the function $f(x) = x^2 + e^{-2x}$?

(a) f is increasing.

(b) f is decreasing.

(c) f is discontinuous.

(d) f has a local minimum.

(e) f has a local maximum.

35. Let A be the area in the first quadrant that lies below both the graphs of $y = 3x^2$ and $y = 3/x$ and to the left of the line $x = k$, where $k > 1$.

(a) Find the area A as a function of k.

(b) When the area is 7, what is k?

(c) If the area A is increasing at the constant rate of 5 square units per second, at what rate is k increasing when $k = 15$?

36. (a) For what number m is the line $y = mx$ tangent to the graph of $y = \ln x$?
(b) Prove that the graph of $y = \ln x$ lies entirely below the line found in part (a).
(c) Use the results of part (b) to show that $e^x \geq x^e$ for $x > 0$.

37. Find the area of the largest rectangle in the fourth quadrant that has three vertices on the coordinate axes and the fourth vertex on the graph of $y = \ln x$.

38. Find the volume of the solid generated by revolving about the y-axis the area under the graph of $y = e^{-x^2}$ from $x = 0$ to the abscissa of the point of inflection of $y = e^{-x^2}$.

39. Let A be the area enclosed by the graph of $y = 1/x$, the x-axis, the line $x = m$, and the line $x = 2m$, where $m > 0$. Which of the following is true about the area A?

(a) Independent of m
(b) Increases as m increases
(c) Decreases as m increases
(d) Decreases as m increases when $m < \frac{1}{2}$; increases as m increases when $m > \frac{1}{2}$
(e) Increases as m increases when $m < \frac{1}{2}$; decreases as m increases when $m > \frac{1}{2}$

40. Which of the following is true about the graph of $y = \ln|x^2 - 1|$ in the interval $(-1, 1)$?
(a) Increasing
(b) Attains a local minimum at $(0, 0)$
(c) Has a range of all real numbers
(d) Concave down
(e) Has an asymptote of $x = 0$

41. Consider $f(x) = (1/x) + \ln x$, defined only on the closed interval $1/e \leq x \leq e$.

(a) Show your reasoning to determine the number(s) x at which f has its absolute maximum and absolute minimum.
(b) For what numbers x is the curve concave up?
(c) Sketch the graph of f over the interval $1/e \leq x \leq e$.

42. Find: $\displaystyle\lim_{x \to 2} \frac{\ln x - \ln 2}{x - 2}$

43. Find: $\displaystyle\lim_{h \to 0} \frac{e^h - 1}{h}$

44. Find: $\displaystyle\lim_{x \to c} \frac{e^x - e^c}{x - c}$

45. In Problem 13 on page 494 suppose that 5 minutes after the thermometer is first brought into the room, it is put into an oven at 200°C. Find the reading 10 minutes after the thermometer is first brought into the room if it remains in the oven.

46. A bank offers 10% interest compounded continuously in an account. Find how much \$1000 will increase to in 1 year and the equivalent rate of interest for 1 year if the compounding is done annually.

47. Sketch the graph of $y = xe^x$.

48. Sketch the graph of $y = x^2 e^{2x}$.

49. Find a function that is equal to its own derivative and such that $f(3) = 10$.

50. Use Newton's method to solve $e^{-x} = x - 4$.

51. Use Newton's method to solve $10x = e^x$.

52. *Calculator Problem.* Use a calculator to evaluate $(3^{0.003})^{1001}$ and $3^{(0.003 \times 1001)}$ by computing the expressions within parentheses first. Did you get the same results? If not, can you explain the difference? What does this tell you about the laws of exponents ((1) on page 457) as applied to calculator arithmetic? Experiment with other values and some of the other laws of exponents.

53. *Calculator Problem.* Use a calculator to evaluate $\log_{10}(2^{0.0001})$ and $0.0001 \times (\log_{10} 2)$ by computing the expressions within parentheses first. Did you get the same results? If not, can you explain the difference? What does this tell you about the laws of logarithms ((3) on page 457) as applied to calculator arithmetic? Experiment with other values and some of the other laws of logarithms.

54. *Calculator Problem.* Verify that $2^{10} \approx 10^3$. By taking $\log_{10}$ of both sides of this approximate equation, show that $\log_{10} 2 \approx 0.3$. (In fact, $\log_{10} 2 = 0.30103\ldots$.) Can you find other powers like 2^{10} that are approximately 10^n for some n? If so, you can approximate other logarithms.

55.*The force exerted by a gas in a cylinder on a piston whose area is A is given by $F = pA$, where p is the force per unit area, or *pressure*. The work W in a displacement of the piston from x_1 to x_2 is

$$W = \int_{x_1}^{x_2} F\,dx = \int_{x_1}^{x_2} pA\,dx = \int_{V_1}^{V_2} p\,dV$$

where dV is the accompanying infinitesimal change of volume of the gas.

(*Problem 55 continues on page 497*)

* Adapted from F. W. Sears, M. W. Zemansky, and H. D. Young, *University Physics* (Reading, Mass.: Addison-Wesley Publishing Co., 1976), p. 131. Reprinted by permission.

55. *(continued)*
(a) During expansion of a gas at constant temperature (isothermal), the pressure depends on the volume according to the relation

$$p = \frac{nRT}{V}$$

where n and R are constants and T is the constant temperature. Calculate the work in expanding the gas isothermally from volume V_1 to volume V_2.

(Problem 55 continues in the next column)

55. *(continued)*
(b) During expansion of a gas at constant entropy (adiabatic), the pressure depends on the volume according to the relation

$$p = \frac{K}{V^{\gamma}}$$

where K and $\gamma \neq 1$, are constants. Calculate the work in expanding the gas adiabatically from V_1 to V_2.

CHALLENGE EXERCISES

1. Prove that $\pi^e < e^\pi$.

2. A function f has the following properties: (i) $f(x + h) = e^h f(x) + e^x f(h)$ for all real numbers x and h; (ii) $f(x)$ has a derivative for all real numbers x; (iii) $f'(0) = 2$.
(a) Show that $f(0) = 0$.
(b) Find $\lim_{x \to 0}[f(x)/x]$.
(c) Prove that there is a real number p such that $f'(x) = f(x) + pe^x$ for all real numbers x; identify p.
(d) Consider the differential equation $y' - y = pe^x$. Multiply both sides by e^{-x} and note that the resulting expression on the left side is the derivative of a product. Use this to solve the differential equation.
(e) Use the solutions to the differential equation in part (d) to find all possible values for $f(x)$ satisfying properties (i), (ii), and (iii).

3. Find a so that the area under the graph of $y = x + (1/x)$ from a to $(a + 1)$ is minimum.

4. (a) Sketch the graph of $y = \frac{1}{2}(e^x + e^{-x})$.
(b) Let R be a point on the graph and let the x-coordinate of R be r $(r \neq 0)$. The tangent line to the graph at R crosses the x-axis at a point Q. Find the coordinates of Q.
(c) If P is the point $(r, 0)$, find the length of PQ as a function of r and the limiting value of this length as r increases without bound.

5. Show that $a \ln(b/a) < b - a < b \ln(b/a)$ if $0 < a < b$.

6. Prove *Bernoulli's inequality* (Jakob Bernoulli): $(1 + x)^p \geq 1 + px$ for $x \geq 0$ and $p > 1$.

7. Show that for any positive integer n, we have

$$\frac{n}{n+1} < \ln\left\{\left(1 + \frac{1}{n}\right)^n\right\} < 1.$$

8. (a) Show that $\lim_{x \to +\infty} xe^{-x} = 0$.
(b) Show that $\lim_{x \to +\infty} \frac{\ln x}{x} = 0$.

9. Use the result of Problem 8 to find the limits.
(a) $\lim_{x \to +\infty} xe^{-2x}$
(b) $\lim_{x \to -\infty} xe^{-2x}$
(c) $\lim_{x \to +\infty} x^2 e^{2x}$
(d) $\lim_{x \to -\infty} x^2 e^{2x}$
(e) $\lim_{x \to +\infty} \frac{\ln 2x}{x^2}$
(f) $\lim_{x \to +\infty} \frac{\ln x}{\sqrt{x}}$

10. Let $m > 0$.
(a) Show that every function $y = mx$ will be less than e^x for all real x beyond some point.
(b) For some m, $y = mx$ will always be less than e^x. Find the set of all such m. Does this set have a largest element?

11. A separable differential equation for y' is a differential equation that can be written in the form $y' = f(x)/g(y)$, where f and g are continuous. Thus,

$$\int g(y)\, dy = \int f(x)\, dx$$

and performing the integration (if possible) will give us a solution. Use this technique to solve parts (a)–(c) below. (You may need to leave your answer in implicitly defined form.)

(Problem 11 continues on page 498)

11. *(continued)*

(a) $\dfrac{y^2}{x}\dfrac{dy}{dx} = 1 + x^2$

(b) $\dfrac{dy}{dx} = y\dfrac{x^2 - 2x + 1}{y + 3}$

(c) $y\dfrac{dy}{dx} = \dfrac{x^2}{y + 4}$; $y = 2$ when $x = 8$

(d) Show that the differential equation

$$y' = \frac{2}{x + y}$$

is not separable.

12. Consider the differential equation $y' + P(x)y = Q(x)$, where P and Q are continuous. Multiply both sides through by

$$\mu(x) = e^{\int P(x)\,dx}$$

and observe that the left side of the resulting equation is the derivative of a product. Solve the original differential equation, giving a general formula.

Inverse Trigonometric Functions; Hyperbolic Functions

Review of Derivatives of Trigonometric Functions; Related Integrals 8.1

Derivatives of Trigonometric Functions

The formulas for the derivatives and most integrals of the trigonometric functions have already been derived. For completeness, we list these formulas again here.

$$\frac{d}{dx}\sin x = \cos x \tag{1}$$

$$\frac{d}{dx}\cos x = -\sin x \tag{2}$$

$$\frac{d}{dx}\tan x = \sec^2 x \tag{3}$$

$$\frac{d}{dx} \cot x = -\csc^2 x \tag{4}$$

$$\frac{d}{dx} \sec x = \sec x \tan x \tag{5}$$

$$\frac{d}{dx} \csc x = -\csc x \cot x \tag{6}$$

Notice that (2), (4), and (6) for the derivatives of $\cos x$, $\cot x$, and $\csc x$ (sometimes referred to as the *cofunctions*) can be obtained from (1), (3), and (5) for the derivatives of $\sin x$, $\tan x$, and $\sec x$ by: (*1*) introducing a minus sign and (*2*) replacing each function by its cofunction. For example, from (5) we know that

$$\frac{d}{dx} \sec x = \sec x \tan x$$

If we replace $\sec x$ by its cofunction $\csc x$ and replace $\tan x$ by its cofunction $\cot x$ and put in a minus sign, we obtain

$$\frac{d}{dx} \csc x = -\csc x \cot x$$

which is (6).

We suggest that (1), (3), and (5) be memorized and that the rule above be applied to get (2), (4), and (6).

Now let's review how the chain rule is used with (1)–(6).

EXAMPLE 1

Find y' if:

(a) $y = \sin e^x$ (b) $y = e^{\tan x}$

(c) $y = \ln(\sec x + \tan x)$ (d) $y = \sec^2(3x + 2)$ (e) $y = x^{\sin x}$

Solution

(a) Let $y = \sin u$ and $u = e^x$. Then

$$y' = \frac{dy}{du}\frac{du}{dx} = (\cos u)(e^x) = e^x \cos e^x$$

(b) Let $y = e^u$ and $u = \tan x$. Then

$$y' = \frac{dy}{du}\frac{du}{dx} = (e^u)(\sec^2 x) = e^{\tan x} \sec^2 x$$

(c) Let $y = \ln u$ and $u = \sec x + \tan x$. Then

$$y' = \frac{dy}{du}\frac{du}{dx} = \left(\frac{1}{u}\right)(\sec x \tan x + \sec^2 x) = \frac{(\sec x + \tan x)(\sec x)}{\sec x + \tan x} = \sec x$$

(d) Let $y = u^2$, $u = \sec v$, and $v = 3x + 2$. Then

$$y' = \frac{dy}{du}\frac{du}{dv}\frac{dv}{dx} = (2u)(\sec v \tan v)(3) = 6\sec^2(3x + 2)\tan(3x + 2)$$

(e) We use logarithmic differentiation:

$$y = x^{\sin x}$$

$$\ln y = \ln x^{\sin x} = \sin x \ln x$$

Taking the derivative of both sides, we have

$$\frac{y'}{y} = \cos x \ln x + (\sin x)\frac{1}{x}$$

$$y' = y\left(\cos x \ln x + (\sin x)\frac{1}{x}\right) = x^{\sin x}\left(\cos x \ln x + (\sin x)\frac{1}{x}\right)$$

Related Integrals

Each of the six differentiation formulas developed thus far leads to a corresponding integration formula:

$$\int \cos x \, dx = \sin x + C \tag{7}$$

$$\int \sin x \, dx = -\cos x + C \tag{8}$$

$$\int \sec^2 x \, dx = \tan x + C \tag{9}$$

$$\int \csc^2 x \, dx = -\cot x + C \tag{10}$$

$$\int \sec x \tan x \, dx = \sec x + C \tag{11}$$

$$\int \csc x \cot x \, dx = -\csc x + C \tag{12}$$

To this list we add four more formulas:

$$\int \tan x \, dx = \ln|\sec x| + C \tag{13}$$

$$\int \sec x \, dx = \ln|\sec x + \tan x| + C \tag{14}$$

$$\int \cot x \, dx = \ln|\sin x| + C \tag{15}$$

$$\int \csc x \, dx = \ln|\csc x - \cot x| + C \tag{16}$$

Formulas (13) and (15) are repetitions of (7) in Section 7.4. We derive (14) by multiplying by $(\sec x + \tan x)/(\sec x + \tan x)$:

$$\int \sec x\,dx = \int \frac{(\sec x)(\sec x + \tan x)}{\sec x + \tan x}\,dx = \int \frac{\sec^2 x + \sec x \tan x}{\sec x + \tan x}\,dx$$

The substitution $u = \sec x + \tan x$ results in $du = (\sec x \tan x + \sec^2 x)\,dx$, so that

$$\int \sec x\,dx = \int \frac{du}{u} = \ln|u| + C = \ln|\sec x + \tan x| + C$$

If you are wondering how we decided to multiply by $(\sec x + \tan x)/(\sec x + \tan x)$ in the first place, look at Example 1, part (c). Also, see Problem 84 for a different formula for $\int \sec x\,dx$.

The derivation of (16) is similar and is left as an exercise (Problem 83).

EXAMPLE 2

Evaluate each integral:

(a) $\displaystyle\int \tan(3x + 1)\,dx$ (b) $\displaystyle\int_0^{\sqrt{\pi}/2} x \sec x^2 \tan x^2\,dx$

Solution

(a) We use the substitution $u = 3x + 1$. Then $du = 3\,dx$, so that

$$\int \tan(3x + 1)\,dx = \frac{1}{3}\int \tan u\,du \underset{\text{By (13)}}{=} \frac{1}{3}\ln|\sec u| + C = \frac{1}{3}\ln|\sec(3x + 1)| + C$$

(b) Let $u = x^2$. Then $du = 2x\,dx$. Also, $u = 0$ when $x = 0$, and $u = \pi/4$ when $x = \sqrt{\pi}/2$. Consequently,

$$\int_0^{\sqrt{\pi}/2} x \sec x^2 \tan x^2\,dx = \frac{1}{2}\int_0^{\pi/4} \sec u \tan u\,du$$

$$\underset{\text{By (11)}}{=} \frac{1}{2}\sec u\Big|_0^{\pi/4} = \frac{1}{\sqrt{2}} - \frac{1}{2} \approx 0.207$$

EXAMPLE 3

Find the volume of the solid of revolution generated by revolving the area under the graph of $y = \sqrt{\cos x}$ from $x = 0$ to $x = \pi/2$ about the x-axis.

Solution

Figure 1 illustrates the situation. We use a disk method here. (The shell method requires that we solve for x in the equation $y = \sqrt{\cos x}$: not an easy task at this stage.) Thus, we partition along the x-axis, obtaining disks

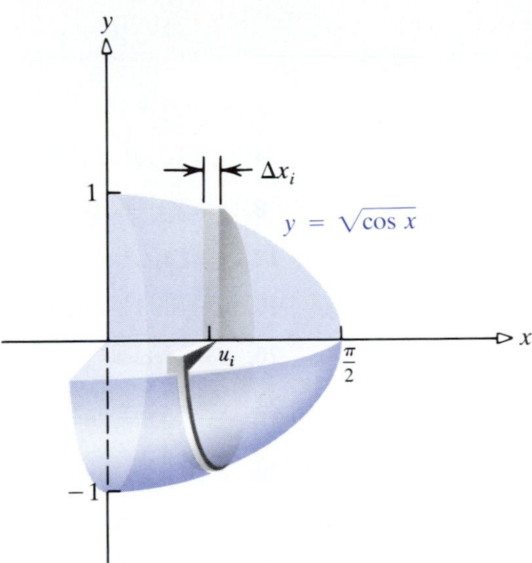

Figure 1

of radius $\sqrt{\cos u_i}$ and width Δx_i. The required volume is

$$V = \pi \int_0^{\pi/2} (\sqrt{\cos x})^2\, dx = \pi \int_0^{\pi/2} \cos x\, dx = \pi \sin x \Big|_0^{\pi/2} = \pi$$

EXAMPLE 4

Compute the arc length of $f(x) = \ln(\cos x)$ from $x = -\pi/4$ to $x = \pi/4$.

Solution

Figure 2 illustrates the situation. Since $f(-x) = f(x)$, the graph of f is symmetric with respect to the y-axis; thus, we need to compute only the arc length from $x = 0$ to $x = \pi/4$ and double the result. Since

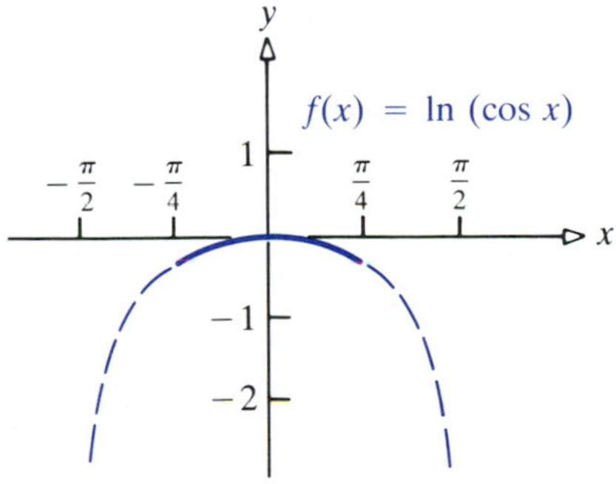

Figure 2

$$f'(x) = -\frac{\sin x}{\cos x} = -\tan x$$

the arc length is given by

$$\begin{aligned} s = \int_a^b \sqrt{1 + (f'(x))^2}\, dx &= \int_{-\pi/4}^{\pi/4} \sqrt{1 + \tan^2 x}\, dx = 2\int_0^{\pi/4} \sqrt{1 + \tan^2 x}\, dx \\ &= 2\int_0^{\pi/4} \sqrt{\sec^2 x}\, dx = 2\int_0^{\pi/4} \sec x\, dx \\ &= 2(\ln|\sec x + \tan x|)\Big|_0^{\pi/4} \\ &= 2[\ln(\sqrt{2} + 1) - \ln(1)] \approx 2(0.8814) = 1.7628 \end{aligned}$$

EXERCISE 8.1

In Problems 1–50 find $y' = dy/dx$.

1. $y = \sec 4x$
2. $y = \cot 5x$
3. $y = \tan 5x$
4. $y = \csc(x^3 + 1)$
5. $y = e^x \sec x$
6. $y = e^x \cot x$
7. $y = e^{-4x} \sin 3x$
8. $y = e^{-2x} \cos 3x$
9. $y = e^{\pi x} \tan \pi x$
10. $y = x^{\pi} \cot \pi x$
11. $y = \sin(x^2)$
12. $y = \cos(x^2)$
13. $y = \sec^4 x - \tan^4 x$
14. $y = \cot^2 x - \csc^2 x$
15. $y = \sec(x^2 + 2x - 1)$
16. $y = \cot(x^3 - 2x + 5)$
17. $y = \csc^2(1 + 2x)$
18. $y = \cot^2(1 + 3x^2)$
19. $y = x^2 \sin 4x$
20. $y = x^2 \cos 4x$
21. $y = \sqrt{\sin(1/x)}$
22. $y = \sqrt{\cos(3/x)}$
23. $y = x \tan x$
24. $y = x \cot x$
25. $y = x^2 \sin^2 x$
26. $y = x \sin \dfrac{1}{x}$
27. $y = \sec x \tan x$
28. $y = \sec^2 2x \tan^3 2x$
29. $y = \dfrac{\sec x}{1 + x}$
30. $y = \dfrac{\tan x}{1 + x}$
31. $x \sec y + y \tan x = x$
32. $\tan(xy) = x$
33. $\sec^2(xy) = 3x$
34. $\csc(x + y) = 2y$
35. $y = \sec^2(x^3 + x)$
36. $y = \cot^2(3x + 1)$
37. $y = e^{\csc^2 x}$
38. $y = \ln(\cot^2 x)$
39. $y = 4^{\cot x}$
40. $y = (\sqrt{3})^{\cos x}$
41. $y = (\cos x)^{\sqrt{3}}$
42. $y = 5^{\csc x}$
43. $y = 6^{\sec x}$
44. $y = 3^{\tan x}$
45. $y = \cos(2^x)$
46. $y = x^{\cos x}$
47. $y = (\cos x)^{\sin x}$
48. $y = \sin(x^{\sqrt{x}})$
49. $y = (\cos x)^x$
50. $y = x^{\ln \sin x}$

In Problems 51–72 evaluate each integral.

51. $\displaystyle\int_0^{\pi/4} \sec^2 x\, dx$
52. $\displaystyle\int_{-1}^{1} \sec x \tan x\, dx$
53. $\displaystyle\int \sec 5x\, dx$
54. $\displaystyle\int \csc 3x\, dx$
55. $\displaystyle\int \tan 2x\, dx$
56. $\displaystyle\int \cot 4x\, dx$
57. $\displaystyle\int \sec 4x \tan 4x\, dx$
58. $\displaystyle\int \csc 3x \cot 3x\, dx$
59. $\displaystyle\int \sec^2 2x\, dx$
60. $\displaystyle\int \csc^2 4x\, dx$
61. $\displaystyle\int \sqrt{\tan x}\, \sec^2 x\, dx$
62. $\displaystyle\int (2 + 3 \cot x)^{3/2} \csc^2 x\, dx$
63. $\displaystyle\int \frac{1}{\sec 5x}\, dx$
64. $\displaystyle\int \frac{1}{\csc 3x}\, dx$
65. $\displaystyle\int \sec(3x - 1)\tan(3x - 1)\, dx$
66. $\displaystyle\int \csc^2 4x\, dx$
67. $\displaystyle\int \frac{\sin x}{\cos^2 x}\, dx$
68. $\displaystyle\int \frac{\cos x}{\sin^2 x}\, dx$
69. $\displaystyle\int_0^{\pi/4} (1 + \sec^2 x)\, dx$
70. $\displaystyle\int_{\pi/4}^{\pi/2} \frac{dx}{\sin^2 x}$
71. $\displaystyle\int \sin x e^{\cos x}\, dx$
72. $\displaystyle\int \sec^2 x e^{\tan x}\, dx$

In Problems 73 and 74 find the area enclosed by the graphs of f and g.

73. $f(x) = \sec^2 x$, $g(x) = \sec x \tan x$; $x = -\pi/3$ and $x = \pi/6$
74. $f(x) = \cos^2 x$, $g(x) = \tan x$; $x = -\pi/4$ and $x = \pi/6$

In Problems 75 and 76 find the arc length.

75. $y = \ln(\sin x)$ from $x = \pi/6$ to $x = \pi/3$

76. $y = \ln(\cos x)$ from $x = 0$ to $x = \pi/4$

In Problems 77 and 78 discuss the graph of each function. Use the outline in Section 4.8 as a guideline.

77. $y = \sqrt[3]{\sin x}$

78. $y = \ln(\tan^2 x)$

79. Find the volume of the solid of revolution generated by revolving the area under the graph of $y = \csc x$ from $\pi/2$ to $3\pi/4$ about the x-axis.

80. Find the volume of the solid of revolution generated by revolving the area under the graph of $y = \sec x$ from 0 to $\pi/4$ about the x-axis.

81. Find the average value of $y = \tan x$ on the interval $[0, \pi/4]$.

82. Find the average value of $y = \sec x$ on the interval $[0, \pi/4]$.

83. Prove that $\int \csc x\, dx = \ln|\csc x - \cot x| + C$. [*Hint:* Multiply and divide by $(\csc x - \cot x)$.]

84. Find y' if $y = \ln[\tan(x/2 + \pi/4)]$. Use the result to show that

$$\int \sec x\, dx = \ln\left[\tan\left(\frac{x}{2} + \frac{\pi}{4}\right)\right] + C \qquad (17)$$

Also, show that the right-hand side of (17) is equal to the right-hand side of (14).

85. Show that $\int \sin x \cos x\, dx$ can be given in three different ways as

$$\int \sin x \cos x\, dx = \tfrac{1}{2}\sin^2 x + C_1 = -\tfrac{1}{2}\cos^2 x + C_2$$

$$= -\tfrac{1}{4}\cos 2x + C_3$$

Find the relationship between the constants C_1 and C_2 and the constants C_1 and C_3.

86. Use the identity below to obtain a formula for $\int \csc x\, dx$.

$$\csc x = \frac{1}{\sin x} = \frac{1}{2\sin(x/2)\cos(x/2)}$$

87. Use the identity below and the result of Problem 86 to obtain (17).

$$\sec x = \frac{1}{\cos x} = \frac{1}{\sin(\pi/2 - x)}$$

88. Use the identities $\cos^2 x = \frac{1}{2}(1 + \cos 2x)$ and $\sin^2 x = \frac{1}{2}(1 - \cos 2x)$ to show that:

(a) $\displaystyle\int \sin^2 x\, dx = \tfrac{1}{2}x - \tfrac{1}{4}\sin 2x + C$

(b) $\displaystyle\int \cos^2 4x\, dx = \tfrac{1}{2}x + \tfrac{1}{16}\sin 8x + C$

(c) $\displaystyle\int \frac{\sin^2\sqrt{x}}{\sqrt{x}}\, dx = \sqrt{x} - \tfrac{1}{2}\sin(2\sqrt{x}) + C$

89. Find the dimensions of the right circular cone of maximum volume having slant height 4 units. (See the figure.)

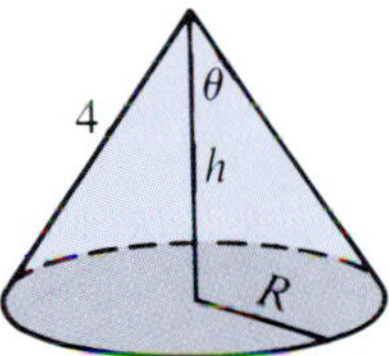

90. A 6 foot fence is erected 4 feet from a house. What is the shortest ladder than can be stood outside the fence and leaned against the wall? What is its angle of inclination θ? (See the figure.)

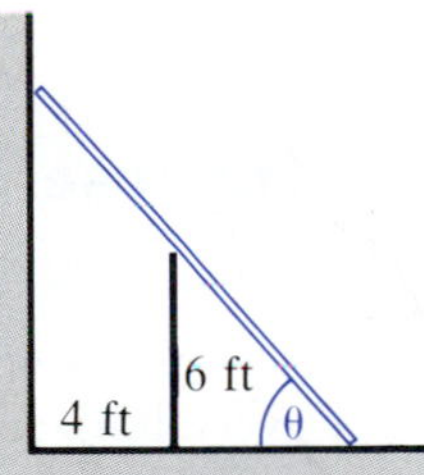

91. The sides of an isosceles triangle (see the figure) are sliding outward with velocity 1 centimeter per minute. At what rate is the area enclosed by the triangle changing when $\theta = 30°$?

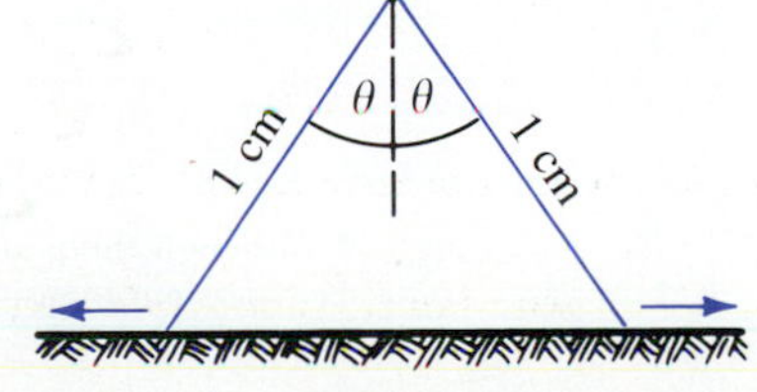

92. A weight hangs on a spring that is 2 meters long when it is stretched out (see the figure). The weight is pulled down and then released. The weight then oscillates up and down, and the length x of the spring when t seconds have elapsed is given by the formula $x = 2 + \cos 2\pi t$. Find:

(a) Length of the spring at times $t = 0, \frac{1}{2}, 1, \frac{3}{2}, \frac{5}{8}$
(b) Velocity of the weight at time $t = \frac{1}{4}$
(c) Acceleration of the weight at time $t = \frac{1}{4}$
(d) Time intervals during which the weight is moving down

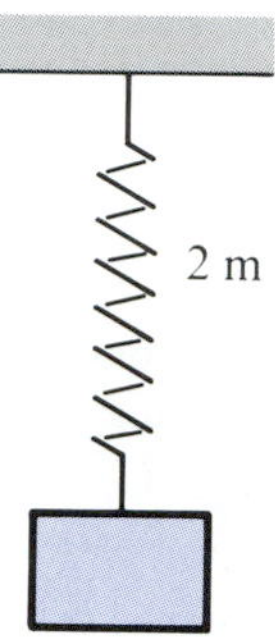

93. The hands of a clock are 2 inches and 3 inches long (see the figure). As the hands move around the clock, they sweep out the triangle OAB. At what rate is the area of the triangle changing at time 12:10?

(Problem 93 continues in the next column)

93. *(continued)*

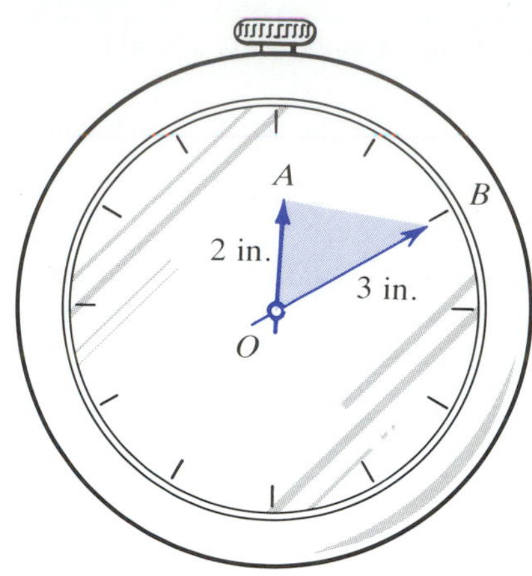

94. A searchlight is following a plane flying at an altitude of 3000 feet in a straight line over the light; the plane's velocity is 500 miles per hour. At what rate is the searchlight turning when the distance between the light and the plane is 5000 feet? (See the figure.)

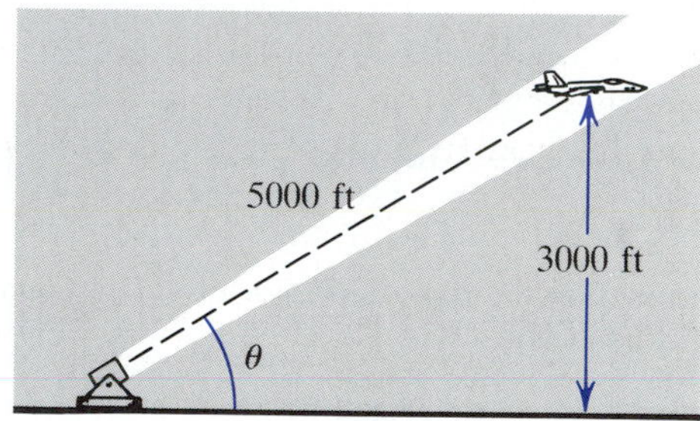

8.2 Inverse Trigonometric Functions

The graph of $y = \sin x$ is shown in Figure 3(a). There we observe that every horizontal line between -1 and 1 inclusive intersects the graph of $y = \sin x$ at infinitely many points. Thus, the function $y = \sin x$ is not one-to-one and consequently has no inverse. However, if we agree to restrict $y = \sin x$ to only that part of its domain for which $-\pi/2 \leq x \leq \pi/2$, we have a one-to-one function. With this restricted domain.* the function $y = \sin x$ has an inverse. To obtain the inverse function, we interchange x and y and solve for y. This value of y is called the *inverse sine of* x and we write

$$y = \sin^{-1} x$$

* Although there are other restrictions that work just as well, mathematicians have agreed to be consistent by using the interval $-\pi/2 \leq x \leq \pi/2$ for $y = \sin x$.

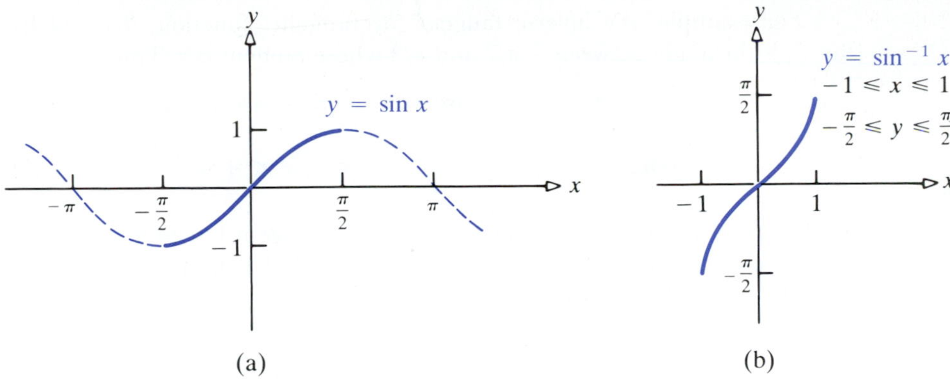

Figure 3

which we read as "*y is the angle whose sine is x*" or "*y is the inverse sine of x.*"*

$$y = \sin^{-1}x \qquad \text{means} \qquad x = \sin y$$

$$\text{where} \quad -\frac{\pi}{2} \leq y \leq \frac{\pi}{2}, \quad -1 \leq x \leq 1 \tag{1}$$

The domain of $y = \sin^{-1}x$ is the closed interval $[-1, 1]$, and the range is the closed interval $[-\pi/2, \pi/2]$. The graph of $y = \sin^{-1}x$ is the reflection about the line $y = x$ of the restricted portion of the graph of $y = \sin x$, as shown in Figure 3(b).

Let's look at an example that reviews the basic idea behind the definition of $y = \sin^{-1}x$.

EXAMPLE 1

Find:

(a) $\sin^{-1}(\sqrt{3}/2)$ (b) $\sin^{-1}(-\sqrt{2}/2)$

Solution

(a) Let $y = \sin^{-1}(\sqrt{3}/2)$. From (1), we have $\sqrt{3}/2 = \sin y$, where $-\pi/2 \leq y \leq \pi/2$. Thus, $y = \pi/3$. [*Note:* $\sin y = \sqrt{3}/2$ has infinitely many solutions, but $y = \pi/3$ is the only solution in $-\pi/2 \leq y \leq \pi/2$.]

(b) Let $y = \sin^{-1}(-\sqrt{2}/2)$. From (1), we have $-\sqrt{2}/2 = \sin y$, where $-\pi/2 \leq y \leq \pi/2$. Thus, $y = -\pi/4$.

Do not misinterpret the use of the -1 in $y = \sin^{-1}x$. This symbolism is used to remind you that it is the inverse of the sine function that is being discussed. If it is desired to discuss the reciprocal of the sine function, then write $y = (\sin x)^{-1}$ or $y = \csc x$.

In a similar manner, we can define inverse functions for the remaining trigonometric functions. In each case we suitably restrict the domain so that the function is one-to-one.

* In some books the inverse sine function is referred to as the *arcsine function*, written arcsin x.

For example, the inverse tangent (arctangent) function, denoted by $\tan^{-1}$, is the angle between $-\pi/2$ and $\pi/2$ whose tangent is x. Thus,

$$\mathbf{y = \tan^{-1} x \quad \text{means} \quad x = \tan y}$$

$$\mathbf{\text{where} \quad -\frac{\pi}{2} < y < \frac{\pi}{2}, \quad -\infty < x < +\infty} \tag{2}$$

See Figures 4(a) and 4(b) for the graphs of the tangent function and its inverse. Note that the values $y = \pm\pi/2$ are excluded from (2), since the tangent is not defined at $\pm\pi/2$.

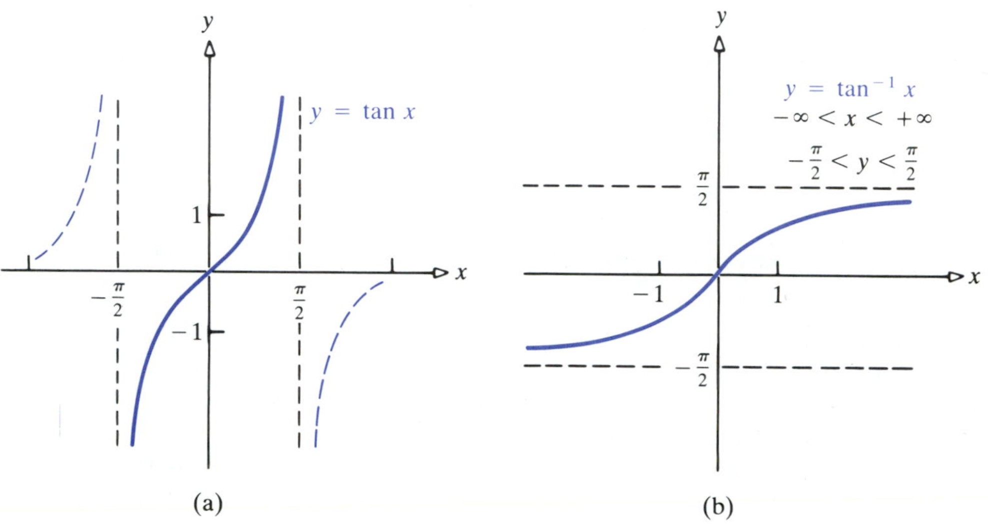

Figure 4

The *inverse secant* (arcsecant) function is denoted by $\sec^{-1}$. Similar to previous definitions,

$$\mathbf{y = \sec^{-1} x \quad \text{means} \quad \sec y = x}$$

$$\mathbf{\text{where} \quad |x| \geq 1, \quad 0 \leq y < \frac{\pi}{2} \quad \text{or} \quad \pi \leq y < \frac{3\pi}{2}} \tag{3}$$

See Figures 5(a) and 5(b). We have restricted y to $0 \leq y < \pi/2$ or $\pi \leq y < 3\pi/2$, instead of using the restrictions $0 \leq y < \pi/2$ or $\pi/2 < y \leq \pi$, in order to make $\tan y \geq 0$. This, in turn, will make the differentiation formula for the inverse secant simpler.

The remaining three inverse trigonometric functions are defined as follows:

Inverse cosine function:

$$y = \cos^{-1} x \quad \text{means} \quad x = \cos y$$

$$\text{where} \quad 0 \leq y \leq \pi \quad \text{and} \quad -1 \leq x \leq 1 \tag{4}$$

Inverse cotangent function:

$$y = \cot^{-1} x \quad \text{means} \quad x = \cot y$$

$$\text{where} \quad 0 < y < \pi \quad \text{and} \quad -\infty < x < +\infty \tag{5}$$

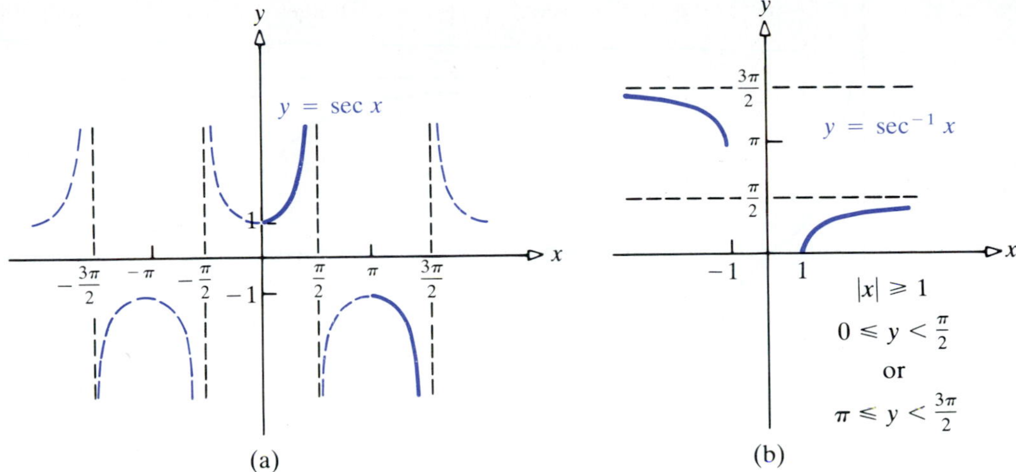

Figure 5

Inverse cosecant function:

$$y = \csc^{-1} x \qquad \text{means} \qquad x = \csc y$$

$$\text{where} \quad -\pi < y \leq -\frac{\pi}{2} \quad \text{or} \quad 0 < y \leq \frac{\pi}{2} \quad \text{and} \quad |x| \geq 1 \qquad \textbf{(6)}$$

See Figures 6, 7, and 8 for these graphs.

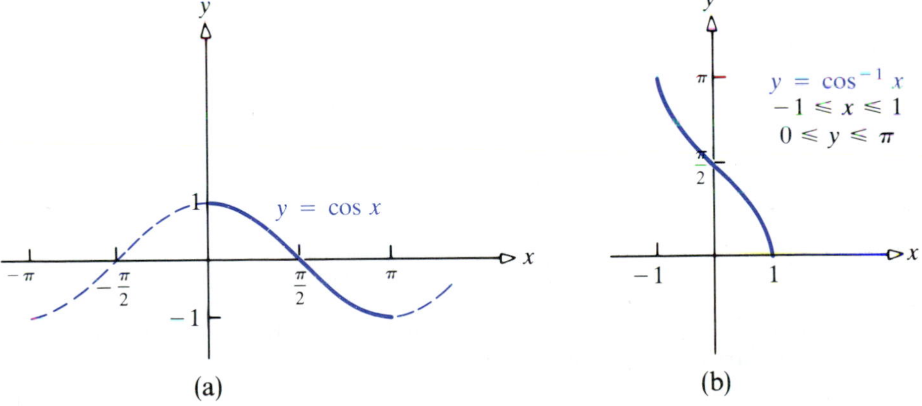

Figure 6

The inverse trigonometric functions satisfy the following three identities:

$$\boldsymbol{\cos^{-1} x = \frac{\pi}{2} - \sin^{-1} x} \qquad \textbf{(7)}$$

$$\boldsymbol{\cot^{-1} x = \frac{\pi}{2} - \tan^{-1} x} \qquad \textbf{(8)}$$

$$\boldsymbol{\csc^{-1} x = \frac{\pi}{2} - \sec^{-1} x} \qquad \textbf{(9)}$$

We shall give only the following restricted proof for (7).

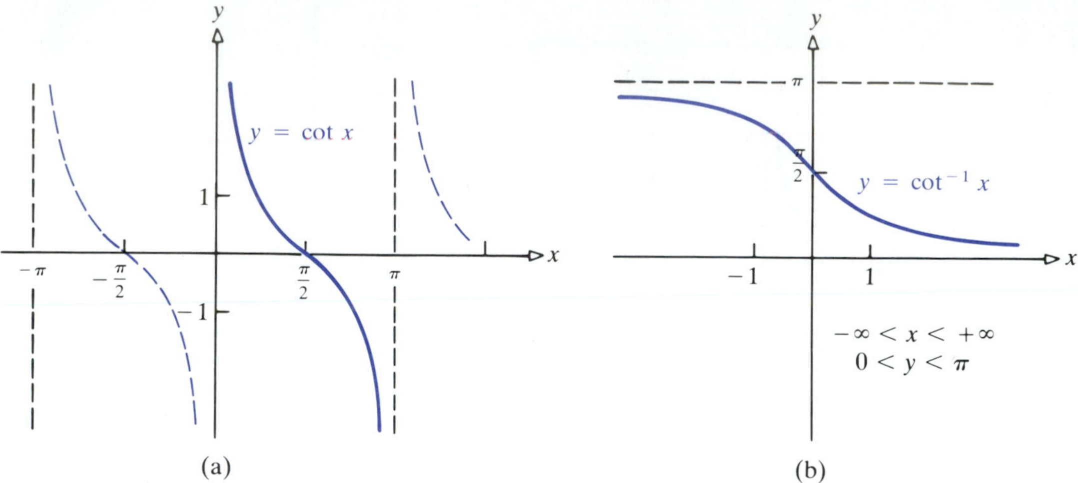

Figure 7

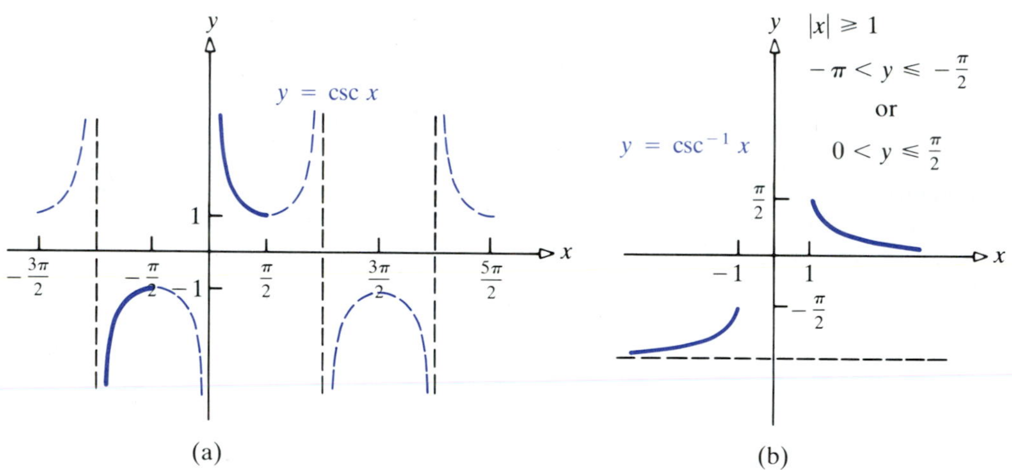

Figure 8

Proof Consider the right triangle shown in Figure 9, in which the angles α and β are acute and complementary. Then

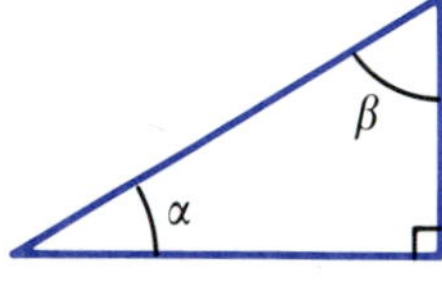

Figure 9

$$\alpha + \beta = \frac{\pi}{2}$$

$$\beta = \frac{\pi}{2} - \alpha \qquad \textbf{(10)}$$

Since $\sin \alpha = \cos \beta$, we may set $\sin \alpha = \cos \beta = x$ to get

$$\alpha = \sin^{-1} x \qquad \text{and} \qquad \beta = \cos^{-1} x \qquad \textbf{(11)}$$

By combining (10) and (11), we get (7).

In Problems 29 and 30 you are asked to give similar derivations of (8) and (9).

EXERCISE 8.2

In Problems 1–20 find the exact value of each expression.

1. $\sin^{-1}\left(\frac{\sqrt{2}}{2}\right)$

2. $\sin^{-1}\left(\frac{-1}{2}\right)$

3. $\sin^{-1}(-1)$

4. $\cot^{-1}(-1)$

5. $\tan^{-1}(0)$

6. $\cos^{-1}(-1)$

7. $\sin^{-1}\left(\frac{\sqrt{2}}{2}\right) + \sin^{-1}\left(-\frac{\sqrt{2}}{2}\right)$

8. $\tan^{-1}\left(\tan\frac{\pi}{4}\right)$

9. $\sin\left[\sin^{-1}\left(\frac{1}{2}\right)\right]$

10. $\cos^{-1}\left[\cos\left(-\frac{\pi}{4}\right)\right]$

11. $\tan^{-1}(\sin 0)$

12. $\tan^{-1}(\cos 0)$

13. $\sin\left[\cos^{-1}\left(\frac{3}{5}\right)\right]$

14. $\sin^{-1}\left(\sin\frac{1}{2}\right)$

15. $\tan(\sec^{-1}5)$

16. $\tan[\sec^{-1}(-3)]$

17. $\sin^{-1}(\sin 2)$

18. $\sin\left[\sin^{-1}\left(\frac{1}{3}\right)\right]$

[*Hint:* For Problems 19 and 20 use the formula for the sine of the sum of two angles.]

19. $\sin\left[\cos^{-1}\left(\frac{1}{3}\right) + \sin^{-1}\left(\frac{2}{3}\right)\right]$

20. $\sin\left[\cos^{-1}\left(\frac{1}{2}\right) + \sin^{-1}\left(\frac{-3}{4}\right)\right]$

21. Show that $\cos(\sin^{-1}x) = \sqrt{1-x^2}$.

22. Show that $\tan(\sin^{-1}x) = x/\sqrt{1-x^2}$.

23. Show that $\sin(2\cos^{-1}x) = 2x\sqrt{1-x^2}$.

24. Show that $\cos(2\sin^{-1}x) = 1 - 2x^2$.

25. Show that $\sin^{-1}(-x) = -\sin^{-1}x$.

26. Show that $\tan^{-1}(-x) = -\tan^{-1}x$.

27. Show that $\cos^{-1}(-x) = \pi - \cos^{-1}x$.

28. Given that $x = \sin^{-1}(\frac{1}{2})$, find $\cos x$, $\tan x$, $\cot x$, $\sec x$, and $\csc x$.

29. Give a restricted proof of (8).

30. Give a restricted proof of (9).

Derivatives of Inverse Trigonometric Functions; Related Integrals 8.3

Derivatives of Inverse Trigonometric Functions

To find the derivative of the function

$$y = \sin^{-1}x \qquad -\frac{\pi}{2} \le y \le \frac{\pi}{2}, \quad -1 \le x \le 1$$

we write $\sin y = x$ and differentiate implicitly with respect to x:

$$\cos y \frac{dy}{dx} = 1$$

$$\frac{dy}{dx} = \frac{1}{\cos y} \qquad \text{provided} \quad y \ne \pm\frac{\pi}{2}$$

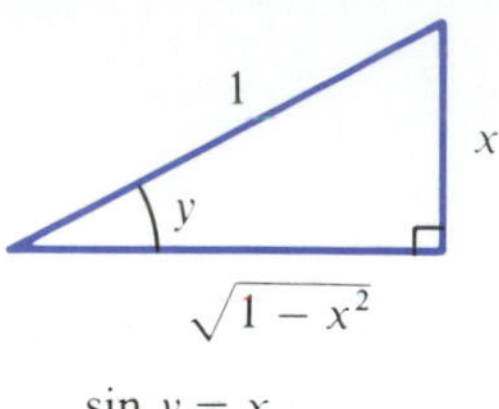

$\sin y = x$
$\cos y = \sqrt{1-x^2}$

Figure 10

Since y is the angle whose sine is x and $-\pi/2 < y < \pi/2$, we have

$$\cos y = \sqrt{1-x^2}$$

In Figure 10 we can "see" this formula for $0 < y < \pi/2$.

[8.3.1] THEOREM / *Derivative of* $y = \sin^{-1}x$.

The derivative of the inverse sine function $y = \sin^{-1}x$ is $dy/dx = 1/\sqrt{1-x^2}$, $-1 < x < 1$. That is,

$$\frac{d}{dx}\sin^{-1}x = \frac{1}{\sqrt{1-x^2}} \qquad -1 < x < 1$$

EXAMPLE 1

Find y' if:

(a) $y = \sin^{-1}(4x^2)$ (b) $y = e^{\sin^{-1}x}$ (c) $y = [\sin^{-1}(3x+5)]^{1/2}$

Solution

(a) We use the chain rule with $y = \sin^{-1}u$ and $u = 4x^2$. Then

$$y' = \frac{dy}{du}\frac{du}{dx} = \left(\frac{1}{\sqrt{1-u^2}}\right)(8x) = \frac{8x}{\sqrt{1-16x^4}}$$

(b) We use the chain rule with $y = e^u$ and $u = \sin^{-1}x$. Then

$$y' = \frac{dy}{du}\frac{du}{dx} = (e^u)\left(\frac{1}{\sqrt{1-x^2}}\right) = \frac{e^{\sin^{-1}x}}{\sqrt{1-x^2}}$$

(c) We use the chain rule with $y = u^{1/2}$, $u = \sin^{-1}v$, and $v = 3x + 5$. Then

$$y' = \frac{dy}{du}\frac{du}{dv}\frac{dv}{dx} = \left(\frac{1}{2u^{1/2}}\right)\left(\frac{1}{\sqrt{1-v^2}}\right)(3)$$

$$= \frac{3}{2[\sin^{-1}(3x+5)]^{1/2}\sqrt{1-(3x+5)^2}}$$

To get a formula for the derivative of the function $y = \tan^{-1}x$, we write $\tan y = x$ and differentiate implicitly with respect to x. Then

$$\sec^2 y\frac{dy}{dx} = 1$$

$$\frac{dy}{dx} = \frac{1}{\sec^2 y}$$

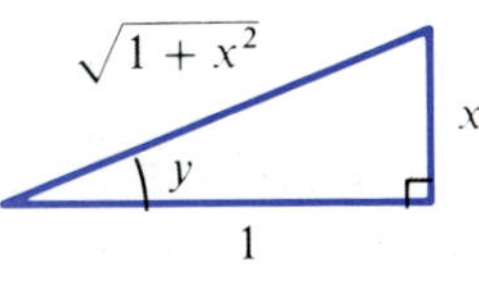

$\tan y = x$
$\sec y = \sqrt{1+x^2}$

Figure 11

Since y is an angle whose tangent is x and $-\pi/2 < y < \pi/2$, we have $\sec y > 0$. It follows that $\sec y = \sqrt{1+x^2}$. Figure 11 illustrates this result. Thus,

$$\sec^2 y = 1 + x^2$$

and we have the following theorem:

[8.3.2] THEOREM / *Derivative of $y = \tan^{-1}x$.*

The derivative of the inverse tangent function $y = \tan^{-1}x$ is $dy/dx = 1/(1 + x^2)$. That is,

$$\frac{d}{dx}\tan^{-1}x = \frac{1}{1+x^2}, \qquad -\frac{\pi}{2} < y < \frac{\pi}{2}$$

EXAMPLE 2

Find y' if:

(a) $y = \tan^{-1}4x$ (b) $y = \sin(\tan^{-1}x)$ (c) $y = \tan^{-1}e^{2x}$

Solution

(a) Set $y = \tan^{-1}u$ and $u = 4x$. Then

$$y' = \frac{dy}{du}\cdot\frac{du}{dx} = \left(\frac{1}{1+u^2}\right)(4) = \frac{4}{1+16x^2}$$

(b) Set $y = \sin u$ and $u = \tan^{-1}x$. Then

$$y' = \frac{dy}{du}\cdot\frac{du}{dx} = (\cos u)\left(\frac{1}{1+x^2}\right) = \frac{\cos(\tan^{-1}x)}{1+x^2}$$

(c) Set $y = \tan^{-1}u$, $u = e^v$, and $v = 2x$. Then

$$y' = \frac{dy}{du}\cdot\frac{du}{dv}\cdot\frac{dv}{dx} = \left(\frac{1}{1+u^2}\right)(e^v)(2) = \frac{2e^{2x}}{1+e^{4x}}$$

Finally, we derive the derivative of $y = \sec^{-1}x$. If we write $x = \sec y$ and differentiate implicitly, we find

$$1 = \sec y \tan y \frac{dy}{dx}$$

By the definition of $y = \sec^{-1}x$, we know that $0 \le y < \pi/2$ or $\pi \le y < 3\pi/2$. We can solve for dy/dx provided $\sec y \ne 0$ and $\tan y \ne 0$. Thus,

$$\frac{dy}{dx} = \frac{1}{\sec y \tan y} \qquad \text{provided } 0 < y < \frac{\pi}{2} \quad \text{or} \quad \pi < y < \frac{3\pi}{2}$$

With these restrictions on y, we have $\tan y > 0$.* Thus, as Figure 12 illustrates,

$$\sec y = x \qquad \tan y = \sqrt{x^2-1}$$

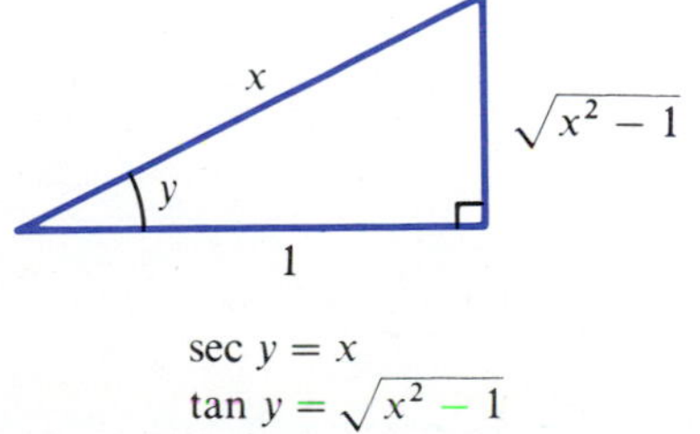

$\sec y = x$
$\tan y = \sqrt{x^2-1}$

Figure 12

Hence, we have the formula

$$\frac{d}{dx}\sec^{-1}x = \frac{1}{x\sqrt{x^2-1}}$$

* Now the reason for the previous restrictions on $y = \sec^{-1}x$ (see (3) in Section 8.2) should be apparent. Any other choice of restrictions would complicate the formula for $(d/dx)\sec^{-1}x$.

[8.3.3] THEOREM / *Derivative of* $y = \sec^{-1}x$.

The derivative of the inverse secant function $y = \sec^{-1}x$ is $dy/dx = 1/(x\sqrt{x^2-1})$, $|x| > 1$; that is,

$$\frac{d}{dx}\sec^{-1}x = \frac{1}{x\sqrt{x^2-1}} \qquad |x| > 1$$

Note that $y = \sec^{-1}x$ is not differentiable when $x = \pm 1$. In fact, as Figure 5 showed, at $x = \pm 1$, the graph of $y = \sec^{-1}x$ has no tangent line.

The three remaining formulas for the derivatives can be obtained by using the identities (7), (8), and (9) in Section 8.2.

Since $\cos^{-1}x = \frac{\pi}{2} - \sin^{-1}x$, $\quad \frac{d}{dx}\cos^{-1}x = \frac{-1}{\sqrt{1-x^2}} \qquad |x| < 1$

Since $\cot^{-1}x = \frac{\pi}{2} - \tan^{-1}x$, $\quad \frac{d}{dx}\cot^{-1}x = \frac{-1}{1+x^2}$

Since $\csc^{-1}x = \frac{\pi}{2} - \sec^{-1}x$, $\quad \frac{d}{dx}\csc^{-1}x = \frac{-1}{x\sqrt{x^2-1}} \qquad |x| > 1$

Related Integrals

Now we turn our attention to the integrals that correspond to each of the derivative formulas just obtained:

$$\int \frac{dx}{\sqrt{1-x^2}} = \sin^{-1}x + C \tag{1}$$

$$\int \frac{dx}{1+x^2} = \tan^{-1}x + C \tag{2}$$

$$\int \frac{dx}{x\sqrt{x^2-1}} = \sec^{-1}x + C \tag{3}$$

EXAMPLE 3

Evaluate each integral:

(a) $\int \frac{dx}{\sqrt{4-x^2}}$ (b) $\int \frac{dx}{9+4x^2}$

(c) $\int \frac{dx}{x^2+x+1}$ (d) $\int \frac{x\,dx}{x^2+x+1}$

Solution

(a) This resembles (1). We rewrite the original problem as follows:

$$\int \frac{dx}{\sqrt{4-x^2}} = \int \frac{dx}{2\sqrt{1-(x/2)^2}}$$

Now we use the substitution $u = x/2$. Then $du = dx/2$, and

$$\int \frac{dx}{\sqrt{4 - x^2}} = \int \frac{2\,du}{2\sqrt{1 - u^2}} = \sin^{-1}u + C = \sin^{-1}\left(\frac{x}{2}\right) + C$$

(b) This resembles (2). We rewrite the original problem as follows:

$$\int \frac{dx}{9 + 4x^2} = \int \frac{dx}{9[1 + (2x/3)^2]}$$

Let $u = 2x/3$. Then $du = 2\,dx/3$, and

$$\int \frac{dx}{9 + 4x^2} = \int \frac{3\,du/2}{9(1 + u^2)} = \frac{1}{6}\tan^{-1}u + C = \frac{1}{6}\tan^{-1}\left(\frac{2x}{3}\right) + C$$

(c) We complete the square to get

$$\int \frac{dx}{x^2 + x + 1} = \int \frac{dx}{(x + \frac{1}{2})^2 + \frac{3}{4}}$$

Let $u = x + \frac{1}{2}$. Then $du = dx$, and

$$\int \frac{dx}{x^2 + x + 1} = \int \frac{du}{u^2 + \frac{3}{4}} = \int \frac{4\,du}{4u^2 + 3} = \int \frac{4\,du}{3[(2u/\sqrt{3})^2 + 1]}$$

Set $2u/\sqrt{3} = v$. Then $2\,du/\sqrt{3} = dv$, and

$$\int \frac{dx}{x^2 + x + 1} = \int \frac{4\sqrt{3}\,dv/2}{3(v^2 + 1)} = \frac{2\sqrt{3}}{3}\tan^{-1}v + C = \frac{2\sqrt{3}}{3}\tan^{-1}\left[\frac{2u}{\sqrt{3}}\right] + C$$

$$= \frac{2\sqrt{3}}{3}\tan^{-1}\left[\frac{2(x + \frac{1}{2})}{\sqrt{3}}\right] + C = \frac{2\sqrt{3}}{3}\tan^{-1}\left[\frac{2x + 1}{\sqrt{3}}\right] + C$$

(d) Note how we force the derivative of the denominator to appear in the numerator:

$$\int \frac{x\,dx}{x^2 + x + 1} = \frac{1}{2}\int \frac{2x\,dx}{x^2 + x + 1} = \frac{1}{2}\int \frac{[(2x + 1) - 1]\,dx}{x^2 + x + 1}$$

$$= \frac{1}{2}\int \frac{(2x + 1)\,dx}{x^2 + x + 1} - \frac{1}{2}\int \frac{dx}{x^2 + x + 1}$$

$$= \frac{1}{2}\ln(x^2 + x + 1) - \frac{\sqrt{3}}{3}\tan^{-1}\left(\frac{2x + 1}{\sqrt{3}}\right) + C$$

↑ See part (c).

The integrals in (1), (2), and (3) have a more general form.

EXAMPLE 4

Verify that

$$\int \frac{dx}{a^2 + x^2} = \frac{1}{a}\tan^{-1}\left(\frac{x}{a}\right) + C \qquad a \neq 0 \tag{4}$$

Solution

$$\int \frac{dx}{a^2 + x^2} = \frac{1}{a^2} \int \frac{dx}{1 + (x/a)^2}$$

We use the substitution $u = x/a$. Then $du = dx/a$, and

$$\int \frac{dx}{a^2 + x^2} = \frac{1}{a^2} \int \frac{a\,du}{1 + u^2} = \frac{1}{a} \tan^{-1} u + C = \frac{1}{a} \tan^{-1} \frac{x}{a} + C$$

In Problems 83 and 84 you are asked to show that

$$\int \frac{dx}{\sqrt{a^2 - x^2}} = \sin^{-1}\left(\frac{x}{a}\right) + C \qquad a > 0 \tag{5}$$

$$\int \frac{dx}{x\sqrt{x^2 - a^2}} = \frac{1}{a} \sec^{-1}\left(\frac{x}{a}\right) + C \qquad a > 0 \tag{6}$$

The next example is reproduced from *Dynamics*,* as an illustration of the occurrence in dynamics of integrals that lead to inverse trigonometric functions.

EXAMPLE 5

Consult Figure 13. The spring-mounted slider moves in the horizontal guide with negligible friction and has a velocity $v_0 > 0$ in the s direction as it crosses the midposition where $s = 0$ and $t = 0$. The two springs together exert a retarding force to the motion of the slider, which gives it an acceleration proportional to the displacement but oppositely directed and equal to $a = -k^2 s$, where k is constant. (The constant is arbitrarily squared for later convenience in the form of the expressions.) Determine the expressions for the displacement s and velocity v as functions of the time t.

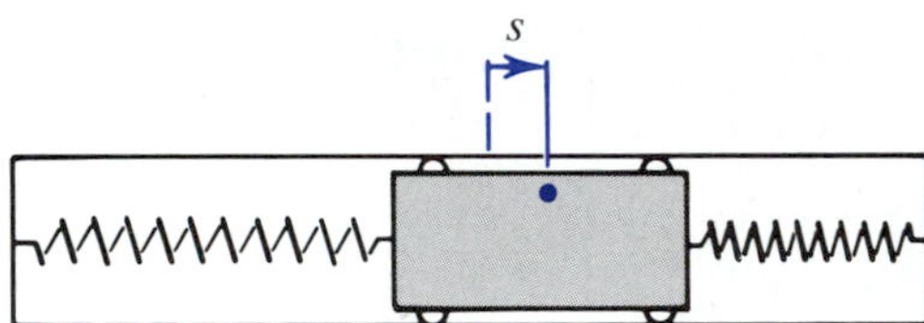

Figure 13

Solution

Since the acceleration is specified in terms of the displacement, the differential relation $v\,dv = a\,ds$ may be integrated. Thus,

$$\int v\,dv = \int -k^2 s\,ds \qquad \text{or} \qquad \frac{v^2}{2} = -\frac{k^2 s^2}{2} + C_1 \qquad \text{a constant}$$

* J. L. Meriam, *Dynamics*, 2d ed. (New York: John Wiley & Sons, 1975), p. 23. Reprinted by permission.

When $s = 0$, $v = v_0$, so that $C_1 = v_0^2/2$, and the velocity becomes

$$v = \pm\sqrt{v_0^2 - k^2 s^2}$$

We will clear up the ambiguity of the sign by imposing the boundary conditions below. This last expression may be integrated by substituting $v = ds/dt$. Thus,

$$\int \frac{ds}{\sqrt{v_0^2 - k^2 s^2}} = \pm \int dt \qquad \text{or} \qquad \frac{1}{k}\sin^{-1}\frac{ks}{v_0} = \pm t + C_2 \qquad \text{a constant}$$

With the requirement that $t = 0$ when $s = 0$, the constant of integration becomes $C_2 = 0$, and we may solve the equation for s, so that

$$s = \pm\frac{v_0}{k}\sin kt$$

The velocity is $v = \dot{s}[= ds/dt]$, which gives

$$v = \pm v_0 \cos kt$$

Imposing the condition $v(0) = v_0 > 0$ leads us to choose the plus sign and we obtain finally

$$s = \frac{v_0}{k}\sin kt \qquad \text{and} \qquad v = v_0 \cos kt$$

EXERCISE 8.3

In Problems 1–38 find $y' = dy/dx$.

1. $y = \sin^{-1}4x$

2. $y = \tan^{-1}5x$

3. $y = \cos^{-1}x^2$

4. $y = \sec^{-1}3x$

5. $y = \csc^{-1}(5x + 1)$

6. $y = \cot^{-1}x^2$

7. $y = \sin^{-1}x + \cos^{-1}x$

8. $y = \tan^{-1}x + \cot^{-1}x$

9. $y = \tan^{-1}\left(\frac{1}{x}\right)$

10. $y = \sin^{-1}\left(\frac{1}{x}\right)$

11. $y = \tan^{-1}\left(\frac{2x - 1}{2x}\right)$

12. $y = \sec^{-1}\sqrt{x}$

13. $y = \cos^{-1}\left(1 - \frac{x}{a}\right)$

14. $y = \frac{-\sqrt{a^2 - x^2}}{x} + \cos^{-1}\left(\frac{x}{a}\right)$

15. $y = \sin^{-1}e^x$

16. $y = \sin^{-1}\left(\frac{x}{a}\right) + \frac{\sqrt{a^2 + x^2}}{x}$

17. $y = \csc^{-1}\sqrt{x}$

18. $y = x(\sin^{-1}x)$

19. $y = \sin^{-1}(1 - x^2)$

20. $y = x\tan^{-1}(x + 1)$

21. $y = (1 + \tan^{-1}x)^2$

22. $y = \sin^{-1}(x^2 + 2x)$

23. $y = \sin^{-1}(1 - x^2)^{1/2}$

24. $y = \sqrt{1 - x^2}\sin^{-1}x$

25. $y = x\cot^{-1}(1 + x^2)$

26. $y = \cos^{-1}x + \frac{1}{1 - x^2}$

27. $y = \sin^{-1}\left(\frac{x - 1}{x + 1}\right)$

28. $y = x \cot^{-1} x + (1 + x)^{3/2}$

29. $y = \sin^{-1}(\cos x)$

30. $y = \tan^{-1}\sqrt{x}$

31. $y = \tan^{-1}(\ln x)$

32. $y = \sin^{-1}(\ln x)$

33. $y = \ln(\tan^{-1} x)$

34. $y = \ln\sqrt{\sin^{-1} x}$

35. $y = \tan^{-1}(\sin x)$

36. $y = \sin(\tan^{-1} x)$

37. $y = x \tan^{-1}\left(\dfrac{x}{a}\right) - \dfrac{1}{2} a \ln(x^2 + a^2)$

38. $y = x \sin^{-1}\left(\dfrac{x}{a}\right) + a \ln\sqrt{a^2 - x^2}$

In Problems 39 and 40 find dy/dx in terms of x and y.

39. $\sin^{-1} y + \cos^{-1} x = y$

40. $\tan^{-1} y = 3x + y$

In Problems 41–64 evaluate each integral.

41. $\displaystyle\int_0^1 \frac{5\,dx}{1 + x^2}$

42. $\displaystyle\int_0^1 \frac{x\,dx}{1 + x^4}$

43. $\displaystyle\int \frac{dx}{x^2 + 25}$

44. $\displaystyle\int \frac{dx}{x\sqrt{x^2 - 4}}$

45. $\displaystyle\int \frac{dx}{\sqrt{9 - x^2}}$

46. $\displaystyle\int_0^1 \frac{dx}{\sqrt{4 - x^2}}$

47. $\displaystyle\int \frac{dx}{\sqrt{16 - 9x^2}}$

48. $\displaystyle\int \frac{dx}{x\sqrt{9x^2 - 16}}$

49. $\displaystyle\int \frac{\cos x}{1 + \sin^2 x}\,dx$

50. $\displaystyle\int_0^1 \frac{e^x}{1 + e^{2x}}\,dx$

51. $\displaystyle\int \frac{\sin x}{\sqrt{4 - \cos^2 x}}\,dx$

52. $\displaystyle\int \frac{\sec^2 x\,dx}{\sqrt{1 - \tan^2 x}}$

53. $\displaystyle\int \frac{8x\,dx}{\sqrt{1 - (2x^2 - 1)^2}}$

54. $\displaystyle\int \frac{5\,dx}{\sqrt{e^{2x} - 16}}$

55. $\displaystyle\int \frac{2\,dx}{3 + 2x + 2x^2}$

56. $\displaystyle\int \frac{3\,dx}{x^2 + 6x + 10}$

57. $\displaystyle\int \frac{x\,dx}{2x^2 + 2x + 3}$

58. $\displaystyle\int \frac{3x\,dx}{x^2 + 6x + 10}$

59. $\displaystyle\int \frac{dx}{\sqrt{2x - x^2}}$

60. $\displaystyle\int_{\sqrt{2}}^2 \frac{dx}{x\sqrt{x^2 - 1}}$

61. $\displaystyle\int \frac{x^2\,dx}{\sqrt{1 - x^6}}$

62. $\displaystyle\int \frac{\cos 2x\,dx}{\sin 2x\sqrt{4 - 4\sin^2 2x}}$

63. $\displaystyle\int \frac{dx}{\sqrt{3x - x^2}}$

64. $\displaystyle\int_2^4 \frac{dx}{\sqrt{8x - x^2}}$

In Problems 65–68 find the area under the graph of each function.

65. Under $y = \dfrac{1}{x^2 + 1}$ from $x = 0$ to $x = \sqrt{3}$

66. Under $y = \dfrac{1}{\sqrt{1 - x^2}}$ from $x = 0$ to $x = \frac{1}{2}$

67. Under $y = \dfrac{1}{5x^2 + 1}$ from $x = 0$ to $x = 1$

68. Under $y = \dfrac{1}{x\sqrt{x^2 - 4}}$ from $x = 3$ to $x = 4$

69. Discuss the graph of the function $f(x) = \cos(\cos^{-1} x)$.

70. If $g(x) = \cos^{-1}(\cos x)$, show that $g'(x) = \dfrac{\sin x}{|\sin x|}$.

71. Discuss the graph of the function $g(x) = \cos^{-1}(\cos x)$.

72. Discuss the graph of the function $k(x) = \cos^{-1}(\sin x)$.

73. Show that $\dfrac{d}{dx}\cot^{-1} x = \dfrac{d}{dx}\tan^{-1}\left(\dfrac{1}{x}\right)$ for all $x \neq 0$.

74. We might try to infer from Problem 73 that $\cot^{-1} x = \tan^{-1}(1/x) + C$ for all $x \neq 0$, where C is a constant. Show, however, that

$$\cot^{-1} x = \begin{cases} \tan^{-1}(1/x) & \text{if } x > 0 \\ \tan^{-1}(1/x) + \pi & \text{if } x < 0 \end{cases}$$

What is the explanation of the incorrect inference?

75. Show that $\dfrac{d}{dx}\tan^{-1}(\cot x) = -1$.

76. A ladder 5 meters long is leaning against a wall. If the lower end of the ladder slides away from the wall at the rate of 0.5 meter per second, at what rate is the inclination of the ladder with respect to the ground changing when the lower end is 4 meters from the wall?

77. A man 2 meters tall walks horizontally at a constant rate of 1 meter per second toward the base of a tower 25 meters high. When the man is 10 meters from the tower, how fast is the angle of elevation changing if that angle is measured from the horizontal to the line joining the top of the man's head to the top of the tower?

78. A picture 4 meters in height is hung on a wall with the lower edge 3 meters above the level of an observer's eye. How far from the wall should the observer stand in order to obtain the most favorable view? (That is, the picture should subtend a maximum angle.)

79. Establish the identity $\sin^{-1}x + \cos^{-1}x = \pi/2$ by showing that the derivative of $y = \sin^{-1}x + \cos^{-1}x$ is 0. Then use the fact that when $x = 0$, $y = \pi/2$.

80. Establish the identity $\tan^{-1}x + \cot^{-1}x = \pi/2$ by showing that the derivative of $y = \tan^{-1}x + \cot^{-1}x$ is 0. Then use the fact that when $x = 1$, $y = \pi/2$.

81. The mean value theorem (4.4.2) guarantees that there is a real number N in $(0, 1)$ such that $f'(N) = f(1) - f(0)$ if f is continuous in $[0, 1]$ and differentiable in $(0, 1)$. Find N if $f(x) = \sin^{-1}x$.

82. A hot air balloon is rising at a speed of 100 meters per minute. What is the rate of change of the angle of elevation of an observer's line of sight if the observer is standing 200 meters from the lift-off point when the balloon is 600 meters high?

83. Verify (5).

84. Verify (6).

85. Find an equation for the tangent line to the graph of $y = \sin^{-1}(x/2)$ at the origin.

86. The region enclosed by the x-axis and the part of the graph of $y = \cos x$ from $x = -\pi/2$ to $x = \pi/2$ is separated into two regions by the line $x = k$. If the area of the region for $-\pi/2 \le x \le k$ is three times the area of the region for $k \le x \le \pi/2$, find k.

87. Use differentiation to verify the integration formula

$$\int \sin^{-1}x\,dx = x\sin^{-1}x + \sqrt{1 - x^2} + C$$

88. Use differentiation to verify the integration formula

$$\int \sqrt{a^2 - x^2}\,dx = \frac{1}{2}x\sqrt{a^2 - x^2} + \frac{1}{2}a^2\sin^{-1}\left(\frac{x}{a}\right) + C$$

Harmonic Motion 8.4

The importance of the sine and cosine functions in applied mathematics derives primarily from the differential equation

$$\frac{d^2x}{dt^2} + \omega^2 x = 0 \qquad \omega > 0 \tag{1}$$

This differential equation occurs in the study of oscillations, such as with a coiled spring or in wave theory (sound waves, radio waves, lightwaves, and so on).

The general solution of the differential equation (1) is

$$x = a\cos\omega t + b\sin\omega t \tag{2}$$

where a and b are constants.

We shall study the derivation of the differential equation (1) as it relates to a coiled spring. Suppose a spring of natural length L is placed in a vertical position, as shown in Figure 14(a). If a body of mass m is attached to the end of the spring, the spring is stretched beyond its natural length and comes to rest in a new equilibrium position of length $L + d$, as shown in Figure 14(b). The force of gravity that acts on the body of mass m is mg, and, by Hooke's law, the resisting force is kd. For equilibrium to occur, we must have

$$mg = kd \tag{3}$$

Suppose we pull the weight down a distance A from this new equilibrium position and then release it (see Fig. 14(c)). We want to find an expression for the displacement of the body from its equilibrum position (length $L + d$) as a function of the time t after the motion begins.

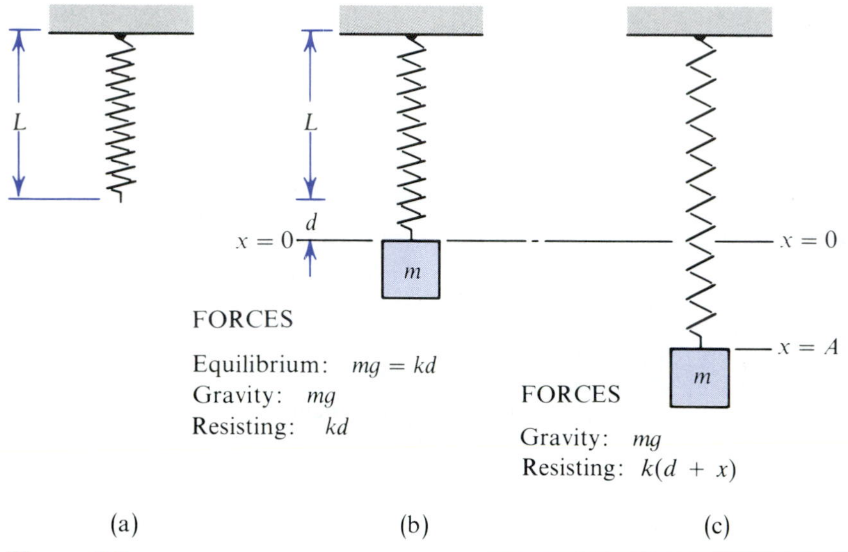

Figure 14

We take the positive direction of x to be downward. If we assume there is no friction in the spring and no air resistance to its motion, the forces acting on the body are

$$\begin{bmatrix}\text{Force due to}\\ \text{gravity}\end{bmatrix} - \begin{bmatrix}\text{Force due to}\\ \text{resisting force}\\ \text{of spring (Hooke's law)}\end{bmatrix} = mg - k(d + x)$$

By Newton's second law of motion, the sum of these forces, F, equals ma, where $a = d^2x/dt^2$ is the acceleration of the body. That is,

$$F = ma$$

or

$$mg - k(d + x) = ma$$

$$mg - kd - kx = m\frac{d^2x}{dt^2} \tag{4}$$

But $mg = kd$, so (4) becomes

$$m \frac{d^2x}{dt^2} = -kx$$

$$m \frac{d^2x}{dt^2} + kx = 0$$

$$\frac{d^2x}{dt^2} + \frac{k}{m} x = 0 \tag{5}$$

The initial conditions require that

$$\text{at} \quad t = 0, \quad x = A \quad \text{and} \quad \frac{dx}{dt} = 0 \tag{6}$$

The first condition arises because the body is pulled down a distance A when it is released. The second condition arises because when $t = 0$, the body has 0 speed.

Since k and m are positive real numbers, we set $k/m = \omega^2$, and the differential equation (5) may be written as

$$\frac{d^2x}{dt^2} + \omega^2 x = 0 \qquad \omega = \sqrt{\frac{k}{m}} \tag{7}$$

As we pointed out in (2), the general solution of this differential equation is

$$x = a \cos \omega t + b \sin \omega t \tag{8}$$

The initial conditions (6) require that at $t = 0$, $x = A$ and $dx/dt = 0$. So we have

$$A = a \cos 0 + b \sin 0 \qquad \text{and} \qquad \frac{dx}{dt} = -a\omega \sin \omega t + b\omega \cos \omega t$$

$$A = a \qquad\qquad 0 = -a\omega \sin 0 + b\omega \cos 0$$

$$0 = b\omega$$

$$0 = b$$

The solution of the differential equation (7) with the initial conditions (6) is therefore

$$x = A \cos \omega t \qquad \omega = \sqrt{\frac{k}{m}}$$

This is an equation describing the motion of a body and is an example of *simple harmonic motion*. The maximum displacement A is called the *amplitude*. The motion is *periodic* with period $T = 2\pi/\omega$. This means that if time is measured in seconds, then every $2\pi/\omega$ seconds the motion repeats itself. The reciprocal $1/T$ of the period T is called the *frequency* $f = \omega/2\pi$, the number of complete cycles per second. In terms of the spring constant k and the mass m of the body, the period T and frequency f are

$$T = 2\pi\sqrt{\frac{m}{k}} \qquad \text{and} \qquad f = \frac{1}{2\pi}\sqrt{\frac{k}{m}}$$

Another Form of the General Solution

Another form of the general solution (2) to the differential equation (1) can be obtained if we write (2) as

$$x = \sqrt{a^2 + b^2}\left(\frac{a}{\sqrt{a^2 + b^2}}\cos \omega t + \frac{b}{\sqrt{a^2 + b^2}}\sin \omega t\right) \tag{9}$$

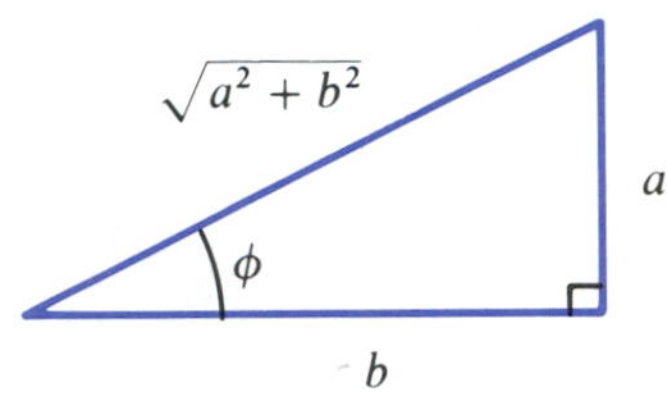

Figure 15

Now we set

$$\sin \phi = \frac{a}{\sqrt{a^2 + b^2}} \qquad \text{and} \qquad \cos \phi = \frac{b}{\sqrt{a^2 + b^2}}$$

(see Fig. 15). An application of the identity

$$\sin(\omega t + \phi) = \sin \omega t \cos \phi + \cos \omega t \sin \phi$$

allows us to write (2) as

$$x = A\sin(\omega t + \phi) \qquad A = \sqrt{a^2 + b^2} \tag{10}$$

This is the equation of simple harmonic motion with amplitude $A = \sqrt{a^2 + b^2}$ and period $2\pi/\omega$. The angle ϕ is referred to as the *phase angle* and the angle ϕ/ω is referred to as the *phase shift* (see Fig. 16).

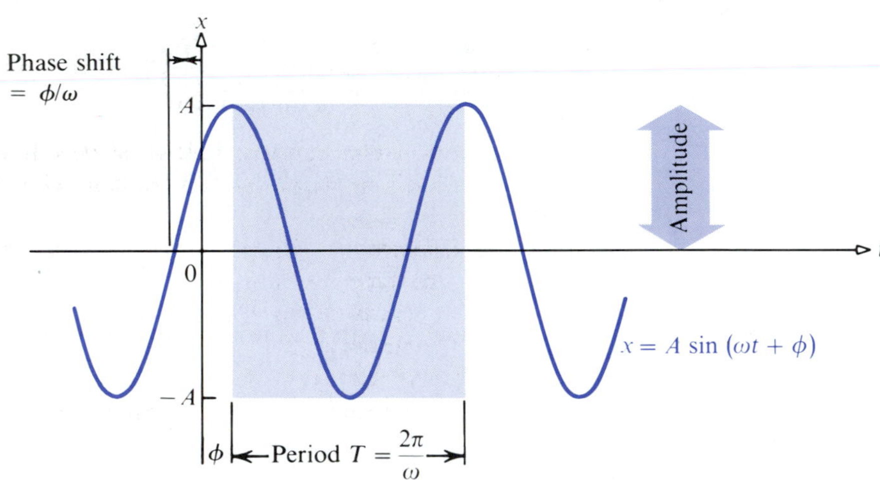

Figure 16

Any motion that obeys the differential equation (1) is referred to as *harmonic oscillation*. Harmonic oscillation occurs not only in elastic bodies but also in vibrating tuning forks, in vibrating strings, inside an atom, and elsewhere. This phenomenon explains why watches keep time and why atoms emit light of a particular frequency.

EXERCISE 8.4

In Problems 1–5 a particle of mass m is attached to one end of a vertical elastic spring whose spring constant is k. The other end of the spring is fixed. If the particle is pulled down a distance A (the amplitude) from the equilibrium position, the displacement of the particle from its equilibrium position at time t is denoted by $x = x(t)$, where at $t = 0$, $x = A$.

1. Suppose that $m = 15$ grams and $x = 12\cos(t/3)$. Find the spring constant.

2. Suppose that $m = 12$ grams and $k = 180$ grams per second per second. Suppose also that, at time $t = 0$ second, the spring is stretched 2 centimeters from its equilibrium position and the particle is at rest. Find $x(t)$. Find the amplitude.

3. Suppose that $m = 1$ gram and $k = 4$ grams per second per second. Suppose also that, at time $t = 0$ second, the spring is compressed 1 centimeter from its equilibrium position. Find an expression for the velocity of the particle.

4. Suppose that $m = 2$ grams and $k = 180$ grams per second per second. Suppose also that, at time $t = 0$ second, the spring is compressed 2 centimeters from its equilibrium position. Find the velocity of the particle at time $t = \pi/6$ second.

5. Suppose that $m = 1$ gram and $k = 0.25$ grams per second per second. Suppose also that, at time $t = 0$ second, the spring is stretched 1 centimeter from its equilibrium position. Find $x(t)$.

6. Consider a simple pendulum whose length in centimeters is l. If the pendulum makes an angle θ with the vertical, its motion (assuming no friction or air resistance) obeys the differential equation

$$\frac{d^2\theta}{dt^2} = -\frac{g}{l}\sin\theta$$

$g \approx 980$ centimeters per second per second

(*Problem 6 continues in the next column*)

6. (*continued*)
If the angle θ is sufficiently small ($<5°$), then $\sin\theta \approx \theta$, and we may replace $\sin\theta$ by θ (θ in radians). Under this assumption, find the equation of the motion, its period, and its frequency. Use the initial conditions that when $t = 0$, $\theta = \theta_0$ and $d\theta/dt = 0$.

7. If the disk in the figure is rotated about the vertical through an angle θ, torsion in the wire will attempt to turn the disk in the opposite direction. The motion (assuming no friction or air resistance) obeys the differential equation

$$\frac{1}{2}mR^2\frac{d^2\theta}{dt^2} = -k\theta$$

where m is the mass of the disk, R is the radius of the disk, and k is the coefficient of torsion of the wire. Find the equation of the motion and its period. Use the initial conditions that when $t = 0$, $\theta = \theta_0$ and $d\theta/dt = 0$.

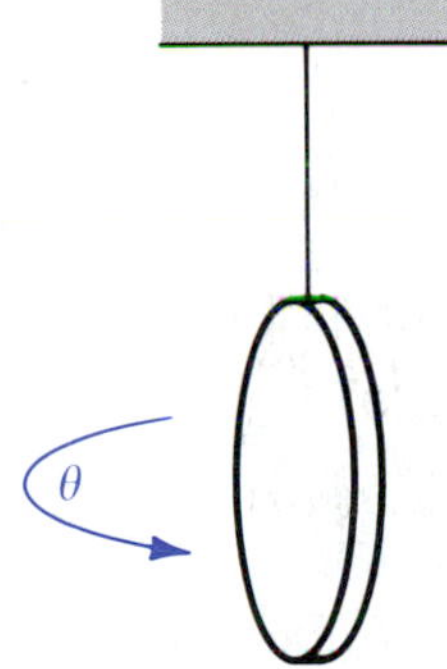

8.5 Hyperbolic Functions

Certain combinations of the functions e^x and e^{-x} occur so frequently in applied mathematics that they warrant special study. This collection of functions, called *hyperbolic functions*, has some properties that bear a close resemblance to those of the trigonometric functions. In fact, hyperbolic functions are so named because they are related to a hyperbola in much the same way that the trigonometric functions (sometimes called *circular functions*) are related to a circle.* Because of this, we give them names such as *hyperbolic sine* (*sinh*) and *hyperbolic cosine* (*cosh*). These functions are defined as

$$\sinh x = \frac{e^x - e^{-x}}{2} \qquad \cosh x = \frac{e^x + e^{-x}}{2} \tag{1}$$

The remaining hyperbolic functions are combinations of the first two; they should be easy to remember, since they are analogous to the relationships among the trigonometric functions.

Hyperbolic tangent: $$\tanh x = \frac{\sinh x}{\cosh x} = \frac{e^x - e^{-x}}{e^x + e^{-x}} \tag{2}$$

Hyperbolic cotangent: $$\coth x = \frac{\cosh x}{\sinh x} = \frac{e^x + e^{-x}}{e^x - e^{-x}} \tag{3}$$

Hyperbolic secant: $$\operatorname{sech} x = \frac{1}{\cosh x} = \frac{2}{e^x + e^{-x}} \tag{4}$$

Hyperbolic cosecant: $$\operatorname{csch} x = \frac{1}{\sinh x} = \frac{2}{e^x - e^{-x}} \tag{5}$$

There are also identities involving the hyperbolic functions that are strikingly reminiscent of the familiar trigonometric identities. Some of the basic ones are:

$$\cosh^2 x - \sinh^2 x = 1 \tag{6}$$

$$\tanh^2 x + \operatorname{sech}^2 x = 1 \tag{7}$$

$$\coth^2 x - \operatorname{csch}^2 x = 1 \tag{8}$$

The first of these is derived directly from the definition, as follows:

$$\cosh^2 x - \sinh^2 x = \left(\frac{e^x + e^{-x}}{2}\right)^2 - \left(\frac{e^x - e^{-x}}{2}\right)^2$$

$$= \frac{e^{2x} + 2e^0 + e^{-2x}}{4} - \frac{e^{2x} - 2e^0 + e^{-2x}}{4} = \frac{2+2}{4} = 1$$

The second identity (7) is obtained from the first by dividing each term by $\cosh^2 x$. Identity (8) is derived similarly.

Numerous other identities may be derived. We list some of them below:

$$\sinh(A + B) = \sinh A \cosh B + \cosh A \sinh B \tag{9}$$

$$\cosh(A + B) = \cosh A \cosh B + \sinh A \sinh B \tag{10}$$

$$\sinh(-A) = -\sinh A \tag{11}$$

$$\cosh(-A) = \cosh A \tag{12}$$

The derivations of these identities are left as exercises (Problems 5–8).

* More on this at the end of the section.

The differentiation formulas for the hyperbolic functions can be obtained readily:

$$\frac{d}{dx}\sinh x = \frac{d}{dx}\left[\frac{1}{2}(e^x - e^{-x})\right] = \frac{e^x + e^{-x}}{2} = \cosh x$$

$$\frac{d}{dx}\cosh x = \frac{d}{dx}\left[\frac{1}{2}(e^x + e^{-x})\right] = \frac{e^x - e^{-x}}{2} = \sinh x$$

Thus,

$$\frac{d}{dx}\sinh x = \cosh x \qquad \frac{d}{dx}\cosh x = \sinh x \tag{13}$$

The formulas for differentiating the other hyperbolic functions are

$$\frac{d}{dx}\tanh x = \operatorname{sech}^2 x \qquad \frac{d}{dx}\coth x = -\operatorname{csch}^2 x \tag{14}$$

$$\frac{d}{dx}\operatorname{sech} x = -\operatorname{sech} x \tanh x \qquad \frac{d}{dx}\operatorname{csch} x = -\operatorname{csch} x \coth x$$

The Graphs of the Hyperbolic Functions

Since $\cosh x = \frac{1}{2}(e^x + e^{-x})$, it follows that, at $x = 0$, $y = \frac{1}{2}(e^0 + e^0) = 1$. Thus, the graph of $y = \cosh x$ has its y-intercept at $(0, 1)$. Since $\cosh(-x) = \cosh x$, the cosh function is an *even function*, and hence its graph is symmetric with respect to the y-axis. Also, since $\frac{1}{2}(e^x + e^{-x})$ is positive for all x, the points on the graph of $y = \cosh x$ are always above the x-axis.

Since $\lim_{x\to+\infty}[\frac{1}{2}(e^x + e^{-x})] = +\infty$, the graph is not bounded above. If $y = \cosh x$, then $dy/dx = \sinh x$. Now, since $\sinh x > 0$ for $x > 0$, $\cosh x$ is increasing for $x > 0$. Also, since $\sinh x < 0$ for $x < 0$, $\cosh x$ is decreasing for $x < 0$. Furthermore, at $x = 0$, $dy/dx = 0$. Hence, there is a horizontal tangent line at the point $(0, 1)$.

Also, $d^2y/dx^2 = \cosh x$, which is always positive. Hence, the point $(0, 1)$ is a local minimum, and the graph is concave upward. Using all this information, we are able to sketch the graph of $\cosh x$, as shown in Figure 17.

A similar analysis will yield enough information to graph $\tanh x$, $\sinh x$, $\coth x$, $\operatorname{sech} x$, and $\operatorname{csch} x$, as shown in Figures 18–22.

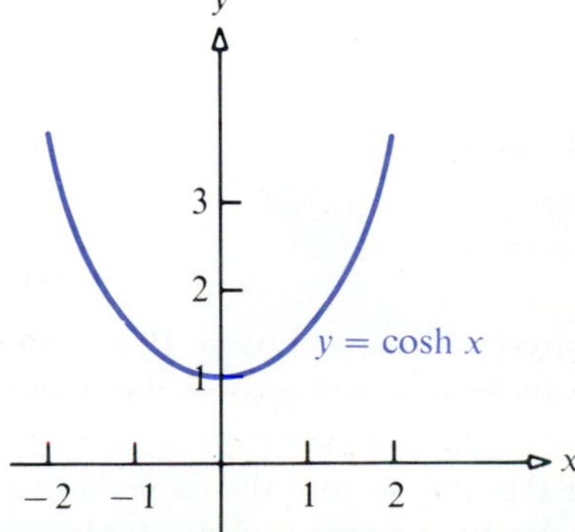

Figure 17

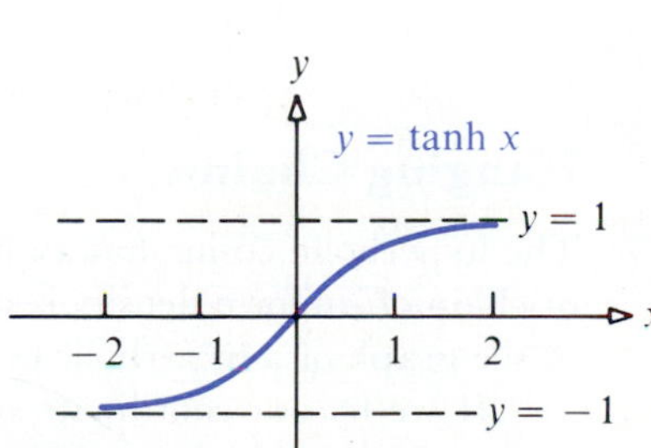

Figure 18

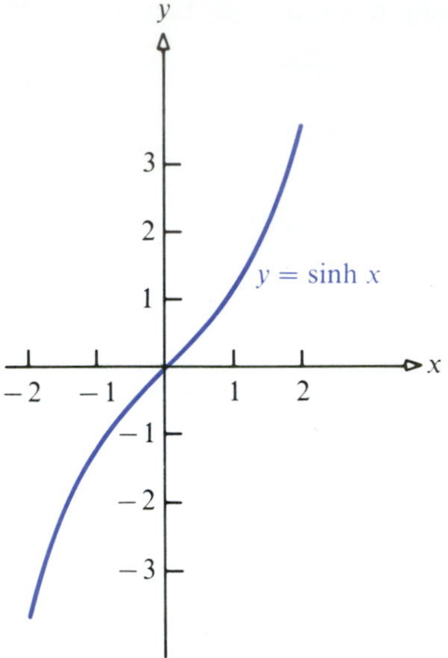

Figure 19

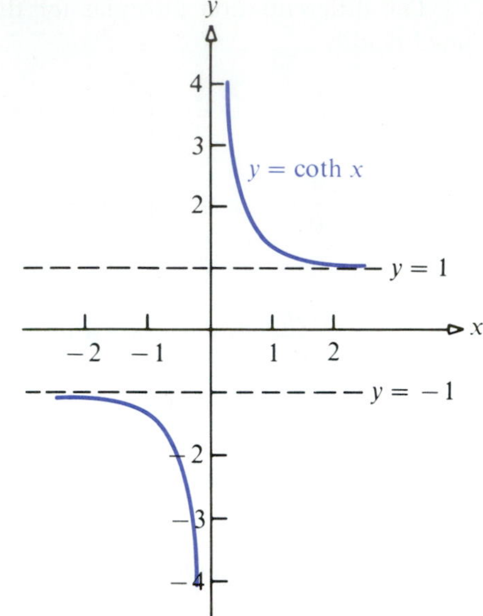

Figure 20

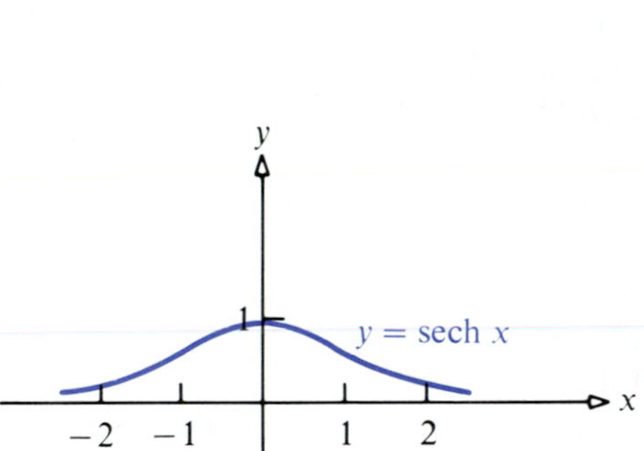

Figure 21

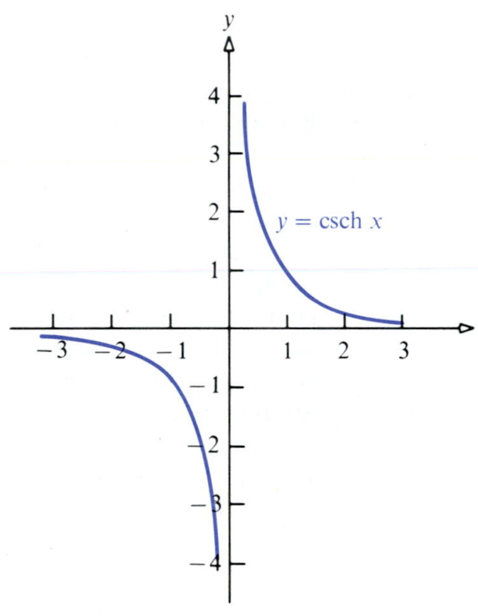

Figure 22

Hanging Chains

The hyperbolic cosine has an interesting physical interpretation. If a cable or chain of uniform density is suspended at its ends, it will assume the shape of the graph of a hyperbolic cosine.

If we fix our coordinate system (as in Fig. 23) so that the cable lies in the xy-plane, with the y-axis vertical and the lowest point of the cable on

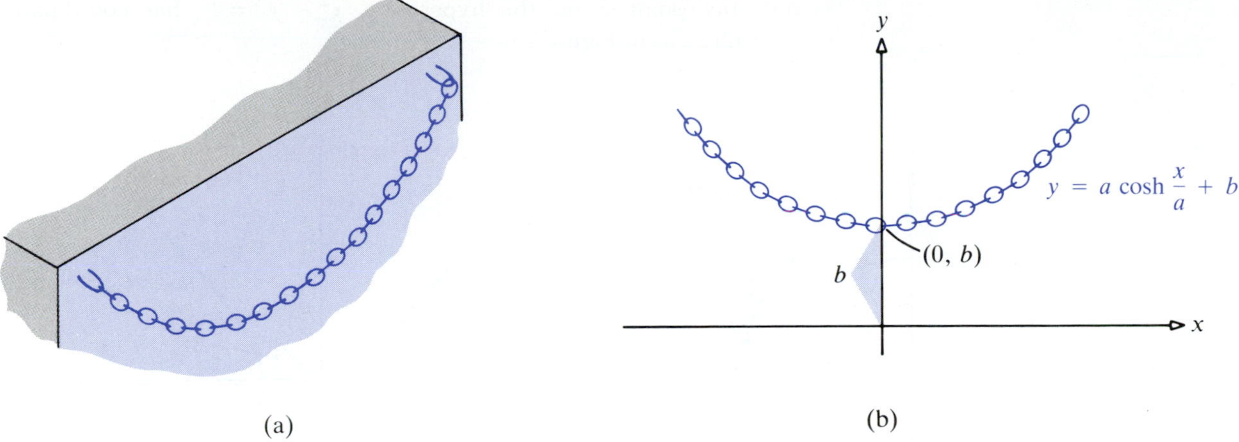

Figure 23

the y-axis at the point $(0, b)$, the shape of the chain will be given by the equation

$$y = a \cosh \frac{x}{a} + b - a \tag{15}$$

where a is a constant that depends on the linear density or weight per unit length of the chain and the tension or horizontal force holding the ends of the chain apart.

The graph of this equation is called a *catenary*.*

Related Integrals

Formulas (13) and (14) give rise to the following integration formulas:

$$\int \sinh x\, dx = \cosh x + C \qquad \int \cosh x\, dx = \sinh x + C$$

$$\int \operatorname{sech}^2 x\, dx = \tanh x + C \qquad \int \operatorname{csch}^2 x\, dx = -\coth x + C$$

$$\int \operatorname{sech} x \tanh x\, dx = -\operatorname{sech} x + C \qquad \int \operatorname{csch} x \coth x\, dx = -\operatorname{csch} x + C \tag{16}$$

Why the Name Hyperbolic?

The trigonometric functions are sometimes referred to as *circular functions* because any point P on the unit circle $x^2 + y^2 = 1$ has coordinates $(\cos t, \sin t)$, as shown in Figure 24. The hyperbolic functions are so named

* From the Latin word *catena*, meaning "chain."

because any point P on the hyperbola $x^2 - y^2 = 1$ has coordinates $(\cosh t, \sinh t)$, as in Figure 25.

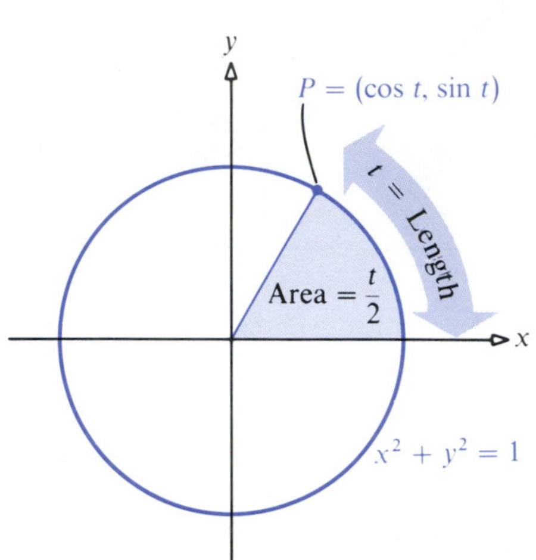

Figure 24

Figure 25

The relationship is even deeper, though. From elementary geometry it is known that the area of the sector shaded in Figure 24 equals $t/2$. And it can be shown that the area shaded in Figure 25 also equals $t/2$.

EXERCISE 8.5

1. Find: $\lim_{x\to 0}\left(\frac{\sinh x}{x}\right)$

2. Find: $\lim_{x\to 0}\left(\frac{\cosh x - 1}{x}\right)$

In Problems 3–10 derive each identity.

3. $\tanh^2 x + \text{sech}^2 x = 1$

4. $\coth^2 x - \text{csch}^2 x = 1$

5. $\sinh(-A) = -\sinh A$

6. $\cosh(-A) = \cosh A$

7. $\sinh(A + B) = \sinh A \cosh B + \cosh A \sinh B$

8. $\cosh(A + B) = \cosh A \cosh B + \sinh A \sinh B$

9. $\cosh 3x = 4\cosh^3 x - 3\cosh x$

10. $\tanh 2x = \frac{2\tanh x}{1 + \tanh^2 x}$

In Problems 11–22 find y'.

11. $y = \sinh 3x$

12. $y = \cosh \frac{x}{2}$

13. $y = \cosh(x^2 + 1)$

14. $y = \cosh(2x^3 - 1)$

15. $y = \sinh x \cosh 4x$

16. $y = \cosh^2 x$

17. $y = \tanh^2 x$

18. $y = e^x \cosh x$

19. $y = \coth \frac{1}{x}$

20. $y = e^x(\cosh x + \sinh x)$

21. $y = \tanh x^2$

22. $y = x^2 \text{sech}\, x$

In Problems 23–30 evaluate each integral.

23. $\int \frac{\sinh\sqrt{x}}{\sqrt{x}}\,dx$

24. $\int \frac{1}{\cosh^2 2x}\,dx$

25. $\int \sinh^2 x\,dx$

26. $\int \cosh^4 8x\,dx$

27. $\int \sinh x \cosh x\,dx$

28. $\int \operatorname{sech}^2 x \tanh x\,dx$

29. $\int \sqrt{\cosh x + 1}\,dx$

30. $\int \cosh^2 x\,dx$

31. Find the area under the graph of $y = \cosh x$ from $x = -1$ to $x = 1$.

32. A cable hangs in the shape of a catenary between two supports at the same height 20 meters apart. The slope of the cable at the right-hand support is $\frac{3}{4}$.

(a) What is the height of the supports above the lowest point in the middle of the cable?

(b) Find the length of the cable.

33. A rope of length L is supported at two points, (c, d) and $(-c, d)$. The middle of the rope is located at $(0, b)$. Find the constant a in (15) for the shape of the rope. [*Hint:* Show that $(d - b + a)^2 - a^2 = L^2/4$.]

34. The famous Gateway Arch to the West in St. Louis, Missouri, is constructed in the shape of a catenary (see the picture). For this arch, the number a in (15) is about 127.7 feet. Find an equation of the arch. (Assume that the coordinate system has been positioned in such a way that $b = a$.)

Gateway Arch to the West, St. Louis, Missouri

35. Find the length of the catenary (15) from the point $(0, b)$ to a point (x, y), $x > 0$.

36. Find the area under the graph of the catenary (15) from $x = 0$ to $x = c$.

37. Find the volume of the solid generated by revolving the area under the graph of the catenary in (15) from $x = 0$ to $x = 1$ about the x-axis.

38. Show that (15) obeys the differential equation $d^2y/dx^2 = (1/a^2)(y + a - b)$.

39. Establish the identity $(\cosh x + \sinh x)^n = \cosh nx + \sinh nx$ for any real number n.

40. Establish (14).

41. Show that the catenary function (15) is a solution of the differential equation

$$\frac{d^2y}{dx^2} = \frac{1}{a}\sqrt{1 + \left(\frac{dy}{dx}\right)^2}$$

Inverse Hyperbolic Functions 8.6

The graph of $y = \sinh x$ is shown in Figure 19 (p. 526). From the graph, we observe that every horizontal line intersects the graph of $y = \sinh x$ in exactly one point. The function $y = \sinh x$ is therefore one-to-one and, hence, has an inverse. We denote the inverse by $\sinh^{-1}$ and define it as

$$y = \sinh^{-1}x \qquad \text{means} \qquad x = \sinh y \tag{1}$$

The domain of $y = \sinh^{-1}x$ is the set of real numbers, and the range is also the set of real numbers.

The graph of $y = \cosh x$ is shown in Figure 17 (p. 525). From the graph, we observe that every horizontal line above 1 intersects the graph of $y = \cosh x$ at two points. Thus, the function is not one-to-one. However, if we restrict $y = \cosh x$ to the nonnegative values of x, we have a one-to-one function that, therefore, has an inverse. We denote the inverse by $\cosh^{-1}$ and define it as

$$y = \cosh^{-1}x \qquad \text{means} \qquad x = \cosh y \qquad y \geq 0 \tag{2}$$

The domain of $y = \cosh^{-1}x$ is $x \geq 1$, and the range is $y \geq 0$.

The other inverse hyperbolic functions are defined similarly.

The inverse hyperbolic functions may be expressed in terms of natural logarithms as follows:

$$\sinh^{-1}x = \ln(x + \sqrt{x^2 + 1}) \qquad \text{for all } x \tag{3}$$

$$\cosh^{-1}x = \ln(x + \sqrt{x^2 - 1}) \qquad x \geq 1 \tag{4}$$

$$\begin{aligned} \tanh^{-1}x &= \frac{1}{2}\ln\left(\frac{1 + x}{1 - x}\right) \qquad -1 < x < 1 \\ \coth^{-1}x &= \frac{1}{2}\ln\left(\frac{x + 1}{x - 1}\right) \qquad |x| > 1 \end{aligned} \tag{5}$$

We will prove (3) here and leave (4) and (5) as exercises (Problems 11 and 12).

Proof Let $y = \sinh^{-1}x$. Then $x = \sinh y = (e^y - e^{-y})/2$. Multiplying by $2e^y$, we get

$$2xe^y = (e^y)^2 - 1$$

$$(e^y)^2 - 2xe^y - 1 = 0$$

This is a quadratic equation in e^y. By solving it for e^y, we find

$$e^y = \tfrac{1}{2}(2x \pm \sqrt{4x^2 + 4}) = x \pm \sqrt{x^2 + 1}$$

Since $e^y > 0$ and $x < \sqrt{x^2 + 1}$ for all x, the minus sign on the right is not possible. As a result,

$$e^y = x + \sqrt{x^2 + 1} \qquad \text{or} \qquad y = \ln(x + \sqrt{x^2 + 1})$$

The derivatives of $y = \sinh^{-1}x$, $y = \cosh^{-1}x$, and $y = \tanh^{-1}x$ are obtained by a straightforward calculation using (3), (4), and (5):

$$\frac{d}{dx}\sinh^{-1}x = \frac{1}{\sqrt{x^2 + 1}} \tag{6}$$

$$\frac{d}{dx}\cosh^{-1}x = \frac{1}{\sqrt{x^2 - 1}} \tag{7}$$

$$\frac{d}{dx}\tanh^{-1}x = \frac{1}{1 - x^2} \tag{8}$$

The differentiation formulas (6)–(8) may, in the usual way, be written as the following integral formulas:

$$\int \frac{dx}{\sqrt{x^2+1}} = \sinh^{-1}x + C \tag{9}$$

$$\int \frac{dx}{\sqrt{x^2-1}} = \cosh^{-1}x + C \tag{10}$$

$$\int \frac{dx}{1-x^2} = \begin{cases} \tanh^{-1}x + C & \text{if } |x| < 1 \quad \text{(11a)} \\ \coth^{-1}x + C & \text{if } |x| > 1 \quad \text{(11b)} \end{cases}$$

The distinction between the two cases $|x| < 1$ and $|x| > 1$ in (11) results from the fact that the inverse hyperbolic tangent is defined for $|x| < 1$, whereas the inverse hyperbolic cotangent is defined for $|x| > 1$.

EXAMPLE 1

$$\int \frac{dx}{\sqrt{9x^2+1}} \underset{u=3x}{\uparrow}= \frac{1}{3}\int \frac{du}{\sqrt{u^2+1}} = \frac{1}{3}\sinh^{-1}3x + C$$

EXAMPLE 2

$$\int_0^{1/3} \frac{dx}{1-x^2} = \tanh^{-1}x\Big|_0^{1/3} = \frac{1}{2}\left[\ln\left|\frac{1+x}{1-x}\right|\right]_0^{1/3} = \frac{1}{2}\ln 2$$

EXAMPLE 3

$$\int_2^4 \frac{dx}{1-x^2} = \coth^{-1}x\Big|_2^4 = \frac{1}{2}\left[\ln\left|\frac{1+x}{1-x}\right|\right]_2^4 = \frac{1}{2}\left[\ln\frac{5}{3} - \ln\frac{3}{1}\right] = \frac{1}{2}\ln 5 - \ln 3$$

EXERCISE 8.6

In Problems 1–10 find y'.

1. $y = \sinh^{-1}3x$
2. $y = \cosh^{-1}4x$
3. $y = \tanh^{-1}(x^2-1)$
4. $y = x\sinh^{-1}x$
5. $y = \cosh^{-1}(2x+1)$
6. $y = x^2\cosh^{-1}x$
7. $y = \tanh^{-1}(\cos 2x)$
8. $y = \cosh^{-1}(\sqrt{x^2-1}), \quad x > \sqrt{2}$
9. $y = \tanh^{-1}(\tan x)$
10. $y = \sinh^{-1}(\sin x)$
11. Show that $\cosh^{-1}x = \ln(x + \sqrt{x^2-1}), \quad x \geq 1.$
12. Show that $\tanh^{-1}x = \frac{1}{2}\ln\left(\frac{1+x}{1-x}\right), \quad -1 < x < 1.$
13. Show that $\frac{d}{dx}\sinh^{-1}x = \frac{1}{\sqrt{x^2+1}}.$
14. Show that $\frac{d}{dx}\cosh^{-1}x = \frac{1}{\sqrt{x^2-1}}.$
15. Show that $\frac{d}{dx}\tanh^{-1}x = \frac{1}{1-x^2}.$
16. Obtain the graph of $y = \sinh^{-1}x$.

17. Obtain the graph of $y = \cosh^{-1}x$.

18. Obtain the graph of $y = \tanh^{-1}x$.

19. Evaluate: $\displaystyle\int_2^3 \frac{1}{\sqrt{x^2-1}}\,dx$

20. Derive the formula $\displaystyle\int \frac{dx}{\sqrt{x^2+a^2}} = \sinh^{-1}\left(\frac{x}{a}\right) + C.$

21. Find the length of the graph of $y = \sqrt{a^2 - x^2}$ in two ways (by calculus and by elementary geometry).

22. Find the minimum value of $y = x - \cosh^{-1}x$.

REVIEW EXERCISES

1. Find y' at $x = \pi/2$, $y = \pi$ for $x \sin y + y \cos x = 0$.

2. Find $\displaystyle\int \frac{1}{\sqrt{x}} \cos^2\sqrt{x}\,dx$.

3. Find y' for $x = \ln(\csc y + \cot y)$.

4. Find $\displaystyle\int \sin^2 3x\,dx$.

5. Write the arc length of $y = \ln(\sec x)$ from $x = 0$ to $x = b$, where $0 < b < \pi/2$, as an integral.

6. Write the arc length of $y = \tan x$ from $x = a$ to $x = b$, where $0 < a < b < \pi/2$, as an integral.

7. Find y' for $y = \ln(\tan^2 x)$.

In Problems 8–12 find the exact value of each expression.

8. $\cos\left[\sin^{-1}\frac{4}{5}\right]$

9. $\sin[\tan^{-1}(-\sqrt{2})]$

10. $\tan[\sec^{-1}(-\sqrt{2})]$

11. $\cos\left[\sin^{-1}\frac{2}{3} + \tan^{-1}\sqrt{2}\right]$

12. $\sin\left[\tan^{-1}\sqrt{3} - \cos^{-1}\frac{\sqrt{2}}{2}\right]$

13. Find $f^{-1}(x)$ if $f(x) = \displaystyle\int_0^x \frac{dt}{\sqrt{1-t^2}}$.

In Problems 14–20 find y'.

14. $y = \sqrt{1-x^2}(\sin^{-1}x)$

15. $y = \sin^{-1}(x-1) + (x-1)\sqrt{2x-x^2}$

16. $y = 2\sqrt{x} - 2\tan^{-1}\sqrt{x}$

17. $y = \dfrac{8x}{x^2+4} - 4\tan^{-1}\dfrac{x}{2} + x$

18. $y = \sin^{-1}(2x-1)$

19. $y = x^2 \tan^{-1}\dfrac{1}{x}$

20. $y = x\tan^{-1}x - \ln\sqrt{1+x^2}$

21. Find: $\displaystyle\int_0^1 \frac{x\,dx}{\sqrt{2-x^4}}$

22. Find: $\displaystyle\int_0^1 \frac{x+1}{x^2+3}\,dx$

23. Find: $\displaystyle\int_{3\sqrt{2}}^6 \frac{dx}{x\sqrt{x^2-9}}$

24. Find: $\displaystyle\int \frac{e^{4x}\,dx}{\sqrt{16-e^{8x}}}$

25. Find the area of the semicircle $y = \sqrt{a^2-x^2}$ by integration.

26. Find the area under the graph of $y = 1/(1+x^2)$ from $x = -r$ to $x = r$ $(r > 0)$. What happens as $r \to +\infty$?

In Problems 27 and 28 a particle of mass m is attached to one end of a vertical elastic spring with spring constant k. The other end of the spring is fixed. If the particle is pulled down a distance A (amplitude), the displacement at time t is given by $x = x(t)$, where at $t = 0$, $x = A$.

27. Let $m = 20$ grams and $x = 12\cos(t/3)$. Find k.

28. Let $m = 24$ grams and $k = 12$ grams per second per second. At $t = 0$ the spring is stretched 2 centimeters from its equilibrium point and is at rest. Find $x(t)$. Find the amplitude.

In Problems 29–34 find y'.

29. $y = x \sinh x$

30. $y = \tanh \frac{x}{2} + \frac{2x}{4 + x^2}$

31. $y = x^2 \tanh^{-1} x$

32. $y = \sqrt{\sinh x}$

33. $y = \cosh e^x$

34. $y = \ln(\sinh^2 x)$

35. Find the area under the graph of $y = \cosh x$ from $x = 0$ to $x = 2$.

36. Find the arc length of $y = \cosh x$ from $(0, 1)$ to $(2, \cosh 2)$.

37. Find y' for $y = \sinh^{-1} e^x$.

38. Find y' for $y = \ln\sqrt{1 - x^2} + x \tanh^{-1} x$.

39. Find the area enclosed by $y = \frac{1}{1 - x^2}$, $x = 0$, $x = \frac{1}{2}$, $y = 0$.

40. Evaluate: $\int_3^5 \frac{dx}{\sqrt{x^2 - 1}}$

CHALLENGE EXERCISES

1. (a) Express the area A of the region in the first quadrant enclosed by the y-axis and the graphs of $y = \tan x$ and $y = k$ for $k > 0$ as a function of k.
(b) What is the value of A when $k = 1$?
(c) If the line $y = k$ is moving upward at the rate of $\frac{1}{10}$ unit per second, at what rate is A changing when $k = 1$?

2. What are the coordinates of the inflection point on the graph of $y = (x + 1)\tan^{-1} x$?

3. Let $f(x) = x \sin^{-1} x$.

(a) Show that f is an even function.
(b) Find $f'(x)$ and $f''(x)$.
(c) Sketch the graph of f.

4. What happens when we try to find the derivative of $f(x) = \sin^{-1}(\cosh x)$? What is the explanation?

5. Since $(d/dx)\tanh^{-1} x = 1/(1 - x^2)$, we might expect that $\int_2^3 dx/(1 - x^2) = \tanh^{-1}(3) - \tanh^{-1}(2)$. Why is this incorrect? What is the correct result?

6. In Section 8.5 we said that the shaded area in Figure 25 is $t/2$. Prove this as follows:

(a) Let A be twice the area outlined in Figure 25, and let (x, y) be the coordinates of P. Explain why $A = x\sqrt{x^2 - 1} - 2\int_1^x \sqrt{u^2 - 1}\, du$.
(b) Differentiate in part (a) to show that $dA/dx = 1/\sqrt{x^2 - 1}$.
(c) Show that $A = \cosh^{-1} x + C$, and explain why $C = 0$.
(d) Why does it follow that $A = t$?

7.* The general equation of simple harmonic motion,

$$x = A \sin(\omega t + \theta_0)$$

can be written in the equivalent form

$$x = B \sin \omega t + C \cos \omega t$$

Find the expressions for the amplitudes B and C in terms of the amplitude A and the initial phase angle θ_0.

8.† A body is vibrating with simple harmonic motion of amplitude 15 centimeters and frequency 4 hertz (cycles per second). Compute:

(a) The maximum values of the acceleration and velocity
(b) The acceleration and velocity when the coordinate is 9 centimeters
(c) The time required to move from the equilibrium position to a point 12 centimeters distant from it

9.† A body of mass 10 grams moves with simple harmonic motion of amplitude 24 centimeters and period 4 seconds. The coordinate is 24 centimeters when $t = 0$. Compute:

(a) The position of the body when $t = 0.5$ second
(b) The magnitude and direction of the force acting on the body when $t = 0.5$ second

(*Problem 9 continues on page 534*)

* Adapted from F. W. Sears, M. W. Zemansky, and H. D. Young, *University Physics* (Reading, Mass.: Addison-Wesley Publishing Co., 1976), p. 206. Reprinted by permission.
† Ibid., p. 208. Reprinted by permission.

9. (*continued*)

(c) The minimum time required for the body to move from its initial position to the point where $x = -12$ centimeters

(d) The velocity of the body when $x = -12$ centimeters

10.* The motion of the piston of an automobile engine is approximately simple harmonic.

(a) If the stroke of an engine (twice the amplitude) is 4 inches and the angular velocity is 3600 revolutions per minute, compute the acceleration of the piston at the end of its stroke.

(b) If the piston weighs 1 pound, what resultant force must be exerted on it at this point?

---

* Ibid., p. 208. Reprinted by permission.

11. (a) Sketch the graph of $y = \mathrm{gd}(x) = \tan^{-1}(\sinh x)$; this is called the *gudermannian of* x (named after Christoph Gudermann).

(b) If $y = \mathrm{gd}(x)$, show that $\cos y = \operatorname{sech} x$ and $\sin y = \tanh x$.

(c) Show that if $y = \mathrm{gd}(x)$, then y satisfies the differential equation $y' = \cos y$.

(d) Use the differential equation of part (c) to obtain the formula $\int \sec\theta\, d\theta = \mathrm{gd}^{-1}\theta + C$. Compare this to (14) in Section 8.1.

Techniques of Integration

9.1 Basic Integration Formulas

In the preceding chapters we developed a collection of basic integration formulas. For convenience, a list of those that are *essential* is given here and inside the covers of this book. If you have not already committed these to memory, then you should do so now.

1. $\displaystyle\int x^\alpha\,dx = \frac{x^{\alpha+1}}{\alpha+1} + C \qquad \alpha \neq -1$

2. $\displaystyle\int \frac{1}{x}\,dx = \ln|x| + C$

3. $\displaystyle\int e^x\,dx = e^x + C$

4. $\displaystyle\int \sin x\,dx = -\cos x + C$

5. $\displaystyle\int \cos x\,dx = \sin x + C$

6. $\displaystyle\int \tan x\,dx = \ln|\sec x| + C$

7. $\int \sec x\,dx = \ln|\sec x + \tan x| + C$

8. $\int \sec^2 x\,dx = \tan x + C$

9. $\int \sec x \tan x\,dx = \sec x + C$

10. $\int \frac{dx}{\sqrt{1 - x^2}} = \sin^{-1} x + C$

11. $\int \frac{dx}{1 + x^2} = \tan^{-1} x + C$

12. $\int \frac{dx}{x\sqrt{x^2 - 1}} = \sec^{-1} x + C$

The next list contains formulas that are *useful*. For short-term use (a test, for example), you may wish to learn them. For long-term or occasional use, you can look them up (inside the covers of this book) or derive them.

13. $\int \cot x\,dx = \ln|\sin x| + C$

14. $\int \csc x\,dx = \ln|\csc x - \cot x| + C$

15. $\int \csc^2 x\,dx = -\cot x + C$

16. $\int \csc x \cot x\,dx = -\csc x + C$

17. $\int \frac{dx}{\sqrt{a^2 - x^2}} = \sin^{-1}\left(\frac{x}{a}\right) + C \qquad a > 0$

18. $\int \frac{dx}{a^2 + x^2} = \frac{1}{a}\tan^{-1}\left(\frac{x}{a}\right) + C \qquad a > 0$

19. $\int \frac{dx}{x\sqrt{x^2 - a^2}} = \frac{1}{a}\sec^{-1}\left(\frac{x}{a}\right) + C \qquad a > 0$

20. $\int a^x\,dx = \frac{a^x}{\ln a} + C \qquad a > 0, \quad a \neq 1$

21. $\int \sinh x\,dx = \cosh x + C$

22. $\int \cosh x\,dx = \sinh x + C$

23. $\int \operatorname{sech}^2 x\,dx = \tanh x + C$

24. $\int \operatorname{csch}^2 x\,dx = -\coth x + C$

25. $\int \operatorname{sech} x \tanh x\,dx = -\operatorname{sech} x + C$

26. $\int \operatorname{csch} x \coth x\,dx = -\operatorname{csch} x + C$

When evaluating an integral, it is best first to compare it with the preceding list. If an expression is identical to one of these, its integral is known. However, quite often it is necessary to use some special techniques in order to express the integrand in a form that is found in the list. For example, one *technique of integration*—the substitution method—has been studied in Chapter 5. We review this technique in the following example:

EXAMPLE 1

Evaluate each integral: (a) $\int \frac{e^x\,dx}{e^x + 4}$ (b) $\int \frac{5\,dx}{\sqrt{1 - 4x^2}}$

Solution

(a) We use the substitution $u = e^x + 4$. Then $du = e^x\,dx$, and

$$\int \frac{e^x}{e^x + 4}\,dx = \int \frac{1}{u}\,du \underset{\uparrow}{=} \ln|u| + C = \ln(e^x + 4) + C$$

Formula 2

(b) We use the substitution $u = 2x$. Then $du = 2\,dx$, and

$$\int \frac{5\,dx}{\sqrt{1-4x^2}} = \left(\frac{5}{2}\right)\int \frac{2\,dx}{\sqrt{1-(2x)^2}} = \left(\frac{5}{2}\right)\int \frac{du}{\sqrt{1-u^2}}$$

$$\underset{\substack{\uparrow \\ \text{Formula 10}}}{=} \frac{5}{2}\sin^{-1}u + C = \frac{5}{2}\sin^{-1}(2x) + C$$

Alternatively, we could have used formula 17 from the list of useful integral formulas:

$$\int \frac{5\,dx}{\sqrt{1-4x^2}} = 5\int \frac{dx}{2\sqrt{\frac{1}{4}-x^2}} \underset{\substack{\uparrow \\ a=\frac{1}{2}}}{=} \frac{5}{2}\sin^{-1}\left(\frac{x}{\frac{1}{2}}\right) + C = \frac{5}{2}\sin^{-1}(2x) + C$$

In the remaining sections of this chapter we detail procedures for identifying the most appropriate substitution to be made to simplify the integral. In addition, we study the techniques of integration referred to as *integration by parts* and *partial fractions*.

Many of the integration formulas that appear inside the covers of this book are derived in the coming sections. A more comprehensive list of integration formulas is found in William H. Beyer, ed., *Standard Mathematical Tables*, 27th ed. (Boca Raton, Fla.: CRC Press, Inc., 1984).

EXERCISE 9.1

In Problems 1–36 evaluate each integral by using the formulas listed in this section.

1. $\int \sqrt{3x+1}\,dx$

2. $\int \sqrt{5x-7}\,dx$

3. $\int x(3x^2-5)^5\,dx$

4. $\int x^2\sqrt{5x^3-2}\,dx$

5. $\int \frac{\sin x}{1+\cos x}\,dx$

6. $\int \frac{\sin x}{1-\cos x}\,dx$

7. $\int \frac{x\,dx}{\sqrt{1-9x^2}}$

8. $\int \frac{x\,dx}{\sqrt{1-x^2}}$

9. $\int \frac{1}{\sqrt{4-x^2}}\,dx$

10. $\int \frac{1}{x\sqrt{9x^2-1}}\,dx$

11. $\int \frac{e^x}{1+2e^x}\,dx$

12. $\int \frac{e^x+1}{e^x}\,dx$

13. $\int \frac{e^{\sqrt{x}}}{\sqrt{x}}\,dx$

14. $\int \frac{e^{\sqrt{3x+1}}}{\sqrt{3x+1}}\,dx$

15. $\int xe^{x^2}\,dx$

16. $\int xe^{-x^2}\,dx$

17. $\int \cos x\, e^{\sin x}\,dx$

18. $\int \sin x\, e^{\cos x}\,dx$

19. $\int \sin x \cos x\,dx$

20. $\int \frac{\sec^2 u}{1+\tan u}\,du$

21. $\int \frac{dx}{x^2+16}$

22. $\int \frac{dx}{x^2+4}$

23. $\int \frac{1+\cos(2x)}{\sin^2(2x)}\,dx$

24. $\int \frac{\sin^2(2x)}{1+\cos(2x)}\,dx$

25. $\int x\csc x^2 \cot x^2\,dx$
[*Hint:* Let $u = x^2$.]

26. $\int \frac{\csc v}{\tan v}\,dv$
[*Hint:* $\cot v = 1/(\tan v)$]

27. $\int \frac{\tan t}{\cos t}\,dt$
[*Hint:* $\sec t = 1/(\cos t)$]

28. $\int \frac{dx}{x\sqrt{x^2 - 4}}$

29. $\int \frac{dx}{x\sqrt{x^2 - 9}}$

30. $\int \frac{dx}{(x + 2)^2 + 1}$

31. $\int \frac{\sin^2(3x)}{1 + \cos(3x)}\, dx$

32. $\int \frac{\cos^2(5x)}{1 + \sin(5x)}\, dx$

33. $\int 5^x\, dx$

34. $\int 9^{5x}\, dx$

35. $\int \frac{\sinh x\, dx}{1 - \cosh x}$

36. $\int \frac{\operatorname{csch}^2 x\, dx}{1 - \coth x}$

37. The *White Lightning* (the first human-powered vehicle to go faster than 55 miles per hour, built by Northrop University students) is hypothesized to accelerate according to the function $a(t) = t^2/(20\sqrt{t^3 + 4})$ meters per second per second. What is its velocity at time t if the initial velocity is 0?

9.2 Integration by Parts

Integration by parts is based on the product rule for differentiation and is an effective and versatile technique for integration.

Recall that if u and v are differentiable functions of x, then

$$\frac{d}{dx}(uv) = v\frac{du}{dx} + u\frac{dv}{dx}$$

Integrating both sides gives

$$uv = \int v\frac{du}{dx}\, dx + \int u\frac{dv}{dx}\, dx$$

$$\int u\frac{dv}{dx}\, dx = uv - \int v\frac{du}{dx}\, dx$$

We may write this in the abbreviated form

$$\int u\, dv = uv - \int v\, du \tag{1}$$

This formula is usually referred to as the *by-parts formula*.

The corresponding formula for definite integrals is

$$\int_a^b u(x)\, dv(x) = u(x)v(x)\Big|_a^b - \int_a^b v(x)\, du(x)$$

where $dv(x) = v'(x)\, dx$ and $du(x) = u'(x)\, dx$.

To apply (1), we separate the integrand into two parts. We call one u and the other dv. We differentiate u to obtain du and integrate dv to obtain v. If we can then integrate $\int v\, du$, the problem is solved. The goal of this procedure, then, is to choose u and dv so that the term $\int v\, du$ is easier to solve than the original problem. As the examples will illustrate, this usually happens when u is simplified by differentiation.

EXAMPLE 1

Evaluate: $\int xe^x\, dx$

Solution

To use the by-parts formula, we choose u and dv so that

$$\int u\,dv = \int xe^x\,dx$$

and $\int v\,du$ is easier to evaluate than $\int u\,dv$. In this example we decide to choose

$$u = x \qquad \text{and} \qquad dv = e^x\,dx$$

As a result of this choice,

$$du = dx \qquad \text{and} \qquad v = \int dv = \int e^x\,dx = e^x$$

Note that we require only a particular antiderivative of dv at this stage; we will add the constant of integration later. Substitution in (1) results in

$$\int \overbrace{x}^{u}\,\overbrace{e^x\,dx}^{dv} = \overbrace{x}^{u}\,\overbrace{e^x}^{v} - \int \overbrace{e^x}^{v}\,\overbrace{dx}^{du} = xe^x - e^x + C = e^x(x-1) + C$$

Let's look once more at Example 1. Suppose we had chosen u and dv differently as

$$u = e^x \qquad \text{and} \qquad dv = x\,dx$$

This choice would have resulted in

$$du = e^x\,dx \qquad \text{and} \qquad v = \frac{x^2}{2}$$

and (1) would have yielded

$$\int xe^x\,dx = \frac{x^2}{2}e^x - \int \frac{x^2 e^x}{2}\,dx$$

As you can see, instead of obtaining an integral that is easier to evaluate, we obtain one that is more complicated than the original. This means that an unwise choice of u and dv has been made.

Unfortunately, there are no general directions for choosing u and dv except that:

1. dx is always a part of dv.
2. It must be possible to integrate dv.
3. u and dv are chosen so that $\int v\,du$ is easier to evaluate than the original integral $\int u\,dv$; this often happens when u is simplified by differentiation.

In making an initial choice for u and dv, a certain amount of trial and error is used. If a selection appears to hold little promise, abandon it and try some other choice. If no choices work, it may be that some other technique of integration should be tried.

Let's look at some more examples.

EXAMPLE 2

Derive the formula* $\int \ln x\, dx = x \ln x - x + C$.

Solution

We choose

$$u = \ln x \qquad \text{and} \qquad dv = dx$$

Then

$$du = \frac{1}{x}\, dx \qquad \text{and} \qquad v = \int dx = x$$

By substituting in (1), we get

$$\int \ln x\, dx = x \ln x - \int (x)\left(\frac{1}{x}\, dx\right) = x \ln x - \int dx = x \ln x - x + C$$

EXAMPLE 3

Evaluate: $\int x \sin x\, dx$

Solution

We choose

$$u = x \qquad \text{and} \qquad dv = \sin x\, dx$$

Then

$$du = dx \qquad \text{and} \qquad v = \int \sin x\, dx = -\cos x$$

By substituting in (1), we get

$$\int x \sin x\, dx = -x \cos x + \int \cos x\, dx = -x \cos x + \sin x + C$$

EXAMPLE 4

Show that † $\int \tan^{-1} x\, dx = x \tan^{-1} x - \frac{1}{2} \ln(1 + x^2) + C$.

Solution

We choose

$$u = \tan^{-1} x \qquad \text{and} \qquad dv = dx$$

Then

$$du = \frac{1}{1 + x^2}\, dx \qquad \text{and} \qquad v = \int dx = x$$

---

* This is formula 118 in the Table of Integrals (inside the covers of this book).

† This is formula 108 in the Table of Integrals (inside the covers of this book).

By substituting in (1), we get

$$\int \tan^{-1}x\, dx = x \tan^{-1}x - \int \frac{x}{1+x^2}\, dx$$

To evaluate the integral $\int [x/(1+x^2)]\, dx$, we use the substitution $t = 1 + x^2$. Then $dt = 2x\, dx$, and

$$\int \frac{x}{1+x^2}\, dx = \frac{1}{2}\int \frac{dt}{t} = \frac{1}{2}\ln|t| = \frac{1}{2}\ln(1+x^2)$$

As a result,

$$\int \tan^{-1}x\, dx = x\tan^{-1}x - \tfrac{1}{2}\ln(1+x^2) + C$$

EXAMPLE 5

Evaluate: $\int_1^2 x \ln x\, dx$

Solution

First, we calculate the indefinite integral $\int x \ln x\, dx$ by using the by-parts formula (1). We choose u and dv as

$$u = \ln x \qquad \text{and} \qquad dv = x\, dx$$

Then

$$du = \frac{1}{x}\, dx \qquad \text{and} \qquad v = \int x\, dx = \frac{x^2}{2}$$

Thus,

$$\int x \ln x\, dx = \frac{x^2}{2}\ln x - \int \frac{x^2}{2}\left(\frac{1}{x}\, dx\right) = \frac{x^2}{2}\ln x - \frac{x^2}{4} + C = \frac{x^2}{2}\left(\ln x - \frac{1}{2}\right) + C$$

It follows that

$$\int_1^2 x \ln x\, dx = \frac{x^2}{2}\left(\ln x - \frac{1}{2}\right)\Bigg|_1^2 = 2\left(\ln 2 - \frac{1}{2}\right) - \frac{1}{2}\left(-\frac{1}{2}\right) = 2\ln 2 - \frac{3}{4} \approx 0.6363$$

Sometimes it may be necessary to integrate by parts more than once to solve a particular problem. The next example is a case in point.

EXAMPLE 6

Evaluate: $\int x^2 e^x\, dx$

Solution

We choose

$$u = x^2 \qquad \text{and} \qquad dv = e^x\, dx$$

Then

$$du = 2x\,dx \qquad \text{and} \qquad v = \int e^x\,dx = e^x$$

Thus,

$$\int x^2 e^x\,dx = x^2 e^x - 2\int x e^x\,dx$$

We must still evaluate $\int x e^x\,dx$. In Example 1 we found that

$$\int x e^x\,dx = x e^x - e^x + C$$

As a result,

$$\int x^2 e^x\,dx = x^2 e^x - 2x e^x + 2e^x - 2C = e^x(x^2 - 2x + 2) + C_1$$

Below, we outline a few of the types of integrals that usually are evaluated by using integration by parts in the manner of Examples 1–6. In the list n is always a positive integer.

I. $\int x^n e^{ax}\,dx \qquad \int x^n \cos ax\,dx \qquad \int x^n \sin ax\,dx$

Here, let $u = x^n$ and $dv =$ what remains.

II. $\int x^n \sin^{-1} x\,dx \qquad \int x^n \cos^{-1} x\,dx \qquad \int x^n \tan^{-1} x\,dx$

Here, let $u = \sin^{-1} x$ or $u = \cos^{-1} x$ or $u = \tan^{-1} x$ and $dv = x^n\,dx$.

III. $\int x^m (\ln x)^n\,dx \qquad m \neq -1$

Here, let $u = (\ln x)^n$ and $dv = x^m\,dx$.

The next examples illustrate a use of integration by parts in quite a different manner from that in any example we have seen so far.

EXAMPLE 7

Show that*

$$\int e^{ax} \cos bx\,dx = \frac{e^{ax}(b \sin bx + a \cos bx)}{a^2 + b^2} + C \qquad b \neq 0 \tag{2}$$

Solution

Let

$$u = e^{ax} \qquad \text{and} \qquad dv = \cos bx\,dx$$

* This is formula 126 in the Table of Integrals (inside the covers of this book).

Then

$$du = ae^{ax}\,dx \qquad \text{and} \qquad v = \frac{1}{b}\sin bx$$

Thus,

$$\int e^{ax}\cos bx\,dx = \frac{e^{ax}\sin bx}{b} - \frac{a}{b}\int e^{ax}\sin bx\,dx \tag{3}$$

Now, apply the by-parts formula (1) to $\int e^{ax}\sin bx\,dx$. We choose

$$u = e^{ax} \qquad \text{and} \qquad dv = \sin bx\,dx$$

Then

$$du = ae^{ax}\,dx \qquad \text{and} \qquad v = -\frac{1}{b}\cos bx$$

Thus,

$$\int e^{ax}\sin bx\,dx = -\left(\frac{1}{b}\right)e^{ax}\cos bx + \frac{a}{b}\int e^{ax}\cos bx\,dx \tag{4}$$

From (3) and (4), we get

$$\int e^{ax}\cos bx\,dx = \frac{1}{b}e^{ax}\sin bx - \frac{a}{b}\left[-\frac{1}{b}e^{ax}\cos bx + \frac{a}{b}\int e^{ax}\cos bx\,dx\right]$$

$$= \frac{1}{b}e^{ax}\sin bx + \frac{a}{b^2}e^{ax}\cos bx - \frac{a^2}{b^2}\int e^{ax}\cos bx\,dx$$

Finally, collecting terms, we get

$$\int e^{ax}\cos bx\,dx + \frac{a^2}{b^2}\int e^{ax}\cos bx\,dx = \frac{1}{b}e^{ax}\sin bx + \frac{a}{b^2}e^{ax}\cos bx$$

$$\left(1 + \frac{a^2}{b^2}\right)\int e^{ax}\cos bx\,dx = \frac{1}{b^2}e^{ax}(b\sin bx + a\cos bx)$$

$$\int e^{ax}\cos bx\,dx = \frac{e^{ax}(b\sin bx + a\cos bx)}{a^2 + b^2} + C$$

As an illustration of (2) for $a = 4$ and $b = 5$, we have

$$\int e^{4x}\cos 5x\,dx = \frac{e^{4x}(5\sin 5x + 4\cos 5x)}{41} + C$$

Let's derive another general formula.

EXAMPLE 8

Derive the formula*

$$\int \sec^n x\,dx = \frac{\sec^{n-2}x\tan x}{n-1} + \frac{n-2}{n-1}\int \sec^{n-2}x\,dx \qquad n > 1 \tag{5}$$

* This is formula 94 in the Table of Integrals (inside the covers of this book).

Solution

We write $\sec^n x = \sec^{n-2} x \sec^2 x$, and choose

$$u = \sec^{n-2} x \qquad \text{and} \qquad dv = \sec^2 x\, dx$$

Then,

$$\begin{aligned} du &= (n-2)\sec^{n-3} x \sec x \tan x\, dx \\ &= (n-2)\sec^{n-2} x \tan x\, dx \qquad \text{and} \qquad v = \tan x \end{aligned}$$

Hence

$$\int \sec^n x\, dx = \sec^{n-2} x \tan x - (n-2)\int \sec^{n-2} x \tan^2 x\, dx$$

We replace $\tan^2 x$ by $\sec^2 x - 1$ to get

$$\begin{aligned} \int \sec^n x\, dx &= \sec^{n-2} x \tan x - (n-2)\int \sec^{n-2} x(\sec^2 x - 1)\, dx \\ &= \sec^{n-2} x \tan x - (n-2)\int \sec^n x\, dx + (n-2)\int \sec^{n-2} x\, dx \end{aligned}$$

By transposing the second term on the right to the left, we get

$$(n-1)\int \sec^n x\, dx = \sec^{n-2} x \tan x + (n-2)\int \sec^{n-2} x\, dx$$

Dividing both sides by $n-1$ produces the result:

$$\int \sec^n x\, dx = \frac{\sec^{n-2} x \tan x}{n-1} + \frac{n-2}{n-1}\int \sec^{n-2} x\, dx$$

Formula (5) is referred to as a *reduction formula* because repeated applications eventually lead to an elementary integral. When n is even, repeated applications lead to $\int \sec^2 x\, dx = \tan x + C$. When n is odd, repeated applications lead to $\int \sec x\, dx = \ln|\sec x + \tan x| + C$.

As an illustration of (5), for $n = 3$, we have

$$\int \sec^3 x\, dx = \frac{\sec x \tan x}{2} + \frac{1}{2}\int \sec x\, dx = \frac{\sec x \tan x}{2} + \frac{1}{2}\ln|\sec x + \tan x| + C \qquad (6)$$

EXERCISE 9.2

In Problems 1–40 use integration by parts to evaluate each integral.

1. $\int xe^{2x}\, dx$ **2.** $\int xe^{-3x}\, dx$ **3.** $\int x \cos x\, dx$ **4.** $\int x \sin(3x)\, dx$

5. $\int \sqrt{x} \ln x\, dx$ **6.** $\int x^{-2} \ln x\, dx$ **7.** $\int \cot^{-1} x\, dx$ **8.** $\int \sin^{-1} x\, dx$

9. $\int (\ln x)^2\,dx$

10. $\int x(\ln x)^2\,dx$

11. $\int e^x \sin x\,dx$

12. $\int_0^{\pi} e^x \cos x\,dx$

13. $\int_0^1 x^2 e^{-x}\,dx$

14. $\int x^2 e^x\,dx$

15. $\int x^2 \sin x\,dx$

16. $\int x^2 \cos x\,dx$

17. $\int x \cos^2 x\,dx$

18. $\int x \sin^2 x\,dx$

19. $\int_0^2 x^2 e^{-3x}\,dx$

20. $\int_0^{\pi/4} x \tan^2 x\,dx$

21. $\int x \sinh x\,dx$

22. $\int \sinh^{-1} x\,dx$

23. $\int x^n \ln x\,dx$ $n \neq -1$, n real

24. $\int x^3 e^{x^2}\,dx$ [*Hint:* Let $u = x^2$, $dv = xe^{x^2}\,dx$.]

25. $\int (\ln x)^3\,dx$

26. $\int (\ln x)^4\,dx$

27. $\int xe^x \cos x\,dx$

28. $\int xe^x \sin x\,dx$

29. $\int x^2(\ln x)^2\,dx$

30. $\int x^3(\ln x)^2\,dx$

31. $\int x^2 \tan^{-1} x\,dx$

32. $\int x \tan^{-1} x\,dx$

33. $\int \sin(\ln x)\,dx$

34. $\int \cos(\ln x)\,dx$

35. $\int \cos x \ln(\sin x)\,dx$

36. $\int 7^x x\,dx$

37. $\int_1^9 \ln\sqrt{x}\,dx$

38. $\int x \csc^2 x\,dx$

39. $\int_1^c x^2(\ln x)^2\,dx$, $c > 0$

40. $\int_0^{\pi/2} (x^2 + x\sin x)\,dx$

In Problems 41–44 evaluate each integral by first making a substitution and then integrating by parts.

41. $\int e^x \ln(2 + e^x)\,dx$

42. $\int e^{4x} \cos e^{2x}\,dx$

43. $\int_1^2 \ln(x + 5)\,dx$

44. $\int \cos x \tan^{-1}(\sin x)\,dx$

45. Derive the formula $\int \ln(x + \sqrt{x^2 + a^2})\,dx = x\ln(x + \sqrt{x^2 + a^2}) - \sqrt{x^2 + a^2} + C.$

46. Derive the formula

$$\int e^{ax} \sin bx\,dx = \frac{e^{ax}(a \sin bx - b \cos bx)}{a^2 + b^2} + C.$$

In Problems 47–52 establish each reduction formula where $n > 1$ is an integer.

47. $$\int x^n \sin^{-1} x\,dx = \frac{x^{n+1}}{n+1}\sin^{-1} x - \frac{1}{n+1}\int \frac{x^{n+1}}{\sqrt{1-x^2}}\,dx$$

48. $$\int x^n \tan^{-1} x\,dx = \frac{x^{n+1}}{n+1}\tan^{-1} x - \frac{1}{n+1}\int \frac{x^{n+1}}{1+x^2}\,dx$$

49. $$\int x^n(ax+b)^{1/2}\,dx = \frac{2x^n(ax+b)^{3/2}}{(2n+3)a} - \frac{2bn}{(2n+3)a}\int x^{n-1}(ax+b)^{1/2}\,dx$$

50. $$\int \frac{dx}{(x^2+1)^{n+1}} = \left(1 - \frac{1}{2n}\right)\int \frac{dx}{(x^2+1)^n} + \frac{x}{2n(x^2+1)^n}$$

51. $$\int \sin^n x\,dx = -\frac{\sin^{n-1} x \cos x}{n} + \frac{n-1}{n}\int \sin^{n-2} x\,dx$$

52. $$\int \sin^n x \cos^m x\,dx = -\frac{\sin^{n-1} x \cos^{m+1} x}{n+m} + \frac{n-1}{n+m}\int \sin^{n-2} x \cos^m x\,dx, \quad m \neq -n$$

53. Suppose $F(x) = \int_0^x tg'(t)\,dt$ for all $x \geq 0$. Show that $F(x) = xg(x) - \int_0^x g(t)\,dt$.

54. (a) Determine $\int x^2 e^{5x}\,dx$.
(b) Using integration by parts, derive a general formula for $\int x^n e^{kx}\,dx$, $k \neq 0$, in which the resulting integrand involves x^{n-1}.

55. When integration by parts is used to find $\int e^x \cosh x\,dx$, what happens? Explain. Find $\int e^x \cosh x\,dx$ without using integration by parts.

56. Try two different choices of u and dv for $\int \frac{\sin x}{x}\,dx$.

Neither will work because $\int \frac{\sin x}{x}\,dx$ has no elementary antiderivative! See Section 9.8.

57. Find the area enclosed by $y = \ln x$, the x-axis, and the line $x = e$.

58. Find the area enclosed by $y = x \sin x$, $y = x$, $x = 0$, and $x = \pi/2$.

In Problems 59 and 60 find the area enclosed by the graphs of f and g.

59. $f(x) = 3 \ln x$ and $g(x) = x \ln x$

60. $f(x) = 4x \ln x$ and $g(x) = x^2 \ln x$

61. Find the volume of the solid generated by revolving the area under the graph of $y = \cos x$ from $x = 0$ to $x = \pi/2$ about the y-axis.

62. Find the volume of the solid generated by revolving the area under the graph of $y = \sin x$ from $x = 0$ to $x = \pi/2$ about the y-axis.

63. Find the volume of the solid generated by revolving the area under the graph of $y = x\sqrt{\sin x}$ from $x = 0$ to $x = \pi/2$ about the x-axis.

64. Show that the following formula is true:

$$\int_0^{\pi/2} \sin^n x\,dx = \int_0^{\pi/2} \cos^n x\,dx = \begin{cases} \dfrac{(n-1)(n-3)\cdots(4)(2)}{n(n-2)\cdots(5)(3)(1)} & \text{if } n > 1 \text{ is odd} \\[2ex] \dfrac{(n-1)(n-3)\cdots(5)(3)(1)}{n(n-2)\cdots(4)(2)}\left(\dfrac{\pi}{2}\right) & \text{if } n > 1 \text{ is even} \end{cases} \tag{7}$$

This is called *Wallis' formula.* [*Hint:* Use the result of Problem 51.]

65. Use Wallis' formula (7) to find:

(a) $\int_0^{\pi/2} \sin^6 x\,dx$ (b) $\int_0^{\pi/2} \sin^5 x\,dx$

(c) $\int_0^{\pi/2} \cos^8 x\,dx$ (d) $\int_0^{\pi/2} \cos^6 x\,dx$

9.3 Trigonometric Integrals

In this section we consider the problem of evaluating certain trigonometric integrals. In studying the techniques here, you should follow the procedures outlined in the examples rather than try to memorize the results.

We confine our discussion to four commonly encountered types of trigonometric integrals:

I. $\int \sin^n x\,dx,\quad \int \cos^n x\,dx$

II. $\int \sin^m x \cos^n x\,dx$

III. $\int \tan^m x \sec^n x\,dx,\quad \int \cot^m x \csc^n x\,dx$

IV. $\int \sin mx \sin nx\,dx,\quad \int \sin mx \cos nx\,dx,\quad \int \cos mx \cos nx\,dx$

Type I

$$\int \sin^n x \, dx \qquad \int \cos^n x \, dx \qquad n \text{ a positive integer}$$

Although we could use integration by parts to obtain a reduction formula for each of these integrals (see Problem 51 in Exercise 9.2), they can also be evaluated in another—usually easier—way.

We consider two cases: (a) n is an odd positive integer, and (b) n is an even positive integer.

(a) When n is an odd positive integer, we begin by writing

$$\int \sin^n x \, dx = \int \sin^{n-1} x \sin x \, dx$$

Since n is odd, $(n - 1)$ is even. Using the fact that $\sin^2 x = 1 - \cos^2 x$, we obtain an integral that can be evaluated by using the substitution $u = \cos x$. The following example illustrates this technique.

EXAMPLE 1

Evaluate: $\int \sin^5 x \, dx$

Solution

$$\int \sin^5 x \, dx = \int \sin^4 x \sin x \, dx = \int (1 - \cos^2 x)^2 \sin x \, dx$$

$$= \int (1 - 2\cos^2 x + \cos^4 x) \sin x \, dx$$

Use the substitution $u = \cos x$. Then $du = -\sin x \, dx$, so that

$$\int \sin^5 x \, dx = -\int (1 - 2u^2 + u^4) \, du = -u + \tfrac{2}{3}u^3 - \tfrac{1}{5}u^5 + C$$

$$= -\cos x + \tfrac{2}{3}\cos^3 x - \tfrac{1}{5}\cos^5 x + C$$

A similar technique may be used to evaluate $\int \cos^n x \, dx$ when n is an odd positive integer. In this case, we write

$$\int \cos^n x \, dx = \int \cos^{n-1} x \cos x \, dx$$

and use the fact that $\cos^2 x = 1 - \sin^2 x$, along with the substitution $u = \sin x$.

(b) To evaluate $\int \sin^n x \, dx$ when n is an even positive integer, we use the identities (half-angle formulas)

$$\sin^2 x = \frac{1 - \cos 2x}{2} \qquad \text{and} \qquad \cos^2 x = \frac{1 + \cos 2x}{2}$$

to obtain an integrand that may be simpler.

EXAMPLE 2

Evaluate: $\int \sin^2 x\, dx$

Solution

$$\int \sin^2 x\, dx = \tfrac{1}{2}\int (1 - \cos 2x)\, dx = \tfrac{1}{2}x - \tfrac{1}{2}\int \cos 2x\, dx = \tfrac{1}{2}x - \tfrac{1}{4}\sin 2x + C$$

EXAMPLE 3

Evaluate: $\int \cos^4 x\, dx$

Solution

$$\begin{aligned}\int \cos^4 x\, dx &= \int (\cos^2 x)^2\, dx = \int \tfrac{1}{4}(1 + \cos 2x)^2\, dx \\ &= \tfrac{1}{4}\int (1 + 2\cos 2x + \cos^2 2x)\, dx\end{aligned}$$

You should be able to supply the necessary details to show that

$$\int \cos^4 x\, dx = \tfrac{3}{8}x + \tfrac{1}{4}\sin 2x + \tfrac{1}{32}\sin 4x + C$$

Type II

$$\int \sin^m x \cos^n x\, dx$$

Again, we consider two cases: (a) at least one of the exponents m or n is an odd positive integer, and (b) both are even positive integers.

Integrals of Type II may be evaluated by using variations of previous techniques.

EXAMPLE 4

Evaluate: $\int \cos^{1/5} x \sin^5 x\, dx$

Solution

Here, $\sin x$ is raised to an odd power and $\cos x$ to the number $\frac{1}{5}$. We write

$$\int \cos^{1/5} x \sin^5 x\, dx = \int \cos^{1/5} x \sin^4 x \sin x\, dx$$

But

$$\sin^4 x = (\sin^2 x)^2 = (1 - \cos^2 x)^2$$

This leads us to make the substitution $u = \cos x$. Then $du = -\sin x\, dx$, and

$$\int \cos^{1/5}x \sin^5 x\, dx = \int \cos^{1/5}x(1 - \cos^2 x)^2 \sin x\, dx$$

$$= \int u^{1/5}(1 - u^2)^2(-du) = -\int (u^{1/5} - 2u^{11/5} + u^{21/5})\, du$$

$$= -\tfrac{5}{6}u^{6/5} + \tfrac{5}{8}u^{16/5} - \tfrac{5}{26}u^{26/5} + C$$

$$= -\tfrac{5}{6}\cos^{6/5}x + \tfrac{5}{8}\cos^{16/5}x - \tfrac{5}{26}\cos^{26/5}x + C$$

Note that if either m or n is an odd positive integer, the other exponent may be *any real number*.

If m and n are both even positive integers, then we may use the identity $\sin^2 x + \cos^2 x = 1$ to obtain a sum of integrals, each one of which involves only powers of either $\sin x$ or $\cos x$. For example,

$$\int \sin^2 x \cos^4 x\, dx = \int (1 - \cos^2 x)\cos^4 x\, dx = \int \cos^4 x\, dx - \int \cos^6 x\, dx$$

The two integrals on the right are now of Type I.

Table 1 summarizes the methods for evaluating $\int \sin^m x \cos^n x\, dx$.

Table 1

Case	Method	Useful Identities
n odd	Substitute $u = \sin x$	$\cos^2 x = 1 - \sin^2 x$
m odd	Substitute $u = \cos x$	$\sin^2 x = 1 - \cos^2 x$
m and n even	Reduce to smaller powers of m or n	$\sin x \cos x = \frac{1}{2}\sin 2x$
		$\sin^2 x = \dfrac{1 - \cos 2x}{2}$
		$\cos^2 x = \dfrac{1 + \cos 2x}{2}$

Type III

$$\int \tan^m x \sec^n x\, dx \qquad \int \cot^m x \csc^n x\, dx$$

This time, we consider three cases: (a) m is an odd positive integer, (b) n is an even positive integer, and (c) m is an even positive integer and n is an odd positive integer.

(a) If m is an odd positive integer, we write $\int \tan^m x \sec^n x\, dx$ as

$$\int \tan^{m-1}x \sec^{n-1}x \sec x \tan x\, dx$$

Since $(m-1)$ is even, we can use the identity $\tan^2 x = \sec^2 x - 1$ to express $\tan^{m-1} x$ in terms of $\sec x$. The substitution $u = \sec x$ will then lead to a simpler integral, since $du = \sec x \tan x\, dx$.

EXAMPLE 5

Evaluate: $\int \tan^3 x \sec^3 x\, dx$

Solution

Here $m = 3$ is odd and

$$\int \tan^3 x \sec^3 x\, dx = \int \tan^2 x \sec^2 x(\sec x \tan x\, dx)$$

$$= \int (\sec^2 x - 1)\sec^2 x(\sec x \tan x\, dx)$$

Let $u = \sec x$, so that $du = \sec x \tan x\, dx$. Then

$$\int \tan^3 x \sec^3 x\, dx = \int (u^2 - 1)u^2\, du = \int (u^4 - u^2)\, du$$

$$= \frac{u^5}{5} - \frac{u^3}{3} + C = \frac{\sec^5 x}{5} - \frac{\sec^3 x}{3} + C$$

(b) If n is an even positive integer, we write $\int \tan^m x \sec^n x\, dx$ as

$$\int \tan^m x \sec^{n-2} x \sec^2 x\, dx$$

We now express $\sec^{n-2} x$ in terms of $\tan x$ by using the identity $\sec^2 x = 1 + \tan^2 x$. The substitution $u = \tan x$ will lead to a simpler integral. (Do you see why?)

EXAMPLE 6

Evaluate: $\int \tan^2 x \sec^4 x\, dx$

Solution

$$\int \tan^2 x \sec^4 x\, dx = \int \tan^2 x \sec^2 x(\sec^2 x)\, dx = \int \tan^2 x(1 + \tan^2 x)\sec^2 x\, dx$$

Let $u = \tan x$, so that $du = \sec^2 x\, dx$. Then

$$\int \tan^2 x \sec^4 x\, dx = \int u^2(1 + u^2)\, du = \int (u^2 + u^4)\, du$$

$$= \frac{u^3}{3} + \frac{u^5}{5} + C = \frac{\tan^3 x}{3} + \frac{\tan^5 x}{5} + C$$

(c) When m is an even positive integer and n is an odd positive integer, we express the integrand of $\int \tan^m x \sec^n x\, dx$ in terms of $\sec x$ by using the identity $\tan^2 x = \sec^2 x - 1$ and then use the technique of integration by parts.

EXAMPLE 7

Evaluate: $\int \tan^2 x \sec x \, dx$

Solution

Here

$$\int \tan^2 x \sec x \, dx = \int (\sec^2 x - 1)\sec x \, dx = \int \sec^3 x \, dx - \int \sec x \, dx$$

By using (6) in Section 9.2 for $\int \sec^3 x \, dx$, we have

$$\int \tan^2 x \sec x \, dx = \tfrac{1}{2} \sec x \tan x - \tfrac{1}{2} \ln|\sec x + \tan x| + C$$

To evaluate integrals of the form $\int \cot^m x \csc^n x \, dx$, we follow the same procedures, except that the identity $\csc^2 x = 1 + \cot^2 x$ is used.

Table 2 summarizes the methods for evaluating $\int \tan^m x \sec^n x \, dx$ $(m \geq 0, \quad n \geq 0)$.

Table 2

Case	Method	Useful Identities
m odd	Substitute $u = \sec x$	$\tan^2 x = \sec^2 x - 1$
n even	Substitute $u = \tan x$	$\sec^2 x = 1 + \tan^2 x$
m even, n odd	Reduce to powers of $\sec x$ alone	$\tan^2 x = \sec^2 x - 1$

Type IV

$$\int \sin mx \sin nx \, dx \qquad \int \sin mx \cos nx \, dx \qquad \int \cos mx \cos nx \, dx$$

Trigonometric integrals of Type IV are handled with the aid of the following trigonometric identities (product-to-sum formulas):

$$2 \sin A \sin B = \cos(A - B) - \cos(A + B)$$

$$2 \sin A \cos B = \sin(A + B) + \sin(A - B)$$

$$2 \cos A \cos B = \cos(A - B) + \cos(A + B)$$

EXAMPLE 8

Evaluate: $\int \sin 3x \sin 2x \, dx$

Solution

We set $A = 3x$ and $B = 2x$. Then

$$\int \sin 3x \sin 2x \, dx = \tfrac{1}{2} \int (\cos x - \cos 5x) \, dx = \tfrac{1}{2} \sin x - \tfrac{1}{10} \sin 5x + C$$

EXERCISE 9.3

In Problems 1–38 evaluate each integral.

1. $\int \sin^2 x \cos x\, dx$

2. $\int \sin^3 x \cos x\, dx$

3. $\int \frac{\sin x\, dx}{\cos^2 x}$

4. $\int \frac{\cos x\, dx}{\sin^4 x}$

5. $\int \sin^3 x\, dx$

6. $\int \cos^3 x\, dx$

7. $\int \sin^2 3x\, dx$

8. $\int \sin^4 x\, dx$

9. $\int \cos^5 x\, dx$

10. $\int \cos^7 x\, dx$

11. $\int \sin^5 3x\, dx$

12. $\int \cos^3 3x\, dx$

13. $\int \sin^3 x \cos^2 x\, dx$

14. $\int \sin^4 x \cos^3 x\, dx$

15. $\int \sin^3 x \cos^5 x\, dx$

16. $\int \sin^3 x \cos^3 x\, dx$

17. $\int \sin^{1/3} x \cos x\, dx$

18. $\int \frac{\sin^2 x}{\cos^3 x}\, dx$

19. $\int \sin^{1/2} x \cos^3 x\, dx$

20. $\int \sin^2 x \cos^4 x\, dx$

21. $\int \sin^2 x \cos^2 x\, dx$

22. $\int \sin^2 3x \cos^2 3x\, dx$

23. $\int \tan^3 x\, dx$

24. $\int \tan^3 x \sec^2 x\, dx$

25. $\int \tan^2 x \sec^2 x\, dx$

26. $\int \tan^4 x \sec^2 x\, dx$

27. $\int \csc^3 x \cot^5 x\, dx$

28. $\int \cot 2x \csc^4 2x\, dx$

29. $\int \tan^2 x \sec^3 x\, dx$

30. $\int \tan^4 x \sec^4 x\, dx$

31. $\int \sin 3x \cos x\, dx$

32. $\int \sin x \cos 3x\, dx$

33. $\int \sin 2x \sin 4x\, dx$

34. $\int \cos x \cos 3x\, dx$

35. $\int \cos 2x \cos x\, dx$

36. $\int \sin 2x \sin x\, dx$

37. $\int \sin \frac{x}{2} \cos \frac{3x}{2}\, dx$

38. $\int \sin \frac{x}{2} \sin \frac{3x}{2}\, dx$

39. Evaluate $\int \sec^n x\, dx$, n a positive even integer, by the use of the identity

$$\sec^n x = \sec^{n-2} x \sec^2 x = (\tan^2 x + 1)^{(n-2)/2} \sec^2 x$$

40. Evaluate $\int \cot^5 x \csc^2 x\, dx$.

41. Find the volume of the solid generated by revolving the area under the graph of $y = \sin x$ from $x = 0$ to $x = \pi$ about the x-axis.

42. Find the volume of the solid generated by revolving the area enclosed by $y = \cos x$, $y = \sin x$, $x = 0$, and $x = \pi/4$ about the x-axis.

43. The acceleration a of a particle at time t is given by $a = \cos^2 t \sin t$ meters per second per second. At $t = 0$ the particle is at the origin and its velocity is 5 meters per second. Find its position at any time t.

9.4 Integration by Trigonometric Substitution: Integrands Containing $\sqrt{a^2 - x^2}$, $\sqrt{a^2 + x^2}$, or $\sqrt{x^2 - a^2}$

When an integrand contains a square root of one of the forms $\sqrt{a^2 - x^2}$, $\sqrt{a^2 + x^2}$, or $\sqrt{x^2 - a^2}$, where $a > 0$, an appropriate trigonometric substitution will eliminate the radical and sometimes transform the integrand into a trigonometric integral like those studied in the previous section. The three cases and the suggested substitutions are listed below.

	Integrand	***Substitution***	***Derived from the Identity***
Case 1:	$\sqrt{a^2 - x^2}$	$x = a \sin \theta$	$1 - \sin^2\theta = \cos^2\theta$
Case 2:	$\sqrt{a^2 + x^2}$	$x = a \tan \theta$	$1 + \tan^2\theta = \sec^2\theta$
Case 3:	$\sqrt{x^2 - a^2}$	$x = a \sec \theta$	$\sec^2\theta - 1 = \tan^2\theta$

The substitutions can be memorized, but you will probably find it easier to draw a right triangle and derive them as needed. Each substitution derives from the Pythagorean theorem. If we place the sides a and x on the triangle appropriately, we can make the third side of the triangle represent any of the cases above. For example, the substitution $x = a \sin \theta$ is used in $\sqrt{a^2 - x^2}$, since it eliminates the radical (see Fig. 1). That is,

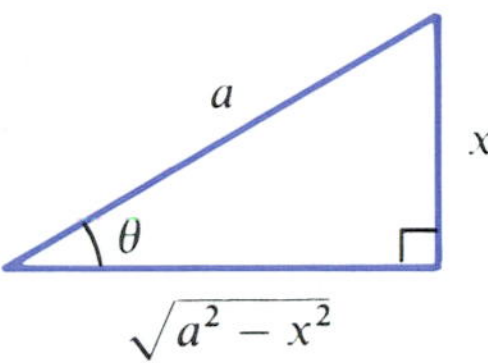

Figure 1

$$\begin{aligned}\sqrt{a^2 - x^2} &= \sqrt{a^2 - a^2\sin^2\theta}\\ &= a\sqrt{1 - \sin^2\theta}\\ &= a\sqrt{\cos^2\theta}\\ &= a|\cos \theta|\end{aligned}$$

Since $x = a \sin \theta$, we have $x/a = \sin \theta$, so that $\theta = \sin^{-1}(x/a)$. From (1) in Section 8.2, we have $-\pi/2 \le \theta \le \pi/2$. With the restriction $-\pi/2 \le \theta \le \pi/2$, $\cos \theta \ge 0$ and, thus, $\sqrt{a^2 - x^2} = a \cos \theta$. Now, if $\sqrt{a^2 - x^2}$ occurs in the denominator of the integrand, we let $x = a \sin \theta$, with the restriction $-\pi/2 < \theta < \pi/2$.

Let's look at an example.

EXAMPLE 1

Evaluate: $\displaystyle\int \frac{dx}{x^2\sqrt{4 - x^2}}$

Solution

Let $x = 2 \sin \theta$, $-\pi/2 < \theta < \pi/2$. Then $dx = 2 \cos \theta \, d\theta$, and

$$\sqrt{4 - x^2} = \sqrt{4 - 4\sin^2\theta} = 2\sqrt{\cos^2\theta} = 2 \cos \theta$$

Therefore,

$$\int \frac{dx}{x^2\sqrt{4 - x^2}} = \int \frac{2 \cos \theta \, d\theta}{(4 \sin^2\theta)(2 \cos \theta)} = \frac{1}{4}\int \csc^2\theta \, d\theta = -\frac{1}{4}\cot \theta + C$$

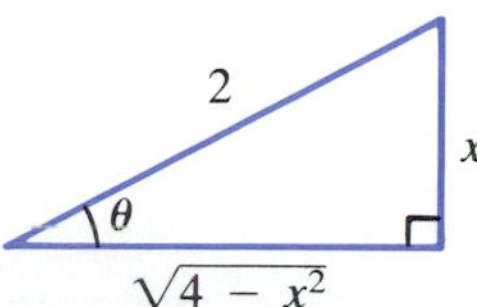

Figure 2

The integral has now been found in terms of θ. In order to express $\cot \theta$ in terms of x, we draw a right triangle, as shown in Figure 2, and use the fact that $\sin \theta = x/2$, $-\pi/2 < \theta < \pi/2$. Using the Pythagorean theorem, we conclude that $\cot \theta = (\sqrt{4 - x^2})/x$. Thus,

$$\int \frac{dx}{x^2\sqrt{4 - x^2}} = -\frac{1}{4}\cot \theta + C = -\frac{1}{4}\frac{\sqrt{4 - x^2}}{x} + C$$

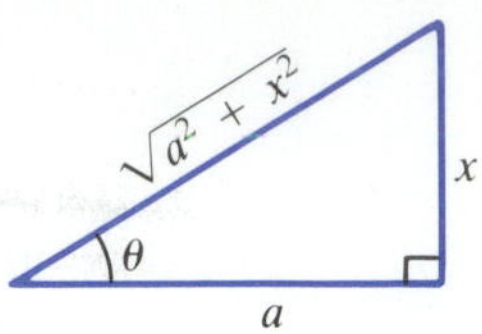

Figure 3

For integrands involving $\sqrt{a^2 + x^2}$, $a > 0$, we use the substitution $x = a \tan \theta$, $-\pi/2 < \theta < \pi/2$ (see Fig. 3). Since $\sec \theta > 0$ when $-\pi/2 < \theta < \pi/2$, we have

$$\sqrt{a^2 + x^2} = \sqrt{a^2 + a^2\tan^2\theta} = a\sqrt{1 + \tan^2\theta} = a\sqrt{\sec^2\theta} = a \sec \theta$$

EXAMPLE 2

Evaluate: $\displaystyle\int \frac{dx}{(x^2 + 9)^{3/2}}$

Solution

The given integrand involves a power of $\sqrt{x^2 + 9}$. Hence, we use the substitution $x = 3 \tan \theta$, $-\pi/2 < \theta < \pi/2$. Then $dx = 3 \sec^2\theta \, d\theta$, and

$$\int \frac{dx}{(x^2 + 9)^{3/2}} = \int \frac{3 \sec^2\theta \, d\theta}{(9 \tan^2\theta + 9)^{3/2}} = \int \frac{3 \sec^2\theta \, d\theta}{27(1 + \tan^2\theta)^{3/2}} = \frac{1}{9}\int \frac{\sec^2\theta \, d\theta}{\sec^3\theta}$$

$$= \frac{1}{9}\int \cos \theta \, d\theta = \frac{1}{9} \sin \theta + C$$

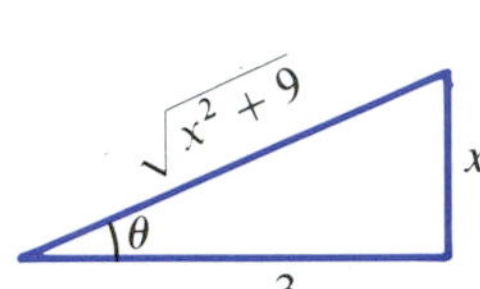

Figure 4

Since $\tan \theta = x/3$, $-\pi/2 < \theta < \pi/2$, from Figure 4 we see that $\sin \theta = x/\sqrt{x^2 + 9}$. Thus,

$$\int \frac{dx}{(x^2 + 9)^{3/2}} = \frac{x}{9\sqrt{x^2 + 9}} + C$$

In evaluating integrals involving $\sqrt{a^2 + x^2}$, $a > 0$, we may also use the substitution $x = a \sinh \theta$. Then $dx = a \cosh \theta \, d\theta$, and

$$\sqrt{a^2 + x^2} = \sqrt{a^2 + a^2\sinh^2\theta} = a\sqrt{1 + \sinh^2\theta} = a\sqrt{\cosh^2\theta} = a \cosh \theta$$

For example,

$$\int \frac{dx}{\sqrt{x^2 + a^2}} = \int \frac{a \cosh \theta}{a \cosh \theta} \, d\theta = \int d\theta = \theta + C = \sinh^{-1}\left(\frac{x}{a}\right) + C$$

$$= \ln\left(\frac{x}{a} + \frac{\sqrt{x^2 + a^2}}{a}\right) + C$$

↑ By (3) in Section 8.6

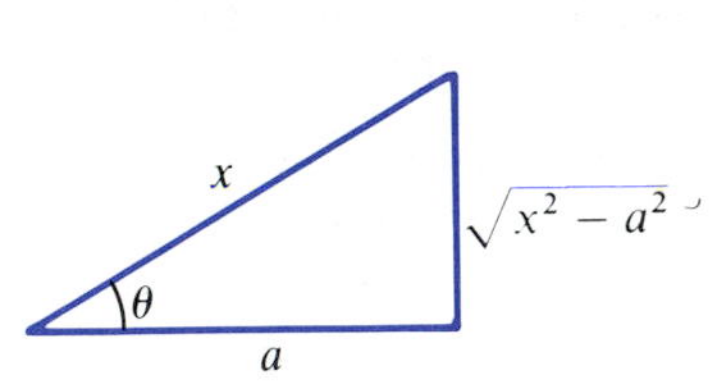

Figure 5

For integrands involving $\sqrt{x^2 - a^2}$, $a > 0$, we may use the substitution $x = a \sec \theta$, $0 < \theta < \pi/2$ or $\pi < \theta < 3\pi/2$.* Since $\tan \theta > 0$ for $0 < \theta < \pi/2$ or $\pi < \theta < 3\pi/2$, we have (see Fig. 5)

$$\sqrt{x^2 - a^2} = \sqrt{a^2(\sec^2\theta - 1)} = a\sqrt{\tan^2\theta} = a \tan \theta$$

EXAMPLE 3

Evaluate: $\displaystyle\int \frac{\sqrt{x^2 - 4}}{x} \, dx$

* We may also use the substitution $x = a \cosh \theta$.

Solution

Let $x = 2 \sec \theta$. Then $dx = 2 \sec \theta \tan \theta \, d\theta$, and

$$\sqrt{x^2 - 4} = \sqrt{4 \sec^2\theta - 4} = 2\sqrt{\sec^2\theta - 1} = 2 \tan \theta$$

Therefore,

$$\int \frac{\sqrt{x^2 - 4}}{x} \, dx = \int \frac{(2 \tan \theta)(2 \sec \theta \tan \theta \, d\theta)}{2 \sec \theta} = 2 \int \tan^2\theta \, d\theta$$

$$= 2 \int (\sec^2\theta - 1) \, d\theta = 2(\tan \theta - \theta) + C$$

$$= \sqrt{x^2 - 4} - 2 \sec^{-1}\left(\frac{x}{2}\right) + C$$

↑ Use Figure 6.

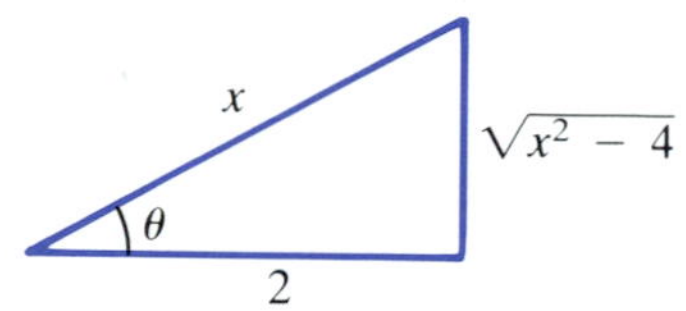

Figure 6

It may be required to complete the square of the expression under the radical before the trigonometric substitution is made. For example, if the integrand contains a square root of the form $\sqrt{x^2 + 2x - 3}$, rewrite it as $\sqrt{(x + 1)^2 - 4}$ and use the substitution $x + 1 = 2 \sec \theta$.

EXAMPLE 4

Find the area enclosed by the ellipse: $\dfrac{x^2}{4} + \dfrac{y^2}{9} = 1$

Solution

The total area (as Fig. 7 illustrates) is four times the area in the first quadrant—namely, $\int_0^2 y \, dx$, where

$$y = 3\sqrt{1 - \frac{x^2}{4}} = \left(\frac{3}{2}\right)\sqrt{4 - x^2}$$

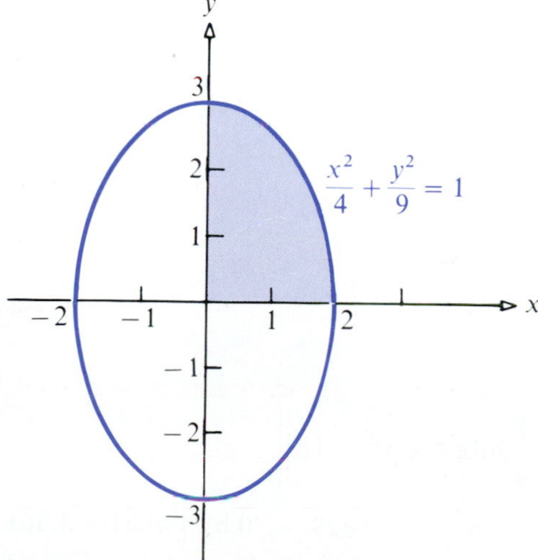

Figure 7

Hence, the total area is

$$A = (4)\left(\frac{3}{2}\right)\int_0^2 \sqrt{4 - x^2}\,dx$$

We make the substitution $x = 2\sin\theta$. Now, when $x = 0$, $2\sin\theta = 0$ and $\theta = 0$; when $x = 2$, $2\sin\theta = 2$ and so $\sin\theta = 1$ and $\theta = \pi/2$. Thus,

$$\begin{aligned}
A &= 6\int_0^2 \sqrt{4 - x^2}\,dx \\
&= 6\int_0^{\pi/2} \sqrt{4 - 4\sin^2\theta}\;2\cos\theta\,d\theta \\
&= 24\int_0^{\pi/2} \cos^2\theta\,d\theta = \frac{24}{2}\int_0^{\pi/2} (1 + \cos 2\theta)\,d\theta \\
&= 12\left(\theta + \frac{\sin 2\theta}{2}\right)\Big|_0^{\pi/2} = 12\left(\frac{\pi}{2} + 0\right) - 12(0) = 6\pi
\end{aligned}$$

In Example 4 the new limits of integration were easy to obtain. This is not always the case.

EXAMPLE 5

Evaluate: $\int_1^3 \sqrt{x^2 - 1}\,dx$

Solution

Here we use the substitution $x = \sec\theta$. Then $dx = \sec\theta\tan\theta\,d\theta$. Since the upper limit $x = 3$ does not result in a nice angle $(\theta = \sec^{-1}3)$, we evaluate the indefinite integral and do not change the limits of integration. Then

$$\begin{aligned}
\int \sqrt{x^2 - 1}\,dx &= \int \tan\theta\sec\theta\tan\theta\,d\theta \\
&= \int \tan^2\theta\sec\theta\,d\theta = \tfrac{1}{2}\sec\theta\tan\theta - \tfrac{1}{2}\ln|\sec\theta + \tan\theta| + C
\end{aligned}$$

See Example 7, Section 9.3.

As Figure 8 illustrates,

$$\sec\theta = x \qquad \tan\theta = \sqrt{x^2 - 1}$$

Hence

$$\begin{aligned}
\int_1^3 \sqrt{x^2 - 1}\,dx &= \left[\left(\frac{x}{2}\right)\sqrt{x^2 - 1} - \frac{1}{2}\ln|x + \sqrt{x^2 - 1}|\right]\Bigg|_1^3 \\
&= \tfrac{3}{2}\sqrt{8} - \tfrac{1}{2}\ln|3 + \sqrt{8}| \approx (1.5)(2.828) - (0.5)(1.763) = 3.361
\end{aligned}$$

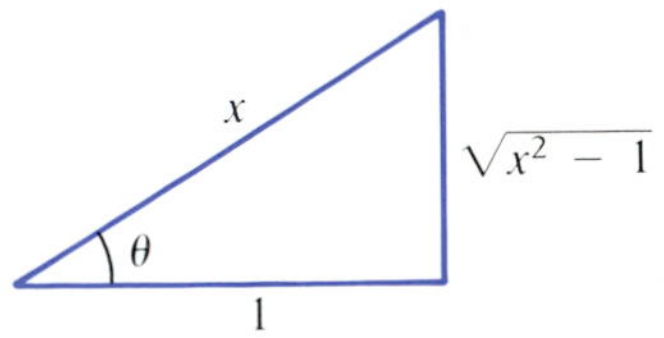

Figure 8

EXERCISE 9.4

In Problems 1–32 evaluate each integral.

1. $\int \frac{dx}{\sqrt{4-x^2}}$
2. $\int x\sqrt{x^2-4}\,dx$
3. $\int \frac{\sqrt{9-x^2}}{x^2}\,dx$
4. $\int \frac{x^2}{\sqrt{9-x^2}}\,dx$
5. $\int \frac{x^2}{\sqrt{x^2-1}}\,dx$
6. $\int \frac{\sqrt{x^2-1}}{x^2}\,dx$
7. $\int \frac{dx}{(x^2+4)^{3/2}}$
8. $\int \frac{x^3}{(x^2+4)^{1/2}}\,dx$
9. $\int \frac{dx}{x^2\sqrt{x^2+4}}$
10. $\int \sqrt{4-x^2}\,dx$
11. $\int \sqrt{16-x^2}\,dx$
12. $\int x^2\sqrt{4-x^2}\,dx$
13. $\int \frac{dx}{(4-x^2)^{3/2}}$
14. $\int \frac{x^2\,dx}{(x^2+9)^{3/2}}$
15. $\int \frac{x^2}{\sqrt{16-x^2}}\,dx$
16. $\int \frac{dx}{x^2\sqrt{x^2-9}}$
17. $\int \frac{\sqrt{x^2-1}}{x}\,dx$
18. $\int \frac{\sqrt{x^2+1}}{x^2}\,dx$
19. $\int \frac{dx}{(x^2-9)^{3/2}}$
20. $\int \frac{x^2+9}{x^6}\,dx$
21. $\int \frac{dx}{\sqrt{1-(x-2)^2}}$
22. $\int \sqrt{4-(x+2)^2}\,dx$
23. $\int \frac{dx}{\sqrt{(x-1)^2-4}}$
24. $\int \frac{dx}{(x-2)\sqrt{(x-2)^2+9}}$
25. $\int \frac{dx}{(x^2-2x+10)^{3/2}}$
26. $\int \frac{dx}{\sqrt{x^2-2x+10}}$
27. $\int \frac{dx}{\sqrt{x^2+2x-3}}$
28. $\int \sqrt{5+4x-x^2}\,dx$
29. $\int_0^2 \frac{x^2\,dx}{(16-x^2)^{3/2}}$
30. $\int_1^2 \frac{dx}{x^2\sqrt{1+9x^2}}$
31. $\int_1^2 \frac{x^3\,dx}{\sqrt{1+4x^2}}$
32. $\int_0^3 \frac{x^2\,dx}{9+x^2}$

33. Evaluate $\int \sqrt{a^2-x^2}\,dx$ and use it to find the area enclosed by the ellipse $\frac{x^2}{a^2}+\frac{y^2}{b^2}=1$.

34. Find the area enclosed by the hyperbola $\frac{x^2}{9}-\frac{y^2}{16}=1$ and the line $x=6$.

35. Derive the formula:

$$\int \frac{dx}{\sqrt{x^2-a^2}} = \ln\left|\frac{x}{a}+\frac{\sqrt{x^2-a^2}}{a}\right| + C$$

36. Evaluate: $\int \sqrt{x^2+a^2}\,dx$
   (a) By using a trigonometric substitution
   (b) By using a hyperbolic substitution

37. Derive the formula for the area of a circle.

38. Find the area under the graph of $y = x^3/\sqrt{9-x^2}$ from $x=0$ to $x=2$.

39. Find the length of the arch of the parabola $y=5x-x^2$ above the x-axis.

40. Find the volume of the solid generated by revolving the area under the graph of $y=1/(x^2+4)$ from $x=0$ to $x=1$ about the x-axis.

In Problems 41 and 42 use integration by parts and then the method of this section to evaluate each integral.

41. $\int x\sin^{-1}x\,dx$
42. $\int x\cos^{-1}x\,dx$

In Problems 43–47 use a trigonometric substitution to derive each formula.

43. $\int \frac{dx}{\sqrt{a^2-x^2}} = \sin^{-1}\left(\frac{x}{a}\right) + C$
44. $\int \frac{dx}{a^2+x^2} = \frac{1}{a}\tan^{-1}\left(\frac{x}{a}\right) + C$

45. $\displaystyle\int \frac{dx}{x\sqrt{x^2-a^2}} = \frac{1}{a}\sec^{-1}\left(\frac{x}{a}\right) + C$

46. $\displaystyle\int \sqrt{x^2-a^2}\,dx = \frac{1}{2}x\sqrt{x^2-a^2} - \frac{1}{2}a^2\ln|x+\sqrt{x^2-a^2}| + C$

47. $\displaystyle\int \frac{dx}{\sqrt{x^2+a^2}} = \ln(x+\sqrt{x^2+a^2}) + C$

48. Find $\int dx/\sqrt{x^2+a^2}$, $a > 0$, by using the substitution $u = \sinh^{-1}(x/a)$. Express your answer in logarithmic form.

9.5 Integrals Involving $ax^2 + bx + c$

Integrals involving $ax^2 + bx + c$, where $a \neq 0$, b and c are real numbers, can often be handled by the technique of completing the square:

$$ax^2 + bx + c = a\left(x^2 + \frac{b}{a}x + \qquad\right) + c$$
$$= a\left(x^2 + \frac{b}{a}x + \frac{b^2}{4a^2}\right) + c - \frac{b^2}{4a} = a\left(x + \frac{b}{2a}\right)^2 + \left(c - \frac{b^2}{4a}\right)$$

The substitution

$$u = x + \frac{b}{2a}$$

reduces the original expression $ax^2 + bx + c$ to the simpler form $au^2 + r$, where $r = c - (b^2/4a)$.

EXAMPLE 1

Evaluate: $\displaystyle\int \frac{dx}{x^2 + 6x + 10}$

Solution

We complete the square by writing

$$x^2 + 6x + 10 = (x^2 + 6x + 9) + 1 = (x+3)^2 + 1$$

Now we can use the substitution $u = x + 3$. Then $du = dx$, and

$$\int \frac{dx}{x^2+6x+10} = \int \frac{dx}{(x+3)^2+1} = \int \frac{du}{u^2+1} = \tan^{-1}u + C = \tan^{-1}(x+3) + C$$

Formula 11

EXAMPLE 2

Evaluate: $\displaystyle\int \frac{(x-1)\,dx}{x^2 + x + 1}$

Solution

Complete the square:

$$x^2 + x + 1 = (x^2 + x + \tfrac{1}{4}) + 1 - \tfrac{1}{4} = (x + \tfrac{1}{2})^2 + \tfrac{3}{4}$$

Now let $u = x + \frac{1}{2}$. Then $du = dx$, $x = u - \frac{1}{2}$, and

$$\int \frac{(x-1)\,dx}{x^2+x+1} = \int \frac{(x-1)\,dx}{(x+\frac{1}{2})^2+\frac{3}{4}} = \int \frac{(u-\frac{3}{2})\,du}{u^2+\frac{3}{4}} = \int \frac{u\,du}{u^2+\frac{3}{4}} - \frac{3}{2}\int \frac{du}{u^2+\frac{3}{4}}$$

$$= \frac{1}{2}\ln\left(u^2+\frac{3}{4}\right) - \sqrt{3}\tan^{-1}\left(\frac{2u}{\sqrt{3}}\right) + C$$

Formulas 2 and 18

$$= \frac{1}{2}\ln(x^2+x+1) - \sqrt{3}\tan^{-1}\left(\frac{2x+1}{\sqrt{3}}\right) + C$$

EXAMPLE 3

Evaluate: $\displaystyle\int \frac{dx}{\sqrt{2x-x^2}}$

Solution

We complete the square by writing

$$-x^2 + 2x = -(x^2 - 2x + 1) + 1 = 1 - (x-1)^2$$

We let $u = x - 1$. Then $du = dx$ and

$$\int \frac{dx}{\sqrt{2x-x^2}} = \int \frac{dx}{\sqrt{1-(x-1)^2}} = \int \frac{du}{\sqrt{1-u^2}} = \sin^{-1}u + C = \sin^{-1}(x-1) + C$$

Formula 10

EXERCISE 9.5

In Problems 1–16 evaluate each integral.

1. $\displaystyle\int \frac{dx}{x^2+4x+5}$
2. $\displaystyle\int \frac{dx}{x^2+2x+5}$
3. $\displaystyle\int \frac{dx}{\sqrt{8+2x-x^2}}$
4. $\displaystyle\int \frac{dx}{\sqrt{5-4x-2x^2}}$
5. $\displaystyle\int \frac{dx}{\sqrt{4x-x^2}}$
6. $\displaystyle\int \frac{dx}{\sqrt{x^2-6x+10}}$
7. $\displaystyle\int \frac{dx}{\sqrt{24-2x-x^2}}$
8. $\displaystyle\int \frac{dx}{\sqrt{9x^2+6x+10}}$
9. $\displaystyle\int \frac{x\,dx}{\sqrt{x^2-2x+5}}$
10. $\displaystyle\int \frac{(x+1)\,dx}{x^2-4x+3}$
11. $\displaystyle\int_1^3 \frac{dx}{\sqrt{x^2-2x+5}}$
12. $\displaystyle\int_{1/2}^1 \frac{x^2\,dx}{\sqrt{2x-x^2}}$
13. $\displaystyle\int \frac{e^x\,dx}{\sqrt{e^{2x}+e^x+1}}$
14. $\displaystyle\int \frac{\cos x\,dx}{\sqrt{\sin^2 x+4\sin x+3}}$
15. $\displaystyle\int \frac{(2x-3)\,dx}{\sqrt{4x-x^2-3}}$
16. $\displaystyle\int \frac{(x+3)\,dx}{\sqrt{x^2+2x+2}}$
17. Show that if $k > 0$, then

$$\int \frac{dx}{\sqrt{(x+h)^2+k}} = \ln[\sqrt{(x+h)^2+k} + x + h] + C \quad \text{for all } x$$

18. Show that if $a > 0$ and $b^2 - 4ac > 0$, then

$$\int \frac{dx}{\sqrt{ax^2+bx+c}} = \frac{1}{\sqrt{a}}\ln\left(\sqrt{ax^2+bx+c} + \sqrt{a}\,x + \frac{b}{2\sqrt{a}}\right) + C \quad \text{for all } x$$

9.6 Integration of Rational Functions by Partial Fractions

In many cases, the integration of *rational functions* can be accomplished by using certain standard integral forms together with the algebraic method of *partial fractions*.

Recall that, by definition, a rational function is the quotient of two polynomials, P and $Q \neq 0$, with no common factors. A rational function in which the polynomial in the numerator is of lower degree than the polynomial in the denominator is called a *proper rational function*; otherwise it is an *improper rational function*. Any improper rational function can be reduced by division to a mixed form, consisting of the sum of a polynomial and a proper rational function.

For example, the rational function $x^2/(x-1)$ is improper. By long division, we obtain

$$\frac{x^2}{x-1} = x + 1 + \frac{1}{x-1} \tag{1}$$

To evaluate

$$\int \frac{x^2}{x-1}\,dx$$

we integrate both sides of (1), obtaining

$$\int \frac{x^2}{x-1}\,dx = \int \left(x + 1 + \frac{1}{x-1}\right) dx = \int (x+1)\,dx + \int \frac{1}{x-1}\,dx$$

$$= \frac{x^2}{2} + x + \ln|x-1| + C$$

Thus, because an improper rational function can always be written as the sum of a polynomial plus a proper rational function, we restrict our discussion to proper rational functions.

In general, a proper rational function P/Q may be written as the sum of simpler fractions, called *partial fractions*. (This is something we will not prove, although this fact will be used in this and the following sections.) The method of integration by partial fractions involves separating the given rational function into a sum of partial fractions. These partial fractions are then integrated by standard integration formulas. The result is that *the integral of every rational function can be expressed in terms of algebraic, logarithmic, and/or inverse trigonometric expressions*.

When a rational function is separated into partial fractions, the result is an *identity*; that is, it is true for all values of the variable for which the expressions involved have meaning. The evaluation of the coefficients of the partial fractions is based on the following theorem from algebra:

[9.6.1] THEOREM

If two polynomials are equal for all values of the variable, then the polynomials have the same degree and the coefficients of like powers of the variable in both polynomials must be equal.

For example, if

$$ax^2 + bx + c = 3x^2 - 5x + 2 \qquad \text{for all } x$$

then

$$a = 3 \qquad b = -5 \qquad c = 2$$

As it turns out, the partial fraction decomposition of P/Q depends on the nature of the factors of the denominator Q. According to the fundamental theorem of algebra, any polynomial in x with real coefficients can be expressed as a product of factors of one or both of the following types:

1. *Linear factors* of the form $ax + b$, where a and b are real numbers
2. *Irreducible quadratic factors* of the form $ax^2 + bx + c$, where a, b, and c are real numbers and $ax^2 + bx + c$ cannot be factored into real linear factors

We begin with the case for which Q contains only nonrepeated linear factors.

Case 1. Q has only nonrepeated linear factors.

The polynomial Q may be written as

$$Q(x) = (x - a_1)(x - a_2) \cdot \cdots \cdot (x - a_n)$$

where none of the real numbers $a_1, a_2, \ldots, a_n$ are the same. In this case, P/Q may be written as the identity

$$\frac{P(x)}{Q(x)} = \frac{A_1}{x - a_1} + \frac{A_2}{x - a_2} + \cdots + \frac{A_n}{x - a_n} \tag{2}$$

in which $A_1, A_2, \ldots, A_n$ are numbers to be found. By integrating both sides of (2), we find

$$\int \frac{P(x)}{Q(x)}\,dx = \int \left(\frac{A_1}{x - a_1} + \cdots + \frac{A_n}{x - a_n}\right) dx = \int \frac{A_1}{x - a_1}\,dx + \cdots + \int \frac{A_n}{x - a_n}\,dx$$

$$= A_1 \ln|x - a_1| + \cdots + A_n \ln|x - a_n| + C$$

The procedure for finding the numbers $A_1, \ldots, A_n$ is illustrated in Example 1.

EXAMPLE 1

Evaluate: $\displaystyle\int \frac{x\,dx}{x^2 - 5x + 6}$

Solution

First we factor $x^2 - 5x + 6 = (x - 2)(x - 3)$. Then we write the identity

$$\frac{x}{(x - 2)(x - 3)} = \frac{A}{x - 2} + \frac{B}{x - 3} \tag{3}$$

Now we multiply both sides by $(x-2)(x-3)$ to clear fractions, giving

$$x = A(x-3) + B(x-2)$$
$$x = (A+B)x - 3A - 2B \tag{4}$$

This is an identity in x. Now, using theorem (9.6.1), we may conclude:

1	$=$	$A+B$
Coefficient of x on left side of (4)		Coefficient of x on right side of (4)
0	$=$	$-3A-2B$
Coefficient of x^0 on left side of (4) (the constant term)		Coefficient of x^0 on right side of (4) (the constant term)

The solution of this system of equations is $A=-2$ and $B=3$. By substituting into (3), we find

$$\frac{x}{(x-2)(x-3)} = \frac{-2}{x-2} + \frac{3}{x-3}$$

so that

$$\int \frac{x}{(x-2)(x-3)}\,dx = \int \frac{-2}{x-2}\,dx + \int \frac{3}{x-3}\,dx$$
$$= -2\ln|x-2| + 3\ln|x-3| + C$$
$$= \ln\left|\frac{(x-3)^3}{(x-2)^2}\right| + C$$

Case 2. Q has repeated linear factors.

If the polynomial Q has a factor $(x-a)^n$, $n \geq 2$ an integer, then in the partial fraction decomposition of P/Q we must allow for the terms

$$\frac{A_1}{x-a} + \frac{A_2}{(x-a)^2} + \cdots + \frac{A_n}{(x-a)^n}$$

where $A_1, \ldots, A_n$ are numbers to be found.

EXAMPLE 2

Evaluate: $\displaystyle\int \frac{dx}{x(x-1)^2}$

Solution

Since x is a distinct linear factor of the denominator Q, by Case 1, we allow for the term A/x in the decomposition of P/Q. Since $(x-1)^2$ is a repeated linear factor, we allow for the terms $B/(x-1)$ and $C/(x-1)^2$ in the decomposition of P/Q. Hence, we write the identity

$$\frac{1}{x(x-1)^2} = \frac{A}{x} + \frac{B}{x-1} + \frac{C}{(x-1)^2} \tag{5}$$

As in Example 1, we clear fractions:

$$1 = A(x-1)^2 + Bx(x-1) + Cx$$

$$1 = (A+B)x^2 + (C-B-2A)x + A$$

By equating coefficients, we obtain the system of equations

$$A + B = 0$$

$$-2A - B + C = 0$$

$$A = 1$$

By solving this system of equations, we find $A=1$, $B=-1$, and $C=1$. Substituting into (5), we get

$$\frac{1}{x(x-1)^2} = \frac{1}{x} + \frac{-1}{x-1} + \frac{1}{(x-1)^2}$$

so that

$$\int \frac{dx}{x(x-1)^2} = \int \frac{dx}{x} - \int \frac{dx}{x-1} + \int \frac{dx}{(x-1)^2}$$

$$= \ln|x| - \ln|x-1| - \frac{1}{x-1} + C = \ln\left|\frac{x}{x-1}\right| - \frac{1}{x-1} + C$$

The numbers to be found in the partial fraction decomposition of P/Q can often be obtained more easily by substituting convenient values of x into the identity obtained after clearing fractions. In Example 2, after clearing fractions, we have the identity

$$1 = A(x-1)^2 + Bx(x-1) + Cx$$

If we let $x=0$, the terms involving B and C drop out, leaving $1 = A(1)$, so that $A=1$. If we let $x=1$, the terms involving A and B drop out, leaving $1 = C(1)$, so that $C=1$. To get B, we can use any x other than $x=0$ or $x=1$ with $A=1$ and $C=1$. Using $x=2$, we get

$$1 = 1(1)^2 + B2(1) + 1(2)$$

$$2B = -2$$

$$B = -1$$

The advantage of this method is evident from the next example.

EXAMPLE 3

Evaluate: $\displaystyle\int \frac{x^3-8}{x^2(x-1)^3}\,dx$

Solution

Due to the x^2 factor, we allow for the terms A/x and B/x^2; and due to the $(x-1)^3$ term, we allow for the terms $C/(x-1)$, $D/(x-1)^2$, and $E/(x-1)^3$.

We write the identity

$$\frac{x^3-8}{x^2(x-1)^3} = \frac{A}{x} + \frac{B}{x^2} + \frac{C}{x-1} + \frac{D}{(x-1)^2} + \frac{E}{(x-1)^3} \tag{6}$$

As usual, we clear fractions and simplify:

$$x^3 - 8 = Ax(x-1)^3 + B(x-1)^3 + Cx^2(x-1)^2 + Dx^2(x-1) + Ex^2 \tag{7}$$

We let $x = 0$. Then

$$-8 = B(-1)$$
$$B = 8$$

We let $x = 1$. Then

$$1 - 8 = E$$
$$E = -7$$

To find the other coefficients, we replace B by 8 and E by -7 in (7) and collect like terms:

$$x^3 - 8 - 8(x-1)^3 + 7x^2 = Ax(x-1)^3 + Cx^2(x-1)^2 + Dx^2(x-1)$$
$$-7x^3 + 31x^2 - 24x = Ax(x-1)^3 + Cx^2(x-1)^2 + Dx^2(x-1)$$
$$-x(x-1)(7x-24) = x(x-1)[A(x-1)^2 + Cx(x-1) + Dx] \tag{8}$$

Notice the factor $x(x-1)$ that appears after simplification of $x^3 - 8 - 8(x-1)^3 + 7x^2$ to $-x(x-1)(7x-24)$ on the left side of (8). This is no coincidence. Although we do not prove it, we assert that this method will always result in a factor of the form $(x - a_1) \cdots (x - a_n)$, which can be canceled from both sides.

By dividing (8) by $x(x-1)$, we obtain

$$-(7x - 24) = A(x-1)^2 + Cx(x-1) + Dx \tag{9}$$

In (9) we let $x = 0$. Then

$$24 = A$$

In (9), we let $x = 1$. Then

$$17 = D$$

In (9), we replace A by 24 and D by 17. Then

$$-(7x - 24) = 24(x-1)^2 + Cx(x-1) + 17x$$

Now we let $x = 2$ (any choice other than 0 or 1 could have been used). Then

$$10 = 24 + 2C + 34$$
$$C = -24$$

Using the values found for A, B, C, D, and E in (6) we have

$$\frac{x^3-8}{x^2(x-1)^3} = \frac{24}{x} + \frac{8}{x^2} - \frac{24}{x-1} + \frac{17}{(x-1)^2} - \frac{7}{(x-1)^3}$$

Thus,

$$\int \frac{x^3-8}{x^2(x-1)^3}\,dx = 24\ln|x| - \frac{8}{x} - 24\ln|x-1| - \frac{17}{x-1} + \frac{7}{2(x-1)^2} + C$$

$$= 24\ln\left|\frac{x}{x-1}\right| - \frac{8}{x} - \frac{17}{x-1} + \frac{7}{2(x-1)^2} + C$$

Although this procedure may seem tedious, it is much faster than solving five equations in five unknowns by equating coefficients in (7).*

We close this section by deriving two useful formulas.

EXAMPLE 4

Derive the formulas:

$$\textbf{(a)}\quad \int \frac{dx}{x^2-a^2} = \frac{1}{2a}\ln\left|\frac{x-a}{x+a}\right| + C \qquad a \neq 0$$

$$\textbf{(b)}\quad \int \frac{dx}{a^2-x^2} = \frac{1}{2a}\ln\left|\frac{x+a}{x-a}\right| + C \qquad a \neq 0 \tag{10}$$

Solution

(a)

$$\frac{1}{x^2-a^2} = \frac{1}{(x-a)(x+a)} = \frac{A}{x-a} + \frac{B}{x+a}$$

By solving for A and B, we find that $A = 1/(2a)$ and $B = -1/(2a)$, so that

$$\int \frac{dx}{x^2-a^2} = \frac{1}{2a}\int \frac{dx}{x-a} - \frac{1}{2a}\int \frac{dx}{x+a} = \frac{1}{2a}\left(\int \frac{dx}{x-a} - \int \frac{dx}{x+a}\right)$$

$$= \frac{1}{2a}(\ln|x-a| - \ln|x+a|) + C = \frac{1}{2a}\ln\left|\frac{x-a}{x+a}\right| + C$$

(b)

$$\int \frac{dx}{a^2-x^2} = -\int \frac{dx}{x^2-a^2} = -\frac{1}{2a}\ln\left|\frac{x-a}{x+a}\right| + C$$

$$= \frac{1}{2a}\ln\left|\frac{x-a}{x+a}\right|^{-1} + C$$

$$= \frac{1}{2a}\ln\left|\frac{x+a}{x-a}\right| + C$$

* The method used in this example is discussed in more detail in H. J. Straight and R. Dowds, *American Mathematical Monthly*, 91, no. 6 (June–July 1984): 365.

EXERCISE 9.6

In Problems 1–4 evaluate each integral by expressing the integrand as the sum of a polynomial plus a proper rational function.

1. $\displaystyle\int \frac{x^2+1}{x+1}\,dx$ **2.** $\displaystyle\int \frac{x^2+4}{x-2}\,dx$ **3.** $\displaystyle\int \frac{x^3+3x-4}{x-2}\,dx$ **4.** $\displaystyle\int \frac{x^3-3x^2+4}{x+3}\,dx$

In Problems 5–22 evaluate each integral.

5. $\displaystyle\int \frac{dx}{(x-2)(x+1)}$ **6.** $\displaystyle\int \frac{dx}{(x+4)(x-1)}$ **7.** $\displaystyle\int \frac{x\,dx}{(x-1)(x-2)}$

8. $\displaystyle\int \frac{3x\,dx}{(x+2)(x-4)}$ **9.** $\displaystyle\int \frac{x^2\,dx}{(x-1)^2(x+1)}$ **10.** $\displaystyle\int \frac{(x+1)\,dx}{x^2(x-2)}$

11. $\displaystyle\int \frac{(x-3)\,dx}{(x+2)(x+1)^2}$ **12.** $\displaystyle\int \frac{(x^2+x)\,dx}{(x+2)(x-1)^2}$ **13.** $\displaystyle\int \frac{x\,dx}{(3x-2)(2x+1)}$

14. $\displaystyle\int \frac{dx}{(2x+3)(4x-1)}$ **15.** $\displaystyle\int \frac{x\,dx}{x^2+2x-3}$ **16.** $\displaystyle\int \frac{x^2-x-8}{(x+1)(x^2+5x+6)}\,dx$

17. $\displaystyle\int \frac{7x+3}{x^3-2x^2-3x}\,dx$ **18.** $\displaystyle\int \frac{x^5+1}{x^6-x^4}\,dx$ **19.** $\displaystyle\int \frac{x^2}{x^3-4x^2+5x-2}\,dx$

20. $\displaystyle\int \frac{x^2+1}{x^3+x^2-5x+3}\,dx$ **21.** $\displaystyle\int \frac{\cos\theta\,d\theta}{\sin^2\theta+\sin\theta-6}$ **22.** $\displaystyle\int \frac{e^t\,dt}{e^{2t}+e^t-2}$

In Problems 23–28, evaluate each integral by first making the proper substitution or integrating by parts and then using partial fraction techniques.

23. $\displaystyle\int \frac{dx}{e^x-e^{-x}}$ **24.** $\displaystyle\int \frac{e^x\,dx}{1-e^{2x}}$ **25.** $\displaystyle\int \frac{e^x\,dx}{e^{2x}+e^x-6}$

26. $\displaystyle\int \ln(1+x^2)\,dx$ **27.** $\displaystyle\int x\tan^{-1}x\,dx$ **28.** $\displaystyle\int \frac{\sin^2 x\cos x\,dx}{\sin^2 x+1}$

In Problems 29–32 evaluate each definite integral using part (a) or (b) of (10).

29. $\displaystyle\int_0^1 \frac{dx}{x^2-9}$ **30.** $\displaystyle\int_2^4 \frac{dx}{x^2-25}$ **31.** $\displaystyle\int_{-2}^3 \frac{dx}{16-x^2}$ **32.** $\displaystyle\int_1^2 \frac{dx}{9-x^2}$

33. Find the area under the graph of $y=4/(x^2-4)$ from $x=3$ to $x=5$.

34. Find the area under the graph of $y=(x-4)/(x+3)^2$ from $x=4$ to $x=6$.

35. Find the volume of the solid generated by revolving the area under the graph of $y=x/(x^2-4)$ from $x=3$ to $x=5$ about the x-axis.

36. Assume that a population grows according to *Verhulst's logistic law of population growth*—that is, $dN/dt = AN - BN^2$, where $N=N(t)$ is the population at time t (in years), and the constants A and B are the *vital coefficients* of the population. Assume that the vital coefficients have the values $A=10^5$ and $B=0.01$. If initially the size of this population was 12×10^5, what will its size be after 10 years? What will the population be after a very long time?

Integration of Rational Functions by Partial Fractions (Continued)

9.7

In this section we treat proper rational functions P/Q, where one of the factors of Q is an irreducible quadratic. A *quadratic polynomial is called irreducible if it cannot be factored into real linear factors.* Thus, $ax^2 + bx + c$ is an irreducible quadratic polynomial if $b^2 - 4ac < 0$. For example, $x^2 + x + 1$ and $x^2 + 4$ are irreducible.

Case 3. Q contains a nonrepeated irreducible quadratic polynomial.

If the polynomial Q contains a nonrepeated irreducible quadratic polynomial $ax^2 + bx + c$, then in the partial fraction decomposition of P/Q we must allow for the term

$$\frac{Ax + B}{ax^2 + bx + c}$$

where A and B are numbers to be found.

EXAMPLE 1

Evaluate: $\displaystyle\int \frac{5x^2 + 3x - 2}{x^3 - 1}\,dx$

Solution

Here, $Q(x) = x^3 - 1 = (x - 1)(x^2 + x + 1)$. Since $(x - 1)$ is a nonrepeated linear factor, by Case 1 we allow for the term $A/(x - 1)$ in the decomposition of P/Q. Since $x^2 + x + 1$ is a nonrepeated irreducible quadratic polynomial, we allow for the term $(Bx + C)/(x^2 + x + 1)$ in the decomposition of P/Q. Hence, we write the identity

$$\frac{5x^2 + 3x - 2}{x^3 - 1} = \frac{A}{x - 1} + \frac{Bx + C}{x^2 + x + 1} \tag{1}$$

We multiply both sides by $x^3 - 1 = (x - 1)(x^2 + x + 1)$ to get

$$5x^2 + 3x - 2 = A(x^2 + x + 1) + (Bx + C)(x - 1) \tag{2}$$

To determine the coefficients A, B, and C, we collect terms:

$$5x^2 + 3x - 2 = (A + B)x^2 + (A - B + C)x + A - C$$

By equating coefficients of like powers of x, we obtain the system of equations

$$A + B = 5$$
$$A - B + C = 3$$
$$A - C = -2$$

A little effort* yields $A = 2$, $B = 3$, and $C = 4$. By now substituting in (1), we get

$$\frac{5x^2 + 3x - 2}{x^3 - 1} = \frac{2}{x - 1} + \frac{3x + 4}{x^2 + x + 1}$$

so that

$$\int \frac{5x^2 + 3x - 2}{x^3 - 1}\,dx = \int \frac{2}{x - 1}\,dx + \int \frac{3x + 4}{x^2 + x + 1}\,dx$$

$$= 2\ln|x - 1| + \int \frac{3x + 4}{x^2 + x + 1}\,dx \qquad \textbf{(3)}$$

For the integral on the right, we complete the square in the denominator and make a substitution (see Section 9.5):

$$\int \frac{3x + 4}{x^2 + x + 1}\,dx = \int \frac{3x + 4}{(x + \frac{1}{2})^2 + \frac{3}{4}}\,dx \underset{u = x + \frac{1}{2}}{\uparrow}= \int \frac{3u + \frac{5}{2}}{u^2 + \frac{3}{4}}\,du = 3\int \frac{u}{u^2 + \frac{3}{4}}\,du + \frac{5}{2}\int \frac{du}{u^2 + \frac{3}{4}}$$

$$= \frac{3}{2}\ln\left(u^2 + \frac{3}{4}\right) + \frac{5}{2}\left(\frac{2}{\sqrt{3}}\right)\tan^{-1}\frac{2}{\sqrt{3}}u + C$$

$$\underset{u = x + \frac{1}{2}}{\uparrow}= \frac{3}{2}\ln(x^2 + x + 1) + \frac{5\sqrt{3}}{3}\tan^{-1}\frac{2x + 1}{\sqrt{3}} + C \qquad \textbf{(4)}$$

Hence, from (3) and (4), we find that

$$\int \frac{5x^2 + 3x - 2}{x^3 - 1}\,dx = 2\ln|x - 1| + \frac{3}{2}\ln(x^2 + x + 1) + \frac{5\sqrt{3}}{3}\tan^{-1}\frac{2x + 1}{\sqrt{3}} + C$$

Note that we could have saved some computation time if, in place of (1), we had written the partial fraction decomposition as

$$\frac{5x^2 + 3x - 2}{x^3 - 1} = \frac{D}{x - 1} + \frac{E(2x + 1) + F}{x^2 + x + 1} \qquad \textbf{(5)}$$

This immediately places the derivative of $x^2 + x + 1$ in the decomposition. Following the usual procedure, we find the numbers D, E, and F to be $D = 2$, $E = \frac{3}{2}$, and $F = \frac{5}{2}$. Integrating (5) saves a few steps.

Case 4. Q has repeated irreducible quadratic polynomials.

If the polynomial Q contains a repeated irreducible quadratic polynomial $(x^2 + bx + c)^n$, $n \geq 2$ an integer, then in the partial fraction decom-

---

* An alternative here is to set $x = 1$ in (2), which results in finding $A = 2$. By resubstituting $A = 2$ into (2) and collecting terms, we get $(3x + 4)(x - 1) = (Bx + C)(x - 1)$, from which we conclude that $B = 3$ and $C = 4$. The effort is somewhat less if the alternative is used.

position of P/Q we must allow for the terms

$$\frac{A_1x + B_1}{x^2 + bx + c} + \frac{A_2x + B_2}{(x^2 + bx + c)^2} + \cdots + \frac{A_nx + B_n}{(x^2 + bx + c)^n}$$

where $A_1, B_1, A_2, B_2, \ldots, A_n, B_n$ are numbers to be found.

EXAMPLE 2

Evaluate: $\displaystyle\int \frac{x^3 + 1}{(x^2 + 4)^2}\, dx$

Solution

The partial fraction decomposition is

$$\frac{x^3 + 1}{(x^2 + 4)^2} = \frac{Ax + B}{x^2 + 4} + \frac{Cx + D}{(x^2 + 4)^2}$$

Upon clearing fractions and combining terms, we arrive at

$$x^3 + 1 = Ax^3 + Bx^2 + (4A + C)x + 4B + D$$

By equating coefficients, we find

$$A = 1 \qquad B = 0 \qquad 4A + C = 0 \qquad 4B + D = 1$$
$$C = -4 \qquad D = 1$$

Therefore,

$$\frac{x^3 + 1}{(x^2 + 4)^2} = \frac{x}{x^2 + 4} + \frac{-4x + 1}{(x^2 + 4)^2}$$

so that

$$\int \frac{x^3 + 1}{(x^2 + 4)^2}\, dx = \int \frac{x}{x^2 + 4}\, dx + \int \frac{-4x + 1}{(x^2 + 4)^2}\, dx$$
$$= \frac{1}{2}\ln(x^2 + 4) + \int \frac{-4x}{(x^2 + 4)^2}\, dx + \int \frac{dx}{(x^2 + 4)^2}$$

In the first integral on the right side, we use the substitution $u = x^2 + 4$. Then $du = 2x\, dx$, and

$$\int \frac{-4x}{(x^2 + 4)^2}\, dx = \int \frac{-2\, du}{u^2} = \frac{2}{u} = \frac{2}{x^2 + 4}$$

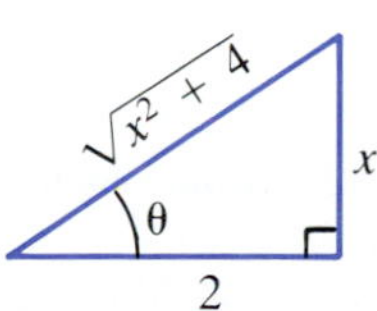

Figure 9

In the second integral on the right, we use the trigonometric substitution $x = 2\tan\theta$. Then $dx = 2\sec^2\theta\, d\theta$, and

$$\int \frac{dx}{(x^2 + 4)^2} = \int \frac{2\sec^2\theta\, d\theta}{16\sec^4\theta} = \frac{1}{8}\int \cos^2\theta\, d\theta = \frac{1}{8}\int \frac{1 + \cos 2\theta}{2}\, d\theta$$
$$= \frac{1}{16}\left(\theta + \frac{1}{2}\sin 2\theta\right) = \frac{1}{16}(\theta + \sin\theta\cos\theta) \underset{\substack{\uparrow \\ \text{From Figure 9}}}{=} \frac{1}{16}\left(\tan^{-1}\frac{x}{2} + \frac{2x}{x^2 + 4}\right)$$

By combining these results, we find

$$\int \frac{x^3+1}{(x^2+4)^2}\,dx = \frac{1}{2}\ln(x^2+4) + \frac{2}{x^2+4} + \frac{1}{16}\tan^{-1}\frac{x}{2} + \frac{x}{8(x^2+4)} + C$$

EXERCISE 9.7

In Problems 1–14 evaluate each integral.

1. $\displaystyle\int \frac{dx}{x(x^2+1)}$

2. $\displaystyle\int \frac{dx}{(x+1)(x^2+4)}$

3. $\displaystyle\int \frac{(2x+1)dx}{x^3-1}$

4. $\displaystyle\int \frac{dx}{x^3-8}$

5. $\displaystyle\int \frac{x+4}{x^2(x^2+4)}\,dx$

6. $\displaystyle\int \frac{10x^2+2x}{(x-1)^2(x^2+2)}\,dx$

7. $\displaystyle\int \frac{x^2+2x+3}{(x+1)(x^2+2x+4)}\,dx$

8. $\displaystyle\int \frac{x^2-11x-18}{x(x^2+3x+3)}\,dx$

9. $\displaystyle\int \frac{x^2+2x+3}{(x^2+4)^2}\,dx$

10. $\displaystyle\int \frac{2x+1}{(x^2+16)^2}\,dx$

11. $\displaystyle\int \frac{x^3\,dx}{(x^2+16)^3}$

12. $\displaystyle\int \frac{x^2\,dx}{(x^2+4)^3}$

13. $\displaystyle\int \frac{\sin\theta\,d\theta}{\cos^3\theta+\cos\theta}$

14. $\displaystyle\int \frac{dt}{e^{2t}+1}$

15. Find the area under the graph of $y = 8/(x^3+1)$ from $x=0$ to $x=2$.

16. Find the volume of the solid generated by revolving the area under the graph of $y = 1/(x^2+1)^2$ from $x=0$ to $x=1$ about the y-axis.

If an integrand is a rational expression of $\sin x$ or $\cos x$ or both, the substitution

$$z = \tan\frac{x}{2} \qquad -\frac{\pi}{2} < \frac{x}{2} < \frac{\pi}{2}$$

or equivalently,

$$\sin x = \frac{2z}{1+z^2} \qquad \cos x = \frac{1-z^2}{1+z^2} \qquad dx = \frac{2\,dz}{1+z^2} \qquad (6)$$

will transform the integrand into a rational function of z.

In Problems 17–32 evaluate each integral using the substitution (6).

17. $\displaystyle\int \frac{dx}{1-\sin x}$

18. $\displaystyle\int \frac{dx}{1+\sin x}$

19. $\displaystyle\int \frac{dx}{1-\cos x}$

20. $\displaystyle\int \frac{dx}{3+2\cos x}$

21. $\displaystyle\int \frac{2\,dx}{\sin x+\cos x}$

22. $\displaystyle\int \frac{dx}{1-\sin x+\cos x}$

23. $\displaystyle\int \frac{\sin x\,dx}{3+\cos x}$

24. $\displaystyle\int \frac{dx}{\tan x-1}$

25. $\displaystyle\int \frac{dx}{\tan x-\sin x}$

26. $\displaystyle\int \frac{\sec x\,dx}{\tan x-2}$

27. $\displaystyle\int \frac{\cot x\,dx}{1+\sin x}$

28. $\displaystyle\int \frac{\sec x\,dx}{1+\sin x}$

29. $\displaystyle\int_0^{\pi/2} \frac{dx}{\sin x+1}$

30. $\displaystyle\int_{\pi/4}^{\pi/3} \frac{\csc x\,dx}{3+4\tan x}$

31. $\displaystyle\int_0^{\pi/2} \frac{\cos x\,dx}{2-\cos x}$

32. $\displaystyle\int_0^{\pi/4} \frac{4\,dx}{\tan x+1}$

33. Show that the two formulas below are equivalent.

$$\int \sec x\,dx = \ln|\sec x + \tan x| + C$$

$$\int \sec x\,dx = \ln\left|\frac{1+\tan(x/2)}{1-\tan(x/2)}\right| + C$$

$$\left[\textit{Hint:}\quad \tan\left(\frac{x}{2}\right) = \frac{\sin(x/2)}{\cos(x/2)} = \frac{\sin^2(x/2)}{\sin(x/2)\cos(x/2)} = \frac{1-\cos x}{\sin x}\right]$$

34. Use the substitution (6) to show that

$$\int \csc x\,dx = \ln\sqrt{\frac{1-\cos x}{1+\cos x}} + C$$

35. Show that the result obtained in Problem 34 is equivalent to the formula

$$\int \csc x\,dx = \ln|\csc x - \cot x| + C$$

36. Compute $\int \cos x\,dx$ using the substitution

$$z = \tan\frac{x}{2}.$$

For Problems 37–50 we first simplify the integrand in the following manner. If the integrand involves fractional powers $x^{p/q}$, $x^{r/s}$, and so forth, the substitution $x = u^n$, where n is the least common denominator of p/q, r/s, and so forth, will transform the integrand into a rational function of u.

37. $\displaystyle\int \frac{x\,dx}{3+\sqrt{x}}$

38. $\displaystyle\int \frac{dx}{\sqrt{x}+2}$

39. $\displaystyle\int \frac{dx}{x-\sqrt[3]{x}}$

40. $\displaystyle\int \frac{x\,dx}{\sqrt[3]{x}-1}$

41. $\displaystyle\int \frac{dx}{\sqrt{x}+\sqrt[3]{x}}$

42. $\displaystyle\int \frac{dx}{\sqrt{x}-\sqrt[3]{x}}$

43. $\displaystyle\int \frac{dx}{\sqrt[3]{2+3x}}$

44. $\displaystyle\int \frac{dx}{\sqrt[4]{1+2x}}$

45. $\displaystyle\int \frac{x\,dx}{(1+x)^{3/4}}$

46. $\displaystyle\int \frac{dx}{(1+x)^{2/3}}$

47. $\displaystyle\int \frac{dx}{\sqrt{x}+1}$

48. $\displaystyle\int \frac{dx}{\sqrt{x}(1+\sqrt[3]{x})^2}$

49. $\displaystyle\int \frac{\sqrt[3]{x}+1}{\sqrt[3]{x}-1}\,dx$

50. $\displaystyle\int \frac{x\,dx}{\sqrt[5]{x+4}}$

51. Find the volume of the solid generated by revolving the area under the graph of $y = \sqrt{x+1} + x$ from $x = 0$ to $x = 3$ about the x-axis.

52. Rework Problem 51 if the revolution is about the y-axis.

Integrals That Are Not Expressible in Terms of Elementary Functions — 9.8

We have seen that the derivatives of all elementary functions are also elementary functions,* but integrals of elementary functions are *not* all expressible (in finite form) in terms of elementary functions. Some examples of integrals of elementary functions that are not expressible in terms of elementary functions are

$$\int e^{x^2}\,dx \qquad \int e^{-x^2}\,dx \qquad \int \frac{\sin x}{x}\,dx$$

$$\int \frac{\cos x}{x}\,dx \qquad \int \frac{e^x\,dx}{x} \qquad \int \frac{dx}{\sqrt{1-x^3}}$$

* The elementary functions are those that we have been discussing, such as polynomials, exponential, logarithmic, trigonometric, inverse trigonometric, hyperbolic, and so on.

An important class of integral forms that are not expressible in terms of elementary functions are the *elliptic integrals*. An elliptic integral has the form

$$\int R(x, \sqrt{P(x)})\, dx$$

where P is a polynomial function of degree 3 or 4 with nonrepeated roots, and R denotes a rational function.

Numerous problems in geometry, mechanics, and other subjects lead to elliptic integrals. For example, the problems of finding the length of an arc of an ellipse and of finding the time of oscillation of a pendulum (for arbitrary angle of oscillation) lead to elliptic integrals.

In the next section we give two techniques for approximating definite integrals. This is especially useful where the integral cannot be expressed in terms of elementary functions.

9.9 Numerical Techniques

So far in this chapter we have dealt mainly with various techniques for evaluating indefinite integrals—that is, finding the antiderivative of a function. One reason for this study is that in order to evaluate a definite integral $\int_a^b f(x)\, dx$ using the second fundamental theorem of calculus, we need to know an antiderivative of the integrand f. When it is not possible to find an antiderivative of the integrand f or when the integrand f is defined by an empirical table of values or by an empirical graph, we turn to numerical techniques to approximate the value of the definite integral. To aid in the following discussion, we shall assume that the integrand f is nonnegative and continuous on the closed interval $[a, b]$.

Most of the methods of approximate integration are based on the fact that a definite integral equals the area under a graph, so that any method of approximating this area will also give an approximation to the integral. Two of the most widely used numerical techniques of approximate integration are the *trapezoidal rule* and *Simpson's rule*. Another useful method is based on the use of *series* (see Chap. 11).

[9.9.1] THEOREM / *Trapezoidal Rule.*

If a function f is continuous on the closed interval $[a, b]$, then

$$\int_a^b f(x)\, dx \approx \left(\frac{1}{2}\right)\left(\frac{b-a}{n}\right)[f(x_0) + 2f(x_1) + 2f(x_2) + \cdots + 2f(x_{n-1}) + f(x_n)] \tag{1}$$

where the closed interval $[a, b]$ has been partitioned into n subintervals $[x_0, x_1], [x_1, x_2], \ldots, [x_{n-1}, x_n]$, each of length $(b - a)/n$.

Proof The trapezoidal rule is based on the idea of representing a definite integral by an area under a graph and approximating this area by a collection of trapezoids obtained by replacing the graph by a set of chords. Suppose that we are to evaluate $\int_a^b f(x)\, dx$ approximately. The integral is then equal to the area enclosed by the graph of f, the x-axis, and the lines $x = a$ and

$x = b$ (see Fig. 10). Now, partition the interval $[a, b]$ into n subintervals

$$[a, x_1], \quad [x_1, x_2], \quad \ldots, \quad [x_{i-1}, x_i], \quad \ldots, \quad [x_{n-1}, b]$$

each of length $\Delta x = (b - a)/n$. The ordinates corresponding to $x_0 = a$, $x_1, x_2, x_{n-1}, \ldots, x_n = b$ are $f(x_0), f(x_1), f(x_2), \ldots, f(x_{n-1}), f(x_n)$. When we join consecutive points on the graph by straight line segments (chords), trapezoids are formed, and the sum of the areas of the trapezoids is taken as an approximation to the area under the graph. Since the area of a trapezoid is equal to half the sum of the length of the parallel sides times the base, we have the following equation for the sum of the areas of the trapezoids:

$$\text{Area} = \tfrac{1}{2}[f(x_0) + f(x_1)]\Delta x + \tfrac{1}{2}[f(x_1) + f(x_2)]\Delta x + \cdots + \tfrac{1}{2}[f(x_{n-1}) + f(x_n)]\Delta x$$

$$= \tfrac{1}{2}\Delta x[f(x_0) + 2f(x_1) + 2f(x_2) + \cdots + 2f(x_{n-1}) + f(x_n)]$$

By setting $\Delta x = (b - a)/n$, we have (1).

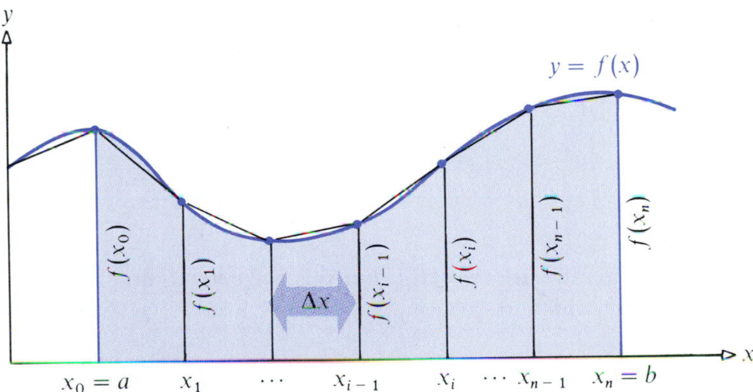

Area of trapezoid in ith subinterval $= \frac{1}{2}[f(x_{i-1}) + f(x_i)]\,\Delta x$

Figure 10

[9.9.2] THEOREM / *Error.*

The difference between the exact value of the integral $\int_a^b f(x)\,dx$ and the approximate value given by the trapezoidal rule is called the *error*. It may be estimated by the formula

$$\textbf{Error} \leq \frac{(b - a)^3 M}{12n^2} \tag{2}$$

where M is the largest value of $|f''(x)|$ on the closed interval $[a, b]$.*

EXAMPLE 1

Use the trapezoidal rule with $n = 4$ to approximate $\displaystyle\int_0^1 \frac{dx}{1 + x^2}$. Estimate the error in using this approximation.

* The verification of this formula, which uses an extension of the mean value theorem, is found in advanced calculus books.

Solution

We partition the interval [0, 1] into four subintervals of equal length—namely, $[0, \frac{1}{4}]$, $[\frac{1}{4}, \frac{1}{2}]$, $[\frac{1}{2}, \frac{3}{4}]$, $[\frac{3}{4}, 1]$. The corresponding values of f are

$$f(0) = \frac{1}{1+0} = 1 \qquad f\left(\frac{1}{4}\right) = \frac{1}{1 + (\frac{1}{4})^2} = \frac{16}{17} \approx 0.94117 \qquad f\left(\frac{1}{2}\right) = \frac{1}{1 + (\frac{1}{2})^2} = \frac{4}{5} = 0.8$$

$$f\left(\frac{3}{4}\right) = \frac{1}{1 + (\frac{3}{4})^2} = \frac{16}{25} = 0.64 \qquad f(1) = \frac{1}{1+1} = \frac{1}{2} = 0.5$$

It is convenient to set up a table, as shown in Table 3. The sum of the entries in the bottom row of the table is 6.26234, so that by the trapezoidal rule (1), we get

$$\int_0^1 \frac{dx}{1+x^2} \approx \left(\frac{1}{8}\right)(6.26234) \approx 0.78279$$

Table 3

	$x = 0$	$x = \frac{1}{4}$	$x = \frac{1}{2}$	$x = \frac{3}{4}$	$x = 1$
$f(x) = 1/(1 + x^2)$	1	0.94117	0.8	0.64	0.5
Factor	$\times 1$	$\times 2$	$\times 2$	$\times 2$	$\times 1$
Product	1	1.88234	1.6	1.28	0.5

We now use (2) to estimate the error. For this, we need to find the maximum value of $|f''(x)|$, which in turn requires that we find $f'''(x)$. Some calculations will lead to

$$f(x) = \frac{1}{1+x^2} \qquad f'(x) = \frac{-2x}{(1+x^2)^2} \qquad f''(x) = \frac{2(3x^2-1)}{(1+x^2)^3}$$

$$f'''(x) = 24x(1-x^2)(1+x^2)^{-4}$$

Because $f'''(x) > 0$ for $0 < x < 1$, f'' has no critical numbers in (0, 1). The largest value for $|f''(x)|$ occurs at the endpoint 0. For $M = |f''(0)| = 2$, $b = 1$, $a = 0$, and $n = 4$, an upper estimate to the error is

$$\text{Error} \le \frac{(1-0)^3(2)}{(12)(4^2)} = \frac{1}{96} \approx 0.0104$$

As a result of Example 1, we have

$$0.78279 - 0.0104 < \int_0^1 \frac{dx}{1+x^2} < 0.78279 + 0.0104$$

Since the exact value of the integral $\int_0^1 dx/(1+x^2)$ is $\tan^{-1}x\big|_0^1 = \pi/4$, we conclude that

$$0.77239 < \frac{\pi}{4} < 0.79319$$

$$3.08956 < \pi < 3.17276$$

In Chapter 11 we give another way of approximating π, which equals 3.14159 correct to five decimal places.

The next example illustrates how the trapezoidal rule is used when only discrete information is known.

EXAMPLE 2

A tree trunk is 140 feet long. At a distance x feet from one end, its sectional area A is given in square feet by the following table at intervals of 20 feet:

x	0	20	40	60	80	100	120	140
A	120	124	128	130	132	136	144	158

Find the approximate volume of the tree trunk.

Solution

From Chapter 6 we know that the volume is

$$V = \int_0^{140} A\,dx$$

Since $n = 7$, $a = 0$, and $b = 140$, by the trapezoidal rule, we find

$$V \approx \tfrac{140}{14}[120 + 2(124) + 2(128) + 2(130) + 2(132) + 2(136) + 2(144) + 158]$$
$$= 18{,}660 \text{ cubic feet}$$

[9.9.3] THEOREM / *Simpson's Rule.*

If a function f is continuous on the closed interval $[a, b]$, then

$$\int_a^b f(x)\,dx \approx \frac{b-a}{3n}[f(x_0) + 4f(x_1) + 2f(x_2) + 4f(x_3) + 2f(x_4) + \cdots + 2f(x_{n-2}) + 4f(x_{n-1}) + f(x_n)] \quad (3)$$

where the closed interval $[a, b]$ has been partitioned into an even number n of subintervals $[x_0, x_1], [x_1, x_2], \ldots, [x_{n-1}, x_n]$, each of length $(b - a)/n$.

Simpson's rule (named after the English mathematician Thomas Simpson, 1710–1761) is obtained by interpreting the definite integral as an area under a graph and by approximating the graph as a collection of parabolic arcs. By using parabolic arcs, instead of chords as in the derivation of the trapezoidal rule, we often get a closer approximation to the area. Thus, as a preliminary to the derivation of Simpson's rule, we need to find a formula for the area under a parabolic arc.

Let the graph in Figure 11 represent the parabola $y = ax^2 + bx + c$. Draw the lines $x = -h$ and $x = h$, and denote the ordinates at $x = -h$, $x = 0$, and $x = h$ by y_0, y_1, and y_2, respectively. The area enclosed by the parabola, the x-axis, and the lines $x = -h$ and $x = h$ is given by

$$\text{Area} = \int_{-h}^{h} y\,dx = \int_{-h}^{h} (ax^2 + bx + c)\,dx = \left(a\frac{x^3}{3} + b\frac{x^2}{2} + cx\right)\bigg|_{-h}^{h}$$

$$= \frac{2}{3}ah^3 + 2ch = \frac{h}{3}(2ah^2 + 6c) \qquad \textbf{(4)}$$

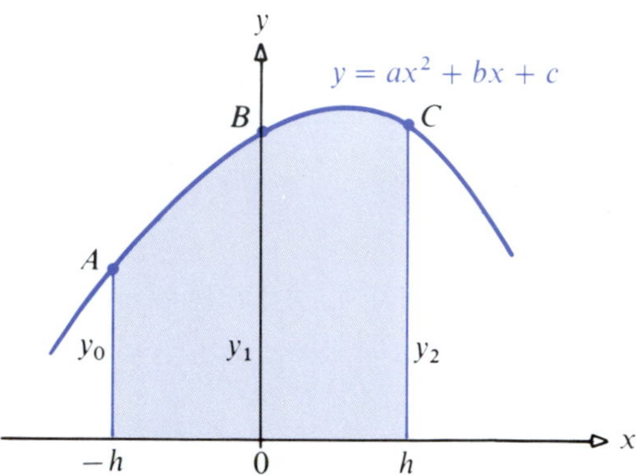

Figure 11

Since the three points $A = (-h, y_0)$, $B = (0, y_1)$, and $C = (h, y_2)$ lie on the parabola, their coordinates satisfy the equation $y = ax^2 + bx + c$, and we have the system of equations

$$\begin{aligned} y_0 &= ah^2 - bh + c \\ y_1 &= c \\ y_2 &= ah^2 + bh + c \end{aligned} \qquad \textbf{(5)}$$

Then

$$y_0 + y_2 = 2ah^2 + 2c$$

and, therefore,

$$y_0 + y_2 + 4y_1 = 2ah^2 + 6c$$

By substituting this in the area formula (4), we get

$$\text{Area} = \frac{h}{3}(y_0 + 4y_1 + y_2) \qquad \textbf{(6)}$$

This formula depends on only the three ordinates and the distance h, and so it is independent of the position of the y-axis. We may state the result as follows: If a parabola with vertical axis is passed through three points

(x_0, y_0), (x_1, y_1), and (x_2, y_2), with the distance h between consecutive abscissas, the area enclosed by the parabola, the x-axis, and the lines $x = x_0$ and $x = x_2$ is given by

$$\text{Area} = \frac{h}{3}(y_0 + 4y_1 + y_2)$$

Proof of Simpson's Rule Let $\int_a^b f(x)\,dx$ be the integral to be approximated; it then equals the area under the graph of $y = f(x)$ from $x = a$ to $x = b$ (see Fig. 12). Partition the closed interval $[a, b]$ into an even number n of subintervals, each of length $\Delta x = (b - a)/n$. The ordinates corresponding to $x_0 = a, x_1, x_2, \ldots, x_{n-1}, x_n = b$ are denoted by $y_0, y_1, y_2, \ldots, y_{n-1}, y_n$. Arrange these ordinates in groups of three, with the last ordinate of each group being the same as the first ordinate of the next group. Through each group of three ordinates, pass an arc of a parabola with vertical axis.

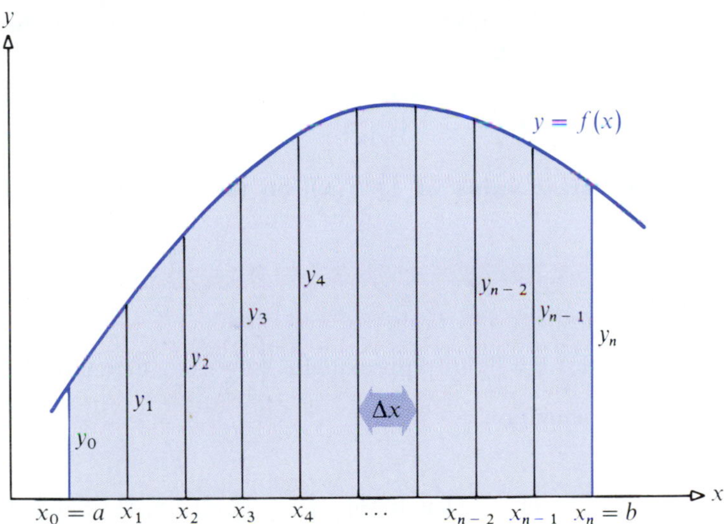

Figure 12

By (6) the area under the graph $y = f(x)$ from $x = x_0$ to $x = x_2$ may be approximated by the area under the parabola passing through (x_0, y_0), (x_1, y_1), and (x_2, y_2)—namely,

$$A_1 = \frac{b - a}{3n}(y_0 + 4y_1 + y_2)$$

Similarly, the area under the graph of $y = f(x)$ from $x = x_2$ to $x = x_4$ may be approximated by the area under the parabola passing through (x_2, y_2), (x_3, y_3), and (x_4, y_4)—namely,

$$A_2 = \frac{b - a}{3n}(y_2 + 4y_3 + y_4)$$

Continuing in this way (by now you should realize why the number of subintervals must be even), we can approximate $\int_a^b f(x)\,dx$ by

$$\int_a^b f(x)\,dx \approx A_1 + A_2 + \cdots + A_n$$

$$= \frac{b-a}{3n}(y_0 + 4y_1 + y_2) + \frac{b-a}{3n}(y_2 + 4y_3 + y_4)$$

$$+ \frac{b-a}{3n}(y_4 + 4y_5 + y_6) + \cdots + \frac{b-a}{3n}(y_{n-2} + 4y_{n-1} + y_n)$$

$$= \frac{b-a}{3n}(y_0 + 4y_1 + 2y_2 + 4y_3 + 2y_4 + \cdots + 2y_{n-2} + 4y_{n-1} + y_n)$$

Thus, we have (3).

[9.9.4] THEOREM / *Error.*

The difference between the exact value of the integral $\int_a^b f(x)\,dx$ and the approximate value given by Simpson's rule is called the *error*. It may be estimated by the formula

$$\text{Error} \le \frac{M(b-a)^5}{180n^4} \tag{7}$$

where M is the greatest value of $|f^{(4)}(x)|$ on the closed interval $[a, b]$.*

EXAMPLE 3

Use Simpson's rule with $n = 4$ to approximate $\int_0^1 \frac{dx}{1+x}$. Estimate the error in using this approximation.

Solution

For $n = 4$, the partition of [0, 1] is $[0, \frac{1}{4}]$, $[\frac{1}{4}, \frac{1}{2}]$, $[\frac{1}{2}, \frac{3}{4}]$, $[\frac{3}{4}, 1]$. Table 4 summarizes the information we need. Thus,

$$\int_0^1 \frac{dx}{1+x} \approx \left(\frac{1}{12}\right)(1 + 3.2 + 1.33333 + 2.28572 + 0.5) = 0.69325$$

Table 4

	$x = 0$	$x = \frac{1}{4}$	$x = \frac{1}{2}$	$x = \frac{3}{4}$	$x = 1$
$f(x) = 1/(1+x)$	1	0.8	0.66666	0.57143	0.5
Factor	×1	×4	×2	×4	×1
Product	1	3.2	1.33333	2.28572	0.5

* The verification of this formula, which uses the extension of the mean value theorem, is found in advanced calculus books.

To estimate the error, we need to find the maximum value of $|f^{(4)}(x)| = |24(1+x)^{-5}|$. On $[0, 1]$, the largest value of $|24(1+x)^{-5}|$ occurs at $x = 0$. For $M = |f^{(4)}(0)| = 24$, $b = 1$, $a = 0$, and $n = 4$, an upper estimate to the error is

$$\text{Error} \leq \frac{24(1-0)^5}{(180)(4^4)} = \frac{1}{1920} \approx 0.00052$$

As a result of Example 3, we have

$$0.69325 - 0.00052 < \int_0^1 \frac{dx}{1+x} < 0.69325 + 0.00052$$

Since the exact value of $\int_0^1 dx/(1+x)$ is $\ln(x+1)|_0^1 = \ln 2$, we conclude that

$$0.69273 < \ln 2 < 0.69377$$

In Chapter 11 we give a different technique for approximating ln 2, which equals 0.69314 correct to five decimal places.

EXERCISE 9.9

In Problems 1–10 use the trapezoidal rule to approximate each integral.

1. $\int_0^4 x^2\,dx;\quad n = 8$

2. $\int_0^3 x^3\,dx;\quad n = 6$

3. $\int_1^2 \frac{dx}{x};\quad n = 4$

4. $\int_0^1 \frac{dx}{1+x};\quad n = 6$

5. $\int_0^\pi \sin x\,dx;\quad n = 6$

6. $\int_{-\pi/2}^{\pi/2} \cos x\,dx;\quad n = 6$

7. $\int_{\pi/2}^{\pi} \frac{\sin x}{x}\,dx;\quad n = 3$

8. $\int_{3\pi/2}^{2\pi} \frac{\cos x}{x}\,dx;\quad n = 3$

9. $\int_0^1 e^{-x^2}\,dx;\quad n = 4$

10. $\int_0^1 e^{x^2}\,dx;\quad n = 4$

In Problems 11–20 use Simpson's rule to approximate each integral.

11. $\int_0^4 x^2\,dx;\quad n = 8$

12. $\int_0^3 x\,dx;\quad n = 6$

13. $\int_1^2 \frac{dx}{x^2};\quad n = 4$

14. $\int_0^2 \sqrt{x}\,dx;\quad n = 6$

15. $\int_0^2 \frac{1}{\sqrt{1+x}}\,dx;\quad n = 6$

16. $\int_0^\pi \sin x\,dx;\quad n = 6$

17. $\int_0^\pi \sqrt{\sin x}\,dx;\quad n = 6$

18. $\int_{-\pi/2}^{\pi/2} \sqrt{\cos x}\,dx;\quad n = 6$

19. $\int_0^1 e^{-x^2}\,dx;\quad n = 4$

20. $\int_0^1 e^{x^2}\,dx;\quad n = 4$

In Problems 21–24 approximate each integral by both the trapezoidal rule and Simpson's rule.

21. $\int_3^6 \frac{x\,dx}{4+x^2}$; use six subintervals

22. $\int_0^1 \frac{dx}{1+x^2}$; use ten subintervals

23. $\int_4^7 \sqrt{9+x^2}\,dx$; use six subintervals

24. $\int_0^{\pi/2} \sqrt{\sin x}\,dx$; use four subintervals

25. Show that $\int_1^2 dx/x = \ln 2$. Then use the trapezoidal rule with $n = 5$ to approximate $\int_1^2 dx/x$ and hence obtain an approximation to ln 2.

26. Use Simpson's rule with $n = 6$ to approximate $\int_1^2 dx/x$ and hence obtain an approximation to ln 2.

27. In the table, S is the area in square meters of the cross section of a railroad cutting, and x meters is the corresponding distance along the line. Use the trapezoidal rule to calculate the number of cubic meters of earth removed to make the cutting from $x = 0$ to $x = 150$. Do not attempt to compute an error, since a function f for the area is not known.

x	0	25	50	75	100	125	150
S	105	118	142	120	110	90	78

28. A series of soundings taken across a river channel is given in the table, where x is the distance from one shore and y is the corresponding depth. Draw the section and find its area by the trapezoidal rule.

x	0	10	20	30	40	50	60	70	80
y	5	10	13.2	15	15.6	12	6	4	0

29. In the table, F is the force in pounds acting on a body in its direction of motion and s is the displacement in feet. Use the trapezoidal rule to calculate the total work done by the force from $s = 0$ to $s = 50$.

s	0	5	10	15	20	25	30	35	40	45	50
F	100	80	66	56	50	45	40	36	33	30	28

30. Use Simpson's rule to find the area enclosed by the pairs of rectangular coordinates in the table, the x-axis, and the lines $x = 2$ and $x = 4.4$.

x	2.0	2.4	2.8	3.2	3.6	4.0	4.4
y	3.03	4.61	5.80	6.59	7.76	8.46	9.19

31. The area of the horizontal section of a reservoir is A square meters at a height x meters from the bottom; corresponding values of x and A are given in the table. Find the volume of water in the reservoir by use of the trapezoidal rule and also by Simpson's rule.

x	0	2.5	5	7.5	10	12.5	15	17.5	20	22.5	25
A	0	2510	3860	4870	5160	5590	5810	6210	6890	7680	8270

32. A gas expands from a volume of 1 cubic inch to 2.5 cubic inches; values of the volume (v) and pressure (p, in pounds per square inch) during the expansion are given in the table. Calculate the total work W done in the expansion by using Simpson's rule $(W = \int_a^b p\,dv)$.

v	1	1.25	1.5	1.75	2	2.25	2.5
p	68.7	55.0	45.8	39.3	34.4	30.5	27.5

33. Use Simpson's rule to approximate the area of the pond pictured in the figure.

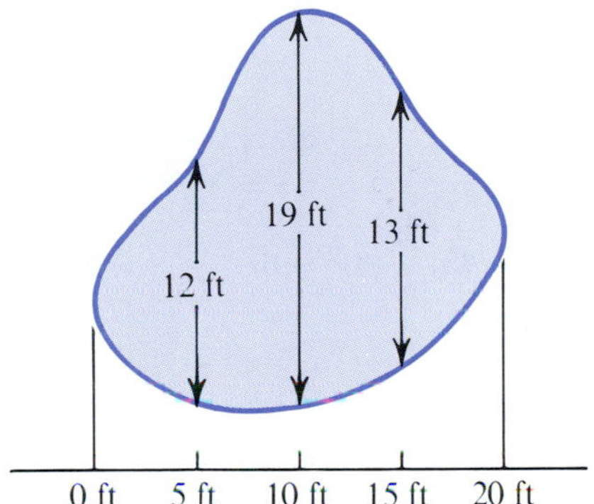

34. Use the trapezoidal rule to find the arc length of the ellipse $9x^2 + 100y^2 = 900$ in the first quadrant from $x = 0$ to $x = 8$. Divide the interval into four equal parts, and obtain an answer to three decimal places.

35. Use (2) to estimate the error that can result in Problem 1.

36. Use (2) to estimate the error that can result in Problem 2.

37. Use (7) to estimate the error that can result in Problem 13.

38. Use (7) to estimate the error that can result in Problem 14.

39. Approximate the arc length of $y = \sin x$ from $x = 0$ to $x = \pi/2$ by using the trapezoidal rule with $n = 3$.

40. Rework Problem 39 by using Simpson's rule with $n = 4$.

41. Approximate the volume of the solid generated by revolving the area under the graph of $y = \sin x$ from $x = 0$ to $x = \pi/2$ about the y-axis by using the trapezoidal rule with $n = 3$.

42. Rework Problem 41 by using Simpson's rule with $n = 4$.

43. For the function below, use the trapezoidal rule with $n = 6$ to approximate $\int_0^\pi f(x)\,dx$.

$$f(x) = \begin{cases} (\sin x)/x & \text{if } x \neq 0 \\ 1 & \text{if } x = 0 \end{cases}$$

44. Use Simpson's rule with $n = 6$ to rework Problem 43.

45. The velocity (v, in meters per second) of a particle at time t is given by the table. Use the trapezoidal rule to approximate the distance traveled from $t = 0$ to $t = 3$.

t	0	0.5	1	1.5	2	2.5	3
v	5.1	5.3	5.6	6.1	6.8	6.7	6.5

46. Use Simpson's rule to rework Problem 45.

REVIEW EXERCISES

In Problems 1–65 evaluate each integral.

1. $\int x^2(x^2 - 2)\,dx$

2. $\int \sqrt{ax + b}\,dx$

3. $\int \dfrac{dx}{x^2 + 4x + 20}$

4. $\int x\sqrt{x + 2}\,dx$

5. $\int \sec^2 2\theta\,d\theta$

6. $\int \dfrac{y^3 + y}{y + 1}\,dy$

7. $\int \dfrac{\sqrt{x}\,dx}{1 + x}$

8. $\int \sec^3\phi \tan\phi\,d\phi$

9. $\int \dfrac{t\,dt}{9 + t^2}$

10. $\int \dfrac{dt}{9 + t^2}$

11. $\int \cot^2\theta\,d\theta$

12. $\int \dfrac{x\,dx}{(1 + x)^4}$

13. $\int e^{\cos x}\sin x\,dx$

14. $\int \sin^3\phi\,d\phi$

15. $\int x\sin 2x\,dx$

16. $\int \frac{(y-2)\,dy}{y^2-4y+2}$

17. $\int (z^2+4)^2\,dz$

18. $\int v\csc^2 v\,dv$

19. $\int t^3\sqrt{2-t}\,dt$

20. $\int x(1-x^2)\,dx$

21. $\int \sin^2 x\cos^3 x\,dx$

22. $\int (4-x^2)^{3/2}\,dx$

23. $\int \frac{dy}{\sqrt{2y+1}}$

24. $\int \frac{3x^2+1}{x^3+2x^2-3x}\,dx$

25. $\int \frac{e^{2t}\,dt}{e^t-2}$

26. $\int \frac{2z+3}{\sqrt{1+2z}}\,dz$

27. $\int \tanh 2v\,dv$

28. $\int \frac{dy}{5+4y+4y^2}$

29. $\int \frac{\sin x+\cos x}{\tan x}\,dx$

30. $\int \frac{e^{2x}\,dx}{e^{2x}+1}$

31. $\int \frac{x\,dx}{x^4-4}$

32. $\int x^3e^{x^2}\,dx$

33. $\int \frac{dx}{x\ln x}$

34. $\int \frac{y^2\,dy}{(y+1)^3}$

35. $\int \tan\left(\frac{1}{4}\pi-\theta\right)d\theta$

36. $\int \frac{\sqrt[4]{x}}{\sqrt{x}+\sqrt[3]{x}}\,dx$

37. $\int a^xb^x\,dx$

38. $\int \sin^2\theta\csc^2 2\theta\,d\theta$

39. $\int x\sec^2 x\,dx$

40. $\int \frac{dx}{\sqrt{16+4x-2x^2}}$

41. $\int \frac{x^3-2x}{x-1}\,dx$

42. $\int \ln(1-y)\,dy$

43. $\int (3y^2-6y)^3(y-1)\,dy$

44. $\int \cos^n\theta\sin\theta\,d\theta$

45. $\int \frac{3x^2\,dx}{1-x}$

46. $\int x^2e^x\,dx$

47. $\int \cos^3 3x\,dx$

48. $\int \frac{dy}{\sqrt{2+3y^2}}$

49. $\int \frac{x^2\,dx}{4-x^2}$

50. $\int \csc^4 x\,dx$

51. $\int x^2\sin^{-1}x\,dx$

52. $\int \frac{\sec^2 z\,dz}{a+b\tan z}$

53. $\int \frac{x+\sqrt{x^2+1}}{x-\sqrt{x^2+1}}\,dx$

54. $\int e^{y/2}\,dy$

55. $\int x\cos^2 x\,dx$

56. $\int \frac{\sec^2 x}{\sqrt{\tan^2 x-6\tan x+10}}\,dx$

57. $\int \frac{dx}{x^2+ax}$

58. $\int \sin^4 y\cos^4 y\,dy$

59. $\int \frac{w^3\,dw}{1-w^2}$

60. $\int \frac{\cos^2 mx\,dx}{\sin^3 mx}$

61. $\int \sin mx\sin nx\,dx \quad m\neq n$

62. $\int \sin mx\cos nx\,dx \quad m\neq n$

63. $\int \cos mx\cos nx\,dx \quad m\neq n$

64. $\int (\ln x)^{-1}\,dx$

65. $\int \frac{x^{3/4}}{x^{1/5}}\,dx$

66. Find $\int dx/\sqrt{x^2-a^2}$, $a>0$, using an appropriate hyperbolic function substitution. Express your answer in logarithmic form.

67. If $\int x^2\cos x\,dx = f(x) - \int 2x\sin x\,dx$, find f.

68. If the graph of $y=f(x)$ contains the point $(0, 2)$, $dy/dx = -x/ye^{x^2}$, and $f(x)>0$ for all x, find f.

69. Let f be a function defined for all $x>-5$ with the following properties: (i) $f''(x) = 1/(3\sqrt{x+5})$ for all x in the domain of f; (ii) the line tangent to the graph of f at $(4, 2)$ has an angle of inclination of 45°. Find an expression for f.

70. The area in the first quadrant enclosed by the graph of $y = \sec x$, $x = \pi/4$, and the axes is rotated about the x-axis. What is the volume of the solid generated?

71. Let A be the area enclosed by the graph of $y = \ln x$, the line $x = e$, and the x-axis.

(a) Find the volume generated by revolving A about the x-axis.

(b) Find the volume generated by revolving A about the y-axis.

72. Let A be the area in the first quadrant enclosed by $y = \sec x$, $y = 2 \sin x$, and the y-axis.

(Problem 72 continues in next column.)

72. *(continued)*

(a) Find A.

(b) Find the volume of the solid formed when A is revolved about the x-axis.

73. Find the length of the graph $y = \ln x$ from $x = \sqrt{3}/3$ to $x = \sqrt{3}$.

74. Let T_n be the approximation to $\int_a^b f(x)\,dx$ given by the trapezoidal rule with n subintervals. Without appealing to the error formula given in the text, prove that $\lim_{n\to+\infty} T_n = \int_a^b f(x)\,dx$.

75. Show that if $f(x) = Ax^3 + Bx^2 + Cx + D$, then Simpson's rule gives the exact value of $\int_a^b f(x)\,dx$.

CHALLENGE EXERCISES

In Problems 1–4 evaluate each integral.

1. $\displaystyle\int \frac{x+1}{x\sqrt{x-2}}\,dx$

2. $\displaystyle\int \sqrt{\frac{a+x}{a-x}}\,dx$

3. $\displaystyle\int \frac{dv}{\sqrt{3v - v^2}}$

4. $\displaystyle\int \frac{(3t+2)\,dt}{t\sqrt{t+1}}$

5. Define $Si(x)$ as

$$Si(x) = \int \frac{\sin x}{x}\,dx$$

$Si(x)$ cannot be expressed in elementary terms; that is, there is no way to express this integral in finite terms involving functions you know so far. However, if we admit $Si(x)$ to the family of "known" functions, then $\int Si(x)\,dx$ is so expressible. Show this.

6. Show that for any positive integer n, we have

$$\int_0^1 e^{x^2}\,dx = e\cdot\left\{1 - \frac{2}{3} + \frac{4}{15} - \frac{8}{105} + \cdots + \frac{(-1)^n 2^n}{(2n+1)(2n-1)\cdots 3\cdot 1}\right\} + (-1)^{n+1}$$

$$\times \frac{2^{n+1}}{(2n+1)(2n-1)\cdots 3\cdot 1}\int_0^1 x^{2n+2}e^{x^2}\,dx$$

7. Starting with the identity

$$f(b) - f(a) = \int_a^b f'(t)\,dt$$

derive the following generalizations of the mean value theorem:

(a) $f(b) - f(a) = f'(a)(b-a) - \displaystyle\int_a^b f''(t)(t-b)\,dt,$

(b) $f(b) - f(a) = f'(a)(b-a) + \dfrac{f''(a)}{2}(b-a)^2 + \displaystyle\int_a^b \frac{f'''(t)}{2}(t-b)^2\,dt.$

8. If $y = f(x)$ has the inverse function given by $x = f^{-1}(y)$, show that

$$\int_a^b f(x)\,dx + \int_{f(a)}^{f(b)} f^{-1}(y)\,dy = bf(b) - af(a)$$

This relationship allows us to choose whichever of the two integrals is the easier to evaluate.

9. *Computer Problem.* Another approach to approximating an integral is called the *Monte Carlo method.* Suppose we have a dart board consisting of a circle inscribed in a square that measures 2 meters on each side (see the figure). Suppose we randomly throw 10,000 darts. (By "randomly" we mean that a dart is equally likely to land anywhere within the square.) If we hit the circle 8567 times, it is reasonable to approximate the area of the circle as

$$\frac{\text{Hits}}{\text{Total}} \times \text{Area of square} = \frac{8567}{10{,}000} \times 4 = 3.4268$$

(Note that the area of the circle is $\pi R^2 = \pi = 3.14159\ldots,$ so that we have an approximation to π.)

It would be rather tedious actually to construct the dart board and throw 10,000 darts, and it would be impossible to throw the darts in a truly random manner. But, fortunately, we can simulate dart throwing on a computer. All that is needed is a random number generator and a few lines of computer coding. Suppose the command RND(1) returns a random

(Problem 9 continues on page 584)

9. (*continued*)

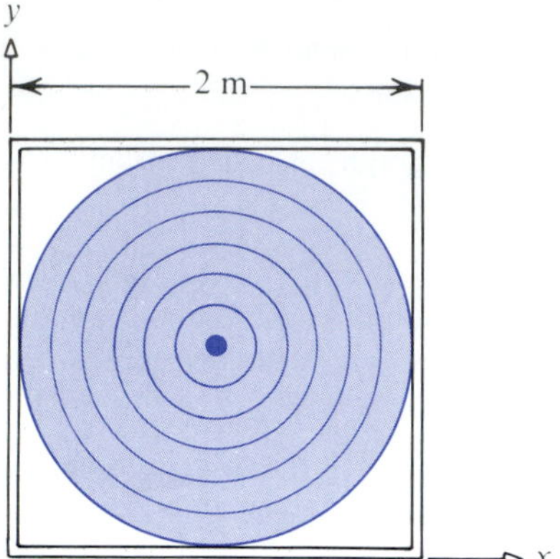

number between 0 and 1. By multiplying by 2, we obtain a random number between 0 and 2. If we call RND(1) 20,000 times, we will obtain 10,000 pairs of numbers, or 10,000 points in our square. Then all we need to do is determine which points are within the circle and keep track of the number of hits. A portion of the computer program in the BASIC language* might be

```
HIT = Ø
FOR I = 1 TO 1Ø,ØØØ
X = 2 * RND(1)
Y = 2 * RND(1)
IF (X - 1) ∧ 2 + (Y - 1) ∧ 2 < 1 THEN
    HIT = HIT + 1
NEXT I
```

Write a program to estimate the area of the circle using the Monte Carlo method.

10. *Computer Problem.* Use the Monte Carlo method (see Problem 9) to approximate the integrals:

(a) $\int_0^1 \frac{dx}{1+x^2}$ (b) $\int_0^1 \frac{dx}{1+x}$ (c) $\int_0^1 e^{x^2}\,dx$

(d) $\int_0^1 f(x)\,dx$, where

$$f(x) = \begin{cases} \dfrac{\sin x}{x} & \text{for} \quad 0 < x \le 1 \\ 1 & \text{for} \quad x = 0 \end{cases}$$

Compare your results in parts (a) and (b) to the approximations of Examples 1 and 3 in Section 9.9.

* The variable names in this program may have to be adjusted to be compatible with your particular system.

11. *Buffon Needle Problem.* A floor pattern consists of lines 1 meter apart. Find the probability that a meter stick thrown at random hits a crack (see the figure).

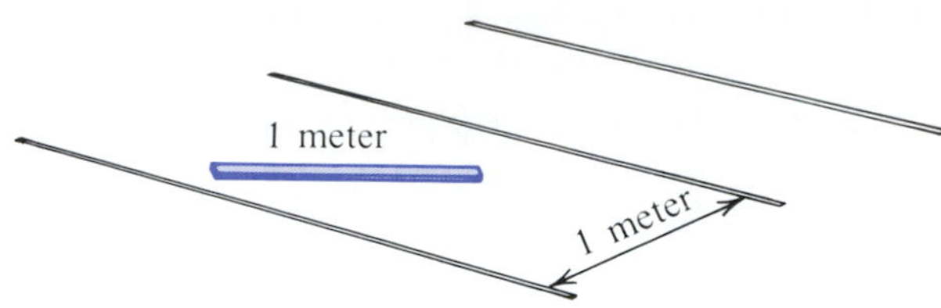

[*Hint:* Let d = Distance from bottom of stick to next higher line and let θ = Angle from horizontal to stick, as indicated:

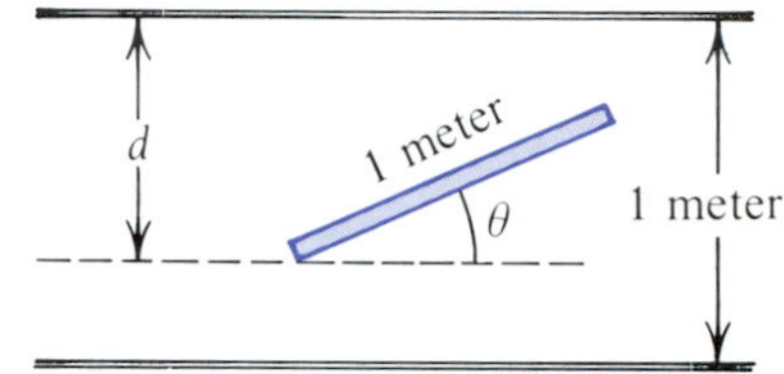

The inequalities $0 \le d \le 1$ and $0 \le \theta \le \pi$ represent all possibilities:

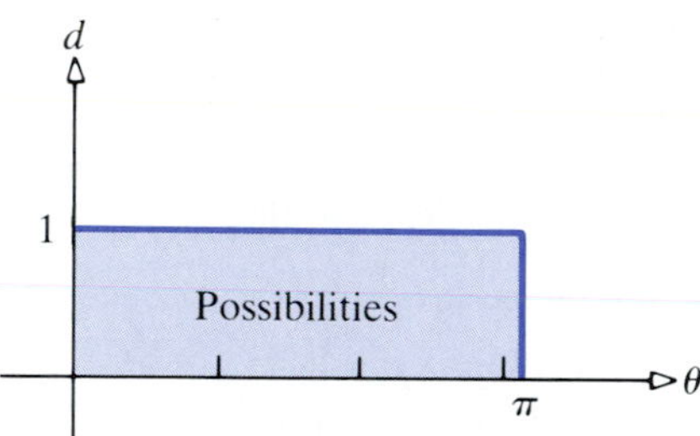

Show that the meter stick hits a crack if and only if $d \le \sin\theta$, so that the shaded area in the figure represents the hits. Deduce that the desired probability is $\int_0^\pi \sin\theta\,d\theta/\pi$.]

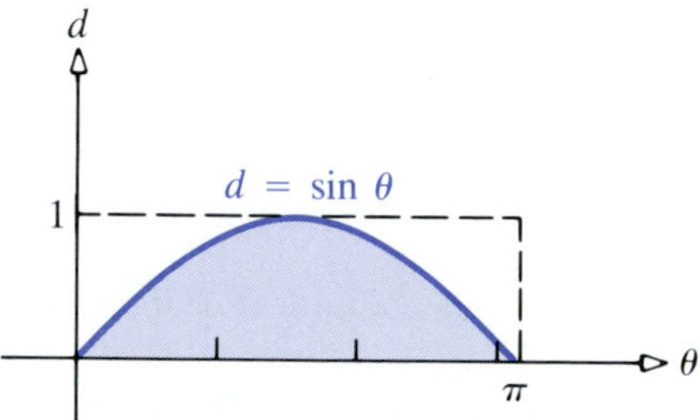

12. *Computer Problem.* Simulate Problem 11 on a computer. Compare the approximation obtained with the exact result obtained in Problem 11.

Indeterminate Forms; Improper Integrals; Taylor Polynomials

10

10.1 The Indeterminate Forms 0/0 and ∞/∞

To evaluate

$$\lim_{x \to c} \frac{f(x)}{g(x)}$$

we usually try first to use the rule

$$\lim_{x \to c} \frac{f(x)}{g(x)} = \frac{\lim_{x \to c} f(x)}{\lim_{x \to c} g(x)} \qquad (1)$$

However, in many cases, this result cannot be applied.

For example, to find

$$\lim_{x \to 2} \frac{x^2 - 4}{x - 2}$$

we do not use (1), since the numerator and the denominator each approach 0. That is,

$$\frac{\lim_{x\to 2}(x^2-4)}{\lim_{x\to 2}(x-2)} \quad \text{leads to} \quad \frac{0}{0}$$

Instead, to find this limit, we rely on algebra to get

$$\lim_{x\to 2}\frac{x^2-4}{x-2} = \lim_{x\to 2}\frac{(x+2)(x-2)}{x-2} = \lim_{x\to 2}(x+2) = 4$$

To find

$$\lim_{x\to +\infty}\frac{3x-2}{x+5}$$

we do not use (1), since the numerator and the denominator each approach infinity. That is,

$$\frac{\lim_{x\to +\infty}(3x-2)}{\lim_{x\to +\infty}(x+5)} \quad \text{leads to} \quad \frac{\infty}{\infty}$$

Instead, to find this limit, we divide the numerator and the denominator by x and find

$$\lim_{x\to +\infty}\frac{3-(2/x)}{1+(5/x)} = 3$$

As another example, to find

$$\lim_{x\to 0}\frac{\sin x}{x}$$

we do not use (1), since it leads to 0/0. Instead, we use a geometric argument (see Section 3.7 in Chap. 3) to show that

$$\lim_{x\to 0}\frac{\sin x}{x} = 1$$

Whenever the application of (1) leads to 0/0 or ∞/∞, we say that $f(x)/g(x)$ is an *indeterminate form*. More precisely:

The quotient $f(x)/g(x)$ is an *indeterminate form at c* if

$$\lim_{x\to c} f(x) = 0 \quad \text{and} \quad \lim_{x\to c} g(x) = 0 \qquad \text{Type } \frac{0}{0} \tag{2}$$

or

$$\lim_{x\to c} f(x) = \pm\infty \quad \text{and} \quad \lim_{x\to c} g(x) = \pm\infty \qquad \text{Type } \frac{\infty}{\infty} \tag{3}$$

L'Hospital's Rule

The word *indeterminate* conveys the idea that the limit cannot be determined without additional work. However, a remarkable theorem, named after the

French mathematician Guillaume Francois de L'Hospital* (1661–1704), provides a simple and general method for finding the limit of an indeterminate form.

[10.1.1] THEOREM / *L'Hospital's Rule*†

Suppose f and g are two functions that are differentiable at every point in an open interval I containing c, except possibly at c itself, and suppose $g'(x) \neq 0$ for all $x \neq c$ in I. Let L denote either a real number or $+\infty$ or $-\infty$, and suppose $f(x)/g(x)$ is the indeterminate form at c of the type $0/0$ or ∞/∞.

$$\text{If} \quad \lim_{x \to c} \frac{f'(x)}{g'(x)} = L \quad \text{then} \quad \lim_{x \to c} \frac{f(x)}{g(x)} = L \tag{4}$$

A partial proof of L'Hospital's rule is given at the end of this section.

Under the conditions stated in the theorem, the limit of a quotient of two functions equals the limit of the quotient of their derivatives. Here are the basic steps used in finding the limits of functions with L'Hospital's rule.

Step 1 Check that the limit $f(x)/g(x)$ is the indeterminate form $0/0$ or ∞/∞. If it is not, do not use L'Hospital's rule.

Step 2 Differentiate f and g separately.

Step 3 Find $\lim_{x \to c} \dfrac{f'(x)}{g'(x)}$. This limit is equal to $\lim_{x \to c} \dfrac{f(x)}{g(x)}$.

EXAMPLE 1 Type $\dfrac{0}{0}$

Use L'Hospital's rule to verify that $\lim_{x \to 0} \dfrac{\sin x}{x} = 1$.

Solution

Let $f(x) = \sin x$ and $g(x) = x$. Since $\lim_{x \to 0} f(x) = \lim_{x \to 0} \sin x = 0$ and $\lim_{x \to 0} g(x) = \lim_{x \to 0} x = 0$, we find that $f(x)/g(x) = (\sin x)/x$ is an indeterminate form at 0 of the type $0/0$. Since

$$\lim_{x \to 0} \frac{f'(x)}{g'(x)} = \lim_{x \to 0} \frac{\cos x}{1} = 1$$

it follows by L'Hospital's rule that

$$\lim_{x \to 0} \frac{\sin x}{x} = 1$$

* Pronounced "lōpital" and sometimes spelled L'Hôpital.

† The result was actually discovered by his teacher, Johann Bernoulli, who sent L'Hospital his proof in a letter in 1694.

In the solution of Example 1 we were careful to determine that the limit of the ratio of the derivatives—namely,

$$\lim_{x\to 0} \frac{\cos x}{1}$$

existed or became infinite *before* applying L'Hospital's rule. However, to avoid complicating things, the usual practice is to proceed directly as follows:

$$\lim_{x\to 0} \frac{\sin x}{x} \underset{\substack{\uparrow \\ \text{Apply} \\ \text{L'Hospital's} \\ \text{rule.}}}{=} \lim_{x\to 0} \frac{\cos x}{1} \underset{\substack{\uparrow \\ \text{Evaluate} \\ \text{limit.}}}{=} 1$$

A word of caution! L'Hospital's rule is applied by taking the derivative of the numerator and the derivative of the denominator *separately*. Be careful not to take the derivative of the quotient.

The next example illustrates that sometimes we may have to apply L'Hospital's rule more than once.

EXAMPLE 2 Type $\frac{0}{0}$

Evaluate: $\lim_{x\to 0} \frac{\sin x - x}{x^2}$

Solution

Since $\lim_{x\to 0}(\sin x - x) = 0$ and $\lim_{x\to 0} x^2 = 0$, we have an indeterminate form at 0 of the type 0/0. Applying L'Hospital's rule, we find

$$\lim_{x\to 0} \frac{\sin x - x}{x^2} = \lim_{x\to 0} \frac{\cos x - 1}{2x}$$

Since $\lim_{x\to 0}(\cos x - 1) = 0$ and $\lim_{x\to 0} 2x = 0$, we still have an indeterminate form at 0 of the type 0/0. Hence, we apply the rule once more. Then

$$\lim_{x\to 0} \frac{\cos x - 1}{2x} = \lim_{x\to 0} \frac{-\sin x}{2} = 0$$

Since the latter limit equals 0, we conclude that

$$\lim_{x\to 0} \frac{\sin x - x}{x^2} = 0$$

The statement (4) is also true if $c = +\infty$ or $c = -\infty$. The next examples illustrate the use of (4) for $c = +\infty$ as it is applied to indeterminate forms at $+\infty$ of the type ∞/∞.

EXAMPLE 3 Type $\frac{\infty}{\infty}$

Evaluate: (a) $\lim_{x\to +\infty} \frac{\ln x}{x}$ (b) $\lim_{x\to +\infty} \frac{x}{e^x}$

Solution

(a) Since $\lim_{x\to+\infty} \ln x = +\infty$ and $\lim_{x\to+\infty} x = +\infty$, we have an indeterminate form at $+\infty$ of the type ∞/∞. By L'Hospital's rule, we have

$$\lim_{x\to+\infty} \frac{\ln x}{x} = \lim_{x\to+\infty} \frac{1/x}{1} = \lim_{x\to+\infty} \frac{1}{x} = 0$$

(b) Since $\lim_{x\to+\infty} x = +\infty$ and $\lim_{x\to+\infty} e^x = +\infty$, we have an indeterminate form at $+\infty$ of the type ∞/∞. Hence, by L'Hospital's rule, we have

$$\lim_{x\to+\infty} \frac{x}{e^x} = \lim_{x\to+\infty} \frac{1}{e^x} = 0$$

The results obtained in this example tell us that x grows faster than $\ln x$, while e^x grows faster than x. In fact, this is true for any *power* of x. That is, for α a positive number,

$$\lim_{x\to+\infty} \frac{\ln x}{x^\alpha} = 0 \qquad \lim_{x\to+\infty} \frac{x^\alpha}{e^x} = 0 \tag{5}$$

You are asked to verify these results in Problems 35 and 37.

EXAMPLE 4 Type $\frac{0}{0}$

Evaluate: $\lim_{x\to-1} \dfrac{3x^2 + 2x - 1}{x^2 + x}$

Solution

Since $\lim_{x\to-1}(3x^2 + 2x - 1) = 0$ and $\lim_{x\to-1}(x^2 + x) = 0$, we have an indeterminate form at -1 of the type 0/0. Hence, we apply L'Hospital's rule:

$$\lim_{x\to-1} \frac{3x^2 + 2x - 1}{x^2 + x} = \lim_{x\to-1} \frac{6x + 2}{2x + 1} = 4$$

Before applying L'Hospital's rule, it is important to check on the indeterminacy of the expression, since a routine and thoughtless application may yield an incorrect result. For instance, in Example 4, note that in obtaining the last equality, we use the fact that the quotient $(6x + 2)/(2x + 1)$ is continuous at $x = -1$, so that a direct substitution yields 4. Be careful not to apply L'Hospital's rule to expressions such as $(6x + 2)/(2x + 1)$. In this case, you would obtain $\frac{6}{2} = 3$, which is a wrong answer.

The next example illustrates how an early simplication may reduce the effort needed to solve the problem.

EXAMPLE 5 Type $\frac{0}{0}$

Evaluate: $\lim_{x\to 0} \dfrac{\tan x - \sin x}{x^2 \tan x}$

Solution

This is an example of an indeterminate form of the type 0/0. However, due to the complicated nature of the denominator, we attempt to simplify using the fact that $\tan x = (\sin x)/(\cos x)$ *before* applying L'Hospital's rule. The result is

$$\frac{\tan x - \sin x}{x^2 \tan x} = \frac{\dfrac{\sin x}{\cos x} - \sin x}{x^2 \dfrac{\sin x}{\cos x}} = \frac{\dfrac{\sin x - \sin x \cos x}{\cos x}}{\dfrac{x^2 \sin x}{\cos x}} = \frac{\sin x(1 - \cos x)}{x^2 \sin x} = \frac{1 - \cos x}{x^2}$$

Now,

$$\lim_{x \to 0} \frac{\tan x - \sin x}{x^2 \tan x} = \lim_{x \to 0} \frac{1 - \cos x}{x^2} \underset{\uparrow}{=} \lim_{x \to 0} \frac{\sin x}{2x} = \frac{1}{2} \cdot \lim_{x \to 0} \frac{\sin x}{x} = \frac{1}{2} \cdot 1 = \frac{1}{2}$$

Apply L'Hospital's rule.

Compare this solution to one that applies L'Hospital's rule to the original expression.

L'Hospital's rule also may be used for one-sided limits.

EXAMPLE 6 Type $\dfrac{\infty}{\infty}$

Evaluate: $\displaystyle\lim_{x \to 0^+} \frac{\cot x}{\ln x}$

Solution

Since $\lim_{x \to 0^+} \cot x = +\infty$ and $\lim_{x \to 0^+} \ln x = -\infty$, we have an indeterminate form at 0 of the type ∞/∞. Hence,

$$\lim_{x \to 0^+} \frac{\cot x}{\ln x} \underset{\uparrow}{=} \lim_{x \to 0^+} \frac{-\csc^2 x}{1/x} = -\lim_{x \to 0^+} \frac{x}{\sin^2 x} \underset{\uparrow}{=} -\lim_{x \to 0^+} \frac{1}{2 \sin x \cos x} = -\infty$$

L'Hospital's rule   L'Hospital's rule

Partial Proof of L'Hospital's Rule

To prove L'Hospital's rule, we use a formula that bears the name of the French mathematician Augustin Cauchy (1789–1857) and is an extension of the mean value theorem (4.4.2).

[10.1.2] THEOREM / ***Cauchy's Mean Value Theorem.***

Let f and g be continuous on the closed interval $[a, b]$ and differentiable on the open interval (a, b). If g' is never 0 on (a, b), there exists a number c in (a, b) such that

$$\frac{f'(c)}{g'(c)} = \frac{f(b) - f(a)}{g(b) - g(a)} \tag{6}$$

Remark: Under the conditions stated, $g(b)$ and $g(a)$ cannot be equal because, if $g(b) = g(a)$, then Rolle's theorem (4.4.1) would assert that g' takes on the value 0 somewhere in (a, b). Since g' is never 0 on (a, b), we conclude that $g(b) \neq g(a)$.

Proof of Cauchy's Mean Value Theorem We prove this result by applying the mean value theorem to the function h defined by

$$h(x) = [g(b) - g(a)][f(x) - f(a)] - [g(x) - g(a)][f(b) - f(a)] \qquad a \leq x \leq b$$

We note that h is continuous on $[a, b]$, h is differentiable on (a, b), and $h(a) = h(b) = 0$. Applying Rolle's theorem to h produces a number c in (a, b) such that $h'(c) = 0$. That is,

$$h'(c) = [g(b) - g(a)]f'(c) - g'(c)[f(b) - f(a)] = 0$$

Thus,

$$\frac{f'(c)}{g'(c)} = \frac{f(b) - f(a)}{g(b) - g(a)}$$

If we let $g(x) = x$ in (6), we obtain the familiar mean value theorem. A partial proof of L'Hospital's rule follows.

Proof of L'Hospital's Rule Suppose that $f(x)/g(x)$ is an indeterminate form at c of the type 0/0, and suppose $\lim_{x\to c}[f'(x)/g'(x)] = L$, where L is a real number. We wish to prove that $\lim_{x\to c}[f(x)/g(x)] = L$. First, we introduce the functions F and G as follows:

$$F(x) = \begin{cases} f(x) & \text{if } x \neq c \\ 0 & \text{if } x = c \end{cases} \qquad G(x) = \begin{cases} g(x) & \text{if } x \neq c \\ 0 & \text{if } x = c \end{cases} \qquad \textbf{(7)}$$

Both F and G are continuous at c, since

$$\lim_{x\to c} F(x) = \lim_{x\to c} f(x) = 0 = F(c) \qquad \lim_{x\to c} G(x) = \lim_{x\to c} g(x) = 0 = G(c)$$

Moreover,

$$F'(x) = f'(x) \qquad G'(x) = g'(x)$$

for all x on the given interval I, except possibly at c. Since the conditions for Cauchy's mean value theorem are met by the functions F and G in either $[x, c]$ or $[c, x]$, there is a number u between c and x such that

$$\frac{F(x) - F(c)}{G(x) - G(c)} = \frac{F'(u)}{G'(u)} = \frac{f'(u)}{g'(u)} \qquad \textbf{(8)}$$

By using (7), this simplifies to

$$\frac{f(x)}{g(x)} = \frac{f'(u)}{g'(u)}$$

Since u is between c and x, it follows from the squeezing theorem (2.5.1) that

$$\lim_{x\to c}\frac{f(x)}{g(x)}=\lim_{u\to c}\frac{f'(u)}{g'(u)}=L$$

A similar argument may be given if L is infinite. The proof when $f(x)/g(x)$ is an indeterminate form at $+\infty$ of the type ∞/∞ is omitted here, but it may be found in books on advanced calculus.

We justify the use of L'Hospital's rule when $c=+\infty$ for an indeterminate form of the type 0/0 by the following argument: In $\lim_{x\to+\infty}[f(x)/g(x)]$, let $x=1/u$. Then, as $x\to+\infty$, $u\to 0^+$, and

$$\lim_{x\to+\infty}\frac{f(x)}{g(x)}=\lim_{u\to 0^+}\frac{f\left(\frac{1}{u}\right)}{g\left(\frac{1}{u}\right)}=\lim_{u\to 0^+}\frac{\frac{d}{du}f\left(\frac{1}{u}\right)}{\frac{d}{du}g\left(\frac{1}{u}\right)}\underset{\text{Chain rule}}{=}\lim_{u\to 0^+}\frac{-\frac{1}{u^2}f'\left(\frac{1}{u}\right)}{-\frac{1}{u^2}g'\left(\frac{1}{u}\right)}\underset{x=1/u}{=}\lim_{x\to+\infty}\frac{f'(x)}{g'(x)}=L$$

A similar argument handles the case $c=-\infty$ (see Problem 73).

EXERCISE 10.1

In Problems 1–62 evaluate each limit.

1. $\lim_{x\to 2}\frac{x^2+x-6}{x^2-3x+2}$

2. $\lim_{x\to -1}\frac{x^2+5x+4}{x^2-4x-5}$

3. $\lim_{x\to 1}\frac{2x^3+5x^2-4x-3}{x^3+x^2-10x+8}$

4. $\lim_{x\to 0}\frac{x^3-3x^2+5x}{x^3-x}$

5. $\lim_{x\to+\infty}\frac{3x^2+7}{4x^2-5}$

6. $\lim_{x\to+\infty}\frac{4x^3+7}{3x^5+6}$

7. $\lim_{x\to 1}\frac{\ln x}{x^2-1}$

8. $\lim_{x\to 1}\frac{\sin \pi x}{x-1}$

9. $\lim_{x\to 0}\frac{e^x-e^{-x}}{\sin x}$

10. $\lim_{x\to 0}\frac{\tan 2x}{\ln(1+x)}$

11. $\lim_{x\to \pi}\frac{1+\cos x}{\sin 2x}$

12. $\lim_{x\to 0}\frac{\ln(1-x)}{e^x-1}$

13. $\lim_{x\to+\infty}\frac{x^2}{e^x}$

14. $\lim_{x\to+\infty}\frac{x^4+x^3}{e^x+1}$

15. $\lim_{x\to 0^+}\frac{\cot x}{\cot 2x}$

16. $\lim_{x\to+\infty}\frac{\ln x}{a^x}$

17. $\lim_{x\to+\infty}\frac{x^2+x-1}{e^x+e^{-x}}$

18. $\lim_{x\to+\infty}\frac{x+\ln x}{x\ln x}$

19. $\lim_{x\to 0}\frac{\sin^2 x}{x}$

20. $\lim_{x\to 0}\frac{\tan x-x}{\sin x}$

21. $\lim_{x\to 0}\frac{e^x+e^{-x}-2}{x^2}$

22. $\lim_{x\to 0}\frac{e^x+e^{-x}-2}{\sin^2 x}$

23. $\lim_{x\to 0}\frac{\cos x-1}{\cos 2x-1}$

24. $\lim_{x\to 0}\frac{e^x-1-\sin x}{1-\cos x}$

25. $\lim_{x\to 0}\frac{\sin x-x}{x^3}$

26. $\lim_{x\to 0}\frac{\tan x-x}{\cos x-1}$

27. $\lim_{x\to 0}\frac{e^x-e^{-x}-2\sin x}{3x^3}$

28. $\lim_{x\to 0} \frac{\tan x - \sin x}{x^3}$

29. $\lim_{x\to +\infty} \frac{\ln(\ln x)}{\ln x}$

30. $\lim_{x\to 1/2^-} \frac{\ln(1-2x)}{\tan \pi x}$

31. $\lim_{x\to 0} \frac{\sin x}{x^3}$

32. $\lim_{x\to 3^-} \frac{\ln x}{\sqrt{3-x}}$

33. $\lim_{x\to 1^-} \frac{\ln(1-x)}{\cot \pi x}$

34. $\lim_{x\to 1^+} \frac{\ln(\ln x)}{1/(1-x)}$

35. $\lim_{x\to +\infty} \frac{\ln x}{x^\alpha}$ $(\alpha > 0)$

36. $\lim_{x\to +\infty} \frac{\ln x}{e^x}$

37. $\lim_{x\to +\infty} \frac{x^\alpha}{e^x}$ $(\alpha > 0)$

38. $\lim_{x\to +\infty} \frac{\cosh x}{x}$

39. $\lim_{x\to 0} \frac{xe^{4x} - x}{1 - \cos 2x}$

40. $\lim_{x\to 0} \frac{x \tan x}{1 - \cos x}$

41. $\lim_{x\to \pi/4} \frac{1 + \cos 4x}{\sec^2 x - 2 \tan x}$

42. $\lim_{x\to 0} \frac{\sec x - 1}{x^2 \sec x}$

43. $\lim_{x\to 0} \frac{2^x - 1}{3^x - 1}$

44. $\lim_{x\to 0} \frac{a^x - b^x}{x}$

45. $\lim_{x\to +\infty} \frac{e^x}{x^5}$

46. $\lim_{x\to 0} \left(\frac{1}{x^6} - \frac{1}{x^3}\right)$

47. $\lim_{x\to \pi/4} \frac{\cos^2 2x}{1 - \tan x}$

48. $\lim_{x\to 0} \frac{\tan^{-1} x}{x}$

49. $\lim_{x\to 0} \frac{e^x - \ln(1+x) - 1}{x^2}$

50. $\lim_{x\to 0} \frac{2x - \sin^{-1} x}{2 \tan^{-1} x - x}$

51. $\lim_{x\to 0} \frac{x - \tan^{-1} x}{x^3}$

52. $\lim_{x\to 0} \frac{\tan 3x}{x \cos x}$

53. $\lim_{x\to 0} \frac{e^{2x} - 1}{x^2 - \sin x}$

54. $\lim_{x\to 0} \frac{\sin 2x - \sin x}{\tan 3x}$

55. $\lim_{x\to 0} \frac{\tan^{-1} x}{\sin^{-1} x}$

56. $\lim_{x\to 0} \frac{\ln(\sec 2x)}{\ln(\sec x)}$

57. $\lim_{x\to 0} \frac{\sinh x}{x}$

58. $\lim_{x\to 0} \frac{\cosh x - 1}{x^2}$

59. $\lim_{x\to 0^+} \frac{x \sin(\frac{1}{x})}{\sin x}$

60. $\lim_{x\to -2} \frac{\ln(x^2 + e^x)}{x + 2}$

61. $\lim_{x\to 0} \frac{\ln(\cosh x)}{x}$

62. $\lim_{x\to 0} \frac{\tanh^{-1} x}{x}$

63. Discuss the behavior of $\frac{\sin x - \tan x}{e^x + e^{-x} - 2}$ as $x \to 0$.

64. Discuss the behavior of $\frac{1 - \cos x - x \sin x}{2 - 2\cos x - \sin^2 x}$ as $x \to 0$.

65. If α and β are positive real numbers, show that $\lim_{x\to +\infty} \frac{(\ln x)^\beta}{x^\alpha} = 0$.

66. If α and β are positive real numbers, show that $\lim_{x\to +\infty} \frac{x^\alpha}{e^{\beta x}} = 0$.

67. Evaluate: $\lim_{x\to 0^+} \frac{\int_0^x \cos t^2\, dt}{\int_0^x e^{t^2}\, dt}$

68. Evaluate: $\lim_{x\to 0^+} \frac{\int_0^x \sin t^3\, dt}{x^4}$

69. If $\alpha, \beta \neq 0$ and $c > 0$ are real numbers, show that $\lim_{x\to c} \frac{x^\alpha - c^\alpha}{x^\beta - c^\beta} = \frac{\alpha}{\beta} c^{\alpha-\beta}$.

70. The equation governing the amount of current I (in amperes) in a simple RL circuit consisting of a resistance R (in ohms), an inductance L (in henrys), and an electromotive force E (in volts) is

$$I = \frac{E}{R}(1 - e^{-Rt/L})$$

Find $\lim_{t\to +\infty} I(t)$ and $\lim_{R\to 0^+} I(t)$.

71. The sum S_n of the first n terms of a geometric series is given by the formula

$$S_n = a + ar + ar^2 + \cdots + ar^{n-1} = \frac{a(r^n - 1)}{r - 1}$$

Evaluate $\lim_{r\to 1} \frac{a(r^n - 1)}{r - 1}$, and compare the result with a geometric series in which $r = 1$.

72. Suppose f is a continuous function. Evaluate
$$\lim_{x\to 0}\left[\frac{1}{x}\int_0^x f(t)\,dt\right].$$

73. Prove L'Hospital's rule when $f(x)/g(x)$ is an indeterminate form at $-\infty$ of the type $0/0$.

10.2 Other Indeterminate Forms ($0 \cdot \infty$, $\infty - \infty$, 0^0, 1^∞, ∞^0)

L'Hospital's rule applies only to indeterminate forms of the types $0/0$ and ∞/∞. By algebraic rearrangement, other indeterminate forms of the types $0 \cdot \infty$, $\infty - \infty$,* 0^0, 1^∞, and ∞^0 often may be transformed into the type $0/0$ or the type ∞/∞, and then L'Hospital's rule may be applied.

Indeterminate Form of Type $0 \cdot \infty$

If a function $F(x) = f(x)g(x)$ results in the indeterminate form $0 \cdot \infty$, we may write the product $f(x)g(x)$ in quotient form as

$$f(x)g(x) = \frac{f(x)}{1/g(x)} \qquad \text{or} \qquad f(x)g(x) = \frac{g(x)}{1/f(x)}$$

which will produce one of the indeterminate forms $0/0$ or ∞/∞. This may then be handled by L'Hospital's rule. The choice of the quotient form is usually dictated by the ease of differentiation. Let's look at two examples.

EXAMPLE 1 Type $0 \cdot \infty$

Evaluate: $\lim_{x\to 0^+} (x \ln x)$

Solution

This is an indeterminate form of the type $0 \cdot \infty$, since $\lim_{x\to 0^+} x = 0$ and $\lim_{x\to 0^+} \ln x = -\infty$. To transform this to type $0/0$ or ∞/∞, we write

$$\lim_{x\to 0^+} (x \ln x) = \lim_{x\to 0^+} \frac{\ln x}{1/x}$$

Since the limit on the right is an indeterminate form of the type ∞/∞, we may apply L'Hospital's rule. Then

$$\lim_{x\to 0^+} (x \ln x) \underset{\text{Algebra}}{=} \lim_{x\to 0^+} \frac{\ln x}{1/x} \underset{\text{L'Hospital's rule}}{=} \lim_{x\to 0^+} \frac{1/x}{-1/x^2} \underset{\text{Algebra}}{=} \lim_{x\to 0^+} (-x) = 0$$

* The indeterminate form of the type $\infty - \infty$ is a convenient notation for any of the following: $(+\infty) - (+\infty)$; $(-\infty) - (-\infty)$; $(+\infty) + (-\infty)$. Note that $(+\infty) + (+\infty) = +\infty$ and $(-\infty) + (-\infty) = -\infty$ are not indeterminate forms.

The idea behind the solution given in Example 1 is to replace a product by a quotient. We chose to replace $x \ln x$ by $(\ln x)/(1/x)$. Why did we make this choice instead of $x/(1/\ln x)$? The answer lies in the subsequent step, where we apply L'Hospital's rule. The choice we made results in a limit that is easy to find; the other choice results in a limit that is more complicated than the original one. Again, experience is the best teacher.

You might be tempted to argue that $0 \cdot \infty$ is 0, since "anything" times 0 is 0. This is not true because $0 \cdot \infty$ is not a product of numbers; it is a statement about limits. The next example illustrates an indeterminate form of the type $0 \cdot \infty$ that yields a nonzero limit.

EXAMPLE 2 Type $0 \cdot \infty$

Evaluate: $\lim_{x \to +\infty} \left(x \sin \frac{1}{x} \right)$

Solution

This is an indeterminate form of the type $0 \cdot \infty$, since $\lim_{x \to +\infty} x = +\infty$ and $\lim_{x \to +\infty} \sin(1/x) = 0$. To transform this, we write

$$\lim_{x \to +\infty} \left(x \sin \frac{1}{x} \right) \underset{\text{Algebra}}{=} \lim_{x \to +\infty} \frac{\sin(1/x)}{1/x} \underset{\substack{\text{Set} \\ t = 1/x \\ x \to +\infty \Rightarrow t \to 0^+}}{=} \lim_{t \to 0^+} \frac{\sin t}{t} = 1$$

Indeterminate Form of the Type $\infty - \infty$

If the limit of a function results in the indeterminate form $\infty - \infty$, it is generally possible to rewrite the function (usually by using algebra or trigonometry) so that it changes to a quotient, producing one of the indeterminate forms $0/0$ or ∞/∞. Let's look at an example.

EXAMPLE 3 Type $\infty - \infty$

Evaluate: $\lim_{x \to 0^+} \left(\frac{1}{x} - \frac{1}{\sin x} \right)$

Solution

This is an indeterminate form of the type $\infty - \infty$, since $\lim_{x \to 0^+} (1/x) = +\infty$ and $\lim_{x \to 0^+} (1/\sin x) = +\infty$. To transform this, we write the difference as a single fraction:

$$\lim_{x \to 0^+} \left(\frac{1}{x} - \frac{1}{\sin x} \right) = \lim_{x \to 0^+} \frac{\sin x - x}{x \sin x}$$

The limit on the right is an indeterminate form of the type $0/0$. By applying

L'Hospital's rule twice, we get

$$\lim_{x \to 0^+} \left(\frac{1}{x} - \frac{1}{\sin x} \right) \underset{\text{Algebra}}{\uparrow=} \lim_{x \to 0^+} \frac{\sin x - x}{x \sin x} \underset{\text{L'Hospital's rule}}{\uparrow=} \lim_{x \to 0^+} \frac{\cos x - 1}{\sin x + x \cos x}$$

$$\underset{\text{L'Hospital's rule}}{\uparrow=} \lim_{x \to 0^+} \frac{-\sin x}{\cos x + \cos x - x \sin x}$$

$$= \lim_{x \to 0^+} \frac{-\sin x}{2 \cos x - x \sin x} = 0$$

Indeterminate Forms of Type 1^∞, 0^0, or ∞^0

A function of the form $f(x)^{g(x)}$ may result in one of the indeterminate forms 1^∞, 0^0, or ∞^0. To handle these, we set $y = f(x)^{g(x)}$ and take logarithms to obtain

$$\ln y = \ln f(x)^{g(x)} = g(x) \ln f(x)$$

The expression on the right is then an indeterminate form of the type $0 \cdot \infty$, which can be evaluated by techniques already discussed.

The four steps listed below are followed when we want to evaluate $\lim_{x \to c} f(x)^{g(x)}$ and this limit results in one of the indeterminate forms 0^0, 1^∞, or ∞^0.

Step 1 Set $y = f(x)^{g(x)}$.

Step 2 Take logarithms: $\ln y = g(x) \ln f(x)$.

Step 3 Evaluate $\lim_{x \to c} \ln y$, if it exists.

Step 4 If $\lim_{x \to c} \ln y = L$, then $\lim_{x \to c} y = e^L$.*

These four steps may be used if $x \to +\infty$ or $x \to -\infty$, or if one-sided limits are involved.

EXAMPLE 4 Type 0^0

Evaluate: $\lim_{x \to 0^+} x^x$

Solution

This expression is of the form 0^0. We follow the steps listed above.

* If $\lim_{x \to c} \ln y = L$, then $\lim_{x \to c} e^{\ln y} = \lim_{x \to c} y = e^L$.

Step 1 Let $y = x^x$.

Step 2 $\ln y = x \ln x$

Step 3 $\lim_{x\to 0^+} \ln y = \lim_{x\to 0^+} x \ln x = 0$ (by Example 1)

Step 4 Since $\lim_{x\to 0^+} \ln y = 0$, $\lim_{x\to 0^+} y = e^0 = 1$, so that $\lim_{x\to 0^+} x^x = 1$.

Caution: Do not stop after showing that $\lim_{x\to c} \ln y = L$ and conclude that $\lim_{x\to c} f(x)^{g(x)} = L$. What we are looking for is $\lim_{x\to c} y$, which equals e^L.

EXAMPLE 5 Type 1^∞

Evaluate: $\lim\limits_{x\to 0^+} (1 + x)^{1/x}$

Solution

This expression is of the form 1^∞. We follow definition (10.2.1).

Step 1 Let $y = (1 + x)^{1/x}$.

Step 2 $\ln y = \ln(1 + x)^{1/x} = (1/x)\ln(1 + x)$

Step 3 $$\lim_{x\to 0^+} \ln y = \lim_{x\to 0^+} \frac{\ln(1 + x)}{x} \underset{\text{L'Hospital's rule}}{=} \lim_{x\to 0^+} \frac{1/(1 + x)}{1} = 1$$

Step 4 Since $\lim_{x\to 0^+} \ln y = 1$, $\lim_{x\to 0^+} y = e^1 = e$, so that $\lim_{x\to 0^+}(1 + x)^{1/x} = e$.

We close this section with some important limits that we shall have occasion to refer to in Chapter 11.

[10.2.1] THEOREM

(a) $\lim\limits_{n\to +\infty} \sqrt[n]{\alpha} = 1$ **α any positive real number**

(b) $\lim\limits_{n\to +\infty} \sqrt[n]{n} = 1$ **(1)**

(c) $\lim\limits_{n\to +\infty} \left(1 + \dfrac{\alpha}{n}\right)^n = e^\alpha$ **α any real number**

Proof

(a) Let $y = \sqrt[n]{\alpha} = \alpha^{1/n}$. Then $\ln y = \ln \alpha^{1/n} = (1/n) \ln \alpha$, so that

$$\lim_{n\to +\infty} \ln y = \lim_{n\to +\infty} \frac{\ln \alpha}{n} = 0$$

Hence,

$$\lim_{n\to +\infty} y = \lim_{n\to +\infty} \sqrt[n]{\alpha} = e^0 = 1$$

(b) Let $y = \sqrt[n]{n} = n^{1/n}$. Then $\ln y = \ln n^{1/n} = (\ln n)/n$, so that

$$\lim_{n\to+\infty} \ln y = \lim_{n\to+\infty} \frac{\ln n}{n} = \lim_{n\to+\infty} \frac{1/n}{1} = 0$$

Hence,

$$\lim_{n\to+\infty} y = \lim_{n\to+\infty} \sqrt[n]{n} = e^0 = 1$$

(c) Let $y = (1 + \alpha/n)^n$. Then $\ln y = n \ln(1 + \alpha/n)$, so that

$$\lim_{n\to+\infty} \ln y = \lim_{n\to+\infty} n \ln\left(1 + \frac{\alpha}{n}\right) = \lim_{n\to+\infty} \frac{\ln(1 + \alpha/n)}{1/n}$$

$$= \lim_{n\to+\infty} \frac{\dfrac{-\alpha/n^2}{1 + \alpha/n}}{-1/n^2} = \lim_{n\to+\infty} \frac{\alpha}{1 + \alpha/n} = \alpha$$

Hence,

$$\lim_{n\to+\infty} y = \lim_{n\to+\infty} \left(1 + \frac{\alpha}{n}\right)^n = e^\alpha$$

EXAMPLE 6

Use the result from Example 4 that $\lim_{x\to 0^+} x^x = 1$ to help sketch the graph of $f(x) = x^x$.

Solution

1. The domain of f is $x > 0$.

2. There are no intercepts.

3. There is no symmetry.

4. There are no horizontal asymptotes, since $\lim_{x\to+\infty} x^x = +\infty$. Since $f(x) = x^x$ is continuous for all $x > 0$, there are no vertical asymptotes.

5. $$\frac{d}{dx} x^x = \frac{d}{dx} e^{x \ln x} = e^{x \ln x}(\ln x + 1) = x^x(\ln x + 1)$$

Thus, $f'(x) = 0$ if $\ln x + 1 = 0$; that is, $\ln x = -1$, or $x = e^{-1}$, so that e^{-1} is the only critical number.

6. $f'(x) > 0$ if $\ln x + 1 > 0$ or $\ln x > -1$ or $x > e^{-1}$. $f'(x) < 0$ if $\ln x + 1 < 0$ or $\ln x < -1$ or $x < e^{-1}$. Therefore, f is increasing on $e^{-1} \le x < +\infty$ and decreasing on $0 < x \le e^{-1}$.

7. By the first derivative test f has a local minimum at $e^{-1} = \dfrac{1}{e}$;

$$f(e^{-1}) = \left(\frac{1}{e}\right)^{1/e} \approx 0.69$$

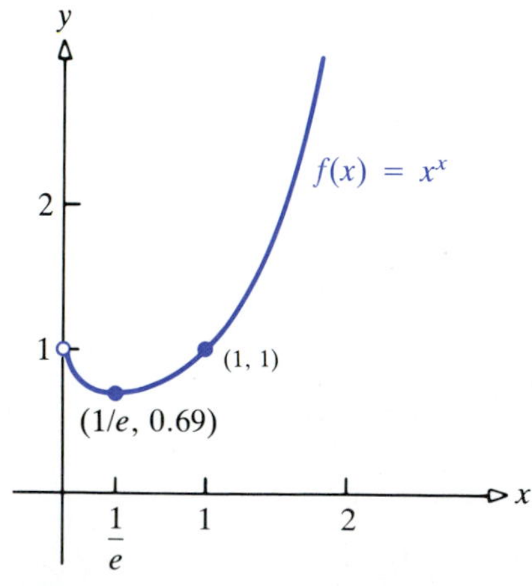

Figure 1

8. Since $f''(x) = x^x(\ln x + 1)^2 + x^x \dfrac{1}{x} > 0$ for all $x > 0$, there are no inflection points and the graph of f is concave up for all $x > 0$.

9. Since $\lim_{x\to 0^+} x^x = 1$, we know the graph approaches the point $(0, 1)$ on the y-axis. See Figure 1.

EXERCISE 10.2

In Problems 1–46 evaluate each limit.

1. $\lim_{x\to 0}(x \cot x)$

2. $\lim_{x\to \pi/2}(1 - \sin x)\tan x$

3. $\lim_{x\to +\infty}[x(e^{1/x} - 1)]$

4. $\lim_{x\to 0^+}(x^2 \ln x)$

5. $\lim_{x\to 0^+}[x \ln(\sin x)]$

6. $\lim_{x\to +\infty}(xe^{-x})$

7. $\lim_{x\to \pi/2}[\tan x \ln(\sin x)]$

8. $\lim_{x\to 0}[\csc x \ln(x + 1)]$

9. $\lim_{x\to 0}(x^2 \csc x \cot x)$

10. $\lim_{x\to \pi/4}(1 - \tan x)\sec 2x$

11. $\lim_{x\to -\infty}(x^2 e^x)$

12. $\lim_{x\to +\infty}(x - 1)e^{-x^2}$

13. $\lim_{x\to 0^+}(x^\alpha \ln x)$, $\alpha > 0$

14. $\lim_{x\to a}(a^2 - x^2)\tan \dfrac{\pi x}{2a}$

15. $\lim_{x\to 1^+}(1 - x)\tan \dfrac{1}{2}\pi x$

16. $\lim_{x\to 0^+}[\sin x \ln(\sin x)]$

17. $\lim_{x\to \pi/2}(\sec x - \tan x)$

18. $\lim_{x\to 0}\left(\csc \dfrac{1}{2}x - \cot \dfrac{1}{2}x\right)$

19. $\lim_{x\to 0}\left(\cot x - \dfrac{1}{x}\right)$

20. $\lim_{x\to 1}\left(\dfrac{1}{\ln x} - \dfrac{x}{\ln x}\right)$

21. $\lim_{x\to 0}\left(\dfrac{1}{x} - \dfrac{1}{e^x - 1}\right)$

22. $\lim_{x\to 1}\left(\dfrac{x}{x - 1} - \dfrac{1}{\ln x}\right)$

23. $\lim_{x\to 1}\left(\dfrac{1}{\ln x} - \dfrac{1}{x - 1}\right)$

24. $\lim_{x\to \pi/2}\left(x \tan x - \dfrac{\pi}{2}\sec x\right)$

25. $\lim_{x\to 0^+}(2x)^{3x}$

26. $\lim_{x\to 0}(1 + x^2)^{1/x^2}$

27. $\lim_{x\to \pi/2}(\sin x)^{\tan x}$

28. $\lim_{x\to 0^+}(\csc x)^{\sin x}$

29. $\lim_{x\to 0^+} x^{\sqrt{x}}$

30. $\lim_{x\to 0} x^{x^2}$

31. $\lim_{x\to 0}\dfrac{3^x - 1}{x}$

32. $\lim_{x\to +\infty} x^{1/x}$

33. $\lim_{x\to 0^+} x^{1/\ln x}$

34. $\lim_{x\to 1}\dfrac{2^{x-1} - x}{x - 1}$

35. $\lim_{x\to +\infty}(\sqrt{x^2 + x + 1} - \sqrt{x^2 - x})$

36. $\lim_{x\to 0}\dfrac{x + 1 - e^x}{x^2}$

37. $\lim_{x\to +\infty}\left(1 + \dfrac{5}{x} + \dfrac{3}{x^2}\right)^x$

38. $\lim_{x\to +\infty}\left(\dfrac{x^2}{x + 1} - \dfrac{x^2}{x - 1}\right)$

39. $\lim_{x\to 0^+}(1 + \sin x)^{\cot x}$

40. $\lim_{x\to 0^+}\left(\dfrac{1}{x}\right)^{\sin x}$

41. $\lim_{x\to +\infty}(1 + x^2)^{1/x}$

42. $\lim_{x\to 0}(\cos x)^{1/x}$

43. $\lim_{x\to 1^-}(1 - x)^{\tan \pi x}$

44. $\lim_{x\to 0}(e^x + x)^{1/x}$

45. $\lim_{x\to 0}\left(\dfrac{\sin x}{x}\right)^{1/x}$

46. $\lim_{x\to \pi/2}(\csc x)^{\tan^2 x}$

47. Show that $\lim_{x\to 0^+}(\cos x + 2 \sin x)^{\cot x} = e^2$.

48. Show that $\lim_{x\to +\infty}\left(\dfrac{x + a}{x - a}\right)^x = e^{2a}$.

49. Find $\lim_{x\to +\infty}\dfrac{P(x)}{e^x}$, where P is a polynomial function.

50. Find $\lim_{x\to +\infty}[\ln(x + 1) - \ln(x - 1)]$.

51. Show that $\lim_{x\to 0^+}\dfrac{e^{-1/x^2}}{x} = 0$.

[*Hint:* Write $\dfrac{e^{-1/x^2}}{x} = \dfrac{1/x}{e^{1/x^2}}$.]

52. If n is an integer, show that $\lim\limits_{x\to 0^+} \dfrac{e^{-1/x^2}}{x^n} = 0.$

53. Show that the function below has a derivative at 0. What is $f'(0)$? [*Hint:* Use Problem 51.]

$$f(x) = \begin{cases} e^{-1/x^2} & \text{if } x \neq 0 \\ 0 & \text{if } x = 0 \end{cases}$$

In Problems 54–57 sketch the graph of each function. Use L'Hospital's rule to determine the asymptotes.

54. $f(x) = \dfrac{\sin 3x}{x\sqrt{4 - x^2}}$

55. $f(x) = x^{\sqrt{x}}$

56. $f(x) = x^{1/x}$

57. $y = \dfrac{1}{x} \tan x, \quad -\dfrac{\pi}{2} < x < \dfrac{\pi}{2}$

58. Show that $\lim_{n\to+\infty} n(\sqrt[n]{x} - 1) = \ln x.$

10.3 Improper Integrals

In previous discussions on the definite integral $\int_a^b f(x)\,dx$, we required that a and b both be numbers and that f be continuous throughout the closed interval $[a, b]$. However, in many situations, one or more of these assumptions is not met. For example, one of the limits of integration might be infinite, or the function f might not be defined at some number in $[a, b]$.

In this section we extend the concept of the definite integral to include:

1. Integrals with infinite intervals of integration, such as

$$\int_1^{+\infty} \frac{1}{\sqrt{x}}\,dx \qquad \int_{-\infty}^{0} \frac{x-3}{x^3-8}\,dx \qquad \int_{-\infty}^{+\infty} \frac{x}{(x^2+1)^2}\,dx$$

2. Integrals in which the integrand becomes infinite within the interval of integration, such as

$$\int_0^1 \frac{1}{\sqrt{x}}\,dx \qquad \int_{-1}^{1} \frac{1}{x^3}\,dx$$

Integrals of these types are called *improper integrals.*

We begin by discussing the situation in which one of the limits of integration is infinite.

[10.3.1] DEFINITION

If f is continuous for all $x \geq a$, then

$$\int_a^{+\infty} f(x)\,dx = \lim_{b\to+\infty}\left[\int_a^b f(x)\,dx\right] \tag{1}$$

provided this limit exists. If the limit exists, the improper integral $\int_a^{+\infty} f(x)\,dx$ is said to *converge.* If the limit does not exist,* the improper integral $\int_a^{+\infty} f(x)\,dx$ is said to *diverge.*

* Remember, a limit exists only when it equals a number.

[10.3.2] DEFINITION

If f is continuous for all $x \le b$, then

$$\int_{-\infty}^{b} f(x)\,dx = \lim_{a\to-\infty}\left[\int_{a}^{b} f(x)\,dx\right] \qquad (2)$$

provided this limit exists. The terms *converge* and *diverge* are defined similarly.

EXAMPLE 1

Determine whether the following improper integrals converge or diverge:

(a) $\int_{1}^{+\infty} \frac{1}{x}\,dx$ (b) $\int_{-\infty}^{0} e^x\,dx$ (c) $\int_{1}^{+\infty} \sin\frac{\pi}{2}x\,dx$

Solution

(a) $$\int_{1}^{+\infty} \frac{1}{x}\,dx = \lim_{b\to+\infty}\left(\int_{1}^{b} \frac{1}{x}\,dx\right) = \lim_{b\to+\infty} \ln x\Big|_{1}^{b} = \lim_{b\to+\infty} \ln b = +\infty$$

Thus, $\int_{1}^{+\infty}(1/x)\,dx$ diverges.

(b) $$\int_{-\infty}^{0} e^x\,dx = \lim_{a\to-\infty}\left(\int_{a}^{0} e^x\,dx\right)$$
$$= \lim_{a\to-\infty} e^x\Big|_{a}^{0} = \lim_{a\to-\infty}(1-e^a)$$
$$= 1 - \lim_{a\to-\infty} e^a = 1 - 0 = 1$$

Thus, $\int_{-\infty}^{0} e^x\,dx$ converges and equals 1.

(c) $$\int_{1}^{+\infty} \sin\frac{\pi}{2}x\,dx = \lim_{b\to+\infty}\left(\int_{1}^{b} \sin\frac{\pi}{2}x\,dx\right) = \lim_{b\to+\infty} \frac{-\cos(\pi/2)x}{\pi/2}\Big|_{1}^{b}$$
$$= \lim_{b\to+\infty}\left(-\frac{2}{\pi}\cos\frac{\pi}{2}b\right) = -\frac{2}{\pi}\lim_{b\to+\infty}\cos\frac{\pi}{2}b$$

This limit does not exist, since as $b \to +\infty$, the value of $\cos(\pi/2)b$ oscillates between -1 and 1. Hence, $\int_{1}^{+\infty}\sin(\pi/2)x\,dx$ diverges.

Geometric Interpretation

If f is continuous and nonnegative for $x \ge a$, then $\int_{a}^{+\infty} f(x)\,dx$ may be interpreted geometrically. For each number $b > a$, the definite integral $\int_{a}^{b} f(x)\,dx$ represents the area under the graph of $y = f(x)$ from a to b, as shown in Figure 2(a). As we let $b \to +\infty$, this area approaches the area under the graph of $y = f(x)$ over the infinite interval $[a, +\infty)$, as shown in Figure 2(b). We define the area under the graph of $y = f(x)$ to the right of a to be $\int_{a}^{+\infty} f(x)\,dx$, provided the improper integral converges. If it diverges, there is no defined area.

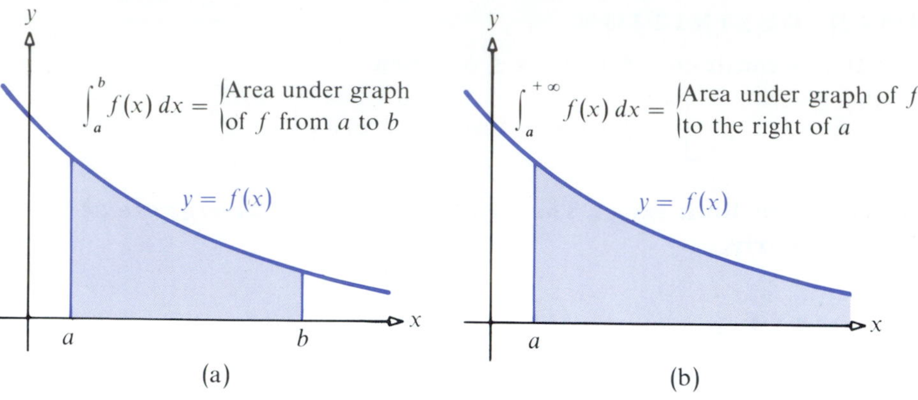

Figure 2

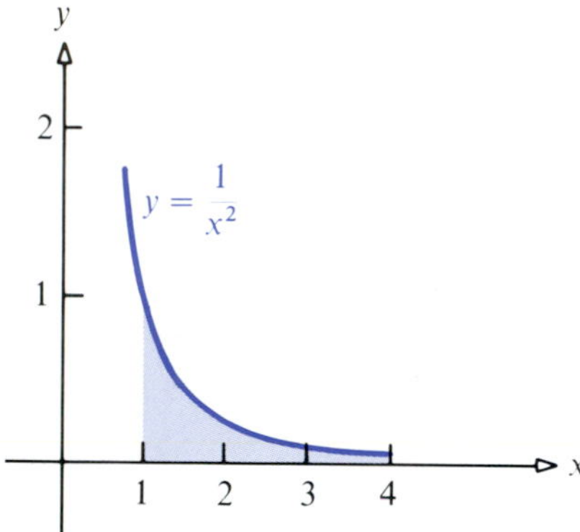

Figure 3

EXAMPLE 2

Find the area under the graph of $y = 1/x^2$ to the right of $x = 1$.

Solution

Look at the graph in Figure 3. Is there a finite number that measures the indicated area? At first, we may be tempted to answer no, but, in fact, it turns out that this area is finite. To find the area, we proceed as follows:

$$\int_1^{+\infty} \frac{1}{x^2}\,dx = \lim_{b\to+\infty}\left(\int_1^b \frac{1}{x^2}\,dx\right) = \lim_{b\to+\infty}\left(-\frac{1}{x}\right)\Bigg|_1^b = \lim_{b\to+\infty}\left(-\frac{1}{b}+1\right) = 1$$

Thus, the area under the graph $f(x) = 1/x^2$ to the right of 1 is 1 square unit.

In Example 1(a) it was shown that $\int_1^{+\infty}(1/x)\,dx$ diverges, whereas in Example 2 it was shown that $\int_1^{+\infty}(1/x^2)\,dx$ converges. Yet, as Figure 4 illustrates, the graphs of $y = 1/x$ and $y = 1/x^2$ are similar on the interval $[1, +\infty)$. The explanation for the difference is that $y = 1/x^2$ approaches 0 more rapidly than $y = 1/x$ as $x \to +\infty$. The difference in the areas under the graphs is large enough to make $\int_1^{+\infty}(1/x)\,dx$ diverge and $\int_1^{+\infty}(1/x^2)\,dx$ converge.

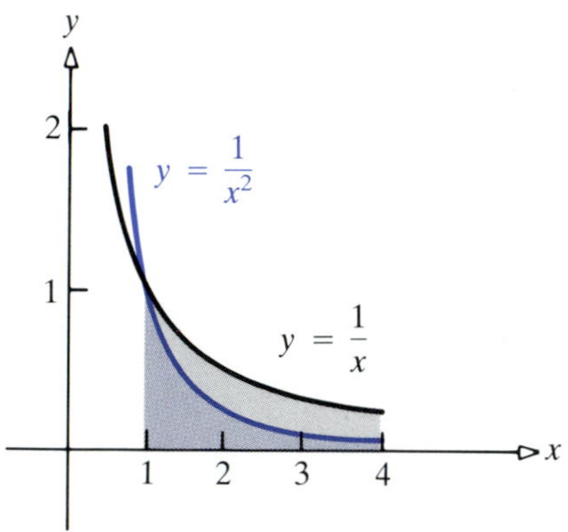

Figure 4

EXAMPLE 3

Find the volume of the solid generated by revolving the area under the graph of $y = 1/x$ to the right of 1 about the x-axis. Use the disk method.

Solution

Figure 5 illustrates the situation. The volume V is

$$V = \int_1^{+\infty} \pi\left(\frac{1}{x}\right)^2 dx = \pi\int_1^{+\infty}\frac{dx}{x^2} = \pi\lim_{b\to+\infty}\left(\int_1^b \frac{dx}{x^2}\right) = \pi\lim_{b\to+\infty}\left(-\frac{1}{x}\right)\Bigg|_1^b$$

$$= \pi\lim_{b\to+\infty}\left(-\frac{1}{b}+1\right) = \pi \text{ cubic units}$$

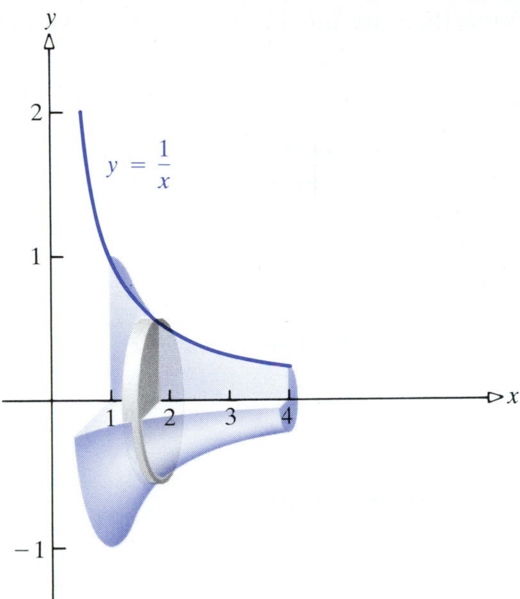

Figure 5

It is interesting to note that the area under the graph of $y = 1/x$ to the right of 1 is not defined, as shown in Example 1(a), whereas the volume obtained by revolving this same area about the x-axis equals π cubic units.

The result obtained in the next example will be put to good use in the following chapter.

[10.3.3] THEOREM

$$\int_1^{+\infty} \frac{dx}{x^p} \quad \textbf{converges if } p > 1 \textbf{ and diverges if } p \leq 1.$$

Proof $p \neq 1$:

$$\int_1^{+\infty} \frac{dx}{x^p} = \lim_{b \to +\infty} \left(\int_1^b \frac{dx}{x^p} \right) = \lim_{b \to +\infty} \left. \frac{x^{-p+1}}{-p+1} \right|_1^b$$

$$= \lim_{b \to +\infty} \frac{b^{-p+1} - 1}{-p+1}$$

$$= \begin{cases} \dfrac{-1}{-p+1} & \text{if } p > 1 \\ +\infty & \text{if } p < 1 \end{cases}$$

$p = 1$:

$$\int_1^{+\infty} \frac{dx}{x^p} = \int_1^{+\infty} \frac{dx}{x} = \lim_{b \to +\infty} \left(\int_1^b \frac{dx}{x} \right)$$

$$= \lim_{b \to +\infty} \ln x \Big|_1^b$$

$$= \lim_{b \to +\infty} \ln b = +\infty$$

In case both limits of integration are infinite, we use the following definition:

[10.3.4] DEFINITION

If f is continuous for all x and if the two improper integrals $\int_{-\infty}^{0} f(x)\,dx$ and $\int_{0}^{+\infty} f(x)\,dx$ *both* converge, then we say that $\int_{-\infty}^{+\infty} f(x)\,dx$ *converges* and we define

$$\int_{-\infty}^{+\infty} f(x)\,dx = \int_{-\infty}^{0} f(x)\,dx + \int_{0}^{+\infty} f(x)\,dx \tag{3}$$

If *either* or *both* of the integrals diverge, then we say that $\int_{0}^{+\infty} f(x)\,dx$ *diverges*.*

EXAMPLE 4

Determine whether $\int_{-\infty}^{+\infty} 4x^3\,dx$ converges or diverges.

Solution

We look first at the improper integral $\int_{-\infty}^{0} 4x^3\,dx$:

$$\int_{-\infty}^{0} 4x^3\,dx = \lim_{a\to-\infty}\left(\int_{a}^{0} 4x^3\,dx\right) = \lim_{a\to-\infty} x^4\Big|_{a}^{0} = \lim_{a\to-\infty}(-a^4) = -\infty$$

There is no need to go on: $\int_{-\infty}^{+\infty} 4x^3\,dx$ is divergent.

A word of caution! Definition (10.3.4) requires that *two* improper integrals each converge in order for $\int_{-\infty}^{+\infty} f(x)\,dx$ to converge. We cannot set $\int_{-\infty}^{+\infty} f(x)\,dx = \lim_{a\to+\infty}[\int_{-a}^{a} f(x)\,dx]$. If we had done this in Example 4, the result would have been $\int_{-a}^{a} 4x^3\,dx = x^4\big|_{-a}^{a} = a^4 - a^4 = 0$, from which we would have incorrectly concluded that $\int_{-\infty}^{+\infty} f(x)\,dx$ converges and equals 0.

Another Type of Improper Integral

[10.3.5] DEFINITION

An integral $\int_{a}^{b} f(x)\,dx$ is also improper when f is continuous on $(a, b]$ but is not defined at a. In this case, we define

$$\int_{a}^{b} f(x)\,dx = \lim_{t\to a^+}\left[\int_{t}^{b} f(x)\,dx\right] \tag{4}$$

provided this limit exists.

[10.3.6] DEFINITION

If f is continuous in $[a, b)$ but is not defined at b, the integral $\int_{a}^{b} f(x)\,dx$ is improper and

$$\int_{a}^{b} f(x)\,dx = \lim_{t\to b^-}\left[\int_{a}^{t} f(x)\,dx\right] \tag{5}$$

provided this limit exists.

---

* In (3) the number 0 is not crucial; any real number can be used.

When the limit exists, the improper integral is said to *converge*; otherwise, it is said to *diverge*.

Figure 6 may help you remember the correct one-sided limit to use. If f is continuous and nonnegative on the interval $(a, b]$, then the improper integral $\int_a^b f(x)\,dx$ may be interpreted geometrically. The integral $\int_t^b f(x)\,dx$ for each number t, $a < t \le b$, represents the area under the graph of $y = f(x)$ over the interval $[t, b]$, as shown in Figure 7(a). As we let $t \to a^+$, this area tends to approach the entire area under the graph of $y = f(x)$ over $(a, b]$, as shown in Figure 7(b). Hence, if $\int_a^b f(x)\,dx$ converges, we shall define its value to be the area under the graph of f from a to b.

(a)
f is continuous on $(a, b]$
f is not defined at a

$a \quad t \quad b$
$t \to a^+$

$$\int_a^b f(x)\,dx = \lim_{t \to a^+} \left[\int_t^b f(x)\,dx\right]$$

(b)
f is continuous on $[a, b)$
f is not defined at b

$a \quad t \quad b$
$t \to b^-$

$$\int_a^b f(x)\,dx = \lim_{t \to b^-} \left[\int_a^t f(x)\,dx\right]$$

Figure 6

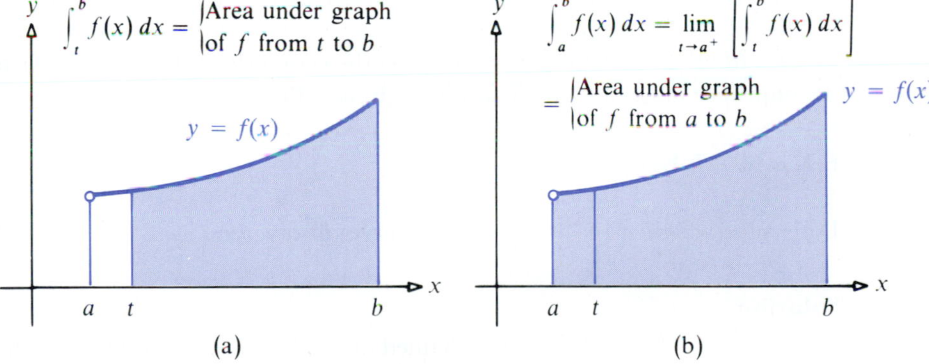

Figure 7

EXAMPLE 5

Determine whether $\displaystyle\int_0^4 \frac{1}{\sqrt{x}}\,dx$ converges or diverges.

Solution

This is an improper integral, since $f(x) = 1/\sqrt{x}$ is not defined at 0. We use (4) to get

$$\int_0^4 \frac{1}{\sqrt{x}}\,dx = \lim_{t \to 0^+} \left(\int_t^4 \frac{1}{\sqrt{x}}\,dx\right) = \lim_{t \to 0^+} \left.\frac{x^{1/2}}{\frac{1}{2}}\right|_t^4 = \lim_{t \to 0^+} (2\sqrt{4} - 2\sqrt{t}) = 4 - \lim_{t \to 0^+} 2\sqrt{t} = 4$$

We conclude that $\int_0^4 (1/\sqrt{x})\,dx$ converges.

EXAMPLE 6

Determine whether $\int_0^{\pi/2} \tan x\, dx$ converges or diverges.

Solution

Since $\tan x$ is not defined at $\pi/2$, $\int_0^{\pi/2} \tan x\, dx$ is an improper integral. Thus, we use (5) to get

$$\int_0^{\pi/2} \tan x\, dx = \lim_{t\to\pi/2^-}\left(\int_0^t \tan x\, dx\right) = \lim_{t\to\pi/2^-} \ln(\sec x)\Big|_0^t$$

$$= \lim_{t\to\pi/2^-} \ln(\sec t) = +\infty$$

We conclude that $\int_0^{\pi/2} \tan x\, dx$ diverges.

Another condition under which $\int_a^b f(x)\, dx$ is an improper integral is when the integrand is not defined at some number c in the open interval (a, b).

[10.3.7] DEFINITION

If f is continuous on $[a, b]$ except at c, $a < c < b$, where f is not defined, the integral $\int_a^b f(x)\, dx$ is improper and

$$\int_a^b f(x)\, dx = \int_a^c f(x)\, dx + \int_c^b f(x)\, dx \tag{6}$$

provided that each of the improper integrals on the right converges.

If either of the improper integrals on the right side of (6) diverges, then the improper integral on the left side of (6) also diverges.

EXAMPLE 7

Determine whether $\int_0^2 \frac{1}{(x-1)^2}\, dx$ converges or diverges.

Solution

Since $f(x) = 1/(x-1)^2$ is not defined at 1, $0 < 1 < 2$, we look at the two improper integrals $\int_0^1 [1/(x-1)^2]\, dx$ and $\int_1^2 [1/(x-1)^2]\, dx$:

$$\int_0^1 \frac{1}{(x-1)^2}\, dx = \lim_{t\to 1^-}\left[\int_0^t \frac{1}{(x-1)^2}\, dx\right] = \lim_{t\to 1^-} \frac{-1}{x-1}\Big|_0^t = \lim_{t\to 1^-}\left(\frac{-1}{t-1} - 1\right) = +\infty$$

Since $\int_0^1 [1/(x-1)^2]\, dx$ diverges, there is no need to investigate the second integral, and we conclude that $\int_0^2 [1/(x-1)^2]\, dx$ diverges.

A word of caution! It is a common mistake to look at an integral like $\int_0^2 dx/(x-1)^2$ and not notice that the integrand is undefined at 1. Then, if you merely plow ahead, using the second fundamental theorem, you get

$$\int_0^2 \frac{dx}{(x-1)^2} = \frac{-1}{x-1}\Big|_0^2 = -1 - 1 = -2$$

which is an incorrect answer. Look before you leap!

EXERCISE 10.3

In Problems 1–8 determine whether each integral is improper. For those that are improper, state the reason.

1. $\int_0^{+\infty} x^2\,dx$
2. $\int_0^5 x^3\,dx$
3. $\int_2^3 \frac{dx}{x-1}$
4. $\int_1^2 \frac{dx}{x-1}$
5. $\int_0^1 \frac{1}{x}\,dx$
6. $\int_{-1}^1 \frac{x}{x^2+1}\,dx$
7. $\int_0^1 \frac{x}{x^2-1}\,dx$
8. $\int_0^{+\infty} e^{-2x}\,dx$

In Problems 9–62 determine whether the given improper integral converges or diverges, and evaluate each convergent improper integral.

9. $\int_{10}^{+\infty} \frac{dx}{x^2}$
10. $\int_0^{+\infty} e^{-x}\,dx$
11. $\int_0^{+\infty} e^{2x}\,dx$
12. $\int_1^{+\infty} \frac{dx}{\sqrt{x}}$
13. $\int_{-\infty}^{-1} \frac{4}{x}\,dx$
14. $\int_3^{+\infty} \frac{dx}{(x-1)^4}$
15. $\int_0^1 \frac{dx}{\sqrt{x}}$
16. $\int_0^1 \frac{dx}{x^3}$
17. $\int_0^1 \frac{dx}{x}$
18. $\int_4^6 \frac{dx}{\sqrt{x-4}}$
19. $\int_0^a \frac{dx}{\sqrt{a-x}}$
20. $\int_1^5 \frac{x\,dx}{\sqrt{5-x}}$
21. $\int_0^{+\infty} xe^{-x^2}\,dx$
22. $\int_0^{+\infty} \cos x\,dx$
23. $\int_0^1 \frac{x\,dx}{(1-x^2)^2}$
24. $\int_0^{+\infty} \frac{x\,dx}{1+x^2}$
25. $\int_0^{+\infty} \frac{x\,dx}{\sqrt{x+1}}$
26. $\int_1^{+\infty} \frac{dx}{x(1+x^2)}$
27. $\int_0^{\pi/4} \tan 2x\,dx$
28. $\int_0^{\pi/2} \csc x\,dx$
29. $\int_{-1}^1 \frac{dx}{\sqrt[3]{x}}$
30. $\int_0^3 \frac{dx}{(x-2)^2}$
31. $\int_0^{2a} \frac{dx}{(x-a)^2}$
32. $\int_a^{3a} \frac{2x\,dx}{(x^2-a^2)^{3/2}}$
33. $\int_{-\infty}^0 \frac{dx}{x^2+4}$
34. $\int_{-\infty}^2 \frac{dx}{\sqrt{4-x}}$
35. $\int_{-\infty}^1 \frac{x\,dx}{\sqrt{2-x}}$
36. $\int_{-\infty}^0 e^x\,dx$
37. $\int_0^{\pi/2} \frac{x\,dx}{\sin x^2}$
38. $\int_0^{+\infty} e^{-x}\sin x\,dx$
39. $\int_{-\infty}^{+\infty} \frac{dx}{x^2+a^2}$
40. $\int_0^4 \frac{2x\,dx}{\sqrt[3]{x^2-4}}$
41. $\int_0^1 \frac{dx}{1-x^2}$
42. $\int_1^2 \frac{dx}{\sqrt{x^2-1}}$
43. $\int_0^a \frac{x^2\,dx}{\sqrt{a^2-x^2}}$
44. $\int_0^3 \frac{x\,dx}{(9-x^2)^{3/2}}$
45. $\int_0^{+\infty} \sin \pi x\,dx$
46. $\int_0^1 \frac{\ln x\,dx}{x}$
47. $\int_1^{+\infty} \frac{dx}{x^{2/3}}$
48. $\int_{-\infty}^{+\infty} \frac{dx}{x^2+4x+5}$
49. $\int_{2a}^{+\infty} \frac{dx}{x^2-a^2}$
50. $\int_2^{+\infty} \frac{dx}{x\sqrt{x^2-1}}$
51. $\int_1^{+\infty} \frac{dx}{x^4+x^2}$
52. $\int_{-\infty}^{+\infty} \frac{dx}{e^x+e^{-x}}$
53. $\int_0^a \frac{dx}{\sqrt{a^2-x^2}}$
54. $\int_a^{2a} \frac{x\,dx}{\sqrt{x^2-a^2}}$
55. $\int_1^2 \frac{dx}{(2-x)^{3/4}}$
56. $\int_0^4 \frac{dx}{\sqrt{8x-x^2}}$
57. $\int_0^{\pi} \frac{1}{1-\cos x}\,dx$
58. $\int_0^{\pi/2} \frac{1}{1-\sin x}\,dx$
59. $\int_{-1}^1 \frac{1}{x^3}\,dx$
60. $\int_0^2 \frac{dx}{x-1}$
61. $\int_0^2 \frac{dx}{(x-1)^{1/3}}$
62. $\int_{-1}^1 \frac{dx}{x^{5/3}}$

63. Find the area, if it exists, between $y = 8a^3/(x^2 + 4a^2)$ and its asymptote.

64. Find the area, if it exists, enclosed by $y = 1/(x + 1)$ and $y = 1/(x + 2)$ on the interval $[0, +\infty)$.

65. Find the area, if it exists, under the graph of $y = 1/(1 + x^2)$ to the right of 0.

66. Find the volume, if it exists, of the solid generated by revolving the area under the graph of $y = e^{-x}$ to the right of 0 about the x-axis.

67. Find the volume, if it exists, of the solid generated by revolving the area under the graph of $y = 1/\sqrt{x}$ to the right of 1 about the x-axis.

68. For what α does $\int_0^1 x^\alpha\, dx$ converge?

69. Show that $\int_0^{+\infty} \sin x\, dx$ and $\int_{-\infty}^0 \sin x\, dx$ each diverge, yet $\lim_{t\to+\infty}(\int_{-t}^t \sin x\, dx) = 0$.

70. Find a function f for which $\int_0^{+\infty} f(x)\, dx$ and $\int_{-\infty}^0 f(x)\, dx$ each diverge, yet $\lim_{t\to+\infty}[\int_{-t}^t f(x)\, dx] = 1$.

In Problems 71–74 use integration by parts and perhaps L'Hospital's rule to evaluate each integral.

71. $\displaystyle\int_0^{+\infty} xe^{-x}\, dx$

72. $\displaystyle\int_0^1 (x \ln x)\, dx$

73. $\displaystyle\int_0^{+\infty} (e^{-x}\cos x)\, dx$

74. $\displaystyle\int_0^{+\infty} (\tan^{-1}x)\, dx$

75. The rate of reaction to a given dose of a drug at time t hours after administration is given by $r(t) = te^{-t^2}$ (measured in appropriate units). Why is it reasonable to define the total *reaction* as the area under the graph of $y = r(t)$ from $t = 0$ to $t = +\infty$? Evaluate the total reaction to the given dose of the drug.

76. The present value (PV) of a capital asset that provides a perpetual stream of revenues that flows continuously at a rate of $R(t)$ dollars per year is given by

$$PV = \int_0^{+\infty} R(t)e^{-rt}\, dt$$

where r is the per annum rate of interest compounded continuously.

(a) Find the present value of an asset if it provides a constant return of \$100 per year and r is 8%.
(b) Find the present value of an asset if it provides a return of $R(t) = 1000 + 80t$ dollars per year and $r = 7\%$.

77. In a problem in electrical theory, the following integral occurs:

$$\int_0^{+\infty} Ri^2\, dt$$

where the current $i = Ie^{-Rt/L}$, t is time, and R, I, and L are constants. Evaluate this integral.

78. The magnetic potential at a point on the axis of a circular coil is given by

$$u = \frac{2\pi NIr}{10} \int_x^{+\infty} \frac{dy}{(r^2 + y^2)^{3/2}}$$

where N, I, r, and x are constants. Evaluate this expression.

79. The field intensity around a long ("infinite") straight wire carrying electric current is given by the integral

$$F = \frac{rIm}{10} \int_{-\infty}^{+\infty} \frac{dy}{(r^2 + y^2)^{3/2}}$$

where r, I, and m are constants. Evaluate this expression.

80. The force of gravitational attraction between two point masses m and M that are r units apart is $F = GmM/r^2$, where G is a constant. Find the work done in moving the mass m along a straight-line path from $r = 1$ unit from M to $r = +\infty$.

81. If n is a positive integer, show that $\int_0^{+\infty} x^n e^{-x}\, dx = n \int_0^{+\infty} x^{n-1} e^{-x}\, dx$.

82. If n is a positive integer, show that $\int_0^{+\infty} x^n e^{-x}\, dx = n!$.

83. Show that $\int_e^{+\infty} dx/[x(\ln x)^p]$ converges if $p > 1$ and diverges if $p \le 1$.

84. Suppose that f and g are continuous on $[1, +\infty)$ and $0 \le g(x) \le f(x)$ for $x \ge 1$. Show that if $\int_1^{+\infty} g(x)\, dx$ diverges then $\int_1^{+\infty} f(x)\, dx$ diverges.

85. Suppose f and g denote continuous functions defined on the interval $[a, +\infty)$, for which $0 \le f(x) \le g(x)$ for $x \ge a$. (a) If $\int_a^{+\infty} f(x)\, dx = +\infty$, show that $\int_a^{+\infty} g(x)\, dx = +\infty$. (b) If $\int_0^{+\infty} g(x)\, dx$ converges, then $\int_0^{+\infty} f(x)\, dx$ converges.

86. Use the result obtained in Problem 85 to show that $\int_0^{+\infty} (1/\sqrt{2 + \sin x})\, dx$ diverges.

87. Use the result obtained in Problem 85 to show that $\int_2^{+\infty} [(\ln x)/\sqrt{x^2 - 1}]\, dx$ diverges.

Laplace transforms are useful in solving a special class of differential equations. The Laplace transform $L\{f(x)\}$ of a function f is defined as

$$L\{f(x)\} = \int_0^{+\infty} e^{-xs} f(t)\, dt \tag{7}$$

In Problems 88–93 compute the Laplace transform of the given function.

88. $f(x) = x$ **89.** $f(x) = \cos x$ **90.** $f(x) = \sin x$ **91.** $f(x) = e^x$

92. $f(x) = e^{ax}$ **93.** $f(x) = 1$

Taylor Polynomials 10.4

Of the functions studied thus far, the polynomial functions

$$P_n(x) = a_0 + a_1 x + a_2 x^2 + \cdots + a_n x^n$$

stand out as the simplest. Such functions are easy to evaluate, since the only operations needed are addition and multiplication. Such is not the case with, for example, the exponential function or the sine function if it is desired to calculate e^π or $\sin 0.1$. Because of the simple nature of polynomials, we are interested in *approximating functions* by a special class of polynomials known as the *Taylor polynomials*.*

We begin by recalling that the linear approximation of f (studied in Section 4.2 of Chap. 4) is a polynomial of degree 1 that serves as an approximation to f. In fact, based on (5) in Section 4.2, the first-degree polynomial $f(c) + f'(c)(x - c)$ approximates f for x close to c. That is,

$$f(x) \approx f(c) + f'(c)(x - c)$$

We define the first-degree polynomial P_1 as

$$P_1(x) = f(c) + f'(c)(x - c)$$

The graph of this approximating polynomial for f is the tangent line to f at $x = c$.

An improvement on this approximation would be a second-degree polynomial whose graph has the same tangent line at c as f and whose second derivative at c equals that of f. Such a polynomial is

$$P_2(x) = f(c) + f'(c)(x - c) + \frac{f''(c)}{2}(x - c)^2$$

because

1. $P_2(c) = f(c)$ — $P_2(x)$ has the same value at c as f.

2. $P_2'(x) = f'(c) + f''(c)(x - c)$
$P_2'(c) = f'(c) + f''(c)(c - c) = f'(c)$ — The slope of the tangent line to P_2 at c is the same as the slope of the tangent line to f at c.

---

* Named for the English mathematician Brook Taylor (1681–1731).

3. $P_2''(x) = f''(c)$
$P_2''(c) = f''(c)$

The second derivative of P_2 at c is the same as that of f at c.

See Figure 8.

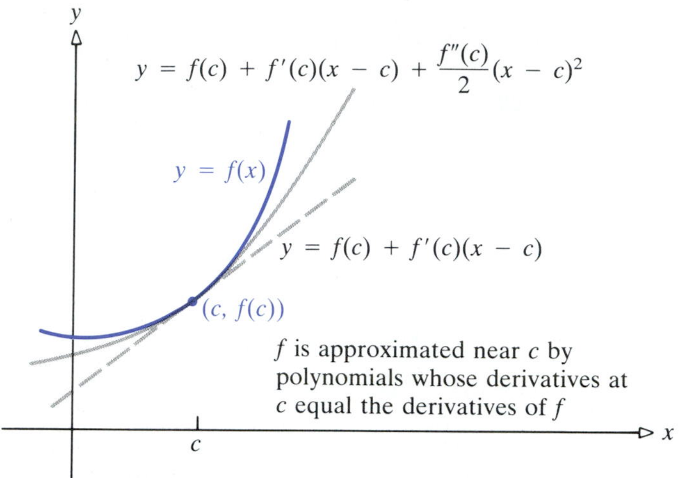

Figure 8

A third-degree polynomial whose first three derivatives at c are equal to those of f at c should provide an even better approximation to f for x near c. As we shall verify, such a polynomial is

$$P_3(x) = f(c) + f'(c)(x - c) + \frac{f''(c)}{2!}(x - c)^2 + \frac{f'''(c)}{3!}(x - c)^3$$

Notice that

1. $P_3(c) = f(c)$
2. $P_3'(x) = f'(c) + f''(c)(x - c) + \frac{f'''(c)}{2}(x - c)^2 \qquad P_3'(c) = f'(c)$
3. $P_3''(x) = f''(c) + f'''(c)(x - c) \qquad P_3''(c) = f''(c)$
4. $P_3'''(x) = f'''(c) \qquad P_3'''(c) = f'''(c)$

If continued, this same reasoning would, after n steps, bring us to a polynomial of degree n:

$$P_n(x) = f(c) + f'(c)(x - c) + \frac{f''(c)}{2!}(x - c)^2 + \cdots + \frac{f^{(n)}(c)}{n!}(x - c)^n \tag{1}$$

This polynomial P_n is called a *Taylor polynomial of degree n for f at c.*

EXAMPLE 1

Write the Taylor polynomial of degree 3 for $f(x) = \sqrt{x}$ at 1.

Solution

We use (1) with $n = 3$, which requires that we calculate the derivatives of f at 1:

$$f(x) = \sqrt{x} \qquad f(1) = 1$$

$$f'(x) = \frac{1}{2\sqrt{x}} \qquad f'(1) = \frac{1}{2}$$

$$f''(x) = -\frac{1}{4x^{3/2}} \qquad f''(1) = -\frac{1}{4}$$

$$f'''(x) = \frac{3}{8x^{5/2}} \qquad f'''(1) = \frac{3}{8}$$

The Taylor polynomial $P_3(x)$ of $f(x) = \sqrt{x}$ at 1 is

$$P_3(x) = f(1) + f'(1)(x-1) + \frac{f''(1)}{2!}(x-1)^2 + \frac{f'''(1)}{3!}(x-1)^3$$

$$= 1 + \frac{x-1}{2} - \frac{(x-1)^2}{8} + \frac{(x-1)^3}{16}$$

EXAMPLE 2

Write the Taylor polynomial of degree 7 for $f(x) = \sin x$ at 0.

Solution

The successive derivatives of $f(x) = \sin x$ at 0 are

$$f(x) = \sin x \qquad f(0) = 0$$

$$f'(x) = \cos x \qquad f'(0) = 1$$

$$f''(x) = -\sin x \qquad f''(0) = 0$$

$$f'''(x) = -\cos x \qquad f'''(0) = -1$$

$$f^{(4)}(x) = \sin x \qquad f^{(4)}(0) = 0$$

The pattern now begins to repeat: even derivatives of f at 0 equal 0, and odd derivatives of f at 0 alternate between 1 and -1. Therefore, the Taylor polynomial of degree 7 is

$$P_7(x) = f(0) + f'(0)x + \frac{f''(0)x^2}{2!} + \frac{f'''(0)x^3}{3!} + \cdots + \frac{f^{(7)}(0)x^7}{7!}$$

$$= x - \frac{x^3}{3!} + \frac{x^5}{5!} - \frac{x^7}{7!}$$

Figure 9 illustrates the graph of P_7 superimposed on the graph of $\sin x$.

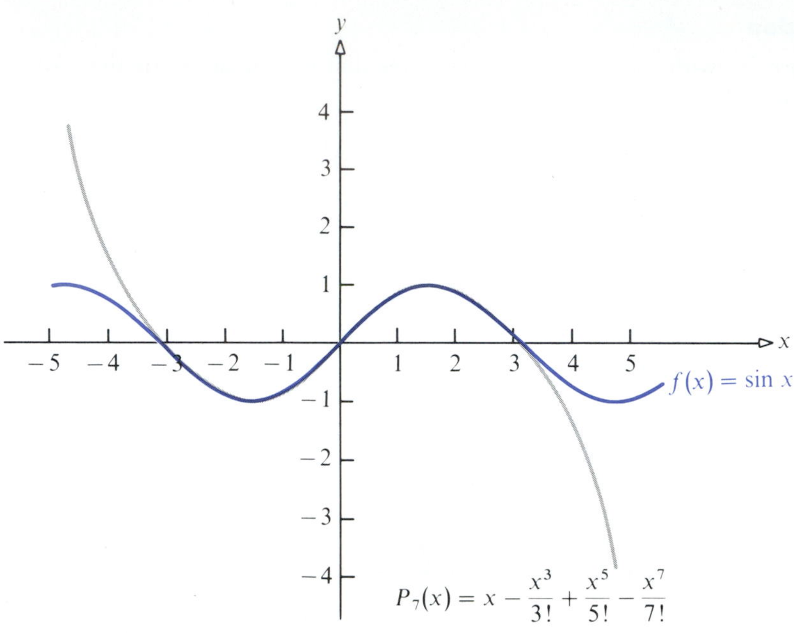

Figure 9

EXAMPLE 3

Write the Taylor polynomial of degree n for $f(x) = e^x$ at 0.

Solution

The successive derivatives of $f(x) = e^x$ at 0 are

$$f(x) = e^x \qquad f(0) = 1$$
$$f'(x) = e^x \qquad f'(0) = 1$$
$$f''(x) = e^x \qquad f''(0) = 1$$

As is evident, the derivative of any order of $f(x) = e^x$ equals 1 at 0. Thus, the Taylor polynomial of degree n of $f(x) = e^x$ at 0 is

$$P_n(x) = f(0) + f'(0)x + \frac{f''(0)x^2}{2!} + \cdots + \frac{f^{(n)}(0)x^n}{n!} = 1 + x + \frac{x^2}{2!} + \cdots + \frac{x^n}{n!}$$

Figure 10 illustrates how good the Taylor polynomials P_1, P_2, P_3 of $f(x) = e^x$ are at approximating e^x near 0.

Lagrange's Form of the Remainder

The accuracy of the polynomial P_n as an approximation to the function f for x close to c is provided by the next result, which is called *Taylor's formula with remainder.*

[10.4.1] THEOREM / *Taylor's Formula with Remainder.*

Let f be a function whose first $(n + 1)$ derivatives exist and are continuous on some interval I containing c. Then, for every x in I,

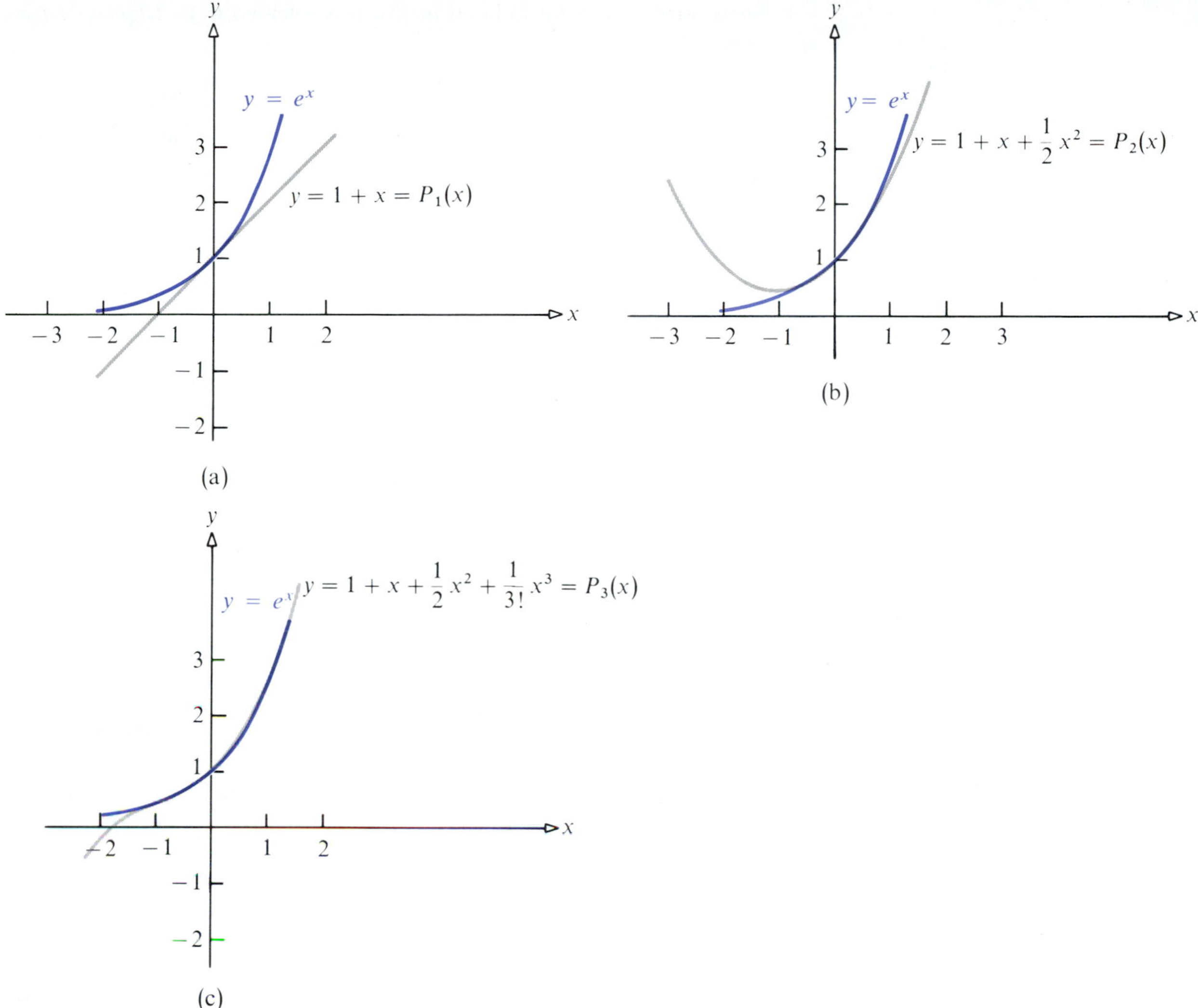

Figure 10

there exists a number u between x and c such that

$$f(x) = f(c) + f'(c)(x - c) + \frac{f''(c)}{2!}(x - c)^2 + \frac{f'''(c)}{3!}(x - c)^3 + \cdots + \frac{f^{(n)}(c)}{n!}(x - c)^n + R_{n+1}(x) \qquad (2)$$

where

$$R_{n+1}(x) = \frac{f^{(n+1)}(u)}{(n+1)!}(x - c)^{n+1} \qquad (3)$$

A proof of theorem (10.4.1) is postponed to the end of the section.

In (2), the sum of the first $n + 1$ terms on the right is the Taylor polynomial of degree n of f at c. The expression R_{n+1} is called the *remainder after*

$n + 1$ *terms* and, as it is formulated in (3), it is referred to as *Lagrange's form of the remainder.*

Since we have

$$P_n(x) = f(c) + f'(c)(x - c) + \frac{f''(c)}{2!}(x - c)^2 + \cdots + \frac{f^{(n)}(c)}{n!}(x - c)^n$$

and

$$R_{n+1}(x) = \frac{f^{(n+1)}(u)}{(n + 1)!}(x - c)^{n+1} \qquad u \text{ between } x \text{ and } c$$

Taylor's formula with remainder asserts that

$$f(x) = P_n(x) + R_{n+1}(x) \tag{4}$$

where P_n is a polynomial of degree n in $(x - c)$, whose first n derivatives at c coincide with those of f at c.

The Taylor polynomial P_n may be considered an approximating polynomial for the function f for x close to c. The remainder R_{n+1} measures the accuracy of this approximation. Thus:

If the numerical value of R_{n+1} is small, then

$$f(x) \approx P_n(x)$$

and the error involved in using the approximation equals $|R_{n+1}(x)|$.

Of course, if we knew the value of $R_{n+1}(x)$, we would know $f(x)$ exactly. The reason we do not know $R_{n+1}(x)$, is that its value depends on the number u and we do not know u. However, we can usually get estimates of the remainder $R_{n+1}(x)$ and thereby determine just how good an approximation we have.

For instance, using the result obtained in Example 1, we can approximate $\sqrt{1.2}$ by $P_3(1.2)$. That is,

$$\sqrt{1.2} \approx P_3(1.2) = 1 + \frac{0.2}{2} - \frac{(0.2)^2}{8} + \frac{(0.2)^3}{16}$$

$$= 1 + 0.1 - 0.005 + 0.0005 = 1.0955$$

The error of this approximation for $\sqrt{1.2}$ is obtained from (3):

$$R_4(x) = \frac{f^{(4)}(u)}{4!}(x - 1)^4$$

where $f^{(4)}(u) = -15/(16u^{7/2})$, $x = 1.2$, and $1 < u < 1.2$. Since $u > 1$, it follows that $1/u < 1$. Therefore, an upper estimate to the error is

$$|R_4(1.2)| = \frac{15}{(4!)(16u^{7/2})}(0.2)^4 < \frac{15(0.2)^4}{384} = 0.0000625$$

By using a hand calculator, we compute the actual error to be

$$|\sqrt{1.2} - 1.0955| \approx |1.0954451 - 1.0955| = 0.0000549$$

EXAMPLE 4

Write the polynomial $f(x) = x^3 - 3x^2 + 6x - 9$ as a polynomial in $(x - 2)$.

Solution

We seek the Taylor polynomial of f at 2:

$$f(x) = x^3 - 3x^2 + 6x - 9 \qquad f(2) = -1$$

$$f'(x) = 3x^2 - 6x + 6 \qquad f'(2) = 6$$

$$f''(x) = 6x - 6 \qquad f''(2) = 6$$

$$f'''(x) = 6 \qquad f'''(2) = 6$$

Since $f^{(4)}(x) = 0$, it follows that $R_4(x) = 0$ and $f(x) = P_3(x)$. That is,

$$f(x) = P_3(x) = -1 + 6(x - 2) + \frac{6(x - 2)^2}{2!} + \frac{6(x - 2)^3}{3!}$$

Thus,

$$x^3 - 3x^2 + 6x - 9 = (x - 2)^3 + 3(x - 2)^2 + 6(x - 2) - 1$$

Example 4 illustrates a general result. If f is a polynomial of degree n, it follows that the remainder R_{n+1} must equal 0, since $f^{(n+1)}(u) = 0$. Hence, a polynomial of degree n is equal to a Taylor polynomial of degree n at c.

Let's look at another example of approximation.

EXAMPLE 5

Approximate sin 1 using a Taylor polynomial of degree 7 for $f(x) = \sin x$ at 0. Estimate the error obtained in using this Taylor polynomial.

Solution

Using the result of Example 2, we have

$$\sin 1 \approx 1 - \frac{(1)^3}{3!} + \frac{(1)^5}{5!} - \frac{(1)^7}{7!}$$

$$\approx 1 - 0.1666666 + 0.0083333 - 0.0001984 = 0.8414683$$

The error obtained in using this Taylor polynomial is

$$|R_8(1)| = \left| \frac{f^{(8)}(u)(1)^8}{8!} \right| = \frac{|\sin u|}{8!}$$

where u is between 0 and 1. Since $|\sin u| \leq 1$, we have

$$|R_8(1)| \leq \frac{1}{8!} \approx 0.0000248$$

Let's look at another example and approximate the number e.

EXAMPLE 6

Use a Taylor polynomial to approximate the number e correct to within 0.0001.

Solution

From Example 3, the Taylor polynomial of degree n at 0 of $f(x) = e^x$ is

$$P_n(x) = 1 + \frac{x}{1!} + \frac{x^2}{2!} + \cdots + \frac{x^n}{n!} \tag{5}$$

and

$$e^x = P_n(x) + R_{n+1}(x) = 1 + \frac{x}{1!} + \frac{x^2}{2!} + \cdots + \frac{x^n}{n!} + \frac{e^u}{(n+1)!}x^{n+1}$$

where u is a number between 0 and x. If we set $x = 1$, we get

$$e^1 = P_n(1) + R_{n+1}(1)$$

$$e = 1 + \frac{1}{1!} + \frac{1}{2!} + \cdots + \frac{1}{n!} + \frac{e^u}{(n+1)!} \qquad 0 < u < 1$$

We need to find an integer n such that the remainder $e^u/(n+1)!$ is less than 0.0001. Our problem in finding such an n is that we do not know exactly what e^u is. However, since we set $x = 1$, we do know that $0 < u < 1$. In addition, since it is common knowledge that $e < 3$, we also know that

$$0 < \frac{e^u}{(n+1)!} < \frac{e}{(n+1)!} < \frac{3}{(n+1)!}$$

The smallest n for which $3/(n+1)! < 0.0001$ is $n = 7$. Thus, for $n = 7$, $e^u/(n+1)! < 0.0001$. Hence, we use $n = 7$ in (5) to find that

$$\begin{aligned} P_7(1) &= 1 + 1 + \frac{1}{2!} + \frac{1}{3!} + \frac{1}{4!} + \frac{1}{5!} + \frac{1}{6!} + \frac{1}{7!} \\ &\approx 1 + 1 + 0.500000 + 0.166667 + 0.041667 + 0.008333 \\ &\quad + 0.001389 + 0.000198 \\ &= 2.718254 \end{aligned}$$

approximates e to within 0.0001. In other words, $|e - 2.718254| < 0.0001$ or $2.718154 < e < 2.718354$.

In approximating e by using a Taylor polynomial of degree n, the error introduced equals $e^u/(n+1)!$. Evidently this error can be made as small as we please by choosing n sufficiently large. However, by taking n larger and larger, we need to add more and more terms to the Taylor polynomial. In fact, we show in the next chapter that e actually equals an infinite sum of numbers. This leads us to the subject of the next chapter—*infinite series.*

Proof of Taylor's Formula with Remainder For a fixed x in I there is a number L (depending on x) for which

$$f(x) = f(c) + \frac{f'(c)}{1!}(x-c) + \frac{f''(c)}{2!}(x-c)^2 + \cdots + \frac{f^{(n)}(c)}{n!}(x-c)^n + \frac{L}{(n+1)!}(x-c)^{n+1} \tag{6}$$

We define the function F so that

$$F(t) = f(x) - f(t) - \frac{f'(t)}{1!}(x-t) - \frac{f''(t)}{2!}(x-t)^2 - \cdots - \frac{f^{(n)}(t)}{n!}(x-t)^n - \frac{L}{(n+1)!}(x-t)^{n+1} \tag{7}$$

The domain of F is $c \le t \le x$ if $x > c$ and $x \le t \le c$ if $x < c$. Clearly, since $f(t), f'(t), \ldots, f^{(n)}(t)$ are each continuous, F is continuous on its domain. Furthermore, F is differentiable and

$$\begin{aligned}\frac{dF}{dt} = F'(t) &= -f'(t) + \left[f'(t) - \frac{f''(t)}{1!}(x-t)\right] \\ &+ \left[\frac{f''(t)}{1!}(x-t) - \frac{f'''(t)}{2!}(x-t)^2\right] + \cdots \\ &+ \left[\frac{f^{(n)}(t)}{(n-1)!}(x-t)^{n-1} - \frac{f^{(n+1)}(t)}{n!}(x-t)^n\right] + \frac{L}{n!}(x-t)^n \\ &= \frac{-f^{(n+1)}(t)}{n!}(x-t)^n + \frac{L}{n!}(x-t)^n \qquad \text{for all } t \text{ between } c \text{ and } x\end{aligned}$$

In view of (6) and (7), we have

$$F(c) = f(x) - f(c) - \frac{f'(c)}{1!}(x-c) - \frac{f''(c)}{2!}(x-c)^2 - \cdots - \frac{f^{(n)}(c)}{n!}(x-c)^n - \frac{L}{(n+1)!}(x-c)^{n+1} = 0$$

and, by direct substitution in (7), we get

$$F(x) = f(x) - f(x) - \frac{f'(x)}{1!}(x-x) - \cdots - \frac{f^{(n)}(x)}{n!}(x-x)^n - \frac{L}{(n+1)!}(x-x)^{n+1} = 0$$

Thus, we may apply Rolle's theorem (4.4.1) to F. Then there is a number u between c and x so that

$$F'(u) = \frac{-f^{(n+1)}(u)}{n!}(x-u)^n + \frac{L(x-u)^n}{n!} = 0$$

Solving this equation for L, we find $L = f^{(n+1)}(u)$. Then, setting $t = c$ and $L = f^{(n+1)}(u)$ in (7) and solving for $f(x)$ gives the desired result.

EXERCISE 10.4

In Problems 1–4 write each polynomial as powers of $(x - 1)$.

1. $f(x) = 3x^2 - 6x + 4$

2. $f(x) = 4x^2 - x + 1$

3. $f(x) = x^3 + x^2 - 8$

4. $f(x) = x^4 + 1$

In Problems 5–22 write the Taylor polynomial of degree n about c for the given c and n. Also, write an expression for the remainder $R_{n+1}(x)$.

5. $f(x) = 3x^3 + 2x^2 - 6x + 5;\quad c = 1,\quad n = 3$

6. $f(x) = 4x^3 - 2x^2 - 4;\quad c = 1,\quad n = 3$

7. $f(x) = 2x^4 - 6x^3 + x;\quad c = -1,\quad n = 4$

8. $f(x) = -3x^4 + 2x^2 - 5;\quad c = -1,\quad n = 4$

9. $f(x) = x^5;\quad c = 2,\quad n = 4$

10. $f(x) = x^6;\quad c = 3,\quad n = 5$

11. $f(x) = \ln x;\quad c = 1,\quad n = 5$

12. $f(x) = \ln(1 + x);\quad c = 0,\quad n = 5$

13. $f(x) = \dfrac{1}{x};\quad c = 1,\quad n = 5$

14. $f(x) = \sqrt[3]{x};\quad c = 1,\quad n = 5$

15. $f(x) = \cos x;\quad c = 0,\quad n = 6$

16. $f(x) = \sin x;\quad c = \dfrac{\pi}{4},\quad n = 7$

17. $f(x) = \cosh x;\quad c = 0,\quad n = 6$

18. $f(x) = e^{-x};\quad c = 0,\quad n = 5$

19. $f(x) = \dfrac{1}{1 - x};\quad c = 0,\quad n = 4$

20. $f(x) = \dfrac{1}{1 + x};\quad c = 0,\quad n = 4$

21. $f(x) = \dfrac{1}{(1 + x)^2};\quad c = 0,\quad n = 3$

22. $f(x) = \dfrac{1}{1 + x^2};\quad c = 0,\quad n = 2$

In Problems 23–32 use a Taylor polynomial to approximate the given number with an error less than that shown.

23. $\ln 1.1$; Error < 0.00001

24. $\ln 1.1$; Error < 0.01

25. $\dfrac{1}{1.1}$; Error < 0.001

26. $\sqrt[3]{0.9}$; Error < 0.001

27. $\cos 1°$; Error < 0.0001
[*Hint:* $1° = 2\pi/360$ radian.]

28. $\sin 46°$; Error < 0.0001

29. e^{-1}; Error < 0.01

30. $e^{2/3}$; Error < 0.01

31. $e^{1/2}$; Error < 0.01

32. e^{-2}; Error < 0.1

In Problems 33–40 write the Taylor polynomial of degree n for each function about c using the indicated values of c and n.

33. $f(x) = x \ln x;\quad n = 4,\quad c = 1$

34. $f(x) = \sqrt{3 + x^2};\quad n = 3,\quad c = 1$

35. $f(x) = e^{-x^2};\quad n = 2,\quad c = 0$

36. $f(x) = \sin x^2;\quad n = 3,\quad c = 0$

37. $f(x) = \tan x;\quad n = 4,\quad c = \dfrac{\pi}{4}$

38. $f(x) = \sec x;\quad n = 3,\quad c = 0$

39. $f(x) = \sqrt{1 + x};\quad n = 3,\quad c = 0$

40. $f(x) = \tan^{-1} x;\quad n = 3,\quad c = 0$

41. The graphs of $y = \sin x$ and $y = \lambda x$ intersect near $x = \pi$ if λ is small. Set $f(x) = \sin x - \lambda x$. Write the Taylor polynomial of degree 2 for f about π and use it to show that an approximate solution of the equation $\sin x = \lambda x$ is $x = \pi/(1 + \lambda)$.

42. Proceed as in Problem 41 and obtain an approximate solution of the equation $\cot x = \lambda x$. [*Hint:* Assume that λ is small and use $n = 2$ and $c = \pi/2$.]

In Problems 43 and 44 use Taylor polynomials to approximate the value of each integral.

43. $\int_0^{0.01} e^{x^2}\,dx$, using a Taylor polynomial of degree 4 for $f(x) = e^{x^2}$ at $c = 0$

44. $\int_{0.1}^{0.2} \sin(x^2)\,dx$, using a Taylor polynomial of degree 4 for $f(x) = \dfrac{e^x}{x}$ at $c = 0$

45. (a) Find: $\lim_{h\to 0} \dfrac{f(x+2h) - 2f(x+h) + f(x)}{h^2}$

(b) Find:

$$\lim_{h\to 0} \frac{f(x+3h) - 3f(x+2h) + 3f(x+h) - f(x)}{h^3}$$

(c) Generalize parts (a) and (b).

46. *Calculator Problem.* The formulas in Problem 45 can be used to approximate derivatives. Approximate $f'(2)$, $f''(2)$, and $f'''(2)$ from the table. The data are for $f(x) = \ln x$. Compare the exact values with your approximations.

x	2.0	2.1	2.2	2.3	2.4
$f(x)$	0.6931	0.7419	0.7885	0.8329	0.8755

REVIEW EXERCISES

In Problems 1–20 find the limit.

1. $\lim_{x\to 0} \dfrac{xe^{3x} - x}{1 - \cos 2x}$

2. $\lim_{x\to \pi/2} \dfrac{\sec^2 x}{\sec^2 3x}$

3. $\lim_{x\to 4} \dfrac{x^2 - 16}{x^2 + x - 20}$

4. $\lim_{x\to 0} \dfrac{e^x - e^{-x}}{\sin x}$

5. $\lim_{x\to 0} \dfrac{\tan x - x}{x - \sin x}$

6. $\lim_{x\to 0} \dfrac{\tan x + \sec x - 1}{\tan x - \sec x + 1}$

7. $\lim_{x\to \theta} \dfrac{\sin x - \sin\theta}{x - \theta}$

8. $\lim_{x\to a} \dfrac{ax - x^2}{a^4 - 2a^3x + 2ax^3 - x^4}$

9. $\lim_{x\to 0} \dfrac{x - \sin x}{x^3}$

10. $\lim_{x\to 0} \dfrac{\tan x - \sin x}{\sin^3 x}$

11. $\lim_{x\to +\infty} (1 + 4x)^{2/x}$

12. $\lim_{x\to 0} \left(\dfrac{1}{x^2} - \dfrac{1}{x^2 \sec x}\right)$

13. $\lim_{x\to 0^+} x \cot \pi x$

14. $\lim_{x\to +\infty} x \sin \dfrac{a}{x}$

15. $\lim_{x\to 1} \left[\dfrac{2}{x^2 - 1} - \dfrac{1}{x - 1}\right]$

16. $\lim_{x\to 0} \left[\dfrac{2}{\sin^2 x} - \dfrac{1}{1 - \cos x}\right]$

17. $\lim_{x\to 0^+} \left(\dfrac{1}{x}\right)^{\tan x}$

18. $\lim_{x\to +\infty} \left(1 + \dfrac{a}{x}\right)^x$

19. $\lim_{x\to 0^+} (\cot x)^x$

20. $\lim_{x\to +\infty} \left(\dfrac{2}{x} + 1\right)^x$

In Problems 21–26 evaluate the integrals.

21. $\int_1^{+\infty} \dfrac{e^{-\sqrt[3]{x}}}{\sqrt[3]{x^2}}\,dx$

22. $\int_1^{+\infty} \dfrac{e^{-\sqrt{x}}}{\sqrt{x}}\,dx$

23. $\int_0^1 \dfrac{\sin\sqrt{x}}{\sqrt{x}}\,dx$

24. $\int_{-a}^{a} \dfrac{dx}{\sqrt{a^2 - x^2}}$

25. $\int_0^1 \dfrac{x\,dx}{\sqrt{1 - x^2}}$

26. $\int_{-\infty}^{0} xe^x\,dx$

27. Show that $\int_0^{\pi/2} \dfrac{\sin x}{\sqrt{\cos x}}\,dx$ converges.

28. Show that $\int_1^{+\infty} \dfrac{\sqrt{1 + x^{1/8}}}{x^{3/4}}\,dx$ diverges.

29. Show that the area enclosed by $y = x^{-2/3}$, $x = 0$, and $x = 1$ is finite.

30. Show that the volume generated when the area enclosed by $y = x^{-2/3}$ $x = 0$, and $x = 1$ is revolved about the x-axis is not finite.

In Problems 31–34 write the Taylor polynomial of degree n about c for each function.

31. $f(x) = e^{2x}$; $n = 4$, $c = 3$

32. $f(x) = \tan x$; $n = 4$, $c = \pi/3$

33. $f(x) = 1/(1 + x)$; $n = 4$, $c = 1$

34. $f(x) = \ln(x + h)$; $n = 6$, $c = 0$

35. Expand $f(x) = x^3 - 2x^2 + 5x - 7$ in powers of $(x - 1)$.

36. Expand $f(x) = 4x^3 - 17x^2 + 11x + 2$ in powers of $(x - 4)$.

37. Expand $f(x) = e^x$ in powers of $(x - 3)$.

38. Expand $f(x) = \ln x$ in powers of $(x - 2)$.

39. Use five terms of the Taylor polynomial to approximate $\sqrt{e}$.

40. Use six terms of the Taylor polynomial to approximate $\cos \pi/3$.

CHALLENGE EXERCISES

1. Explain why L'Hospital's rule does not apply to $\lim_{x\to 0} \frac{x^2 \sin(1/x)}{\sin x}$.

2. (a) Prove that the area in the first quadrant enclosed by the graph of $y = e^{-x}$ and the x-axis is divided into two equal parts by the line $x = \ln 2$.

(b) If the two equal areas described in part (a) are rotated about the x-axis, are the resulting volumes equal? If so, prove it. If not, determine which one is larger and by how much.

3. Find constants a, b, c, and d so that

$$\lim_{x\to 0} \frac{\sin ax + bx + cx^2 + dx^3}{x^5} = \frac{4}{15}.$$

4. Find the length of the curve $y = \sqrt{x - x^2} - \sin^{-1}\sqrt{x}$.

5. Evaluate: $\int_{-\infty}^{a} e^{(x - e^x)}\, dx$

6. Evaluate: $\int_{-\infty}^{+\infty} e^{(x - e^x)}\, dx$

7. A *probability density function* is a function f whose domain is the set of all real numbers such that (i) $f(x) \geq 0$ for all x and (ii) $\int_{-\infty}^{+\infty} f(x)\, dx = 1$. Show that the function below is a probability density function. (This is called the *uniform density function*.)

$$f(x) = \begin{cases} 0 & \text{if } x < a \\ \dfrac{1}{b - a} & \text{if } a \leq x \leq b \\ 0 & \text{if } x > b \end{cases}$$

8. Show that the following function is a probability density

(Problem 8 continues in the next column)

8. *(continued)*

function for $a > 0$. (This is called the *exponential density function.*)

$$f(x) = \begin{cases} \dfrac{1}{a} e^{-x/a} & \text{if } x \geq 0 \\ 0 & \text{if } x < 0 \end{cases}$$

9. The *expected value* or *mean* μ associated with the density function f is defined by

$$\mu = \int_{-\infty}^{+\infty} xf(x)\, dx$$

We can think of the expected value as a weighted average of various probabilities. Calculate the expected value associated with the probability density function given in Problem 7.

10. Calculate the expected value associated with the probability density function given in Problem 8.

11. The *variance* σ^2 associated with the probability density function f is defined by

$$\sigma^2 = \int_{-\infty}^{+\infty} (x - \mu)^2 f(x)\, dx$$

We can think of the variance as the average square of the deviation from the mean. We define the *standard deviation* σ associated with the function f to be the square root of the variance σ^2. Calculate the variance and standard deviation associated with the probability density function given in Problem 7.

12. Calculate the variance and standard deviation associated with the probability density function given in Problem 8.

11

Infinite Series

Historical Perspectives

Both Newton and Leibniz realized that the foundation of the theory they had helped to create was poorly and inadequately laid. Newton especially seemed to be aware of the seriousness of the difficulties and inconsistencies in his efforts to explain the logical basis of calculus. Although there were mathematicians in the eighteenth century who made attempts to supply the missing rigor for the subject, most of the mathematical talent of the day was busy at the task of expanding and refining the methodology of calculus rather than worrying about the solidity of its logical foundation. Much of this expanded methodology depended directly on the use of infinite series. The missing rigor was finally supplied in the nineteenth century—largely in response to inconsistencies and paradoxes in the theory of infinite series that could no longer be ignored if the development of calculus was to proceed.

Among the most important contributors to the development of calculus methods were the three Swiss mathematicians Jakob Bernoulli (1654–1705), his brother Johann (1667–1748), and Leonhard Euler (1707–1783). Much of their work involved the representation of functions by infinite series for the purposes of integration and differentiation. In particular, this method was the standard way of treating the trigonometric, exponential, and logarithmic functions at the time.

Both of the Bernoulli brothers were in regular correspondence with Leibniz, and each made a major contribution in interpreting and completing the details in Leibniz's sketchy papers on calculus. A famous dispute developed between the ambitious brothers when Johann presented some results that Jakob had communicated to him as if he had discovered them himself. When Jakob learned of this, he reciprocated and took credit for some of Johann's work. Perhaps justice was served in the end though, because, as we mentioned in Chapter 10, one of Johann's nicest discoveries is now known as L'Hospital's rule—credit for it wrongly going to Johann's benefactor, Guillaume L'Hospital.

Euler was one of the most prolific mathematicians ever (his collected works fill more than 70 volumes!) and one of the most capable. He had the good fortune to be educated in mathematics by Johann Bernoulli. Euler was noted for his remarkable memory, and although he was totally blind for the last 17 years of his life, he employed a secretary to record his discoveries and over 400 of his research papers were written during those years. The range of Euler's work covered all the mathematics known in his day. In 1755, he wrote the first reasonably complete textbook on differential calculus, which he followed in 1768–1770 with a three-volume text on integral calculus. Through these popular books, his numerous research papers, and correspondence, Euler exerted an inestimable influence on the development of calculus.

Throughout most of the eighteenth century, calculus was regarded as essentially an extension of algebra, expanded through the use and manipulation of infinite series. However, even Euler, who accomplished this manipulation with greater success than anyone, had no method for analyzing the convergence or divergence of a series, and confusion about the proper role of infinite series abounded.

The present-day concepts of convergence and divergence were not clearly defined until the early nineteenth century in the works of the great French mathematician Augustin Cauchy and the Czechoslovakian Bernhard Bolzano. Although others made important contributions, these two men were the first to put forth the concept of the sum of a series being the limit of its sequence of partial sums. The fact that over 150 years elapsed between Newton's use of the binomial theorem to expand a function into an infinite series and integrate it term-by-term, and an acceptable definition of the sum of a series, indicates how hard-won the concepts of convergence and divergence were.

Sequences 11.1

The study of infinite sums of numbers has important applications in physics and engineering since it provides an alternate way of representing functions. In particular, *infinite series* may be used to approximate numbers such as e, π, ln 2, and so on. The theory of infinite series is developed through the use of a special kind of function called a *sequence*.

[11.1.1] DEFINITION / *Sequence.*

A *sequence* is a function whose domain is the set of positive integers.

Let us *see* what this means. Figure 1(a) is the graph of the function $f(x) = 1/x$ for $x > 0$. If all of the graph was erased *except* the points $(1, 1)$, $(2, \frac{1}{2})$, $(3, \frac{1}{3})$, and so on, as shown in Figure 1(b), then these points would form the graph of a sequence. In the graph of a sequence, there is *one point* for every positive integer.

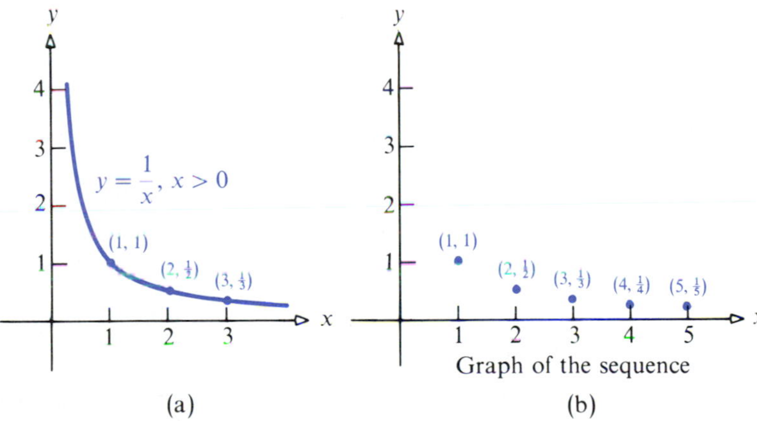

Figure 1

A sequence is often represented by listing its values in order. For example, for the above sequence we write

$$f(1), \quad f(2), \quad f(3), \quad f(4), \quad f(5), \quad \ldots$$

or

$$1, \quad \tfrac{1}{2}, \quad \tfrac{1}{3}, \quad \tfrac{1}{4}, \quad \tfrac{1}{5}, \quad \ldots$$

The list never ends, as indicated by the three dots at the end. The numbers in the list are called the *terms* of the sequence.

It is common to use a letter with a subscript, such as s_1, to stand for the first term, s_2 for the second term, s_3 for the third, and so on. Thus, for the above sequence, we have

$$s_1 = f(1) = 1$$
$$s_2 = f(2) = \tfrac{1}{2}$$
$$s_3 = f(3) = \tfrac{1}{3}$$
$$\vdots$$
$$s_n = f(n) = \frac{1}{n}$$
$$\vdots$$

It is easy to find any term of this particular sequence because it has the simple rule $s_n = f(n) = 1/n$. In general, if a rule for the nth term of a sequence is known, then any term of that sequence may be found.

EXAMPLE 1

The nth term of a sequence is

$$b_n = g(n) = \frac{2n - 1}{n^3}$$

Write the first three terms.

Solution

$$b_1 = g(1) = \frac{2(1) - 1}{1^3} = 1$$

$$b_2 = g(2) = \frac{2(2) - 1}{2^3} = \frac{3}{8}$$

$$b_3 = g(3) = \frac{2(3) - 1}{3^3} = \frac{5}{27}$$

When a rule for the nth term of a sequence is known, there is yet another advantage; namely, we may shorten our notation. For example, we may write $\{1/n\}$ for the sequence $1, \tfrac{1}{2}, \tfrac{1}{3}, \tfrac{1}{4}, \tfrac{1}{5}, \tfrac{1}{6}, \ldots$. In words, $\{1/n\}$ is the sequence whose nth term is $1/n$. Similarly, $\{2n/(n + 1)\}$ is the sequence whose nth term is $2n/(n + 1)$—namely,

$$\frac{2(1)}{1 + 1}, \quad \frac{2(2)}{2 + 1}, \quad \frac{2(3)}{3 + 1}, \quad \ldots, \quad \frac{2n}{n + 1}, \quad \ldots$$

or

$$1, \quad \frac{4}{3}, \quad \frac{3}{2}, \quad \ldots, \quad \frac{2n}{n + 1}, \quad \ldots$$

EXAMPLE 2

Write the first five terms of the sequence

$$\{(-1)^n(\tfrac{1}{2})^n\}$$

Solution

If c_n is used to denote the nth term of this sequence, then

$$c_n = (-1)^n(\tfrac{1}{2})^n$$

Thus,

$$c_1 = (-1)^1(\tfrac{1}{2})^1 = -\tfrac{1}{2}$$
$$c_2 = (-1)^2(\tfrac{1}{2})^2 = \tfrac{1}{4}$$
$$c_3 = (-1)^3(\tfrac{1}{2})^3 = -\tfrac{1}{8}$$
$$c_4 = \tfrac{1}{16}$$
$$c_5 = -\tfrac{1}{32}$$

In the sequence given in Example 2 the signs of the terms *alternate* because the factor $(-1)^n$ equals -1 when n is odd and equals 1 when n is even. Factors that have just the reverse behavior are $(-1)^{n+1}$ and $(-1)^{n-1}$. These equal 1 when n is odd and equal -1 when n is even. For example, the first five terms of the sequence $\{(-1)^{n+1}(\tfrac{1}{2})^n\}$ are $\tfrac{1}{2}, -\tfrac{1}{4}, \tfrac{1}{8}, -\tfrac{1}{16}, \tfrac{1}{32}$. Compare these with the first five terms of the sequence $\{c_n\} = \{(-1)^n(\tfrac{1}{2})^n\}$, listed in the solution to Example 2.

The rule defining a sequence is often expressed by an explicit formula for its nth term in terms of the index n. Sometimes, though, a sequence is indicated by an observed pattern in the first few terms so that a natural choice for the nth term is suggested. Example 3 will give you the general idea.

EXAMPLE 3

(a) $e, \dfrac{e^2}{2}, \dfrac{e^3}{3}, \ldots$ $\qquad a_n = \dfrac{e^n}{n}$

(b) $1, \tfrac{1}{3}, \tfrac{1}{9}, \tfrac{1}{27}, \ldots$ $\qquad b_n = (\tfrac{1}{3})^{n-1}$

(c) $1, 3, 5, 7, \ldots$ $\qquad c_n = 2n - 1$

(d) $1, 4, 9, 16, 25, \ldots$ $\qquad d_n = n^2$

(e) $\tfrac{2}{2}, \tfrac{4}{3}, \tfrac{6}{4}, \tfrac{8}{5}, \ldots$ $\qquad e_n = \dfrac{2n}{n+1}$

(f) $1, -\tfrac{1}{2}, \tfrac{1}{3}, -\tfrac{1}{4}, \tfrac{1}{5}, \ldots$ $\qquad f_n = \dfrac{(-1)^{n+1}}{n}$

(g) $1, \tfrac{1}{2}, 1, \tfrac{1}{4}, 1, \tfrac{1}{6}, \ldots$ $\qquad g_n = \begin{cases} 1 & \text{if } n \text{ is odd} \\ \dfrac{1}{n} & \text{if } n \text{ is even} \end{cases}$

(h) $0, -1, -1, -2, -2, -3, -3, \ldots$ $\qquad h_n = \begin{cases} -\dfrac{n-1}{2} & \text{if } n \text{ is odd} \\ -\dfrac{n}{2} & \text{if } n \text{ is even} \end{cases}$

The graphs of sequences (e)–(h) are given in Figure 2 on page 626.

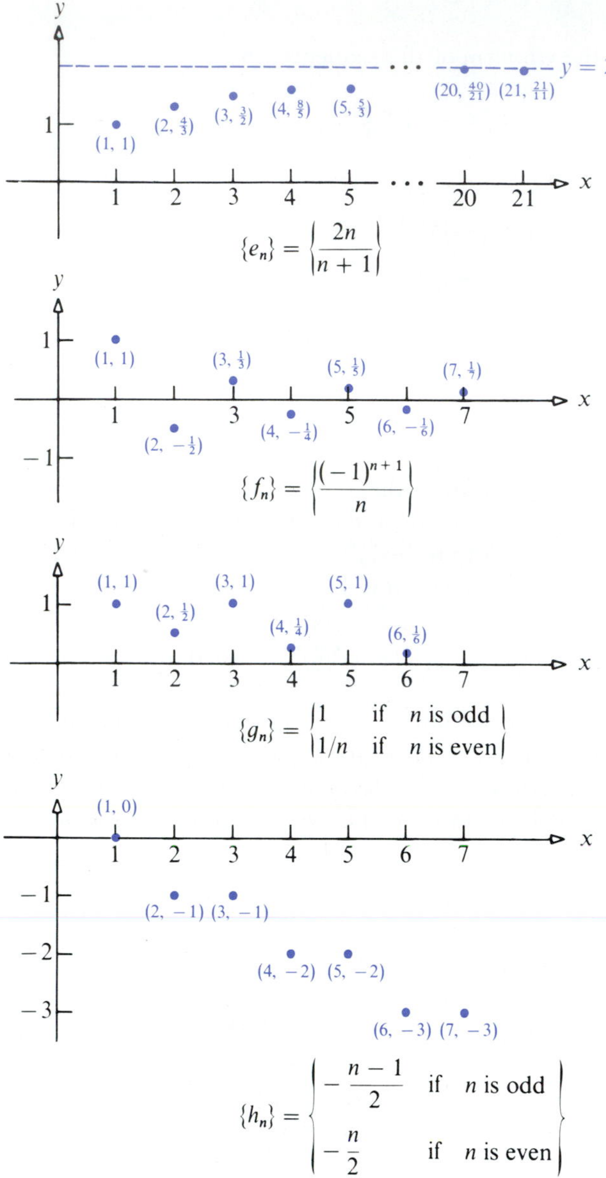

Figure 2

Convergent Sequences

Consider the sequence $\{e_n\} = \{2n/(n+1)\}$, whose graph is given in Figure 2. As we look to the right in this graph, we see that the points get closer and closer to the line $y = 2$. In fact, if we could look far enough to the right, the points would eventually *appear* to lie on the line $y = 2$ (although in actuality they would always be some very small distance below). This is because the nth term $2n/(n+1)$ becomes a closer and closer approximation to 2 as n grows larger. In fact, as the following table suggests, $2n/(n+1)$ can be made as close as we please to 2 by making n sufficiently large:

n	1	9	99	999	9,999	999,999
$2n/(n+1)$	1	1.8	1.98	1.998	1.9998	1.999998

We describe this behavior by saying that the sequence $\{2n/(n+1)\}$ *converges to* 2. In this case, we write

$$\lim_{n\to+\infty} \frac{2n}{n+1} = 2$$

We now state the definition of convergence.

[11.1.2] DEFINITION / ***Limit of a Sequence.***

Let L be a real number and let $\{s_n\}$ be a sequence. The statement $\{s_n\}$ *converges to L* means that for any given $\varepsilon > 0$, there exists a positive integer N such that

$$|s_n - L| < \varepsilon \qquad \text{for all integers} \quad n > N$$

We use the notation $\lim_{n\to+\infty} s_n = L$ to mean that $\{s_n\}$ converges to L. The real number L is called the *limit* of the sequence $\{s_n\}$.

In general, if a sequence converges to some number, the sequence is said to be *convergent*; otherwise, it is said to be *divergent*.

Figure 3 on page 628 provides a geometric interpretation of the statement

$$\lim_{n\to+\infty} s_n = L$$

The figure also shows that the convergence (or divergence) of a sequence is in no way affected by the beginning terms of the sequence. In fact, we may ignore the beginning terms of a sequence in making a determination of convergence or divergence.

Many of the familiar properties of limits of functions also apply to limits of sequences. We state, without proof, some of these results.

[11.1.3] THEOREM / ***Properties of Convergent Sequences.***

If $\{s_n\}$ and $\{t_n\}$ are convergent sequences and if c is a number, then:

(a) $\lim_{n\to+\infty} (cs_n) = c \lim_{n\to+\infty} s_n$

(b) $\lim_{n\to+\infty} (s_n \pm t_n) = \lim_{n\to+\infty} s_n \pm \lim_{n\to+\infty} t_n$

(c) $\lim_{n\to+\infty} (s_n t_n) = \left(\lim_{n\to+\infty} s_n\right)\left(\lim_{n\to+\infty} t_n\right)$

(d) $\lim_{n\to+\infty} \dfrac{s_n}{t_n} = \dfrac{\lim_{n\to+\infty} s_n}{\lim_{n\to+\infty} t_n}$ **provided** $\lim_{n\to+\infty} t_n \neq 0$

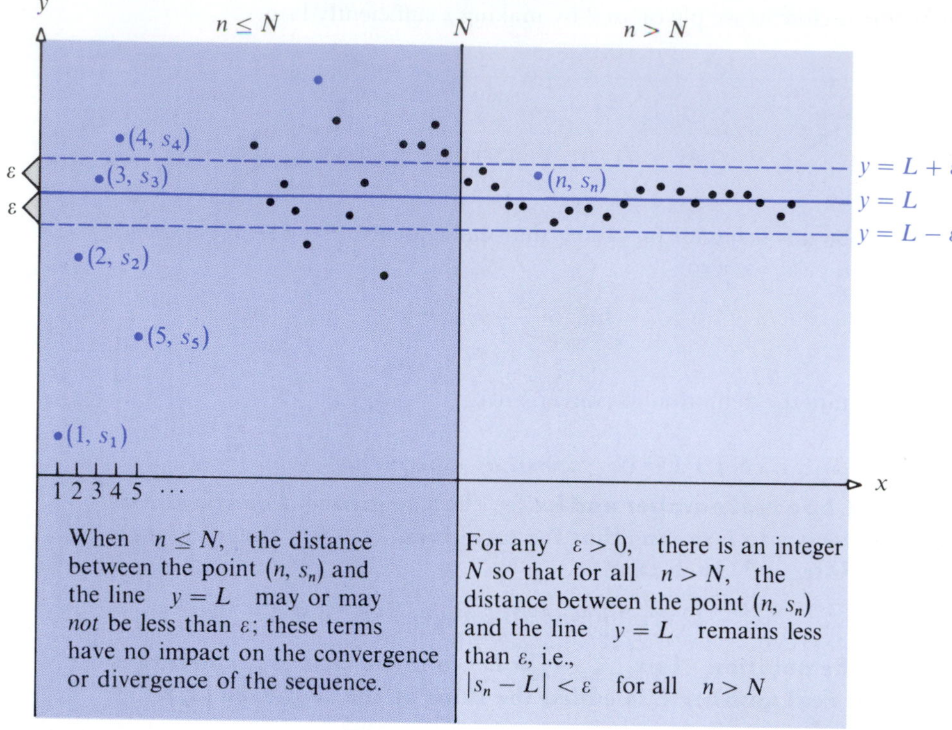

Figure 3

EXAMPLE 4

Determine whether the sequence $\left\{\dfrac{3n^2 + 5n - 2}{6n^2 - 6n + 5}\right\}$ converges or diverges. If it converges, determine the limit of the sequence.

Solution

We find the limit of the nth term by dividing the numerator and denominator by n^2:

$$\lim_{n\to+\infty} \frac{3n^2 + 5n - 2}{6n^2 - 6n + 5} = \lim_{n\to+\infty} \frac{3 + 5/n - 2/n^2}{6 - 6/n + 5/n^2}$$

$$= \frac{\lim_{n\to+\infty}(3 + 5/n - 2/n^2)}{\lim_{n\to+\infty}(6 - 6/n + 5/n^2)} = \frac{(3 + 0 - 0)}{(6 - 0 + 0)} = \frac{1}{2}$$

↑ Use theorem (11.1.3(d)).

The sequence $\{(3n^2 + 5n - 2)/(6n^2 - 6n + 5)\}$ converges and the limit is $\frac{1}{2}$.

EXAMPLE 5

Determine whether the sequence $\{2^n\}$ converges or diverges. If it converges, determine the limit.

Solution

Since $\lim_{n\to+\infty} 2^n = +\infty$, the sequence $\{2^n\}$ diverges.

Recall that a sequence $\{s_n\}$ is a function whose domain is the set of positive integers. A function f with the two properties: (*1*) its domain is the set of positive real numbers, and (*2*) $f(n) = s_n$ is called a *related function* of the sequence $\{s_n\}$.

EXAMPLE 6

(a) If $\{s_n\} = \{1/n\}$, a related function is $f(x) = 1/x$.
(b) If $\{s_n\} = \{n/e^n\}$, a related function is $f(x) = x/e^x$.

There is a connection between the convergence of certain sequences $\{s_n\}$ and the behavior at infinity of a related function f of $\{s_n\}$. The following result, which we state without proof, explains this connection.

[11.1.4] THEOREM

Let $\{s_n\}$ be a sequence and let f be a related function of $\{s_n\}$. Suppose L is a real number.

$$\textbf{If} \quad \lim_{x\to+\infty} f(x) = L \quad \textbf{then} \quad \lim_{n\to+\infty} s_n = L$$

Because of this result, it may be possible to use L'Hospital's rule to evaluate the limit of a sequence $\{s_n\}$, provided there is a related function f that meets the necessary requirements. We demonstrate such a use of L'Hospital's rule in the next example.

EXAMPLE 7

Determine whether the sequence $\{n/e^n\}$ converges or diverges. If it converges, determine the limit.

Solution

We cannot use theorem (11.1.3) for this sequence, since both the numerator and the denominator approach infinity as $n \to +\infty$. Furthermore, L'Hospital's rule cannot be used, since it applies to functions of a real variable, not to sequences. However, in view of theorem (11.1.4), we can look at the related function $f(x) = x/e^x$, and apply L'Hospital's rule to it. Thus,

$$\lim_{x\to+\infty} f(x) = \lim_{x\to+\infty} \frac{x}{e^x} = \lim_{x\to+\infty} \frac{1}{e^x} = 0$$

Based on theorem (11.1.4), we conclude that the sequence $\{n/e^n\}$ converges and its limit is 0.

EXAMPLE 8

Determine whether the sequence $\{1/n + \sin n\pi\}$ converges or diverges. If it converges, determine the limit.

Solution

For this sequence, we list the terms—namely,

$$1, \ \tfrac{1}{2}, \ \tfrac{1}{3}, \ \tfrac{1}{4}, \ \tfrac{1}{5}, \ \ldots$$

As $n \to +\infty$, the terms converge. The limit of the sequence is 0.

We observe that the technique used in Example 7 will not work with the sequence in Example 8. A related function f for the sequence $\{1/n + \sin n\pi\}$ is $f(x) = 1/x + \sin \pi x$, where x is a real number. As $x \to +\infty$, the values of f oscillate back and forth between -1 and 1. Therefore, $\lim_{x \to +\infty} f(x)$ does not exist and so theorem (11.1.4) cannot be used.

EXAMPLE 9

Determine whether the sequence $\{1 + (-1)^n\}$ converges or diverges. If it converges, tell the limit.

Solution

We list the terms of the sequence—namely,

$$0, \ 2, \ 0, \ 2, \ 0, \ 2, \ldots$$

Since the terms oscillate from 0 to 2, there is no single number L that the terms continually approach. We conclude that the sequence diverges.

This example illustrates an important point: Sequences may diverge without tending to $+\infty$ or to $-\infty$. The sequence discussed in Example 9 has another important characteristic—its value is never more than 2 nor less than 0.

Bounded Sequences

If every point of the graph of a sequence $\{s_n\}$ lies between the horizontal lines $y = -K$ and $y = K$, we say that $\{s_n\}$ is *bounded by K*. Figure 4 illustrates the idea.

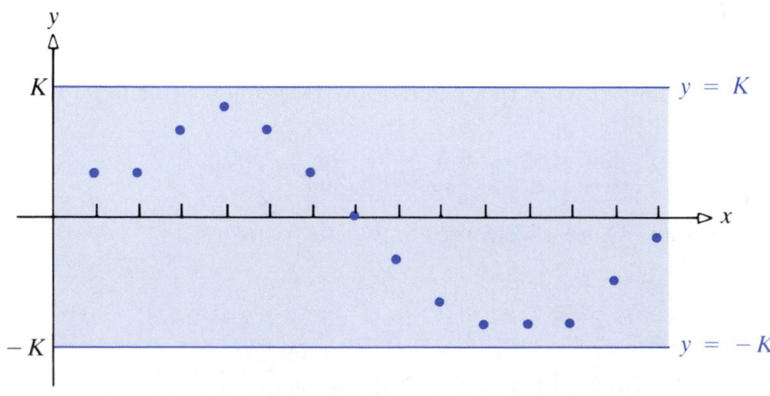

Figure 4

[11.1.5] DEFINITION / ***Bounded Sequence.***

A sequence $\{s_n\}$ is *bounded* if and only if there exists a positive number K such that

$$|s_n| \leq K \qquad \text{for all integers } n$$

The sequence $\{1/n\}$ is bounded, since $|1/n| \leq 1$ for all n. The sequence $\{n^2\}$ is *unbounded*, since for any choice of K, there is an integer n so that $n^2 > K$.

[11.1.6] THEOREM

Any convergent sequence is bounded.

Proof Let $\{s_n\}$ denote a convergent sequence. There is a number L so that $\lim_{n\to+\infty} s_n = L$. We use definition (11.1.2) and, for simplicity, we take $\varepsilon = 1$. Then we know that there exists a positive integer N such that

$$|s_n - L| < 1 \qquad \text{for all} \quad n > N$$

This means that

$$|s_n| < 1 + |L| \qquad \text{for all} \quad n > N$$

If we choose K to be the largest number in the finite collection

$$|s_1|, \quad |s_2|, \quad |s_3|, \quad \ldots, \quad |s_N|, \quad 1 + |L|$$

it will follow that $|s_n| \leq K$ for *all* n. Consequently, the sequence $\{s_n\}$ is bounded.

The contrapositive of theorem (11.1.6) provides a test for divergent sequences.

[11.1.7] THEOREM / ***Test for Divergence.***

If a sequence is unbounded, then it diverges.

For example, the sequences $\{n - 1/n\}$, $\{n/3\}$, and $\{e^n\}$ are unbounded and are therefore divergent.

Warning: The converse of theorem (11.1.7) is not true. Bounded sequences may converge or they may diverge. Example 9 demonstrates a bounded sequence that diverges.

Monotonic Sequences

[11.1.8] DEFINITION / ***Monotonic Sequence.***

A sequence $\{s_n\}$ is said to be:

(a) ***Increasing*** **if and only if** $s_n < s_{n+1}$ **for each n**
(b) ***Nondecreasing*** **if and only if** $s_n \leq s_{n+1}$ **for each n**
(c) ***Decreasing*** **if and only if** $s_n > s_{n+1}$ **for each n**
(d) ***Nonincreasing*** **if and only if** $s_n \geq s_{n+1}$ **for each n**

A sequence that has one of these properties is called *monotonic*.

Table 1 lists three ways to show whether a given sequence $\{s_n\}$ is increasing or decreasing.

Table 1

	To Show $\{s_n\}$ Is Decreasing	To Show $\{s_n\}$ Is Increasing
Algebraic Difference	Show that $s_{n+1} - s_n < 0$ for all n	Show that $s_{n+1} - s_n > 0$ for all n
Algebraic Ratio	If $s_n > 0$ for all n, show that $\dfrac{s_{n+1}}{s_n} < 1$ for all n	If $s_n > 0$ for all n, show that $\dfrac{s_{n+1}}{s_n} > 1$ for all n
Derivative	Show that the derivative of the related real function f, defined for $x > 0$ and for which $s_n = f(n)$ for all n, is negative for all $x > 0$	Show that the derivative of the related real function f, defined for $x > 0$ and for which $s_n = f(n)$ for all n, is positive for all $x > 0$

EXAMPLE 10

Show that each of the following sequences is monotonic by telling whether it is increasing, nondecreasing, decreasing, or nonincreasing.

(a) $\{s_n\} = \left\{\dfrac{n}{n+1}\right\}$ (b) $\{s_n\} = \left\{\dfrac{e^n}{n!}\right\}$ (c) $\{s_n\} = \{\ln n\}$

Solution

(a) We choose the algebraic difference test. Since $s_n = n/(n+1)$ and $s_{n+1} = (n+1)/(n+2)$, we find

$$s_{n+1} - s_n = \frac{n+1}{n+2} - \frac{n}{n+1} = \frac{n^2 + 2n + 1 - n^2 - 2n}{(n+2)(n+1)}$$

$$= \frac{1}{(n+2)(n+1)} > 0 \qquad \text{for all } n$$

Hence, $\{n/(n+1)\}$ is an increasing sequence.

(b) The presence of a factorial usually means the algebraic ratio test applies. Since $s_n = e^n/n!$ and $s_{n+1} = e^{n+1}/(n+1)!$ we find

$$\frac{s_{n+1}}{s_n} = \frac{e^{n+1}/(n+1)!}{e^n/n!} = \left(\frac{e^{n+1}}{e^n}\right)\frac{n!}{(n+1)!} = \frac{e}{n+1} < 1 \qquad \text{for all } n > 1$$

Hence, after the first term, $\{e^n/n!\}$ is a decreasing sequence.

(c) We use the derivative test. The related function f of the sequence $\{\ln n\}$ is $f(x) = \ln x$. Since $(d/dx)(\ln x) = 1/x > 0$ for all $x > 0$, it follows that f is an increasing function and hence $\{\ln n\}$ is an increasing sequence.

Not all sequences are monotonic. For example, the sequences $\{s_n\} = \{\sin(\pi/2)n\}$ and $\{t_n\} = \{1 + (-1)^n/n^2\}$ are not monotonic (see Figs. 5 and 6). Although both s_n and t_n are bounded sequences, s_n diverges and t_n converges.

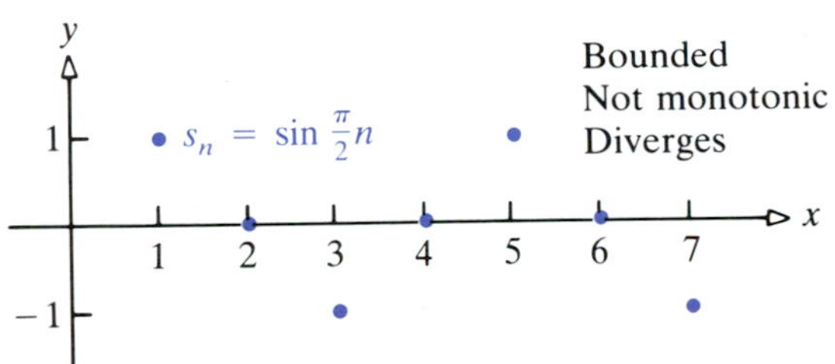

Figure 5

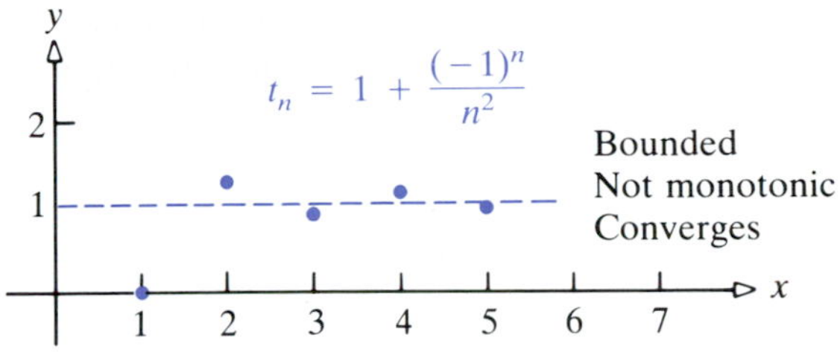

$t_n = 1 + \frac{(-1)^n}{n^2}$

Bounded
Not monotonic
Converges

Figure 6

The sequence $\{n\}$ is monotonic, not bounded, and diverges.

Thus, we have examples of monotonic sequences that diverge and bounded sequences that diverge. However, when a sequence is both monotonic and bounded, it will always converge. We take this property as an axiom.

[11.1.9] AXIOM

A bounded monotonic sequence converges.

Figure 7 provides an illustration for axiom (11.1.9).

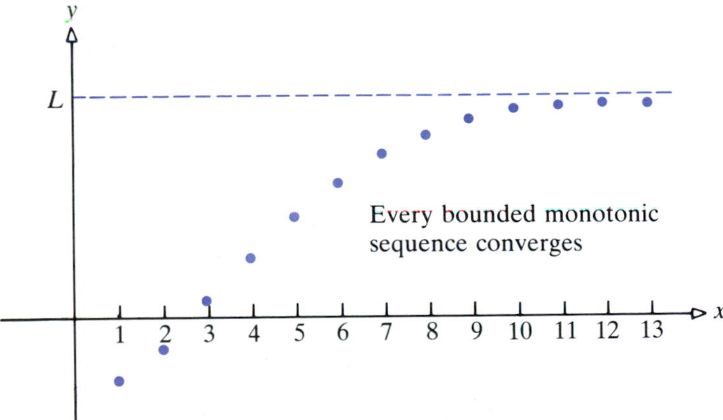

Figure 7

In using axiom (11.1.9) to show that a sequence $\{s_n\}$ converges, it is necessary to show only that $\{s_n\}$ is either (*1*) increasing (or nondecreasing) and bounded from above, that is, for some M, $s_n \leq M$ for all n; or (*2*) decreasing (or nonincreasing) and bounded from below, that is, for some m, $s_n \geq m$ for all n.

EXAMPLE 11

Determine whether the sequence $\{s_n\} = \left\{\dfrac{2^n}{n!}\right\}$ converges or diverges.

Solution

We shall show that the sequence $\{2^n/n!\}$ is both bounded and monotonic. To show that it is monotonic, we examine the algebraic ratio s_{n+1}/s_n.

$$\frac{s_{n+1}}{s_n} = \frac{2^{n+1}/(n+1)!}{2^n/n!} = \frac{2^{n+1}n!}{(n+1)!2^n} = \frac{2}{n+1} \leq 1 \qquad \text{for all} \quad n \geq 1$$

Thus, $s_{n+1} \leq s_n$ for all n, so the sequence $\{s_n\}$ is nonincreasing. Next, each term of the sequence is positive, so that $s_n \geq 0$ for all n. Thus, the sequence $\{s_n\}$ is bounded from below. Hence, $\{s_n\}$ converges.

The use of axiom (11.1.9) to show that a sequence $\{s_n\}$ converges does not provide an answer to the question, What is $\lim_{n\to+\infty} s_n$? Its use merely answers the question, Does $\{s_n\}$ converge? This does not handicap us, since in most cases we are interested in knowing only whether a sequence converges or diverges.

Summary

To determine whether a sequence converges:

1. Look at the first few terms of the sequence to see if a trend is developing. For example, the first five terms of the sequence $\{1 + (-1)^n/n^2\}$ are $1 - 1$, $1 + \frac{1}{4}$, $1 - \frac{1}{9}$, $1 + \frac{1}{16}$, and $1 - \frac{1}{25}$. The pattern suggests that the sequence converges to 1.
2. Calculate the limit of the nth term using any available limit technique, including a related real function and L'Hospital's rule. For example, for the sequence $\{\ln n/n\}$, we examine the limit of $f(x) = \ln x/x$ as $x \to +\infty$:

$$\lim_{x\to+\infty} f(x) = \lim_{x\to+\infty} \frac{\ln x}{x} = \lim_{x\to+\infty} \frac{1/x}{1} = 0$$

   Thus, the sequence $\{\ln n/n\}$ converges.
3. Show that the sequence is both monotonic and bounded.

EXERCISE 11.1

In Problems 1–10 the nth term of a sequence is given. Write the first four terms of the sequence.

1. $s_n = \dfrac{1}{n}$ **2.** $s_n = \dfrac{2}{n^2}$ **3.** $s_n = \ln n$ **4.** $s_n = (-1)^{n+1}$

5. $s_n = \dfrac{(-1)^{n+1}}{2n+1}$ **6.** $s_n = \dfrac{n}{\ln(n+1)}$ **7.** $s_n = \dfrac{1-(-1)^n}{2}$

8. $s_n = \begin{cases} n^2 + n & \text{if } n \text{ is even} \\ 4n + 1 & \text{if } n \text{ is odd} \end{cases}$ **9.** $s_n = \dfrac{n!}{2^n}$ **10.** $s_n = \dfrac{n!}{n^2}$

In Problems 11–18 the first few terms of a sequence are provided. Find an expression for the nth term of each sequence, assuming the pattern continues for all n.

11. 2, 4, 6, 8, 10, . . .
12. 1, 3, 5, 7, 9, . . .
13. 2, 4, 8, 16, 32, . . .
14. 1, 4, 9, 16, 25, 36, . . .
15. $\frac{1}{2}, -\frac{1}{3}, \frac{1}{4}, -\frac{1}{5}, \frac{1}{6}, \ldots$
16. 1, −2, 3, −4, 5, . . .
17. $\frac{1}{2}, \frac{2}{3}, \frac{3}{4}, \frac{4}{5}, \ldots$
18. $\frac{1}{2}, \frac{4}{3}, \frac{9}{4}, \frac{16}{5}, \ldots$

In Problems 19–50 tell whether the given sequence converges or diverges. If it converges, determine the limit.

19. $\left\{\frac{5}{n}\right\}$
20. $\left\{\frac{n}{5}\right\}$
21. $\left\{2 - \frac{4}{n}\right\}$
22. $\left\{\left(4 + \frac{2}{n}\right)^2\right\}$
23. $\left\{1 + \frac{(-1)^n}{n}\right\}$
24. $\left\{\frac{1 + (-1)^n}{n}\right\}$
25. $\{(0.5)^n\}$
26. $\left\{\left(\frac{1}{3}\right)^n\right\}$
27. $\{1 + (-1)^n\}$
28. $\left\{\frac{(-1)^n}{2n}\right\}$
29. $\left\{\frac{n + (-1)^n}{n}\right\}$
30. $\left\{\left(-\frac{1}{3}\right)^n\right\}$
31. $\left\{\left(1 - \frac{1}{n}\right)\left(1 - \frac{1}{n^2}\right)\right\}$
32. $\left\{\left(1 - \frac{1}{n}\right)\left(1 - \frac{1}{n^2}\right)\left(1 - \frac{1}{n^3}\right)\right\}$
33. $\left\{\frac{n^2}{2n + 1} - \frac{n^2}{2n - 1}\right\}$
34. $\left\{\frac{6n^4 - 5}{7n^4 + 3}\right\}$
35. $\left\{\frac{n}{\sqrt{n^2 + 1}}\right\}$
36. $\left\{\frac{\sqrt{n} + 2}{\sqrt{n} + 5}\right\}$
37. $\left\{2 - \frac{1}{2^n}\right\}$
38. $\left\{\frac{n^2}{3^n}\right\}$
39. $\left\{\frac{1}{ne^{-n}}\right\}$
40. $\left\{\frac{n}{e^n}\right\}$
41. $\left\{\cos\left(n\pi + \frac{\pi}{2}\right)\right\}$
42. $\left\{\sin \frac{n\pi}{2}\right\}$
43. $\left\{n \sin \frac{1}{n}\right\}$
44. $\left\{\frac{\sqrt{n}}{e^n}\right\}$
45. $\{\ln n - \ln(n + 1)\}$
46. $\left\{\frac{\ln(n + 1)}{n + 1}\right\}$
47. $\{\tan^{-1} n\}$
48. $\left\{\frac{n^2 \tan^{-1} n}{n^2 + 1}\right\}$
49. $\left\{\frac{n + \sin n}{n + \cos 4n}\right\}$
50. $\left\{\frac{n^2}{2n + 1} \sin \frac{1}{n}\right\}$

In Problems 51–60 determine whether the given sequence is monotonic. If it is, tell whether it is increasing, decreasing, nonincreasing, or nondecreasing.

51. $\left\{1 - \frac{1}{n}\right\}$
52. $\left\{\frac{2n + 1}{n}\right\}$
53. $\left\{\frac{n!}{3^n}\right\}$
54. $\left\{\frac{n!}{n^2}\right\}$
55. $\{ne^{-n}\}$
56. $\left\{\frac{\ln n}{\sqrt{n}}\right\}$
57. $\left\{\frac{3}{n}\right\}$
58. $\{\sqrt{n}\}$
59. $\left\{\frac{\sqrt{n + 1}}{n}\right\}$
60. $\left\{\left(\frac{1}{3}\right)^n\right\}$

In Problems 61–82 the nth term of a sequence is given. Tell whether the sequence converges or diverges. Give a reason for your decision.

61. $s_n = \frac{3}{n}$
62. $s_n = \frac{(-1)^n}{n}$
63. $s_n = \frac{3^n + 1}{4^n}$
64. $s_n = \frac{(n - 1)^2}{e^n}$
65. $s_n = \sqrt{n}$
66. $s_n = 1 - \left(\frac{1}{2}\right)^n$

67. $s_n = (0.8)^n$

68. $s_n = \dfrac{n+1}{n^2}$

69. $s_n = \dfrac{\ln(n+1)}{\sqrt{n}}$

70. $s_n = \ln\left(\dfrac{n+1}{3n}\right)$

71. $s_n = \cos n\pi$

72. $s_n = ne^{-n}$

73. $s_n = \dfrac{\sqrt{n+1}}{n}$

74. $s_n = \dfrac{\sin n}{n}$

75. $s_n = \dfrac{(-3)^n}{n^{100}}$

76 $s_n = \dfrac{2^n}{(2)(4)(6)\cdots(2n)}$

77. $s_n = \dfrac{3^{n+1}}{(3)(6)(9)\cdots(3n)}$

78. $s_n = (-1)^n\sqrt{n}$

79. $s_n = \sqrt{8-(1/n)}$

80. $s_n = \dfrac{\sqrt{2n+7}}{n+7}$

81. $s_n = \dfrac{5^n}{(n+1)^2}$

82. $s_n = 1 + \dfrac{1}{n^2 + n\cos n + 1}$

[*Hint:* Show that the derivative of $1/(x^2 + x\cos x + 1)$ is negative for $x > 1$.]

83. For what integers n is $1/n$ within $1/100$ of $1/50$?

84. For what integers n is $1/n$ within $1/100$ of 0?

85. For what integers n is $(n-1)/n$ within $1/100$ of 0.9?

86. For what integers n is $(n-1)/n$ within $1/100$ of 1.1?

87. Prove that if $0 < r < 1$, then $\lim_{n\to+\infty} r^n = 0$. [*Hint:* Let $r = 1/(1+p)$, where $p > 0$; then, by the binomial theorem, $r^n = 1/(1+p)^n = 1/[1 + np + n(n-1)p^2/2 + \cdots + p^n] < 1/np$.]

88. Use the result of Problem 87 to prove that if $-1 < r < 0$, then $\lim_{n\to+\infty} r^n = 0$.

89. Prove that if $r > 1$, then $\lim_{n\to+\infty} r^n = +\infty$. [*Hint:* Let $r = 1 + p$, where $p > 0$; then by the binomial theorem, $r^n = (1+p)^n = 1 + np + n(n-1)p^2/2 + \cdots + p^n > np$.]

90. Use the result of Problem 89 to show that if $r < -1$, then $\lim_{n\to+\infty} r^n$ does not exist. [*Hint:* r^n oscillates between positive and negative values.]

91. Prove that if the sequence $\{s_n\}$ is convergent and $\lim_{n\to+\infty} s_n = L$, then the sequence $\{s_n^2\}$ is also convergent and $\lim_{n\to+\infty} s_n^2 = L^2$.

92. A sequence $\{s_n\}$ is said to be a *Cauchy sequence* if and only if for each $\varepsilon > 0$, there exists a positive integer N such that

$$|s_n - s_m| < \varepsilon \qquad \text{for all} \quad n, m > N$$

Show that every convergent sequence is a Cauchy sequence.

93. Prove that if $\lim_{n\to+\infty} s_n = L$, then $\lim_{n\to+\infty}|s_n|$ exists and $\lim_{n\to+\infty}|s_n| = |L|$. Is the converse true?

94. Let $a_1 > 0$ and $b_1 > 0$ be two real numbers for which $a_1 > b_1$. Define sequences $\{a_n\}$ and $\{b_n\}$ as

$$a_{n+1} = \frac{a_n + b_n}{2} \qquad b_{n+1} = \sqrt{a_n b_n}$$

(a) Show that $b_n < b_{n+1} < a_1$ for all n.
(b) Show that $b_1 < a_{n+1} < a_n$ for all n.
(c) Show that $0 < a_{n+1} - b_{n+1} < \dfrac{a_1 - b_1}{2^n}$
(d) Show that $\lim_{n\to+\infty} a_n$ and $\lim_{n\to+\infty} b_n$ each exist and are equal.

95. The famous *Fibonacci sequence* $\{u_n\}$ is defined recursively by the formula

$$u_{n+2} = u_{n+1} + u_n \qquad u_1 = 1, \quad u_2 = 1$$

(a) Write the first eight terms of the Fibonacci sequence.
(b) Verify that a closed form representation for u_n is given by

$$u_n = \frac{(1+\sqrt{5})^n - (1-\sqrt{5})^n}{2^n\sqrt{5}}$$

[*Hint:* Show that $u_1 = 1$, $u_2 = 1$, and $u_{n+2} = u_{n+1} + u_n$.]

Infinite Series **11.2**

Is it logical for the sum of an infinite collection of nonzero numbers to be finite? Consider Figure 8. The square in the figure has each side of length 1 unit, and hence its area is 1 square unit. When we divide the box into two boxes of equal area, each half has an area of $\frac{1}{2}$ square unit. If we continue this process indefinitely, we obtain a decomposition of the area 1 into boxes of area $\frac{1}{2}, \frac{1}{4}, \frac{1}{8}, \frac{1}{16}$, and so forth. Therefore,

$$1 = \tfrac{1}{2} + \tfrac{1}{4} + \tfrac{1}{8} + \tfrac{1}{16} + \cdots$$

Surprised?

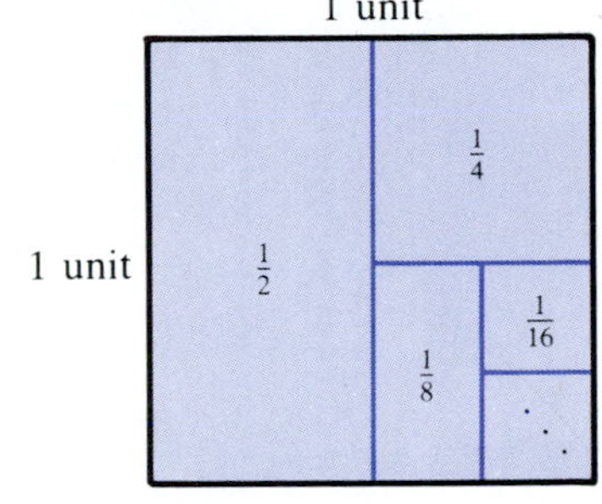

Figure 8

Let's look at this result from a different point of view by starting with the infinite sum

$$\tfrac{1}{2} + \tfrac{1}{4} + \tfrac{1}{8} + \tfrac{1}{16} + \cdots \qquad \textbf{(1)}$$

One way we might add this up is by the use of *partial sums* in order to see whether a trend develops. The first five partial sums are

$$\tfrac{1}{2} = 0.5$$
$$\tfrac{1}{2} + \tfrac{1}{4} = \tfrac{3}{4} = 0.75$$
$$\tfrac{1}{2} + \tfrac{1}{4} + \tfrac{1}{8} = \tfrac{3}{4} + \tfrac{1}{8} = \tfrac{7}{8} = 0.875$$
$$\tfrac{1}{2} + \tfrac{1}{4} + \tfrac{1}{8} + \tfrac{1}{16} = \tfrac{7}{8} + \tfrac{1}{16} = \tfrac{15}{16} = 0.9375$$
$$\tfrac{1}{2} + \tfrac{1}{4} + \tfrac{1}{8} + \tfrac{1}{16} + \tfrac{1}{32} = \tfrac{15}{16} + \tfrac{1}{32} = \tfrac{31}{32} = 0.96875$$

Each of these sums uses more of the terms in (1) and each sum seems to be getting closer to 1.

[11.2.1] DEFINITION / ***Infinite Series.***

If $a_1, a_2, \ldots, a_n, \ldots$ is some infinite collection of numbers, the expression

$$\sum_{k=1}^{+\infty} a_k = a_1 + a_2 + \cdots + a_n + \cdots$$

is called an *infinite series* or, simply, a *series*. The numbers $a_1, a_2, \ldots, a_n, \ldots$ are called *terms*, and the number a_n is called the *nth term* or the *general term*. As before, the symbol $\sum$ stands for *summation*; k is the *index of summation*.

The index of summation is not fixed for a given series. For example, the series in (1), which may be written as

$$\frac{1}{2} + \frac{1}{2^2} + \frac{1}{2^3} + \cdots + \frac{1}{2^n} + \cdots = \sum_{k=1}^{+\infty} \frac{1}{2^k}$$

also may be written in any of the following equivalent ways:

$$\sum_{k=1}^{+\infty} \frac{1}{2}\left(\frac{1}{2^{k-1}}\right) \qquad \sum_{k=0}^{+\infty} \frac{1}{2^{k+1}} \qquad \sum_{k=4}^{+\infty} \frac{1}{2^{k-3}}$$

In most of our work, the index of summation will begin at 1.

To define a sum of an infinite series $\sum_{k=1}^{+\infty} a_k$, we make use of the sequence $\{S_n\}$ defined by

$$S_1 = a_1$$

$$S_2 = a_1 + a_2 = \sum_{k=1}^{2} a_k$$

$$\vdots$$

$$S_n = a_1 + a_2 + \cdots + a_n = \sum_{k=1}^{n} a_k$$

This sequence $\{S_n\}$ is called the *sequence of partial sums* of the series $\sum_{k=1}^{+\infty} a_k$.

For example, consider again the series

$$\sum_{k=1}^{+\infty} \frac{1}{2^k} = \frac{1}{2} + \frac{1}{2^2} + \frac{1}{2^3} + \frac{1}{2^4} + \cdots = \frac{1}{2} + \frac{1}{4} + \frac{1}{8} + \frac{1}{16} + \cdots$$

As it turns out, the partial sums can each be written as 1 minus a power of $\frac{1}{2}$, as follows:

$$S_1 = \frac{1}{2} = 1 - \frac{1}{2}$$

$$S_2 = \frac{1}{2} + \frac{1}{4} = \frac{3}{4} = 1 - \frac{1}{4} = 1 - \frac{1}{2^2}$$

$$S_3 = \frac{1}{2} + \frac{1}{4} + \frac{1}{8} = \frac{7}{8} = 1 - \frac{1}{8} = 1 - \frac{1}{2^3}$$

$$S_4 = \frac{15}{16} = 1 - \frac{1}{16} = 1 - \frac{1}{2^4}$$

$$S_5 = \frac{31}{32} = 1 - \frac{1}{2^5}$$

$$\vdots$$

$$S_n = 1 - \frac{1}{2^n}$$

The nth partial sum is $S_n = 1 - (1/2^n)$, and, as n gets larger and larger, the sequence $\{S_n\}$ of partial sums approaches a limit—namely,

$$\lim_{n\to+\infty} S_n = \lim_{n\to+\infty} \left(1 - \frac{1}{2^n}\right) = \lim_{n\to+\infty} 1 - \lim_{n\to+\infty} \frac{1}{2^n} = 1 - 0 = 1$$

We agree to call this limit the *sum of the series*, and we write

$$\sum_{k=1}^{+\infty} \frac{1}{2^k} = \frac{1}{2} + \frac{1}{4} + \frac{1}{8} + \frac{1}{16} + \cdots = 1 \tag{2}$$

[11.2.2] DEFINITION / ***Convergence and Divergence.***

If the sequence $\{S_n\}$ of partial sums of an infinite series $\sum_{k=1}^{+\infty} a_k$ has a limit S, then the series *converges* and is said to have the *sum*

S. That is, if $\lim_{n\to+\infty} S_n = S$, then

$$\sum_{k=1}^{+\infty} a_k = a_1 + a_2 + \cdots + a_n + \cdots = S$$

An infinite series *diverges* if the sequence of partial sums diverges.

Remember that if an infinite series does not converge, then the symbol $a_1 + a_2 + \cdots = \sum_{k=1}^{+\infty} a_k$ is *not* the name of a number.

Let's look at some examples.

EXAMPLE 1

Show that:

$$\sum_{k=1}^{+\infty} \frac{1}{k(k+1)} = \frac{1}{1\cdot 2} + \frac{1}{2\cdot 3} + \frac{1}{3\cdot 4} + \cdots = \frac{1}{2} + \frac{1}{6} + \frac{1}{12} + \cdots = 1$$

Solution

We begin with the sequence $\{S_n\}$ of partial sums—namely,

$$S_1 = \frac{1}{1\cdot 2}$$

$$S_2 = \frac{1}{1\cdot 2} + \frac{1}{2\cdot 3}$$

$$S_3 = \frac{1}{1\cdot 2} + \frac{1}{2\cdot 3} + \frac{1}{3\cdot 4}$$

$$\vdots$$

$$S_n = \frac{1}{1\cdot 2} + \frac{1}{2\cdot 3} + \frac{1}{3\cdot 4} + \cdots + \frac{1}{n(n+1)}$$

To find the sum of the infinite series requires that we find the limit of the sequence $\{S_n\}$. This is usually difficult because to find $\lim_{n\to+\infty} S_n$ requires that S_n be expressed as a function of n. In this example, though, since

$$\frac{1}{n(n+1)} = \frac{1}{n} - \frac{1}{n+1}$$

we can express S_n as

$$S_n = \left(\frac{1}{1} - \frac{1}{2}\right) + \left(\frac{1}{2} - \frac{1}{3}\right) + \cdots + \left(\frac{1}{n-1} - \frac{1}{n}\right) + \left(\frac{1}{n} - \frac{1}{n+1}\right)$$

Note that all the terms except the first and last cancel,* so that

$$S_n = 1 - \frac{1}{n+1}$$

---

* Expressions in which the middle terms cancel in this manner are sometimes referred to as *telescoping*.

It follows that

$$\lim_{n\to+\infty} S_n = \lim_{n\to+\infty}\left(1 - \frac{1}{n+1}\right) = 1$$

Thus, $\sum_{k=1}^{+\infty} 1/[k(k+1)]$ converges, and its sum is 1.

EXAMPLE 2

Show that the series $\sum_{k=1}^{+\infty}(-1)^k = -1 + 1 - 1 + \cdots$ diverges.

Solution

The sequence $\{S_n\}$ of partial sums for this series is

$$\begin{aligned}
S_1 &= -1\\
S_2 &= -1 + 1 = 0\\
S_3 &= -1 + 1 - 1 = -1\\
S_4 &= -1 + 1 - 1 + 1 = 0\\
&\vdots\\
S_n &= \begin{cases} -1 & \text{if } n \text{ is odd}\\ 0 & \text{if } n \text{ is even}\end{cases}
\end{aligned}$$

Since the sequence $\{S_n\}$ diverges, the infinite series diverges.

EXAMPLE 3

Show that the series $\sum_{k=1}^{+\infty} k = 1 + 2 + 3 + \cdots$ diverges.

Solution

The sequence $\{S_n\}$ of partial sums is

$$\begin{aligned}
S_1 &= 1\\
S_2 &= 1 + 2\\
S_3 &= 1 + 2 + 3\\
&\vdots\\
S_n &= 1 + 2 + 3 + \cdots + n
\end{aligned}$$

We seek to express S_n in a way that will make it easy to find $\lim_{n\to+\infty} S_n$. Using equation (9), Section 5.1, we write

$$S_n = 1 + 2 + \cdots + n = \frac{n(n+1)}{2}$$

Since $\lim_{n\to+\infty} S_n = \lim_{n\to+\infty}[n(n+1)/2] = +\infty$, $\{S_n\}$ diverges. Hence, the series $\sum_{k=1}^{+\infty} k$ diverges.

Geometric Series

[11.2.3] DEFINITION / *Geometric Series.*

A special infinite series that is useful in many applied problems is the *geometric series*:

$$\sum_{k=0}^{+\infty} ar^k = \sum_{k=1}^{+\infty} ar^{k-1} = a + ar + ar^2 + \cdots + ar^{n-1} + \cdots \qquad a \neq 0 \tag{3}$$

In this series, the ratio of each term to its predecessor is r, where r is some fixed real number.

To determine the conditions for convergence of the geometric series, we examine the nth partial sum:

$$S_n = a + ar + ar^2 + \cdots + ar^{n-1} \tag{4}$$

If $r = 1$, the series becomes $\sum_{k=1}^{+\infty} a = a + a + \cdots + a + \cdots$ and the nth partial sum is

$$S_n = a + a + \cdots + a = na$$

In this case, $\{S_n\}$ diverges (since $a \neq 0$), so that for $r = 1$, the geometric series diverges.

If $r = -1$, the series is $\sum_{k=1}^{+\infty} a(-1)^{k-1} = a - a + a - a + \cdots$ and the nth partial sum is

$$S_n = \begin{cases} 0 & \text{if } n \text{ is even} \\ a & \text{if } n \text{ is odd} \end{cases}$$

Since $\{S_n\}$ diverges, we conclude that for $r = -1$, the geometric series diverges.

Suppose $r \neq 1$ and $r \neq -1$. To see what happens for other choices of r, we multiply both sides of (4) by r, obtaining

$$rS_n = ar + ar^2 + \cdots + ar^n$$

By subtracting this from the expression for S_n in (4), we have

$$\begin{aligned} S_n - rS_n &= (a + ar + ar^2 + \cdots + ar^{n-1}) - (ar + ar^2 + \cdots + ar^{n-1} + ar^n) \\ &= a - ar^n \\ S_n(1 - r) &= a(1 - r^n) \end{aligned}$$

Since $r \neq 1$, we can express the nth partial sum of the geometric series as

$$S_n = \frac{a(1 - r^n)}{1 - r} = \frac{a - ar^n}{1 - r} = \frac{a}{1 - r} - \frac{ar^n}{1 - r}$$

Now,

$$\lim_{n\to+\infty} S_n = \lim_{n\to+\infty}\left[\frac{a}{1 - r} - \frac{ar^n}{1 - r}\right] = \frac{a}{1 - r} - \frac{a}{1 - r}\lim_{n\to+\infty} r^n$$

Combining the results obtained in Problems 87 and 88 of Exercise 11.1, we conclude that:

$$\text{For} \quad -1 < r < 1, \quad \lim_{n\to+\infty} r^n = 0 \quad \text{so that} \quad \lim_{n\to+\infty} S_n = \frac{a}{1-r}$$

Hence, for $-1 < r < 1$, the geometric series converges and its sum is $a/(1-r)$.

Combining the results obtained in Problems 89 and 90 of Exercise 11.1, we have:

$$\text{For} \quad r > 1 \quad \text{or} \quad r < -1, \quad \lim_{n\to+\infty} r^n \text{ does not exist}$$

Hence, for $r > 1$ or $r < -1$, the geometric series diverges.

We summarize these results in the following theorem:

[11.2.4] THEOREM

(a) If $-1 < r < 1$, the geometric series $\sum_{k=1}^{+\infty} ar^{k-1}$ converges, and its sum is

$$\sum_{k=1}^{+\infty} ar^{k-1} = \frac{a}{1-r}$$

(b) If $|r| \geq 1$, the geometric series $\sum_{k=1}^{+\infty} ar^{k-1}$ diverges.

EXAMPLE 4

Determine whether each geometric series converges or diverges. If it converges, find its sum.

(a) $\sum_{k=1}^{+\infty} 8\left(\frac{2}{5}\right)^{k-1}$ (b) $\sum_{k=1}^{+\infty} \left(-\frac{5}{9}\right)^{k-1}$ (c) $\sum_{k=1}^{+\infty} 3\left(\frac{3}{2}\right)^{k-1}$ (d) $\sum_{k=1}^{+\infty} \frac{1}{2^k}$

Solution

(a) We have $a = 8$, $r = \frac{2}{5}$. Since $|r| < 1$, the series converges and

$$\sum_{k=1}^{+\infty} 8\left(\frac{2}{5}\right)^{k-1} = \frac{8}{1-\frac{2}{5}} = 8\left(\frac{5}{3}\right) = \frac{40}{3}$$

(b) We have $a = 1$, $r = -\frac{5}{9}$. Since $|r| < 1$, the series converges and

$$\sum_{k=1}^{+\infty} \left(-\frac{5}{9}\right)^{k-1} = \frac{1}{1+\frac{5}{9}} = \frac{9}{14}$$

(c) We have $a = 3$, $r = \frac{3}{2}$. Since $|r| > 1$, the series diverges.

(d) The given series is not exactly in the form $\sum_{k=1}^{+\infty} ar^{k-1}$. However, since

$$\sum_{k=1}^{+\infty} \frac{1}{2^k} = \sum_{k=1}^{+\infty} \left(\frac{1}{2}\right)^k = \sum_{k=1}^{+\infty} \frac{1}{2}\left(\frac{1}{2}\right)^{k-1}$$

we conclude that $\sum_{k=1}^{+\infty}(1/2^k)$ is a geometric series with $a=\frac{1}{2}$, $r=\frac{1}{2}$. Hence, the series converges and its sum is

$$\sum_{k=1}^{+\infty}\frac{1}{2^k}=\frac{\frac{1}{2}}{1-\frac{1}{2}}=1$$

which, of course, verifies the solution given earlier in (2).

Geometric series can be used to express a nonterminating repeating decimal as a rational number.

EXAMPLE 5

Express the repeating decimal 0.3333. . . as a quotient of two integers.

Solution

We write the infinite decimal 0.3333. . . as an infinite series:

$$0.333\ldots = 0.3 + 0.03 + 0.003 + 0.0003 + \cdots = \frac{3}{10}+\frac{3}{10^2}+\frac{3}{10^3}+\cdots$$

$$=\sum_{k=1}^{+\infty}\frac{3}{10^k}=\sum_{k=1}^{+\infty}\frac{3}{10}\left(\frac{1}{10}\right)^{k-1}$$

This is a geometric series with $a=\frac{3}{10}$ and $r=\frac{1}{10}$. Since $|r|<1$, the series converges and its sum is

$$\sum_{k=1}^{+\infty}\frac{3}{10^k}=\frac{\frac{3}{10}}{1-\frac{1}{10}}=\frac{3}{9}=\frac{1}{3}$$

Hence, $0.3333\ldots=\frac{1}{3}$.

EXAMPLE 6

A ball is dropped from a height of 12 meters. Each time it strikes the ground, it bounces back to a height three-fourths the distance from which it fell. Find the total distance traveled by the ball before it comes to rest. (See Fig. 9.)

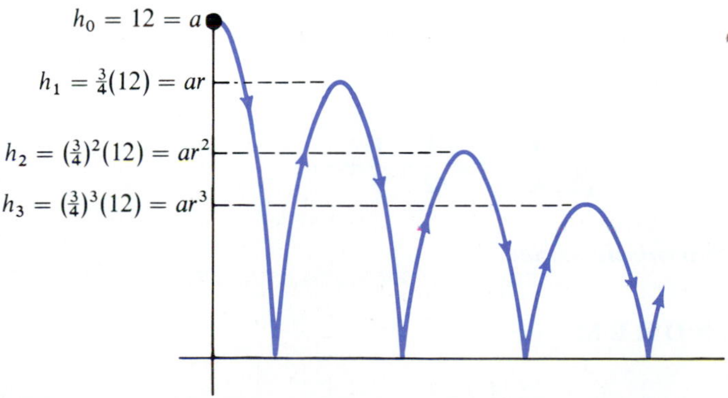

Figure 9

Solution

Let h_n denote the height the ball reaches after the nth bounce. Then

$$\begin{aligned}
h_0 &= 12\\
h_1 &= \tfrac{3}{4}(12)\\
h_2 &= \tfrac{3}{4}[\tfrac{3}{4}(12)] = (\tfrac{3}{4})^2(12)\\
&\vdots\\
h_k &= (\tfrac{3}{4})^k(12)
\end{aligned}$$

After the first bounce, the ball travels up a distance $h_1 = \frac{3}{4}(12)$ and then the same distance back down. Between the first and second bounce, the total distance traveled is therefore $h_1 + h_1 = 2h_1$. The *total* distance H traveled by the ball is

$$\begin{aligned}
H = h_0 + 2h_1 + 2h_2 + 2h_3 + \cdots &= h_0 + \sum_{k=1}^{+\infty} 2h_k\\
&= 12 + \sum_{k=1}^{+\infty} 2\left[12\left(\frac{3}{4}\right)^k\right]\\
&= 12 + \sum_{k=1}^{+\infty} 24\left[\frac{3}{4}\left(\frac{3}{4}\right)^{k-1}\right]\\
&= 12 + \sum_{k=1}^{+\infty} 18\left(\frac{3}{4}\right)^{k-1}
\end{aligned}$$

The last expression is a geometric series with $a = 18$, $r = \frac{3}{4}$. Hence, it converges and

$$H = 12 + \sum_{k=1}^{+\infty} 18\left(\frac{3}{4}\right)^{k-1} = 12 + \frac{18}{1 - \frac{3}{4}} = 84 \text{ meters}$$

The ball travels a total distance of 84 meters.

Harmonic Series

[11.2.5] DEFINITION / *Harmonic Series.*

The infinite series

$$\sum_{k=1}^{+\infty} \frac{1}{k} = 1 + \frac{1}{2} + \frac{1}{3} + \cdots$$

is called the *harmonic series.*

[11.2.6] THEOREM

The harmonic series $\sum_{k=1}^{+\infty} \frac{1}{k}$ diverges.

Proof To show this, we look at the partial sums whose index is a power of 2:

$$S_1 = 1 > 1(\tfrac{1}{2})$$

$$S_2 = 1 + \tfrac{1}{2} > \tfrac{1}{2} + \tfrac{1}{2} = 2(\tfrac{1}{2})$$

↑ Replace 1 by $\frac{1}{2}$ which is smaller.

$$S_4 = 1 + \tfrac{1}{2} + (\tfrac{1}{3} + \tfrac{1}{4}) > 2(\tfrac{1}{2}) + (\tfrac{1}{4} + \tfrac{1}{4}) = 3(\tfrac{1}{2})$$

↑ $1 + \frac{1}{2} > 2(\frac{1}{2})$ and $\frac{1}{3} > \frac{1}{4}$

$$S_8 = 1 + \tfrac{1}{2} + (\tfrac{1}{3} + \tfrac{1}{4}) + (\tfrac{1}{5} + \tfrac{1}{6} + \tfrac{1}{7} + \tfrac{1}{8}) > 3(\tfrac{1}{2}) + (\tfrac{1}{8} + \tfrac{1}{8} + \tfrac{1}{8} + \tfrac{1}{8}) = 4(\tfrac{1}{2})$$

↑ $\frac{1}{5} > \frac{1}{8}, \frac{1}{6} > \frac{1}{8}, \frac{1}{7} > \frac{1}{8}$

$$\vdots$$

$$S_{2^{n-1}} > n(\tfrac{1}{2})$$

We conclude that $\{S_{2^{n-1}}\}$ is unbounded and, hence, so is $\{S_n\}$,* so that $\{S_n\}$ diverges. Thus, the harmonic series $\sum_{k=1}^{+\infty}(1/k)$ diverges.

Summary

1. A series converges if and only if its sequence of partial sums converges.
2. The geometric series $\sum_{k=1}^{+\infty} ar^{k-1}$, $a \neq 0$, converges for $|r| < 1$ and diverges for $|r| \geq 1$. If it converges, its sum is $a/(1 - r)$.
3. The harmonic series $\sum_{k=1}^{+\infty}(1/k)$ diverges.

Application in Biology†

In this application we study the rate of occurrence of *retinoblastoma*, a rare type of eye cancer in children, which, at the turn of this century, was nearly always fatal.

To begin, we need a term from biology. An *allele* (*allelomorph*) is a gene that gives rise to one of a pair of contrasting characteristics, such as smooth or rough, tall or short. Each person normally has two such genes for each characteristic. An individual may have two "tall" genes, two "short" genes, or one of each. In reproduction, each parent gives one of the two types to the child.

The tendency to develop retinoblastoma apparently depends on a single dominant allele—say, A. If the corresponding normal allele is represented by a, the mutation rate from a to A in each generation is approximately $m = 0.00002 = 2 \times 10^{-5}$. In this example, we ignore the very unlikely possibility of mutation from A to a. With the medical care available in the early 1950's, approximately 70% of those affected with the disease survived, although they usually became blind in one or both eyes. Let us assume that survivors reproduce at about half the normal rate. (This assumption is based on scientific guesswork.) Then the productive proportion of persons affected

---

* Do you see why? The sequence of partial sums is increasing, so that once we know that $\{S_{2^{n-1}}\}$ is unbounded, we also know that $\{S_n\}$ itself is unbounded.

† Adapted from J. V. Neel and W. J. Schull, *Human Heredity*, 3rd impression (Chicago: University of Chicago Press, 1958), pp. 333–334. Reprinted by permission.

with the disease is $r = 0.35$. This rate is remarkable, considering that in around 1900, r was approximately 0.

Starting with 0 inherited cases in an early generation, for the nth consecutive generation, we obtain a rate of

m due to mutation in the nth generation

mr due to mutation in the $(n-1)$st generation

mr^2 due to mutation in the $(n-2)$nd generation

$\vdots$

mr^n due to mutation in the zero (original) generation

Hence, the total rate of occurrence of the disease in the nth generation is

$$p_n = m + mr + \cdots + mr^n = \frac{m(1 - r^{n+1})}{1 - r}$$

from which

$$p = \lim_{n\to+\infty} p_n = \frac{m}{1-r} = 3.08 \times 10^{-5}$$

Thus, the total rate of persons affected with the disease will be slightly more than 50% higher than the mutation rate.

Observe that if $r = 0$, then $p = m$. Thus, retinoblastoma has become more frequent with better medical care. As medical care improves, the rate of occurrence of the disease can be expected to become even greater. Neel and Schull point out that, with improved medical care, the frequency of the gene A increases rapidly at first, then more slowly, until an equilibrium point is reached. This equilibrium point is closely approximated after about eight generations.

EXERCISE 11.2

In Problems 1–4 find the fourth partial sum of each series.

1. $\sum_{k=1}^{+\infty} \left(\frac{3}{4}\right)^{k-1}$ **2.** $\sum_{k=1}^{+\infty} \frac{(-1)^{k+1}}{3^{k-1}}$ **3.** $\sum_{k=1}^{+\infty} k$ **4.** $\sum_{k=1}^{+\infty} \ln k$

In Problems 5–14 determine whether each geometric series converges or diverges. If it converges, find its sum.

5. $1 + \frac{1}{3} + \frac{1}{9} + \cdots + \left(\frac{1}{3}\right)^n + \cdots$

6. $1 + \frac{1}{4} + \frac{1}{16} + \cdots + \left(\frac{1}{4}\right)^n + \cdots$

7. $1 + 2 + 4 + \cdots + 2^n + \cdots$

8. $1 - \frac{1}{2} + \frac{1}{4} - \frac{1}{8} + \cdots + \frac{(-1)^{n-1}}{2^{n-1}} + \cdots$

9. $\left(\frac{1}{7}\right)^2 + \left(\frac{1}{7}\right)^3 + \cdots + \left(\frac{1}{7}\right)^n + \cdots$

10. $\left(\frac{3}{4}\right)^5 + \left(\frac{3}{4}\right)^6 + \cdots + \left(\frac{3}{4}\right)^n + \cdots$

11. $\sum_{k=1}^{+\infty} (\sqrt{2})^{k-1}$ **12.** $\sum_{k=1}^{+\infty} (0.33)^{k-1}$ **13.** $\sum_{k=1}^{+\infty} 7\left(\frac{1}{3}\right)^k$ **14.** $\sum_{k=1}^{+\infty} \left(\frac{7}{4}\right)^k$

In Problems 15–34 use the examples and theorems of this section to determine whether each series converges or diverges. If it converges, find its sum.

15. $\sum_{k=1}^{+\infty} \frac{1}{100^k}$

16. $\sum_{k=0}^{+\infty} e^{-k}$

17. $\sum_{k=1}^{+\infty} \frac{10}{2^k}$

18. $\sum_{k=1}^{+\infty} \frac{2^k 3^k}{4^k}$

19. $\sum_{k=1}^{+\infty} (2^{1/3})^k$

20. $\sum_{k=1}^{+\infty} (3^{1/2})^k$

21. $\sum_{k=1}^{+\infty} \left(\frac{1}{k+2} - \frac{1}{k+3}\right)$

22. $\sum_{k=1}^{+\infty} \left[\frac{1}{k^2} - \frac{1}{(k+1)^2}\right]$

23. $\sum_{k=1}^{+\infty} \frac{3^k}{7^k}$

24. $\sum_{k=1}^{+\infty} \frac{4^{k+1}}{7^k}$

25. $\sum_{k=1}^{+\infty} 2^{-k} 3^{k+1}$

26. $\sum_{k=1}^{+\infty} 3^{1-k} 2^{1+k}$

27. $\sum_{k=1}^{+\infty} \left(-\frac{1}{3}\right)^k$

28. $\sum_{k=1}^{+\infty} \frac{\pi}{3^k}$

29. $\sum_{k=1}^{+\infty} \left(\frac{1}{3^{k+1}} - \frac{1}{3^k}\right)$

30. $\sum_{k=1}^{+\infty} \left(-\frac{3}{2}\right)^k$

31. $\sum_{k=1}^{+\infty} \left(\sin\frac{1}{k} - \sin\frac{1}{k+1}\right)$

32. $\sum_{k=1}^{+\infty} \left(\tan\frac{1}{k} - \tan\frac{1}{k+1}\right)$

33. $\sum_{k=1}^{+\infty} \frac{1}{4k^2 - 1}$
$\left[\textit{Hint:} \quad \frac{1}{4k^2-1} = \frac{1}{2}\left(\frac{1}{2k-1} - \frac{1}{2k+1}\right)\right]$

34. $\sum_{k=1}^{+\infty} \frac{1}{k(k+1)(k+2)}$ $\left[\textit{Hint:} \quad \frac{1}{k(k+1)(k+2)} = \frac{1}{2}\left(\frac{1}{k(k+1)} - \frac{1}{(k+1)(k+2)}\right)\right]$

In Problems 35–38 express each repeating decimal as a rational number by using a geometric series.

35. 0.555. . .

36. 0.999. . .

37. 4.28555. . .
[*Hint:* Write the number as 4.28 + 0.00555. . . .]

38. 7.162162. . .

In Problems 39 and 40 use a geometric series to prove the given statement.

39. $\frac{x}{x-1} = \sum_{k=1}^{+\infty} \frac{1}{x^{k-1}}$; for $|x| > 1$

40. $\frac{1}{1+x} = \sum_{k=0}^{+\infty} (-1)^k x^k$; for $|x| < 1$

41. A ball is dropped from a height of 18 feet. Each time it strikes the ground, it bounces back to two-thirds of the previous height. Find the total distance traveled by the ball before it comes to rest.

42. A rich man promises to give you \$1000 on January 1, 1990. Each day thereafter he will give you $\frac{9}{10}$ of what he gave you the previous day. What is the total amount you will receive? What is the first date on which the amount you receive is less than 1¢?

43. A coin-flipping game involves two people who successively flip a coin. The first person to obtain a head is the winner. In probability, it turns out that the person who flips first has the probability of winning given by the series below. Find this number.

$$\frac{1}{2} + \frac{1}{8} + \frac{1}{32} + \cdots + \frac{1}{2^{2n-1}} + \cdots$$

44. Use a hand calculator to find the smallest number n for which $\sum_{k=1}^{n} (1/k) \geq 3$.

45. Answer Problem 44 for $\sum_{k=1}^{n} (1/k) \geq 4$.

46. Show that the series $\sum_{k=1}^{+\infty} \ln\frac{k}{k+1}$ diverges.

47. Show that the series $\sum_{k=1}^{+\infty} \frac{\sqrt{k+1} - \sqrt{k}}{\sqrt{k(k+1)}}$ converges and has the sum 1.

48. Zeno's paradox is about a race between Achilles and a tortoise, in which the tortoise is allowed a certain lead at the start of the race. Zeno claimed the tortoise must win such a race. He reasoned that in order for Achilles to overtake the tortoise, at some time he must cover $\frac{1}{2}$ of the distance that originally separated them. Then, when he covers $\frac{1}{2}$ of the new distance separating them, he will still have $\frac{1}{4}$ of the original distance remaining. And so on. Therefore, by Zeno's reasoning, Achilles never catches the tortoise. Use a series argument to explain this paradox. Assume that the difference in speed between Achilles and the tortoise is v meters per second.

49. Show that: $\sum_{k=1}^{+\infty} \frac{1}{k(k+2)} = \frac{3}{4}$

50. Show that: $\sum_{k=1}^{+\infty} \frac{1}{k(k+1)(k+2)} = \frac{1}{4}$

51. Show that: $\sum_{k=1}^{+\infty} \frac{1}{k(k+1)(k+2)(k+3)} = \frac{1}{18}$

52. Show that: $\sum_{k=1}^{+\infty} \frac{1}{k(k+1)(k+2)\cdots(k+\alpha)} = \frac{1}{\alpha}\left(\frac{1}{\alpha!}\right)$
$\alpha \geq 1$, an integer

11.3 Properties of Series; The Integral Test

In Section 11.2 we determined the convergence or divergence of a series by finding a single compact expression for the sequence $\{S_n\}$ of partial sums as a function of n, and then examining $\lim_{n\to+\infty} S_n$. For most series, however, this is not possible. Thus, obtaining the sum of a convergent infinite series is often very difficult. However, in many applications involving series, the significant point is to know whether or not the series converges. Knowing the sum of a convergent series, though desirable, is not always necessary. We are therefore interested in finding tests for convergence or divergence. To be useful, these tests should depend only on a knowledge of the form of the general term of the series. Our first result states that the limit of the nth term of a series must equal 0 if the series is to converge.

[11.3.1] THEOREM

If the series $\sum_{k=1}^{+\infty} a_k$ converges, then $\lim_{n\to+\infty} a_n = 0$.

Proof The nth partial sum of $\sum_{k=1}^{+\infty} a_k$ is $S_n = \sum_{k=1}^{n} a_k$. Since $S_{n-1} = \sum_{k=1}^{n-1} a_k$, it follows that

$$a_n = S_n - S_{n-1}$$

But $\sum_{k=1}^{+\infty} a_k$ converges. Hence, the sequence $\{S_n\}$ has a limit—say, S. Since $\lim_{n\to+\infty} S_n = S$ and $\lim_{n\to+\infty} S_{n-1} = S$, we find that

$$\lim_{n\to+\infty} a_n = \lim_{n\to+\infty} (S_n - S_{n-1}) = \lim_{n\to+\infty} S_n - \lim_{n\to+\infty} S_{n-1} = S - S = 0$$

Theorem (11.3.1) provides a *necessary condition* for convergence, since the condition $\lim_{n\to+\infty} a_n = 0$ necessarily holds for *any convergent series*. However, the condition is *not a sufficient condition* for convergence, since there are many divergent series $\sum_{k=1}^{+\infty} a_k$ for which $\lim_{n\to+\infty} a_n = 0$.

For example, consider the harmonic series $\sum_{k=1}^{+\infty}(1/k)$. The limit of the nth term is $\lim_{n\to+\infty}(1/n) = 0$, and yet the series diverges. For the geometric series $\sum_{k=1}^{+\infty}(1/2^{k-1})$, the limit of the nth term is $\lim_{n\to+\infty}(1/2^{n-1}) = 0$, and the series converges $(r = \frac{1}{2})$.

The contrapositive of theorem (11.3.1) provides a most useful test for divergence.

[11.3.2] THEOREM | *Test for Divergence.*

If the nth term of an infinite series $\sum_{k=1}^{+\infty} a_k$ does not approach 0 as $n \to +\infty$, the series is divergent.

EXAMPLE 1

(a) $\sum_{k=1}^{+\infty} 87$ diverges, since $\lim_{n \to +\infty} 87 = 87 \neq 0$.

(b) $\sum_{k=1}^{+\infty} k$ diverges, since $\lim_{n \to +\infty} n = +\infty \neq 0$.

(c) $\sum_{k=1}^{+\infty} (-1)^k$ diverges, since $\lim_{n \to +\infty} (-1)^n$ does not exist.

(d) $\sum_{k=1}^{+\infty} 2^k$ diverges, since $\lim_{n \to +\infty} 2^n = +\infty \neq 0$.

Properties of Series

Next we discuss some properties of convergent and divergent series. Knowing these properties can sometimes be helpful in determining the convergence or divergence of a given series.

[11.3.3] THEOREM

If two infinite series are identical after a certain term, then either both converge or both diverge. If both series converge, they will not necessarily have the same sum.

To verify this statement, consider the two series

$$\sum_{k=1}^{+\infty} a_k = a_1 + a_2 + \cdots + a_p + a_{p+1} + \cdots + a_n + \cdots$$

$$\sum_{k=1}^{+\infty} b_k = b_1 + b_2 + \cdots + b_p + a_{p+1} + \cdots + a_n + \cdots$$

in which after the first p terms of each series the remaining terms are identical. The sequence $\{S_n\}$ of partial sums of $\sum_{k=1}^{+\infty} a_k$ and the sequence $\{T_n\}$ of partial sums of $\sum_{k=1}^{+\infty} b_k$ are given by

$$S_n = a_1 + a_2 + \cdots + a_p + a_{p+1} + \cdots + a_n$$

$$T_n = b_1 + b_2 + \cdots + b_p + a_{p+1} + \cdots + a_n$$

Now,

$$S_n - T_n = (a_1 + \cdots + a_p) - (b_1 + \cdots + b_p)$$

$$S_n = T_n + (a_1 + \cdots + a_p) - (b_1 + \cdots + b_p)$$

$$\lim_{n \to +\infty} S_n = \lim_{n \to +\infty} T_n + (a_1 + \cdots + a_p) - (b_1 + \cdots + b_p)$$

Consequently, either both limits exist (both series converge) or both limits do not exist (both series diverge). No other possibility can occur.

[11.3.4] THEOREM

If $\sum_{k=1}^{+\infty} a_k = S$ and $\sum_{k=1}^{+\infty} b_k = T$ are two convergent series, then the series $\sum_{k=1}^{+\infty}(a_k + b_k)$ and $\sum_{k=1}^{+\infty}(a_k - b_k)$ are also convergent. Moreover,

$$\sum_{k=1}^{+\infty} (a_k + b_k) = \sum_{k=1}^{+\infty} a_k + \sum_{k=1}^{+\infty} b_k = S + T$$

$$\sum_{k=1}^{+\infty} (a_k - b_k) = \sum_{k=1}^{+\infty} a_k - \sum_{k=1}^{+\infty} b_k = S - T$$

The proof is left as an exercise (Problem 45).

[11.3.5] THEOREM

Let c be a nonzero constant. If $\sum_{k=1}^{+\infty} a_k = S$ is a convergent series, then the series $\sum_{k=1}^{+\infty} ca_k$ is also convergent. Moreover,

$$\sum_{k=1}^{+\infty} ca_k = c\sum_{k=1}^{+\infty} a_k = cS$$

If $\sum_{k=1}^{+\infty} a_k$ is a divergent series, then the series $\sum_{k=1}^{+\infty} ca_k$ is also divergent.

The proof is left as an exercise (Problem 46).

Let's look at an example to illustrate these properties.

EXAMPLE 2

Determine whether each series converges or diverges. If it converges, find its sum.

(a) $\sum_{k=4}^{+\infty} \frac{1}{k}$ (b) $\sum_{k=1}^{+\infty} \frac{2}{k}$ (c) $\sum_{k=1}^{+\infty} \left(\frac{1}{2^{k-1}} + \frac{1}{3^{k-1}}\right)$

Solution

(a) Except for the first three terms, the series $\sum_{k=4}^{+\infty}(1/k)$ is the harmonic series, which diverges. By theorem (11.3.3), it follows that $\sum_{k=4}^{+\infty}(1/k)$ also diverges.

(b) Except for a factor of 2, this is a harmonic series, which diverges. Hence, by theorem (11.3.5), $\sum_{k=1}^{+\infty}(2/k)$ diverges.

(c) The series $\sum_{k=1}^{+\infty}(1/2^{k-1})$ is a convergent geometric series, as is the series $\sum_{k=1}^{+\infty}(1/3^{k-1})$. Hence, by theorem (11.3.4), $\sum_{k=1}^{+\infty}[(1/2^{k-1}) + (1/3^{k-1})]$ converges, and its sum is

$$\sum_{k=1}^{+\infty} \left(\frac{1}{2^{k-1}} + \frac{1}{3^{k-1}}\right) = \sum_{k=1}^{+\infty} \frac{1}{2^{k-1}} + \sum_{k=1}^{+\infty} \frac{1}{3^{k-1}}$$

$$= \frac{1}{1-\frac{1}{2}} + \frac{1}{1-\frac{1}{3}} = 2 + \frac{3}{2} = 3.5$$

Series of Positive Terms

As we mentioned earlier, the definition of convergence of an infinite series requires that we know the sequence $\{S_n\}$ of partial sums of the series. In many cases it is difficult to obtain a formula for $\{S_n\}$, and therefore it is desirable to have tests for convergence that bypass the sequence of partial sums. For series whose terms are all positive, it is possible to construct tests for convergence that require the use of the *nth term of the series* and do not depend on a knowledge of the form of the sequence $\{S_n\}$ of partial sums.

Consider an infinite series

$$\sum_{k=1}^{+\infty} a_k = a_1 + a_2 + \cdots + a_n + \cdots$$

in which each term is positive. Because each term is positive, the sequence $\{S_n\}$ of partial sums will be increasing.* If the sequence of partial sums is bounded (from above), it follows from axiom (11.1.9) that it will converge. Hence, we have the *general convergence test*:

[11.3.6] THEOREM / *General Convergence Test.*

An infinite series of positive terms will converge if and only if its sequence of partial sums is bounded. The sum of such an infinite series will not exceed an upper bound.

We use this fact in developing standard types of tests for convergence of positive series; the first one we discuss is the *integral test.*

The Integral Test

[11.3.7] THEOREM / *Integral Test.*

Let f be a continuous, positive, decreasing function defined for all real numbers $x \geq 1$. Let $a_k = f(k)$ for all positive integers k. Then the series $\sum_{k=1}^{+\infty} a_k$ converges if and only if the improper integral $\int_1^{+\infty} f(x)\,dx$ converges.

The proof of this result is given at the end of the section.

The integral test is usually applied when the nth term of the series has the form of a continuous, positive, decreasing function whose antiderivative can be readily found.

EXAMPLE 3

Determine whether the series $\displaystyle\sum_{k=1}^{+\infty} \frac{4}{k^2+1}$ converges or diverges.

---

* Do you see why? $S_{n+1} = a_{n+1} + S_n > S_n$

Solution

The function f defined by $f(x) = 4/(x^2 + 1)$ is continuous, positive, and decreasing for $x \geq 1$. Thus, we can use the integral test:

$$\int_1^{+\infty} \frac{4}{x^2+1}\,dx = \lim_{b\to+\infty} \int_1^b \frac{4}{x^2+1}\,dx = \lim_{b\to+\infty} 4\tan^{-1}x\Big|_1^b$$

$$= \lim_{b\to+\infty} (4\tan^{-1}b - 4\tan^{-1}1) = 4\cdot\frac{\pi}{2} - 4\cdot\frac{\pi}{4} = \pi$$

Since the improper integral converges, we conclude that the series also converges.

Caution: In Example 3 the fact that $\int_1^{+\infty} 4/(x^2+1)\,dx = \pi$ does not mean that the sum of the series is π.

EXAMPLE 4

Determine whether the series $\sum_{k=1}^{+\infty} \frac{2k}{k^2+1}$ converges or diverges.

Solution

The function f defined by $f(x) = 2x/(x^2+1)$ is continuous, positive, and decreasing for $x \geq 1$. Thus, we can use the integral test:

$$\int_1^{+\infty} \frac{2x}{x^2+1}\,dx = \lim_{b\to+\infty} \int_1^b \frac{2x}{x^2+1}\,dx = \lim_{b\to+\infty} \ln(x^2+1)\Big|_1^b$$

$$= \lim_{b\to+\infty} [\ln(b^2+1) - \ln 2] = +\infty$$

Since the improper integral diverges, the series also diverges.

EXAMPLE 5

Determine whether the series $\sum_{k=2}^{+\infty} \frac{1}{k(\ln k)^2}$ converges or diverges.

Solution

The function $f(x) = 1/[x(\ln x)^2]$ is continuous, positive, and decreasing for all $x \geq 2$. Using the integral test, we proceed to investigate the improper integral $\int_2^{+\infty} dx/[x(\ln x)^2]$. In $\int dx/[x(\ln x)^2]$ make the substitution $u = \ln x$, $du = dx/x$. Then

$$\int \frac{dx}{x(\ln x)^2} = \int \frac{du}{u^2} = \frac{u^{-1}}{-1} + C = \frac{-1}{\ln x} + C$$

Hence,

$$\int_2^{+\infty} \frac{dx}{x(\ln x)^2} = \lim_{b\to+\infty} \frac{-1}{\ln x}\Big|_2^b = \left(\lim_{b\to+\infty} \frac{-1}{\ln b}\right) + \frac{1}{\ln 2} = \frac{1}{\ln 2}$$

Hence, $\sum_{k=2}^{+\infty} 1/[k(\ln k)^2]$ converges.

p-Series

[11.3.8] DEFINITION / *p-Series.*

A *p-series* is an infinite series of the form

$$\sum_{k=1}^{+\infty} \frac{1}{k^p} = 1 + \frac{1}{2^p} + \frac{1}{3^p} + \cdots + \frac{1}{n^p} + \cdots$$

where p is a positive real number.

A p-series is sometimes referred to as a *hyperharmonic series*, since the harmonic series is a special case of a p-series when $p = 1$.

Some examples of p-series follow:

$$p = 1: \quad \sum_{k=1}^{+\infty} \frac{1}{k} = 1 + \frac{1}{2} + \frac{1}{3} + \cdots + \frac{1}{n} + \cdots$$

$$p = \frac{1}{2}: \quad \sum_{k=1}^{+\infty} \frac{1}{k^{1/2}} = 1 + \frac{1}{2^{1/2}} + \frac{1}{3^{1/2}} + \cdots + \frac{1}{n^{1/2}} + \cdots$$

$$p = 3: \quad \sum_{k=1}^{+\infty} \frac{1}{k^3} = 1 + \frac{1}{2^3} + \frac{1}{3^3} + \cdots + \frac{1}{n^3} + \cdots$$

The following theorem establishes the values of p for which the p-series converges and diverges.

[11.3.9] THEOREM / *Convergence, Divergence of the p-Series.*

The *p*-series

$$\sum_{k=1}^{+\infty} \frac{1}{k^p} = 1 + \frac{1}{2^p} + \frac{1}{3^p} + \cdots + \frac{1}{n^p} + \cdots$$

converges if $p > 1$ and diverges if $0 < p \le 1$.

Proof We use the integral test. The function f defined by $f(x) = 1/x^p$, $p > 0$, is continuous, positive, and decreasing for $x \ge 1$. Hence, the series $\sum_{k=1}^{+\infty}(1/k^p)$ converges if and only if the improper integral $\int_1^{+\infty}(1/x^p)\,dx$ converges. Now,

$$\int_1^{+\infty} \frac{1}{x^p}\,dx = \int_1^{+\infty} x^{-p}\,dx = \lim_{b\to+\infty} \int_1^b x^{-p}\,dx$$

$$= \begin{cases} \displaystyle\lim_{b\to+\infty} \left.\frac{x^{-p+1}}{-p+1}\right|_1^b & \text{if } p \ne 1 \\ \displaystyle\lim_{b\to+\infty} \ln x\Big|_1^b & \text{if } p = 1 \end{cases}$$

$$= \begin{cases} \displaystyle\lim_{b\to+\infty} \frac{b^{1-p} - 1}{1 - p} & \begin{cases} \text{converges} & \text{if } p > 1 \quad (1 - p < 0) \\ \text{diverges} & \text{if } p < 1 \quad (1 - p > 0) \end{cases} \\ \displaystyle\lim_{b\to+\infty} \ln b & \text{diverges} \quad (p = 1) \end{cases}$$

Hence, $\sum_{k=1}^{+\infty}(1/k^p)$ converges if $p > 1$ and diverges if $0 < p \le 1$. ■

EXAMPLE 6

(a) The series

$$\sum_{k=1}^{+\infty} \frac{1}{k^3} = 1 + \frac{1}{2^3} + \frac{1}{3^3} + \cdots + \frac{1}{n^3} + \cdots$$

converges, since it is a p-series where $p = 3$.

(b) The series

$$\sum_{k=1}^{+\infty} \frac{1}{\sqrt{k}} = 1 + \frac{1}{\sqrt{2}} + \frac{1}{\sqrt{3}} + \cdots + \frac{1}{\sqrt{n}} + \cdots$$

diverges, since it is a p-series where $p = \frac{1}{2}$.

Theorem (11.3.9) has an interesting corollary that establishes a pair of bounds for the sum of a convergent p-series.

[11.3.10] COROLLARY

If $p > 1$, then

$$\frac{1}{p-1} < \sum_{k=1}^{+\infty} \frac{1}{k^p} < 1 + \frac{1}{p-1}$$

Proof To establish corollary (11.3.10), we refer to the shaded areas in Figures 10 and 11. By combining the inequalities given in these figures, we obtain

$$\int_1^{+\infty} \frac{dx}{x^p} < \sum_{k=1}^{+\infty} \frac{1}{k^p} < 1 + \int_1^{+\infty} \frac{dx}{x^p}$$

Since $p > 1$, we have

$$\int_1^{+\infty} \frac{dx}{x^p} = \lim_{b \to +\infty} \int_1^b \frac{dx}{x^p} = \frac{1}{p-1}$$

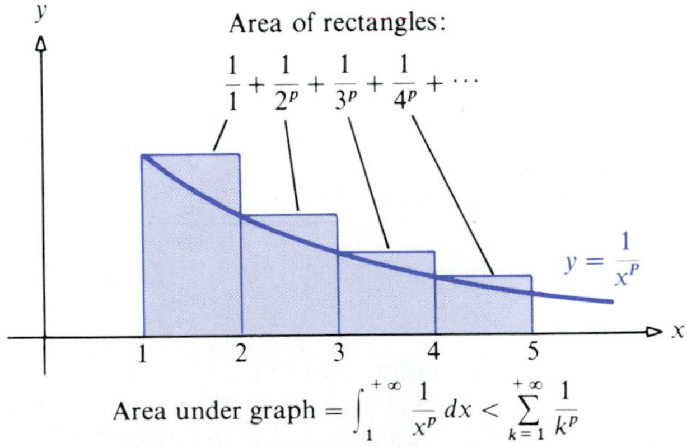

Figure 10

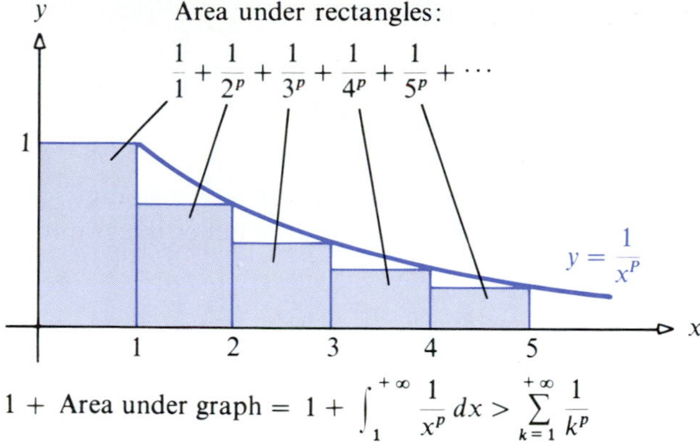

Figure 11

and, therefore, it follows that

$$\frac{1}{p-1} < \sum_{k=1}^{+\infty} \frac{1}{k^p} < 1 + \frac{1}{p-1}$$

For example, based on corollary (11.3.10), we know that

$$p = 2: \qquad 1 < \sum_{k=1}^{+\infty} \frac{1}{k^2} < 2$$

$$p = 3: \qquad \frac{1}{2} < \sum_{k=1}^{+\infty} \frac{1}{k^3} < \frac{3}{2}$$

The actual sum of most convergent p-series remains an open question.*

Proof of the Integral Test The proof is motivated by Figures 12 and 13. Suppose $\int_1^{+\infty} f(x)\,dx$ converges. Then $\lim_{n\to+\infty} \int_1^n f(x)\,dx$ exists and equals

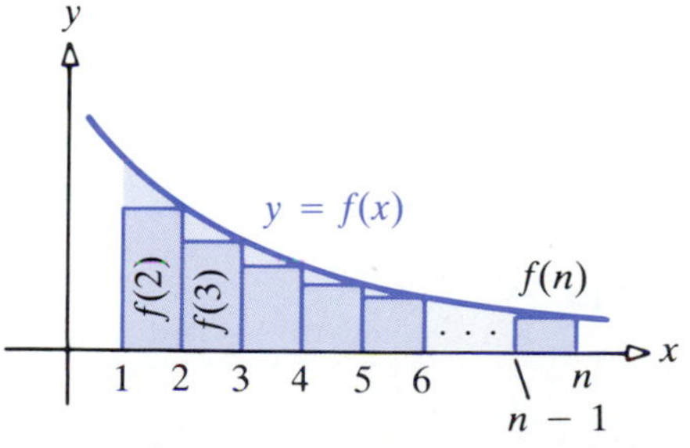

Figure 12

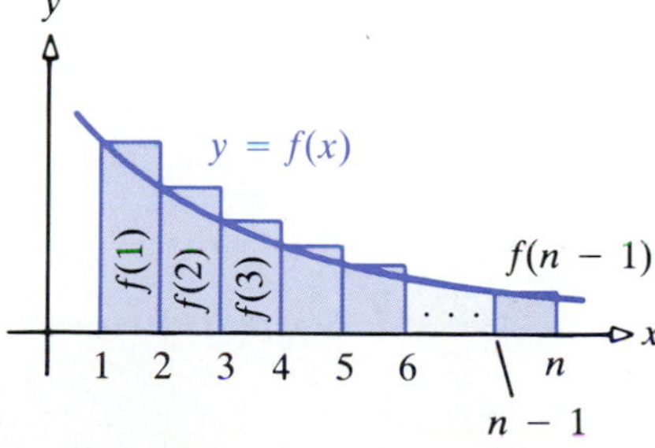

Figure 13

* Since $\sum_{k=1}^{+\infty}(1/k^p)$ converges for $p > 1$, it has a sum. In particular, when $p > 1$ is an integer, it has a sum. However, this exact sum has been found only for p any positive even integer, and the exact sum remains unknown for p a positive odd integer. In 1752, Euler was the first to show that $\sum_{k=1}^{+\infty}(1/k^2) = \pi^2/6$.

some number L. Since $f(x) > 0$, we must have

$$\int_1^n f(x)\,dx < L \qquad \text{for any } n$$

But f is decreasing on the interval $[1, n]$. Thus, the $(n-1)$ rectangles indicated in Figure 12 have a total area less than the area under the graph of $y = f(x)$ from 1 to n. That is,

$$f(2) + f(3) + \cdots + f(n) < \int_1^n f(x)\,dx < L$$

Since $a_1 = f(1)$, $a_2 = f(2)$, $\ldots$, $a_n = f(n)$, upon adding $f(1) = a_1$ to both sides, we have

$$a_1 + a_2 + a_3 + \cdots + a_n = f(1) + f(2) + \cdots + f(n) < f(1) + \int_1^n f(x)\,dx < f(1) + L$$

This means that the partial sums of $\sum_{k=1}^{+\infty} a_k$ are bounded and, hence, $\sum_{k=1}^{+\infty} a_k$ converges.

Now suppose $\int_1^{+\infty} f(x)\,dx$ diverges. Since f is decreasing on $[1, n]$, the $(n-1)$ rectangles indicated in Figure 13 have a total area greater than the area under the graph of $y = f(x)$ from 1 to n. That is,

$$f(1) + f(2) + \cdots + f(n-1) > \int_1^n f(x)\,dx$$

But $\int_1^n f(x)\,dx \to +\infty$ as $n \to +\infty$. Hence, the nth partial sum of $\sum_{k=1}^{+\infty} a_k$ diverges so that $\sum_{k=1}^{+\infty} a_k$ diverges.

EXERCISE 11.3

In Problems 1–10 determine whether the given p-series converges or diverges.

1. $\sum_{k=1}^{+\infty} \frac{1}{k^2}$

2. $\sum_{k=1}^{+\infty} \frac{1}{k^4}$

3. $\sum_{k=1}^{+\infty} \frac{1}{k^{1/3}}$

4. $\sum_{k=1}^{+\infty} \frac{1}{k^{2/3}}$

5. $1 + \frac{1}{2\sqrt{2}} + \frac{1}{3\sqrt{3}} + \frac{1}{4\sqrt{4}} + \cdots$

6. $1 + \frac{1}{\sqrt[3]{2}} + \frac{1}{\sqrt[3]{3}} + \frac{1}{\sqrt[3]{4}} + \cdots$

7. $1 + \frac{1}{4\sqrt{2}} + \frac{1}{9\sqrt{3}} + \frac{1}{16\sqrt{4}} + \cdots$

8. $1 + \frac{1}{8} + \frac{1}{27} + \frac{1}{64} + \cdots$

9. $\sum_{k=1}^{+\infty} \frac{1}{k^{0.1}}$

10. $\sum_{k=1}^{+\infty} \frac{1}{k^{\pi}}$

In Problems 11–16 use the test for divergence (11.3.2) to show that each series diverges.

11. $\sum_{k=1}^{+\infty} 16$

12. $\sum_{k=1}^{+\infty} \frac{k+9}{k}$

13. $\sum_{k=1}^{+\infty} \ln k$

14. $\sum_{k=1}^{+\infty} \frac{k^2+3}{\sqrt{k}}$

15. $\sum_{k=1}^{+\infty} \frac{k^2}{k^2+4}$

16. $\sum_{k=1}^{+\infty} \frac{e^k}{k}$

In Problems 17–30 test each series for convergence or divergence by using the integral test.

17. $\sum_{k=1}^{+\infty} \frac{1}{k^{1.01}}$ **18.** $\sum_{k=1}^{+\infty} \frac{1}{k^{0.9}}$ **19.** $\sum_{k=1}^{+\infty} \frac{\ln k}{k}$ **20.** $\sum_{k=1}^{+\infty} \frac{1}{\sqrt{k^2+25}}$ **21.** $\sum_{k=1}^{+\infty} ke^{-k}$

22. $\sum_{k=1}^{+\infty} ke^{-k^2}$ **23.** $\sum_{k=1}^{+\infty} \frac{1}{k^2+1}$ **24.** $\sum_{k=1}^{+\infty} \frac{1}{k(k+1)}$ **25.** $\sum_{k=2}^{+\infty} \frac{1}{k(\ln k)}$ **26.** $\sum_{k=2}^{+\infty} \frac{1}{k(\ln k)^3}$

27. $\sum_{k=2}^{+\infty} \frac{1}{k\sqrt{k^2-1}}$ **28.** $\sum_{k=1}^{+\infty} \ln \frac{k+2}{k}$ **29.** $\sum_{k=1}^{+\infty} \frac{\ln k}{k^2}$ **30.** $\sum_{k=2}^{+\infty} \frac{1}{k\sqrt{\ln k}}$

In Problems 31–42 determine whether the given series converges or diverges.

31. $\sum_{k=1}^{+\infty} \frac{10}{k}$ **32.** $\sum_{k=1}^{+\infty} \frac{2}{1+k}$ **33.** $\sum_{k=1}^{+\infty} \frac{k^2+1}{4k+1}$ **34.** $\sum_{k=1}^{+\infty} \frac{k^3}{k^3+3}$

35. $\sum_{k=1}^{+\infty} \left(k+\frac{1}{k}\right)$ **36.** $\sum_{k=1}^{+\infty} \left(\frac{1}{3^k}-\frac{1}{4^k}\right)$ **37.** $\sum_{k=1}^{+\infty} \left(\frac{1}{3k}-\frac{1}{4k}\right)$ **38.** $\sum_{k=1}^{+\infty} \left(k-\frac{10}{k}\right)$

39. $\sum_{k=1}^{+\infty} \sin \frac{\pi}{2} k$ **40.** $\sum_{k=1}^{+\infty} \sec \pi k$ **41.** $\sum_{k=3}^{+\infty} \frac{k+1}{k-2}$ **42.** $\sum_{k=5}^{+\infty} \frac{2k^5+3}{k^5-4k^4}$

43. Use the integral test to prove that the series $\sum_{k=2}^{+\infty} 1/[k(\ln k)^p]$ is convergent if and only if $p > 1$.

44. Use the integral test to prove that the series $\sum_{k=3}^{+\infty} 1/[k(\ln k)[\ln(\ln k)^p]]$ is convergent if and only if $p > 1$.

45. Prove that the sum of two convergent series is a convergent series.

46. Prove that if $\sum_{k=1}^{+\infty} a_k = S$, then $\sum_{k=1}^{+\infty} ca_k = cS$ for any real number c.

47. Suppose $\sum_{k=N+1}^{+\infty} a_k = S$ and suppose $a_1 + a_2 + \cdots + a_N = K$. Prove that $\sum_{k=1}^{+\infty} a_k$ converges and its sum is $S + K$.

48. Let $S = 1 + 2 + 4 + 8 + \cdots$. Then

$$2S = 2 + 4 + 8 + 16 + \cdots$$

$$= -1 + (1 + 2 + 4 + \cdots) = -1 + S$$

Therefore,

$$S = 1 + 2 + 4 + 8 + \cdots = -1$$

What did we do wrong here?

49. Find examples to show that the series $\sum_{k=1}^{+\infty}(a_k + b_k)$ and $\sum_{k=1}^{+\infty}(a_k - b_k)$ may converge or diverge if $\sum_{k=1}^{+\infty} a_k$ and $\sum_{k=1}^{+\infty} b_k$ each diverge.

50. If $\sum_{k=1}^{+\infty} a_k$ converges and $\sum_{k=1}^{+\infty} b_k$ diverges, prove that $\sum_{k=1}^{+\infty}(a_k + b_k)$ diverges.

Comparison Tests 11.4

Series may be tested for convergence or divergence by comparing them with series whose behavior is known. Suppose $\sum_{k=1}^{+\infty} b_k$ is a series of positive terms that is known to converge, and suppose $\sum_{k=1}^{+\infty} a_k$ is the series of positive terms to be tested. If S_n is the nth partial sum of $\sum_{k=1}^{+\infty} b_k$ and if B is its sum, then we have $S_n < B$. If term-by-term $a_k \le b_k$, that is, if

$$a_1 \le b_1, \quad a_2 \le b_2, \quad \ldots, \quad a_n \le b_n, \quad \ldots$$

then it follows that

$$\left(n\text{th partial sum of } \sum_{k=1}^{+\infty} a_k\right) \le S_n < B$$

and hence, by theorem (11.3.6), $\sum_{k=1}^{+\infty} a_k$ must also converge. Thus, we have the first comparison test.

[11.4.1] THEOREM / *Comparison Test I: Convergence.*

A series $\sum_{k=1}^{+\infty} a_k$ of positive terms is convergent if each of its terms is less than or equal to the corresponding term of a known convergent series $\sum_{k=1}^{+\infty} b_k$ of positive terms.

It is appropriate to recall at this point that the early terms in a series have no effect on the convergence or divergence of the series. In fact, comparison test I is true if $0 < a_n \le b_n$ for all $n \ge N$, where N is some suitably selected integer. We use this in the next example by ignoring the first term.

EXAMPLE 1

Prove that the series given below converges.

$$\sum_{k=1}^{+\infty} \frac{1}{k^k} = 1 + \frac{1}{2^2} + \frac{1}{3^3} + \cdots + \frac{1}{n^n} + \cdots$$

Solution

We have already seen that the geometric series $\sum_{k=1}^{+\infty}(1/2^k)$ converges to 1. Therefore, since $1/n^n \le 1/2^n$ for $n \ge 2$,* each term of the given series is less than or equal to each corresponding term of a convergent series (except for the first term). Therefore, by comparison test I, we conclude that $\sum_{k=1}^{+\infty}(1/k^k)$ converges.

In Example 1 you may ask, How did you know that the given series should be compared to $\sum_{k=1}^{+\infty}(1/2^k)$? The answer is that only practice and experience will guide you.

Now, let's look at a comparison test for divergence. Let $\sum_{k=1}^{+\infty} c_k$ be a known divergent series of positive terms. If S_n is the nth partial sum $\sum_{k=1}^{+\infty} c_k$, then we know that $\{S_n\}$ is unbounded. Suppose $\sum_{k=1}^{+\infty} a_k$ is the series of positive terms to be tested, and suppose that term-by-term we have

$$a_1 \ge c_1, \quad a_2 \ge c_2, \quad \ldots, \quad a_n \ge c_n, \quad \ldots$$

Then the nth partial sum of $\sum_{k=1}^{+\infty} a_k$ is greater than or equal to the nth partial sum of $\sum_{k=1}^{+\infty} c_k$, which is unbounded. Hence, the nth partial sum of $\sum_{k=1}^{+\infty} a_k$ is unbounded, so that $\sum_{k=1}^{+\infty} a_k$ diverges. Thus, we have a comparison test for divergence.

[11.4.2] THEOREM / *Comparison Test II: Divergence.*

A series $\sum_{k=1}^{+\infty} a_k$ of positive terms is divergent if each of its terms is greater than or equal to the corresponding term of a known divergent series $\sum_{k=1}^{+\infty} c_k$ of positive terms.

---

* Do you see why? $n \ge 2 \Rightarrow n^n \ge 2^n \Rightarrow 1/n^n \le 1/2^n$

EXAMPLE 2

Show that the following series diverges: $\displaystyle\sum_{k=1}^{+\infty} \frac{k+3}{k(k+2)}$

Solution

We choose to compare the given series to the harmonic series $\sum_{k=1}^{+\infty}(1/k)$, which diverges.

$$\frac{n+3}{n(n+2)} = \underbrace{\left(\frac{n+3}{n+2}\right)}_{\substack{\uparrow \\ \text{This number is always} \\ \text{greater than 1.}}}\left(\frac{1}{n}\right) > \frac{1}{n}$$

It follows from theorem (11.4.2) that $\sum_{k=1}^{+\infty}(k+3)/[k(k+2)]$ diverges.

EXAMPLE 3

Use comparison tests to show that the p-series:

(a) Converges if $p > 1$ (b) Diverges if $0 < p \leq 1$

Solution

(a) We group the terms of the series as follows:

$$\begin{aligned} 1 + \frac{1}{2^p} + \frac{1}{3^p} + \cdots = 1 &+ \left(\frac{1}{2^p} + \frac{1}{3^p}\right) + \left(\frac{1}{4^p} + \frac{1}{5^p} + \frac{1}{6^p} + \frac{1}{7^p}\right) \\ &+ \left(\frac{1}{8^p} + \frac{1}{9^p} + \frac{1}{10^p} + \cdots + \frac{1}{15^p}\right) \\ &+ \left(\frac{1}{16^p} + \cdots + \frac{1}{31^p}\right) + \cdots \end{aligned}$$

in which each group contains twice as many terms as the preceding group. Since $p > 1$, we have

$$\frac{1}{2^p} + \frac{1}{3^p} < \frac{1}{2^p} + \frac{1}{2^p} = \frac{2}{2^p} = \frac{1}{2^{p-1}}$$

$$\frac{1}{4^p} + \frac{1}{5^p} + \frac{1}{6^p} + \frac{1}{7^p} < \frac{1}{4^p} + \frac{1}{4^p} + \frac{1}{4^p} + \frac{1}{4^p} = \frac{4}{4^p} = \frac{1}{4^{p-1}} = \frac{1}{(2^{p-1})^2}$$

$$\frac{1}{8^p} + \cdots + \frac{1}{15^p} < \frac{1}{8^p} + \cdots + \frac{1}{8^p} = \frac{8}{8^p} = \frac{1}{8^{p-1}} = \frac{1}{(2^{p-1})^3}$$

and so on. Therefore,

$$1 + \frac{1}{2^p} + \frac{1}{3^p} + \cdots < 1 + \frac{1}{2^{p-1}} + \frac{1}{(2^{p-1})^2} + \frac{1}{(2^{p-1})^3} + \cdots$$

The series on the right is a geometric series whose ratio is $1/2^{p-1}$. This ratio is less than 1, since $p > 1$. It therefore converges. Hence, by comparison test I, the p-series $\sum_{k=1}^{+\infty}(1/k^p)$ converges if $p > 1$.

(b) If $p = 1$, the p-series is the harmonic series, which has been shown to be divergent. Suppose $p < 1$. Then each term $1/n^p$ of the p-series is greater than or equal to the corresponding term $1/n$ of the harmonic series. That is,

$$\frac{1}{n^p} \geq \frac{1}{n} \qquad \text{for all} \quad n \geq 1$$

Since the harmonic series diverges, it follows from comparison test II that the p-series $\sum_{k=1}^{+\infty}(1/k^p)$ diverges if $0 < p < 1$.

Comparison tests I and II are algebraic tests that require certain inequalities to hold. The next comparison test is analytic; it requires that certain conditions on the limit of a ratio occur. You may find this test the easiest to use when a comparison test is called for.

[11.4.3] THEOREM / *Limit Comparison Test.*

Suppose $\sum_{k=1}^{+\infty} a_k$ and $\sum_{k=1}^{+\infty} b_k$ are each positive series. If L is a positive real number and if $\lim_{n\to+\infty}(a_n/b_n) = L$, then both series converge or both diverge.

Proof If $\lim_{n\to+\infty}(a_n/b_n) = L > 0$, there is a number N so that

$$\frac{L}{2} < \frac{a_n}{b_n} < \frac{3L}{2} \qquad \text{for all} \quad n > N^*$$

Since $b_n > 0$, this is the same as

$$\frac{L}{2} b_n < a_n < \frac{3L}{2} b_n \qquad \text{for all} \quad n > N$$

If $\sum_{k=1}^{+\infty} a_k$ converges, then so does $\sum_{k=1}^{+\infty}(L/2)b_k$ and hence, by theorem (11.3.5), so does

$$\sum_{k=1}^{+\infty} b_k = \frac{2}{L}\sum_{k=1}^{+\infty} \frac{L}{2} b_k$$

If $\sum_{k=1}^{+\infty} b_k$ converges, then so does $\sum_{k=1}^{+\infty}(3L/2)b_k$ and, by comparison test I, so does $\sum_{k=1}^{+\infty} a_k$. Thus, $\sum_{k=1}^{+\infty} a_k$ and $\sum_{k=1}^{+\infty} b_k$ converge together. Consequently, they also diverge together.

The limit comparison test is quite versatile for comparing algebraically complex series to a p-series. The correct choice of the p-series to use in this comparison is obtained by disregarding all terms but the highest powers of n in both the numerator and the denominator of the series to be tested.

---

* To see this, use definition (11.1.2) with $\varepsilon = L/2$.

For example:

1. To test $\sum_{k=1}^{+\infty} \frac{1}{3k^2 + 5k + 2}$, use the p-series $\sum_{k=1}^{+\infty} \frac{1}{k^2}$, because for large n,

$$\frac{1}{3n^2 + 5n + 2} \approx \left(\frac{1}{3}\right)\frac{1}{n^2}$$

2. To test $\sum_{k=1}^{+\infty} \frac{2k^2 + 5}{3k^3 - 5k^2 + 2}$, use the p-series $\sum_{k=1}^{+\infty} \frac{1}{k}$, because for large n,

$$\frac{2n^2 + 5}{3n^3 - 5n^2 + 2} \approx \frac{2n^2}{3n^3} = \left(\frac{2}{3}\right)\frac{1}{n}$$

3. To test $\sum_{k=1}^{+\infty} \frac{\sqrt{3k + 1}}{\sqrt{4k^2 - 2k + 1}}$, use the p-series $\sum_{k=1}^{+\infty} \frac{1}{k^{1/2}}$, because for large n,

$$\frac{\sqrt{3n + 1}}{\sqrt{4n^2 - 2n + 1}} \approx \frac{\sqrt{3}\sqrt{n}}{2\sqrt{n^2}} = \left(\frac{\sqrt{3}}{2}\right)\frac{1}{n^{1/2}}$$

EXAMPLE 4

Test the series $\sum_{k=1}^{+\infty}[1/(2k^{3/2} + 5)]$ for convergence or divergence, using the limit comparison test.

Solution

We compare this series to the convergent p-series $\sum_{k=1}^{+\infty}(1/k^{3/2})$. Thus, we evaluate

$$\lim_{n \to +\infty} \frac{\frac{1}{2n^{3/2} + 5}}{\frac{1}{n^{3/2}}} = \lim_{n \to +\infty} \frac{n^{3/2}}{2n^{3/2} + 5} = \lim_{n \to +\infty} \frac{1}{2 + \frac{5}{n^{3/2}}} = \frac{1}{2}$$

By the limit comparison test, since $\sum_{k=1}^{+\infty}(1/k^{3/2})$ converges, $\sum_{k=1}^{+\infty}[1/(2k^{3/2} + 5)]$ does also.

EXAMPLE 5

Test the series below for convergence or divergence using the limit comparison test.

$$\sum_{k=1}^{+\infty} \frac{3\sqrt{k} + 2}{\sqrt{k^3 - 3k^2 + 1}}$$

Solution

We choose for (limit) comparison the p-series $\sum_{k=1}^{+\infty}(1/k)$, since

$$\frac{3\sqrt{n} + 2}{\sqrt{n^3 - 3n^2 + 1}} \approx \frac{3\sqrt{n}}{n^{3/2}} \approx \frac{3}{n}$$

Then

$$\lim_{n\to+\infty} \frac{\dfrac{3\sqrt{n}+2}{\sqrt{n^3-3n^2+1}}}{\dfrac{1}{n}} = \lim_{n\to+\infty} \frac{(3\sqrt{n}+2)n}{\sqrt{n^3-3n^2+1}} = \lim_{n\to+\infty} \frac{3n^{3/2}+2n}{\sqrt{n^3-3n^2+1}}$$

$$= \lim_{n\to+\infty} \frac{3+\dfrac{2}{n^{1/2}}}{\sqrt{1-\dfrac{3}{n}+\dfrac{1}{n^3}}} = 3$$

By the limit comparison test, since $\sum_{k=1}^{+\infty}(1/k)$ diverges, the series to be tested does also.

Effective use of the comparison tests for convergence and divergence requires a prior knowledge of convergent and divergent series to compare against. We list below some series we have already encountered that are useful for this purpose.

	Convergent	***Divergent***
1. The geometric series $\sum_{k=1}^{+\infty} ar^{k-1}$	$\lvert r\rvert < 1$	$r \geq 1$ or $r \leq -1$
2. The harmonic series $\sum_{k=1}^{+\infty} (1/k)$		Divergent
3. The p-series $\sum_{k=1}^{+\infty} (1/k^p)$	$p > 1$	$0 < p \leq 1$
4. The series $\sum_{k=1}^{+\infty} (1/k^k)$	Convergent	

EXERCISE 11.4

In Problems 1–6 use comparison test I or II to test each series for convergence or divergence.

1. Test $\sum_{k=1}^{+\infty} \dfrac{1}{k(k+1)}$ by comparing it with $\sum_{k=1}^{+\infty} \dfrac{1}{k^2}$.

2. Test $\sum_{k=1}^{+\infty} \dfrac{1}{(k+2)^2}$ by comparing it with $\sum_{k=1}^{+\infty} \dfrac{1}{k^2}$.

3. Test $\sum_{k=2}^{+\infty} \dfrac{\sqrt{k}}{k-1}$ by comparing it with $\sum_{k=2}^{+\infty} \dfrac{1}{\sqrt{k}}$.

4. Test $\sum_{k=1}^{+\infty} \dfrac{1}{(2k-1)(2^k)}$ by comparing it with $\sum_{k=1}^{+\infty} \dfrac{1}{2^k}$.

5. Test $\sum_{k=2}^{+\infty} \dfrac{1}{\sqrt{k(k-1)}}$ by comparing it with $\sum_{k=2}^{+\infty} \dfrac{1}{k}$.

6. Test $\sum_{k=1}^{+\infty} \dfrac{1}{k(k+1)(k+2)}$ by comparing it with $\sum_{k=1}^{+\infty} \dfrac{1}{k^3}$.

In Problems 7–22 use the limit comparison test to test each series for convergence or divergence.

7. $\sum_{k=1}^{+\infty} \dfrac{1}{(k+1)(k+2)}$

8. $\sum_{k=1}^{+\infty} \dfrac{1}{k^2+1}$

9. $\sum_{k=1}^{+\infty} \dfrac{1}{\sqrt{k^2+1}}$

10. $\sum_{k=1}^{+\infty} \dfrac{\sqrt{k}}{k+4}$

11. $\sum_{k=1}^{+\infty} \dfrac{3\sqrt{k}+2}{2k^2+5}$

12. $\sum_{k=2}^{+\infty} \dfrac{3\sqrt{k}+2}{2k-3}$

13. $\sum_{k=2}^{+\infty} \dfrac{1}{k\sqrt{k^2-1}}$

14. $\sum_{k=1}^{+\infty} \dfrac{k}{(2k-1)^2}$

15. $\sum_{k=1}^{+\infty} \frac{3k+4}{k2^k}$

16. $\sum_{k=1}^{+\infty} \frac{1}{2^k+1}$

17. $\sum_{k=1}^{+\infty} \frac{5}{3^k+2}$

18. $\sum_{k=2}^{+\infty} \frac{k-1}{k2^k}$

19. $\sum_{k=1}^{+\infty} \frac{k+5}{k^{k+1}}$

20. $\sum_{k=1}^{+\infty} \frac{5}{k^k+1}$

21. $\sum_{k=1}^{+\infty} \sin\frac{1}{k}$

22. $\sum_{k=1}^{+\infty} \tan\frac{1}{k}$

In Problems 23 and 24 determine whether each series converges or diverges.

23. $\sum_{k=2}^{+\infty} \frac{\ln(k+1)}{k^3-4}$

24. $\sum_{k=2}^{+\infty} \frac{\ln k}{k+3}$

25. Prove that any series of the form $\sum_{k=1}^{+\infty}(d_k/10^k)$, where the d_k are digits $(0, 1, 2, \ldots, 9)$, converges.

26. Prove that if $\sum_{k=1}^{+\infty} a_k$ is a positive series to be tested and if $\sum_{k=1}^{+\infty} c_k$ is a known convergent, positive series and if $\lim_{n\to+\infty}(a_n/c_n)$ exists, then the series $\sum_{k=1}^{+\infty} a_k$ converges.

27. Prove that if $\sum_{k=1}^{+\infty} a_k$ is a positive series to be tested and if $\sum_{k=1}^{+\infty} d_k$ is a known divergent, positive series and if $\lim_{n\to+\infty}(a_n/d_n)$ exists and is nonzero, then the series $\sum_{k=1}^{+\infty} a_k$ diverges. Also, if $\lim_{n\to+\infty}(a_n/d_n) = +\infty$, then $\sum_{k=1}^{+\infty} a_k$ diverges.

Alternating Series; Absolute Convergence 11.5

Alternating Series

In Sections 11.3 and 11.4 we studied series with positive terms; we now discuss series that have both positive and negative terms. Even though the behavior of such series is markedly different from that of series with positive terms, we shall find that we can make good use of the latter to clarify the former. By far the most frequently encountered series with both positive and negative terms is the *alternating series*, in which the terms are alternately positive and negative.

[11.5.1] DEFINITION / ***Alternating Series.***

An *alternating series* is a series either of the form

$$\sum_{k=1}^{+\infty} (-1)^{k+1} a_k = a_1 - a_2 + a_3 - a_4 + \cdots$$

or of the form

$$\sum_{k=1}^{+\infty} (-1)^k a_k = -a_1 + a_2 - a_3 + a_4 - \cdots$$

where each number a_k is positive for every k.

For example, the series

$$1 - \frac{1}{2} + \frac{1}{3} - \frac{1}{4} + \cdots = \sum_{k=1}^{+\infty} \frac{(-1)^{k+1}}{k}$$

and

$$-1 + \frac{1}{3!} - \frac{1}{5!} + \frac{1}{7!} - \cdots = \sum_{k=1}^{+\infty} \frac{(-1)^k}{(2k-1)!}$$

are alternating series.

A test for convergence of alternating series is given next.

[11.5.2] THEOREM / ***Alternating Series Test.***

If the numbers a_k of the alternating series

$$\sum_{k=1}^{+\infty} (-1)^{k+1} a_k = a_1 - a_2 + a_3 - a_4 + \cdots \qquad a_k > 0$$

satisfy the following two conditions:

(a) The a_k tend to 0—that is, $\lim_{n\to+\infty} a_n = 0$
(b) The a_k are nonincreasing—that is,
$a_1 \geq a_2 \geq a_3 \geq \cdots \geq a_n \geq a_{n+1} \geq \cdots$

then the alternating series converges.

Proof We first consider the partial sums with an even number of terms:

$$\begin{aligned} S_{2n} &= a_1 - a_2 + a_3 - a_4 + \cdots + a_{2n-1} - a_{2n} \\ &= (a_1 - a_2) + (a_3 - a_4) + \cdots + (a_{2n-1} - a_{2n}) \end{aligned}$$

Since $a_1 \geq a_2$, $a_3 \geq a_4$, ..., $a_{2n-1} \geq a_{2n}$, the difference shown within each pair of parentheses is either 0 or positive. Hence, the sequence of partial sums S_{2n} is increasing—that is,

$$S_2 \leq S_4 \leq S_6 \leq \cdots \leq S_{2n} \leq S_{2n+2} \leq \cdots \qquad \textbf{(1)}$$

For the partial sums with an odd number of terms, we group the terms a little differently and write

$$\begin{aligned} S_{2n+1} &= a_1 - a_2 + a_3 - a_4 + a_5 - \cdots - a_{2n} + a_{2n+1} \\ &= a_1 - (a_2 - a_3) - (a_4 - a_5) - \cdots - (a_{2n} - a_{2n+1}) \end{aligned}$$

Here again, the difference within each pair of parentheses is either positive or 0, and it is to be subtracted from the previous term. Hence, the sequence of partial sums S_{2n+1} is decreasing—that is,

$$S_1 \geq S_3 \geq S_5 \geq \cdots \geq S_{2n-1} \geq S_{2n+1} \geq \cdots \qquad \textbf{(2)}$$

Figure 14 on page 665 illustrates what happens.

Finally, we note that

$$S_{2n+1} - S_{2n} = a_{2n+1} > 0 \qquad \text{so that} \qquad S_{2n} < S_{2n+1} \qquad \textbf{(3)}$$

We combine (1), (2), and (3) as

$$S_2 \leq S_4 \leq S_6 \leq \cdots \leq S_{2n} < S_{2n+1} \leq \cdots \leq S_5 \leq S_3 \leq S_1$$

The sequence $S_2, S_4, S_6, \ldots$ is increasing and bounded above by S_1. By axiom (11.1.9), this sequence has a limit, which we call T. Similarly, the sequence $S_1, S_3, S_5, \ldots$ is decreasing and bounded below, so it has a limit, which we call S. Furthermore, by (3),

$$S - T = \lim_{n\to+\infty} S_{2n+1} - \lim_{n\to+\infty} S_{2n} = \lim_{n\to+\infty} (S_{2n+1} - S_{2n}) = \lim_{n\to+\infty} a_{2n+1}$$

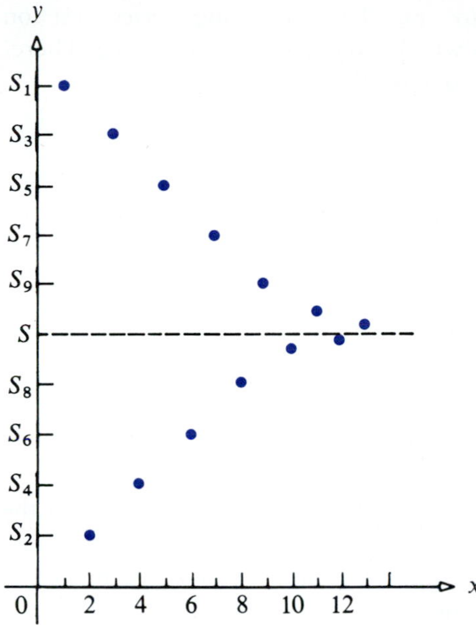

Figure 14

Therefore, since $\lim_{n\to+\infty} a_{2n+1} = 0$, we have $S - T = 0$. This means that $S = T$ and, hence, the sequences $\{S_{2n}\}$ and $\{S_{2n+1}\}$ both converge to S. It follows that

$$\lim_{n\to+\infty} S_n = S$$

and so the alternating series converges.

Let's now look at an example of how theorem (11.5.2) is used.

EXAMPLE 1

Test the alternating harmonic series given below for convergence or divergence.

$$\sum_{k=1}^{+\infty} \frac{(-1)^{k+1}}{k} = 1 - \frac{1}{2} + \frac{1}{3} - \frac{1}{4} + \cdots$$

Solution

Since

$$\lim_{n\to+\infty} a_n = \lim_{n\to+\infty} \frac{1}{n} = 0 \qquad \text{and} \qquad a_{n+1} = \frac{1}{n+1} < \frac{1}{n} = a_n$$

the conditions of the test are satisfied. Hence, the series converges.*

* We shall discover in Section 11.9 that the sum of this alternating harmonic series is ln 2.

As Example 1 illustrates, to use the alternating series test on $\sum_{k=1}^{+\infty}(-1)^k a_k$, we need to determine whether the a_k are nonincreasing. Therefore, we reproduce here three ways to make this verification:

1. Show that $a_{n+1} - a_n \leq 0$ for all n.
2. Show that $a_{n+1}/a_n \leq 1$ for all n.
3. Show that the derivative of the related function representing a_n is always negative or 0.

EXAMPLE 2

We can show that the a_k in the series $\sum_{k=1}^{+\infty}[(-1)^{k+1}/k]$ are decreasing by observing that the derivative of the related function $f(x) = 1/x$, namely, $f'(x) = -1/x^2$, is negative for all $x \geq 1$.

In applying the alternating series test, it is always best to check on the limit of the nth term first. If it is not 0, the test for divergence tells us immediately that the series diverges. If it equals 0, then we should continue on to determine whether the a_n are nonincreasing. For example, in testing the alternating series

$$\sum_{k=2}^{+\infty}(-1)^k \frac{k}{k-1} = 2 - \frac{3}{2} + \frac{4}{3} - \frac{5}{4} + \cdots$$

we find out that $\lim_{n\to+\infty} a_n = \lim_{n\to+\infty}[n/(n-1)] = 1$. There is no need to look further. By the test for divergence, the series diverges.

Estimating the Sum of an Alternating Series

An important characteristic of a convergent alternating series is that it is possible to estimate quite easily the error made by taking a partial sum as an approximation to the sum of the series.

[11.5.3] THEOREM / *Error Estimate.*

For an alternating series $\sum_{k=1}^{+\infty}(-1)^{k+1} a_k$ satisfying the conditions of Theorem 11.5.2, the error E_n made in taking the sum of the first n terms as an approximation to the sum S of the series is numerically less than or equal to the $(n+1)$st term of the series. That is,

$$|E_n| \leq a_{n+1}$$

Proof If n is even, $S_n < S$, and

$$0 < S - S_n = a_{n+1} - (a_{n+2} - a_{n+3}) - \cdots \leq a_{n+1}$$

If n is odd, $S_n > S$, and

$$0 < S_n - S = a_{n+1} - (a_{n+2} - a_{n+3}) - \cdots \leq a_{n+1}$$

In either case,

$$|E_n| = |S - S_n| \leq a_{n+1}$$

We have seen in Example 1 that the alternating harmonic series converges. Table 2 provides estimates of the sum of this series and the maximum error associated with the estimate using $n = 3, 9, 99$, and 999 terms. Observe the slow rate of convergence, as evidenced by the rather small change in the maximum error. For example, using an additional 900 terms reduces the estimated error by merely $|E_{999} - E_{99}| = 0.009$.

Table 2

Number of Terms	Estimate of the Sum	Maximum Error
$n = 3$	$1 - \frac{1}{2} + \frac{1}{3} \approx 0.83333$	$\lvert E_3\rvert \le \frac{1}{4} = 0.25$
$n = 9$	$\sum_{k=1}^{9} \frac{(-1)^{k+1}}{k} \approx 0.74563$	$\lvert E_9\rvert \le \frac{1}{10} = 0.1$
$n = 99$	$\sum_{k=1}^{99} \frac{(-1)^{k+1}}{k} \approx 0.7$	$\lvert E_{99}\rvert \le \frac{1}{100} = 0.01$
$n = 999$	$\sum_{k=1}^{999} \frac{(-1)^{k+1}}{k} \approx 0.694$	$\lvert E_{999}\rvert \le \frac{1}{1000} = 0.001$

EXAMPLE 3

Approximate the sum of the alternating series below to within 0.0001.

$$\sum_{k=0}^{+\infty} \frac{(-1)^k}{(2k)!} = 1 - \frac{1}{2!} + \frac{1}{4!} - \frac{1}{6!} + \frac{1}{8!} - \cdots$$

Solution

We leave it to you to verify that the series actually converges. The fifth term of the series, $1/8! \approx 0.000025$, is the first term less than or equal to 0.0001, and it represents an upper estimate to the error due to using the sum of the first four terms as an approximation to the sum. Hence,

$$\sum_{k=0}^{+\infty} \frac{(-1)^k}{(2k)!} \approx 1 - \frac{1}{2!} + \frac{1}{4!} - \frac{1}{6!} = 1 - \frac{1}{2} + \frac{1}{24} - \frac{1}{720} \approx 0.54028$$

accurate to within 0.0001.*

Absolute Convergence

In order to discuss ways of determining the convergence or divergence of a series $\sum_{k=1}^{+\infty} a_k$ in which the terms a_k are sometimes positive and sometimes

* You will see in Section 11.8 that the actual sum of this series is cos 1. Since it turns out that other values of cos x can be similarly computed, this example illustrates one way to obtain a table of trigonometric values.

negative (not necessarily alternating), we require the following definition:

[11.5.4] DEFINITION / *Absolute Convergence.*

A series $\sum_{k=1}^{+\infty} a_k$ is *absolutely convergent* if

$$\sum_{k=1}^{+\infty} |a_k| = |a_1| + |a_2| + \cdots + |a_n| + \cdots \quad \textbf{is convergent}$$

The next result gives a test for convergence of a series $\sum_{k=1}^{+\infty} a_k$, in which the terms a_k are sometimes positive and sometimes negative.

[11.5.5] THEOREM / *Absolute Convergence Test.*

If a series $\sum_{k=1}^{+\infty} a_k$ is absolutely convergent, then it is convergent.

Proof For each n,

$$-|a_n| \le a_n \le |a_n|$$

Hence, by adding $|a_n|$, we have

$$0 \le a_n + |a_n| \le 2|a_n|$$

If $\sum_{k=1}^{+\infty}|a_k|$ converges, then $\sum_{k=1}^{+\infty} 2|a_k| = 2\sum_{k=1}^{+\infty}|a_k|$ converges. Thus, by comparison test I, $\sum_{k=1}^{+\infty}(a_k+|a_k|)$ converges. But $a_n=(a_n+|a_n|)-|a_n|$. Since $\sum_{k=1}^{+\infty} a_k$ is the difference of two convergent series, it also converges. ∎

EXAMPLE 4

The series

$$1 - \frac{1}{2} - \frac{1}{4} + \frac{1}{8} - \frac{1}{16} - \frac{1}{32} + \frac{1}{64} + \cdots$$

converges absolutely, since $1 + \frac{1}{2} + \frac{1}{4} + \cdots = \sum_{k=1}^{+\infty}(\frac{1}{2})^{k-1}$, a geometric series with $r = \frac{1}{2}$, converges. ∎

EXAMPLE 5

Test the series below for convergence or divergence.

$$\sum_{k=1}^{+\infty} \frac{\sin k}{k^2} = \frac{\sin 1}{1^2} + \frac{\sin 2}{2^2} + \frac{\sin 3}{3^2} + \cdots$$

Solution

This series is not a series of positive terms, nor is it an alternating series. To use theorem (11.5.5), we investigate the series $\sum_{k=1}^{+\infty}|(\sin k)/k^2|$. Since

$$\left|\frac{\sin n}{n^2}\right| \le \frac{1}{n^2}$$

for all n, and since $\sum_{k=1}^{+\infty}(1/k^2)$ is a convergent p-series, we conclude by comparison test I that $\sum_{k=1}^{+\infty}|(\sin k)/k^2|$ is convergent. Hence, $\sum_{k=1}^{+\infty}[(\sin k)/k^2]$ is absolutely convergent (by definition) and, by theorem (11.5.5), it follows that $\sum_{k=1}^{+\infty}[(\sin k)/k^2]$ is convergent.

The converse of theorem (11.5.5) is not necessarily true; that is, there are convergent series that are not absolutely convergent. For instance, we have shown that the alternating harmonic series

$$\sum_{k=1}^{+\infty}\frac{(-1)^{k+1}}{k}=1-\frac{1}{2}+\frac{1}{3}-\frac{1}{4}+\frac{1}{5}-\cdots$$

converges. The series of absolute values yields the harmonic series $\sum_{k=1}^{+\infty}(1/k)$, which is divergent.

[11.5.6] DEFINITION / *Conditional Convergence.*

A series that is convergent without being absolutely convergent is called *conditionally convergent.*

EXAMPLE 6

Determine the numbers p for which the series $\sum_{k=1}^{+\infty}[(-1)^{k+1}/k^p]$ is absolutely convergent, conditionally convergent, and divergent.

Solution

We begin by testing the series for absolute convergence. Therefore, we consider the series of absolute values—namely, $\sum_{k=1}^{+\infty}(1/k^p)$. This is the p-series, which converges if $p>1$ and diverges if $p\leq 1$. Hence, $\sum_{k=1}^{+\infty}[(-1)^{k+1}/k^p]$ is absolutely convergent if $p>1$.

It remains to determine what happens when $p\leq 1$. By using the alternating series test, we find:

$$\lim_{n\to+\infty}\frac{1}{n^p}=0 \quad \text{if} \quad 0<p\leq 1; \qquad \lim_{n\to+\infty}\frac{1}{n^p}=1 \quad \text{if} \quad p=0; \qquad \lim_{n\to+\infty}\frac{1}{n^p}=+\infty \quad \text{if} \quad p<0$$

Consequently, $\sum_{k=1}^{+\infty}[(-1)^{k+1}/k^p]$ diverges if $p\leq 0$. To verify part (b) of the alternating series test, when $0<p\leq 1$, we observe that

$$\frac{d}{dx}\frac{1}{x^p}=\frac{-p}{x^{p+1}}<0 \quad \text{for all } x, \text{ when} \quad 0<p\leq 1$$

Thus, $\sum_{k=1}^{+\infty}[(-1)^{k+1}/k^p]$ is conditionally convergent if $0<p\leq 1$.

Figure 15 on page 670 summarizes the preceding discussion. As illustrated, a series $\sum_{k=1}^{+\infty}a_k$ is either convergent or divergent; and, if it is convergent, it is either absolutely convergent or conditionally convergent.

The flowchart in Figure 16 may be helpful in determining whether a given series is absolutely convergent, conditionally convergent, or divergent.

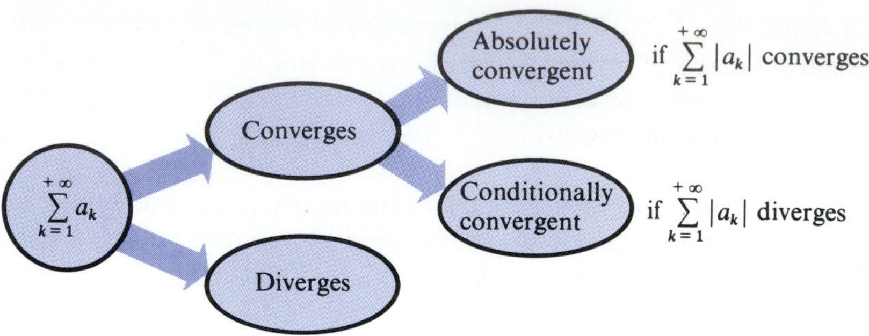

Figure 15

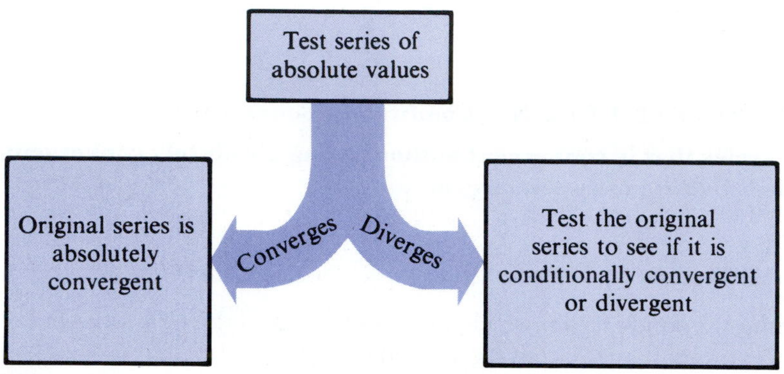

Figure 16

Some Properties

Absolute convergence of a series is stronger than conditional convergence, and absolutely convergent series possess properties that conditionally convergent series do not. We state some of these properties without a proof:

1. If a series is absolutely convergent, any rearrangement of its terms will result in a series that is also absolutely convergent to the same sum.
2. If a series is conditionally convergent, its terms may be rearranged to form a new series that converges to any sum we like. In fact, they may be rearranged to form a divergent series.
3. If two series $\sum_{k=1}^{+\infty} a_k$ and $\sum_{k=1}^{+\infty} b_k$ are absolutely convergent, the series $\sum_{k=1}^{+\infty} (a_k + b_k)$ is also absolutely convergent. Moreover,

$$\sum_{k=1}^{+\infty} (a_k + b_k) = \sum_{k=1}^{+\infty} a_k + \sum_{k=1}^{+\infty} b_k$$

4. If two series $\sum_{k=1}^{+\infty} a_k$ and $\sum_{k=1}^{+\infty} b_k$ are absolutely convergent, the series $\sum_{k=1}^{+\infty} c_k$, where $c_k = a_1 b_{k-1} + a_2 b_{k-2} + \cdots + a_{k-1} b_1$, is also absolutely convergent. Here

$$\sum_{k=1}^{+\infty} c_k = a_1 b_1 + (a_1 b_2 + a_2 b_1) + (a_1 b_3 + a_2 b_2 + a_3 b_1) + \cdots$$

is formed by multiplying the terms of one series by the terms of the other series. Its sum will be the product of the sums of the original series.

5. If a series converges absolutely, then the series consisting of just the positive terms converges, as does the series consisting of just the negative terms.
6. If a series is conditionally convergent, then the series consisting of just the positive terms diverges, as does the series consisting of just the negative terms.

EXERCISE 11.5

In Problems 1–8 test each alternating series for convergence or divergence by using the alternating series test (11.5.2).

1. $\sum_{k=1}^{+\infty} (-1)^{k+1} \frac{1}{k^2}$ **2.** $\sum_{k=1}^{+\infty} (-1)^{k+1} \frac{1}{2\sqrt{k}}$ **3.** $\sum_{k=2}^{+\infty} (-1)^{k} \frac{1}{k \ln k}$ **4.** $\sum_{k=1}^{+\infty} (-1)^{k+1} \frac{k+1}{k^2}$

5. $\sum_{k=1}^{+\infty} (-1)^{k+1} \frac{k+1}{k}$ **6.** $\sum_{k=1}^{+\infty} (-1)^{k+1} \frac{k^2}{5k^2+2}$ **7.** $\sum_{k=1}^{+\infty} \frac{(-1)^{k+1}}{(k+1)2^k}$ **8.** $\sum_{k=0}^{+\infty} (-1)^{k} \frac{1}{k!}$

In Problems 9–12 approximate the sum of each series accurate to within 0.001.

9. $\sum_{k=0}^{+\infty} (-1)^{k} \frac{1}{k!}$ **10.** $\sum_{k=0}^{+\infty} (-1)^{k} \frac{1}{2k+1}\left(\frac{1}{3}\right)^{2k+1}$ **11.** $\sum_{k=1}^{+\infty} (-1)^{k+1} \frac{1}{k^4}$ **12.** $\sum_{k=0}^{+\infty} (-1)^{k} \frac{1}{k!}\left(\frac{1}{2}\right)^{k}$

In Problems 13–18 find a positive integer n such that S_n approximates the sum of the given convergent alternating series to within 0.0001; that is, find n such that $|S - S_n| < 10^{-4}$.

13. $\sum_{k=1}^{+\infty} (-1)^{k+1} \frac{1}{k^2}$ **14.** $\sum_{k=0}^{+\infty} (-1)^{k} \frac{1}{k!}$ **15.** $\sum_{k=0}^{+\infty} (-1)^{k} \frac{1}{k!}\left(\frac{1}{3}\right)^{k}$ **16.** $\sum_{k=1}^{+\infty} (-1)^{k+1} \frac{1}{k^k}$

17. $\sum_{k=1}^{+\infty} (-1)^{k+1} \frac{1}{k}$ **18.** $\sum_{k=1}^{+\infty} (-1)^{k+1} \frac{1}{\sqrt{k}}$

In Problems 19–38 determine whether each series is conditionally convergent, absolutely convergent, or divergent.

19. $\sum_{k=1}^{+\infty} \frac{(-1)^{k+1}}{2k}$ **20.** $\sum_{k=1}^{+\infty} \frac{(-1)^{k+1}}{3k-4}$ **21.** $\sum_{k=1}^{+\infty} \frac{(-1)^{k+1}}{k(k+1)}$ **22.** $\sum_{k=1}^{+\infty} \frac{(-1)^{k+1}\ln k}{k^3}$

23. $\sum_{k=1}^{+\infty} \frac{(-1)^{k+1}}{k\sqrt{k+3}}$ **24.** $\sum_{k=0}^{+\infty} \frac{(-1)^{k+1}}{\sqrt{k^2+1}}$ **25.** $\sum_{k=1}^{+\infty} \frac{(-1)^{k+1}\sqrt{k}}{k^2+1}$ **26.** $\sum_{k=1}^{+\infty} \frac{(-1)^{k+1}\sqrt{k}}{k+1}$

27. $\sum_{k=1}^{+\infty} (-1)^{k+1}\left(\frac{1}{5}\right)^{k}$ **28.** $\sum_{k=1}^{+\infty} \frac{(-1)^{k+1}}{e^k}$ **29.** $\sum_{k=2}^{+\infty} \frac{(-1)^{k}}{k \ln k}$ **30.** $\sum_{k=2}^{+\infty} \frac{(-1)^{k}}{k(\ln k)^2}$

31. $\sum_{k=1}^{+\infty} (-1)^{k+1} \frac{\sin k}{k^2+1}$ **32.** $\sum_{k=1}^{+\infty} (-1)^{k+1} \frac{\cos k}{k^2}$ **33.** $\sum_{k=1}^{+\infty} (-1)^{k+1} \frac{e^k}{k}$ **34.** $\sum_{k=1}^{+\infty} (-1)^{k+1} \frac{1}{ke^k}$

35. $\sum_{k=1}^{+\infty} \frac{(-1)^{k+1}\tan^{-1}k}{k}$ **36.** $\sum_{k=1}^{+\infty} \frac{(-1)^{k+1}2^k}{k^2}$ **37.** $\sum_{k=1}^{+\infty} (-1)^{k+1} \frac{\ln k}{k}$ **38.** $\sum_{k=1}^{+\infty} \frac{(-1)^{k+1}}{k^{1/k}}$

39. Show that the series $\sum_{k=1}^{+\infty}[(-1)^{k+1}/k^{1/3}]$ is conditionally convergent.

40. Show that the positive terms of $\sum_{k=1}^{+\infty}[(-1)^{k+1}/k]$ diverge.

41. Show that the negative terms of $\sum_{k=1}^{+\infty}[(-1)^{k+1}/k]$ diverge.

42. Show that the series $\sum_{k=1}^{+\infty}[(-1)^{k+1}/k]$ can be rearranged so that the resulting series converges to 0.

43. Show that the series $\sum_{k=1}^{+\infty}[(-1)^{k+1}/k]$ can be rearranged so that the resulting series converges to 2.

44. Show that the series $\sum_{k=1}^{+\infty}[(-1)^{k+1}/k]$ can be rearranged so that the resulting series diverges.

45. Show that the series

$$e^{-x}\cos x + e^{-2x}\cos 2x + e^{-3x}\cos 3x + \cdots$$

is absolutely convergent for all positive values of x. [*Hint:* Use the fact that $|\cos\theta| \le 1$.]

46. What can you say about the convergence of the series

$$1 + r\cos\theta + r^2\cos 2\theta + r^3\cos 3\theta + \cdots$$

47. What is wrong with the following argument?

$$A = 1 - \tfrac{1}{2} + \tfrac{1}{3} - \tfrac{1}{4} + \tfrac{1}{5} - \tfrac{1}{6} + \tfrac{1}{7} - \tfrac{1}{8} + \cdots$$

$$(\tfrac{1}{2})A = \tfrac{1}{2} - \tfrac{1}{4} + \tfrac{1}{6} - \tfrac{1}{8} + \cdots$$

Thus,

$$A + (\tfrac{1}{2})A = 1 + \tfrac{1}{3} - \tfrac{1}{2} + \tfrac{1}{5} + \tfrac{1}{7} - \tfrac{1}{4} + \cdots$$

The series on the right is a rearrangement of the terms of the series A. Hence, its sum is A, so that

$$A + (\tfrac{1}{2})A = A$$

$$A = 0$$

But

$$A = (1 - \tfrac{1}{2}) + (\tfrac{1}{3} - \tfrac{1}{4}) + (\tfrac{1}{5} - \tfrac{1}{6}) + \cdots > 0???$$

In Problems 48–50 you are asked to consider an incorrect argument given by Jakob Bernoulli to prove that $\sum_{k=1}^{+\infty}\{1/[k(k+1)]\} = \frac{1}{2} + \frac{1}{6} + \frac{1}{12} + \cdots = 1$. Bernoulli's argument went as follows: Let $N = 1 + \frac{1}{2} + \frac{1}{3} + \frac{1}{4} + \cdots$. Then $N - 1 = \frac{1}{2} + \frac{1}{3} + \frac{1}{4} + \frac{1}{5} + \cdots$. Now, subtract term-by-term to get $N - (N-1) = (1 - \frac{1}{2}) + (\frac{1}{2} - \frac{1}{3}) + (\frac{1}{3} - \frac{1}{4}) + \cdots$, or $1 = \frac{1}{2} + \frac{1}{6} + \frac{1}{12} + \cdots$.

48. What is wrong with Bernoulli's argument?

49. In general, what can be said about the convergence or divergence of a series formed by taking the term-by-term difference (or sum) of two divergent series? Support your answer with examples.

50. Although the method is wrong, Bernoulli's conclusion is correct; that is, it is true that $1 = \frac{1}{2} + \frac{1}{6} + \frac{1}{12} + \cdots$. Prove it! [*Hint:* Look at the partial sums using the form that the series was in just after our term-by-term subtraction above.]

11.6 Ratio Test; Root Test

Ratio Test

One of the most practical tests for convergence of a series makes use of the ratio of consecutive terms.

[11.6.1] THEOREM / ***Ratio Test.***

Let $\sum_{k=1}^{+\infty} a_k$ be a series of nonzero terms.

(a) If $\displaystyle\lim_{n\to+\infty}\left|\frac{a_{n+1}}{a_n}\right| < 1$, then $\sum_{k=1}^{+\infty} a_k$ converges.

(b) If $\displaystyle\lim_{n\to+\infty}\left|\frac{a_{n+1}}{a_n}\right| > 1$ or if $\displaystyle\lim_{n\to+\infty}\left|\frac{a_{n+1}}{a_n}\right| = +\infty$, then $\sum_{k=1}^{+\infty} a_k$ diverges.

(c) If $\displaystyle\lim_{n\to+\infty}\left|\frac{a_{n+1}}{a_n}\right| = 1$, the test fails to indicate either convergence or divergence.

A proof is given at the end of this section. Let's look at some examples of how theorem (11.6.1) is applied.

EXAMPLE 1

Use the ratio test to test each series for convergence or divergence.

(a) $\sum_{k=1}^{+\infty} \frac{k}{4^k}$ (b) $\sum_{k=1}^{+\infty} \frac{2^k}{k}$ (c) $\sum_{k=1}^{+\infty} \frac{3k+1}{k^2}$ (d) $\sum_{k=0}^{+\infty} \frac{1}{k!}$

Solution

(a) For $\sum_{k=1}^{+\infty}(k/4^k)$, we have $a_{n+1} = (n+1)/4^{n+1}$ and $a_n = n/4^n$. We form their ratio:

$$\left|\frac{a_{n+1}}{a_n}\right| = \frac{(n+1)/4^{n+1}}{n/4^n} = \frac{n+1}{4n} = \frac{1+(1/n)}{4}$$

Since

$$\lim_{n\to+\infty}\left|\frac{a_{n+1}}{a_n}\right| = \lim_{n\to+\infty}\frac{1+(1/n)}{4} = \frac{1}{4} < 1$$

the series converges.

(b) For $\sum_{k=1}^{+\infty}(2^k/k)$, we have $a_{n+1} = 2^{n+1}/(n+1)$ and $a_n = 2^n/n$, and their ratio is

$$\left|\frac{a_{n+1}}{a_n}\right| = \frac{2^{n+1}}{n+1}\left(\frac{n}{2^n}\right) = \frac{2n}{n+1} = \frac{2}{1+(1/n)}$$

Then

$$\lim_{n\to+\infty}\left|\frac{a_{n+1}}{a_n}\right| = \lim_{n\to+\infty}\frac{2}{1+(1/n)} = 2$$

The series diverges.

(c) For $\sum_{k=1}^{+\infty}[(3k+1)/k^2]$, we have $a_{n+1} = (3n+4)/(n+1)^2$ and $a_n = (3n+1)/n^2$. Their ratio is

$$\left|\frac{a_{n+1}}{a_n}\right| = \left[\frac{3n+4}{(n+1)^2}\right]\left(\frac{n^2}{3n+1}\right) = \left(\frac{3n+4}{3n+1}\right)\left(\frac{n^2}{n^2+2n+1}\right)$$

$$= \left[\frac{3+(4/n)}{3+(1/n)}\right]\left[\frac{1}{1+(2/n)+(1/n^2)}\right]$$

Then

$$\lim_{n\to+\infty}\left|\frac{a_{n+1}}{a_n}\right| = \lim_{n\to+\infty}\left\{\left[\frac{3+(4/n)}{3+(1/n)}\right]\left[\frac{1}{1+(2/n)+(1/n^2)}\right]\right\} = 1$$

The ratio test gives no information about this series. [You can show that the series diverges by comparing it to the harmonic series $\sum_{k=1}^{+\infty}(1/k)$.]

(d) $\sum_{k=1}^{+\infty}(1/k!)$, we have $a_{n+1} = 1/(n+1)!$ and $a_n = 1/n!$, so

$$\left|\frac{a_{n+1}}{a_n}\right| = \frac{n!}{(n+1)!} = \frac{1}{n+1}$$

Then

$$\lim_{n\to+\infty}\left|\frac{a_{n+1}}{a_n}\right| = \lim_{n\to+\infty}\frac{1}{n+1} = 0$$

Hence, $\sum_{k=1}^{+\infty}(1/k!)$ converges.*

As Example 1 illustrates, the ratio test is especially useful in handling series containing factorials and/or powers.

EXAMPLE 2

Determine whether the series $\sum_{k=1}^{+\infty}(k!/k^k)$ converges or diverges.

Solution

Since $a_n = n!/n^n$ and $a_{n+1} = (n+1)!/(n+1)^{n+1}$, we find that

$$\left|\frac{a_{n+1}}{a_n}\right| = \frac{(n+1)!/(n+1)^{n+1}}{n!/n^n} = \frac{(n+1)!n^n}{n!(n+1)^{n+1}} = \frac{n^n}{(n+1)^n} = \frac{1}{[(n+1)/n]^n}$$

$$= \frac{1}{[1+(1/n)]^n}$$

Hence,

$$\lim_{n\to+\infty}\left|\frac{a_{n+1}}{a_n}\right| = \lim_{n\to+\infty}\frac{1}{[1+(1/n)]^n} = \frac{1}{\lim_{n\to+\infty}[1+(1/n)]^n} \underset{\uparrow}{=} \frac{1}{e}$$

By theorem (7.6.1)

Since $1/e < 1$, the series converges.

Caution: Here we have found that the ratio $|a_{n+1}/a_n|$ converges to $1/e$; this does not mean that $\sum_{k=1}^{+\infty}(k!/k^k)$ converges to $1/e$. In fact, it is not known what the sum of the series is; all we know is that it converges.

We conclude our discussion of the ratio test with some observations:

1. To test $\sum_{k=1}^{+\infty} a_k$ for convergence, it is important to check whether the *limit of the ratio* $|a_{n+1}/a_n|$—not the ratio itself—is less than 1. For example, the ratio $|a_{n+1}/a_n|$ for the harmonic series $\sum_{k=1}^{+\infty}(1/k)$, which diverges, is $n/(n+1) < 1$, but its limit is not less than 1.
2. For divergence, it is sufficient that the ratio $|a_{n+1}/a_n|$ itself is greater than 1 for all n.

* We shall discover in Section 11.9 that $\sum_{k=1}^{+\infty}(1/k!) = e$.

3. The ratio test is not conclusive when $\lim_{n\to+\infty}|a_{n+1}/a_n| = 1$. It is also not conclusive when $\lim_{n\to+\infty}|a_{n+1}/a_n| \neq +\infty$ does not exist.
4. If the general term of an infinite series involves the index n, either exponentially or factorially, the ratio test often provides an answer to the question of convergence or divergence.

Root Test

The root test works well for series of nonzero terms whose nth term involves an nth power.

[11.6.2] THEOREM / *Root Test.*

Let a_k be a series of nonzero terms.

(a) If $\lim_{n\to+\infty} \sqrt[n]{|a_n|} < 1$, then $\sum_{k=1}^{+\infty} a_k$ converges.
(b) If $\lim_{n\to+\infty} \sqrt[n]{|a_n|} > 1$, then $\sum_{k=1}^{+\infty} a_k$ diverges.
(c) If $\lim_{n\to+\infty} \sqrt[n]{|a_n|} = 1$, the test is inconclusive.

The proof of this result is similar to the proof given at the end of this section for the ratio test; we leave it as an exercise (Problem 36).

EXAMPLE 3

Determine whether the series $\sum_{k=1}^{+\infty} \frac{e^k}{k^k}$ converges or diverges.

Solution

$$\lim_{n\to+\infty} \sqrt[n]{|a_n|} = \lim_{n\to+\infty} \sqrt[n]{\frac{e^n}{n^n}} = \lim_{n\to+\infty} \frac{e}{n} = 0 < 1$$

Thus, the series converges.

EXAMPLE 4

Determine whether the series $\sum_{k=1}^{+\infty} \left(\frac{8k+3}{5k-2}\right)^k$ converges or diverges.

Solution

$$\lim_{n\to+\infty} \sqrt[n]{\left(\frac{8n+3}{5n-2}\right)^n} = \lim_{n\to+\infty} \frac{8n+3}{5n-2} = \frac{8}{5} > 1$$

Thus, the series diverges.

Proof of the Ratio Test Let

$$\lambda = \lim_{n\to+\infty} \left|\frac{a_{n+1}}{a_n}\right|$$

***Part (a):* $0 \le \lambda < 1$.** Let r be any number such that $\lambda < r < 1$. Since $\lim_{n\to+\infty}|a_{n+1}/a_n| = \lambda < r$, by the definition of limit, we can find a number

N so that for $n > N$, the ratio $|a_{n+1}/a_n|$ will differ from λ by as little as we please, and will therefore be less than r. Then

$$\left|\frac{a_{N+1}}{a_N}\right| < r \quad \text{or} \quad |a_{N+1}| < r|a_N|$$

$$\left|\frac{a_{N+2}}{a_{N+1}}\right| < r \quad \text{or} \quad |a_{N+2}| < r|a_{N+1}| < r^2|a_N|$$

$$\left|\frac{a_{N+3}}{a_{N+2}}\right| < r \quad \text{or} \quad |a_{N+3}| < r|a_{N+2}| < r^3|a_N|$$

Thus, each term of the series $|a_{N+1}| + |a_{N+2}| + \cdots$ is less than the corresponding term of the geometric series $|a_N|r + |a_N|r^2 + |a_N|r^3 + \cdots$, which is convergent, since $|r| < 1$. By comparison test I, the series $|a_{N+1}| + |a_{N+2}| + \cdots$ is also convergent and, hence, the series $\sum_{k=1}^{+\infty}|a_k|$ is convergent. By theorem (11.5.5), the series $\sum_{k=1}^{+\infty}a_k$ converges.

***Part (b):* $1 < \lambda \le +\infty$.** Suppose $\lambda < +\infty$. Let r be any number such that $1 < r < \lambda$. Since $\lim_{n\to+\infty}|a_{n+1}/a_n| = \lambda$, we can find a number N so that for $n > N$, the ratio $|a_{n+1}/a_n|$ will differ from λ by as little as we please, and will therefore be greater than r. That is, for all $n > N$, we have $|a_{n+1}/a_n| > r > 1$ so that $|a_{n+1}| > |a_n|$. After the Nth term, the terms are positive and increasing. Hence, $\lim_{n\to+\infty} a_n \ne 0$ and therefore by the test for divergence, the series diverges. The proof when $\lambda = +\infty$ is left as an exercise (Problem 40).

***Part (c):* $\lambda = 1$.** To show that the test fails for $\lambda = 1$, we merely check two series—one that diverges and another that converges—to show that no conclusion can be drawn. Consider $\sum_{k=1}^{+\infty}(1/k)$ and $\sum_{k=1}^{+\infty}(1/k^2)$. The first is the harmonic series and we know it diverges; the second is a p-series with $p > 1$ and it converges. It is left to you to show that $\lambda = 1$ in each case. (See Problems 31 and 32.)

EXERCISE 11.6

In Problems 1–18 test for convergence or divergence by using the ratio test.

1. $\displaystyle\sum_{k=1}^{+\infty} \frac{(2k-1)(2k+1)}{2^k}$ **2.** $\displaystyle\sum_{k=1}^{+\infty} \frac{1}{(2k+1)2^k}$ **3.** $\displaystyle\sum_{k=1}^{+\infty} k\left(\frac{2}{3}\right)^k$ **4.** $\displaystyle\sum_{k=1}^{+\infty} \frac{(k+1)!}{3^k}$

5. $\displaystyle\sum_{k=1}^{+\infty} \frac{10^k}{(2k)!}$ **6.** $\displaystyle\sum_{k=1}^{+\infty} \frac{5^k}{k^2}$ **7.** $\displaystyle\sum_{k=1}^{+\infty} \frac{k}{(2k-2)!}$ **8.** $\displaystyle\sum_{k=1}^{+\infty} \frac{(2k)!}{5^k 3^{k-1}}$

9. $\displaystyle\sum_{k=1}^{+\infty} \frac{2^k}{k(k+1)}$ **10.** $\displaystyle\sum_{k=1}^{+\infty} \frac{k!}{k^2(k+1)^2}$ **11.** $\displaystyle\sum_{k=1}^{+\infty} \frac{k^3}{k!}$ **12.** $\displaystyle\sum_{k=1}^{+\infty} \frac{k!}{k^{k+1}}$

13. $\displaystyle\sum_{k=2}^{+\infty} \frac{3^{k-1}}{k2^k}$ **14.** $\displaystyle\sum_{k=1}^{+\infty} \frac{k(k+2)}{3^k}$ **15.** $\displaystyle\sum_{k=1}^{+\infty} \frac{k}{e^k}$ **16.** $\displaystyle\sum_{k=1}^{+\infty} \frac{e^k}{k^3}$

17. $\displaystyle\sum_{k=1}^{+\infty} k2^k$ **18.** $\displaystyle\sum_{k=1}^{+\infty} \frac{1}{k}2^k$

In Problems 19–30 test for convergence or divergence by using the root test.

19. $\sum_{k=1}^{+\infty}\left(\frac{2k+1}{5k+1}\right)^k$

20. $\sum_{k=1}^{+\infty}\left(\frac{3k-1}{2k+1}\right)^k$

21. $\sum_{k=1}^{+\infty}\left(\frac{k}{5}\right)^k$

22. $\sum_{k=1}^{+\infty}\left(\frac{\sqrt{4k^2+1}}{k}\right)^k$

23. $\sum_{k=1}^{+\infty}\left(\frac{\ln k}{k}\right)^k$

24. $\sum_{k=1}^{+\infty}\left(\frac{k^2+1}{k}\right)^k$

25. $\sum_{k=1}^{+\infty}\left(\frac{\sqrt{k^2+1}}{3k}\right)^k$

26. $\sum_{k=1}^{+\infty}\frac{\pi^{2k}}{k^k}$

27. $\sum_{k=1}^{+\infty}\frac{k^2}{2^k}$

28. $\sum_{k=1}^{+\infty}\frac{k^3}{3^k}$

29. $\sum_{k=1}^{+\infty}\left(\frac{1}{\ln k}\right)^k$

30. $\sum_{k=1}^{+\infty}\frac{k}{3^k}$

31. For the series $\sum_{k=1}^{+\infty}(1/k)$, show that $\lim_{n\to+\infty}(a_{n+1}/a_n) = 1$, yet $\sum_{k=1}^{+\infty}(1/k)$ diverges.

32. For the series $\sum_{k=1}^{+\infty}(1/k^2)$, show that $\lim_{n\to+\infty}(a_{n+1}/a_n) = 1$, yet $\sum_{k=1}^{+\infty}(1/k^2)$ converges.

33. Prove that $\lim_{n\to+\infty}(n!/n^n) = 0$, where n denotes a positive integer.

34. Show that the root test is inconclusive for $\sum_{k=1}^{+\infty}(1/k)$ and for $\sum_{k=1}^{+\infty}(1/k^2)$.

35. Use the ratio test to find the positive numbers x for which the series $\sum_{k=1}^{+\infty}(x^k/k^2)$ converges or diverges.

36. Prove theorem (11.6.2).

37. For a series $\sum_{k=1}^{+\infty} a_k$, show that if the ratio $\left|\frac{a_{n+1}}{a_n}\right| > 1$ for all n, then the series diverges.

38. Give an example of a convergent series $\sum_{k=1}^{+\infty} a_k$ for which $\lim_{n\to+\infty}\left|\frac{a_{n+1}}{a_n}\right| \neq +\infty$ does not exist.

39. Give an example of a divergent series $\sum_{k=1}^{+\infty} a_k$ for which $\lim_{n\to+\infty}\left|\frac{a_{n+1}}{a_n}\right| \neq +\infty$ does not exist.

40. Prove that $\sum_{k=1}^{+\infty} a_k$ diverges if $\lim_{n\to+\infty}\left|\frac{a_{n+1}}{a_n}\right| = +\infty$

41. Determine whether the following series is convergent or divergent:

$$\frac{1}{3} - \frac{2^3}{3^2} + \frac{3^3}{3^3} - \frac{4^3}{3^4} + \cdots + \frac{(-1)^{n-1}n^3}{3^n} + \cdots$$

Summary of Tests — 11.7

The preceding sections contained many tests for convergence or divergence of a series. In the exercises following these sections, each time a series was to be tested, the test to be applied was also cited or the section in which the question appeared virtually gave the test away. Such advantages, unfortunately, are not of the real world. Therefore, in this section we summarize the tests already encountered and give some clues as to what test, in general, gives the best chance of working.

We begin with Table 3 (page 678), a list of the tests we've encountered, and Table 4 (page 679), a list of the important series we've analyzed.

The following outline offers one method of attack for determining the convergence or divergence of a series. Obviously, there are other methods, equally good, that you can devise for yourself.

1. Check to see if the given series is either a geometric series or a p-series.
2. Take the limit of the nth term of the series to see if the divergence test applies.

3. If the series is positive and meets the conditions of the integral test, the integral test may be tried.
4. If the series is positive and the nth term is a quotient of sums or differences of powers of n, the limit comparison test with an appropriate p-series will usually do the job.

Table 3 Tests for Convergence and Divergence of Series

Name	Description	Comment
Test for divergence (11.3.2)	$\sum_{k=1}^{+\infty} a_k$ diverges if $\lim_{n\to+\infty} a_n \neq 0$	Test gives no information about convergence of $\sum_{k=1}^{+\infty} a_k$ if $\lim_{n\to+\infty} a_n = 0$
Integral test for series of positive terms (11.3.7)	$\sum_{k=1}^{+\infty} a_k$ converges (diverges) if $\int_1^{+\infty} f(x)\,dx$ converges (diverges) where $f(n) = a_n$ for all n and f is continuous, positive, and decreasing for $x \geq 1$	Good to use if f is easy to integrate
Comparison test I: convergence for series of positive terms (11.4.1)	$\sum_{k=1}^{+\infty} a_k$ converges if $0 \leq a_k \leq b_k$ and if $\sum_{k=1}^{+\infty} b_k$ converges	You must choose a series $\sum_{k=1}^{+\infty} b_k$ of positive terms whose convergence is known
Comparison test II: divergence for series of positive terms (11.4.2)	$\sum_{k=1}^{+\infty} a_k$ diverges if $a_k \geq c_k \geq 0$ and if $\sum_{k=1}^{+\infty} c_k$ diverges	You must choose a series $\sum_{k=1}^{+\infty} c_k$ of positive terms whose divergence is known
Limit comparison test for series of positive terms (11.4.3)	$\sum_{k=1}^{+\infty} a_k$ converges (diverges) if $\sum_{k=1}^{+\infty} b_k$ converges (diverges) and if $\lim_{n\to+\infty}(a_n/b_n) = L > 0$	You must choose a series $\sum_{k=1}^{+\infty} b_k$ of positive terms whose convergence (divergence) is known
Alternating series test (11.5.2)	$\sum_{k=1}^{+\infty}(-1)^{k+1}a_k$ converges if a_k are decreasing and if $\lim_{n\to+\infty} a_n = 0$ $(a_k > 0)$	Refer to theorem (11.5.3) for the procedure to estimate the error
Absolute convergence test (11.5.5)	If $\sum_{k=1}^{+\infty}\lvert a_k\rvert$ converges, then $\sum_{k=1}^{+\infty} a_k$ converges	
Ratio test (11.6.1)	$\sum_{k=1}^{+\infty} a_k$ converges if $\lim_{n\to+\infty}\lvert a_{n+1}/a_n\rvert < 1$ $\sum_{k=1}^{+\infty} a_k$ diverges if $\lim_{n\to+\infty}\lvert a_{n+1}/a_n\rvert > 1$	Good to use if a_n consists of factorials The test gives no information if $\lim_{n\to+\infty}\lvert a_{n+1}/a_n\rvert = 1$
Root test (11.6.2)	$\sum_{k=1}^{+\infty} a_k$ converges if $\lim_{n\to+\infty}\sqrt[n]{\lvert a_n\rvert} < 1$ $\sum_{k=1}^{+\infty} a_k$ diverges if $\lim_{n\to+\infty}\sqrt[n]{\lvert a_n\rvert} > 1$	Good to use if a_n involves nth powers The test gives no information if $\lim_{n\to+\infty}\sqrt[n]{\lvert a_n\rvert} = 1$

Table 4 Important Series

Name	Description	Comment
Geometric series (11.2.4)	$\sum_{k=1}^{+\infty} ar^{k-1} = a + ar + ar^2 + \cdots, \quad a \neq 0$	Converges to $a/(1-r)$ if $\lvert r\rvert < 1$; Diverges otherwise
Harmonic series (11.2.5)	$\sum_{k=1}^{+\infty} \frac{1}{k} = 1 + \frac{1}{2} + \frac{1}{3} + \cdots$	Diverges
p-series (11.3.9)	$\sum_{k=1}^{+\infty} \frac{1}{k^p} = 1 + \frac{1}{2^p} + \frac{1}{3^p} + \cdots$	Converges if $p > 1$; Diverges if $0 < p \leq 1$
k-to-the-k series	$\sum_{k=1}^{+\infty} \frac{1}{k^k} = 1 + \frac{1}{2^2} + \frac{1}{3^3} + \frac{1}{4^4} + \cdots$	Converges
Factorial series	$\sum_{k=0}^{+\infty} \frac{1}{k!} = 1 + 1 + \frac{1}{2} + \frac{1}{6} + \frac{1}{24} + \cdots$	Converges
Alternating harmonic series	$\sum_{k=1}^{+\infty} \frac{(-1)^{k+1}}{k} = 1 - \frac{1}{2} + \frac{1}{3} - \frac{1}{4} + \cdots$	Converges

5. If the series is positive and the preceding attempts fail to provide a conclusion, then try comparison test I or II.
6. *Series with negative terms.* For alternating series that meet the proper conditions, use the alternating series test (it is sometimes better to apply the test for absolute convergence first). For other series containing negative terms, always test for absolute convergence first.
7. If the series has terms involving products, factorials, or powers, the ratio test is a good choice.
8. If the series has an nth term that involves an nth power, try the root test.

EXERCISE 11.7

Test each series for convergence (absolute or conditional) or divergence. Use any test you wish.

1. $\sum_{k=1}^{+\infty} \frac{9k^3 + 5k^2}{k^{5/2} + 4}$

2. $\sum_{k=1}^{+\infty} \frac{(-1)^{k+1}}{\sqrt{2k+1}}$

3. $6 + 2 + \frac{2}{3} + \frac{2}{9} + \frac{2}{27} + \cdots$

4. $1 + \frac{1 \cdot 2}{1 \cdot 3} + \frac{1 \cdot 2 \cdot 3}{1 \cdot 3 \cdot 5} + \frac{1 \cdot 2 \cdot 3 \cdot 4}{1 \cdot 3 \cdot 5 \cdot 7} + \cdots$

5. $\sum_{k=1}^{+\infty} \frac{1}{k^2} \sin \frac{\pi}{k}$

6. $\sum_{k=1}^{+\infty} \frac{3k+2}{k^3+1}$

7. $1 + \frac{2^2+1}{2^3+1} + \frac{3^2+1}{3^3+1} + \frac{4^2+1}{4^3+1} + \cdots$

8. $2 + \frac{3}{2} \cdot \frac{1}{4} + \frac{4}{3} \cdot \frac{1}{4^2} + \frac{5}{4} \cdot \frac{1}{4^3} + \cdots$

9. $\sum_{k=1}^{+\infty} \frac{k+4}{k\sqrt{3k-2}}$

10. $\sum_{k=1}^{+\infty} \frac{\sin k}{k^3}$

11. $\sum_{k=1}^{+\infty} \frac{3^{2k-1}}{k^2 + 2k}$

12. $\sum_{k=1}^{+\infty} \frac{5^k}{k!}$

13. $\sum_{k=1}^{+\infty} \left(1 + \frac{2}{k}\right)^k$

14. $\sum_{k=1}^{+\infty} \frac{k^2 + 4}{e^k}$

15. $\frac{2}{3} - \frac{3}{4} \cdot \frac{1}{2} + \frac{4}{5} \cdot \frac{1}{3} - \frac{5}{6} \cdot \frac{1}{4} + \cdots$

16. $\frac{1}{2} - \frac{4}{2^3 + 1} + \frac{9}{3^3 + 1} - \frac{16}{4^3 + 1} + \cdots$

17. $\sum_{k=1}^{+\infty} \frac{k!}{(2k)!}$

18. $\sum_{k=1}^{+\infty} k^3 e^{-k^4}$

19. $1 + \frac{1 \cdot 3}{2!} + \frac{1 \cdot 3 \cdot 5}{3!} + \frac{1 \cdot 3 \cdot 5 \cdot 7}{4!} + \cdots$

20. $\frac{1}{\sqrt{1 \cdot 2 \cdot 3}} + \frac{1}{\sqrt{2 \cdot 3 \cdot 4}} + \frac{1}{\sqrt{3 \cdot 4 \cdot 5}} + \cdots$

21. $\sum_{k=4}^{+\infty} \left(\frac{1}{k-3} - \frac{1}{k}\right)$

22. $\sum_{k=1}^{+\infty} \frac{1}{\sqrt{k} + 100}$

23. $\sum_{k=1}^{+\infty} \frac{k^2 + 5k}{3 + 5k^2}$

24. $\sum_{k=1}^{+\infty} \frac{1}{\sqrt[3]{k^4 + 4}}$

25. $\sum_{k=1}^{+\infty} \frac{1}{11}\left(\frac{-3}{2}\right)^k$

26. $\sum_{k=1}^{+\infty} \frac{1}{\sqrt{k^3 + 1}}$

27. $1 - \frac{2!}{1 \cdot 3} + \frac{3!}{1 \cdot 3 \cdot 5} - \frac{4!}{1 \cdot 3 \cdot 5 \cdot 7} + \cdots$

28. $\frac{1}{3} - \frac{2}{4} + \frac{3}{5} - \frac{4}{6} + \cdots$

29. $\sum_{k=1}^{+\infty} (-1)^k \frac{k^3}{k^3 + e^k}$

30. $\sum_{k=1}^{+\infty} \frac{k(-4)^{3k}}{5^k}$

31. $\sum_{k=1}^{+\infty} \left(-\frac{1}{k}\right)^k$

32. $\sum_{k=1}^{+\infty} \frac{5}{2^k + 1}$

33. $\sum_{k=1}^{+\infty} \frac{\ln k}{2k^3 - 1}$

34. $\sum_{k=1}^{+\infty} (\sqrt{k+1} - \sqrt{k})$

35. $\sum_{k=1}^{+\infty} \sin^3\left(\frac{1}{k}\right)$

36. $\sum_{k=1}^{+\infty} e^{-k^2}$

37. $\frac{\sin\sqrt{1}}{1^{3/2}} - \frac{\sin\sqrt{2}}{2^{3/2}} + \frac{\sin\sqrt{3}}{3^{3/2}} - \cdots$

38. $\sum_{k=2}^{+\infty} \frac{(-1)^{k-1}}{k(\ln k)^3}$

39. $\sum_{k=1}^{+\infty} \frac{1}{(2k)^k}$

40. $\sum_{k=1}^{+\infty} \left(\frac{\ln k}{1000}\right)^k$

41. $\sum_{k=2}^{+\infty} \ln \frac{k}{k+1}$

42. $\sum_{k=2}^{+\infty} \frac{\ln(2k+1)}{\sqrt{k^2 - 2\sqrt{k^3 - 2k - 3}}}$

43. $\sum_{k=1}^{+\infty} \frac{\sqrt{k}}{\sqrt{k^3 - k + 1}\,\ln(2k+1)}$

44. $\sum_{k=1}^{+\infty} \frac{1}{\cosh^2 k}$

45. $\sum_{k=1}^{+\infty} \frac{\tan^{-1} k}{k^2}$

46. $\sum_{k=2}^{+\infty} \frac{1}{\sqrt{k}(\ln k)^4}$

47. $\sum_{k=2}^{+\infty} \frac{(\ln k)^2}{k^{5/2}}$

11.8 Power Series

In this section we study series whose terms are variable.

[11.8.1] DEFINITION / ***Power Series.***

If x is a variable, then a series of the form

$$\sum_{k=0}^{+\infty} a_k x^k = a_0 + a_1 x + a_2 x^2 + \cdots$$

where the coefficients a_0, a_1, $a_2, \ldots$ are constants, is called a

power series in x.* **A series of the form**

$$\sum_{k=0}^{+\infty} a_k(x-c)^k = a_0 + a_1(x-c) + a_2(x-c)^2 + \cdots$$

where c is a constant, is called a *power series in* $(x - c)$.

For a particular x, a power series in x reduces to a series of real numbers like the series studied thus far. For example, $\sum_{k=1}^{+\infty}(x^k/k)$ is a power series in x. It certainly converges (to 0) if $x = 0$. If $x = 1$, it becomes the harmonic series $\sum_{k=1}^{+\infty}(1/k)$, which is divergent. If $x = -1$, it becomes the alternating harmonic series $\sum_{k=1}^{+\infty}[(-1)^k/k]$, which is convergent.

To find all numbers x for which a power series in x is convergent, we often rely on the ratio test, since in a power series, x is raised to a power.

EXAMPLE 1

Find all x for which the following power series are convergent.

(a) $\displaystyle\sum_{k=0}^{+\infty} \frac{x^k}{k!} = 1 + x + \frac{x^2}{2!} + \frac{x^3}{3!} + \cdots$ (b) $\displaystyle\sum_{k=0}^{+\infty} \frac{kx^k}{4^k} = \frac{x}{4} + \frac{2x^2}{4^2} + \frac{3x^3}{4^3} + \cdots$

(c) $\displaystyle\sum_{k=0}^{+\infty} k!x^k = 1 + x + 2!x^2 + 3!x^3 + \cdots$

Solution

(a) To test the series $\sum_{k=0}^{+\infty}(x^k/k!)$ for convergence, we apply the ratio test. For the series under investigation, we take

$$b_n = \frac{x^n}{n!} \qquad \text{and} \qquad b_{n+1} = \frac{x^{n+1}}{(n+1)!}$$

Then

$$\lim_{n\to+\infty}\left|\frac{b_{n+1}}{b_n}\right| = \lim_{n\to+\infty}\left|\frac{x^{n+1}/(n+1)!}{x^n/n!}\right| = \lim_{n\to+\infty}\frac{|x|^{n+1}n!}{(n+1)!|x|^n}$$

$$= \lim_{n\to+\infty}\frac{|x|}{n+1} = 0$$

Since the limit is less than 1 for every x, it follows from the ratio test that the given series is absolutely convergent for all real numbers.†

(b) For $\sum_{k=0}^{+\infty}(kx^k/4^k)$, $b_n = nx^n/4^n$ and $b_{n+1} = (n+1)x^{n+1}/4^{n+1}$. Then

$$\lim_{n\to+\infty}\left|\frac{b_{n+1}}{b_n}\right| = \lim_{n\to+\infty}\frac{(n+1)|x|^{n+1}(4^n)}{4^{n+1}(n|x|^n)} = \lim_{n\to+\infty}\frac{(n+1)|x|}{4n} = \frac{|x|}{4}$$

* For power series, we agree that if $x = 0$, the series equals a_0.

† This suggests the following interesting result: Since $\sum_{k=0}^{+\infty}(x^k/k!)$ converges absolutely for every x, the limit of the nth term tends to 0. Thus, we have $\lim_{n\to+\infty}(x^n/n!) = 0$ for every x.

By the ratio test, the series will converge if $|x|/4 < 1$ or $|x| < 4$. It will diverge if $|x|/4 > 1$ or $|x| > 4$. The ratio test gives no information when $|x|/4 = 1$. However, we may check these values directly by replacing x by 4 and -4. For $x = 4$, the series becomes

$$\sum_{k=0}^{+\infty} \frac{k4^k}{4^k} = \sum_{k=1}^{+\infty} k = 1 + 2 + \cdots$$

which diverges. For $x = -4$, we obtain

$$\sum_{k=1}^{+\infty} \frac{k(-4)^k}{4^k} = \sum_{k=1}^{+\infty} \frac{(-1)^k k(4^k)}{4^k} = \sum_{k=1}^{+\infty} (-1)^k k = -1 + 2 - 3 + \cdots$$

which also diverges. Thus, the series $\sum_{k=0}^{+\infty}(kx^k/4^k)$ converges absolutely for $-4 < x < 4$ and diverges for $|x| \geq 4$.

(c) For $\sum_{k=0}^{+\infty} k!x^k$, $b_n = n!x^n$ and $b_{n+1} = (n+1)!x^{n+1}$. Thus,

$$\lim_{n\to+\infty} \left|\frac{b_{n+1}}{b_n}\right| = \lim_{n\to+\infty} \frac{(n+1)!|x|^{n+1}}{n!|x|^n} = \lim_{n\to+\infty} (n+1)|x|$$

$$= \begin{cases} 0 & \text{if } x = 0 \\ +\infty & \text{if } x \neq 0 \end{cases}$$

Thus, this power series converges for only $x = 0$. For any other x, it diverges.

The results obtained in Example 1 are typical of all power series. In order to prove this, we need the following preliminary result.

[11.8.2] THEOREM

(a) If the power series $\sum_{k=0}^{+\infty} a_k x^k$ converges for a number $x_0 \neq 0$, then it converges absolutely for all numbers x such that $|x| < |x_0|$.

(b) If the power series $\sum_{k=0}^{+\infty} a_k x^k$ diverges for a number x_1, then it diverges for all numbers x such that $|x| > |x_1|$.

Proof of Part (a) Assume that $\sum_{k=0}^{+\infty} a_k x_0^k$ converges. Then

$$\lim_{n\to+\infty} a_n x_0^n = 0$$

Thus, there is a positive integer N such that $|a_n x_0^n| < 1$ for all $n \geq N$. For any number x such that $|x| < |x_0|$,

$$|a_n x^n| = \left|\frac{a_n x^n x_0^n}{x_0^n}\right| = |a_n x_0^n|\left|\frac{x}{x_0}\right|^n < \left|\frac{x}{x_0}\right|^n \qquad \text{for } n \geq N$$

But the series $\sum_{k=0}^{+\infty}|x/x_0|^k$ is a convergent geometric series because $|x/x_0| < 1$. Therefore, by comparison test I, the series $\sum_{k=0}^{+\infty}|a_k x^k|$ converges, and so the power series $\sum_{k=0}^{+\infty} a_k x^k$ converges absolutely for all numbers x such that $|x| < |x_0|$.

Proof of Part (b) If the series converges for some number x such that $|x| > |x_1|$, it must converge for x_1 (by part (a)), which is contrary to the hypothesis. Therefore, the series diverges for all x such that $|x| > |x_1|$.

From theorem (11.8.2) we conclude:

[11.8.3] THEOREM

For a power series $\sum_{k=0}^{+\infty} a_k x^k$, exactly one of the following is true:

1. **The series converges for only $x = 0$.**
2. **The series converges absolutely for all x.**
3. **There is a positive number R such that the series converges absolutely for all x for which $|x| < R$ and diverges for all x for which $|x| > R$.**

[11.8.4] DEFINITION / *Interval of Convergence; Radius of Convergence.*

The set of all numbers x for which a power series converges is called the *interval of convergence* of the power series. If, in theorem (11.8.3), possibility 3 holds, then the number R is called the *radius of convergence* of the power series. If possibility 1 holds, then $R = 0$; if possibility 2 holds, then $R = +\infty$.

As Example 1 illustrates, the ratio test is the most useful method for obtaining the interval of convergence and the radius of convergence of a power series. However, no conclusion may be drawn about convergence or divergence at the endpoints of the interval of convergence. At an endpoint, a power series may be absolutely convergent, conditionally convergent, or divergent.

Let's return to Example 1. In Example 1(a) the series converges absolutely for all x. In this case, the radius of convergence is $+\infty$, and the interval of convergence is $(-\infty, +\infty)$. In Example 1(b) the series converges absolutely for $|x| < 4$ and diverges for $|x| \geq 4$. In this case, 4 is the radius of convergence, and the open interval $(-4, 4)$ is the interval of convergence (see Fig. 17). In Example 1(c) the series converges absolutely for only $x = 0$. In this case the radius of convergence is 0.

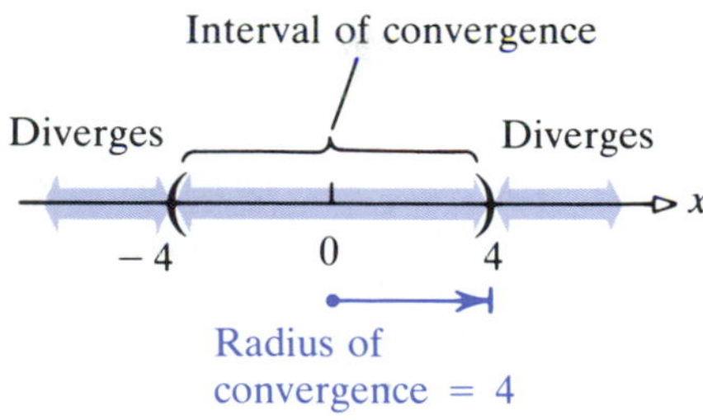

Figure 17

EXAMPLE 2

Find the interval of convergence of: $\sum_{k=1}^{+\infty} \frac{x^{2k}}{k}$

Solution

By applying the ratio test, we obtain

$$\lim_{n\to+\infty}\left|\frac{x^{2n+2}/(n+1)}{x^{2n}/n}\right| = \lim_{n\to+\infty}\frac{n}{n+1}x^2 = x^2 \lim_{n\to+\infty}\frac{n}{n+1} = x^2$$

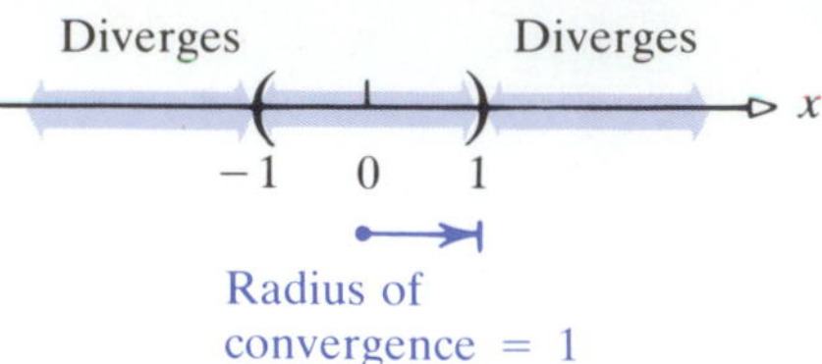

Figure 18

For the series to converge, we require $x^2 < 1$; that is, $-1 < x < 1$. At the endpoints, we find that for $x = -1$ or for $x = 1$, the series reduces to the harmonic series $\sum_{k=1}^{+\infty}(1/k)$, which diverges. Consequently, the interval of convergence is $-1 < x < 1$ (see Fig. 18).

To find the interval of convergence of a power series in $(x - c)$, namely, $\sum_{k=0}^{+\infty} a_k(x - c)^k$, we let $z = x - c$. The result is a power series in z: $\sum_{k=0}^{+\infty} a_k z^k$. The interval of convergence of the latter series may be used to find the interval of convergence of the former. For example, if $\sum_{k=0}^{+\infty} a_k z^k$ has $(-R, R]$ for its interval of convergence, then $-R < z \leq R$, so that when we replace z by $(x - c)$, we get $-R < (x - c) \leq R$ or $c - R < x \leq c + R$. Thus, the interval of convergence is $(c - R, c + R]$ and R is the radius of convergence.

EXAMPLE 3

Find the interval of convergence of: $\displaystyle\sum_{k=0}^{+\infty} (-1)^k \frac{(x-2)^k}{k+1}$

Solution

By the ratio test, we obtain

$$\lim_{n \to +\infty} \left| \frac{(-1)^{n+1}(x-2)^{n+1}/(n+2)}{(-1)^n(x-2)^n/(n+1)} \right| = \lim_{n \to +\infty} \left| \frac{(n+1)(x-2)}{n+2} \right| = |x - 2|$$

For the series to converge, we require $|x - 2| < 1$; that is, $-1 < x - 2 < 1$ or $1 < x < 3$. The endpoints $x = 1$ and $x = 3$ must be checked separately. If $x = 1$, the series becomes

$$1 + \frac{1}{2} + \frac{1}{3} + \cdots + \frac{1}{n+1} + \cdots$$

which is the divergent harmonic series. If $x = 3$, we get

$$1 - \frac{1}{2} + \frac{1}{3} - \frac{1}{4} + \cdots + \frac{(-1)^n}{n+1} + \cdots$$

the alternating harmonic series, which converges. Hence, the interval of convergence is $(1, 3]$, as shown in Figure 19.

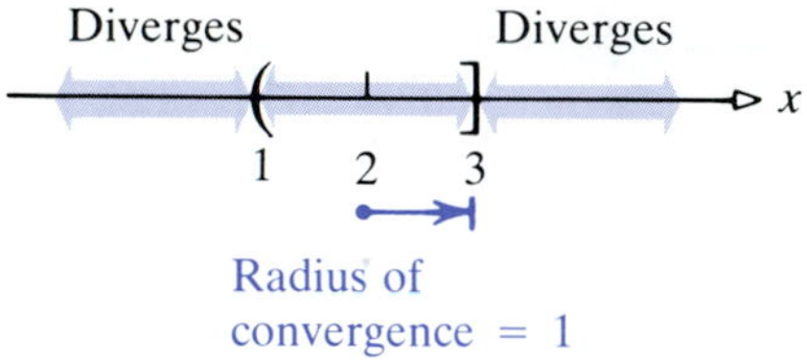

Diverges Diverges x

1 2 3

Radius of convergence = 1

Figure 19

EXERCISE 11.8

In Problems 1–20 find the interval of convergence.

1. $\displaystyle\sum_{k=0}^{+\infty} \frac{x^k}{k+5}$
2. $\displaystyle\sum_{k=0}^{+\infty} \frac{x^k}{1+k^2}$
3. $\displaystyle\sum_{k=0}^{+\infty} \frac{k^2x^k}{3^k}$
4. $\displaystyle\sum_{k=0}^{+\infty} \frac{2^kx^k}{3^k}$
5. $\displaystyle\sum_{k=0}^{+\infty} \frac{kx^k}{2k+1}$
6. $\displaystyle\sum_{k=0}^{+\infty} (6x)^k$
7. $\displaystyle\sum_{k=0}^{+\infty} \frac{k(2x)^k}{3^k}$
8. $\displaystyle\sum_{k=0}^{+\infty} (-1)^k \frac{(2x)^k}{k!}$

9. $\sum_{k=0}^{+\infty} \frac{(-1)^k}{(2k+1)!} x^{2k+1}$ **10.** $\sum_{k=1}^{+\infty} (kx)^k$ **11.** $\sum_{k=1}^{+\infty} \frac{kx^k}{\ln(k+1)}$ **12.** $\sum_{k=0}^{+\infty} (-1)^k \frac{(x-3)^{2k}}{9^k}$

13. $\sum_{k=1}^{+\infty} \frac{k(k+1)x^k}{4^k}$ **14.** $\sum_{k=1}^{+\infty} \frac{(-1)^k(x-5)^k}{k(k+1)}$ **15.** $\sum_{k=1}^{+\infty} \frac{x^k}{\ln(k+1)}$ **16.** $\sum_{k=0}^{+\infty} \frac{x^k}{e^k}$

17. $\sum_{k=0}^{+\infty} (x-3)^k$ **18.** $\sum_{k=0}^{+\infty} \frac{(x+1)^k}{k!}$ **19.** $\sum_{k=0}^{+\infty} (-1)^k \frac{(x-1)^{4k}}{k!}$ **20.** $\sum_{k=1}^{+\infty} \frac{(x+1)^k}{k(k+1)(k+2)}$

In Problems 21 and 22 find the radius of convergence.

21. $\sum_{k=1}^{+\infty} \frac{k^k x^k}{k!}$ **22.** $\sum_{k=0}^{+\infty} \frac{3^k(x-2)^k}{k!}$

In Problems 23–28 find all x for which the series converges.

23. $\sum_{k=1}^{+\infty} \frac{1}{kx^k}$ **24.** $\sum_{k=1}^{+\infty} \frac{(\sin x)^k}{2k+1}$ **25.** $\sum_{k=1}^{+\infty} \frac{\sin k\pi x}{k^2}$ **26.** $\sum_{k=1}^{+\infty} \frac{1}{k(x-4)^k}$

27. $\sum_{k=0}^{+\infty} (-1)^k 2^k (\sin x)^k, \quad -\pi/2 \le x \le \pi/2$ **28.** $\sum_{k=1}^{+\infty} (-1)^k (\tan x)^k, \quad -\pi/2 < x < \pi/2$

29. If $\sum_{k=1}^{+\infty} a_k x^k$ converges for $x = 3$, what, if anything, can be said about the convergence at $x = 2$? Can anything be said about the convergence at $x = 5$?

30. If $\sum_{k=1}^{+\infty} a_k (x-2)^k$ converges for $x = 6$, at what other numbers x must the series necessarily converge?

31. If the series $\sum_{k=1}^{+\infty} a_k x^k$ converges for $x = 6$ and diverges for $x = -8$, what, if anything, can be said about the truth of the following statements?

(a) The series converges for $x = 2$.
(b) The series diverges for $x = 7$.
(c) The series is absolutely convergent for $x = 6$.
(d) The series converges for $x = -6$.
(e) The series diverges for $x = 10$.
(f) The series is absolutely convergent for $x = 4$.

32. If the series $\sum_{k=1}^{+\infty} a_k 3^k$ converges, show that the series $\sum_{k=1}^{+\infty} ka_k 2^k$ also converges.

33. If $R > 0$ is the radius of convergence of $\sum_{k=1}^{+\infty} a_k x^k$, show that $\lim_{n\to+\infty} |a_{n+1}/a_n| = 1/R$, provided this limit exists.

34. If R is the radius of convergence of $\sum_{k=1}^{+\infty} a_k x^k$, show that the radius of convergence of $\sum_{k=1}^{+\infty} a_k x^{2k}$ is $\sqrt{R}$.

35. Prove that if a power series converges absolutely at one endpoint of its interval of convergence, then the power series is absolutely convergent at the other endpoint.

Functions Represented by Power Series; Taylor and Maclaurin Series 11.9

Functions Represented by Power Series

A power series $\sum_{k=0}^{+\infty} a_k x^k$ defines a function f whose domain is the interval of convergence of the power series. If I is the interval of convergence of $\sum_{k=0}^{+\infty} a_k x^k$, we say that f is *represented by* the power series $\sum_{k=0}^{+\infty} a_k x^k$ on I by writing

$$f(x) = a_0 + a_1 x + a_2 x^2 + \cdots + a_n x^n + \cdots$$

for all x in I.*

* As before, we define $f(0) = a_0$.

For example, if f is represented by the power series $\sum_{k=0}^{+\infty} a_k x^k$ whose interval of convergence is I and if x_0 is a number in I, then $f(x_0)$ may be found by finding the sum of the series

$$f(x_0) = a_0 + a_1 x_0 + a_2 x_0^2 + \cdots + a_n x_0^n + \cdots$$

Thus, a power series representation for a function f may be used to calculate functional values.

The function f represented by a power series has properties similar to a polynomial. We state three of these properties without proof.

[11.9.1] THEOREM / ***Properties of Power Series.***

Let $\sum_{k=0}^{+\infty} a_k x^k$ be a power series in x having a nonzero radius of convergence R. Define the function f to be

$$f(x) = a_0 + a_1 x + a_2 x^2 + \cdots + a_n x^n + \cdots$$

Then at every number x within the interval of convergence of the given series:

(a) f is continuous:

$$\lim_{x \to x_0} \left(\sum_{k=0}^{+\infty} a_k x^k \right) = \sum_{k=0}^{+\infty} \left(\lim_{x \to x_0} a_k x^k \right) = \sum_{k=0}^{+\infty} a_k x_0^k$$

(b) f is differentiable:

$$\frac{d}{dx} \left(\sum_{k=0}^{+\infty} a_k x^k \right) = \sum_{k=0}^{+\infty} \left(\frac{d}{dx} a_k x^k \right) = \sum_{k=1}^{+\infty} k a_k x^{k-1}$$

(c) f can be integrated:

$$\int_0^x \left(\sum_{k=0}^{+\infty} a_k t^k \right) dt = \sum_{k=0}^{+\infty} \left(\int_0^x a_k t^k \, dt \right) = \sum_{k=0}^{+\infty} \frac{a_k x^{k+1}}{k+1}$$

Parts (b) and (c) state that a power series can be differentiated or integrated term-by-term and that the resulting series represents the derivative and integral, respectively, of the function represented by the power series at every number x within the interval of convergence of the original power series. Moreover, it can be shown that the power series obtained by differentiating or integrating a power series whose radius of convergence is R, both converge and have the same radius of convergence R.

Caution: Read the above statement carefully! It states that the *radii of convergence of* $\sum_{k=0}^{+\infty} a_k x^k$, $\sum_{k=1}^{+\infty} k a_k x^{k-1}$, and $\sum_{k=0}^{+\infty} [a_k x^{k+1}/(k+1)]$ are the same; it does not imply that the *intervals of convergence* are the same. For example, the interval of convergence of $\sum_{k=1}^{+\infty} (x^k/k)$ is $[-1, 1)$, whereas the interval of convergence of its derivative $\sum_{k=1}^{+\infty} x^{k-1}$ is $(-1, 1)$. Endpoints must always be checked separately.

As the following example illustrates, theorem (11.9.1) provides a tool for obtaining new power series representations from existing ones.

EXAMPLE 1

The geometric series $1 + x + x^2 + \cdots$ converges for $-1 < x < 1$ and has the sum $1/(1 - x)$. That is,

$$\frac{1}{1-x} = 1 + x + x^2 + \cdots + x^n + \cdots \qquad -1 < x < 1$$

An application of part (b) of theorem (11.9.1) reveals that

$$\frac{1}{(1-x)^2} = 1 + 2x + 3x^2 + \cdots + nx^{n-1} + \cdots$$

whose interval of convergence is $-1 < x < 1$.

The next example is a little more subtle.

EXAMPLE 2

Use the geometric series $1 + x + x^2 + \cdots$ to find a power series representation for $\ln[1/(1 - x)]$.

Solution

The geometric series $1 + x + x^2 + \cdots$ converges for $-1 < x < 1$ and has the sum $1/(1 - x)$. Thus,

$$\frac{1}{1-x} = 1 + x + x^2 + \cdots + x^n + \cdots \qquad -1 < x < 1 \qquad \textbf{(1)}$$

By part (c) of theorem (11.9.1) we may integrate (1), obtaining

$$\int_0^x \frac{1}{1-t}\,dt = x + \frac{x^2}{2} + \frac{x^3}{3} + \cdots + \frac{x^{n+1}}{n+1} + \cdots$$

But

$$\int_0^x \frac{1}{1-t}\,dt = -\ln(1-t)\Big|_0^x = -\ln(1-x) = \ln\frac{1}{1-x}$$

Hence,

$$\ln\frac{1}{1-x} = x + \frac{x^2}{2} + \frac{x^3}{3} + \cdots + \frac{x^{n+1}}{n+1} + \cdots \qquad \textbf{(2)}$$

It is easy to verify that the interval of convergence of the power series in (2) is $-1 \le x < 1$. In fact, with $x = -1$ we obtain

$$\ln\tfrac{1}{2} = -1 + \tfrac{1}{2} - \tfrac{1}{3} + \tfrac{1}{4} - \cdots$$

Since $\ln\frac{1}{2} = -\ln 2$, we have found that the sum of the alternating harmonic series is ln 2; that is,

$$\mathbf{\ln 2 = 1 - \tfrac{1}{2} + \tfrac{1}{3} - \tfrac{1}{4} + \cdots} \qquad \textbf{(3)}$$

EXAMPLE 3 *Gregory's Series.*

Develop the series

$$\tan^{-1}x = x - \frac{x^3}{3} + \frac{x^5}{5} - \frac{x^7}{7} + \cdots + (-1)^n \frac{x^{2n+1}}{2n+1} + \cdots \qquad -1 \le x \le 1 \tag{4}$$

This is called *Gregory's series.*

Solution

In the geometric series (1), replace x by $-x^2$. The result is

$$\frac{1}{1+x^2} = 1 - x^2 + x^4 - x^6 + \cdots \qquad -1 < x < 1 \tag{5}$$

Hence, by part (c) of theorem (11.9.1), we may integrate (5) to obtain

$$\int_0^x \frac{dt}{1+t^2} = \int_0^x (1 - t^2 + t^4 - \cdots)\, dt$$

Thus, we have

$$\tan^{-1}x = x - \frac{x^3}{3} + \frac{x^5}{5} - \frac{x^7}{7} + \cdots + (-1)^n \frac{x^{2n+1}}{2n+1} + \cdots$$

The radius of convergence of the series above is $R = 1$. To find the interval of convergence, we test $x = -1$ and $x = 1$. For $x = -1$, we get

$$-1 + \tfrac{1}{3} - \tfrac{1}{5} + \tfrac{1}{7} - \cdots$$

For $x = 1$, we get

$$1 - \tfrac{1}{3} + \tfrac{1}{5} - \tfrac{1}{7} + \cdots$$

Each of these is a form of the alternating harmonic series, which converges. Hence, the interval of convergence of (4) is $-1 \le x \le 1$.

As the preceding examples illustrate, it is often possible to obtain a power series representation for a given function by starting with a known series and differentiating, integrating, or substituting in it. We now give a more direct method for finding a power series representation for a large class of functions.

Taylor and Maclaurin Series

Consider the power series

$$\sum_{k=0}^{+\infty} a_k(x-c)^k = a_0 + a_1(x-c) + a_2(x-c)^2 + \cdots + a_n(x-c)^n + \cdots \tag{6}$$

Suppose its interval of convergence is $(c-R, c+R)$, $0 < R \le +\infty$. For a given function f, we show how to obtain a power series representation. We begin by assuming* that for $c - R < x < c + R$, the function f

---

* The conditions under which this assumption is justified are considered later.

is actually represented by the power series (6). That is,

$$f(x) = \sum_{k=0}^{+\infty} a_k(x-c)^k = a_0 + a_1(x-c) + a_2(x-c)^2 + \cdots + a_n(x-c)^n + \cdots \tag{7}$$

We may obtain the coefficients $a_0, a_1, \ldots$ in terms of f and its derivatives by using part (b) of theorem (11.9.1).

$$f'(x) = \sum_{k=1}^{+\infty} ka_k(x-c)^{k-1} = a_1 + 2a_2(x-c) + 3a_3(x-c)^2 + \cdots$$

$$f''(x) = \sum_{k=2}^{+\infty} k(k-1)a_k(x-c)^{k-2} = 2a_2 + 6a_3(x-c) + 12a_4(x-c)^2 + \cdots$$

$$f'''(x) = \sum_{k=3}^{+\infty} k(k-1)(k-2)a_k(x-c)^{k-3} = 6a_3 + 24a_4(x-c) + \cdots$$

and for any positive integer n,

$$\begin{aligned} f^{(n)}(x) &= \sum_{k=n}^{+\infty} k(k-1)(k-2)\cdots(k-n+1)a_k(x-c)^{k-n} \\ &= n!a_n + (n+1)!a_{n+1}(x-c) + \cdots \end{aligned}$$

Now we set $x = c$ in each of these equations:

$$\begin{array}{ll} f(c) = a_0 & a_0 = f(c) \\ f'(c) = a_1 & a_1 = f'(c) \\ f''(c) = 2a_2 & a_2 = \dfrac{f''(c)}{2!} \\ f'''(c) = 3!a_3 & a_3 = \dfrac{f'''(c)}{3!} \\ \vdots & \vdots \\ f^{(n)}(c) = n!a_n & a_n = \dfrac{f^{(n)}(c)}{n!} \end{array}$$

Hence, by substituting in (7), we obtain

$$f(x) = f(c) + f'(c)(x-c) + \frac{f''(c)}{2!}(x-c)^2 + \cdots + \frac{f^{(n)}(c)}{n!}(x-c)^n + \cdots$$

And we have proved the following result:

[11.9.2] THEOREM / *Taylor Series.*

If a function f can be represented by the power series $\sum_{k=0}^{+\infty} a_k(x-c)^k$ whose radius of convergence is R, then

$$f(x) = f(c) + f'(c)(x-c) + \frac{f''(c)(x-c)^2}{2!} + \cdots + \frac{f^{(n)}(c)(x-c)^n}{n!} + \cdots \tag{8}$$

for all x such that $c - R < x < c + R$. This series representation for f is called a *Taylor series in* $(x - c)$ of the function f.

We have shown that if f has a series representation of the form (7) then it must be the Taylor series (8). That is, there is only one power series in $(x - c)$ that defines a function f.

[11.9.3] DEFINITION / *Maclaurin Series.*

When $c = 0$, the Taylor series becomes

$$f(x) = f(0) + f'(0)x + \frac{f''(0)x^2}{2!} + \cdots + \frac{f^{(n)}(0)x^n}{n!} + \cdots \qquad \textbf{(9)}$$

This series representation of f is called a *Maclaurin series.*

Note that in a Maclaurin series all the derivatives that appear are evaluated at 0, and the interval of convergence of this series has its center at 0.

The Taylor series in $(x - c)$ of a function f is usually referred to as the *Taylor expansion of f about c*; the Maclaurin series of a function f is the *Taylor expansion of f about* 0.

EXAMPLE 4

Assuming that $f(x) = e^x$ can be represented by a power series in x, find its Maclaurin series.

Solution

To find the Maclaurin series of f, we must evaluate f and its derivatives at 0:

$$\begin{array}{ll} f(x) = e^x & f(0) = 1 \\ f'(x) = e^x & f'(0) = 1 \\ f''(x) = e^x & f''(0) = 1 \\ \vdots & \vdots \end{array}$$

Hence, by (9), we find

$$f(x) = e^x = 1 + x + \frac{x^2}{2!} + \frac{x^3}{3!} + \cdots + \frac{x^n}{n!} + \cdots = \sum_{k=0}^{+\infty} \frac{x^k}{k!}$$

We have shown in Example 4 that if e^x can be represented by a power series, then the power series takes the form $\sum_{k=0}^{+\infty}(x^k/k!)$. In Example 1, part (a), in Section 11.8 (p. 681), we showed that this series converges absolutely for all x. What we have not demonstrated yet, but soon will, is that this series actually converges to e^x for all x.

The conditions on a function f that guarantee that its power series representation actually converges to f are obtained as follows: We begin with the observation that the $(n + 1)$st partial sum of the Taylor series in $(x - c)$ of f is the Taylor polynomial $P_n(x)$ of degree n for f at c (see (1) in Section 10.4). As a result,

$$P_n(x) = f(x) - R_{n+1}(x)$$

where the remainder R_{n+1} is

$$R_{n+1}(x) = \frac{f^{(n+1)}(u)(x-c)^{n+1}}{(n+1)!} \tag{10}$$

for some u between c and x. (The number u depends on n.)

The next result uses this remainder R_{n+1} to establish sufficient conditions for the power series representation of a function f to actually converge to f.

[11.9.4] THEOREM / *Convergence of Taylor Series.*

If a function f has derivatives of all orders in some interval $(c-R, c+R)$ and if

$$\lim_{n\to+\infty} R_{n+1}(x) = 0$$

for all x in $(c-R, c+R)$, then

$$f(x) = \sum_{k=0}^{+\infty} \frac{f^{(k)}(c)}{k!}(x-c)^k$$

$$= f(c) + f'(c)(x-c) + \frac{f''(c)}{2!}(x-c)^2 + \cdots + \frac{f^{(n)}(c)}{n!}(x-c)^n + \cdots$$

for all x in $(c-R, c+R)$.

Proof If f has derivatives of all orders in the interval $(c-R, c+R)$, then $P_n(x) = f(x) - R_{n+1}(x)$ is the $(n+1)$st partial sum of the infinite series $\sum_{k=0}^{+\infty}[f^{(k)}(c)/k!](x-c)^k$. If we also have $\lim_{n\to+\infty} R_{n+1}(x) = 0$, then

$$\lim_{n\to+\infty} P_n(x) = \lim_{n\to+\infty} [f(x) - R_{n+1}(x)] = f(x)$$

and the series $\sum_{k=0}^{+\infty}[f^{(k)}(c)/k!](x-c)^k$ converges to f.

Although at first glance it may appear that the theorem is easy to use, in practice it is not always easy to show that $\lim_{n\to+\infty} R_{n+1}(x) = 0$. One of the reasons for this is that the term $f^{(n+1)}(u)$ that appears in R_{n+1} depends on n, making the limit difficult to evaluate. Let's look at an example that illustrates this.

EXAMPLE 5

Prove that the series developed in Example 4 converges to e^x for every number x. That is, prove that

$$e^x = 1 + \frac{x}{1!} + \frac{x^2}{2!} + \frac{x^3}{3!} + \cdots + \frac{x^n}{n!} + \cdots \qquad \text{for all } x \tag{11}$$

Solution

In order to prove that (11) is actually true for every x, we have to verify that $\lim_{n\to+\infty} R_{n+1}(x) = 0$. Since $f^{(n+1)}(x) = e^x$,

$$R_{n+1}(x) = \frac{x^{n+1}}{(n+1)!}e^u \qquad \text{where } u \text{ is between 0 and } x$$

To show that $\lim_{n\to+\infty} R_{n+1}(x) = 0$, we consider two cases: (a) $x > 0$ and (b) $x < 0$.

(a) When $x > 0$, we have $0 < u < x$, so that $e^u < e^x$ and, hence, for every positive integer n,

$$0 < R_{n+1}(x) = \frac{x^{n+1}e^u}{(n+1)!} < \frac{e^x x^{n+1}}{(n+1)!}$$

Now, by the ratio test (11.6.1), the series $\sum_{k=0}^{+\infty}[x^{k+1}/(k+1)!]$ converges for all x. Therefore, from theorem (11.3.1), it follows that $\lim_{n\to+\infty}[x^{n+1}/(n+1)!] = 0$. Hence,

$$\lim_{n\to+\infty} \frac{e^x x^{n+1}}{(n+1)!} = e^x \lim_{n\to+\infty} \frac{x^{n+1}}{(n+1)!} = 0$$

so that, by the squeezing theorem (2.5.1), $\lim_{n\to+\infty} R_{n+1}(x) = 0$.

(b) When $x < 0$, we have $x < u < 0$ and $0 < e^u < 1$, so that

$$|R_{n+1}(x)| = \frac{|x|^{n+1}}{(n+1)!} e^u < \frac{|x|^{n+1}}{(n+1)!}$$

The right member approaches 0 as $n \to +\infty$, so that by the squeezing theorem, $\lim_{n\to+\infty} R_{n+1}(x) = 0$. Thus, by theorem (11.9.4), the Maclaurin series representation (11) for e^x is true for all x, and we can write

$$e^x = 1 + x + \frac{x^2}{2!} + \frac{x^3}{3!} + \cdots + \frac{x^n}{n!} + \cdots = \sum_{k=0}^{+\infty} \frac{x^k}{k!} \qquad \text{for all } x \tag{12}$$

In particular, if $x = 1$, we find

$$e = 1 + 1 + \frac{1}{2!} + \frac{1}{3!} + \cdots + \frac{1}{n!} + \cdots = \sum_{k=0}^{+\infty} \frac{1}{k!} \tag{13}$$

There are functions whose Taylor series converge for all x in some interval—but to some other function. An example is

$$f(x) = \begin{cases} e^{-1/x^2} & \text{if } x \neq 0 \\ 0 & \text{if } x = 0 \end{cases} \tag{14}$$

With the aid of L'Hospital's rule (10.1.1), we can show that $f'(0) = 0$ (see Problem 53 in Exercise 10.2, p. 600). Although it is much more difficult, it also can be shown that f has derivatives of all orders at 0 and that $f^{(n)}(0) = 0$ for all $n \geq 1$. Hence, the Maclaurin series for f is

$$0 + 0(x) + 0(x)^2 + \cdots + 0(x^n) + \cdots$$

This series converges for every x and its sum is obviously the zero function—not f. Consequently, the series has $f(x)$ as its sum only when $x = 0$ and for no other x. This illustrates that convergence of the Taylor series of a function is not sufficient to ensure that the series converges to the function;

the remainder R_{n+1} must tend to 0 *as* $n \to +\infty$. In fact, for the function f defined by (14), it follows from the above discussion that R_{n+1} does not approach 0 as $n \to +\infty$.

EXERCISE 11.9

In Problems 1–18 use theorem (11.9.2) to find the Taylor expansion of each function about the given number c, under the assumption that such an expansion exists.

1. $f(x) = \sin x; \quad c = 0$

2. $f(x) = \cos x; \quad c = 0$

3. $f(x) = e^x; \quad c = 1$

4. $f(x) = e^{2x}; \quad c = -1$

5. $f(x) = \sin x; \quad c = \pi/4$

6. $f(x) = \cos x; \quad c = \pi/3$

7. $f(x) = \dfrac{1}{x}; \quad c = 1$

8. $f(x) = \dfrac{1}{\sqrt{x}}; \quad c = 4$

9. $f(x) = \dfrac{1}{(1+x)^2}; \quad c = 0$

10. $f(x) = \dfrac{1}{1+x^2}; \quad c = 0$

11. $f(x) = \dfrac{1}{1-3x}; \quad c = 0$

12. $f(x) = \dfrac{1}{1+2x^3}; \quad c = 0$

13. $f(x) = \ln(1+x); \quad c = 0$

14. $f(x) = \ln x; \quad c = 1$

15. $f(x) = 3x^3 + 2x^2 + 5x - 6; \quad c = 0$

16. $f(x) = 4x^4 - 2x^3 - x; \quad c = 0$

17. $f(x) = 3x^3 + 2x^2 + 5x - 6; \quad c = 1$

18. $f(x) = 4x^4 - 2x^3 + x; \quad c = 1$

19. In the geometric series (1), replace x by x^2 to obtain the power series representation for $1/(1 - x^2)$. What is its interval of convergence?

20. Integrate the power series of Problem 19 to obtain the power series for $\frac{1}{2}\ln[(1+x)/(1-x)]$. What is its interval of convergence?

21. Use the power series found in Problem 20 to get an approximation for ln 2 accurate to within 0.001.

22. Use Gregory's series (4) to obtain a series for $\pi/4$. How many terms of this series are required to obtain an approximation for π accurate to within 0.0001?

23. By letting $x = 1$ in Gregory's series (4) we have

$$\tan^{-1} 1 = 1 - \frac{1}{3} + \frac{1}{5} - \cdots$$

It seems possible to use it to obtain an approximation to π. However, since this series converges very slowly (see Problem 22), it is next to useless for this purpose. A more rapidly convergent series is obtained by using the identity

$$\tan^{-1} 1 = \tan^{-1}\tfrac{1}{2} + \tan^{-1}\tfrac{1}{3}$$

Use $x = \frac{1}{2}$ and $x = \frac{1}{3}$ in Gregory's series, together with this identity, to get an approximation for π accurate to within 0.0001.

24. Leibniz derived the following formula for $\pi/4$: $\pi/4 = 1 - \frac{1}{3} + \frac{1}{5} - \frac{1}{7} + \frac{1}{9} - \cdots$.

(a) Evaluate: $\displaystyle\int_0^1 \frac{1}{1+x^2}\,dx$

(b) Expand the integrand in part (a) into a power series and integrate it term-by-term to get Leibniz's formula.

(c) Find the sum of the first 10 terms in the above series. Does it appear that Leibniz's formula is a useful one for approximating π?

25. Euler believed that $\frac{1}{2} = 1 - 1 + 1 - 1 + 1 - 1 + \cdots$. He based his argument to support this equation on his belief in the identification of a series and the values of the function from which it was derived.

(a) Write the Maclaurin series expansion for $1/(1+x)$. Do this without calculating any derivatives.

(b) Evaluate both sides of the equation you derived in part (a) at $x = 1$ to arrive at the formula above.

(c) Criticize the procedure in part (b).

26. (a) Write the first three nonzero terms and the general term of the Taylor series expansion about $x = 0$ of $f(x) = 3\sin(x/2)$.

(b) What is the interval of convergence for the series found in part (a)?

(c) What is the minimum number of terms of the series in part (a) that are *necessary* to approximate f on the interval $(-2, 2)$ with an error not exceeding 0.1?

11.10 Other Important Series

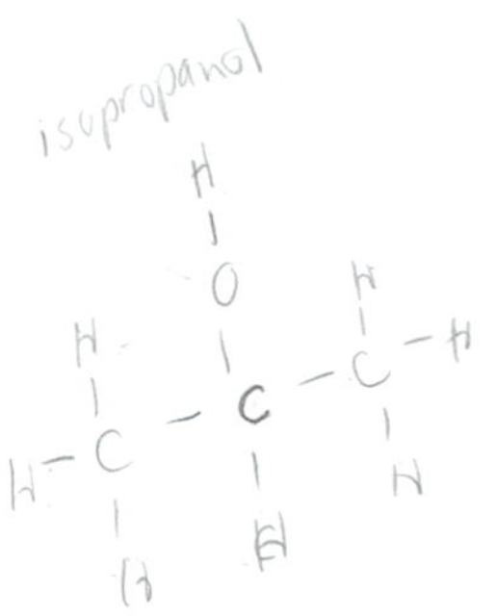

In Section 11.9 we saw that the Maclaurin series representation for $f(x) = e^x$,

$$e^x = 1 + x + \frac{x^2}{2!} + \frac{x^3}{3!} + \cdots + \frac{x^n}{n!} + \cdots \qquad \textbf{(1)}$$

converges and equals e^x for all real numbers x.

In this section we develop the Maclaurin series representation for $\sin x$, $\cos x$, $\cosh x$, and some other functions.

Series for sin *x*

For $f(x) = \sin x$, the value of f and its derivatives at 0 are

$$\begin{aligned} f(x) &= \sin x & f(0) &= 0 \\ f'(x) &= \cos x & f'(0) &= 1 \\ f''(x) &= -\sin x & f''(0) &= 0 \\ f'''(x) &= -\cos x & f'''(0) &= -1 \end{aligned}$$

Successive derivatives follow this same pattern, so that for $n=0, 1, 2, 3, \ldots,$ we have

$$\begin{aligned} f^{(2n)}(x) &= (-1)^n \sin x & f^{(2n)}(0) &= 0 \\ f^{(2n+1)}(x) &= (-1)^n \cos x & f^{(2n+1)}(0) &= (-1)^n \end{aligned}$$

By Taylor's formula with remainder (10.4.1), we find that

$$\sin x = x - \frac{x^3}{3!} + \frac{x^5}{5!} - \frac{x^7}{7!} + \cdots + (-1)^n \frac{x^{2n+1}}{(2n+1)!} + R_{2n+3}(x) \qquad \textbf{(2)}$$

At this point, we know that if $\sin x$ can be represented by a power series in x, then it is given by (2). To prove that the series on the right of (2) actually converges to $\sin x$ for all x, we need to show that

$$\lim_{n \to +\infty} R_{2n+3}(x) = \lim_{n \to +\infty} (-1)^{n+1} \frac{f^{(2n+3)}(u) x^{2n+3}}{(2n+3)!} = 0 \qquad \text{for all } x$$

where u is between 0 and x. Since $|f^{(2n+3)}(u)| = |\cos u| \le 1$ for every number u, we see that

$$|R_{2n+3}(x)| = \frac{|f^{(2n+3)}(u)|}{(2n+3)!} |x|^{2n+3} \le \frac{|x|^{2n+3}}{(2n+3)!}$$

But by the ratio test (11.6.1), the series $\sum_{k=0}^{+\infty} [|x|^{2k+3}/(2k+3)!]$ converges for all x. Hence, it follows by theorem (11.3.1) that

$$\lim_{n \to +\infty} \left[\frac{|x|^{2n+3}}{(2n+3)!} \right] = 0$$

so that by the squeezing theorem (2.5.1), $\lim_{n\to+\infty}|R_{2n+3}(x)| = 0$. Therefore, $\lim_{n\to+\infty} R_{2n+3}(x) = 0$. Thus, by theorem (11.9.4), the series converges to $\sin x$ for all x and we may write

$$\sin x = x - \frac{x^3}{3!} + \frac{x^5}{5!} - \cdots + (-1)^n \frac{x^{2n+1}}{(2n+1)!} + \cdots \qquad \text{for all } x \qquad (3)$$

From the discussion so far, it is clear that the task of finding the Taylor series representation of a function by taking successive derivatives and then showing that $\lim_{n\to+\infty} R_{2n+3}(x) = 0$ is a difficult one. Consequently, to minimize this difficulty, we make use of the series developed thus far and properties of theorem (11.9.1). For example, we can obtain the Taylor series for $f(x) = \cos x$ quite easily by merely differentiating the series for $\sin x$.

Series for cos *x*

By using part (b) of theorem (11.9.1), we can obtain the series for $\cos x$ by differentiating the series (3) term-by-term. The result is

$$\cos x = 1 - \frac{x^2}{2!} + \frac{x^4}{4!} - \cdots + (-1)^n \frac{x^{2n}}{(2n)!} + \cdots \qquad \text{for all } x \qquad (4)$$

We could have derived (4) in the same way we obtained (3). It could also have been derived by integrating (3) term-by-term.

Calculation of Logarithms

In Example 2 of Section 11.9 we discovered that

$$\ln \frac{1}{1-x} = x + \frac{x^2}{2} + \frac{x^3}{3} + \cdots + \frac{x^n}{n} + \cdots \qquad -1 \le x < 1 \qquad (5)$$

It might appear that we could use (5) to compute logarithms of numbers. Unfortunately, though, this series converges only for $-1 \le x < 1$. In addition, unless x is close to 0, the series converges so slowly that too many terms would be required for practical use. (Recall the rate of convergence of the alternating harmonic series, p. 667.)

We now develop a useful formula for computing logarithms. If we multiply (5) by (-1) and use the fact that $-\ln(1/A) = \ln(1/A)^{-1} = \ln A$, we find that

$$\ln(1-x) = -x - \frac{x^2}{2} - \frac{x^3}{3} - \cdots \qquad (6)$$

In (6), we replace x by $-x$. The result is

$$\ln(1+x) = x - \frac{x^2}{2} + \frac{x^3}{3} - \cdots \qquad (7)$$

Now, we subtract (6) from (7) to get

$$\ln \frac{1+x}{1-x} = 2\left(x + \frac{x^3}{3} + \frac{x^5}{5} + \cdots\right) \qquad (8)$$

This series converges for $-1 < x < 1$.

If N is a positive integer, we let $x = 1/(2N+1)$. Then $0 < x < 1$. A little computation shows that

$$\frac{1+x}{1-x} = \frac{N+1}{N}$$

By substituting in (8) and using some properties of logarithms, we get

$$\ln(N+1) = \ln N + 2\left[\frac{1}{2N+1} + \frac{1}{3}\left(\frac{1}{2N+1}\right)^3 + \frac{1}{5}\left(\frac{1}{2N+1}\right)^5 + \cdots\right] \qquad \textbf{(9)}$$

This series converges quite rapidly for all positive integers N.

For example, if $N = 1$, we find

$$\ln 2 = 2\left[\frac{1}{3} + \frac{1}{3}\left(\frac{1}{3}\right)^3 + \frac{1}{5}\left(\frac{1}{3}\right)^5 + \cdots\right]$$

$$\approx 2\left[\frac{1}{3} + \frac{1}{3}\left(\frac{1}{3}\right)^3 + \frac{1}{5}\left(\frac{1}{3}\right)^5\right] \approx 0.693004 \qquad \textbf{(10)}$$

The use of three terms of this series achieves accuracy to within 0.0002.

To get ln 3, we set $N = 2$ in (9) and use (10). By continuing in this way, a table of natural logarithms may be written.

Other Series

We can use the series arrived at in this section to obtain other power series representations. For example, if in (1), we replace x by $-x$, we obtain

$$e^{-x} = 1 - x + \frac{x^2}{2!} - \frac{x^3}{3!} + \cdots + (-1)^n\frac{x^n}{n!} + \cdots \qquad \text{for all } x \qquad \textbf{(11)}$$

Since

$$\cosh x = \frac{e^x + e^{-x}}{2}$$

its power series representation can be found by adding corresponding terms in (1) and (11) and then dividing by 2. The result is

$$\cosh x = 1 + \frac{x^2}{2!} + \frac{x^4}{4!} + \frac{x^6}{6!} + \cdots + \frac{x^{2n}}{(2n)!} + \cdots \qquad \text{for all } x \qquad \textbf{(12)}$$

We can obtain the series representation for $e^x\cos x$ by multiplying the series (1) by the series (4). That is,

$$e^x\cos x = \left(1 + x + \frac{x^2}{2!} + \frac{x^3}{3!} + \frac{x^4}{4!} + \cdots\right)\left(1 - \frac{x^2}{2!} + \frac{x^4}{4!} - \cdots\right)$$

For the sake of illustration, we list the terms up to and including the fifth power:

$$e^x \cos x = 1\left(1 - \frac{x^2}{2!} + \frac{x^4}{4!}\right) + x\left(1 - \frac{x^2}{2!} + \frac{x^4}{4!}\right) + \frac{x^2}{2!}\left(1 - \frac{x^2}{2!}\right)$$

$$+ \frac{x^3}{3!}\left(1 - \frac{x^2}{2!}\right) + \frac{x^4}{4!}(1) + \frac{x^5}{5!}(1) + \cdots$$

$$= \left(1 - \frac{x^2}{2} + \frac{x^4}{24}\right) + \left(x - \frac{x^3}{2} + \frac{x^5}{24}\right) + \left(\frac{x^2}{2} - \frac{x^4}{4}\right)$$

$$+ \left(\frac{x^3}{6} - \frac{x^5}{12}\right) + \frac{x^4}{24} + \frac{x^5}{120} + \cdots$$

$$= 1 + x + \left(-\frac{1}{2} + \frac{1}{2}\right)x^2 + \left(-\frac{1}{2} + \frac{1}{6}\right)x^3$$

$$+ \left(\frac{1}{24} - \frac{1}{4} + \frac{1}{24}\right)x^4 + \left(\frac{1}{24} - \frac{1}{12} + \frac{1}{120}\right)x^5 + \cdots$$

$$= 1 + x - \frac{1}{3}x^3 - \frac{1}{6}x^4 - \frac{1}{30}x^5 + \cdots$$

Numerical Techniques

As mentioned in Section 9.11, power series are sometimes useful for approximating definite integrals when the Second Fundamental Theorem of Calculus cannot be applied. Let's look at an illustration.

EXAMPLE 1

Using power series, approximate $\int_0^{1/2} e^{-x^2}\,dx$ accurate to within 0.001.

Solution

In (11), we replace x by x^2. The power series for e^{-x^2} is, therefore,

$$e^{-x^2} = 1 - x^2 + \frac{x^4}{2!} - \frac{x^6}{3!} + \frac{x^8}{4!} - \cdots$$

Then, by part (c) of theorem (11.9.1), we have

$$\int_0^{1/2} e^{-x^2}\,dx = \int_0^{1/2} \left(1 - x^2 + \frac{x^4}{2!} - \frac{x^6}{3!} + \frac{x^8}{4!} - \cdots\right) dx$$

$$= \left(x - \frac{x^3}{3} + \frac{x^5}{2!5} - \frac{x^7}{3!7} + \cdots\right)\Bigg|_0^{1/2}$$

$$= \frac{1}{2} - \frac{1}{3(2)^3} + \frac{1}{2!5(2)^5} - \frac{1}{3!7(2)^7} + \cdots$$

$$= 0.5 - 0.041666 + 0.003125 - 0.00019 + \cdots$$

Since this is an alternating series, the error due to using the first three terms as an approximation is less than 0.00019 (see theorem (11.5.3)). Thus, by summing the first three terms, we have

$$\int_0^{1/2} e^{-x^2}\,dx \approx 0.46146$$

to within 0.001.

Note that only three terms were needed to obtain the desired accuracy. To get this same accuracy by Simpson's rule or the trapezoidal rule would have required a partition of $[0, \frac{1}{2}]$ so fine that even the use of a computer would be more time-consuming and, of course, more costly than merely adding three terms. Thus, efficiency of technique is an important consideration for solving numerical problems.

We close this section with an illustration of one procedure for actually finding the sum of an infinite series.

EXAMPLE 2

Find the exact sum of the series below.

$$\frac{1}{1!3} + \frac{1}{2!4} + \frac{1}{3!5} + \cdots + \frac{1}{n!(n+2)} + \cdots$$

Solution

We start with

$$e^x = 1 + x + \frac{x^2}{2!} + \frac{x^3}{3!} + \cdots$$

We multiply this by x and integrate from 0 to 1 to get

$$\int_0^1 xe^x\,dx = \int_0^1 \left(x + x^2 + \frac{x^3}{2!} + \frac{x^4}{3!} + \cdots\right) dx = \frac{1}{2} + \left(\frac{1}{1!3} + \frac{1}{2!4} + \frac{1}{3!5} + \cdots\right)$$

Since $\int_0^1 xe^x\,dx = 1$ (use integration by parts), we find that

$$\frac{1}{1!3} + \frac{1}{2!4} + \frac{1}{3!5} + \cdots + \frac{1}{n!(n+2)} + \cdots = 1 - \frac{1}{2} = \frac{1}{2}$$

EXERCISE 11.10

In Problems 1–8 find a Maclaurin series that converges to each function.

1. e^{2x} **2.** $\sin^3 x$ **3.** $\ln(1 + x^2)$ **4.** e^{-x^2}

5. $\sinh x$ **6.** $\sin x^2$ **7.** $e^x \sin x$ **8.** $e^{-x}\cos x$

In Problems 9 and 10 use a Maclaurin series for f to get one for g. The first four nonzero terms will be sufficient.

9. $f(x) = \dfrac{1}{\sqrt{1 - x^2}}$; $g(x) = \sin^{-1} x$

10. $f(x) = \tan x$; $g(x) = \ln(\cos x)$

In Problems 11–16 approximate each expression accurate to within 0.0001 by using power series.

11. $\int_0^1 \sin x^2\,dx$

12. $\int_0^1 e^{-x^2}\,dx$

13. $\cos 0.1$

14. $\sin 0.1$

15. $\int_0^1 \frac{\sin x}{x}\,dx$

16. $\int_0^{0.1} \frac{\ln(1+x)}{x}\,dx$

In Problems 17–20 find each limit by means of series.

17. $\lim_{x\to 0} \frac{\sin x - x}{x^2}$

18. $\lim_{x\to 0} \frac{\sin x}{x}$

19. $\lim_{x\to 0} \frac{x \sin x}{1-\cos x}$

20. $\lim_{h\to 0} \frac{e^h - 1}{h}$

21. Find the exact sum of the infinite series below.

$$\frac{x^3}{1(3)} - \frac{x^5}{3(5)} + \frac{x^7}{5(7)} - \frac{x^9}{7(9)} + \cdots \quad \text{for} \quad x = 1$$

22. Obtain the series (4) for $\cos x$ by integrating the series (3) for $\sin x$.

23. Obtain the series (4) for $\cos x$ in the same way as that of the series (3) for $\sin x$.

24. Use the result (9) to show that $\ln 3 \approx 1.09861$.

25. Use series to approximate $\int_0^{0.16}[(\sin x)/\sqrt{x}]\,dx$ to within 10^{-3}.

11.11 The Binomial Series

In algebra we study the binomial theorem, which states that if m is a positive integer and a, b are real numbers, then

$$(a+b)^m = a^m + ma^{m-1}b + \cdots + b^m = \sum_{k=0}^{m} \binom{m}{k} a^{m-k}b^k$$

where

$$\binom{m}{k} = \frac{m(m-1)\cdots(m-k+1)}{k!}$$

In particular, for $a = 1$, $b = x$, and m a positive integer, this becomes

$$(1+x)^m = 1 + mx + \frac{m(m-1)}{2}x^2 + \cdots + mx^{m-1} + x^m \tag{1}$$

In this section we obtain a Maclaurin series for the function $f(x) = (1+x)^m$, where *m is any real number*.

We begin by investigating the derivatives of f at 0:

$$\begin{aligned}
f(x) &= (1+x)^m & f(0) &= 1 \\
f'(x) &= m(1+x)^{m-1} & f'(0) &= m \\
f''(x) &= m(m-1)(1+x)^{m-2} & f''(0) &= m(m-1) \\
&\vdots & &\vdots \\
f^{(n)}(x) &= m(m-1)(m-2)\cdots & f^{(n)}(0) &= m(m-1)(m-2)\cdots \\
&\quad (m-n+1)(1+x)^{m-n} & &\quad (m-n+1)
\end{aligned}$$

The Maclaurin series is

$$(1+x)^m = 1 + mx + \frac{m(m-1)}{2!}x^2 + \cdots + \binom{m}{n}x^n + \cdots = \sum_{k=0}^{+\infty}\binom{m}{k}x^k \quad \textbf{(2)}$$

where

$$\binom{m}{k} = \frac{m(m-1)\cdots(m-k+1)}{k!} \qquad k \neq 0^* \quad \textbf{(3)}$$

Prompted by (1), we call the series given by (2) the *binomial series*. The expression in (3) is called the *binomial coefficient of* x^k.

The following result, which we state without proof, gives the conditions under which the infinite series (2) actually converges to $(1+x)^m$:

[11.11.1] THEOREM

$$(1+x)^m = \sum_{k=0}^{+\infty}\binom{m}{k}x^k$$

(a) for all x if m is a nonnegative integer
(b) for $-1 < x < 1$ if $m \le -1$
(c) for $-1 < x \le 1$ if $-1 < m < 0$
(d) for $-1 \le x \le 1$ if $m > 0$, m not an integer

EXAMPLE 1

Represent $\sqrt{1+x}$ in a Maclaurin series and find the interval of convergence.

Solution

We use the binomial series with $m = \frac{1}{2}$. The result is

$$(1+x)^{1/2} = 1 + \frac{1}{2}x + \frac{(\frac{1}{2})(-\frac{1}{2})}{2!}x^2 + \frac{\frac{1}{2}(-\frac{1}{2})(-\frac{3}{2})}{3!}x^3 + \cdots$$

By part (d) of theorem (11.11.1), the series converges for $-1 \le x \le 1$.

EXAMPLE 2

Represent the function $f(x) = \sin^{-1}x$ by a Maclaurin series and find the interval of convergence.

Solution

We observe that

$$\sin^{-1}x = \int_0^x \frac{dt}{\sqrt{1-t^2}} = \int_0^x (1-t^2)^{-1/2}\,dt$$

* We define $\binom{m}{0} = 1$ for any m.

By expanding the integrand in a binomial series, we find

$$(1-t^2)^{-1/2} = 1 + \frac{t^2}{2} + \left(\frac{1}{2}\right)\left(\frac{3}{2}\right)\left(\frac{t^4}{2!}\right) + \left(\frac{1}{2}\right)\left(\frac{3}{2}\right)\left(\frac{5}{2}\right)\left(\frac{t^6}{3!}\right) + \cdots$$

which is valid for $-1 < t \leq 1$. Integrating the right side term-by-term, we obtain

$$\sin^{-1} x = x + \left(\frac{1}{2}\right)\left(\frac{x^3}{3}\right) + \left(\frac{1}{2}\right)\left(\frac{3}{4}\right)\left(\frac{x^5}{5}\right) + \left(\frac{1}{2}\right)\left(\frac{3}{4}\right)\left(\frac{5}{6}\right)\left(\frac{x^7}{7}\right) + \cdots \qquad -1 < x \leq 1$$

Summary

Table 5 summarizes information about the series discussed in the previous sections.

Table 5 Important Series

Name	Description	Comment
Taylor series	$f(x) = \sum_{k=0}^{n} \frac{f^{(k)}(c)(x-c)^k}{k!} + R_{n+1}(x)$	Converges to $f(x)$ if $\lim_{n\to+\infty} R_{n+1}(x) = 0$
Remainder formula	$R_{n+1}(x) = \frac{f^{(n+1)}(u)(x-c)^{n+1}}{(n+1)!}$	u between c and x
$\frac{1}{1-x}$	$\frac{1}{1-x} = \sum_{k=0}^{+\infty} x^k$	Converges to $1/(1-x)$ for $-1 < x < 1$
e^x	$e^x = \sum_{k=0}^{+\infty} \frac{x^k}{k!}$	Converges to e^x for all x
$\sin x$	$\sin x = \sum_{k=0}^{+\infty} \frac{(-1)^k x^{2k+1}}{(2k+1)!}$	Converges to $\sin x$ for all x
$\cos x$	$\cos x = \sum_{k=0}^{+\infty} \frac{(-1)^k x^{2k}}{(2k)!}$	Converges to $\cos x$ for all x
$\ln(1+x)$	$\ln(1+x) = \sum_{k=0}^{+\infty} \frac{(-1)^k x^{k+1}}{k+1}$	Converges to $\ln(1+x)$ for $-1 < x \leq 1$
$\ln(N+1)$	$\ln(N+1) = \ln N + 2\sum_{k=1}^{+\infty} \frac{1}{2k-1}\left(\frac{1}{2N+1}\right)^{2k-1}$	N is a positive integer; converges for all N
$(1+x)^m$	$(1+x)^m = \sum_{k=0}^{+\infty} \binom{m}{k} x^k$	For convergence information, see theorem (11.11.1)

EXERCISE 11.11

In Problems 1–6 use a binomial series to represent each function. Determine the interval of convergence.

1. $\sqrt{1+x^2}$ **2.** $\frac{1}{\sqrt{1-x}}$ **3.** $(1+x)^{1/5}$ **4.** $\frac{1}{1+x^2}$ **5.** $\frac{1}{(1+x)^{3/4}}$ **6.** $\frac{2x}{\sqrt{1-x}}$

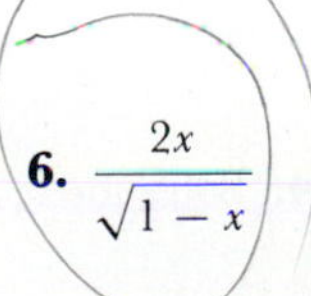

In Problems 7–10 approximate each definite integral to within 0.001.

7. $\int_0^{0.2} \sqrt[3]{1+x^4}\,dx$

8. $\int_0^{1/2} \sqrt[3]{1+x}\,dx$

9. $\int_0^{1/2} \frac{1}{\sqrt[3]{1+x^2}}\,dx$

10. $\int_0^{1/2} \frac{1}{\sqrt{1+x^3}}\,dx$

11. Show that $[1+(1/n)]^n < e$ for all $n > 0$.

12. Prove that $(1+x)^m = \sum_{k=0}^{+\infty} \binom{m}{k} x^k$ when m is a nonnegative integer by showing that $R_{n+1}(x) \to 0$ as $n \to +\infty$.

13. Show that the series $\sum_{k=0}^{+\infty} \binom{m}{k} x^k$ converges absolutely for $|x| < 1$ and diverges for $|x| > 1$. [*Hint:* Use the ratio test.]

REVIEW EXERCISES

In Problems 1–22 test each series for convergence (absolute or conditional) or divergence.

1. $\sum_{k=1}^{+\infty} \sin \frac{(-1)^k}{k^2}$

2. $\sum_{k=1}^{+\infty} \sin \frac{(-1)^k}{k}$

3. $\sum_{k=1}^{+\infty} \frac{\sin^{-1}\frac{1}{k}}{\frac{1}{k}}$

4. $\sum_{k=2}^{+\infty} \frac{\ln k}{k}$

5. $\sum_{k=1}^{+\infty} \ln\left(1+\frac{1}{k}\right)$

6. $\sum_{k=1}^{+\infty} \left(1-\cos\frac{(-1)^k}{k}\right)$

7. $\sum_{k=1}^{+\infty} \frac{1}{(k+a)^2}$, a real, not a negative integer

8. $\sum_{k=2}^{+\infty} \frac{(-1)^k}{\sqrt[p]{k^3+1}}$, $p > 0$

9. $\sum_{k=1}^{+\infty} \left(\sin\frac{1}{k} - \frac{1}{k}\right)$

10. $\sum_{k=1}^{+\infty} \frac{1}{\left(1+\frac{k^2+1}{k^2}\right)^k}$

11. $\sum_{k=1}^{+\infty} \frac{2\cdot 4\cdot 6\cdots(2k)}{1\cdot 3\cdot 5\cdots(2k-1)}$

12. $\sum_{k=1}^{+\infty} \left(\frac{1}{k^2}\right)^{\cos^2(k\pi/4)}$

13. $\sum_{k=1}^{+\infty} \frac{1}{k\ln\left(1+\frac{1}{k}\right)}$

14. $\sum_{k=1}^{+\infty} \frac{k^2}{(1+k^3)\ln\sqrt[3]{1+k^3}}$

15. $\sum_{k=1}^{+\infty} \frac{k^a}{2^k}$, a real

16. $\sum_{k=1}^{+\infty} \frac{k!}{k^k}$

17. $\sum_{k=1}^{+\infty} [\sqrt{k+1}-\sqrt{k}]$

18. $\sum_{k=2}^{+\infty} (-1)^k k^{(1-k)/k}$

19. $\sum_{k=1}^{+\infty} \frac{\left(1+\frac{1}{k^2}\right)^{k2}}{2^k}$

20. $\sum_{k=1}^{+\infty} \frac{2^k + k^{2001}}{3^k}$

21. $\sum_{k=2}^{+\infty} \frac{(-1)^k}{(\ln k)^{\ln k}}$

22. $\sum_{k=1}^{+\infty} c_k$, where $c_k = \begin{cases} \frac{1}{a^k} & \text{if } k \text{ is even} \\ \frac{-1}{b^k} & \text{if } k \text{ is odd} \end{cases}$ $a, b > 1$

23. Find the sum of the geometric series

$$(\sqrt{2}+1) + 1 + (\sqrt{2}-1) + \cdots$$

24. Solve for x: $\frac{x}{2+2x} = x + x^2 + x^3 + \cdots$

25. Show that the sum of any convergent geometric series whose first term and common ratio are rational is rational.

26. Suppose $\sum_{k=1}^{+\infty} a_k$ converges. Assume $a_n \neq 0$ for all n.

Show that $\sum_{k=1}^{+\infty} \frac{a_k}{1+a_k}$ converges.

27. Show that a convergent sequence $\{a_n\}$ cannot have two distinct limits.

In Problems 28–36 determine the interval of convergence.

28. $\sum_{k=0}^{+\infty} x^{k^2}$

29. $\sum_{k=1}^{+\infty} \frac{(x-3)^{3k-1}}{k^2}$

30. $\sum_{k=1}^{+\infty} \frac{x^k}{\sqrt[3]{k}}$

31. $\sum_{k=0}^{+\infty} (-1)^k \frac{1}{k!(k+1)!}\left(\frac{x}{2}\right)^{2k+1}$

32. $\sum_{k=0}^{+\infty} \frac{2^k}{k!} x^k$

33. $\sum_{k=1}^{+\infty} \frac{k^k}{(k!)^2} x^k$

34. $\sum_{k=1}^{+\infty} \frac{k^a}{a^k}(x-a)^k, \quad a \neq 0$

35. $\sum_{k=0}^{+\infty} \frac{(k!)^2}{(2k)!}(x-1)^k$

36. $\sum_{k=0}^{+\infty} \frac{\sqrt{k!}}{(2k)!} x^k$

37. Suppose $\sum_{k=0}^{+\infty} a_k x^k$ converges for $|x| < R$ and that $\lim_{k\to+\infty} \left|\frac{a_{k+1}}{a_k}\right|$ exists. Show that $\sum_{k=1}^{+\infty} k a_k x^{k-1}$ and $\sum_{k=0}^{+\infty} \frac{a_k}{k+1} x^{k+1}$ also converge for $|x| < R$.

38. Prove property 3 on page 670.

39. Prove property 5 on page 671.

40. Prove property 6 on page 671.

CHALLENGE EXERCISES

1. Show that the following series converges:
$$1 + \frac{2}{2^2} + \frac{3}{3^3} + \frac{1}{4^4} + \frac{2}{5^5} + \frac{3}{6^6} + \cdots$$

2. Show that the series $\sum_{k=1}^{+\infty}(1/k^{1+1/k})$ diverges.

3. $\sum_{k=1}^{+\infty} \frac{k^2-3k-2}{k^2(k+1)^2}$; find the sum of the series if it converges.

4. $1 - 1 - \frac{1}{2} + \frac{1}{3} + \frac{1}{3} - \frac{1}{9} - \frac{1}{4} + \frac{1}{27} + \frac{1}{5} - \frac{1}{81} - \cdots$; find the sum of the series if it converges.

5. Show that: $\sum_{k=1}^{+\infty} \frac{k}{(k+1)!} = 1$

6. Find the interval of convergence of $\sum_{k=1}^{+\infty} k^x$.

7. Find the interval of convergence of the series: $\sum_{k=1}^{+\infty} \frac{(x-2)^k}{k(3^k)}$

8. Let $a_k = (-1)^{k+1} \int_0^{\pi/k} \sin kx\, dx$.

(a) Evaluate a_k.

(b) Show that the infinite series $\sum_{k=1}^{+\infty} a_k$ converges.

(c) Show that $1 \leq \sum_{k=1}^{+\infty} a_k \leq \frac{3}{2}$.

9. Consider the finite sum $S_n = \sum_{k=1}^{n}[1/(1+k^2)]$.

(a) By comparing S_n with an appropriate integral, prove that $S_n \leq \tan^{-1} n$ for $n \geq 1$.

(b) Use part (a) to deduce that $\sum_{k=1}^{+\infty}[1/(1+k^2)]$ exists.

(c) Prove that $\pi/4 \leq \sum_{k=1}^{+\infty}[1/(1+k^2)] \leq \pi/2$.

10. Let $s_n = 1/1! + 1/2! + \cdots + 1/n!, \quad n = 1, 2, 3, \ldots$

(a) Show that $n! \geq 2^{n-1}$.

(b) Show that $0 < s_n \leq 1 + \frac{1}{2} + (\frac{1}{2})^2 + \cdots + (\frac{1}{2})^{n-1}$.

(c) Show that $0 < s_n < s_{n+1} < 2$. Hence, conclude that there exists $S = \lim_{n\to+\infty} s_n \leq 2$.

(d) Let $t_n = [1 + (1/n)]^n$. Show that $t_n = 1 + 1 + (1/2!)[1-(1/n)] + (1/3!)[1-(1/n)][1-(2/n)] + \cdots + (1/n!)[1-(1/n)][1-(2/n)] \cdots [1-(n-1)/n] < s_n + 1$.

(e) Show that $0 < t_n < t_{n+1} < 3$. Hence, conclude that there exists $e = \lim_{n\to+\infty} t_n \leq 3$.

11. Prove that $\{(3^n + 5^n)^{1/n}\}$ converges.

12. From the fact that $\sin t \leq t$ for all $t \geq 0$, use integration repeatedly to prove
$$1 - \frac{x^2}{2!} \leq \cos x \leq 1 - \frac{x^2}{2!} + \frac{x^4}{4!} \quad \text{for all } x \geq 0$$

13. Find an elementary expression for $\sum_{k=1}^{+\infty}[x^{k+1}/k(k+1)]$. [*Hint:* Integrate the series for $\ln[1/(1-x)]$.]

An interesting relationship between the nth partial sum of the harmonic series and $\ln n$ was discovered by Euler. In particular, he showed that

$$\gamma = \lim_{n\to+\infty}\left(1 + \frac{1}{2} + \frac{1}{3} + \cdots + \frac{1}{n} - \ln n\right)$$

exists and is approximately equal to 0.5772. *Euler's number*, as γ is called, appears in many interesting areas of mathematics. For example, it is involved in the evaluation of the exponential integral, $\int_x^{+\infty}(e^{-t}/t)\,dt$, which is important in applied mathematics. It is also related to two special functions—the *gamma function* and *Riemann's zeta function* (see Problem 17). Surprisingly, it is still unknown whether Euler's number is rational or irrational. Problems 14–16 concern Euler's number.

14. The harmonic series diverges quite slowly. For example, the partial sums S_{10}, S_{20}, S_{50}, and S_{100} have approximate values 2.92897, 3.59774, 4.49921, and 5.18738, respectively. In fact, the sum of the first million terms of the harmonic series is about 14.4. With this in mind, what would you conjecture about the rate of convergence of the limit defining γ? Test your conjecture by calculating approximate values for γ by using the partial sums given above.

***15.** Let $C_n = 1 + \frac{1}{2} + \cdots + \frac{1}{n} - \ln n$.

(a) By consulting the figure, show that if $f(x) = 1/x$, then

$$\int_1^n f(x)\,dx \le f(1) + \cdots + f(n-1)$$

Show further that this reduces to $0 < 1/n \le C_n$.

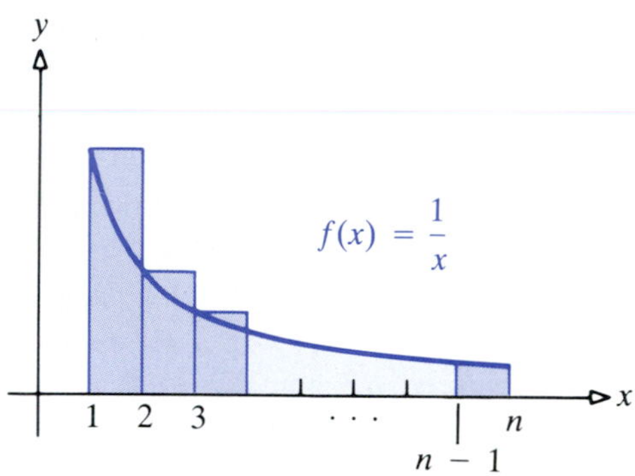

(b) By consulting the figure, show that if $f(x) = 1/x$, then

$$\int_n^{n+1} f(x)\,dx > f(n+1)$$

Compute $C_n - C_{n+1}$ and deduce that $\{C_n\}$ is a decreasing sequence bounded from below by 0 and that it converges. Thus, Euler's number γ exists.

(*Problem 15 continues in the next column*)

* Based on an article by R. Boas, *American Mathematics Monthly* (April 1977).

15. (*continued*)

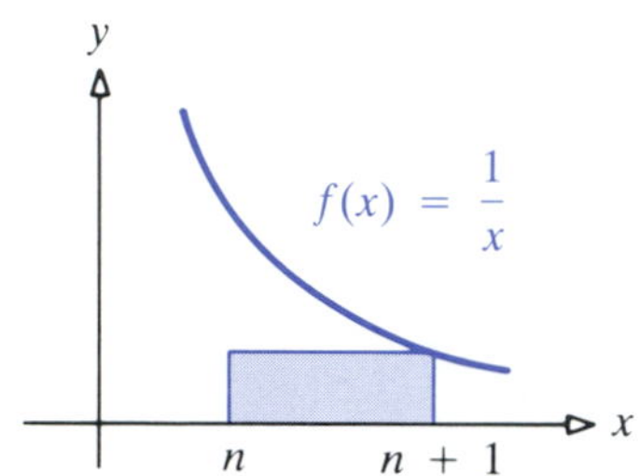

(c) Since $\lim_{n\to+\infty} C_n = \gamma$, we have the approximation

$$S_n = 1 + \frac{1}{2} + \frac{1}{3} + \cdots + \frac{1}{n} \approx \gamma + \ln n$$

In fact, this approximation is very good. It can be used to estimate the growth of S_n. Suppose we want to find the smallest n for which $S_n \ge M$. We can solve $M = \gamma + \ln n$ for n and round the result to an integer. Solve this for $M = 3$ and $M = 4$. Repeat for $M = 10$ and $M = 20$.

16. Use the value of Euler's number given above and your calculator to estimate

$$1 + \frac{1}{2} + \frac{1}{3} + \cdots + \frac{1}{1{,}000{,}000{,}000}$$

17. *Riemann's zeta function* is defined as

$$\zeta(s) = 1 + \frac{1}{2^s} + \frac{1}{3^s} + \cdots \qquad \text{for } s > 1$$

As mentioned in the footnote on page 655, Euler showed that $\zeta(2) = \pi^2/6$. He also found the value of the zeta function for many other even values of s. As of now, no one knows the value of the zeta function for odd values of s. However, it is not too difficult to approximate these values, as this problem demonstrates.

(a) Use your calculator to find $\sum_{k=1}^{10}(1/k^3)$.

(b) Use integrals, in a way analogous to their use in the integral test, to set upper and lower bounds on $\sum_{k=1}^{+\infty}(1/k^3)$.

(c) What can you conclude about $\zeta(3)$?

18. Let x be a fixed positive number and define a sequence by $a_{n+1} = (\frac{1}{2})[a_n + (x/a_n)]$, where a_1 is chosen an arbitrary positive number.

(a) Show that this sequence converges to a limit $\sqrt{x}$.
(b) *Calculator Problem.* Use this sequence to approximate $\sqrt{28}$. How accurate is a_3? a_6?

19. Let n be positive integer and $|x| < 1$. Evaluate

$$\sum_{k=0}^{+\infty}\left(1 + \frac{1}{2^k} + \cdots + \frac{1}{n^k}\right)x^k$$

20. Show that a real number has a repeating decimal if and only if it is rational.

21. Solve the differential equation $y'(x) = y(x)$ by supposing $y(x)$ has a Maclaurin expansion

$$y(x) = \sum_{k=0}^{+\infty} a_k x^k$$

substituting, and obtaining a formula for a_k. Check your answer by solving $y' = y$ by methods you know.

22. Consider the differential equation

$$(1 + x^2)y'' - 4xy' + 6y = 0$$

Assuming there is a solution $y(x) = \sum_{k=0}^{+\infty} a_k x^k$, substitute and obtain a formula for a_k. Note that you have no other method of solving this. Your answer should have the form

$$y(x) = a_0(1 - 3x^2) + a_1(x - \tfrac{1}{3}x^3) \qquad a_0, a_1 \text{ real}$$

23. Let a power series $s(x)$ be convergent for $|x| < R$. Assume we have $s(x) = \sum_{k=0}^{+\infty} a_k x^k$ with partial sums $s_n(x) = \sum_{k=0}^{n} a_k x^k$. Suppose for any $\varepsilon > 0$ there is a number N so that $n > N$ guarantees that $|s(x) - s_n(x)| < \varepsilon/3$ for all $|x| < R$. Show that $s(x)$ is continuous for all $|x| < R$.

24. Let a_n be a sequence of nonzero real numbers and define a new sequence p_n from a_n as

$$p_n = a_1 \cdots a_n = \prod_{k=1}^{n} a_k$$

If $\lim_{n\to+\infty} p_n$ exists and is nonzero,* we say that p is the infinite product of a_n and call the p_n the partial products. In this case, we write $p = \prod_{k=1}^{+\infty} a_k$ and say that $\prod_{k=1}^{+\infty} a_k$ converges. If $\prod_{k=1}^{+\infty} a_k$ does not converge, we say that it diverges.

(*Problem 24 continues in the next column*)

* One can allow 0, but for simplicity we avoid 0.

24. (*continued*)

(a) Give an example of an infinite product that diverges to 0.
(b) Show that $\prod_{k=1}^{+\infty}(1 + a_k)$ converges if and only if $\sum_{k=1}^{+\infty} a_k$ converges. [*Hint:* You will want to consider $\lim_{k\to+\infty}\{[\ln(1 + a_k)]/a_k\}$.]
(c) Show that $\prod_{p\text{ prime}}[1 - (1/p^s)]$ converges if $s > 1$. (Here, the product is taken over prime numbers p.)
(d) It turns out that $1/\zeta(s) = \prod_{p\text{ prime}}[1 - (1/p^s)]$ for $s > 1$. Verify this formally.

†25. (a) By considering graphs like those shown earlier in Problem 15, show that if f is decreasing, positive, and continuous, then

$$f(n+1) + \cdots + f(m) \le \int_n^m f(x)\,dx \le f(n) + \cdots + f(m-1)$$

(b) Under the assumption of part (a), prove that if $\sum_{k=1}^{+\infty} f(k)$ converges, then

$$\sum_{k=n+1}^{+\infty} f(k) \le \int_n^{+\infty} f(x)\,dx \le \sum_{k=n}^{+\infty} f(k)$$

(c) Let $f(x) = 1/x^2$. Use the inequality in part (b) to determine *exactly* how many terms of the series $\sum_{k=1}^{+\infty}(1/k^2)$ one must take in order to have $|\text{Error}| < (\frac{1}{2})10^{-2}$. To have $|\text{Error}| < (\frac{1}{2})10^{-10}$.

‡26. Let $\{a_n\}$ be a sequence that decreases to 0. Define

$$R_n = \sum_{k=n+1}^{+\infty}(-1)^{k+1}a_k \qquad \Delta a_k = a_k - a_{k+1}$$

Suppose that the sequence $\{\Delta a_k\}$ decreases.

(a) Show that the series $\sum_{k=1}^{+\infty}(-1)^{k+1}\Delta a_k$ is a convergent alternating series.
(b) Derive

$$|R_n| = \frac{a_n}{2} + \frac{1}{2}\sum_{k=1}^{+\infty}(-1)^k \Delta a_{k+n-1}$$

Show that $\sum_{k=1}^{+\infty}(-1)^k\Delta a_{k+n-1} < 0$. Deduce $|R_n| < \frac{1}{2}a_n$.

(c) Derive

$$|R_n| = \frac{a_{n+1}}{2} + \frac{1}{2}\sum_{k=1}^{+\infty}(-1)^{k+1}\Delta a_{k+n}$$

Deduce $a_{n+1}/2 < |R_n|$.

† Based on an article by R. Boas, *American Mathematics Monthly* (April 1977).

‡ Based on R. Johnsonbaugh, "Summing an Alternating Series," *American Mathematics Monthly* (1979).

12 Conics

12.1 Introduction

The word *conic* is derived from the word *cone*. A cone can be generated as follows: Let a and g be two distinct lines that intersect at a point V. Rotate the line g around the line a, maintaining the angle between a and g. The resulting surface is called a *cone*. The line a is called the *axis* of the cone; the line g is called a *generator* of the cone. The point V where the axis and generator intersect is called the *vertex* of the cone. See Figure 1.

The curves that result by intersecting a cone with a plane are called *conic sections* or, simply, *conics*. The most interesting conics occur when the intersecting plane does *not* pass through the vertex. These conics are either *circles* (the intersecting plane is perpendicular to the axis of the cone), *ellipses* (the plane is tipped from the perpendicular position), *hyperbolas* (the plane is parallel to the axis), or *parabolas* (the plane is parallel to a generator). Figure 2 shows all these possibilities.

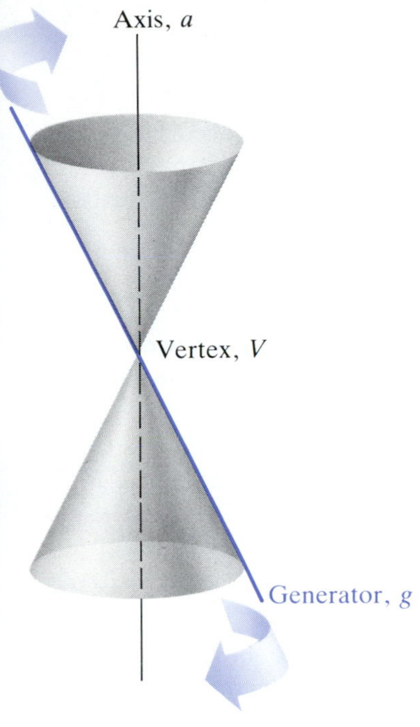

Figure 1

Conics of less interest occur when the intersecting plane passes through the vertex of the cone. These conics, which may be either a point, a line, or a pair of intersecting lines, are called *degenerate conics*.

Historically, many of the interesting properties of conics were discovered and studied by Apollonius around 200 BC. Because they have many current uses, conics are still studied. For example, parabolic reflectors are used in radar, solar energy devices, and reflecting telescopes. Artificial satellites circle the earth in approximately elliptical orbits. Elliptic surfaces are useful for reflecting light beams and sound waves from one place to another. Hyperbolic curves are sometimes used to locate the positions of ships at sea.

Apollonius used the methods of *euclidean geometry* to study conics, but we shall use the more powerful methods of *analytic geometry*. First, we define the conics in terms of distances from certain fixed lines and points. Then we use cartesian coordinates and algebra to derive the equations of these conics.

We could use many words to describe the shape, size, and position of a conic in the *xy*-plane. But, to paraphrase an old saying, "An equation is worth a thousand words." As this chapter shows, the equation of a conic contains complete information concerning the shape and position of that conic in the *xy*-plane.

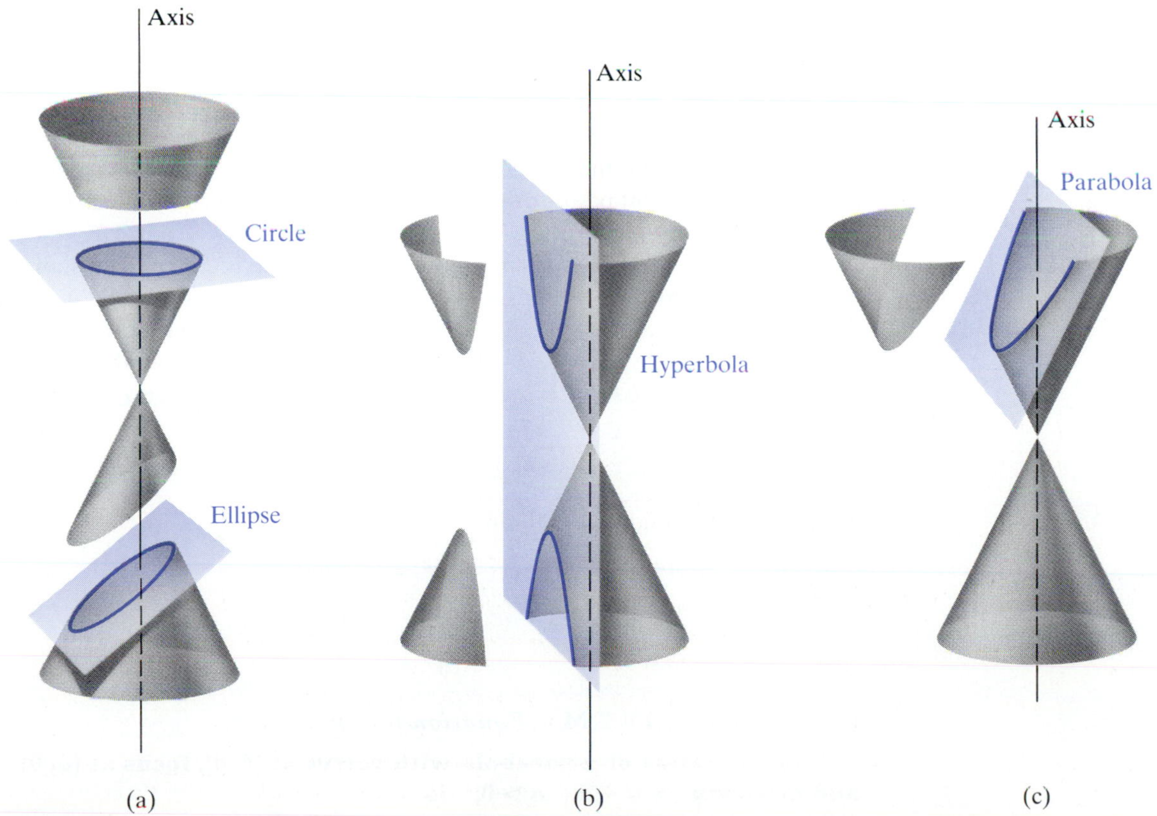

Figure 2

12.2 The Parabola

We begin our study of conics with the *parabola*. The parabola occurs naturally as the path of a projectile (such as a thrown ball or an artillery shell). It also occurs in technical applications, such as the parabolic reflector, the parabolic arch, and the parabolic suspension cable.

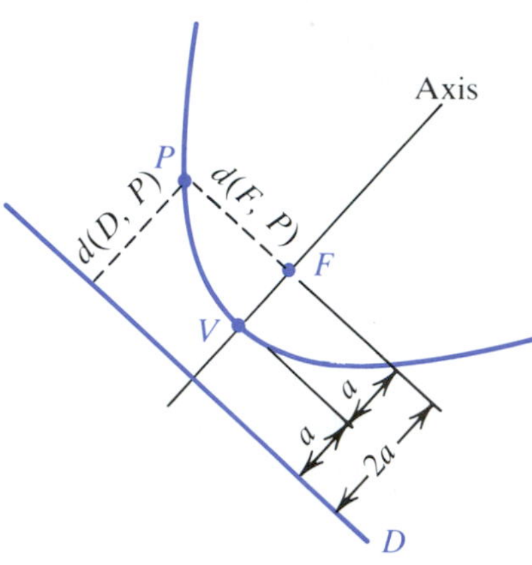

Figure 3

[12.2.1] DEFINITION / *Parabola.*

A *parabola* is the collection of all points P in a plane whose distance from a fixed point F is equal to its distance from a fixed line D.

The point F is called the *focus* of the parabola, and the line D is called the *directrix*. If we denote the distance between two points F and P by $d(F, P)$ and the distance between the line D and a point P by $d(D, P)$, then the parabola with focus F and directrix D is the set of all points P such that

$$d(F, P) = d(D, P) \tag{1}$$

See Figure 3.

The line passing through the focus F perpendicular to the directrix D is called the *axis* of the parabola. The intersection V of the axis with the parabola is called the *vertex*. This point is midway between the focus and the directrix and lies on the parabola. Usually (and for the purpose of this section) the distance from the focus F to the directrix D is denoted by $2a$, where a is a positive real number. As a result, the distance from the focus to the vertex of a parabola is a (refer to Fig. 3).

To obtain the equation of a parabola, we position the axes of our coordinate systems so that the vertex V, focus F, and directrix D are conveniently located. A convenient position for the vertex V is $(0, 0)$. Then the focus F can be placed on the x-axis or the y-axis.

Suppose we begin by placing F on the *positive* x-axis. Since the distance from the focus to the vertex is a, the focus F is located at $(a, 0)$, $a > 0$. Also, because the vertex V is equidistant from F and the directrix, the directrix must be given by $x = -a$. If $P = (x, y)$ is any point on the parabola, then by (1), $d(F, P) = d(D, P)$, and by using the distance formula, we find

$$\sqrt{(x-a)^2 + y^2} = |x + a|$$

By squaring both sides, we obtain

$$(x-a)^2 + y^2 = (x+a)^2$$
$$x^2 - 2ax + a^2 + y^2 = x^2 + 2ax + a^2$$
$$y^2 = 4ax$$

[12.2.2] THEOREM / *Equation of a Parabola.*

The equation of a parabola with vertex at $(0, 0)$, focus at $(a, 0)$, and directrix $x = -a$, $a > 0$, is

$$y^2 = 4ax \tag{2}$$

Conversely, any equation of the form (2) is a parabola with vertex at (0, 0), focus at $(a, 0)$, and directrix $x = -a$, $a > 0$.

EXAMPLE 1

(a) The parabola with vertex at (0, 0) and focus at (2, 0) has the equation $y^2 = 8x$.

(b) The equation $y^2 = 16x$ represents a parabola with vertex (0, 0), focus (4, 0), and directrix $x = -4$.

EXAMPLE 2

Find the equation of a parabola with vertex at the origin, focus at (3, 0), and directrix the line $x = -3$. Graph the parabola.

Solution

Method I Using definition (12.2.1)

$$d(F, P) = d(D, P)$$

$$\sqrt{(x-3)^2 + y^2} = |x + 3|$$

$$(x-3)^2 + y^2 = (x+3)^2$$

$$y^2 = 12x$$

Method II Using theorem (12.2.2) Here, $a = 3$ and the parabola is of the type described by theorem (12.2.2). Thus, the form of the equation is

$$y^2 = 4ax$$

With $a = 3$,

$$y^2 = 12x$$

See Figure 4 for the graph.

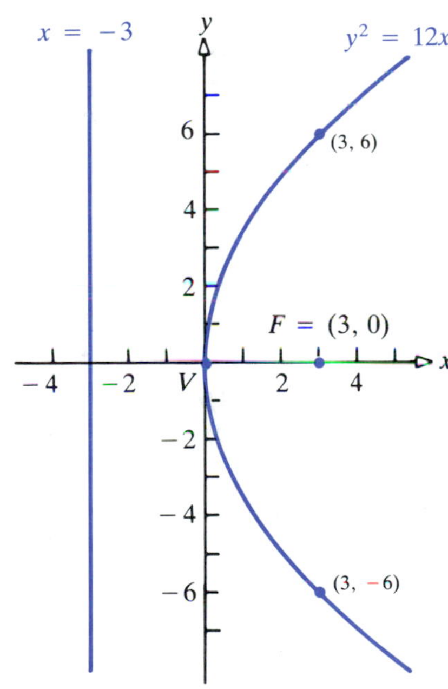

Figure 4

The parabola $y^2 = 4ax$ is symmetric with respect to the x-axis, since, when y is replaced by $-y$, the relationship between x and y is unchanged.

If the focus F is located on the negative x-axis, negative y-axis, or positive y-axis, a different parabola and, hence, a different equation results. The four possible parabolas with vertex at (0, 0) and focus on a coordinate axis are illustrated in Figure 5. In each case, the distance from the focus to the vertex is denoted by a, where a is positive. The four equations are given in Table 1.

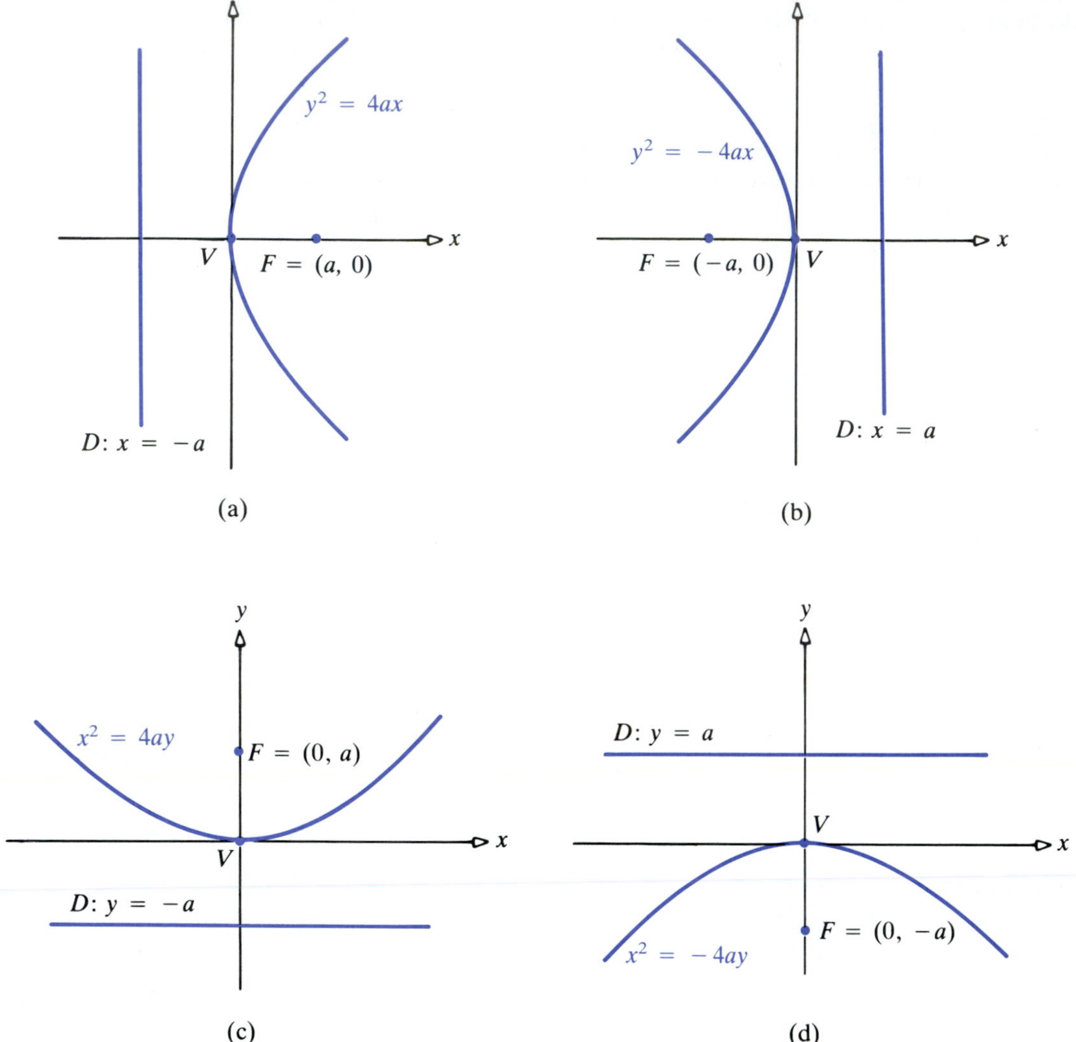

Figure 5

Table 1

Equation	Vertex	Focus	Directrix	Graph	Symmetry
$y^2 = 4ax$	(0, 0)	$(a, 0)$	$x = -a$	Figure 5(a)	x-axis
$y^2 = -4ax$	(0, 0)	$(-a, 0)$	$x = a$	Figure 5(b)	x-axis
$x^2 = 4ay$	(0, 0)	$(0, a)$	$y = -a$	Figure 5(c)	y-axis
$x^2 = -4ay$	(0, 0)	$(0, -a)$	$y = a$	Figure 5(d)	y-axis

EXAMPLE 3

Find the vertex, focus, and directrix of the parabola $x^2 = -16y$.

Solution

First, we observe that this type of parabola has vertex at (0, 0), focus on the negative y-axis, and directrix parallel to the x-axis. By comparing the given equation $x^2 = -16y$ to $x^2 = -4ay$, we find that $a = 4$. Thus, the focus is at (0, −4) and the directrix is $y = 4$. See Figure 6 for a graph.

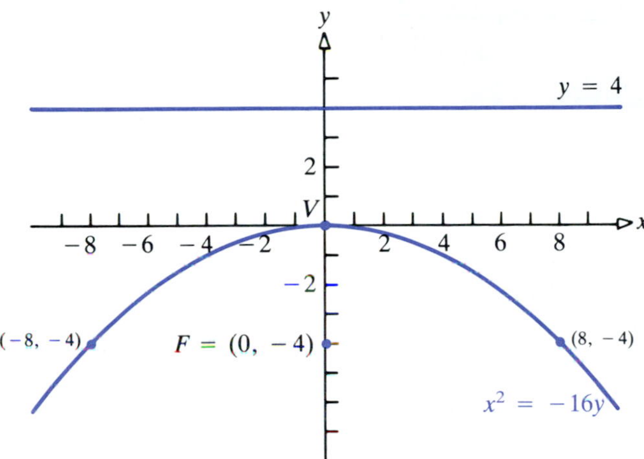

Figure 6

EXAMPLE 4

Find an equation for the tangent line to the parabola $y^2 = -8x$ at the point (−2, 4).

Solution

The slope of the tangent line is obtained by computing the derivative y'. By implicit differentiation, we find

$$2yy' = -8$$

$$y' = \frac{-4}{y}$$

At $y = 4$, we find $y' = -1$. The slope of the tangent line to $y^2 = -8x$ at (−2, 4) is therefore −1 and an equation of this tangent line is

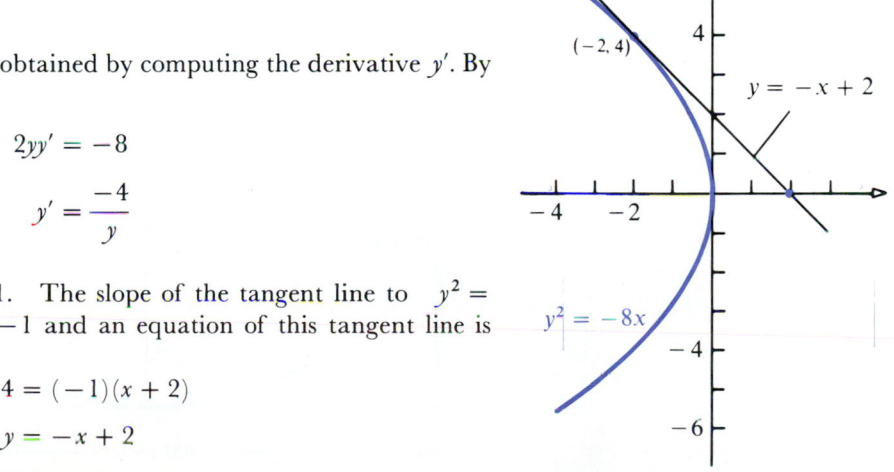

$$y - 4 = (-1)(x + 2)$$

$$y = -x + 2$$

See Figure 7.

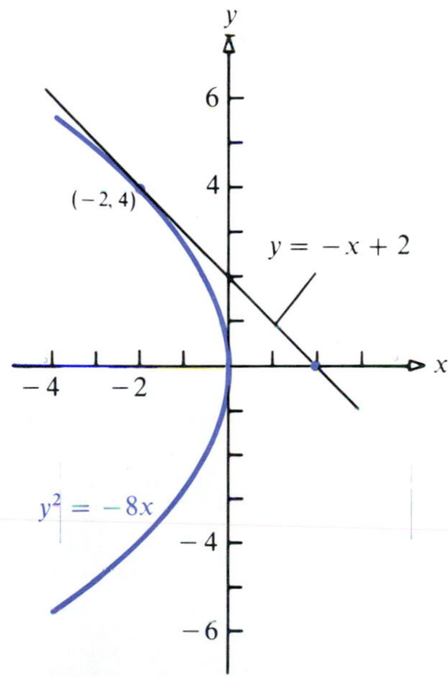

Figure 7

Suspension Property

In a suspension bridge the main cables are parabolic. The reason for this is that if the total weight of a bridge is uniformly distributed along its length, the only cable shape that will bear the load evenly is that of a parabola.

EXAMPLE 5

A suspension bridge with weight uniformly distributed along its length has twin towers that extend 100 meters above the road surface and are 400 meters apart. The cables are parabolic and are tangent to the road surface at the center of the bridge. Find the height of the cables at a point 100 meters from the center. (The road is assumed horizontal.)

Solution

We begin by drawing Figure 8. The cable has the shape of a parabola with vertex at (0, 0) and focus along the positive y-axis. The equation of this parabola is $x^2 = 4ay$. Since the cable is 100 meters high when $x = 200$, we find

$$(200)^2 = 4a(100)$$
$$a = 100$$

To find the height of the cable when $x = 100$, we solve for y in $x^2 = 400y$, obtaining $y = 25$. The cable is 25 meters high at 100 meters from the center of the bridge.

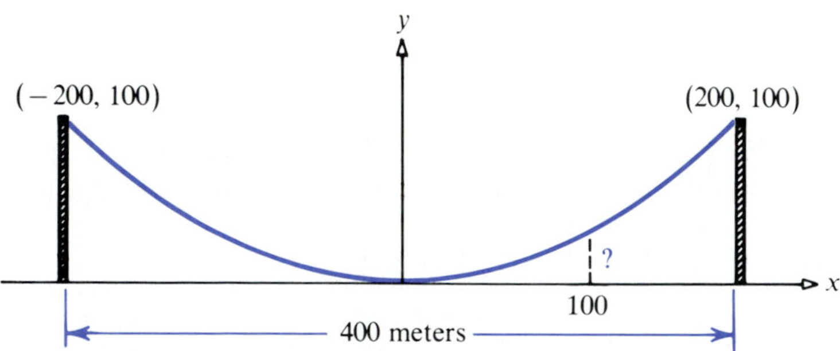

Figure 8

Reflecting Property

Another important property of a parabola is its application to reflecting. Suppose a mirror has the shape of a *paraboloid of revolution*—that is, a surface formed by rotating a parabola about its axis. If a light is placed at the focus of the mirror, all the rays emanating from it will be reflected off the mirror in lines parallel to the axis. This property is the principle behind the design of automobile headlights, where the bulb is placed at the focus (see Fig. 9).

Conversely, when rays of light emanating from a distant source (like the sun) strike a parabolic mirror, they will be reflected to a single point—the focus. This latter property is the principle behind some solar energy devices and most reflecting telescopes. You are asked to prove this property in Problem 40.

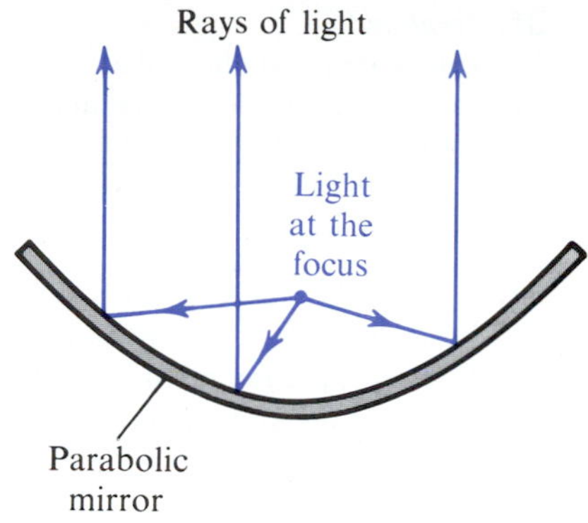

Figure 9

EXERCISE 12.2

In Problems 1–16 find the equation of the parabola(s) having the stated properties.

1. Focus at $(1, 0)$; vertex at $(0, 0)$

2. Focus at $(0, 2)$; vertex at $(0, 0)$

3. Focus at $(-2, 0)$; vertex at $(0, 0)$

4. Focus at $(-3, 0)$; vertex at $(0, 0)$

5. Focus at $(2, 0)$; directrix $x = -2$

6. Focus at $(0, -3)$; directrix $y = 3$

7. Focus at $(0, 3)$; directrix $y = -3$

8. Focus at $(-4, 0)$; directrix $x = 4$

9. Vertex at $(0, 0)$; directrix $x = 2$

10. Vertex at $(0, 0)$; directrix $y = 3$

11. Vertex at $(0, 0)$; $a = 3$; directrix to the right of $(0, 0)$ and parallel to the y-axis.

12. Vertex at $(0, 0)$; $a = 4$; directrix below $(0, 0)$ and parallel to the x-axis

13. Vertex at $(0, 0)$; axis coinciding with the x-axis; containing the point $(2, 8)$

14. Vertex at $(0, 0)$; axis coinciding with the y-axis; containing the point $(4, -4)$

15. Vertex at $(0, 0)$; $a = 3$; axis coinciding with the negative x-axis

16. Vertex at $(0, 0)$; $a = 2$; axis coinciding with the negative y-axis

In Problems 17–24 find the vertex, focus, and directrix of each parabola. Graph each parabola.

17. $y^2 = 8x$

18. $x^2 = 4y$

19. $x^2 = -12y$

20. $y^2 = -4x$

21. $y^2 = -16x$

22. $x^2 = 16y$

23. $x^2 = 8y$

24. $y^2 = 8x$

25. Find an equation of the tangent line to the parabola $x^2 = -8y$ at the point $(4, -2)$.

26. Find an equation of the tangent line to the parabola $y^2 = -12x$ at the point $(-3, 6)$.

27. Find the area of the region in the first quadrant enclosed by the parabola $x^2 = 8y$, the y-axis, and the line $y = 2$.

28. Find the area of the region enclosed by the parabola $y^2 = 4x$, the x-axis, and the line $x = 4$.

29. Find the volume of the solid of revolution obtained by rotating the region described in Problem 27 about the x-axis.

30. Find the volume of the solid of revolution obtained by rotating the region described in Problem 28 about the y-axis.

31. A suspension bridge with weight uniformly distributed along its length has twin towers that extend 75 meters above the road surface and are 300 meters apart. The

(Problem 31 continues on page 714)

31. *(continued)*
cables are parabolic and are suspended from the tops of the towers. The cables are tangent to the road surface at the center of the bridge. Find the height of the cables at a point 80 meters from the center. Assume the road is horizontal.

32. A parabolic arch has a span of 120 feet and a maximum height of 25 feet. Choose suitable rectangular axes and find the equation of the parabola. Then calculate the height of the arch at points 10 feet, 20 feet, and 40 feet from the center.

33. A radar antenna is constructed so that any cross section through its axis is a parabola. Suppose the receiver is located at the focus. Find the location of this receiver if the antenna is 5 feet across at the opening and is 1 foot deep.

34. If a parabolic reflector has a diameter of 8 inches and is 6 inches deep, how far from the vertex of the parabola should a light be placed so that the rays may be reflected parallel to the axis?

35. Show that the area of a parabolic segment cut off by a chord perpendicular to the axis of the parabola is equal to two-thirds of the area of the circumscribed rectangle.

36. A trough 10 feet long with a vertical parabolic cross section 4 feet deep and 4 feet across the top is filled with water (see the figure). Find the work done in pumping out the trough.

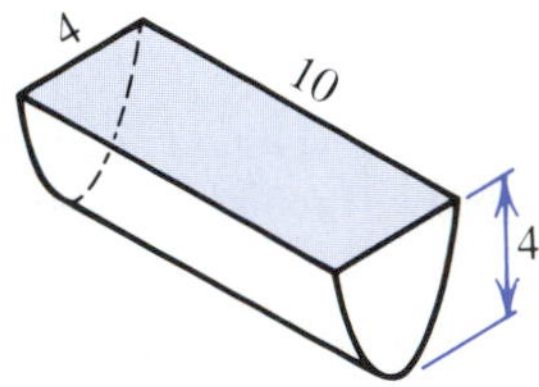

37. A plate in the form of a parabolic segment cut off by a chord perpendicular to the axis is immersed vertically in water. The vertex is at the surface and the axis is vertical. The plate is 20 feet deep and 12 feet broad (see the figure). Find the force on one face of the segment in tons. (The mass density ρ of water is 1.94 slugs per cubic foot; 1 ton = 2000 pounds.)

(Problem 37 continues in the next column)

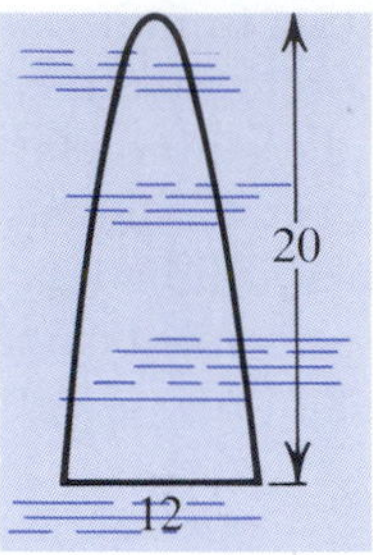

38. An arched window with base width $2b$ and height h is to be set into a wall. The arch is to be either an arc of a parabola or a half-cycle of a cosine curve. Of these two window designs, which has the greater area? Justify your answer.

39. Prove that an equation of the tangent line at any point (x_0, y_0) of the parabola $y^2 = 4ax$ is $y_0 y = 2a(x + x_0)$.

40. In the figure show that the angle of incidence α equals the angle of reflection β.

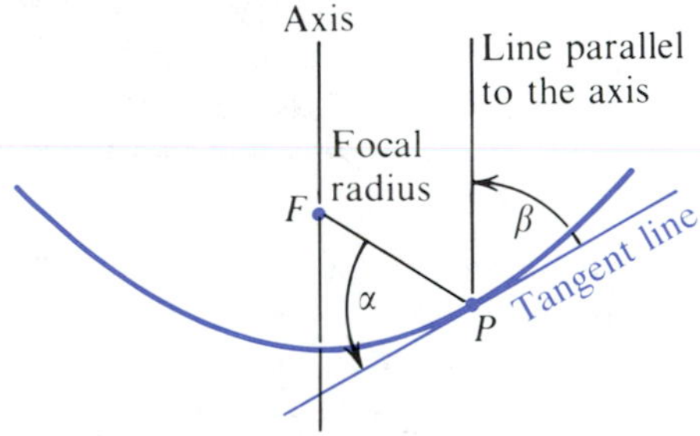

The tangent line makes equal angles with the focal radius and the line parallel to the axis.

41. Show that an equation of the form

$$Ax^2 + Ey = 0 \qquad A \neq 0, \quad E \neq 0$$

is the equation of a parabola with vertex at $(0, 0)$ and axis the y-axis. Find its focus and directrix.

42. Show that an equation of the form

$$Cy^2 + Dx = 0 \qquad C \neq 0, \quad D \neq 0$$

is the equation of a parabola with vertex at $(0, 0)$ and axis the x-axis. Find its focus and directrix.

The Ellipse

12.3

An *ellipse* is defined as follows:

[12.3.1] DEFINITION / *Ellipse*.

An *ellipse* is the collection of all points in the plane, the sum of whose distances from two fixed points, called the *foci*, is a constant.

An ellipse can be drawn with the help of two thumbtacks, a pencil, and some string. With the thumbtacks, fix two points (the foci) on a piece of paper and attach the end of a piece of string to each thumbtack. The length of the string is the constant referred to in the definition. Loop the string around a pencil and make the string taut. Keeping the string taut, move the pencil about the two thumbtacks. The pencil will then trace out an ellipse as shown in Figure 10.

In Figure 10 the foci are F and F'. The line joining the foci is called the *major axis*. The midpoint of the foci is called the *center* of the ellipse. The line through the center and perpendicular to the major axis is called the *minor axis*. The points of intersection V and V' of the ellipse with the major axis are the *vertices* of the ellipse.

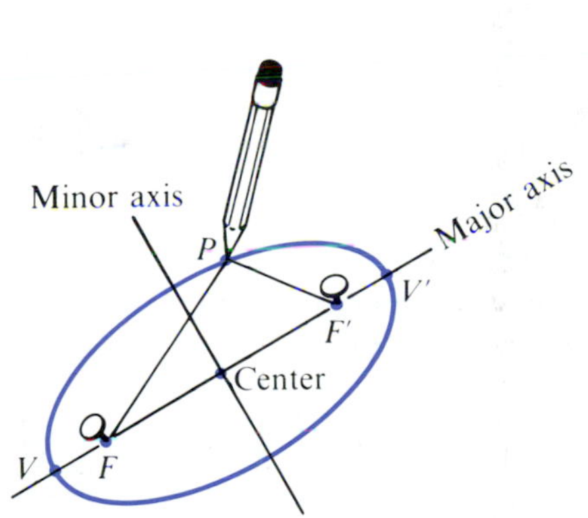

Figure 10

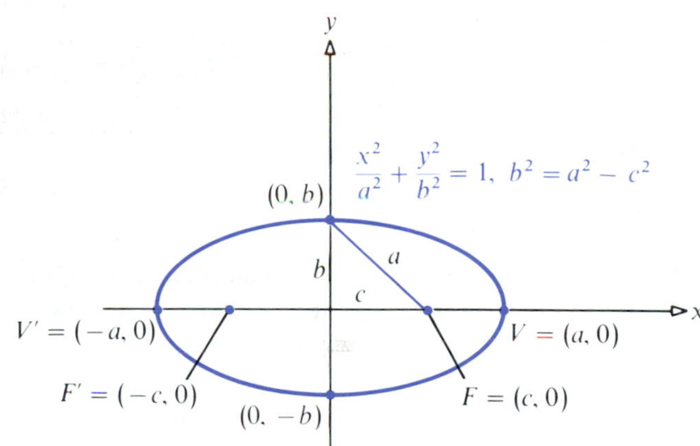

Figure 11

For convenience in obtaining the standard equation of an ellipse, we place the center at (0, 0) and use the x-axis as the major axis (see Fig. 11). If one focus is at $F = (c, 0)$, $c > 0$, the other focus must be at $F' = (-c, 0)$. Now we let the constant sum of the distances of a point $P = (x, y)$ on the ellipse from the foci F and F' be denoted by $2a$. Then, by definition (12.3.1), we have

$$d(F, P) + d(F', P) = 2a$$

By the distance formula,

$$\sqrt{(x-c)^2+y^2}+\sqrt{(x+c)^2+y^2}=2a$$

$$\sqrt{(x-c)^2+y^2}=2a-\sqrt{(x+c)^2+y^2}$$

$$(x-c)^2+y^2=4a^2-4a\sqrt{(x+c)^2+y^2}+(x+c)^2+y^2$$

$$x^2-2cx+c^2+y^2=4a^2-4a\sqrt{(x+c)^2+y^2}+x^2+2cx+c^2+y^2$$

$$4a\sqrt{(x+c)^2+y^2}=4a^2+4cx$$

$$a\sqrt{(x+c)^2+y^2}=a^2+cx$$

Square both sides again and simplify:

$$a^2[(x+c)^2+y^2]=(a^2+cx)^2$$

$$a^2[x^2+2cx+c^2+y^2]=a^4+2a^2cx+c^2x^2$$

$$a^2x^2+a^2c^2+a^2y^2=a^4+c^2x^2$$

$$(a^2-c^2)x^2+a^2y^2=a^4-a^2c^2$$

$$(a^2-c^2)x^2+a^2y^2=a^2(a^2-c^2)$$

To get points on the ellipse off the x-axis, we must have $a > c$ (see Fig. 11). Since $a > c$, it follows that $a^2 > c^2$, so that $a^2 - c^2 > 0$. As a result, we may set $b^2 = a^2 - c^2$, $b > 0$, to get

$$b^2x^2+a^2y^2=a^2b^2$$

or

$$\frac{x^2}{a^2}+\frac{y^2}{b^2}=1 \qquad \text{where} \quad b^2=a^2-c^2, \quad a>b$$

[12.3.2] THEOREM / *Equation of an Ellipse.*

The standard equation of an ellipse with center at (0, 0) and foci at (c, 0) and ($-c$, 0) is

$$\frac{x^2}{a^2}+\frac{y^2}{b^2}=1 \qquad \text{where} \quad b^2=a^2-c^2, \qquad a>b \tag{1}$$

Conversely, any equation of the form (1) is an ellipse with center at (0, 0), foci at (c, 0) and ($-c$, 0), and major axis along the x-axis.

The graph of the ellipse defined by (1) is symmetric with respect to the x-axis, the y-axis, and the origin.

By setting $y = 0$, we find that the vertices of the ellipse obey the equation $x^2/a^2 = 1$, and, hence, the vertices are at $V = (a, 0)$ and $V' = (-a, 0)$, as shown in Figure 11. If we set $x = 0$, we obtain the y-intercepts of the ellipse, $(0, b)$ and $(0, -b)$, which are often called the *covertices* of the ellipse.

We agree that the length of the major axis is the distance from V to V', and the length of the minor axis is the distance between the two y-intercepts.

Thus,

Length of major axis is $2a$; length of minor axis is $2b$.

Because $b^2 + c^2 = a^2$, the major axis is longer than the minor axis.

[12.3.3] THEOREM / *Equation of an Ellipse.*

If we position the ellipse so that its center is at (0, 0) and the y-axis is the major axis, then its foci are at $(0, c)$ and $(0, -c)$ and its equation is

$$\frac{x^2}{b^2} + \frac{y^2}{a^2} = 1 \quad \text{where} \quad b^2 = a^2 - c^2, \quad a > b \tag{2}$$

See Figure 12.

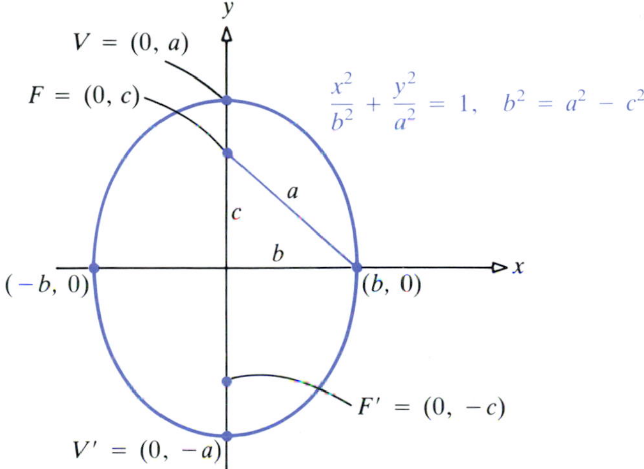

Figure 12

The next example illustrates the difference between equations (1) and (2).

EXAMPLE 1

Discuss and graph the equation $9x^2 + 16y^2 = 144$.

Solution

To get the equation in standard form, we divide both sides by 144:

$$\frac{x^2}{16} + \frac{y^2}{9} = 1$$

Since the denominator of the x^2 term is larger than the denominator of the y^2 term, this equation is of the form (1). Thus, this is the equation of an ellipse whose major axis is the x-axis, with $a^2 = 16$, $b^2 = 9$, and $c^2 = a^2 - b^2 = 7$. The foci are at $(\sqrt{7}, 0)$ and $(-\sqrt{7}, 0)$; the vertices are at $(4, 0)$ and $(-4, 0)$. The graph is given in Figure 13.

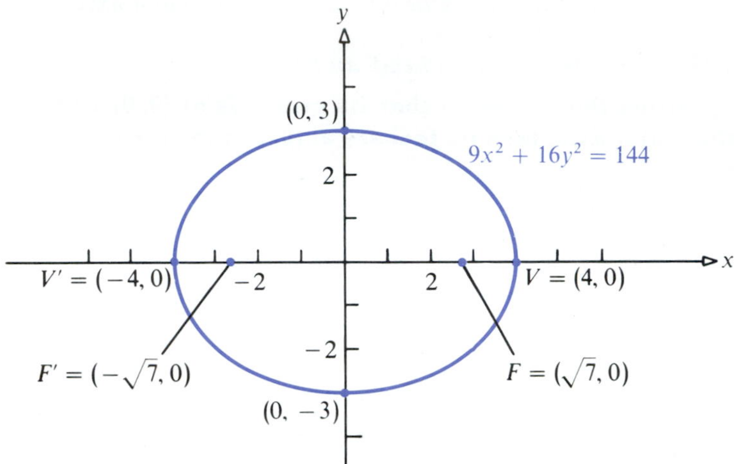

Figure 13

EXAMPLE 2

Find the equation of the ellipse with center at $(0, 0)$, a focus at $(2, 0)$, and a vertex at $(-3, 0)$.

Solution

The second focus is at $(-2, 0)$ and the other vertex is at $(3, 0)$. Since the major axis is along the x-axis, the form of the equation of the ellipse is

$$\frac{x^2}{a^2} + \frac{y^2}{b^2} = 1 \qquad \text{where} \quad b^2 = a^2 - c^2, \quad a > b$$

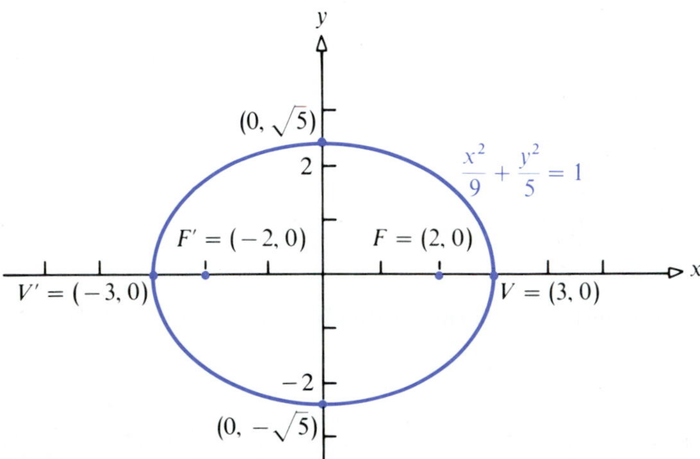

Figure 14

Since a focus is at (2, 0) and a vertex is at (3, 0), we know that $c = 2$ and $a = 3$. As a result, $b^2 = a^2 - c^2 = 9 - 4 = 5$. The equation of the ellipse is

$$\frac{x^2}{9} + \frac{y^2}{5} = 1$$

See Figure 14.

EXAMPLE 3

Find the equation of the ellipse with center at (0, 0), length of major axis 8, and a focus at (0, 2).

Solution

The second focus is at $(0, -2)$, and the major axis is the y-axis. The form of the equation of this ellipse is

$$\frac{x^2}{b^2} + \frac{y^2}{a^2} = 1 \qquad \text{where} \quad b^2 = a^2 - c^2, \quad a > b$$

Since the length of the major axis is $2a = 8$, we have $a = 4$. In addition, $c = 2$. Thus, $b^2 = 16 - 4 = 12$ and the equation is

$$\frac{x^2}{12} + \frac{y^2}{16} = 1$$

See Figure 15.

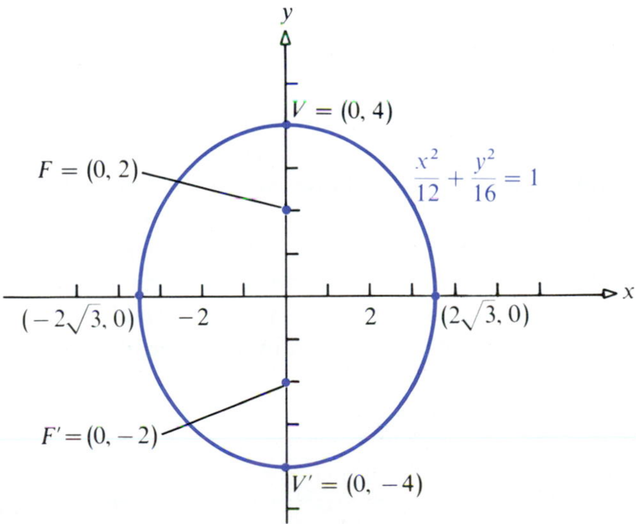

Figure 15

A circle can be thought of as a special kind of ellipse. Indeed, if the two foci of an ellipse move toward the center, then $c \to 0$. Consequently, $b \to a$ (since $b = \sqrt{a^2 - c^2}$). Thus, the length of the major axis becomes equal to the length of the minor axis, and the equation becomes

$$\frac{x^2}{a^2} + \frac{y^2}{a^2} = 1 \qquad \text{or} \qquad x^2 + y^2 = a^2$$

a circle of radius a and center at $(0, 0)$.

EXAMPLE 4

Find an equation for the tangent line to the ellipse $4x^2 + y^2 = 5$ at the point $(1, -1)$.

Solution

We need to compute the derivative y' at $(1, -1)$. By implicit differentiation, we have

$$8x + 2yy' = 0$$

$$y' = \frac{-4x}{y}$$

At $(1, -1)$, the slope of the tangent line is $-4/(-1) = 4$. An equation of the tangent line is

$$y + 1 = 4(x - 1)$$

$$y = 4x - 5$$

Applications

The ellipse has many uses in science and engineering. For example, the orbits of the planets and some comets about the sun are elliptical, with the sun positioned at a focus. If a source of light (or sound) is placed at one focus of an ellipse, the rays reflect off the ellipse and meet at the other focus (see Fig. 16). This is because the tangent lines to an ellipse make equal angles with the focal radii (see Problem 26). This property of an ellipse is used in so-called whispering galleries, which are rooms designed with ceilings that are elliptical arcs. A person standing at a focus of the ellipse can whisper and be heard by another person standing at the other focus because all the sound rays emanating from the first person reflect off the ceiling to the other person.

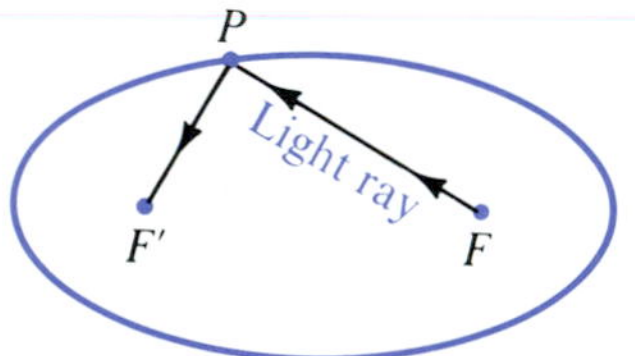

Figure 16

Semielliptical arches are often used in stone and concrete bridges. Elliptical gears are used in some types of machinery when variable rates of motion are required. And springs are often formed in the shape of an ellipse or semiellipse.

EXERCISE 12.3

In Problems 1–8 find the equation of the ellipse having the stated properties.

1. Center at $(0, 0)$; focus at $(4, 0)$; vertex at $(6, 0)$

2. Center at $(0, 0)$; focus at $(3, 0)$; vertex at $(5, 0)$

3. Center at $(0, 0)$; focus at $(0, 2)$; vertex at $(0, -3)$

4. Center at $(0, 0)$; focus at $(0, -3)$; vertex at $(0, 4)$

5. Foci at $(3, 0)$ and $(-3, 0)$; vertex at $(5, 0)$

6. Focus at $(0, -2)$; vertices at $(0, 3)$ and $(0, -3)$

7. Foci at $(4, 0)$ and $(-4, 0)$; length of minor axis is 2

8. Vertices at $(0, 5)$ and $(0, -5)$; length of minor axis is 4

In Problems 9–16 discuss and graph each equation.

9. $\dfrac{x^2}{4} + \dfrac{y^2}{9} = 1$ **10.** $\dfrac{x^2}{9} + \dfrac{y^2}{4} = 1$ **11.** $x^2 + 4y^2 = 16$ **12.** $9x^2 + y^2 = 18$

13. $4x^2 + y^2 = 4$ **14.** $9x^2 + 4y^2 = 36$ **15.** $x^2 + y^2 = 16$ **16.** $x^2 + y^2 = 4$

17. Find an equation for the tangent line to the ellipse $\dfrac{x^2}{4} + \dfrac{y^2}{3} = 1$ at the point $(1, \frac{3}{2})$.

18. Find an equation for the tangent line to the ellipse $x^2 + \dfrac{y^2}{4} = 1$ at the point $(\frac{1}{2}, \sqrt{3})$.

19. The *eccentricity* e of an ellipse is defined as the number c/a. Since $c < a$, it follows that $e < 1$. Describe the general shape of an ellipse whose eccentricity is:
(a) Close to 0 (b) Equal to $\frac{1}{2}$ (c) Close to 1

20. The orbit of the earth is an ellipse with the sun at one focus. If the *semimajor axis* (half the major axis) is approximately 92 million miles long and the eccentricity is $\frac{1}{60}$, find the greatest and shortest distances of the earth from the sun.

21. In order to support a bridge, an arch in the shape of the upper half of an ellipse is built (see the figure). Write an equation of the ellipse if the bridge is to span a river 20 meters wide and the center of the arch is 8 meters above the center of the river. Let the x-axis coincide with the water level and the y-axis pass through the center of the arch.

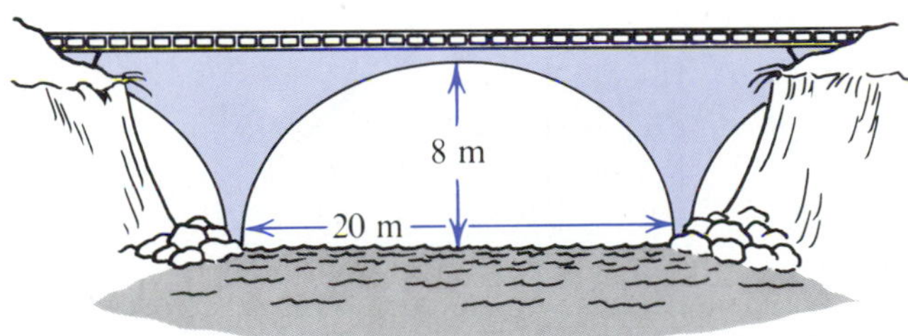

22. The arch of a bridge is a semiellipse with horizontal major axis. The span is 30 feet, and the top of the arch is 10 feet above the major axis. The roadway is horizontal and is 2 feet above the top of the arch. Find the vertical distance from the roadway to the arch at 5 foot intervals along the roadway starting at the middle of the bridge.

23. An arch in the form of half an ellipse is 40 feet wide and 15 feet high at the center. Find the height of the arch at intervals of 10 feet along its width starting at the middle of the bridge.

24. The Colosseum in Rome is in the form of an ellipse 615 feet long and 510 feet wide. Find the area enclosed by the Colosseum.

25. Show that the area enclosed by the ellipse $\dfrac{x^2}{a^2} + \dfrac{y^2}{b^2} = 1$ is πab.

26. Show that the tangent line to an ellipse makes equal angles with the focal radii (see the figure).

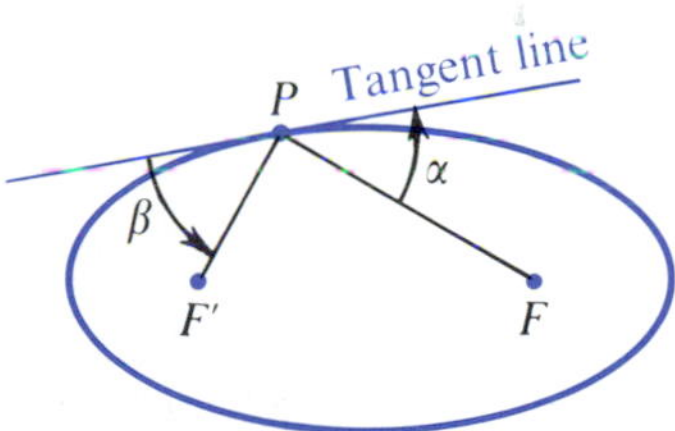

The tangent line makes equal angles with the focal radii.

27. Prove that an equation for the tangent line at any point (x_0, y_0) on the ellipse $\dfrac{x^2}{a^2} + \dfrac{y^2}{b^2} = 1$ is $\dfrac{xx_0}{a^2} + \dfrac{yy_0}{b^2} = 1$.

28. Show that an equation of the form

$$Ax^2 + Cy^2 + F = 0 \qquad A \neq 0, \quad C \neq 0, \quad F \neq 0$$

where A and C are of the same sign and F is of opposite sign:
(a) Is the equation of an ellipse with center at (0, 0) if $A \neq C$
(b) Is the equation of a circle with center at (0, 0) if $A = C$

12.4 The Hyperbola

A hyperbola is defined as follows:

[12.4.1] DEFINITION / *Hyperbola.*

A *hyperbola* is the collection of all points in the plane, the difference of whose distances from two fixed points, called the *foci*, is a constant.

In Figure 17 the line joining the foci is called the *transverse axis*. The midpoint of the foci is called the *center* of the hyperbola. The line through the center and perpendicular to the transverse axis is called the *conjugate axis*. The hyperbola consists of two separate curves, called *branches*, which are symmetric with respect to the conjugate axis, the transverse axis, and the center. The points of intersection of these branches with the transverse axis are the *vertices* of the hyperbola.

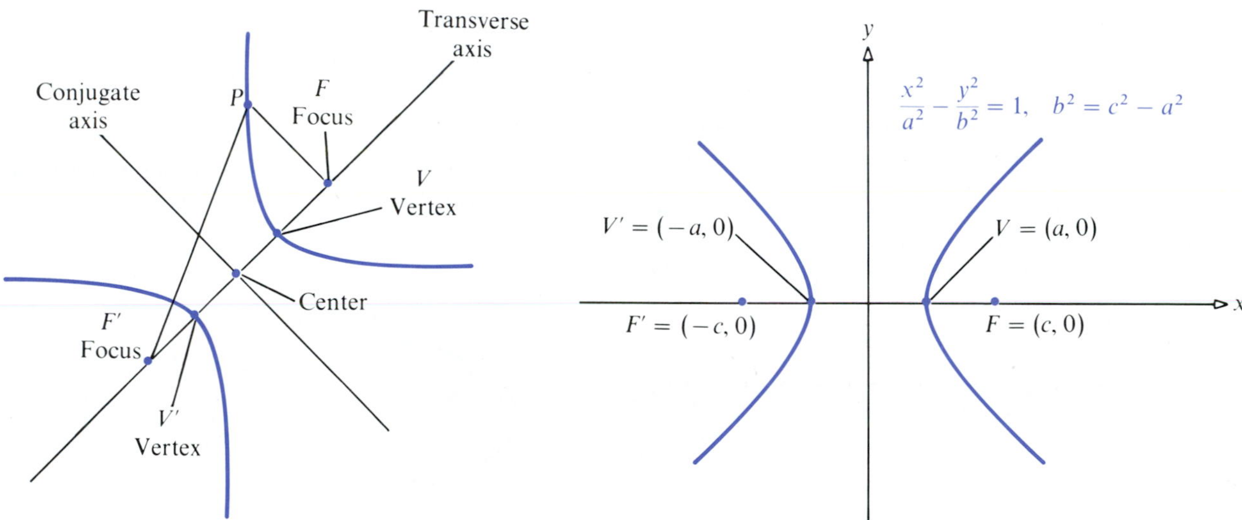

Figure 17 **Figure 18**

For convenience in obtaining the standard equation of a hyperbola, we place the center at $(0, 0)$ and use the x-axis as the transverse axis. If one focus is at $F = (c, 0)$, $c > 0$, the other focus must be at $F' = (-c, 0)$, as shown in Figure 18. Then, if we let the constant difference of the distances from a point $P = (x, y)$ on the hyperbola to the foci F and F' be denoted by $\pm 2a$,* by the definition, we have

$$d(F', P) - d(F, P) = \pm 2a$$

* If P is on the right branch, the plus sign is used; if P is on the left branch, the minus sign is used.

By the distance formula,

$$\sqrt{(x+c)^2+y^2} - \sqrt{(x-c)^2+y^2} = \pm 2a$$
$$\sqrt{(x+c)^2+y^2} = \pm 2a + \sqrt{(x-c)^2+y^2}$$
$$(x+c)^2+y^2 = 4a^2 \pm 4a\sqrt{(x-c)^2+y^2} + (x-c)^2+y^2$$
$$x^2+2cx+c^2+y^2 = 4a^2 \pm 4a\sqrt{(x-c)^2+y^2} + x^2-2cx+c^2+y^2$$
$$4cx-4a^2 = \pm 4a\sqrt{(x-c)^2+y^2}$$
$$cx-a^2 = \pm a\sqrt{(x-c)^2+y^2}$$

We square both sides again to get

$$(cx-a^2)^2 = a^2[(x-c)^2+y^2]$$
$$c^2x^2-2ca^2x+a^4 = a^2[x^2-2cx+c^2+y^2]$$
$$c^2x^2+a^4 = a^2x^2+a^2c^2+a^2y^2$$
$$(c^2-a^2)x^2-a^2y^2 = a^2c^2-a^4$$
$$(c^2-a^2)x^2-a^2y^2 = a^2(c^2-a^2)$$

To get points on the hyperbola off the x-axis, we must have $a < c$ (refer to Fig. 18). Since $0 < a < c$, it follows that $a^2 < c^2$, so that $c^2 - a^2 > 0$. By setting $b^2 = c^2 - a^2$, $b > 0$, we get

$$b^2x^2 - a^2y^2 = a^2(b^2)$$
$$\frac{x^2}{a^2} - \frac{y^2}{b^2} = 1 \qquad \text{where} \quad b^2 = c^2 - a^2$$

[12.4.2] THEOREM / *Equation of a Hyperbola.*

The standard equation of a hyperbola with center at (0, 0) and foci at (c, 0) and ($-c$, 0) is

$$\frac{x^2}{a^2} - \frac{y^2}{b^2} = 1 \qquad \textbf{where} \quad b^2 = c^2 - a^2 \tag{1}$$

Conversely, any equation of the form (1) is a hyperbola with center at (0, 0), foci at (c, 0) and ($-c$, 0), and transverse axis along the x-axis.

The graph is symmetric with respect to the x-axis, the y-axis, and the origin.

By setting $y = 0$, we find that the vertices of the hyperbola obey the equation $x^2/a^2 = 1$ or $x^2 = a^2$, and, hence, the vertices are at $V = (a, 0)$ and $V' = (-a, 0)$. By setting $x = 0$, we obtain the equation $y^2/b^2 = -1$, which has no solution. Thus, this hyperbola never crosses the y-axis; in fact, there are no points on this hyperbola for which $|x| < a$.

EXAMPLE 1

Find the equation of the hyperbola with center at (0, 0), a focus at (3, 0), and a vertex at (2, 0). Graph this hyperbola.

Solution

The second focus is at $(-3, 0)$ and the second vertex is at $(-2, 0)$. Since the transverse axis is the x-axis, the form of the equation of the hyperbola is

$$\frac{x^2}{a^2} - \frac{y^2}{b^2} = 1$$

Since $a = 2$ and $c = 3$, we have $b^2 = c^2 - a^2 = 9 - 4 = 5$. The equation of this hyperbola is

$$\frac{x^2}{4} - \frac{y^2}{5} = 1$$

See Figure 19.

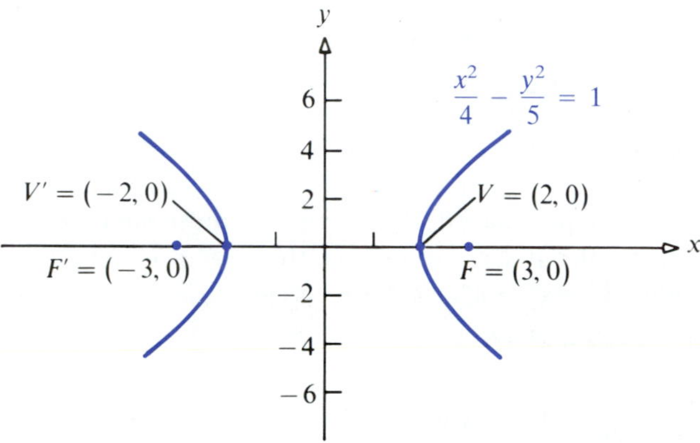

Figure 19

EXAMPLE 2

Discuss and graph the equation $9x^2 - 16y^2 = 144$.

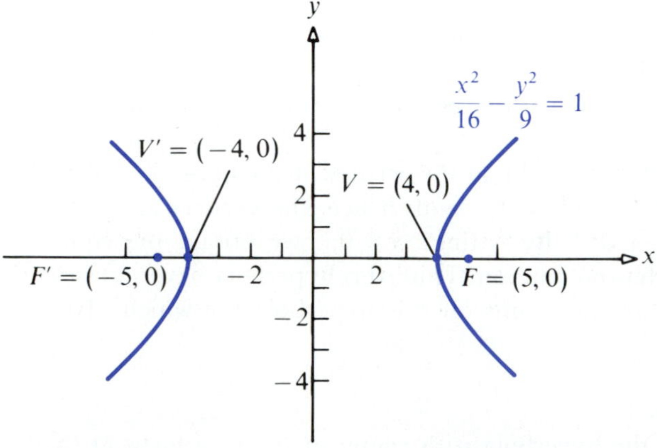

Figure 20

Solution

To get the equation in standard form, we divide both sides by 144:

$$\frac{x^2}{16} - \frac{y^2}{9} = 1$$

This is the equation of a hyperbola with center at $(0, 0)$, whose transverse axis is the x-axis. Also, $a^2 = 16$, $b^2 = 9$, and $c^2 = a^2 + b^2 = 25$. The foci are at $(5, 0)$ and $(-5, 0)$; the vertices are at $(4, 0)$ and $(-4, 0)$; see Figure 20.

[12.4.3] THEOREM / *Equation of a Hyperbola.*

If we position the hyperbola so that (0, 0) is its center and the y-axis is the transverse axis (see Fig. 21), then its foci are at $(0, c)$ and $(0, -c)$ and its equation is

$$\frac{y^2}{a^2} - \frac{x^2}{b^2} = 1 \qquad \text{where} \quad b^2 = c^2 - a^2 \tag{2}$$

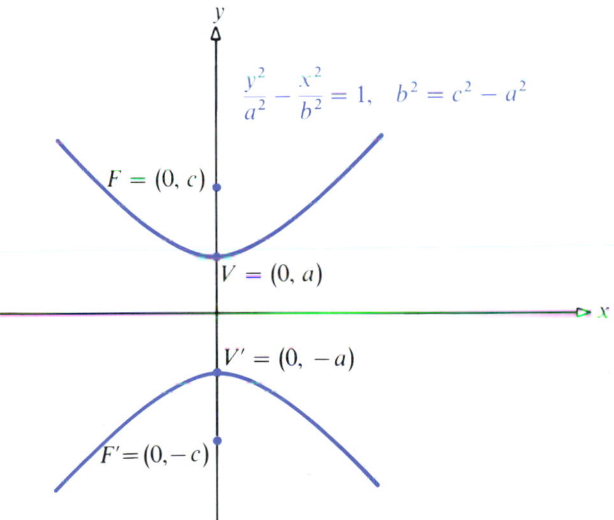

Figure 21

Asymptotes

Hyperbolas have one feature that parabolas and ellipses do not possess—hyperbolas have *asymptotes*. Recall that an asymptote is a line with the property that the distance from an asymptote to points on the graph gets arbitrarily close to 0 as the points on the graph recede indefinitely far from the origin.

[12.4.4] THEOREM / *Asymptotes of a Hyperbola.*

The hyperbola

$$\frac{x^2}{a^2} - \frac{y^2}{b^2} = 1$$

has the two asymptotes

$$y = \frac{b}{a}x \qquad \text{and} \qquad y = -\frac{b}{a}x$$

See Figure 22. A portion of the proof of this statement follows.

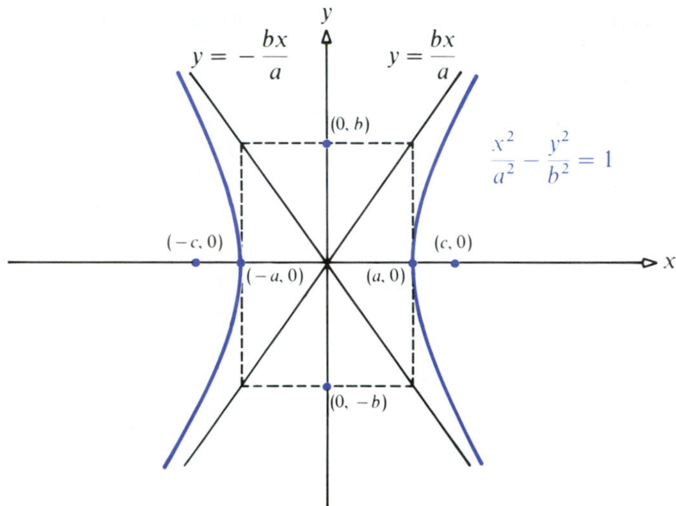

Figure 22

Consider the hyperbola

$$\frac{x^2}{a^2} - \frac{y^2}{b^2} = 1$$

If we solve for y, we get

$$y^2 = \frac{b^2}{a^2}(x^2 - a^2)$$

$$y = \pm\frac{b}{a}\sqrt{x^2 - a^2}$$

Hence, the two branches of the hyperbola are given by

$$y_1 = \frac{b}{a}\sqrt{x^2 - a^2} \qquad \text{and} \qquad y_2 = -\frac{b}{a}\sqrt{x^2 - a^2}$$

To prove that the line $y = (b/a)x$ is an asymptote to $y_1 = (b/a)\sqrt{x^2 - a^2}$, we must show that the vertical distance between the line $y = (b/a)x$ and y_1 tends to 0 as x gets large. That is,

$$\lim_{x \to +\infty} (y_1 - y) = \lim_{x \to +\infty} \left(\frac{b}{a}\sqrt{x^2 - a^2} - \frac{b}{a}x\right) = 0$$

To show this, we first note that

$$\frac{b}{a}\sqrt{x^2 - a^2} - \frac{b}{a}x = \frac{b}{a}(\sqrt{x^2 - a^2} - x) = \frac{b}{a}\left(\frac{\sqrt{x^2 - a^2} + x}{\sqrt{x^2 - a^2} + x}\right)(\sqrt{x^2 - a^2} - x)$$

$$= \frac{b}{a}\left(\frac{-a^2}{\sqrt{x^2 - a^2} + x}\right) = \frac{-ab}{\sqrt{x^2 - a^2} + x}$$

By taking the limit, we get

$$\lim_{x \to +\infty}\left(\frac{b}{a}\sqrt{x^2 - a^2} - \frac{b}{a}x\right) = \lim_{x \to +\infty}\left(\frac{-ab}{\sqrt{x^2 - a^2} + x}\right) \underset{\substack{\uparrow \\ \sqrt{x^2 - a^2} \approx x}}{=} \lim_{x \to +\infty}\left(\frac{-ab}{2x}\right) = 0$$

Similarly, we can show that $y = (b/a)x$ is also an asymptote to

$$y_2 = -\left(\frac{b}{a}\right)\sqrt{x^2 - a^2} \qquad \text{as} \quad x \to -\infty$$

and that $y = -(b/a)x$ is an asymptote to

$$y_1 = \left(\frac{b}{a}\right)\sqrt{x^2 - a^2} \qquad \text{as} \quad x \to -\infty$$

and to

$$y_2 = -\left(\frac{b}{a}\right)\sqrt{x^2 - a^2} \qquad \text{as} \quad x \to +\infty$$

(See Problems 28 and 29.)

The asymptotes can be used as a guide for approximating the graph of a hyperbola. This is accomplished by locating the vertices $(a, 0)$ and $(-a, 0)$ of the hyperbola and the two points $(0, b)$ and $(0, -b)$. These four points are used to form a rectangle (see Fig. 22). The diagonals of this rectangle have slope b/a and $-b/a$, and their extensions are the asymptotes $y = (b/a)x$ and $y = -(b/a)x$. The graph of the hyperbola can now be obtaining by using the rectangle and the asymptotes as guides.

[12.4.5] THEOREM / *Asymptotes of a Hyperbola.*

For the hyperbola

$$\frac{y^2}{a^2} - \frac{x^2}{b^2} = 1$$

the asymptotes are the lines

$$y = \frac{a}{b}x \qquad \text{and} \qquad y = -\frac{a}{b}x$$

EXAMPLE 3

Discuss the equation $4y^2 - x^2 = 4$ and graph it by using the asymptotes as guides.

Solution

To get the equation in standard form, we divide both sides by 4:

$$y^2 - \frac{x^2}{4} = 1$$

This is the equation of a hyperbola with center at $(0, 0)$ and the y-axis as transverse axis. Furthermore, $a^2 = 1$, $b^2 = 4$, and $c^2 = a^2 + b^2 = 5$. The foci are at $(0, \sqrt{5})$ and $(0, -\sqrt{5})$; the vertices are at $(0, 1)$ and $(0, -1)$. We form the rectangle containing the vertices $(0, 1)$ and $(0, -1)$ and the points $(2, 0)$ and $(-2, 0)$. The extension of the diagonals of this rectangle are the asymptotes $y = x/2$ and $y = -x/2$ (see Fig. 23).

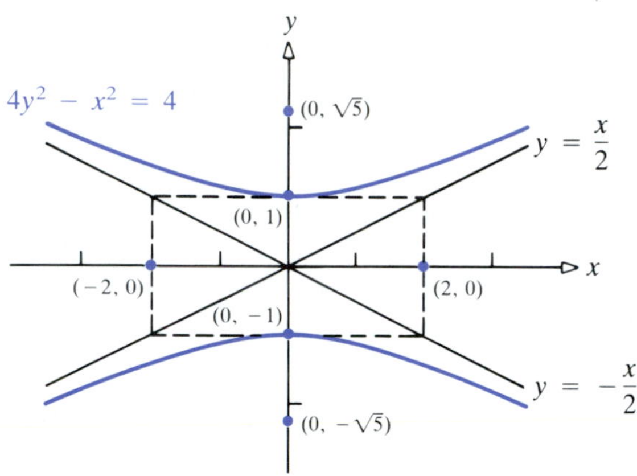

Figure 23

EXERCISE 12.4

In Problems 1–8 find the equation of the hyperbola that has the stated properties.

1. Center at $(0, 0)$; focus at $(6, 0)$; vertex at $(4, 0)$

2. Center at $(0, 0)$; focus at $(5, 0)$; vertex at $(3, 0)$

3. Center at $(0, 0)$; focus at $(0, -3)$; vertex at $(0, 2)$

4. Center at $(0, 0)$; focus at $(0, 4)$; vertex at $(0, -3)$

5. Foci at $(5, 0)$ and $(-5, 0)$; vertex at $(3, 0)$

6. Focus at $(0, -3)$; vertices at $(0, 2)$ and $(0, -2)$

7. Foci at $(4, 0)$ and $(-4, 0)$; asymptote $y = 2x$

8. Vertices at $(0, 5)$ and $(0, -5)$; asymptote $y = 3x$

In Problems 9–16 discuss and graph each equation.

9. $\frac{x^2}{4} - \frac{y^2}{9} = 1$

10. $\frac{y^2}{9} - \frac{x^2}{4} = 1$

11. $x^2 - 4y^2 = 16$

12. $y^2 - 9x^2 = 18$

13. $y^2 - 4x^2 = 4$

14. $9x^2 - y^2 = 36$

15. $x^2 - y^2 = 16$

16. $y^2 - x^2 = 16$

17. The *eccentricity* e of a hyperbola is defined as the number c/a. Since $c > a$, it follows that $e > 1$. Describe the general shape of a hyperbola whose eccentricity is close to 1. What is the shape if e is very large?

18. Use the result of Problem 17 to calculate the eccentricity of the hyperbolas in Problems 9–16.

19. A hyperbola for which $a = b$ is called an *equilateral hyperbola*. Find the eccentricity of an equilateral hyperbola.

20. Two hyperbolas that have the same set of asymptotes are called *conjugate*. Show that the hyperbolas

$$\frac{x^2}{4} - y^2 = 1 \quad \text{and} \quad y^2 - \frac{x^2}{4} = 1$$

are conjugate. Graph each hyperbola.

21. Show that all of the hyperbolas

$$\frac{x^2}{\cos^2\alpha} - \frac{y^2}{\sin^2\alpha} = 1$$

have their foci at $(\pm 1, 0)$ for all values of α.

22. Find the equation of the hyperbola that has the foci of the ellipse $4x^2 + 9y^2 = 36$ for vertices and the vertices of the ellipse as foci.

23. Find the equation of the asymptotes of the hyperbolas:

(a) $16x^2 - 25y^2 = 400$ (b) $9y^2 - x^2 = 18$

24. Prove that the product of the perpendicular distances of any point on a hyperbola from its asymptotes is constant.

25. Find equations for the tangent and the normal to each of the following hyperbolas at the point indicated:

(a) $4x^2 - 16y^2 = 48$; $(4, -1)$
(b) $x^2 - y^2 = 5$; $(3, -2)$

26. Prove that an ellipse and a hyperbola that have the same foci intersect at right angles.

27. *Cooling Tower* (see Chap. 6, p. 407). Find the equation of the hyperbola with center $(0, 0)$, one vertex at $(147, 0)$, and passing through the point $(155, 123)$. (A calculator will be very helpful for this exercise.)

28. Show that $y = (b/a)x$ is an asymptote to $y_2 = -(b/a)\sqrt{x^2 - a^2}$ as $x \to -\infty$

29. Show that $y = -(b/a)x$ is an asymptote to

$$y_1 = \frac{b}{a}\sqrt{x^2 - a^2} \quad \text{as} \quad x \to -\infty$$

and

$$y_2 = -\frac{b}{a}\sqrt{x^2 - a^2} \quad \text{as} \quad x \to +\infty.$$

30. Show that the graph of an equation of the form

$$Ax^2 + Cy^2 + F = 0 \qquad A \neq 0, \quad C \neq 0, \quad F \neq 0$$

where A and C are of opposite sign, is a hyperbola with center at $(0, 0)$.

31. Prove that an equation of the tangent line at any point (x_0, y_0) of the hyperbola $(x^2/a^2) - (y^2/b^2) = 1$ is $(x_0x/a^2) - (y_0y/b^2) = 1$.

12.5 Translation and Rotation of Axes

In this section we show that the graph of a general second-degree polynomial in two variables—that is,

$$Ax^2 + Bxy + Cy^2 + Dx + Ey + F = 0 \tag{1}$$

is a conic. We assume that A, B, and C are not all simultaneously 0, since, if they were, (1) would represent a line,* provided D and E are not both 0.

We begin with the situation for which $B = 0$ in (1). In this case, the xy term is absent, so that (1) has the form

$$Ax^2 + Cy^2 + Dx + Ey + F = 0 \tag{2}$$

The approach is to complete the squares by grouping together the terms involving x and those involving y.

* It may happen that (1) represents a degenerate conic. For example, $x^2 - y^2 = 0$ is the equation of two lines, $x - y = 0$ and $x + y = 0$. The equation $x^2 + y^2 + 1 = 0$ has no graph. We shall not consider such equations here.

EXAMPLE 1

Complete the squares involving x and y in the polynomial equation below, and identify the conic represented by this equation.

$$x^2 + 2y^2 - 4x - 12y + 20 = 0$$

Solution

First, we group the terms:

$$(x^2 - 4x) + 2(y^2 - 6y) = -20$$

By adding 4 to $(x^2 - 4x)$ and 9 to $(y^2 - 6y)$, we get

$$(x^2 - 4x + 4) + 2(y^2 - 6y + 9) = -20 + 4 + 2(9)$$

or

$$(x - 2)^2 + 2(y - 3)^2 = 2$$

By dividing by 2, we find

$$\frac{(x - 2)^2}{2} + (y - 3)^2 = 1$$

Next, we replace $(x - 2)$ by x' and $(y - 3)$ by y'. Then

$$\frac{x'^2}{2} + y'^2 = 1$$

This is an ellipse, with center at $(x', y') = (0, 0)$ and vertices at $x' = \pm\sqrt{2}$, $y' = 0$. As Figure 24 illustrates, the substitutions $x' = x - 2$ and $y' = y - 3$ enable us to identify the point (2, 3) in the xy-coordinate system as the origin (0, 0) in the $x'y'$-coordinate system.

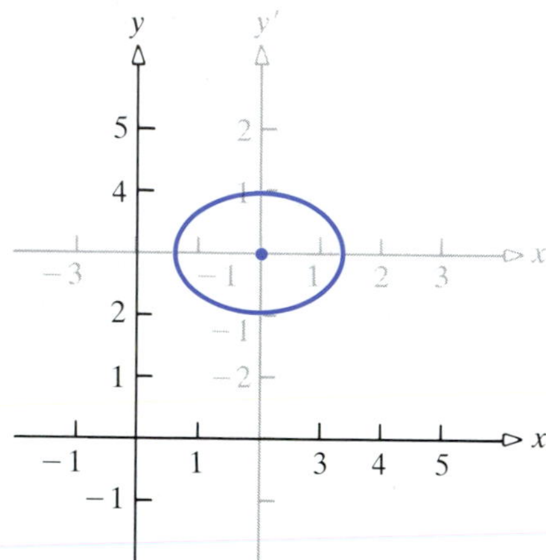

Figure 24

Translation of Axes

In general, if we wish the point (h, k) in the xy-coordinate system to be identified as the origin of the $x'y'$-coordinate system, we use the substitutions

$$\mathbf{x' = x - h \quad \text{and} \quad y' = y - k}$$

This procedure is referred to as a *translation of axes*. The result of a translation of axes is that new axes are obtained that are parallel to the old ones (see Fig. 25).

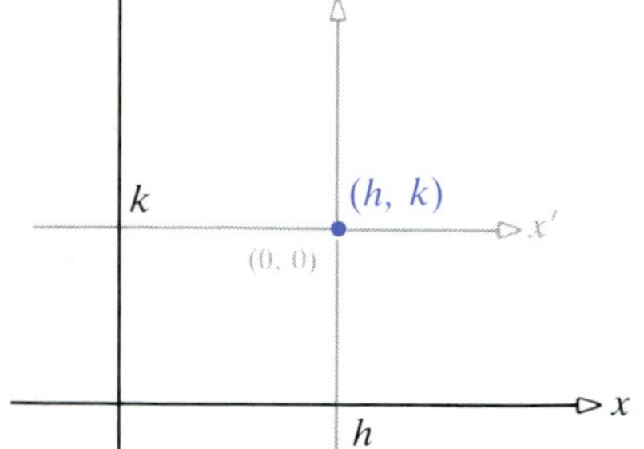

Figure 25

EXAMPLE 2

By a translation of axes, determine the nature of the conic represented by the equation

$$2x^2 - y^2 + 8x + 4y + 3 = 0$$

Graph the equation in the xy-coordinate system.

Solution

First, we group terms:

$$2(x^2 + 4x) - (y^2 - 4y) = -3$$

Next, we complete the squares:

$$2(x^2 + 4x + 4) - (y^2 - 4y + 4) = -3 + 8 - 4$$
$$2(x + 2)^2 - (y - 2)^2 = 1$$

Finally, we translate the axes by using the substitutions $x' = x + 2$ and $y' = y - 2$:

$$2x'^2 - y'^2 = 1$$
$$\frac{x'^2}{\frac{1}{2}} - y'^2 = 1$$

This is the equation of a hyperbola with center at $(0, 0)$ and vertices at $(\pm 1/\sqrt{2}, 0)$ of the $x'y'$-coordinate system. The origin of the $x'y'$ system is the point $(-2, 2)$ of the xy system (see Fig. 26).

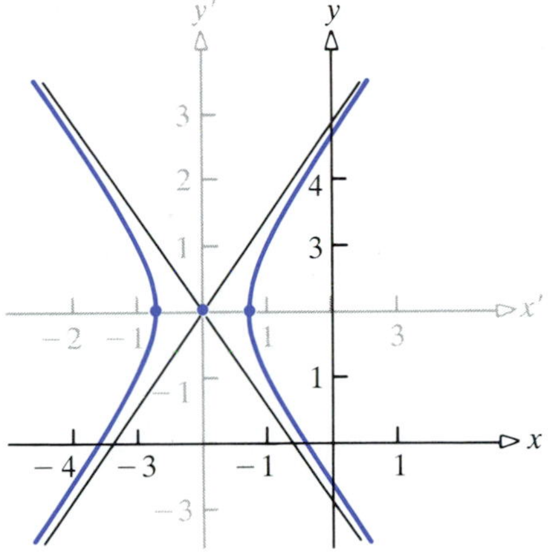

Figure 26

Rotation of Axes

Next, we consider the more difficult case where $B \neq 0$ in the equation

$$Ax^2 + Bxy + Cy^2 + Dx + Ey + F = 0$$

Determination of the nature of this conic requires a *rotation of axes*. In a rotation of axes, the origin remains fixed while the x-axis and y-axis are rotated counterclockwise through a positive acute angle θ, becoming the x''-axis and y''-axis, respectively.

Using Figure 27, we obtain the relationship between the coordinates of a point $P = (x'', y'')$ in the new coordinate system and the coordinates of this same point $P = (x, y)$ in the old coordinate system. Let r denote the distance from the origin O to the point P and let α denote the angle between

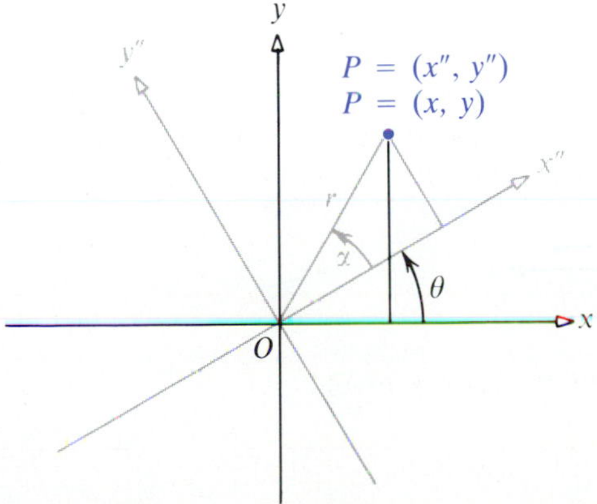

Figure 27

the positive x''-axis and the line OP. Then

$$x'' = r\cos\alpha \qquad x = r\cos(\alpha + \theta)$$

$$y'' = r\sin\alpha \qquad y = r\sin(\alpha + \theta)$$

We want to express x and y in terms of x'', y'', and θ. To accomplish this, we use some identities from trigonometry:

$$\mathbf{x} = r\cos(\alpha + \theta) = r(\cos\alpha\cos\theta - \sin\alpha\sin\theta) = \mathbf{x''\cos\theta - y''\sin\theta}$$

$$\mathbf{y} = r\sin(\alpha + \theta) = r(\sin\alpha\cos\theta + \cos\alpha\sin\theta) = \mathbf{y''\cos\theta + x''\sin\theta} \qquad \textbf{(3)}$$

Although these equations may appear to be cumbersome, they are not difficult to use in practice.

EXAMPLE 3

Transform the equation $xy = 1$ by rotating the axes counterclockwise through an angle of $\pi/4$.

Solution

Replace θ by $\pi/4$ in (3). Then

$$x = \frac{1}{\sqrt{2}}x'' - \frac{1}{\sqrt{2}}y'' \qquad \text{and} \qquad y = \frac{1}{\sqrt{2}}y'' + \frac{1}{\sqrt{2}}x''$$

Substitute these into $xy = 1$:

$$\frac{1}{\sqrt{2}}(x'' - y'')\frac{1}{\sqrt{2}}(y'' + x'') = 1$$

$$\frac{1}{2}(x''^2 - y''^2) = 1$$

$$\frac{x''^2}{2} - \frac{y''^2}{2} = 1$$

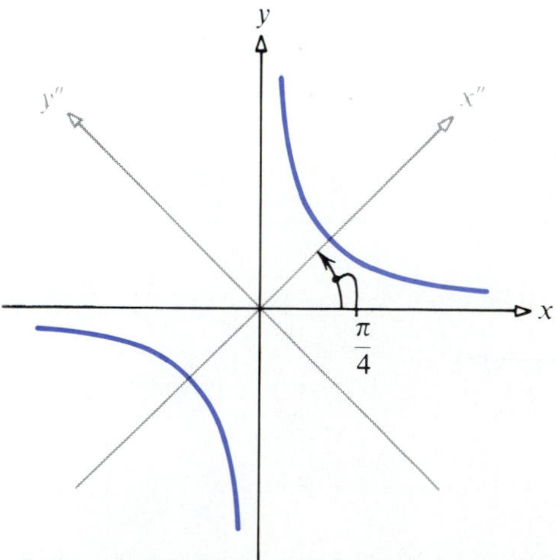

Figure 28

This is the equation of a hyperbola with center at the origin of the $x''y''$-coordinate system and transverse axis along the x''-axis (see Fig. 28).

As Example 3 illustrates, a rotation counterclockwise through an appropriate positive acute angle θ will result in a second-degree equation involving x'' and y'', but with no $x''y''$ term. Then the methods already discussed (including possibly a translation of axes) may be used to determine the nature of the conic and its graph.

To derive a formula for determining the appropriate angle θ, we begin with the equation

$$Ax^2 + Bxy + Cy^2 + Dx + Ey + F = 0 \tag{4}$$

We rotate the axes counterclockwise through an angle θ by using (3):

$$A(x''\cos\theta - y''\sin\theta)^2 + B(x''\cos\theta - y''\sin\theta)(y''\cos\theta + x''\sin\theta)$$
$$+ C(y''\cos\theta + x''\sin\theta)^2 + D(x''\cos\theta - y''\sin\theta) + E(y''\cos\theta + x''\sin\theta) + F = 0$$

By expanding the expression and collecting terms, we obtain an equation of the form

$$A''x''^2 + B''x''y'' + C''y''^2 + D''x'' + E''y'' + F = 0 \tag{5}$$

where

$$A'' = A\cos^2\theta + B\sin\theta\cos\theta + C\sin^2\theta$$
$$B'' = 2(C - A)\sin\theta\cos\theta + B(\cos^2\theta - \sin^2\theta)$$
$$C'' = A\sin^2\theta - B\sin\theta\cos\theta + C\cos^2\theta$$
$$D'' = D\cos\theta + E\sin\theta$$
$$E'' = -D\sin\theta + E\cos\theta$$

Since we want the $x''y''$ term to be absent, we set $B'' = 0$:

$$B'' = 2(C - A)\sin\theta\cos\theta + B(\cos^2\theta - \sin^2\theta) = 0$$

Using the double-angle formulas, we get

$$(C - A)\sin 2\theta + B\cos 2\theta = 0$$

Since we are assuming that $B \neq 0$, we find

$$\cot 2\theta = \frac{A - C}{B}$$

The following theorem summarizes this discussion.

[12.5.1] THEOREM / *Elimination of the xy Term.*

To transfer the equation

$$Ax^2 + Bxy + Cy^2 + Dx + Ey + F = 0$$

into one in x'' and y'', without the $x''y''$ term, rotate the axes counterclockwise through the angle θ, $0 < \theta < \pi/2$, that obeys the

equation

$$\cot 2\theta = \frac{A - C}{B} \tag{6}$$

For example, to eliminate the xy term from the equation $xy = 1$, we note that $A = 0$, $B = 1$, and $C = 0$, so that

$$\cot 2\theta = 0$$

$$2\theta = \frac{\pi}{2}$$

$$\theta = \frac{\pi}{4}$$

This was the angle that worked successfully in Example 3.

EXAMPLE 4

By a rotation of axes, determine the nature of the conic with the equation

$$x^2 + 10\sqrt{3}xy + 11y^2 = 4 \tag{7}$$

Solution

Here $A = 1$, $B = 10\sqrt{3}$, and $C = 11$, so that the angle of rotation needed to eliminate the xy term is given by

$$\cot 2\theta = \frac{A - C}{B} = \frac{1 - 11}{10\sqrt{3}} = \frac{-1}{\sqrt{3}}$$

This implies that

$$2\theta = \frac{2\pi}{3}$$

$$\theta = \frac{\pi}{3}$$

By using this angle in the rotation formulas (3), we obtain

$$x = \frac{1}{2}x'' - \frac{\sqrt{3}}{2}y'' \qquad \text{and} \qquad y = \frac{1}{2}y'' + \frac{\sqrt{3}}{2}x''$$

By substituting for x and y in (7), we find that

$$\left(\frac{1}{2}x'' - \frac{\sqrt{3}}{2}y''\right)^2 + 10\sqrt{3}\left(\frac{1}{2}x'' - \frac{\sqrt{3}}{2}y''\right)\left(\frac{1}{2}y'' + \frac{\sqrt{3}}{2}x''\right) + 11\left(\frac{1}{2}y'' + \frac{\sqrt{3}}{2}x''\right)^2 = 4$$

By expanding and collecting terms, we obtain

$$16x''^2 - 4y''^2 = 4$$

$$\frac{x''^2}{\frac{1}{4}} - y''^2 = 1$$

This is the equation of a hyperbola with center at the origin of the $x''y''$-coordinate system and transverse axis along the x''-axis (see Fig. 29).

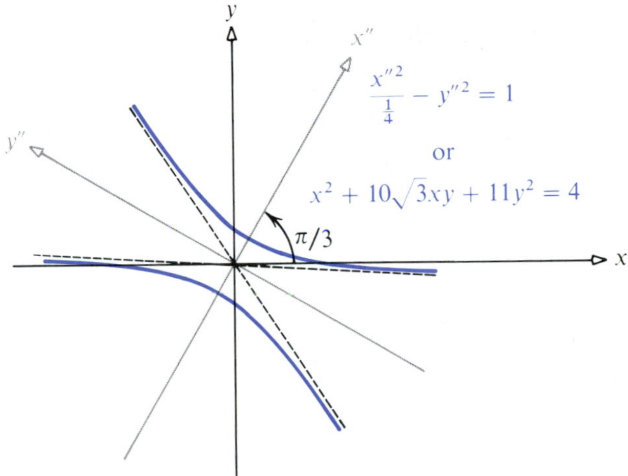

Figure 29

It may happen that an equation of the form (4) will require a rotation of axes to eliminate the xy term, and will then require a translation of axes. In any case, an equation of the form (4) is a conic.

We conclude this section with the following result.

[12.5.2] THEOREM

Consider the second-degree equation

$$Ax^2 + Bxy + Cy^2 + Dx + Ey + F = 0$$

1. **If $B^2 - 4AC > 0$, then the graph is a hyperbola.**
2. **If $B^2 - 4AC < 0$, then the graph is an ellipse.**
3. **If $B^2 - 4AC = 0$, then the graph is a parabola.**

The quantity $B^2 - 4AC$ is called the *discriminant*.

The proof of this result is outlined in Problems 26 and 27.

EXERCISE 12.5

In Problems 1–12 use a translation of axes to determine the nature of each conic. Graph each equation in the xy-coordinate system.

1. $x^2 + 4y^2 - 2x = 0$

2. $4x^2 + y^2 - 4y = 0$

3. $4x^2 - 4y^2 + 8y - 5 = 0$

4. $x^2 - y^2 - 2x + 2y - 1 = 0$

5. $4x^2 - y - 16x + 13 = 0$

6. $y^2 + 4x - 4y = 0$

7. $9x^2 + 16y^2 + 54x - 32y - 47 = 0$

8. $4x^2 + 9y^2 + 24x + 18y + 9 = 0$

9. $25y^2 - 9x^2 - 54x - 100y + 10 = 0$

10. $4y^2 - 8x - x^2 + 32y + 49 = 0$

11. $4x^2 - 3y^2 + 8x + 12y - 20 = 0$

12. $3x^2 + 4y^2 - 12x + 8y + 4 = 0$

In Problems 13–18 determine the nature of each conic. Use a rotation of axes to graph each equation in the xy-coordinate system.

13. $xy = 4$

14. $xy = -4$

15. $3x^2 + 5\sqrt{3}xy + 8y^2 = 10$

16. $x^2 + 4\sqrt{3}xy + 5y^2 = 10$

17. $2x^2 - 2\sqrt{3}xy = 5$

18. $y^2 - \sqrt{3}xy = 4$

In Problems 19–22 use theorem (12.5.2) to determine whether the given second-degree equation is a parabola, ellipse, or hyperbola.

19. $9x^2 - 6xy + y^2 - x + y - 2 = 0$

20. $6x^2 + 11xy + 3y^2 + 11x - y - 10 = 0$

21. $10x^2 + 12xy + 4y^2 - x - y = 0$

22. $9x^2 + 12xy + 4y^2 - x - y = 0$

23. Show that under a rotation of axes the sum $A + C$ is invariant. That is, show that $A'' + C'' = A + C$ for any angle θ.

24. Show that under a rotation of axes the expression $B^2 - 4AC$ is invariant. Refer to Problem 23.

25. Determine the conic given by the equation $17x^2 - 12xy + 8y^2 - 80 = 0$ by rotating the coordinate axes through an angle θ such that $\cos\theta = \sqrt{5}/5$ and $\sin\theta = 2\sqrt{5}/5$.

26. By examining the standard forms presented in this chapter, explain why (except for degenerate cases) the equation $Ax^2 + Cy^2 + Dx + Ey + F = 0$ (A and C not both 0) represents an ellipse if $AC > 0$ (a circle if $A = C$), a parabola if $AC = 0$, and a hyperbola if $AC < 0$.

27. When a rotation of axes eliminates the xy term from the equation $Ax^2 + Bxy + Cy^2 + Dx + Ey + F = 0$, a new equation $A''x''^2 + B''x''y'' + C''y''^2 + D''x'' + E''y'' + F = 0$ is obtained, with $B'' = 0$. According to Problem 24, $B''^2 - 4A''C'' = B^2 - 4AC$. Use this fact, together with Problem 26, to explain why (except for degenerate cases) the graph is a hyperbola, parabola, or ellipse, depending on whether $B^2 - 4AC$ is positive, 0, or negative, respectively.

REVIEW EXERCISES

1. Write the equation of the parabola with focus at (0, 3) and directrix $y = -3$.

2. Write the equation of the parabola with focus at (4, 0) and vertex (0, 0).

3. Find the arc length of the parabola $x^2 = 4ay$ from $x = 0$ to $x = b$.

4. Show that the graph of $x^{1/2} + y^{1/2} = a^{1/2}$ is part of a parabola.

5. Find the equation of the parabola with focus (2, 4) and directrix $y = -3$.

6. Find the equation of the parabola with focus at the origin and directrix $x = 5$.

7. A parabolic arch with a vertical axis is 25 feet across the bottom and 15 feet tall. What is the horizontal distance across the arch at a distance 6 feet below the vertex?

8. The cable for a suspension bridge hangs in a parabolic arch above a roadway 500 feet long. The height of the towers at the ends of the roadway is 80 feet. The cable is 20 feet above the roadway in its middle. How high is the cable above the roadway at a point 100 feet from the middle?

9. The base of a solid is the sector of the parabola $y = 4x^2$ below $y = 1$. Find the volume of the solid if cross sections taken perpendicular to the y-axis

(Problem 9 continues on page 737)

9. (*continued*) are:

(a) Rectangles of height 3
(b) Equilateral triangles with two vertices on the parabola

10. Given the parabola $y^2 = 4ax$, $a > 0$, show that the length of the chord through the focus perpendicular to the axis is $4a$. (This chord is called the *latus rectum* of the parabola.)

11. Write the equation of the ellipse with center at (0, 0), focus at (2, 0), and vertex at (3, 0).

12. Write the equation of the ellipse with center at (0, 0), focus at (2, 0), and semiminor axis of 3.

13. Write the equation of the graph such that the distance of each point from the point (3, 0) is $\frac{3}{5}$ of its distance from the line $x = \frac{25}{3}$.

14. Write the equation of the graph such that the distance of each point from the point (4, 0) is $\frac{4}{5}$ of its distance from the line $x = \frac{25}{4}$.

15. Find the equation for the tangent line to $x^2 + 4y^2 = 4$ at the point $(1, \sqrt{3}/2)$.

16. The base of a solid coincides with the ellipse $4x^2 + 9y^2 = 36$. Find the volume of the solid if cross sections taken perpendicular to the x-axis are:

(a) Semicircles
(b) Squares

17. A bridge is built in the shape of the upper half of an ellipse. The bottom of the bridge has to be 36 meters wide, and it has to be 12 meters high in the middle. What is the equation for the arch?

18. Find the equation of the ellipse having its foci at the points $(-1, 0)$ and $(7, 0)$ and having a major axis of 10.

19. Find the equation of the ellipse having its foci at the points (0, 0) and (0, 4) and having a major axis of 6.

20. Find the equation of the ellipse that passes through the point $(4, \frac{12}{5})$ and has its foci at the points $(-3, 0)$ and $(3, 0)$.

21. Find the coordinates of the foci of the hyperbola $(y^2/9) - (x^2/16) = 1$.

22. Find the vertices of the hyperbola $x^2 = y^2 - 4y + 8$.

23. Find the center of the hyperbola $4x^2 - 5y^2 - 16x + 10y + 31 = 0$.

24. Write the equation of the hyperbola with vertices $(0, \pm 12)$ and asymptotes of $y = \pm\frac{3}{4}x$.

25. Find the equations of the asymptotes of the hyperbola $(x^2/25) - (y^2/36) = 1$.

26. Write the equation of the graph such that each point has a distance from the point (5, 0) equal to $\frac{5}{3}$ of its distance from the line $x = \frac{9}{5}$.

27. Write the equation of the graph such that each point has a distance from the point (5, 0) equal to $\frac{5}{2}$ of its distance from the line $x = \frac{9}{5}$.

28. Find the equation of the hyperbola having its foci at the points (0, 0) and (3, 0) and the difference of the distances of any point on it from the foci equal to 2.

29. The foci of a hyperbola are at the points $(-5, 2)$ and $(5, 2)$, and the difference of the distances of any point on it from the foci is 4. Find the equation of the hyperbola.

30. Find the locus of a point that has the property that its distance from the point (4, 0) is twice the distance from the line $x = 1$.

In Problems 31–35 identify each equation.

31. $5x^2 - 2y^2 + 50x + 16y + 83 = 0$

32. $4x^2 + 9y^2 - 4x + 6y + 9 = 0$

33. $y^2 - 2y - 2x - 5 = 0$

34. $y^2 - 4xy + 3x^2 + 4x = 0$

35. $2x^2 + 4xy + 4y^2 + x + 4y - 5 = 0$

36. Find all the intercepts of $x^2 + y^2 - 6x - 2y + 1 = 0$.

37. For what values of c will the graph of $7x - 4y + c = 0$ be tangent to the graph of $3x^2 - y^2 + x = 0$?

38. Sketch the graph of $x^2 + 2xy + 2y^2 = 8$.

39. Write the equations of the asymptotes to the graph of $2x^2 + 3xy - 2y^2 + x + 2y + 2 = 0$.

40. Sketch the graph of $y^2 - 2xy + 1 = 0$.

CHALLENGE EXERCISES

1. Let A be the area in the first quadrant enclosed by the graphs of $(x^2/9) + (y^2/81) = 1$ and $3x + y = 9$.

(a) Set up—but do not evaluate—an integral representing the area A. Express the integrand as a function of a single variable.

(b) Set up—but do not evaluate—an integral representing the volume of the solid generated when A is rotated about the x-axis. Express the integrand as a function of a single variable.

(c) Set up—but do not evaluate—an integral representing the volume of the solid generated when A is rotated about the y-axis. Express the integrand as a function of a single variable.

2. The points A and B on the line segment AP are free to move on the x-axis and y-axis, respectively (see the figure). If $\overline{AP} = a$ and $\overline{BP} = b$, show that P traces out an ellipse as A and B move on their respective axes.

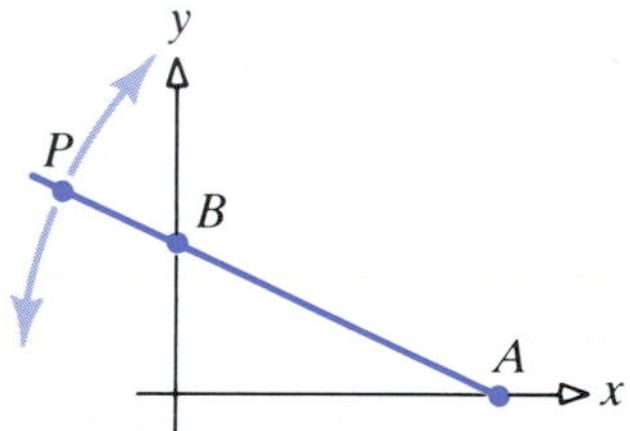

3. Given an ellipse, we may choose a coordinate system so that an equation of the ellipse is given by $(x^2/a^2) + (y^2/b^2) = 1$, where $a > b > 0$. Let F be the focus $(c, 0)$, where $c > 0$. The corresponding *directrix* D is the line $x = a/e$, where e is the eccentricity. Show that the ellipse may be described as the set of points $P = (x, y)$ satisfying the condition $d(F, P)/d(D, P) = e$. (The same condition characterizes the ellipse if F is $(-c, 0)$ and D is the line $x = -a/e$, as you can verify. Hence, the ellipse has a directrix corresponding to each focus.)

4. Rework Problem 3 for an arbitrary hyperbola placed in a coordinate system so that its equation is $(x^2/a^2) - (y^2/b^2) = 1$, where $a > 0$ and $b > 0$.

5. In view of Problems 3 and 4 (and the definition of a parabola in terms of its focus F and directrix D), explain why the condition $d(F, P)/d(D, P) = e$ characterizes every conic (except for degenerate cases), where the conic is an ellipse, parabola, or hyperbola depending on whether $e < 1$, $e = 1$, or $e > 1$, respectively.

6. Show that any chord through the focus of the parabola $y^2 = 4ax$, $a > 0$, has length $2a + x_1 + x_2$, where (x_1, y_1) and (x_2, y_2) are the endpoints of the chord.

7. At each endpoint of the latus rectum of the parabola $y^2 = 4ax$, a tangent line is drawn. Show that these tangent lines intersect at right angles on the directrix of the parabola.

8. Given the ellipse $(x^2/a^2) + (y^2/b^2) = 1$, show that the length of the chord through either focus perpendicular to the line joining the vertices is $2b^2/a$. (Each of these chords is a *latus rectum* of the ellipse.)

9. Rework Problem 8 for the hyperbola $(x^2/a^2) - (y^2/b^2) = 1$.

10. A tangent line is drawn at each endpoint of the latus rectum through the focus $(c, 0)$ of the ellipse $(x^2/a^2) + (y^2/b^2) = 1$. Show that these tangent lines intersect on the directrix $x = a/e$ and have slopes $\pm e$.

11. Rework Problem 10 for the hyperbola $(x^2/a^2) - (y^2/b^2) = 1$.

12. Two lines perpendicular to the asymptotes are drawn through the focus $(c, 0)$ of the hyperbola given by $(x^2/a^2) - (y^2/b^2) = 1$. Prove that each line intersects an asymptote on the directrix of the hyperbola.

13. Find the two parabolas with axes parallel to the coordinate axes having the common tangent line $y = x$ at $P = (1, 1)$; the focus of one parabola is at $(0, 1)$ and the focus of the other parabola is at $(1, 0)$.

Polar Coordinates; Parametric Equations

13

13.1 Polar Coordinates

Until now we have used a system of rectangular coordinates to locate a point in the plane. In this section we describe another system called *polar coordinates*. As we shall soon discover, there are many situations where polar coordinates offer certain advantages over rectangular coordinates.

In a rectangular coordinate system a point in the plane is represented by a pair of numbers (x, y), where x and y represent the signed distance of the point from the y- and x-axis, respectively. In a polar coordinate system a point is represented by a pair of numbers (r, θ), where r is the distance from the origin (called the *pole*) to the point, and θ is an angle between the positive x-axis (called the *polar axis*) and a ray from the origin through the point (see Fig. 1 on page 740).

Just as a rectangular grid is used to locate points given by rectangular coordinates (see Fig. 2(a) on page 740), we can use a grid consisting of

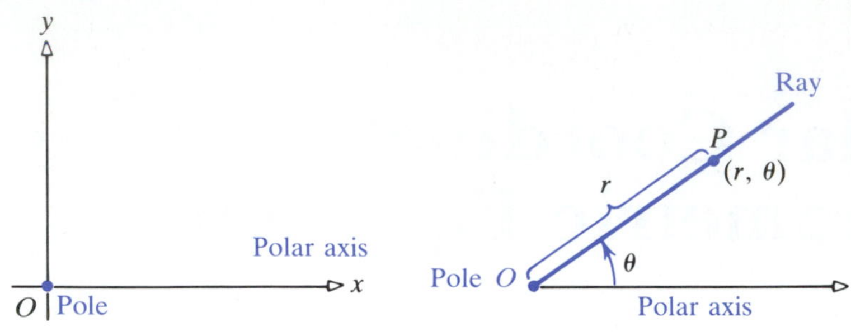

Figure 1

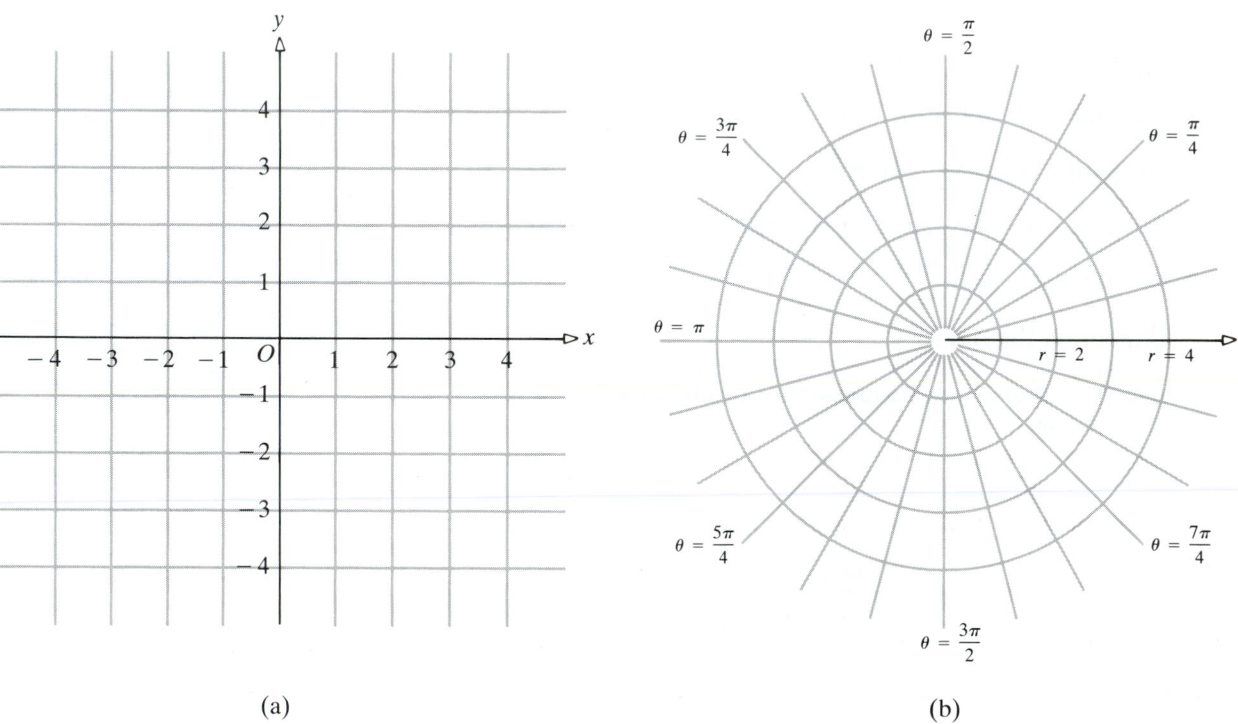

(a) (b)

Figure 2

concentric circles with center at the pole and rays with vertex at the pole to locate points given by polar coordinates (see Fig. 2(b)).

For example, in polar coordinates the point $P = (3, 2\pi/3)$ is located by first drawing the angle with radian measure $2\pi/3$, with vertex at the pole and initial side along the polar axis. Then the point on the terminal side that is 3 units from the pole is the point P, as shown in Figure 3(a). We could also have located this point P by using the polar coordinates $(3, -4\pi/3)$, as shown in Figure 3(b). In fact, the polar coordinates $(3, 8\pi/3)$ also define this same point P, as shown in Figure 3(c). This point P may also be located by the polar coordinates $(-3, -\pi/3)$ because when the first member of the ordered pair of polar coordinates is negative, we follow the convention that this means *to reflect the point about the origin.* In Figure 3(d) we first measure out 3 units along the ray that makes an angle of $-\pi/3$ radians with the polar axis. Then, following this convention, we obtain point P. In Figure 3(e) we follow the same procedure, using the polar coordinates $(-3, 5\pi/3)$ to obtain point P.

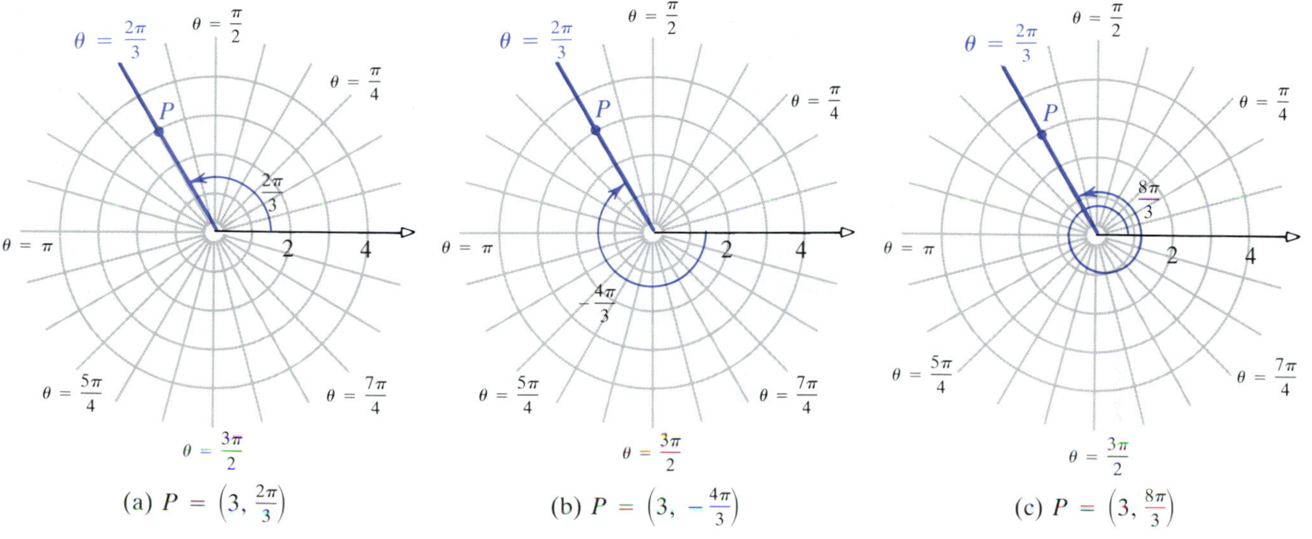

(a) $P = \left(3, \frac{2\pi}{3}\right)$ (b) $P = \left(3, -\frac{4\pi}{3}\right)$ (c) $P = \left(3, \frac{8\pi}{3}\right)$

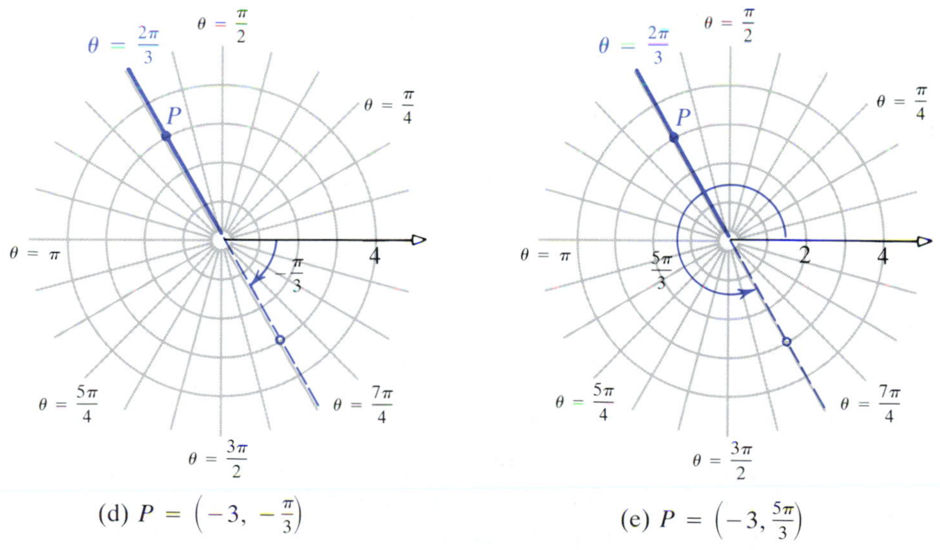

(d) $P = \left(-3, -\frac{\pi}{3}\right)$ (e) $P = \left(-3, \frac{5\pi}{3}\right)$

Figure 3

We can now see a major difference between a rectangular coordinate system and a polar coordinate system. In the former, each point has exactly one pair of rectangular coordinates; in the latter, a point may have infinitely many polar coordinates.

In general, if (r, θ) are polar coordinates of a point, then that same point can be represented by

$$(r, \theta + 2n\pi) \qquad \text{or} \qquad (-r, \theta + (2n+1)\pi) \qquad n \text{ an integer}$$

The pole itself has polar coordinates of the form $(0, \theta)$, where θ can be any number.

Points (r, θ), where r is negative, deserve special attention. The location of the point, instead of being on the terminal side of the angle θ, is on the ray from the pole extending in the direction opposite to the terminal side and at a distance $|r|$ from the pole. Again, refer to Figure 3(d) and 3(e).

To summarize, every pair of numbers (r, θ) determines a unique point P such that $|OP| = |r|$. Furthermore, θ is the radian measure of an angle having its initial side along the polar axis and its terminal side along OP if $r > 0$, and along OP extended through the origin in the opposite direction if $r < 0$. If $r = 0$, then the point $(0, \theta)$ is the pole for any θ.

Let's look at another example.

EXAMPLE 1

Plot the point whose polar coordinates are $(-2, -\pi/4)$, and find three other polar coordinates of the same point such that:

(a) $r > 0$ and $0 < \theta < 2\pi$ (b) $r > 0$ and $-2\pi < \theta < 0$
(c) $r < 0$ and $0 < \theta < 2\pi$

Solution

The point $(-2, -\pi/4)$ is located by first drawing the direction indicated by the angle $-\pi/4$. Then we locate P on the extension of the terminal side of θ, 2 units from the pole (see Fig. 4). Figure 5 illustrates the answers to parts (a), (b), and (c).

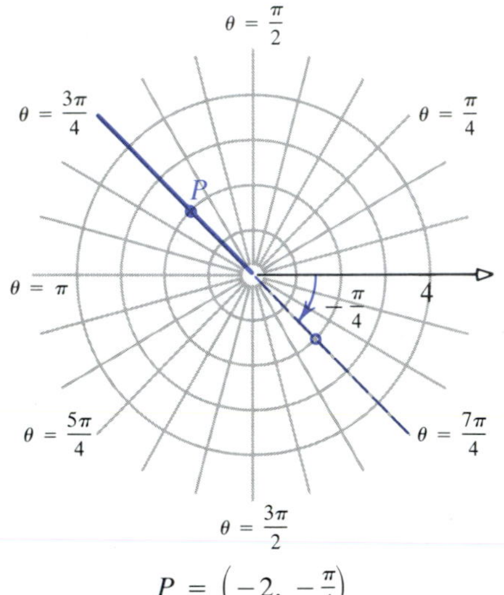

$P = \left(-2, -\frac{\pi}{4}\right)$

Figure 4

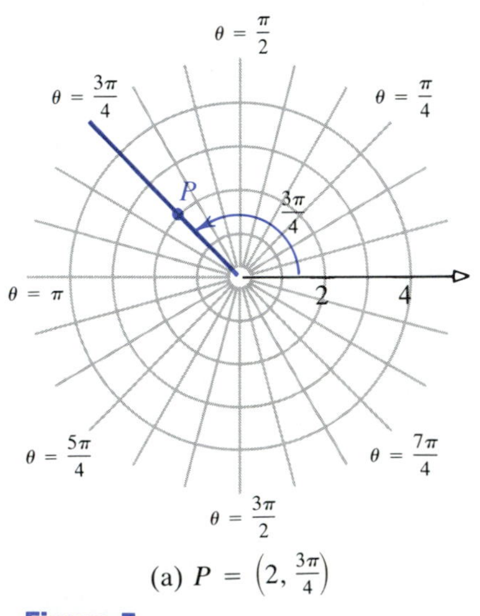

(a) $P = \left(2, \frac{3\pi}{4}\right)$

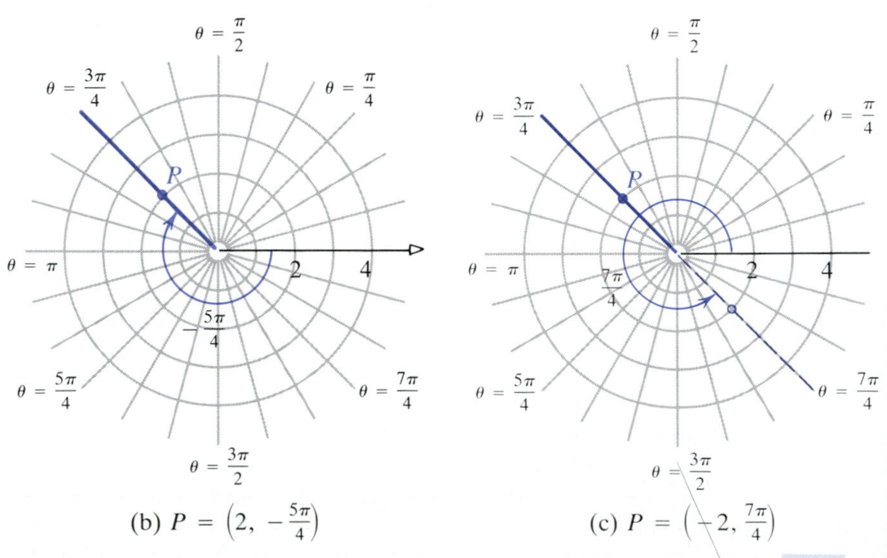

(b) $P = \left(2, -\frac{5\pi}{4}\right)$ (c) $P = \left(-2, \frac{7\pi}{4}\right)$

Figure 5

Relationship Between Polar and Rectangular Coordinates of a Point

It is frequently convenient and of great advantage to be able to transform coordinates or equations in rectangular form into polar form, or vice versa. To do this, we recall that the origin in rectangular coordinates is the pole in polar coordinates and that the positive x-axis in rectangular coordinates is the polar axis in polar coordinates. The y-axis is represented by the ray $\theta = \pi/2$. See Figure 6.

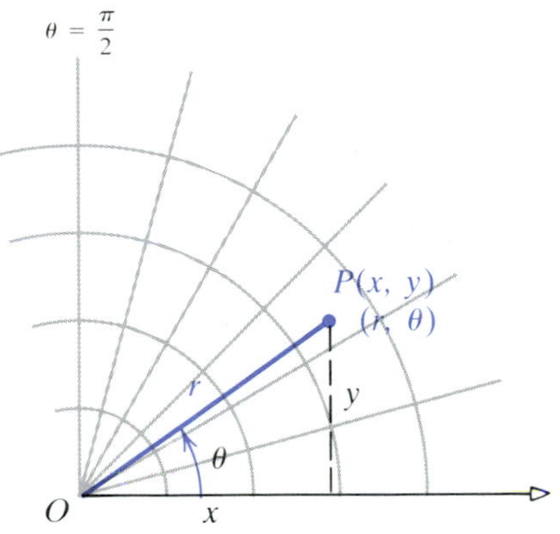

Figure 6

Suppose a point P has the polar coordinates (r, θ). We seek the rectangular coordinates (x, y) of P. If $r > 0$, then the point P is on the terminal side of θ and

$$\cos\theta = \frac{x}{r} \qquad \sin\theta = \frac{y}{r}$$

If $r < 0$, then the point $P = (r, \theta)$ can be represented as $(-r, \pi + \theta)$, where $-r > 0$. Then we have

$$\cos(\pi + \theta) = -\cos\theta = \frac{x}{-r} \qquad \sin(\pi + \theta) = -\sin\theta = \frac{y}{-r}$$

Thus, whether $r > 0$ or $r < 0$, we have

$$x = r\cos\theta \qquad y = r\sin\theta \tag{1}$$

If $r = 0$, then P is the pole and these relationships hold.

Now suppose P has the rectangular coordinates (x, y). We seek a representation (r, θ) for P in polar coordinates. Refer again to Figure 6. Since (x, y) lies on a circle with radius r and center at $(0, 0)$, we have $x^2 + y^2 = r^2$. Using this fact and the definition of $\tan\theta$, it follows that if $x \neq 0$, then

$$r^2 = x^2 + y^2 \qquad \tan\theta = \frac{y}{x}$$

If $x = 0$, then the point P is on the y-axis. In this event, $r = y$, $\theta = \pi/2$ yields the desired result. Of course, we could have obtained these results directly from (1).

[13.1.1] THEOREM

If P is any point in the plane with polar coordinates (r, θ) and rectangular coordinates (x, y), then r, θ and x, y are related by the equations

$$x = r\cos\theta \qquad y = r\sin\theta \tag{2}$$

$$r^2 = x^2 + y^2 \qquad \tan\theta = \frac{y}{x} \quad \text{if } x \neq 0 \tag{3}$$

$$r = y \qquad \theta = \frac{\pi}{2} \quad \text{if } x = 0$$

We use (2) to find the rectangular coordinates of a point given in polar coordinates. We use (3) to find the polar coordinates of a point given in rectangular coordinates. Care must be taken when using (3). The solution for

θ in $\tan\theta = y/x$, $x \neq 0$, will yield two values of θ, $0 \leq \theta < 2\pi$. If the value of θ that corresponds to the quadrant where P is located is chosen, then r must equal the square root of $x^2 + y^2$. If the other value of θ is chosen, then r must equal the negative of the square root of $x^2 + y^2$.

EXAMPLE 2

Find the rectangular coordinates of a point whose polar coordinates are:

(a) $(4, \pi/3)$ (b) $(-2, 3\pi/4)$ (c) $(-3, -5\pi/6)$

Solution

We use equations (2).

(a)
$$x = 4\cos\frac{\pi}{3} = 4\left(\frac{1}{2}\right) = 2$$
$$y = 4\sin\frac{\pi}{3} = 4\left(\frac{\sqrt{3}}{2}\right) = 2\sqrt{3}$$

The rectangular coordinates are $(2, 2\sqrt{3})$.

(b)
$$x = -2\cos\frac{3\pi}{4} = -2\left(-\frac{\sqrt{2}}{2}\right) = \sqrt{2}$$
$$y = -2\sin\frac{3\pi}{4} = -2\left(\frac{\sqrt{2}}{2}\right) = -\sqrt{2}$$

The rectangular coordinates are $(\sqrt{2}, -\sqrt{2})$.

(c)
$$x = -3\cos\left(-\frac{5\pi}{6}\right) = -3\left(-\frac{\sqrt{3}}{2}\right) = \frac{3\sqrt{3}}{2}$$
$$y = -3\sin\left(-\frac{5\pi}{6}\right) = -3\left(-\frac{1}{2}\right) = \frac{3}{2}$$

The rectangular coordinates are $(3\sqrt{3}/2, 3/2)$.

EXAMPLE 3

Find the polar coordinates of the point P whose rectangular coordinates are $(4, -4)$.

Solution

From (3), we find

$$r^2 = (4)^2 + (-4)^2 = 32$$
$$r = \pm\sqrt{32} = \pm 4\sqrt{2}$$
$$\tan\theta = -\frac{4}{4} = -1 \qquad \text{and} \qquad \theta = \frac{3\pi}{4} \quad \text{or} \quad \frac{7\pi}{4}$$

Since $(4, -4)$ is located in the fourth quadrant, it follows that

$$r = -4\sqrt{2} \qquad \text{when} \qquad \theta = \frac{3\pi}{4}$$

$$r = 4\sqrt{2} \qquad \text{when} \qquad \theta = \frac{7\pi}{4}$$

We have thus found two (equivalent) polar coordinates for the point P—namely, $(-4\sqrt{2}, 3\pi/4)$ and $(4\sqrt{2}, 7\pi/4)$. See Figures 7 and 8.

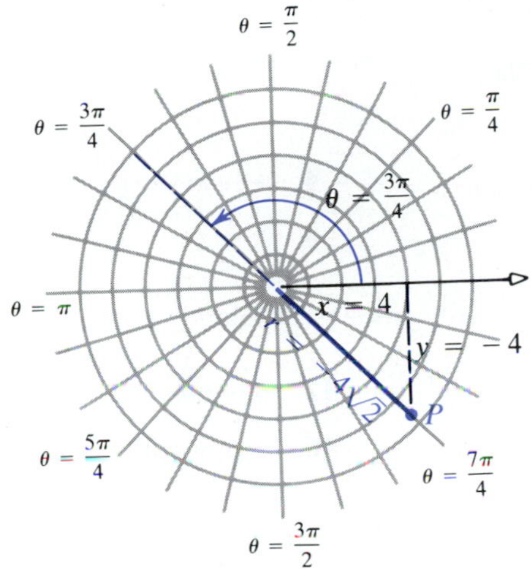

Figure 7

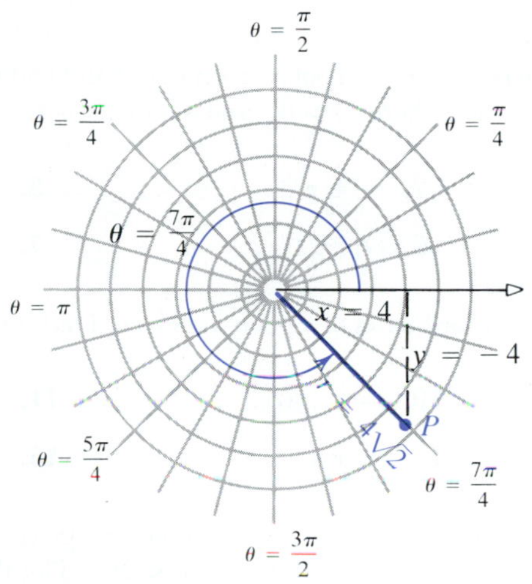

Figure 8

EXAMPLE 4

Transform the equation $r = 2 \cos \theta$ from polar coordinates to rectangular coordinates and identify the graph.

Solution

Multiply both sides of the equation by r to obtain

$$r^2 = 2(r \cos \theta)$$

By replacing r^2 by $x^2 + y^2$ and $r \cos \theta$ by x, we get

$$x^2 + y^2 = 2x$$

This equation can be written as $x^2 - 2x + 1 + y^2 = 1$ or equivalently as $(x - 1)^2 + y^2 = 1$, which is the equation of a circle in rectangular coordinates with center at $(1, 0)$ and radius 1.

EXAMPLE 5

Transform the equation $2xy = a^2$ from rectangular coordinates to polar coordinates.

Solution

By substituting $x = r\cos\theta$ and $y = r\sin\theta$ in the given equation, we get

$$2r^2 \sin\theta\cos\theta = a^2$$

Since $2\sin\theta\cos\theta = \sin 2\theta$, the equation becomes

$$r^2 \sin 2\theta = a^2$$

EXERCISE 13.1

In Problems 1–8 plot each point (given in polar coordinates) and find two other polar coordinates of each point—one where $r > 0$ and another where $r < 0$.

1. $(4, \pi/3)$ **2.** $(-4, \pi/3)$ **3.** $(-4, -\pi/3)$ **4.** $(4, -\pi/3)$

5. $(\sqrt{2}, \pi/4)$ **6.** $(7, 7\pi/4)$ **7.** $(6, 4\pi/3)$ **8.** $(5, \pi/2)$

In Problems 9–16 the polar coordinates of a point are given. Find the rectangular coordinates of each point.

9. $(6, \pi/6)$ **10.** $(-6, \pi/6)$ **11.** $(-6, -\pi/6)$ **12.** $(6, -\pi/6)$

13. $(5, \pi/2)$ **14.** $(8, \pi/4)$ **15.** $(2\sqrt{2}, -\pi/4)$ **16.** $(-5, -\pi/3)$

In Problems 17–22 the rectangular coordinates of a point are given. Find polar coordinates of each point for which $r > 0$ and $0 \le \theta < 2\pi$. Plot the point.

17. $(2, -2)$ **18.** $(-2, 2)$ **19.** $(-2, -2\sqrt{3})$ **20.** $(0, -3)$

21. $(-\sqrt{3}, 1)$ **22.** $(3\sqrt{2}, -3\sqrt{2})$

In Problems 23–30 the letters x, y represent rectangular coordinates. Write each equation in terms of the polar coordinates r, θ.

23. $\dfrac{x^2}{4} + \dfrac{y^2}{9} = 1$ **24.** $x - 4y + 4 = 0$ **25.** $x^2 + y^2 - 4x = 0$ **26.** $y = -6$

27. $x^2 = 1 - 4y$ **28.** $y^2 = 1 - 4x$ **29.** $xy = 1$ **30.** $x^2 + y^2 - 2x + 4y = 0$

In Problems 31–42 the letters r, θ represent polar coordinates. Write each equation in terms of the rectangular coordinates x, y.

31. $r = \cos\theta$ **32.** $r = 2 + \cos\theta$ **33.** $r^2 = \sin\theta$ **34.** $r^2 = 1 - \sin\theta$

35. $r = \dfrac{4}{1 - \cos\theta}$ **36.** $r = \dfrac{4}{4 - \cos\theta}$ **37.** $r^2 = \theta$ **38.** $\theta = -\dfrac{\pi}{4}$

39. $r = 2$ **40.** $r = -5$ **41.** $\tan\theta = 4$ **42.** $\cot\theta = 3$

43. Show that the formula for the distance d between two points $P_1 = (r_1, \theta_1)$ and $P_2 = (r_2, \theta_2)$ is

$$d = \sqrt{r_1^2 + r_2^2 - 2r_1 r_2 \cos(\theta_2 - \theta_1)}$$

Graphs in Polar Coordinates

13.2

An equation whose variables are polar coordinates is called a *polar equation.* The *graph of a polar equation* is the set of all points whose polar coordinates satisfy the equation.

One method we use to graph polar equations is to convert the equation to rectangular coordinates. In the discussion that follows, (x, y) represents the rectangular coordinates of a point P, and (r, θ) represents the polar coordinates of the point P.

EXAMPLE 1

Identify and graph the equation $r \sin \theta = 2$.

Solution

Since $y = r \sin \theta$, we can write the equation as

$$y = 2$$

We conclude that the graph of $r \sin \theta = 2$ is a horizontal line 2 units above the pole. See Figure 9.

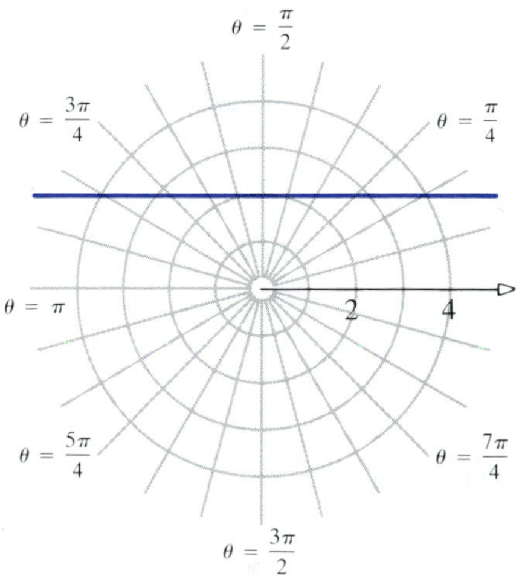

Figure 9

EXAMPLE 2

Identify and graph the equation $r \cos \theta = -3$.

Solution

Since $x = r \cos \theta$, we can write the equation as

$$x = -3$$

We conclude that the graph of $r \cos \theta = -3$ is a vertical line 3 units to the left of the pole. See Figure 10.

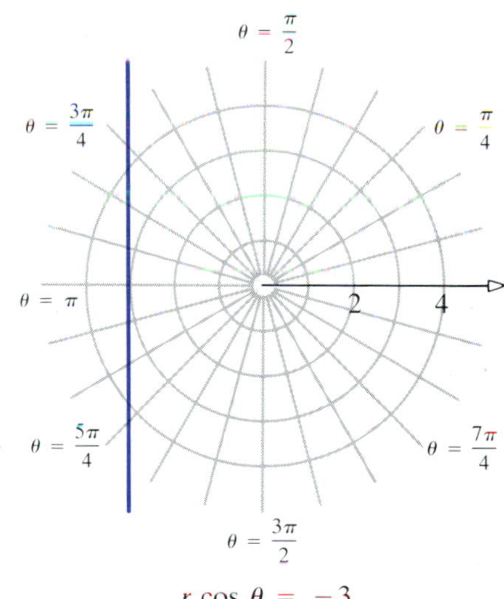

Figure 10

Based on Examples 1 and 2, we are led to the following results. The proofs are left as exercises (Problems 45 and 46).

[13.2.1] THEOREM / *Equations of Horizontal and Vertical Lines.*

If a is a real number, then the graph of the equation

$$r \sin \theta = a$$

is a horizontal line a units above the pole if $a > 0$, and $|a|$ units below it if $a < 0$. The graph of the equation

$$r \cos \theta = a$$

is a vertical line a units to the right of the pole if $a > 0$, and $|a|$ units to the left of the pole if $a < 0$.

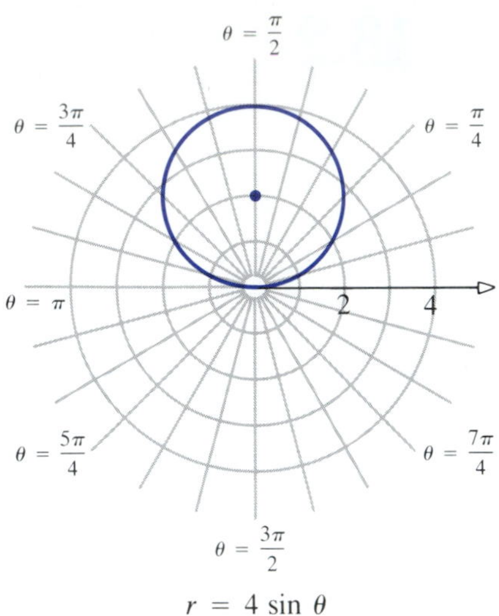

Figure 11

EXAMPLE 3

Identify and graph the equation $r = 4 \sin \theta$.

Solution

To transform to rectangular coordinates, we multiply each side by r:

$$r^2 = 4r \sin \theta$$

Now we use the facts that $r^2 = x^2 + y^2$ and $y = r \sin \theta$. Then,

$$x^2 + y^2 = 4y$$
$$x^2 + (y^2 - 4y) = 0$$
$$x^2 + (y - 2)^2 = 4$$

This is the equation of a circle with center at the rectangular coordinates (0, 2) and radius 2. See Figure 11.

EXAMPLE 4

Identify and graph the equation $r = -2 \cos \theta$.

Solution

We proceed as in Example 3.

$$r^2 = -2r \cos \theta$$
$$x^2 + y^2 = -2x$$
$$x^2 + 2x + y^2 = 0$$
$$(x + 1)^2 + y^2 = 1$$

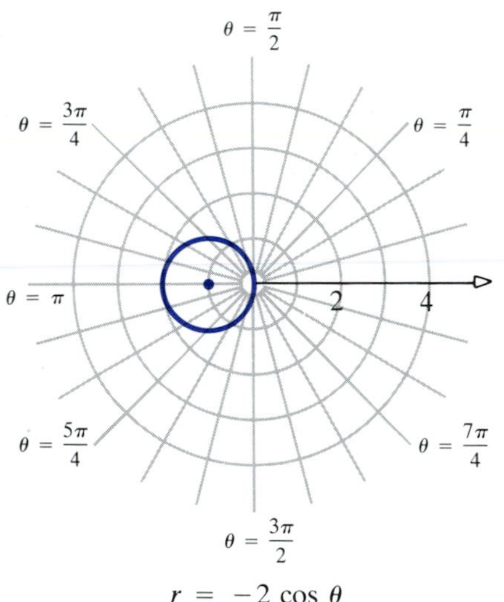

Figure 12

This is the equation of a circle with center at the rectangular coordinates $(-1, 0)$ and radius 1. See Figure 12.

Based on Examples 3 and 4, we are led to the following results. The proofs are left as exercises (Problems 47–50).

[13.2.2] THEOREM

Let $a > 0$ be a positive real number.

Equation	**Description**
(a) $r = 2a \sin \theta$	**Circle: radius a; center at $(0, a)$**
(b) $r = -2a \sin \theta$	**Circle: radius a; center at $(0, -a)$**
(c) $r = 2a \cos \theta$	**Circle: radius a; center at $(a, 0)$**
(d) $r = -2a \cos \theta$	**Circle: radius a; center at $(-a, 0)$**

Each circle passes through the pole.

EXAMPLE 5

Identify and graph the equations:

(a) $r = 3$ (b) $\theta = \frac{\pi}{4}$ (c) $\theta = -\frac{\pi}{6}$

Solution

(a) The points (r, θ) that obey the equation $r = 3$ are those that lie 3 units from the pole. Thus, the graph of $r = 3$ is a circle with center at the pole and radius 3. See Figure 13(a).

(b) The points (r, θ) that obey the equation $\theta = \pi/4$ lie on a line through the pole whose slope is $\tan \theta = \tan \pi/4 = 1$. See Figure 13(b).

(c) The graph of the equation $\theta = -\pi/6$ is a line through the pole whose slope is $\tan \theta = \tan(-\pi/6) = -\sqrt{3}/3$. See Figure 13(c).

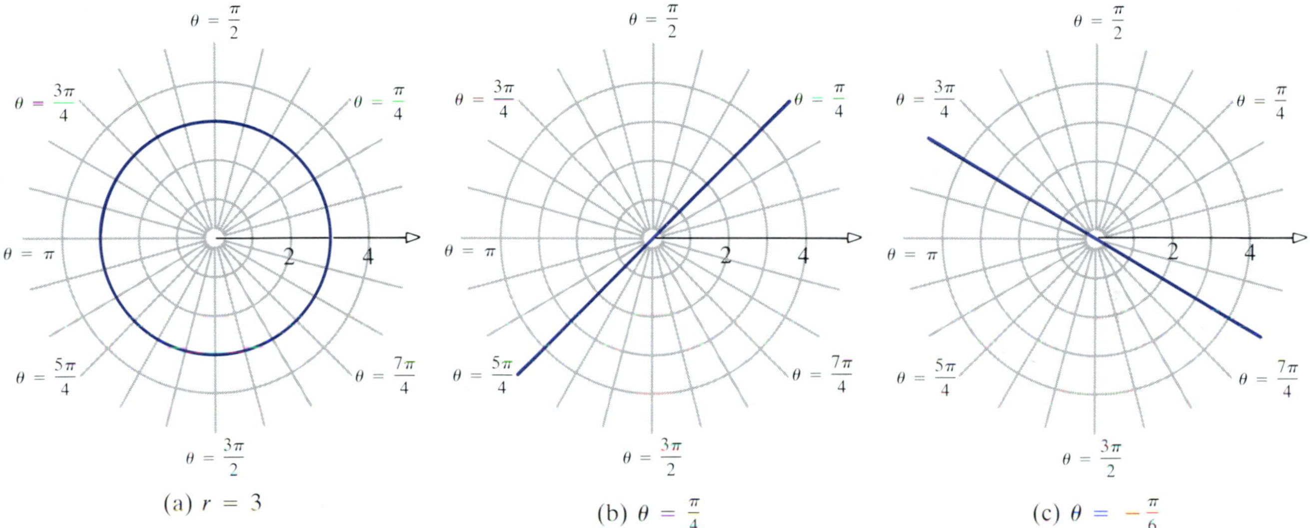

Figure 13

Symmetry

In polar coordinates, the points (r, θ) and $(r, -\theta)$ are symmetric with respect to the polar axis (see Fig. 14(a)). The points (r, θ) and $(r, \pi - \theta)$ are symmetric with respect to the line $\theta = \pi/2$ (see Fig. 14(b)). The points (r, θ) and $(-r, \theta)$ are symmetric with respect to the pole (see Fig. 14(c)). The following tests are a consequence of these observations.

[13.2.3] THEOREM / *Tests for Symmetry.*

Symmetry with respect to the polar axis (x-axis): In a polar equation, replace θ by $-\theta$. If an equivalent equation results, the graph is symmetric with respect to the polar axis.

Symmetry with respect to the line $\theta = \pi/2$ (y-axis): In a polar equation, replace θ by $\pi - \theta$. If an equivalent equation results, the graph is symmetric with respect to the line $\theta = \pi/2$.

Symmetry with respect to the pole (origin): In a polar equation, replace r by $-r$. If an equivalent equation results, the graph is symmetric with respect to the pole.

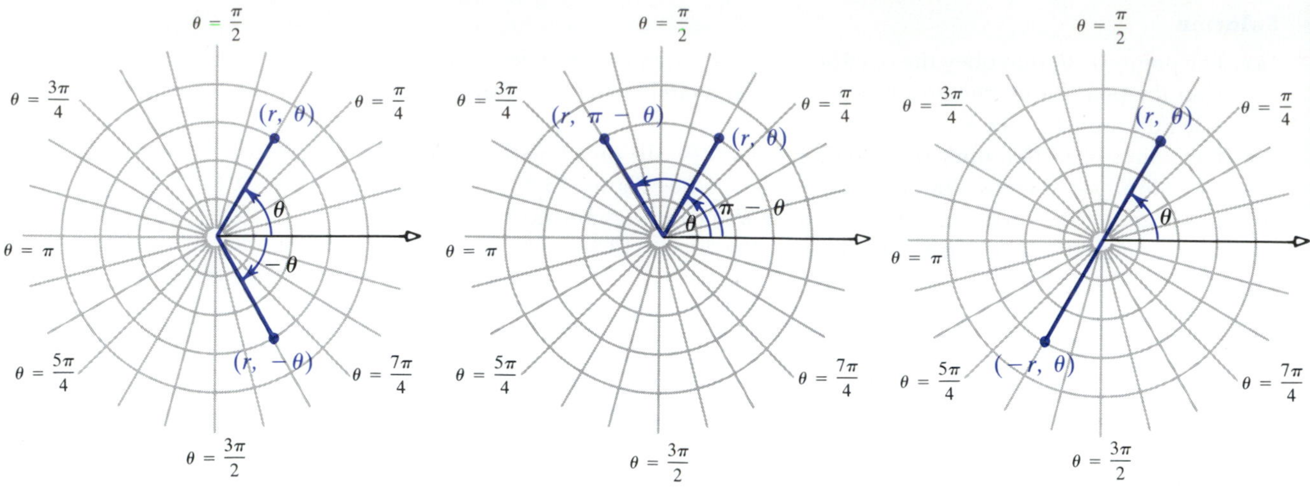

(a) Point symmetric with respect to the polar axis

(b) Point symmetric with respect to the line $\theta = \frac{\pi}{2}$

(c) Point symmetric with respect to the pole

Figure 14

The three tests for symmetry are *sufficient* conditions for symmetry, but they are not *necessary* conditions; that is, an equation may fail the above tests and yet its graph may be symmetric with respect to the polar axis, the line $\theta = \pi/2$, or the pole. A case in point is the graph of $r = \sin 2\theta$, which turns out to be symmetric with respect to the polar axis, the line $\theta = \pi/2$, and the pole, yet all the tests given above fail.

EXAMPLE 6

Graph the equation $r = 1 + 2\cos\theta$.

Solution

First we check for symmetry:

With respect to the polar axis: Replace θ by $-\theta$. The result is

$$r = 1 + 2\cos(-\theta) = 1 + 2\cos\theta$$

Thus, the graph is symmetric with respect to the polar axis.

With respect to the line $\theta = \pi/2$: Replace θ by $\pi - \theta$. The result is

$$r = 1 + 2\cos(\pi - \theta) = 1 - 2\cos\theta$$

The test fails, so the graph may or may not be symmetric with respect to the line $\theta = \pi/2$.

With respect to the pole: Replace r by $-r$. The test fails, so the graph may or may not be symmetric with respect to the pole.

Next, we identify points on the graph of $r = 1 + 2\cos\theta$ by letting θ vary from 0 to $\pi/6$ to $\pi/3$ to $\pi/2$, and so on and calculating the corresponding value of r (Table 1). Due to the periodicity of $\cos\theta$, we need to consider only values of θ from 0 to 2π, since for other values of θ, the same r values will be found.

Table 1

θ	0	$\frac{\pi}{6}$	$\frac{\pi}{3}$	$\frac{\pi}{2}$	$\frac{2\pi}{3}$	$\frac{5\pi}{6}$	π	$\frac{7\pi}{6}$	$\frac{4\pi}{3}$	$\frac{3\pi}{2}$	$\frac{5\pi}{3}$	$\frac{11\pi}{6}$	2π
$r = 1 + 2\cos\theta$	3	$1+\sqrt{3}$	2	1	0	$1-\sqrt{3}$	-1	$1-\sqrt{3}$	0	1	2	$1+\sqrt{3}$	3

Next, we plot the points (r, θ) obtained in Table 1. See Figure 15(a). Now we trace out the graph, beginning at the point $(3, 0)$. We note that as θ varies from 0 to $\pi/6$, the value of r varies from 3 to $1+\sqrt{3} \approx 2.7$. At $\theta = \pi/3$, $r = 2$; at $\theta = \pi/2$, $r = 1$; until at $\theta = 2\pi/3$, we have $r = 0$. This flow is shown in Figure 15(b). The flow of points as θ continues

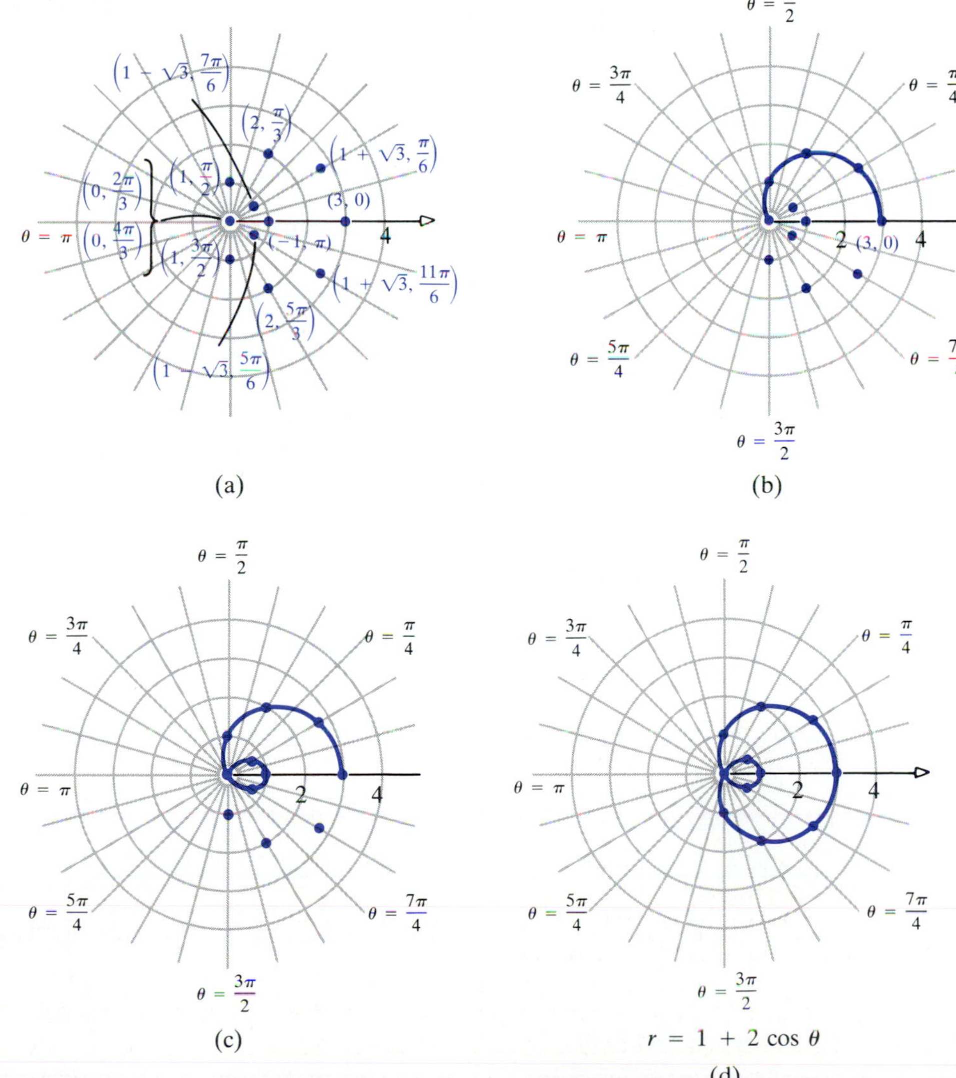

(a)

(b)

(c)

$r = 1 + 2\cos\theta$

(d)

Figure 15

from $2\pi/3$ to π to $4\pi/3$ is shown in Figure 15(c). Finally, after θ varies from $4\pi/3$ to $3\pi/2$ to $5\pi/3$ to 2π, we obtain the complete graph in Figure 15(d).

The curve in Figure 15(d) is an example of a *limaçon with inner loop*. Limaçons with inner loop are characterized by equations of the form

$$r = a + b\cos\theta \qquad r = a + b\sin\theta$$

$$r = a - b\cos\theta \qquad r = a - b\sin\theta$$

where $a > 0$, $b > 0$, and $a < b$.

EXAMPLE 7

Graph the equation $r = 1 - \sin\theta$.

Solution

We check for symmetry first.

Polar axis: Replace θ by $-\theta$. The result is

$$r = 1 - \sin(-\theta) = 1 + \sin\theta$$

The test fails, so the graph may or may not be symmetric with respect to the polar axis.

The line $\theta = \pi/2$: Replace θ by $\pi - \theta$. The result is

$$r = 1 - \sin(\pi - \theta) = 1 - \sin\theta$$

Thus, the graph is symmetric with respect to the line $\theta = \pi/2$.

The pole: Replace r by $-r$. The test fails, so the graph may or may not be symmetric with respect to the pole.

Next, we identify points on the graph by letting the angle θ vary from 0 to $\pi/6$ to $\pi/3$ to $\pi/2$ and so on, and calculating the corresponding values of r (Table 2). Due to the periodicity of $\sin\theta$, we need to consider only values from 0 to 2π.

We plot these points in Figure 16 and begin the flow of points on the graph with the point $(1, 0)$.

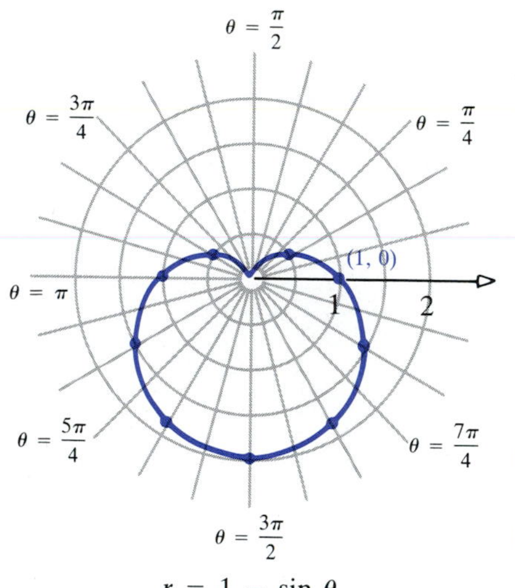

Figure 16

Table 2

θ	0	$\frac{\pi}{6}$	$\frac{\pi}{3}$	$\frac{\pi}{2}$	$\frac{2\pi}{3}$	$\frac{5\pi}{6}$	π	$\frac{7\pi}{6}$	$\frac{4\pi}{3}$	$\frac{3\pi}{2}$	$\frac{5\pi}{3}$	$\frac{11\pi}{6}$	2π
$r = 1 - \sin\theta$	1	$1-\frac{1}{2}=\frac{1}{2}$	$1-\frac{\sqrt{3}}{2}$	0	$1-\frac{\sqrt{3}}{2}$	$1-\frac{1}{2}=\frac{1}{2}$	1	$1+\frac{1}{2}=\frac{3}{2}$	$1+\frac{\sqrt{3}}{2}$	$1+1=2$	$1+\frac{\sqrt{3}}{2}$	$1+\frac{1}{2}=\frac{3}{2}$	1

The curve in Figure 16 is an example of a *cardioid* (a heart-shaped curve). Cardioids are characterized by equations of the form

$$r = a(1 + \cos\theta) \qquad r = a(1 + \sin\theta)$$

$$r = a(1 - \cos\theta) \qquad r = a(1 - \sin\theta)$$

where $a > 0$. The graphs are heart-shaped.

EXAMPLE 8

Graph the equation $r = 2\cos 2\theta$.

Solution

We check for symmetry.

Polar axis: If we replace θ by $-\theta$, the result is

$$r = 2\cos 2(-\theta) = 2\cos 2\theta$$

Thus, the graph is symmetric with respect to the polar axis.

The line $\theta = \pi/2$: If we replace θ by $\pi - \theta$, we obtain

$$r = 2\cos 2(\pi - \theta) = 2\cos 2\theta$$

Thus, the graph is symmetric with respect to the line $\theta = \pi/2$. Consequently, since the graph is symmetric with respect to both the polar axis and the line $\theta = \pi/2$, it must be symmetric with respect to the pole.

Next, we construct Table 3. Since $\cos 2\theta$ has period π, we consider first values of θ from 0 to π. We plot and connect these points in Figure 17(a).

The graph is not complete yet. To obtain the remaining points, continue Table 3 for $\theta = \pi$ to $\theta = 2\pi$ or use the symmetry with respect to the polar axis. The graph of $r = 2\cos 2\theta$ is shown in Figure 17(b).

Table 3

θ	0	$\frac{\pi}{6}$	$\frac{\pi}{4}$	$\frac{\pi}{3}$	$\frac{\pi}{2}$	$\frac{2\pi}{3}$	$\frac{3\pi}{4}$	$\frac{5\pi}{6}$	π
$r = 2\cos 2\theta$	2	1	0	-1	-2	-1	0	1	2

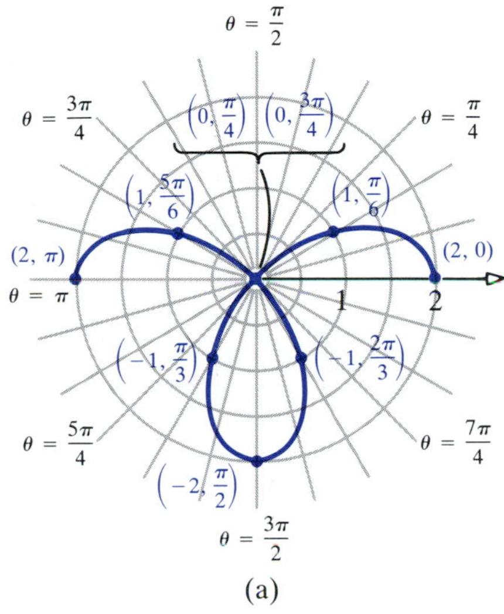

(a)

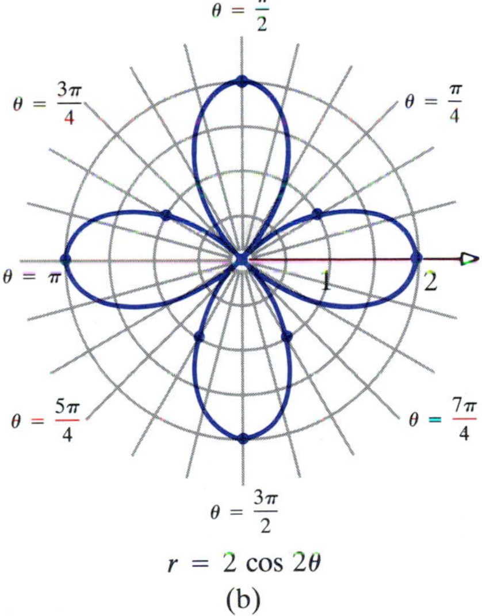

$r = 2\cos 2\theta$

(b)

Figure 17

The curve in Figure 17(b) is called a *rose with four petals*. Roses are characterized by equations of the form

$$r = a\cos n\theta \qquad r = a\sin n\theta \qquad a > 0$$

The graphs are rose-shaped. If n is even, the rose has $2n$ petals; if n is odd, the rose has n petals.

EXAMPLE 9

Graph the equation $r^2 = 4 \sin 2\theta$.

Solution

We leave it for you to verify that the graph is symmetric with respect to the pole. Table 4 provides points on the graph for values of $\theta = 0$ through $\theta = \pi$. (Note that there are no points on the graph for $\pi/2 < \theta < \pi$ (quadrant II), since $\sin 2\theta < 0$ for such values.) The points in Table 4 are plotted in Figure 18(a). The remaining points on the graph may be obtained either by using symmetry or by continuing Table 4 for $\pi \leq \theta \leq 3\pi/2$. Figure 18(b) shows the final graph.

Table 4

θ	0	$\frac{\pi}{6}$	$\frac{\pi}{4}$	$\frac{\pi}{3}$	$\frac{\pi}{2}$
$r^2 = 4 \sin 2\theta$	0	$2\sqrt{3}$	4	$2\sqrt{3}$	0

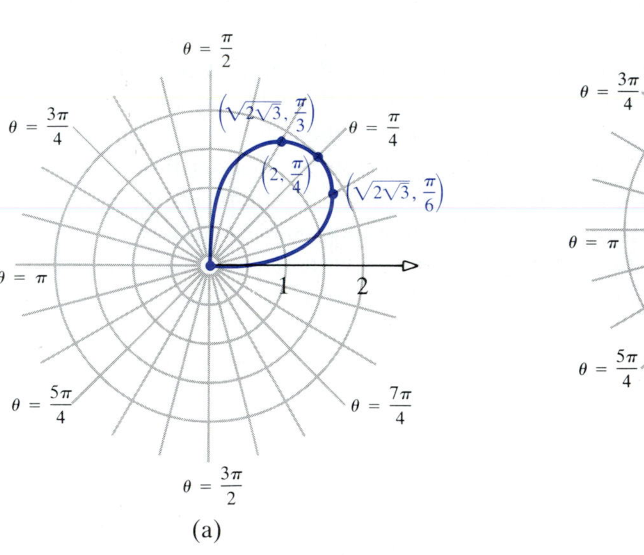

(a)

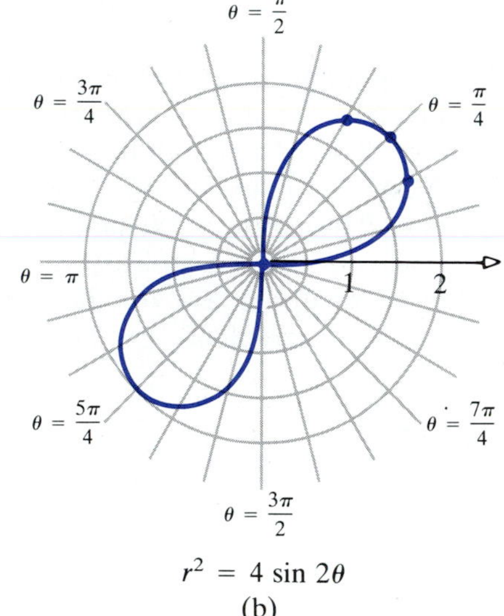

(b)

Figure 18

The curve in Figure 18(b) is an example of a *lemniscate*. Lemniscates are characterized by equations of the form

$$r^2 = a \sin 2\theta \qquad r^2 = a \cos 2\theta \qquad a \neq 0$$

The graphs are propeller-shaped.

EXAMPLE 10

Sketch the graph of $r = e^{\theta/5}$.

Solution

The tests for symmetry with respect to the pole, the polar axis, and the line $\theta = \pi/2$ fail. Furthermore, there is no number θ for which $r = 0$. Hence, the graph does not pass through the pole. We observe that r is positive for all θ, that r increases as θ increases, that $r \to 0$ as $\theta \to -\infty$, and that $r \to +\infty$ as $\theta \to +\infty$. With the help of a calculator, we get the values in Table 5. See Figure 19 for the graph. This is called a *logarithmic spiral*, since its equation may be written as $\theta = 5 \ln r$ and it spirals infinitely often both in approaching the pole and in receding from it.

Table 5

θ	0	$\frac{\pi}{4}$	$\frac{\pi}{2}$	π	$\frac{3\pi}{2}$	2π	$-\frac{\pi}{4}$	$-\frac{\pi}{2}$	$-\pi$	$-\frac{3\pi}{2}$
$r = e^{\theta/5}$	1	1.17	1.37	1.87	2.57	3.51	0.85	0.73	0.53	0.39

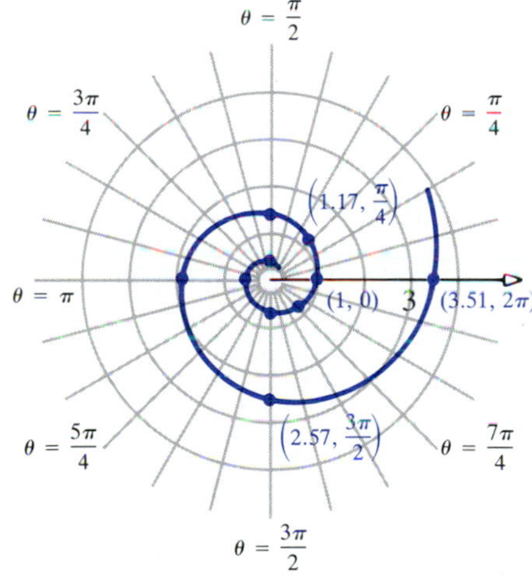

Figure 19

EXERCISE 13.2

In Problems 1–16 identify and graph each polar equation by converting to a rectangular equation.

1. $r = 4$ **2.** $r = 2$ **3.** $\theta = \pi/3$ **4.** $\theta = -\pi/4$

5. $r \sin \theta = 4$ **6.** $r \cos \theta = 4$ **7.** $r \cos \theta = -2$ **8.** $r \sin \theta = -2$

9. $r = 2 \cos \theta$ **10.** $r = 2 \sin \theta$ **11.** $r = -4 \sin \theta$ **12.** $r = -4 \cos \theta$

13. $r \sec \theta = 4$ **14.** $r \csc \theta = 8$ **15.** $r \csc \theta = -2$ **16.** $r \sec \theta = -4$

In Problems 17–32 graph each polar equation. Be sure to test for symmetry.

17. $r = 1 + 2 \sin \theta$ **18.** $r = 1 - 2 \sin \theta$ **19.** $r = 2 - 3 \cos \theta$ **20.** $r = 2 + 4 \cos \theta$

21. $r = 2 + 2\cos\theta$

22. $r = 1 + \sin\theta$

23. $r = 3 - 3\sin\theta$

24. $r = 2 - 2\cos\theta$

25. $r = 3\cos 2\theta$

26. $r = 2\sin 2\theta$

27. $r = 4\sin 3\theta$

28. $r = 3\cos 4\theta$

29. $r^2 = 9\cos 2\theta$

30. $r^2 = \sin 2\theta$

31. $r = 2^\theta$

32. $r = 3^\theta$

In Problems 33–42 graph each polar equation.

33. $r = \dfrac{2}{1 - \cos\theta}$ (parabola)

34. $r = \dfrac{2}{1 - 2\cos\theta}$ (hyperbola)

35. $r = \dfrac{1}{3 - 2\cos\theta}$ (ellipse)

36. $r = \dfrac{1}{1 - \cos\theta}$ (parabola)

37. $r = \theta, \quad \theta \geq 0$ (spiral of Archimedes)

38. $r = 3/\theta, \quad \theta > 0$ (reciprocal spiral)

39. $r = \csc\theta - 2, \quad 0 < \theta < \pi$ (conchoid)

40. $r = \sin\theta\tan\theta$ (cissoid)

41. $r = \tan\theta$ (kappa curve)

42. $r = \cos(\theta/2)$

43. Show that $r = 4(\cos\theta + 1)$ and $r = 4(\cos\theta - 1)$ have the same graph.

44. Show that $r = 5(\sin\theta + 1)$ and $r = 5(\sin\theta - 1)$ have the same graph.

45. Show that the graph of the equation $r\sin\theta = a$ is a horizontal line a units above the pole if $a > 0$, and $|a|$ units below it if $a < 0$.

46. Show that the graph of the equation $r\cos\theta = a$ is a vertical line a units to the right of the pole if $a > 0$, and $|a|$ units to the left if $a < 0$.

47. Show that the graph of the equation $r = 2a\sin\theta$, $a > 0$, is a circle of radius a with center at the rectangular coordinates $(0, a)$.

48. Show that the graph of the equation $r = -2a\sin\theta$, $a > 0$, is a circle of radius a with center at the rectangular coordinates $(0, -a)$.

49. Show that the graph of the equation $r = 2a\cos\theta$, $a > 0$, is a circle of radius a with center at the rectangular coordinates $(a, 0)$.

50. Show that the graph of the equation $r = -2a\cos\theta$, $a > 0$, is a circle of radius a with center at the rectangular coordinates $(-a, 0)$.

51. Let e denote a positive number. This number e is called the *eccentricity*. (Do not confuse this e with the number e that is the base of the natural logarithm function.) Let d denote the distance from a focus to the directrix of a conic. Position the conic so that its focus is at the pole and its directrix is the line $r\cos\theta = -d$, $d > 0$.

(*Problem 51 continues in the next column*)

51. (*continued*)
Show that the polar equation of this conic is

$$r = \frac{ed}{1 - e\cos\theta}$$

where the conic is:

(a) An ellipse if $e < 1$
(b) A parabola if $e = 1$
(c) A hyperbola if $e > 1$

52. The planet Mercury travels around the sun in an elliptical orbit given approximately by

$$r = \frac{(3.442)10^7}{1 - 0.206\cos\theta}$$

where r is measured in miles and the sun is at the pole. Find the distance from Mercury to the sun at *aphelion* (greatest distance from the sun) and at *perihelion* (shortest distance from the sun).

53. Kepler showed that a line joining a planet to the sun sweeps out equal areas in space in equal intervals of time (see the figure). Use this information to determine whether a planet travels faster or slower at aphelion than at perihelion (refer to Problem 52).

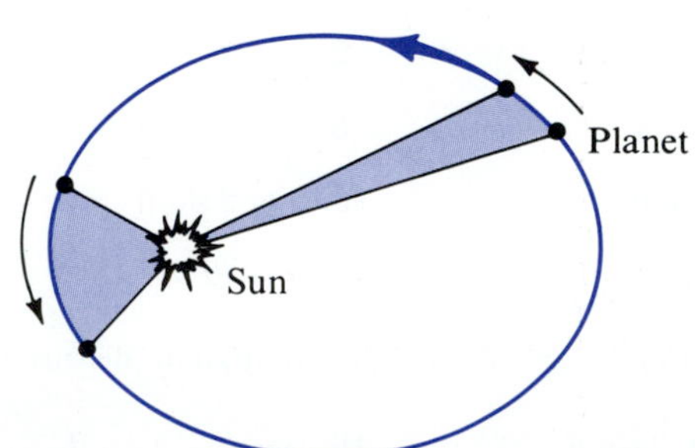

Area in Polar Coordinates

13.3

In this section we consider the problem of finding the area of a region in the plane enclosed by the graph of a polar equation and two rays that have the pole as a common vertex. The technique used is analogous to the one introduced in Chapter 5. Here, however, instead of using the formula for the area of a rectangle, we use the fact that for a circle of radius r, a sector of central angle θ (measured in radians) has area

$$A = \tfrac{1}{2}r^2\theta$$

(see Fig. 20).

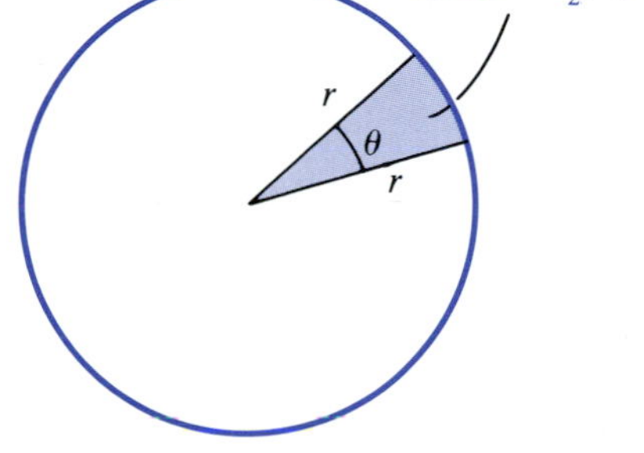

Figure 20

The formula for area in polar coordinates is developed as follows: In Figure 21, $r = f(\theta)$ denotes a continuous, nonnegative function defined on $\alpha \le \theta \le \beta$, and A is the area enclosed by the graph of the equation $r = f(\theta)$ and by the rays $\theta = \alpha$ and $\theta = \beta$, where $0 \le \alpha < \beta \le 2\pi$. It will prove helpful to think of the area A as being "swept out" by rays, beginning with the ray $\theta = \alpha$ and continuing up to the ray $\theta = \beta$.

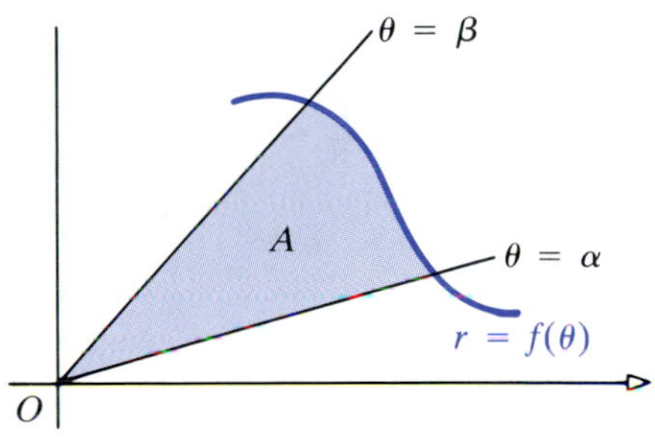

Figure 21

Now refer to Figure 22, where we have partitioned $[\alpha, \beta]$ into n subintervals

$$[\alpha, \theta_1], \quad [\theta_1, \theta_2], \quad \ldots, \quad [\theta_{i-1}, \theta_i], \quad \ldots, \quad [\theta_{n-1}, \beta]$$

We denote the measure of the ith subinterval by $\Delta\theta_i = \theta_i - \theta_{i-1}$, and we select an angle θ_i^* in $[\theta_{i-1}, \theta_i]$. The quantity $\frac{1}{2}[f(\theta_i^*)]^2\Delta\theta_i$ is the area of the circular sector with radius $r = f(\theta_i^*)$ and central angle $\Delta\theta_i = \theta_i - \theta_{i-1}$ and gives an approximation to the area OPQ. The sum of the areas of these sectors

$$\sum_{i=1}^{n} \frac{1}{2}[f(\theta_i^*)]^2\Delta\theta_i$$

is an estimate of the area A. As the length of each subinterval gets smaller and smaller—that is, as the norm of the partition approaches 0—the sum $\sum_{i=1}^{n} \frac{1}{2}[f(\theta_i^*)]^2\Delta\theta_i$ becomes a better and better approximation to the area

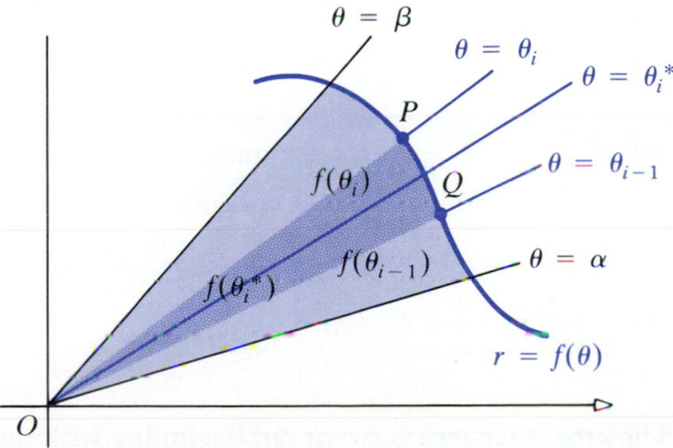

Figure 22

we seek. However, this sum is a Riemann sum. Hence, since f is continuous, as the norm of the partition approaches 0, its limit is the definite integral $\int_\alpha^\beta \frac{1}{2}[f(\theta)]^2\,d\theta$. This suggests the following result:

[13.3.1] THEOREM / *Area in Polar Coordinates.*

If $r = f(\theta)$ is continuous and nonnegative on $\alpha \le \theta \le \beta$, $0 \le \alpha < \beta \le 2\pi$, the area A enclosed by the graph of the equation $r = f(\theta)$ and the rays $\theta = \alpha$ and $\theta = \beta$ is given by

$$A = \int_\alpha^\beta \frac{1}{2}[f(\theta)]^2\,d\theta = \int_\alpha^\beta \frac{1}{2}r^2\,d\theta \tag{1}$$

In order to use (1), it is necessary to first draw an accurate graph of the equation $r = f(\theta)$, $\alpha \le \theta \le \beta$. In doing this, draw rays, indicating the start $(\theta = \alpha)$ and finish $(\theta = \beta)$, that sweep out the area to be found. These rays then determine the limits of integration to use in (1).

EXAMPLE 1

Find the area in the first quadrant enclosed by the graph of the equation $r = 1 - \sin\theta$.

Solution

We drew the graph of $r = 1 - \sin\theta$, a cardioid, in Example 7 of Section 13.2. Its graph is reproduced in Figure 23, with the area to be found shaded. As the figure illustrates, the area we seek is swept out beginning with the ray $\theta = 0$ and ending with the ray $\theta = \pi/2$. Thus, our limits of integration are 0 and $\pi/2$.

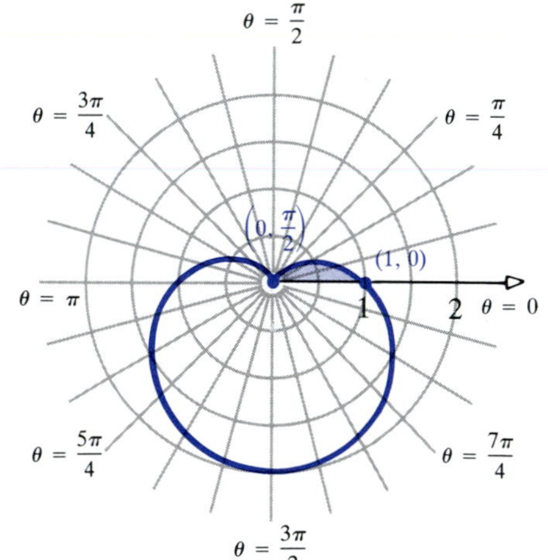

Figure 23

$$A = \int_\alpha^\beta \frac{1}{2}r^2\,d\theta = \int_0^{\pi/2} \frac{1}{2}(1 - \sin\theta)^2\,d\theta = \frac{1}{2}\int_0^{\pi/2} (1 - 2\sin\theta + \sin^2\theta)\,d\theta$$

Using the identity $\sin^2\theta = \frac{1}{2}(1 - \cos 2\theta)$, we find

$$\begin{aligned} A &= \frac{1}{2}\int_0^{\pi/2} \left[1 - 2\sin\theta + \frac{1}{2}(1 - \cos 2\theta)\right]d\theta \\ &= \frac{1}{2}\int_0^{\pi/2} \left[\frac{3}{2} - 2\sin\theta - \frac{1}{2}\cos 2\theta\right]d\theta \\ &= \frac{1}{2}\left[\frac{3}{2}\theta + 2\cos\theta - \frac{1}{4}\sin 2\theta\right]\Bigg|_0^{\pi/2} = \frac{3\pi - 8}{8} \text{ square units} \end{aligned}$$

EXAMPLE 2

Find the area enclosed by the cardioid $r = 1 - \sin\theta$.

Solution

Look again at Figure 23. The area A we seek is swept out beginning with the ray $\theta = 0$ and ending with the ray $\theta = 2\pi$. Thus, we proceed, as in

Example 1, to obtain

$$A = \int_0^{2\pi} \frac{1}{2}(1 - \sin\theta)^2\, d\theta$$

$$= \frac{1}{2}\left[\frac{3}{2}\theta + 2\cos\theta - \frac{1}{4}\sin 2\theta\right]\Bigg|_0^{2\pi} = \frac{3\pi}{2} \text{ square units}$$

Another way of finding the area enclosed by the cardioid $r = 1 - \sin\theta$ is to take advantage of the symmetry with respect to the line $\theta = \pi/2$. Since the area to the left of the line $\theta = \pi/2$ is half of the area A we seek, and since it is swept out by the rays $\theta = \pi/2$ and $\theta = 3\pi/2$, we can compute the area A as follows:

$$A = 2\int_{\pi/2}^{3\pi/2} \frac{1}{2}(1 - \sin\theta)^2\, d\theta$$

We leave it for you to verify that, in fact, this yields the same answer: $3\pi/2$.

EXAMPLE 3

Find the area enclosed by the graph of $r = 2\cos 3\theta$, a rose with three petals.

Solution

Figure 24 depicts the area A we seek. Using symmetry, we see that the area in the first quadrant represents one-sixth of the area A. Since the area in the first quadrant is swept out by the rays $\theta = 0$ and $\theta = \pi/6$, the area A is given by:

$$A = 6\int_0^{\pi/6} \frac{1}{2}(4\cos^2 3\theta)\, d\theta = 12\int_0^{\pi/6} \frac{1 + \cos 6\theta}{2}\, d\theta = 12\left(\frac{\theta}{2} + \frac{\sin 6\theta}{12}\right)\Bigg|_0^{\pi/6}$$

$$= 12\left(\frac{\pi}{12}\right) = \pi \text{ square units}$$

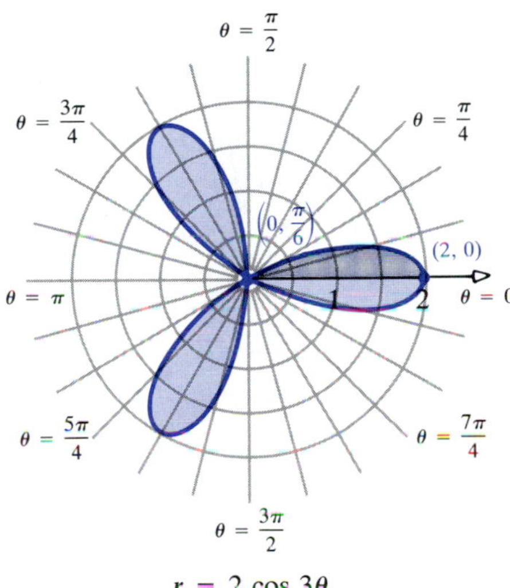

Figure 24

EXAMPLE 4

Find the area enclosed by the graph of the equation $r = 2 + \cos\theta$.

Solution

The area A we seek is shown in Figure 25. Since the graph is symmetric with respect to the polar axis, it is necessary to calculate only the area enclosed by $r = 2 + \cos\theta$ and swept out by the rays $\theta = 0$ and $\theta = \pi$, and then multiply this result by 2. Thus, the area A we seek is

$$A = 2\int_0^{\pi} \frac{1}{2}(2 + \cos\theta)^2\, d\theta = \int_0^{\pi} (4 + 4\cos\theta + \cos^2\theta)\, d\theta$$

$$= \int_0^{\pi} \left[4 + 4\cos\theta + \frac{1}{2}(1 + \cos 2\theta)\right] d\theta$$

$$= \left(4\theta + 4\sin\theta + \frac{\theta}{2} + \frac{1}{4}\sin 2\theta\right)\Bigg|_0^{\pi} = \frac{9\pi}{2} \text{ square units}$$

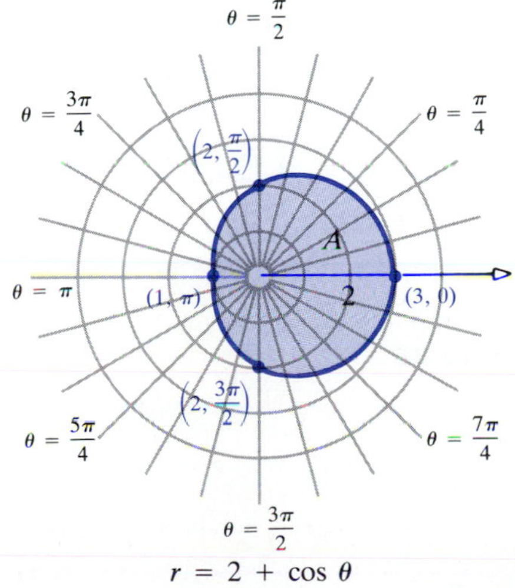

Figure 25

EXAMPLE 5

Find the area that lies outside the cardioid $r = 1 + \cos\theta$ and inside the circle $r = 3\cos\theta$.

Solution

First we graph each equation using the same coordinate system (see Fig. 26(a)). To complete the figure, we need to know the points of intersection of the two graphs. To find them, we equate the given expressions for r:

$$1 + \cos\theta = 3\cos\theta$$
$$2\cos\theta = 1$$
$$\cos\theta = \frac{1}{2}$$
$$\theta = -\frac{\pi}{3} \qquad \theta = \frac{\pi}{3}$$

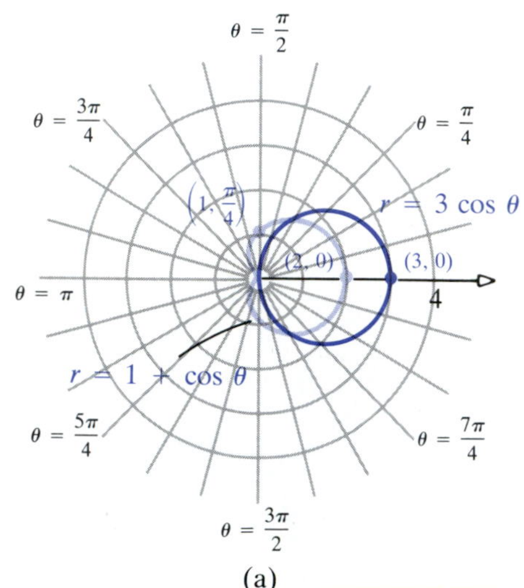

(a)

Thus, the graphs intersect at the points $(3/2, -\pi/3)$ and $(3/2, \pi/3)$. The area A we seek is shown as the shaded portion in Figure 26(b). Notice that the area is the difference between the area enclosed by the circle $r = 3\cos\theta$ swept out by the rays $\theta = -\pi/3$ and $\theta = \pi/3$, and the area enclosed by the cardioid $r = 1 + \cos\theta$ swept out by the same rays. Thus, the area A we seek is

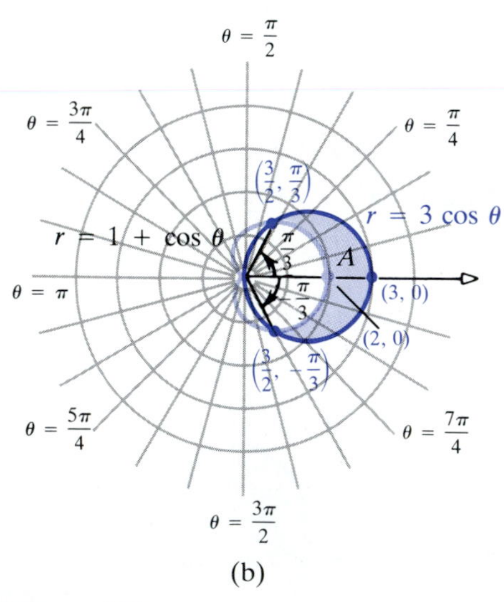

(b)

Figure 26

$$\begin{aligned}
A &= \int_{-\pi/3}^{\pi/3} \frac{1}{2}(3\cos\theta)^2\, d\theta - \int_{-\pi/3}^{\pi/3} \frac{1}{2}(1 + \cos\theta)^2\, d\theta \\
&= \frac{1}{2}\int_{-\pi/3}^{\pi/3} [9\cos^2\theta - 1 - 2\cos\theta - \cos^2\theta]\, d\theta \\
&= \frac{1}{2}\int_{-\pi/3}^{\pi/3} (8\cos^2\theta - 2\cos\theta - 1)\, d\theta \\
&= \frac{1}{2}\int_{-\pi/3}^{\pi/3} \left[8\left(\frac{1 + \cos 2\theta}{2}\right) - 2\cos\theta - 1\right] d\theta \\
&= \frac{1}{2}\int_{-\pi/3}^{\pi/3} (3 + 4\cos 2\theta - 2\cos\theta)\, d\theta \\
&= \frac{1}{2}[3\theta + 2\sin 2\theta - 2\sin\theta]\Big|_{-\pi/3}^{\pi/3} = \pi \text{ square units}
\end{aligned}$$

Warning: Notice in Figure 26(a) that the circle and the cardioid, in fact, intersect in three points. Yet we found only two points of intersection when we equated the expressions for r. The reason we failed to detect the pole as a point of intersection is that, on $r = 1 + \cos\theta$, it has coordinates $(0, \pi)$ and, on $r = 3\cos\theta$, it has coordinates $(0, \pi/2)$ and $(0, 3\pi/2)$.

We see, then, that it is important to make rough sketches of the graphs when we are looking for the points of intersection of two polar graphs. This is the only way to find out whether they have points in common other than those found by solving their equations simultaneously. Since the pole presents particular difficulties, always set $r = 0$ in both equations to determine whether the graphs go through the pole.

EXERCISE 13.3

In each of the following problems draw a figure and indicate the area to be found.

In Problems 1–4 find the area enclosed by the given polar equation and swept out by the given rays.

1. $r = 3\cos\theta;\quad \theta = 0,\quad \theta = \pi/3$

2. $r = 3\sin\theta;\quad \theta = 0,\quad \theta = \pi/4$

3. $r = a\theta;\quad \theta = 0,\quad \theta = 2\pi$

4. $r = e^{a\theta};\quad \theta = 0,\quad \theta = \pi/2$

In Problems 5–8 find the area enclosed by each polar equation.

5. $r = 1 + \cos\theta$

6. $r = 2 - \sec\theta$

7. $r = 2\sin^2(\theta/2)$

8. $r = 3(1 - \sin\theta)$

In Problems 9–12 find the area of one loop of the given polar equation.

9. $r = 4\cos 2\theta$

10. $r = 2\sin 3\theta$

11. $r^2 = 4\cos 2\theta$

12. $r = a^2\cos 2\theta$

In Problems 13–16 find the indicated area.

13. Inside $r = 2\sin\theta$; outside $r = 1$

14. Inside $r = 4\cos\theta$; outside $r = 2$

15. Inside $r = \sin\theta$; outside $r = 1 - \cos\theta$

16. Inside $r^2 = 4\cos 2\theta$; outside $r = \sqrt{2}$

17. Find the area of the small loop of the limaçon $r = 1 + 2\cos\theta$.

18. Find the area of the small loop of the graph of $r = 1 + 2\sin 2\theta$.

19. Find the area inside the circle $r = 8\cos\theta$ and to the right of the line $r = 2\sec\theta$.

20. Find the area inside the circle $r = 10\sin\theta$ and above the line $r = 2\csc\theta$.

21. Find the area outside the circle $r = 3$ and inside the cardioid $r = 2 + 2\cos\theta$.

22. Find the area inside the circle $r = \sin\theta$ and outside the cardioid $r = 1 + \cos\theta$.

23. Find the area common to the circle $r = \cos\theta$ and the cardioid $r = 1 - \cos\theta$.

24. Find the area common to the circles $r = \cos\theta$ and $r = \sin\theta$.

25. Find the area common to the inside of the cardioid $r = 1 + \sin\theta$ and the outside of the cardioid $r = 1 + \cos\theta$.

26. Find the area enclosed by the loop of the strophoid $r = \sec\theta - 2\cos\theta,\quad -\pi/2 < \theta < \pi/2$.

27. Find the area enclosed by the lines $\theta = 0$ and $\theta = 1$ and $r = e^{-\theta},\quad 0 \le \theta \le 1$.

28. Find the area enclosed by the lines $\theta = 0$ and $\theta = 1$ and $r = e^{\theta},\quad 0 \le \theta \le 1$.

29. Find the area enclosed by the lines $\theta = 1$ and $\theta = \pi$ and $r = 1/\theta,\quad 1 \le \theta \le \pi$.

30. Find the area inside the lemniscate $r^2 = 8\cos 2\theta$ and outside the circle $r = 2$.

31. Find the area inside the outer loop but outside the inner loop of $r = 1 + 2\sin\theta$.

13.4 Parametric Equations

Equations of the form $y = f(x)$, f a function, have graphs that are cut no more than once by any vertical line. To study more complicated graphs for which the above rule may not hold, we need a different method. One method is to use a pair of equations and a third variable.

[13.4.1] DEFINITION / *Parametric Equations; Plane Curve.*

Let $x(t)$ and $y(t)$ be two functions defined on some interval I and set

$$x = x(t) \qquad \text{and} \qquad y = y(t) \tag{1}$$

To each number t in I, there corresponds a value of x and a value of y that in turn determine the rectangular coordinates of a point in the xy-plane. The collection of all such points is called a *curve in the plane* or, simply, a *plane curve*. Equations (1) are called *parametric equations* of the curve, and the variable t is called a *parameter* of the curve.

EXAMPLE 1

Consider the parametric equations

$$x = 4t \qquad y = t^2$$

Table 6

t	-2	-1	0	1	2	3
$x = 4t$	-8	-4	0	4	8	12
$y = t^2$	4	1	0	1	4	9
(x, y)	$(-8, 4)$	$(-4, 1)$	$(0, 0)$	$(4, 1)$	$(8, 4)$	$(12, 9)$

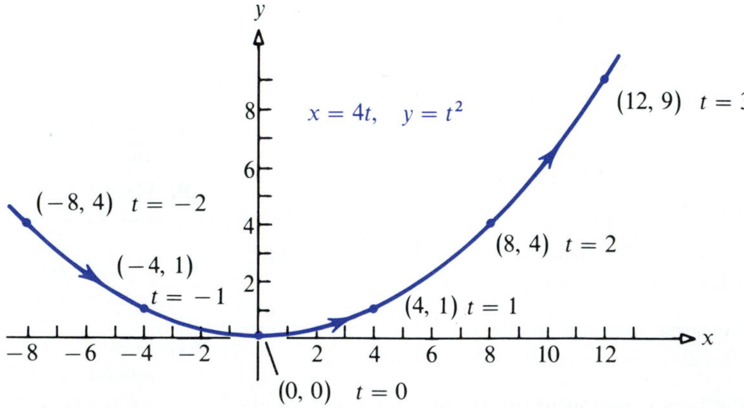

Figure 27

Here the parameter is t. For each number t, there corresponds a value for x and a value for y. When, say, $t = 3$, then $x = 12$ and $y = 9$. When $t = -1$, then $x = -4$ and $y = 1$. We can set up a table listing various choices for t and the corresponding values of x and y, as shown in Table 6. Figure 27 illustrates the curve whose parametric equations are $x = 4t$ and $y = t^2$. The arrows on the graph indicate the direction, or *orientation*, of the curve for increasing values of the parameter t.

This curve is familiar. In fact, we can identify it readily, once we eliminate the parameter t from the two equations $x = 4t$ and $y = t^2$. We do this as follows: In the equation $x = 4t$, we solve for t to get $t = x/4$. Then we replace t in the other equation, $y = t^2$, by $t = x/4$. The result is the single equation $y = x^2/16$, which is a parabola. We shall refer to such an equation as the *rectangular equation* of the curve in order to distinguish it from the parametric equations.

EXAMPLE 2

Find the rectangular equation of the curve whose parametric equations are

$$x = R\cos t \qquad y = R\sin t \qquad R > 0$$

Graph this curve, indicating its orientation.

Solution

We can eliminate the parameter t by using a familiar trigonometric identity:

$$\cos^2 t + \sin^2 t = \left(\frac{x}{R}\right)^2 + \left(\frac{y}{R}\right)^2 = 1$$

$$x^2 + y^2 = R^2$$

The curve is a circle with center at the origin and radius R. As the parameter t increases, the points (x, y) on the circle are traced out in the counterclockwise direction. See Figure 28.

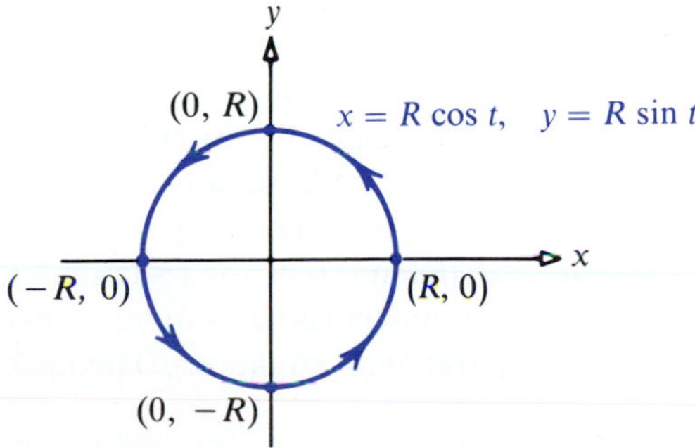

Figure 28

Some observations should be made concerning Example 2. The parameter t has no restrictions placed on it, so it is assumed to vary from $-\infty$ to

$+\infty$. Thus, the graph in Figure 28 is actually being repeated each time t increases by 2π. If we wanted the curve to consist of exactly one revolution in the counterclockwise direction, we could write

$$x = R\cos t \qquad y = R\sin t \qquad 0 \le t \le 2\pi$$

In this case, the curve starts at $(R, 0)$, $t = 0$, and ends at $(R, 0)$, $t = 2\pi$.

If we wanted the curve to consist of exactly three revolutions in the counterclockwise direction, we could write

$$x = R\cos t \qquad y = R\sin t \qquad -2\pi \le t \le 4\pi$$

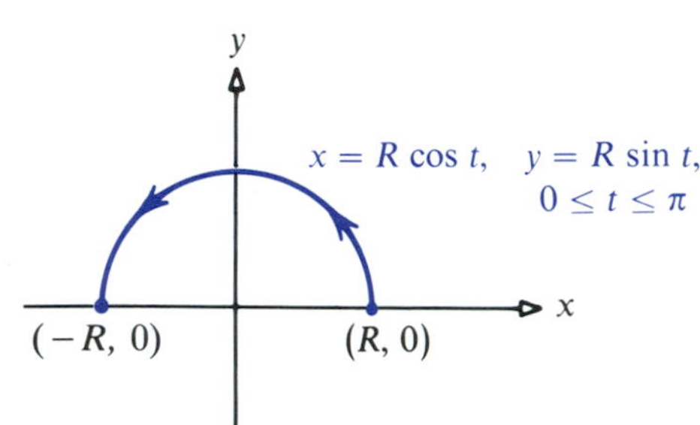

Figure 29

Here the choice of the t interval is completely arbitrary. We could just as well have used $0 \le t \le 6\pi$ or $2\pi \le t \le 8\pi$ or any interval of length 6π.

If we wanted the curve to consist of the upper semicircle of radius R with a counterclockwise orientation, we could write

$$x = R\cos t \qquad y = R\sin t \qquad 0 \le t \le \pi$$

See Figure 29.

If we wanted the curve to consist of a right semicircle of radius R with a *clockwise* orientation, we could use the parametric equations

$$x = R\sin t \qquad y = R\cos t \qquad 0 \le t \le \pi$$

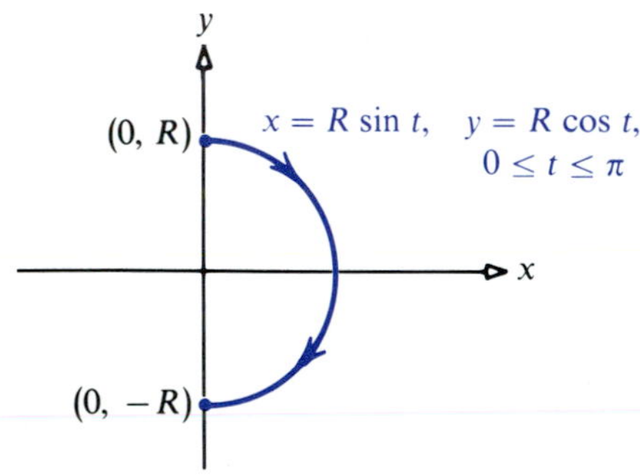

Figure 30

See Figure 30. Note that the parametric equations are not unique. Other choices that also give the graph in Figure 30 are

$$x = R\sin t \qquad y = -R\cos(\pi - t) \qquad 0 \le t \le \pi$$

or

$$x = R\sin 3t \qquad y = R\cos 3t \qquad 0 \le t \le \pi/3$$

and infinitely many others.

These observations should convince you of the flexibility and potential usefulness of using parametric equations to define curves. However, in the examples we chose, it was easy to eliminate the parameter t to get the rectangular equation. This may not always be the case. Furthermore, in our examples, the graph of the rectangular equation (obtained by the elimination of the parameter) and the graph of the curve defined by the parametric equations (obtained by locating the values of x and y corresponding to each t) were identical. But it sometimes happens that the graph of the rectangular equation obtained by the elimination of the parameter contains more points than the graph of the curve defined by the parametric equations. The next example illustrates this.

EXAMPLE 3

Find the rectangular equation of the curve whose parametic equations are

$$x = \cos 2t \qquad y = \sin t$$

Solution

We can eliminate the parameter t from these equations by using the trigonometric identity $\sin^2 t = \frac{1}{2}(1 - \cos 2t)$. Since $y = \sin t$ and $x = \cos 2t$, it follows that

$$y^2 = \frac{1}{2}(1 - x)$$

so the curve seems to be the parabola $y^2 = \frac{1}{2}(1 - x)$, as shown in Figure 31(a). However, the curve described by the parametric equations does not consist of all points on this parabola. Since $x = \cos 2t$, we have $-1 \leq x \leq 1$, and also, since $y = \sin t$, we have $-1 \leq y \leq 1$. Thus, the curve described by the given parametric equations is only part of the parabola, as shown in Figure 31(b).

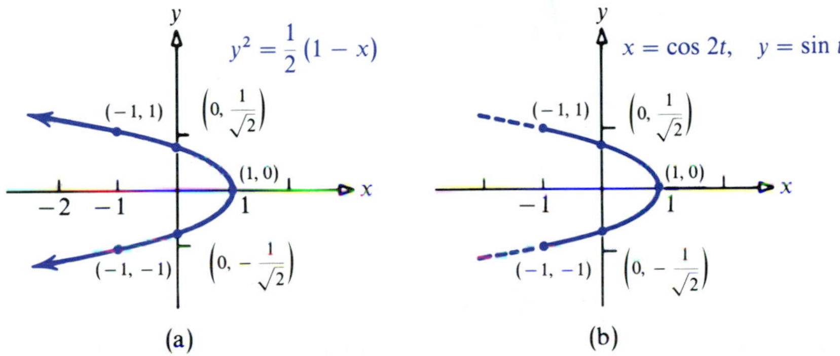

Figure 31

Time as a Parameter

If we use the parameter t to define time, then the parametric equations $x = x(t)$ and $y = y(t)$ specify how the x- and y-coordinates of a moving point vary with time.

One advantage in using parametric equations, rather than a rectangular equation, to describe the motion of an object is that we are able to specify not only *where* the object travels but also *when* it gets to any given place. The rectangular equation tells us only the path along which the object travels.

EXAMPLE 4

Describe the motion of an object that moves so that at time t it has coordinates

$$x = 3\cos t \qquad y = 4\sin t \qquad 0 \leq t \leq 2\pi$$

Solution

We eliminate the parameter t by using the familiar identity $\sin^2 t + \cos^2 t = 1$. The result is

$$\frac{x^2}{9} + \frac{y^2}{16} = 1$$

so the path is an ellipse. When $t = 0$, the object is at $(3, 0)$. As t increases, the object moves around the ellipse in a counterclockwise direction, reaching $(0, 4)$ when $t = \pi/2$; $(-3, 0)$ when $t = \pi$; $(0, -4)$ when $t = 3\pi/2$; and getting back to the starting point when $t = 2\pi$. See Figure 32.

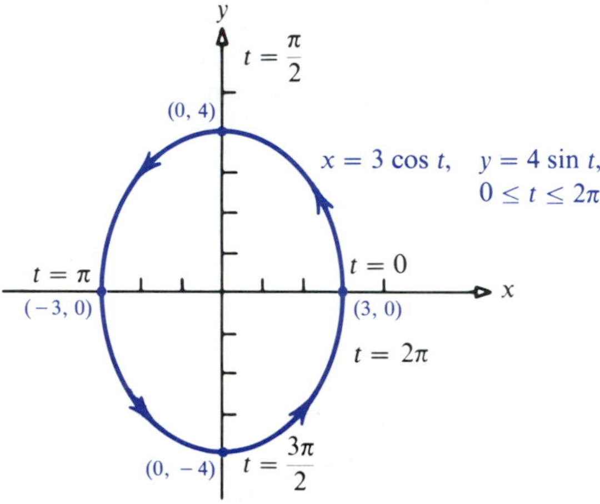

Figure 32

Finding Parametric Equations

If the rectangular equation of a curve is known, a variety of parametric equations can be used to represent it.

EXAMPLE 5

Consider the ellipse

$$\frac{x^2}{4} + y^2 = 1$$

Let the parameter t denote time (in seconds). The position of an object, whose motion around this ellipse begins at $(2, 0)$, is counterclockwise, and requires 1 second for a complete revolution, is given by the parametric equations

$$x = 2 \cos 2\pi t \qquad y = \sin 2\pi t \qquad 0 \le t \le 1$$

The position of an object, whose motion around the ellipse begins at $(0, 1)$, is clockwise, and requires 2 seconds for a complete revolution, is given by

$$x = 2 \sin \pi t \qquad y = \cos \pi t \qquad 0 \le t \le 2$$

EXAMPLE 6

Find parametric equations for the cardioid $r = 2 - 2 \sin \theta$.

Solution

Parametric equations for $r = 2 - 2 \sin \theta$ can be found by substituting for

r in the relationships

$$x = r\cos\theta = (2 - 2\sin\theta)\cos\theta = 2\cos\theta(1 - \sin\theta)$$

$$y = r\sin\theta = (2 - 2\sin\theta)\sin\theta = 2\sin\theta(1 - \sin\theta)$$

Here the parameter θ is unrestricted. If we restrict θ to $0 \le \theta < 2\pi$, then the points on the cardioid are traversed exactly once as θ varies from 0 to 2π. See Figure 33.

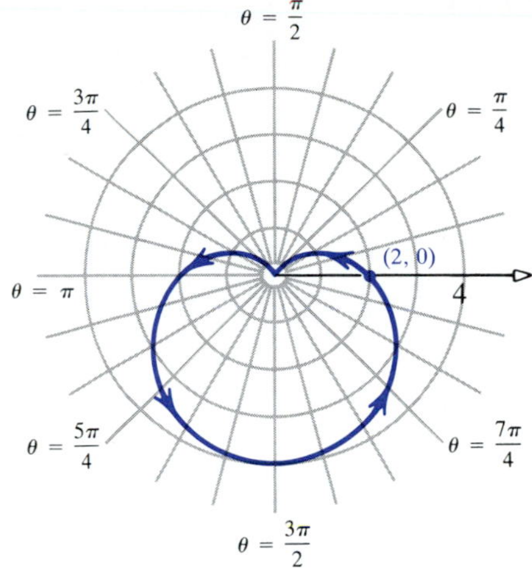

Figure 33

The cycloid, which is discussed next, provides an example of how the parametric equations of a curve can be developed. In this case the parametric equations provide a more natural way to write the equation than rectangular equations would.

The Cycloid

Suppose that a circle rolls along a horizontal line without slipping. As the circle rolls along the line, a point P on the circle will trace out a curve called a *cycloid* (see Fig. 34). An attempt to derive the equation of the cycloid in rectangular coordinates will soon demonstrate how complicated the task is. However, the derivation in terms of parametric equations is relatively easy.

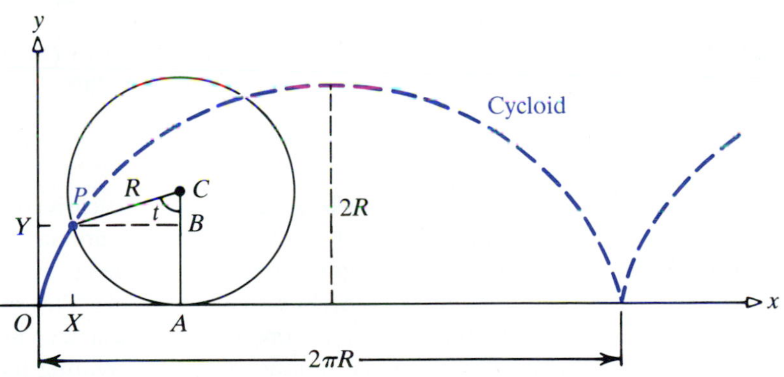

Figure 34

We begin with a circle of radius R and take the fixed line on which the circle rolls as the x-axis. Let the origin be one of the points at which the point P comes in contact with the x-axis. Figure 34 illustrates the position of this point P after the circle has rolled somewhat. The angle t (in radians) measures the angle through which the circle has rolled.

Since we require no slippage, it follows that

$$\text{Arc } AP = |OA|$$

Therefore,

$$Rt = |OA|$$

The x-coordinate of the point P is

$$|OX| = |OA| - |XA| = Rt - R\sin t = R(t - \sin t)$$

The y-coordinate of the point P is equal to

$$|OY| = |AC| - |BC| = R - R\cos t = R(1 - \cos t)$$

[13.4.2] THEOREM / *Equations of a Cycloid.*

The parametric equations of the cycloid are

$$\mathbf{x = R(t - \sin t) \qquad y = R(1 - \cos t)} \tag{2}$$

Applications to Mechanics

If, in the equations of the cycloid (2), we let $R < 0$, we obtain an inverted cycloid (see Fig. 35(a)). The inverted cycloid arises as a result of some remarkable applications in the field of mechanics. We shall mention two of them—the *brachistochrone* and the *tautochrone*.*

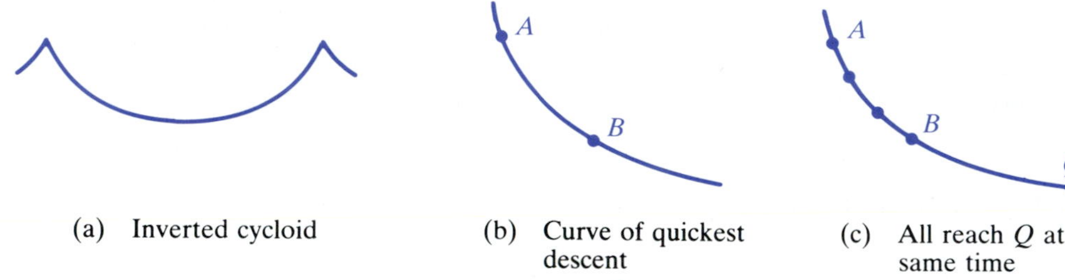

(a) Inverted cycloid (b) Curve of quickest descent (c) All reach Q at same time

Figure 35

Brachistochrone. This is the curve of quickest descent. If a particle is constrained to follow some path from one point A to a lower point B (not on the same vertical line) and is acted upon only by gravity, the time needed to make the descent is least if the path is an inverted cycloid (see Fig. 35(b)). This remarkable discovery, which is attributed to many famous mathematicians (including Johann Bernoulli and Pascal), was a significant step in creating the branch of mathematics known as the *calculus of variations*.

Cycloid Cycloid Cycloid

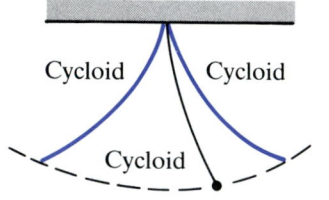

A flexible pendulum constrained by cycloids swings in a cycloid

Figure 36

Tautochrone. Let Q be the lowest point on an inverted cycloid. If several particles placed at various positions on an inverted cycloid simultaneously begin to slide down the cycloid, they will reach the point Q at the same time, as indicated in Figure 35(c). The tautochrone property of the cycloid was used by the Dutch mathematician, physicist, and astronomer Christiaan Huygens (1629–1695) to construct a pendulum clock with a bob that swings along a cycloid (see Fig. 36). In Huygens's clock, the bob was made to swing along a cycloid by suspending the bob on a thin wire constrained by two plates shaped like cycloids. In a clock of this design, the period of the pendulum is independent of its amplitude.

* In Greek, *brachistochrone* means "the shortest time," and *tautochrone* means "equal time."

EXERCISE 13.4

In Problems 1–24 sketch the graph of the curve defined by the given parametric equations and show its orientation. Find the rectangular equation of each curve.

1. $x = 2t + 1, \quad y = t + 2, \quad -\infty < t < +\infty$
2. $x = t - 2, \quad y = 3t + 1, \quad -\infty < t < +\infty$
3. $x = 2t + 1, \quad y = t + 2, \quad 0 \le t \le 2$
4. $x = t - 2, \quad y = 3t + 1, \quad 0 \le t \le 2$
5. $x = \sin t, \quad y = \cos t, \quad 0 \le t \le 2\pi$
6. $x = \cos t, \quad y = \sin t, \quad 0 \le t \le \pi$
7. $x = 2 \sin t - 3, \quad y = 2 \cos t + 1, \quad 0 \le t \le \pi$
8. $x = 4 \cos t + 1, \quad y = 4 \sin t - 3, \quad 0 \le t \le \pi$
9. $x = 2 \sin t, \quad y = 3 \cos t, \quad 0 \le t \le 2\pi$
10. $x = 4 \cos t, \quad y = 3 \sin t, \quad 0 \le t \le 2\pi$
11. $x = 3, \quad y = 2t, \quad -\infty < t < +\infty$
12. $x = 4t + 1, \quad y = 2t, \quad -\infty < t < +\infty$
13. $x = 2, \quad y = t^2 + 4, \quad 0 \le t < +\infty$
14. $x = t + 3, \quad y = t^3, \quad -4 \le t \le 4$
15. $x = t + 5, \quad y = \sqrt{t}, \quad t \ge 0$
16. $x = 2t^2, \quad y = 2t^3, \quad 0 \le t \le 3$
17. $x = t^{1/2} + 1, \quad y = t^{3/2}, \quad t \ge 1$
18. $x = 2e^t, \quad y = 1 - e^t, \quad t \ge 0$
19. $x = e^t, \quad y = t, \quad -\infty < t < +\infty$
20. $x = t, \quad y = 1/t, \quad -\infty < t < +\infty, \quad t \ne 0$
21. $x = \sec t, \quad y = \tan t, \quad -\pi/2 < t < \pi/2$
22. $x = 3 \sinh t, \quad y = 2 \cosh t, \quad -\infty < t < +\infty$
23. $x = 3 \sin^2 t - 2, \quad y = 2 \cos t, \quad 0 \le t \le \pi$
24. $x = 1 + 2 \sin^2 t, \quad y = 2 - \cos t, \quad 0 \le t \le 2\pi$

In Problems 25–28 find two different pairs of parametric equations for each rectangular equation.

25. $y = 4x^3$
26. $y = 2x^2$
27. $x = \frac{1}{3}\sqrt{y} - 3$
28. $x = 3y^3 - 2\sqrt{y} + 5y + 2$

In Problems 29 and 30 use the rectangular equation $(x^2/9) + (y^2/4) = 1$. Find its parametric equations under the given conditions on the motion of a particle along this curve.

29. The motion begins at (3, 0), is counterclockwise, and requires 3 seconds for a complete revolution.
30. The motion begins at (3, 0), is clockwise, and requires 3 seconds for a complete revolution.

In Problems 31–36 eliminate the parameter to obtain the rectangular equation.

31. $x = \dfrac{1}{t^2}, \quad y = \dfrac{2}{t^2 + 1}$
32. $x = \dfrac{3t}{\sqrt{t^2 + 1}}, \quad y = \dfrac{3}{\sqrt{t^2 + 1}}$
33. $x = \dfrac{4}{\sqrt{4 - t^2}}, \quad y = \dfrac{4t}{\sqrt{4 - t^2}}$
34. $x = \sqrt{t - 3}, \quad y = \sqrt{t + 1}$
35. $x = \sin \theta - 2, \quad y = 4 - 2 \cos \theta$
36. $x = 2 + \tan \theta, \quad y = 3 - 2 \sec \theta$

37. Sketch and compare the graphs of the following parametric equations.

    (a) $x = t, \quad y = t^2$ (b) $x = \sqrt{t}, \quad y = t$
    (c) $x = e^t, \quad y = e^{2t}$ (d) $x = \cos t, \quad y = 1 - \sin^2 t$

38. Sketch and compare the graphs of the following parametric equations.

    (a) $x = \sec t, \quad y = \tan t$ (b) $x = t, \quad y = \sqrt{t^2 - 1}$
    (c) $x = \sqrt{t + 1}, \quad y = \sqrt{t}$

39. The position of a projectile (air resistance neglected) at the end of t seconds, fired with an initial velocity v_0 feet per second and at angle θ from the horizontal, is given by the parametric equations

$$x = (v_0 \cos \theta)t \qquad y = (v_0 \sin \theta)t - 16t^2$$

(a) By eliminating the parameter t, show that the trajectory is the arc of a parabola from the initial point to the point of impact.

(b) The speed v at any time t is given by

$$v = \left[\left(\frac{dx}{dt}\right)^2 + \left(\frac{dy}{dt}\right)^2\right]^{1/2}$$

Find v when $t = 1$ and $t = 2$.

(c) Show that the projectile hits the ground when $t = \frac{1}{16}v_0 \sin \theta$. Find how far the projectile has traveled horizontally at that time.

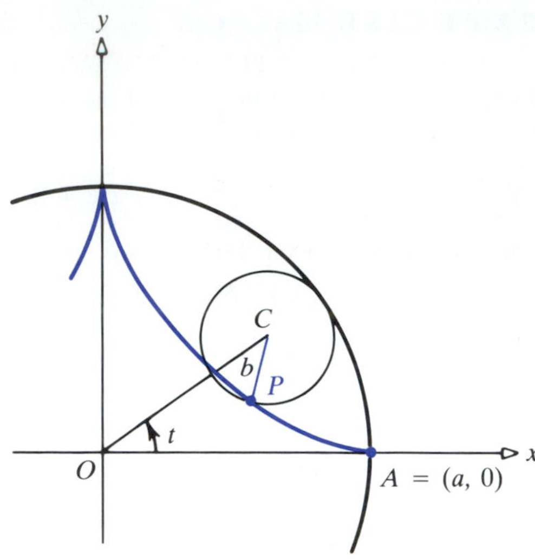

Figure for Problems 40 and 41

40. Let a circle C of radius b roll, without slipping, inside a fixed circle with radius a, $a > b$. A fixed point P on the circle C traces out a curve called a *hypocycloid*. If $A = (a, 0)$ is the initial position of the tracing point P and if t denotes the angle from the positive x-axis to the line segment from the origin to the center of C, show that the parametric equations of the hypocycloid are

$$x = (a - b)\cos t + b \cos \frac{a - b}{b} t$$

$$y = (a - b)\sin t - b \sin \frac{a - b}{b} t \qquad 0 \le t \le 2\pi$$

41. The figure shows a hypocycloid with $a = 4b$. Show that the rectangular equation of the hypocycloid with $a = 4b$ is $x^{2/3} + y^{2/3} = a^{2/3}$.

42. If the circle C of Problem 40 rolls on the outside of the second circle, find the parametric equations of the curve traced by P. This curve is called an *epicycloid*.

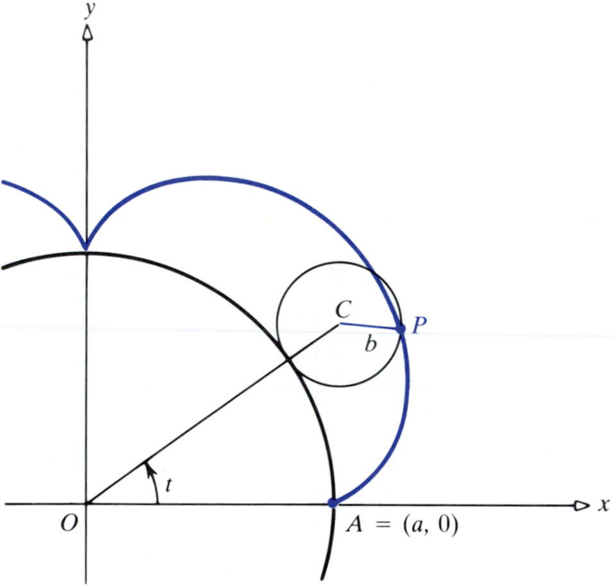

Figure for Problem 42

13.5 Tangent Lines; Arc Length

Tangent Lines

In Chapter 3 we saw that for a differentiable function $y = f(x)$, the derivative $dy/dx = f'(x)$ equals the slope of the tangent line to the graph of f at a point (x, y). We now seek a formula for the slope of the tangent line to a curve if the curve is defined by parametric equations. We begin with a definition.

[13.5.1] DEFINITION / *Smooth Curve.*

Let C denote a curve defined by the parametric equations

$$x = x(t) \qquad y = y(t) \qquad a \le t \le b$$

where $x(t)$ and $y(t)$ are each continuous and differentiable on $[a, b]$. If, in addition, dx/dt and dy/dt are each continuous and never simultaneously 0 on $[a, b]$ (except possibly at the endpoints $t = a$, $t = b$), the curve C is called *smooth*.

The reason for the name *smooth* is that the graph of a smooth curve has no corners or cusps.

It can be shown that a smooth curve $x = x(t)$, $y = y(t)$, for which dx/dt is never 0, can be represented by a rectangular equation $y = f(x)$. If $(x(t), y(t))$ is a point on the curve, we can write

$$y = y(t) = f(x(t))$$

By the chain rule,

$$\frac{dy}{dt} = \frac{dy}{dx}\frac{dx}{dt}$$

Since $dx/dt \neq 0$, we have

$$\frac{dy}{dx} = \frac{dy/dt}{dx/dt}$$

[13.5.2] THEOREM

For a smooth curve C: $x = x(t)$, $y = y(t)$, $a \le t \le b$, the slope of the tangent line to C at a point $(x(t), y(t))$ is given by the formula

$$\frac{dy}{dx} = \frac{dy/dt}{dx/dt} \tag{1}$$

provided $dx/dt \neq 0$.

Equation (1) does not apply if $dx/dt = 0$. However, at a number t where $dx/dt = 0$ and $dy/dt \neq 0$, there is generally a *vertical tangent line*. The behavior of the graph at points for which dx/dt and dy/dt are simultaneously 0 can be determined only by more detailed analysis. Such points, called *singular points*, do not occur for smooth curves.

EXAMPLE 1

Find an equation of the tangent line to the curve with parametric equations $x = 3t^2$, $y = 2t$ at $t = 1$.

Solution

By (1),

$$\frac{dy}{dx} = \frac{dy/dt}{dx/dt} = \frac{2}{6t} = \frac{1}{3t}$$

At $t = 1$, the slope of the tangent line is $\frac{1}{3}$ and $x = 3$ and $y = 2$. Therefore, an equation of the tangent line is $y - 2 = \frac{1}{3}(x - 3)$ or $y = x/3 + 1$.

For the parametric equations in Example 1, $dx/dt = 0$ at $t = 0$. Since $dy/dt = 2 \neq 0$ at $t = 0$, we expect a vertical tangent line at $(0, 0)$. See Figure 37.

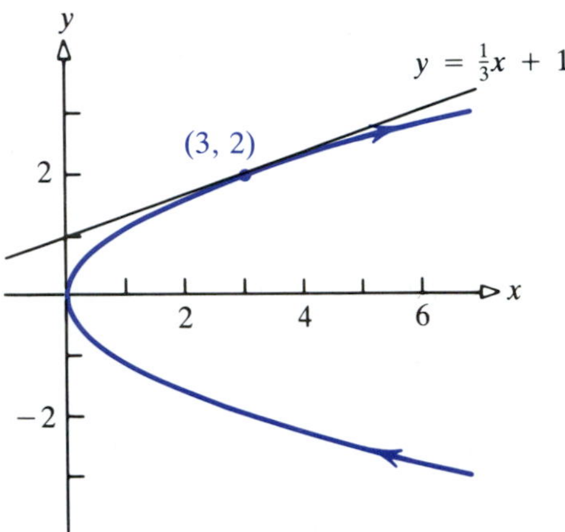

Figure 37

EXAMPLE 2

Show that the slope of the tangent line to the cycloid

$$x = R(t - \sin t) \qquad y = R(1 - \cos t) \qquad 0 \leq t \leq 2\pi$$

is $\cot(t/2)$.

Solution

The derivatives dx/dt and dy/dt are

$$\frac{dx}{dt} = R - R\cos t \qquad \frac{dy}{dt} = R\sin t$$

Hence, the slope of the tangent line is given by

$$\frac{dy}{dx} = \frac{dy/dt}{dx/dt} = \frac{R\sin t}{R(1 - \cos t)}$$

$$= \frac{\sin t}{1 - \cos t} = \frac{\sin 2(t/2)}{1 - \cos 2(t/2)} = \frac{2\sin(t/2)\cos(t/2)}{2\sin^2(t/2)} = \cot\frac{t}{2}$$

EXAMPLE 3

Determine where the cardioid $r = 2 - 2\sin\theta$ has vertical tangent lines.

Solution

We begin by writing parametric equations for the cardioid, using θ as a parameter. Then

$$x = (2 - 2\sin\theta)\cos\theta = 2(1 - \sin\theta)\cos\theta$$

$$y = (2 - 2\sin\theta)\sin\theta = 2(1 - \sin\theta)\sin\theta$$

The derivatives $dx/d\theta$ and $dy/d\theta$ are

$$\frac{dx}{d\theta} = 2(-\cos\theta)\cos\theta + 2(1 - \sin\theta)(-\sin\theta)$$

$$= 2(-\cos^2\theta + \sin^2\theta - \sin\theta) = 2(2\sin^2\theta - \sin\theta - 1)$$

$$\frac{dy}{d\theta} = 2(-\cos\theta)\sin\theta + 2(1 - \sin\theta)\cos\theta = 2[\cos\theta(1 - 2\sin\theta)]$$

To find where the vertical tangent lines are, we let $dx/d\theta = 0$ and solve for θ. Then,

$$2\sin^2\theta - \sin\theta - 1 = 0$$

$$(2\sin\theta + 1)(\sin\theta - 1) = 0$$

$$\sin\theta = -\frac{1}{2} \qquad \text{or} \qquad \sin\theta = 1$$

$$\theta = \frac{7\pi}{6} \qquad \frac{11\pi}{6} \qquad \text{or} \qquad \frac{\pi}{2}$$

At $\theta = 7\pi/6$ and $\theta = 11\pi/6$, we have $dx/d\theta = 0$ and $dy/d\theta \neq 0$. Thus, at $(3, 7\pi/6)$ and at $(3, 11\pi/6)$, the cardioid has a vertical tangent line. At $\theta = \pi/2$, we have $dx/d\theta = 0$ and $dy/d\theta = 0$. Thus, the point $(0, \pi/2)$ is a singular point. (Although we shall not prove it, the cardioid has a vertical tangent line at $(0, \pi/2)$ as well.) See Figure 38.

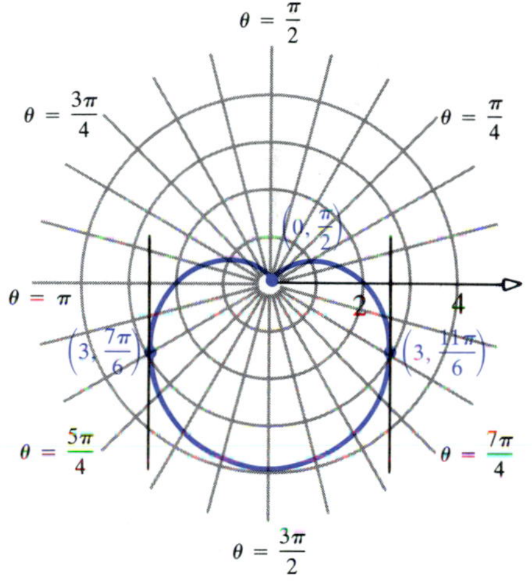

Figure 38

Arc Length

In Section 6.5 in Chapter 6 we developed the formula (5) for the arc length of the graph of a function $y = f(x)$. For some graphs, such as the one in Figure 39, this formula cannot be used, since such a graph is not the graph of a function. In this section we develop a formula for the arc length of a smooth curve.

Figure 39

[13.5.3] THEOREM / ***Formula for Arc Length.***

For a smooth curve C,

$$x = x(t) \qquad y = y(t) \qquad a \le t \le b$$

the length s of C from $t = a$ to $t = b$ is given by the formula

$$s = \int_a^b \sqrt{\left(\frac{dx}{dt}\right)^2 + \left(\frac{dy}{dt}\right)^2}\, dt \tag{2}$$

We provide a partial proof at the end of the section.

EXAMPLE 4

Find the length of the curve defined by the parametric equations

$$x = t^3 + 2 \qquad y = 2t^{9/2}$$

from the point where $t = 0$ to the point where $t = 3$.

Solution

We have $dx/dt = 3t^2$ and $dy/dt = 9t^{7/2}$. Therefore, by (2), the length s of the curve from $t = 0$ to $t = 3$ is

$$s = \int_0^3 \sqrt{(3t^2)^2 + (9t^{7/2})^2}\, dt = \int_0^3 \sqrt{9t^4 + 81t^7}\, dt = \int_0^3 3t^2\sqrt{1 + 9t^3}\, dt$$

Make the substitution $u = 1 + 9t^3$. Then $du = 27t^2\, dt$, and

$$s = \int_0^3 3t^2\sqrt{1 + 9t^3}\, dt$$

$$= \int_1^{244} \sqrt{u}\left(\frac{du}{9}\right) = \frac{1}{9}\left(\frac{u^{3/2}}{\frac{3}{2}}\right)\Bigg|_1^{244} = \frac{2}{27}(244\sqrt{244} - 1) \approx 282.3$$

EXAMPLE 5

Use theorem (13.5.3) to verify the familiar formula $s = R\theta$ for the length s of the arc of a circle of radius R subtended by a central angle of θ radians.

Solution

We set up our coordinate system so that the circle has its center at the origin and the central angle θ has its initial side along the positive x-axis (see Fig. 40). We can represent the circle by the parametric equations

$$x = R\cos t \qquad y = R\sin t \qquad 0 \le t \le 2\pi$$

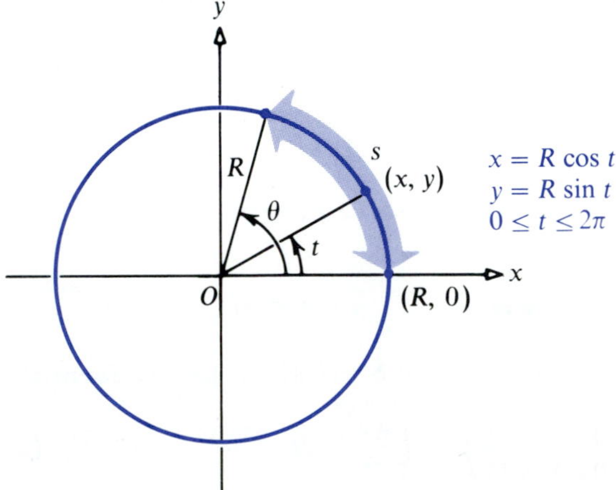

Figure 40

in which the parameter t is a central angle. We seek the length s of the arc from $t = 0$ to $t = \theta$. By (2), we find

$$s = \int_0^\theta \sqrt{\left(\frac{dx}{dt}\right)^2 + \left(\frac{dy}{dt}\right)^2}\, dt = \int_0^\theta \sqrt{R^2\sin^2 t + R^2\cos^2 t}\, dt = \int_0^\theta R\, dt = R\theta$$

In case the circular arc is a full circle—that is, if $\theta = 2\pi$—we get $s = 2\pi R$, the formula for the circumference of the circle.

EXAMPLE 6

Find the length of one arch of the cycloid

$$x = R(t - \sin t) \qquad y = R(1 - \cos t) \qquad R > 0$$

Solution

We obtain one arch of the cycloid when t changes from 0 to 2π. By using (2) with $dx/dt = R - R\cos t$ and $dy/dt = R\sin t$, we get

$$s = \int_0^{2\pi} \sqrt{R^2(1 - \cos t)^2 + R^2\sin^2 t}\, dt$$

We simplify the radical before going any further. Since $\sin^2 t + \cos^2 t = 1$, we have

$$\sqrt{R^2(1 - \cos t)^2 + R^2\sin^2 t} = \sqrt{R^2(2 - 2\cos t)} = R\sqrt{2}\sqrt{1 - \cos t}$$

By using the half-angle identity $1 - \cos t = 2\sin^2(t/2)$ and remembering that $\sin(t/2) \geq 0$ if $0 \leq t \leq 2\pi$, we get

$$R\sqrt{2}\sqrt{1 - \cos t} = 2R\sin\frac{t}{2}$$

Thus, the arc length s from $t = 0$ to $t = 2\pi$ is

$$s = \int_0^{2\pi} 2R\sin\frac{t}{2}\, dt = -4R\cos\frac{t}{2}\bigg|_0^{2\pi} = 8R$$

Arc Length of a Rectangular Equation $y = f(x)$

We verify next that (2) reduces to (5) from Section 6.5 (p. 425) in the case where the smooth curve C is the graph of a function. If the smooth curve C, defined by the parametric equations

$$x = x(t) \qquad y = y(t) \qquad a \leq t \leq b$$

is the graph of a function $y = f(x)$, then we can use x as the parameter t and write the parametric equations for C as

$$x = t \qquad y = f(t) \qquad a \leq t \leq b$$

Since $dx/dt = 1$ and $dy/dt = dy/dx = f'(x)$, (2) takes the form

$$s = \int_a^b \sqrt{\left(\frac{dx}{dt}\right)^2 + \left(\frac{dy}{dt}\right)^2}\, dt = \int_a^b \sqrt{1 + [f'(x)]^2}\, dx$$

which is the same as (5) from Section 6.5.

Arc Length of the Graph of a Polar Equation $r = f(\theta)$

Suppose a curve C is given by the polar equation $r = f(\theta)$, where f is continuous and has a continuous derivative on $\alpha \le \theta \le \beta$. Using θ as a parameter, we write parametric equations for the curve C as

$$x = r\cos\theta = f(\theta)\cos\theta \qquad y = r\sin\theta = f(\theta)\sin\theta$$

Thus,

$$\frac{dx}{d\theta} = -f(\theta)\sin\theta + f'(\theta)\cos\theta \qquad \frac{dy}{d\theta} = f(\theta)\cos\theta + f'(\theta)\sin\theta$$

After simplification, we obtain

$$\left(\frac{dx}{d\theta}\right)^2 + \left(\frac{dy}{d\theta}\right)^2 = [f(\theta)]^2 + [f'(\theta)]^2$$

[13.5.4] THEOREM | ***Arc Length of the Graph of a Polar Equation.***

If a curve C is defined by the polar equation $r = f(\theta)$, $\alpha \le \theta \le \beta$, and if $f'(\theta)$ is continuous on $[\alpha, \beta]$, then the arc length s of C from $\theta = \alpha$ to $\theta = \beta$ is

$$s = \int_\alpha^\beta \sqrt{[f(\theta)]^2 + [f'(\theta)]^2}\, d\theta = \int_\alpha^\beta \sqrt{r^2 + \left(\frac{dr}{d\theta}\right)^2}\, d\theta \qquad (3)$$

EXAMPLE 7

Find the arc length of the logarithmic spiral

$$r = f(\theta) = e^{3\theta} \qquad 0 \le \theta \le 2$$

Solution

By applying (3) with $f(\theta) = e^{3\theta}$ and $f'(\theta) = 3e^{3\theta}$, we get

$$s = \int_0^2 \sqrt{(e^{3\theta})^2 + (3e^{3\theta})^2}\, d\theta = \int_0^2 \sqrt{10e^{6\theta}}\, d\theta$$

$$= \int_0^2 \sqrt{10}e^{3\theta}\, d\theta = \frac{\sqrt{10}}{3} e^{3\theta}\Big|_0^2 = \frac{\sqrt{10}}{3}(e^6 - 1) \approx 424.197$$

The Differential of Arc Length

For a smooth curve C defined by the parametric equations

$$x = x(t) \qquad y = y(t) \qquad a \le t \le b$$

the arc length s along C from a to b is given by (2). The arc length s along C from a to some variable t will, in general, be a function of t given by

$$s = s(t) = \int_a^t \sqrt{\left(\frac{dx}{du}\right)^2 + \left(\frac{dy}{du}\right)^2}\, du$$

where, for convenience, we use u as an artificial variable.

Differentiating the above expression with respect to t gives us

$$\frac{ds}{dt} = \sqrt{\left(\frac{dx}{dt}\right)^2 + \left(\frac{dy}{dt}\right)^2} \tag{4}$$

so that

$$\left(\frac{ds}{dt}\right)^2 = \left(\frac{dx}{dt}\right)^2 + \left(\frac{dy}{dt}\right)^2$$

In terms of differentials, we may write this as

$$(ds)^2 = (dx)^2 + (dy)^2$$
$$ds = \sqrt{(dx)^2 + (dy)^2} \tag{5}$$

Let's interpret (5) geometrically. As Figure 41 illustrates, the differential $ds = \sqrt{(dx)^2 + (dy)^2}$ is the length of the hypotenuse of a right triangle with sides of lengths dx and dy. The hypotenuse is along the tangent line to the curve because dy/dx is the slope of the tangent line.

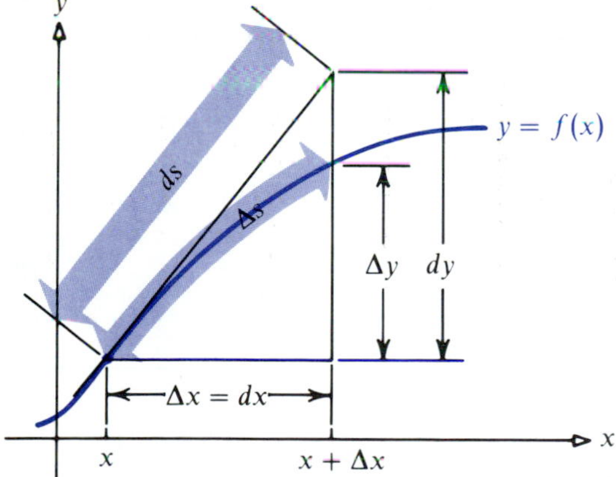

Figure 41

The differential ds may be used to approximate the arc length Δs between two nearby points. This approximating tool is particularly useful when the definite integral (2) is difficult, or perhaps impossible, to evaluate.

EXAMPLE 8

Use the differential ds to approximate the arc length along the curve $x = t^2$, $y = t^3$ from $t = 1$ to $t = 1.1$.

Solution

In preparation for using (5), we calculate the differentials

$$dx = 2t\,dt \qquad \text{and} \qquad dy = 3t^2\,dt$$

The differential $dt = 0.1$, so that the values of dx and dy at $t = 1$ are

$$dx = 2(1)(0.1) = 0.2 \qquad dy = 3(1)^2(0.1) = 0.3$$

Hence, an approximation to the arc length s is

$$ds = \sqrt{(dx)^2 + (dy)^2} = \sqrt{(0.2)^2 + (0.3)^2} = \sqrt{0.13} \approx 0.36$$

Partial Proof of Theorem (13.5.3) The procedure is much like the one we used in Chapter 6. First, we partition the closed interval $[a, b]$ into n subintervals,

$$[a, t_1], \quad [t_1, t_2], \quad \ldots, \quad [t_{i-1}, t_i], \quad \ldots, \quad [t_{n-1}, b]$$

and we let $\Delta t_i = t_i - t_{i-1}$. Corresponding to each number $a, t_1, t_2, \ldots, t_{n-1}, b$, there is a succession of points $P_0, P_1, P_2, \ldots, P_n$, on the curve (see Fig. 42). We join each point to its successor by a line segment. The sum of the lengths of these line segments provides an approximation to the length of the curve from $t = a$ to $t = b$. This sum may be written as

$$d(P_0, P_1) + d(P_1, P_2) + \cdots + d(P_{n-1}, P_n) = \sum_{i=1}^{n} d(P_{i-1}, P_i) \tag{6}$$

where $d(P_{i-1}, P_i)$ is the length of the line segment joining P_{i-1} and P_i. By the formula for the distance between two points, it follows that the length of each line segment is

$$d(P_{i-1}, P_i) = \sqrt{[x(t_i) - x(t_{i-1})]^2 + [y(t_i) - y(t_{i-1})]^2} \tag{7}$$

By using (6), we can write the sum of the lengths of the line segments as

$$\sum_{i=1}^{n} \sqrt{[x(t_i) - x(t_{i-1})]^2 + [y(t_i) - y(t_{i-1})]^2} \tag{8}$$

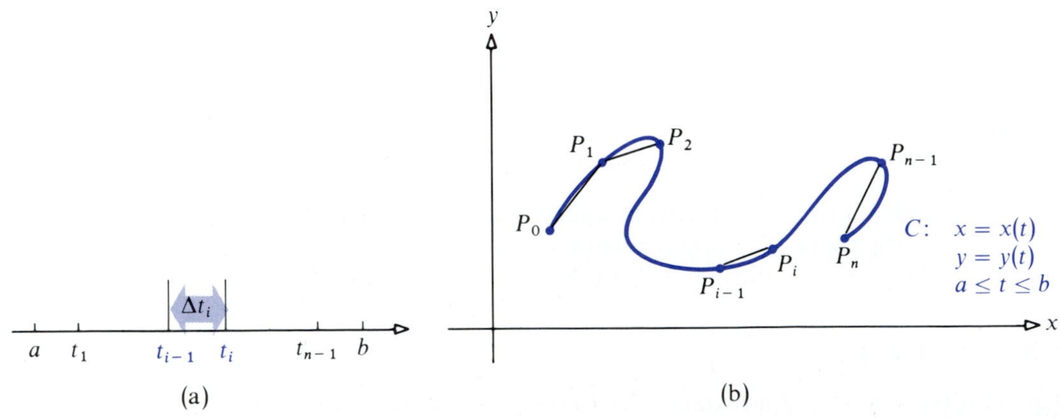

Figure 42

Since we assume that each of the functions $x(t)$ and $y(t)$ has a continuous derivative on $[a, b]$, it follows that $x(t)$ and $y(t)$ have derivatives on each subinterval $[t_{i-1}, t_i]$. When we apply the mean value theorem (4.4.2) to $x(t)$ and $y(t)$ on $[t_{i-1}, t_i]$, the result is that there exist numbers u_i and v_i in each open interval (t_{i-1}, t_i) such that

$$x(t_i) - x(t_{i-1}) = \left[\frac{dx}{dt}(u_i)\right]\Delta t_i \qquad y(t_i) - y(t_{i-1}) = \left[\frac{dy}{dt}(v_i)\right]\Delta t_i \tag{9}$$

Thus, the sum of the lengths of the line segments in (8) becomes

$$\sum_{i=1}^{n} \sqrt{\left[\frac{dx}{dt}(u_i)\right]^2 + \left[\frac{dy}{dt}(v_i)\right]^2}\,\Delta t_i \tag{10}$$

This sum is not a Riemann sum, since the numbers u_i and v_i are not necessarily equal. However, there is a result (usually given in advanced calculus) that states that the limit of the sum in (10) as the norm of the partition approaches 0 is a definite integral.* But as the length of each subinterval gets smaller and smaller—that is, as the norm of the partition approaches 0—it is plausible to define the limit of this sum as the length of the curve from $t = a$ to $t = b$, namely,

$$s = \int_a^b \sqrt{\left(\frac{dx}{dt}\right)^2 + \left(\frac{dy}{dt}\right)^2}\, dt$$

EXERCISE 13.5

In Problems 1–20 find a rectangular equation of the tangent line to each curve at the given number.

1. $x = 2t^2, \quad y = t \quad$ at $\quad t = 2$
2. $x = t, \quad y = 3t^2 \quad$ at $\quad t = -2$
3. $x = 3t, \quad y = 2t^2 - 1 \quad$ at $\quad t = 1$
4. $x = 2t, \quad y = t^2 - 2 \quad$ at $\quad t = 2$
5. $x = \sqrt{t}, \quad y = 1/t \quad$ at $\quad t = 4$
6. $x = 2/t^2, \quad y = 1/t \quad$ at $\quad t = 1$
7. $x = \dfrac{t}{t+2}, \quad y = \dfrac{4}{t+2} \quad$ at $\quad t = 0$
8. $x = \dfrac{t^2}{1+t}, \quad y = \dfrac{1}{1+t} \quad$ at $\quad t = 0$
9. $x = e^{-t}, \quad y = e^{-t} \quad$ at $\quad t = 0$
10. $x = e^{2t}, \quad y = e^{t} \quad$ at $\quad t = 0$
11. $x = \sin t, \quad y = \cos t \quad$ at $\quad t = \pi/4$
12. $x = \sin^2 t, \quad y = \cos t \quad$ at $\quad t = \pi/4$
13. $x = 4 \sin t, \quad y = 3 \cos t \quad$ at $\quad t = \pi/3$
14. $x = 2 \sin t - 1, \quad y = \cos t + 2 \quad$ at $\quad t = \pi/6$
15. $r = 2 \cos 3\theta \quad$ at $\quad \theta = \pi/6$
16. $r = 3 \sin 3\theta \quad$ at $\quad \theta = \pi/3$
17. $r = 2 + \cos \theta \quad$ at $\quad \theta = \pi/4$
18. $r = 3 - \sin \theta \quad$ at $\quad \theta = \pi/6$
19. $r = 4 + 5 \sin \theta \quad$ at $\quad \theta = \pi/4$
20. $r = 1 - 2 \cos \theta \quad$ at $\quad \theta = \pi/4$

* This is where the continuity of the derivatives is used—to guarantee the existence of the definite integral.

In Problems 21–36 find all points on the curve at which the tangent line is vertical.

21. $x = 2t^2, \quad y = t$

22. $x = 4\sqrt{t}, \quad y = 2t$

23. $x = \sin t, \quad y = \cos t$

24. $x = 3 \sin t, \quad y = \cos t$

25. $r = 3 - 3 \cos \theta$

26. $r = 1 + \cos \theta$

27. $r = 3 + \sin \theta$

28. $r = 2 + \cos \theta$

29. $r = 2 \cos 3\theta$

30. $r = 3 \sin 2\theta$

31. $r = 3 \cos \theta$

32. $r = 4 \sin \theta$

33. $r = e^\theta$

34. $r = \ln \theta$

35. $r = 2 \sec \theta$

36. $r = 3 \csc \theta$

In Problems 37–48 find the arc length of each curve on the given interval.

37. $x(t) = t^3, \quad y(t) = t^2, \quad 0 \leq t \leq 2$

38. $x(t) = 3t^2 + 1, \quad y(t) = t^3 - 1, \quad 0 \leq t \leq 2$

39. $x(t) = t - 1, \quad y(t) = \frac{1}{2}t^2, \quad 0 \leq t \leq 2$

40. $x(t) = t^2, \quad y(t) = 2t, \quad 1 \leq t \leq 3$

41. $x(t) = 4 \sin t, \quad y(t) = 4 \cos t, \quad -\pi/2 \leq t \leq \pi/2$

42. $x(t) = 6 \sin t, \quad y(t) = 6 \cos t, \quad -\pi/2 \leq t \leq \pi/2$

43. $x(t) = 2 \sin t - 1, \quad y(t) = 2 \cos t + 1, \quad 0 \leq t \leq 2\pi$

44. $x(t) = e^t \sin t, \quad y(t) = e^t \cos t, \quad 0 \leq t \leq \pi$

45. $r = f(\theta) = e^{\theta/2}, \quad 0 \leq \theta \leq 2$

46. $r = f(\theta) = e^{2\theta}, \quad 0 \leq \theta \leq 2$

47. $r = f(\theta) = \cos^2(\theta/2), \quad 0 \leq \theta \leq \pi$

48. $r = f(\theta) = \sin^2(\theta/2), \quad 0 \leq \theta \leq \pi$

49. Find the arc length of the spiral $r = \theta, \quad 0 \leq \theta \leq 2\pi$.

50. Find the perimeter of the cardioid $r = f(\theta) = 1 - \cos \theta, \quad -\pi \leq \theta \leq \pi$.

51. Find the arc length of one arch of the four-cusped hypocycloid $x = b \sin^3 t, \quad y = b \cos^3 t, \quad 0 \leq t \leq \pi/2$.

52. Find the entire arc length of the curve $r = a \sin^3(\theta/3)$.

In Problems 53–58 find the distance a particle travels along the given path during the indicated time.

53. $x(t) = 3t, \quad y(t) = t^2 - 3, \quad 0 \leq t \leq 2$

54. $x(t) = t^2, \quad y(t) = 3t, \quad 0 \leq t \leq 2$

55. $x(t) = (t^2/2) + 1, \quad y(t) = \frac{1}{3}(2t + 3)^{3/2}, \quad 0 \leq t \leq 2$

56. $x(t) = a \cos t, \quad y(t) = a \sin t, \quad 0 \leq t \leq \pi$

57. $x(t) = \cos 2t, \quad y(t) = \sin^2 t, \quad 0 \leq t \leq \pi/2$

58. $x(t) = 1/t, \quad y(t) = \ln t, \quad 1 \leq t \leq 2$

In Problems 59–70 find dy/dx and d^2y/dx^2. [*Hint:* Use the chain rule to calculate d^2y/dx^2.]

59. $x = e^t \cos t, \quad y = e^t \sin t$

60. $x = 2 \sin t, \quad y = \cos 2t$

61. $x = t + 1/t, \quad y = 4 + t$

62. $x = t^2 + 1, \quad y = \sqrt{t}$

63. $x = a(\cos \theta + \theta \sin \theta), \quad y = a(\sin \theta - \theta \cos \theta)$

64. $x = \cos^3 \theta, \quad y = \sin^3 \theta$

65. $x = 3t^2, \quad y = 2t$

66. $x = t + \dfrac{1}{t}, \quad y = t - \dfrac{1}{t}$

67. $x = 2 \cos 2\theta, \quad y = \sin \theta$

68. $x = 1 + e^{-t}, \quad y = e^{3t}$

69. $x = \cot^2 \theta, \quad y = \cot \theta$

70. $x = \sin t, \quad y = \sec^2 t$

In Problems 71–74 use the differential ds to approximate the indicated arc length.

71. $x = t^{1/3}$, $y = t^2$; from $t = 1$ to $t = 1.1$

72. $x = \sqrt{t}$, $y = t^3$; from $t = 1$ to $t = 1.2$

73. $x = a \sin t$, $y = b \cos t$; from $t = 0$ to $t = 0.1$

74. $x = e^{at}$, $y = e^{bt}$; from $t = 0$ to $t = 0.2$

75. Find the points on the graph of $x = t^2 + 2$, $y = t^3 - 4t$, where the tangent line is horizontal and where it is vertical. Find an equation of the tangent line(s) at the point $(6, 0)$.

76. Find the length of the circumference of a circle of radius 1 by inscribing regular polygons in the circle, finding their perimeters, and then allowing the number of sides of the polygons to tend to infinity.

Surface Area of a Solid of Revolution

13.6

We begin this discussion with the simple situation illustrated in Figure 43(a). If the line segment of length L is revolved about the axis of revolution, we obtain a frustum of a right circular cone, where r and R are the base radii and L is the slant height, as shown in Figure 43(b). The surface area is

$$\text{Surface area} = \frac{2\pi(R+r)L}{2}$$

$$= 2\pi\frac{R+r}{2}L = 2\pi\ (\text{Average radius})(\text{Slant height}) \qquad \textbf{(1)}$$

We shall use this equation to obtain a general formula for the surface area of a solid of revolution.

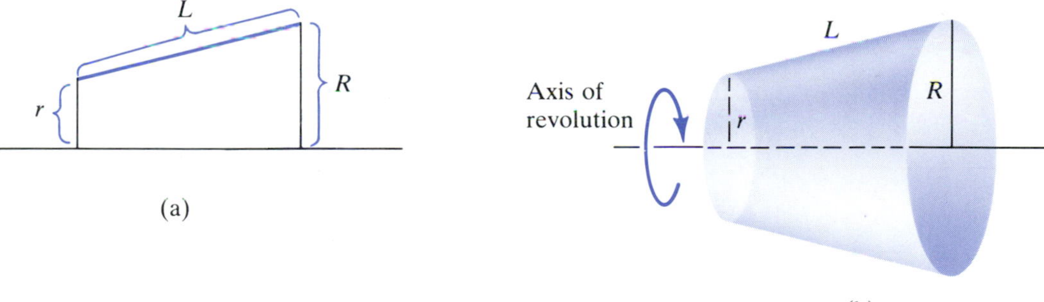

Figure 43

Let C denote a smooth curve defined by the parametric equations $x = x(t)$, $y = y(t)$, $a \le t \le b$, where $y = y(t) \ge 0$ on $[a, b]$. Revolve C about the x-axis to obtain a solid of revolution. We wish to find the surface area of this solid of revolution.

We begin by partitioning the interval $a \le t \le b$ into n subintervals $[a, t_1], [t_1, t_2], \ldots, [t_{i-1}, t_i], \ldots, [t_{n-1}, b]$. We denote the length of the ith subinterval by Δt_i. Corresponding to each number $a, t_1, t_2, \ldots, t_{i-1}, t_i, \ldots, t_{n-1}, b$, there is a succession of points $P_0, P_1, \ldots, P_{i-1}, P_i, \ldots, P_{n-1}, P_n$ on the curve C. We join each point to its successor by a line segment and concentrate on the line segment joining the points P_{i-1} and P_i. See Figure 44(a).

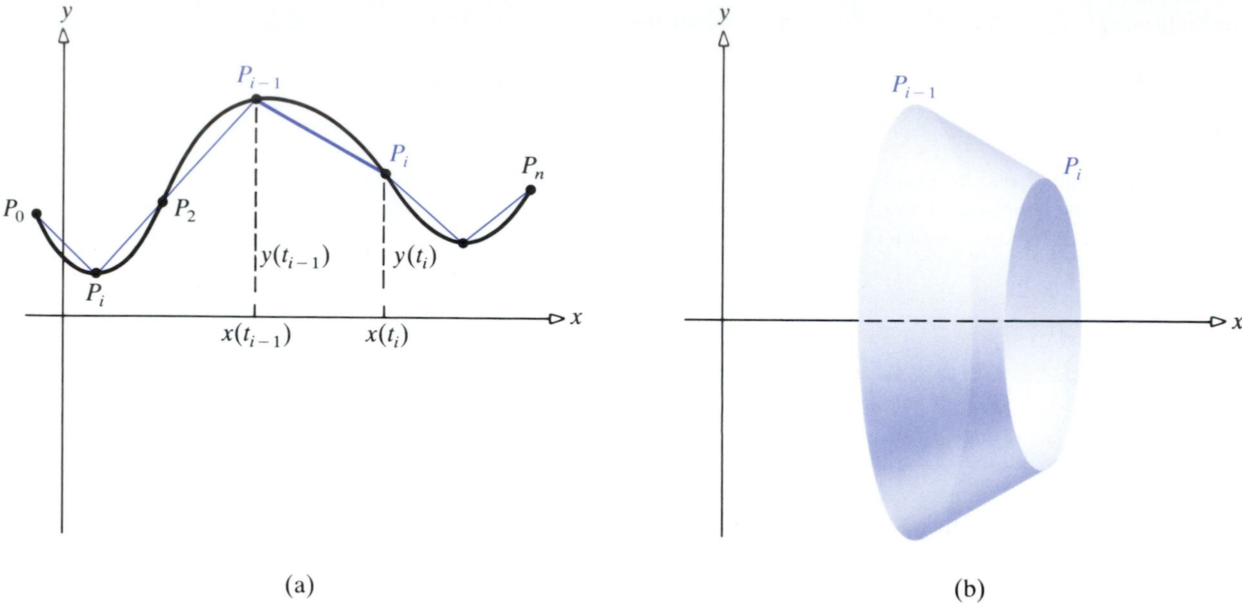

Figure 44

When this segment, of length $d(P_{i-1}, P_i)$, is revolved about the x-axis, it generates a frustum of a right circular cone whose surface area is

$$\Delta S_i \underset{\uparrow \atop \text{By (1)}}{=} 2\pi\left[\frac{y(t_{i-1}) + y(t_i)}{2}\right][d(P_{i-1}, P_i)] \tag{2}$$

Looking back at the development for arc length in Section 13.5, we can combine (7) and (9) to get

$$d(P_{i-1}, P_i) = \sqrt{\left[\frac{dx}{dt}(u_i)\right]^2 + \left[\frac{dy}{dt}(v_i)\right]^2}\,\Delta t_i \tag{3}$$

where u_i and v_i are numbers in the ith subinterval. By using (3) in (2), we write the area of the surface generated by the sum of the line segments as

$$\sum_{i=1}^{n} \Delta S_i = \sum_{i=1}^{n} 2\pi\left[\frac{y(t_{i-1}) + y(t_i)}{2}\right]\sqrt{\left[\frac{dx}{dt}(u_i)\right]^2 + \left[\frac{dy}{dt}(v_i)\right]^2}\,\Delta t_i$$

This sum provides an approximation to the surface area we seek. By using an argument similar to the one used in developing the formula for arc length, it is plausible to give the following definition:

[13.6.1] DEFINITION / ***Surface Area of a Solid of Revolution.***

The surface area S of the solid of revolution generated by revolving the smooth curve C: $x = x(t)$, $y = y(t)$, $a \le t \le b$,

where $y(t) \geq 0$, about the x-axis is

$$S = \int_a^b 2\pi y \sqrt{\left(\frac{dx}{dt}\right)^2 + \left(\frac{dy}{dt}\right)^2}\, dt \tag{4}$$

EXAMPLE 1

Find the surface area that is generated by revolving the curve $x = 2t^3$, $y = 3t^2$, $0 \leq t \leq 1$, about the x-axis.

Solution

By using (4) with $dx/dt = 6t^2$, $dy/dt = 6t$, we get

$$S = 2\pi \int_0^1 3t^2\sqrt{36t^4 + 36t^2}\, dt = 36\pi \int_0^1 t^3\sqrt{t^2+1}\, dt$$

$$= \frac{36\pi}{2}\int_1^2 (u-1)\sqrt{u}\, du = 18\pi\left(\frac{2}{5}u^{5/2} - \frac{2}{3}u^{3/2}\right)\Bigg|_1^2 = \frac{24\pi}{5}(\sqrt{2}+1)$$

Set $u = t^2 + 1$.

To aid in memorizing (4), regard the integrand as the product of the slant height $\sqrt{(dx/dt)^2 + (dy/dt)^2}$ and the circumference $2\pi y$ of the circle traced by a point (x, y) on the corresponding subarc. Remember, too, that the limits of integration are parameter values, not x values.

[13.6.2] THEOREM

The surface area S of a solid of revolution obtained by revolving the smooth curve C: $x = x(t)$, $y = y(t)$, $a \leq t \leq b$, where $x = x(t) \geq 0$, about the y-axis is

$$S = \int_a^b 2\pi x \sqrt{\left(\frac{dx}{dt}\right)^2 + \left(\frac{dy}{dt}\right)^2}\, dt \tag{5}$$

Surface Area of a Solid of Revolution Using a Rectangular Equation $y = f(x)$

Suppose we have a smooth curve C that is defined by the rectangular equation $y = f(x)$, $a \leq x \leq b$, where $f(x) \geq 0$ on $[a, b]$. A set of parametric equations for this curve is $x = t$ and $y = f(t)$. Then $dx/dt = 1$, $dy/dt = f'(t) = f'(x)$, and $dt = dx$. The surface area S of the solid of revolution obtained by revolving C about the x-axis is

$$S = \int_a^b 2\pi y\sqrt{1 + [f'(x)]^2}\, dx \tag{6}$$

EXAMPLE 2

Find the surface area of the solid generated by revolving $y = \sqrt{x}$ about the x-axis from $x = 0$ to $x = 1$.

Solution

Here we use (6):

$$S = \int_0^1 2\pi\sqrt{x}\sqrt{1 + \frac{1}{4x}}\,dx = 2\pi \int_0^1 \frac{1}{2}\sqrt{4x+1}\,dx$$

$$\underset{\uparrow}{=} \pi \int_1^5 \sqrt{u}\left(\frac{du}{4}\right) = \frac{\pi}{4}\left(\frac{u^{3/2}}{\frac{3}{2}}\right)\bigg|_1^5 = \frac{\pi}{6}(5\sqrt{5} - 1)$$

Set $u = 4x + 1$.

Surface Area of a Solid of Revolution Using a Polar Equation $r = f(\theta)$

Suppose a smooth curve C is given by the polar equation $r = f(\theta)$, $\alpha \le \theta \le \beta$. One set of parametric equations for this curve is

$$x(\theta) = f(\theta)\cos\theta \qquad y(\theta) = f(\theta)\sin\theta$$

Then

$$\frac{dx}{d\theta} = f'(\theta)\cos\theta - f(\theta)\sin\theta \qquad \frac{dy}{d\theta} = f'(\theta)\sin\theta + f(\theta)\cos\theta$$

$$\left(\frac{dx}{d\theta}\right)^2 + \left(\frac{dy}{d\theta}\right)^2 = f(\theta)^2 + f'(\theta)^2$$

So, the surface area S of the solid of revolution obtained by revolving C about the polar axis is

$$S = 2\pi \int_\alpha^\beta f(\theta)\sin\theta\sqrt{[f(\theta)]^2 + [f'(\theta)]^2}\,d\theta \qquad (7)$$

EXAMPLE 3

Find the surface area of the solid generated by revolving the arc of the circle $r = a$, $0 \le \theta \le \pi/4$, about the polar axis.

Solution

By using (7) with $r = f(\theta) = a$, we find

$$S = 2\pi \int_0^{\pi/4} a\sin\theta\sqrt{a^2}\,d\theta = 2\pi a^2 \int_0^{\pi/4} \sin\theta\,d\theta$$

$$= 2\pi a^2(-\cos\theta)\bigg|_0^{\pi/4} = 2\pi a^2\left(\frac{-\sqrt{2}}{2} + 1\right) = \pi a^2(2 - \sqrt{2})$$

EXERCISE 13.6

In Problems 1–14 find the surface area of the solid generated by revolving the given curve about the x-axis.

1. $x = 3t^2, \quad y = 6t, \quad 0 \le t \le 1$

2. $x = t^2, \quad y = 2t, \quad 0 \le t \le 3$

3. $x = a\cos^3\theta, \quad y = a\sin^3\theta, \quad 0 \le \theta \le \pi/2$

4. $x = a(t - \sin t), \quad y = a(1 - \cos t), \quad 0 \le t \le \pi$

5. $y = x^3, \quad 0 \le x \le 1$

6. $y = \dfrac{x^4}{8} + \dfrac{1}{4x^2}, \quad 1 \le x \le 2$

7. $y = \dfrac{a}{2}(e^{x/a} + e^{-x/a}), \quad 0 \le x \le a$

8. $x = \dfrac{1}{4}y^2, \quad 0 \le y \le 2$

9. $y = \sqrt{a^2 - x^2}, \quad -a \le x \le a$

10. $y = e^{-x}, \quad 0 \le x \le 1$

11. $y = e^x, \quad 0 \le x \le 1$

12. $y = e^{-x}, \quad 0 \le x < +\infty$

13. $x^{2/3} + y^{2/3} = a^{2/3}, \quad 0 \le x \le a, \quad y \ge 0$

14. $y = \dfrac{1}{x}, \quad 1 \le x \le 2$

In Problems 15 and 16 find the surface area of the solid generated by revolving the given curve about the y-axis.

15. $x = 3t^2, \quad y = 2t^3, \quad 0 \le t \le 1$

16. $x = 2t + 1, \quad y = t^2 + 3, \quad 0 \le t \le 3$

In Problems 17–20 find the surface area of the solid generated by revolving the given curve about the polar axis.

17. $r = \sin\theta, \quad 0 \le \theta \le \pi/2$

18. $r = 1 + \cos\theta, \quad 0 \le \theta \le \pi$

19. $r = e^{\theta}, \quad 0 \le \theta \le \pi$

20. $r = 2a\cos\theta, \quad 0 \le \theta \le \pi/2$

21. Show that if the curve $y = 1/x, \quad x > 1,$ is revolved about the x-axis, then the volume of the solid generated is finite, but its surface area is infinite.

22. Prove that the surface area of a right circular cone of altitude h and base b is $\pi b\sqrt{h^2 + b^2}$.

23. When an arc of a catenary $y = \cosh x, \quad a \le x \le b,$ is rotated about the x-axis, it generates a surface called a *catenoid*, which has the least area of all surfaces generated by rotating curves having the same endpoints. What is its area?

24. Develop a formula for the surface area of a sphere of radius R.

25. A sphere of radius R has a hole of radius $a < R$ drilled through it. The axis of the hole coincides with a diameter of the sphere. Find the surface area of that part of the sphere that remains.

26. A plug is made to repair the hole in the sphere in Problems 25. What is its surface area?

REVIEW EXERCISES

In Problems 1–8 convert the given polar equations to cartesian equations.

1. $r = 4\sin 2\theta$

2. $r = e^{\theta/2}$

3. $r = \dfrac{1}{1 + 2\cos\theta}$

4. $r = a - \sin\theta$

5. $r = \sec\theta + 2$

6. $r^2 = 4\cos 2\theta$

7. $r = \sin 3\theta$

8. $r = \theta$

In Problems 9–13 convert the given cartesian equation to a polar equation.

9. $x^2 + y^2 = x$

10. $(x^2 + y^2)^2 = x^2 - y^2$

11. $y^2 = (x^2 + y^2)\cos^2[(x^2 + y^2)^{1/2}]$

12. $\dfrac{x^2}{a^2} + \dfrac{y^2}{b^2} = 1$

13. $y = a$

In Problems 14–27 sketch the graph of each polar equation.

14. $r = \dfrac{2}{1 - \cos\theta}$

15. $r^2 = 4\cos 2\theta$

16. $r = \dfrac{1}{2} - \sin\theta$

17. $r = \dfrac{4}{1 - 2\cos\theta}$

18. $r^2 = 4\sin 2\theta$

19. $r = 1 - \sin\theta$

20. $r = \dfrac{1}{1 - \frac{1}{6}\cos\theta}$

21. $r = \dfrac{1}{1 + 2\cos\theta}$

22. $r = 2 - \sin\theta$

23. $r = \sec\theta + 2$

24. $r = \dfrac{2}{1 + 3\cos\theta}$

25. $r^2 = 1 + \sin^2\theta$

26. $r^2 = 1 - \sin^2\theta$

27. $r = \dfrac{3}{1 - \sin\theta}$

28. Find the area enclosed by the right branch of the graph of $r = \sec\theta + 2$ and the lines $\theta = \pm\pi/4$.

29. Find the area enclosed by the graph of $r^2 = 4\cos 2\theta$ but outside the circle $r = \sqrt{2}$.

30. Find the area of the region common to the graphs of $r = \cos\theta$ and $r = 1 - \cos\theta$.

In Problems 31–37 find the rectangular equation of the curve with the given parametric equations.

31. $x = \cosh t, \quad y = \sinh$

32. $x = e^t, \quad y = e^{-t}$

33. $x = \sinh^{-1} t, \quad y = \sqrt{t^2 + 1}$

34. $x = \dfrac{t^2}{1 + t}, \quad y = \dfrac{t}{1 + t}$

35. $x = \sin 2t, \quad y = \cos t$

36. $x = \tan t, \quad y = \frac{1}{3}(\sec^2 t + 1)$

37. $x = e^t, \quad y = \frac{1}{2}e^{2t} - \frac{1}{4}t$

38. Find parametric equations for the ellipse $\dfrac{x^2}{a^2} + \dfrac{y^2}{b^2} = 1.$

39. Show that the ellipse in Problem 38 has infinitely many parametric equations.

40. Find parametric equations for the hyperbola $\dfrac{x^2}{a^2} - \dfrac{y^2}{b^2} = 1.$

In Problems 41–44 find the arc lengths indicated.

41. Arc length of the curve of Problem 33 from $t = 0$ to $t = 1$

42. Arc length of the curve of Problem 36 from $t = 0$ to $t = \pi/4$

43. Arc length of the curve of Problem 37 from $t = 0$ to $t = 2$

44. Arc length of the curve $x = \frac{1}{2}y^2 - \frac{1}{4}\ln y$ from $y = 1$ to $y = 2$

45. Find the area of the surface generated by revolving the curve of Problem 33 from $t = 0$ to $t = 1$ about the x-axis.

CHALLENGE EXERCISES

1. Find an expression for d^2y/dx^2 if $x = f(t)$, $y = g(t)$, where f and g have second derivatives.

2. Find d^2y/dx^2 if $x = a\cos^3\theta$, $y = a\sin^3\theta$.

3. Express $r^2 = \cos 2\theta$ in rectangular coordinates free of radicals.

4. In our earlier discussion of polar equations for conic sections, we have had one of the foci at the pole. Write the general equation for a conic in polar coordinates where we do not have a focus at a pole. Suppose we are given a focus F with polar coordinates (r_1, θ_1) and a directrix D with equation $r\cos(\theta+\theta_0) = -d$, where $d > 0$. Let the eccentricity be e.

5. Find parametric equations for the circle $x^2 + y^2 = R^2$, using as parameter the slope m of the line through $(-R, 0)$ and the general point $P = (x, y)$ on the circle.

6. Find parametric equations for the parabola $y = x^2$ using as parameter the slope m of the line joining the point $(1, 1)$ to the general point $P = (x, y)$ on the parabola.

7. Show that the area enclosed by the graph of $r\theta = a$ and any two radii r_1 and r_2 is proportional to the difference of the radii, $r_1 - r_2$.

8. Find the point on the curve $x = \frac{4}{3}t^3 + 3t^2$, $y = t^3 - 4t^2$, such that the length of the curve from $(0, 0)$ to (x, y) is $(80\sqrt{2} - 40)/3$.

9. Show that a smooth curve $(x(t), y(t))$ for t in $[a, b]$ can be parameterized by its arc length; that is, there are functions $f(s)$ and $g(s)$ so that the curve is given by $(f(s), g(s))$ for s in $[0, L]$, where L is the length of the curve, and where for each s_0 in $[0, L]$, the length of the portion of the curve given by $(f(s), g(s))$ for s in $[0, s_0]$ is s_0.

10. Show that the surface area of the solid generated by revolving the first-quadrant arc of $(x^2/a^2) + (y^2/b^2) = 1$ about the x-axis is

$$S = \pi b^2 + \frac{\pi ab}{e}\sin^{-1}e$$

where e is the eccentricity of the ellipse. (See Problem 19 in Exercise 12.3, p. 721.)

11. Show that $r = a\cos\theta + b\sin\theta$ is the equation of a circle. Find the center and radius of the circle.

12. Prove that the area of the triangle with vertices $(0, 0)$, (r_1, θ_1), (r_2, θ_2) is

$$A = \tfrac{1}{2}r_1r_2\sin(\theta_2 - \theta_1) \qquad 0 \le \theta_1 < \theta_2 \le \pi$$

13. Find the surface area of a parabolic reflector of a searchlight, generated by revolving an arc of a parabola about its axis, if the searchlight is 1 meter across and $\frac{1}{4}$ meter deep.

14. Consider the equation $r = \sin n\theta$. Discuss the possibilities for the graph when n is a positive integer.

Graphing Polar and Parametric Curves on a Graphing Calculator*

In this vignette, short programs for the Casio graphing calculator are used to graph the curves associated with polar and parametric equations. Programs are stored on the Casio using Mode 2, the program writing and editing mode.

* Gregory D. Foley
The Ohio State University

EXAMPLE 1

Graph $r = 6\sin 2.5\theta$.

Solution

This can be accomplished on the Casio by storing and executing the program given below. The portions of the program in **boldface** type may vary from problem to problem. Notice that the variable T (used for the angle θ) takes on values from 0 to 4π in 0.1 increments. The functional value R is computed for each of these values of T. These polar coordinates (R, T) are converted to rectangular coordinates; the point is plotted; then it is connected by a line segment to its immediate predecessor (except for the first time through the loop when there is no preceding point). The Line command is given in boldface type because in some cases you may wish to delete this step to avoid connecting points. The Range parameters are chosen so that the entire graph will appear on the screen and fit the 3:2 aspect ratio of the Casio screen. The line with the Range command can be completely deleted so that the Range can be changed without editing the program. Note that a Range statement within a program will cause the graphics screen to be cleared each time the program is executed.

In case you have not programmed the Casio before, here is how you enter and execute the program: Press [MODE] [2]. The line

Prog 01234567989

should appear at the the bottom of the screen. The blinking 0 indicates that Program Storage Register 0 is the selected program address. The left and right cursor keys can be used to select a different program storage address. With 0 still blinking, press [EXE]. You should see a blinking cursor on an otherwise blank screen, indicating that the Casio is ready to accept program statements. Key in the following, pressing [EXE] after each line *except the last*; the EXE key acts as a carriage return key on a typewriter or computer when the Casio is in Mode 2.

```
0 → T
Range −9, 9, 1, −6, 6, 1
Lbl 1
6 sin 2.5T → R
Rec(R, T)
Plot I,J
Line
T + 0.1 → T
T ≤ 4π ⇒ Goto 1
```

After you key in the last line, press [MODE] [1] to return to RUN mode. Check to see that your Casio is set in *radian mode*. (Why is this necessary?) Then press [Prog] [0] [EXE] to execute the program.

When the program is executed, the Casio flashes back and forth between the text window and the graphics window. At the end of execution, the Casio will display a number (in this case, 12.56637061) in the text window, and the graph will be stored in the graphics window. The G ↔ T command will reveal the graph shown in the following figure.

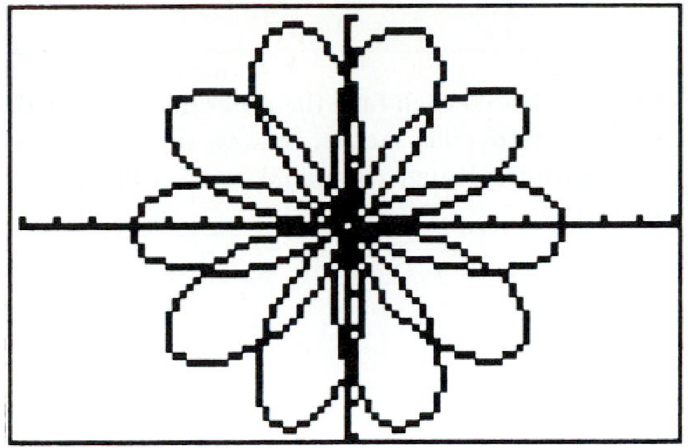

The graph of $r = 6 \sin 2.5\theta$ as it appears on the Casio using a viewing rectangle of $[-9, 9] \times [-6, 6]$

EXAMPLE 2

Graph $r = 4 \cos 3\theta$.

Solution

Edit the program used in Example 1 to read as indicated below. Be sure your Casio is set in radian mode; then execute the program.

```
0 → T
Range −6, 6, 1, −4, 4, 1
Lbl 1
4 cos 3T → R
Rec(R, T)
Plot I,J
Line
T + .1 → T
T ≤ π ⇒ Goto 1
```

EXAMPLE 3

Graph the curve given by the parametric equations

$$x = 3 \cos t \qquad y = 4 \sin t \qquad 0 \le t \le 2\pi$$

Solution

Edit the program used in Example 1 to read as indicated below. Once again, be sure your Casio is set in radian mode, and execute the program.

```
0 → T
Range −6, 6, 1, −4, 4, 1
Lbl 1
3 cos T → X
4 sin T → Y
Plot X,Y
Line
T + .1 → T
T ≤ 2π ⇒ Goto 1
```

Compare your graph with Figure 32 of Chapter 13.

EXERCISES

1–16. Modify the programs given in this vignette to produce the following figures from Chapter 13: Figures 9–12, 15–19, 23–27, 31, and 33. *Note:* A Range statement within a program will cause the graphics screen to be cleared each time the program is executed. So, in order to overlay curves, remove the line in the program with the Range command, and set the Range when in Mode 1.

Vectors; Analytic Geometry in Space

14

Rectangular Coordinates in Space 14.1

In Chapter 1 we established a correspondence between the points on a line and the real numbers. We showed that each point in a plane can be associated with an ordered pair of real numbers. Here we show that each point in space can be associated with an *ordered triple* of real numbers.

We begin by selecting a fixed point called the *origin*. Through the origin, we draw three mutually perpendicular lines; these are called the *coordinate axes*, and they are usually labeled the *x-axis*, *y-axis*, and *z-axis*. On each of the three lines we choose one direction as positive and select an appropriate scale, as indicated in Figure 1 on page 792.

As indicated in Figure 2, we position the positive z-axis so that the system is *right-handed*. This conforms to the so-called *right-hand rule*, which asserts that if the index finger of the right hand points in the direction of the positive

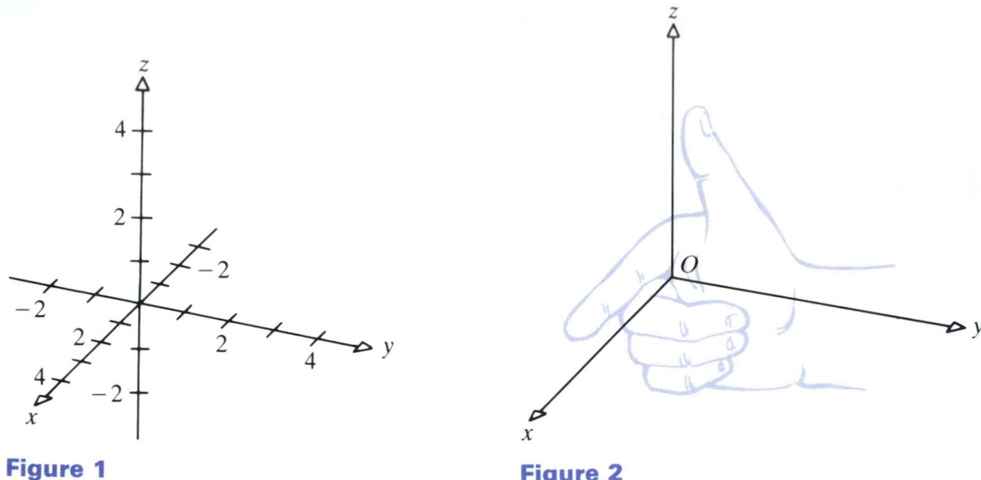

Figure 1

Figure 2

x-axis and the middle finger points in the direction of the positive y-axis, then the thumb will point in the direction of the positive z-axis.*

Just as we did in one and two dimensions, we assign coordinates to each point P in three dimensions. Specifically, we identify a point P with an ordered triple of real numbers (x, y, z), and we refer to it as "the point (x, y, z)." Thus, "the point $(3, 5, 7)$" is the point for which $x = 3,\quad y = 5,\quad z = 7$; that is, starting from the origin, we reach P by moving 3 units along the positive x-axis, then 5 units in the direction of the positive y-axis, and finally 7 units in the direction of the positive z-axis. Figure 3 illustrates the location of the point $(3, 5, 7)$, as well as the points $(3, 5, 0)$ and $(0, 5, 0)$. Observe

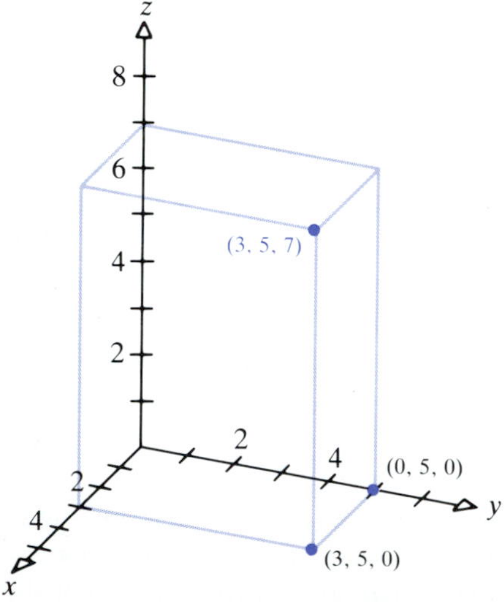

Figure 3

* Although there are left-handed systems and left-handed rules, we shall adopt the usual convention and use only a right-handed system.

that any point on the x-axis will have the form $(x, 0, 0)$. Similarly, $(0, y, 0)$ and $(0, 0, z)$ represent points on the y-axis and z-axis, respectively.

In addition, all points of the form $(x, y, 0)$ constitute a plane called the xy-plane. This plane is perpendicular to the z-axis. Similarly, the points $(0, y, z)$ form the yz-plane, which is perpendicular to the x-axis; and the points $(x, 0, z)$ form the xz-plane, which is perpendicular to the y-axis. See Fig. 4(a). Figure 4(b) illustrates that points of the form (x, y, z), where $z = 5$, lie in a plane parallel to the xy-plane. Similarly, points (x, y, z), where $y = 7$, lie in a plane parallel to the xz-plane.

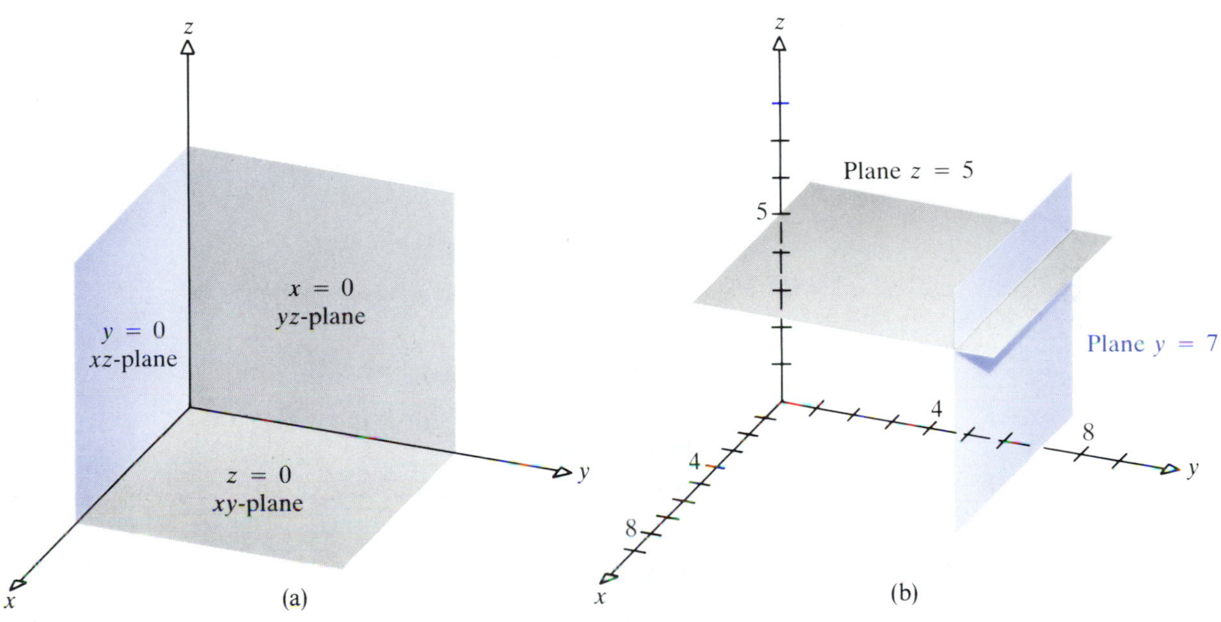

Figure 4

Distance in Space

To derive a formula for the distance $|P_1P_2|$ between two points $P_1 = (x_1, y_1, z_1)$ and $P_2 = (x_2, y_2, z_2)$, we apply the Pythagorean theorem twice. As Figure 5 on page 794 illustrates, we utilize the point $A = (x_2, y_2, z_1)$. The first application of the Pythagorean theorem involves observing that the triangle P_1AP_2 is a right triangle in which the side of length $|P_1P_2|$ is the hypotenuse. As a result,

$$|P_1P_2| = \sqrt{|P_1A|^2 + |AP_2|^2} \quad \textbf{(1)}$$

The points P_1 and A lie in a plane parallel to the xy-plane. Thus, we can use the formula for distance in two dimensions and obtain

$$|P_1A| = \sqrt{(x_2 - x_1)^2 + (y_2 - y_1)^2} \quad \textbf{(2)}$$

The points P_2 and A lie along a line parallel to the z-axis, so that $|AP_2| = |z_2 - z_1|$ and $|AP_2|^2 = (z_2 - z_1)^2$. This fact, together with (1) and (2), gives a formula for the distance between two points in space.

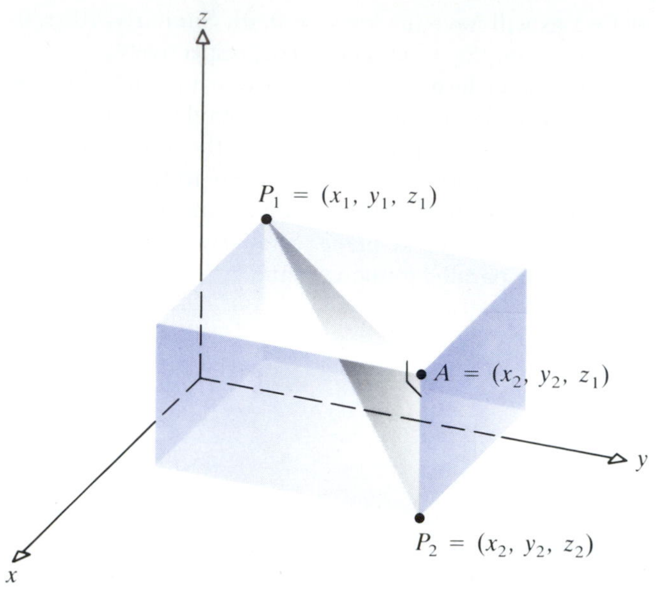

Figure 5

[14.1.1] THEOREM / *Distance Formula.*

The distance $|P_1P_2|$ from $P_1=(x_1, y_1, z_1)$ to $P_2=(x_2, y_2, z_2)$ is given by the formula

$$|P_1P_2| = \sqrt{(x_2 - x_1)^2 + (y_2 - y_1)^2 + (z_2 - z_1)^2} \tag{3}$$

EXAMPLE 1

If $P_1 = (1, 3, -2)$ and $P_2 = (2, -1, -3)$, the distance $|P_1P_2|$ from P_1 to P_2 is

$$|P_1P_2| = \sqrt{(2-1)^2 + (-1-3)^2 + (-3+2)^2}$$
$$= \sqrt{1 + 16 + 1} = \sqrt{18} = 3\sqrt{2}$$

The Sphere

In space the collection of all points that are the same distance from some fixed point is called a *sphere* (see Fig. 6). The constant distance is called the *radius*, and the fixed point is the *center* of the sphere. Any point $P = (x, y, z)$ on a sphere of radius R and center at the point $P_0 = (x_0, y_0, z_0)$ obeys $|PP_0| = R$. By the distance formula (3), the equation of this sphere is

$$\sqrt{(x - x_0)^2 + (y - y_0)^2 + (z - z_0)^2} = R$$

Squaring both sides, we obtain the following result:

[14.1.2] THEOREM / *Equation of a Sphere.*

The equation of a sphere with center at (x_0, y_0, z_0) and radius R is

$$(x - x_0)^2 + (y - y_0)^2 + (z - z_0)^2 = R^2$$

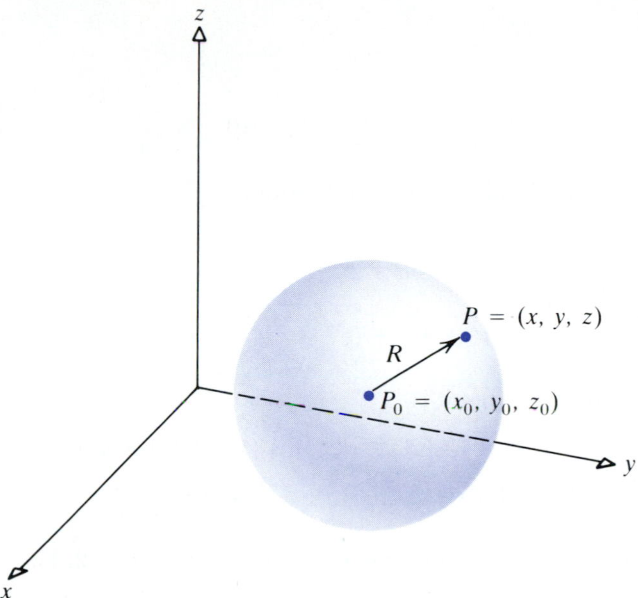

Figure 6

EXAMPLE 2

The equation of the sphere illustrated in Figure 7, with radius 2 and center $(-1, 2, 0)$, is

$$(x + 1)^2 + (y - 2)^2 + z^2 = 4$$

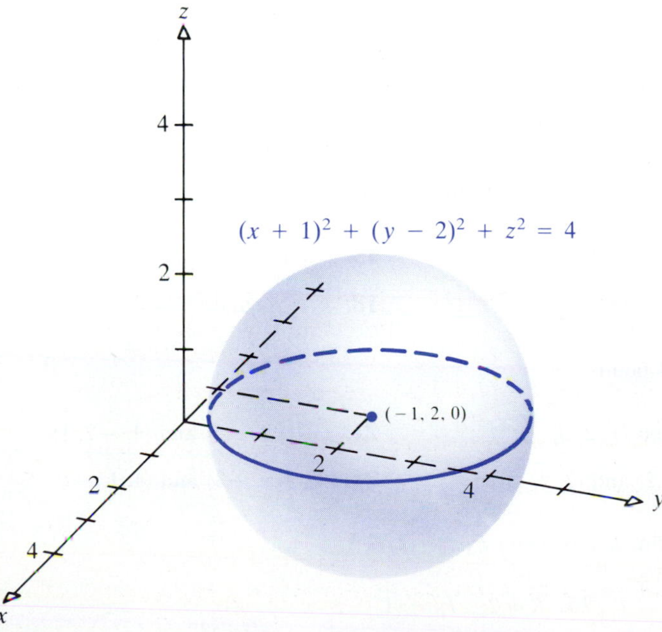

Figure 7

EXAMPLE 3

Show that

$$x^2 + y^2 + z^2 + 2x + 4y - 2z = 10$$

is the equation of a sphere. Find its center and radius.

Solution

Rewrite the expression as

$$(x^2 + 2x) + (y^2 + 4y) + (z^2 - 2z) = 10$$

and complete the squares. The result is

$$(x^2 + 2x + 1) + (y^2 + 4y + 4) + (z^2 - 2z + 1) = 10 + 1 + 4 + 1$$
$$(x + 1)^2 + (y + 2)^2 + (z - 1)^2 = 16$$

This is the equation of a sphere with radius 4 and center at $(-1, -2, 1)$.

EXERCISE 14.1

In Problems 1–6 plot each point in a three-dimensional coordinate system.

1. $(1, 1, 1)$ **2.** $(0, 0, 1)$ **3.** $(0, 2, 5)$

4. $(-1, 5, 0)$ **5.** $(-3, 1, 0)$ **6.** $(4, -1, -3)$

In Problems 7–12 opposite vertices of a rectangular box whose edges are parallel to the coordinate axes are given. List the coordinates of the other six vertices of the box.

7. $(0, 0, 0)$; $(2, 1, 3)$ **8.** $(0, 0, 0)$; $(4, 2, 2)$ **9.** $(1, 2, 3)$; $(3, 4, 5)$

10. $(5, 6, 1)$; $(3, 8, 2)$ **11.** $(-1, 0, 2)$; $(4, 2, 5)$ **12.** $(-2, -3, 0)$; $(-6, 7, 1)$

In Problems 13–18 describe in words the set of all points (x, y, z) that satisfy the given conditions.

13. $y = 3$ **14.** $z = -3$ **15.** $x = 0$

16. $x = 1$ and $y = 0$ **17.** $z = 5$ **18.** $x = y$ and $z = 0$

In Problems 19–24 find the distance between each pair of points.

19. $(1, 3, 0)$ and $(4, 1, 2)$ **20.** $(3, 2, 1)$ and $(1, 2, 3)$ **21.** $(-1, 2, -3)$ and $(4, -2, 1)$

22. $(-2, 1, 3)$ and $(4, 0, -3)$ **23.** $(4, -2, -2)$ and $(3, 2, 1)$ **24.** $(2, -3, -3)$ and $(4, 1, -1)$

In Problems 25–28 find the equation of a sphere with radius R and center P_0.

25. $R = 1$; $P_0 = (3, 1, 1)$ **26.** $R = 2$; $P_0 = (1, 2, 2)$

27. $R = 3$; $P_0 = (-1, 1, 2)$ **28.** $R = 1$; $P_0 = (-3, 1, -1)$

In Problems 29–34 find the radius and center of each sphere.

29. $x^2 + y^2 + z^2 + 2x - 2y = 2$

30. $x^2 + y^2 + z^2 + 2x - 2z = -1$

31. $x^2 + y^2 + z^2 - 4x + 4y + 2z = 0$

32. $x^2 + y^2 + z^2 - 4x = 0$

33. $2x^2 + 2y^2 + 2z^2 - 8x + 4z = -1$

34. $3x^2 + 3y^2 + 3z^2 + 6x - 6y = 3$

In Problems 35–38 write the equation of the sphere described.

35. The endpoints of a diameter are $(-2, 0, 4)$ and $(2, 6, 8)$.

36. The endpoints of a diameter are $(1, 3, 6)$ and $(-3, 1, 4)$.

37. The center is at $(-3, 2, 1)$, and the sphere passes through the point $(4, -1, 3)$.

38. The center is at $(0, -3, 4)$, and the sphere passes through the point $(2, 1, 1)$.

39. Show that the points $(-2, 6, 0)$, $(4, 9, 1)$, and $(-3, 2, 18)$ are the vertices of a right triangle.

40. Show that the points $(2, 2, 2)$, $(0, 1, 2)$, $(-1, 3, 3)$, and $(3, 0, 1)$ are the vertices of a parallelogram.

41. Show that the points $(2, 2, 2)$, $(2, 0, 1)$, $(4, 1, -1)$, and $(4, 3, 0)$ are the vertices of a rectangle.

42. Find an equation of the sphere that passes through the vertices of the right triangle cited in Problem 39 and that has a diameter along the hypotenuse of this triangle.

43. Show that the points $(2, 4, 2)$, $(2, 1, 5)$, and $(5, 1, 2)$ are the vertices of an equilateral triangle.

Introduction to Vectors 14.2

Roughly speaking, a *vector** is a quantity that has both *magnitude and direction*. A vector is usually portrayed by an arrow of length equal to the magnitude of the vector and pointing in the appropriate direction.

We've encountered vectors earlier in this book. For example, *velocity* is a vector. The velocity of an airplane can be represented by an arrow that points in the direction of movement (see Fig. 8). The length of the arrow represents the speed of the airplane. If the airplane speeds up, the length of the arrow increases. If the direction of the airplane changes, the direction of the arrow changes.

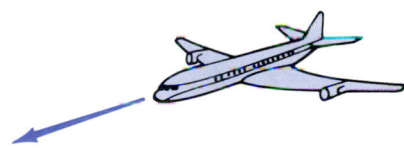

Figure 8

Force and *acceleration* are also vectors.† If a force **F** is exerted on an object of mass m, the force causes the object to accelerate *in the direction of the force*. Thus, force and acceleration are quantities that have both magnitude and direction, and hence they are vectors. On the other hand, mass has magnitude but no direction, and hence it is not a vector. Figure 9 illustrates the situation just described.

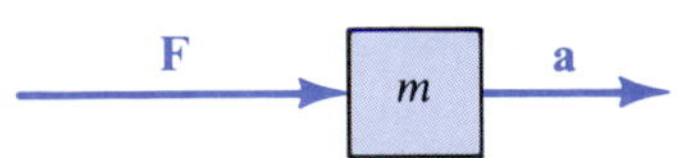

Figure 9

Vectors are closely related to the geometric idea of a *directed line segment*. Consider two points P and Q. If P and Q are distinct points, there is exactly one line containing both P and Q. The points that are on the part of the line that joins P and Q—including P and Q—make up the *line segment from P to Q*. If we order the points P and Q—say, as PQ—then we have a *directed*

* From the Latin *vehere*: "to carry."

† Boldface letters will be used to denote vectors, in order to distinguish them from numbers. For handwritten work, an arrow placed over a letter may be used to denote a vector.

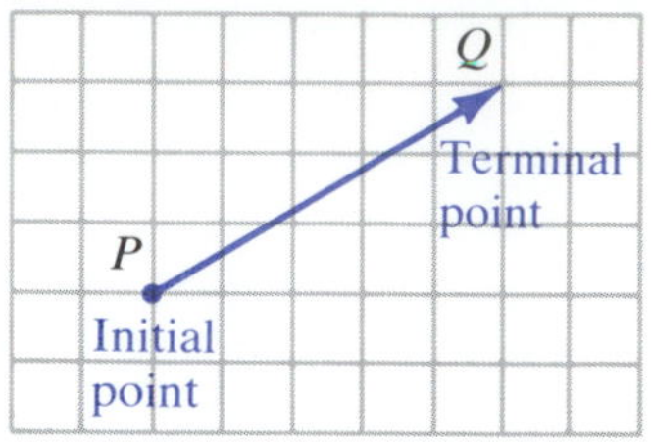

Directed line segment $\overrightarrow{PQ}$

Figure 10

line segment from P to Q, which we denote by $\overrightarrow{PQ}$. For the directed line segment $\overrightarrow{PQ}$, we call P the *initial point* and Q the *terminal point* (see Fig. 10).

The magnitude of the directed line segment $\overrightarrow{PQ}$ is the distance from the point P to the point Q; that is, it is the length of the line segment. The direction of $\overrightarrow{PQ}$ is from P to Q. If a vector $\mathbf{v}$ has the same magnitude and the same direction as the directed line segment $\overrightarrow{PQ}$, then we write

$$\mathbf{v} = \overrightarrow{PQ}$$

The vector $\mathbf{v}$ whose magnitude is 0 is called the *zero vector* $\mathbf{0}$.

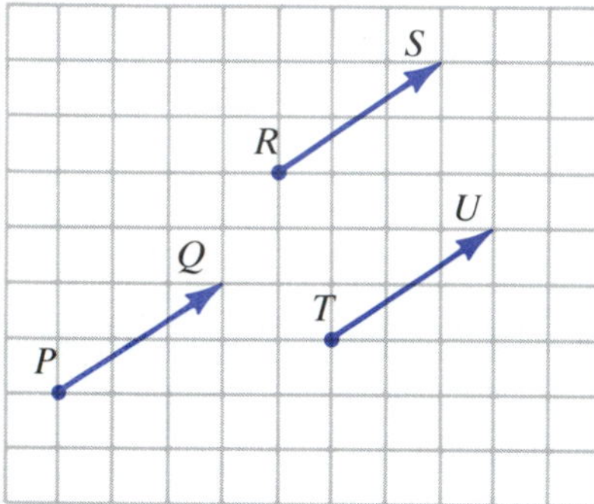

Figure 11

[14.2.1] DEFINITION / *Equal Vectors.*

Two vectors v and w are *equal*, written

$$\mathbf{v} = \mathbf{w}$$

if they have the same magnitude and the same direction.

For example, the vectors shown in Figure 11 have the same magnitude and the same direction and so are equal, even though they have different initial points and different terminal points.

As a result, we find it useful to think of a vector as simply an arrow, keeping in mind that two arrows (vectors) are equal if they have the same direction and the same magnitude (length).

Adding Vectors

[14.2.2] DEFINITION / *Sum of Two Vectors.*

We define the *sum* v + w of two vectors as follows: We position the vectors v and w so that the terminal point of v coincides with the initial point of w. The vector v + w is then represented by the arrow directed from the initial point of v to the terminal point of w.

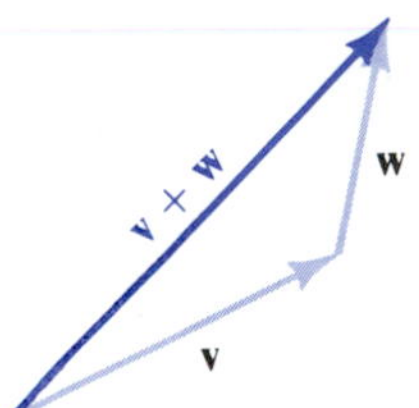

Figure 12

Figure 12 illustrates this idea.

We said earlier that forces are vectors. But how do we know that forces "add" the same way vectors do? Well, physicists tell us they do, and laboratory experiments bear it out. Thus, if $\mathbf{F}_1$ and $\mathbf{F}_2$ are two forces simultaneously acting on an object, the vector sum $\mathbf{F}_1 + \mathbf{F}_2$ is equal to the force that produces the same effect as that obtained when the force $\mathbf{F}_1$ is applied followed by the force $\mathbf{F}_2$. The force $\mathbf{F}_1 + \mathbf{F}_2$ is sometimes called the *resultant* of $\mathbf{F}_1$ and $\mathbf{F}_2$ (see Fig. 13).

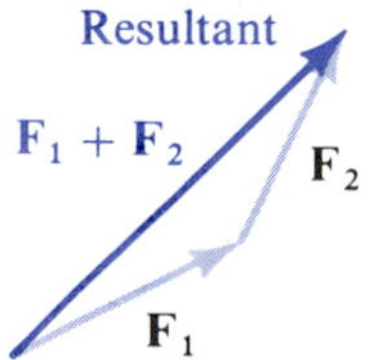

Figure 13

Another physical example of a vector is velocity. For example, if $\mathbf{w}$ is a vector describing the velocity of the wind relative to the earth and $\mathbf{v}$ is a vector describing the velocity of an airplane in the air, then $\mathbf{w} + \mathbf{v}$ is the vector describing the velocity of the airplane relative to the earth (see Fig. 14).

Vector addition is *commutative*; that is, if $\mathbf{v}$ and $\mathbf{w}$ are any two vectors, then

$$\mathbf{v} + \mathbf{w} = \mathbf{w} + \mathbf{v} \tag{1}$$

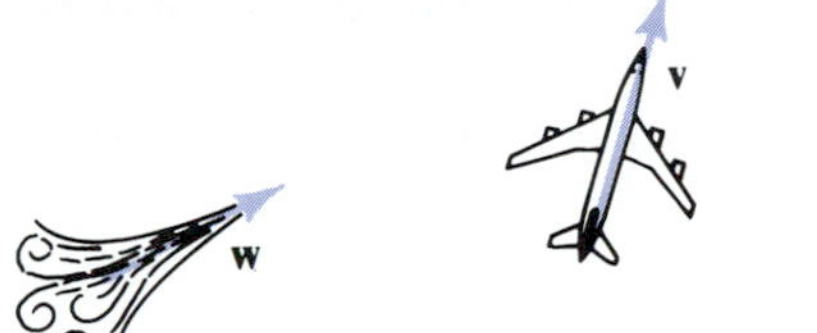

(a) Velocity of wind relative to earth (b) Velocity of airplane relative to air (c) Resultant equals velocity of airplane relative to earth

Figure 14

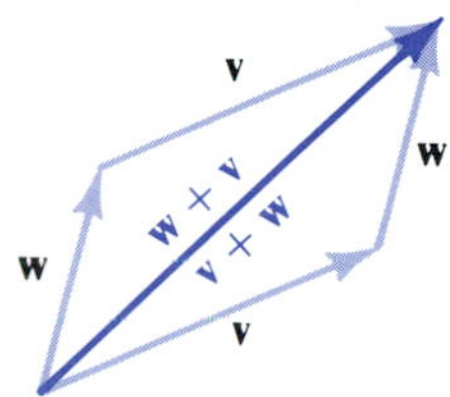

Figure 15

Figure 15 illustrates the validity of this fact. (Observe that (1) is another way of saying that opposite sides of a parallelogram are equal and parallel.)

Vector addition is also *associative*; that is, if **u**, **v**, and **w** are vectors, then

$$(\mathbf{u} + \mathbf{v}) + \mathbf{w} = \mathbf{u} + (\mathbf{v} + \mathbf{w})$$

See Figure 16.

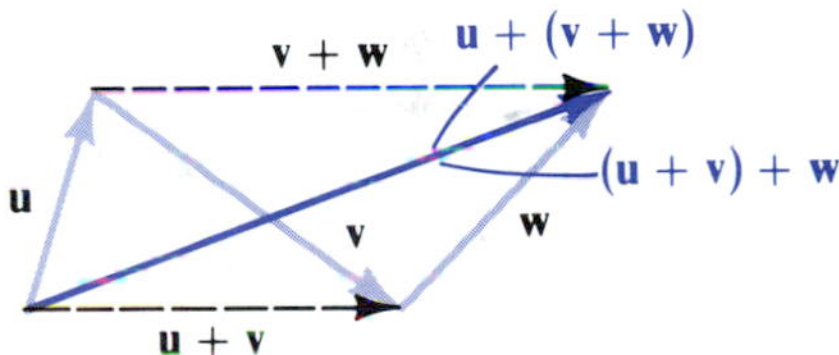

Figure 16

The zero vector has the property that

$$\mathbf{v} + \mathbf{0} = \mathbf{0} + \mathbf{v} = \mathbf{v}$$

for any vector **v**.

If **v** is a vector, then $-\mathbf{v}$ is the vector that has the same magnitude as **v**, but whose direction is opposite to **v**. See Figure 17. Furthermore,

$$\mathbf{v} + (-\mathbf{v}) = \mathbf{0}$$

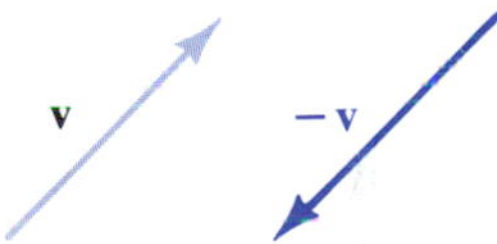

Figure 17

[14.2.3] DEFINITION / *Difference of Two Vectors.*

If v and w are two vectors, we define the *difference* v − w as

$$\mathbf{v} - \mathbf{w} = \mathbf{v} + (-\mathbf{w})$$

Figure 18 illustrates that to find the difference $\mathbf{v} - \mathbf{w}$, we need to produce the vector that, when added to **w**, gives **v**.

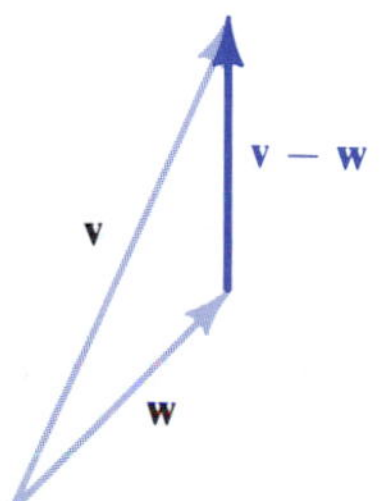

Figure 18

Multiplying Vectors by Numbers

We use the word *scalar* to mean a real number. Scalars are quantities that have only *magnitude*. In physics, the quantities *mass*, *time*, *density*, *temperature*, and *speed* are examples of scalars.

[14.2.4] DEFINITION / *Product* $\alpha\mathbf{v}$.

If α is a scalar and v is a vector, the *product* $\alpha\mathbf{v}$ is defined as follows:

1. **If $\alpha > 0$, the product $\alpha\mathbf{v}$ is the vector having magnitude α times the magnitude of v and having the same direction as v.**

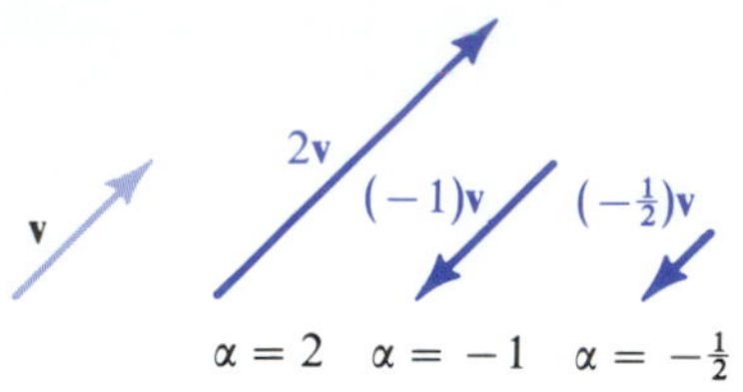

Figure 19

2. If $\alpha < 0$, the product $\alpha\mathbf{v}$ is the vector having magnitude $|\alpha|$ times the magnitude of v and having the opposite direction as v.

3. If $\alpha = 0$ or if $\mathbf{v} = \mathbf{0}$, then $\alpha\mathbf{v} = \mathbf{0}$.

See Figure 19.

For example, if **a** is the acceleration of an object of mass m due to a force **F** being exerted on it, then by Newton's second law of motion,

$$\mathbf{F} = m\mathbf{a}$$

Here, $m\mathbf{a}$ is the product of the scalar m and the vector **a**.

The product of a scalar and a vector has the following properties:

$$(0)\mathbf{v} = \mathbf{0} \qquad (1)\mathbf{v} = \mathbf{v} \qquad (-1)\mathbf{v} = -\mathbf{v}$$

$$(\alpha + \beta)\mathbf{v} = \alpha\mathbf{v} + \beta\mathbf{v} \qquad \alpha(\mathbf{v} + \mathbf{w}) = \alpha\mathbf{v} + \alpha\mathbf{w} \qquad \alpha(\beta\mathbf{v}) = (\alpha\beta)\mathbf{v}$$

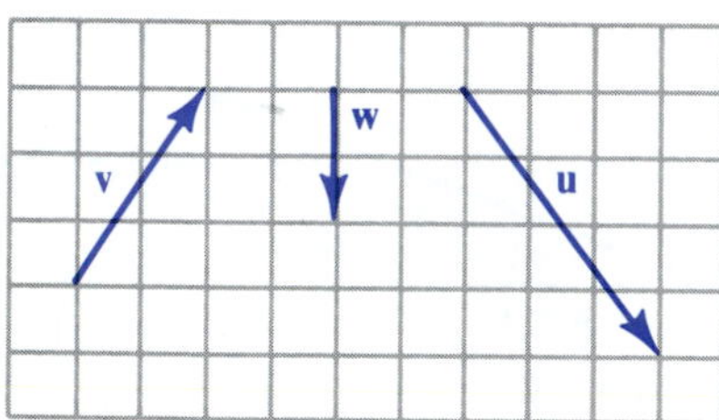

Figure 20

EXAMPLE 1

Given the vectors **v**, **w**, and **u** in Figure 20, construct:

(a) $\mathbf{v} - \mathbf{w}$ (b) $2\mathbf{v} - \mathbf{w} + \mathbf{u}$ (c) $\frac{2}{3}\mathbf{u} + \frac{1}{2}(\mathbf{v} - \mathbf{w})$

Solution

The solutions are illustrated in Figure 21.

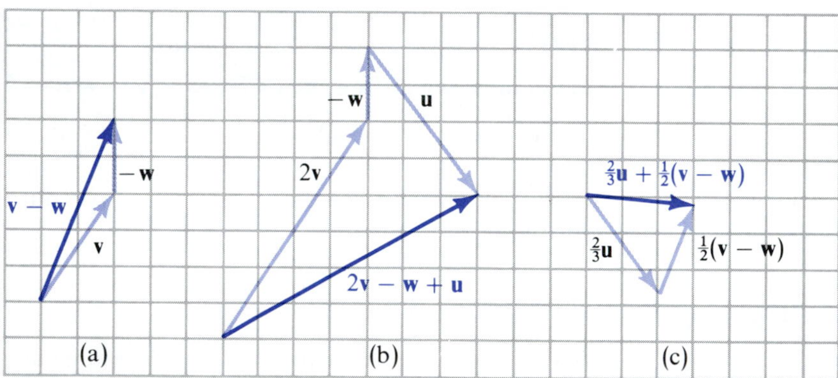

Figure 21

In the above discussion we did not concern ourselves with *dimensionality*. In fact, everything we have said thus far is true in two-, three-, and n-dimensional space. However, there are some vector concepts that relate directly to the dimension. For example, the cross product, which will be taken up in Section 14.5, is used only in three-dimensional space.

EXERCISE 14.2

1. State which of the following are scalars and which are vectors.

(a) Volume (b) Speed (c) Force
(d) Work (e) Mass (f) Distance
(g) Age (h) Velocity

2. Given the vectors **v**, **w**, and **u** in the figure, construct:

(a) $2\mathbf{v}$ (b) $-2\mathbf{v}$
(c) $\mathbf{v} + \mathbf{w}$ (d) $\mathbf{v} - \mathbf{w}$
(e) $\mathbf{w} - \mathbf{v}$ (f) $\mathbf{v} - 2\mathbf{w}$
(g) $\mathbf{v} - \mathbf{w} + 3\mathbf{u}$ (h) $2\mathbf{u} - \frac{1}{3}(\mathbf{v} - \mathbf{w})$

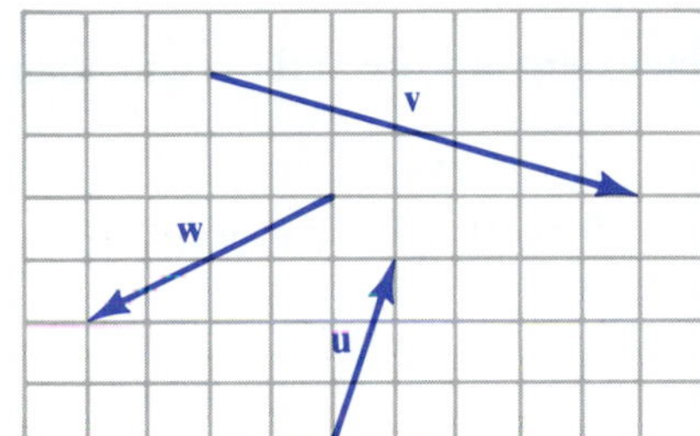

In Problems 3–10 use the accompanying figure.

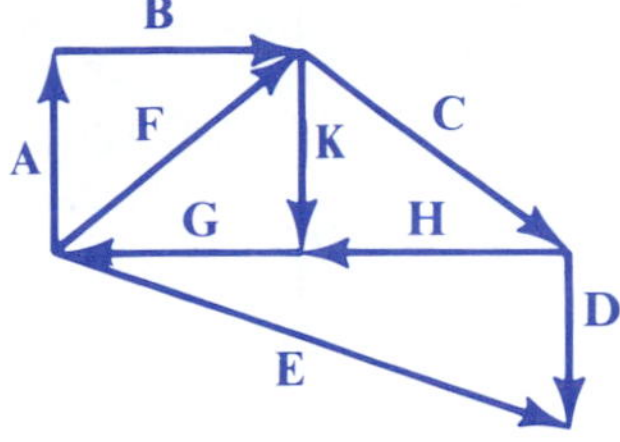

3. Find the vector **x** if $\mathbf{x} + \mathbf{B} = \mathbf{F}$.

4. Find the vector **x** if $\mathbf{x} + \mathbf{D} = \mathbf{E}$.

5. Write **C** in terms of **E**, **D**, and **F**.

6. Write **G** in terms of **C**, **D**, **E**, and **K**.

7. Write **E** in terms of **G**, **H**, and **D**.

8. Write **E** in terms of **A**, **B**, **C**, and **D**.

9. What is $\mathbf{A} + \mathbf{B} + \mathbf{K} + \mathbf{G}$?

10. What is $\mathbf{A} + \mathbf{B} + \mathbf{C} + \mathbf{H} + \mathbf{G}$?

11. Let **v** and **w** be nonzero vectors represented by arrows with the same initial point. Let the terminal points of **v** and **w** be P and Q, respectively. Let **u** denote the vector represented by an arrow from the initial point of **v** to the midpoint of the directed line segment $\overrightarrow{PQ}$. Write **u** in terms of **v** and **w**.

12. Find nonzero scalars α and β such that

$$\alpha\mathbf{v} + \beta(\mathbf{v} - \mathbf{w}) + 4(\mathbf{v} + \mathbf{w}) = \mathbf{0}$$

for every pair of vectors **v** and **w**.

Vectors in the Plane and in Space 14.3

In the preceding section, we relied on geometry to provide insight into vectors. In this section, we look at vectors algebraically, taking advantage of a rectangular coordinate system. This will require that we distinguish between vectors in the plane and vectors in space.

If **v** is a vector in the plane and if its initial point is at the origin and its terminal point is at (v_1, v_2), then we may represent **v** by the ordered pair

$$\mathbf{v} = \langle v_1, v_2 \rangle$$

The numbers v_1 and v_2 are called the *components* of **v**.

If **v** is a vector in space and if its initial point is at the origin and its terminal point is at (v_1, v_2, v_3), then we may represent **v** by the ordered triplet

$$\mathbf{v} = \langle v_1, v_2, v_3 \rangle$$

The numbers v_1, v_2, and v_3 are called the *components* of **v**. See Figure 22.

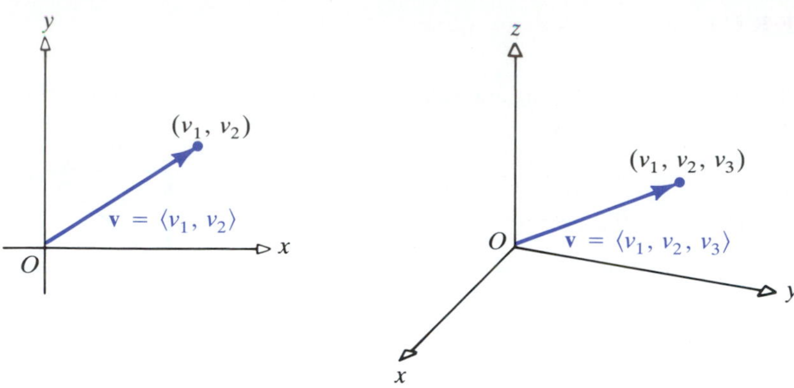

Figure 22

If the initial point and the terminal point of **v** are at the origin, then **v** is the zero vector. As a result,

$$\mathbf{0} = \langle 0, 0 \rangle \text{ in the plane} \qquad \mathbf{0} = \langle 0, 0, 0 \rangle \text{ in space}$$

Position Vectors

[14.3.1] DEFINITION / *Position Vector.*

A vector whose initial point is at the origin is called a *position vector*.

The next result states that any vector whose initial point is not at the origin is equal to a unique position vector.

[14.3.2] THEOREM

Suppose v is a vector with initial point P_1, not necessarily the origin, and terminal point P_2, so that $\mathbf{v} = \overrightarrow{P_1P_2}$. Then v is equal to

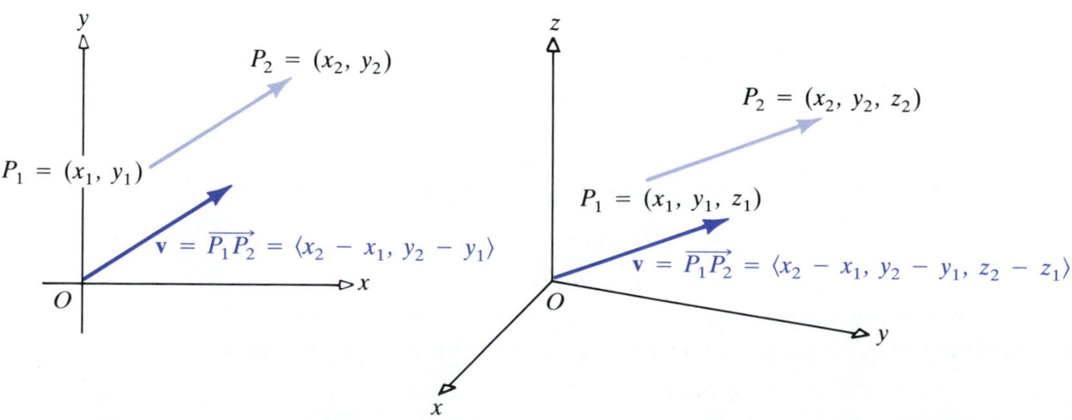

Figure 23

the position vector whose components are:

In the plane: **If** $P_1 = (x_1, y_1)$ **and** $P_2 = (x_2, y_2)$, **then**

$$\mathbf{v} = \langle x_2 - x_1, y_2 - y_1 \rangle \tag{1}$$

In space: **If** $P_1 = (x_1, y_1, z_1)$ **and** $P_2 = (x_2, y_2, z_2)$, **then**

$$\mathbf{v} = \langle x_2 - x_1, y_2 - y_1, z_2 - z_1 \rangle \tag{2}$$

See Figure 23.

EXAMPLE 1

Find the position vector of the vector $\mathbf{v} = \overrightarrow{P_1P_2}$ if

(a) $P_1 = (-1, 2)$, $P_2 = (4, 6)$
(b) $P_1 = (3, 4, 2)$, $P_2 = (7, -8, 5)$

Solution

(a) By (1), the position vector equal to **v** is

$$\mathbf{v} = \langle 4 - (-1), 6 - 2 \rangle = \langle 5, 4 \rangle$$

See Figure 24.

(b) By (2), the position vector equal to **v** is

$$\mathbf{v} = \langle 7 - 3, -8 - 4, 5 - 2 \rangle = \langle 4, -12, 3 \rangle$$

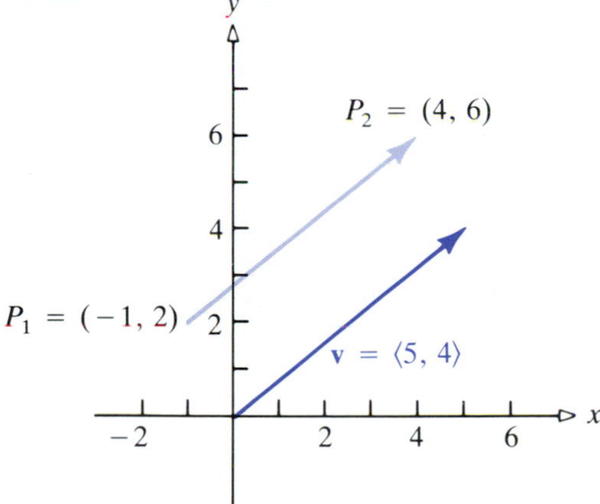

Figure 24

Two position vectors **v** and **w** are equal if and only if the terminal point of **v** is the same as the terminal point of **w**. This leads to the following result:

[14.3.3] THEOREM / ***Equality of Vectors.***

Two position vectors in the plane $\mathbf{v} = \langle v_1, v_2 \rangle$ **and** $\mathbf{w} = \langle w_1, w_2 \rangle$ **are equal if and only if**

$$v_1 = w_1 \quad \textbf{and} \quad v_2 = w_2$$

Two position vectors in space $\mathbf{v} = \langle v_1, v_2, v_3\rangle$ and $\mathbf{w} = \langle w_1, w_2, w_3\rangle$ are equal if and only if

$$v_1 = w_1 \qquad v_2 = w_2 \qquad v_3 = w_3$$

Because of theorem (14.3.3), we can replace any vector (directed line segment) by a unique position vector, and vice versa. This flexibility is what underlies the power of using vectors. Unless otherwise specified, from now on the term *vector* will mean the unique position vector equal to it.

Operations with Vectors

Next, we define how to add and subtract vectors and how to form the product of a scalar and a vector using components.

[14.3.4] DEFINITION

If $\mathbf{v} = \langle v_1, v_2\rangle$ and $\mathbf{w} = \langle w_1, w_2\rangle$ are two vectors in the plane and if α is a scalar, then

$$\mathbf{v} + \mathbf{w} = \langle v_1 + w_1, v_2 + w_2\rangle$$

$$\mathbf{v} - \mathbf{w} = \langle v_1 - w_1, v_2 - w_2\rangle$$

$$\alpha\mathbf{v} = \langle \alpha v_1, \alpha v_2\rangle$$

If $\mathbf{v} = \langle v_1, v_2, v_3\rangle$ and $\mathbf{w} = \langle w_1, w_2, w_3\rangle$ are two vectors in space and if α is a scalar, then

$$\mathbf{v} + \mathbf{w} = \langle v_1 + w_1, v_2 + w_2, v_3 + w_3\rangle$$

$$\mathbf{v} - \mathbf{w} = \langle v_1 - w_1, v_2 - w_2, v_3 - w_3\rangle$$

$$\alpha\mathbf{v} = \langle \alpha v_1, \alpha v_2, \alpha v_3\rangle$$

These definitions are compatible with the geometric ones given in Section 14.2. See Figure 25 for a justification in the plane.

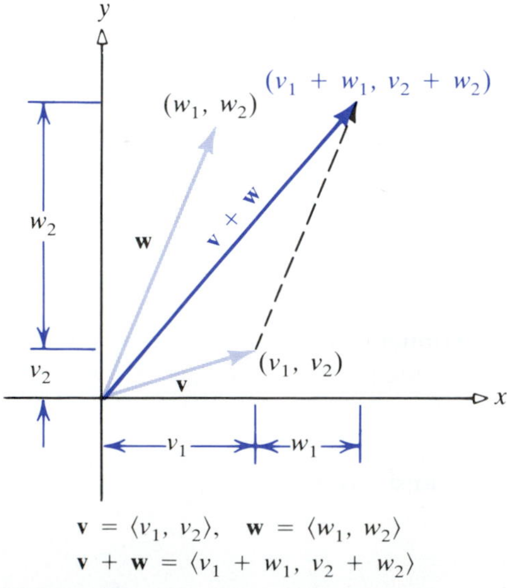

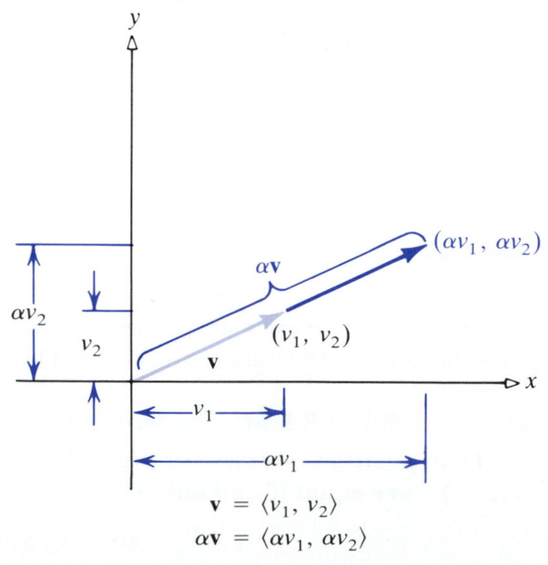

Figure 25

EXAMPLE 2

If $\mathbf{v} = \langle 2, -3 \rangle$ and $\mathbf{w} = \langle -1, 2 \rangle$, find:

(a) $\mathbf{v} + \mathbf{w}$ (b) $2\mathbf{v}$ (c) $2\mathbf{v} - \mathbf{w}$

Solution

(a) $\mathbf{v} + \mathbf{w} = \langle 2, -3 \rangle + \langle -1, 2 \rangle = \langle 1, -1 \rangle$
(b) $2\mathbf{v} = 2\langle 2, -3 \rangle = \langle 4, -6 \rangle$
(c) $2\mathbf{v} - \mathbf{w} = \langle 4, -6 \rangle - \langle -1, 2 \rangle = \langle 5, -8 \rangle$

EXAMPLE 3

If $\mathbf{v} = \langle 2, 3, -1 \rangle$ and $\mathbf{w} = \langle -1, -2, 4 \rangle$, find:

(a) $\mathbf{v} - \mathbf{w}$ (b) $\frac{1}{2}\mathbf{w}$ (c) $2\mathbf{v} + 3\mathbf{w}$

Solution

(a) $\mathbf{v} - \mathbf{w} = \langle 2, 3, -1 \rangle - \langle -1, -2, 4 \rangle = \langle 3, 5, -5 \rangle$
(b) $\frac{1}{2}\mathbf{w} = \frac{1}{2}\langle -1, -2, 4 \rangle = \langle -\frac{1}{2}, -1, 2 \rangle$
(c) $2\mathbf{v} + 3\mathbf{w} = \langle 4, 6, -2 \rangle + \langle -3, -6, 12 \rangle = \langle 1, 0, 10 \rangle$

Properties of Vectors

[14.3.5] THEOREM

If u, v, and w are vectors and if α and β are scalars, then

(a) $\mathbf{u} + \mathbf{v} = \mathbf{v} + \mathbf{u}$
(b) $\mathbf{u} + (\mathbf{v} + \mathbf{w}) = (\mathbf{u} + \mathbf{v}) + \mathbf{w}$
(c) $\mathbf{v} + \mathbf{0} = \mathbf{0} + \mathbf{v} = \mathbf{v}$
(d) $\mathbf{v} + (-\mathbf{v}) = \mathbf{0}$
(e) $\alpha(\mathbf{u} + \mathbf{v}) = \alpha\mathbf{u} + \alpha\mathbf{v}$
(f) $(\alpha + \beta)\mathbf{v} = \alpha\mathbf{v} + \beta\mathbf{v}$
(g) $\alpha(\beta\mathbf{v}) = (\alpha\beta)\mathbf{v}$
(h) $1\mathbf{v} = \mathbf{v}$

We saw some of these properties in the previous section where a geometric justification was given. For example, look back to Figure 16 for a justification of property (b).

Proof We shall prove only property (a) for vectors in the plane and property (h) for vectors in space. The remaining proofs are left as exercises (Problems 73 and 74).

(a) If $\mathbf{u} = \langle u_1, u_2 \rangle$ and if $\mathbf{v} = \langle v_1, v_2 \rangle$, then

$$\begin{aligned}\mathbf{u} + \mathbf{v} &= \langle u_1, u_2 \rangle + \langle v_1, v_2 \rangle = \langle u_1 + v_1, u_2 + v_2 \rangle \\ &= \langle v_1 + u_1, v_2 + u_2 \rangle = \langle v_1, v_2 \rangle + \langle u_1, u_2 \rangle = \mathbf{v} + \mathbf{u}\end{aligned}$$

(h) If $\mathbf{v} = \langle v_1, v_2, v_3 \rangle$, then

$$1\mathbf{v} = \langle 1v_1, 1v_2, 1v_3 \rangle = \langle v_1, v_2, v_3 \rangle = \mathbf{v}$$

Parallel Vectors

[14.3.6] DEFINITION / *Parallel Vectors.*

Let v and w be two vectors. If there is a nonzero scalar α such that $\mathbf{v} = \alpha\mathbf{w}$, then v and w are called *parallel vectors*.

EXAMPLE 4

(a) The vectors

$$\mathbf{v} = \langle 2, 3, -1 \rangle \qquad \text{and} \qquad \mathbf{w} = \langle 4, 6, -2 \rangle$$

are parallel, since $\mathbf{v} = \frac{1}{2}\mathbf{w}$. In this case, **v** and **w** have the same direction.

(b) The vectors

$$\mathbf{v} = \langle 1, 2, 3 \rangle \qquad \text{and} \qquad \mathbf{w} = \langle -3, -6, -9 \rangle$$

are parallel, since $\mathbf{v} = -\frac{1}{3}\mathbf{w}$. In this case, **v** and **w** have opposite direction.

Magnitude of a Vector

[14.3.7] DEFINITION / *Magnitude of a Vector.*

The *magnitude* or *length* of a vector v, denoted by the symbol $\|\mathbf{v}\|$, is the distance of the terminal point P of v from the origin O. Sometimes $\|\mathbf{v}\|$ is called the *norm* of the vector v.

[14.3.8] THEOREM

If $\mathbf{v} = \langle v_1, v_2 \rangle$ is a vector in the plane, the magnitude of v is

$$\|\mathbf{v}\| = \sqrt{v_1^2 + v_2^2}$$

If $\mathbf{v} = \langle v_1, v_2, v_3 \rangle$ is a vector in space, the magnitude of v is

$$\|\mathbf{v}\| = \sqrt{v_1^2 + v_2^2 + v_3^2}$$

The proof is a direct consequence of the distance formula.

EXAMPLE 5

Find the magnitude of **v** if:

(a) $\mathbf{v} = \langle 3, -4 \rangle$ (b) $\mathbf{v} = \langle 2, 3, -1 \rangle$

Solution

(a) $\|\mathbf{v}\| = \sqrt{3^2 + (-4)^2} = \sqrt{9 + 16} = 5$

(b) $\|\mathbf{v}\| = \sqrt{2^2 + 3^2 + (-1)^2} = \sqrt{4 + 9 + 1} = \sqrt{14}$

[14.3.9] THEOREM

If α is a scalar and v is a vector, then

$$\|\alpha\mathbf{v}\| = |\alpha|\,\|\mathbf{v}\|$$

Proof We shall give a proof for a vector **v** in the plane. The proof for a vector **v** in space is left as an exercise (Problem 75). Suppose $\mathbf{v} = \langle v_1, v_2 \rangle$. Then $\alpha\mathbf{v} = \langle \alpha v_1, \alpha v_2 \rangle$ and

$$\begin{aligned} \|\alpha\mathbf{v}\| &= \sqrt{(\alpha v_1)^2 + (\alpha v_2)^2} = \sqrt{\alpha^2 v_1^2 + \alpha^2 v_2^2} \\ &= \sqrt{\alpha^2(v_1^2 + v_2^2)} = \sqrt{\alpha^2}\sqrt{v_1^2 + v_2^2} = |\alpha|\|v\| \end{aligned}$$

Unit Vectors

[14.3.10] DEFINITION / *Unit Vector.*

A *unit vector* v is one for which $\|\mathbf{v}\| = 1$.

In many applications it is useful to be able to find a unit vector that has the same direction as a given vector **v**.

[14.3.11] THEOREM

For any nonzero vector v, the vector

$$\mathbf{u} = \frac{\mathbf{v}}{\|\mathbf{v}\|}$$

is a unit vector that has the same direction as v.

Proof Since $1/\|\mathbf{v}\|$ is a positive scalar, it follows that **u** has the same direction as **v**. Furthermore, by theorem (14.3.9), we have

$$\|\mathbf{u}\| = \left\|\frac{\mathbf{v}}{\|\mathbf{v}\|}\right\| = \frac{\|\mathbf{v}\|}{\|\mathbf{v}\|} = 1$$

so that **u** is a unit vector.

The process of multiplying a nonzero vector **v** by $1/\|\mathbf{v}\|$ to obtain a unit vector that has the same direction as **v** is called *normalizing* **v**.

EXAMPLE 6

Find a unit vector **u** that has the same direction as:

(a) $\mathbf{v} = \langle 3, -4 \rangle$ (b) $\mathbf{v} = \langle -1, 2, -2 \rangle$

Solution

(a) Since $\mathbf{v} = \langle 3, -4 \rangle$, then $\|\mathbf{v}\| = \sqrt{9 + 16} = 5$. The unit vector **u** in the same direction as **v** is

$$\mathbf{u} = \frac{\mathbf{v}}{\|\mathbf{v}\|} = \frac{\mathbf{v}}{5} = \left\langle \frac{3}{5}, \frac{-4}{5} \right\rangle$$

(b) Since $\mathbf{v} = \langle -1, 2, -2 \rangle$, then $\|\mathbf{v}\| = \sqrt{1 + 4 + 4} = 3$. The unit vector **u** in the same direction as **v** is

$$\mathbf{u} = \frac{\mathbf{v}}{\|\mathbf{v}\|} = \frac{\mathbf{v}}{3} = \left\langle \frac{-1}{3}, \frac{2}{3}, \frac{-2}{3} \right\rangle$$

Standard Basis Vectors

The unit vectors that are directed along the positive coordinate axes are called *standard basis vectors*. In the plane, the standard basis vectors are

$$\mathbf{i} = \langle 1, 0 \rangle \qquad \mathbf{j} = \langle 0, 1 \rangle$$

In space, the standard basis vectors are

$$\mathbf{i} = \langle 1, 0, 0 \rangle \qquad \mathbf{j} = \langle 0, 1, 0 \rangle \qquad \mathbf{k} = \langle 0, 0, 1 \rangle$$

See Figures 26(a) and 26(b) for an illustration.

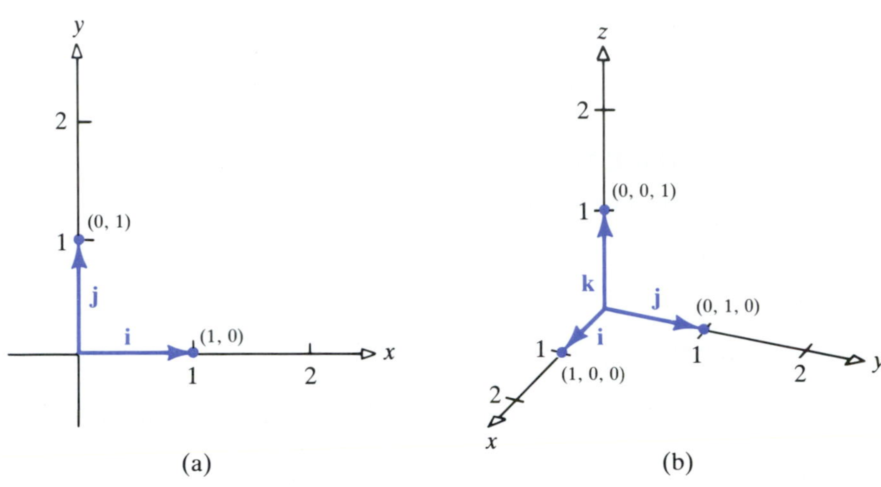

Figure 26

The importance of these vectors lies in the next result.

[14.3.12] THEOREM

Every vector $\mathbf{v} = \langle v_1, v_2 \rangle$ in the plane can be written in terms of the vectors i and j as

$$\mathbf{v} = v_1\mathbf{i} + v_2\mathbf{j}$$

Every vector $\mathbf{v} = \langle v_1, v_2, v_3 \rangle$ in space can be written in terms of the vectors i, j, and k as

$$\mathbf{v} = v_1\mathbf{i} + v_2\mathbf{j} + v_3\mathbf{k}$$

Proof We shall prove this result only for vectors in the plane. The proof for vectors in space is left as an exercise (Problem 76).

$$\mathbf{v} = \langle v_1, v_2 \rangle = \langle v_1, 0 \rangle + \langle 0, v_2 \rangle = v_1\langle 1, 0 \rangle + v_2\langle 0, 1 \rangle = v_1\mathbf{i} + v_2\mathbf{j}$$

EXAMPLE 7

(a) $\langle 4, 3 \rangle = 4\mathbf{i} + 3\mathbf{j}$
(b) $\langle 2, -3 \rangle = 2\mathbf{i} + (-3)\mathbf{j} = 2\mathbf{i} - 3\mathbf{j}$

(c) $\langle 1, -3, 2\rangle = \mathbf{i} - 3\mathbf{j} + 2\mathbf{k}$
(d) $\langle 0, 3, 4\rangle = 0\mathbf{i} + 3\mathbf{j} + 4\mathbf{k} = 3\mathbf{j} + 4\mathbf{k}$

EXAMPLE 8

If $\mathbf{v} = 2\mathbf{i} - 3\mathbf{j}$ and $\mathbf{w} = -\mathbf{i} + 2\mathbf{j}$, find:

(a) $2\mathbf{v} - \mathbf{w}$ (b) $\|\mathbf{v}\|$ (c) $\|-3\mathbf{v} + 2\mathbf{w}\|$

Solution

(a) $2\mathbf{v} - \mathbf{w} = 2(2\mathbf{i} - 3\mathbf{j}) - (-\mathbf{i} + 2\mathbf{j}) = (4\mathbf{i} - 6\mathbf{j}) - (-\mathbf{i} + 2\mathbf{j}) = 5\mathbf{i} - 8\mathbf{j}$

(b) $\|\mathbf{v}\| = \sqrt{2^2 + (-3)^2} = \sqrt{4 + 9} = \sqrt{13}$

(c)
$$\begin{aligned}-3\mathbf{v} + 2\mathbf{w} &= -3(2\mathbf{i} - 3\mathbf{j}) + 2(-\mathbf{i} + 2\mathbf{j}) = (-6\mathbf{i} + 9\mathbf{j}) + (-2\mathbf{i} + 4\mathbf{j})\\ &= -8\mathbf{i} + 13\mathbf{j}\end{aligned}$$
$$\|-3\mathbf{v} + 2\mathbf{w}\| = \sqrt{(-8)^2 + (13)^2} = \sqrt{64 + 169} = \sqrt{233}$$

EXAMPLE 9

If $\mathbf{v} = 2\mathbf{i} + 3\mathbf{j} - \mathbf{k}$ and $\mathbf{w} = -\mathbf{i} - 2\mathbf{j} + 4\mathbf{k}$, find:

(a) $\|\mathbf{v}\|$ (b) $2\mathbf{v} - 3\mathbf{w}$

Solution

(a) $\|\mathbf{v}\| = \sqrt{2^2 + 3^2 + (-1)^2} = \sqrt{4 + 9 + 1} = \sqrt{14}$
(b)
$$\begin{aligned}2\mathbf{v} - 3\mathbf{w} &= 2(2\mathbf{i} + 3\mathbf{j} - \mathbf{k}) - 3(-\mathbf{i} - 2\mathbf{j} + 4\mathbf{k})\\ &= (4\mathbf{i} + 6\mathbf{j} - 2\mathbf{k}) - (-3\mathbf{i} - 6\mathbf{j} + 12\mathbf{k}) = 7\mathbf{i} + 12\mathbf{j} - 14\mathbf{k}\end{aligned}$$

EXAMPLE 10

An airplane has an air speed of 400 kilometers per hour in an easterly direction. The wind velocity is 80 kilometers per hour in a southeasterly direction.

(a) Find a unit vector having southeast as direction.
(b) Find a vector 80 units long having the same direction as the vector found in part (a).
(c) Use parts (a) and (b) to find the actual speed of the airplane relative to the ground.

Solution

We set up a coordinate system in which the direction north is along the positive y-axis. Then the direction east is along the positive x-axis (see Fig. 27). Using a scale of 1 unit = 1 kilometer per hour, we set

$\mathbf{v}_a$ = Velocity of the airplane in the air = $400\mathbf{i}$

$\mathbf{v}_g$ = Velocity of the airplane relative to the ground

$\mathbf{v}_w$ = Velocity of the wind

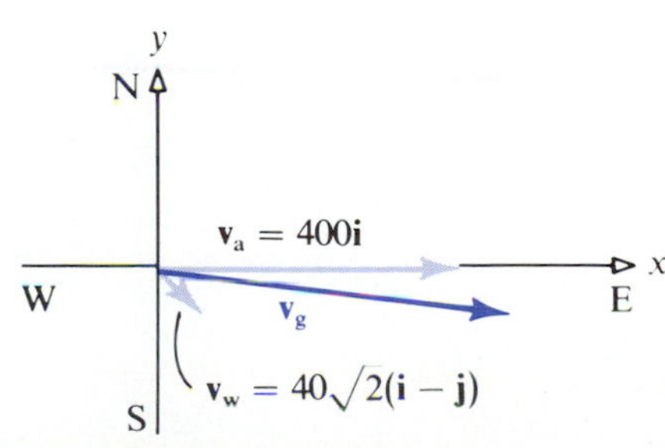

Figure 27

(a) A vector having southeast as direction is $\mathbf{i} - \mathbf{j}$. The corresponding unit vector in this direction is

$$\frac{\mathbf{i} - \mathbf{j}}{\|\mathbf{i} - \mathbf{j}\|} = \frac{\mathbf{i} - \mathbf{j}}{\sqrt{1+1}} = \frac{1}{\sqrt{2}}(\mathbf{i} - \mathbf{j})$$

(b) The velocity of the wind $\mathbf{v}_w$ is a vector in the direction of $(1/\sqrt{2})(\mathbf{i} - \mathbf{j})$ with magnitude 80. Hence,

$$\mathbf{v}_w = 80\frac{1}{\sqrt{2}}(\mathbf{i} - \mathbf{j}) = 40\sqrt{2}(\mathbf{i} - \mathbf{j})$$

(c) The velocity $\mathbf{v}_g$ of the airplane relative to the ground is

$$\mathbf{v}_g = \mathbf{v}_a + \mathbf{v}_w = 400\mathbf{i} + 40\sqrt{2}(\mathbf{i} - \mathbf{j}) = (400 + 40\sqrt{2})\mathbf{i} - 40\sqrt{2}\mathbf{j}$$

The actual speed of the airplane relative to the ground is

$$\|\mathbf{v}_g\| = \sqrt{(400 + 40\sqrt{2})^2 + (40\sqrt{2})^2} \approx 460.06 \text{ kilometers per hour}$$

↑ Use a calculator.

In Section 14.4 we find the direction of the airplane relative to the ground. Those of you who are well versed in trigonometry can find the direction now, but as you will see, the method we use in Section 14.4 is much easier.

Vectors in *n* Dimensions

In a space of n dimensions, the rectangular coordinates of a point are given by an ordered n-tuple $(v_1, v_2, \ldots, v_n)$ of real numbers. A position vector $\mathbf{v}$ in n-space whose initial point is at the origin and whose terminal point is at $(v_1, v_2, \ldots, v_n)$ is defined as

$$\mathbf{v} = \langle v_1, v_2, \ldots, v_n \rangle$$

Here the numbers $v_1, v_2, \ldots, v_n$ are the *components of* $\mathbf{v}$. In n-space there are n coordinate axes and, along each positive axis, we can define standard basis vectors

$$\mathbf{u}_1 = \langle 1, 0, \ldots, 0 \rangle, \quad \mathbf{u}_2 = \langle 0, 1, \ldots, 0 \rangle, \quad \ldots, \quad \mathbf{u}_n = \langle 0, 0, \ldots, 1 \rangle$$

Any vector $v = \langle v_1, v_2, \ldots, v_n \rangle$ in n-space can be written as

$$\mathbf{v} = v_1\mathbf{u}_1 + v_2\mathbf{u}_2 + \cdots + v_n\mathbf{u}_n$$

We add two vectors in n-space by adding respective components. To multiply a scalar times a vector, we multiply each component by the scalar.

The magnitude of the vector $\mathbf{v} = v_1\mathbf{u}_1 + v_2\mathbf{u}_2 + \cdots + v_n\mathbf{u}_n$ is

$$\|\mathbf{v}\| = \sqrt{v_1^2 + v_2^2 + \cdots + v_n^2}$$

EXERCISE 14.3

In Problems 1–12 find the position vector of $\overrightarrow{P_1P_2}$.

1. $P_1 = (2, 3)$; $P_2 = (6, -2)$

2. $P_1 = (4, -1)$; $P_2 = (-3, 2)$

3. $P_1 = (0, 5)$; $P_2 = (-1, 6)$

4. $P_1 = (-2, 0)$; $P_2 = (3, 2)$

5. $P_1 = (0, 0)$; $P_2 = (x, y)$

6. $P_1 = (0, 0)$; $P_2 = (0, y)$

7. $P_1 = (6, 2, 1)$; $P_2 = (3, 0, 2)$

8. $P_1 = (4, 7, 0)$; $P_2 = (0, 5, 6)$

9. $P_1 = (-1, 0, 1)$; $P_2 = (2, 0, 0)$

10. $P_1 = (6, 2, 2)$; $P_2 = (2, 6, 2)$

11. $P_1 = (0, 0, 0)$; $P_2 = (x, y, z)$

12. $P_1 = (0, 0, 0)$; $P_2 = (0, y, 0)$

In Problems 13–18 use $\mathbf{v} = 2\mathbf{i} - 3\mathbf{j}$ and $\mathbf{w} = \mathbf{i} + 2\mathbf{j}$ to find the indicated quantity.

13. $2\mathbf{v} - \mathbf{w}$

14. $\mathbf{v} + 5\mathbf{w}$

15. $\frac{1}{3}\mathbf{v} + \frac{1}{2}\mathbf{w}$

16. $\frac{2}{3}\mathbf{v} - \frac{1}{2}\mathbf{w}$

17. $\|\mathbf{v} - \mathbf{w}\|$

18. $\|\mathbf{v} + \mathbf{w}\|$

In Problems 19–28 use $\mathbf{u} = \mathbf{i} - 2\mathbf{j} + 3\mathbf{k}$, $\mathbf{v} = 3\mathbf{i} + \mathbf{j} - \mathbf{k}$, and $\mathbf{w} = 6\mathbf{i} + \mathbf{j} + \mathbf{k}$ to find the indicated quantity.

19. $3\mathbf{v} - 2\mathbf{w}$

20. $-2\mathbf{v} + \mathbf{w}$

21. $\|\mathbf{v} - \mathbf{w}\|$

22. $\|2\mathbf{v} + \mathbf{w}\|$

23. $2\mathbf{u} - 3\mathbf{v} + 4\mathbf{w}$

24. $\dfrac{\mathbf{u} + \mathbf{v}}{\|\mathbf{w}\|}$

25. $\|5\mathbf{u} - \mathbf{v} + \mathbf{w}\|$

26. $\|\mathbf{u}\| + \|\mathbf{v}\| + \|\mathbf{w}\|$

27. $\|\mathbf{u} + \mathbf{v} + \mathbf{w}\|$

28. $\dfrac{\mathbf{u}}{\|\mathbf{v} + \mathbf{w}\|}$

In Problems 29–40 find the magnitude of $\mathbf{v}$.

29. $\mathbf{v} = 4\mathbf{i} - 3\mathbf{j}$

30. $\mathbf{v} = \mathbf{i} - 12\mathbf{j}$

31. $\mathbf{v} = \mathbf{i} + \mathbf{j}$

32. $\mathbf{v} = \mathbf{i} - \mathbf{j}$

33. $\mathbf{v} = a\mathbf{i} - a\mathbf{j}$

34. $\mathbf{v} = (\cos\theta)\mathbf{i} + (\sin\theta)\mathbf{j}$

35. $\mathbf{v} = 4\mathbf{i} + 2\mathbf{j} - \mathbf{k}$

36. $\mathbf{v} = \mathbf{i} - \mathbf{j} + \mathbf{k}$

37. $\mathbf{v} = \mathbf{i} + \mathbf{j} + \mathbf{k}$

38. $\mathbf{v} = 2\mathbf{i} - \mathbf{k}$

39. $\mathbf{v} = a\mathbf{i} + a\mathbf{j} + a\mathbf{k}$, $a > 0$

40. $\mathbf{v} = (\cos\theta)\mathbf{i} + (\sin\theta)\mathbf{j} + \mathbf{k}$

In Problems 41–52 find a unit vector in the same direction as $\mathbf{v}$. What is the unit vector in the opposite direction?

41. $\mathbf{v} = 5\mathbf{i} - 12\mathbf{j}$

42. $\mathbf{v} = 3\mathbf{i} + 4\mathbf{j}$

43. $\mathbf{v} = 2\mathbf{i} + \mathbf{j}$

44. $\mathbf{v} = \mathbf{i} - \mathbf{j}$

45. $\mathbf{v} = \dfrac{1}{2}\mathbf{i} + \dfrac{\sqrt{3}}{2}\mathbf{j}$

46. $\mathbf{v} = \dfrac{\sqrt{2}}{2}(\mathbf{i} - \mathbf{j})$

47. $\mathbf{v} = \mathbf{i} + \mathbf{j} + \mathbf{k}$

48. $\mathbf{v} = \mathbf{i} - \mathbf{j} - \mathbf{k}$

49. $\mathbf{v} = \cos\theta\mathbf{i} + \sin\theta\mathbf{j} + 2\sqrt{2}\,\mathbf{k}$

50. $\mathbf{v} = \sin\theta\mathbf{i} + \cos\theta\mathbf{j} + \mathbf{k}$

51. $\mathbf{v} = 3\mathbf{i} - 4\mathbf{j} + 12\mathbf{k}$

52. $\mathbf{v} = \dfrac{1}{2}\mathbf{i} - \dfrac{1}{2}\mathbf{j} + \dfrac{\sqrt{2}}{2}\mathbf{k}$

53. Find a vector whose magnitude is 4 that is parallel to $2\mathbf{i} - 3\mathbf{j} + 4\mathbf{k}$.

54. Find a vector whose magnitude is 2 that is parallel to $-\mathbf{i} + \mathbf{j} + 2\mathbf{k}$.

55. Find α so that the vectors $2\mathbf{i} + \mathbf{j} - \mathbf{k}$ and $2\alpha\mathbf{i} + \mathbf{j} + \mathbf{k}$ have the same magnitude.

56. Find α so that $\|\alpha\mathbf{i} + (\alpha + 1)\mathbf{j} + 2\mathbf{k}\| = 3$.

57. Find all numbers x so that $\|\mathbf{v} + \mathbf{w}\| = 5$, where $\mathbf{v} = 2\mathbf{i} - \mathbf{j}$ and $\mathbf{w} = x\mathbf{i} + 3\mathbf{j}$.

58. Find all numbers x so that the vector represented by $\overrightarrow{P_1P_2}$ with $P_1 = (-3, 1)$ and $P_2 = (x, 4)$ has length 5.

In Problems 59–64 find all vectors $\mathbf{v} = a\mathbf{i} + b\mathbf{j}$ that have the desired properties.

59. Magnitude is 4; makes an angle of 30° with the positive x-axis.

60. Magnitude is 2; makes an angle of 45° with the positive x-axis.

61. Magnitude is 5; **i** component is twice the **j** component.

62. Magnitude is 2; **i** component equals **j** component.

63. **j** component is 1; makes an angle of 135° with the positive x-axis.

64. **i** component is -3; makes an angle of 210° with the positive x-axis.

65. An airplane has an air speed of 500 kilometers per hour in an easterly direction. If the wind velocity is 60 kilometers per hour in a northwesterly direction, find the speed of the airplane relative to the ground.

66. An airplane, after 1 hour of flying, arrives at a point 200 miles due south of the departure point. If, during the flight, there was a steady wind of 20 miles per hour from the northwest, what was the airplane's average air speed?

67. Forces of 2 and 5 pounds act at an angle of 30° on an object. Determine graphically what force is necessary to balance them.

68. Forces of 1 and 2 pounds act at an angle of 135° on an object. Determine graphically what force is necessary to balance them.

69. Represent graphically:

(a) A velocity of 20 kilometers per hour in a direction 60° west of north

(b) A velocity of 30 kilometers per hour in a direction 30° north of east

70. Forces $\mathbf{F}_1$, $\mathbf{F}_2$, $\mathbf{F}_3$, $\mathbf{F}_4$ act as shown in the figure on an object Q. What force is needed to prevent Q from moving?

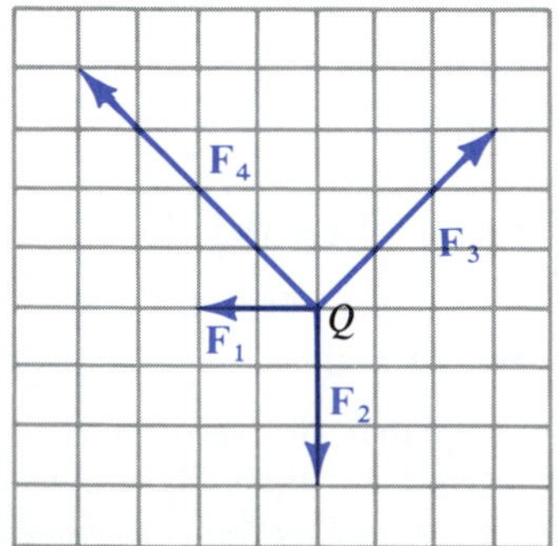

71. An airplane travels in a northeasterly direction at 250 kilometers per hour relative to the ground, due to the fact that there is an easterly wind of 50 kilometers per hour relative to the ground. How fast would the plane be going if there were no wind?

72. A woman traveling east at 8 miles per hour finds that the wind appears to be coming from the north. Upon doubling her speed, she finds that the wind appears to be coming from the northeast. Find the velocity of the wind.

73. Prove theorem (14.3.5), parts (b)–(h), for vectors in the plane.

74. Prove theorem (14.3.5), parts (a)–(g), for vectors in space.

75. Prove theorem (14.3.9) for a vector in space.

76. Prove theorem (14.3.12) for vectors in space.

77. Let A, B, C, and D be the vertices of a tetrahedron (triangular pyramid) in space. Let $\mathbf{b} = \overrightarrow{AB}$, $\mathbf{c} = \overrightarrow{AC}$, and $\mathbf{d} = \overrightarrow{AD}$. Express the directed edges $\overrightarrow{BC}$, $\overrightarrow{BD}$, and $\overrightarrow{CD}$ of the tetrahedron in terms of the vectors **b**, **c**, and **d**.

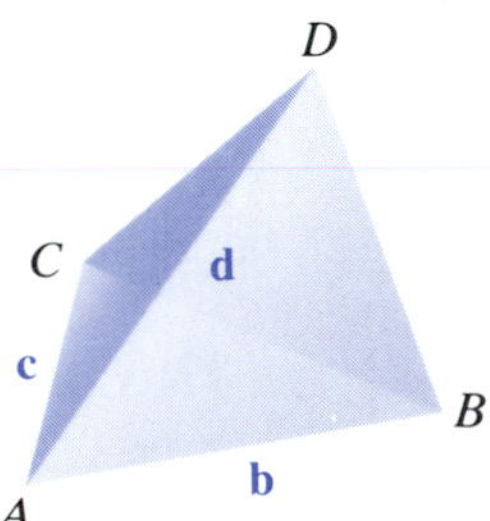

78. Let A, B, C, D, E, F, G, and H be the vertices of a parallelepiped in space, whose faces are the parallelograms $ABCD$, $ABFE$, $AEHD$, and so forth. Let $\mathbf{b} = \overrightarrow{AB}$, $\mathbf{e} = \overrightarrow{AE}$, and $\mathbf{d} = \overrightarrow{AD}$. Express the vectors $\overrightarrow{AC}$, $\overrightarrow{AF}$, $\overrightarrow{AG}$, $\overrightarrow{FG}$, and $\overrightarrow{EG}$ in terms of **b**, **e**, and **d**.

The Dot Product 14.4

We shall discuss two different definitions for the product of two vectors. In this section we discuss the dot product; in Section 14.5 we discuss the cross product.

[14.4.1] DEFINITION / *Dot Product.*

If $\mathbf{v} = v_1\mathbf{i} + v_2\mathbf{j}$ and $\mathbf{w} = w_1\mathbf{i} + w_2\mathbf{j}$ are two vectors in the plane, the *dot product* $\mathbf{v} \cdot \mathbf{w}$ is defined as

$$\mathbf{v} \cdot \mathbf{w} = v_1w_1 + v_2w_2$$

If $\mathbf{v} = v_1\mathbf{i} + v_2\mathbf{j} + v_3\mathbf{k}$ and $\mathbf{w} = w_1\mathbf{i} + w_2\mathbf{j} + w_3\mathbf{k}$ are two vectors in space, the *dot product* $\mathbf{v} \cdot \mathbf{w}$ is defined as

$$\mathbf{v} \cdot \mathbf{w} = v_1w_1 + v_2w_2 + v_3w_3$$

EXAMPLE 1

(a) If $\mathbf{v} = 2\mathbf{i} - 3\mathbf{j}$ and $\mathbf{w} = \mathbf{i} + \mathbf{j}$, then

$$\mathbf{v} \cdot \mathbf{w} = (2)(1) + (-3)(1) = 2 - 3 = -1$$

(b) If $\mathbf{v} = 2\mathbf{i} - \mathbf{j} + \mathbf{k}$ and $\mathbf{w} = 4\mathbf{i} + 2\mathbf{j} - \mathbf{k}$, then

$$\mathbf{v} \cdot \mathbf{w} = (2)(4) + (-1)(2) + (1)(-1) = 8 - 2 - 1 = 5$$

$$\mathbf{v} \cdot \mathbf{v} = 4 + 1 + 1 = 6$$

$$\mathbf{w} \cdot \mathbf{w} = 16 + 4 + 1 = 21$$

Note that the dot product $\mathbf{v} \cdot \mathbf{w}$ of two vectors is a real number. Because of this, the dot product is sometimes referred to as the *scalar product.*

We now list some algebraic properties of the dot product.

[14.4.2] THEOREM / *Properties of the Dot Product.*

If u, v, and w are vectors and α is any scalar, then

(a) $\mathbf{u} \cdot \mathbf{v} = \mathbf{v} \cdot \mathbf{u}$ **(b)** $\mathbf{u} \cdot (\mathbf{v} + \mathbf{w}) = \mathbf{u} \cdot \mathbf{v} + \mathbf{u} \cdot \mathbf{w}$
(c) $\alpha(\mathbf{u} \cdot \mathbf{v}) = (\alpha\mathbf{u}) \cdot \mathbf{v}$ **(d)** $\mathbf{0} \cdot \mathbf{v} = 0$
(e) $\mathbf{v} \cdot \mathbf{v} = \|\mathbf{v}\|^2$

Proof We shall prove only properties (a) and (e) for vectors in space. The remaining proofs are left as exercises (Problems 47–50).

(a) If $\mathbf{u} = u_1\mathbf{i} + u_2\mathbf{j} + u_3\mathbf{k}$ and $\mathbf{v} = v_1\mathbf{i} + v_2\mathbf{j} + v_3\mathbf{k}$, then

$$\mathbf{u} \cdot \mathbf{v} = u_1v_1 + u_2v_2 + u_3v_3 = v_1u_1 + v_2u_2 + v_3u_3 = \mathbf{v} \cdot \mathbf{u}$$

(e) If $\mathbf{v} = v_1\mathbf{i} + v_2\mathbf{j} + v_3\mathbf{k}$, then

$$\mathbf{v} \cdot \mathbf{v} = v_1v_1 + v_2v_2 + v_3v_3 = v_1^2 + v_2^2 + v_3^2 = \|\mathbf{v}\|^2$$

As a result of property (e) in theorem (14.4.2), we conclude that

$$\mathbf{v} \cdot \mathbf{v} \geq 0 \quad \text{and} \quad \mathbf{v} \cdot \mathbf{v} = 0 \quad \text{if and only if} \quad \mathbf{v} = \mathbf{0}$$

Angle Between Vectors

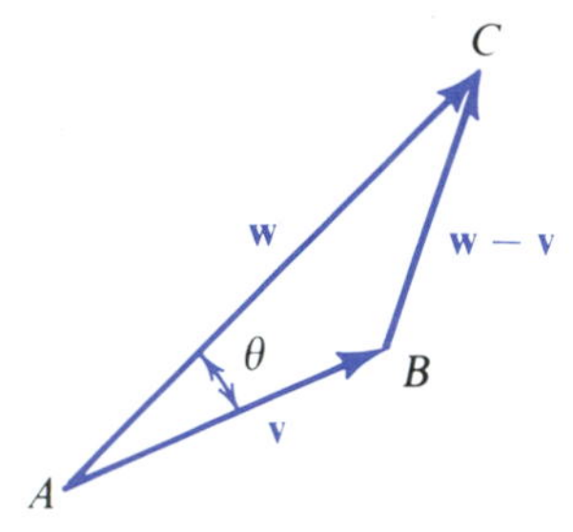

Figure 28

The dot product may be used to calculate the angle between two nonzero vectors. Consider triangle ABC in Figure 28. Let θ denote the angle at vertex A, and let the vectors $\mathbf{v}$ and $\mathbf{w}$ be represented by the directed line segments $\overrightarrow{AB}$ and $\overrightarrow{AC}$, respectively. Then the vector $\mathbf{w} - \mathbf{v}$ can be represented by the directed line segment $\overrightarrow{BC}$. The sides of the triangle have lengths $\|\mathbf{v}\|$, $\|\mathbf{w}\|$, and $\|\mathbf{w} - \mathbf{v}\|$. If we use the law of cosines,* we find

$$\|\mathbf{w} - \mathbf{v}\|^2 = \|\mathbf{v}\|^2 + \|\mathbf{w}\|^2 - 2\|\mathbf{v}\|\|\mathbf{w}\|\cos\theta \tag{1}$$

Now we use property (e) of theorem (14.4.2) to rewrite (1) as

$$(\mathbf{w} - \mathbf{v}) \cdot (\mathbf{w} - \mathbf{v}) = \mathbf{v} \cdot \mathbf{v} + \mathbf{w} \cdot \mathbf{w} - 2\|\mathbf{v}\|\|\mathbf{w}\|\cos\theta$$

A double application of property (b) of theorem (14.4.2) to the term $(\mathbf{w} - \mathbf{v}) \cdot (\mathbf{w} - \mathbf{v})$ yields

$$\begin{aligned}(\mathbf{w} - \mathbf{v}) \cdot (\mathbf{w} - \mathbf{v}) &= \mathbf{w} \cdot (\mathbf{w} - \mathbf{v}) - \mathbf{v} \cdot (\mathbf{w} - \mathbf{v}) = \mathbf{w} \cdot \mathbf{w} - \mathbf{w} \cdot \mathbf{v} - \mathbf{v} \cdot \mathbf{w} + \mathbf{v} \cdot \mathbf{v} \\ &\underset{\uparrow}{=} \mathbf{w} \cdot \mathbf{w} - \mathbf{v} \cdot \mathbf{w} - \mathbf{v} \cdot \mathbf{w} + \mathbf{v} \cdot \mathbf{v} \underset{\uparrow}{=} \|\mathbf{w}\|^2 - 2\mathbf{v} \cdot \mathbf{w} + \|\mathbf{v}\|^2\end{aligned} \tag{2}$$

By theorem (14.4.2(a)) By theorem (14.4.2(e))

Since $\|\mathbf{w} - \mathbf{v}\|^2 = (\mathbf{w} - \mathbf{v}) \cdot (\mathbf{w} - \mathbf{v})$, we may combine (1) and (2) to get

$$-2\|\mathbf{v}\|\|\mathbf{w}\|\cos\theta = -2(\mathbf{v} \cdot \mathbf{w})$$

$$\mathbf{v} \cdot \mathbf{w} = \|\mathbf{v}\|\|\mathbf{w}\|\cos\theta$$

So, we have proved the following result:

[14.4.3] THEOREM / *Angle Between Vectors.*

The angle θ, $0 \leq \theta \leq \pi$, between two nonzero vectors v and w is obtained by the formula

$$\cos\theta = \frac{\mathbf{v} \cdot \mathbf{w}}{\|\mathbf{v}\|\|\mathbf{w}\|} \tag{3}$$

EXAMPLE 2

Find the angle θ between the vectors $\mathbf{v} = 2\mathbf{i} - \mathbf{j} + \mathbf{k}$ and $\mathbf{w} = -\mathbf{i} + \mathbf{j}$.

Solution

$$\mathbf{v} \cdot \mathbf{w} = -2 - 1 = -3 \qquad \|\mathbf{v}\| = \sqrt{4 + 1 + 1} = \sqrt{6} \qquad \|\mathbf{w}\| = \sqrt{1 + 1} = \sqrt{2}$$

* The law of cosines states that $c^2 = a^2 + b^2 - 2ab\cos\theta$, where a, b, and c are the lengths of the sides of a triangle and θ is the angle opposite the side of length c.

Thus,

$$\cos\theta = \frac{\mathbf{v}\cdot\mathbf{w}}{\|\mathbf{v}\|\|\mathbf{w}\|} = \frac{-3}{\sqrt{6}\sqrt{2}} = \frac{-3}{\sqrt{12}} = \frac{-\sqrt{3}}{2}$$

Since $0 \le \theta \le \pi$, the angle θ between $\mathbf{v}$ and $\mathbf{w}$ is $5\pi/6$ radians (see Fig. 29).

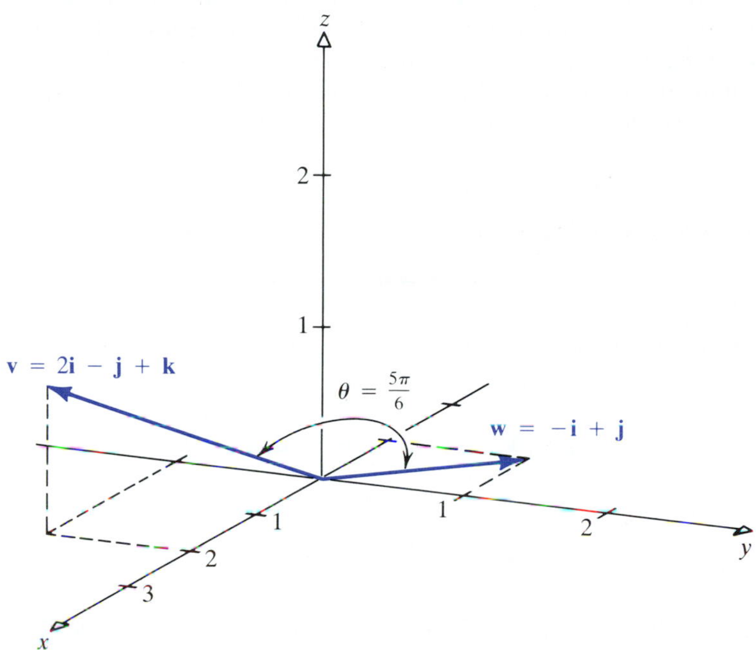

Figure 29

EXAMPLE 3

An airplane has an air speed of 400 kilometers per hour in an easterly direction. The wind velocity is 80 kilometers per hour in a southeasterly direction. Find the direction of the airplane relative to the ground.

Solution

This is the information supplied in Example 10 of Section 14.3. There, we found that the velocity of the airplane relative to the ground was

$$\mathbf{v}_g = (400 + 40\sqrt{2})\mathbf{i} - 40\sqrt{2}\mathbf{j}$$

The angle θ between $\mathbf{v}_g$ and the vector $\mathbf{v}_a = 400\mathbf{i}$ (the velocity of the airplane in the air) obeys

$$\cos\theta = \frac{\mathbf{v}_g\cdot\mathbf{v}_a}{\|\mathbf{v}_g\|\|\mathbf{v}_a\|} = \frac{(400+40\sqrt{2})(400)}{(460.06)(400)} \approx 0.9924$$

Use a calculator.

$$\theta \approx 7.06^\circ$$

The direction of the plane relative to the ground is 7.06° south of east.

Orthogonal Vectors

If the angle θ between two nonzero vectors $\mathbf{v}$ and $\mathbf{w}$ is 0 or π, the vectors $\mathbf{v}$ and $\mathbf{w}$ are parallel. If the angle θ between two nonzero vectors $\mathbf{v}$ and $\mathbf{w}$ is $\pi/2$, the vectors $\mathbf{v}$ and $\mathbf{w}$ are *orthogonal*.*

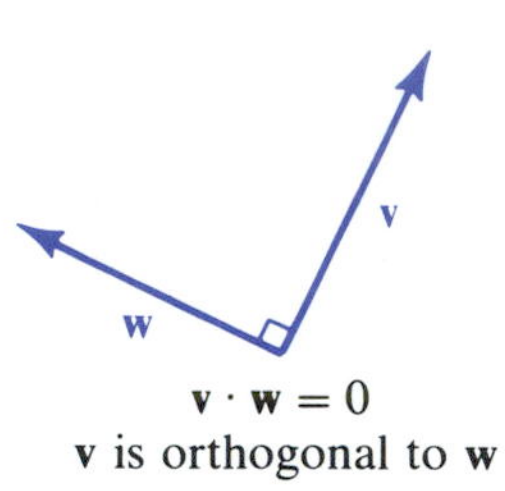

$\mathbf{v}\cdot\mathbf{w}=0$
$\mathbf{v}$ is orthogonal to $\mathbf{w}$

Figure 30

If follows from (3) that if $\mathbf{v}$ and $\mathbf{w}$ are orthogonal, then $\mathbf{v}\cdot\mathbf{w}=0$, since $\cos(\pi/2)=0$.

On the other hand, if $\mathbf{v}\cdot\mathbf{w}=0$, then either $\mathbf{v}=\mathbf{0}$ or $\mathbf{w}=\mathbf{0}$ or $\cos\theta=0$. In the latter case, $\theta=\pi/2$ and $\mathbf{v}$ and $\mathbf{w}$ are orthogonal. If $\mathbf{v}$ or $\mathbf{w}$ is the zero vector, then since the zero vector has no specific direction, we adopt the convention that the zero vector is orthogonal to every vector. See Figure 30.

[14.4.4] THEOREM / *Orthogonal Vectors.*

Two vectors v and w are orthogonal if and only if

$$\mathbf{v}\cdot\mathbf{w}=0$$

EXAMPLE 4

The vectors

$$\mathbf{v}=2\mathbf{i}-\mathbf{j}+5\mathbf{k} \qquad \text{and} \qquad \mathbf{w}=3\mathbf{i}+\mathbf{j}-\mathbf{k}$$

are orthogonal, since

$$\mathbf{v}\cdot\mathbf{w}=6-1-5=0$$

The standard basis vectors $\mathbf{i}$ and $\mathbf{j}$ in the plane are orthogonal, since

$$\mathbf{i}\cdot\mathbf{j}=1\cdot 0+0\cdot 1=0$$

The standard basis vectors $\mathbf{i}$, $\mathbf{j}$, and $\mathbf{k}$ in space are mutually orthogonal, since

$$\mathbf{i}\cdot\mathbf{j}=0 \qquad \mathbf{j}\cdot\mathbf{k}=0 \qquad \mathbf{k}\cdot\mathbf{i}=0$$

EXAMPLE 5

Find a scalar α so that the vectors $\mathbf{v}=2\alpha\mathbf{i}+\mathbf{j}-\mathbf{k}$ and $\mathbf{w}=\mathbf{i}-\alpha\mathbf{j}+\mathbf{k}$ are orthogonal.

Solution

It is required that $\mathbf{v}\cdot\mathbf{w}=0$, so

$$\mathbf{v}\cdot\mathbf{w}=2\alpha-\alpha-1=0$$
$$\alpha=1$$

* *Orthogonal*, *perpendicular*, and *normal* are terms that mean "meet at a right angle." It is customary to refer to two vectors as *orthogonal*, two lines as *perpendicular*, and a line and a plane or a vector and a plane as *normal*.

Cauchy–Schwarz Inequality

For any angle θ we have $|\cos\theta| \le 1$. Thus, for nonzero vectors $\mathbf{v}$ and $\mathbf{w}$, it follows from (3) that

$$\frac{|\mathbf{v}\cdot\mathbf{w}|}{\|\mathbf{v}\|\,\|\mathbf{w}\|} \le 1$$

Hence, for any pair of nonzero vectors $\mathbf{v}$ and $\mathbf{w}$ (the result is trivially true for $\mathbf{v}=\mathbf{0}$ or $\mathbf{w}=\mathbf{0}$), $|\mathbf{v}\cdot\mathbf{w}| \le \|\mathbf{v}\|\,\|\mathbf{w}\|$.

[14.4.5] THEOREM / ***Cauchy–Schwarz Inequality.***

If v and w are vectors, then

$$|\mathbf{v}\cdot\mathbf{w}| \le \|\mathbf{v}\|\|\mathbf{w}\| \qquad \mathbf{(4)}$$

The Cauchy–Schwarz inequality is used to prove the triangle inequality.

[14.4.6] THEOREM / ***Triangle Inequality.***

If v and w are vectors, then

$$\|\mathbf{v}+\mathbf{w}\| \le \|\mathbf{v}\| + \|\mathbf{w}\|$$

Proof We use the Cauchy–Schwarz inequality and several properties of the dot product:

$$\|\mathbf{v}+\mathbf{w}\|^2 = (\mathbf{v}+\mathbf{w})\cdot(\mathbf{v}+\mathbf{w}) = \mathbf{v}\cdot(\mathbf{v}+\mathbf{w}) + \mathbf{w}\cdot(\mathbf{v}+\mathbf{w})$$

By theorem (14.4.2(e)) By theorem (14.4.2(b))

$$= (\mathbf{v}\cdot\mathbf{v}) + (\mathbf{v}\cdot\mathbf{w}) + (\mathbf{w}\cdot\mathbf{v}) + (\mathbf{w}\cdot\mathbf{w}) = \|\mathbf{v}\|^2 + 2(\mathbf{v}\cdot\mathbf{w}) + \|\mathbf{w}\|^2$$

By theorem (14.4.2(b)) By theorem (14.4.2(a), (e))

$$\le \|\mathbf{v}\|^2 + 2|\mathbf{v}\cdot\mathbf{w}| + \|\mathbf{w}\|^2 \le \|\mathbf{v}\|^2 + 2\|\mathbf{v}\|\|\mathbf{w}\| + \|\mathbf{w}\|^2 = (\|\mathbf{v}\| + \|\mathbf{w}\|)^2$$

By (4)

By taking square roots, we find

$$\|\mathbf{v}+\mathbf{w}\| \le \|\mathbf{v}\| + \|\mathbf{w}\|$$

Direction Angles of Vectors in Space

A nonzero vector $\mathbf{v}$ in space can be described by specifying its three *direction angles* α, β, γ and its magnitude. These direction angles are defined as

α = Angle between $\mathbf{v}$ and the positive x-axis, $0 \le \alpha \le \pi$

β = Angle between $\mathbf{v}$ and the positive y-axis, $0 \le \beta \le \pi$

γ = Angle between $\mathbf{v}$ and the positive z-axis, $0 \le \gamma \le \pi$

See Figure 31 on page 818.

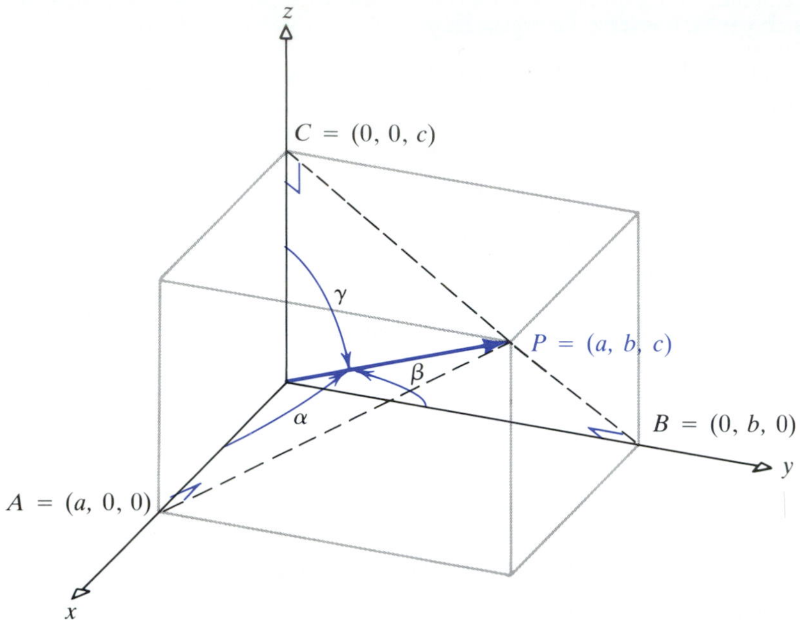

$0 \leqslant \alpha \leqslant \pi, \quad 0 \leqslant \beta \leqslant \pi, \quad 0 \leqslant \gamma \leqslant \pi$

Figure 31

Our first problem will be to find an expression for α, β, and γ in terms of the components of a vector. Let $\mathbf{v} = v_1\mathbf{i} + v_2\mathbf{j} + v_3\mathbf{k}$ denote a nonzero vector. The angle α between $\mathbf{v}$ and the positive x-axis obeys

$$\cos \alpha = \frac{\mathbf{v} \cdot \mathbf{i}}{\|\mathbf{v}\|\,\|\mathbf{i}\|} = \frac{v_1}{\|\mathbf{v}\|}$$

Similarly,

$$\cos \beta = \frac{v_2}{\|\mathbf{v}\|} \qquad \cos \gamma = \frac{v_3}{\|\mathbf{v}\|}$$

Since $\|\mathbf{v}\| = \sqrt{v_1^2 + v_2^2 + v_3^2}$, we have the following result:

[14.4.7] THEOREM

If $\mathbf{v} = v_1\mathbf{i} + v_2\mathbf{j} + v_3\mathbf{k}$ is a nonzero vector in space, the direction angles α, β, γ obey

$$\cos \alpha = \frac{v_1}{\sqrt{v_1^2 + v_2^2 + v_3^2}}, \ \cos \beta = \frac{v_2}{\sqrt{v_1^2 + v_2^2 + v_3^2}}, \ \cos \gamma = \frac{v_3}{\sqrt{v_1^2 + v_2^2 + v_3^2}} \qquad (5)$$

The numbers $\cos \alpha$, $\cos \beta$, and $\cos \gamma$ are called the *direction cosines* of the vector $\mathbf{v}$. They play the same role in three dimensions as slope does in two dimensions.

EXAMPLE 6

Find the direction cosines of $\mathbf{v} = -3\mathbf{i} + 2\mathbf{j} - 6\mathbf{k}$.

Solution

$$\|\mathbf{v}\| = \sqrt{(-3)^2 + 2^2 + (-6)^2} = \sqrt{49} = 7$$

By using (5), we get

$$\cos\alpha = \frac{-3}{7} \qquad \cos\beta = \frac{2}{7} \qquad \cos\gamma = \frac{-6}{7}$$

[14.4.8] THEOREM

If α, β, γ are the direction angles of a nonzero vector v in space, then

$$\cos^2\alpha + \cos^2\beta + \cos^2\gamma = 1 \tag{6}$$

The proof is a direct consequence of (5).

Based on (6), when two direction cosines are known, the third is determined up to its sign. Thus, knowing two direction cosines is not sufficient to uniquely determine the direction of a vector in space.

EXAMPLE 7

The vector **v** makes an angle of $\alpha = \pi/3$ with the positive x-axis, an angle of $\beta = \pi/3$ with the positive y-axis, and an acute angle γ with the positive z-axis. Find γ.

Solution

By (6), we have

$$\cos^2\left(\frac{\pi}{3}\right) + \cos^2\left(\frac{\pi}{3}\right) + \cos^2\gamma = 1$$

$$\left(\frac{1}{2}\right)^2 + \left(\frac{1}{2}\right)^2 + \cos^2\gamma = 1$$

$$\cos^2\gamma = \frac{1}{2}$$

$$\cos\gamma = \frac{\sqrt{2}}{2} \qquad \text{or} \qquad \cos\gamma = -\frac{\sqrt{2}}{2}$$

$$\gamma = \frac{\pi}{4} \qquad \text{or} \qquad \gamma = \frac{3\pi}{4}$$

Since we are requiring that γ be acute, the answer is $\gamma = \pi/4$.

The direction cosines of a vector give information about only the direction of the vector; they provide no information about its magnitude. For example, *any* vector parallel to the xy-plane and making an angle of $\pi/4$ radian with the positive x- and y-axes has direction cosines

$$\cos\alpha = \frac{\sqrt{2}}{2} \qquad \cos\beta = \frac{\sqrt{2}}{2} \qquad \cos\gamma = 0$$

However, if the direction angles *and* the magnitude of a vector are known, then the vector is uniquely determined.

EXAMPLE 8

Show that any nonzero vector $\mathbf{v}$ in space can be written in terms of its magnitude and direction cosines as

$$\mathbf{v} = \|\mathbf{v}\|[(\cos\alpha)\mathbf{i} + (\cos\beta)\mathbf{j} + (\cos\gamma)\mathbf{k}] \tag{7}$$

Solution

Let $\mathbf{v} = v_1\mathbf{i} + v_2\mathbf{j} + v_3\mathbf{k}$. From (5), we see that

$$v_1 = \|\mathbf{v}\|\cos\alpha \qquad v_2 = \|\mathbf{v}\|\cos\beta \qquad v_3 = \|\mathbf{v}\|\cos\gamma$$

Thus,

$$\begin{aligned}\mathbf{v} = v_1\mathbf{i} + v_2\mathbf{j} + v_3\mathbf{k} &= \|\mathbf{v}\|(\cos\alpha)\mathbf{i} + \|\mathbf{v}\|(\cos\beta)\mathbf{j} + \|\mathbf{v}\|(\cos\gamma)\mathbf{k}\\ &= \|\mathbf{v}\|[(\cos\alpha)\mathbf{i} + (\cos\beta)\mathbf{j} + (\cos\gamma)\mathbf{k}]\end{aligned}$$

The result shown in (7) tells us that the direction cosines of a vector $\mathbf{v}$ are also the components of the unit vector in the direction of $\mathbf{v}$.

Projection of a Vector

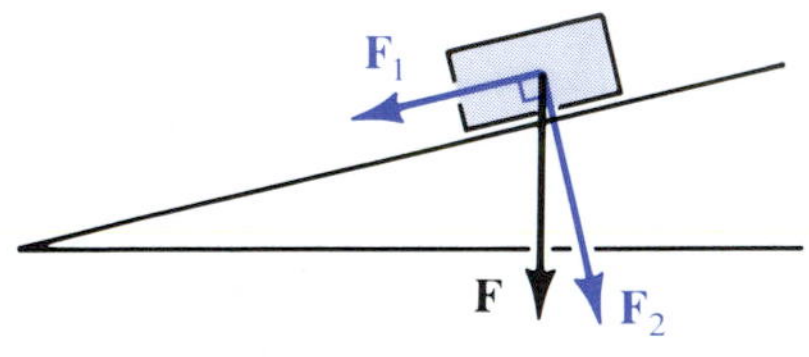

Figure 32

In many physical applications it is often necessary to find "how much" of a vector is applied along a given direction. In Figure 32 the force $\mathbf{F}$ due to gravity is pulling down on the block. To study the effect of gravity on the block, it is necessary to determine how much of $\mathbf{F}$ is actually pulling the block down the incline ($\mathbf{F}_1$) and how much is pulling the block against the incline ($\mathbf{F}_2$). Knowing the composition of $\mathbf{F}$ allows us to determine when friction is overcome and when the block will slide down the incline.

Suppose $\mathbf{v} \neq 0$ and $\mathbf{w} \neq 0$ are two vectors with the same initial point P. We seek to decompose $\mathbf{v}$ into two vectors: one, $\mathbf{v}_1$, that is parallel to $\mathbf{w}$ and the other, $\mathbf{v}_2$, that is orthogonal to $\mathbf{w}$. The vector $\mathbf{v}_1$ is called the *vector projection of* $\mathbf{v}$ *along* $\mathbf{w}$ and is denoted by $\text{proj}_{\mathbf{w}}\mathbf{v}$. See Figure 33.

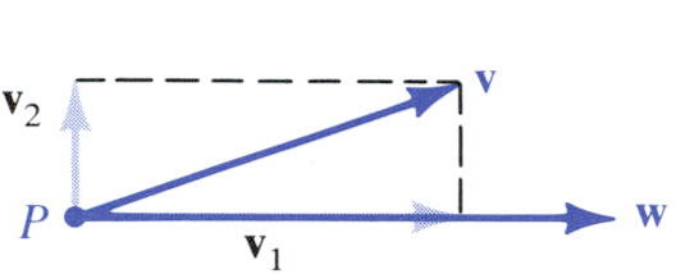

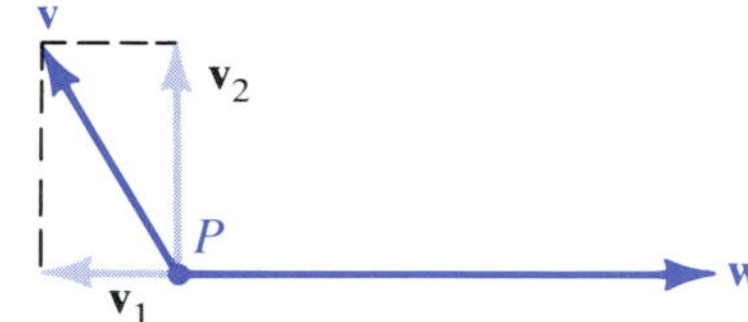

Figure 33

The vector $\mathbf{v}_1$ is obtained as follows: From the terminal point of $\mathbf{v}$ drop a perpendicular to the line through $\mathbf{w}$. The vector $\mathbf{v}_1$ is the vector from P to the foot of this perpendicular; the vector $\mathbf{v}_2$ is merely $\mathbf{v}_2 = \mathbf{v} - \mathbf{v}_1$. Note that

$\mathbf{v} = \mathbf{v}_1 + \mathbf{v}_2$, $\mathbf{v}_1$ is parallel to $\mathbf{w}$, and $\mathbf{v}_2$ is orthogonal to $\mathbf{w}$.

This is the decomposition of $\mathbf{v}$ we wanted.

We seek a formula for $\mathbf{v}_1$ that is based on a knowledge of the vectors $\mathbf{v}$ and $\mathbf{w}$. Since $\mathbf{v} = \mathbf{v}_1 + \mathbf{v}_2$, we have

$$\mathbf{v} \cdot \mathbf{w} = (\mathbf{v}_1 + \mathbf{v}_2) \cdot \mathbf{w} = \mathbf{v}_1 \cdot \mathbf{w} + \mathbf{v}_2 \cdot \mathbf{w} \tag{8}$$

Since $\mathbf{v}_2$ is orthogonal to $\mathbf{w}$, we have $\mathbf{v}_2 \cdot \mathbf{w} = 0$. Since $\mathbf{v}_1$ is parallel to $\mathbf{w}$, we have $\mathbf{v}_1 = \alpha\mathbf{w}$ for some scalar α. Thus, (8) can be written as

$$\mathbf{v} \cdot \mathbf{w} = \alpha\mathbf{w} \cdot \mathbf{w} = \alpha\|\mathbf{w}\|^2$$

$$\alpha = \frac{\mathbf{v} \cdot \mathbf{w}}{\|\mathbf{w}\|^2}$$

Thus,

$$\mathbf{v}_1 = \alpha\mathbf{w} = \frac{\mathbf{v} \cdot \mathbf{w}}{\|\mathbf{w}\|^2}\mathbf{w}$$

[14.4.9] THEOREM / ***Vector Projection.***

If $\mathbf{v} \neq \mathbf{0}$ and $\mathbf{w} \neq \mathbf{0}$ are two vectors, the vector projection of v along w is

$$\text{proj}_{\mathbf{w}}\mathbf{v} = \frac{\mathbf{v} \cdot \mathbf{w}}{\|\mathbf{w}\|^2}\mathbf{w}$$

The decomposition of $\mathbf{v}$ into $\mathbf{v}_1$ and $\mathbf{v}_2$, where $\mathbf{v}_1$ is parallel to $\mathbf{w}$ and $\mathbf{v}_2$ is perpendicular to $\mathbf{w}$, is

$$\mathbf{v}_1 = \text{proj}_{\mathbf{w}}\mathbf{v} = \frac{\mathbf{v} \cdot \mathbf{w}}{\|\mathbf{w}\|^2}\mathbf{w} \qquad \mathbf{v}_2 = \mathbf{v} - \mathbf{v}_1 \tag{9}$$

EXAMPLE 9

Find the vector projection of $\mathbf{v} = 2\mathbf{i} - \mathbf{j} + \mathbf{k}$ along $\mathbf{w} = \mathbf{i} + \mathbf{j} + \mathbf{k}$. Decompose $\mathbf{v}$ into two vectors $\mathbf{v}_1$ and $\mathbf{v}_2$, where $\mathbf{v}_1$ is parallel to $\mathbf{w}$ and $\mathbf{v}_2$ is orthogonal to $\mathbf{w}$.

Solution

We use (9):

$$\mathbf{v}_1 = \text{proj}_{\mathbf{w}}\mathbf{v} = \frac{\mathbf{v} \cdot \mathbf{w}}{\|\mathbf{w}\|^2}\mathbf{w} = \frac{2 - 1 + 1}{(\sqrt{3})^2}\mathbf{w} = \frac{2}{3}\mathbf{w} = \frac{2}{3}(\mathbf{i} + \mathbf{j} + \mathbf{k})$$

$$\mathbf{v}_2 = \mathbf{v} - \mathbf{v}_1 = (2\mathbf{i} - \mathbf{j} + \mathbf{k}) - \frac{2}{3}(\mathbf{i} + \mathbf{j} + \mathbf{k}) = \frac{4}{3}\mathbf{i} - \frac{5}{3}\mathbf{j} + \frac{1}{3}\mathbf{k}$$

Work Done by a Constant Force

In Chapter 6 the work W done by a constant force $\mathbf{F}$ in moving an object from a point A to a point B was defined as

$$W = (\text{Magnitude of force})(\text{Distance}) = \|\mathbf{F}\|\|\overrightarrow{AB}\|$$

In this definition, it was assumed that the force $\mathbf{F}$ was applied along the line of motion $\overrightarrow{AB}$. If the constant force $\mathbf{F}$ is not along the line of motion, but instead is at an angle θ to the direction of motion (see Figure 34), then the *work W done by* $\mathbf{F}$ in moving an object from A to B is defined as

$$W = \mathbf{F} \cdot \overrightarrow{AB} \tag{10}$$

This definition is compatible with the force times distance definition from Chapter 6, since

$$\begin{aligned} W &= (\text{Amount of force in direction of } \overrightarrow{AB})(\text{Distance}) \\ &= \|\text{proj}_{\mathbf{AB}}\mathbf{F}\|\|\overrightarrow{AB}\| = \frac{\mathbf{F} \cdot \overrightarrow{AB}}{\|\overrightarrow{AB}\|^2}\|\overrightarrow{AB}\|\|\overrightarrow{AB}\| = \mathbf{F} \cdot \overrightarrow{AB} \end{aligned}$$

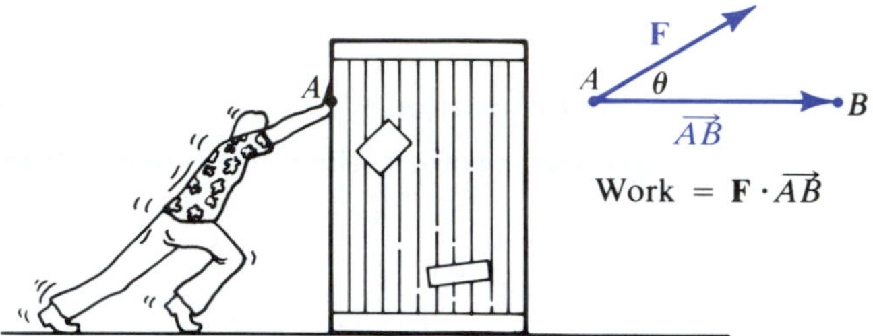

Figure 34

EXAMPLE 10

Find the work done by a force of 2 newtons acting in the direction $\mathbf{i} + \mathbf{j} + \mathbf{k}$ in moving an object 1 meter from $(0, 0, 0)$ to $(1, 0, 0)$.

Solution

First, we must express the force $\mathbf{F}$ as a vector. From (5), the direction cosines of the force are

$$\cos \alpha = \frac{1}{\sqrt{3}} \qquad \cos \beta = \frac{1}{\sqrt{3}} \qquad \cos \gamma = \frac{1}{\sqrt{3}}$$

Since the force has magnitude 2, by (7) it can be written as

$$\mathbf{F} = \frac{2}{\sqrt{3}}(\mathbf{i} + \mathbf{j} + \mathbf{k})$$

The line of motion of the object is along $\overrightarrow{AB} = \mathbf{i}$. The work W is therefore

$$W = \mathbf{F} \cdot \overrightarrow{AB} = \frac{2}{\sqrt{3}}(\mathbf{i} + \mathbf{j} + \mathbf{k}) \cdot \mathbf{i} = \frac{2}{\sqrt{3}} \text{ joules}$$

EXAMPLE 11

Figure 35(a) shows a person pushing on a lawn mower handle with a force of 30 pounds. How much work is done in moving the lawn mower a distance of 75 feet if the handle makes an angle of 60° with the ground?

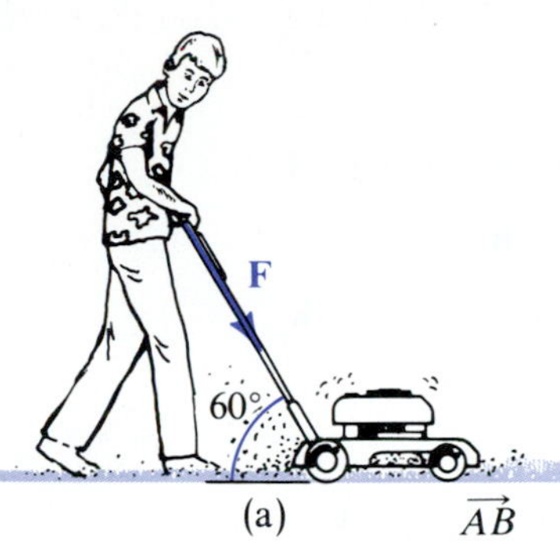

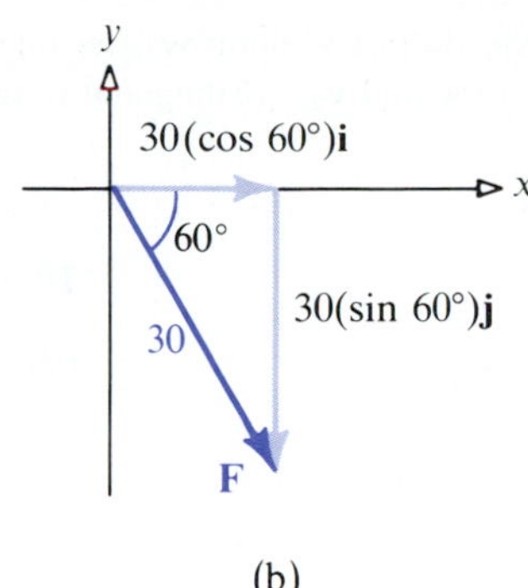

Figure 35

Solution

We choose to set up our coordinates in such a way that the lawn mower is moved from (0, 0) to (75, 0). Thus, the motion is along $\overrightarrow{AB} = 75\mathbf{i}$. As a result of (7), the force vector **F**, as shown in Figure 35(b), is

$$\mathbf{F} = 30(\cos 60°)\mathbf{i} - 30(\sin 60°)\mathbf{j} = 15\mathbf{i} - 15\sqrt{3}\mathbf{j}$$

By (10), the work W done is

$$W = \mathbf{F} \cdot \overrightarrow{AB} = (15\mathbf{i} - 15\sqrt{3}\mathbf{j}) \cdot 75\mathbf{i} = 1125 \text{ foot-pounds}$$

Vectors in *n* Dimensions

A nonzero vector **v** in n-space will have *n direction angles* $\alpha_1, \alpha_2, \ldots, \alpha_n$, each defined as the angle α_i, $0 \le \alpha_i \le \pi$, $i = 1, 2, \ldots n$, between **v** and the corresponding positive axis.

If $\mathbf{v} = v_1\mathbf{u}_1 + v_2\mathbf{u}_2 + \cdots + v_n\mathbf{u}_n$ is a nonzero vector and $\alpha_1, \alpha_2, \ldots, \alpha_n$ are its direction angles, then the direction cosines of **v** are

$$\cos \alpha_1 = \frac{v_1}{\|\mathbf{v}\|}, \quad \cos \alpha_2 = \frac{v_2}{\|\mathbf{v}\|}, \quad \ldots, \quad \cos \alpha_n = \frac{v_n}{\|\mathbf{v}\|}$$

and

$$\cos^2\alpha_1 + \cos^2\alpha_2 + \cdots + \cos^2\alpha_n = 1$$

EXERCISE 14.4

In Problems 1–6 find the dot product $\mathbf{v} \cdot \mathbf{w}$, and find the cosine of the angle θ between **v** and **w**.

1. $\mathbf{v} = 2\mathbf{i} - 3\mathbf{j} + \mathbf{k}$, $\mathbf{w} = \mathbf{i} - \mathbf{j} + \mathbf{k}$

2. $\mathbf{v} = -3\mathbf{i} + 2\mathbf{j} - \mathbf{k}$, $\mathbf{w} = 2\mathbf{i} + \mathbf{j} - \mathbf{k}$

3. $\mathbf{v} = \mathbf{i} - \mathbf{j}$, $\mathbf{w} = \mathbf{j} + \mathbf{k}$

4. $\mathbf{v} = \mathbf{j} - \mathbf{k}$, $\mathbf{w} = \mathbf{i} + \mathbf{k}$

5. $\mathbf{v} = 3\mathbf{i} + \mathbf{j} - \mathbf{k}$, $\mathbf{w} = -2\mathbf{i} - \mathbf{j} + \mathbf{k}$

6. $\mathbf{v} = \mathbf{i} - 3\mathbf{j} + 4\mathbf{k}$, $\mathbf{w} = 4\mathbf{i} - \mathbf{j} + 3\mathbf{k}$

In Problems 7–12 find the vector projection of $\mathbf{v}$ along $\mathbf{w}$. Decompose $\mathbf{v}$ into two vectors $\mathbf{v}_1$ and $\mathbf{v}_2$, where $\mathbf{v}_1$ is parallel to $\mathbf{w}$ and $\mathbf{v}_2$ is orthogonal to $\mathbf{w}$.

7. $\mathbf{v} = 2\mathbf{i} - 3\mathbf{j} + \mathbf{k}, \quad \mathbf{w} = \mathbf{i} - \mathbf{j} + \mathbf{k}$

8. $\mathbf{v} = -3\mathbf{i} + 2\mathbf{j} - \mathbf{k}, \quad \mathbf{w} = 2\mathbf{i} + \mathbf{j} - \mathbf{k}$

9. $\mathbf{v} = \mathbf{i} - \mathbf{j}, \quad \mathbf{w} = \mathbf{j} + \mathbf{k}$

10. $\mathbf{v} = \mathbf{j} - \mathbf{k}, \quad \mathbf{w} = \mathbf{i} + \mathbf{k}$

11. $\mathbf{v} = 3\mathbf{i} + \mathbf{j} - \mathbf{k}, \quad \mathbf{w} = -2\mathbf{i} - \mathbf{j} + \mathbf{k}$

12. $\mathbf{v} = \mathbf{i} - 3\mathbf{j} + 4\mathbf{k}, \quad \mathbf{w} = 4\mathbf{i} - \mathbf{j} + 3\mathbf{k}$

In Problems 13–16 find a real number α so that the vectors $\mathbf{v}$ and $\mathbf{w}$ are orthogonal.

13. $\mathbf{v} = 2\alpha\mathbf{i} + \mathbf{j} - \mathbf{k}, \quad \mathbf{w} = \mathbf{i} - \mathbf{j} + \mathbf{k}$

14. $\mathbf{v} = \mathbf{i} + 2\alpha\mathbf{j} - \mathbf{k}, \quad \mathbf{w} = \mathbf{i} - \mathbf{j} + \mathbf{k}$

15. $\mathbf{v} = \alpha\mathbf{i} + \mathbf{j} + \mathbf{k}, \quad \mathbf{w} = \mathbf{i} + \alpha\mathbf{j} + 4\mathbf{k}$

16. $\mathbf{v} = \mathbf{i} - \alpha\mathbf{j} + 2\mathbf{k}, \quad \mathbf{w} = 2\alpha\mathbf{i} + \mathbf{j} + \mathbf{k}$

17. Find a real number α so that the angle θ between the vectors $\mathbf{v} = \alpha\mathbf{i} + \mathbf{j} + \mathbf{k}$ and $\mathbf{w} = \mathbf{i} + \alpha\mathbf{j} + \mathbf{k}$ is $\pi/3$.

18. Find a real number α so that the angle θ between the vectors $\mathbf{v} = \alpha\mathbf{i} - \mathbf{j} + \mathbf{k}$ and $\mathbf{w} = \mathbf{i} - \mathbf{j} + \alpha\mathbf{k}$ is $\pi/3$.

In Problems 19–22 find a vector $\mathbf{v}$ in space that has the given magnitude and direction angles.

19. $\|\mathbf{v}\| = 3, \quad \alpha = \pi/3, \quad \beta = \pi/4, \quad 0 < \gamma < \pi/2$

20. $\|\mathbf{v}\| = 3$, direction angles are equal, $\mathbf{v}$ has positive components

21. $\|\mathbf{v}\| = 2, \quad \alpha = \pi/4, \quad \pi/2 < \beta < \pi, \quad \gamma = \pi/3$

22. $\|\mathbf{v}\| = \frac{1}{2}, \quad \cos\alpha > 0, \quad \cos\beta = \frac{1}{4}, \quad \cos\gamma = \sqrt{\frac{7}{8}}$

23. Show that the vector projection of $\mathbf{v}$ along $\mathbf{i}$ is $(\mathbf{v}\cdot\mathbf{i})\mathbf{i}$. In fact, show that we can always write a vector $\mathbf{v}$ as

$$\mathbf{v} = (\mathbf{v}\cdot\mathbf{i})\mathbf{i} + (\mathbf{v}\cdot\mathbf{j})\mathbf{j} + (\mathbf{v}\cdot\mathbf{k})\mathbf{k}$$

24. Find all numbers α and β so that the vectors $2\alpha\mathbf{i} - 2\mathbf{j} + \mathbf{k}$ and $\beta\mathbf{i} + 2\mathbf{j} + 2\mathbf{k}$ are orthogonal and have the same magnitude.

25. If $\|\mathbf{v}\| = 2$, $\|\mathbf{w}\| = 6$, and the angle between $\mathbf{v}$ and $\mathbf{w}$ is $\pi/6$, find $\|\mathbf{v} + \mathbf{w}\|$ and $\|\mathbf{v} - \mathbf{w}\|$.

26. A wagon is pulled horizontally by exerting a force of 20 pounds on the handle at an angle of 30° with the horizontal. How much work is done in moving the wagon 100 feet?

27. Find the work done in moving an object along a vector $\mathbf{u} = 3\mathbf{i} + 2\mathbf{j} - 5\mathbf{k}$ if the applied force is $\mathbf{F} = 2\mathbf{i} - \mathbf{j} - \mathbf{k}$.

28. Find the work done by a force of 3 newtons acting in the direction $2\mathbf{i} + \mathbf{j} + 2\mathbf{k}$ in moving an object 2 meters from $(0, 0, 0)$ to $(0, 2, 0)$.

29. Find the work done by a force of 1 newton acting in the direction $2\mathbf{i} + 2\mathbf{j} + \mathbf{k}$ in moving an object 3 meters from $(0, 0, 0)$ to $(1, 2, 2)$.

30. Find the acute angle that a constant unit force vector makes with the positive x-axis if the work done by the force in moving a particle from $(0, 0)$ to $(4, 0)$ equals 2.

31. Find all numbers z so that the triangle with vertices $A = (1, -1, 0)$, $B = (-2, 2, 1)$, and $C = (0, 2, z)$ is a right triangle with right angle at C.

32. An airplane has an air speed of 500 kilometers per hour in a northerly direction. The wind velocity is 60 kilometers per hour in a southeasterly direction. Find the actual speed and direction of the plane relative to the ground.

33. A stream 1 kilometer wide has a constant current of 5 kilometers per hour. At what angle to the shore should a person head a boat, which is capable of maintaining a constant speed of 15 kilometers per hour, in order to reach a point directly opposite?

34. A river is 500 meters wide and has a current of 1 kilometer per hour. If a person can swim at a rate of 2 kilometers per hour, at what angle to the shore should she swim if she wishes to cross the river to a point directly opposite? How long will it take to swim across the river?

35. An airplane travels 200 miles due west and then 150 miles 60° north of west. Determine the resultant displacement.

36. Prove the *polarization identity*:
$\|\mathbf{u} + \mathbf{v}\|^2 - \|\mathbf{u} - \mathbf{v}\|^2 = 4(\mathbf{u} \cdot \mathbf{v})$

37. (a) If $\mathbf{u}$ and $\mathbf{v}$ have the same magnitude, then show that $\mathbf{u} + \mathbf{v}$ and $\mathbf{u} - \mathbf{v}$ are orthogonal.
(b) Use this to prove that an angle inscribed in a semicircle is a right angle (see the figure).

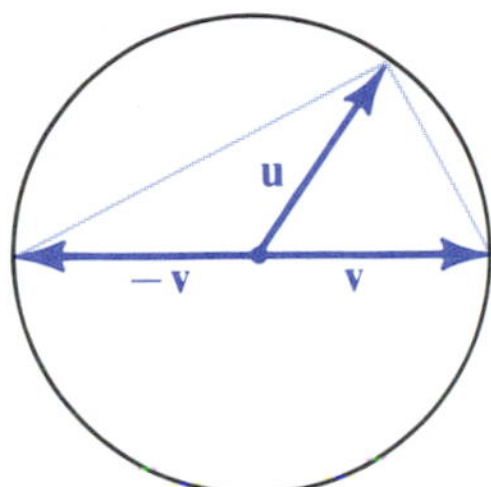

38. Let $\mathbf{r}_1$ and $\mathbf{r}_2$ be unit vectors in the plane making angles α and β with the positive x-axis.

(a) Prove that $\mathbf{r}_1 = (\cos \alpha)\mathbf{i} + (\sin \alpha)\mathbf{j}$ and $\mathbf{r}_2 = (\cos \beta)\mathbf{i} + (\sin \beta)\mathbf{j}$.
(b) By considering $\mathbf{r}_1 \cdot \mathbf{r}_2$, prove the trigonometric formulas below:

$$\cos(\alpha - \beta) = \cos \alpha \cos \beta + \sin \alpha \sin \beta$$

$$\cos(\alpha + \beta) = \cos \alpha \cos \beta - \sin \alpha \sin \beta$$

39. Let $\mathbf{v}$ and $\mathbf{w}$ denote nonzero vectors. Show that the vector $\mathbf{v} - \alpha\mathbf{w}$ is orthogonal to $\mathbf{w}$ if $\alpha = \dfrac{\mathbf{v} \cdot \mathbf{w}}{\|\mathbf{w}\|^2}$.

40. Let $\mathbf{v}$ and $\mathbf{w}$ denote nonzero vectors. Show that the vectors $\|\mathbf{w}\|\mathbf{v} + \|\mathbf{v}\|\mathbf{w}$ and $\|\mathbf{w}\|\mathbf{v} - \|\mathbf{v}\|\mathbf{w}$ are orthogonal.

41. In the definition of work given in (10), what is the work done if $\mathbf{F}$ is orthogonal to $\overrightarrow{AB}$?

42. Let $\mathbf{v} = v_1\mathbf{i} + v_2\mathbf{j} + v_3\mathbf{k}$ and $\mathbf{w} = w_1\mathbf{i} + w_2\mathbf{j} + w_3\mathbf{k}$ be two nonzero vectors that are not parallel. Find a vector $\mathbf{u} \neq \mathbf{0}$ that is orthogonal to both $\mathbf{v}$ and $\mathbf{w}$.

43. Let $\mathbf{v}$ and $\mathbf{w}$ denote nonzero vectors, and let $\mathbf{u} = \|\mathbf{w}\|\mathbf{v} + \|\mathbf{v}\|\mathbf{w}$. Show that the angle between $\mathbf{v}$ and $\mathbf{u}$ equals the angle between $\mathbf{w}$ and $\mathbf{u}$.

44. Let $\mathbf{w}$ be a nonzero vector, and let $\mathbf{u}$ denote a unit vector. Prove that the unit vector $\mathbf{u}$ making $\mathbf{w} \cdot \mathbf{u}$ a maximum is the unit vector pointing in the same direction as $\mathbf{w}$.

45. Let $\mathbf{u}$ and $\mathbf{w}$ be two unit vectors, and let θ be the angle between them. Find $\frac{1}{2}\|\mathbf{u} - \mathbf{w}\|$ in terms of θ.

46. Under what conditions is the Cauchy–Schwarz inequality an equality?

47. Prove part (b) of theorem (14.4.2) for vectors in the plane and in space.

48. Prove part (c) of theorem (14.4.2) for vectors in the plane and in space.

49. Prove part (d) of theorem (14.4.2) for vectors in the plane and in space.

50. Prove parts (a) and (e) of theorem (14.4.2) for vectors in the plane.

51. If $\mathbf{v}$ is a vector for which $\mathbf{v} \cdot \mathbf{i} = 0$, $\mathbf{v} \cdot \mathbf{j} = 0$, and $\mathbf{v} \cdot \mathbf{k} = 0$, find $\mathbf{v}$.

52. Solve for $\mathbf{x}$ in terms of α, $\mathbf{a}$, $\mathbf{b}$, and $\mathbf{c}$, if

$$\alpha\mathbf{x} + (\mathbf{x} \cdot \mathbf{b})\mathbf{a} = \mathbf{c} \qquad \alpha \neq 0, \quad \alpha + \mathbf{a} \cdot \mathbf{b} \neq 0$$

[*Hint:* First find $\mathbf{x} \cdot \mathbf{b}$, then $\mathbf{x}$.]

53. Show that:

(a) There is no vector for which $\alpha = \pi/6$ and $\beta = \pi/4$.
(b) If α and β are positive acute direction angles of a vector, then $\alpha + \beta \geq \pi/2$.

54. (a) Prove that the vectors $\mathbf{u} = 3\mathbf{i} + \mathbf{j} - 2\mathbf{k}$, $\mathbf{v} = -\mathbf{i} + 3\mathbf{j} + 4\mathbf{k}$, and $\mathbf{w} = 4\mathbf{i} - 2\mathbf{j} - 6\mathbf{k}$ can form the sides of a triangle.
(b) Find the lengths of the medians of the triangle.

55. Let θ be the angle between the nonzero vectors $\mathbf{u} = u_1\mathbf{i} + u_2\mathbf{j} + u_3\mathbf{k}$ and $\mathbf{v} = v_1\mathbf{i} + v_2\mathbf{j} + v_3\mathbf{k}$. Show that

$$\sin^2\theta = \frac{(u_2v_3 - u_3v_2)^2 + (u_1v_3 - u_3v_1)^2 + (u_1v_2 - u_2v_1)^2}{\|\mathbf{u}\|^2\|\mathbf{v}\|^2}$$

[*Hint:* Begin with $\sin^2\theta = 1 - \cos^2\theta$.]

56. Let (r, θ) be any point of the polar coordinate curve $r = f(\theta)$, where $r \geq 0$, and let

$$\mathbf{u}_r = (\cos \theta)\mathbf{i} + (\sin \theta)\mathbf{j} \qquad \mathbf{u}_\theta = (-\sin \theta)\mathbf{i} + (\cos \theta)\mathbf{j}$$

(a) Show that $\mathbf{u}_r$ and $\mathbf{u}_\theta$ are unit vectors.
(b) Explain why $\mathbf{u}_r$ and $\mathbf{u}_\theta$ are orthogonal.
(c) Show that $\mathbf{u}_r$ has the same direction as the ray from the origin to (r, θ) and that $\mathbf{u}_\theta$ is 90° counterclockwise from $\mathbf{u}_r$.

57. Let $\mathbf{u}_1$, $\mathbf{u}_2$, and $\mathbf{u}_3$ be noncoplanar vectors. Let $\mathbf{w}_1 = \mathbf{u}_1/\|\mathbf{u}_1\|$, $\mathbf{v}_2 = \mathbf{u}_2 - (\mathbf{u}_2 \cdot \mathbf{w}_1)\mathbf{w}_1$, $\mathbf{w}_2 = \mathbf{v}_2/\|\mathbf{v}_2\|$, $\mathbf{v}_3 = \mathbf{u}_3 - (\mathbf{u}_3 \cdot \mathbf{w}_1)\mathbf{w}_1 - (\mathbf{u}_3 \cdot \mathbf{w}_2)\mathbf{w}_2$, and $\mathbf{w}_3 = \mathbf{v}_3/\|\mathbf{v}_3\|$. Show that $\mathbf{w}_1$, $\mathbf{w}_2$, and $\mathbf{w}_3$ are mutually orthogonal unit vectors. (This is the *Gram–Schmidt orthogonalization process*.)

58. Use the procedure of Problem 57 to transform $-\mathbf{i}+\mathbf{j}$, $2\mathbf{i}+\mathbf{k}$, and $3\mathbf{i}-\mathbf{j}+2\mathbf{k}$ into a set of mutually orthogonal unit vectors.

59. Find an example of three nonzero vectors **a**, **b**, and **c** for which $\mathbf{b} \neq \mathbf{c}$, neither **b** nor **c** is orthogonal to **a**, and $\mathbf{a} \cdot \mathbf{b} = \mathbf{a} \cdot \mathbf{c}$.

60. Provide the details of the following outline of an alternate proof of the Cauchy–Schwarz inequality: If **u** and **v** denote nonzero vectors, first show that

$$0 \leq \|\alpha\mathbf{u} + \beta\mathbf{v}\|^2 = \alpha^2\|\mathbf{u}\|^2 + 2\alpha\beta\mathbf{u}\cdot\mathbf{v} + \beta^2\|\mathbf{v}\|^2$$

for any scalars α and β. Then set $\alpha = \|\mathbf{v}\|$ and $\beta = -\|\mathbf{u}\|$ to get the desired result.

61. The dot product of two vectors $\mathbf{v} = v_1\mathbf{u}_1 + v_2\mathbf{u}_2 + \cdots + v_n\mathbf{u}_n$ and $\mathbf{w} = w_1\mathbf{u}_1 + w_2\mathbf{u}_2 + \cdots + w_n\mathbf{u}_n$ in n-space is defined as

$$\mathbf{v}\cdot\mathbf{w} = v_1w_1 + v_2w_2 + \cdots + v_nw_n$$

Show that the five properties listed in theorem (14.4.2) hold.

14.5 The Cross Product

The cross product of two vectors is of special interest for those studying physics, particularly mechanics and electricity. It is used, for example, to describe angular velocity, torque, and angular momentum. In geometry, the cross product of two vectors turns out to be a vector that is orthogonal to each of the two vectors.

We begin with an algebraic definition of the cross product and list some algebraic properties that follow from that definition. Later in the section we look at some of the geometric and physical properties of the cross product. As mentioned earlier, the cross product is only defined for vectors in space.

[14.5.1] DEFINITION / *Cross Product.*

If $\mathbf{v} = v_1\mathbf{i} + v_2\mathbf{j} + v_3\mathbf{k}$ and $\mathbf{w} = w_1\mathbf{i} + w_2\mathbf{j} + w_3\mathbf{k}$ are two vectors in space, the *cross product* $\mathbf{v} \times \mathbf{w}$ is defined as the vector

$$\mathbf{v} \times \mathbf{w} = (v_2w_3 - v_3w_2)\mathbf{i} - (v_1w_3 - v_3w_1)\mathbf{j} + (v_1w_2 - v_2w_1)\mathbf{k} \quad (1)$$

Note that the cross product $\mathbf{v} \times \mathbf{w}$ of two vectors is itself a vector. Because of this, the cross product is sometimes referred to as the *vector product*.

EXAMPLE 1

If $\mathbf{v} = 2\mathbf{i} - \mathbf{j} + \mathbf{k}$ and $\mathbf{w} = 4\mathbf{i} + 2\mathbf{j} - \mathbf{k}$, then

$$\begin{aligned}\mathbf{v} \times \mathbf{w} &= [(-1)(-1) - (1)(2)]\mathbf{i} - [(2)(-1) - (1)(4)]\mathbf{j} + [(2)(2) - (-1)(4)]\mathbf{k} \\ &= -\mathbf{i} + 6\mathbf{j} + 8\mathbf{k}\end{aligned}$$

Determinants may be used as an aid in remembering (1). A *second-order determinant* is defined as

$$\begin{vmatrix} v_1 & v_2 \\ w_1 & w_2 \end{vmatrix} = v_1w_2 - v_2w_1$$

A *third-order determinant* is defined as

$$\begin{vmatrix} A & B & C \\ v_1 & v_2 & v_3 \\ w_1 & w_2 & w_3 \end{vmatrix} = \begin{vmatrix} v_2 & v_3 \\ w_2 & w_3 \end{vmatrix} A - \begin{vmatrix} v_1 & v_3 \\ w_1 & w_3 \end{vmatrix} B + \begin{vmatrix} v_1 & v_2 \\ w_1 & w_2 \end{vmatrix} C \quad (2)$$

EXAMPLE 2

$$\begin{vmatrix} 2 & 3 \\ -1 & 2 \end{vmatrix} = (2)(2) - (-1)(3) = 4 + 3 = 7$$

and

$$\begin{vmatrix} A & B & C \\ 2 & 3 & 1 \\ -1 & 2 & -1 \end{vmatrix} = \begin{vmatrix} 3 & 1 \\ 2 & -1 \end{vmatrix} A - \begin{vmatrix} 2 & 1 \\ -1 & -1 \end{vmatrix} B + \begin{vmatrix} 2 & 3 \\ -1 & 2 \end{vmatrix} C$$

$$= (-3 - 2)A - (-2 + 1)B + (4 + 3)C = -5A + B + 7C$$

By using determinants, the cross product of the vectors $\mathbf{v} = v_1\mathbf{i} + v_2\mathbf{j} + v_3\mathbf{k}$ and $\mathbf{w} = w_1\mathbf{i} + w_2\mathbf{j} + w_3\mathbf{k}$ may be written symbolically as

$$\mathbf{v} \times \mathbf{w} = \begin{vmatrix} \mathbf{i} & \mathbf{j} & \mathbf{k} \\ v_1 & v_2 & v_3 \\ w_1 & w_2 & w_3 \end{vmatrix} = \begin{vmatrix} v_2 & v_3 \\ w_2 & w_3 \end{vmatrix} \mathbf{i} - \begin{vmatrix} v_1 & v_3 \\ w_1 & w_3 \end{vmatrix} \mathbf{j} + \begin{vmatrix} v_1 & v_2 \\ w_1 & w_2 \end{vmatrix} \mathbf{k}$$

$$= (v_2w_3 - v_3w_2)\mathbf{i} - (v_1w_3 - v_3w_1)\mathbf{j} + (v_1w_2 - v_2w_1)\mathbf{k} \qquad (3)$$

EXAMPLE 3

If $\mathbf{v} = 2\mathbf{i} - \mathbf{j} + \mathbf{k}$ and $\mathbf{w} = 4\mathbf{i} + 2\mathbf{j} - \mathbf{k}$, find: (a) $\mathbf{v} \times \mathbf{w}$ (b) $\mathbf{w} \times \mathbf{v}$

Solution

(a) $$\mathbf{v} \times \mathbf{w} = \begin{vmatrix} \mathbf{i} & \mathbf{j} & \mathbf{k} \\ 2 & -1 & 1 \\ 4 & 2 & -1 \end{vmatrix} = (1 - 2)\mathbf{i} - (-2 - 4)\mathbf{j} + (4 + 4)\mathbf{k}$$

$$= -\mathbf{i} + 6\mathbf{j} + 8\mathbf{k}$$

(b) $$\mathbf{w} \times \mathbf{v} = \begin{vmatrix} \mathbf{i} & \mathbf{j} & \mathbf{k} \\ 4 & 2 & -1 \\ 2 & -1 & 1 \end{vmatrix} = \mathbf{i} - 6\mathbf{j} - 8\mathbf{k}$$

The cross product has some interesting algebraic and geometric properties. We first list the algebraic properties.

[14.5.2] THEOREM / ***Properties of the Cross Product.***

If u, v, and w are vectors and if α is a scalar, then:

(a) $\mathbf{v} \times \mathbf{v} = \mathbf{0}$

(b) $\alpha(\mathbf{v} \times \mathbf{w}) = (\alpha\mathbf{v}) \times \mathbf{w} = \mathbf{v} \times (\alpha\mathbf{w})$

(c) $\mathbf{v} \times \mathbf{w} = -(\mathbf{w} \times \mathbf{v})$

(d) $\mathbf{v} \times (\mathbf{w} + \mathbf{u}) = (\mathbf{v} \times \mathbf{w}) + (\mathbf{v} \times \mathbf{u})$

(e) $\|\mathbf{v} \times \mathbf{w}\|^2 = \|\mathbf{v}\|^2\|\mathbf{w}\|^2 - (\mathbf{v} \cdot \mathbf{w})^2$

We derive parts (a), (c), and (e) below and leave the proofs of parts (b) and (d) as exercises (Problems 23 and 24).

Proof

(a) If $\mathbf{v} = v_1\mathbf{i} + v_2\mathbf{j} + v_3\mathbf{k}$, then

$$\mathbf{v} \times \mathbf{v} = (v_2v_3 - v_2v_3)\mathbf{i} - (v_1v_3 - v_1v_3)\mathbf{j} + (v_1v_2 - v_1v_2)\mathbf{k} = \mathbf{0}$$

(c) If $\mathbf{v} = v_1\mathbf{i} + v_2\mathbf{j} + v_3\mathbf{k}$ and $\mathbf{w} = w_1\mathbf{i} + w_2\mathbf{j} + w_3\mathbf{k}$, then

$$\mathbf{v} \times \mathbf{w} = (v_2w_3 - w_2v_3)\mathbf{i} - (v_1w_3 - w_1v_3)\mathbf{j} + (v_1w_2 - w_1v_2)\mathbf{k}$$

and

$$\mathbf{w} \times \mathbf{v} = (w_2v_3 - v_2w_3)\mathbf{i} - (w_1v_3 - v_1w_3)\mathbf{j} + (w_1v_2 - v_1w_2)\mathbf{k}$$

Consequently, $\mathbf{v} \times \mathbf{w} = -(\mathbf{w} \times \mathbf{v})$.

(e) If $\mathbf{v} = v_1\mathbf{i} + v_2\mathbf{j} + v_3\mathbf{k}$ and $\mathbf{w} = w_1\mathbf{i} + w_2\mathbf{j} + w_3\mathbf{k}$, then

$$\begin{aligned}\|\mathbf{v} \times \mathbf{w}\|^2 &= (v_2w_3 - w_2v_3)^2 + (v_1w_3 - w_1v_3)^2 + (v_1w_2 - w_1v_2)^2 \\ &= (v_2w_3)^2 - 2v_2w_2v_3w_3 + (w_2v_3)^2 + (v_1w_3)^2 \\ &\quad - 2v_1w_1v_3w_3 + (w_1v_3)^2 + (v_1w_2)^2 \\ &\quad - 2v_1w_1v_2w_2 + (w_1v_2)^2\end{aligned}$$

$$\begin{aligned}\|\mathbf{v}\|^2\|\mathbf{w}\|^2 - (\mathbf{v} \cdot \mathbf{w})^2 &= (v_1^2 + v_2^2 + v_3^2)(w_1^2 + w_2^2 + w_3^2) \\ &\quad - (v_1w_1 + v_2w_2 + v_3w_3)^2 \\ &= (v_1w_2)^2 + (v_1w_3)^2 + (v_2w_1)^2 + (v_2w_3)^2 \\ &\quad + (v_3w_1)^2 + (v_3w_2)^2 - 2v_1v_2w_1w_2 \\ &\quad - 2v_1v_3w_1w_3 - 2v_2v_3w_2w_3\end{aligned}$$

Reordering the terms on the right in both equations gives the same result.

We shall find the cross products of the unit vectors **i**, **j**, and **k** to be particularly useful:

$$\mathbf{i} \times \mathbf{j} = \begin{vmatrix} \mathbf{i} & \mathbf{j} & \mathbf{k} \\ 1 & 0 & 0 \\ 0 & 1 & 0 \end{vmatrix} = 0\mathbf{i} + 0\mathbf{j} + \mathbf{k} = \mathbf{k}$$

Similarly,

$$\mathbf{j} \times \mathbf{k} = \begin{vmatrix} \mathbf{i} & \mathbf{j} & \mathbf{k} \\ 0 & 1 & 0 \\ 0 & 0 & 1 \end{vmatrix} = \mathbf{i} \qquad \mathbf{k} \times \mathbf{i} = \begin{vmatrix} \mathbf{i} & \mathbf{j} & \mathbf{k} \\ 0 & 0 & 1 \\ 1 & 0 & 0 \end{vmatrix} = \mathbf{j}$$

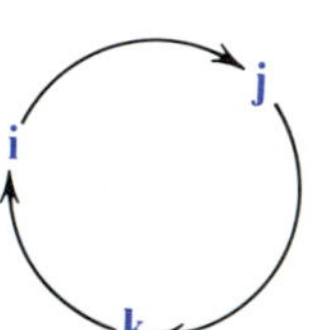

Figure 36

Note the cyclic pattern for the cross products of **i**, **j**, and **k**:

$$\mathbf{i} \times \mathbf{j} = \mathbf{k} \qquad \mathbf{j} \times \mathbf{k} = \mathbf{i} \qquad \mathbf{k} \times \mathbf{i} = \mathbf{j}$$

Figure 36 may be helpful in remembering this pattern.

We list the remaining cross products involving **i**, **j**, and **k**; they are a consequence of properties (a) and (c) of theorem (14.5.2).

$$\mathbf{j} \times \mathbf{i} = -\mathbf{k} \qquad \mathbf{k} \times \mathbf{j} = -\mathbf{i} \qquad \mathbf{i} \times \mathbf{k} = -\mathbf{j}$$

$$\mathbf{i} \times \mathbf{i} = \mathbf{0} \qquad \mathbf{j} \times \mathbf{j} = \mathbf{0} \qquad \mathbf{k} \times \mathbf{k} = \mathbf{0}$$

Geometric Properties of the Cross Product

The geometric properties of the cross product are summarized in the following theorem.

[14.5.3] THEOREM

If v and w are vectors, then:

(a) $\mathbf{v} \times \mathbf{w}$ **is orthogonal to both v and w**
(b) $\|\mathbf{v} \times \mathbf{w}\| = \|\mathbf{v}\|\|\mathbf{w}\|\sin\theta$, **where** θ **is the angle between** $\mathbf{v} \neq \mathbf{0}$ **and** $\mathbf{w} \neq \mathbf{0}$
(c) $\|\mathbf{v} \times \mathbf{w}\|$ **is the area of the parallelogram having** $\mathbf{v} \neq \mathbf{0}$ **and** $\mathbf{w} \neq \mathbf{0}$ **as adjacent sides**
(d) $\mathbf{v} \times \mathbf{w} = \mathbf{0}$ **if and only if v and w are parallel and** $\mathbf{v} \neq \mathbf{0}, \mathbf{w} \neq \mathbf{0}$

Proof of Part (a) Let $\mathbf{v} = v_1\mathbf{i} + v_2\mathbf{j} + v_3\mathbf{k}$ and $\mathbf{w} = w_1\mathbf{i} + w_2\mathbf{j} + w_3\mathbf{k}$. Then

$$\mathbf{v} \times \mathbf{w} = (v_2w_3 - w_2v_3)\mathbf{i} - (v_1w_3 - w_1v_3)\mathbf{j} + (v_1w_2 - w_1v_2)\mathbf{k}$$

so that

$$\mathbf{v} \cdot (\mathbf{v} \times \mathbf{w}) = v_1v_2w_3 - v_1w_2v_3 - v_2v_1w_3 + v_2w_1v_3 + v_3v_1w_2 - v_3w_1v_2 = 0$$

and

$$\mathbf{w} \cdot (\mathbf{v} \times \mathbf{w}) = w_1v_2w_3 - w_1w_2v_3 - w_2v_1w_3 + w_2w_1v_3 + w_3v_1w_2 - w_3w_1v_2 = 0$$

Thus, $\mathbf{v} \times \mathbf{w}$ is orthogonal to $\mathbf{v}$, and $\mathbf{v} \times \mathbf{w}$ is orthogonal to $\mathbf{w}$.

The vector $\mathbf{v} \times \mathbf{w}$ is normal to the plane containing $\mathbf{v}$ and $\mathbf{w}$. As Figure 37 illustrates, there are two vectors that are normal to the plane containing $\mathbf{v}$ and $\mathbf{w}$. If the plane is represented by this page, one such vector is directed up and the other is directed down. Which of these is the direction of $\mathbf{v} \times \mathbf{w}$? It can be shown that the direction of $\mathbf{v} \times \mathbf{w}$ is determined by the thumb of the right hand when the other fingers of the right hand are cupped so that they point in the direction of the angle θ from $\mathbf{v}$ to $\mathbf{w}$ (see Fig. 38).

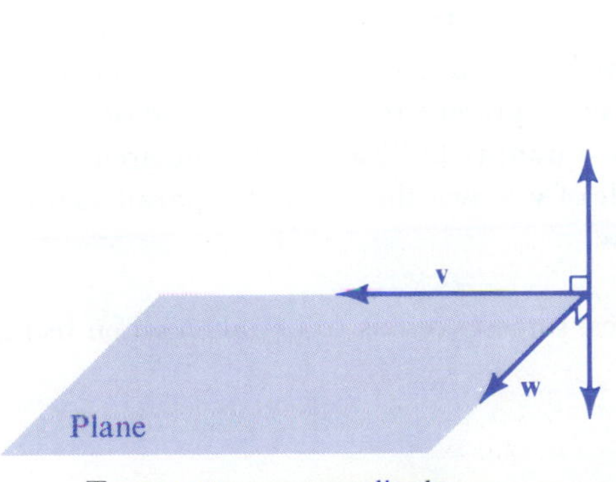

Figure 37

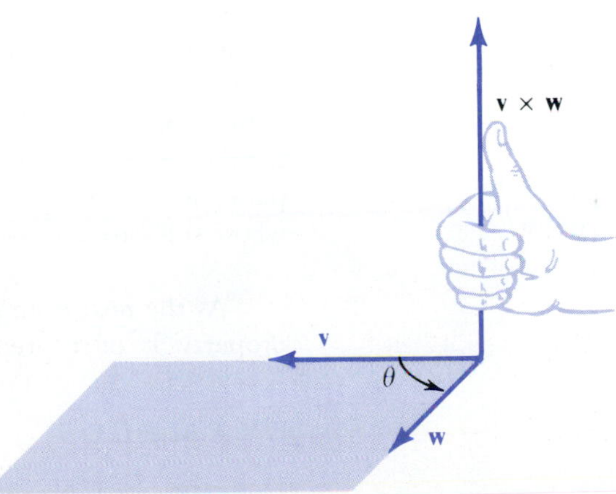

Figure 38

As in Section 14.1, this is usually referred to as the *right-hand rule* and the coordinate system we have been using is called *right-handed.*

EXAMPLE 4

Find a vector orthogonal to each of the vectors $\mathbf{v} = 2\mathbf{i} - \mathbf{j} + \mathbf{k}$ and $\mathbf{w} = 4\mathbf{i} + 2\mathbf{j} - \mathbf{k}$.

Solution

Based on the preceding discussion, such a vector is

$$\mathbf{v} \times \mathbf{w} \underset{\text{By Example 3}}{\uparrow{=}} -\mathbf{i} + 6\mathbf{j} + 8\mathbf{k}$$

This may be verified as follows:

$$\mathbf{v} \cdot (\mathbf{v} \times \mathbf{w}) = (2\mathbf{i} - \mathbf{j} + \mathbf{k}) \cdot (-\mathbf{i} + 6\mathbf{j} + 8\mathbf{k}) = -2 - 6 + 8 = 0$$

and

$$\mathbf{w} \cdot (\mathbf{v} \times \mathbf{w}) = (4\mathbf{i} + 2\mathbf{j} - \mathbf{k}) \cdot (-\mathbf{i} + 6\mathbf{j} + 8\mathbf{k}) = -4 + 12 - 8 = 0$$

Hence, $\mathbf{v} \times \mathbf{w}$ is orthogonal to both $\mathbf{v}$ and $\mathbf{w}$.

Proof of Part (b) of Theorem (14.5.3) Let θ be the angle between the nonzero vectors $\mathbf{v}$ and $\mathbf{w}$. Then, by part (e) of theorem (14.5.2), we have

$$\|\mathbf{v} \times \mathbf{w}\|^2 = \|\mathbf{v}\|^2\|\mathbf{w}\|^2 - (\mathbf{v} \cdot \mathbf{w})^2$$

Since $\mathbf{v} \cdot \mathbf{w} = \|\mathbf{v}\|\|\mathbf{w}\|\cos\theta$, we find

$$\|\mathbf{v} \times \mathbf{w}\|^2 = \|\mathbf{v}\|^2\|\mathbf{w}\|^2 - \|\mathbf{v}\|^2\|\mathbf{w}\|^2\cos^2\theta = \|\mathbf{v}\|^2\|\mathbf{w}\|^2(1 - \cos^2\theta)$$
$$= \|\mathbf{v}\|^2\|\mathbf{w}\|^2\sin^2\theta$$

By taking square roots, we have the result

$$\|\mathbf{v} \times \mathbf{w}\| = \|\mathbf{v}\|\|\mathbf{w}\|\sin\theta$$

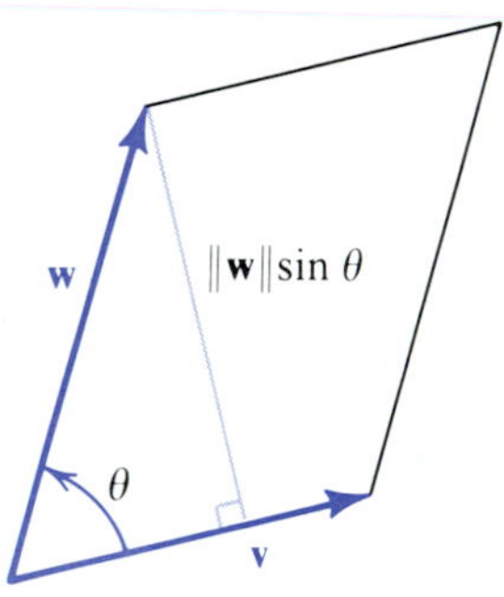

Figure 39
Area = Base × Height
$= \|\mathbf{v}\|\|\mathbf{w}\|\sin\theta$

Proof of Part (c) of Theorem (14.5.2) Part (b) of theorem (14.5.3) is used to prove part (c). If $\mathbf{v}$ and $\mathbf{w}$ are nonzero vectors and θ is the angle between $\mathbf{v}$ and $\mathbf{w}$, then $\|\mathbf{v}\|$ and $\|\mathbf{w}\|$ represent the lengths of the sides of a parallelogram (see Fig. 39) and the quantity $\|\mathbf{v}\|\|\mathbf{w}\|\sin\theta$ is the area of the parallelogram. Thus, the magnitude of $\mathbf{v} \times \mathbf{w}$ is the area of the parallelogram whose sides are the vectors $\mathbf{v}$ and $\mathbf{w}$.

As the next example illustrates, some care has to be taken when using property (c) of theorem (14.5.3).

EXAMPLE 5

Find the area of the parallelogram whose vertices are $P_1 = (0, 0, 0)$, $P_2 = (-1, 2, 4)$, $P_3 = (2, -1, 4)$, and $P_4 = (1, 1, 8)$.

Solution

The area of the parallelogram is $\|\mathbf{v} \times \mathbf{w}\|$, where $\mathbf{v}$ and $\mathbf{w}$ are two adjacent sides of the parallelogram. We may pick

$$\mathbf{v} = \overrightarrow{P_1P_2} = -\mathbf{i} + 2\mathbf{j} + 4\mathbf{k} \qquad \text{and} \qquad \mathbf{w} = \overrightarrow{P_1P_3} = 2\mathbf{i} - \mathbf{j} + 4\mathbf{k}$$

Be careful! Not all pairs of vertices give a side. For example, $\overrightarrow{P_1P_4}$ is not a side; it is a diagonal of the parallelogram, since $\overrightarrow{P_1P_2} + \overrightarrow{P_1P_3} = \overrightarrow{P_1P_4}$. Now

$$\mathbf{v} \times \mathbf{w} = (-\mathbf{i} + 2\mathbf{j} + 4\mathbf{k}) \times (2\mathbf{i} - \mathbf{j} + 4\mathbf{k}) = 12\mathbf{i} + 12\mathbf{j} - 3\mathbf{k}$$

So, the area of the parallelogram is

$$\|\mathbf{v} \times \mathbf{w}\| = \sqrt{144 + 144 + 9} = \sqrt{297} = 3\sqrt{33} \approx 17.23 \text{ square units}$$

Proof of Part (d) of Theorem (14.5.3) If $\mathbf{v}$ and $\mathbf{w}$ are parallel vectors $(\sin\theta = 0)$, it follows from part (b) of theorem (14.5.3) that $\|\mathbf{v} \times \mathbf{w}\| = 0$ or $\mathbf{v} \times \mathbf{w} = \mathbf{0}$. On the other hand, if $\mathbf{v} \times \mathbf{w} = \mathbf{0}$, then

$$\|\mathbf{v}\| = 0 \qquad \text{or} \qquad \|\mathbf{w}\| = 0 \qquad \text{or} \qquad \sin\theta = 0$$

In the latter case, $\theta = 0$ or $\theta = \pi$ and the vectors $\mathbf{v}$ and $\mathbf{w}$ are parallel. If $\mathbf{v}$ or $\mathbf{w}$ is the zero vector, then, since the zero vector has no specific direction, we adopt the convention that the zero vector is parallel to every vector.* Thus, we have the result we seek.

We use part (d) of theorem (14.5.3) as a criterion for parallel vectors. For example, the vectors $\mathbf{v} = 2\mathbf{i} + \mathbf{j} - \mathbf{k}$ and $\mathbf{w} = -4\mathbf{i} - 2\mathbf{j} + 2\mathbf{k}$ are parallel, since

$$\mathbf{v} \times \mathbf{w} = \begin{vmatrix} \mathbf{i} & \mathbf{j} & \mathbf{k} \\ 2 & 1 & -1 \\ -4 & -2 & 2 \end{vmatrix} = \begin{vmatrix} 1 & -1 \\ -2 & 2 \end{vmatrix}\mathbf{i} - \begin{vmatrix} 2 & -1 \\ -4 & 2 \end{vmatrix}\mathbf{j} + \begin{vmatrix} 2 & 1 \\ -4 & -2 \end{vmatrix}\mathbf{k}$$

$$= 0\mathbf{i} + 0\mathbf{j} + 0\mathbf{k} = \mathbf{0}$$

For these vectors $\mathbf{v}$ and $\mathbf{w}$, we note that $\mathbf{v} = -2\mathbf{w}$. Thus, the magnitude of $\mathbf{v}$ is twice that of $\mathbf{w}$, and $\mathbf{v}$ and $\mathbf{w}$ are opposite in direction.

Angular Velocity

Figure 40 illustrates a rigid body that is rotating about a fixed axis l with a constant angular speed ω.† The *angular velocity* $\boldsymbol{\omega}$ is defined as the vector of magnitude ω whose direction is parallel to the axis l, so that if the fingers of the right hand are wrapped about l in the direction of the rotation, the thumb will point in the direction of $\boldsymbol{\omega}$.

---

* With this convention, the zero vector is both parallel and perpendicular to every vector. This apparent contradiction will cause us no trouble, however, because the angle between two vectors is never applied when one of the vectors is the zero vector.

† Think of a phonograph record (rigid body) on a spindle (axis) rotating at $\omega = 45$ rpm. Its angular speed ω is $\omega = (45 \text{ revolutions/minute})(2\pi \text{ radians/revolution}) = 90\pi$ radians/minute.

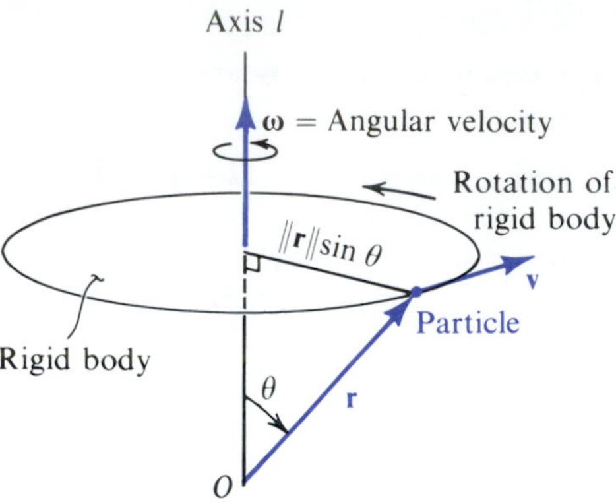

Figure 40

We are interested in obtaining a formula for the velocity $\mathbf{v}$ of a particle on the rigid body. To fix our ideas, let the origin O be on the axis l of rotation and let $\mathbf{r}$ be the position vector of the particle. Since the motion is circular and $\|\mathbf{r}\|\sin\theta$ equals the distance of the particle from the axis l, it follows that

$$\|\mathbf{v}\| = \omega\|\mathbf{r}\|\sin\theta$$

Next, since the velocity $\mathbf{v}$ is necessarily perpendicular to both $\mathbf{r}$ and $\boldsymbol{\omega}$, it follows that $\mathbf{v}$ is parallel to $\boldsymbol{\omega} \times \mathbf{r}$. By the right-hand rule used to define $\boldsymbol{\omega}$, $\mathbf{v}$ is in the direction of $\boldsymbol{\omega} \times \mathbf{r}$, rather than $\mathbf{r} \times \boldsymbol{\omega}$. So, by part (b) of theorem (14.5.3),

$$\|\boldsymbol{\omega} \times \mathbf{r}\| = \|\boldsymbol{\omega}\|\,\|\mathbf{r}\|\sin\theta = \|\mathbf{v}\|$$

It follows that

$$\mathbf{v} = \boldsymbol{\omega} \times \mathbf{r} \tag{4}$$

EXAMPLE 6

A rigid body rotates with constant angular speed ω radians per second about a line parallel to $2\mathbf{i} + \mathbf{j} + 2\mathbf{k}$. Find the speed of a particle on this body at the instant it passes through the point $(1, 3, 5)$. Assume the distance is in meters.

Solution

The vector $2\mathbf{i} + \mathbf{j} + 2\mathbf{k}$ is parallel to the axis l of rotation of the rigid body and hence is parallel to $\boldsymbol{\omega}$. Since $\frac{2}{3}\mathbf{i} + \frac{1}{3}\mathbf{j} + \frac{2}{3}\mathbf{k}$ is a unit vector parallel to l, we have $\boldsymbol{\omega} = \pm\omega(\frac{2}{3}\mathbf{i} + \frac{1}{3}\mathbf{j} + \frac{2}{3}\mathbf{k})$. (We use a $\pm$ sign, since the problem gives no information about the direction of $\boldsymbol{\omega}$.)

The position vector of the particle is $\mathbf{r} = \mathbf{i} + 3\mathbf{j} + 5\mathbf{k}$. Hence, from (4), the velocity $\mathbf{v}$ of the particle is

$$\mathbf{v} = \boldsymbol{\omega} \times \mathbf{r} = \pm\omega\begin{vmatrix} \mathbf{i} & \mathbf{j} & \mathbf{k} \\ \frac{2}{3} & \frac{1}{3} & \frac{2}{3} \\ 1 & 3 & 5 \end{vmatrix} = \pm\omega(-\tfrac{1}{3}\mathbf{i} - \tfrac{8}{3}\mathbf{j} + \tfrac{5}{3}\mathbf{k})$$

The speed v of the particle is

$$v = \|\mathbf{v}\| = \omega\sqrt{\tfrac{1}{9} + \tfrac{64}{9} + \tfrac{25}{9}} = \sqrt{10}\,\omega \text{ meters per second}$$

Triple Scalar Product

If $\mathbf{u}$, $\mathbf{v}$, and $\mathbf{w}$ are three vectors in space, the dot product of $\mathbf{u}$ and $\mathbf{v} \times \mathbf{w}$—namely,

$$\mathbf{u} \cdot (\mathbf{v} \times \mathbf{w})$$

is called the *triple scalar product.*

The triple scalar product is not too difficult to compute if we use determinants.

[14.5.4] THEOREM

If $\mathbf{u} = u_1\mathbf{i} + u_2\mathbf{j} + u_3\mathbf{k}$, $\mathbf{v} = v_1\mathbf{i} + v_2\mathbf{j} + v_3\mathbf{k}$, and $\mathbf{w} = w_1\mathbf{i} + w_2\mathbf{j} + w_3\mathbf{k}$ are three vectors, the triple scalar product $\mathbf{u} \cdot (\mathbf{v} \times \mathbf{w})$ is given by

$$\mathbf{u} \cdot (\mathbf{v} \times \mathbf{w}) = \begin{vmatrix} u_1 & u_2 & u_3 \\ v_1 & v_2 & v_3 \\ w_1 & w_2 & w_3 \end{vmatrix} \tag{5}$$

We leave the proof as an exercise (Problem 44).

EXAMPLE 7

Find $\mathbf{u} \cdot (\mathbf{v} \times \mathbf{w})$ if $\mathbf{u} = \mathbf{i} + \mathbf{j} + \mathbf{k}$, $\mathbf{v} = 2\mathbf{i} - 3\mathbf{j} + 2\mathbf{k}$, and $\mathbf{w} = \mathbf{i} - 2\mathbf{j} - \mathbf{k}$.

Solution

Using (5) we obtain

$$\mathbf{u} \cdot (\mathbf{v} \times \mathbf{w}) = \begin{vmatrix} 1 & 1 & 1 \\ 2 & -3 & 2 \\ 1 & -2 & -1 \end{vmatrix} = \begin{vmatrix} -3 & 2 \\ -2 & -1 \end{vmatrix}(1) - \begin{vmatrix} 2 & 2 \\ 1 & -1 \end{vmatrix}(1) + \begin{vmatrix} 2 & -3 \\ 1 & -2 \end{vmatrix}(1)$$
$$= 7 - (-4) + (-1) = 10$$

The triple scalar product has an interesting geometric application. Consider the parallepiped shown in Figure 41 on page 834 that has the vectors $\mathbf{u}$, $\mathbf{v}$, and $\mathbf{w}$ as adjacent sides. The area of the base parallelogram formed by $\mathbf{v}$ and $\mathbf{w}$ has area $\|\mathbf{v} \times \mathbf{w}\|$. If θ is the angle between $\mathbf{u}$ and $\mathbf{v} \times \mathbf{w}$, then the height of the parallepiped will be $\|\mathbf{u}\||\cos\theta|$. (We use $|\cos\theta|$ in case $\cos\theta < 0$, which will occur if $\pi/2 < \theta < \pi$.) The volume V of the parallepiped is

$$V = (\text{Height})(\text{Area of base}) = \|\mathbf{u}\||\cos\theta|\|\mathbf{v} \times \mathbf{w}\| = |\mathbf{u} \cdot (\mathbf{v} \times \mathbf{w})|$$

[14.5.5] THEOREM

The volume V of a parallepiped whose adjacent sides are the vectors u, v, and w is

$$V = |\mathbf{u} \cdot (\mathbf{v} \times \mathbf{w})|$$

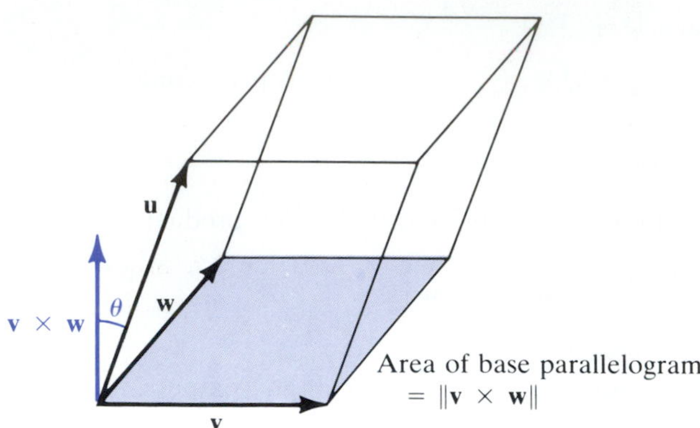

Figure 41

EXERCISE 14.5

In Problems 1–10 compute the cross product $\mathbf{v} \times \mathbf{w}$. Check your answer by showing that $\mathbf{v}$ and $\mathbf{w}$ are each orthogonal to $\mathbf{v} \times \mathbf{w}$.

1. $\mathbf{v} = 2\mathbf{i} + \mathbf{j} - \mathbf{k}, \quad \mathbf{w} = \mathbf{i} - \mathbf{j} + \mathbf{k}$

2. $\mathbf{v} = 4\mathbf{i} - \mathbf{j} + 2\mathbf{k}, \quad \mathbf{w} = 2\mathbf{i} + \mathbf{j} + \mathbf{k}$

3. $\mathbf{v} = \mathbf{i} + \mathbf{j}, \quad \mathbf{w} = \mathbf{i} - \mathbf{j}$

4. $\mathbf{v} = \mathbf{j} - \mathbf{k}, \quad \mathbf{w} = \mathbf{i} - \mathbf{j}$

5. $\mathbf{v} = 3\mathbf{i} - 2\mathbf{j} + \mathbf{k}, \quad \mathbf{w} = \mathbf{i} + \mathbf{j}$

6. $\mathbf{v} = 2\mathbf{i} - \mathbf{j}, \quad \mathbf{w} = \mathbf{i} + \mathbf{j} - 3\mathbf{k}$

7. $\mathbf{v} = -\mathbf{i} + 8\mathbf{j} + 3\mathbf{k}, \quad \mathbf{w} = 7\mathbf{i} + 2\mathbf{j}$

8. $\mathbf{v} = 2\mathbf{j} - \mathbf{k}, \quad \mathbf{w} = -3\mathbf{i} + \mathbf{j} + \mathbf{k}$

9. $\mathbf{v} = 2\mathbf{i} + 3\mathbf{j} - 4\mathbf{k}, \quad \mathbf{w} = -\mathbf{i} + \mathbf{j} - 4\mathbf{k}$

10. $\mathbf{v} = (\cos\theta)\mathbf{i} + (\sin\theta)\mathbf{j}, \quad \mathbf{w} = (\sin\theta)\mathbf{i} + (\cos\theta)\mathbf{j}$

In Problems 11–16 find the area of the parallelogram with one corner at P and sides PQ and PR.

11. $P = (1, -3, 7)$; $Q = (2, 1, 1)$; $R = (6, -1, 2)$

12. $P = (0, 1, 1)$; $Q = (2, 0, -4)$; $R = (-3, -2, 1)$

13. $P = (-2, 1, 6)$; $Q = (2, 1, -7)$; $R = (4, 1, 1)$

14. $P = (0, 0, 3)$; $Q = (2, -5, 3)$; $R = (1, 1, -2)$

15. $P = (1, 1, -6)$; $Q = (5, -3, 0)$; $R = (-2, 4, 1)$

16. $P = (-4, 6, 3)$; $Q = (1, 1, -5)$; $R = (2, 2, 2)$

In Problems 17–20 find the area of the parallelogram whose vertices are P_1, P_2, P_3, and P_4.

17. $P_1 = (0, 0, 0)$; $P_2 = (1, 2, 3)$; $P_3 = (2, -1, 1)$; $P_4 = (3, 1, 4)$

18. $P_1 = (0, 0, 0)$; $P_2 = (-1, 2, 0)$; $P_3 = (2, 3, -4)$; $P_4 = (1, 5, -4)$

19. $P_1 = (1, 2, -1)$; $P_2 = (4, 2, -3)$; $P_3 = (6, -5, 2)$; $P_4 = (9, -5, 0)$

20. $P_1 = (-1, 1, 1)$; $P_2 = (-1, 2, 2)$; $P_3 = (-3, 4, -5)$; $P_4 = (-3, 5, -4)$

21. Show that the area of the triangle whose vertices are the endpoints of the vectors $\mathbf{u}$, $\mathbf{v}$, and $\mathbf{w}$ is

$$A = \tfrac{1}{2}\|(\mathbf{v} - \mathbf{u}) \times (\mathbf{w} - \mathbf{u})\|$$

22. Use the result of Problem 21 to find the area of the triangle with vertices $(0, 0, 0)$, $(2, 3, -2)$, and $(-1, 1, 4)$.

23. Derive part (b) of theorem (14.5.2).

24. Derive part (d) of theorem (14.5.2).

25. Give an example to show that the cross product is not associative; that is, find vectors $\mathbf{u}$, $\mathbf{v}$, and $\mathbf{w}$ so that $\mathbf{u} \times (\mathbf{v} \times \mathbf{w}) \neq (\mathbf{u} \times \mathbf{v}) \times \mathbf{w}$.

26. If $\mathbf{v} \times \mathbf{w} = \mathbf{0}$ and $\mathbf{v} \cdot \mathbf{w} = 0$, can you draw any conclusions about $\mathbf{v}$ or $\mathbf{w}$?

27. Find a unit vector normal to the plane containing $\mathbf{v} = 2\mathbf{i} - 6\mathbf{j} - 3\mathbf{k}$ and $\mathbf{w} = 4\mathbf{i} + 3\mathbf{j} - \mathbf{k}$.

28. Find a unit vector normal to the plane containing $\mathbf{v} = \mathbf{i} + \mathbf{j} - 2\mathbf{k}$ and $\mathbf{w} = 3\mathbf{i} + 2\mathbf{j} - \mathbf{k}$.

29. A rigid body rotates about an axis through the origin with a constant angular speed of 30 radians per second. If the angular velocity $\boldsymbol{\omega}$ points in the direction of $\mathbf{i} + \mathbf{j} + \mathbf{k}$, find the speed of a particle at the instant it passes through the point $(-1, 2, 3)$. Assume the distance scale is in meters.

30. For the rigid body in Problem 29, at what points will the speed be 60 meters per second?

31. A rigid body rotates with constant angular speed ω about a line parallel to $3\mathbf{i} + \mathbf{j} - 2\mathbf{k}$. Find the speed of a particle at the instant that it passes through the point $(4, 4, 0)$. Assume that distance is measured in meters. Find ω if the speed of the particle at $(4, 4, 0)$ is $8\sqrt{14}$ meters per second.

32. Show that if $\mathbf{u}$ and $\mathbf{v}$ are orthogonal vectors, then $\|\mathbf{u} \times \mathbf{v}\| = \|\mathbf{u}\|\|\mathbf{v}\|$. If, in addition, $\mathbf{u}$ and $\mathbf{v}$ are unit vectors, show that $\mathbf{u} \times \mathbf{v}$ is, too.

33. Show that three nonzero vectors $\mathbf{u}$, $\mathbf{v}$, and $\mathbf{w}$ lie in the same plane if and only if $\mathbf{u} \cdot (\mathbf{v} \times \mathbf{w}) = 0$.

34. Show that $\mathbf{u} \cdot (\mathbf{v} \times \mathbf{w}) = (\mathbf{u} \times \mathbf{v}) \cdot \mathbf{w}$.

35. Show that $\mathbf{u} \cdot (\mathbf{v} \times \mathbf{w}) = \mathbf{v} \cdot (\mathbf{w} \times \mathbf{u}) = \mathbf{w} \cdot (\mathbf{u} \times \mathbf{v})$.

36. If $\mathbf{u}$, $\mathbf{v}$, and $\mathbf{w}$ are three vectors, the expression $\mathbf{u} \times (\mathbf{v} \times \mathbf{w})$ is called the *triple vector product*. Show that

$$\mathbf{u} \times (\mathbf{v} \times \mathbf{w}) = (\mathbf{u} \cdot \mathbf{w})\mathbf{v} - (\mathbf{u} \cdot \mathbf{v})\mathbf{w}$$

37. Show that

$$\mathbf{u} \times (\mathbf{v} \times \mathbf{w}) + \mathbf{v} \times (\mathbf{w} \times \mathbf{u}) + \mathbf{w} \times (\mathbf{u} \times \mathbf{v}) = \mathbf{0}$$

38. If $\mathbf{v}$, $\mathbf{w}$, $\mathbf{u}$ and $\mathbf{v}'$, $\mathbf{w}'$, $\mathbf{u}'$ are such that

$$\mathbf{v}' \cdot \mathbf{v} = \mathbf{w}' \cdot \mathbf{w} = \mathbf{u}' \cdot \mathbf{u} = 1$$

$$\mathbf{v}' \cdot \mathbf{w} = \mathbf{v}' \cdot \mathbf{u} = \mathbf{w}' \cdot \mathbf{v} = \mathbf{w}' \cdot \mathbf{u} = \mathbf{u}' \cdot \mathbf{v} = \mathbf{u}' \cdot \mathbf{w} = 0$$

prove that

$$\mathbf{v}' = \frac{\mathbf{w} \times \mathbf{u}}{\mathbf{v} \cdot \mathbf{w} \times \mathbf{u}} \qquad \mathbf{w}' = \frac{\mathbf{u} \times \mathbf{v}}{\mathbf{v} \cdot \mathbf{w} \times \mathbf{u}} \qquad \mathbf{u}' = \frac{\mathbf{v} \times \mathbf{w}}{\mathbf{v} \cdot \mathbf{w} \times \mathbf{u}}$$

39. Prove Lagrange's identity:
$(\mathbf{a} \times \mathbf{b}) \cdot (\mathbf{c} \times \mathbf{d}) = (\mathbf{a} \cdot \mathbf{c})(\mathbf{b} \cdot \mathbf{d}) - (\mathbf{a} \cdot \mathbf{d})(\mathbf{b} \cdot \mathbf{c})$

40. Prove that:
$(\mathbf{a} \times \mathbf{b}) \times (\mathbf{c} \times \mathbf{d}) = [\mathbf{a} \cdot (\mathbf{b} \times \mathbf{d})]\mathbf{c} - [\mathbf{a} \cdot (\mathbf{b} \times \mathbf{c})]\mathbf{d}$

In Problems 41 and 42 use vector methods to prove each statement.

41. The diagonals of a parallelogram are perpendicular if and only if the parallelogram is a rhombus.

42. The altitudes of a triangle meet at one point; the medians of a triangle meet at one point.

43. Solve for $\mathbf{x}$ in terms of α, $\mathbf{a}$, and $\mathbf{b}$ if

$$\alpha\mathbf{x} + \mathbf{x} \times \mathbf{a} = \mathbf{b} \qquad \alpha \neq 0$$

[*Hint:* First find $\mathbf{x} \cdot \mathbf{a}$, then $\mathbf{x} \times \mathbf{a}$.]

44. Prove theorem (14.5.4).

Lines in Space 14.6

We now take up the question of finding the vector equation of a line in space. Let $P_1 = (x_1, y_1, z_1)$ and $P_2 = (x_2, y_2, z_2)$ be two distinct points on a line L. The vector $\mathbf{D}$ represented by the directed line segment $\overrightarrow{P_1P_2}$ is a nonzero vector parallel to L (see Fig. 42).

Let $\mathbf{r}_1$ denote the position vector of P_1 and let $\mathbf{r}$ denote the position vector of any point $P = (x, y, z)$. If we insist that P is on L, then the vector $\mathbf{r} - \mathbf{r}_1$ must be parallel to the vector $\mathbf{D}$, so that

$$\mathbf{r} - \mathbf{r}_1 = t\mathbf{D} \qquad \text{for some scalar } t$$

$$\mathbf{r} = \mathbf{r}_1 + t\mathbf{D}$$

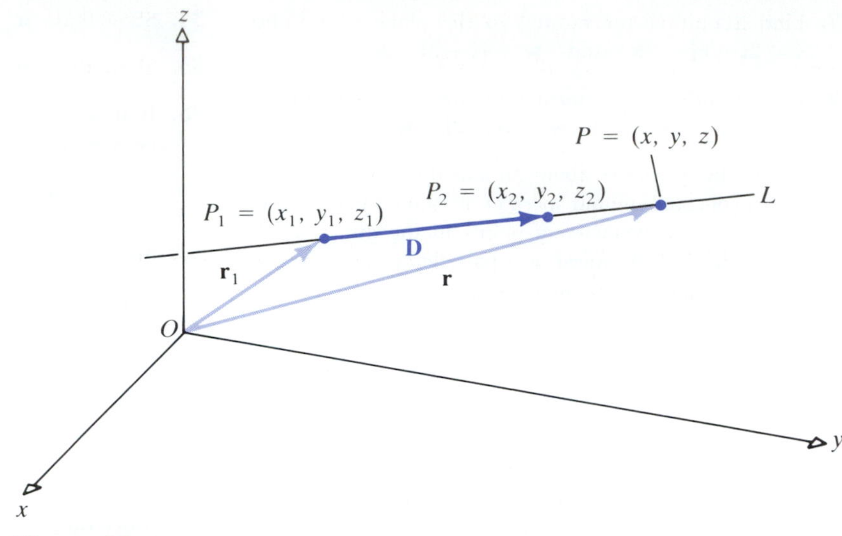

Figure 42

[14.6.1] THEOREM / *Vector Equation of a Line.*

The vector equation of a line L parallel to $\mathbf{D} \neq \mathbf{0}$ and passing through the point $P_1 = (x_1, y_1, z_1)$ is

$$\mathbf{r} = \mathbf{r}_1 + t\mathbf{D} \tag{1}$$

where r_1 is the position vector of P_1 and r is the position vector of any point P on L.

In the vector equation (1), the vector $\mathbf{r}_1$ is known (it is the position vector of the point P_1 on L) and the vector $\mathbf{D}$ is known (it is any nonzero vector parallel to L). To locate any point P on L, we simply assign an appropriate number to the scalar t. For each number t, we obtain the position vector of a point on L.

EXAMPLE 1

Find the vector equation of a line L passing through the two points $(1, 2, -1)$ and $(4, 3, -2)$.

Solution

The vector $\mathbf{D}$ in the direction from $(1, 2, -1)$ to $(4, 3, -2)$ is

$$\mathbf{D} = 3\mathbf{i} + \mathbf{j} - \mathbf{k}$$

The vector equation of L is

$$\mathbf{r} = \mathbf{r}_1 + t\mathbf{D} = \mathbf{i} + 2\mathbf{j} - \mathbf{k} + t(3\mathbf{i} + \mathbf{j} - \mathbf{k}) = (1 + 3t)\mathbf{i} + (2 + t)\mathbf{j} + (-1 - t)\mathbf{k}$$

Parametric Equations of a Line

The vector equation (1) of a line L parallel to $\mathbf{D} = a\mathbf{i} + b\mathbf{j} + c\mathbf{k}$ and passing through the point $P_1 = (x_1, y_1, z_1)$ may be written as

$$\mathbf{r} = \mathbf{r}_1 + t\mathbf{D}$$

$$x\mathbf{i} + y\mathbf{j} + z\mathbf{k} = (x_1\mathbf{i} + y_1\mathbf{j} + z_1\mathbf{k}) + t(a\mathbf{i} + b\mathbf{j} + c\mathbf{k})$$

Since two vectors are equal if and only if corresponding components are equal, we find

$$x = x_1 + at \qquad y = y_1 + bt \qquad z = z_1 + ct \tag{2}$$

These equations are called *parametric equations of the line L*, and the variable t is a *parameter*. By assigning values to the parameter t, we obtain points (x, y, z) on L. For example, when $t = 0$, we obtain the point (x_1, y_1, z_1) on L.

EXAMPLE 2

Find the parametric equations of a line L containing the point $(2, -3, 1)$ and parallel to $4\mathbf{i} + \frac{3}{4}\mathbf{j} - \mathbf{k}$.

Solution

Set $(x_1, y_1, z_1) = (2, -3, 1)$ and $a\mathbf{i} + b\mathbf{j} + c\mathbf{k} = 4\mathbf{i} + \frac{3}{4}\mathbf{j} - \mathbf{k}$. Then, by substituting in (2), the parametric equations of L are

$$x = 2 + 4t \qquad y = -3 + \tfrac{3}{4}t \qquad z = 1 - t$$

Symmetric Equations of a Line

In the parametric equations (2), if the numbers a, b, and c (the components of the vector $\mathbf{D}$) are each nonzero, we may solve for t, obtaining

$$t = \frac{x - x_1}{a} = \frac{y - y_1}{b} = \frac{z - z_1}{c}$$

By dropping the t, we have

$$\frac{x - x_1}{a} = \frac{y - y_1}{b} = \frac{z - z_1}{c} \tag{3}$$

These equations are referred to as *symmetric equations of the line L*.

EXAMPLE 3

Find symmetric equations of the line L passing through the point $(1, -1, 2)$ and parallel to $5\mathbf{i} - 2\mathbf{j} + 3\mathbf{k}$.

Solution

By using (3) with $a = 5$, $b = -2$, $c = 3$, $x_1 = 1$, $y_1 = -1$, and $z_1 = 2$, we get

$$\frac{x - 1}{5} = \frac{y + 1}{-2} = \frac{z - 2}{3}$$

Symmetric equations of a line are not unique. For example, since the vector $-10\mathbf{i} + 4\mathbf{j} - 6\mathbf{k}$ is parallel to $5\mathbf{i} - 2\mathbf{j} + 3\mathbf{k}$, we can write symmetric equations of the line described in Example 3 as

$$\frac{x-1}{-10} = \frac{y+1}{4} = \frac{z-2}{-6}$$

EXAMPLE 4

Find symmetric equations for the line L containing the points

$$P_1 = (3, -2, 1) \qquad \text{and} \qquad P_2 = (1, -5, 2)$$

Solution

In order to use (3) we must find a vector $\mathbf{D}$ parallel to L. Since P_1 and P_2 are distinct points lying on L, the directed line segment $\overrightarrow{P_1P_2}$ can be used for $\mathbf{D}$. Therefore, $\mathbf{D} = -2\mathbf{i} - 3\mathbf{j} + \mathbf{k}$. If we use $P_1 = (3, -2, 1)$ in (3), we obtain the symmetric equations

$$\frac{x-3}{-2} = \frac{y+2}{-3} = \frac{z-1}{1}$$

If we use $P_2 = (1, -5, 2)$, instead of P_1, we get another form of the symmetric equations—namely,

$$\frac{x-1}{-2} = \frac{y+5}{-3} = \frac{z-2}{1}$$

Either representation is correct.

In (2), if one of the numbers a, b, or c equals 0, we may still write symmetric equations for L. For example, if $a = 0$, but $b \neq 0$ and $c \neq 0$, we rearrange (2) as

$$x = x_1 \qquad \frac{y-y_1}{b} = \frac{z-z_1}{c}$$

and call these *symmetric equations of the line L*. This particular line lies in the plane $x = x_1$.

EXAMPLE 5

Find the symmetric equations of the line L that contains the point $(5, -2, 3)$ and is parallel to $3\mathbf{j} - 2\mathbf{k}$.

Solution

In this case, $a = 0$, $b = 3$, and $c = -2$ in (2). Hence, the symmetric equations of L are

$$x = 5 \qquad \frac{y+2}{3} = \frac{z-3}{-2}$$

EXAMPLE 6

Let

$$\frac{x-6}{3} = \frac{y+2}{1} = \frac{z+3}{-2}$$

be symmetric equations of the line L. Find a vector parallel to L and find two points on L.

Solution

The denominators give the numbers $a = 3$, $b = 1$, and $c = -2$ in (3). Therefore, a vector parallel to L is $3\mathbf{i} + \mathbf{j} - 2\mathbf{k}$. We also know from (3) that $P_1 = (6, -2, -3)$ lies on L. To obtain a second point, we arbitrarily assign x a value—say, $x = 0$—to obtain

$$\frac{0-6}{3} = \frac{y+2}{1} = \frac{z+3}{-2}$$

from which we find $y = -4$, $z = 1$. Hence, another point on L is $(0, -4, 1)$.

EXAMPLE 7

Find symmetric equations of the line passing through the point $(1, 0, -1)$ that is perpendicular to each of the lines

$$\frac{x+1}{-1} = \frac{y-2}{2} = \frac{z+3}{1} \qquad \text{and} \qquad \frac{x}{3} = \frac{y-1}{1} = \frac{z+2}{-2}$$

Solution

The two lines are parallel, respectively, to the vectors

$$\mathbf{v} = -\mathbf{i} + 2\mathbf{j} + \mathbf{k} \qquad \text{and} \qquad \mathbf{w} = 3\mathbf{i} + \mathbf{j} - 2\mathbf{k}$$

A vector orthogonal to both $\mathbf{v}$ and $\mathbf{w}$ is

$$\mathbf{v} \times \mathbf{w} = \begin{vmatrix} \mathbf{i} & \mathbf{j} & \mathbf{k} \\ -1 & 2 & 1 \\ 3 & 1 & -2 \end{vmatrix} = -5\mathbf{i} + \mathbf{j} - 7\mathbf{k}$$

The line parallel to this vector and passing through the point $(1, 0, -1)$ is given by the symmetric equations

$$\frac{x-1}{-5} = \frac{y}{1} = \frac{z+1}{-7}$$

Skew Lines

In space, two distinct lines intersect, are parallel, or are *skew*. They *intersect* when they have exactly one point in common. They are *parallel* when they lie in a plane and do not intersect. They are *skew* if they do not intersect and are not parallel. See Figure 43 on page 840.

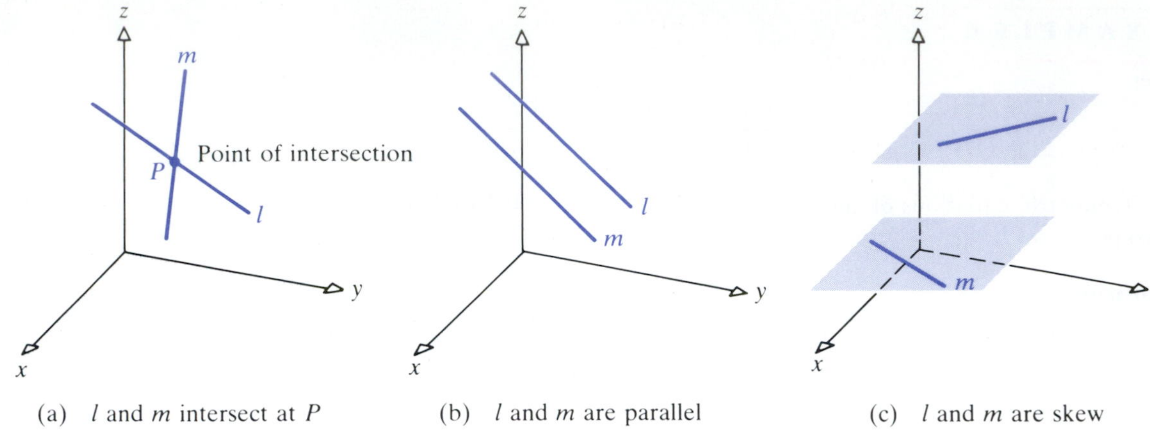

(a) l and m intersect at P (b) l and m are parallel (c) l and m are skew

Figure 43

As an example of skew lines, think of the path of two airplanes—one at an altitude of 1000 meters traveling north and the other at an altitude of 3000 meters traveling east. The next example illustrates a method for determining whether a given pair of lines are skew, parallel, or intersecting.

EXAMPLE 8

Determine whether the lines given below intersect, are parallel, or are skew.

$$l: \quad \mathbf{r} = (3\mathbf{i} + 2\mathbf{j} + \mathbf{k}) + t(\mathbf{i} - \mathbf{j} + \mathbf{k}) = (3 + t)\mathbf{i} + (2 - t)\mathbf{j} + (1 + t)\mathbf{k}$$

$$m: \quad \mathbf{R} = (5\mathbf{i} + 6\mathbf{j} + \mathbf{k}) + T(\mathbf{i} - 4\mathbf{j} + 2\mathbf{k}) = (5 + T)\mathbf{i} + (6 - 4T)\mathbf{j} + (1 + 2T)\mathbf{k}$$

Solution

The line l is parallel to the vector $\mathbf{i} - \mathbf{j} + \mathbf{k}$ and the line m is parallel to the vector $\mathbf{i} - 4\mathbf{j} + 2\mathbf{k}$. Since these vectors are not parallel, l and m either intersect or are skew. To see which is the case, we suppose they intersect. In order for this to occur, some value of the parameter t and some value of the parameter T must give a point (position vector $\mathbf{r}$) on l that is the same as a point (position vector $\mathbf{R}$) on m. We set $\mathbf{r} = \mathbf{R}$, obtaining

$$(3 + t)\mathbf{i} + (2 - t)\mathbf{j} + (1 + t)\mathbf{k} = (5 + T)\mathbf{i} + (6 - 4T)\mathbf{j} + (1 + 2T)\mathbf{k}$$

Then, by equating the components of each vector, we obtain

$$3 + t = 5 + T \qquad 2 - t = 6 - 4T \qquad 1 + t = 1 + 2T$$

From the last of these conditions, $t = 2T$. Using this in the first two conditions, we find

$$3 + 2T = 5 + T \qquad 2 - 2T = 6 - 4T$$

$$T = 2 \qquad\qquad T = 2$$

Thus, $t = 4$ and $T = 2$. The point of intersection can be found by setting $t = 4$ in l or $T = 2$ in m. The result is $\mathbf{r} = \mathbf{R} = 7\mathbf{i} - 2\mathbf{j} + 5\mathbf{k}$. The point of intersection is $(7, -2, 5)$.

The next example illustrates the situation when two lines are skew.

EXAMPLE 9

Show that the lines given below are skew.

$$l: \quad \mathbf{r} = (3\mathbf{i} + 2\mathbf{j} + \mathbf{k}) + t(\mathbf{i} - \mathbf{j} + \mathbf{k})$$

$$m: \quad \mathbf{R} = (5\mathbf{i} + 6\mathbf{j} + \mathbf{k}) + T(\mathbf{i} - \mathbf{j} + 2\mathbf{k})$$

Solution

The line l is parallel to the vector $\mathbf{i} - \mathbf{j} + \mathbf{k}$ and the line m is parallel to the vector $\mathbf{i} - \mathbf{j} + 2\mathbf{k}$. Since these vectors are not parallel, l and m either intersect or are skew. As in Example 8, we set $\mathbf{r} = \mathbf{R}$, obtaining

$$3 + t = 5 + T \qquad 2 - t = 6 - T \qquad 1 + t = 1 + 2T$$

From the last of these, $t = 2T$. Then

$$3 + 2T = 5 + T \qquad 2 - 2T = 6 - T$$

$$T = 2 \qquad\qquad T = -4$$

Thus, l and m have at least two points in common. This is impossible, since l and m are distinct lines. (Why?) Consequently, l and m are skew.

EXERCISE 14.6

In Problems 1–8 find the vector equation, parametric equations, and symmetric equations of the line described.

1. Passing through the point $(1, 2, 3)$ and parallel to $2\mathbf{i} - \mathbf{j} + \mathbf{k}$
2. Passing through the point $(-1, 1, 5)$ and parallel to $\mathbf{i} + \mathbf{j} - \mathbf{k}$
3. Passing through the point $(4, -1, 6)$ and parallel to $\mathbf{i} + \mathbf{j}$
4. Passing through the point $(3, 2, -1)$ and parallel to $\mathbf{j} - \mathbf{k}$
5. Passing through the points $(1, -1, 3)$ and $(4, 2, 1)$
6. Passing through the points $(-2, 3, 0)$ and $(1, -1, 2)$
7. Passing through the points $(0, 0, 1)$ and $(2, 3, 1)$
8. Passing through the points $(0, 0, 0)$ and $(1, 2, -1)$

In Problems 9–12 determine the point of intersection and the cosine of the angle between the lines l and m.

9. l: $\mathbf{r} = (2 - t)\mathbf{i} + (4 + 2t)\mathbf{j} + (-5 + t)\mathbf{k}$
   m: $\mathbf{R} = (4 - T)\mathbf{i} + (3 + T)\mathbf{j} + (-13 + 3T)\mathbf{k}$
10. l: $\mathbf{r} = (5 + 2t)\mathbf{i} + (6 - t)\mathbf{j} + 2t\mathbf{k}$
    m: $\mathbf{R} = (7 + 3T)\mathbf{i} + (5 - 2T)\mathbf{j} + (2 - T)\mathbf{k}$
11. l: $\mathbf{r} = t\mathbf{i} + (1 + 2t)\mathbf{j} + (-3 + t)\mathbf{k}$
    m: $\mathbf{R} = (-3 + T)\mathbf{i} + (1 - 4T)\mathbf{j} + (2 - 7T)\mathbf{k}$
12. l: $\mathbf{r} = (2 - 3t)\mathbf{i} + 6t\mathbf{j} + (-2 + 5t)\mathbf{k}$
    m: $\mathbf{R} = -2T\mathbf{i} + (1 + T)\mathbf{j} + 2T\mathbf{k}$

In Problems 13–16 find symmetric equations of the line passing through the point P that is perpendicular to the two directions given.

13. $P = (0, 0, 0)$; $\mathbf{a} = 2\mathbf{i} + 3\mathbf{j} - 2\mathbf{k}$; $\mathbf{b} = 3\mathbf{i} + \mathbf{j} + 2\mathbf{k}$
14. $P = (0, 0, 0)$; $\mathbf{a} = 4\mathbf{i} + 2\mathbf{j} + \mathbf{k}$; $\mathbf{b} = 5\mathbf{i} + 3\mathbf{j} + \mathbf{k}$
15. $P = (1, 2, -1)$; $\mathbf{a} = 2\mathbf{i} + 4\mathbf{j} - 2\mathbf{k}$; $\mathbf{b} = -3\mathbf{i} - 2\mathbf{j} + \mathbf{k}$
16. $P = (-1, 3, 2)$; $\mathbf{a} = -2\mathbf{i} + 2\mathbf{j} - 3\mathbf{k}$; $\mathbf{b} = 4\mathbf{i} + 2\mathbf{j} + \mathbf{k}$

In Problems 17–27 determine whether the lines l and m intersect, are parallel, or are skew.

17. l: $\mathbf{r} = (2 - 3t)\mathbf{i} + 6t\mathbf{j} + (-2 + 6t)\mathbf{k}$
m: $\mathbf{R} = (6 - T)\mathbf{i} + (2 + 2T)\mathbf{j} + (5 + 2T)\mathbf{k}$

18. l: $\mathbf{r} = (3 + 3t)\mathbf{i} + 6t\mathbf{j} + (3 - 2t)\mathbf{k}$
m: $\mathbf{R} = (3 - 2T)\mathbf{i} + 4T\mathbf{j} + (3 + 7T)\mathbf{k}$

19. l: $\mathbf{r} = (4 + t)\mathbf{i} + (3 - t)\mathbf{j} + 6t\mathbf{k}$
m: $\mathbf{R} = (4 + T)\mathbf{i} + (3 - T)\mathbf{j} + (2 - 2T)\mathbf{k}$

20. l: $\mathbf{r} = (2 - 2t)\mathbf{i} + (7 + 8t)\mathbf{j} - 6t\mathbf{k}$
m: $\mathbf{R} = (6 + T)\mathbf{i} + (-5 - 4T)\mathbf{j} + 3T\mathbf{k}$

21. l: $\mathbf{r} = (5 + 2t)\mathbf{i} + (6 - t)\mathbf{j} + (8 - t)\mathbf{k}$
m: $\mathbf{R} = (4 + T)\mathbf{i} + T\mathbf{j} + (2 + T)\mathbf{k}$

22. l: $\mathbf{r} = (2 - t)\mathbf{i} + t\mathbf{j} + (1 - t)\mathbf{k}$
m: $\mathbf{R} = (6 + T)\mathbf{i} + (-4 + T)\mathbf{j} + T\mathbf{k}$

23. l: $\dfrac{x-3}{2} = \dfrac{y+2}{3} = \dfrac{z-1}{4}$
m: $\dfrac{x+4}{-4} = \dfrac{y-3}{-6} = \dfrac{z+4}{-8}$

24. l: $\dfrac{x}{3} = \dfrac{y-2}{4} = \dfrac{z+4}{1}$
m: $\dfrac{x-6}{3} \neq \dfrac{y+2}{4} = \dfrac{z-3}{2}$

25. l: $\dfrac{x+1}{5} = \dfrac{y-2}{4} = \dfrac{z-3}{-3}$
m: $\dfrac{x+1}{6} = \dfrac{y-2}{3} = \dfrac{z+3}{2}$

26. l: $\dfrac{x+5}{6} = \dfrac{y-2}{3} = \dfrac{z+4}{-1}$
m: $\dfrac{x}{1} = \dfrac{y-2}{3} = \dfrac{z-8}{2}$

27. l: $\dfrac{x+5}{6} = \dfrac{y-2}{3} = \dfrac{z+4}{7}$
m: $\dfrac{x+5}{6} = \dfrac{y+1}{3} = \dfrac{z-2}{7}$

28. Write the parametric equations of a line passing through the point $(-1, 5, 6)$ and parallel to the line

$$\frac{x+1}{5} = \frac{y-2}{4} = \frac{z-3}{-3}$$

29. Write the parametric equations of a line passing through the point $(1, -2, -3)$ and parallel to the line

$$\frac{x+1}{6} = \frac{y+2}{2} = \frac{z}{-1}$$

30. What property can you assign to the lines l and m, given below, if $a_1b_1 + a_2b_2 + a_3b_3 = 0$?

$$l: \quad \frac{x-x_1}{a_1} = \frac{y-y_1}{a_2} = \frac{z-z_1}{a_3}$$

$$m: \quad \frac{x-x_1}{b_1} = \frac{y-y_1}{b_2} = \frac{z-z_1}{b_3}$$

31. What property can you assign to the lines l and m, given below, if $a_1/b_1 = a_2/b_2 = a_3/b_3$ and $(x_1, y_1, z_1) \neq (x_2, y_2, z_2)$?

$$l: \quad \frac{x-x_1}{a_1} = \frac{y-y_1}{a_2} = \frac{z-z_1}{a_3}$$

$$m: \quad \frac{x-x_2}{b_1} = \frac{y-y_2}{b_2} = \frac{z-z_2}{b_3}$$

32. Let

$$\frac{x-4}{2} = \frac{y+1}{-1} = \frac{z-2}{2}$$

be symmetric equations of the line L. Find a vector parallel to L, and find two points on L.

33. Let

$$x + 1 = y + 3 = \frac{z+4}{2}$$

be symmetric equations of the line L. Find a vector parallel to L, and find two points on L.

34. Find symmetric equations of the line passing through the centers of the spheres

$$x^2 + y^2 + z^2 - 2x - 4y + 4z = 8$$

and

$$x^2 + y^2 + z^2 + 2x + 6y + 4z = 20$$

35. Two unidentified flying objects are at $(t, -t, 1 - t)$ and $(t - 3, 2t, 4t - 1)$ at time t.

(a) Describe the paths of the objects.
(b) Find where the paths intersect (or determine that they don't). Will the objects collide?
(c) Find the acute angle between the paths.

36. Find the line perpendicular to the lines

$$l: \quad x = 1 - t, \quad y = t, \quad z = 2t - 1$$

and

$$m: \quad x = t + 1, \quad y = -t, \quad z = t - 1$$

at their point of intersection. Why is this line parallel to the xy-plane?

Planes

14.7

Recall that the vector equation of a line is determined once a point on the line and a vector parallel to the line are known. To determine the vector equation of a plane, we use a point on the plane and a vector perpendicular to the plane. A vector that is perpendicular to a plane is called a *normal* to the plane.

Suppose the point $P_1 = (x_1, y_1, z_1)$ is on a plane and suppose the nonzero vector $\mathbf{N}$ is normal to the plane (see Fig. 44). Let $P = (x, y, z)$ be any point on the plane. Denote by $\mathbf{r}_1$ and $\mathbf{r}$ the position of the points P_1 and P, respectively. Since we are insisting that P is a point on the plane, the vector $\mathbf{r} - \mathbf{r}_1$ will always be orthogonal to the vector $\mathbf{N}$; that is, $(\mathbf{r} - \mathbf{r}_1) \cdot \mathbf{N} = 0$.

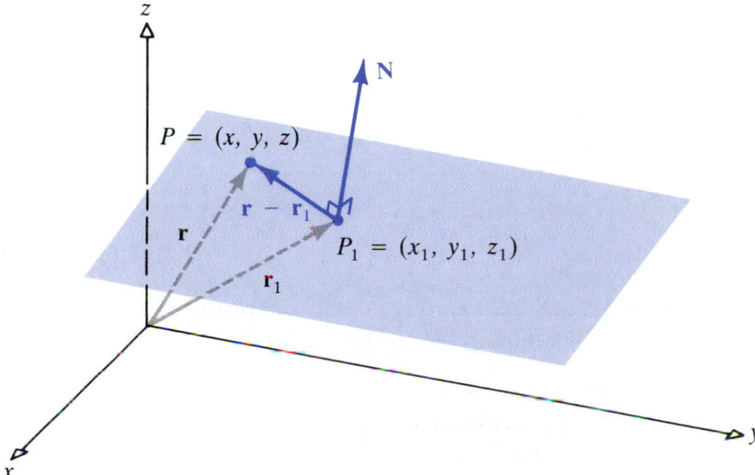

Figure 44

[14.7.1] THEOREM / *Vector Equation of a Plane.*

The vector equation of the plane containing the point P_1 is

$$(\mathbf{r} - \mathbf{r}_1) \cdot \mathbf{N} = 0 \tag{1}$$

where $\mathbf{r}_1$ denotes the position vector of P_1, $\mathbf{r}$ is the position vector of any point P on the plane, and $\mathbf{N}$ is a normal to the plane.

If, in the above, the coordinates of P_1 are (x_1, y_1, z_1) and the vector $\mathbf{N} = A\mathbf{i} + B\mathbf{j} + C\mathbf{k}$, then

$$[(x - x_1)\mathbf{i} + (y - y_1)\mathbf{j} + (z - z_1)\mathbf{k}] \cdot (A\mathbf{i} + B\mathbf{j} + C\mathbf{k}) = 0$$

$$A(x - x_1) + B(y - y_1) + C(z - z_1) = 0 \tag{2}$$

or

$$Ax + By + Cz = D$$

where $D = Ax_1 + By_1 + Cz_1$. This equation is called the *general equation of a plane*. Notice that the coefficients of x, y, and z in the general equation are the components of the normal vector **N**. The converse is also true.

[14.7.2] THEOREM / *General Equation of a Plane.*

Any equation of the form

$$Ax + By + Cz = D \tag{3}$$

with at least one of the numbers A, B, C not 0, is the equation of a plane, and the vector $A\mathbf{i} + B\mathbf{j} + C\mathbf{k}$ is a normal to this plane.

EXAMPLE 1

Find the general equation of the plane containing the point $(1, 2, -1)$ if the vector $2\mathbf{i} + 3\mathbf{j} - 4\mathbf{k}$ is a normal to it.

Solution

Here a point in the plane is $(x_1, y_1, z_1) = (1, 2, -1)$ and a vector normal to the plane is $\mathbf{N} = 2\mathbf{i} + 3\mathbf{j} - 4\mathbf{k}$. Using (2) the equation of the plane is

$$2(x - 1) + 3(y - 2) - 4(z + 1) = 0$$

which simplifies to $2x + 3y - 4z = 12$. Figure 45 depicts the plane given by $2x + 3y - 4z = 12$. Note the use of the intercepts to locate three points on the plane; these are useful for graphing it.

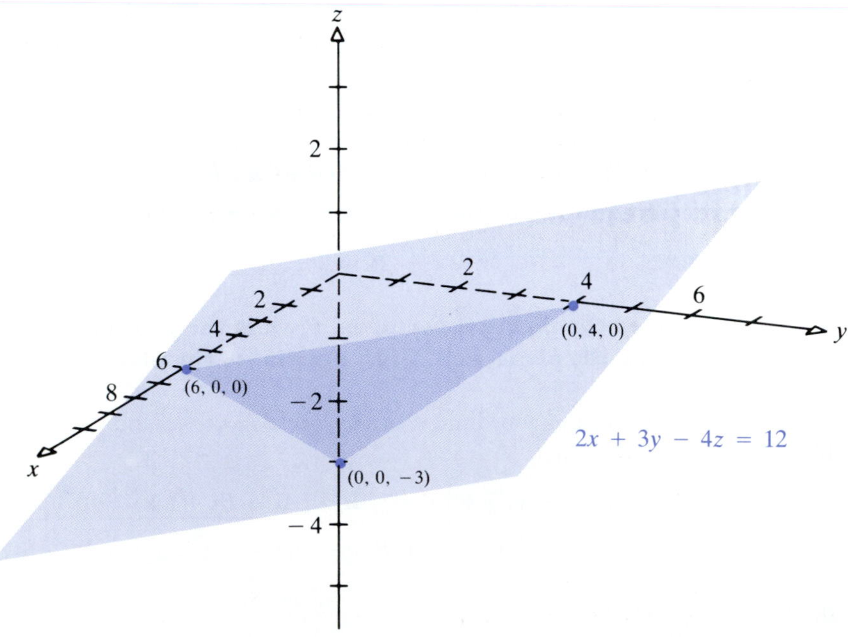

Figure 45

The next example illustrates the use of (3) to find the equation of the plane through three noncollinear points.

EXAMPLE 2

Find the general equation of the plane determined by the points $P_1 = (1, -1, 2)$, $P_2 = (3, 0, 0)$, and $P_3 = (4, 2, 1)$.

Solution

The vectors $\mathbf{v} = \overrightarrow{P_1P_2} = 2\mathbf{i} + \mathbf{j} - 2\mathbf{k}$ and $\mathbf{w} = \overrightarrow{P_1P_3} = 3\mathbf{i} + 3\mathbf{j} - \mathbf{k}$ lie in the plane. The vector

$$\mathbf{N} = \mathbf{v} \times \mathbf{w} = \begin{vmatrix} \mathbf{i} & \mathbf{j} & \mathbf{k} \\ 2 & 1 & -2 \\ 3 & 3 & -1 \end{vmatrix} = \begin{vmatrix} 1 & -2 \\ 3 & -1 \end{vmatrix}\mathbf{i} - \begin{vmatrix} 2 & -2 \\ 3 & -1 \end{vmatrix}\mathbf{j} + \begin{vmatrix} 2 & 1 \\ 3 & 3 \end{vmatrix}\mathbf{k} = 5\mathbf{i} - 4\mathbf{j} + 3\mathbf{k}$$

is orthogonal to both $\mathbf{v}$ and $\mathbf{w}$ and, thus, is normal to the plane. By using this normal vector and the point $P_1 = (1, -1, 2)$ (we could have used P_2 or P_3), we obtain the equation

$$5(x - 1) - 4(y + 1) + 3(z - 2) = 0$$

$$5x - 4y + 3z = 15$$

Parallel Planes; Angle Between Planes

Two distinct planes are *parallel* if they have parallel normals. If two planes have parallel normals and a point in common, they are *identical*. For example, the planes

$$p_1\colon\ 2x + 3y - z = 3 \qquad p_2\colon\ 2x + 3y - z = 4$$

have parallel normals. Since these planes cannot have any points in common (why?), the plane p_1 and p_2 are parallel. The plane p_3: $4x + 6y - 2z = 6$ is identical to the plane p_1. (Why?)

If two planes have nonparallel normals, they *intersect*. In this case, the *angle θ between the planes* is defined as the nonobtuse angle between the normals (see Fig. 46). Depending on the choice of normals, there are actually two angles between the normals. Since these angles are always supplementary, there is no confusion in agreeing to name the nonobtuse angle as the angle between the planes. This is the angle whose cosine is nonnegative.

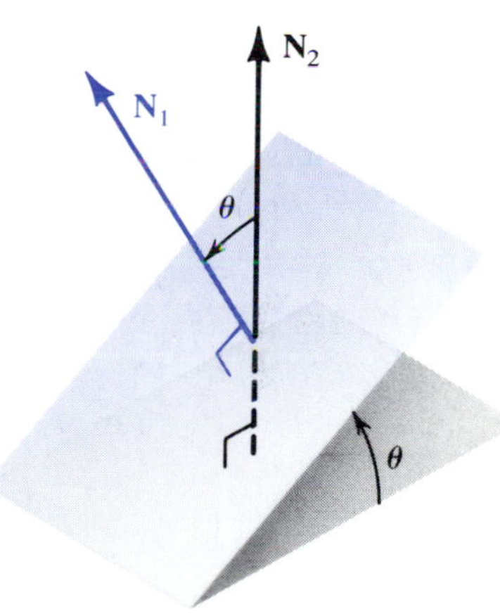

Figure 46

[14.7.3] THEOREM / *Angle Between Planes.*

If $\mathbf{N}_1$ and $\mathbf{N}_2$ are normals of two intersecting planes, the angle θ between these planes obeys

$$\cos\theta = \frac{|\mathbf{N}_1 \cdot \mathbf{N}_2|}{\|\mathbf{N}_1\|\|\mathbf{N}_2\|} \qquad 0 \le \theta \le \frac{\pi}{2} \tag{4}$$

EXAMPLE 3

Find the angle between the planes

$$p_1\colon\ 2x - 3y + 6z = 5 \qquad p_2\colon\ 3x + y + 2z = 6$$

Solution

First, we note that the normals $\mathbf{N}_1$ and $\mathbf{N}_2$ to p_1 and p_2, respectively, can be chosen to be

$$\mathbf{N}_1 = 2\mathbf{i} - 3\mathbf{j} + 6\mathbf{k} \qquad \mathbf{N}_2 = 3\mathbf{i} + \mathbf{j} + 2\mathbf{k}$$

Since the normals are not parallel, the planes intersect. The cosine of the angle θ between p_1 and p_2 is

$$\cos\theta = \frac{|\mathbf{N}_1 \cdot \mathbf{N}_2|}{\|\mathbf{N}_1\|\|\mathbf{N}_2\|} = \frac{|6 - 3 + 12|}{\sqrt{4 + 9 + 36}\sqrt{9 + 1 + 4}} = \frac{15}{7\sqrt{14}} \approx 0.5727$$

Thus, $\theta \approx 0.96$ radian.

In Example 3, notice that we take the absolute value of $\mathbf{N}_1 \cdot \mathbf{N}_2$ to ensure that the nonobtuse angle between p_1 and p_2 is being calculated.

Distance from a Point to a Plane

To find the distance from a point P_0 to a plane, we use projections. As Figure 47 illustrates, if P_1 is any point on the plane itself, the distance from P_0 to the plane may be found by calculating the magnitude of the projection of $\overrightarrow{P_1P_0}$ along the normal $\mathbf{N}$ of the plane.

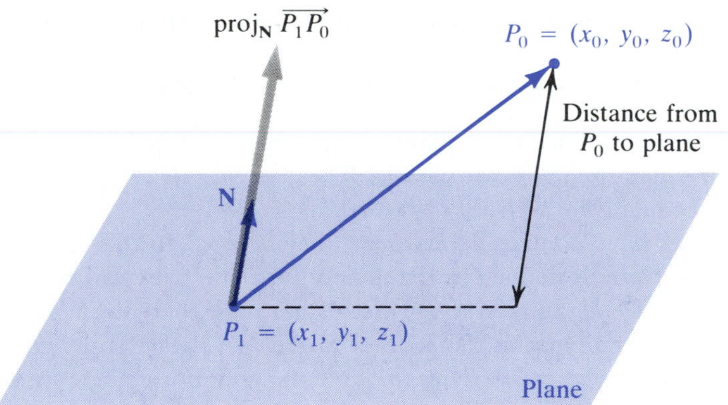

Figure 47

Let the plane be given by the equation

$$Ax + By + Cz = D$$

The distance from the point $P_0 = (x_0, y_0, z_0)$ to the plane is given by

$$\begin{bmatrix}\text{Distance from} \\ P_0 \text{ to plane}\end{bmatrix} = \|\text{proj}_{\mathbf{N}}\overrightarrow{P_1P_0}\| = \frac{|\overrightarrow{P_1P_0} \cdot \mathbf{N}|}{\|\mathbf{N}\|}$$

where $P_1 = (x_1, y_1, z_1)$ is a point on the plane. But $\mathbf{N} = A\mathbf{i} + B\mathbf{j} + C\mathbf{k}$

and $\overrightarrow{P_1P_0} = (x_0 - x_1)\mathbf{i} + (y_0 - y_1)\mathbf{j} + (z_0 - z_1)\mathbf{k}$. Hence,

$$\begin{bmatrix}\text{Distance from}\\ P_0 \text{ to plane}\end{bmatrix} = \frac{|A(x_0 - x_1) + B(y_0 - y_1) + C(z_0 - z_1)|}{\sqrt{A^2 + B^2 + C^2}}$$

$$= \frac{|(Ax_0 + By_0 + Cz_0) - (Ax_1 + By_1 + Cz_1)|}{\sqrt{A^2 + B^2 + C^2}} \underset{\uparrow}{=} \frac{|Ax_0 + By_0 + Cz_0 - D|}{\sqrt{A^2 + B^2 + C^2}}$$

P_1 is on the plane, so $Ax_1 + By_1 + Cz_1 = D$.

[14.7.4] THEOREM / *Distance from a Point to a Plane.*

The distance from the point $P_0 = (x_0, y_0, z_0)$ to the plane $Ax + By + Cz = D$ is

$$\frac{|Ax_0 + By_0 + Cz_0 - D|}{\sqrt{A^2 + B^2 + C^2}} \tag{5}$$

EXAMPLE 4

Find the distance from the point $(2, 3, -1)$ to the plane $x + 4y + z = 5$.

Solution

Here $A = 1$, $B = 4$, $C = 1$, $D = 5$, and $P_0 = (2, 3, -1)$. Thus, the distance is

$$\frac{|(1)(2) + (4)(3) + (1)(-1) - 5|}{\sqrt{1 + 16 + 1}} = \frac{8}{3\sqrt{2}} = \frac{4\sqrt{2}}{3}$$

EXERCISE 14.7

In Problems 1–16 find the general equation of the plane whose properties are stated.

1. Parallel to the xy-plane and 4 units above it
2. Parallel to the yz-plane and 2 units to the right of it
3. Parallel to the xz-plane and containing the point $(1, -2, 3)$
4. Parallel to the xy-plane and containing the point $(2, -3, 4)$
5. Containing the point $(1, -1, 2)$ and normal to $2\mathbf{i} - \mathbf{j} + \mathbf{k}$
6. Containing the point $(-3, 2, 1)$ and normal to $\mathbf{i} + \mathbf{j} - 2\mathbf{k}$
7. Containing the point $(0, 1, 5)$ and normal to $2\mathbf{i} + \mathbf{j} + 3\mathbf{k}$
8. Containing the point $(1, 0, -2)$ and normal to $-3\mathbf{i} + \mathbf{j} + \mathbf{k}$
9. Containing the point $(0, 5, -2)$ and parallel to the plane $x + 2y - z = 6$
10. Containing the point $(1, -2, 0)$ and parallel to the plane $2x - y + 3z = 10$
11. Containing the point $(10, 3, -4)$ and parallel to the plane $x - y + 3z = 5$
12. Containing the point $(-1, 2, 3)$ and parallel to the plane $2x + 3y - 4z = 8$
13. Containing the point $(2, 3, -1)$ and normal to the line $\dfrac{x-1}{2} = \dfrac{y-3}{5} = \dfrac{z+1}{-2}$
14. Containing the point $(-1, 2, 3)$ and normal to the line $\dfrac{x+5}{3} = \dfrac{y+2}{4} = \dfrac{z-4}{4}$

15. Containing the point $(0, 1, -2)$ and normal to the line $\dfrac{x-1}{2} = \dfrac{y+3}{3} = \dfrac{z-2}{2}$

16. Containing the point $(-1, 0, 4)$ and normal to the line $\dfrac{x+2}{3} = \dfrac{y+4}{2} = \dfrac{z-1}{1}$

17. Determine which of the points below lie on the plane $3x - 2y = 0$.

$(2, 3, 0) \quad (5, -1, 2) \quad (6, 0, 3) \quad (0, 0, 0)$

18. Determine which of the points below lie on the plane $2x + y - z = 2$.

$(0, 1, -1) \quad (1, 1, 1) \quad (3, 2, 1) \quad (-1, 1, 1)$

19. Find parametric equations of the line passing through the point $(1, 2, -1)$ and normal to the plane $2x - y + z = 6$.

20. Find parametric equations of the line passing through the point $(2, 3, -1)$ and normal to the plane $x + y - z = 3$.

21. Find the point of intersection of the plane $2x + y - z = 5$ and the line $\dfrac{x-1}{2} = \dfrac{y+3}{4} = \dfrac{z-1}{1}$.

22. Find the point of intersection of the plane $x + y - 2z = 8$ and the line $\dfrac{x+1}{2} = \dfrac{y-3}{1} = \dfrac{z-4}{-2}$.

23. Find the point of intersection of the plane $2x + 3y + z = 5$ and the line $\dfrac{x-3}{1} = \dfrac{y+4}{2} = \dfrac{z-1}{2}$.

24. Find the point of intersection of the plane $x + y - z = 3$ and the line $\dfrac{x+2}{2} = \dfrac{y-3}{1} = \dfrac{z}{2}$.

In Problems 25–30 find the general equation of the plane determined by the points P_1, P_2, and P_3.

25. $P_1 = (0, 0, 0)$; $P_2 = (1, 2, -1)$; $P_3 = (-1, 1, 0)$

26. $P_1 = (0, 0, 0)$; $P_2 = (3, -1, 2)$; $P_3 = (-3, 1, 0)$

27. $P_1 = (1, 2, 1)$; $P_2 = (3, 2, 2)$; $P_3 = (4, -1, -1)$

28. $P_1 = (-1, 2, 0)$; $P_2 = (3, 4, -1)$; $P_3 = (-2, -1, 0)$

29. $P_1 = (6, 8, -2)$; $P_2 = (4, -1, 0)$; $P_3 = (1, 0, 0)$

30. $P_1 = (-3, -4, 0)$; $P_2 = (6, -7, 2)$; $P_3 = (0, 0, 1)$

In Problems 31–38 determine whether the two planes intersect, are parallel, or are identical. If they intersect, find the cosine of the angle between them.

31. p_1: $2x - y + z = 2$; p_2: $x + y + z = 3$

32. p_1: $x + y - z = 5$; p_2: $2x + 3y - 4z = 1$

33. p_1: $2x - 3y + z = 1$; p_2: $2x - 3y + 4z = 2$

34. p_1: $x - y = 2$; p_2: $y - z = 2$

35. p_1: $2x - y + z = 3$; p_2: $4x - y + 6z = 7$

36. p_1: $x + y - z = 1$; p_2: $-2x - 2y + 2z = -2$

37. p_1: $x - 2y + z = 1$; p_2: $3x - 6y + 3z = 3$

38. p_1: $x + y - z = 1$; p_2: $x + y - z = 2$

In Problems 39–42 find the distance from the point to the plane.

39. $(1, 2, -1)$; $2x - y + z = 1$

40. $(-1, 3, -2)$; $x + 2y - 3z = 4$

41. an $-1, 1)$; $-x + y - 3z = 6$

42. $(-2, 1, 1)$; $-3x + 2y + z = 1$

43. Find equations for the planes containing the origin that are perpendicular to $x - 2y - z = 0$ and make an angle of 60° with the positive y-axis.

44. Explain why the set of points (x, y, z) equidistant from the points $(1, 3, 0)$ and $(-1, 1, 2)$ is a plane. Then find its equation in two ways, as follows:

(*Problem 44 continues in the next column*)

44. (*continued*)

(a) Use the distance formula to equate the distances between (x, y, z) and the given points, simplifying the result to obtain an equation of the plane.

(b) Name a point of the plane and a vector normal to the plane, and use the answers to find an equation of the plane.

45. Find symmetric equations of the line of intersection of the planes $2x + y - z = 6$ and $x - y + 3z = 4$.

46. Find symmetric equations of the line that contains $(2, 0, -3)$, is perpendicular to $\mathbf{i} + 2\mathbf{j} - \mathbf{k}$, and is parallel to $2x + 3y - z = 1$.

47. Find the point of intersection of the line through the points $(0, 2, -2)$ and $(2, 1, -3)$ and the plane through $(0, 4, -2)$, $(1, 3, -2)$, and $(2, 2, -3)$.

48. Find an equation of the plane parallel to the line $\mathbf{r} = 2\mathbf{i} + t(-\mathbf{i} + \mathbf{j} + 2\mathbf{k})$ and containing the points $(2, 2, -1)$ and $(1, 0, 1)$.

Quadric Surfaces 14.8

Earlier, we showed that the graph of a second-degree equation in two variables,

$$Ax^2 + Bxy + Cy^2 + Dx + Ey + F = 0$$

is, except for degenerate cases, a conic section—that is, a parabola, ellipse, circle, or hyperbola. (You may want to review Chap. 12.)

The graph of a second-degree equation in three variables,

$$Ax^2 + By^2 + Cz^2 + Dxy + Exz + Fyz + Gx + Hy + Iz + J = 0$$

is called a *quadric surface*, and, except for degenerate cases,* there are nine distinct types. As in the case of conics, the equations we use to identify the nine quadric surfaces are the so-called standard equations. For those that have centers, the center will be at the origin; for those that are symmetric, the axis of symmetry will be a coordinate axis.

In the equations that follow we characterize each quadric surface by citing its *intercepts* (points at which the coordinate axes are crossed), its *traces* (the intersection of the surface with the coordinate planes), and its *sections* (the intersection of the surface with other planes, usually taken parallel to the coordinate planes). The letters a, b, and c denote positive numbers.

[14.8.1] DEFINITION / ***Equation of an Ellipsoid.***

$$\frac{x^2}{a^2} + \frac{y^2}{b^2} + \frac{z^2}{c^2} = 1 \qquad \textbf{(Fig. 48)}$$

The intercepts are the points $(\pm a, 0, 0)$, $(0, \pm b, 0)$, and $(0, 0, \pm c)$.

The traces are all ellipses. For example, the trace in the xy-plane, obtained by setting $z = 0$ in the equation, is the ellipse $(x^2/a^2) + (y^2/b^2) = 1$. All sections parallel to the coordinate planes are ellipses.

A computer graph of the ellipsoid $x^2 + (y^2/9) + (z^2/4) = 1$ is provided in Figure 48(b).†

---

* For example, we exclude the equation $x^2 + y^2 + z^2 = 0$, whose graph is the point $(0, 0, 0)$. We also exclude the equation $x^2 + y^2 + 3z^2 = -2$, whose graph consists of no points.

† See the preface for a description of the techniques used to create the computer-generated graphs.

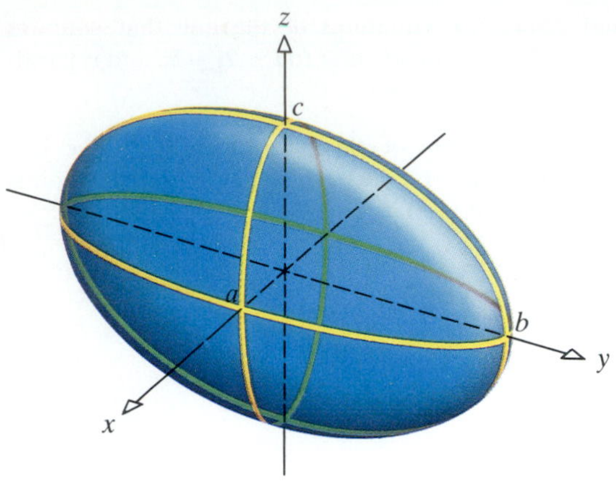

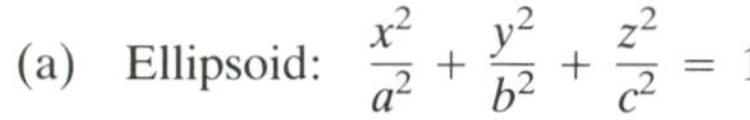
(a) Ellipsoid: $\dfrac{x^2}{a^2} + \dfrac{y^2}{b^2} + \dfrac{z^2}{c^2} = 1$

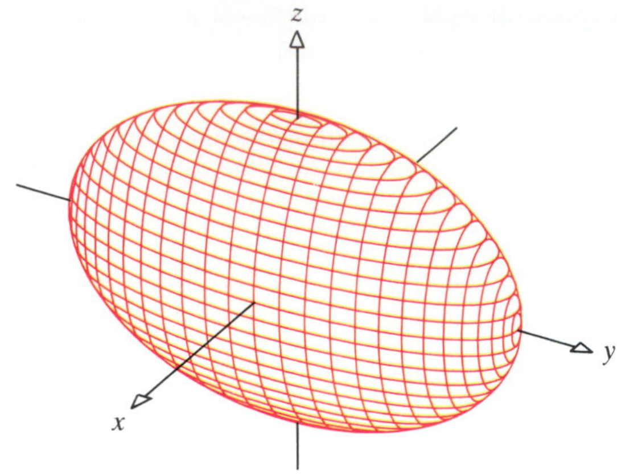

(b) Ellipsoid: $x^2 + \dfrac{y^2}{9} + \dfrac{z^2}{4} = 1$

Figure 48

[14.8.2] DEFINITION / ***Equation of an Elliptic Cone.***

$$z^2 = \frac{x^2}{a^2} + \frac{y^2}{b^2} \qquad \textbf{(Fig. 49)}$$

The lone intercept is the origin (0, 0, 0), and it is referred to as the *vertex* of the cone.

The trace in the xy-plane is the origin; the trace in the yz-plane consists of the pair of intersecting lines $z = \pm y/b$; the trace in the xz-plane consists of the pair of intersecting lines $z = \pm x/a$.

Sections parallel to the xy-plane are ellipses.

If $a = b$, the elliptic cone becomes a *circular cone*. A computer graph of the circular cone $z^2 = 3x^2 + 3y^2$ is provided in Figure 49(b).

[14.8.3] DEFINITION / ***Equation of an Elliptic Paraboloid.***

$$z = \frac{x^2}{a^2} + \frac{y^2}{b^2} \qquad \textbf{(Fig. 50)}$$

The lone intercept is the origin (0, 0, 0), and it is referred to as the *vertex* of the elliptic paraboloid.

The trace in the xy-plane is the origin; the trace in the yz-plane is the parabola $z = y^2/b^2$; the trace in the xz-plane is the parabola $z = x^2/a^2$.

Sections parallel to the xy-plane are ellipses; sections parallel to the other coordinate planes are parabolas.

Observe that $z \geq 0$, so that this surface (except for the origin) lies above the xy-plane.

If $a = b$, the surface is called a *paraboloid of revolution*. A computer graph of the paraboloid of revolution $z = \frac{1}{2}(x^2 + y^2)$ is provided in Figure 50(b).

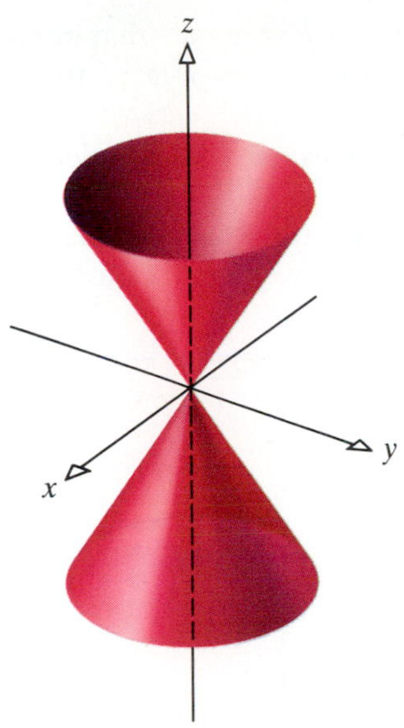

(a) Elliptic cone: $z^2 = \frac{x^2}{a^2} + \frac{y^2}{b^2}$

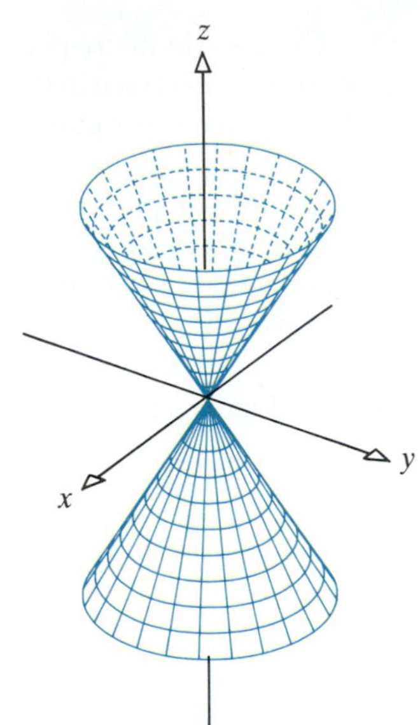

(b) Circular cone: $z^2 = 3x^2 + 3y^2$

Figure 49

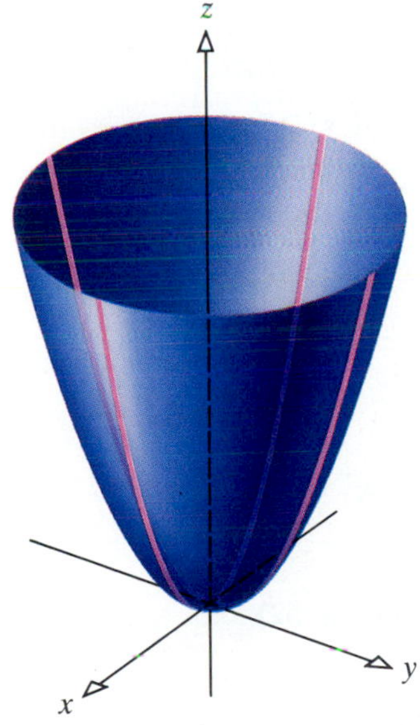

(a) Elliptic paraboloid: $z = \frac{x^2}{a^2} + \frac{y^2}{b^2}$

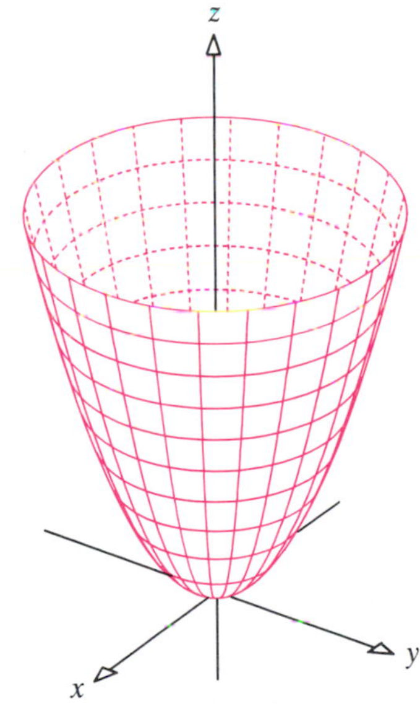

(b) Paraboloid of revolution: $z = \frac{1}{2}(x^2 + y^2)$

Figure 50

[14.8.4] DEFINITION / ***Equation of a Hyperbolic Paraboloid.***

$$z = \frac{y^2}{b^2} - \frac{x^2}{a^2} \qquad \textbf{(Fig. 51)}$$

The lone intercept is at the origin.

The trace in the xy-plane is the pair of lines $y/b = \pm x/a$ that intersect at the origin; the trace in the yz-plane is the parabola $z = y^2/b^2$; the trace in the xz-plane is the parabola $z = -x^2/a^2$.

Sections parallel to the xy-plane are hyperbolas; sections parallel to the other coordinate planes are parabolas.

Note that the origin is a minimum point for the trace in the yz-plane and is a maximum point for the trace in the xy-plane. Such a point is called a *saddle point* of the surface.

A computer graph of the hyperbolic paraboloid $z = y^2 - x^2$ is provided in Figure 51(b).

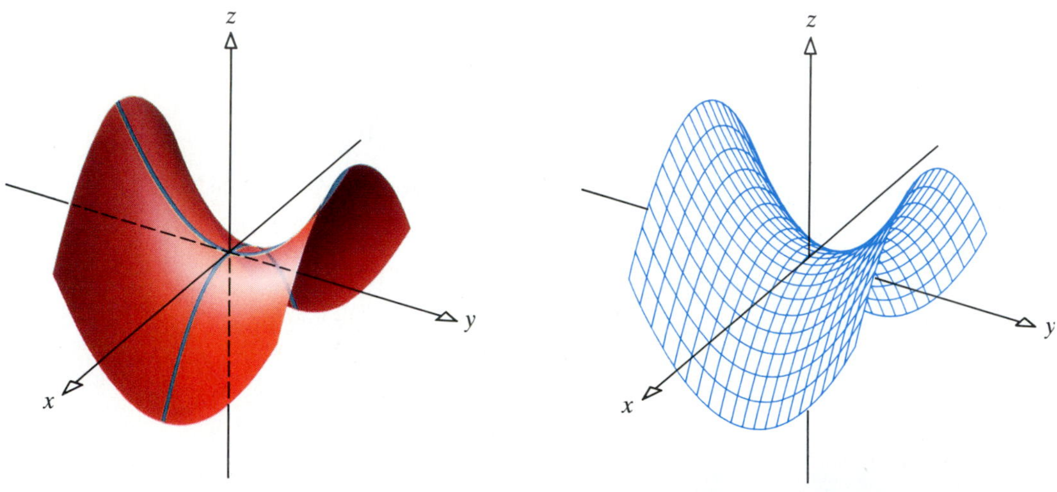

(a) Hyperbolic paraboloid: $z = \dfrac{y^2}{b^2} - \dfrac{x^2}{a^2}$ (b) Hyperbolic paraboloid: $z = y^2 - x^2$

Figure 51

[14.8.5] DEFINITION / *Equation of a Hyperboloid of One Sheet.*

$$\frac{x^2}{a^2} + \frac{y^2}{b^2} - \frac{z^2}{c^2} = 1 \qquad \textbf{(Fig. 52)}$$

The intercepts are at $(\pm a, 0, 0)$ and $(0, \pm b, 0)$.

The trace in the xy-plane is the ellipse $(x^2/a^2) + (y^2/b^2) = 1$; the trace in the yz-plane is the hyperbola $(y^2/b^2) - (z^2/c^2) = 1$; the trace in the xz-plane is the hyperbola $(x^2/a^2) - (z^2/c^2) = 1$.

Sections parallel to the xy-plane are ellipses; sections parallel to the other coordinate planes are hyperbolas. Here z is referred to as the *axis* of the hyperboloid of one sheet.

If $a = b$, the surface is called a *hyperboloid of revolution*. A computer graph of the hyperboloid of revolution $x^2 + y^2 - (z^2/3) = 1$ is provided in Figure 52(b).

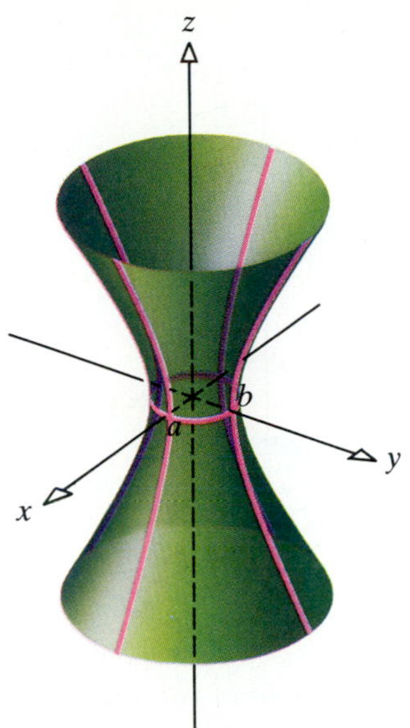

(a) Hyperboloid of one sheet:
$$\frac{x^2}{a^2} + \frac{y^2}{b^2} - \frac{z^2}{c^2} = 1$$

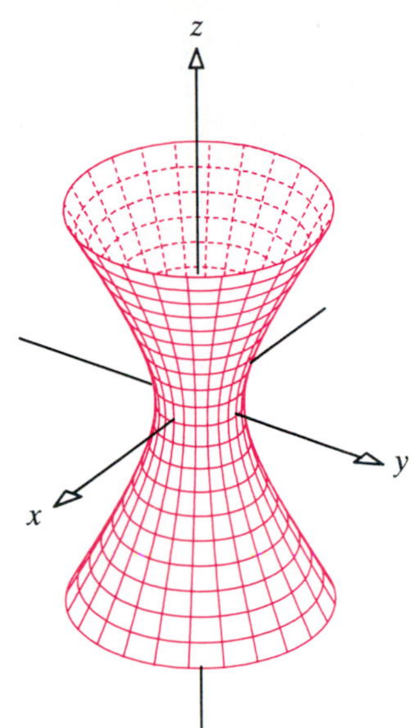

(b) Hyperboloid of revolution:
$$x^2 + y^2 - \frac{z^2}{3} = 1$$

Figure 52

[14.8.6] DEFINITION / ***Equation of a Hyperboloid of Two Sheets.***

$$\frac{x^2}{a^2} + \frac{y^2}{b^2} - \frac{z^2}{c^2} = -1 \qquad \textbf{(Fig. 53)}$$

The intercepts are the two points $(0, 0, \pm c)$.

There is no trace in the xy-plane; the trace in the yz-plane is the hyperbola $(z^2/c^2) - (y^2/b^2) = 1$; and in the xz-plane the trace is $(z^2/c^2) - (x^2/a^2) = 1$, a hyperbola.

Sections parallel to the xy-plane are ellipses; sections parallel to the other coordinate axes are hyperbolas. Here z is referred to as the *axis* of the hyperboloid of two sheets.

Note that this surface consists of two parts, one for which $z \geq c$, and the other for which $z \leq -c$.

A computer graph of the hyperboloid of two sheets $x^2 + y^2 - (z^2/3) = -1$ is provided in Figure 53(b).

The three remaining quadric surfaces are called *cylinders*. Their standard equations are characterized by the fact that one of the variables is missing from the equation. In naming these surfaces, we have chosen z as the missing variable. As a result, z is unrestricted, and the cylinder will be unbounded in the z direction.

Cylinders are surfaces that are generated by a line moving along a curve while remaining parallel to a fixed line. We therefore describe cylinders in terms of how they are generated.

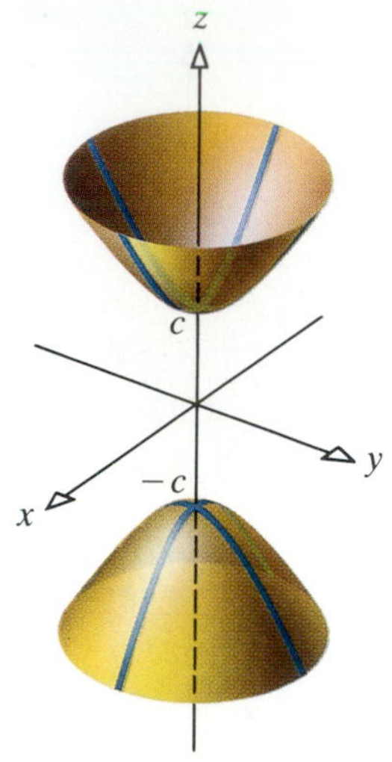

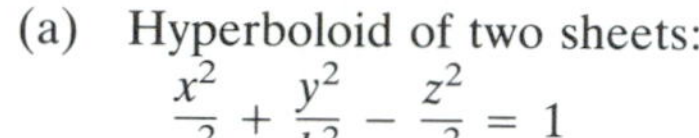

(a) Hyperboloid of two sheets: $\dfrac{x^2}{a^2} + \dfrac{y^2}{b^2} - \dfrac{z^2}{c^2} = 1$

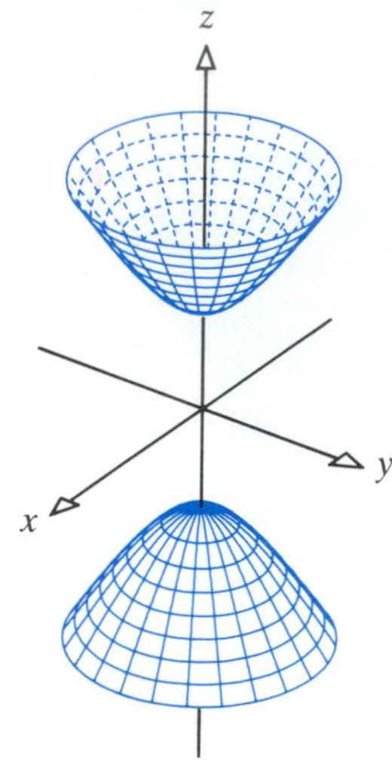

(b) Hyperboloid of two sheets: $x^2 + y^2 - \dfrac{z^2}{3} = -1$

Figure 53

[14.8.7] DEFINITION / ***Equation of a Parabolic Cylinder.***

$$x^2 = 4ay \qquad \textbf{(Fig. 54)}$$

This surface is generated by a line parallel to the z-axis moving along the parabola $x^2 = 4ay$. This surface is easily visualized by taking a piece of paper and folding it so that two opposite edges trace out a portion of a parabola.

A computer graph of the parabolic cylinder $x^2 = 4y$ is provided in Figure 54(b).

[14.8.8] DEFINITION / ***Equation of an Elliptic Cylinder.***

$$\frac{x^2}{a^2} + \frac{y^2}{b^2} = 1 \qquad \textbf{(Fig. 55)}$$

This surface is generated by a line parallel to the z-axis moving along the ellipse $(x^2/a^2) + (y^2/b^2) = 1$. This surface can be visualized by taking a piece of paper and attaching the two opposite edges in the shape of an ellipse.

A computer graph of the elliptic cylinder $(x^2/4) + (y^2/9) = 1$ is provided in Figure 55(b).

[14.8.9] DEFINITION / ***Equation of a Hyperbolic Cylinder.***

$$\frac{x^2}{a^2} - \frac{y^2}{b^2} = 1 \qquad \textbf{(Fig. 56)}$$

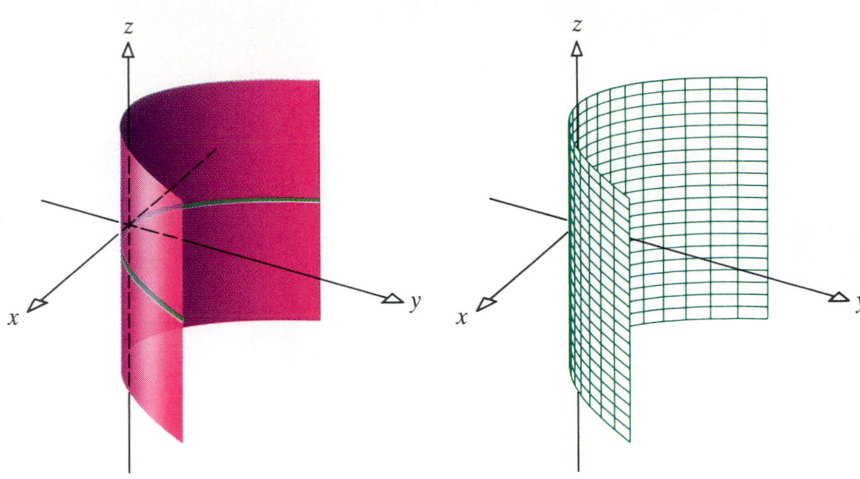

(a) Parabolic cylinder: $x^2 = 4ay$ (b) Parabolic cylinder: $x^2 = 4y$

Figure 54

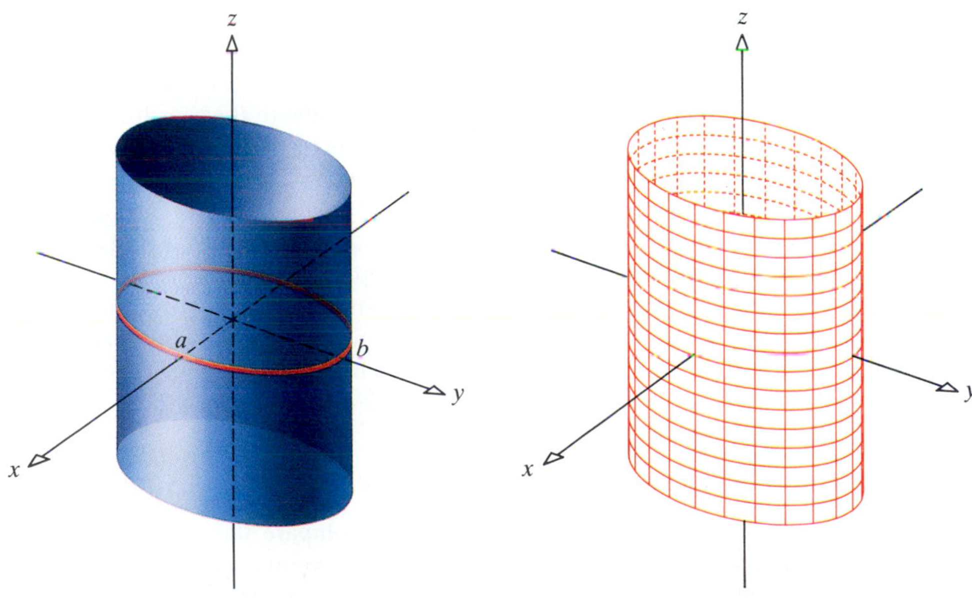

(a) Elliptic cylinder: $\frac{x^2}{a^2} + \frac{y^2}{b^2} = 1$ (b) Elliptic cylinder: $\frac{x^2}{4} + \frac{y^2}{9} = 1$

Figure 55

This surface is generated by a line parallel to the z-axis moving along the hyperbola $(x^2/a^2) - (y^2/b^2) = 1$. This surface consists of two parts, each of which may be visualized by folding a piece of paper so that two opposite edges trace out a portion of a hyperbola.

A computer graph of the hyperbolic cylinder $(x^2/9) - (y^2/4) = 1$ is provided in Figure 56(b).

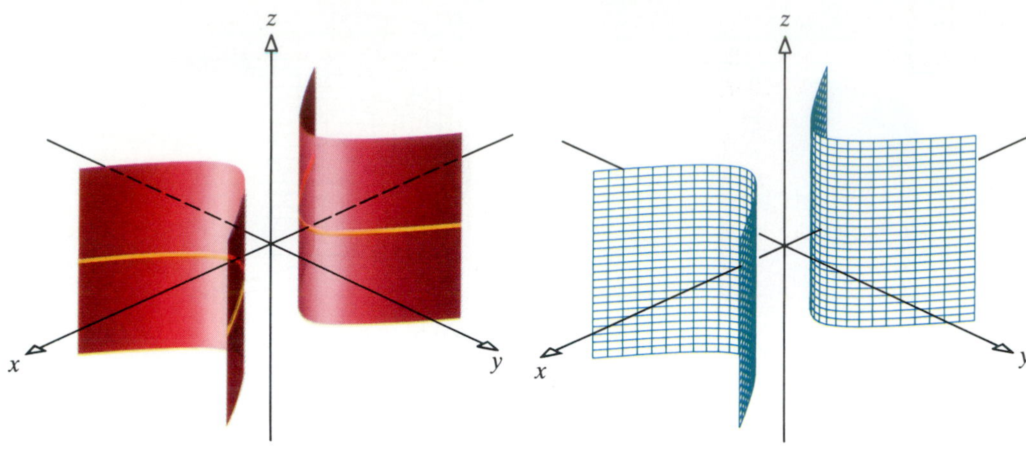

(a) Hyperbolic cylinder: $\dfrac{x^2}{a^2} - \dfrac{y^2}{b^2} = 1$ (b) Hyperbolic cylinder: $\dfrac{x^2}{9} - \dfrac{y^2}{4} = 1$

Figure 56

The quadric surfaces classified here are in standard position, with centers (or vertex) at the origin and symmetries with respect to the coordinate axes. When the center (or vertex) of a quadric surface is not located at the origin, but there is symmetry with respect to lines parallel to the coordinate axes, a simple translation of axes may be applied to place the equation in standard form. The correct translation is obtained by completing squares. If the symmetry of the quadric surface is with respect to lines that are not parallel to the coordinate axes, a rotation of axes, perhaps followed by a translation of axes, is required to place the equation in standard form.

Using rotations and translations, we can show that our list of nine basic equations provides an exhaustive classification of quadrics. Rotations in space are somewhat involved and we shall not discuss them here. However, the idea is quite similar to that of rotations in the plane (see Section 12.5).

Identifying and Sketching Quadric Surfaces

To identify and sketch a quadric surface, compare the given equation to the ones discussed in this section. Once you have matched it, the nature of the graph is apparent. Here are some suggestions to help you make the comparison:

1. Remember that the nine types of quadric surfaces listed are given in standard position. An interchange of variables does not affect the classification. For example, the equation $z = (x^2/4) + (y^2/9)$ is an elliptic paraboloid and its graph lies above the xy-plane. The equation $y = (x^2/9) + (z^2/4)$ is also an elliptic paraboloid. (Do you see why?) Its graph lies to the right of the xz-plane.
2. Cylinders are characterized by a variable missing from the equation. For example, the equation $(y^2/4) + (z^2/9) = 1$ is an elliptic cylinder whose graph is perpendicular to the yz-plane.

3. The technique of completing the square is sometimes required to identify the given quadric surface. In this case, a translation of axes will help to get the correct location of the graph.

EXAMPLE 1

Identify the surface given by each equation and sketch its graph.

(a) $4x^2 - 18y^2 + 9z^2 = 36$ (b) $4y^2 + 9z^2 - 36x = 0$
(c) $x^2 + z^2 - 4x = 0$ (d) $4y^2 + 9z^2 - 36x - 8y + 36z + 148 = 0$

Solution

(a) Divide both sides of the equation by 36 to get

$$\frac{x^2}{9} - \frac{y^2}{2} + \frac{z^2}{4} = 1$$

This equation resembles the one given in definition (14.8.5). We conclude that the surface is a hyperboloid of one sheet whose axis is the y-axis (see Fig. 57).

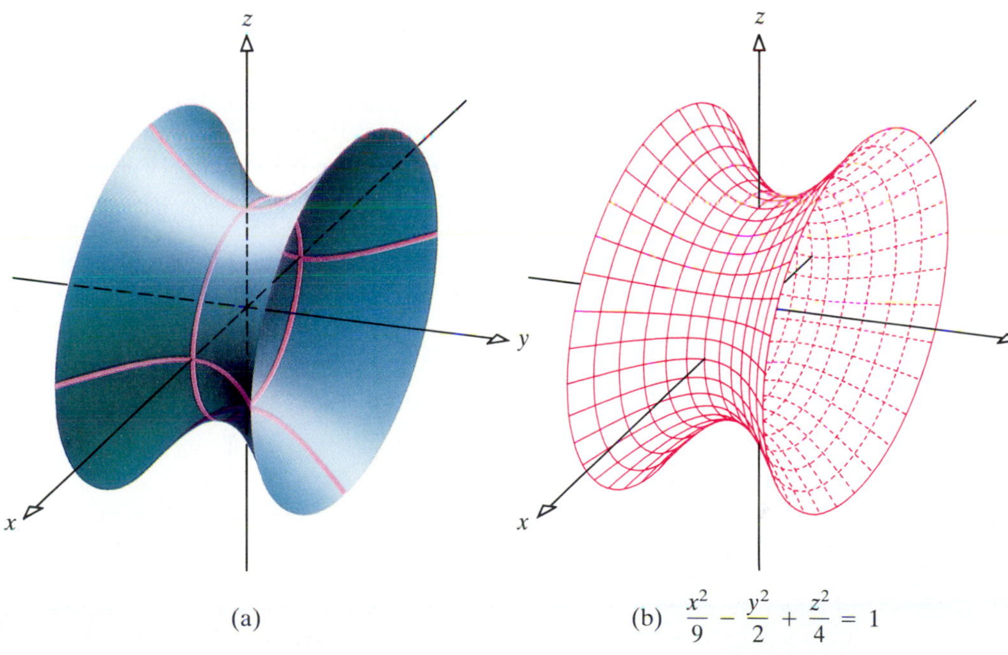

Figure 57

(b) The equation can be written as

$$\frac{y^2}{9} + \frac{z^2}{4} = x$$

which resembles the one in definition (14.8.3). Its graph is therefore an elliptic paraboloid with vertex at the origin whose axis is the x-axis (see Fig. 58).

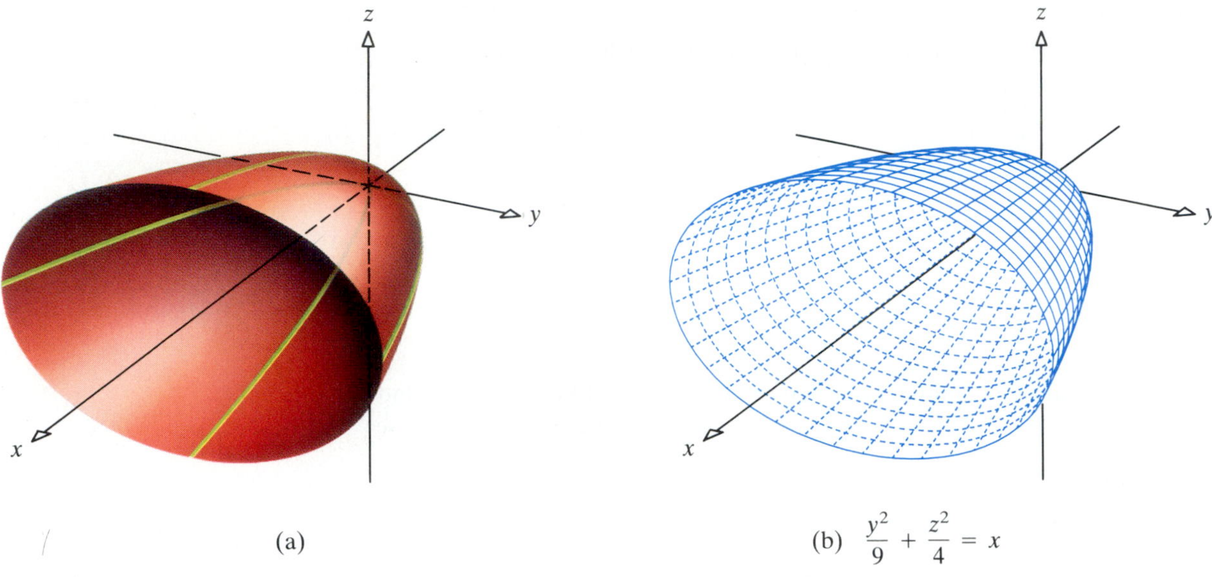

(a) (b) $\frac{y^2}{9} + \frac{z^2}{4} = x$

Figure 58

(c) Since the y variable is missing, this surface is a cylinder. By completing the square in x, the equation can be written as $(x - 2)^2 + z^2 = 4$. This is the equation of a circular cylinder with center axis parallel to the y-axis passing through $(2, 0, 0)$ and of radius 2. Its graph is perpendicular to the xz-plane (see Fig. 59).

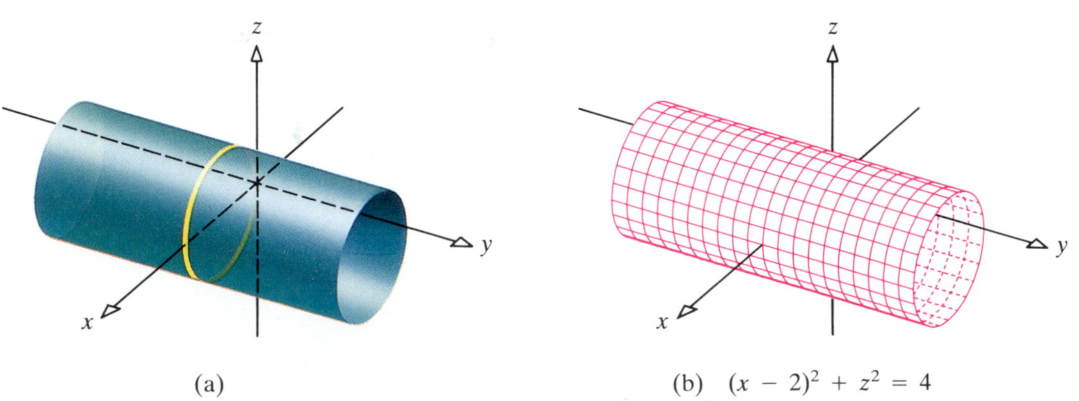

(a) (b) $(x - 2)^2 + z^2 = 4$

Figure 59

(d) We rewrite the equation and complete the squares:

$$\begin{aligned}4y^2 - 8y + 9z^2 + 36z &= 36x - 148\\4(y^2 - 2y) + 9(z^2 + 4z) &= 36x - 148\\4(y-1)^2 + 9(z+2)^2 &= 36x - 148 + 40\\4(y-1)^2 + 9(z+2)^2 &= 36(x-3)\\\frac{(y-1)^2}{9} + \frac{(z+2)^2}{4} &= x - 3\end{aligned}$$

We recognize this as an elliptic paraboloid with vertex at $(3, 1, -2)$ whose axis is the x-axis (see Fig. 60 and compare it to Fig. 58).

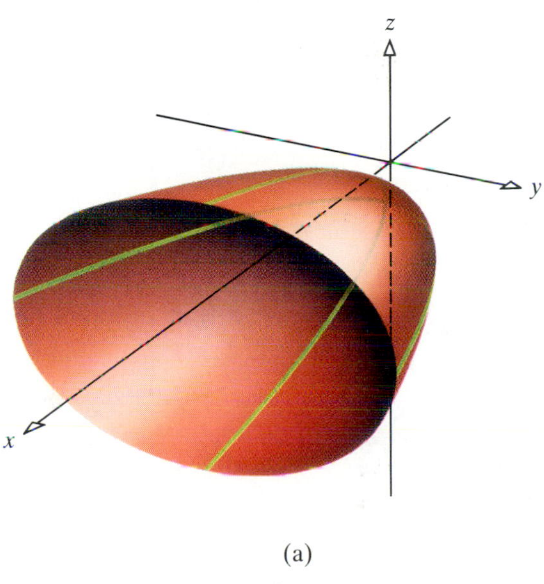

(a)

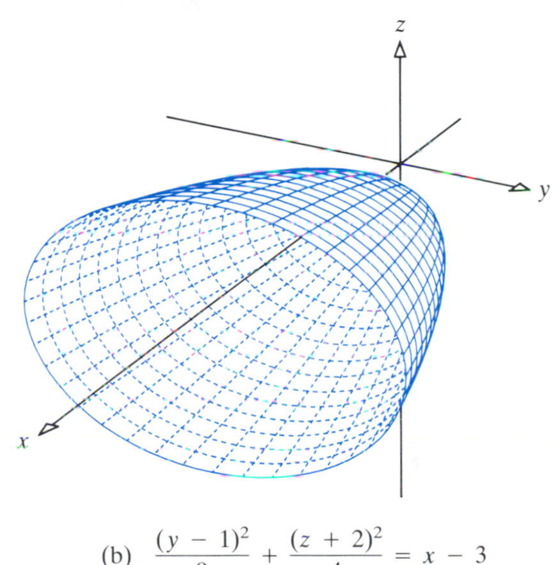

(b) $\dfrac{(y-1)^2}{9} + \dfrac{(z+2)^2}{4} = x - 3$

Figure 60

EXERCISE 14.8

In Problems 1–22 identify and sketch each quadric surface.

1. $z = x^2 + y^2$
2. $z = x^2 - y^2$
3. $4x^2 + y^2 + 4z^2 = 4$
4. $2x^2 + y^2 + z^2 = 1$
5. $z^2 = x^2 + 2y^2$
6. $x^2 + 2y^2 - z^2 = 1$
7. $x = 4z^2$
8. $x^2 + y^2 = 1$
9. $x^2 + 2y^2 - z^2 = -4$
10. $z = x^2 - 2y^2$
11. $y^2 - x^2 = 4$
12. $2x = y^2$
13. $z = x^2 + 2y^2$
14. $z^2 = 3x^2 + 4y^2$
15. $4y^2 - x^2 = 1$
16. $4x^2 + y^2 - z^2 = 1$
17. $x^2 + 4x + y^2 = z^2 + 4z$
18. $x^2 + y^2 + 2y = z - 1$
19. $y = 4x^2 + 9z^2$
20. $x + 4 = 2y^2$
21. $x^2 - y^2 - 4z^2 + 4y + 8z - 9 = 0$
22. $4x^2 + y^2 + 4z^2 - 16x - 4y - 8z + 20 = 0$

23. Explain why the graph of $xy = 1$ is a cylinder. Sketch it. Is the graph a quadric surface?

24. Explain why the graph of $z = \sin y$ is a cylinder. Sketch it. Is the graph a quadric surface?

Figures A–J are graphs of quadric surfaces. In Problems 25–30 match each equation with its graph.

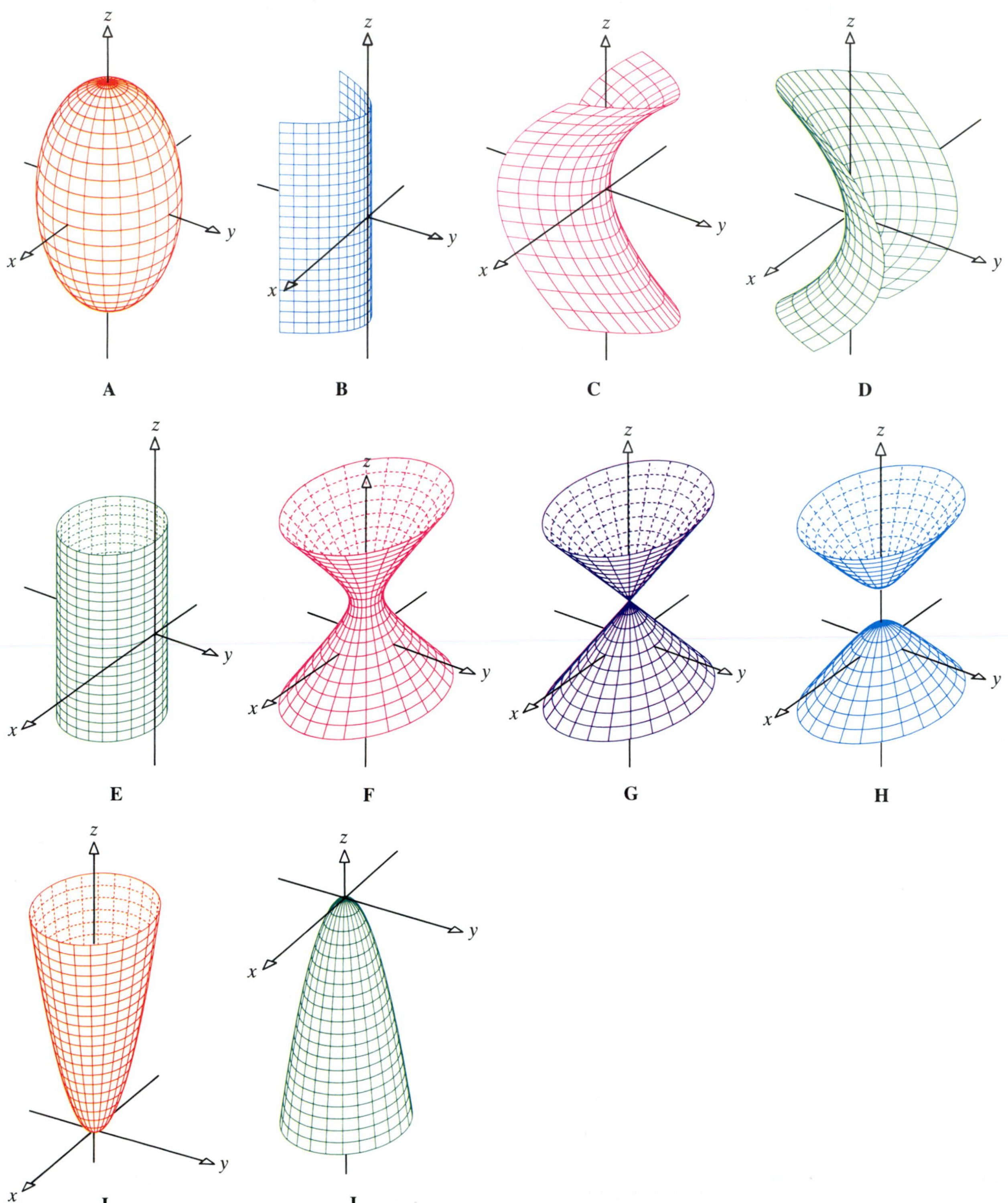

25. $3x^2 + 4y^2 + z = 0$ **26.** $3x^2 + 4y^2 + 4y = 0$ **27.** $3x^2 + 2y^2 - (z - 2)^2 + 1 = 0$

28. $z^2 - 4x^2 = 3y$ **29.** $x^2 + 2y^2 - z^2 + 4z = 4$ **30.** $3x^2 + 3y^2 + z^2 = 1$

31. Figures K–O are computer graphs of quadric surfaces. Match each graph to one of the following:

(a) $z = 4y^2 - x^2$ (b) $2z = x^2 + 4y^2$ (c) $2x^2 + y^2 - z^2 = 1$
(d) $2x^2 + y^2 + 3z^2 = 1$ (e) $y^2 = 4x$

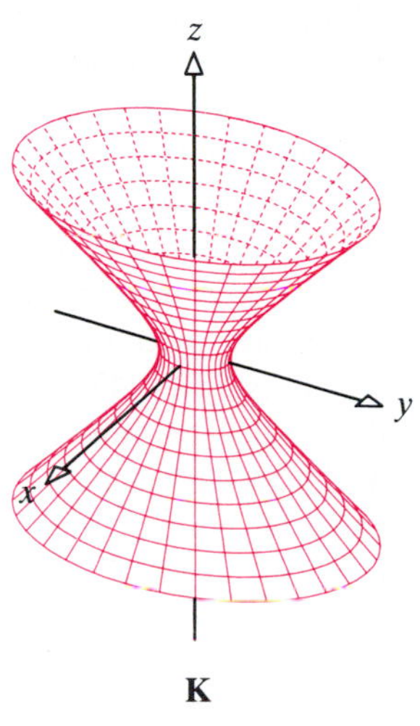

K

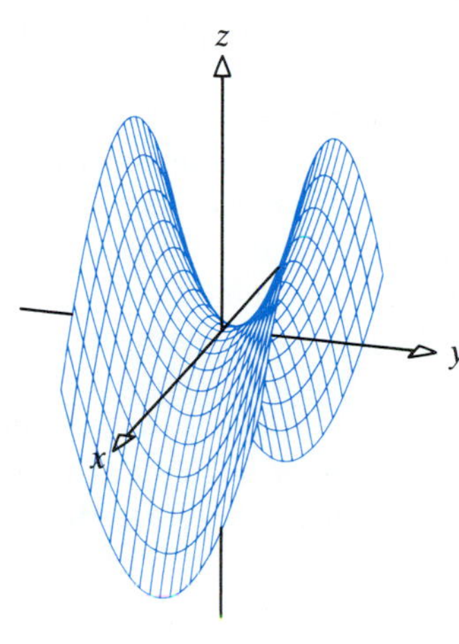

L

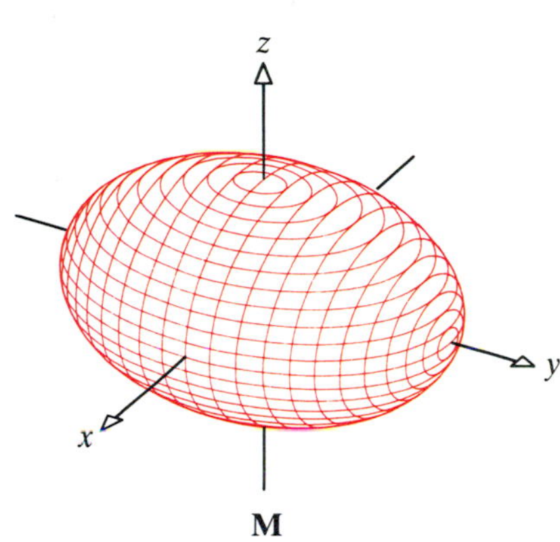

M

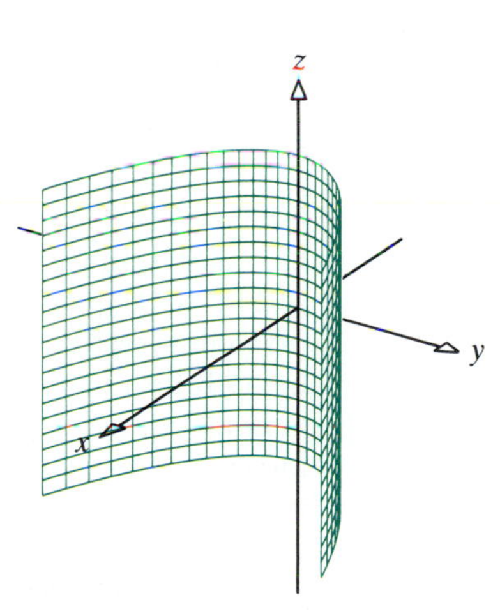

N

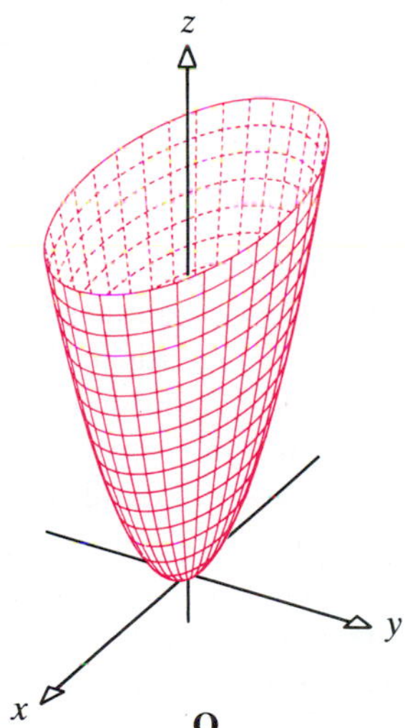

O

32. Sketch $z = xy$. (This surface is a hyperbolic paraboloid rotated 45° about the z-axis.)

33. Show that through each point on the hyperboloid of one sheet

$$\frac{x^2}{a^2} + \frac{y^2}{b^2} - \frac{z^2}{c^2} = 1$$

there are two lines lying entirely on the surface.

Historical Perspectives

Most mathematical concepts have undergone many changes and interpretations after their introduction into mathematics, and this process is especially apparent in the development of the concept of vector. Vector analysis is a fairly young branch of calculus; it was developed in the late nineteenth century. However, the concept of a vector quantity—that is, a quantity with both magnitude and direction—was undoubtedly present in much mathematical thought long before it became formalized. The *parallelogram law for addition* was known to Aristotle and was stated explicitly by Galileo, but the first extensive use of vectors came about in attempts to visualize *complex numbers*. Two-dimensional vectors and vector algebra form an essential part of the framework of the geometric representation of complex numbers introduced by the French mathematician Argand in 1813 and later by the great German mathematician and physicist Karl Friedrich Gauss (1777–1855).

Gauss was born in Brunswick, Germany, to working-class parents. He received his education through the generosity of Duke Wilhelm, after showing great promise in elementary school. He worked at the University of Göttingen from 1807 until his death, rarely leaving the city of Göttingen even for short periods of time. In mathematics he made fundamental discoveries in algebra, differential geometry, complex analysis, potential theory, number theory, and noneuclidean geometry. Gauss's accomplishments were extraordinary not only in mathematics but also in physics, where he did first-rate work in the study of optics, electricity and magnetism, and astronomy. Even though he published only a fraction of his work, Gauss was widely known and acclaimed to be the greatest mathematician since Newton. The awe-inspiring scope of his life's work has led many to consider him the last of the "universal geniuses." His work on complex numbers did much to popularize the geometric interpretation of complex numbers discussed in Problems 1–3.

The success of the two-dimensional interpretation of complex numbers motivated the Irish mathematician Rowan Hamilton (1805–1865) to search for an analogous three-dimensional "complex" algebra. Hamilton was born in Dublin, Ireland, and ranks second only to Newton among British mathematicians. Unlike Newton, Hamilton's genius was apparent at an early age. When he was 8 years old, he is said to have been literate in six languages: Latin, Hebrew, Greek, Italian, French, and, of course, English. As a 23-year-old undergraduate, he was appointed Professor of Astronomy at Trinity College. Today, he is perhaps most famous for *Hamilton's principle* in physics, but his discovery of four-dimensional quantities, which he called *quaternions*, was an accomplishment with equally important implications in the motivation that it provided for the introduction of vectors into calculus later on. Hamilton's quaternions consisted of a sum of a scalar (real number) part and a vector part, although he did not treat the vector part separately (see Problems 4 and 5).

The German mathematician Grassmann and others also worked on the problem of extending or generalizing the complex numbers to higher dimensions, but it was Hamilton's quaternions that found the most immediate application and eventually led to vector analysis. James Clerk Maxwell (1831–1879), the Scottish mathematical physicist, used and developed

Hamilton's theory of quaternions, distinguishing the scalar and vector parts of the quantities. His famous work on electromagnetic theory, based on what came to be known as *Maxwell's equations*, was very influential in making vectors and vector methods in analysis well known and in motivating their further development.

However, it was in the independent work of the American physicist and mathematician Josiah Willard Gibbs (1839–1903) and the Englishman Oliver Heaviside (1850–1925) that the distinction between scalar and vector was drawn once and for all, and that the concept of vector was finally divorced from the theory of quaternions. Both men introduced the vector cross product and placed the scalar product in proper perspective. Modern vector analysis is usually considered to have begun with Gibbs and Heaviside; their developments of the subject proceed from a starting point not too dissimilar from the treatment in this chapter. In the early 1880's, while Professor of Mathematical Physics at Yale University, Gibbs produced some notes on vector analysis for his students' use. These notes formed the basis for a textbook in vector analysis that was published in 1901 by one of his former students. Gibbs's work on vector analysis was among the first major American contributions to mathematics.

Throughout the early development of the theory of vectors, the geometric interpretation, as opposed to the algebraic interpretation, was the prevalent viewpoint. The basic vector operations were defined geometrically and the **i**, **j**, **k** notation was not used except when attempts were made to evade vector arguments with the aid of a rectangular coordinate system. Of course, today we recognize that the real advantages of vector analysis are realized through the combination of geometry and algebra inherent in the uses of vector algebra, or linear algebra, to simplify and clarify many of its applications.

HISTORICAL EXERCISES

1. A *complex number* z can be thought of as a number of the form $z = a + bi$, where a and b are real numbers and $i^2 = -1$. (This use of the symbol i should not be confused with the vector **i**.) The number a is called the *real part of* z, and b is called the *imaginary part of* z. The *Argand diagram* of $z = a + bi$ is shown in the figure. The quantity $r = \sqrt{a^2 + b^2}$ is called the *modulus* (length) of the complex number z. To add two complex numbers $a + bi$ and $c + di$, we simply add the corresponding real and imaginary parts.

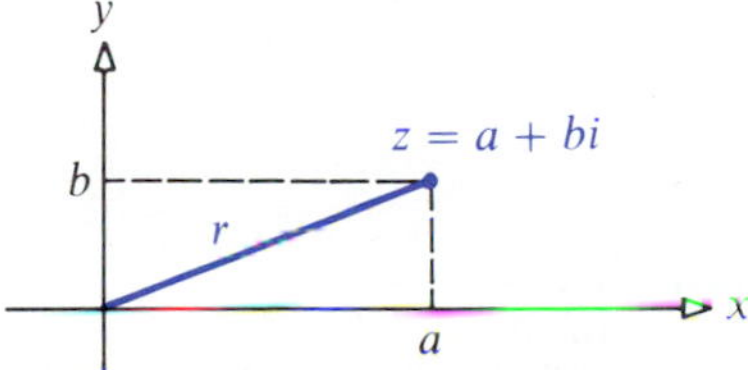

(*Problem 1 continues in the next column*)

1. (*continued*)

(a) Use the above definition to add z_1 to z_2, where $z_1 = 3 + 2i$ and $z_2 = -1 + 2i$.

(b) Sketch the Argand diagrams for z_1, z_2, and the sum $z_1 + z_2$.

2. To multiply two complex numbers $a + bi$ and $c + di$, we multiply as though we were multiplying two binomials, use the fact that $i^2 = -1$, and collect the real and imaginary parts.

(a) Use the above definition to multiply z_1 and z_2, as given in Problem 1, part (a).

(b) Find the product of $2 + 3i$ and $2 - 3i$.

(c) What can you say, in general, about the product of a complex number $z = a + bi$ and its *complex conjugate* $\bar{z} = a - bi$? Interpret this result geometrically.

(d) What vector operation is suggested by your geometric interpretation in part (c)?

3. Recall from Problem 2, part (c), that the conjugate of $z = a + bi$ is $\bar{z} = a - bi$.

(a) Define a new operation on complex numbers—say, $\diamond$—as follows: $z \diamond w = (z)(\bar{w})$. Let $z = a + bi$ and $w = c + di$, and write the expression for $z \diamond w$. What kind of number is $z \diamond w$?

(b) Show that the operation $\diamond$ satisfies all the properties of the dot product (inner product) of two vectors listed in theorem (14.4.2) except (a). (This is the reason that $z \diamond w$ is called the *inner product* of the complex numbers z and w.)

4. Hamilton was the first person to think of complex numbers as ordered pairs of real numbers (a, b), where a is the real part and b is the imaginary part of the complex number. Since $(a + bi) + (c + di) = (a + c) + (b + d)i$, we can write this in ordered-pair notation as

$$(a, b) + (c, d) = (a + c, \quad b + d)$$

(a) Write the definition of complex multiplication in ordered-pair notation.

(*Problem 4 continues in the next column*)

4. (*continued*)

(b) Show that complex multiplication is a commutative operation. That is, show that $(a, b) \times (c, d) = (c, d) \times (a, b)$ for all complex numbers (a, b) and (c, d).

5. Hamilton's *quaternions* can be represented as ordered four-tuples, or quadruples. Multiplication of quaternions can then be defined (in a way analogous to complex multiplication, as in Problem 4) as

$$(a, b, c, d) \times (e, f, g, h)$$

$$= (ae - bf - cg - dh, \ af + be + ch - dg,$$

$$ag - bh + ce + df, \quad ah + bg - cf + de)$$

(a) Verify that quaternion multiplication is *not* commutative by providing a counterexample.

(b) Is there an identity for quaternion multiplication? That is, is there a quaternion that when multiplied by any other quaternion Q will produce Q as the result? If so, what is the identity?

REVIEW EXERCISES

1. Find the magnitude of $\mathbf{i} + 2\mathbf{j} + 3\mathbf{k}$.

2. Find the distance between $(1, 2, 3)$ and $(7, 5, 1)$.

3. Find the distance between $(1, 2, 3)$ and $(3, -2, -5)$.

4. Write the equation of the sphere with center at $(1, 2, 3)$ and radius of 4.

5. Find the radius and center of the sphere

$$x^2 + y^2 + z^2 - 4x + 8y = 5$$

6. If $\mathbf{v} = \langle -2, -1, 3\rangle$ and $\mathbf{w} = \langle 5, 4, -2\rangle$; find $3\mathbf{v} - 2\mathbf{w}$.

7. If $\mathbf{v} = 2\mathbf{i} + 3\mathbf{j} - \mathbf{k}$ and $\mathbf{w} = -\mathbf{i} - 2\mathbf{j} + 3\mathbf{k}$; find $\|\mathbf{v} + \mathbf{w}\|$.

8. Find a unit vector in the opposite direction of $4\mathbf{i} + 12\mathbf{j} - 3\mathbf{k}$.

9. Write a space vector with a magnitude of 25 whose projection on the xy-plane is $12\mathbf{i} + 16\mathbf{j}$.

10. Find $\mathbf{u}$ and $\mathbf{v}$ if

$$\mathbf{u} - 3\mathbf{v} = 2\mathbf{i} + \mathbf{j} - \mathbf{k}$$

$$2\mathbf{u} + \mathbf{v} = \mathbf{i} + \mathbf{j}$$

11. Given $\mathbf{u} = \mathbf{i} - 2\mathbf{j}$ and $\mathbf{v} = 3\mathbf{i} + \mathbf{j}$, find vectors $\mathbf{w}_1$ and $\mathbf{w}_2$ such that $\mathbf{u} = \mathbf{w}_1 - \mathbf{w}_2$, $\mathbf{v}$ is parallel to $\mathbf{w}_1$, and $\mathbf{u}$ is orthogonal to $\mathbf{w}_2$.

12. If θ is the angle between the nonzero vectors $\mathbf{v}$ and $\mathbf{w}$, $0 \le \theta \le \pi$, show that

$$\sin^2\theta = \frac{\|\mathbf{v}\|^2\|\mathbf{w}\|^2 - (\mathbf{v} \cdot \mathbf{w})^2}{\|\mathbf{v}\|^2\|\mathbf{w}\|^2}$$

13. Find the total work done by gravity when a particle of mass m moves once around the triangle shown in the figure. (The force of gravity is mg, where $g \doteq 9.80$ meters per second per second.)

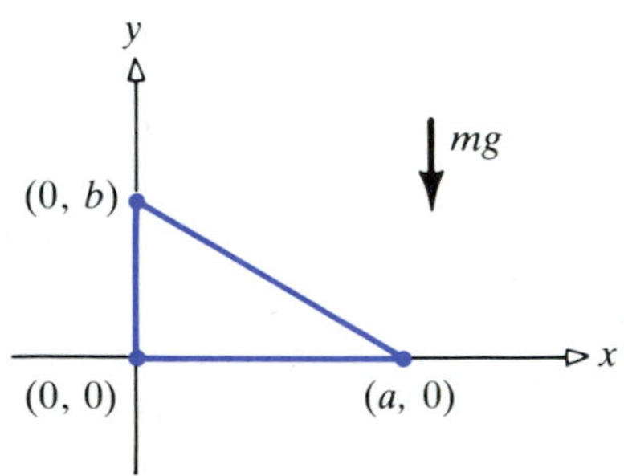

14. Three forces are applied to an object in mutually orthogonal directions: 12 units along the x-axis, 16 units along the y-axis, and 15 units along the z-axis. What is the resultant force?

15. Find the angle between the vectors $\mathbf{u} = \mathbf{i} + 2\mathbf{j} + 3\mathbf{k}$ and $\mathbf{v} = 2\mathbf{i} - 3\mathbf{j} - \mathbf{k}$.

16. Find the angle A of the triangle ABC with vertices $A = (1, 0, 1)$, $B = (2, -1, 1)$, $C = (-2, 1, 0)$.

17. Find the angle between the vectors $5\mathbf{i} + 2\mathbf{j} - 3\mathbf{k}$ and $-\mathbf{i} + 2\mathbf{j} + 3\mathbf{k}$.

18. Show that the volume of the tetrahedron with adjacent edges $\mathbf{u}$, $\mathbf{v}$, and $\mathbf{w}$ is $\frac{1}{6}|\mathbf{u} \cdot (\mathbf{v} \times \mathbf{w})|$.

19. Find the volume of the parallelpiped determined by $\overrightarrow{AB}$, $\overrightarrow{AC}$, and $\overrightarrow{AD}$, where $A = (0, 1, 0)$, $B = (1, 1, 4)$, $C = (2, 0, 1)$, and $D = (3, 5, 8)$.

20. A rigid body rotates with constant angular speed of ω radians per second about a line parallel to $2\mathbf{i} + 3\mathbf{j} + 6\mathbf{k}$. Find the speed of a particle on this body at the instant it passes through the point $(1, 2, 3)$. Assume the distance is in meters.

21. Find the line perpendicular to the lines

$$l: \quad x = 1 - t, \quad y = t, \quad z = 2t - 1$$

$$m: \quad x = t + 1, \quad y = -t, \quad z = t - 1$$

22. The paths of two objects are described by $\langle t, -t, 1 - t\rangle$ and $\langle t - 3, 2t, 4t - 1\rangle$ at time t. Will the paths intersect? If so, where? Will the objects collide? If so, when?

23. Write the equation of a line passing through $(1, 2, 3)$ and parallel to $\mathbf{v} = 2\mathbf{i} - \mathbf{j} - 4\mathbf{k}$.

24. Find the scalar α so that $\mathbf{v} = 2\mathbf{i} - \mathbf{j} - 4\mathbf{k}$ and $\mathbf{w} = \alpha\mathbf{i} + \alpha\mathbf{j} + 3\mathbf{k}$ are orthogonal.

25. Find an equation of the sphere with $(2, 2, -3)$ and $(-2, 6, 5)$ as endpoints of a diameter.

26. Find all points (x, y, z) satisfying the system of equations

$$5x + 2y - z = 3$$

$$2x + y + z = 1$$

27. Find an equation of the plane with nonzero intercepts $(a, 0, 0)$, $(0, b, 0)$, $(0, 0, c)$.

28. Find an equation of the plane tangent to the sphere $(x - 1)^2 + (y + 2)^2 + (z - 2)^2 = 6$ at the point $(2, -1, 0)$.

29. Find an equation of the sphere with center at $(-2, 1, 5)$ and tangent to $x + 4y - z = 7$.

30. Find an equation of the plane containing the lines

$$l: \quad x = t + 1, \quad y = 3t, \quad z = t - 1$$

$$m: \quad \frac{x + 3}{2} = \frac{y - 3}{3} = z$$

31. Find equations for the spheres with radius $\sqrt{26}$ that are tangent to $3x + y - 4z = 20$ at the point $(1, 1, -4)$.

32. Find symmetric equations of the line normal to the plane containing the lines

$$x - 2 = \frac{y + 1}{2} = \frac{z - 1}{2} \quad \text{and} \quad x + 1 = \frac{y + 8}{3} = \frac{z}{-3}$$

at their point of intersection.

33. Determine whether the expressions below represent a vector (**V**) or a scalar (S), or are merely nonsense (N).

(a) $(\mathbf{V}_1 \cdot \mathbf{V}_2) \cdot \mathbf{V}_3$ (b) $(\mathbf{V}_1 \cdot \mathbf{V}_2) \cdot 5$
(c) $(\mathbf{V}_1 \times \mathbf{V}_2) \times \mathbf{V}_3$ (d) $(\mathbf{V}_1 \times \mathbf{V}_2) \cdot \mathbf{V}_3$
(e) $(\mathbf{V}_1 \cdot \mathbf{V}_2) \times \mathbf{V}_3$ (f) $\dfrac{\mathbf{V}_1 \times \mathbf{V}_2}{\|\mathbf{V}_1\|}$, $\mathbf{V}_1 \neq \mathbf{0}$

34. Suppose that $\mathbf{v} = c_1\mathbf{u}_1 + c_2\mathbf{u}_2$ is a vector in the plane. Find c_1 and c_2 in terms of $\mathbf{u}_1$, $\mathbf{u}_2$, and $\mathbf{v}$. What are c_1 and c_2 in case $\mathbf{u}_1$ and $\mathbf{u}_2$ are unit orthogonal vectors?

35. Show that the graph of

$$Ax^2 + Ay^2 + Az^2 + Dx + Ey + Fz = G, \quad A > 0$$

is a sphere, a point, or the empty set. When is it a sphere, and what are the center and radius?

36. Find an equation of the sphere passing through $(3, 0, 0)$, $(0, 0, 1)$, $(-1, -3, 1)$, and $(2, 0, 1)$.

37. Identify the following quadric surfaces:

(a) $\dfrac{x^2}{4} - \dfrac{y^2}{9} = 1$ (b) $x^2 + y^2 - \dfrac{z^2}{4} = -1$

(c) $z = \dfrac{y^2}{9} - \dfrac{x^2}{4}$ (d) $z = \dfrac{x^2}{4} + \dfrac{y^2}{9}$

(e) $z^2 = \dfrac{x^2}{4} + \dfrac{y^2}{9}$ (f) $\dfrac{x^2}{4} + \dfrac{y^2}{9} + z^2 = 1$

(g) $\dfrac{x^2}{4} + \dfrac{y^2}{9} - z^2 = 1$ (h) $y^2 = 9x$

(i) $\dfrac{x^2}{4} + \dfrac{y^2}{9} = 1$

CHALLENGE EXERCISES

1. We proved in the text that the distance between the point $P_0 = (x_0, y_0, z_0)$ and the plane $Ax + By + Cz = D$ is

$$d = \frac{|Ax_0 + By_0 + Cz_0 - D|}{\sqrt{A^2 + B^2 + C^2}}$$

Derive this formula differently, as follows:

(a) Show that the line through P_0 normal to the plane has the parametric equations $x = x_0 + At$, $y = y_0 + Bt$, $z = z_0 + Ct$.

(b) If $P = (x, y, z)$ is the point of intersection of the line in part (a) with the plane, show that $x - x_0 = At$, $y - y_0 = Bt$, $z - z_0 = Ct$, where

$$t = \frac{-(Ax_0 + By_0 + Cz_0) + D}{A^2 + B^2 + C^2}$$

(c) Explain why d is the distance between P_0 and P, and use the distance formula to finish the proof.

2. Prove that the distance between the parallel planes $Ax + By + Cz = D_1$ and $Ax + By + Cz = D_2$ is

$$d = \frac{|D_2 - D_1|}{\sqrt{A^2 + B^2 + C^2}}$$

3. Suppose that the cross product of $\mathbf{v} = v_1\mathbf{i} + v_2\mathbf{j} + v_3\mathbf{k}$ and $\mathbf{w} = w_1\mathbf{i} + w_2\mathbf{j} + w_3\mathbf{k}$ has not yet been defined, and we seek a vector $\mathbf{u} = x\mathbf{i} + y\mathbf{j} + z\mathbf{k}$ orthogonal to both $\mathbf{v}$ and $\mathbf{w}$.

(a) Explain why (x, y, z) satisfies the system of equations $v_1x + v_2y + v_3z = 0$, $w_1x + w_2y + w_3z = 0$.

(b) By eliminating first y and then x from the system, show that

$$(v_1w_2 - w_1v_2)x = (v_2w_3 - w_2v_3)z$$

and

$$(v_1w_2 - w_1v_2)y = -(v_1w_3 - w_1v_3)z$$

(c) Use the result in part (b) to show that one solution of the system is

$$x = v_2w_3 - w_2v_3$$

$$y = -(v_1w_3 - w_1v_3)$$

$$z = v_1w_2 - w_1v_2$$

Compare $u = x\mathbf{i} + y\mathbf{j} + z\mathbf{k}$ (for this choice of x, y, z) with the definition of $\mathbf{v} \times \mathbf{w}$ in Section 14.5.

4. Let points $P_1 = (x_1, y_1, z_1)$ and $P_2 = (x_2, y_2, z_2)$ have respective position vectors $\mathbf{r}_1$ and $\mathbf{r}_2$. Show that the point $P = (x, y, z)$ that divides the segment P_1P_2 in the ratio $t = \overline{P_1P}/\overline{P_1P_2}$ has position vector $\mathbf{r} = \mathbf{r}_1 + t(\mathbf{r}_2 - \mathbf{r}_1)$. Thus, conclude that

$$x = x_1 + t(x_2 - x_1)$$

$$y = y_1 + t(y_2 - y_1)$$

$$z = z_1 + t(z_2 - z_1)$$

5. Let $\mathbf{u}$ and $\mathbf{v}$ be nonparallel vectors in the plane. Show that for each vector $\mathbf{w}$ there exist unique constants s and t so that $\mathbf{w} = s\mathbf{u} + t\mathbf{v}$. Interpret your result geometrically.

6. Let two skew lines have respective direction vectors $\mathbf{M}$ and $\mathbf{N}$. Let A and B be points on the respective lines, and let $\mathbf{w} = \overrightarrow{AB}$. Show that the distance between the two lines is the magnitude of the vector projection of $\mathbf{w}$ on $\mathbf{M} \times \mathbf{N}$:

$$|\text{proj}_{\mathbf{M}\times\mathbf{N}}\mathbf{w}| = \frac{|\mathbf{w} \cdot (\mathbf{M} \times \mathbf{N})|}{\|\mathbf{M} \times \mathbf{N}\|}$$

(The shortest distance is measured along the common perpendicular to the two lines.)

7. Find the distance between the lines

$$\frac{x-3}{2} = \frac{y}{3} = z \quad \text{and} \quad x = \frac{y+1}{-2} = \frac{z-2}{-1}$$

8. What is the minimum distance between the skew lines

$$x - 1 = y - 2 = z + 6$$

and

$$\frac{x-1}{2} = \frac{y+2}{-3} = z - 10$$

Locate the points on each line at which the distance is minimum.

9. Let points A, B, and C determine a plane p. If $\mathbf{u} = \overrightarrow{OA}$, $\mathbf{v} = \overrightarrow{OB}$, and $\mathbf{w} = \overrightarrow{OC}$, prove that the vector $\mathbf{u} \times \mathbf{v} + \mathbf{v} \times \mathbf{w} + \mathbf{w} \times \mathbf{u}$ is normal to p.

10. Show that $\mathbf{w} = \|\mathbf{v}\|\mathbf{u} + \|\mathbf{u}\|\mathbf{v}$ bisects the angle between $\mathbf{u}$ and $\mathbf{v}$.

11. Let $\mathbf{u}$ and $\mathbf{v}$ be fixed vectors and define $g(t) = \|\mathbf{u} - t\mathbf{v}\|^2$. Find the minimum value of g and deduce the Cauchy–Schwarz inequality $|\mathbf{u} \cdot \mathbf{v}| \leq \|\mathbf{u}\|\|\mathbf{v}\|$.

12. Show that two vectors in the plane are parallel if and only if their projections on two fixed mutually orthogonal vectors are proportional.

13. Show that $(\mathbf{r} - \mathbf{b}) \cdot (\mathbf{r} + \mathbf{b}) = 0$ is a vector equation of the sphere, with $\mathbf{b}$ and $-\mathbf{b}$ as endpoints of a diameter. Show that $(\mathbf{r} - \mathbf{r}_0 - \mathbf{b}) \cdot (\mathbf{r} - \mathbf{r}_0 + \mathbf{b}) = 0$ is a vector equation of the sphere with center $\mathbf{r}_0$ and radius $\|\mathbf{b}\|$.

14. Let A, B, C, and D be the vertices of a square in the plane with center at Q. Show that

$$\overrightarrow{OQ} = \frac{\overrightarrow{OA} + \overrightarrow{OB} + \overrightarrow{OC} + \overrightarrow{OD}}{4}$$

15. Sketch the set of points in the plane with position vector $\mathbf{r}$ satisfying $\mathbf{r} \cdot (\mathbf{i} + \mathbf{j}) \geq 0$.

16. Refer to Problem 57, page 825. Generalize the Gram–Schmidt orthogonalization process to vectors in n dimensions.

15

Vector Functions

15.1 Vector Functions and Their Derivatives

Until now, we have dealt only with functions whose domain and range are both sets of real numbers. Such functions may be appropriately described as *real functions of a real variable* or, simply, *real functions.* In this chapter, we discuss functions whose domain is a set of real numbers, but whose range consists of vectors in the plane or vectors in space. Such functions are called *vector-valued functions of a real variable* or, simply, *vector functions.* We denote a vector function by $\mathbf{r} = \mathbf{r}(t)$, where t is a real number defined on some interval $a \leq t \leq b$ and $\mathbf{r}$ is a position vector.

[15.1.1] DEFINITION / ***Vector Function.***

A vector function in the plane has as components two real functions $x = x(t)$ and $y = y(t)$, each defined on $a \leq t \leq b$, and

takes the form

$$\mathbf{r} = \mathbf{r}(t) = x(t)\mathbf{i} + y(t)\mathbf{j}$$

A vector function in space has as components three real functions $x = x(t)$, $y = y(t)$, **and** $z = z(t)$, **each defined on** $a \le t \le b$, **and takes the form**

$$\mathbf{r} = \mathbf{r}(t) = x(t)\mathbf{i} + y(t)\mathbf{j} + z(t)\mathbf{k}$$

As we did with vectors and scalars, we shall write the symbol for a vector function using boldface type and for a real function using lightface italic type.

In definition (15.1.1), the domain of the vector function $\mathbf{r} = \mathbf{r}(t)$ was specified as $a \le t \le b$. As with real functions, if no domain is specified for a vector function, it is assumed that the domain consists of all real numbers for which each of the component functions are defined.

Graph of a Vector Function

In general, the graph of a vector function $\mathbf{r}=\mathbf{r}(t)$ with domain $a \le t \le b$ is a curve in the sense that, as t varies over the interval $a \le t \le b$, the tip of the vector $\mathbf{r}=\mathbf{r}(t)$, whose initial point is the origin, traces out the curve. The components of the vector function are the *parametric equations* of the curve. We shall refer to the curve C: $\mathbf{r}(t) = x(t)\mathbf{i} + y(t)\mathbf{j}$, $a \le t \le b$, as a *curve in the plane* or as a *plane curve*. The curve C: $\mathbf{r}(t) = x(t)\mathbf{i} + y(t)\mathbf{j} + z(t)\mathbf{k}$, $a \le t \le b$, will be referred to as a *curve in space* or as a *space curve* (see Fig. 1).

The curve traced out by a vector function $\mathbf{r} = \mathbf{r}(t)$, $a \le t \le b$, has a *direction*, or *orientation*, at each point. We shall take as the *positive direction* along the curve, the direction in which the vector $\mathbf{r} = \mathbf{r}(t)$ moves as t increases.

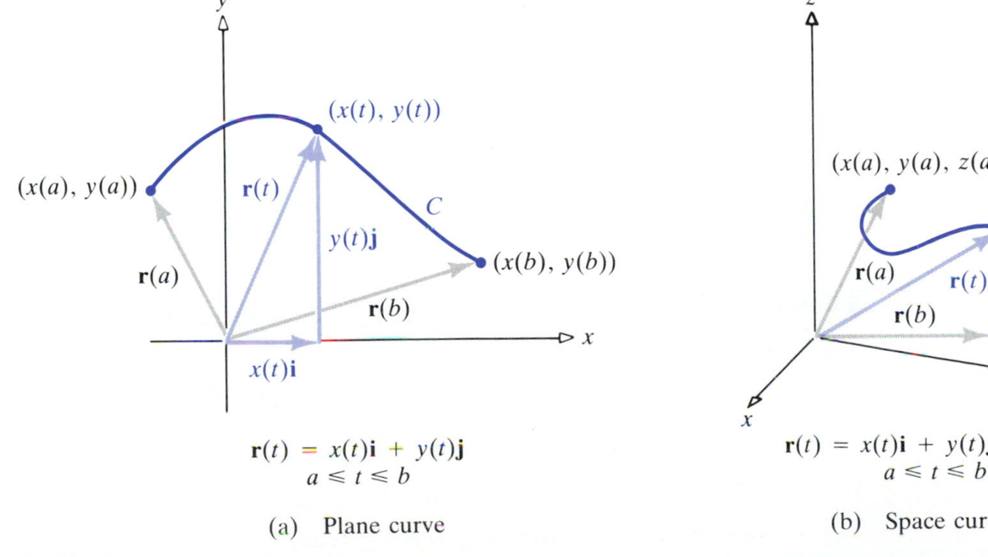

Figure 1

EXAMPLE 1

The vector function

$$\mathbf{r}(t) = \cos t\mathbf{i} + \sin t\mathbf{j} \qquad 0 \le t \le 2\pi$$

traces out a circle with center at the origin and radius equal to 1 unit. The positive direction along the circle is counterclockwise, beginning at $\mathbf{r}(0) = \mathbf{i}$. See Figure 2.

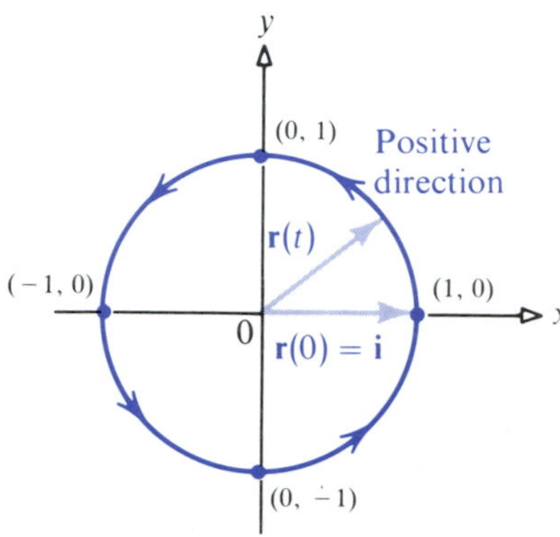

$\mathbf{r}(t) = (\cos t)\mathbf{i} + (\sin t)\mathbf{j}, \quad 0 \le t \le 2\pi$
$\|\mathbf{r}(t)\| = \sqrt{\cos^2 t + \sin^2 t} = 1 \quad$ for all t

Figure 2

EXAMPLE 2

The vector function

$$\mathbf{r}(t) = \sin t\mathbf{i} + \cos t\mathbf{j} \qquad 0 \le t \le \pi$$

traces out a semicircle with center at the origin and radius equal to 1 unit. The positive direction along the semicircle is clockwise, beginning at $\mathbf{r}(0) = \mathbf{j}$ and ending at $\mathbf{r}(\pi) = -\mathbf{j}$. See Figure 3.

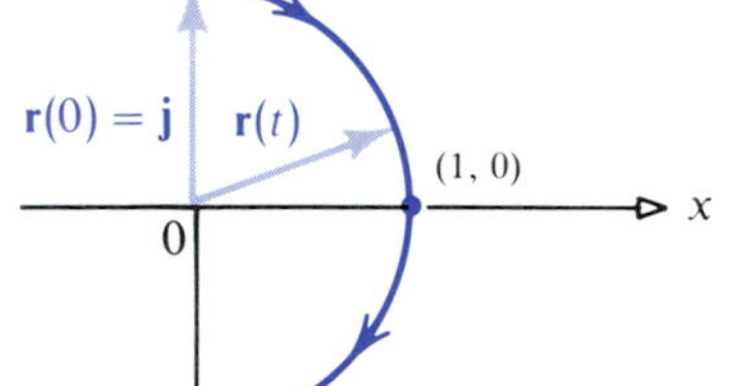

$\mathbf{r}(t) = (\sin t)\mathbf{i} + (\cos t)\mathbf{j}, \quad 0 \le t \le \pi$

Figure 3

EXAMPLE 3

Discuss the curve traced out by the vector function

$$\mathbf{r}(t) = (2 + 3t)\mathbf{i} + (3 - t)\mathbf{j} + 2t\mathbf{k}$$

Solution

The components of this vector function,

$$x(t) = 2 + 3t \qquad y(t) = 3 - t \qquad z(t) = 2t$$

are the parametric equations of a line in space that passes through the point (2, 3, 0). Symmetric equations of this line are

$$\frac{x - 2}{3} = \frac{y - 3}{-1} = \frac{z}{2}$$

Since $\mathbf{r}(1) = 5\mathbf{i} + 2\mathbf{j} + 2\mathbf{k}$ and $\mathbf{r}(0) = 2\mathbf{i} + 3\mathbf{j}$, the positive direction of this line is $\mathbf{r}(1) - \mathbf{r}(0) = 3\mathbf{i} - \mathbf{j} + 2\mathbf{k}$.

EXAMPLE 4

Discuss the curve traced out by

$$\mathbf{r}(t) = a \cos t\mathbf{i} + a \sin t\mathbf{j} + bt\mathbf{k} \qquad t \geq 0$$

where a and b are positive constants.

Solution

The parametric equations of this curve are

$$x = a \cos t \qquad y = a \sin t \qquad z = bt$$

For any t, a point (x, y, z) on this curve lies on the right circular cylinder

$$x^2 + y^2 = a^2$$

For $t = 0$, the point $(a, 0, 0)$ is on the curve. As t increases, the vector $\mathbf{r} = \mathbf{r}(t)$ starts at $(a, 0, 0)$ and winds around the circular cylinder, one revolution for every increase of 2π in t. See Figure 4.

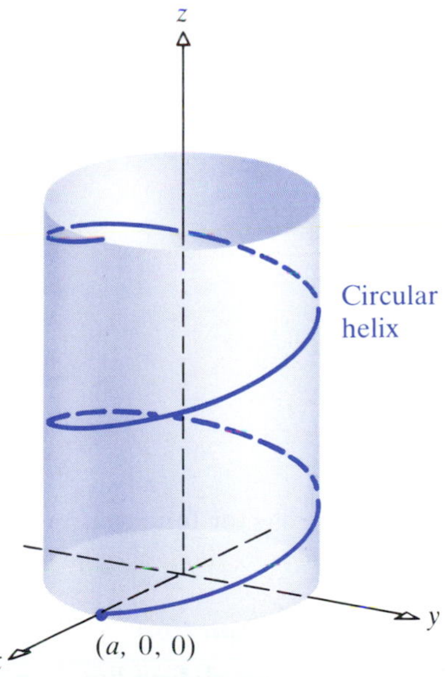

Figure 4

The space curve shown in Figure 4 is called a *circular helix* and may be viewed as a coiled spring, much like a Slinky® toy. One of the many occurrences in nature of a helix is in the model of deoxyribonucleic acid (DNA), which consists of a *double helix*—one helix intertwined with another helix (see Fig. 5).

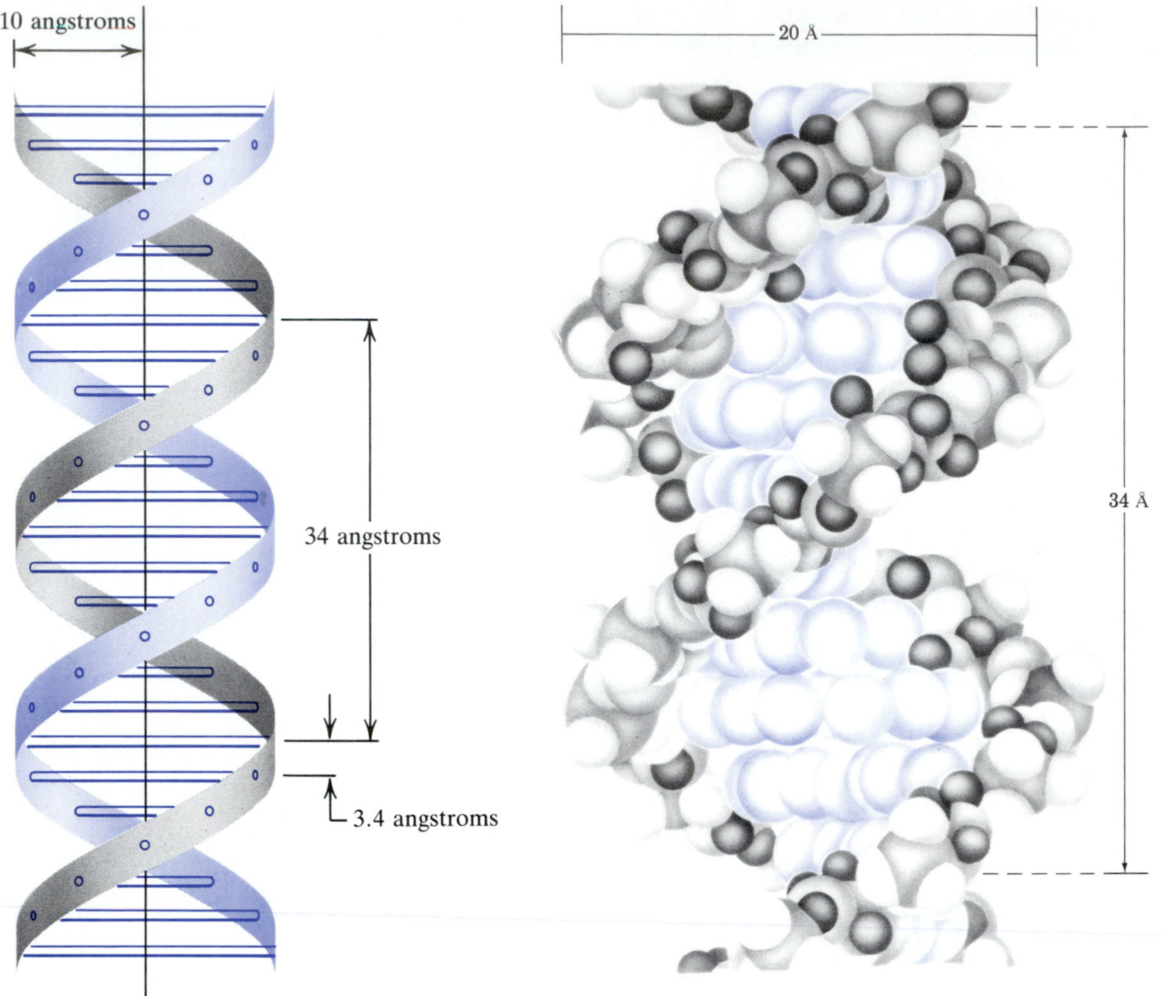

Figure 5

Limit, Continuity, and Derivative

The notions of limit, continuity, and derivative for vector functions are extensions of these same ideas for real functions.

[15.1.2] DEFINITION / ***Limit of a Vector Function.***

For a vector function $\mathbf{r}(t) = x(t)\mathbf{i} + y(t)\mathbf{j}$, **we define** $\lim_{t\to t_0} \mathbf{r}(t)$ **as**

$$\lim_{t\to t_0} \mathbf{r}(t) = \left[\lim_{t\to t_0} x(t)\right]\mathbf{i} + \left[\lim_{t\to t_0} y(t)\right]\mathbf{j}$$

provided $\lim_{t\to t_0} x(t)$ **and** $\lim_{t\to t_0} y(t)$ **both exist.**

For a vector function $\mathbf{r}(t) = x(t)\mathbf{i} + y(t)\mathbf{j} + z(t)\mathbf{k}$, **we define**

$\lim_{t\to t_0} \mathbf{r}(t)$ as

$$\lim_{t\to t_0} \mathbf{r}(t) = \left[\lim_{t\to t_0} x(t)\right]\mathbf{i} + \left[\lim_{t\to t_0} y(t)\right]\mathbf{j} + \left[\lim_{t\to t_0} z(t)\right]\mathbf{k}$$

provided $\lim_{t\to t_0} x(t)$, $\lim_{t\to t_0} y(t)$, and $\lim_{t\to t_0} z(t)$ exist.

Thus, we find the limit of a vector function by finding the limits of its component functions.

[15.1.3] DEFINITION / ***Continuous Vector Function.***

A vector function $\mathbf{r} = r(t)$ is *continuous* at t_0 if:

1. **$\mathbf{r}(t_0)$ exists**
2. **$\lim_{t\to t_0} \mathbf{r}(t)$ exists**
3. **$\lim_{t\to t_0} \mathbf{r}(t) = \mathbf{r}(t_0)$**

Definitions (15.1.2) and (15.1.3) lead to the following result:

[15.1.4] THEOREM

A vector function is continuous at t_0 if and only if each of its component functions is continuous at t_0.

EXAMPLE 5

For the vector function

$$\mathbf{r}(t) = t^2\mathbf{i} + (1 + t)\mathbf{j} + \sin t\mathbf{k}$$

find $\lim_{t\to 0} \mathbf{r}(t)$. Determine whether $\mathbf{r}$ is continuous at 0.

Solution

$$\begin{aligned}\lim_{t\to 0} \mathbf{r}(t) &= \lim_{t\to 0}[t^2\mathbf{i} + (1 + t)\mathbf{j} + \sin t\mathbf{k}] = \lim_{t\to 0} t^2\mathbf{i} + \lim_{t\to 0}(1 + t)\mathbf{j} + \lim_{t\to 0} \sin t\mathbf{k} \\ &= (0)\mathbf{i} + (1)\mathbf{j} + (0)\mathbf{k} = \mathbf{j}\end{aligned}$$

This vector function is continuous at 0 because $\mathbf{r}(0) = \mathbf{j}$.

The definition of continuity on an interval for vector functions follows the pattern developed in Chapter 2 for real functions.

Our main interest, however, is with the derivative. The definition of a derivative for vector functions is analogous to the definition for real functions.

[15.1.5] DEFINITION / ***Derivative of a Vector Function.***

The *derivative* of the vector function $\mathbf{r} = \mathbf{r}(t)$, denoted by $\mathbf{r}'(t)$, is defined by

$$\mathbf{r}'(t) = \lim_{h\to 0} \frac{\mathbf{r}(t + h) - \mathbf{r}(t)}{h}$$

provided this limit exists. In this event, we say $\mathbf{r}(t)$ is *differentiable*.

The next result is most useful. It states that the derivative of a vector function can be found by differentiating its components.

[15.1.6] THEOREM

If $\mathbf{r}(t) = x(t)\mathbf{i} + y(t)\mathbf{j}$ and if $x = x(t)$ and $y = y(t)$ are each differentiable, then

$$\mathbf{r}'(t) = x'(t)\mathbf{i} + y'(t)\mathbf{j}$$

If $\mathbf{r}(t) = x(t)\mathbf{i} + y(t)\mathbf{j} + z(t)\mathbf{k}$ and if $x = x(t)$, $y = y(t)$, and $z = z(t)$ are each differentiable, then

$$\mathbf{r}'(t) = x'(t)\mathbf{i} + y'(t)\mathbf{j} + z'(t)\mathbf{k}$$

Proof We shall prove this result for a vector function in space.

$$\begin{aligned}
\mathbf{r}'(t) &= \lim_{h\to 0} \frac{\mathbf{r}(t+h) - \mathbf{r}(t)}{h} \\
&= \lim_{h\to 0} \frac{[x(t+h) - x(t)]\mathbf{i} + [y(t+h) - y(t)]\mathbf{j} + [z(t+h) - z(t)]\mathbf{k}}{h} \\
&= \lim_{h\to 0}\left[\frac{x(t+h) - x(t)}{h}\right]\mathbf{i} + \lim_{h\to 0}\left[\frac{y(t+h) - y(t)}{h}\right]\mathbf{j} + \lim_{h\to 0}\left[\frac{z(t+h) - z(t)}{h}\right]\mathbf{k} \\
&= x'(t)\mathbf{i} + y'(t)\mathbf{j} + z'(t)\mathbf{k}
\end{aligned}$$

In some situations, we shall prefer to use the Leibniz notation, which for a differentiable vector function $\mathbf{r} = \mathbf{r}(t)$, has the form:

$$\frac{d\mathbf{r}}{dt} = \frac{dx}{dt}\mathbf{i} + \frac{dy}{dt}\mathbf{j} \qquad \text{in the plane}$$

$$\frac{d\mathbf{r}}{dt} = \frac{dx}{dt}\mathbf{i} + \frac{dy}{dt}\mathbf{j} + \frac{dz}{dt}\mathbf{k} \qquad \text{in space}$$

EXAMPLE 6

Find the derivative $\mathbf{r}'(t)$ of each vector function.

(a) $\mathbf{r}(t) = 2 \sin t\mathbf{i} + 3 \cos t\mathbf{j}$ (b) $\mathbf{r}(t) = t^2\mathbf{i} + (1 + t)\mathbf{j} + \sin t\mathbf{k}$

Solution

(a) $\dfrac{d\mathbf{r}}{dt} = \mathbf{r}'(t) = 2 \cos t\mathbf{i} - 3 \sin t\mathbf{j}$

(b) $\dfrac{d\mathbf{r}}{dt} = \mathbf{r}'(t) = 2t\mathbf{i} + \mathbf{j} + \cos t\mathbf{k}$

Higher-Order Derivatives

If it is possible to take the derivative of $\mathbf{r}'(t)$, we obtain the second derivative $\mathbf{r}''(t)$. Continuing in this fashion, we may compute $\mathbf{r}'''(t)$, $\mathbf{r}^{(4)}(t)$, and so on, provided these derivatives exist.

EXAMPLE 7

Find $\mathbf{r}'(t)$ and $\mathbf{r}''(t)$ for $\mathbf{r}(t) = e^t\mathbf{i} + \ln t\mathbf{j} - \cos t\mathbf{k}$.

Solution

$$\mathbf{r}'(t) = e^t\mathbf{i} + \frac{1}{t}\mathbf{j} + \sin t\mathbf{k}$$

$$\mathbf{r}''(t) = e^t\mathbf{i} - \frac{1}{t^2}\mathbf{j} + \cos t\mathbf{k}$$

Differentiation Formulas

The manner in which the derivative of a vector function is defined is quite similar to the way the derivative of a real function is defined. This similarity extends to the formulas for vector differentiation.

[15.1.7] THEOREM / *Derivative Formulas.*

If $u(t)$ is a differentiable scalar function and if $\mathbf{f}(t)$ and $\mathbf{g}(t)$ are differentiable vector functions, we have the following formulas:

(a) $[\mathbf{f}(t) + \mathbf{g}(t)]' = \mathbf{f}'(t) + \mathbf{g}'(t)$

(b) $[u(t)\mathbf{f}(t)]' = u'(t)\mathbf{f}(t) + u(t)\mathbf{f}'(t)$

(c) $[\mathbf{f}(t) \cdot \mathbf{g}(t)]' = \mathbf{f}'(t) \cdot \mathbf{g}(t) + \mathbf{f}(t) \cdot \mathbf{g}'(t)$

(d) $[\mathbf{f}(t) \times \mathbf{g}(t)]' = \mathbf{f}'(t) \times \mathbf{g}(t) + \mathbf{f}(t) \times \mathbf{g}'(t)$

We verify part (b) here and leave the proofs of parts (a), (c), and (d) as exercises (Problems 58–60).

Proof of Part (b) Let $\mathbf{f}(t) = f_1(t)\mathbf{i} + f_2(t)\mathbf{j} + f_3(t)\mathbf{k}$. Then

$$u(t)\mathbf{f}(t) = u(t)f_1(t)\mathbf{i} + u(t)f_2(t)\mathbf{j} + u(t)f_3(t)\mathbf{k}$$

By using the rule for differentiating the product of two real functions, we obtain

$$\begin{aligned}[u(t)\mathbf{f}(t)]' &= [u(t)f_1(t)]'\mathbf{i} + [u(t)f_2(t)]'\mathbf{j} + [u(t)f_3(t)]'\mathbf{k} \\ &= [u'(t)f_1(t) + u(t)f'_1(t)]\mathbf{i} + [u'(t)f_2(t) + u(t)f'_2(t)]\mathbf{j} + [u'(t)f_3(t) + u(t)f'_3(t)]\mathbf{k} \\ &= u'(t)[f_1(t)\mathbf{i} + f_2(t)\mathbf{j} + f_3(t)\mathbf{k}] + u(t)[f'_1(t)\mathbf{i} + f'_2(t)\mathbf{j} + f'_3(t)\mathbf{k}] \\ &= u'(t)\mathbf{f}(t) + u(t)\mathbf{f}'(t)\end{aligned}$$

EXAMPLE 8

Find the derivative of $\mathbf{f}(t) \times \mathbf{g}(t)$ for

$$\mathbf{f}(t) = \cos t\mathbf{i} + \sin t\mathbf{j} + t\mathbf{k} \qquad \mathbf{g}(t) = t\mathbf{i} + \ln t\mathbf{j} + \mathbf{k}$$

Solution

$$\mathbf{f}'(t) = -\sin t\mathbf{i} + \cos t\mathbf{j} + \mathbf{k} \qquad \mathbf{g}'(t) = \mathbf{i} + \frac{1}{t}\mathbf{j}$$

$$
\begin{aligned}
[\mathbf{f}(t) \times \mathbf{g}(t)]' &= \mathbf{f}'(t) \times \mathbf{g}(t) + \mathbf{f}(t) \times \mathbf{g}'(t) \\
&= (-\sin t\mathbf{i} + \cos t\mathbf{j} + \mathbf{k}) \times (t\mathbf{i} + \ln t\mathbf{j} + \mathbf{k}) \\
&\quad + (\cos t\mathbf{i} + \sin t\mathbf{j} + t\mathbf{k}) \times \left(\mathbf{i} \times \frac{1}{t}\mathbf{j}\right) \\
&= \begin{vmatrix} \mathbf{i} & \mathbf{j} & \mathbf{k} \\ -\sin t & \cos t & 1 \\ t & \ln t & 1 \end{vmatrix} + \begin{vmatrix} \mathbf{i} & \mathbf{j} & \mathbf{k} \\ \cos t & \sin t & t \\ 1 & \dfrac{1}{t} & 0 \end{vmatrix} \\
&= (\cos t - \ln t)\mathbf{i} - (-\sin t - t)\mathbf{j} + [(-\sin t)(\ln t) - t\cos t]\mathbf{k} \\
&\quad + (-1)\mathbf{i} - (-t)\mathbf{j} + \left(\frac{1}{t}\cos t - \sin t\right)\mathbf{k} \\
&= (\cos t - \ln t - 1)\mathbf{i} + (\sin t + 2t)\mathbf{j} \\
&\quad + \left[\frac{1}{t}\cos t - \sin t - (\sin t)(\ln t) - t\cos t\right]\mathbf{k}
\end{aligned}
$$

In using part (d) of theorem (15.1.7), care must be taken. Remember that the cross product is not commutative, so order is important.

The solution to Example 8 could have been obtained by first taking the cross product and then differentiating. Which operation to perform first in a particular situation is determined by convenience and ease of computation.

The Chain Rule

If $\mathbf{r} = \mathbf{r}(u)$ is a vector function and if the parameter u is a function of another parameter t—say, $u = f(t)$—then the composite vector function $\mathbf{r} = \mathbf{r}(u) = \mathbf{r}(f(t))$ is a vector function of the parameter t. To find the derivative of a composite vector function, we use the chain rule.

[15.1.8] THEOREM / *Chain Rule.*

If $\mathbf{r} = \mathbf{r}(u)$ is a differentiable vector function and if $u = f(t)$ is a differentiable real function, then $\mathbf{r} = \mathbf{r}(u) = \mathbf{r}(f(t))$ is a differential vector function of t and

$$\frac{d\mathbf{r}}{dt} = \frac{d\mathbf{r}}{du}\frac{du}{dt}$$

EXERCISE 15.1

In Problems 1–8 find the value of each vector function at t.

1. $\mathbf{r}(t) = t^2\mathbf{i} - 2t\mathbf{j}, \quad t = 1$
2. $\mathbf{r}(t) = t^3\mathbf{i} + 2t\mathbf{j}, \quad t = 2$
3. $\mathbf{r}(t) = \sin t\mathbf{i} - \cos t\mathbf{j}, \quad t = \pi/4$
4. $\mathbf{r}(t) = \tan t\mathbf{i} + \cos 2t\mathbf{j}, \quad t = 0$
5. $\mathbf{r}(t) = e^{2t}\mathbf{i} + t^2\mathbf{j} - 2\mathbf{k}, \quad t = 0$
6. $\mathbf{r}(t) = \ln t\mathbf{i} - \mathbf{j} + t^3\mathbf{k}, \quad t = 1$

7. $\mathbf{r}(t) = t\mathbf{i} - \cos\frac{\pi t}{4}\mathbf{j} + 2t\mathbf{k}, \quad t = 4$

8. $\mathbf{r}(t) = \sin\frac{3\pi t}{4}\mathbf{i} + 3\mathbf{j} - 3t^2\mathbf{k}, \quad t = 1$

In Problems 9–20 a vector function defining a curve in the plane is given. Graph each vector function, indicating its orientation.

9. $\mathbf{r}(t) = 2t\mathbf{i} + t^2\mathbf{j}, \quad t \ge 0$

10. $\mathbf{r}(t) = t^2\mathbf{i} - 4t\mathbf{j}, \quad t \le 0$

11. $\mathbf{r}(t) = t\mathbf{i}, \quad -1 \le t \le 1$

12. $\mathbf{r}(t) = t\mathbf{j}, \quad -1 \le t \le 1$

13. $\mathbf{r}(t) = t\mathbf{i} + t\mathbf{j}$

14. $\mathbf{r}(t) = 3t\mathbf{i} + 2t\mathbf{j}$

15. $\mathbf{r}(t) = 3t\mathbf{i} - 2t\mathbf{j}$

16. $\mathbf{r}(t) = t\mathbf{i} + t^2\mathbf{j}$

17. $\mathbf{r}(t) = \cos t\mathbf{i} - \sin t\mathbf{j}, \quad 0 \le t \le \pi/2$

18. $\mathbf{r}(t) = \cos t\mathbf{i} + \sin t\mathbf{j}, \quad 0 \le t \le \pi/2$

19. $\mathbf{r}(t) = \sin^2 t\mathbf{i} + \cos^2 t\mathbf{j}, \quad 0 \le t \le \pi/2$

20. $\mathbf{r}(t) = \sin^2 t\mathbf{i} - \cos^2 t\mathbf{j}, \quad 0 \le t \le \pi/2$

In Problems 21–34 find $\mathbf{r}'(t)$ and $\mathbf{r}''(t)$.

21. $\mathbf{r}(t) = 4t^2\mathbf{i} - 2t^3\mathbf{j}$

22. $\mathbf{r}(t) = 8t\mathbf{i} + 4t^3\mathbf{j}$

23. $\mathbf{r}(t) = 4\sqrt{t}\mathbf{i} + 2e^t\mathbf{j}$

24. $\mathbf{r}(t) = e^{3t}\mathbf{i} + \sqrt[3]{t}\mathbf{j}$

25. $\mathbf{r}(t) = \sin 2t\mathbf{i} - \cos 2t\mathbf{j}$

26. $\mathbf{r}(t) = \sin^2 t\mathbf{i} - \cos^2 t\mathbf{j}$

27. $\mathbf{r}(t) = \mathbf{i} + t\mathbf{j} + t^2\mathbf{k}$

28. $\mathbf{r}(t) = \mathbf{i} - \mathbf{j} + t\mathbf{k}$

29. $\mathbf{r}(t) = t^2\mathbf{i} + t^3\mathbf{j} - t\mathbf{k}$

30. $\mathbf{r}(t) = (1 + t)\mathbf{i} - 3t^2\mathbf{j} + t\mathbf{k}$

31. $\mathbf{r}(t) = e^t\cos t\mathbf{i} + e^t\sin t\mathbf{j} + t\mathbf{k}$

32. $\mathbf{r}(t) = e^{-t}\cos t\mathbf{i} + e^{-t}\sin t\mathbf{j} - t\mathbf{k}$

33. $\mathbf{r}(t) = (t - t^3)\mathbf{i} + (t + t^3)\mathbf{j} - t\mathbf{k}$

34. $\mathbf{r}(t) = (t^2 - t)\mathbf{i} + (t^2 + t)\mathbf{j} + t\mathbf{k}$

In Problems 35–42 a vector function defining a curve in the plane is given. Graph each curve, indicating its orientation. Include in the sketch the vectors $\mathbf{r}(0)$ and $\mathbf{r}'(0)$. Draw $\mathbf{r}'(0)$ so its initial point is at the tip of $\mathbf{r}(0)$.

35. $\mathbf{r}(t) = t\mathbf{i} + t^2\mathbf{j}$

36. $\mathbf{r}(t) = 2t^2\mathbf{i} - t\mathbf{j}$

37. $\mathbf{r}(t) = t\mathbf{i} + e^t\mathbf{j}$

38. $\mathbf{r}(t) = t\mathbf{i} + \ln(1 + t)\mathbf{j}$

39. $\mathbf{r}(t) = 3\sin t\mathbf{i} - 3\cos t\mathbf{j}$

40. $\mathbf{r}(t) = 4\sin t\mathbf{i} + 4\cos t\mathbf{j}$

41. $\mathbf{r}(t) = 2\cos t\mathbf{i} - 3\sin t\mathbf{j}$

42. $\mathbf{r}(t) = -\cos t\mathbf{i} + 2\sin t\mathbf{j}$

In Problems 43–48 find $[\mathbf{f}(t) \cdot \mathbf{g}(t)]'$.

43. $\mathbf{f}(t) = t^2\mathbf{i} - t\mathbf{j}, \quad \mathbf{g}(t) = t\mathbf{i} + t^2\mathbf{j}$

44. $\mathbf{f}(t) = t^3\mathbf{i} + t\mathbf{j}, \quad \mathbf{g}(t) = t\mathbf{i} - 2t\mathbf{j}$

45. $\mathbf{f}(t) = e^t\mathbf{i} + e^{-t}\mathbf{j}, \quad \mathbf{g}(t) = t\mathbf{i} - t^2\mathbf{j}$

46. $\mathbf{f}(t) = t\mathbf{i} + 4\sqrt{t}\mathbf{j}, \quad \mathbf{g}(t) = 2t\mathbf{i} - t^2\mathbf{j}$

47. $\mathbf{f}(t) = \sin\omega t\mathbf{i} + \cos\omega t\mathbf{j}, \quad \mathbf{g}(t) = \mathbf{i} + \mathbf{j}$

48. $\mathbf{f}(t) = \sin^2 t\mathbf{i} - \cos^2 t\mathbf{j}, \quad \mathbf{g}(t) = \mathbf{i} - \mathbf{j}$

In Problems 49–54 find $[\mathbf{f}(t) \cdot \mathbf{g}(t)]'$ and $[\mathbf{f}(t) \times \mathbf{g}(t)]'$.

49. $\mathbf{f}(t) = 2t\mathbf{i} + t^2\mathbf{j} - 5\mathbf{k}, \quad \mathbf{g}(t) = t^2\mathbf{i} + 2t\mathbf{j} + \mathbf{k}$

50. $\mathbf{f}(t) = t^3\mathbf{i} - t^2\mathbf{j} + t\mathbf{k}, \quad \mathbf{g}(t) = t\mathbf{i} - t^2\mathbf{j} + t^3\mathbf{k}$

51. $\mathbf{f}(t) = \cos 2t\mathbf{i} + \sin 2t\mathbf{j} + \mathbf{k}, \quad \mathbf{g}(t) = \cos t\mathbf{i} + \sin t\mathbf{j} + \mathbf{k}$

52. $\mathbf{f}(t) = \tan t\mathbf{i} + t\mathbf{j} + \mathbf{k}, \quad \mathbf{g}(t) = \sec t\mathbf{i} + 3t\mathbf{j} - \mathbf{k}$

53. $\mathbf{f}(t) = e^{2t}\mathbf{i} + e^{-2t}\mathbf{j} + t\mathbf{k}, \quad \mathbf{g}(t) = e^{-t}\mathbf{i} + e^{-2t}\mathbf{j} - t\mathbf{k}$

54. $\mathbf{f}(t) = \ln t\mathbf{i} + t\mathbf{j} - \ln(t + 1)\mathbf{k}, \quad \mathbf{g}(t) = \frac{1}{t}\mathbf{i} + t^2\mathbf{j} - t\mathbf{k}$

55. For $\mathbf{f}(t) = \sin t\mathbf{i} - \cos t\mathbf{j}$, show that $\mathbf{f}(t)$ and $\mathbf{f}''(t)$ are parallel.

56. For $\mathbf{f}(t) = e^{3t}\mathbf{i} + e^{-3t}\mathbf{j}$, show that $\mathbf{f}(t)$ and $\mathbf{f}''(t)$ are parallel.

57. Given the vectors $\mathbf{u} = \cos \omega t\mathbf{i} + \sin \omega t\mathbf{j}$ and $\mathbf{v} = t\mathbf{i} + t^2\mathbf{j} + t^3\mathbf{k}$, find:

(a) $\dfrac{d\mathbf{u}}{dt}$ and $\left\|\dfrac{d\mathbf{u}}{dt}\right\|$ (b) $\dfrac{d^2\mathbf{v}}{dt^2}$ and $\left\|\dfrac{d^2\mathbf{v}}{dt^2}\right\|$

58. Prove part (a) of theorem (15.1.7): If $\mathbf{f}(t)$ and $\mathbf{g}(t)$ are differentiable vector functions, show that

$$[\mathbf{f}(t) + \mathbf{g}(t)]' = \mathbf{f}'(t) + \mathbf{g}'(t)$$

59. Prove part (c) of theorem (15.1.7): If $\mathbf{f}(t)$ and $\mathbf{g}(t)$ are differentiable vector functions, show that

$$[\mathbf{f}(t) \cdot \mathbf{g}(t)]' = \mathbf{f}'(t) \cdot \mathbf{g}(t) + \mathbf{f}(t) \cdot \mathbf{g}'(t)$$

60. Prove part (d) of theorem (15.1.7): If $\mathbf{f}(t)$ and $\mathbf{g}(t)$ are differentiable vector functions, show that

$$[\mathbf{f}(t) \times \mathbf{g}(t)]' = \mathbf{f}'(t) \times \mathbf{g}(t) + \mathbf{f}(t) \times \mathbf{g}'(t)$$

61. If $\mathbf{f}(t)$ and $\mathbf{g}(t)$ are differentiable vector functions, show that

$$[\mathbf{f}(t) \times \mathbf{g}(t)]' = -[\mathbf{g}(t) \times \mathbf{f}(t)]'$$

62. If $\mathbf{r}(t)$ is a twice differentiable vector function, show that

$$[\mathbf{r}(t) \times \mathbf{r}'(t)]' = \mathbf{r}(t) \times \mathbf{r}''(t)$$

63. If $\mathbf{r}(t)$ is a differentiable vector function, show that $\|\mathbf{r}(t)\|$ is constant if and only if $\mathbf{r}(t) \cdot \mathbf{r}'(t) = 0$ for all t.

64. Show that

$$\lim_{t\to t_0} \mathbf{r}(t) = \mathbf{L} \quad \text{if and only if} \quad \lim_{t\to t_0} \|\mathbf{r}(t) - \mathbf{L}\| = 0$$

65. If $\mathbf{r} = \mathbf{r}(t)$ is a differentiable vector function, show that

$$\frac{d}{dt}\|\mathbf{r}(t)\| = \frac{\mathbf{r}(t) \cdot \mathbf{r}'(t)}{\|\mathbf{r}(t)\|}$$

66. If $\mathbf{r} = \mathbf{r}(t)$ is a differentiable vector function, show that

$$\frac{d}{dt}\frac{\mathbf{r}(t)}{\|\mathbf{r}(t)\|} = \frac{\mathbf{r}'(t)}{\|\mathbf{r}(t)\|} - \frac{\mathbf{r}(t) \cdot \mathbf{r}'(t)}{\|\mathbf{r}(t)\|^3}\mathbf{r}(t)$$

67. The position vector of a particle at time t is $\mathbf{r} = (1 - t^3)\mathbf{i} + t^2\mathbf{j}$. Eliminate the parameter to find y as a function of x, and sketch the path of the particle. Indicate the direction of travel.

68. Rework Problem 67 if $\mathbf{r} = t^2\mathbf{i} + (4t^2 - t^4)\mathbf{j}$.

69. Find an equation in x and y of the graph of the vector function $\mathbf{r} = \sin^2 t\mathbf{i} + \tan t\mathbf{j}$, $-\pi/2 < t < \pi/2$, and sketch the graph. Indicate the positive direction.

70. Find an equation in x and y of the graph of the vector function $\mathbf{r} = e^t\mathbf{i} + e^{-t}\mathbf{j}$ and sketch the graph. Indicate the direction of travel.

15.2 Unit Tangent and Normal Vectors; Arc Length

Geometric Interpretation of the Derivative of a Vector Function

We have already seen that a vector function $\mathbf{r} = \mathbf{r}(t)$ that is defined and differentiable on some interval traces out a curve as t ranges over the interval. For a given number t_0, $\mathbf{r} = \mathbf{r}(t_0)$ is the position vector of a point P_0 on the curve. If $h \neq 0$ is an increment of t_0, the vector $\mathbf{r}(t_0 + h)$ is also the position vector of a point on the curve (see Fig. 6).

The vector

$$\frac{\mathbf{r}(t_0 + h) - \mathbf{r}(t_0)}{h} = \frac{1}{h}[\mathbf{r}(t_0 + h) - \mathbf{r}(t_0)]$$

represents a vector parallel to $\mathbf{r}(t_0 + h) - \mathbf{r}(t_0)$. If we allow h to approach 0, the limiting position of this vector is a vector tangent to the curve at P_0. But

$$\mathbf{r}'(t_0) = \lim_{h\to 0} \frac{\mathbf{r}(t_0 + h) - \mathbf{r}(t_0)}{h}$$

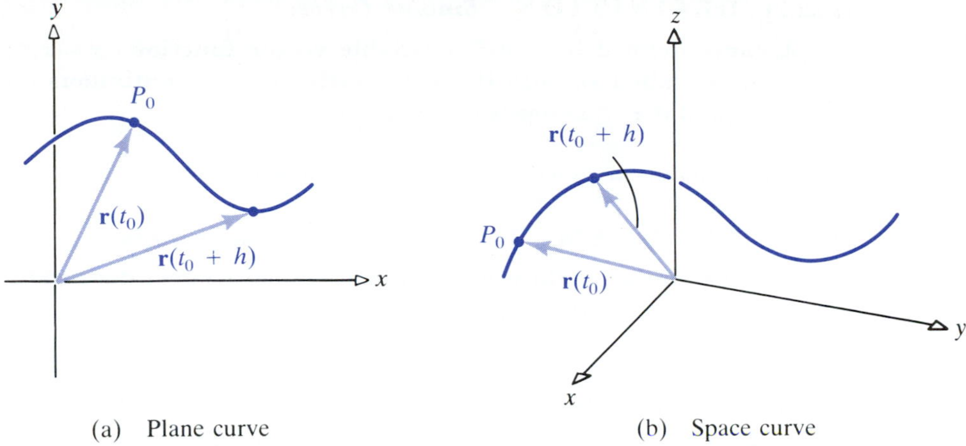

(a) Plane curve (b) Space curve

Figure 6

If this limit is not $\mathbf{0}$ at P_0, then the direction of $\mathbf{r}'(t_0)$ defines the *direction of the tangent* line to the curve at P_0. This leads us to make the following definition:

[15.2.1] DEFINITION / *Tangent Vector.*

Let $\mathbf{r} = \mathbf{r}(t)$ denote a differentiable vector function defined on $a \leq t \leq b$, and let P_0 be the point on the graph of $\mathbf{r} = \mathbf{r}(t)$ corresponding to $t = t_0$. If $\mathbf{r}'(t_0) \neq \mathbf{0}$, then the vector $\mathbf{r}'(t_0)$ is the *tangent vector* to the graph of $\mathbf{r} = \mathbf{r}(t)$ at t_0. If the tangent vector $\mathbf{r}'(t_0)$ is placed so its initial point is at P_0, then the line containing P_0 in the direction of $\mathbf{r}'(t_0)$ is the *tangent line* to the graph of $\mathbf{r} = \mathbf{r}(t)$ at t_0.

Figure 7 illustrates these definitions.

Note that there is no tangent vector defined if $\mathbf{r}'(t_0) = \mathbf{0}$.

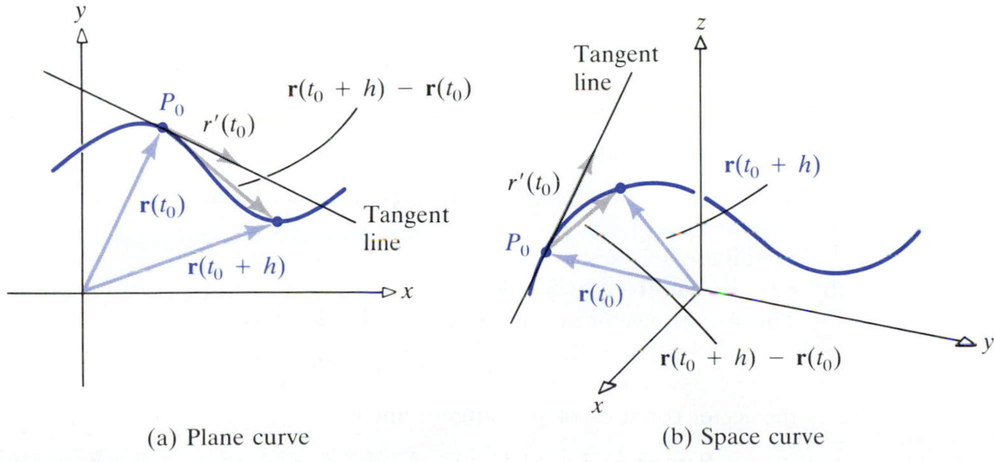

(a) Plane curve (b) Space curve

Figure 7

[15.2.2] DEFINITION / *Smooth Curve.*

A curve defined by a differentiable vector function $\mathbf{r} = \mathbf{r}(t)$, $a \le t \le b$, is called *smooth* if its derivative $\mathbf{r}'(t)$ is continuous on $a \le t \le b$ and if $\mathbf{r}'(t)$ is never $\mathbf{0}$ for $a \le t \le b$.

Thus, smooth curves have a tangent vector at every point.

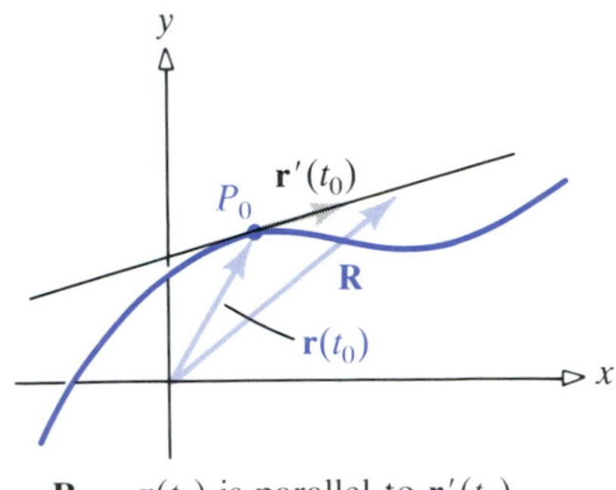

$\mathbf{R} - \mathbf{r}(t_0)$ is parallel to $\mathbf{r}'(t_0)$

Figure 8

[15.2.3] THEOREM / *Vector Equation of Tangent Line.*

The vector equation of the tangent line to a smooth curve defined by $\mathbf{r} = \mathbf{r}(t)$, $a \le t \le b$, at t_0 is given by

$$\mathbf{R}(u) = \mathbf{r}(t_0) + u\mathbf{r}'(t_0)$$

where $\mathbf{R}(u)$ is the position vector of any point on the tangent line and, to avoid confusion, we use u as the parameter of the line.

Proof See Figure 8. Suppose $\mathbf{R}$ is the position vector of any point on the tangent. The direction of the tangent line at t_0 is $\mathbf{r}'(t_0)$. For any u, the vector $\mathbf{R} - \mathbf{r}(t_0)$ is parallel to $\mathbf{r}'(t_0)$. Thus, the equation of the tangent line is

$$\mathbf{R}(u) - \mathbf{r}(t_0) = u\mathbf{r}'(t_0)$$

or

$$\mathbf{R}(u) = \mathbf{r}(t_0) + u\mathbf{r}'(t_0)$$

where u is the parameter of the line.

EXAMPLE 1

Find the tangent vector:

(a) To the plane curve $\mathbf{r}(t) = 2t^2\mathbf{i} + 4t\mathbf{j}$ at $t = 1$
(b) To the helix $\mathbf{r}(t) = \cos t\mathbf{i} + \sin t\mathbf{j} + t\mathbf{k}$ at $t = \pi$

What is the vector equation of the tangent line at these points?

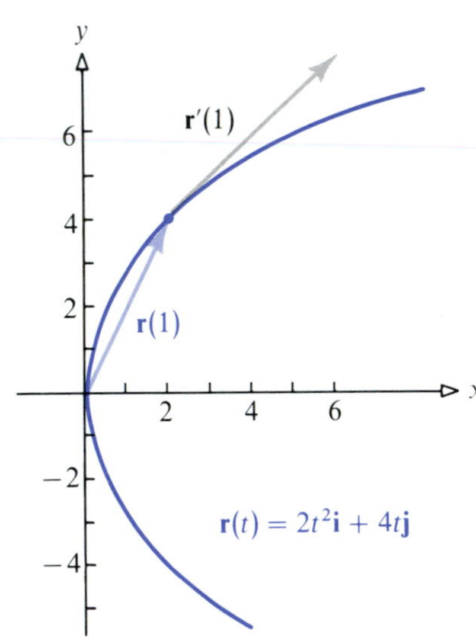

Figure 9

Solution

(a) $\mathbf{r}'(t) = 4t\mathbf{i} + 4\mathbf{j}$
The tangent vector at 1 is $\mathbf{r}'(1) = 4\mathbf{i} + 4\mathbf{j}$. Since

$$\mathbf{r}(1) = 2\mathbf{i} + 4\mathbf{j}$$

the vector equation of the tangent line is

$$\mathbf{R}(u) = (2\mathbf{i} + 4\mathbf{j}) + u(4\mathbf{i} + 4\mathbf{j}) = (2 + 4u)\mathbf{i} + (4 + 4u)\mathbf{j}$$

See Figure 9.

(b) $\mathbf{r}'(t) = -\sin t\mathbf{i} + \cos t\mathbf{j} + \mathbf{k}$
The tangent vector at π is $\mathbf{r}'(\pi) = -\mathbf{j} + \mathbf{k}$. Since

$$\mathbf{r}(\pi) = -\mathbf{i} + \pi\mathbf{k}$$

the vector equation of the tangent line is

$$\mathbf{R}(u) = (-\mathbf{i} + \pi\mathbf{k}) + u(-\mathbf{j} + \mathbf{k}) = -\mathbf{i} - u\mathbf{j} + (\pi + u)\mathbf{k}$$

See Figure 10.

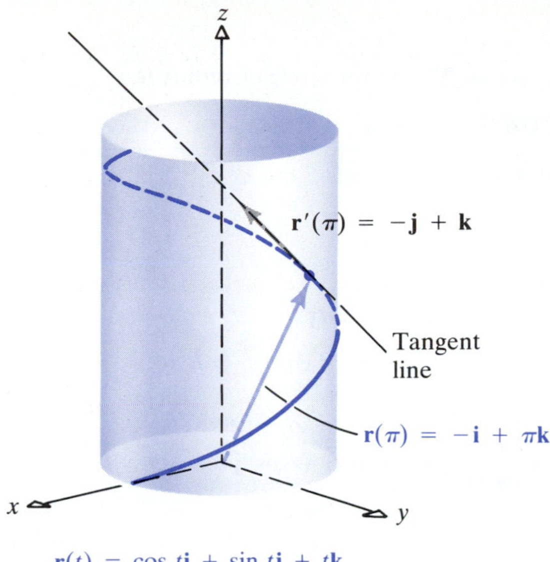

Figure 10

EXAMPLE 2

For the helix

$$\mathbf{r}(t) = \cos t\mathbf{i} + \sin t\mathbf{j} + t\mathbf{k} \qquad 0 \leq t \leq 2\pi$$

show that the acute angle between the tangent vector and the z-axis is $\pi/4$ radian.

Solution

The tangent vector at any point P on the helix is given by

$$\mathbf{r}'(t) = -\sin t\mathbf{i} + \cos t\mathbf{j} + \mathbf{k}$$

The direction of the z-axis is $\mathbf{k}$. The cosine of the acute angle θ between $\mathbf{r}'(t)$ and $\mathbf{k}$ is

$$\cos\theta = \frac{\mathbf{r}'(t)\cdot\mathbf{k}}{\|\mathbf{r}'(t)\|\,\|\mathbf{k}\|} \underset{\|\mathbf{k}\|=1}{\uparrow}= \frac{1}{\sqrt{\sin^2 t + \cos^2 t + 1}} = \frac{1}{\sqrt{2}} = \frac{\sqrt{2}}{2}$$

Hence, $\theta = \pi/4$ radian.

Unit Tangent and Normal Vectors

[15.2.4] DEFINITION / *Unit Tangent Vector.*

For a smooth curve C defined by the vector function $\mathbf{r} = \mathbf{r}(t)$, $a \leq t \leq b$, the *unit tangent vector* $\mathbf{T}(t)$ to C at t is defined as

$$\mathbf{T}(t) = \frac{\mathbf{r}'(t)}{\|\mathbf{r}'(t)\|} \tag{1}$$

EXAMPLE 3

Show that the unit tangent vector $\mathbf{T}(t)$ to the circle of radius R,

$$\mathbf{r}(t) = R\cos t\mathbf{i} + R\sin t\mathbf{j} \qquad 0 \le t \le 2\pi$$

is everywhere orthogonal to $\mathbf{r}(t)$.

Solution

We calculate $\mathbf{r}'(t)$ and $\|\mathbf{r}'(t)\|$:

$$\begin{aligned}\mathbf{r}'(t) &= -R\sin t\mathbf{i} + R\cos t\mathbf{j}\\ \|\mathbf{r}'(t)\| &= \sqrt{(-R\sin t)^2 + (R\cos t)^2} = R\end{aligned}$$

From (1), we have

$$\mathbf{T}(t) = \frac{\mathbf{r}'(t)}{\|\mathbf{r}'(t)\|} = -\sin t\mathbf{i} + \cos t\mathbf{j}$$

It follows that

$$\begin{aligned}\mathbf{T}(t)\cdot\mathbf{r}(t) &= (-\sin t\mathbf{i} + \cos t\mathbf{j})\cdot(R\cos t\mathbf{i} + R\sin t\mathbf{j})\\ &= -R\sin t\cos t + R\sin t\cos t = 0\end{aligned}$$

for all t. Hence, $\mathbf{T}(t)$ is everywhere orthogonal to $\mathbf{r}(t)$ (see Fig. 11).

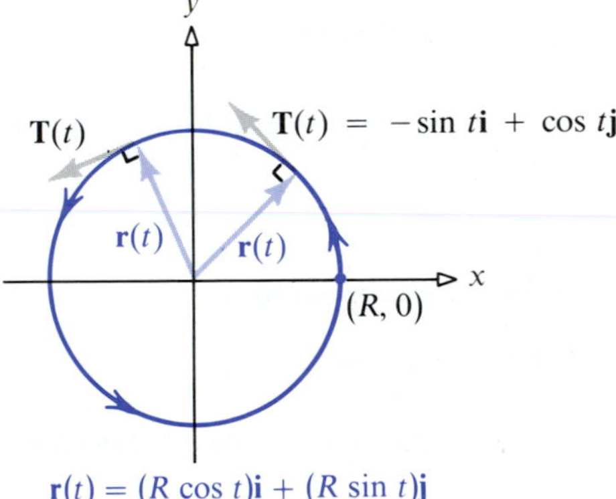

Figure 11

Suppose a smooth curve C is traced out by a twice differentiable vector function $\mathbf{r} = \mathbf{r}(t)$. Then the unit tangent vector $\mathbf{T}(t)$ is itself differentiable. Furthermore, since $\mathbf{T}(t)$ is a unit vector, we know that

$$\mathbf{T}(t)\cdot\mathbf{T}(t) = 1 \qquad \text{for all } t$$

If we differentiate this expression, we find that

$$\begin{aligned}\frac{d}{dt}[\mathbf{T}(t)\cdot\mathbf{T}(t)] &= \frac{d}{dt}1\\ \mathbf{T}'(t)\cdot\mathbf{T}(t) + \mathbf{T}(t)\cdot\mathbf{T}'(t) &= 0\\ \mathbf{T}(t)\cdot\mathbf{T}'(t) &= 0\end{aligned}$$

Thus, $\mathbf{T}'(t)$ is a vector that is orthogonal to $\mathbf{T}(t)$ at every point of C. Moreover, $\mathbf{T}'(t)/\|\mathbf{T}'(t)\|$ is a unit vector orthogonal to $\mathbf{T}(t)$ at every point of C.

[15.2.5] DEFINITION / *Principal Unit Normal Vector.*

For a smooth curve C defined by a twice differentiable vector function $\mathbf{r} = \mathbf{r}(t)$, $a \leq t \leq b$, the *principal unit normal vector* $\mathbf{N}(t)$ to C at t is defined as

$$\mathbf{N}(t) = \frac{\mathbf{T}'(t)}{\|\mathbf{T}'(t)\|}$$

If C is a plane curve, there are two unit vectors that are orthogonal to the unit tangent vector $\mathbf{T}$. If C is a curve in space, there are infinitely many such vectors. Definition (15.2.5) merely identifies one such vector and gives it a name. In fact, for a plane curve C, the principal unit normal vector $\mathbf{N}$ is the vector orthogonal to $\mathbf{T}$ that points toward the concave side of C.

To summarize, the unit tangent vector $\mathbf{T}(t)$ points in the direction of the tangent line; the principal unit normal vector $\mathbf{N}(t)$ is orthogonal to $\mathbf{T}(t)$ at every point.

EXAMPLE 4

Find the principal unit normal vector $\mathbf{N}(t)$ of the circle discussed in Example 3.

Solution

In Example 3 we found that the unit tangent vector $\mathbf{T}(t)$ is given by

$$\mathbf{T}(t) = -\sin t\mathbf{i} + \cos t\mathbf{j}$$

Hence, the unit principal normal vector $\mathbf{N}(t)$ is

$$\mathbf{N}(t) = \frac{\mathbf{T}'(t)}{\|\mathbf{T}'(t)\|} = \frac{-\cos t\mathbf{i} - \sin t\mathbf{j}}{\sqrt{(-\cos t)^2 + (-\sin t)^2}} = -\cos t\mathbf{i} - \sin t\mathbf{j}$$

For this circle, the unit principal normal vector $\mathbf{N}(t) = -\mathbf{r}(t)/\|\mathbf{r}(t)\|$. That is, $\mathbf{N}(t)$ is a unit vector opposite in direction to the vector $\mathbf{r}(t)$ and hence is directed toward the center of the circle. See Figure 12.

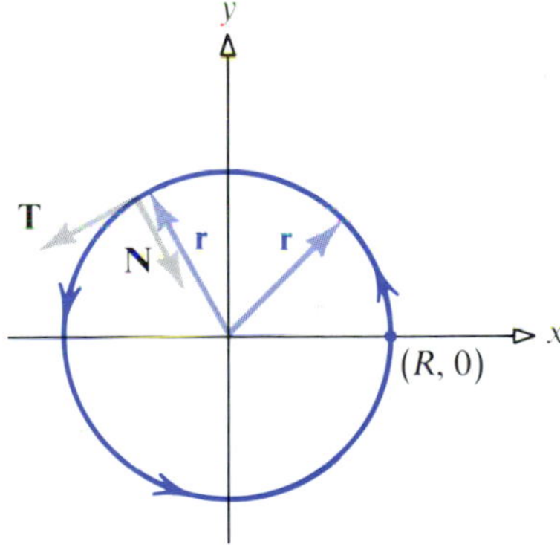

$\mathbf{r}(t) = (R\cos t)\mathbf{i} + (R\sin t)\mathbf{j}$

Figure 12

EXAMPLE 5

Show that the principal unit normal vector $\mathbf{N}(t)$ of the helix

$$\mathbf{r}(t) = \cos t\mathbf{i} + \sin t\mathbf{j} + t\mathbf{k}$$

is orthogonal to the z-axis.

Solution

$$\mathbf{r}'(t) = -\sin t\mathbf{i} + \cos t\mathbf{j} + \mathbf{k}$$

$$\|\mathbf{r}'(t)\| = \sqrt{(-\sin t)^2 + (\cos t)^2 + 1} = \sqrt{2}$$

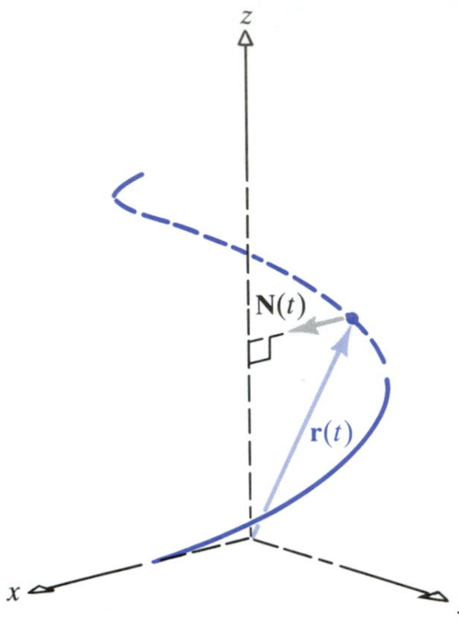

Figure 13

$$\mathbf{T}(t) = \frac{\mathbf{r}'(t)}{\|\mathbf{r}'(t)\|} = \frac{-\sin t\mathbf{i} + \cos t\mathbf{j} + \mathbf{k}}{\sqrt{2}}$$

$$\mathbf{T}'(t) = \frac{-\cos t\mathbf{i} - \sin t\mathbf{j}}{\sqrt{2}}$$

$$\|\mathbf{T}'(t)\| = \frac{1}{\sqrt{2}}$$

$$\mathbf{N}(t) = \frac{\mathbf{T}'(t)}{\|\mathbf{T}'(t)\|} = -\cos t\mathbf{i} - \sin t\mathbf{j}$$

Since the direction of the z-axis is $\mathbf{k}$, it follows that $\mathbf{N}(t) \cdot \mathbf{k} = 0$ for all t, and, hence, $\mathbf{N}(t)$ is orthogonal to the direction of the z-axis. See Figure 13.

Arc Length

In Chapter 13, we derived a formula for the arc length of a smooth plane curve. We state below, without proof, the extension of this result for a smooth curve defined by a vector function.

[15.2.6] THEOREM / ***Arc Length of a Vector Function.***

If a smooth curve C is defined by the vector function $\mathbf{r} = \mathbf{r}(t)$, $a \le t \le b$, the arc length s along C from $t = a$ to $t = b$ is

$$s = \int_a^b \|\mathbf{r}'(t)\|\, dt \tag{2}$$

For a smooth curve C in the plane defined by

$$\mathbf{r}(t) = x(t)\mathbf{i} + y(t)\mathbf{j} \qquad a \le t \le b$$

we have

$$\mathbf{r}'(t) = \frac{dx}{dt}\mathbf{i} + \frac{dy}{dt}\mathbf{j} \qquad \|\mathbf{r}'(t)\| = \sqrt{\left(\frac{dx}{dt}\right)^2 + \left(\frac{dy}{dt}\right)^2}$$

so that (2) becomes

$$s = \int_a^b \sqrt{\left(\frac{dx}{dt}\right)^2 + \left(\frac{dy}{dt}\right)^2}\, dt$$

which is in agreement with (2) in Section 13.5.

For a smooth curve C in space, defined by

$$\mathbf{r}(t) = x(t)\mathbf{i} + y(t)\mathbf{j} + z(t)\mathbf{k} \qquad a \le t \le b$$

we have

$$\mathbf{r}'(t) = \frac{dx}{dt}\mathbf{i} + \frac{dy}{dt}\mathbf{j} + \frac{dz}{dt}\mathbf{k} \qquad \|\mathbf{r}'(t)\| = \sqrt{\left(\frac{dx}{dt}\right)^2 + \left(\frac{dy}{dt}\right)^2 + \left(\frac{dz}{dt}\right)^2}$$

so that (2) becomes

$$s = \int_a^b \sqrt{\left(\frac{dx}{dt}\right)^2 + \left(\frac{dy}{dt}\right)^2 + \left(\frac{dz}{dt}\right)^2}\, dt$$

EXAMPLE 6

Find the arc length of:

(a) The circle $\mathbf{r}(t) = R \cos t\mathbf{i} + R \sin t\mathbf{j}$ from $t = 0$ to $t = 2\pi$
(b) The circular helix $\mathbf{r}(t) = R \cos t\mathbf{i} + R \sin t\mathbf{j} + t\mathbf{k}$ from $t = 0$ to $t = 2\pi$

Solution

(a)

$$\mathbf{r}'(t) = -R \sin t\mathbf{i} + R \cos t\mathbf{j}$$

$$\|\mathbf{r}'(t)\| = \sqrt{(-R \sin t)^2 + (R \cos t)^2} = R$$

By using (2), we find the arc length s to be

$$s = \int_0^{2\pi} R\, dt = 2\pi R$$

which is the familiar formula for the circumference of a circle.

(b)

$$\mathbf{r}'(t) = -R \sin t\mathbf{i} + R \cos t\mathbf{j} + \mathbf{k}$$

$$\|\mathbf{r}'(t)\| = \sqrt{(-R \sin t)^2 + (R \cos t)^2 + 1^2} = \sqrt{R^2 + 1}$$

Hence, the desired arc length s is

$$s = \int_0^{2\pi} \sqrt{R^2 + 1}\, dt = 2\pi\sqrt{R^2 + 1}$$

The arc length s of a smooth curve C defined by $\mathbf{r} = \mathbf{r}(t)$, $a \le t \le b$, from a to an arbitrary parameter value t is

$$s(t) = \int_a^t \|\mathbf{r}'(u)\|\, du$$

Thus,

$$\frac{ds}{dt} = \|\mathbf{r}'(t)\| \tag{3}$$

In particular, if $\mathbf{r} = \mathbf{r}(t) = x(t)\mathbf{i} + y(t)\mathbf{j}$ is a plane curve, then

$$\frac{ds}{dt} = \sqrt{\left(\frac{dx}{dt}\right)^2 + \left(\frac{dy}{dt}\right)^2} \tag{4}$$

which agrees with (4) in Section 13.5 (p. 777).

If $\mathbf{r} = \mathbf{r}(t) = x(t)\mathbf{i} + y(t)\mathbf{j} + z(t)\mathbf{k}$ is a curve in space, then

$$\frac{ds}{dt} = \sqrt{\left(\frac{dx}{dt}\right)^2 + \left(\frac{dy}{dt}\right)^2 + \left(\frac{dz}{dt}\right)^2} \tag{5}$$

We'll have occasion to refer to (3), (4), and (5) a little later.

EXERCISE 15.2

In Problems 1–16 find the tangent vector to each curve at t. Write the vector equation of the tangent line at t.

1. $\mathbf{r}(t) = t\mathbf{i} + t^2\mathbf{j}, \quad t = 1$

2. $\mathbf{r}(t) = t\mathbf{i} - t^3\mathbf{j}, \quad t = -1$

3. $\mathbf{r}(t) = (t^2 + 1)\mathbf{i} + (1 - t)\mathbf{j}, \quad t = 1$

4. $\mathbf{r}(t) = t^3\mathbf{i} + 3t\mathbf{j}, \quad t = 1$

5. $\mathbf{r}(t) = \dfrac{2t}{t+1}\mathbf{i} - \dfrac{t^2}{t+1}\mathbf{j}, \quad t = 1$

6. $\mathbf{r}(t) = \sqrt{t}\mathbf{i} + \frac{1}{2}t\mathbf{j}, \quad t = 4$

7. $\mathbf{r}(t) = e^t\mathbf{i} + e^{-t}\mathbf{j}, \quad t = 0$

8. $\mathbf{r}(t) = e^{2t}\mathbf{i} + e^{-t}\mathbf{j}, \quad t = 0$

9. $\mathbf{r}(t) = (1 - 3t)\mathbf{i} + 2t\mathbf{j} - (5 + t)\mathbf{k}, \quad t = 0$

10. $\mathbf{r}(t) = (2 + t)\mathbf{i} + (2 - t)\mathbf{j} + 3t\mathbf{k}, \quad t = 0$

11. $\mathbf{r}(t) = \cos 2t\mathbf{i} + \sin 2t\mathbf{j} - 5\mathbf{k}, \quad t = \pi/4$

12. $\mathbf{r}(t) = 3\mathbf{i} + \cos t\mathbf{j} + \sin t\mathbf{k}, \quad t = \pi/6$

13. $\mathbf{r}(t) = 2\cos t\mathbf{i} + \mathbf{j} + 2\sin t\mathbf{k}, \quad t = \pi/6$

14. $\mathbf{r}(t) = 2\cos 2t\mathbf{i} + 2\sin 2t\mathbf{j} + 5\mathbf{k}, \quad t = \pi/2$

15. $\mathbf{r}(t) = e^t\cos t\mathbf{i} + e^t\sin t\mathbf{j} + e^t\mathbf{k}, \quad t = 0$

16. $\mathbf{r}(t) = e^{-t}\cos t\mathbf{i} + e^{-t}\sin t\mathbf{j} - e^{-t}\mathbf{k}, \quad t = 0$

In Problems 17–32 find the unit tangent vector and the unit normal vector at t.

17. $\mathbf{r}(t) = t\mathbf{i} + t^2\mathbf{j}, \quad t = 1$

18. $\mathbf{r}(t) = t\mathbf{i} - t^3\mathbf{j}, \quad t = -1$

19. $\mathbf{r}(t) = (t^2 + 1)\mathbf{i} + (1 - t)\mathbf{j}, \quad t = 1$

20. $\mathbf{r}(t) = t^3\mathbf{i} + 3t\mathbf{j}, \quad t = 1$

21. $\mathbf{r}(t) = \dfrac{2t}{t+1}\mathbf{i} - \dfrac{t^2}{t+1}\mathbf{j}, \quad t = 1$

22. $\mathbf{r}(t) = \sqrt{t}\mathbf{i} + \frac{1}{2}t\mathbf{j}, \quad t = 4$

23. $\mathbf{r}(t) = e^t\mathbf{i} + e^{-t}\mathbf{j}, \quad t = 0$

24. $\mathbf{r}(t) = e^{2t}\mathbf{i} + e^{-t}\mathbf{j}, \quad t = 0$

25. $\mathbf{r}(t) = (1 - 3t)\mathbf{i} + 2t\mathbf{j} - (5 + t)\mathbf{k}, \quad t = 0$

26. $\mathbf{r}(t) = (2 + t)\mathbf{i} + (2 - t)\mathbf{j} + 3t\mathbf{k}, \quad t = 0$

27. $\mathbf{r}(t) = \cos 2t\mathbf{i} + \sin 2t\mathbf{j} - 5\mathbf{k}, \quad t = \pi/4$

28. $\mathbf{r}(t) = 3\mathbf{i} + \cos t\mathbf{j} + \sin t\mathbf{k}, \quad t = \pi/6$

29. $\mathbf{r}(t) = 2\cos t\mathbf{i} + \mathbf{j} + 2\sin t\mathbf{k}, \quad t = \pi/6$

30. $\mathbf{r}(t) = 2\cos 2t\mathbf{i} + 2\sin 2t\mathbf{j} + 5\mathbf{k}, \quad t = \pi/2$

31. $\mathbf{r}(t) = e^t\cos t\mathbf{i} + e^t\sin t\mathbf{j} + e^t\mathbf{k}, \quad t = 0$

32. $\mathbf{r}(t) = e^{-t}\cos t\mathbf{i} + e^{-t}\sin t\mathbf{j} - e^{-t}\mathbf{k}, \quad t = 0$

In Problems 33–42 find the arc length s of each vector function.

33. $\mathbf{r}(t) = t\mathbf{i} + (t^{2/3} + 1)\mathbf{j}$ from $t = 1$ to $t = 8$

34. $\mathbf{r}(t) = t\mathbf{i} + t^{3/2}\mathbf{j}$ from $t = 0$ to $t = 4$

35. $\mathbf{r}(t) = 8t\mathbf{i} + \dfrac{t^6 + 2}{t^2}\mathbf{j}$ from $t = 1$ to $t = 2$

36. $\mathbf{r}(t) = t\mathbf{i} + (1 - t^{2/3})^{3/2}\mathbf{j}$ from $t = \frac{1}{8}$ to $t = 1$

37. $\mathbf{r}(t) = t\mathbf{i} + 2t\mathbf{j} + t\mathbf{k}$ from $t = 0$ to $t = 1$

38. $\mathbf{r}(t) = 2t\mathbf{i} + t\mathbf{j} + 3t\mathbf{k}$ from $t = 1$ to $t = 3$

39. $\mathbf{r}(t) = \sin 2t\mathbf{i} + \cos 2t\mathbf{j} + t\mathbf{k}$ from $t = 0$ to $t = \pi$

40. $\mathbf{r}(t) = \sin t\mathbf{i} + \cos t\mathbf{j} + bt\mathbf{k}$ from $t = 0$ to $t = 2\pi$

41. $\mathbf{r}(t) = e^t\mathbf{i} + e^{-t}\mathbf{j} + \sqrt{2}\, t\mathbf{k}$ from $t = 0$ to $t = 1$

42. $\mathbf{r}(t) = \cos^3 t\mathbf{i} + \sin^3 t\mathbf{j} + \mathbf{k}$ from $t = 0$ to $t = \pi/2$

43. Find all points on the curve traced out by

$$\mathbf{r}(t) = t^2\mathbf{i} + (t^2 - 1)\mathbf{j} - t\mathbf{k}$$

at which $\mathbf{r}(t)$ and its tangent line are orthogonal.

44. Show that at any point, the tangent vectors to the helix

$$\mathbf{r}(t) = \alpha \cos t\mathbf{i} + \alpha \sin t\mathbf{j} + \beta t\mathbf{k} \qquad \alpha, \beta \text{ real numbers}$$

intersect the direction of the z-axis at a constant angle.

If two curves intersect, the angle between their tangent lines at the point of intersection is called the *angle between the curves.* In Problems 45 and 46 find the angle between the curves $\mathbf{r}_1(t)$ and $\mathbf{r}_2(T)$ at the point of intersection indicated.

45. $\mathbf{r}_1(t) = t^2\mathbf{i} + \sin \pi t\mathbf{j} + \mathbf{k}$ at $(1, 0, 1)$
$\mathbf{r}_2(T) = \mathbf{i} + T\mathbf{j} + (1 + T)\mathbf{k}$

46. $\mathbf{r}_1(t) = (e^t - 1)\mathbf{i} - \cos \pi t\mathbf{j} + t\mathbf{k}$ at $(0, -1, 0)$
$\mathbf{r}_2(T) = (1 - T)\mathbf{i} - \mathbf{j} + (T - 1)\mathbf{k}$

47. Find the length of $\mathbf{r}(t) = e^t\cos t\mathbf{i} + e^t\sin t\mathbf{j} + e^t\mathbf{k}$, $0 \le t \le 2\pi$.

48. Find the length of $\mathbf{r}(t) = t^2\mathbf{i} - 2\sqrt{2}t\mathbf{j} + (t^2 - 1)\mathbf{k}$, $0 \le t \le 1$.

49. Evaluate $\int_0^3 \|t^2\mathbf{i} - 2t\mathbf{j} + 2\mathbf{k}\|\, dt$.

Motion along a Curve 15.3

An important physical application of vector functions and their derivatives is found in the study of the motion of an object. If we think of the mass of the object as being concentrated at the object's center of gravity, then the object may be represented as a point. In this case, we refer to the object as a *particle.* As the particle moves through space, its coordinates x, y, and z are each functions of time t; its position vector at time t is given by the vector function:

$$\text{Position:} \quad \mathbf{r}(t) = x(t)\mathbf{i} + y(t)\mathbf{j} + z(t)\mathbf{k}$$

[15.3.1] DEFINITION / *Velocity; Acceleration; Speed.*

The *velocity*, *acceleration*, and *speed* of a particle whose motion is along a smooth curve traced out by a twice differentiable vector function $\mathbf{r}(t)$ are defined as:

$$\textbf{Velocity:} \quad \mathbf{v}(t) = \mathbf{r}'(t) = \frac{dx}{dt}\mathbf{i} + \frac{dy}{dt}\mathbf{j} + \frac{dz}{dt}\mathbf{k}$$

$$\textbf{Acceleration:} \quad \mathbf{a}(t) = \mathbf{r}''(t) = \frac{d^2x}{dt^2}\mathbf{i} + \frac{d^2y}{dt^2}\mathbf{j} + \frac{d^2z}{dt^2}\mathbf{k}$$

$$\textbf{Speed:} \quad v(t) = \|\mathbf{v}(t)\| = \|\mathbf{r}'(t)\| = \sqrt{\left(\frac{dx}{dt}\right)^2 + \left(\frac{dy}{dt}\right)^2 + \left(\frac{dz}{dt}\right)^2}$$

These definitions have analogous counterparts for particles whose position vector at time t is given by a vector function in the plane. Figure 14 illustrates the position vector and velocity vector of a particle moving along a plane curve.

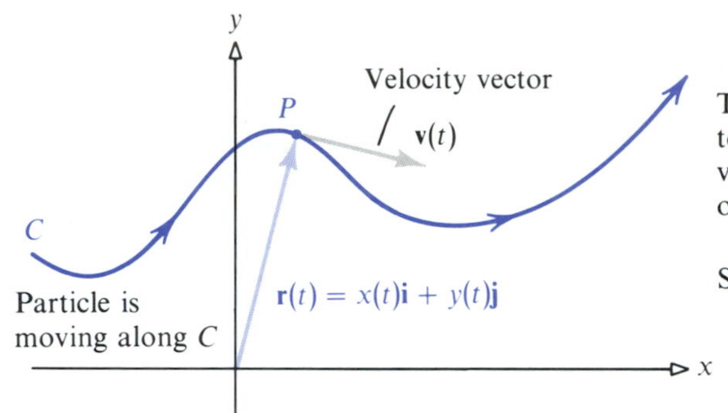

The velocity vector **v** is tangent to C at P. The magnitude of the velocity vector equals the speed of the object:

$$\text{Speed} = \sqrt{\left(\frac{dx}{dt}\right)^2 + \left(\frac{dy}{dt}\right)^2}$$

Figure 14

EXAMPLE 1

Find the velocity, acceleration, and speed of a particle whose motion is along:

(a) The plane curve $\mathbf{r}(t) = (\frac{1}{2}t^2 + t)\mathbf{i} + t^3\mathbf{j}, \quad 0 \le t \le 2$
(b) The space curve $\mathbf{r}(t) = t\mathbf{i} + t^2\mathbf{j} + t^3\mathbf{k}, \quad 0 \le t \le 2$

For each curve, sketch the path of motion of the particle and illustrate $\mathbf{v}(1)$ and $\mathbf{a}(1)$.

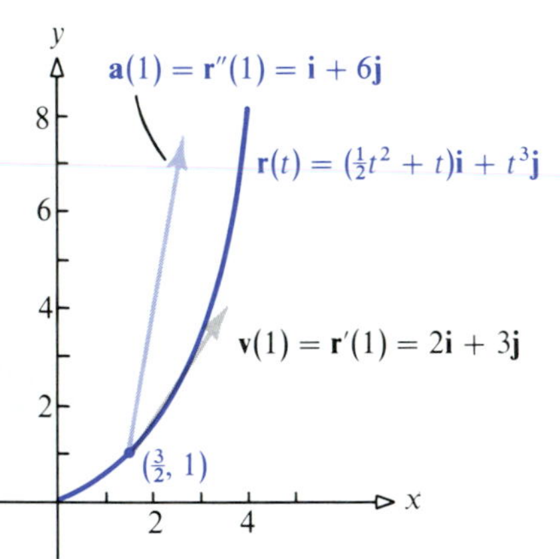

Figure 15

Solution

(a) The velocity, acceleration, and speed are

$$\mathbf{v}(t) = \mathbf{r}'(t) = (t + 1)\mathbf{i} + 3t^2\mathbf{j}$$
$$\mathbf{a}(t) = \mathbf{r}''(t) = \mathbf{i} + 6t\mathbf{j}$$
$$v(t) = \|\mathbf{v}(t)\| = \sqrt{9t^4 + t^2 + 2t + 1}$$

For $t = 1$, we have

$$\mathbf{v}(1) = \mathbf{r}'(1) = 2\mathbf{i} + 3\mathbf{j} \qquad \text{and} \qquad \mathbf{a}(1) = \mathbf{r}''(1) = \mathbf{i} + 6\mathbf{j}$$

Figure 15 illustrates the path of motion, $\mathbf{v}(1)$, and $\mathbf{a}(1)$.

(b) The velocity, acceleration, and speed are

$$\mathbf{v}(t) = \mathbf{r}'(t) = \mathbf{i} + 2t\mathbf{j} + 3t^2\mathbf{k}$$
$$\mathbf{a}(t) = \mathbf{r}''(t) = 2\mathbf{j} + 6t\mathbf{k}$$
$$v(t) = \|\mathbf{v}(t)\| = \sqrt{1 + 4t^2 + 9t^4}$$

When $t = 1$, we have

$$\mathbf{v}(1) = \mathbf{i} + 2\mathbf{j} + 3\mathbf{k} \qquad \text{and} \qquad \mathbf{a}(1) = 2\mathbf{j} + 6\mathbf{k}$$

Figure 16 illustrates the graph of $\mathbf{r}(t)$, $\mathbf{v}(1)$, and $\mathbf{a}(1)$.

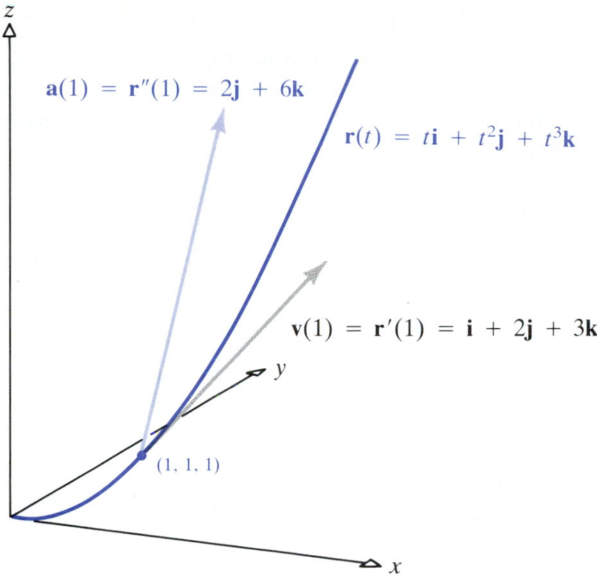

Figure 16

EXAMPLE 2

Find the force acting on an object of mass m whose motion is along the elliptic path

$$\mathbf{r}(t) = \alpha \cos \omega t\mathbf{i} + \beta \sin \omega t\mathbf{j} \qquad 0 \leq t \leq 2\pi$$

Solution

To find the force, we must first find the acceleration of the particle and then apply Newton's law, $\mathbf{F} = m\mathbf{a}$.

$$\mathbf{v}(t) = -\alpha\omega \sin \omega t\mathbf{i} + \beta\omega \cos \omega t\mathbf{j}$$
$$\mathbf{a}(t) = -\alpha\omega^2\cos \omega t\mathbf{i} - \beta\omega^2\sin \omega t\mathbf{j} = -\omega^2\mathbf{r}(t)$$

Hence, by Newton's law,

$$\mathbf{F}(t) = m\mathbf{a}(t) = -m\omega^2\mathbf{r}(t)$$

Thus, the direction of the force vector $\mathbf{F}(t)$ is opposite to that of the position vector $\mathbf{r}(t)$ at any t (see Fig. 17).

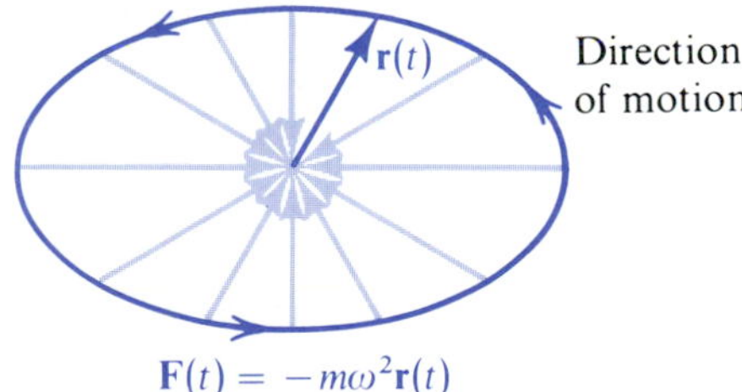

Figure 17

The next example will be referred to frequently, so study it carefully.

EXAMPLE 3

Find the position vector of a particle that moves counterclockwise along a circle of radius R with a constant speed v_0. Find the velocity and acceleration of this particle. Also find the magnitude of the acceleration.

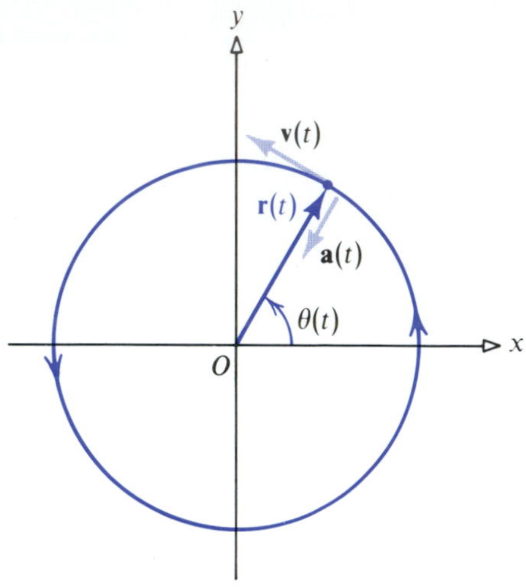

Figure 18

Solution

For convenience, we place the circle so that it lies in the xy-plane, with its center at the origin. Furthermore, at time $t = 0$, we assume the particle is on the positive x-axis. After an arbitrary time t, the particle has moved counterclockwise along the circle. As Figure 18 illustrates, if $\mathbf{r}(t)$ is the position vector of the particle at time t and if $\theta(t)$ is the angle between $\mathbf{r}(t)$ and the positive x-axis, it follows that

$$\mathbf{r}(t) = R\cos\theta(t)\mathbf{i} + R\sin\theta(t)\mathbf{j} \tag{1}$$

The velocity $\mathbf{v}(t)$ is

$$\mathbf{v}(t) = \frac{d\mathbf{r}}{dt} = -R\sin\theta(t)\frac{d\theta}{dt}\mathbf{i} + R\cos\theta(t)\frac{d\theta}{dt}\mathbf{j}$$

so that the speed is

$$v(t) = \|\mathbf{v}(t)\| = \sqrt{R^2\sin^2\theta\left(\frac{d\theta}{dt}\right)^2 + R^2\cos^2\theta\left(\frac{d\theta}{dt}\right)^2} = \sqrt{R^2\left(\frac{d\theta}{dt}\right)^2} = R\left|\frac{d\theta}{dt}\right|$$

But the speed $v(t)$ is a constant. Thus, we let $v(t) = v_0$. Furthermore, since θ is increasing, $d\theta/dt > 0$. Hence,

$$v_0 = R\frac{d\theta}{dt}$$

$$\frac{d\theta}{dt} = \frac{v_0}{R}$$

But the quantity v_0/R is the angular speed ω. Thus,

$$\frac{d\theta}{dt} = \omega$$

Therefore, $\theta(t) = \omega t + \theta_0$. But the initial conditions given are that at $t = 0$, $\theta = 0$. Hence, $\theta(t) = \omega t$.* Using this in (1), we find that the position vector $\mathbf{r}(t)$ of the particle is given by

$$\mathbf{r}(t) = R\cos\omega t\mathbf{i} + R\sin\omega t\mathbf{j}$$

The velocity $\mathbf{v}(t)$ and acceleration $\mathbf{a}(t)$ are

$$\mathbf{v}(t) = -R\omega\sin\omega t\mathbf{i} + R\omega\cos\omega t\mathbf{j}$$

$$\mathbf{a}(t) = -R\omega^2\cos\omega t\mathbf{i} - R\omega^2\sin\omega t\mathbf{j} = -\omega^2\mathbf{r}(t) \tag{2}$$

The magnitude of the acceleration is

$$\|\mathbf{a}(t)\| = \omega^2\|\mathbf{r}(t)\| = \omega^2 R = \frac{v_0^2}{R} \tag{3}$$

---

* The function $\theta(t)$ may be found another way. Since the speed of the particle about the circle is constant, the rate at which the angle θ changes per unit time $(d\theta/dt)$, the *angular speed*, is a constant ω. That is, $d\theta/dt = \omega$. Solving this differential equation with the initial condition that $\theta = 0$ when $t = 0$, we get $\theta = \omega t$.

We observe that although the speed of the particle in Example 3 is constant, its velocity is not. Furthermore, the direction of the acceleration is opposite that of the position vector **r** and, hence, is directed toward the center of the circle. By Newton's law $\mathbf{F} = m\mathbf{a}$, the force vector **F** is also directed toward the center of the circle. Such a force is called a *centripetal force*.

The results obtained in Example 3 may be used to calculate the speed required of a satellite to maintain a near-earth circular orbit* (the gravitational attractions of other bodies are ignored). Let R be the distance of a satellite from the center of the earth. From (3), the magnitude of the acceleration of the satellite whose motion is circular is

$$\|\mathbf{a}(t)\| = \frac{v_0^2}{R}$$

But the magnitude of the acceleration of the satellite must equal g, the acceleration of gravity for earth.† Hence,

$$\frac{v_0^2}{R} = g$$

$$v_0 = \sqrt{gR}$$

[15.3.2] THEOREM / ***Near-Earth Orbits.***

The speed v_0 required to maintain a near-earth circular orbit is

$$v_0 = \sqrt{gR} \tag{4}$$

where R is the distance of the satellite from the center of the earth and g is the acceleration of gravity.

For example, the speed required of a communications satellite whose circular orbit is to be 4500 miles from the center of the earth is

$$v_0 = \sqrt{(79{,}036)(4500)} \approx 18{,}859 \text{ miles per hour}$$

Let's look at another use of (3).

EXAMPLE 4

A motorcycle with mass 150 kilograms is driven at a constant speed of 120 kilometers per hour on a circular track whose radius is 100 meters. To keep the motorcycle from skidding, what frictional force must be exerted by the tires on the track?

Solution

By Newton's law, the force **F** required to keep an object of mass m traveling along a curve traced out by $\mathbf{r} = \mathbf{r}(t)$ is $\mathbf{F} = m\mathbf{a}$. The magnitude of

---

* Near-earth orbits are above 100 miles (out of the earth's atmosphere) up to an altitude of approximately 15,000 miles.

† Although the acceleration of gravity at such altitudes is somewhat less than $g \approx 32.2$ feet per second per second $\approx 79{,}036$ miles per hour per hour, we shall ignore this discrepancy in our calculations.

the frictional force exerted by the tires must therefore equal

$$\|\mathbf{F}\| = m\|\mathbf{a}\|$$

But the motion is circular. Hence, from (3), we find

$$\|\mathbf{F}\| = m\left(\frac{v_0^2}{R}\right) = 150 \text{ kg}\left(\frac{120^2 \text{ km}^2/\text{hr}^2}{100 \text{ m}}\right) = (150)(144)\left(\frac{1000^2}{3600^2}\right)\frac{\text{kg m}}{\text{sec}^2} \approx 1667 \text{ newtons}$$

Tangential and Normal Components of the Acceleration

The acceleration vector $\mathbf{a}(t)$ of a particle whose motion is along a smooth curve traced out by a twice differentiable vector function $\mathbf{r} = \mathbf{r}(t)$ has been defined as $\mathbf{a}(t) = \mathbf{r}''(t)$. We show now that the acceleration vector $\mathbf{a}(t)$ lies in the plane determined by the unit tangent vector $\mathbf{T}(t)$ and the principal unit normal vector $\mathbf{N}(t)$, and we express $\mathbf{a}(t)$ in terms of $\mathbf{T}(t)$ and $\mathbf{N}(t)$.

Using the fact that $\mathbf{T}(t) = \mathbf{r}'(t)/\|\mathbf{r}'(t)\|$, we can express the velocity vector $\mathbf{v}(t)$ in terms of $\mathbf{T}(t)$:

$$\mathbf{v}(t) = \mathbf{r}'(t) = \|\mathbf{r}'(t)\|\mathbf{T}(t) = \|\mathbf{v}(t)\|\mathbf{T}(t) = v(t)\mathbf{T}(t) \tag{5}$$

where $v(t) = \|\mathbf{v}(t)\|$ is the speed of the particle. Thus, we have the following result:

[15.3.3] THEOREM

The velocity vector $\mathbf{v}(t)$ of a particle whose motion is along a smooth curve traced out by a twice differentiable vector function $\mathbf{r} = \mathbf{r}(t)$ is directed along the tangent to the curve.

By differentiating (5), we can get an expression for the acceleration vector:

$$\mathbf{a}(t) = \mathbf{r}''(t) = \frac{d}{dt}[v(t)\mathbf{T}(t)] = v'(t)\mathbf{T}(t) + v(t)\mathbf{T}'(t)$$

Since $\mathbf{N}(t) = \mathbf{T}'(t)/\|\mathbf{T}'(t)\|$, we can write the above expression as

$$\mathbf{a}(t) = v'(t)\mathbf{T}(t) + v(t)\|\mathbf{T}'(t)\|\mathbf{N}(t) \tag{6}$$

Thus we have the following result:

[15.3.4] THEOREM

The acceleration vector $\mathbf{a}(t)$ of a particle whose motion is along a smooth curve traced out by a twice differentiable vector function $\mathbf{r} = \mathbf{r}(t)$ lies in the plane of the unit tangent vector $\mathbf{T}(t)$ and the principal unit normal vector $\mathbf{N}(t)$ of the curve.

See Figure 19 for an illustration.

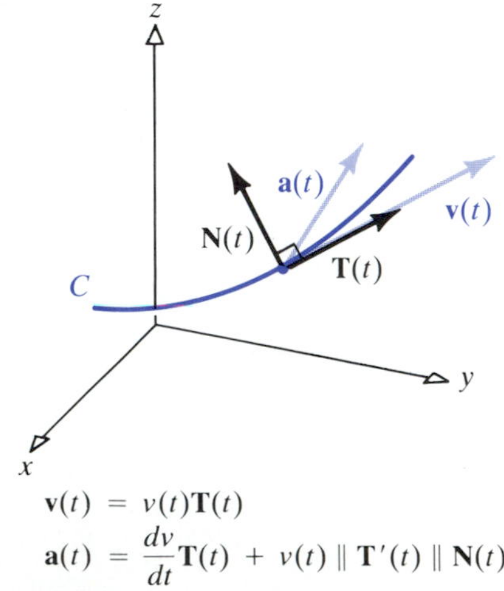

$\mathbf{v}(t) = v(t)\mathbf{T}(t)$

$\mathbf{a}(t) = \frac{dv}{dt}\mathbf{T}(t) + v(t)\,\|\,\mathbf{T}'(t)\,\|\,\mathbf{N}(t)$

Figure 19

The *tangential component*, denoted by $a_{\mathbf{T}}$, and the *normal component*, denoted by $a_{\mathbf{N}}$, of the acceleration vector $\mathbf{a}$ are

$$a_{\mathbf{T}} = v'(t) = \frac{dv}{dt} \qquad a_{\mathbf{N}} = v(t)\|\mathbf{T}'(t)\| \tag{7}$$

Although the formula for $a_{\mathbf{T}}$ is easy to use, the one for $a_{\mathbf{N}}$ is usually not easy to calculate directly. As an alternative, we obtain a formula for $a_{\mathbf{N}}$ that requires knowing $\mathbf{a}(t)$ and $a_{\mathbf{T}}$. To derive it, we combine (6) and (7) and write

$$\mathbf{a}(t) = a_{\mathbf{T}}\mathbf{T}(t) + a_{\mathbf{N}}\mathbf{N}(t)$$

Then

$$\begin{aligned} \mathbf{a}(t) \cdot \mathbf{a}(t) &= (a_{\mathbf{T}}\mathbf{T} + a_{\mathbf{N}}\mathbf{N}) \cdot (a_{\mathbf{T}}\mathbf{T} + a_{\mathbf{N}}\mathbf{N}) \\ \|\mathbf{a}(t)\|^2 &= a_{\mathbf{T}}^2\|\mathbf{T}\|^2 + 2a_{\mathbf{T}}a_{\mathbf{N}}\mathbf{T} \cdot \mathbf{N} + a_{\mathbf{N}}^2\|\mathbf{N}\|^2 \end{aligned}$$

Since $\mathbf{T} \cdot \mathbf{N} = 0$ and $\|\mathbf{N}\| = \|\mathbf{T}\| = 1$, we get

$$\|\mathbf{a}(t)\|^2 = a_{\mathbf{T}}^2 + a_{\mathbf{N}}^2$$

Hence, we find that

$$a_{\mathbf{N}} = \sqrt{\|\mathbf{a}(t)\|^2 - a_{\mathbf{T}}^2} \tag{8}$$

Formula (8) is usually easier to use in practice than the formula for $a_{\mathbf{N}}$ in (7).

EXAMPLE 5

Find the tangential and normal components of the acceleration of a particle whose motion is along the ellipse

$$\mathbf{r}(t) = 3 \sin t\mathbf{i} + 4 \cos t\mathbf{j}$$

Graph the ellipse, showing $\mathbf{a}$, $a_{\mathbf{T}}$, and $a_{\mathbf{N}}$ when $t = \pi/4$.

Solution

First,

$$\begin{aligned} \mathbf{v}(t) &= \mathbf{r}'(t) = 3 \cos t\mathbf{i} - 4 \sin t\mathbf{j} \\ v(t) &= \|\mathbf{v}(t)\| = \sqrt{(3 \cos t)^2 + (-4 \sin t)^2} = \sqrt{9 \cos^2 t + 16 \sin^2 t} \\ \mathbf{a}(t) &= \mathbf{r}''(t) = -3 \sin t\mathbf{i} - 4 \cos t\mathbf{j} \end{aligned}$$

The tangential component of the acceleration is

$$a_{\mathbf{T}} = \frac{dv}{dt} = \frac{-18 \cos t \sin t + 32 \sin t \cos t}{2\sqrt{9 \cos^2 t + 16 \sin^2 t}} = \frac{7 \sin t \cos t}{\sqrt{9 \cos^2 t + 16 \sin^2 t}}$$

To find the normal component of the acceleration, we find $\|\mathbf{a}(t)\|$ and then

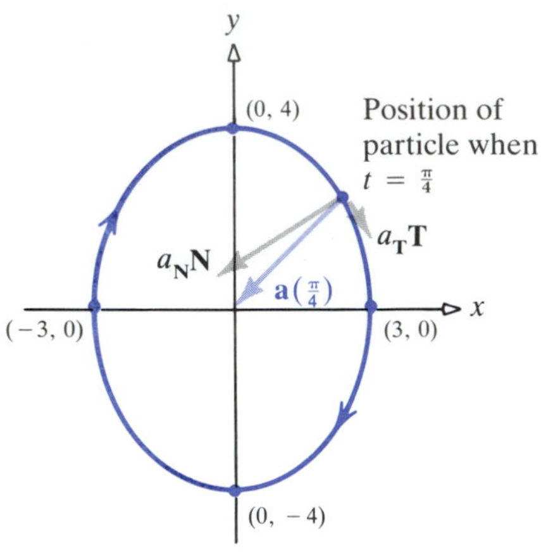

Figure 20

use (8):

$$\|\mathbf{a}(t)\| = \sqrt{(-3\sin t)^2 + (-4\cos t)^2} = \sqrt{9\sin^2 t + 16\cos^2 t}$$

$$a_{\mathbf{N}} = \sqrt{\|\mathbf{a}\|^2 - a_{\mathbf{T}}^2} = \sqrt{9\sin^2 t + 16\cos^2 t - \frac{49\sin^2 t\cos^2 t}{9\cos^2 t + 16\sin^2 t}}$$

$$= \sqrt{\frac{144(\sin^4 t + 2\sin^2 t\cos^2 t + \cos^4 t)}{9\cos^2 t + 16\sin^2 t}} = \frac{12}{\sqrt{9\cos^2 t + 16\sin^2 t}}$$

At $t = \pi/4$, we have

$$\mathbf{a}(\pi/4) = \frac{-3}{\sqrt{2}}\mathbf{i} - \frac{4}{\sqrt{2}}\mathbf{j} \qquad a_{\mathbf{T}} = \frac{7\sqrt{2}}{10} \qquad a_{\mathbf{N}} = \frac{12\sqrt{2}}{5}$$

The motion of the particle on the ellipse is clockwise, starting at $\mathbf{r}(0) = 4\mathbf{j}$. When $t = \pi/4$, the particle is at the position $\mathbf{r}(\pi/4) = (3/\sqrt{2})\mathbf{i} + (4/\sqrt{2})\mathbf{j}$. See Figure 20.

EXAMPLE 6

Find the tangential and normal components of the acceleration of a particle whose motion is along the helix

$$\mathbf{r}(t) = \cos t\mathbf{i} + \sin t\mathbf{j} + t\mathbf{k}$$

Solution

$$\mathbf{v}(t) = \mathbf{r}'(t) = -\sin t\mathbf{i} + \cos t\mathbf{j} + \mathbf{k}$$

$$v(t) = \|\mathbf{v}(t)\| = \sqrt{(-\sin t)^2 + (\cos t)^2 + 1} = \sqrt{2}$$

The tangential component of the acceleration

$$a_{\mathbf{T}} = \frac{dv}{dt} = 0$$

To find the normal component, we first find $\mathbf{a}(t)$ and then use (8):

$$\mathbf{a}(t) = \frac{d\mathbf{v}}{dt} = -\cos t\mathbf{i} - \sin t\mathbf{j}$$

$$\|\mathbf{a}(t)\| = \sqrt{(-\cos t)^2 + (-\sin t)^2} = 1$$

$$a_{\mathbf{N}} = \sqrt{\|\mathbf{a}\|^2 - a_{\mathbf{T}}^2} = 1$$

Application: The UVW-Axes in Orbital Mechanics

A spacecraft in orbit about the earth follows a curve in space, which is modeled as the graph of a vector function of one variable—time, t. The spacecraft carries a gyroscope whose three axes remain fixed in direction throughout time and are aligned with the $\mathbf{i}$, $\mathbf{j}$, $\mathbf{k}$ unit vectors of the xyz-coordinate system (see Fig. 21). The xyz-coordinate system has its origin at the center of the earth and provides the same basic frame of reference for both the earthbound observers and the spacecraft.

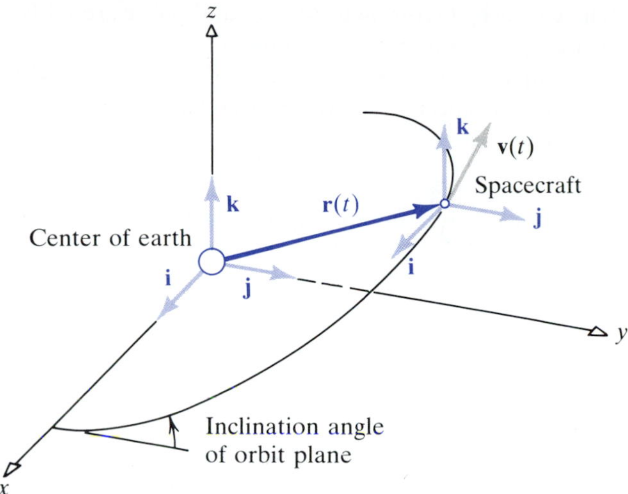

Figure 21

The position vector from the origin is given by the function

$$\mathbf{r}(t) = x(t)\mathbf{i} + y(t)\mathbf{j} + z(t)\mathbf{k}$$

The velocity is the vector tangent to the orbit expressed by $\mathbf{v}(t) = \mathbf{r}'(t)$.

If the earth is taken to have all its mass concentrated at the center, then the laws of Kepler and Newton tell us that any orbit remains in a *fixed plane* containing that center. Furthermore, a closed orbit describes an ellipse (or circle) in that plane with the earth's center placed at *one of the foci*; the other focus remains vacant.

The UVW system of axes is an orthogonal set of unit vectors $\mathbf{i}_U(t)$, $\mathbf{i}_V(t)$, $\mathbf{i}_W(t)$, which are "attached" to the spacecraft's orbit at each point. This set of vectors is particularly useful in manned spacecraft, since it corresponds to the axes of *yaw*, *roll*, and *pitch* customarily used in airplanes flying over the surface of the earth. We define these time-dependent unit vectors as follows:

$$\mathbf{i}_U = \frac{\mathbf{r}}{\|\mathbf{r}\|} \qquad \mathbf{i}_W = \frac{\mathbf{r} \times \mathbf{v}}{\|\mathbf{r} \times \mathbf{v}\|} \qquad \mathbf{i}_V = \mathbf{i}_W \times \mathbf{i}_U \tag{9}$$

Note that $\mathbf{i}_U(t)$ lies in the orbit plane and points radially *outward* (see Fig. 22). The vector $\mathbf{i}_W(t)$ is *normal* to the orbit plane, since both $\mathbf{r}(t)$ and $\mathbf{v}(t)$ are

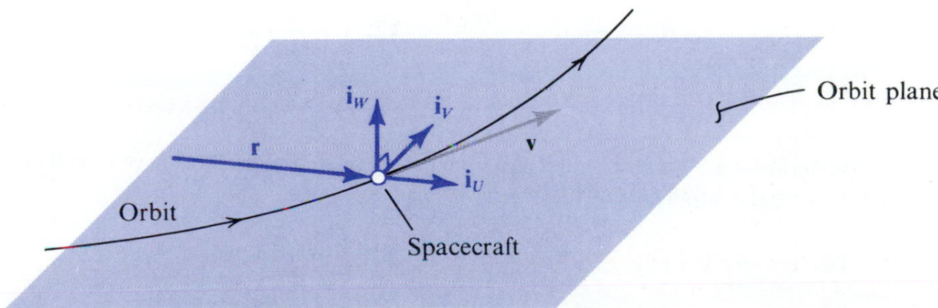

Figure 22

vectors in that plane. The vector $\mathbf{i}_V(t)$ completes the set and points, roughly, in the direction of the velocity; it lies in the orbit plane.

Now, consider a circular orbit intersecting the x-axis, inclined to the xy-plane at 30°. Suppose that its radius is a and the motion has angular speed ω. Then its position vector is

$$\mathbf{r}(t) = a\left(\cos \omega t\mathbf{i} + \frac{\sqrt{3}}{2}\sin \omega t\mathbf{j} + \frac{1}{2}\sin \omega t\mathbf{k}\right)$$

Therefore,

$$\mathbf{v}(t) = \mathbf{r}'(t) = a\omega\left(-\sin \omega t\mathbf{i} + \frac{\sqrt{3}}{2}\cos \omega t\mathbf{j} + \frac{1}{2}\cos \omega t\mathbf{k}\right)$$

In this case, using (9), we obtain

$$\mathbf{i}_U = \cos \omega t\mathbf{i} + \frac{\sqrt{3}}{2}\sin \omega t\mathbf{j} + \frac{1}{2}\sin \omega t\mathbf{k}$$

$$\mathbf{i}_W = -\frac{1}{2}\mathbf{j} + \frac{\sqrt{3}}{2}\mathbf{k}$$

$$\mathbf{i}_V = -\sin \omega t\mathbf{i} + \frac{\sqrt{3}}{2}\cos \omega t\mathbf{j} + \frac{1}{2}\cos \omega t\mathbf{k}$$

Thus, we have the directions of the UVW-axes expressed in terms of the onboard gyroscope axes.

EXERCISE 15.3

In Problems 1–14 sketch the motion of a particle traveling along the curve traced out by $\mathbf{r} = \mathbf{r}(t)$. Include in the sketch the vectors $\mathbf{v}(0)$ and $\mathbf{a}(0)$.

1. $\mathbf{r}(t) = t\mathbf{i} + t^2\mathbf{j}$

2. $\mathbf{r}(t) = t\mathbf{i} - t^3\mathbf{j}$

3. $\mathbf{r}(t) = (t^2 + 1)\mathbf{i} + (1 - t)\mathbf{j}$

4. $\mathbf{r}(t) = t^3\mathbf{i} + 3t\mathbf{j}$

5. $\mathbf{r}(t) = \dfrac{2t}{t+1}\mathbf{i} - \dfrac{t^2}{t+1}\mathbf{j},\ t \geq 0$

6. $\mathbf{r}(t) = \sqrt{t}\mathbf{i} + \frac{1}{2}t\mathbf{j}$

7. $\mathbf{r}(t) = e^t\mathbf{i} + e^{-t}\mathbf{j}$

8. $\mathbf{r}(t) = e^{2t}\mathbf{i} + e^{-t}\mathbf{j}$

9. $\mathbf{r}(t) = t\mathbf{i} + e^t\mathbf{j}$

10. $\mathbf{r}(t) = t\mathbf{i} + \ln(1 + t)\mathbf{j}$

11. $\mathbf{r}(t) = 3\sin t\mathbf{i} - 3\cos t\mathbf{j}$

12. $\mathbf{r}(t) = 4\sin t\mathbf{i} + 4\cos t\mathbf{j}$

13. $\mathbf{r}(t) = 2\cos t\mathbf{i} - 3\sin t\mathbf{j}$

14. $\mathbf{r}(t) = -\cos t\mathbf{i} + 2\sin t\mathbf{j}$

In Problems 15–30 find the velocity $\mathbf{v}(t)$, acceleration $\mathbf{a}(t)$, and speed $v(t)$ of a particle traveling along the curve traced out by $\mathbf{r} = \mathbf{r}(t)$. Also, find the tangential and normal components of the acceleration.

15. $\mathbf{r}(t) = 2t\mathbf{i} + (t + 1)\mathbf{j}$

16. $\mathbf{r}(t) = e^t\mathbf{i} + e^{2t}\mathbf{j}$

17. $\mathbf{r}(t) = 2\cos t\mathbf{i} + 3\sin t\mathbf{j}$

18. $\mathbf{r}(t) = e^{-t}\mathbf{i} + e^{-2t}\mathbf{j}$

19. $\mathbf{r}(t) = (1 - 3t)\mathbf{i} + 2t\mathbf{j} - (5 + t)\mathbf{k}$

20. $\mathbf{r}(t) = (2 + t)\mathbf{i} + (2 - t)\mathbf{j} + 3t\mathbf{k}$

21. $\mathbf{r}(t) = \cos 2t\mathbf{i} + \sin 2t\mathbf{j} - 5\mathbf{k}$

22. $\mathbf{r}(t) = 3\mathbf{i} + \cos t\mathbf{j} + \sin t\mathbf{k}$

23. $\mathbf{r}(t) = \dfrac{t^2}{t+1}\mathbf{i} + \dfrac{1}{t+1}\mathbf{j} + \dfrac{t}{t+1}\mathbf{k}$

24. $\mathbf{r}(t) = t\mathbf{i} + \sin t\mathbf{j} + \cos t\mathbf{k}$

25. $\mathbf{r}(t) = t\mathbf{i} + t^2\mathbf{j} + t^3\mathbf{k}$

26. $\mathbf{r}(t) = a\cos t\mathbf{i} + b\sin t\mathbf{j} + ct\mathbf{k}, \quad a > 0, \quad b > 0, \quad c > 0$

27. $\mathbf{r}(t) = e^t\cos t\mathbf{i} + e^t\sin t\mathbf{j} + e^t\mathbf{k}$

28. $\mathbf{r}(t) = e^{-t}\cos t\mathbf{i} + e^{-t}\sin t\mathbf{j} - e^{-t}\mathbf{k}$

29. $\mathbf{r}(t) = \ln t\mathbf{i} + \sqrt{t}\mathbf{j} + t^{3/2}\mathbf{k}, \quad t > 0$

30. $\mathbf{r}(t) = \cosh t\mathbf{i} + \sinh t\mathbf{j} + t\mathbf{k}$

31. Show that if a particle moves along a path at constant speed, the tangential component of the acceleration is 0.

32. Suppose that the function $\mathbf{r}(t) = e^t\mathbf{i} + e^{-t}\mathbf{j}$ gives the position vector of a particle at time t.

(a) Show that the force on the particle is directed away from the origin.
(b) What is the minimum speed of the particle and where does it occur?
(c) Find the tangential and normal components of acceleration at the point found in part (b).
(d) The answers in part (c) are $a_{\mathbf{T}} = 0$ and $a_{\mathbf{N}} = \|\mathbf{a}(t)\| = \sqrt{2}$. How could these have been predicted?

33. Show that if the speed of a particle along a curve is constant, then the velocity and acceleration vectors are orthogonal.

34. If the motion of a particle is along the curve traced out by $\mathbf{r}(t) = \alpha\cosh t\mathbf{i} + \beta\sinh t\mathbf{j}, \quad \alpha > 0, \quad \beta > 0$, show that the force acting on the particle is parallel to $\mathbf{r}(t)$.

In Problems 35–43 use the idea expressed by (3) and (4).

35. A girl is spinning a bucket of water in a horizontal plane at the end of a rope of length L.

(a) If she triples the speed of the bucket, how many times as hard must she pull on the rope?
(b) If, instead, she doubles the length of the rope but keeps the speed of the bucket the same, will she have to pull more or less? How much?

36. Find the distance of a satellite from the surface of the earth if it moves around the earth in a circular orbit at a constant speed of 18,630 miles per hour. (The radius of the earth is approximately 4000 miles.)

37. Find the speed of a satellite that moves in a circular orbit around the earth at a height of 100 miles.

38. At the Fermi National Accelerator Laboratory in Batavia, Illinois, protons are accelerated along a circular route of radius 1 kilometer. Find the magnitude of the force necessary to give a proton of mass m a constant speed of 280,000 kilometers per second.

39. A weather satellite orbits the earth in a circle every $1\frac{1}{2}$ hours. How high is it above the earth?

40. Some communications satellites remain stationary above a fixed point on the equator of the earth's surface. How high are such satellites and what is their common velocity? (Assume the earth turns once every 24 hours.)

41. If, in Example 4 in this section, the motorcycle is driven at a speed that is 10% faster, by how much is the frictional force of the tires increased?

42. If, in Example 4 in this section, the radius of the circular track is halved, how much slower should the motorcycle be driven so as not to increase the frictional force?

43. A race car of mass 1000 kilograms is driven at a constant speed of 200 kilometers per hour around a circular track whose radius is 75 meters. What frictional force must be exerted by the tires on the track to keep the car from skidding?

44. A particle moves on the circle $x^2 + y^2 = 1$ so that at time $t \geq 0$ the position is given by the vector

$$\mathbf{r}(t) = \frac{1 - t^2}{1 + t^2}\mathbf{i} + \frac{2t}{1 + t^2}\mathbf{j}$$

(a) Find the velocity vector.
(b) Is the particle ever at rest? Justify your answer.
(c) Give the coordinates of the point that the particle approaches as t increases without bound.

45. The *power* expended by a force $\mathbf{F}(t)$ acting on an object of mass m with velocity $\mathbf{v}(t)$ is the dot product

$$\text{Power} = \mathbf{F}(t) \cdot \mathbf{v}(t)$$

The *kinetic energy* of an object of mass m and velocity $\mathbf{v}(t)$ equals one-half the product of its mass times the square of its speed:

$$\text{Kinetic energy} = \tfrac{1}{2}m\|\mathbf{v}(t)\|^2$$

Show that the power of an object equals the instantaneous rate of change of its kinetic energy with respect to time.

46. In classical mechanics, the *momentum* $\mathbf{p}(t)$ of an object of mass m at time t is defined as $\mathbf{p}(t) = m\mathbf{v}(t)$, where $\mathbf{v}(t)$ is the velocity of the object at time t. Show that force equals the instantaneous rate of change of momentum with respect to time.

47. The *torque* τ* produced by a force $\mathbf{F}(t)$ acting on an object whose position at time t is $\mathbf{r}(t)$ is defined as $\tau(t) = \mathbf{r}(t) \times \mathbf{F}(t)$. The torque measures the twist imparted on the object by the force. The *angular momentum* $\mathbf{L}$ of an object of mass m and velocity $\mathbf{v}(t)$ whose position at time t is $\mathbf{r}(t)$ is $\mathbf{L}(t) = \mathbf{r}(t) \times m\mathbf{v}(t)$. Show that the instantaneous rate of change of angular momentum with respect to time equals the torque.

48. A central force $\mathbf{F}(t)$ is one whose direction is proportional to the position vector $\mathbf{r}(t)$ of the object it acts upon; that is, $\mathbf{F}(t) = u(t)\mathbf{r}(t)$, where $u(t)$ is a scalar function. Show that a central force produces zero torque.

49. A particle moves on a disk from the center directly toward the edge. If the disk is revolving in the counterclockwise direction at a constant angular speed ω, the position of the particle at time t is $\mathbf{r}(t) = t\mathbf{R}(t)$ where $\mathbf{R}(t) = \cos \omega t\mathbf{i} + \sin \omega t\mathbf{j}$ is the position vector of any point on the disk.

(a) Show that the velocity $\mathbf{v}$ of the particle is

$$\mathbf{v} = \cos \omega t\mathbf{i} + \sin \omega t\mathbf{j} + t\mathbf{v}_d$$

where $\mathbf{v}_d = \mathbf{R}'(t)$ is the velocity of the rotating disk.

(b) Also show that the acceleration $\mathbf{a}$ of the particle is

$$\mathbf{a} = 2\mathbf{v}_d + t\mathbf{a}_d$$

where $\mathbf{a}_d = \mathbf{R}''(t)$ is the acceleration of the ro-

(*Problem 49 continues in the next column*)

* Greek letter tau.

49. (*continued*)

tating disk. The extra term $2\mathbf{v}_d$ is the so-called *Coriolis acceleration*, which results from the interaction of the rotation of the disk and the motion of the particles on the disk.

50. Refer to Problem 49.

(a) Find the velocity and acceleration of a particle revolving on a rotating disk according to

$$\mathbf{r}(t) = t^2 \cos \omega t\mathbf{i} + t^2 \sin \omega t\mathbf{j}$$

(b) What is the Coriolis acceleration?

51. The position vector of a particle at time $t \geq 0$ is $\mathbf{r} = e^{-t}\cos t\mathbf{i} + e^{-t}\sin t\mathbf{j}$. Show that the path of the particle is part of the spiral $r = e^{-\theta}$ (in polar coordinates). Sketch the path and indicate the direction of travel.

52. In Problem 51 show that the angle between the position vector and the velocity vector is always 135°.

53. Suppose a particle moves along the curve $\mathbf{r}(t) = 4 \cos t\mathbf{i} - 2 \cos 2t\mathbf{j}$.

(a) Show that the particle oscillates on an arc of a parabola.

(b) Sketch the path.

(c) Find the acceleration $\mathbf{a}$ at points of zero velocity.

54. Find the maximum magnitude of the force acting on a particle of mass m whose motion is along the curve $\mathbf{r}(t) = 4 \cos t\mathbf{i} - 2 \cos 2t\mathbf{j}$.

55. If $\mathbf{r}'(t) = \mathbf{b} \times \mathbf{r}(t)$ for all t, where $\mathbf{b}$ is a constant vector, show that the acceleration $\mathbf{a}(t)$ is perpendicular to $\mathbf{b}$ and that the speed is constant.

56. A particle moves along the path $y = 3x^2 - x^3$ with the horizontal component of the velocity identically equal to $\frac{1}{3}$. Find the acceleration at the points where the velocity $\mathbf{v}$ is horizontal. Sketch the path and indicate $\mathbf{v}$ and $\mathbf{a}$ at these points.

15.4 Curvature

Arc Length As Parameter

We have seen that different curves may be defined by choosing different parameters. For example, the upper half of a unit circle may be defined as

$$\mathbf{r}(t) = t\mathbf{i} + \sqrt{1 - t^2}\mathbf{j} \qquad -1 \leq t \leq 1$$

where the parameter t represents position on the x-axis, or it can be defined as

$$\mathbf{r}(t) = \cos t\mathbf{i} + \sin t\mathbf{j} \qquad 0 \leq t \leq \pi$$

where the parameter t is arc length as measured along the circle itself.

Just as time is the preferred parameter for motion along a curve, arc length along the curve is the preferred parameter for discussing a curve's geometric properties. The next result provides a way for telling whether the parameter used represents arc length.

[15.4.1] THEOREM

Suppose a smooth curve C is defined by the vector function

$$\mathbf{r} = \mathbf{r}(t) \qquad a \le t \le b$$

The parameter t is the arc length along C if and only if

$$\|\mathbf{r}'(t)\| = 1 \qquad \text{for all } t$$

Proof The arc length function $s(t)$ along the curve C defined by $\mathbf{r} = \mathbf{r}(t)$, $a \le t \le b$, from a to an arbitrary value of t is given by

$$s(t) = \int_a^t \|\mathbf{r}'(u)\|\, du \tag{1}$$

If $\|\mathbf{r}'(t)\| = 1$ for all t, then the arc length s is

$$s(t) = \int_a^t du = t - a$$

That is, the parameter t is the arc length s measured along C.

If the parameter t is arc length s, then we may differentiate (1) to obtain

$$\frac{ds}{dt} = 1 = \|\mathbf{r}'(t)\| \qquad \text{for all } t$$

EXAMPLE 1

Determine whether the parameter used in defining the curves below is arc length.

(a) C: $\mathbf{r}(t) = 2\sin\frac{t}{2}\mathbf{i} + 2\cos\frac{t}{2}\mathbf{j}$, $0 \le t \le 2\pi$

(b) C: $\mathbf{r}(t) = 3t^2\mathbf{i} - 4t\mathbf{j}$, $0 \le t \le 2$

Solution

(a) $$\mathbf{r}'(t) = \cos\frac{t}{2}\mathbf{i} - \sin\frac{t}{2}\mathbf{j}$$

$$\|\mathbf{r}'(t)\| = \left(\cos^2\frac{t}{2}\right) + \left(\sin^2\frac{t}{2}\right) = 1 \qquad \text{for all } t$$

The parameter t is arc length as measured along C.

(b) $$\mathbf{r}'(t) = 6t\mathbf{i} - 4\mathbf{j}$$

$$\|\mathbf{r}'(t)\| = \sqrt{36t^2 + 16}$$

Since $\|\mathbf{r}'(t)\| \ne 1$, the parameter t does not measure arc length along C.

Figure 23

Curvature

Look at the curve shown in Figure 23. At the point P the curve bends sharply; at Q the curve hardly bends at all. *Curvature* is a measure of how sharply a curve bends. For example, for the curve in Figure 23, the curvature at P is greater than the curvature at Q.

The bending of a curve at a point P may be quantified by using the unit tangent vector $\mathbf{T}(t)$ at P. If we move a little away from P and there is little or no change in the direction of $\mathbf{T}(t)$, then the curve has not bent very much. If we move a little away from P and the direction of $\mathbf{T}(t)$ changes dramatically, then the curve has bent quite a lot. See Figure 24. Thus, we measure the bend of a curve by evaluating the magnitude of the rate of change of the tangent vector with respect to arc length. This discussion is the basis for the following definition.

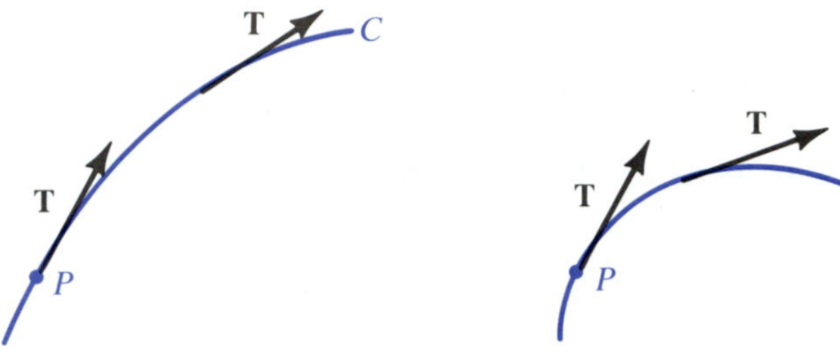

Figure 24

[15.4.2] DEFINITION / *Curvature.*

Suppose a smooth curve C is defined by the twice differentiable vector function $\mathbf{r} = \mathbf{r}(s)$, where the parameter s is the arc length along C. The *curvature* $\kappa = \kappa(s)$* at a point P on C is

$$\kappa = \left\| \frac{d\mathbf{T}}{ds} \right\|$$

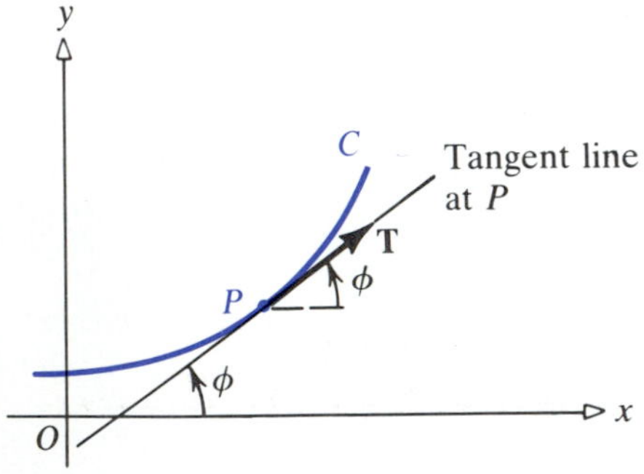

Curvature in the Plane

Figure 25

For a plane curve C, the curvature κ at a point P of C can be interpreted by using the inclination ϕ of the tangent line at P to the positive x-axis. See Figure 25. The components of the unit tangent vector $\mathbf{T}$ can be expressed in terms of ϕ as

$$\mathbf{T} = \cos\phi\mathbf{i} + \sin\phi\mathbf{j}$$

We use the chain rule to find $d\mathbf{T}/ds$:

$$\frac{d\mathbf{T}}{ds} = \frac{d\mathbf{T}}{d\phi}\frac{d\phi}{ds} = (-\sin\phi\mathbf{i} + \cos\phi\mathbf{j})\frac{d\phi}{ds}$$

* The Greek letter κ (kappa).

The curvature κ is

$$\kappa = \left\|\frac{d\mathbf{T}}{ds}\right\| = \left|\frac{d\phi}{ds}\right| \sqrt{(-\sin\phi)^2 + (\cos\phi)^2} = \left|\frac{d\phi}{ds}\right|$$

[15.4.3] THEOREM / ***Curvature of a Plane Curve.***

The *curvature* κ of a plane curve C at a point (x, y) on C is

$$\kappa = \left|\frac{d\phi}{ds}\right|$$

where ϕ is the inclination to the positive x-axis of the tangent line to C at (x, y) and s is the arc length as measured along the curve C.

Thus, κ is the absolute value of the rate of change of ϕ with respect to s.

We will use theorem (15.4.3) to calculate the curvature at any point on a straight line, but before we do this, let's guess the answer. A straight line does not bend at all. Hence, its curvature must be 0.

[15.4.4] THEOREM

The curvature of a straight line is everywhere 0.

Proof In the case of a straight line, $\Delta\phi = 0$ everywhere on the line, since the tangent line coincides with the line itself. Then $d\phi/ds = 0$ and, therefore, by theorem (15.4.3), $\kappa = |0| = 0$ at all points of the line.

Now let's use theorem (15.4.3) to calculate the curvature at any point on a circle. Again, let's see if we can guess the answer. A circle is a curve that bends uniformly. Hence, we suspect that the curvature of a circle is a constant. What the constant is may surprise you!

[15.4.5] THEOREM

The curvature of a circle is the same at every point on it, and it is equal to the reciprocal of the radius. That is, if R is the radius of a circle, then the curvature κ at any point on the circle is

$$\kappa = \frac{1}{R}$$

Proof We position the circle so that its center is at the origin, as shown in Figure 26. At any point P on the circle, the inclination ϕ to the positive x-axis of the tangent line is

$$\phi = \frac{\pi}{2} + \theta$$

where θ is a central angle. The arc length s is $s = R\theta$. Hence,

$$\phi = \frac{\pi}{2} + \frac{s}{R}$$

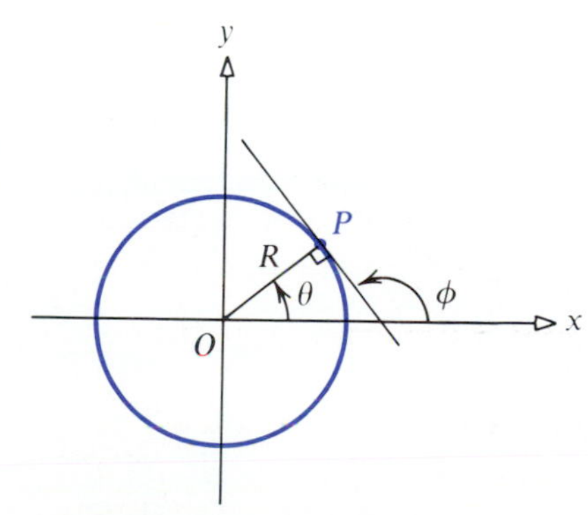

Figure 26

By differentiating ϕ with respect to s, we find that

$$\kappa = \left|\frac{d\phi}{ds}\right| = \frac{1}{R}$$

Let's now derive a formula for the curvature of the graph of a twice differentiable function $y = f(x)$ at a given point $P = (x, y)$ on the graph. By using theorem (15.4.3) and the chain rule, we have

$$\kappa = \left|\frac{d\phi}{ds}\right| = \left|\left(\frac{d\phi}{dx}\right)\left(\frac{dx}{ds}\right)\right| = \left|\frac{d\phi/dx}{ds/dx}\right| \tag{2}$$

To obtain an expression for $d\phi/dx$, we use the fact that the slope of the tangent line to the graph of $y = f(x)$ is

$$\tan\phi = \frac{dy}{dx}$$

Then, upon differentiating each side with respect to x, we obtain

$$\sec^2\phi \frac{d\phi}{dx} = y''$$

$$(1 + \tan^2\phi)\frac{d\phi}{dx} = y''$$

$$\frac{d\phi}{dx} = \frac{y''}{1 + (y')^2} \tag{3}$$

Next we seek an expression for ds/dx. The arc length s of $y = f(x)$ from a to an arbitrary x is

$$s = \int_a^x \sqrt{1 + [f'(u)]^2}\, du$$

from which

$$\frac{ds}{dx} = \sqrt{1 + [f'(x)]^2} = \sqrt{1 + y'^2} \tag{4}$$

By substituting the results (3) and (4) into (2), we obtain

$$\kappa = \left|\frac{y''/[1 + (y')^2]}{\sqrt{1 + (y')^2}}\right| = \frac{|y''|}{[1 + (y')^2]^{3/2}}$$

[15.4.6] THEOREM / ***Curvature of $y = f(x)$.***

The curvature κ of the graph of a twice differentiable function $y = f(x)$ at a point (x, y) on its graph is given by

$$\kappa = \frac{|y''|}{[1 + (y')^2]^{3/2}} \tag{5}$$

where y' and y'' are to be evaluated at (x, y).

EXAMPLE 2

Find the curvature of the parabola $y = \frac{1}{4}x^2$ at the point (2, 1). Compare this to the curvature of the graph of $y = (x^3 + 4)/12$ at (2, 1).

Solution

From the equation $y = \frac{1}{4}x^2$, we find that $y' = \frac{1}{2}x$ and $y'' = \frac{1}{2}$. At the point (2, 1), $y' = 1$, $y'' = \frac{1}{2}$. By using this information in (5), we obtain

$$\kappa = \frac{\frac{1}{2}}{(1 + 1)^{3/2}} = \frac{1}{4\sqrt{2}} \approx 0.177$$

For $y = (x^3 + 4)/12$, we find $y' = \frac{1}{4}x^2$ and $y'' = \frac{1}{2}x$. At (2, 1) these become $y' = 1$ and $y'' = 1$. The curvature κ of this graph at (2, 1) is

$$\kappa = \frac{1}{(1 + 1)^{3/2}} = \frac{1}{2\sqrt{2}} \approx 0.354$$

Figure 27 illustrates the graphs of these two functions. Note that they have the same tangent line at (2, 1). The tangent line to the graph of $y = \frac{1}{4}x^2$ at (2, 1) turns more slowly $(\kappa \approx 0.177)$ than does the tangent line to the graph of $y = (x^3 + 4)/12$ at (2, 1) $(\kappa \approx 0.354)$.

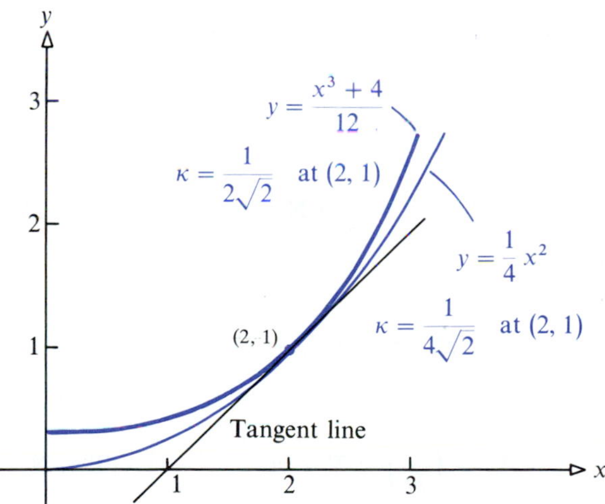

Figure 27

A Curvature Formula for Curves in Space

To use the definition $\kappa = \|d\mathbf{T}/ds\|$ effectively to compute the curvature κ of a space curve C requires that the curve be expressed with arc length s as a parameter. Since this is usually not the case, it is helpful to have a formula for the curvature κ when the curve is expressed by a parameter other than arc length. We now give such a formula.

[15.4.7] THEOREM / *Curvature in Space.*

If a smooth curve C in space is defined by the twice differentiable vector function $\mathbf{r} = \mathbf{r}(t)$, its curvature κ is given by the formula

$$\kappa = \frac{\|\mathbf{r}'(t) \times \mathbf{r}''(t)\|}{\|\mathbf{r}'(t)\|^3} \tag{6}$$

Proof Since $\|\mathbf{r}'(t)\| = ds/dt$, we have

$$\mathbf{T}(t) = \frac{\mathbf{r}'(t)}{\|\mathbf{r}'(t)\|} \quad \text{or} \quad \mathbf{r}'(t) = \|\mathbf{r}'(t)\|\mathbf{T}(t) = \frac{ds}{dt}\mathbf{T}(t) \tag{7}$$

Now,

$$\mathbf{r}''(t) = \frac{d}{dt}\left(\frac{ds}{dt}\mathbf{T}(t)\right) = \frac{d^2s}{dt^2}\mathbf{T}(t) + \frac{ds}{dt}\mathbf{T}'(t) \tag{8}$$

But

$$\mathbf{T}'(t) = \frac{d\mathbf{T}}{ds}\frac{ds}{dt} \quad \text{and} \quad \mathbf{N}(t) = \frac{\mathbf{T}'(t)}{\|\mathbf{T}'(t)\|}$$

Thus,

$$\mathbf{T}'(t) = \|\mathbf{T}'(t)\|\mathbf{N}(t) = \left\|\frac{d\mathbf{T}}{ds}\right\|\frac{ds}{dt}\mathbf{N}(t) \underset{\kappa = \|d\mathbf{T}/ds\|}{\uparrow{=}} \kappa\frac{ds}{dt}\mathbf{N}(t) \tag{9}$$

Using the expression above for $\mathbf{T}'(t)$ in (8), we get

$$\mathbf{r}''(t) = \frac{d^2s}{dt^2}\mathbf{T}(t) + \kappa\left(\frac{ds}{dt}\right)^2\mathbf{N}(t) \tag{10}$$

From (7) and (10), we get

$$\begin{aligned}\mathbf{r}'(t) \times \mathbf{r}''(t) &= \frac{ds}{dt}\mathbf{T} \times \left[\frac{d^2s}{dt^2}\mathbf{T} + \left(\frac{ds}{dt}\right)^2\kappa\mathbf{N}\right] \\ &= \frac{ds}{dt}\left[\frac{d^2}{dt^2}(\mathbf{T} \times \mathbf{T}) + \left(\frac{ds}{dt}\right)^2\kappa(\mathbf{T} \times \mathbf{N})\right] \underset{\mathbf{T} \times \mathbf{T} = \mathbf{0}}{\uparrow{=}} \left(\frac{ds}{dt}\right)^3\kappa(\mathbf{T} \times \mathbf{N})\end{aligned}$$

Hence,

$$\|\mathbf{r}'(t) \times \mathbf{r}''(t)\| = \left(\frac{ds}{dt}\right)^3\kappa\|\mathbf{T} \times \mathbf{N}\| \tag{11}$$

Since $\mathbf{T}$ and $\mathbf{N}$ are orthogonal unit vectors, $\mathbf{T} \times \mathbf{N}$ is also a unit vector (see Problem 32 in Exercise 14.5). From (7), it follows that $\|\mathbf{r}'(t)\| = ds/dt$. Using these facts and solving (11) for κ give the required formula.

EXAMPLE 3

Find the curvature κ of the curve $\mathbf{r}(t) = t\mathbf{i} + t^2\mathbf{j} + t^3\mathbf{k}.$

Solution

$$\mathbf{r}'(t) = \mathbf{i} + 2t\mathbf{j} + 3t^2\mathbf{k}$$

$$\|\mathbf{r}'(t)\| = \sqrt{1 + 4t^2 + 9t^4}$$

Since $\|\mathbf{r}'(t)\| \neq 1$, we know that the parameter t is not arc length. As a result, we proceed to calculate $\mathbf{r}''(t)$ in order to use theorem [15.4.7].

$$\mathbf{r}''(t) = 2\mathbf{j} + 6t\mathbf{k}$$

$$\mathbf{r}'(t) \times \mathbf{r}''(t) = \begin{vmatrix} \mathbf{i} & \mathbf{j} & \mathbf{k} \\ 1 & 2t & 3t^2 \\ 0 & 2 & 6t \end{vmatrix} = 6t^2\mathbf{i} - 6t\mathbf{j} + 2\mathbf{k}$$

$$\|\mathbf{r}'(t) \times \mathbf{r}''(t)\| = \sqrt{36t^4 + 36t^2 + 4} = 2\sqrt{9t^4 + 9t^2 + 1}$$

Thus,

$$\kappa = \frac{2\sqrt{9t^4 + 9t^2 + 1}}{(1 + 4t^2 + 9t^4)^{3/2}}$$

Equation (6) can be used to find the curvature of a plane curve C provided a **k**-component of 0 is appended to the vector function defining C. Thus, to use (6) for a plane curve C, the vector function defining C is written as $\mathbf{r}(t) = x(t)\mathbf{i} + y(t)\mathbf{j} + 0\mathbf{k}$.

EXAMPLE 4

Find the curvature κ of the plane curve:

$$\mathbf{r}(t) = 3 \sin t\mathbf{i} + 4 \cos t\mathbf{j}, \qquad 0 \leq t \leq 2\pi$$

(a) At $t = 0$ (b) At $t = \pi/2$

Solution

We express the plane curve as a curve in space with a 0 **k**-component so that (6) can be used. Then,

$$\mathbf{r}(t) = 3 \sin t\mathbf{i} + 4 \cos t\mathbf{j} + 0\mathbf{k}$$

$$\mathbf{r}'(t) = 3 \cos t\mathbf{i} - 4 \sin t\mathbf{j}$$

$$\|\mathbf{r}'(t)\| = \sqrt{9 \cos^2 t + 16 \sin^2 t} = \sqrt{9 + 7 \sin^2 t}$$

$$\mathbf{r}''(t) = -3 \sin t\mathbf{i} - 4 \cos t\mathbf{j}$$

$$\mathbf{r}'(t) \times \mathbf{r}''(t) = \begin{vmatrix} \mathbf{i} & \mathbf{j} & \mathbf{k} \\ 3 \cos t & -4 \sin t & 0 \\ -3 \sin t & -4 \cos t & 0 \end{vmatrix} = 0\mathbf{i} + 0\mathbf{j} - 12\mathbf{k}$$

$$\kappa = \frac{\|\mathbf{r}'(t) \times \mathbf{r}''(t)\|}{\|\mathbf{r}'(t)\|^3} = \frac{12}{(9 + 7 \sin^2 t)^{3/2}} \tag{12}$$

(a) At $t = 0$, $\kappa = \dfrac{12}{9^{3/2}} = \dfrac{4}{9}$.

(b) At $t = \dfrac{\pi}{2}$, $\kappa = \dfrac{12}{16^{3/2}} = \dfrac{12}{64} = \dfrac{3}{16}$.

Osculating Circle

Associated with each point P of a twice differentiable smooth curve C at which $\kappa \neq 0$, there is a circle, called the *osculating circle*, with the following characteristics:

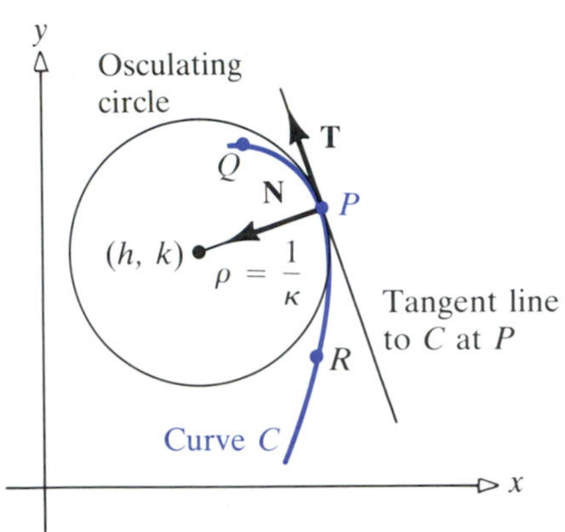

Figure 28

1. The osculating circle passes through P.
2. The tangent line to C at P and the tangent line to the osculating circle at P are the same.
3. The radius of the osculating circle is $\rho = 1/\kappa$.
4. The center of the osculating circle lies on a line in the direction of the principal unit normal $\mathbf{N}$ to C at P.

Figure 28 illustrates these characteristics.

The osculating circle of a curve C is sometimes called the *circle of curvature*. Its radius $\rho = 1/\kappa$ is called the *radius of curvature*, and its center is called the *center of curvature*.

Since the osculating circle and the curve share a common tangent at each point and have the same curvature at each point, sometimes the osculating circle is used as an approximation to the curve C. In fact, the osculating circle at P is the limiting circle that results from the circle passing through three points, P, Q, and R, on C as Q and R are made to approach P.

EXAMPLE 5

Find the radius of the osculating circle of the curve defined by

$$\mathbf{r}(t) = 3 \sin t\mathbf{i} + 4 \cos t\mathbf{j}$$

(a) At $t = 0$ (b) At $t = \pi/2$

Solution

This is the same curve discussed in Example 4. Using (12), the radius ρ of the osculating circle is

$$\rho = \frac{1}{\kappa} = \frac{(9 + 7 \sin^2 t)^{3/2}}{12}$$

(a) At $t = 0$, $\rho = 9/4$.
(b) At $t = \pi/2$, $\rho = 16/3$.

Figures 29(a) and 29(b) illustrate the curve and its osculating circles.

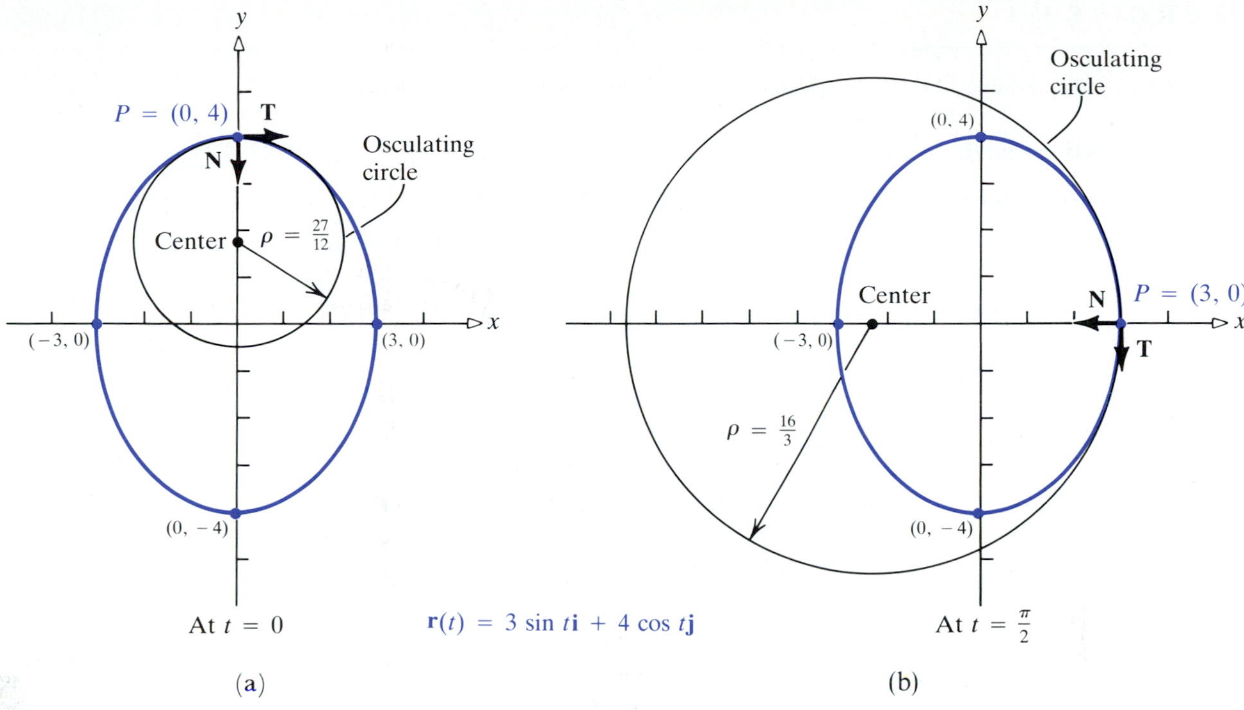

Figure 29

Tangential and Normal Components of Acceleration

In the proof of theorem (15.4.7), we show in (10) that the acceleration $\mathbf{a}(t) = \mathbf{r}''(t)$ of a particle whose motion is along a curve C traced out by a twice differentiable vector function $\mathbf{r} = \mathbf{r}(t)$ is

$$\mathbf{a}(t) = \frac{d^2s}{dt}\mathbf{T}(t) + \left(\frac{ds}{dt}\right)^2 \kappa\mathbf{N}(t)$$

where $v(t) = ds/dt$ is the speed of the particle, κ is the curvature, $\mathbf{T}$ is the unit tangent vector, and $\mathbf{N}$ is the unit principal normal vector. Thus, we may write

$$\mathbf{a}(t) = \frac{dv}{dt}\mathbf{T}(t) + v^2\kappa\mathbf{N}(t)$$

The tangential and normal components of the acceleration are

$$a_{\mathbf{T}} = \frac{dv}{dt} \qquad a_{\mathbf{N}} = v^2\kappa$$

This formula for $a_{\mathbf{N}}$ offers a better alternative for finding $a_{\mathbf{N}}$ than does (7) from Section 15.3.

EXERCISE 15.4

In Problems 1–10 determine whether the parameter used to define each curve is arc length.

1. $\mathbf{r}(t) = 4\cos t\mathbf{i} - 4\sin t\mathbf{j}, \quad 0 \le t \le 2\pi$

2. $\mathbf{r}(t) = \sin 3t\mathbf{i} + \cos 3t\mathbf{j}, \quad 0 \le t \le 2\pi$

3. $\mathbf{r}(t) = t^2\mathbf{i} + t\mathbf{j}, \quad 0 \le t \le 4$

4. $\mathbf{r}(t) = t\mathbf{i} + t^3\mathbf{j}, \quad 0 \le t \le 2$

5. $\mathbf{r}(t) = (2t+1)\mathbf{i} + (3t-2)\mathbf{j}, \quad 0 \le t \le 5$

6. $\mathbf{r}(t) = \mathbf{i} + t\mathbf{j}, \quad 0 \le t \le 1$

7. $\mathbf{r}(t) = \left(\frac{2}{\sqrt{13}}t + 1\right)\mathbf{i} + \left(\frac{3}{\sqrt{13}}t - 2\right)\mathbf{j}, \quad 0 \le t \le 5\sqrt{13}$

8. $\mathbf{r}(t) = \mathbf{i} + t^2\mathbf{j}, \quad 0 \le t \le 1$

9. $\mathbf{r}(t) = \sin t\mathbf{i} + \cos t\mathbf{j} + t\mathbf{k}$

10. $\mathbf{r}(t) = a\sin t\mathbf{i} + a\cos t\mathbf{j} + \sqrt{1-b^2}\,t\mathbf{k}$

In Problems 11–24 find the curvature of each curve.

11. $y = x^2$, at $(1, 1)$

12. $y = 2x - x^2$, at $(1, 1)$

13. $y = x^2 - x^3$, at $x = 1$

14. $y = x^3$, at $x = 2$

15. $y = \sqrt{x}$, at $x = 2$

16. $y = 1/x$, at $(1, 1)$

17. $y = x^{-3/2}$, at $x = 1$

18. $y = 1/\sqrt{x}$, at $(1, 1)$

19. $x^2 - y^2 = a^2$, at $x = 2a$

20. $4x^2 + 9y^2 = 36$, at $(0, 2)$

21. $y = \cos x$, at $x = 0$

22. $y = \sec x - 1$, at $x = \frac{1}{4}\pi$

23. $y = e^x$, at $(0, 1)$

24. $y = \ln(x+1)$, at $x = 2$

In Problems 25–34 find the curvature $\kappa = \kappa(t)$ of the curve C defined by the vector function.

25. $\mathbf{r}(t) = t^2\mathbf{i} + (2/t)\mathbf{j}$

26. $\mathbf{r}(t) = 2t\mathbf{i} + t^3\mathbf{j}$

27. $\mathbf{r}(t) = t\mathbf{i} + 2t\mathbf{j} + t\mathbf{k}$

28. $\mathbf{r}(t) = 2t\mathbf{i} + t\mathbf{j} + 3t\mathbf{k}$

29. $\mathbf{r}(t) = \sin 2t\mathbf{i} + \cos 2t\mathbf{j} + t\mathbf{k}$

30. $\mathbf{r}(t) = \sin t\mathbf{i} + \cos t\mathbf{j} + bt\mathbf{k}$

31. $\mathbf{r}(t) = e^t\mathbf{i} + e^{-t}\mathbf{j} + \sqrt{2}\,t\mathbf{k}$

32. $\mathbf{r}(t) = \cos^3 t\mathbf{i} + \sin^3 t\mathbf{j} + \mathbf{k}$

33. $\mathbf{r}(t) = a(3t - t^3)\mathbf{i} + 3at^2\mathbf{j} + a(3t + t^3)\mathbf{k}$

34. $\mathbf{r}(t) = 4a\cos^3 t\mathbf{i} + 4a\sin^3 t\mathbf{j} + 3a\cos 2t\mathbf{k}$

In Problems 35–44 find the radius of curvature at the point P of the curve C defined by the vector function.

35. $\mathbf{r}(t) = 3t^2\mathbf{i} + (3t - t^3)\mathbf{j}$, at $t = 1$

36. $\mathbf{r}(t) = \sin t\mathbf{i} + \cos 2t\mathbf{j}$, at $t = \pi/4$

37. $\mathbf{r}(t) = t\mathbf{i} + t^2\mathbf{j} + t^3\mathbf{k}$, at $t = 1$

38. $\mathbf{r}(t) = 2\sin t\mathbf{i} + 2\cos t\mathbf{j} + 3t\mathbf{k}$, at $t = 0$

39. $\mathbf{r}(t) = \sin 2t\mathbf{i} + \cos 2t\mathbf{j} + t\mathbf{k}$, at $t = \pi/4$

40. $\mathbf{r}(t) = \sin t\mathbf{i} + \cos t\mathbf{j} + bt\mathbf{k}$, at $t = \pi/4$

41. $\mathbf{r}(t) = e^t\mathbf{i} + e^{-t}\mathbf{j} + \sqrt{2}\,t\mathbf{k}$, at $t = 0$

42. $\mathbf{r}(t) = \cos^3 t\mathbf{i} + \sin^3 t\mathbf{j} + \mathbf{k}$, at $t = \pi/3$

43. $\mathbf{r}(t) = a(3t - t^3)\mathbf{i} + 3at^2\mathbf{j} + a(3t + t^3)\mathbf{k}$, at $t = 1$

44. $\mathbf{r}(t) = 4a\cos^3 t\mathbf{i} + 4a\sin^3 t\mathbf{j} + 3a\cos 2t\mathbf{k}$, at $t = \pi/2$

In Problems 45–50 find the radius of curvature of each curve.

45. $y = x^3 - 6x$, at $(1, -5)$

46. $y = 1/x^2$, at $(-1, 1)$

47. $y = \sin x$, at $(\pi/2, 1)$

48. $y = e^{-x}$, at $(0, 1)$

49. $x^2 + xy + y^2 = 3$, at $(1, 1)$

50. $y^2 - y + x = 0$, at $(0, 0)$

51. Show that the radius of curvature of the parabola $y = ax^2 + bx + c$ is a minimum at its vertex.

52. Show that the radii of curvature at the ends of the axes of the ellipse $b^2x^2 + a^2y^2 = a^2b^2$ are b^2/a and a^2/b.

53. Find the point at which the curvature is maximum on the curve $y = \ln x$.

54. Find the point at which the curvature is maximum on the curve $y = e^x$.

55. Find the point at which the curvature is maximum on the curve $y = \frac{1}{3}x^3$.

56. Show that the formula for the curvature of a polar curve $r = f(\theta)$ is

$$\kappa = \frac{|r^2 + 2(dr/d\theta)^2 - r(d^2r/d\theta^2)|}{[r^2 + (dr/d\theta)^2]^{3/2}}$$

In Problems 57–62 use the result of Problem 56 to find the curvature of each polar curve.

57. $r = 2\cos 2\theta$, at $\theta = \pi/12$

58. $r = e^{a\theta}$, at $\theta = \pi/2$

59. $r = a\theta$, at $\theta = 1$, $a > 0$

60. $r = 1 - \cos\theta$, at $\theta = 0$

61. $r = 3 - 2\sin\theta$, at $\theta = \pi/6$

62. $r = 2 + 3\cos\theta$, at $\theta = \pi/3$

63. Show that the curvature of the catenary $y = a\cosh(x/a)$ at any point is a/y^2.

64. What is the curvature at a point of inflection of a plane curve?

65. Find the curvature of the cissoid $y^2(2 - x) = x^3$ at the point $(1, 1)$.

66. Find the curvature of the cycloid $x = \theta - \sin\theta$, $y = 1 - \cos\theta$ at the highest point of an arch.

67. Find the curvature of the curve $\mathbf{r}(t) = (1 - t^3)\mathbf{i} + t^2\mathbf{j}$. Where is the curvature undefined? Do you see any geometric reason for this?

68. Find the curvature of the spiral $\mathbf{r}(t) = e^{-t}\cos t\mathbf{i} + e^{-t}\sin t\mathbf{j}$. How does it behave when $t \to +\infty$? Do you see any geometric reason for this?

69. Find the curvature κ of the curve $\mathbf{r}(t) = 2a\cos t\mathbf{i} + 2a\sin t\mathbf{j} + bt^2\mathbf{k}$.

70. Suppose $\mathbf{r} = \mathbf{r}(t)$ is the position vector of a particle

(*Problem 70 continues in the next column*)

70. (*continued*)
at time t. If the normal component of acceleration equals 0 at any time t, explain why the motion of the particle must be in a straight line.

71. We have shown that if $\mathbf{r}(t) = x(t)\mathbf{i} + y(t)\mathbf{j} + z(t)\mathbf{k}$ is the position vector of a particle at time t, the normal component of acceleration is $a_{\mathbf{N}} = \kappa v^2$, where $v = ds/dt$ is speed. Use this fact to find $a_{\mathbf{N}}$ when the motion of a particle is given by $\mathbf{r}(t) = \cos t\mathbf{i} + \sin t\mathbf{j} + t\mathbf{k}$.

72. Use the formula

$$\mathbf{F} = m\mathbf{a} = m\frac{dv}{dt}\mathbf{T} + mkv^2\mathbf{N}$$

to discuss the forces on a passenger in a car. What force corresponds to the push against the seat experienced by the passenger when the accelerator is depressed? What force corresponds to the push against the door experienced by the passenger when the car is going around a curve? How does this latter force vary with the curvature of the road? How does it vary with the speed of the car?

In Problems 73–76 use the fact that the graph of a twice differentiable function $y = f(x)$ can be described as the space curve $\mathbf{r} = \mathbf{r}(t) = t\mathbf{i} + f(t)\mathbf{j} + 0\mathbf{k}$ to prove each result.

73. $\dfrac{dt}{ds} = \dfrac{1}{\sqrt{1 + f'^2}}$

74. $\mathbf{T} = \dfrac{\mathbf{i} + f'\mathbf{j}}{\sqrt{1 + f'^2}}$

75. $\dfrac{d\mathbf{T}}{ds} = \dfrac{f''}{(1 + f'^2)^2}(-f'\mathbf{i} + \mathbf{j})$

76. $\kappa = \dfrac{|f''|}{(1 + f'^2)^{3/2}}$

77. Find $a > 0$ so that $\mathbf{r}(t) = a\cos t\mathbf{i} + a\sin t\mathbf{j} + t\mathbf{k}$ has maximum curvature.

78. Compare the solutions of Problems 5 and 7. Then show how to change the parameter t of the line

$$\mathbf{r}(t) = (at + b)\mathbf{i} + (ct + d)\mathbf{j}$$

where either $a \neq 0$ or $c \neq 0$, to one that is arc length as measured along the line.

79. Suppose a smooth curve C is defined by a twice differentiable vector function $\mathbf{r} = \mathbf{r}(t)$, $a \le t \le b$. If the curvature $\kappa \neq 0$ at a point P on C, show that the position vector $\mathbf{C}$ of the center of the osculating circle at P obeys

$$\mathbf{C}(t) = \mathbf{r}(t) + \rho\mathbf{N}(t)$$

where $\mathbf{N}$ is the principal unit normal vector to C at P, and $\rho = 1/\kappa$.

80. Use the result of Problem 79 to find the center and radius of the osculating circle for the helix

$$\mathbf{r}(t) = a \sin t\mathbf{i} + a \cos t\mathbf{j} + a^2t\mathbf{k}$$

(a) At $t = \pi/2$ (b) At $t = \pi$

81. If a particle moves along the graph of $y = f(x)$, show that $a_{\mathbf{N}} = 0$ at a point of inflection of the graph.

82. A particle moves along the graph of $y = \frac{1}{2}x^2$ with constant speed $v(t) = 2$. Find $a_{\mathbf{T}}$ and $a_{\mathbf{N}}$ in terms of x alone at a general point $(x, \frac{1}{2}x^2)$ on the graph.

Use the following discussion for Problems 83–87. We have discussed the unit tangent vector $\mathbf{T}(t)$ and the unit normal vector $\mathbf{N}(t)$ for a smooth curve C in space defined by a twice differentiable vector function. Suppose C is defined by the vector function

$$\mathbf{r} = \mathbf{r}(s) \qquad a \le s \le b$$

and s is arc length as measured along C. Since C is a space curve, we can define a vector $\mathbf{B}(s) = \mathbf{T}(s) \times \mathbf{N}(s)$, called the *binormal vector*.

83. Show that the three vectors $\mathbf{T}$, $\mathbf{N}$, and $\mathbf{B}$ form a collection of mutually orthogonal unit vectors at each point on C.

84. Show that $\dfrac{d\mathbf{T}}{ds} = \kappa(s)\mathbf{N}(s)$.

85. Show that $d\mathbf{B}/ds$ is orthogonal to both $\mathbf{B}(s)$ and $\mathbf{T}(s)$.

86. If the *torsion* $\tau(s)$ of C is defined by the equation $d\mathbf{B}/ds = -\tau\mathbf{N}$, show that $(d\mathbf{N}/ds) = \tau\mathbf{B} - \kappa\mathbf{T}$.

87. Calculate κ, $\mathbf{T}$, $\mathbf{N}$, and $\mathbf{B}$ for

$$\mathbf{r}(s) = \frac{1}{\sqrt{2}}\left[\sin s\mathbf{i} + \cos s\mathbf{j} + s\mathbf{k}\right].$$

15.5 Integrals of Vector Functions

The integration of a vector function $\mathbf{r} = \mathbf{r}(t)$ is performed by integrating each component. That is, if $\mathbf{r}(t) = x(t)\mathbf{i} + y(t)\mathbf{j} + z(t)\mathbf{k}$, then

$$\int \mathbf{r}(t)\,dt = \left[\int x(t)\,dt\right]\mathbf{i} + \left[\int y(t)\,dt\right]\mathbf{j} + \left[\int z(t)\,dt\right]\mathbf{k}$$

In actually doing integration problems, remember to insert the vector constant of integration. For example,

$$\int \left(\sin t\mathbf{i} + \cos t\mathbf{j} + 2\mathbf{k}\right) dt = \left(\int \sin t\,dt\right)\mathbf{i} + \left(\int \cos t\,dt\right)\mathbf{j} + \left(\int 2\,dt\right)\mathbf{k}$$

$$= (-\cos t + c_1)\mathbf{i} + (\sin t + c_2)\mathbf{j} + (2t + c_3)\mathbf{k}$$

The next example illustrates how to determine a vector function when its derivative and an initial condition are given.

EXAMPLE 1

Find $\mathbf{r}(t)$ if $\mathbf{r}'(t) = 2t\mathbf{i} + e^t\mathbf{j} + e^{-t}\mathbf{k}$ and $\mathbf{r}(0) = \mathbf{i} - \mathbf{j} + \mathbf{k}$.

Solution

$$\mathbf{r}(t) = \int \mathbf{r}'(t)dt = \int (2t\mathbf{i} + e^t\mathbf{j} + e^{-t}\mathbf{k})\,dt$$

$$= \left(\int 2t\,dt\right)\mathbf{i} + \left(\int e^t\,dt\right)\mathbf{j} + \left(\int e^{-t}\,dt\right)\mathbf{k}$$

$$= (t^2 + c_1)\mathbf{i} + (e^t + c_2)\mathbf{j} + (c_3 - e^{-t})\mathbf{k}$$

By applying the condition that $\mathbf{r}(0) = \mathbf{i} - \mathbf{j} + \mathbf{k}$, we find

$$\mathbf{i} - \mathbf{j} + \mathbf{k} = c_1\mathbf{i} + (1 + c_2)\mathbf{j} + (c_3 - 1)\mathbf{k}$$

so

$$1 = c_1 \qquad -1 = 1 + c_2 \qquad 1 = c_3 - 1$$
$$c_1 = 1 \qquad c_2 = -2 \qquad c_3 = 2$$

Hence, the vector function $\mathbf{r}$ is

$$\mathbf{r}(t) = (t^2 + 1)\mathbf{i} + (e^t - 2)\mathbf{j} + (2 - e^{-t})\mathbf{k}$$

Newton's Second Law of Motion

From Newton's second law of motion, $\mathbf{F}(t) = m\mathbf{a}(t)$, once the force $\mathbf{F}(t)$ acting on a particle of known constant mass m is given, the acceleration vector $\mathbf{a}(t)$ is determined. Since the velocity vector $\mathbf{v}(t)$ is the integral of $\mathbf{a}(t)$—that is,

$$\mathbf{v}(t) = \int \mathbf{a}(t)\,dt$$

we can calculate the velocity and speed of the particle, provided a boundary condition on $\mathbf{v}(t)$ is known. Furthermore, since

$$\mathbf{r}(t) = \int \mathbf{v}(t)\,dt$$

we can determine the path of the particle when a boundary condition on $\mathbf{r}(t)$ is known.

Projectile Problem

We use these facts to find the path of a projectile fired at an inclination θ to the horizontal and with initial speed v_0. For convenience, we select the starting point as the origin, the x-axis as horizontal, and the y-axis as positive upward (see Fig. 30).

When time $t = 0$, the position of the projectile is the origin, and the initial velocity vector $\mathbf{v}(0)$ has magnitude v_0 and inclination θ. Consequently,

$$\mathbf{r}(0) = \mathbf{0} \qquad \text{and} \qquad \mathbf{v}(0) = v_0 \cos\theta\mathbf{i} + v_0 \sin\theta\mathbf{j} \tag{1}$$

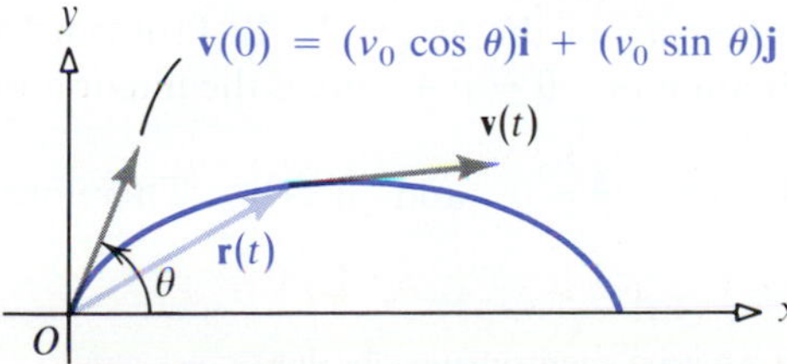

Figure 30

We let $\mathbf{r}(t)$ represent the position vector of the projectile after time t. We wish to find an expression for $\mathbf{r}(t)$ in terms of v_0 and θ. We assume that the only force acting on the projectile is the force $\mathbf{F} = -mg\mathbf{j}$,* where g is the acceleration of gravity and m is the mass of the projectile.

From Newton's second law of motion, the acceleration $\mathbf{a}(t)$ of the projectile obeys

$$m\mathbf{a}(t) = -mg\mathbf{j} \qquad \text{or} \qquad \mathbf{r}''(t) = -g\mathbf{j}$$

Integrating both sides with respect to t gives

$$\mathbf{v}(t) = -gt\mathbf{j} + \mathbf{c}$$

where $\mathbf{c}$ is a constant vector. When $t = 0$, $\mathbf{v}(0) = \mathbf{c}$. Hence, from (1), the velocity vector of the projectile is

$$\mathbf{v}(t) = v_0 \cos\theta\mathbf{i} + v_0 \sin\theta\mathbf{j} - gt\mathbf{j}$$

Integrating both sides with respect to t gives

$$\mathbf{r}(t) = (v_0 \cos\theta)t\mathbf{i} + (v_0 \sin\theta)t\mathbf{j} - \tfrac{1}{2}gt^2\mathbf{j} + \mathbf{d}$$

where $\mathbf{d}$ is a constant vector. When $t = 0$, the position of the projectile is at the origin. Hence, $\mathbf{d} = \mathbf{0}$. Thus, the position vector of the projectile at time t is

$$\mathbf{r}(t) = (v_0 \cos\theta)t\mathbf{i} + (v_0 \sin\theta)t\mathbf{j} - \tfrac{1}{2}gt^2\mathbf{j}$$

The parametric equations of the motion of the projectile are

$$x = (v_0 \cos\theta)t \qquad y = -\tfrac{1}{2}gt^2 + (v_0 \sin\theta)t \tag{2}$$

By eliminating t from these equations, we find

$$y = \frac{-g}{2v_0^2 \cos^2\theta}x^2 + x\tan\theta \tag{3}$$

which is the equation of a parabola.

The x-intercepts of this parabola are found by setting $y = 0$. Hence, we find that

$$x = 0 \qquad \text{and} \qquad x = \frac{2v_0^2 \cos^2\theta \tan\theta}{g} = \frac{v_0^2 \sin 2\theta}{g} \tag{4}$$

The number $(v_0^2 \sin 2\theta)/g$ is called the *range* of the projectile. In Problem 15 you are asked to show that an inclination of $\theta = \pi/4$ gives the maximum range.

The projectile hits the ground when $y = 0$ and $t > 0$. Therefore,

* This is called the *flat earth approximation*. It is a good approximation for short-range missiles.

we solve (2) for t:

$$0 = -\tfrac{1}{2}gt^2 + (v_0 \sin\theta)t$$
$$\tfrac{1}{2}gt = v_0 \sin\theta$$
$$t = \frac{2v_0 \sin\theta}{g}$$

The projectile is in the air for $(2v_0 \sin\theta)/g$ seconds.

The projectile reaches its maximum height when $dy/dt = 0$ or $t = (v_0 \sin\theta)/g$ and, therefore, the maximum height is

$$y = -\tfrac{1}{2}g\left(\frac{v_0^2 \sin^2\theta}{g^2}\right) + (v_0 \sin\theta)\left(\frac{v_0 \sin\theta}{g}\right) = \frac{v_0^2 \sin^2\theta}{2g} \qquad (5)$$

The formulas for the range of a projectile and its maximum height as given by (4) and (5) are valid only for the initial conditions specified in (1). If different initial conditions are given, (4) and (5) will be different. In such instances, merely follow the same pattern of solution, adjusting where needed for different initial conditions.

EXERCISE 15.5

In Problems 1–4 evaluate each integral.

1. $\int (\sin t\mathbf{i} - \cos t\mathbf{j} + t\mathbf{k})\,dt$

2. $\int (t^2\mathbf{i} - t\mathbf{j} + e^t\mathbf{k})\,dt$

3. $\int (\cos t\mathbf{i} + \sin t\mathbf{j} - \mathbf{k})\,dt$

4. $\int (e^t\mathbf{i} - \sqrt{t}\mathbf{j} + t^2\mathbf{k})\,dt$

In Problems 5–8 find the velocity, speed, and position of a particle having the given acceleration, initial velocity, and initial position.

5. $\mathbf{a} = -32\mathbf{k}$, $\mathbf{v}(0) = \mathbf{0}$, $\mathbf{r}(0) = \mathbf{0}$

6. $\mathbf{a} = -32\mathbf{k}$, $\mathbf{v}(0) = \mathbf{i} + \mathbf{j}$, $\mathbf{r}(0) = \mathbf{0}$

7. $\mathbf{a} = \cos t\mathbf{i} + \sin t\mathbf{j}$, $\mathbf{v}(0) = \mathbf{i}$, $\mathbf{r}(0) = \mathbf{j}$

8. $\mathbf{a} = \cos t\mathbf{i} + \sin t\mathbf{j}$, $\mathbf{v}(0) = \mathbf{j}$, $\mathbf{r}(0) = \mathbf{i}$

9. Find $\mathbf{r}(t)$ if $\mathbf{r}'(t) = e^t\mathbf{i} - \ln t\mathbf{j} + 2t\mathbf{k}$ and $\mathbf{r}(1) = \mathbf{j} + \mathbf{k}$.

10. Find $\mathbf{r}(t)$ if $\mathbf{r}'(t) = t\mathbf{i} + e^{-t}\mathbf{j} - (1/t)\mathbf{k}$ and $\mathbf{r}(1) = \mathbf{i} - \mathbf{j} + 2\mathbf{k}$.

In Problems 11–17 assume $g = 9.8$ meters per second per second and neglect air resistance.

11. A projectile was fired at an angle of 30° with the horizontal with an initial speed of 520 meters per second. What were its range, the time of flight, and the greatest height reached?

12. A projectile was fired with an initial speed of 200 meters per second at an inclination of 60° with the horizontal. What were its range, the time of flight, and the greatest height reached?

13. A projectile was fired with an initial speed of 100 meters per second at an angle of inclination of $\tan^{-1}(\frac{5}{12})$. Find the equations of the path and the range.

14. A projectile was fired with an initial speed of 120 meters per second at an angle of inclination of $\tan^{-1}(\frac{3}{4})$. Find the equations of the path and the range.

15. Show that the maximum range of the projectile of (4) occurs when $\theta = \pi/4$ and has the value v_0^2/g.

16. Show that the speed of the projectile of (2) is least when the projectile is at its highest point.

17. A projectile is fired with a speed of 500 meters per second at an inclination of 45° from a point 30 meters above level ground. Find the point where the projectile will strike the ground.

18. A force whose magnitude is 5 and whose direction is along the positive x-axis is continuously applied to an object of mass $m = 1$ kilogram. If at $t = 0$, (*Problem 18 continues in the next column*)

18. (*continued*) the position of the object is the origin and if its velocity is 3 meters per second in the direction of the positive y-axis, find:

(a) The speed and velocity after t seconds
(b) The position after the force has been applied for t seconds
(c) The path of the object

19. An object of mass m leaves the point (1, 2) with initial velocity $\mathbf{v}_0 = 3\mathbf{i} + 4\mathbf{j}$. Thereafter, it is subjected only to the force $\mathbf{F} = m(-\mathbf{i} - \mathbf{j})/\sqrt{2}$. Find the formula for the position of the object at any time $t > 0$.

20. An projectile is propelled horizontally at a height of 3 meters above the ground in order to hit a target 1 meter high that is 30 meters away (see the figure). What should the initial velocity of the projectile be?

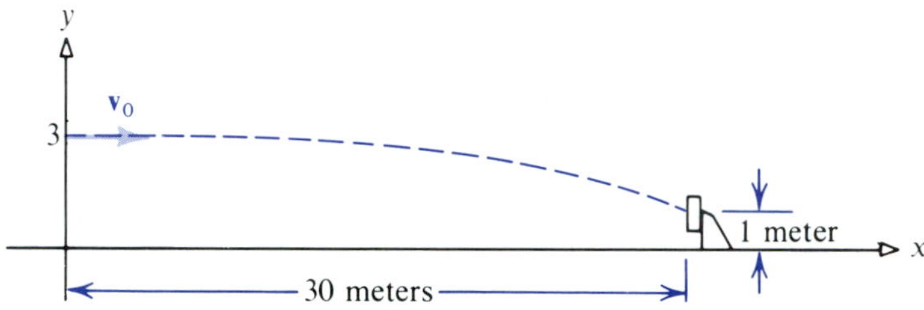

21. *Calculator Problem.* In a Metro Conference basketball game on January 21, 1980, between Florida State University and Virginia Tech, a record was set. Les Henson, who is 6 feet 6 inches tall, made a basket from $89\frac{1}{4}$ feet downcourt, to win the game for Virginia Tech by a score of 79 to 77. Assuming he released the ball at a height of 6 feet 6 inches and threw it at an angle of 45° (to maximize distance), with what initial velocity was the ball tossed? See the accompanying figure. (Assume $g = 32$ feet per second per second.)

22. *Calculator Problem.* A baseball is hit at an angle of 45° to the horizontal from an initial height of 3 feet. If the ball just clears the vines in front of the bleachers in Wrigley Field, which are 10 feet high and a distance of 400 feet from the plate, what was the initial speed of the ball? How long did it take the ball to reach the vines?

23. *Calculator Problem.* A certain outfielder throws a baseball at an angle of 45° to the horizontal from an initial height of 6 feet. If he can throw the ball with an initial velocity of 100 feet per second, what is the maximum distance he can be from home plate to ensure that the ball reaches home plate on the fly? How long is the ball in flight? (The answers reveal how fast a runner must be to get from third base to home plate on a fly ball and beat the throw to score a run.)

24. A gun, lifted at an angle θ_0 to the horizontal, is aimed at an elevated target, which is released at the moment the gun is fired. No matter what the initial speed v_0 of the bullet, show that it will always hit the falling target. Use the accompanying figure as a guide.

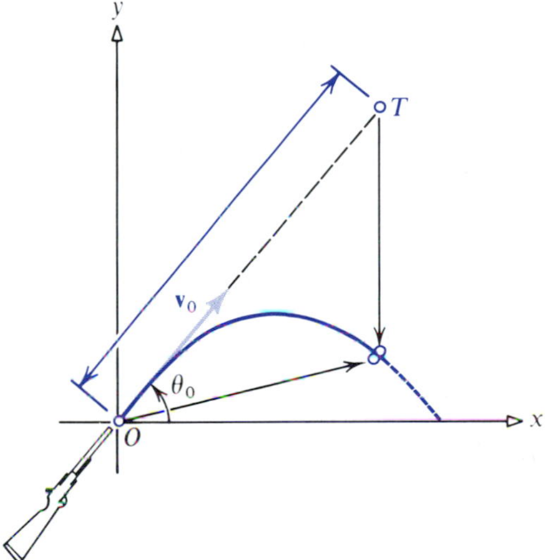

25. A plane is flying at an elevation of 4.0 kilometers with a constant horizontal speed of 400 kilometers per hour toward a point directly above its target. At what angle of sight α should a package be released in order to strike the target? Use the accompanying figure as a guide. [*Hint:* $g \approx 127{,}008$ kilometers per hour per hour.]

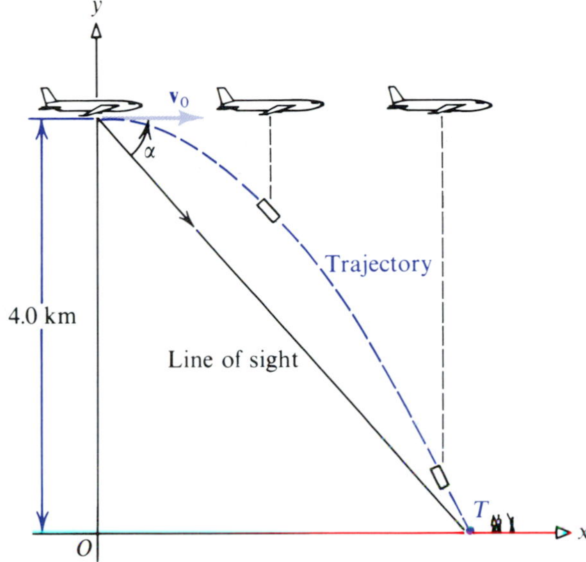

26. Write a vector equation for the curve $\mathbf{r}(t) = 2t\mathbf{i} + (2t-1)\mathbf{j} + t\mathbf{k}$, $0 \le t \le 2$, using arc length s as a parameter. [*Hint:* For each t, $0 \le t \le 2$, calculate the length $s(t)$ of the curve from 0 to t.]

27. Rework Problem 26 for a coil of the helix $\mathbf{r}(t) = \cos t\mathbf{i} + \sin t\mathbf{j} + t\mathbf{k}$, $0 \le t \le 2\pi$.

28. Show that a particle subject to no outside forces is either stationary or moves with constant speed along a straight line.

29. Show that if $\mathbf{r}'(t) = \mathbf{0}$ for all t on some interval I, then $\mathbf{r}(t) = \mathbf{c}$, a constant vector, for all t in I.

30. Show that if $\mathbf{f}'(t) = \mathbf{g}'(t)$ for all t in some interval I, then $\mathbf{f}(t) = \mathbf{g}(t) + \mathbf{c}$ for all t in I.

31. If $\mathbf{c}$ is a constant vector, show that

$$\int_a^b [\mathbf{c} \cdot \mathbf{r}(t)]\, dt = \mathbf{c} \cdot \int_a^b \mathbf{r}(t)\, dt$$

32. Use the result of Problem 31 to show that

$$\left\| \int_a^b \mathbf{r}(t)\, dt \right\| \le \int_a^b \|\mathbf{r}(t)\|\, dt$$

[*Hint:* Set $\mathbf{c} = \int_a^b \mathbf{r}(t)\, dt$ and evaluate $\|\mathbf{c}\|^2$.]

Kepler's Laws of Planetary Motion 15.6

In the sixteenth century, Copernicus (1473–1543) conjectured the *heliocentric theory of planetary motion*—that is, that the planets travel in circular orbits with the sun as center. This theory agreed better with observation than the

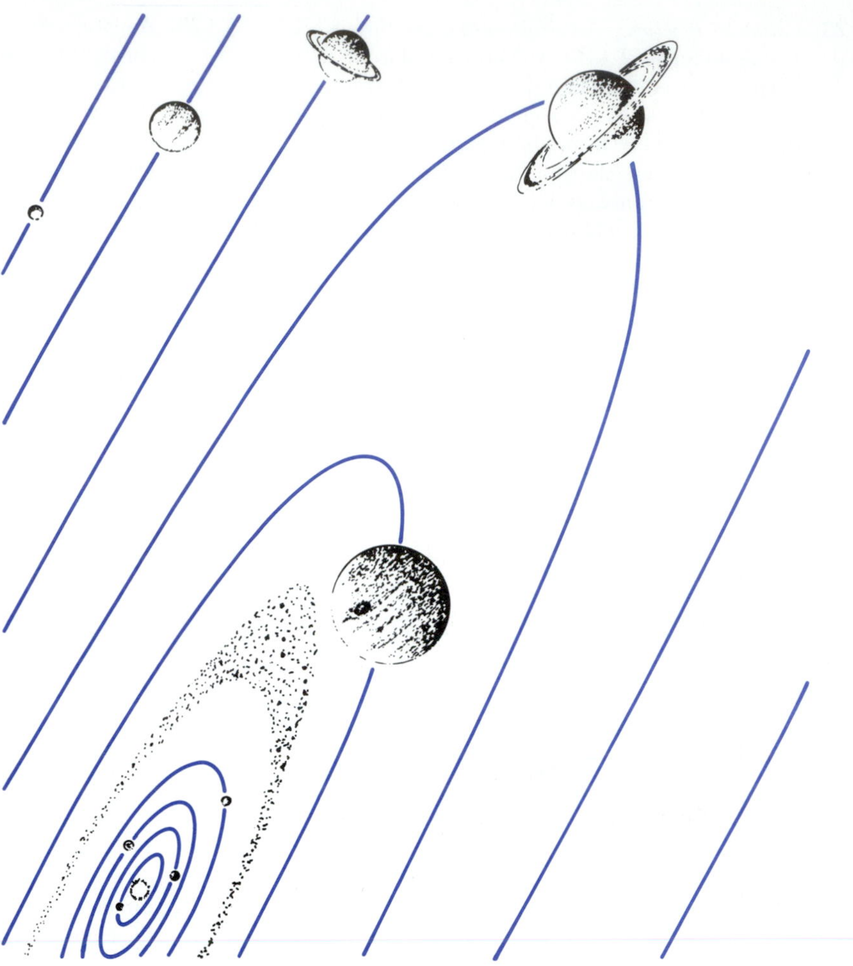

earlier *geocentric theory*—that the sun, moon, and planets travel in circular paths around the earth.

However, Copernicus' theory still did not explain some observable facts. Early in the seventeenth century, Kepler, working with the observations of the Danish astronomer Tycho Brahe (1546–1601), stated three laws to explain the orbits of planets:

1. The orbit of each planet is an ellipse with the sun at a focus (1609).
2. The speed of a planet is such that a line joining the sun to the planet sweeps out an area at a constant rate (1609).
3. The square of the period of revolution of the planet is proportional to the cube of the length of the semimajor axis of the planet's elliptical orbit (1619).

Although Kepler's laws of planetary motion agreed well with observable facts, a sound footing for this theory was not obtained until Newton used calculus to organize a theory that explained the motion of both planetary bodies and the earth. Here, we develop the first and second of Kepler's laws with the aid of Newton's ideas and vector calculus.

We position our coordinate system so that the sun is at the origin and let $\mathbf{r}(t)$ denote the position vector of a planet in orbit about the sun (see Fig. 31). Then $\mathbf{r}$ is subject to two laws:

Newton's second law of motion: $\mathbf{F} = m\mathbf{r}''(t)$ **(1)**

Newton's inverse square law: $\mathbf{F} = \dfrac{-GmM}{\|\mathbf{r}\|^2}\dfrac{\mathbf{r}}{\|\mathbf{r}\|}$ **(2)**

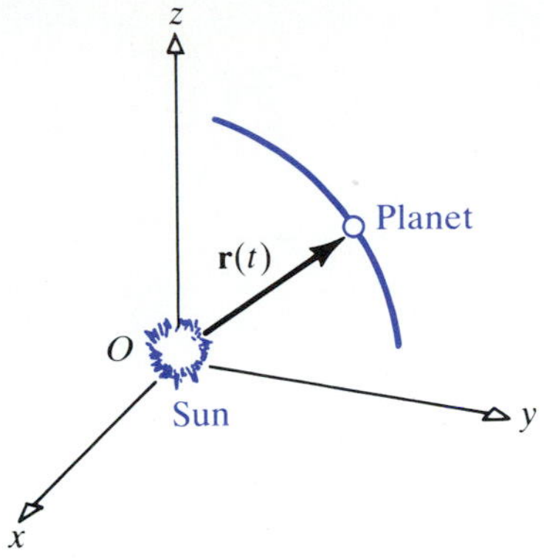

Figure 31

The first of these is the familiar law stating that the force acting on an object is proportional to the acceleration of the object, its mass being the constant of proportionality. Equation (2) states that the force $\mathbf{F}$ of attraction between two objects of mass m and M is inversely proportional to the square of the distance between them and is directed along a line joining them, G ($\approx 6.67 \times 10^{-11}$ newton-meter^2 per kilogram^2) being the constant of proportionality.

From (1) and (2), we may write

$$\mathbf{r}''(t) = \frac{-GM}{\|\mathbf{r}\|^3}\mathbf{r}$$

If $\mathbf{v}(t)$ is the velocity of the planet, it follows that

$$\frac{d\mathbf{v}}{dt} = \frac{-GM}{\|\mathbf{r}\|^3}\mathbf{r} \qquad \textbf{(3)}$$

Then we can conclude that $d\mathbf{v}/dt$ and $\mathbf{r}$ are parallel, so that

$$\mathbf{r} \times \frac{d\mathbf{v}}{dt} = \mathbf{0}$$

Because of this,

$$\frac{d}{dt}(\mathbf{r} \times \mathbf{v}) = \frac{d\mathbf{r}}{dt} \times \mathbf{v} + \mathbf{r} \times \frac{d\mathbf{v}}{dt} = \mathbf{v} \times \mathbf{v} + \mathbf{0} = \mathbf{0}$$

Hence, it follows by integrating that

$$\mathbf{r} \times \mathbf{v} = \mathbf{D} \qquad \textbf{(4)}$$

where $\mathbf{D}$ is a constant vector.

Kepler's second law may now be verified. As Figure 32 illustrates, if A is the area swept out by the vector $\mathbf{r}$ in time t, then

$$\Delta A = \text{Area swept out by } \mathbf{r} \text{ in time } \Delta t \approx \frac{1}{2}\|\mathbf{r} \times \Delta\mathbf{r}\|$$

$$\frac{\Delta A}{\Delta t} \approx \frac{1}{2}\left\|\mathbf{r} \times \frac{\Delta\mathbf{r}}{\Delta t}\right\|$$

$$\frac{dA}{dt} = \frac{1}{2}\left\|\mathbf{r} \times \frac{d\mathbf{r}}{dt}\right\| = \frac{1}{2}\|\mathbf{r} \times \mathbf{v}\| \underset{\text{By (4)}}{\uparrow}{=} \frac{1}{2}\|\mathbf{D}\| = \text{Constant}$$

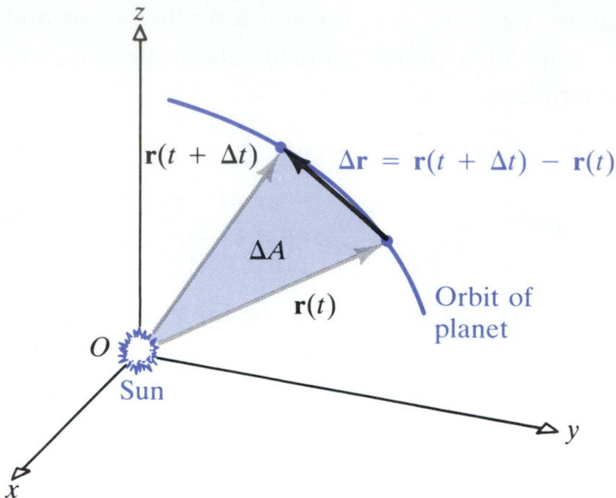

Figure 32

That is:

[15.6.1] THEOREM / *Kepler's Second Law.*

The speed of the planet is such that vectorial area is swept out at a constant rate.

To put it another way, in equal amounts of time, equal vectorial areas are swept out (see Fig. 33).

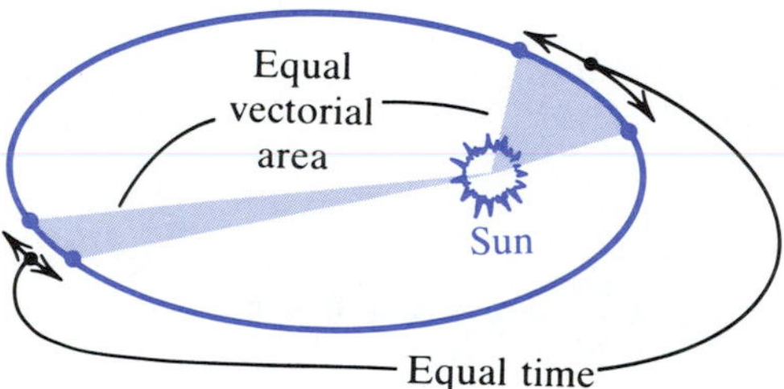

Figure 33

To obtain Kepler's first law, we return to (3) and (4). Then, using the formula for triple cross products, $\mathbf{r} \times (\mathbf{r} \times \mathbf{v}) = (\mathbf{r} \cdot \mathbf{v})\mathbf{r} - (\mathbf{r} \cdot \mathbf{r})\mathbf{v}$, we find that

$$\frac{d\mathbf{v}}{dt} \times \mathbf{D} = \frac{-GM}{\|\mathbf{r}\|^3} \mathbf{r} \times (\mathbf{r} \times \mathbf{v}) = \frac{-GM}{\|\mathbf{r}\|^3} [(\mathbf{r} \cdot \mathbf{v})\mathbf{r} - (\mathbf{r} \cdot \mathbf{r})\mathbf{v}]$$

But

$$\frac{d}{dt}(\mathbf{r} \cdot \mathbf{r}) = \frac{d\mathbf{r}}{dt} \cdot \mathbf{r} + \mathbf{r} \cdot \frac{d\mathbf{r}}{dt} = 2\mathbf{r} \cdot \frac{d\mathbf{r}}{dt} = 2\mathbf{r} \cdot \mathbf{v}$$

and

$$\frac{d}{dt}(\mathbf{r} \cdot \mathbf{r}) = \frac{d}{dt}\|\mathbf{r}\|^2 = 2\|\mathbf{r}\| \frac{d}{dt}\|\mathbf{r}\|$$

Hence,

$$\|\mathbf{r}\|\frac{d}{dt}\|\mathbf{r}\| = \mathbf{r}\cdot\mathbf{v}$$

so that

$$\frac{d\mathbf{v}}{dt}\times\mathbf{D} = \frac{-GM}{\|\mathbf{r}\|^3}\left(\|\mathbf{r}\|\frac{d}{dt}\|\mathbf{r}\|\mathbf{r} - \|\mathbf{r}\|^2\mathbf{v}\right) = GM\left(\frac{\mathbf{v}}{\|\mathbf{r}\|} - \frac{\mathbf{r}}{\|\mathbf{r}\|^2}\frac{d}{dt}\|\mathbf{r}\|\right)$$

But

$$\frac{d}{dt}\frac{\mathbf{r}}{\|\mathbf{r}\|} = \frac{d\mathbf{r}/dt}{\|\mathbf{r}\|} - \frac{\mathbf{r}}{\|\mathbf{r}\|^2}\frac{d}{dt}\|\mathbf{r}\| = \frac{\mathbf{v}}{\|\mathbf{r}\|} - \frac{\mathbf{r}}{\|\mathbf{r}\|^2}\frac{d}{dt}\|\mathbf{r}\|$$

Hence,

$$\frac{d\mathbf{v}}{dt}\times\mathbf{D} = GM\frac{d}{dt}\frac{\mathbf{r}}{\|\mathbf{r}\|}$$

Integrating yields

$$\mathbf{v}\times\mathbf{D} = GM\frac{\mathbf{r}}{\|\mathbf{r}\|} + \mathbf{H} \tag{5}$$

where **H** is a constant vector.

If we then find the dot product of both sides of (5) by **r**, the result is

$$\mathbf{r}\cdot(\mathbf{v}\times\mathbf{D}) = GM\|\mathbf{r}\| + \mathbf{r}\cdot\mathbf{H}$$

But $\mathbf{A}\cdot(\mathbf{B}\times\mathbf{C}) = (\mathbf{A}\times\mathbf{B})\cdot\mathbf{C}$; hence,

$$(\mathbf{r}\times\mathbf{v})\cdot\mathbf{D} = GM\|\mathbf{r}\| + \mathbf{r}\cdot\mathbf{H}$$

By (4), we have

$$\|\mathbf{D}\|^2 = GM\|\mathbf{r}\| + \mathbf{r}\cdot\mathbf{H}$$

If θ is the angle between **r** and **H**, it follows that

$$\|\mathbf{D}\|^2 = GM\|\mathbf{r}\| + \|\mathbf{r}\|\|\mathbf{H}\|\cos\theta$$

Solving for $\|\mathbf{r}\|$, we obtain

$$\|\mathbf{r}\| = \frac{\|\mathbf{D}\|^2}{GM + \|\mathbf{H}\|\cos\theta} = \frac{\|\mathbf{D}\|^2}{GM}\left(\frac{1}{1 + e\cos\theta}\right) \tag{6}$$

where $e = \|\mathbf{H}\|/GM$. This is the equation of a conic: a hyperbola if $e > 1$, a parabola if $e = 1$, an ellipse if $e < 1$. Since the planets have stable orbits, the only possibility is an ellipse. Thus, we have Kepler's first law of planetary motion:

[15.6.2] THEOREM / ***Kepler's First Law.***

The orbit of each planet is an ellipse with the sun at a focus.

EXERCISE 15.6

1. If one revolution of Jupiter requires $5\sqrt{5}$ years, what is the distance of Jupiter from the sun? Assume that orbits are circles and that the average distance of the earth from the sun is 9.3×10^7 miles.

2. Mercury has a "year" of 88 days. How far is Mercury from the sun?

3. The mean distance of Pluto from the sun is 39.5 times that of the earth. What is the "year" of Pluto?

4. From (6), deduce that:

$$\text{Length of semimajor axis} = \frac{\|\mathbf{D}\|^2}{GM}\left(\frac{1}{1-e^2}\right)$$

$$\text{Length of semiminor axis} = \frac{\|\mathbf{D}\|^2}{GM}\left(\frac{1}{\sqrt{1-e^2}}\right)$$

$$\text{Area of ellipse} = \frac{\pi\|\mathbf{D}\|^4}{G^2M^2}\left[\frac{1}{(1-e^2)^{3/2}}\right]$$

REVIEW EXERCISES

In Problems 1–5 graph the given curve. Also find the unit tangent vector at 0 if possible.

1. $\mathbf{r}(t) = \cosh t\mathbf{i} + \sinh t\mathbf{j}$

2. $\mathbf{r}(t) = \sinh^{-1}t\mathbf{i} + \sqrt{1+t^2}\mathbf{j}$

3. $\mathbf{r}(t) = a\cos t\mathbf{i} + b\sin t\mathbf{j}, \quad a > 0, \quad b > 0$

4. $\mathbf{r}(t) = t\mathbf{i} + e^t\mathbf{j}$

5. $\mathbf{r}(t) = \cos^3 t\mathbf{i} + \sin^3 t\mathbf{j}$

In Problems 6–8 give the unit tangent and the principal unit normal vectors at $t = 0$.

6. $\mathbf{r}(t) = e^t\mathbf{i} + e^{-t}\mathbf{j} + \mathbf{k}$

7. $\mathbf{r}(t) = t\mathbf{i} + 2t\mathbf{j} + \sqrt{1-5t^2}\mathbf{k}, \quad 0 \le t \le \sqrt{5}/5$

8. $\mathbf{r}(t) = e^t\sin t\mathbf{i} + e^t\cos t\mathbf{j} + e^t\mathbf{k}$

9. For Problems 6–8 name surfaces upon which the given curves lie.

10. Using the result of Problem 9, graph the curves in Problems 6 and 7.

11. Using the result of Problem 9 describe the graph of the curve in Problem 8.

12. Suppose the motion of a particle in space is given by $\mathbf{r}(t)$. Show that the tangent vector is always orthogonal to $\mathbf{r}(t)$ if $\mathbf{r}(t)$ lies on a sphere.

13. Suppose a curve is given by $\mathbf{r}(t)$. Find two nonparallel vectors that are orthogonal to $\mathbf{T}(t)$ other than the principal normal. Assume $\mathbf{T}(t) \ne \mathbf{0} \ne \mathbf{N}(t)$.

14. Find the arc length of the curve in Problem 5 if $0 \le t \le 2\pi$.

15. Find the arc length of the curve in Problem 7.

16. Find the arc length of the curve

$$\mathbf{r}(t) = \cos t\mathbf{i} + \sin t\mathbf{j} + \frac{t}{5}\mathbf{k} \qquad 0 \le t \le 2\pi$$

17–24. For appropriate t, find the velocity, acceleration, speed, and the tangential and normal components of the acceleration of the curves in Problems 1–8.

25. Use the figure to show that the coordinates (h, k) of the center of curvature of $y = f(x)$ are

$$h = x - \rho\sin\phi \qquad k = y + \rho\cos\phi$$

where ρ is the radius of curvature. Show that

$$\sin\phi = \frac{y'}{\sqrt{1+(y')^2}} \qquad \cos\phi = \frac{1}{\sqrt{1+(y')^2}}$$

so

$$h = x - \frac{y'[1+(y')^2]}{y''} \qquad k = y + \frac{1+(y')^2}{y''}$$

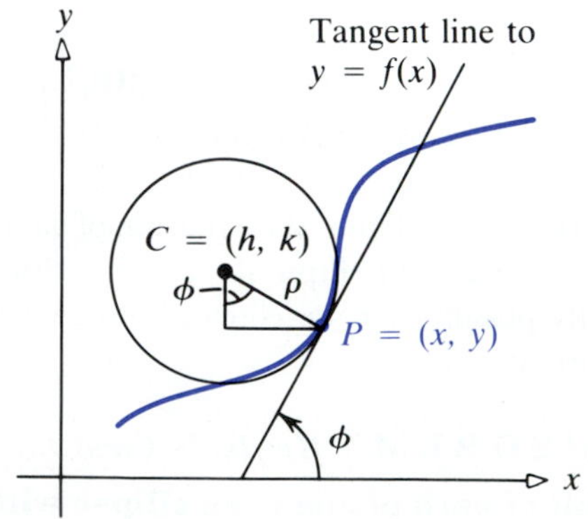

In Problems 26–30 use the result obtained in Problem 25.

26. Find the center of curvature of $y = x^2$ at $x = 1$.

27. Find the center of curvature of $y = \sin x$ at $x = \pi/2$.

28. Find the center of curvature of $y = x/(x+1)$ at $(0, 0)$.

29. Find the center of curvature of $x^3 + y^3 = 4xy$ at $(2, 2)$.

30. Find the center of curvature of $xy = 4$ at $x = 2$.

In Problems 31–37 find the osculating circle at the point indicated. It might be helpful to view curves in parametric form.

31. On $y = \cosh x$ at $x = 0$

32. On $y = \ln \sec x$ at $x = \pi/4$

33. On $y = \ln \sin x$ at $x = \pi/2$

34. On the curve of Problem 2 at $t = 0$

35. On the curve of Problem 5 at $t = \pi/3$

36. On the curve of Problem 7 at $t = 0$

37. On the curve of Problem 8 at $t = 0$

38. Find the minimum curvature of the curve of Problem 5.

39. Find the speed, velocity, and position at time t for a particle whose acceleration at time t is given by $\mathbf{a}(t) = \tan t \sec t\mathbf{i} + \cos t\mathbf{j}$. Assume the initial velocity and position to be $\mathbf{0}$.

40. A projectile is fired in the uphill direction of a rise that makes a 30° angle with the horizontal. Suppose the projectile is fired at an angle of inclination of 45° with the horizontal with an initial speed of 100 feet per second. Find the point of impact.

(*Problem 40 continues in the next column*)

40. (*continued*)

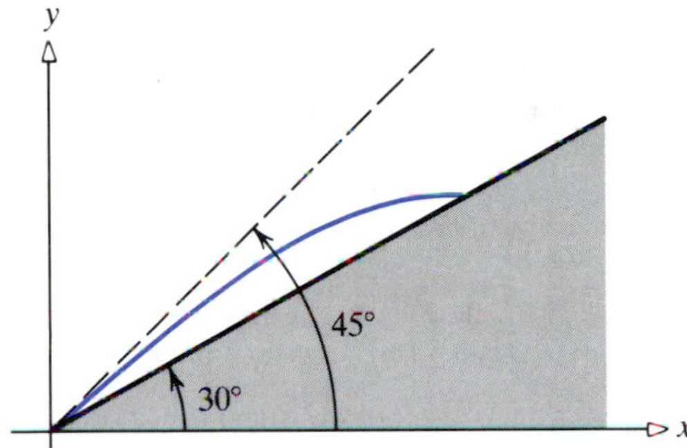

41. Suppose $\mathbf{r}(t)$ is a differentiable vector function with a nonzero tangent vector at t_0. Find the direction cosines for the tangent vector at t_0.

CHALLENGE EXERCISES

1. As a point P moves along a curve C, the center of curvature corresponding to P traces out a curve C_1 called the *evolute* of C; conversely, C is the *involute* of C_1. Show that parametric equations of the evolute of $y = \frac{1}{2}x^2$ are $h = -x^3$, $k = \frac{3}{2}x^2 + 1$. Then eliminate the parameter x to obtain

$$h^2 = \frac{8}{27}(k-1)^3$$

2. Refer to Problem 1. Find parametric equations and a rectangular equation for the evolute of $x = a\cos\theta$, $y = b\sin\theta$.

3. Show that if $\mathbf{r}'(t) \cdot \mathbf{r}(t) = 0$ for all t, then the motion $\mathbf{r}(t)$ is on the surface of a sphere centered at the origin. What can be said about a motion for which $\mathbf{r}''(t) \cdot \mathbf{r}'(t) = 0$ for all t? What can be said about a motion for which $\mathbf{r}'(t) \times \mathbf{r}(t) = \mathbf{0}$ for all t?

4. Let $\mathbf{u}_r = \cos\theta\mathbf{i} + \sin\theta\mathbf{j}$ and $\mathbf{u}_\theta = -\sin\theta\mathbf{i} + \cos\theta\mathbf{j}$.

(a) Show that $\mathbf{u}_r$ and $\mathbf{u}_\theta$ are unit orthogonal vectors and that $d\mathbf{u}_r/d\theta = \mathbf{u}_\theta$ and $d\mathbf{u}_\theta/d\theta = -\mathbf{u}_r$.

(b) Suppose that $\mathbf{r}(t)$ is the position vector of a plane curve and that the tip of $\mathbf{r}(t)$ has polar coordinates $(r(t), \theta(t))$. Show that $\mathbf{r}(t) = \|\mathbf{r}(t)\|\mathbf{u}_r = r(t)\mathbf{u}_r(t)$.

(*Problem 4 continues on page 922*)

4. *(continued)*

(c) Use the chain rule to show that

$$\mathbf{v}(t) = \frac{dr}{dt}\mathbf{u}_r + r\frac{d\theta}{dt}\mathbf{u}_\theta = r'\mathbf{u}_r + r\theta'\mathbf{u}_\theta.$$

(d) Show that $\mathbf{a}(t) = [r'' - r(\theta')^2]\mathbf{u}_r + [r\theta'' + 2r'\theta']\mathbf{u}_\theta$.

5. Express $\mathbf{v}$ and $\mathbf{a}$ in terms of $\mathbf{u}_r$ and $\mathbf{u}_\theta$ for a motion along the polar coordinate curve $r = 2 + \cos t$, $\theta = 2t$.

6. Find polar coordinate equations for $\mathbf{r}(t) = e^{2t}\cos t\mathbf{i} + e^{2t}\sin t\mathbf{j}$ and express $\mathbf{v}$ and $\mathbf{a}$ in terms of $\mathbf{u}_r$ and $\mathbf{u}_\theta$.

7. (a) For a particle moving under the influence of a central force directed toward the origin, show that $r\theta'' + 2r'\theta' = 0$.

(b) Multiply this equation by r and integrate to obtain $r^2\theta' = c$, c a constant.

(c) Using the expression for area in polar coordinates, deduce Kepler's second law for *any* central force.

8. If $\mathbf{c}$ is a constant vector and $\mathbf{r}(t)$ is continuous on a closed interval $[a, b]$, prove that

$$\int_a^b [\mathbf{c} \times \mathbf{r}(t)]\, dt = \mathbf{c} \times \int_a^b \mathbf{r}(t)\, dt$$

9. Estimate, correct to five decimal places, the length of the curve $\mathbf{r}(t) = \frac{1}{3}t^3\mathbf{i} + (t-1)\mathbf{j} + 2\mathbf{k}$, $0 \le t \le \frac{1}{2}$, by expanding the integrand in a power series.

10. Use Simpson's rule, theorem (9.9.3) in Section 9.9, with $n = 4$ to approximate the length of the curve $\mathbf{r}(t) = t^2\mathbf{i} + t^3\mathbf{j} + (2t+3)\mathbf{k}$, $0 \le t \le 2$.

In Problems 11–13 we extend the idea of a parametric curve to a parametric surface. A line with coordinate t can be used to parameterize a curve; we think of the curve as being "one-dimensional," having arc length as its dimension. Similarly, a plane region with coordinates (t_1, t_2) can be used to parameterize a surface; we think of the surface as being "two-dimensional," having arc length as measured with respect to t_1 for fixed values of t_2 as one dimension and arc length as measured with respect to t_2 for fixed values of t_1 as the other dimension. We do not have any formal definition of dimension here, but we can intuitively think of it as being the number of variables used in a parameterization.

11. As (t_1, t_2) is allowed to take values in the square, $0 \le t_1 \le 2\pi$ and $0 \le t_2 \le 2\pi$, the vectors

$$a \sin t_1 \cos t_2\mathbf{i} + b \sin t_1 \sin t_2\mathbf{j} + c \cos t_1\mathbf{k}$$

will fill out a surface. What is the surface?

12. Find a parameterization for the surface

$$\frac{x^2}{a^2} + \frac{y^2}{b^2} - \frac{z^2}{c^2} = 1$$

13. The components $a \sin t_1 \cos t_2$ and $b \sin t_1 \sin t_2$ in Problem 11 depend on t_1 and t_2. More generally, components used in a parameterization of a surface can depend on two variables t_1 and t_2 and are examples of

(Problem 13 continues in the next column)

13. *(continued)*

what is called a function of two variables. If we write $x(t_1, t_2) = a \sin t_1 \cos t_2$, $y(t_1, t_2) = b \sin t_1 \sin t_2$ and $z(t_1, t_2) = c \cos t_1$, then the surface of Problem 11 is given by

$$\mathbf{r}(t_1, t_2) = x(t_1, t_2)\mathbf{i} + y(t_1, t_2)\mathbf{j} + z(t_1, t_2)\mathbf{k}$$

Now fix $t_2 = t_2^*$. As t_1 goes from 0 to 2π, what happens to $\mathbf{r}(t_1, t_2^*)$? Fix $t_1 = t_1^*$. As t_2 goes from 0 to 2π, what happens to $\mathbf{r}(t_1^*, t_2)$?

Functions of Several Variables

16

Functions of Two or More Variables and Their Graphs 16.1

Many important problems in physics, engineering, and other sciences require functions of more than one variable. For example, to calculate the volume V of water stored in a cylindrical tank, we need to know both the height h of the tank and its radius R. The volume $V = \pi R^2 h$ depends on both h and R, and we say that V is a *function of two variables*, R and h. In function notation, we write

$$V = f(R, h) = \pi R^2 h$$

For example, if $R = 10$ centimeters and $h = 3$ centimeters, then

$$V = f(10, 3) = \pi(10^2)(3) = 300\pi \text{ cubic centimeters}$$

[16.1.1] DEFINITION / *Function of Two Variables.*

A *function f of two variables x* and *y* is a rule that assigns a unique real number $z = f(x, y)$ to each point (x, y) in some subset D of the xy-plane.

In the equation $z = f(x, y)$ we refer to z as the *dependent* variable, and we refer to x and y as the *independent* variables. The set D in the definition is called the *domain* of the function f. It is the set of points in the xy-plane for which the function is defined. The *range* of f consists of all real numbers $f(x, y)$ where (x, y) is in D. Figure 1 illustrates one way of depicting $z = f(x, y)$. We shall have occasion to use this kind of "picture" of f later on.

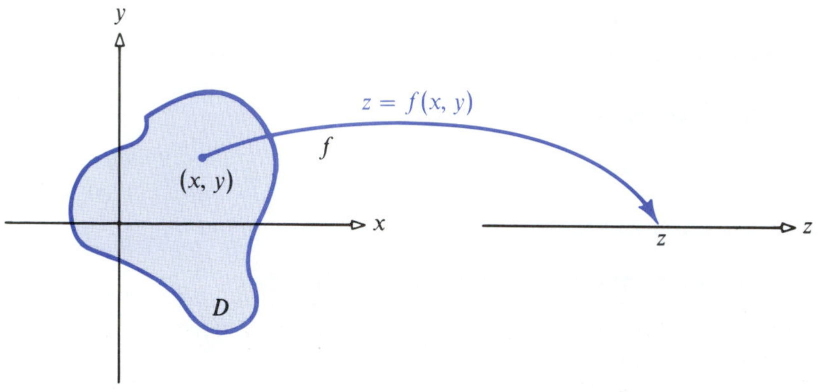

Figure 1

EXAMPLE 1

Let $f(x, y) = \sqrt{x} + x\sqrt{y}$. Find:

(a) $f(0, 0)$ (b) $f(1, 4)$ (c) $f(a^2, 9b^2)$, $a > 0$, $b > 0$

(d) $f(x + \Delta x, y)$ (e) $f(x, y + \Delta y)$

Solution

(a) $f(0, 0) = \sqrt{0} + 0\sqrt{0} = 0$ (b) $f(1, 4) = \sqrt{1} + 1\sqrt{4} = 1 + 2 = 3$

(c) $f(a^2, 9b^2) = \sqrt{a^2} + a^2\sqrt{9b^2} = a + 3a^2b$

(d) $f(x + \Delta x, y) = \sqrt{x + \Delta x} + (x + \Delta x)\sqrt{y}$

(e) $f(x, y + \Delta y) = \sqrt{x} + x\sqrt{y + \Delta y}$

[16.1.2] DEFINITION / *Function of Three Variables.*

A *function f of three variables x, y,* and *z* is a rule that assigns a unique real number $w = f(x, y, z)$ to point (x, y, z) in some subset D of space.

We refer to w as the *dependent* variable and x, y, and z as the *independent* variables. The domain D is some collection of points in space. Figure 2 illustrates a way of depicting $w = f(x, y, z)$.

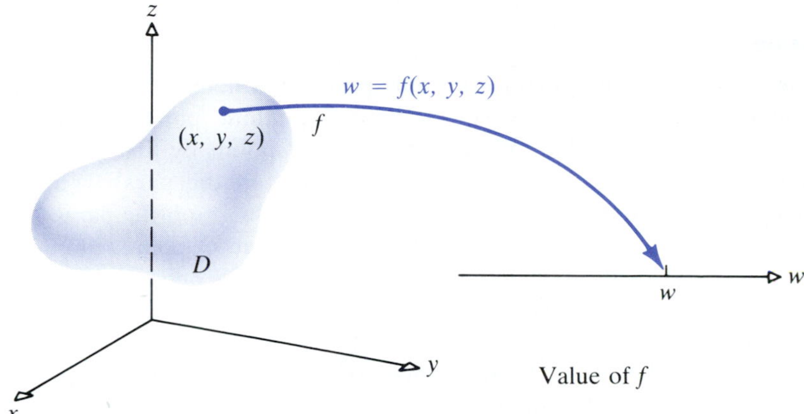

D is the domain of $w = f(x, y, z)$

Figure 2

Functions of n variables can also be defined. A *function f of n variables* $x_1, x_2, \ldots, x_n$ is a rule that assigns a unique real number $z = f(x_1, x_2, \ldots, x_n)$ to each point $(x_1, x_2, \ldots, x_n)$ in a certain subset D of n-dimensional space. Here z is the dependent variable and $x_1, x_2, \ldots, x_n$ are the n independent variables.

Collectively, functions of two or more variables are referred to as *functions of several variables*. As with functions of a single variable, a function of several variables is usually given by a formula, and, unless otherwise stated, the domain is taken to be the largest set of points for which this rule or formula makes sense.

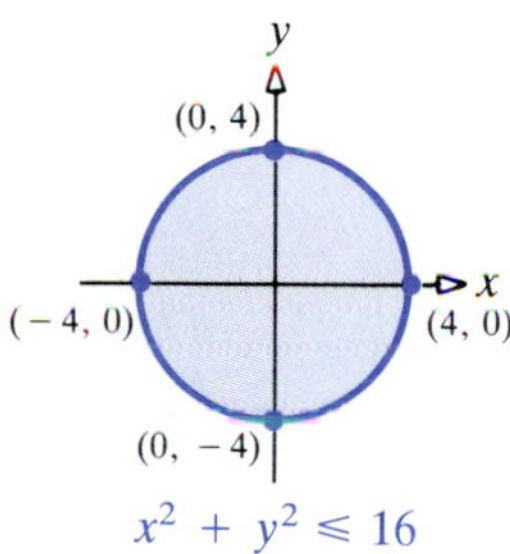

Figure 3
Domain of $f(x, y) = \sqrt{16 - x^2 - y^2}$

EXAMPLE 2

Describe the domain of each function and graph the domain.

(a) $z = f(x, y) = \sqrt{16 - x^2 - y^2}$ (b) $w = f(x, y, z) = \dfrac{1}{x^2 + y^2 + z^2}$

Solution

(a) The expression inside the square root must be nonnegative. Thus, the domain of f consists of all points in the plane for which

$$16 - x^2 - y^2 \geq 0$$

That is, the domain consists of all points inside and on the circle $x^2 + y^2 = 16$. The shaded portion of Figure 3 illustrates the domain.

(b) The denominator $x^2 + y^2 + z^2$ equals 0 only if $x = y = z = 0$. Thus, the domain of f consists of all points in space except $(0, 0, 0)$. See Figure 4. Note the use of an open circle at $(0, 0, 0)$ to indicate that this point is excluded.

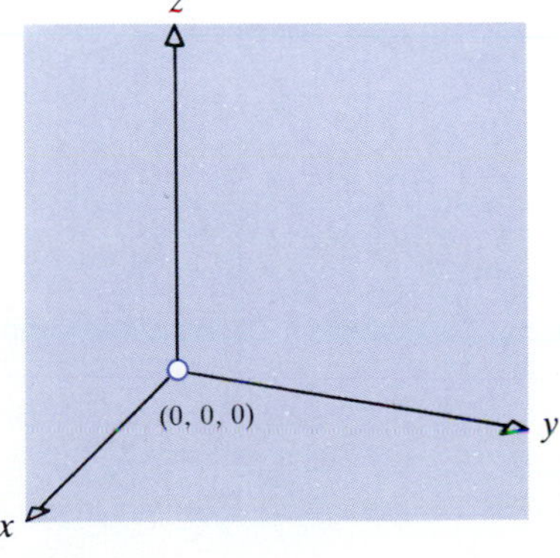

Figure 4
Domain of $f(x, y, z) = \dfrac{1}{x^2 + y^2 + z^2}$

EXAMPLE 3

Describe the domain of the function $f(x, y) = \ln(y^2 - 4x)$ and graph the domain.

Solution

The logarithm function is defined for only positive numbers. Therefore, the domain of $f(x, y) = \ln(y^2 - 4x)$ is limited to those points for which $y^2 - 4x > 0$ or $y^2 > 4x$. We graph this set of points by first graphing the parabola $y^2 = 4x$. We indicate that points on the parabola itself are not part of the domain by using a dashed curve to graph the parabola. The parabola $y^2 = 4x$ divides the plane into two sets of points: those for which $y^2 < 4x$ and those for which $y^2 > 4x$. To determine which points are "inside" and which are "outside" the parabola, we merely choose a point not on the curve $y^2 = 4x$ and test it. For example, the point $(2, 0)$ belongs to the set $y^2 < 4x$, since $0^2 < (4)(2)$. The set of points for which $y^2 > 4x$, the domain of f, is shaded in Figure 5.

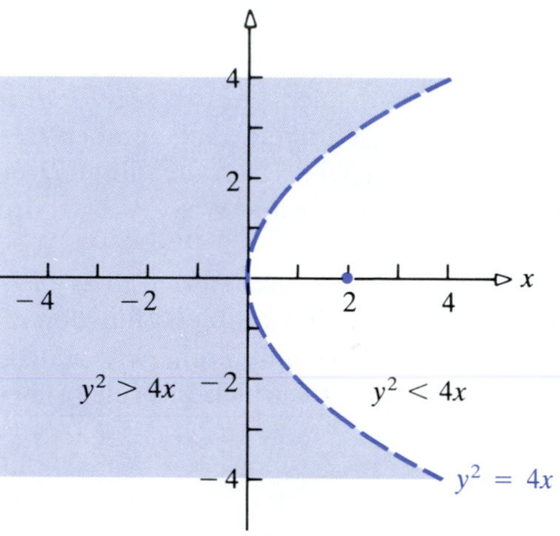

Domain of $f(x, y) = \ln(y^2 - 4x)$

Figure 5

EXAMPLE 4

Describe the domain of the function $w = f(x, y, z) = \sqrt{x^2 + y^2 + z^2 - 1}$ and graph the domain.

Solution

Since square roots of negative numbers are not permitted, the domain of this function consists of all points for which $x^2 + y^2 + z^2 - 1 \geq 0$. Therefore, the domain consists of all points outside and on the unit sphere (see Fig. 6).

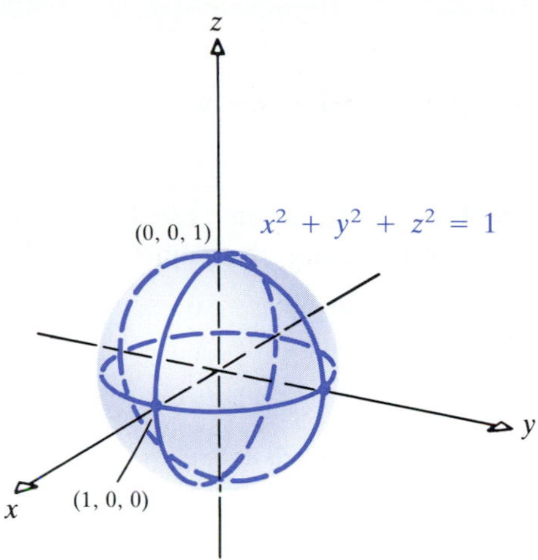

Figure 6

In practical problems the domain often is determined by physical considerations. Table 1 lists a few examples.

Table 1

Description	Formula	Domain
Volume V of a cylindrical tank of height h and radius R	$V(R, h) = \pi R^2 h$	$R > 0, \quad h > 0$
Area of a rectangle with sides x and y	$A(x, y) = xy$	$x > 0, \quad y > 0$
Surface area A of a closed box with sides of length x, y, and z	$A(x, y, z) = 2xy + 2yz + 2xz$	$x > 0, \quad y > 0, \quad z > 0$
Volume of a rectangular parallelepiped with sides x, y, and z	$V(x, y, z) = xyz$	$x > 0, \quad y > 0, \quad z > 0$
Magnitude of force of attraction between two bodies, one located at the origin and the other at (x, y, z), of masses m and M; G is a constant	$F(x, y, z) = \dfrac{GmM}{x^2 + y^2 + z^2}$	$(x, y, z) \neq (0, 0, 0)$

Graphing Functions of Two Variables

The *graph* of a function $z = f(x, y)$ of two variables is called a *surface* and consists of all points (x, y, z) for which $z = f(x, y)$ and (x, y) is in the domain of f.

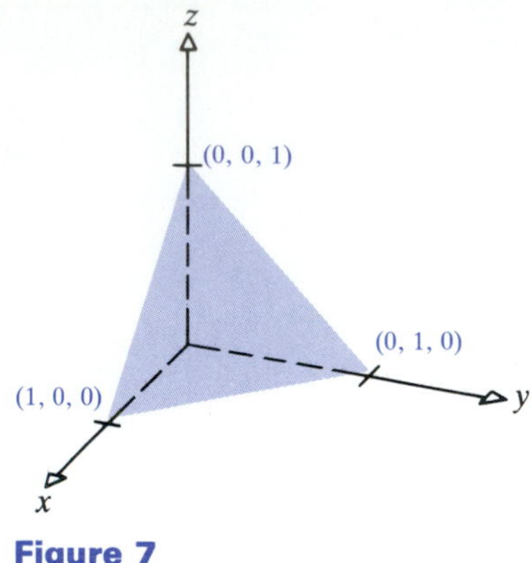

Figure 7

EXAMPLE 5

Describe the graph of the function $z = f(x, y) = 1 - x - y$.

Solution

In Chapter 14 we showed that the graph of the equation $z = 1 - x - y$, or $x + y + z = 1$, is a plane whose intercepts are the points $(1, 0, 0)$, $(0, 1, 0)$, and $(0, 0, 1)$. See Figure 7.

EXAMPLE 6

Describe the graph of the function $z = f(x, y) = x^2 + 4y^2$.

Solution

The equation $z = x^2 + 4y^2$ is the equation of an elliptic paraboloid (a quadric surface*) whose vertex is at the origin (see Fig. 8).

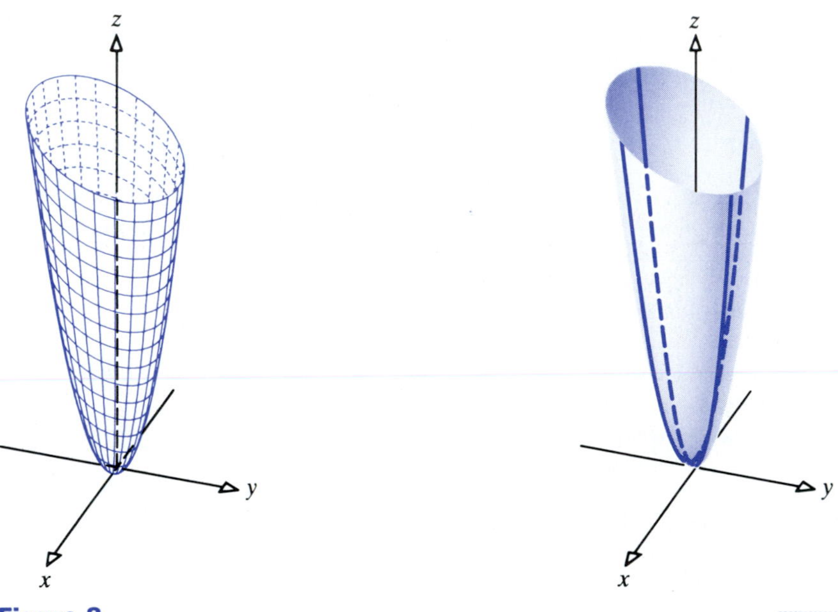

Figure 8

EXAMPLE 7

Describe the graph of the function $z = f(x, y) = \sqrt{x^2 + y^2}$.

Solution

The equation $z = \sqrt{x^2 + y^2}$, or equivalently, $z^2 = x^2 + y^2$, $z \geq 0$, is the equation of part of a circular cone whose vertex is at the origin. Since z must be nonnegative, the graph of the function consists of only the upper portion of the cone (see Fig. 9).

* These surfaces are discussed in detail in Section 14.8.

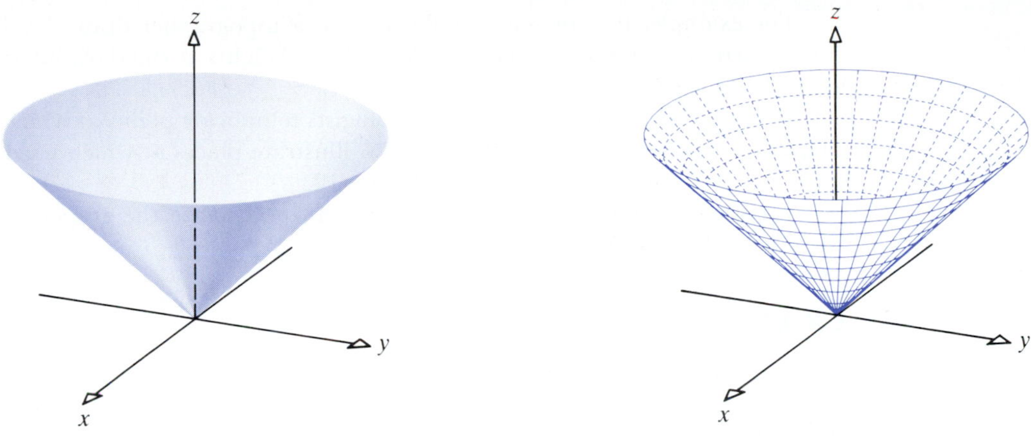

Figure 9

Level Curves

Ordinarily, the graph of a function of two variables is difficult to draw. In practice, such as in *topography* (map-making), the idea of the graph of $z = f(x, y)$ is conveyed by drawing properly labeled curves on which z is fixed. Such curves are of two types: level curves and contour lines. A *level curve* is a curve in the xy-plane for which $f(x, y) = c$, where c is a number in the range of f (see Fig. 10). A *contour line* is the curve resulting from the intersection of the graph of f with a plane parallel to the xy-plane. The corresponding level curve is the projection of the contour line on the xy-plane. If several level curves of a surface are drawn, the surface may be visualized by viewing them together, in much the same way that a movie film results from a collection of still pictures.

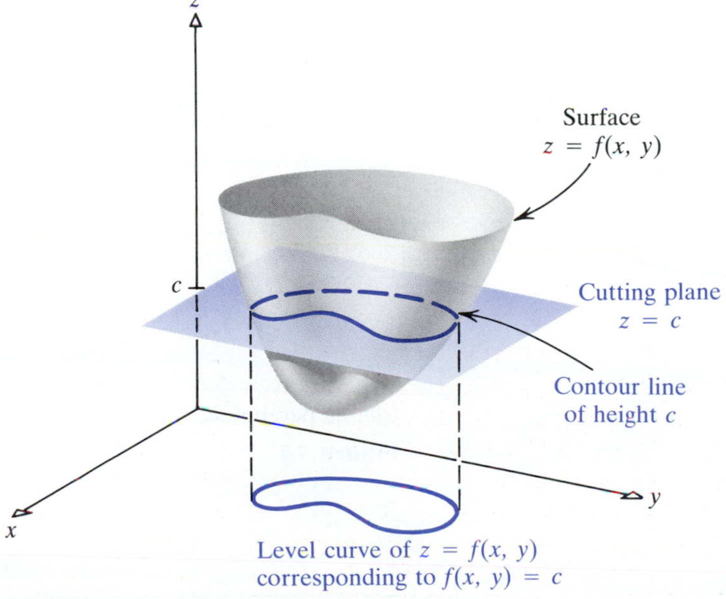

Figure 10

For example, to represent a hilly terrain, a topographer draws level curves corresponding to contour lines for various heights at standard intervals. See Figure 11.

Level curves are also used by meteorologists to indicate points at which barometric pressure is fixed (*isobars*) and to illustrate places at which wind speed remains constant (*isolines*).* See Figure 12.

We shall use the technique of level curves to visualize the graphs of functions of two variables.

EXAMPLE 8

Sketch the level curves of the function $z = f(x, y) = x^2 + 4y^2$.

Solution

Since $z \geq 0$, the level curves of f obey

$$x^2 + 4y^2 = c \qquad c \geq 0$$

For $c = 0$, this is a point—the origin. For $c > 0$, the level curves are concentric ellipses. Figure 13 shows the level curves for $c = 0$, $c = 1$, $c = 4$, and $c = 16$. A computer graph of $z = x^2 + 4y^2$ is shown in Figure 14.

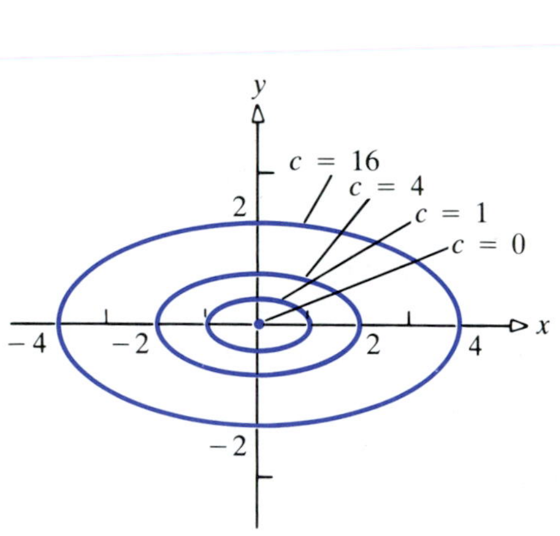

Level curves are concentric ellipses: $x^2 + 4y^2 = c$

Figure 13

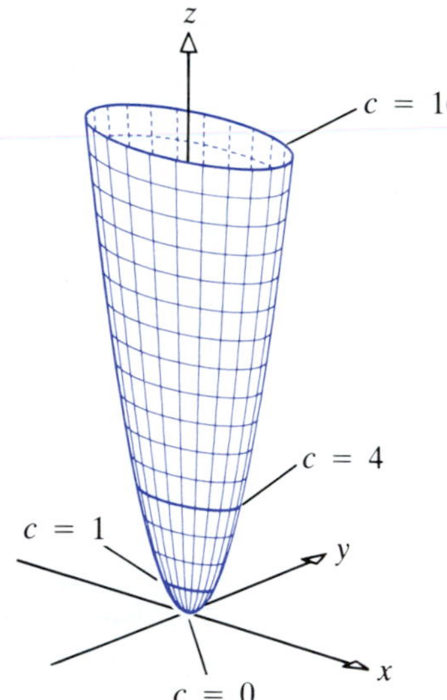

Elliptic paraboloid: $z = x^2 + 4y^2$

Figure 14

* The prefix *iso* is Greek, meaning "same."

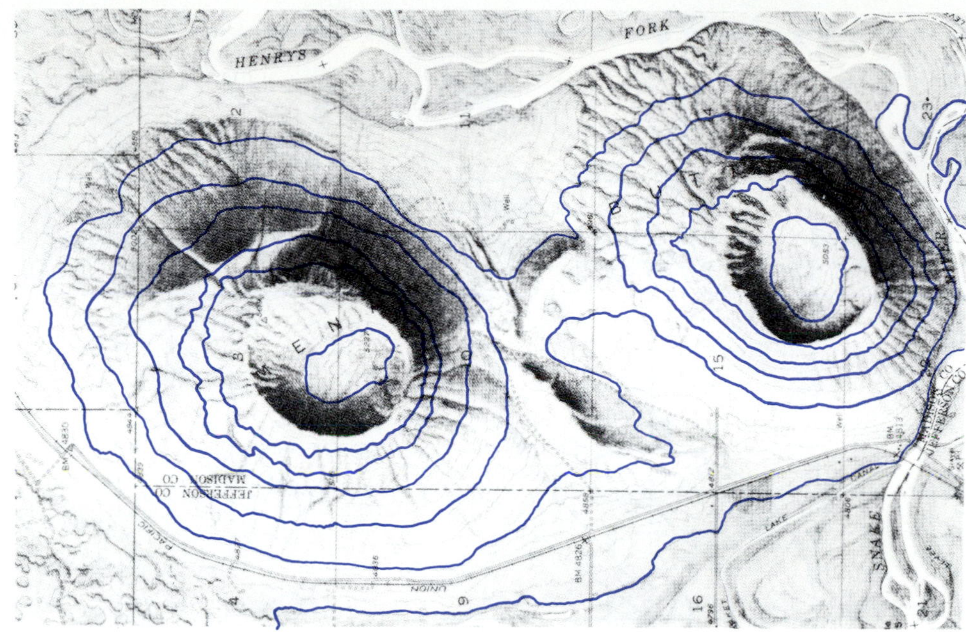

Figure 11
Menan Buttes, Idaho. Contour intervals are shown in color for each 100 foot change in altitude. The crater on the left is about 100 feet deep, and the crater on the right is about 150 feet deep. (Courtesy U.S. Geological Survey.)

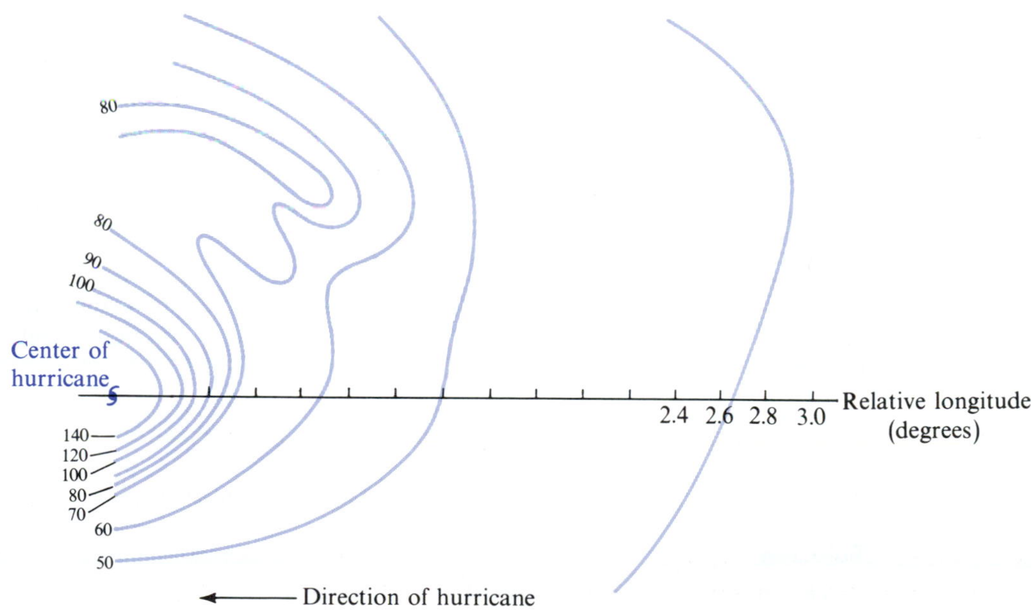

Isolines of wind speeds in rear quadrants of Pacific hurricane Ava

Figure 12
Pacific hurricane Ava, June 6, 1973. (Courtesy U.S. Department of Commerce, National Oceanic and Atmospheric Administration.)

EXAMPLE 9

Sketch the level curves of the function $z = f(x, y) = e^{x^2+y^2}$.

Solution

Since $z \geq 1$, the level curves obey $e^{x^2+y^2} = c$ or $x^2 + y^2 = \ln c$, $c \geq 1$. The level curves are concentric circles if $c > 1$. For $c = 1$, the level curve reduces to a single point at the origin. Figure 15 illustrates the level curves of f. A computer graph of $z = e^{x^2+y^2}$ is given in Figure 16.

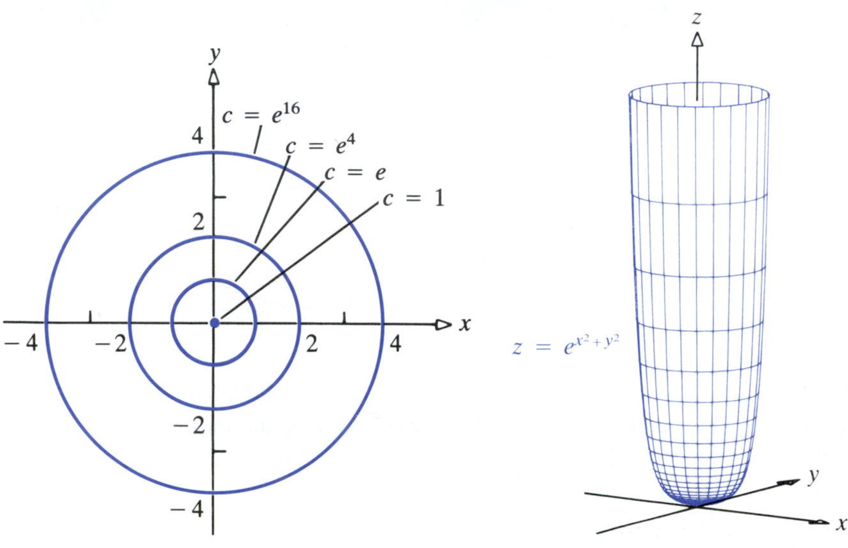

Figure 15 **Figure 16**

Level Surfaces

The graph of a function $w = f(x, y, z)$ of three variables consists of all points (x, y, z, w) for which $w = f(x, y, z)$ and (x, y, z) is in the domain of f. We cannot draw the graph of such a function, since it requires four dimensions. However, we may try to visualize such graphs by examining *level surfaces*—that is, the surfaces obtained by setting w equal to a constant. Although level surfaces are usually difficult to sketch, they do provide useful information. We shall limit ourselves to some simple examples.

EXAMPLE 10

Sketch the level surfaces of the function $w = f(x, y, z) = x^2 + y^2 + z^2$.

Solution

Since $w \geq 0$, the level surfaces obey

$$x^2 + y^2 + z^2 = c \qquad c \geq 0$$

These consist of concentric spheres if $c > 0$ and the origin if $c = 0$ (Fig. 17).

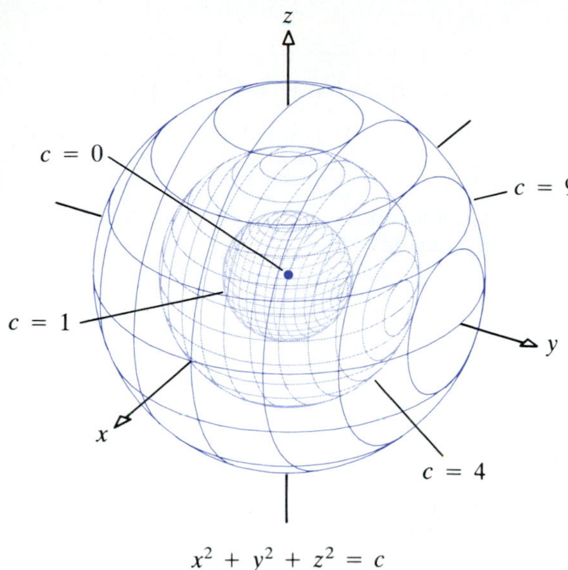

Figure 17

EXAMPLE 11

Sketch the level surfaces of the function $w = f(x, y, z) = 2x + 3y + z$.

Solution

The level surfaces obey

$$2x + 3y + z = c$$

This is a collection of parallel planes, each plane having the vector $\mathbf{N} = 2\mathbf{i} + 3\mathbf{j} + \mathbf{k}$ as normal (Fig. 18).

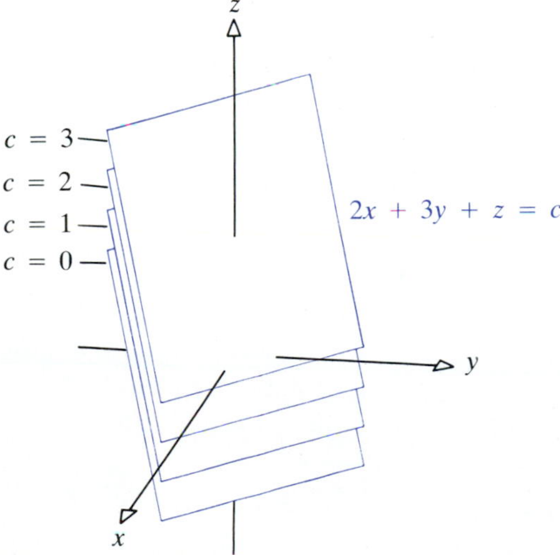

Figure 18

EXERCISE 16.1

1. Let $f(x, y) = 3x + 2y + xy$. Find:
(a) $f(1, 0)$ (b) $f(0, 1)$ (c) $f(2, 1)$
(d) $f(x + \Delta x, y)$ (e) $f(x, y + \Delta y)$

2. Let $f(x, y) = x^2y + x + 1$. Find:
(a) $f(0, 0)$ (b) $f(0, 1)$ (c) $f(2, 1)$
(d) $f(x + \Delta x, y)$ (e) $f(x, y + \Delta y)$

3. Let $f(x, y) = \sqrt{xy} + x$. Find:
(a) $f(0, 0)$ (b) $f(0, 1)$
(c) $f(a^2, t^2)$, $a > 0$, $t > 0$
(d) $f(x + \Delta x, y)$ (e) $f(x, y + \Delta y)$

4. Let $f(x, y) = e^{x+y}$. Find:
(a) $f(0, 0)$ (b) $f(1, -1)$
(c) $f(x + \Delta x, y)$ (d) $f(x, y + \Delta y)$

5. Let $f(x, y, z) = \dfrac{3xy + z}{x^2 + y^2 - z^2}$. Find:
(a) $f(1, 1, 1)$ (b) $f(0, 0, 1)$
(c) $f(0, 1, 0)$ (d) $f(\sin t, \cos t, 0)$

6. Let $f(x, y, z) = \dfrac{xyz}{x^2 + y^2 + z^2}$. Find:
(a) $f(0, 1, 1)$ (b) $f(1, -1, 1)$
(c) $f(a, a, a)$ $a \neq 0$ (d) $f(\sin t, \cos t, a)$

7. Let $f(x, y, z, t) = x^2 + y^2 + z^2 - t$. Find:
(a) $f(0, 0, 0, 5)$ (b) $f(1, 1, 1, 0)$ (c) $f(a, a, a, a^2)$

8. Let $f(x, y, z, t) = e^t \sin x \sin y \sin z$. Find:
(a) $f(\pi/4, \pi/6, \pi/3, 0)$ (b) $f(3\pi/2, \pi/2, \pi/2, 1)$
(c) $f(0, \pi, 2\pi, 3\pi)$

9. Let $f(x, y) = 3xy - x^2$, $x = x(t) = \sqrt{t}$, $y = y(t) = t^2$. Find:
(a) $f(x(0), y(1))$ (b) $f(x(1), y(0))$
(c) $f(x(4), y(2))$

10. Let $f(x, y) = x^2 + xy + y^2$, $x(t) = t$, $y(t) = t^2$. Find:
(a) $f(x(0), y(0))$ (b) $f(x(1), y(1))$
(c) $f(x(2), y(2))$

11. Let $F(x, y) = e^{x^2+y^2}$, $g(x) = x^2$, $h(y) = 2y - 1$. Find $F(g(x), h(y))$.

12. Let $g(x, y) = x \cos(xy)$, $u(x, y) = x^2y$, $v(x, y) = y^2x$. Find $g(u(x, y), v(x, y))$.

In Problems 13–26 describe the domain of each function and graph the domain. Use a solid curve to indicate that the boundary is included and a dashed curve to indicate that the boundary is excluded.

13. $z = f(x, y) = \dfrac{\sqrt{x}}{\sqrt{y}}$

14. $z = f(x, y) = \sqrt{x}\sqrt{y}$

15. $z = f(x, y) = \sqrt{xy}$

16. $z = f(x, y) = \dfrac{xy}{x^2 + y^2}$

17. $z = f(x, y) = e^x \sin y$

18. $z = f(x, y) = \dfrac{\ln x}{\ln y}$

19. $z = f(x, y) = \sqrt{\dfrac{x^2 + y^2}{x^2 - y^2}}$

20. $z = f(x, y) = \sqrt{\dfrac{x^2 + y^2}{xy}}$

21. $z = f(x, y) = \dfrac{5}{\sqrt{2y + x^2}}$

22. $z = f(x, y) = \ln(x^2 - y^2)$

23. $z = f(x, y) = \dfrac{x}{\sqrt{x^2 + y^2 - 4}}$

24. $z = f(x, y) = \dfrac{y}{\sqrt{9 - x^2 - y^2}}$

25. $z = f(x, y) = e^{x^2+y^2}\sin(xy)$

26. $z = f(x, y) = \tan^{-1}(x^2 + y^2)$

In Problems 27–30 describe the domain of each function.

27. $w = f(x, y, z) = \dfrac{x^2 + y^2}{z^2}$

28. $w = f(x, y, z) = e^z \ln(x^2 + y^2)$

29. $w = f(x, y, z) = \dfrac{z \sin x}{\cos y}$

30. $w = f(x, y, z) = \dfrac{xyz}{\sqrt{x^2 + y^2 + z^2}}$

In Problems 31–40 sketch a graph of each surface.

31. $z = f(x, y) = 3 - x - y$

32. $z = f(x, y) = 2 + x - y$

33. $z = f(x, y) = x^2 + y^2$

34. $z = f(x, y) = x^2 - y^2$

35. $z = f(x, y) = \sqrt{4 - x^2 - y^2}$

36. $z = f(x, y) = \sqrt{x^2 + y^2 - 4}$

37. $z = f(x, y) = \sin y$

38. $z = f(x, y) = \cos x$

39. $z = f(x, y) = 4 - x^2 - y^2$

40. $z = f(x, y) = x^2 + y^2 - 4$

For the functions given in Problems 41–48 sketch the level curves corresponding to the given values of c.

41. $z = f(x, y) = x^2 - y^2;\quad c = 0, 1, 4, 9$

42. $z = f(x, y) = 2x^2 + y^2;\quad c = 0, 1, 4, 9$

43. $z = f(x, y) = \sqrt{1 - x^2 - y^2};\quad c = 0, \frac{1}{4}, \frac{1}{9}$

44. $z = f(x, y) = \sqrt{x^2 + y^2 - 4};\quad c = 0, 1, 4, 9$

45. $z = f(x, y) = x^2 - 2y;\quad c = 0, 1, 4, 9$

46. $z = f(x, y) = y^2 - x;\quad c = 0, 1, 4$

47. $z = f(x, y) = x + \sin y;\quad c = 0, 2, 4, 8$

48. $z = f(x, y) = y - \ln x;\quad c = 1, 2, 4$

In Problems 49–54 describe the level surfaces associated with f.

49. $f(x, y, z) = x^2 + y^2 + z^2$

50. $f(x, y, z) = x + y - z$

51. $f(x, y, z) = z - 2x - 2y$

52. $f(x, y, z) = 4 - x^2 - y^2$

53. $f(x, y, z) = x^2 + y^2$

54. $f(x, y, z) = z$

55. Sketch on the same set of axes the level curves of $f(x, y) = x^2 - y^2$ and $g(x, y) = xy$ for $c = \pm 1, \pm 2, \pm 3$.

56. Refer to Problem 55. Show that at each point $P_0 \neq (0, 0)$, the level curve of $f(x, y) = x^2 - y^2$ through P_0 is perpendicular to the level curve of $g(x, y) = xy$ through P_0. The two families of level curves are said to be *orthogonal*.

57. Write the equation for the surface area of an open box as a function of its length x, width y, and depth z.

58. The cost of the bottom and top of a cylindrical tank is \$300 per square meter and the cost of the sides is \$500 per square meter. Write the total cost of constructing such a tank as a function of the radius R and height h (both in meters).

59. Write the total cost of constructing an open rectangular box if the cost per square centimeter of the material to be used for the bottom is \$4, for two of the opposite sides is \$3, and for the remaining pair of opposite sides is \$2.

60. Repeat Problem 59 for a closed rectangular box that has a top made of material costing \$5 per square centimeter.

61. The formula

$$V(x, y) = \frac{9}{\sqrt{4 - (x^2 + y^2)}}$$

(*Problem 61 continues in the next column*)

61. (*continued*)
gives the electrical potential V (in volts) at a point (x, y) in the xy-plane. Draw the equipotential curves (level curves) for $V = 36, 18, 9$. Describe the surface $z = V(x, y)$.

62. The temperature T (in °C) at any point (x, y) of a flat plate situated in the xy-plane is $T = 60 - 2x^2 - 3y^2$. Draw the isothermal curves (level curves) for $T = 60, 54, 48, 6, 0$. Describe the surface $z = T(x, y)$.

63. The strength E of an electric field at a point (x, y, z) due to an infinitely long charged wire lying along the x-axis is

$$E(x, y, z) = \frac{3}{\sqrt{y^2 + z^2}}$$

Describe the level surfaces of E.

64. The magnitude F of the force of attraction between two bodies, one located at the origin and the other at the point $(x, y, z) \neq (0, 0, 0)$, of masses m and M is given by

$$F = \frac{GmM}{x^2 + y^2 + z^2}$$

where G (equal to 6.67×10^{-11} in SI units) is a positive constant. Describe the level surfaces of F.

16.2 Limits and Continuity

In Chapter 2 we defined $\lim_{x \to x_0} f(x)$ as follows: Let f be a function defined on an open interval containing x_0, except possibly for x_0 itself. Then $\lim_{x \to x_0} f(x) = L$ if, for any given number $\varepsilon > 0$, there is a number $\delta > 0$ so that

$$\text{whenever} \quad 0 < |x - x_0| < \delta \quad \text{then} \quad |f(x) - L| < \varepsilon$$

In this definition, the phrase "whenever $0 < |x - x_0| < \delta$" means "for all x within a distance δ of x_0." To define limits or functions of two or more variables, we need the following concepts:

[16.2.1] DEFINITION / ***Neighborhood; Disk; Ball.***

Let P_0 be a point in the plane or in space. A *δ-neighborhood* of P_0 consists of all points P that lie within a distance δ of P_0—that is, all points P for which $d(P, P_0) < \delta$. In the plane, a δ-neighborhood of P_0 is called a *disk of radius δ, centered at P_0*; in space, a δ-neighborhood of P_0 is called a *ball of radius δ, centered at P_0*.

Now let's define the limit of a function of two variables.

[16.2.2] DEFINITION / ***Limit of a Function of Two Variables.***

Let f be a function of two variables defined at each point in some disk with center at (x_0, y_0), except possibly for (x_0, y_0) itself. Then the *limit of $f(x, y)$ as (x, y) approaches (x_0, y_0)* is L if, for any given number $\varepsilon > 0$, there is a number $\delta > 0$ so that

$$\text{whenever} \quad 0 < \sqrt{(x - x_0)^2 + (y - y_0)^2} < \delta \quad \text{then} \quad |f(x, y) - L| < \varepsilon$$

In this case we say that the limit exists and write

$$\lim_{(x, y) \to (x_0, y_0)} f(x, y) = L.$$

See Figures 19 and 20 for graphic interpretations of the definition.

[16.2.3] DEFINITION / ***Limit of a Function of Three Variables.***

Let f be a function of three variables defined at each point in some ball with center at (x_0, y_0, z_0), except possibly for (x_0, y_0, z_0) itself. Then the *limit of $f(x, y, z)$ as (x, y, z) approaches (x_0, y_0, z_0)* is L if, for any given number $\varepsilon > 0$, there is a number $\delta > 0$ so that whenever $0 < \sqrt{(x - x_0)^2 + (y - y_0)^2 + (z - z_0)^2} < \delta$ then $|f(x, y, z) - L| < \varepsilon$. In this case we say that the limit exists and write

$$\lim_{(x, y, z) \to (x_0, y_0, z_0)} f(x, y, z) = L$$

See Figure 21.

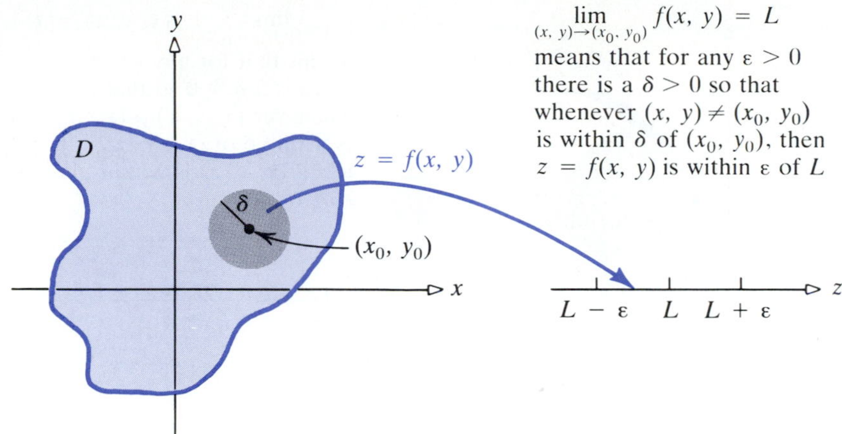

y
D
δ
(x_0, y_0)
x
$z = f(x, y)$
$\lim_{(x,y)\to(x_0,y_0)} f(x, y) = L$
means that for any $\varepsilon > 0$
there is a $\delta > 0$ so that
whenever $(x, y) \neq (x_0, y_0)$
is within δ of (x_0, y_0), then
$z = f(x, y)$ is within ε of L
$L - \varepsilon$ L $L + \varepsilon$
z

Figure 19

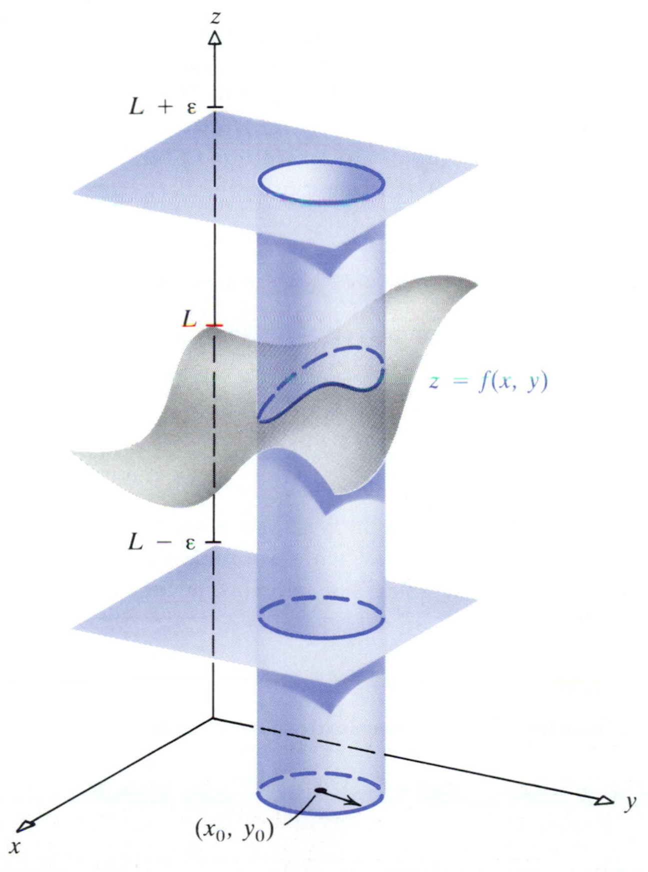

Figure 20

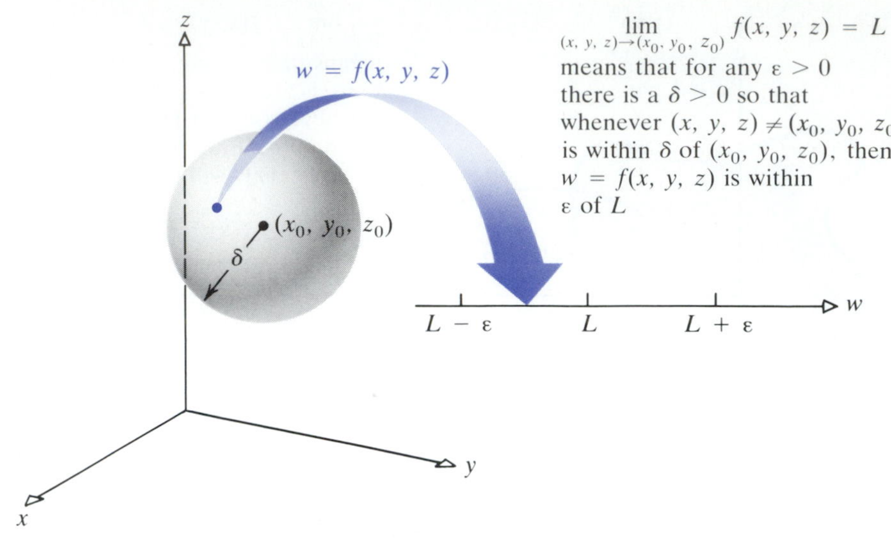

z
$w = f(x, y, z)$
$\lim_{(x, y, z)\to(x_0, y_0, z_0)} f(x, y, z) = L$
means that for any $\varepsilon > 0$
there is a $\delta > 0$ so that
whenever $(x, y, z) \neq (x_0, y_0, z_0)$
is within δ of (x_0, y_0, z_0), then
$w = f(x, y, z)$ is within
ε of L
(x_0, y_0, z_0)
δ
$L - \varepsilon$
L
$L + \varepsilon$
w
y
x

Figure 21

Note the similarity between definitions (16.2.2) and (16.2.3). In fact, limits of functions of more than three variables are defined in a like way. Thus, suppose $f(P)$ denotes a function of two or more variables defined at each point P in some neighborhood of P_0, except possibly at P_0 itself. Then

$$\lim_{P\to P_0} f(P) = L$$

if, for any given $\varepsilon > 0$, there is a number $\delta > 0$ so that

$$\text{whenever} \quad 0 < d(P, P_0) < \delta \quad \text{then} \quad |f(P) - L| < \varepsilon$$

Algebra of Limits

The algebra of limits for functions of several variables is analogous to that for functions of one variable. We state below, without proof, algebraic properties of limits of functions of two variables. The same properties are true for functions of three or more variables.

[16.2.4] THEOREM | *Properties of Limits.*

Let f and g be functions of two variables for which

$$\lim_{(x, y)\to(x_0, y_0)} f(x, y) = L \qquad \text{and} \qquad \lim_{(x, y)\to(x_0, y_0)} g(x, y) = M$$

where L and M are two real numbers.

Limit of a Sum:

$$\lim_{(x, y)\to(x_0, y_0)} [f(x, y) + g(x, y)] = \lim_{(x, y)\to(x_0, y_0)} f(x, y) + \lim_{(x, y)\to(x_0, y_0)} g(x, y) = L + M$$

Limit of a Difference:

$$\lim_{(x,y)\to(x_0,y_0)} [f(x,y) - g(x,y)] = \lim_{(x,y)\to(x_0,y_0)} f(x,y) - \lim_{(x,y)\to(x_0,y_0)} g(x,y) = L - M$$

If k is any real number, then

$$\lim_{(x,y)\to(x_0,y_0)} [kf(x,y)] = k \lim_{(x,y)\to(x_0,y_0)} f(x,y) = kL$$

Limit of a Product:

$$\lim_{(x,y)\to(x_0,y_0)} [f(x,y)g(x,y)] = \left[\lim_{(x,y)\to(x_0,y_0)} f(x,y)\right]\left[\lim_{(x,y)\to(x_0,y_0)} g(x,y)\right] = LM$$

Limit of a Quotient: **If** $\lim_{(x,y)\to(x_0,y_0)} g(x,y) = M \neq 0$,

$$\lim_{(x,y)\to(x_0,y_0)} \frac{f(x,y)}{g(x,y)} = \frac{\lim_{(x,y)\to(x_0,y_0)} f(x,y)}{\lim_{(x,y)\to(x_0,y_0)} g(x,y)} = \frac{L}{M}$$

Although we shall not prove it, the following result is a consequence of definition (16.2.2).

[16.2.5] THEOREM

$$\lim_{(x,y)\to(x_0,y_0)} c = c, \quad \text{where } c \text{ is a constant}$$

$$\lim_{(x,y)\to(x_0,y_0)} f(x) = \lim_{x\to x_0} f(x)$$

$$\lim_{(x,y)\to(x_0,y_0)} g(y) = \lim_{y\to y_0} g(y)$$

Similar results are true for limits of functions of more than two variables.

We often use theorems (16.2.4) and (16.2.5) to find limits.

EXAMPLE 1

Find each limit.

(a) $\displaystyle\lim_{(x,y)\to(1,2)} (3x^2 + 2xy + y^2)$ (b) $\displaystyle\lim_{(x,y)\to(1,2)} \frac{xy}{x^2 + y^2}$

Solution

(a)
$$\begin{aligned}\lim_{(x,y)\to(1,2)} (3x^2 + 2xy + y^2) &= \lim_{(x,y)\to(1,2)} 3x^2 + \lim_{(x,y)\to(1,2)} 2xy + \lim_{(x,y)\to(1,2)} y^2 \\ &= 3\left(\lim_{x\to 1} x\right)^2 + 2\left(\lim_{x\to 1} x\right)\left(\lim_{y\to 2} y\right) + \left(\lim_{y\to 2} y\right)^2 \\ &= 3 + 4 + 4 = 11\end{aligned}$$

(b)
$$\lim_{(x,y)\to(1,2)} \frac{xy}{x^2 + y^2} = \frac{\lim_{(x,y)\to(1,2)} xy}{\lim_{(x,y)\to(1,2)} (x^2 + y^2)} = \frac{\left(\lim_{x\to 1} x\right)\left(\lim_{y\to 2} y\right)}{\lim_{x\to 1} x^2 + \lim_{y\to 2} y^2} = \frac{2}{1+4} = \frac{2}{5}$$

A criterion for the existence of the limit of a function of one variable was that the left-hand limit equals the right-hand limit. More simply, this says that the limit must be the same no matter how the point x_0 is approached. Of course, on a line, there are only two ways to approach x_0—from the left or from the right. For $P_0 = (x_0, y_0)$ or for $P_0 = (x_0, y_0, z_0)$, however, there are infinitely many ways to approach P_0—namely, along any curve that passes through P_0. See Figure 22. Thus, in order for the limit to exist, it is necessary that the function approach this limit if we approach P_0 along any curve that contains P_0. To put it another way:

[16.2.6] THEOREM

If $\lim_{P \to P_0} f(P)$ is computed along two different curves of approach to P_0 and if two different answers are obtained, then $\lim_{P \to P_0} f(P)$ does not exist.

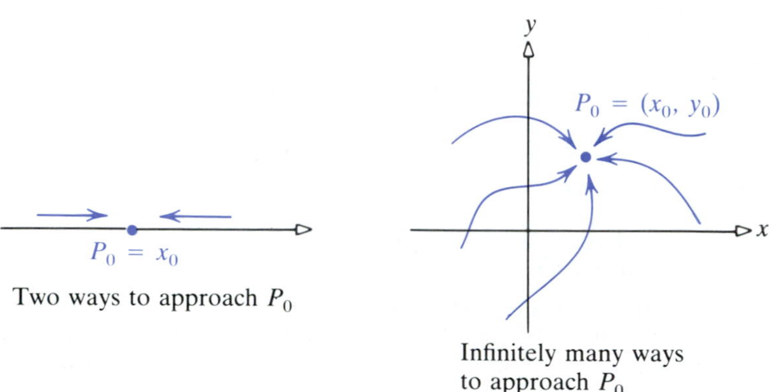

Figure 22

Let's look at an example that illustrates how theorem (16.2.6) is used.

EXAMPLE 2

Show that the following limit does not exist: $\lim_{(x,y)\to(0,0)} \dfrac{xy}{x^2 + y^2}$

Solution

We seek two curves of approach to (0, 0) that result in different limits. Suppose we evaluate the limit using the line $y = 2x$. Then

$$\text{Along } y = 2x: \quad \lim_{(x,y)\to(0,0)} \frac{xy}{x^2 + y^2} \underset{\substack{\uparrow \\ y = 2x}}{=} \lim_{x\to 0} \frac{2x^2}{x^2 + 4x^2} = \lim_{x\to 0} \frac{2x^2}{5x^2} = \frac{2}{5}$$

Using the line $y = -x$, we find

$$\text{Along } y = -x: \quad \lim_{(x,y)\to(0,0)} \frac{xy}{x^2 + y^2} \underset{\substack{\uparrow \\ y = -x}}{=} \lim_{x\to 0} \frac{-x^2}{x^2 + x^2} = \lim_{x\to 0} -\frac{x^2}{2x^2} = -\frac{1}{2}$$

Since two different answers are obtained, we conclude that the limit does not exist. See Figure 23.

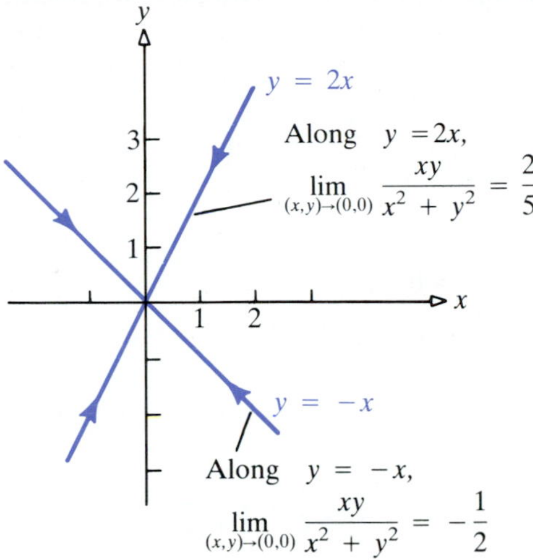

Figure 23

Be careful! Even if we obtain the same limit along two curves of approach (or three, or even several hundred), we would not be able to draw any conclusion about the existence of the limit of the function. For instance, in Example 2, using the two curves of approach $y = 2x$ and $x = 2y$ would have resulted in the same value for the limit, and yet the limit does not exist.

EXAMPLE 3

Show that the following limit does not exist: $\lim_{(x,y)\to(0,0)} \frac{x^2y}{x^4+y^2}$

Solution

We seek two curves of approach to (0, 0) that result in different limits. Based on our experience with the function in Example 2, we will try to save time by simultaneously testing all nonvertical lines of approach to (0, 0). Thus, we use $y = mx$. The result is

$$\lim_{(x,y)\to(0,0)} \frac{x^2y}{x^4+y^2} \underset{y=mx}{\uparrow}= \lim_{x\to0} \frac{x^2mx}{x^4+m^2x^2} = \lim_{x\to0} \frac{mx}{x^2+m^2} = 0$$

Along every nonvertical line through (0, 0), the limit is 0. But we still cannot conclude that the limit is 0. Suppose we approach (0, 0) along $y = x^2$. Then

$$\lim_{(x,y)\to(0,0)} \frac{x^2y}{x^4+y^2} \underset{y=x^2}{\uparrow}= \lim_{x\to0} \frac{x^4}{x^4+x^4} = \frac{1}{2}$$

Since two different curves of approach result in different limiting values, the given limit does not exist.

To show that a limit does exist may require the use of an ε, δ argument.

EXAMPLE 4

Use an ε, δ argument to show that $\lim_{(x,y)\to(0,0)} \frac{2xy^2}{x^2+y^2} = 0$.

Solution

Given any $\varepsilon > 0$, we need to find $\delta > 0$ so that

$$\text{whenever} \quad 0 < \sqrt{x^2+y^2} < \delta \quad \text{then} \quad |(2xy^2)/(x^2+y^2)| < \varepsilon$$

Now, $y^2 \le x^2 + y^2$, so that $y^2/(x^2+y^2) \le 1$ for all (x, y) not equal to $(0, 0)$. Then,

$$\left|\frac{2xy^2}{x^2+y^2}\right| = \frac{2|x|y^2}{x^2+y^2} \le 2|x| = 2\sqrt{x^2} \le 2\sqrt{x^2+y^2}$$

Given $\varepsilon > 0$, choose $\delta \le \varepsilon/2$. Then whenever $0 < \sqrt{x^2+y^2} < \delta$, we have

$$\left|\frac{2xy^2}{x^2+y^2}\right| \le 2\sqrt{x^2+y^2} < 2\delta \le \varepsilon$$

That is,

$$\lim_{(x,y)\to(0,0)} \frac{2xy^2}{x^2+y^2} = 0$$

Continuous Functions

When we introduced continuity for a function of one variable, we did so by defining what was meant by a function f being continuous at a point x_0, where f is defined on an open interval containing x_0. Then we extended the definition to include continuity on intervals: open, closed, or neither. For functions of two or three variables, we also begin by defining continuity at a point in the plane or in space. However, to extend the definition of continuity of such functions, we need some terminology that characterizes properties of sets of points in the plane and in space.

[16.2.7] DEFINITION / ***Interior Point; Boundary Point; Open Set; Closed Set.***

Let S be a set of points. A point P_0 is called an *interior point* of S if there is a δ-neighborhood of P_0 that lies entirely in S. A point P_0 of S is called a *boundary point* of S if every δ-neighborhood of P_0 contains points both in S and not in S. A set S is called *open* if every point of S is an interior point. A set S is called *closed* if it contains all its boundary points. A set S that contains some, but not all, of its boundary points is neither open nor closed.

Figures 24 and 25 illustrate these definitions in the plane.

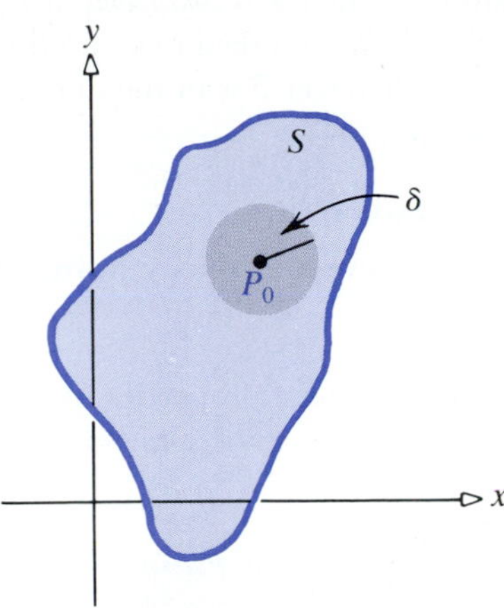

There is a δ-neighborhood of P_0 that lies entirely in S, so P_0 is an interior point of S

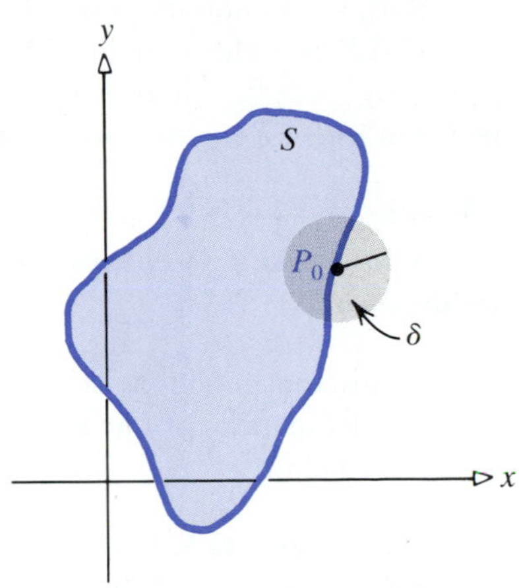

Every δ-neighborhood of P_0 contains points both in S and not in S, so P_0 is a boundary point of S

Figure 24

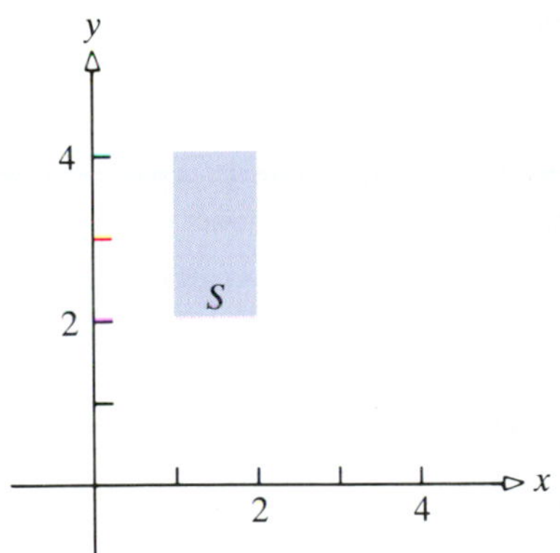

$1 < x < 2, \quad 2 < y < 4$
Every point in S is an interior point, so S is open

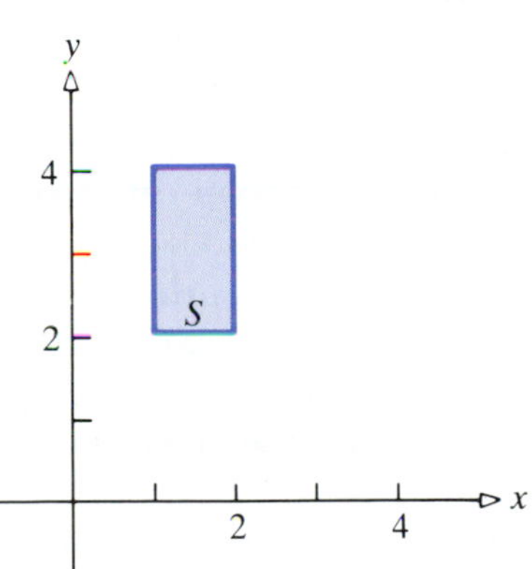

$1 \leqslant x \leqslant 2, \quad 2 \leqslant y \leqslant 4$
S contains all its boundary points, so S is closed

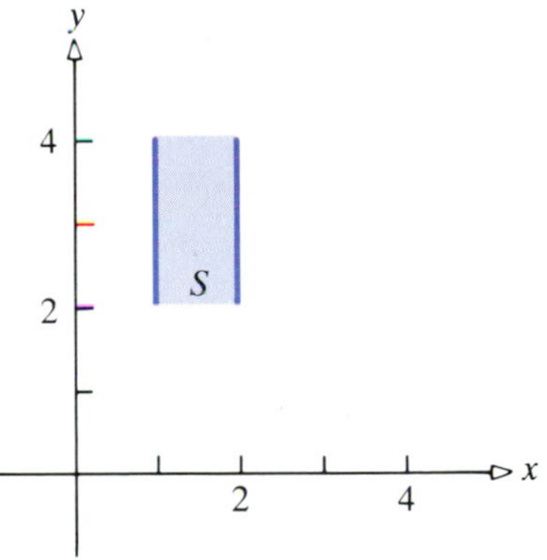

$1 < x < 2, \quad 2 \leqslant y \leqslant 4$
S contains some, but not all, its boundary points, so S is neither open nor closed

Figure 25

[16.2.8] DEFINITION / ***Continuous at P_0.***

Let f be a function defined on an open set. If

1. **P_0 is in this open set; that is, $f(P_0)$ is defined**
2. **$\lim_{P \to P_0} f(P)$ exists**
3. **$\lim_{P \to P_0} f(P) = f(P_0)$**

then f is *continuous* at P_0.

If any one of these three conditions is not met, then we say that f is *discontinuous* at P_0.

If P_0 is a boundary point of the domain of f, then f is *continuous at* P_0, provided P_0 is in the domain and $\lim_{P \to P_0} f(P) = L$ is taken to mean that given any $\varepsilon > 0$, there is a $\delta > 0$ so that whenever P is in the domain of f and $d(P, P_0) < \delta$, then $d(f(P_0), L) < \varepsilon$.

[16.2.9] DEFINITION / *Continuous Function.*

A function f is *continuous* if it is continuous at every point of its domain.

Continuous functions of two or three variables have properties similar to those of functions of one variable. We state without proof some of these properties below.

[16.2.10] THEOREM / *Properties of Continuous Functions.*

1. **The sum, difference, product, and composite of two continuous functions are continuous.**
2. **The quotient of two functions that are continuous at P_0 is also continuous at P_0, provided the denominator is not 0 at P_0.**
3. **If f is a function of one variable continuous at b in its domain, then the function $g(x, y) = f(x)$ is continuous at (b, y) for all y, and the function $h(x, y) = f(y)$ is continuous at (x, b) for all x.**
4. **If h is a continuous function of one variable and g is a continuous function of two variables, then their composite $f(x, y) = h(g(x, y))$ is a continuous function of two variables.**
5. **If $f(x, y)$ is continuous at (x_0, y_0), then f is continuous in each variable separately;* that is,**

$$\lim_{x \to x_0}\left[\lim_{y \to y_0} f(x, y)\right] = \lim_{y \to y_0}\left[\lim_{x \to x_0} f(x, y)\right] = f(x_0, y_0)$$

EXAMPLE 5

Discuss the continuity of:

(a) $f(x, y) = \dfrac{x^2 + y^2}{x^2 - y^2}$ (b) $g(x, y) = \dfrac{x^2 + 2xy - y^2}{x^2 + y^2 + 2}$

Solution

(a) As a consequence of property 2 above, any rational function in two variables is continuous wherever its denominator is nonzero. Thus,

$$f(x, y) = \frac{x^2 + y^2}{x^2 - y^2}$$

is continuous at all (x_0, y_0), where $x_0 \neq y_0$ and $x_0 \neq -y_0$.

* The converse is false. It is possible for a function of two variables to be continuous in each variable separately and yet fail to be continuous in both variables simultaneously.

(b) The function

$$g(x, y) = \frac{x^2 + 2xy - y^2}{x^2 + y^2 + 2}$$

is continuous everywhere, since its denominator is never 0.

EXAMPLE 6

The function

$$f(x, y) = \tan^{-1}\left(\frac{x^2 + 2xy - y^2}{x^2 + y^2 + 2}\right)$$

is continuous everywhere, since it is the composite of the inverse tangent function, which is continuous, and the rational function discussed in part (b) of Example 5.

We can obtain the limits of continuous functions by direct substitution. For example,

$$\lim_{(x, y)\to(0, 0)} \tan^{-1}\left(\frac{x^2 + 2xy - y^2}{x^2 + y^2 + 2}\right) = \tan^{-1}(0) = 0$$

EXERCISE 16.2

In Problems 1–20 find each limit.

1. $\lim_{(x, y)\to(1, 2)} (x^2 + xy - y^2)$

2. $\lim_{(x, y)\to(-1, 3)} (x^2 + y^2 - 3x)$

3. $\lim_{(x, y)\to(\pi/2, \pi)} (\sin x \cos y)$

4. $\lim_{(x, y)\to(\pi, \pi/4)} [\sin(x + y)\cos(x - y)]$

5. $\lim_{(x, y)\to(0, 0)} e^{x^2 - y^2}$

6. $\lim_{(x, y)\to(1, 0)} \ln(2x^2 + 3y^2)$

7. $\lim_{(x, y, z)\to(1, -1, 2)} (3x^2y + y^2z)$

8. $\lim_{(x, y, z)\to(0, 1, -1)} (x^2 - y^2z^2)$

9. $\lim_{(x, y)\to(3, 2)} \frac{x^2 - y^2}{x - y}$

10. $\lim_{(x, y)\to(2, 2)} \frac{4x^2 - y^2}{2x - y}$

11. $\lim_{(x, y)\to(1, 5)} \frac{4x - xy}{x(4y - y^2)}$

12. $\lim_{(x, y)\to(2, 2)} \frac{x^2 + 2xy + y^2 - 9}{x + y - 3}$

13. $\lim_{(x, y, z)\to(2, 2, -1)} \frac{x^2 - y^2 - 2yz - z^2}{x - y - z}$

14. $\lim_{(x, y, z)\to(2, 1, 2)} \frac{x^2 + xy - xz}{x^2 + 2xy + y^2 - z^2}$

15. $\lim_{(x, y)\to(\pi, 0)} \frac{\sin x \cos y}{x}$

16. $\lim_{(x, y)\to(\pi, \pi)} \frac{\sin y(1 - \cos x)}{xy}$

17. $\lim_{(x, y)\to(0, 0)} \frac{e^x - 1}{e^y}$

18. $\lim_{(x, y)\to(0, 0)} \frac{e^x \sin y - \sin y}{e^y}$

19. $\lim_{(x, y)\to(0, 0)} \frac{e^{x^2 + y^2}\cos x^2}{\cos y^2}$

20. $\lim_{(x, y)\to(0, 0)} \frac{e^{x^2 y}}{\cos 2x}$

In Problems 21–28 calculate each limit:

(a) Along the x-axis (b) Along the y-axis (c) Along $y = x$
(d) Along $y = 3x$ (e) Along $y = x^2$

21. $\lim_{(x, y)\to(0, 0)} \frac{3xy}{2x^2 + y^2}$

22. $\lim_{(x, y)\to(0, 0)} \frac{2xy}{x^2 + 3y^2}$

23. $\lim_{(x, y)\to(0, 0)} \frac{xy^2}{x^2 + y^4}$

24. $\lim_{(x, y)\to(0, 0)} \frac{2x^2y}{3x^4 + y^2}$

25. $\lim_{(x,y)\to(0,0)} \frac{3x^2y^2}{x^4+y^4}$

26. $\lim_{(x,y)\to(0,0)} \frac{x^2}{x^2+y^2}$

27. $\lim_{(x,y)\to(0,0)} \frac{x^2+xy}{x^2+y^2}$

28. $\lim_{(x,y)\to(0,0)} \frac{(x-y)^2}{x^2+y^2}$

In Problems 29–32 find: $\lim_{(x,y,z)\to(0,0,0)} \frac{2xyz}{x^4+y^2+z^2}$ along the indicated paths.

29. Along the line $x = t, \quad y = t, \quad z = t$

30. Along the line $x = 2t, \quad y = 3t, \quad z = 4t$

31. Along the curve $x = t, \quad y = t^2, \quad z = t^2$

32. Along the line $x = at, \quad y = bt, \quad z = ct,$ where $a^2 + b^2 + c^2 > 0$

In Problems 33–36 show that each limit does not exist.

33. $\lim_{(x,y)\to(0,0)} \frac{2x^2+y^2}{x^2+y^2}$

34. $\lim_{(x,y)\to(0,0)} \frac{2xy}{x^2+y^2}$

35. $\lim_{(x,y)\to(0,0)} \frac{x^4-y^2}{x^2+y^2}$

36. $\lim_{(x,y)\to(0,0)} \frac{x^2+y^4}{x^2+y^2}$

37. Discuss the continuity of the function

$$f(x, y) = \frac{xy^2}{x^2+y^2}$$

Is it possible to define $f(0, 0)$ so that f is continuous there?

38. Discuss the continuity of the function

$$f(x, y) = \frac{x^2y}{x^2+y^2}$$

Is it possible to define $f(0, 0)$ so that f is continuous there?

39. Discuss the continuity of the function

$$f(x, y) = \frac{2x^2+y^2}{x^2+y^2}$$

Is it possible to define $f(0, 0)$ so that f is continuous there?

40. Discuss the continuity of the function

$$f(x, y) = \frac{x^4-y^2}{x^2+y^2}$$

Is it possible to define $f(0, 0)$ so that f is continuous there?

41. Use an ε, δ argument to show that

$$\lim_{(x,y)\to(0,0)} \frac{\sin(x^2+y^2)}{x^2+y^2} = 1$$

42. Use an ε, δ argument to show that

$$\lim_{(x,y)\to(0,0)} \frac{2x^2y}{x^2+y^2} = 0$$

43. Show that the function

$$f(x, y) = \begin{cases} \dfrac{3xy}{x^2+y^2} & \text{if} \quad (x, y) \neq (0, 0) \\ 0 & \text{if} \quad (x, y) = (0, 0) \end{cases}$$

is not continuous at $(0, 0)$.

44. Show that the function

$$f(x, y) = \begin{cases} \dfrac{\sin xy}{x^2+y^2} & \text{if} \quad (x, y) \neq (0, 0) \\ 1 & \text{if} \quad (x, y) = (0, 0) \end{cases}$$

is not continuous at $(0, 0)$.

16.3 Partial Derivatives

Suppose $z = f(x, y)$ is a function of two variables x and y. If we hold y fixed—say, at $y = y_0$—then the function $z = f(x, y) = f(x, y_0)$ is now a function of the single variable x. If f is differentiable at $x = x_0$, it is called the *partial derivative of f with respect to x at* (x_0, y_0) and is denoted by $f_x(x_0, y_0)$.

Similarly, if we hold x fixed at $x = x_0$ in $z = f(x, y)$, then $z = f(x_0, y)$ becomes a function of the single variable y. If f is differentiable at $y = y_0$, it is called the *partial derivative of f with respect to y at* (x_0, y_0) and is denoted by $f_y(x_0, y_0)$.

Just as the derivative of a function $y = f(x)$ is itself a function, so too the partial derivatives f_x and f_y of $z = f(x, y)$, in general, are functions of two variables. We define the partial derivatives of $z = f(x, y)$ as follows:

[16.3.1] DEFINITION / ***Partial Derivatives.***

Let $z = f(x, y)$ denote a function of two variables and let (x, y) be any interior point* of the domain of f. The partial derivatives of f with respect to x and y are functions f_x and f_y defined as follows:

$$f_x(x, y) = \lim_{\Delta x \to 0} \frac{f(x + \Delta x, y) - f(x, y)}{\Delta x}$$

$$f_y(x, y) = \lim_{\Delta y \to 0} \frac{f(x, y + \Delta y) - f(x, y)}{\Delta y} \qquad (1)$$

provided these limits exist.

Observe the similarity between the above definitions and the definition of a derivative given in Chapter 3. Observe also that in $f_x(x, y)$, an increment Δx is given to x, while y is fixed; in $f_y(x, y)$, an increment Δy is given to y, while x is fixed.

The usual rules for finding derivatives (sum, difference, product, quotient, and chain) may be used to find partial derivatives; just remember that to find $f_x(x, y)$, y is treated as a constant while differentiating f with respect to x; and to find $f_y(x, y)$, x is treated as a constant while differentiating f with respect to y.

EXAMPLE 1

Find $f_x(x, y)$ and $f_y(x, y)$ for:

(a) $f(x, y) = 3x^2y + 2x - 3y$ (b) $f(x, y) = x \sin y + y \sin x$

Solution

(a) To find $f_x(x, y)$, treat y as a constant in $f(x, y) = 3x^2y + 2x - 3y$ and differentiate with respect to x. The result is

$$f_x(x, y) = 6xy + 2$$

To find $f_y(x, y)$, treat x as a constant and differentiate with respect to y.

* Although partial derivatives may be defined at boundary points of the domain of f, these are more difficult to handle and will not be discussed in this book.

The result is

$$f_y(x, y) = 3x^2 - 3$$

(b) $f_x(x, y) = \sin y + y \cos x; \quad f_y(x, y) = x \cos y + \sin x$

EXAMPLE 2

Find $f_x(\pi/3, \pi/6)$ and $f_y(\pi/3, \pi/6)$ if $f(x, y) = x \sin y + y \sin x$.

Solution

$$f_x(x, y) = \sin y + y \cos x$$
$$f_x(\pi/3, \pi/6) = \sin \pi/6 + \pi/6 \cos \pi/3 = 1/2 + \pi/12$$
$$f_y(x, y) = x \cos y + \sin x$$
$$f_y(\pi/3, \pi/6) = \pi/3 \cos \pi/6 + \sin \pi/3 = \pi\sqrt{3}/6 + \sqrt{3}/2$$

Another Notation

There is another notation used for the partial derivatives $f_x(x, y)$ and $f_y(x, y)$ of a function $z = f(x, y)$, which we introduce here:

$$f_x(x, y) = \frac{\partial f}{\partial x} = \frac{\partial z}{\partial x} \qquad f_y(x, y) = \frac{\partial f}{\partial y} = \frac{\partial z}{\partial y}$$

The symbols $\partial/\partial x$ and $\partial/\partial y$ denote operations on a function to obtain the partial derivatives with respect to x in the case of $\partial/\partial x$ and with respect to y in the case of $\partial/\partial y$.

EXAMPLE 3

$$\frac{\partial}{\partial x}(e^x \cos y) = e^x \cos y \qquad \frac{\partial}{\partial y}(e^x \cos y) = -e^x \sin y$$

Geometric Interpretation

For a geometric interpretation of the partial derivatives of $z = f(x, y)$, look at the graph of the surface $z = f(x, y)$ shown in Figure 26. In computing $f_x(x, y)$, we hold y fixed—say, at $y = y_0$—and then differentiate with respect to x. But holding y fixed at y_0 is equivalent to intersecting the surface $z = f(x, y)$ with the plane $y = y_0$, and the result is the curve $z = f(x, y_0)$. Thus, the partial derivative f_x is the slope of the tangent line to this curve.

[16.3.2] THEOREM

The partial derivative $f_x(x_0, y_0)$ equals the slope of the tangent line to the curve of intersection of the surface $z = f(x, y)$ and the plane $y = y_0$ at the point (x_0, y_0, z_0) on the surface.

The partial derivative $f_y(x_0, y_0)$ equals the slope of the tangent line to the curve of intersection of the surface $z = f(x, y)$ and the plane $x = x_0$ at the point (x_0, y_0, z_0) on the surface.

See Figure 27.

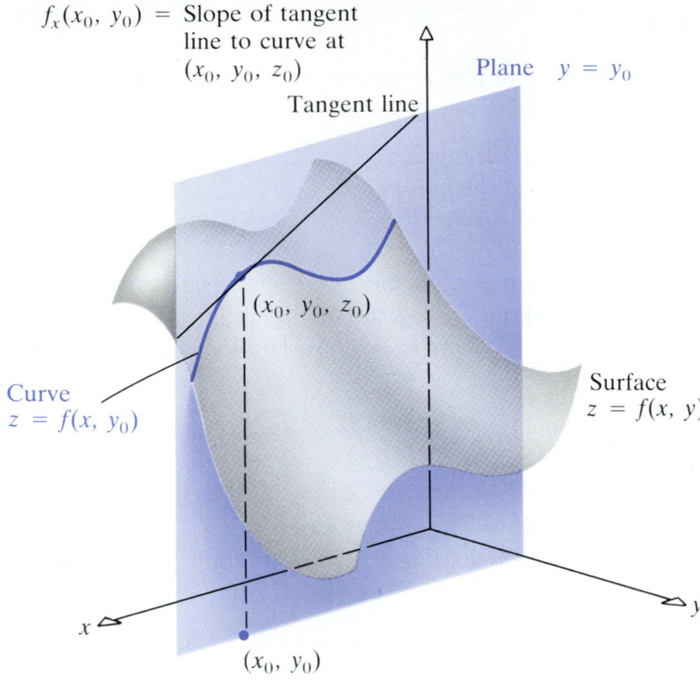
$f_x(x_0, y_0)$ = Slope of tangent line to curve at (x_0, y_0, z_0)
Plane $y = y_0$
Tangent line
(x_0, y_0, z_0)
Curve $z = f(x, y_0)$
Surface $z = f(x, y)$
x
y
(x_0, y_0)

Figure 26

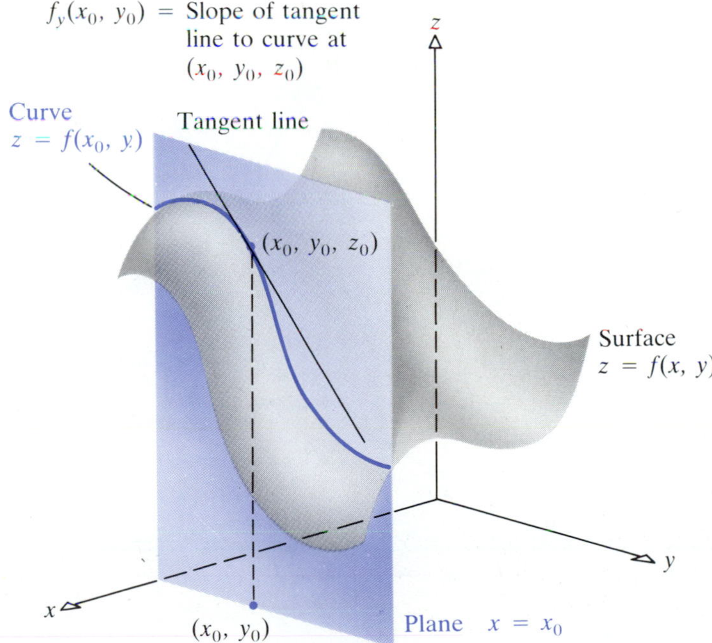
$f_y(x_0, y_0)$ = Slope of tangent line to curve at (x_0, y_0, z_0)
z
Curve $z = f(x_0, y)$
Tangent line
(x_0, y_0, z_0)
Surface $z = f(x, y)$
y
x
(x_0, y_0)
Plane $x = x_0$

Figure 27

EXAMPLE 4

Find an equation of the tangent line to the curve of intersection of the surface $z = f(x, y) = 16 - x^2 - y^2$:

(a) With the plane $y = 2$ at the point $(1, 2, 11)$
(b) With the plane $x = 1$ at the point $(1, 2, 11)$

Solution

(a) The slope of the tangent line to the curve of intersection of $z = 16 - x^2 - y^2$ and the plane $y = 2$ at any point is $f_x(x, y) = -2x$. At the point $(1, 2, 11)$, the slope is -2. An equation of this tangent line is

$$z - 11 = -2(x - 1) \qquad y = 2$$

(b) The slope of the tangent line to the curve of intersection of $z = 16 - x^2 - y^2$ and the plane $x = 1$ at any point is $f_y(x, y) = -2y$. At the point $(1, 2, 11)$, the slope is -4. An equation of this tangent line is

$$z - 11 = -4(y - 2) \qquad x = 1$$

Figure 28 illustrates these solutions.

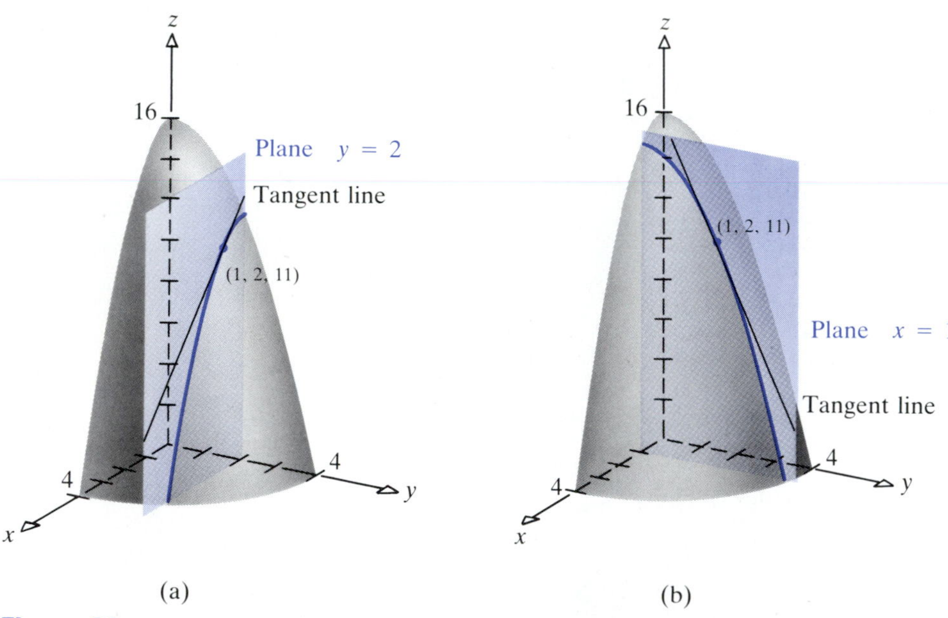

Figure 28

Rate of Change

As we noted earlier, the definition of the partial derivatives of $z = f(x, y)$ is a generalization of the definition of the derivative of a function of one variable. It is not surprising, therefore, to see similarities in the interpretations. For example, $f_x(x, y)$ equals the rate of change of f in a direction parallel to the x-axis (while y is held fixed), as shown in Figure 29.

Similarly, $f_y(x, y)$ equals the rate of change of f in a direction parallel to the y-axis (x is held fixed), as shown in Figure 30.

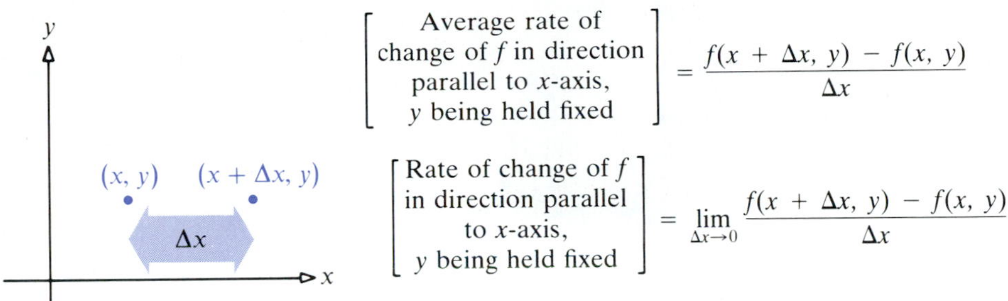

Figure 29

y

$\bullet(x, y + \Delta y)$

Δy

$\bullet(x, y)$

x

$$\left[\begin{array}{c}\text{Average rate of}\\ \text{change of } f \text{ in direction}\\ \text{parallel to } y\text{-axis,}\\ x \text{ being held fixed}\end{array}\right] = \frac{f(x, y + \Delta y) - f(x, y)}{\Delta y}$$

$$\left[\begin{array}{c}\text{Rate of change of } f\\ \text{in direction parallel}\\ \text{to } y\text{-axis,}\\ x \text{ being held fixed}\end{array}\right] = \lim_{\Delta y \to 0} \frac{f(x, y + \Delta y) - f(x, y)}{\Delta y}$$

Figure 30

EXAMPLE 5

A particle moves along the path obtained from the intersection of the sphere $x^2 + y^2 + z^2 = 14$ and the plane $y = 1$. At what rate is z changing with respect to x when the particle is at $P = (2, 1, 3)$?

Solution

The path of intersection of the sphere and the plane $y = 1$ is a circle (see Fig. 31). The point P lies on the upper hemisphere, since the z-coordinate is positive. Therefore,

$$z = f(x, y) = \sqrt{14 - x^2 - y^2}$$

The rate at which z changes with respect to x is $f_x(x, y)$. From the above equation, we obtain

$$f_x(x, y) = \frac{\partial}{\partial x}(14 - x^2 - y^2)^{1/2} = \frac{1}{2}(14 - x^2 - y^2)^{-1/2}(-2x)$$

$$= \frac{-x}{\sqrt{14 - x^2 - y^2}}$$

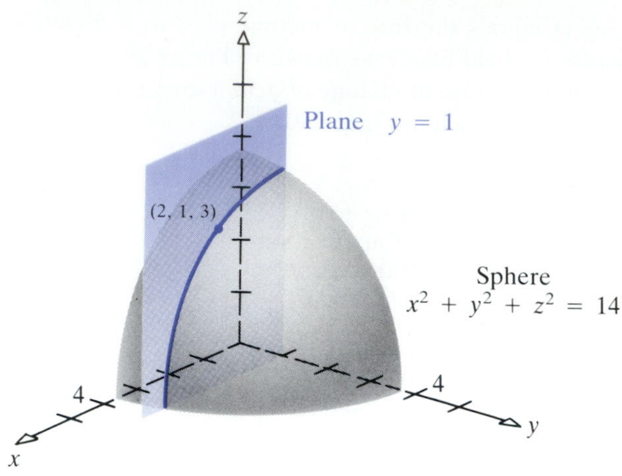

Figure 31

At the point (2, 1, 3), the rate at which z is changing with respect to x is

$$f_x(2, 1) = \frac{-2}{\sqrt{14 - 4 - 1}} = -\frac{2}{3}$$

EXAMPLE 6

The temperature (in °C) of a metal plate, located in the xy-plane, at any point (x, y) is given by the formula $T = 24(x^2 + y^2)^2$. Find the rate of change of T in a direction parallel to the x-axis at the point $(1, -2)$.

Solution

The rate of change of temperature in a direction parallel to the x-axis is given by

$$T_x(x, y) = 2(24)(x^2 + y^2)(2x)$$

At $(1, -2)$, we find $T_x(1, -2) = 96(5) = 480°\text{C}$. We interpret $T_x(1, -2) = 480°\text{C}$ to mean that as one moves in a horizontal direction (y fixed) away from the point $(1, -2)$, the temperature of the plate is increasing at the rate of 480°C per unit of distance moved.

Higher-Order Partial Derivatives

For a function $z = f(x, y)$ of two variables for which the limits (1) exist, there are two *first-order partial derivatives*: $f_x(x, y)$ and $f_y(x, y)$. If it is possible to differentiate each of these partially with respect to x or y, there will result four *second-order partial derivatives*—namely,

$$f_{xx}(x, y) = \frac{\partial}{\partial x} f_x(x, y) = \frac{\partial}{\partial x}\frac{\partial z}{\partial x} = \frac{\partial^2 z}{\partial x^2} \qquad f_{xy}(x, y) = \frac{\partial}{\partial y} f_x(x, y) = \frac{\partial}{\partial y}\frac{\partial z}{\partial x} = \frac{\partial^2 z}{\partial y\, \partial x}$$

$$f_{yx}(x, y) = \frac{\partial}{\partial x} f_y(x, y) = \frac{\partial}{\partial x}\frac{\partial z}{\partial y} = \frac{\partial^2 z}{\partial x\, \partial y} \qquad f_{yy}(x, y) = \frac{\partial}{\partial y} f_y(x, y) = \frac{\partial}{\partial y}\frac{\partial z}{\partial y} = \frac{\partial^2 z}{\partial y^2}$$

The two second-order partial derivatives

$$\frac{\partial^2 z}{\partial x\, \partial y} = f_{yx}(x, y) \qquad \text{and} \qquad \frac{\partial^2 z}{\partial y\, \partial x} = f_{xy}(x, y)$$

are called *mixed partials. Be careful*! Observe the differences in the above two equations. The notation $\partial^2 z/\partial x\, \partial y = f_{yx}$ means that first we should differentiate f partially with respect to y and then differentiate the result partially with respect to x—in that order! On the other hand, $\partial^2 z/\partial y\, \partial x = f_{xy}$ means we should differentiate with respect to x and then with respect to y.

EXAMPLE 7

Find all second-order partial derivatives of $z = f(x, y) = x \ln y + ye^x$.

Solution

$$f_x = \ln y + ye^x \qquad f_y = \frac{x}{y} + e^x$$

Therefore,

$$f_{xx} = \frac{\partial}{\partial x}(f_x) = \frac{\partial}{\partial x}(\ln y + ye^x) = ye^x \qquad f_{xy} = \frac{\partial}{\partial y}(f_x) = \frac{\partial}{\partial y}(\ln y + ye^x) = \frac{1}{y} + e^x$$

$$f_{yx} = \frac{\partial}{\partial x}(f_y) = \frac{\partial}{\partial x}\left(\frac{x}{y} + e^x\right) = \frac{1}{y} + e^x \qquad f_{yy} = \frac{\partial}{\partial y}(f_y) = \frac{\partial}{\partial y}\left(\frac{x}{y} + e^x\right) = \frac{-x}{y^2}$$

Note in Example 7 that $f_{xy} = f_{yx}$ for all (x, y). As it turns out, this will be the case for most functions we encounter. The conditions under which the equality of the mixed partials holds are given next.

[16.3.3] THEOREM / *Equality of Mixed Partials.*

Let $z = f(x, y)$ denote a function of two variables whose domain is D. Let (x_0, y_0) be an interior point of D. If the partial derivatives f_x, f_y, f_{xy}, and f_{yx} exist at each point of some disk centered at (x_0, y_0), and if f_{xy} and f_{yx} are continuous at (x_0, y_0), then $f_{xy}(x_0, y_0) = f_{yx}(x_0, y_0)$.

A proof of this result may be found in most advanced calculus texts. For two examples of a function for which the mixed partials are not equal, see Problems 72 and 73.

Continuity and Partial Derivatives

In Chapter 3 we proved the important result that if a function f of one variable has a derivative at x_0, then f is continuous at x_0. However, if a function f of more than one variable has partial derivatives at a point P_0, it may fail to be continuous at P_0.

EXAMPLE 8

Show that the function

$$f(x, y) = \begin{cases} \dfrac{xy}{x^2 + y^2} & \text{if} \quad (x, y) \neq (0, 0) \\ 0 & \text{if} \quad (x, y) = (0, 0) \end{cases}$$

has partial derivatives at (0, 0) but is not continuous at that point.

Solution

We use definition (16.3.1):

$$f_x(0, 0) = \lim_{\Delta x \to 0} \frac{f(0 + \Delta x, 0) - f(0, 0)}{\Delta x}$$

$$f_y(0, 0) = \lim_{\Delta y \to 0} \frac{f(0, 0 + \Delta y) - f(0, 0)}{\Delta y}$$

Since $f(0 + \Delta x, 0) = f(\Delta x, 0) = 0/(\Delta x)^2 = 0$, $f(0, 0 + \Delta y) = f(0, \Delta y) = 0$, and $f(0, 0) = 0$, it follows that

$$f_x(0, 0) = \lim_{\Delta x \to 0} 0 = 0 \qquad f_y(0, 0) = \lim_{\Delta y \to 0} 0 = 0$$

Thus, f has partial derivatives at (0, 0).

We saw in Example 2 in Section 16.2 that

$$\lim_{(x, y) \to (0, 0)} \frac{xy}{x^2 + y^2}$$

does not exist. As a result, f is not continuous at (0, 0). Figure 32 shows a computer graph of this function.

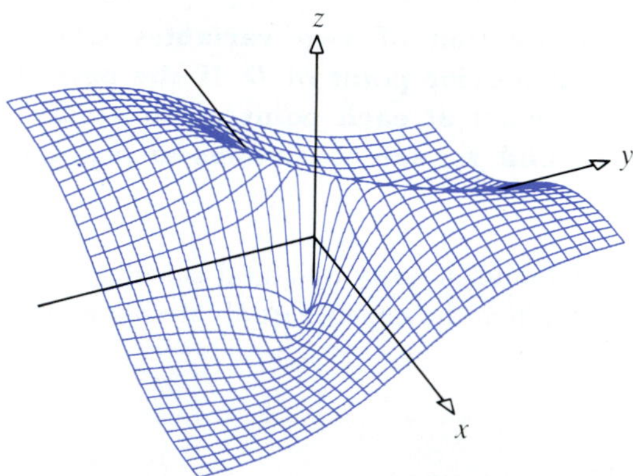

$$f(x, y) = \begin{cases} \dfrac{xy}{x^2 + y^2} & \text{if } (x, y) \neq (0, 0) \\ 0 & \text{if } (x, y) = (0, 0) \end{cases}$$

Figure 32

Functions of Three or More Variables

The idea of partial differentiation may be extended to functions of three or more variables. Thus, if $w = f(x, y, z)$ is a function of three variables, there will be three partial derivatives: the partial derivative with respect to x is f_x; the partial derivative with respect to y is f_y; and the partial derivative with respect to z is f_z. Each of these is calculated by differentiating with respect to the indicated variable, while holding the other two fixed. The function f_x equals the rate of change of w with respect to x (y and z are fixed); the function f_y equals the rate of change of w with respect to y (x and z are fixed); and the function f_z equals the rate of change of w with respect to z (x and y are fixed).

EXAMPLE 9

If $w = f(x, y, z) = 5x^2yz^3$, then

$$f_x = 10xyz^3 \qquad f_y = 5x^2z^3 \qquad f_z = 15x^2yz^2$$

We may also write the partial derivatives of $w = f(x, y, z)$ as

$$f_x = \frac{\partial w}{\partial x} \qquad f_y = \frac{\partial w}{\partial y} \qquad f_z = \frac{\partial w}{\partial z}$$

EXAMPLE 10

The area of a triangle is given by $A = \frac{1}{2}ab \sin \gamma$. See Figure 33.

(a) Find the rate of change of A with respect to a when b and γ are held fixed.
(b) Find the rate of change of A with respect to γ when a and b are held fixed.
(c) Find the rate of change of b with respect to a when A and γ are held fixed.
(d) Calculate each of the above if $a = 20$, $b = 30$, and $\gamma = \pi/6$.

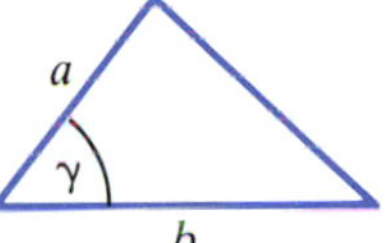

Figure 33

Solution

(a) $\dfrac{\partial A}{\partial a} = \dfrac{1}{2} b \sin \gamma$ (b) $\dfrac{\partial A}{\partial \gamma} = \dfrac{1}{2} ab \cos \gamma$

(c) $b = \dfrac{2A}{a \sin \gamma}$; $\dfrac{\partial b}{\partial a} = -\dfrac{2A}{a^2 \sin \gamma} = -\dfrac{2(\frac{1}{2}ab \sin \gamma)}{a^2 \sin \gamma} = -\dfrac{b}{a}$

(d) $\dfrac{\partial A}{\partial a} = \dfrac{1}{2}(30)\left(\sin \dfrac{\pi}{6}\right) = \dfrac{15}{2}$; $\dfrac{\partial A}{\partial \gamma} = \dfrac{1}{2}(20)(30)\left(\cos \dfrac{\pi}{6}\right) = 150\sqrt{3}$;

$\dfrac{\partial b}{\partial a} = -\dfrac{3}{2}$

Higher-order partial derivatives of $w = f(x, y, z)$ are defined in an analogous way to those of a function of two variables. For example,

$$f_{zy} = \frac{\partial}{\partial y}\left(\frac{\partial w}{\partial z}\right) = \frac{\partial^2 w}{\partial y\, \partial z} \qquad f_{xz} = \frac{\partial}{\partial z}\left(\frac{\partial w}{\partial x}\right) = \frac{\partial^2 w}{\partial z\, \partial x}$$

As in the case of functions of two variables, for most functions $w = f(x, y, z)$, the mixed partials have the property that $f_{xy} = f_{yx}$, $f_{xz} = f_{zx}$, and $f_{yz} = f_{zy}$.

If $z = f(x_1, x_2, \ldots, x_n)$ is a function of n variables, there are n partial derivatives of f. The partial derivative of f with respect to x_1, $\partial f/\partial x_1$, is found by differentiating f with respect to x_1, while holding the remaining variables fixed. The other partial derivatives $\partial f/\partial x_2, \partial f/\partial x_3, \ldots, \partial f/\partial x_n$ are defined similarly.

EXAMPLE 11

If $z = f(x_1, x_2, \ldots, x_n) = x_1^2 + x_2^2 + \cdots + x_n^2$, then

$$\frac{\partial f}{\partial x_1} = 2x_1, \quad \frac{\partial f}{\partial x_2} = 2x_2, \quad \ldots, \quad \frac{\partial f}{\partial x_n} = 2x_n$$

Collectively, we can write these partial derivatives as

$$\frac{\partial f}{\partial x_i} = 2x_i \qquad i = 1, 2, \ldots, n$$

EXERCISE 16.3

In Problems 1–14 find f_x and f_y.

1. $f(x, y) = x^2y + 6y^2$

2. $f(x, y) = 3x^2 + 6xy^3$

3. $f(x, y) = x^3/y^3$

4. $f(x, y) = \dfrac{x + y}{y}$

5. $f(x, y) = e^y\cos x + e^x\sin y$

6. $f(x, y) = x^2\cos y + y^2\sin x$

7. $f(x, y) = e^{2x+3y}$

8. $f(x, y) = \cos(x^2y^3)$

9. $f(x, y) = \ln\sqrt{x^2 + y^2}$

10. $f(x, y) = \tan^{-1}(y/x)$

11. $f(x, y) = \sin^2(2xy)$

12. $f(x, y) = e^{(x^2+y^2)^{1/2}}$

13. $f(x, y) = x^y, \quad x > 0$

14. $f(x, y) = \sin[\ln(x^2 + y^2)]$

In Problems 15–22 compute f_{xx}, f_{xy}, f_{yx}, and f_{yy}. Check to verify that $f_{xy} = f_{yx}$.

15. $f(x, y) = 6x^2 - 8xy + 9y^2$

16. $f(x, y) = x^3/y^3$

17. $f(x, y) = \ln(x^3 + y^2)$

18. $f(x, y) = \ln(y/x)$

19. $f(x, y) = e^{2x+3y}$

20. $f(x, y) = \tan^{-1}(y/x)$

21. $f(x, y) = \cos(x^2y^3)$

22. $f(x, y) = \sin^2(xy)$

In Problems 23–34 compute f_x, f_y, and f_z.

23. $f(x, y, z) = xy + yz + xz$

24. $f(x, y, z) = xe^y + ye^z + ze^x$

25. $f(x, y, z) = xy\sin z - yz\sin x$

26. $f(x, y, z) = 1/\sqrt{x^2 + y^2 + z^2}$

27. $f(x, y, z) = z\tan^{-1}(y/x)$

28. $f(x, y, z) = e^{xyz}$

29. $f(x, y, z) = (x + y)^z$

30. $f(x, y, z) = x^{y+z}$

31. $f(x, y, z) = x^{yz}$

32. $f(x, y, z) = \sin[\ln(x^2 + y^2 + z^2)]$

33. $f(x, y, z) = \tan^{-1}[(xy)/z]$

34. $f(x, y, z) = e^{x^2+y^2}\ln z$

In Problems 35 and 36 use definition (16.3.1) to calculate $f_x(0, 0)$ and $f_y(0, 0)$.

35. $f(x, y) = \begin{cases} \dfrac{x^3 + y^3}{x^2 + y^2} & \text{if} \quad (x, y) \neq (0, 0) \\ 0 & \text{if} \quad (x, y) = (0, 0) \end{cases}$

36. $f(x, y) = \begin{cases} \dfrac{x^2 y^2}{x^2 + 4y^3} & \text{if} \quad (x, y) \neq (0, 0) \\ 0 & \text{if} \quad (x, y) = (0, 0) \end{cases}$

In Problems 37–42 find an equation of the tangent line to the curve of intersection of the surface with the plane at the point indicated.

37. $z = x^2 + y^2, \quad y = 2, \quad \text{at } (1, 2, 5)$

38. $z = x^2 - y^2, \quad x = 3, \quad \text{at } (3, 1, 8)$

39. $z = \sqrt{1 - x^2 - y^2}, \quad x = 0, \quad \text{at } (0, 1/2, \sqrt{3}/2)$

40. $z = \sqrt{16 - x^2 - y^2}, \quad y = 2, \quad \text{at } (\sqrt{3}, 2, 3)$

41. $z = \sqrt{x^2 - y^2}, \quad x = 4, \quad \text{at } (4, 1, \sqrt{15})$

42. $z = e^x \ln y, \quad y = e, \quad \text{at } (0, e, 1)$

43. Find the rate of change of $z = \ln\sqrt{x^2 + y^2}$ at $(3, 4, \ln 5)$:

(a) In a direction parallel to the x-axis
(b) In a direction parallel to the y-axis

44. Find the rate of change of $z = e^y \sin x$ at $(\pi/3, 0, \sqrt{3}/2)$:

(a) In a direction parallel to the x-axis
(b) In a direction parallel to the y-axis

45. The temperature distribution T of a heated plate located in the xy-plane is given by

$$T = T(x, y) = \left(\frac{100}{\ln 2}\right) \ln(x^2 + y^2) \qquad \text{for} \quad 1 \leq x^2 + y^2 \leq 9$$

(a) Show that $T = 0$ if $x^2 + y^2 = 1$; and $T = 200$ if $x^2 + y^2 = 4$.
(b) Find the rate of change of T in a direction parallel to the x-axis at the point $(1, 0)$ and at the point $(0, 1)$.
(c) Find the rate of change of T in a direction parallel to the y-axis at the point $(2, 0)$ and at the point $(0, 2)$

46. Rework parts (b) and (c) of Problem 45 if the temperature distribution is $T = T(x, y) = 100/\sqrt{x^2 + y^2}$.

47. Find $\partial x/\partial r$, $\partial x/\partial \theta$, $\partial y/\partial r$, and $\partial y/\partial \theta$ if $x = r\cos\theta$, $y = r\sin\theta$.

48. Find $\partial r/\partial x$, $\partial \theta/\partial x$, $\partial r/\partial y$, and $\partial \theta/\partial y$ by solving for r and θ in $x = r\cos\theta$, $y = r\sin\theta$.

49. Show that $\partial u/\partial x = \partial v/\partial y$ and $\partial u/\partial y = -\partial v/\partial x$ for $u = e^x \cos y$, $v = e^x \sin y$.

50. Rework Problem 49 for $u = \ln\sqrt{x^2 + y^2}$, $v = \tan^{-1}(y/x)$.

51. If $u = x^2 + 4y^2$, show that $x(\partial u/\partial x) + y(\partial u/\partial y) = 2u$.

52. If $u = xy^2$, show that $x(\partial u/\partial x) + y(\partial u/\partial y) = 3u$.

53. If $w = x^2 + y^2 - 3yz$, show that $x(\partial w/\partial x) + y(\partial w/\partial y) + z(\partial w/\partial z) = 2w$.

54. If $w = (xz + y^2)/yz$, show that $x(\partial w/\partial x) + y(\partial w/\partial y) + z(\partial w/\partial z) = 0$.

55. If $z = \cos(x + y) + \cos(x - y)$, show that $\partial^2 z/\partial x^2 - \partial^2 z/\partial y^2 = 0$.

56. If $z = \sin(x - y) + \ln(x + y)$, show that $\partial^2 z/\partial x^2 = \partial^2 z/\partial y^2$.

A function $z = f(x, y)$ that obeys the partial differential equation $\partial^2 z/\partial x^2 + \partial^2 z/\partial y^2 = 0$ is called *harmonic*.* In Problems 57–60 show that each function is harmonic.

57. $z = \ln\sqrt{x^2 + y^2}$

58. $z = e^{ax} \sin ay$

59. $z = \tan^{-1}(y/x)$

60. $z = e^{ax} \cos ay$

61. Show that $z = e^{x^2 + y^2}/(x^2 + y^2)$ obeys the partial differential equation $y(\partial z/\partial x) = x(\partial z/\partial y)$.

62. Show that $z = x^2 y^2/(x + y)$ obeys the partial differential equation $x(\partial z/\partial x) + y(\partial z/\partial y) = 3z$.

---

* This equation is referred to as *Laplace's equation*. It is of great importance in many branches of mathematical physics, such as the flow of fluids, heat, elasticity, electricity, and so forth.

63. The ideal gas law $PV = nrT$ is used to describe the relationship between pressure P, volume V, and temperature T of a confined gas, where n is the number of moles of the gas and r is the universal gas constant. Show that $(\partial V/\partial T)(\partial T/\partial P)(\partial P/\partial V) = -1$.

64. The function

$$z = f(x, y, r) = \frac{1 + (1 - x)y}{1 + r} - 1$$

describes the net gain or loss of money invested, where x = annual marginal tax rate, y = annual effective yield on an investment, and r = annual inflation rate.

(a) Find the annual net gain or loss if money is invested at an effective yield of 6% when the marginal tax rate is 25% and the inflation rate is 10%; that is, find $f(0.25, 0.06, 0.1)$.

(b) Find $f_x(x, 0.04, 0.05)$, which equals the rate of change of gain (or loss) of money with respect to the marginal tax rate when the effective yield is 4% and the inflation rate is 5%.

(*Problem 64 continues in the next column*)

64. (*continued*)

(c) Find $f_y(0.25, y, 0.05)$, which equals the rate of change of gain (or loss) of money with respect to the effective yield when the marginal tax rate is 25% and the inflation rate is 5%.

(d) Find $f_r(0.25, 0.04, r)$, which equals the rate of change of gain (or loss) of money with respect to the inflation rate when the marginal tax rate is 25% and the effective yield is 4%.

65. The speed of sound v in a gas depends on the pressure p and density d of the gas according to the formula $v(p, d) = k\sqrt{p/d}$, where k is some constant. Find the rate of change of velocity with respect to p and with respect to d.

66. Suppose a vibrating string is governed by the equation $f(x, t) = 2\cos 5t \sin x$, where x is the horizontal distance of a point on the string, t is time, and $f(x, t)$ is the amplitude of the point. Verify that $\partial^2 f/\partial t^2 = 25(\partial^2 f/\partial x^2)$ at all points (x, t).

In Problems 67–70 find f_{xy} and f_{yx} and show that $f_{xy} = f_{yx}$.

67. $f(x, y) = \dfrac{x}{x^2 + y^2}$

68. $f(x, y) = \tan^{-1}\left(\dfrac{y}{x}\right)$

69. $f(x, y) = e^{x \ln y}$

70. $f(x, y) = x \sin x \cos y$

71. (a) Consider two coordinate systems as given in the figure. Let $f(x, y) = 3x^2 + 4y^3$. Evaluate $f_x(1, 6)$ and $f_y(1, 6)$.

(b) Let (a, b) be the $x'y'$-coordinates of $(1, 6)$. Let $\bar{f}(x', y') = f(x, y)$. Evaluate $\bar{f}_{x'}(a, b)$ and $\bar{f}_{y'}(a, b)$.

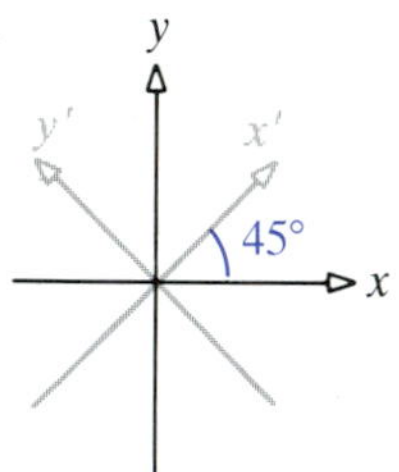

72. Define the function

$$f(x, y) = \begin{cases} \dfrac{xy^3}{x^2 + y^2} & \text{if } (x, y) \neq (0, 0) \\ 0 & \text{if } (x, y) = (0, 0) \end{cases}$$

(a) Find f_x and f_y. [*Hint:* $f_x(x, y)$, $(x, y) \neq (0, 0)$, can be found by differentiating. To find $f_x(0, 0)$, use the definition of the partial derivative.]

(b) Show that $f_{xy}(0, 0) \neq f_{yx}(0, 0)$.

73. Let

$$f(x, y) = \begin{cases} \dfrac{xy(x^2 - y^2)}{x^2 + y^2} & \text{if } (x, y) \neq (0, 0) \\ 0 & \text{if } (x, y) = (0, 0) \end{cases}$$

Use the definition of partial derivative as the limit of a difference quotient to show that:

(a) $f_x(0, y) = -y$ (b) $f_y(x, 0) = x$
(c) $f_{xy}(0, 0) = -1$ (d) $f_{yx}(0, 0) = 1$

(This provides another example of a function whose mixed second partial derivatives f_{xy} and f_{yx} are different.)

74. If you are told that f is a function of two variables whose partial derivatives are $f_x(x, y) = 3x - y$ and $f_y(x, y) = x - 3y$, should you believe it? Explain.

75. Use definition (16.3.1) to show that the function $z = \sqrt{x^2 + y^2}$ does not have partial derivatives at $(0, 0)$. By discussing the graph of the function, give a geometric reason why this should be so.

76. Find the first partial derivatives of $w = x^{yz}$.

77. Find an equation of the tangent line to the curve at $(1, -2, 1)$ cut from the surface $z = 4x^2 - y^2 + 1$ by the plane $x = 1$. By the plane $y = -2$.

78. Interpret $f_x(1, -\frac{1}{3})$ and $f_y(1, -\frac{1}{3})$ geometrically if $z = f(x, y) = 4x^2 + 9y^2 - 12$.

79. If $u = z \tan^{-1}(x/y)$, show that $(\partial^2 u/\partial x^2) + (\partial^2 u/\partial y^2) + (\partial^2 u/\partial z^2) = 0$.

80. Show that $u = e^{-\alpha^2 t}\sin \alpha x$ satisfies the equation $\partial u/\partial t = \partial^2 u/\partial x^2$ for all values of the constant α.

81. Show that $xf_x + yf_y + zf_z = 0$ for $f(x, y, z) = e^{x/y} + e^{y/z} + e^{z/x}$.

82. Find a in terms of b and c so that $f(t, x, y) = e^{at}\sin bx \cos cy$ satisfies $f_t = f_{xx} + f_{yy}$.

83. Suppose $u(x, y)$ and $v(x, y)$ have continuous second partial derivatives, and $u_x = v_y$, $u_y = -v_x$. Show that u and v are harmonic functions.

84. Show that $f(x, y, z) = (x^2 + y^2 + z^2)^{-1/2}$ satisfies the three-dimensional Laplace equation

$$f_{xx} + f_{yy} + f_{zz} = 0$$

85. Show that $f(x, t) = \cos(x + ct)$ satisfies the one-dimensional wave equation $f_{tt} = c^2 f_{xx}$, where c is a constant.

86. Suppose a thin metal rod extends along the x-axis from $x = 0$ to $x = 20$, and for each x, $0 \le x \le 20$, the temperature of the rod at time $t \ge 0$ is $T(t, x) = 40e^{-\lambda t}\sin(\pi x/20)$, where $\lambda > 0$ is a constant.

(a) Show that $T_t = -\lambda T$, $T_{xx} = -(\pi^2/400)\,T$, and $T_t = (1/k^2)\,T_{xx}$ for some k.

(b) Graph the initial temperature distribution, $y = T(0, x)$, $0 \le x \le 20$.

(c) At what point(s) on the rod is the rate of cooling one-half the maximum rate of cooling?

87. Find f_x and f_y if $f(x, y) = \int_x^y \ln(\cos \sqrt{t})\, dt$.

16.4 Differentiability

In the case of a differentiable function of one variable, the differential provides an approximation to the change in that function. That is, the change in $y = f(x)$ from x_0 to $(x_0 + \Delta x)$—namely,

$$\Delta y = f(x_0 + \Delta x) - f(x_0)$$

may be approximated by the differential

$$dy = f'(x_0)\Delta x$$

for $\Delta x \approx 0$ (see Chap. 4).

The idea behind approximating Δy by dy can also be viewed in the following manner: Define η (Greek letter eta) as

$$\eta = \frac{\Delta y}{\Delta x} - f'(x_0) \qquad \textbf{(1)}$$

By multiplying by Δx and solving for Δy, we find that

$$\boldsymbol{\Delta y = f'(x_0)\Delta x + \eta\Delta x} \qquad \textbf{(2)}$$

where η is a function depending on Δx, for which $\lim_{\Delta x \to 0} \eta = 0$.

Figure 34 gives a geometric interpretation of (2). The term $\eta\Delta x$ represents the difference between Δy and $dy = f'(x)\Delta x$. It is evident from Figure 34 that $\eta\Delta x \to 0$ as $\Delta x \to 0$. But Figure 34 does not show something that (1) does show—that $\eta \to 0$ as $\Delta x \to 0$. (Do you see why? Since f is differentiable, it follows that $\Delta y/\Delta x \to f'(x_0)$ as $\Delta x \to 0$.)

With this as background, we extend the notion of differentiability to functions of two variables.

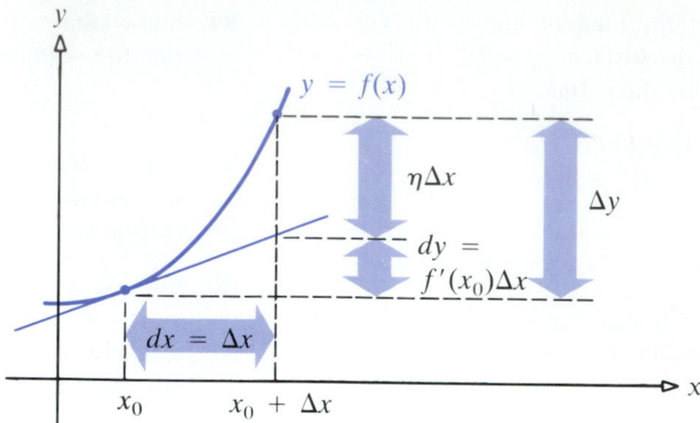

Figure 34

Differentiability of $z = f(x, y)$

Let $z = f(x, y)$ denote a function of two variables whose domain is D. Let (x_0, y_0) be an interior point of D, and let Δx and Δy (which represent changes in x and in y, respectively) be chosen so that the point $(x_0 + \Delta x, y_0 + \Delta y)$ is also in D. The *change in* z, denoted by Δz, is defined as

$$\Delta z = f(x_0 + \Delta x, y_0 + \Delta y) - f(x_0, y_0) \tag{3}$$

See Figure 35.

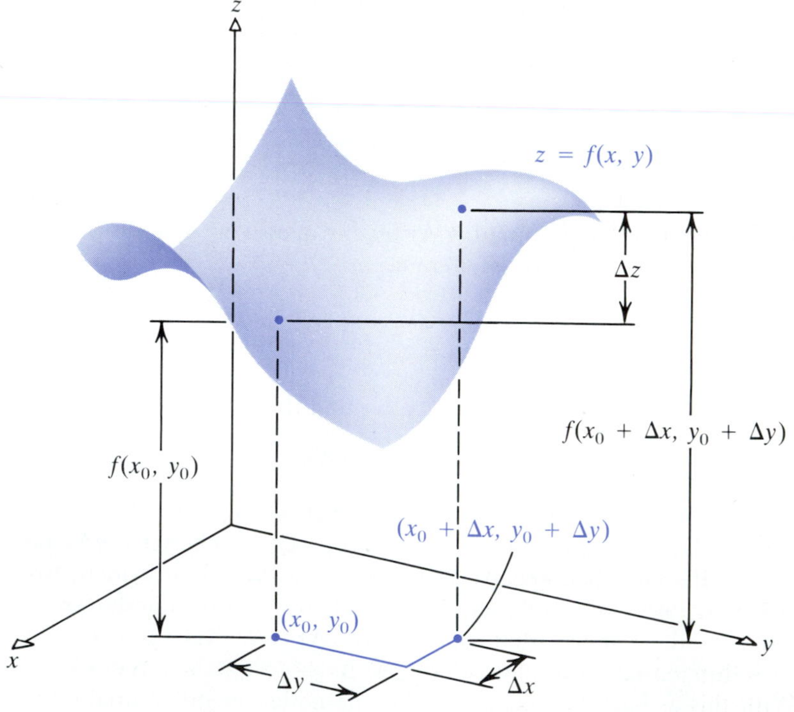

Figure 35

EXAMPLE 1

Find the change Δz in $z = f(x, y) = x^2y - 1$ from (x_0, y_0) to $(x_0 + \Delta x, y_0 + \Delta y)$. Use the answer to calculate the change in z from $(1, 2)$ to $(1.1, 1.9)$.

Solution

From (3) we find

$$\begin{aligned}\Delta z &= f(x_0 + \Delta x, y_0 + \Delta y) - f(x_0, y_0)\\ &= (x_0 + \Delta x)^2(y_0 + \Delta y) - 1 - (x_0^2 y_0 - 1)\\ &= x_0^2\Delta y + 2x_0 y_0 \Delta x + 2x_0 \Delta x \Delta y + y_0(\Delta x)^2 + (\Delta x)^2 \Delta y\end{aligned}$$

The change in z from $(1, 2)$ to $(1.1, 1.9)$ may be computed by substituting $x_0 = 1$, $y_0 = 2$, $\Delta x = 0.1$, and $\Delta y = -0.1$ in the above expression. The change is found to be $\Delta z = 0.299$. This answer could also have been found by computing $f(1.1, 1.9) - f(1, 2)$.

We use an equation similar to (2) to define *differentiability* of a function of two variables:

[16.4.1] DEFINITION / *Differentiability of $z = f(x, y)$ at (x_0, y_0).*

Let $z = f(x, y)$ denote a function of two variables whose domain is D. Let (x_0, y_0) be an interior point of D and let Δx and Δy be chosen so that the point $(x_0 + \Delta x, y_0 + \Delta y)$ is also in D. Suppose also that $f_x(x_0, y_0)$ and $f_y(x_0, y_0)$ exist. If the change Δz from (x_0, y_0) to $(x_0 + \Delta x, y_0 + \Delta y)$ can be expressed in the form

$$\Delta z = f_x(x_0, y_0)\Delta x + f_y(x_0, y_0)\Delta y + \eta_1 \Delta x + \eta_2 \Delta y \tag{4}$$

where η_1 and η_2 are each functions of Δx and Δy such that

$$\lim_{(\Delta x, \Delta y)\to(0,0)} \eta_1 = 0 \quad \text{and} \quad \lim_{(\Delta x, \Delta y)\to(0,0)} \eta_2 = 0$$

then $z = f(x, y)$ is said to be *differentiable at* (x_0, y_0).

If the domain D of a function $z = f(x, y)$ is an open set and if $z = f(x, y)$ is differentiable at every point of D, then we say *f is differentiable on D*. If the domain D of a function is the xy-plane and if the function is differentiable at every point in the xy-plane, then we say f is *everywhere differentiable* or, simply, is *differentiable*.

Let's use the definition to show that the function given in Example 1 is differentiable.

EXAMPLE 2

Show that $z = f(x, y) = x^2y - 1$ is differentiable.

Solution

The domain of f is the xy-plane. Let (x, y) denote any point. We need to find η_1 and η_2 so that

$$\Delta z - f_x(x, y)\Delta x - f_y(x, y)\Delta y = \eta_1 \Delta x + \eta_2 \Delta y \tag{5}$$

where

$$\lim_{(\Delta x, \Delta y)\to(0,0)} \eta_1 = 0 \qquad \text{and} \qquad \lim_{(\Delta x, \Delta y)\to(0,0)} \eta_2 = 0$$

Since $f(x, y) = x^2y - 1$, it follows that

$$f_x(x, y) = 2xy \qquad \text{and} \qquad f_y(x, y) = x^2$$

By using these facts and the calculation of Δz obtained in Example 1, we find that

$$\begin{aligned}\Delta z - f_x(x, y)\Delta x - f_y(x, y)\Delta y &= x^2\Delta y + 2xy\Delta x + 2x\Delta x\Delta y + y(\Delta x)^2 \\ &\quad + (\Delta x)^2\Delta y - 2xy\Delta x - x^2\Delta y \\ &= 2x\Delta x\Delta y + y(\Delta x)^2 + (\Delta x)^2\Delta y\end{aligned}$$

There are several ways in which we can write the right-hand side of the above equation so that it conforms to (5). One way is the following:

$$\Delta z - f_x(x, y)\Delta x - f_y(x, y)\Delta y = (2x\Delta y + y\Delta x)\Delta x + (\Delta x)^2\Delta y$$

where $\eta_1 = 2x\Delta y + y\Delta x$ and $\eta_2 = (\Delta x)^2$. It is now clear that η_1 and η_2 both approach 0 as $(\Delta x, \Delta y)$ approaches $(0, 0)$; thus, the function is differentiable.

We emphasize that *the choice of η_1 and η_2 in (4) is not necessarily unique.* In Example 2, there are several other convenient and natural ways to write the equation, including the following:

1. $\eta_1\Delta x + \eta_2\Delta y = (y\Delta x)\Delta x + [2x\Delta x + (\Delta x)^2]\Delta y$
2. $\eta_1\Delta x + \eta_2\Delta y = (y\Delta x + \Delta x\Delta y)\Delta x + (2x\Delta x)\Delta y$
3. $\eta_1\Delta x + \eta_2\Delta y = (2x\Delta y + y\Delta x + \Delta x\Delta y)\Delta x + (0)\Delta y$

In each case (as you may easily verify), η_1 and η_2 are functions that approach 0 as Δx and Δy approach 0.

Later in this section we give an example of a function $z = f(x, y)$ that is *not* differentiable. For now, we proceed to give conditions on a function $z = f(x, y)$ that will guarantee that it is differentiable.

[16.4.2] THEOREM

Let $z = f(x, y)$ denote a function of two variables whose domain is D. Let (x_0, y_0) be an interior point of D. If the partial derivatives f_x and f_y exist at each point of some disk centered at (x_0, y_0), and if f_x and f_y are each continuous at (x_0, y_0), then f is differentiable at (x_0, y_0).

We defer the proof of this result to the end of the section.

EXAMPLE 3

Show that $z = f(x, y) = x^2y - 1$ is differentiable.

Solution

We'll use theorem (16.4.2). The domain D of f is the xy-plane. The partial derivatives

$$f_x(x, y) = 2xy \qquad f_y(x, y) = x^2$$

exist and are continuous at every point of D. Consequently, f is differentiable.

Differentials

[16.4.3] DEFINITION / *Total Differential.*

Let $z = f(x, y)$ denote a function of two variables whose domain is D. Let (x_0, y_0) be an interior point of D and let Δx and Δy be chosen so that the point $(x_0 + \Delta x, y_0 + \Delta y)$ is also in D. If f is differentiable at the point (x_0, y_0), we define the differentials dx and dy as

$$dx = \Delta x \qquad dy = \Delta y$$

The differential dz, also called the *total differential of* $z = f(x, y)$, is defined as

$$dz = f_x(x_0, y_0)\, dx + f_y(x_0, y_0)\, dy \tag{6}$$

If $z = f(x, y)$ is a differentiable function, a comparison of (6) with (4) yields the expression

$$\Delta z = dz + \eta_1 \Delta x + \eta_2 \Delta y \tag{7}$$

where $\lim_{(\Delta x, \Delta y) \to (0, 0)} \eta_1 = 0$ and $\lim_{(\Delta x, \Delta y) \to (0, 0)} \eta_2 = 0$. When $dx(= \Delta x)$ and $dy(= \Delta y)$ are close to 0—and hence η_1 and η_2 are also close to 0—we see from (7) that dz is, therefore, approximately equal to Δz. Since dz is usually easier to calculate than Δz, we make use of the fact that when Δx and Δy are close to 0, then dz is an approximation to Δz. Of course, in using dz as an approximation to Δz, the error that results equals the expression $\eta_1 \Delta x + \eta_2 \Delta y$.

EXAMPLE 4

For the function $z = f(x, y) = x^2 y - 1$ use the total differential dz to approximate the change in z from $(1, 2)$ to $(1.1, 1.9)$.

Solution

We showed in Example 3 that f is differentiable. We use (6) with $f_x(x, y) = 2xy$ and $f_y(x, y) = x^2$, and find the total differential dz at any point (x, y) to be

$$dz = 2xy\, dx + x^2\, dy$$

At $(1, 2)$, $dz = 4\, dx + dy$. By using $dx = 0.1$ and $dy = -0.1$, we estimate the change in z to be

$$\Delta z \approx dz = 4(0.1) + (-0.1) = 0.3$$

The actual change in z was computed in Example 1 to be 0.299, so the use of differentials resulted in an error of 0.001. Hence, by (7), we see that, in this case, $\eta_1 \Delta x + \eta_2 \Delta y = -0.001$.

EXAMPLE 5

A cola company requires cans in the shape of a right circular cylinder of height 10 centimeters and radius 3 centimeters. If the manufacturer of the cans claims a percentage error of no more than 0.2% in the height and no more than 0.1% in the radius, approximately what is the maximum error in the volume?

Solution

The volume V of a right circular cylinder of height h and radius R is $V = \pi R^2 h$. By (6), the total differential dV is

$$dV = \frac{\partial V}{\partial R}\, dR + \frac{\partial V}{\partial h}\, dh = 2\pi Rh\, dR + \pi R^2\, dh$$

The relative error in R is $|\Delta R|/R = |dR|/R = 0.001$, and the relative error in h is $|\Delta h|/h = |dh|/h = 0.002$. The relative error in the volume is

$$\frac{|\Delta V|}{V} \approx \frac{|dV|}{V} = \frac{|2\pi Rh\, dR + \pi R^2\, dh|}{\pi R^2 h} \le 2\frac{|dR|}{R} + \frac{|dh|}{h} = 2(0.001) + 0.002 = 0.004$$

The maximum percentage error in the volume is therefore approximately 0.4%, so the actual volume of the container is about $90\pi \pm (0.004)(90\pi)$ and lies between $89.64\pi \approx 281.612$ and $90.36\pi \approx 283.874$ cubic centimeters.

Differentiability and Continuity

We showed in Chapter 3 that differentiable functions of a single variable are necessarily continuous. We now extend this result to functions of two variables.

[16.4.4] THEOREM

Let $z = f(x, y)$ denote a function of two variables whose domain is D. Let (x_0, y_0) be an interior point of D. If f is differentiable at (x_0, y_0), then f is continuous at (x_0, y_0).

Proof Since $z = f(x, y)$ is differentiable at (x_0, y_0), we can express Δz by

$$\Delta z = f_x(x_0, y_0)\Delta x + f_y(x_0, y_0)\Delta y + \eta_1 \Delta x + \eta_2 \Delta y$$

where $\lim_{(\Delta x, \Delta y) \to (0,0)} \eta_1 = 0$ and $\lim_{(\Delta x, \Delta y) \to (0,0)} \eta_2 = 0$. We write Δz as

$$\Delta z = [f_x(x_0, y_0) + \eta_1]\Delta x + [f_y(x_0, y_0) + \eta_2]\Delta y$$

and set $\Delta x = x - x_0$ and $\Delta y = y - y_0$. Then $(\Delta x, \Delta y) \to (0, 0)$ is equivalent to $(x, y) \to (x_0, y_0)$, so that

$$\lim_{(x,y) \to (x_0, y_0)} \Delta z = \lim_{(x,y) \to (x_0, y_0)} \{[f_x(x_0, y_0) + \eta_1](x - x_0) + [f_y(x_0, y_0) + \eta_2](y - y_0)\} = 0$$

Since $\Delta z = f(x, y) - f(x_0, y_0)$, then $\lim_{(x,y) \to (x_0, y_0)} \Delta z = 0$ is equivalent to $\lim_{(x,y) \to (x_0, y_0)} f(x, y) = f(x_0, y_0)$. Hence, it follows that f is continuous at (x_0, y_0).

For a function $y = f(x)$, the notions of derivative and differentiability are equivalent, so that differentiability (existence of a derivative) implies continuity for functions of a single variable. However, for functions of two variables, even though differentiability implies continuity, the existence of partial derivatives at a point does not necessarily result in continuity at that point. Refer back to Example 8 in Section 16.3. The function

$$f(x, y) = \begin{cases} \dfrac{xy}{x^2 + y^2} & \text{if } (x, y) \neq (0, 0) \\ 0 & \text{if } (x, y) = (0, 0) \end{cases}$$

has partial derivatives at $(0, 0)$ but is not continuous at $(0, 0)$. We use the contrapositive of theorem (16.4.4), a function that is not continuous at (x_0, y_0) is not differentiable at (x_0, y_0). Thus, we have found a function that has partial derivatives at $(0, 0)$ but is not differentiable at $(0, 0)$. We conclude that for a function of two variables, the notions of partial derivative and differentiable are quite different.

The following corollary is a consequence of theorems (16.4.2) and (16.4.4).

[16.4.5] COROLLARY

Let $z = f(x, y)$ denote a function of two variables whose domain is D. Let (x_0, y_0) be an interior point of D. If the partial derivatives f_x and f_y exist at each point of some disk centered at (x_0, y_0), and if f_x and f_y are each continuous at (x_0, y_0), then f is continuous at (x_0, y_0).

Although the precise formulations are given as theorems, the following summary may be helpful:

1. Continuity of f_x and $f_y \Rightarrow$ Differentiability of f.
2. Differentiability of $f \Rightarrow$ Continuity of f.
3. Continuity of f_x and $f_y \Rightarrow$ Continuity of f.
4. Existence of f_x and f_y does not necessarily mean f is differentiable.
5. Existence of f_x and f_y does not necessarily mean f is continuous.

Differentials for Functions of Three or More Variables

Under suitable conditions the definitions and theorems given for functions of two variables extend to functions of three or more variables. Thus, if $w = f(x, y, z)$ is a function of three variables, we define f to be differentiable at a point (x, y, z) if the change Δw in w can be expressed in the form

$$\Delta w = f_x(x, y, z)\Delta x + f_y(x, y, z)\Delta y + f_z(x, y, z)\Delta z + \eta_1\Delta x + \eta_2\Delta y + \eta_3\Delta z$$

where η_1, η_2, and η_3 are each functions of Δx, Δy, and Δz such that

$$\lim_{(\Delta x, \Delta y, \Delta z)\to(0,0,0)} \eta_1 = 0 \qquad \lim_{(\Delta x, \Delta y, \Delta z)\to(0,0,0)} \eta_2 = 0 \qquad \lim_{(\Delta x, \Delta y, \Delta z)\to(0,0,0)} \eta_3 = 0$$

If $w = f(x, y, z)$ is differentiable at a point (x, y, z), the *total differential* dw is defined as

$$dw = f_x(x, y, z)\,dx + f_y(x, y, z)\,dy + f_z(x, y, z)\,dz$$

where $dx = \Delta x$, $dy = \Delta y$, and $dz = \Delta z$.

EXAMPLE 6

If $w = f(x, y, z) = 3x^2\sin^2 y \cos z$, find the total differential dw.

Solution

$$dw = 6x \sin^2 y \cos z\, dx + 6x^2 \sin y \cos y \cos z\, dy - 3x^2 \sin^2 y \sin z\, dz$$

It can be shown that if $w = f(x, y, z)$ is defined within a ball centered at (x_0, y_0, z_0), and if f_x, f_y, and f_z exist in this ball and are continuous at (x_0, y_0, z_0), then f is differentiable at (x_0, y_0, z_0).

The remarks above extend to functions of more than three variables in a completely analogous way.

Application in Astrophysics

The luminosity L (total power output in watts) of a star is given by the formula

$$L = 4\pi R^2 \sigma T^4$$

where R is its radius (in meters), T is its effective surface temperature (in degrees Kelvin),* and σ is the Stefan–Boltzmann constant. Our sun presently has $L_0 = 3.90 \times 10^{26}$ watts, $R_0 = 6.94 \times 10^8$ meters, and $T_0 = 4800$ K. Suppose another billion years of evolution are expected to result in the changes $\Delta R = +0.08 \times 10^8$ meters and $\Delta T = +100$ K. What will be the resulting percent increase in luminosity?

* 0 K = −273°C; 100 K = −173°C

We begin with $L = 4\pi R^2 \sigma T^4$ and take the natural log of each side. Then

$$\ln L = \ln 4\pi + 2 \ln R + \ln \sigma + 4 \ln T$$

Then, since σ is a constant,

$$\frac{dL}{L} = 2\frac{dR}{R} + 4\frac{dT}{T}$$

The relative error in luminosity is, therefore,

$$\frac{\Delta L}{L} \approx \frac{dL}{L} = 2\frac{dR}{R} + 4\frac{dT}{T} \approx \frac{2(0.08) \times 10^8}{6.94 \times 10^8} + \frac{4(100)}{4800}$$

$$= 2(0.0115) + 4(0.0208) = 0.106$$

The percent increase in luminosity is approximately 10.6%.

Incidentally, a reasonable, though rough, guess at how this would affect the earth's temperature is

$$\Delta T_e \approx (\tfrac{1}{4})(0.106)\, T_e = (0.0265)(290 \text{ K}) = +7.69 \text{ K}$$

Such a change in temperature would be enough to modify the earth's climate.

Proof of Theorem (16.4.2) The proof depends on the mean value theorem for derivatives. Let Δx and Δy be changes, not both 0, in x and in y, respectively, so that the point $(x_0 + \Delta x, y_0 + \Delta y)$ lies in some disk centered at (x_0, y_0). The change in z is

$$\begin{aligned}\Delta z &= f(x_0 + \Delta x, y_0 + \Delta y) - f(x_0, y_0)\\ &= f(x_0 + \Delta x, y_0 + \Delta y) - f(x_0, y_0 + \Delta y) + f(x_0, y_0 + \Delta y) - f(x_0, y_0)\end{aligned} \tag{8}$$

The expression $f(x, y_0 + \Delta y)$ is a function of x alone, and its (partial) derivative $f_x(x, y_0 + \Delta y)$ exists in the disk centered at (x_0, y_0). Hence, by the mean value theorem, there is a u between x_0 and $(x_0 + \Delta x)$, so that

$$f(x_0 + \Delta x, y_0 + \Delta y) - f(x_0, y_0 + \Delta y) = f_x(u, y_0 + \Delta y)\Delta x \tag{9}$$

Similarly, the expression $f(x_0, y)$ is a function of y alone, and its derivative $f_y(x_0, y)$ exists in the disk centered at (x_0, y_0). Hence, by the mean value theorem, there is a v between y_0 and $(y_0 + \Delta y)$, so that

$$f(x_0, y_0 + \Delta y) - f(x_0, y_0) = f_y(x_0, v)\Delta y \tag{10}$$

By substituting (9) and (10) in (8), we obtain

$$\Delta z = f_x(u, y_0 + \Delta y)\Delta x + f_y(x_0, v)\Delta y \tag{11}$$

We introduce the functions η_1 and η_2 defined by

$$\eta_1 = f_x(u, y_0 + \Delta y) - f_x(x_0, y_0) \qquad \text{and} \qquad \eta_2 = f_y(x_0, v) - f_y(x_0, y_0) \tag{12}$$

and observe that η_1 and η_2 have the desired property that

$$\lim_{(\Delta x, \Delta y)\to(0,0)} \eta_1 = \lim_{(\Delta x, \Delta y)\to(0,0)} [f_x(u, y_0 + \Delta y) - f_x(x_0, y_0)]$$
$$= f_x(x_0, y_0) - f_x(x_0, y_0) = 0$$
$$\lim_{(\Delta x, \Delta y)\to(0,0)} \eta_2 = \lim_{(\Delta x, \Delta y)\to(0,0)} [f_y(x_0, v) - f_y(x_0, y_0)]$$
$$= f_y(x_0, y_0) - f_y(x_0, y_0) = 0$$

since f_x and f_y are continuous at (x_0, y_0) and $u \to x_0$, $v \to y_0$ as $(\Delta x, \Delta y) \to (0, 0)$. As a result of (12), we may write (11) as

$$\Delta z = f_x(x_0, y_0)\Delta x + f_y(x_0, y_0)\Delta y + \eta_1 \Delta x + \eta_2 \Delta y$$

If we compare the above expression to (4), we conclude that f is differentiable at (x_0, y_0).

EXERCISE 16.4

In Problems 1–14 find the total differential of each function.

1. $z = x^2 + y^2$

2. $z = 2x^2 + xy - y^2$

3. $z = x \sin y + y \sin x$

4. $z = \tan^{-1}(y/x)$

5. $z = e^x \cos y + e^{-x} \sin y$

6. $z = \ln(y/x)$

7. $z = \ln(x^2 + y^2)$

8. $z = e^{xy}$

9. $w = x^2y + y^2z + z^2x$

10. $w = xyz$

11. $w = xe^{yz} + ye^{xz} + ze^{xy}$

12. $w = \ln(x^2 + y^2 + z^2)$

13. $w = e^t(\ln xy + \ln xz + \ln yz)$

14. $w = \dfrac{xyzt}{x + y + z}$

In Problems 15–18 show that the function $z = f(x, y)$ is differentiable at any point (x, y) in its domain by: (a) finding Δz; (b) finding η_1 and η_2 so that (4) holds; and (c) showing that $\lim_{(\Delta x, \Delta y)\to(0,0)} \eta_1 = 0$ and $\lim_{(\Delta x, \Delta y)\to(0,0)} \eta_2 = 0$.

15. $z = f(x, y) = xy^2 - 2xy$

16. $z = f(x, y) = 3x^2 + y^2$

17. $z = f(x, y) = y^2/x$

18. $z = f(x, y) = 2x/y$

19. Use differentials to estimate the change in $z = x^2 + y^2$ from $(1, 3)$ to $(1.1, 3.2)$.

20. Use differentials to estimate the change in $z = 2x^2 + xy - y^2$ from $(2, -1)$ to $(2.1, -1.1)$.

21. Use differentials to estimate the change in $z = e^x \ln(xy)$ from $(1, 2)$ to $(0.9, 2.1)$.

22. Use differentials to estimate the change in $z = xy/(x + y)$ from $(-1, 2)$ to $(-0.9, 1.9)$.

23. Use differentials to estimate $(\sqrt[4]{16.01})(\sqrt[5]{32.1})$.

24. Use differentials to estimate $(2.01)^6/\sqrt{3.89}$.

25. Using differentials, estimate the change in the volume of a right circular cylinder if the height changes from 2 to 2.1 centimeters and the radius changes from 0.5 to 0.51 centimeter.

26. For the data in Problem 25, what is the estimated change in surface area? Assume that the cylinder is closed at both the top and the bottom.

27. Estimate the increase in area of a triangle if its base is increased from 2 to 2.05 centimeters and its altitude is increased from 5 to 5.1 centimeters.

28. If the base of a triangle is increased from 5 to 5.1 centimeters and its altitude is decreased from 10 to 9.8 centimeters, how is the area affected?

29. If $x = r\cos\theta$, $y = r\sin\theta$, show that $x\,dy - y\,dx = r^2\,d\theta$.

30. The specific gravity of an object is defined as $s = a/(a - w)$, where a is the weight of the object in air and w is its weight in water. If a is found to be 6 pounds with a possible error of 1% and w is 5 pounds with a possible error of 2%, what is the maximum error in the specific gravity?

31. In a parallel circuit the total resistance R due to two resistances R_1 and R_2 obeys $1/R = (1/R_1) + (1/R_2)$. If $R_1 = 50$ ohms with a possible error of 1.2% and $R_2 = 75$ ohms with a possible error of 1%, what is the maximum error in the total resistance?

32. A tank consists of a hemisphere mounted on a cylinder of the same radius (see the figure). The height and radius of the cylinder were measured as 14 meters and 5 meters, respectively. However, the device used to make this measurement was found to be in error by 1%. What is the maximum error in the volume of the tank?

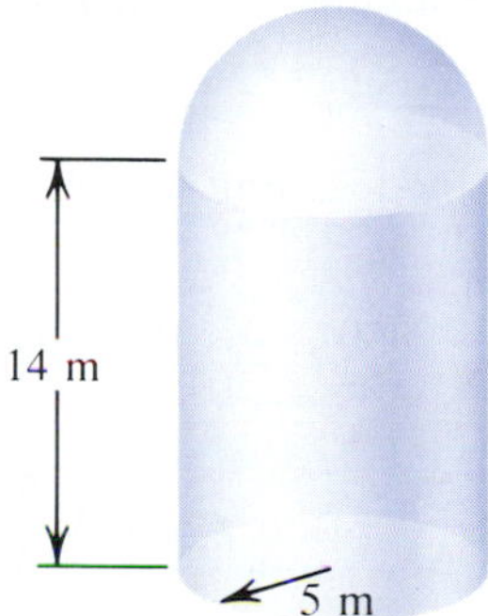

33. The index of refraction is defined as $\mu = (\sin i)/(\sin r)$, where i is the angle of incidence and r is the angle of refraction. If $i = 30°$ and $r = 60°$, and each is subject to a possible error of 2%, approximately what is the maximum relative error for μ?

34. The equation $PV = kT$, where k is a constant, relates the pressure P, volume V, and temperature T of a confined ideal gas. If $P = 0.1$ gram per square millimeter, $V = 12$ cubic millimeters, and $T = 32°$C, aproximate the change in P if V and T change to 15 cubic millimeters and 29°C, respectively.

(*Problem 34 continues in the next column*)

34. (*continued*)

35. Two sides and the included angle of a triangle are measured by a ruler and protractor, which are subject to errors of 2% and 3%, respectively. The area of the triangle is then computed from the formula $A = \frac{1}{2}bc\sin\alpha$.

(a) Show that: $\dfrac{dA}{A} = \dfrac{db}{b} + \dfrac{dc}{c} + \cot\alpha\,d\alpha$

(b) If $\alpha = \pi/4$, approximately what is the maximum error in the computation of A?

36. Show that the function below has partial derivatives at (0, 0), but that f is not continuous at (0, 0) and, hence, is not differentiable at that point.

$$f(x, y) = \begin{cases} \dfrac{2xy}{x^2 + y^2} & \text{if } (x, y) \neq (0, 0) \\ 0 & \text{if } (x, y) = (0, 0) \end{cases}$$

37. Show that the function below has partial derivatives at (0, 0), but that f is not continuous at (0, 0) and, hence, is not differentiable at that point.

$$f(x, y) = \begin{cases} \dfrac{xy(1 + y^2)}{x^2 + y^2} & \text{if } (x, y) \neq (0, 0) \\ 0 & \text{if } (x, y) = (0, 0) \end{cases}$$

38. For the function below, show that $f_x(1, 1)$ and $f_y(1, 1)$ each exist, but that f is not differentiable at (1, 1).

$$f(x, y) = \begin{cases} \dfrac{xy - 1}{x^2 + y^2 - 2} & \text{if } x^2 + y^2 \neq 2 \\ \frac{1}{2} & \text{if } x^2 + y^2 = 2 \end{cases}$$

39. For the function below, show that $f_x(0, 0)$ and $f_y(0, 0)$ each exist, but that f is not differentiable at (0, 0).

$$f(x, y) = \begin{cases} \dfrac{x^2y^2}{x^4 + y^4} & \text{if } (x, y) \neq (0, 0) \\ 0 & \text{if } (x, y) = (0, 0) \end{cases}$$

Chain Rules 16.5

For a function $z = f(x, y)$ of two variables, there are several versions of the chain rule. The first one we discuss occurs when the two independent variables x and y are each functions of a single variable t—say, $x = x(t)$ and $y = y(t)$. Then the composite function $z = f((x(t), y(t))$ is a function of a single variable t. The first chain rule gives a formula for finding dz/dt.

[16.5.1] THEOREM / *Chain Rule.*

If $x = x(t)$ and $y = y(t)$ are each differentiable at t, and if $z = f(x, y)$ is differentiable at the point $(x(t), y(t))$, then $z = f(x(t), y(t))$ is differentiable at t. Moreover,

$$\frac{dz}{dt} = \frac{\partial z}{\partial x}\frac{dx}{dt} + \frac{\partial z}{\partial y}\frac{dy}{dt} \tag{1}$$

Proof Since $dz/dt = \lim_{\Delta t \to 0}(\Delta z/\Delta t)$, we seek an expression for Δz. Since z is differentiable at the point $(x, y) = (x(t), y(t))$, we can write Δz as

$$\Delta z = \frac{\partial z}{\partial x}\Delta x + \frac{\partial z}{\partial y}\Delta y + \eta_1 \Delta x + \eta_2 \Delta y$$

where η_1 and η_2 are functions of Δx and Δy and $\lim_{(\Delta x, \Delta y)\to(0,0)} \eta_1 = 0$ and $\lim_{(\Delta x, \Delta y)\to(0,0)} \eta_2 = 0$. Now,

$$\frac{dz}{dt} = \lim_{\Delta t \to 0}\frac{\Delta z}{\Delta t} = \lim_{\Delta t \to 0}\left[\frac{\partial z}{\partial x}\frac{\Delta x}{\Delta t} + \frac{\partial z}{\partial y}\frac{\Delta y}{\Delta t} + \eta_1\frac{\Delta x}{\Delta t} + \eta_2\frac{\Delta y}{\Delta t}\right]$$

In the right-hand expression, $\partial z/\partial x$ and $\partial z/\partial y$ are evaluated at $(x(t), y(t))$ and hence do not depend on Δt. However, since

$$\lim_{\Delta t \to 0}\Delta x = \lim_{\Delta t \to 0}\left[\frac{\Delta x}{\Delta t}\cdot \Delta t\right] = \frac{dx}{dt}\cdot 0 = 0 \qquad \text{and} \qquad \lim_{\Delta t \to 0}\Delta y = 0$$

it follows that as $\Delta t \to 0$, then $(\Delta x, \Delta y) \to (0, 0)$ so that $\eta_1 \to 0$ and $\eta_2 \to 0$. Putting all this together, we have

$$\begin{aligned}
\frac{dz}{dt} &= \frac{\partial z}{\partial x}\lim_{\Delta t \to 0}\frac{\Delta x}{\Delta t} + \frac{\partial z}{\partial y}\lim_{\Delta t \to 0}\frac{\Delta y}{\Delta t} + \lim_{\Delta t \to 0}\eta_1 \cdot \lim_{\Delta t \to 0}\frac{\Delta x}{\Delta t} + \lim_{\Delta t \to 0}\eta_2 \cdot \lim_{\Delta t \to 0}\frac{\Delta y}{\Delta t} \\
&= \frac{\partial z}{\partial x}\frac{dx}{dt} + \frac{\partial z}{\partial y}\frac{dy}{dt} + 0 \cdot \frac{dx}{dt} + 0 \cdot \frac{dy}{dt} \\
&= \frac{\partial z}{\partial x}\frac{dx}{dt} + \frac{\partial z}{\partial y}\frac{dy}{dt}
\end{aligned}$$

The schematic given in Figure 36 may prove helpful in remembering the form of (1).

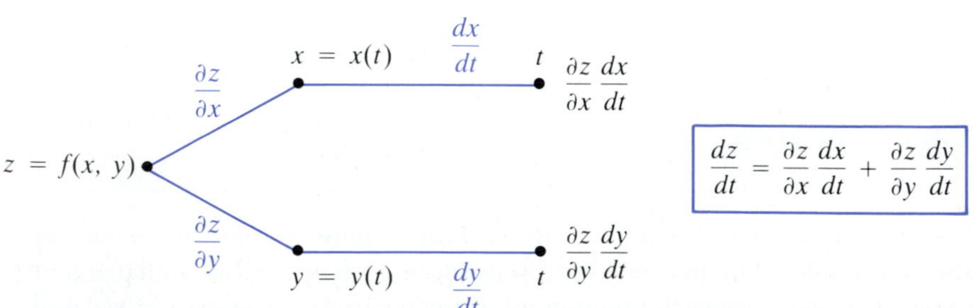

Figure 36

EXAMPLE 1

If $z = x^2y - y^2x$ and $x = t$, $y = t^2$, find dz/dt.

Solution

We use theorem (16.5.1):

$$\frac{dz}{dt} = \frac{\partial z}{\partial x}\frac{dx}{dt} + \frac{\partial z}{\partial y}\frac{dy}{dt} = (2xy - y^2)(1) + (x^2 - 2xy)(2t)$$

Since $x = t$ and $y = t^2$, we have

$$\frac{dz}{dt} = (2t^3 - t^4) + (t^2 - 2t^3)(2t) = -5t^4 + 4t^3$$

We could also have solved Example 1 by forming the composite first and then differentiating:

$$z = x^2y - y^2x = t^4 - t^5$$

$$\frac{dz}{dt} = 4t^3 - 5t^4$$

In this case the alternative solution is easier. This will not always be the case, though.

EXAMPLE 2

Suppose $z = e^x \sin y$ and $x = e^t$, $y = (\pi/3)e^{-t}$. Find dz/dt when $t = 0$.

Solution

We use theorem (16.5.1):

$$\frac{dz}{dt} = \frac{\partial z}{\partial x}\frac{dx}{dt} + \frac{\partial z}{\partial y}\frac{dy}{dt} = (e^x \sin y)e^t + (e^x \cos y)\left(-\frac{\pi}{3}e^{-t}\right)$$

When $t = 0$, we have $x = 1$ and $y = \pi/3$. Thus, when $t = 0$,

$$\frac{dz}{dt} = \left(e \sin\frac{\pi}{3}\right)(1) + \left(e \cos\frac{\pi}{3}\right)\left(-\frac{\pi}{3}\right) = \frac{e\sqrt{3}}{2} - \frac{\pi e}{6} = \frac{e}{6}(3\sqrt{3} - \pi)$$

There is another version of the chain rule when $z = f(x, y)$ and x and y are each functions of two independent variables u and v—say, $x = g(u, v)$ and $y = h(u, v)$. Then the composite function $z = f(x, y) = f(g(u, v), h(u, v))$ is a function of the two variables u and v. This chain rule gives formulas for finding $\partial z/\partial u$ and $\partial z/\partial v$.

[16.5.2] THEOREM / *Chain Rule.*

Let $z = f(g(u, v), h(u, v))$ be the composite of $z = f(x, y)$ and $x = g(u, v)$, $y = h(u, v)$. If g and h are each continuous and have continuous first-order partial derivatives at a point (u, v),

which is an interior point of the domains of both g and h, and if f is differentiable in some disk centered at the point $(x, y) = (g(u, v), h(u, v))$, then

$$\frac{\partial z}{\partial u} = \frac{\partial z}{\partial x}\frac{\partial x}{\partial u} + \frac{\partial z}{\partial y}\frac{\partial y}{\partial u} \qquad \frac{\partial z}{\partial v} = \frac{\partial z}{\partial x}\frac{\partial x}{\partial v} + \frac{\partial z}{\partial y}\frac{\partial y}{\partial v} \tag{2}$$

Proof To find $\partial z/\partial u$, we hold v fixed. Then $x = g(u, v)$ and $y = h(u, v)$ are functions of u alone so we can use theorem (16.5.1). In doing so, dz/dt is replaced by $\partial z/\partial u$ and dx/dt, dy/dt by $\partial x/\partial u$, $\partial y/\partial u$. A similar argument is used for $\partial z/\partial v$.

The schematics given in Figures 37 and 38 may prove helpful in remembering the form of (2).

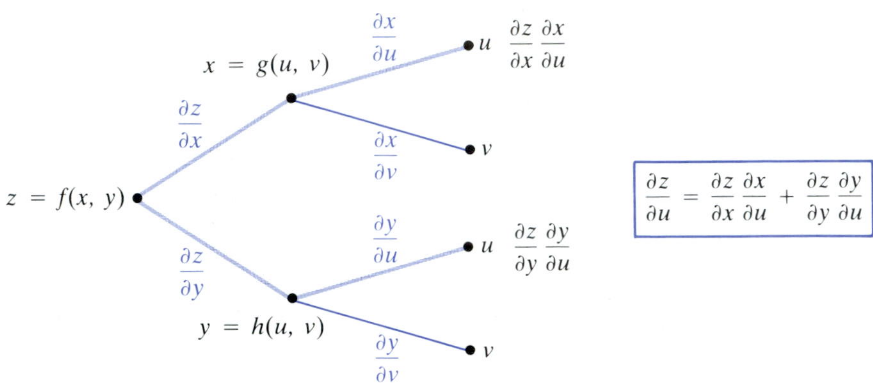

Figure 37

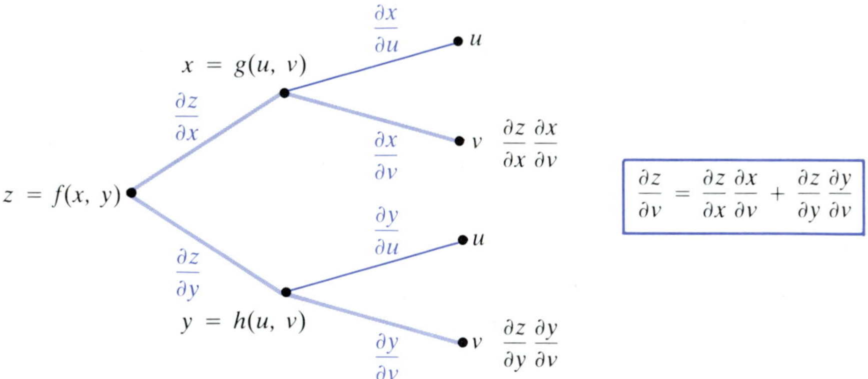

Figure 38

EXAMPLE 3

If $z = x^2 + xy - y^2$ and $x = e^{2u+v}$, $y = \ln(v/u)$, find $\partial z/\partial u$ and $\partial z/\partial v$.

Solution

$$\frac{\partial z}{\partial u} = \frac{\partial z}{\partial x}\frac{\partial x}{\partial u} + \frac{\partial z}{\partial y}\frac{\partial y}{\partial u} = (2x + y)(2e^{2u+v}) + (x - 2y)\left(\frac{-v/u^2}{v/u}\right)$$

$$= \left[4e^{2u+v} + 2\ln\left(\frac{v}{u}\right)\right]e^{2u+v} + \left[e^{2u+v} - 2\ln\left(\frac{v}{u}\right)\right]\left(\frac{-1}{u}\right)$$

$$\frac{\partial z}{\partial v} = \frac{\partial z}{\partial x}\frac{\partial x}{\partial v} + \frac{\partial z}{\partial y}\frac{\partial y}{\partial v} = (2x + y)(e^{2u+v}) + (x - 2y)\left(\frac{1/u}{v/u}\right)$$

$$= \left[2e^{2u+v} + \ln\left(\frac{v}{u}\right)\right]e^{2u+v} + \left[e^{2u+v} - 2\ln\left(\frac{v}{u}\right)\right]\frac{1}{v}$$

As Example 3 illustrates, when $\partial z/\partial u$ and $\partial z/\partial v$ are found, they should be expressed as functions of u and v alone.

The symmetry of the chain rule formula (2) remains even when the number of variables is increased. Thus, if $z = f(x_1, x_2, \ldots, x_m)$ is a differentiable function, and if $x_1 = g_1(u_1, u_2, \ldots, u_n)$, $x_2 = g_2(u_1, u_2, \ldots, u_n)$, $\ldots$, $x_m = g_m(u_1, u_2, \ldots, u_n)$ each possess continuous first-order partial derivatives, then the composite function $z = f(g_1, g_2, \ldots, g_m)$ is a function of $u_1, u_2, \ldots, u_n$, and

$$\begin{aligned}
\frac{\partial z}{\partial u_1} &= \frac{\partial z}{\partial x_1}\frac{\partial x_1}{\partial u_1} + \frac{\partial z}{\partial x_2}\frac{\partial x_2}{\partial u_1} + \cdots + \frac{\partial z}{\partial x_m}\frac{\partial x_m}{\partial u_1} \\
\frac{\partial z}{\partial u_2} &= \frac{\partial z}{\partial x_1}\frac{\partial x_1}{\partial u_2} + \frac{\partial z}{\partial x_2}\frac{\partial x_2}{\partial u_2} + \cdots + \frac{\partial z}{\partial x_m}\frac{\partial x_m}{\partial u_2} \\
&\vdots \\
\frac{\partial z}{\partial u_n} &= \frac{\partial z}{\partial x_1}\frac{\partial x_1}{\partial u_n} + \frac{\partial z}{\partial x_2}\frac{\partial x_2}{\partial u_n} + \cdots + \frac{\partial z}{\partial x_m}\frac{\partial x_m}{\partial u_n}
\end{aligned} \tag{3}$$

More compactly, we may write

$$\frac{\partial z}{\partial u_i} = \sum_{j=1}^{m} \frac{\partial z}{\partial x_j}\frac{\partial x_j}{\partial u_i} \qquad i = 1, 2, \ldots, n$$

Let's look at an example to see how (3) is used in practice.

EXAMPLE 4

If $f(x, y, z) = x^2 + y^2 + z^2$ and $x = uvwt$, $y = e^{u+v+w+t}$, $z = u + 2v + 3w + 4t$, find $\partial f/\partial u$, $\partial f/\partial v$, $\partial f/\partial w$, and $\partial f/\partial t$.

Solution

We use (3):

$$\frac{\partial f}{\partial u} = \frac{\partial f}{\partial x}\frac{\partial x}{\partial u} + \frac{\partial f}{\partial y}\frac{\partial y}{\partial u} + \frac{\partial f}{\partial z}\frac{\partial z}{\partial u}$$

$$= (2x)(vwt) + (2y)(e^{u+v+w+t}) + (2z)(1)$$

$$= 2uv^2w^2t^2 + 2e^{2(u+v+w+t)} + 2(u + 2v + 3w + 4t)$$

$$\begin{aligned}\frac{\partial f}{\partial v} &= \frac{\partial f}{\partial x}\frac{\partial x}{\partial v} + \frac{\partial f}{\partial y}\frac{\partial y}{\partial v} + \frac{\partial f}{\partial z}\frac{\partial z}{\partial v} \\ &= (2x)(uwt) + (2y)(e^{u+v+w+t}) + (2z)(2) \\ &= 2u^2vw^2t^2 + 2e^{2(u+v+w+t)} + 4(u+2v+3w+4t)\end{aligned}$$

$$\begin{aligned}\frac{\partial f}{\partial w} &= \frac{\partial f}{\partial x}\frac{\partial x}{\partial w} + \frac{\partial f}{\partial y}\frac{\partial y}{\partial w} + \frac{\partial f}{\partial z}\frac{\partial z}{\partial w} \\ &= (2x)(uvt) + (2y)(e^{u+v+w+t}) + (2z)(3) \\ &= 2u^2v^2wt^2 + 2e^{2(u+v+w+t)} + 6(u+2v+3w+4t)\end{aligned}$$

$$\begin{aligned}\frac{\partial f}{\partial t} &= \frac{\partial f}{\partial x}\frac{\partial x}{\partial t} + \frac{\partial f}{\partial y}\frac{\partial y}{\partial t} + \frac{\partial f}{\partial z}\frac{\partial z}{\partial t} \\ &= (2x)(uvw) + (2y)(e^{u+v+w+t}) + (2z)(4) \\ &= 2u^2v^2w^2t + 2e^{2(u+v+w+t)} + 8(u+2v+3w+4t)\end{aligned}$$

Again, note that the partial derivatives of f in Example 4 are expressed in terms of u, v, w, and t alone.

EXAMPLE 5

Let $p = f(v-w, v-u, u-w)$ be a differentiable function. Show that

$$\frac{\partial p}{\partial u} + \frac{\partial p}{\partial v} + \frac{\partial p}{\partial w} = 0$$

Solution

If we let $x = v - w$, $y = v - u$, and $z = u - w$, then $p = f(x, y, z)$ and we can use (3):

$$\begin{aligned}\frac{\partial p}{\partial u} &= \frac{\partial p}{\partial x}\frac{\partial x}{\partial u} + \frac{\partial p}{\partial y}\frac{\partial y}{\partial u} + \frac{\partial p}{\partial z}\frac{\partial z}{\partial u} \\ &= \frac{\partial p}{\partial x}(0) + \frac{\partial p}{\partial y}(-1) + \frac{\partial p}{\partial z}(1) = -\frac{\partial p}{\partial y} + \frac{\partial p}{\partial z}\end{aligned}$$

$$\begin{aligned}\frac{\partial p}{\partial v} &= \frac{\partial p}{\partial x}\frac{\partial x}{\partial v} + \frac{\partial p}{\partial y}\frac{\partial y}{\partial v} + \frac{\partial p}{\partial z}\frac{\partial z}{\partial v} \\ &= \frac{\partial p}{\partial x}(1) + \frac{\partial p}{\partial y}(1) + \frac{\partial p}{\partial z}(0) = \frac{\partial p}{\partial x} + \frac{\partial p}{\partial y}\end{aligned}$$

$$\begin{aligned}\frac{\partial p}{\partial w} &= \frac{\partial p}{\partial x}\frac{\partial x}{\partial w} + \frac{\partial p}{\partial y}\frac{\partial y}{\partial w} + \frac{\partial p}{\partial z}\frac{\partial z}{\partial w} \\ &= \frac{\partial p}{\partial x}(-1) + \frac{\partial p}{\partial y}(0) + \frac{\partial p}{\partial z}(-1) = -\frac{\partial p}{\partial x} - \frac{\partial p}{\partial z}\end{aligned}$$

But

$$\left(-\frac{\partial p}{\partial y} + \frac{\partial p}{\partial z}\right) + \left(\frac{\partial p}{\partial x} + \frac{\partial p}{\partial y}\right) + \left(-\frac{\partial p}{\partial x} - \frac{\partial p}{\partial z}\right) = 0$$

Therefore,

$$\frac{\partial p}{\partial u} + \frac{\partial p}{\partial v} + \frac{\partial p}{\partial w} = 0$$

Theorem (16.5.1) can be extended. If each of the functions $x_1, x_2, \ldots, x_m$ in (3) is a differentiable function of a single variable—say, t—then, after composition, the function $w = f(x_1, x_2, \ldots, x_m)$ is a differentiable function of the one variable t. In this case, (3) becomes

$$\frac{dw}{dt} = \frac{\partial w}{\partial x_1}\frac{dx_1}{dt} + \frac{\partial w}{\partial x_2}\frac{dx_2}{dt} + \cdots + \frac{\partial w}{\partial x_m}\frac{dx_m}{dt} \tag{4}$$

where each of the partial derivatives $\partial w/\partial x_1, \ldots, \partial w/\partial x_m$ is expressed in terms of t.

EXAMPLE 6

If $w = x^2y + y^2z$ and $x = t$, $y = t^2$, $z = t^3$, then w is a function of t and

$$\begin{aligned}\frac{dw}{dt} &= \frac{\partial w}{\partial x}\frac{dx}{dt} + \frac{\partial w}{\partial y}\frac{dy}{dt} + \frac{\partial w}{\partial z}\frac{dz}{dt}\\ &= (2xy)(1) + (x^2 + 2yz)(2t) + (y^2)(3t^2)\\ &= 2t^3 + (t^2 + 2t^5)(2t) + (t^4)(3t^2) = 7t^6 + 4t^3\end{aligned}$$

In Example 6, w can easily be found in terms of t—namely,

$$w = x^2y + y^2z = (t)^2t^2 + (t^2)^2t^3 = t^4 + t^7 = t^7 + t^4$$

Therefore,

$$\frac{dz}{dt} = 7t^6 + 4t^3$$

Of course, (4) was introduced so that we can deal with cases where w cannot be so easily expressed in terms of t.

Implicit Differentiation

If a differentiable function $y = f(x)$ is defined implicitly by the equation $F(x, y) = 0$, we can find the derivative dy/dx by applying the chain rule. (Recall that in Chap. 3 we used implicit differentiation to get dy/dx.) If $y = f(x)$ is a function defined by the equation $F(x, y) = 0$, it follows that $F(x, f(x)) \equiv 0$. For convenience, we set

$$z = F(u, y) \qquad u = x \qquad y = f(x)$$

Then

$$\frac{dz}{dx} = \frac{\partial F}{\partial u}\cdot\frac{du}{dx} + \frac{\partial F}{\partial y}\cdot\frac{dy}{dx} \tag{5}$$

But the composite $z = F(u, y) = F(x, f(x)) \equiv 0$, so $dz/dx = 0$. Since $u = x$, then $du/dx = 1$, and (5) becomes

$$\frac{\partial F}{\partial x}(1) + \frac{\partial F}{\partial y} \cdot \frac{dy}{dx} = 0$$

And, if $\partial F/\partial y \neq 0$, then

$$\frac{dy}{dx} = -\frac{F_x}{F_y} \tag{6}$$

EXAMPLE 7

For $F(x, y) = x^2y + y^2 - 2x = 0$, find dy/dx by using (6).

Solution

$$F_x = \frac{\partial F}{\partial x} = 2xy - 2 \qquad F_y = \frac{\partial F}{\partial y} = x^2 + 2y$$

Hence, by (6), if $x^2 + 2y \neq 0$,

$$\frac{dy}{dx} = \frac{-F_x}{F_y} = -\frac{2xy - 2}{x^2 + 2y} = \frac{2(1 - xy)}{x^2 + 2y}$$

You may wish to compare the above method with the method of implicit differentiation used in Chapter 3.

If a differentiable function $z = f(x, y)$ is defined implicitly by the equation $F(x, y, z) = 0$, we can find the partial derivatives $\partial z/\partial x$ and $\partial z/\partial y$ by applying the chain rule. For convenience, we set $w = F(u, v, z)$ and $u = x$, $v = y$, and $z = f(x, y)$. Since the composite function $w = F(x, y, f(x, y)) \equiv 0$, it follows that $\partial w/\partial x = 0$ and $\partial w/\partial y = 0$. For $\partial w/\partial x$, we obtain

$$\frac{\partial w}{\partial x} = \frac{\partial F}{\partial u} \cdot \frac{\partial u}{\partial x} + \frac{\partial F}{\partial v} \cdot \frac{\partial v}{\partial x} + \frac{\partial F}{\partial z} \cdot \frac{\partial z}{\partial x} = 0$$

Since $u = x$ and $v = y$, we have

$$\frac{\partial F}{\partial x}(1) + \frac{\partial F}{\partial y}(0) + \frac{\partial F}{\partial z} \cdot \frac{\partial z}{\partial x} = 0$$

and if $\partial F/\partial z \neq 0$, it follows that

$$\frac{\partial z}{\partial x} = -\frac{\partial F/\partial x}{\partial F/\partial z} = -\frac{F_x(x, y, z)}{F_z(x, y, z)} \tag{7}$$

In a similar way we can show that

$$\frac{\partial z}{\partial y} = -\frac{F_y(x, y, z)}{F_z(x, y, z)} \tag{8}$$

EXAMPLE 8

For $F(x, y, z) = x^2z^2 + y^2 - z^2 + 6yz - 10 = 0$, find $\partial z/\partial x$ and $\partial z/\partial y$ by using (7) and (8).

Solution

$$F_x = \frac{\partial F}{\partial x} = 2xz^2 \qquad F_y = 2y + 6z \qquad F_z = 2zx^2 - 2z + 6y$$

Thus, if $F_z = 2zx^2 - 2z + 6y \neq 0$, then

$$\frac{\partial z}{\partial x} = -\frac{2xz^2}{2zx^2 - 2z + 6y} = \frac{-xz^2}{zx^2 - z + 3y}$$

$$\frac{\partial z}{\partial y} = -\frac{2y + 6z}{2zx^2 - 2z + 6y} = -\frac{y + 3z}{zx^2 - z + 3y}$$

EXERCISE 16.5

In Problems 1–10 find dz/dt by using the chain rule.

1. $z = x^2 + y^2$, where $x = \sin t$; $y = \cos 2t$

2. $z = x^2 - y^2$, where $x = \sin 2t$, $y = \cos t$

3. $z = x^2 + y^2$, where $x = te^t$, $y = te^{-t}$

4. $z = x^2 - y^2$, where $x = te^{-t}$, $y = t^2e^{-t}$

5. $z = e^x \sin y$, where $x = \sqrt{t}$, $y = \pi t$

6. $z = e^{x/y}$, where $x = \sqrt{t}$; $y = t^3 + 1$

7. $z = e^{x/y}$, where $x = te^t$, $y = e^{t^2}$

8. $z = \ln(xy)$, where $x = t^5$, $y = \sqrt{t+1}$

9. $z = e^{x^2+y^2}$, where $x = \sin 2t$, $y = \cos t$

10. $z = e^{x^2-y^2}$, where $x = \sin 2t$, $y = \cos 2t$

In Problems 11–22 find $\partial z/\partial u$ and $\partial z/\partial v$ by using the chain rule.

11. $z = x^2 + y^2$, $x = ue^v$, $y = ve^u$

12. $z = x^2 - y^2$, $x = u \ln v$, $y = v \ln u$

13. $z = e^x \sin y$, $x = u^2 v$, $y = \ln(uv)$

14. $z = \frac{1}{y} \ln x$, $x = \sqrt{uv}$, $y = \frac{v}{u}$

15. $z = se^r$, $r = u^2 + v^2$, $s = \frac{v}{u}$

16. $z = \sqrt{s^2 + r^2}$, $s = \ln(uv)$, $r = \sqrt{uv}$

17. $z = xy^2w^3$, $x = 2u + v$, $y = 5u - 3v$, $w = 2u + 3v$

18. $z = x^2 - y^2 + w$, $x = e^{u+v}$, $y = uv$, $w = \frac{v}{u}$

19. $z = \ln(x^2 + y^2)$, $x = \frac{v^2}{u}$, $y = \frac{u}{v^2}$

20. $z = x \sin y - y \sin x$, $x = u^2 v$, $y = uv^2$

21. $z = x^2 + y^2$, $x = \sin(u - v)$, $y = \cos(u + v)$

22. $z = e^x + y$, $x = \tan^{-1}\left(\frac{u}{v}\right)$, $y = \ln(u + v)$

In Problems 23–32 find dw/dt.

23. $w = x^2 + y^2 - z^2$, $x = te^t$, $y = te^{-t}$, $z = e^{2t}$

24. $w = x^2 - y^2 - z^2$, $x = te^{-t}$, $y = t^2e^{-t}$, $z = e^{-t}$

25. $w = e^x \sin y \cos z$, $x = \sqrt{t}$, $y = \pi t$, $z = \frac{t}{2}$

26. $w = \ln(xyz)$, $x = t^5$, $y = \sqrt{t+1}$, $z = t^2$

27. $w = z \ln\left(\frac{x}{y}\right)$, $x = te^t$, $y = e^{t^2}$, $z = e^{2t}$

28. $w = ze^{x/y}$, $x = \sqrt{t}$, $y = t^3 + 1$, $z = e^t$

29. $w = x^2yz$, $x = \sin t$, $y = \cos t$, $z = e^t$

30. $w = \sqrt{xyz}$, $x = e^t$, $y = te^t$, $z = t^2e^t$

31. $w = \dfrac{xy}{x^2 + y^2}$, $x = \sin t$, $y = \cos t$

32. $w = y \ln x + xy + \tan y$, $x = \dfrac{t}{t+1}$, $y = t^3 - t$

In Problems 33–38 find dy/dx by using (6).

33. $F(x, y) = x^2y - y^2x + xy - 5 = 0$

34. $F(x, y) = x^3y^2 - xy + x^2y - 10 = 0$

35. $F(x, y) = x \sin y + y \sin x - 2 = 0$

36. $F(x, y) = xe^y + ye^x - xy = 0$

37. $F(x, y) = x^{1/3} + y^{1/3} - 1 = 0$

38. $F(x, y) = x^{2/3} + y^{2/3} - 1 = 0$

In Problems 39–42 find $\partial z/\partial x$ and $\partial z/\partial y$ by using (7) and (8).

39. $F(x, y, z) = xz + 3yz^2 + x^2y^3 - 5z = 0$

40. $F(x, y, z) = x^2z + y^2z + x^3y - 10z = 0$

41. $F(x, y, z) = \sin z + y \cos z + xyz - 10 = 0$

42. $F(x, y, z) = xe^{yz} + ye^{xz} + xyz = 0$

43. If $z = f(x, y)$, $x = r\cos\theta$, $y = r\sin\theta$, show that

$$\left(\frac{\partial z}{\partial r}\right)^2 + \frac{1}{r^2}\left(\frac{\partial z}{\partial \theta}\right)^2 = \left(\frac{\partial z}{\partial x}\right)^2 + \left(\frac{\partial z}{\partial y}\right)^2$$

44. If $z = f(x, y)$ and $x = u\cos\theta - v\sin\theta$, $y = u\sin\theta + v\cos\theta$, with θ a constant, show that $(\partial f/\partial u)^2 + (\partial f/\partial v)^2 = (\partial f/\partial x)^2 + (\partial f/\partial y)^2$.

45. If $z = f(u - v, v - u)$, show that $(\partial z/\partial u) + (\partial z/\partial v) = 0$. [*Hint:* Let $x = u - v$, $y = v - u$.]

46. If $z = vf(u^2 - v^2)$, show that $v(\partial z/\partial u) + u(\partial z/\partial v) = uz/v$.

47. If $w = f(u)$ and $u = \sqrt{x^2 + y^2 + z^2}$, show that $(\partial w/\partial x)^2 + (\partial w/\partial y)^2 + (\partial w/\partial z)^2 = (dw/du)^2$.

48. If $z = f(y/x)$, show that $x(\partial z/\partial x) + y(\partial z/\partial y) = 0$.

49. Show that if $z = f(u/v, v/w)$, then $u(\partial z/\partial u) + v(\partial z/\partial v) + w(\partial z/\partial w) = 0$.

50. Let $z = f(y + ax) + g(y - ax)$, $a \neq 0$. Show that z satisfies the wave equation $\partial^2 z/\partial x^2 = a^2(\partial^2 z/\partial y^2)$.

51. If $z = f(x, y)$ and $x = g(u, v)$, $y = h(u, v)$, find expressions for $\partial^2 z/\partial u^2$, $\partial^2 z/\partial u\,\partial v$, and $\partial^2 z/\partial v^2$.

52. A certain confined gas obeys the ideal gas law $PV = 20T$. If the temperature of the gas is increasing at the rate of 5°C per second and if, when the temperature is 80°C, the pressure is 10 newtons per square meter and is decreasing at the rate of 2 newtons per square meter per second, find the rate of change of the volume.

53. Prove that if $F(x, y, z) = 0$ is differentiable, then $(\partial z/\partial x)(\partial x/\partial y)(\partial y/\partial z) = -1$.

In Problems 54 and 55 find $\partial w/\partial x$, $\partial w/\partial y$, and $\partial w/\partial z$.

54. $w = (2x + 3y)^{4z}$

55. $w = (2x)^{3y+4z}$

56. Let $y = f(a, b)$, $a = h(s, t)$, $b = k(s, t)$. When $s = 1$ and $t = 3$, we know that

$$\frac{\partial h}{\partial s} = 4 \qquad \frac{\partial k}{\partial s} = -3 \qquad \frac{\partial h}{\partial t} = 1 \qquad \frac{\partial k}{\partial t} = -5$$

Also,

$$h(1, 3) = 6 \qquad k(1, 3) = 2 \qquad f_a(6, 2) = 7 \qquad f_b(6, 2) = 2$$

What are $\partial y/\partial s$ and $\partial y/\partial t$ at (1, 3)?

57. Suppose we denote the expression $(\partial^2/\partial x^2) + (\partial^2/\partial y^2)$ by Δ. If $z = f(x, y)$, show that

$$\Delta f = \frac{\partial^2 f}{\partial r^2} + \frac{1}{r}\left(\frac{\partial f}{\partial r}\right) + \frac{1}{r^2}\left(\frac{\partial^2 f}{\partial \theta^2}\right)$$

where $x = r\cos\theta$ and $y = r\sin\theta$.

58. Suppose that $F(x, y)$ has continuous second-order partial derivatives and $F(x, y) = 0$ defines y as a function of x. Show that

$$\frac{d^2y}{dx^2} = -\frac{F_y^2F_{xx} - 2F_xF_yF_{xy} + F_x^2F_{yy}}{F_y^3} \quad \text{where } F_y \neq 0$$

59. Use the result of Problem 58 to find d^2y/dx^2 if $x^3 + 3xy - y^3 = 6$.

REVIEW EXERCISES

In Problems 1–3 sketch each cylindrical surface.

1. $y = \sin x$

2. $y = \ln z$

3. $z = e^x$

4. Find $\partial z/\partial y$ and $\partial z/\partial x$ for $z = x^y$.

5. Show that $x(\partial z/\partial x) + y(\partial z/\partial y) = 5z$ for $z = x^3y^2 - 2xy^4 + 3x^2y^3$.

6. Show that $x(\partial z/\partial x) + y(\partial z/\partial y) = z$ for $z = (xy)/(x + y)$.

7. Find $(\partial z/\partial x)^2 + (\partial z/\partial y)^2$ for $z = e^x \sin y + e^y \sin x$.

8. For the ellipsoid $(x^2/24) + (y^2/12) + (z^2/6) = 1$, find the slope of the curve in the section made by the plane $y = 1$, where $x = 4$ and z is positive.

9. The volume V of a gas varies directly with the temperature T and inversely with the pressure P. Find $\partial V/\partial T$ and $\partial V/\partial P$.

10. Let $f(x, y) = (3xy^2)/(x^2 + y^4)$. Show that

$$\lim_{(x, y)\to(0, 0)} f(x, y) = 0$$

for all linear approaches, $y = mx$, to the origin. Find $\lim_{(x, y)\to(0, 0)} f(x, y)$ along the parabola $x = y^2$. What does this show?

11. Where is $z = x^2 + y^2$ continuous?

12. Show that $\displaystyle\lim_{(x, y, z)\to(0, 0, 0)} \frac{4xy}{x^2 + y^2 + z^2}$ does not exist.

13. The function $z = \dfrac{xy}{x^2 + y^2}$ is continuous everywhere except at the origin. Can it be defined there to make it continuous?

In Problems 14–17 find all second-order partial derivatives.

14. $w = e^{xyz}$

15. $w = ze^{xy}$

16. $F(x, y, z) = e^x \sin y + e^y \sin z$

17. $w = z \tan^{-1}(y/x)$

18. If $f(x, y) = \sqrt{x^2 - y^2}$, find $f_x(2, 1)$ and $f_y(2, -1)$.

19. If $F(x, y) = e^x \sin y$, find $F_x(0, \pi/6)$ and $F_y(0, \pi/6)$.

20. Show that there does not exist a function $f(x, y)$ such that $f_x(x, y) = 2x - y$ and $f_y(x, y) = x - 2y$.

21. Let

$$f(x,y) = \begin{cases} \dfrac{\sin(x^2 + y^2)}{x^2 + y^2} & \text{if } (x, y) \neq (0, 0) \\ 1 & \text{if } (x, y) = (0, 0) \end{cases}$$

Is f continuous at $(0, 0)$? Why or why not?

22. Let

$$f(x, y) = \begin{cases} \dfrac{\sin(x^2 - y^2)}{x^2 + y^2} & \text{if } (x, y) \neq (0, 0) \\ 1 & \text{if } (x, y) = (0, 0) \end{cases}$$

Is f continuous at $(0, 0)$? Why or why not?

23. Show that $xf_x + yf_y = 0$ for $f(x, y) = \sin^{-1}(y/x)$.

24. Show that $x^2f_{xx} + 2xyf_{xy} + y^2f_{yy} = 0$ for $f(x, y) = xy/(x + y)$.

25. Find f_x and f_y if $f(x, y) = \sqrt{\dfrac{x + 2y}{3x - y}}$. (Use logarithmic differentiation.)

26. Find f_x and f_y if $f(x, y) = \dfrac{(x + y)(x - y)}{(2x + 3y)(3x - 2y)}$.

27. Find $\partial a/\partial b$, $\partial a/\partial c$, $\partial a/\partial A$, $\partial c/\partial A$ for the law of cosines: $a^2 = b^2 + c^2 - 2bc \cos A$.

In Problems 28–31 find the total differential.

28. $z = x\sqrt{1-y^2}$

29. $z = \sin^{-1}(x/y)$

30. $w = ze^{xy}$

31. $w = \ln(xyz)$

32. Use the differential of $f(x, y) = y^2\cos x$ to find an approximate value of $f(0.05, 1.98)$. (Compare your answer with a calculator result.)

33. The electrical resistance R of a wire is proportional to the length L of the wire and inversely proportional to the square of its diameter D. If L has a 1% error and D has a 2% error, what is the maximum percentage error in the computations of R?

34. If $x = r\cos\theta$ and $y = r\sin\theta$, show that

$$\begin{vmatrix} \dfrac{\partial x}{\partial r} & \dfrac{\partial x}{\partial \theta} \\ \dfrac{\partial y}{\partial r} & \dfrac{\partial y}{\partial \theta} \end{vmatrix} = r \quad \text{and} \quad \begin{vmatrix} \dfrac{\partial r}{\partial x} & \dfrac{\partial r}{\partial y} \\ \dfrac{\partial \theta}{\partial x} & \dfrac{\partial \theta}{\partial y} \end{vmatrix} = \frac{1}{r}$$

35. Find $\partial z/\partial x$ and $\partial z/\partial y$ for

$$F(x, y, z) = x^2 + y^2 - 2xyz = 0$$

36. Find $\partial z/\partial x$ and $\partial z/\partial y$ for

$$F(x, y, z) = 2x\sin y + 2y\sin x + 2xyz = 0$$

37. Find dz/dt for $z = \sin(xy) - x\sin y$ if $x = e^t$, $y = te^t$.

38. Find $\dfrac{dw}{dt}$ for $w = \dfrac{x}{y} + \dfrac{y}{z} + \dfrac{z}{x}$ if $x = \dfrac{1}{t}$, $y = \dfrac{1}{t^2}$, $z = \dfrac{1}{t^3}$.

39. Find $\partial u/\partial r$ and $\partial u/\partial s$ in terms of r and s for $u = xy + yz - xz$, $x = r + s$, $y = rs$, $z = s$.

40. Find $\partial u/\partial r$ and $\partial u/\partial s$ in terms of r and s for $u = \sqrt{x^2+y^2+z^2}$, $x = r\cos s$, $y = r\sin s$, $z = \sqrt{r^2+s^2}$.

CHALLENGE EXERCISES

1. Describe the set of points (x, y, z) satisfying the conditions $x^2 + y^2 + z^2 < 1$, $x^2 + y^2 < z^2$, and $z > 0$.

2. Sketch the surfaces $x^2 + y^2 + z^2 = 4$ and $z = \frac{1}{3}(x^2 + y^2)$ and their curve of intersection above the xy-plane. What is the length of this curve?

3. Find f_x and f_y at $(0, 0)$ if

$$f(x, y) = \begin{cases} e^{-1/(x^2+y^2)} & \text{if } (x, y) \neq (0, 0) \\ 0 & \text{if } (x, y) = (0, 0) \end{cases}$$

4. Calculate f_x, f_y, f_{xx}, f_{yy}, and f_{xy} for $f(x, y) = (xy)^{xy}$. What is the domain of f?

5. Show that $f(r, \theta) = r^n\sin n\theta$ satisfies

$$f_{rr} + (1/r)f_r + (1/r^2)f_{\theta\theta} = 0$$

(This is Laplace's equation in polar coordinates.)

6. Show that the function below has first partial derivatives at all points.

$$f(x, y) = \begin{cases} \dfrac{x^3 - y^3}{x^2 + y^2} & \text{if } (x, y) \neq (0, 0) \\ 0 & \text{if } (x, y) = (0, 0) \end{cases}$$

7. Let $u = r^m\cos m\theta$. Show that

$$\frac{\partial^2 u}{\partial r^2} + \frac{1}{r^2}\left(\frac{\partial^2 u}{\partial \theta^2}\right) + \frac{1}{r}\left(\frac{\partial u}{\partial r}\right) = 0 \quad \text{for all } m$$

8. Find the total differential of $f(x, y, z) = x^{y^z}$ and $g(x, y, z) = (x^y)^z$.

9. Find symmetric equations of the tangent lines at $(x_0, y_0, f(x_0, y_0))$ to the curve of intersection of $z = f(x, y)$ and $y = y_0$, and the curve of intersection of $z = f(x, y)$ and $x = x_0$. Write an equation of the plane determined by these two lines. What is the geometric relationship of this plane to the surface $z = f(x, y)$?

10. Let

$$f(t, x) = \int_0^{x/2\sqrt{\lambda t}} e^{-u^2}\, du$$

Show that $f_t = \lambda f_{xx}$. [*Hint:* Use the chain rule to show that $(d/dx)\int_0^{g(x)} h(u)\, du = h(g(x))g'(x)$.]

Finding Partial Derivatives on the HP-28*

This vignette extends the one-step method for calculating derivatives, which was introduced in the vignette at the end of Chapter 3, Finding Derivatives on the HP-28, to partial derivatives of functions of two or more variables. Partial derivatives are found in essentially the same way as ordinary derivatives on the HP. It is helpful, however, to store the original function so that it can be reused to compute the partial derivative for each independent variable. We begin by revisiting Problem 1 from Exercise 16.3.

EXAMPLE 1

Find f_x and f_y for $f(x, y) = x^2y + 6y^2$.

Solution

Turn on the HP. Clear the stack by pressing [ ] [CLEAR], and purge the memories X and Y by pressing ['] [X] [ ] [PURGE] ['] [Y] [ ] [PURGE]. Then store the expression for the function in F:

['] [X] [ ] [^] [2] [×] [Y] [+] [6] [×] [Y] [ ] [^] [2] [ENTER] ['] [F] [STO]

To find f_x, key in

['] [F] [ENTER] ['] [X] [ENTER] [ ] [d/dx]

To find f_y, key in

['] [F] [ENTER] ['] [Y] [ENTER] [ ] [d/dx]

The stack should look like this:

```
2:          '2*X*Y'
1:   'X^2+6*(2*Y)'
```

So $f_x = 2xy$ and $f_y = x^2 + 12y$.

Often we are interested in the value of a partial derivative at a particular point. Such a value can be found using the HP. Instead of purging X and Y in Example 1, we could have stored values for these variables. For instance, suppose you wished to evaluate the first partial derivative of the function f with respect to x at $(x, y) = (1, 2)$—that is, to find $f_x(1, 2)$. Simply store 1 in X and 2 in Y by keying in

[1] [ENTER] ['] [X] [STO] [2] [ENTER] ['] [Y] [STO]

and then evaluate $f_x(1, 2)$ by pressing

['] [F] [ENTER] ['] [X] [ENTER] [ ] [d/dx]

The answer should be 4 because $f_x = 2xy$ implies that $f_x(1, 2) = 2(1)(2) = 4$.

* Gregory D. Foley
The Ohio State University

This shows why it is important to purge the independent variables if you want to find a general partial derivative: the HP will automatically assign within the partial derivative expressions any values stored for the variables. To see whether you have previously stored a value for a particular variable, check your USER menu. If a variable is listed there, you have assigned it a value. To check the value of any variable, use the VISIT command. In many cases it may be easier just to purge the variable or to store the value or algebraic expression desired.

Next we consider how to compute higher-order partial derivatives using the HP-28. In this case, we save the intermediate answers; that is, we store the first-order partial derivatives as variables. Problem 15 from Exercise 16.3 serves as the example.

EXAMPLE 2

Compute f_{xx}, f_{xy}, f_{yx}, and f_{yy} for $f(x, y) = 6x^2 - 8xy + 9y^2$.

Solution

As before, you should clear the stack and purge the memories X and Y if they appear on your USER menu. Then, store the expression for the function in F:

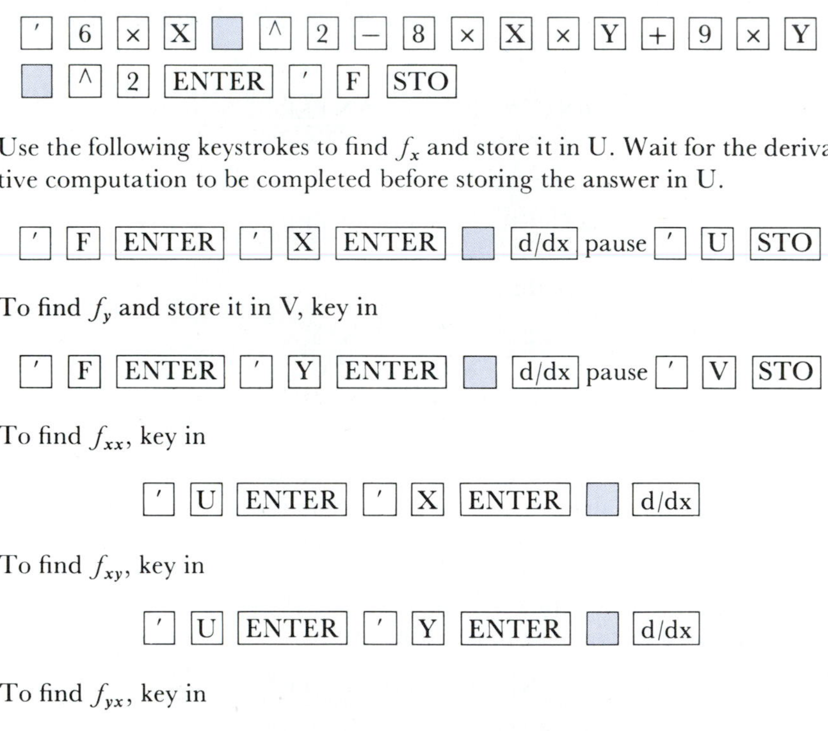

[′] [6] [×] [X] [ ] [^] [2] [−] [8] [×] [X] [×] [Y] [+] [9] [×] [Y] [ ] [^] [2] [ENTER] [′] [F] [STO]

Use the following keystrokes to find f_x and store it in U. Wait for the derivative computation to be completed before storing the answer in U.

[′] [F] [ENTER] [′] [X] [ENTER] [ ] [d/dx] pause [′] [U] [STO]

To find f_y and store it in V, key in

[′] [F] [ENTER] [′] [Y] [ENTER] [ ] [d/dx] pause [′] [V] [STO]

To find f_{xx}, key in

[′] [U] [ENTER] [′] [X] [ENTER] [ ] [d/dx]

To find f_{xy}, key in

[′] [U] [ENTER] [′] [Y] [ENTER] [ ] [d/dx]

To find f_{yx}, key in

[′] [V] [ENTER] [′] [X] [ENTER] [ ] [d/dx]

To find f_{yy}, key in

[′] [V] [ENTER] [′] [Y] [ENTER] [ ] [d/dx]

The stack should look like this:

```
4:   12
3:   -8
2:   -8
1:   18
```

So $f_{xx} = 12$, $f_{xy} = f_{yx} = -8$, and $f_{yy} = 18$.

Even complicated chain rule problems, like those found in Section 16.5, can be solved on the HP-28, although the HP-28C does not have enough memory to solve many such problems. The following example can be done on either the HP-28C or the HP-28S.

EXAMPLE 3

Find dz/dt given that $z = e^u \sin v$, $u = \sqrt{t}$, and $v = \pi t$.

Solution

Purge T. Store 'EXP(U)*SIN(V)' in Z; store '√T' in U, and store 'π*T' in V. Then to find dz/dt, key in

['] [Z] [ENTER] ['] [T] [ENTER] [ ] [d/dx]

The stack should look like this:

```
1:  'INV(2*√T)*EXP(√T)*
    SIN(π*T)+EXP(√T)*(
    COS(π*T)*π)'
```

So $dz/dt = \dfrac{1}{2\sqrt{t}} e^{\sqrt{t}} \sin \pi t + \pi e^{\sqrt{t}} \cos \pi t.$

EXERCISES

1–34. Solve Problems 1–34 from Exercise 16.3 using the HP-28.

35–66. Solve Problems 1–32 from Exercise 16.5 using the HP-28.

17 Directional Derivative, Gradient, and Extrema

17.1 Directional Derivative; Gradient

Recall that the partial derivatives of a function $z = f(x, y)$ at (x_0, y_0) have been defined by

$$f_x(x_0, y_0) = \lim_{\Delta x \to 0} \frac{f(x_0 + \Delta x, y_0) - f(x_0, y_0)}{\Delta x}$$

$$f_y(x_0, y_0) = \lim_{\Delta y \to 0} \frac{f(x_0, y_0 + \Delta y) - f(x_0, y_0)}{\Delta y}$$

where (x_0, y_0) is an interior point of the domain of f. The partial derivative $f_x(x_0, y_0)$ equals the rate of change of f at (x_0, y_0) in a *direction parallel to the x-axis*; similarly, $f_y(x_0, y_0)$ equals the rate of change of f at (x_0, y_0) in a *direction parallel to the y-axis*. The *directional derivative*, a generalization of the partial derivative, equals the rate of change of f at (x_0, y_0) in *any chosen direction in the xy-plane*.

To lay the groundwork for the definition of the directional derivative, we first let $P_0 = (x_0, y_0)$ denote an interior point in the domain of a function $z = f(x, y)$. Then we let $\mathbf{u} = (\cos\theta)\mathbf{i} + (\sin\theta)\mathbf{j}$ denote a unit vector with initial point at P_0 so that $\mathbf{u}$ makes an angle θ with the positive x-axis (see Fig. 1). Finally, we let L denote a directed line segment with initial point P_0 in the direction of $\mathbf{u}$. We choose the point $P = (x, y)$ on L, different from P_0, so that the directed line segment $\overrightarrow{P_0P}$ is in the domain of f. If $\overrightarrow{P_0P} = t\mathbf{u}$, then the coordinates of P are $(x_0 + t\cos\theta, y_0 + t\sin\theta)$.

See Figure 2.

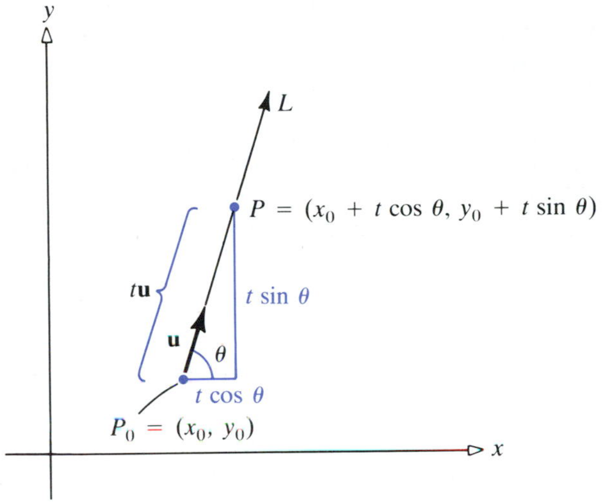

Figure 1

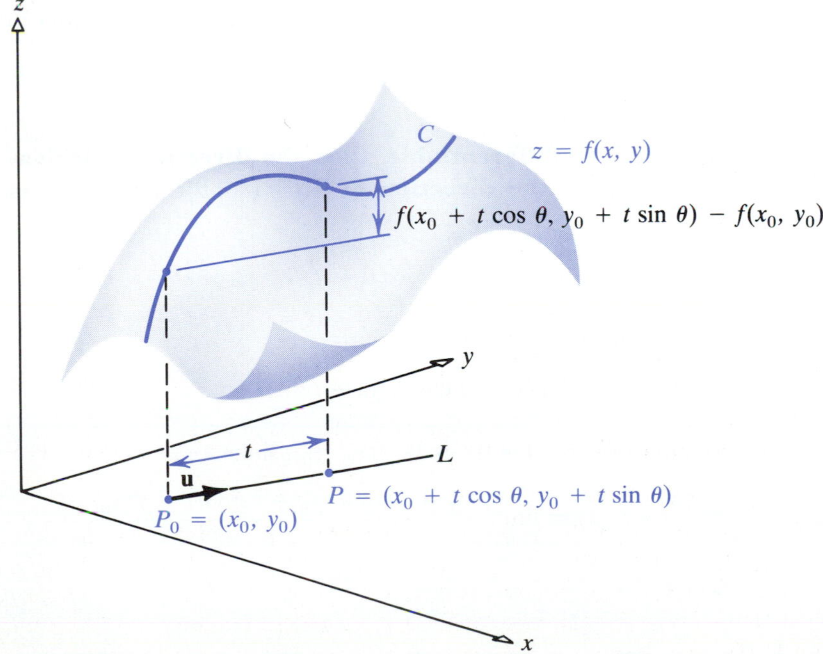

Figure 2

The average rate of change of $z = f(x, y)$ from P_0 to P is

$$\frac{f(x_0 + t\cos\theta,\ y_0 + t\sin\theta) - f(x_0, y_0)}{t}$$

By taking the limit as $t \to 0$, we obtain the rate of change of f at (x_0, y_0) in the direction of **u**, provided this limit exists. This limit, denoted by $D_{\mathbf{u}}f(x_0, y_0)$, is the *directional derivative of f at (x_0, y_0) in the direction of* **u**.

[17.1.1] DEFINITION / ***Directional Derivative.***

Let $z = f(x, y)$ denote a function of two variables whose domain is D, and let (x_0, y_0) be an interior point of D. Let $\mathbf{u} = (\cos\theta)\mathbf{i} + (\sin\theta)\mathbf{j}$ be a unit vector. The *directional derivative of f at (x_0, y_0) in the direction of* u is defined as

$$D_{\mathbf{u}}f(x_0, y_0) = \lim_{t\to 0}\frac{f(x_0 + t\cos\theta,\ y_0 + t\sin\theta) - f(x_0, y_0)}{t} \tag{1}$$

provided this limit exists.

As (1) implies, the directional derivative of f in the direction of **u** is a number. In fact, it is easy to show that the partial derivatives of f at (x_0, y_0) are special cases of the directional derivative $D_{\mathbf{u}}f(x_0, y_0)$. If we set $\theta = 0$, then $\mathbf{u} = (\cos 0)\mathbf{i} + (\sin 0)\mathbf{j} = \mathbf{i}$ is a vector parallel to the x-axis. From (1), the directional derivative of f at (x_0, y_0) in this direction is

$$D_{\mathbf{u}}f(x_0, y_0) = D_{\mathbf{i}}f(x_0, y_0) = \lim_{t\to 0}\frac{f(x_0 + t,\ y_0) - f(x_0, y_0)}{t} = f_x(x_0, y_0)$$

Similarly, if $\theta = \pi/2$, then $\mathbf{u} = \mathbf{j}$ and $D_{\mathbf{j}}f(x_0, y_0) = f_y(x_0, y_0)$.

When $z = f(x, y)$ is differentiable, a simple formula can be used to calculate directional derivatives.

[17.1.2] THEOREM

If $z = f(x, y)$ is differentiable, then the directional derivative of f at (x_0, y_0) in the direction of $\mathbf{u} = (\cos\theta)\mathbf{i} + (\sin\theta)\mathbf{j}$ is given by

$$D_{\mathbf{u}}f(x_0, y_0) = f_x(x_0, y_0)\cos\theta + f_y(x_0, y_0)\sin\theta \tag{2}$$

Proof We define the function g as

$$g(t) = f(x_0 + t\cos\theta,\ y_0 + t\sin\theta)$$

Then the derivative of g at $t = 0$ is $D_{\mathbf{u}}f(x_0, y_0)$, since

$$g'(0) = \lim_{t\to 0}\frac{g(t) - g(0)}{t} = \lim_{t\to 0}\frac{f(x_0 + t\cos\theta,\ y_0 + t\sin\theta) - f(x_0, y_0)}{t}$$

$$= D_{\mathbf{u}}f(x_0, y_0)$$

↑ By (1)

We apply the chain rule to

$$g(t) = f(x, y) \qquad x = x_0 + t\cos\theta \qquad y = y_0 + t\sin\theta$$

to get

$$g'(t) = \frac{\partial f}{\partial x}\frac{dx}{dt} + \frac{\partial f}{\partial y}\frac{dy}{dt} = \frac{\partial f}{\partial x}\cos\theta + \frac{\partial f}{\partial y}\sin\theta$$

At $t = 0$, $x = x_0$ and $y = y_0$, so

$$D_{\mathbf{u}}f(x_0, y_0) = g'(0) = f_x(x_0, y_0)\cos\theta + f_y(x_0, y_0)\sin\theta$$

EXAMPLE 1

Find the directional derivative $D_{\mathbf{u}}f(x, y)$ of $f(x, y) = x^2y + y^2$ in the direction of

$$\mathbf{u} = \left(\cos\frac{\pi}{4}\right)\mathbf{i} + \left(\sin\frac{\pi}{4}\right)\mathbf{j}$$

What is $D_{\mathbf{u}}f(1, 2)$?

Solution

The partial derivatives of f are

$$f_x(x, y) = 2xy \qquad f_y(x, y) = x^2 + 2y$$

Hence, by (2),

$$D_{\mathbf{u}}f(x, y) = (2xy)\cos\frac{\pi}{4} + (x^2 + 2y)\sin\frac{\pi}{4} = \sqrt{2}xy + \frac{\sqrt{2}}{2}(x^2 + 2y)$$

$$D_{\mathbf{u}}f(1, 2) = 2\sqrt{2} + \frac{\sqrt{2}}{2}(1 + 4) = \frac{9\sqrt{2}}{2}$$

The directional derivative of a function f in the direction of $\mathbf{a}$, where $\mathbf{a}$ is any nonzero vector, is defined as the directional derivative of f in the direction of $\mathbf{u}$, where $\mathbf{u} = \mathbf{a}/\|\mathbf{a}\|$, with the unit vector having the same direction as $\mathbf{a}$.

EXAMPLE 2

Find the directional derivative of $f(x, y) = x\sin y$ at $(2, \pi/3)$ in the direction of $\mathbf{a} = 3\mathbf{i} + 4\mathbf{j}$.

Solution

The vector $\mathbf{a}$ is not a unit vector. However, the unit vector $\mathbf{u}$ in the direction of $\mathbf{a}$ is

$$\mathbf{u} = \frac{\mathbf{a}}{\|\mathbf{a}\|} = \frac{3}{5}\mathbf{i} + \frac{4}{5}\mathbf{j}$$

The partial derivatives of f are

$$f_x(x, y) = \sin y \qquad f_y(x, y) = x \cos y$$

At $(2, \pi/3)$,

$$f_x\left(2, \frac{\pi}{3}\right) = \frac{\sqrt{3}}{2} \qquad f_y\left(2, \frac{\pi}{3}\right) = 1$$

The directional derivative of f at $(2, \pi/3)$ in the direction of $\mathbf{a} = 3\mathbf{i} + 4\mathbf{j}$ is

$$D_{\mathbf{a}}f\left(2, \frac{\pi}{3}\right) = \left(\frac{\sqrt{3}}{2}\right)\left(\frac{3}{5}\right) + (1)\left(\frac{4}{5}\right) = \frac{3\sqrt{3} + 8}{10}$$

Geometric Interpretation

The partial derivative $f_y(x_0, y_0)$ equals the slope of the tangent line to the curve of intersection of the surface $z = f(x, y)$ and the plane $x = x_0$. The directional derivative $D_{\mathbf{u}}f(x_0, y_0)$ of f at (x_0, y_0) in the direction of $\mathbf{u}$ equals the slope of the tangent line to the curve of intersection of the surface $z = f(x, y)$ and a plane: the plane through a line L in the direction of $\mathbf{u}$ and perpendicular to the xy-plane. Let's see why.

Look at Figure 3. The point $P_0 = (x_0, y_0)$ gives rise to a point $(x_0, y_0, f(x_0, y_0))$ on the surface. As we move along the line L in the direction of $\mathbf{u}$, we trace out points on a curve C that lies on the surface. This curve C is the intersection of the surface with the plane containing L that is perpendicular to the xy-plane. The point $P = (x_0 + t\cos\theta, y_0 + t\sin\theta)$ on L

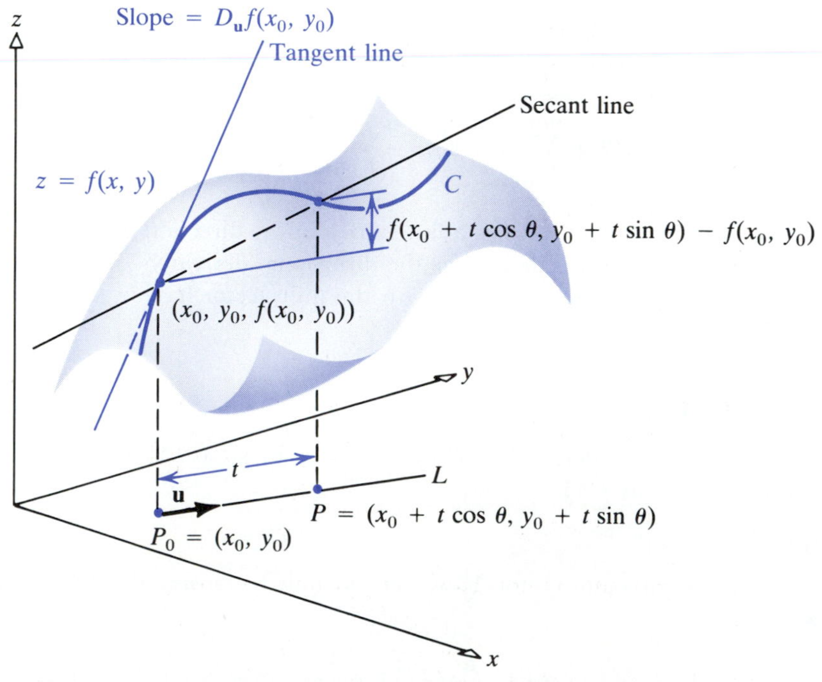

Figure 3

also gives rise to a point on C. The quantity

$$\frac{f(x_0 + t\cos\theta,\ y_0 + t\sin\theta) - f(x_0, y_0)}{t}$$

represents the slope of the secant line joining the two points on C. Its limit as $t \to 0$ is the slope of the tangent line to C at P_0.

[17.1.3] THEOREM

Let $z = f(x, y)$ denote a differentiable function. The directional derivative $D_{\mathbf{u}}f(x_0, y_0)$ of f at (x_0, y_0) in the direction of u equals the slope of the tangent line to the curve C at the point $(x_0, y_0, f(x_0, y_0))$ on the surface, where C is the intersection of the surface with the plane perpendicular to the xy-plane and containing u.

Gradient

[17.1.4] DEFINITION / *Gradient of f, ∇f.*

Let $z = f(x, y)$ denote a differentiable function. The vector $\nabla f(x, y)$, read "del f" and called the *gradient of f*, is defined* as

$$\nabla f(x, y) = f_x(x, y)\mathbf{i} + f_y(x, y)\mathbf{j} \tag{3}$$

The gradient of f and the directional derivative of f are related.

[17.1.5] THEOREM

The dot product of the vector $\nabla f(x, y)$ and the unit vector $\mathbf{u} = (\cos\theta)\mathbf{i} + (\sin\theta)\mathbf{j}$ is a scalar that is equal to the directional derivative $D_{\mathbf{u}}f(x, y)$. That is,

$$\nabla f(x, y) \cdot \mathbf{u} = D_{\mathbf{u}}f(x, y)$$

Proof We use (3) and theorem (17.1.2):

$$\nabla f(x, y) \cdot \mathbf{u} = f_x(x, y)\cos\theta + f_y(x, y)\sin\theta = D_{\mathbf{u}}f(x, y)$$

EXAMPLE 3

Find the gradient of $f(x, y) = x^3y$ at the point (2, 1) and use it to find the directional derivative of f at (2, 1) in the direction from (2, 1) to (3, 5).

Solution

The gradient of f at (x, y) is

$$\nabla f(x, y) = f_x(x, y)\mathbf{i} + f_y(x, y)\mathbf{j} = 3x^2y\mathbf{i} + x^3\mathbf{j}$$

* The abbreviation "del f" is short for "delta f," the symbol ∇ being the inverted Greek letter delta. In some books, ∇f is written as "grad f."

The gradient of f at $(2, 1)$ is

$$\nabla f(2, 1) = 12\mathbf{i} + 8\mathbf{j}$$

The unit vector $\mathbf{u}$ from $(2, 1)$ to $(3, 5)$ is

$$\mathbf{u} = \frac{\mathbf{i} + 4\mathbf{j}}{\sqrt{17}}$$

Using theorem (17.1.5), we find that the directional derivative of f at $(2, 1)$ in the direction of $\mathbf{u}$ is

$$D_{\mathbf{u}}(2, 1) = \nabla f(2, 1) \cdot \mathbf{u} = (12\mathbf{i} + 8\mathbf{j}) \cdot \frac{\mathbf{i} + 4\mathbf{j}}{\sqrt{17}} = \frac{44}{\sqrt{17}}$$

Just as the differential operator $D = d/dx$ is used to denote the operation of differentiation, the *vector differential operator* ∇ may be used to symbolize the operation

$$\nabla = \mathbf{i}\frac{\partial}{\partial x} + \mathbf{j}\frac{\partial}{\partial y}$$

It is important to remember that ∇ is an operator and has no meaning except when it operates on a function $z = f(x, y)$.

Properties of the Gradient

As we shall see, the gradient of a differentiable function $z = f(x, y)$ provides information about the direction to move so that the value of z increases most rapidly.

[17.1.6] THEOREM | ***Properties of the Gradient.***

Let $z = f(x, y)$ denote a function that is differentiable at (x_0, y_0).

(a) If $\nabla f(x_0, y_0) = 0$, then $D_{\mathbf{u}} f(x_0, y_0) = 0$ for all directions u.
(b) If $\nabla f(x_0, y_0) \neq 0$, then the directional derivative of f at (x_0, y_0) with the largest value is the one in the direction of $\nabla f(x_0, y_0)$. This largest value is $\|\nabla f(x_0, y_0)\|$.
(c) If $\nabla f(x_0, y_0) \neq 0$, then the directional derivative of f at (x_0, y_0) with the smallest value is the one in the direction of $-\nabla f(x_0, y_0)$. This smallest value is $-\|\nabla f(x_0, y_0)\|$.

Proof We use theorem (17.1.5).
(a) If $\nabla f(x_0, y_0) = 0$, then for any direction $\mathbf{u}$, we have

$$D_{\mathbf{u}} f(x_0, y_0) = \nabla f(x_0, y_0) \cdot \mathbf{u} = \mathbf{0} \cdot \mathbf{u} = 0$$

(b) and (c) Suppose $\nabla f(x_0, y_0) \neq 0$. We let ϕ denote the angle between the unit vector $\mathbf{u}$ and the gradient $\nabla f(x_0, y_0)$. Then

$$\begin{aligned} D_{\mathbf{u}} f(x_0, y_0) &= \nabla f(x_0, y_0) \cdot \mathbf{u} = \|\nabla f(x_0, y_0)\| \|\mathbf{u}\| \cos \phi \\ &= \|\nabla f(x_0, y_0)\| \cos \phi \end{aligned} \tag{4}$$

The value of $D_{\mathbf{u}}f(x_0, y_0)$ is largest when $\cos\phi = 1$—that is, when the angle ϕ between $\mathbf{u}$ and ∇f is 0. Hence, the directional derivative $D_{\mathbf{u}}f(x_0, y_0)$ at (x_0, y_0) is largest when the direction of $\mathbf{u}$ is the same as that of ∇f at (x_0, y_0). It follows from (4) that if $\mathbf{u}$ has the same direction as $\nabla f(x_0, y_0)$, the largest value of the directional derivative $D_{\mathbf{u}}f(x_0, y_0)$ is $\|\nabla f(x_0, y_0)\|$. Similarly, the smallest value of $D_{\mathbf{u}}f(x_0, y_0)$ occurs when $\cos\phi = -1$—that is, when $\mathbf{u}$ and ∇f are opposite in direction $(\phi = \pi)$. Under these conditions, the smallest value is $-\|\nabla f(x_0, y_0)\|$.

EXAMPLE 4

Find the direction for which the directional derivative of $f(x, y) = x^2 - xy + y^2$ at $(1, -2)$ is largest and find that largest value.

Solution

From theorem (17.1.6), we need to calculate the gradient of f at $(1, -2)$:

$$\nabla f(x, y) = (2x - y)\mathbf{i} + (2y - x)\mathbf{j} \qquad \nabla f(1, -2) = 4\mathbf{i} - 5\mathbf{j}$$

The direction for which $D_{\mathbf{u}}f(1, -2)$ is largest is $4\mathbf{i} - 5\mathbf{j}$. The largest value of the directional derivative at $(1, -2)$ equals the magnitude of the gradient—namely, $\|\nabla f(1, -2)\| = \sqrt{16 + 25} = \sqrt{41}$.

The directional derivative $D_{\mathbf{u}}f(x_0, y_0)$ equals the rate of change of f at (x_0, y_0) in the direction of $\mathbf{u}$. When $\mathbf{u}$ has the same direction as $\nabla f(x_0, y_0)$, the rate at which f changes at (x_0, y_0) is largest. It follows that $f(x, y)$ will increase most rapidly in this direction. Thus, we may restate theorem (17.1.6) in the following way:

[17.1.7] THEOREM

The value of $z = f(x, y)$ at (x_0, y_0) increases most rapidly in the direction of $\nabla f(x_0, y_0)$ and decreases most rapidly in the direction of $-\nabla f(x_0, y_0)$. The rate of change of z at (x_0, y_0) is 0 for directions orthogonal to $\nabla f(x_0, y_0)$.

The second statement is a consequence of (4). The rate of change of z in the direction of $\mathbf{u}$ is also when $D_{\mathbf{u}}f(x_0, y_0) = 0$. This occurs when $\cos\phi = 0$, that is, when $\phi = \pi/2$. In other words, this occurs for directions orthogonal to $\nabla f(x_0, y_0)$.

EXAMPLE 5

A metal plate is situated on the xy-plane in such a way that the temperature T at any point $P = (x, y)$ is inversely proportional to the distance of P from $(0, 0)$. If the temperature at $(-3, 4)$ equals 50°C, in what direction will the temperature at $(-3, 4)$ increase the fastest? In what direction is the rate of change of T at $(-3, 4)$ equal to 0?

Solution

The temperature T at any point (x, y) is given by

$$T(x, y) = \frac{k}{\sqrt{x^2 + y^2}}$$

where k is the constant of proportionality. Since $T = 50$ when $(x, y) = (-3, 4)$, we find that $k = T\sqrt{x^2 + y^2} = 50(5) = 250$. The gradient of T is

$$\nabla T(x, y) = T_x(x, y)\mathbf{i} + T_y(x, y)\mathbf{j} = \frac{-250x}{(x^2 + y^2)^{3/2}}\mathbf{i} + \frac{-250y}{(x^2 + y^2)^{3/2}}\mathbf{j}$$

$$\nabla T(-3, 4) = 6\mathbf{i} - 8\mathbf{j}$$

By theorem (17.1.7), the direction along which the temperature increases the fastest is $6\mathbf{i} - 8\mathbf{j}$. The temperature decreases the fastest in the opposite direction—that is, in the direction of $-6\mathbf{i} + 8\mathbf{j}$. Along either direction perpendicular to $6\mathbf{i} - 8\mathbf{j}$, the rate of change of the temperature is 0. Such a direction is therefore a tangent to an isothermal curve (a curve of constant temperature).

Let's look at another interpretation of theorem (17.1.7): Suppose the surface $z = f(x, y)$ represents a mountain whose elevation is z (see Fig. 4). The level curves of f correspond to contours along which the elevation remains fixed—that is, along which there is no increase or decrease in the value of z. The direction along which the elevation increases most rapidly is perpendicular to the level curves. In other words, the most direct route to the summit is in the direction of ∇f, which is perpendicular to the level curves.* The shortest route down the mountain is in the direction of $-\nabla f$, which is also perpendicular to the level curves. A stream of water would flow this way down the mountain.

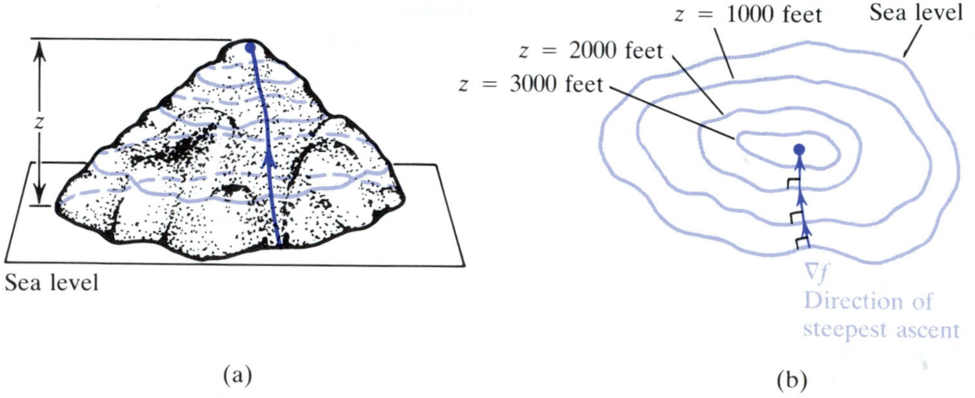

Figure 4

* The *grade* of a mountain (or hill) is a measure of its steepness—hence, the name *gradient* given to ∇f, the direction of steepest ascent.

We formalize this discussion in the following theorem:

[17.1.8] THEOREM

Let $z = f(x, y)$ denote a function that is differentiable at a point $P_0 = (x_0, y_0)$. If $\nabla f(x_0, y_0) \neq 0$, then $\nabla f(x_0, y_0)$ is normal to the level curve of f at P_0.

Proof See Figure 5. Let $f(x, y) = k$ be the level curve through P_0. Suppose this level curve is represented parametrically by $x = x(t)$ and $y = y(t)$, with $x(t_0) = x_0$ and $y(t_0) = y_0$. Then

$$f(x(t), y(t)) = k$$

By differentiating with respect to t, we have

$$f_x(x(t), y(t))x'(t) + f_y(x(t), y(t))y'(t) = 0$$

or, equivalently,

$$\nabla f(x, y) \cdot [x'(t)\mathbf{i} + y'(t)\mathbf{j}] = 0$$

In particular, when $t = t_0$, this equation says that $\nabla f(x_0, y_0)$ is normal to the level curve of f at (x_0, y_0).

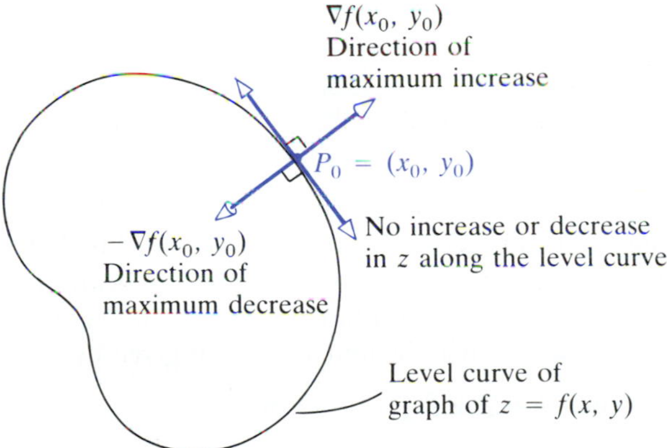

Figure 5

EXAMPLE 6

For the function $f(x, y) = \sqrt{x^2 + y^2}$, sketch the level curve passing through the point (3, 4) and sketch the gradient at this point.

Solution

Since $z = \sqrt{x^2 + y^2}$ is a cone, the level curves of this surface are circles. Because $f(3, 4) = \sqrt{9 + 16} = 5$, the level curve through (3, 4) is the circle $x^2 + y^2 = 25$. Since

$$\nabla f(x, y) = \frac{x}{\sqrt{x^2 + y^2}}\mathbf{i} + \frac{y}{\sqrt{x^2 + y^2}}\mathbf{j}$$

the gradient at (3, 4) is

$$\nabla f(3, 4) = \tfrac{3}{5}\mathbf{i} + \tfrac{4}{5}\mathbf{j}$$

See Figure 6.

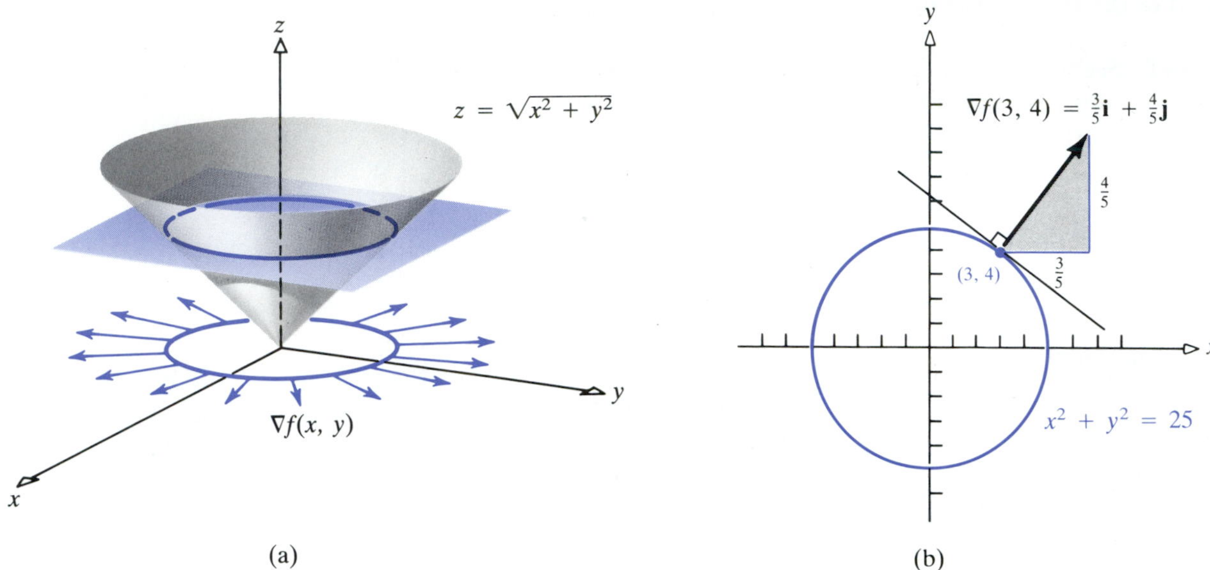

Figure 6

Functions of Three Variables

We conclude this section with an informal extension of the directional derivative and the gradient to functions of three variables. We begin with the directional derivative $D_{\mathbf{u}}f(x_0, y_0, z_0)$ of a function $w = f(x, y, z)$ at (x_0, y_0, z_0) in the direction of $\mathbf{u}$ in which the unit vector $\mathbf{u}$ is given by

$$\mathbf{u} = (\cos \alpha)\mathbf{i} + (\cos \beta)\mathbf{j} + (\cos \gamma)\mathbf{k} \tag{5}$$

where α, β, and γ are the direction angles of $\mathbf{u}$ and $\cos^2\alpha + \cos^2\beta + \cos^2\gamma = 1$. Then

$$D_{\mathbf{u}}f(x_0, y_0, z_0) = f_x(x_0, y_0, z_0)\cos \alpha + f_y(x_0, y_0, z_0)\cos \beta + f_z(x_0, y_0, z_0)\cos \gamma \tag{6}$$

is the rate of change of f at (x_0, y_0, z_0) in the direction of $\mathbf{u}$.

The gradient of f at (x_0, y_0, z_0) is the vector

$$\nabla f(x_0, y_0, z_0) = f_x(x_0, y_0, z_0)\mathbf{i} + f_y(x_0, y_0, z_0)\mathbf{j} + f_z(x_0, y_0, z_0)\mathbf{k} \tag{7}$$

From (5), (6), and (7), we have the formula

$$D_{\mathbf{u}}f(x_0, y_0, z_0) = \nabla f(x_0, y_0, z_0) \cdot \mathbf{u} = \|\nabla f(x_0, y_0, z_0)\|\cos \phi$$

where ϕ is the angle between $\nabla f(x_0, y_0, z_0)$ and $\mathbf{u}$.

As we saw with functions of two variables, the gradient $\nabla f(x_0, y_0, z_0)$ is the direction along which the value of f increases most rapidly, and $-\nabla f(x_0, y_0, z_0)$ is the direction along which the value of f decreases most rapidly. For example, suppose $T = T(x, y, z)$ is the temperature of a homogeneous body at the point (x, y, z). At the point (x_0, y_0, z_0) on the body, heat will flow in the direction of greatest decrease in temperature—namely, $-\nabla T(x_0, y_0, z_0)$—and this direction is perpendicular to the level surface through (x_0, y_0, z_0).*

We now state the analog of theorem (17.1.8) to functions of three variables.

[17.1.9] THEOREM

Let $w = f(x, y, z)$ denote a function that is differentiable at a point $P_0 = (x_0, y_0, z_0)$. If $\nabla f(x_0, y_0, z_0) \neq 0$, the gradient $\nabla f(x_0, y_0, z_0)$ is normal to the level surface S of f through P_0.

Proof Suppose that $f(x, y, z) = k$ is the equation of the level surface. Suppose C is any curve that lies on S. We let C be defined by the parametric equations $x = x(t)$, $y = y(t)$, $z = z(t)$. Suppose that $x_0 = x(t_0)$, $y_0 = y(t_0)$, $z_0 = z(t_0)$. Since C lies on S, we have $f(x(t), y(t), z(t)) = k$ for all t. Thus, we have

$$f_x(x(t), y(t), z(t))x'(t) + f_y(x(t), y(t), z(t))y'(t) + f_z(x(t), y(t), z(t))z'(t) = 0$$

or, equivalently,

$$\nabla f(x, y, z) \cdot [x'(t)\mathbf{i} + y'(t)\mathbf{j} + z'(t)\mathbf{k}] = 0$$

In particular,

$$\nabla f(x_0, y_0, z_0) \cdot [x'(t_0)\mathbf{i} + y'(t_0)\mathbf{j} + z'(t_0)\mathbf{k}] = 0$$

Since C is an arbitrary curve on S, it follows that $\nabla f(x_0, y_0, z_0)$ is orthogonal to the tangent vector of every curve lying on S that passes through (x_0, y_0, z_0). Thus, ∇f is normal to S at P_0. ∎

EXERCISE 17.1

In Problems 1–12 find the directional derivative of each function at the indicated point and in the indicated direction.

1. $f(x, y) = xy^2 + x^2$ at $(-1, 2)$ in the direction $\theta = \pi/3$

2. $f(x, y) = 3xy + y^2$ at $(2, 1)$ in the direction $\theta = \pi/4$

3. $f(x, y) = 2xy - y^2$ at $(-1, 3)$ in the direction $\theta = 2\pi/3$

4. $f(x, y) = 2xy + x^2$ at $(0, 3)$ in the direction $\theta = 4\pi/3$

5. $f(x, y) = xe^y + ye^x$ at $(0, 0)$ in the direction $\theta = \pi/6$

6. $f(x, y) = x \ln y$ at $(5, 1)$ in the direction $\theta = \pi/4$

* Recall that level surfaces were defined in Section 16.1.

7. $f(x, y) = \tan^{-1}(y/x)$ at $(1, 1)$ in the direction of $\mathbf{u} = (3\mathbf{i} - 4\mathbf{j})/5$

8. $f(x, y) = \ln\sqrt{x^2 + y^2}$ at $(3, 4)$ in the direction of $\mathbf{u} = (5\mathbf{i} + 12\mathbf{j})/13$

9. $f(x, y, z) = z \tan^{-1}(y/x)$ at $(1, 1, 3)$ in the direction of $\mathbf{a} = \mathbf{i} + \mathbf{j} - \mathbf{k}$

10. $f(x, y, z) = \sqrt{x^2 + y^2 + z^2}$ at $(3, 4, 0)$ in the direction of $\mathbf{a} = \mathbf{i} - \mathbf{j} + \mathbf{k}$

11. $f(x, y, z) = xe^{yz}$ at $(1, 0, 1)$ in the direction of $\mathbf{a} = 2\mathbf{i} + \mathbf{j}$

12. $f(x, y, z) = z \ln(x/y)$ at $(1, 1, 2)$ in the direction of $\mathbf{a} = \mathbf{j} + \mathbf{k}$

In Problems 13–28 find the gradient of f at the given point P. Use it to find the directional derivative of f in the direction from P to Q.

13. $f(x, y) = xy^2 + x^2$; $P = (1, 2)$, $Q = (2, 4)$

14. $f(x, y) = 2xy + x^2$; $P = (-1, 1)$, $Q = (1, 2)$

15. $f(x, y) = 2xy + x^2$; $P = (0, 3)$, $Q = (4, 1)$

16. $f(x, y) = 3xy + y^2$; $P = (2, 1)$, $Q = (4, 1)$

17. $f(x, y) = \tan^{-1}(y/x)$; $P = (1, 0)$, $Q = (4, \pi)$

18. $f(x, y) = \ln\sqrt{x^2 + y^2}$; $P = (3, 4)$, $Q = (0, 5)$

19. $f(x, y) = x^2e^y$; $P = (2, 0)$, $Q = (3, 0)$

20. $f(x, y) = e^{x^2+y^2}$; $P = (1, 2)$, $Q = (2, 3)$

21. $f(x, y) = \dfrac{x}{x^2 + y^2}$; $P = (1, 2)$, $Q = (2, 3)$

22. $f(x, y) = \sqrt{x^2 + y^2}$; $P = (3, 4)$, $Q = (0, 5)$

23. $f(x, y, z) = x^2y - xyz^2$; $P = (0, 1, 2)$, $Q = (1, 4, 3)$

24. $f(x, y, z) = x^2y + y^2z + z^2x$; $P = (1, 2, -1)$, $Q = (2, 0, 1)$

25. $f(x, y, z) = \sqrt{x^2 + y^2 + z^2}$; $P = (1, 2, 2)$, $Q = (2, 1, 1)$

26. $f(x, y, z) = x\sqrt{y} + y\sqrt{z}$; $P = (0, 4, 9)$, $Q = (1, 9, 4)$

27. $f(x, y, z) = xe^{yz}$; $P = (1, 0, 1)$, $Q = (2, 2, 1)$

28. $f(x, y, z) = z \ln(x/y)$; $P = (1, 1, 2)$, $Q = (2, 2, 1)$

In Problems 29–34 find the direction at P along which each function increases most rapidly. Find the rate of increase in this direction.

29. $z = xy^2 + x^2$; $P = (-1, 2)$

30. $z = 3xy + y^2$; $P = (2, 1)$

31. $z = xe^y + ye^x$; $P = (0, 0)$

32. $z = x \ln y$; $P = (5, 1)$

33. $w = z \tan^{-1}(y/x)$; $P = (1, 1, 3)$

34. $w = \sqrt{x^2 + y^2 + z^2}$; $P = (3, 4, 0)$

In Problems 35–40 sketch the level curve of f that passes through the point P, and sketch ∇f at P.

35. $f(x, y) = x^2 + y^2$: $P = (3, 4)$

36. $f(x, y) = x^2 - y^2$; $P = (2, -1)$

37. $f(x, y) = x^2 - 4y^2$; $P = (3, \sqrt{5}/2)$

38. $f(x, y) = x^2 + 4y^2$; $P = (-2, 0)$

39. $f(x, y) = x^2y$; $P = (3, \frac{1}{9})$

40. $f(x, y) = xy$; $P = (1, 1)$

41. A metal plate is situated on the xy-plane in such a way that the temperature T at any point (x, y) is given by $T = e^x \sin y + e^y \sin x$. What is the rate of change in temperature at $(0, 0)$ in the direction of $3\mathbf{i} - 4\mathbf{j}$? At $(0, 0)$, in what direction is the rate of change of temperature the greatest? In what direction is it the least? In what directions is it 0?

42. Rework Problem 41 if $T = \ln\sqrt{1 - (x^2 + y^2)}$.

43. The electrical potential V at any point (x, y) is given by $V = \ln\sqrt{x^2 + y^2}$. Find the rate of change of potential V at any point $(x, y) \neq (0, 0)$:

(a) In a direction toward $(0, 0)$

(*Problem 43 continues on page 997*)

43. *(continued)*
(b) In the two directions orthogonal to a direction toward $(0, 0)$
(c) In what direction is the rate of change in potential V largest?
(d) In what direction is the rate of change smallest?

44. The surface of a hill may be represented by the equation $z = 8 - 2x^2 - y^2$. If a freshwater spring is located at the point $(1, 2, 2)$, in what direction will the water flow?

45. Suppose that you are climbing a mountain in the shape of the surface $x^2 + y^2 - 5x + z = 0$. (The x-axis points east, the y-axis north, and the z-axis up.)

(a) Your route is such that as you pass through the point $(1, 1, 3)$ you are heading northeast. At what rate (with respect to distance) is your altitude changing at that point?
(b) Another climber at $(1, 2, 0)$ wants to move upward as quickly as possible. In what direction should he start? At what rate is his altitude changing when he starts?
(c) A third climber at $(2, 1, 5)$ wants to remain at the same altitude. In what direction(s) may she go?

46. Suppose that the temperature at each point of the coordinate plane is $T = 3x^2 + 4y^2 + 5$ (in degrees Fahrenheit).

(a) If we leave the point $(3, 4)$ heading for the point $(4, 3)$, how fast (in degrees per unit of distance) is the temperature changing as we leave?
(b) After reaching $(4, 3)$, in what direction should we go if we want to cool off as fast as possible? How fast is the temperature changing as we leave $(4, 3)$?
(c) Suppose that we want to move away from $(4, 3)$ along a path of constant temperature. What is an equation of our path and what is the temperature?
(d) A person at the origin may go in *any* direction and will experience the same rate of change of temperature as he leaves. Why? What is the rate?

47. Suppose that $z = xy^2$. In what direction(s) may we go from the point $(-1, 1)$ if we want the rate of change of z to be 2?

48. Find the direction through the point $(2, 1)$ in which the function $u = 4x^2 + 9y^2$ has the maximum rate of change. Show that this direction is that of the normal to the curve $4x^2 + 9y^2 = 25$ at the point $(2, 1)$. Also, find the value of this maximum rate of change.

49. Show that the level curves of $f(x, y) = x^2 - y^2$ are orthogonal to the level curves of $h(x, y) = xy$ for all $(x_0, y_0) \neq (0, 0)$.

50. Find a unit vector $\mathbf{u}$ that is normal to the level curve of $f(x, y) = 4x^2y$ through $P = (1, -2)$ at P.

51. Find a unit vector $\mathbf{u}$ that is normal to the level curve of $f(x, y) = 2x^2 + y^2 + 1$ through $P = (1, 1)$ at P.

52. Under the hypotheses of theorem (17.1.8), show that $D_{\mathbf{u}}f(x_0, y_0) = 0$ in the directions orthogonal to that of $\nabla f(x_0, y_0)$.

53. If $u = f(x, y)$ and $v = g(x, y)$ are differentiable, show that:

(a) $\nabla(ku) = k\nabla u$, k a constant
(b) $\nabla(u + v) = \nabla u + \nabla v$
(c) $\nabla(uv) = u\nabla v + v\nabla u$
(d) $\nabla\left(\dfrac{u}{v}\right) = \dfrac{v\nabla u - u\nabla v}{v^2}$
(e) $\nabla u^\alpha = \alpha u^{\alpha-1}\nabla u$, α a real number

54. Show that for a nonzero vector $\mathbf{a} = a_1\mathbf{i} + a_2\mathbf{j}$ and a differentiable function $z = f(x, y)$,

$$D_{\mathbf{u}}f(x, y) = \frac{a_1(\partial f/\partial x) + a_2(\partial f/\partial y)}{\sqrt{a_1^2 + a_2^2}}$$

where $\mathbf{u} = \mathbf{a}/\|\mathbf{a}\|$.

55. Suppose that the electric potential (voltage V) at each point in space is $V = e^{xyz}$ and that electric charges move in the direction of greatest *potential drop* (most rapid decrease of potential). In what direction does a charge at the point $(1, -1, 2)$ move? How fast does the potential change as the charge leaves this point?

56. A hill is in the shape of the surface $z = 5 - x^2 - 3y^2$. Standing at the point $(1, 1, 1)$, we may ski in the direction $\mathbf{i}$ or $\mathbf{j}$. Use partial derivatives to determine which direction is best if we want the descent to start out as slowly as possible.

58. The temperature at each point of region $x^2 + y^2 + z^2 \leq 9$ is $T = \sqrt{9 - x^2 - y^2 - z^2}$. If we start at the point $(0, 1, 2)$ and move across the region in a straight path ending at $(2, 1, 2)$, find the (instantaneous) rate of change of T at an arbitrary point on the path.

58. Let $F(x, y) = 0$ be the equation of a curve in the xy-plane, where F is differentiable. If (x_0, y_0) is a point on the curve, show that $\nabla F(x_0, y_0)$ is normal to the curve at (x_0, y_0).

59. Use Problem 58 to show that the tangent to the curve $F(x, y) = 0$ at (x_0, y_0) is given by $a(x - x_0) + b(y - y_0) = 0$, where $a = F_x(x_0, y_0)$ and $b = F_y(x_0, y_0)$. (Assume that a and b are not both 0.)

60. Use Problem 59 to find the following tangents to the hyperbola $x^2 - y^2 = 16$:

(a) The tangent at $(5, 3)$; check by the methods of single variable calculus, finding dy/dx at $(5, 3)$ by implicit differentiation

(b) The tangent at $(4, 0)$; note that this tangent is vertical, which requires special treatment in single variable calculus

61. Assuming that $b \neq 0$ in Problem 59, show that the slope of the tangent to the curve $F(x, y) = 0$ at (x_0, y_0) is $m = -F_x(x_0, y_0)/F_y(x_0, y_0)$. (This is a proof of (6) in Section 16.5.)

62. The temperature at any point (x, y) of a rectangular plate lying in the xy-plane is given by $T = x \sin 2y$. Find the rate of change of temperature at the point $(1, \pi/4)$ in the direction making an angle of $\pi/6$ with the x-axis.

63. If $f(x, y, z) = z^3 + 3xz - y^2$, find the directional derivatives of f at $(1, 2, 1)$ in the direction of the line $x - 1 = y - 2 = z - 1$.

64. Find the directional derivatives of $f(x, y, z) = 2xy + xz - z^2$ at $(1, -1, 0)$ in the direction of the line of intersection of the planes $2x + y - z = 0$ and $x - 2z = 0$.

65. Let $\mathbf{r} = x\mathbf{i} + y\mathbf{j}$ and $r = \|\mathbf{r}\| = \sqrt{x^2 + y^2}$.

(a) Show that: $\nabla r^n = nr^{n-2}\mathbf{r}$

(b) Show that: $\nabla g(r) = g'(r)\dfrac{\mathbf{r}}{r}$

(c) Show that: $D_{\mathbf{u}}[g(r)] = \dfrac{g'(r)}{r}(\mathbf{u} \cdot \mathbf{r})$

(d) Show that: $\nabla\left(\dfrac{x}{r}\right) = \dfrac{\mathbf{i}}{r} - \left(\dfrac{x}{r}\right)\left(\dfrac{\mathbf{r}}{r^2}\right)$

17.2 Tangent Planes

The idea behind the tangent plane to a surface at a point P_0 on the surface is based on the definition of a tangent line to a smooth curve.

[17.2.1] DEFINITION / *Tangent Plane.*

Let $P_0 = (x_0, y_0, z_0)$ denote a point on a surface S, and let C denote a smooth curve that passes through P_0 and lies entirely in S. If the tangent lines of any such curve at P_0 all lie in the same plane, then this plane is called the *tangent plane to S at P_0*.

Figure 7 illustrates the definition.

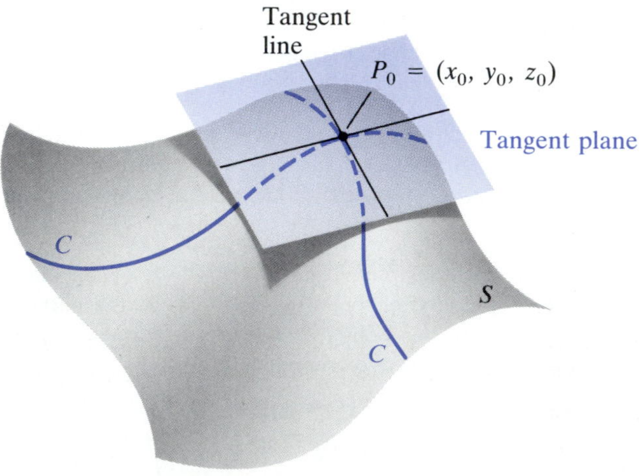

Figure 7

Consider a surface S defined by an equation of the form $F(x, y, z) = 0$ and let $P_0 = (x_0, y_0, z_0)$ be a point on S. Suppose F is differentiable at P_0 and $\nabla F(x_0, y_0, z_0) \neq \mathbf{0}$. For any smooth curve C that passes through P_0 and lies in S, the tangent lines to C at P_0 will be orthogonal to the vector $\nabla F(x_0, y_0, z_0)$ and will therefore all lie in the same plane. As a result, the surface S will have a tangent plane at P_0. See Figure 8.

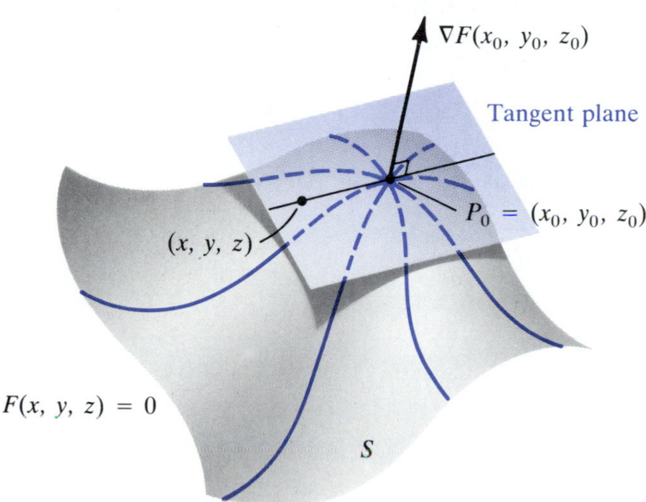

Figure 8

Suppose (x, y, z) is any point on the tangent plane to S at P_0. Since $\nabla F(x_0, y_0, z_0)$ is normal to the tangent plane at P_0, a vector equation of this tangent plane is

$$\nabla F(x_0, y_0, z_0) \cdot [(x - x_0)\mathbf{i} + (y - y_0)\mathbf{j} + (z - z_0)\mathbf{k}] = 0$$

Since

$$\nabla F(x_0, y_0, z_0) = F_x(x_0, y_0, z_0)\mathbf{i} + F_y(x_0, y_0, z_0)\mathbf{j} + F_z(x_0, y_0, z_0)\mathbf{k}$$

an equation of the tangent plane is

$$F_x(x_0, y_0, z_0)(x - x_0) + F_y(x_0, y_0, z_0)(y - y_0) + F_z(x_0, y_0, z_0)(z - z_0) = 0$$

The preceding discussion leads to the following result:

[17.2.2] THEOREM / *Equation of Tangent Plane.*

Let F denote a function that is differentiable at a point $P_0 = (x_0, y_0, z_0)$ on the surface $F(x, y, z) = 0$. If $\nabla F(x_0, y_0, z_0) \neq 0$, then the surface has a tangent plane at P_0. Moreover, an equation of the tangent plane at P_0 is

$$F_x(x_0, y_0, z_0)(x - x_0) + F_y(x_0, y_0, z_0)(y - y_0) + F_z(x_0, y_0, z_0)(z - z_0) = 0 \qquad \textbf{(1)}$$

EXAMPLE 1

Find an equation of the tangent plane to the surface $x^2 + y^2 - z^2 - 24 = 0$ at the point $(3, -4, 1)$.

Solution

Here, $F(x, y, z) = x^2 + y^2 - z^2 - 24$, so that

$$F_x(x, y, z) = 2x \qquad F_y(x, y, z) = 2y \qquad F_z(x, y, z) = -2z$$

At the point $(3, -4, 1)$, we have

$$F_x(3, -4, 1) = 6 \qquad F_y(3, -4, 1) = -8 \qquad F_z(3, -4, 1) = -2$$

An equation of the tangent plane is, therefore,

$$6(x - 3) - 8(y + 4) - 2(z - 1) = 0$$
$$3x - 4y - z - 24 = 0$$

See Figure 9 for an illustration.

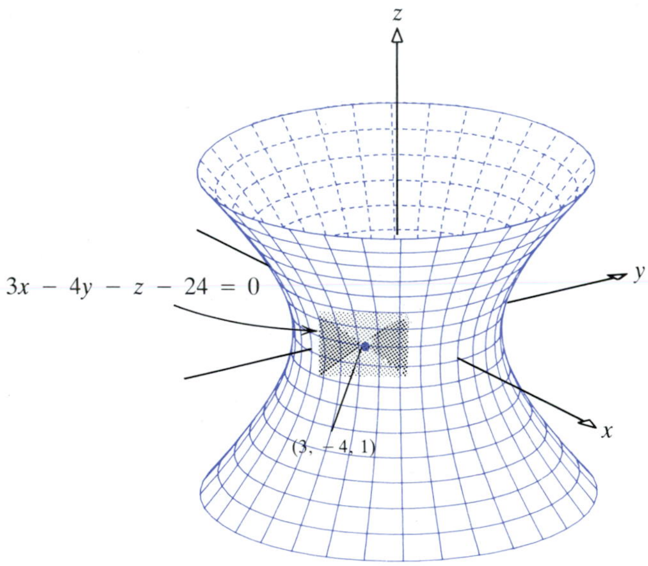

Figure 9

Normal Line

The line normal to the tangent plane of a surface $F(x, y, z) = 0$ at a point $P_0 = (x_0, y_0, z_0)$ is called the *normal line to the surface at* P_0. Since the gradient $\nabla F(x_0, y_0, z_0)$ is normal to the tangent plane at P_0, it follows that the normal line is in the direction of the gradient. By using the notation of Section 14.6, we can write the vector equation for the normal line as

$$\mathbf{r}(t) = \mathbf{r}_0 + t\nabla F(x_0, y_0, z_0)$$

where $\mathbf{r}_0 = x_0\mathbf{i} + y_0\mathbf{j} + z_0\mathbf{k}$ is the position vector of P_0.

The corresponding set of parametric equations is

$$x = x_0 + at \qquad y = y_0 + bt \qquad z = z_0 + ct$$

where $a = F_x(x_0, y_0, z_0)$, $b = F_y(x_0, y_0, z_0)$, $c = F_z(x_0, y_0, z_0)$. If $abc \neq 0$, the symmetric equations

$$\frac{x - x_0}{a} = \frac{y - y_0}{b} = \frac{z - z_0}{c} \tag{2}$$

can also be used.

EXAMPLE 2

Find symmetric equations of the normal line to the surface $x^2 + y^2 - z^2 - 24 = 0$ at the point $(3, -4, 1)$.

Solution

Using the result obtained in Example 1, we find the equation of the tangent plane:

$$3x - 4y - z - 24 = 0$$

Consequently, the symmetric equations of the normal line at $(3, -4, 1)$ are

$$\frac{x - 3}{3} = \frac{y + 4}{-4} = \frac{z - 1}{-1}$$

Surfaces Given by $z = f(x, y)$

To obtain an equation of the tangent plane to the surface $z = f(x, y)$ at (x_0, y_0, z_0), we write the equation of the surface as

$$F(x, y, z) = z - f(x, y) = 0$$

Then

$$F_x = -f_x \qquad F_y = -f_y \qquad F_z = 1$$

so that by theorem (17.2.2), an equation of the tangent plane to $z = f(x, y)$ at (x_0, y_0, z_0) is

$$-f_x(x_0, y_0)(x - x_0) - f_y(x_0, y_0)(y - y_0) + (z - z_0) = 0$$

or

$$z - z_0 = f_x(x_0, y_0)(x - x_0) + f_y(x_0, y_0)(y - y_0) \tag{3}$$

The parametric equations for the normal line are

$$x = x(t) = x_0 + tf_x(x_0, y_0)$$
$$y = y(t) = y_0 + tf_y(x_0, y_0)$$
$$z = z(t) = z_0 - t$$

In Examples 1 and 2, the same results could have been obtained by solving the equation $x^2 + y^2 - z^2 - 24 = 0$ for z and using the function

$$z = f(x, y) = \sqrt{x^2 + y^2 - 24}$$

whose graph is the part of the surface lying above the xy-plane. This would yield $f_x(3, -4) = 3$, $f_y(3, -4) = -4$, and $z = 3x - 4y - 24$ as an equation for the tangent plane at $(3, -4, 1)$. The parametric equations for the normal line would be

$$x = 3 + 3t \qquad y = -4 - 4t \qquad z = 1 - t$$

EXERCISE 17.2

In Problems 1–14 find an equation of the tangent plane and an equation of the normal line to each surface at the point indicated.

1. $x^2 + y^2 + z^2 = 14$ at $(1, -2, 3)$

2. $x^2 - y^2 + z^2 = 4$ at $(-1, 1, 2)$

3. $2x^2 + 3y^2 + z^2 = 12$ at $(2, 1, -1)$

4. $4x^2 + y^2 + 2z^2 = 7$ at $(1, 1, -1)$

5. $z^2 = x^2 + 3y^2$ at $(1, -1, -2)$

6. $x^2 + y^2 - 2z^2 = -13$ at $(2, 1, 3)$

7. $z = x^2 + y^2$ at $(-2, 1, 5)$

8. $z = 2x^2 - 3y^2$ at $(2, 1, 5)$

9. $2x^2 + y^2 = z$ at $(1, 0, 2)$

10. $2x^2 - y^2 = z$ at $(1, 0, 2)$

11. $z = e^x \cos y$ at $(0, \pi/2, 0)$

12. $z = \ln(x^2 + y^2)$ at $(1, -1, \ln 2)$

13. $x^{2/3} + y^{2/3} + z^{2/3} = 9$ at $(1, 8, -8)$

14. $x^{1/2} + y^{1/2} + z^{1/2} = 6$ at $(1, 4, 9)$

In Problems 15–20 determine those point(s) on the surface at which the tangent plane is parallel to the xy-plane.

15. $z = 6x - 4y - x^2 - 2y^2$

16. $z = 4x - 2y + x^2 + y^2$

17. $z = x^2 + 2xy + y^2$

18. $z = x^2 - 3xy + y^2$

19. $z = 2x^4 - y^2 - x^2 - 2y$

20. $z = x^2 + y^4 - 4y^2 - 2x$

21. Two surfaces are said to be *tangent at a common point* P_0 if each has the same tangent plane at P_0. Show that the surfaces $x^2 + z^2 + 4y = 0$ and $x^2 + y^2 + z^2 - 6z + 7 = 0$ are tangent at the point $(0, -1, 2)$.

22. Two surfaces are *orthogonal at a common point* P_0 if their normal lines at P_0 are orthogonal. Show that the surfaces $x^2 + y^2 + z^2 = 4$ and $x^2 + y^2 - z^2 = 0$ are orthogonal at $(0, \sqrt{2}, \sqrt{2})$. In fact, show that the surfaces are orthogonal at every point of intersection.

23. Prove that the normal lines of a sphere of radius a given by $x^2 + y^2 + z^2 = a^2$ pass through the center of the sphere.

24. Write equations of the tangent plane and normal line to $x^2 - 3y^2 - 4z^2 = 2$ at $(3, 1, 1)$.

25. Write equations of the tangent plane and normal line to $x^2 + 2y^2 - 5z^2 = 4$ at $(1, 2, 1)$.

26. Let $P_0 = (x_0, y_0, z_0)$ be a point on the sphere $x^2 + y^2 + z^2 = R^2$. Show that the normal line to the sphere at P_0 passes through the point $(-x_0, -y_0, -z_0)$.

27. Find equations for the tangent line to the curve of intersection of the surfaces $x \sin yz = 1$, $ze^{y^2-x^2} = \pi/2$ at the point $(1, 1, \pi/2)$.

28. Find the points on the surface $(x^2/9) + (y^2/4) + (z^2/36) = 1$ where the tangent plane is parallel to the plane $2x - 3y + z = 4$.

29. An Eskimo whose igloo is in the shape of the surface $z = 4 - x^2 - y^2$ wants to drill a hole perpendicular to the igloo at the point $(1, 1, 2)$ so that a flat solar panel can be bolted to the igloo. In what direction should the Eskimo drill?

30. Show that the tangent plane to the cylinder $x^2 + y^2 = a^2$ at the point (x_0, y_0, z_0) is given by the equation $x_0x + y_0y = a^2$.

31. Show that the tangent plane to the cone $(z^2/c) = (x^2/a^2) + (y^2/b^2)$ at the point $(x_0, y_0, z_0) \neq (0, 0, 0)$ is given by the equation $(z_0z/c^2) = (x_0x/a^2) + (y_0y/b^2)$. Why do we restrict the point of tangency to be different from the origin?

32. Show that at any point, the sum of the intercepts on the coordinate axes of the tangent plane to the surface $x^{1/2} + y^{1/2} + z^{1/2} = a^{1/2}$ is a constant.

33. Find equations of the tangent plane and normal line to $\sin(x + y) + \sin(y + z) = 1$ at $(0, \pi/2, \pi/2)$.

Extrema of Functions of Two Variables 17.3

We saw in Chapter 4 that an important application of the derivative is to find the extrema (maxima and minima) of a function of a single variable. In this section we find that partial derivatives are used in a similar way to find the extrema of a function of two variables.

[17.3.1] DEFINITION / ***Local Maximum; Local Minimum.***

Let $z = f(x, y)$ denote a function with domain D and let (x_0, y_0) be an interior point of D. We say that f has a *local maximum* at (x_0, y_0) if for all points (x, y) within some disk centered at (x_0, y_0), we have

$$f(x_0, y_0) \geq f(x, y)$$

We say that f has a *local minimum* at (x_0, y_0) if for all points (x, y) within some disk centered at (x_0, y_0) we have

$$f(x_0, y_0) \leq f(x, y)$$

[17.3.2] DEFINITION / ***Absolute Maximum; Absolute Minimum.***

Let $z = f(x, y)$ denote a function with domain D. We say that f has an *absolute maximum* at (x_0, y_0) if $f(x_0, y_0) \geq f(x, y)$ for all points (x, y) in the domain D of f. We say that f has an *absolute minimum* at (x_0, y_0) if $f(x_0, y_0) \leq f(x, y)$ for all points (x, y) in the domain D of f.

Collectively, we refer to the local maxima and local minima of f as *local extrema* and to the absolute maximum and absolute minimum of f as *absolute extrema.*

Figure 10 illustrates these definitions.

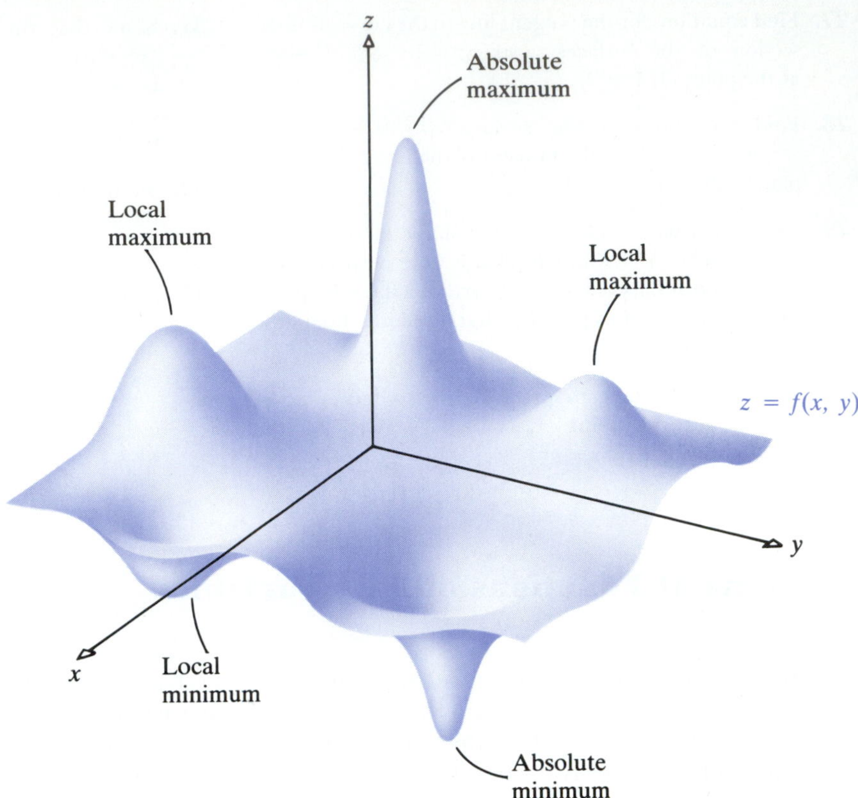

Figure 10

Local Extrema

To develop a procedure for locating local extrema, we first let $z = f(x, y)$ denote a function of two variables that is defined and continuous on an open set D. Suppose f has a local maximum at the point (x_0, y_0) in D. Then the curve that results from the intersection of the surface $z = f(x, y)$ and any plane through $(x_0, y_0, 0)$ that is perpendicular to the xy-plane will have a local maximum at (x_0, y_0), as shown in Figure 11. In particular, the curve that results from intersecting $z = f(x, y)$ with the plane $x = x_0$ has this property. This means that if $f_y(x_0, y_0)$ exists, then $f_y(x_0, y_0) = 0$. By a similar argument, we must also have $f_x(x_0, y_0) = 0$. This leads us to formulate the following necessary condition for local extrema:

[17.3.3] THEOREM / *A Necessary Condition for Local Extrema.*

Let $z = f(x, y)$ denote a function of two variables defined and continuous on an open set containing the point (x_0, y_0). Suppose $f_x(x_0, y_0)$ and $f_y(x_0, y_0)$ each exist. If f has a local extremum at (x_0, y_0), then

$$f_x(x_0, y_0) = 0 \qquad \text{and} \qquad f_y(x_0, y_0) = 0$$

From this theorem we see that local extrema of a function occur at those points at which the partial derivatives exist and are 0 simultaneously.

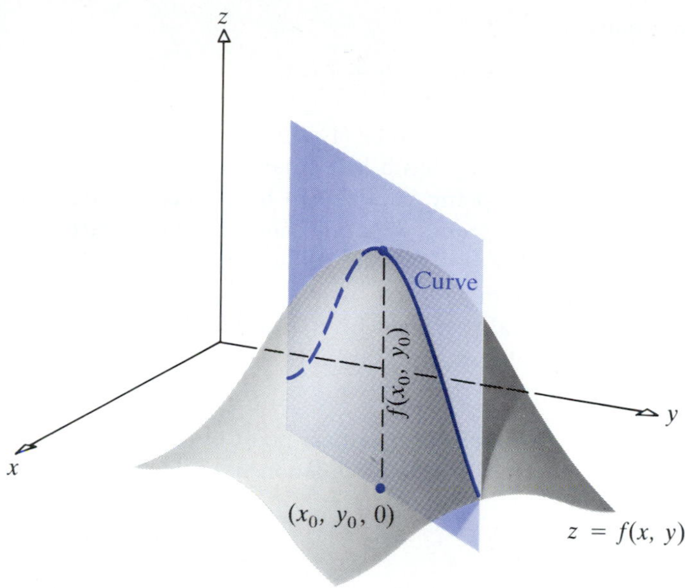

Figure 11

It can also happen that the local extrema of a function occur at points at which one or both of the partial derivatives fail to exist. This leads to the following definition:

[17.3.4] DEFINITION / *Critical Point.*

Let $z = f(x, y)$ denote a function of two variables whose domain is an open set D containing the point (x_0, y_0). The point (x_0, y_0) is a critical point of f if

(a) $f_x(x_0, y_0) = f_y(x_0, y_0) = 0$, or
(b) $f_x(x_0, y_0)$ or $f_y(x_0, y_0)$ does not exist

We can now restate theorem (17.3.3) as follows:

[17.3.5] THEOREM

Let $z = f(x, y)$ denote a function of two variables whose domain is an open set D. If f has a local extremum at (x_0, y_0), then (x_0, y_0) is a critical point of f.

EXAMPLE 1

Locate all local maxima and local minima for

$$z = f(x, y) = x^2 + y^2 - 2x + 4y$$

Solution

The partial derivatives of f are

$$f_x = 2x - 2 \qquad \text{and} \qquad f_y = 2y + 4$$

Since both of these exist for all x and y, the only critical points are those for which each partial derivative equals 0. Thus, the critical points of f obey

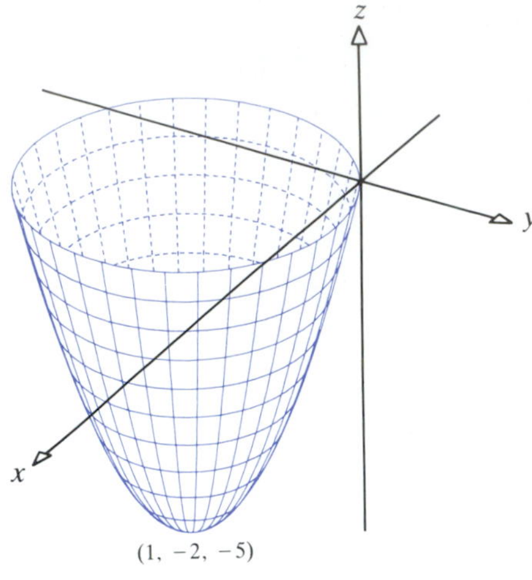

Figure 12

the system of equations

$$2x - 2 = 0 \qquad 2y + 4 = 0$$

Solving, we find the only critical point to be $(1, -2)$.

Whether the function actually has a local maximum or a local minimum or neither at the critical point $(1, -2)$ can be verified by algebraic means. For this particular function, we may complete the squares, obtaining $z = (x - 1)^2 + (y + 2)^2 - 5$, from which we conclude that at $(1, -2)$ there is a local minimum and that $z = -5$ is in fact the absolute minimum. This conclusion is also evident from a graph of the surface (an elliptic paraboloid, as seen in Fig. 12).

As with functions of one variable, a critical point of a function of two variables may not be a point at which the function has a local extremum.

EXAMPLE 2

Consider the hyperbolic paraboloid $z = f(x, y) = y^2 - x^2$. Show that $(0, 0)$ is the only critical point and that neither a local maximum nor a local minimum occurs at $(0, 0)$.

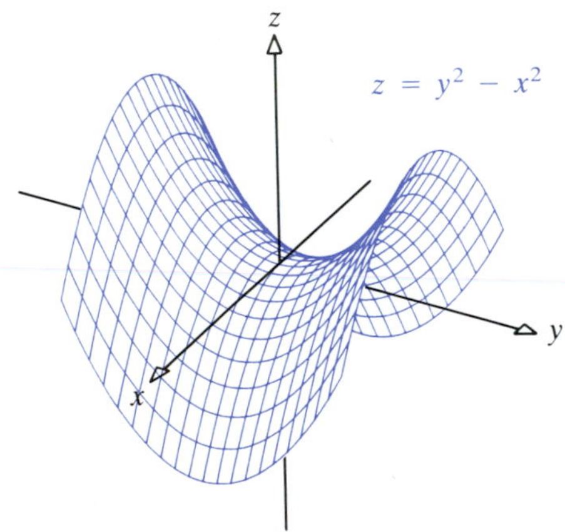

Figure 13

Solution

We have

$$f_x(x, y) = -2x \qquad \text{and} \qquad f_y(x, y) = 2y$$

Both equal 0 at $(0, 0)$; this is the only critical point. Next, we observe that $f(0, 0) = 0$, $f(x, 0) = -x^2 < 0$ for all $x \neq 0$, and $f(0, y) = y^2 > 0$ for all $y \neq 0$. See Figure 13.

If we consider the values of $f(x, 0)$ along the x-axis, we have a function of x that attains a maximum value of 0 at the origin. However, if we consider the values of $f(0, y)$ along the y-axis, we have a function that attains a minimum value at the origin. In other words, at the critical point $(0, 0)$, the function appears to have a maximum when viewed in one direction and to have a minimum when viewed in another direction and, thus, has neither a maximum nor a minimum.

We say that the function f has a *saddle point* on its graph at $(x_0, y_0, f(x_0, y_0))$ if (x_0, y_0) is a critical point and f does not have a local extremum at (x_0, y_0). Thus, the function $z = f(x, y) = y^2 - x^2$ of Example 2 has a saddle point at $(0, 0, 0)$.

In Examples 1 and 2, we were able to determine whether the critical point is a local maximum, a local minimum, or a saddle point by using algebraic and/or geometric arguments. For most functions, such arguments cannot be used. The following test, which we state without proof, provides an analytic method for determining the character of critical points. This test is analogous to the second derivative test for functions of one variable. A proof may be found in texts on advanced calculus.

[17.3.6] THEOREM / *Second Derivative Test.*

Let $z = f(x, y)$ denote a function of two variables for which the first- and second-order partial derivatives are continuous in some disk containing the point (x_0, y_0). Suppose that $f_x(x_0, y_0) = 0$ and $f_y(x_0, y_0) = 0$. Let

$$A = f_{xx}(x_0, y_0)$$

$$B = f_{xy}(x_0, y_0)$$

$$C = f_{yy}(x_0, y_0)$$

$$D = D(x_0, y_0) = [f_{xx}(x_0, y_0)][f_{yy}(x_0, y_0)] - [f_{xy}(x_0, y_0)]^2 = AC - B^2$$

(a) If $D > 0$ and $f_{xx}(x_0, y_0) > 0$, then f has a local minimum at (x_0, y_0).
(b) If $D > 0$ and $f_{xx}(x_0, y_0) < 0$, then f has a local maximum at (x_0, y_0).
(c) If $D < 0$, then f has a saddle point at $(x_0, y_0, f(x_0, y_0))$.
(d) If $D = 0$, then the test gives no information.

EXAMPLE 3

Find all local maxima and local minima for

$$z = f(x, y) = x^2 + xy + y^2 - 6x + 6$$

Solution

The critical points of f obey the equations

$$f_x(x, y) = 2x + y - 6 = 0 \qquad f_y(x, y) = x + 2y = 0$$

Solving this system of equations, we find $x = 4$ and $y = -2$ so that $(4, -2)$ is the only critical point. The second-order partial derivatives of f are

$$f_{xx}(x, y) = 2 \qquad f_{yy}(x, y) = 2 \qquad f_{xy}(x, y) = 1$$

At the critical point $(4, -2)$, we have

$$f_{xx}(4, -2) = 2 \qquad f_{yy}(4, -2) = 2 \qquad f_{xy}(4, -2) = 1$$

$$D = [f_{xx}(4, -2)][f_{yy}(4, -2)] - [f_{xy}(4, -2)]^2 = 4 - 1 = 3 > 0$$

Since $f_{xx}(4, -2) = 2 > 0$ and $D > 0$, it follows from theorem (17.3.6) that f has a local minimum at $(4, -2)$. The value of f at this local minimum is $z = f(4, -2) = -6$.

To find the critical points of a function $z = f(x, y)$ requires that we be able to solve the system of equations $f_x = 0$, $f_y = 0$. If this results in a system of linear equations, techniques for finding the solution are abundant. However, if the system is nonlinear, no general method is available. In this event, various manipulations such as substitution and addition/subtraction are used. Care must be taken because extraneous roots are sometimes introduced.

EXAMPLE 4

Find all local maxima and local minima for

$$z = f(x, y) = x^3 + y^2 + 2xy - 4x - 3y + 5$$

Solution

The critical points obey the equations

$$f_x(x, y) = 3x^2 + 2y - 4 = 0 \qquad f_y(x, y) = 2y + 2x - 3 = 0$$

We use a substitution method to solve the system. In $f_y(x, y) = 0$, we solve for y, obtaining $y = \frac{1}{2}(3 - 2x)$, and substitute for y in $f_x(x, y) = 0$. The result is

$$3x^2 + 2[\tfrac{1}{2}(3 - 2x)] - 4 = 0$$
$$3x^2 - 2x - 1 = 0$$
$$(3x + 1)(x - 1) = 0$$
$$x = -\tfrac{1}{3} \qquad \text{or} \qquad x = 1$$

When $x = -\frac{1}{3}$,

$$y = \tfrac{1}{2}(3 - 2x) = \tfrac{1}{2}(3 + \tfrac{2}{3}) = \tfrac{11}{6}$$

When $x = 1$,

$$y = \tfrac{1}{2}(3 - 2x) = \tfrac{1}{2}$$

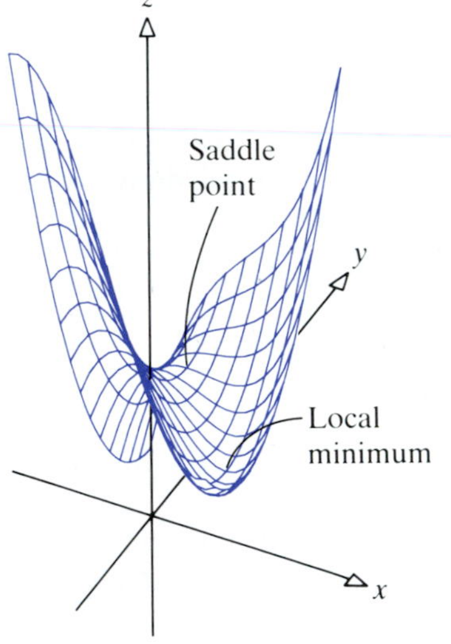

$z = x^3 + y^2 + 2xy - 4x - 3y + 5$

Figure 14

Thus, the critical points are $(-\frac{1}{3}, \frac{11}{6})$ and $(1, \frac{1}{2})$.* The second-order partial derivatives of f are

$$f_{xx}(x, y) = 6x \qquad f_{xy}(x, y) = 2 \qquad f_{yy}(x, y) = 2$$

At $(-\frac{1}{3}, \frac{11}{6})$, $\quad f_{xx}(-\frac{1}{3}, \frac{11}{6}) = -2 \qquad f_{xy}(-\frac{1}{3}, \frac{11}{6}) = 2 \qquad f_{yy}(-\frac{1}{3}, \frac{11}{6}) = 2$

$$D(-\tfrac{1}{3}, \tfrac{11}{6}) = -4 - 4 = -8 < 0$$

Thus, f has a saddle point at $(-\frac{1}{3}, \frac{11}{6}, \frac{317}{108})$.
At $(1, \frac{1}{2})$

$$D(1, \tfrac{1}{2}) = 12 - 4 = 8 > 0 \qquad f_{xx}(1, \tfrac{1}{2}) = 12 > 0$$

Hence, f has a local minimum at $(1, \frac{1}{2})$. A graph of

$$z = f(x, y) = x^3 + y^2 + 2xy - 4x - 3y + 5$$

is provided in Figure 14.

---

* You should check to make certain that neither of these roots is extraneous.

Absolute Extrema

In the case of a function of one variable, the extreme value theorem stated that if f is continuous on a closed interval $[a, b]$, then f has an absolute maximum and an absolute minimum. For functions of two variables there is a similar result. If D is the domain of a function f of two variables, we have said that D is closed if it contains all its boundary points. We say that D is *bounded* if D can be enclosed by some disk. With this definition, we can state the following analog of the extreme value theorem:

[17.3.7] THEOREM / ***Extreme Value Theorem.***

Let $z = f(x, y)$ denote a function of two variables. If f is a continuous function defined on a closed, bounded set D, then f has an absolute maximum and an absolute minimum on D.

To find the absolute extrema of a function meeting the criteria of theorem (17.3.7), we note that if f has an extreme value at (x_0, y_0), then (x_0, y_0) either is a critical point of f or is a boundary point of D. Thus, we have the following test:

[17.3.8] THEOREM / ***Test for Absolute Maximum and Absolute Minimum.***

Let $z = f(x, y)$ denote a function of two variables. If f is a continuous function defined on a closed, bounded set D, the absolute maximum and the absolute minimum of f are, respectively, the largest and smallest values found among the following:

(a) The values of f at the critical points of f in D.
(b) The extreme values of f on the boundary of D.

EXAMPLE 5

Find the absolute maximum and the absolute minimum of

$$z = f(x, y) = 2x - 2xy + y^2$$

whose domain is the rectangle $0 \leq x \leq 4, \quad 0 \leq y \leq 3$.

Solution

Since f is continuous on its domain and since its domain is a closed, bounded set, we know by theorem (17.3.7) that f has an absolute maximum and an absolute minimum on its domain. The critical points of f obey

$$f_x(x, y) = 2 - 2y = 0 \qquad f_y(x, y) = -2x + 2y = 0$$

Solving, we find that the only critical point is $(1, 1)$. The value of f at $(1, 1)$ is

$$f(1, 1) = 1$$

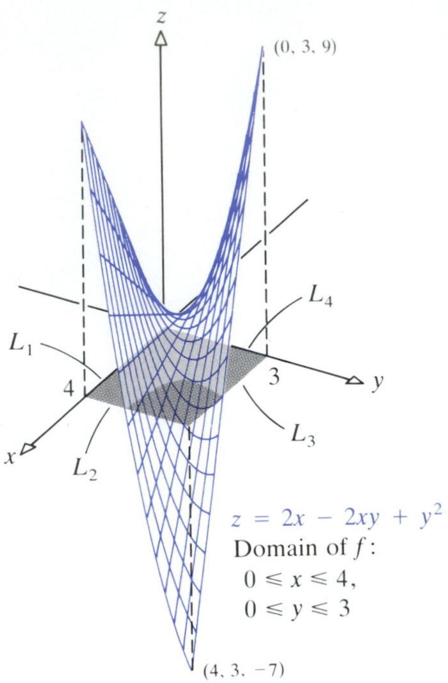

Figure 15

The boundary of the domain of f consists of four line segments (see Fig. 15). We evaluate f on each one:

On L_1: $0 \leq x \leq 4$, $y = 0$, so $f(x, y) = f(x, 0) = 2x$. Here f is increasing on $0 \leq x \leq 4$, so its extreme values will occur at the endpoints: $x = 0$ and $x = 4$. At $x = 0$, $f(0, 0) = 0$; at $x = 4$, $f(4, 0) = 8$.

On L_2: $x = 4$, $0 \leq y \leq 3$, so $f(x, y) = f(4, y) = 8 - 8y + y^2$. Using techniques from Chapter 4, we find that the critical numbers of f obey $-8 + 2y = 0$ or $y = 4$. Since this is not a value in $0 \leq y \leq 3$, the extreme values occur at the endpoints: $y = 0$, $y = 3$. At $y = 0$, $f(4, 0) = 8$; at $y = 3$, $f(4, 3) = -7$.

On L_3: $0 \leq x \leq 4$, $y = 3$, so $f(x, y) = f(x, 3) = 2x - 6x + 9 = -4x + 9$. Here f is decreasing on $0 \leq x \leq 4$, so its extreme values will occur at the endpoints: $x = 0$, $x = 4$. At $x = 0$, $f(0, 3) = 9$; at $x = 4$, $f(4, 3) = -7$.

On L_4: $x = 0$, $0 \leq y \leq 3$, so $f(x, y) = f(0, y) = y^2$. On $0 \leq y \leq 3$, f is increasing, so its extreme values occur at the endpoints: $y = 0$, $y = 3$. At $y = 0$, $f(0, 0) = 0$; at $y = 3$, $f(0, 3) = 9$.

We list below the values of f at the critical point $(1, 1)$ and at the extreme values on the boundary:

Point	(1, 1)	(0, 0)	(4, 0)	(4, 3)	(0, 3)
Value	1	0	8	−7	9

The absolute maximum of f on $0 \leq x \leq 4$, $0 \leq y \leq 3$ is $f(0, 3) = 9$; the absolute minimum is $f(4, 3) = -7$. See Figure 15.

EXAMPLE 6

A manufacturer wishes to make an open rectangular box of volume $V = 500$ cubic centimeters using the least possible amount of material. Find the dimensions of the box. See Figure 16.

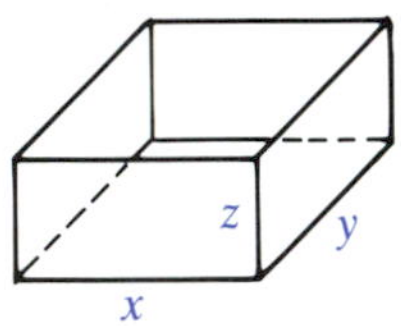

Figure 16

Solution

Let x and y be the dimensions of the base of the box, and let z be the height of the box. Then

$$V = 500 = xyz \qquad x > 0, \quad y > 0, \quad z > 0 \tag{1}$$

The surface area S—that is, the amount of material used—is

$$S = xy + 2yz + 2zx \tag{2}$$

Now we solve for z in (1) and substitute into (2):

$$S = xy + \frac{1000}{x} + \frac{1000}{y}$$

This is the function to be minimized. Differentiation yields

$$S_x = y - \frac{1000}{x^2} \qquad S_y = x - \frac{1000}{y^2}$$

To determine the critical points, we set these derivatives equal to 0:

$$S_x = y - \frac{1000}{x^2} = 0 \qquad S_y = x - \frac{1000}{y^2} = 0 \tag{3}$$

and we find

$$y = \frac{1000}{x^2} \qquad \text{and} \qquad x = \frac{1000}{y^2}$$

By solving these simultaneously (using substitution), we get

$$x = y = (1000)^{1/3} = 10$$

From (3), it follows that

$$S_{xx} = \frac{2000}{x^3} \qquad S_{xy} = 1 \qquad S_{yy} = \frac{2000}{y^3}$$

and, hence, at $(x, y) = (10, 10)$,

$$S_{yy} = S_{xx} = 2 \qquad D = S_{xx}S_{yy} - (S_{xy})^2 = 3$$

The conditions for a local minimum have been met and S has a local minimum at (10, 10). What we are seeking, though, is the point where S attains its absolute minimum. Since the domain of S is not bounded, we have no guarantee that S has an absolute minimum. But the physical considerations of the problem require that the absolute minimum exists. Therefore, it must be the case that the local minimum is also the absolute minimum. Consequently, we know that the dimensions (in centimeters) of the open box of fixed volume that uses the least amount of material are

$$x = 10 \qquad y = 10 \qquad z = \frac{500}{100} = 5$$

In Example 6, the conditions of theorem (17.3.7) were not met, since the domain of S is not bounded. In such cases, to show that an absolute minimum occurs where a local minimum has been found (and that an absolute maximum occurs where a local maximum has been found) can be extremely difficult. Fortunately, in applied problems such as Example 6, it is often possible to argue from physical or geometric considerations either that an absolute minimum must exist and that it must occur at a local minimum or, similarly, that an absolute maximum must occur where one of the local maxima has been found.

The theory just discussed can be generalized to functions of more than two variables. If, for example, $w = f(x, y, z)$, the critical points of f occur when f_x, f_y, and f_z are simultaneously 0, but it is more difficult to determine their nature. Fortunately, in practice, the correct answer can often be anticipated on physical grounds.

EXERCISE 17.3

In Problems 1–6 find all the critical points for each function.

1. $f(x, y) = x^4 - 2x^2 + y^2 + 5$

2. $f(x, y) = x^2 - y^2 + 6x - 2y + 4$

3. $f(x, y) = 4xy - x^4 - y^4 + 2$

4. $f(x, y) = x^3 + 6xy + 3y^2 + 3$

5. $f(x, y) = x^4 + y^4$

6. $f(x, y) = xy + \dfrac{2}{x} + \dfrac{4}{y}$

In Problems 7–26 find all local maxima, local minima, and saddle points for each function.

7. $z = x^2 + y^2 - 2x + 4y + 2$

8. $z = x^2 + y^2 - 4x + 2y - 4$

9. $z = x^2 + 4y^2 - 4x + 8y - 1$

10. $z = 2x^2 - y^2 + 4x - 4y + 8$

11. $z = x^3 - 6xy + y^3$

12. $z = x^2 - 3xy - y^2$

13. $z = x^2 + 3xy - y^2 + 4y - 6x$

14. $z = 2x^2 + xy + y^2 - 2x + 3y + 6$

15. $z = x^3 + 3xy + y^3$

16. $z = x^3 - 3xy - y^3$

17. $z = x^3 + x^2y + y^2$

18. $z = 3y^3 - x^2y + x$

19. $z = \dfrac{y}{x + y}$

20. $z = \dfrac{x}{x + y}$

21. $z = \cos x + \cos y$

22. $z = x^2 + 4 - 4x \cos y$

23. $z = y^2 - 6y \cos x + 6$

24. $z = x^2 - 6x \sin y + 2$

25. $z = e^{xy}$

26. $z = xye^{-(x+y)}$

In Problems 27–32 find the absolute maximum and absolute minimum of f on R.

27. $z = f(x, y) = x^2 - 4xy + 4y$
R: $0 \le x \le 2$, $0 \le y \le 1$

28. $z = f(x, y) = x^2 - 4xy + 4y$
R: $0 \le x \le 3$, $0 \le y \le 2$

29. $z = f(x, y) = x^2 + 2x + y^2 - 2y$
R: $x^2 + y^2 \le 4$

30. $z = f(x, y) = x^2 + 4x + y^2 - 4y$
R: $x^2 + y^2 \le 9$

31. $z = f(x, y) = 4x - 3y + 5$; R is the closed triangular region whose vertices are $(0, 0)$, $(0, 4)$, and $(5, 4)$

32. $z = f(x, y) = 3xy - 3x - 6y + 1$; R is the closed triangular region whose vertices are $(0, 0)$, $(5, 0)$, and $(0, 3)$

33. Find the highest point of the paraboloid $x^2 + y^2 - 6x + 4y + z + 12 = 0$ as follows:

(a) Find the vertex of the paraboloid without using calculus.

(b) Express z as a function of x and y and use calculus to find its maximum value.

34. A certain mountain is in the shape of the surface $z = 2xy - 2x^2 - y^2 - 8x + 6y + 4$. (The unit of distance is 1000 feet.) If sea level is the xy-plane, how high is the mountain?

35. An open rectangular box has a fixed surface area. Find the dimensions that make its volume a maximum.

36. Rework Example 6 if the box is closed.

37. Rework Problem 35 if the box is closed.

38. Find the dimensions of an open rectangular box having a volume of 72 cubic meters if the cost per square meter of the material to be used is \$4 for the bottom, \$3 for one of the sides, and \$2 for the three remaining sides, and the total cost is to be minimized.

39. The U.S. Post Office regulations require that the combined (sum) length and girth of a parcel post package being sent to a post office in the United States may not exceed 84 inches.

(Problem 39 continues on page 1013)

39. *(continued)*

(a) Find the length, width, and height of the rectangular box of maximum volume that can be mailed, subject to the 84 inch restriction.

(b) Find the dimensions of a circular tube of maximum volume that can be mailed, subject to the 84 inch restriction.

40. The reaction to an injection of x units of a certain drug, t hours after the injection, is given by

$$y = x^2(a - x)te^{-t} \qquad a \text{ constant}$$

Find the values of x and t, if any, that will maximize y.

41. An open irrigation channel is to be made into a symmetric form with three straight sides, as indicated in the illustration. If the perimeter is of length L, find a channel design that will allow the maximum possible flow. Is this design preferable to an open semicircular channel?

42. The cost of material for the sides of a rectangular shipping container is a cents per square foot; the cost of the top and bottom material is $\frac{3}{2}a$ cents per square foot. If the volume is to be $\frac{3}{2}$ cubic feet, what dimensions of the container will minimize its cost?

43. A fuel reservoir at D (see the figure) is to service plants located at A, B, and C, as shown. The cost, in thousands of dollars, of connecting the plants to D is determined by the formula

$$F = 6x^2 + 6y^2 - 4x - 6y + 5$$

Find the location of D that will minimize this cost.

(Problem 43 continues in the next column)

43. *(continued)*

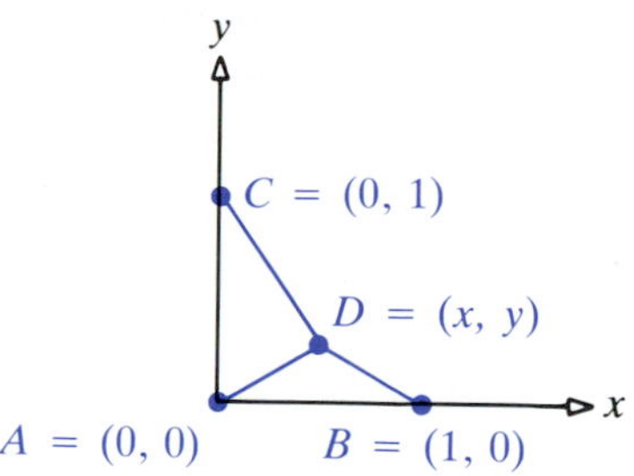

44. A metal detector is used to locate an underground pipe. When several meter readings of the detector are compared, it is found that the reading at an arbitrary point (x, y) is given by the formula

$$M = y(x - x^2) - y^2 \text{ volts} \qquad x \geq 0, \quad y \geq 0$$

Find the point (x, y) where the reading is largest.

45. A steel manufacturer produces two grades of steel, x tons of grade A and y tons of grade B. The cost C and revenue R are given in dollars by the formulas

$$C = \tfrac{1}{20}x^2 + 700x + y^2 - 150y - \tfrac{1}{2}xy$$

$$R = 2700x - \tfrac{3}{20}x^2 + 1000y - y^2 + \tfrac{1}{2}xy + 10{,}000$$

If $P = \text{Profit} = R - C$, find the production (in tons) of grades A and B that maximizes the manufacturer's profit.

46. Find the point in the plane $3x + 2y + z = 14$ that is nearest the origin.

47. Let $g(x, y) = Ax^2 + 2Bxy + Cy^2$, where A, B, and C are constants. Show that:

(a) If $AC - B^2 > 0$, $A \neq 0$, then $g(x, y)$ has a maximum at $(0, 0)$ if $A < 0$, or a minimum if $A > 0$.

(b) If $AC - B^2 < 0$, then $g(x, y)$ has both positive and negative values at points (x, y) near $(0, 0)$. Thus, $(0, 0, 0)$ is a saddle point for g.

The Method of Lagrange Multipliers 17.4

We begin with an example.

EXAMPLE 1

At what point in the first quadrant on the hyperbola $xy = 4$ is the value of $z = 12x + 3y$ a minimum? What is this minimum?

Solution

In this problem we are asked to find the minimum of $z = 12x + 3y$ subject to the condition that x and y obey the equation $xy = 4$. We can express this as a minimum problem in one variable by setting $y = 4/x$ in the expression $z = 12x + 3y$, obtaining

$$z = 12x + \frac{12}{x}$$

The critical numbers obey

$$\frac{dz}{dx} = 12 - \frac{12}{x^2} = 0$$

$$\frac{12}{x^2} = 12$$

$$x = 1 \qquad \text{or} \qquad x = -1$$

We ignore $x = -1$, since the point $(-1, -4)$ is not in the first quadrant. Since $d^2z/dx^2 = 24/x^3 > 0$ for $x = 1$, we conclude that when $x = 1$ (and, therefore, $y = 4/x = 4$), the value of z is a minimum. Thus, $z = 12x + 3y$ is a minimum at the point $(1, 4)$ on the hyperbola, and this minimum value is $z = 24$.

In this example we found the minimum value of $z = 12x + 3y$ subject to the condition that $xy = 4$. We solved the problem by eliminating the variable y and treating the problem as a typical minimum problem in one variable. Sometimes, though, it is not easy—and is perhaps even impossible—to eliminate a variable. In such cases, the *method of Lagrange multipliers* may be used.

The purpose of the method of *Lagrange multipliers* is to maximize (or minimize) a function subject to an auxiliary condition (or constraint) on the domain of the function. Specifically, for two variables: $z = f(x, y)$ is to be maximized (or minimized) subject to the constraint $g(x, y) = 0$.

Geometric Argument

Suppose we wish to minimize the function $z = f(x, y) = 12x + 3y$, subject to the constraint $g(x, y) = xy - 4 = 0$. The graph of $g(x, y) = xy - 4 = 0$ is a curve in the xy-plane. We seek the minimum value of $f(x, y) = 12x + 3y$ for values (x, y) on the curve $xy - 4 = 0$. Now let's look at the level curves of $z = 12x + 3y$. See Figure 17. On a given level curve, the value of z is fixed, and we seek the smallest such value when x and y are restricted to the hyperbola $xy = 4$. We observe in Figure 17 that the minimum value of z occurs precisely at the point of tangency of the line $12x + 3y = 24$ with the hyperbola $xy = 4$. At this point of tangency, the two curves have a common tangent and, hence, the same normal line. Since the gradient gives the direction of the normal line to a curve, it follows that ∇f is parallel to ∇g. That is, there is a scalar $\lambda \neq 0$ so that

$$\nabla f = \lambda \nabla g \qquad \text{where} \quad g(x, y) = 0$$

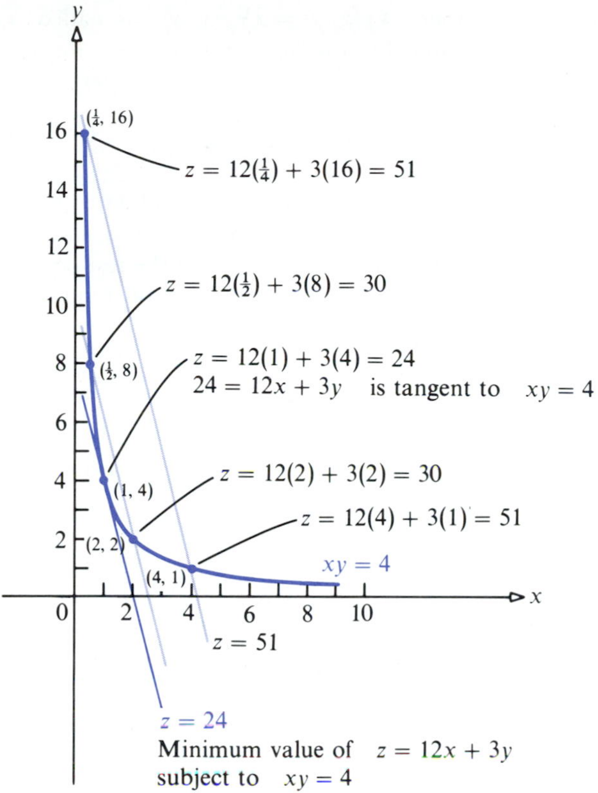

Figure 17

We now state and prove a theorem that suggests a method for solving problems of the type under discussion.

[17.4.1] THEOREM

Let f and g be functions of two variables with continuous partial derivatives at every point of some open set containing the curve $g(x, y) = 0$. Suppose that f, when restricted to points on the curve $g(x, y) = 0$, has a local extremum at the point (x_0, y_0). Suppose also that $\nabla g(x_0, y_0) \neq 0$. Then there is a number λ such that

$$\nabla f(x_0, y_0) = \lambda \nabla g(x_0, y_0) \tag{1}$$

Proof Let C be the curve $g(x, y) = 0$. Suppose that $\mathbf{r}(t) = x(t)\mathbf{i} + y(t)\mathbf{j}$ is a parametric equation in vector form for C. Define $s(t) = f(x(t), y(t))$ and choose t_0 so that $x(t_0) = x_0$, $y(t_0) = y_0$. We are assuming that $s(t)$ has either a local maximum or a local minimum at t_0. Thus, $s'(t_0) = 0$. Then, by the chain rule,

$$0 = s'(t_0) = f_x(x_0, y_0)x'(t_0) + f_y(x_0, y_0)y'(t_0) = \nabla f(x_0, y_0) \cdot \mathbf{r}'(t_0)$$

If $\nabla f(x_0, y_0)$ is nonzero, it is orthogonal to $\mathbf{r}'(t_0)$, which is a tangent vector to C at (x_0, y_0). By theorem (17.1.8), the vector $\nabla g(x_0, y_0)$ is also orthogonal to $\mathbf{r}'(t_0)$. If $\nabla f(x_0, y_0) = \mathbf{0}$, then let $\lambda = 0$. Thus, in every case, there is a number λ such that $\nabla f(x_0, y_0) = \lambda \nabla g(x_0, y_0)$. ∎

The number λ in the equations $\nabla f(x, y) = \lambda \nabla g(x, y)$ is called a *Lagrange multiplier*. These equations and the equation $g(x, y) = 0$ form a system of three equations in the three unknowns, x, y, and λ—namely,

$$f_x(x, y) = \lambda g_x(x, y) \qquad f_y(x, y) = \lambda g_y(x, y) \qquad g(x, y) = 0 \tag{2}$$

We call a solution (x_0, y_0) of (2) a *test point*, since it is a candidate for the desired extrema. The maximum and minimum values, if they exist, may be found by choosing the largest and smallest values of $f(x, y)$ at the test points.

Let's rework Example 1, this time using the method of Lagrange multipliers.

EXAMPLE 2

Find the minimum value of $z = f(x, y) = 12x + 3y$ subject to the condition $xy = 4$, $x > 0$, $y > 0$.

Solution

The test points obey the equations

$$\nabla f(x, y) = \lambda \nabla g(x, y) \qquad g(x, y) = xy - 4 = 0$$

where λ is a number that is not known. The gradients of f and g are

$$\nabla f = 12\mathbf{i} + 3\mathbf{j} \qquad \nabla g = y\mathbf{i} + x\mathbf{j}$$

so that

$$12 = \lambda y \qquad 3 = \lambda x \qquad xy - 4 = 0$$

Eliminating λ from the first two equations results in $3y = 12x$, or $y = 4x$, so the third equation becomes

$$4x^2 - 4 = 0$$
$$x^2 = 1$$
$$x = 1 \quad \text{or} \quad x = -1$$

We ignore $x = -1$, since $x > 0$ is the domain of the function. When $x = 1$, we have $y = 4$, so the only test point is $(1, 4)$. The corresponding minimum value of $z = 12x + 3y$ is $z = 24$.

EXAMPLE 3

Find the maximum and minimum values of $z = f(x, y) = 3x - y + 1$ subject to the condition $3x^2 + y^2 = 9$.

Solution

The test points satisfy the equations

$$\nabla f = \lambda \nabla g \qquad g(x, y) = 3x^2 + y^2 - 9 = 0$$

where λ is to be determined. These equations are

$$3 = \lambda 6x \qquad -1 = \lambda 2y \qquad 3x^2 + y^2 - 9 = 0$$

From the first two equations, we find that $x = 1/(2\lambda)$ and $y = -1/(2\lambda)$, so the third equation may be written as

$$\frac{3}{4\lambda^2} + \frac{1}{4\lambda^2} = 9$$

$$\frac{1}{\lambda^2} = 9$$

$$\lambda = \pm\tfrac{1}{3}$$

Hence, $x = \pm\frac{3}{2}$, $y = \mp\frac{3}{2}$, and the test points are $(\frac{3}{2}, -\frac{3}{2})$ and $(-\frac{3}{2}, \frac{3}{2})$. The corresponding values of z are $z = \frac{9}{2} + \frac{3}{2} + 1 = 7$ and $z = -\frac{9}{2} - \frac{3}{2} + 1 = -5$. Thus, the maximum value of z on the ellipse is 7 and the minimum value is -5.

The next example shows how a Lagrange multiplier may be used to find the absolute extrema of a function whose domain is closed and bounded.

EXAMPLE 4

Find the absolute maximum and minimum of the function $f(x, y) = x^2 + y^2 + 4x - 4y + 3$ on $x^2 + y^2 \leq 2$.

Solution

We first look for critical points:

$$f_x = 2x + 4 = 0 \qquad f_y = 2y - 4 = 0$$

$$x = -2 \qquad y = 2$$

But the inequality $x^2 + y^2 \leq 2$ is not satisfied at $x = -2$, $y = 2$; the point $(-2, 2)$ lies outside the domain of the function. Thus, the function has no critical points. The extrema must occur on the boundary of the domain—that is, on the curve $x^2 + y^2 = 2$. We apply the Lagrange method with the constraint $g(x, y) = x^2 + y^2 - 2 = 0$. The test point conditions are

$$f_x = 2x + 4 = 2x\lambda \qquad f_y = 2y - 4 = 2y\lambda \qquad x^2 + y^2 = 2$$

Eliminating λ from the first two equations yields $y = -x$. By substituting in the third equation and solving, we get two test points: $(1, -1)$ and $(-1, 1)$. Thus, the extreme values are

$$f(1, -1) = 13 \qquad \text{Maximum value}$$

$$f(-1, 1) = -3 \qquad \text{Minimum value}$$

See Figure 18.

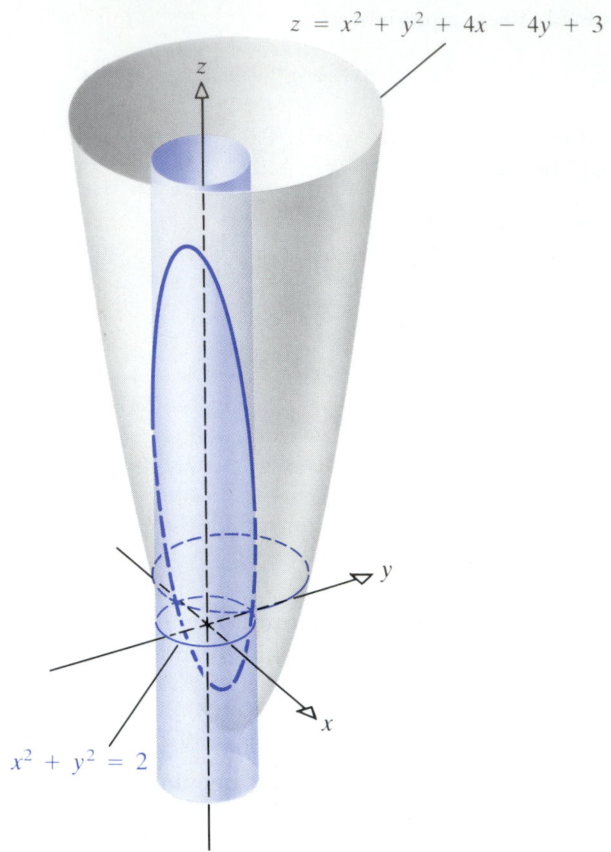

Figure 18

Functions of Three Variables

For functions of three variables, the method of Lagrange multipliers works in the same way. The extreme values of $w = f(x, y, z)$ subject to the condition $g(x, y, z) = 0$, if they exist, occur at the solutions (x, y, z) of the system of equations

$$\nabla f(x, y, z) = \lambda \nabla g(x, y, z) \qquad g(x, y, z) = 0 \tag{2}$$

Note that (2) is a system of four equations in the four unknowns x, y, z, and λ.

EXAMPLE 5

A box that is open on top is to have a fixed volume of 12 cubic meters. The material used to make the bottom of the box costs \$3 per square meter, while the material used for the sides costs \$1 per square meter. What should the dimensions be so that the cost is a minimum?

Solution

Let x be the width of the box, y the depth of the box, and z the height of the box. Then xy is the area of the bottom of the box. We seek to minimize

$$C = 3xy + 2xz + 2yz$$

subject to the condition $g(x, y, z) = xyz - 12 = 0$. The test points satisfy

$$\nabla C = \lambda \nabla g \qquad g(x, y, z) = xyz - 12 = 0$$

That is,

$$3y + 2z = \lambda yz \qquad 3x + 2z = \lambda xz \qquad 2x + 2y = \lambda xy \qquad xyz - 12 = 0 \quad \textbf{(3)}$$

Since $x > 0$, $y > 0$, and $z > 0$, we can solve the first three equations for λ, obtaining

$$\lambda = \frac{3}{z} + \frac{2}{y} \qquad \lambda = \frac{3}{z} + \frac{2}{x} \qquad \lambda = \frac{2}{y} + \frac{2}{x}$$

From these, we find that

$$y = x \qquad \text{and} \qquad z = \tfrac{3}{2}x$$

Hence, from the last equation in (3), we get

$$\tfrac{3}{2}x^3 = 12$$
$$x = 2$$

The test point is therefore (2, 2, 3). The dimensions of the box are 2 meters × 2 meters × 3 meters, and the minimum cost is \$36.

Several Constraints

The method of Lagrange multipliers may also be used for problems involving several constraints. In the two-constraint situation, we are asked to find the extreme values of a function $z = f(x, y)$ subject to the two constraints $g(x, y) = 0$ and $h(x, y) = 0$. The extreme values of f, if they exist, will occur at solutions (x, y) of the system of equations

$$\nabla f(x, y) = \lambda_1 \nabla g(x, y) + \lambda_2 \nabla h(x, y) \qquad g(x, y) = 0 \quad h(x, y) = 0 \quad \textbf{(4)}$$

where λ_1 and λ_2 are two numbers called *Lagrange multipliers*. Note that (4) is a system of four equations in the four unknowns x, y, λ_1, and λ_2.

Similar remarks hold for the three-variable case, as the next example illustrates.

EXAMPLE 6

Find the points on the intersection of the ellipsoid $x^2 + y^2 + 9z^2 = 25$ and the plane $x + 3y - 2z = 0$ that are the farthest from the origin. Also, find the points that are the closest to the origin.

Solution

See Figure 19. Let C be the curve resulting from the intersection of the ellipsoid and the plane, and let $P = (x, y, z)$ be a point on C. The function we wish to maximize and minimize is the distance from P to the origin—namely, $d = f(x, y, z) = \sqrt{x^2 + y^2 + z^2}$. The constraints under which this is to be done are

$$g(x, y, z) = x^2 + y^2 + 9z^2 - 25 = 0 \qquad \text{and} \qquad h(x, y, z) = x + 3y - 2z = 0$$

Set

$$\nabla f(x, y, z) = \lambda_1 \nabla g(x, y, z) + \lambda_2 \nabla h(x, y, z)$$

where λ_1 and λ_2 are to be determined. The desired values of x, y, and z are solutions of

$$\frac{x}{\sqrt{x^2 + y^2 + z^2}} = \lambda_1(2x) + \lambda_2(1)$$

$$\frac{y}{\sqrt{x^2 + y^2 + z^2}} = \lambda_1(2y) + \lambda_2(3)$$

$$\frac{z}{\sqrt{x^2 + y^2 + z^2}} = \lambda_1(18z) + \lambda_2(-2)$$

Elimination of λ_1 and λ_2 from these equations gives $3xz - yz = 0$. By using this equation together with the equations of the plane and the ellipsoid, we obtain the following four points:

$$\left(\frac{-15}{\sqrt{10}}, \frac{5}{\sqrt{10}}, 0\right), \qquad \left(\frac{15}{\sqrt{10}}, \frac{-5}{\sqrt{10}}, 0\right),$$

$$\left(\frac{5}{\sqrt{235}}, \frac{15}{\sqrt{235}}, \frac{25}{\sqrt{235}}\right), \qquad \left(\frac{-5}{\sqrt{235}}, \frac{-15}{\sqrt{235}}, \frac{-25}{\sqrt{235}}\right)$$

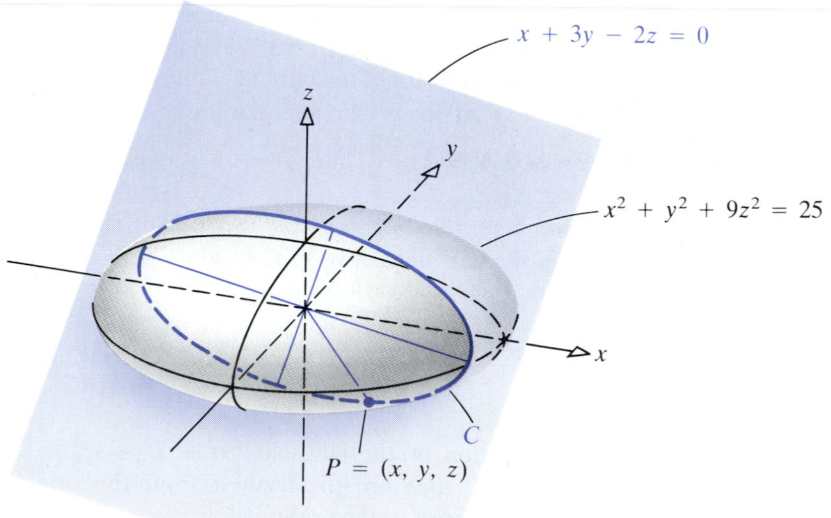

Figure 19

The distance of each of these from $(0, 0, 0)$ is

$$d = 5 \qquad d = 5 \qquad d = \sqrt{\tfrac{175}{47}} \qquad d = \sqrt{\tfrac{175}{47}}$$

respectively. Thus, the points $(-15/\sqrt{10}, 5/\sqrt{10}, 0)$ and $(15/\sqrt{10}, -5/\sqrt{10}, 0)$ are farthest from the origin, while the points $(5/\sqrt{235}, 15/\sqrt{235}, 25/\sqrt{235})$ and $(-5/\sqrt{235}, -15/\sqrt{235}, -25/\sqrt{235})$ are closest. Since the intersection of the ellipsoid and the plane is an ellipse, the points we have found are the ends of the axes of this ellipse.

EXERCISE 17.4

In Problems 1–8 use the method of Lagrange multipliers to find the maximum and minimum values of f subject to the given condition. Tell where the extreme values occur.

1. $f(x, y) = 3xy$; $g(x, y) = x^2 + 2y^2 - 4 = 0$

2. $f(x, y) = x - 2y^2$; $g(x, y) = x^2 + y^2 - 1 = 0$

3. $f(x, y) = x^2 + 4y^3$; $g(x, y) = x^2 + 2y^2 - 2 = 0$

4. $f(x, y) = 2xy$; $g(x, y) = x^2 + y^2 - 2 = 0$

5. $f(x, y) = xy$; $g(x, y) = 9x^2 + 4y^2 - 36 = 0$

6. $f(x, y) = x^2 - 4xy + 4y^2$; $g(x, y) = x^2 + y^2 - 4 = 0$

7. $f(x, y, z) = 4x - 3y + 2z$; $g(x, y, z) = x^2 + y^2 - 6z = 0$

8. $f(x, y, z) = x^2 + 2y^2 + z^2$; $g(x, y, z) = 2x - 3y + z - 6 = 0$

9. Find the point on the line $x - 3y = 6$ that is closest to the origin.

10. Find the point on the plane $2x + y - 3z = 6$ that is closest to the origin.

11. At which points on the ellipse $x^2 + 2y^2 = 2$ is the product xy a maximum?

12. Find the dimensions of a box that is open at the top, so that the volume is a maximum when the surface area is fixed at 24 square centimeters.

13. A box that is open at the top and has a volume of 12 cubic meters is to be made from material costing \$1 per square meter. What dimensions minimize the cost?

14. Suppose that $T = T(x, y, z) = 100x^2yz$ is the temperature (in degrees Celsius) at any point (x, y, z) on the sphere given by $x^2 + y^2 + z^2 = 1$. Find the points on the sphere where the temperature is greatest and least. What is the temperature at these points?

15. A farmer has 340 meters of fencing for enclosing two separate fields, one of which is to be a rectangle twice as long as it is wide and the other a square. The square field must contain at least 100 square meters, and the rectangular one must contain at least 800 square meters.

(a) If x is the width of the rectangular field, what are the maximum and minimum possible values of x?

(b) What is the greatest number of square meters that can be enclosed in the two fields? Justify your answer.

16. The surface $xyz = -1$ is cut by the plane $x + y + z = 1$, resulting in a curve C. Find the points on C that are nearest to the origin and farthest from it.

17. Find the maximum and minimum values of the function $f(x, y, z) = x^2 + y^2 + z^2$ subject to $z^2 = x^2 + y^2$ and $x + y - z + 1 = 0$.

18. Find the minimum value of $w = x^2 + y^2 + z^2$ subject to the constraints $2x + y + 2z = 9$, $5x + 5y + 7z = 29$.

19. You found in Problem 37 of Exercise 17.3 that the closed rectangular box of fixed surface area and maximum volume is a cube. Use the method of Lagrange multipliers to confirm this fact.

20. It is shown in single variable calculus that a cylindrical can of fixed surface area and maximum volume has an altitude equal to the diameter of its base. Use the method of Lagrange multipliers to confirm this fact.

REVIEW EXERCISES

In Problems 1–10 find the directional derivatives of the given function at the indicated point in the direction of **a**.

1. $f(x, y) = \ln(\sec(x^2 + y^2))$ at $(\sqrt{\pi/8}, \sqrt{\pi/8})$; $\mathbf{a} = \mathbf{i} + \mathbf{j}$

2. $f(x, y, z) = \ln(\ln(x^2 + y^2 + z^2))$ at $(\sqrt{e}, 0, 0)$; $\mathbf{a} = \mathbf{i} + 2\mathbf{j} + 3\mathbf{k}$

3. $f(x, y) = \dfrac{x^2 - y^2}{x^2 + y^2}$ at $(1, 1)$; $\mathbf{a} = \mathbf{i} - \mathbf{j}$

4. $f(x, y, z) = \ln(xyz)$ at $(1, 1, 3)$; $\mathbf{a} = \mathbf{i} + \mathbf{j} + 3\mathbf{k}$

5. $f(x, y, z) = (2x + y + z)^2 + xyz$ at $(1, 1, 1)$; $\mathbf{a} = -\mathbf{i} - \mathbf{j} - \mathbf{k}$

6. $f(x, y, z) = \dfrac{x}{\sqrt{x^2 + 2y^2 + 3z^2}}$ at $(1, 2, 1)$; $\mathbf{a} = -\mathbf{i} + \mathbf{j} + \mathbf{k}$

7. $f(x, y, z) = ye^{-x}(x^2 + y^2 + z^2 + 1)$ at $(0, 0, 0)$; $\mathbf{a} = 2\mathbf{i} + \mathbf{j} + 2\mathbf{k}$

8. $f(x, y, z) = \sinh x \cosh(y + z)$ at $(1, 1, 1)$; $\mathbf{a} = 2\mathbf{i} - \mathbf{j}$

9. $f(x, y) = y \sin^{-1} x$ at $(0, 1)$; $\mathbf{a} = 3\mathbf{i} + \mathbf{j}$

10. $f(x, y) = \begin{cases} \dfrac{\sin(x^2 + y^2)}{x^2 + y^2} & \text{if } (x, y) \neq (0, 0) \\ 1 & \text{if } (x, y) = (0, 0) \end{cases}$
at $(0, 0)$; $\mathbf{a} = \mathbf{i} + \mathbf{j}$

11. Consider the tangent plane to the graph of a differentiable function $z = f(x, y)$ at a point (x_0, y_0). Consider a tangent vector $\mathbf{v}_1$ on that plane whose $\mathbf{j}$ component is 0 and a tangent vector $\mathbf{v}_2$ on that plane whose $\mathbf{i}$ component is 0. Must $\mathbf{v}_1$ and $\mathbf{v}_2$ be at right angles? See the figure.

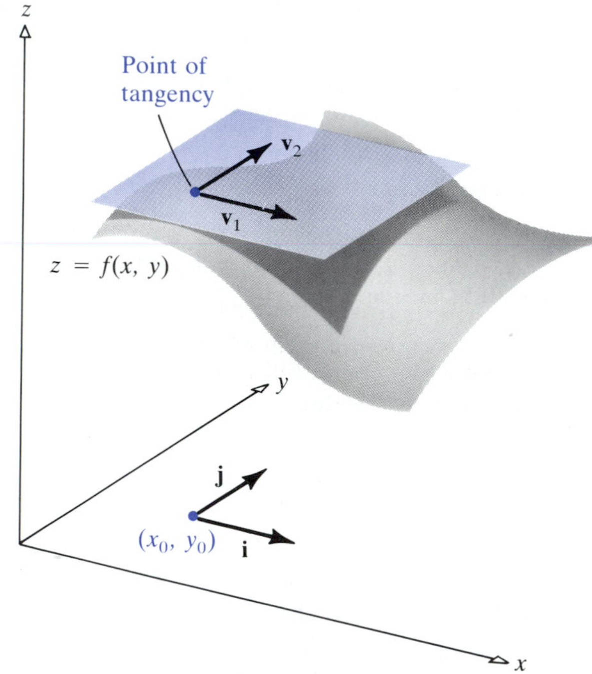

12. Suppose $z = f(x, y)$ has directional derivatives in all directions. Need f be differentiable?

13–22. Find the normal lines and tangent planes to the surfaces at the given points in Problems 1–10. For Problems 1, 3, 9, and 10, $z = f(x, y)$ defines the surface; for Problems 2 and 4–8, $f(x, y, z) = 0$ defines the surface.

23. Show that the surfaces $x^2 + y^2 + z^2 = 1$ and $x^2 + y^2 - z^2 = 1$ are tangent at every point of intersection. [Refer to Problem 21, Exercise 17.2.]

In Problems 24–27 find all local extrema and saddle points.

24. $f(x, y) = c\sqrt{1 - (x^2/a^2) - (y^2/b^2)}$, $c > 0$

25. $f(x, y) = x^2 + xy + y^2 + 6x$

26. $f(x, y) = x^3 - y^3 + 3xy$

27. $f(x, y) = \sin x \cdot \sin y$

28. A flat metal plate has the shape of the ellipse

$$\frac{x^2}{3} + \frac{y^2}{6} \le 1$$

If the plate is heated so that the temperature $T(x, y)$ is given by

$$T(x, y) = x^2 + y^2$$

find the hottest and coldest points on the plate and give the temperature at each of these places.

29. Find the volume of the largest rectangular solid that can be inscribed in the interior of the surface

$$\frac{x^2}{a^2} + \frac{y^2}{b^2} + \frac{z^2}{c^2} = 1$$

and whose sides are parallel to the axis.

30. Find the points on the surface $xyz = 1$ closest to the origin.

31. Maximize the volume of a rectangular solid that has three faces on the coordinate planes and one vertex on the plane $(x/a) + (y/b) + (z/c) = 1$ where $a, b, c > 0$.

32. Let the point $P = (x_0, y_0, z_0)$ be given. Find the minimum distance from P to a plane with equation $Ax + By + Cz = D$ using techniques of this chapter.

33. At what points on the union of the two curves $x^2 + y^2 = 1$ and $x^3 + y^3 = 1$ is the function $f(x, y) = x^4 + y^4 + 4$ a maximum? At what point is it a minimum?

34. The base of a rectangular box costs five times as much as do the other five sides. Find the proportions of the dimensions for the cheapest possible box of volume V.

35. Find the extreme values of $f(x, y, z) = xyz$ on the surface $x^2 + y^2 + z^2 = 1$.

36. Minimize $x^4 + y^4 + z^4$ subject to the constraint $Ax + By + Cz = D$.

37. Find the extreme values of $f(x, y, z) = x^2 + y^2 + z^2$ on the intersection of the cone $x^2 + y^2 - z^2 = 0$ and the plane $x + y - z = 1$.

38. Find the points of intersection of the plane $x + y + z = 1$ and the hyperboloid $x^2 + y^2 - z^2 = 1$ nearest the origin.

39. Let a circle of radius R be given. Find the triangle of largest perimeter that can be inscribed in the circle.

40. Find the extreme values of $f(x, y, z) = xyz$ subject to the two constraints $x^2 + y^2 = 1$ and $y = 3z$.

CHALLENGE EXERCISES

1. Find the distance between the skew lines

L_1: $x = t - 6, \quad y = t, \quad z = 2t$

and

L_2: $x = t, \quad y = t, \quad z = -t$

2. Show that if the sum of the sines of the angles of a triangle is a maximum, then the triangle must be equilateral.

3. Find the point of the paraboloid $z = 2 - x^2 - y^2$ that is closest to the point $(1, 1, 2)$.

4. A scientist plots the points $(x_1, y_1), \ldots, (x_n, y_n)$ from her experimental data. Her theory tells her that the points should lie on a straight line, but they do not—experimental error, perhaps. She is looking for the line $y = mx + b$ that "best" fits the data. The

(*Problem 4 continues in the next column*)

4. (*continued*)

most often used criterion is the least squares fit, for which m and b are chosen to minimize

$$\sum_{k=1}^{n} (mx_k + b - y_k)^2$$

Show that the minimizing values of m and b are

$$m = \frac{n \sum_{k=1}^{n} x_k y_k - \sum_{k=1}^{n} x_k \sum_{k=1}^{n} y_k}{n \sum_{k=1}^{n} x_k^2 - \left(\sum_{k=1}^{n} x_k\right)^2}$$

$$b = \frac{1}{n}\left(\sum_{k=1}^{n} y_k - m \sum_{k=1}^{n} x_k\right)$$

5. Find the least squares estimate $y = mx + b$ for the points $(0, 2)$, $(1, 1)$, $(2, 2)$, $(3, 4)$, and $(4, 4)$. Plot the points and draw the line. See Problem 4.

6. Find equations of the tangent plane and normal line to $(yz)^{xz} = 16$ at $(2, 1, 2)$.

7. Show that if $\nabla f(x, y) = c(x\mathbf{i} + y\mathbf{j})$, where c is a constant, then $f(x, y)$ is constant on any circle of radius k, centered at $(0, 0)$. [*Hint:* Set $x = k \cos \theta$, $y = k \sin \theta$, and calculate $df/d\theta$.]

8. Suppose that $z^2 + xz - y + w^2 = 0$ and $x^2 + y^2 + z^2 + w^2 = 18$. Determine z and w as functions of x and y. Find $D_{\mathbf{u}}z$ at $x = -2$, $y = 3$, $z = 1$, $w = 2$ in the direction from $(0, 1)$ to $(2, -3)$.

9. What points of the surface $xy - z^2 - 6y + 36 = 0$ are closest to the origin?

10. Let $z = f(x, y)$ have continuous second-order partial derivatives. If $\mathbf{u} = u_1\mathbf{i} + u_2\mathbf{j}$ is a unit vector, we have a directional derivative $D_{\mathbf{u}}f(x, y) = g(x, y)$. If $g(x, y)$ is differentiable and if $\mathbf{v} = v_1\mathbf{i} + v_2\mathbf{j}$ is a unit vector, we have a directional derivative $D_{\mathbf{v}}g(x, y)$. We can view this second quantity as a second-order directional derivative for $z = f(x, y)$. Compute it in terms of f_{xx}, f_{xy}, and f_{yy}, showing that it has the value

$$u_1v_1f_{xx} + (u_1v_2 + u_2v_1)f_{xy} + u_2v_2f_{yy}$$

11. In Problem 10, if we set $A = f_{xx}(x_0, y_0)$, $B = f_{xy}(x_0, y_0)$, and $C = f_{yy}(x_0, y_0)$, then we will have a bilinear expression in the components u_1 and u_2 of $\mathbf{u}$, and v_1 and v_2 of $\mathbf{v}$.

(*Problem 11 continues in the next column*)

11. (*continued*)

(a) In particular, if $\mathbf{u} = \mathbf{v}$, show that the second-order directional derivative is given at (x_0, y_0) by

$$Au_1^2 + 2Bu_1u_2 + Cu_2^2$$

(b) Let $f_x(x_0, y_0) = 0 = f_y(x_0, y_0)$. Show that if $A > 0$ and $AC - B^2 > 0$, then $z = f(x, y)$ has a local minimum at (x_0, y_0); if $A < 0$ and $AC - B^2 > 0$, then $z = f(x, y)$ has a local maximum at (x_0, y_0); and if $AC - B^2 < 0$, then $z = f(x, y)$ has a saddle point at (x_0, y_0).

(c) Show by example that if $AC - B^2 = 0$, we may have any of the above three behaviors of $z = f(x, y)$.

12. Recall the discussion of a parameterized surface in the Challenge Exercises at the end of Chapter 13. Let the parameterization be given by $\mathbf{r}(t_1, t_2)$ and let $\mathbf{r}^* = \mathbf{r}(t_1^*, t_2^*)$ lie on the surface (t_1^*, t_2^* being fixed numbers). If $\mathbf{r}^* = (x^*, y^*, z^*)$, show that the equation of the tangent plane to the surface at $\mathbf{r}^*$ is given by

$$\begin{vmatrix} x - x^* & y - y^* & z - z^* \\ \dfrac{\partial}{\partial t_1}x(t_1, t_2) & \dfrac{\partial}{\partial t_1}y(t_1, t_2) & \dfrac{\partial}{\partial t_1}z(t_1, t_2) \\ \dfrac{\partial}{\partial t_2}x(t_1, t_2) & \dfrac{\partial}{\partial t_2}y(t_1, t_2) & \dfrac{\partial}{\partial t_2}z(t_1, t_2) \end{vmatrix} = 0$$

provided no tangent vector is 0.

Graphical Solutions to a Constrained Optimization Problem*

Some Background from Statistics

In this section we will consider the makeup of a stock portfolio containing two stocks. Because stock prices fluctuate, we will need to make a decision based on the expected return, the intensity of the fluctuation of the stock, and the joint variation of the two stocks. The concepts that we will use from statistics are mean, variance, and covariance. If we have a stock X exhibiting values measured over a period of time $x_1, x_2, \ldots, x_n$, then we define the sample mean to be

$$E(X) = \frac{1}{n}\sum_{i=1}^{n} x_i$$

The sample variance is defined to be

$$V(X) = E(X^2) - E(X)^2$$

* L. Carl Leinbach
Gettysburg College

where X^2 is the random variable obtained by squaring the observed values of X. If we have a second stock Y, the joint variation of the two stocks is called the covariance of XY and is defined to be

$$C(X, Y) = E(XY) - E(X)E(Y)$$

The exact meaning and use of these measures are the topic of a course in statistics. The use of them in the following example is superficial and done only as a method of quantifying an investor's expectations and the risk involved in a joint investment.

Selecting a Stock Portfolio

A stockbroker has determined to construct a portfolio for a client that consists of two stocks, X and Y. The client has budgeted \$50 for investing in this particular portfolio. The broker needs to choose the portfolio that will maximize the return on the stocks and minimize the risk involved in such a choice. If the broker chooses x shares of X and y shares of Y, then the return can be estimated as

$$R(x, y) = m_1 x + m_2 y$$

where m_1 and m_2 are the sample means of X and Y, respectively. The commonly used measures of risk are the variance and covariance of the stocks. Thus, we define the measure of risk to be

$$V(x, y) = v_1 x^2 + cxy + v_2 y^2$$

where v_1 and v_2 are the variances of X and Y and c is the covariance.

Another factor that must be taken into account is the nature of the client. Any observer of investors knows that some individuals are very conservative (adverse to risk) and others are much more liberal (accepting of risk). Thus, the broker needs to assign a value to the investor's acceptance of risk. Values close to 0 indicate a more liberal investor, and larger values indicate a more conservative investing attitude. This value is given as a risk factor f. Thus, the investor is seeking values for x and y that will maximize

$$F(x, y) = R(x, y) - fV(x, y)$$

In order to make our discussion more specific, let's assign some values to the means, variances, and covariances of the stocks. We must also make a judgment about the investor. Stock X has been reasonably stable, exhibiting an average return of 5 per share with a variance of 4. Stock Y, on the other hand, has been less stable, with an average return of 10 per share and a variance of 100. The covariance of the two stocks is estimated at 5. Our investor is moderately conservative when it comes to money matters, so we assign a value of .1 to f. This yields the function

$$\begin{aligned} F(x, y) &= 5x + 10y - .1(4x^2 + 5xy + 100y^2) \\ &= 5x + 10y - .4x^2 - .5xy - 10y^2 \end{aligned}$$

If we look at the graph of $F(x, y)$ over the region $-2.5 \le x, y \le 2.5$ we see that the graph has the shape of a parabolic cylinder that is slanted upward. (Looking at it over a larger region we see that it is actually a paraboloid and that this is just one small slice of the surface.)

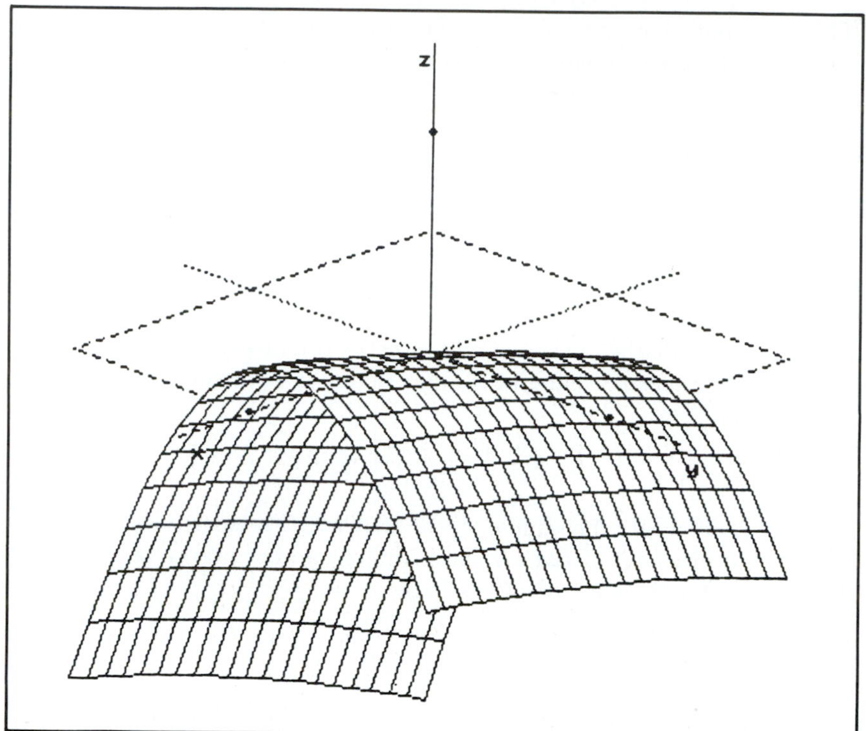

The highest point on the graph appears to lie on the boundary, $x = 2.5$. This, however, does not take into account the fact that the investor has set the size of the investment in this portfolio at \$50. If the current price of stock X is 20 and of stock Y is 30, then the maximum value is subject to the constraint

$$20x + 30y = 50$$

Recalling that the constrained maximum occurs when the gradient of F is normal to the surface represented by the constraint, we will use a program in MicroCalc that depicts the direction fields for the gradient and tangent vectors to the surface $z = F(x, y)$ at several points (x, y) and also show the projection of the intersection of the constraint surface with the graph of $F(x, y)$ on the xy-plane. The following figure is the graph of the direction fields and the projection over the region $0 \le x, \quad y \le 2.5$.

In the lower right corner of the graph the direction field of the gradient appears to be perpendicular to the projection. Thus, we assume that the maximum occurs in this region. We move the window to this region as in the following figure and zoom in to obtain a magnified view of the direction field.

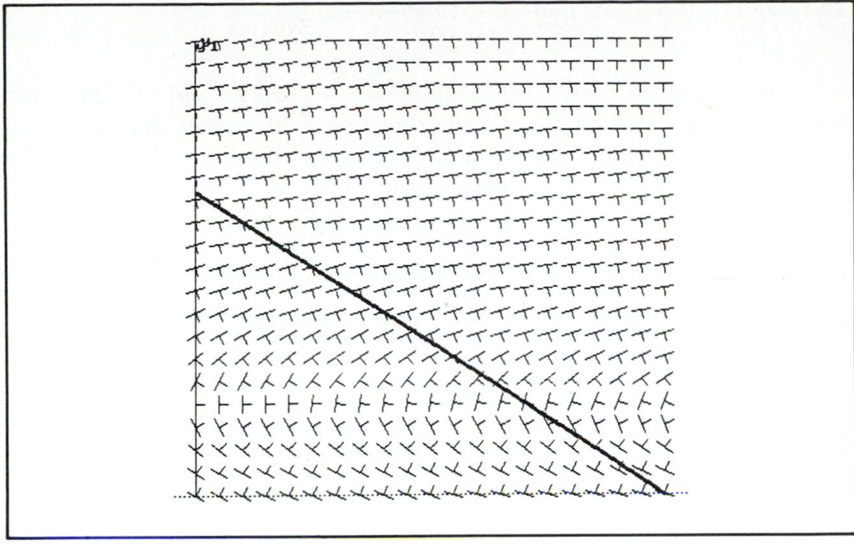
y1

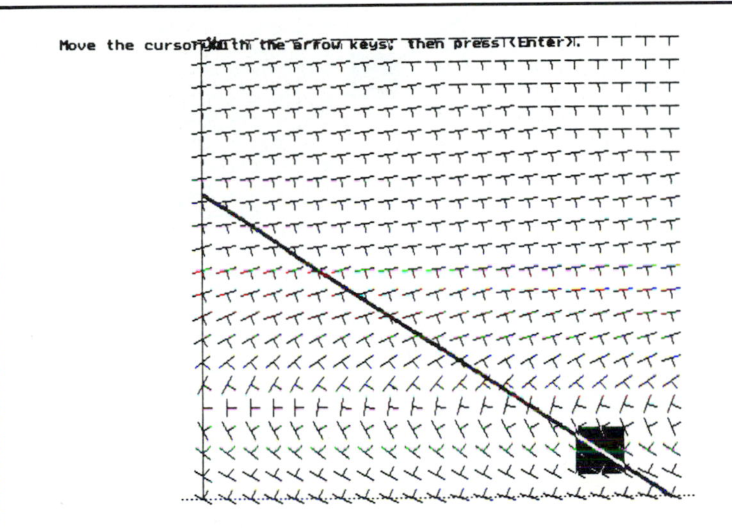
Move the cursor with the arrow keys, then press <Enter>.

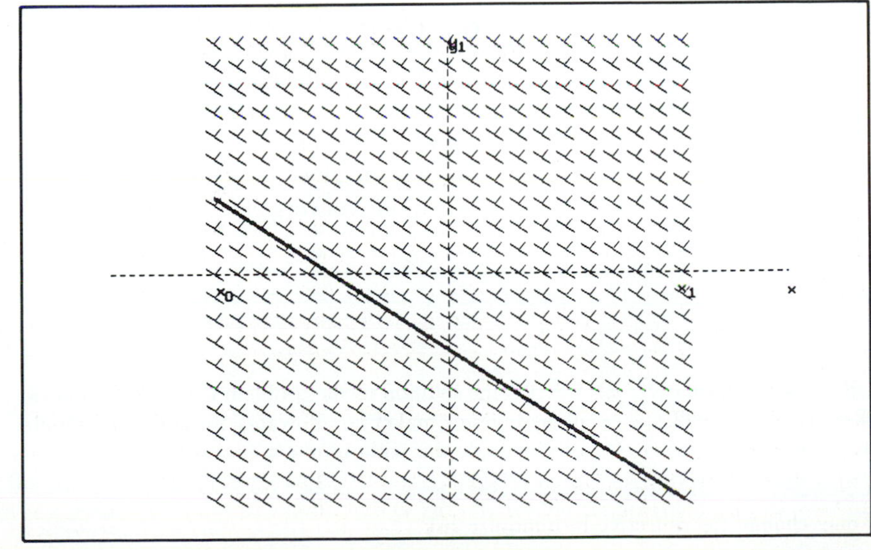
y1
x0
x1
x

The magnified view is the region $2 \le x \le 2.25$ and $0.125 \le y \le 0.375$. Further examination of the direction field reveals that the maximum value for the constrained F may be realized when x lies between 2.1500 and 2.1525 and when y lies between 0.23125 and 0.233750 (see the following figure).

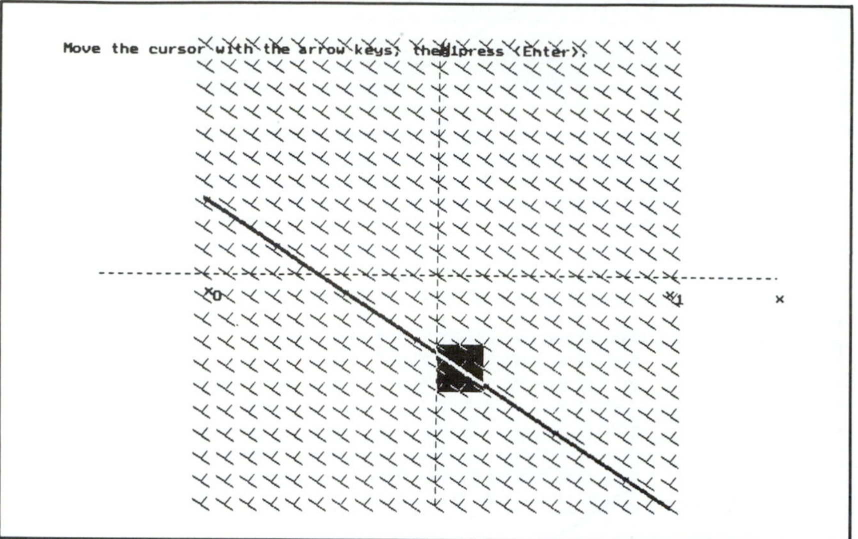

These estimates can be made more accurate by exiting the graphics program and activating the LaGrange multiplier program that uses the extension of Newton's method to solve the system of equations for x, y, and t (the Lagrange multiplier):

$$\frac{\partial F}{\partial x} - t\frac{\partial G}{\partial x} = 0$$

$$\frac{\partial F}{\partial y} - t\frac{\partial G}{\partial y} = 0$$

$$G - C = 0$$

The center of the viewing window is used as the starting values for x and y. After two iterations the method converges to $x = 2.18596$, $y = 0.20936$, and $t = 0.15733$. The corresponding value of F is 10.444889. Thus, although the graphical investigation did not give completely accurate results, it provided a good estimate for rapid convergence of the numerical method.

EXERCISES

1. Repeat the above investigation for an investor who has a very liberal attitude toward risk—say, $f = 0.02$. Do the same for a very conservative investor—say, $f = 0.5$.

2. What is the solution to the problem if $f = 0$—that is, if the investor makes no allowance for risk? You should be able to do this without much analysis.

3. An extremely conservative investor may change the objective to minimize risk. What is the solution in this case?

Multiple Integrals

18

The definite integral of a function of a single variable can be extended to functions of two or more variables. Integrals of a function of two or more variables are called *multiple integrals*. Specifically, the integral of a function of two variables is called a *double integral*, and integrals of a function of three variables are called *triple integrals*. For consistency, we shall call the integral of a function of one variable a *single integral*.

One of the many uses of multiple integration is to compute areas and volumes of regions more general than those considered in Chapter 6.

18.1 The Double Integral

Since the development of multiple integrals parallels that of a single integral, we begin by reviewing the definition of a single integral (see Chap. 5).

In the definition of a single integral $\int_a^b f(x)\,dx$, we required that f be defined on a closed (finite) interval $a \le x \le b$. Let's review the process that led to defining $\int_a^b f(x)\,dx$.

1. We partitioned $[a, b]$ into n subintervals whose lengths were Δx_1, $\Delta x_2, \ldots, \Delta x_n$.
2. In each subinterval we chose a number u_i.
3. We then evaluated f at u_i and formed the Riemann sums

$$\sum_{i=1}^{n} f(u_i)\Delta x_i$$

4. We took the limit as the norm $||P||$ of the partition (the length of the largest subinterval) approached 0. If this limit exists, then we defined

$$\int_a^b f(x)\,dx = \lim_{||P|| \to 0} \sum_{i=1}^{n} f(u_i)\Delta x_i$$

For a function f of two variables, the double integral of f is defined in a similar way. We begin by considering a function f of two variables defined over a closed rectangular region R given by

$$R: \quad a \leq x \leq b, \quad c \leq y \leq d$$

We divide the rectangular region R into subrectangles by drawing lines parallel to the coordinate axes and let n be the total number of subrectangles constructed. We now have a *partition* P of R into n subrectangles R_1, $R_2, \ldots, R_n$ (see Fig. 1).

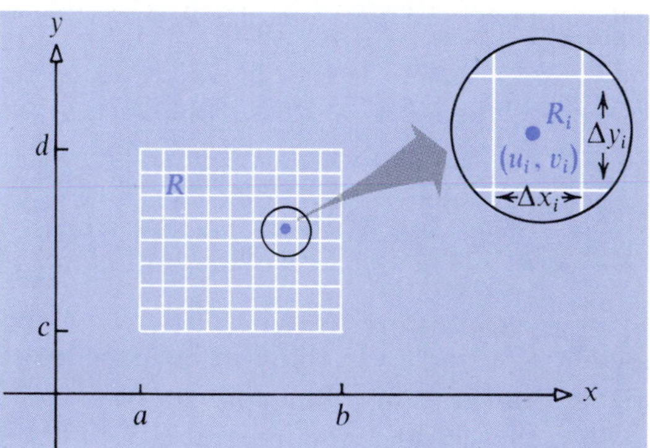

Figure 1

The *norm* of the partition P, denoted by $||P||$, is defined as the length of the longest diagonal of the rectangles R_i, $i = 1, 2, \ldots, n$. If Δx_i denotes the length of the rectangle R_i and Δy_i is its width, then its area is $\Delta A_i = \Delta x_i \Delta y_i$. In each rectangle R_i, we arbitrarily select a point (u_i, v_i) and evaluate the function f there. Then we can form the sum

$$\sum_{i=1}^{n} f(u_i, v_i)\Delta A_i \tag{1}$$

This sum is called a *Riemann sum* of f for the partition P and depends on both the choice of a partition and the choice of the points (u_i, v_i). However, if all such sums can be made as close as we please to a number I by choosing partitions whose norms are sufficiently close to 0, then I is defined as the limit of these sums as $\|P\|$ approaches 0.

[18.1.1] DEFINITION

Let f be a function defined on a closed rectangular region R. If there is a number I with the property that for any $\varepsilon > 0$, there is a $\delta > 0$ such that

$$\left| \sum_{i=1}^{n} f(u_i, v_i)\Delta A_i - I \right| < \varepsilon$$

for every partition P for which $\|P\| < \delta$ and for any choice of (u_i, v_i) in R_i, then I is called the limit* of the sums of the form $\sum_{i=1}^{n} f(u_i, v_i)\Delta A_i$, and we write

$$I = \lim_{\|P\| \to 0} \sum_{i=1}^{n} f(u_i, v_i)\Delta A_i$$

The definition asserts that the Riemann sums of f for the partition P can be made as close as we please to I if the norm of the partition P is sufficiently close to 0. We now give a name to this number I, provided it exists.

[18.1.2] DEFINITION / *Double Integral.*

Let f be a function defined on a closed rectangular region R. Then the *double integral of f over R*, denoted by $\iint_R f(x, y)\, dA$, is defined by

$$\iint_R f(x, y)\, dA = \lim_{\|P\| \to 0} \sum_{i=1}^{n} f(u_i, v_i)\Delta A_i$$

provided this limit exists. In this case, f is said to be *integrable* on R.

Other symbols for the double integral of f over R are

$$\iint_R f(x, y)\, dx\, dy \qquad \text{and} \qquad \iint_R f(x, y)\, dy\, dx$$

EXAMPLE 1

Let $f(x, y) = xy$ be a function defined over the unit square having its lower left corner at $(1, 0)$ and its upper right corner at $(2, 1)$. Use a Riemann sum to approximate the double integral of f over this region by partitioning the unit square into four congruent subsquares and using the lower left corner of each as (u_i, v_i).

* The proof that such a number, if it exists, is unique is similar to the proof given in Appendix I.

Solution

Figure 2 illustrates the situation. By using (1), in which $\Delta A_i = (\frac{1}{2})(\frac{1}{2}) = \frac{1}{4}$ for $i = 1, 2, 3, 4,$ the Riemann sum is

$$\sum_{i=1}^{4} f(u_i, v_i)\Delta A_i = f(1, 0)\Delta A_1 + f(\tfrac{3}{2}, 0)\Delta A_2 + f(1, \tfrac{1}{2})\Delta A_3 + f(\tfrac{3}{2}, \tfrac{1}{2})\Delta A_4$$

$$= 0 \cdot \tfrac{1}{4} + 0 \cdot \tfrac{1}{4} + \tfrac{1}{2} \cdot \tfrac{1}{4} + \tfrac{3}{4} \cdot \tfrac{1}{4} = (\tfrac{1}{2} + \tfrac{3}{4}) \cdot \tfrac{1}{4} = \tfrac{5}{16}$$

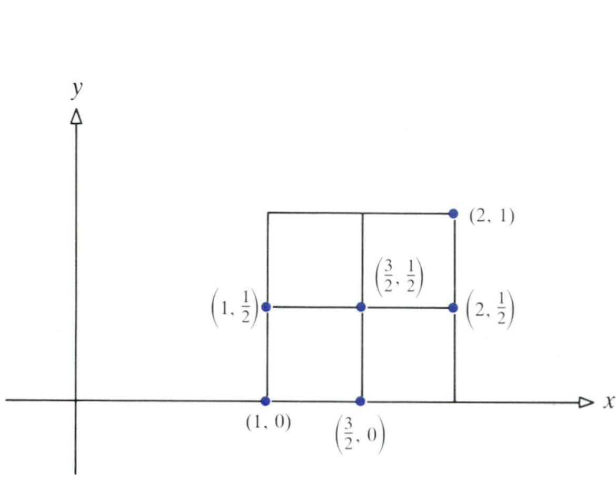

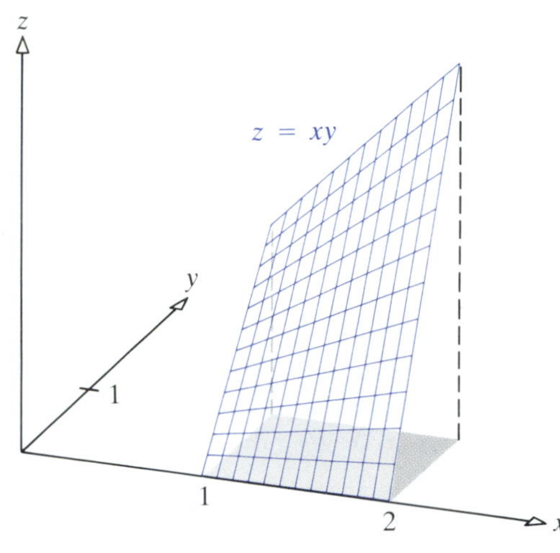

Figure 2

EXAMPLE 2

In Example 1 use the upper right corners as (u_i, v_i) to approximate the double integral.

Solution

By using (1), we get the Riemann sum

$$\sum_{i=1}^{4} f(u_i, v_i)\Delta A_i = f(\tfrac{3}{2}, \tfrac{1}{2})\Delta A_1 + f(2, \tfrac{1}{2})\Delta A_2 + f(\tfrac{3}{2}, 1)\Delta A_3 + f(2, 1)\Delta A_4$$

$$= \tfrac{3}{4} \cdot \tfrac{1}{4} + 1 \cdot \tfrac{1}{4} + \tfrac{3}{2} \cdot \tfrac{1}{4} + 2 \cdot \tfrac{1}{4} = \tfrac{21}{16}$$

Evaluating $\iint_R f(x, y)\, dA$: R a Closed Rectangular Region

The evaluation of a double integral of a function f of two variables whose domain is a closed rectangle is equivalent to the evaluation of a pair of definite integrals in which one of the integrations is performed *partially*. Partial integration is merely the reverse of partial differentiation. The symbol $\int_a^b f(x, y)\, dx$ is an instruction to hold y fixed and integrate with respect to x. The result will be a function of y alone. Similarly, $\int_c^d f(x, y)\, dy$ is an instruction to hold x fixed and integrate with respect to y. The result here is a function of x alone.

EXAMPLE 3

Evaluate:

(a) $\int_1^2 2x^2y\,dx$ (b) $\int_0^4 2x^2y\,dy$

Solution

(a) Hold y fixed and integrate with respect to x:

$$\int_1^2 2x^2y\,dx = 2y\int_1^2 x^2\,dx = 2y\frac{x^3}{3}\bigg|_1^2 = 2y\cdot\frac{7}{3} = \frac{14y}{3}$$

(b) Hold x fixed and integrate with respect to y:

$$\int_0^4 2x^2y\,dy = 2x^2\int_0^4 y\,dy = 2x^2\frac{y^2}{2}\bigg|_0^4 = 16x^2$$

EXAMPLE 4

Evaluate:

(a) $\int_0^4\left[\int_1^2 2x^2y\,dx\right]dy$ (b) $\int_1^2\left[\int_0^4 2x^2y\,dy\right]dx$

Solution

(a) $$\int_0^4\left[\int_1^2 2x^2y\,dx\right]dy = \int_0^4 \frac{14y}{3}\,dy = \frac{7y^2}{3}\bigg|_0^4 = \frac{112}{3}$$

From Example 3, part (a)

(b) $$\int_1^2\left[\int_0^4 2x^2y\,dy\right]dx = \int_1^2 16x^2\,dx = \frac{16x^3}{3}\bigg|_1^2 = \frac{112}{3}$$

From Example 3, part (b)

Integrals of the form

$$\int_a^b\left[\int_c^d f(x, y)\,dy\right]dx \qquad \text{and} \qquad \int_c^d\left[\int_a^b f(x, y)\,dx\right]dy$$

are called *iterated integrals*. In the iterated integral on the left, the function f is integrated partially with respect to y from c to d, resulting in a function of x that is integrated from a to b. In the iterated integral on the right, the function f is integrated partially with respect to x from a to b, resulting in a function of y that is integrated from c to d. The following result, called Fubini's theorem,* provides a practical way to evaluate a double integral. Although we shall not prove this theorem, we will provide a geometric justification under more general conditions in the next section.

* Named for the Italian mathematician Guido Fubini (1879–1943).

[18.1.3] THEOREM / *Fubini's Theorem.*

Let $z = f(x, y)$ denote a function of two variables defined on the closed rectangular region R given by $a \le x \le b$, $c \le y \le d$. If f is continuous on R, then

$$\iint_R f(x, y)\, dA = \int_a^b \left[\int_c^d f(x, y)\, dy \right] dx = \int_c^d \left[\int_a^b f(x, y)\, dx \right] dy \qquad (2)$$

Note that theorem (18.1.3) states that either form of the iterated integral in (2) may be used to evaluate a double integral over a closed rectangular region.

EXAMPLE 5

Evaluate $\iint_R xy\, dA$ if R is the closed rectangular region $1 \le x \le 2$, $0 \le y \le 1$.

Solution

Based on theorem (18.1.3), the double integral equals either of the iterated integrals

$$\int_1^2 \left[\int_0^1 xy\, dy \right] dx \qquad \text{or} \qquad \int_0^1 \left[\int_1^2 xy\, dx \right] dy$$

We evaluate the left integral; you are encouraged to evaluate the right one.

$$\iint_R xy\, dA = \int_1^2 \left[\int_0^1 xy\, dy \right] dx = \int_1^2 x \frac{y^2}{2} \Bigg|_0^1 dx = \int_1^2 \frac{1}{2} x\, dx = \frac{1}{4} x^2 \Bigg|_1^2 = \frac{3}{4}$$

The Double Integral over a More General Region

We begin by restating two definitions given earlier.

[18.1.4] DEFINITION / *Closed; Bounded.*

A set R of points in the xy-plane is said to be *closed* if R contains all its boundary points. A set R of points in the xy-plane is said to be *bounded* if R can be enclosed within some suitably large rectangle whose sides are parallel to the coordinate axes.*

Suppose f is a function of two variables defined on a closed, bounded region R. Then we can enclose R in a rectangle. Now we partition the rectangle into subrectangles by drawing lines parallel to the coordinate axes and discard any rectangle thus formed that does not lie entirely in R (see Fig. 3). Let n be the number of rectangles that remain. Now we may proceed as we

---

* This definition of a bounded set is equivalent to the one given in Section 17.3 (p. 1009). If a set of points can be enclosed within some suitably large disk, it can also be enclosed within some suitably large rectangle, and vice versa.

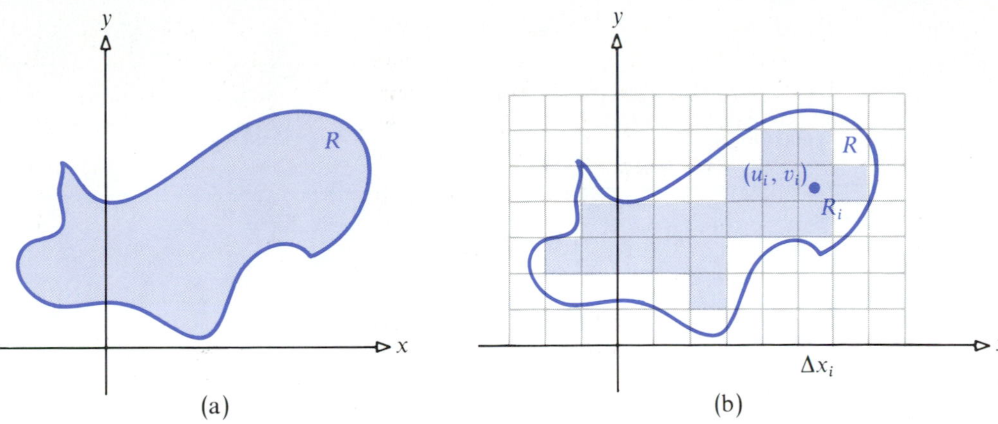

Figure 3

did earlier to define the double integral of a function f defined on the region R as

$$\iint_R f(x, y)\, dA = \lim_{\|P\| \to 0} \sum_{i=1}^{n} f(u_i, v_i)\Delta x_i \Delta y_i \tag{3}$$

Two comments are in order: First, it is apparent (at least intuitively) that the areas of the discarded rectangles will approach 0 as the norm of the partition approaches 0, so that having discarded them will not affect the limit of the sum. Second, if the limit of the sum exists for rectangular partitions, it can be shown that the manner in which R is partitioned does not matter, as long as each subregion obtained has an area.

We state without proof a condition under which the limit is (3) exists. The similarity of this result and theorem (5.3.4) for a single integral is apparent.

[18.1.5] THEOREM

If $z = f(x, y)$ is a continuous function defined on a closed, bounded region R, then f is integrable on R.

Geometric Interpretation

If a function f of a single variable is continuous and nonnegative on a closed interval $[a, b]$, then the single integral $\int_a^b f(x)\, dx$ equals the area under the graph of f from $x = a$ to $x = b$. Similarly, suppose $z = f(x, y)$ is a continuous and nonnegative function defined on a closed, bounded region R. The graph of f is then a surface lying above the xy-plane. Figure 4 shows a typical rectangle that results from partitioning R. The area of this rectangle is $\Delta x_i \Delta y_i$. The value $f(u_i, v_i)$ is the height of a rectangular solid whose base has the area $\Delta x_i \Delta y_i$. The volume of this rectangular solid is the product $f(u_i, v_i)\Delta x_i \Delta y_i$. The sum $\sum_{i=1}^{n} f(u_i, v_i)\Delta x_i \Delta y_i$ over all the rectangles of the partition will approximate the volume of the solid under the graph of the surface $z = f(x, y)$ and above R. But this sum is also an approximation

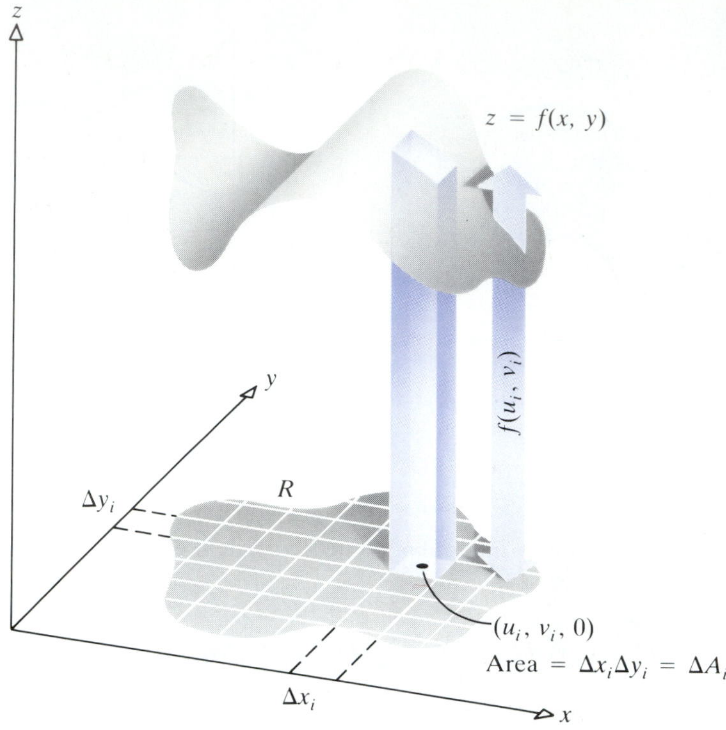

Figure 4

to the double integral $\iint_R f(x, y)\, dA$. As a result, the volume of the solid is equal to the value of the double integral.*

[18.1.6] THEOREM

Let f denote a continuous, nonnegative function defined on a closed, bounded region R. The volume V under the surface $z = f(x, y)$ and over the region R is given by

$$V = \iint_R f(x, y)\, dA \tag{4}$$

In particular, if $f(x, y) = 1$ in (4), the double integral $\iint_R dA$ represents the *area* of the region R, since, when the height of the solid is 1, the area and the volume are numerically equal.

EXAMPLE 6

Find the volume V under $z = f(x, y) = x^2 + y^2$ and over the rectangular region $0 \le x \le 2$, $0 \le y \le 1$.

* It can be proved that this formula for volume is consistent with the formulas for volume given in Chapter 6.

Solution

Figure 5 illustrates the volume we seek. From theorems (18.1.3) and (18.1.6), the volume V is

$$V = \iint_R (x^2 + y^2)\, dA = \int_0^2 \left[\int_0^1 (x^2 + y^2)\, dy\right] dx = \int_0^2 \left(x^2 y + \frac{y^3}{3}\right)\Bigg|_0^1 dx$$

$$= \int_0^2 \left(x^2 + \frac{1}{3}\right) dx = \left(\frac{x^3}{3} + \frac{x}{3}\right)\Bigg|_0^2 = \frac{10}{3} \text{ cubic units}$$

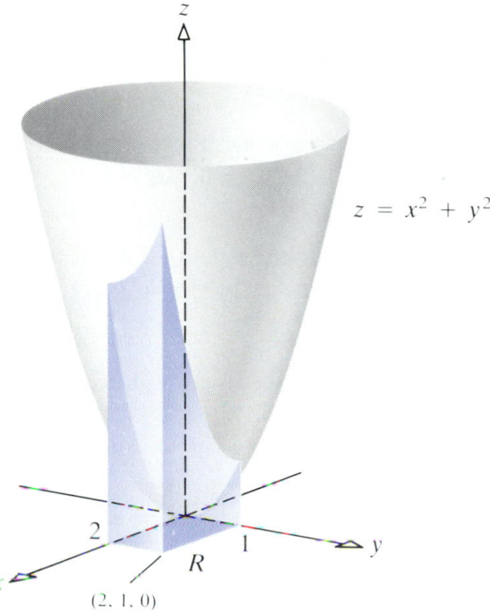

Figure 5

Properties of Double Integrals

We close this section by listing some properties of the double integral. The similarity between these properties and those of a single integral is apparent. The proofs of these properties are analogous to the ones given in Chapter 5, and so are omitted.

[18.1.7] THEOREM / ***Properties of Double Integrals.***

Suppose f and g are continuous functions of two variables defined on a closed, bounded region R. Then

$$\iint_R [f(x, y) + g(x, y)]\, dA = \iint_R f(x, y)\, dA + \iint_R g(x, y)\, dA \qquad (5)$$

$$\iint_R cf(x, y)\, dA = c \iint_R f(x, y)\, dA \qquad c \text{ a constant} \qquad (6)$$

If R consists of two subregions, R_1 and R_2, that have no points in common except for points lying on portions of their common boundary, then

$$\iint_R f(x, y)\,dA = \iint_{R_1} f(x, y)\,dA + \iint_{R_2} f(x, y)\,dA \tag{7}$$

If $f(x, y) \geq 0$ throughout R, then $\iint_R f(x, y)\,dA \geq 0$.

EXERCISE 18.1

In Problems 1–4 approximate $\iint_R f(x, y)\,dA$ by a Riemann sum. Use the indicated partition of R and choose (u_i, v_i) as the lower left corner of each subregion.

1. $f(x, y) = x^2 + y^2$; $1 \leq x \leq 3$, $2 \leq y \leq 4$; $x_0 = 1$, $x_1 = 2$, $x_2 = 3$, $y_0 = 2$, $y_1 = 3$, $y_2 = 4$

2. $f(x, y) = x^2 - y^2$; $0 \leq x \leq 4$, $1 \leq y \leq 3$; $x_0 = 0$, $x_1 = 1$, $x_2 = 2$, $x_3 = 3$, $x_4 = 4$, $y_0 = 1$, $y_1 = 2$, $y_2 = 3$

3. $f(x, y) = 2xy - y^2$; $0 \leq x \leq 4$, $0 \leq y \leq 2$; $x_0 = 0$, $x_1 = 2$, $x_2 = 4$, $y_0 = 0$, $y_1 = 1$, $y_2 = 2$

4. $f(x, y) = 2x^2y + x$; $-1 \leq x \leq 1$, $-2 \leq y \leq 0$; $x_0 = -1$, $x_1 = 0$, $x_2 = 1$, $y_0 = -2$, $y_1 = -1$, $y_2 = 0$

5. Find the Riemann sum for Example 1 if R is divided into nine equal squares and the lower left corners are taken as (u_i, v_i).

6. Rework Problem 5 using the upper right corners for (u_i, v_i).

In Problems 7–16 evaluate each iterated integral.

7. $\int_0^1 \left[\int_0^2 x^2y\,dy\right] dx$

8. $\int_0^3 \left[\int_0^2 3xy\,dy\right] dx$

9. $\int_{-1}^1 \left[\int_0^1 3x^2y^2\,dx\right] dy$

10. $\int_0^2 \left[\int_{-1}^1 2xy^2\,dx\right] dy$

11. $\int_0^{\pi/4} \left[\int_0^2 x\cos y\,dx\right] dy$

12. $\int_0^{\pi/3} \left[\int_0^3 x^2\sin y\,dx\right] dy$

13. $\int_0^1 \left[\int_0^{\pi/2} e^x\cos y\,dy\right] dx$

14. $\int_0^2 \left[\int_0^{\pi/2} e^y\sin x\,dx\right] dy$

15. $\int_1^2 \left[\int_{-\pi/2}^{\pi/2} \frac{\sin y}{x}\,dy\right] dx$

16. $\int_1^2 \left[\int_0^{\pi/2} \frac{\cos y}{x}\,dy\right] dx$

In Problems 17–24 find the volume under $z = f(x, y)$ and over the given rectangular region.

17. $f(x, y) = x + 2y$; $0 \leq x \leq 1$, $0 \leq y \leq 2$

18. $f(x, y) = 2x + 3y$; $0 \leq x \leq 2$, $0 \leq y \leq 3$

19. $f(x, y) = \sin x$; $0 \leq x \leq \pi/2$, $0 \leq y \leq 1$

20. $f(x, y) = \cos y$; $0 \leq x \leq 1$, $0 \leq y \leq \pi/2$

21. $f(x, y) = \dfrac{xy}{x^2 + y^2}$; $1 \leq x \leq 2$, $1 \leq y \leq 2$

22. $f(x, y) = \dfrac{y^2}{x^2}$; $1 \leq x \leq 2$, $1 \leq y \leq 2$

23. $f(x, y) = e^{x+y}$; $0 \leq x \leq 2$, $0 \leq y \leq 1$

24. $f(x, y) = e^{x-y}$; $0 \leq x \leq 2$, $1 \leq y \leq 2$

25. Let R be the region enclosed by $y = 1$, $y = -1$, $x = 0$, and $x = 1$; let R_1 and R_2 be the subregions of R in the first and fourth quadrants, respectively. Suppose $f(x, y)$ is continuous on R and

$$\iint_R 3f(x, y)\,dA - 2\iint_{R_1} f(x, y)\,dA = \iint_{R_2} f(x, y)\,dA$$

$$\iint_{R_2} 5f(x, y)\,dA - 2\iint_{R_1} f(x, y)\,dA = 18$$

Find $\iint_R f(x, y)\,dA$.

26. Using the properties of double integrals, prove that if the functions f and g are integrable on R and $f(x, y) \le g(x, y)$ for all (x, y) in R, then $\iint_R f(x, y)\,dA \le \iint_R g(x, y)\,dA$.

Evaluation of $\iint_R f(x, y)\,dx\,dy$ over Nonrectangular Regions 18.2

Until now, we have been evaluating double integrals of a function f defined on a closed rectangle. In this section, we see how iteration can be used to evaluate double integrals over more general regions, which we classify as x-simple and y-simple regions.

[18.2.1] DEFINITION / *x-Simple Region.*

Let R denote a closed, bounded region. Then R is an x-simple region if R has a boundary that consists of two smooth curves, $y = g_1(x)$ and $y = g_2(x)$, $g_1(x) \le g_2(x)$, $a \le x \le b$, and possibly a part of the vertical lines $x = a$ and $x = b$.

Figure 6 illustrates some typical x-simple regions.

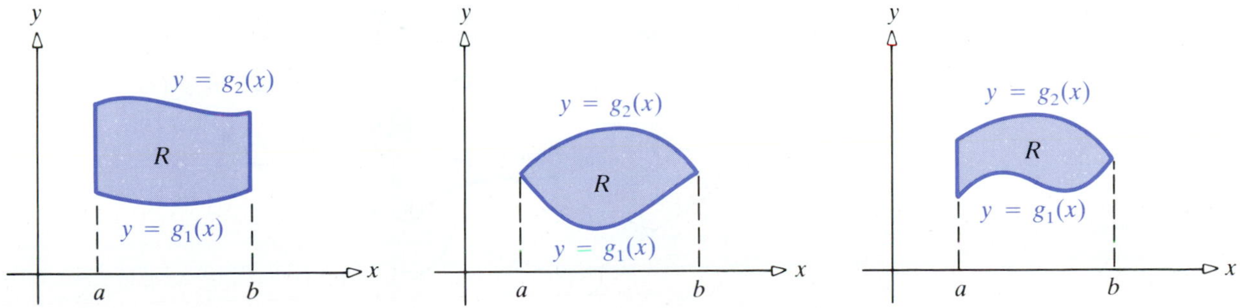

Figure 6

The double integral of a function defined on closed, bounded region that is x-simple can be expressed as an iterated integral.

[18.2.2] THEOREM / *Fubini's Theorem for x-Simple Regions.*

If f is a continuous function defined on a closed, bounded region R that is x-simple, the double integral of f on R is given by

$$\iint_R f(x, y)\, dA = \int_a^b \left[\int_{g_1(x)}^{g_2(x)} f(x, y)\, dy\right] dx$$

We give an intuitive (geometric) argument* for theorem (18.2.2) when $f(x, y) \geq 0$ on R. At a number x_i in the closed interval $[a, b]$, let $A(x_i)$ denote the area of the intersection of the plane $x = x_i$ and the solid under the surface $z = f(x, y)$ and over R (see Fig. 7). By the method developed in Chapter 6, we find the volume V of this solid

$$V = \int_a^b A(x)\, dx$$

But this volume V is also given by the double integral of f over R. Thus,

$$V = \iint_R f(x, y)\, dA = \int_a^b A(x)\, dx \qquad \textbf{(1)}$$

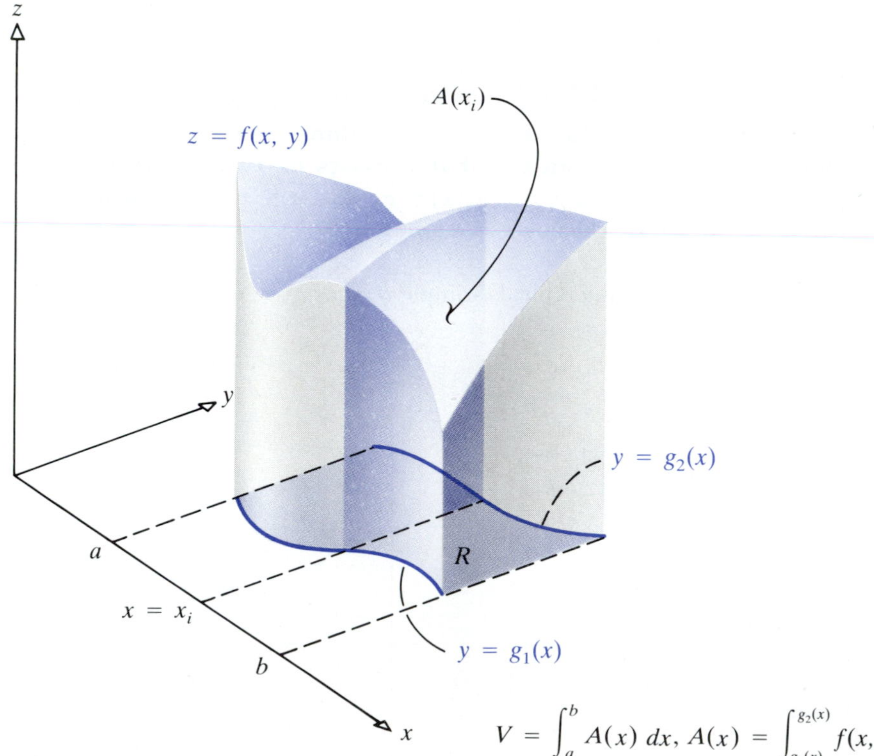

Figure 7

* An analytic proof is given in most books in advanced calculus.

But $A(x_i)$ is the area of the plane region under the surface $z = f(x_i, y)$ from $y = g_1(x_i)$ to $y = g_2(x_i)$. That is,

$$A(x) = \int_{g_1(x)}^{g_2(x)} f(x, y)\,dy \tag{2}$$

where x is held fixed. By substituting (2) for $A(x)$ in (1), we find

$$\iint_R f(x, y)\,dA = \int_a^b \left[\int_{g_1(x)}^{g_2(x)} f(x, y)\,dy \right] dx$$

EXAMPLE 1

Evaluate $\iint_R xy\,dA$ if R is the region enclosed by

$$y = x^2 \quad \text{and} \quad y = \sqrt{x}$$

Solution

We begin by graphing the region R (see Fig. 8). Note that R is a closed, bounded region that is x-simple, where $g_1(x) = x^2$, $g_2(x) = \sqrt{x}$, $0 \le x \le 1$. Thus, by theorem (18.2.2),

$$\iint_R xy\,dA = \int_0^1 \left[\int_{x^2}^{\sqrt{x}} xy\,dy \right] dx = \int_0^1 x \frac{y^2}{2}\Big|_{x^2}^{\sqrt{x}} dx$$

$$= \int_0^1 \frac{x}{2}(x - x^4)\,dx = \int_0^1 \frac{1}{2}(x^2 - x^5)\,dx = \frac{1}{2}\left[\frac{x^3}{3} - \frac{x^6}{6}\right]\Big|_0^1 = \frac{1}{12}$$

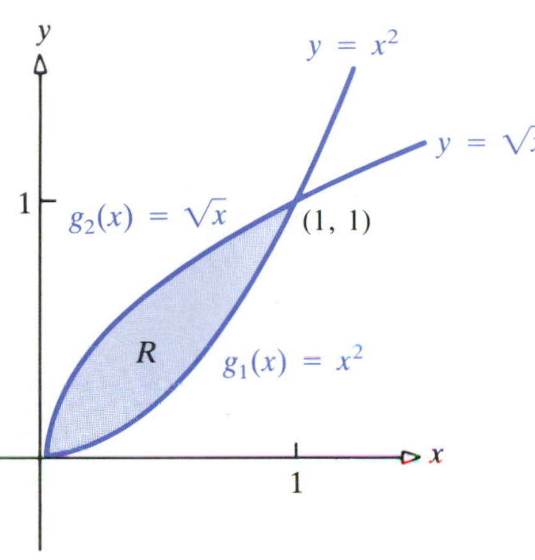

Figure 8

We define y-simple regions as follows:

[18.2.3] DEFINITION / *y-Simple Region.*

Let R denote a closed, bounded region. Then R is a y-simple region if R has a boundary that consists of two smooth curves, $x = h_1(y)$ and $x = h_2(y)$, $h_1(y) \le h_2(y)$, $c \le y \le d$, and possibly a part of the horizontal lines $y = c$ and $y = d$.

Figure 9 illustrates some typical y-simple regions.

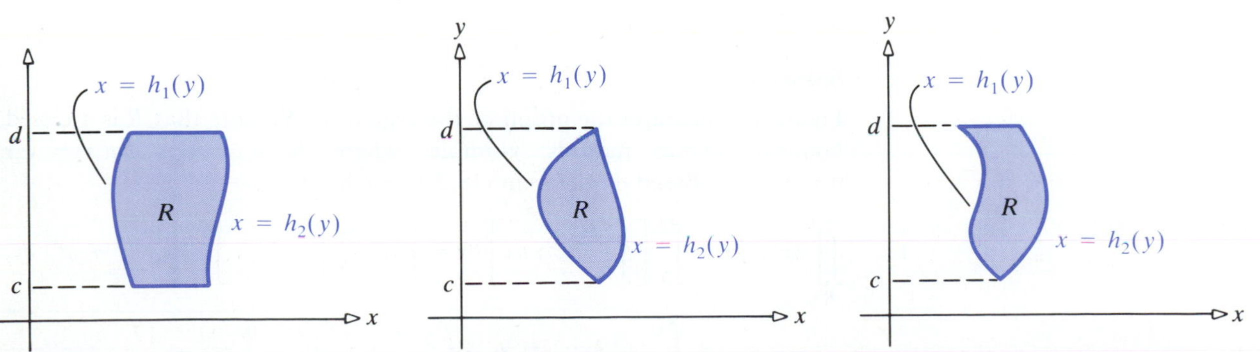

Figure 9

The double integral of a function defined on a closed, bounded region that is y-simple can be expressed as an iterated integral.

[18.2.4] THEOREM / *Fubini's Theorem for y-Simple Regions.*

If f is a continuous function defined on a closed, bounded region R that is y-simple, the double integral of f on R is given by

$$\iint_R f(x, y)\, dA = \int_c^d \left[\int_{h_1(y)}^{h_2(y)} f(x, y)\, dx \right] dy$$

Figure 10 provides justification for this result.

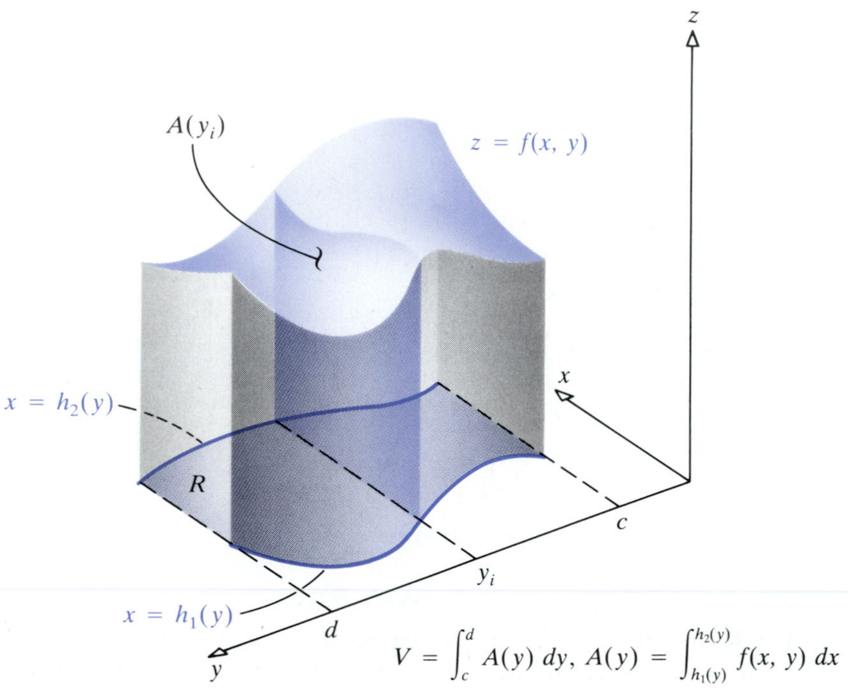

Figure 10

EXAMPLE 2

Evaluate $\iint_R 3x^2y\, dA$ if R is the region enclosed by $x = \sqrt{y}$, $x = -y$, and the line $y = 1$.

Solution

Figure 11 illustrates the graph of the region R. We note that R is a closed, bounded region that is y-simple, where $h_1(y) = -y$, $h_2(y) = \sqrt{y}$, $0 \le y \le 1$. Based on theorem (18.2.4), we have

$$\iint_R 3x^2y\, dA = \int_0^1 \left[\int_{-y}^{\sqrt{y}} 3x^2y\, dx \right] dy = \int_0^1 3y \frac{x^3}{3} \Bigg|_{-y}^{\sqrt{y}} dy = \int_0^1 y(y^{3/2} + y^3)\, dy$$

$$= \int_0^1 (y^{5/2} + y^4)\, dy = \left(\frac{y^{7/2}}{7/2} + \frac{y^5}{5} \right) \Bigg|_0^1 = \frac{2}{7} + \frac{1}{5} = \frac{17}{35}$$

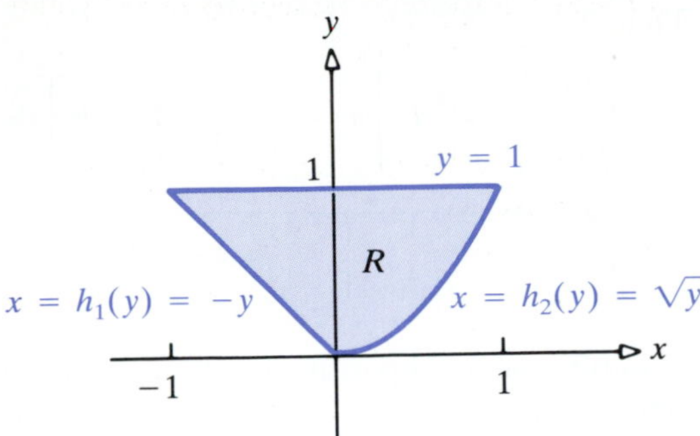

Figure 11

When the region R is both x-simple and y-simple, a *choice* of the order of integration is available. In some cases, integration in one order requires simpler techniques than is required by the opposite order. The next example illustrates that one order of integration is impossible, whereas the opposite is straightforward.

EXAMPLE 3

By changing the order of integration, evaluate

$$\int_0^1 \left[\int_{2x}^2 e^{y^2}\,dy\right] dx$$

Solution

We observe that the integration cannot be done in the order indicated, since e^{y^2} has no elementary antiderivative. Thus, we are forced to change the order of the integration. To accomplish this, we need to determine the region R on which the integration is performed. We infer from the given iterated integral that R is x-simple and R is enclosed by $y = 2x$ and $y = 2$, $0 \le x \le 1$. In Figure 12 we see that region R is also y-simple and is enclosed by $x = 0$

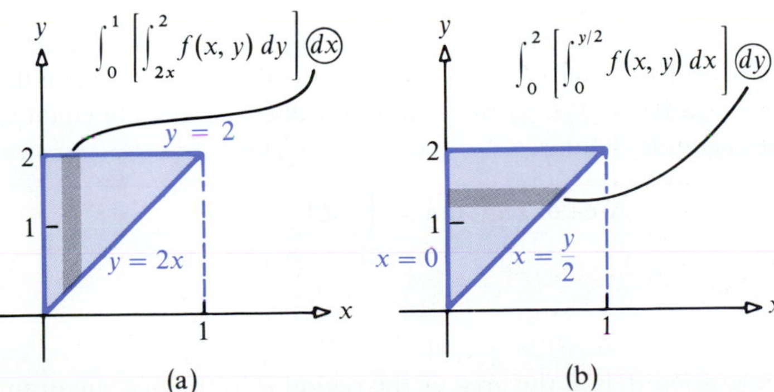

Figure 12

and $x = y/2$, $0 \le y \le 2$. Thus,

$$\int_0^1 \left[\int_{2x}^2 e^{y^2}\,dy\right] dx = \iint_R e^{y^2}\,dA = \int_0^2 \left[\int_0^{y/2} e^{y^2}\,dx\right] dy = \int_0^2 e^{y^2}\left[\int_0^{y/2} dx\right] dy$$

$$= \int_0^2 \frac{ye^{y^2}}{2}\,dy = \frac{1}{4}e^{y^2}\Big|_0^2 = \frac{1}{4}(e^4 - 1)$$

If a region R is neither x-simple nor y-simple, it may be possible to partition R into subregions, each of which is x-simple or y-simple. In this situation, we use (7) on page 1038 to evaluate the double integral. Figure 13 illustrates how this may be done.

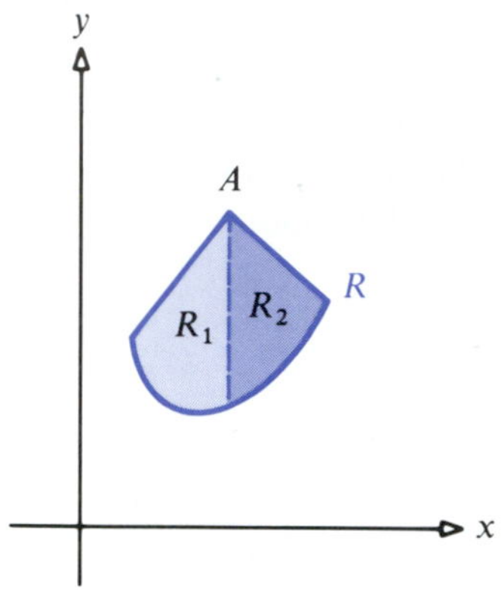

(a) R is not y-simple because of the corner at A; draw a vertical line through A, decomposing R into two subregions R_1 and R_2, each being x-simple

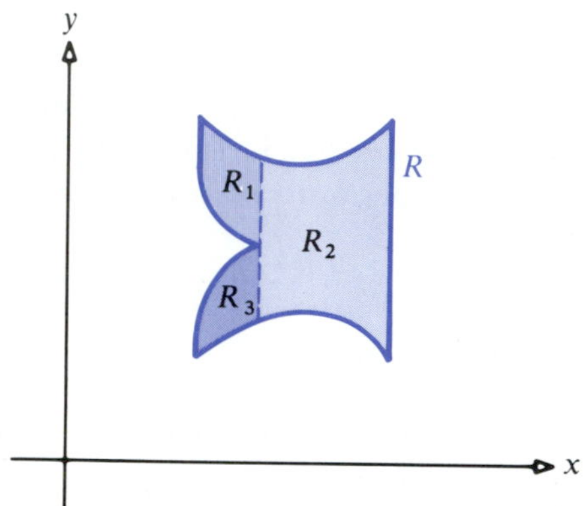

(b) R is neither x-simple nor y-simple
$R = R_1 \cup R_2 \cup R_3$
R_1, R_2, R_3 are x-simple

$$\iint_R f(x, y)\,dA = \iint_{R_1} f(x, y)\,dA + \iint_{R_2} f(x, y)\,dA + \iint_{R_3} f(x, y)\,dA$$

Figure 13

Area and Volume as Double Integrals

Double integrals can be used to find the area of a closed, bounded region R. This is accomplished by letting $f(x, y) = 1$ in theorem (18.1.6). Then the volume under the surface $f(x, y) = 1$ and over R is numerically equal to the area of the region R. That is,

$$\text{Area of region } R = \iint_R dA$$

EXAMPLE 4

Use double integration to find the area of the region R in the first quadrant enclosed by the parabola $y = 6x - x^2$ and the line $y = 4x - 8$.

Solution

Figure 14 illustrates the region R. Note that R is neither x-simple nor y-simple. It is not x-simple because of the corner at $(2, 0)$; it is not y-simple because of the corner at $(4, 8)$. Because of this, we decompose R into regions that have smooth boundaries. We choose to split R with a vertical line through the point $(2, 0)$, thus obtaining two regions R_1 and R_2, each of which we may treat as x-simple.* The desired area is

$$\iint_R dA = \iint_{R_1} dA + \iint_{R_2} dA = \int_0^2 \left[\int_0^{6x-x^2} dy\right] dx + \int_2^4 \left[\int_{4x-8}^{6x-x^2} dy\right] dx$$

$$= \int_0^2 (6x - x^2)\, dx + \int_2^4 (-x^2 + 2x + 8)\, dx = \tfrac{28}{3} + \tfrac{28}{3} = \tfrac{56}{3}$$

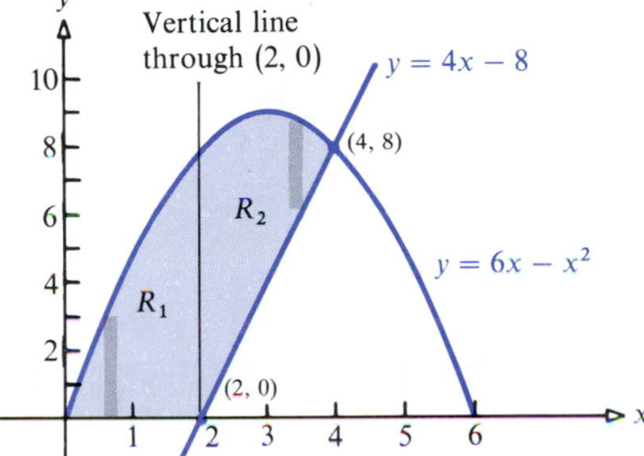

Figure 14

The volumes of certain solids can be expressed as double integrals that may then be evaluated as iterated integrals.

EXAMPLE 5

Find the volume of the solid enclosed in the first octant by the plane $x + y + z = 1$.

Solution

Figure 15 illustrates the solid and the region R in the xy-plane. We see that R is enclosed by the coordinate axes $x = 0$ and $y = 0$ and by the line $x + y = 1$ (or $y = 1 - x$). We shall treat R as an x-simple region so that y will vary according to $0 \le y \le 1 - x$ and x will vary according to

---

* In Problem 41 you are asked to do this example by splitting R with a horizontal line through the point $(4, 8)$.

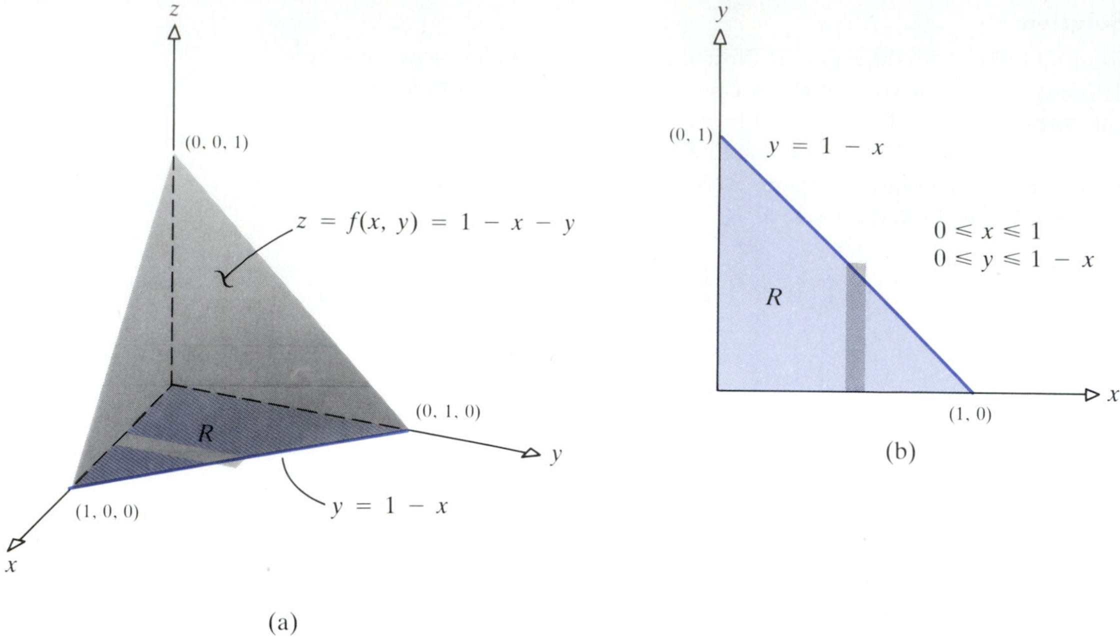

Figure 15

$0 \le x \le 1$. Thus, the volume V of the solid is

$$V = \iint_R z\, dA = \int_0^1 \left[\int_0^{1-x} (1 - x - y)\, dy\right] dx = \int_0^1 \left[\left((1-x)y - \frac{y^2}{2}\right)\Big|_0^{1-x}\right] dx$$

$$= \int_0^1 \left[(1-x)^2 - \frac{(1-x)^2}{2}\right] dx = \frac{1}{2}\int_0^1 (1-x)^2\, dx$$

$$= \frac{1}{2}\left(x - x^2 + \frac{x^3}{3}\right)\Big|_0^1 = \frac{1}{6} \text{ cubic unit}$$

EXAMPLE 6

Find the volume of the solid under the graph of the elliptic paraboloid $z = 8 - 2x^2 - y^2$ and above the first quadrant of the xy-plane.

Solution

The solid $z = 8 - 2x^2 - y^2,\ z \ge 0,$ is illustrated in Figure 16(a). We treat the region R as x-simple, as shown in Figure 16(b). Then,

$$V = \iint_R (8 - 2x^2 - y^2)\, dA = \int_0^2 \left[\int_0^{\sqrt{8-2x^2}} (8 - 2x^2 - y^2)\, dy\right] dx$$

The inside integral is

$$\int_0^{\sqrt{8-2x^2}} (8 - 2x^2 - y^2)\, dy = \left[(8 - 2x^2)y - \frac{y^3}{3}\right]\Big|_0^{\sqrt{8-2x^2}} = \frac{2}{3}(8 - 2x^2)^{3/2}$$

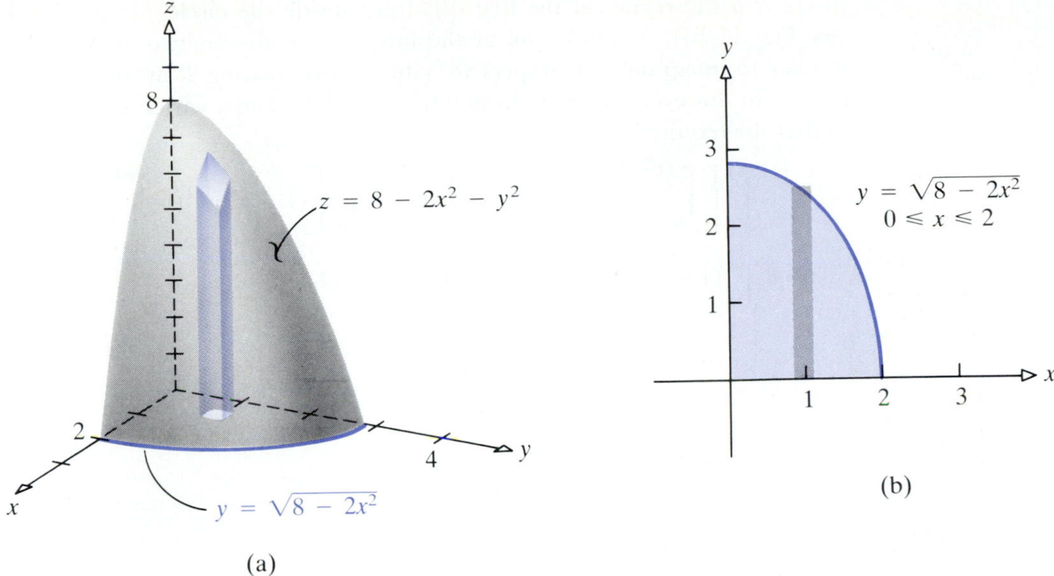

Figure 16

The required volume is therefore

$$V = \frac{2}{3}\int_0^2 (8 - 2x^2)^{3/2}\, dx$$

If we use the substitution $x = 2 \sin \theta$, $dx = 2 \cos \theta\, d\theta$, then we obtain

$$V = \frac{2}{3}\int_0^{\pi/2} 16\sqrt{2}(\cos^3\theta)(2 \cos \theta\, d\theta) = \frac{64\sqrt{2}}{3}\int_0^{\pi/2} \cos^4\theta\, d\theta$$

$$= \frac{64\sqrt{2}}{3}\left[\frac{(3)(1)}{(4)(2)}\frac{\pi}{2}\right] = 4\sqrt{2}\pi \text{ cubic units.}$$

↑ By Wallis' formula ((7) in Section 9.2)

EXAMPLE 7

Compute the volume common to the two cylinders $x^2 + y^2 = 1$ and $x^2 + z^2 = 1$.

Solution

Each cylinder has radius 1; their axes are perpendicular to each other and lie on the z-axis and y-axis, respectively. Figure 17(a) shows the portion of the solid lying in the first octant. We compute the volume of this portion of the solid. Then, since the volume in each octant is the same, the required volume will be eight times as much as the volume in the first octant; that is,

$$V = 8 \iint_R z\, dx\, dy = 8 \iint_R (1 - x^2)^{1/2}\, dx\, dy$$

where R is the region in the first quadrant inside the circle $x^2 + y^2 = 1$ (see Fig. 17(b)). A quick look at the integrand will convince us that it is simpler to integrate with respect to y first, thus treating R as an x-simple region. In this event, y varies from 0 to $(1 - x^2)^{1/2}$ and x varies from 0 to 1, so that the required volume is

$$V = 8\int_0^1 \left[\int_0^{\sqrt{1-x^2}} (1 - x^2)^{1/2}\, dy\right] dx = 8\int_0^1 \left[(1 - x^2)^{1/2}(y)\Big|_0^{\sqrt{1-x^2}}\right] dx$$

$$= 8\int_0^1 (1 - x^2)\, dx = 8\left(x - \frac{x^3}{3}\right)\Big|_0^1 = \frac{16}{3} \text{ cubic units}$$

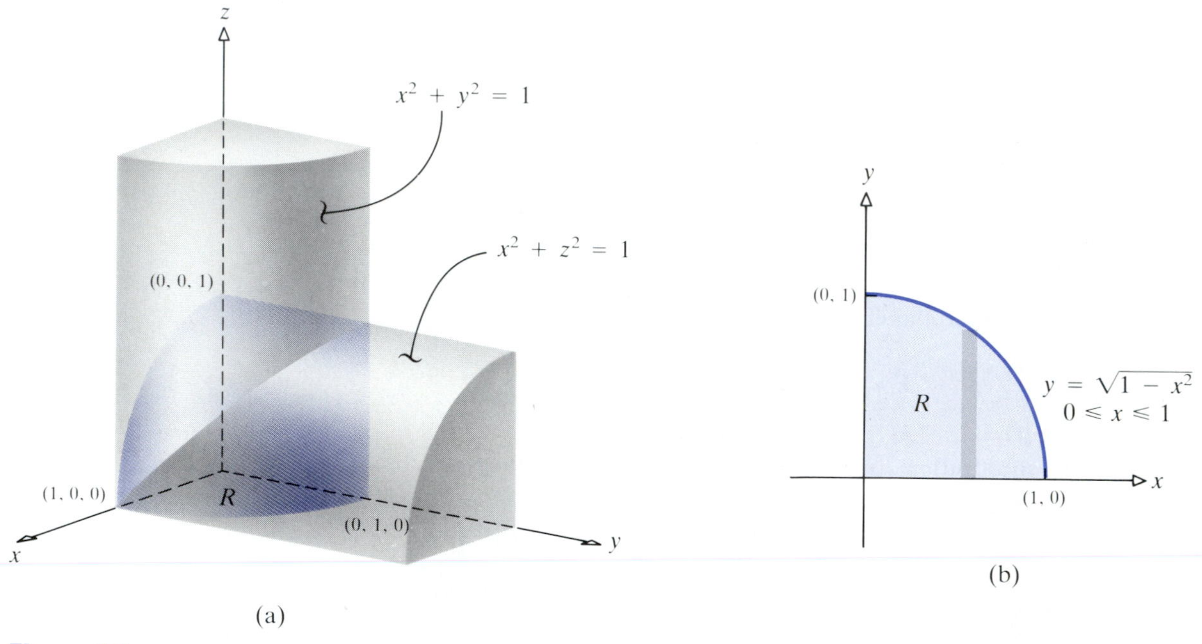

Figure 17

EXERCISE 18.2

In working the following problems, start by drawing an accurate figure illustrating the region R used in the integration. Pay particular attention to the boundary, since it determines the correct limits of integration.

In Problems 1–18 evaluate each iterated integral.

1. $\int_0^1 \left[\int_{x^2}^{\sqrt{x}} dy\right] dx$

2. $\int_0^1 \left[\int_{y^2}^{\sqrt{y}} x\, dx\right] dy$

3. $\int_{-1}^2 \left[\int_{y^2}^{y+2} dx\right] dy$

4. $\int_0^1 \left[\int_x^{2x} y\, dy\right] dx$

5. $\int_0^1 \left[\int_1^{e^y} \frac{y}{x}\, dx\right] dy$

6. $\int_0^1 \left[\int_0^{x^2} xe^y\, dy\right] dx$

7. $\int_0^2 \left[\int_y^{2y} xy\, dx\right] dy$

8. $\int_2^4 \left[\int_1^{y^2} \frac{y}{x^2}\, dx\right] dy$

9. $\int_0^1 \left[\int_{x^2}^x \sqrt{xy}\, dy\right] dx$

10. $\int_0^1 \left[\int_{x^2}^{x} \sqrt{x}\,dy\right] dx$

11. $\int_0^a \left[\int_{a-x}^{\sqrt{a^2-x^2}} y\,dy\right] dx$

12. $\int_0^1 \left[\int_x^1 \frac{dy}{x^2+1}\right] dx$

13. $\int_0^{\sqrt{\pi}} \left[\int_0^{x^2} x \sin y\,dy\right] dx$

14. $\int_0^1 \left[\int_y^{\sqrt{y}} (x^2+y^2)\,dx\right] dy$

15. $\int_1^2 \left[\int_0^{\ln x} xe^y\,dy\right] dx$

16. $\int_0^1 \left[\int_{x^5}^{x^2} (x+y)\,dy\right] dx$

17. $\int_2^3 \left[\int_0^{1/y} \ln y\,dx\right] dy$

18. $\int_0^{\pi} \left[\int_{\pi-y}^{\pi+y} \sin(x+y)\,dx\right] dy$

In Problems 19–22 evaluate each double integral.

19. $\iint_R (x+y)\,dA$; R is the region enclosed by $y = x^2$ and $y^2 = 8x$

20. $\iint_R (x^2 - y^2)\,dA$; R is the region enclosed by $y = x$ and $y = x^2$

21. $\iint_R y^2\,dA$; R is the region enclosed by $y = 2 - x$ and $y = x^2$

22. $\iint_R xy\,dA$; R is the region enclosed by $y^2 = x + 1$ and $y = 1 - x$

In Problems 23–32 use double integration to find the area of each region.

23. Enclosed by $y = x^3$ and $y = x^2$

24. Enclosed by $y = 2\sqrt{x}$ and $y = x^2/4$

25. Enclosed by $y = x^2 - 9$ and $y = 9 - x^2$

26. Enclosed by $x^2 + y^2 = 16$ and $y^2 = 6x$

27. Enclosed by $y = x^2$ and $x + y - 2 = 0$

28. Enclosed by the triangle with vertices (0, 0), (5, 0), and (2, 3)

29. Enclosed by $y = 1/\sqrt{x-1}$, $y = 0$, $x = 2$, and $x = 5$

30. Enclosed by $y = x^{3/2}$ and $y = x$

31. Enclosed by the line $x + y = 3$ and the hyperbola $xy = 2$

32. Enclosed by the hyperbola $xy = \sqrt{3}$ and the circle $x^2 + y^2 = 4$, in the first quadrant only

In Problems 33–40 find the volume of each solid.

33. The tetrahedron enclosed by the plane $x + 2y + z = 2$ and the coordinate planes

34. The elliptic paraboloid $4x^2 + 9y^2 = 36z$ between the planes $z = 0$ and $z = 1$

35. Below the paraboloid $z = x^2 + y^2$ and above the square in the xy-plane enclosed by the lines $x = \pm 1$ and $y = \pm 1$

36. Enclosed by the cylinder $z = 9 - y^2$ and the planes $x = 0$, $x = 4$, and $z = 0$

37. Enclosed by the surfaces $z = 4 - x^2 - y^2$, $z = 0$, and $x = 0$ (front half only)

38. Enclosed by the surfaces $x^2 + y^2 = 1$, $z = x$, and $z = x^2$

39. In the first octant below $y = z^2$, above $z = 0$, and enclosed on the sides by $y = x^2$ and $y = 4$

40. Enclosed by $y = e^x$ and the planes $x = 0$, $x = 1$, $z = 0$, and $z = y$, in the first octant only

41. Find the area of the region R enclosed by $y = 6x - x^2$ and the x-axis above the line $y = 4x - 8$ by splitting R with a horizontal line through the point (4, 8). (Refer to Fig. 14.)

42. Use double integration to find the area in the first quadrant enclosed by the parabola $x^2 = 9y$, the y-axis, and the circle $x^2 + y^2 = 10$.

In Problems 43–48 change the order of integration of each iterated integral. Do not integrate.

43. $\int_0^1 \left[\int_0^{x} f(x, y)\,dy\right] dx$

44. $\int_0^2 \left[\int_{y^2}^{2y} f(x, y)\,dx\right] dy$

45. $\int_0^a \left[\int_0^{\sqrt{a^2-y^2}} f(x, y)\,dx\right] dy$

46. $\int_0^{2\sqrt[3]{2}} \left[\int_{x^2/4}^{\sqrt{x}} f(x, y)\, dy \right] dx$ **47.** $\int_0^4 \left[\int_{y^2-2y}^{2y} f(x, y)\, dx \right] dy$ **48.** $\int_2^5 \left[\int_{x^2-6x+9}^{x-1} f(x, y)\, dy \right] dx$

In Problems 49–60 change the order of integration of the given iterated integral and then evaluate it.

49. $\int_0^1 \left[\int_x^{3x} xy\, dy \right] dx$ **50.** $\int_0^1 \left[\int_y^{\sqrt{y}} x^2y\, dx \right] dy$ **51.** $\int_0^1 \left[\int_{1-x}^{2-x} (x + y)\, dy \right] dx$

52. $\int_0^{\sqrt{\pi/2}} \left[\int_y^{\sqrt{\pi/2}} \sin x^2\, dx \right] dy$ **53.** $\int_0^{1/2} \left[\int_{2x}^1 e^{y^2}\, dy \right] dx$ **54.** $\int_0^1 \left[\int_y^1 \frac{\sin x}{x}\, dx \right] dy$

55. $\int_0^1 \left[\int_0^{\sqrt{1-x^2}} \frac{dy}{\sqrt{1 - y^2}} \right] dx$ **56.** $\int_1^5 \left[\int_{-\sqrt{y-1}}^{\sqrt{y-1}} y\, dx \right] dy$ **57.** $\int_0^1 \left[\int_y^1 \sqrt{2 + x^2}\, dx \right] dy$

58. $\int_0^1 \left[\int_{\sqrt[3]{x}}^1 \sqrt{1 + y^4}\, dy \right] dx$ **59.** $\int_0^1 \left[\int_{\sqrt{y}}^1 e^{y/x}\, dx \right] dy$ **60.** $\int_0^1 \left[\int_{\sin^{-1} y}^{\pi/2} e^{\cos x}\, dx \right] dy$

61. Evaluate $\iint_R xy\, dA$, where R is the region enclosed by $y = 4 - x^2$ and $y = x^2$ to the right of the y-axis.

62. Evaluate $\iint_R xy\, dA$, where R is the region enclosed by $x = 4 - y^2$ and $x = y^2$ above the x-axis.

63. Find the volume of the solid enclosed by a portion of the parabolic cylinder $z = x^2/2$ and the four planes $y = 0$, $y = x$, $x = 2$, and $z = 0$.

64. Find the volume of the solid enclosed by the coordinate planes, the plane $y = 3$, and the surface $z = y + 1 - x^2$.

65. Find the volume of the solid enclosed by the surfaces $z = xe^y$, $z = 0$, $y = 0$, $x = 1$, and $y = x^2$.

66. Find the volume of the solid below the graph of $z = x^2 + y^2$ and inside $x^2 + y^2 = 1$.

In Problems 67–70 each of the iterated integrals represents the volume of a solid. Describe the solid.

67. $\int_0^1 \left[\int_0^x \sqrt{1 - x^2}\, dy \right] dx$ **68.** $\int_0^2 \left[\int_{y/2}^{\sqrt{y}} (x + y)\, dx \right] dy$ **69.** $\int_0^\pi \left[\int_0^{\sin x} 3\, dy \right] dx$ **70.** $\int_0^2 \left[\int_{-2}^2 2\, dy \right] dx$

71. Evaluate:

(a) $\iint_R x\, dA$; R is the region enclosed by the circle of radius 1 centered at the origin

(b) $4 \iint_{R_1} x\, dA$; R_1 is the region in the first quadrant enclosed by the circle of radius 1 centered at the origin (This shows that although the region we are integrating over is symmetric about the x-axis and the y-axis, we must also take into consideration the behavior of the function on this region before we use symmetry.)

72. Evaluate:

(a) $\iint_R xy^2\, dA$; R is the square enclosed by the lines $x = -1$, $x = 1$, $y = -1$, and $y = 1$

(b) $4 \iint_{R_1} xy^2\, dA$; R_1 is the square enclosed by the lines $x = 0$, $x = 1$, $y = 0$, and $y = 1$

18.3 Double Integrals Using Polar Coordinates

Some regions (such as circles and cardioids) over which a double integral is to be evaluated are more readily described using polar coordinates rather than rectangular coordinates. In such cases, it may be easier to evaluate the double integral by utilizing polar coordinates.

We start with a closed, bounded region R enclosed by the rays $\theta = \alpha$, $\theta = \beta$ $(0 \le \alpha < \beta < 2\pi)$, and the smooth curves $r = r_1(\theta)$, $r = r_2(\theta)$

$(r_1 \leq r_2)$, as shown in Figure 18. We proceed to partition the two intervals $[\alpha, \beta]$ and $[r_1, r_2]$ as depicted in Figure 19. The result is a collection of circular segments. Let n denote the number of such circular segments that lie entirely in R. We now have a partition P of the region whose norm $\|P\|$ is taken as the length of the longest diagonal of the circular segments.

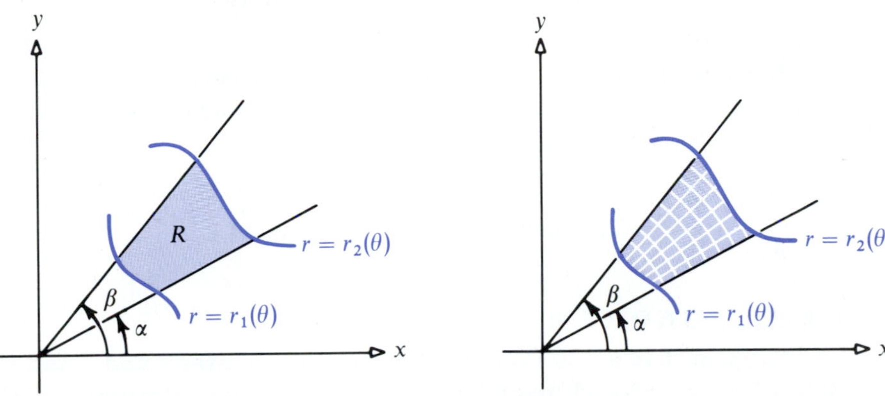

Figure 18

Figure 19

As has been the usual practice, we calculate the area ΔA_i of a typical circular segment R_i (see Fig. 20). Because this area ΔA_i is the difference of the areas of two circular sectors, we have

$$\begin{aligned}\Delta A_i &= \tfrac{1}{2}r_i^2\Delta\theta_i - \tfrac{1}{2}r_{i-1}^2\Delta\theta_i = \tfrac{1}{2}(r_i^2 - r_{i-1}^2)\Delta\theta_i \\ &= \tfrac{1}{2}(r_i + r_{i-1})(r_i - r_{i-1})\Delta\theta_i \underset{\uparrow}{=} \tfrac{1}{2}(r_i + r_{i-1})\Delta r_i\Delta\theta_i\end{aligned}$$

$$\Delta r_i = r_i - r_{i-1}$$

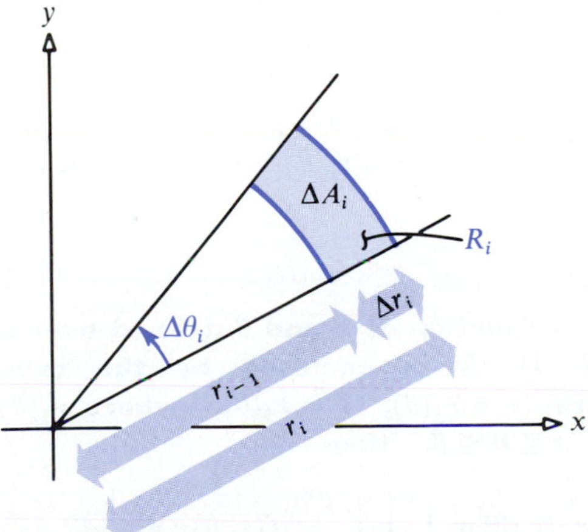

Figure 20

Now we pick a point $(\bar{r}_i, \bar{\theta}_i)$ in the circular segment R_i in such a way that $\bar{r}_i = \frac{1}{2}(r_i + r_{i-1})$. Then

$$\Delta A_i = \bar{r}_i \Delta r_i \Delta \theta_i \tag{1}$$

Let f be a function of the polar coordinates r and θ, and form the sum

$$\sum_{i=1}^{n} f(\bar{r}_i, \bar{\theta}_i)\Delta A_i = \sum_{i=1}^{n} f(\bar{r}_i, \bar{\theta}_i)\bar{r}_i \Delta r_i \Delta \theta_i \tag{2}$$

It can be shown that if f is continuous on the region R, then the limit of the sum in (2) as the norm $\|P\|$ approaches 0 exists and equals the double integral of f over R. That is,

$$\iint_R f(r, \theta)\, dA = \lim_{\|P\| \to 0} \sum_{i=1}^{n} f(\bar{r}_i, \bar{\theta}_i)\Delta A_i = \lim_{\|P\| \to 0} \sum_{i=1}^{n} f(\bar{r}_i, \bar{\theta}_i)\bar{r}_i \Delta r_i \Delta \theta_i$$

[18.3.1] THEOREM

Suppose f is a function of the polar coordinates r and θ and is defined over a closed, bounded region R. If f is continuous on R, then the double integral of f over R using polar coordinates is given by

$$\iint_R f(r, \theta)\, dA = \lim_{\|P\| \to 0} \sum_{i=1}^{n} f(\bar{r}_i, \bar{\theta}_i)\bar{r}_i \Delta r_i \Delta \theta_i = \iint_R f(r, \theta) r\, dr\, d\theta$$

It is important to remember that the integrand contains a factor of r. This is because in polar coordinates, the differential dA of area A (see (1)) is given by

$$dA = r\, dr\, d\theta$$

Evaluation of Double Integrals Using Polar Coordinates

As with double integrals in rectangular coordinates, certain regions in polar coordinates will lead to iterated integrals. Look back at Figure 18, where the region R has as fixed bounds the rays $\theta = \alpha$, $\theta = \beta$, $\alpha \le \beta$, and has as variable bounds the polar equations $r = r_1(\theta)$, $r = r_2(\theta)$, where $r_1 \le r_2$.

[18.3.2] THEOREM

Let f denote a continuous function of r and θ defined over a closed, bounded region R. If R is enclosed by the rays $\theta = \alpha$, $\theta = \beta$, and the curves $r = r_1(\theta)$, $r = r_2(\theta)$, where $r_1(\theta)$ and $r_2(\theta)$ are continuous for $\alpha \le \theta \le \beta$, then

$$\iint_R f(r, \theta)\, dA = \iint_R f(r, \theta) r\, dr\, d\theta = \int_{\alpha}^{\beta} \left[\int_{r_1(\theta)}^{r_2(\theta)} f(r, \theta) r\, dr \right] d\theta$$

As before, in evaluating the inside integral $\int_{r_1(\theta)}^{r_2(\theta)} f(r, \theta) r\, dr$, we treat θ as a constant and integrate partially with respect to r.

EXAMPLE 1

Evaluate $\iint_R \cos\theta\, dA$, where R is the region enclosed by $\theta = 0$, $\theta = \pi/4$, and $r = 4\cos\theta$.

Solution

Figure 21 illustrates the region R. We have

$$\begin{aligned}
\iint_R \cos\theta\, dA &= \int_0^{\pi/4} \left[\int_0^{4\cos\theta} \cos\theta\, r\, dr\right] d\theta = \int_0^{\pi/4} \cos\theta \left.\frac{r^2}{2}\right|_0^{4\cos\theta} d\theta \\
&= \int_0^{\pi/4} 8\cos^3\theta\, d\theta = \int_0^{\pi/4} 8(1 - \sin^2\theta)\cos\theta\, d\theta \\
&= 8\left[\sin\theta - \frac{\sin^3\theta}{3}\right]\Bigg|_0^{\pi/4} = 8\left[\frac{\sqrt{2}}{2} - \frac{\sqrt{2}}{12}\right] = \frac{10\sqrt{2}}{3}
\end{aligned}$$

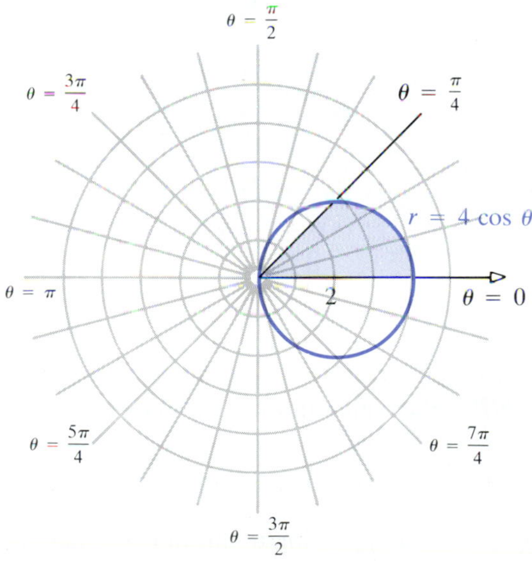

Figure 21

When the integrand of a double integral $\iint_R f(x, y)\, dx\, dy$ is expressed as a function of $x^2 + y^2$, it is sometimes easier to perform the integration in polar coordinates than in rectangular coordinates.

EXAMPLE 2

Evaluate the iterated integral: $$\int_0^2 \left[\int_0^{\sqrt{4-x^2}} (x^2 + y^2)\, dy\right] dx$$

Solution

The region R is enclosed by the graphs of $y = \sqrt{4 - x^2}$ (a portion of a circle of radius 2) and $0 \le x \le 2$ (see Fig. 22). In polar coordinates this region R is simply $0 \le r \le 2$, $0 \le \theta \le \pi/2$. The integrand $x^2 + y^2$ becomes r^2 and the differential of area $dy\,dx$ becomes $r\,dr\,d\theta$. By changing the limits of integration, we find

$$\int_0^2 \left[\int_0^{\sqrt{4-x^2}} (x^2 + y^2)\,dy \right] dx = \int_0^{\pi/2} \left[\int_0^2 r^2 r\,dr \right] d\theta$$

$$= \int_0^{\pi/2} \left[\frac{r^4}{4} \Big|_0^2 \right] d\theta = 4\left(\frac{\pi}{2}\right) = 2\pi$$

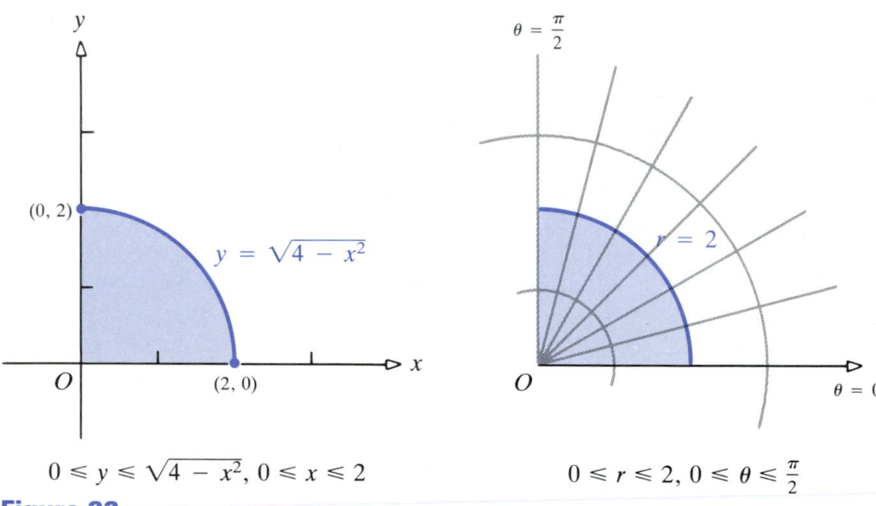

Figure 22

EXAMPLE 3

Evaluate $\iint_R e^{x^2+y^2}\,dx\,dy$, where R is the region in the first quadrant inside the circle $x^2 + y^2 = a^2$.

Solution

Without using polar coordinates, we can approximate this integral only by numerical techniques. However, a change to polar coordinates results in

$$\iint_R e^{x^2+y^2}\,dx\,dy = \iint_R e^{r^2} r\,dr\,d\theta = \int_0^{\pi/2} \left[\int_0^a e^{r^2} r\,dr \right] d\theta$$

$$= \frac{1}{2} \int_0^{\pi/2} (e^{a^2} - 1)\,d\theta = \left(\frac{\pi}{4}\right)(e^{a^2} - 1)$$

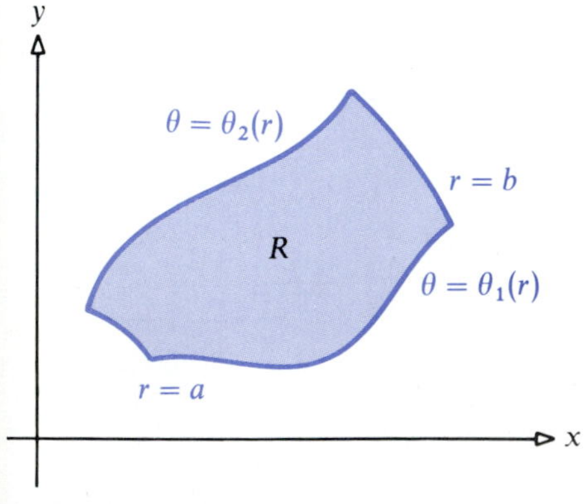

Figure 23

There is another type of region R that lends itself to iteration in polar coordinates. This is the area enclosed by the two circles $r = a$, $r = b$, $0 \le a < b$, and by the graphs of two curves $\theta = \theta_1(r)$, $\theta = \theta_2(r)$, where $0 \le \theta_1 \le \theta_2 \le 2\pi$ (see Fig. 23).

[18.3.3] THEOREM

Let f denote a continuous function of r and θ defined over a closed, bounded region R. If R is enclosed by the circles $r = a$, $r = b$, and by the curves $\theta = \theta_1(r)$, $\theta = \theta_2(r)$, where $\theta_1(r)$ and $\theta_2(r)$ are continuous for $a \le r \le b$ and $\theta_1(r) \le \theta_2(r)$, then

$$\iint\limits_R f(r, \theta)\, dA = \iint\limits_R f(r, \theta) r\, dr\, d\theta = \int_a^b \left[\int_{\theta_1(r)}^{\theta_2(r)} f(r, \theta)\, d\theta \right] r\, dr$$

In evaluating the inside integral $\int_{\theta_1(r)}^{\theta_2(r)} f(r, \theta)\, d\theta$, we treat r as a constant and integrate partially with respect to θ.

Geometric Interpretation

Provided $f(r, \theta) \ge 0$ and f is continuous over a region R, the double integral $\iint_R f(r, \theta)\, dA$ represents the volume of the solid under the graph of the surface $z = f(r, \theta)$ and above the region R in the xy-plane.

An intuitive argument follows: Suppose R is partitioned into n circular segments. As we did earlier, we select a point $(\bar{r}_i, \bar{\theta}_i)$, $\bar{r}_i = \frac{1}{2}(r_i + r_{i-1})$, in the ith circular segment R_i. See Figure 24. The quantity $f(\bar{r}_i, \bar{\theta}_i)$ measures the altitude of a solid whose base is R_i. The volume of this solid is $f(\bar{r}_i, \bar{\theta}_i)\Delta A_i$. The sum of all such solids corresponding to the n circular

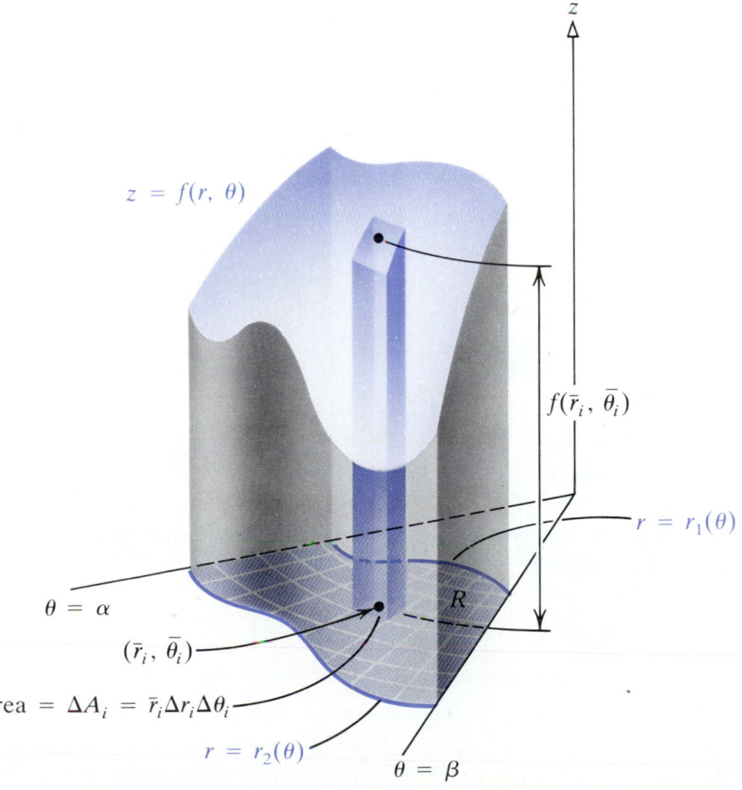

Figure 24

segments is

$$\sum_{i=1}^{n} f(\bar{r}_i, \bar{\theta}_i)\Delta A_i = \sum_{i=1}^{n} f(\bar{r}_i, \bar{\theta}_i)\bar{r}_i \Delta r_i \Delta\theta_i$$

[18.3.4] THEOREM

The volume V of the solid under the surface $z = f(r, \theta)$ and above the region R in the xy-plane is

$$V = \lim_{\|P\| \to 0} \sum_{i=1}^{n} f(\bar{r}_i, \bar{\theta}_i)\bar{r}_i \Delta r_i \Delta\theta_i = \iint_R f(r, \theta) r\, dr\, d\theta \tag{3}$$

EXAMPLE 4

Find the volume of the solid under the surface $z = x^2 + y^2$, above the xy-plane, and inside the cylinder $x^2 + y^2 = 2y$.

Solution

The plane $x = 0$ divides the solid into two parts with equal volume. Hence, we may evaluate the volume for $x \geq 0$ and double the result. Let $x = r\cos\theta$, $y = r\sin\theta$. Then $x^2 + y^2 = 2y$ becomes $r = 2\sin\theta$, and the restrictions on r and θ become $0 \leq \theta \leq \pi/2$ and $0 \leq r \leq 2\sin\theta$ (see Fig. 25). The required volume V is

$$V = 2\int_0^{\pi/2} \left[\int_0^{2\sin\theta} f(r, \theta) r\, dr\right] d\theta$$

The surface $z = f(r, \theta)$ under which the volume lies is the paraboloid of

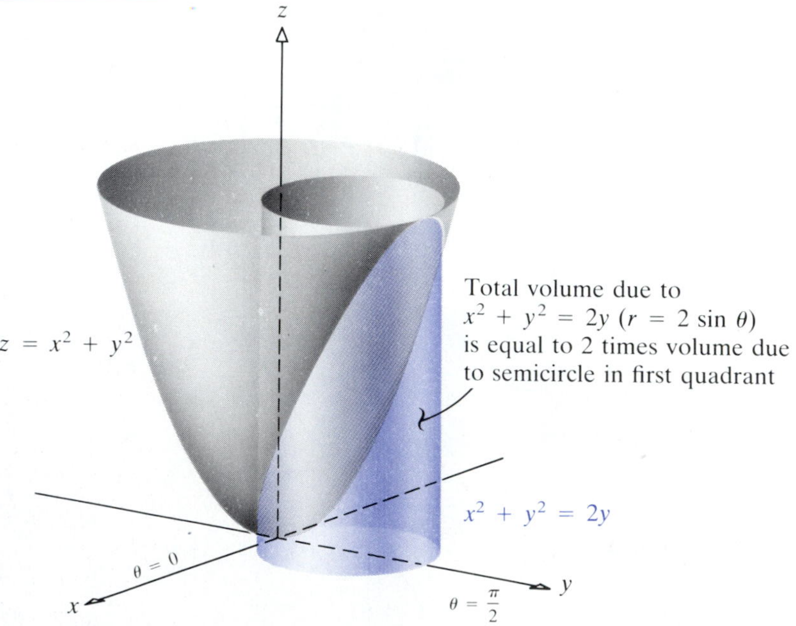

Figure 25

revolution $z = x^2 + y^2 = r^2$. Thus,

$$V = 2\int_0^{\pi/2}\left[\int_0^{2\sin\theta} r^2 r\,dr\right]d\theta = 8\int_0^{\pi/2}\sin^4\theta\,d\theta = \frac{3\pi}{2}$$

If $f(r, \theta) = 1$, then (3) represents the volume of a solid whose height is constantly 1. As a result, the value of (3) in this case gives the area A of the region R.

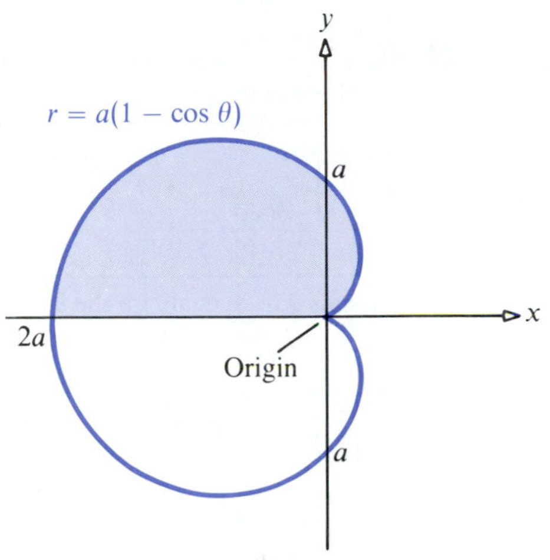

Figure 26

EXAMPLE 5

Use double integration to find the area enclosed by the cardioid $r = a(1 - \cos\theta)$.

Solution

We see that the area we wish to calculate is twice the shaded area in Figure 26; hence, the required area is

$$A = 2\int_0^{\pi}\left[\int_0^{a(1-\cos\theta)} (1)r\,dr\right]d\theta = 2\int_0^{\pi}\frac{a^2}{2}(1 - 2\cos\theta + \cos^2\theta)\,d\theta = 3\pi\frac{a^2}{2}$$

EXERCISE 18.3

In Problems 1–4 evaluate the iterated integral.

1. $\int_0^{\pi/4}\left[\int_0^{\cos\theta} r\,dr\right]d\theta$

2. $\int_0^{\pi/3}\left[\int_0^{\sin\theta} r\,dr\right]d\theta$

3. $\int_{-\pi/4}^{\pi/4}\left[\int_0^{4\sin\theta} \cos\theta\, r\,dr\right]d\theta$

4. $\int_{-\pi/3}^{5\pi/3}\left[\int_0^{2\cos\theta} \sin\theta\, r\,dr\right]d\theta$

In Problems 5–10 evaluate the double integral $\iint_R f(r, \theta)\,dA$. Graph the region R.

5. $\iint_R 3\sin\theta\, r\,dr\,d\theta$; $0 \le r \le 2$, $0 \le \theta \le \pi/2$

6. $\iint_R 4\cos\theta\, r\,dr\,d\theta$; $0 \le r \le 3$, $0 \le \theta \le \pi/2$

7. $\iint_R 2r\sin\theta\, r\,dr\,d\theta$; $0 \le r \le 1$, $0 \le \theta \le \pi/2$

8. $\iint_R 3r\cos\theta\, r\,dr\,d\theta$; $0 \le r \le 2$, $0 \le \theta \le \pi/4$

9. $\iint_R 3r^2\sin\theta\, r\,dr\,d\theta$; $0 \le r \le 2$, $-\pi/2 \le \theta \le \pi/2$

10. $\iint_R 2r^2\cos\theta\, r\,dr\,d\theta$; $0 \le r \le 4$, $-\pi/2 \le \theta \le \pi/2$

In Problems 11–24 evaluate each double integral by changing to polar coordinates.

11. $\int_0^1\left[\int_0^{\sqrt{1-x^2}} dy\right]dx$

12. $\int_0^3\left[\int_0^y \sqrt{x^2 + y^2}\,dx\right]dy$

13. $\int_0^4\left[\int_0^{\sqrt{4y-y^2}} (x^2 + y^2)\,dx\right]dy$

14. $\int_0^2\left[\int_0^{\sqrt{4-x^2}} \sqrt{4 - x^2 - y^2}\,dy\right]dx$

15. $\int_0^1\left[\int_0^{\sqrt{1-x^2}} \cos(x^2 + y^2)\,dy\right]dx$

16. $\int_0^1\left[\int_0^{\sqrt{1-y^2}} e^{\sqrt{x^2+y^2}}\,dx\right]dy$

17. $\iint_R e^{-(x^2+y^2)}\,dx\,dy$; R is the region in the first quadrant enclosed by the circles $x^2+y^2=1$ and $x^2+y^2=4$

18. $\iint_R (y/\sqrt{x^2+y^2})\,dx\,dy$; R is the region in the first quadrant inside the circle $x^2+y^2=a^2$

19. $\iint_R x\,dx\,dy$; R is the region enclosed by the circle $x^2+y^2=x$

20. $\iint_R y^2\,dx\,dy$; R is the region enclosed by the circle $x^2+y^2=2y$

21. $\iint_R \sqrt{x^2+y^2}\,dx\,dy$; R is the region enclosed by $r=3+\cos\theta$

22. $\iint_R (x^2+y^2)\,dx\,dy$; R is the region enclosed by $r=2(1+\sin\theta)$

23. $\displaystyle\int_{-2}^{2}\left[\int_{-\sqrt{4-y^2}}^{\sqrt{4-y^2}}(x^2+y^2)^2\,dx\right]dy$

24. $\displaystyle\int_{0}^{2}\left[\int_{\sqrt{2x-x^2}}^{\sqrt{4x-x^2}}dy\right]dx+\int_{2}^{4}\left[\int_{0}^{\sqrt{4x-x^2}}dy\right]dx$

In Problems 25–28 evaluate the iterated integral in polar coordinates, and sketch the region of integration.

25. $\displaystyle\int_{\pi/2}^{\pi}\left[\int_{0}^{1}r\cos\theta\,dr\right]d\theta$

26. $\displaystyle\int_{1}^{2}\left[\int_{0}^{\pi}re^{r}\,d\theta\right]dr$

27. $\displaystyle\int_{1/2\pi}^{1/\pi}\left[\int_{1/r}^{2\pi}d\theta\right]dr$

28. $\displaystyle\int_{0}^{1}\left[\int_{0}^{\cos^{-1}(r/2)}r\sin\theta\,d\theta\right]dr$

In Problems 29–38 use double integrals in polar coordinates.

29. Find the volume of the solid cut from the sphere $x^2+y^2+z^2=a^2$ by the cylinder $x^2+y^2=ay$.

30. Find the volume of the solid enclosed by the paraboloid $x^2+y^2=az$, the xy-plane, and the cylinder $x^2+y^2=a^2$.

31. Find the volume of the solid enclosed by the ellipsoid $x^2+y^2+4z^2=4$.

32. Find the volume of the solid cut from the ellipsoid $x^2+y^2+4z^2=4$ by the cylinder $x^2+y^2=1$.

33. Find the area enclosed by one leaf of the rose $r=\sin 3\theta$.

34. Find the area enclosed by one loop of $r^2=9\sin 2\theta$.

35. Find the area that lies inside the circle $r=4\cos\theta$ but outside the circle $r=\cos\theta$.

36. Find the area that lies inside the circle $r=1$ but outside the cardioid $r=1+\cos\theta$.

37. Find the area that lies inside the cardioid $r=1+\cos\theta$ but outside the circle $r=\frac{1}{2}$.

38. Find the area that lies inside the limaçon $r=3-\cos\theta$ but outside the circle $r=5\cos\theta$.

In Problems 39 and 40 replace the given iterated integral(s) in polar coordinates with iterated integral(s) in cartesian coordinates. Do not evaluate the integrals.

39. $\displaystyle\int_{0}^{1/\sqrt{2}}\left[\int_{0}^{\sin^{-1}r}r\,d\theta\right]dr$

40. $\displaystyle\int_{0}^{1}\left[\int_{\cos^{-1}r}^{\pi/2}r^2\sin\theta\,d\theta\right]dr$

In Problems 41 and 42 reverse the order of integration in the indicated iterated integrals in polar coordinates.

41. $\displaystyle\int_{0}^{\sqrt{2}/2}\left[\int_{\sin^{-1}r}^{\cos^{-1}r}f(r,\theta)\,d\theta\right]dr$

42. $\displaystyle\int_{\sqrt{2}/2}^{1}\left[\int_{\cos^{-1}r}^{\sin^{-1}r}f(r,\theta)\,d\theta\right]dr$

18.4 Center of Mass; Moment of Inertia

Mass of a Lamina

In many practical situations it is convenient to regard thin sheets of material, such as copper stripping, as two-dimensional. A *lamina* is a plane area that represents a two-dimensional distribution of matter. If the material is of constant mass density,* the lamina is called *homogeneous*. The mass m of a homogeneous lamina is ρA, where A is the area of the lamina and ρ is its constant mass density.

In general, however, substances are not homogeneous and so the mass density is variable. Suppose a lamina is represented by a certain closed, bounded region R of the xy-plane and its mass density $\rho = \rho(x, y)$ varies continuously over R. To find the total mass m of such a lamina, we may use double integration.

First, partition the region R into n rectangles. See Figure 27. Then, in a representative rectangle R_i of area ΔA_i, choose a point (u_i, v_i). An approximation to the mass due to the ith rectangle is

$$[\text{Mass density}] \times [\text{Area}] = \rho(u_i, v_i)\Delta A_i$$

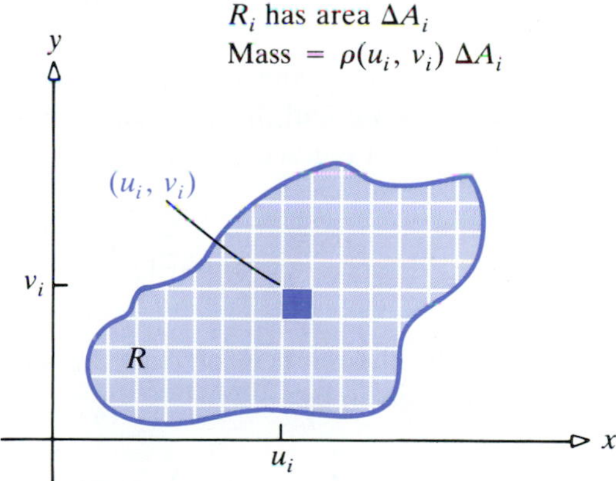

Figure 27

By adding up all the masses, the total mass of the lamina may be approximated by

$$\sum_{i=1}^{n} \rho(u_i, v_i)\Delta A_i$$

* The mass density of a two-dimensional material is defined as the mass per unit area of the material. In SI units, mass density is measured in kilograms per square meter (kg/m^2); in U.S. customary units, it is measured in slugs per square foot ($slug/ft^2$).

If the norm $\|P\|$ of the partition is allowed to approach 0, the total mass m of the lamina is given by a double integral:

$$m = \lim_{\|P\| \to 0} \sum_{i=1}^{n} \rho(u_i, v_i)\Delta A_i = \iint_R \rho(x, y)\, dA \quad \textbf{(1)}$$

Observe that if the mass density ρ is constant, we have the familiar formula

$$m = \iint_R \rho\, dA = \rho \iint_R dA = \rho A$$

If a particle of mass m is located a distance l from a fixed axis a, the *moment of mass* M_a* with respect to this axis is defined as

$$M_a = ml$$

Consider a lamina of variable mass density ρ represented by a certain region R of the xy-plane. Refer again to Figure 27. Partition R into n rectangles and concentrate on the ith rectangle R_i, whose area is ΔA_i. Pick a point (u_i, v_i) in R_i. The moment of mass M_x with respect to the x-axis of the ith rectangle may be approximated by

$$[\text{Mass}] \times [\text{Distance from } x\text{-axis}] = [\rho(u_i, v_i)\Delta A_i](v_i)$$

By adding up all these moments and taking the limit as the norm $\|P\|$ approaches 0, we find the moment of mass M_x with respect to the x-axis of the entire lamina is given by

$$M_x = \lim_{\|P\| \to 0} \sum_{i=1}^{n} v_i \rho(u_i, v_i)\Delta A_i = \iint_R y\rho(x, y)\, dA \quad \textbf{(2)}$$

Similarly, the moment of mass M_y with respect to the y-axis of the entire lamina is given by

$$M_y = \lim_{\|P\| \to 0} \sum_{i=1}^{n} u_i \rho(u_i, v_i)\Delta A_i = \iint_R x\rho(x, y)\, dA \quad \textbf{(3)}$$

[18.4.1] THEOREM / ***Mass of a Lamina.***

Suppose a lamina of variable mass density $\rho = \rho(x, y)$ is represented by a closed, bounded region R of the xy-plane. If ρ is continuous on R, then

$$\textbf{Mass of lamina} = m = \iint_R \rho(x, y)\, dA$$

---

* Sometimes M_a is referred to as the *first moment* of the particle about the axis a.

Moment of mass with respect to the x-axis $= M_x = \iint_R y\rho(x, y)\, dA$

Moment of mass with respect to the y-axis $= M_y = \iint_R x\rho(x, y)\, dA$

Center of Mass

The *center of mass* of a lamina is defined as the point $(\bar{x}, \bar{y})$ whose coordinates satisfy the equations

$$m\bar{x} = M_y \qquad m\bar{y} = M_x \tag{4}$$

From theorem (18.4.1), we conclude that

$$\bar{x} = \frac{\iint_R x\rho(x, y)\, dA}{\iint_R \rho(x, y)\, dA} \qquad \bar{y} = \frac{\iint_R y\rho(x, y)\, dA}{\iint_R \rho(x, y)\, dA}$$

We distinguish here between the *center of mass* (or *center of gravity*) of a lamina, which is defined in (4), and the *centroid* of a lamina. The latter is a purely geometric property of the lamina and coincides with the center of mass in the case of a homogeneous body. To see the distinction, a square lamina always has its centroid at the (geometric) center, but the same square with variable density will almost always have its center of mass off-center.

For a physical interpretation of the center of mass of a lamina, consider a flat piece of corrugated cardboard cut in the shape of the lamina and weighted appropriately to reflect the mass density of the lamina. If a piece of string is attached to the cardboard at the center of mass $(\bar{x}, \bar{y})$, then the cardboard, when suspended from the string, will lie in a horizontal position.

EXAMPLE 1

Find the center of mass of a lamina in the shape of an isosceles right triangle, if the mass density ρ is directly proportional to the square of the distance from the vertex opposite the hypotenuse.

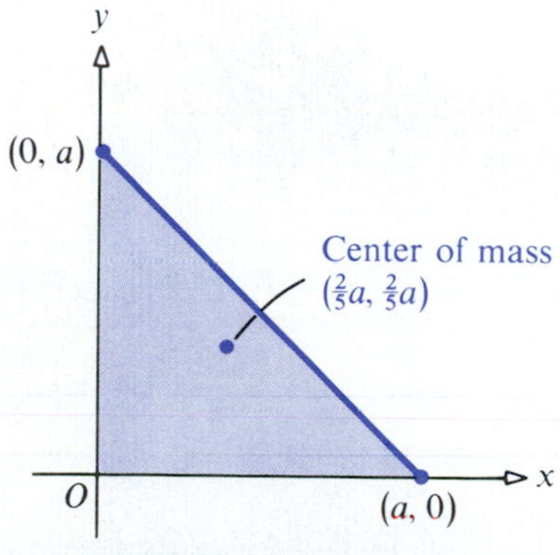

Figure 28

Solution

We situate the lamina in the xy-plane in such a way that the vertex opposite the hypotenuse is at the origin and the two equal sides—say, of length a—lie along the positive coordinate axes (see Fig. 28).

The mass density of the lamina is then

$$\rho(x, y) = k(x^2 + y^2)$$

where k is a constant. From theorem (18.4.1), the mass m of this lamina is

$$m = \iint_R \rho(x, y)\, dA = \iint_R k(x^2 + y^2)\, dx\, dy = k \int_0^a \left[\int_0^{a-x} (x^2 + y^2)\, dy \right] dx$$

$$= k \int_0^a \left[\left(x^2 y + \frac{y^3}{3} \right) \Big|_0^{a-x} \right] dx = k \int_0^a \frac{1}{3}(a^3 - 3a^2 x + 6ax^2 - 4x^3)\, dx$$

$$= \frac{k}{3}\left(a^4 - \frac{3a^4}{2} + 2a^4 - a^4 \right) = \frac{ka^4}{6}$$

Due to the symmetric character of the region R and the mass density, the center of mass $(\bar{x}, \bar{y})$ must lie on the line $y = x$. Consequently, if we find $\bar{x}$, we also know $\bar{y}$. Using theorem (18.4.1), we find

$$M_y = \iint_R x\rho(x, y)\, dA = \iint_R xk(x^2 + y^2)\, dx\, dy = k \int_0^a \left[\int_0^{a-x} (x^3 + xy^2)\, dy \right] dx$$

$$= \frac{k}{3} \int_0^a (a^3 x - 3a^2 x^2 + 6ax^3 - 4x^4)\, dx = \frac{ka^5}{15}$$

From (4), we conclude that

$$\bar{x} = \frac{M_y}{m} = \frac{ka^5/15}{ka^4/6} = \frac{2}{5} a$$

The center of mass of the lamina is $(\frac{2}{5}a, \frac{2}{5}a)$.

EXAMPLE 2

Find the center of mass of a lamina in the shape of a region R in the xy-plane that lies outside the circle $x^2 + y^2 = a^2$ and inside the circle $x^2 + y^2 = 2ax$, if the mass density ρ is inversely proportional to the distance from the origin.

Solution

Figure 29 illustrates the region R. Since the distance of a point (x, y) from the origin is $\sqrt{x^2 + y^2}$, the mass density of the lamina at any point (x, y) in R is

$$\rho(x, y) = \frac{k}{\sqrt{x^2 + y^2}}$$

where k is a constant. The mass m of the lamina is given by

$$m = \iint_R \rho(x, y) dA = k \iint_R \frac{dx\, dy}{\sqrt{x^2 + y^2}}$$

Due to the character of the region R and the integrand, it appears that integration in polar coordinates is preferable. By transforming to polar coordinates,

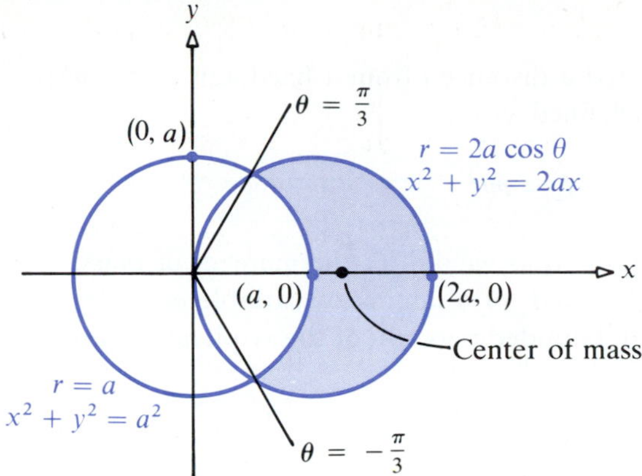

Figure 29

we have

$$\begin{aligned} m &= k \iint_R \frac{1}{r}\, r\, dr\, d\theta = k \int_{-\pi/3}^{\pi/3} \left[\int_a^{2a\cos\theta} dr \right] d\theta \\ &= k \int_{-\pi/3}^{\pi/3} (2a\cos\theta - a)\, d\theta = k(2a\sin\theta - a\theta)\Big|_{-\pi/3}^{\pi/3} \\ &= k\left(2a\sqrt{3} - a\frac{2\pi}{3}\right) = \frac{2ka}{3}(3\sqrt{3} - \pi) \end{aligned}$$

Due to the symmetric character of the mass density and the region R, it follows that the center of mass lies on the x-axis. Consequently, $\bar{y} = 0$. To find $\bar{x}$, we first need M_y.

$$\begin{aligned} M_y &= \iint_R x\rho(x, y)\, dx\, dy = k \iint_R r\cos\theta\, dr\, d\theta = k \int_{-\pi/3}^{\pi/3} \left[\int_a^{2a\cos\theta} r\cos\theta\, dr \right] d\theta \\ &= \frac{k}{2} \int_{-\pi/3}^{\pi/3} \cos\theta(4a^2\cos^2\theta - a^2)\, d\theta = \frac{k}{2} \int_{-\pi/3}^{\pi/3} (3a^2\cos\theta - 4a^2\sin^2\theta\cos\theta)\, d\theta \\ &= \frac{k}{2}\left(3a^2\sin\theta - \frac{4a^2\sin^3\theta}{3}\right)\Big|_{-\pi/3}^{\pi/3} = ka^2\sqrt{3} \end{aligned}$$

Thus,

$$\begin{aligned} \bar{x} &= \frac{M_y}{m} = \frac{ka^2\sqrt{3}}{(2ka/3)(3\sqrt{3} - \pi)} \\ &= \frac{3a\sqrt{3}}{2(3\sqrt{3} - \pi)} \approx 1.26a \end{aligned}$$

The center of mass is approximately $(1.26a, 0)$.

Moment of Inertia

If a particle of mass m is located a distance l from a fixed axis a, its *moment of inertia* I_a* about this axis is defined as

$$I_a = ml^2$$

Proceeding in the same fashion as we did for the moment of mass, the moment of inertia I_x about the x-axis of a lamina of variable mass density $\rho(x, y)$ represented by a closed, bounded region R of the xy-plane is given by

$$I_x = \lim_{\|P\| \to 0} \sum_{i=1}^{n} \overbrace{[\rho(u_i, v_i)\Delta A_i]}^{\text{Mass}}\overbrace{(v_i^2)}^{\text{Square of distance}} = \iint_R y^2 \rho(x, y)\, dA \tag{5}$$

Similarly, the moment of inertia I_y about the y-axis is given by

$$I_y = \lim_{\|P\| \to 0} \sum_{i=1}^{n} \overbrace{[\rho(u_i, v_i)\Delta A_i]}^{\text{Mass}}\overbrace{(u_i^2)}^{\text{Square of distance}} = \iint_R x^2 \rho(x, y)\, dA \tag{6}$$

and the moment of inertia I_o about the origin (z-axis) is given by

$$I_o = \lim_{\|P\| \to 0} \sum_{i=1}^{n} \overbrace{[\rho(u_i, v_i)\Delta A_i]}^{\text{Mass}}\overbrace{(u_i^2 + v_i^2)}^{\text{Square of distance}} = \iint_R (x^2 + y^2)\rho(x, y)\, dA \tag{7}$$

The moment of inertia I_o about the origin is sometimes referred to as the *polar moment of inertia* or the *polar second moment*.

A consequence of (5), (6), and (7) is that

$$I_o = I_x + I_y \tag{8}$$

The polar moment is frequently used to find the moments I_x and I_y when symmetry is present—that is, when $I_x = I_y$.

EXAMPLE 3

Find the polar moment of inertia of a homogeneous lamina of mass density ρ in the shape of a region R in the xy-plane enclosed by the circle $x^2 + y^2 = a^2$. Use the polar moment to find the moments of inertia I_x and I_y of this lamina.

---

* Sometimes I_a is referred to as the *second moment* of the particle about the axis a.

Solution

We shall use polar coordinates. From (7), we have

$$I_o = \iint_R (x^2 + y^2)\rho\, dA = \rho \iint_R r^2 r\, dr\, d\theta$$

$$= \rho \int_0^{2\pi} \left[\int_0^a r^3\, dr \right] d\theta = \frac{\rho}{4} \int_0^{2\pi} a^4\, d\theta = \frac{\pi a^4 \rho}{2}$$

By symmetry, it is evident that $I_x = I_y$. From (8), we conclude that

$$I_x = I_y = \frac{\pi a^4 \rho}{4}$$

In dynamics, the moment of inertia of a lamina occurs in connection with the study of rotational motion. If a rigid body (lamina) is rotated about an axis a with angular velocity ω, its kinetic energy K is given by

$$K = \tfrac{1}{2} I_a \omega^2 \qquad (9)$$

For example, suppose the lamina in Example 3 has mass density $\rho = 20$ kilograms per square meter and is of radius $a = 1$ meter. If it is rotated about the origin at a constant angular velocity of 2π radians per second, its kinetic energy is

$$K = \left(\frac{1}{2}\right)\left(\frac{\pi a^4 \rho}{2}\right)(4\pi^2) = 20\pi^3 \text{ kg-m}^2/\text{sec}^2 \approx 620 \text{ joules}$$

Observe the similarity between (9) and the formula for the kinetic energy K of a particle of mass m moving in a straight line with speed v—namely,

$$K = \tfrac{1}{2} m v^2$$

EXERCISE 18.4

In Problems 1–12 use double integration to find the mass and center of mass of each lamina for the given mass density ρ.

1. Lamina in the shape of a rectangle enclosed by the lines $x = 2$, $y = 4$, and the coordinate axes; $\rho = 3x^2y$

2. Lamina in the shape of a rectangle enclosed by the lines $x = 1$, $y = 2$, and the coordinate axes; $\rho = 2x^2y^2$

3. Lamina in the shape of a region in the first quadrant enclosed by $y^2 = x$, $x = 1$, and the x-axis; $\rho = 2x + 3y$

4. Lamina in the shape of a region in the first quadrant enclosed by $y^2 = 4x$, $x = 1$, and the x-axis; $\rho = x + 1$

5. Lamina in the shape of a region enclosed by $y^2 = x$ and $y = x$; ρ is proportional to the distance from the y-axis

6. Lamina in the shape of a region enclosed by $y = \sin x$, $x = 0$, $x = \pi$, and $y = 0$; ρ is proportional to the distance from the x-axis

7. Lamina in the shape of a triangle enclosed by the lines $2x + 3y = 6$, $x = 0$, and $y = 0$; ρ is proportional to the sum of the distances from the coordinate axes

8. Lamina in the shape of a triangle enclosed by the lines $3x + 4y = 12$, $x = 0$, and $y = 0$; ρ is proportional to the product of the distances from the coordinate axes

9. Lamina in the shape of the region inside the cardioid $r = 1 + \sin\theta$; ρ is proportional to the distance from the pole

10. Lamina in the shape of the region enclosed by one leaf of the rose $r = \cos 2\theta$; ρ is proportional to the distance from the pole

11. Lamina in the shape of the region inside the graph of $r = 2a\sin\theta$ and outside the graph of $r = a$; ρ is inversely proportional to the distance from the pole

12. Lamina in the shape of the region outside the limaçon $r = 2 - \cos\theta$ and inside the circle $r = 4\cos\theta$; ρ is inversely proportional to the distance from the pole

In Problems 13–20 use double integration to find the moment of inertia about the indicated axis for each homogeneous lamina of mass density ρ.

13. Lamina in the shape of a triangle enclosed by the lines $2x + 3y = 6$, $x = 0$, $y = 0$; about the x-axis

14. Rework Problem 13 for the moment of inertia about the y-axis.

15. Lamina in the shape of a rectangle enclosed by the lines $x = a$, $y = b$, $x = 0$, $y = 0$; about the x-axis; $a > 0$, $b > 0$

16. Rework Problem 15 for the moment of inertia about the y-axis.

17. Lamina in the shape of the region enclosed by $y = x^2$ and $y = 2 - x^2$; about the y-axis

18. Lamina in the shape of the region enclosed by the loop of $y^2 = x^2(4 - x)$; about the y-axis

19. Lamina in the shape of the ellipse $b^2x^2 + a^2y^2 = a^2b^2$; about the x-axis

20. Lamina in the shape of the region enclosed by $x^{2/3} + y^{2/3} = a^{2/3}$; about the x-axis

21. Find the mass of a circular washer with inner radius a and outer radius b, if its mass density is inversely proportional to the square of the distance from the center.

22. Rework Problem 21 if the mass density is inversely proportional to the distance from the center.

23. Find the mass and center of mass of a lamina in the shape of the region enclosed on the left by the line $x = a$ $(a > 0)$ and on the right by the circle $r = 2a\cos\theta$, if its mass density is inversely proportional to the distance from the y-axis.

24. Find the mass and center of mass of a lamina in the shape of the smaller region cut from the circle $r = 6$ by the line $r\cos\theta = 3$, if its mass density is $\rho = \cos^2\theta$.

25. Show that the center of mass of a rectangular homogeneous lamina lies at the intersection of its diagonals.

26. Show that the center of mass of a triangular homogeneous lamina lies at the point of intersection of its medians.

27. Find the center of mass of the lamina inside $r = 4\cos\theta$ and outside $r = 2\sqrt{3}$, if the mass density is inversely proportional to the distance from the origin.

28. Find the moment of inertia about the x-axis of the lamina inside the limaçon $r = 3 + 2\cos\theta$ and outside $r = 2$, if the mass density is inversely proportional to the square of the distance from the origin.

29. A homogeneous lamina is in the shape of a right triangle of base b and altitude h. Show that its moment of inertia about the base is $\frac{1}{6}mh^2$, where m is the mass.

30. Find the mass of the lamina enclosed by $y = x^2$ and $y = x^3$; $\rho = \sqrt{xy}$ at any point (x, y).

31. Find the center of mass and the mass of the lamina enclosed by $x = y - 2$ and $x = -y^2$; $\rho = x^2$ at any point (x, y).

32. Find the center of mass of the lamina enclosed by $x = 0$, $y = x^2$, and $x - 2y + 1 = 0$; $\rho = 2x + 8y + 2$ at any point (x, y).

18.5 Area of a Surface

In Chapter 13 we used a single integral to calculate the surface area of a solid of revolution—a surface obtained by revolving the graph of a continuous function $y = f(x)$ about an axis. In this section we show how double integration can be used to calculate the area of the part of a surface $z = f(x, y)$

(not necessarily a solid of revolution) that lies above a closed, bounded region R of the xy-plane.

We begin with the simplest case in which the surface is a plane $z = ax + by + c$ and R is a rectangle with sides of lengths Δx and Δy. See Figure 30. We let S be the area of that part of the plane that lies over the rectangle.

[18.5.1] THEOREM

Let R be a closed rectangular region in the xy-plane. If R has sides of length Δx and Δy, then the surface area S of that part of the plane $z = ax + by + c$ that projects onto the region R is given by

$$S = \sqrt{a^2 + b^2 + 1}\,\Delta x \Delta y \tag{1}$$

Proof As suggested by Figure 30, the surface is a parallelogram. What we want then is the area of this parallelogram. The vector marked $\mathbf{u}$ in the figure has $(x_0, y_0, ax_0 + by_0 + c)$ as the initial point and $(x_0 + \Delta x, y_0, a(x_0 + \Delta x) + by_0 + c)$ as the terminal point. Hence,

$$\mathbf{u} = \Delta x\mathbf{i} + a\Delta x\mathbf{k} = \Delta x(\mathbf{i} + a\mathbf{k})$$

The vector marked $\mathbf{v}$ has $(x_0, y_0, ax_0 + by_0 + c)$ as the initial point and $(x_0, y_0 + \Delta y, ax_0 + b(y_0 + \Delta y) + c)$ as the terminal point. Hence,

$$\mathbf{v} = \Delta y\mathbf{j} + b\Delta y\mathbf{k} = \Delta y(\mathbf{j} + b\mathbf{k})$$

The area of the parallelogram is

$$\begin{aligned}\|\mathbf{u} \times \mathbf{v}\| &= \|\Delta x(\mathbf{i} + a\mathbf{k}) \times \Delta y(\mathbf{j} + b\mathbf{k})\| = \|(\mathbf{i} + a\mathbf{k}) \times (\mathbf{j} + b\mathbf{k})\|\Delta x\Delta y \\ &= \|-a\mathbf{i} - b\mathbf{j} + \mathbf{k}\|\Delta x\Delta y = \sqrt{a^2 + b^2 + 1}\,\Delta x\Delta y\end{aligned}$$

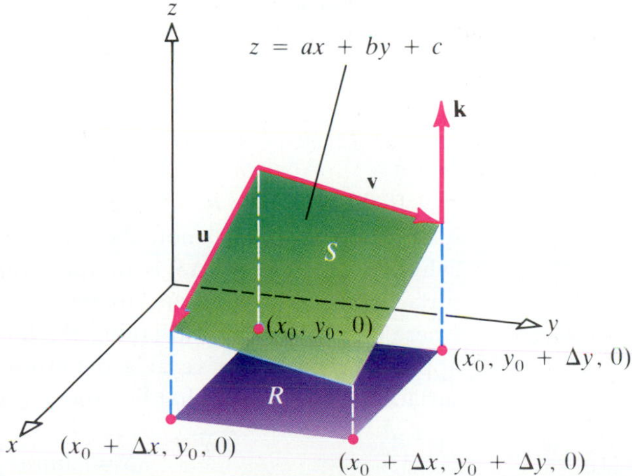

Figure 30

We are now prepared to derive a formula for the surface area of the graph of a nonnegative function $z = f(x, y)$ whose graph is not necessarily a plane. We assume that f has continuous first-order partial derivatives at every point of a closed, bounded region R.

Look at Figure 31(a). Using lines parallel to the x- and y-axes, we partition R into n rectangles and concentrate on the ith rectangle R_i having sides of lengths Δx_i and Δy_i. We pick a point (u_i, v_i) in R_i and at the corresponding point $P_i = (u_i, v_i, f(u_i, v_i))$ on the surface, we construct the tangent plane to the surface (see Fig. 31(b)). The equation of this tangent plane can be written as

$$z = f_x(u_i, v_i)(x - u_i) + f_y(u_i, v_i)(y - v_i) + f(u_i, v_i) \tag{2}$$

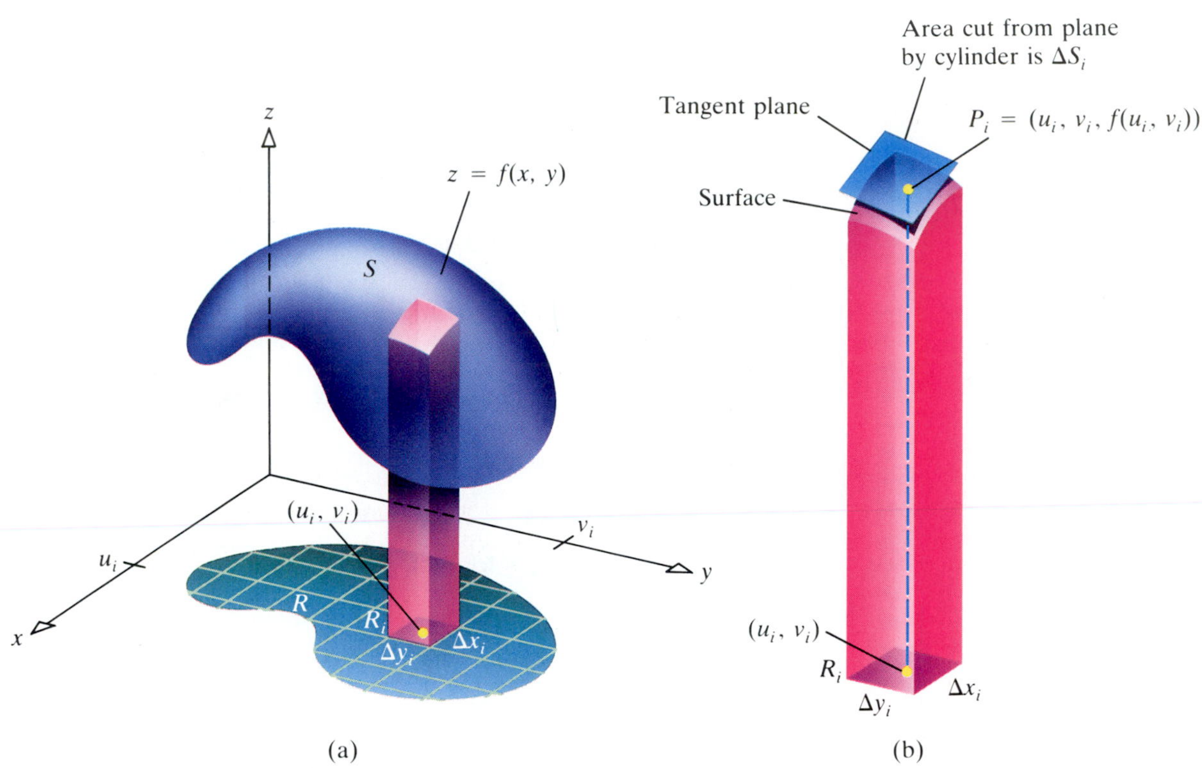

Figure 31

If the rectangle R_i is projected upward parallel to the z-axis, the result is a cylinder. Let ΔS_i denote the area cut from the tangent plane at P_i by the cylinder. The number ΔS_i is an approximation to the area of the part of the surface that lies above the ith rectangle R_i. By adding up all these parts (there are n of them) and taking the limit as the norm $\|P\|$ of the partition approaches 0, we arrive at a definition for the area S of the part of the surface $z = f(x, y)$ that lies above the region R—namely,

$$S = \lim_{\|P\| \to 0} \sum_{i=1}^{n} \Delta S_i \tag{3}$$

provided the limit exists.

To obtain a formula for ΔS_i, we use the equation of the tangent plane (2) with (1). The result is

$$\Delta S_i = \sqrt{[f_x(u_i, v_i)]^2 + [f_y(u_i, v_i)]^2 + 1}\ \Delta x_i \Delta y_i \qquad \textbf{(4)}$$

Now substitute this expression for ΔS_i in (3) to obtain

$$S = \lim_{||P|| \to 0} \sum_{i=1}^{n} \sqrt{[f_x(u_i, v_i)]^2 + [f_y(u_i, v_i)]^2 + 1}\ \Delta x_i \Delta y_i \qquad \textbf{(5)}$$

Since the first-order partial derivatives of f are continuous on R, S is a double integral.

[18.5.2] THEOREM / ***Area of a Surface.***

Let $z = f(x, y)$ denote a function of two variables defined on a closed, bounded region R. If f_x and f_y are continuous on R, then the area S of the part of the surface that lies over R is given by

$$S = \iint_R \sqrt{[f_x(x, y)]^2 + [f_y(x, y)]^2 + 1}\ dx\, dy \qquad \textbf{(6)}$$

EXAMPLE 1

Find the surface area of the graph of $z = f(x, y) = \frac{2}{3}(x^{3/2} + y^{3/2})$ that lies above the rectangle enclosed by $x = 0$, $x = 1$, $y = 0$, and $y = 2$.

Solution

Figure 32 illustrates the situation.

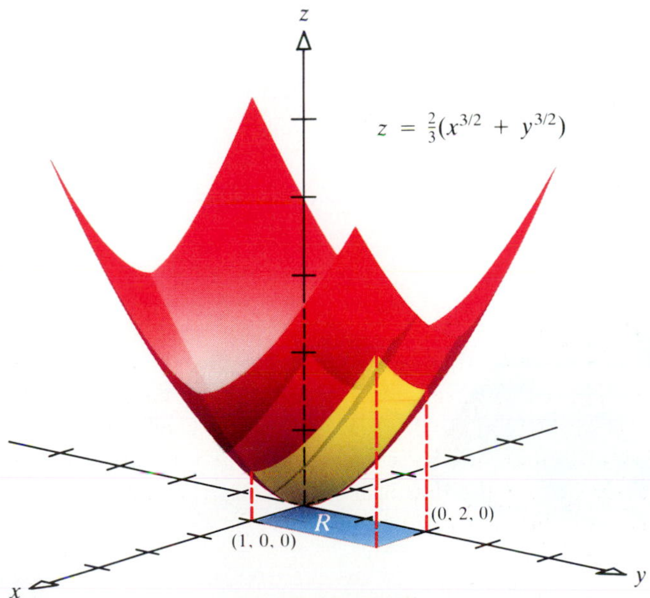

Figure 32

Since $f_x(x, y) = x^{1/2}$ and $f_y(x, y) = y^{1/2}$, it follows from theorem (18.5.2) that

$$S = \iint_R \sqrt{[f_x(x, y)]^2 + [f_y(x, y)]^2 + 1}\, dx\, dy = \iint_R \sqrt{x + y + 1}\, dx\, dy$$

$$= \int_0^1 \left[\int_0^2 \sqrt{x + y + 1}\, dy\right] dx = \int_0^1 \left[\tfrac{2}{3}(x + y + 1)^{3/2}\Big|_0^2\right] dx$$

$$= \tfrac{2}{3}\int_0^1 [(x + 3)^{3/2} - (x + 1)^{3/2}]\, dx = (\tfrac{2}{3})(\tfrac{2}{5})[(x + 3)^{5/2} - (x + 1)^{5/2}]\Big|_0^1$$

$$= \tfrac{4}{15}(32 - 9\sqrt{3} - 4\sqrt{2} + 1) = \tfrac{4}{15}(33 - 9\sqrt{3} - 4\sqrt{2})$$

$$\approx 3.135 \text{ square units}$$

EXAMPLE 2

Find the surface area of the part of the paraboloid $z = f(x, y) = 1 - x^2 - y^2$ that lies above the xy-plane.

Solution

Figure 33 shows the graph of the part of the surface $z = 1 - x^2 - y^2$ that lies above the xy-plane and its projection onto the xy-plane. The region R is enclosed by the circle $x^2 + y^2 = 1$. Since $f_x(x, y) = -2x$ and $f_y(x, y) = -2y$, it follows that the desired surface area S is given by

$$S = \iint_R \sqrt{(-2x)^2 + (-2y)^2 + 1}\, dx\, dy$$

$$= \iint_R \sqrt{4(x^2 + y^2) + 1}\, dx\, dy$$

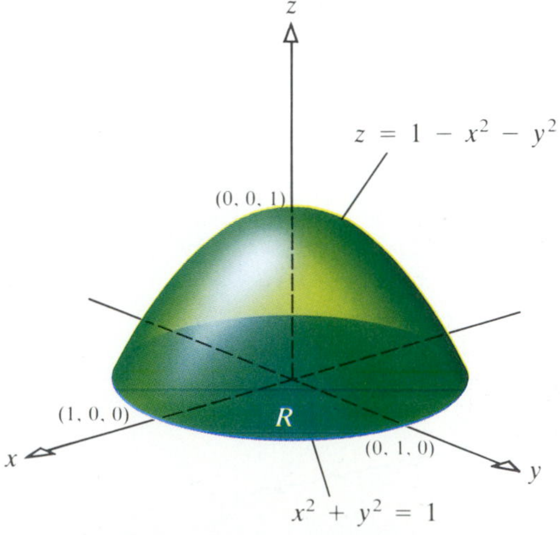

Figure 33

17. (*continued*)
ten as

$$S = \iint_R \sec \gamma \, dx \, dy$$

where γ is the positive acute angle between the normal $\mathbf{n} = A\mathbf{i} + B\mathbf{j} + C\mathbf{k}$ to the plane and $\mathbf{k}$.

18. Let F denote a function of three variables that possesses continuous first-order partial derivatives at each point of its domain. Suppose also that $F_z(x, y, z)$ is never 0. If S is the area of the part of the surface $F(x, y, z) = 0$, that lies over the closed, bounded region R, show that

$$S = \iint_R \frac{\sqrt{[F_x(x, y, z)]^2 + [F_y(x, y, z)]^2 + [F_z(x, y, z)]^2}}{|F_z(x, y, z)|} \, dx \, dy$$

The Triple Integral 18.6

The triple integral of a function of three variables is defined in a way analogous to the definition of the double integral.

We begin by considering a function f of three variables defined over a box-shaped region S of three-dimensional space. We divide the region S into rectangular boxes by drawing planes parallel to the coordinate planes. Let n be the total number of boxes thus constructed. We now have a partition P of S into n boxes $S_1, S_2, \ldots, S_n$ (see Fig. 35).

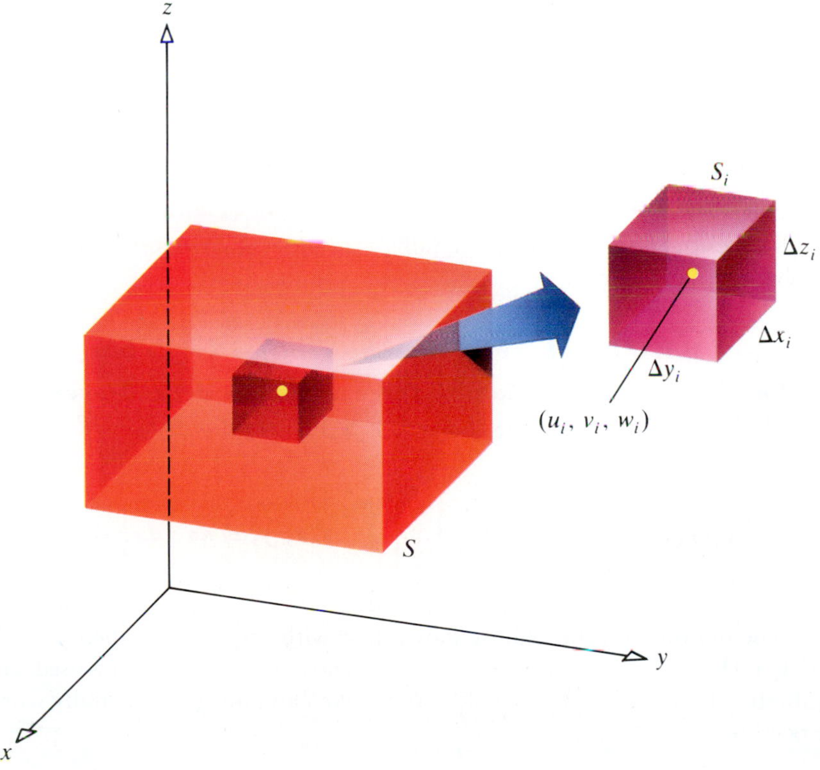

Figure 35

The norm of the partition P, denoted by $\|P\|$, is defined as the length of the longest diagonal of the boxes S_i. If Δx_i, Δy_i, and Δz_i denote the length, width, and height of the ith box, respectively, then its volume is $\Delta V_i = \Delta x_i \Delta y_i \Delta z_i$. In each box S_i we arbitrarily select a point (u_i, v_i, w_i) and evaluate the function f there. Then we form the sum

$$\sum_{i=1}^{n} f(u_i, v_i, w_i)\,\Delta V_i$$

All sums of this form are referred to as *Riemann sums* of f for the partition P. If all such sums approach a limit as the norm of the partition approaches 0, this limit is called the *triple integral of f over S*.

[18.6.1] DEFINITION / *Triple Integral.*

Let f denote a function of three variables defined over a closed, bounded, box-shaped region S. The triple integral of f over S is defined as

$$\iiint_S f(x, y, z)\, dV = \lim_{\|P\| \to 0} \sum_{i=1}^{n} f(u_i, v_i, w_i)\Delta V_i$$

provided this limit exists.

Other symbols for the triple integral of f over S are

$$\iiint_S f(x, y, z)\, dx\, dy\, dz \qquad \iiint_S f(x, y, z)\, dx\, dz\, dy$$

and so on.

Triple integrals of many functions can be evaluated using iteration involving three integrals.

[18.6.2] THEOREM / *Fubini's Theorem for Triple Integrals.*

Let f denote a function of three variables defined over the closed box S: $x_1 \le x \le x_2$, $y_1 \le y \le y_2$, $z_1 \le z \le z_2$. If f is continuous on S, then

$$\iiint_S f(x, y, z)\, dV = \int_{x_1}^{x_2} \left\{ \int_{y_1}^{y_2} \left[\int_{z_1}^{z_2} f(x, y, z)\, dz \right] dy \right\} dx \qquad (1)$$

The iteration on the right is shown first with respect to z, then y, and finally x. However, the same result is obtained no matter what order is used. In evaluating $\int_{z_1}^{z_2} f(x, y, z)\, dz$, we hold x and y fixed and integrate partially with respect to z.

EXAMPLE 1

Evaluate $\iiint_S 4xyz\, dV$ if S is the closed box $0 \le x \le 1$, $0 \le y \le 2$, $0 \le z \le 3$.

Solution

Based on (1), we find

$$\iiint_S 4xyz\, dV = \int_0^1 \left\{ \int_0^2 \left[\int_0^3 4xyz\, dz \right] dy \right\} dx = \int_0^1 \left\{ \int_0^2 \left[4xy \frac{z^2}{2} \Big|_0^3 \right] dy \right\} dx$$

$$= \int_0^1 \left\{ \int_0^2 18xy\, dy \right\} dx = \int_0^1 18x \frac{y^2}{2} \Big|_0^2 dx = \int_0^1 36x\, dx$$

$$= 18x^2 \Big|_0^1 = 18$$

The Triple Integral over a More General Region

Consider a region S in space enclosed on the top by the surface $z = z_2(x, y)$ and on the bottom by the surface $z = z_1(x, y)$, where $z_1 \le z_2$ and z_1 and z_2 are continuous functions. The sides of S are cylinders whose intersection with the xy-plane forms the boundary of a closed, bounded region R (see Fig. 36). We shall call such a region S *xy-simple*, since the projection of S

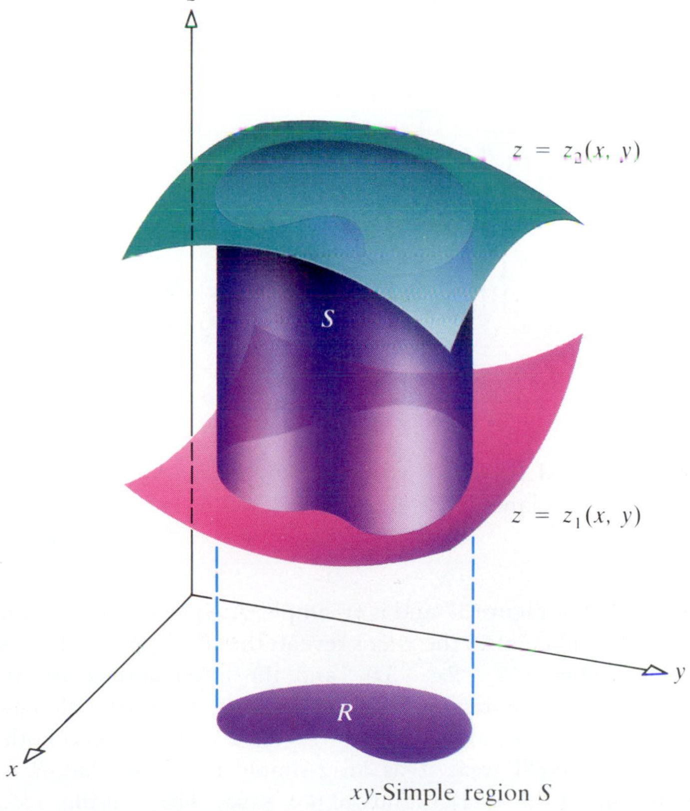

Figure 36

onto the xy-plane results in a closed, bounded region R and S lies between two surfaces. The next result, which we state without proof, shows how to evaluate triple integrals over regions that are xy-simple.

[18.6.3] THEOREM

Let S denote an xy-simple region that lies between the two surfaces $z = z_1(x, y)$ and $z = z_2(x, y)$, $z_1 \le z_2$, and whose projection onto the xy-plane is a closed, bounded region R. If f is a function of three variables that is continuous on S, then

$$\iiint_S f(x, y, z)\, dV = \iint_R \left[\int_{z_1(x,y)}^{z_2(x,y)} f(x, y, z)\, dz\right] dA \tag{2}$$

Depending on the order of integration used to evaluate this double integral, there are two iterated forms of the triple integral (2). If the region R is x-simple, then

$$\begin{aligned}\iiint_S f(x, y, z)\, dV &= \iint_R \left[\int_{z_1(x,y)}^{z_2(x,y)} f(x, y, z)\, dz\right] dA \\ &= \int_a^b \left\{\int_{y_1(x)}^{y_2(x)} \left[\int_{z_1(x,y)}^{z_2(x,y)} f(x, y, z)\, dz\right] dy\right\} dx \end{aligned} \tag{3}$$

If the region R is y-simple, then

$$\begin{aligned}\iiint_S f(x, y, z)\, dV &= \iint_R \left[\int_{z_1(x,y)}^{z_2(x,y)} f(x, y, z)\, dz\right] dA \\ &= \int_c^d \left\{\int_{x_1(y)}^{x_2(y)} \left[\int_{z_1(x,y)}^{z_2(x,y)} f(x, y, z)\, dz\right] dx\right\} dy \end{aligned} \tag{4}$$

EXAMPLE 2

Evaluate the triple integral $\iiint_S 4x\, dV$ over the tetrahedron formed by the coordinate planes and the plane $2x + 3y + 4z = 12$.

Solution

The region S is pictured in Figure 37 and is xy-simple. A typical plane section of S from a plane perpendicular to the x-axis reveals that the upper surface is the plane $z = z_2(x, y) = \frac{1}{4}(12 - 2x - 3y)$ and the lower surface is the plane $z = z_1(x, y) = 0$. The region R in the xy-plane is the triangle enclosed by the x-axis, the y-axis, and the line $y = \frac{1}{3}(12 - 2x)$ and is both x-simple and y-simple. We'll treat it as an x-simple region so that R is given by $y = 0$, $y = \frac{1}{3}(12 - 2x)$, and $0 \le x \le 6$. Then, using (3),

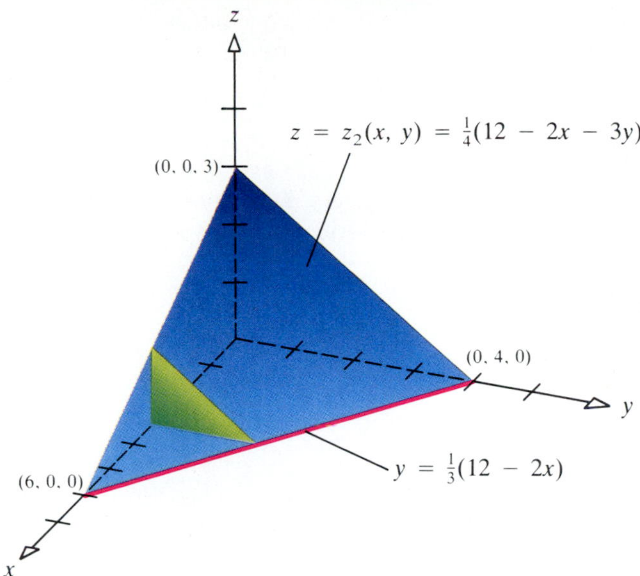

Figure 37

we find

$$\iiint_S 4x\, dV = \int_0^6 \left\{ \int_0^{(1/3)(12-2x)} \left[\int_0^{(1/4)(12-2x-3y)} 4x\, dz \right] dy \right\} dx$$

$$= \int_0^6 \left[\int_0^{(1/3)(12-2x)} (12x - 2x^2 - 3xy)\, dy \right] dx$$

$$= \int_0^6 (24x - 8x^2 + \tfrac{2}{3}x^3)\, dx = 72$$

EXAMPLE 3

Express $\iiint_S f(x, y, z)\, dV$ as an iterated integral if S is the region in the first octant that is enclosed by the paraboloid $z = 16 - 4x^2 - y^2$ and the xy-plane.

Solution

Figure 38 illustrates the region S, which is xy-simple. The upper surface is the paraboloid $z = z_2(x, y) = 16 - 4x^2 - y^2$ and the lower surface is the plane $z = z_1(x, y) = 0$. The region R in the xy-plane is enclosed by the x-axis, the y-axis, and one-fourth of the ellipse $4x^2 + y^2 = 16$ and is both x-simple and y-simple. We use (3):

$$\iiint_S f(x, y, z)\, dV = \int_0^2 \left\{ \int_0^{\sqrt{16-4x^2}} \left[\int_0^{16-4x^2-y^2} f(x, y, z)\, dz \right] dy \right\} dx$$

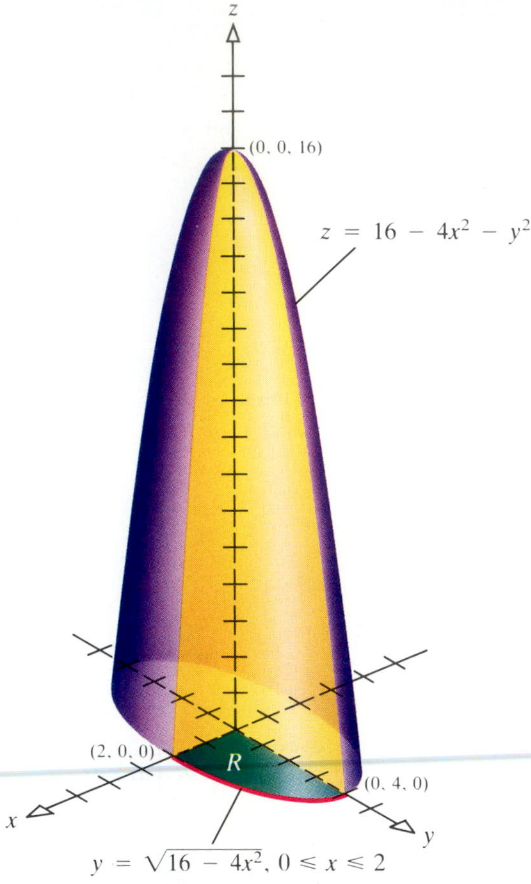

Figure 38

If $f(x, y, z) = 1$ on S, then (3) or (4) may be used to calculate the volume of the region S. That is,

$$\text{Volume of } S = \iiint_S dV$$

EXAMPLE 4

Find the volume V of the tetrahedron formed by the coordinate planes and the plane $2x + 3y + 4z = 12$.

Solution

Refer back to Example 2 and Figure 37. Then,

$$V = \iiint_S dV = \int_0^6 \left\{ \int_0^{(1/3)(12-2x)} \left[\int_0^{(1/4)(12-2x-3y)} dz \right] dy \right\} dx$$

$$= \frac{1}{4} \int_0^6 \left[\int_0^{(1/3)(12-2x)} (12 - 2x - 3y)\, dy \right] dx = \frac{1}{6} \int_0^6 (36 - 12x + x^2)\, dx = 12 \text{ cubic units}$$

Mass; Center of Mass; Moment of Inertia

Another useful application of triple integrals is in finding mass, centers of mass, and moments of inertia. The formulas are obtained in a manner much like those involving double integrals, so we merely state them here.

The mass m of a body of continuous mass density $\rho = \rho(x, y, z)$ in a region S of volume V is

$$m = \iiint\limits_S \rho\, dV$$

The center of mass $(\bar{x}, \bar{y}, \bar{z})$ of this body obeys

$$m\bar{x} = \iiint\limits_S x\rho\, dV \qquad m\bar{y} = \iiint\limits_S y\rho\, dV \qquad m\bar{z} = \iiint\limits_S z\rho\, dV$$

The moment of inertia about an axis is

$$I = \iiint\limits_S r^2\rho\, dV$$

where r is the distance from the point (x, y, z) of the body to the axis about which the moment is to be calculated.

EXAMPLE 5

Find the mass of the body in the shape of a tetrahedron cut from the first octant by the plane $x + y + z = 1$, if the mass density is proportional to the distance from the yz-plane. Locate its center of mass.

Solution

See Figure 39. The region S describing the body is xy-simple. It is enclosed by the surfaces $z = z_1(x, y) = 0$ and $z = z_2(x, y) = 1 - x - y$, and its projection onto the xy-plane is the closed, bounded x-simple region R defined by $y = 0$, $y = 1 - x$, $0 \le x \le 1$. Since the mass density of the body is $\rho = kx$, k a constant, the mass m is given by

$$m = \iiint\limits_S kx\, dV = k\int_0^1 \left\{\int_0^{1-x}\left[\int_0^{1-x-y} x\, dz\right] dy\right\} dx = \frac{k}{24}$$

The center of mass $(\bar{x}, \bar{y}, \bar{z})$ is

$$\bar{x} = \frac{\iiint_S kx^2\, dV}{m} = \frac{k\iiint_S x^2\, dV}{k/24} = 24\int_0^1 \left\{\int_0^{1-x}\left[\int_0^{1-x-y} x^2\, dz\right] dy\right\} dx = \frac{2}{5}$$

$$\bar{y} = \frac{\iiint_S kxy\, dV}{m} = 24\int_0^1 \left\{\int_0^{1-x}\left[\int_0^{1-x-y} xy\, dz\right] dy\right\} dx = \frac{1}{5}$$

$$\bar{z} = \frac{\iiint_S kxz\, dV}{m} = 24\int_0^1 \left\{\int_0^{1-x}\left[\int_0^{1-x-y} xz\, dz\right] dy\right\} dx = \frac{1}{5}$$

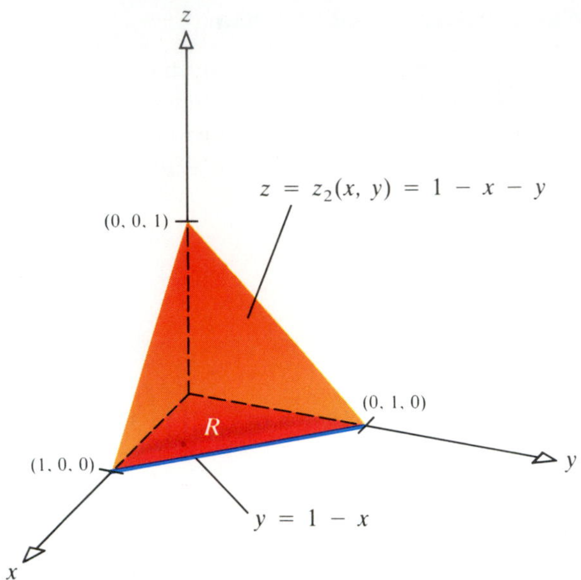

Figure 39

EXAMPLE 6

Find the moment of inertia about the z-axis of the homogeneous body of mass density ρ in the first octant enclosed by the surface $z = 4xy$ and the planes $z = 0$, $x = 3$, $y = 2$.

Solution

The moment of inertia I_z about the z-axis is

$$I_z = \iiint_S \rho(x^2 + y^2)\, dV = \rho \int_0^3 \left\{ \int_0^2 \left[\int_0^{4xy} (x^2 + y^2)\, dz \right] dy \right\} dx = 234\rho$$

xz-Simple and yz-Simple Regions

We end this section by showing that for certain regions in space, triple integrals are best evaluated by integrating first with respect to x or y rather than z.

If a region S in space is enclosed by the two surfaces $y = y_1(x, z)$ and $y = y_2(x, z)$, where $y_1 \leq y_2$ and y_1, y_2 are continuous functions, and if the projection of S onto the xz-plane is a closed, bounded region R, then we call S an *xz-simple region* (see Fig. 40).

[18.6.4] THEOREM

If f is a function of three variables that is continuous on such an xz-simple region S, then

$$\iiint_S f(x, y, z)\, dV = \iint_R \left[\int_{y_1(x,z)}^{y_2(x,z)} f(x, y, z)\, dy \right] dx\, dz$$

If a region S in space is enclosed by the two surfaces $x = x_1(y, z)$ and $x = x_2(y, z)$, where $x_1 \leq x_2$ and x_1, x_2 are continuous functions, and if the projection of S onto the yz-plane is a closed, bounded region R, then we call S a *yz-simple region* (see Fig. 41).

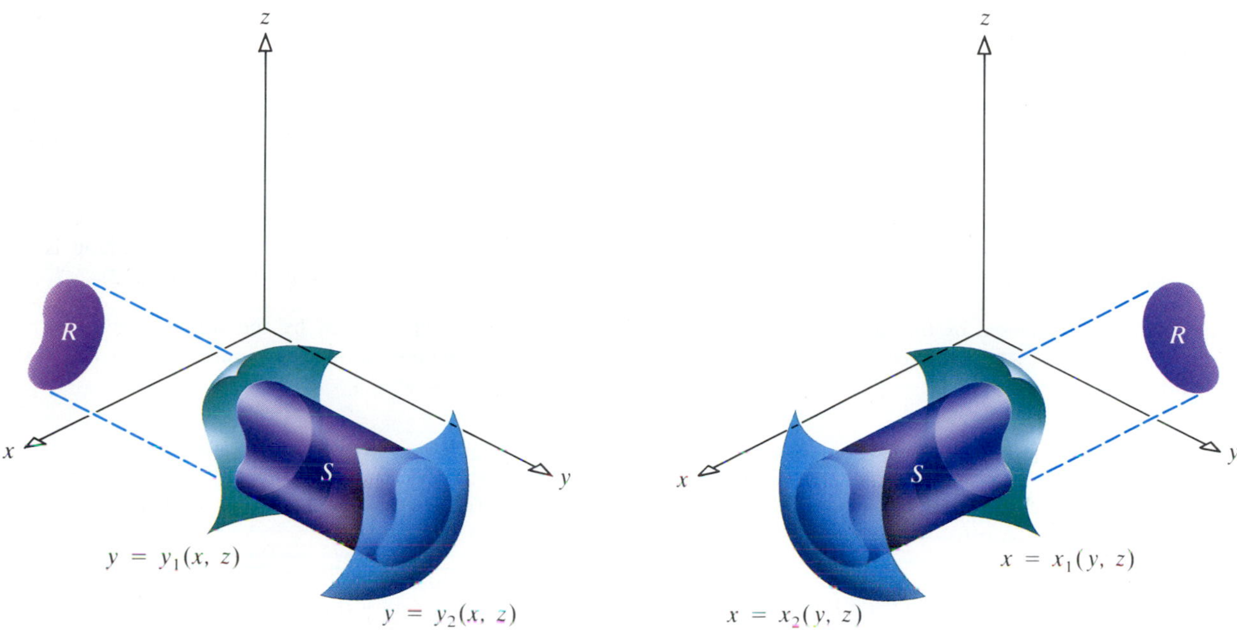

Figure 40

Figure 41

[18.6.5] THEOREM

If f is a function of three variables that is continuous on such a *yz*-simple region S, then

$$\iiint_S f(x, y, z)\, dV = \iint_R \left[\int_{x_1(y,z)}^{x_2(y,z)} f(x, y, z)\, dx \right] dy\, dz$$

EXERCISE 18.6

In Problems 1–6 evaluate the triple integral of f over the closed box S.

1. $\iiint_S xy^2\, dV$; S: $0 \leq x \leq 1,\ \ 0 \leq y \leq 1,\ \ 0 \leq z \leq 4$

2. $\iiint_S x^2yz\, dV$; S: $0 \leq x \leq 2,\ \ 0 \leq y \leq 1,\ \ 0 \leq z \leq 2$

3. $\iiint_S (x^2 + y^2 + z^2)\, dV$; S: $0 \leq x \leq 1,\ \ 0 \leq y \leq 1,\ \ 0 \leq z \leq 1$

4. $\iiint_S (x^2 - y^2 + z^2)\, dV$; S: $0 \leq x \leq 1,\ \ 0 \leq y \leq 2,\ \ 0 \leq z \leq 1$

5. $\iiint_S e^z \sin x \cos y\, dV$; S: $0 \leq x \leq \pi/2,\ \ 0 \leq y \leq \pi/2,\ \ 0 \leq z \leq 1$

6. $\iiint_S e^{-z} \cos x \cos y\, dV$; S: $0 \leq x \leq \pi/2,\ \ 0 \leq y \leq \pi/2,\ \ 0 \leq z \leq 1$

In Problems 7–10 evaluate each iterated triple integral.

7. $\int_0^2 \left\{ \int_0^{2-3x} \left[\int_0^{x+y} x\, dz \right] dy \right\} dx$

8. $\int_0^1 \left\{ \int_0^{4-x} \left[\int_0^{2x+y} z\, dz \right] dy \right\} dx$

9. $\int_0^3 \left\{ \int_z^{z+2} \left[\int_y^{y+z} 2x\, dx \right] dy \right\} dz$

10. $\int_0^{\pi/2} \left\{ \int_y^{\pi/2} \left[\int_0^{xy} \sin \frac{z}{y}\, dz \right] dx \right\} dy$

In Problems 11–16 evaluate $\iiint_S xyz\, dV$.

11. S is the solid enclosed by the surfaces $z_1 = 0$, $z = 2 - x - y$, whose projection on the xy-plane is the rectangle $0 \le x \le 1$, $0 \le y \le 2$.

12. S is the solid enclosed by the surfaces $z_1 = 0$, $z = 2 - x^2 - y^2$, whose projection onto the xy-plane is the rectangle $0 \le x \le 2$, $0 \le y \le 1$.

13. S is the solid enclosed by the surfaces $z = 0$, $z = xy$, whose projection onto the xy-plane is the triangle with vertices $(0, 0)$, $(0, 1)$, and $(1, 0)$.

14. S is the solid enclosed by the surfaces $z = 0$, $z = x^2 + y^2$, whose projection onto the xy-plane is the triangle with vertices $(0, 0)$, $(1, 0)$, and $(0, 2)$.

15. S is the solid enclosed by the surfaces $z = 1 - x - y$, $z = 3 - x - y$, whose projection onto the xy-plane is the circle $x^2 + y^2 = 1$.

16. S is the solid enclosed by the surfaces $z = 0$, $z = x^2 + y$, whose projection onto the xy-plane is the circle $x^2 + y^2 = 4$.

In Problems 17 and 18 express $\iiint_S f(x, y, z)\, dV$ as an iterated integral in six different ways for each region S.

17. S is the region enclosed by the coordinate planes and the plane $x + 2y + 3z = 6$.

18. S is the region enclosed by the coordinate planes and the plane $x + y + z = 3$.

In Problems 19–22 evaluate each triple integral.

19. $\iiint_S x\, dV$, if S is the region enclosed by the tetrahedron having vertices at $(0, 0, 0)$, $(1, 1, 0)$, $(1, 0, 0)$, $(1, 0, 1)$

20. $\iiint_S (x^2 + z^2)\, dV$, if S is the same region as in Problem 19

21. $\iiint_S (xy + 3y)\, dV$, if S is the region enclosed by the cylinder $x^2 + y^2 = 9$ and the planes $x + z = 3$, $y = 0$, and $z = 0$

22. $\iiint_S xyz\, dV$, if S is the region enclosed by the cylinders $x^2 + y^2 = 1$ and $x^2 + z^2 = 1$

In Problems 23–28 use triple integration to find the volume of the indicated region.

23. Enclosed by $z = 4 - y^2$, $z = 9 - x$, $x = 0$, and $z = 0$

24. Enclosed by $z = 0$, $z = 1 - x^2$, and $z = 1 - y^2$

25. Enclosed by $y^2 = z$, $x = 0$, and $x = y - z$

26. Enclosed by $z = x^2 + y^2$ and $z = 16 - x^2 - y^2$

27. Enclosed by $z = x^2 + y^2$ and $z = 2 - x$

28. Enclosed by $z^2 = 4x$ and $x^2 + y^2 = 2x$

29. Find the mass of a body in the shape of a right circular cylinder of height h and radius a, if its mass density is proportional to the square of the distance from the axis of the cylinder.

30. Find the mass of a body in the shape of a cube of edge a if its mass density is proportional to the square of the distance from one corner.

31. Find the mass of a body in the shape of a tetrahedron cut from the first octant by the plane $x + y + z = 1$, if the mass density is proportional to the product of the distances from the three coordinate planes.

(*Problem 31 continues in the next column*)

31. (*continued*)

32. Find the moments of inertia I_x and I_y for the solid region enclosed by the hemisphere $z = \sqrt{9 - x^2 - y^2}$ and lying above the xy-plane if the mass density is proportional to the distance from the xy-plane.

33. Set up, but do not evaluate, the integral of the function $f(x, y, z) = x^2yz$ over the region enclosed by the cone $3x^2 + 3y^2 = z^2$, $z \ge 0$, and the plane $z = 3$.

Triple Integrals Using Cylindrical Coordinates 18.7

We have seen that in many instances the evaluation of a double integral is more easily accomplished using polar coordinates rather than rectangular coordinates. For triple integrals, we give two alternatives to integration in rectangular coordinates: one utilizes *cylindrical coordinates* and the other uses *spherical coordinates* (introduced in Section 18.8).

If the rectangular coordinates of a point P in three-dimensional space are (x, y, z) and if the polar coordinates for the projection of P onto the xy-plane are (r, θ), then P may be located by the ordered triple (r, θ, z), called the *cylindrical coordinates of P*.*

Figure 42 illustrates the role of r, θ, and z in locating a point P whose cylindrical coordinates are (r, θ, z). The algebraic relationship of the cylindrical coordinates (r, θ, z) and the rectangular coordinates (x, y, z) of a point P are given by the formulas

$$x = r\cos\theta \qquad y = r\sin\theta \qquad z = z \tag{1}$$

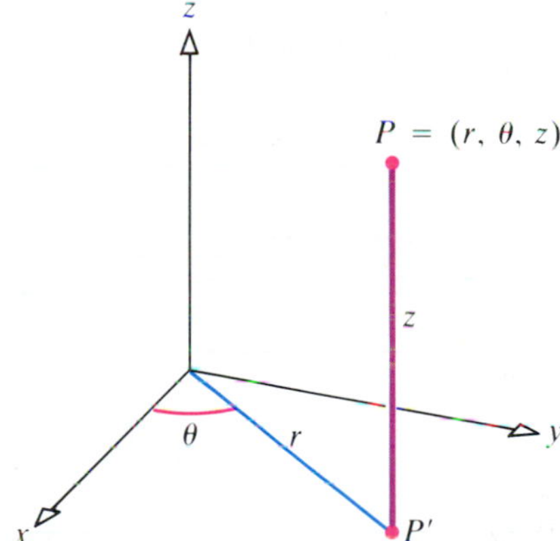

Figure 42

EXAMPLE 1

If the cylindrical coordinates of a point P are $(6, \pi/3, -2)$, find the rectangular coordinates for P.

Solution

From (1), we find that

$$x = 6\cos\frac{\pi}{3} = 3 \qquad y = 6\sin\frac{\pi}{3} = 3\sqrt{3} \qquad z = -2$$

---

* The reason for the name *cylindrical* coordinates is that the surfaces $r = a$ constant are cylinders.

Table 1 is a list of several equations in rectangular coordinates and their respective equations in cylindrical coordinates. Figure 43 illustrates the graph of each equation.

Table 1

Surface	Rectangular	Cylindrical
(a) Plane	$y = x \tan k$	$\theta = k$
(b) Plane	$z = k$	$z = k$
(c) Cylinder	$x^2 + y^2 = a^2$	$r = a$
(d) Sphere	$x^2 + y^2 + z^2 = R^2$	$r^2 + z^2 = R^2$
(e) Circular cone	$x^2 + y^2 = a^2z^2$	$r = az$
(f) Circular paraboloid	$x^2 + y^2 = az$	$r^2 = az$

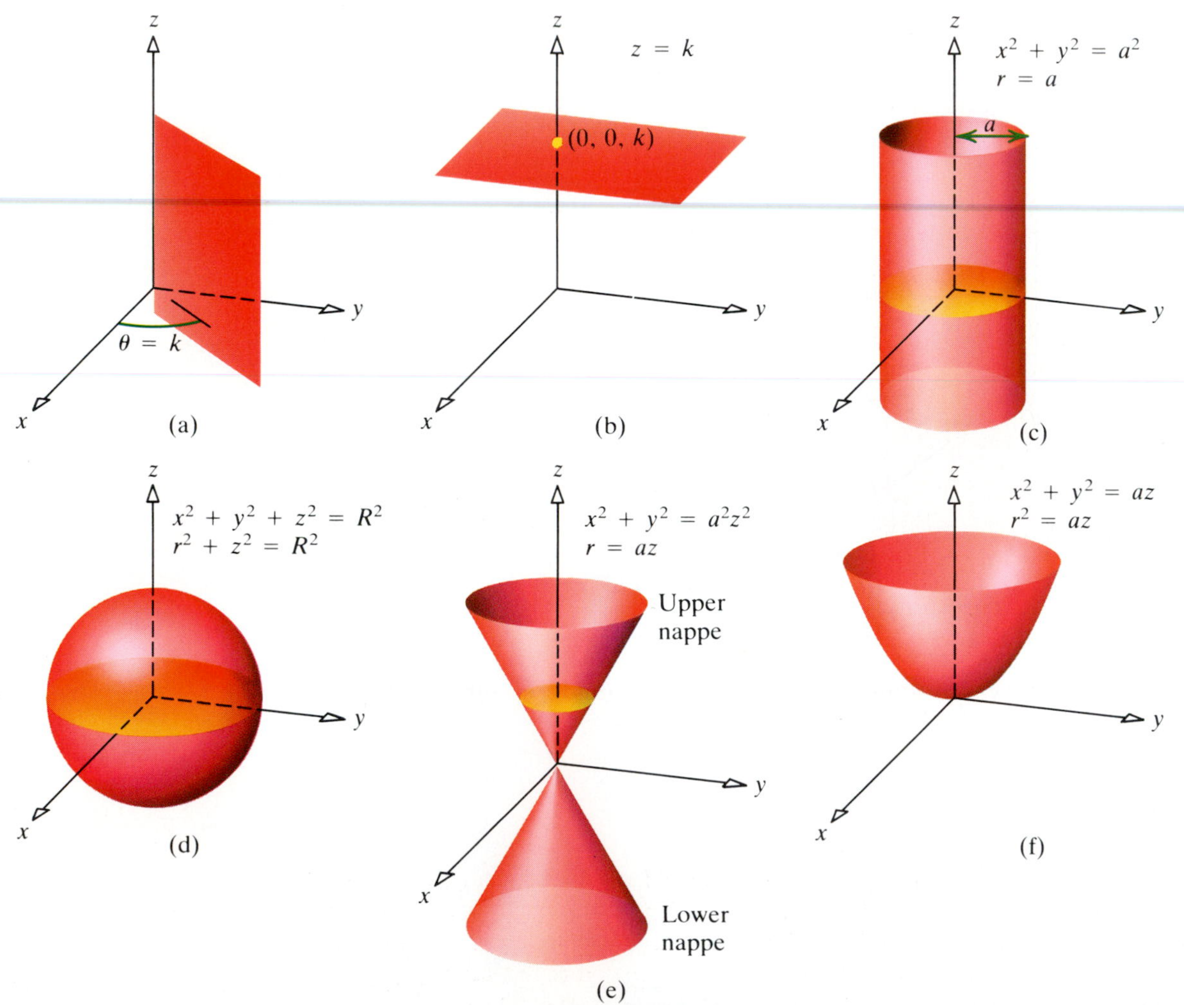

Figure 43

For triple integral problems in which there is symmetry about an axis, particularly problems concerned with cylinders and cones, it may be easier to evaluate the triple integral by utilizing cylindrical coordinates.

The solid regions most easily described by cylindrical coordinates are cylindrical wedges of the form

$$r_1 \leq r \leq r_2 \qquad \alpha \leq \theta \leq \beta \qquad z_1 \leq z \leq z_2$$

Such a region is enclosed by two planes $(\theta = \alpha$ and $\theta = \beta)$ passing through the z-axis, two concentric circular cylinders $(r = r_1$ and $r = r_2)$ centered about the z-axis, and two parallel planes $(z = z_1$ and $z = z_2)$ perpendicular to the z-axis. See Figure 44. The volume V is given by

$$V = \bar{r}\Delta r\Delta\theta\Delta z$$

where

$$\bar{r} = \frac{r_1 + r_2}{2} \qquad \Delta r = r_2 - r_1 \qquad \Delta\theta = \beta - \alpha \qquad \Delta z = z_2 - z_1$$

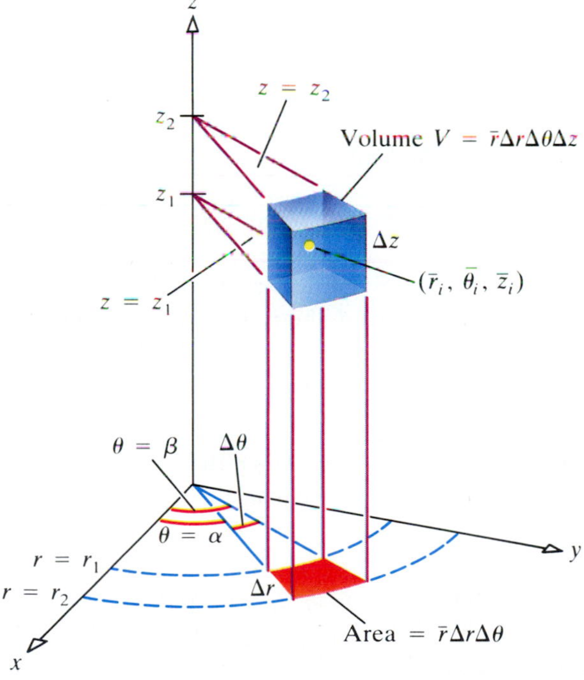

Figure 44

To evaluate a triple integral over a solid region S, we divide S into a number of wedge-shaped subregions by passing planes and cylinders through S. The shaded portion of Figure 45 illustrates a typical subregion, denoted S_i. We let n denote the number of such subregions that lie entirely in S. We now have a partition P of the region whose norm $\|P\|$ is taken as the length of the longest diagonal of these subregions. The volume ΔV_i of a typical

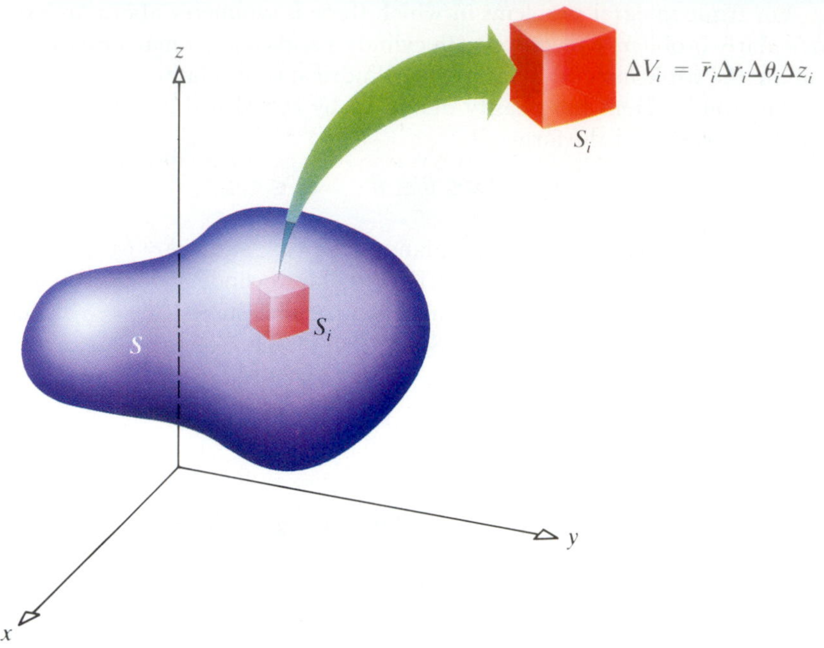

Figure 45

subregion S_i is given by

$$\Delta V_i = \bar{r}_i \Delta r_i \Delta\theta_i \Delta z_i$$

where Δr_i, $\Delta\theta_i$, and Δz_i are the dimensions of the subregion S_i and $\bar{r}_i$ is the r-coordinate of a point in its interior.

Now we let f be a function of the cylindrical coordinates r, θ, and z. We pick a point $(\bar{r}_i, \bar{\theta}_i, \bar{z}_i)$ in S_i and evaluate f there. Then we form the sum

$$\sum_{i=1}^{n} f(\bar{r}_i, \bar{\theta}_i, \bar{z}_i)\bar{r}_i \Delta r_i \Delta\theta_i \Delta z_i$$

It can be shown that if f is continuous on S, then the limit of the sum above as the norm $\|P\|$ approaches 0 exists and equals the triple integral of f over S. That is,

$$\iiint_S f(r, \theta, z)\, dV = \lim_{\|P\| \to 0} \sum_{i=1}^{n} f(\bar{r}_i, \bar{\theta}_i, \bar{z}_i)\bar{r}_i \Delta r_i \Delta\theta_i \Delta z_i$$

$$= \iiint_S f(r, \theta, z) r\, dr\, d\theta\, dz$$

It is important to remember that the integrand contains a factor of r. This is because in cylindrical coordinates the differential dV of volume is given by

$$dV = r\, dr\, d\theta\, dz$$

Sometimes we can evaluate a triple integral by using iterated integrals.

[18.7.1] THEOREM

Suppose a solid S is enclosed by the surfaces $z = z_1(r, \theta)$ and $z = z_2(r, \theta)$, where $z_1 \le z_2$ and z_1, z_2 are continuous functions. Suppose the projection of S onto the $r\theta$-plane is a closed, bounded region R. If f is a continuous function of three variables on S, then

$$\iiint_S f(r, \theta, z)\, dV = \iint_R \left[\int_{z_1(r,\theta)}^{z_2(r,\theta)} f(r, \theta, z)\, dz\right] r\, dr\, d\theta$$

Here the inside integral $\int_{z_1(r,\theta)}^{z_2(r,\theta)} f(r, \theta, z)\, dz$ is evaluated by holding r and θ fixed and integrating partially with respect to z.

If the region R is enclosed by the rays $\theta = \alpha$, $\theta = \beta$, and curves $r = r_1(\theta)$, $r = r_2(\theta)$, where $r_1 \le r_2$, and r_1, r_2 are continuous for $\alpha \le \theta \le \beta$, then

$$\iiint_S f(r, \theta, z) r\, dr\, d\theta\, dz = \int_\alpha^\beta \left\{\int_{r_1(\theta)}^{r_2(\theta)} \left[\int_{z_1(r,\theta)}^{z_2(r,\theta)} f(r, \theta, z)\, dz\right] r\, dr\right\} d\theta \tag{2}$$

Five other formulations for iteration are also possible; the choice of which to use depends on S.

EXAMPLE 2

Find the volume of the region enclosed by the sphere $x^2 + y^2 + z^2 = 4$ and the cylinder $(x - 1)^2 + y^2 = 1$. Figure 46 illustrates the situation.

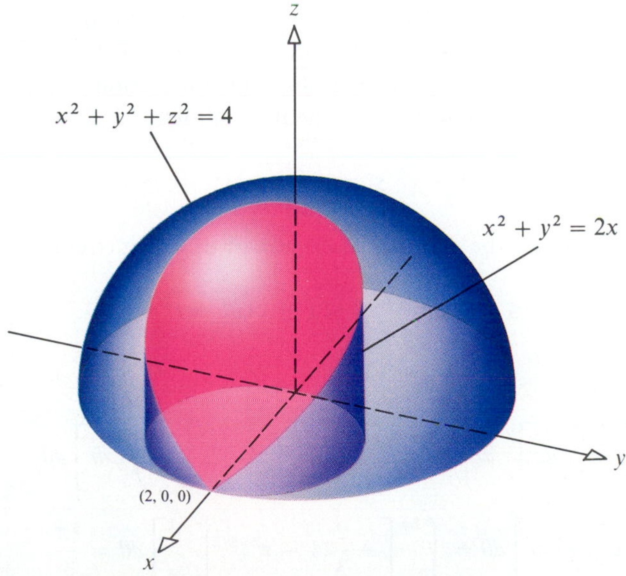

Figure 46

Solution

We use cylindrical coordinates because the variables x and y appear in the equations for the sphere and the cylinder only in combinations of $x^2 + y^2$. The equation of the sphere in cylindrical coordinates is $r^2 + z^2 = 4$ and that of the cylinder is $r = 2\cos\theta$. Since both the region and the integral are symmetric about the half plane $\theta = 0$ and the plane $z = 0$, we have

$$V = 4\int_0^{\pi/2}\left\{\int_0^{2\cos\theta}\left[\int_0^{\sqrt{4-r^2}} r\,dz\right]dr\right\}d\theta = 4\int_0^{\pi/2}\left[\int_0^{2\cos\theta}(4-r^2)^{1/2}r\,dr\right]d\theta$$

$$= -2\int_0^{\pi/2} \tfrac{2}{3}(4-r^2)^{3/2}\Big|_0^{2\cos\theta}\,d\theta = \tfrac{4}{3}\int_0^{\pi/2}[8-(4-4\cos^2\theta)^{3/2}]\,d\theta$$

$$= \tfrac{32}{3}\int_0^{\pi/2}(1-\sin^3\theta)\,d\theta = \tfrac{16}{9}(3\pi - 4)$$

The next example illustrates a triple integral stated in rectangular coordinates that is more easily evaluated using cylindrical coordinates.

EXAMPLE 3

Use cylindrical coordinates to evaluate

$$\int_{-1}^{1}\left\{\int_{-\sqrt{1-x^2}}^{\sqrt{1-x^2}}\left[\int_0^{2\sqrt{1-x^2-y^2}} dz\right]dy\right\}dx$$

Solution

From the limits of integration on z, we see that the integral is the volume of the upper half of the ellipsoid $4x^2 + 4y^2 + z^2 = 4$, $z \geq 0$. From the x and y limits of integration, the projection R in the xy-plane is enclosed by the circle $x^2 + y^2 = 1$. See Figure 47. The region S of integration and its projection on the xy-plane can be described by the inequalities

$$0 \leq z \leq 2\sqrt{1-x^2-y^2} \qquad -\sqrt{1-x^2} \leq y \leq \sqrt{1-x^2} \qquad -1 \leq x \leq 1$$

In cylindrical coordinates the region S is described by the inequalities

$$0 \leq r \leq 1 \qquad 0 \leq \theta \leq 2\pi \qquad 0 \leq z \leq 2\sqrt{1-r^2}$$

Thus,

$$V = \int_0^{2\pi}\left\{\int_0^1\left[\int_0^{2\sqrt{1-r^2}} dz\right]r\,dr\right\}d\theta = \int_0^{2\pi}\left[\int_0^1\left(z\Big|_{z=0}^{z=2\sqrt{1-r^2}}\right)r\,dr\right]d\theta$$

$$= \int_0^{2\pi}\left[\int_0^1 2r\sqrt{1-r^2}\,dr\right]d\theta = \int_0^{2\pi}\left[-\frac{2}{3}(1-r^2)^{3/2}\Big|_{r=0}^{r=1}\right]d\theta = \frac{4\pi}{3}$$

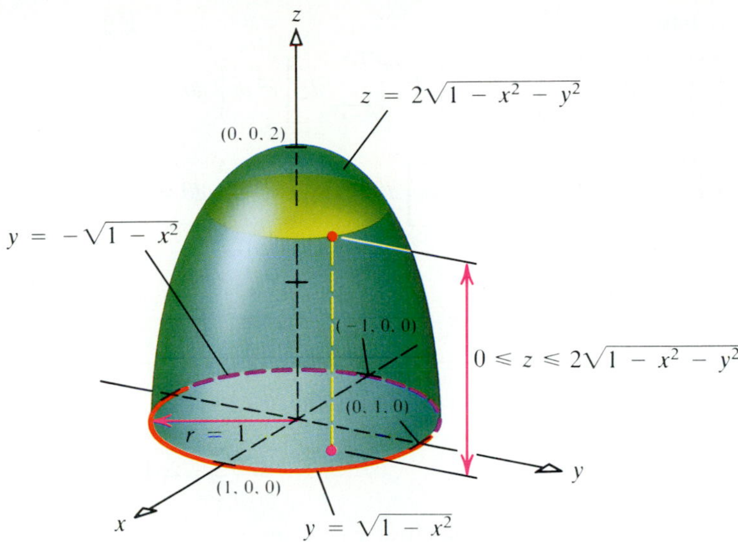

Figure 47

EXAMPLE 4

Find the moment of inertia about the z-axis of a homogeneous body of mass density ρ in the shape of a region enclosed by the paraboloid $z = 1 - x^2 - y^2$ and the xy-plane.

Solution

In cylindrical coordinates, the region S is $0 \leq \theta \leq 2\pi$, $0 \leq r \leq 1$, and $z = 0$, $z = 1 - r^2$. The moment of inertia I_z about the z-axis is

$$I_z = \iiint_S r^2 \rho \, dV = \rho \iiint_S r^2 r \, dr \, d\theta \, dz = \rho \int_0^{2\pi} \left\{ \int_0^1 \left[\int_0^{1-r^2} dz \right] r^3 \, dr \right\} d\theta$$

$$= \rho \int_0^{2\pi} \left[\int_0^1 r^3 (1 - r^2) \, dr \right] d\theta = \rho \int_0^{2\pi} \frac{1}{12} \, d\theta = \rho \frac{\pi}{6}$$

EXAMPLE 5

Find the moment of inertia of a homogeneous body in the shape of a sphere about a diameter.

Solution

Let ρ denote the constant mass density of the body and let a be the radius of the sphere. Situate the sphere so that its center is at the origin. The equation of the sphere is then $x^2 + y^2 + z^2 = a^2$, or $r^2 + z^2 = a^2$. Since the moment of inertia about a diameter is given by the moment of inertia about the z-axis, we have

$$I_z = \iiint_S \rho r^2\, dV = \rho \int_0^{2\pi} \left\{ \int_0^a \left[\int_{-\sqrt{a^2-r^2}}^{\sqrt{a^2-r^2}} dz \right] r^3\, dr \right\} d\theta$$

$$= \rho \int_0^{2\pi} \left[\int_0^a (r^3) 2\sqrt{a^2 - r^2}\, dr \right] d\theta = 2\rho \int_0^{2\pi} \frac{2a^5}{15}\, d\theta = \frac{8\pi\rho a^5}{15}$$

EXERCISE 18.7

In Problems 1–4 find the cylindrical coordinates of the point with the given rectangular coordinates.

1. $(-\sqrt{3}, -1, -5)$ **2.** $(-1, \sqrt{3}, 4)$ **3.** $(1, 1, \sqrt{2})$ **4.** $(2, -2, 4)$

In Problems 5–8 find the rectangular coordinates of the point with the given cylindrical coordinates.

5. $(2, \pi/6, -5)$ **6.** $(4, \pi/3, 3)$ **7.** $(4, \pi/6, 2)$ **8.** $(2, \pi/2, 0)$

In Problems 9 and 10 evaluate each iterated integral.

9. $\int_{\pi/6}^{\pi/2} \left\{ \int_0^3 \left[\int_0^{r \sin\theta} r \csc^3\theta\, dz \right] dr \right\} d\theta$

10. $\int_0^{\pi/3} \left\{ \int_0^{\sin\theta} \left[\int_0^{r \sin\theta} r\, dz \right] dr \right\} d\theta$

In Problems 11–16 use triple integration in cylindrical coordinates.

11. Find the mass of a homogeneous body in the shape of a sphere of radius a.

12. Find the mass of a body in the shape of a sphere of radius a, if the mass density is proportional to the square of the distance from the center.

13. Find the center of mass of a homogeneous body in the shape of a region enclosed by the surface $x^2 + y^2 = 4z$ and the plane $z = 2$.

14. Find the center of mass of a homogeneous body in the shape of a region in the first octant enclosed by the surface $z = xy$ and the cylinder $x^2 + y^2 = 4$.

15. Find the center of mass of a homogeneous body in the shape of a region enclosed by the inside of the sphere $x^2 + y^2 + z^2 = 12$ and above the paraboloid $z = x^2 + y^2$.

16. Find the center of mass of a homogeneous body in the shape of a region enclosed by the paraboloid $z = x^2 + y^2$ and the plane $z = 4$.

In Problems 17 and 18 each integral is given in cylindrical coordinates. Express each integral in rectangular coordinates. Do not evaluate.

17. $\int_0^{2\pi} \left\{ \int_0^4 \left[\int_{-r}^{\sqrt{16-r^2}} z^2 r^5 \cos^4\theta\, dz \right] dr \right\} d\theta$

18. $\int_0^{\pi/2} \left\{ \int_0^2 \left[\int_{-r^2}^9 z^2 r^4 \sin\theta\, dz \right] dr \right\} d\theta$

19. Use cylindrical coordinates to find the mass of the homogeneous solid bounded on the sides by $x^2 + y^2 = 1$, on the bottom by the xy-plane, and on the top by $x^2 + y^2 + z^2 = 2$.

Triple Integrals Using Spherical Coordinates 18.8

Spherical Coordinates

Suppose (x, y, z) are the rectangular coordinates of a point P (different from the origin) in three-dimensional space. We define the numbers ρ, θ, and ϕ by

$\rho = |OP|$, distance from O to P

θ = Angle between positive x-axis and OP', where P' is the projection of P onto the xy-plane

ϕ = Angle between positive z-axis and OP, $0 \le \phi \le \pi$

As Figure 48(a) illustrates, the point P may be located by the ordered triple (ρ, θ, ϕ), called the *spherical* coordinates of* P. The algebraic relationship of the spherical coordinates (ρ, θ, ϕ) and the rectangular coordinates (x, y, z) of a point P are obtained from Figure 48(b). There, we conclude that

$$x = |OP'|\cos\theta \qquad y = |OP'|\sin\theta$$

Since $|OP'| = |QP| = \rho \sin\phi$ and $|OQ| = z = \rho\cos\phi$, it follows that

$$\boldsymbol{x = \rho \sin\phi \cos\theta \qquad y = \rho\sin\phi\sin\theta \qquad z = \rho\cos\phi} \tag{1}$$

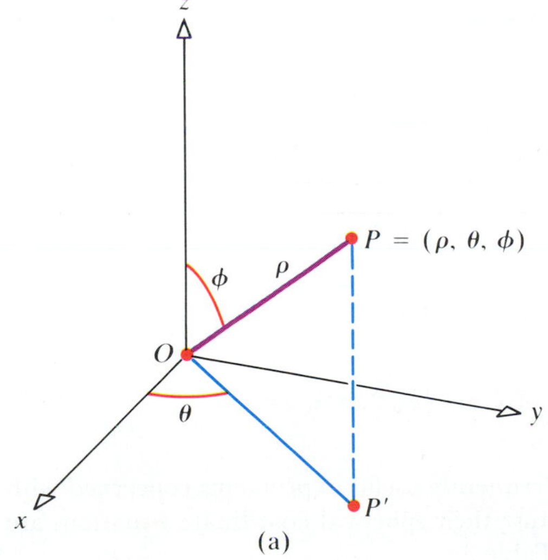

(a)

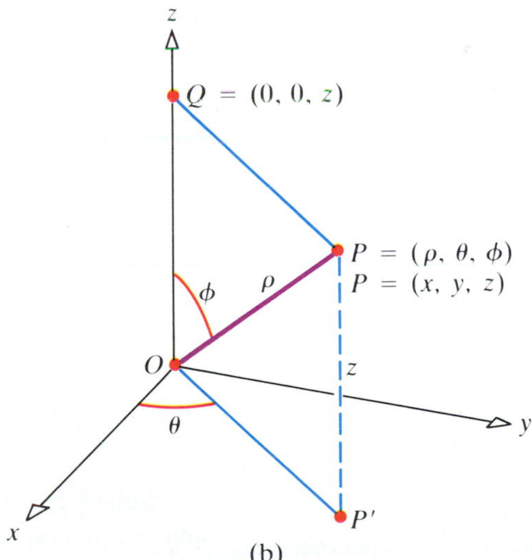

(b)

Figure 48

* The reason for the name *spherical* coordinates is that the surfaces ρ = a constant are spheres.

Based on (1), we have

$$\rho = \sqrt{x^2 + y^2 + z^2} \qquad \tan\theta = \frac{y}{x} \qquad \cos\phi = \frac{z}{\rho} \tag{2}$$

EXAMPLE 1

If the spherical coordinates of a point P are $(4, \pi/6, 2\pi/3)$, find the rectangular and cylindrical coordinates for P.

Solution

From (1), we find that

$$x = 4\sin\frac{2\pi}{3}\cos\frac{\pi}{6} = 4\left(\frac{\sqrt{3}}{2}\right)\left(\frac{\sqrt{3}}{2}\right) = 3$$

$$y = 4\sin\frac{2\pi}{3}\sin\frac{\pi}{6} = 4\left(\frac{\sqrt{3}}{2}\right)\left(\frac{1}{2}\right) = \sqrt{3}$$

$$z = 4\cos\frac{2\pi}{3} = 4\left(-\frac{1}{2}\right) = -2$$

Thus, the rectangular coordinates of P are $(3, \sqrt{3}, -2)$. Since $r^2 = x^2 + y^2 = 9 + 3 = 12$, the cylindrical coordinates of P are $(2\sqrt{3}, \pi/6, -2)$.

EXAMPLE 2

If the rectangular coordinates of a point P are $(1, \sqrt{3}, -2)$, find the spherical coordinates of P.

Solution

From (2), we find that

$$\rho = \sqrt{x^2 + y^2 + z^2} = \sqrt{1 + 3 + 4} = \sqrt{8} = 2\sqrt{2}$$

$$\tan\theta = \sqrt{3} \qquad \theta = \frac{\pi}{3} \qquad \cos\phi = \frac{-1}{\sqrt{2}} \qquad \phi = \frac{3\pi}{4}$$

Thus, the spherical coordinates of P are $(2\sqrt{2}, \pi/3, 3\pi/4)$.

Table 2

Surface	Equation
(a) Sphere	$\rho = a$
(b) Half cone	$\phi = a$
(c) Half plane	$\theta = a$

Spherical coordinates are frequently useful in problems concerned with spheres, planes, and cones because their spherical coordinate equations are relatively simple, as shown by Table 2.

The surface $\phi = a$, $0 < a < \pi$, $a \neq \pi/2$, requires some explanation. The surface is a half cone, which is generated by revolving any ray emanating from the origin at an angle a about the positive z-axis. The various surfaces are illustrated in Figure 49.

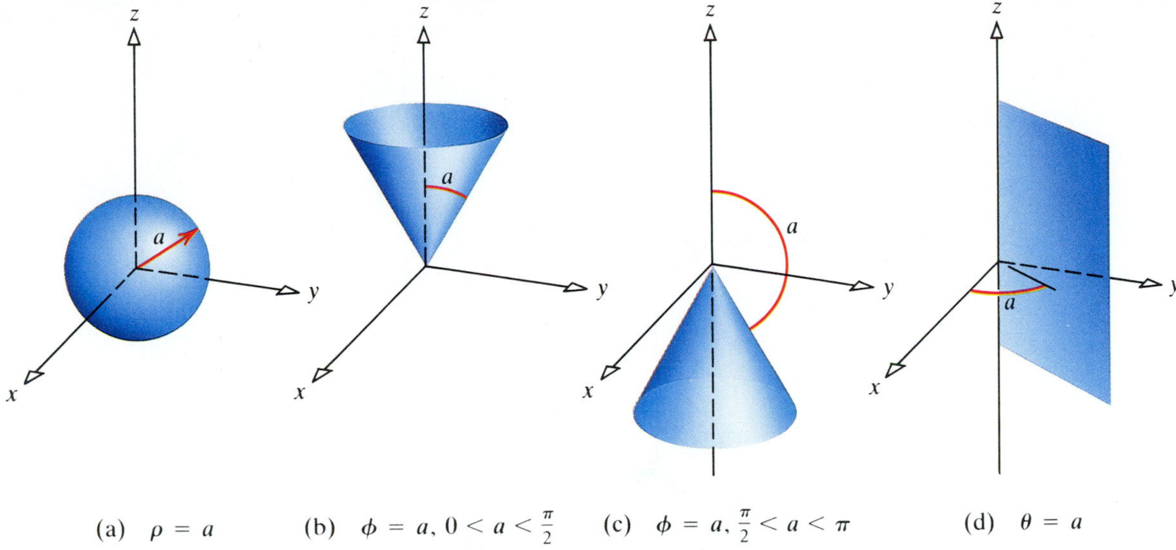

(a) $\rho = a$ (b) $\phi = a, 0 < a < \frac{\pi}{2}$ (c) $\phi = a, \frac{\pi}{2} < a < \pi$ (d) $\theta = a$

Figure 49

Triple Integrals

Let (ρ, θ, ϕ) be the spherical coordinates of a point. We partition a region S by drawing concentric spheres (ρ a constant), planes through the z-axis (θ a constant), and circular cones with vertex at the origin and axis along the z-axis (ϕ a constant). We let n denote the number of such subregions that lie entirely in S. We now have a partition of S whose norm $\|P\|$ is taken as the length of the longest diagonal of these subregions.

We proceed to calculate the volume ΔV_i of a typical subregion S_i (see Fig. 50). By treating the subregion as though it were a rectangular parallelepiped, we obtain the following approximation to ΔV_i:

$$\Delta V_i = (\bar{\rho}_i \Delta\phi_i)(\Delta\rho_i)(\bar{\rho}_i \sin \bar{\phi}_i \Delta\theta_i) = \bar{\rho}_i^2 \sin \bar{\phi}_i \Delta\rho_i \Delta\theta_i \Delta\phi_i$$

where $(\bar{\rho}_i, \bar{\theta}_i, \bar{\phi}_i)$ is some point in S_i. Now we let f be a function in the spherical coordinates ρ, θ, and ϕ. We pick a point $(\bar{\rho}_i, \bar{\theta}_i, \bar{\phi}_i)$ in each S_i and evaluate f there. Then we form the sum

$$\sum_{i=1}^{n} f(\bar{\rho}_i, \bar{\theta}_i, \bar{\phi}_i)\bar{\rho}_i^2 \sin \bar{\phi}_i \Delta\rho_i \Delta\theta_i \Delta\phi_i$$

It can be shown that if f is continuous on S, then the limit of the above sum, as the norm $\|P\|$ approaches 0, exists and equals the triple integral of f over S. That is,

$$\iiint_S f(\rho, \theta, \phi)\, dV = \lim_{\|P\| \to 0} \sum_{i=1}^{n} f(\bar{\rho}_i, \bar{\theta}_i, \bar{\phi}_i)\bar{\rho}_i^2 \sin \bar{\phi}_i \Delta\rho_i \Delta\theta_i \Delta\phi_i$$

$$= \iiint_S f(\rho, \theta, \phi)\rho^2 \sin \phi \, d\rho \, d\theta \, d\phi \qquad \textbf{(3)}$$

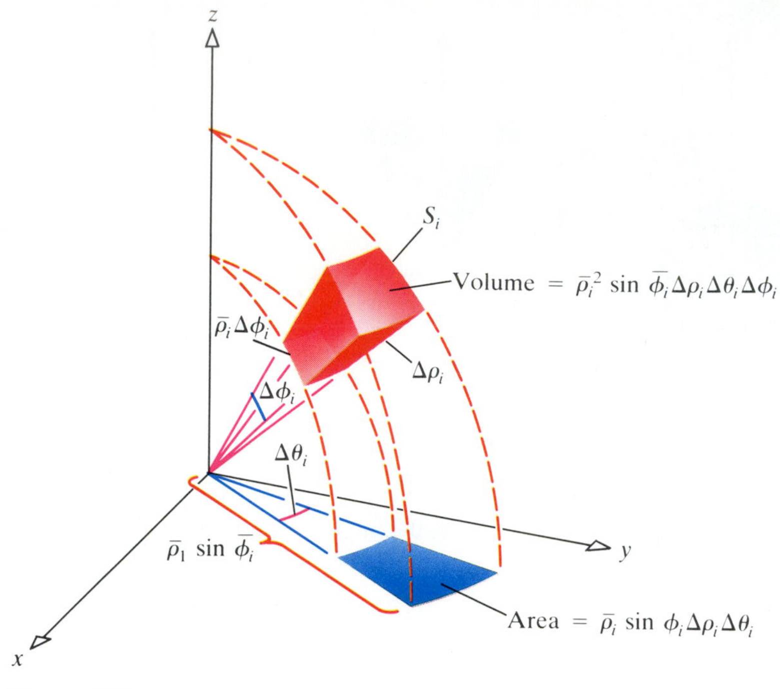

Figure 50

It is important to remember that the integrand contains the factor $\rho^2 \sin \phi$. This is because in spherical coordinates the differential dV of volume is given by

$$dV = \rho^2 \sin \phi \; d\rho \; d\theta \; d\phi$$

EXAMPLE 3

Find the volume that is cut from a sphere whose radius is 1 by a cone that makes an angle of 30° with the positive z-axis.

Solution

From Figure 51, we see that the limits on the variables of integration ρ, θ, and ϕ are

$$0 \le \rho \le 1 \qquad 0 \le \theta \le 2\pi \qquad 0 \le \phi \le \pi/6$$

Thus,

$$\begin{aligned} V &= \int_0^{\pi/6} \left\{ \int_0^{2\pi} \left[\int_0^1 \rho^2 \sin \phi \; d\rho \right] d\theta \right\} d\phi \\ &= \frac{2\pi}{3}\left(1 - \frac{\sqrt{3}}{2}\right) \text{ cubic units} \end{aligned}$$

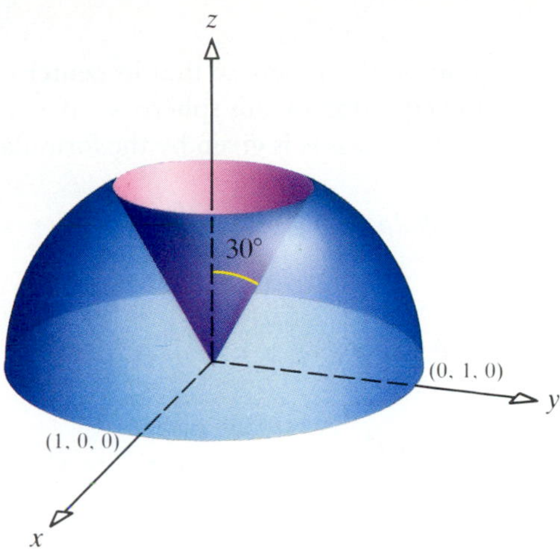

Figure 51

EXAMPLE 4

Use spherical coordinates to evaluate the integral

$$\iiint_S z\,dx\,dy\,dz$$

where S is defined by the inequalities

$$\sqrt{x^2+y^2} \le z \qquad x^2+y^2+z^2 \le 1 \qquad z \ge 0$$

Solution

S can be described in spherical coordinates as follows: $z \ge 0$ corresponds to $\phi \le \pi/2$, and $x^2+y^2+z^2 \le 1$ corresponds to $\rho \le 1$ (it is the upper half of the hemispherical solid $x^2+y^2+z^2 \le 1$). For the surface $z = \sqrt{x^2+y^2}$ (which is a half cone), we get $x^2+y^2 = z^2$. The cross section of this cone with the xz-plane is the pair of lines $z^2 = x^2$ or $z = \pm x$, which form an angle $\phi = \pi/4$ with the positive z-axis. Since $z = \rho\cos\phi$, our integral becomes

$$\iiint_S z\,dx\,dy\,dz = \iiint_S \rho\cos\phi\,\rho^2\sin\phi\,d\rho\,d\theta\,d\phi = \int_0^{2\pi}\left\{\int_0^{\pi/4}\sin\phi\cos\phi\left[\int_0^1 \rho^3\,d\rho\right]d\phi\right\}d\theta$$

$$= \frac{1}{4}\int_0^{2\pi}\left[\int_0^{\pi/4}\frac{1}{2}\sin 2\phi\,d\phi\right]d\theta = \frac{1}{16}\int_0^{2\pi} d\theta = \frac{\pi}{8}$$

EXAMPLE 5

Find the mass of a body in the shape of a sphere, if the mass density δ* is proportional to the distance from the center.

* Since we are using ρ as a spherical coordinate here, we will use δ as the symbol for mass density in order to avoid confusion.

Solution

Let a be the radius of the sphere, and position the sphere so that its center is at the origin. In spherical coordinates the equation of this sphere is $\rho = a$. The density δ of the sphere is $\delta = k\rho$. The mass m is given by the formula

$$m = \iiint_S \delta\, dV = k \iiint_S \rho\rho^2 \sin\phi\, d\rho\, d\phi\, d\theta$$

$$= k \int_0^{2\pi} \left\{ \int_0^{\pi} \left[\int_0^a \rho^3 \sin\phi\, d\rho \right] d\phi \right\} d\theta$$

$$= \frac{ka^4}{4} \int_0^{2\pi} \left[\int_0^{\pi} \sin\phi\, d\phi \right] d\theta = \frac{ka^4}{4} \int_0^{2\pi} 2\, d\theta = k\pi a^4$$

Spherical Coordinates in Navigation

If we assume that the surface of the earth is spherical, there is a simple relationship between the spherical coordinates we have defined and the system of latitude and longitude measurements used in geography. The origin is placed at the center of the earth and the z-axis is chosen to be the diameter through the North and South Poles. See Figure 52(a). The equator is then the great circle in the xy-plane. The x-axis is chosen so that the xz-plane will pass through the Greenwich Observatory in London. The longitude for a point P on the surface of the earth is then the angle we have called θ, except that degree measure is used and east and west are measured from the great circle through the poles and Greenwich. London itself thus has longitude 0°.

The latitude for a point on the surface of the earth north of the equator is $90° - \phi$; thus, latitude is measured from the equator rather than from the North Pole. For a point on the earth south of the equator, the latitude is $\phi - 90°$. Points on the equator have $\phi = 90°$ and latitude 0°. For the

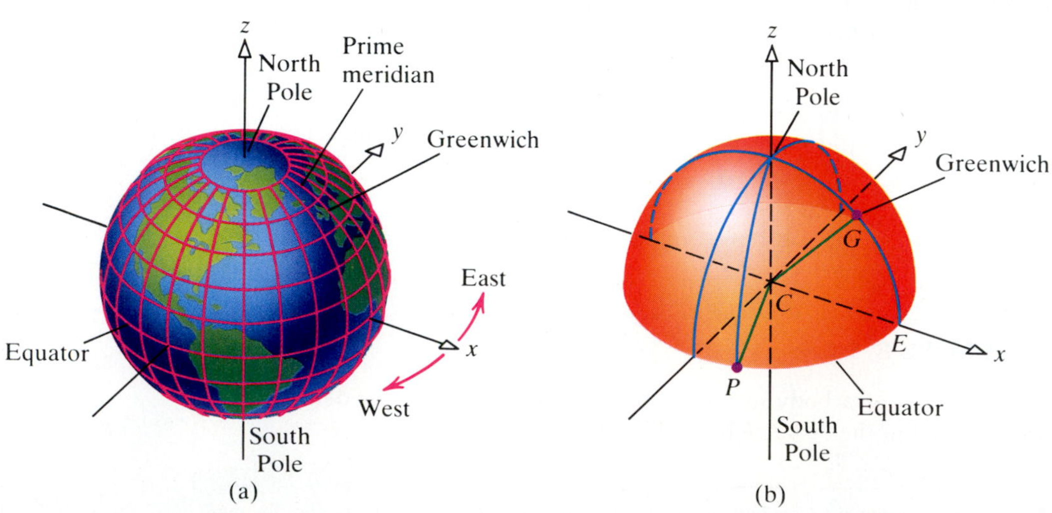

Figure 52

North Pole, we write $\phi = 0°$ and north latitude 90°; for the South Pole, we write $\phi = 180°$ and south latitude 90°.

EXAMPLE 6

Greenwich has a longitude of 0° and an approximate latitude of $51\frac{1}{2}°$ N. If we use $R_e = 4000$ miles for the radius of the earth and use degrees instead of radians, the spherical coordinates for Greenwich are $(4000, 0°, 38\frac{1}{2}°)$. See Figure 52(b), where the angle $ECG = 51\frac{1}{2}°$.

EXERCISE 18.8

In Problems 1–8 find the spherical coordinates of the point with the given rectangular coordinates.

1. $(-\sqrt{2}, -\sqrt{2}, 2\sqrt{3})$ **2.** $(-1, \sqrt{3}, 2)$ **3.** $(1, 1, \sqrt{2})$ **4.** $(1, -\sqrt{3}, -2)$

5. (1, 2, 3) (Use a calculator to find the answer to the nearest 0.01).

6. $(1, -1, \sqrt{2})$ **7.** $(0, 3\sqrt{3}, 3)$ **8.** $(-5\sqrt{3}, 5, 0)$

In Problems 9 and 10 fill in the missing coordinates. Assume that the three ordered triples represent the rectangular, cylindrical, and spherical coordinates of some point in space.

9. (x, y, z) (r, θ, z) (ρ, θ, ϕ)
$(-1, \quad, \quad)$ $(\quad, \quad, -2\sqrt{3}/3)$ $(\quad, 4\pi/3, \quad)$

10. (x, y, z) (r, θ, z) (ρ, θ, ϕ)
$(\quad, \quad, 2\sqrt{2})$ $(\sqrt{8}, \quad, \quad)$ $(\quad, \pi/6, \quad)$

In Problems 11–14 evaluate each iterated integral.

11. $\int_0^{\pi/2} \left\{ \int_0^{\sin\phi} \left[\int_0^{\pi/4} \rho^2 \sin\phi \, d\theta \right] d\rho \right\} d\phi$

12. $\int_0^{\pi/2} \left\{ \int_0^{\pi/2} \left[\int_0^{\sin\phi} \rho^2 \sin\phi \cos\phi \, d\rho \right] d\theta \right\} d\phi$

13. $\int_0^{2\pi} \left\{ \int_0^{\pi/4} \left[\int_0^{\sec\phi} \rho^2 \sin^2\phi \, d\rho \right] d\phi \right\} d\theta$

14. $\int_0^{\pi/4} \left\{ \int_0^{\cos\phi} \left[\int_0^{2\pi} \rho^2 \sin\phi \, d\theta \right] d\rho \right\} d\phi$

In Problems 15–18 use triple integration in spherical coordinates.

15. Find the mass of a homogeneous body in the shape of a sphere of radius a.

16. Find the mass of a body in the shape of a sphere of radius a, if the mass density is proportional to the square of the distance from the center.

17. Find the center of mass of a body in the shape of a hemisphere of radius a, if the mass density is proportional to the distance from the center.

18. Find the center of mass of a body in the shape of a hemisphere, if the mass density is proportional to the distance from the axis of symmetry.

In Problems 19 and 20 use either cylindrical or spherical coordinates to evaluate each triple integral.

19. $\int_0^2 \left\{ \int_0^2 \left[\int_0^{\sqrt{4-x^2}} \sqrt{x^2+y^2} \, dy \right] dx \right\} dz$

20. $\int_0^2 \left\{ \int_0^{\sqrt{4-y^2}} \left[\int_0^{\sqrt{4-x^2-y^2}} \frac{2z}{\sqrt{x^2+y^2}} \, dz \right] dx \right\} dy$

21. The magnitude of the resultant gravitational force of a solid hemisphere of radius a and constant mass density δ on a unit mass particle situated at the center of the base of the hemisphere is given by the triple integral

$$F = k\delta \iiint_V \frac{\cos\phi}{\rho^2}\, dV$$

where the center of the sphere is at the origin and spherical coordinates are used. Evaluate the integral.

22. Use spherical coordinates to integrate $f(x, y, z) = \sqrt{x^2 + y^2 + z^2}$ over the region above the cone $z = -\sqrt{3x^2 + 3y^2}$ and inside the sphere $x^2 + y^2 + z^2 = 4$.

In Problems 23 and 24 set up the triple integral $\iiint_S f(x, y, z)\, dV$ for S in rectangular, cylindrical, and spherical coordinates.

23. S is the solid sphere of radius a with center at the origin.

24. S is the region inside the cylinder $x^2 + y^2 = 4$ and inside the sphere $x^2 + y^2 + z^2 = 9$.

The integrals in Problems 25 and 26 are given in spherical coordinates. Express each in rectangular coordinates. Do not evaluate.

25. $\displaystyle\int_0^{\pi} \left\{ \int_{3\pi/4}^{\pi} \left[\int_0^4 \rho^5 \cos\theta \sin^2\phi \, d\rho \right] d\phi \right\} d\theta$

26. $\displaystyle\int_{\pi/2}^{3\pi/2} \left\{ \int_{\pi/2}^{\pi} \left[\int_0^2 \frac{\rho^4 \sin\phi \cos\phi}{\rho^2 + 3}\, d\rho \right] d\phi \right\} d\theta$

27. The integral below is given in cylindrical coordinates. Express it in spherical coordinates. Do not evaluate.

$$\int_0^{\pi} \left\{ \int_0^3 \left[\int_0^{\sqrt{9-z^2}} r^2 \sin\theta \, dr \right] dz \right\} d\theta$$

28. A solid occupies the region $\sqrt{x^2 + y^2} \le z \le 1$ and has density $\delta(x, y, z) = z\sqrt{x^2 + y^2 + z^2}$. Determine its mass.

29. Find the volume of the solid enclosed on the outside by the sphere $\rho = 2$ and on the inside by the surface $\rho = 1 + \cos\phi$.

30. Use spherical coordinates to find the centroid of the hemisphere of radius a, whose base is on the xy-plane.

31. Find the center of mass of a homogeneous solid enclosed from above by the sphere $x^2 + y^2 + z^2 = 9$ and from below by the half cone $z = \sqrt{x^2 + y^2}$.

32. Find the center of mass of a homogeneous solid in the shape of the wedge $x^2 + y^2 = 16$, $z = 2y$, $y \ge 0$, $z \ge 0$.

33. Use spherical coordinates to find the mass of the solid in the first octant between the spheres $x^2 + y^2 + z^2 = a^2$ and $x^2 + y^2 + z^2 = b^2$ $(a > b)$ if the density at any point is inversely proportional to its distance from the origin.

18.9 Jacobians

In evaluating a single integral, $\int_a^b f(x)\, dx$, one technique we used often was substitution: If we let $x = g(u)$, then $dx = g'(u)\, du$ and

$$\int_a^b f(x)\, dx = \int_c^d f(g(u))g'(u)\, du$$

where $a = g(c)$ and $b = g(d)$. In making this substitution, we changed the variable from x to u and introduced the factor $g'(u)$ in the integrand.

In working with a double integral, $\iint_R f(x, y)\,dx\,dy$, we sometimes changed the variables x and y to polar coordinates r and θ by letting $x = r\cos\theta$, $y = r\sin\theta$, and $dx\,dy = r\,dr\,d\theta$. That is,

$$\iint\limits_R f(x, y)\,dx\,dy = \iint\limits_{R^\#} f(r\cos\theta, r\sin\theta)r\,dr\,d\theta$$

where $R^\#$ is the region R expressed in polar coordinates. In changing the variables, we introduced the factor r into the integrand on the right.

In working with a triple integral, $\iiint_S f(x, y, z)\,dx\,dy\,dz$, we sometimes changed the variables x, y, and z to cylindrical coordinates r, θ, and z by letting $x = r\cos\theta$, $y = r\sin\theta$, and $dx\,dy\,dz = r\,dr\,d\theta\,dz$. That is,

$$\iiint\limits_S f(x, y, z)\,dx\,dy\,dz = \iiint\limits_{S^\#} f(r\cos\theta, r\sin\theta, z)r\,dr\,d\theta\,dz$$

where $S^\#$ is the region S expressed in cylindrical coordinates. Here, a factor of r is introduced into the integrand on the right. In changing the variables of a triple integral to spherical coordinates, we found

$$\iiint\limits_S f(x, y, z)\,dx\,dy\,dz = \iiint\limits_{S^\#} f(\rho\sin\phi\cos\theta, \rho\sin\phi\sin\theta, \rho\cos\phi)\rho^2\sin\phi\,d\rho\,d\theta\,d\phi$$

where $S^\#$ is the region S expressed in spherical coordinates. Here, a factor of $\rho^2\sin\phi$ is introduced into the integrand on the right.

In general, when the variables of an integral are changed, a factor, called the *Jacobian*, is introduced into the resulting integrand.

Jacobian in Two Variables

[18.9.1] DEFINITION / *Jacobian; Two Variables.*

For a change of variables from x, y to u, v given by

$$x = g_1(u, v) \qquad y = g_2(u, v)$$

the Jacobian of x and y with respect to u and v is

$$\frac{\partial(x, y)}{\partial(u, v)} = \begin{vmatrix} \dfrac{\partial x}{\partial u} & \dfrac{\partial x}{\partial v} \\[2ex] \dfrac{\partial y}{\partial u} & \dfrac{\partial y}{\partial v} \end{vmatrix} = \frac{\partial x}{\partial u}\frac{\partial y}{\partial v} - \frac{\partial x}{\partial v}\frac{\partial y}{\partial u}$$

EXAMPLE 1

Show that in changing from rectangular coordinates (x, y) to polar coordinates (r, θ), the Jacobian of x, y with respect to r, θ is r.

Solution

The change of variables is given by

$$x = r\cos\theta \qquad y = r\sin\theta$$

The partial derivatives are

$$\frac{\partial x}{\partial r} = \cos\theta \qquad \frac{\partial x}{\partial \theta} = -r\sin\theta \qquad \frac{\partial y}{\partial r} = \sin\theta \qquad \frac{\partial y}{\partial \theta} = r\cos\theta$$

The Jacobian of x, y with respect to r, θ is

$$\frac{\partial(x, y)}{\partial(r, \theta)} = \begin{vmatrix} \frac{\partial x}{\partial r} & \frac{\partial x}{\partial \theta} \\ \frac{\partial y}{\partial r} & \frac{\partial y}{\partial \theta} \end{vmatrix} = \begin{vmatrix} \cos\theta & -r\sin\theta \\ \sin\theta & r\cos\theta \end{vmatrix}$$

$$= \cos\theta \cdot r\cos\theta - \sin\theta(-r\sin\theta) = r\cos^2\theta + r\sin^2\theta = r$$

In changing variables to evaluate a double integral, we make use of the following result, which we state without proof:

[18.9.2] THEOREM

Suppose R is a closed, bounded region of the xy-plane and $R^{\#}$ is a closed, bounded region of the uv-plane, such that each point in R corresponds to one and only one point in $R^{\#}$ using the change of variables $x = g_1(u, v)$, $y = g_2(u, v)$. If the following are true:

(a) f is continuous on R,
(b) g_1 and g_2 have continuous partial derivatives on $R^{\#}$, and
(c) the Jacobian $\partial(x, y)/\partial(u, v) \neq 0$ on $R^{\#}$,

then

$$\iint_R f(x, y)\,dx\,dy = \iint_{R^{\#}} f(g_1(u, v), g_2(u, v)) \left|\frac{\partial(x, y)}{\partial(u, v)}\right| du\,dv \qquad (1)$$

Notice in (1) that the absolute value of the Jacobian is introduced as a factor on the right side.

Changing variables can simplify the integration of $\iint_R f(x, y)\,dx\,dy$ in two ways: either by simplifying the region R or by simplifying the integrand $f(x, y)$.

EXAMPLE 2

Evaluate $\iint_R y\,dx\,dy$, where R is the region enclosed by the lines $2x + y = 0$, $2x + y = 3$, $x - y = 0$, and $x - y = 2$.

Solution

Figure 53(a) illustrates the region R. If we let $u = 2x + y$ and $v = x - y$, the parallelogram region R will transform into a rectangular region

$R^{\#}$, since now $R^{\#}$ is the region enclosed by $u = 0$, $u = 3$, $v = 0$, $v = 2$ (see Fig. 53(b)).

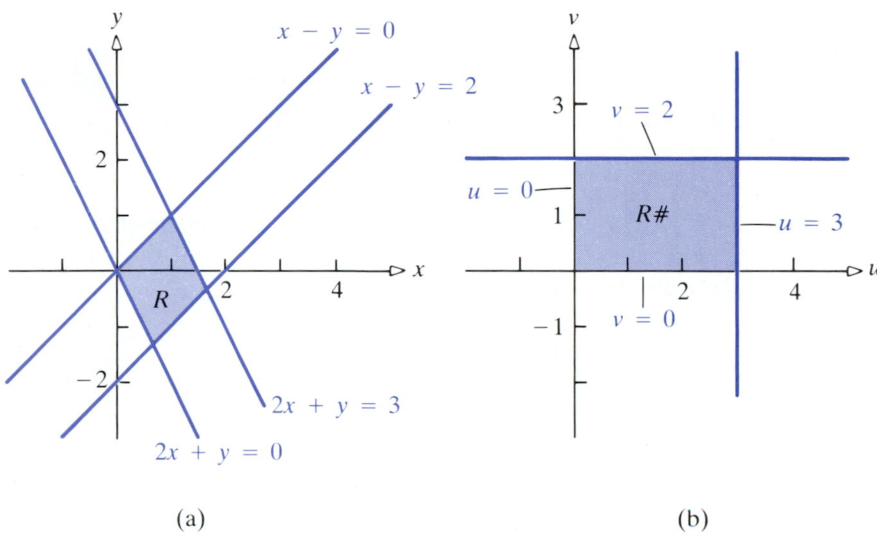

Figure 53

To transform the integral, we need to express x, y in terms of u, v. Since

$$u = 2x + y \qquad v = x - y$$

we can solve for x and y:

$$x = \frac{u + v}{3} \qquad y = \frac{u - 2v}{3}$$

The Jacobian is

$$\frac{\partial(x, y)}{\partial(u, v)} = \begin{vmatrix} \dfrac{\partial x}{\partial u} & \dfrac{\partial x}{\partial v} \\ \dfrac{\partial y}{\partial u} & \dfrac{\partial y}{\partial v} \end{vmatrix} = \begin{vmatrix} \dfrac{1}{3} & \dfrac{1}{3} \\ \dfrac{1}{3} & -\dfrac{2}{3} \end{vmatrix} = \frac{-2}{9} - \frac{1}{9} = -\frac{1}{3}$$

Now we use (1):

$$\iint_R y\,dx\,dy = \iint_{R^{\#}} \frac{(u - 2v)}{3} \cdot \left|\frac{\partial(x, y)}{\partial(u, v)}\right| du\,dv = \int_0^2 \left[\int_0^3 \frac{1}{9}(u - 2v)\,du\right] dv$$

$$= \frac{1}{9}\int_0^2 \left(\frac{u^2}{2} - 2uv\right)\Bigg|_0^3 dv = \frac{1}{9}\int_0^2 \left(\frac{9}{2} - 6v\right) dv$$

$$= \frac{1}{9}\left(\frac{9v}{2} - 3v^2\right)\Bigg|_0^2 = \frac{1}{9}(9 - 12) = -\frac{1}{3}$$

EXAMPLE 3

Evaluate

$$\iint_R (x + y)\sin(x - y)\, dx\, dy$$

where R is the region enclosed by $y=x$, $y=x-2$, $y=-x$, $y=-x+1$.

Solution

The integrand $(x + y)\sin(x - y)$ suggests the following change of variables: $u = x + y$, $v = x - y$. The region R in the xy-plane and the region $R^{\#}$ in the uv-plane are shown in Figure 54. Solving

$$u = x + y \qquad v = x - y$$

for x and y, we obtain

$$x = \frac{u + v}{2} \qquad y = \frac{u - v}{2}$$

The Jacobian is

$$\frac{\partial(x, y)}{\partial(u, v)} = \begin{vmatrix} \frac{1}{2} & \frac{1}{2} \\ \frac{1}{2} & -\frac{1}{2} \end{vmatrix} = -\frac{1}{4} - \frac{1}{4} = -\frac{1}{2}$$

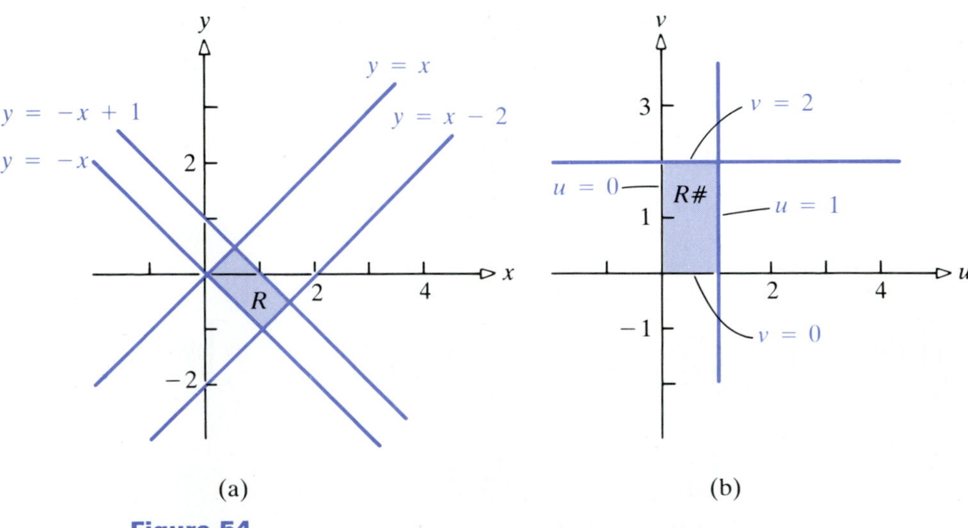

Figure 54

The integral, under this change of variables, becomes

$$\iint_R (x+y)\sin(x-y)\,dx\,dy = \iint_{R^\#} u\sin v\left|\frac{\partial(x,y)}{\partial(u,v)}\right| du\,dv = \frac{1}{2}\int_0^1\left[\int_0^2 u\sin v\,dv\right]du$$

$$= \frac{1}{2}\int_0^1 u(-\cos v)\Big|_0^2\,du = \frac{1}{2}(1-\cos 2)\int_0^1 u\,du$$

$$= \frac{1}{4}(1-\cos 2)$$

Jacobian in Three Variables

[18.9.3] DEFINITION / *Jacobian; Three Variables.*

For a change of variables from x, y, z to u, v, w given by

$$x = g_1(u,v,w) \qquad y = g_2(u,v,w) \qquad z = g_3(u,v,w)$$

the Jacobian of x, y, z with respect to u, v, w is given by

$$\frac{\partial(x,y,z)}{\partial(u,v,w)} = \begin{vmatrix} \dfrac{\partial x}{\partial u} & \dfrac{\partial x}{\partial v} & \dfrac{\partial x}{\partial w} \\ \dfrac{\partial y}{\partial u} & \dfrac{\partial y}{\partial v} & \dfrac{\partial y}{\partial w} \\ \dfrac{\partial z}{\partial u} & \dfrac{\partial z}{\partial v} & \dfrac{\partial z}{\partial w} \end{vmatrix}$$

In changing the variables of a triple integral, we use the following result:

[18.9.4] THEOREM

Suppose S is a closed, bounded region of xyz space and $S^\#$ is a closed, bounded region of uvw space, such that each point in S corresponds with one and only one point in $S^\#$ using the change of variables $x = g_1(u,v,w)$, $y = g_2(u,v,w)$ $z = g_3(u,v,w)$. If the following are true:

(a) f is continuous on S,
(b) g_1, g_2, g_3 have continuous partial derivatives on $S^\#$, and
(c) the Jacobian $\partial(x,y,z)/\partial(u,v,w) \neq 0$ on $S^\#$,

then

$$\iiint_S f(x,y,z)\,dx\,dy\,dz$$

$$= \iiint_{S^\#} f(g_1(u,v,w), g_2(u,v,w), g_3(u,v,w))\left|\frac{\partial(x,y,z)}{\partial(u,v,w)}\right| du\,dv\,dw$$

EXAMPLE 4

Show that in changing from rectangular coordinates (x, y, z) to spherical coordinates (ρ, θ, ϕ), the triple integral of f over S becomes

$$\iiint_S f(x, y, z)\, dx\, dy\, dz$$

$$= \iiint_{S^\#} f(\rho \sin \phi \cos \theta, \rho \sin \phi \sin \theta, \rho \cos \phi)\rho^2 \sin \phi\, d\rho\, d\theta\, d\phi$$

Solution

The equations for changing variables from rectangular coordinates (x, y, z) to spherical coordinates (ρ, θ, ϕ) are

$$x = \rho \sin \phi \cos \theta \qquad y = \rho \sin \phi \sin \theta \qquad z = \rho \cos \phi$$

The Jacobian $\partial(x, y, z)/\partial(\rho, \theta, \phi)$ is

$$\frac{\partial(x, y, z)}{\partial(\rho, \theta, \phi)} = \begin{vmatrix} \frac{\partial x}{\partial \rho} & \frac{\partial x}{\partial \theta} & \frac{\partial x}{\partial \phi} \\ \frac{\partial y}{\partial \rho} & \frac{\partial y}{\partial \theta} & \frac{\partial y}{\partial \phi} \\ \frac{\partial z}{\partial \rho} & \frac{\partial z}{\partial \theta} & \frac{\partial z}{\partial \phi} \end{vmatrix} = \begin{vmatrix} \sin \phi \cos \theta & -\rho \sin \phi \sin \theta & \rho \cos \phi \cos \theta \\ \sin \phi \sin \theta & \rho \sin \phi \cos \theta & \rho \cos \phi \sin \theta \\ \cos \phi & 0 & -\rho \sin \phi \end{vmatrix}$$

$$= \sin \phi \cos \theta(-\rho^2 \sin^2 \phi \cos \theta)$$

$$+ \rho \sin \phi \sin \theta(-\rho \sin^2 \phi \sin \theta - \rho \cos^2 \phi \sin \theta)$$

$$+ \rho \cos \phi \cos \theta(-\rho \sin \phi \cos \theta \cos \phi)$$

$$= -\rho^2[\sin^3 \phi \cos^2 \theta + \sin \phi \sin^2 \theta + \sin \phi \cos^2 \phi \cos^2 \theta]$$

$$= -\rho^2 \sin \phi[\sin^2 \phi \cos^2 \theta + \sin^2 \theta + \cos^2 \phi \cos^2 \theta] = -\rho^2 \sin \phi$$

Thus,

$$\iiint_S f(x, y, z)\, dx\, dy\, dz$$

$$= \iiint_{S^\#} f(\rho \sin \phi \cos \theta, \rho \sin \phi \sin \theta, \rho \cos \phi) \left| \frac{\partial(x, y, z)}{\partial(\rho, \theta, \phi)} \right| d\rho\, d\theta\, d\phi$$

$$= \iiint_{S^\#} f(\rho \sin \phi \cos \theta, \rho \sin \phi \sin \theta, \rho \cos \phi)\rho^2 \sin \phi\, d\rho\, d\theta\, d\phi$$ ■

EXERCISE 18.9

In Problems 1–8 find the Jacobian $\partial(x, y)/\partial(u, v)$.

1. $x = u + v, \quad y = u - v$

2. $x = u^2 - v^2, \quad y = u^2 + v^2$

3. $x = u + a, \quad y = v + b$

4. $x = au + bv, \quad y = cu + dv$

5. $x = u/v, \quad y = u + v$

6. $x = e^u \cos v, \quad y = e^u \sin v$

7. $x = ve^u, \quad y = ue^{-v}$

8. $x = u \cos v, \quad y = v \sin u$

In Problems 9 and 10 find the Jacobian $\partial(x, y, z)/\partial(u, v, w)$.

9. $x = u + v + w, \quad y = u + v - w, \quad z = u - v - w$

10. $x = u, \quad y = v^2, \quad z = w^3$

In Problems 11–20 evaluate each integral using the given change of variables.

11. $\iint_R x\,dx\,dy$, where R is the region enclosed by $2x - y = 0, \quad 2x - y = 4, \quad x + y = 0, \quad x + y = 3;$ $u = 2x - y, \quad v = x + y$

12. $\iint_R (x^2 + y^2)\,dx\,dy$, where R is the region enclosed by $2x - y = 1, \quad 2x - y = 3, \quad x + y = 1, \quad x + y = 2;$ $u = 2x - y, \quad v = x + y$

13. $\iint_R xy\,dx\,dy$, where R is the triangular region whose vertices are $(-1, 1)$, $(1, 1)$, and $(0, 0)$; $u = x + y$, $v = x - y$

14. $\iint_R x^2y\,dx\,dy$, where R is the triangular region whose vertices are $(-1, 1)$, $(1, 1)$, and $(0, 0)$; $u = x + y$, $v = x - y$

15. $\iint_R (x + y)\sin(x - y)\,dx\,dy$, where R is the triangular region whose vertices are $(-1, 1)$, $(1, 1)$, and $(0, 0)$; $u = x + y, \quad v = x - y$

16. $\iint_R (x + y)e^{x-y}\,dx\,dy$, where R is the triangular region whose vertices are $(-1, 1)$, $(1, 1)$, and $(0, 0)$; $u = x + y, \quad v = x - y$

17. $\iint_R xy\,dx\,dy$, where R is the region enclosed by the ellipse $9x^2 + 16y^2 = 144; \quad u = x/4, \quad v = y/3$

18. $\iint_R (x^2 - y)\,dx\,dy$, where R is the region enclosed by the ellipse $9x^2 + 16y^2 = 144; \quad u = x/4, \quad v = y/3$

19. $\iiint_S xy\,dx\,dy\,dz$, where S is the solid enclosed by the ellipsoid $(x^2/4) + (y^2/4) + (z^2/9) = 1; \quad u = x/2, \quad v = y/2,$ $w = z/3$

20. $\iiint_S x^2y\,dx\,dy\,dz$, where S is the solid enclosed by the ellipsoid $(x^2/9) + (y^2/4) + z^2 = 1; \quad u = x/3, \quad v = y/2,$ $w = z$

In Problems 21–24 evaluate each integral.

21. $\iint_R x^2y\,dx\,dy$, where R is the region enclosed by the lines $2x - 3y = 0, \quad 2x - 3y = 1, \quad x + 2y = 0, \quad x + 2y = 3$

22. $\iint_R x\,dx\,dy$, where R is the region enclosed by the ellipse $(x^2/9) + (y^2/16) = 1$

23. $\iint_R x \cos xy\,dx\,dy$, where R is the region enclosed by $xy = 1, \quad xy = 3, \quad x = 1, \quad x = 3$

24. $\iint_R ye^{xy}\,dx\,dy$, where R is the region enclosed by $xy = 1, \quad xy = 3, \quad x = 1, \quad x = 3$

REVIEW EXERCISES

1. Evaluate: $\displaystyle\int_0^1 \left[\int_y^1 ye^{-x^3}\,dx\right] dy$

2. Evaluate: $\displaystyle\int_0^1 \left[\int_0^{\sqrt{1+x^2}} \frac{1}{x^2 + y^2 + 1}\,dy\right] dx$

3. Evaluate: $\displaystyle\int_0^{\pi/4} \left[\int_0^{\tan x} \sec x\,dy\right] dx$

4. Evaluate: $\displaystyle\int_0^1 \left[\int_{-x}^{x} e^{x+y}\,dy\right] dx$

5. Let R be the square region $-2 \le x \le 2$, $-2 \le y \le 2$ in the xy-plane, and suppose that $f(x, y) = xy$.

(a) Using a partition of R into four squares of equal area and evaluating f at the midpoint of each square, compute the corresponding Riemann sum of f over R.

(b) Using the same partition as in part (a), but evaluating f at an appropriate point of each square, determine the largest possible Riemann sum and the smallest. What is the average of these values?

(c) What is the actual value of $\iint_R f(x, y)\, dA$?

6. Evaluate $\iint_R x \sin y^3\, dA$, where R is the triangle with vertices $(0, 0)$, $(0, 2)$, and $(2, 2)$.

7. Reverse the order of integration in the iterated integral $\int_1^e [\int_0^{\ln x} (dy/x)]\, dx$ and evaluate it both ways.

8. Reverse the order of integration in the iterated integral $\int_0^{\pi/2} [\int_0^{\cos x} \sin x\, dy]\, dx$ and evaluate it both ways.

9. Find the volume of the tetrahedron enclosed by the coordinate planes and the plane $(x/a) + (y/b) + (z/c) = 1$. (Assume that a, b, and c are positive.)

10. Use multiple integration to show that the volume of the ellipsoid $(x^2/a^2) + (y^2/b^2) + (z^2/c^2) = 1$ is $\frac{4}{3}\pi abc$. (Assume that a, b, and c are positive.) What does this formula reduce to if $a = b = c$?

11. Find the volume of the solid in the first octant enclosed by the coordinate planes $a^2 y = b(a^2 - x^2)$ and $a^2 z = c(a^2 - x^2)$.

12. Use a double integral to find the area enclosed by the parabola $y^2 = 16x$ and the line $y = 4x - 8$.

13. Use a double integral to find the area of the circle $x^2 + y^2 = r^2$.

14. Use a double integral to calculate the first-quadrant area enclosed by $y^2 = x^3$ and $y = x$.

15. Use a double integral and polar coordinates to find the area of the circle $r = 2 \cos \theta$.

16. Find the first-quadrant area outside $r = 2a$ and inside $r = 4a \cos \theta$.

17. The mass density at each point of a circular washer of inner radius a and outer radius b is inversely proportional to the square of the distance from the center.

(a) Find the mass. Then discuss its behavior as $a \to 0$. (Note that the mass density becomes infinite as we approach the center of the washer.)

(b) Find the moment of inertia of the washer about its center, and show that (unlike the mass) it remains finite as $a \to 0$.

18. Find the mass of a square lamina of side a, if the mass density is proportional to the distance from a fixed vertex to the square.

19. Find the center of mass of the homogeneous body enclosed by $bx^2 = a^2 y$ and $ay = bx$, $a > 0, b > 0$.

20. Find the center of mass of the homogeneous hemispherical shell $0 \le a \le r \le b$, $0 \le \phi \le \pi/2$.

21. Derive the formula $S = \pi a \sqrt{a^2 + h^2}$ for the lateral surface area of a right circular cone of base radius a and altitude h.

22. Show that the area of the first-octant portion of the plane $(x/a) + (y/b) + (z/c) = 1$ (where a, b, and c are positive) is $S = \frac{1}{2}\sqrt{a^2b^2 + b^2c^2 + c^2a^2}$.

23. Find the surface area of the paraboloid $z = x^2 + y^2$ below the plane $z = 1$.

24. Find the area of the cone $x^2 + y^2 = 3z^2$ inside the cylinder $x^2 + y^2 = 4y$.

25. Describe the solid S whose volume is given by the iterated integral $\int_0^1 \{\int_{y^2}^1 [\int_0^{1-x} dz]\, dx\}\, dy$. Set up the other five iterated integrals for the volume of S.

26. Describe the solid S whose volume is given by the iterated integral $\int_0^1 \{\int_0^{x^2} [\int_0^y dz]\, dy\}\, dx$. Set up the other five iterated integrals for the volume of S.

27. Show that: $\int_a^b [\int_a^y f(x)\, dx]\, dy = \int_a^b (b - x) f(x)\, dx$

28. Show that:

$$\int_a^b \left\{ \int_a^z \left[\int_a^y f(x)\, dx \right] dy \right\} dz = \int_a^b \frac{(b - x)^2}{2} f(x)\, dx$$

29. Find the volume in the first octant of the ellipsoid $(x^2/a^2) + (y^2/b^2) + (z^2/c^2) = 1$.

30. Find the volume in the first octant enclosed by $y = 0$, $z = 0$, $x + y = 2$, $x + 2y = 6$, and $y^2 + z^2 = 4$.

31. Use spherical coordinates with triple integrals to find the volume between the spheres $x^2 + y^2 + z^2 = 4$ and $x^2 + y^2 + z^2 = 1$.

32. Find the volume cut from the sphere $\rho = a$ by the cone $\phi = \alpha$.

33. Use cylindrical coordinates to evaluate

$$\int_0^2 \left\{ \int_0^{\sqrt{4-x^2}} \left[\int_0^{\sqrt{4-x^2-y^2}} \frac{z}{\sqrt{x^2 + y^2}}\, dz \right] dy \right\} dx$$

34. Find the center of mass of the areas enclosed by $4x = y^2$, $y = 0$, $x = 4$.

35. Use cylindrical coordinates to find the volume in the first octant inside the cylinder $r = 1$ and the plane $3x + 2y + 6z = 6$.

36. Use cylindrical coordinates to evaluate

$$\int_0^a \left\{ \int_0^{\sqrt{a^2-x^2}} \left[\int_0^{\sqrt{a^2-x^2-y^2}} dz \right] dy \right\} dx$$

37. Use spherical coordinates to evaluate the expression in Problem 36.

38. Evaluate $\iint_R e^{\sqrt{x}}\, dA$, where R is the region enclosed by $y = x$, $x = 1$, $y = 0$.

39. Evaluate $\iint_R (x + y)^3\, dx\, dy$, where the sides of R are lines of the form $x + y = a$, $x + y = b$, $x - 2y = c$, and $x - 2y = d$. Introduce $u = x + y$ and $v = x - 2y$. Assume $a < b$ and $c < d$.

40. Evaluate $\iint_R xy\, dx\, dy$, where R is the region enclosed by $2x + y = 0$, $2x + y = 3$, $x - y = 0$, $x - y = 2$.

In Problems 41–45 evaluate the given integral by first transforming into an equivalent integral in cylindrical or spherical coordinates (whichever is more convenient).

41. $\displaystyle\int_{-2}^{2} \left\{ \int_{-\sqrt{4-y^2}}^{\sqrt{4-y^2}} \left[\int_0^{\sqrt{16-x^2-y^2}} dz \right] dx \right\} dy$

42. $\displaystyle\int_0^{1} \left\{ \int_{-\sqrt{1-x^2}}^{\sqrt{1-x^2}} \left[\int_{\sqrt{x^2+y^2}}^{\sqrt{2-x^2-y^2}} dz \right] dy \right\} dx$

43. $\displaystyle\int_{-\sqrt{2}}^{\sqrt{2}} \left\{ \int_{-\sqrt{2-x^2}}^{\sqrt{2-x^2}} \left[\int_0^{4} (x^2 + y^2)z\, dz \right] dy \right\} dx$

44. $\displaystyle\int_0^{1} \left\{ \int_0^{1} \left[\int_0^{\sqrt{1-x^2}} dy \right] dx \right\} dz$

45. $\displaystyle\int_0^{a} \left\{ \int_0^{\sqrt{a^2-x^2}} \left[\int_{h\sqrt{x^2+y^2}/a}^{\sqrt{a^2-x^2-y^2}} dz \right] dy \right\} dx$

46. Evaluate the following by either transforming to a more convenient coordinate system or using a different order of integration.

(a) $\displaystyle\int_0^{1} \left[\int_0^{\sqrt{1-x^2}} (x^2 + y^2)^{3/2}\, dy \right] dx$

(b) $\displaystyle\int_{-1}^{1} \left\{ \int_{-\sqrt{1-y^2}}^{\sqrt{1-y^2}} \left[\int_{\sqrt{x^2+y^2}}^{\sqrt{2-x^2-y^2}} dz \right] dx \right\} dy$

(c) $\displaystyle\int_0^{1} \left[\int_y^{1} (x^2 + 1)^{2/3}\, dx \right] dy$

CHALLENGE EXERCISES

1. Suppose that $f(x, y)$ is integrable over a rectangular region R in the xy-plane, and let P be a partition of R into n subrectangles of equal area ΔA. Evaluating f at the midpoint (u_i, v_i) of the ith subrectangle $(i = 1, 2, \ldots, n)$, let $M(P)$ be the average of these n functional values.

(a) Show that $M(P) = (1/A) \sum_{i=1}^{n} f(u_i, v_i)\Delta A$, where A is the area of R.

(b) Explain why

$$\lim_{\|P\|\to 0} M(P) = (1/A) \iint_R f(x, y)\, dA$$

(This is called the *average value of f over R*.)

2. By analogy with Problem 1 we define the average value of f over a region R that is not necessarily rectangular to be the number $(1/A) \iint_R f(x, y)\, dA$.

(a) In single variable calculus the average value of a function f over the interval $a \le x \le b$ is defined to be the number $[1/(b - a)] \int_a^b f(x)\, dx$. In what sense is this a special case of the above definition of the average value of f over R?

(b) Let $f(x, y, z)$ be integrable over the region S in space. What definition would you give for the average value of f over S?

3. Suppose that a homogeneous lamina of mass density 1 occupies a region R in the xy-plane. Show that the average value of $f(x, y) = x$ over R (as defined in Problem 2) is $\bar{x}$, the first coordinate of the center of mass of the lamina. What is the average value of $g(x, y) = y$ over R?

4. To calculate $\int_{-\infty}^{+\infty} e^{-x^2}\,dx$, let $I_a = \int_0^a e^{-x^2}\,dx$.

(a) Show that:

$$I_a^2 = \int_0^a e^{-x^2}\,dx \int_0^a e^{-y^2}\,dy = \int_0^a \int_0^a e^{-(x^2+y^2)}\,dx\,dy$$

(b) Let $J_a = \iint_R e^{-(x^2+y^2)}\,dA$, where R is the quarter circle $0 \le \theta \le \pi/2$, $0 \le r \le a$. Show that

$$|I_a^2 - J_a| < \frac{(4-\pi)a^2}{4} e^{-a^2}$$

(c) Evaluate J_a.

(d) Show that

$$\lim_{a \to -\infty} J_a = \pi/4 \qquad \text{and} \qquad \lim_{a \to +\infty} |I_a^2 - J_a| = 0$$

Thus, show that $\int_{-\infty}^{+\infty} e^{-x^2}\,dx = \sqrt{\pi}$. This integral is of special importance in statistics.

***5.** (a) Prove that the moment of inertia of the thin flat plate in the figure, about the z-axis, equals the sum of its moments of inertia about the x- and y-axes.

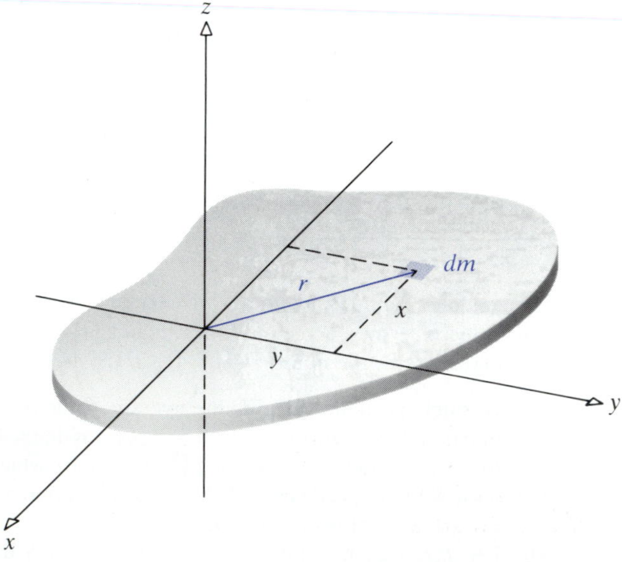

(*Problem 5 continues in the next column*)

* Adapted from F. W. Sears, M. W. Zemansky, and H. D. Young, *University Physics* (Reading, Mass.: Addison-Wesley Publishing Co., 1976), p. 177. Reprinted by permission.

5. (*continued*)

(b) Given that the moment of inertia of a disk about an axis through its center and perpendicular to its plane is $mR^2/2$ (where m is mass and R is the radius), use the relation from part (a) to find its moment of inertia about a diameter.

(c) Derive the result of part (b) by direct integration of the defining equation $I = \int r^2\,dm$.

(d) What is the moment of inertia of a disk about an axis tangent to its edge?

6. Evaluate $\int_1^2 \{\int_0^1 [1/(x^2\sqrt{x^2+y^2})]\,dy\}\,dx$. Use polar coordinates.

7. Use a double integral to derive the formula for the surface area of a sphere.

8. The center of a sphere of radius R is on the surface of a right cylinder of base radius $R/2$. Find the surface area of the sphere inside the cylinder.

9. Find the volume enclosed on the top by the sphere $x^2 + y^2 + z^2 = 5$ and on the bottom by the paraboloid $x^2 + y^2 = 4z$.

10. Show that:

(a) $$4\int_0^4 \left\{\int_0^{\sqrt{16-x^2}} \left[\int_{(x^2+y^2)/4}^4 dz\right] dy\right\} dx$$

(b) $$4\int_0^4 \left\{\int_0^{2\sqrt{z}} \left[\int_0^{\sqrt{4z-x^2}} dy\right] dx\right\} dz$$

(c) $$4\int_0^4 \left\{\int_{y^2/4}^4 \left[\int_0^{\sqrt{4z-y^2}} dx\right] dz\right\} dy$$

represent the same volume; do not evaluate.

Topics in Vector Calculus

19

Vector Fields 19.1

In Chapter 15 we introduced the idea of a vector function, $\mathbf{r} = \mathbf{r}(t)$, $a \leq t \leq b$. This type of function associates a vector $\mathbf{r}$ to a real number. A *vector field* is a function that associates a vector to a point in the plane or to a point in space.

[19.1.1] DEFINITION / *Vector Field.*

Let $P = P(x, y)$ and $Q = Q(x, y)$ denote functions of two variables defined on a set R. A *vector field* F *over* R is defined as the function

$$\mathbf{F} = \mathbf{F}(x, y) = P(x, y)\mathbf{i} + Q(x, y)\mathbf{j}$$

Let $P = P(x, y, z)$, $Q = Q(x, y, z)$ $R = R(x, y, z)$ denote functions of three variables defined on a set S. A *vector field* F *over*

***S* is defined as the function**

$$\mathbf{F} = \mathbf{F}(x, y, z) = P(x, y, z)\mathbf{i} + Q(x, y, z)\mathbf{j} + R(x, y, z)\mathbf{k}$$

The graph of a vector field **F** consists of vectors. If $\mathbf{F} = \mathbf{F}(x, y)$, the graph of **F** consists of vectors in the plane with initial point at (x, y). If $\mathbf{F} = \mathbf{F}(x, y, z)$, the graph of **F** consists of vectors in space with initial point at (x, y, z). See Figures 1(a) and 1(b).

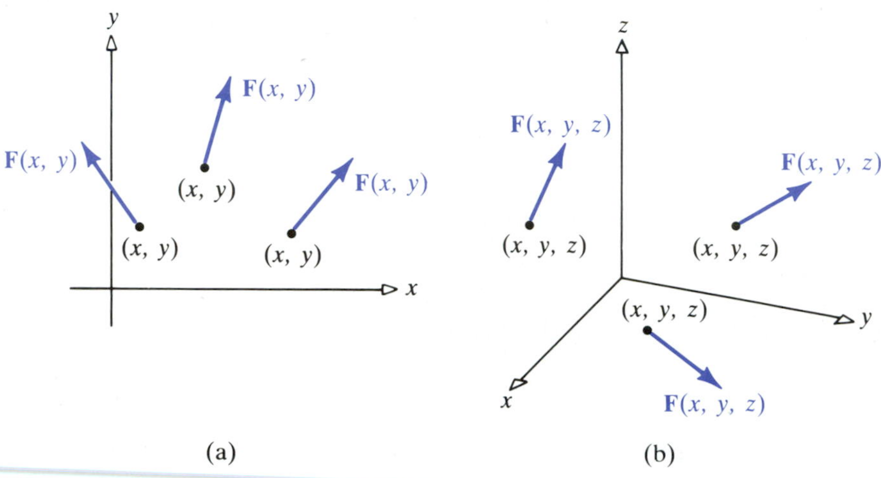

Figure 1

EXAMPLE 1

Describe the graph of the vector field $\mathbf{F} = \mathbf{F}(x, y) = -y\mathbf{i} + x\mathbf{j}$.

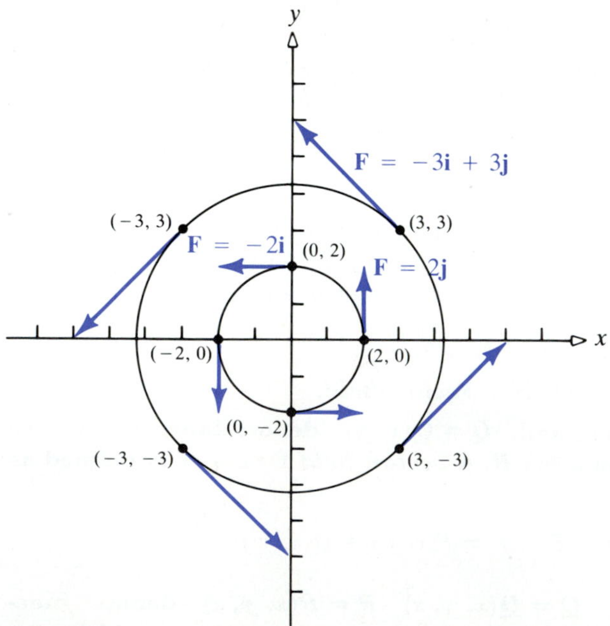

Figure 2

Solution

Figure 2 illustrates some vectors in the field. Notice that each vector in the field is tangent to a circle, and the direction of the vectors indicates that the field is rotating counterclockwise. This field might represent the motion of wheels spinning on an axle, with each vector equal to the velocity at a point of a wheel. In this case, **F** would be called a *velocity field.*

Another example of a velocity field arises in describing the flow of a fluid through a pipe or the flow of blood through an artery. At each point inside the pipe (or artery), a vector is defined that represents the velocity of the flow (see Fig. 3). The length of each arrow represents the speed of the flow.

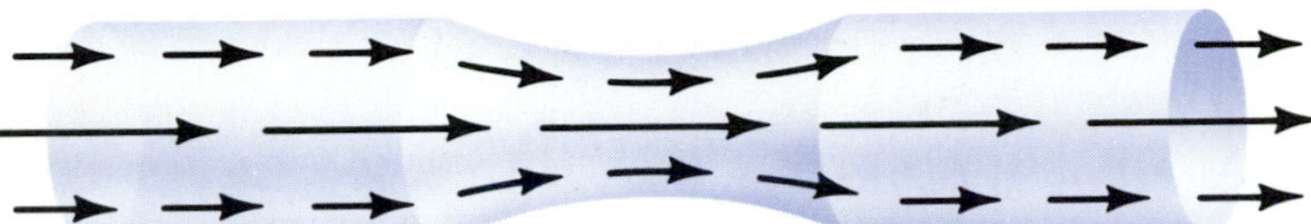

Figure 3

EXAMPLE 2 Newton's Law of Gravitation

A *gravitational field* is defined as the vector field

$$\mathbf{F} = \mathbf{F}(x, y, z) = \frac{-mMG}{x^2 + y^2 + z^2}\mathbf{u}$$

where m and M are the masses of two particles, one located at $(0, 0, 0)$ and the other at (x, y, z); G is the gravitational constant; and $\mathbf{u}$ is a unit vector from $(0, 0, 0)$ to (x, y, z). Here **F** is the force of attraction of the two particles. The force vectors **F** are directed from the particle at (x, y, z) toward the particle at $(0, 0, 0)$. See Figure 4.

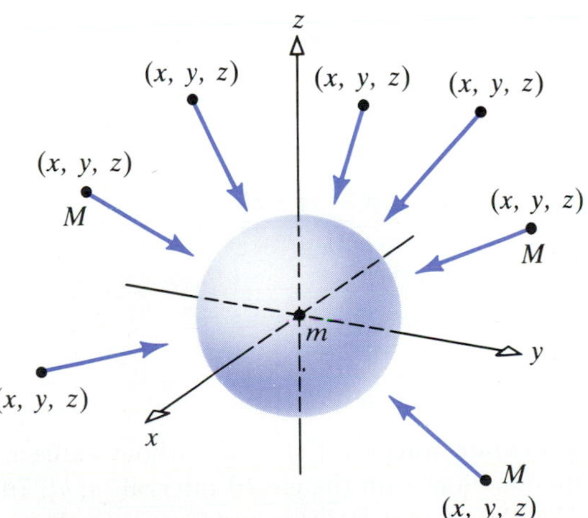

Figure 4

EXAMPLE 3 Coulomb's Law

An *electric force field* is defined as the vector field

$$\mathbf{F} = \mathbf{F}(x, y, z) = \frac{\varepsilon q Q}{x^2 + y^2 + z^2}\mathbf{u}$$

where ε is a constant that depends on the units being used; q and Q are the charges of two particles, one located at $(0, 0, 0)$ and the other at (x, y, z); and $\mathbf{u}$ is the unit vector from $(0, 0, 0)$ to (x, y, z). Here $\mathbf{F}$ is the electric force exerted by the charge at (x, y, z) on the charge at $(0, 0, 0)$. For like charges, we have $qQ > 0$ and the force $\mathbf{F}$ is *repulsive*. For unlike charges, we have $qQ < 0$ and the force $\mathbf{F}$ is *attractive*.

The gradient of a function f is also an example of a vector field, since

$$\nabla f(x, y) = f_x(x, y)\mathbf{i} + f_y(x, y)\mathbf{j}$$

and

$$\nabla f(x, y, z) = f_x(x, y, z)\mathbf{i} + f_y(x, y, z)\mathbf{j} + f_z(x, y, z)\mathbf{k}$$

[19.1.2] DEFINITION / *Conservative Vector Fields; Potential Function.*

A vector field F is called *conservative* if F is the gradient of some function that has continuous first order partial derivatives. That is, *F* is conservative if

$$\mathbf{F} = \nabla f$$

The function f is called the *potential function for* F.

Not all vector fields are conservative. Those that are have important characteristics that we will discuss in Sections 19.3 and 19.4. We'll also provide an easy way to tell whether or not a vector field is conservative.

EXERCISE 19.1

In Problems 1–10 describe the graph of each vector field.

1. $\mathbf{F} = \mathbf{F}(x, y) = x\mathbf{i} + y\mathbf{j}$
2. $\mathbf{F} = \mathbf{F}(x, y) = x\mathbf{i} - y\mathbf{j}$
3. $\mathbf{F} = \mathbf{F}(x, y) = \mathbf{i} + x\mathbf{j}$
4. $\mathbf{F} = \mathbf{F}(x, y) = y\mathbf{i} - \mathbf{j}$
5. $\mathbf{F} = \mathbf{F}(x, y) = \mathbf{i}$
6. $\mathbf{F} = \mathbf{F}(x, y) = \mathbf{i} + \mathbf{j}$
7. $\mathbf{F} = \mathbf{F}(x, y, z) = z\mathbf{k}$
8. $\mathbf{F} = \mathbf{F}(x, y, z) = x\mathbf{i}$
9. $\mathbf{F} = \mathbf{F}(x, y, z) = \dfrac{x\mathbf{i} + y\mathbf{j} + z\mathbf{k}}{\sqrt{x^2 + y^2 + z^2}}$
10. $\mathbf{F} = \mathbf{F}(x, y, z) = -\dfrac{x\mathbf{i} + y\mathbf{j} + z\mathbf{k}}{\sqrt{x^2 + y^2 + z^2}}$

19.2 Line Integrals

In Chapter 5 we defined the definite integral $\int_a^b f(x)\,dx$, whose value is determined by the values of the function f on the closed interval $[a, b]$. In Chapter 18 we discussed the double integral $\iint_R f(x, y)\,dA$, whose value is determined by the values of f on a closed, bounded region R, and the triple integral $\iiint_S f(x, y, z)\, dV$, whose value is determined by the values of f on a closed, bounded region S of space. In this section we define a *line integral*,* whose value is determined by the values of a function f along a curve C.

Line Integrals in the Plane

Let C be a smooth curve whose parametric equations are given by

$$x = x(t) \qquad y = y(t) \qquad a \le t \le b$$

See Figure 5. As t increases from a to b, the corresponding points $(x(t), y(t))$ trace out the curve C from the point $A = (x(a), y(a))$ to the point $B = (x(b), y(b))$; that is, the orientation of C is from A to B. Let $f(x, y)$ be a function that is defined on some region containing the curve C.

Partition the closed interval $[a, b]$ into n subintervals,

$$[a, t_1], \quad [t_1, t_2], \quad \ldots, \quad [t_{i-1}, t_i], \quad \ldots, \quad [t_{n-1}, b]$$

and denote the length of each subinterval by $\Delta t_1, \Delta t_2, \ldots, \Delta t_n$. Corresponding to each number $a = t_0, t_1, \ldots, t_n = b$ of the partition, there is a succession of points $P_0, P_1, \ldots, P_n$ on the curve C. These points divide the curve into n subarcs of lengths $\Delta s_1, \Delta s_2, \Delta s_3, \ldots, \Delta s_n$ (see Fig. 6).

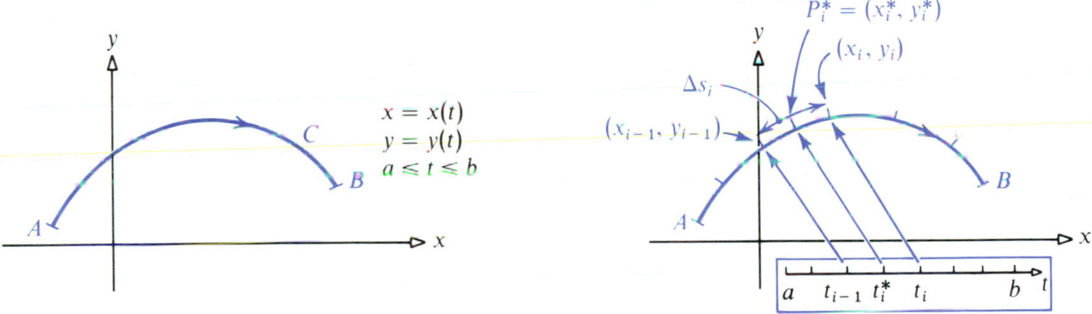

Figure 5 **Figure 6**

Let $P_i^* = (x_i^*, y_i^*)$ be an arbitrary point on the ith subarc, corresponding to the number t_i^*, $t_{i-1} \le t_i^* \le t_i$. We define the *norm* $\|\Delta\|$ of the partition to be the largest arc length Δs_i. Now we form the sum

$$\sum_{i=1}^{n} f(x_i^*, y_i^*)\Delta s_i$$

This sum depends on both the choice of a partition and the choice of P_i^*. However, if all such sums can be made as close as we please to a number L by choosing partitions whose norms are sufficiently close to 0, then L is the limit of these sums as $\|\Delta\|$ approaches 0.

---

* Such an integral might better be called a *curve integral*, but we shall follow the usual practice and use the term *line integral*.

[19.2.1] DEFINITION / ***Line Integral of f along C.***

Let C be a smooth curve defined by the parametric equations $x = x(t)$, $y = y(t)$, $a \leq t \leq b$. Let $f(x, y)$ be a function defined on some region containing C. The line integral of f along C from $t = a$ to $t = b$ is

$$\int_C f(x, y)\, ds = \lim_{\|\Delta\| \to 0} \sum_{i=1}^{n} f(x_i^*, y_i^*)\Delta s_i$$

provided this limit exists.

Evaluation of Line Integrals

We state without proof a method for calculating the value of a line integral when the function f is continuous on some region that contains C.

[19.2.2] THEOREM

Let C be a smooth curve defined by the parametric equations

$$x = x(t) \qquad y = y(t) \qquad a \leq t \leq b$$

Let $f(x, y)$ be a function that is continuous on some region that contains the curve C. Then the line integral of f along C from $t = a$ to $t = b$ exists and is given by

$$\int_C f(x, y)\, ds = \int_a^b f(x(t), y(t))\sqrt{\left(\frac{dx}{dt}\right)^2 + \left(\frac{dy}{dt}\right)^2}\, dt \tag{1}$$

This result expresses the fact that under certain conditions a line integral is equal to a definite integral. When this is the case, the definite integral may be evaluated by techniques already studied.

Although (1) may seem complicated, it is simply the result of substituting the parametric equations $x = x(t)$, $y = y(t)$ of C for x and y in f and using the arc length formula for ds developed in Chapter 13. Here is a concrete example to illustrate the technique.

EXAMPLE 1

Evaluate $\int_C y\, ds$, if C is the curve given by the parametric equations $x = t$, $y = \sqrt{t}$, $2 \leq t \leq 6$.

Solution

The curve C is part of a parabola. The element ds of arc length along C is given by

$$ds = \sqrt{\left(\frac{dx}{dt}\right)^2 + \left(\frac{dy}{dt}\right)^2}\, dt$$

where

$$\frac{dx}{dt} = 1 \qquad \text{and} \qquad \frac{dy}{dt} = \frac{1}{2\sqrt{t}}$$

Thus,

$$ds = \sqrt{1 + \frac{1}{4t}}\,dt = \sqrt{\frac{4t+1}{4t}}\,dt = \frac{\sqrt{4t+1}}{2\sqrt{t}}\,dt$$

By applying (1), we obtain

$$\int_C y\,ds \underset{\substack{\uparrow \\ y=\sqrt{t}}}{=} \int_2^6 \sqrt{t}\,\frac{\sqrt{4t+1}}{2\sqrt{t}}\,dt = \frac{1}{2}\int_2^6 \sqrt{4t+1}\,dt = \left(\frac{1}{8}\right)\frac{(4t+1)^{3/2}}{\frac{3}{2}}\Bigg|_2^6 = \frac{49}{6}$$

It can be shown that the value of the line integral along a curve C *does not* depend on the parametric representation of the curve; all parameterizations with the same orientation satisfying the assumptions made on $x = x(t)$ and $y = y(t)$ in theorem (19.2.2) will give the same value. The next example illustrates this phenomenon.

EXAMPLE 2

Evaluate $\int_C (x^2 + y)\,ds$, if C is the line segment from $(0, 0)$ to $(1, 2)$ and C is parameterized as:

(a) $x = t, \quad y = 2t, \quad 0 \le t \le 1$
(b) $x = \sin t, \quad y = 2\sin t, \quad 0 \le t \le \pi/2$

Solution

We note that the two parameterizations of C are merely different ways of representing a part of the line $y = 2x$ from $(0, 0)$ to $(1, 2)$.

(a) Since $dx/dt = 1$, $dy/dt = 2$, the differential ds of arc length is given by $ds = \sqrt{1^2 + 2^2}\,dt = \sqrt{5}\,dt$, so that

$$\int_C (x^2 + y)\,ds \underset{\substack{\uparrow \\ x=t \\ y=2t}}{=} \int_0^1 (t^2 + 2t)\sqrt{5}\,dt = \sqrt{5}\left(\frac{t^3}{3} + t^2\right)\Bigg|_0^1 = \frac{4\sqrt{5}}{3}$$

(b) With $dx/dt = \cos t$, $dy/dt = 2\cos t$, the differential ds of arc length is $ds = \sqrt{\cos^2 t + 4\cos^2 t}\,dt = \sqrt{5\cos^2 t}\,dt = \sqrt{5}\cos t\,dt$. Thus,

$$\int_C (x^2 + y)\,ds \underset{\substack{\uparrow \\ x=\sin t \\ y=2\sin t}}{=} \int_0^{\pi/2} (\sin^2 t + 2\sin t)\sqrt{5}\cos t\,dt$$

$$= \sqrt{5}\int_0^{\pi/2} (\sin^2 t + 2\sin t)\cos t\,dt$$

$$= \sqrt{5}\left(\frac{\sin^3 t}{3} + \sin^2 t\right)\Bigg|_0^{\pi/2} = \frac{4\sqrt{5}}{3}$$

Applications

An easily understood physical application of a line integral is the problem of finding the mass of a long thin piece of wire of variable density, whose shape is described by the curve C in Figure 6. Suppose the linear density (mass per unit length) at the point (x, y) is $\rho(x, y)$. Within each subinterval $[t_{i-1}, t_i]$ of $[a, b]$, we arbitrarily choose a number t_i^*. Then the mass of the corresponding short piece of wire between (x_{i-1}, y_{i-1}) and (x_i, y_i) can be approximated by $\Delta m_i = \rho(x_i^*, y_i^*)\Delta s_i$, where Δs_i is the length of the short piece and (x_i^*, y_i^*) is the point corresponding to t_i^*. The sum

$$\sum_{i=1}^{n} \rho(x_i^*, y_i^*)\Delta s_i$$

over all the subintervals is an approximation of the mass m of the wire. This approximation can be made as close as we please to m by taking the subintervals to be sufficiently small and sufficiently numerous (that is, by taking the limit as $\|\Delta\|$ approaches 0). Thus,

$$m = \int_C \rho(x, y)\, ds \tag{2}$$

is the mass of the wire.

EXAMPLE 3

Find the mass of a thin piece of wire in the shape of a semicircle $x = 2\cos t$, $y = 2\sin t$, $0 \le t \le \pi$, if the linear density of the wire is $\rho(x, y) = y + 2$.

Solution

The mass of the wire is given by

$$\begin{aligned}\text{Mass} &= \int_C \rho(x, y)\, ds = \int_C (y + 2)\, ds \\ &= \int_0^\pi (2\sin t + 2)\sqrt{(-2\sin t)^2 + (2\cos t)^2}\, dt \\ &= \int_0^\pi 2(2\sin t + 2)\, dt \\ &= 4[-\cos t + t]\Big|_0^\pi = 4[\pi + 2]\end{aligned}$$

If $z = f(x, y) \ge 0$, then $\int_C f(x, y)\, ds$ equals the lateral surface area of the cylinder that lies above the xy-plane and below the surface $z = f(x, y)$ and is formed by lines parallel to the z-axis and intersecting C. See Figure 7.

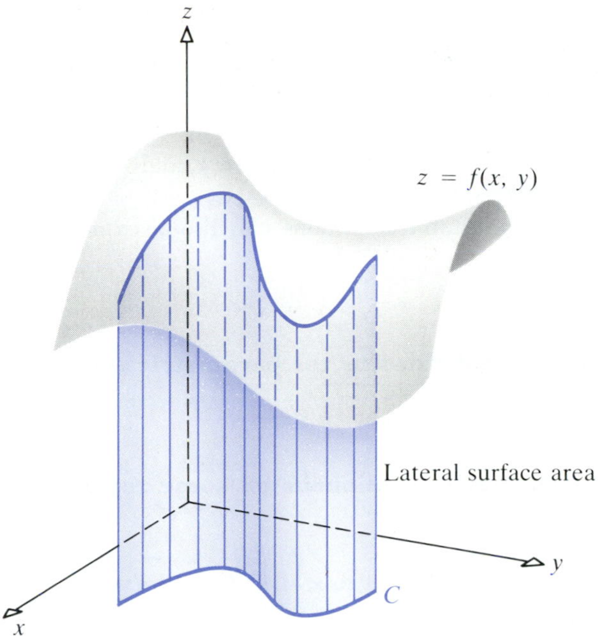

Figure 7

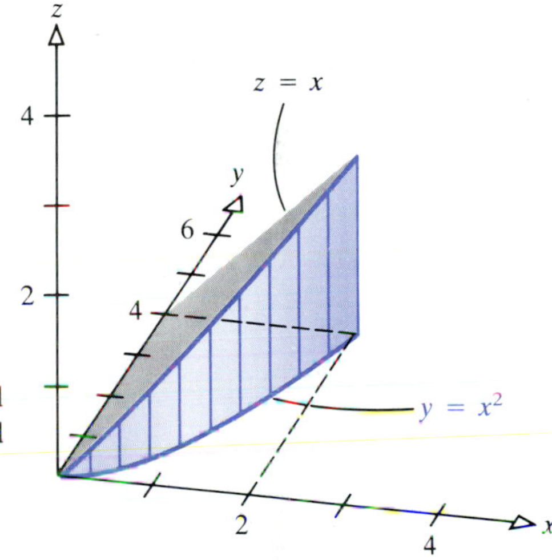

Figure 8

EXAMPLE 4

Find the lateral surface area of the cylinder that lies above the xy-plane and below the surface $z = x$ and is formed by lines parallel to the z-axis and intersecting the curve $y = x^2$, $0 \le x \le 2$.

Solution

Figure 8 illustrates the situation. Since $ds = \sqrt{1 + y'^2}\,dx$, we have

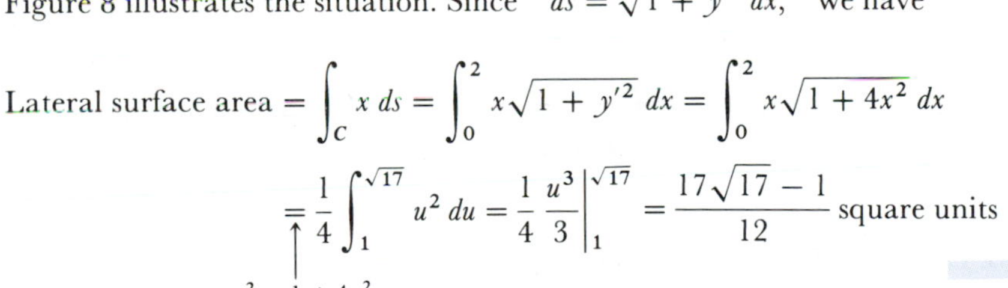

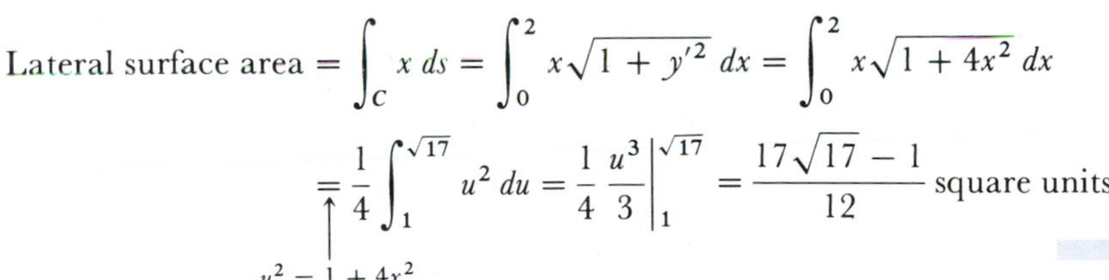

$$\text{Lateral surface area} = \int_C x\,ds = \int_0^2 x\sqrt{1 + y'^2}\,dx = \int_0^2 x\sqrt{1 + 4x^2}\,dx$$

$$\underset{u^2 = 1 + 4x^2}{=} \frac{1}{4}\int_1^{\sqrt{17}} u^2\,du = \frac{1}{4}\,\frac{u^3}{3}\bigg|_1^{\sqrt{17}} = \frac{17\sqrt{17} - 1}{12} \text{ square units}$$

Other Types of Line Integrals

If we replace Δs_i by Δx_i or Δy_i in definition (19.2.1), we obtain two other types of line integrals:

$$\int_C f(x, y)\,dx = \lim_{\|\Delta\| \to 0} \sum_{i=1}^{n} f(x_i^*, y_i^*)\,\Delta x_i \qquad \textbf{(3)}$$

and

$$\int_C f(x, y)\, dy = \lim_{\|\Delta\| \to 0} \sum_{i=1}^{n} f(x_i^*, y_i^*) \Delta y_i \tag{4}$$

We refer to (3) as the *line integral of f along C with respect to x*, and (4) is called the *line integral of f along C with respect to y*.

These line integrals, under suitable conditions, can also be expressed as a single integral.

[19.2.3] THEOREM

Let C be a smooth curve defined by the parametric equations

$$x = x(t) \qquad y = y(t) \qquad a \le t \le b$$

Let $f(x, y)$ be a function that is continuous on some region that contains the curve C. Then:

Line integral of f along C with respect to x $= \int_C f(x, y)\, dx$

$$= \int_a^b f(x(t), y(t))x'(t)\, dt$$

Line integral of f along C with respect to y $= \int_C f(x, y)\, dy$

$$= \int_a^b f(x(t), y(t))y'(t)\, dt$$

In general, each of these line integrals will have a different value.

EXAMPLE 5

Evaluate $\int_C (x - 3y)\, dx$ and $\int_C (x - 3y)\, dy$, if C is the part of the parabola $x = y^2$ that joins the points $(1, 1)$ and $(4, 2)$.

Solution

We can write the parametric equations of the curve C as $x = t^2$, $y = t$, $1 \le t \le 2$. Hence, $dx = 2t\, dt$, $dy = dt$, and

$$\int_C (x - 3y)\, dx = \int_1^2 (t^2 - 3t)2t\, dt = -\frac{13}{2}$$

$$\int_C (x - 3y)\, dy = \int_1^2 (t^2 - 3t)\, dt = -\frac{13}{6}$$

It is not always necessary to use parametric equations for C to evaluate the line integral of some function f. Let's see how (3) and (4) look if the curve C is given by a rectangular equation.

If a smooth curve C is given by the rectangular equation $y = g(x)$, $a \leq x \leq b$, a parameterization of C is

$$x = t \qquad y = g(t) \qquad a \leq t \leq b$$

where $dx = dt$ and $dy = g'(t)\,dt$. Under these conditions, (3) and (4) have the form

$$\int_C f(x, y)\,dx = \int_a^b f(t, g(t))\,dt = \int_a^b f(x, g(x))\,dx \tag{5}$$

$$\int_C f(x, y)\,dy = \int_a^b f(t, g(t))g'(t)\,dt = \int_a^b f(x, g(x))g'(x)\,dx \tag{6}$$

For example, the curve C in Example 5 has the rectangular equation $y = \sqrt{x}$, $1 \leq x \leq 4$. If we use (5) and (6) to evaluate $\int_C (x - 3y)\,dx$ and $\int_C (x - 3y)\,dy$, the computation is a little easier than the solution we gave in Example 5.

$$\int_C (x - 3y)\,dx \underset{\substack{\uparrow \\ y = \sqrt{x}}}{=} \int_1^4 (x - 3\sqrt{x})\,dx = -\frac{13}{2}$$

$$\int_C (x - 3y)\,dy \underset{\substack{\uparrow \\ y = \sqrt{x} \\ dy = \frac{1}{2\sqrt{x}}dx}}{=} \int_1^4 (x - 3\sqrt{x})\left(\frac{1}{2\sqrt{x}}\right)dx = -\frac{13}{6}$$

Line Integrals of the Form $\int_C (P\,dx + Q\,dy)$

[19.2.4] DEFINITION / $\int_C(P\,dx + Q\,dy)$; *C Smooth.*

Let C denote a smooth curve and let $P = P(x, y)$ and $Q = Q(x, y)$ be functions of two variables that are continuous on some region containing C. The *line integral of $P\,dx + Q\,dy$ along C* is defined as

$$\int_C (P\,dx + Q\,dy) = \int_C P(x, y)\,dx + \int_C Q(x, y)\,dy \tag{7}$$

EXAMPLE 6

Evaluate the line integral $\int_C (y^2\,dx - x^2\,dy)$ along:

C_1: The parabola $x = t$, $y = t^2$ joining the two points (0, 0) and (2, 4)

C_2: The line $x = t$, $y = 2t$ joining the two points (0, 0) and (2, 4)

Solution

Along C_1, we have

$$\int_{C_1} (y^2\,dx - x^2\,dy) = \int_0^2 [t^4\,dt - t^2(2t\,dt)] = \int_0^2 (t^4 - 2t^3)\,dt = -\frac{8}{5}$$

Along C_2, we have

$$\int_{C_2} (y^2\,dx - x^2\,dy) = \int_0^2 [4t^2\,dt - t^2(2\,dt)] = \int_0^2 2t^2\,dt = \frac{16}{3}$$

We observe that the value of the line integral in Example 6 depends on the curve C over which the integration takes place. In the next section we investigate conditions under which the value of the integral is independent of path—that is, where the value of the integral depends only on the endpoints of the curve.

The line integral $\int_C (P\,dx + Q\,dy)$ can be written compactly using vectors. Thus, if $\mathbf{F} = \mathbf{F}(x, y) = P(x, y)\mathbf{i} + Q(x, y)\mathbf{j}$ is a vector field that is continuous on some region containing the smooth curve C: $\mathbf{r} = \mathbf{r}(t) = x(t)\mathbf{i} + y(t)\mathbf{j}$, $a \le t \le b$, then $d\mathbf{r} = dx\mathbf{i} + dy\mathbf{j}$ and

$$\int_C \mathbf{F} \cdot d\mathbf{r} = \int_C (P\,dx + Q\,dy)$$

Orientation

In evaluating a line integral over a curve C, the orientation of C plays a role. If C is a smooth curve, we let $-C$ denote the same curve but with reverse orientation. Then

$$\int_C (P\,dx + Q\,dy) = -\int_{-C} (P\,dx + Q\,dy)$$

Thus, a reversal of orientation alters the value of the line integral by a factor of -1 (see Fig. 9).

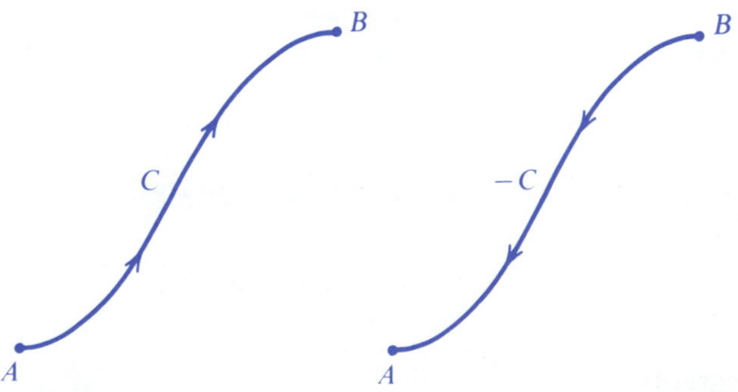

Figure 9

$$\int_C (P\,dx + Q\,dy) = -\int_{-C} (P\,dx + Q\,dy)$$

Piecewise Smooth Curves

In definition (19.2.4), it is assumed that the curve C is a smooth curve. That is, the two parametric equations $x(t)$ and $y(t)$ defining C have continuous derivatives dx/dt and dy/dt not simultaneously 0 on $a \le t \le b$. If a curve C consists of a finite number of smooth curves—say, $C_1, C_2, \ldots, C_n$—that are joined together end to end (see Fig. 10), then we say that C is *piecewise smooth.*

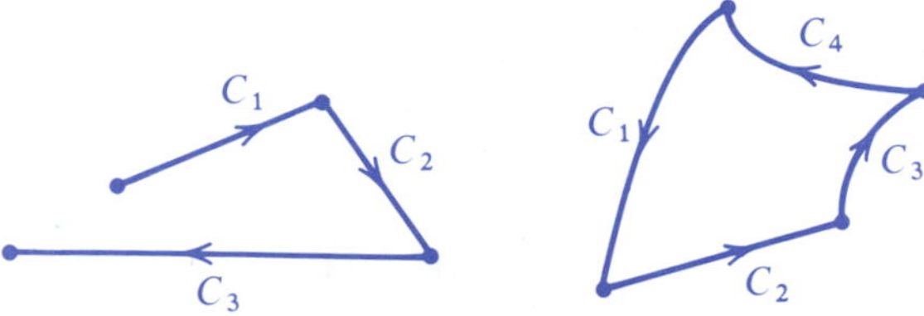

Two piecewise smooth curves

Figure 10

[19.2.5] DEFINITION / $\int_C(P\,dx + Q\,dy)$; *C* Piecewise Smooth.

Let C denote a piecewise smooth curve, and let $P = P(x, y)$ and $Q = Q(x, y)$ be functions of two variables that are continuous on some region containing C. The line integral of $P\,dx + Q\,dy$ along C is defined as

$$\int_C (P\,dx + Q\,dy)$$

$$= \int_{C_1} (P\,dx + Q\,dy) + \int_{C_2} (P\,dx + Q\,dy) + \cdots + \int_{C_n} (P\,dx + Q\,dy)$$

where C consists of the smooth curves $C_1, C_2, \ldots, C_n$.

EXAMPLE 7

Evaluate $\int_C(xy\,dx + x^2\,dy)$ along the piecewise smooth curve C illustrated in Figure 11.

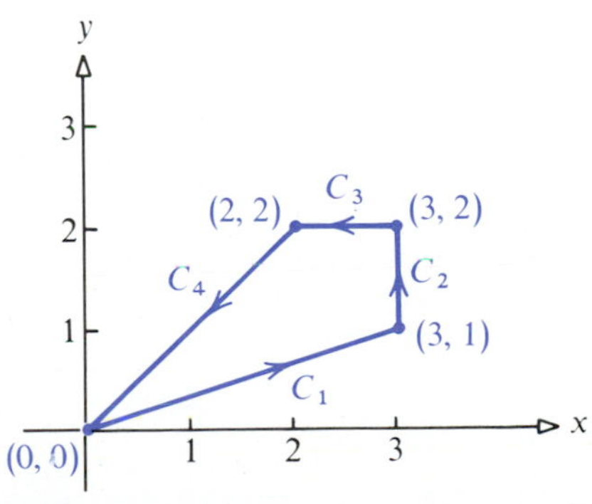

Figure 11

Solution

The values of the line integral along each of the smooth curves C_1, C_2, C_3, and C_4 are

$$C_1\colon\quad y = \frac{x}{3} \qquad dy = \frac{dx}{3} \qquad 0 \le x \le 3$$

$$\int_{C_1} (xy\,dx + x^2\,dy) = \int_0^3 \left[x\left(\frac{x}{3}\right) dx + x^2\,\frac{dx}{3}\right] = 6$$

C_2: $x = 3$ $\quad dx = 0$ $\quad 1 \le y \le 2$

$$\int_{C_2} (xy\,dx + x^2\,dy) = \int_1^2 9\,dy = 9$$

C_3: $y = 2$ $\quad dy = 0$ $\quad$ Watch the orientation here: x varies from 3 to 2.

$$\int_{C_3} (xy\,dx + x^2\,dy) = \int_3^2 2x\,dx = -5$$

C_4: $y = x$ $\quad dy = dx$ $\quad$ Watch the orientation here: x varies from 2 to 0.

$$\int_{C_4} (xy\,dx + x^2\,dy) = \int_2^0 (x^2\,dx + x^2\,dx) = -\frac{16}{3}$$

Thus,

$$\int_C (xy\,dx + x^2\,dy) = 6 + 9 - 5 - \frac{16}{3} = \frac{14}{3}$$

Line Integrals in Space

The definition of a line integral in space is analogous to that in the plane, so we will not give the details here. Instead, we state without proof the main results we need to evaluate line integrals in space.

[19.2.6] THEOREM

Suppose C is a smooth curve defined by the parametric equations

$$x = x(t) \qquad y = y(t) \qquad z = z(t) \qquad a \le t \le b$$

If $P(x, y, z)$, $Q(x, y, z)$, $R(x, y, z)$ are functions that are continuous on some region containing C, then

$$\int_C P(x, y, z)\,dx = \int_a^b P(x(t), y(t), z(t))x'(t)\,dt$$

$$\int_C Q(x, y, z)\,dy = \int_a^b Q(x(t), y(t), z(t))y'(t)\,dt$$

$$\int_C R(x, y, z)\,dz = \int_a^b R(x(t), y(t), z(t))z'(t)\,dt$$

$$\int_C (P\,dx + Q\,dy + R\,dz) = \int_C P\,dx + \int_C Q\,dy + \int_C R\,dz$$

The line integral $\int_C (P\,dx + Q\,dy + R\,dz)$ can be written compactly using vectors. Thus, if $\mathbf{F} = \mathbf{F}(x, y, z) = P(x, y, z)\mathbf{i} + Q(x, y, z)\mathbf{j} + R(x, y, z)\mathbf{k}$ is a vector field that is continuous on some region containing the smooth curve C: $\mathbf{r} = \mathbf{r}(t) = x(t)\mathbf{i} + y(t)\mathbf{j} + z(t)\mathbf{k}$, $a \le t \le b$, then

$$\int_C \mathbf{F} \cdot d\mathbf{r} = \int_C (P\,dx + Q\,dy + R\,dz)$$

A curve C in space is *piecewise smooth* if it consists of a finite number of smooth curves $C_1, C_2, \ldots, C_n$ that are joined together end to end. If C: $\mathbf{r} = \mathbf{r}(t)$ is piecewise smooth and if $\mathbf{F} = \mathbf{F}(x, y, z)$ is a vector field that is continuous on some region containing C, then

$$\int_C \mathbf{F} \cdot d\mathbf{r} = \int_{C_1} \mathbf{F} \cdot d\mathbf{r} + \int_{C_2} \mathbf{F} \cdot d\mathbf{r} + \cdots + \int_{C_n} \mathbf{F} \cdot d\mathbf{r}$$

where C consists of the smooth curves $C_1, C_2, \ldots, C_n$.

EXAMPLE 8

Evaluate $\int_C \mathbf{F} \cdot d\mathbf{r}$ if $\mathbf{F}(x, y, z) = xy^2\mathbf{i} + x^2z\mathbf{j} - (y - x)\mathbf{k}$ and the curve C is $\mathbf{r}(t) = t\mathbf{i} + t^2\mathbf{j} + t^3\mathbf{k}$, $0 \le t \le 1$.

Solution

The parametric equations of the curve C are

$$x = t \qquad y = t^2 \qquad z = t^3$$

Hence,

$$\mathbf{F} = xy^2\mathbf{i} + x^2z\mathbf{j} - (y - x)\mathbf{k} = t^5\mathbf{i} + t^5\mathbf{j} - (t^2 - t)\mathbf{k}$$

and

$$d\mathbf{r} = \frac{d\mathbf{r}}{dt}\, dt = (\mathbf{i} + 2t\mathbf{j} + 3t^2\mathbf{k})\, dt$$

Therefore,

$$\begin{aligned} \mathbf{F} \cdot d\mathbf{r} &= [t^5\mathbf{i} + t^5\mathbf{j} - (t^2 - t)\mathbf{k}] \cdot (\mathbf{i} + 2t\mathbf{j} + 3t^2\mathbf{k})\, dt \\ &= [t^5 + 2t^6 - 3t^2(t^2 - t)]\, dt = (2t^6 + t^5 - 3t^4 + 3t^3)\, dt \end{aligned}$$

so that

$$\int_C \mathbf{F} \cdot d\mathbf{r} = \int_0^1 (2t^6 + t^5 - 3t^4 + 3t^3)\, dt = \frac{253}{420}$$

EXERCISE 19.2

In Problems 1–4 evaluate each line integral for the given curve C.

1. $\int_C (x + y^2)\, ds$; C: $x = t$, $y = t$, $0 \le t \le 1$

2. $\int_C (x + y^2)\, ds$; C: $x = \sin t$, $y = \sin t$, $0 \le t \le \pi$

3. $\int_C x^2y\, ds$; C: $x = \cos t$, $y = \cos t$, $0 \le t \le \pi/2$

4. $\int_C x\, ds$; C: $x = t$, $y = 3t^2$, $0 \le t \le 1$

5. Evaluate $\int_C (x^2y\, dx + xy\, dy)$ along the curve C: $x^2 + y^2 = 1$ from $(1, 0)$ to $(0, 1)$, using the following parameterizations:

(a) $x = \cos t$, $y = \sin t$, $0 \le t \le \pi/2$
(b) $y = \sqrt{1 - x^2}$, $0 \le x \le 1$

6. Evaluate $\int_C [y\, dx + (x - 16y)\, dy]$ along each of the given curves from $(2, 0)$ to $(0, 4)$.

(a) The straight line joining the two points
(b) The parabola $y = 4 - x^2$
(c) The straight line from $(2, 0)$ to $(2, 2)$, followed by the straight line from $(2, 2)$ to $(0, 4)$

7. Evaluate $\int_C[(x+2y)\,dx+(2x+y)\,dy]$, where:

(a) C is the curve $y=x^2$ from $(0, 0)$ to $(1, 1)$
(b) C is the curve $y=x^3$ from $(0, 0)$ to $(1, 1)$
(c) C is the curve $x=\cos t$, $y=\sin t$, $0\le t\le \pi/2$ from $(1, 0)$ to $(0, 1)$
(d) C is the curve $x=1-t$, $y=t$, $0\le t\le 1$

8. Evaluate $\int_C(yz\,dx+xz\,dy+xy\,dz)$, where C consists of line segments connecting the points $(0, 0, 0)$, $(1, 0, 0)$, $(1, 1, 0)$, and $(1, 1, 1)$, in that order.

9. Evaluate $\int_C[y^2\,dx+(xy-x^2)\,dy]$ from $(0, 0)$ to $(1, 3)$, where:

(a) C is the line $y=3x$
(b) C is the parabola $y^2=9x$

10. Evaluate $\int_C[(x^2+y^2)\,dx+3x^2y\,dy]$, where C is the parabola $y=x^2$ from $(-2, 4)$ to $(2, 4)$.

11. Evaluate $\int_C[(x\cos y)\,dx-(y\sin x)\,dy]$, where C consists of line segments connecting the points $(0, 0)$, $(1, 0)$, $(1, 1)$, and $(0, 1)$, in that order.

12. Evaluate $\int_C(z\,dx+x\,dy+y\,dz)$, where C is the circular helix $x=a\cos t$, $y=a\sin t$, $z=t$, $0\le t\le 2\pi$.

13. Let C_1 be the curve $x=\cos\theta$, $y=\sin\theta$, $0\le\theta\le 4\pi$, and let C_2 be the curve $x=\cos t^2$, $y=\sin t^2$, $0\le t\le 2\sqrt{\pi}$. For $P=x^3y$, $Q=y^2x$, show that $\int_{C_1}(P\,dx+Q\,dy)=\int_{C_2}(P\,dx+Q\,dy)$. Explain why this is so.

14. Evaluate $\int_C[ye^{xy}\,dx+\sin x\,dy+(xy/z)\,dz]$, where C is the curve $x=t$, $y=t^2$, $z=t^3$, $1\le t\le 3$.

15. Evaluate $\int_C[(yz/x)\,dx+e^y\,dy+\sin z\,dz]$, where C is the curve $x=t^3$, $y=t$, $z=t^2$, $2\le t\le 3$.

16. Evaluate $\int_C(xy\,dx+x^2z\,dy+xyz\,dz)$, where C is the curve $x=e^t$, $y=e^{-t}$, $z=t^2$, $0\le t\le 1$.

17. Evaluate the line integral below, where C is the curve $x=\cos t$, $y=\sin t$, $-\pi\le t\le\pi$.

$$\int_C\left(\frac{y\,dx}{\sqrt{x^2+y^2}}+\frac{x\,dy}{\sqrt{x^2+y^2}}\right)$$

18. Evaluate the line integral below, where C is the curve $x^2-y^2=16$ from $(4, 0)$ to $(5, 3)$.

$$\int_C\left(\frac{-y}{x\sqrt{x^2-y^2}}\,dx+\frac{1}{\sqrt{x^2-y^2}}\,dy\right)$$

19. Evaluate $\int_C[yz\,dx+(y+zx)\,dy+xz\,dz]$, where C consists of line segments connecting the points $(0, 0, 0)$, $(1, 0, 0)$, $(1, 2, 0)$, $(1, 2, 1)$, and $(0, 0, 1)$, in that order.

20. Evaluate $\int_C 2y\,ds$, where C is the curve $y=\frac{1}{2}x^3$ joining $(0, 0)$ to $(2, 4)$.

21. Evaluate $\int_C[xz\,dx+(y+z)\,dy+x\,dz]$, where C is the curve $x=e^t$, $y=e^{-t}$, $z=e^{2t}$ from $t=0$ to $t=1$.

22. Evaluate $\int_C[1/(x^2+y^2+z^2)]\,ds$, where C is the helix $x=a\cos t$, $y=a\sin t$, $z=bt$, $0\le t\le 1$.

23. Evaluate $\int_C(y^n\,dx+x^n\,dy)$, where C is the ellipse $x=a\sin t$, $y=b\cos t$, $0\le t\le 2\pi$.

24. Evaluate $\int_C y^2\,ds$, where C is the first arch of the cycloid $x=a(t-\sin t)$, $y=a(1-\cos t)$.

25. Evaluate $\int_C\sqrt{x^2+y^2}\,ds$, where C is the curve $x=a(\cos t+t\sin t)$, $y=a(\sin t-t\cos t)$, $0\le t\le 2\pi$.

26. Evaluate $\int_C(x+y)\,ds$, where C is the curve $x=t$, $y=3t^2/\sqrt{2}$, $0\le t\le 1$.

27. Evaluate $\int_C\mathbf{F}\cdot d\mathbf{r}$ if $\mathbf{F}(x, y)=(x+2y)\mathbf{i}+(2x+y)\mathbf{j}$ and C is $\mathbf{r}(t)=t\mathbf{i}+t^2\mathbf{j}$, $0\le t\le 1$.

28. Evaluate $\int_C\mathbf{F}\cdot d\mathbf{r}$ if $\mathbf{F}(x, y)=x^2\mathbf{i}+xy\mathbf{j}$ and C is $\mathbf{r}(t)=(\cos t)\mathbf{i}+(\sin t)\mathbf{j}$, $0\le t\le\pi$.

29. Evaluate $\int_C\mathbf{F}\cdot d\mathbf{r}$ if $\mathbf{F}(x, y, z)=xy\mathbf{i}+x^2z\mathbf{j}+xyz\mathbf{k}$ and C is $\mathbf{r}(t)=e^t\mathbf{i}+e^{-t}\mathbf{j}+t^2\mathbf{k}$, $0\le t\le 1$.

30. Evaluate $\int_C\mathbf{F}\cdot d\mathbf{r}$ if $\mathbf{F}(x, y, z)=xy\mathbf{i}-y\mathbf{j}+z\mathbf{k}$ and:

(a) C is the line segment from $(0, 0, 0)$ to $(1, 0, 0)$
(b) C is the line segment from $(1, 0, 0)$ to $(1, 2, 0)$
(c) C is the line segment from $(1, 2, 0)$ to $(1, 2, 3)$
(d) C is the line segment from $(0, 0, 0)$ to $(1, 2, 3)$

In Problems 31–34 find the lateral surface area of the cylinder that lies above the xy-plane and below the surface $z=f(x, y)$ and is formed by lines parallel to the z-axis and intersecting C.

31. $z=f(x, y)=1-x^2$; C: $x=\sin t$, $y=\cos t$, $0\le t\le 2\pi$

32. $z=f(x, y)=1-y^2$; C: $x=\sin t$, $y=\cos t$, $0\le t\le 2\pi$

33. $z=f(x, y)=2xy$; C: $y=\sqrt{1-x^2}$, $0\le x\le 1$

34. $z=f(x, y)=x+y$; C: $y=\sqrt{1-x^2}$, $0\le x\le 1$

In Problems 35 and 36 find the mass of a thin wire in the shape of the curve C whose linear density is ρ.

35. C: $y = x^2$, $1 \le x \le 2$; $\rho = \rho(x, y) = 4x$

36. C: $x = 3 \sin t$, $y = 3 \cos t$, $0 \le t \le \pi/2$; $\rho = \rho(x, y) = x + y + 1$

37. Write a formula similar to (2) for the mass of a thin wire in the shape of a space curve C whose linear density is $\rho = \rho(x, y, z)$.

38. Use the formula developed in Problem 37 to find the mass of a thin wire in the shape of the helix C: $x = 2 \cos t$, $y = 2 \sin t$, $z = 5t$, $0 \le t \le 2\pi$, whose linear density is $\rho(x, y, z) = k$, a constant.

39. Use the formula developed in Problem 37 to find the mass of a thin wire in the shape of a helix C: $x = 3 \cos t$, $y = 3 \sin t$, $z = 4t$, $0 \le t \le 2\pi$, whose linear density is $\rho(x, y, z) = 2 + z$.

40. Use the formula developed in Problem 37 to find the mass of a thin wire in the shape of a helix C: $x = 2 \cos t$, $y = 2 \sin t$, $z = 5t$, $0 \le t \le 2\pi$, if the density is proportional to the square of the distance from the origin.

41. Derive the formula $A = 2\pi Rh$ for the surface area of a right circular cylinder using the line integral $\int_C f(x, y)\, ds$, where $z = f(x, y) = h$, $h > 0$, and C is the circle $x^2 + y^2 = R^2$.

Independence of Path in the Plane

19.3

Recall Example 6 of Section 19.2 in which we evaluated a line integral along two different curves with common endpoints and obtained two different values for the integral. By way of contrast, consider the following example, in which $\int_C (P\, dx + Q\, dy)$ turns out to be the same for several different curves connecting two given points.

EXAMPLE 1

Evaluate $\int_C (2xy\, dx + x^2\, dy)$ if:

(a) C consists of line segments from (3, 1) to (5, 1) and from (5, 1) to (5, 6); see Figure 12(a)
(b) C is the line segment from (3, 1) to (5, 6); see Figure 12(b)
(c) C is a part of the parabola $x = 2t + 1$, $y = 2t^2 - t$, $1 \le t \le 2$; see Figure 12(c).

Solution

(a) The curve C consists of two line segments whose equations are

$$C_1: \quad 3 \le x \le 5 \qquad y = 1$$

$$C_2: \quad x = 5 \qquad 1 \le y \le 6$$

The line integral along C is the sum of the line integrals along C_1 and along C_2. Along C_1 we have $y = 1$, so that $dy = 0$. Hence,

$$\int_{C_1} (2xy\, dx + x^2\, dy) = \int_3^5 (2x\, dx + 0) = x^2 \Big|_3^5 = 16$$

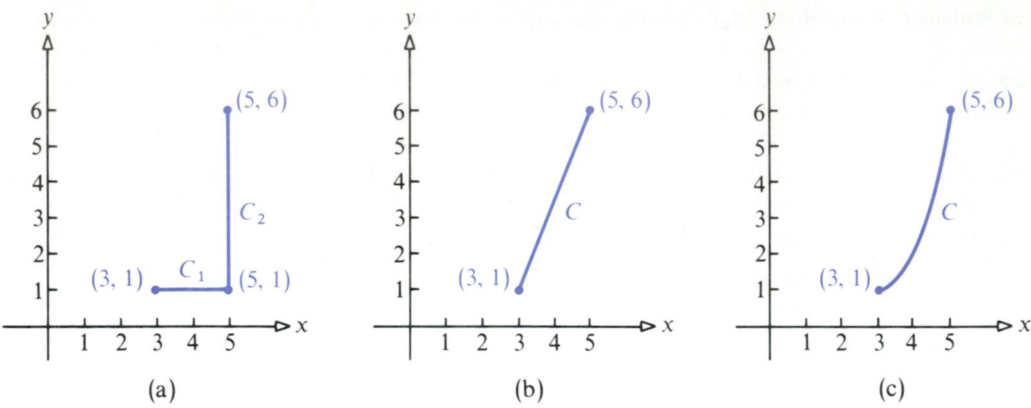

Figure 12

Along C_2 we have $x = 5$, so that $dx = 0$. Hence,

$$\int_{C_2} (2xy\,dx + x^2\,dy) = \int_1^6 (0 + 25\,dy) = 25y\Big|_1^6 = 125$$

Consequently, the line integral along C equals $16 + 125 = 141$.

(b) The curve C is a line segment whose equation is $y = (\frac{5}{2})x - \frac{13}{2}$, $3 \le x \le 5$. By using the fact that $dy = (\frac{5}{2})\,dx$, we find

$$\int_C (2xy\,dx + x^2\,dy) = \int_3^5 \left[2x\left(\frac{5x}{2} - \frac{13}{2}\right)dx + x^2\left(\frac{5}{2}\right)dx\right]$$

$$= \int_3^5 \left[\left(\frac{15}{2}\right)x^2 - 13x\right]dx = \left[\left(\frac{5}{2}\right)x^3 - \left(\frac{13}{2}\right)x^2\right]\Bigg|_3^5 = 141$$

(c) The curve C is part of a parabola. Since $dx = 2\,dt$ and $dy = (4t - 1)\,dt$, we find

$$\int_C (2xy\,dx + x^2\,dy) = \int_1^2 [2(2t + 1)(2t^2 - t)2\,dt + (2t + 1)^2(4t - 1)\,dt]$$

$$= (8t^4 + 4t^3 - 2t^2 - t)\Big|_1^2 = 141$$

In Example 1, we evaluated the line integral $\int_C (2xy\,dx + x^2\,dy)$ along three different curves and arrived at the same value in each instance. The next result demonstrates that this happens for a line integral $\int_C \mathbf{F} \cdot d\mathbf{r}$, whenever the vector field $\mathbf{F}$ is conservative—that is, whenever $\mathbf{F}$ is the gradient of some function f with continuous first-order partial derivatives. In our example, $\mathbf{F}(x, y) = 2xy\mathbf{i} + x^2\mathbf{j}$ is conservative, since $\mathbf{F}$ is the gradient of the function $f(x, y) = x^2y$ $[\nabla f(x, y) = 2xy\mathbf{i} + x^2\mathbf{j} = \mathbf{F}(x, y)]$.

[19.3.1] THEOREM / *Fundamental Theorem of Line Integrals.*

Let $\mathbf{F} = \mathbf{F}(x, y) = P(x, y)\mathbf{i} + Q(x, y)\mathbf{j}$ denote a vector field, where $P(x, y)$ and $Q(x, y)$ are continuous on some open region R

containing the points (x_0, y_0) and (x_1, y_1). Let C denote a piecewise smooth curve that lies entirely in R beginning at the point (x_0, y_0) and ending at the point (x_1, y_1). If F is conservative on R—that is, if F is the gradient of some function f so that

$$\mathbf{F}(x, y) = \nabla f(x, y)$$

throughout R—then

$$\int_C (P\,dx + Q\,dy) = \int_{(x_0, y_0)}^{(x_1, y_1)} (P\,dx + Q\,dy)$$

$$= f(x, y)\Big|_{(x_0, y_0)}^{(x_1, y_1)} = f(x_1, y_1) - f(x_0, y_0) \tag{1}$$

$$\int_C \mathbf{F} \cdot d\mathbf{r} = \int_C \nabla f \cdot d\mathbf{r} = \int_{(x_0, y_0)}^{(x_1, y_1)} \mathbf{F} \cdot d\mathbf{r}$$

$$= f(x, y)\Big|_{(x_0, y_0)}^{(x_1, y_1)} = f(x_1, y_1) - f(x_0, y_0) \tag{1'}$$

Proof We give a proof for a smooth curve C. (The proof for a piecewise smooth curve may be obtained in the same manner by considering one piece at a time.) Let $x = x(t)$, $y = y(t)$, $a \le t \le b$ be the parametric equations of the curve C. The initial point and endpoint of C may be written as

$$(x_0, y_0) = (x(a), y(a)) \qquad \text{and} \qquad (x_1, y_1) = (x(b), y(b))$$

Since $\mathbf{F} = P\mathbf{i} + Q\mathbf{j} = \nabla f = (\partial f/\partial x)\mathbf{i} + (\partial f/\partial y)\mathbf{j}$, we find that

$$\int_C (P\,dx + Q\,dy) = \int_C \left(\frac{\partial f}{\partial x}\,dx + \frac{\partial f}{\partial y}\,dy\right) = \int_a^b \left(\frac{\partial f}{\partial x}\frac{dx}{dt} + \frac{\partial f}{\partial y}\frac{dy}{dt}\right) dt$$

$$\underset{\uparrow}{=} \int_a^b \frac{d}{dt}[f(x(t), y(t))]\,dt = f(x(t), y(t))\Big|_{t=a}^{t=b}$$

By chain rule (4) in Section 16.5

$$= f(x(b), y(b)) - f(x(a), y(a)) = f(x_1, y_1) - f(x_0, y_0)$$

Some observations about (1) and (1′) are in order. The right-hand side depends only on the endpoints (x_0, y_0) and (x_1, y_1) of the curve C. As a result, the value of the line integral will not change if C is replaced by any other piecewise smooth curve in R, so long as the new curve also connects (x_0, y_0) to (x_1, y_1) (see Fig. 13). As a result, line integrals $\int_C \mathbf{F} \cdot d\mathbf{r}$ for which the conditions of theorem (19.3.1) are met are said to be *independent of the path*. In other words, for a conservative vector field, $\int_C \mathbf{F} \cdot d\mathbf{r}$ will have the same value no matter what path C has taken from $A = (x_0, y_0)$ to $B = (x_1, y_1)$.

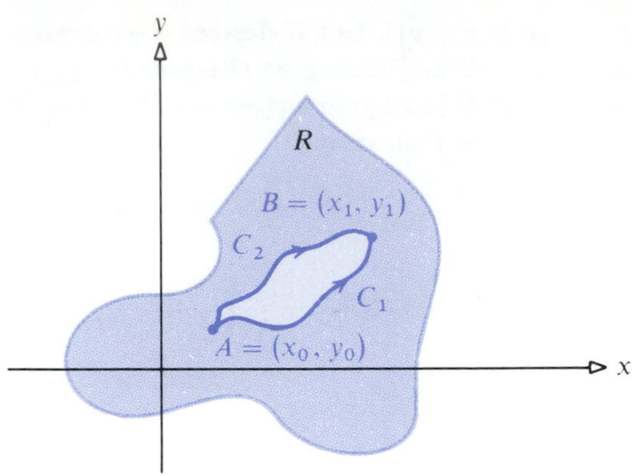

Figure 13
If $\mathbf{F} = P\mathbf{i} + Q\mathbf{j} = \nabla f$, then
(1) $\int_C (P\,dx + Q\,dy) = f(x_1, y_1) - f(x_0, y_0)$ for any curve C joining A to B
(2) $\int_{C_1} (P\,dx + Q\,dy) = \int_{C_2} (P\,dx + Q\,dy)$ for any curves C_1 and C_2 joining A to B

EXAMPLE 2

The vector field $\mathbf{F} = \mathbf{F}(x, y) = (2xy + 24x)\mathbf{i} + (x^2 + 16)\mathbf{j}$ is conservative, since $\mathbf{F}$ is the gradient of $f(x, y) = x^2y + 12x^2 + 16y$. Use this fact to evaluate

$$\int_C [(2xy + 24x)\,dx + (x^2 + 16)\,dy]$$

where C is any path joining the point $(1, 1)$ to $(2, 4)$.

Solution

We use two methods to evaluate the given line integral.

Method I This method utilizes the potential function $f(x, y) = x^2y + 12x^2 + 16y$ whose gradient is

$$\nabla f = (2xy + 24x)\mathbf{i} + (x^2 + 16)\mathbf{j} = \mathbf{F}(x, y)$$

Thus, by (1),

$$\int_C [(2xy + 24x)\,dx + (x^2 + 16)\,dy] = f(x, y)\Big|_{(1,1)}^{(2,4)}$$
$$= f(2, 4) - f(1, 1) = 99$$

Method II This method uses the fact that the given line integral is independent of path, so that we can evaluate the line integral along *any* path joining $(1, 1)$ and $(2, 4)$. We choose the path shown in Figure 14, since it makes the integration easy. Then

$$\int_C [(2xy + 24x)\,dx + (x^2 + 16)\,dy] = \int_{C_1} [(2xy + 24x)\,dx + (x^2 + 16)\,dy]$$
$$+ \int_{C_2} [(2xy + 24x)\,dx + (x^2 + 16)\,dy]$$
$$= \int_1^2 (2x + 24x)\,dx + \int_1^4 (4 + 16)\,dy$$
$$= 39 + 60 = 99$$

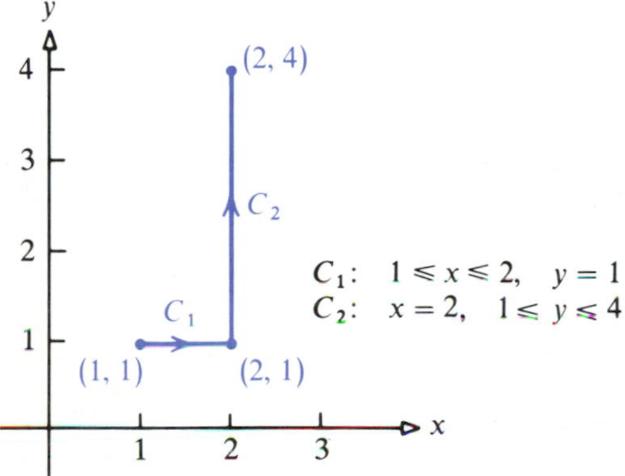

Figure 14

In general, if it is required to evaluate the line integral $\int_C \mathbf{F} \cdot d\mathbf{r} = \int_C (P\,dx + Q\,dy)$, where $\mathbf{F} = P\mathbf{i} + Q\mathbf{j}$ is conservative so that $\mathbf{F}$ is the gradient of some function f, then either of the methods outlined in Example 2 may be used.

Method I Use (1) or (1′) directly by evaluating f at the endpoints (x_0, y_0) and (x_1, y_1) of C. That is,

$$\int_C \mathbf{F} \cdot d\mathbf{r} = \int_C (P\,dx + Q\,dy) = f(x_1, y_1) - f(x_0, y_0)$$

Method II Use the fact that $\int_C (P\,dx + Q\,dy)$ is independent of path and select some suitable path joining the endpoints of C to evaluate the line integral.

Some care must be exercised when Method II is used. For example, suppose we are asked to evaluate the line integral

$$\int_C \frac{-y\,dx + x\,dy}{x^2}$$

along the curve C joining the points $(1, -1)$ and $(4, 2)$. As you can verify, the vector function $\mathbf{F} = \mathbf{F}(x, y) = (-y\mathbf{i} + x\mathbf{j})/x^2$ is conservative, since $\mathbf{F}$ is the gradient of the function $f(x, y) = y/x$. In using Method II to evaluate this line integral, you may not, for example, choose a path (such as $x = y^2$) that intersects the y-axis $(x = 0)$, since $P = -y/x^2$ and $Q = 1/x$ are not continuous if $x = 0$. However, any path that lies entirely in the first and fourth quadrants can be used (see Fig. 15).

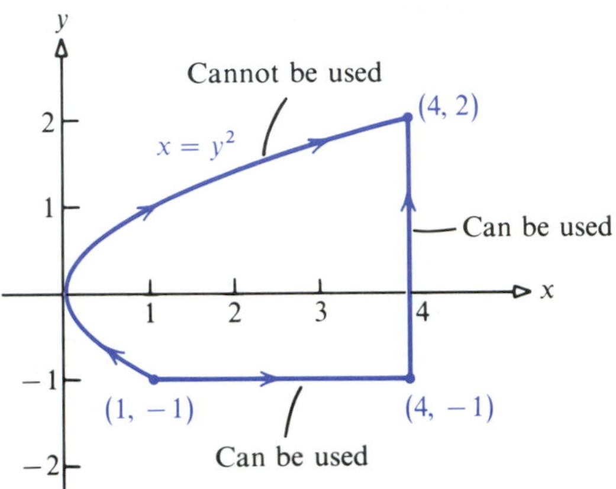

Figure 15

Line Integrals of Conservative Fields over Closed Curves

A special case of theorem (19.3.1) occurs when C is a *closed curve*—that is, when $(x_0, y_0) = (x_1, y_1)$. Then $f(x_1, y_1) = f(x_0, y_0)$ and the value of the line integral is 0.

[19.3.2] COROLLARY

Consider a line integral $\int_C \mathbf{F} \cdot d\mathbf{r}$. Suppose F is a conservative vector field on some open region *R*. If *C* is a closed, piecewise smooth curve that lies entirely in *R*, then

$$\int_C \mathbf{F} \cdot d\mathbf{r} = \int_C (P\,dx + Q\,dy) = 0$$

EXAMPLE 3

The vector field $\mathbf{F} = \mathbf{F}(x, y) = \dfrac{-y\mathbf{i} + x\mathbf{j}}{x^2 + y^2}$ is conservative on any region R that does not contain any points on the y-axis, since $\mathbf{F}$ is the gradient of $f(x, y) = \tan^{-1} y/x$. Based on corollary (19.3.2), $\int_C \mathbf{F} \cdot d\mathbf{r} = 0$ along any closed, piecewise smooth curve C that does not cross or touch the y-axis $(x = 0)$.

Converse of the Fundamental Theorem on Line Integrals

Based on the fundamental theorem of line integrals (19.3.1), if, on an open region R, the vector field $\mathbf{F}$ is conservative, so that F is the gradient of some function f—that is, if $\nabla f = \mathbf{F}$ on R—then $\int_C \mathbf{F} \cdot d\mathbf{r}$ is independent of the path taken in R. If R is *connected*—that is, if any two points in R can be joined by a piecewise smooth curve that lies entirely in R—then the converse of theorem (19.3.1) is true. Figure 16(a) illustrates the idea of a connected set. In Figure 16(a), any two points of R can be joined by a piecewise smooth curve that lies in R, so R is connected. In Figure 16(b), R is not connected, since the points A and B cannot be joined by a curve *that lies entirely in R*.

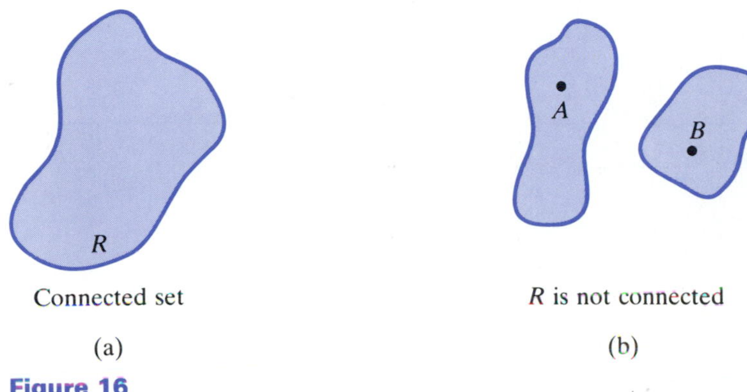

Figure 16

[19.3.3] THEOREM

Let $\mathbf{F} = \mathbf{F}(x, y) = P(x, y)\mathbf{i} + Q(x, y)\mathbf{j}$ denote a vector field, where $P(x, y)$ and $Q(x, y)$ are continuous on some open, connected region R containing the point (x_0, y_0). Let (x, y) be any point in R. If the line integral

$$\int_C \mathbf{F} \cdot d\mathbf{r} = \int_C (P\,dx + Q\,dy)$$

has the same value for every piecewise smooth curve C in R joining (x_0, y_0) and (x, y), then F is conservative on R.

Proof Suppose the line integral $\int_C (P\,dx + Q\,dy)$ has the same value for every piecewise smooth curve C in R that joins (x_0, y_0) to (x, y). Then the line integral $\int_C (P\,dx + Q\,dy)$ is independent of path. Hence, if C is any piecewise smooth curve in R joining a fixed point (x_0, y_0) to an arbitrary point (x, y), then the line integral $\int_C (P\,dx + Q\,dy)$ will define a function f that depends only on (x, y) and not on C. Thus, we may define

$$f(x, y) = \int_{(x_0, y_0)}^{(x, y)} (P\,dx + Q\,dy) \tag{2}$$

Any piecewise smooth path in R joining (x_0, y_0) and (x, y) is allowed. We shall use two paths, one in which a horizontal line segment is used, and another in which a vertical line segment is used.* As Figure 17 illustrates,

C_1: Consists of a piecewise smooth curve joining (x_0, y_0) to (x_1, y) plus a horizontal line segment joining (x_1, y) to (x, y)

C_2: Consists of a piecewise smooth curve joining (x_0, y_0) to (x, y_1) plus a vertical segment joining (x, y_1) to (x, y)

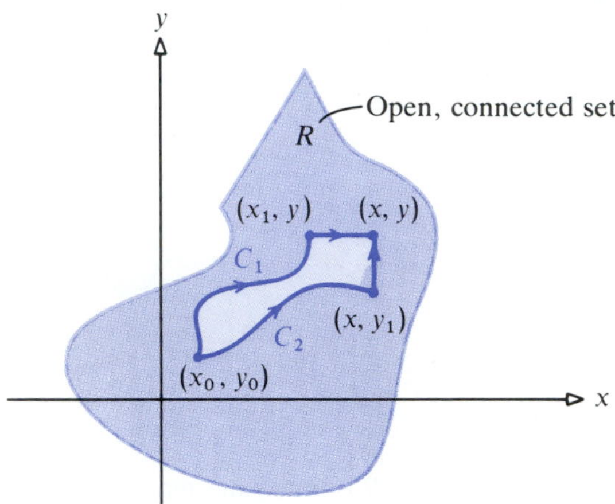

Figure 17

For C_1, (2) takes the form

$$f(x, y) = \int_{(x_0, y_0)}^{(x_1, y)} (P\,dx + Q\,dy) + \int_{(x_1, y)}^{(x, y)} (P\,dx + Q\,dy) \tag{3}$$

The first integral on the right in (3) is a function of y alone, since x_0 and x_1 are constants. Hence, its partial derivative with respect to x equals 0. That is,

$$\frac{\partial}{\partial x} \int_{(x_0, y_0)}^{(x_1, y)} (P\,dx + Q\,dy) = 0 \tag{4}$$

Consider now the second integral on the right in (3). Since the value of y is the same in both the lower and upper limits of integration, it follows that $dy = 0$. Thus,

$$\frac{\partial}{\partial x} \int_{(x_1, y)}^{(x, y)} (P\,dx + Q\,dy) = \frac{\partial}{\partial x} \int_{x_1}^{x} P(x, y)\,dx = P(x, y) \tag{5}$$

Combining (4) and (5), we find from (3) that

$$\frac{\partial}{\partial x} f(x, y) = P(x, y) \tag{6}$$

* Such line segments must exist, since R is open and connected.

For C_2, (2) takes the form

$$f(x, y) = \int_{(x_0, y_0)}^{(x, y_1)} (P\,dx + Q\,dy) + \int_{(x, y_1)}^{(x, y)} (P\,dx + Q\,dy) \tag{7}$$

The first integral on the right in (7) is a function of x alone, since y_0 and y_1 are constants. Hence, its partial derivative with respect to y equals 0. That is,

$$\frac{\partial}{\partial y} \int_{(x_0, y_0)}^{(x, y_1)} (P\,dx + Q\,dy) = 0 \tag{8}$$

Since the value of x is the same in both the lower and upper limits of integration of the second integral on the right in (7), it follows that $dx = 0$. Thus,

$$\frac{\partial}{\partial y} \int_{(x, y_1)}^{(x, y)} (P\,dx + Q\,dy) = \frac{\partial}{\partial y} \int_{y_1}^{y} Q(x, y)\,dy = Q(x, y) \tag{9}$$

Combining (8) and (9), we find from (7) that

$$\frac{\partial}{\partial y} f(x, y) = Q(x, y) \tag{10}$$

By combining (6) and (10), we find that

$$\nabla f = \frac{\partial f}{\partial x}\mathbf{i} + \frac{\partial f}{\partial y}\mathbf{j} = P(x, y)\mathbf{i} + Q(x, y)\mathbf{j} = \mathbf{F}$$

That is, the vector field $\mathbf{F}$ is conservative. ■

Based on Theorems (19.3.1) and (19.3.3), we have the following result:

[19.3.4] THEOREM

Suppose a vector field F is continuous on an open, connected region R. Then $\int_C \mathbf{F} \cdot d\mathbf{r}$ is independent of path in R if and only if F is conservative on R.

Reconstructing a Function from Its Gradient; Finding the Potential Function of a Conservative Vector Field

Suppose $\mathbf{F} = \mathbf{F}(x, y)$ is a conservative vector field. Then there is a potential function f so that $\nabla f = \mathbf{F}$. The proof of theorem (19.3.3) provides a way of finding the potential function f when its gradient $\nabla f = \mathbf{F}$ is known. This is called *reconstructing a function from its gradient.*

EXAMPLE 4

If it is known that the vector field

$$\mathbf{F} = \mathbf{F}(x, y) = (6xy + y^3)\mathbf{i} + (3x^2 + 3xy^2)\mathbf{j}$$

is conservative, find the potential function f of $\mathbf{F}$.

Solution

We seek the function $f(x, y)$ for which

$$\nabla f = \mathbf{F} = (6xy + y^3)\mathbf{i} + (3x^2 + 3xy^2)\mathbf{j}$$

Let

$$P(x, y) = 6xy + y^3 \qquad \text{and} \qquad Q(x, y) = 3x^2 + 3xy^2$$

Since

$$\nabla f = \frac{\partial f}{\partial x}\mathbf{i} + \frac{\partial f}{\partial y}\mathbf{j} = P(x, y)\mathbf{i} + Q(x, y)\mathbf{j}$$

we have

$$\frac{\partial f}{\partial x} = 6xy + y^3 \qquad \frac{\partial f}{\partial y} = 3x^2 + 3xy^2 \tag{11}$$

We integrate the left equation in (11) partially with respect to x to obtain

$$f(x, y) = 3x^2y + xy^3 + h(y) \tag{12}$$

where the "constant of integration," $h(y)$, is a function of y. We differentiate (12) with respect to y and combine the result with the right equation in (11):

$$\frac{\partial f}{\partial y} = 3x^2 + 3xy^2 + h'(y) = 3x^2 + 3xy^2$$

Now we solve for $h'(y)$:

$$h'(y) = 0$$

$$h(y) = K, \quad \text{a constant}$$

Then, from (12) the potential function f of $\mathbf{F}$ is

$$f(x, y) = 3x^2y + xy^3 + K$$

We can verify that f is, in fact, a potential function of $\mathbf{F}$ by calculating the gradient of f:

$$\nabla f = \frac{\partial f}{\partial x}\mathbf{i} + \frac{\partial f}{\partial y}\mathbf{j} = (6xy + y^3)\mathbf{i} + (3x^2 + 3xy^2)\mathbf{j} = \mathbf{F}(x, y)$$

We observe that if we had chosen to integrate the right equation in (11) partially with respect to y, we would have obtained

$$f(x, y) = 3x^2y + xy^3 + k(x)$$

where the "constant of integration," $k(x)$, is a function of x. Then

$$\frac{\partial f}{\partial x} = 6xy + y^3 = 6xy + y^3 + k'(x)$$

so that $k'(x) = 0$ or $k(x) = K$, a constant. Thus,

$$f(x, y) = 3x^2y + xy^3 + K$$

as before.

As the solution to Example 4 illustrates, if we know that a vector field **F** is conservative, then we can find its potential function f. But how can we tell whether a vector field **F** is conservative?

The forthcoming theorem (19.3.7) provides the answer. But first we need the following definitions:

[19.3.5] DEFINITION / ***Closed Curve; Simple Curve.***

Let $\mathbf{r} = \mathbf{r}(t)$, $a \le t \le b$, denote a piecewise smooth curve C. Then C is *closed* if $\mathbf{r}(a) = \mathbf{r}(b)$—that is, if the initial point and terminal point of C coincide. The curve C is called *simple* if it does not intersect itself for $a < t < b$.

Figure 18 illustrates some of the possibilities.

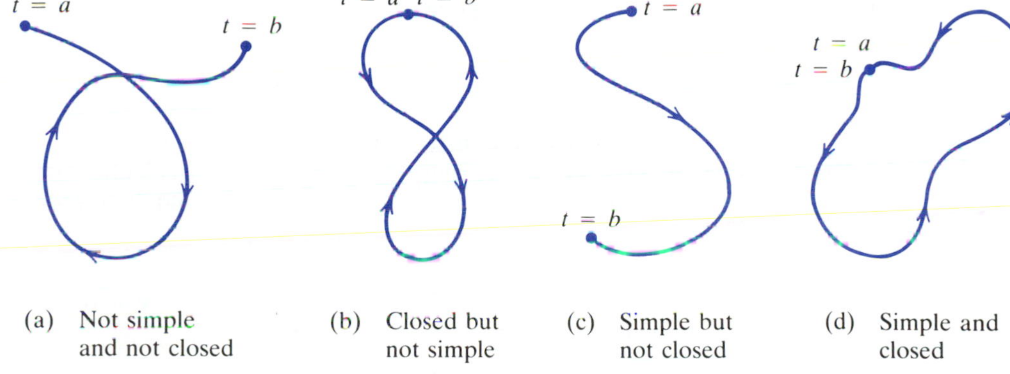

Figure 18

[19.3.6] DEFINITION

Let R denote a set of points in the plane. Then R is called *simply connected* if its boundary consists of a single simple closed curve and the interior of any simple closed curve in R contains only points of R.

Intuitively, a set R is not simply connected if it has "holes" or if it consists of two separate pieces. Figure 19 illustrates some of the possibilities.

[19.3.7] THEOREM

Let $\mathbf{F} = \mathbf{F}(x, y) = P(x, y)\mathbf{i} + Q(x, y)\mathbf{j}$ denote a vector field, where P and Q are continuous on some simply connected, open region R. Suppose $\partial P/\partial y$ and $\partial Q/\partial x$ are also continuous on R. Then $\mathbf{F}(x, y) = P(x, y)\mathbf{i} + Q(x, y)\mathbf{j}$ is conservative on R if and

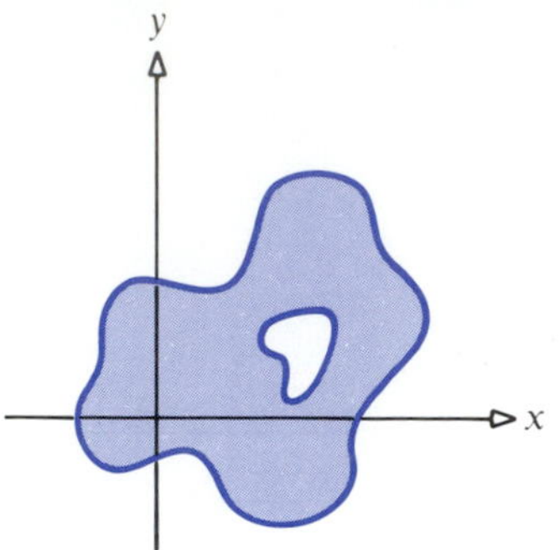

(a) Not simply connected, has "holes"; boundary consists of two simple closed curves; interior of a simple closed curve contains points not in set

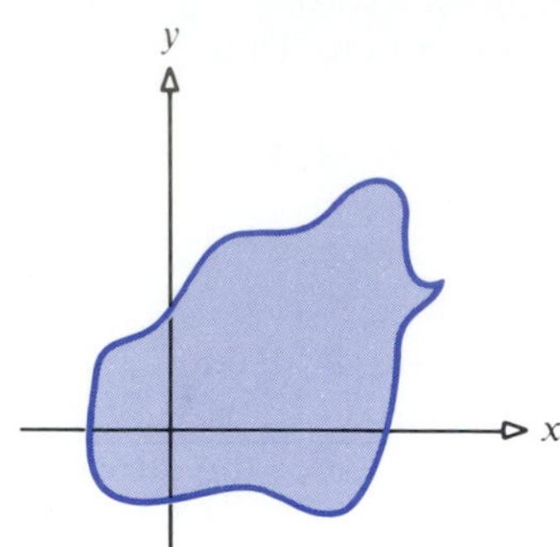

(b) Simply connected; boundary consists of one simple closed curve

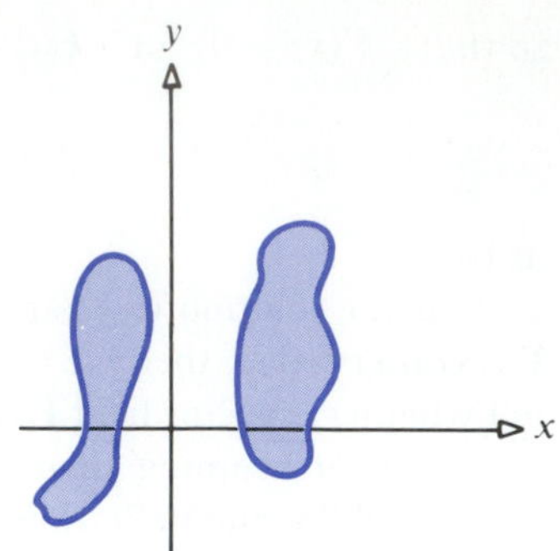

(c) Not simply connected; consists of two pieces; boundary consists of two simple closed curves

Figure 19

only if

$$\frac{\partial P}{\partial y} = \frac{\partial Q}{\partial x}$$

throughout *R*.

We shall prove only that if $\mathbf{F}$ is conservative, then $\partial P/\partial y = \partial Q/\partial x$. The proof of the converse is generally given in advanced calculus and is omitted here. However, the proof of the converse in a special case is given in Section 19.5.

Proof Suppose that $\mathbf{F} = P\mathbf{i} + Q\mathbf{j}$ is conservative. Then $\mathbf{F}$ is the gradient of some function f. That is,

$$\nabla f = \frac{\partial f}{\partial x}\mathbf{i} + \frac{\partial f}{\partial y}\mathbf{j} = P\mathbf{i} + Q\mathbf{j}$$

Then

$$\frac{\partial f}{\partial x} = P \qquad \text{and} \qquad \frac{\partial f}{\partial y} = Q$$

so that

$$\frac{\partial^2 f}{\partial y\,\partial x} = \frac{\partial P}{\partial y} \qquad \text{and} \qquad \frac{\partial^2 f}{\partial x\,\partial y} = \frac{\partial Q}{\partial x}$$

Since we assume that $\partial P/\partial y$ and $\partial Q/\partial x$ are continuous in $\mathbf{R}$, we apply theorem (16.3.3), which states that under these conditions, the mixed partials are equal. As a result,

$$\frac{\partial P}{\partial y} = \frac{\partial Q}{\partial x}$$

EXAMPLE 5

(a) The vector field $\mathbf{F} = 2xy\mathbf{i} + (x^2 + 1)\mathbf{j}$ is conservative on the entire plane, since

$$\frac{\partial}{\partial x}(x^2 + 1) = 2x \qquad \text{and} \qquad \frac{\partial}{\partial y}(2xy) = 2x$$

are equal for any choice of (x, y).

(b) The vector field $\mathbf{F} = (x/y^2)\mathbf{i} - (x^2/y^3)\mathbf{j}$ is conservative on any region not containing points on the x-axis $(y = 0)$, since

$$\frac{\partial}{\partial x}\frac{x^2}{(-y^3)} = \frac{-2x}{y^3} \qquad \text{and} \qquad \frac{\partial}{\partial y}\frac{x}{y^2} = \frac{-2x}{y^3}$$

are equal, provided $y \neq 0$.

(c) The line integral $\displaystyle\int_C \mathbf{F} \cdot d\mathbf{r} = \int_C \left[\frac{x}{y^2}\,dx - \frac{x^2}{y^3}\,dy\right]$ is independent of path in any simply connected region not containing points on the x-axis, since $\mathbf{F}$ is conservative on such a region.

EXAMPLE 6

(a) Show that the line integral $\int_C \mathbf{F} \cdot d\mathbf{r} = \int_C [(2xy + 24x)\,dx + (x^2 + 16)\,dy]$ is independent of path.

(b) Find a function f such that $\nabla f = (2xy + 24x)\mathbf{i} + (x^2 + 16)\mathbf{j}$.

(c) Evaluate $\int_C [(2xy + 24x)\,dx + (x^2 + 16)\,dy]$, where C is any piecewise smooth curve joining $(0, 1)$ to $(1, 2)$.

Solution

(a) Let $P(x, y) = 2xy + 24x$ and $Q(x, y) = x^2 + 16$. Then $\partial P/\partial y = 2x$ and $\partial Q/\partial x = 2x$. Since P, Q, $\partial P/\partial y$, and $\partial Q/\partial x$ are continuous in the entire plane and since $\partial P/\partial y = \partial Q/\partial x$, it follows from theorem (19.3.7) that $\mathbf{F} = \mathbf{F}(x, y) = (2xy + 24x)\mathbf{i} + (x^2 + 16)\mathbf{j}$ is conservative. By theorem (19.3.1), the integral $\int_C \mathbf{F} \cdot d\mathbf{r}$ is independent of path.

(b) Since $\mathbf{F} = \mathbf{F}(x, y) = (2xy + 24x)\mathbf{i} + (x^2 + 16)\mathbf{j}$ is conservative, $\mathbf{F}$ has a potential function for which

$$\nabla f = \mathbf{F} = (2xy + 24x)\mathbf{i} + (x^2 + 16)\mathbf{j}$$

Since

$$\nabla f = \frac{\partial f}{\partial x}\mathbf{i} + \frac{\partial f}{\partial y}\mathbf{j}$$

we have

$$\frac{\partial f}{\partial x} = 2xy + 24x \qquad \text{and} \qquad \frac{\partial f}{\partial y} = x^2 + 16$$

By integrating the first of these partial derivatives partially with respect to x, we obtain

$$f(x, y) = \int (2xy + 24x)\, dx = x^2y + 12x^2 + h(y)$$

where the "constant of integration," $h(y)$, is actually a function of y. If we now differentiate f with respect to y, we get

$$\frac{\partial f}{\partial y} = x^2 + h'(y)$$

But $\partial f/\partial y = Q = x^2 + 16$; thus, $h'(y) = 16$. By integrating with respect to y, we obtain

$$h(y) = 16y + K$$

where K is a constant. Thus,

$$f(x, y) = x^2y + 12x^2 + 16y + K$$

(c) Since $\int_C \mathbf{F} \cdot d\mathbf{r}$ is independent of path, we may use the potential function $f(x, y) = x^2y + 12x^2 + 16y$ of $\mathbf{F}$ to evaluate it.

$$\int_C \mathbf{F} \cdot d\mathbf{r} = \int_C \nabla f \cdot d\mathbf{r} = f(x, y)\Big|_{(0,1)}^{(1,2)} = f(1, 2) - f(0, 1) = 30$$

EXAMPLE 7

(a) Show that the line integral

$$\int_C \mathbf{F} \cdot d\mathbf{r} = \int_C [(y \cos x + 2xe^y)\, dx + (\sin x + x^2e^y + 4)\, dy]$$

is independent of path.

(b) Find a function f such that

$$\nabla f = (y \cos x + 2xe^y)\mathbf{i} + (\sin x + x^2e^y + 4)\mathbf{j}.$$

Solution

(a) Let $P(x, y) = y \cos x + 2xe^y$ and $Q(x, y) = \sin x + x^2e^y + 4$. Then

$$\frac{\partial P}{\partial y} = \cos x + 2xe^y = \frac{\partial Q}{\partial x}$$

Since P, Q, $\partial P/\partial y$, and $\partial Q/\partial x$ are continuous in the entire plane, and since $\partial P/\partial y = \partial Q/\partial x$, it follows from theorem (19.3.7) that $\mathbf{F} = \mathbf{F}(x, y) = (y \cos x + 2xe^y)\mathbf{i} + (\sin x + x^2e^y + 4)\mathbf{j}$ is conservative. By theorem (19.3.1), the integral $\int_C \mathbf{F} \cdot d\mathbf{r}$ is independent of path.

(b) Since $\mathbf{F}$ is conservative, $\mathbf{F}$ has a potential function f for which

$$\nabla f = \mathbf{F} = (y \cos x + 2xe^y)\mathbf{i} + (\sin x + x^2e^y + 4)\mathbf{j}$$

Since $\nabla f = (\partial f/\partial x)\mathbf{i} + (\partial f/\partial y)\mathbf{j}$, we have

$$\frac{\partial f}{\partial x} = y \cos x + 2xe^y \qquad \text{and} \qquad \frac{\partial f}{\partial y} = \sin x + x^2 e^y + 4 \tag{13}$$

We integrate the second of these functions partially with respect to y to obtain

$$f(x, y) = y \sin x + x^2 e^y + 4y + k(x) \tag{14}$$

in which the "constant of integration" is denoted by the function $k(x)$. Now we differentiate with respect to x to get

$$\frac{\partial f}{\partial x} = y \cos x + 2xe^y + k'(x)$$

We equate this with the expression for $\partial f/\partial x$ found in (13):

$$y \cos x + 2xe^y + k'(x) = y \cos x + 2xe^y$$
$$k'(x) = 0$$

So

$$k(x) = K \qquad \text{where } K \text{ is a constant}$$

By substituting for $k(x)$ in (14), we obtain

$$f(x, y) = y \sin x + x^2 e^y + 4y + K$$

EXAMPLE 8

Determine whether $\int_C (x^2 y\, dx + xy^2\, dy)$ is independent of path anywhere in the plane.

Solution

Let $P = x^2 y$ and $Q = xy^2$. Then $\partial P/\partial y = x^2$ and $\partial Q/\partial x = y^2$. Since these two functions are not equal (except on the graph of the equation $x^2 = y^2$, which is not an open set), the line integral is not independent of path anywhere in the plane.

Summary

Let $\mathbf{F} = \mathbf{F}(x, y) = P(x, y)\mathbf{i} + Q(x, y)\mathbf{j}$ denote a vector field, where P and Q are two functions that are continuous on an open, connected region R containing a piecewise smooth curve C.

If $\mathbf{F} = P\mathbf{i} + Q\mathbf{j}$ is conservative, then $\int_C \mathbf{F} \cdot d\mathbf{r} = \int_C (P\, dx + Q\, dy)$ is independent of path.

If $\int_C (P\, dx + Q\, dy)$ is independent of path and if C is a closed curve in R, then $\int_C \mathbf{F} \cdot d\mathbf{r} = \int_C (P\, dx + Q\, dy) = 0$.

If $\int_C(P\,dx + Q\,dy)$ is independent of path, then $\mathbf{F} = P\mathbf{i} + Q\mathbf{j}$ is conservative.

If $\partial P/\partial y$ and $\partial Q/\partial x$ are also continuous on R and if R is simply connected, then $\mathbf{F} = P\mathbf{i} + Q\mathbf{j}$ is conservative if and only if $\partial P/\partial y = \partial Q/\partial x$.

EXERCISE 19.3

In Problems 1–4 use theorem (19.3.7) to determine whether **F** is a conservative vector field.

1. $\mathbf{F}(x, y) = x^2\mathbf{i} + y^2\mathbf{j}$

2. $\mathbf{F}(x, y) = xy\mathbf{i} + xy\mathbf{j}$

3. $\mathbf{F}(x, y) = xe^y\mathbf{i} + \frac{1}{2}x^2e^y\mathbf{j}$

4. $\mathbf{F}(x, y) = (x^2 + y^2)\mathbf{i} + (2xy - \sin y)\mathbf{j}$

In Problems 5–10:

(a) Show that the line integral $\int_C(P\,dx + Q\,dy)$ is independent of path.
(b) Find a function f such that $\nabla f = P(x, y)\mathbf{i} + Q(x, y)\mathbf{j}$.
(c) Evaluate the integral $\int_C(P\,dx + Q\,dy)$.

5. $\int_C(x\,dx + y\,dy)$; C is any curve joining the points $(1, 3)$ and $(2, 5)$

6. $\int_C[2xy\,dx + (x^2 + 1)\,dy]$; C is any curve joining the points $(1, -4)$ and $(-2, 3)$

7. $\int_C[(x^2 + 3y)\,dx + 3x\,dy]$; C is any curve joining the points $(1, 2)$ and $(-3, 5)$

8. $\int_C[(2x + y + 1)\,dx + (x + 3y + 2)\,dy]$; C is any curve joining the points $(0, 0)$ and $(1, 2)$

9. $\int_C[(4x^3 + 20xy^3 - 3y^4)\,dx + (30x^2y^2 - 12xy^3 + 5y^4)\,dy]$; C is any curve joining the points $(0, 0)$ and $(1, 1)$

10. $\int_C[(2yx^{-1})\,dx + (\ln x^2)\,dy]$; C is any curve in the first quadrant joining the points $(1, 1)$ and $(5, 5)$

In Problems 11–22 show that each line integral $\int_C \mathbf{F}\cdot d\mathbf{r} = \int_C(P\,dx + Q\,dy)$ is independent of path in the entire plane. Find a potential function f for **F**.

11. $\displaystyle\int_C (3x^2y^2\,dx + 2x^3y\,dy)$

12. $\displaystyle\int_C [(2x + y)\,dx + (x - 2y)\,dy]$

13. $\displaystyle\int_C [(x + 3y)\,dx + 3x\,dy]$

14. $\displaystyle\int_C [(2x + y)\,dx + (2y + x)\,dy]$

15. $\displaystyle\int_C [(2xy - y^2)\,dx + (x^2 - 2xy)\,dy]$

16. $\displaystyle\int_C [y^2\,dx + (2yx - e^y)\,dy]$

17. $\displaystyle\int_C [(x^2 - x + y^2)\,dx - (ye^y - 2xy)\,dy]$

18. $\displaystyle\int_C [(3x^2y + xy^2 + e^x)\,dx + (x^3 + x^2y + \sin y)\,dy]$

19. $\displaystyle\int_C [(y\cos x - 2\sin y)\,dx - (2x\cos y - \sin x)\,dy]$

20. $\displaystyle\int_C [(e^x\sin y + 2y\sin x)\,dx + (e^x\cos y - 2\cos x)\,dy]$

21. $\displaystyle\int_C [(2x + y\cos x)\,dx + \sin x\,dy]$

22. $\displaystyle\int_C [(\cos y - \cos x)\,dx + (e^y - x\sin y)\,dy]$

23. Show that $\int_C \left(\frac{-y}{x^2+y^2}\,dx + \frac{x}{x^2+y^2}\,dy\right)$ is independent of path in the rectangle R whose vertices are $(1/a, -a)$, $(a, -a)$, $(1/a, -1/a)$, and $(a, -1/a)$, $a>1$. Find the potential function f of $\mathbf{F}$.

24. Given the constant vector $\mathbf{c}$, show that $\nabla(\mathbf{c}\cdot\mathbf{r}) = \mathbf{c}$, where $\mathbf{r}=x\mathbf{i}+y\mathbf{j}+z\mathbf{k}$. Use the result to prove that $\int_C \mathbf{c}\cdot d\mathbf{r} = 0$, where C is any circle in space.

25. Suppose that $\mathbf{F}(x, y)$ is a force field directed toward the origin with magnitude inversely proportional to the square of the distance from the origin. (Such "inverse square law" forces are common in nature.) Show that $\mathbf{F}$ is conservative, and find a potential function.

26. Suppose f and g are differentiable functions of one variable. Show that $\int_C[f(x)\,dx+g(y)\,dy]=0$, where C is any circle in the coordinate plane.

Work 19.4

One important physical application of line integrals is to the concept of work. Here is a review of our development to date.

If an object is pushed along a straight line segment of length s by a constant force of magnitude F acting in the direction of the motion, then the work W done is

$$W = Fs \tag{1}$$

If the force acts in the opposite direction, then the work is $-Fs$.

In Chapter 6 we generalized (1) to the case where the force F is variable. There we showed that the work W done by a variable force $F = F(x)$ acting in the direction of the motion of an object as that object moves along a straight line from $x = a$ to $x = b$ is

$$W = \int_a^b F(x)\,dx \tag{2}$$

Next, in Chapter 14, we took up the question of the work W done by a constant force vector $\mathbf{F}$ acting on an object as that object moves in the direction of the vector $\mathbf{r}$ for the distance $\|\mathbf{r}\|$, and we found that

$$W = \mathbf{F}\cdot\mathbf{r} \tag{3}$$

The purpose of this section is to generalize (2) and (3) to give a definition for the work done by a vector field $\mathbf{F}$ acting on an object as that object moves along a smooth curve C from the point A on C to the point B on C.

Let

$$\mathbf{F} = \mathbf{F}(x, y) = P(x, y)\mathbf{i} + Q(x, y)\mathbf{j}$$

denote the vector force field exerted on an object at the point (x, y) in some open, connected set R. We assume P and Q have continuous first-order partial derivatives at each point in R. Let C be a smooth curve lying entirely in R and defined by the vector function $\mathbf{r} = \mathbf{r}(t) = x(t)\mathbf{i} + y(t)\mathbf{j}$, $a \le t \le b$.

We proceed as usual to partition the closed interval $[a, b]$ into n subintervals,

$$[a, t_1],\quad [t_1, t_2],\quad \ldots,\quad [t_{i-1}, t_i],\quad \ldots,\quad [t_{n-1}, b]$$

and denote the length of each subinterval by $\Delta t_1, \Delta t_2, \ldots, \Delta t_n$. Corresponding to each number $a = t_0, t_1, t_2, \ldots, t_n = b$ of the partition, there is a sequence of points $P_0, P_1, \ldots, P_n$ on the curve C. These points subdivide the curve into n subarcs of lengths $\Delta s_1, \Delta s_2, \ldots, \Delta s_n$. The norm $\|\Delta\|$ of this partition is the largest of the arc lengths Δs_i, $i = 1, 2, \ldots, n$. Let $P_i^* = (x_i^*, y_i^*)$ be some point on the ith subarc $\widehat{P_{i-1}P_i}$ of C. If the norm $\|\Delta\|$ is small enough, then the work W_i done by the force $\mathbf{F}$ in moving an object along the arc $\widehat{P_{i-1}P_i}$ can be approximated by the work done by the constant force

$$\mathbf{F}_i^* = \mathbf{F}(x_i^*, y_i^*)$$

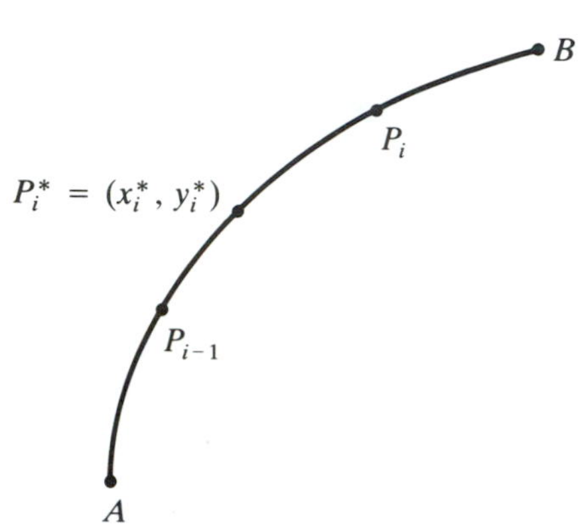

Figure 20

in moving the object along the directed line segment $\Delta\mathbf{r}_i = \overrightarrow{P_{i-1}P_i}$ (see Fig. 20). That is, by (3),

$$W_i \approx \mathbf{F}(x_i^*, y_i^*) \cdot \Delta\mathbf{r}_i$$

The total work W done by $\mathbf{F}$ along C from the point A: $\mathbf{r} = \mathbf{r}(a)$ to the point B: $\mathbf{r} = \mathbf{r}(b)$ is $W = \sum_{i=1}^{n} W_i$. This leads to the following definition:

[19.4.1] DEFINITION / *Work.*

Let $\mathbf{F} = \mathbf{F}(x, y) = P(x, y)\mathbf{i} + Q(x, y)\mathbf{j}$ denote a vector field, where P and Q are continuous with continuous partial derivatives at each point of an open, connected set R. Let C: $\mathbf{r} = \mathbf{r}(t)$, $a \le t \le b$, be a piecewise smooth curve lying entirely in R. The work W done by the force $\mathbf{F}$ in moving an object along C is defined as

$$W = \lim_{\|\Delta\| \to 0} \sum_{i=1}^{n} \mathbf{F}(x_i^*, y_i^*) \cdot \Delta\mathbf{r}_i = \int_C \mathbf{F} \cdot d\mathbf{r} = \int_C (P\,dx + Q\,dy)$$

EXAMPLE 1

Find the work done by the force $\mathbf{F}(x, y) = -y\mathbf{i} + x\mathbf{j}$ in moving an object along the half circle

$$C: \quad \mathbf{r}(t) = (\cos t)\mathbf{i} + (\sin t)\mathbf{j} \qquad 0 \le t \le \pi$$

Solution

On C we have $x(t) = \cos t$, $y(t) = \sin t$. Hence,

$$\mathbf{F}(x(t), y(t)) = (-\sin t)\mathbf{i} + (\cos t)\mathbf{j}$$

$$d\mathbf{r}(t) = [(-\sin t)\mathbf{i} + (\cos t)\mathbf{j}]\,dt$$

$$\mathbf{F} \cdot d\mathbf{r} = (\sin^2 t + \cos^2 t)\,dt = dt$$

Therefore,

$$W = \int_C \mathbf{F} \cdot d\mathbf{r} = \int_0^{\pi} dt = \pi$$

Application to Kinetic Energy

The work W done by a force $\mathbf{F}$ in moving an object along a piecewise smooth curve C defined by $\mathbf{r} = \mathbf{r}(t), \quad a \le t \le b,$ is

$$W = \int_C \mathbf{F} \cdot d\mathbf{r} = \int_a^b \mathbf{F} \cdot \mathbf{r}'(t)\, dt \tag{4}$$

By utilizing Newton's second law of motion, $\mathbf{F} = m\mathbf{r}''(t)$, where $\mathbf{r}''(t)$ is the acceleration of the object, we find that (4) takes the form

$$\begin{aligned} W &= \int_a^b m\mathbf{r}''(t) \cdot \mathbf{r}'(t)\, dt = \frac{m}{2}\int_a^b \frac{d}{dt}[\mathbf{r}'(t) \cdot \mathbf{r}'(t)]\, dt = \frac{m}{2}\int_a^b \frac{d}{dt}\|\mathbf{r}'(t)\|^2\, dt \\ &= \frac{m}{2}\|\mathbf{r}'(t)\|^2\Big|_a^b \end{aligned} \tag{5}$$

Recall that the *kinetic energy* K of an object of mass m that moves along a curve with velocity $\mathbf{v}$ is

$$K = \frac{1}{2}m\|\mathbf{v}\|^2 \underset{\substack{\uparrow \\ \mathbf{v} = \mathbf{r}'(t)}}{=} \frac{m}{2}\|\mathbf{r}'(t)\|^2 \tag{6}$$

By combining (5) and (6), we obtain the result

$$W = K(b) - K(a)$$

In other words,

$$\begin{bmatrix} \text{Work done by} \\ \text{a force } \mathbf{F} \text{ in} \\ \text{moving an object} \\ \text{from } A \text{ to } B \end{bmatrix} = \begin{bmatrix} \text{Change in} \\ \text{kinetic energy} \\ \text{from } A \text{ to } B \end{bmatrix} \tag{7}$$

If the force $\mathbf{F}$ is conservative and if the curve C is closed, then by virtue of corollary (19.3.2), it follows that

$$\text{Work} = \int_C \mathbf{F} \cdot d\mathbf{r} = 0$$

Based on (7), we have the following result:

[19.4.2] THEOREM

In a conservative field of force, the work W done by $\mathbf{F}$ in moving an object along a closed path is 0, and the object returns to its original position with the same kinetic energy it started with.

Physicists paraphrase this by saying that, in a conservative field of force, work is a function of position and not path.

Conservation of Energy

For a conservative field of force, the principle of conservation of energy holds. Suppose an object of mass m moves along a smooth curve C defined by $\mathbf{r} = \mathbf{r}(t)$ in a conservative field of force $\mathbf{F} = \mathbf{F}(x, y)$. The *potential energy* $U = U(x, y)$ of the object due to $\mathbf{F}$ is that function U for which*

$$\nabla U = -\mathbf{F}$$

The principle of conservation of energy may be stated as follows:

[19.4.3] THEOREM

In a conservative field of force the sum of the potential and kinetic energies of an object is constant.

The proof of this statement is relatively straightforward.

Proof The potential energy $U = U(x, y)$ of an object moving along a curve C in a conservative field of force $\mathbf{F}$ obeys

$$\nabla U = -\mathbf{F}(x, y)$$

If the object moves along the curve C: $\mathbf{r} = \mathbf{r}(t)$, its kinetic energy is

$$\tfrac{1}{2}m\|\mathbf{r}'(t)\|^2 = \tfrac{1}{2}m[\mathbf{r}'(t) \cdot \mathbf{r}'(t)]$$

Let $E = E(t)$ equal the sum of the potential and kinetic energies of the object at time t. Then

$$E(t) = U(x, y) + \frac{1}{2}m\|\mathbf{r}'(t)\|^2 = U(x, y) + \left(\frac{m}{2}\right)[\mathbf{r}'(t) \cdot \mathbf{r}'(t)]$$

We show that E is constant by showing that $E'(t) = 0$:

$$E'(t) \underset{\substack{\uparrow \\ \text{Chain rule}}}{=} \left(\frac{\partial U}{\partial x}\right)\left(\frac{dx}{dt}\right) + \left(\frac{\partial U}{\partial y}\right)\left(\frac{dy}{dt}\right) + m[\mathbf{r}'(t) \cdot \mathbf{r}''(t)] \tag{8}$$

But

$$\mathbf{r}'(t) = \frac{dx}{dt}\mathbf{i} + \frac{dy}{dt}\mathbf{j} \quad \text{and} \quad \frac{\partial U}{\partial x}\mathbf{i} + \frac{\partial U}{\partial y}\mathbf{j} = \nabla U = -\mathbf{F} \underset{\substack{\uparrow \\ \text{Newton's law}}}{=} -m\mathbf{r}''(t) \tag{9}$$

By combining (8) and (9), we find

$$E'(t) = -m[\mathbf{r}''(t) \cdot \mathbf{r}'(t)] + m[\mathbf{r}'(t) \cdot \mathbf{r}''(t)] = 0$$

---

* The existence of $U = U(x, y)$ is guaranteed, since $\mathbf{F}$ is conservative.

EXERCISE 19.4

In Problems 1–10 calculate the work done by the field of force $\mathbf{F}$ in moving an object along each curve between the indicated points.

1. $\mathbf{F} = y\mathbf{i} + x\mathbf{j}$ along $\mathbf{r}(t) = t\mathbf{i} + t^2\mathbf{j}$ from $t = 0$ to $t = 1$
2. $\mathbf{F} = xy\mathbf{i} + y^2\mathbf{j}$ along $\mathbf{r}(t) = t\mathbf{i} + t^2\mathbf{j}$ from $t = 0$ to $t = 1$
3. $\mathbf{F} = (x - 2y)\mathbf{i} + xy\mathbf{j}$ along $\mathbf{r}(t) = (3\cos t)\mathbf{i} + (2\sin t)\mathbf{j}$ from $t = 0$ to $t = \pi/2$
4. $\mathbf{F} = x\mathbf{i} - y\mathbf{j}$ along $\mathbf{r}(t) = (\cos t)\mathbf{i} + (\sin t)\mathbf{j}$ from $t = 0$ to $t = 2\pi$
5. $\mathbf{F} = x^2y\mathbf{i} + (x^2 - y^2)\mathbf{j}$ along $y = 2x^2$ from $(0, 0)$ to $(1, 2)$
6. $\mathbf{F} = (y\sin x)\mathbf{i} - (x\cos y)\mathbf{j}$ along $y = x$ from $(0, 0)$ to $(1, 1)$
7. $\mathbf{F} = (y - x^2)\mathbf{i} + x\mathbf{j}$ along the upper half of the circle $x^2 + y^2 = 1$ from $(1, 0)$ to $(-1, 0)$
8. $\mathbf{F} = (y - x^2)\mathbf{i} + x\mathbf{j}$ along the line segments joining. $(1, 0)$ to $(1, 1)$ to $(-1, 1)$ to $(-1, 0)$
9. $\mathbf{F} = y^3\mathbf{i} + x^3\mathbf{j}$ along the ellipse $\mathbf{r}(t) = (a\cos t)\mathbf{i} + (b\sin t)\mathbf{j}$ from $t = 0$ to $t = 2\pi$
10. $\mathbf{F} = -y^2\mathbf{i} + x^2\mathbf{j}$ along the upper half of the ellipse $(x^2/a^2) + (y^2/b^2) = 1$ from $(a, 0)$ to $(-a, 0)$

11. Verify that the force field $\mathbf{F} = y\mathbf{i} - x\mathbf{j}$ is nonconservative. Show that the integral $\int_C \mathbf{F} \cdot d\mathbf{r}$ is dependent on the path of integration by taking two paths in which the starting point is the origin $(0, 0)$ and the endpoint is $(1, 1)$. For one path, take the line $x = y$. For the other path, take the x-axis out to the point $(1, 0)$ and then the line $x = 1$ up to the point $(1, 1)$.

12. A particle moves in a clockwise direction from the origin to the point $(2a, 0)$ along the upper half of the circle $(x - a)^2 + y^2 = a^2$ and is acted upon by a force with constant magnitude 3 and with direction $\mathbf{i} + \mathbf{j}$. Find the work done by this force.

13. A particle moves along the curve $\mathbf{r}(t) = 64\sqrt{3}\, t\mathbf{i} + (64t - 16t^2)\mathbf{j}$ from $t = 0$ to $t = 4$ and is acted upon by a force whose magnitude is directly proportional to the speed of the particle and whose direction is opposite to that of the velocity. Find the work done by this force.

14. A particle moves from the point $(0, 0)$ to the point $(1, 0)$ along the curve $y = ax(1 - x)$ and is acted upon by a force given by $\mathbf{F}(x, y) = (y^2 + 1)\mathbf{i} + (x + y)\mathbf{j}$. Find a so that the work done is a minimum.

15. The repelling force between a charged particle P at the origin and an oppositely charged particle Q at (x, y) is

$$\mathbf{F}(x, y) = \frac{x}{(x^2 + y^2)^{3/2}}\mathbf{i} + \frac{y}{(x^2 + y^2)^{3/2}}\mathbf{j}$$

Find the work done by $\mathbf{F}$ as Q moves along the line segment from $(1, 0)$ to $(-1, 2)$.

16. Suppose sea level is the xy-plane and the z-axis points upward. Then the gravitational force on an object of mass m near sea level may be taken to be $\mathbf{F} = mg\mathbf{k}$, where g is the acceleration due to gravity. Show that the work done by gravity on an object moving from (x_1, y_1, z_1) to (x_2, y_2, z_2) along any path is $W = mg(z_2 - z_1)$.

17. Suppose an object travels through a force field along a path that is always normal to the force. Show that the work done in moving the object from one point to another is 0. Why does it follow that the gravitational field of the earth does no work on a satellite in circular orbit?

Green's Theorem in the Plane 19.5

*Green's theorem** relates the value of a line integral along a simple closed curve C to a certain double integral over the region R enclosed by C. Green's

* Named in honor of the Englishman George Green (1793–1841), who first used it in 1828. Although the theorem was actually discovered earlier by Gauss and Lagrange, it became widely known as *Green's theorem* because Green used it in applications to electricity and magnetism, fluid flow, and other areas of physics.

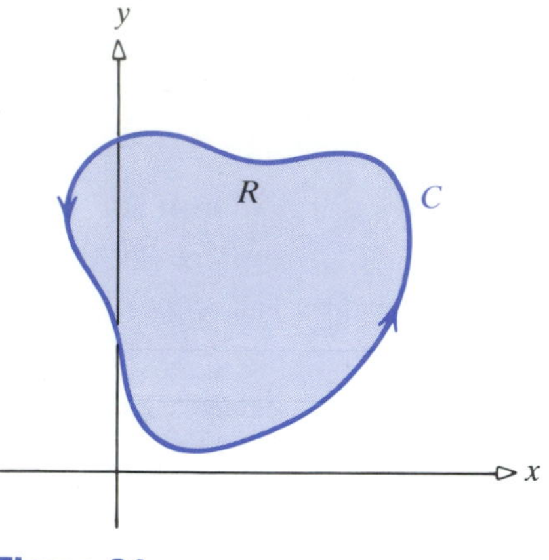

Figure 21

theorem may be stated in several ways, but the most common form is the following:

$$\oint_C (P\,dx + Q\,dy) = \iint_R \left(\frac{\partial Q}{\partial x} - \frac{\partial P}{\partial y}\right) dx\,dy \tag{1}$$

The curve C is the boundary of the region R, and the symbol $\oint$ indicates that the simple closed curve C is to be traversed in a counterclockwise direction; that is, the region R will always lie to the left as one moves along the curve (see Fig. 21).

The result (1) requires two assumptions:

1. The functions P and Q possess continuous first-order partial derivatives at each point in the region R.*
2. The curve C may be any piecewise smooth, simple closed curve, and the region R is both simply connected and closed.

We now give a rather general formulation of Green's theorem, but we will confine our proof to regions that are both x-simple and y-simple (see Section 18.2).

[19.5.1] THEOREM / *Green's Theorem.*

Let C denote a piecewise smooth, simple closed curve, and let R be the region consisting of C and its interior. Let P and Q denote functions that possess continuous first-order partial derivatives on some open region that contains R. Then

$$\oint_C (P\,dx + Q\,dy) = \iint_R \left(\frac{\partial Q}{\partial x} - \frac{\partial P}{\partial y}\right) dx\,dy$$

where the line integral is taken around C in the counterclockwise direction.

Proof for regions that are both x-Simple and y-Simple. It is sufficient to prove that

$$\oint_C P(x, y)\,dx = -\iint_R \frac{\partial P}{\partial y}\,dx\,dy \tag{2}$$

and

$$\oint_C Q(x, y)\,dy = \iint_R \frac{\partial Q}{\partial x}\,dx\,dy \tag{3}$$

since the result will then follow from adding these two equations. Suppose the region R is both x-simple and y-simple. Then, since it is x-simple (see

* Somewhat less stringent conditions will still work.

Fig. 22), the boundary of R consists of two smooth curves, C_1: $y = g_1(x)$ and C_2: $y = g_2(x)$, $g_1(x) \leq g_2(x)$, $a \leq x \leq b$.

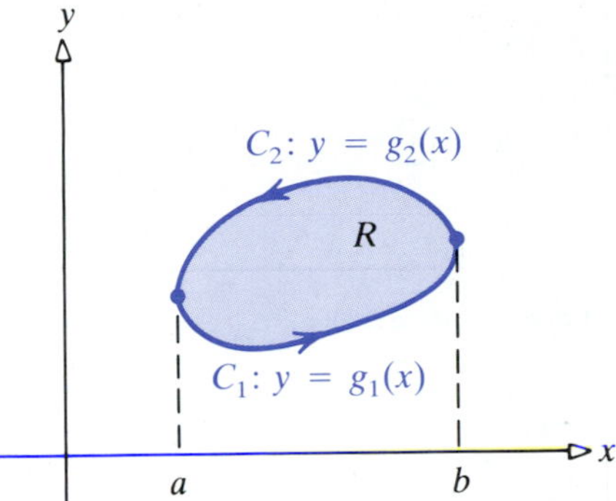

Figure 22

To establish (2) we begin with the double integral on the right side of (2). Since R is x-simple, the double integral of $-\partial P/\partial y$ over R is given by

$$-\iint_R \frac{\partial P}{\partial y}\, dx\, dy = -\int_a^b \left[\int_{g_1(x)}^{g_2(x)} \frac{\partial P}{\partial y}\, dy\right] dx = -\int_a^b \left[P(x, y)\Big|_{g_1(x)}^{g_2(x)}\right] dx$$

$$= -\int_a^b [P(x, g_2(x)) - P(x, g_1(x))]\, dx$$

$$= \int_a^b [P(x, g_1(x)) - P(x, g_2(x))]\, dx$$

$$= \int_a^b P(x, g_1(x))\, dx - \int_a^b P(x, g_2(x))\, dx$$

$$= \int_{C_1} P(x, y)\, dx - \int_{C_2} P(x, y)\, dx$$

$$= \int_{C_1} P(x, y)\, dx + \int_{-C_2} P(x, y)\, dx = \oint_C P(x, y)\, dx$$

Thus, we have established (2). The proof of (3) uses the fact that the region R is also y-simple. It is obtained in a similar manner and is left as an exercise (Problem 29).

EXAMPLE 1

Use Green's theorem to evaluate the line integral

$$\oint_C [(-2xy + y^2)\, dx + x^2\, dy]$$

where C is the boundary of the region R enclosed by $y = 4x$ and $y = 2x^2$.

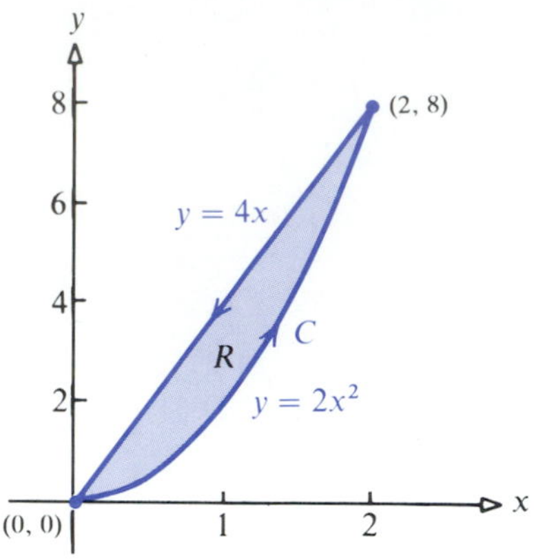

Figure 23

Solution

Figure 23 illustrates the curve C and the region R. Let

$$P(x, y) = -2xy + y^2 \qquad \text{and} \qquad Q(x, y) = x^2$$

By applying Green's theorem, we find

$$\oint_C [(-2xy + y^2)\,dx + x^2\,dy] = \iint_R \left[\frac{\partial}{\partial x}(x^2) - \frac{\partial}{\partial y}(-2xy + y^2)\right] dx\,dy$$

$$= \iint_R (4x - 2y)\,dx\,dy$$

$$\underset{\substack{\uparrow \\ R \text{ is } x\text{-simple.}}}{=} \int_0^2 \left[\int_{2x^2}^{4x} (4x - 2y)\,dy\right] dx$$

$$= \int_0^2 \left[(4xy - y^2)\Big|_{2x^2}^{4x}\right] dx$$

$$= \int_0^2 [(16x^2 - 16x^2) - (8x^3 - 4x^4)]\,dx$$

$$= \left(-2x^4 + \frac{4}{5}x^5\right)\Big|_0^2 = -32 + \frac{128}{5}$$

$$= \frac{-32}{5}$$

We can verify the solution found in Example 1 by evaluating the line integral directly:

$$\oint_C [(-2xy + y^2)\,dx + x^2\,dy] = \int_{C_1:\ y=2x^2} [(-2xy + y^2)\,dx + x^2\,dy]$$

$$+ \int_{C_2:\ y=4x} [(-2xy + y^2)\,dx + x^2\,dy]$$

$$= \int_0^2 [(-4x^3 + 4x^4)\,dx + 4x^3\,dx]$$

$$+ \int_2^0 [(-8x^2 + 16x^2)\,dx + 4x^2\,dx]$$

$$= 4\frac{x^5}{5}\Big|_0^2 + 12\frac{x^3}{3}\Big|_2^0 = \frac{128}{5} - 32$$

$$= \frac{-32}{5}$$

The next example illustrates how Green's theorem is used to evaluate line integrals that may not be easily evaluated by the techniques of Section 19.2.

EXAMPLE 2

Use Green's theorem to evaluate the line integral

$$\oint_C [(e^{-x^2} + y^2)\,dx + (\ln y - x^2)\,dy]$$

where C is the square illustrated in Figure 24.

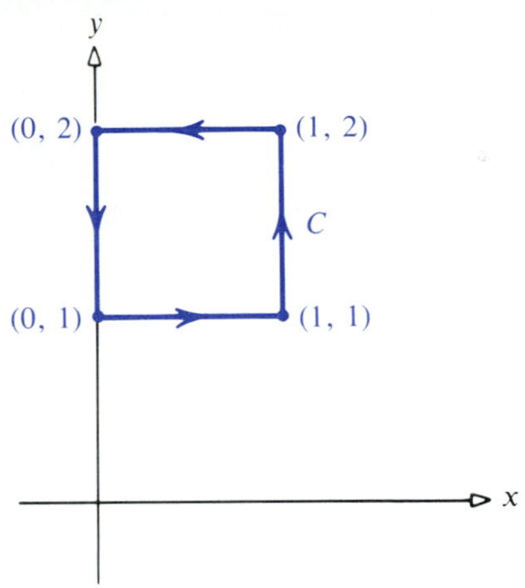

Figure 24

Solution

Because e^{-x^2} appears in the integrand of the line integral, prior techniques may be useless. However, an application of Green's theorem produces

$$\begin{aligned}\oint_C [(e^{-x^2} + y^2)\,dx + (\ln y - x^2)]\,dy &= \iint_R (-2x - 2y)\,dx\,dy \\ &= -2\int_0^1 \left[\int_1^2 (x+y)\,dy\right] dx \\ &= -2\int_0^1 \left[\left(xy + \frac{y^2}{2}\right)\Bigg|_1^2\right] dx \\ &= -2\int_0^1 \left(x + \frac{3}{2}\right) dx \\ &= -2\left(\frac{x^2}{2} + \frac{3x}{2}\right)\Bigg|_0^1 = -4\end{aligned}$$

Application of Green's Theorem to Area

Green's theorem can be used to express the area A of a region R enclosed by a piecewise smooth, simple closed curve C as a line integral.

[19.5.2] THEOREM

If C denotes a piecewise smooth, simple closed curve C, the area A of the region R enclosed by C is given by

$$A = \frac{1}{2}\oint_C (-y\,dx + x\,dy)$$

Proof In Green's theorem, put $P = P(x, y) = -y$ and $Q = Q(x, y) = x$. Then

$$\oint_C (-y\,dx + x\,dy) = \iint_R \left[\frac{\partial}{\partial x}(x) - \frac{\partial}{\partial y}(-y)\right] dx\,dy = \iint_R 2\,dx\,dy = 2A$$

In Problem 30 you are asked to show that the area A can also be expressed by the formulas

$$A = \oint_C x\,dy \qquad \text{and} \qquad A = \oint_C (-y)\,dx \tag{4}$$

EXAMPLE 3

Use theorem (19.5.2) to find the area of the region enclosed by the ellipse $(x^2/a^2) + (y^2/b^2) = 1$.

Solution

We use the parametric equations of the ellipse: $x = a \cos t$, $y = b \sin t$, $0 \le t \le 2\pi$. We thus have $dx = -a \sin t\, dt$ and $dy = b \cos t\, dt$. Then

$$\begin{aligned} A &= \tfrac{1}{2} \oint_C (-y\, dx + x\, dy) \\ &= \tfrac{1}{2} \int_0^{2\pi} [-b \sin t(-a \sin t\, dt) + a \cos t(b \cos t\, dt)] \\ &= \tfrac{1}{2} \int_0^{2\pi} ab(\sin^2 t + \cos^2 t)\, dt = \tfrac{1}{2}(ab) \int_0^{2\pi} dt = \tfrac{1}{2} ab(2\pi) = \pi ab \end{aligned}$$

Application of Green's Theorem to Conservative Fields of Force

Suppose that $\mathbf{F} = \mathbf{F}(x, y) = P(x, y)\mathbf{i} + Q(x, y)\mathbf{j}$ denotes a conservative field of force for which the components P and Q possess continuous first-order partial derivatives on an open, connected region R. It follows from theorem (19.3.7) that $\partial P/\partial y = \partial Q/\partial x$ throughout R. We shall now prove for a special case that the converse is true, provided R is simply connected.

[19.5.3] THEOREM

Let $\mathbf{F} = P(x, y)\mathbf{i} + Q(x, y)\mathbf{j}$ denote a field of force whose components P and Q possess continuous first-order partial derivatives throughout an open, simply connected region R. If $\partial P/\partial y = \partial Q/\partial x$ throughout R, then $\mathbf{F}$ is conservative.

To show that $\mathbf{F}$ is conservative, we need only show that for any curve C in R, $\int_C \mathbf{F} \cdot d\mathbf{r} = \int_C (P\, dx + Q\, dy)$ is independent of path; that is, that it has the same value for any two piecewise smooth curves joining a fixed point A in R to an arbitrary point (x, y) in R.

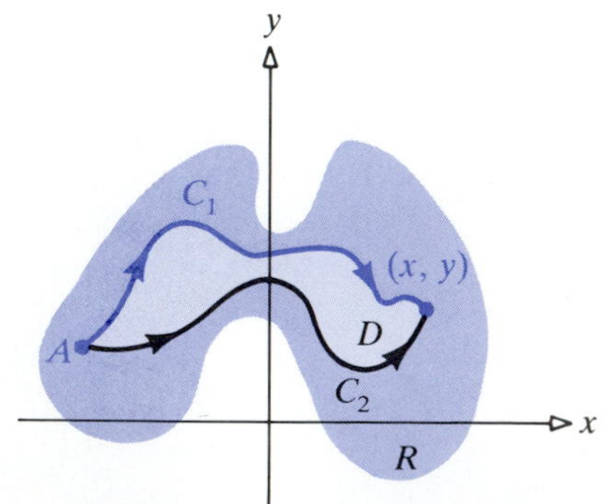

Figure 25

Proof for a Special Case Let C_1 and C_2 be two piecewise smooth curves in R joining A to (x, y). We assume these curves do not intersect or coincide except at A and (x, y).* The curves C_1 and C_2 will therefore form the boundary of a region D that is a subset of R, since R is simply connected (see Fig. 25).

* This assumption may be relaxed so that the curves C_1 and C_2 may sometimes intersect or even coincide in R.

We apply Green's theorem on this region D to get

$$\iint_D \left(\frac{\partial Q}{\partial x} - \frac{\partial P}{\partial y}\right) dx\, dy = \int_{C_2 \cup -C_1} (P\, dx + Q\, dy)$$
$$= \int_{C_2} (P\, dx + Q\, dy) + \int_{-C_1} (P\, dx + Q\, dy)$$
$$= \int_{C_2} (P\, dx + Q\, dy) - \int_{C_1} (P\, dx + Q\, dy)$$

But $\partial P/\partial y = \partial Q/\partial x$ throughout R. So,

$$\iint_D \left(\frac{\partial Q}{\partial x} - \frac{\partial P}{\partial y}\right) dx\, dy = 0$$

Hence, we conclude that $\int_{C_1}(P\, dx + Q\, dy) = \int_{C_2}(P\, dx + Q\, dy)$. Thus, the line integral is independent of path, and **F** is conservative.

Green's Theorem for Multiply Connected Regions

For Green's theorem to hold, we assume that the piecewise smooth, simple closed curve C and its interior form the region R over which the double integral is evaluated. We now state the formulation of Green's theorem for *multiply connected regions.*

[19.5.4] THEOREM

Let $C_1, C_2, \ldots, C_n$ be n piecewise smooth, simple closed curves for which:

1. **No two of the curves intersect.**
2. **Each of the curves lies in the interior of a piecewise smooth, simple closed curve C.**
3. **Any curve C_i is exterior to every curve C_j, $i \neq j$, $i, j = 1, 2, \ldots, n$.**

Let R denote the region consisting of the interior of C and its boundary, less the interior of each of the curves $C_1, C_2, \ldots, C_n$. Let P and Q denote functions that possess continuous first-order partial derivatives throughout R. Then

$$\iint_R \left(\frac{\partial Q}{\partial x} - \frac{\partial P}{\partial y}\right) dx\, dy = \oint_C (P\, dx + Q\, dy) - \sum_{i=1}^{n} \oint_{C_i} (P\, dx + Q\, dy)$$

Figure 26(a) illustrates a multiply connected region.

We give the idea of the proof for $n = 1$: In Figure 26(b), curve C_1 is a circle contained within the larger circle C. We construct line segments $\overline{AB}$ and $\overline{ED}$ that join the curves C and C_1. This decomposes R into two regions R_1 and R_2, with boundaries Γ_1 and Γ_2, respectively. Since R_1 and R_2 each satisfy the

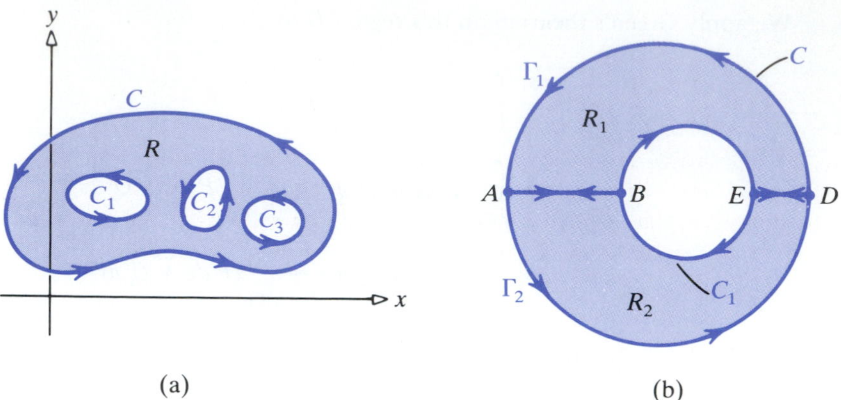

Figure 26

conditions of Green's theorem, we find

$$\iint_R \left(\frac{\partial Q}{\partial x} - \frac{\partial P}{\partial y}\right) dx\, dy = \iint_{R_1} \left(\frac{\partial Q}{\partial x} - \frac{\partial P}{\partial y}\right) dx\, dy + \iint_{R_2} \left(\frac{\partial Q}{\partial x} - \frac{\partial P}{\partial y}\right) dx\, dy$$

$$= \oint_{\Gamma_1} (P\, dx + Q\, dy) + \oint_{\Gamma_2} (P\, dx + Q\, dy)$$

$$= \int_{\widehat{DA}} (P\, dx + Q\, dy) + \int_{\overline{AB}} (P\, dx + Q\, dy) + \int_{\widehat{BE}} (P\, dx + Q\, dy)$$

$$+ \int_{\overline{ED}} (P\, dx + Q\, dy) + \int_{\overline{DE}} (P\, dx + Q\, dy) + \int_{\widehat{EB}} (P\, dx + Q\, dy)$$

$$+ \int_{\overline{BA}} (P\, dx + Q\, dy) + \int_{\widehat{AD}} (P\, dx + Q\, dy)$$

Along the line segments $\overline{AB}$ and $\overline{ED}$,

$$\int_{\overline{AB}} (P\, dx + Q\, dy) = -\int_{\overline{BA}} (P\, dx + Q\, dy)$$

and

$$\int_{\overline{ED}} (P\, dx + Q\, dy) = -\int_{\overline{DE}} (P\, dx + Q\, dy)$$

Hence,

$$\iint_R \left(\frac{\partial Q}{\partial x} - \frac{\partial P}{\partial y}\right) dx\, dy = \oint_C (P\, dx + Q\, dy) + \oint_{-C_1} (P\, dx + Q\, dy)$$

$$= \oint_C (P\, dx + Q\, dy) - \oint_{C_1} (P\, dx + Q\, dy) \quad ■$$

The next result is a consequence of theorem (19.5.4).

[19.5.5] THEOREM

Let P and Q denote functions that possess continuous first-order partial derivatives throughout an open, connected set R. Let C_1 and C_2 denote two piecewise smooth, simple closed curves, each lying entirely in R, for which C_2 is not in the exterior of C_1. If $\partial P/\partial y = \partial Q/\partial x$ throughout R, then

$$\int_{C_1} (P\,dx + Q\,dy) = \int_{C_2} (P\,dx + Q\,dy)$$

Figure 27 illustrates a set R satisfying the conditions of theorem (19.5.5). This theorem may be restated informally as:

If $\partial P/\partial y = \partial Q/\partial x$ throughout R, then the value of a line integral along a piecewise smooth, simple closed curve equals the value of the same line integral along any other piecewise smooth, simple closed curve that can be obtained from the original curve by a continuous deformation, with all of the intermediate curves of the deformation being entirely in R.

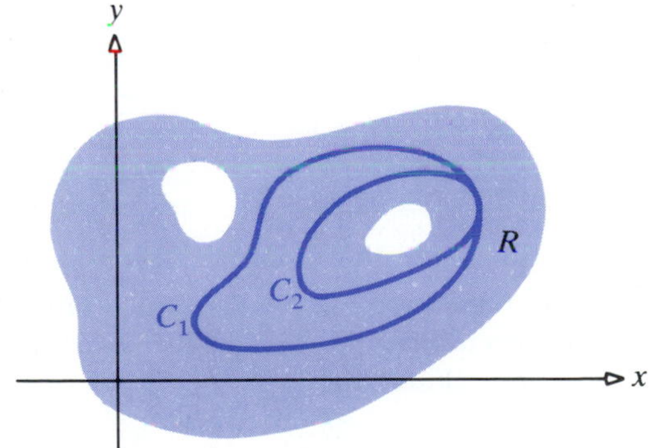

Figure 27

Let's look at an example.

EXAMPLE 4

Evaluate: $\oint_C \left(\frac{-y}{x^2 + y^2}\,dx + \frac{x}{x^2 + y^2}\,dy \right)$

(a) If C is the curve $x^{2/3} + y^{2/3} = 1$

(b) If C is any piecewise smooth, simple closed curve not containing the origin in its interior

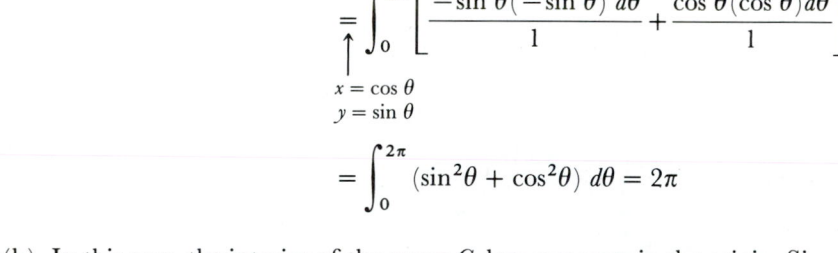

Solution

We first note that for

$$P = \frac{-y}{x^2 + y^2} \qquad \text{and} \qquad Q = \frac{x}{x^2 + y^2}$$

we have

$$\frac{\partial P}{\partial y} = \frac{y^2 - x^2}{(x^2 + y^2)^2} \qquad \text{and} \qquad \frac{\partial Q}{\partial x} = \frac{y^2 - x^2}{(x^2 + y^2)^2}$$

Next, we note that, except at (0, 0), $\partial P/\partial y$ and $\partial Q/\partial x$ are continuous.

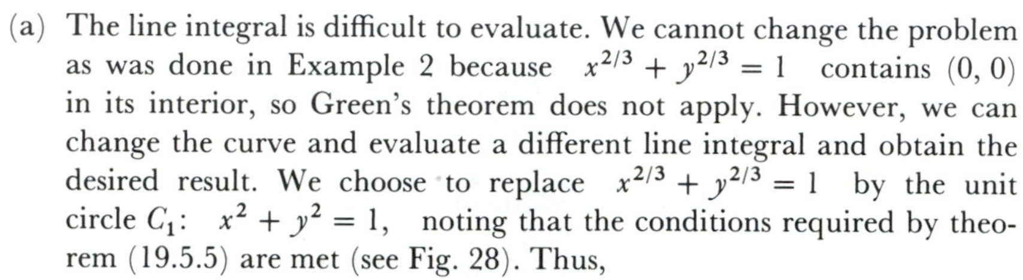

(a) The line integral is difficult to evaluate. We cannot change the problem as was done in Example 2 because $x^{2/3} + y^{2/3} = 1$ contains (0, 0) in its interior, so Green's theorem does not apply. However, we can change the curve and evaluate a different line integral and obtain the desired result. We choose to replace $x^{2/3} + y^{2/3} = 1$ by the unit circle C_1: $x^2 + y^2 = 1$, noting that the conditions required by theorem (19.5.5) are met (see Fig. 28). Thus,

Figure 28

$$\oint_C \left(\frac{-y}{x^2 + y^2}\,dx + \frac{x}{x^2 + y^2}\,dy\right) = \oint_{C_1} \left(\frac{-y}{x^2 + y^2}\,dx + \frac{x}{x^2 + y^2}\,dy\right)$$

$$\underset{\substack{\uparrow \\ x = \cos\theta \\ y = \sin\theta}}{=} \int_0^{2\pi} \left[\frac{-\sin\theta(-\sin\theta)\,d\theta}{1} + \frac{\cos\theta(\cos\theta)\,d\theta}{1}\right]$$

$$= \int_0^{2\pi} (\sin^2\theta + \cos^2\theta)\,d\theta = 2\pi$$

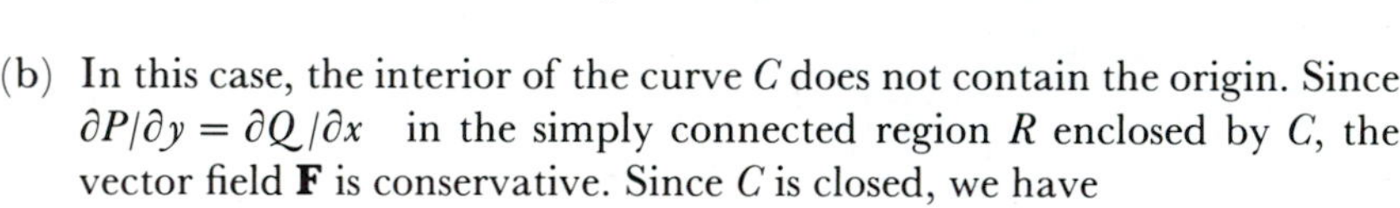

(b) In this case, the interior of the curve C does not contain the origin. Since $\partial P/\partial y = \partial Q/\partial x$ in the simply connected region R enclosed by C, the vector field $\mathbf{F}$ is conservative. Since C is closed, we have

$$\oint_C \left(\frac{-y}{x^2 + y^2}\,dx + \frac{x}{x^2 + y^2}\,dy\right) = 0$$

EXERCISE 19.5

In Problems 1–6 use Green's theorem to evaluate the given line integral. In each case, assume that C is the perimeter of the rectangle with vertices (0, 0), (4, 0), (4, 3), (0, 3), and that the orientation of C is counterclockwise.

1. $\oint_C y\,dx$

2. $\oint_C x\,dy$

3. $\oint_C [xy\,dx + (x + y)\,dy]$

4. $\oint_C (3y\,dx - 2x\,dy)$

5. $\oint_C (xy^2\,dx + x^2y\,dy)$

6. $\oint_C (y \sin xy\,dx + x \sin xy\,dy)$

In Problems 7–10 use Green's theorem to evaluate the given line integral. In each case, assume that C is the unit circle $x^2 + y^2 = 1$, and that the orientation of C is counterclockwise.

7. $\oint_C (-x^2y\,dx + y^2x\,dy)$

8. $\oint_C [y(x^2 + y^2)\,dx - x(x^2 + y^2)\,dy]$

9. $\oint_C [(x^2 - y^3)\,dx + (x^2 + y^2)\,dy]$

10. $\oint_C [(xy^3 + \sin x)\,dx + (x^2y^2 + 4x)\,dy]$

In Problems 11–16 use Green's theorem to evaluate the indicated integral. In each case, C is to be traversed in the counterclockwise direction.

11. $\oint_C [(x^2 + y)\,dx + (x - y^2)\,dy]$, where C is the boundary of the region R enclosed by $y = x^{3/2}$, the x-axis, and $x = 1$

12. $\oint_C [(4x^2 - 8y^2)\,dx + (y - 6xy)\,dy]$, where C is the boundary of the region R enclosed by $y = \sqrt{x}$ and $y = x^2$

13. $\oint_C [(x^3 - x^2y)\,dx + xy^2\,dy]$, where C is the boundary of the region R enclosed by $y = x^2$ and $x = y^2$

14. $\oint_C [(x^2 - y^2)\,dx + xy\,dy]$, where C is the boundary of the region R enclosed by $x = y^2$ and $x = 1$

15. $\oint_C [(1/y)\,dx + (1/x)\,dy]$, where C is the boundary of the region R enclosed by $y = 1$, $x = 9$, and $y = \sqrt{x}$

16. $\oint_C (x^2y\,dx - y^2x\,dy)$, where C is the boundary of the region R enclosed by $y = \sqrt{a^2 - x^2}$ and $y = 0$

In Problems 17–20 use Green's theorem to find the area of the region described.

17. Enclosed by $y = x^2$ and $y = x + 2$

18. Enclosed by $y = x^2 - 1$ and $y = 0$

19. Under one arch of the cycloid

$\mathbf{r}(t) = (2\pi t - \sin 2\pi t)\mathbf{i} + (1 - \cos 2\pi t)\mathbf{j}, \quad 0 \le t \le 1$

20. Enclosed by the hypocycloid $x^{2/3} + y^{2/3} = 1$ [*Hint:* Use the parametric equations $x = \cos^3 t$, $y = \sin^3 t$.]

State why each of the line integrals in Problems 21–24 is equal to 0.

21. $\oint_C (xe^{x^2+y^2}\,dx + ye^{x^2+y^2}\,dy)$, where C is any piecewise smooth, simple closed curve in the plane

22. $\oint_C [e^{xy}(xy + 1)\,dx + x^2e^{xy}\,dy]$, where C is any piecewise smooth, simple closed curve in the plane

23. $\oint_C (e^x\sin y\,dx + e^x\cos y\,dy)$, where C is any piecewise smooth, simple closed curve in the plane

24. $\oint_C \left[\frac{x\,dx}{(x^2 + y^2)^{1/2}} + \frac{y\,dy}{(x^2 + y^2)^{1/2}}\right]$, where C is any piecewise smooth, simple closed curve not containing the origin in its interior

In Problems 25–28 evaluate $\oint_C \frac{y\,dx - x\,dy}{x^2 + y^2}$ about the indicated curve C.

25. C: $x^2 + y^2 = 4$

26. C: $(x - 2)^2 + (y + 3)^2 = 1$

27. C: $x^{2/3} + y^{2/3} = 1$

28. C: $3x^2 + 6y^2 = 4$

29. Complete the proof of Green's theorem by establishing (3).

30. Use theorem (19.5.2) to obtain (4).

19.6 Surface Integrals

Surface integrals are similar to line integrals, but the integration takes place on a surface instead of along a curve.

Let Σ be a surface given by $z = f(x, y)$, where f is continuous and has continuous first-order partial derivatives on a closed, bounded region R. See Figure 29. Following our usual practice, we partition R into n rectangles and concentrate on the ith rectangle R_i, whose area is ΔA_i. Then we pick a point (u_i, v_i) in each rectangle R_i and erect a vertical column on each of these rectangles to intersect the tangent plane at $(u_i, v_i, f(u_i, v_i))$ on Σ in an area $\Delta\sigma_i$. The number $\Delta\sigma_i$ is an approximation to the area of the part of the surface that lies above the ith rectangle R_i.

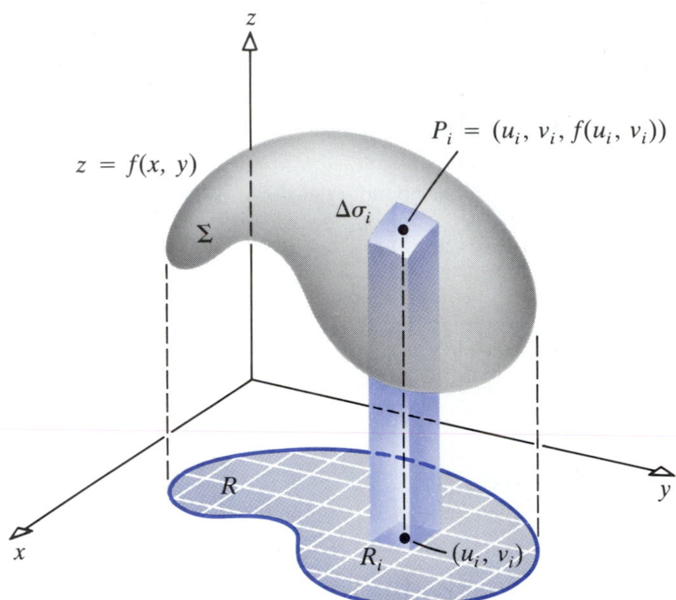

Figure 29

Let $\phi(x, y, z)$ be a function that is defined over a region of space containing the surface Σ. Evaluate ϕ at the point $(u_i, v_i, f(u_i, v_i))$ on Σ, and form the product $\phi(u_i, v_i, f(u_i, v_i))\,\Delta\sigma_i$. Then form the sum

$$\sum_{i=1}^{n} \phi(u_i, v_i, f(u_i, v_i))\Delta\sigma_i$$

Since

$$\Delta\sigma_i = \sqrt{[f_x(u_i, v_i)]^2 + [f_y(u_i, v_i)]^2 + 1}\;\Delta x_i \Delta y_i$$

(refer to (5) in Section 18.5), we are led to the following definition:

[19.6.1] DEFINITION / *Surface Integral of ϕ over Σ.*

Let Σ be a surface given by $z = f(x, y)$, where f is continuous and has continuous first-order partial derivatives on a closed,

bounded region R. Let $\phi = \phi(x, y, z)$ be a function defined on some region containing Σ. The *surface integral of ϕ over Σ* is defined as

$$\iint_\Sigma \phi(x, y, z)\, d\sigma = \lim_{\|P\|\to 0} \sum_{i=1}^{n} \phi(u_i, v_i, f(u_i, v_i))\Delta\sigma_i$$
$$= \lim_{\|P\|\to 0} \sum_{i=1}^{n} \phi(u_i, v_i, f(u_i, v_i))\sqrt{[f_x(u_i, v_i)]^2 + [f_y(u_i, v_i)]^2 + 1}\, \Delta x_i \Delta y_i$$

provided this limit exists.

We have a method for evaluating surface integrals when ϕ is continuous on some region that contains Σ.

[19.6.2] THEOREM

Let Σ be a surface given by $z = f(x, y)$, where f is continuous and has continuous first-order partial derivatives on a closed, bounded region R. Let $\phi = \phi(x, y, z)$ denote a function that is continuous on some region containing Σ. Then the surface integral of ϕ over Σ is given by

$$\iint_\Sigma \phi(x, y, z)\, d\sigma = \iint_R \phi(x, y, f(x, y))\sqrt{[f_x(x, y)]^2 + [f_y(x, y)]^2 + 1}\, dx\, dy \tag{1}$$

Thus, under certain conditions, a surface integral over Σ equals a double integral over the region R on which Σ is defined.

EXAMPLE 1

Evaluate $\iint_\Sigma x\, d\sigma$, where Σ is the surface given by $z = f(x, y) = x^2 + y$, whose domain is $0 \le x \le 1$, $0 \le y \le 4$.

Solution

For $f(x, y) = x^2 + y$, we find $f_x(x, y) = 2x$ and $f_y(x, y) = 1$, so that

$$\sqrt{f_x^2 + f_y^2 + 1} = \sqrt{4x^2 + 1 + 1} = \sqrt{4x^2 + 2}$$

Thus, we use (1), where R is the closed rectangle $0 \le x \le 1$, $0 \le y \le 4$.

$$\iint_\Sigma x\, d\sigma = \iint_R x\sqrt{4x^2 + 2}\, dx\, dy = \int_0^1 \left[\int_0^4 x\sqrt{4x^2 + 2}\, dy\right] dx$$
$$= 4\int_0^1 x\sqrt{4x^2 + 2}\, dx = 4\frac{2}{3}\frac{1}{8}(4x^2 + 2)^{3/2}\Big|_0^1 = \frac{1}{3}(6\sqrt{6} - 2\sqrt{2})$$

EXAMPLE 2

Evaluate $\iint_\Sigma (x^2 + y^2)\, d\sigma$, where Σ is the part of the surface of the paraboloid $z = f(x, y) = 1 - x^2 - y^2$ that lies above the xy-plane.

Solution

For $f(x, y) = 1 - x^2 - y^2$, we find that

$$\sqrt{f_x^2 + f_y^2 + 1} = \sqrt{4(x^2 + y^2) + 1}.$$

Using this fact, we get

$$\iint_\Sigma (x^2 + y^2)\, d\sigma = \iint_R (x^2 + y^2)\sqrt{4(x^2 + y^2) + 1}\, dx\, dy$$

where R is the region enclosed by the circle $x^2 + y^2 = 1$. See Figure 30. Because the integrand is a function of $x^2 + y^2$, we choose to use polar coordinates to evaluate the double integral. The result is

$$\iint_\Sigma (x^2 + y^2)\, d\sigma = \int_0^{2\pi} \left[\int_0^1 r^2\sqrt{4r^2 + 1}\, r\, dr \right] d\theta$$

$$\underset{\text{Set } u = \sqrt{4r^2+1}.}{=} \int_0^{2\pi} \left[\int_1^{\sqrt{5}} \frac{u^4 - u^2}{16}\, du \right] d\theta = \frac{1}{16} \int_0^{2\pi} \frac{50\sqrt{5} + 2}{15}\, d\theta \approx 0.948\pi$$

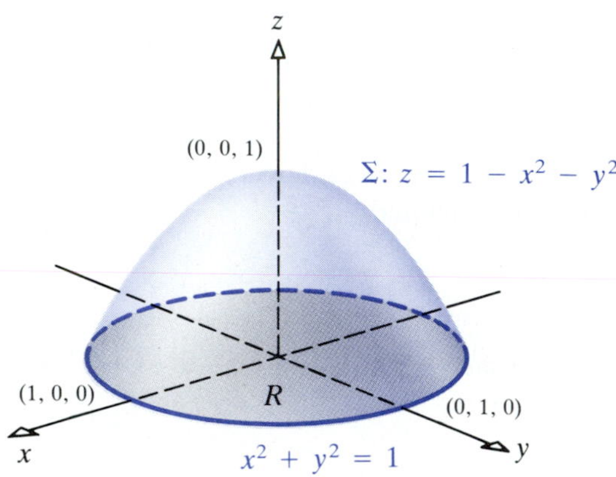

Figure 30

Applications

If $\phi(x, y, z) = 1$ on Σ, the surface integral of ϕ over Σ is

$$\iint_\Sigma d\sigma = \iint_R \sqrt{[f_x(x, y)]^2 + [f_y(x, y)]^2 + 1}\, dx\, dy$$

which is the formula for the surface area of Σ. Refer to theorem (18.5.2).

If the surface Σ is thought of as a lamina with variable mass density $\rho = \rho(x, y, z)$, the mass M of Σ is given by

$$M = \text{Mass of } \Sigma = \iint_\Sigma \rho(x, y, z)\, d\sigma$$

EXAMPLE 3

A lamina is in the shape of the cone $z = 6 - \sqrt{x^2 + y^2}$ that lies between the planes $z = 2$ and $z = 5$. If the mass density of the lamina is $\rho(x, y, z) = \sqrt{x^2 + y^2}$, find the mass M of the lamina.

Solution

For $z = f(x, y) = 6 - \sqrt{x^2 + y^2}$, we have

$$f_x(x, y) = \frac{-x}{\sqrt{x^2 + y^2}} \qquad f_y(x, y) = \frac{-y}{\sqrt{x^2 + y^2}}$$

$$\sqrt{[f_x(x, y)]^2 + f_y(x, y)]^2 + 1} = \sqrt{\frac{x^2}{x^2 + y^2} + \frac{y^2}{x^2 + y^2} + 1} = \sqrt{2}$$

Thus,

$$M = \text{Mass} = \iint_{\Sigma} \rho \, d\sigma = \iint_{R} \sqrt{x^2 + y^2}\,\sqrt{2}\,dx\,dy$$

where R is the region enclosed by the two circles $x^2 + y^2 = 16$ and $x^2 + y^2 = 1$. See Figure 31. We'll use polar coordinates to evaluate the double integral. Then R is the region enclosed by $r = 4$, $r = 1$, $0 \le \theta \le 2\pi$.

$$M = \sqrt{2} \int_0^{2\pi} \left[\int_1^4 r^2 \, dr \right] d\theta = \sqrt{2} \int_0^{2\pi} \frac{r^3}{3} \bigg|_1^4 d\theta$$

$$= (\sqrt{2})\left(\frac{63}{3}\right)(2\pi) = 42\sqrt{2}\pi$$

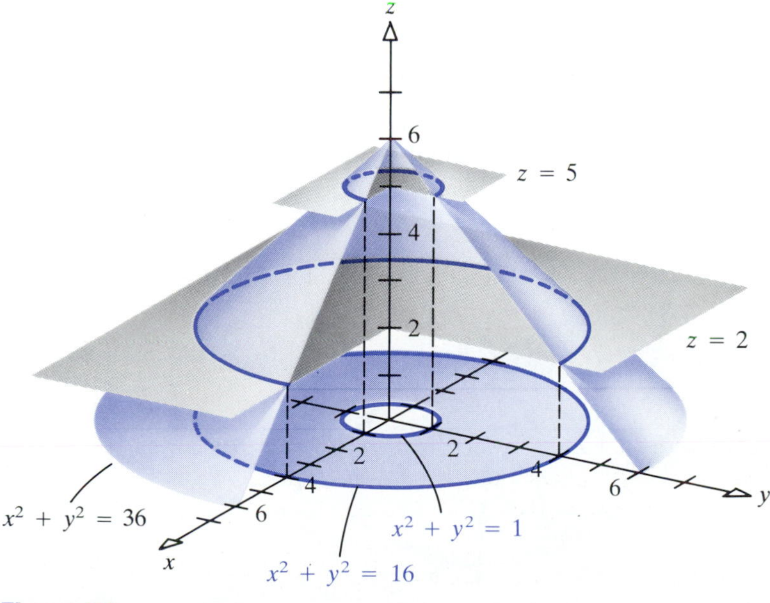

Figure 31

Other Types of Surface Integrals

[19.6.3] THEOREM

If a surface Σ is given by the equation $y = h(x, z)$, where h is continuous and has continuous first-order partial derivatives h_x and h_z on some closed, bounded region of the xz-plane, and if $\phi = \phi(x, y, z)$ is continuous on some region that contains Σ, then

$$\iint_{\Sigma} \phi(x, y, z)\, d\sigma = \iint_{R} \phi(x, h(x, z), z)\sqrt{[h_x(x, z)]^2 + [h_z(x, z)]^2 + 1}\, dx\, dz$$

We omit the proof.

[19.6.4] THEOREM

If a surface Σ is given by the equation $x = g(y, z)$, where g is continuous and has continuous first-order partial derivatives g_y and g_z on some closed, bounded region of the yz-plane, and if $\phi = \phi(x, y, z)$ is continuous on some region that contains Σ, then

$$\iint_{\Sigma} \phi(x, y, z)\, d\sigma = \iint_{R} \phi(g(y, z), y, z)\sqrt{[g_y(y, z)]^2 + [g_z(y, z)]^2 + 1}\, dy\, dz$$

We omit the proof.

Oriented Surfaces

Let Σ denote a surface that has a tangent plane at each point, except possibly at boundary points. At each point of Σ, there are two unit normal vectors, $\mathbf{n}$ and $-\mathbf{n}$ (see Fig. 32). Suppose it is possible to choose a unit normal vector $\mathbf{n}$ at each point of a surface so that $\mathbf{n}$ varies continuously over the surface and, as $\mathbf{n}$ moves around any closed curve on Σ, $\mathbf{n}$ returns to its original direction (see Fig. 33). Such a surface is said to be *oriented*, with the vector $\mathbf{n}$ providing its *orientation*. Thus, an orientable surface has two possible orientations. An example of a nonorientable surface is the Möbius strip (see Fig. 34).

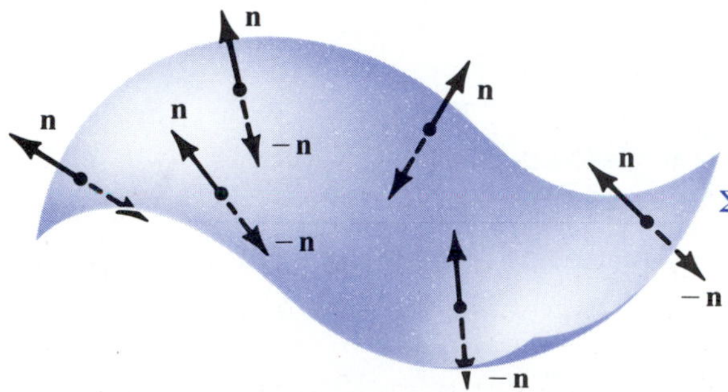

Figure 32

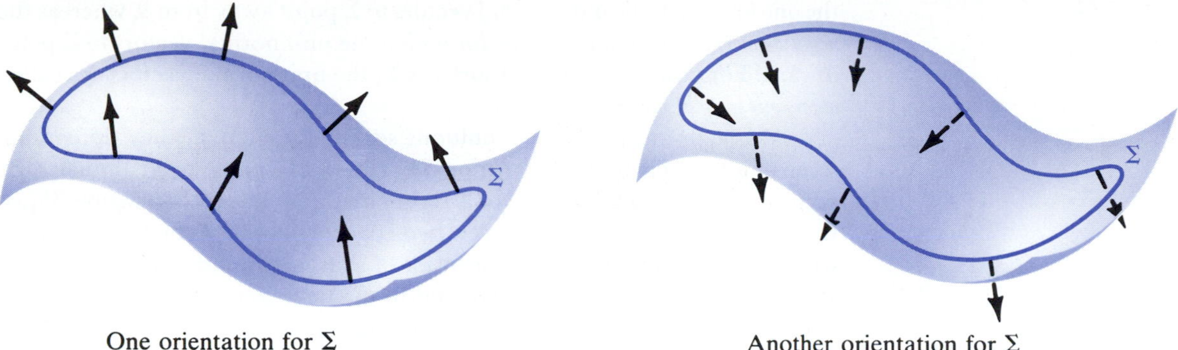

One orientation for Σ Another orientation for Σ

Figure 33

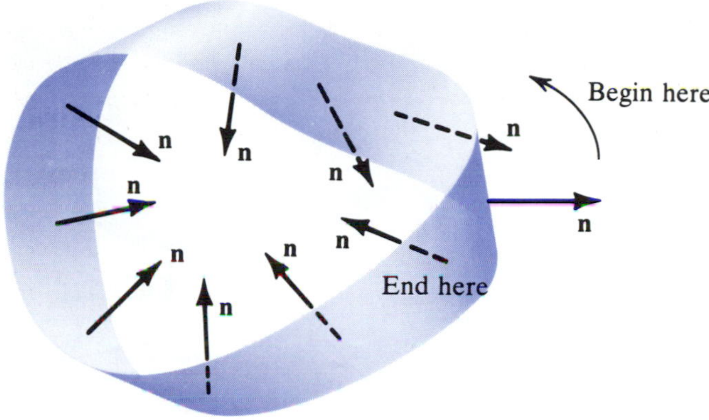

Figure 34
Möbius strip; a nonoriented surface

For an oriented surface Σ defined by $z = f(x, y)$, we can write the equation of the surface as

$$F(x, y, z) = z - f(x, y) = 0$$

The two unit vectors normal to Σ are

$$\mathbf{n} = \frac{\nabla F}{\|\nabla F\|} = \frac{-f_x(x, y)\mathbf{i} - f_y(x, y)\mathbf{j} + \mathbf{k}}{\sqrt{[f_x(x, y)]^2 + [f_y(x, y)]^2 + 1}} \tag{2}$$

and

$$-\mathbf{n} = \frac{-\nabla F}{\|\nabla F\|} = \frac{f_x(x, y)\mathbf{i} + f_y(x, v)\mathbf{j} - \mathbf{k}}{\sqrt{[f_x(x, y)]^2 + [f_y(x, y)]^2 + 1}} \tag{3}$$

Since the $\mathbf{k}$ component of $\mathbf{n}$ is positive, it will point in the upward direction, and so $\mathbf{n}$ is called the *upward unit normal vector* of Σ. The vector $-\mathbf{n}$ is called the *downward unit normal vector* of Σ.

For a closed surface Σ—that is, a surface that forms the boundary of a closed, bounded region S in space—we agree that the *positive orientation* of Σ is the one for which the unit normal vectors to Σ point away from S, whereas the *negative orientation* of Σ is the one for which the unit normal vectors to Σ point inward. For positively oriented surfaces Σ, the unit normal vectors are called *outer unit normal vectors*.

Although at first these definitions seem easy enough, they do contain some hidden difficulties. The region S enclosed by Σ may be a familiar solid such as a cube, a solid sphere, or a right circular cylinder. See Figure 35(a). But S might be the region between two concentric spheres or the region left when a sphere is removed from the inside of a large ellipsoid. See Figure 35(b). We shall limit our discussion here and in the next section to simple types of regions S and leave the general discussion to courses such as metric topology.

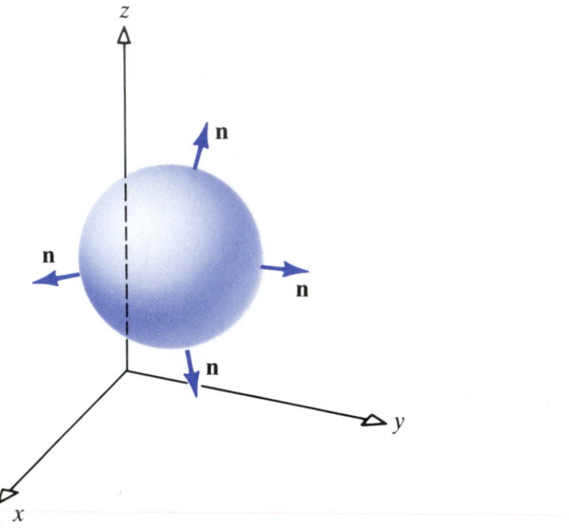

(a) Σ is a sphere
S is the interior of the sphere
Σ is positively oriented

(b) S is the region left when a sphere is removed from an ellipsoid
Σ is positively oriented

Figure 35

Regions Enclosed by *xy*-Simple Surfaces

Consider the positively oriented *xy*-simple surface Σ that forms the boundary of the closed, bounded region S depicted in Figure 36. There we see that Σ actually consists of three surfaces, Σ_1, Σ_2, and Σ_3. Note that the outer unit normals $\mathbf{n}_1$, $\mathbf{n}_2$, and $\mathbf{n}_3$ of the three surfaces each point away from the region S enclosed by $\Sigma = \Sigma_1 \cup \Sigma_2 \cup \Sigma_3$.

We assume the surfaces Σ_1 and Σ_2 are defined by the equations

$$\Sigma_2: \quad z = f_2(x, y) \qquad \Sigma_1: \quad z = f_1(x, y)$$

where $f_1(x, y) < f_2(x, y)$ on R, f_1 and f_2 are continuous on R, and f_1, f_2 possess continuous partial derivatives on R. The surface Σ_3 is the portion of a cylindrical surface between Σ_1 and Σ_2 formed by lines parallel to the z-axis and along the boundary of R. The region S is enclosed by Σ_2 on the top, Σ_1 on the bottom, and the lateral surface Σ_3 on the sides.

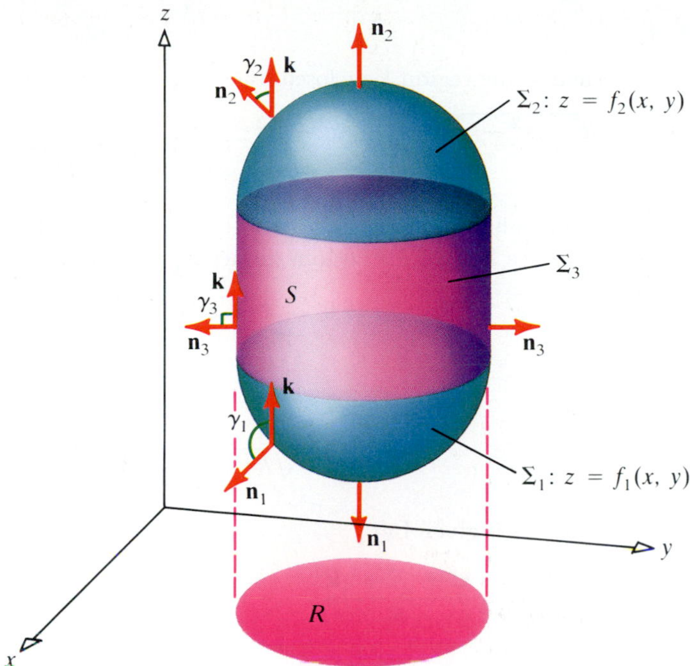

Figure 36

We seek expressions for the $\mathbf{k}$ component of the outer unit normals to Σ. For the top surface Σ_2: $z = f_2(x, y)$, the outer unit normal $\mathbf{n}_2$ is the same as the upward unit normal to Σ_2. Thus, from (2), we have

$$\mathbf{n}_2 = \frac{-f_{2_x}(x, y)\mathbf{i} - f_{2_y}(x, y)\mathbf{j} + \mathbf{k}}{\sqrt{[f_{2_x}(x, y)]^2 + [f_{2_y}(x, y)]^2 + 1}}$$

The $\mathbf{k}$ component of $\mathbf{n}_2$ is given by the direction cosine, $\cos \gamma_2$:

$$\cos \gamma_2 = \mathbf{n}_2 \cdot \mathbf{k} = \frac{1}{\sqrt{[f_{2_x}(x, y)]^2 + [f_{2_y}(x, y)]^2 + 1}} \tag{4}$$

For the bottom surface Σ_1: $z = f_1(x, y)$, the outer unit normal $\mathbf{n}_1$, equals the downward unit normal to Σ_1. Thus, from (3), we have

$$\mathbf{n}_1 = \frac{f_{1_x}(x, y)\mathbf{i} + f_{1_y}(x, y)\mathbf{j} - \mathbf{k}}{\sqrt{[f_{1_x}(x, y)]^2 + [f_{1_y}(x, y)]^2 + 1}}$$

The $\mathbf{k}$ component of $\mathbf{n}_1$ is given by the direction cosine, $\cos \gamma_1$:

$$\cos \gamma_1 = \mathbf{n}_1 \cdot \mathbf{k} = \frac{-1}{\sqrt{[f_{1_x}(x, y)]^2 + [f_{1_y}(x, y)]^2 + 1}} \tag{5}$$

For the lateral surface Σ_3, the outer unit normal $\mathbf{n}_3$ is orthogonal to $\mathbf{k}$ at each point on Σ_3. The $\mathbf{k}$ component of $\mathbf{n}_3$, $\cos \gamma_3$, is therefore 0.

EXAMPLE 4

Find the outer unit normal to the region S enclosed by

$$z = \sqrt{R^2 - x^2 - y^2} \qquad 0 \le x^2 + y^2 \le R^2$$

Solution

The region S is the interior of a hemisphere with center at (0, 0, 0) and radius R (see Fig. 37). The top surface $\boldsymbol{\Sigma}_2$ and the bottom surface $\boldsymbol{\Sigma}_1$ are defined by

$$\boldsymbol{\Sigma}_2\!: \ z = \sqrt{R^2 - x^2 - y^2} \qquad \boldsymbol{\Sigma}_1\!: \ z = 0 \qquad 0 \le x^2 + y^2 \le R^2$$

Since for $\boldsymbol{\Sigma}_2$

$$\frac{\partial z}{\partial x} = \frac{-x}{\sqrt{R^2 - x^2 - y^2}} = \frac{-x}{z} \qquad \text{and} \qquad \frac{\partial z}{\partial y} = \frac{-y}{\sqrt{R^2 - x^2 - y^2}} = \frac{-y}{z}$$

the outer unit normal $\mathbf{n}_2$ of $\boldsymbol{\Sigma}_2$ is given by

$$\mathbf{n}_2 = \frac{x\mathbf{i} + y\mathbf{i} + z\mathbf{k}}{R}$$

The outer unit normal $\mathbf{n}_1$ of $\boldsymbol{\Sigma}_1$ is $-\mathbf{k}$.

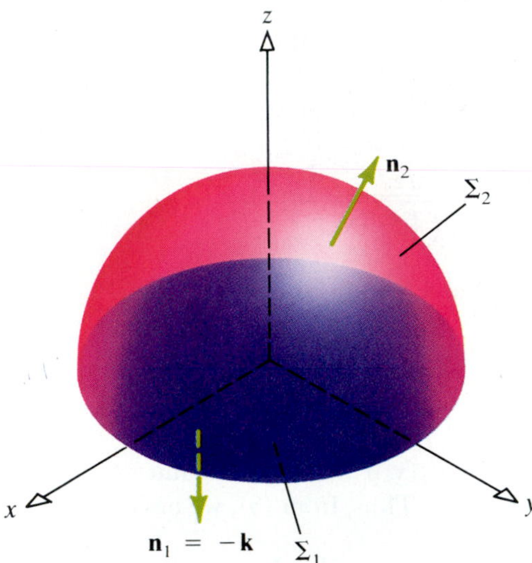

Figure 37

Surface Integrals of Vector Fields; Flux

Suppose $\mathbf{F} = \mathbf{F}(x, y, z)$ is a vector field that represents the velocity of a fluid, such as air or water. Let $\rho = \rho(x, y, z)$ denote the variable mass density of the fluid. Then $\rho\mathbf{F}$ equals the *rate of flow* (mass per unit time) per unit area. Let $\boldsymbol{\Sigma}$ be an oriented surface immersed in the fluid that does not impede its flow. If $\mathbf{n}$ is a unit normal vector of $\boldsymbol{\Sigma}$, then $\rho\mathbf{F} \cdot \mathbf{n}$ equals the rate of flow of fluid in the direction of $\mathbf{n}$. We seek a formula for the total mass of fluid that flows across Σ in unit time.

As Figure 38 illustrates, the mass of fluid crossing a patch of surface with area $\Delta\sigma_i$ per unit time in the direction of the unit normal $\mathbf{n}_i$ is approximately*

$$\rho\mathbf{F}\cdot\mathbf{n}_i\Delta\sigma_i$$

where $\mathbf{F}$ is evaluated at some point on the patch of surface. The total mass of fluid flowing across the surface Σ per unit time in the direction of $\mathbf{n}$ is therefore

$$\frac{\text{Total mass of fluid}}{\text{Unit time}} = \lim_{\|\Delta\|\to 0}\sum_{i=1}^{n}\rho\mathbf{F}\cdot\mathbf{n}_i\Delta\sigma_i = \iint_{\Sigma}\rho\mathbf{F}\cdot\mathbf{n}\,d\sigma \tag{6}$$

The surface integral $\iint_\Sigma \rho\mathbf{F}\cdot\mathbf{n}\,d\sigma$ in (6) is called the *flux of* $\mathbf{F}$ across Σ.

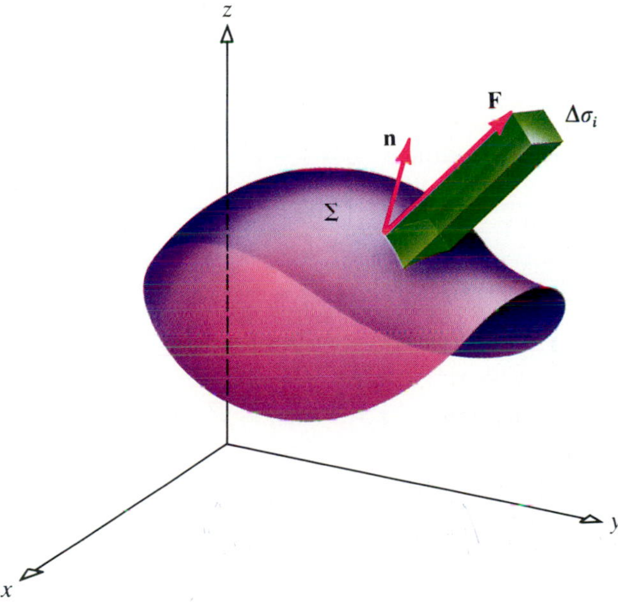

Figure 38

EXAMPLE 5

Find the total mass of fluid of constant mass density ρ flowing across the paraboloid $z = 4 - x^2 - y^2$, $z \geq 0$, per unit time in the direction of the upward unit normal, if the velocity of the fluid at any point on the paraboloid is $\mathbf{F} = \mathbf{F}(x, y, z) = x\mathbf{i} + y\mathbf{j} + 2z\mathbf{k}$ (see Fig. 39).

* The velocity $\mathbf{F}$ is decomposed into a component parallel to the patch and a component normal to it. The parallel component carries no fluid across the surface. Hence, the mass carried across per unit time is the normal component of $\rho\mathbf{F}$.

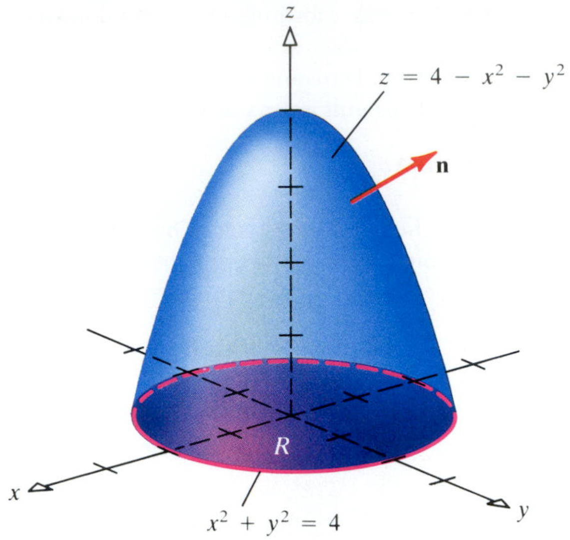

Figure 39

Solution

The upward unit normal $\mathbf{n}$ to the surface $z = 4 - x^2 - y^2$ is

$$\mathbf{n} = \frac{2x\mathbf{i} + 2y\mathbf{j} + \mathbf{k}}{\sqrt{4x^2 + 4y^2 + 1}}$$

Thus,

$$\mathbf{F} \cdot \mathbf{n} = \frac{2x^2 + 2y^2 + 2z}{\sqrt{4x^2 + 4y^2 + 1}}$$

The total mass of fluid flowing across $z = 4 - x^2 - y^2$ per unit time in the direction of $\mathbf{n}$ is therefore,

$$\iint_{\Sigma} \rho\mathbf{F} \cdot \mathbf{n}\, d\sigma = \rho \iint_R \frac{2x^2 + 2y^2 + 2z}{\sqrt{4x^2 + 4y^2 + 1}} \sqrt{4x^2 + 4y^2 + 1}\, dx\, dy$$

$$\underset{\substack{\uparrow \\ 2z = 8 - 2x^2 - 2y^2}}{=} 8\rho \iint_R dx\, dy \underset{\substack{\uparrow \\ R \text{ is a disk} \\ \text{of radius 2.}}}{=} 8\rho\pi(2)^2 = 32\,\rho\pi$$

EXAMPLE 6

Find the total mass of fluid of constant mass density ρ flowing across the cube enclosed by the planes $x = 0$, $x = 1$, $y = 0$, $y = 1$, $z = 0$, and $z = 1$ per unit time, in the direction of the outer unit normal if the velocity of the fluid at any point on the cube is $\mathbf{F} = \mathbf{F}(x, y, z) = 4xz\mathbf{i} - y^2\mathbf{j} + yz\mathbf{k}$. See Figure 40.

Solution

The total mass of fluid flowing across the cube per unit time in the direction

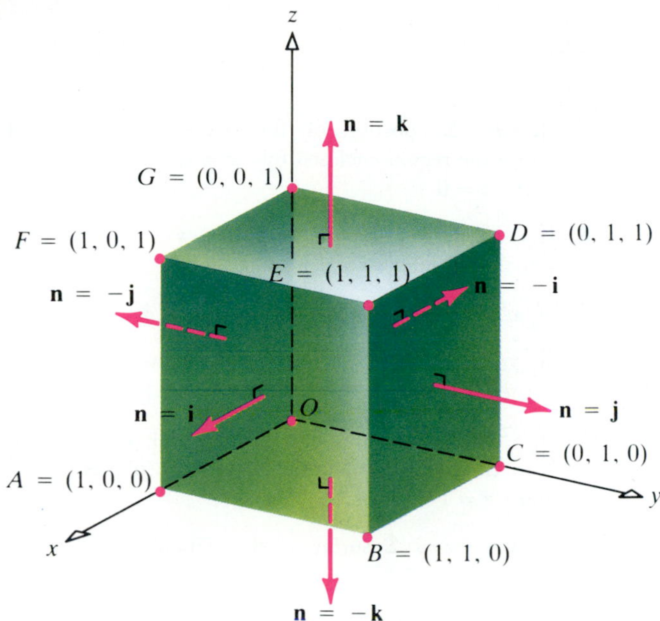

Figure 40

of the outer unit normal $\mathbf{n}$ is

$$\iint_{\Sigma} \rho \mathbf{F} \cdot \mathbf{n} \, d\sigma = \rho \iint_{\Sigma} \mathbf{F} \cdot \mathbf{n} \, d\sigma$$

We decompose $\boldsymbol{\Sigma}$ into its six faces and calculate the surface integral over each one in Table 1. From the table, we can see that the total mass of fluid is $0 + 0 + 0 + 2\rho - \rho + \frac{1}{2}\rho = \frac{3}{2}\rho$.

Table 1

	Face		$\mathbf{n}$	$\mathbf{F}$	$\mathbf{F} \cdot \mathbf{n}$	$\rho \iint \mathbf{F} \cdot \mathbf{n} \, d\sigma$
Σ_1	$ABCO$	$z = 0$	$-\mathbf{k}$	$-y^2\mathbf{j}$	0	0
Σ_2	$OAFG$	$y = 0$	$-\mathbf{j}$	$4xz\mathbf{i}$	0	0
Σ_3	$OCDG$	$x = 0$	$-\mathbf{i}$	$-y^2\mathbf{j} + yz\mathbf{k}$	0	0
Σ_4	$ABEF$	$x = 1$	$\mathbf{i}$	$4z\mathbf{i} - y^2\mathbf{j} + yz\mathbf{k}$	$4z$	$\rho \int_0^1 \int_0^1 4z \, dy \, dz = 2\rho$
Σ_5	$BCDE$	$y = 1$	$\mathbf{j}$	$4xz\mathbf{i} - \mathbf{j} + z\mathbf{k}$	-1	$-\rho \int_0^1 \int_0^1 dx \, dz = -\rho$
Σ_6	$DEFG$	$z = 1$	$\mathbf{k}$	$4x\mathbf{i} - y^2\mathbf{j} + y\mathbf{k}$	y	$\rho \int_0^1 \int_0^1 y \, dx \, dy = \frac{1}{2}\rho$

EXERCISE 19.6

In Problems 1–12 evaluate each surface integral.

1. $\iint_\Sigma (x^2 + y^2)\, d\sigma$; Σ: $z = f(x, y) = 2x + 3y + 5$, $0 \le x \le 1$, $0 \le y \le 1$

2. $\iint_\Sigma x^2y^2\, d\sigma$; Σ: $z = f(x, y) = 3x + 4y + 8$, $0 \le x \le 1$, $0 \le y \le 1$

3. $\iint_\Sigma y\, d\sigma$; Σ: $z = f(x, y) = x + 5y^2$, $0 \le x \le 2$, $0 \le y \le 1$

4. $\iint_\Sigma 4x\, d\sigma$; Σ: $z = f(x, y) = 2x^2 + y$, $0 \le x \le 1$, $0 \le y \le 2$

5. $\iint_\Sigma (x^2 + y^2 + z)\, d\sigma$; Σ: $z = x + y + 1$, $0 \le x \le 1$, $0 \le y \le 1$

6. $\iint_\Sigma (x + y + z)\, d\sigma$; Σ: portion of the plane $z = x - 3y$ above the region enclosed by $y = 0$, $x = 1$, and $x = 3y$ in the xy-plane

7. $\iint_\Sigma x\, d\sigma$; Σ: portion of the plane $x + y + 2z = 4$ above the region enclosed by $x = 1$, $y = 1$, $x = 0$, and $y = 0$

8. $\iint_\Sigma y\, d\sigma$; Σ: portion of the plane $x + y + z = 2$ inside the cylinder $x^2 + y^2 = 1$

9. $\iint_\Sigma x^2z\, d\sigma$; Σ: $x^2 + y^2 = 1$, $0 \le z \le 1$

10. $\iint_\Sigma yz\, d\sigma$; Σ: first-octant part of the plane $x + 2y + 3z = 6$

11. $\iint_\Sigma x^2\, d\sigma$; Σ: piece of the cylinder $x^2 + y^2 = 1$ that is in the first octant and between the planes $z = 0$ and $z = 1$

12. $\iint_\Sigma (x + y)z\, d\sigma$; Σ: surface of the unit cube

In Problems 13–20 find the total mass of fluid of constant mass density ρ flowing out of the cube in Figure 40 per unit time in the direction of the outer unit normal, if the velocity of the fluid at any point on the cube is given by $\mathbf{F}$.

13. $\mathbf{F} = x\mathbf{i}$

14. $\mathbf{F} = y\mathbf{i}$

15. $\mathbf{F} = z\mathbf{i}$

16. $\mathbf{F} = x\mathbf{i} + y\mathbf{j}$

17. $\mathbf{F} = x\mathbf{i} + y\mathbf{j} + z\mathbf{k}$

18. $\mathbf{F} = z^2\mathbf{i}$

19. $\mathbf{F} = x^2\mathbf{i} + y^2\mathbf{j} + z^2\mathbf{k}$

20. $\mathbf{F} = x^2\mathbf{i} + y^2\mathbf{j}$

In Problems 21–24 evaluate $\iint_\Sigma \mathbf{F} \cdot \mathbf{n}\, d\sigma$, where $\mathbf{n}$ is the upward unit normal of Σ.

21. $\mathbf{F} = x\mathbf{i} + y\mathbf{j} + z\mathbf{k}$; Σ: the upper half of the hemispherical region $x^2 + y^2 + z^2 = 1$, $z \ge 0$

22. $\mathbf{F} = -y\mathbf{i} + x\mathbf{j} + z\mathbf{k}$; Σ: same as in Problem 21

23. $\mathbf{F} = (x + y)\mathbf{i} + (2x - z)\mathbf{j} + y\mathbf{k}$; Σ: the tetrahedron formed by the coordinate planes and the plane $z + 2x + 2y = 8$

24. $\mathbf{F} = 2x\mathbf{i} - x^2\mathbf{j} + (z - 2x + 2y)\mathbf{k}$; Σ: the tetrahedron formed by the coordinate planes and the plane $2x + 2y + z = 6$

25. Find the mass of a lamina in the shape of a cone $z = \sqrt{x^2 + y^2}$, $2 \le z \le 4$, if the mass density of the lamina is $\rho = \rho(x, y, z) = 8 - z$.

26. Find the mass of a lamina in the shape of a cone $z = \sqrt{x^2 + y^2}$, $2 \le z \le 4$, if the mass density of the lamina is $\rho = \rho(x, y, z) = \sqrt{x^2 + y^2}$.

27. Find the total mass of fluid of constant mass density flowing across the paraboloid $z = x^2 + y^2$, $z \le 5$, per unit time in the direction of the upward unit normal, if the velocity of the fluid at any point on the paraboloid is $\mathbf{F} = \mathbf{F}(x, y, z) = -x\mathbf{i} - y\mathbf{j} - z\mathbf{k}$.

28. Find the total mass of fluid of constant mass density flowing across the paraboloid $z = 9 - x^2 - y^2$, $z \ge 0$, in the direction of the upward unit normal, if the velocity of the fluid at any point on the paraboloid is $\mathbf{F} = \mathbf{F}(x, y, z) = x\mathbf{i} + y\mathbf{j} + 3\mathbf{k}$.

29. If $\mathbf{n}$ is the upward unit normal to a surface Σ: $z = f(x, y)$ and $\mathbf{F} = P\mathbf{i} + Q\mathbf{j} + R\mathbf{k}$, show that

$$\iint_\Sigma \mathbf{F} \cdot \mathbf{n}\, d\sigma = \iint_D \left(-P\frac{\partial f}{\partial x} - Q\frac{\partial f}{\partial y} + R\right) dx\, dy$$

30. Explain why the area of a smooth surface Σ is $A = \iint_\Sigma d\sigma$. When the surface is the graph of $z = f(x, y)$ over the region R in the xy-plane, show that this formula reduces to

$$A = \iint_R \sqrt{\left(\frac{\partial z}{\partial x}\right)^2 + \left(\frac{\partial z}{\partial y}\right)^2 + 1}\, dx\, dy$$

as given in Chapter 18.

31. The sphere $x^2 + y^2 + z^2 = a^2$ is covered by a thin material with mass density at each point proportional to the distance from the xy-plane. Find the mass of the material.

The Divergence Theorem 19.7

Recall that Green's theorem expresses a relationship between a certain double integral extended over a plane region and a line integral taken around its boundary. There are two ways to generalize this result to space. One of these, known as the *divergence theorem* (or *Gauss' theorem*), is the subject of this section, and the other, known as *Stokes' theorem*, is the subject of the next section.

[19.7.1] DEFINITION / ***Divergence of* F.**

Let $\mathbf{F} = \mathbf{F}(x, y, z) = P(x, y, z)\mathbf{i} + Q(x, y, z)\mathbf{j} + R(x, y, z)\mathbf{k}$ denote a vector field, where $\partial P/\partial x, \partial Q/\partial y, \partial R/\partial z$ each exists. The *divergence of* F, denoted by div F, is defined as the function

$$\text{div}\,\mathbf{F} = \nabla \cdot \mathbf{F} = \frac{\partial P}{\partial x} + \frac{\partial Q}{\partial y} + \frac{\partial R}{\partial z}$$

EXAMPLE 1

The divergence of

$$\mathbf{F}(x, y, z) = x^2yz^2\mathbf{i} + (2xz + y^3)\mathbf{j} + x^2y^3z\mathbf{k}$$

is

$$\text{div}\,\mathbf{F} = \frac{\partial}{\partial x}(x^2yz^2) + \frac{\partial}{\partial y}(2xz + y^3) + \frac{\partial}{\partial z}(x^2y^3z) = 2xyz^2 + 3y^2 + x^2y^3$$

[19.7.2] THEOREM / ***Divergence Theorem.***

Let Σ be a positively oriented surface with outer unit normal n that encloses a closed, bounded region S of space. Let $\mathbf{F} = \mathbf{F}(x, y, z) = P(x, y, z)\mathbf{i} + Q(x, y, z)\mathbf{j} + R(x, y, z)\mathbf{k}$ denote a vector field for which P, Q, and R have continuous first-order partial derivatives on an open set containing S. Then

$$\iiint_S \text{div}\,\mathbf{F}\,dV = \iint_\Sigma \mathbf{F} \cdot \mathbf{n}\,d\sigma$$

If $\mathbf{n} = \cos\alpha\,\mathbf{i} + \cos\beta\,\mathbf{j} + \cos\gamma\,\mathbf{k}$, then

$$\iiint_S \left(\frac{\partial P}{\partial x} + \frac{\partial Q}{\partial y} + \frac{\partial R}{\partial z}\right) dV = \iint_\Sigma (P\cos\alpha + Q\cos\beta + R\cos\gamma)\,d\sigma$$

It is beyond the scope of this book to prove theorem (19.7.2) for the general conditions on the region S. However, we can give the idea behind the proof. The basis of the proof is to verify each of the three equations

$$\iiint_S \frac{\partial P}{\partial x}\,dV = \iint_\Sigma P\cos\alpha\,d\sigma \qquad \textbf{(1)}$$

$$\iiint_S \frac{\partial Q}{\partial y}\, dV = \iint_\Sigma Q \cos \beta \, d\sigma \tag{2}$$

$$\iiint_S \frac{\partial R}{\partial z}\, dV = \iint_\Sigma R \cos \gamma \, d\sigma \tag{3}$$

Once these are verified, theorem (19.7.2) follows when we add these three equations. We shall give a proof for (3) for a special version of the region S.

Suppose the surface Σ that encloses S is a positively oriented, xy-simple surface (see Fig. 41). Then Σ actually consists of three surfaces, Σ_1, Σ_2, and Σ_3, with the outer unit normals $\mathbf{n}_1$, $\mathbf{n}_2$, and $\mathbf{n}_3$, respectively.

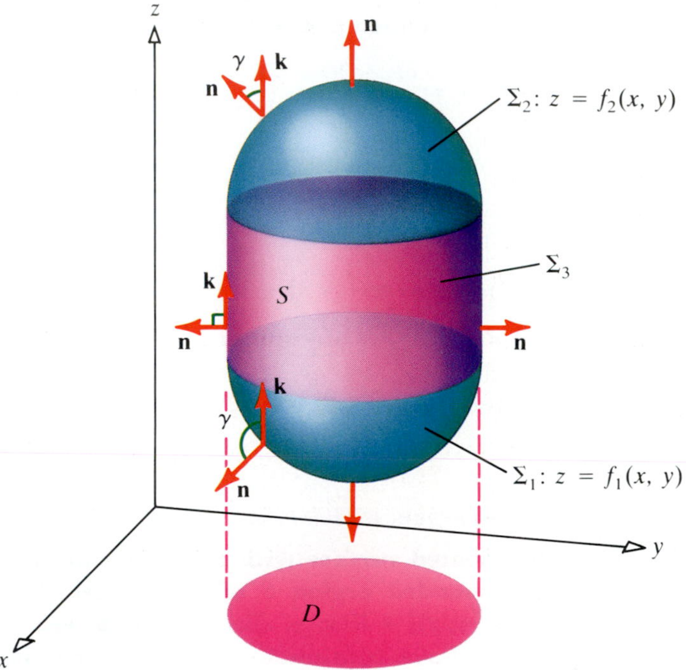

Figure 41

We assume that surfaces Σ_1 and Σ_2 are defined by the equations

$$\Sigma_2: \quad z = f_2(x, y) \qquad \Sigma_1: \quad z = f_1(x, y)$$

where $f_1(x, y) < f_2(x, y)$ on D, f_1 and f_2 are continuous on D, and f_1, f_2 have continuous partial derivatives on D. The surface Σ_3 is the portion of a cylindrical surface between Σ_1 and Σ_2 formed by lines parallel to the z-axis and along the boundary of D. The region S is enclosed by Σ_2 on the top, Σ_1 on the bottom, and the lateral surface Σ_3 on the sides.

We shall now show that

$$\iiint_S \frac{\partial R}{\partial z}\, dV = \iint_\Sigma R \cos \gamma \, d\sigma$$

We start with the right side, using the fact that $\Sigma = \Sigma_1 \cup \Sigma_2 \cup \Sigma_3$:

$$\iint_{\Sigma} R\cos\gamma\, d\sigma = \iint_{\Sigma_1} R\cos\gamma\, d\sigma + \iint_{\Sigma_2} R\cos\gamma\, d\sigma + \iint_{\Sigma_3} R\cos\gamma\, d\sigma \tag{4}$$

On the lateral surface Σ_3, $\gamma = \pi/2$ so that $\cos\gamma = 0$. Therefore,

$$\iint_{\Sigma_3} R\cos\gamma\, d\sigma = 0 \tag{5}$$

On the top surface Σ_2: $z = f_2(x, y)$, $\cos\gamma$ is given by (4) on page 1163. Thus,

$$\iint_{\Sigma_2} R\cos\gamma\, d\sigma = \iint_{\Sigma_2} R(x, y, f_2(x, y)) \frac{1}{\sqrt{[f_{2_x}(x, y)]^2 + [f_{2_y}(x, y)]^2 + 1}}\, d\sigma$$

$$\underset{\uparrow}{=} \iint_{D} R(x, y, f_2(x, y)) \frac{\sqrt{[f_{2_x}(x, y)]^2 + [f_{2_y}(x, y)]^2 + 1}\, dx\, dy}{\sqrt{[f_{2_x}(x, y)]^2 + [f_{2_y}(x, y)]^2 + 1}}$$

By theorem (19.6.2)

$$= \iint_{D} R(x, y, f_2(x, y))\, dx\, dy \tag{6}$$

On the bottom surface Σ_1: $z = f_1(x, y)$, $\cos\gamma$ is given by (5) on page 1163. Thus,

$$\iint_{\Sigma_1} R\cos\gamma\, d\sigma = \iint_{\Sigma_1} R(x, y, f_1(x, y)) \frac{-1}{\sqrt{[f_{1_x}(x, y)]^2 + [f_{1_y}(x, y)]^2 + 1}}\, d\sigma$$

$$\underset{\uparrow}{=} -\iint_{D} R(x, y, f_1(x, y)) \cdot \frac{\sqrt{[f_{1_x}(x, y)]^2 + [f_{1_y}(x, y)]^2 + 1}\, dx\, dy}{\sqrt{[f_{1_x}(x, y)]^2 + [f_{1_y}(x, y)]^2 + 1}}$$

By theorem (19.6.2)

$$= -\iint_{D} R(x, y, f_1(x, y))\, dx\, dy \tag{7}$$

Thus, based on (4), (5), (6), and (7), we have

$$\iint_{\Sigma} R\cos\gamma\, d\sigma = \iint_{D} R(x, y, f_2(x, y))\, dx\, dy - \iint_{D} R(x, y, f_1(x, y))\, dx\, dy$$

$$= \iint_{D} [R(x, y, f_2(x, y)) - R(x, y, f_1(x, y))]\, dx\, dy$$

$$= \iint_{D} \left[\int_{f_1(x, y)}^{f_2(x, y)} \frac{\partial R}{\partial z}\, dz\right] dx\, dy = \iiint_{S} \frac{\partial R}{\partial z}\, dV$$

which proves (3).

EXAMPLE 2

Redo Example 6 of Section 19.6 using the divergence theorem.

Solution

Suppose S is the solid cube, Σ is its surface, and $\mathbf{F} = 4xz\mathbf{i} - y^2\mathbf{j} + yz\mathbf{k}$. Since $\operatorname{div} \mathbf{F} = 4z - 2y + y = 4z - y$, an application of the divergence theorem yields

$$\iint_{\Sigma} \rho\mathbf{F} \cdot \mathbf{n}\, d\sigma = \rho \iint_{\Sigma} \mathbf{F} \cdot \mathbf{n}\, d\sigma = \rho \iiint_{S} \operatorname{div} \mathbf{F}\, dV$$

$$= \rho \int_0^1 \left\{ \int_0^1 \left[\int_0^1 (4z - y)\, dz \right] dy \right\} dx$$

$$= \rho \int_0^1 \left[\int_0^1 (2 - y)\, dy \right] dx = \rho \int_0^1 \frac{3}{2}\, dx = \tfrac{3}{2}\rho$$

EXAMPLE 3

Let Σ be the surface of a cylindrical solid S whose boundary is $x^2 + y^2 = 4$, $z = 0$, and $z = 1$. Let $\mathbf{F} = x^3\mathbf{i} + y^3\mathbf{j} + z^2\mathbf{k}$ and let $\mathbf{n}$ be the outer unit normal to Σ (see Fig. 42). Use the divergence theorem to evaluate $\iint_\Sigma \mathbf{F} \cdot \mathbf{n}\, d\sigma$.

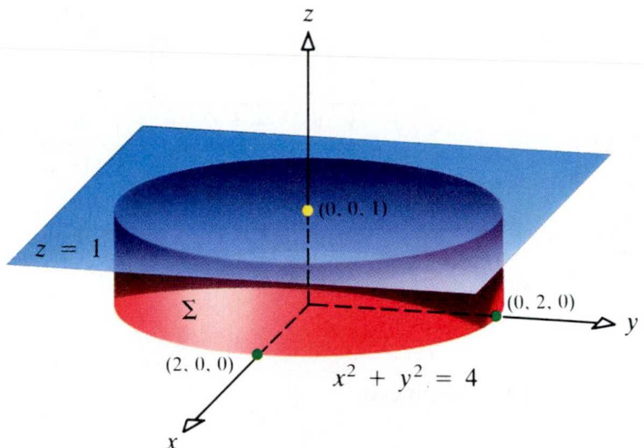

Figure 42

Solution

Since $\operatorname{div} \mathbf{F} = 3x^2 + 3y^2 + 2z$, an application of the divergence theorem yields

$$\iint_{\Sigma} \mathbf{F} \cdot \mathbf{n}\, d\sigma = \iiint_{S} (3x^2 + 3y^2 + 2z)\, dV$$

By using cylindrical coordinates to evaluate the triple integral, we obtain

$$\iint_{\Sigma} \mathbf{F} \cdot \mathbf{n}\, d\sigma = \int_0^{2\pi} \left\{ \int_0^2 \left[\int_0^1 (3r^2 + 2z) r\, dz \right] dr \right\} d\theta$$

$$= \int_0^{2\pi} \left[\int_0^2 (3r^3 z + z^2 r) \Big|_{z=0}^{z=1} dr \right] d\theta$$

$$= \int_0^{2\pi} \left[\int_0^2 (3r^3 + r)\, dr \right] d\theta$$

$$= \int_0^{2\pi} \left(\frac{3r^4}{4} + \frac{r^2}{2} \right) \Big|_{r=0}^{r=2} d\theta = \int_0^{2\pi} 14\, d\theta = 28\pi$$

EXAMPLE 4

Let $\mathbf{F}(x, y, z) = x^3\mathbf{i} + y^3\mathbf{j} + z^3\mathbf{k}$. Let $\mathbf{\Sigma}$ be the surface of the hemispherical region S, enclosed by $z = \sqrt{4 - x^2 - y^2}$ and $z = 0$. Let $\mathbf{n}$ be the outer unit normal to $\mathbf{\Sigma}$ (see Fig. 43). Use the divergence theorem to evaluate $\iint_{\Sigma} \mathbf{F} \cdot \mathbf{n}\, d\sigma$.

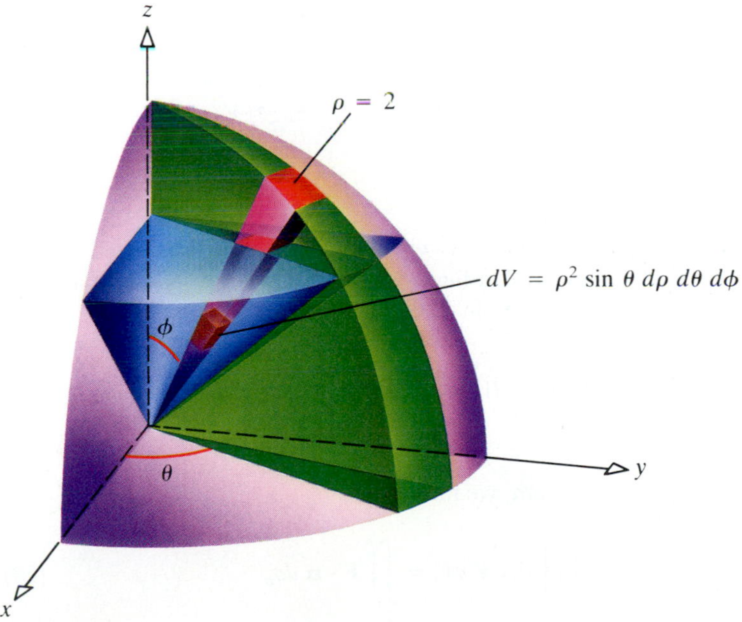

Figure 43
Note that only the first octant is shown.

Solution

Since $\operatorname{div} \mathbf{F} = 3x^2 + 3y^2 + 3z^2 = 3(x^2 + y^2 + z^2)$, an application of the divergence theorem yields

$$\iint_{\Sigma} \mathbf{F} \cdot \mathbf{n} \, d\sigma = 3 \iiint_{S} (x^2 + y^2 + z^2) \, dV$$

We use spherical coordinates to evaluate the triple integral, obtaining

$$\iint_{\Sigma} \mathbf{F} \cdot \mathbf{n} \, d\sigma = 3 \int_{\phi=0}^{\pi/2} \left\{ \int_{\theta=0}^{2\pi} \left[\int_{\rho=0}^{2} (\rho^2)\rho^2 \sin \phi \, d\rho \right] d\theta \right\} d\phi = \frac{6\pi(2^5)}{5} = \frac{192\pi}{5}$$

Interpretation of div F

Recall that the law of the mean for a single integral states that if a function f of one variable is continuous on a closed interval $[a, b]$, then there is a number c in (a, b) such that

$$\int_a^b f(x) \, dx = f(c)L$$

where $L = b - a$ is the length of $[a, b]$.

In an analogous way, there is a law of the mean for triple integrals that asserts that if f is a continuous function of three variables defined on a simply connected, closed, bounded region S, then there is a point (x^*, y^*, z^*) in S such that

$$\iiint_{S} f(x, y, z) \, dV = f(x^*, y^*, z^*)V \tag{8}$$

where V is the volume of S.

Let $\mathbf{F} = \mathbf{F}(x, y, z)$ denote a continuous vector field defined on and within a spherical region S_a of radius a with center at (x_1, y_1, z_1). Based on (8), there is a point (x^*, y^*, z^*) in S_a for which

$$\iiint_{S_a} \text{div } \mathbf{F} \, dV_a = \text{div } \mathbf{F}(x^*, y^*, z^*) V_a \tag{9}$$

Using the divergence theorem, we find that

$$\iiint_{S_a} \text{div } \mathbf{F} \, dV_a = \iint_{\Sigma_a} \mathbf{F} \cdot \mathbf{n} \, d\sigma_a \tag{10}$$

where Σ_a is the sphere of radius a with center at (x_1, y_1, z_1). Combining (9) and (10), we have

$$\text{div } \mathbf{F}(x^*, y^*, z^*) = \frac{\iint_{\Sigma_a} \mathbf{F} \cdot \mathbf{n} \, d\sigma_a}{V_a} \tag{11}$$

The ratio on the right side of (11) is the flux of **F** per unit volume across Σ_a. If we let the radius $a \to 0$, then $(x^*, y^*, z^*) \to (x_1, y_1, z_1)$, so that

$$\text{div } \mathbf{F}(x_1, y_1, z_1) = \lim_{a \to 0} \left(\frac{\iint_{\Sigma_a} \mathbf{F} \cdot \mathbf{n} \, d\sigma_a}{V_a} \right) \tag{12}$$

In other words,

div $\mathbf{F}(x_1, y_1, z_1)$
= Limiting value of the flux of F per unit volume across Σ_a

If **F** is the velocity of a steady fluid flow and if $\text{div } \mathbf{F}(x_1, y_1, z_1) > 0$, then the net flow is out of V_a and the point (x_1, y_1, z_1) is called a *source*. If $\text{div } \mathbf{F}(x_1, y_1, z_1) < 0$, then the net flow is into V_a and the point (x_1, y_1, z_1) is called a *sink*. In terms of this vocabulary, the divergence theorem states that the net flow out of a volume V equals the sum of the sources and sinks in V.

If there are no sources or sinks in a region, then $\text{div } \mathbf{F} = 0$, and we call **F** a *solenoidal* vector field. For such fields, the equation

$$\text{div } \mathbf{F} = \frac{\partial P}{\partial x} + \frac{\partial Q}{\partial y} + \frac{\partial R}{\partial z} = 0$$

is referred to as the *equation of continuity*.

The result given in (12) can be taken as a starting point for defining the divergence of **F**, and all the properties may be derived from it, including the divergence theorem. In fact, this was the original formulation of the theorem as given by Gauss. Using (12) as a definition has the advantage of being coordinate-free.

We close this section with an example that is important in the study of electrostatic fields, *Coulomb's law of electrostatic attraction* (*gravitation*).

EXAMPLE 5

Let $\mathbf{F} = \mathbf{F}(x, y, z)$ denote an electric force field

$$\mathbf{F} = \mathbf{F}(x, y, z) = \frac{\varepsilon q Q}{x^2 + y^2 + z^2} \mathbf{u}$$

where ε is a constant that depends on the units being used; q and Q are the charges of two particles, one located at $(0, 0, 0)$ and the other at (x, y, z); and $\mathbf{u}$ is the unit vector from $(0, 0, 0)$ to (x, y, z). Let Σ denote a positively oriented surface that encloses a region S of space. Show that the following statements are true under the assumptions of the divergence theorem.

(a) If neither Σ nor its interior contains the point $(0, 0, 0)$, then $\iint_\Sigma \mathbf{F} \cdot \mathbf{n} \, d\sigma = 0$, that is, flux $\mathbf{F}$ over $\Sigma = 0$.
(b) If the interior of Σ contains the point $(0, 0, 0)$, then $\iint_\Sigma \mathbf{F} \cdot \mathbf{n} \, d\sigma = 4\pi\varepsilon q Q$, that is, flux $\mathbf{F}$ over $\Sigma = 4\pi\varepsilon q Q$.

Solution

First we note that $\mathbf{u} = \dfrac{x\mathbf{i} + y\mathbf{j} + z\mathbf{k}}{\sqrt{x^2 + y^2 + z^2}}$, so that

$$\mathbf{F} = \mathbf{F}(x, y, z) = \varepsilon q Q \frac{x\mathbf{i} + y\mathbf{j} + z\mathbf{k}}{(x^2 + y^2 + z^2)^{3/2}}$$

The electric field $\mathbf{F}$ is continuous everywhere except at $(0, 0, 0)$.

(a) Here, Σ is a surface that forms the boundary of a region S satisfying the assumptions of the divergence theorem. Furthermore, since

$$\frac{\partial}{\partial x}\left[\frac{x}{(x^2+y^2+z^2)^{3/2}}\right] + \frac{\partial}{\partial y}\left[\frac{y}{(x^2+y^2+z^2)^{3/2}}\right] + \frac{\partial}{\partial z}\left[\frac{z}{(x^2+y^2+z^2)^{3/2}}\right]$$

$$= \frac{(x^2+y^2+z^2)^{3/2} - x(\frac{3}{2})(2x)\sqrt{x^2+y^2+z^2}}{(x^2+y^2+z^2)^3}$$

$$+ \frac{(x^2+y^2+z^2)^{3/2} - y(\frac{3}{2})(2y)\sqrt{x^2+y^2+z^2}}{(x^2+y^2+z^2)^3}$$

$$+ \frac{(x^2+y^2+z^2)^{3/2} - z(\frac{3}{2})(2z)\sqrt{x^2+y^2+z^2}}{(x^2+y^2+z^2)^3}$$

$$= \frac{3(x^2+y^2+z^2)^{3/2} - 3\sqrt{x^2+y^2+z^2}\,(x^2+y^2+z^2)}{(x^2+y^2+z^2)^3} = 0$$

it follows that $\operatorname{div} \mathbf{F} = 0$. Therefore, by the divergence theorem,

$$\iint_{\Sigma} \mathbf{F} \cdot \mathbf{n}\, d\sigma = \iiint_{S} \operatorname{div} \mathbf{F}\, dV = 0$$

(b) We note that since $\mathbf{F}$ is not continuous at the origin, the divergence theorem cannot be applied. However, the following argument will help us prove part (b). Let S be the closed region enclosed by two separate surfaces: the surface Σ and a sphere Σ_1 of radius a, with center at $(0, 0, 0)$, as shown in Figure 44. The outer surface is Σ and the inner surface is Σ_1. Now $\mathbf{F}$ is continuous throughout S. Therefore, we can apply the divergence theorem to get

$$\iiint_{S} \operatorname{div} \mathbf{F}\, dV = \iint_{\Sigma} \mathbf{F} \cdot \mathbf{n}\, d\sigma + \iint_{\Sigma_1} \mathbf{F} \cdot \mathbf{n}\, d\sigma$$

From part (a), we know that $\iiint_S \operatorname{div} \mathbf{F}\, dV = 0$. Thus,

$$\iint_{\Sigma} \mathbf{F} \cdot \mathbf{n}\, d\sigma = -\iint_{\Sigma_1} \mathbf{F} \cdot \mathbf{n}\, d\sigma$$

On the inner surface Σ_1, a sphere of radius a, the outer unit normal is

$$\mathbf{n} = -\frac{x\mathbf{i} + y\mathbf{j} + z\mathbf{k}}{a}$$

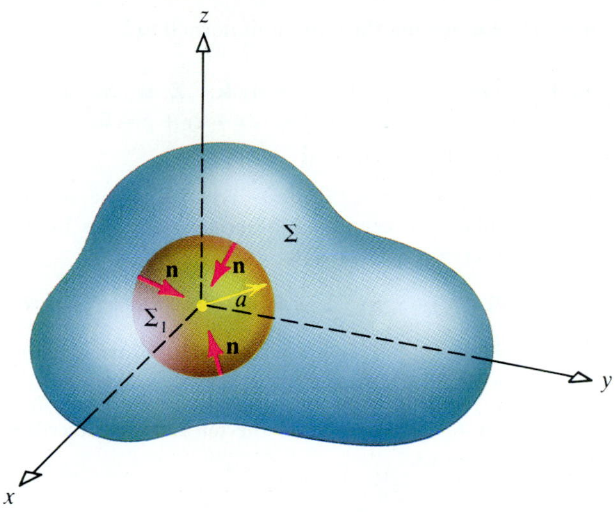

Figure 44

Hence,

$$\mathbf{F}\cdot\mathbf{n} = -\varepsilon q Q \frac{x^2+y^2+z^2}{a^4} = -\frac{\varepsilon q Q}{a^2}$$

Thus,

$$\iint_{\Sigma} \mathbf{F}\cdot\mathbf{n}\,d\sigma = -\iint_{\Sigma_1}\left(-\frac{\varepsilon q Q}{a^2}\right) d\sigma = \frac{\varepsilon q Q}{a^2}\iint_{\Sigma_1} d\sigma \underset{\uparrow}{=} \frac{\varepsilon q Q}{a^2}(4\pi a^2) = 4\pi\varepsilon q Q$$

Surface area of the sphere Σ_1 is $4\pi a^2$.

Summary

Green's theorem: $\iint_R [(\partial Q/\partial x) - (\partial P/\partial y)]\,dx\,dy = \oint_C (P\,dx + Q\,dy)$ relates a certain double integral over a region R enclosed by the curve C to a certain line integral over C, the (one-dimensional) boundary of R.

The divergence theorem: $\iiint_S \operatorname{div}\mathbf{F}\,dV = \iint_\Sigma \mathbf{F}\cdot n\,d\sigma$ relates a certain triple integral over a region S enclosed by Σ to a certain surface integral over Σ, the (two-dimensional) boundary of S.

EXERCISE 19.7

In Problems 1–4 find the divergence of **F**.

1. $\mathbf{F}(x, y, z) = x^2\mathbf{i} + y^2\mathbf{j} + z^2\mathbf{k}$

2. $\mathbf{F}(x, y, z) = x\mathbf{i} + xy\mathbf{j} + xyz\mathbf{k}$

3. $\mathbf{F}(x, y, z) = (x + \cos x)\mathbf{i} + (y + y\sin x)\mathbf{j} + 2z\mathbf{k}$

4. $\mathbf{F}(x, y, z) = xye^z\mathbf{i} + x^2e^z\mathbf{j} + x^2ye^z\mathbf{k}$

In Problems 5–14 use the divergence theorem to evaluate $\iint_\Sigma \mathbf{F}\cdot\mathbf{n}\,d\sigma$. Here $\mathbf{n}$ denotes the outer unit normal to Σ.

5. $\mathbf{F} = (2xy + 2z)\mathbf{i} + (y^2 + 1)\mathbf{j} - (x + y)\mathbf{k}$; Σ is the surface of the region enclosed by $x + y + z = 4$, $x = 0$, $y = 0$, and $z = 0$

6. $\mathbf{F} = (2xy + z)\mathbf{i} + y^2\mathbf{j} - (x + 4y)\mathbf{k}$; Σ is the surface of the region enclosed by $2x + 2y + z = 6$, $x = 0$, $y = 0$, and $z = 0$

7. $\mathbf{F} = x^2\mathbf{i} + y^2\mathbf{j} + z^2\mathbf{k}$; Σ is the surface of the region enclosed by $x = 0$, $x = 1$, $y = 0$, $y = 1$, $z = 0$, and $z = 1$

8. $\mathbf{F} = (x - y)\mathbf{i} + (y - z)\mathbf{j} + (x - y)\mathbf{k}$; Σ is the surface of a cube with center at the origin and faces in the planes $x = \pm 1$, $y = \pm 1$, and $z = \pm 1$

9. $\mathbf{F} = x\mathbf{i} + y\mathbf{j} + z\mathbf{k}$; Σ is the sphere $x^2 + y^2 + z^2 = 1$

10. $\mathbf{F} = 2x\mathbf{i} + 2y\mathbf{j} + 2z\mathbf{k}$; Σ is the sphere $x^2 + y^2 + z^2 = 2$

11. $\mathbf{F} = x^2\mathbf{i} + 2y\mathbf{j} + 4z^2\mathbf{k}$; Σ is the surface of the cylinder $x^2 + y^2 \le 4$, $0 \le z \le 2$

12. $\mathbf{F} = x\mathbf{i} + 2y^2\mathbf{j} + 3z^2\mathbf{k}$; Σ is the surface of the cylinder $x^2 + y^2 \le 9$, $0 \le z \le 1$

13. $\mathbf{F} = (x + \cos x)\mathbf{i} + (y + y\sin x)\mathbf{j} + 2z\mathbf{k}$; Σ is the tetrahedron with vertices $(0, 0, 0)$, $(1, 0, 0)$, $(0, 1, 0)$, and $(0, 0, 1)$

14. $\mathbf{F} = yz\mathbf{i} + xz\mathbf{j} + xy\mathbf{k}$; Σ is the tetrahedron described in Problem 13

15. Let $\mathbf{F} = x^2\mathbf{i} + y^2\mathbf{j} + z^2\mathbf{k}$ and $\mathbf{n} = (\cos\alpha)\mathbf{i} + (\cos\beta)\mathbf{j} + (\cos\gamma)\mathbf{k}$. Use the divergence theorem to show that $\iint_\Sigma \mathbf{F}\cdot\mathbf{n}\,d\sigma = 8\pi q^4/3$, if Σ is the surface $x^2 + y^2 + z^2 = 2qz$.

16. What is the value of the integral in Problem 15, if Σ is the surface of the cube enclosed by $x = 0$, $x = q$, $y = 0$, $y = q$, $z = 0$, and $z = q$?

17. Let $\mathbf{F} = x\mathbf{i} + y\mathbf{j} + z\mathbf{k}$; let Σ be the surface of a region T obeying the divergence theorem; and let $\mathbf{n}$ be the outer unit normal to Σ. Show that the volume V of T is given by the formula $V = \frac{1}{3}\iint_\Sigma \mathbf{F}\cdot\mathbf{n}\,d\sigma$.

In Problems 18–20 use the expression given in Problem 17 to find each volume.

18. A rectangular parallelepiped with sides of length a, b, and c

19. A right circular cone with height h and base radius R [*Hint:* The calculation is simplified with the cone oriented as shown in the figure.]

20. A sphere of radius R

21. Let $\mathbf{F} = 3x\mathbf{i} + 4y\mathbf{j} + (7z + 2x)\mathbf{k}$ and $\mathbf{G} = 2x\mathbf{i} + 3y\mathbf{j} + (9z + 6y)\mathbf{k}$. Let Σ be the surface of a region T obeying the assumption of the divergence theorem. Prove that $\iint_\Sigma \mathbf{F}\cdot\mathbf{n}\,d\sigma = \iint_\Sigma \mathbf{G}\cdot\mathbf{n}\,d\sigma$.

22. Let f and g be two scalar functions. Let Σ be the surface of a region T obeying the assumption of the divergence theorem. Prove that $\iint_\Sigma (f\nabla g\cdot\mathbf{n})\,d\sigma = \iiint_T (f\nabla^2 g + \nabla f\cdot\nabla g)\,dV$. [*Hint:* Let $\mathbf{F} = f\nabla g$ in the divergence theorem and use the fact that $\nabla^2 g = \nabla\cdot\nabla g = g_{xx} + g_{yy} + g_{zz}$.]

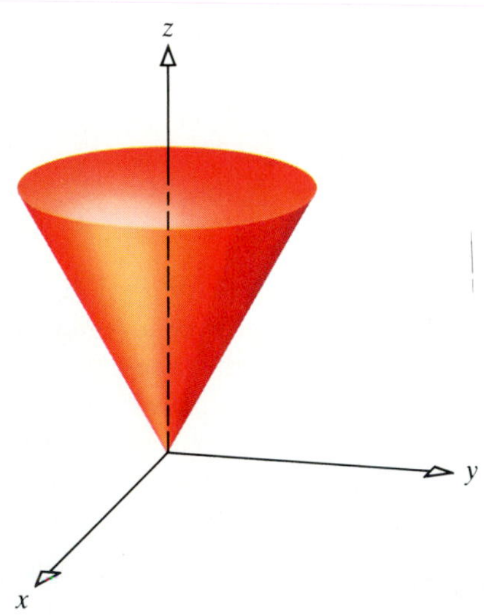

In Problems 23 and 24 verify the divergence theorem.

23. $\mathbf{F}(x, y, z) = x\mathbf{i} + y\mathbf{j} + z\mathbf{k}$; Σ is the surface of the sphere $x^2 + y^2 + z^2 = 100$

24. $\mathbf{F}(x, y, z) = x\mathbf{i} + y\mathbf{i} + z\mathbf{k}$; Σ is the closed cylindrical surface $x^2 + y^2 = 1$ between $z = 0$ and $z = 2$

25. Prove that if $\mathbf{F}$ is a constant vector field and Σ is the surface of a region T obeying the assumption of the divergence theorem, then $\iint_\Sigma \mathbf{F} \cdot \mathbf{n}\, d\sigma = 0$.

26. Suppose the velocity of a fluid flow in space is constant. Show that the flux through any closed surface is 0. What is the physical interpretation?

27. Let $\mathbf{F}(x, y, z) = (1/\rho)(x\mathbf{i} + y\mathbf{j} + z\mathbf{k})$, where $\rho = \sqrt{x^2 + y^2 + z^2}$. Show that $\operatorname{div} \mathbf{F} = 2/\rho$.

28. Use Problem 27 and the divergence theorem to compute

$$\iiint_S \frac{dV}{\sqrt{x^2 + y^2 + z^2}}$$

where S is the region inside the sphere $x^2 + y^2 + z^2 = 1$. Check by evaluating the triple integral directly.

Stokes' Theorem 19.8

To get started, we need the following definition:

[19.8.1] DEFINITION / *Curl of* F.

Let $\mathbf{F} = \mathbf{F}(x, y, z) = P(x, y, z)\mathbf{i} + Q(x, y, z)\mathbf{j} + R(x, y, z)\mathbf{k}$ denote a vector field, where P, Q, and R have first-order partial derivatives. The curl of F, denoted by curl F, is defined as

$$\operatorname{curl} \mathbf{F} = \nabla \times \mathbf{F} = \left(\frac{\partial R}{\partial y} - \frac{\partial Q}{\partial z}\right)\mathbf{i} - \left(\frac{\partial R}{\partial x} - \frac{\partial P}{\partial z}\right)\mathbf{j} + \left(\frac{\partial Q}{\partial x} - \frac{\partial P}{\partial y}\right)\mathbf{k}$$

Rather than memorize this definition, you can write the curl in the form of the symbolic determinant:

$$\operatorname{curl} \mathbf{F} = \begin{vmatrix} \mathbf{i} & \mathbf{j} & \mathbf{k} \\ \dfrac{\partial}{\partial x} & \dfrac{\partial}{\partial y} & \dfrac{\partial}{\partial z} \\ P & Q & R \end{vmatrix}$$

EXAMPLE 1

Find curl $\mathbf{F}$ if $\mathbf{F} = x^2y\mathbf{i} - 2xz\mathbf{j} + 2yz\mathbf{k}$.

Solution

$$\operatorname{curl} \mathbf{F} = \begin{vmatrix} \mathbf{i} & \mathbf{j} & \mathbf{k} \\ \dfrac{\partial}{\partial x} & \dfrac{\partial}{\partial y} & \dfrac{\partial}{\partial z} \\ x^2y & -2xz & 2yz \end{vmatrix}$$

$$= \left[\frac{\partial}{\partial y}(2yz) - \frac{\partial}{\partial z}(-2xz)\right]\mathbf{i} - \left[\frac{\partial}{\partial x}(2yz) - \frac{\partial}{\partial z}(x^2y)\right]\mathbf{j} + \left[\frac{\partial}{\partial x}(-2xz) - \frac{\partial}{\partial y}(x^2y)\right]\mathbf{k}$$

$$= (2z + 2x)\mathbf{i} - (0 - 0)\mathbf{j} + (-2z - x^2)\mathbf{k} = (2z + 2x)\mathbf{i} + (-2z - x^2)\mathbf{k}$$

Now we are ready to introduce *Stokes' theorem*,* which may be stated in vector form as

$$\oint_C \mathbf{F} \cdot d\mathbf{r} = \iint_{\Sigma} \operatorname{curl} \mathbf{F} \cdot \mathbf{n}\, d\sigma \qquad \text{under certain conditions on } \Sigma \qquad \textbf{(1)}$$

We briefly discuss the conditions on Σ. The surface Σ must be *smooth*; that is, if the surface is defined by $z = f(x, y)$, then f has continuous first-order partial derivatives. The surface Σ must also be *simply connected*. By this we mean that there are no "holes" in Σ. Finally, we require the surface Σ to be *orientable*. Recall that this means it is possible to assign a direction to the unit normal $\mathbf{n}$ at each point of Σ, which we take as the positive direction of $\mathbf{n}$. The positively oriented vector $\mathbf{n}$ must vary continuously, and, as it moves around any closed curve on the surface, $\mathbf{n}$ must return to its original direction.

The curve C of the line integral in (1) is a piecewise smooth, simple closed curve that forms a boundary of the surface Σ. The positively oriented surface Σ induces a *positive orientation* on C in the sense that if you walk around C in the positive direction with your head pointing in the same direction as the unit normal $\mathbf{n}$ to Σ, then the surface will always be to your left. See Figure 45.

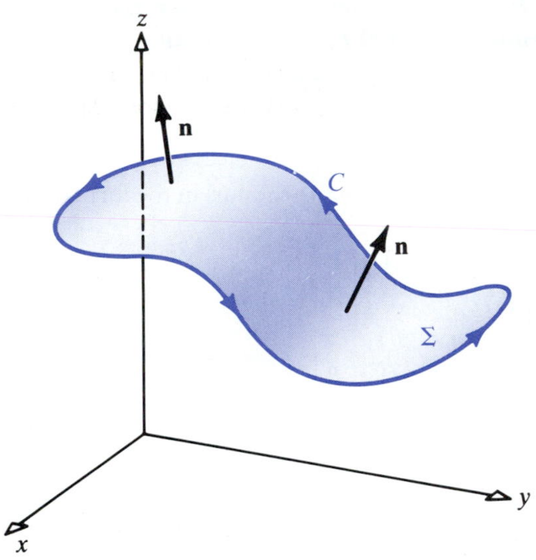

Figure 45

We now give a precise formulation of Stokes' theorem.

[19.8.2] THEOREM / *Stokes' Theorem.*

Let Σ denote a smooth, simply connected orientable surface bounded by a piecewise smooth, simple closed curve C. Let $\mathbf{F} =$

* Named after the English scientist George Gabriel Stokes (1819–1903).

$\mathbf{F}(x, y, z) = P(x, y, z)\mathbf{i} + Q(x, y, z)\mathbf{j} + R(x, y, z)\mathbf{k}$ denote a vector field, where P, Q, and R have continuous first-order partial derivatives throughout a region S containing Σ and C. Let n denote the positive unit normal to Σ, and let C be positively oriented, as described above. Then

$$\oint_C \mathbf{F} \cdot d\mathbf{r} = \iint_\Sigma \text{curl } \mathbf{F} \cdot \mathbf{n} \, d\sigma$$

This statement can also be written as

$$\oint_C (P\,dx + Q\,dy + R\,dz)$$

$$= \iint_\Sigma \left[\left(\frac{\partial R}{\partial y} - \frac{\partial Q}{\partial z} \right) dy\,dz - \left(\frac{\partial R}{\partial x} - \frac{\partial P}{\partial z} \right) dz\,dx + \left(\frac{\partial Q}{\partial x} - \frac{\partial P}{\partial y} \right) dx\,dy \right]$$

The statement of Stokes' theorem as given above is quite general, and its proof is beyond the scope of this book. Let's look at some examples that use Stokes' theorem.

EXAMPLE 2

Verify Stokes' theorem for $\mathbf{F} = y\mathbf{i} - x\mathbf{j}$, where Σ is the paraboloid $z = x^2 + y^2$, with the circle $x^2 + y^2 = 1$, $z = 1$, as its boundary (see Fig. 46).

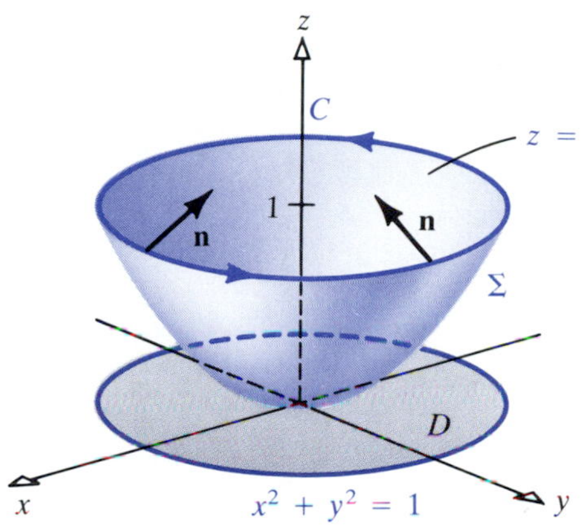

Figure 46

Solution

A set of parametric equations for C are $x = \cos t$, $y = \sin t$, $z = 1$. We use these to evaluate the line integral $\oint_C \mathbf{F} \cdot d\mathbf{r}$.

$$\oint_C \mathbf{F} \cdot d\mathbf{r} = \oint_C (y\,dx - x\,dy) = \int_0^{2\pi} [\sin t(-\sin t)\,dt - \cos t \cos t\,dt]$$

$$= -\int_0^{2\pi} [\sin^2 t + \cos^2 t]\,dt = -2\pi$$

To evaluate the surface integral, $\iint_\Sigma \text{curl } \mathbf{F} \cdot \mathbf{n}\,d\sigma$, we compute curl $\mathbf{F}$ and $\mathbf{n}$:

$$\text{curl } \mathbf{F} = \begin{vmatrix} \mathbf{i} & \mathbf{j} & \mathbf{k} \\ \dfrac{\partial}{\partial x} & \dfrac{\partial}{\partial y} & \dfrac{\partial}{\partial z} \\ y & -x & 0 \end{vmatrix} = 0\mathbf{i} - 0\mathbf{j} - 2\mathbf{k} = -2\mathbf{k}$$

$$\mathbf{n} = \frac{-2x\mathbf{i} - 2y\mathbf{j} + \mathbf{k}}{\sqrt{4x^2 + 4y^2 + 1}}$$

$$\text{curl } \mathbf{F} \cdot \mathbf{n} = \frac{-2}{\sqrt{4x^2 + 4y^2 + 1}}$$

On Σ: $z = x^2 + y^2$, we have

$$\frac{\partial z}{\partial x} = 2x \qquad \frac{\partial z}{\partial y} = 2y \qquad \sqrt{\left(\frac{\partial z}{\partial x}\right)^2 + \left(\frac{\partial z}{\partial y}\right)^2 + 1} = \sqrt{4x^2 + 4y^2 + 1}$$

Thus,

$$\begin{aligned}\iint_\Sigma \operatorname{curl} \mathbf{F} \cdot \mathbf{n}\, d\sigma &= \iint_\Sigma \frac{-2}{\sqrt{4x^2 + 4y^2 + 1}}\, d\sigma \\ &= \iint_D \frac{-2}{\sqrt{4x^2 + 4y^2 + 1}} \sqrt{4x^2 + 4y^2 + 1}\, dx\, dy\end{aligned}$$

where D is the interior of the circle $x^2 + y^2 = 1$. Thus,

$$\iint_\Sigma \operatorname{curl} \mathbf{F} \cdot \mathbf{n}\, d\sigma = -2 \iint_D dx\, dy = -2\pi$$

EXAMPLE 3

Use Stokes' theorem to evaluate the surface integral $\iint_\Sigma \operatorname{curl} \mathbf{F} \cdot \mathbf{n}\, d\sigma$, where $\mathbf{F} = \mathbf{F}(x, y, z) = y\mathbf{i} + x\mathbf{j} + z\mathbf{k}$ and Σ is the surface enclosed by $z = 5 - (x^2 + y^2)$, $z \geq 1$.

Solution

Figure 47 illustrates the situation. The boundary curve C is a circle $x^2 + y^2 = 4$ that lies in the plane $z = 1$. The vector form of C is

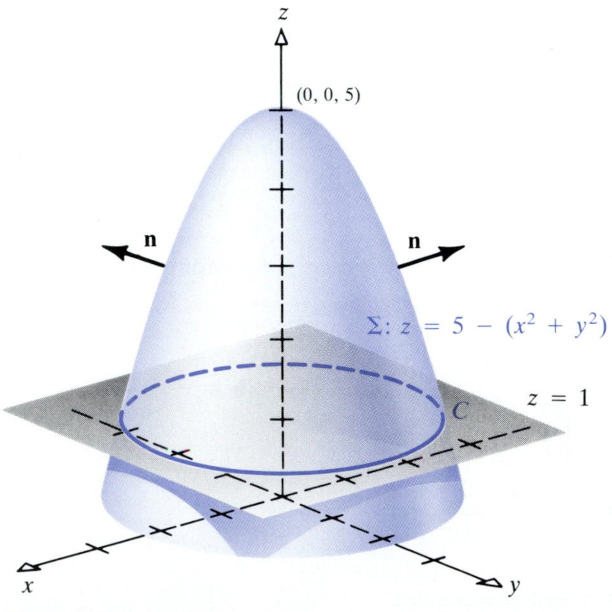

Figure 47

$\mathbf{r} = \mathbf{r}(t) = 2\cos t\mathbf{i} + 2\sin t\mathbf{j} + \mathbf{k}$. Then,

$$d\mathbf{r} = (-2\sin t\mathbf{i} + 2\cos t\mathbf{j})\,dt$$

$$\mathbf{F} = y\mathbf{i} + x\mathbf{j} + z\mathbf{k} = \sin t\mathbf{i} + \cos t\mathbf{j} + \mathbf{k}$$

$$\mathbf{F}\cdot d\mathbf{r} = (-2\sin^2 t + 2\cos^2 t)\,dt = 2\cos 2t\,dt$$

By Stokes' theorem, we have

$$\iint_{\Sigma} \operatorname{curl}\mathbf{F}\cdot\mathbf{n}\,d\sigma = \oint_C \mathbf{F}\cdot d\mathbf{r} = \int_0^{2\pi} 2\cos 2t\,dt = \sin 2t\Big|_0^{2\pi} = 0$$

In Example 3, we calculated the surface integral $\iint_\Sigma \operatorname{curl}\mathbf{F}\cdot\mathbf{n}\,d\sigma$ by using the line integral of $\mathbf{F}$ on the boundary C of the surface Σ. This means that for any surface Σ_1 with the same orientation and the same boundary curve as Σ, the surface integral $\iint_{\Sigma_1} \operatorname{curl}\mathbf{F}\cdot\mathbf{n}_1\,d\sigma_1$ will have the same value as $\iint_\Sigma \operatorname{curl}\mathbf{F}\cdot\mathbf{n}\,d\sigma$.

Stokes' theorem can be used to avoid calculation of a surface integral by substituting an equivalent line integral. However, it is probably more frequently used in the opposite direction—that is, in cases when the line integral is not easily evaluated, but the quantity $\operatorname{curl}\mathbf{F}\cdot\mathbf{n}$ is of relatively simple form on some open surface with boundary C. This fact is illustrated in the next example.

EXAMPLE 4

Evaluate $I = \oint_C[(e^{-x^2/2} - yz)\,dx + (e^{-y^2/2} + xz + 2x)\,dy + (e^{-z^2/2} + 5)\,dz]$, where C is the circle $x = \cos t$, $y = \sin t$, $z = 2$, $0 \le t \le 2\pi$.

Solution

If we let $\mathbf{F} = P\mathbf{i} + Q\mathbf{j} + R\mathbf{k}$, with $P = e^{-x^2/2} - yz$, $Q = e^{-y^2/2} + xz + 2x$, and $R = e^{-z^2/2} + 5$, it can be verified that

$$\operatorname{curl}\mathbf{F} = -x\mathbf{i} - y\mathbf{j} + (2 + 2z)\mathbf{k}$$

To apply Stokes' theorem, we take Σ to be a plane region enclosed by the circle C in the plane $z = 2$ so that $\mathbf{n} = \mathbf{k}$. Hence, $\operatorname{curl}\mathbf{F}\cdot\mathbf{n} = 6$ on Σ, and we may write

$$I = \iint_\Sigma \operatorname{curl}\mathbf{F}\cdot\mathbf{n}\,d\sigma = 6\iint_\Sigma dx\,dy = 6\pi$$

Application of Stokes' Theorem to Conservative Fields of Force

A vector field $\mathbf{F} = \mathbf{F}(x, y) = P(x, y, z)\mathbf{i} + Q(x, y, z)\mathbf{j} + R(x, y, z)\mathbf{k}$ is *conservative* if $\mathbf{F}$ is the gradient of some function $f = f(x, y, z)$. It is an extension of earlier proofs to show that $\mathbf{F}$ is conservative if and only if $\int_C \mathbf{F}\cdot d\mathbf{r}$ is independent of path, where C is a piecewise smooth curve in space and

$\mathbf{F} = \mathbf{F}(x, y, z)$ is defined on some open, connected region of space. In particular, if C is simple and closed, then $\mathbf{F}$ is conservative if and only if $\oint_C \mathbf{F} \cdot d\mathbf{r} = 0$ for every piecewise smooth, simple closed curve C.

Stokes' theorem can be used in an easy way to tell whether a vector field $\mathbf{F}$ in space is conservative.

[19.8.3] THEOREM / *Conservative Vector Fields in Space.*

Let $\mathbf{F} = P(x, y, z)\mathbf{i} + Q(x, y, z)\mathbf{j} + R(x, y, z)\mathbf{k}$ denote a vector field whose components P, Q, and R possess continuous first-order partial derivatives throughout a region S whose boundary is a surface Σ satisfying the conditions of Stokes' theorem. Then $\mathbf{F}$ is conservative if and only if $\text{curl } \mathbf{F} = \mathbf{0}$ throughout S.

Proof We shall give only an outline of the proof. Suppose $\text{curl } \mathbf{F} = \mathbf{0}$. Then, by Stokes' theorem,

$$\oint_C \mathbf{F} \cdot d\mathbf{r} = \iint_\Sigma \text{curl } \mathbf{F} \cdot \mathbf{n}\, d\sigma = 0$$

for any closed curve C. That is, $\mathbf{F}$ is conservative.

Conversely, suppose $\mathbf{F}$ is conservative. Then, $\oint_C \mathbf{F} \cdot d\mathbf{r} = 0$ around every closed path C. Assume that $\text{curl } \mathbf{F} \neq \mathbf{0}$ at some point P. Then, since curl $\mathbf{F}$ is continuous, there will be a region with P as an interior point where $\text{curl } \mathbf{F} \neq \mathbf{0}$. Let Σ be a surface contained in this region whose normal $\mathbf{n}$ at each point has the same direction as curl $\mathbf{F}$, that is, $\text{curl } \mathbf{F} = a\mathbf{n}$, where a is a positive constant. If C is a boundary of Σ, then, by Stokes' theorem,

$$\oint_C \mathbf{F} \cdot d\mathbf{r} = \iint_\Sigma \text{curl } \mathbf{F} \cdot \mathbf{n}\, d\sigma = a \iint_\Sigma \mathbf{n} \cdot \mathbf{n}\, d\sigma > 0$$

which contradicts the hypothesis that $\oint_C \mathbf{F} \cdot d\mathbf{r} = 0$. Hence, we conclude that $\text{curl } \mathbf{F} = \mathbf{0}$. ∎

Under the hypothesis of theorem (19.8.3), we may now list the following equivalent statements for a vector field $\mathbf{F} = P\mathbf{i} + Q\mathbf{j} + R\mathbf{k}$ in space; these are parallel to those listed in the Summary at the end of Section 19.3 for a vector field in the plane.

1. **$\mathbf{F}$ is conservative.**
2. **$\mathbf{F}$ is the gradient of some function.**
3. **The work done by $\mathbf{F}$ in moving an object of mass m from a point A to a point B in Σ is independent of the path chosen from A to B.**
4. **The work done by $\mathbf{F}$ in moving an object of mass m along any closed, piecewise smooth curve C in Σ is 0.**
5. **$\text{curl } \mathbf{F} = \mathbf{0}$.**

Of all these equivalent conditions, the easiest to establish is statement 5, as the next example illustrates.

EXAMPLE 5

Show that $\mathbf{F} = (\frac{3}{5}y^5 + 2z^2)\mathbf{i} + 3xy^4\mathbf{j} + 4xz\mathbf{k}$ is conservative.

Solution

$$\text{curl } \mathbf{F} = \begin{vmatrix} \mathbf{i} & \mathbf{j} & \mathbf{k} \\ \dfrac{\partial}{\partial x} & \dfrac{\partial}{\partial y} & \dfrac{\partial}{\partial z} \\ \frac{3}{5}y^5 + 2z^2 & 3xy^4 & 4xz \end{vmatrix} = (0-0)\mathbf{i} - (4z - 4z)\mathbf{j} + (3y^4 - 3y^4)\mathbf{k} = \mathbf{0}$$

Thus, $\mathbf{F}$ is conservative.

Interpretation of curl F

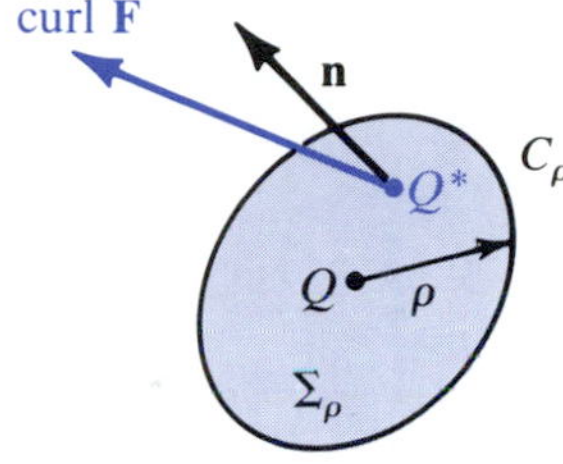

Figure 48

Just as the divergence theorem gives a new interpretation for the divergence of $\mathbf{F}$, we may use Stokes' theorem to give an interpretation for the curl of a vector. Let $Q = (x_1, y_1, z_1)$ be the center of a circular disk Σ_ρ of radius ρ and let C_ρ be the boundary of Σ_ρ (see Fig. 48). There is a law of the mean for double integrals which, when combined with Stokes' theorem, gives us

$$\oint_{C_\rho} \mathbf{F} \cdot d\mathbf{r} = \iint_{\Sigma_\rho} \text{curl } \mathbf{F} \cdot \mathbf{n}\, d\sigma = (\text{curl } \mathbf{F} \cdot \mathbf{n})_{Q^*}(\pi\rho^2)$$

where $(\text{curl } \mathbf{F} \cdot \mathbf{n})_{Q^*}$ denotes the value of curl $\mathbf{F} \cdot \mathbf{n}$ evaluated at a suitably chosen point $Q^* = (x^*, y^*, z^*)$ in Σ_ρ, and $\pi\rho^2$ is the area of Σ_ρ. Thus,

$$(\text{curl } \mathbf{F} \cdot \mathbf{n})_{Q^*} = \frac{1}{\pi\rho^2} \oint_{C_\rho} \mathbf{F} \cdot d\mathbf{r}$$

Now, if ρ is allowed to approach 0, then $Q^* \to Q$, and hence,

$$(\text{curl } \mathbf{F} \cdot \mathbf{n})_Q = \lim_{\rho \to 0} \frac{1}{\pi\rho^2} \oint_{C_\rho} \mathbf{F} \cdot d\mathbf{r} \tag{2}$$

In the case where $\mathbf{F}$ is the velocity of a fluid, the integral $\oint_{C_\rho} \mathbf{F} \cdot d\mathbf{r}$ in (2) is referred to as the *circulation*, or *whirling tendency*, around C_ρ, and it measures the extent to which the corresponding fluid rotates around the circle C_ρ in the given direction. Equation (2) therefore states that the component of curl $\mathbf{F}$ at (x_1, y_1, z_1) in the direction of $\mathbf{n}$ is the limiting ratio of circulation to area for a circle about (x_1, y_1, z_1) with $\mathbf{n}$ as a normal. Thus,

Circulation per unit of area at $(x_1, y_1, z_1) = (\text{curl } \mathbf{F} \cdot \mathbf{n})_Q$

The left-hand side of this equation is a maximum at Q when $\mathbf{n}$ has the same direction as curl $\mathbf{F}$. Suppose that a small paddle wheel of radius ρ is introduced into the fluid at Q with its axle directed along $\mathbf{n}$. The rate of spin of the paddle wheel will be affected by the circulation of the fluid around C_ρ. The wheel will spin fastest when the circulation integral is maximized; that is, it will spin fastest when the axle of the paddle wheel is in the direction of curl $\mathbf{F}$ (see Fig. 49).

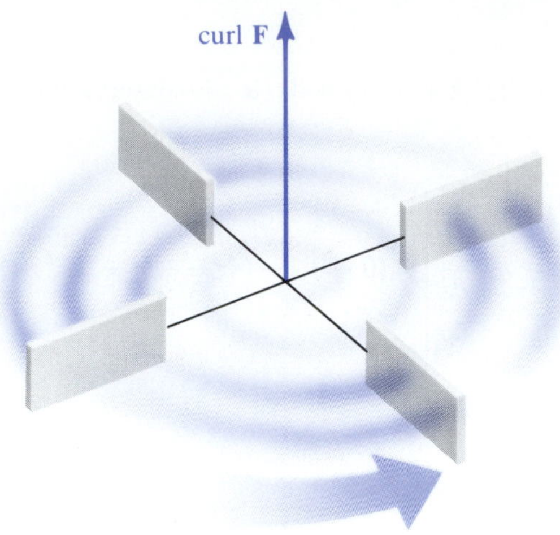

Figure 49

We now give another interpretation of the curl based on motion. Knowledge of angular velocity (discussed in Chap. 14) will be helpful here.

[19.8.4] THEOREM

Let F be the velocity vector of a fluid rotating about a fixed axis and let ω be the constant angular velocity. Then

$$\text{curl } \mathbf{F} = 2\boldsymbol{\omega}$$

Thus, if a fluid experiences a rotation, the curl of the velocity vector is a constant vector equal to twice the angular velocity vector.

Proof Since the motion of a fluid is simply a rotation about a given axis fixed in space, then we may represent the angular velocity by a constant vector

$$\boldsymbol{\omega} = \omega_1\mathbf{i} + \omega_2\mathbf{j} + \omega_3\mathbf{k}$$

where the length of $\boldsymbol{\omega}$ is the angular speed and the direction of $\boldsymbol{\omega}$ is along the direction of the axis of rotation, in accordance with the right-hand rule. If we take the origin of a rectangular coordinate system on the axis and denote the position vector of a point (x, y, z) by $\mathbf{r} = x\mathbf{i} + y\mathbf{j} + z\mathbf{k}$, then the velocity vector $\mathbf{F}$ is

$$\mathbf{F} = \boldsymbol{\omega} \times \mathbf{r}$$

and

$$\text{curl } \mathbf{F} = \begin{vmatrix} \mathbf{i} & \mathbf{j} & \mathbf{k} \\ \dfrac{\partial}{\partial x} & \dfrac{\partial}{\partial y} & \dfrac{\partial}{\partial z} \\ \omega_2 z - \omega_3 y & \omega_3 x - \omega_1 z & \omega_1 y - \omega_2 x \end{vmatrix}$$

Since $\boldsymbol{\omega}$ is constant, straightforward computation shows that

$$\text{curl } \mathbf{F} = 2\omega_1\mathbf{i} + 2\omega_2\mathbf{j} + 2\omega_3\mathbf{k} = 2\boldsymbol{\omega}$$ ■

Summary

Green's theorem: $\iint_R [(\partial Q/\partial x) - (\partial P/\partial y)]\, dx\, dy = \oint_C (P\, dx + Q\, dy)$ relates a certain double integral over a region R enclosed by the curve C to a certain line integral over C.

Stokes' theorem: $\iint_\Sigma \text{curl}\, \mathbf{F} \cdot \mathbf{n}\, d\sigma = \oint_C \mathbf{F} \cdot d\mathbf{r}$ relates a certain surface integral over Σ to a certain line integral over a boundary curve C of Σ.

EXERCISE 19.8

In Problems 1–12 find the curl of $\mathbf{F}$.

1. $\mathbf{F}(x, y) = x\mathbf{i} + y\mathbf{j}$

2. $\mathbf{F}(x, y) = y\mathbf{i} + x\mathbf{j}$

3. $\mathbf{F}(x, y, z) = xyz\mathbf{i} + xz\mathbf{j} + z\mathbf{k}$

4. $\mathbf{F}(x, y, z) = 4x\mathbf{i} - y\mathbf{j} - 2z\mathbf{k}$

5. $\mathbf{F}(x, y, z) = 3xyz^2\mathbf{i} + (y^2 \sin z)\mathbf{j} + xe^{2z}\mathbf{k}$

6. $\mathbf{F}(x, y, z) = yz\mathbf{i} + z^2x\mathbf{j} + yz\mathbf{k}$

7. $\mathbf{F}(x, y, z) = \dfrac{x\mathbf{i}}{x^2 + y^2 + z^2} + \dfrac{y\mathbf{j}}{x^2 + y^2 + z^2} + \dfrac{z\mathbf{k}}{x^2 + y^2 + z^2}$

8. $\mathbf{F}(x, y, z) = e^x\mathbf{i} + x^2y\mathbf{j} + e^z\mathbf{k}$

9. $\mathbf{F}(x, y, z) = (\cos x)\mathbf{i} + (\sin y)\mathbf{j} + e^{xz}\mathbf{k}$

10. $\mathbf{F}(x, y, z) = (\sin xy)\mathbf{i} + (\cos xy^2)\mathbf{j} + x\mathbf{k}$

11. $\mathbf{F}(x, y, z) = (x + y)\mathbf{i} + (y + z)\mathbf{j} + (z + x)\mathbf{k}$

12. $\mathbf{F}(x, y, z) = (y + z)\mathbf{i} + (z + x)\mathbf{j} + (z + y + x)\mathbf{k}$

13. Show that $\text{div}(\text{curl}\, \mathbf{F}) = 0$, where $\mathbf{F} = P\mathbf{i} + Q\mathbf{j} + R\mathbf{k}$ and P, Q, and R are continuously twice differentiable functions of x, y, and z.

14. Show that $\text{curl}(\mathbf{F} + \mathbf{G}) = \text{curl}\, \mathbf{F} + \text{curl}\, \mathbf{G}$ and $\text{curl}\, c\mathbf{F} = c(\text{curl}\, \mathbf{F})$, c a constant.

15. Determine which of the following forces in space are conservative.

(a) $\mathbf{F} = x\mathbf{i} + y\mathbf{i}$
(b) $\mathbf{F} = y\mathbf{i} + x\mathbf{j}$
(c) $\mathbf{F} = y\mathbf{i} - x\mathbf{j}$
(d) $\mathbf{F} = xy\mathbf{i} + yz\mathbf{j} + zx\mathbf{k}$
(e) $\mathbf{F} = yz\mathbf{i} + zx\mathbf{j} + xy\mathbf{k}$

16. Find the value of the constant c so that the following forces in space are conservative:

(a) $\mathbf{F} = xy\mathbf{i} + cx^2\mathbf{j}$
(b) $\mathbf{F} = \dfrac{z}{y}\mathbf{i} + c\dfrac{xz}{y^2}\mathbf{j} + \dfrac{x}{y}\mathbf{k}, \quad y \neq 0$

In Problems 17–20 verify Stokes' theorem for the given $\mathbf{F}$ and Σ.

17. $\mathbf{F} = y\mathbf{i} - x\mathbf{j}$; Σ is the hemisphere $z = \sqrt{1 - x^2 - y^2}$

18. $\mathbf{F} = z\mathbf{i} + x\mathbf{j} + y\mathbf{k}$; Σ is the hemisphere $z = \sqrt{1 - x^2 - y^2}$

19. $\mathbf{F} = (z - y)\mathbf{i} + (z + x)\mathbf{j} - (x + y)\mathbf{k}$; Σ is the portion of the paraboloid $z = 1 - x^2 - y^2$ that lies above the plane $z = 0$

20. $\mathbf{F} = y\mathbf{i} + z\mathbf{j} + x\mathbf{k}$; Σ is the portion of the paraboloid $z = 1 - x^2 - y^2$, $z \geq 0$

In Problems 21–24 apply Stokes' theorem to evaluate the given line integral, and verify your answer by a direct calculation of the line integral.

21. $\oint_C [(y + z)\, dx + (z + x)\, dy + (x + y)\, dz]$; C is the circle $x^2 + y^2 + z^2 = 1$, $x + y + z = 0$

22. $\oint_C [(y - z)\, dx + (z - x)\, dy + (x - y)\, dz]$; C is the ellipse $x^2 + y^2 = 1$, $x + z = 1$

23. $\oint_C [x\, dx + (x + y)\, dy + (x + y + z)\, dz]$; C is the curve $x = 2\cos t$, $y = 2\sin t$, $z = 2$, $0 \leq t \leq 2\pi$

24. $\oint_C (y^2\, dx + z^2\, dy + x^2\, dz)$; C is the triangle with vertices $(1, 0, 0)$, $(0, 1, 0)$, and $(0, 0, 1)$

25. Let $\mathbf{F} = 2xy^2z\mathbf{i} + 2x^2yz\mathbf{j} + (x^2y^2 - 2z)\mathbf{k}$. Show that $\mathbf{F}$ is conservative.

26. Show that $\mathbf{F} = y\mathbf{i} - x\mathbf{j} + z\mathbf{k}$ is not a conservative field. Nevertheless, there are certain paths C for which $\oint_C \mathbf{F} \cdot d\mathbf{r} = 0$. Find one.

27. A particle is moved from the origin to the point (a, b, c) in a field of force $\mathbf{F} = (x + y)\mathbf{i} + (x - z)\mathbf{j} + (z - y)\mathbf{k}$. Show that the work done depends on only a, b, and c, and find this value.

28. Verify Stokes' theorem when $\mathbf{F}(x, y, z) = y^2\mathbf{i} + x\mathbf{j} - xz\mathbf{k}$ and Σ is the surface $z = 1 - x^2 - y^2$, $z \geq 0$.

29. If

$$\mathbf{F}(x, y, z) = z\mathbf{i} + x\mathbf{j} + y\mathbf{k}$$

calculate $\iint_\Sigma (\text{curl } \mathbf{F} \cdot \mathbf{n})\, d\sigma$, where Σ is the hemisphere $z = \sqrt{1 - x^2 - y^2}$.

30. Rework Problem 29 where Σ is the circular region $x^2 + y^2 \leq 1$, $z = 0$.

31. By direct calculation show that $\int_C (z\,dx + x\,dy + y\,dz) = \sqrt{3}\pi$, where C is the curve given by $x + y + z = 0$, $x^2 + y^2 + z^2 = 1$. Obtain the same result by using Stokes' theorem.

REVIEW EXERCISES

1. Evaluate $\displaystyle\int_C \left(\frac{dx}{y} + \frac{dy}{x}\right)$, where C is the arc of the parabola $y = x^2$ from $(1, 1)$ to $(2, 4)$.

2. Using methods of Section 19.2, find $\int_C [y \cos xy\, dx + x \cos xy\, dy]$, where C is the curve $x = t^2$, $y = t^3$, $0 \leq t \leq 1$.

3. Confirm that the integral in Problem 2 is independent of path, and compute it by following the right-angle path from $(0, 0)$ to $(1, 0)$ to $(1, 1)$, using methods of Section 19.2.

4. Find a function f whose gradient is $(y \cos xy)\mathbf{i} + (x \cos xy)\mathbf{j}$ and explain why Problem 2 can be worked by writing

$$\int_C [y \cos xy\, dx + x \cos xy\, dy] = f(1, 1) - f(0, 0)$$

5. Using methods of Section 19.2, find $\int_C [(yz - y - z)\, dx + (xz - x - z)\, dy + (xy - x - y)\, dz]$, where C is the twisted cubic $x = t$, $y = t^2$, $z = t^3$, $0 \leq t \leq 1$.

6. Find a function $f(x, y, z)$ whose gradient is $(yz - y - z)\mathbf{i} + (xz - x - z)\mathbf{j} + (xy - x - y)\mathbf{k}$, and use it to confirm the answer to Problem 5.

7. Find $\int_C [e^x \sin y\, dx + e^x \cos y\, dy]$, where C is the arc of the parabola $y = x^2$ from $(0, 0)$ to $(1, 1)$.

8. Show that if the work done by a force $\mathbf{F}(x, y)$ in moving an object along any piecewise smooth, closed curve in an open set S in the plane is 0, then the work done by $\mathbf{F}$ in moving an object from a point A to a point B in S is independent of the piecewise smooth path chosen.

9. Evaluate $\int_C [x^2y^3\, dx - xy^4\, dy]$, where C is the arc of the parabola $y^2 = x$ from $(0, 0)$ to $(1, 1)$.

10. Evaluate $\int_C x^2y^2\, ds$, where C is the upper branch of the unit circle.

11. Let f and g have continuous partial derivatives in a plane region S, and let C be a piecewise smooth curve in S going from A to B. Show that

$$\int_C f\nabla g \cdot d\mathbf{r} = f(B)g(B) - f(A)g(A) - \int_C g\nabla f \cdot d\mathbf{r}$$

12. Parameterize the unit circle by $x(t) = \cos t$, $y(t) = \sin t$, and let $\mathbf{F}(x, y) = \dfrac{y}{(x^2 + y^2)^{3/2}}\mathbf{i} - \dfrac{x}{(x^2 + y^2)^{3/2}}\mathbf{j}$. Compute the work done by going around the unit circle against $\mathbf{F}$ from $t = 0$ to $t = 2\pi$.

13. Find the work done by the force $\mathbf{F} = y \sin x\mathbf{i} + \sin x\mathbf{j}$ as a particle moves along the curve $y = \sin x$ from $x = 0$ to $x = 2\pi$.

14. Find the work done by the force

$$\mathbf{F} = \frac{x}{x^2 + y^2}\mathbf{i} + \frac{y}{x^2 + y^2}\mathbf{j}$$

as a particle moves along the curve $t \cos t\,\mathbf{i} + t \sin t\,\mathbf{j}$ from $(\pi, 0)$ to $(2\pi, 0)$.

15. Evaluate $\int_C (y^2\, dx - x^2\, dy)$, where C is the square with vertices $(0, 0)$, $(1, 0)$, $(1, 1)$, and $(0, 1)$ traversed counterclockwise. Do not use Green's theorem.

16. Rework Problem 15 using Green's theorem.

17. Evaluate $\int_C [(x - y)\, dx + (x + y)\, dy]$, where C is the ellipse $x = 2 \cos t$, $y = 3 \sin t$, $0 \leq t \leq 2\pi$, without using Green's theorem.

18. Rework Problem 17 using Green's theorem.

19. Consider the integral $\int_C \frac{y\,dx - x\,dy}{(x^2 + y^2)^{3/2}}$. For exactly which simple closed, piecewise smooth curves C can Green's theorem be applied to transform this line integral to a double integral?

20. Evaluate $\iint_\Sigma x\,d\sigma$, where Σ is the surface $x^2 + y^2 = k^2$, $-1 \le z \le 1$; $k > 0$.

21. Evaluate $\iint_\Sigma (xz\cos\alpha + yz\cos\beta + x^2\cos\gamma)\,d\sigma$, where Σ is the upper half of the unit sphere together with the plane $z = 0$ and the cosines are the direction cosines for the outer unit normal to Σ.

22. Evaluate $\iint_\Sigma (x\cos\alpha + y\cos\beta)\,d\sigma$, where Σ is the positive octant of the unit sphere and the cosines are the direction cosines to the outer unit normal to Σ.

23. Evaluate $\iint_\Sigma (1/\sqrt{1 + 4z})\,d\sigma$, where Σ is the surface $x^2 + y^2 = z$, $x^2 + y^2 \le 1$.

24. Evaluate $\iint_\Sigma z\,d\sigma$, where Σ is the surface $z = 9 - x - y$, $x^2 + y^2 \le 9$.

25. Evaluate $\iint_\Sigma \mathbf{F} \cdot \mathbf{n}\,d\sigma$, where Σ is the upper half of the unit sphere and where $\mathbf{F} = \mathbf{n}$.

26. Evaluate $\iint_\Sigma \cos x\,d\sigma$, where Σ is the portion of the plane $x = y + z$, $x \le \pi$, $y, z \ge 0$.

27. Evaluate $\iint_\Sigma \mathbf{F} \cdot \mathbf{n}\,d\sigma$, where Σ is a level surface of a function $f(x, y, z)$ with outer unit normal $\mathbf{n}$, $\mathbf{F} = \nabla f$, and f satisfies the equation $(\partial f/\partial x)^2 + (\partial f/\partial y)^2 + (\partial f/\partial z)^2 = 1$. Your answer should depend on Σ.

28. Evaluate $\iint_\Sigma \mathbf{F} \cdot \mathbf{n}\,d\sigma$, where Σ is the surface $x^2 + y^2 = 1$, $0 \le z \le 1$, and $\mathbf{F} = x^2\mathbf{i} + y\mathbf{j} - z\mathbf{k}$.

29. Compute the integral $\iiint_T \operatorname{div} \mathbf{F}\,dV$, where T is the unit ball, $|\mathbf{r}| \le 1$, and $\mathbf{F} = |\mathbf{r}|^a\mathbf{r}$.

30. In Problem 22 in Exercise 19.7, the result to be demonstrated is sometimes known as Green's first identity. Under the same assumptions on T and Σ, show Green's second identity: $\iiint_T (f\nabla^2 g - g\nabla^2 f)\,dV = \iint_\Sigma (f\nabla g - g\nabla f) \cdot \mathbf{n}\,d\sigma$.

In Problems 31–32, assume that the hypotheses of the divergence theorem hold for Σ and T.

31. Show that if f satisfies the Laplace equation $(f_{xx} + f_{yy} + f_{zz} = 0)$ in a closed, bounded region T with boundary Σ with outer unit normal $\mathbf{n}$, then $\iint_\Sigma \nabla f \cdot \mathbf{n}\,d\sigma = 0$.

32. Show that for any vector field $\mathbf{F}$ with appropriate derivatives on a closed, bounded region T with orientable boundary Σ with outer unit normal $\mathbf{n}$, we have

$$\iint_\Sigma (\operatorname{curl} \mathbf{F}) \cdot \mathbf{n}\,d\sigma = 0.$$

33. Prove that no twice differentiable vector function exists whose curl is $x\mathbf{i} + y\mathbf{j} + z\mathbf{k}$.

34. Let $\mathbf{F}(x, y, z) = x^3\mathbf{i} + y^3\mathbf{j} + z^3\mathbf{k}$ be the velocity of a fluid flow in space, where the mass density of the fluid is 1.

(a) Find the flux through the sphere $x^2 + y^2 + z^2 = 1$.
(b) Find the circulation around the circle $x^2 + y^2 = 1$ in the xy-plane.

35. Suppose

$$\mathbf{F}(x, y, z) = P(x, y, z)\mathbf{i} + Q(x, y, z)\mathbf{j} + R(x, y, z)\mathbf{k}$$

is a vector field with continuous and differentiable components in a simply connected region of space. Use Stokes' theorem to show that $\mathbf{F}$ is a gradient field

(Problem 35 continues in the next column)

35. *(continued)*
if $Q_x = P_y$, $R_y = Q_z$, and $P_z = R_x$. Why does it follow that $\int_C \mathbf{F} \cdot d\mathbf{r}$ is independent of path when these conditions hold?

36. Use Problem 35 to show that $\int_C [(yz - y - z)\,dx + (xz - x - z)\,dy + (xy - x - y)\,dz]$ is independent of path. (See Problems 5 and 6.)

37. Find $\int_C [(x - y)\,dx + (y - z)\,dy + (z - x)\,dz]$, where C is the boundary of the first-octant portion of the plane $x + y + z = 1$ (traversed counterclockwise when viewed from above).

38. Rework Problem 37 using Stokes' theorem.

39. Let $\mathbf{F}(x, y, z)$ be a vector function with continuous and differentiable components, and let Σ be a sphere with outer unit normal $\mathbf{n}$. Use Stokes' theorem to show that $\iint_\Sigma \operatorname{curl} \mathbf{F} \cdot \mathbf{n}\,d\sigma = 0$.

40. Let C be a closed curve contained in an open region in space, and let f and g have continuous partial derivatives in that region. Show that $\int_C (f\nabla g) \cdot d\mathbf{r} = -\int_C (g\nabla f) \cdot d\mathbf{r}$.

41. Let C be a smooth simple closed curve lying on an orientable surface Σ and let f and g have continuous partial derivatives on Σ. Show that $\int_C (f\nabla g) \cdot d\mathbf{r} = \iint_\Sigma (\nabla f \times \nabla g) \cdot \mathbf{n}\,d\sigma$. Assume that the portion of Σ bounded by C is smooth and simply connected.

CHALLENGE EXERCISES

1. A homogeneous lamina occupies a region in the xy-plane with boundary C (oriented counterclockwise). Show that its center of mass has the coordinates given below, where A is the area of the region.

$$\bar{x} = \frac{1}{2A}\int_C x^2\,dy \qquad \bar{y} = \frac{-1}{2A}\int_C y^2\,dx$$

2. In Problem 1 show that the moment of inertia about the origin is

$$I_o = \frac{1}{3}\int_C (x^3\,dy - y^3\,dx)$$

(Assume that the mass density of the lamina is 1.)

3. Use Problem 1 to find the center of mass of a homogeneous lamina enclosed by the coordinate axes and the hypocycloid $x = a\cos^3 t$, $y = a\sin^3 t$, $0 \le t \le \pi/2$.

4. Find the center of mass of a homogeneous wire in the shape of the hypocycloid $x = a\cos^3 t$, $y = a\sin^3 t$, $0 \le t \le \pi/2$.

5. Find the center of mass of the upper half of the sphere $x^2 + y^2 + z^2 = a^2$ covered by a thin material with mass density at each point proportional to the distance from the xy-plane.

6. Suppose the line integral of $f(x, y)$ along the curve C exists. Use the mean value theorem to establish the formula of theorem (19.2.2). Apply it to the quantity $S(t) = \int_a^t \sqrt{x'(u)^2 + y'(u)^2}\,du$; $S'(t_0)$ is the length of the portion of the curve traced out as t moves from a to t_0.

7. It follows from Problem 6 that the value of a line integral of a function along a smooth curve, given the hypotheses of theorem (19.2.2), is independent of the parameterization. Why? In particular, if x and y are differentiable functions of the arc length as one moves along the curve, describe what form the integral in theorem (19.2.2) takes if we use s as the parameter.

8. We have seen two kinds of line integrals: (*1*) $\int_C P\,dx + Q\,dy$ and (*2*) $\int_C f\,ds$. These are closely connected.

(a) If we parameterize C by arc length, show how an integral (*1*) can be written in the form (*2*).

(b) If arc length can be expressed as a function of x and y, show how an integral (*2*) can be written in the form (*1*). In particular, show that the integral

$$\int_C (x + y^2)\,ds = \int_C \frac{x^2 + xy^2}{\sqrt{x^2 + y^2}}\,dx + \frac{xy + y^3}{\sqrt{x^2 + y^2}}\,dy$$

where C is the curve $x = t$, $y = t$ for $0 \le t \le 1$.

(*Problem 8 continues in the next column*)

8. (*continued*)

(c) More generally, given a differentiable function $\eta(x, y)$, we can construct a line integral (*3*) $\int_C f(x, y)\,d\eta(x, y)$ obtaining an integral of f along C with respect to η. Write this integral in the form (*1*). Suppose we can parameterize C as $x = x(\eta)$, $y = y(\eta)$, and we are given an integral in the form (*1*). Express it in the form (*3*).

(d) Do part (c) with $r = \eta(x, y) = \sqrt{x^2 + y^2}$ and with $\theta = \eta(x, y) = \tan^{-1}(y/x)$.

9. Let R be a closed, simply connected region in the xy-plane, and let R' be a closed, simply connected region in the st-plane. Suppose $x = x(s, t)$ and $y = y(s, t)$ are differentiable and that, as (s, t) is allowed to vary throughout R', (x, y) varies throughout R in a one-to-one correspondence. Suppose further that R has boundary C, which is a closed, piecewise smooth curve, and R' has boundary C', which is a closed, piecewise smooth curve. Finally, suppose that as (s, t) moves along C', $(x(s, t), y(s, t))$ moves along C. Show that

$$\iint_R dx\,dy = \iint_{R'} |(x_s y_t - x_t y_s)|\,ds\,dt$$

by writing $\iint_R dx\,dy = \frac{1}{2}\int_C (x\,dy - y\,dx)$, writing the latter integral in terms of s and t, and applying Green's theorem to the resulting integral (which will have the form $\int_{C'} (P\,ds + Q\,dt)$.

10. Apply the result from Problem 9 to show that for R and R' as above in the xy-plane and the $r\theta$-plane, respectively, we have

$$\iint_R dx\,dy = \iint_{R'} r\,dr\,d\theta$$

11. *Transformation of a Double Integral.* Let $\phi(x, y)$ be continuous and let $P(x, y) = -\int \phi(x, y)\,dy$ and $Q(x, y) = \int \phi(x, y)\,dx$. Generalize the result of Problem 9 using P and Q to show that

$$\iint_R \phi(x, y)\,dx\,dy$$

$$= \iint_{R'} \phi(x(s, t), y(x, s))|x_s y_t - x_t y_s|\,ds\,dt$$

Go through a similar process as in Problem 9, using Green's theorem on the integral $\int_C P\,dx + Q\,dy$.

20 Differential Equations

You have already had some contact with the subject of differential equations through the consideration of problems arising from certain topics in physics. For example, in Section 4.9 in Chapter 4, simple differential equations dealing with the motion of a particle were solved by antidifferentiation. In this chapter we present a more general treatment of the subject, including some important applications.

20.1 Classification of Ordinary Differential Equations

An *ordinary differential equation* is an equation involving x, y, and various derivatives of y. Some examples of ordinary differential equations are given on the next page.

1. $\frac{dy}{dx} + y + 5 = 0$

2. $\frac{x - y}{x + y} + \frac{y - x}{x - y}\frac{dy}{dx} = 0$

3. $7\frac{d^2y}{dx^2} - 5\frac{dy}{dx} + 4x^3y = 0$

4. $\frac{dy}{dx} + \frac{d^3y}{dx^3} = 1$

5. $\left(\frac{dy}{dx}\right)^3 = 8\frac{d^2y}{dx^2}$

6. $5x^2 = 6y^3\left(\frac{dy}{dx}\right)^2 + 18$

The *order* of a differential equation is the order of the highest-order derivative of y appearing in the equation. For the equations above, 1, 2, and 6 are first-order, 3 and 5 are second-order, and 4 is third-order.

The exponent of the highest-order derivative in a differential equation is called the *degree* of the equation. For example, equations 1–5 above are of degree 1, whereas 6 is of degree 2. Being able to recognize the order and degree of a differential equation is an important step in finding its solution. (In this chapter we discuss only differential equations of degree 1.)

An important class of differential equations consists of those called *linear differential equations*. If $P_1, P_2, \ldots, P_n$ and Q are functions of one variable, then

$$\frac{d^ny}{dx^n} + P_1(x)\frac{d^{n-1}y}{dx^{n-1}} + \cdots + P_{n-1}(x)\frac{dy}{dx} + P_n(x)y = Q(x) \tag{1}$$

is called a *linear differential equation of order n*. The term *linear* is used to describe these differential equations, since they arise from certain kinds of functions that in linear algebra are called *linear transformations*. The equation

$$3\frac{d^2y}{dx^2} + 8\frac{dy}{dx} - 5x^3y = 0$$

is an example of a second-order linear differential equation.

If in (1), $Q(x) = 0$ for every x, then the differential equation is called a *homogeneous linear differential equation of order n*.

A function $y = f(x)$ is called a *solution* of a given differential equation on an interval I if the equation is satisfied for all $x \in I$ when $f(x)$ and its derivatives are substituted into the equation.

EXAMPLE 1

Show that $y = -4x^2$ is a solution of the differential equation

$$dy/dx = -8x$$

Solution

Differentiation of y yields immediately $dy/dx = -8x$, so y is a solution of the given differential equation.

EXAMPLE 2

Show that $y = e^{5x}$ is a solution of the differential equation $(dy/dx) + 3y = 8e^{5x}$.

Solution

Notice that $dy/dx = 5e^{5x}$, so that

$$\frac{dy}{dx} + 3y = 5e^{5x} + 3e^{5x} = 8e^{5x}$$

Thus, y is a solution of the given equation.

EXAMPLE 3

Show that $y = \left(4 + \frac{x}{2}\right)e^x + 3e^{-x}$ is a solution of the differential equation

$$\frac{d^2y}{dx^2} - y = e^x.$$

Solution

We observe that

$$\frac{dy}{dx} = \frac{1}{2}e^x + \left(4 + \frac{x}{2}\right)e^x - 3e^{-x} \qquad \frac{d^2y}{dx^2} = \left(5 + \frac{x}{2}\right)e^x + 3e^{-x}$$

Therefore,

$$\left(5 + \frac{x}{2}\right)e^x + 3e^{-x} - \left(4 + \frac{x}{2}\right)e^x - 3e^{-x} = e^x$$

A *general solution* of a differential equation is a set of all solutions of the differential equation given in the form of an equation with arbitrary constants. The number of arbitrary constants in the general solution agrees with the order of the differential equation. For example, it can be shown that every solution of $(d^2y/dx^2) - 12x^2 = 0$ can be written in the form $y = x^4 + C_1x + C_2$, where C_1 and C_2 are constants.

As another example of the use of this terminology, consider the second-order differential equation

$$\frac{d^2y}{dx^2} + y = 0 \tag{2}$$

It can be verified that a general solution is $y = C_1 \sin x + C_2 \cos x$. Equation (2) has a general solution involving two arbitrary constants.

Any solution of a differential equation that can be obtained from the general solution by assigning values to the arbitrary constants is called a *particular solution*. A particular solution of (2) is obtained by choosing values for the arbitrary constants. So, for example, $C_1 = 6$ and $C_2 = 9$ yield the particular solution

$$y = 6 \sin x + 9 \cos x$$

EXAMPLE 4

The function $f(x) = x^4 + 5$ is a particular solution of the differential equation $(d^2y/dx^2) - 12x^2 = 0$.

The main problem in working with differential equations is to find the general solution. Once this is known, it is relatively simple to determine the arbitrary constants (that is, find a particular solution) if *initial conditions* are specified. Sometimes the term *boundary conditions* is used.

Suppose we are required to find a solution for $(dy/dx) - x^2 = 0$ that has the value 5 when x has the value 3. Then we must find a differentiable function f such that $f(3) = 5$ and $f'(x) - x^2 = 0$ for all x in the domain of f. By integrating $f'(x) = x^2$, we find that $\frac{1}{3}x^3 + C$ is the general solution for this equation. We can determine a particular solution $f(x)$ by giving C a value such that f satisfies the given conditions. Substituting 3 for x and 5 for $f(x)$ into the general solution, we obtain

$$f(3) = 5 = \tfrac{1}{3}(3)^3 + C$$

$$5 = 9 + C$$

$$C = -4$$

The required particular solution is $f(x) = \frac{1}{3}x^3 - 4$.

In other words, by specifying a value for the arbitrary constant in the general solution of this first-order differential equation, we can find a particular solution that satisfies a specified initial condition. In this case, the solution $f(x) = \frac{1}{3}x^3 - 4$ satisfies $f(3) = 5$.

EXAMPLE 5

If $y = e^{-x}(x + 1)$, show that y is the particular solution of $(dy/dx) + y = e^{-x}$, such that $y = 1$ when $x = 0$.

Solution

Differentiating y, we find that

$$\frac{dy}{dx} = -xe^{-x}$$

so that

$$\frac{dy}{dx} + y = \underbrace{-xe^{-x}}_{\frac{dy}{dx}} + \underbrace{e^{-x}(x + 1)}_{y} = e^{-x}$$

Therefore, y satisfies the differential equation. Since $y = e^{-0}(0 + 1) = 1$, it follows that y also satisfies the initial condition $y = 1$ when $x = 0$.

Often the solution of a differential equation is given implicitly. In such situations, implicit differentiation can be used to verify that the solution actually satisfies the differential equation. The following example illustrates the technique involved.

EXAMPLE 6

Show that $x^2 - x^3y + 3y^4 = C$ is an implicit solution of

$$\frac{dy}{dx} = \frac{3x^2y - 2x}{12y^3 - x^3} \qquad (3)$$

Solution

If we differentiate $x^2 - x^3y + 3y^4 = C$ implicitly with respect to x, we obtain

$$2x - 3x^2y - x^3\frac{dy}{dx} + 12y^3\frac{dy}{dx} = 0$$

from which we find

$$\frac{dy}{dx} = \frac{3x^2y - 2x}{12y^3 - x^3}$$

In Example 6 we see that the solution is given by $F(x, y) = C$, where $F(x, y) = x^2 - x^3y + 3y^4$. More generally, if $F(x, y) = C$ defines implicitly some function that solves a differential equation, then we refer to $F(x, y) = C$ as an *implicit solution* of the equation. For example, $x^2 + y^2 = 4$ is an implicit solution of the differential equation

$$y^2\frac{dy}{dx} + xy = 0$$

since, upon differentiating $x^2 + y^2 = 4$ implicitly, we have

$$\frac{dy}{dx} = -\frac{x}{y}$$

and so

$$y^2\frac{dy}{dx} + xy = y^2\left(-\frac{x}{y}\right) + xy = -xy + xy = 0$$

EXERCISE 20.1

In Problems 1–10 state the order and degree of each equation and whether the equation is linear or nonlinear. If it is linear, state whether it is homogeneous or nonhomogeneous.

1. $\frac{dy}{dx} + x^2y = xe^x$

2. $\frac{d^3y}{dx^3} + 4\frac{d^2y}{dx^2} - 5\frac{dy}{dx} + 3y = \sin x$

3. $\frac{d^4y}{dx^4} + 3\frac{d^2y}{dx^2} + 5y = 0$

4. $\frac{d^2y}{dx^2} + y \sin x = 0$

5. $\frac{d^2y}{dx^2} + x \sin y = 0$

6. $\frac{d^6x}{dt^6} + \frac{d^4x}{dt^4} + \frac{d^3x}{dt^3} + x = t$

7. $\left(\dfrac{dr}{ds}\right)^3 = \dfrac{d^2r}{ds^2} + 1$

8. $\dfrac{d^2y}{ds^2} + 3s\dfrac{dy}{ds} = y$

9. $x(y'')^3 + (y')^4 - y = 0$

10. $\dfrac{dy}{dx} = 1 - xy + y^2$

In Problems 11–20 verify that the given function is a solution of the differential equation.

11. $y = e^x + 3e^{-x}$, $\dfrac{d^2y}{dx^2} - y = 0$

12. $y = 5 \sin x + 2 \cos x$, $\dfrac{d^2y}{dx^2} + y = 0$

13. $y = \sin 2x$, $\dfrac{d^2y}{dx^2} + 4y = 0$

14. $y = \cos 2x$, $\dfrac{d^2y}{dx^2} + 4y = 0$

15. $y = \dfrac{3}{1 - x^3}$, $\dfrac{dy}{dx} = x^2y^2$

16. $y = 1 + Ce^{-(x^2)/2}$, $\dfrac{dy}{dx} + xy = x$

17. $y = C_1e^{ax} + C_2e^{-ax}$, $\dfrac{d^2y}{dx^2} - a^2y = 0$

18. $y = \sinh x$, $\dfrac{d^2y}{dx^2} = y$

19. $y = \dfrac{a^2kt}{1 + akt}$, $\dfrac{dy}{dt} = k(a - y)^2$

20. $y = \ln(C - e^{-x})$, $\dfrac{dy}{dx} = e^{-(x+y)}$

In Problems 21–24 use implicit differentiation to show that the given equation is a solution of the differential equation.

21. $y^2 + 2xy - x^2 = C$, $\dfrac{dy}{dx} = \dfrac{x - y}{x + y}$

22. $x^2 - y^2 = Cx^2y^2$, $\dfrac{dy}{dx} = \dfrac{y^3}{x^3}$ [*Hint:* Solve for C first.]

23. $x^2 - y^2 = Cx$, $\dfrac{dy}{dx} = \dfrac{x^2 + y^2}{2xy}$ [*Hint:* Solve for C first.]

24. $x^2y + 12x^2 + 16y = 9$, $\dfrac{dy}{dx} = -\dfrac{2xy + 24x}{x^2 + 16}$

In Problems 25–28 verify that y is the particular solution that satisfies the initial condition.

25. $\dfrac{dy}{dx} = \dfrac{y}{x}$; $y = 3$ when $x = 1$, $y = 3x$

26. $\dfrac{dy}{dx} = 3y$; $y = 2$ when $x = 0$, $y = 2e^{3x}$

27. $\dfrac{dy}{dx} = y^2$; $y = 1$, when $x = 0$, $y = \dfrac{1}{1 - x}$

28. $x\dfrac{dy}{dx} + y = x^2$; $y = 4$ when $x = 3$, $y = \dfrac{x^2}{3} + \dfrac{3}{x}$

29. Each of the following is a solution to the differential equation $(d^2y/dx^2) + y = 0$. Show that each can be written in the form $A \sin x + B \cos x$, where A and B are arbitrary constants.

(a) $C_1 \sin(x + C_2)$ (b) $C_1 \cos(x + C_2)$

30. Show that each of the following solutions to $(d^2y/dx^2) + y = 0$ written with two arbitrary constants could be written in terms of a single arbitrary constant.

(a) $C_1 \sin x + 3C_2 \sin x$
(b) $C_1 \sin x + C_2 \cos(x - \frac{9}{2}\pi)$
(c) $C_1 \sin x + C_2 \sin(-x)$

31. Find the values of n so that $y = e^{nx}$ is a solution of $y'' + y' - 6y = 0$.

32. Find a differential equation that has the function defined by $y = e^x + e^{-x}$ as a solution.

Separation of Variables in First-Order Equations

20.2

We shall limit our attention in this section to first-order differential equations, which can be written in *derivative form* as

$$\frac{dy}{dx} = f(x, y) \tag{1}$$

or in *differential form* as

$$M(x, y)\,dx + N(x, y)\,dy = 0 \tag{2}$$

Separable Equations

We start with a procedure for solving a first-order differential equation that is separable.

[20.2.1] DEFINITION / ***Separable Equations.***

A first-order differential equation is said to be *separable* if it can be written in the form

$$M(x)\,dx + N(y)\,dy = 0 \tag{3}$$

where M is a continuous function of x alone and N is a continuous function of y alone.

The following steps are used to solve separable differential equations:

1. Express the given equation in the differential form

$$M(x)\,dx + N(y)\,dy = 0 \qquad \text{or} \qquad M(x)\,dx = -N(y)\,dy$$

2. Integrate to obtain the general solution

$$\int M(x)\,dx + \int N(y)\,dy = C \qquad \text{or} \qquad \int M(x)\,dx = -\int N(y)\,dy + C$$

Some examples will help clarify the idea.

EXAMPLE 1

Solve: $y^2 \dfrac{dy}{dx} = x^3$

Solution

We can write the given equation in the form

$$y^2\,dy = x^3\,dx$$

thus separating the variables. Integration then gives

$$\frac{y^3}{3} = \frac{x^4}{4} + C$$

An implicit solution is therefore

$$4y^3 = 3x^4 + 12C \qquad \text{or} \qquad 4y^3 = 3x^4 + K$$

where $K = 12C$ is an arbitrary constant.

EXAMPLE 2

Solve $\dfrac{dy}{dx} = \dfrac{3y}{x - 1}$ and find the particular solution for which $y = 1$ when $x = 3$.

Solution

If $y \neq 0$, we can separate the variables and write

$$\frac{dy}{y} = \frac{3\,dx}{x - 1}$$

After integrating, we have

$$\ln|y| = C_1 + 3\ln|x - 1|$$

Then by using properties of the natural logarithm function, we find

$$e^{\ln|y|} = e^{[c_1 + \ln|x-1|^3]}$$

$$|y| = e^{c_1}|x - 1|^3$$

Exactly one of the following situations holds:

$$y = e^{c_1}(x - 1)^3 \qquad \text{or} \qquad y = -e^{c_1}(x - 1)^3$$

In either case there is a constant C such that $C = e^{c_1}$ or $C = -e^{c_1}$, whichever is applicable, and we conclude that a general solution for the original differential equation is

$$y = C(x - 1)^3 \tag{4}$$

Even though this equation was derived under the assumption that $y \neq 0$, note that in this case the function defined by $y = 0$ satisfies the given differential equation and can be obtained from (4) by letting $C = 0$.

To find the desired particular solution, set $y = 1$ and $x = 3$ in (4) to obtain $y = \frac{1}{8}(x - 1)^3$.

Homogeneous Equations

Some first-order differential equations that are not separable can be made separable by a change of variable. This is true for differential equations of the form $(dy/dx) = f(x, y)$, where f is a *homogeneous function*.

[20.2.2] DEFINITION / *Homogeneous of Degree k.*

The function $f(x, y)$ is said to be *homogeneous of degree k in x and y* if and only if

$$f(tx, ty) = t^k f(x, y)$$

where k is a real number.

EXAMPLE 3

(a) $f(x, y) = 3x^2 - xy + y^2$ is a homogeneous function of degree 2, since

$$\begin{aligned} f(tx, ty) &= 3(tx)^2 - txty + (ty)^2 = t^2 3x^2 - t^2xy + t^2y^2 \\ &= t^2(3x^2 - xy + y^2) = t^2 f(x, y) \end{aligned}$$

(b) $f(x, y) = \sqrt{x + 4y}$ is a homogeneous function of degree $\frac{1}{2}$, since

$$f(tx, ty) = \sqrt{tx + 4ty} = \sqrt{t(x + 4y)} = t^{1/2}\sqrt{x + 4y} = t^{1/2} f(x, y)$$

(c) $f(x, y) = x/\sqrt{x^2 - y^2}$ is a homogeneous function of degree 0, since

$$f(tx, ty) = \frac{tx}{\sqrt{(tx)^2 - (ty)^2}} = \frac{tx}{\sqrt{t^2(x^2 - y^2)}} = \frac{tx}{t\sqrt{x^2 - y^2}} = t^0 f(x, y)$$

(d) $f(x, y) = x - y^2$ is not a homogeneous function, since

$$f(tx, ty) = tx - (ty)^2 = t(x - ty^2) \neq t(x - y^2) \qquad \text{for all } t$$

[20.2.3] DEFINITION / *Homogeneous Differential Equation.*

A differential equation of the form

$$M(x, y)\, dx + N(x, y)\, dy = 0$$

is said to be *homogeneous* if M and N are homogeneous functions of the same degree.

To solve homogeneous differential equations by the method of separation of variables, we use the following change of variable theorem:

[20.2.4] THEOREM / *Change of Variable for Homogeneous Equations.*

If $M(x, y)\, dx + N(x, y)\, dy = 0$ is homogeneous, then it can be transformed into an equation whose variables are separable by the substitution

$$y = vx$$

where v is a differentiable function of x.

Proof Let $y = vx$. Then $dy = v\, dx + x\, dv$ and, by substitution,

$$M(x, y)\, dx + N(x, y)\, dy = M(x, vx)\, dx + N(x, vx)(v\, dx + x\, dv) = 0$$

If M and N are homogeneous of degree n, it follows that

$$x^n M(1, v)\,dx + x^n N(1, v)(v\,dx + x\,dv) = 0$$

$$M(1, v)\,dx + N(1, v)v\,dx + N(1, v)x\,dv = 0$$

$$[M(1, v) + vN(1, v)]\,dx + N(1, v)x\,dv = 0$$

$$\frac{dx}{x} + \frac{N(1, v)}{M(1, v) + vN(1, v)}\,dv = 0$$

provided no denominators are 0. Moreover, if v is a solution to this separable differential equation, we can reverse the steps to show that $y = vx$ is a solution to the given homogeneous differential equation.

We hasten to point out that the preceding formula should not be memorized; rather the procedure should be worked through each time.

Let's look at an example.

EXAMPLE 4

Solve the differential equation $(x^2 - 3y^2)\,dx + 2xy\,dy = 0$.

Solution

Both $x^2 - 3y^2$ and $2xy$ are homogeneous of degree 2. If we let $y = xv$ and $dy = x\,dv + v\,dx$, it follows that

$$(x^2 - 3(xv)^2)\,dx + 2x(xv)\underbrace{(x\,dv + v\,dx)}_{dy} = 0$$

$$(x^2 - x^2v^2)\,dx + 2x^3v\,dv = 0$$

$$x^2(1 - v^2)\,dx + 2x^3v\,dv = 0$$

Dividing by x^2 and separating variables, we have for $x \neq 0$, $v \neq \pm 1$

$$(1 - v^2)\,dx + 2xv\,dv = 0$$

$$\frac{2v\,dv}{v^2 - 1} = \frac{dx}{x}$$

Integrating, we find

$$\ln|v^2 - 1| = \ln|x| + C_1$$

Since C_1 is a constant, we may write $C_1 = \ln C_2$. Hence,

$$|v^2 - 1| = C_2|x|$$

where C_2 is an arbitrary constant. Now we replace v by y/x:

$$\left|\frac{y^2}{x^2} - 1\right| = C_2|x|$$

$$|y^2 - x^2| = C_2|x|x^2$$

$$y^2 - x^2 = \pm C_2x^3$$

Since C_2 may be either positive or negative, the solution may be written in the simple form

$$y^2 - x^2 = Cx^3$$

where C is an arbitrary constant.

EXERCISE 20.2

In Problems 1–14 solve each differential equation.

1. $\frac{dy}{dx} = xy^{1/2}$

2. $\frac{dy}{dx} = \frac{x^2}{1 + y^2}$

3. $\frac{dy}{dx} = \frac{\sin^2 y}{1 - x^2}$

4. $\cos y \frac{dy}{dx} = \frac{\sin y}{x}$

5. $\frac{dy}{dx} = e^{y-x}$

6. $\frac{dy}{dx} = \left(\frac{y-1}{x-1}\right)^2$

7. $x\frac{dy}{dx} + 2y = 5$

8. $\frac{dy}{dx} - xy^2 = x$

9. $\frac{dy}{dx} = \cos x \frac{dy}{dx} - y \sin x$

10. $\frac{dy}{dx} = y + \sec x \frac{dy}{dx}$

11. $\frac{dy}{dx} + xy = x$

12. $(3x + 1)\, dx + e^{x+y}\, dy = 0$

13. $\ln x \frac{dx}{dy} = \frac{x}{y}$

14. $\frac{dy}{dx} = \frac{x+2}{2-y}$

In Problems 15–24 obtain the particular solution of the differential equation. (Use the results obtained in Problems 1–10.)

15. $\frac{dy}{dx} = xy^{1/2}$; $y = 1$ when $x = 2$

16. $\frac{dy}{dx} = \frac{x^2}{1 + y^2}$; $y = -1$ when $x = 0$

17. $\frac{dy}{dx} = \frac{\sin^2 y}{1 - x^2}$; $y = \frac{\pi}{4}$ when $x = \frac{1}{2}$

18. $\cos y \frac{dy}{dx} = \frac{\sin y}{x}$; $y = \frac{\pi}{3}$ when $x = -1$

19. $\frac{dy}{dx} = e^{y-x}$; $y = 0$ when $x = 0$

20. $\frac{dy}{dx} = \left(\frac{y-1}{x-1}\right)^2$; $y = 2$ when $x = 0$

21. $x\frac{dy}{dx} + 2y = 5$; $y = 1$ when $x = -1$

22. $\frac{dy}{dx} - xy^2 = x$; $y = 1$ when $x = 2$

23. $\frac{dy}{dx} = \cos x \frac{dy}{dx} - y \sin x$; $y = \frac{1}{2}$ when $x = \pi$

24. $\frac{dy}{dx} = y + \sec x \frac{dy}{dx}$; $y = 2$ when $x = \frac{\pi}{4}$

In Problems 25–34 determine whether the function is homogeneous and, if so, determine the degree.

25. $f(x, y) = 2x^2 - 3xy - y^2$

26. $f(x, y) = x^3 - xy^2 + y^3$

27. $f(x, y) = x^3 - xy + y^3$

28. $f(x, y) = x^2 - xy^2 + y^2$

29. $f(x, y) = 2x + \sqrt{x^2 + y^2}$

30. $f(x, y) = \sqrt{x + y}$

31. $f(x, y) = \tan(3x/y)$

32. $f(x, y) = e^{x/y}$

33. $f(x, y) = \ln(x/y)$

34. $f(x, y) = x \ln x - x \ln y$

In Problems 35–48 solve the homogeneous differential equation.

35. $(x - y)\, dx + x\, dy = 0$

36. $(x + y)\, dx + x\, dy = 0$

37. $(x^2 + y^2)\, dx + (x^2 - xy)\, dy = 0$

38. $xy\,dx + (x^2 + y^2)\,dy = 0$

39. $\dfrac{dy}{dx} = \dfrac{y - x}{y + x}$

40. $\dfrac{dy}{dx} = \dfrac{x + 2y}{2x + y}$

41. $x(x^2 - y^2)\dfrac{dy}{dx} - y(x^2 + y^2) = 0$

42. $x^2\dfrac{dy}{dx} = x^2 - xy - 2y^2$

43. $\dfrac{dy}{dx} = \dfrac{2xy}{x^2 + y^2}$

44. $\dfrac{dy}{dx} = \dfrac{x + 2y}{2x - y}$

45. $\dfrac{dy}{dx} = \dfrac{2xy}{x^2 + y^2}$; $y = 1$ when $x = 2$

46. $x^2\dfrac{dy}{dx} = x^2 - xy - 2y^2$; $y = 1$ when $x = -1$

47. $x(x^2 - y^2)\dfrac{dy}{dx} - y(x^2 + y^2) = 0$; $y = -2$ when $x = -1$

48. $\dfrac{dy}{dx} = \dfrac{x + 2y - 3}{2x - y + 1}$; $y = 4$ when $x = 3$ [*Hint:* Let $x = u + h$ and $y = v + k$, where h and k are constants chosen so as to eliminate the constant terms on the right side.]

49. If we assume a simplified mathematical model, the rate of growth of a fad at any time t among a population is proportional to the product xy, where x is the number who have adopted the fad and y is the number who have not adopted the fad at time t (assuming that all the members of the population are in contact with each other). Suppose that on a certain day ($t = 0$) 2 members from a club of 30 members begin wearing a new style of clothing. The next day one more member adopts the clothing style.

(a) Express the number of members x in the club who have adopted the style as a function of t days.

(b) In approximately how many days will half of the club members have adopted the style?

50. From physics it is known that the rate of flow of water from a tank through an orifice of area A square feet is given approximately by $kA\sqrt{2gh}$, where $k = 0.6$, $g = 32$ feet per second per second, and h is the height of the water level in feet above the orifice. Suppose a cylindrical tank with height 6 feet and diameter 3 feet is filled with water. At time $t = 0$ a valve is opened and the water begins draining through a hole of diameter 1 inch in the bottom of the tank.

(a) Express the height of the water in the tank as a function of time t.

(b) How much time is required to empty the tank?

20.3 Exact Differential Equations

We know that when a first-order differential equation can be put in the form

$$P(x)\,dx + Q(y)\,dy = 0$$

a set of solutions can be determined by integration—that is, by finding a function whose differential is $P(x)\,dx + Q(y)\,dy$. This idea can be extended to some differential equations of the form

$$M(x, y)\,dx + N(x, y)\,dy = 0 \qquad \textbf{(1)}$$

in which separation of variables may not be possible.

[20.3.1] DEFINITION / *Exact Differential Equation.*

The differential equation

$$\boldsymbol{M(x, y)\,dx + N(x, y)\,dy = 0} \qquad \textbf{(2)}$$

is said to be an *exact differential equation* in a simply connected, open set R if there is a function f of two variables with continuous partial derivatives such that

$$\frac{\partial f}{\partial x} = M \quad \text{and} \quad \frac{\partial f}{\partial y} = N \tag{3}$$

for all (x, y) in R.

Note that the exactness of (2) is equivalent to finding a (potential) function whose gradient is $M(x, y)\mathbf{i} + N(x, y)\mathbf{j}$, and so by theorem (19.3.7) we have the following theorem:

[20.3.2] THEOREM / *Test for Exactness.*

The differential equation

$$M(x, y)\,dx + N(x, y)\,dy = 0$$

is exact at every point (x, y) in the plane if and only if

$$\frac{\partial M}{\partial y} = \frac{\partial N}{\partial x}$$

If $M(x, y)\,dx + N(x, y)\,dy$ is exact, then it can be written in the equivalent form $f_x\,dx + f_y\,dy = 0$, where f is such that $f_x = M$ and $f_y = N$. But $f_x\,dx + f_y\,dy = df$, the total differential of f, and so this equation is equivalent to the equation $df = 0$, whose solution is $f(x, y) = C$. Thus, solving an exact differential equation is exactly the same as finding a (potential) function whose gradient is $M\mathbf{i} + N\mathbf{j}$. A method for determining f is described in Section 19.2.

EXAMPLE 1

Show that the differential equation

$$(2x + 3x^2y)\,dx + (x^3 + 2y - 3y^2)\,dy = 0$$

is exact, and find the general solution.

Solution

Let $M = 2x + 3x^2y$ and $N = x^3 + 2y - 3y^2$, and compute

$$\frac{\partial M}{\partial y} = 3x^2 \quad \text{and} \quad \frac{\partial N}{\partial x} = 3x^2$$

We conclude that the differential equation is exact. Let $f(x, y) = C$ be the required implicit solution. Then

$$df = \frac{\partial f}{\partial x}\,dx + \frac{\partial f}{\partial y}\,dy = (2x + 3x^2y)\,dx + (x^3 + 2y - 3y^2)\,dy$$

As a result,

$$\frac{\partial f}{\partial x} = M = 2x + 3x^2y \tag{4}$$

$$\frac{\partial f}{\partial y} = N = x^3 + 2y - 3y^2 \tag{5}$$

We shall find f from (4) by integrating both sides with respect to x while holding y constant. The result is

$$f(x, y) = x^2 + x^3y + B(y) \tag{6}$$

where the usual arbitrary constant of integration is now a function $B(y)$ of y alone, which is as yet unknown. To determine $B(y)$, we use the fact that the function f in (6) must also satisfy (5). Hence,

$$\frac{\partial f}{\partial y} = x^3 + B'(y) \qquad \text{and} \qquad \frac{\partial f}{\partial y} = x^3 + 2y - 3y^2$$

Thus,

$$B'(y) = 2y - 3y^2$$
$$B(y) = y^2 - y^3 + C_1$$

Replacing $B(y)$ in (6), we find

$$f(x, y) = x^2 + x^3y + y^2 - y^3 + C_1$$

Hence, an implicit solution of the given differential equation is

$$x^2 + x^3y + y^2 - y^3 = C$$

where the constant C_1 has been absorbed by the constant C.

EXAMPLE 2

Show that the differential equation

$$(\cos y - \cos x)\, dx + (e^y - x \sin y)\, dy = 0$$

is exact, and find the particular solution such that $y(\pi) = 0$, that is, $y = 0$ when $x = \pi$.

Solution

Let $M = \cos y - \cos x$ and $N = e^y - x \sin y$, and compute

$$\frac{\partial M}{\partial y} = -\sin y \qquad \text{and} \qquad \frac{\partial N}{\partial x} = -\sin y$$

We conclude that the differential equation is exact. Let $f(x, y) = C$ be the required solution. Then

$$df = \frac{\partial f}{\partial x}\, dx + \frac{\partial f}{\partial y}\, dy = (\cos y - \cos x)\, dx + (e^y - x \sin y)\, dy$$

Hence,

$$\frac{\partial f}{\partial x} = M = \cos y - \cos x \tag{7}$$

$$\frac{\partial f}{\partial y} = N = e^y - x \sin y \tag{8}$$

We shall find f from (7) by integrating both sides with respect to x while holding y constant. The result is

$$f(x, y) = x \cos y - \sin x + B(y) \tag{9}$$

where B is a function of y alone, which is yet unknown. From (8) and (9), we find

$$\frac{\partial f}{\partial y} = -x \sin y + B'(y) \qquad \text{and} \qquad \frac{\partial f}{\partial y} = e^y - x \sin y$$

Hence,

$$B'(y) = e^y$$
$$B(y) = e^y + C_1$$

Thus,

$$f(x, y) = x \cos y - \sin x + e^y + C_1$$

and the general solution is

$$x \cos y - \sin x + e^y = C \tag{10}$$

where, again, the constant C_1 has been absorbed by the constant C. We now determine C in (10) so that the condition $y = 0$ when $x = \pi$ is satisfied:

$$\pi \cos 0 - \sin \pi + e^0 = C$$
$$C = 1 + \pi$$

The required particular solution is therefore

$$x \cos y - \sin x + e^y = 1 + \pi$$

EXAMPLE 3

The equation

$$-y\,dx + x\,dy = 0$$

is not exact, since

$$\frac{\partial}{\partial y}(-y) \neq \frac{\partial}{\partial x}(x)$$

We can, however, convert this into an exact equation by multiplying by $1/x^2$. The differential equation

$$-\frac{y}{x^2}\,dx+\frac{1}{x}\,dy=0$$

is exact because

$$\frac{\partial}{\partial y}\left(-\frac{y}{x^2}\right)=\frac{\partial}{\partial x}\left(\frac{1}{x}\right)=-\frac{1}{x^2}$$

A term (such as $1/x^2$ in Example 3) that converts a nonexact equation into an exact equation is one example of what is known as an *integrating factor*, a term that converts a differential equation into a form that can be solved by a simple integration. Every differential equation with a solution of the form $u(x, y)=C$ has an infinite number of integrating factors that make the problem exact. However, it is in general as difficult to find an integrating factor that makes the problem exact as it is to solve the original equation. Problems 24 and 25 consider two special cases when an integrating factor can be obtained quite readily.

EXERCISE 20.3

In Problems 1–18 show that the given equation is exact and solve.

1. $(4x-2y+5)\,dx+(2y-2x)\,dy=0$

2. $(3x^2+3xy^2)\,dx+(3x^2y-3y^2+2y)\,dy=0$

3. $(a^2-2xy-y^2)\,dx-(x+y)^2\,dy=0$

4. $(2ax+by+g)\,dx+(2e^y+bx+h)\,dy=0$

5. $\frac{1}{y}\,dx-\frac{x}{y^2}\,dy=0$

6. $\frac{y\,dx-x\,dy}{x^2}=0$

7. $(x-1)^{-1}\,y\,dx+[\ln(2x-2)+y^{-1}]\,dy=0$

8. $2xy^{-1}\,dy+(2\ln(5y)+x^{-1})\,dx=0$

9. $(x+3)^{-1}\cos y\,dx-[\sin y\ln(5x+15)-y-1]\,dy=0$

10. $p^2\sec 2\theta\tan 2\theta\,d\theta+p(\sec 2\theta+2)\,dp=0$

11. $\cos(x+y^2)\,dx+2y\cos(x+y^2)\,dy=0$

12. $(\sin 2\theta-2p\cos 2\theta)\,dp+(2p\cos 2\theta+2p^2\sin 2\theta)\,d\theta=0$

13. $e^{2x}(dy+2y\,dx)=x^2\,dx$

14. $e^{x^2}(dy+2xy\,dx)=3x^2\,dx$

15. $\left[\frac{1}{x+y}+y^2\right]dx+\left[\frac{1}{x+y}+2xy\right]dy=0$

16. $\frac{y^2-2x^2}{xy^2-x^3}\,dx+\frac{2y^2-x^2}{y^3-x^2y}\,dy=0$

17. $2y^3\sin 2x\,dx-3y^2\cos 2x\,dy=0$

18. $\frac{3y^2\,dx}{x^2+3x}+\left(2y\ln\frac{5x}{x+3}+3\sin y\right)dy=0$

In Problems 19–22 each of the differential equations is exact. Find the particular solution.

19. $(1+y^2+xy^2)\,dx+(x^2y+y+2xy)\,dy=0;\quad y(1)=1$

20. $(3x^2y^{-1}+2x)\,dx+(y^2-x^3y^{-2})\,dy=0;\quad y(3)=3$

21. $[2xy-\sin(x)]\,dx=(2y-x^2)\,dy;\quad y(0)=1$

22. $y[y+\sin(x)]\,dx-\left[\cos(x)-2xy+\frac{1}{1+y^2}\right]dy=0;$ $y(0)=1$

23. Show that any differential equation in which the variables can be separated is exact.

24. Suppose the equation $M(x, y)\,dx + N(x, y)\,dy = 0$ has the property that $(N_x - M_y)/M$ is a function of y only and that

$$u(y) = e^{\left[\int \frac{N_x - M_y}{M}\,dy\right]}$$

Show that $u(y)[M(x, y)\,dx + N(x, y)\,dy] = 0$ is exact.

25. Suppose the equation $M(x, y)\,dx + N(x, y)\,dy = 0$ has the property that $(M_y - N_x)/N$ is a function of x only and that

$$u(x) = e^{\left[\int \frac{M_y - N_x}{N}\,dx\right]}$$

Show that $u(x)[M(x, y)\,dx + N(x, y)\,dy] = 0$ is exact.

In Problems 26–32:

(a) Change the given equation to an exact equation by using one of the techniques described in Problems 24 and 25.
(b) Solve the resulting equation and verify that the solution also satisfies the given equation.

26. $2y\,dx + x\,dy = 0$

27. $y\,dx + 3x\,dy = 0$

28. $(x^2 + y^2 + 1)\,dx + (x^2 - 2xy)\,dy = 0$

29. $4x^2y\,dx + (x^3 + y)\,dy = 0$

30. $y\,dx + (x^2y - x)\,dy = 0$

31. $(\cos y + x)\,dx + x \sin y\,dy = 0$

32. $(x^2 - x \sin y)\,dx + x^2 \cos y\,dy = 0$

First-Order Linear Differential Equations 20.4

In this section we introduce integrating factors to solve one of the most important classes of first-order differential equations—first-order *linear* differential equations.

[20.4.1] DEFINITION / *First-Order Linear Differential Equation.*

A first-order differential equation is said to be *linear* if it can be written in the form

$$\frac{dy}{dx} + P(x)y = Q(x) \qquad (1)$$

where P and Q are continuous functions of x.

The differential equation

$$\frac{dy}{dx} - \frac{3}{x}y = x^2$$

is an example of a first-order linear differential equation.

If in (1) $Q = 0$, then we can write the equation in the separable form

$$\frac{dy}{y} + P(x)\,dx = 0$$

which can be solved to give

$$\ln|y| + \int P(x)\,dx = C$$

Taking the exponential of both sides, we get

$$|y|e^{\int P(x)\,dx} = e^C$$

Note that by the product rule and the chain rule, we have

$$\frac{d}{dx}\left(ye^{\int P(x)\,dx}\right) = \frac{dy}{dx}e^{\int P(x)\,dx} + P(x)ye^{\int P(x)\,dx} \tag{2}$$

Thus, if $Q(x)$ is not the zero function, we can multiply (1) through by $e^{\int P(x)\,dx}$ to obtain

$$\frac{dy}{dx}e^{\int P(x)\,dx} + P(x)ye^{\int P(x)\,dx} = Q(x)e^{\int P(x)\,dx}$$

which from (2) is equivalent to

$$\frac{d}{dx}\left(ye^{\int P(x)\,dx}\right) = Q(x)e^{\int P(x)\,dx}$$

Integrating, we obtain,

$$ye^{\int P(x)\,dx} = \int Q(x)e^{\int P(x)\,dx}\,dx + C$$

or, equivalently,

$$y = e^{-\int P(x)\,dx}\left[\int Q(x)e^{\int P(x)\,dx} + C\right]$$

The expression $e^{\int P(x)\,dx}$ by which we multiplied (1) in order to obtain (2) is called an integrating factor for the differential equation.

[20.4.2] THEOREM / *Solution of a First-Order Linear Differential Equation.*

The general solution of $y' + P(x)y = Q(x)$ is

$$ye^{\int P(x)\,dx} = \int Q(x)e^{\int P(x)\,dx}\,dx + C \tag{3}$$

Rather than trying to memorize (3), just remember that multiplying by the integrating factor $e^{\int P(x)\,dx}$ transforms the left side of the differential equation into the derivative of the product $ye^{\int P(x)\,dx}$.

EXAMPLE 1

Solve the linear differential equation

$$\frac{dy}{dx} + \frac{3y}{x} = x^2 \tag{4}$$

Solution

Here $P(x) = 3/x$ and $Q(x) = x^2$. From (2), an integrating factor for (4) is of the form

$$e^{\int P(x)\,dx} = e^{\int(3/x)\,dx} = e^{3\ln|x| + C} = e^C|x|^3 \tag{5}$$

For convenience we shall let $C = 0$. It follows that for $x > 0$, x^3 is an integrating factor for (4), and for $x < 0$, $-x^3$ is an integrating factor. In either case multiplication of each member of (4) by x^3 then yields

$$x^3\frac{dy}{dx} + 3x^2y = x^5$$

$$\frac{d}{dx}(x^3y) = x^5$$

$$x^3y = \int x^5\,dx = \frac{x^6}{6} + C$$

Hence, a general solution of (4) is

$$y = \frac{x^3}{6} + \frac{C}{x^3}$$

Extensions of Linear Differential Equations

The equation

$$\frac{dy}{dx} + P(x)y = Q(x)y^n \qquad n \neq 0, \quad n \neq 1 \tag{6}$$

where P and Q are functions of x only, is called *Bernoulli's equation*, after James Bernoulli, who studied it in 1695.

To solve (6), we multiply by y^{-n} to get

$$y^{-n}\frac{dy}{dx} + P(x)y^{-n+1} = Q(x) \tag{7}$$

We make the substitution $v = y^{-n+1}$. Then $(dv/dx) = (1 - n)y^{-n}(dy/dx)$ and $y^{-n}(dy/dx) = (1/(1 - n))(dv/dx)$, where $n \neq 1$. Substituting into (7), we obtain the equation

$$\frac{1}{1 - n}\frac{dv}{dx} + P(x)v = Q(x)$$

or

$$\frac{dv}{dx} + (1 - n)P(x)v = (1 - n)Q(x) \tag{8}$$

which is a linear equation in x and v. The solution method for a Bernoulli equation should be clear; namely, transform an equation of the form (6) into the linear equation (8) in terms of x and v. Find the general solution of (8) and then rewrite the general solution in terms of the original variables x and y.

EXAMPLE 2

Determine the general solution of the differential equation

$$\frac{dy}{dx} + \frac{1}{x}y = (\ln x)y^2$$

Solution

This is a Bernoulli equation with $P(x) = 1/x$, $Q(x) = \ln x$, and $n = 2$. Multiply the differential equation by y^{-2} to obtain

$$y^{-2}\frac{dy}{dx} + \frac{1}{x}y^{-1} = \ln(x)$$

Now let $v = y^{-1}$. Then $(dv/dx) = -y^{-2}(dy/dx)$ or $-(dv/dx) = y^{-2}(dy/dx)$. Substituting these expressions into our new equation yields

$$-\frac{dv}{dx} + \frac{1}{x}v = \ln(x)$$

$$\frac{dv}{dx} - \frac{1}{x}v = -\ln(x)$$

which is linear. Using the method for solving linear equations, we multiply by

$$e^{\int(-1/x)\,dx} = e^{-\ln x} = e^{\ln(x^{-1})} = \frac{1}{x}$$

This produces

$$\frac{1}{x}\frac{dv}{dx} - \frac{1}{x^2}v = \frac{-1}{x}\ln(x)$$

$$\frac{d}{dx}\left(\frac{1}{x}v\right) = \frac{-1}{x}\ln(x)$$

Integrating this equation yields

$$\frac{1}{x}v = -\int \frac{1}{x}\ln(x)\,dx + C = -\frac{1}{2}(\ln x)^2 + C$$

$$v = Cx - \frac{1}{2}x(\ln x)^2$$

Finally, since $v = y^{-1} = 1/y$ we have

$$\frac{1}{y} = Cx - \frac{1}{2}x(\ln x)^2$$

$$y = \frac{1}{Cx - \frac{1}{2}x(\ln x)^2}$$

EXAMPLE 3

Determine the general solution of the differential equation

$$\frac{dy}{dx} = 4y + 2e^x y^{1/2}$$

Solution

First, we write the equation in the standard form

$$y' - 4y = 2e^x y^{1/2}$$

This is a Bernoulli equation with $P(x) = -4$, $Q(x) = 2e^x$, and $n = \frac{1}{2}$. We multiply the equation by $y^{-1/2}$ to obtain

$$y^{-1/2}y' - 4y^{1/2} = 2e^x$$

and let $v = y^{1/2}$. Then $v' = \frac{1}{2}y^{-1/2}y'$ or $2v' = y^{-1/2}y'$. Substituting into our new equation, we have

$$2v' - 4v = 2e^x$$

$$v' - 2v = e^x$$

which is linear in x and v. We multiply this equation by $e^{\int -2\,dx} = e^{-2x}$. This yields the equation

$$e^{-2x}v' - 2e^{-2x}v = e^{-x}$$

$$(e^{-2x}v)' = e^{-x}$$

Integrating, we have

$$e^{-2x}v = -e^{-x} + C$$

$$v = Ce^{2x} - e^x$$

Finally, since $v = y^{1/2}$,

$$y^{1/2} = Ce^{2x} - e^x$$

$$y = (Ce^{2x} - e^x)^2$$

is the general solution of the given differential equation.

EXERCISE 20.4

In Problems 1–22 solve the given equation after finding the required integrating factor.

1. $(2x + y)\,dx - x\,dy = 0$

2. $\dfrac{dy}{dx} + \dfrac{2y}{x} = x^2 + 1$

3. $x\dfrac{dy}{dx} - y = x^2e^x$

4. $\dfrac{dy}{dx} + 2xy = 2x$

5. $\dfrac{dy}{dx} + \dfrac{1}{x}y = 3x$

6. $\dfrac{dy}{dx} + \dfrac{y}{x} = x^2$

7. $\frac{dy}{dx} - 2y = e^{-x}$

8. $\frac{dy}{dx} - \frac{y}{x} = x^{3/2}$

9. $\frac{dr}{d\theta} + \frac{4r}{\theta} = \theta$

10. $\frac{dy}{dx} + e^x y = e^x$

11. $\frac{dy}{dx} - \frac{2y}{x+1} = 3(x+1)^2$

12. $\frac{dy}{dx} + y \cot x = \csc^2 x$

13. $\frac{dy}{dx} + y \tan x = \cos^2 x$

14. $\frac{dy}{dx} - y \csc x = \sin 2x$

15. $(1 + x^2)\frac{dy}{dx} + xy = x^3$

16. $\cos x \frac{dy}{dx} + y = \sec x$

17. $(x^2 + 2y^2)\, dx + xy\, dy = 0$

18. $(3x^2y^2 - e^y x^4)\frac{dy}{dx} = 2xy^3$

19. $\frac{dy}{dx} = \frac{y}{x - y^2}$

20. $\frac{dy}{dx} + y \tan x = \cos x$

21. $\frac{dy}{dx} + 2xy = y$

22. $dx + (2x - y^2)\, dy = 0$

In Problems 23–26 find the particular solution satisfying the given condition.

23. $\frac{dy}{dx} + y = e^{-x}$; $y = 5$ when $x = 0$

24. $\frac{dy}{dx} + \frac{2y}{x} = \frac{4}{x}$; $y = 6$ when $x = 1$

25. $\frac{dy}{dx} + \frac{y}{x} = e^x$; $y = e^{-1}$ when $x = -1$

26. $\frac{dy}{dx} + y \cot x = 2 \cos x$; $y = 3$ when $x = \frac{\pi}{2}$

In Problems 27–32 solve the Bernoulli equation.

27. $dy + x^{-1}y\, dx = 3x^2y^2\, dx$

28. $dy + y\, dx = 2xy^2e^x\, dx$

29. $dx + 2xy^{-1}\, dy = 2x^2y^2\, dy$

30. $2\frac{dy}{dx} - yx^{-1} = 5x^3y^3$

31. $dx - 2xy\, dy = 6x^3y^2e^{-2y^2}\, dy$

32. $3\frac{dy}{dx} + 3x^{-1}y = 2x^2y^4$

20.5 Applications

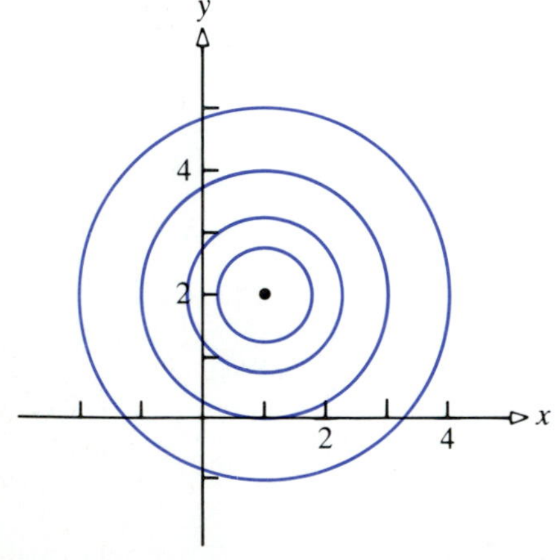

Figure 1

As discussed previously, many important problems in science, engineering, the social sciences, and business can be formulated mathematically as differential equations. A variety of applications in these fields have been treated in Chapters 4 and 7. These included rectilinear motion, exponential growth and decay, continuous compounding of interest, and Newton's law of cooling. In this section we consider a few additional applications of first-order differential equations, including some geometric problems.

A Geometric Application: Orthogonal Trajectories

Consider the one-parameter family of circles

$$(x-1)^2 + (y-2)^2 = C \qquad C \ge 0 \tag{1}$$

with center at the point (1, 2) and (arbitrary) radius $\sqrt{C}$ (see Fig. 1). If we differentiate (1) with respect to x, we get

$$2(x-1)+2(y-2)y'=0$$

$$y'=-\frac{x-1}{y-2} \tag{2}$$

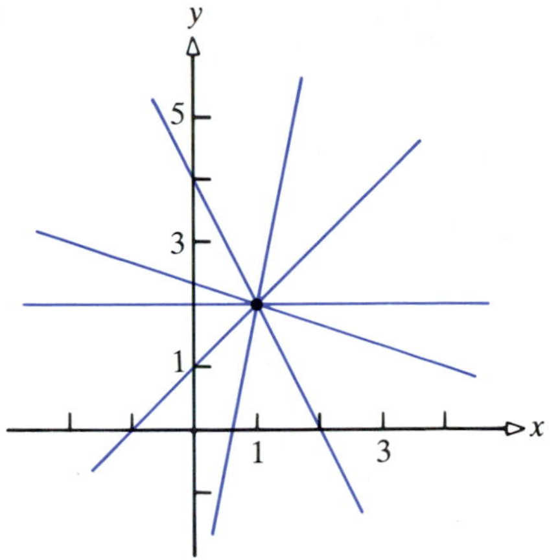

Figure 2

This is the differential equation of the family. Note also that if we choose a specific point (x, y), $(y \neq 2)$, on one of the circles, then (2) gives the slope of the tangent line at (x, y).

Now consider the family of straight lines passing through the point $(1, 2)$ (see Fig. 2):

$$y-2=K(x-1) \tag{3}$$

The differential equation for this family is

$$y'=\frac{y-2}{x-1} \tag{4}$$

Comparing (2) and (4), we see that the right side of (2) is the negative reciprocal of the right side of (4). From this, we conclude that if (x, y) is a point of intersection of one of the circles in (1) and one of the lines in (3), then the line and the circle are perpendicular (orthogonal) to each other at this point.

A curve that intersects each member of a given family of curves at right angles is called an *orthogonal trajectory* of the given family. Each line in the family (3) is an orthogonal trajectory of the family of circles (1), and conversely. In general, if $F(x, y, C)=0$ and $G(x, y, K)=0$ are one-parameter families of curves, such that each member of one family intersects the members of the other family orthogonally, then the two families are said to be *orthogonal trajectories* of each other.

Orthogonal trajectories occur naturally in a variety of physical applications. For example, iron filings sprinkled on a pane of glass over a bar magnet arrange themselves along a family of curved lines indicating the direction of magnetic force. A family of orthogonal projections can be determined by locating all the points in the plane of the glass with the same potential energy.

A procedure for finding a family $G(x, y, K)=0$ of orthogonal trajectories for a given family $F(x, y, C)=0$ is given below.

Step 1 Determine the differential equation for the given family.

Step 2 Replace y' in this differential equation by $-1/y'$; the resulting equation is the differential equation for the family of orthogonal trajectories.

Step 3 Determine the general solution of the equation obtained in Step 2.

EXAMPLE 1

Determine the family of orthogonal trajectories for the one-parameter family $y^2=Cx^3$.

Solution

Differentiate the given equation with respect to x to obtain

$$2yy' = 3Cx^2$$

Eliminating C between the two equations, we have

$$2yy' = \frac{3y^2x^2}{x^3}$$

$$y' = \frac{3y}{2x}$$

Now the differential equation for the family of orthogonal trajectories is

$$\frac{-1}{y'} = \frac{3y}{2x}$$

$$y' = -\frac{2x}{3y}$$

This is a separable equation that can be written as

$$2x\,dx + 3y\,dy = 0$$

The general solution is

$$x^2 + \frac{3}{2}y^2 = K$$

Thus, the orthogonal trajectories for the family $y^2 = Cx^3$ is a family of ellipses $x^2 + \frac{3}{2}y^2 = K$.

Growth and Decay

Unlimited exponential growth and exponential decay (e.g., radioactive decay) are represented mathematically by the first-order differential equation

$$\frac{dy}{dt} = ky$$

where k is a constant. These models were discussed in Chapter 7. Note that the differential equation can be treated either as a separable equation or as a linear equation. The general solution is

$$y(t) = Ce^{kt} \qquad \text{or} \qquad y(t) = y(0)e^{kt}$$

It is easy to verify that

$$\lim_{t\to+\infty} y(t) = +\infty \quad \text{if} \quad k > 0 \qquad \text{and} \qquad \lim_{t\to+\infty} y(t) = 0 \quad \text{if} \quad k < 0$$

In contrast to unlimited growth or decay to 0, there are many situations in which the growth or decay is limited by some natural barrier. For example, a new company entering a market in which the total sales in the market is $\$M$ might expect its annual sales to grow, but the sales cannot exceed $\$M$. A possible model for the company to use to predict the growth of its sales is to assume that the rate of growth is proportional to the difference between its sales at time t and the upper limit $\$M$. If $y = y(t)$ denotes the sales at time t, then the mathematical representation of this problem is

$$\frac{dy}{dt} = k(M - y) \tag{5}$$

where k is a positive constant.

The differential equation (5) also expresses Newton's law of heating, with $y(t)$ denoting the temperature of an object at time t units after it is placed in a warmer surrounding medium that is at the constant temperature M. Of course, the temperature of the object will never exceed M. See also Newton's law of cooling, which is treated in Chapter 7. Also, social scientists have used this differential equation to model the learning of certain specific skills—for example, a person learning to type.

Equation (5) can be treated either as a separable equation or as a linear equation. In either case, the general solution is found to be

$$y(t) = M + Ce^{-kt}$$

If we assume that $y(0) = R \geq 0$, then we get $R = M + C$ or $C = R - M$, and

$$y(t) = M + (R - M)e^{-kt}$$

See Figure 3.

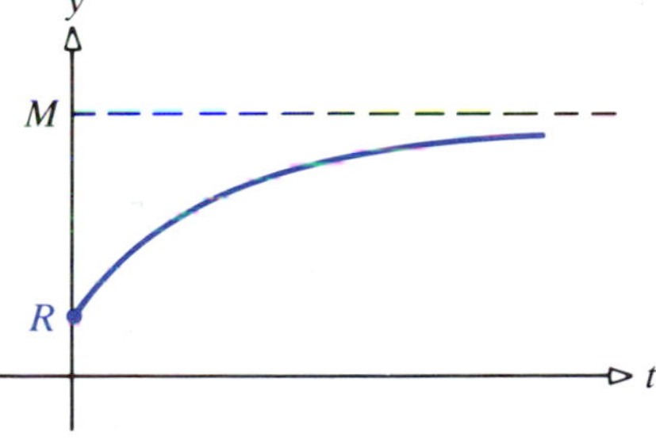

Figure 3

In the special case $R = 0$, we have

$$y(t) = M(1 - e^{-kt})$$

The function $y = y(t)$ is increasing, its graph is concave downward, and $\lim_{t \to +\infty} y(t) = M$.

EXAMPLE 2

The annual sales of a new company are expected to grow at a rate proportional to the difference between the sales at time t and an upper limit of \$5 million. Suppose the sales are 0 initially and are \$1 million after 4 years of operation. Determine the annual sales at any time t. How long will it take for the annual sales to reach \$4 million?

Solution

Using the general discussion above, we find that the sales at time t are given by

$$y(t) = 5(1 - e^{-kt})$$

We use the boundary condition $y(4) = 1$ to determine the value of the constant k:

$$1 = 5(1 - e^{-4k})$$
$$e^{-4k} = 0.8$$
$$k = \frac{-\ln(0.8)}{4} \approx 0.056$$

Thus,

$$y(t) = 5(1 - e^{-0.056t})$$

To find out how long it will take for sales to reach \$4 million, we solve

$$4 = 5(1 - e^{-0.056t})$$
$$-0.056t = \ln(0.2)$$
$$t = \frac{\ln(0.2)}{-0.056} \approx 29 \text{ years}$$

Logistic Equations

In the mid-nineteenth century the Belgian mathematical biologist P. F. Verhulst used the differential equation

$$\frac{dy}{dt} = ky(M - y) \tag{6}$$

where k and M are positive constants, to predict the human population of various countries. This equation is now known as the *logistic equation* and its solutions are called *logistic functions*. Rewriting the equation as

$$\frac{dy}{dt} = kMy - ky^2$$

we can interpret the first term on the right side, kMy, as a growth rate and the second term, $-ky^2$, as an "inhibition" or "competition" term that has the effect of retarding growth.

This differential equation has been useful in predicting the growth of certain types of bacteria. It also provides a reasonable model for describing the spread of an epidemic caused by introducing an infected individual into a population. In this instance, if M denotes the size of the population and $y = y(t)$ denotes the number of infected individuals at time t, then the differential equation states that the spread (rate of growth) of the disease is proportional to the product of the number of infected individuals and the number of individuals who have not been infected. Social scientists have used this differential equation to study the spread of information throughout a population, and economists have used it to predict such things as growth of sales, effects of advertising, and company growth.

Equation (6) can be treated either as a separable equation or as a Bernoulli equation. Using either solution method, the general solution is found to be

$$y(t) = \frac{M}{1 + Ce^{-kMt}}$$

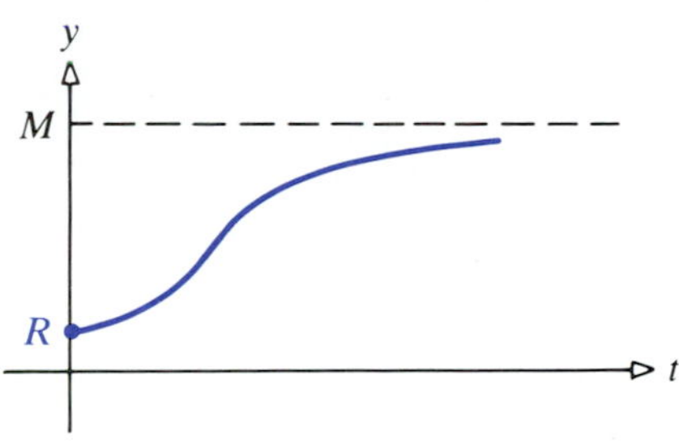

Figure 4

If we assume that $y(0) = R,\ \ 0 \le R < M,$ then $R = M/(1 + C)$ or $C = (M - R)/R$ and

$$y(t) = \frac{RM}{R + (M - R)e^{-kMt}}$$

Note that $\lim_{t \to +\infty} y(t) = M.$ The graph of $y = y(t)$ is given in Figure 4.

EXAMPLE 3

An influenza epidemic is spreading throughout a population of 50,000 people at a rate that is proportional to the product of the number of infected people and the number of well people. Suppose that 100 people were infected initially and 1000 were infected after 10 days. Determine the number of infected people at any time t. How long will it take to have half the population infected?

Solution

By the general result above, we have

$$y(t) = \frac{100(50{,}000)}{100 + 49{,}900e^{-50{,}000kt}} = \frac{50{,}000}{1 + 499e^{-50{,}000kt}}$$

where k can be determined from the boundary condition $y(10) = 1000.$ We have

$$1000 = \frac{50{,}000}{1 + 499e^{-500{,}000k}}$$

$$499e^{-500{,}000k} = 49$$

$$-500{,}000k = \ln\left(\frac{49}{499}\right)$$

$$k \approx 0.0000046$$

Thus,

$$y(t) = \frac{50{,}000}{1 + 499e^{-0.23t}}$$

For $y(t) = 25{,}000$ we get

$$e^{-0.23t} = \frac{1}{499}$$

$$t = \frac{\ln(1/499)}{-0.23} \approx 27 \text{ days}$$

Mechanics: Freely Falling Bodies

The motion of a rigid body falling freely toward the earth was treated in Chapter 4. It was shown that if air resistance and small changes in acceleration with altitude are neglected, then the position $s = s(t)$ at time t of a body of mass m is governed by the equation

$$ma = -mg$$

or

$$a = -g \tag{7}$$

where $a = d^2s/dt^2$ is the acceleration of the body and g is the gravitational constant. Hence, we have chosen a coordinate system in which the positive direction is up. This accounts for the minus sign in (7).

Equation (7) is a second-order equation in s—namely,

$$\frac{d^2s}{dt^2} = -g \tag{8}$$

If we let $v = ds/dt$ denote the velocity, then (8) is a first-order equation in v,

$$\frac{dv}{dt} = -g$$

The general solutions of these equations are

$$v(t) = -gt + C_1 \qquad \text{and} \qquad s(t) = -\tfrac{1}{2}gt^2 + C_1t + C_2$$

If v_0 and s_0 denote the velocity and position of the body at time $t = 0$, respectively, then

$$v(t) = -gt + v_0 \qquad \text{and} \qquad s(t) = -\tfrac{1}{2}gt^2 + v_0t + s_0$$

A more realistic model of the motion of a freely falling body takes into account the effects of air resistance. A common assumption is that air resistance exerts a force that is proportional to the speed and that opposes the motion of the body. With this assumption, it can be shown that the motion is described by the differential equation

$$m\frac{d^2s}{dt^2} = -mg - kv \tag{9}$$

where $k > 0$ is the constant of proportionality. This constant depends on the particular body. For example, a sheet of paper will have a much larger constant of proportionality than a solid lead sphere.

If we let $v = ds/dt$, then (9) becomes

$$m\frac{dv}{dt} = -mg - kv$$

$$\frac{dv}{dt} + \frac{k}{m}v = -g$$

This is a first-order linear nonhomogeneous equation. The general solution is

$$v(t) = -\frac{mg}{k} + C_1 e^{-(k/m)t}$$

If we assume that the body has an initial velocity v_0—that is, $v(0) = v_0$—then

$$v(t) = -\frac{mg}{k} + \left(v_0 + \frac{mg}{k}\right)e^{-(k/m)t} \tag{10}$$

Note that $\lim_{t\to+\infty} v(t) = -mg/k$; the limiting velocity of a freely falling body depends only on the mass of the body and the constant of proportionality k. This limit is called the *terminal velocity* of the body.

EXAMPLE 4

A paratrooper and his parachute weigh 192 pounds. At the instant his parachute opens, he is falling at 150 feet per second. Assume that air resistance is 200 pounds when the velocity is 20 feet per second. How fast is the paratrooper falling 3 seconds after his parachute opens? What is his terminal velocity?

Solution

In the U.S. customary system of units, length is measured in feet, weight in pounds, and time in seconds. In this system, $g \approx 32$ feet per second per second. The mass of the paratrooper and his parachute is $m = W/g = 192/32 = 6$ slugs. Using the fact that air resistance is 200 pounds when the velocity is 20 feet per second, we have

$$200 = k \cdot 20$$

$$k = 10$$

Thus, from (10), the velocity of the paratrooper at any time t after his parachute opens is

$$v(t) = -\frac{192}{10} + \left(-150 + \frac{192}{10}\right)e^{-(10/6)t} = -19.2 - 130.8e^{-(10/6)t}$$

Therefore, after 3 seconds, the paratrooper's velocity is

$$v(3) = -19.2 - 130.8e^{-5.5} \approx -20.08 \text{ feet per second}$$

His terminal velocity is $\lim_{t\to+\infty} v(t) = -mg/k = -19.2$ feet per second.

Mixtures

EXAMPLE 5

A large tank contains 81 gallons of brine in which 20 pounds of salt are dissolved. Brine containing 3 pounds of dissolved salt per gallon runs into the tank at the rate of 5 gallons per minute. The mixture, kept uniform by stirring, runs out of the tank at the rate of 2 gallons per minute. How much salt is in the tank at the end of 37 minutes?

Solution

Let $y(t)$ be the amount of salt in the tank at any time t. Then $y'(t)$ is the rate of change of the amount of salt in the tank at time t. Clearly,

$$y'(t) = \text{Rate in} - \text{Rate out}$$

where "rate in" is the rate at which salt runs into the tank at time t, and "rate out" is the rate at which salt runs out of the tank at time t. Here

$$\text{Rate in} = (3 \text{ pounds per gallon})(5 \text{ gallons per minute}) = 15 \text{ pounds per minute}$$

That is, 15 pounds of salt per minute flow into the tank. To compute the rate out, we first find the concentration of salt at time t—that is, the amount of salt per gallon of brine at time t. Since

$$\text{Concentration} = \frac{\text{Pounds of salt in tank at time } t}{\text{Gallons of brine in tank at time } t} = \frac{y(t)}{81 + (5 - 2)t}$$

we have

$$\begin{aligned}\text{Rate out} &= \left[\frac{y(t)}{81 + 3t} \text{ pounds per gallon}\right](2 \text{ gallons per minute}) \\ &= \frac{2y(t)}{81 + 3t} \text{ pounds per minute}\end{aligned}$$

Thus, the differential equation describing this mixture problem is

$$y' = 15 - \frac{2y}{81 + 3t} \tag{11}$$

We also have the boundary condition that $y = 20$ when $t = 0$. The differential equation in (11) is linear and can be written in the form

$$y' + \frac{2}{81 + 3t}y = 15$$

By multiplying both sides by the integrating factor

$$e^{\int[2/(81+3t)]\,dt} = e^{(2/3)\int[dt/(27+t)]} = e^{(2/3)[\ln(27+t)]} = (27 + t)^{2/3}$$

we obtain

$$\frac{d}{dt}[y(t)(27 + t)^{2/3}] = 15(27 + t)^{2/3}$$

By integrating both sides with respect to t, we obtain

$$y(t)(27 + t)^{2/3} = 15(27 + t)^{5/3}\frac{3}{5} + C$$

$$y(t) = 9(27 + t) + C(27 + t)^{-2/3}$$

By using the boundary condition, we find that

$$20 = 9(27) + C(27)^{-2/3}$$

and so $C = -2007$. Thus,

$$y(t) = 9(27 + t) - 2007(27 + t)^{-2/3}$$

and the amount of salt in the tank at the end of 37 minutes is

$$y(37) = 9(27 + 37) - 2007(27 + 37)^{-2/3} = 450.6 \text{ pounds}$$

EXERCISE 20.5

In Problems 1–6 determine the orthogonal trajectories of the given family.

1. $x^2 + y^2 = 2Cx$ **2.** $xy = C$ **3.** $y^2 = 2(C - x)$ **4.** $y = Cx^2$

5. All parabolas with vertex at (1, 2) and a vertical axis [*Hint:* First determine the one-parameter family.]

6. All circles tangent to the x-axis at (3, 0)

7. Show that the family of parabolas $y^2 = 4C(x + C)$ is self-orthogonal.

8. Show that the family of ellipses

$$\frac{x^2}{C^2} + \frac{y^2}{C^2 - 4} = 1$$

is self-orthogonal.

9. The annual sales of a new company are expected to grow at a rate that is proportional to the difference between the sales and an upper limit of \$20 million. The sales are \$0 initially and are \$3 million for the second year.

(a) Determine the annual sales at any time t.
(b) What will the sales be during the eighth year of operations?
(c) How long will it take for sales to reach \$12 million?

10. A culture of bacteria is growing in a medium that can support a maximum of 1000 bacteria. The rate of growth of the population at time t is proportional to the difference between 1000 and the number present at time t. The culture contains 100 bacteria initially, and after 1 hour there are 150 bacteria.

(a) How many bacteria will there be after 5 hours?
(b) When will the culture contain 995 bacteria?

11. The number of words per minute W that a person can type increases with practice. Assume that the rate of change of W is proportional to the difference between W and an upper limit of 150—that is, proportional to (*Problem 11 continues in the next column*)

11. (*continued*) $150 - W$. Assume also that a beginner cannot type at all—that is, $W = 0$ when $t = 0$—and that he can type 30 words per minute after 10 hours of practice.

(a) How many words per minute can he type after 25 hours of practice?
(b) How many hours of practice will be required in order for him to type 100 words per minute?

12. Newton's law of cooling states that the rate of decrease of the temperature u of an object immersed in a medium of constant temperature T is proportional to the difference $u - T$. The differential equation that governs the temperature of the object at time t is

$$\frac{du}{dt} = k(u - T)$$

where k is a negative constant.

(a) Determine the general solution of this differential equation.
(b) Determine the particular solution that satisfies the initial condition $u(0) = u_0$.

13. A metal bar with initial temperature 25°C is dropped into a container of boiling water (100°C). After 5 seconds, the temperature of the bar is 35°C.

(a) What will the temperature of the bar be after 1 minute?
(b) How long will it take for the temperature of the bar to be within 0.5°C of the temperature of the boiling water?

14. A pie is removed from a 350°F oven to cool in a room with temperature 72°F. Five minutes later the temperature of the pie is 200°F.

(a) What is the temperature of the pie after 15 minutes?
(b) How long will it take for the pie to be 100°F and thus ready to eat?

15. A drug is injected into a patient's bloodstream at a constant rate r (milligrams per second). Simultaneously, the drug is removed from the bloodstream at a rate proportional to the amount $y(t)$ present at time t.

(a) Derive the differential equation that models this problem.
(b) Determine the general solution of the differential equation in part (a) and give the particular solution that satisfies the initial condition $y(0) = 0$.

16. A rumor spreads through a population of 5000 people at a rate that is proportional to the product of the number of people who have heard the rumor and the number who have not heard it. Suppose that 100 people initiated the rumor and 500 have heard it after 3 days.

(a) How many people will have heard the rumor after 8 days?
(b) How long will it take for half the people to hear the rumor?

17. A flu virus is spreading through a college campus of 10,000 students at a rate that is proportional to the product of the number of infected students and the number of well students. Assume 10 students were infected initially and 200 students are infected after 10 days.

(a) How many students will be infected after 5 days?
(b) How long will it take for 75% of the students to be infected?

18. The developers of a planned community assume that the population $P(t)$ of the community will be governed by the logistic equation

$$\frac{dP}{dt} = P(10^{-1} - 10^{-5}P)$$

where t is measured in months. The initial population will be $P(0) = 1500$.

(a) What is the limiting value of the population?
(b) When will the population equal one-half of the limiting value?

19. Consider the logistic equation (6). Show that dy/dt is increasing if $y < M/2$ and is decreasing if $y > M/2$. From this it follows that the growth rate is a maximum when $y = M/2$.

20. Using (10), determine the position $s(t)$ of a freely falling body at any time t. Assume that the initial position is $s(0) = s_0$. Also, give the expressions for $v(t)$ and $s(t)$ in the special case $v(0) = s(0) = 0$.

21. A body of mass 2 kilograms is dropped from rest from a height of 2000 meters. As it falls, the air resistance is equal to $\frac{1}{2}v$, where v is the velocity measured in meters per second.

(a) Determine the velocity and the distance fallen at time t.
(b) What is the terminal velocity of the object?
(c) With what velocity does the object strike the earth?

22. A skydiver and her parachute weigh 160 pounds. She free-falls from rest from a height of 10,000 feet for 20 seconds. Assume that there is no air resistance. After her parachute opens, the air resistance is four times her velocity.

(a) How fast will the skydiver be falling 4 seconds after her parachute opens?
(b) How long will it take for the skydiver to hit the ground?
(c) What is her velocity when she lands?

23. A skydiver and his parachute weigh 160 pounds. He free-falls from rest from a height of 10,000 feet for 20 seconds. Assume that the air resistance during the free-fall is $\frac{1}{2}v$ and that the air resistance after his chute opens is $5v$.

(a) What is the skydiver's velocity when his parachute opens?
(b) How long will it take for the skydiver to hit the ground?
(c) What is his velocity when he lands?
[*Hint:* There are two distinct differential equations that govern the velocity and position of the skydiver: one for the free-fall period and the other for the period after his parachute opens.]

24. An object of mass 100 grams is thrown vertically upward with an initial velocity of 200 centimeters per second. The air resistance is $200v$.

(a) How long does it take for the object to reach its maximum height and what is the maximum height?
(b) How long does it take for the object to hit the ground and what is its impact velocity?

25. The equation governing the amount of electrical charge q (in coulombs) of an RC circuit consisting of a resistance R, a capacitance C (in farads), an electromotive force E, and no inductance is

$$\frac{dq}{dt} + \frac{1}{RC}q = \frac{E}{R}$$

(*Problem 25 continues on page 1223*)

25. (*continued*)
where t is the time in seconds. Solve the differential equation, assuming E, R, and C are constants and $q = 0$ when $t = 0$.

26. The equation governing the amount of current I (in amperes) in an RL circuit consisting of a resistance R (in ohms), an inductance L (in henrys), and an electromotive force $E_0 \sin \omega t$ is given by Kirchhoff's second law,

$$L\frac{dI}{dt} + RI = E_0 \sin \omega t$$

Find I as a function of t if $I = I_0$ when $t = 0$.

27. The basic equation governing the amount of current I (in amperes) in a simple RL circuit consisting of a resistance R (in ohms), an inductance L (in henrys), and an electromotive force E (in volts) is

$$\frac{dI}{dt} + \frac{R}{L}I = \frac{E}{L}$$

where t is the time in seconds. Solve the differential equation, assuming that E, R, and L are constants and $I = 0$ when $t = 0$.

28. A large tank contains 40 gallons of brine in which 10 pounds of salt are dissolved. Brine containing 2 pounds of dissolved salt per gallon runs into the tank at the rate of 4 gallons per minute. The mixture, kept uniform by stirring, runs out of the tank at the rate of 3 gallons per minute.

(a) How much salt is in the tank at any time t?

(b) Find the amount of salt in the tank at the end of 1 hour.

29. A tank initially contains 10 gallons of pure water. Starting at time $t = 0$, brine containing 3 pounds of salt per gallon flows into the tank at the rate of 2 gallons per minute. The mixture is kept uniform by stirring, and the well-stirred mixture flows out of the tank at the same rate as the inflow. How much salt is in the tank after 5 minutes? How much salt is in the tank after a very long time?

Second-Order Homogeneous Linear Differential Equations

20.6

In this and the following section we will discuss some solution methods for certain types of second-order linear differential equations. The theory and methods presented here can be extended to linear differential equations of arbitrary order n.

[20.6.1] DEFINITION / ***Linear Differential Equation of Order Two.***

A second-order linear differential equation is of the form

$$\frac{d^2y}{dx^2} + b(x)\frac{dy}{dx} + c(x)y = Q(x) \qquad \textbf{(1)}$$

where b, c, and Q are continuous on some interval I.

If $Q(x) = 0$ in (1), then we have

$$\frac{d^2y}{dx^2} + b(x)\frac{dy}{dx} + c(x)y = 0 \qquad \textbf{(2)}$$

and (2) is called *homogeneous*. (Note that the use of the term *homogeneous* in the case of these second-order equations indicates something quite different from the use of this term in the first-order equations considered in Section 20.2.) If $Q(x) \neq 0$, then (1) is said to be *nonhomogeneous*. We shall restrict our work to equations in which b and c are constant functions.

In this section we discuss equations that are homogeneous. In Section 20.7 we will discuss equations that are nonhomogeneous.

We begin with a definition of linear independence. We say the functions y_1 and y_2 are *linearly independent* if the only solution to the equation

$$c_1 y_1 + c_2 y_2 = 0$$

is the trivial one, $c_1 = c_2 = 0$. Otherwise, this set of functions is *linearly dependent*. For example, $y_1(x) = x$ and $y_2(x) = 5x$ are linearly dependent because $c_1 x + c_2(5x) = 0$ has nonzero solutions $c_1 = -5$ and $c_2 = 1$.

The following theorem points out the importance of linear independence in obtaining the general solution of a second-order linear differential equation with constant coefficients.

[20.6.2] THEOREM / *General Solution of a Second-Order Linear Differential Equation.*

If y_1 and y_2 are two linearly independent solutions of the differential equation

$$y'' + by' + cy = 0 \tag{3}$$

then

$$y_c = c_1 y_1 + c_2 y_2 \tag{4}$$

where c_1 and c_2 are constants, is the general solution of (3).

[20.6.3] THEOREM

If $y_c = c_1 y_1 + c_2 y_2$ is the general solution of (3), where y_1 and y_2 are linearly independent and y_p defines *any* particular solution of (1), then *every* solution of (1) can be expressed in the form

$$y = y_c + y_p$$

Thus, to find the complete solution of (1), first find two linearly independent solutions y_1 and y_2 of (3). Then add to $c_1 y_1 + c_2 y_2$ any particular solution y_p of (1).

Let us assume that (3) has a solution of the form

$$y(x) = e^{mx}$$

where m is a constant, and substitute $y(x) = e^{mx}$, $y'(x) = me^{mx}$, and $y''(x) = m^2 e^{mx}$ into the differential equation. Then we have

$$m^2 e^{mx} + bme^{mx} + ce^{mx} = 0$$

$$e^{mx}(m^2 + bm + c) = 0$$

Since $e^{mx} > 0$ for all x, we conclude that $y(x) = e^{mx}$ is a solution of (3) if and only if

$$m^2 + bm + c = 0 \tag{5}$$

This equation is called the *characteristic equation* for the differential equation (3). It is a quadratic in m and its two roots, m_1 and m_2, called the *characteristic roots*, can be found either by factoring the characteristic equation or by the quadratic formula

$$m_1 = \frac{-b + \sqrt{b^2 - 4c}}{2} \qquad m_2 = \frac{-b - \sqrt{b^2 - 4c}}{2}$$

There are exactly three cases to consider.

Case 1. $b^2 - 4c > 0$: m_1 and m_2 are real numbers, $m_1 \neq m_2$

In this case, $y_1(x) = e^{m_1 x}$ and $y_2(x) = e^{m_2 x}$ are linearly independent solutions of (4), and the general solution of the equation is

$$y(x) = c_1 e^{m_1 x} + c_2 e^{m_2 x}$$

EXAMPLE 1

Determine the general solution of the differential equation

$$y'' - 2y' - 8y = 0$$

and give the particular solution that satisfies the initial conditions $y(0) = 2$, $y'(0) = -1$.

Solution

The characteristic equation for the differential equation is

$$m^2 - 2m - 8 = 0$$

$$(m - 4)(m + 2) = 0$$

Thus, the characteristic roots are $m_1 = 4$ and $m_2 = -2$. The functions

$$y_1(x) = e^{4x} \qquad y_2(x) = e^{-2x}$$

are linearly independent solutions of the differential equation, and the general solution is

$$y(x) = c_1 e^{4x} + c_2 e^{-2x}$$

Applying the given initial conditions, we get the pair of equations

$$c_1 + c_2 = 2$$

$$4c_1 - 2c_2 = -1$$

The solution is $c_1 = \frac{1}{2}$, $c_2 = \frac{3}{2}$. Therefore, the particular solution is

$$y(x) = \tfrac{1}{2}e^{4x} + \tfrac{3}{2}e^{-2x}$$

Case 2. $b^2 - 4c = 0$: $m_1 = m_2 = r$, a real number

In this case, the characteristic equation must be

$$(m - r)^2 = m^2 - 2mr + r^2 = 0$$

and the corresponding differential equation is

$$y'' - 2ry' + r^2 y = 0$$

We know that $y_1(x) = e^{rx}$ is one solution of the differential equation, and so we need to find a second solution y_2 that is independent of y_1. To motivate the result here, consider the special case $r = 0$. With $r = 0$, the differential equation is

$$y'' = 0$$

and the characteristic equation is

$$m^2 = (m - 0)^2 = 0$$

Clearly, the general solution of the differential equation is

$$y(x) = c_1 + c_2 x$$

which we can interpret as being

$$y(x) = c_1 e^{0x} + c_2 x e^{0x}$$

Thus, we are led to consider $y_2(x) = xe^{rx}$ as a possible solution of the original equation. It can be verified that y_2 does indeed satisfy the differential equation, and y_2 is independent of y_1. Therefore,

$$y(x) = c_1 e^{rx} + c_2 x e^{rx}$$

is the general solution of the differential equation in this case.

EXAMPLE 2

Determine the general solution of the differential equation

$$y'' + 6y' + 9y = 0$$

Solution

The characteristic equation is

$$m^2 + 6m + 9 = 0$$

$$(m + 3)^2 = 0$$

Thus, $m_1 = m_2 = -3$; $y_1(x) = e^{-3x}$ and $y_2(x) = xe^{-3x}$ are independent solutions, and

$$y(x) = c_1 e^{-3x} + c_2 x e^{-3x}$$

is the general solution.

Case 3. $b^2 - 4c < 0$: $m_1 = \lambda + i\omega$, $m_2 = \lambda - i\omega$, $i^2 = -1$

In this case, the functions $z_1(x) = e^{(\lambda + i\omega)x}$ and $z_2(x) = e^{(\lambda - i\omega)x}$ are independent solutions of the differential equation, and the general solution is

$$z(x) = K_1 e^{(\lambda + i\omega)x} + K_2 e^{(\lambda - i\omega)x} = e^{\lambda x}(K_1 e^{i\omega x} + K_2 e^{-i\omega x})$$

K_1 and K_2 are arbitrary constants.

However, the functions represented here are complex-valued and we are seeking real-valued solutions. We can use Euler's formula

$$e^{i\theta} = \cos\theta + i \cdot \sin\theta$$

to construct real-valued functions from z_1 and z_2. In particular,

$$z_1(x) = e^{\lambda x} e^{i\omega x} = e^{\lambda x}(\cos\omega x + i \cdot \sin\omega x)$$

$$z_2(x) = e^{\lambda x} e^{-i\omega x} = e^{\lambda x}(\cos\omega x - i \cdot \sin\omega x)$$

It follows that

$$y_1(x) = \frac{1}{2}[z_1(x) + z_2(x)] = e^{\lambda x}\cos\omega x$$

and

$$y_2(x) = \frac{1}{2i}[z_1(x) - z_2(x)] = e^{\lambda x}\sin\omega x$$

are real-valued solutions of the differential equation. The solutions y_1 and y_2 are linearly independent and the general solution of the equation is

$$y(x) = c_1 e^{\lambda x}\cos\omega x + c_2 e^{\lambda x}\sin\omega x = e^{\lambda x}[c_1 \cos\omega x + c_2 \sin\omega x]$$

EXAMPLE 3

Determine the general solution of the differential equation

$$y'' + 2y' + 5y = 0$$

Solution

The characteristic equation is

$$m^2 + 2m + 5 = 0$$

and the characteristic roots are $m_1 = -1 + 2i$ and $m_2 = -1 - 2i$. Thus,

$$y_1(x) = e^{-x}\cos 2x \qquad y_2(x) = e^{-x}\sin 2x$$

are independent solutions and

$$y(x) = e^{-x}[c_1 \cos 2x + c_2 \sin 2x]$$

is the general solution.

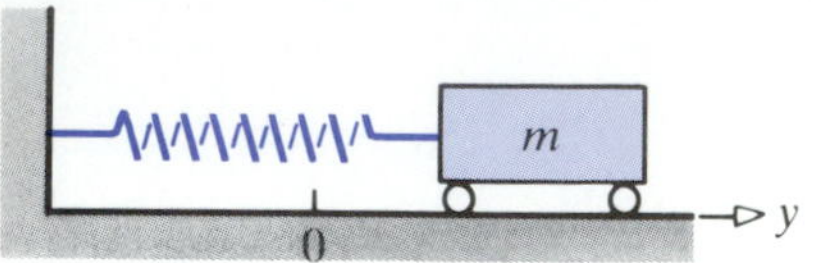

Figure 5

Vibrating Mechanical Systems

Suppose we have a simple mechanical system consisting of a mass m connected to a wall by a spring as illustrated in Figure 5. The mass moves on a smooth surface, and $y = y(t)$ denotes its position at time t. We assume that the spring exerts a force (restoration) that is proportional to the displacement, and that the surface exerts a resistance to the motion of the mass (damping) that is proportional to the speed. For example, the resistance could be caused by friction. From Newton's law of motion, $F = ma$ the motion of the mass after some displacement from the equilibrium position is

$$m\frac{d^2y}{dt^2} = -b\frac{dy}{dt} - cy$$

or

$$\frac{d^2y}{dt^2} + \frac{b}{m}\frac{dy}{dt} + \frac{c}{m}y = 0 \tag{6}$$

where $b > 0$ and $c > 0$ are the damping and restoring constants, respectively. If, in addition, we assume there is an external force applied to the system—say, mq—that is dependent on time only, then the motion of the mass is given by

$$\frac{d^2y}{dt^2} + \frac{b}{m}\frac{dy}{dt} + \frac{c}{m}y = q(t) \tag{7}$$

Note that (6) is a second-order linear homogeneous equation with constant coefficients and so the results of this section can be applied. Equation (7) is a second-order linear nonhomogeneous equation, and it will be treated at the end of the next section.

We begin the analysis of the system by considering the case where there is no damping and no external force $(b = 0,\ \ q = 0)$. In this case, the system is in a state of *free vibration*. From (6), the motion of the system is described by

$$\frac{d^2y}{dt^2} + k^2y = 0 \qquad k^2 = \frac{c}{m} \tag{8}$$

The general solution of this equation is given by

$$y(t) = c_1\cos(kt) + c_2\sin(kt) \tag{9}$$

If we let $R = (c_1^2 + c_2^2)^{1/2}$ and let θ be the angle such that $c_1 = R\cos\theta$, $c_2 = R\sin\theta$, then the general solution (9) can be written in the form

$$y(t) = R\cos(kt - \theta)$$

It is clear from this representation that the solution is a periodic function of period $2\pi/k$ and frequency k. These numbers are called the *natural period* and the *natural frequency* of the system. The numbers R and θ are called the *amplitude* and the *phase shift*, respectively. The motion of the system is usually called *simple harmonic motion* (see Section 8.4 in Chapter 8). See Figure 6.

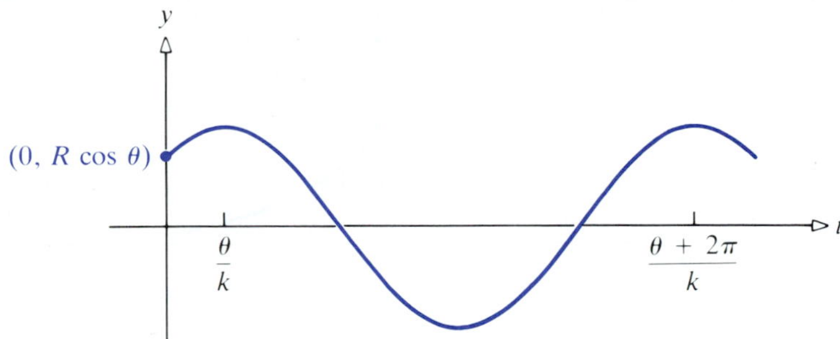

Figure 6

If we include the effects of damping, then (6) describes the motion, and the system is said to be in a state of *damped free vibration*. The characteristic equation corresponding to (6) is

$$s^2 + \frac{b}{m}s + \frac{c}{m} = 0 \qquad \text{or} \qquad ms^2 + bs + c = 0$$

and the characteristic roots are

$$s_1, s_2 = \frac{-b \pm \sqrt{b^2 - 4cm}}{2m} = \frac{-b}{2m} + \left[\frac{b^2}{4m^2} - \frac{c}{m}\right]^{1/2}$$

It is convenient to define new constants in terms of b, c, and m. Let $p = b/2m$ and $k^2 = c/m$. The positive constant p is called the *damping factor* and k is the *frequency*. In terms of these constants, the characteristic roots are

$$s_1, s_2 = -p \pm (p^2 - k^2)^{1/2}$$

The three possible cases together with the forms of the solutions are given here:

1. $p^2 - k^2 > 0$ $\qquad y(t) = e^{-pt}(c_1 e^{rt} + c_2 e^{-rt})$
$\qquad r = (p^2 - k^2)^{1/2}, \quad p > r$

2. $p^2 - k^2 = 0$ $\qquad y(t) = e^{-pt}(c_1 + c_2 t)$

3. $p^2 - k^2 < 0$ $\qquad y(t) = e^{-pt}(c_1 \cos \omega t + c_2 \sin \omega t)$
$\qquad \omega^2 = k^2 - p^2$

Note that in all three cases $\lim_{t \to +\infty} y(t) = 0$, regardless of the values of the constants c_1 and c_2. In Cases 1 and 2 there is no oscillatory behavior. These cases are known as *overdamping* and *critical damping*, respectively. Depending on the initial conditions, the graph of a solution will be one of the curves illustrated in Figure 7. Case (3) is called *underdamped* motion. By letting $R = (c_1^2 + c_2^2)^{1/2}$, $\cos \theta = c_1/R$ and $\sin \theta = c_2/R$, as in the case of free vibration, we get

$$y(t) = Re^{-pt}\cos(\omega t - \theta)$$

The graph of a solution will look like Figure 8.

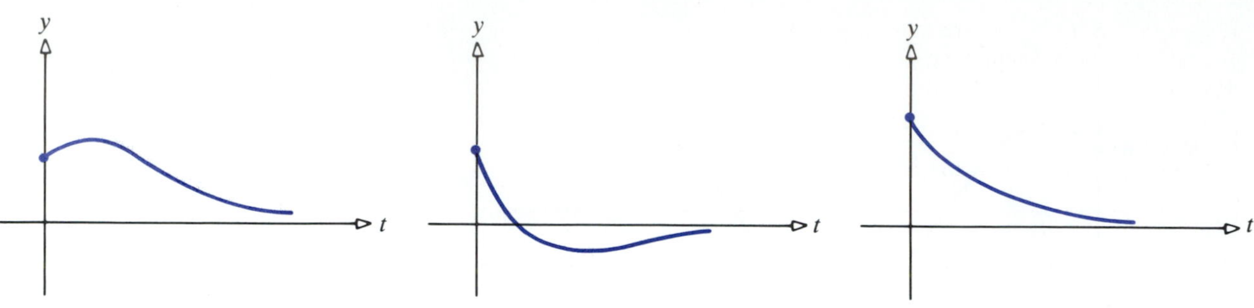

Figure 7

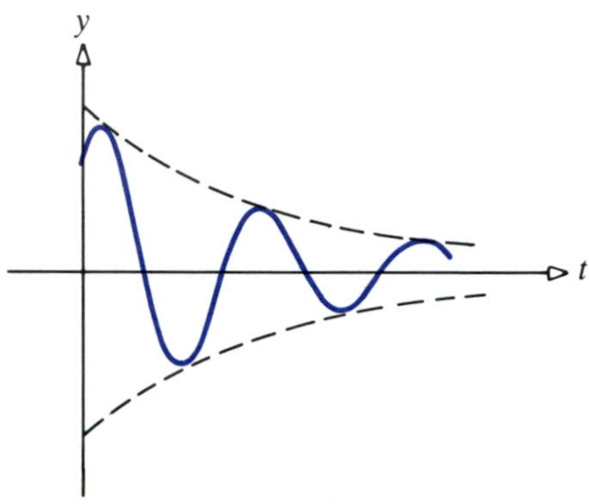

Figure 8

EXERCISE 20.6

In Problems 1–10 determine the general solution of each differential equation.

1. $y'' - 4y' + 3y = 0$
2. $y'' + 2y' = 0$
3. $y'' + 2y' + y = 0$
4. $y'' - 4y = 0$
5. $2y'' + 3y' - 2y = 0$
6. $y'' + 4y = 0$
7. $y'' - 4y' + 8y = 0$
8. $y'' + 4y' + 4y = 0$
9. $y'' + y' + y = 0$
10. $6y'' - 7y' - 3y = 0$

In Problems 11–14 determine the particular solution.

11. $y'' - y' - 6y = 0; \quad y(0) = 1, \quad y'(0) = 3$
12. $y'' + 9y = 0; \quad y(0) = 1, \quad y'(0) = -6$
13. $y'' - 6y' + 9y = 0; \quad y(1) = 0, \quad y'(1) = 2e$
14. $y'' - 4y' + 5y = 0; \quad y(\pi/2) = 0, \quad y'(\pi/2) = -1$

The method for solving second-order linear homogeneous differential equations with constant coefficients can be extended to equations of arbitrary order n. In particular, consider the nth-order equation

$$y^{(n)} + a_{n-1}y^{(n-1)} + a_{n-2}y^{(n-2)} + \cdots + a_1 y' + a_0 y = 0$$

where $a_0, a_1, \ldots, a_{n-1}$ are constants. The characteristic equation is

$$m^n + a_{n-1}m^{n-1} + a_{n-2}m^{n-2} + \cdots + a_1 m + a_0 = 0$$

If r is a root of the characteristic equation, then $y = e^{rx}$ is a solution of the differential equation. Roots of multiplicity greater than 1 and complex conjugate roots are handled in the same manner as the second-order case. In Problems 15–20 determine the general solution of each differential equation.

15. $y''' + 2y'' - y' - 2y = 0$ [*Hint:* $r = 1$ is a root of the characteristic equation.]

16. $y^{(4)} - y = 0$

17. $y''' - y'' + 9y' - 9y = 0$ [*Hint:* $r = 1$ is a root of the characteristic equation.]

18. $y''' + y'' - 2y' = 0$

19. $y^{(4)} - 4y''' + 5y'' - 4y' + 4y = 0$ [*Hint:* $r = 2$ is a double root of the characteristic equation.]

20. $y^{(4)} - 8y' = 0$

Suppose $y = c_1e^{rx} + c_2e^{sx}$ is the general solution of a second-order linear homogeneous differential equation with constant coefficients. Then $y_1(x) = e^{rx}$ and $y_2(x) = e^{sx}$ are independent solutions of the equation, r and s are the roots of the characteristic equation, and the characteristic equation must be $m^2 - (r + s)m + rs = 0$. Therefore, the differential equation is

$$y'' - (r + s)y' + rsy = 0$$

The case of repeated roots and complex conjugate roots can be handled in a similar manner. In Problems 21–29, determine a second-order linear homogeneous differential equation with constant coefficients that has the given general solution.

21. $y(x) = c_1e^{-2x} + c_2e^{4x}$

22. $y(x) = c_1 \cos 3x + c_2 \sin 3x$

23. $y(x) = c_1 + c_2e^x$

24. $y(x) = c_1e^{-2x} + c_2xe^{-2x}$

25. $y(x) = e^{2x}[c_1 \cos 3x + c_2 \sin 3x]$

26. $y(x) = c_1e^{5x} + c_2e^{-5x}$

27. $y(x) = c_1 + c_2x$

28. $y(x) = c_1e^x + c_2xe^x$

29. $y(x) = c_1e^x\cos x + c_2e^x\sin x$

30. A second-order linear differential equation having the special form

$$x^2y'' + \beta xy' + \gamma y = 0$$

is called an *Euler equation*. Show that the change of independent variable $z = \ln x$ transforms an

(*Problem 30 continues in the next column*)

30. (*continued*)
Euler equation into a linear equation with constant coefficients. [*Hint:* If $z = \ln x$, then $(dy/dx) = (dy/dz)(dz/dx) = (1/x)(dy/dz)$. Now differentiate this equation with respect to x to find d^2y/dx^2 in terms of the derivatives of y with respect to z. Then substitute into the differential equation.]

In Problems 31–35 use the result in Problem 30 to determine the general solution of each Euler equation.

31. $x^2y'' + 4xy' + 2y = 0$

32. $x^2y'' - 2xy' + 2y = 0$

33. $x^2y'' - 4xy' + 6y = 0$

34. $x^2y'' + xy' + y = 0$

35. $x^2y'' + 5xy' + 13y = 0$

36. Show that the general solution of $y'' - y = 0$ can be written as

$$y(x) = c_1 \cosh x + c_2 \sinh x$$

37. Suppose that $y = y_1(x)$ and $y = y_2(x)$ are nontrivial solutions of the second-order linear homogeneous differential equation

$$y'' + p_1(x)y' + p_2(x)y = 0 \qquad x \in I$$

The function $W(x) = y_1(x)y_2'(x) - y_2(x)y_1'(x)$ is

(*Problem 37 continues in the next column*)

37. (*continued*)
called the *Wronskian* of y_1 and y_2. The Wronskian can also be represented by the 2×2 determinant

$$W(x) = \begin{vmatrix} y_1(x) & y_2(x) \\ y_1'(x) & y_2'(x) \end{vmatrix}$$

(a) Show that if y_1 is a constant multiple of y_2 on the interval I—that is, if $y_1(x) = ky_2(x)$ for all $x \in I$—then $W(x) \equiv 0$ on I.

(*Problem 37 continues on page 1232*)

37. *(continued)*

(b) Show that if there is a point $a \in I$ such that $W(a) = 0$, then $y_1(x) = ky_2(x)$ for all $x \in I$, k a constant. It follows from this that $W(x) \equiv 0$ on I.

It follows from parts (a) amd (b) that $W(x) \equiv 0$ on I if and only if y_1 and y_2 are linearly dependent on I.

38. Consider the second-order linear homogeneous equation with constant coefficients

$$y'' + by' + cy = 0$$

Then $P(m) = m^2 + bm + c = 0$ is the characteristic equation.

(a) Show that if P has negative real roots, or if P has complex conjugate roots with negative real parts then $\lim_{x\to+\infty} y(x) = 0$ for all solutions $y = y(x)$ of the differential equation.

(b) Give a necessary and sufficient condition for all solutions of the differential equation to be bounded on $[0, +\infty)$.

39. Consider the nonhomogeneous equation $y'' + by' + cy = ke^{rx}$, where k, c, b, and r are constants. The characteristic equation for the corresponding homogeneous equation is $P(m) = m^2 + bm + c = 0$.

(a) Show that if $P(r) \neq 0$, then $y = ke^{rx}/P(r)$ is a solution of the given equation.

(Problem 39 continues in the next column)

39. *(continued)*

(b) Show that if $P(r) = 0$ and $P'(r) \neq 0$, then $y = kxe^{rx}/P'(r)$ is a solution of the given equation.

(c) Show that if $P(r) = P'(r) = 0$, then $y = kx^2e^{rx}/P''(r)$ is a solution of the given equation.

40. Determine the solutions of (8) satisfying the following initial conditions:

(a) $y(0) = 1$, $y'(0) = 0$
(b) $y(0) = 0$, $y'(0) = 1$
(c) $y(0) = 1$, $y'(0) = 1$

41. Equation (6) can be written $y'' + 2py' + k^2y = 0$, where $p = b/2m$ and $k^2 = c/m$. In each of the cases (i) $p = 1$, $k^2 = 2$; (ii) $p = 1$, $k^2 = 1$; and (iii) $p = 1$, $k^2 = \frac{3}{4}$, graph the solution that satisfies the initial conditions $y(0) = 1$, $y'(0) = -1$.

42. In the case of underdamping, show that if t_1 and t_2 are successive relative maxima of a solution $y = y(t)$, then $y(t_2)/y(t_1) = e^{-\left[\frac{-2\pi p}{\omega}\right]}$ where $p = b/2m$, $k^2 = \dfrac{c}{m}$, and $\omega^2 = k^2 - p^2$.

43. Show that if the damping constant b in (6) is such that the motion is either critically damped or overdamped, then the mass passes through 0 at most once.

44. In the case of critical damping, determine a condition on $y'(0)$ such that a solution satisfying $y(0) = 1$ will pass through 0.

20.7 Linear Nonhomogeneous Differential Equations

In this section we develop two methods for finding a particular solution of a given nonhomogeneous equation

$$\frac{d^2y}{dx^2} + p_1(x)\frac{dy}{dx} + p_2(x)y = Q(x), \quad Q(x) \neq 0$$

The first method is called *variation of parameters*. This is a general method in the sense that it can be applied to any nonhomogeneous equation. However, it depends on knowing the general solution of the homogeneous equation. The second method is called *undetermined coefficients*. This method is essentially restricted to nonhomogeneous equations with constant coefficients in which the function Q has a particular form. When it can be used, the method of undetermined coefficients is usually simpler than variation of parameters. We shall develop both methods in the context of nonhomogeneous equations with constant coefficients. That is, we shall restrict our attention to equations of the form

$$y'' + by' + cy = Q(x) \tag{1}$$

where b and c are constants and $Q \neq 0$ is a continuous function on some interval I.

Variation of Parameters

Suppose that $y = y_1(x)$ and $y = y_2(x)$ are linearly independent solutions of

$$y'' + by' + cy = 0 \tag{2}$$

Then

$$y(x) = c_1 y_1(x) + c_2 y_2(x)$$

is the general solution of (2). We replace the arbitrary constants c_1 and c_2 by the functions $u = u(x)$ and $v = v(x)$, which are to be determined so that

$$y_p(x) = y_1(x)u(x) + y_2(x)v(x)$$

is a particular solution of the nonhomogeneous equation (1). The replacement of the "parameters" c_1, c_2 by the "variables" u, v is the basis of the term *variation of parameters*. Since there are two unknowns u and v to be determined, we shall impose two conditions on these unknowns. One condition is that y_p should solve the differential equation (1). The second condition is at our disposal, and we shall choose it in a manner that will simplify the calculations.

Differentiating y_p, we have

$$y_p' = y_1'u + y_1u' + y_2'v + y_2v' \tag{3}$$

We choose as the second condition on u, v that

$$y_1u' + y_2v' = 0 \tag{4}$$

This condition is chosen because it simplifies the first derivative y_p' considerably and because it will lead to a simple pair of "linear" equations in the unknowns u and v. With this, condition (3) is reduced to

$$y_p' = y_1'u + y_2'v \tag{5}$$

and

$$y_p'' = y_1'u' + y_1''u + y_2'v' + y_2''v$$

Now substitute y_p, y_p' (given by (5)), and y_p'' into the left side of the differential equation (1). This gives

$$\begin{aligned}(y_1'u' + y_1''u + y_2'v' + y_2''v) + b(y_1'u + y_2'v) + c(y_1u + y_2v)\\ = (y_1'' + by_1' + cy_1)u + (y_2'' + by_2' + cy_2)v + y_1'u' + y_2'v'\end{aligned}$$

Since y_1 and y_2 are solutions of (2), it follows that

$$y_p'' + by_p' + cy_p = y_1'u' + y_2'v'$$

The condition that y_p should satisfy (1) is

$$y_1'u' + y_2'v' = Q(x) \tag{6}$$

Equations (4) and (6) constitute a system of two ("linear") equations in the two unknowns u and v:

$$\begin{aligned} y_1u' + y_2v' &= 0 \\ y_1'u' + y_2'v' &= Q(x) \end{aligned} \tag{7}$$

Of course, this system actually involves u' and v', but if we can solve for these, then we can simply integrate to find u and v. Solving (7) for u' and v', we find

$$u' = \frac{-y_2Q}{y_1y_2' - y_2y_1'} \qquad v' = \frac{y_1Q}{y_1y_2' - y_2y_1'}$$

The expression $W(x) = y_1(x)y_2'(x) - y_2(x)y_1'(x)$ is the *Wronskian* discussed in Problem 37 in Exercise 20.6.

Integrating u' and v', we have

$$u(x) = \int \frac{-y_2(x)Q(x)}{W(x)}\,dx \qquad v(x) = \int \frac{y_1(x)Q(x)}{W(x)}\,dx$$

and

$$y_p(x) = y_1(x)\int \frac{-y_2(x)Q(x)}{W(x)}\,dx + y_2(x)\int \frac{y_1(x)Q(x)}{W(x)}\,dx$$

is a particular solution of the nonhomogeneous equation (1).

EXAMPLE 1

Use the method of variation of parameters to find a particular solution of

$$y'' - y' - 6y = 2e^{-2x}$$

Also, give the general solution of this equation.

Solution

The functions $y_1(x) = e^{3x}$ and $y_2(x) = e^{-2x}$ are linearly independent solutions of the equation $y'' - y' - 6y = 0$. Substituting these functions into the system (7), we have

$$\begin{aligned} e^{3x}u' + e^{-2x}v' &= 0 \\ 3e^{3x}u' - 2e^{-2x}v' &= 2e^{-2x} \end{aligned}$$

The solution of this system is

$$u' = \frac{-e^{-2x}2e^{-2x}}{-5e^{x}} = \frac{2}{5}e^{-5x}$$

$$v' = \frac{e^{3x}2e^{-2x}}{-5e^{x}} = -\frac{2}{5}$$

Therefore,

$$u(x) = \int \frac{2}{5}e^{-5x}\,dx = -\frac{2}{25}e^{-5x} \qquad v(x) = \int -\frac{2}{5}\,dx = -\frac{2}{5}x$$

[*Note:* Since we are seeking only one function u and one function v, we do not include the arbitrary constants in the integration steps.] Now

$$y_p(x) = e^{3x}\left(\frac{-2}{25}e^{-5x}\right) + e^{-2x}\left(-\frac{2}{5}x\right) = -\frac{2}{25}e^{-2x} - \frac{2}{5}xe^{-2x}$$

and the general solution is

$$y(x) = c_1e^{3x} + c_2e^{-2x} - \frac{2}{25}e^{-2x} - \frac{2}{5}xe^{-2x} = c_1e^{3x} + \left(c_2 - \frac{2}{25}\right)e^{-2x} - \frac{2}{5}xe^{-2x}$$

Since $c_2 - \frac{2}{25}$ is an arbitrary constant, we simply relabel it c_2 and write the general solution as

$$y(x) = c_1e^{3x} + c_2e^{-2x} - \frac{2}{5}xe^{-2x} \tag{8}$$

Undetermined Coefficients

In contrast to the method of variation of parameters, this method is restricted to nonhomogeneous equations with constant coefficients in which the nonhomogeneous term $Q = Q(x)$ has a special form. To motivate the method, consider the differential expression on the left side of (1):

$$y'' + by' + cy \tag{9}$$

If we substitute $y(x) = Ae^{rx}$ into this expression, we get

$$Ar^2e^{rx} + Abre^{rx} + Ace^{rx} = (Ar^2 + Abr + Ac)e^{rx} = Ke^{rx}$$

That is, substituting Ae^{rx} into $y'' + by' + cy$ produces a constant multiple of e^{rx}. We can use this fact to "guess" the form of a particular solution y_p of a nonhomogeneous differential equation in which the nonhomogeneous term $Q(x) = ae^{rx}$ for some constant a.

EXAMPLE 2

Determine a particular solution y_p of the nonhomogeneous equation

$$y'' - 2y' - 8y = 10e^{3x}$$

Solution

If we substitute $y_p(x) = Ae^{3x}$ into the left side of the equation, then we know that we will get the expression Ke^{3x}, where K is a constant that involves the unknown coefficient A and the known constants 3, -2, and -8. We want to determine A so that $K = 10$. In this particular case,

$$y_p = Ae^{3x} \qquad y'_p = 3Ae^{3x} \qquad y''_p = 9Ae^{3x}$$

and

$$y''_p - 2y'_p - 8y_p = 9Ae^{3x} - 6Ae^{3x} - 8Ae^{3x} = -5Ae^{3x}$$

Now we set $-5A = 10$ and get $A = -2$. Therefore,

$$y_p(x) = -2e^{3x}$$

is a particular solution of the nonhomogeneous equation.

In the same manner, it can be verified that if $y(x) = A \cdot \cos px + B \cdot \sin px$ is substituted into (9), then the result will have the form

$$K \cdot \cos px + M \cdot \sin px$$

where K and M are constants, and the result of substituting $y(x) = Ae^{rx}\cos px + Be^{rx}\sin px$ into (9) will be an expression of the form

$$Ke^{rx}\cos px + Me^{rx}\sin px$$

In each of these cases, K and M are constants that involve A and B, the numbers p and r, and the coefficients b and c.

Based on these observations, if we are given the differential equation

$$y'' + by' + cy = Q(x)$$

where

$$Q(x) = \begin{cases} ae^{rx} \\ a \cdot \cos px \quad \text{or} \quad a \cdot \sin px \\ ae^{rx}\cos px \quad \text{or} \quad ae^{rx}\sin px \end{cases}$$

then we assume that a particular solution y_p has the form

$$y_p(x) = \begin{cases} Ae^{rx} \\ A \cdot \cos px + B \cdot \sin px \\ Ae^{rx}\cos px + Be^{rx}\sin px \end{cases}$$

respectively, and we attempt to determine values for the undetermined coefficients A, or A and B. Note that $Q(x) = ae^{0x} = a$, a constant, is included in this analysis, and in this case $y_p = A$. The next example gives another illustration of the method. There is also the possibility that the assumed form for y_p might not be correct. We explain and illustrate this complication after the example.

EXAMPLE 3

Determine a particular solution of the nonhomogeneous equation

$$y'' + 2y' + y = 3e^x \cos x$$

Solution

Since $Q(x) = 3e^x \cos x$, we assume that y_p will have the form

$$y_p(x) = Ae^x \cos x + Be^x \sin x$$

Then

$$\begin{aligned} y_p' &= Ae^x \cos x - Ae^x \sin x + Be^x \sin x + Be^x \cos x \\ &= (A + B)e^x \cos x + (B - A)e^x \sin x \end{aligned}$$

and

$$\begin{aligned} y_p'' &= (A + B)e^x \cos x - (A + B)e^x \sin x + (B - A)e^x \sin x + (B - A)e^x \cos x \\ &= 2Be^x \cos x - 2Ae^x \sin x \end{aligned}$$

Substituting y_p and its derivatives into the differential equation, we have

$$\begin{aligned} &2Be^x \cos x - 2Ae^x \sin x + 2(A + B)e^x \cos x \\ &\qquad + 2(B - A)e^x \sin x + Ae^x \cos x + Be^x \sin x \\ &\qquad = (3A + 4B)e^x \cos x + (-4A + 3B)e^x \sin x = 3e^x \cos x \end{aligned}$$

Equating the coefficients of corresponding terms yields

$$\begin{aligned} 3A + 4B &= 3 \\ -4A + 3B &= 0 \end{aligned}$$

This system of equations has the unique solution $A = \frac{9}{25}$, $B = \frac{12}{25}$ and

$$y_p(x) = \tfrac{9}{25}e^x \cos x + \tfrac{12}{25}e^x \sin x$$

is a particular solution of the differential equation.

We now illustrate a complication that can arise in trying to use the method of undetermined coefficients. Consider the differential equation in Example 1:

$$y'' - y' - 6y = 2e^{-2x} \tag{10}$$

Since $Q(x) = 2e^{-2x}$ we assume that a particular solution y_p has the form

$$y_p(x) = Ae^{-2x}$$

Then $y_p' = -2Ae^{-2x}$ and $y_p'' = 4Ae^{-2x}$. Substituting y_p and its derivatives into the differential equation, we have

$$y_p'' - y_p' - 6y_p = 4Ae^{-2x} - (-2Ae^{-2x}) - 6Ae^{-2x} = (4 + 2 - 6)Ae^{-2x} = 2e^{-2x}$$

or

$$0 \cdot Ae^{-2x} = 2e^{-2x}$$

So we want to solve

$$0 \cdot A = 2$$

which is impossible. The problem is that our assumed solution y_p is a solution of the homogeneous equation. From Example 1, note that a particular solution has the form $y_p(x) = Axe^{-2x}$. Substituting

$$y_p = Axe^{-2x} \qquad y_p' = Ae^{-2x} - 2Axe^{-2x} \qquad y_p'' = -4Ae^{-2x} + 4Axe^{-2x}$$

into the differential equation (10), we have

$$-4Ae^{-2x} + 4Axe^{-2x} - (Ae^{-2x} - 2Axe^{-2x}) - 6Axe^{-2x} = 2e^{-2x}$$

or

$$-5Ae^{-2x} = 2e^{-2x}$$

Thus,

$$-5A = 2 \qquad \text{or} \qquad A = -\tfrac{2}{5}$$

Therefore, $y_p(x) = \frac{-2}{5}xe^{-2x}$ is a particular solution of (10), as established in Example 1.

In a similar manner, it can be verified that the form of a particular solution of

$$y'' + 4y' + 4y = e^{-2x}$$

is $y_p(x) = Ax^2e^{-2x}$, since $y_1(x) = Ae^{-2x}$ and $y_2(x) = Axe^{-2x}$ are solutions of the homogeneous equation. Also, the differential equation

$$y'' + 4y = \cos 2x$$

will have a particular solution of the form $y_p = Ax \cdot \cos(2x) + Bx \cdot \sin(2x)$, since both $\sin 2x$ and $\cos 2x$ are solutions of the homogeneous equation.

The point of this discussion is that in using the method of undetermined coefficients, we assume the form of a particular solution y_p based on the form of the nonhomogeneous term Q. But before proceeding with the method, we must first check to see whether y_p is a solution of the homogeneous equation. If it is, then we multiply y_p by the smallest power of x such that the new y_p is not a solution of the homogeneous equation. It is advisable to solve the homogeneous equation before attempting to solve the nonhomogeneous equation.

Table 1 summarizes the method of undetermined coefficients.

The method of undetermined coefficients can also be used if the function Q has one of the forms ax^ne^{rx}, $ax^n\cos px$, $ax^n\sin px$, $ax^ne^{rx}\cos px$, or $ax^ne^{rx}\sin px$, where n is a positive integer. This will be considered in the exercises.

Table 1 A particular solution of $y'' + by' + cy = Q(x)$

$Q(x)$	$y_p(x)$
ax^n, where n is a positive integer	$P(x)$, a general polynomial of degree m
ae^{rx}	$Ax^k e^{rx}$
$a \cdot \cos px$ or $a \cdot \sin px$	$Ax^m \cos px + Bx^m \sin px$
$ae^{rx}\cos px$ or $ae^{rx}\sin px$	$Ax^m e^{rx}\cos px + Bx^m e^{rx}\sin px$

where k and m are the smallest nonnegative integers such that y_p is *not* a solution of the homogeneous equation. [*Note:* $k = 0, 1$, or 2; $m = 0$ or 1.)

Forced Vibrations

We conclude this section by considering the effect of an externally applied force on the spring–mass system introduced at the end of Section 20.6. As indicated there, the differential equation that describes the motion of the mass is

$$\frac{d^2y}{dt^2} + \frac{b}{m}\frac{dy}{dt} + \frac{c}{m}y = Q(t) \tag{11}$$

and the system is said to be in a state of *forced vibration.*

As a particular example, suppose that there is no damping $(b = 0)$ and that the applied force is periodic and proportional to $\cos ht$. Then the motion is given by

$$\frac{d^2y}{dt^2} + k^2 y = r \cos ht \tag{12}$$

where $k^2 = c/m$ and r is the constant of proportionality. It can be verified using the methods of this section that the general solution of (12) is

$$y(t) = c_1 \cos kt + c_2 \sin kt + \frac{r}{k^2 - h^2} \cos ht$$

provided $h^2 \neq k^2$. As shown in Section 20.6, the general solution can also be written in the form

$$y(t) = R \cdot \cos(kt - \theta) + \frac{r}{k^2 - h^2} \cos ht \tag{13}$$

From this representation of the general solution, we can see that the motion of the mass is the sum of two periodic functions of different frequencies and, in general, different amplitudes. A typical graph of a solution $y = y(t)$ is given in Figure 9.

An interesting phenomenon known as *resonance* occurs when the period of the applied force equals the natural period of the system. From (13), note that if h is "close to" k, then the amplitude of y is large. If $h = k$ in (12), then

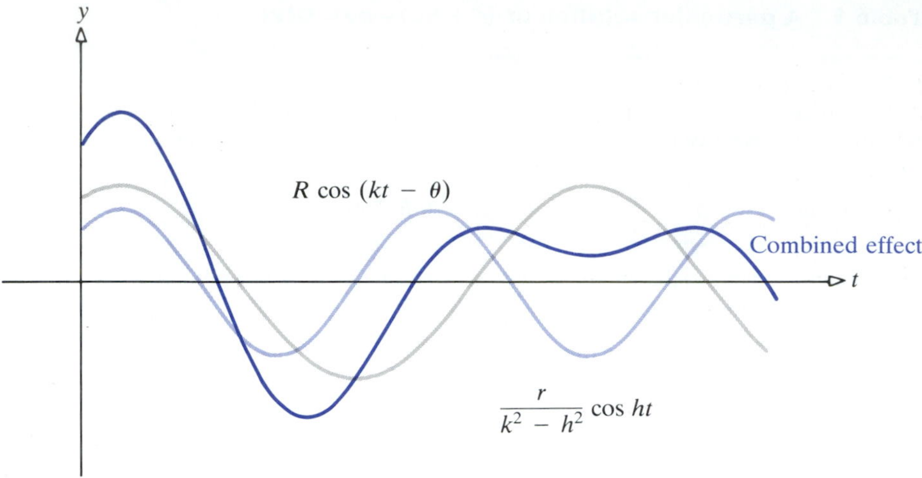

Figure 9

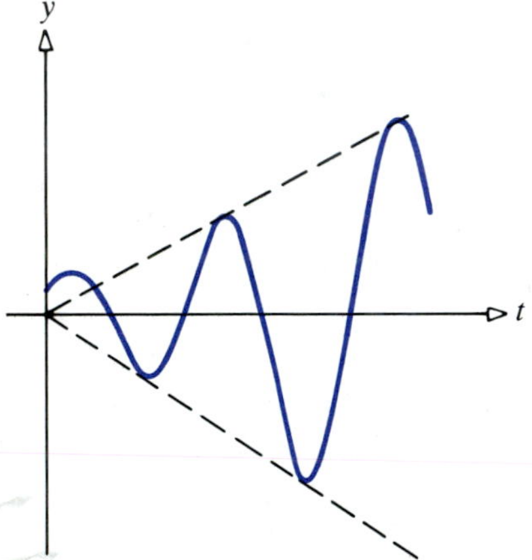

Figure 10

the general solution of the resulting differential equation is

$$y(t) = c_1 \cos kt + c_2 \sin kt + \frac{r}{2k} t \cdot \sin kt$$

and it is clear that the motion is unbounded as $t \to +\infty$. The graph of a typical solution in this case is given in Figure 10.

In general, the motion of the spring–mass system with damping and forcing function $r \cdot \cos ht$ is described by

$$\frac{d^2y}{dt^2} + \frac{b}{m}\frac{dy}{dt} + \frac{c}{m} y = r \cdot \cos ht \tag{14}$$

The general solution of (14) has the form

$$y(t) = c_1 e^{s_1 t} + c_2 e^{s_2 t} + A \cdot \cos(ht - \theta)$$

where s_1 and s_2 are the characteristic roots for the homogeneous equation of (14), and A and θ are constants that are determined explicitly from the coefficients m, b, c, r, and h. As shown in the case of damped free vibrations, the function

$$y_c(t) = c_1 e^{s_1 t} + c_2 e^{s_2 t}$$

has the limit 0 as $t \to +\infty$. Therefore, as $t \to +\infty$,

$$y(t) \to y_p(t) = A \cdot \cos(ht - \theta)$$

For this reason, $y_c(t)$ is called the *transient solution* and $y_p(t)$ is called the *steady state solution*.

EXERCISE 20.7

In Problems 1–8 the general solution of the homogeneous equation is given. Use the method of variation of parameters to find a particular solution of the nonhomogeneous equation. Also, give the general solution of each equation.

1. $y'' + y = \sec^3 x; \quad y = c_1 \cos x + c_2 \sin x$

2. $y'' + 4y = x; \quad y = c_1 \cos 2x + c_2 \sin 2x$

3. $y'' + 4y' + 4y = \frac{1}{x^2}e^{-2x}; \quad y = c_1 e^{-2x} + c_2 x e^{-2x}$

4. $y'' - y' - 2y = -3e^{2x}; \quad y = c_1 e^{-x} + c_2 e^{2x}$

5. $y'' - 2y' + y = xe^x; \quad y = c_1 e^x + c_2 x e^x$

6. $y'' - y = \sinh x; \quad y = c_1 e^x + c_2 e^{-x}$

7. $y'' - 5y' + 6y = e^x \cos x; \quad y = c_1 e^{2x} + c_2 e^{3x}$

8. $y'' + 4y = 3 \csc 2x; \quad y = c_1 \cos 2x + c_2 \sin 2x$

In Problems 9–20 use the method of undetermined coefficients to find a particular solution of each nonhomogeneous equation. Also, give the general solution of each equation.

9. $y'' - 3y' + 2y = \sin x$

10. $y'' - 2y' - 3y = 10e^{4x}$

11. $y'' - 6y' + 9y = 4e^{-3x}$

12. $y'' - y = e^x \cos 2x$

13. $y'' + 4y = 2$ [Note: $2 = 2e^{0x}$. Thus, $y_p = Ae^{0x} = A$.]

14. $y'' + 2y' = \cos 2x$

15. $y'' - y' - 6y = 2e^{2x} + 3e^{4x}$

16. $y'' - 4y' + 3y = 3e^x$

17. $y'' - y' - 2y = -3e^{-2x}$

18. $y'' + 4y' + 4y = xe^{-2x}$

19. $y'' + 4y = \cos 2x$

20. $y'' - y' - 2y = 2e^{-x} + e^{2x}$

The method of undetermined coefficients can also be used when the nonhomogeneous term Q has the form $Q(x) = ax^k e^{rx}$, where a is a constant and k is a positive integer. In this case, it can be verified that the equation has a particular solution of the form

$$y_p = (A_k x^k + A_{k-1}x^{k-1} + \cdots + A_1 x + A_0)e^{rx}$$

where $A_k, A_{k-1}, \ldots, A_0$ are the undetermined coefficients, provided $y = e^{rx}$ is not a solution of the homogeneous equation. If $y = e^{rx}$ is a solution of the homogeneous equation, then $y_p = x^n(A_k x^k + \cdots + A_1 x + A_0)e^{rx}$, where n is the smallest positive integer such that $y = x^n e^{rx}$ is not a solution of the homogeneous equation. In the same manner, the method of undetermined coefficients can be used when the nonhomogeneous term Q has one of the following forms: $ax^k \cos px$, $ax^k \sin px$, $ax^k e^{rx} \cos px$, $ax^k e^{rx} \sin px$. In Problems 21–28 use the method of undetermined coefficients to find a particular solution of each equation. Also, give the general solution of each equation.

21. $y'' - 2y' + 2y = xe^x$

22. $y'' - y = x^2 e^{2x}$

23. $y'' - 5y' + 6y = xe^{2x}$

24. $y'' - 2y' + y = xe^x$

25. $y'' - y' - 2y = x \sin 2x$

26. $y'' - y = x^2 \cos x$

27. $y'' + y = x \cos x$

28. $y'' + y' = xe^x \sin x$

In Problems 29–33 determine the general solution of each differential equation.

29. $y'' + y = x^2 + \tan x$

30. $y'' + 2y = 2 + e^x$

31. $y'' - 4y' + 3y = (1 + e^{-x})^{-1}$

32. $y'' + 2y' + 2y = x^2 + \sin x$

33. $y'' - 4y' + 4y = xe^{2x} + \frac{1}{x^2}e^{2x}$

34. Verify that the solution of (11) is given by (13).

35. Show that the solution of (12) that satisfies the initial conditions $y(0) = y'(0) = 0$ can be written in the form

$$y(t) = \frac{2r}{k^2 - h^2} \sin\left(\frac{k-h}{2}\right)t \sin\left(\frac{k+h}{2}\right)t$$

36. Determine the general solution of

$$y'' + k^2 y = r \cos kt$$

20.8 Power Series Methods

In the preceding sections we considered some of the basic theory and solution methods for various types of differential equations. It should be obvious that we have not exhausted all the possibilities for first-order or higher-order equations. Indeed, if we simply write down a differential equation at random, it is almost certain that it will not fall into any of the classes we have covered, and it is equally certain that we will not be able to solve the equation exactly. For this reason, in any detailed study of differential equations, considerable emphasis is placed on methods for approximating solutions. We shall introduce one class of approximation methods here—namely, power series methods. The methods that will be developed in this section depend heavily on the treatment of power series in Chapter 11.

Using power series methods, we assume that a solution $y = y(x)$ of a given differential equation has a power series expansion of the form

$$\sum_{n=0}^{+\infty} a_n x^n \tag{1}$$

We calculate corresponding power series for the terms in the differential equation, as well as the power series for y', y'', and so on. These power series are then substituted into the differential equation, producing a relationship between the coefficients. This relationship will specify each coefficient a_n in terms of one or more of the lower-order coefficients. We will need to use the fact that power series representations are unique in the sense that

$$\sum_{n=0}^{+\infty} a_n x^n = \sum_{n=0}^{+\infty} b_n x^n \tag{2}$$

for all x on some open interval containing $x = 0$ if and only if $a_n = b_n$ for all n.

The next two examples illustrate this method.

EXAMPLE 1

Determine a power series solution of the form $\sum_{n=0}^{+\infty} a_n x^n$ for the differential equation

$$y' - 2y = e^{-x}$$

Solution

Assume that

$$y(x) = \sum_{n=0}^{+\infty} a_n x^n$$

Then

$$y'(x) = \sum_{n=1}^{+\infty} n a_n x^{n-1}$$

Also,

$$e^{-x} = \sum_{n=0}^{+\infty} \frac{(-1)^n}{n!} x^n$$

Substituting these power series into the differential equation, we have

$$\sum_{n=1}^{+\infty} na_n x^{n-1} - 2 \sum_{n=0}^{+\infty} a_n x^n = \sum_{n=0}^{+\infty} \frac{(-1)^n}{n!} x^n \tag{3}$$

The next step is to reduce this equation to a relationship between the coefficients. One way to derive this relationship is to write out the terms of the power series on both sides of the equation

$$a_1 + 2a_2 x + 3a_3 x^2 + 4a_4 x^3 + \cdots - 2(a_0 + a_1 x + a_2 x^2 + a_3 x^3 + \cdots)$$
$$= 1 - x + \frac{1}{2!} x^2 - \frac{1}{3!} x^3 + \cdots$$

and combine the coefficients of corresponding powers of x. This gives

$$(a_1 - 2a_0) + (2a_2 - 2a_1)x + (3a_3 - 2a_2)x^2 + (4a_4 - 2a_3)x^3 + \cdots$$
$$= 1 - x + \frac{1}{2} x^2 - \frac{1}{6} x^3 + \cdots$$

Now, using the uniqueness of power series representations noted above, the coefficients of corresponding powers of x must be equal, and we have

$$a_1 - 2a_0 = 1 \quad \text{or} \quad a_1 = 2a_0 + 1$$

$$2a_2 - 2a_1 = -1 \quad \text{or} \quad a_2 = \frac{2a_1 - 1}{2} = 2a_0 + \frac{1}{2}$$

$$3a_3 - 2a_2 = \frac{1}{2} \quad \text{or} \quad a_3 = \frac{2a_2 + \frac{1}{2}}{3} = \frac{4}{3} a_0 + \frac{1}{2}$$

$$4a_4 - 2a_3 = -\frac{1}{6} \quad \text{or} \quad a_4 = \frac{2a_3 - \frac{1}{6}}{4} = \frac{2}{3} a_0 + \frac{5}{24}$$

and so on. Thus,

$$y(x) = a_0 + (2a_0 + 1)x + (2a_0 + \tfrac{1}{2})x^2 + (\tfrac{4}{3}a_0 + \tfrac{1}{2})x^3 + (\tfrac{2}{3}a_0 + \tfrac{5}{24})x^4 + \cdots$$
$$= a_0(1 + 2x + 2x^2 + \tfrac{4}{3}x^3 + \tfrac{2}{3}x^4 + \cdots) + x + \tfrac{1}{2}x^2 + \tfrac{1}{2}x^3 + \tfrac{5}{24}x^4 + \cdots$$

where a_0 is arbitrary. This is the power series representation of the general solution. If we had carried the calculations far enough, we could have recognized $y(x) = a_0 e^{2x} + \frac{1}{3}e^{2x} - \frac{1}{3}e^{-x} = (a_0 + \frac{1}{3})e^{2x} - \frac{1}{3}e^{-x}$. Since the differential equation is a first-order linear equation, we could use the methods given in Section 20.6 to show that the general solution is $y(x) = Ce^{2x} - \frac{1}{3}e^{-x}$, which agrees with the expression above when $C = a_0 + \frac{1}{3}$.

In calculating the coefficients in Example 1, notice the pattern that is developed. From this pattern it is easy to see the general relationship between the coefficients—namely,

$$(n+1)a_{n+1} - 2a_n = \frac{(-1)^n}{n!} \qquad n = 0, 1, 2, \ldots \tag{4}$$

This formula is called a *recursion formula*, and it can be used to calculate each of the coefficients a_n, $n \geq 1$, in terms of a_0, as we did in Example 1.

A more direct method for obtaining the recursion formula is through a formal manipulation of the power series. Changing the index of summation from n to $n + 1$ in the first power series on the left side of (3) and multiplying the second power series by -2, we have

$$\sum_{n=0}^{\infty} (n+1)a_{n+1}x^n + \sum_{n=0}^{\infty} -2a_{nx}{}^n = \sum_{n=0}^{\infty} \frac{(-1)^n}{n!} x^n$$

Now, adding the two series on the left

$$\sum_{n=0}^{\infty} [(n+1)a_{n+1} - 2a_n]x^n = \sum_{n=0}^{n} \frac{(-1)^n}{n!} x^n$$

and equating coefficients, we obtain (4) immediately. This formal approach is used in the next example.

EXAMPLE 2

Determine a power series representation of the general solution of the differential equation

$$\frac{d^2y}{dx^2} + x\frac{dy}{dx} + 2y = 0$$

Also, determine a power series representation of the solution that satisfies the boundary conditions $y(0) = 2$, $y'(0) = -3$.

Solution

Assume that $y(x) = \sum_{n=0}^{+\infty} a_n x^n$. Then

$$y'(x) = \sum_{n=1}^{+\infty} na_n x^{n-1}$$

$$y''(x) = \sum_{n=2}^{+\infty} n(n-1)a_n x^{n-2}$$

Substituting these power series into the differential equation, we have

$$\sum_{n=2}^{+\infty} n(n-1)a_n x^{n-2} + x\sum_{n=1}^{+\infty} na_n x^{n-1} + 2\sum_{n=0}^{+\infty} a_n x^n = 0$$

or

$$\sum_{n=2}^{+\infty} n(n-1)a_n x^{n-2} + \sum_{n=1}^{+\infty} na_n x^n + \sum_{n=0}^{+\infty} 2a_n x^n = 0$$

Next we adjust the summation indexes so that x^n appears in each series. In this case we need only to replace n by $n + 2$ in the first series to get

$$\sum_{n=0}^{+\infty} (n+2)(n+1)a_{n+2}x^n + \sum_{n=1}^{+\infty} na_n x^n + \sum_{n=0}^{+\infty} 2a_n x^n = 0$$

Now, the first and third series start with $n = 0$, whereas the second series starts with $n = 1$. We write out the $n = 0$ term of the first and third series separately. This gives

$$2 \cdot 1a_2 + 2a_0 + \sum_{n=1}^{+\infty} (n+2)(n+1)a_{n+2}x^n + \sum_{n=1}^{+\infty} na_n x^n + \sum_{n=1}^{+\infty} 2a_n x^n = 0$$

$$2 \cdot 1a_2 + 2a_0 + \sum_{n=1}^{+\infty} [(n+2)(n+1)a_{n+2} + (n+2)a_n]x^n = 0$$

Therefore, equating the coefficents of corresponding powers of x (the coefficients on the right side are all 0), we have

$$2a_2 + 2a_0 = 0 \tag{5}$$

and

$$(n+2)(n+1)a_{n+2} + (n+2)a_n = 0 \qquad n = 1, 2, 3, \ldots \tag{6}$$

From (5) we get

$$a_2 = -a_0$$

and from (6) we get

$$a_{n+2} = \frac{-a_n}{(n+1)} \qquad n = 1, 2, 3, \ldots$$

The latter equation is the recursion formula. We can use this formula to obtain all of the coefficients in terms of a_0 and a_1. In particular, we have

$$a_3 = \frac{-a_1}{2} \qquad a_4 = \frac{-a_2}{3} = \frac{a_0}{3}$$

$$a_5 = \frac{-a_3}{4} = \frac{a_1}{2 \cdot 4} \qquad a_6 = \frac{-a_4}{5} = \frac{-a_0}{3 \cdot 5}$$

$$a_7 = \frac{-a_5}{6} = \frac{-a_1}{2 \cdot 4 \cdot 6} \qquad a_8 = \frac{-a_6}{7} = \frac{a_0}{3 \cdot 5 \cdot 7}$$

and so on. Since a_0 and a_1 can be chosen arbitrarily, the power series representation for the general solution is

$$y(x) = a_0\left(1 - x^2 + \frac{x^4}{3} - \frac{x^6}{3 \cdot 5} + \frac{x^8}{3 \cdot 5 \cdot 7} - \cdots\right) + a_1\left(x - \frac{x^3}{2} + \frac{x^5}{2 \cdot 4} - \frac{x^7}{2 \cdot 4 \cdot 6} + \cdots\right)$$

To solve the boundary-value problem, note that the conditions $y(0) = 2$ and $y'(0) = -3$ imply that $a_0 = 2$ and $a_1 = -3$. Thus, the solution of the problem is

$$y(x) = 2\left(1 - x^2 + \frac{x^4}{3} - \frac{x^6}{3 \cdot 5} + \cdots\right) - 3\left(x - \frac{x^3}{2} + \frac{x^5}{2 \cdot 4} + \cdots\right)$$

$$= 2 - 3x - 2x^2 + \frac{3}{2}x^3 + \frac{2}{3}x^4 - \frac{3}{8}x^5 - \frac{2}{15}x^6 + \cdots$$

A second type of series solution method involves a differential equation with initial conditions and makes use of Taylor's theorem as given in Section 11.9.

EXAMPLE 3

Use Taylor's theorem to find the series solution of

$$y'' = x^2y + e^xy'$$

given the initial conditions $y(0) = 1$, $y'(0) = 1$. Then use the first five terms of the series to approximate values of y for $0 \le x \le 1$.

Solution

Here we assume that the solution of the initial-value problem has the Maclaurin series expansion

$$y(x) = \sum_{n=0}^{+\infty} \frac{y^{(n)}(0)}{n!} x^n = y(0) + y'(0)x + \frac{y''(0)}{2!}x^2 + \frac{y'''(0)}{3!}x + \cdots$$

From the initial conditions, $y(0) = 1$, $y'(0) = 1$, and from the differential equation

$$y''(0) = 0^2 \cdot 1 + e^0 \cdot 1 = 1$$

Differentiating the differential equation with respect to x, we have

$$y''' = x^2y' + 2xy + e^xy'' + e^xy'$$

Then

$$y'''(0) = 0^2 \cdot 1 + 2 \cdot 0 \cdot 1 + e^0 \cdot 1 + e^0 \cdot 1 = 2$$

$$y^{(4)} = x^2y'' + 4xy' + 2y + e^xy''' + 2e^xy'' + e^xy'$$

$$y^{(4)}(0) = 0^2 \cdot 1 + 4 \cdot 0 \cdot 1 + 2 \cdot 1 + e^0 \cdot 2 + 2 \cdot e^0 \cdot 1 + e^0 \cdot 1 = 7$$

and so on. Thus,

$$y(x) = 1 + x + \frac{1}{2!}x^2 + \frac{2}{3!}x^3 + \frac{7}{4!}x^4 + \frac{23}{5!}x^5 + \frac{78}{6!}x^6 + \frac{325}{7!}x^7 + \frac{1511}{8!}x^8 + \cdots \quad \textbf{(7)}$$

Using this series, we compute Tables 2 and 3 for values of y in the interval $0 \le x \le 1$. In Table 2 we use the first five terms and in Table 3 we use the first nine terms. In comparing the y-values in Table 2 with those in Table 3, we observe that the farther we move from the center of conver-

Table 2 Five terms

x	0.0	0.1	0.2	0.3	0.4	0.5	0.6	0.7	0.8	0.9	1.0
y	1.0	1.1054	1.2231	1.3564	1.5088	1.6849	1.8898	2.1294	2.4101	2.7394	3.125

Table 3 Nine terms

x	0.0	0.1	0.2	0.3	0.4	0.5	0.6	0.7	0.8	0.9	1.0
y	1.0	1.1054	1.2232	1.3569	1.5113	1.6932	1.9071	2.1818	2.4715	2.9571	3.5270

gence of the infinite series, the less accurate the estimate will be. This confirms what we already know about Taylor series: that for a fixed number of terms, the farther we move from the center of convergence (in this case $c = 0$), the less accurate the estimate will be.

EXERCISE 20.8

In Problems 1–10 use power series to solve the differential equation.

1. $y'' + y = 0$
2. $y'' + xy = 0$
3. $y'' + x^2y' + xy = 0$
4. $y''' + y = 0$
5. $(x^2 + 2)y'' - 3xy' + 4y = 0$
6. $y'' + x^2y = 0$
7. $y'' + 3xy' + 3y = 0$
8. $y' + 3xy = 0$
9. $y''' - 2xy = 0$
10. $(1 + x^2)y'' - 4xy' + 6y = 0$

In Problems 11–20 use power series to solve the initial-value problem. In each problem, compute the coefficients out to the fifth power of the x-term.

11. $y'' + xy' + y = 0;\quad y(0) = 1,\quad y'(0) = 0$
12. $y'' - (\sin x)y = 0;\quad y(0) = 0,\quad y'(0) = 1,\quad y''(0) = 0$
13. $y'' + y' + e^x y = 0;\quad y(0) = 2,\quad y'(0) = 1$
14. $y^{(4)} - \ln(1 + x)y = 0;\quad y(0) = 1,\quad y'(0) = 1,\quad y''(0) = 0,\quad y'''(0) = 0$
15. $y'' + (3 + x)y = 0;\quad y(0) = 1,\quad y'(0) = 0$
16. $y'' + x^2y = 0;\quad y(0) = 0,\quad y'(0) = 2$
17. $y'' - 2xy' + y = 0;\quad y(0) = 2,\quad y'(0) = 1$
18. $y''' - 3x^2y' + 2xy = 0;\quad y(0) = 1,\quad y'(0) = 1,\quad y''(0) = 0$
19. $y''' + 4y'' + 2y' - x^3y = 0;\quad y(0) = 1,\quad y'(0) = 1,\quad y''(0) = 0$
20. $y'' + (\cos x)y = 0;\quad y(0) = 0,\quad y'(0) = 1$

REVIEW EXERCISES

In Problems 1–10, (a) identify each differential equations as separable, homogeneous, exact, or linear, and (b) solve each differential equation.

1. $x(y^2 + 1)\,dx + y(x^2 - x)\,dy = 0$
2. $(2y + e^{2x})\,dx + (2x + e^{2y})\,dy = 0$

3. $x\dfrac{dy}{dx} + y = x^6$

4. $\dfrac{dy}{dx} - 2y = e^{3x}$

5. $(2xy^3 + y^2\cos x - 2x)\,dx + (3x^2y^2 + 2y\sin x)\,dy = 0$

6. $\dfrac{dy}{dx} + x^2y - y = x^2 - 1$

7. $y\sin x\,dx - \cos x\,dy = 0$

8. $(2x\sin y - \ln y)\,dx + \left(x^2\cos y - \dfrac{x}{y} + 3y^2\right)dy = 0$

9. $xy' - 2y = x + 1$

10. $(x - y\tan(y/x))\,dx + (x\tan(y/x))\,dy = 0$

In Problems 11–20 find the particular solution.

11. $xy' + 2y = 6, \quad y(2) = 8$

12. $xy\,dy = (y^2 + x\sqrt{x^2 + y^2})\,dx, \quad y(1) = 1$

13. $y' + \dfrac{y}{x} = 2, \quad y(1) = 0$

14. $(x^2 + xy + y^2)\,dx - x^2\,dy = 0, \quad y(1) = 1$

15. $\dfrac{dy}{dx} = \dfrac{y - 1}{x + 3}, \quad y(-1) = 0$

16. $(1 - \cos x)y' + y\sin x = 0, \quad y(\pi/4) = 1$

17. $\dfrac{dy}{dx} - \dfrac{2y}{x} = x^2\cos x, \quad y\left(\dfrac{\pi}{2}\right) = 3$

18. $\left(\dfrac{2x - 1}{y}\right)dx + x\left(\dfrac{1 - x}{y^2}\right)dy = 0, \quad y(-1) = 3$

19. $y^2\sin x\,dx + \left(\dfrac{1}{x} - \dfrac{y}{x}\right)dy = 0, \quad y(\pi) = 1$

20. $(x^2 + y^2)\,dx - xy\,dy = 0, \quad y(1) = 2$

In Problems 21–30 determine the general solution of the given second-order differential equation.

21. $y'' + 9y = 0$

22. $y'' + 3y' + y = 0$

23. $y'' + 3y + 3 = 0$

24. $y'' + 2y' - 35y = 0$

25. $y'' - 9y = 0$

26. $2y'' + 4y' + 8y = 0$

27. $4y'' - 7y' = 0$

28. $y'' + 3y = 0$

29. $3y'' + 4y = 0$

30. $2y'' + 3y' + 8y = 0$

In Problems 31–36 find the particular solution.

31. $y'' - 4y' + 13y = 0, \quad y(0) = 2, \quad y'(0) = -5$

32. $y'' + 4y' + 4y = 0; \quad y(0) = 3, \quad y'(0) = 6$

33. $y'' + y' = 0, \quad y(0) = 2, \quad y'(0) = 1$

34. $y'' - 6y' + 25y = 0; \quad y(0) = y'(0) = -3$

35. $y'' - 4y' + 4y = 0, \quad y(1) = 1, \quad y'(1) = 1$

36. $y'' + \sqrt{10}y' + \frac{5}{2}y = 0; \quad y(-2) = 10, \quad y'(-2) = 0$

In Problems 37–42 use the method of undetermined coefficients to solve the given differential equation.

37. $y'' - y' - 6y = e^x\sin x$

38. $y'' + 2y' + 4y = \cos 4x$

39. $y'' + y = x + e^x$

40. $y'' - 4y' + 4y = xe^{2x}$

41. $y'' + y = 3 - 4\cos x$

42. $y'' + y = 8\cos^3 x$

In Problems 43–46 use the method of variation of parameters to solve the given differential equation.

43. $y'' - 3y' + 2y = e^x/(e^x + 1)$

44. $y'' + y = \tan^2 x$

45. $y'' + y = \sec^3 x$

46. $y'' - 4y' + 4y = x^2e^{2x}$

In Problems 47 and 48 find the series solution to the differential equation.

47. $(x-2)y' + y = 0$

48. $x^2y'' + y' - 2y = 0$

49. Assuming that a spherical drop of water evaporates at a rate proportional to its surface area, find the radius as a function of time.

50. A cylindrical tank has a leak in the bottom, and water flows out at a rate proportional to the pressure at the bottom. If the tank loses 2% of its water in 24 hours, when will it be half empty?

51. Suppose that the food supply available will support a maximum number M of bacteria, and that the rate of growth of the bacteria is proportional to the difference between the maximum number and the number present. Find an expression for the number of bacteria as a function of time.

CHALLENGE EXERCISES

1. *Age of the Earth's Crust.* Uranium has a half-life of 4.5×10^9 years. The decomposition sequence is very complicated, producing a very large number of intermediate radioactive products, but the final product is an isotope of lead with an atomic weight of 206, called uranium lead. Assuming for simplicity that the change from uranium to lead is direct, prove that $u = u_0e^{-kt}$, $l = u_0(1 - e^{-kt})$, where u and l denote the number of uranium and uranium lead atoms, respectively, present at time t.

We can measure the ratio $r = l/u$ in a rock and, if we assume that all of the uranium lead came from decomposition of the uranium originally present, we can obtain a lower bound for the age of the rock, and consequently a lower bound for the age of the earth's crust. Prove that this lower bound is given by

$$t = \frac{1}{k}\ln(1+r) = \frac{1}{k}\left(r - \frac{r^2}{2} + \frac{r^3}{3} - \cdots\right) \approx \frac{r}{k}$$

In a certain rock, a chemical determination gave $l/u = 0.054$. Using these data, show that the rock is about 350 million years old.

2. A room 150 feet by 50 feet by 20 feet receives fresh air at the rate of 5000 cubic feet per minute. If the fresh air contains 0.04% carbon dioxide and the air in the room initially contained 0.3% carbon dioxide, find the percentage of carbon dioxide after 1 hour. What is the percentage after 2 hours?

3. Near the earth's surface the attraction due to gravity is practically constant, but according to Newton's law the force of attraction exerted by the earth on a given body

(*Problem 3 continues in the next column*)

3. (*continued*)
is k/x^2, where x is the distance of the body from the center of the earth. Show that if a rocket is shot upward from the earth's surface with initial velocity v_0, and if air resistance is neglected, then

$$v^2 = v_0^2 - 2Rg + 2R^2g/x,$$

where $R = 3960$ miles, is the radius of the earth. [*Hint:* Observe that

$$\frac{d^2x}{dt^2} = \frac{dv}{dt} = \frac{dv}{dx}\frac{dx}{dt} = v\frac{dv}{dx}$$

4. *Escape Velocity.* For the rocket of Problem 3 find the smallest value of v_0 necessary to keep it going indefinitely (neglecting the attractive forces due to celestial bodies other than the earth). This is called the *escape velocity* for the earth.

5. A body suspended by a vibrating spring is acted on by a damping force that is proportional to speed. The distance x of the body below its equilibrium position satisfies the differential equation $x'' + Bx' + 9x = 0$. For what values of B will the motion be oscillatory? Assume $B > 0$.

6. According to Newton's law, the rate at which a body cools is proportional to the difference between the temperature of the body and that of the surrounding medium. If a certain steel bar has a temperature of 1230°C and cools to 1030°C in 10 minutes when the surrounding temperature is 30°C, how long will it take for the bar to cool to 80°C?

7. If the water in a tank runs out through a small hole in the bottom, then the rate of flow is proportional to the square root of the height of the water in the tank. Prove that if the tank is a cylinder with its axis vertical, then the time required for three-fourths of the water to run out is equal to the time required for the remaining one-fourth of the water to run out.

8. A country has in circulation \$3 billion of paper currency. Each day about \$10 million comes into the banks and the same amount is paid out. The government decides to issue new currency, and whenever the old-style currency comes into the bank, it is destroyed and replaced by the new currency. How long will it take for the currency in circulation to become 95% new?

Appendix I Theorems, Proofs, and Definitions

Limit Theorems, Proofs, and Definitions I.1

Uniqueness of Limit

The limit of a function, if it exists, is unique; that is, a function cannot approach two different limits at the same time.

THEOREM

If $\lim_{x\to c} f(x) = L_1$ and $\lim_{x\to c} f(x) = L_2$, then $L_1 = L_2$.

Proof Let us assume that $L_1 \neq L_2$. We will show that this assumption leads to a contradiction. Since $\lim_{x\to c} f(x) = L_1$, by the definition of limit, for any given $\varepsilon > 0$, there is a $\delta_1 > 0$ such that for x in the domain of f,

$$|f(x) - L_1| < \varepsilon \quad \text{whenever} \quad 0 < |x - c| < \delta_1 \tag{1}$$

Similarly, since $\lim_{x\to c} f(x) = L_2$, by the definition of limit, for any given $\varepsilon > 0$, there is a $\delta_2 > 0$ such that for x in the domain of f,

$$|f(x) - L_2| < \varepsilon \qquad \text{whenever} \qquad 0 < |x - c| < \delta_2 \tag{2}$$

Now, we know that $L_1 - L_2 = L_1 - f(x) + f(x) - L_2$, so, by applying the triangle inequality, we have

$$|L_1 - L_2| = |L_1 - f(x) + f(x) - L_2| \le |L_1 - f(x)| + |f(x) - L_2| \tag{3}$$

From (1), (2), and (3), we conclude that for any given $\varepsilon > 0$, there exist δ_1 and δ_2 such that whenever $0 < |x - c| < \delta_1$ and $0 < |x - c| < \delta_2$, we have

$$|L_1 - L_2| < \varepsilon + \varepsilon = 2\varepsilon \tag{4}$$

But, if we let $\varepsilon = \frac{1}{2}|L_1 - L_2| > 0$, then, from (4), we have

$$|L_1 - L_2| < 2\varepsilon = |L_1 - L_2|$$

which is a contradiction. Hence, $L_1 = L_2$.

Algebra of Limits

THEOREM

Let f and g be two functions for which $\lim_{x\to c} f(x) = L$ and $\lim_{x\to c} g(x) = M$, L and M being two real numbers.

[2.2.1] ***Limit of a Sum.***

$$\lim_{x\to c}[f(x) + g(x)] = \lim_{x\to c} f(x) + \lim_{x\to c} g(x) = L + M$$

[2.2.3] ***Limit of a Constant Times a Function.*** **If k is any real number,**

$$\lim_{x\to c}[kf(x)] = k \lim_{x\to c} f(x) = kL$$

[2.2.4] ***Limit of a Product.***

$$\lim_{x\to c}[f(x)g(x)] = [\lim_{x\to c} f(x)][\lim_{x\to c} g(x)] = LM$$

Proof of Theorem (2.2.1) We wish to show that for any $\varepsilon > 0$ there must exist a $\delta > 0$ such that

$$|f(x) + g(x) - (L + M)| < \varepsilon \qquad \text{whenever} \qquad 0 < |x - c| < \delta$$

Since $\lim_{x\to c} f(x) = L$, by the definition of limit, given $\varepsilon/2 > 0$, there is a $\delta_1 > 0$ such that

$$|f(x) - L| < \frac{\varepsilon}{2} \qquad \text{whenever} \qquad 0 < |x - c| < \delta_1$$

Since $\lim_{x\to c} g(x) = M$, for this same choice of $\varepsilon/2$ there is a $\delta_2 > 0$

such that

$$|g(x) - M| < \frac{\varepsilon}{2} \qquad \text{whenever} \qquad 0 < |x - c| < \delta_2$$

Now let δ be the smaller of δ_1 and δ_2. Then $\delta \le \delta_1$ and $\delta \le \delta_2$; and, by using this δ, we can state the following:

$$|f(x) - L| < \frac{\varepsilon}{2} \qquad \text{whenever} \qquad 0 < |x - c| < \delta$$

$$|g(x) - M| < \frac{\varepsilon}{2} \qquad \text{whenever} \qquad 0 < |x - c| < \delta$$

Therefore, by the triangle inequality,

$$\begin{aligned} |f(x) + g(x) - (L + M)| &= |f(x) - L + g(x) - M| \\ &\le |f(x) - L| + |g(x) - M| \le \frac{\varepsilon}{2} + \frac{\varepsilon}{2} = \varepsilon \end{aligned}$$

whenever $0 < |x - c| < \delta$. Hence, $\lim_{x \to c}[f(x) + g(x)] = L + M$.

Proof of Theorem (2.2.3) If the constant $k = 0$, the result is obvious. For $k \ne 0$, we look at

$$|kf(x) - kL| = |k[f(x) - L]| = |k||f(x) - L|$$

Since $\lim_{x \to c} f(x) = L$, given $\varepsilon/|k| > 0$, there is a $\delta > 0$ so that whenever $0 < |x - c| < \delta$, we have $|f(x) - L| < \varepsilon/|k|$. Hence, for any $\varepsilon > 0$, there is a $\delta > 0$ such that whenever $0 < |x - c| < \delta$, we have

$$|kf(x) - kL| = |k||f(x) - L| < |k|\frac{\varepsilon}{|k|} = \varepsilon$$

Proof of Theorem (2.2.4) We look at $|f(x)g(x) - LM|$ and see that we should subtract and add $f(x)M$ to get terms involving $g(x) - M$ and $f(x) - L$:

$$\begin{aligned} |f(x)g(x) - LM| &= |f(x)g(x) - f(x)M + f(x)M - LM| \\ &= |f(x)[g(x) - M] + [f(x) - L]M| \\ &\le |f(x)||g(x) - M| + |f(x) - L||M| \end{aligned}$$

Since $\lim_{x \to c} f(x) = L$ and $\lim_{x \to c} g(x) = M$, we know that there is a $\delta_1 > 0$ such that if $0 < |x - c| < \delta_1$, then

$$|f(x) - L| < 1 \qquad \text{from which} \qquad |f(x)| < 1 + |L| \tag{5}$$

Given $\varepsilon > 0$, there is a δ_2 such that if $0 < |x - c| < \delta_2$, then

$$|g(x) - M| < \frac{\varepsilon}{1 + |L| + |M|} \tag{6}$$

Given $\varepsilon > 0$, there is a δ_3 such that if $0 < |x - c| < \delta_3$, then

$$|f(x) - L| < \frac{\varepsilon}{1 + |L| + |M|} \tag{7}$$

So we choose $\delta = \min(\delta_1, \delta_2, \delta_3)$. Then, if $0 < |x - c| < \delta$, we may combine (5), (6), and (7) to get

$$|f(x)g(x) - LM| < [1 + |L|]\frac{\varepsilon}{1 + |L| + |M|} + |M|\frac{\varepsilon}{1 + |L| + |M|} = \varepsilon$$

[2.5.1] THEOREM / ***Squeezing Theorem.***

Let f, g, and h be functions such that $f(x) \leq g(x) \leq h(x)$ for all numbers x in some open interval containing c, except possibly at c. If $\lim_{x \to c} f(x) = L$ and $\lim_{x \to c} h(x) = L$, then $\lim_{x \to c} g(x) = L$.

Proof Let $\varepsilon > 0$. Since $\lim_{x \to c} f(x) = \lim_{x \to c} h(x) = L$, there exist positive numbers δ_1 and δ_2 such that for x in the open interval,

$$|f(x) - L| < \varepsilon \quad \text{whenever} \quad 0 < |x - c| < \delta_1$$

$$|h(x) - L| < \varepsilon \quad \text{whenever} \quad 0 < |x - c| < \delta_2$$

We choose δ to be the smaller of δ_1 and δ_2. Then $0 < |x - c| < \delta$ implies that $|f(x) - L| < \varepsilon$ and $|h(x) - L| < \varepsilon$. In other words, $0 < |x - c| < \delta$ implies that $L - \varepsilon < f(x) < L + \varepsilon$ and $L - \varepsilon < h(x) < L + \varepsilon$. Since $f(x) \leq g(x) \leq h(x)$ for all $x \neq c$ in the open interval, it follows that if $0 < |x - c| < \delta$ and x is in the open interval, we have

$$L - \varepsilon < f(x) \leq g(x) \leq h(x) < L + \varepsilon$$

Thus, for any given $\varepsilon > 0$, there is some $\delta > 0$ such that when $0 < |x - c| < \delta$ and x is in the open interval, then $L - \varepsilon < g(x) < L + \varepsilon$. Hence, $\lim_{x \to c} g(x) = L$.

THEOREM

Let a function f be defined on some interval containing c and let f be continuous at c. If $f(c) > 0$, then there is a subinterval containing c throughout which $f(x) > 0$.

Proof Since f is continuous at c, we know that $\lim_{x \to c} f(x) = f(c) > 0$. Then, given any $\varepsilon > 0$, and, in particular, $\varepsilon = \frac{1}{2}f(c)$, there is a $\delta > 0$ such that whenever $|x - c| < \delta$, we have

$$|f(x) - f(c)| < \varepsilon = \tfrac{1}{2}f(c)$$

For x in the interval $c - \delta < x < c + \delta$, it follows that

$$-\tfrac{1}{2}f(c) < f(x) - f(c) < \tfrac{1}{2}f(c)$$

$$\tfrac{1}{2}f(c) < \quad f(x) \quad < \tfrac{3}{2}f(c)$$

That is, $f(x) > \frac{1}{2}f(c) > 0$ in the interval $c - \delta < x < c + \delta$.

Limit Definitions

DEFINITION / ***Left-Hand Limit.***

Let f be a function defined on the open interval (a, c). The *limit of $f(x)$ as x approaches c from the left* is L, written

$$\lim_{x \to c^-} f(x) = L$$

if for any given number $\varepsilon > 0$, a positive number δ exists such that

$$|f(x) - L| < \varepsilon \qquad \text{whenever} \qquad c - \delta < x < c \quad \text{and } x \text{ is in the domain of } f$$

See Figure 1 for a geometric interpretation.

DEFINITION / ***Right-Hand Limit.***

Let f be a function defined on the open interval (c, b). The *limit of $f(x)$ as x approaches c from the right* is L, written

$$\lim_{x \to c^+} f(x) = L$$

if for any given number $\varepsilon > 0$, a positive number δ exists such that

$$|f(x) - L| < \varepsilon \qquad \text{whenever} \qquad c < x < c + \delta \quad \text{and } x \text{ is in the domain of } f$$

See Figure 2 for an illustration.

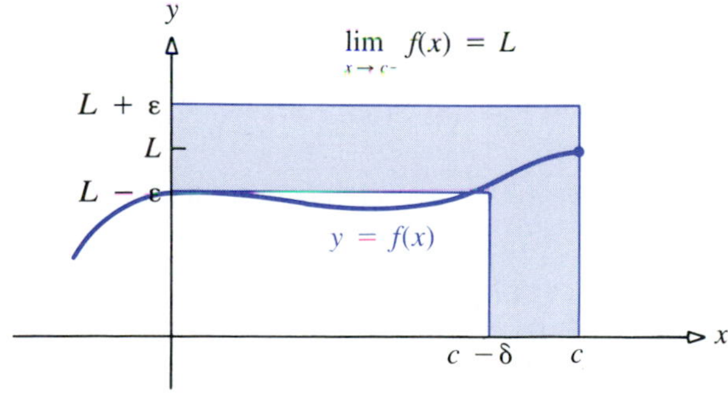

Figure 1

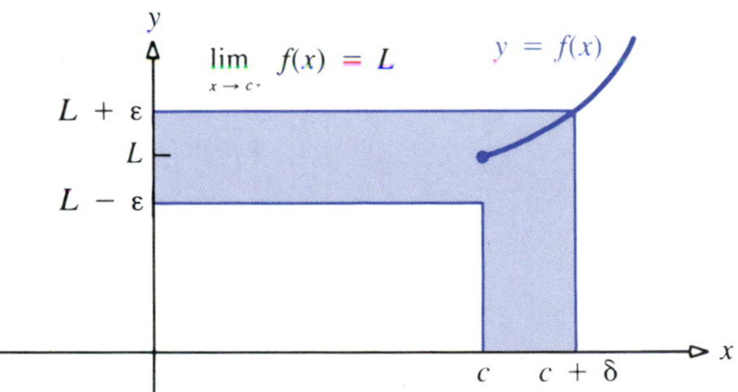

Figure 2

Derivative Theorems and Proofs **I.2**

[3.7.1] THEOREM / ***Chain Rule.***

If f and g are differentiable functions, the composite function $f \circ g$ is also differentiable. Moreover, if $y = f(u)$ and $u = g(x)$,

then the derivative of $y = (f \circ g)(x)$ is

$$\frac{dy}{dx} = \frac{dy}{du}\frac{du}{dx}$$

Proof If $y = f(u)$ and $u = g(x)$, the composite function is $y = (f \circ g)(x) = f(g(x))$. First, the function $y = f(u)$ is differentiable, so that

$$\lim_{\Delta u \to 0} \frac{\Delta y}{\Delta u} = \frac{dy}{du}$$

Define a new function t to be

$$t = \begin{cases} 0 & \text{if } \Delta u = 0 \\ \dfrac{\Delta y}{\Delta u} - \dfrac{dy}{du} & \text{if } \Delta u \neq 0 \end{cases}$$

For $\Delta u \neq 0$,

$$\frac{\Delta y}{\Delta u} = \frac{dy}{du} + t$$

$$\Delta y = \frac{dy}{du}\Delta u + t\,\Delta u \tag{1}$$

Since for $\Delta u = 0$, we have $\Delta y = 0$ (y is a continuous function of u), (1) is valid whether $\Delta u = 0$ or $\Delta u \neq 0$. Division by Δx yields

$$\frac{\Delta y}{\Delta x} = \frac{dy}{dx}\frac{\Delta u}{\Delta x} + t\frac{\Delta u}{\Delta x}$$

Take the limit as $\Delta x \to 0$. Then

$$\lim_{\Delta x \to 0} \frac{\Delta y}{\Delta x} = \lim_{\Delta x \to 0} \frac{dy}{du}\frac{\Delta u}{\Delta x} + \lim_{\Delta x \to 0} t\frac{\Delta u}{\Delta x} = \frac{dy}{du}\frac{du}{dx} + \lim_{\Delta x \to 0} t\frac{du}{dx}$$

Now when $\Delta x \to 0$, so does $\Delta u \to 0$, since u is a continuous function of x. Thus,

$$\lim_{\Delta x \to 0} t = \lim_{\Delta u \to 0} t \underset{\text{By (1)}}{\uparrow}= \lim_{\Delta u \to 0} \left(\frac{\Delta y}{\Delta u} - \frac{dy}{du}\right) = 0$$

Hence,

$$\frac{dy}{dx} = \lim_{\Delta x \to 0} \frac{\Delta y}{\Delta x} = \frac{dy}{du}\frac{du}{dx} + \lim_{\Delta u \to 0} t\frac{du}{dx} = \frac{dy}{du}\frac{du}{dx} + 0\frac{du}{dx} = \frac{dy}{du}\frac{du}{dx}$$

Inverse Functions

THEOREM / *Continuity of an Inverse Function.*

If f denotes a continuous one-to-one function defined on the open interval (a, b), then its inverse $g = f^{-1}$ is continuous.

Proof If f is continuous, then, being one-to-one, f is either increasing on (a, b) or decreasing on (a, b). (This proof is left to you.) Suppose f is increasing on (a, b). Let $y_0 = f(x_0)$. We need to show that g is continuous at y_0, given that f is continuous at x_0. The number $g(y_0) = x_0$ is in the open interval (a, b). Choose $\varepsilon > 0$ sufficiently small so that $g(y_0) - \varepsilon$ and $g(y_0) + \varepsilon$ are also in (a, b). We wish to find $\delta > 0$ so that:

$$\text{If} \quad y_0 - \delta < y < y_0 + \delta \quad \text{then} \quad g(y_0) - \varepsilon < g(y) < g(y_0) + \varepsilon$$

This can be done by choosing δ to obey

$$f[g(y_0) - \varepsilon] < y_0 - \delta \qquad \text{and} \qquad y_0 + \delta < f[g(y_0) + \varepsilon]$$

Then, if $y_0 - \delta < y < y_0 + \delta$, we have

$$f[g(y_0) - \varepsilon] < y < f[g(y_0) + \varepsilon]$$

But g is also increasing, since f is. Hence,

$$g(y_0) - \varepsilon < g(y) < g(y_0) + \varepsilon$$

We can handle the case where f is decreasing on (a, b) in a similar way.

[3.8.1] THEOREM | ***Derivative of an Inverse Function.***

Let $y = f(x)$ and $x = g(y)$ be inverse functions. Assume that f is differentiable on an open interval containing x_0 and that $y_0 = f(x_0)$. If $f'(x_0) \neq 0$, then g is differentiable at y_0, and

$$g'(y_0) = \frac{1}{f'(x_0)}$$

Proof Since f and g are inverses of one another, we have $f(x) = y$ if and only if $x = g(y)$. Hence, we have the following identity, where $g(y_0) = x_0$:

$$\frac{g(y) - g(y_0)}{y - y_0} = \frac{x - x_0}{f(x) - f(x_0)} = \frac{1}{\dfrac{f(x) - f(x_0)}{x - x_0}}$$

By the previous theorem, the continuity of f at x_0 implies the continuity of g at y_0; hence, $y \to y_0$ as $x \to x_0$. By taking the limits of both sides of the above identity, we have

$$g'(y_0) = \lim_{x \to x_0} \frac{g(y) - g(y_0)}{y - y_0} = \frac{1}{\displaystyle\lim_{x \to x_0} \frac{f(x) - f(x_0)}{x - x_0}} = \frac{1}{f'(x_0)}$$

which completes the proof.

Theorem (3.8.1) and its proof are illustrated in Figure 3.

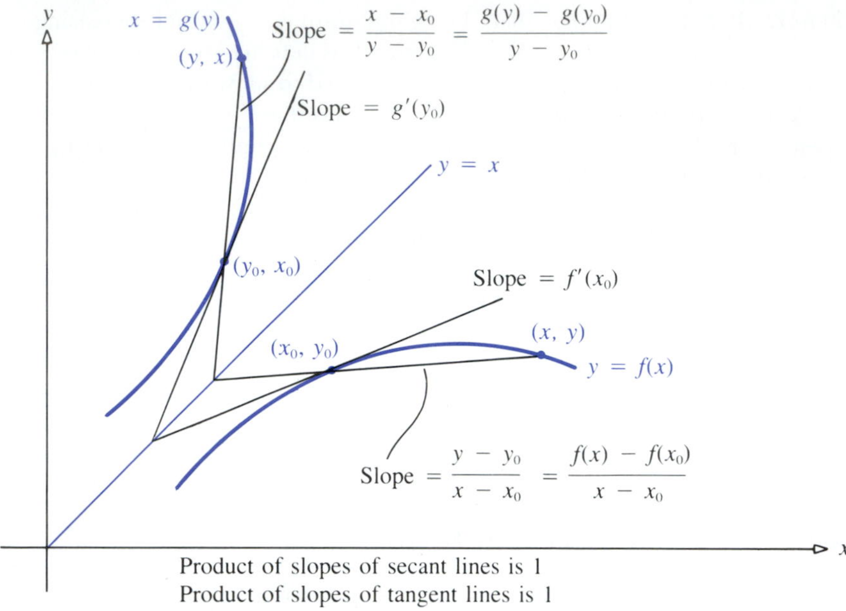

Figure 3

I.3 Integral Theorems and Proofs

[5.4.1] THEOREM

If f is continuous on a closed interval containing the numbers a, b, and c, then

$$\int_a^b f(x)\,dx + \int_b^c f(x)\,dx = \int_a^c f(x)\,dx$$

no matter what the order of the numbers a, b, and c.

Proof, Part 1 Assume that $a < b < c$. Let $\varepsilon > 0$. Since f is continuous on $[a, b]$ and on $[b, c]$, there is a $\delta_1 > 0$ such that

$$\left|\sum_{i=1}^{k} f(u_i)\Delta x_i - \int_a^b f(x)\,dx\right| < \frac{\varepsilon}{2} \tag{1}$$

for every Riemann sum $\sum_{i=1}^{k} f(u_i)\Delta x_i$ for f on $[a, b]$ whose partition P_1 of $[a, b]$ has norm $\|P_1\| < \delta_1$. There is also a $\delta_2 > 0$ such that

$$\left|\sum_{i=1}^{j} f(u_i)\Delta x_i - \int_b^c f(x)\,dx\right| < \frac{\varepsilon}{2} \tag{2}$$

for every Riemann sum $\sum_{i=1}^{j} f(u_i)\Delta x_i$ for f on $[b, c]$ whose partition P_2 of $[b, c]$ has norm $\|P_2\| < \delta_2$.

Let $\delta = \min\{\delta_1, \delta_2\}$. Then (1) and (2) hold with δ replacing δ_1 and δ_2. If we now add (1) and (2), term by term, and if $\|P_1\| < \delta$ and

$\|P_2\| < \delta$, we have

$$\left|\sum_{i=1}^{k} f(u_i)\,\Delta x_i - \int_a^b f(x)\,dx\right| + \left|\sum_{i=1}^{j} f(u_i)\,\Delta x_i - \int_b^c f(x)\,dx\right| < \frac{\varepsilon}{2} + \frac{\varepsilon}{2} = \varepsilon$$

By using the triangle inequality, this result implies that

$$\left|\sum_{i=1}^{k} f(u_i)\,\Delta x_i - \int_a^b f(x)\,dx + \sum_{i=1}^{j} f(u_i)\,\Delta x_i - \int_b^c f(x)\,dx\right| < \varepsilon \qquad \textbf{(3)}$$

if $\|P_1\| < \delta$ and $\|P_2\| < \delta$.

Denote $P_1 \cup P_2$ by P^*, a partition of $[a, c]$ having b as a point of division. Then

$$\sum_{i=1}^{k} f(u_i)\,\Delta x_i + \sum_{i=1}^{j} f(u_i)\,\Delta x_i = \sum_{i=1}^{n} f(u_i)\,\Delta x_i$$

is a Riemann sum for f on P^*. Since $\|P^*\| < \delta$ implies that $\|P_1\| < \delta$ and $\|P_2\| < \delta$, it follows from (3) that

$$\left|\sum_{i=1}^{n} f(u_i)\,\Delta x_i - \left[\int_a^b f(x)\,dx + \int_b^c f(x)\,dx\right]\right| < \varepsilon$$

for every Riemann sum $\sum_{i=1}^{n} f(u_i)\,\Delta x_i$ for f on $[a, c]$ whose partition P^* of $[a, c]$ has b for a point of division and norm $\|P^*\| < \delta$. Therefore,

$$\int_a^c f(x)\,dx = \int_a^b f(x)\,dx + \int_b^c f(x)\,dx$$

Part 2 There are six possible orderings for the points a, b, and c:

$$a < b < c \qquad a < c < b \qquad b < a < c$$

$$b < c < a \qquad c < a < b \qquad c < b < a$$

In part 1 we showed that the theorem is true for the order $a < b < c$. Now consider any other order—say, $b < c < a$. From part 1, we have

$$\int_b^c f(x)\,dx + \int_c^a f(x)\,dx = \int_b^a f(x)\,dx \qquad \textbf{(4)}$$

However,

$$\int_c^a f(x)\,dx = -\int_a^c f(x)\,dx \qquad \int_b^a f(x)\,dx = -\int_a^b f(x)\,dx$$

By substituting this in (4), we obtain

$$\int_b^c f(x)\,dx - \int_a^c f(x)\,dx = -\int_a^b f(x)\,dx$$

$$\int_a^b f(x)\,dx + \int_b^c f(x)\,dx = \int_a^c f(x)\,dx$$

which proves the theorem for the order $b < c < a$.

The proofs for the remaining four orders are similar. ∎

[5.4.2] THEOREM

If a function f is continuous on a closed interval $[a, b]$ and if m and M denote the absolute minimum and absolute maximum of f on $[a, b]$, respectively, then

$$m(b-a) \leq \int_a^b f(x)\, dx \leq M(b-a)$$

Proof We shall prove here only that $m(b-a) \leq \int_a^b f(x)\, dx$. We assume the contrary—namely, that

$$m(b-a) > \int_a^b f(x)\, dx \tag{5}$$

We show that this assumption leads to a contradiction, and therefore that the theorem must be true. Since f is continuous on $[a, b]$,

$$\lim_{\|P\| \to 0} \sum_{i=1}^{n} f(u_i)\Delta x_i = \int_a^b f(x)\, dx$$

By (5), $m(b-a) - \int_a^b f(x)\, dx > 0$. Let

$$\varepsilon = m(b-a) - \int_a^b f(x)\, dx \tag{6}$$

Then there is a $\delta > 0$ such that for all partitions P of $[a, b]$ with $\|P\| < \delta$, we have

$$\sum_{i=1}^{n} f(u_i)\Delta x_i - \int_a^b f(x)\, dx < \varepsilon$$

which is equivalent to

$$\int_a^b f(x)\, dx - \varepsilon < \sum_{i=1}^{n} f(u_i)\Delta x_i < \int_a^b f(x)\, dx + \varepsilon$$

By (6), the second inequality can be written as

$$\sum_{i=1}^{n} f(u_i)\Delta x_i < \int_a^b f(x)\, dx + \varepsilon = \int_a^b f(x)\, dx + m(b-a) - \int_a^b f(x)\, dx = m(b-a)$$

Consequently,

$$\sum_{i=1}^{n} f(u_i)\Delta x_i < m(b-a) = \sum_{i=1}^{n} m\Delta x_i$$

which implies that for every partition P of $[a, b]$ with $\|P\| < \delta$,

$$f(u_i) < m$$

for some u_i in $[a, b]$. But this result is impossible because m is the absolute minimum of f on $[a, b]$. Thus, the assumption $m(b-a) > \int_a^b f(x)\, dx$ is false, and it follows that

$$m(b-a) \leq \int_a^b f(x)\, dx$$

The proof that $\int_a^b f(x)\, dx \leq M(b-a)$ is similar. ■

Appendix II
Formulas

The Binomial Theorem II.1

BINOMIAL THEOREM

If n is a positive integer, then

$$(x+y)^n = \binom{n}{0}x^n + \binom{n}{1}x^{n-1}y + \binom{n}{2}x^{n-2}y^2 + \cdots + \binom{n}{k}x^{n-k}y^k + \cdots + \binom{n}{n}y^n$$

where $\binom{n}{r} = \dfrac{n!}{r!(n-r)!}$

Observe that the powers of x begin at n and decrease by 1, while the powers of y begin with 0 and increase by 1. Also, in $(x + y)^n$, the coefficient of $x^{n-k}y^k$ is $\binom{n}{k}$.

II.2 Algebra

(1) ***Quadratic Formula:*** The solutions of the quadratic equation $ax^2 + bx + c = 0 \quad (a \neq 0)$ are

$$x = \frac{-b \pm \sqrt{b^2 - 4ac}}{2a}$$

provided $b^2 - 4ac \geq 0$. Otherwise, if $b^2 - 4ac < 0$, there are no real solutions.

(2) ***Exponents:***

$$a^m \cdot a^n = a^{m+n} \qquad (ab)^n = a^n b^n$$

$$\frac{a^m}{a^n} = a^{m-n}, \quad a \neq 0 \qquad a^{-n} = \frac{1}{a^n}, \quad a \neq 0$$

$$(a^m)^n = a^{mn} \qquad a^{p/q} = \sqrt[q]{a^p} = (\sqrt[q]{a})^p, \quad a > 0$$

(3) ***Logarithms:***

$$\log_a x + \log_a y = \log_a xy \qquad \log_a 1 = 0$$

$$\log_a x - \log_a y = \log_a \frac{x}{y} \qquad \log_a a = 1$$

$$r \log_a x = \log_a x^r \qquad a^{\log_a x} = x$$

$$\log_a x = \frac{\log_b x}{\log_b a}$$

(4) ***Determinants:***

$$\begin{vmatrix} a & b \\ c & d \end{vmatrix} = ad - bc$$

$$\begin{vmatrix} a_1 & b_1 & c_1 \\ a_2 & b_2 & c_2 \\ a_3 & b_3 & c_3 \end{vmatrix} = a_1 \begin{vmatrix} b_2 & c_2 \\ b_3 & c_3 \end{vmatrix} - b_1 \begin{vmatrix} a_2 & c_2 \\ a_3 & c_3 \end{vmatrix} + c_1 \begin{vmatrix} a_2 & b_2 \\ a_3 & b_3 \end{vmatrix}$$

II.3 Geometry

(1) ***Triangle:***

Area $= \frac{1}{2}bh$

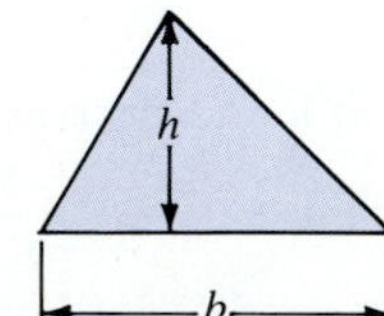

(2) ***Parallelogram:***

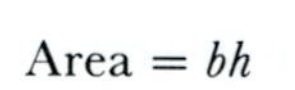

Area $= bh$

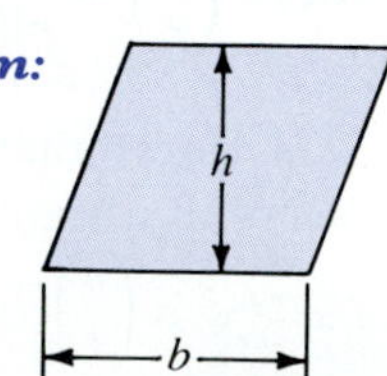

(3) ***Trapezoid:***

$\text{Area} = \frac{1}{2}h(a + b)$

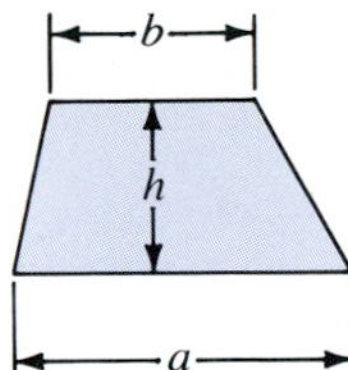

(4) ***Circle:***

$\text{Area} = \pi R^2$

$\text{Circumference} = 2\pi R$

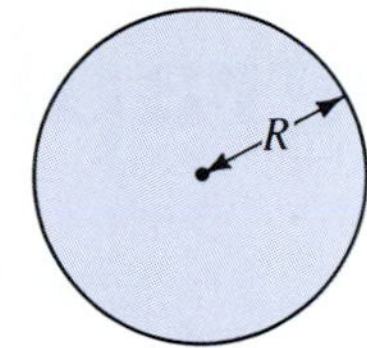

(5) ***Sector of Circle:***

$\text{Area} = \frac{1}{2}R^2\theta$

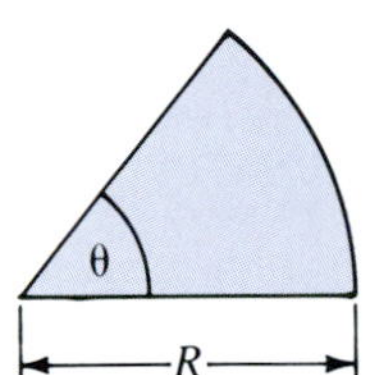

(6) ***Ellipse:***

$\text{Area} = \pi ab$

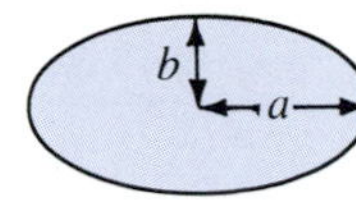

(7) ***Right Circular Cylinder***

$\text{Volume} = \pi R^2 h$

$\text{Lateral surface} = 2\pi Rh$

$\text{Total surface} = 2\pi R(R + h)$

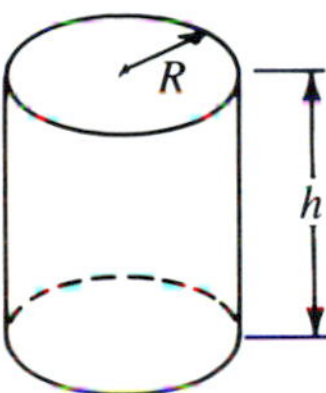

(8) ***Right Circular Cone:***

$\text{Volume} = \frac{1}{3}\pi R^2 h$

$\text{Lateral surface} = \pi Rs$

$\text{Total surface} = \pi R(R + s)$

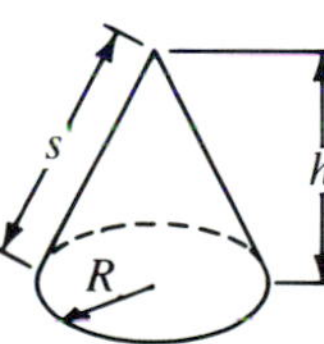

(9) ***Sphere:***

$\text{Volume} = \frac{4}{3}\pi R^3$

$\text{Surface} = 4\pi R^2$

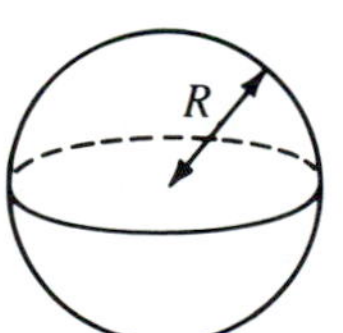

(10) ***Frustum of a Right Circular Cone:***

$\text{Volume} = \frac{1}{3}\pi h(R_1^2 + R_1R_2 + R_2^2)$

$\text{Lateral surface} = \pi s(R_1 + R_2)$

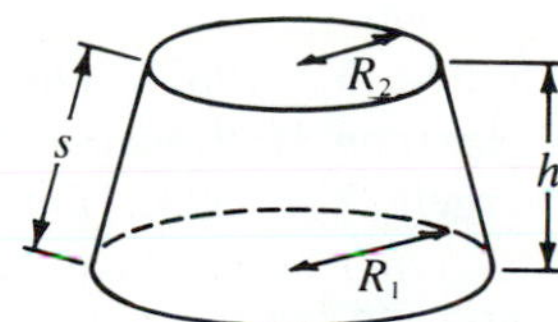

Appendix III Bibliography

Demana, F., & Waits, B. K. (1988). The Ohio State University calculator and computer precalculus project: The mathematics of tomorrow today! *The AMATYC Review*, *10*(1), 46–55.

Demana, F., & Waits, B. K. (1988). Pitfalls in graphical computation, or why a single graph isn't enough. *College Mathematics Journal*, *19*, 236–241.

Foley, G. D. (1987). Future shock: Hand-held computers. *The AMATYC Review*, *9*(1), 53–57.

Foley, G. D. (1987). Reader reflections: Zoom revisited. *Mathematics Teacher*, *80*, 606.

Nievergelt, Y. (1987). The chip with the college education: The HP-28C. *American Mathematical Monthly*, *94*, 895–902.

Nievergelt, Y. (1988). The HP-28S brings computations and theory back together in the classroom. *Notices of the American Mathematical Society*, *35*, 799–804.

Potter, D. (1987, August). Seven ounces of pocket graphics: Casio's new fx-8000G and friends. *Pico: The Magazine of Portable Computing*, pp. 22–23.

Potter, D. (1987, August). Casio's fx-8000G goes to high school. *Pico: The Magazine of Portable Computing*, pp. 23–24.

Shumway, R. J. (1988). Graphics calculators: Skills versus concepts. In J. de Lang & M. Doorman (Eds.), *Senior Secondary Mathematics Education* (pp. 136–140). Utrecht, The Netherlands: University Press.

Small, D., Hosack, J., & Lane, K. (1986). Computer algebra systems in undergraduate instruction. *College Mathematics Journal*, *17*, 423–433.

Smith, D. A., Porter, G. J., Leinbach, L. C., & Wenger, R. H. (Eds.) (1988). *Computers and mathematics: The use of computers in undergraduate instruction* [MAA Notes No. 9]. Washington, DC: Mathematical Association of America.

Steen, L. A. (1987, October 14). Who still does math with paper and pencil? *The Chronicle of Higher Education*, p. A48.

Steen, L. A. (Ed.). (1988). *Calculus for a new century: A pump, not a filter* [MAA Notes No. 8]. Washington, DC: Mathematical Association of America.

Tucker, T. (1987, January–February). Calculators with a college education? *Focus: The Newsletter of the Mathematical Association of America*, pp. 1, 5.

Waits, B. K., & Demana, F. (1988). Manipulative algebra—the culprit or the scapegoat? *Mathematics Teacher*, *81*, 322–334.

Appendix IV Answers to Odd-Numbered Problems

CHAPTER 1

Exercise 1.1

1. Rational **3.** Rational **5.** Irrational **7.** Rational **9.** Irrational **11.** $>$ **13.** $=$ **15.** $x \leq -1$
17. $x \geq -1$ **19.** $x \leq -4$ **21.** $x \geq \frac{9}{5}$ **23.** $x \leq 2$ or $x \geq 3$ **25.** $-4 < x < -3$ **27.** $x = 5$ or $x = -5$
29. $-3 \leq x \leq 3$ **31.** $-1 < x < 7$ **33.** $0 \leq x \leq 4$ **35.** $x < -1$ or $x > 7$ **37.** $-\infty < x < +\infty$
39. $2.99 < x < 3.01$ **41.** $-\frac{7}{4} \leq x \leq \frac{9}{4}$ **43.** $x < 0$ or $x > \frac{1}{3}$ **45.** $\frac{8}{5} \leq x < 2$ **47.** $-4 < x < 3$
49. $x < 1$ or $x > 3$ **51.** $x < 0$ or $1 < x < 3$ **53.** $x < -\frac{1}{2}$ or $x > \frac{1}{2}$ **55.** $x \geq 0$ **57.** $x \leq 1$ or $x \geq 2$
59. $\frac{71}{33}$ **61.** $\frac{104,209}{49,500}$ **63.** 1 **65.** $-\frac{1}{2} < x < \frac{7}{2}$
67. $|xy - 6| = |y(x-2) + 2(y-3)| \leq |y||x-2| + 2|y-3| < (\frac{31}{10})(\frac{1}{5}) + (2)(\frac{1}{10}) = \frac{41}{50}$
69. $|x - y| = |(x-a) + (a-y)| \leq |x-a| + |a-y| < \frac{1}{3} + \frac{1}{3} = \frac{2}{3}$

71. $a < b \Rightarrow a + a < a + b \Rightarrow a < \dfrac{a+b}{2}$

$a < b \Rightarrow a + b < b + b \Rightarrow \dfrac{a+b}{2} < b$

Let m = arithmetic mean

$D(a, m) = \left| a - \left(\dfrac{a+b}{2}\right) \right|$

$\left|\dfrac{a}{2} - \dfrac{b}{2}\right| = \left|\dfrac{b}{2} - \dfrac{a}{2}\right|$

$= \left| b - \left(\dfrac{a+b}{2}\right)\right| = D(b, m)$

73. $\dfrac{1}{h} = \left(\dfrac{a+b}{2}\right)\left(\dfrac{1}{ab}\right)$; $h = \dfrac{ab}{(a+b)/2}$

$a < \dfrac{a+b}{2} < b$

$\dfrac{a}{ab} < \left(\dfrac{a+b}{2}\right)\left(\dfrac{1}{ab}\right) < \dfrac{b}{ab}$

$\dfrac{1}{b} < \dfrac{1}{h} < \dfrac{1}{a}$

$a < h < b$

$\dfrac{(\sqrt{ab})^2}{\dfrac{a+b}{2}} = \dfrac{ab}{\dfrac{a+b}{2}} = \dfrac{2ab}{a+b} = h$

75. P = Perimeter

For a circle, $P = 2\pi R \Rightarrow R = \dfrac{P}{2\pi} \Rightarrow A = \pi R^2 = \dfrac{P^2}{4\pi}$

For a square, $P = 4s \Rightarrow s = \dfrac{P}{4} \Rightarrow A = s^2 = \dfrac{P^2}{16} < \dfrac{P^2}{4\pi}$

77. $b - a \geq 0$ and $-c > 0 \Rightarrow -c(b-a) \geq 0 \Rightarrow ac \geq bc$ **79.** $b - a \geq 0$ and $b + a > 0 \Rightarrow (b-a)(b+a) \geq 0 \Rightarrow b^2 \geq a^2$

Exercise 1.2

1. 5 **3.** $\sqrt{4.88} \approx 2.209$

5.

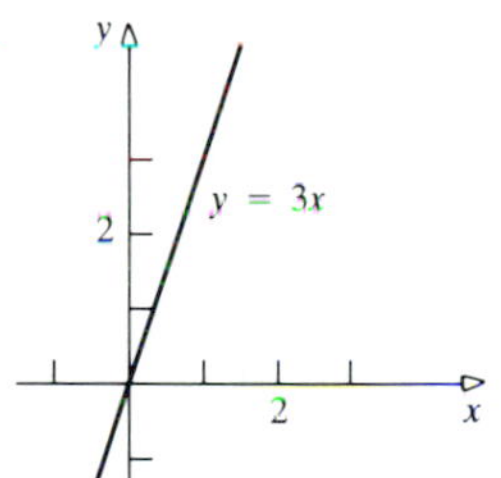

7.

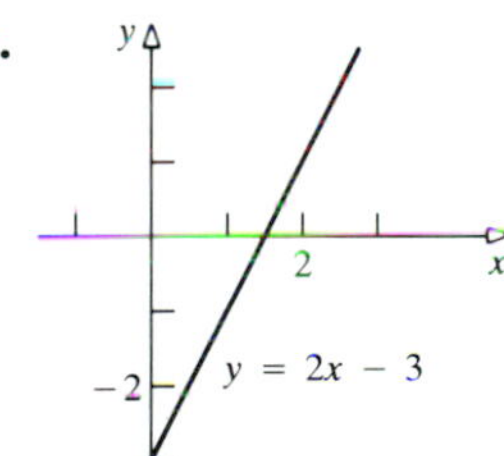

9.

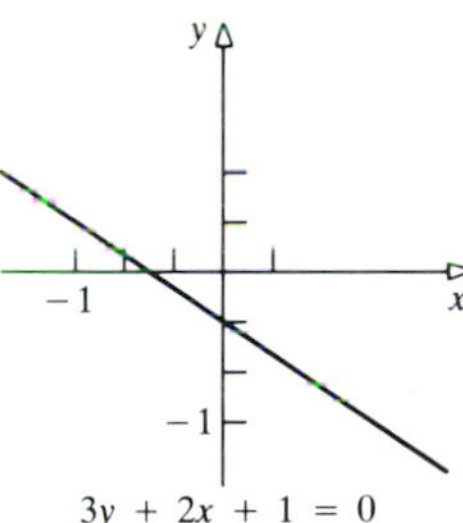

11.

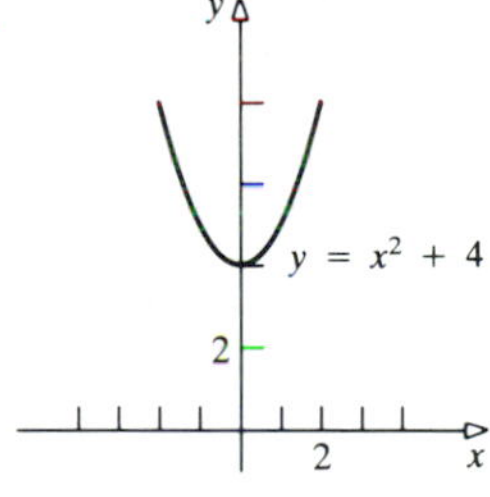

13. $x^2 + y^2 - 4x + 6y - 3 = 0$ **15.** $x^2 + y^2 - 2x + 4y + 4 = 0$ **17.** $(1, 0)$; 2 **19.** $(-2, 3)$; 4 **21.** $(4, 4)$

23. $(\frac{1}{2}, \frac{1}{2})$ **25.**

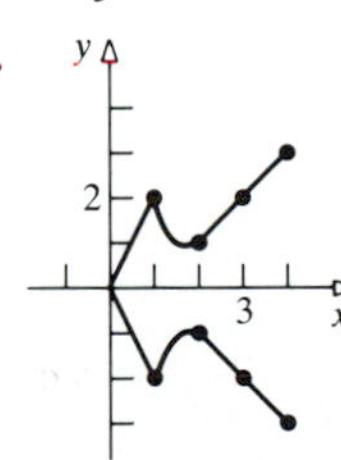

27.

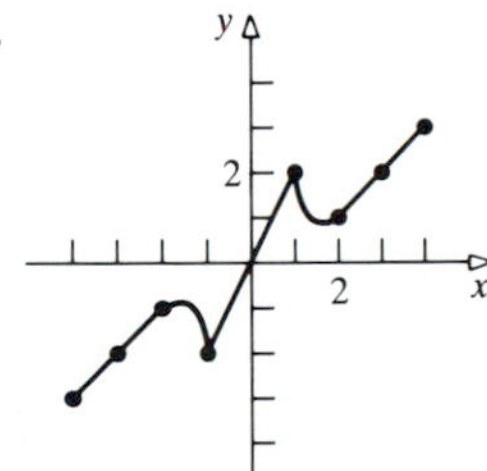

29. No symmetry **31.** Origin

33. y-axis **35.** Origin **37.** $x^2 + y^2 - 2x + 4y + 3 = 0$ **39.** $x^2 + y^2 - 4x - 6y + 9 = 0$

41. $x^2 + y^2 + 6x + 4y - 36 = 0$ **43.** (a) $(-a/2, -b/2)$; $\frac{1}{2}\sqrt{a^2 + b^2}$ (b) $a^2 + b^2 - 4c > 0$

45. $x + y - 2 = 0$ **47.** $5, \sqrt{73}, 2\sqrt{13}$ **49.** $6, 4, 2\sqrt{13}$; right angle triangle

51. $2\sqrt{17}, \sqrt{34}, \sqrt{34}$; both **53.** $(240°, 0.16)$

55. $\triangle P_1RP_2$ similar to $\triangle P_1QP$, where $R = (x_2, y_1)$ and $Q = (x, y_1)$

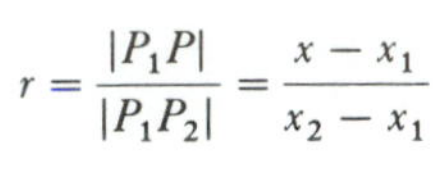

$$r = \frac{|P_1P|}{|P_1P_2|} = \frac{x - x_1}{x_2 - x_1}$$

$$x = (1 - r)x_1 + rx_2$$

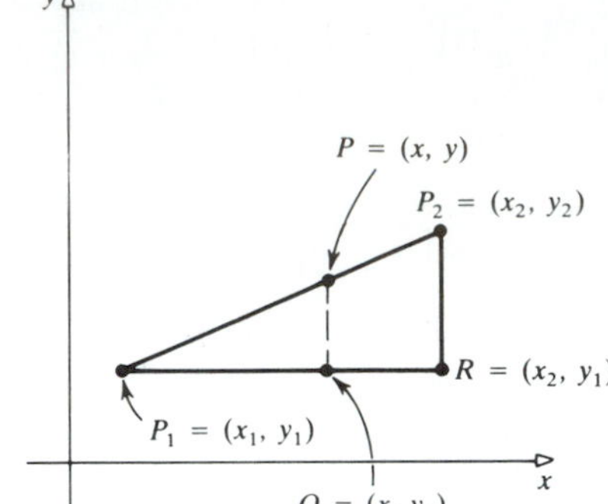

57. P_2 **59.** (9, 8)

Exercise 1.3

1. $2x - y + 7 = 0$ **3.** $2x + 3y + 1 = 0$ **5.** $x - 2y + 5 = 0$ **7.** $3x + y - 3 = 0$ **9.** $x - 2y - 2 = 0$
11. $x - 1 = 0$ **13.** $\frac{3}{2}$; -3 **15.** $-\frac{1}{2}$; 2 **17.** Undefined; no y-intercept **19.** $2x - y = 0$ **21.** $-\frac{2}{3}$
23. 2 **25.** 0 **27.** Parallel **29.** Intersecting; $(0, -2)$ **31.** Intersecting; $(-\frac{2}{3}, \frac{8}{3})$ **33.** Yes, same line
35. Line from $(-3, -4)$ to $(0, 2)$ has slope 2; line from $(0, 2)$ to $(6, -1)$ has slope $-\frac{1}{2}$
37. $\frac{17}{4}$ **39.** $-\frac{23}{5}$
41. The first and fourth lines are parallel, as are the second and third lines. The slopes of the first set are negative reciprocals of the slopes of the second set.
43. $m_{AB} = 2$, $m_{AC} = -\frac{1}{2}$; $|AB| = \sqrt{245}$, $|BC| = \sqrt{325}$, $|AC| = \sqrt{80} \Rightarrow |AB|^2 + |AC|^2 = |BC|^2$
45. $15a + 8b - 102 = 0$; $15a + 8b + 34 = 0$ **47.** $(x - 1)^2 + (y + 2)^2 = 16$ **49.** $x - y + 5 = 0$
51. $3x - 2y - 4 = 0$ **53.** $x + 3y - 9 = 0$ **55.** \$50,000 at 14%; \$50,000 at 10%
57. $°C = \frac{5}{9}(°F - 32)$; $°C = 21.11$

Exercise 1.4

1. (a) 7 (b) -8 (c) -2 (d) $3x + 4$ (e) $3x + 3\Delta x - 2$ (f) $(3/x) - 2$
3. (a) -15 (b) 5.7 (c) 4.41 (d) 25 (e) $4 + 4\Delta x + (\Delta x)^2$ **5.** $f(0) = 3$; $f(2) = 4$ **7.** Positive **9.** 3
11. $-3 < x < 6$ **13.** $-3 \le x \le 10$ **15.** Yes **17.** Yes **19.** No **21.** Yes **23.** No **25.** Yes
27. Yes **29.** $\mathbb{R}$ **31.** $x \ge 1$ **33.** $\mathbb{R}$ **35.** $x \ne 2$ **37.** $x > 0$
39. $-\infty < x < 5$

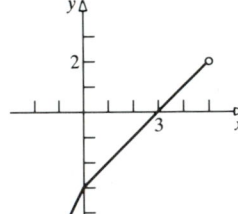

41. $-2 \le x < +\infty$

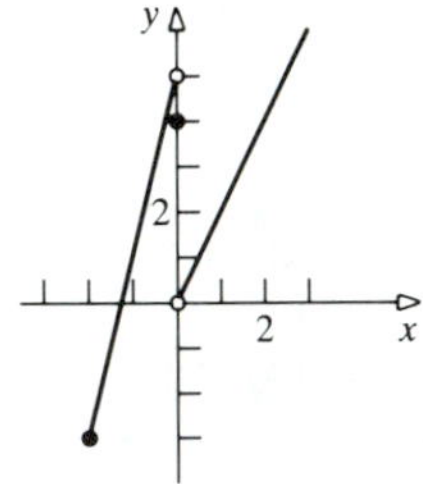

43. $\mathbb{R}$

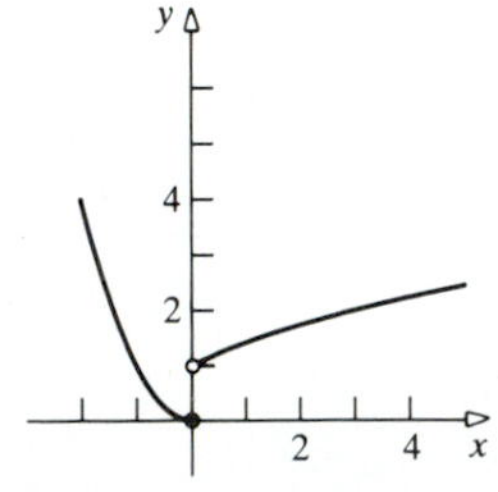

45. $\mathbb{R}$

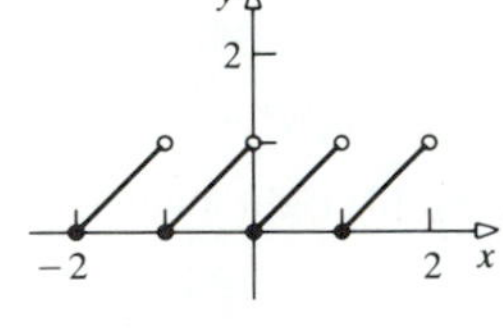

47. $\mathbb{R}$

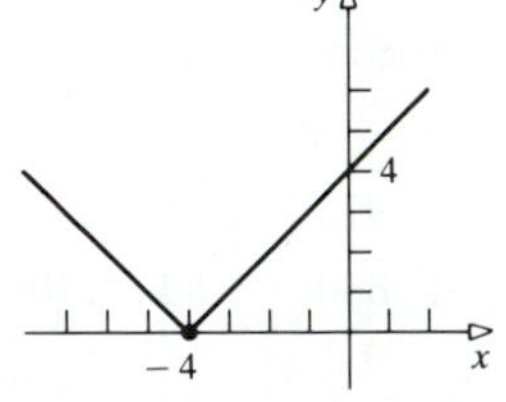

49.

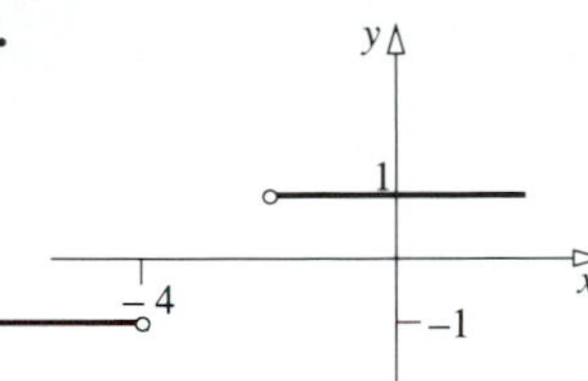

51.

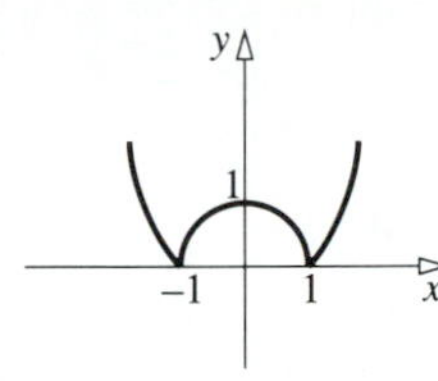

53. (a) $2x + 2\Delta x + 5$ (b) $2\Delta x$ (c) 2

55. (a) $x^2 + 2x\Delta x + (\Delta x)^2 + 3x + 3\Delta x + 4$ (b) $2x\Delta x + (\Delta x)^2 + 3\Delta x$ (c) $2x + (\Delta x) + 3$

57. $\dfrac{\sqrt{x+\Delta x} - \sqrt{x}}{\Delta x} = \dfrac{\sqrt{x+\Delta x} - \sqrt{x}}{\Delta x} \dfrac{\sqrt{x+\Delta x} + \sqrt{x}}{\sqrt{x+\Delta x} + \sqrt{x}} = \dfrac{1}{\sqrt{x+\Delta x} + \sqrt{x}}$ **59.** -1

61. (a) 15.1 m; 14.07 m; 12.94 m; 11.72 m (b) 2.02 sec (c) The rock falls faster on the earth.

(d) $H(x) = \begin{cases} 20 - 4.9x^2 & \text{if } 0 \le x \le 10\sqrt{2}/7 \approx 2.02 \\ 0 & \text{if } x > 10\sqrt{2}/7 \approx 2.02 \end{cases}$

63. $A(x) = \frac{1}{2}x(3000 - 2x); 0 < x < 1500$ **65.** $A = (7 - 2x)(11 - 2x); 0 < x < \frac{7}{2}; 0 < A < 77$

67. $V = 4x(100 - x)(8 - x)$ **69.** $A = [(100 - 2x)/(\pi + 2)^2](4x + \pi x + 50\pi)$

71. $T - 70 = [-415/2(10^6)][e - 29(10^6)]$ where T is temperature and e is elasticity

73. (a) 1 (b) -1 (c) 1 (d) -1 (e) 1 (f) -1 (g) 1 (h) -1 **75.** Yes **77.** Yes **79.** No **81.** Yes

Exercise 1.5

1. (a) $2x^2 + x - 1$ (b) $-2x^2 + x - 1$ (c) $2x^3 - 2x^2$ (d) $(x - 1)/2x^2$; $\mathbb{R}, \mathbb{R}, \mathbb{R}, \mathbb{R}, x \neq 0$

3. (a) $\sqrt{x+1} + x + 1$ (b) $\sqrt{x+1} - x - 1$ (c) $(x+1)^{3/2}$ (d) $1/\sqrt{x+1}$; $x \ge -1, \mathbb{R}, x \ge -1, x \ge -1, x > -1$

5. (a) $(2/x) + 1$ (b) -1 (c) $(1/x)[(1/x) + 1]$ (d) $1/(1 + x)$; $x \neq 0, x \neq 0, x \neq 0, x \neq 0, x \neq 0, -1$ **7.** Yes

9. No

11. No

13. $x \neq -2$

15. $x \neq -2, 2$

17.

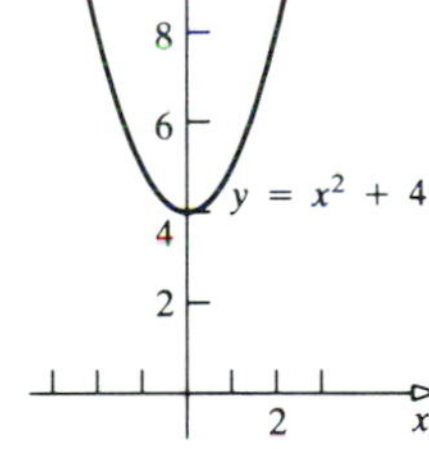

19.

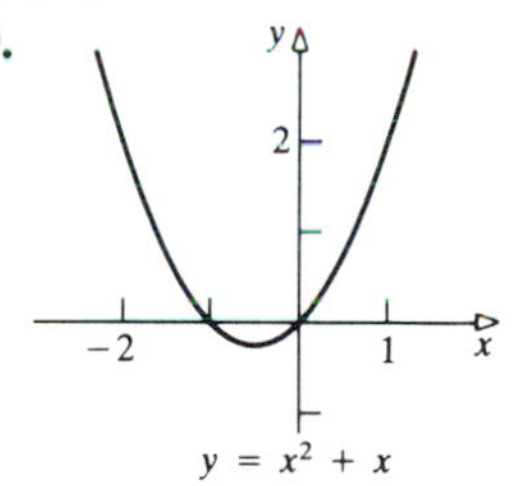

21.

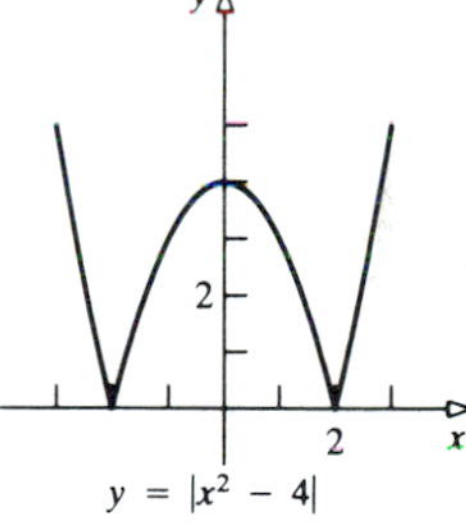

23. Odd **25.** Even

27. Neither **29.** Use x and x^3.

31. Use x^2 and x^4. **33.** Use x and x^3.

35. $a = 0, c = 0, b$ any number

37. $g(x) = 5 - \frac{7}{2}x$

39.

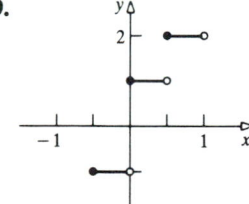

41. (a)

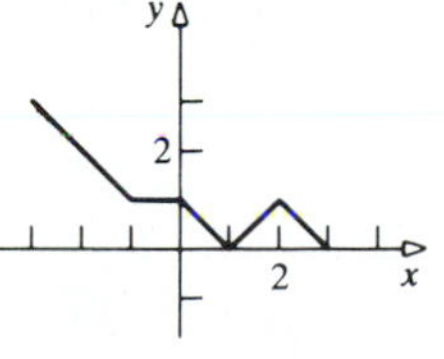

(b)

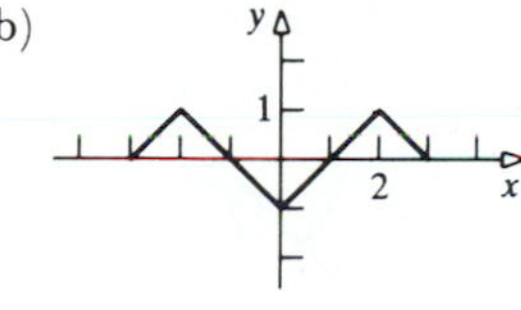

(c)

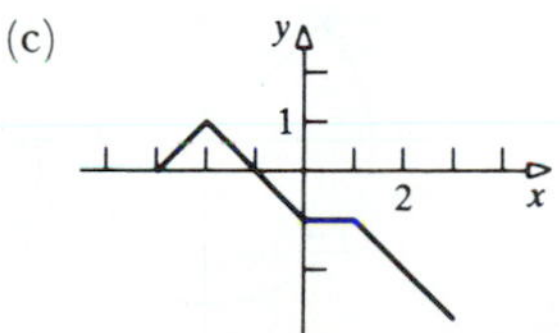

(d)

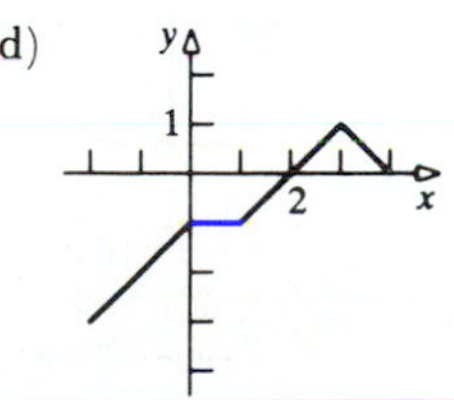

Exercise 1.6

1. (a) $3x^2 + 1$ (b) $(3x + 1)^2$ (c) x^4 (d) $9x + 4$

3. (a) $\sqrt{x^2 - 1}$ (b) $x - 1$ (c) $(x^2 - 1)^2 - 1 = x^4 - 2x^2$ (d) $\sqrt[4]{x}$

5. (a) $(1 - x)/(1 + x)$ (b) $(x + 1)/(x - 1)$ (c) x (d) $-1/x$

7. (a) $(8/x^2)(6 - x)$ (b) $2/\sqrt{3x^4 - 2x^2}$ (c) $\sqrt{2}\sqrt[4]{x}$ (d) $3(3x^4 - 2x^2)^4 - 2(3x^4 - 2x^2)^2$

9. $f(x) = \sqrt{x}$; $g(x) = x^2 + x - 1$ **11.** $f(x) = x^7$; $g(x) = x^2 - 1$ **13.** $f(x) = 1/x^2$; $g(x) = 3x - 5$ **15.** $g(x) = \sqrt[3]{x}$

17. (a) $4x - 3$ (b) $9 - 12x + 4x^2$ **19.** 2; 41 **21.** $p = -1$

23. (a) Neither (b) Odd (c) Even (d) Odd (e) Even (f) Even

Exercise 1.7

1. One-to-one **3.** Not one-to-one **5.** One-to-one **7.** Not one-to-one

9.

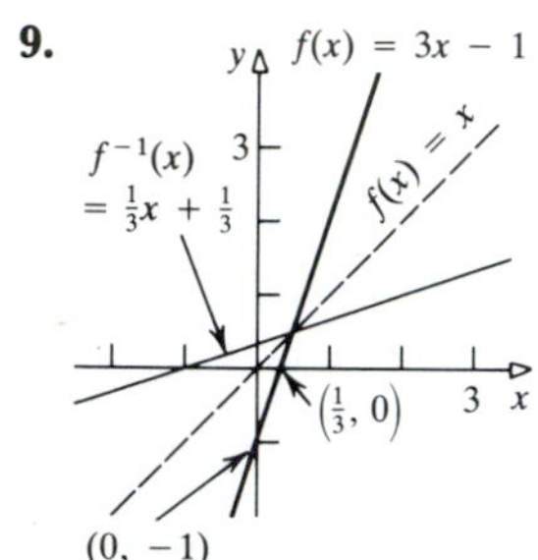

11.

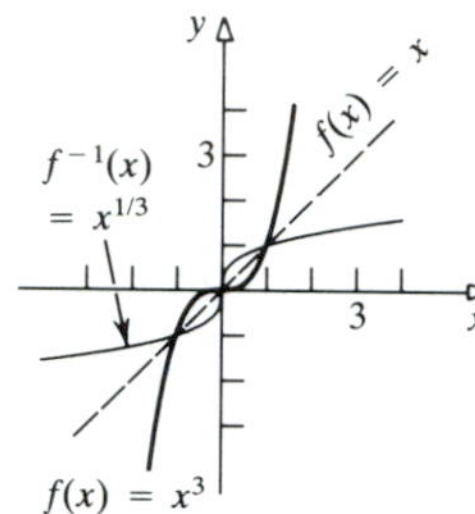

13.

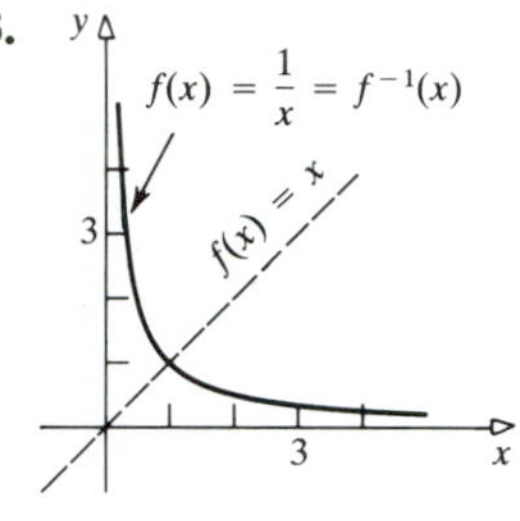

15.

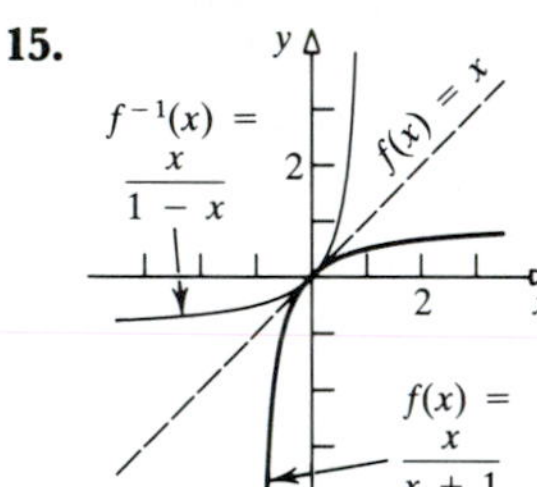

17. $3x_1 + 5 = 3x_2 + 5 \Rightarrow x_1 = x_2$; $(x - 5)/3$; all real numbers

19. $\dfrac{2x_1 + 3}{5x_1 - 6} = \dfrac{2x_2 + 3}{5x_2 - 6} \Rightarrow x_1 = x_2$; $(3 + 6x)/(5x - 2)$; $x \neq \frac{2}{5}$ **21.** $\dfrac{1}{x_1} = \dfrac{1}{x_2} \Rightarrow x_1 = x_2$; $1/x$; $x \neq 0$ **23.** First quadrant

25. $(f \circ g)(x) = f(g(x)) = \frac{9}{5}[\frac{5}{9}(x - 32)] + 32 = x$ **27.** $f^{-1}(x) = \sqrt[5]{x + 1}$ **29.** n is odd.

31. $(f \circ g) \circ (f \circ g)^{-1}(x) = x$ and $f(g(g^{-1}(f^{-1}(x)))) = (f \circ g) \circ (g^{-1} \circ f^{-1})(x) = x$; hence $(f \circ g)^{-1} = g^{-1} \circ f^{-1}$

Exercise 1.8

1. (a) $\pi/4 \approx 0.785$ (b) $-\pi/2 \approx -1.57$ (c) $3\pi/4 \approx 2.355$ (d) $\pi \approx 3.14$ (e) $-7\pi/6 \approx -3.66$ (f) $5\pi/3 \approx 5.23$ (g) $-4\pi \approx -12.56$ (h) $5\pi/2 \approx 7.85$ (i) $\pi/18 \approx 0.17$ (j) $-4\pi/9 \approx -1.40$ (k) $\pi/90 \approx 0.03$ (l) $2\pi/3 \approx 2.09$

3. (a)

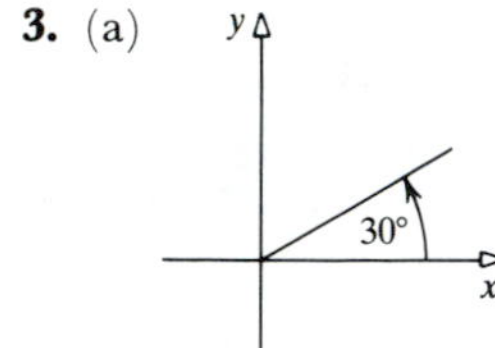

(b)

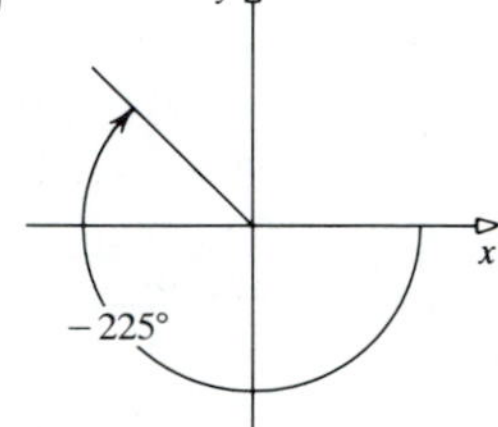

(c)

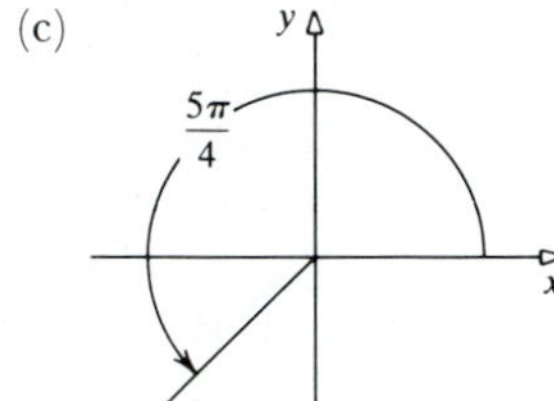

(d)

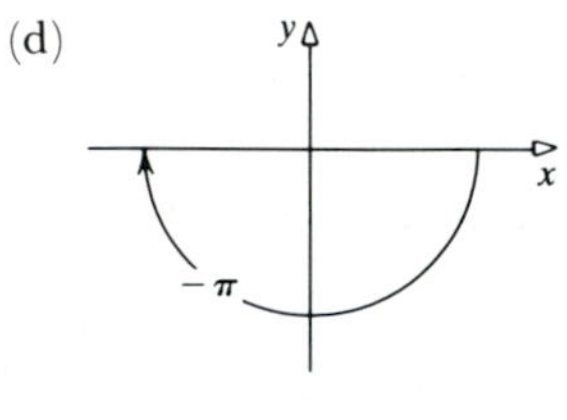

(e) 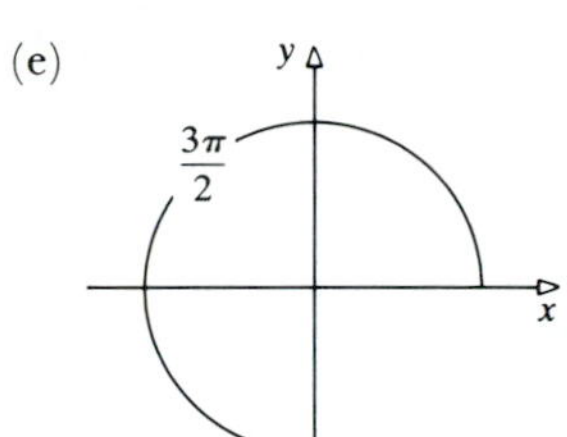

(f) 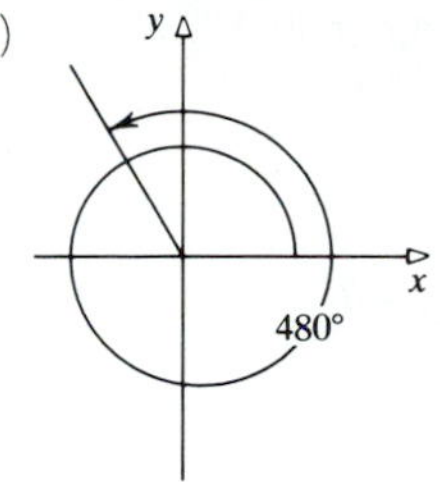

(g) 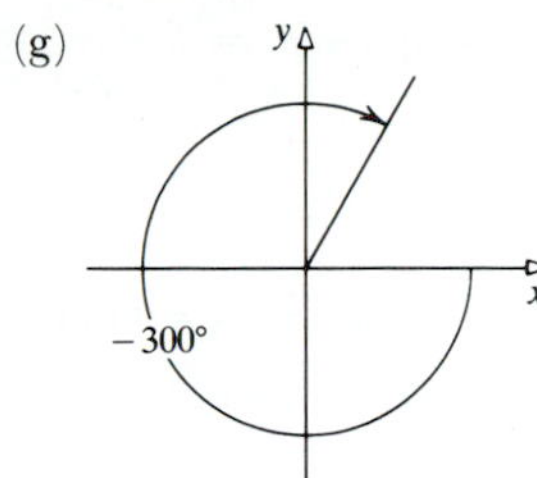

(h) 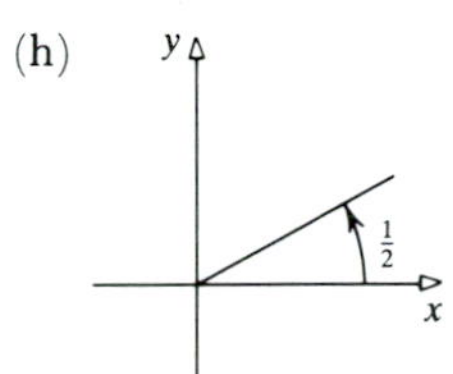

(i) 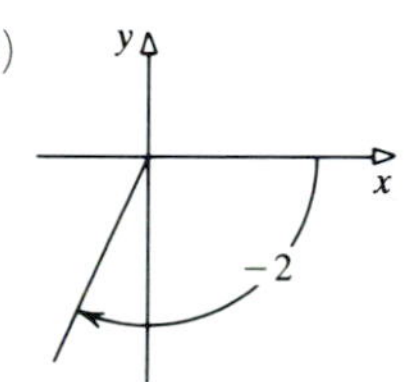

5.

Angle	sin x	cos x	tan x	cot x	sec x	csc x
(a) $x = 0$	0	1	0	Undefined	1	Undefined
(b) $x = \pi/2$	1	0	Undefined	0	Undefined	1
(c) $x = \pi$	0	−1	0	Undefined	−1	Undefined
(d) $x = 3\pi/2$	−1	0	Undefined	0	Undefined	−1

7. (a) $\pi/6, 5\pi/6$ (b) $\pi/4, 7\pi/4$ (c) $0, \pi$ (d) $4\pi/3, 5\pi/3$ (e) $2\pi/3, 4\pi/3$ (f) $3\pi/4, 7\pi/4$ (g) $5\pi/6, 11\pi/6$ (h) $\pi/6, 11\pi/6$ (i) $\pi/4, 3\pi/4$ (j) $\pi/4, 5\pi/4$ (k) 0 (l) $3\pi/2$

9. (a) $\cos\theta = -\frac{4}{5}$, $\tan\theta = -\frac{3}{4}$, $\cot\theta = -\frac{4}{3}$, $\sec\theta = -\frac{5}{4}$, $\csc\theta = \frac{5}{3}$
(b) $\sin\theta = -\frac{12}{13}$, $\tan\theta = \frac{12}{5}$, $\cot\theta = \frac{5}{12}$, $\sec\theta = -\frac{13}{5}$, $\csc\theta = -\frac{13}{12}$
(c) $\sin\theta = \frac{12}{13}$, $\cos\theta = \frac{5}{13}$, $\cot\theta = \frac{5}{12}$, $\sec\theta = \frac{13}{5}$, $\csc\theta = \frac{13}{12}$
(d) $\cos\theta = 2\sqrt{2}/3$, $\tan\theta = -1/2\sqrt{2}$, $\cot\theta = -2\sqrt{2}$, $\sec\theta = 3/2\sqrt{2}$, $\csc\theta = -3$
(e) $\sin\theta = -\sqrt{21}/5$, $\tan\theta = -\sqrt{21}/2$, $\cot\theta = -2/\sqrt{21}$, $\sec\theta = \frac{5}{2}$, $\csc\theta = -5/\sqrt{21}$
(f) $\sin\theta = 1/\sqrt{101}$, $\cos\theta = -10/\sqrt{101}$, $\cot\theta = -10$, $\sec\theta = -\sqrt{101}/10$, $\csc\theta = \sqrt{101}$

11. (a) $2\pi/3$ (b) $4\pi/3$ (c) $2\pi/3$ (d) $4\pi/3$ **13.** $-\sin\theta$ **15.** $-\sin\theta$ **17.** $-\cos\theta$ **19.** $-\cos\theta$

21. $\sec\alpha - \cos\alpha = \dfrac{1}{\cos\alpha} - \cos\alpha = \dfrac{1-\cos^2\alpha}{\cos\alpha} = \dfrac{\sin^2\alpha}{\cos\alpha} = \dfrac{\sin\alpha}{\cos\alpha}\sin\alpha = \tan\alpha\sin\alpha$

23. $\cos^4(2x) - \sin^4(2x) = [\cos^2(2x) - \sin^2(2x)][\cos^2(2x) + \sin^2(2x)] = \cos^2(2x) - \sin^2(2x) = \cos(4x)$

25. $\tan x + \tan y = \dfrac{\sin x}{\cos x} + \dfrac{\sin y}{\cos y} = \dfrac{\sin x\cos y + \cos x\sin y}{\cos x\cos y} = \dfrac{\sin(x+y)}{\cos x\cos y}$

27. $\cot\theta - \tan\theta = \dfrac{\cos\theta}{\sin\theta} - \dfrac{\sin\theta}{\cos\theta} = \dfrac{\cos^2\theta - \sin^2\theta}{\sin\theta\cos\theta} = \dfrac{\cos(2\theta)}{\frac{1}{2}\sin(2\theta)} = 2\cot(2\theta)$

29. $\cos(7\alpha) + \cos(5\alpha) = 2\cos[\frac{1}{2}(7\alpha+5\alpha)]\cos[\frac{1}{2}(7\alpha-5\alpha)] = 2\cos(6\alpha)\cos\alpha$

31. $\sin^2 x - \sin^2 y = (\sin x + \sin y)(\sin x - \sin y) = [2\sin\frac{1}{2}(x+y)\cos\frac{1}{2}(x-y)][2\cos\frac{1}{2}(x+y)\sin\frac{1}{2}(x-y)]$
$= [2\sin\frac{1}{2}(x+y)\cos\frac{1}{2}(x+y)][2\sin\frac{1}{2}(x-y)\cos\frac{1}{2}(x-y)] = \sin(x+y)\sin(x-y)$

33. Period $= 2\pi$ **35.** Period $= \pi$ **37.** Period $= 4\pi$ **39.** Period $= \pi/2$ **41.** Period $= 4\pi$ **43.** Period $= 2\pi$

45. Period $= 8\pi$ **47.** Period $= 2$ **49.** $\dfrac{\pi}{4} + n\pi$ **51.** $\pm\dfrac{\pi}{18} + \dfrac{2n\pi}{3}$ **53.** $\pm\dfrac{\pi}{6} + 2n\pi$; $\pm\dfrac{5\pi}{6} + 2n\pi$

55. $2n\pi$; $\pm\dfrac{\pi}{3} + 2n\pi$ **57.** $\dfrac{(3n-1)\pi}{6}$

59. (a) $(f+g)(x+p) = f(x+p) + g(x+p) = f(x) + g(x) = (f+g)(x)$
(b) $(f \cdot g)(x+p) = f(x+p) \cdot g(x+p) = f(x) \cdot g(x) = (f \cdot g)(x)$
(c) $\dfrac{f}{g}(x+p) = \dfrac{f(x+p)}{g(x+p)} = \dfrac{f(x)}{g(x)} = (f/g)(x)$

61. $\tan[\theta_2 - (\pi/2)] = -1/\tan\theta_2 = -1/m_2 = m_1 = \tan\theta_1 \Rightarrow \theta_2 - \theta_1 = \pi/2$ **63.** $\frac{1}{3}$ **65.** $\frac{7}{4}$ **67.** 3 **69.** $\frac{11}{7}$

71. 3 **73.** 5 **75.** 18°, 72° **77.** 45°, 45°, 90°

79. (a) $\frac{1}{2}, -2, \frac{1}{2}, -2$
(b) AB and CD, BC and AD are parallel; AB is perpendicular to BC and AD; BC is perpendicular to CD and AB
(c) Undefined (d) (2, 4) (e) $2\sqrt{5}$ (f) (3, 2)

81. If θ is the included angle and a is the side of the triangle, then the area $= a^2 \sin\dfrac{\theta}{2}\cos\dfrac{\theta}{2} = \dfrac{1}{2}a^2 \sin\theta$.

83. $s = L\sin\theta$; $\cos\theta = \dfrac{L-h}{L} = 1 - \dfrac{h}{L}$; $\dfrac{h}{L} = 1 - \cos\theta$

$$h = L(1-\cos\theta) = s\,\frac{1-\cos\theta}{\sin\theta} = s\tan\frac{\theta}{2}$$

$$h = s\,\frac{1-\sin\alpha}{\cos\alpha} = s\cot\frac{\alpha}{2}$$

85. $L^2 = R^2 + x^2 - 2Rx\cos\theta$; $x^2 - 2Rx\cos\theta + R^2 - L^2 = 0$

$$x = \frac{2R\cos\theta + \sqrt{4R^2\cos^2\theta - 4R^2 + 4L^2}}{2} = R\cos\theta + \sqrt{R^2\cos^2\theta - R^2 + L^2}$$

Review Exercises

1. $-\frac{14}{3} < x < -\frac{10}{3}$, $x \neq -4$ **3.** $-2 < x < 2$ **5.** (2, 4)

7. (a) $\sqrt{5}$ (b) $(-\frac{1}{2}, -1)$ (c) $y = 2x$ (d) $y = -\frac{1}{2}x - \frac{5}{4}$ **9.** (a) 5 (b) $(\frac{9}{2}, 3)$ (c) $y = 3$ (d) $x = \frac{9}{2}$

11. (a) Center: $(-\frac{1}{2}, -1)$; radius: $\sqrt{5}/2$; $(x+\frac{1}{2})^2 + (y+1)^2 = \frac{5}{4}$ (b) $(x+\frac{3}{2})^2 + (y+\frac{1}{2})^2 = \frac{5}{4}$

13. (a) Center: $(\frac{9}{2}, 3)$; radius: $\frac{5}{2}$; $(x-\frac{9}{2})^2 + (y-3)^2 = \frac{25}{4}$ (b) $(x-7)^2 + (y-3)^2 = \frac{25}{4}$

15. $2 \pm 4\sqrt{3}$ **17.** $\dfrac{\sqrt{2-\sqrt{2}}}{2}, \dfrac{\sqrt{2+\sqrt{2}}}{2}, \dfrac{\sqrt{2+\sqrt{2+\sqrt{2}}}}{2}$ **19.** $2x - 3y + 5 = 0$ **21.** (1, 4)

23. $\dfrac{-1}{\sqrt{4-x-\Delta x} + \sqrt{4-x}}$

25. (a) f is one-to-one by the horizontal line test and hence has an inverse.
(b) Domain of $f = \mathbb{R}$
Domain of f^{-1} = Range of $f = -2, -1 < x < 0, 0 < x$
(c)

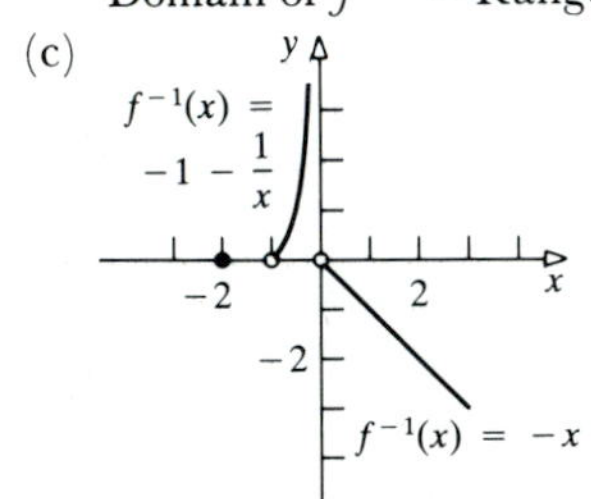

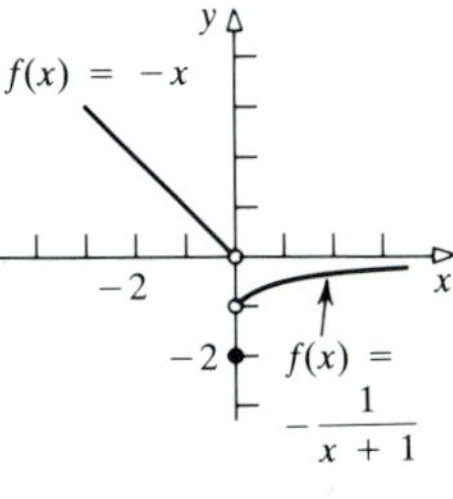

27. (a) $(-\infty, -1) \cup (-1, 1) \cup (1, +\infty)$ (b) $(-\infty, -1) \cup (-1, 1]$

29. There are infinitely many such functions. For example, let x_0 be rational, where $-1 \le x_0 \le 1$. Then one infinite family is defined by

$$f(x) = \begin{cases} \sqrt{1-x^2} & \text{if } x \ne x_0 \\ -\sqrt{1-x_0^2} & \text{if } x = x_0 \end{cases}$$

31. (a) $(f \circ f)(x_0) = f(f(x_0)) = f(x_0) = x_0$ (b) $-3, 1$
(c) Two fixed points if $(d-a)^2 + 4bc > 0$ and $c \ne 0$; one if $(d-a)^2 + 4bc = 0$; none if $(d-a)^2 + 4bc < 0$.
If $c = 0$, $x = \dfrac{b}{d-a} \Rightarrow$ fixed point.

33. $y = -\dfrac{h}{b}x + h$

35. Assume that $\sqrt{2} = m/n$, where m and n are integers, not both even. Then $m^2 = 2n^2 \Rightarrow m^2$ even $\Rightarrow m$ even; hence n must be odd. Let $m = 2a$, $n = 2b+1$. We get $4a^2 = 2(4b^2 + 4b + 1)$, or $2a^2 = 4b^2 + 4b + 1$. Since the left-hand side of this equation must be an even number whereas the right-hand side is an odd number, we have a contradiction. Therefore, $\sqrt{2}$ is irrational.

37. (a) $|x-a| = \begin{cases} x-a & \text{if } x-a \ge 0 \\ a-x & \text{if } x-a < 0 \end{cases}$

$|a-x| = \begin{cases} a-x & \text{if } a-x > 0 \\ x-a & \text{if } a-x \le 0 \end{cases}$

But $x - a \ge 0$ if and only if $x \ge a$ or, equivalently, if and only if $a - x \le 0$; $x - a < 0$ if and only if $x < a$ or, equivalently, if and only if $a - x > 0$.

(b) $|ax| = \begin{cases} ax & \text{if } ax \ge 0 \\ -ax & \text{if } ax < 0 \end{cases}$

$$ax > 0 \Rightarrow \begin{cases} a > 0 \text{ and } x > 0 \\ \text{or} \\ a < 0 \text{ and } x < 0 \end{cases} \Rightarrow \begin{cases} a = |a| \text{ and } x = |x| \\ \text{or} \\ |a| = -a \text{ and } |x| = -x \end{cases}$$

In both cases, $|a| \cdot |x| = ax$ and so $|ax| = |a| \cdot |x|$.

$$ax < 0 \Rightarrow \begin{cases} a > 0 \text{ and } x < 0 \\ \text{or} \\ a < 0 \text{ and } x > 0 \end{cases} \Rightarrow \begin{cases} |a| = a \text{ and } |x| = -x \\ \text{or} \\ |a| = -a \text{ and } |x| = x \end{cases}$$

In both cases, $|a| \cdot |x| = -ax = |ax|$.

(c) $|x| = \begin{cases} x & \text{if } x \ge 0 \\ -x & \text{if } x < 0 \end{cases}$

If $x \ge 0$, then $|x| = x \ge -|x|$. If $x < 0$, then $-|x| = x \le |x|$. Thus, we always have $-|x| \le x \le |x|$.

(d) $\left|\dfrac{x}{a}\right| = \begin{cases} \dfrac{x}{a} & \text{if } \dfrac{x}{a} \ge 0 \\ -\dfrac{x}{a} & \text{if } \dfrac{x}{a} < 0 \end{cases}$

$$\text{If } \frac{x}{a} \ge 0, \text{ then } \begin{cases} x \ge 0 \text{ and } a \ge 0 \\ \text{or} \\ x < 0 \text{ and } a < 0 \end{cases} \Rightarrow \begin{cases} x = |x| \text{ and } a = |a| \\ \text{or} \\ x = -|x| \text{ and } a = -|a|. \end{cases}$$

In both cases, $\dfrac{x}{a} = \dfrac{|x|}{|a|}$ and so $\left|\dfrac{x}{a}\right| = \dfrac{|x|}{|a|}$.

$$\text{If } \frac{x}{a} < 0, \text{ then } \begin{cases} x < 0 \text{ and } a > 0 \\ \text{or} \\ x > 0 \text{ and } a < 0 \end{cases}$$

$$\Rightarrow \begin{cases} x = -|x| \text{ and } a = |a| \\ \text{or} \\ x = |x| \text{ and } a = -|a| \end{cases}$$

$$\Rightarrow -\frac{x}{a} = -\frac{(-|x|)}{|a|} \text{ or } -\frac{x}{a} = -\frac{|x|}{(-|a|)}$$

$$\Rightarrow \left|\frac{x}{a}\right| = -\frac{x}{a} = \frac{|x|}{|a|}$$

39. (a) $a - b < c < a + b \Leftrightarrow -b < c - a < b \Leftrightarrow |c - a| < b$
(b) To disallow $a = c$, we can say $c - a \neq 0$ or, equivalently, $|c - a| > 0$. Thus, $a - b < c < a + b$ and $a \neq c \Leftrightarrow 0 < |c - a| < b$.

Challenge Exercises

1. In order to find the minimum distance between points P and Q on L, it will suffice to find the minimum of the square distance. We have for $Q = (x, y)$ and assuming $B \neq 0$:

$$|PQ|^2 = (x - x_1)^2 + (y - y_1)^2 = (x - x_1)^2 + \left(\frac{C}{B} - \frac{A}{B}x - y_1\right)^2$$

$$= \left(1 + \frac{A^2}{B^2}\right)\left(x^2 + 2\frac{ABy_1 - B^2x_1 - AC}{A^2 + B^2}x + \frac{B^2x_1^2 + C^2 + B^2y_1^2 - 2BCy_1}{A^2 + B^2}\right)$$

$$= \left(1 + \frac{A^2}{B^2}\right)\left(x + \frac{ABy_1 - B^2x_1 - AC}{A^2 + B^2}\right)^2 + \frac{(Ax_1 + By_1 - C)^2}{A^2 + B^2}$$

$$\therefore \quad |PQ|^2 \geq \frac{(Ax_1 + By_1 - C)^2}{A^2 + B^2} \Rightarrow |PQ| \geq \frac{|Ax_1 + By_1 - C|}{\sqrt{A^2 + B^2}}$$

If $B = 0$, then the line has equation $x = \dfrac{C}{A}$ and the distance from (x_1, y_1) to L is then

$$\left|x_1 - \frac{C}{A}\right| = \frac{|Ax_1 + By_1 - C|}{\sqrt{A^2 + B^2}}$$

3. The closest point on T to P is Q, since all other points on T lie more than $|PQ|$ units away from P, being outside of C. By 2, L is $\perp$ to T.

5. Let $m \neq 0$. Since $y = mx$ and $y = \dfrac{x}{m}$ are inverse functions, the lines $y = mx$ and $y = \dfrac{x}{m}$ are symmetric about the lines $y = \pm x$ and so make the same angles with these lines. The lines $y = \pm x$ contain the centers of all circles to which the lines $y = mx$ and $y = \dfrac{x}{m}$ are both tangent. Thus all centers have coordinates $(h, \pm h)$ and have distances to the tangent lines given by

$$\frac{|mh \mp h|}{\sqrt{m^2 + 1}}$$

The circles have equations

$$(*) \qquad (x - h)^2 + (y \mp h)^2 = \frac{h^2(m \mp 1)^2}{m^2 + 1}$$

If $m = 0$, then we have the case where the lines are the x- and y-axes. In this case, the centers are again along the lines $y = \pm x$. The equations are $(x - h)^2 + (y \mp h)^2 = h^2$ so $(*)$ actually holds for $m = 0$, too.

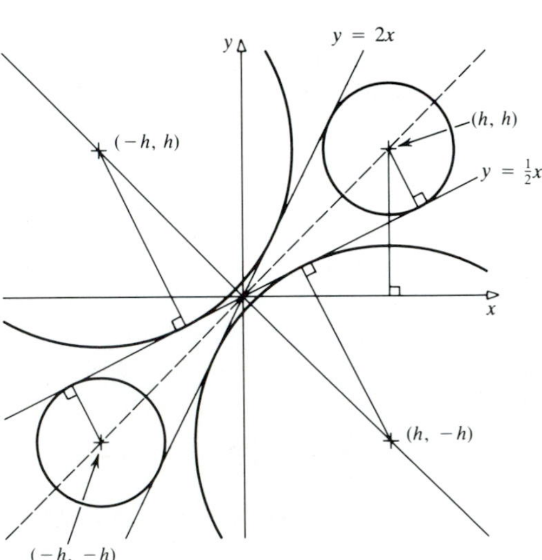

7. Let P and Q be as given. The line L determined by O and P has equation $y - \frac{y_1}{x_1}x = 0$. The length (see the diagram)

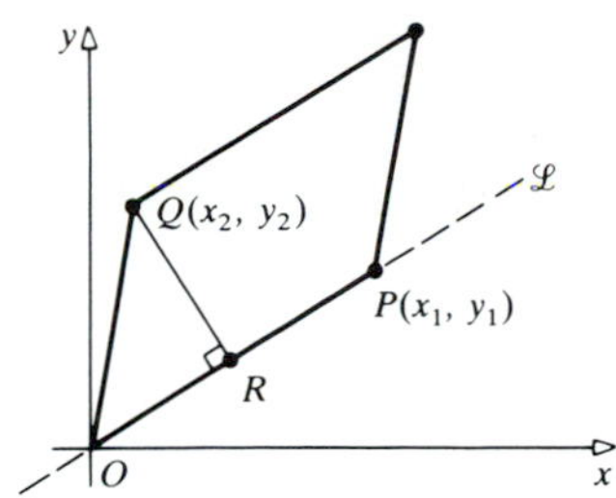

$$|QR| = \frac{\left|\frac{y_1}{x_1}x_2 - y_2\right|}{\sqrt{\left(\frac{y_1}{x_1}\right)^2 + (-1)^2}} = \frac{|y_1x_2 - y_2x_1|}{\sqrt{x_1^2 + y_1^2}}$$

is the altitude. We have

$$A = Bh = |OP||QR| = \sqrt{x_1^2 + y_1^2} \cdot \frac{|x_1y_2 - x_2y_1|}{\sqrt{x_1^2 + y_1^2}} = |x_1y_2 - x_2y_1|$$

9. By challenge exercise 8, $x_1y_2 - x_2y_1 = 0$ means that the parallelogram determined by O, P, and Q has area O and this forces O, P, and Q to be collinear.

Conversely if O, P, and Q are collinear, then we consider two cases. In the first case, the line L containing O, P, and Q is vertical. This forces $x_1 = x_2 = 0$, hence $x_1y_2 - x_2y_1 = 0$. In the second case the line L is not vertical and so $x_1 \neq 0 \neq x_2$. The slope m of L is given by

$$m = \frac{y_1}{x_1} = \frac{y_2}{x_2}$$

Thus cross-multiplying gives $x_1y_2 = x_2y_1$ hence $x_1y_2 - x_2y_1 = 0$.

11. We require $1 + x \neq 0 \Rightarrow x \neq -1$; $1 + \frac{1}{1+x} \neq 0 \Rightarrow 1 + x + 1 \neq 0 \Rightarrow x \neq -2$;

$1 + \cfrac{1}{1 + \cfrac{1}{1+x}} \neq 0 \Rightarrow 1 + \frac{1}{1+x} \neq -1 \Rightarrow x \neq -\frac{3}{2}$. Domain: $(-\infty, -2) \cup (-2, -\frac{3}{2}) \cup (-\frac{3}{2}, -1) \cup (-1, +\infty)$.

13. We have for real x in the domain of $f(x)$: $f(x) = \frac{f(x) + f(-x)}{2} + \frac{f(x) - f(-x)}{2}$

We claim $g_1(x) = \frac{f(x) + f(-x)}{2}$ is even: $g_1(-x) = \frac{f(-x) + f(-(-x))}{2} = \frac{f(x) + f(-x)}{2} = g_1(x)$

We claim $g_2(x) = \frac{f(x) - f(-x)}{2}$ is odd: $g_2(-x) = \frac{f(-x) - f(-(-x))}{2} = \frac{f(-x) - f(x)}{2} = -g_2(x)$

Thus $f(x) = g_1(x) + g_2(x)$, $g_1(x)$ even and $g_2(x)$ odd. Suppose $f(x) = h_1(x) + h_2(x)$ for $h_1(x)$ even and $h_2(x)$ odd. Then $g_1(x) + g_2(x) = h_1(x) + h_2(x) \Rightarrow g_1(x) - h_1(x) = h_2(x) - g_2(x)$. Now $g_1 - h_1$ is even since g_1 and h_1 are: $g_1(-x) - h_1(-x) = g_1(x) - h_1(x)$ and similarly $h_2 - g_2$ is odd since g_2 and h_2 are. Since $g_1 - h_1 = h_2 - g_2$, each of these functions is actually both even and odd. For $F(x)$ to be both we have $F(x) = F(-x) = -F(x) \Rightarrow F(x) = 0$. Thus $g_1 - h_1 = 0 = h_2 - g_2$ and so $g_1 = h_1$ and $h_2 = g_2$.

15. In the figure, $2y + x = 200$, so $y = 100 - \frac{x}{2}$. The area is

$$A = 100x - \frac{x^2}{2} = -\tfrac{1}{2}(x^2 - 200x + 10{,}000) + 5{,}000 = -\tfrac{1}{2}(x - 100)^2 + 5{,}000$$

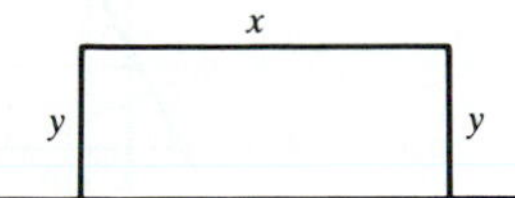

A is maximum when $x = 100$ and $y = 50$. A has a maximum of 5,000 m^2.

CHAPTER 2

Exercise 2.1

1.

x	0.9	0.99	0.999	1.001	1.01	1.1
$f(x)$	1.8	1.98	1.998	2.002	2.02	2.2

$\lim_{x\to 1} 2x = 2$

3.

x	0.1	0.01	0.001	−0.001	−0.01	−0.1
$f(x)$	2.01	2.0001	2.000001	2.000001	2.0001	2.01

$\lim_{x\to 0}(x^2 + 2) = 2$

5.

x	−2.5	−2.9	−2.99	−3.01	−3.1	−3.5
$f(x)$	−5.5	−5.9	−5.99	−6.01	−6.1	−6.5

$\lim_{x\to -3} \dfrac{x^2 - 9}{x + 3} = -6$

7.

x	−0.2	−0.1	−0.01	0.01	0.1	0.2
$f(x)$	1.01355	1.00335	1.00003	1.00003	1.00335	1.01355

$\lim_{x\to 0} \dfrac{\tan x}{x} = 1$

9.

x	−0.2	−0.1	−0.01	0.01	0.1	0.2
$f(x)$	−0.249	−0.111	−0.0101	0.0099	0.0907	0.165

$\lim_{x\to 0} \dfrac{\sin x}{1 + \tan x} = 0$

11. Exists

13. Exists

15. Does not exist

17. Does not exist

19. $\lim_{x\to 2} f(x)$ exists

$\lim_{x\to 2} f(x) = 9$

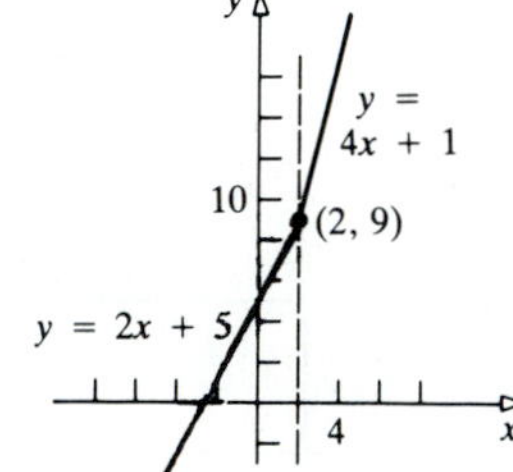

21. $\lim_{x\to 1} f(x)$ exists

$\lim_{x\to 1} f(x) = 2$

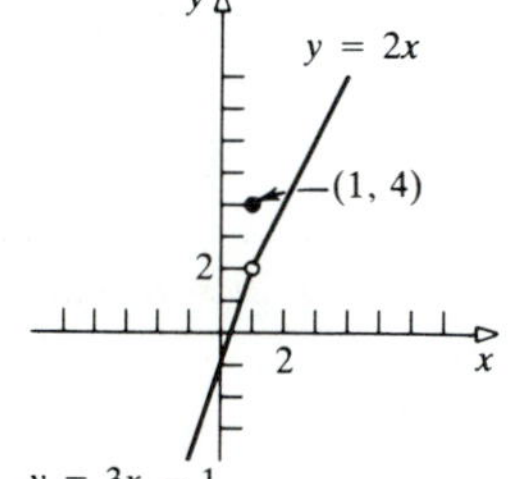

23. $\lim_{x\to1} f(x)$ exists

$\lim_{x\to1} f(x) = 2$

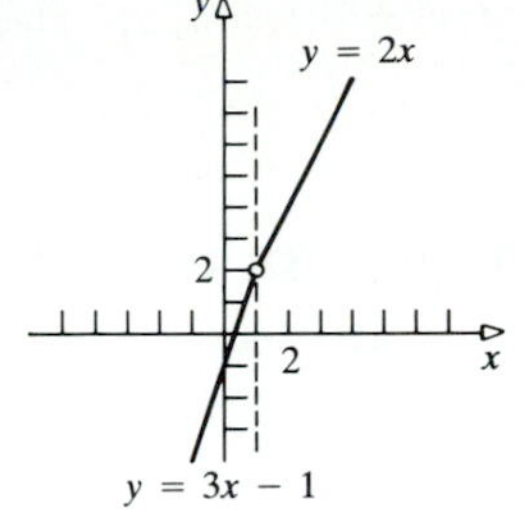

25. Since there is no single number that the values of f are close to when x is close to 0, $\lim_{x\to0} f(x)$ does not exist.

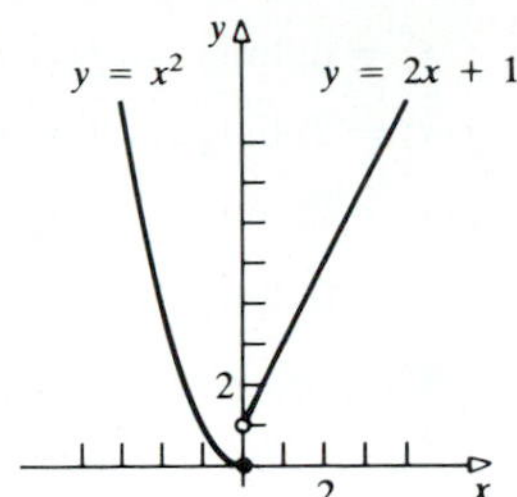

27. -3 **29.** -1 **31.** (a) 15 (b) $y = 12x - 12$ (c)

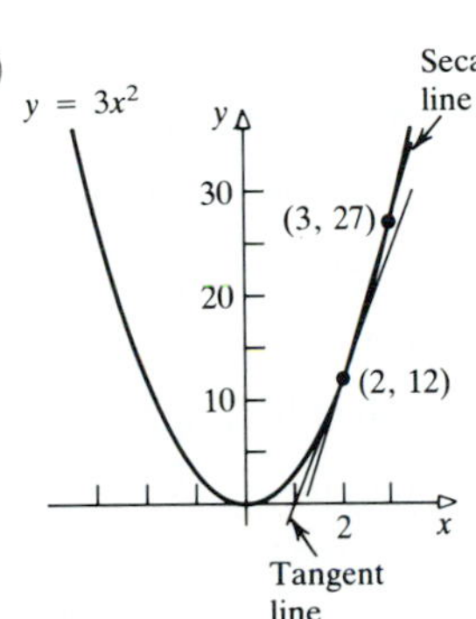

33. (a) $m_{\text{sec}} = 2 + (h/2)$ (b) 1.75; 2.25 (c)

h	-0.5	-0.1	-0.001	0.001	0.1	0.5
m_{sec}	1.75	1.95	1.9995	2.0005	2.05	2.25

(d) 2 (e) $y = 2x - 3$ (f)

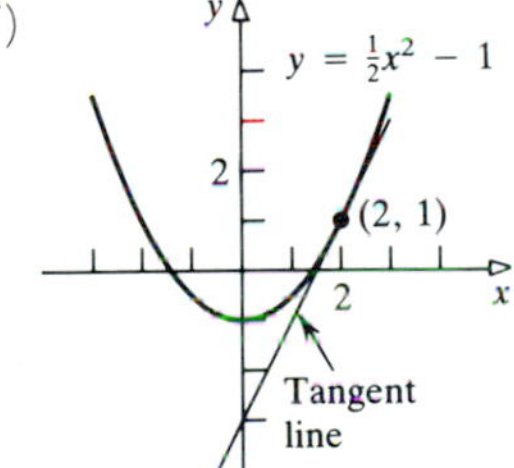

Exercise 2.2

1. 14 **3.** 5 **5.** 12 **7.** $-\frac{31}{8}$ **9.** 6 **11.** 4 **13.** 4 **15.** 0 **17.** 4 **19.** $\frac{13}{4}$ **21.** 4
23. $\frac{12}{5}$ **25.** -4 **27.** $-\frac{1}{7}$ **29.** $\frac{2}{3}$ **31.** 2 **33.** 12 **35.** $\sqrt{2}/4$ **37.** $-1/x^2$ **39.** $1/2\sqrt{x}$ **41.** $\frac{-1}{16}$
43. $\frac{1}{6}$ **45.** 6 **47.** 10 **49.** 8 **51.** 4 **53.** $6x + 4$ **55.** $-2/x^2$ **57.** -6 **59.** na^{n-1} **61.** m/n

63. $(a + b)/2$ **65.**

x	3.9	3.99	3.999	4.001	4.01	4.1
$f(x)$	0.2516	0.2502	0.2500	0.2500	0.2498	0.2485

$\lim_{x\to4} f(x) = 0.25$

67. 0

Exercise 2.3

1. 5 **3.** -19 **5.** $\frac{3}{2}$ **7.** 6 **9.** 3 **11.** 1 **13.** 0 **15.** 1 **17.** 0 **19.** 0
21. $\lim_{x\to0^-} f(x) = 0$; $\lim_{x\to0^+} f(x) = 0$; yes, $\lim_{x\to0} f(x) = 0$

23. $\lim_{x\to 3^-} f(x) = 6$; $\lim_{x\to 3^+} f(x) = 6$; yes, $\lim_{x\to 3} f(x) = 6$ **25.** $\lim_{x\to 1^-} f(x) = -1$; $\lim_{x\to 1^+} f(x) = 2$; no

27. $\lim_{x\to 1^-} f(x) = 2$; $\lim_{x\to 1^+} f(x) = 2$; yes, $\lim_{x\to 1} f(x) = 2$

29. $\lim_{x\to 1^-} f(x) = 2$; $\lim_{x\to 1^+} f(x) = 2$; yes, $\lim_{x\to 1} f(x) = 2$

31. $\lim_{x\to 3^-} f(x) = 0$; $\lim_{x\to 3^+} f(x) = 0$; yes, $\lim_{x\to 3} f(x) = 0$ **33.** $\sqrt{5}$ **35.** 0 **37.** $\sqrt{5}$ **39.** 3 **41.** Yes, 0

43. Yes, 0 **45.** $f(x) = x/(x-c)$, $g(x) = -c/(x-c)$, $\lim_{x\to c}[f(x) + g(x)] = 1$ **47.** $f(x) = |x|/x$; $c = 0$

Exercise 2.4

1. Continuous **3.** Continuous **5.** Continuous **7.** Continuous **9.** Discontinuous **11.** Discontinuous

13. Continuous **15.** Continuous **17.** Discontinuous **19.** $f(2) = 4$ **21.** $f(1) = 2$ **23.** All real numbers

25. All real numbers **27.** All real numbers, except $x = 2$ **29.** All real numbers, except $x = 2$ and $x = -2$

31. All real numbers **33.** All real numbers **35.** All real numbers, except $x \geq 2$ **37.** $(0, 9)$ and $(9, +\infty)$

39. $(-\infty, \frac{2}{3})$ and $(\frac{2}{3}, +\infty)$ **41.** Yes, since $\lim_{x\to 0} f(x) = f(0) = \sqrt{15}$

43. No, since $\lim_{x\to 2^-} f(x) = \sqrt{9} = 3$ and $f(2) = \sqrt{5}$ **45.** Continuous on its domain

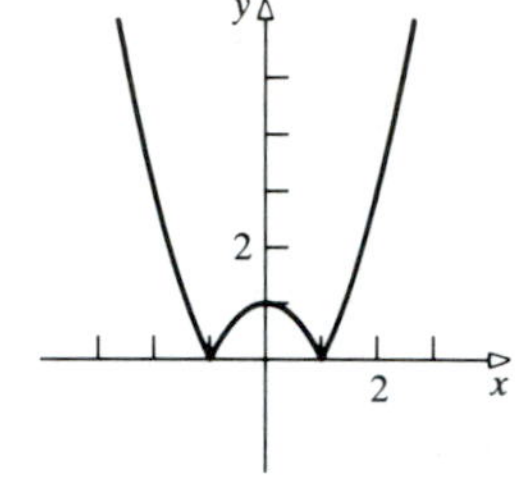

47. $A = 1$; $B = -1$ **49.** f has a zero. **51.** Theorem gives no information. **53.** f has a zero.

55. Let $\delta > 0$; $f(1+\delta) > 0$ and $f(2-\delta) < 0$. Then f is continuous on $(1, 2)$. Therefore there exists $c \in (1, 2)$ such that $f(c) = 0$.

57. An example is $f(x) = x^2 - \frac{1}{4}$. **59.** $k = \frac{1}{6}$ **61.** $\lim_{x\to 0^-} |x|/x = -1$, $\lim_{x\to 0^+} |x|/x = 1$

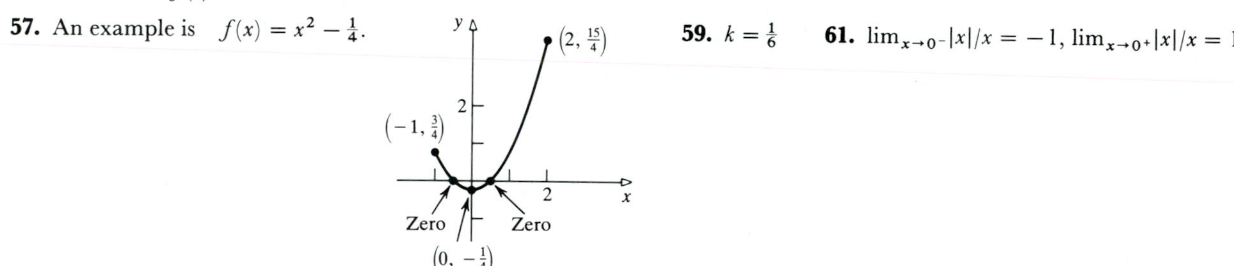

63. Take $f(x) = x$, $g(x) = x^2 - 4$, $c = 2$ or -2.

Exercise 2.5

1. 0 **3.** sin 1 **5.** $(1 + \sqrt{3})/2$ **7.** 6 **9.** 7 **11.** 1 **13.** 0 **15.** 1 **17.** $\frac{1}{2}$

19. $\lim\limits_{x\to 0} \dfrac{\sin ax}{\sin bx} = \lim\limits_{x\to 0} \dfrac{ax(\sin ax)/ax}{bx(\sin bx)/bx} = \lim\limits_{x\to 0} \dfrac{ax}{bx} = \dfrac{a}{b}$ $(a, b \neq 0)$

21. $\lim\limits_{x\to 0} \dfrac{\sin ax}{bx} = \lim\limits_{x\to 0} \dfrac{ax(\sin ax)/ax}{bx} = \lim\limits_{x\to 0} \dfrac{ax}{bx} = \dfrac{a}{b}$ $(a, b \neq 0)$ **23.** π

25. Since $\lim_{x\to 0}(1 - x^2) = 1$ and $\lim_{x\to 0} \cos x = 1$ the result follows.

27. Since $0 \leq f(x) \leq 1$ for every x and $x^2 > 0$ for $x \neq 0$, then on multiplying the inequality by x^2, we have $0 \leq x^2 f(x) \leq x^2$. Since $\lim_{x\to 0} x^2 = 0$, then $\lim_{x\to 0} x^2 f(x) = 0$.

29. $x \neq 0$, $|\sin(1/x)| \leq 1$
$x \neq 0$, $|x^n \sin(1/x)| = |x^n||\sin(1/x)| \leq |x^n|$, $-|x^n| < x^n \sin(1/x) < |x^n|$
Since $\lim_{x\to 0}(-|x^n|) = 0$ and $\lim_{x\to 0}|x^n| = 0$, then $\lim_{x\to 0}[x^n \sin(1/x)] = 0$.

31. Yes

33. $\lim\limits_{x\to x_0} \sin x = \lim\limits_{h\to 0} \sin(x_0 + h) \underset{\substack{\uparrow \\ \text{by (15)} \\ \text{sec. 1.8}}}{=} \lim\limits_{h\to 0} \sin x_0 \cos h + \cos x_0 \sin h = \sin x_0 \lim\limits_{h\to 0} \cos h + \cos x_0 \lim\limits_{h\to 0} \sin h = \sin x_0$. Use a similar argument for the cosine function.

Exercise 2.6

1. (a) $\delta = 0.025$ (b) $\delta = 0.0025$ (c) $\delta = 0.00025$ (d) $\delta = \varepsilon/4$ **3.** (a) $\delta = 0.1$ (b) $\delta = 0.01$ (c) $\delta = \varepsilon$
5. $\delta < 0.005$ **7.** $\delta < \frac{1}{12}$ **9.** Take $\delta = \varepsilon/3$. If $0 < |x - 2| < \delta$, then $|3x - 6| < \varepsilon$.
11. Take $\delta = \varepsilon/2$. If $0 < |x| < \delta$, then $|(2x + 5) - 5| < \varepsilon$.
13. Take $\delta = \varepsilon/5$. If $0 < |x + 3| < \delta$, then $|(-5x + 2) - 17| < \varepsilon$.
15. Take $\delta \leq \min(1, \varepsilon/3)$. If $0 < |x - 2| < \delta$, then $|x| < 3$ and $|x^2 - 2x| < \varepsilon$.
17. Take $\delta \leq \min(1, 2\varepsilon/7)$. If $0 < |x - 1| < \delta$, then $|1/(3 - x)| < 1$ and $|(1 + 2x)/(3 - x) - \frac{3}{2}| < \varepsilon$.
19. Take $\delta = \varepsilon^3$. If $0 < |x| < \delta$, then $|\sqrt[3]{x}| < \varepsilon$.
21. Suppose $\varepsilon = 1$. There is a $\delta > 0$ so that whenever $0 < |x - 3| < \delta$, then $|(3x - 1) - 12| < 1 \Rightarrow 4 < x < \frac{14}{3}$. For any $\delta < 1$—say, $\delta = \frac{1}{2}$—we have $0 < |x - 3| < \delta \Rightarrow 2.5 < x < 3.5$. This is impossible.
23. $\left|\dfrac{1}{x^2 + 9} - \dfrac{1}{18}\right| = \dfrac{|18 - (x^2 + 9)|}{18(x^2 + 9)} = \dfrac{|x + 3||x - 3|}{18(x^2 + 9)} \leq \dfrac{7|x - 3|}{18(13)}$
Given any $\varepsilon > 0$, choose δ to be the smaller of 1 or $\frac{234}{7}\varepsilon$.
25. Suppose $\varepsilon = 0.1$. There is a $\delta > 0$ so that whenever $0 < |x - 1| < \delta$, then $|x^2 - 1.31| < 0.1 \Rightarrow 1.1 < x < \sqrt{1.41}$. For any $\delta < 0.2$—say, $\delta = 0.01$—we have $0 < |x - 1| < 0.01 \Rightarrow 0.99 < x < 1.01$. This is impossible.
27. Take $\delta = \sqrt{\varepsilon}$. If $0 < |x| < \delta$, then $|(4 - x^2) - 4| < \varepsilon$.

Historical Perspectives

1. $y = (-\sqrt{3}/3)x + (4\sqrt{3}/3)$; $m_{L_2} = \sqrt{3}$; $m_{L_1} = -\sqrt{3}/3$ **3.** $y = 4x - 4$ **5.** $y = (2y_0/x_0)x - y_0$
7. $m = \frac{1}{2}$; $y = \frac{1}{2}x$ **9.** $y = -\frac{3}{4}x + \frac{25}{4}$ **11.** $y = 6x - 9$ **13.** $y = 12x - 16$

Review Exercises

1.

x	-0.1	-0.01	-0.001	0.1	0.01	0.001
$\dfrac{1 - \cos x}{x^2}$	0.49958	0.50000	0.50000	0.49958	0.50000	0.50000

$\lim\limits_{x\to 0} \dfrac{1 - \cos x}{x^2} = \dfrac{1}{2}$ **3.** $\lim\limits_{x\to 2} f(x)$ does not exist.

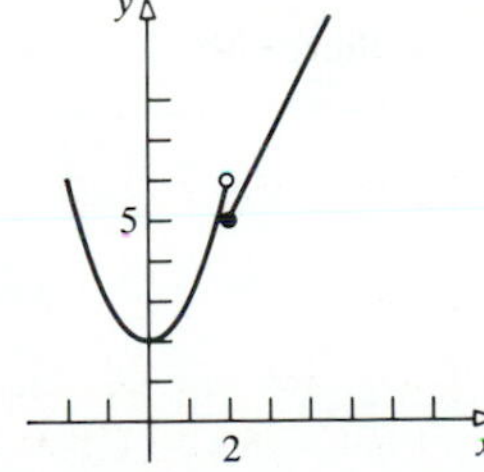

5. -1 **7.** $-3/x^2$ **9.** 1 **11.** 27 **13.** 4 **15.** $\frac{25}{2}$ **17.** 0 **19.** $-\frac{1}{4}$ **21.** Does not exist

23. 1 **25.** -1 **27.** 8 **29.** $\lim_{x\to 3^-} f(x) = 10$; $\lim_{x\to 3^+} f(x) = 10$; $\lim_{x\to 3} f(x) = 10$ **31.** Discontinuous

33. Discontinuous **35.** Continuous **37.** Continuous

39. One possibility: $f(x) = \begin{cases} x & \text{if } x \neq 0 \\ 1 & \text{if } x = 0 \end{cases}$

This function does not contradict the intermediate value theorem, since it is not continuous on $[-1, 1]$.

41. $(-\infty, -3) \cup (-3, -2) \cup (-2, +\infty)$

43. $f(3) = -30$, $f(4) = 42$; therefore, there exists a value c in $[3, 4]$ such that $f(c) = 0$.

45. $\lim\limits_{x\to 2} \dfrac{x^2 - 2x}{x - 2} = 2$, but $\lim\limits_{x\to 2} \dfrac{x^2}{x-2} - \lim\limits_{x\to 2} \dfrac{2x}{x-2}$, limit does not exist

47. Solve $y = x^3$ and $y = 1 - x^2$ simultaneously to get $x^3 + x^2 - 1 = 0 = f(x)$. $f(0) = -1$, $f(1) = 1$; $f(0) < 0 < f(1)$. Therefore, there exists a value c on $[0, 1]$ such that $f(c) = 0$.

49. $\delta \le 0.005$ **51.** Take $\delta \le \min(1, 26\varepsilon)$. If $0 < |x - 2| < \delta$, then $|(x + 2)/(x^2 + 9)| < \frac{1}{2}$ and $\left|\dfrac{1}{x^2+9} - \dfrac{1}{13}\right| < \varepsilon$.

53. $\left|\dfrac{2x}{x+3} - 1\right| < \varepsilon$; $\left|\dfrac{2x - (x+3)}{x+3}\right| < \varepsilon$; $\left|\dfrac{x-3}{x+3}\right| < \varepsilon$

Let $\delta = 1$, $2 < x < 4$, $5 < |x + 3| < 7$

$\left|\dfrac{x-3}{x+3}\right| < \dfrac{|x-3|}{5} < \varepsilon$; $|x - 3| < 5\varepsilon$

Therefore, let $\delta = \min\{1, 5\varepsilon\}$.

55. 1 **57.** $\frac{1}{4}$ **59.** 0

Challenge Exercises

1. $|f(x) - L| = |4x^3 + 3x^2 - 24x + 22 - 5| < \varepsilon$

Let $0 < |x - 1| < \delta = 1$.

$$|4(x-1)^3 + 15x^2 - 36x + 21| = |4(x-1)^3 + 15(x-1)^2 - 6(x-1)|$$
$$\le 4|x-1|^3 + 15|x-1|^2 + 6|x-1| < 4\delta + 15\delta + 6\delta = 25\delta$$

Therefore, let $\delta = \min\{1, \varepsilon/25\}$.

3. The existence of the limit means that $f(a + h) - f(a) \to 0$ as $h \to 0$. Therefore, $\lim_{h\to 0} f(a + h) = f(a)$, which by definition means that $f(x)$ is continuous at $x = a$.

5. To show that there is at least one number $c \in [0, 1]$ such that $f(c) = c$, let $g(x) = f(x) - x$. Then f and x are continuous on $[0, 1] \Rightarrow g$ is continuous on $[0, 1]$; $g(0) = f(0) - 0 = f(0) \Rightarrow 0 \le g(0) \le 1$; $g(1) = f(1) - 1 \Rightarrow -1 \le f(1) - 1 \le 0 \Rightarrow -1 \le g(1) \le 0$. Thus, 0 is an intermediate value to $g(0)$ and $g(1)$. Then, by the intermediate value theorem, there is a $c \in (0, 1)$ such that $g(c) = 0$. Hence, $f(c) - c = 0$ or $f(c) = c$.

7. Let $L \ge 0$ be any arbitrary nonnegative real number. Let $\varepsilon > 0$ be given. Then choosing δ so that $\delta = 1/(L + \varepsilon)$, we have that for all x with $0 < x < \delta = 1/(L + \varepsilon)$, and $1/x > L + \varepsilon$, or $1/x - L > \varepsilon$. This implies that $\lim_{x\to 0^+} 1/x \neq L$ for any arbitrarily chosen $L > 0$. Hence, L cannot be the limit. Similarly, letting $L < 0$ be any arbitrary negative real number, we have $\lim_{x\to 0^-} 1/x \neq L$. This proves the result.

9. f is continuous at 0, and discontinuous everywhere else.

CHAPTER 3

Exercise 3.1

1. 4, 16; $6t + 4$ m/sec **3.** 61, 60.1, 60.01; 60 cm/sec **5.** 100, 68, -28 ft/sec; 6.25 sec; 3.125 sec

7. $\frac{1}{2}, \frac{4}{3}, 4(\sqrt{2} - 1)$ km/hr **9.** (a) 7.91 sec (b) -126.62 ft/sec (c) -32 ft/sec (d) -253.24 ft/sec

11. (a) 0, 2, -2 (b) $y = -1, y = 2x - 2, y = -2x - 2$ (c)

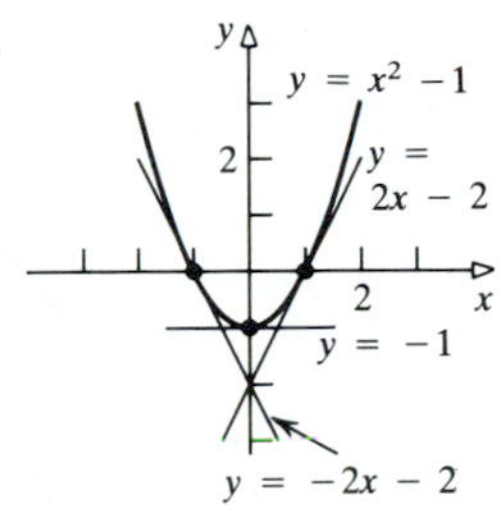

13. (a) $-1, -\frac{1}{4}, -\frac{1}{9}$ (b) $y = -x + 2, y = -\frac{1}{4}x + 1, y = -\frac{1}{9}x + \frac{2}{3}$ (c)

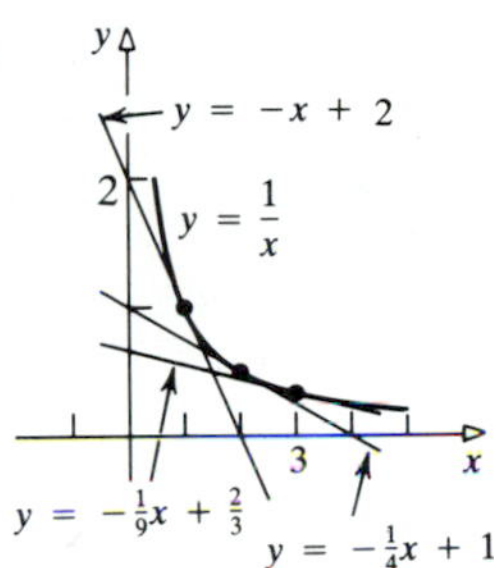

15. (a) 0, 3, 3 (b) $y = 0, y = 3x - 2, y = 3x + 2$ (c)

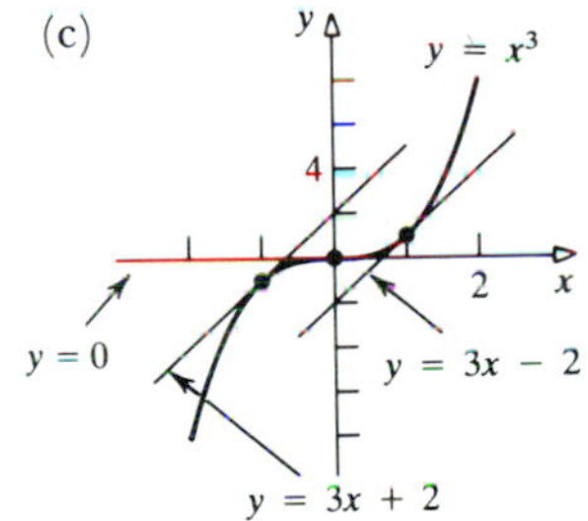

17. $y = 6x - 9$

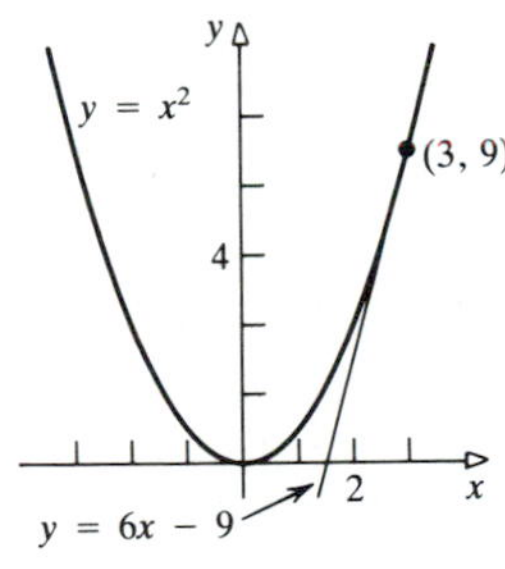

19. $y = 4x$

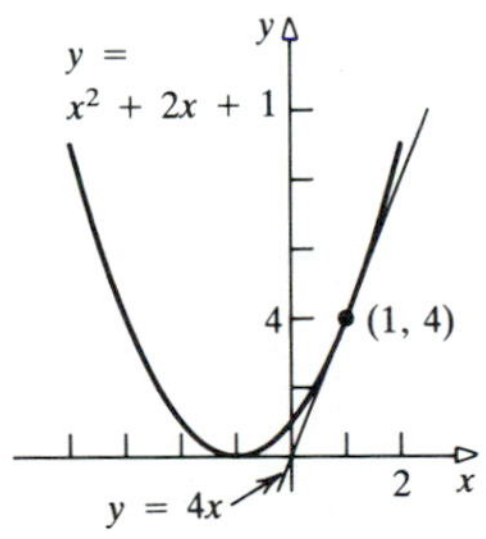

21. $y = -x + 2$

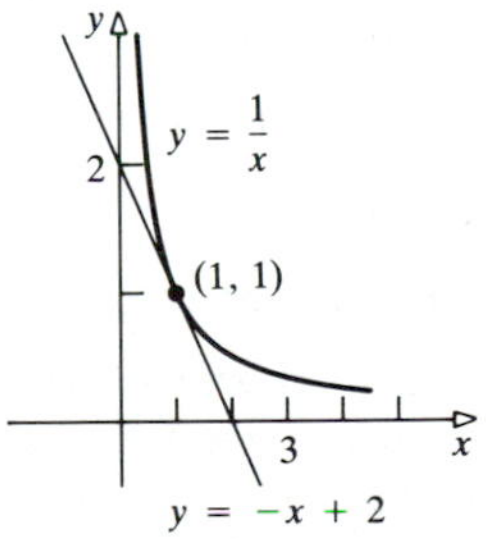

23. $y = -\frac{1}{36}x + \frac{7}{36}$ **25.** $y = -\frac{1}{2}x + \frac{3}{2}$ **27.** No **29.** $f'(2) = -3$ **31.** (a) 5 (b) 5 (c) 5

33. (a) 0 (b) $\frac{7}{16}$ (c) $\dfrac{2c}{c+3} - \dfrac{c^2}{(c+3)^2}$ **35.** 0.59

37. $-\frac{7}{2}$; thus, the average rate of disintegration of protein into amino acids from 0 to 2 hours is $-\frac{7}{2}$ gram/hour.

Exercise 3.2

1. 2 **3.** 0 **5.** -5 **7.** $\frac{1}{4}$ **9.** -7 **11.** 2 **13.** $2x$ **15.** $6x+1$ **17.** 0 **19.** $5/2\sqrt{x}$ **21.** m
23. $2/(x+1)^2$ **25.** $-10x/(1+x^2)^2$ **27.** x^2, 2 **29.** x^2, 1 **31.** $\sin x$, $\pi/6$ **33.** $2x^2-x$, 2
35. $(-\sqrt{2}, 4)$, $(\sqrt{2}, 4)$ **37.** (a) $-2, 4$ (b) 0, 2, 6

39. Yes

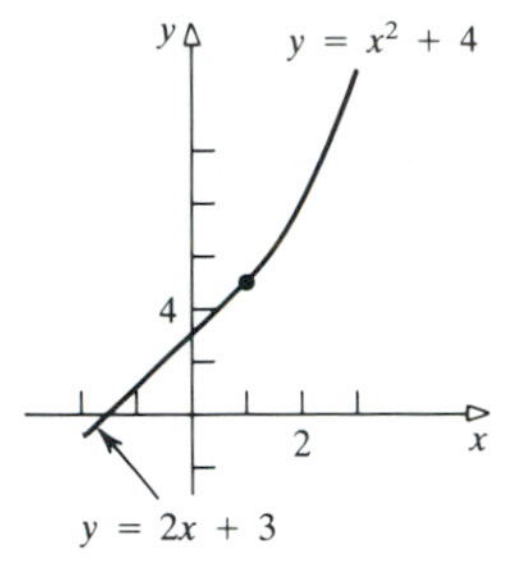

41. No

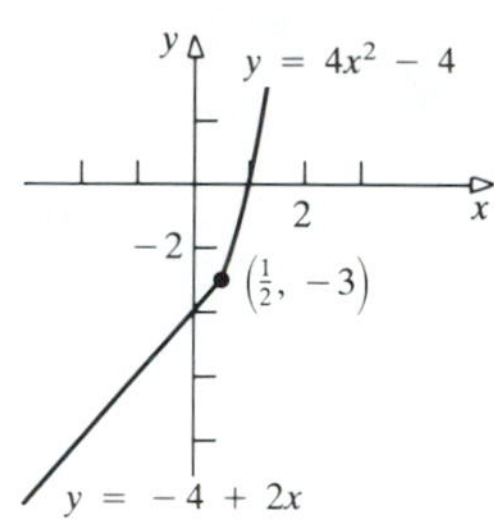

43. No

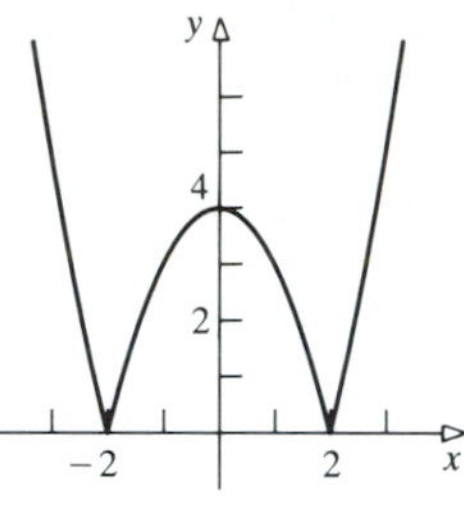

45. No

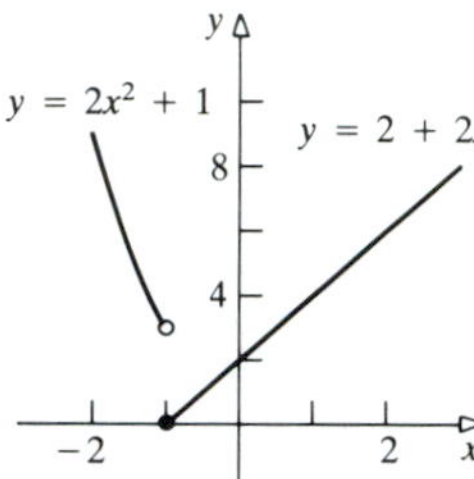

47. (a) Yes (b) Yes (c)

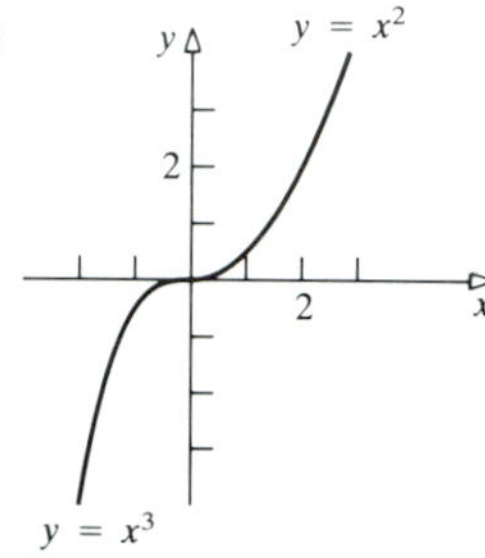

49. 75 ft/sec; 0; no

51. $\frac{4}{3}$ **53.**

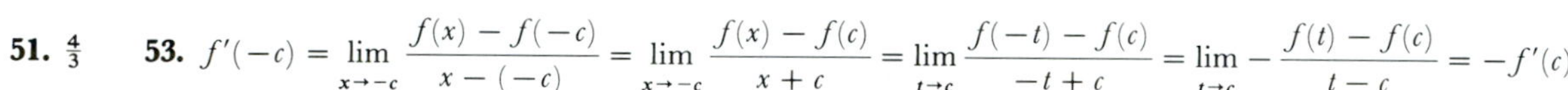

$$f'(-c) = \lim_{x\to -c} \frac{f(x)-f(-c)}{x-(-c)} = \lim_{x\to -c} \frac{f(x)-f(c)}{x+c} = \lim_{t\to c} \frac{f(-t)-f(c)}{-t+c} = \lim_{t\to c} -\frac{f(t)-f(c)}{t-c} = -f'(c)$$

55. $f'(x) = \lim_{h\to 0} \frac{f(x+h)-f(x)}{h} = \lim_{h\to 0} \frac{f(x)f(h)-f(x)}{h} = f(x)\lim_{h\to 0} \frac{f(h)-1}{h} = f(x)\lim_{h\to 0} \frac{f(h)-f(0)}{h-0} = f(x)f'(0)$

57. $(\frac{3}{2}, \frac{27}{8})$ **59.** $-I' = kI$ $(k > 0)$

61. (a) $\frac{4\pi}{3}[3R^2\Delta R + 3R(\Delta R)^2 + (\Delta R)^3]$ (b) $\frac{4\pi}{3}[3R^2 + 3R(\Delta R) + (\Delta R)^2]$ (c) $4\pi R^2$

63. $\frac{1}{2}$

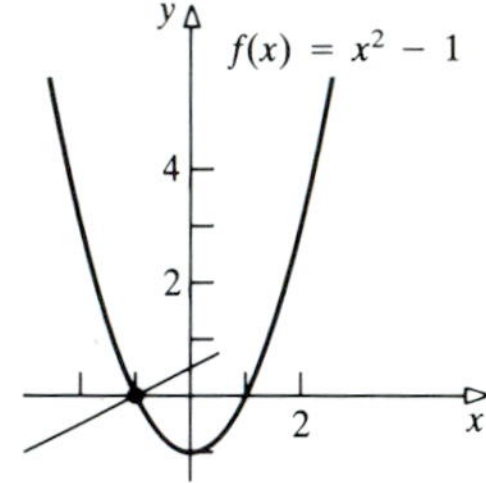

65. $-\frac{1}{5}$

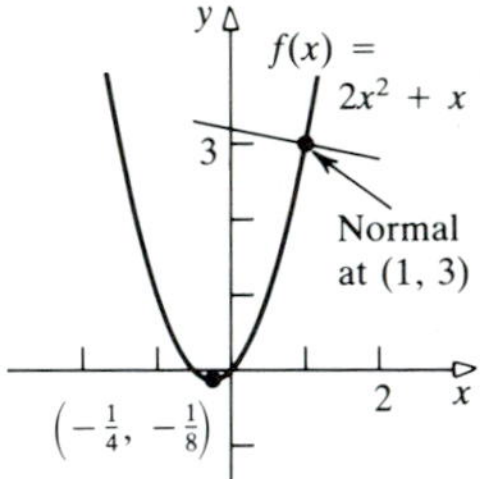

67. -4

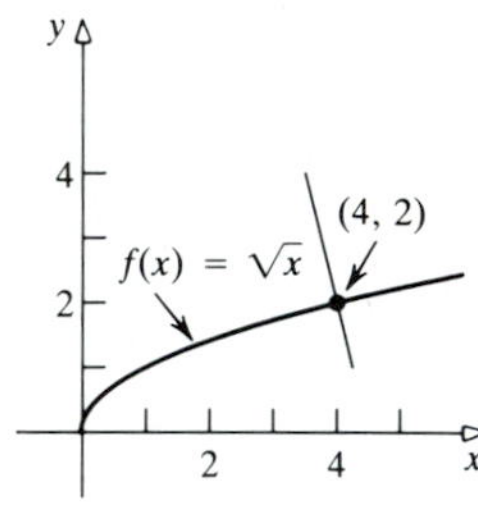

Exercise 3.3

1. 0 **3.** $\sqrt{2}$ **5.** $45x^{14}$ **7.** 3 **9.** $2x+3$ **11.** $40x^4-5$ **13.** $\frac{4}{3}x^3-3$ **15.** $3\pi x^2+3x$ **17.** $\frac{5}{3}x^4$
19. $\frac{7}{5}x^6-\frac{6}{5}x$ **21.** $\frac{7}{9}x^6-\frac{5}{9}$ **23.** $2ax+b$ **25.** $\sqrt{3}$ **27.** $2\pi R$ **29.** $4\pi R^2$ **31.** 3; $y=3x-1$ **33.** 2
35. $-1, 1$ **37.** $y-2=5(x-1)$ **39.** $y=3x-9$; $9x-3y+5=0$ **41.** -1, 74 m/sec
43. This is the derivative of $5x^8$ at $\frac{1}{2}$, which equals $40(\frac{1}{2})^7 = \frac{5}{16}$.

45. $F(x) = f(x) - g(x) = f(x) + (-1)g(x)$; $F'(x) = \dfrac{d}{dx}f(x) + \dfrac{d}{dx}(-1)g(x) = f'(x) + (-1)g'(x) = f'(x) - g'(x)$

47. $f'(c) = \lim\limits_{x\to c} \dfrac{x^n - c^n}{x - c} = \lim\limits_{x\to c} \dfrac{(x-c)(x^{n-1} + cx^{n-2} + \cdots + c^{n-1})}{x - c} = nc^{n-1}$

Exercise 3.4

1. $5x^4 + 3x^2 - 2x$ **3.** $18x^2 + 6x - 10$ **5.** $16t^7 + 10t^4 - 4t^3 - 1$ **7.** $-3t^{-4}$ **9.** $\dfrac{-40}{x^5} - \dfrac{6}{x^3}$

11. $\dfrac{2}{(s+1)^2}$ **13.** $\dfrac{12x^2 + 32x + 6}{(3x+4)^2}$ **15.** $3 - \dfrac{1}{3t^2}$ **17.** $\dfrac{-4}{(1+2u)^2}$ **19.** $9x^2 + \dfrac{2}{3x^3}$ **21.** $\dfrac{-1}{t^2} + \dfrac{2}{t^3} - \dfrac{3}{t^4}$

23. $\dfrac{-3w^2}{(w^3 - 1)^2}$ **25.** $\frac{5}{4}$; $y = \frac{5}{4}x - \frac{3}{4}$ **27.** $0, -2$

29. (a) $y' = 2x(3x-2) + x^2(3) = 9x^2 - 4x$ (b) $y = 3x^3 - 2x^2$; $y' = 9x^2 - 4x$ (c) They are equal.

31. -2 units/m **33.** $\dfrac{d}{dx}[f(x)g(x)h(x)] = f(x)\dfrac{d}{dx}[g(x)h(x)] + g(x)h(x)\dfrac{d}{dx}[f(x)]$; now use (1)

35. $6x^5 - 5x^4 + 20x^3 - 18x^2 + 2x - 5$ **37.** $9x^2(x^3 + 1)^2$

39. $\dfrac{3}{x^4}\left(1 - \dfrac{1}{x}\right)\left(1 - \dfrac{1}{x^2}\right) + \dfrac{2}{x^3}\left(1 - \dfrac{1}{x}\right)\left(1 - \dfrac{1}{x^3}\right) + \dfrac{1}{x^2}\left(1 - \dfrac{1}{x^2}\right)\left(1 - \dfrac{1}{x^3}\right)$

41. If $k(x) = \dfrac{1}{f(x)\cdot g(x)}$, then using the quotient rule we have

$$k' = \frac{(f\cdot g)\cdot 0 - 1\cdot(fg' + gf')}{f^2g^2} = -\frac{1}{fg}\left(\frac{fg' + gf'}{fg}\right) = -k\left(\frac{f'}{f} + \frac{g'}{g}\right)$$

Exercise 3.5

1. 2; 0 **3.** $6x + 1$; 6 **5.** $1 - (1/x^2)$; $2/x^3$ **7.** $\dfrac{1}{(t+1)^2}$; $\dfrac{-2}{(t+1)^3}$ **9.** $\dfrac{x^2 + 2x}{(x+1)^2}$; $\dfrac{2}{(x+1)^3}$

11. $\dfrac{-12}{(3x+5)^2}$; $\dfrac{72}{(3x+5)^3}$

13. (a) $y' = 20x^3 - 6x^2 + 1$; $y'' = 60x^2 - 12x$ (b) $y' = \dfrac{-1}{x^2}$; $y'' = \dfrac{2}{x^3}$

(c) $y' = 2(x^3 + 5) + 3x^2(2x + 1) = 8x^3 + 3x^2 + 10$; $y'' = 24x^2 + 6x$

(d) $y' = \dfrac{(x)(2) - (1)(2x - 5)}{x^2} = \dfrac{5}{x^2}$; $y'' = \dfrac{-10}{x^3}$

15. (a) $y' = 12x^2 - 6x + 1$; $y'' = 24x - 6$; $y''' = 24$ (b) $y' = 3ax^2 + 2bx + c$; $y'' = 6ax + 2b$; $y''' = 6a$

17. 0 **19.** 0 **21.** $7! = 5040$ **23.** $32t + 20$; 32 **25.** $9.8t + 4$; 9.8 **27.** $x^2g''(x) + 4xg'(x) + 2g(x)$

29. (a) $-5 + 2t$ (b) $t = \frac{5}{2}$ **31.** (a) $-2t/(t^2 - 1)^2$ (b) $t = 0$

33. (a) $-9.8t + 39.2$ m/sec (b) 4 sec (c) 78.4 m (d) -9.8 m/sec^2 (e) 8 sec (f) -39.2 m/sec (g) 156.8 m

35. $F'(x) = f'(x)g(x) + f(x)g'(x)$;
$F''(x) = f''(x)g(x) + f'(x)g'(x) + f'(x)g'(x) + f(x)g''(x) = f''(x)g(x) + 2f'(x)g'(x) + f(x)g''(x)$

37. $f(x) = \dfrac{1}{x}$; $f'(x) = \dfrac{-1}{x^2}$; $f''(x) = \dfrac{2}{x^3}$; $f'''(x) = \dfrac{-2(3)}{x^4}$; $f^{(n)}(x) = \dfrac{(-1)^n n(n-1)(n-2)\cdots(3)(2)(1)}{x^{n+1}}$

39. $V = \dfrac{-(1 + t^2)}{(t^2 - 1)^2}$; $a = \dfrac{2t(t^2 + 3)}{(t^2 - 1)^3}$

Exercise 3.6

1. $3 \cos x + 2 \sin x$ **3.** $\cos^2 x - \sin^2 x$ **5.** $\sec x \tan^2 x + \sec^3 x$ **7.** $\dfrac{-x \csc^2 x - \cot x}{x^2}$ **9.** $2x \tan x + x^2 \sec^2 x$

11. $x \sec^2 x + \tan x - 3 \sec x \tan x$ **13.** $\dfrac{-1}{1 - \cos x}$ **15.** $\dfrac{x \cos x + \cos x - \sin x}{(1 + x)^2}$ **17.** $\dfrac{-2}{(\sin x - \cos x)^2}$

19. $\dfrac{x \tan^2 x - x + \sec x \tan x - \tan x}{(1 + x \sin x)^2}$ **21.** $-\csc x \cot^2 x - \csc^3 x$ **23.** $\dfrac{2 \sec^2 x}{(1 - \tan x)^2}$ **25.** $-\sin x$

27. $\sec x \tan^2 x + \sec^3 x$ **29.** $2 \cos x - x \sin x$ **31.** $-2 \sin x + 3 \cos x$ **33.** $2 \sin x + 4x \cos x - x^2 \sin x$

35. $-a \sin x - b \cos x$ **37.** -1 **39.** $\dfrac{-2}{2 + \sqrt{3}}$ **41.** $y = x$ **43.** $3\sqrt{3}x + 6y = 3 + \sqrt{3}\pi$ **45.** $y = x$

47. $y = \sqrt{2}$ **49.** $(-1)^{n/2} \sin x$ if n is even; $(-1)^{(n-1)/2} \cos x$ if n is odd **51.** Derivative of $\cos x$ at $\pi/2$, which is -1

53. $y' = A \cos t - B \sin t$; $y'' = -A \sin t - B \cos t = -(A \sin t + B \cos t) = -y$

55.
$$\frac{d}{dx} \sin x = \lim_{h \to 0} \frac{\sin(x + h) - \sin x}{h} = \lim_{h \to 0} 2\left\{\frac{\cos[(2x + h)/2]\sin(h/2)}{h}\right\}$$
$$= 2 \lim_{h \to 0} \cos\left(\frac{2x + h}{2}\right) \cdot \lim_{h \to 0} \frac{\sin(h/2)}{2(h/2)} = 2(\cos x)(\tfrac{1}{2}) = \cos x$$

57. $f'(0) = \lim_{x \to 0} \dfrac{f(x) - f(0)}{x - 0} = \lim_{x \to 0} \dfrac{\cos x - 1}{x}$ **59.** $v = (-\pi/2)\sin 4\pi t$; $a = -2\pi^2 \cos 4\pi t$

Exercise 3.7

1. $15x^2(x^3 + 1)^4$ **3.** $\dfrac{2x}{(x^2 + 2)^2}$ **5.** $-\dfrac{2(x + 1)}{x^3}$ **7.** $\dfrac{-30}{x^7}\left(\dfrac{1}{x^6} - 1\right)^4$ **9.** $6(3x + 5)$ **11.** $-18(6x - 5)^{-4}$

13. $8x(x^2 + 5)^3$ **15.** $7(t^5 - t^2 + t)^6(5t^4 - 2t + 1)$ **17.** $3\left(x - \dfrac{1}{x}\right)^2\left(1 + \dfrac{1}{x^2}\right)$ **19.** $\dfrac{3z^2}{(z + 1)^4}$ **21.** $2 \tan x \sec^2 x$

23. $4 \sin t \cos t$ **25.** $\dfrac{-6x(3x^2 + 1)(x^3 + x + 2)}{(x^3 - 1)^3}$ **27.** $2x(x^2 + 4)(2x^3 - 1)^2(13x^3 + 36x - 2)$

29. $4[5x + (3x + 6x^2)^3]^3[5 + 9(3x + 6x^2)^2(1 + 4x)]$ **31.** $\dfrac{-(4x^3 - 2)}{(x^4 - 2x + 1)^2}$ **33.** $\dfrac{-10}{(x - 1)^{11}}$ **35.** $4 \cos 4x$

37. $5 \sec^2 5x$ **39.** $6x \cos(3x^2 + 4)$ **41.** $24 \sin 3x \cos 3x$ **43.** $4(x + 1)\cos(x^2 + 2x - 1)$

45. $-6 \csc^2(1 + 3x)\cot(1 + 3x)$ **47.** $2x(\sin 4x + 2x \cos 4x)$ **49.** $(-1/x^2)\cos(1/x)$ **51.** $2x \sin x(\sin x + x \cos x)$

53. $\dfrac{-4x(1 - x^2)^{10}(5x^2 + 6)}{(1 + x^2)^2}$ **55.** $-12x(1 + \cos^3 x^2)(\cos^2 x^2)(\sin x^2)$

57. (a) $y' = (2u)(3x^2) = 6x^2(x^3 + 1)$ (b) $y' = 2(x^3 + 1)(3x^2) = 6x^2(x^3 + 1)$
(c) $y = x^6 + 2x^3 + 1$; $y' = 6x^5 + 6x^2 = 6x^2(x^3 + 1)$ (d) All are equal.

59. $y = 0$ **61.** $y = \frac{1}{27}(-7x + 16)$ **63.** $y = 2x + 1$ **65.** $\dfrac{-576(48 + x^4)^2}{x^{13}}$ **67.** $\dfrac{-64}{x^5}$

69. $6 \tan^2 2x \sec^2 2x$ **71.** $-5x^3[5x^5 \cos(x^5) + 4 \sin(x^5)]$ **73.** $2^n n!$

75. $f(x) = -f(-x)$; $f'(x) = -\dfrac{d}{dx} f(-x) = (-1)(-1)f'(-x) = f'(-x)$ **77.** $2xf'(x^2 + 1)$

79. $\dfrac{-2}{(x - 1)^2} f'\left(\dfrac{x + 1}{x - 1}\right)$ **81.** $\cos x f'(\sin x)$ **83.** $\sin^2 x f''(\cos x) - \cos x f'(\cos x)$ **85.** -12 **87.** 78

89. $3\text{ m/sec};\ -6(2-t)\text{ m/sec}^2$ **91.** $\dfrac{-20\pi}{9}\sin\dfrac{\pi}{6}t$ **93.** $\dfrac{d}{dx}\cos x = \dfrac{d}{dx}\sin\left(\dfrac{\pi}{2}-x\right) = (-1)\cos\left(\dfrac{\pi}{2}-x\right) = -\sin x$

95. $f'(x) = \lim\limits_{h\to 0}\dfrac{f(x+h)-f(x)}{h} = \lim\limits_{h\to 0}\dfrac{f(x)f(h)-f(x)}{h} = f(x)\lim\limits_{h\to 0}\dfrac{f(h)-1}{h} = f(x)\lim\limits_{h\to 0}\dfrac{[1+hg(h)]-1}{h}$

$= f(x)\lim\limits_{h\to 0}\dfrac{hg(h)}{h} = f(x)\lim\limits_{h\to 0} g(h) = f(x)$

97. 2

Exercise 3.8

1. $-x/y$ **3.** $-2y/x$ **5.** $\dfrac{2x-y}{x+2y}$ **7.** $\dfrac{4y-2x}{2y-4x-1}$ **9.** $-2x/y^2$ **11.** $\dfrac{1-12x^2}{6y^2}$ **13.** $(y/x)^3$

15. $-(y/x)^2$ **17.** $\dfrac{6x(x^2+y)^2}{1-3(x^2+y)^2}$ **19.** y/x **21.** $\dfrac{-x(y^2-1)^2}{y}$ **23.** $\dfrac{\sin y}{1-x\cos y}$

25. $\dfrac{-[1+y+2\cos(2x+3y)]}{x+3\cos(2x+3y)}$ **27.** $\dfrac{\sec^2(x-y)}{1+\sec^2(x-y)}$ **29.** $\dfrac{\cos(x+y)-\sin(x-y)}{1-\cos(x+y)-\sin(x-y)}$

31. $\dfrac{1-2y^3\sqrt{t}(1+t)}{2(1+3xy^2)\sqrt{t}(1+t)}$, where $t=\tan^2(xy^3+y)$ **33.** $\dfrac{y^2\sin x+\cos(x+y)}{2y\cos x-\cos(x+y)}$ **35.** $-x/y;\ \dfrac{-y^2+x^2}{y^3}$

37. $\dfrac{-y(1+2x)}{x(1+x)};\ \dfrac{2y(3x^2+3x+1)}{x^2(1+x)^2}$ **39.** $-\frac{1}{2};\ y=-\frac{1}{2}x+\frac{5}{2}$

41. $m_{\tan}=-x/y$; slope of OP is y/x; product of slopes is -1 **43.** (a) $\dfrac{-(y+1)}{4y+x}$ (b) $y=-\frac{1}{3}x+\frac{5}{3}$ (c) $(6,-3)$

45. $-\frac{1}{2}$ **47.** 2 **49.** $\frac{1}{2};\ \frac{1}{5}$ **51.** $\dfrac{dv}{ds}=g\dfrac{dt}{ds}=\dfrac{g}{ds/dt}=\dfrac{g}{v}=\dfrac{g}{gt}=\dfrac{1}{t}$ **53.** $\frac{126}{129}$ radian/sec

1. $\frac{2}{3}x^{-1/3}$ **3.** $\frac{2}{3}x^{-1/3}$ **5.** $\dfrac{-1}{x^{3/2}}-\dfrac{1}{x^{4/3}}+\dfrac{6}{x^{5/2}}-\dfrac{6}{x^{7/4}}$ **7.** $\dfrac{1}{3x^{2/3}}+\dfrac{1}{3x^{4/3}}$ **9.** $\dfrac{-1}{2\sqrt{3x^3}}$ **11.** $\dfrac{3x^2}{2(x^3-1)^{1/2}}$

13. $\dfrac{\cos x}{2\sqrt{\sin x}}$ **15.** $\dfrac{\sec\sqrt{x}\tan\sqrt{x}}{2\sqrt{x}}$ **17.** $\dfrac{2x^2-1}{\sqrt{x^2-1}}$ **19.** $\dfrac{5x^3+2}{2(x^3+1)^{1/2}}$ **21.** $\dfrac{-x}{(3-x^2)^{1/2}}-\dfrac{x}{(4-x^2)^{1/2}}$

23. $\dfrac{x}{\sqrt{x^2+1}}$ **25.** $\dfrac{-1}{(x+1)^{1/2}(x-1)^{3/2}}$ **27.** $\dfrac{|x|(32x+3)}{2\sqrt{x(8x+1)}}$ **29.** $\dfrac{1}{6x^{2/3}\sqrt{4+\sqrt[3]{x}}}$ **31.** $\dfrac{3x}{|x|}$ **33.** $\dfrac{2(2x-1)}{|2x-1|}$

35. $\dfrac{-\sin x\cos x}{|\cos x|}$ **37.** $\dfrac{x\cos|x|}{|x|}$ **39.** $\dfrac{-|y|}{y\cos|y|}$ **41.** $\dfrac{3x^2}{2}(\cos x)^{1/2}(2\cos x-x\sin x)$

43. $\dfrac{-x}{\sqrt{x^2+1}}\sin\sqrt{x^2+1}\cos(\cos\sqrt{x^2+1})$ **45.** $\dfrac{8x^2+x-6}{(x^2-3)^{1/2}(6x+1)^{2/3}}$ **47.** $\dfrac{18x^3+32x^2+3}{2(3x+4)^{3/2}(2x^3-1)^{1/3}}$

49. $\dfrac{18x^4+17x^2+2}{6\sqrt[3]{(2x^3+x)^2}\sqrt[4]{(x^2+1)^3}}$ **51.** $\dfrac{x}{\sqrt{x^2+1}};\ \dfrac{1}{(x^2+1)^{3/2}}$ **53.** $y=-\frac{1}{4}x+1$

55. $y'=-y^{1/3}/x^{1/3}$ **57.** $\frac{1}{4}$

Exercise 3.10

1. 1.154 **3.** 0.2128 **5.** 3.13496 **7.** 1.157 **9.** 1.495 **11.** 3.3166 (start with $c_1 = 3$)
13. 1.9129 (start with $c_1 = 1$) **15.** 1.02 ft **17.** 1.02

Review Exercises

1. $an(ax+b)^{n-1}$ **3.** $\dfrac{2-3x}{2\sqrt{1-x}}$ **5.** $\dfrac{x^2+2x}{(x+1)^2}$ **7.** $\dfrac{1}{(a^2-r^2)^{3/2}}$ **9.** $3x(x^2+4)^{1/2}$ **11.** $\dfrac{-6-2x}{x^3}$

13. $\dfrac{-2z}{(z^2+1)^2}$ **15.** $\dfrac{x(x^2-2)}{(x^2-1)^{3/2}}$ **17.** $\dfrac{1}{a^2p^2\sqrt{p^2-1}}$ **19.** $x(5x^2+2a^2)(a^2+x^2)^{1/2}$ **21.** $\dfrac{1}{2\sqrt{x+2}}$

23. $\dfrac{9}{(y^2+9)^{3/2}}$ **25.** $\dfrac{-x}{\sqrt{1-x^2}}$ **27.** $\dfrac{1-2z}{(1-z+z^2)^2}$ **29.** $\dfrac{-3x^2}{2\sqrt{1-x^3}}$ **31.** $\frac{3}{2}\sqrt{1+u}$ **33.** $\dfrac{-2x}{(x-1)^3}$

35. $1-\sqrt{a/x}$ **37.** $2x\cos 2x + \sin 2x$ **39.** $(\sec u)(\sec u + \tan u)$ **41.** $\dfrac{\cos z}{2\sqrt{1+\sin z}}$

43. $\dfrac{1}{8(1+(1+(1+x)^{1/2})^{1/2})^{1/2}(1+(1+x)^{1/2})^{1/2}(1+x)^{1/2}}$ **45.** $-\left(\dfrac{6003}{2}\sin x\cos^2 x\right)\dfrac{(1+\sqrt{1+\cos^3 x})^{2000}}{\sqrt{1+\cos^3 x}}$

47. $\dfrac{1}{1+5y^4}$ **49.** $\frac{1}{2}$ **51.** $[2, +\infty)$ **53.** $\dfrac{1-y\sec^2(xy)}{x\sec^2(xy)}$

55. (a) $1, -1/2$ (b) $y = 45x - 65$ (c) $y = -4x^3 + 3x - 1$ **57.** $\dfrac{n!}{(1-x)^{n+1}}$ **59.** (a) $t = 1, 4$ minutes (b) 6

61. Tangent line: $y = 2\sqrt{3}(x-1)$ Normal line: $y = \dfrac{\sqrt{3}}{3}\left(\dfrac{-x}{2}+7\right)$

Challenge Exercises

1. $\dfrac{f^2g' + g^2f'}{(f+g)^2}$

3. (a) $f'_1 f_2 \cdots f_n + f_1 f'_2 f_3 \cdots f_n + \cdots + f_1 f_2 \cdots f_{n-1} f'_n$
(b) $\dfrac{-f'_1}{f_1^2 f_2 \cdots f_n} + \dfrac{-f'_2}{f_1 f_2^2 f_3 \cdots f_n} + \cdots + \dfrac{-f'_n}{f_1 f_2 \cdots f_{n-1} f_n^2}$

5. If $y = x^n$ and $n = 2k+1$, where $k \geq 0$, then $y = x^{2k+1}$ and $y' = (2k+1)x^{2k}$. At $(1, 1)$ $m_T = 2k+1$; at $(-1, -1)$ $m_T = 2k+1$. Therefore, the tangent lines are parallel.

7. $y' = nx^{n-1} \Rightarrow m_T$ at $(1, 1) = n$, and the equation of the tangent line is $y = nx + (1-n)$. Therefore, the y-intercept is $1-n$.

9. (a) $xy = c_1$ and $-x^2 + y^2 = c_2$ give $xy' + y = 0$ and $-2x + 2yy' = 0$. Thus, $y' = -y/x$ and $y' = x/y$, respectively, and the product of the slopes of the tangent lines at a point (x, y) is -1. Hence, the tangent lines are perpendicular.

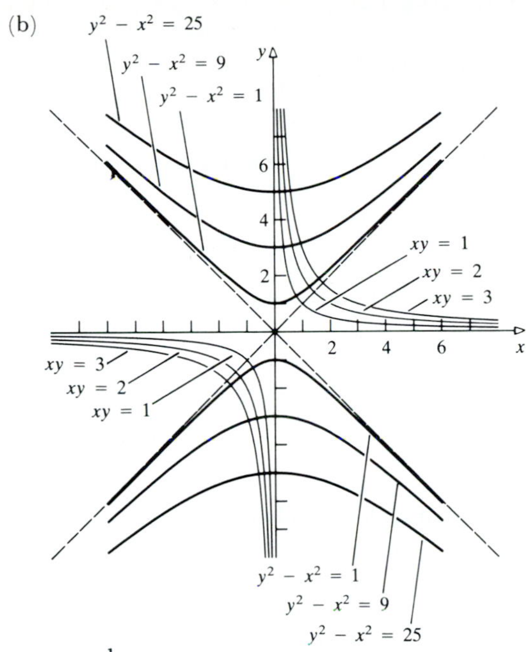

11. (a) $c = \dfrac{1}{2A}$ (b) $1, -1$ (c) $(0, (4AB - 1)/4A)$ **13.** (a) $4x + 2$ (b) $-2 \sin x$ (c) $f^*(x) = 2f'(x)$

15. Let p be a period of f so that $f(x + p) = f(x)$. Then
$f'(x + p) = \lim_{h \to 0} \dfrac{f(x + p + h) - f(x + p)}{h} = \lim_{h \to 0} \dfrac{f(x + h) - f(x)}{h} = f'(x)$. Thus, f' is periodic with period p.

17. $Q = (\frac{9}{4}, \frac{81}{16})$ **19.** $(\frac{13}{4}, \frac{3}{2})$

21. $x \neq 0 \Rightarrow f'(x) = 2x \sin \dfrac{1}{x} - \cos \dfrac{1}{x}$ $f'(0) = \lim_{h \to 0} \dfrac{f(0 + h) - f(0)}{h} = \lim_{h \to 0} \dfrac{f(h)}{h} = \lim_{h \to 0} \dfrac{h^2 \sin \frac{1}{h}}{h} = \lim_{h \to 0} h \sin \dfrac{1}{h}$

For real $h \neq 0$, $h \geq h \sin(1/h) \geq -h$, and so, by the squeezing theorem, $\lim_{h \to 0} h \sin(1/h) = 0$. Therefore, $f'(0) = 0$ and f has a derivative at 0. But $\lim_{x \to 0} f'(x) = \lim_{x \to 0}[2x \sin(1/x) - \cos(1/x)]$. Now, $\lim_{x \to 0} 2x \sin(1/x) = 0$, $\lim_{x \to 0} \cos(1/x)$ does not exist. Therefore, $f'(x)$ is not continuous at 0.

23. $F'(x) = mx = f(x)$

25. $D = f_1 f_4 - f_3 f_2$; $D' = f'_1 f_4 + f_1 f'_4 - f'_3 f_2 - f_3 f'_2 = f'_1 f_4 - f_3 f'_2 + f_1 f'_4 - f'_3 f_2 = \begin{vmatrix} f'_1 & f'_2 \\ f_3 & f_4 \end{vmatrix} + \begin{vmatrix} f_1 & f_2 \\ f'_3 & f'_4 \end{vmatrix}$

CHAPTER 4

Exercise 4.1

1. $-\frac{8}{3}$ **3.** -1 **5.** 40 **7.** $5\pi^2$ **9.** 900 cm³/sec **11.** $-8/\sqrt{2009}$ cm/min **13.** $-\frac{3}{4}$ m²/min
15. $\frac{1}{30}$ m/min **17.** $50\sqrt{2}$ ft/sec **19.** $(-\sqrt{3}/3, -\sqrt{3}/9)$, $(\sqrt{3}/3, \sqrt{3}/9)$, $(0, 0)$
21. $(60\sqrt{14})/7$ ft/sec; at sunset the sun's rays are horizontal, and at its highest point the ball is level with the top of the dome.
23. 8500π m/min **25.** $4/\pi$ m/min **27.** $3 - (9\pi/32)$ m³/min
29. (a) $-3/2\sqrt{55}$ m/sec (b) $-1/2\sqrt{3}$ m/sec (c) $-3/2\sqrt{7}$ m/sec **31.** $4\sqrt{10}/7$ m/sec **33.** $2\sqrt{3}\pi/45$ cm²/min
35. 1.75 kg/cm²/min **37.** $2\sqrt{5}/5$ m/sec **39.** $107/\sqrt{241}$ m/sec **41.** $-3/40$ radian/sec **43.** -4.8 lb/sec

Exercise 4.2

1. $(3x^2 - 2)\,dx$ **3.** $\dfrac{-x^2 + 2x - 6}{(x^2 + 2x - 8)^2}\,dx$ **5.** $(6\cos 2x + 1)\,dx$ **7.** $-y/x;\ -x/y$ **9.** $-x/y;\ -y/x$

11. $\dfrac{x(x - 2y)}{x^2 - y^2};\ \dfrac{x^2 - y^2}{x(x - 2y)}$ **13.** $\dfrac{2}{3\cos 3y};\ \dfrac{3\cos 3y}{2}$ **15.** $\dfrac{1}{2\sqrt{x}}\,dx$ **17.** $(3x^2 + 1)\,dx$ **19.** $f(x) \approx 2x - 3$

21. $f(x) \approx \frac{1}{4}x + 1$ **23.** $f(x) \approx \dfrac{\sqrt{3}}{2}x + \dfrac{6 - \pi\sqrt{3}}{12}$

25. (a) $5.916\overline{6}$ (b) 5.12 (c) 0.9 if $x = 1$; 0.9129 if $x = 1.21$ (d) 0.4849 **27.** (a) 0.006 (b) 0.00125

29. 2π cm^2 **31.** 3.6π m^3 **33.** 6% **35.** 8π cm^3 **37.** 30 m; 0.9% **39.** 72 min

41. (a) $dy = f'(x)\,dx = y'\,dx;\ dy' = f''(x)\,dx = y''\,dx$ (b) $dy'' = f'''(x)\,dx = y'''\,dx;\ d(y'^2) = 2f'(x)f''(x)\,dx$

Exercise 4.3

1. Local maximum at x_2, local and absolute minima at x_3, local maximum at x_5, local minimum at x_7, absolute maximum at x_8.

3. -1 **5.** 3 **7.** 0, 2 **9.** 0, 1, -1 **11.** 0 **13.** 0 **15.** $-1, 1, -\sqrt{2}/2, \sqrt{2}/2$ **17.** 0, 2

19. -3, 1, 0 **21.** 4, 3 **23.** $3, -3, 3\sqrt{3}, -3\sqrt{3}$ **25.** $0, \pi, 2\pi$ **27.** $\pm\pi/8, \pm 3\pi/8$ **29.** 1 **31.** 1

33. 15, -1 **35.** 1, -8 **37.** 16, -4 **39.** 9, 0 **41.** 1, 0 **43.** 4, 2 **45.** $\frac{1}{2}, -\frac{1}{2}$ **47.** $0, -\frac{1}{2}$

49. $128\sqrt[3]{2}$, 0 **51.** $\frac{1}{3}, -1$ **53.** $\sqrt[3]{18}/3\sqrt{3} \approx 0.504$, 0 **55.** 9, 1 **57.** 8, 0 **59.** 40 mph

61. $A = 6, B = 0, C = 2$ **63.** Reverse the inequalities given in the proof of theorem (4.3.4).

65. If f has a local minimum at c, then either $f'(c) = 0$ or $f'(c)$ does not exist (Problem 63). Therefore, $g'(c) = 0$ or does not exist. Since the graph of g is symmetric to the graph of f with respect to the x-axis, g has a local maximum at c.

Exercise 4.4

1. $\frac{3}{2}$ **3.** 1 **5.** $-\sqrt{3}/3$ **7.** $-\sqrt{3}/3, \sqrt{3}/3$ **9.** $-1, 0, 1$ **11.** $\pi/4, 3\pi/4$ **13.** $f(-2) \neq f(1)$

15. $f'(0)$ does not exist. **17.** $\frac{1}{2}$ **19.** 1 **21.** $\frac{7}{3}$ **23.** $\sqrt{3}$ **25.** $(\frac{14}{9})^3$ **27.** $f'(0)$ does not exist.

29. $\dfrac{f(b) - f(a)}{b - a}$ = slope of line; f is not differentiable at $x = 0$

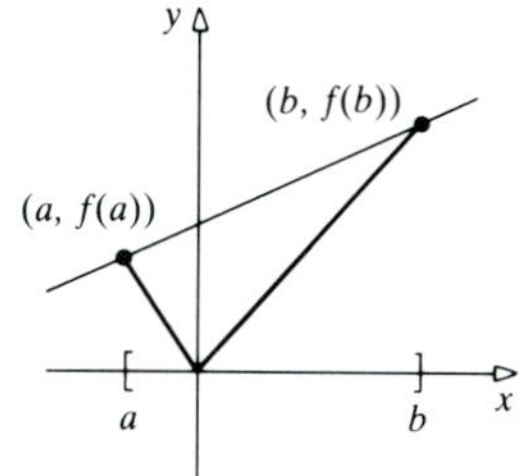

31. By the mean value theorem, $(\sqrt{66} - \sqrt{64})/(66 - 64) = 1/2\sqrt{c}$, $64 < c < 66$. But $8 = \sqrt{64} < \sqrt{c} < \sqrt{66} \Rightarrow 1/\sqrt{66} < 1/\sqrt{c} < \frac{1}{8}$. Since $1/\sqrt{66} > 1/\sqrt{81} = \frac{1}{9}$, then $\frac{1}{9} < 1/\sqrt{c} < \frac{1}{8} \Rightarrow \frac{1}{9} < \sqrt{66} - 8 < \frac{1}{8}$.

33. $f(0) = f(1) = 0$; $f'(x) = (x - 1)\cos x + \sin x = 0$ for some x, $0 < x < 1$. Divide by $\cos x$.

35. Assume the contrary. Then there are at least three roots x_1, x_2, x_3. Set $f(x) = x^n + ax + b$. Apply Rolle's theorem to f on $[x_1, x_2]$ and on $[x_2, x_3]$, obtaining an impossible result.

37. Follow the outline given for the solution to Problem 35.

39. If x_1, x_2, x_3 are the three numbers, apply Rolle's theorem to f on $[x_1, x_2]$ and again on $[x_2, x_3]$.

41. Average acceleration $= (60 - 40)/\frac{1}{3} = 60$ mi/hr^2 = acceleration at some time between 4:00 and 4:20 PM by the mean value theorem.

Exercise 4.5

1. $f'(x) = 6x^2 - 12x + 6 = 6(x-1)^2 \geq 0$ **3.** $f'(x) = 1/(x+1)^2 > 0$

5. (a) $\frac{1}{2}$ (b) Decreasing on $(-\infty, \frac{1}{2}]$, increasing on $[\frac{1}{2}, +\infty)$; local minimum at $\frac{1}{2}$

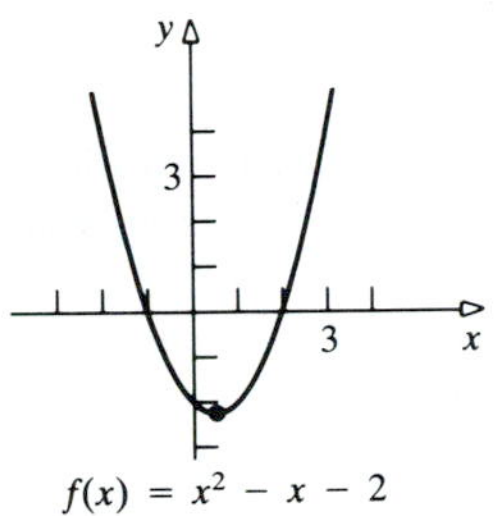

$f(x) = x^2 - x - 2$

7. (a) 1 (b) Increasing on $(-\infty, 1]$, decreasing on $[1, +\infty)$; local maximum at 1

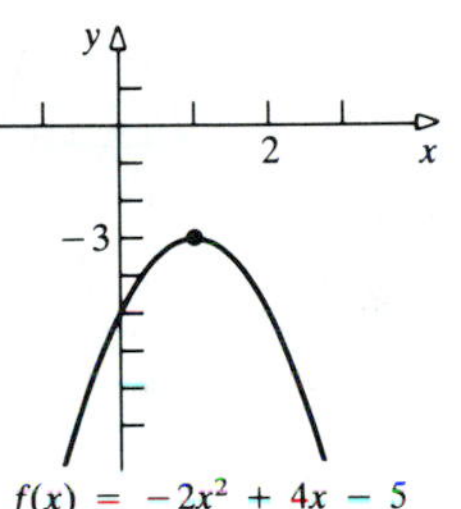

$f(x) = -2x^2 + 4x - 5$

9. (a) $-1, 0$ (b) Increasing on $(-\infty, -1]$, decreasing on $[-1, 0]$, increasing on $[0, +\infty)$; local maximum at -1, local minimum at 0

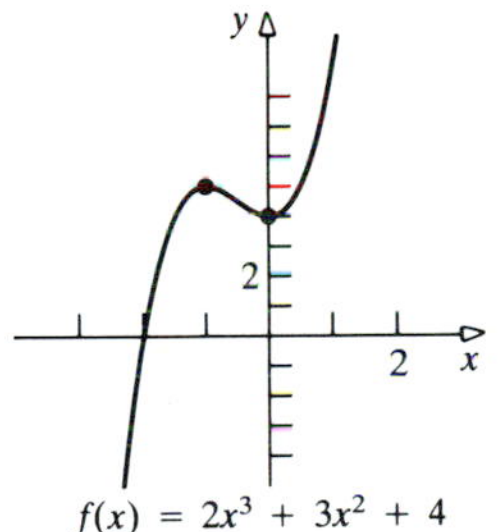

$f(x) = 2x^3 + 3x^2 + 4$

11. (a) -2 (b) Increasing for all x; no local extrema

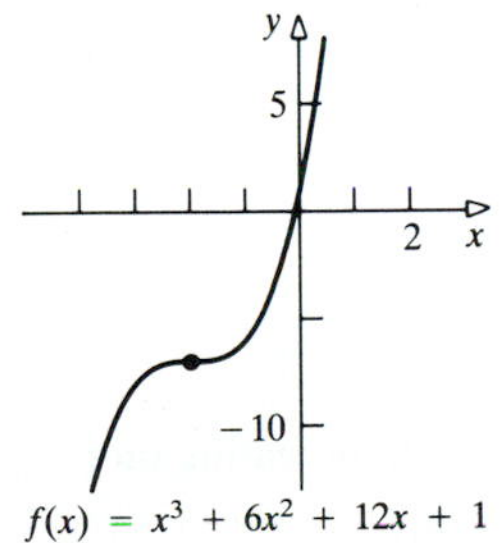

$f(x) = x^3 + 6x^2 + 12x + 1$

13. (a) 0, 3 (b) Decreasing on $(-\infty, 3]$, increasing on $[3, +\infty)$; local minimum at 3

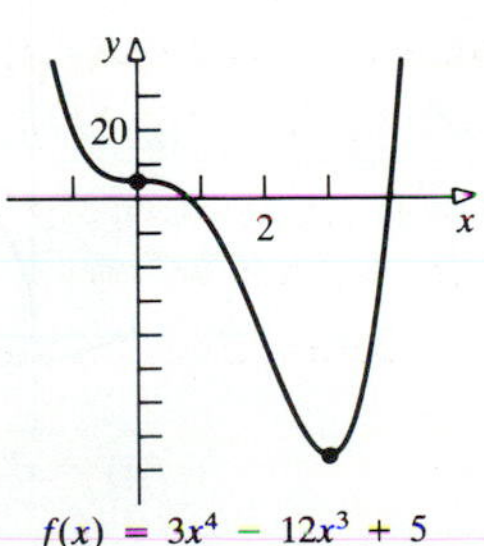

$f(x) = 3x^4 - 12x^3 + 5$

15. (a) 0, 1 (b) Decreasing on $(-\infty, 1]$, increasing on $[1, +\infty)$; local minimum at 1

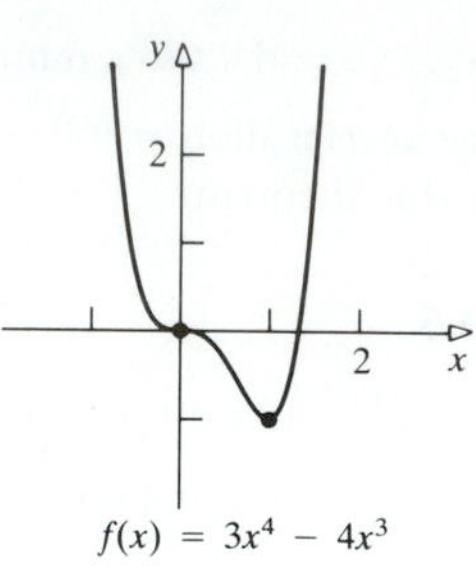

$f(x) = 3x^4 - 4x^3$

17. (a) -3, 1 (b) Increasing on $(-\infty, -3]$, decreasing on $[-3, 1]$, increasing on $[1, +\infty)$; local maximum at -3, local minimum at 1

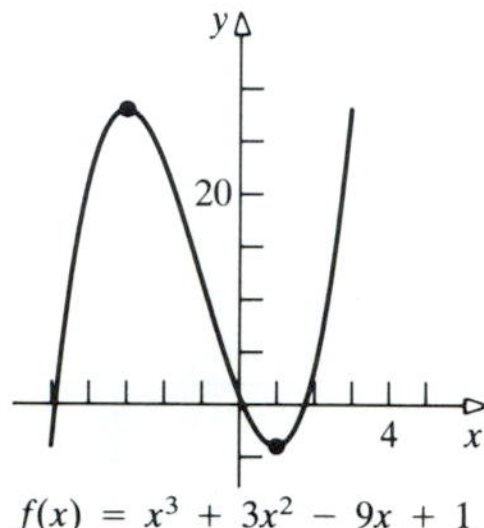

$f(x) = x^3 + 3x^2 - 9x + 1$

19. (a) $-\frac{2}{3}$, 0 (b) Increasing on $(-\infty, -\frac{2}{3}]$, decreasing on $[-\frac{2}{3}, 0]$, increasing on $[0, +\infty)$; local maximum at $-\frac{2}{3}$, local minimum at 0

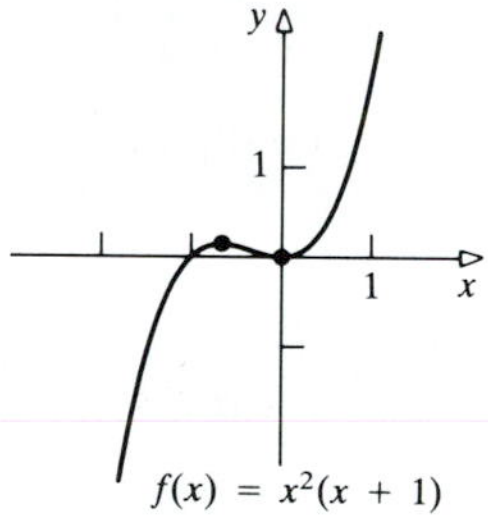

$f(x) = x^2(x + 1)$

21. (a) 0 (b) Decreasing on $(-\infty, 0]$, increasing on $[0, +\infty)$; local minimum at 0

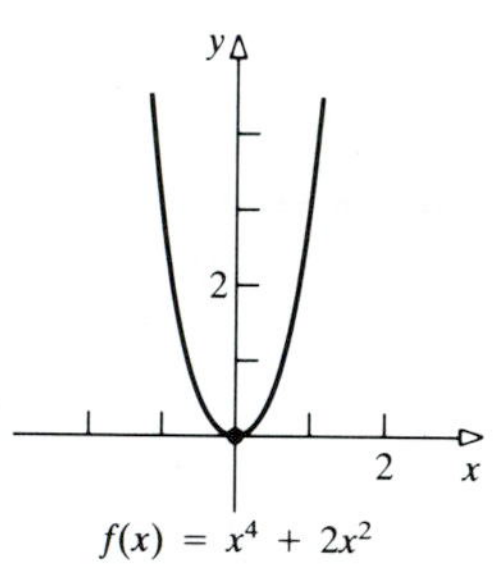

$f(x) = x^4 + 2x^2$

23. (a) $-\frac{1}{8}$, 0 (b) Decreasing on $(-\infty, -\frac{1}{8}]$, increasing on $[-\frac{1}{8}, +\infty)$; local minimum at $-\frac{1}{8}$

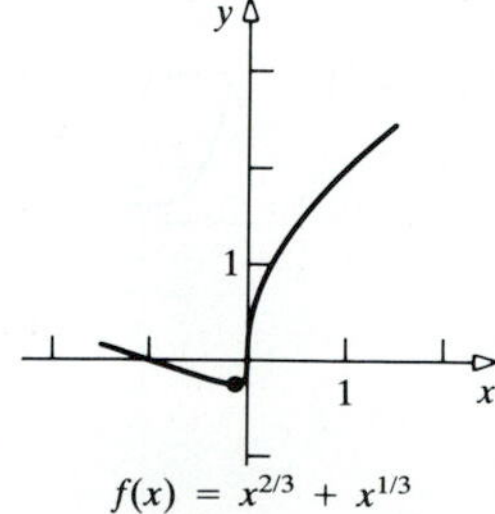

$f(x) = x^{2/3} + x^{1/3}$

25. (a) 0, 4 (b) Increasing on $(-\infty, 0]$, decreasing on $[0, 4]$, increasing on $[4, +\infty)$; local maximum at 0, local minimum at 4

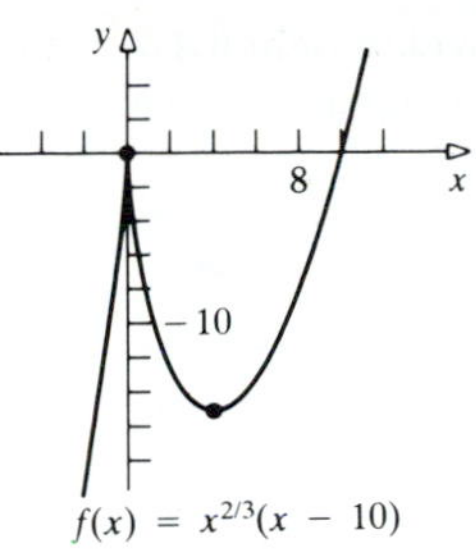

$f(x) = x^{2/3}(x - 10)$

27. (a) $-1, 0, 1$ (b) Decreasing on $(-\infty, -1]$, increasing on $[-1, 0]$, decreasing on $[0, 1]$, increasing on $[1, +\infty)$; local minima at -1 and 1, local maximum at 0

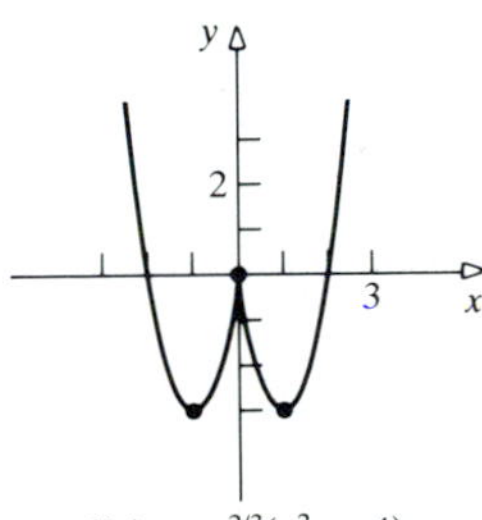

$f(x) = x^{2/3}(x^2 - 4)$

29. (a) $-1, 0, 1$ (b) Decreasing on $(-\infty, -1]$, increasing on $[-1, 0]$, decreasing on $[0, 1]$, increasing on $[1, +\infty)$; local minima at -1 and 1, local maximum at 0

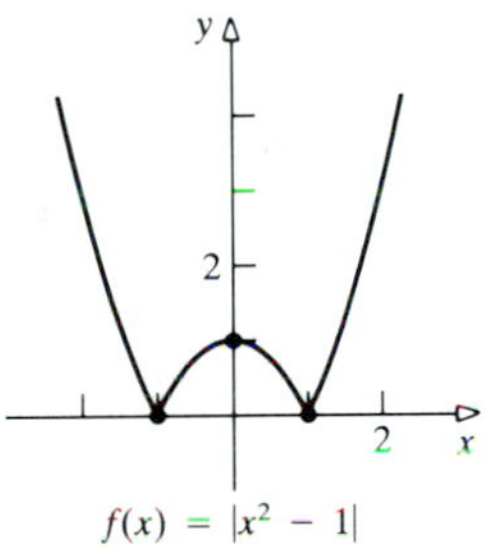

$f(x) = |x^2 - 1|$

31. (a) $(2k + 1)\pi/2$ (b) Decreasing on $[(4k + 1)\pi/2, (4k + 3)\pi/2]$, increasing on $[(4k + 3)\pi/2, (4k + 5)\pi/2]$; local maxima at $(4k + 1)\pi/2$, local minima at $(4k + 3)\pi/2$, k an integer

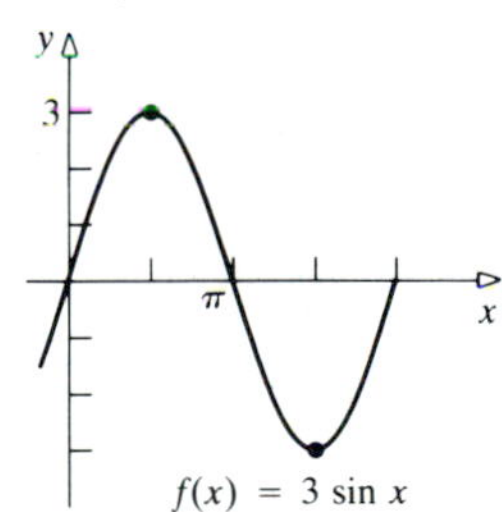

$f(x) = 3 \sin x$

33. (a) $\tan x = -\frac{1}{2}$, or $x \approx -0.4636$ radian $\pm k\pi$, k an integer
(b) Increasing on $[-0.4636 + 2k\pi, -0.4636 + (2k + 1)\pi]$,
decreasing on $[-0.4636 + (2k + 1)\pi, -0.4636 + (2k + 2)\pi]$;
local minima at $-0.4636 + 2k\pi$,
local maxima at $-0.4636 + (2k + 1)\pi$

(c)

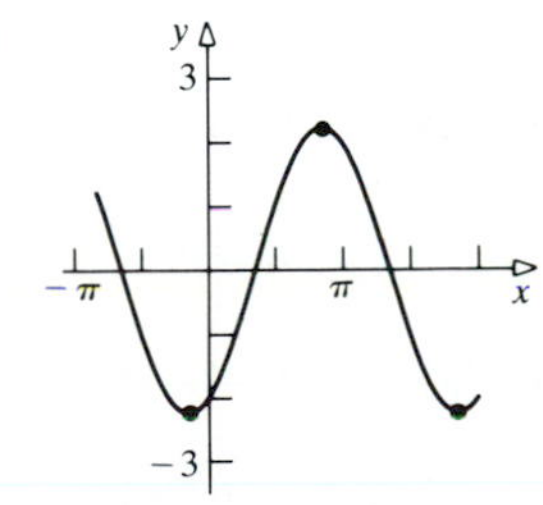

$f(x) = \sin x - 2 \cos x$

35. Moves to right if $t > 1$, to left if $t < 1$; reverses direction if $t = 1$; velocity increases for all t

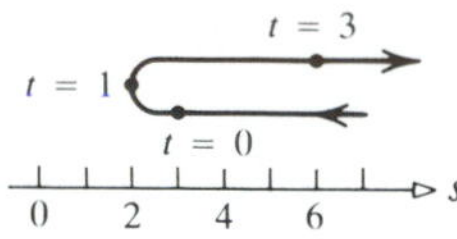

37. Moves to right if $t < -3$ or $t > 1$; moves to left if $-3 < t < 1$; reverses direction if $t = -3$, $t = 1$; velocity increasing if $t \geq -1$; velocity decreasing if $t \leq -1$

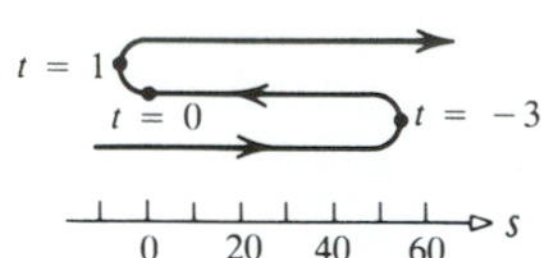

39. Always moves to right; velocity always decreasing

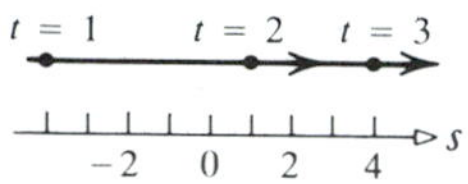

41. Moves to right if $0 < t < \pi/6$ or $\pi/2 < t < 2\pi/3$; moves to left if $\pi/6 < t < \pi/2$; reverses direction at $t = \pi/6$, $t = \pi/2$; velocity decreasing if $0 \leq t \leq \pi/3$; velocity increasing if $\pi/3 \leq t \leq 2\pi/3$

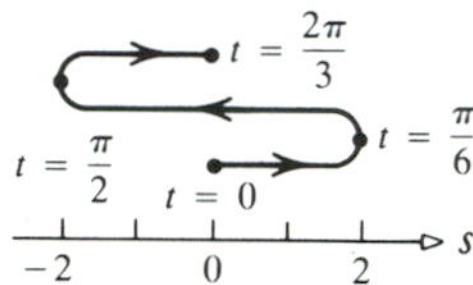

43. $f'(x) = 2ax + b = 0$ if $x = -b/2a$; $a < 0$: $f'(x) = 2a[x + b/2a] > 0$ if $x < -b/2a$; $f'(x) < 0$ if $x > -b/2a$; the case where $a > 0$ is similar.

45. $a = -\frac{7}{8}$, $b = \frac{21}{4}$, $c = 0$, $d = 5$

47. f is decreasing on $a \leq x \leq c$ since $f'(x) < 0$ for $a < x < c$; f is increasing on $c \leq x \leq b$ since $f'(x) > 0$ for $c < x < b$; for all x in $[a, b]$, $f(c) \leq f(x)$; f has local minimum at c.

49. $f'(x) = 1/3x^{2/3} > 0$ if $x \neq 0$; 0 is the only critical number; f is increasing for all x.

51. $\frac{1}{3}$ **53.** No **55.** $\left(\sin\frac{x}{2}\right)^2 \leq \left(\frac{x}{2}\right)^2$ and $\sin^2\frac{x}{2} = \frac{1 - \cos x}{2}$; therefore, $\frac{1 - \cos x}{2} \leq \frac{x^2}{4} \Rightarrow \cos x \geq 1 - \frac{x^2}{2}$

Exercise 4.6

1. Concave up for all x; no inflection points

3. Concave up on $[3, +\infty)$; concave down on $(-\infty, 3]$; inflection point at 3

5. Concave up on $(-\infty, 0]$ and $[2, +\infty)$; concave down on $[0, 2]$; inflection points at 0 and 2

7. Concave up on $(0, +\infty)$; concave down on $(-\infty, 0)$; no inflection points

9. Concave up on $(-\infty, 0]$; concave down on $[0, +\infty)$; inflection point at 0

11. Concave up on $(-\infty, 0)$ and $(0, 3]$; concave down on $[3, +\infty)$; inflection point at 3

13. No local extrema; concave down on $(-\infty, 1]$; concave up on $[1, +\infty)$; inflection point at 1

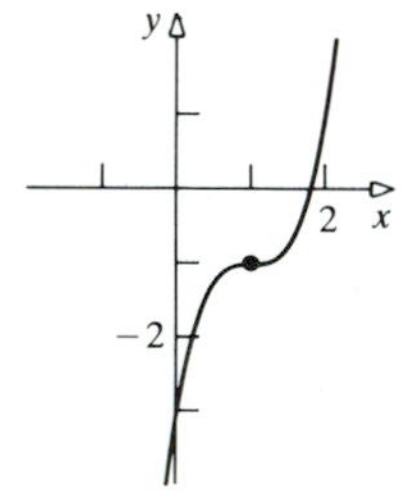

15. Local maximum at 3; local minimum at 2; concave up on $(-\infty, \frac{5}{2}]$; concave down on $[\frac{5}{2}, +\infty)$; inflection point at $\frac{5}{2}$

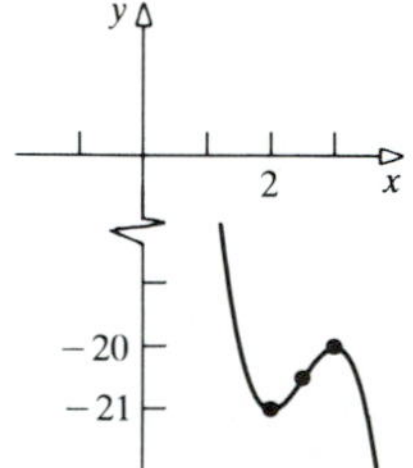

17. Local minimum at 1; concave up for all x; no inflection points

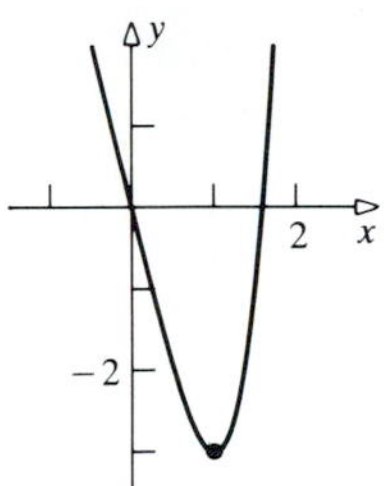

19. Local maximum at 4; local minimum at 0; concave up on $(-\infty, 3]$; concave down on $[3, +\infty)$; inflection point at 3

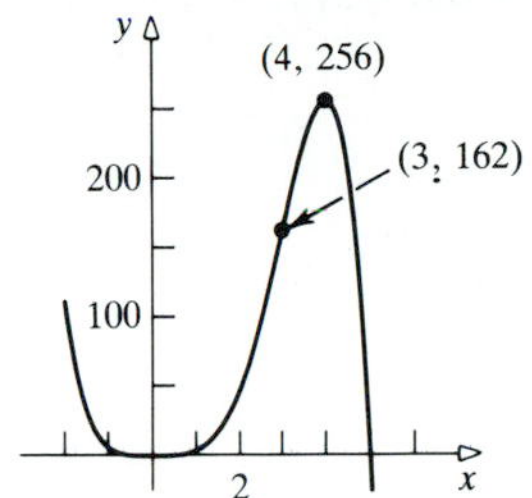

21. Local maximum at -2; local minimum at 2; concave down on $(-\infty, -\sqrt{2}]$; concave up on $[-\sqrt{2}, 0]$; concave down on $[0, \sqrt{2}]$; concave up on $[\sqrt{2}, +\infty)$; inflection points at $-\sqrt{2}, 0, \sqrt{2}$

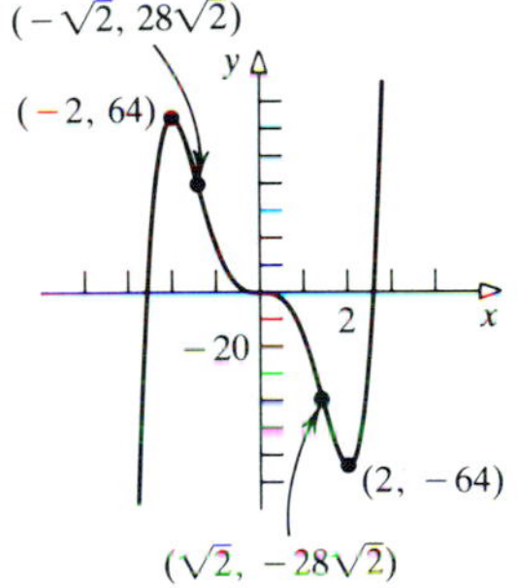

23. Local minimum at $\frac{1}{8}$; concave up on $(-\infty, -\frac{1}{4}]$; concave down on $[-\frac{1}{4}, 0]$; concave up on $[0, +\infty)$; inflection points at $-\frac{1}{4}$, 0

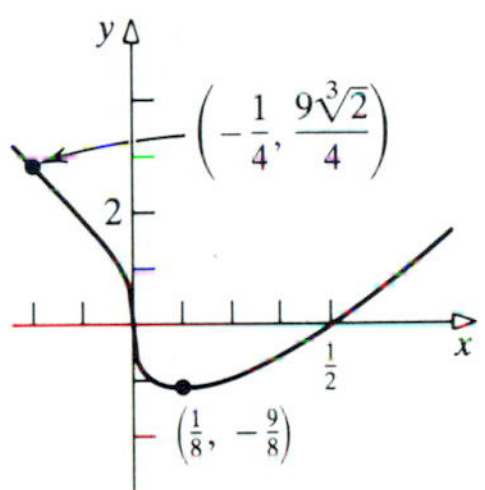

25. Local maximum at 0; local minimum at 4; concave down on $(-\infty, -2]$; concave up on $[-2, +\infty)$; inflection point at -2

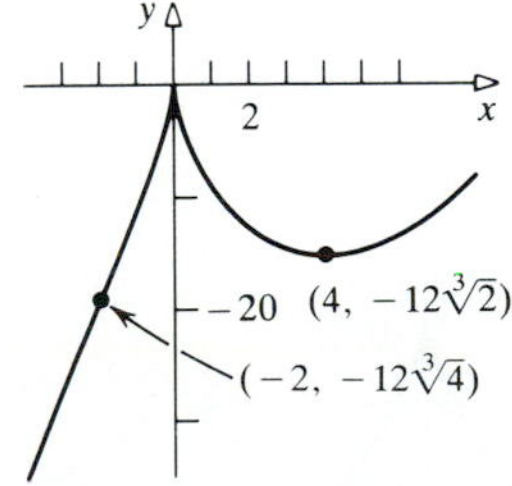

27. Local maximum at 0; local minima at $-\sqrt{2}$ and $\sqrt{2}$; concave up for all x; no inflection points

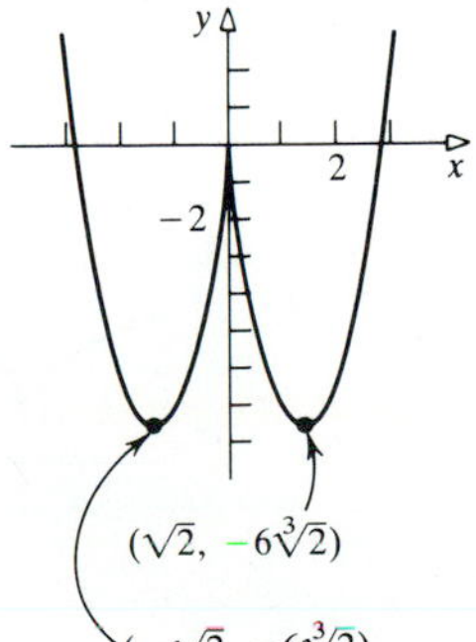

29. Local minimum at 0;
concave down on $(-\infty, -\sqrt{3}/3]$;
concave up on $[-\sqrt{3}/3, \sqrt{3}/3]$;
concave down on $[\sqrt{3}/3, +\infty)$;
inflection points at $-\sqrt{3}/3$ and $\sqrt{3}/3$

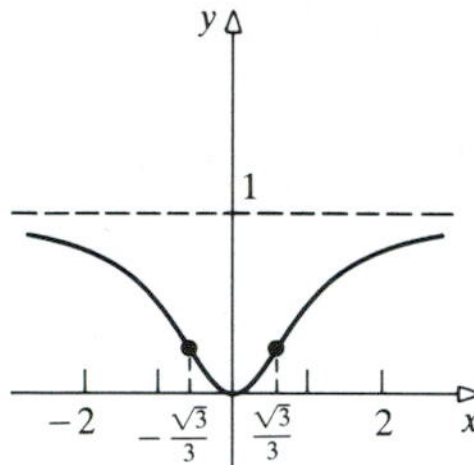

31. Local maximum at 1;
concave down on $[0, (3 + 2\sqrt{3})/3]$;
concave up on $[(3 + 2\sqrt{3})/3, +\infty)$;
inflection point at $(3 + 2\sqrt{3})/3$

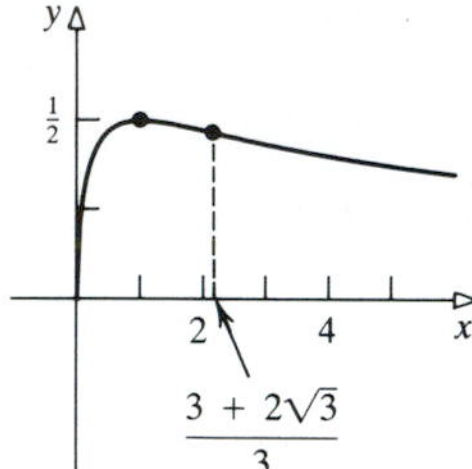

33. Local maximum at $\frac{1}{2}$;
concave down on $[0, 1]$;
no inflection points

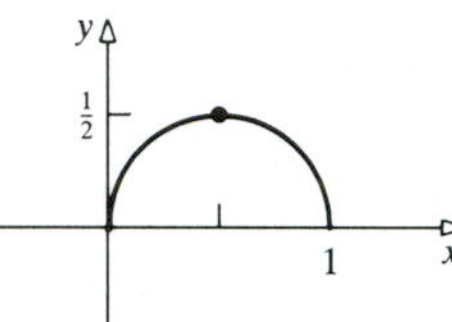

35. Local maximum at $-\frac{2}{3}$;
local minimum at 0;
concave down on $(-\infty, -\frac{1}{3})$;
concave up on $(-\frac{1}{3}, +\infty)$;
no inflection points

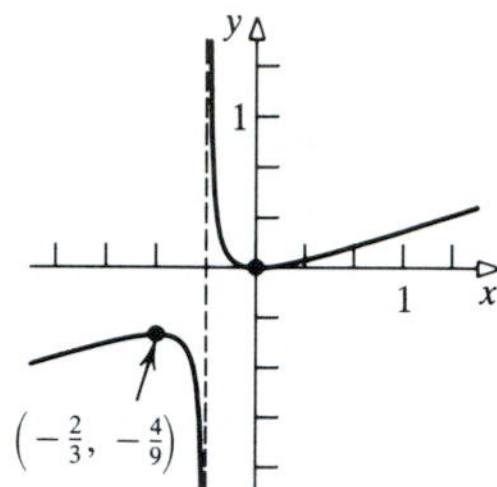

37. Local minima at $k\pi$, k an integer;
local maxima at $(2k + 1)(\pi/2)$;

concave up on $\left(\frac{4k-1}{4}\right)\pi < x < \left(\frac{4k+1}{4}\right)\pi$;

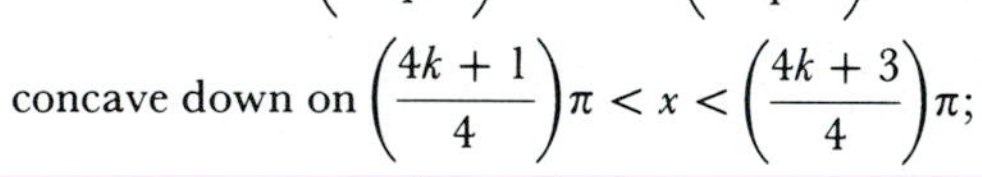

concave down on $\left(\frac{4k+1}{4}\right)\pi < x < \left(\frac{4k+3}{4}\right)\pi$;

inflection points at $\left(\frac{2k+1}{4}\right)\pi$

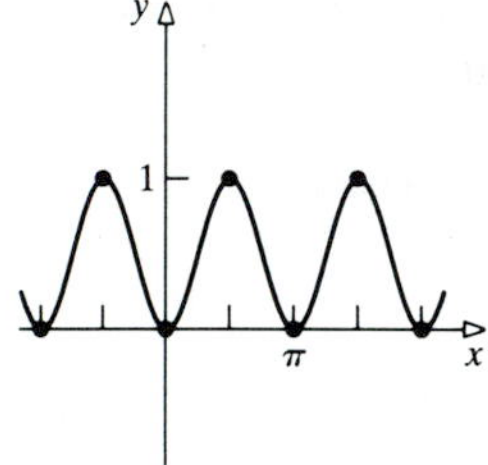

39. Local minimum at $\pi/3$;
local maximum at $5\pi/3$;
concave up on $0 \leq x \leq \pi$;
concave down on $\pi \leq x \leq 2\pi$;
inflection point at π

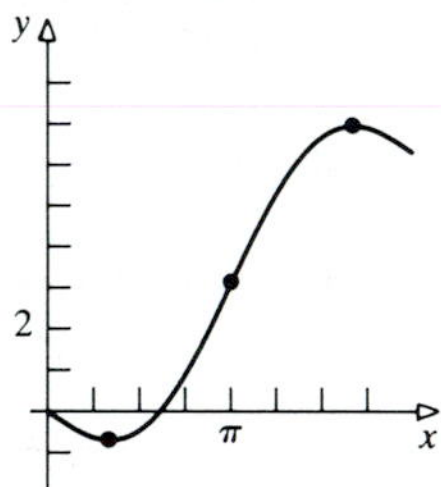

41.

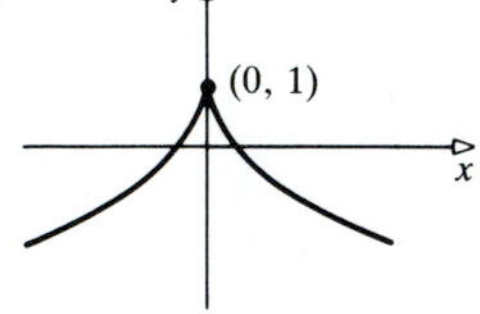

43.

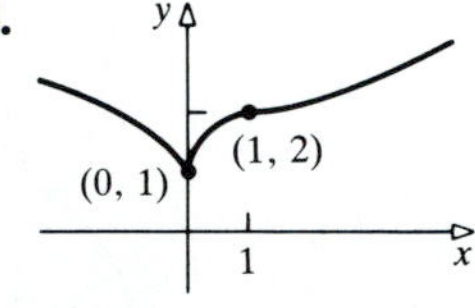

45.

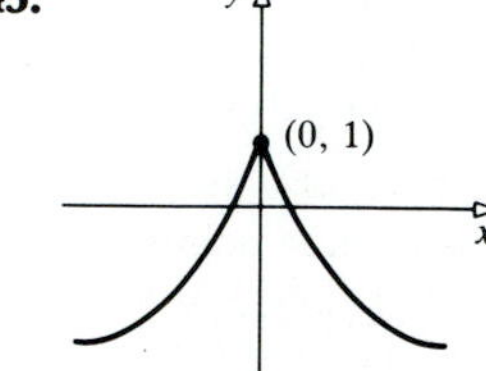

47.

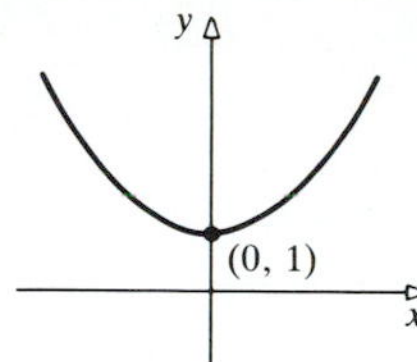

49.

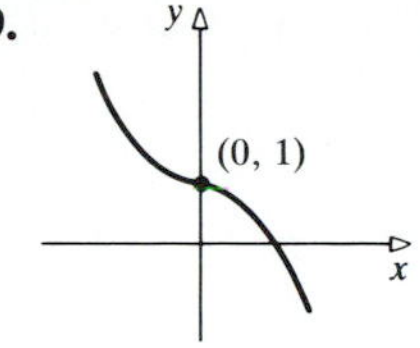

51.

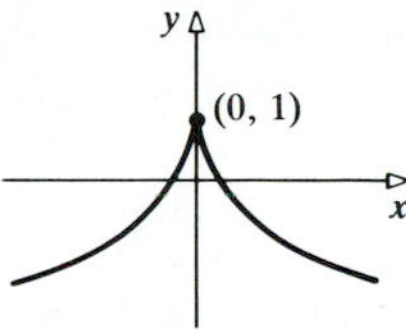

53. $a = -3, b = 9$

55. $f(x) = x^2 - 8x + 21$; f has a local minimum at $(4, 5)$; f is concave up for all x; therefore, $(4, 5)$ is an absolute minimum and $f(x) > 0$ for all x.

57. $f''(x) = 2a \neq 0$ for any x; $a > 0$; $a < 0$

59. Local maximum at $\pi/3$, local minimum at $4\pi/3$; inflection points at $5\pi/6$, $11\pi/6$

61. If P is of degree n, then P' is of degree $n - 1$ and P'' is of degree $n - 2$. A polynomial of degree n has at most n roots.

63. $f''(x) = n(n-1)(x - a)^{n-2}$; if n is even, so is $n - 2$, and so $f''(x) \geq 0$ for all x.

65. Proof is similar to part (a) of theorem (4.6.2) to and including (1). If f'' is negative throughout (a, b), then from theorem (4.5.2), f' is decreasing on (a, b). For $x_1 > c$, $f'(x_1) < f'(c)$ so (1) may be written as $f(x) < f(c) + f'(c)(x - c)$. For $x_1 < c$, $f'(x_1) > f'(c)$, so (1) may be written $f(x) < f(c) + f'(c)(x - c)$. Therefore, $f(x)$ lies below the tangent line to $(c, f(c))$ throughout (a, b) for c arbitrary in (a, b). Thus, f is concave down on $[a, b]$.

Exercise 4.7

1. 1 **3.** 2 **5.** 3 **7.** $-\infty$ **9.** $+\infty$ **11.** $\frac{1}{3}$ **13.** $-\infty$ **15.** $+\infty$ **17.** $+\infty$ **19.** $\sqrt{5}/5$
21. -1 **23.** 0 **25.** $-\infty$ **27.** 0 **29.** $+\infty$ **31.** $y = 3$; $x = 0$ **33.** $y = 0$; $x = 1$ **35.** $y = 3$; $x = -1$
37. $y = 0$; $x = -1$; $x = 1$ **39.** $y = 1$; no vertical **41.** No horizontal; $x = 0$ **43.** No horizontal; no vertical

45. $y = a/c$; $x = -d/c$ **47.** $\lim_{x\to 0}|f'(x)| = \lim_{x\to 0}\left|\frac{1}{3x^{2/3}}\right| = +\infty$ **49.** $\lim_{x\to -4^+}\left|\frac{1}{2\sqrt{x+4}}\right| = +\infty$

51. $\lim_{x\to 3}\left|\frac{2}{3(x-3)^{1/3}}\right| = +\infty$

53. For a rational function, discontinuities occur when the denominator becomes 0. If the numerator is not 0 (i.e., the function is irreducible), then the fraction becomes infinitely large (+ or −) at the discontinuities.

55. For $x^2 + y^2 = 1$, $\frac{dy}{dx} = -\frac{x}{y}$; $y = 0$ for $x = \pm 1$; $\lim_{x\to\pm 1}|f'(x)| = +\infty$. Tangent lines are vertical.

Exercise 4.8

1.
1. Domain: $x \neq 2$
2. y-intercept: $-\frac{1}{2}$
3. No symmetries
4. Horizontal asymptote: $y = 0$; vertical asymptote: $x = 2$
5. No critical number
6. f decreasing on $(-\infty, 2)$, $(2, +\infty)$
7. No local extreme points
8. f concave up on $(2, +\infty)$; f concave down on $(-\infty, 2)$
9. No inflection points
10.

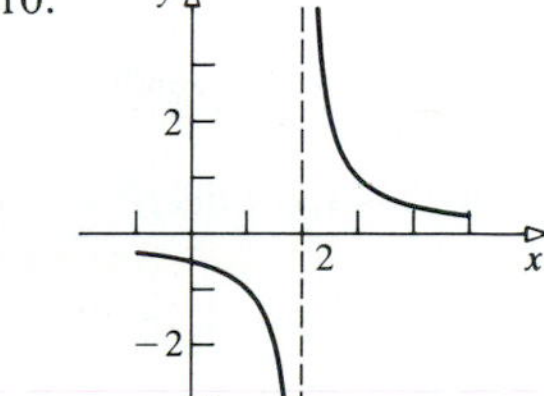

3. 1. Domain: $x \neq 1$
2. Intercept: $(0, 0)$
3. No symmetries
4. Horizontal asymptote: $y = 1$; vertical asymptote: $x = 1$
5. No critical number
6. f decreasing on $(-\infty, 1)$, $(1, +\infty)$
7. No local extreme points
8. f concave up on $(1, +\infty)$; f concave down on $(-\infty, 1)$
9. No inflection points

10. 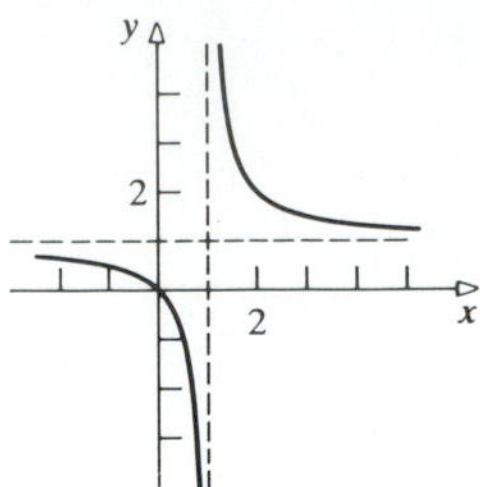

5. 1. Domain: $x \neq \pm 2$
2. No x-intercepts; y-intercept: $(0, -\frac{1}{2})$
3. Symmetric about y-axis
4. Horizontal asymptote: $y = 0$; vertical asymptotes: $x = \pm 2$
5. Critical number: $x = 0$
6. f increasing on $(-\infty, -1)$, $(-1, 0)$; f decreasing on $(0, 1)$, $(1, +\infty)$
7. Local maximum at 0
8. f concave up on $(-\infty, -2)$ and $(2, +\infty)$; f concave down on $(-2, 2)$
9. No inflection points

10. 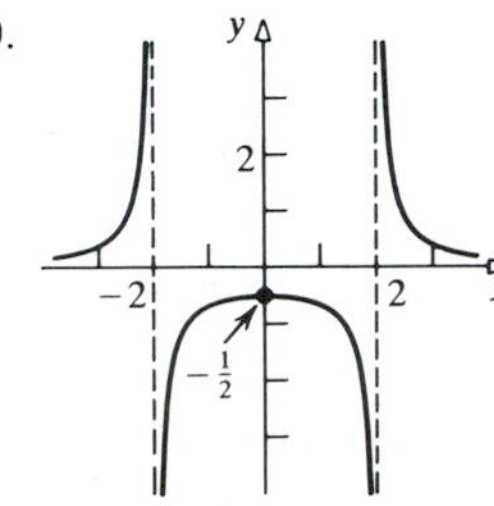

7. 1. Domain: $x \neq -1$
2. x-intercept: $(\frac{1}{2}, 0)$; y-intercept: $(0, -1)$
3. No symmetry
4. Horizontal asymptote: $y = 2$; vertical asymptote: $x = -1$
5. No critical numbers
6. f increasing on $(-\infty, -1)$, $(-1, +\infty)$
7. No local extrema
8. f concave up on $(-\infty, -1)$; f concave down on $(-1, +\infty)$
9. No inflection points

10. 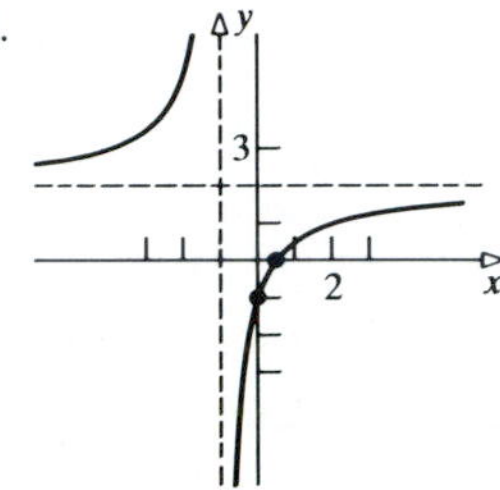

9. 1. Domain: $\mathbb{R}$
2. x-intercept: $(0, 0)$; y-intercept: $(0, 0)$
3. Symmetric about origin
4. Horizontal asymptote: $y = 0$
5. Critical numbers: $x = \pm 1$
6. f decreasing on $(-\infty, -1)$, $(1, +\infty)$; f increasing on $(-1, 1)$
7. Local maximum at 1; local minimum at -1
8. f concave down on $(-\infty, -\sqrt{3}]$ and $[0, \sqrt{3}]$;
   f concave up on $[-\sqrt{3}, 0]$ and $[\sqrt{3}, +\infty)$
9. Inflection points at $0, \pm\sqrt{3}$

10. 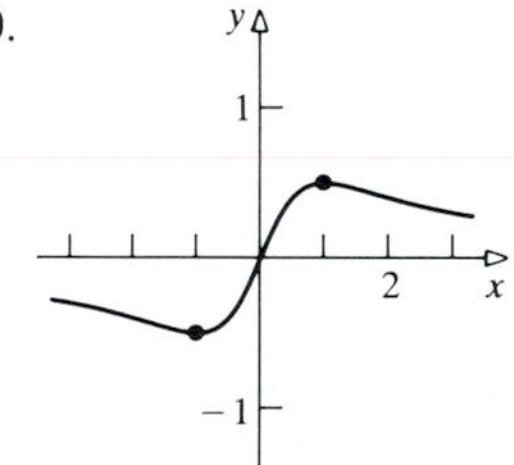

11. 1. Domain: $x \neq \pm 4$
2. No x-intercepts; y-intercept: $(0, -\frac{1}{2})$
3. Symmetric about y-axis
4. Horizontal asymptote: $y = 0$; vertical asymptotes: $x = \pm 4$
5. Critical number: $x = 0$
6. Increasing $(-\infty, -4)$, $(-4, 0)$; decreasing $(0, 4)$, $(4, +\infty)$
7. Local maximum at 0
8. Concave up on $(-\infty, -4)$ and $(4, +\infty)$; concave down on $(-4, 4)$
9. No inflection points

10. 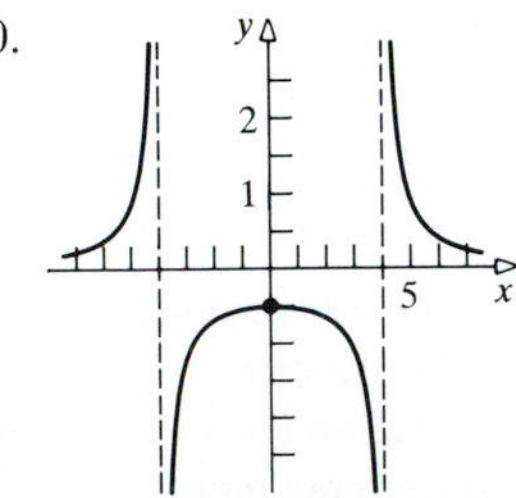

13. 1. Domain: $x \neq 0$
2. No intercepts
3. Symmetry about origin
4. Vertical asymptote: $x = 0$; oblique asymptote: $y = x/2$
5. Critical numbers: $x = \pm 1$
6. f increasing on $(-\infty, -1)$, $(1, +\infty)$; f decreasing on $(-1, 0)$, $(0, 1)$

7. Local maximum at $x = -1$; local minimum at $x = 1$
8. f concave up on $(0, +\infty)$; f concave down on $(-\infty, 0)$
9. No inflection points
10.

15. 1. Domain: $x \neq 0$
2. No intercepts
3. Symmetric with respect to y-axis
4. Vertical asymptote: $x = 0$
5. Critical numbers: $x = \pm 1$
6. f increasing on $(-1, 0)$, $(1, +\infty)$; f decreasing on $(-\infty, -1)$, $(0, 1)$
7. Local minima at $x = -1, 1$
8. f concave up in entire domain
9. No inflection points
10.

17. 1. Domain: $x \neq 2$
2. x-intercepts: $(-\sqrt{3}, 0)$, $(\sqrt{3}, 0)$; y-intercept: $(0, \frac{3}{4})$
3. No symmetries
4. Vertical asymptote: $x = 2$; oblique asymptote: $y = \frac{1}{2}x + 1$
5. Critical numbers: 1, 3
6. f increasing on $(-\infty, 1]$, $[3, +\infty)$; f decreasing on $[1, 2)$, $(2, 3]$
7. Local maximum at $x = 1$; local minimum at $x = 3$
8. f concave up on $(2, +\infty)$; f concave down on $(-\infty, 2)$
9. No inflection points
10.

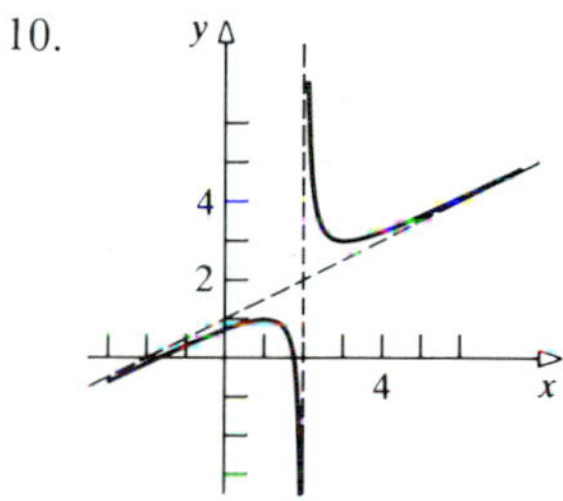

19. 1. Domain: $x \neq 2$
2. y-intercept: $(0, -2)$
3. No symmetries
4. Vertical asymptote: $x = 2$; oblique asymptote: $y = x$
5. Critical numbers: 0, 4
6. f increasing on $(-\infty, 0)$, $(4, +\infty)$; f decreasing on $(0, 4)$
7. Local maximum at $x = 0$; local minimum at $x = 4$
8. f concave up on $(2, +\infty)$; f concave down on $(-\infty, 2)$
9. No inflection points
10.

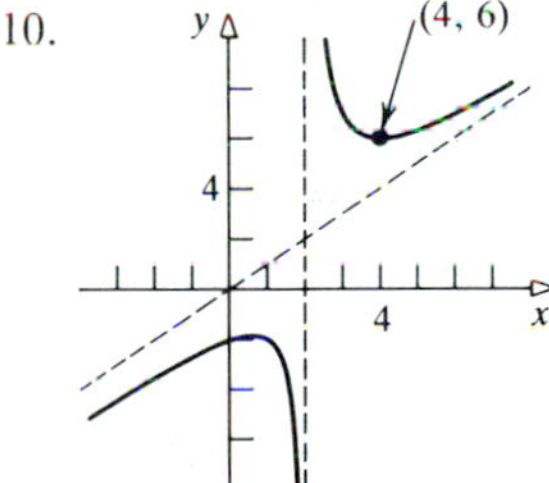

21. 1. Domain: $x \neq 0$
2. No intercepts
3. Symmetric about origin
4. Vertical asymptote: $x = 0$; oblique asymptote: $y = x$
5. Critical numbers: $x = \pm\sqrt{2}$
6. f increasing on $(-\infty, -\sqrt{2}]$, $[\sqrt{2}, +\infty)$; f decreasing on $[-\sqrt{2}, 0)$, $(0, \sqrt{2}]$
7. Local maximum at $x = -\sqrt{2}$; local minimum at $x = \sqrt{2}$
8. f concave up on $(0, +\infty)$; f concave down on $(-\infty, 0)$
9. No inflection points
10.

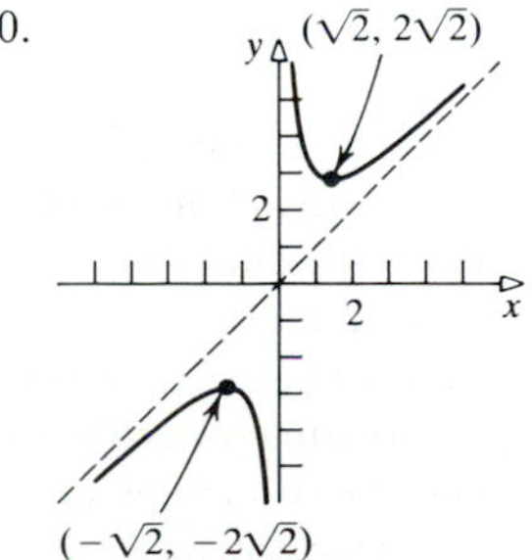

23. 1. Domain: $x \neq 0$
2. No intercepts
3. Symmetric about origin
4. Vertical asymptote: $x = 0$
5. Critical numbers: $x = -1, 1$

6. f increasing on $(-\infty, -1]$, $[1, +\infty)$; f decreasing on $[-1, 0)$, $(0, 1]$
7. Local maximum at -1; local minimum at 1
8. f concave up on $(0, +\infty)$; f concave down on $(-\infty, 0)$
9. No inflection points

10. 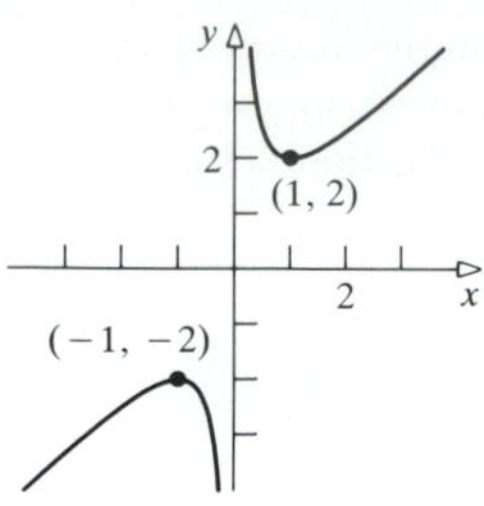

25.
1. Domain: $x \neq -3$
2. x-intercept: $(0, 0)$; y-intercept: $(0, 0)$
3. No symmetry
4. Vertical asymptote: $x = -3$
5. Critical numbers: $x = 0, -6$
6. f increasing on $(-\infty, -6]$, $[0, +\infty)$; f decreasing on $[-6, -3)$, $(-3, 0]$
7. Local maximum at -6; local minimum at 0
8. f concave down on $(-\infty, -3)$; f concave up on $(-3, +\infty)$
9. No inflection points

10. 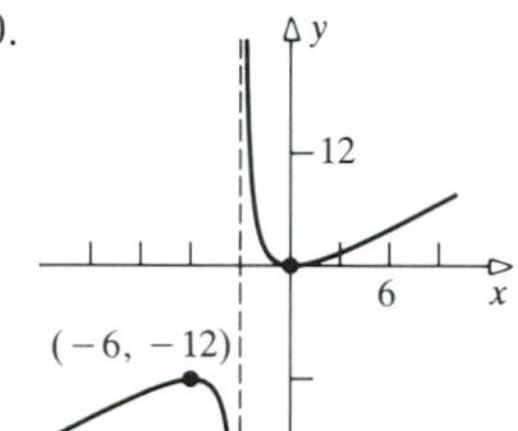

27.
1. Domain: $x \neq 1$
2. x-intercept: $(0, 0)$; y-intercept: $(0, 0)$
3. No symmetry
4. Horizontal asymptote: $y = 0$; vertical asymptote: $x = 1$
5. Critical numbers: $x = -2, 0$
6. f increasing on $[-2, 0]$; f decreasing on $(-\infty, -2]$, $[0, 1)$, $(1, +\infty)$
7. Local maximum at 0; local minimum at -2
8. f concave down on $(-\infty, (-4 - 3\sqrt{2})/2]$ and $[(-4 + 3\sqrt{2})/2, 1)$; f concave up on $[(-4 - 3\sqrt{2})/2, (-4 + 3\sqrt{2})/2]$ and $(1, +\infty)$
9. Inflection points: $x = (-4 \pm 3\sqrt{2})/2$

10. 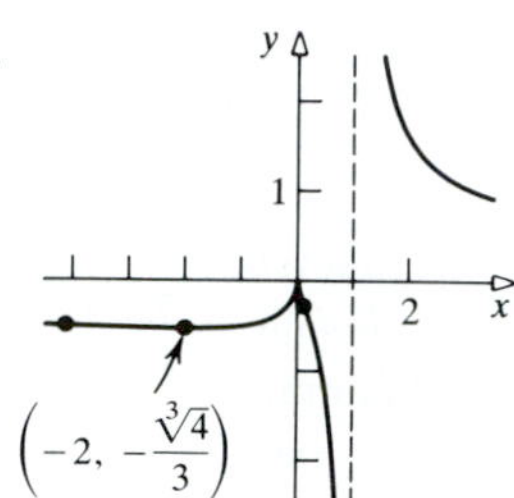

29.
1. Domain: $x \neq 0$
2. No x-intercept; no y-intercept
3. No symmetry
4. Horizontal asymptote: $y = 1$; vertical asymptote: $x = 0$
5. Critical number: $x = -2$
6. f increasing on $[-2, 0)$; f decreasing on $(-\infty, -2]$, $(0, +\infty)$
7. Local minimum at -2
8. f concave down on $(-\infty, -3]$; f concave up on $[-3, 0)$ and $(0, +\infty)$
9. Inflection point at -3

10. 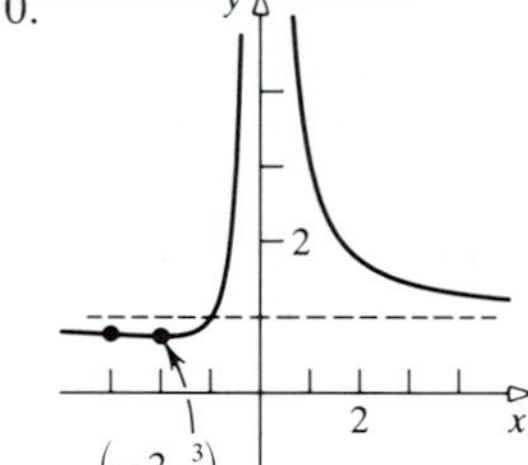

31.
1. Domain: $(-\infty, 3]$
2. x-intercept: $(3, 0)$; y-intercept: $(0, \sqrt{3})$
3. No symmetries
4. No asymptotes
5. Critical number: $x = 3$
6. f decreasing on entire domain
7. No local extreme points
8. f concave down on entire domain
9. No points of inflection

10. 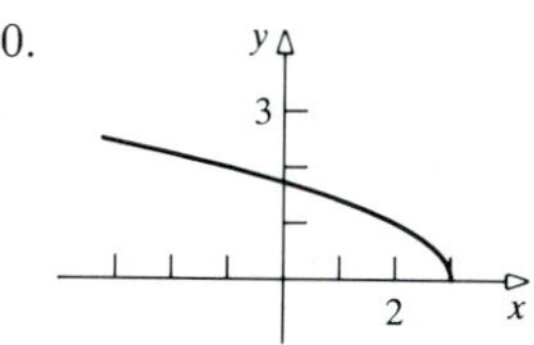

33.
1. Domain: $[0, +\infty)$
2. Intercept: $(0, 0)$
3. No symmetries
4. No asymptotes
5. Critical number: 0

6. f increasing on entire domain
7. No local extreme points
8. f concave down on entire domain
9. No points of inflection
10.

35. 1. Domain: $(-\infty, 2]$
2. x-intercepts: $(0, 0)$, $(2, 0)$; y-intercept: $(0, 0)$
3. No symmetries
4. No asymptotes
5. Critical numbers: $\frac{4}{3}$, 2
6. f increasing on $(-\infty, \frac{4}{3}]$; f decreasing on $[\frac{4}{3}, 2]$
7. Local maximum at $\frac{4}{3}$
8. f concave down on entire domain
9. No inflection point
10.

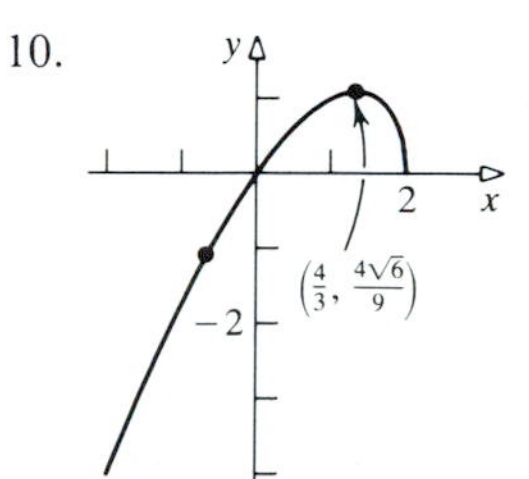

37. 1. Domain: $(-1, +\infty)$
2. Intercept: $(0, 0)$
3. No symmetries
4. Vertical asymptote: $x = -1$
5. Critical number: 0
6. f increasing on $[0, +\infty)$; f decreasing on $(-1, 0]$
7. Local minimum at 0
8. f concave up on entire domain
9. No inflection points
10.

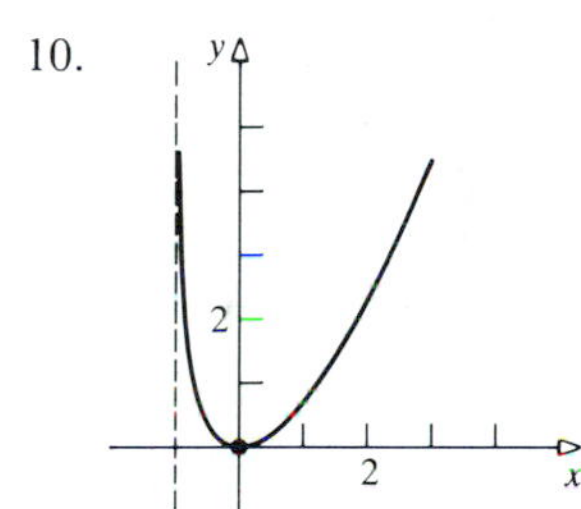

39. 1. Domain: $x \neq -1$, $x \neq 2$
2. y-intercept: $(0, -\frac{1}{2})$
3. No symmetries
4. Horizontal asymptote: $y = 0$; vertical asymptotes: $x = -1, 2$
5. Critical number: $\frac{1}{2}$
6. f increasing on $(-\infty, -1)$, $(-1, \frac{1}{2})$; f decreasing on $(\frac{1}{2}, 2)$, $(2, +\infty)$
7. Local maximum at $\frac{1}{2}$
8. f concave up on $(-\infty, -1)$, $(2, +\infty)$; f concave down on $(-1, 2)$
9. No inflection points
10.

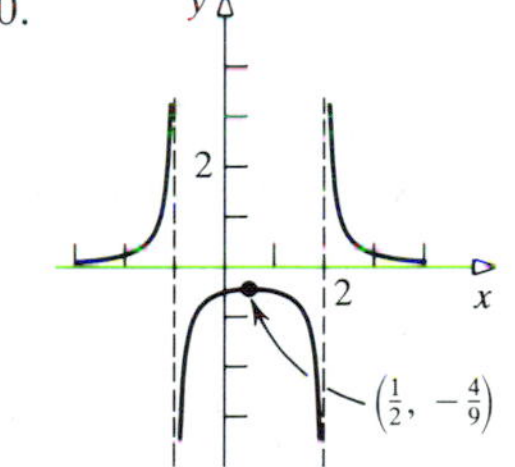

41. 1. Domain: $x \neq 0$, $x \neq -4$
2. No intercepts
3. No symmetries
4. Horizontal asymptote: $y = 0$; vertical asymptotes: $x = 0, -4$
5. Critical number: $-\frac{8}{3}$
6. f increasing on $[-\frac{8}{3}, 0)$; f decreasing on $(-\infty, -4)$, $(-4, -\frac{8}{3}]$, $(0, +\infty)$
7. Local minimum at $-\frac{8}{3}$
8. f concave up on $(-4, 0)$, $(0, +\infty)$; f concave down on $(-\infty, -4)$
9. No inflection points
10.

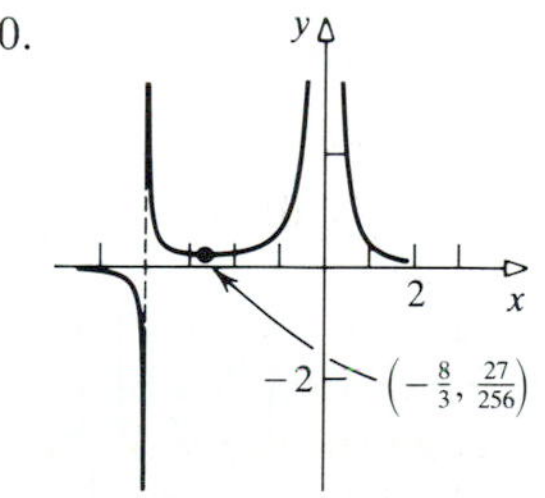

43. 1. Domain: $\mathbb{R}$
2. x-intercepts: $(-1, 0)$, $(1, 0)$; y-intercept: $(0, 1)$
3. Symmetry with respect to y-axis
4. No asymptotes
5. Critical numbers: 0, ± 1
6. f increasing on $[-1, 0]$, $[1, +\infty)$; f decreasing on $(-\infty, -1]$, $[0, 1]$

7. Local minima at ± 1; local maximum at 0
8. f concave down on $[-1, 1]$; f concave up on $(-\infty, -1]$, $[1, +\infty)$
9. No inflection points
10. 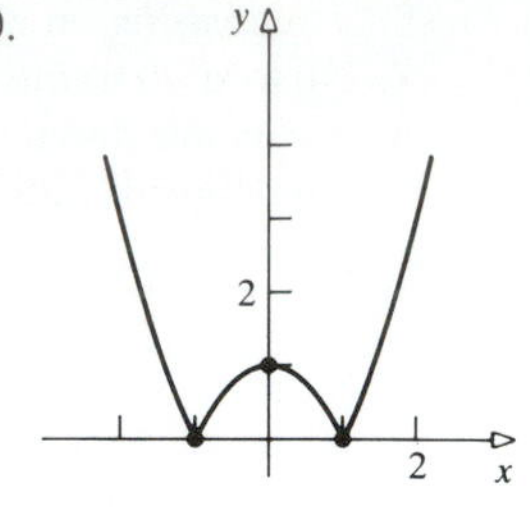

45. 1. Domain: $\mathbb{R}$
2. x-intercepts: $(-8, 0)$, $(-1, 0)$; y-intercept: $(0, 2)$
3. No symmetries
4. No asymptotes
5. Critical numbers: $0, -\frac{27}{8}$
6. f increasing on $[-\frac{27}{8}, +\infty)$; f decreasing on $(-\infty, -\frac{27}{8}]$
7. Local minimum at $-\frac{27}{8}$
8. f concave up on $[-27, 0]$; f concave down on $(-\infty, -27]$, $[0, +\infty)$
9. Inflection points: $-27, 0$
10. 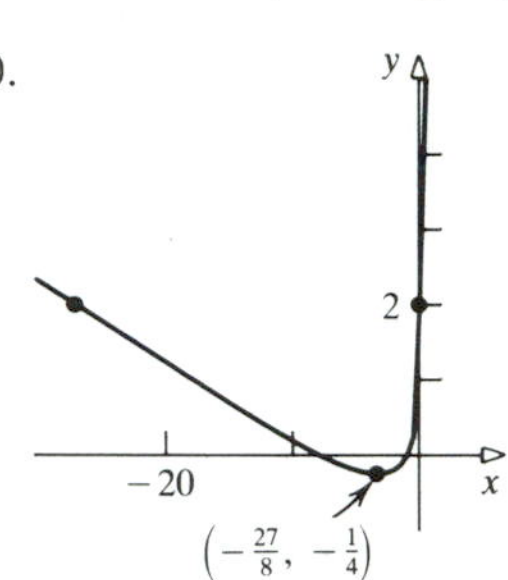

47. 1. Domain: $\mathbb{R}$
2. x-intercepts: $\dfrac{\pi}{4}, \dfrac{5\pi}{4}$, on $[0, 2\pi]$; y-intercept: -1
3. No symmetries
4. No asymptotes
5. Critical numbers: $\dfrac{3\pi}{4}, \dfrac{7\pi}{4}$ on $[0, 2\pi]$
6. f increasing on $\left[0, \dfrac{3\pi}{4}\right], \left[\dfrac{7\pi}{4}, 2\pi\right]$; f decreasing on $\left[\dfrac{3\pi}{4}, \dfrac{7\pi}{4}\right]$
7. Local maximum at $\dfrac{3\pi}{4}$; local minimum $\dfrac{7\pi}{4}$
8. f concave up on $\left[0, \dfrac{\pi}{4}\right], \left[\dfrac{5\pi}{4}, 2\pi\right]$; f concave down on $\left[\dfrac{\pi}{4}, \dfrac{5\pi}{4}\right]$
9. Inflection points: $\dfrac{\pi}{4}, \dfrac{5\pi}{4}$
10. 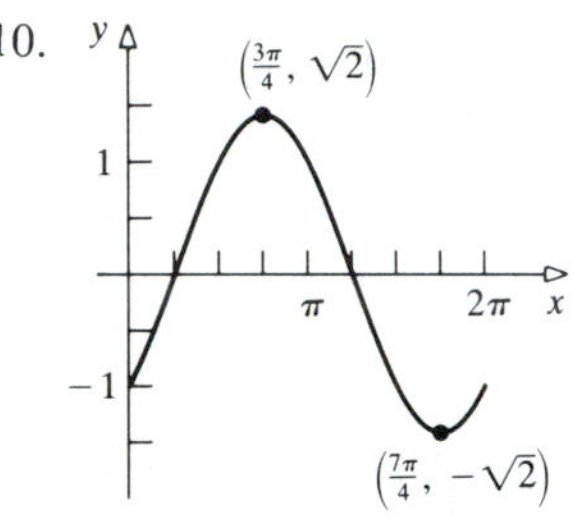

49. 1. Domain: $\mathbb{R}$ 
2. x-intercepts: .905 radians, 5.379 radians; y-intercept: -1
3. Symmetric with respect to the y-axis
4. No asymptotes
5. Critical numbers: $0, \dfrac{2\pi}{3}, \pi, \dfrac{4\pi}{3}, 2\pi$, for one period
6. f increasing on $\left[0, \dfrac{2\pi}{3}\right], \left[\pi, \dfrac{4\pi}{3}\right]$; f decreasing on $\left[\dfrac{2\pi}{3}, \pi\right], \left[\dfrac{4\pi}{3}, 2\pi\right]$ for one period
7. Local maximum at $\dfrac{2\pi}{3}, \dfrac{4\pi}{3}$; local minimum at $0, \pi, 2\pi$
8. 9. See graph
10. 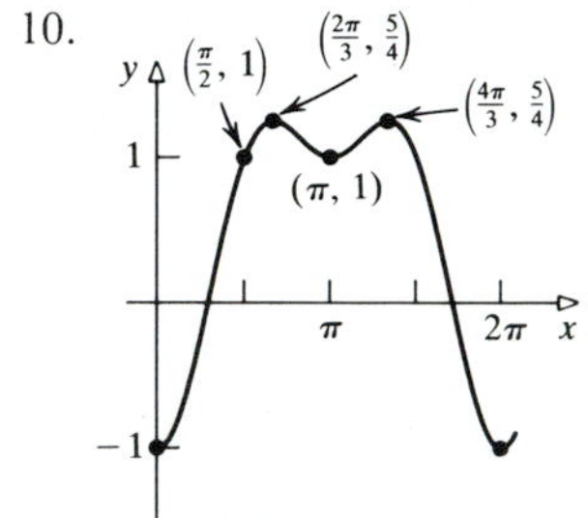

51. 1. Domain: all $x \neq \dfrac{\pi}{2} + k\pi$ k integer
2. x-intercepts: $0, \pi, 2\pi$; y-intercept: 0
3. Symmetric with respect to the origin

4. Vertical asymptotes: $\frac{\pi}{2}, \frac{3\pi}{2}$
5. No critical numbers
6. f is decreasing on $\left[0, \frac{\pi}{2}\right), \left(\frac{\pi}{2}, \frac{3\pi}{2}\right), \left(\frac{3\pi}{2}, 2\pi\right]$
7. No local extremum
8. f concave up on $\left(\frac{\pi}{2}, \pi\right], \left(\frac{3\pi}{2}, 2\pi\right]$; f concave down on $\left[0, \frac{\pi}{2}\right), \left[\pi, \frac{3\pi}{2}\right)$
9. Inflection points: $0, \pi, 2\pi$
10. 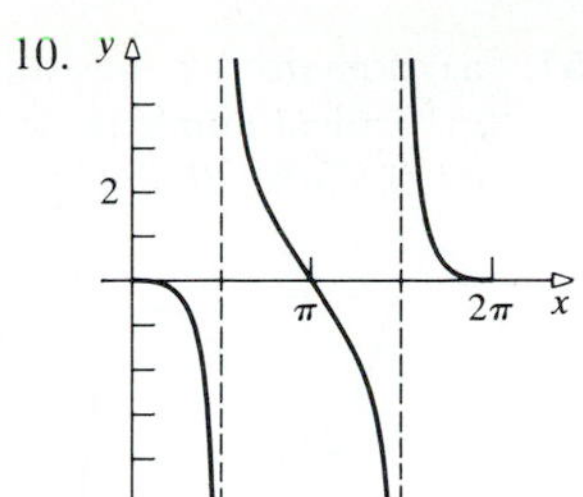

53. 1. Domain: $\mathbb{R}$
2. Intercept: (0, 0)
3. Symmetric about origin
4. No asymptotes
5. Critical numbers: $\frac{\pi}{3} + k\pi, \frac{2\pi}{3} + k\pi$
6. f increasing on $\left[-\frac{\pi}{3} + k\pi, \frac{\pi}{3} + k\pi\right]$; f decreasing on $\left[\frac{\pi}{3} + k\pi, \frac{2\pi}{3} + k\pi\right]$
7. Local maxima at $\frac{\pi}{3} + k\pi$; local minima at $\frac{2\pi}{3} + k\pi$
8. f concave up on $\left[\frac{\pi}{2} + k\pi, \pi + k\pi\right]$; f concave down on $\left[k\pi, \frac{\pi}{2} + k\pi\right]$
9. Inflection points at $\frac{k\pi}{2}$
10. 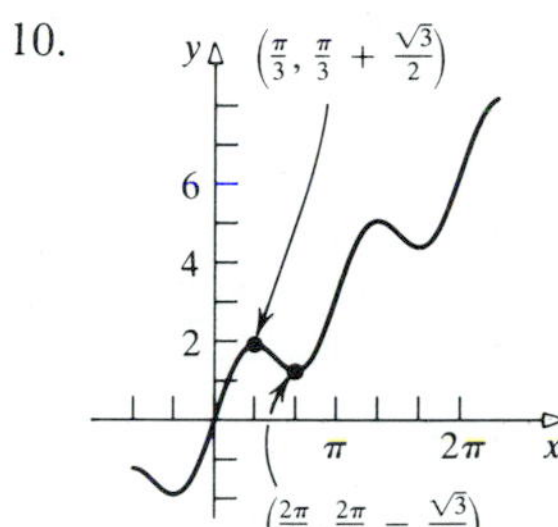

55. 1. Domain: $\mathbb{R}$
2. x-intercept: ≈ 1.0299; y-intercept: $(0, -2)$
3. No symmetries
4. No asymptotes
5. Critical numbers: $-\frac{\pi}{6} + 2k\pi, \frac{7\pi}{6} + 2k\pi$, k integer
6. f increasing on $\left[-\frac{\pi}{6} + 2k\pi, \frac{7\pi}{6} + 2k\pi\right]$; f decreasing on $\left[\frac{7\pi}{6} + 2k\pi, \frac{11\pi}{6} + 2k\pi\right]$
7. Local maxima at $\frac{7\pi}{6} + 2k\pi$; local minima at $\frac{11\pi}{6} + 2k\pi$
8. f concave up on $\left[-\frac{\pi}{2} + 2k\pi, \frac{\pi}{2} + 2k\pi\right]$; f concave down on $\left[\frac{\pi}{2} + 2k\pi, \frac{3\pi}{2} + 2k\pi\right]$
9. Inflection points: $(2k + 1)\frac{\pi}{2}$
10. 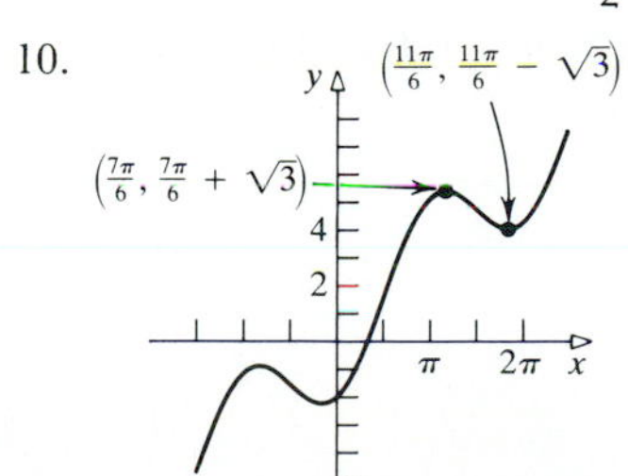

57. (a) domain: $(-\infty, 6]$ (b) intercepts: (0, 0) and (6, 0)
(c) critical numbers: 4; vertical tangent at 6, horizontal tangent at 4
(d) No inflection point at $x = 8$ (8 is not in domain of f); f is concave down on the domain

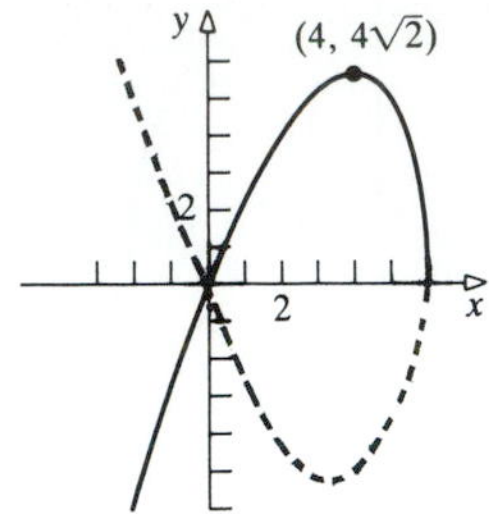

59. 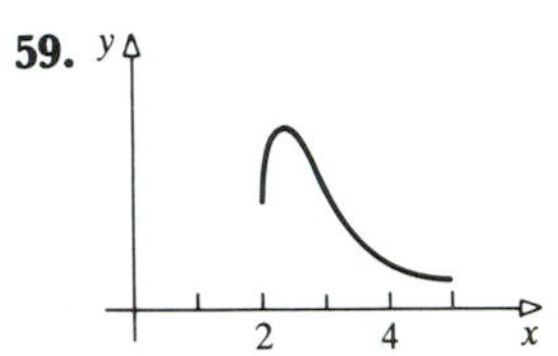

61. 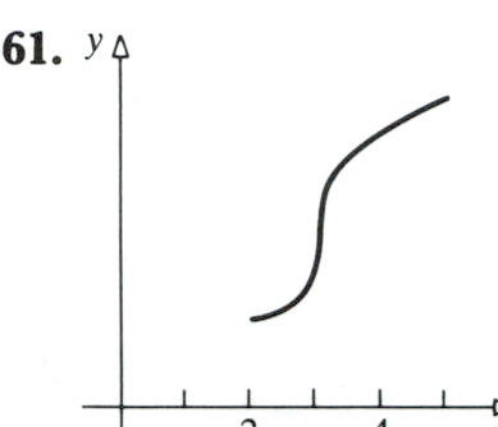

63. 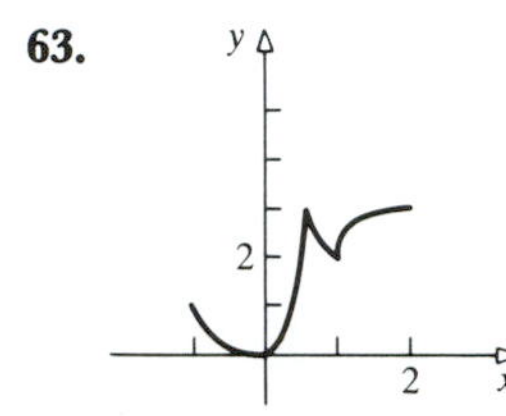

Exercise 4.9

1. 1,125,000 m^2 **3.** $L/4$ by $L/4$ m **5.** 16,666,667 m^2 **7.** Base $= L/3$; legs $= L/3$

9. Height = 4 cm; base = 16 by 16 cm **11.** $10\sqrt[3]{2}$ by $10\sqrt[3]{2}$ by $10\sqrt[3]{2}$ cm

13. Height $= 50/\sqrt[3]{5\pi}$ cm; radius $= 4\sqrt[3]{25/\pi}$ cm **15.** \$21.00 **17.** (1, 1) **19.** $\approx(1.784, 0.818)$

21. $dF/d\theta = -mc(c \cos\theta - \sin\theta)/(c \sin\theta + \cos\theta)^2 = 0$ when $\tan\theta = c$

23. (a) 40 mph (b) 51.64 mph (c) 53.39 mph **25.** 1.98 km from box **27.** $\sqrt{p^2 + 4qr}$

29. Width $= 2\sqrt{3}$; depth $= 4\sqrt{15}/3$ **31.** $5\sqrt{5}$ m **33.** $\pi/2$ **35.** Base-to-height ratio is 2 to 1

37. Diameter-to-height ratio is $2/(1 + \pi)$ **39.** $16\sqrt{\pi}$ cm; $16\sqrt{\pi + 4}$ cm

41. Circle of length 35 cm; circle of length $35\pi/(4 + \pi)$ cm **43.** Diameter $= 20/(\pi + 4)$ m; height $= 10/(\pi + 4)$ m

45. $y = -\left(\dfrac{b}{a}\right)^{1/3} x + b^{1/3}(a^{2/3} + b^{2/3})$

Exercise 4.10

1. $\frac{2}{3}x^6 + C$ **3.** $2x^{5/2} + C$ **5.** $(-2/x) + C$ **7.** $\frac{2}{3}x^{3/2} + C$ **9.** $x^4 - x^3 + x + C$ **11.** $3x^3 - 6x^2 + 4x + C$

13. $2x^{3/2} - 4x^{1/2} + C$ **15.** $\frac{1}{6}x^2 + \frac{7}{3}x + C$ **17.** $x^2 - 3\sin x + C$ **19.** $y = x^3 - x^2 + x + 1$

21. $v = t^3 - t^2 + t + 4$ **23.** $s = \frac{1}{4}t^4 - (1/t) + \frac{11}{4}$ **25.** $y = \frac{1}{2}x^2 + 2\cos x - 2$ **27.** $s = -16t^2 + 128t$

29. $s = \frac{1}{2}t^3 + 18t + 2$ **31.** $F(x) = x\cos x + \sin x + 1$ **33.** $10\sqrt{3}$ m/sec **35.** 12.25 m **37.** 13.13 m/sec

39. 8 g-cm/sec^2 **41.** 13.84 m/sec

Review Exercises

1. $\frac{4}{5}$ cm^2/min **3.** $\dfrac{1150}{\sqrt{13}}$ mph **5.** 4%

7. (a) $dN/dt = kPN - kN^2$ is the differential equation.

(b) $\dfrac{d^2N}{dt^2} = kP\dfrac{dN}{dt} - 2kN\dfrac{dN}{dt} = k(P - 2N)\dfrac{dN}{dt} = k(P - 2N)\cdot kN(P - N) = k^2N(P - 2N)(P - N)$. Therefore, $d^2N/dt^2 = 0$ when $N = 0$, $N = P$, or $N = \frac{1}{2}P$. For $t > 0$, $N(t) > 0$ and $P - N(t) > 0$, and so we have $N''(t) > 0$ if $N(t) < \frac{1}{2}P$ and $N''(t) < 0$ if $N(t) > \frac{1}{2}P$. Hence, there is an inflection point when half the town is infected.

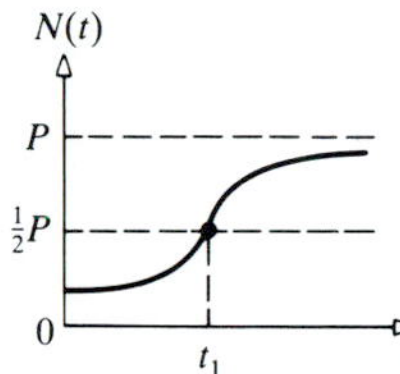

9. $5 + \frac{4}{75} \approx 5.053$ **11.** $\{0, 1\}$

13. f is continuous on $[0, 2]$ and differentiable on $(0, 2)$, and $f(0) = f(2) = 0$. The tangent is horizontal at $x = \frac{2}{3}$.

15. The mean value theorem applies, and $c = \frac{9}{4}$.

17. (a) Domain: $\mathbb{R}$

(b) x-intercepts: $(-2, 0)$, $(0, 0)$, $(1, 0)$; y-intercepts: $(0, 0)$

(c) No symmetries

(d) No asymptotes

(e) Critical numbers: $\dfrac{-(1 + \sqrt{7})}{3}, \dfrac{-(1 - \sqrt{7})}{3}$

(f) f increasing on $\left[\dfrac{-(1 + \sqrt{7})}{3}, \dfrac{-(1 - \sqrt{7})}{3}\right]$; f decreasing on $\left(-\infty, \dfrac{-(1 + \sqrt{7})}{3}\right] \cup \left[\dfrac{-(1 - \sqrt{7})}{3}, +\infty\right)$

(g) Local maximum at $\dfrac{-(1 - \sqrt{7})}{3}$; local minimum at $\dfrac{-(1 + \sqrt{7})}{3}$

(h) f concave up on $(-\infty, -\frac{1}{3}]$; f concave down on $[-\frac{1}{3}, +\infty)$.

(i) Inflection point at $x = -\frac{1}{3}$

(j)

19.

21. $f(1) = f(3) = 0$. f is continuous on $[1, 3]$ and differentiable on $(1, 3)$, since f is a polynomial. Thus, by Rolle's theorem, there is a $c \in (1, 3)$ such that $f'(c) = 0$. $f'(x) = (x^2 - 4x + 3)(2x + 1) + (2x - 4)(x^2 + x + 1)$, a polynomial. $f'(1) = -6$; $f'(3) = 26$. Since $-6 < 0 < 26$ and f' is continuous, the intermediate value theorem guarantees that there exists a $c \in (1, 3)$ such that $f'(c) = 0$.

23. f is concave up on $(-\infty, +\infty)$. **25.** (d) **27.** (b)

29.

x	y	f'	f''
-3	2		0
-1	5	0	
2	-4	0	
6	-1		0

$\lim_{x\to-\infty} f = 1$

$\lim_{x\to+\infty} f = 0$

$\lim_{x\to 0^-} f = -\infty$

$\lim_{x\to 0^+} f = +\infty$

$f'' > 0$ if $x \in (-\infty, -3) \cup (0, 6)$

$f'' < 0$ if $x \in (-3, 0) \cup (6, +\infty)$

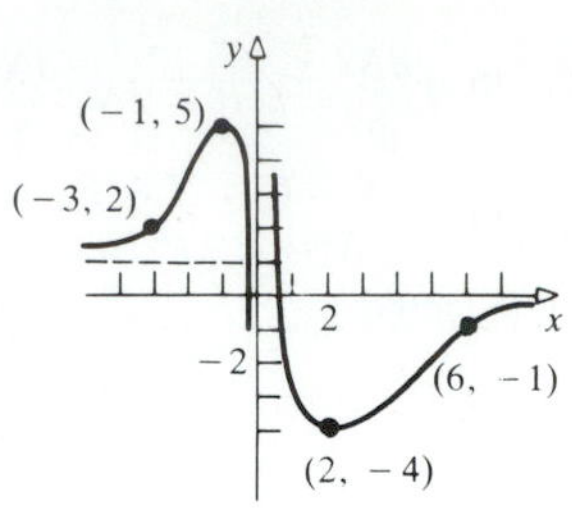

31. $\frac{105}{16}$ **33.** Horizontal asymptote: $y = 0$
Vertical asymptotes: $x = -2$, $x = 2$

35. Domain: $[3, +\infty)$; range: $[0, +\infty)$
Intercept: $(3, 0)$
Concave up on $[4, +\infty)$; concave down on $[3, 4]$
Inflection point at $(4, 4)$

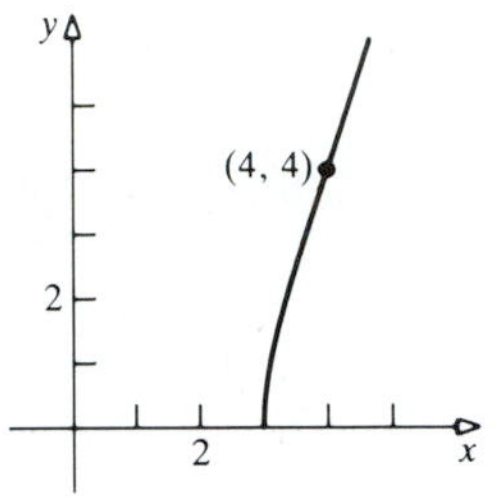

37. $(2, 2)$ **39.** $x^4 - 3x^3 + 5x^2 - 3x + C$ **41.** 6.2 cm/sec **43.** 1075

45. $x = 0$ is the only critical number.
If n is odd, f is increasing on its entire domain.
If n is even, f is increasing on one side of $x = 0$ and decreasing on the other side, depending on the sign of $ad - bc$.

Challenge Exercises

1. $s(3) = 15.7$ m, and the ball is moving downward. The shadow of the ball is moving downward on the wall at 49 m/sec, and the ball is 15.7 m above the ground.

3. $-9\sqrt{3}$ m^3/min **5.** $13\frac{1}{5}$ ft/sec

7. (a) If $f(x) = \dfrac{ax^{2n} + b}{cx^n + d}$, then $\operatorname{dom} f = \left\{x\colon x^n \neq -\dfrac{d}{c}\right\}$, and

$$f'(x) = \frac{(cx^n + d)(2anx^{2n-1}) - (ax^{2n} + b)(cnx^{n-1})}{(cx^n + d)^2} = \frac{acnx^{3n-1} + 2adnx^{2n-1} - bcnx^{n-1}}{(cx^n + d)^2}$$

or $f'(x) = \dfrac{nx^{n-1}[acx^{2n} + 2adx^n - bc]}{(cx^n + d)^2}$. The only critical numbers are when $x = 0$ or when $acx^{2n} + 2adx^n - bc = 0$, which is a quadratic in x^n. There are two possibilities, since

$$x^n = \frac{-ad \pm \sqrt{a^2d^2 + abc^2}}{ac} = \frac{-d \pm \sqrt{d^2 + bc^2/a}}{c}:$$

(i) If n is odd, then the equation $acx^{2n} + 2adx^n - bc = 0$ has two solutions for x^n—say, r_1 and r_2—and hence two solutions for x, $\sqrt[n]{r_1}$ and $\sqrt[n]{r_2}$.

(ii) If n is even, then the equation has two solutions for x^n—say, r_1 and r_2. If r_1 and r_2 are both positive, then x can be $\pm\sqrt{r_1}$ and $\pm\sqrt{r_2}$. Hence, there are at most five critical numbers.

(b) Consider the function $f(x) = \dfrac{3x^4 - 3}{3x^2 - 5}$. Here $a = 3$, $b = -3$, $c = 3$, $d = -5$, and $n = 2$. Hence, there are critical numbers for $x^2 = \dfrac{5 \pm \sqrt{25 - 9}}{3} = \dfrac{5 \pm 4}{3}$, or 1/3 and 3. Thus, $x = \pm\sqrt{3}$ and $\pm\sqrt{3}/3$. The critical numbers are $x = 0$, $\pm\sqrt{3}/3$, and $\pm\sqrt{3}$.

9. (a) Car A (b) At $t = 0$ and $t = (\sqrt{33} - 1)/4 \approx 1.186$ (c) $t = \frac{2}{3}$ (d) $t = \frac{2}{3}$

CHAPTER 5

Exercise 5.1

1. $1 + \sqrt[3]{2} + \sqrt[3]{3} + \sqrt[3]{4} + \sqrt[3]{5} + \sqrt[3]{6} + \sqrt[3]{7}$ **3.** $2 + 4 + 8 + 16 + 32$ **5.** $\frac{1}{1} + \frac{1}{3} + \frac{1}{5} + \frac{1}{7} + \frac{1}{9} + \frac{1}{11}$

7. $0 + 3 + 8 + 15 + \cdots + (n^2 - 1)$ **9.** $-\frac{1}{2} + \frac{1}{3} - \frac{1}{4} + \frac{1}{5} - \cdots + (-1)^n \dfrac{1}{n+1}$

11. (a) $\sum_{i=0}^{n-1} (i+1)^2$ (b) $\sum_{i=1}^{n} i^2$ **13.** (a) $\sum_{i=0}^{n} 2^i$ (b) $\sum_{i=1}^{n+1} 2^{i-1}$ **15.** (a) $\sum_{i=0}^{n} \dfrac{1}{2^i}$ (b) $\sum_{i=1}^{n+1} \dfrac{1}{2^{i-1}}$

17. (a) $\sum_{i=0}^{n-1} (i+1)(i+2)$ (b) $\sum_{i=1}^{n} i(i+1)$ **19.** (a) $\sum_{i=0}^{n} x^{2i}$ (b) $\sum_{i=1}^{n+1} x^{2(i-1)}$ **21.** $n(n+1)(n+2)$

23. $n^2(n+3)(n-1)/4$ **25.** $(n+1)^2/4n^2$ **27.** $(1 - n^2)/6n^2$

29. $\sum_{i=1}^{n} [(i+1)^2 - i^2] = \sum_{i=1}^{n} (2i+1) = 2\sum_{i=1}^{n} i + \sum_{i=1}^{n} 1 = \dfrac{2n(n+1)}{2} + n = n^2 + 2n = (n+1)^2 - 1$

31. $\sum_{i=1}^{n} \left(\dfrac{1}{i} - \dfrac{1}{i+1}\right) = \left(1 - \dfrac{1}{2}\right) + \left(\dfrac{1}{2} - \dfrac{1}{3}\right) + \cdots + \left(\dfrac{1}{n} - \dfrac{1}{n+1}\right) = 1 - \dfrac{1}{n+1}$ **33.** $\frac{1}{4}$ **35.** 1

37. $\sum_{i=1}^{n} i^3 - \sum_{i=1}^{n} (i-1)^3 = 3\sum_{i=1}^{n} i^2 - 3\sum_{i=1}^{n} i + \sum_{i=1}^{n} 1$

$$\frac{n^2(n+1)^2}{4} - \frac{(n-1)^2(n^2)}{4} = 3\sum_{i=1}^{n} i^2 - \frac{3n(n+1)}{2} + n$$

Regroup to get $\sum_{i=1}^{n} i^2 = \dfrac{n(n+1)(2n+1)}{6}$.

39.
$$\sum_{i=1}^{n} \cos ix = \frac{1}{2\sin\frac{1}{2}x}\left(\sum_{i=1}^{n}\left[\sin\left(i + \frac{1}{2}\right)x - \sin\left(i - \frac{1}{2}\right)x\right]\right)$$
$$= \frac{1}{2\sin\frac{1}{2}x}\left[\sin\left(n + \frac{1}{2}\right)x - \sin\frac{1}{2}x\right] \quad \text{(using result of Problem 38)}$$
$$= \frac{1}{2\sin\frac{1}{2}x}\left[\sin\left(\frac{n}{2} - \frac{n+1}{2}\right)x + \sin\left(\frac{n}{2} + \frac{n+1}{2}\right)x\right]$$

For the second part, use $\sin a \cos b = \frac{1}{2}[\sin(a - b) + \sin(a + b)]$

$$\sum_{i=1}^{n} \cos ix = \frac{1}{2\sin\frac{1}{2}x}\left[\sin\frac{nx}{2}\cos\frac{(n+1)x}{2}\right]$$

41. $\dfrac{r^{n+1} - r}{r - 1}$

Exercise 5.2

1. $[1, 2]$, $[2, 3]$, $[3, 4]$ **3.** $[-1, -\frac{1}{2}]$, $[-\frac{1}{2}, 0]$, $[0, \frac{1}{2}]$, $[\frac{1}{2}, 1]$, $[1, \frac{3}{2}]$, $[\frac{3}{2}, 2]$, $[2, \frac{5}{2}]$, $[\frac{5}{2}, 3]$, $[3, \frac{7}{2}]$, $[\frac{7}{2}, 4]$

5. $s_4 = 48$, $S_4 = 56$; $s_8 = 50$, $S_8 = 54$ **7.** $s_4 = \frac{7}{4}$, $S_4 = \frac{15}{4}$; $s_8 = \frac{35}{16}$, $S_8 = \frac{51}{16}$

9. $s_4 = \frac{77}{60}$, $S_4 = \frac{25}{12}$; $s_8 = \frac{3601}{2520}$, $S_8 = \frac{4609}{2520}$

11. (a)

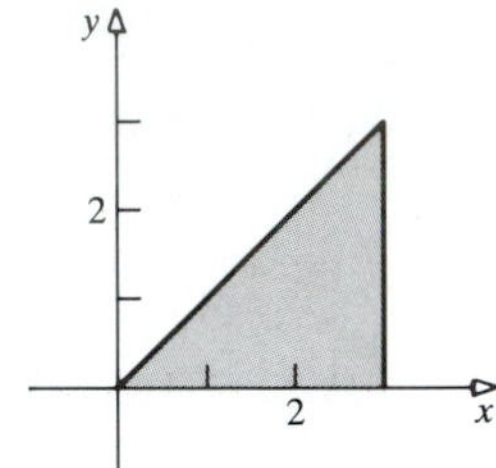

(b) $\left[0, \frac{3}{n}\right], \left[\frac{3}{n}, \frac{6}{n}\right], \left[\frac{6}{n}, \frac{9}{n}\right], \ldots, \left[\frac{3(n-1)}{n}, 3\right]$

(c) $s_n = \sum_{i=1}^{n} \left[(i-1)\frac{3}{n}\right]\left(\frac{3}{n}\right)$, since $c_i = (i-1)\frac{3}{n}$ and $\Delta x = \frac{3}{n}$

(d) $S_n = \sum_{i=1}^{n} \left[i\left(\frac{3}{n}\right)\right]\left(\frac{3}{n}\right)$, since $c_i = i\left(\frac{3}{n}\right)$ and $\Delta x = \frac{3}{n}$

(e)

n	5	10	50	100
s_n	$\frac{36}{10}$	$\frac{81}{20}$	$\frac{441}{100}$	$\frac{891}{200}$
S_n	$\frac{54}{10}$	$\frac{99}{20}$	$\frac{459}{100}$	$\frac{909}{200}$

(f) $$\lim_{n\to+\infty} s_n = \lim_{n\to+\infty} \sum_{i=1}^{n} \left[(i-1)\frac{3}{n}\right]\left(\frac{3}{n}\right)$$
$$= \lim_{n\to+\infty} \frac{9}{n^2}\sum_{i=1}^{n}(i-1) = \lim_{n\to+\infty} \frac{9}{n^2}\,\frac{n(n-1)}{2} = \frac{9}{2}$$
$$\lim_{n\to+\infty} S_n = \lim_{n\to+\infty} \sum_{i=1}^{n} \left[i\left(\frac{3}{n}\right)\right]\left(\frac{3}{n}\right) = \lim_{n\to+\infty} \frac{9}{n^2}\sum_{i=1}^{n} i = \lim_{n\to+\infty} \frac{9}{n^2}\,\frac{n(n+1)}{2} = \frac{9}{2}$$

13. 150 **15.** 20 **17.** $\frac{8}{3}$ **19.** $\frac{16}{3}$ **21.** 4 **23.** 10 **25.** 20 **27.** $\frac{8}{3}$ **29.** $\frac{16}{3}$ **31.** 4 **33.** 10

35. Let the vertices of the right triangle be $(0, 0)$, $(0, H)$, and $(B, 0)$. The equation of the line determined by $(0, H)$ and $(B, 0)$ is then $y = -(H/B)x + H$. Choose $\Delta x = B/n$ and right-hand endpoints; that is, $x_i = i\Delta x$. Then,

$$A = \lim_{n\to+\infty} \sum_{i=1}^{n} f(x_i)\Delta x = \lim_{n\to+\infty} \sum_{i=1}^{n} \left[-\frac{H}{B}(i\Delta x) + H\right]\Delta x = \lim_{n\to+\infty} \left[-\frac{BH}{2}\left(\frac{n+1}{n}\right) + BH\right] = \frac{1}{2}BH$$

37. $m_i = \frac{1}{2}\left[\frac{3(i-1)}{n} + \frac{3i}{n}\right] = \frac{6i-3}{2n}$; $\lim_{n\to+\infty} \sum_{i=1}^{n} f(m_i)\Delta x = \lim_{n\to+\infty} \sum_{i=1}^{n} \frac{(6i-3)^2}{4n^2}\left(\frac{3}{n}\right) = 9$

39. n equal subintervals give $\Delta x = \frac{b-a}{n}$, and the ith subinterval is $\left[a + (i-1)\left(\frac{b-a}{n}\right), a + i\left(\frac{b-a}{n}\right)\right]$.

$$s_n = \sum_{i=1}^{n} \left[a + (i-1)\left(\frac{b-a}{n}\right)\right]\left(\frac{b-a}{n}\right) = \frac{b-a}{n}\,a\sum_{i=1}^{n} 1 + \left(\frac{b-a}{n}\right)^2 \sum_{i=1}^{n}(i-1) = \frac{b^2-a^2}{2} - \frac{(b-a)^2}{2n}$$

Similarly, $S_n = \sum_{i=1}^{n} \left[a + i\left(\frac{b-a}{n}\right)\right]\left(\frac{b-a}{n}\right) = \frac{b^2-a^2}{2} + \frac{(b-a)^2}{2n}$. Clearly, $s_n < \frac{b^2-a^2}{2} < S_n$.

Exercise 5.3

1. 2 **3.** $\frac{11}{4}$ **5.** $-\frac{7}{2}$ **7.** $-\frac{128}{3}$ **9.** $\frac{21}{2}$ **11.** $(1/n^{3/2})\sum_{i=1}^{n}\sqrt{i}$ **13.** $\sum_{i=1}^{n}[1/(2n+i)]$

15. $6n\sum_{i=1}^{n}[1/(n+3i)^2]$

17. $\int_0^4 f(x)\,dx$ equals the sum of four rectangles with bases the intervals [0, 1], [1, 2], [2, 3], [3, 4]; namely, $(0+1+2+3)\cdot 1 = 6$

19. $\int_a^b k\,dx = \lim_{n\to+\infty}\sum_{i=1}^{n} k\frac{b-a}{n} = k(b-a)\lim_{n\to+\infty}\frac{1}{n}\sum_{i=1}^{n} 1 = k(b-a)$

Exercise 5.4

1. 5 **3.** $\frac{255}{8}$ **5.** $\frac{7}{2}$ **7.** 1 **9.** $\frac{2}{3}$ **11.** $\frac{1}{2}$ **13.** $-\frac{1}{15}$ **15.** $\frac{5}{3}$ **17.** -3 **19.** $\frac{12}{7}$ **21.** $\frac{20}{3}$

23. $(a/5)+b$ **25.** $[(b^3-a^3)+6(b^2-a^2)+12(b-a)]/3$ **27.** 9 **29.** $\frac{13}{3}$ **31.** $\frac{7}{3}$ **33.** $-\frac{2}{3}$ **35.** $\frac{43}{6}$

37. $\frac{124}{3}$ **39.** $\sqrt{x^2+1}$ **41.** $(3+t^2)^{3/2}$ **43.** $f(x)$ **45.** $6x^2\sqrt{4x^6+1}$ **47.** $5x^4\sec x^5$ **49.** $-3x^2(x^6-5)^{10}$

51. 160 **53.** $\sqrt{3}$ **55.** $2\sqrt{3}/3$ **57.** 12; 32 **59.** $\pi\sqrt{2}/8$; $\pi/4$ **61.** 1; $\sqrt{2}$

63. 47.5; 35.5; 41.5; $\int_1^5 x^2\,dx = \frac{124}{3}$ **65.** 9 **67.** $-(x-1)^2$

69. $F'(x) = G'(x) = f(x)$; therefore, $F(x) = G(x) + C$. If $x = d$, then $F(d) = \int_c^d f(t)\,dt = G(d) + C = \int_d^d f(t)\,dt + C = C$; therefore, $F(x) - G(x) = \int_c^d f(t)\,dt$

71. $\frac{1}{2}[f(x)]^2$ is an antiderivative of $f(x)f'(x)$. Thus, $\int_a^b f(x)f'(x)\,dx = \frac{1}{2}[f(x)]^2\Big|_a^b$

Exercise 5.5

1. $\int_3^{11} f(x)\,dx - \int_7^{11} f(x)\,dx = \int_3^{11} f(x)\,dx + \int_{11}^{7} f(x)\,dx = \int_3^7 f(x)\,dx$

3. $\int_0^4 f(x)\,dx - \int_6^4 f(x)\,dx = \int_0^4 f(x)\,dx + \int_4^6 f(x)\,dx = \int_0^6 f(x)\,dx$ **5.** 3 **7.** 31 **9.** 17

11. On [0, 1], $x \ge x^3$. Apply (6).

13. Let $F' = f$. Then kF is an antiderivative of kf. Thus, $\int_a^b kf(x)\,dx = kF(x)\Big|_a^b = kF(b) - kF(a) = kF(x)\Big|_a^b = k\int_a^b f(x)\,dx.$

15. If $f(x) \le g(x)$ on $[a, b]$, then $g(x) - f(x) \ge 0$ and is continuous on $[a, b]$. By (5), $\int_a^b [g(x) - f(x)]\,dx \ge 0$ or $\int_a^b f(x)\,dx \le \int_a^b g(x)\,dx.$

17. $\frac{11}{6}$ **19.** $\frac{1}{3}$ **21.** $\frac{2}{3}$ **23.** 9 **25.** -26 **27.** $2/\pi$ **29.** $\frac{5}{2}g \approx 24.5$ m/sec; $\frac{10}{3}g \approx 32.7$ m/sec **31.** 37.5°C

33. 12 m/sec **35.** $1000 + \frac{3}{2} - \frac{2}{3}\sqrt{3}$ kg/m^3

37. By theorem (5.4.3), there is u in $[a, b]$ so that $f'(u) = \frac{1}{b-a}\int_a^b f'(x)\,dx = \frac{f(b)-f(a)}{b-a}$ = Average slope.

Exercise 5.6

1. $6x + C$ **3.** $\frac{3}{2}x^2 + C$ **5.** $-\frac{1}{3}t^{-3} + C$ **7.** $\frac{1}{3}x^3 + 2x + C$ **9.** $\frac{1}{8}x^8 + x + C$ **11.** $x^4 - x^3 + \frac{5}{2}x^2 - 2x + C$
13. $2z^{3/2} + \frac{1}{2}z^2 + C$ **15.** $\frac{1}{2}x^2 + \frac{1}{x} + C$ **17.** $\frac{1}{3}u^3 - \frac{1}{2}u^2 + C$ **19.** $\frac{3}{4}x^4 - \frac{1}{x} + C$ **21.** $\frac{1}{2}t^2 + 2t + C$
23. $\frac{1}{12}(2x + 1)^6 + C$ **25.** $\frac{1}{3}(x^2 - 9)^{3/2} + C$ **27.** $\sqrt{1 + x^2} + C$ **29.** $\frac{2}{5}(x + 3)^{5/2} - 2(x + 3)^{3/2} + C$
31. $-\frac{1}{3}\cos 3x + C$ **33.** $-\frac{1}{2}\cos x^2 + C$ **35.** $\frac{2}{7}(x + 1)^{7/2} - \frac{4}{5}(x + 1)^{5/2} + \frac{2}{3}(x + 1)^{3/2} + C$
37. $\frac{2}{3}(x + 1)^{3/2} - 2(x + 1)^{1/2} + C$ **39.** $\frac{2}{3}(s - 5)^{3/2} + C$ **41.** $\dfrac{-2}{3(1 + \sqrt{x})^3} + C$ **43.** $\frac{2}{5}(x + 1)^{5/2} - 4(x + 1)^{3/2} + C$
45. $\frac{4}{9}(4 + t\sqrt{t})^{3/2} + C$ **47.** $\frac{1}{12}[(z^2 + 1)^4 - 3]^{3/2} + C$ **49.** $\frac{1}{6}\sin^6 x + C$ **51.** $-\frac{1}{5}\cos^5 x + C$
53. $\frac{1}{3}\cos(1 - x^3) + C$ **55.** $-\frac{1}{2}\cos(2x + 5) + C$ **57.** $-\frac{2}{21}$ **59.** $-\frac{1}{28}$ **61.** $-\frac{232}{5}$ **63.** $\frac{16}{3}$ **65.** $\frac{300}{7}$
67. $2 - \sqrt{3}$ **69.** 0 **71.** $\frac{1}{3}(\sqrt{3} - 1)$ **73.** 0 **75.** $\dfrac{\sqrt{185} - 5}{3}$ **77.** $b = \sqrt[3]{2}$ **79.** 1 **81.** $\frac{38}{3}$ **83.** $\frac{2}{3}$
85. (a) $\frac{1}{3}(x + 1)^3 + C_1$ (b) $\frac{1}{3}x^3 + x^2 + x + C_2$; $C_2 = C_1 + \frac{1}{3}$
87. $\int x\sqrt{x}\,dx = \frac{2}{5}x^{5/2} + C_1$; $\int x\,dx = \frac{1}{2}x^2 + C_2$; $\int \sqrt{x}\,dx = \frac{2}{3}x^{3/2} + C_3$
89. $\int \dfrac{x^2 - 1}{x - 1}\,dx = \frac{1}{2}x^2 + x + C_1$; $\int (x^2 - 1)\,dx = \frac{1}{3}x^3 - x + C_2$; $\int (x - 1)\,dx = \frac{1}{2}x^2 - x + C_3$ **91.** $\frac{1}{2}$
93. Since $\sin\theta = \cos(\pi/2 - \theta)$, let $\phi = \pi/2 - \theta$. **95.** 8
97. $\int_{-a}^{a} f(x)\,dx = \int_{-a}^{0} f(x)\,dx + \int_{0}^{a} f(x)\,dx = -\int_{-a}^{0} f(-x)\,dx + \int_{0}^{a} f(x)\,dx = \int_{a}^{0} f(t)\,dt + \int_{0}^{a} f(x)\,dx = 0$
(In the last step, let $x = -t$.)

Historical Perspectives

1. (a) 20 ft (b)

3. (a) Area $ABE = \frac{1}{2}(t_0)(kt_0) = kt_0^2/2$; Area $ABCD = (t_0)(kt_0/2) = kt_0^2/2$ (b) Area $ABE = \int_0^{t_0} kt\,dt = \frac{1}{2}kt^2\Big|_0^{t_0} = \frac{1}{2}kt_0^2$
5. Area $A = \int_a^b [f(x) - g(x)]\,dx = \int_a^b M[h(x) - k(x)]\,dx = M\int_a^b [h(x) - k(x)]\,dx = M(\text{Area } B)$
7. (a) Area $R_1 = ab$; Area $R_2 = AB$; $ka = A$; $kb = B$; Area R_1/Area $R_2 = ab/AB = ab/A(kb) = a/A(A/a) = a^2/A^2$
(b) Area $T_1 = \frac{1}{2}bh$; Area $T_2 = \frac{1}{2}BH$; $kb = B$; $kh = H$; Area T_1/Area $T_2 = bh/BH = bh/B(kh) = b/Bk = b/B(B/b) = b^2/B^2$

Review Exercises

1. $\dfrac{n}{b(n + 1)}(a + bx)^{(n+1)/n} + C$ **3.** $\dfrac{1}{14}\left(2\sqrt{x^2 + 3} - \dfrac{4}{x} + 9\right)^7 + C$ **5.** $\dfrac{1}{3}x^3 + x - \dfrac{1}{2(x^2 + 1)} + C$
7. $\dfrac{1}{9}\left[\dfrac{1}{(2 - 3x)^2} - \dfrac{1}{2 - 3x}\right] + C$ **9.** -2 **11.** $\frac{1}{12}$ **13.** $\frac{2}{3}$ **15.** $\frac{16}{3}$

17. No. Let $f(x) = \begin{cases} 1 & \text{if } x \text{ is rational} \\ -1 & \text{if } x \text{ is irrational} \end{cases}$ on $[0, 1]$. We have $|f(x)| = 1$ on $[0, 1]$. Clearly, $|f(x)|$ is integrable on $[0, 1]$. However, any Riemann sum in which all selected points are irrational will have a sum of -1, whatever the partition; and if the selected points are rational, it will have a sum of 1, whatever the partition.

19. The given integrals are equal for $n = 0, 1, 2, \ldots$. **21.** $\frac{1}{2}$ **23.** $F'(c)$ **25.** (c) **27.** $\dfrac{-x}{\sqrt{1-x^2}}$ **29.** $a = 4$

31. 4 **33.** (c) **35.** Area $= 1 - \dfrac{1}{r}$; $\lim\limits_{r\to+\infty} A(r) = 1$ **37.** $-\frac{9}{4}(\frac{1}{4})^{1/3} \le$ Integral ≤ 54 **39.** -2

41. Suppose that $f(x) \ne 0$ for all $x \in [a, b]$. Then, either $f(x) > 0$ for all $x \in [a, b] \Rightarrow \int_a^b f(x)\,dx > 0$ or $f(x) < 0$ for all $x \in [a, b] \Rightarrow \int_a^b f(x)\,dx < 0$. Conclusion: We cannot have $f(x) \ne 0$ for all $x \in [a, b]$, and so there is at least one number $u \in [a, b]$ such that $f(u) = 0$.

43. Let $[a, b]$ be $[0, 1]$ and let c be irrational. Any regular partition will have division points of the form k/n, which are rational. Thus, c will never itself be a division point, and so no partition of either $[0, c]$ or $[c, 1]$ will ever be induced. Recall the theorem that for integrable $f(x)$ on an interval $[a, b]$, we have

$$\int_a^b f(x)\,dx = \int_a^c f(x)\,dx + \int_c^b f(x)\,dx$$

Each integral is a limit of Riemann sums over a partition of its respective interval. Thus, the theorem would be affected.

Challenge Exercises

1. (a) Let x and Δx be rational. Let $L_b(x) = \log_b x$

$$\lim_{\Delta x\to 0} \frac{\log_b(x+\Delta x) - \log_b x}{\Delta x} = \lim_{\Delta x\to 0} \frac{1}{\Delta x} \log_b\left(\frac{x+\Delta x}{x}\right) = \lim_{\Delta x\to 0} \log_b\left(\frac{x+\Delta x}{x}\right)^{1/\Delta x} = \lim_{\Delta x\to 0} \log_b\left(1 + \frac{\Delta x}{x}\right)^{(x/\Delta x)(1/x)}$$

$$= \lim_{\Delta x\to 0} \frac{1}{x} \log_b\left(1 + \frac{\Delta x}{x}\right)^{x/\Delta x}$$

(b) If $\log_b' x$ exists, then for rational x,

$$\log_b' x = \lim_{\Delta x\to 0} \frac{\log_b(x+\Delta x) - \log_b x}{\Delta x} = \lim_{\substack{\Delta x\to 0\\ \Delta x \text{ rational}}} \frac{\log_b(x+\Delta x) - \log_b x}{\Delta x} = \lim_{\substack{\Delta x\to 0\\ \Delta x \text{ rational}}} \frac{1}{x} \log_b\left(1 + \frac{\Delta x}{x}\right)^{x/\Delta x}$$

$$= \frac{1}{x} \log_b\left[\lim_{\substack{\Delta x\to 0\\ \Delta x \text{ rational}}} \left(1 + \frac{\Delta x}{x}\right)^{x/\Delta x}\right] = \frac{1}{x} \log_b e$$

Assuming that $\log_b' x$ is continuous, we deduce that $\log_b' x = \dfrac{1}{x} \log_b e$ for real x. The choice of b that gives $\log_b' x = \dfrac{1}{x}$ is $b = e$.

3. (a) 0

(b) $F'(x) = \dfrac{1}{ax} \cdot (ax)' = \dfrac{1}{x} = L'(x)$. $F' = L' \Rightarrow F - L$ is constant. $F(1) = L(a)$; $L(1) = 0$. Therefore, $F - L = L(a)$, and $F(x) = L(x) + L(a)$ for $x > 0$. Set $x = b$: $L(ab) = L(a) + L(b)$.

(c) Let $F(x) = \displaystyle\int_1^{x^r} \frac{dt}{t}$; $F'(x) = \dfrac{1}{x^r} \cdot rx^{r-1} = \dfrac{r}{x} = r \cdot \dfrac{1}{x} = r \cdot L'(x)$. Therefore, $F' = rL'$, and $F - rL$ is constant. But $F(1) = 0$, and $rL(1) = r \cdot 0 = 0$. Therefore, $F = rL$, and $L(x^r) = rL(x)$ for $x > 0$. Set $x = a$.

(d) In part (c) let $r = -1$.

(e) $(-\infty, +\infty)$

5. (a) We have, for rational x and Δx,

$$\frac{L(x+\Delta x)-L(x)}{\Delta x}=\frac{1}{x}L\left[\left(1+\frac{\Delta x}{x}\right)^{x/\Delta x}\right]$$

Thus, for rational x,

$$\lim_{\substack{\Delta x\to 0\\ \Delta x\text{ rational}}}\frac{L(x+\Delta x)-L(x)}{\Delta x}=\frac{1}{x}\lim_{\substack{\Delta x\to 0\\ \Delta x\text{ rational}}}L\left[\left(1+\frac{\Delta x}{x}\right)^{x/\Delta x}\right]$$

But

$$\lim_{\substack{\Delta x\to 0\\ \Delta x\text{ rational}}}\frac{L(x+\Delta x)-L(x)}{\Delta x}=\lim_{\Delta x\to 0}\frac{L(x+\Delta x)-L(x)}{\Delta x}=\frac{1}{x}$$

Thus,

$$\lim_{\substack{\Delta x\to 0\\ \Delta x\text{ rational}}}L\left[\left(1+\frac{\Delta x}{x}\right)^{x/\Delta x}\right]=1$$

Now, $E=L^{-1}$ is continuous, and so

$$\lim_{\substack{\Delta x\to 0\\ \Delta x\text{ rational}}}E\left[L\left(1+\frac{\Delta x}{x}\right)^{x/\Delta x}\right]=E(1)$$

But $E=L^{-1}$, and so $\displaystyle\lim_{\substack{\Delta x\to 0\\ \Delta x\text{ rational}}}\left(1+\frac{\Delta x}{x}\right)^{x/\Delta x}=E(1).$

(b) $E(1)\approx 2.71828$ (c) $L(e)=1$. Take $b=e$.

7. $\dfrac{\pi}{4}-\dfrac{1}{2}$ **9.** $f'(x)$ is continuous and $f(0)=0$.

11. (a) (a) An extension of theorem (5.5.1)
(b) Mean value theorem for integrals
(c) Mean value theorem for derivatives; combine sums
(d) Triangle inequality
(e) $|f'(x)|\leq M$; u_i, t_i in (x_{i-1}, x_i)
(f) $\Delta x_i=(b-a)/n$

(b) The actual error = 0.4; the estimate = 0.8 (c) The actual error = $0.048\overline{33}$; the estimate = 0.2

CHAPTER 6

Exercise Set 6.1

1. $A=\frac{1}{2}$

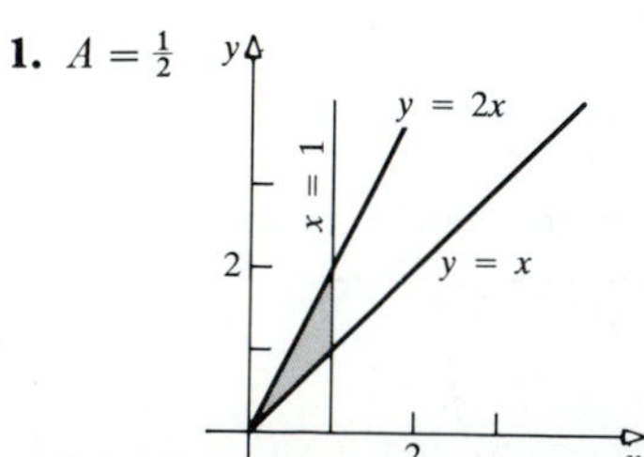

3. $A=\frac{1}{6}$

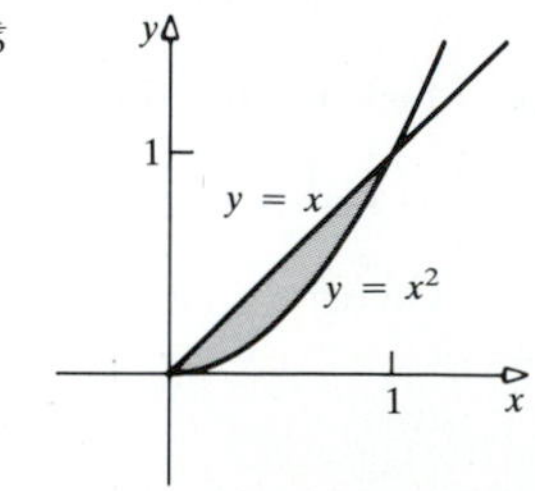

5. $A=\frac{1}{6}$

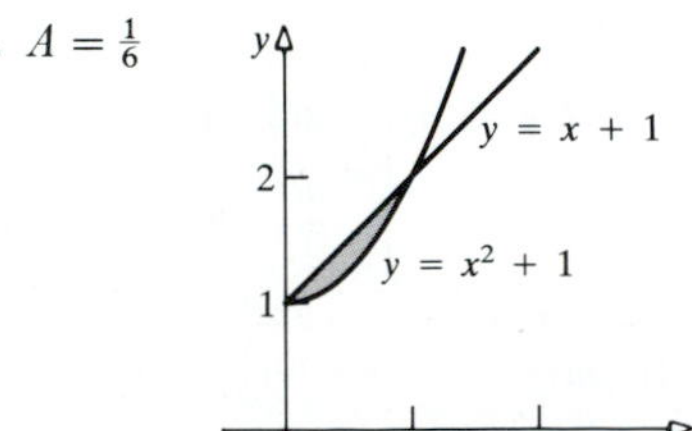

7. $A = \frac{5}{12}$

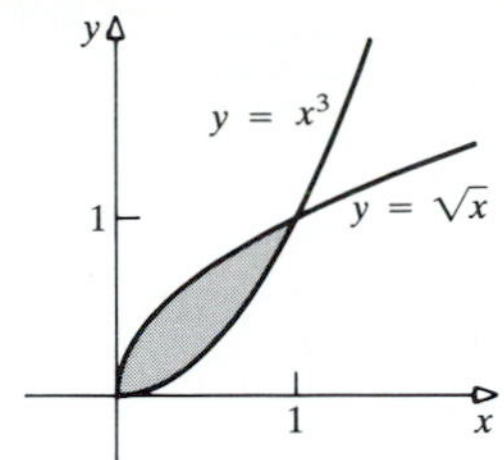

9. $A = \frac{4}{15}$

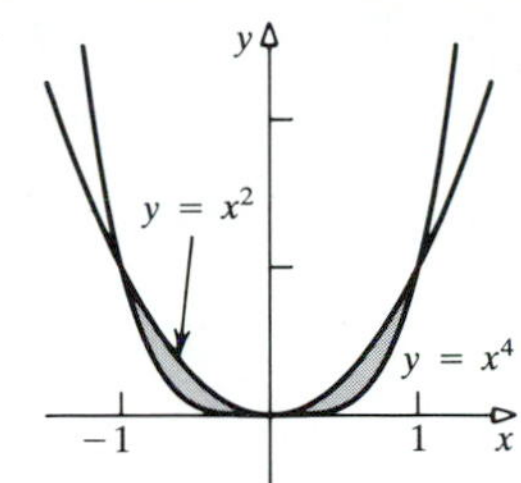

11. $A = \frac{8}{3}$

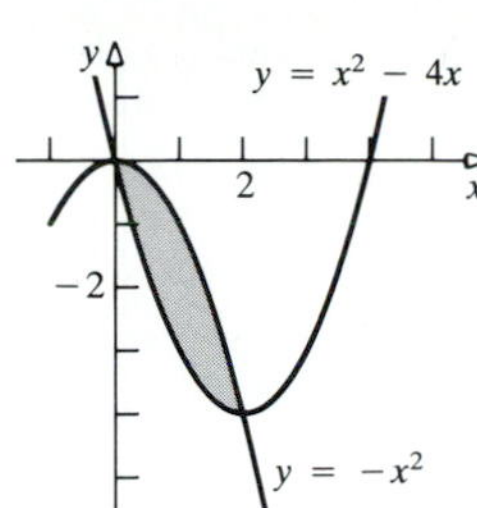

13. $A = \frac{9}{2}$

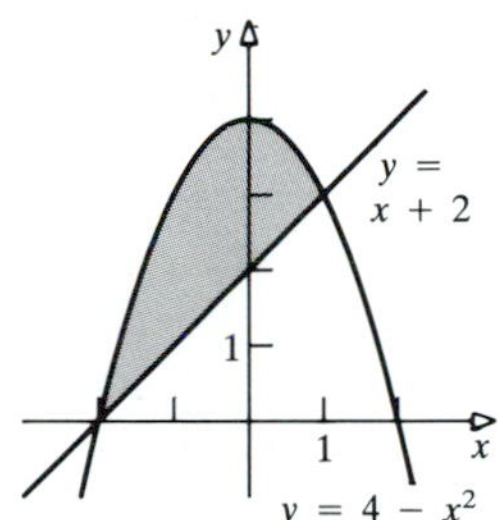

15. $A = 8$

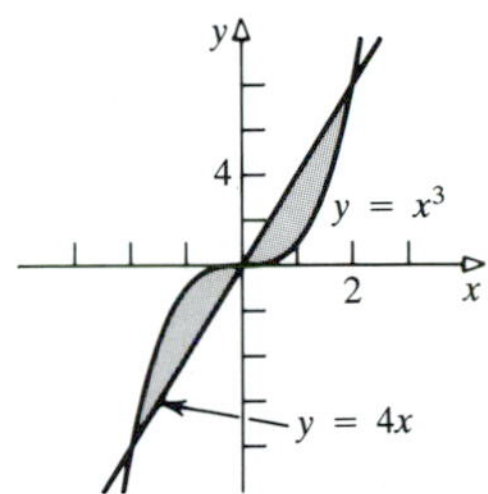

17. $A = \frac{1}{3}$

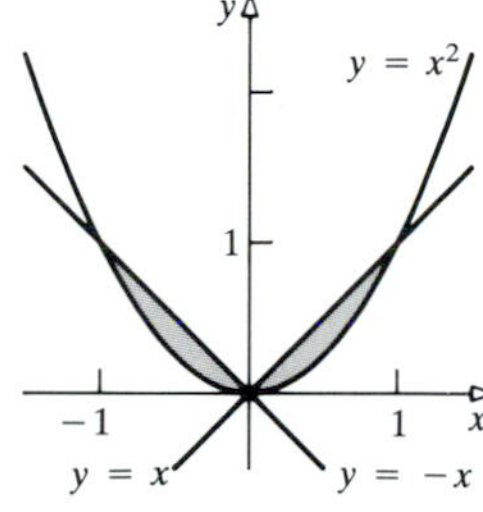

19. $A = 12$

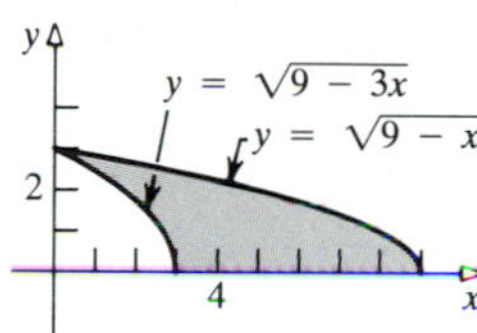

21. $A = \frac{125}{6}$

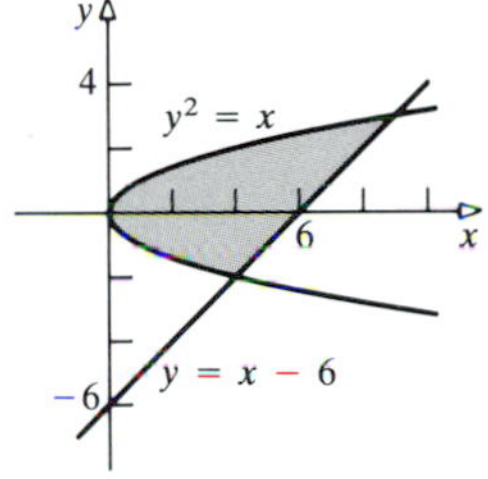

23. $A = \dfrac{\pi}{6} - \dfrac{1}{2}$

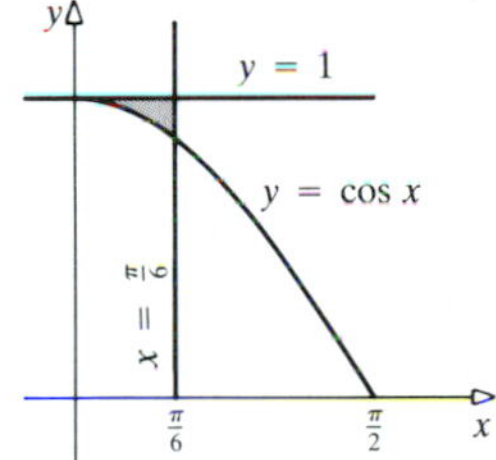

25. $4(\sqrt{2} - 1)$

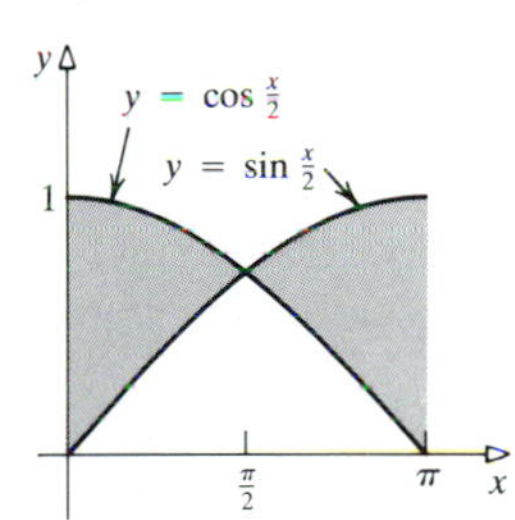

27. $A = 2\sqrt{2}/3$

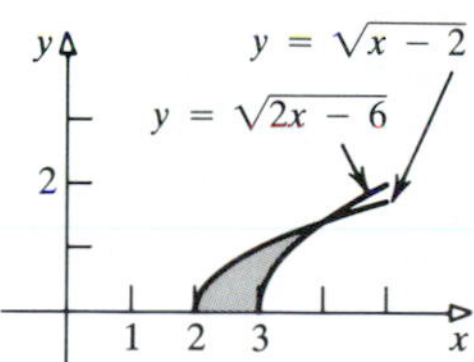

29. $A = \frac{9}{2}$ **31.** $A = \frac{125}{24}$ **33.** $A = \frac{1}{6}$ **35.** $\sqrt[3]{\frac{1}{2}}$

37. Shaded area $= \frac{9}{2}$; area of parallelogram $= \frac{27}{4}$ **39.** $\frac{1}{2}$

Exercise Set 6.2

1. $V = 32\pi/5$

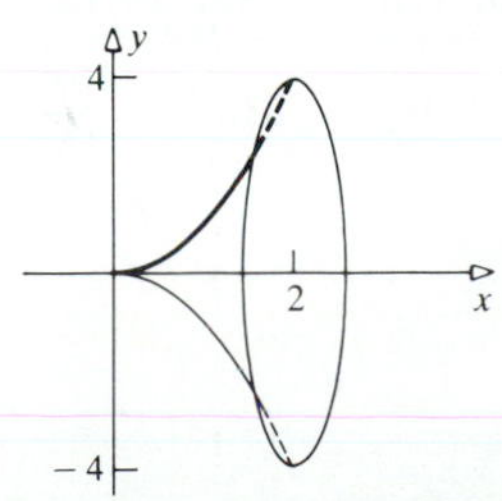

3. $V = 128\pi/7$

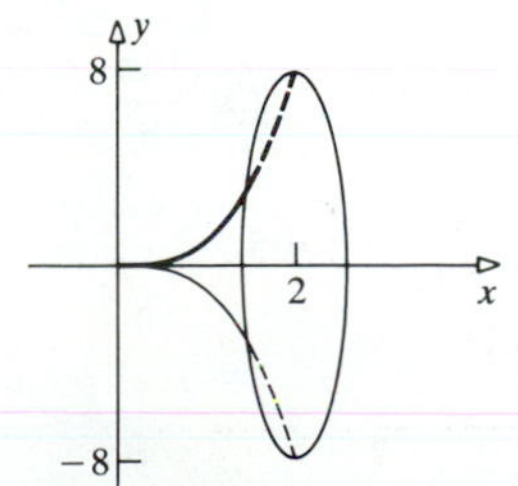

5. $V = 30\pi$

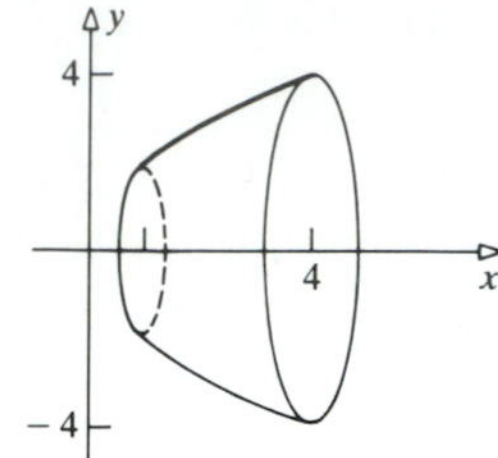

7. $V = \pi/2$

9. $V = 15\pi/2$

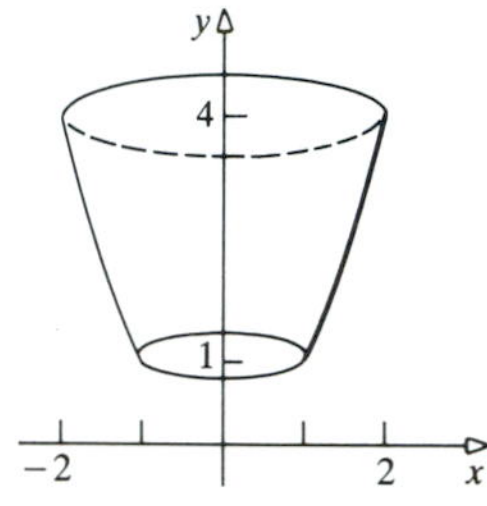

11. $V = 3\pi/4$

13. $V = 2\pi(\tan 1 - 1) \approx 3.502$

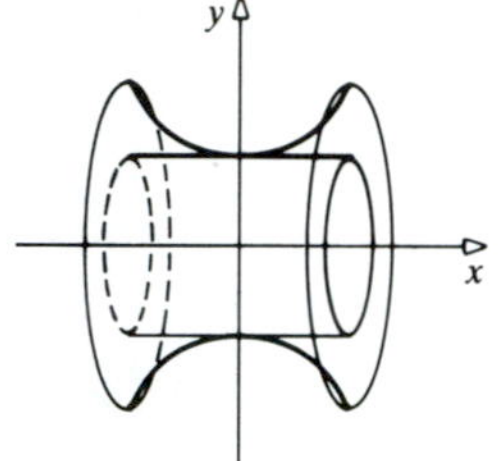

15. $V = 20\pi/7$

17. $V = 2\pi/15$

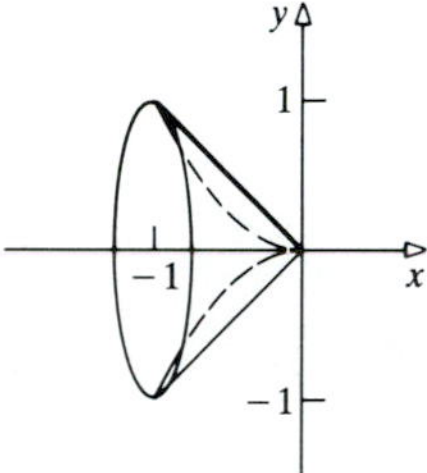

19. $V = 2\pi/15$

21. $V = 128\pi/5$ **23.** $V = 64\pi/5$ **25.** $V = 32\pi$ **27.** $V = \pi/6$ **29.** $V = 256\pi/15$ **31.** $V = \dfrac{363\pi}{64}$

33. $V = \pi\left(2 - \dfrac{\pi}{4}\right)$ **35.** $V = \dfrac{\pi}{6}$ **37.** (a) $\dfrac{128\pi}{5}$ (b) 8π (c) $\dfrac{256\pi}{15}$ (d) $\dfrac{40\pi}{3}$

39. (a) $\dfrac{2\pi}{15}$ (b) $\dfrac{\pi}{6}$ (c) $\dfrac{5\pi}{6}$ (d) $\dfrac{\pi}{5}$ **41.** (a) $\dfrac{\pi k^5}{30}$ (b) $\dfrac{\pi k^4}{6}$ (c) 5

Exercise 6.3

1. 32π **3.** $334\pi/5$ **5.** $\pi/10$ **7.** $768\pi/7$ **9.** $4\pi/5$ **11.** $2\pi/15$ **13.** 128π **15.** $4\pi/15$ **17.** $308\pi/3$

19. 9π **21.** $1177\pi/10$ **23.** $\dfrac{16\pi}{3}$ **25.** $\dfrac{17\pi}{6}$ **27.** $\dfrac{\pi}{6}$ **29.** $V \approx 53{,}135{,}993$ ft^3 **31.** (a) $\dfrac{16\pi}{15}$ (b) $\dfrac{8\pi}{3}$

33. (a) 16π (b) $\dfrac{512\pi}{3}$ **35.** $\dfrac{x(x+2)}{\sqrt{\pi}}$

Exercise 6.4

1. 4 **3.** $\frac{16}{3}$ **5.** $\pi/40$ **7.** $\frac{256{,}000}{3}$ m^3 **9.** $37\pi/1320$ **11.** $V = \pi \int_0^h \left(\frac{R}{h}x\right)^2 dx = \frac{\pi R^2}{h^2}\frac{x^3}{3}\Big|_0^h = \frac{\pi}{3}R^2h$

13. $\pi h^2\left(R - \frac{h}{3}\right)$ **15.** $\frac{250\sqrt{3}}{9}$

Exercise 6.5

1. $2\sqrt{10}$ **3.** $\sqrt{13}$ **5.** $\frac{1}{27}(80\sqrt{10} - 13\sqrt{13})$ **7.** $\frac{1}{27}(80\sqrt{10} - 8)$ **9.** $\frac{2}{3}(2\sqrt{2} - 1)$ **11.** 45 **13.** 21

15. $\frac{33}{16}$ **17.** $\frac{1}{27}(13\sqrt{13} - 8)$ **19.** $\frac{2}{27}(10\sqrt{10} - 1)$ **21.** $6a$ **23.** $(31\sqrt{31} - 13\sqrt{13})/27$

25. $\frac{8}{243}(82^{3/2} - 1) + 4\sqrt{37}$ **27.** $\frac{17}{12}$ **29.** $\int_0^2 \sqrt{1 + 4x^2}\,dx$ **31.** $5\int_0^4 \frac{dx}{\sqrt{25 - x^2}}$

33. $\frac{1}{a}\int_0^{a/2} \sqrt{\frac{a^4 - (a^2 - b^2)x^2}{a^2 - x^2}}\,dx$

35. (a) $\int_{x_2}^{x_1} \frac{1}{\sqrt{1 - x^2}}\,dx$ (b) $2\pi - \int_{x_2}^{x_1} \frac{1}{\sqrt{1 - x^2}}\,dx$ (c) $\int_0^{y_1} \frac{1}{\sqrt{1 - y^2}}\,dy + \frac{\pi}{2} + \int_0^{x_2} \frac{1}{\sqrt{1 - x^2}}\,dx$

Exercise 6.6

1. $\int_0^1 \frac{3}{2}x\,dx = \frac{3}{4}$ J **3.** $\frac{10}{3}$ J **5.** 10 ft **7.** $72\pi\rho g \approx 2{,}216{,}708$ J **9.** $\frac{64}{3}\pi\rho g \approx 656{,}802$ J

11. $36\pi\rho g \approx 1{,}108{,}354$ J **13.** $10{,}500\rho g \approx 656{,}250$ ft-lb **15.** 1.363×10^8 J **17.** $-m/2a$ **19.** 1500 J

21. 4950 ft-lb **23.** 209,210 in.-lb **25.** $\frac{112\pi\rho g}{3}$

Exercise 6.7

1. $36\rho g \approx 352{,}800$ N **3.** $360\rho g \approx 22{,}488$ lb **5.** $\frac{14}{3}\rho g \approx 45{,}733$ N **7.** $\frac{16}{3}\rho g \approx 52{,}267$ N **9.** $\rho g/4 \approx 2450$ N

11. $\frac{128}{3}\rho g \approx 2{,}555.39$ lb

Exercise 6.8

1. 0, 29, $(\frac{29}{15}, 0)$ **3.** 62, 29, $(\frac{29}{15}, \frac{62}{15})$ **5.** $(\frac{9}{4}, \frac{27}{10})$ **7.** $(\frac{7}{8}, \frac{19}{8})$ **9.** $(\frac{8}{5}, \frac{16}{7})$ **11.** $(\frac{13}{6}, \frac{7}{12})$ **13.** $(\frac{43}{65}, \frac{59}{65})$

15. $(-\frac{1}{2}, \frac{12}{5})$ **17.** $\left(0, \frac{2\pi + 3\sqrt{3}}{8(3\sqrt{3} - \pi)}\right)$ **19.** 2π, 0, $\left(0, \frac{2\pi}{4 + \pi}\right)$ (origin taken at center of square)

21. $\frac{b^2c}{3}, \frac{abc}{3}, \left(\frac{a}{3}, \frac{b}{3}\right)$ **23.** $\frac{h^2a}{10}, \frac{ha^2}{4}, \left(\frac{3a}{4}, \frac{3h}{10}\right)$ **25.** $72\pi^2$ **27.** $\frac{32\pi}{15}$ **29.** $\frac{\pi R^2 h}{3}$

Review Exercises

1. $\frac{\pi}{6} + \frac{\sqrt{3}}{2} - 1$ **3.** 72 **5.** $-1 + \sqrt{13}$ **7.** 4 **9.** $\frac{a^2}{6}$ **11.** $\frac{256\pi}{5}$ **13.** $\frac{\pi^2}{2}$ **15.** $\frac{128\pi}{15}$ **17.** $\frac{32\pi}{5}$

19. $\frac{5\pi}{6}$ **21.** $\frac{3456\pi}{35}$ **23.** $\frac{4\pi}{3}(64 - 7\sqrt{7})$ **25.** $\frac{10\pi}{3}$ **27.** $\frac{209}{6}$ **29.** 8 **31.** 3,360,000 ft-lb

33. 6.4 joules **35.** $12\rho g \approx 750$ lb **37.** (a) $2{,}560{,}000\,\rho g$ lb $\approx 1.6 \times 10^8$ lb (b) $1{,}395{,}000\,\rho g$ lb $\approx 8.72 \times 10^7$ lb

39. $\bar{x} = \frac{836}{217}$, $\bar{y} = \frac{80}{31}$ **41.** 72π

Challenge Exercises

1. Given a region to the right of the y-axis, suppose that we subdivide the x-axis for $a \le x \le b$ in considering A. Using the shell method, we find that the volume of the solid generated by revolving about the y-axis is given by $V = 2\pi \int_a^b xh(x)\,dx$, where $h(c)$ is the "height" of the region at $x = c$. If we revolve about the line $x = -k$, $k > 0$, the volume generated would be found by replacing x by $x + k$ in the formula. This gives
$2\pi \int_a^b (x+k)h(x)\,dx = 2\pi \int_a^b xh(x)\,dx + 2k\pi \int_a^b h(x)\,dx = V + 2k\pi A$, since the area of the region is $A = \int_a^b h(x)\,dx$.

3. $\frac{16}{3}r^3$ cubic units 5. $\frac{4}{3}$ square units

7. We will use $\sin\frac{A}{2} = \sqrt{\frac{1-\cos A}{2}}$. By Problem 6, $2\pi \approx 2^{n+1}\sin(45/2^{n-2})$. Thus, $\pi \approx 2^n \sin(45/2^{n-2})$.
Replacing n by $n+2$, we get $\pi \approx 2^{n+2}\sin(45/2^n)$. Let $a_n = \sin(45/2^n)$ for $n = 0, 1, \ldots$. Then $a_0 = \sin 45° = 1/\sqrt{2}$ and

$$a_{n+1} = \sin(45/2^{n+1}) = \sin\left(\frac{45/2^n}{2}\right) = \sqrt{\frac{1-\cos(45/2^n)}{2}} = \sqrt{\frac{1-\sqrt{1-\sin^2(45/2^n)}}{2}} = \sqrt{\frac{1-\sqrt{1-a_n^2}}{2}}$$

n	0	1	2	3	4	5	6
a_n	0.707107	0.382683	0.195090	0.098017	0.049068	0.024541	0.012272
$2^{n+2}a_n$	2.8284	3.061467	3.121445	3.136548	3.140331	3.14127	3.141514

9. $F(x) = \frac{1}{4}x^2$, $0 \le x \le 2$

CHAPTER 7

Exercise 7.2

1. $a + b$ 3. $\frac{b}{2}$ 5. $-(2a + b)$ 7. $b - a$ 9.

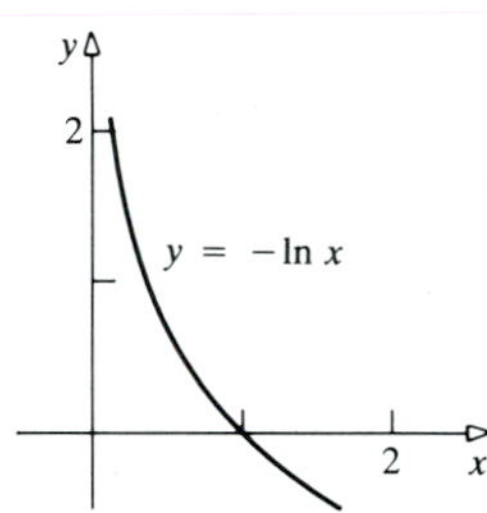

11.

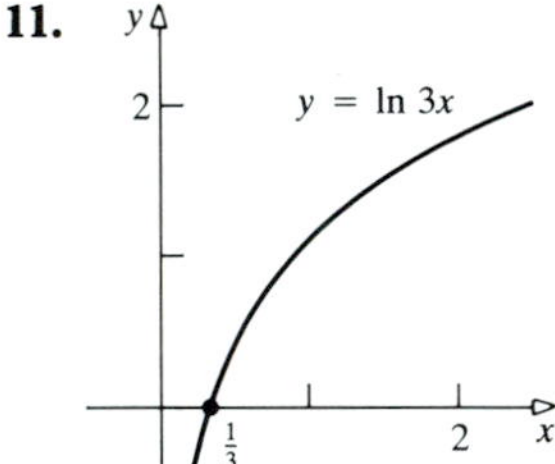

13. $1/x$ 15. $\dfrac{2x^2}{x^2+4} + \ln(x^2+4)$ 17. $\dfrac{x^2}{x^2+1} + \dfrac{1}{2}\ln(x^2+1)$ 19. $\dfrac{1}{1-x^2}$ 21. $\dfrac{1}{x\ln x}$ 23. $\dfrac{1}{x(x^2+1)}$

25. $\dfrac{2x^4 - 5x^2 + 1}{x(x^4-1)}$ 27. $\cot x$ 29. $\dfrac{1}{2x(\ln x)^{1/2}}$ 31. $-x\tan 2x + \frac{1}{2}\ln(\cos 2x)$ 33. $\dfrac{1}{\sqrt{x^2+4}}$ 35. $\dfrac{1}{\sqrt{x^2+a^2}}$

37. $4x(x^2+1)(2x^3-1)^3(8x^3+6x-1)$

39. $(x^3+1)(x-1)(x^4+5)\left(\dfrac{3x^2}{x^3+1} + \dfrac{1}{x-1} + \dfrac{4x^3}{x^4+5}\right)$ 41. $\dfrac{x^2(x^3+1)}{\sqrt{x^2+1}}\left(\dfrac{2}{x} + \dfrac{3x^2}{x^3+1} - \dfrac{x}{x^2+1}\right)$

43. $\dfrac{x\cos x}{(x^2+1)^3\sin x}\left(\dfrac{1}{x} - \tan x - \dfrac{6x}{x^2+1} - \cot x\right)$ 45. $-\dfrac{y(x\ln y + y)}{x(y\ln x + x)}$ 47. y/x 49. $y = 5x - 1$

51. $t > 1 \Rightarrow t > \sqrt{t} \Rightarrow 1/t < 1/\sqrt{t} \Rightarrow \ln x = \int_1^x dt/t < \int_1^x dt/\sqrt{t} = 2(\sqrt{x} - 1)$, $x > 1$ 53. 0

55. $f'(x) > 0$ for $x > 0$, $f'(x) < 0$ for $x < 0$, and $f(0) = 0$. Hence, $-x + \ln(1 + x)^{1+x} \geq 0$ for $x > -1$ and $\ln(1 + x)^{1+x} \geq x$.

57. $9!/x$

59.
1. Domain: $(-2, 2)$
2. x-intercepts: $(-\sqrt{3}, 0)$, $(\sqrt{3}, 0)$; y-intercept: $(0, 2 \ln 2)$
3. Symmetric about y-axis
4. Vertical asymptotes: $x = \pm 2$
5. Critical number: $x = 0$
6. f increasing on $(-2, 0]$; f decreasing on $[0, 2)$
7. Local maximum at $x = 0$
8. f concave down on entire domain
9. No inflection points
10.

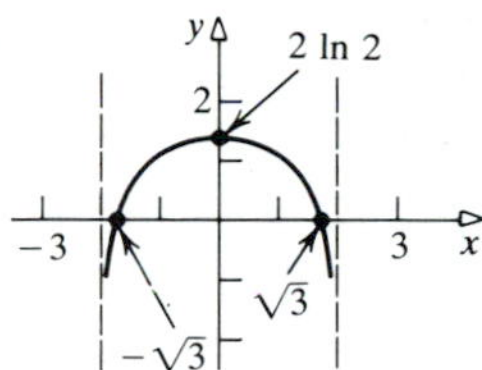

61.
1. Domain: 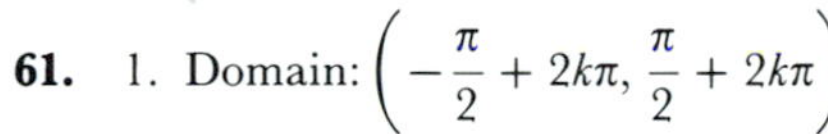 $\left(-\dfrac{\pi}{2} + 2k\pi, \dfrac{\pi}{2} + 2k\pi\right)$
2. Intercepts: $(2k\pi, 0)$
3. Symmetry about y-axis
4. Vertical asymptotes: $\dfrac{\pi}{2} + k\pi$
5. Critical numbers: $2k\pi$
6. f increasing on $\left(-\dfrac{\pi}{2} + 2k\pi, 2k\pi\right]$; f decreasing on $\left[2k\pi, \dfrac{\pi}{2} + 2k\pi\right)$
7. Local maxima at $2k\pi$
8. f concave down on entire domain
9. No inflection points
10.

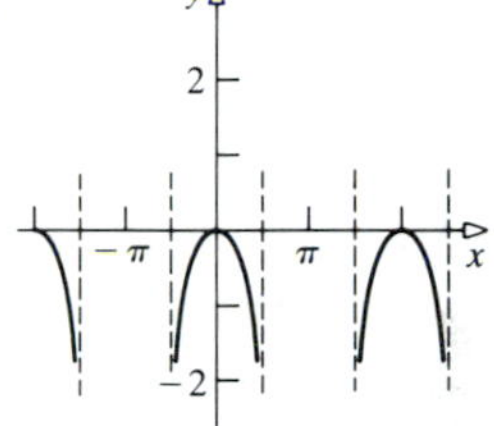

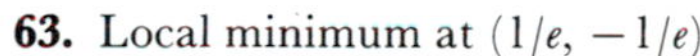

63. Local minimum at $(1/e, -1/e)$

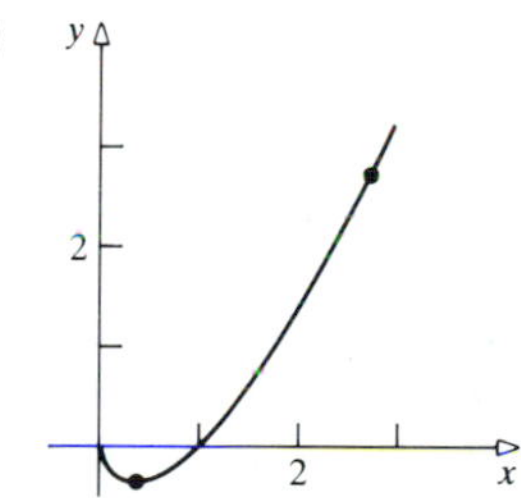

65. 2 **67.** e **69.** $1/\sqrt{e} \approx 0.607$

Exercise 7.3

1. $x - x^2$ **3.** $1/x$ **5.** $-x^2$ **7.** xe^x **9.** $\ln 2$ **11.** $x = \pm e^{-5/2}$ **13.** $x = 0$ or $x = \ln 2$ **15.** $+\infty$

17. 0 **19.** 1 **21.** 1 **23.** $15e^{3x}$ **25.** $\dfrac{e^x - e^{-x}}{2}$ **27.** $e^{-3x}\left(\dfrac{1}{x} - 3 \ln 2x\right)$ **29.** $e^x(\sin x + \cos x)$

31. $e^{ax}(b \cos bx + a \sin bx)$ **33.** $\dfrac{xe^{\sqrt{x^2-9}}}{\sqrt{x^2 - 9}}$ **35.** $\dfrac{-e^{1/x}}{x^2}$ **37.** $\dfrac{2ae^{ax}}{(e^{ax} + 1)^2}$ **39.** $\dfrac{e^{x+y}}{1 - e^{x+y}}$ **41.** $\dfrac{1 + ye^x}{1 - e^x}$

43. $\dfrac{9900e^{-x}}{(1 + 99e^{-x})^2}$ **45.** $-2xe^{x^2}\sin e^{x^2}$ **47.** $e^x\cot e^x$ **49.** $ye^x\ln y$ **51.** $\dfrac{e^y\sin x - e^x\sin y}{e^x\cos y + e^y\cos x}$ **53.** $a^n e^{ax}$

55. $y'' = 4e^{2x}$ **57.** $y'' = 4Ae^{2x} + 4Be^{-2x}$ **59.** $y' = 2Ae^{2x} + 3Be^{3x}$; $y'' = 4Ae^{2x} + 9Be^{3x}$

61. $e^{-at}[(A\omega - Ba)\cos \omega t - (B\omega + Aa)\sin \omega t]$

63. No local extrema, no points of inflection, and concave up everywhere

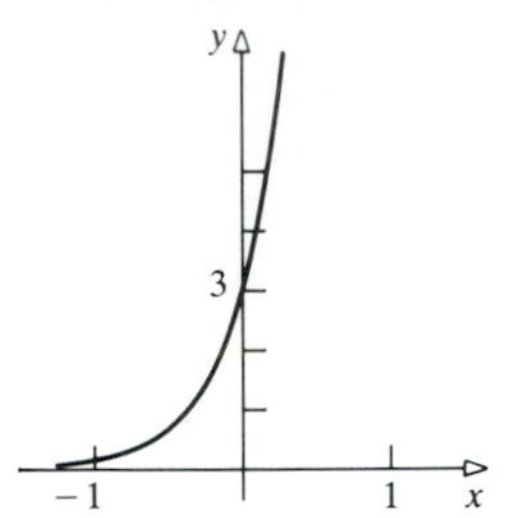

65. Local maximum at $x = 0$, points of inflection at $x = \sqrt{2}/2$ and $x = -\sqrt{2}/2$, concave up on $(-\infty, -\sqrt{2}/2]$, concave down on $[-\sqrt{2}/2, \sqrt{2}/2]$, and concave up on $[\sqrt{2}/2, +\infty)$

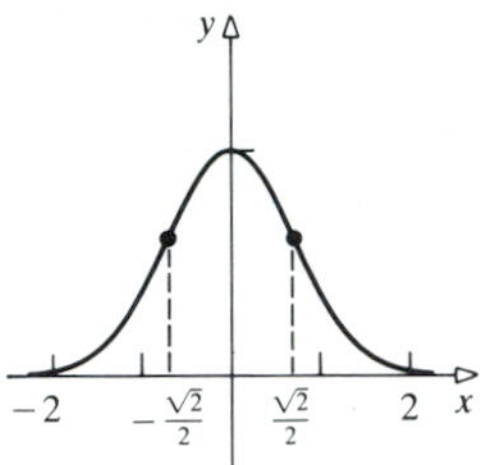

67. No local extrema, an inflection point at $x = -\frac{1}{2}$, concave down on $(-\infty, -\frac{1}{2}]$, and concave up on $[-\frac{1}{2}, 0)$ and $(0, +\infty)$

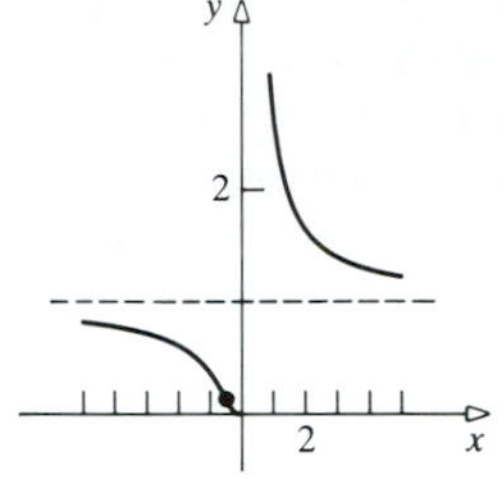

69. Absolute maximum = 1 and occurs at $x = 0$. Absolute minimum = $e - 3$ and occurs at $x = 1$.

71. (a) $-2(\sin x)e^{\cos x}$; $-2e^{\cos x}(\cos x - \sin^2 x)$ (b) $-\frac{5}{2}$ **73.** $di/dt = -Ri/L$

75. $f(0) = 0$; therefore, $f(x) > f(0) = 0$ if $x > 0$. Also note that $f'(x) > 0$ when $x > 0$.

77. From Problem 76, $e^x > x^2$ for all $x > 0$. Hence, $0 \le x/e^x < x/x^2 = 1/x$. Now apply the squeezing theorem to get $0 \le \lim_{x\to+\infty} x/e^x < \lim_{x\to+\infty} 1/x = 0$. Therefore $\lim_{x\to+\infty} x/e^x = 0$.

79. $c = -\ln 3$ **81.** $f(-1) = -1 + e^{-1} < 0$; $f(0) = 1$. Therefore, $f(c) = 0$ for $-1 < c < 0$. **83.** ≈ 0.56714

85. Tangent line at $x = x_0$; $y - e^{x_0} = e^{x_0}(x - x_0)$. Solving for $y = 0$, we get $x = x_0 - 1$.

87. Using the squeezing theorem and $1 < x^2 e^{1/x^2}$ for $x \neq 0$, we get $f'(x) = \dfrac{2}{x^3} e^{-1/x^2}$; for $x = 0$, $f'(x) = \lim_{h\to 0} \dfrac{1}{he^{1/h^2}}$, where for $h < 0$, $0 > \dfrac{1}{he^{1/h^2}} > h$, for $h > 0$, $0 < \dfrac{1}{he^{1/h^2}} < h$ Hence, $f'(0) = 0$.

89. $11e^{10}$ square units per second **91.** $(\sqrt{2}/2, e^{-1/2})$ and $(-\sqrt{2}/2, e^{-1/2})$

93.

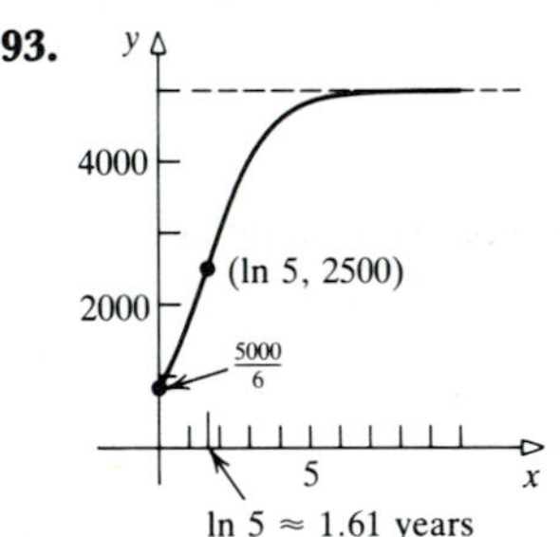

95.
$$\ln(e^\alpha/e^\beta) = \ln e^\alpha - \ln e^\beta = \alpha - \beta = \ln e^{\alpha-\beta}$$
$$\ln(e^\alpha/e^\beta) = \ln e^{\alpha-\beta}$$
$$e^\alpha/e^\beta = e^{\alpha-\beta}$$

97. The implicit equation does not represent a differentiable function $y = f(x)$, since $e^{x+y} > x + y$ for all real x and y.

99. (a) By theorem (5.4.2)
(b) By integration
(c) $e^{1/(x+1)} < 1 + \dfrac{1}{x} < e^{1/x}$, raise to the power x, $e^{x/(1+x)} < \left(1 + \dfrac{1}{x}\right)^x < e$
(d) By squeezing theorem (2.5.1)
(e) From part (c), $e^{x/(x+1)} < \left(1 + \dfrac{1}{x}\right)^x$ raised to the power $(x + 1)/x$

Exercise 7.4

1. $\frac{1}{3}e^{3x} + C$ **3.** $\frac{1}{3}e^{3x+1} + C$ **5.** $\frac{1}{2}x^2 + \frac{1}{7}e^{7x} + C$ **7.** $\frac{1}{3}(\ln 2 + 1)$ **9.** $-\frac{1}{5}\ln|1 - 5x| + C$
11. $-\frac{5}{2}\ln|1 - x^2| + C$ **13.** $\frac{1}{3}\ln(e^6 + e^3 + 1)$ **15.** $(e - 1)/(2e^2)$ **17.** $3e^{\sqrt[3]{x}} + C$ **19.** $-\ln(6 + e^{-x}) + C$
21. $2\sqrt{1 + e^x} + C$ **23.** $\dfrac{1}{\ln 2} - \dfrac{1}{\ln 3}$ **25.** $\frac{1}{4}(\ln 5x)^4 + C$ **27.** $\sin(1) - \sin e^{-\pi}$ **29.** $\ln 2 - 2$
31. $-4\ln(1 + e^{-x}) + C$ **33.** $\frac{1}{2}\ln|2\sin x - 1| + C$ **35.** $\frac{1}{2}$ **37.** $\ln 2$ **39.** $\frac{1}{3}(e^6 - 3e^2 + 2)$
41. $(\pi/2)(1 - e^{-4})$ **43.** $e - 1$

Exercise 7.5

1. $\sqrt{2}x^{\sqrt{2}-1}$ **3.** $2\sqrt{2}x(1 + x^2)^{\sqrt{2}-1}$ **5.** $(\ln \frac{1}{2})(\frac{1}{2})^x$ **7.** $1/(x \ln 2)$ **9.** $2x/[(\ln 2)(1 + x^2)]$
11. $2^{-x}[\cos x - (\ln 2)\sin x]$ **13.** $(3x)^x[1 + \ln(3x)]$ **15.** $2(\ln x)x^{\ln x - 1}$ **17.** $\dfrac{(3x)^{\sqrt{x}}}{2\sqrt{x}}(\ln 3x + 2)$
19. $e^x x^{e^x}(\ln x + \frac{1}{x})$ **21.** $2^{\sin x}(\cos x)(\ln 2)$ **23.** $x^{\sin x}\left[\dfrac{\sin x}{x} + (\ln x)(\cos x)\right]$ **25.** $(\sin x)^{\cos x + 1}[\cot^2 x - \ln(\sin x)]$
27. $\dfrac{1 - xy\ln 2}{x^2 \ln 2}$ **29.** $\dfrac{1}{\ln 3}\left(\dfrac{2x}{1 + x^2}\right)$ **31.** $\dfrac{2^x}{\ln 2} + C$ **33.** $\dfrac{2^{3x+5}}{3\ln 2} + C$ **35.** $\dfrac{1}{3\pi + 1}(2^{3\pi+1} - 1)$ **37.** $(\ln 3)(\ln 2)$
39. $\ln 2$ **41.** By (7), $\log_b a = \dfrac{\ln a}{\ln b} = \dfrac{1}{(\ln b)/(\ln a)} = \dfrac{1}{\log_a b}$ **43.**
$$e^{x_1} < a < e^{x_2}$$
$$\ln e^{x_1} < \ln a < \ln e^{x_2} \text{ (ln is increasing)}$$
$$x_1 < \ln a < x_2$$

45.
$$y = f(x)^{g(x)}$$
$$\ln y = g(x)\ln f(x)$$
$$\frac{y'}{y} = g(x)\frac{f'(x)}{f(x)} + \ln f(x)g'(x)$$
$$\text{or } y' = g(x)f(x)^{g(x)-1}f'(x) + f(x)^{g(x)}\ln f(x)g'(x)$$

47. $(xy)^\alpha = e^{\alpha \ln xy} = e^{\alpha(\ln x + \ln y)} = e^{\alpha \ln x} \cdot e^{\alpha \ln y} = x^\alpha \cdot y^\alpha$

Exercise 7.6

1. e^2 **3.** $e^{1/3}$ **5.** \$530.92; \$531.99; $6\frac{1}{4}\%$ is better **7.** \$941.76; \$942.18 **9.** 83.18 months; 84.2 months

Exercise 7.7

1. 7.679 g **3.** Approximately 9950 years ago **5.** 6944 **7.** $7500e^2$
9. (a) $10{,}000e^{(\ln 5)t/10}$ (b) 250,000 (c) 4.3 min **11.** 4.6×10^6 years **13.** 16.5°C
15. (a) 625.53 mb (b) 303.42 mb (c) 429.03 mb (d) 348.63 mb (e) 57,396 m

Review Exercises

1. $v' = \dfrac{2y}{y^2 + 1}$ **3.** $y' = \dfrac{1}{a}x^{(1/a)-1} + a^{1/x}(\ln a)\left(\dfrac{-1}{x^2}\right)$ **5.** $y' = 2\cot 2x$ **7.** $g'(x) = \dfrac{2(x - 1)}{x(x - 2)}$ **9.** $y' = \dfrac{-4x}{x^4 - 1}$
11. $w' = \dfrac{-1}{2\sqrt{x}\sqrt{x + a}}$ **13.** $f'(x) = \dfrac{1}{a\ln b}\left[\dfrac{1}{x} - \dfrac{2x}{3(x^2 + a^2)}\right]$ **15.** $f'(x) = \log_{10} e$ **17.** $-\frac{1}{2}(1 + 3\ln 2)$
19. $\ln|e^x - 1| + \ln|1 - e^{-x}| + C$ **21.** $\dfrac{(\ln 15)^2}{2\ln 3}$ **23.** $\ln\ln\ln x + C$ **25.** $5(\ln 3)(\ln(\log_3 x))$

27. (a) Proof: $f(x) = \ln ax \Rightarrow f'(x) = \frac{1}{x}$; $g(x) = \ln x \Rightarrow g'(x) = \frac{1}{x}$; $f'(x) = g'(x) \Rightarrow f(x) = g(x) + C$; let $C = f(1) - g(1) = \ln a - \ln 1 = \ln a$. Therefore, $\ln ax = \ln x + \ln a$. Let $x = b$ to get $\ln(ab) = \ln b + \ln a$.
(b) From part (a), $\ln ax - \ln x = \ln a$ for all $x > 0$.

29. Domain: $f = (0, +\infty)$; vertical asymptote at $x = 0$; no horizontal asymptotes; local minimum at $(1, 1)$; f concave up on $(0, +\infty)$

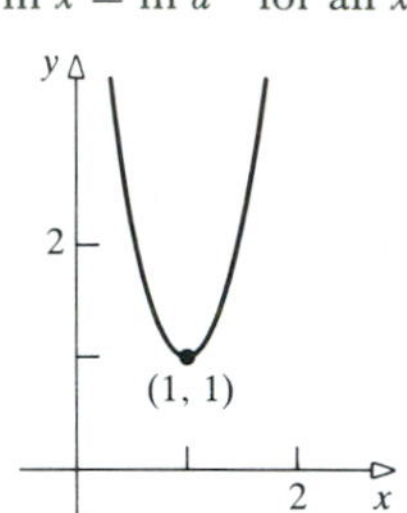

31. $f(x) = e^{1-x}$

33. (a) $P(x) = x^2/(b^2e^{b^2})$ (b) $V = \dfrac{\pi b^2}{2e^{b^2}}$
(c) $V'(b) = \pi b e^{-b^2}(1 - b^2) \Rightarrow$ critical number at $b = 0, \pm 1$. There are local maxima at $b = \pm 1$ and a local minimum at $b = 0$. $V_{\max} = V(1) = \dfrac{\pi}{2e}$.

35. (a) $A = 1 + 3 \ln k$ (b) $k = e^2$ (c) 25 units/sec **37.** $A_{\max} = 1/e$ **39.** (a)

41. (a) $f'(x) = \dfrac{x-1}{x^2} \Rightarrow$ critical number at $x = 1$. $f\left(\dfrac{1}{e}\right) = e - 1$, maximum value. $f(1) = 1$, minimum value. $f(e) = \dfrac{1}{e} + 1$.
(b) $[1/e, 2]$
(c)

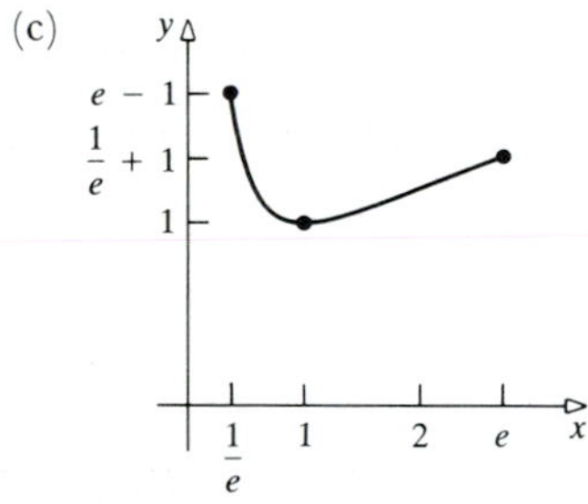

43. 1 **45.** $\approx 104.8°C$ **47.** Local minimum at $(-1, -e^{-1})$; inflection point at $(-2, -2e^{-2})$

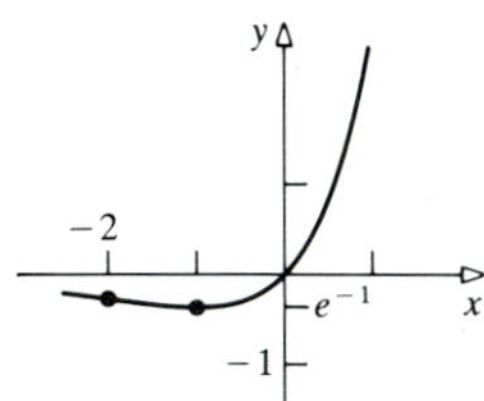

49. $y = 10e^{x-3}$ **51.** $\approx 3.57715, 0.11183$

53. $\log_{10}(2^{0.0001}) \approx 0.0000301$ and $0.0001 \log_{10} 2 \approx 0.000030103$, implying round-off error.

55. (a) $nRT \ln(V_2/V_1)$ (b) $[k/(1-\gamma)](V_2^{1-\gamma} - V_1^{1-\gamma})$

Challenge Exercises

1. Let $f(x) = (\ln x)/x$; $f'(x) = (1 - \ln x)/x^2 \Rightarrow$ critical number at $x = e$. f has a maximum value at $x = e$; $f(e) = 1/e$. Hence, for all $x > 0$, $(\ln x)/x \le 1/e \Rightarrow e \ln x \le x \Rightarrow \ln x^e \le x \Rightarrow x^e \le e^x$. Let $x = \pi \in (0, +\infty)$, and since $\pi \ne e$, $\pi^e < e^\pi$.

3. $(\sqrt{5}-1)/2$

5. Let $y = f(x) = \ln x$ on $[a, b]$; $a > 0$. Since $f(x)$ is continuous on $[a, b]$ and differentiable on (a, b), there is a $c \in (a, b)$ such that $f'(c) = \dfrac{f(b)-f(a)}{b-a}$. $f'(x) = \dfrac{1}{x} \Rightarrow \dfrac{1}{c} = \dfrac{\ln b - \ln a}{b-a} \Rightarrow \dfrac{1}{c}(b-a) = \ln\left(\dfrac{b}{a}\right) \Rightarrow (b-a) = c\ln\dfrac{b}{a}$ and $b > a \Rightarrow \dfrac{b}{a} > 1$. $a < c < b$ and $\dfrac{b}{a} > 1 \Rightarrow \ln\dfrac{b}{a} > 0 \Rightarrow a\ln\dfrac{b}{a} < c\ln\dfrac{b}{a} < b\ln\dfrac{b}{a} \Rightarrow a\ln\dfrac{b}{a} < (b-a) < b\ln\dfrac{b}{a}$.

7. For $n > 0$, we have by the mean value theorem, $\dfrac{\ln(n+1)-\ln n}{(n+1)-n} = \dfrac{1}{c}$, for some $c \in (n, n+1)$. Thus, $\ln\left(\dfrac{n+1}{n}\right) = \dfrac{1}{c}$. Also, $\dfrac{1}{n+1} < \dfrac{1}{c} < \dfrac{1}{n}$, and so $\dfrac{1}{n+1} < \ln\left(\dfrac{n+1}{n}\right) < \dfrac{1}{n}$. Then, $\dfrac{n}{n+1} < \ln\left[\left(1+\dfrac{1}{n}\right)^n\right] < 1$.

9. (a) 0 (b) $-\infty$ (c) $+\infty$ (d) 0 (e) 0 (f) 0

11. (a) $y = (\frac{3}{2}x^2 + \frac{3}{4}x^4 + C)^{1/3}$ (b) $y + 3\ln y = \frac{1}{3}x^3 - x^2 + x + C$ (c) $\frac{1}{3}y^3 + 2y^2 = \frac{1}{3}x^3 - 160$

(d) Suppose that the given differential equation is separable. Then it has the form $y' = \dfrac{f(x)}{g(y)}$, where f and g are continuous. Thus, for some such f and g, we have $\dfrac{2}{x+y} = \dfrac{f(x)}{g(y)}$. Hence, $2g(y) = (x+y)f(x)$ for all (x, y). We fix y. The left-hand side is constant, and so independent of x. In fact, $f(x) = \dfrac{2g(y)}{x+y}$, and so $f(x)$ is differentiable. We can thus differentiate the equation $2g(y) = (x+y)f(x)$, to get $0 = (x+y)f'(x) + f(x)$. Then, $f(x) = -(x+y)f'(x)$. Now, $f(x)$ is not constant, since $\dfrac{f(x)}{g(y)} = \dfrac{2}{x+y}$ has different values for different values of x; in fact, $f'(x) = \dfrac{-2g(y)}{(x+y)^2}$ is never 0. Thus, we can divide, obtaining $-\dfrac{f(x)}{f'(x)} = x + y$. Hence, for all real y, $x + y = -\dfrac{f(x)}{f'(x)}$. But $-\dfrac{f(x)}{f'(x)}$ is independent of y; it has the same value for a given x whatever y is. But this is not true of the other side of the equation, $x + y$. This is a contradiction and, therefore, the original differential equation is not separable.

CHAPTER 8

Exercise 8.1

1. $4\sec 4x\tan 4x$ **3.** $5\sec^2 5x$ **5.** $e^x(\sec x)(1+\tan x)$ **7.** $e^{-4x}(3\cos 3x - 4\sin 3x)$ **9.** $\pi e^{\pi x}(\sec^2\pi x + \tan\pi x)$
11. $2x\cos(x^2)$ **13.** $4\sec^2 x\tan x$ **15.** $2(x+1)\sec(x^2+2x-1)\tan(x^2+2x-1)$ **17.** $-4\csc^2(1+2x)\cot(1+2x)$
19. $2x(2x\cos 4x + \sin 4x)$ **21.** $-\dfrac{\cos(1/x)}{2x^2\sqrt{\sin(1/x)}}$ **23.** $x\sec^2 x + \tan x$ **25.** $2x\sin x(x\cos x + \sin x)$
27. $\sec x(\sec^2 x + \tan^2 x)$ **29.** $\dfrac{(1+x)\sec x\tan x - \sec x}{(1+x)^2}$ **31.** $\dfrac{1-\sec y - y\sec^2 x}{x\sec y\tan y + \tan x}$ **33.** $\dfrac{3-2y\sec^2(xy)\tan(xy)}{2x\sec^2(xy)\tan(xy)}$
35. $2(3x^2+1)\sec^2(x^3+x)\tan(x^3+x)$ **37.** $-2\csc^2 x\cot x e^{\csc^2 x}$ **39.** $-4^{\cot x}(\ln 4)(\csc^2 x)$
41. $-\sqrt{3}\sin x(\cos x)^{\sqrt{3}-1}$ **43.** $6^{\sec x}(\ln 6)(\sec x\tan x)$ **45.** $-\sin(2^x)2^x\ln 2$
47. $(\cos x)^{\sin x}[\cos x\ln(\cos x) - \sin x\tan x]$ **49.** $(\cos x)^x[\ln\cos(x) - x\tan x]$ **51.** 1 **53.** $\frac{1}{5}\ln|\sec 5x + \tan 5x| + C$
55. $\frac{1}{2}\ln|\sec 2x| + C$ **57.** $\frac{1}{4}\sec 4x + C$ **59.** $\frac{1}{2}\tan 2x + C$ **61.** $\frac{2}{3}(\tan x)^{3/2} + C$ **63.** $\frac{1}{5}\sin 5x + C$
65. $\frac{1}{3}\sec(3x-1) + C$ **67.** $\sec x + C$ **69.** $(\pi+4)/4$ **71.** $-e^{\cos x} + C$ **73.** $(2\sqrt{3})/3 + 2$
75. $-\ln[\sqrt{3}(2-\sqrt{3})]$

77. 1. Domain: $\mathbb{R}$
2. x-intercepts: $(k\pi, 0)$
3. Asymptotes: none
4. Symmetric about the origin
5. Critical numbers: $(2k+1)\frac{\pi}{2}$, $k\pi$
6. f increasing on $\left[-\frac{\pi}{2}+2k\pi, \frac{\pi}{2}+2k\pi\right]$; f decreasing on $\left[\frac{\pi}{2}+2k\pi, \frac{3\pi}{2}+2k\pi\right]$
7. Local maxima at $\frac{\pi}{2}+2k\pi$; local minima at $\frac{3\pi}{2}+2k\pi$
8. f concave up on $[\pi+2k\pi, 2\pi+2k\pi]$; f concave down on $[2k\pi, \pi+2k\pi]$
9. Inflection points: $k\pi$
10.

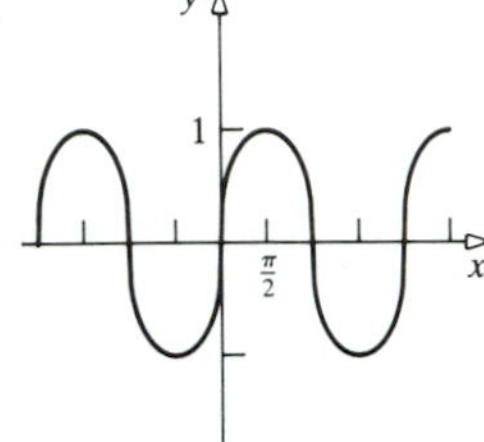

79. π **81.** $(2/\pi)\ln 2$ **83.** $\int \csc x\, dx = \int \frac{\csc^2 x - \csc x \cot x}{\csc x - \cot x}\, dx = \ln|\csc x - \cot x| + C$

85. $C_2 = \frac{1}{2} + C_1$; $C_3 = C_1 + \frac{1}{4}$ **87.** $\ln\left|\tan\left(\frac{x}{2}+\frac{\pi}{4}\right)\right| + C$ **89.** $R = \frac{4\sqrt{6}}{3}$; $h = \frac{4\sqrt{3}}{3}$ **91.** $\frac{\sqrt{3}}{3}$ cm^2/min

93. $\left(\frac{11\pi}{120}\right)\cos\left(\frac{11\pi}{36}\right) \approx 0.165$ in.^{2}/min

Exercise 8.2

1. $\pi/4$ **3.** $-\pi/2$ **5.** 0 **7.** 0 **9.** $\frac{1}{2}$ **11.** 0 **13.** $\frac{4}{5}$ **15.** $2\sqrt{6}$ **17.** $\pi - 2$ **19.** $(2+\sqrt{40})/9$
29. $\alpha + \beta = \pi/2$; $\cot\alpha = \tan\beta = x$; $\alpha = \cot^{-1}x$, $\beta = \tan^{-1}x$; therefore, $\cot^{-1}x + \tan^{-1}x = \pi/2$

Exercise 8.3

1. $4/\sqrt{1-16x^2}$ **3.** $-2x/\sqrt{1-x^4}$ **5.** $-5/(5x+1)\sqrt{(5x+1)^2-1}$ **7.** 0 **9.** $-1/(1+x^2)$
11. $2/(8x^2-4x+1)$ **13.** $\frac{1}{\sqrt{2ax-x^2}}$ **15.** $\frac{e^x}{\sqrt{1-e^{2x}}}$ **17.** $-1/2x\sqrt{x-1}$ **19.** $-2x/\sqrt{2x^2-x^4}$
21. $2(1+\tan^{-1}x)/(1+x^2)$ **23.** $-x/\sqrt{x^2-x^4}$ **25.** $-2x^2/(x^4+2x^2+2) + \cot^{-1}(1+x^2)$ **27.** $1/[\sqrt{x}(x+1)]$
29. $(-\sin x)/\sqrt{\sin^2 x}$ **31.** $1/x[1+(\ln x)^2]$ **33.** $1/[(1+x^2)\tan^{-1}x]$ **35.** $(\cos x)/(1+\sin^2 x)$ **37.** $\tan^{-1}(x/a)$
39. $\sqrt{1-y^2}/[\sqrt{1-x^2}(1-\sqrt{1-y^2})]$ **41.** $5\pi/4$ **43.** $\frac{1}{5}\tan^{-1}(x/5)+C$ **45.** $\sin^{-1}(x/3)+C$
47. $\frac{1}{3}\sin^{-1}(3x/4)+C$ **49.** $\tan^{-1}(\sin x)+C$ **51.** $-\sin^{-1}(\cos x/2)+C$ **53.** $2\sin^{-1}(2x^2-1)+C$
55. $\frac{2}{\sqrt{5}}\tan^{-1}\frac{2x+1}{\sqrt{5}}+C$ **57.** $\frac{1}{4}\ln(2x^2+2x+3) - \frac{1}{2\sqrt{5}}\tan^{-1}\frac{2x+1}{\sqrt{5}}$ **59.** $\sin^{-1}(x-1)+C$ **61.** $\frac{1}{3}\sin^{-1}x^3+C$
63. $\sin^{-1}\frac{1}{3}(2x-3)+C$ **65.** $\pi/3$ **67.** $(1/\sqrt{5})\tan^{-1}\sqrt{5}$
69. The domain of f is $[-1, 1]$; within the domain $f(x) = x$.

71. $g(x) = \begin{cases} x & \text{if } 0 \le x \le \pi \\ 2\pi - x & \text{if } \pi \le x \le 2\pi \end{cases}$ The graph repeats every 2π.

73. $\frac{d}{dx}\tan^{-1}\left(\frac{1}{x}\right) = \frac{-1/x^2}{1+(1/x)^2} = \frac{-1}{1+x^2} = \frac{d}{dx}\cot^{-1}x$ **75.** $\frac{d}{dx}\tan^{-1}(\cot x) = \frac{-\csc^2 x}{1+\cot^2 x} = \frac{-\csc^2 x}{\csc^2 x} = -1$

77. 0.0366 radian/sec **79.** $y' = \frac{1}{\sqrt{1-x^2}} - \frac{1}{\sqrt{1-x^2}} = 0 \Rightarrow \sin^{-1}x + \cos^{-1}x = C \Rightarrow$ if $x = 0$, then $C = \pi/2$

81. $N = \sqrt{1 - \frac{4}{\pi^2}} \approx 0.77$ **83.** $\int \frac{dx}{\sqrt{a^2 - x^2}} = \frac{1}{a}\int \frac{dx}{\sqrt{1-(x/a)^2}} \underset{u = x/a}{=} \frac{1}{a}\int \frac{a\,du}{\sqrt{1-u^2}} = \sin^{-1}\left(\frac{x}{a}\right) + C$ **85.** $y = x/2$

87. $\frac{d}{dx}(x\sin^{-1}x + \sqrt{1-x^2} + C) = \sin^{-1}x + \frac{x}{\sqrt{1-x^2}} - \frac{x}{\sqrt{1-x^2}} = \sin^{-1}x$

Exercise 8.4

1. $\frac{5}{3}$ **3.** $2\sin(2t)$ **5.** $\cos(t/2)$ **7.** $\theta(t) = \theta_0\cos[\sqrt{2k/m}(t/R)]$; $T = 2\pi R\sqrt{m/2k}$

Exercise 8.5

1. 1 **3.** $\tanh^2 x + \operatorname{sech}^2 x = \frac{\sinh^2 x}{\cosh^2 x} + \frac{1}{\cosh^2 x} = \frac{1 + \sinh^2 x}{\cosh^2 x} = \frac{\cosh^2 x}{\cosh^2 x} = 1$

5. $\sinh(-A) = \frac{e^{-A} - e^{A}}{2} = -\frac{e^{A} - e^{-A}}{2} = -\sinh A$

7. $\sinh A\cosh B + \cosh A\sinh B = \left(\frac{e^A - e^{-A}}{2}\right)\left(\frac{e^B + e^{-B}}{2}\right) + \left(\frac{e^A + e^{-A}}{2}\right)\left(\frac{e^B - e^{-B}}{2}\right) = \frac{e^{A+B} - e^{-(A+B)}}{2} = \sinh(A+B)$

9. $\cosh 3x = \cosh(2x + x) = \cosh 2x\cosh x + \sinh 2x\sinh x$
$= \cosh x(\cosh^2 x + \sinh^2 x) + \sinh x(2\sinh x\cosh x)$
$= 2\cosh^3 x - \cosh x + 2\cosh^3 x - 2\cosh x = 4\cosh^3 x - 3\cosh x$

11. $3\cosh 3x$ **13.** $2x\sinh(x^2+1)$ **15.** $\cosh x\cosh 4x + 4\sinh x\sinh 4x$ **17.** $2\tanh x\operatorname{sech}^2 x$

19. $[\operatorname{csch}^2(1/x)]/x^2$ **21.** $2x\operatorname{sech}^2 x^2$ **23.** $2\cosh\sqrt{x} + C$ **25.** $\frac{1}{4}\sinh 2x - \frac{1}{2}x + C$ **27.** $\frac{1}{2}\sinh^2 x + C$

29. $2\sqrt{2}\sinh\left(\frac{x}{2}\right) + C$ **31.** $e - (1/e)$ **33.** $\frac{L^2}{8(d-b)} - \frac{d-b}{2}$ **35.** $a\sinh\left(\frac{x}{a}\right)$

37. $\pi\left[\frac{a^3}{4}\sinh\left(\frac{2}{a}\right) + \frac{a^2}{2} + 2a^2(b-a)\sinh\left(\frac{1}{a}\right) + (b-a)^2\right]$

39. $(\cosh x + \sinh x)^n = \left(\frac{e^x + e^{-x} + e^x - e^{-x}}{2}\right)^n = e^{nx} = \cosh(nx) + \sinh(nx)$

Exercise 8.6

1. $3/\sqrt{9x^2+1}$ **3.** $2/(2x - x^3)$ **5.** $1/\sqrt{x^2+x}$ **7.** $-2\csc 2x$ **9.** $(\sec^2 x)/(1 - \tan^2 x)$

11. Let $y = \cosh^{-1}x$ for $x \ge 1$; then $x = \cosh y = (e^y + e^{-y})/2$ or $2x = e^y + e^{-y}$; $e^{2y} - 2xe^y + 1 = 0$;
$e^y = \frac{2x \pm \sqrt{4x^2 - 4(1)}}{2(1)} = x \pm \sqrt{x^2 - 1}$. The only possible solution is $e^y = x + \sqrt{x^2 - 1}$, so
$\cosh^{-1}x = y = \ln(x + \sqrt{x^2 - 1})$ for $x \ge 1$.

13. $y = \sinh^{-1}x$
$x = \sinh y$
$1 = (\cosh y)(y')$
$$y' = \frac{1}{\cosh y} = \frac{1}{\sqrt{\sinh^2 y + 1}} = \frac{1}{\sqrt{x^2+1}}$$

15. $y = \tanh^{-1}x$
$x = \tanh y$
$1 = (\operatorname{sech}^2 y)(y')$
$$y' = \frac{1}{1 - \tanh^2 y} = \frac{1}{1 - x^2}$$

17.

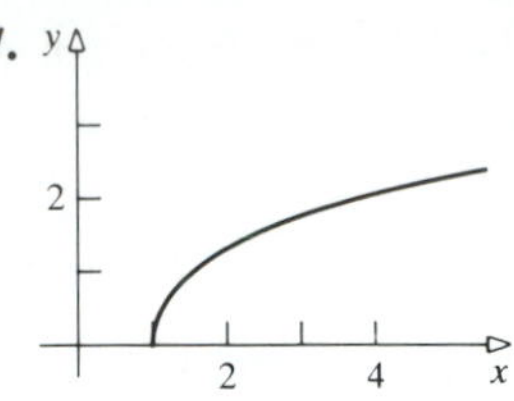

19. $\ln\left(\frac{3 + \sqrt{8}}{2 + \sqrt{3}}\right)$

21. Arc length $= \pi a$. By geometry, the arc length $= \frac{1}{2} \times$ circumference of circle of radius a, or πa.

Review Exercises

1. -2 **3.** $-\sin y$ **5.** $\int_0^b \sec x\, dx$ **7.** $2 \csc x \sec x$ **9.** $-\sqrt{\frac{2}{3}}$ **11.** $\frac{\sqrt{5} - 2\sqrt{2}}{3\sqrt{3}}$ **13.** $\sin x, -\frac{\pi}{2} < x < \frac{\pi}{2}$

15. $2\sqrt{2x - x^2}$ **17.** $\frac{(x^2-4)^2}{(x^2+4)^2}$ **19.** $-\frac{x^2}{x^2+1} + 2x \tan^{-1}\frac{1}{x}$ **21.** $\frac{\pi}{8}$ **23.** $\frac{\pi}{36}$ **25.** $\frac{\pi a^2}{2}$ **27.** $k = \frac{20}{9}$

29. $x \cosh x + \sinh x$ **31.** $\frac{x^2}{1-x^2} + 2x \tanh^{-1}x, |x| < 1$ **33.** $e^x \sinh e^x$ **35.** $\sinh 2$ **37.** $\frac{e^x}{\sqrt{1 + e^{2x}}}$

39. $\tanh^{-1}\frac{1}{2} \approx 0.55$

Challenge Exercises

1. (a) $A = k \tan^{-1}k - \frac{1}{2}\ln(k^2+1)$ (b) $\frac{\pi}{4} - \frac{1}{2}\ln 2$ (c) $\frac{\pi}{40}$ square units/sec

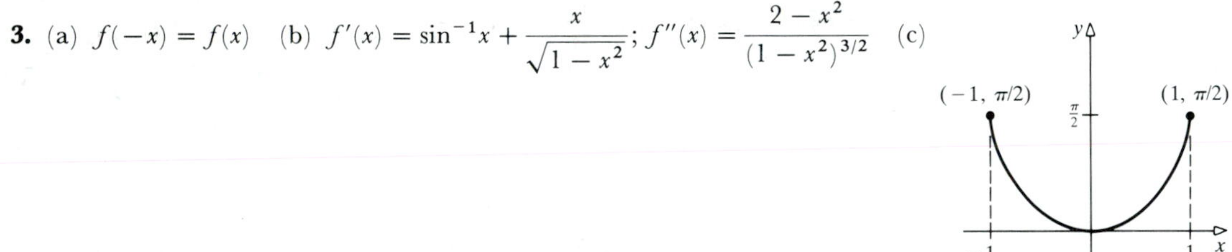

3. (a) $f(-x) = f(x)$ (b) $f'(x) = \sin^{-1}x + \frac{x}{\sqrt{1-x^2}}$; $f''(x) = \frac{2-x^2}{(1-x^2)^{3/2}}$ (c)

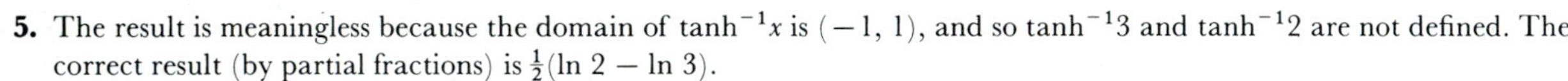

5. The result is meaningless because the domain of $\tanh^{-1}x$ is $(-1, 1)$, and so $\tanh^{-1}3$ and $\tanh^{-1}2$ are not defined. The correct result (by partial fractions) is $\frac{1}{2}(\ln 2 - \ln 3)$.

7. $B = A \cos \theta_0$; $C = A \sin \theta_0$

9. (a) $+12\sqrt{2}$ cm (b) Direction is downward; magnitude $= 30\pi^2\sqrt{2}$ dynes (c) $\frac{4}{3}$ sec (d) $-6\pi\sqrt{3}$ cm/sec

11. (a)

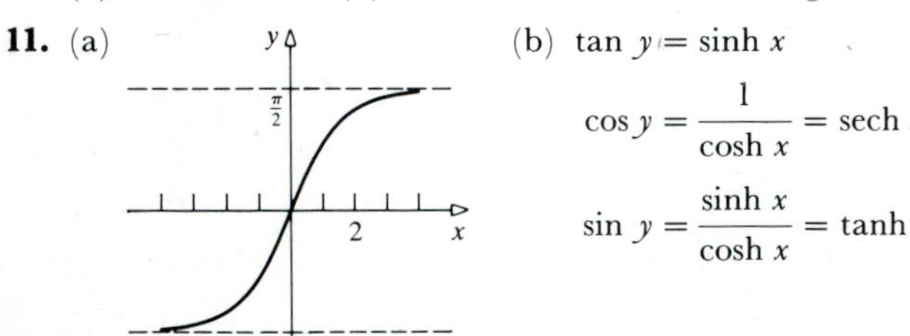

(b) $\tan y = \sinh x$
$$\cos y = \frac{1}{\cosh x} = \operatorname{sech} x$$
$$\sin y = \frac{\sinh x}{\cosh x} = \tanh x$$

(c) $y' = \frac{\cosh x}{1 + \sinh^2 x} = \frac{\cosh x}{\cosh^2 x} = \operatorname{sech} x = \cos y$ (d) $y' = \cos y$; $\int \sec \theta\, d\theta = \int dx = x + C = \operatorname{gd}^{-1}(\theta) + C$

CHAPTER 9

Exercise 9.1

1. $\frac{2}{9}(3x+1)^{3/2}+C$ **3.** $\frac{1}{36}(3x^2-5)^6+C$ **5.** $-\ln|1+\cos x|+C$ **7.** $-\frac{1}{9}(1-9x^2)^{1/2}+C$
9. $\sin^{-1}(x/2)+C$ **11.** $\frac{1}{2}\ln(1+2e^x)+C$ **13.** $2e^{\sqrt{x}}+C$ **15.** $\frac{1}{2}e^{x^2}+C$ **17.** $e^{\sin x}+C$ **19.** $\frac{1}{2}\sin^2 x+C$
21. $\frac{1}{4}\tan^{-1}(x/4)+C$ **23.** $-\frac{1}{2}(\cot 2x+\csc 2x)+C$ **25.** $-\frac{1}{2}\csc x^2+C$ **27.** $\sec t+C$ **29.** $\frac{1}{3}\sec^{-1}(x/3)+C$
31. $x-\frac{1}{3}\sin(3x)+C$ **33.** $(5^x/\ln 5)+C$ **35.** $-\ln|1-\cosh x|+C$ **37.** $v(t)=\frac{1}{30}(t^3+4)^{1/2}-\frac{1}{15}$

Exercise 9.2

1. $(e^{2x}/4)(2x-1)+C$ **3.** $x\sin x+\cos x+C$ **5.** $(2x^{3/2}/3)(\ln x-\frac{2}{3})+C$ **7.** $x\cot^{-1}x+\frac{1}{2}\ln(1+x^2)+C$
9. $x(\ln x)^2-2x\ln x+2x+C$ **11.** $(e^x/2)(\sin x-\cos x)+C$ **13.** $2-5e^{-1}$ **15.** $2x\sin x+(2-x^2)\cos x+C$
17. $(x^2/4)+[(x\sin 2x)/4]+[(\cos 2x)/8]+C$ **19.** $\frac{2}{27}(1-25e^{-6})$ **21.** $x\cosh x-\sinh x+C$
23. $x^{n+1}[(\ln x)/(n+1)-1/(n+1)^2]+C$ **25.** $x[(\ln x)^3-3(\ln x)^2+6\ln x-6]+C$
27. $(e^x/2)[x(\cos x+\sin x)-\sin x]+C$ **29.** $(x^3/3)[(\ln x)^2-\frac{2}{3}\ln x+\frac{2}{9}]+C$
31. $(x^3/3)\tan^{-1}x-(x^2/6)+\frac{1}{6}\ln(1+x^2)+C$ **33.** $(x/2)[\sin(\ln x)-\cos(\ln x)]+C$ **35.** $(\sin x)[\ln(\sin x)-1]+C$
37. $\frac{1}{2}(9\ln 9-8)$ **39.** $(c^3/3)[(\ln c)^2-\frac{2}{3}\ln c+\frac{2}{9}]-\frac{2}{27}$ **41.** $(2+e^x)[\ln(2+e^x)-1]+C$ **43.** $7\ln 7-6\ln 6-1$
45. Let $u=\ln(x+\sqrt{x^2+a^2})$, $dv=dx$. **47.** Let $u=\sin^{-1}x$, $dv=x^n\,dx$. **49.** Let $u=x^n$, $dv=(ax+b)^{1/2}\,dx$.
51. Let $u=\sin^{n-1}x$, $dv=\sin x\,dx$.
53. Use integration by parts, with $u=t$ and $dv=g'(t)\,dt$, to get $tg(t)\Big]_0^x-\int_0^x g(t)\,dt=xg(x)-\int_0^x g(t)\,dt$
55. An identity in x results; $\frac{1}{4}(2x+e^{2x})+C$. **57.** 1 **59.** $\frac{9}{2}\ln 3-4$ **61.** $\pi(\pi-2)$ **63.** $\pi(\pi-2)$
65. (a) $5\pi/32$ (b) $\frac{8}{15}$ (c) $35\pi/256$ (d) $5\pi/32$

Exercise 9.3

1. $\frac{1}{3}\sin^3x+C$ **3.** $\sec x+C$ **5.** $-\cos x+\frac{1}{3}\cos^3x+C$ **7.** $\frac{1}{2}[x-(\sin 6x)/6]+C$
9. $\sin x-\frac{2}{3}\sin^3x+\frac{1}{5}\sin^5x+C$ **11.** $-\frac{1}{3}\cos 3x+\frac{2}{9}\cos^3 3x-\frac{1}{15}\cos^5 3x+C$ **13.** $-\frac{1}{3}\cos^3x+\frac{1}{5}\cos^5x+C$
15. $-\frac{1}{6}\cos^6x+\frac{1}{8}\cos^8x+C$ **17.** $\frac{3}{4}\sin^{4/3}x+C$ **19.** $\frac{2}{3}\sin^{3/2}x-\frac{2}{7}\sin^{7/2}x+C$ **21.** $\frac{1}{8}[x-(\sin 4x)/4]+C$
23. $(\tan^2x)/2+\ln|\cos x|+C$ **25.** $(\tan^3x)/3+C$ **27.** $(-\csc^7x)/7+(2\csc^5x)/5-(\csc^3x)/3+C$
29. $\frac{1}{4}\tan x\sec^3x-\frac{1}{8}\sec x\tan x-\frac{1}{8}\ln|\sec x+\tan x|+C$ **31.** $-\frac{1}{4}\cos 2x-\frac{1}{8}\cos 4x+C$
33. $\frac{1}{4}\sin 2x-\frac{1}{12}\sin 6x+C$ **35.** $\frac{1}{2}\sin x+\frac{1}{6}\sin 3x+C$ **37.** $\frac{1}{2}\cos x-\frac{1}{4}\cos 2x+C$
39. $\tan x+\frac{k-1}{3}\tan^3x+\frac{(k-1)(k-2)}{2\cdot 5}\tan^5x+\cdots+\frac{k-1}{2k-3}\tan^{2k-3}x+\frac{1}{2k-1}\tan^{2k-1}x+C$ **41.** $\pi^2/2$
43. $s(t)=\frac{16}{3}t-\frac{1}{3}\sin t+\frac{1}{9}\sin^3t$

Exercise 9.4

1. $\sin^{-1}(x/2)+C$ **3.** $(-\sqrt{9-x^2}/x)-\sin^{-1}(x/3)+C$ **5.** $(x/2)\sqrt{x^2-1}+\frac{1}{2}\ln|x+\sqrt{x^2-1}|+C$
7. $(x/4\sqrt{x^2+4})+C$ **9.** $(-\sqrt{x^2+4}/4x)+C$ **11.** $\frac{1}{2}[x\sqrt{16-x^2}+16\sin^{-1}(x/4)]+C$ **13.** $(x/4\sqrt{4-x^2})+C$
15. $(-x/2)\sqrt{16-x^2}+8\sin^{-1}(x/4)+C$ **17.** $\sqrt{x^2-1}-\sec^{-1}x+C$ **19.** $(-x/9\sqrt{x^2-9})+C$
21. $\sin^{-1}(x-2)+C$ **23.** $\ln|(x-1)+\sqrt{(x-1)^2-4}|+C$ **25.** $[(x-1)/9\sqrt{x^2-2x+10}]+C$

27. $\ln|(x+1)+\sqrt{(x+1)^2-4}|+C$ **29.** $(\sqrt{3}/3)-(\pi/6)$ **31.** $(7\sqrt{17}-\sqrt{5})/24$ **33.** πab
35. Let $x = a\sec\theta$. **37.** $A = 4\int_0^R \sqrt{R^2-x^2}\,dx = \pi R^2$ **39.** 13.9 **41.** $\frac{1}{4}[(2x^2-1)\sin^{-1}x + x\sqrt{1-x^2}]+C$
43. Let $x = a\sin\theta$. **45.** Let $x = a\sec\theta$. **47.** Let $x = a\tan\theta$.

Exercise 9.5

1. $\tan^{-1}(x+2)+C$ **3.** $\sin^{-1}\left(\frac{x-1}{3}\right)+C$ **5.** $\sin^{-1}\left(\frac{x-2}{2}\right)+C$ **7.** $\sin^{-1}\left(\frac{x+1}{5}\right)+C$
9. $\sqrt{x^2-2x+5}+\ln\left|\frac{\sqrt{x^2-2x+5}}{2}+\frac{x-1}{2}\right|+C$ **11.** $\ln|\sqrt{2}+1|$ **13.** $\ln(\sqrt{e^{2x}+e^x+1}+e^x+\frac{1}{2})+C$
15. $-2\sqrt{4x-x^2-3}+\sin^{-1}(x-2)+C$ **17.** substitute $x+h=\sqrt{k}\tan\theta$

Exercise 9.6

1. $(x^2/2)-x+2\ln|x+1|+C$ **3.** $(x^3/3)+x^2+7x+10\ln|x-2|+C$ **5.** $\frac{1}{3}\ln|(x-2)/(x+1)|+C$
7. $\ln[(x-2)^2/|x-1|]+C$ **9.** $\frac{1}{4}\ln|(x+1)(x-1)^3|-\left[\frac{1}{2(x-1)}\right]+C$
11. $5\ln|(x+1)/(x+2)|+[4/(x+1)]+C$ **13.** $\frac{2}{21}\ln|3x-2|+\frac{1}{14}\ln|2x+1|+C$ **15.** $\frac{1}{4}\ln|(x+3)^3(x-1)|+C$
17. $\ln|(x-3)^2/x(x+1)|+C$ **19.** $1/(x-1)+\ln|(x-2)^4/(x-1)^3|+C$
21. Let $u=\sin\theta$; $\frac{1}{5}\ln|(\sin\theta-2)/(\sin\theta+3)|+C$. **23.** $\frac{1}{2}\ln\left|\frac{e^{-x}-1}{e^{-x}+1}\right|+C$ **25.** $\frac{1}{5}\ln\left|\frac{e^x-2}{e^x+3}\right|+C$
27. $\frac{x^2+1}{2}\tan^{-1}x-\frac{x}{2}+C$ **29.** $(-\ln 2)/6$ **31.** $(\ln 21)/8$ **33.** $\ln\frac{15}{7}$ **35.** $\pi(\frac{1}{8}\ln\frac{15}{7}+\frac{19}{105})$

Exercise 9.7

1. $\ln|x/\sqrt{x^2+1}|+C$ **3.** $\ln(|x-1|/\sqrt{x^2+x+1})+(1/\sqrt{3})\tan^{-1}[(2x+1)/\sqrt{3}]+C$
5. $\frac{1}{8}\ln|x^2/(x^2+4)|-(1/x)-\frac{1}{2}\tan^{-1}(x/2)+C$ **7.** $\frac{1}{3}\ln|(x+1)^2\sqrt{x^2+2x+4}|+C$
9. $\frac{7}{16}\tan^{-1}(x/2)-1/(x^2+4)-x/8(x^2+4)+C$ **11.** $-\frac{(x^2+8)}{2(x^2+16)^2}+C$ **13.** $\frac{1}{2}\ln|(1+\cos^2\theta)/(\cos^2\theta)|+C$
15. 8.72 **17.** $\frac{2}{1-\tan(x/2)}+C$ **19.** $-\cot(x/2)+C$ **21.** $\sqrt{2}\ln\left|\frac{\tan(x/2)-1+\sqrt{2}}{\tan(x/2)-1-\sqrt{2}}\right|+C$
23. $-\ln|3+\cos x|+C$ **25.** $-\frac{1}{4}\left(\tan\frac{x}{2}\right)^{-2}-\frac{1}{2}\ln\left|\tan\frac{x}{2}\right|+C$ **27.** $\ln\left|\frac{2\tan(x/2)}{[1+\tan(x/2)]^2}\right|+C$ **29.** 1
31. $\pi\left(\frac{4}{3\sqrt{3}}-\frac{1}{2}\right)$ **33.**

$$\sec x+\tan x=\frac{1+\sin x}{\cos x}=\frac{1+2\sin(x/2)\cos(x/2)}{\cos^2(x/2)-\sin^2(x/2)}=\frac{\sec^2(x/2)+2\tan(x/2)}{1-\tan^2(x/2)}$$
$$=\frac{1+\tan^2(x/2)+2\tan(x/2)}{[1+\tan(x/2)][1-\tan(x/2)]}=\frac{1+\tan(x/2)}{1-\tan(x/2)}$$

35. $\csc x-\cot x=\frac{1-\cos x}{\sin x}=\frac{1-\cos x}{\sqrt{(1-\cos x)(1+\cos x)}}=\sqrt{\frac{1-\cos x}{1+\cos x}}$ **37.** $\frac{2}{3}x^{3/2}-3x+18\sqrt{x}-54\ln|\sqrt{x}+3|+C$
39. $\frac{3}{2}\ln|x^{2/3}-1|+C$ **41.** $2\sqrt{x}-3\sqrt[3]{x}+6\sqrt[6]{x}-6\ln|\sqrt[6]{x}+1|+C$ **43.** $\frac{1}{2}(2+3x)^{2/3}+C$
45. $\frac{4}{5}(1+x)^{5/4}-4(1+x)^{1/4}+C$ **47.** $2[\sqrt{x}-\ln|\sqrt{x}+1|]+C$ **49.** $x+3x^{2/3}+6x^{1/3}+6\ln|x^{1/3}-1|+C$
51. $959\pi/30$

Exercise 9.9

1. 21.5 **3.** 0.6970 **5.** 1.9541 **7.** 0.4832 **9.** 0.74298 **11.** 21.333 **13.** 0.5004 **15.** 1.4642 **17.** 2.3351 **19.** 0.7468 **21.** 0.5622; 0.5620 **23.** 18.8396; 18.8371 **25.** 0.6956 **27.** 16,787.5 m^3 **29.** 2500 **31.** 131,787.5 m^3; 132,625 m^3 **33.** 230 **35.** 0.1667 **37.** 0.0026 **39.** 1.9101 **41.** 6.4287 **43.** 1.8446 **45.** 18.15

Review Exercises

1. $\frac{1}{5}x^5 - \frac{2}{3}x^3 + C$ **3.** $\frac{1}{4}\tan^{-1}\left(\frac{x+2}{4}\right) + C$ **5.** $\frac{1}{2}\tan 2\theta + C$ **7.** $2\sqrt{x} - 2\tan^{-1}\sqrt{x} + C$ **9.** $\frac{1}{2}\ln(9 + t^2) + C$

11. $-\cot\theta - \theta + C$ **13.** $-e^{\cos x} + C$ **15.** $-\frac{x}{2}\cos 2x + \frac{1}{4}\sin 2x + C$ **17.** $\frac{1}{5}z^5 + \frac{8}{3}z^3 + 16z + C$

19. $\frac{2}{9}(2-t)^{9/2} - \frac{12}{7}(2-t)^{7/2} + \frac{24}{5}(2-t)^{5/2} - \frac{16}{3}(2-t)^{3/2} + C$ **21.** $\frac{1}{3}\sin^3 x - \frac{1}{5}\sin^5 x + C$ **23.** $\sqrt{2y+1} + C$

25. $e^t + 2\ln|e^t - 2| + C$ **27.** $\frac{1}{2}\ln(\cosh 2v) + C$ **29.** $\sin x + \cos x + \ln|\csc x - \cot x| + C$ **31.** $\frac{1}{8}\ln\left|\frac{x^2-2}{x^2+2}\right| + C$

33. $\ln|\ln x| + C$ **35.** $\ln\left|\cos\left(\frac{\pi}{4} - \theta\right)\right| + C$ **37.** $\frac{(ab)^x}{\ln(ab)} + C$ **39.** $x\tan x + \ln|\cos x| + C$

41. $\frac{1}{3}x^3 + \frac{1}{2}x^2 - x - \ln|x-1| + C$ **43.** $\frac{1}{24}(3y^2 - 6y)^4 + C$ **45.** $-\frac{3}{2}x^2 - 3x - 3\ln|1-x| + C$

47. $\frac{1}{3}\sin 3x - \frac{1}{9}\sin^3 3x + C$ **49.** $-x - \ln\left|\frac{x-2}{x+2}\right| + C$ **51.** $\frac{1}{3}x^3\sin^{-1}x + \frac{1}{3}\sqrt{1-x^2} - \frac{1}{9}(1-x^2)^{3/2} + C$

53. $-\frac{2}{3}x^3 - x - \frac{2}{3}(x^2+1)^{3/2} + C$ **55.** $\frac{1}{4}x^2 + \frac{1}{4}x\sin 2x + \frac{1}{8}\cos 2x + C$ **57.** $\frac{1}{a}\ln\left|\frac{x}{x+a}\right| + C$

59. $-\frac{1}{2}w^2 - \frac{1}{2}\ln|w^2 - 1| + C$ **61.** $\frac{1}{2(m-n)}\sin(m-n)x - \frac{1}{2(m+n)}\sin(m+n)x + C$

63. $\frac{1}{2(m-n)}\sin(m-n)x + \frac{1}{2(m+n)}\sin(m+n)x + C$ **65.** $\frac{20}{31}x^{31/20} + C$ **67.** $f(x) = x^2\sin x + C$

69. $f(x) = \frac{4}{9}(x+5)^{3/2} - x - 6$ **71.** (a) $\pi(e-2)$ (b) $\frac{\pi}{2}(e^2+1)$ **73.** $\frac{6-2\sqrt{3}}{3} - \ln(2\sqrt{3} - 3)$

75. The maximum error possible in using Simpson's rule is less than or equal to $[M(b-a)^5]/180n^4$, where M is the maximum value of $|f^{(4)}(x)|$ for $x \in [a, b]$. If $f(x) = Ax^3 + Bx^2 + Cx + D$, then $f^{(4)}(x) = 0$ and $M = 0$. The maximum error is 0, therefore, and Simpson's rule is exact.

Challenge Exercises

1. $2\sqrt{x-2} + \sqrt{2}\tan^{-1}(\sqrt{(x-2)/2}) + C$ **3.** $\sin^{-1}\frac{2v-3}{3} + C$

5. $\int Si(x)\,dx = xSi(x) - \int x\frac{d}{dx}[Si(x)]\,dx$ (using integration by parts) $= xSi(x) - \int x\frac{\sin x}{x}\,dx = xSi(x) + \cos x + C$

7. (a) Using integration by parts, let $u = f'(t)$ and $dv = dt$.

$$\int_a^b f'(t)\,dt = tf'(t)\Big|_a^b - \int_a^b tf''(t)\,dt = bf'(b) - af'(a) - \int_a^b tf''(t)\,dt = bf'(a) - bf'(a) + bf'(b) - af'(a) - \int_a^b tf''(t)\,dt = f'(a)(b-a) + b(f'(b) - f'(a)) - \int_a^b tf''(t)\,dt = f'(a)(b-a) + b\int_a^b f''(t)\,dt - \int_a^b tf''(t)\,dt = f'(a)(b-a) - \int_a^b (t-b)f''(t)\,dt$$

(b) Follow a procedure similar to the one in part (a).

11. If d and θ are as described in the problem, then $\sin\theta$ is the vertical distance that the stick displaces. Hence, the stick will hit a crack if and only if this is at least d—that is, if and only if $d \le \sin\theta$. Thus, the probability of hitting a crack is the probability that $d \le \sin\theta$ for $0 \le d \le 1$, $0 \le \theta \le \pi$, which is the proportion of the shaded area to the rectangular area, $\dfrac{\int_0^\pi \sin\theta\, d\theta}{\pi} = \dfrac{2}{\pi}$

CHAPTER 10

Exercise 10.1

1. 5 **3.** $-\frac{12}{5}$ **5.** $\frac{3}{4}$ **7.** $\frac{1}{2}$ **9.** 2 **11.** 0 **13.** 0 **15.** 2 **17.** 0 **19.** 0 **21.** 1 **23.** $\frac{1}{4}$
25. $-\frac{1}{6}$ **27.** $\frac{2}{9}$ **29.** 0 **31.** $+\infty$ **33.** 0 **35.** 0 **37.** 0 **39.** 2 **41.** 2
43. $(\ln 2)/(\ln 3)$ **45.** $+\infty$ **47.** 0 **49.** 1 **51.** $\frac{1}{3}$ **53.** -2 **55.** 1 **57.** 1

59. Limit does not exist **61.** 0 **63.** 0 **65.** $\displaystyle\lim_{x\to+\infty}\frac{(\ln x)^\beta}{x^\alpha} = \lim_{x\to+\infty}\frac{\beta(\ln x)^{\beta-1}(1/x)}{\alpha x^{\alpha-1}} = \lim_{x\to+\infty}\frac{\beta(\ln x)^{\beta-1}}{\alpha x^\alpha} = \cdots = 0$

67. 1 **69.** $\displaystyle\lim_{x\to c}\frac{x^\alpha - c^\alpha}{x^\beta - c^\beta} = \lim_{x\to c}\frac{\alpha x^{\alpha-1}}{\beta x^{\beta-1}} = \frac{\alpha}{\beta}c^{\alpha-\beta}$ **71.** an

73. Let $x = 1/u$. Then $x \to -\infty$ implies $u \to 0^-$: $\displaystyle\lim_{x\to-\infty}\frac{f(x)}{g(x)} = \lim_{u\to 0^-}\frac{f(1/u)}{g(1/u)} = \lim_{u\to 0^-}\frac{(-1/u^2)f'(1/u)}{(-1/u^2)g'(1/u)} = \lim_{x\to-\infty}\frac{f'(x)}{g'(x)}$

Exercise 10.2

1. 1 **3.** 1 **5.** 0 **7.** 0 **9.** 1 **11.** 0 **13.** 0 **15.** $\frac{2}{\pi}$ **17.** 0 **19.** 0 **21.** $\frac{1}{2}$ **23.** $\frac{1}{2}$ **25.** 1
27. 1 **29.** 1 **31.** $\ln 3$ **33.** e **35.** 1 **37.** e^5 **39.** e **41.** 1 **43.** 1 **45.** 1

47. $y = (\cos x + 2\sin x)^{\cot x}$; $\displaystyle\lim_{x\to0^+}\ln y = 2$; $\displaystyle\lim_{x\to0^+} y = e^2$ **49.** 0 **51.** $\displaystyle\lim_{x\to0^+}\frac{e^{-1/x^2}}{x} = \lim_{x\to0^+}\frac{1/x}{e^{1/x^2}} = \lim_{x\to0^+}\frac{x}{2e^{1/x^2}} = 0$

53. $\displaystyle f'(0) = \lim_{h\to0}\frac{e^{-1/h^2} - 0}{h} = 0$

55. 1. Domain: $x > 0$
2. y-intercept: none. Note $\displaystyle\lim_{x\to0^+} x^{\sqrt{x}} = 1$
3. No symmetries
4. No asymptotes
5. Critical number: e^{-2}
6. f increasing on $(e^{-2}, +\infty)$; f decreasing on $(0, e^{-2})$
7. Local minimum at e^{-2}
8. f concave up on entire domain
9. No inflection points
10.

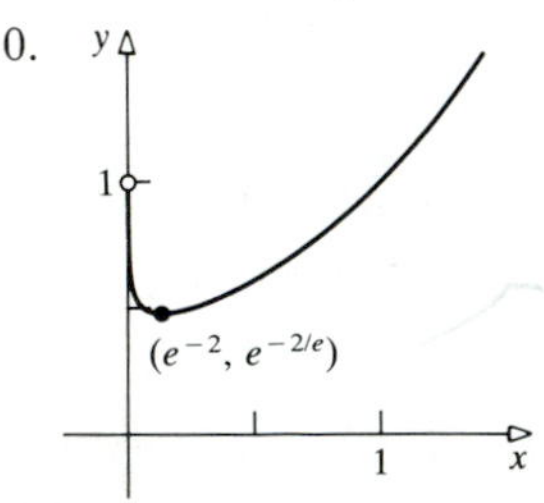

57. $y = \dfrac{1}{x}\tan x$

$\displaystyle\lim_{x\to0}\frac{\tan x}{x} = \lim_{x\to0}\frac{\sec^2 x}{1} = 1$

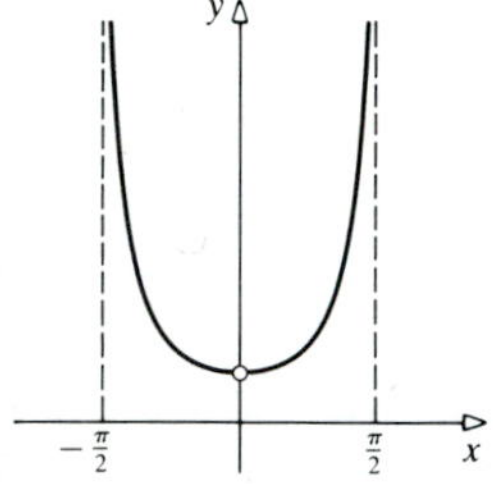

Exercise 10.3

1. Improper; upper limit is $+\infty$ **3.** Not improper **5.** Improper; $1/x$ not defined at 0

7. Improper; integrand not defined at 1 **9.** Converges; $\frac{1}{10}$ **11.** Diverges **13.** Diverges **15.** Converges; 2

17. Diverges **19.** Converges; $2\sqrt{a}$ **21.** Converges; $\frac{1}{2}$ **23.** Diverges **25.** Diverges **27.** Diverges

29. Converges; 0 **31.** Diverges **33.** Converges; $\pi/4$ **35.** Diverges **37.** Diverges **39.** Converges; π/a

41. Diverges **43.** Converges; $\pi a^2/4$ **45.** Diverges **47.** Diverges **49.** Converges; $(1/2a)\ln 3$

51. Converges; $1 - (\pi/4)$ **53.** Converges; $\pi/2$ **55.** Converges; 4 **57.** Diverges **59.** Diverges

61. Converges; 0 **63.** $4a^2\pi$ **65.** $\pi/2$ **67.** Does not exist

69. $\int_0^{+\infty} \sin x\, dx = \lim_{b\to+\infty} (-\cos b + 1)$, which does not exist; $\lim_{t\to+\infty} \int_{-t}^{t} \sin x\, dx = \lim_{t\to+\infty} [\cos(-t) - \cos t] = 0$ **71.** 1

73. $\frac{1}{2}$ **75.** $\frac{1}{2}$ **77.** $\frac{1}{2}LI^2$ **79.** $Im/5r$ **81.** Use by-parts formula with $u = x^n$, $dv = e^{-x}\, dx$. **83.** Let $u = \ln x$.

85. (a) $g(x) \geq f(x) \geq$ for $x \geq a \Rightarrow \int_a^b g(x)\, dx \geq \int_a^b f(x)\, dx \Rightarrow \lim_{b\to+\infty} \int_a^b g(x)\, dx \geq \lim_{b\to+\infty} \int_a^b f(x)\, dx \Rightarrow \int_a^{+\infty} g(x)\, dx \geq$ $\int_a^{+\infty} f(x)\, dx$ but $\int_a^{+\infty} f(x)\, dx$ diverges. Thus, so does $\int_a^{+\infty} g(x)\, dx$.

(b) Let $\int_a^{+\infty} g(x)\, dx = L$.

$$\int_a^b [g(x) - f(x)]\, dx = L - \int_a^b f(x)\, dx \geq 0$$

or $0 \leq \int_a^b f(x)\, dx \leq L$. Hence, $\int_a^b f(x)\, dx$ converges.

87. Since $\dfrac{\ln x}{\sqrt{x^2-1}} > \dfrac{1}{\sqrt{x^2-1}}$ for $x > e$ and $\int_2^{+\infty} \dfrac{dx}{\sqrt{x^2-1}}$ diverges, then so does $\int_2^{+\infty} \dfrac{\ln x\, dx}{\sqrt{x^2-1}}$. **89.** $s/(s^2+1)$

91. $1/(s-1)$ **93.** $1/s$

Exercise 10.4

1. $1 + 3(x-1)^2$ **3.** $-6 + 5(x-1) + 4(x-1)^2 + (x-1)^3$

5. $P_3(x) = 4 + 7(x-1) + 11(x-1)^2 + 3(x-1)^3$; $R_4(x) = 0$

7. $P_4(x) = 7 - 25(x+1) + 30(x+1)^2 - 14(x+1)^3 + 2(x+1)^4$; $R_5(x) = 0$

9. $P_4(x) = 32 + 80(x-2) + 80(x-2)^2 + 40(x-2)^3 + 10(x-2)^4$; $R_5(x) = (x-2)^5$

11. $P_5(x) = (x-1) - \frac{1}{2}(x-1)^2 + \frac{1}{3}(x-1)^3 - \frac{1}{4}(x-1)^4 + \frac{1}{5}(x-1)^5$; $R_6(x) = -(x-1)^6/6u^6$

13. $P_5(x) = 1 - (x-1) + (x-1)^2 - (x-1)^3 + (x-1)^4 - (x-1)^5$; $R_6(x) = (x-1)^6/u^7$

15. $P_6(x) = 1 - (x^2/2!) + (x^4/4!) - (x^6/6!)$; $R_7(x) = (x^7/7!)\sin u$

17. $P_6(x) = 1 + (x^2/2!) + (x^4/4!) + (x^6/6!)$; $R_7(x) = (x^7/7!)\sinh u$

19. $P_4(x) = 1 + x + x^2 + x^3 + x^4$; $R_5(x) = x^5/(1-u)^6$ **21.** $P_3(x) = 1 - 2x + 3x^2 - 4x^3$; $R_4(x) = 5x^4/(1+u)^6$

23. 0.095308 ($n = 4$) **25.** 0.91 ($n = 2$) **27.** 0.999847 ($n = 2$) **29.** 0.375 ($n = 4$) **31.** 1.6458 ($n = 3$)

33. $P_4(x) = (x-1) + \frac{1}{2}(x-1)^2 - \frac{1}{6}(x-1)^3 + \frac{1}{12}(x-1)^4$ **35.** $P_2(x) = 1 - x^2$

37. $P_4(x) = 1 + 2[x - (\pi/4)] + 2[x - (\pi/4)]^2 + \frac{8}{3}[x - (\pi/4)]^3 + \frac{10}{3}[x - (\pi/4)]^4$ **39.** $P_3(x) = 1 + \frac{1}{2}x - \frac{1}{8}x^2 + \frac{1}{16}x^3$

41. $f(x) \approx -\lambda\pi - (1+\lambda)(x-\pi) = 0$ **43.** 0.0100003 **45.** (a) Differentiate with respect to h; $f''(x)$ (b) $f'''(x)$

Review Exercises

1. $\frac{3}{2}$ **3.** $\frac{8}{9}$ **5.** 2 **7.** $\cos\theta$ **9.** $\frac{1}{6}$ **11.** 1 **13.** $\frac{1}{\pi}$ **15.** $-\frac{1}{2}$ **17.** 1 **19.** 1 **21.** $3/e$

23. $2(1-\cos 1)$ **25.** 1

27. $\int_0^{\pi/2} \frac{\sin x}{\sqrt{\cos x}}\,dx = \lim_{b\to\pi/2^-} \int_0^b \frac{\sin x}{\sqrt{\cos x}}\,dx = \lim_{b\to\pi/2^-} -2\sqrt{\cos x}\Big]_0^b = \lim_{b\to\pi/2^-}(-2\sqrt{\cos b}+2) = 2$

29. $A = \int_0^1 x^{-2/3}\,dx = \lim_{b\to 0^+}\int_b^1 x^{-2/3}\,dx = \lim_{b\to 0^+} 3x^{1/3}\Big]_b^1 = \lim_{b\to 0^+} 3(1-b^{1/3}) = 3$

31. $P_4(x) = e^6[1+2(x-3)+2(x-3)^2+\frac{4}{3}(x-3)^3+\frac{2}{3}(x-3)^4]$

33. $P_4(x) = \frac{1}{2} - \frac{(x-1)}{2^2} + \frac{(x-1)^2}{2^3} - \frac{(x-1)^3}{2^4} + \frac{(x-1)^4}{2^5}$ **35.** $P_3(x) = -3+4(x-1)+(x-1)^2+(x-1)^3$

37. $P_n(x) = e^3\left[1+(x-3)+\frac{(x-3)^2}{2}+\frac{(x-3)^3}{3!}+\cdots+\frac{(x-3)^n}{n!}\right]$ **39.** $\sqrt{e}\approx 1.6484$

Challenge Exercises

1. In order to apply L'Hospital's rule to $\lim_{x\to 0}\frac{f(x)}{g(x)}$, we must have $\lim_{x\to 0}\frac{f'(x)}{g'(x)} = L$, where L is a real number or $+\infty$ or $-\infty$. In this case, $\lim_{x\to 0}\frac{f'(x)}{g'(x)} = \lim_{x\to 0}\frac{2x\sin(1/x)+x^2\cos(1/x)(-1/x^2)}{\cos(x)} = \lim_{x\to 0}\frac{2x\sin(1/x)-\cos(1/x)}{\cos(x)}$, which fails to exist due to $\cos(1/x)$ oscillating between -1 and $+1$.

3. $a=2, b=-2, c=0, d=\frac{4}{3}$ **5.** $1-e^{-ea}$

7. $\int_{-\infty}^{+\infty} f(x)\,dx = \int_{-\infty}^{a} 0\,dx + \int_a^b \frac{1}{b-a}\,dx + \int_b^{+\infty} 0\,dx = \int_a^b \frac{1}{b-a}\,dx = \frac{1}{b-a}x\Big|_a^b = \frac{1}{b-a}(b-a) = 1$

9. $\mu = \frac{b+a}{2}$ **11.** $\sigma^2 = \frac{(b-a)^2}{12}$; $\sigma = \frac{(b-a)}{2\sqrt{3}}$

CHAPTER 11

Exercise 11.1

1. $1, \frac{1}{2}, \frac{1}{3}, \frac{1}{4}$ **3.** 0, ln 2, ln 3, ln 4 **5.** $\frac{1}{3}, -\frac{1}{5}, \frac{1}{7}, -\frac{1}{9}$ **7.** 1, 0, 1, 0 **9.** $\frac{1}{2}, \frac{1}{2}, \frac{3}{4}, \frac{3}{2}$ **11.** $s_n = 2n$ **13.** $s_n = 2^n$

15. $s_n = \frac{(-1)^{n+1}}{n+1}$ **17.** $s_n = \frac{n}{n+1}$ **19.** Convergent; 0 **21.** Convergent; 2 **23.** Convergent; 1

25. Convergent; 0 **27.** Divergent **29.** Convergent; 1 **31.** Convergent; 1 **33.** Convergent; $-\frac{1}{2}$

35. Convergent; 1 **37.** Convergent; 2 **39.** Divergent **41.** Convergent; 0 **43.** Convergent; 1

45. Convergent; 0 **47.** Convergent; $\pi/2$ **49.** Convergent; 1 **51.** Increasing **53.** Nonmonotonic

55. Decreasing **57.** Decreasing **59.** Decreasing **61.** Convergent; $\lim_{n\to+\infty} s_n = 0$

63. Convergent; $\lim_{n\to+\infty} s_n = 0$ **65.** Divergent; $\lim_{n\to+\infty} s_n = +\infty$ **67.** Convergent; $\lim_{n\to+\infty} s_n = 0$

69. Convergent; $\lim_{n\to+\infty} s_n = 0$ **71.** Divergent; oscillates from -1 to 1 **73.** Convergent; $\lim_{n\to+\infty} s_n = 0$

75. Divergent; $\lim_{n\to+\infty} s_n$ does not exist. **77.** Convergent; $\lim_{n\to+\infty} s_n = 0$ **79.** Convergent; $\lim_{n\to+\infty} s_n = 2\sqrt{2}$

81. Divergent; $\lim_{n\to+\infty} s_n = +\infty$ **83.** $34 \le n \le 99$ **85.** 10, 11 **87.** $0 < r^n < 1/np$; apply the squeezing theorem

89. $r^n > np$; $\lim_{n\to+\infty} np = +\infty$

91. $|s_n^2 - L^2| = |s_n - L||s_n + L| \le |s_n - L|(|s_n| + |L|)$; use the fact that a convergent sequence is bounded.

93. $||s_n| - |L|| \le |s_n - L|$; the converse is false; take $\{s_n\} = \{(-1)^n\}$.

95. (a) 1, 1, 2, 3, 5, 8, 13, 21 (b) Use $(1 + \sqrt{5})^2 = 2[2 + (1 + \sqrt{5})]$; $(1 - \sqrt{5})^2 = 2[2 + (1 - \sqrt{5})]$.

Exercise 11.2

1. $S_4 = 1 + \frac{3}{4} + \frac{9}{16} + \frac{27}{64} = \frac{175}{64}$ **3.** $S_4 = 1 + 2 + 3 + 4 = 10$ **5.** Converges; $\frac{3}{2}$ **7.** Diverges **9.** Converges; $\frac{1}{42}$
11. Diverges **13.** Converges; $\frac{7}{2}$ **15.** Converges; $\frac{1}{99}$ **17.** Converges; 10 **19.** Diverges **21.** Converges; $\frac{1}{3}$
23. Converges; $\frac{3}{4}$ **25.** Diverges **27.** Converges; $-\frac{1}{4}$ **29.** Converges; $-\frac{1}{3}$ **31.** Converges; sin 1
33. Converges; $\frac{1}{2}$ **35.** $\frac{5}{9}$ **37.** $\frac{3857}{900}$ **39.** $\left|\frac{1}{x}\right| < 1$; $\sum_{k=1}^{+\infty} \frac{1}{x^{k-1}} = \frac{1}{1 - (1/x)} = \frac{x}{x - 1}$ **41.** 90 ft **43.** $\frac{2}{3}$ **45.** 31

47. $\frac{\sqrt{k+1} - \sqrt{k}}{\sqrt{k(k+1)}} = \frac{1}{\sqrt{k}} - \frac{1}{\sqrt{k+1}}$, so Sum $= \left(1 - \frac{1}{\sqrt{2}}\right) + \left(\frac{1}{\sqrt{2}} - \frac{1}{\sqrt{3}}\right) + \cdots = 1$. **49.** $\frac{1}{k(k+2)} = \frac{1}{2}\left(\frac{1}{k} - \frac{1}{k+2}\right)$

51. $\frac{1}{k(k+1)(k+2)(k+3)} = \frac{1}{6}\left(\frac{1}{k} - \frac{1}{k+3}\right) - \frac{1}{2}\left(\frac{1}{k+1} - \frac{1}{k+2}\right)$

Exercise 11.3

1. Converges **3.** Diverges **5.** Converges **7.** Converges **9.** Diverges **11.** $\lim_{n \to +\infty} a_n = 16 \ne 0$

13. $\lim_{n \to +\infty} \ln n = +\infty \ne 0$ **15.** $\lim_{n \to +\infty} \frac{n^2}{n^2 + 4} = 1 \ne 0$ **17.** Convergent **19.** Divergent **21.** Convergent

23. Convergent **25.** Divergent **27.** Convergent **29.** Convergent

31. Diverges (a multiple of the harmonic series) **33.** Diverges $\left(\lim_{n \to +\infty} a_n = +\infty \ne 0\right)$

35. Diverges $\left(\lim_{n \to +\infty} a_n = +\infty \ne 0\right)$ **37.** Diverges $\left(\frac{1}{3k} - \frac{1}{4k} = \frac{1}{12k}\text{, a multiple of the harmonic series}\right)$

39. Diverges (sequence of partial sums = 1, 1, 0, 0, 1, 1, . . .) **41.** Diverges $\left(\lim_{n \to +\infty} a_n = 1 \ne 0\right)$

43. $\int_2^{+\infty} \frac{dx}{x(\ln x)^p} = \int_{\ln 2}^{+\infty} \frac{du}{u^p}$ is convergent if and only if $p > 1$.

45. S_n is the nth partial sum of $\sum_{k=1}^{+\infty} a_k$; T_n is the nth partial sum of $\sum_{k=1}^{+\infty} b_k$; $S_n + T_n$ is the nth partial sum of $\sum_{k=1}^{+\infty} (a_k + b_k)$.

47. The nth partial sum of $\sum_{k=1}^{+\infty} a_k$ is $S_n + K$ if $n \ge N$, where S_n is the nth partial sum of $\sum_{k=N+1}^{+\infty} a_k$.

49. Many examples are possible: let $a_k = k + \frac{1}{k^2}$, $b_k = k$. Then $\sum_{k=1}^{+\infty} (a_k + b_k) = \sum_{k=1}^{+\infty} \left(2k + \frac{1}{k^2}\right)$, which is divergent, but $\sum_{k=1}^{+\infty} (a_k - b_k) = \sum_{k=1}^{+\infty} \frac{1}{k^2}$, which is convergent.

Exercise 11.4

1. Convergent **3.** Divergent **5.** Divergent **7.** Convergent **9.** Divergent **11.** Convergent
13. Convergent **15.** Convergent **17.** Convergent **19.** Convergent **21.** Divergent **23.** Convergent

25. Since $\frac{d_k}{10^k} < \frac{10}{10^k} = \frac{1}{10^{k-1}}$, and $\sum_{k=1}^{+\infty} \frac{1}{10^{k-1}}$ converges $\left(\text{a geometric series with ratio } \frac{1}{10}\right)$, so does the original series.

27. $\lim_{n\to+\infty} \frac{a_n}{d_n} = p > 0 \Rightarrow$ for some $\varepsilon > 0$; $\frac{a_n}{d_n} \geq \varepsilon$ for all $n >$ some $N \Rightarrow a_n \geq \varepsilon d_n$; use comparison test II;

$\lim_{n\to+\infty} \frac{a_n}{d_n} = +\infty \Rightarrow \frac{a_n}{d_n} > 1$ for all $n >$ some $N \Rightarrow a_n > d_n$; use comparison test II.

Exercise 11.5

1. Converges **3.** Converges **5.** Diverges **7.** Converges **9.** 0.368 **11.** 0.947 **13.** 99 **15.** 4 **17.** 9999 **19.** Conditionally convergent **21.** Absolutely convergent **23.** Absolutely convergent **25.** Absolutely convergent **27.** Absolutely convergent **29.** Conditionally convergent **31.** Absolutely convergent **33.** Divergent **35.** Conditionally convergent **37.** Conditionally convergent **39.** $\sum_{k=1}^{+\infty} \frac{1}{k^{1/3}}$ diverges.

41. This is a form of the harmonic series.

43. Begin with enough positive terms so that the sum exceeds 2; then add in just enough of the negative terms for the sum to be less than 2; continuing this process gives a series that converges to 2.

45. $|e^{-kx}\cos kx| \leq (e^{-x})^k = (1/e^x)^k$, a convergent geometric series

47. A rearrangement of the conditionally convergent series A does not have the sum A.

49. The results of Problem 49 in Exercise 11.3 tell us that the term-by-term sum or difference of two divergent series may either converge or diverge.

Exercise 11.6

1. Convergent **3.** Convergent **5.** Convergent **7.** Convergent **9.** Divergent **11.** Convergent **13.** Divergent **15.** Convergent **17.** Divergent **19.** Convergent **21.** Divergent **23.** Convergent **25.** Convergent **27.** Convergent **29.** Convergent **31.** $\lim_{n\to+\infty} \frac{1/(n+1)}{1/n} = 1$

33. $\sum_{k=1}^{+\infty} \frac{k!}{k^k}$ converges by Example 2; thus, $\lim_{n\to+\infty} \frac{n!}{n^n} = 0$. **35.** $0 < x \leq 1$

37. If $\left|\frac{a_{n+1}}{a_n}\right| > 1$ for all n, then $|a_{n+1}| > |a_n|$, so the sequence $|a_n|$ is increasing and $\lim_{n\to+\infty} a_n \neq 0$.

39. Let $a_{2k} = 2$ and $a_{2k+1} = 1$. Then $\frac{a_{2k+1}}{a_{2k}} = \frac{1}{2}$ and $\frac{a_{2k+2}}{a_{2k+1}} = 2$, so $\lim_{n\to+\infty} \left|\frac{a_{n+1}}{a_n}\right| \neq +\infty$ but does not exist.

41. It is absolutely convergent by the ratio test.

Exercise 11.7

1. Divergent **3.** Convergent **5.** Convergent **7.** Divergent **9.** Divergent **11.** Divergent **13.** Divergent **15.** Conditionally convergent **17.** Convergent **19.** Divergent **21.** Convergent **23.** Divergent **25.** Divergent **27.** Absolutely convergent **29.** Absolutely convergent **31.** Absolutely convergent **33.** Convergent **35.** Convergent **37.** Absolutely convergent **39.** Convergent **41.** Divergent **43.** Divergent **45.** Convergent **47.** Convergent

Exercise 11.8

1. $-1 \leq x < 1$ **3.** $-3 < x < 3$ **5.** $-1 < x < 1$ **7.** $-\frac{3}{2} < x < \frac{3}{2}$ **9.** $-\infty < x < +\infty$ **11.** $-1 < x < 1$ **13.** $-4 < x < 4$ **15.** $-1 \leq x < 1$ **17.** $2 < x < 4$ **19.** $-\infty < x < +\infty$ **21.** $1/e$ **23.** $x > 1$ or $x \leq -1$ **25.** $-\infty < x < +\infty$ **27.** $-\pi/6 < x < \pi/6$ **29.** Convergent at $x = 2$; no

31. (a) True (b) Cannot say (c) Cannot say (d) Cannot say (e) True (f) True

33. $\lim_{n\to+\infty}\left|\frac{a_{n+1}}{a_n}\right| = \rho$; $|x| < R$; $\lim_{n\to+\infty}\left|\frac{a_{n+1}x^{n+1}}{a_nx^n}\right| = |x|\rho < 1$; $\rho = \frac{1}{R}$

35. If either $\sum_{k=1}^{+\infty}|a_kx_0^k|$ or $\sum_{k=1}^{+\infty}|a_k(-x_0)^k|$ converges, so does the other.

Exercise 11.9

1. $\sum_{k=0}^{+\infty}\frac{(-1)^kx^{2k+1}}{(2k+1)!}$ **3.** $\sum_{k=0}^{+\infty}\frac{e(x-1)^k}{k!}$ **5.** $\sum_{k=0}^{+\infty}\frac{\sqrt{2}}{2}(-1)^k\left[\frac{(x-\pi/4)^{2k+1}}{(2k+1)!}+\frac{(x-\pi/4)^{2k}}{(2k)!}\right]$ **7.** $\sum_{k=0}^{+\infty}(-1)^k(x-1)^k$

9. $\sum_{k=0}^{+\infty}(-1)^k(k+1)x^k$ **11.** $\sum_{k=0}^{+\infty}(3x)^k$ **13.** $\sum_{k=1}^{+\infty}\frac{(-1)^{k+1}x^k}{k}$ **15.** $-6+5x+2x^2+3x^3$

17. $4+18(x-1)+11(x-1)^2+3(x-1)^3$ **19.** $\sum_{k=0}^{+\infty}x^{2k}$; $-1 < x < 1$ **21.** 0.693 **23.** 3.1415

25. (a) $1/(1+x) = 1 - x + x^2 - x^3 + \cdots$ (b) $\frac{1}{2} = 1 - 1 + 1 - 1 + \cdots$ (c) Series in part (a) diverges if $x = 1$.

Exercise 11.10

1. $\sum_{k=0}^{+\infty}\frac{(2x)^k}{k!}$ **3.** $\sum_{k=1}^{+\infty}\frac{(-1)^{k+1}x^{2k}}{k}$ **5.** $\sum_{k=0}^{+\infty}\frac{x^{2k+1}}{(2k+1)!}$ **7.** $x + x^2 + \frac{1}{3}x^3 - \frac{1}{30}x^5 + \cdots$

9. Integrate f; $x + \frac{1}{6}x^3 + \frac{3}{40}x^5 + \frac{5}{112}x^7$ **11.** 0.3095 **13.** 0.995 **15.** 0.946 **17.** 0 **19.** 2

21. $\frac{1}{4}(\pi - 2)$ (integrate $x\tan^{-1}x$) **23.** $f(0) = 1$; $f'(0) = 0$; $f''(0) = -1$; $f'''(0) = 0$; etc. **25.** 0.0427

Exercise 11.11

1. $\sum_{k=0}^{+\infty}\binom{1/2}{k}x^{2k}$; $-1 \le x \le 1$ **3.** $\sum_{k=0}^{+\infty}\binom{1/5}{k}x^k$; $-1 \le x \le 1$ **5.** $\sum_{k=0}^{+\infty}\binom{-3/4}{k}x^k$; $-1 < x \le 1$ **7.** 0.2 **9.** 0.487

11. Use the binomial theorem to expand $(1 + 1/n)^n$. Each term is no larger than the corresponding terms of $\sum_{k=0}^{+\infty}(1/k!)$, which converges to e.

13. $\lim_{n\to+\infty}\frac{a_{n+1}}{a_n} = \lim_{n\to+\infty}\frac{|m-n|}{n+1}|x| = |x|$

Review Exercises

1. Absolutely convergent **3.** Divergent **5.** Divergent **7.** Convergent **9.** Absolutely convergent

11. Divergent **13.** Divergent **15.** Absolutely convergent **17.** Divergent **19.** Absolutely convergent

21. Absolutely convergent **23.** $\frac{3\sqrt{2}+4}{2}$

25. If $a = \frac{m}{n}$ and $r = \frac{p}{q}$ (m, n, p, q integers), then Sum $= \frac{m/n}{1-(p/q)} = \frac{mq}{nq-np}$, which is rational.

27. $|M - L| = |a_n - L - a_n + M| \le |a_n - L| + |a_n - M|$. Then each expression on the right can be made less than $\frac{1}{2}|M-L|$, so $|M - L| < \frac{1}{2}|M-L| + \frac{1}{2}|M-L|$, a contradiction.

29. $[2, 4]$ **31.** $(-\infty, +\infty)$ **33.** $(-\infty, +\infty)$ **35.** $(-3, 5)$

37. Use the ratio test and the fact that $\frac{(k+1)a_{k+1}}{ka_k}$ and $\frac{a_{k+1}/(k+2)}{a_k/(k+1)}$ both approach $\frac{a_{k+1}}{a_k}$.

39. Let $b_k = a_k$ if $a_k \ge 0$, and 0 otherwise. Then $b_k \le |a_k|$ and converges by the comparison test. A similar argument applies to $c_k = -a_k$ for $a_k < 0$, and 0 otherwise. Then Σb_k and $\Sigma(-c_k)$ are the sums of the positive and negative terms, respectively.

Challenge Exercises

1. Compare with the convergent series $\sum_{k=1}^{+\infty} \frac{k!}{k^k}$ **3.** Converges; -1

5. Differentiate the Maclaurin series for $(e^x - 1)/x$ and put $x = 1$. **7.** $[-1, 5)$

9. (a) $S_n \le \int_0^n \frac{dx}{1+x^2} = \tan^{-1} n$ (b) S_n is bounded and monotonic; $S_n \le \pi/2$.

(c) $S_n \ge \int_1^n \frac{dx}{1+x^2} \ge \pi/4$; from part (b), $\pi/4 \le \lim_{n\to+\infty} S_n \le \pi/2$

11. $a_n = 3(1 + (5/3)^n)^{1/n}$, so $\ln a_n/3 = \frac{1}{n}\ln(1 + (5/3)^n)$. Apply L'Hospital's rule to show that $\ln a_n/3$ converges; therefore a_n also converges.

13. $x + (1-x)\ln(1-x)$

15. (a) $\ln n \le 1 + \frac{1}{2} + \cdots + 1/(n-1) = C_n - 1/n + \ln n \Rightarrow 1/n \le C_n$

(b) $C_n - C_{n+1} = \ln(n+1) - \ln n - \frac{1}{n+1} > \frac{1}{n+1} - \frac{1}{n+1} = 0$; use axiom (11.1.9); then $\lim_{n\to+\infty} C_n = \gamma$.

(c) 11; 31; 12,367; 2.724×10^8

17. (a) 1.197532 (b) Use $f(x) = 1/x^3$; $\frac{1}{2} - \frac{1}{n^2}\left(\frac{1}{2} - \frac{1}{n}\right) < S_n < \frac{3}{2} - \frac{1}{2n^2}$ (c) $\frac{1}{2} < \zeta(3) < \frac{3}{2}$

19. $\frac{1}{1-x} + \frac{2}{2-x} + \frac{3}{3-x} + \cdots + \frac{n}{n-x}$

21. $y' = y$ implies $\sum_{n=0}^{+\infty} a_n x^n = \sum_{n=1}^{+\infty} n a_n x^{n-1}$, implying $a_0 = a_1$, $a_1 = 2a_2$, $a_2 = 3a_3, \ldots$. Thus, $a_2 = \frac{1}{2}a_0$, $a_3 = \frac{1}{6}a_0, \ldots, a_n = (1/n!)\, a_0$, so $y = a_0 \sum \frac{x^n}{n!} = a_0 e^x$.

23. $|S(x) - S(x_0)| \le |S(x) - S_n(x)| + |S_n(x) - S_n(x_0)| + |S_n(x_0) - S(x_0)|$. The first term can be made less than $\varepsilon/3$, the middle term approaches 0 (since S_n a polynomial $\Rightarrow$ S_n continuous), and the last term can be made less than $\varepsilon/3$ by setting $x = x_0$. Thus, by choosing $|x - x_0| < \delta$, $|S(x) - S(x_0)| < \varepsilon/3 + \varepsilon/3 + \varepsilon/3$, implying that S is continuous.

25. (c) 200 terms; 2×10^{10} terms

CHAPTER 12

Exercise 12.2

1. $y^2 = 4x$ **3.** $y^2 = -8x$ **5.** $y^2 = 8x$ **7.** $x^2 = 12y$ **9.** $y^2 = -8x$ **11.** $y^2 = -12x$ **13.** $y^2 = 32x$

15. $y^2 = -12x$

17. Vertex $(0, 0)$; focus $(2, 0)$; directrix $x = -2$

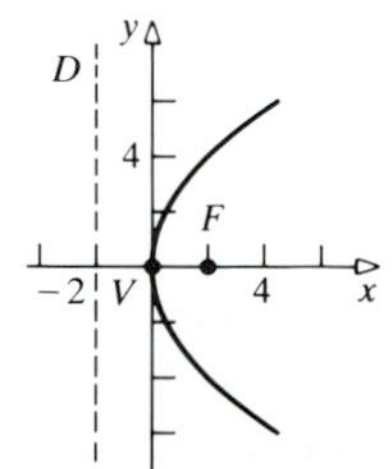

19. Vertex $(0, 0)$; focus $(0, -3)$; directrix $y = 3$

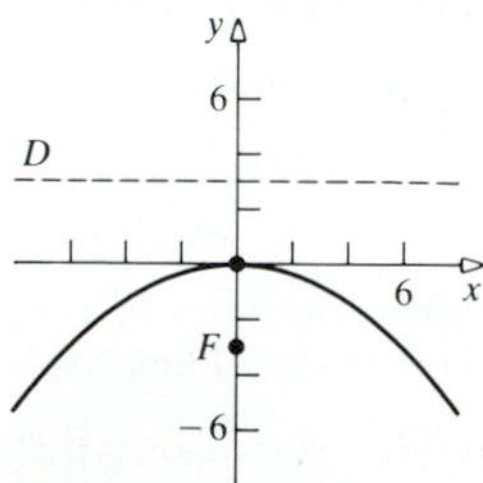

21. Vertex $(0, 0)$; focus $(-4, 0)$; directrix $x = 4$

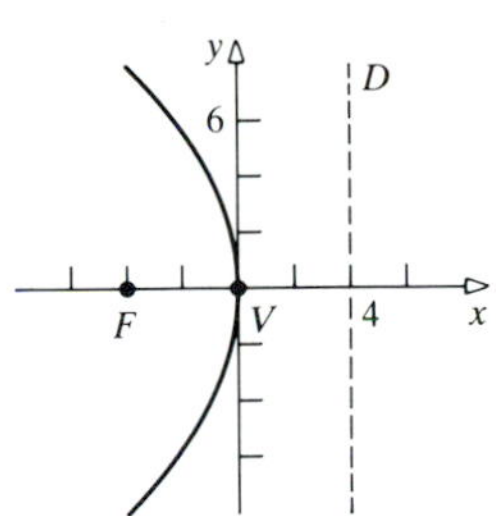

23. Vertex $(0, 0)$; focus $(0, 2)$; directrix $y = -2$

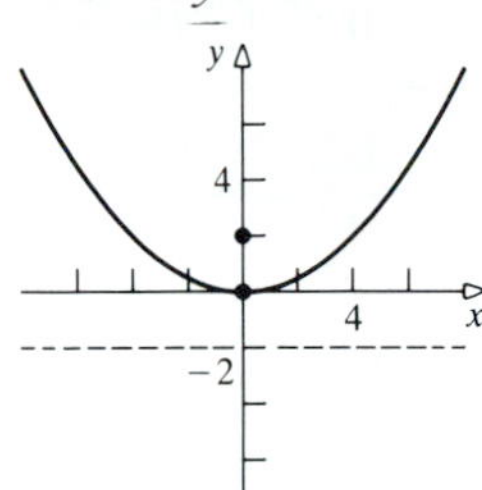

25. $y = -x + 2$ **27.** $\frac{16}{3}$ **29.** $64\pi/5$ **31.** $\frac{64}{3}$ m **33.** $\frac{25}{16}$ ft from the vertex

35. Area of rectangle $= 2x_0 \dfrac{(x_0^2)}{4a} = \dfrac{x_0^3}{2a}$

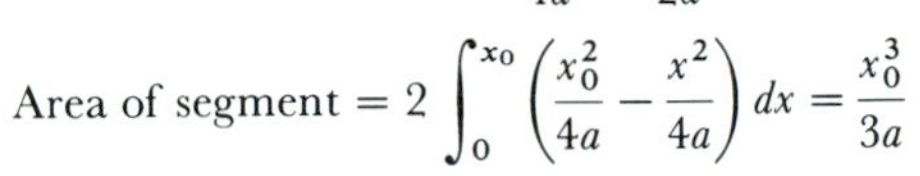

Area of segment $= 2\displaystyle\int_0^{x_0} \left(\frac{x_0^2}{4a} - \frac{x^2}{4a}\right) dx = \frac{x_0^3}{3a}$

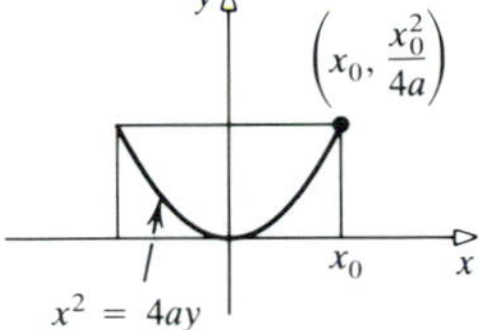

37. 60 tons **39.** $2yy' = 4a$; $y - y_0 = \dfrac{4a}{2y_0}(x - x_0)$; $y_0 y = 2a(x + x_0)$

41. The given equation can be put into the form $x^2 = 4(-E/4A)y$, which is the form of a parabola with vertex $(0, 0)$ and axis the y-axis. The focus is at $(0, -E/4A)$ and directrix $y = E/4A$.

Exercise 12.3

1. $\dfrac{x^2}{36} + \dfrac{y^2}{20} = 1$ **3.** $\dfrac{x^2}{5} + \dfrac{y^2}{9} = 1$ **5.** $\dfrac{x^2}{25} + \dfrac{y^2}{16} = 1$ **7.** $\dfrac{x^2}{17} + y^2 = 1$

9. Major axis = y-axis; $a^2 = 9$; $b^2 = 4$; $c^2 = 5$; foci at $(0, \sqrt{5})$, $(0, -\sqrt{5})$; vertices at $(0, 3)$, $(0, -3)$

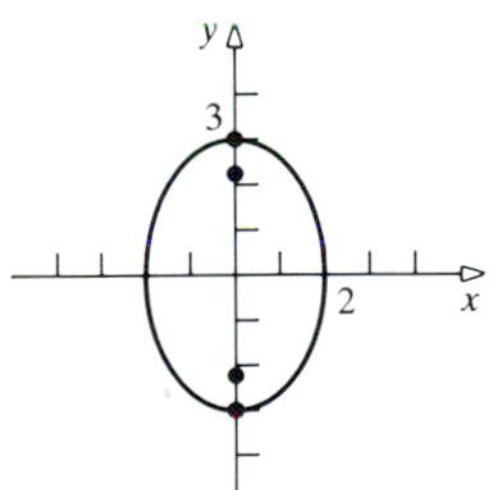

11. Major axis = x-axis; $a^2 = 16$; $b^2 = 4$; $c^2 = 12$; foci at $(2\sqrt{3}, 0)$, $(-2\sqrt{3}, 0)$; vertices at $(4, 0)$, $(-4, 0)$

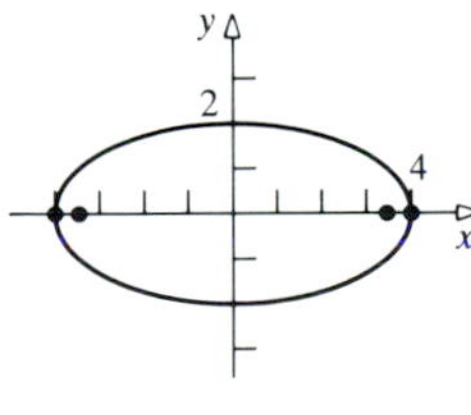

13. Major axis = y-axis; $a^2 = 4$; $b^2 = 1$; $c^2 = 3$;

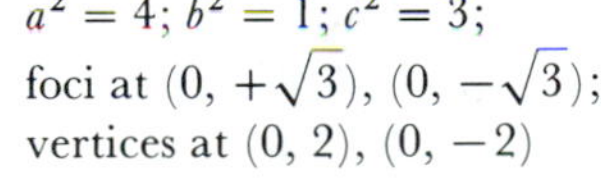

foci at $(0, +\sqrt{3})$, $(0, -\sqrt{3})$; vertices at $(0, 2)$, $(0, -2)$

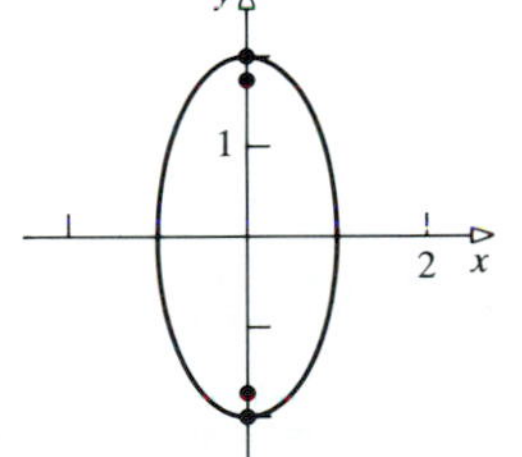

15. Circle with $R = 4$

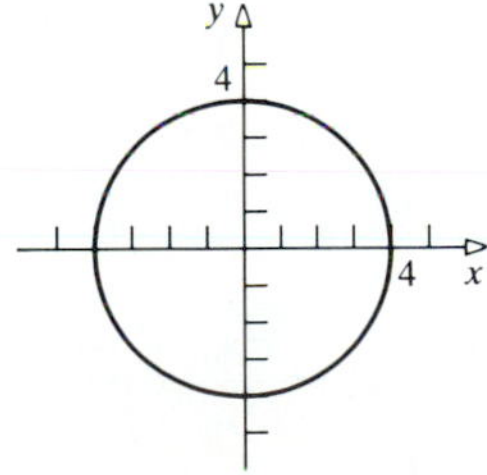

17. $y = -\frac{1}{2}x + 2$

19. (a) Circular (b) Elliptical (c) Flat (elongated) ellipse **21.** $\dfrac{x^2}{100} + \dfrac{y^2}{64} = 1$

23. $\frac{x^2}{400} + \frac{y^2}{225} = 1$

Interval	± 20	± 10	0
Height	0	$15\sqrt{3}/2$	15

25. $A = 4\int_0^a \frac{b}{a}\sqrt{a^2 - x^2}\,dx = \pi ab$

27. $\frac{2x}{a^2} + \frac{2yy'}{b^2} = 0;\ y' = \frac{-b^2}{a^2}\frac{x_0}{y_0};\ y - y_0 = \frac{-b^2}{a^2}\frac{x_0}{y_0}(x - x_0);\ \frac{xx_0}{a^2} + \frac{yy_0}{b^2} = 1$

Exercise 12.4

1. $\frac{x^2}{16} - \frac{y^2}{20} = 1$ **3.** $\frac{y^2}{4} - \frac{x^2}{5} = 1$ **5.** $\frac{x^2}{9} - \frac{y^2}{16} = 1$ **7.** $\frac{5x^2}{16} - \frac{5y^2}{64} = 1$

9. Hyperbola;
x-axis is transverse axis;
center is $(0, 0)$;
vertices at $(\pm 2, 0)$;
foci at $(\pm\sqrt{13}, 0)$;
asymptotes are $y = \pm\frac{3}{2}x$

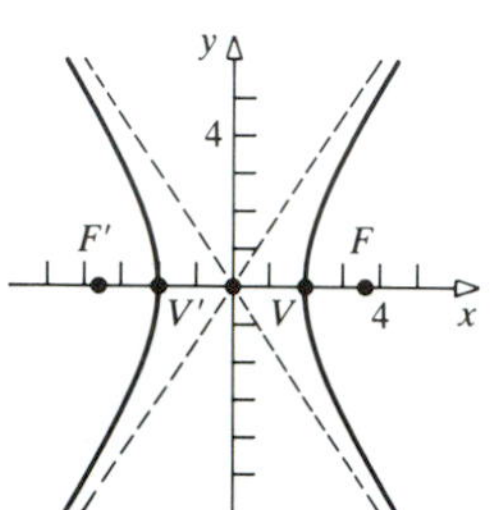

11. Hyperbola;
x-axis is transverse axis;
center is $(0, 0)$;
vertices at $(\pm 4, 0)$;
foci at $(\pm 2\sqrt{5}, 0)$;
asymptotes are $y = \pm\frac{1}{2}x$

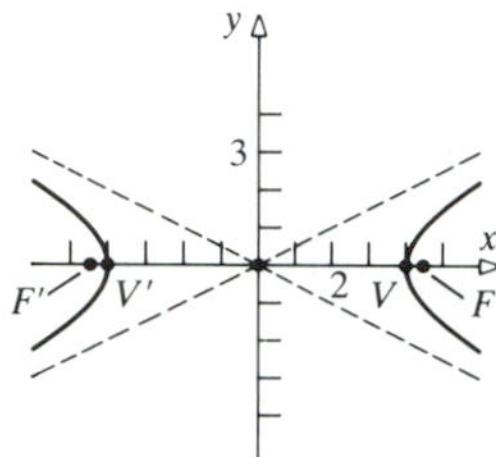

13. Hyperbola;
y-axis is transverse axis;
center is $(0, 0)$;
vertices at $(0, \pm 2)$;
foci at $(0, \pm\sqrt{5})$;
asymptotes are $y = \pm 2x$

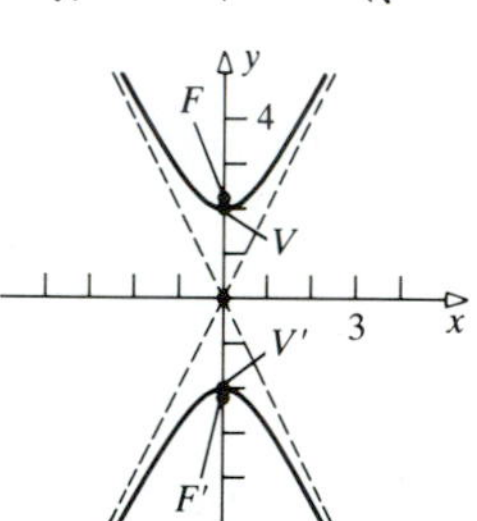

15. Hyperbola;
x-axis is transverse axis;
center is $(0, 0)$;
vertices at $(\pm 4, 0)$;
foci at $(\pm 4\sqrt{2}, 0)$;
asymptotes are $y = \pm x$

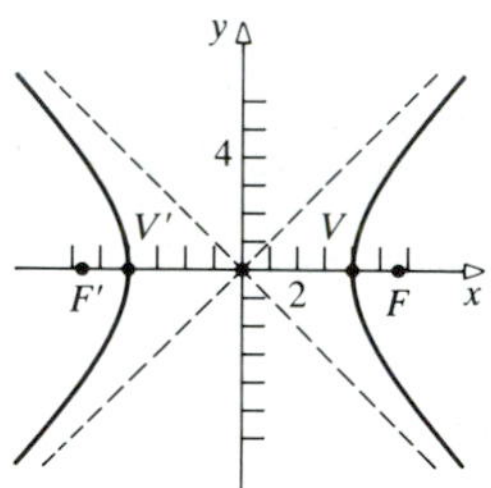

17. If e is close to 1, then c is close to a and the hyperbola is very narrow. If e is very large, the focus is farther from the vertex and the hyperbola is very broad or wide.

19. $e = \sqrt{2}$ **21.** $a^2 = \cos^2\alpha$, $b^2 = \sin^2\alpha$; so $c^2 = a^2 + b^2 = \cos^2\alpha + \sin^2\alpha = 1$ or $c = \pm 1$, so foci are $(\pm 1, 0)$.

23. (a) $y = \pm\frac{4}{5}x$ (b) $y = \pm\frac{1}{3}x$

25. (a) Tangent: $y = -x + 3$; normal: $y = x - 5$ (b) Tangent: $y = -\frac{3}{2}x + \frac{5}{2}$; normal: $y = \frac{2}{3}x - 4$

27. $\frac{x^2}{21{,}609} - \frac{y^2}{135{,}316} = 1$

29.
$$\lim_{x\to-\infty}(y_1 - y) = \lim_{x\to-\infty}\left[\frac{b}{a}\sqrt{x^2 - a^2} - \left(-\frac{b}{a}x\right)\right] = \frac{b}{a}\lim_{x\to-\infty}[\sqrt{x^2 - a^2} + x]$$
$$= \frac{b}{a}\lim_{x\to-\infty}(\sqrt{x^2 - a^2} + x)\frac{\sqrt{x^2 - a^2} - x}{\sqrt{x^2 - a^2} - x} = \frac{b}{a}\lim_{x\to-\infty}\frac{-a^2}{\sqrt{x^2 - a^2} - x} = -ab\lim_{x\to-\infty}\frac{1}{-2x} = 0;$$
$$\lim_{x\to+\infty}(y_2 - y) = \lim_{x\to+\infty}\left[-\frac{b}{a}\sqrt{x^2 - a^2} - \left(-\frac{b}{a}x\right)\right] = -\frac{b}{a}\lim_{x\to+\infty}[\sqrt{x^2 - a^2} - x] = -\frac{b}{a}\lim_{x\to+\infty}\frac{-a^2}{\sqrt{x^2 - a^2} + x}$$
$$= ab\lim_{x\to+\infty}\frac{1}{2x} = 0$$

31. By implicit differentiation, $y' = \frac{b^2x_0}{a^2y_0}$, so the tangent line has equation $y - y_0 = \frac{b^2x_0}{a^2y_0}(x - x_0)$ or $\frac{x_0x}{a^2} - \frac{y_0y}{b^2} = \frac{x_0^2}{a^2} - \frac{y_0^2}{b^2}$. But since (x_0, y_0) is on the hyperbola, the right side $= 1$.

Exercise 12.5

1. Ellipse;
center at $(1, 0)$;
$(x-1)^2 + \dfrac{y^2}{\frac{1}{4}} = 1$

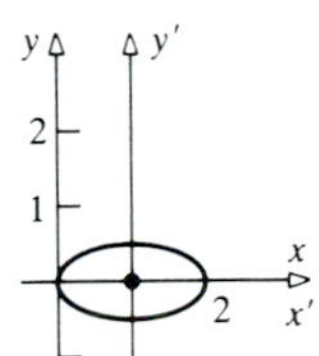

3. Hyperbola;
center at $(0, 1)$;
$\dfrac{x^2}{\frac{1}{4}} - \dfrac{(y-1)^2}{\frac{1}{4}} = 1$

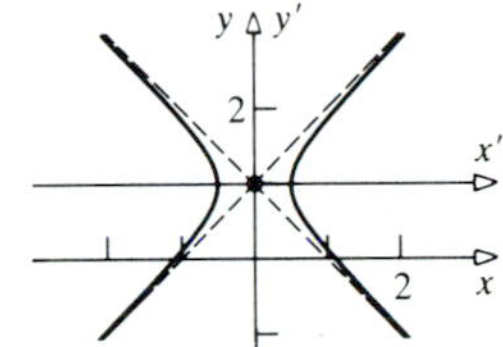

5. Parabola;
vertex at $(2, -3)$;
$(x-2)^2 = \frac{1}{4}(y+3)$

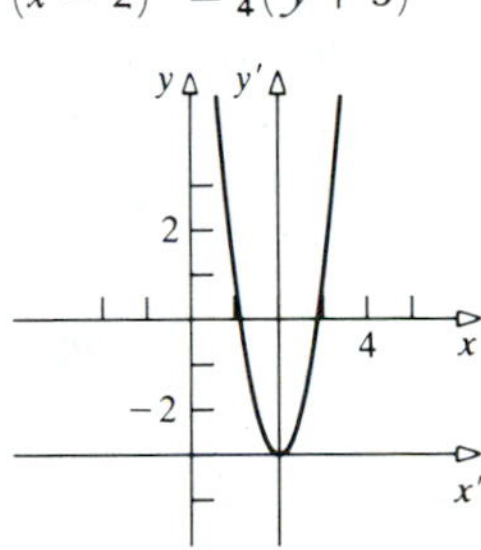

7. Ellipse;
center at $(-3, 1)$;
$\dfrac{(x+3)^2}{16} + \dfrac{(y-1)^2}{9} = 1$

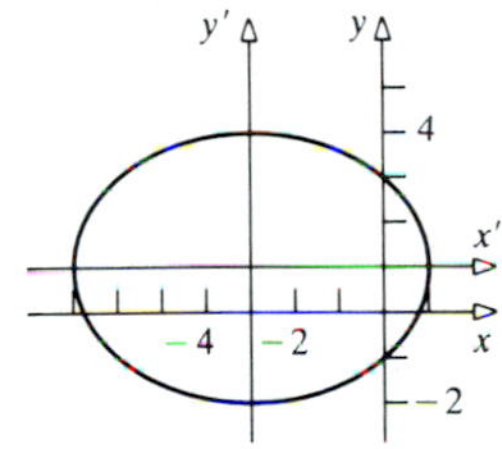

9. Hyperbola;
center at $(-3, 2)$;
$\dfrac{25(y-2)^2}{9} - (x+3)^2 = 1$

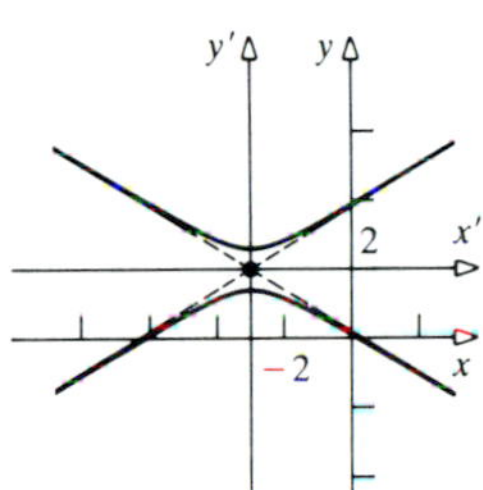

11. Hyperbola;
center at $(-1, 2)$;
$\dfrac{(x+1)^2}{3} - \dfrac{(y-2)^2}{4} = 1$

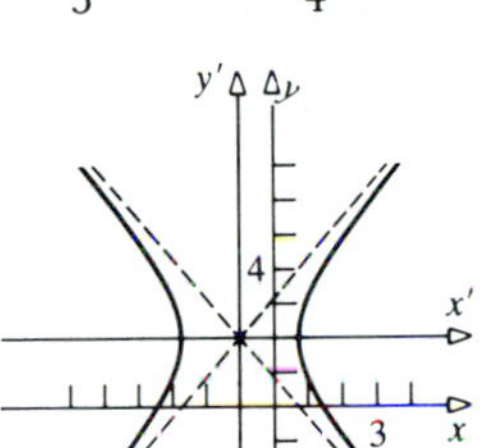

13. Hyperbola;
center at $(0, 0)$;
$\dfrac{x''^2}{8} - \dfrac{y''^2}{8} = 1$;
rotation $= \pi/4$

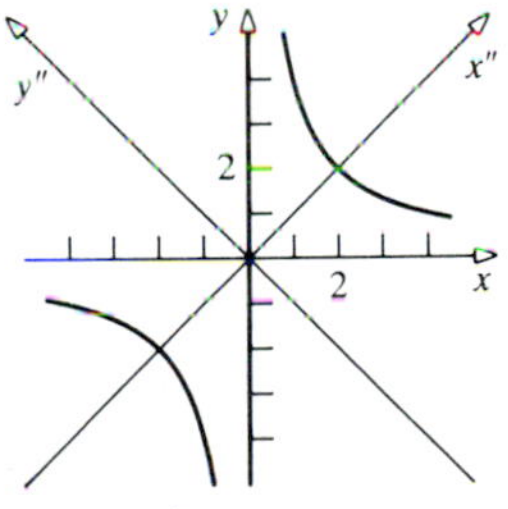

15. Ellipse;
center at $(0, 0)$;
$\dfrac{21x''^2}{20} + \dfrac{y''^2}{20} = 1$;
rotation $= \pi/3$

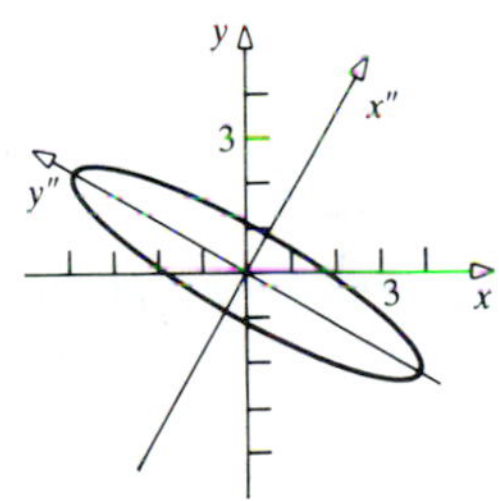

17. Hyperbola;
center at $(0, 0)$;
$\dfrac{3y''^2}{5} - \dfrac{x''^2}{5} = 1$;
rotation $= \pi/3$

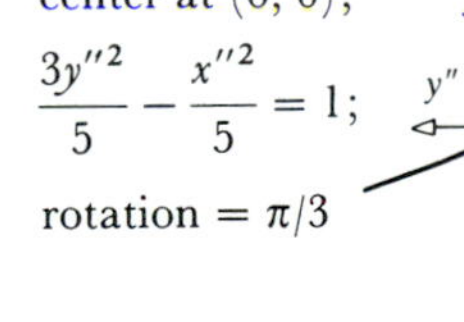

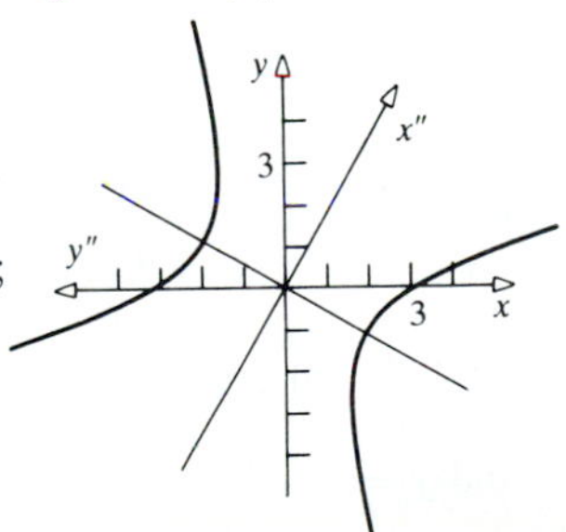

19. Parabola **21.** Ellipse

23. $A'' + C'' = A\cos^2\theta + B\sin\theta\cos\theta + C\sin^2\theta + A\sin^2\theta - B\sin\theta\cos\theta + C\cos^2\theta$
$= A(\cos^2\theta + \sin^2\theta) + C(\sin^2\theta + \cos^2\theta) = A + C$

25. $\dfrac{(x'')^2}{4^2} + \dfrac{(y'')^2}{2^2} = 1$; an ellipse

27. If $B^2 - 4AC = B'' - 4A''C'' = -4A''C'' > 0$, then $A''C'' < 0$ and the graph is a hyperbola. If $B^2 - 4AC = B'' - 4A''C'' = -4A''C'' < 0$, then $A''C'' > 0$ and the graph is an ellipse. If $B^2 - 4AC = B'' - 4A''C'' = 0$, then the graph is a parabola.

Review Exercises

1. $x^2 = 12y$ **3.** $\frac{b}{2}\sqrt{1 + \frac{b^2}{4a^2}} + a\sinh^{-1}\frac{b}{2a}$ or $\frac{b}{2}\sqrt{1 + \frac{b^2}{4a^2}} + a\ln\left[\frac{b}{2a} + \sqrt{\frac{b^2}{4a^2} + 1}\right]$ **5.** $(x - 2)^2 = 14(y - \frac{1}{2})$

7. $5\sqrt{10}$ ft **9.** (a) 2 cubic units (b) $\sqrt{3}/8$ cubic units **11.** $\frac{x^2}{9} + \frac{y^2}{5} = 1$ **13.** $\frac{x^2}{25} + \frac{y^2}{16} = 1$

15. $\sqrt{3}x + 6y = 4\sqrt{3}$ **17.** $\frac{x^2}{324} + \frac{y^2}{144} = 1,\ y \geq 0$ **19.** $\frac{x^2}{5} + \frac{(y-2)^2}{9} = 1$ **21.** $(0, \pm 5)$ **23.** $(2, 1)$ **25.** $y = \pm$

27. $\frac{(x - \frac{25}{21})^2}{\frac{1024}{441}} - \frac{y^2}{\frac{256}{21}} = 1$ **29.** $\frac{x^2}{4} - \frac{(y-2)^2}{21} = 1$ **31.** Hyperbola **33.** Parabola **35.** Ellipse

37. Points of tangency are $(1, 2)$, $(-\frac{4}{3}, -2)$, so $c = 1, \frac{4}{3}$.

39. $y'' = x'' + \frac{2}{\sqrt{10}}$ and $y'' = -x''$, or (in xy-plane) $2x - y + 1 = 0$ and $x + 2y = 0$.

Challenge Exercises

1. (a) $\int_0^3 (3\sqrt{9 - x^2} + 3x - 9)\,dx$ (b) $18\pi\int_0^3 (3x - x^2)\,dx$ or $2\pi\int_0^9 y\left(\sqrt{9 - \frac{y^2}{9}} + \frac{y}{3} - 3\right)dy$

(c) $2\pi\int_0^3 x(3\sqrt{9 - x^2} + 3x - 9)\,dx$ or $2\pi\int_0^9 \left(y - \frac{y^2}{9}\right)dy$

3. The condition is equivalent to $\frac{\sqrt{(x-c)^2 + y^2}}{\left|x - \frac{a}{e}\right|} = e$ or $(x - c)^2 + y^2 = e^2\left(x - \frac{a}{e}\right)^2$. Expanding and simplifying yield $x^2(1 - e^2) + 2x(ae - c) + y^2 = a^2 - c^2$. Then $ae = c$ and $a^2 - c^2 = b^2$ implies $x^2\left(1 - \frac{c^2}{a^2}\right) + y^2 = b^2$ or $\frac{x^2}{a^2} + \frac{y^2}{b^2} = 1$, an ellipse.

5. Problems 3 and 4 have shown that the definition of $e = c/a$ led to the standard equation of the ellipse $(e < 1)$ and the hyperbola $(e > 1)$. For the parabola, $d(F, P) = d(D, P)$ so $d(F, P)/d(D, P) = e = 1$

7. Implicit differentiation yields $y' = 2a/y$. Since the ends of the latus rectum are at $(a, \pm 2a)$, the slopes of the tangents there are 1 and -1, so the lines are perpendicular. Their equations are $y = x + a$ and $y = -x - a$, which intersect at $(-a, 0)$, which lies on the directrix.

9. The required chord's endpoints are $\left(c, \pm b\sqrt{\frac{c^2}{a^2} - 1}\right)$, so its length is $2b\sqrt{\frac{c^2 - a^2}{a^2}} = 2b\sqrt{\frac{b^2}{a^2}} = \frac{2b^2}{a}$.

11. Implicit differentiation yields $y' = \frac{xb^2}{ya^2}$. Since the endpoints of the latus rectum are $\left(c, \pm\frac{b^2}{a}\right)$, the slopes of the tangents there are $\pm\frac{cb^2}{ab^2} = \pm e$, and since their equations are $y = ex - a$ and $y = -ex + a$, they intersect at $\left(\frac{a}{e}, 0\right)$, which is on the directrix.

13. The two parabolas are $x^2 = 2(y - \frac{1}{2})$ and $y^2 = 2(x - \frac{1}{2})$.

CHAPTER 13

Exercise 13.1

1. $(4, \pi/3)$; $(4, 7\pi/3)$; $(-4, -2\pi/3)$

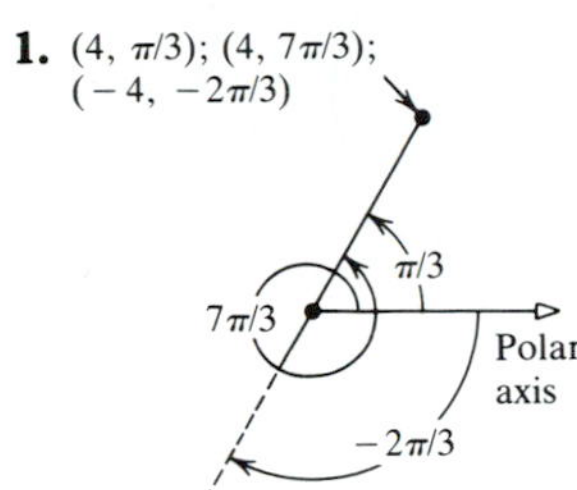

3. $(-4, -\pi/3)$; $(-4, -7\pi/3)$; $(4, 2\pi/3)$

2π/3
Polar
axis
−7π/3
−π/3

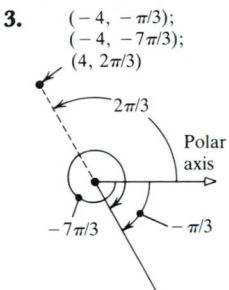

5. $(\sqrt{2}, \pi/4)$; $(\sqrt{2}, 9\pi/4)$; $(-\sqrt{2}, 5\pi/4)$

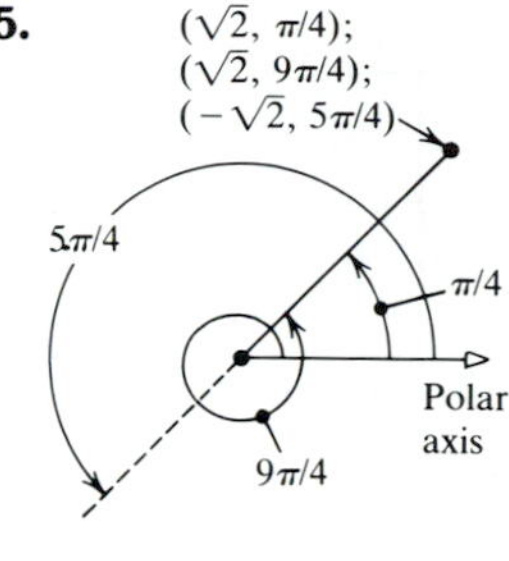

7.

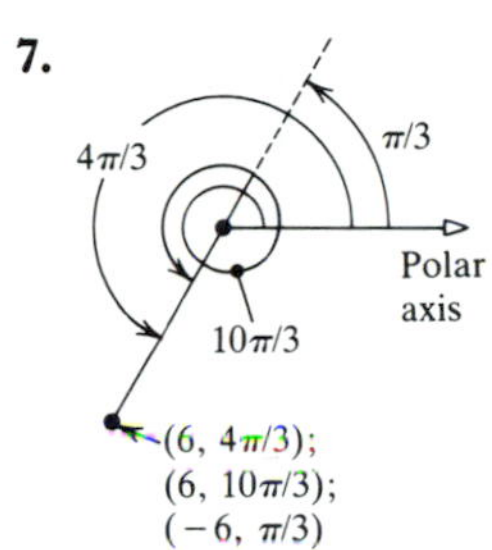

$(6, 4\pi/3)$; $(6, 10\pi/3)$; $(-6, \pi/3)$

9. $(3\sqrt{3}, 3)$ **11.** $(-3\sqrt{3}, 3)$ **13.** $(0, 5)$ **15.** $(2, -2)$

17.

7π/4
Polar
axis
$(2\sqrt{2}, 7\pi/4)$

19.

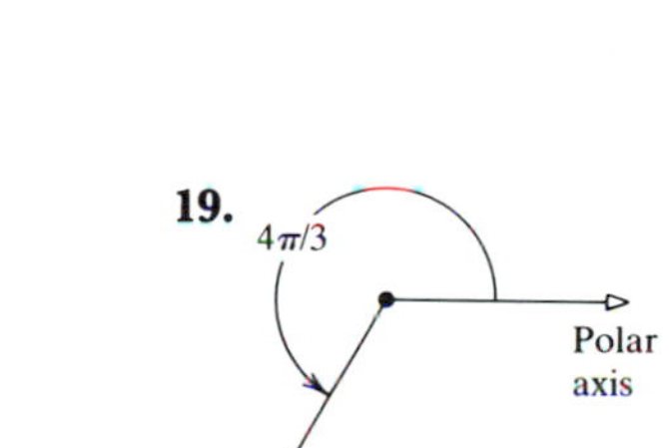

21. $(2, 5\pi/6)$

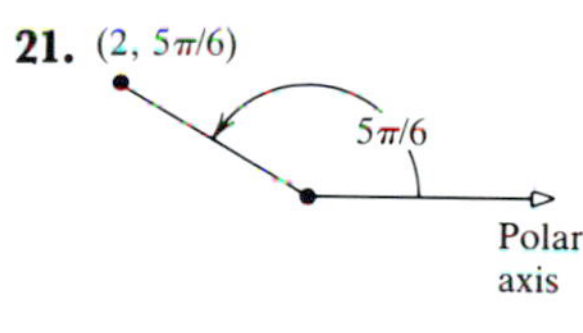

23. $r^2\left(\dfrac{\cos^2\theta}{4} + \dfrac{\sin^2\theta}{9}\right) = 1$ **25.** $r = 4\cos\theta$ **27.** $r^2\cos^2\theta + 4r\sin\theta - 1 = 0$ **29.** $r^2\sin 2\theta = 2$

31. $x^2 + y^2 - x = 0$ **33.** $(x^2 + y^2)^{3/2} - y = 0$ **35.** $y^2 = 8(x + 2)$ **37.** $y = x\tan(x^2 + y^2)$

39. $x^2 + y^2 = 4$ **41.** $y = 4x$ **43.** Apply the law of cosines to the triangle OP_1P_2.

Exercise 13.2

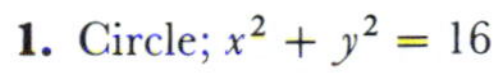

1. Circle; $x^2 + y^2 = 16$

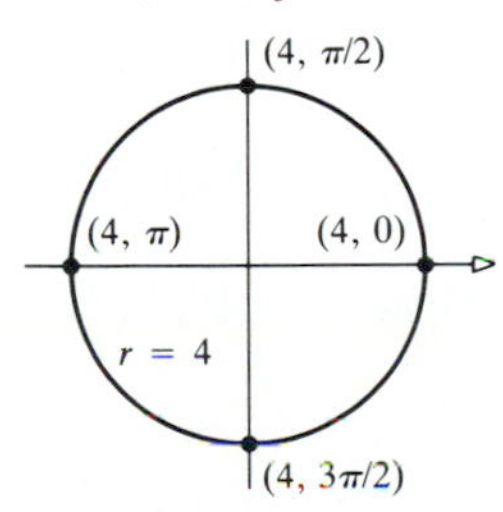

3. Line; $y = \sqrt{3}x$

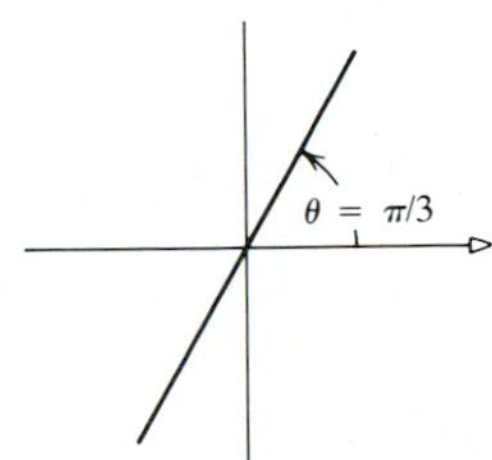

5. Horizontal line; $y = 4$

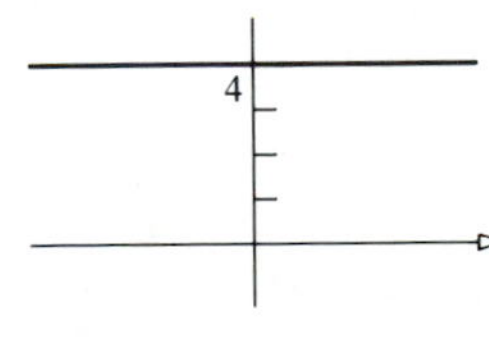

7. Vertical line; $x = -2$

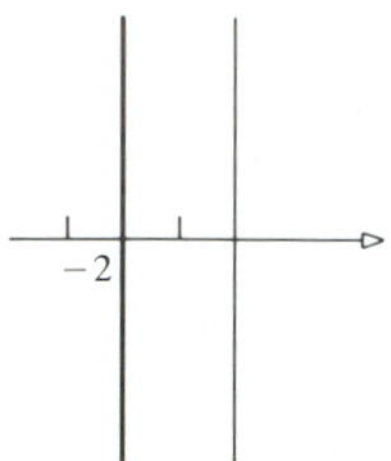

9. Circle; $(x - 1)^2 + y^2 = 1$

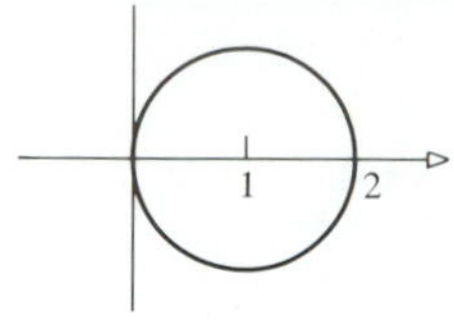

11. Circle; $x^2 + (y + 2)^2 = 4$

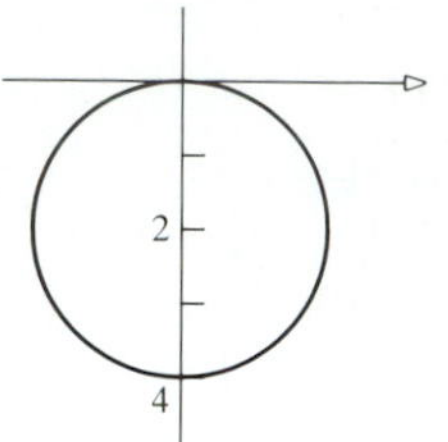

13. Circle; $(x - 2)^2 + y^2 = 4$ (origin excluded)

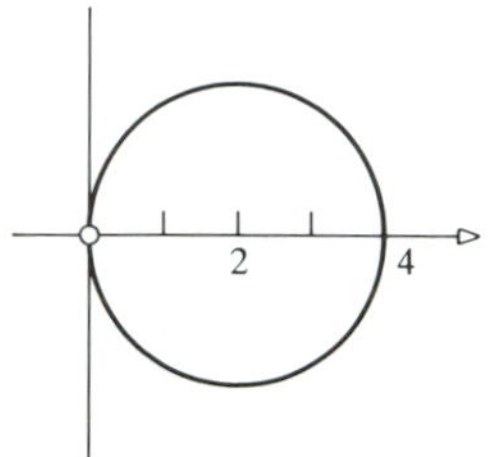

15. Circle; $x^2 + (y + 1)^2 = 1$ (origin excluded)

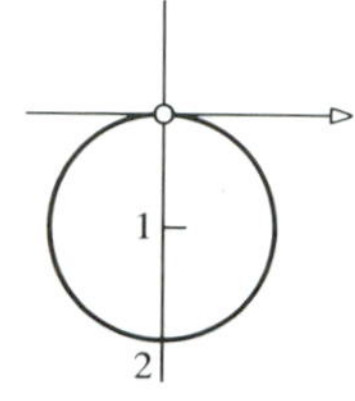

17. Symmetry: $\theta = \dfrac{\pi}{2}$

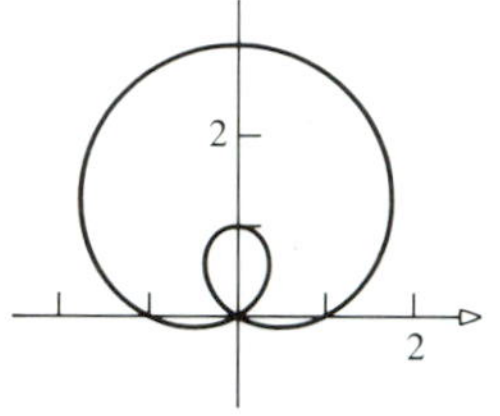

19. Symmetry: polar axis

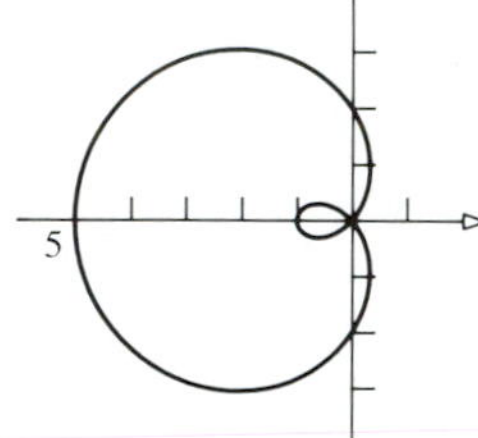

21. Symmetry: polar axis

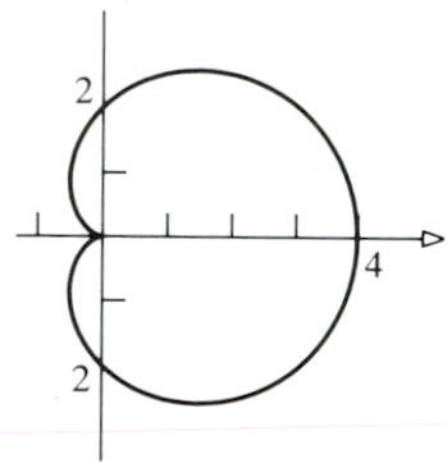

23. Symmetry: $\theta = \dfrac{\pi}{2}$

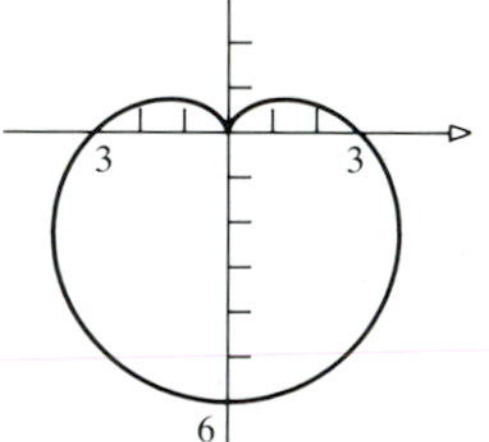

25. Symmetry: polar axis, $\theta = \dfrac{\pi}{2}$, origin (pole)

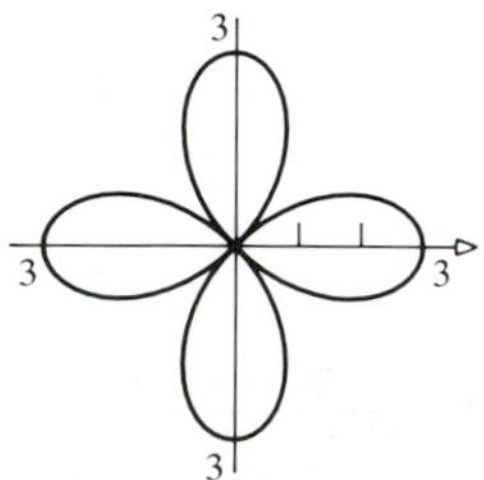

27. Symmetry: $\theta = \dfrac{\pi}{2}$

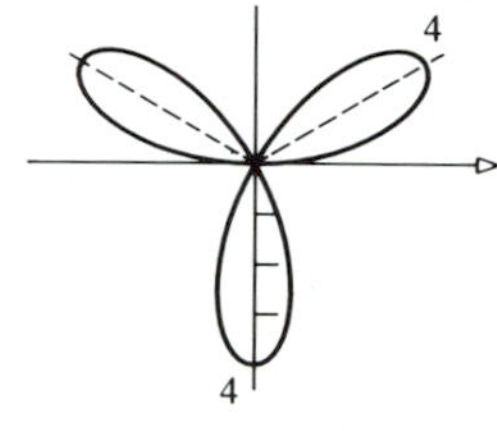

29. Symmetry: polar axis, $\theta = \dfrac{\pi}{2}$, origin (pole)

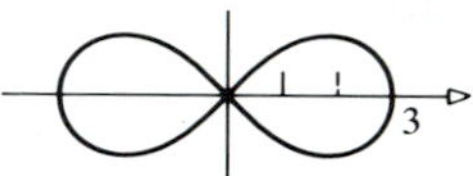

31. Symmetry: none

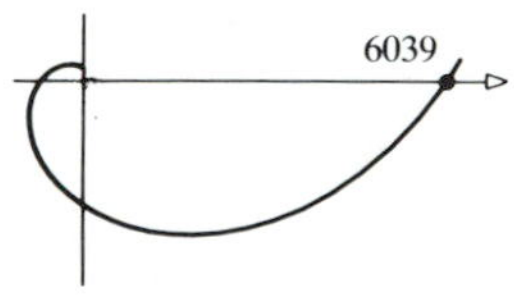

33.

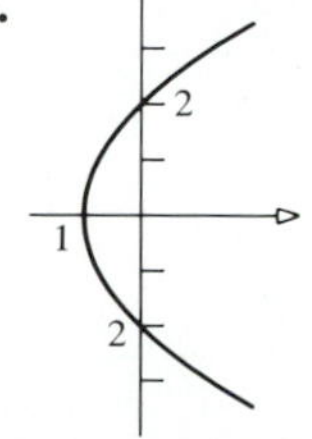

35.

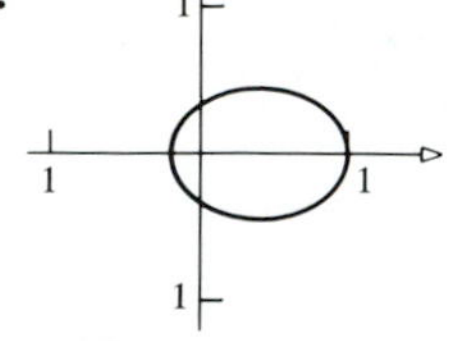

37.

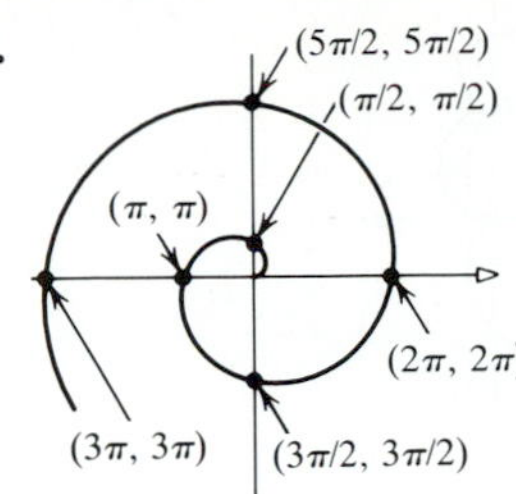

39.

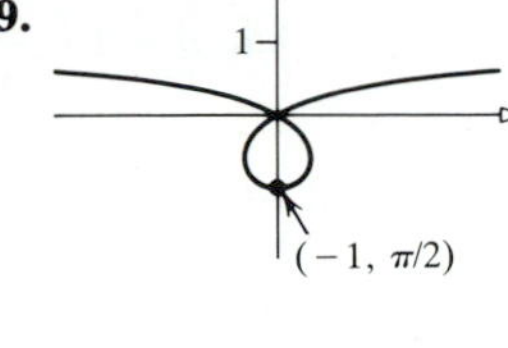

41.

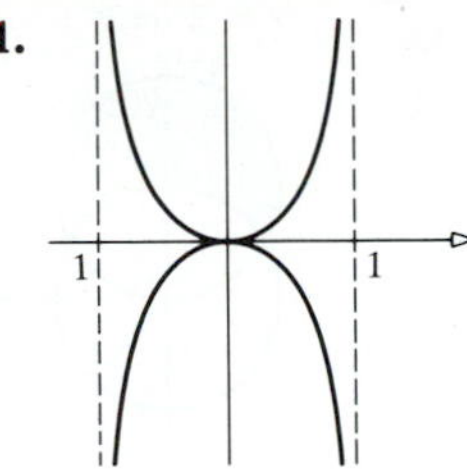

43. The point (r_0, θ_0) can also be named $(-r_0, \pi + \theta_0)$. If (r_0, θ_0) satisfies $r = 4 + 4\cos\theta$, then $r_0 = 4 + 4\cos\theta_0$. But $-4 + 4\cos(\pi + \theta_0) = -4 - 4\cos\theta_0 = -(r_0)$, so $(-r_0, \pi + \theta_0)$ satisfies $r = -4 + 4\cos\theta$. By a similar argument in the other direction, we know that both graphs contain the same points.

45. $r\sin\theta = y$, so the equation is $y = a$.

47. $r = 2a\sin\theta$ is equivalent to $r^2 = 2ar\sin\theta$ or $x^2 + y^2 = 2ay$, which is $x^2 + y^2 - 2ay + a^2 = a^2$ or $(x - 0)^2 + (y - a)^2 = a^2$, a circle with center $(0, a)$ and radius $\sqrt{a^2} = a$.

49. Use an argument similar to that in Problem 47.

51. $\dfrac{|FP|}{|DP|} = e$ where the conic is (a) an ellipse if $e < 1$; (b) a parabola if $e = 1$; and (c) a hyperbola if $e > 1$;

$|FP| = r$, $|DP| = d + r\cos\theta$, $r = e(d + r\cos\theta)$; $r = \dfrac{ed}{1 - e\cos\theta}$

53. Since the line joining the planet to the sun is longer at aphelion, the planet must travel slowest there to sweep out equal areas.

Exercise 13.3

1. $A = \dfrac{3\pi}{4} + \dfrac{9\sqrt{3}}{16}$

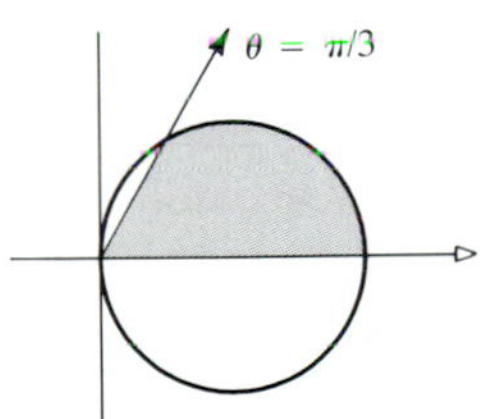

3. $A = \dfrac{4a^2\pi^3}{3}$

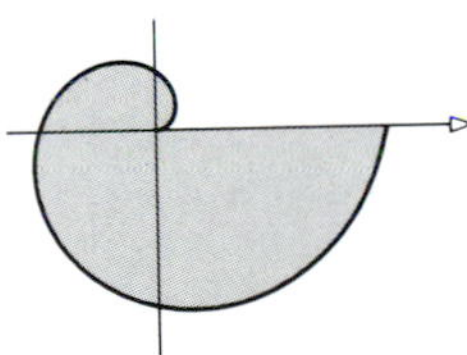

5. $A = \dfrac{3\pi}{2}$

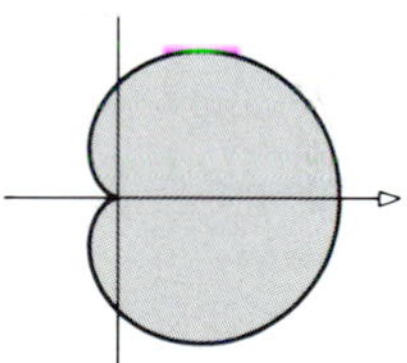

7. $A = \dfrac{3\pi}{2}$

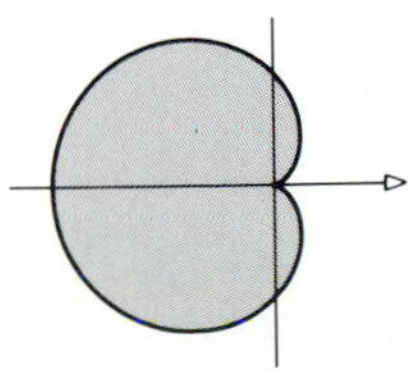

9. $A = 2\pi$

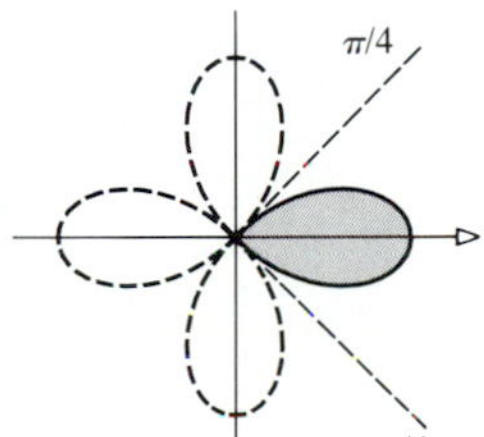

11. $A = 2$

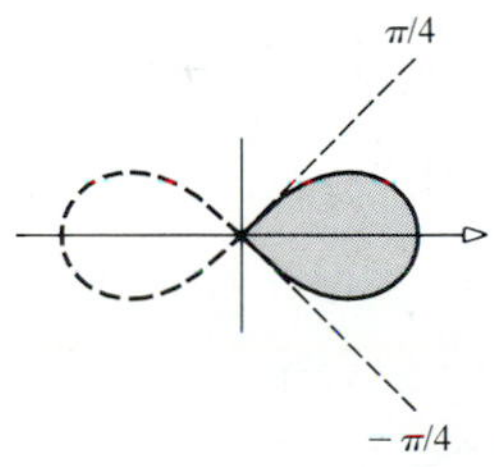

13. $A = \dfrac{\pi}{3} + \dfrac{\sqrt{3}}{2}$

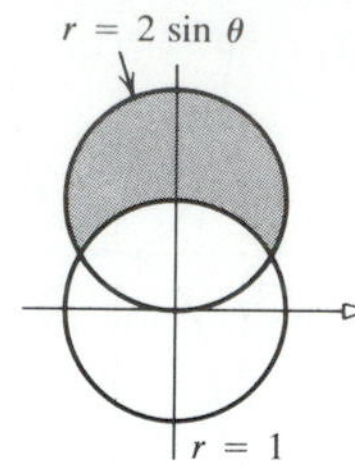

15. $A = 1 - \dfrac{\pi}{4}$

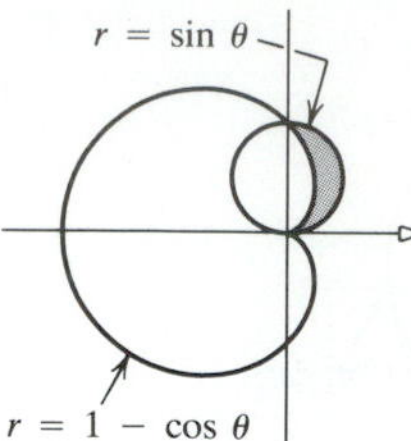

17. $A = \pi - \dfrac{3\sqrt{3}}{2}$

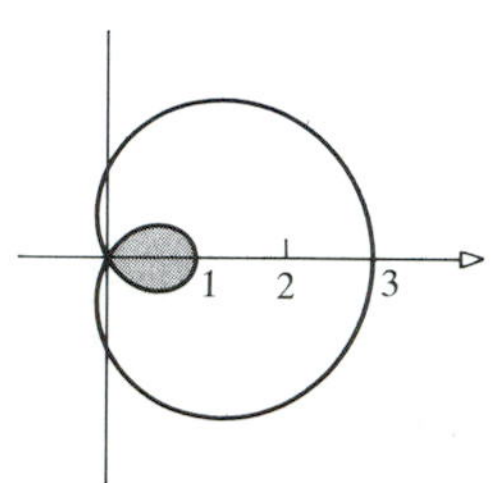

19. $A = \dfrac{32\pi}{3} + 4\sqrt{3}$

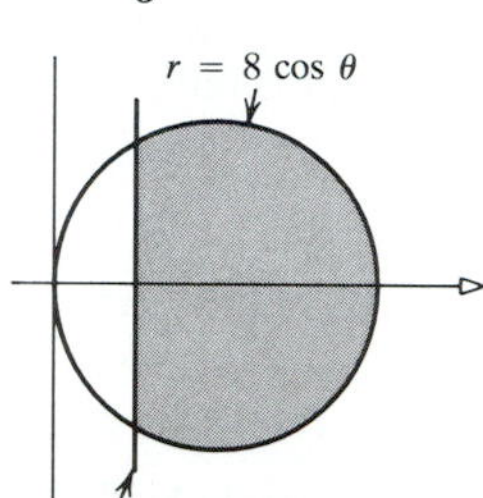

21. $A = \dfrac{9\sqrt{3}}{2} - \pi$

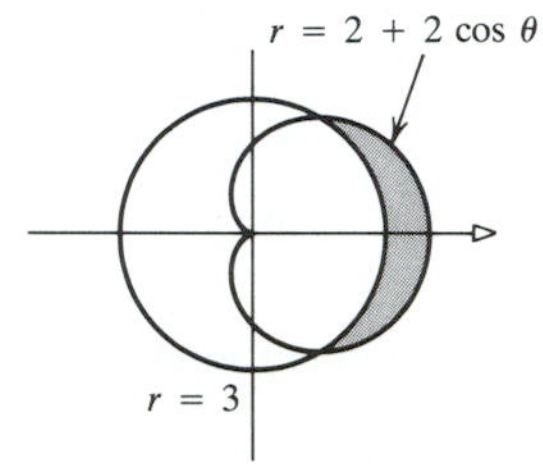

23. $A = \dfrac{7\pi}{12} - \sqrt{3}$

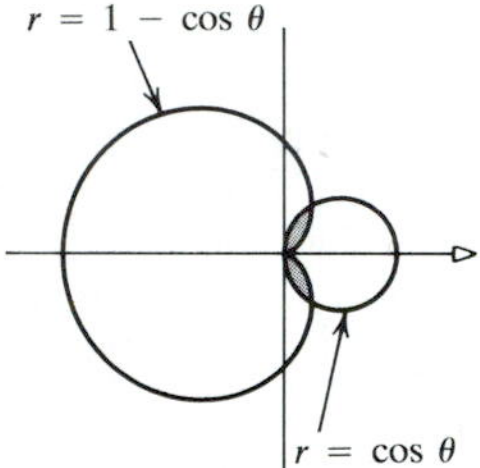

25. $A = 2\sqrt{2}$

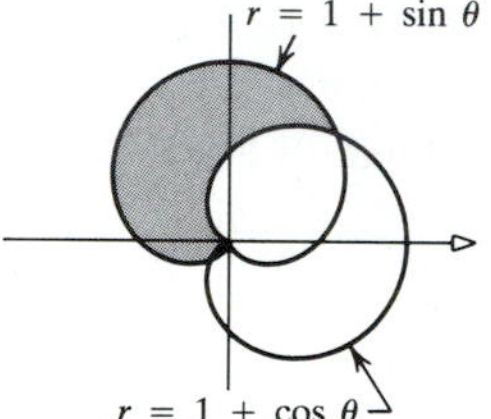

27. $(e^2 - 1)/4e^2$ **29.** $(\pi - 1)/2\pi$ **31.** $\pi + 3\sqrt{3}$

Exercise 13.4

1. $x - 2y + 3 = 0$

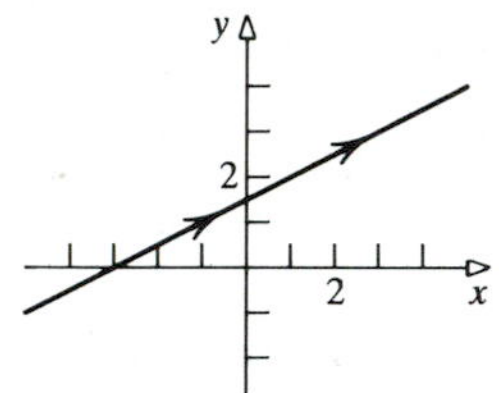

3. $x - 2y + 3 = 0,\ 1 \le x \le 5$

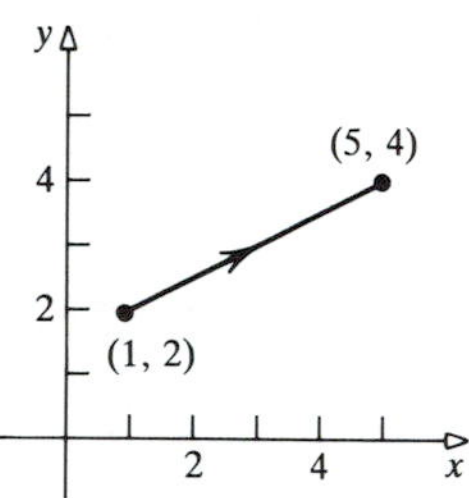

5. $x^2 + y^2 = 1$

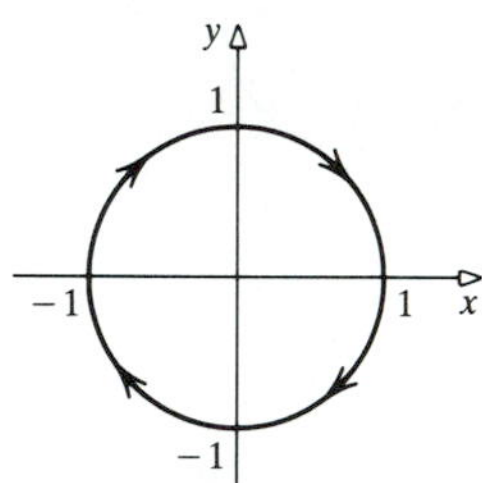

7. $(x + 3)^2 + (y - 1)^2 = 4,\ x \ge -3$

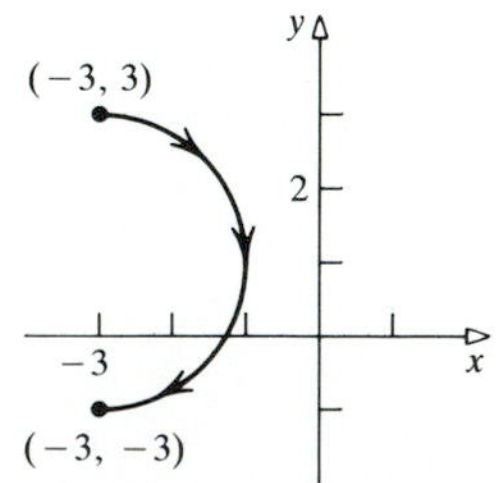

9. $\dfrac{x^2}{4} + \dfrac{y^2}{9} = 1$

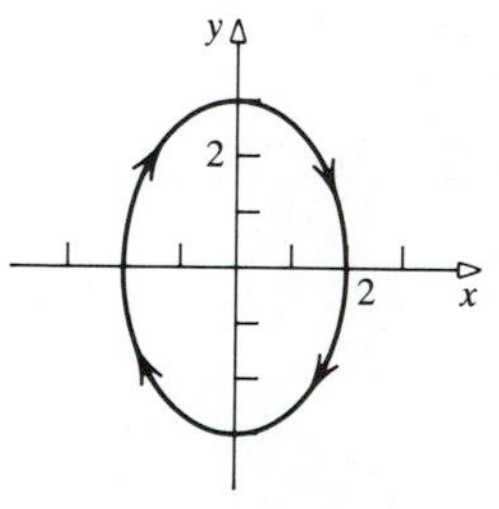

11. $x = 3$

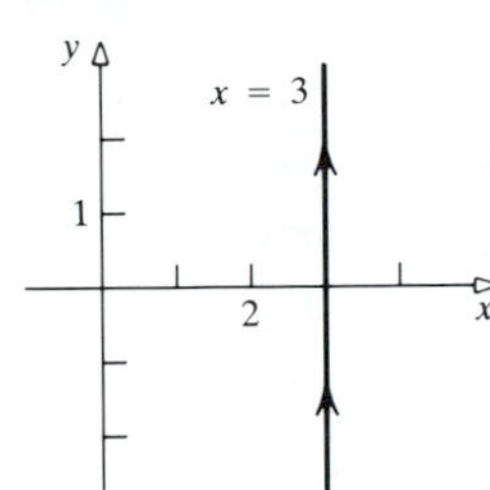

13. $x = 2;\ y \geq 4$

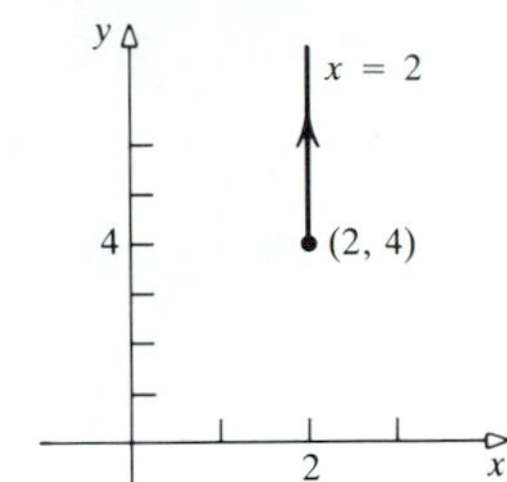

15. $y^2 = x - 5;\ x \geq 5,\ y \geq 0$

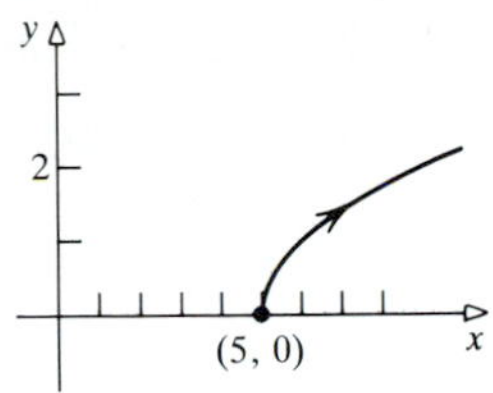

17. $y = (x - 1)^3;$
$x \geq 2,\ y \geq 1$

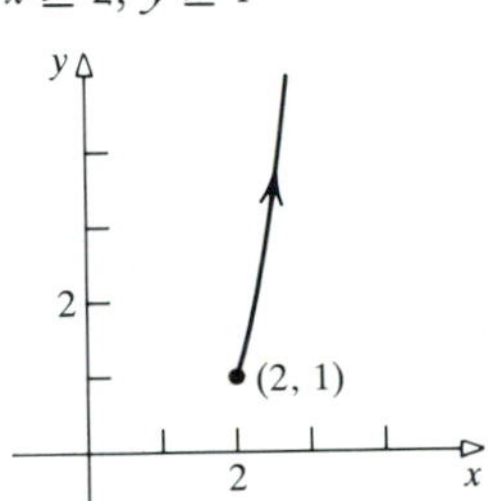

19. $y = \ln x$

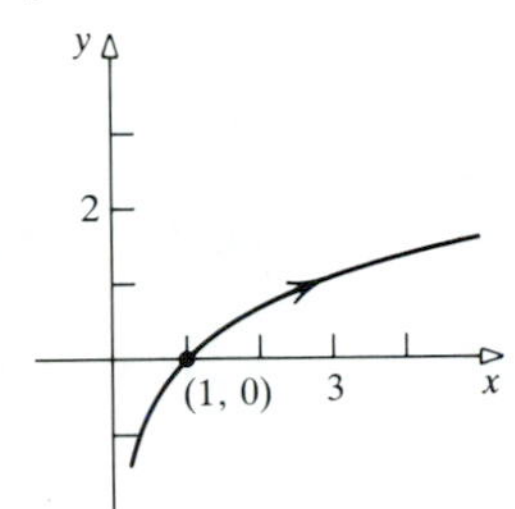

21. $x^2 - y^2 = 1;\ x \geq 1$

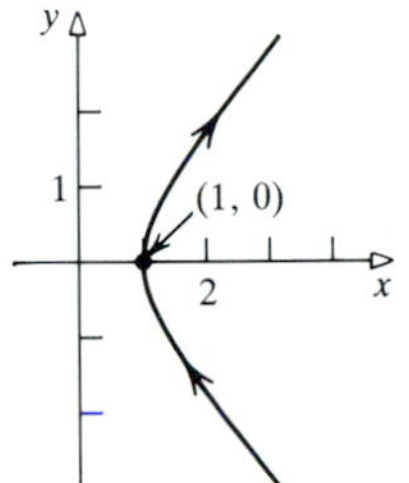

23. $3y^2 = -4(x - 1),\ -2 \leq x \leq 1$

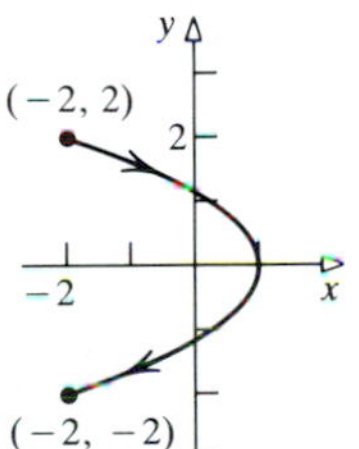

25. (*1*) $x = t$, $y = 4t^3$ (*2*) $x = t^{1/3}$, $y = 4t$

27. (*1*) $x = t - 3$, $y = 9t^2$ (*2*) $x = \frac{1}{3}t^{1/2} - 3$, $y = t, \quad t \geq 0$

29. $x = 3\cos(2\pi t/3)$
$y = 2\sin(2\pi t/3)$

31. $y = 2x/(x + 1),\ x > 0$

33. $\dfrac{x^2}{4} - \dfrac{y^2}{16} = 1,\ x \geq 2$

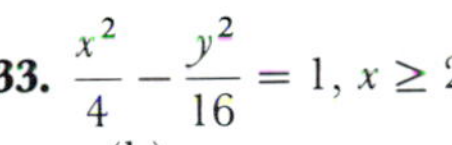

35. $(x + 2)^2 + [(y - 4)^2/4] = 1$

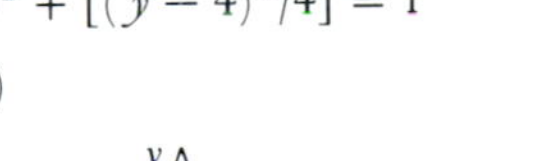

37. (a)

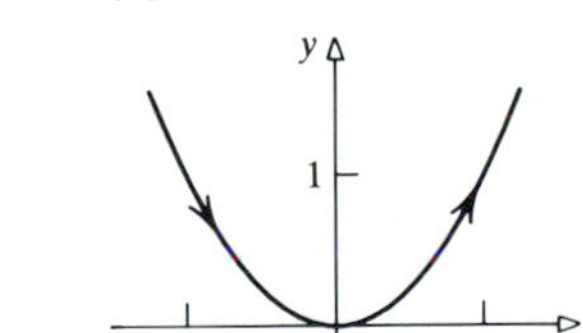

(b)

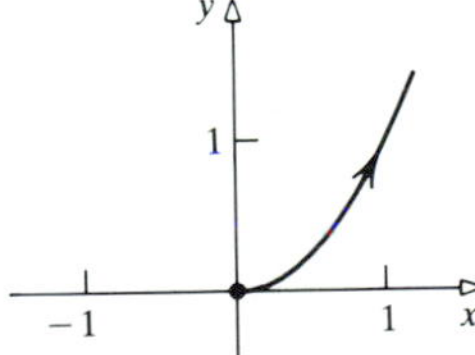

(c)

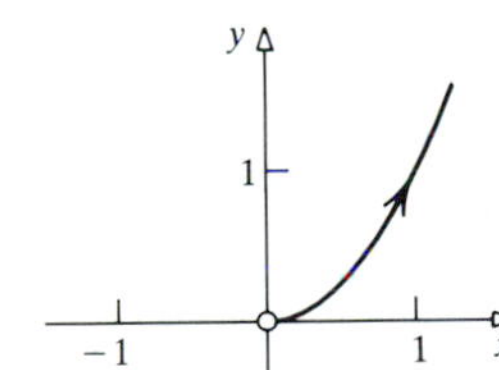

(d)

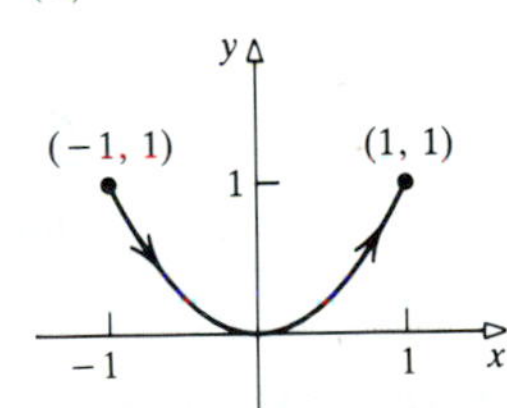

39. (a) $y = (\tan\theta)x - \dfrac{16}{v_0^2\cos^2\theta}x^2$ (b) $v(1) = (v_0^2 - 64v_0\sin\theta + 1024)^{1/2};\ v(2) = (v_0^2 - 128v_0\sin\theta + 4096)^{1/2}$

(c) $y = 0 \Rightarrow t = \dfrac{v_0\sin\theta}{16};\ x = \dfrac{v_0^2\sin 2\theta}{32}$

41. $x = 3b\cos t + b\cos 3t = a\cos^3 t;\ y = 3b\sin t - b\sin 3t = a\sin^3 t;\ (x/a)^{2/3} + (y/a)^{2/3} = 1$

Exercise 13.5

1. $x - 8y + 8 = 0$ **3.** $4x - 3y - 9 = 0$ **5.** $x + 4y - 3 = 0$ **7.** $2x + y - 2 = 0$ **9.** $x - y = 0$
11. $x + y - \sqrt{2} = 0$ **13.** $3\sqrt{3}x + 4y - 24 = 0$ **15.** $x - \sqrt{3}y = 0$ **17.** $\sqrt{2}x + (1 + \sqrt{2})y - 2\sqrt{2} - \frac{9}{2} = 0$
19. $(2\sqrt{2} + 5)x + 2\sqrt{2}y - 20\sqrt{2} - \frac{57}{2} = 0$ **21.** $(0, 0)$ **23.** $(\pm 1, 0)$ **25.** $(6, \pi), \left(\frac{3}{2}, \frac{\pi}{3}\right), \left(\frac{3}{2}, \frac{5\pi}{3}\right)$
27. $\left(\frac{9 + \sqrt{17}}{4}, \sin^{-1}\left(\frac{-3 + \sqrt{17}}{4}\right)\right), \left(\frac{9 + \sqrt{17}}{4}, \pi - \sin^{-1}\left(\frac{-3 + \sqrt{17}}{4}\right)\right)$ or $(3.28, 16.31°)$, $(3.28, 163.69°)$
29. $(2, 0)$, $(0, \pi/2)$, $(-1.84, 52.24°)$, $(1.84, 127.76°)$ **31.** $(3, 0), \left(0, \frac{\pi}{2}\right)$
33. $\left(e^{\pi/4}, \frac{\pi}{4}\right), \left(e^{5\pi/4}, \frac{5\pi}{4}\right), \left(e^{9\pi/4}, \frac{9\pi}{4}\right), \ldots (e^{(4n+1)\pi/4}, (4n + 1)\pi/4), \ldots$ **35.** Every point on the curve
37. $\frac{8}{27}(10\sqrt{10} - 1)$ **39.** $\sqrt{5} + \frac{1}{2}\ln(2 + \sqrt{5})$ **41.** 4π **43.** 4π **45.** $\sqrt{5}(e - 1)$ **47.** 2
49. $\pi\sqrt{1 + 4\pi^2} + \frac{1}{2}\ln(2\pi + \sqrt{1 + 4\pi^2})$ **51.** $3b/2$ **53.** $5 + \frac{9}{4}\ln 3$ **55.** $\frac{1}{2}(3\sqrt{11} - \sqrt{3}) + \ln\frac{3 + \sqrt{11}}{1 + \sqrt{3}}$
57. $\sqrt{5}$ **59.** $\frac{dy}{dx} = \frac{\sin t + \cos t}{\cos t - \sin t}; \frac{d^2y}{dx^2} = \frac{2}{e^t(\cos t - \sin t)^3}$ **61.** $\frac{dy}{dx} = \frac{t^2}{t^2 - 1}; \frac{d^2y}{dx^2} = \frac{-2t^3}{(t^2 - 1)^3}$
63. $\frac{dy}{dx} = \tan\theta; \frac{d^2y}{dx^2} = \frac{1}{a\theta\cos^3\theta}$ **65.** $\frac{dy}{dx} = 1/3t; \frac{d^2y}{dx^2} = -1/18t^3$ **67.** $\frac{dy}{dx} = -1/(8\sin\theta); \frac{d^2y}{dx^2} = -1/(64\sin^3\theta)$
69. $\frac{dy}{dx} = \frac{1}{2}\tan\theta; \frac{d^2y}{dx^2} = -\frac{1}{4}\tan^3\theta$ **71.** $\sqrt{0.0411} \approx 0.2028$ **73.** $a/10$
75. Horizontal: $(\frac{10}{3}, \pm 16/3\sqrt{3})$;
vertical: $(2, 0)$; $y = 2(x - 6)$, $y = -2(x - 6)$

Exercise 13.6

1. $24\pi(2\sqrt{2} - 1)$ **3.** $6\pi a^2/5$ **5.** $(\pi/27)(10\sqrt{10} - 1)$ **7.** $(\pi a^2/4)(e^2 - e^{-2}) + \pi a^2$ **9.** $4\pi a^2$
11. $\pi[e\sqrt{1 + e^2} + \ln(e + \sqrt{1 + e^2}) - \sqrt{2} - \ln(1 + \sqrt{2})]$ **13.** $6\pi a^2/5$ **15.** $(24\pi/5)(\sqrt{2} + 1)$ **17.** $\pi^2/2$
19. $(2\sqrt{2}\pi/5)(1 + e^{2\pi})$ **21.** $S = 2\pi\int_1^{+\infty}\frac{\sqrt{1 + x^4}}{x^3}\,dx$ diverges; $V = \pi\int_1^{+\infty}\frac{1}{x^2}\,dx = \pi$ converges
23. $(\pi/2)[\sinh 2b - \sinh 2a + 2(b - a)]$ **25.** $4\pi R\sqrt{R^2 - a^2}$

Review Exercises

1. $(x^2 + y^2)^3 = 64x^2y^2$ **3.** $3x^2 - y^2 - 4x + 1 = 0$ **5.** $(x^2 + y^2) = 4x^2/(x - 1)^2$ **7.** $(x^2 + y^2)^2 = 3x^2y - y^3$
9. $r^2 = r\cos\theta$ or $r = \cos\theta$ (since this still contains $r = 0$) **11.** $r^2\sin^2\theta = r^2\cos^2 r$ or $\sin^2\theta = \cos^2 r$
13. $r\sin\theta = a$ or $r = a\csc\theta$

15.

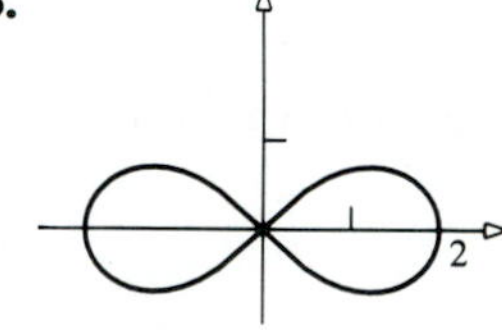

17.

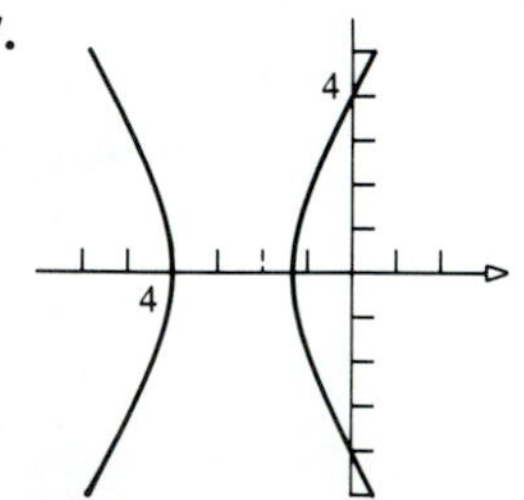

19.
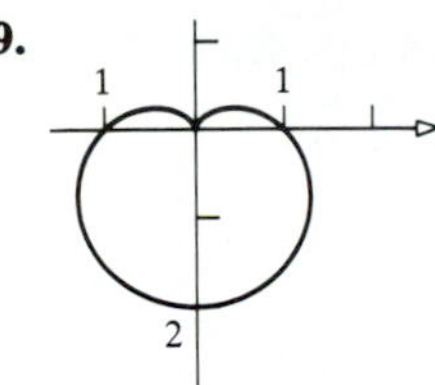

21.
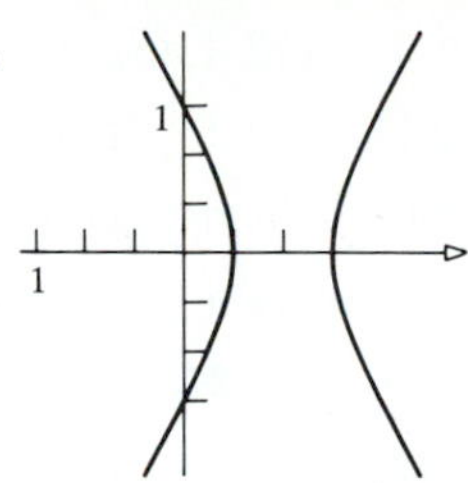

23.
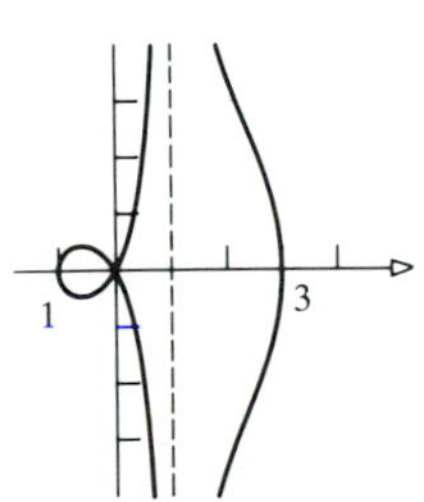

25.
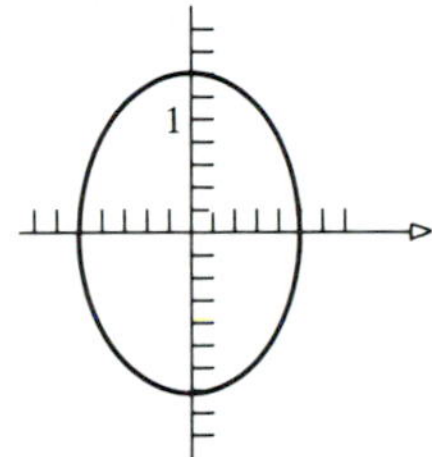

27.
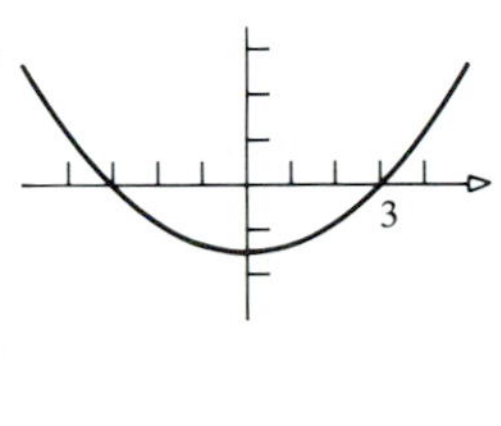

29. $2\sqrt{3} - 2\pi/3$ **31.** $x^2 - y^2 = 1$ **33.** $y = \cosh x$ **35.** $x^2 = 4y^2 - 4y^4$ **37.** $y = \frac{1}{2}x^2 - \frac{1}{4}\ln x$

39. Any set of equations of the form $x = a\cos\omega t$, $y = b\sin\omega t$ (ω a constant) will parameterize the ellipse.

41. 1 **43.** $\frac{1}{2}e^4$ **45.** $\pi\sqrt{2} + \pi\ln(1 + \sqrt{2})$

Challenge Exercises

1. $\dfrac{f'g'' - f''g'}{(f')^3}$ **3.** $(x^2 + y^2)^2 = x^2 - y^2$ **5.** $\begin{cases} x = R\left(\dfrac{1 - m^2}{1 + m^2}\right) \\ y = R\left(\dfrac{2m}{1 + m^2}\right) \end{cases}$

7. $\text{Area} = \displaystyle\int_{\theta_1}^{\theta_2} \frac{1}{2}\left(\frac{a}{\theta}\right)^2 d\theta = \frac{a^2}{2\theta_1} - \frac{a^2}{2\theta_2} = \frac{a}{2}\left(\frac{a}{\theta_1} - \frac{a}{\theta_2}\right) = \frac{a}{2}(r_1 - r_2)$

9. Let $s = S(t) = \displaystyle\int_0^t \sqrt{[x'(u)]^2 + [y'(u)]^2}\, du$; so $S(t)$ is the length of the curve from 0 to t. Now $S'(t) = \sqrt{[x'(t)]^2 + [y'(t)]^2} > 0$, since the curve is smooth. Thus, S is increasing and so has an inverse function $S^{-1}(s) = t$. Let $f(s) = x(S^{-1}(s))$, $g(s) = y(S^{-1}(s))$. Then

$$\int_0^s \sqrt{[f'(v)]^2 + [g'(v)]^2}\, dv = \int_0^s \sqrt{[x'(S^{-1}(u))]^2 + [y'(S^{-1}(u))]^2}\, \frac{d}{du} S^{-1}(u)\, du = \int_0^s \frac{\sqrt{[x'(S^{-1}(u))]^2 + [y'(S^{-1}(u))]^2}}{\sqrt{[x'(S^{-1}(u))]^2 + [y'(S^{-1}(u))]^2}}\, du =$$

$\displaystyle\int_0^s du = S$, as required.

11. $(a/2, b/2)$; $\frac{1}{2}\sqrt{a^2 + b^2}$ **13.** $(\pi/6)(2\sqrt{2} - 1)$

CHAPTER 14

Exercise 14.1

1.
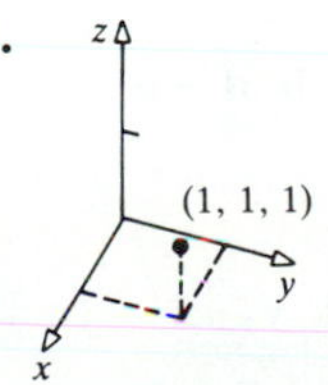

3.
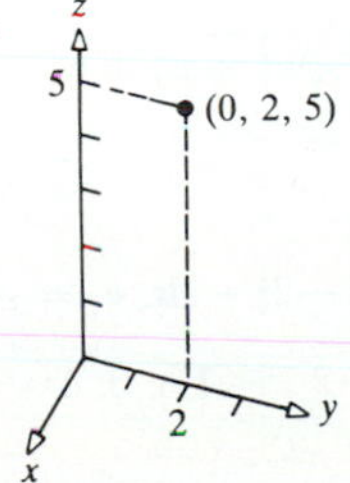

5.
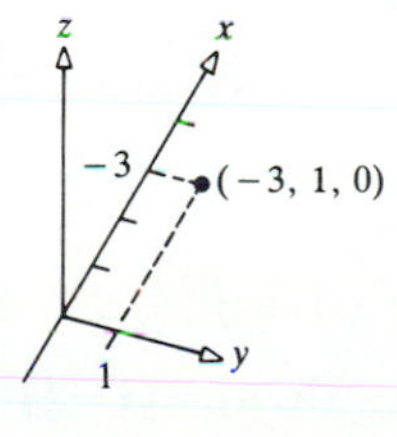

7. $(0, 0, 3)$; $(0, 1, 0)$; $(2, 0, 0)$; $(0, 1, 3)$; $(2, 1, 0)$; $(2, 0, 3)$ **9.** $(1, 2, 5)$; $(1, 4, 3)$; $(3, 2, 3)$; $(1, 4, 5)$; $(3, 4, 3)$; $(3, 2, 5)$
11. $(-1, 0, 5)$; $(-1, 2, 2)$; $(4, 0, 2)$; $(-1, 2, 5)$; $(4, 0, 5)$; $(4, 2, 2)$
13. Plane parallel to xz-plane, 3 units to the right of it **15.** The yz-plane
17. Plane parallel to xy-plane, 5 units above it **19.** $\sqrt{17}$ **21.** $\sqrt{57}$ **23.** $\sqrt{26}$
25. $(x-3)^2 + (y-1)^2 + (z-1)^2 = 1$ **27.** $(x+1)^2 + (y-1)^2 + (z-2)^2 = 9$ **29.** 2; $(-1, 1, 0)$
31. 3; $(2, -2, -1)$ **33.** $3/\sqrt{2}$; $(2, 0, -1)$ **35.** $x^2 + (y-3)^2 + (z-6)^2 = 17$
37. $(x+3)^2 + (y-2)^2 + (z-1)^2 = 62$ **39.** Use Pythagorean theorem and distance formula.
41. Use Pythagorean theorem and distance formula. **43.** Each side has length $3\sqrt{2}$.

Exercise 14.2

1. All are scalars except (c) and (h). **3.** $\mathbf{A}$ **5.** $-\mathbf{F} + \mathbf{E} - \mathbf{D}$ **7.** $-\mathbf{G} - \mathbf{H} + \mathbf{D}$ **9.** $\mathbf{0}$ **11.** $\frac{1}{2}(\mathbf{v} + \mathbf{w})$

Exercise 14.3

1. $4\mathbf{i} - 5\mathbf{j}$ **3.** $-\mathbf{i} + \mathbf{j}$ **5.** $x\mathbf{i} + y\mathbf{j}$ **7.** $-3\mathbf{i} - 2\mathbf{j} + \mathbf{k}$ **9.** $3\mathbf{i} - \mathbf{k}$ **11.** $x\mathbf{i} + y\mathbf{j} + z\mathbf{k}$ **13.** $3\mathbf{i} - 8\mathbf{j}$
15. $\frac{7}{6}\mathbf{i}$ **17.** $\sqrt{26}$ **19.** $-3\mathbf{i} + \mathbf{j} - 5\mathbf{k}$ **21.** $\sqrt{13}$ **23.** $17\mathbf{i} - 3\mathbf{j} + 13\mathbf{k}$ **25.** $\sqrt{453}$ **27.** $\sqrt{109}$ **29.** 5
31. $\sqrt{2}$ **33.** $\sqrt{2}\,|a|$ **35.** $\sqrt{21}$ **37.** $\sqrt{3}$ **39.** $\sqrt{3}\,a$ **41.** $\frac{5}{13}\mathbf{i} - \frac{12}{13}\mathbf{j}$; $\frac{-5}{13}\mathbf{i} + \frac{12}{13}\mathbf{j}$
43. $(2/\sqrt{5})\mathbf{i} + (1/\sqrt{5})\mathbf{j}$; $(-2/\sqrt{5})\mathbf{i} - (1/\sqrt{5})\mathbf{j}$ **45.** $\frac{1}{2}\mathbf{i} + (\sqrt{3}/2)\mathbf{j}$; $-\frac{1}{2}\mathbf{i} - (\sqrt{3}/2)\mathbf{j}$
47. $\frac{1}{\sqrt{3}}\mathbf{i} + \frac{1}{\sqrt{3}}\mathbf{j} + \frac{1}{\sqrt{3}}\mathbf{k}$; $\frac{-1}{\sqrt{3}}\mathbf{i} - \frac{1}{\sqrt{3}}\mathbf{j} - \frac{1}{\sqrt{3}}\mathbf{k}$ **49.** $\frac{\cos\theta}{3}\mathbf{i} + \frac{\sin\theta}{3}\mathbf{j} + \frac{2\sqrt{2}}{3}\mathbf{k}$; $\frac{-\cos\theta}{3}\mathbf{i} - \frac{\sin\theta}{3}\mathbf{j} - \frac{2\sqrt{2}}{3}\mathbf{k}$
51. $\frac{3}{13}\mathbf{i} - \frac{4}{13}\mathbf{j} + \frac{12}{13}\mathbf{k}$; $\frac{-3}{13}\mathbf{i} + \frac{4}{13}\mathbf{j} - \frac{12}{13}\mathbf{k}$ **53.** $\pm[(-8/\sqrt{29})\mathbf{i} + (12/\sqrt{29})\mathbf{j} - (16/\sqrt{29})\mathbf{k}]$ **55.** $\alpha = \pm 1$
57. $-2 \pm \sqrt{21}$ **59.** $\mathbf{v} = 2\sqrt{3}\mathbf{i} \pm 2\mathbf{j}$ **61.** $\mathbf{v} = 2\sqrt{5}\mathbf{i} + \sqrt{5}\mathbf{j}$; $\mathbf{v} = -2\sqrt{5}\mathbf{i} - \sqrt{5}\mathbf{j}$ **63.** $\mathbf{v} = -\mathbf{i} + \mathbf{j}$
65. 459.54 km/hr **67.**

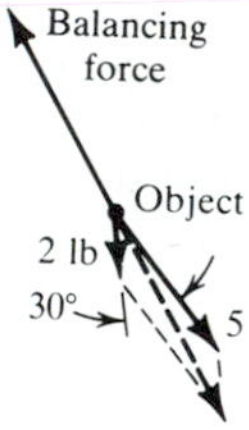

69. (a)

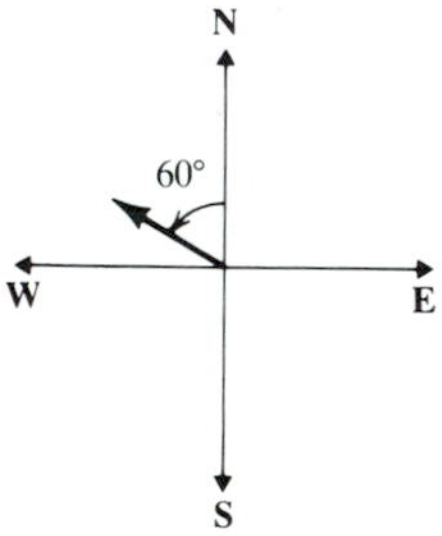

(b)

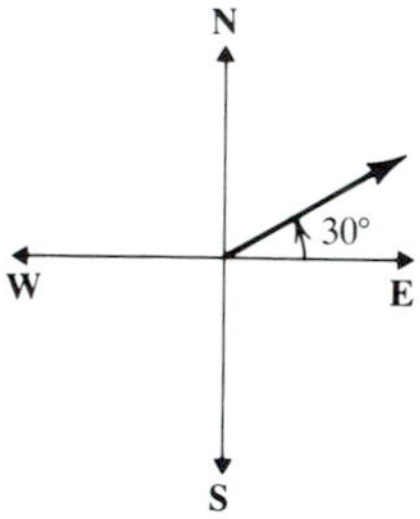

71. 217.54 km/hr

73. (b) $\mathbf{u} + (\mathbf{v} + \mathbf{w}) = \langle u_1, u_2\rangle + (\langle v_1, v_2\rangle + \langle w_1, w_2\rangle) = \langle u_1, u_2\rangle + \langle v_1 + w_1, v_2 + w_2\rangle =$
$\langle u_1 + (v_1 + w_1), u_2 + (v_2 + w_2)\rangle = \langle (u_1 + v_1) + w_1, (u_2 + v_2) + w_2\rangle = \langle u_1 + v_1, u_2 + v_2\rangle + \langle w_1, w_2\rangle =$
$(\mathbf{u} + \mathbf{v}) + \mathbf{w}$
(d) $\mathbf{u} + (-\mathbf{u}) = \langle u_1, u_2\rangle + [-\langle u_1, u_2\rangle] = \langle u_1, u_2\rangle + \langle -u_1, -u_2\rangle = \langle 0, 0\rangle = \mathbf{0}$
(f) $(\alpha + \beta)\mathbf{v} = (\alpha + \beta)\langle v_1, v_2\rangle = \langle (\alpha + \beta)v_1, (\alpha + \beta)v_2\rangle = \langle \alpha v_1 + \beta v_1, \alpha v_2 + \beta v_2\rangle = \langle \alpha v_1, \alpha v_2\rangle + \langle \beta v_1, \beta v_2\rangle =$
$\alpha\mathbf{v} + \beta\mathbf{v}$

75. $\|\alpha\mathbf{v}\| = \|\langle \alpha v_1, \alpha v_2, \alpha v_3\rangle\| = \sqrt{\alpha^2 v_1^2 + \alpha^2 v_2^2 + \alpha^2 v_3^2} = \sqrt{\alpha^2}\sqrt{v_1^2 + v_2^2 + v_3^2} = |\alpha|\,\|\mathbf{v}\|$ **77.** $\mathbf{c} - \mathbf{b}$; $\mathbf{d} - \mathbf{b}$; $\mathbf{d} - \mathbf{c}$

Exercise 14.4

1. 6; $\sqrt{\frac{6}{7}}$ **3.** -1; $-\frac{1}{2}$ **5.** -8; $-8/\sqrt{66}$ **7.** $\mathbf{v}_1 = 2\mathbf{i} - 2\mathbf{j} + 2\mathbf{k}$, $\mathbf{v}_2 = -\mathbf{j} - \mathbf{k}$ **9.** $\mathbf{v}_1 = -\frac{1}{2}\mathbf{j} - \frac{1}{2}\mathbf{k}$, $\mathbf{v}_2 = \mathbf{i} - \frac{1}{2}\mathbf{j} + \frac{1}{2}\mathbf{k}$
11. $\mathbf{v}_1 = \frac{8}{3}\mathbf{i} + \frac{4}{3}\mathbf{j} - \frac{4}{3}\mathbf{k}$, $\mathbf{v}_2 = \frac{1}{3}\mathbf{i} - \frac{1}{3}\mathbf{j} + \frac{1}{3}\mathbf{k}$ **13.** 1 **15.** -2 **17.** $0, 4$ **19.** $\frac{3}{2}\mathbf{i} + (3/\sqrt{2})\mathbf{j} + \frac{3}{2}\mathbf{k}$

21. $\sqrt{2}\mathbf{i} - \mathbf{j} + \mathbf{k}$ **23.** Remember, $\|\mathbf{i}\| = 1$. **25.** $2\sqrt{10 + 3\sqrt{3}}; 2\sqrt{10 - 3\sqrt{3}}$ **27.** 9 **29.** $\frac{8}{3}$ joules

31. $-1, 2$ **33.** Approx. 70.5° **35.** $-275\mathbf{i} + 75\sqrt{3}\mathbf{j}$ **37.** (a) $(\mathbf{u} + \mathbf{v}) \cdot (\mathbf{u} - \mathbf{v}) = \|\mathbf{u}\|^2 - \|\mathbf{v}\|^2 = 0$

39. $(\mathbf{v} - \alpha\mathbf{w}) \cdot \mathbf{w} = \mathbf{v} \cdot \mathbf{w} - \alpha\|\mathbf{w}\|^2 = 0$ **41.** 0 **43.** Use theorems (14.4.2) and (14.4.3). **45.** $\sin(\theta/2)$

47.
$$\begin{aligned}\mathbf{u} \cdot (\mathbf{v} + \mathbf{w}) &= (u_1\mathbf{i} + u_2\mathbf{j} + u_3\mathbf{k}) \cdot [(v_1 + w_1)\mathbf{i} + (v_2 + w_2)\mathbf{j} + (v_3 + w_3)\mathbf{k}] \\ &= u_1(v_1 + w_1) + u_2(v_2 + w_2) + u_3(v_3 + w_3) \\ &= (u_1v_1 + u_2v_2 + u_3v_3) + (u_1w_1 + u_2w_2 + u_3w_3) \\ &= \mathbf{u} \cdot \mathbf{v} + \mathbf{u} \cdot \mathbf{w}\end{aligned}$$

49. $\mathbf{0} \cdot \mathbf{v} = 0 \cdot v_1 + 0 \cdot v_2 + 0 \cdot v_3 = 0$ **51.** $\mathbf{v} = \mathbf{0}$

53. (a) $\cos^2(\pi/6) + \cos^2(\pi/4) > 1$ (b) $\cos^2\beta + \cos^2\gamma = \sin^2\alpha \Rightarrow \sin\alpha \geq \cos\beta \Rightarrow \pi/4 \leq \alpha, \beta \leq \pi/2$

55. $\sin^2\theta = 1 - \cos^2\theta = \dfrac{\|\mathbf{u}\|^2\|\mathbf{v}\|^2 - (\mathbf{u} \cdot \mathbf{v})^2}{\|\mathbf{u}\|^2\|\mathbf{v}\|^2} = \dfrac{(u_1^2 + u_2^2 + u_3^2)(v_1^2 + v_2^2 + v_3^2) - (u_1v_1 + u_2v_2 + u_3v_3)^2}{\|\mathbf{u}\|^2\|\mathbf{v}\|^2}$. Expand the numerator and show that it equals $(u_2v_3 - u_3v_2)^2 + (u_1v_3 - u_3v_1)^2 + (u_1v_2 - u_2v_1)^2$.

57. $\|\mathbf{w}_1\| = \left\|\dfrac{\mathbf{u}_1}{\|\mathbf{u}_1\|}\right\| = 1$; $\mathbf{w}_1 \cdot \mathbf{w}_2 = \mathbf{w}_1 \cdot \dfrac{[\mathbf{u}_2 - (\mathbf{u}_2 \cdot \mathbf{w}_1)\mathbf{w}_1]}{\|\mathbf{v}_2\|} = \dfrac{\mathbf{w}_1 \cdot \mathbf{u}_2 - \mathbf{u}_2 \cdot \mathbf{w}_1}{\|\mathbf{v}_2\|} = 0$

59. One set is: $\mathbf{a} = 2\mathbf{i} + \mathbf{j} - \mathbf{k}$, $\mathbf{b} = 3\mathbf{i} + 2\mathbf{j} + \mathbf{k}$, $\mathbf{c} = \mathbf{i} + 3\mathbf{j} - 2\mathbf{k}$

61. The proofs follow the pattern of those for vectors in the plane and in space.

Exercise 14.5

1. $-3(\mathbf{j} + \mathbf{k})$ **3.** $-2\mathbf{k}$ **5.** $-\mathbf{i} + \mathbf{j} + 5\mathbf{k}$ **7.** $-6\mathbf{i} + 21\mathbf{j} - 58\mathbf{k}$ **9.** $-8\mathbf{i} + 12\mathbf{j} + 5\mathbf{k}$ **11.** $\sqrt{1013}$

13. 58 **15.** $46\sqrt{2}$ **17.** $5\sqrt{3}$ **19.** $\sqrt{998}$ **21.** Use theorem (14.5.3(c)).

23.
$$\begin{aligned}\alpha(\mathbf{v} \times \mathbf{w}) &= \alpha[(v_2w_3 - v_3w_2)\mathbf{i} - (v_1w_3 - v_3w_1)\mathbf{j} + (v_1w_2 - v_2w_1)\mathbf{k}] \\ &= [(\alpha v_2)w_3 - (\alpha v_3)w_2]\mathbf{i} - [(\alpha v_1)w_3 - (\alpha v_3)w_1]\mathbf{j} + [(\alpha v_1)w_2 - (\alpha v_2)w_1]\mathbf{k} \\ &= (\alpha\mathbf{v}) \times \mathbf{w}\end{aligned}$$

25. $\mathbf{i} \times (\mathbf{i} \times \mathbf{j}) = -\mathbf{j}$; $(\mathbf{i} \times \mathbf{i}) \times \mathbf{j} = \mathbf{0}$ **27.** $\pm(\frac{3}{7}\mathbf{i} - \frac{2}{7}\mathbf{j} + \frac{6}{7}\mathbf{k})$ **29.** $10\sqrt{78}$ m/sec

31. $8\sqrt{\frac{3}{14}}\,\omega$ m/sec; $\omega = 14/\sqrt{3}$ rad/sec

33. If $\mathbf{u}$, $\mathbf{v}$ and $\mathbf{w}$ lie in plane p, then $\mathbf{v} \times \mathbf{w}$ is normal to p and orthogonal to all vectors in p, so $\mathbf{u} \cdot (\mathbf{v} \times \mathbf{w}) = 0$. If $\mathbf{u} \cdot (\mathbf{v} \times \mathbf{w}) = 0$, then $\mathbf{u} \perp (\mathbf{v} \times \mathbf{w})$. But then $\mathbf{u}$, $\mathbf{v}$, and $\mathbf{w}$ are all orthogonal to $\mathbf{v} \times \mathbf{w}$, so they lie in one plane.

35. Use the representation of the triple scalar product as a determinant and properties of determinants.

37. Use Problem 36 and commutativity of the scalar product.

39. $(\mathbf{a} \times \mathbf{b}) \cdot (\mathbf{c} \times \mathbf{d}) = \mathbf{a} \cdot [\mathbf{b} \times (\mathbf{c} \times \mathbf{d})] = \mathbf{a} \cdot [(\mathbf{b} \cdot \mathbf{d})\mathbf{c} - (\mathbf{b} \cdot \mathbf{c})\mathbf{d}] = (\mathbf{a} \cdot \mathbf{c})(\mathbf{b} \cdot \mathbf{d}) - (\mathbf{a} \cdot \mathbf{d})(\mathbf{b} \cdot \mathbf{c})$

41. $(\mathbf{u} - \mathbf{v}) \cdot (\mathbf{u} + \mathbf{v}) = 0$ if and only if $\mathbf{u} \cdot \mathbf{u} - \mathbf{v} \cdot \mathbf{v} = 0$ or $\|\mathbf{u}\| = \|\mathbf{v}\|$

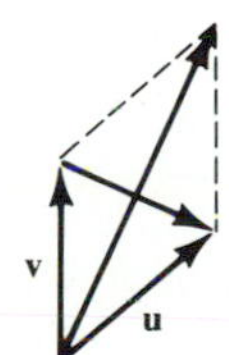

43. $\dfrac{1}{\alpha^2 + \mathbf{a} \cdot \mathbf{a}}\left[\alpha\mathbf{b} - (\mathbf{b} \times \mathbf{a}) + \dfrac{\mathbf{b} \cdot \mathbf{a}}{\alpha}\mathbf{a}\right]$

Exercise 14.6

1. $\mathbf{r} = (1 + 2t)\mathbf{i} + (2 - t)\mathbf{j} + (3 + t)\mathbf{k}$; $x = 1 + 2t$, $y = 2 - t$, $z = 3 + t$; $(x - 1)/2 = (y - 2)/(-1) = (z - 3)/1$

3. $\mathbf{r} = (4 + t)\mathbf{i} + (-1 + t)\mathbf{j} + 6\mathbf{k}$; $x = 4 + t$, $y = -1 + t$, $z = 6$; $x - 4 = y + 1$, $z = 6$

5. $\mathbf{r} = (1 + 3t)\mathbf{i} + (-1 + 3t)\mathbf{j} + (3 - 2t)\mathbf{k}$; $x = 1 + 3t$, $y = -1 + 3t$, $z = 3 - 2t$; $(x - 1)/3 = (y + 1)/3 = (z - 3)/(-2)$

7. $\mathbf{r} = 2t\mathbf{i} + 3t\mathbf{j} + \mathbf{k}$; $x = 2t$, $y = 3t$, $z = 1$; $x/2 = y/3$, $z = 1$ **9.** $(1, 6, -4)$; $\sqrt{\frac{6}{11}}$ **11.** $(-2, -3, -5)$; $-7/(3\sqrt{11})$
13. $x/8 = y/(-10) = z/(-7)$ **15.** $x = 1$, $y - 2 = (z + 1)/2$ **17.** Parallel **19.** Intersect at $(\frac{17}{4}, \frac{11}{4}, \frac{3}{2})$
21. Intersect at $(\frac{25}{3}, \frac{13}{3}, \frac{19}{3})$ **23.** Parallel **25.** Skew **27.** Parallel **29.** $x = 1 + 6t$, $y = -2 + 2t$, $z = -3 - t$
31. Parallel **33.** $\mathbf{i} + \mathbf{j} + 2\mathbf{k}$; $(-1, -3, -4)$, $(1, -1, 0)$
35. (a) Two straight lines, one through $(0, 0, 1)$ in the direction $\mathbf{i} - \mathbf{j} - \mathbf{k}$, the other through $(-3, 0, -1)$ along $\mathbf{i} + 2\mathbf{j} + 4\mathbf{k}$.
(b) They intersect at $(-2, 2, 3)$. They do not collide, since they reach this point at different times (the first at $t = -2$, the second at $t = 1$).
(c) $50.95°$

Exercise 14.7

1. $z = 4$ **3.** $y = -2$ **5.** $2x - y + z = 5$ **7.** $2x + y + 3z = 16$ **9.** $x + 2y - z = 12$
11. $x - y + 3z = -5$ **13.** $2x + 5y - 2z = 21$ **15.** $2x + 3y + 2z = -1$ **17.** $(2, 3, 0)$ and $(0, 0, 0)$
19. $x = 1 + 2t$, $y = 2 - t$, $z = -1 + t$ **21.** $(3, 1, 2)$ **23.** $(4, -2, 3)$ **25.** $x + y + 3z = 0$
27. $3x + 7y - 6z = 11$ **29.** $2x + 6y + 29z = 2$ **31.** Intersect; $\sqrt{2}/3$ **33.** Intersect; $17/\sqrt{406}$
35. Intersect; $5\sqrt{3}/\sqrt{106}$ **37.** Identical **39.** $2/\sqrt{6}$ **41.** $12/\sqrt{11}$
43. $\left(1 + \frac{\sqrt{2}}{2}\right)x + y + \left(-1 + \frac{\sqrt{2}}{2}\right)z = 0$; $\left(1 - \frac{\sqrt{2}}{2}\right)x + y + \left(-1 - \frac{\sqrt{2}}{2}\right)z = 0$ **45.** $\frac{x - \frac{10}{3}}{-2} = \frac{y + \frac{2}{3}}{7} = \frac{z}{3}$
47. $(4, 0, -4)$

Exercise 14.8

1. Elliptic paraboloid

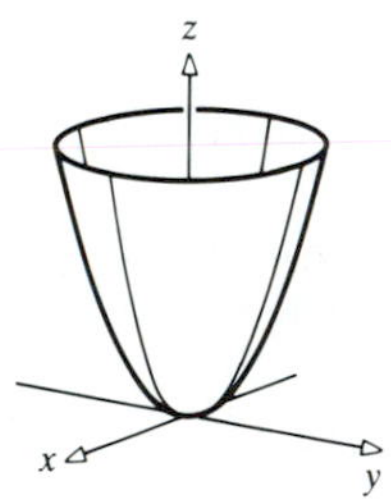

3. Ellipsoid

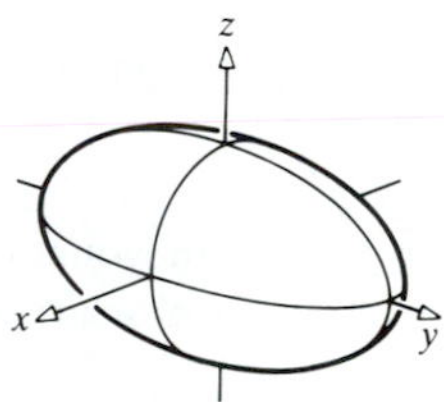

5. Elliptic cone

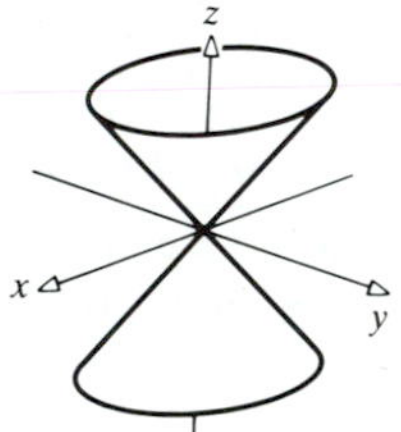

7. Parabolic cylinder

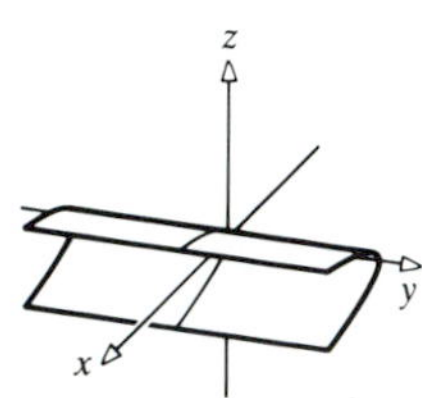

9. Hyperboloid of two sheets

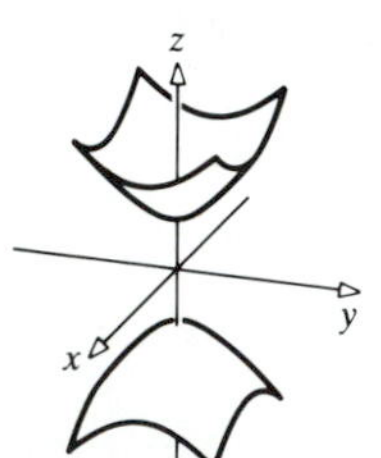

11. Hyperbolic cylinder

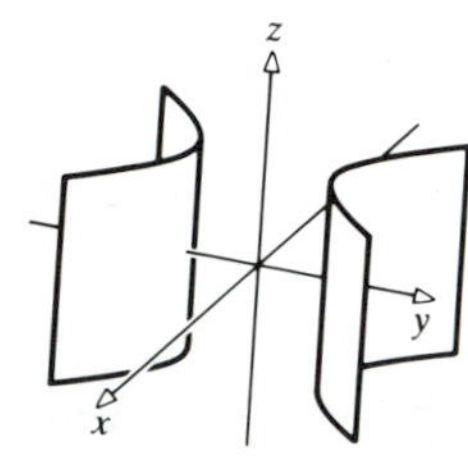

13. Elliptic paraboloid

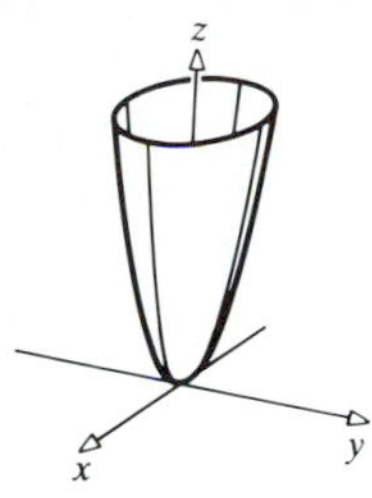

15. Hyperbolic cylinder

17. Elliptic cone

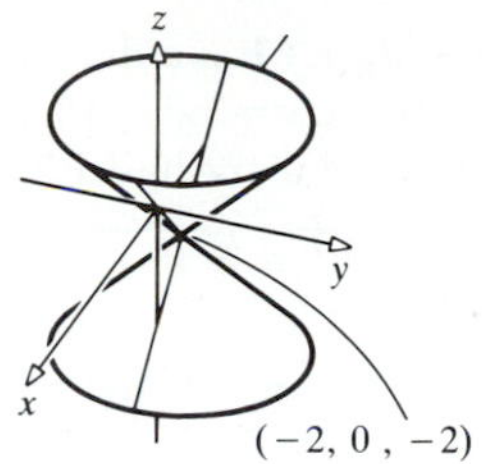

19. Elliptic paraboloid

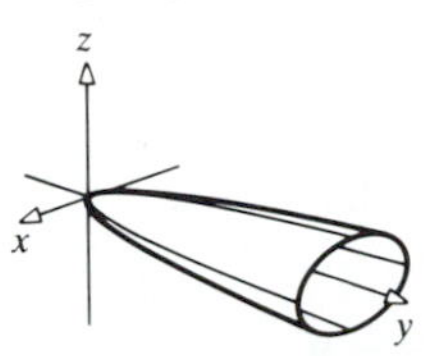

21. Hyperboloid of two sheets

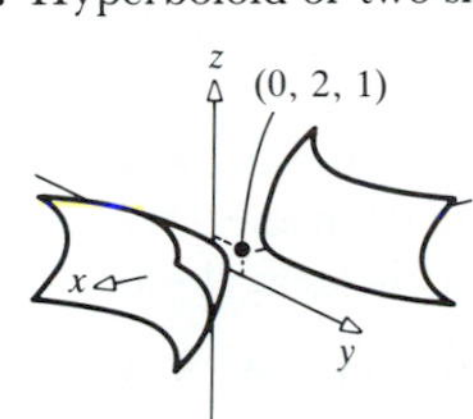

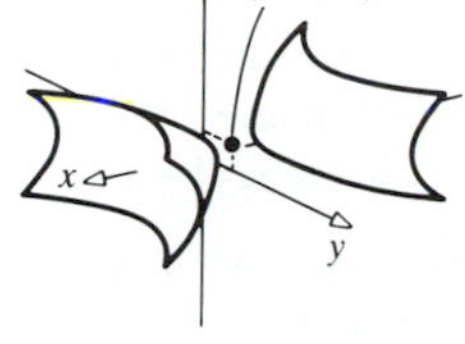

23. z is missing; yes

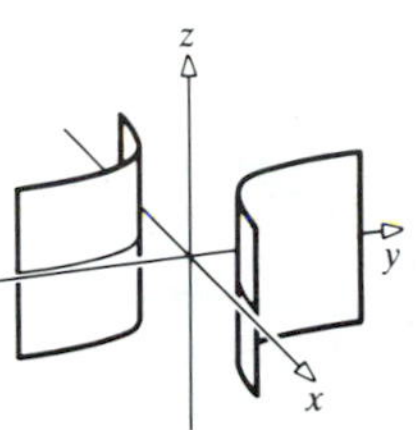

25. **J** **27.** **H** **29.** **G** **31.** (a) **L** (b) **O** (c) **K** (d) **M** (e) **N**

Historical Exercises

1. (a) $2 + 4i$
(b)

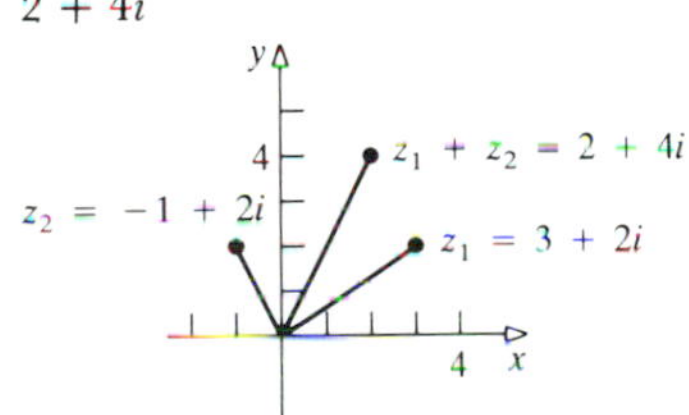

3. (a) $z \diamond w = (ac + bd) + (bc - ad)i$; complex
(b) Use properties of the conjugate. For example, $v \diamond v = v \cdot \bar{v} = a^2 + b^2 = r^2 = \|v\|^2$

5. (a) $(0, 1, 0, 0) \times (0, 0, 1, 0) \neq (0, 0, 1, 0) \times (0, 1, 0, 0)$ (b) Yes; $(1, 0, 0, 0)$

Review Exercises

1. $\sqrt{14}$ **3.** $2\sqrt{21}$ **5.** Center: $(2, -4, 0)$; $r = 5$ **7.** $\sqrt{6}$ **9.** $12\mathbf{i} + 16\mathbf{j} \pm 15\mathbf{k}$

11. $\mathbf{w}_1 = 15\mathbf{i} + 5\mathbf{j}$; $\mathbf{w}_2 = 14\mathbf{i} + 7\mathbf{j}$ **13.** Distance $= \mathbf{AB} + \mathbf{BC} + \mathbf{CA} = \mathbf{0}$, so work $= 0$. **15.** $2\pi/3$ or $120°$

17. $115.7°$ or 2.02 radians **19.** 32 cubic units **21.** $x = 1 + 3t$, $y = 3t$, $z = -1$ **23.** $\dfrac{x-1}{2} = \dfrac{y-2}{-1} = \dfrac{z-3}{-4}$

25. $x^2 + (y-4)^2 + (z-1)^2 = 24$ **27.** $\dfrac{x}{a} + \dfrac{y}{b} + \dfrac{z}{c} = 1$ **29.** $(x+2)^2 + (y-1)^2 + (z-5)^2 = \dfrac{50}{9}$

31. $(x-4)^2 + (y-2)^2 + (z+8)^2 = 26$ or $(x+2)^2 + y^2 + z^2 = 26$

33. (a) N (b) S (c) V (d) S (e) N (f) V

35. Complete the squares; center $\left(\dfrac{-D}{2A}, \dfrac{-E}{2A}, \dfrac{-F}{2A}\right)$, radius $\sqrt{D^2 + E^2 + F^2 + 4AG}/2A$

37. (a) Hyperbolic cylinder (b) Hyperboloid of two sheets (c) Hyperbolic paraboloid (d) Elliptic paraboloid (e) Elliptic cone (f) Ellipsoid (g) Hyperboloid of one sheet (h) Parabolic cylinder (i) Elliptic cylinder

Challenge Exercises

1. (a) $A\mathbf{i} + B\mathbf{j} + C\mathbf{k}$ is a direction vector for the line.
   (b) Substitute $x = x_0 + At$, $y = y_0 + Bt$, $z = z_0 + Ct$ into the equation for the plane and solve for t.
   (c) P is in the plane and $\overrightarrow{PP_0}$ is normal to the plane. Then
   $$d = \sqrt{A^2t^2 + B^2t^2 + C^2t^2} = |t|\sqrt{A^2 + B^2 + C^2} = \frac{|Ax_0 + By_0 + Cz_0 - D|}{\sqrt{A^2 + B^2 + C^2}}.$$

3. (a) Because $\mathbf{v} \cdot \mathbf{u} = 0$ and $\mathbf{w} \cdot \mathbf{u} = 0$ (b) Solve simultaneously (c) Substitute $z = v_1w_2 - v_2w_1$ in (b)

5. $\mathbf{u} = u_1\mathbf{i} + u_2\mathbf{j}$, $\mathbf{v} = v_1\mathbf{i} + v_2\mathbf{j}$, $u_1v_2 - v_1u_2 \neq 0$; if $\mathbf{w} = w_1\mathbf{i} + w_2\mathbf{j}$, then $s = \dfrac{v_2w_1 - v_1w_2}{u_1v_2 - v_1u_2}$, $t = \dfrac{u_1w_2 - u_2w_1}{u_1v_2 - v_1u_2}$.

7. $14/\sqrt{59}$

9. 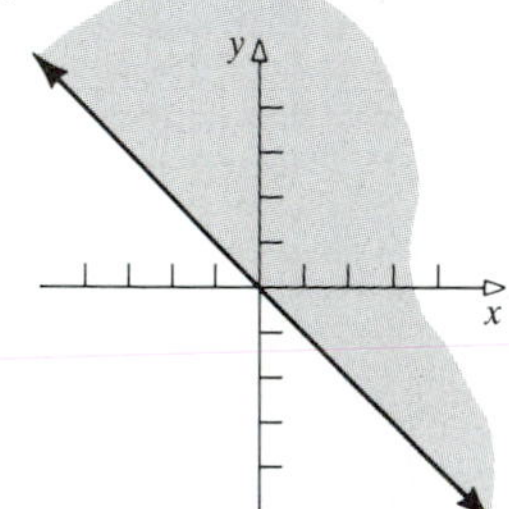

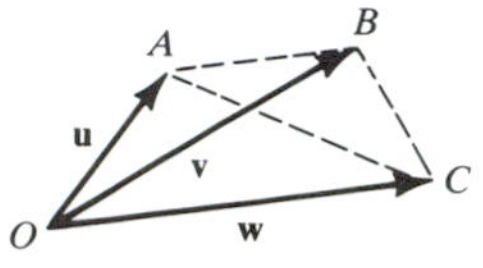

$\overrightarrow{AB} = \mathbf{v} - \mathbf{u}$ and $\overrightarrow{AC} = \mathbf{w} - \mathbf{u}$ are in the plane; $\overrightarrow{AB} \times \overrightarrow{AC}$ is normal to the plane. (We assume that A, B, C do not all fall on a line.) Use theorem (14.5.2), part (d) to evaluate:
$$\overrightarrow{AB} \times \overrightarrow{AC} = (\mathbf{v} - \mathbf{u}) \times (\mathbf{w} - \mathbf{u}) = (\mathbf{v} - \mathbf{u}) \times \mathbf{w} - (\mathbf{v} - \mathbf{u}) \times \mathbf{u}$$
$$= \mathbf{v} \times \mathbf{w} - \mathbf{u} \times \mathbf{w} - \mathbf{v} \times \mathbf{u} + \mathbf{u} \times \mathbf{u} = \mathbf{u} \times \mathbf{v} + \mathbf{v} \times \mathbf{w} + \mathbf{w} \times \mathbf{u}$$

11. $g(t) = \|\mathbf{u}\|^2 - 2t\mathbf{u} \cdot \mathbf{v} + t^2\|\mathbf{v}\|^2$. Finding $g'(t)$ and setting it to 0 gives $t_0 = (\mathbf{u} \cdot \mathbf{v})/\|\mathbf{v}\|^2$ and $g(t_0) = \|\mathbf{u}\|^2 - (\mathbf{u} \cdot \mathbf{v})^2/\|\mathbf{v}\|^2$. But $g(t) \geq 0$, so $\|\mathbf{u}\|^2 \geq (\mathbf{u} \cdot \mathbf{v})^2/\|\mathbf{v}\|^2$, $\|\mathbf{u}\|^2\|\mathbf{v}\|^2 \geq (\mathbf{u} \cdot \mathbf{v})^2$, and since $\|\mathbf{u}\|$ and $\|\mathbf{v}\|$ are positive, $\|\mathbf{u}\|\|\mathbf{v}\| \geq |\mathbf{u} \cdot \mathbf{v}|$.

13. $(\mathbf{r} - \mathbf{b}) \cdot (\mathbf{r} + \mathbf{b}) = 0 \Rightarrow \|\mathbf{r}\| = \|\mathbf{b}\|$, a sphere of radius $\|\mathbf{b}\|$, center at (0, 0, 0). Similarly, $(\mathbf{r} - \mathbf{r}_0 - \mathbf{b}) \cdot (\mathbf{r} - \mathbf{r}_0 + \mathbf{b}) = 0 \Rightarrow \|\mathbf{r} - \mathbf{r}_0\| = \|\mathbf{b}\| \Rightarrow$ a sphere centered at $\mathbf{r}_0$.

15.

CHAPTER 15

Exercise 15.1

1. $\mathbf{r}(1) = \mathbf{i} - 2\mathbf{j}$ 3. $\mathbf{r}\left(\frac{\pi}{4}\right) = \frac{\sqrt{2}}{2}\mathbf{i} - \frac{\sqrt{2}}{2}\mathbf{j}$ 5. $\mathbf{r}(0) = \mathbf{i} - 2\mathbf{k}$ 7. $\mathbf{r}(4) = 4\mathbf{i} + \mathbf{j} + 8\mathbf{k}$

9. 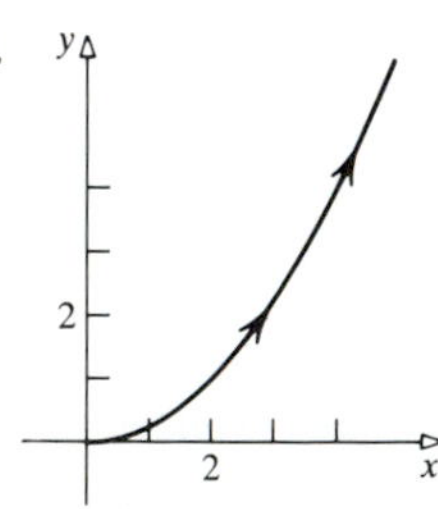

11. 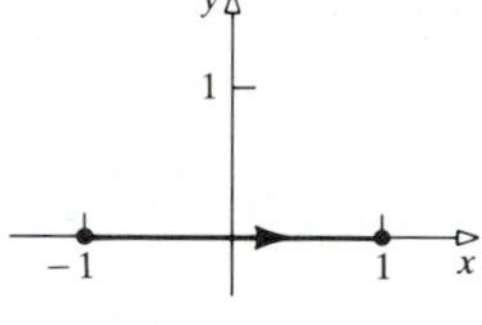

13. 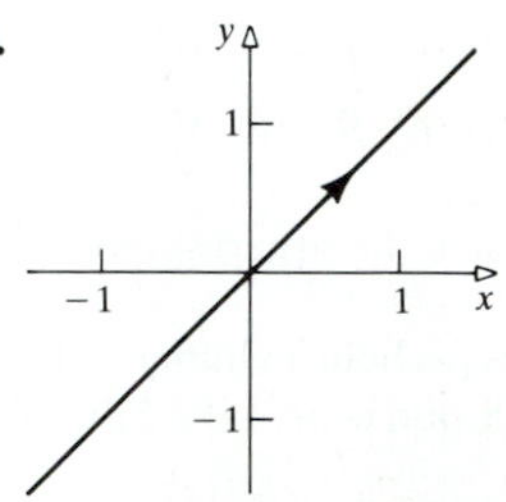

15.

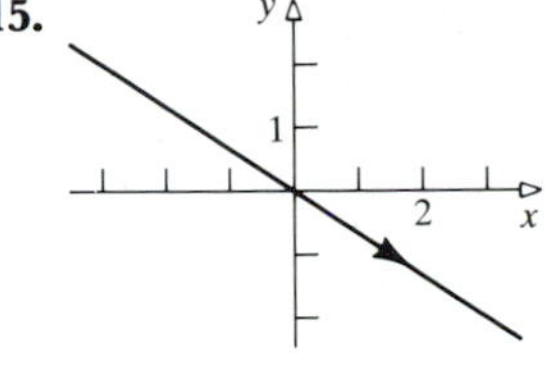

17.

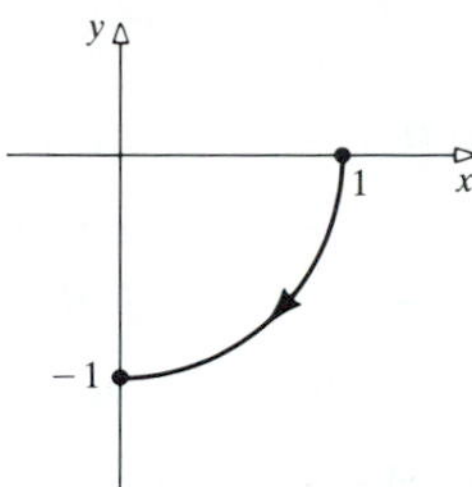

19.

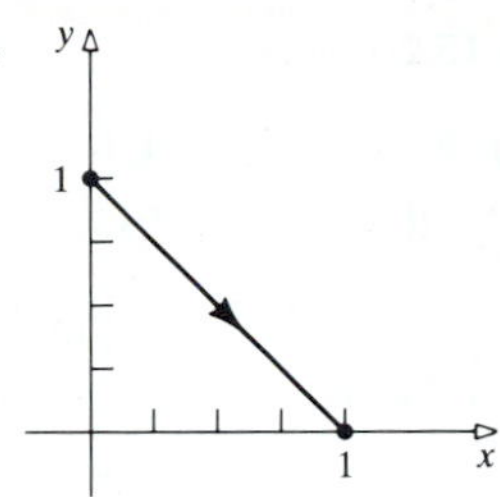

21. $\mathbf{r}'(t) = 8t\mathbf{i} - 6t^2\mathbf{j}$, $\mathbf{r}''(t) = 8\mathbf{i} - 12t\mathbf{j}$ **23.** $\mathbf{r}'(t) = \dfrac{2}{\sqrt{t}}\mathbf{i} + 2e^t\mathbf{j}$, $\mathbf{r}''(t) = \dfrac{-1}{t\sqrt{t}}\mathbf{i} + 2e^t\mathbf{j}$

25. $\mathbf{r}'(t) = 2\cos 2t\mathbf{i} + 2\sin 2t\mathbf{j}$, $\mathbf{r}''(t) = -4\sin 2t\mathbf{i} + 4\cos 2t\mathbf{j}$ **27.** $\mathbf{r}'(t) = \mathbf{j} + 2t\mathbf{k}$, $\mathbf{r}''(t) = 2\mathbf{k}$

29. $\mathbf{r}'(t) = 2t\mathbf{i} + 3t^2\mathbf{j} - \mathbf{k}$; $\mathbf{r}''(t) = 2\mathbf{i} + 6t\mathbf{j}$

31. $\mathbf{r}'(t) = e^t(\cos t - \sin t)\mathbf{i} + e^t(\sin t + \cos t)\mathbf{j} + \mathbf{k}$; $\mathbf{r}''(t) = -2e^t\sin t\mathbf{i} + 2e^t\cos t\mathbf{j}$

33. $\mathbf{r}'(t) = (1 - 3t^2)\mathbf{i} + (1 + 3t^2)\mathbf{j} - \mathbf{k}$; $\mathbf{r}''(t) = -6t\mathbf{i} + 6t\mathbf{j}$

35. $\mathbf{r}(0) = \mathbf{0}$
$\mathbf{r}'(t) = \mathbf{i} + 2t\mathbf{j}$
$\mathbf{r}'(0) = \mathbf{i}$

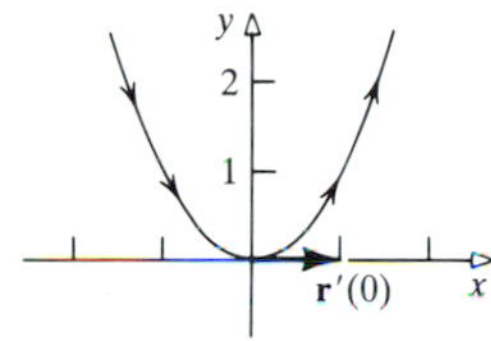

37. $\mathbf{r}(0) = \mathbf{j}$
$\mathbf{r}'(t) = \mathbf{i} + e^t\mathbf{j}$
$\mathbf{r}'(0) = \mathbf{i} + \mathbf{j}$

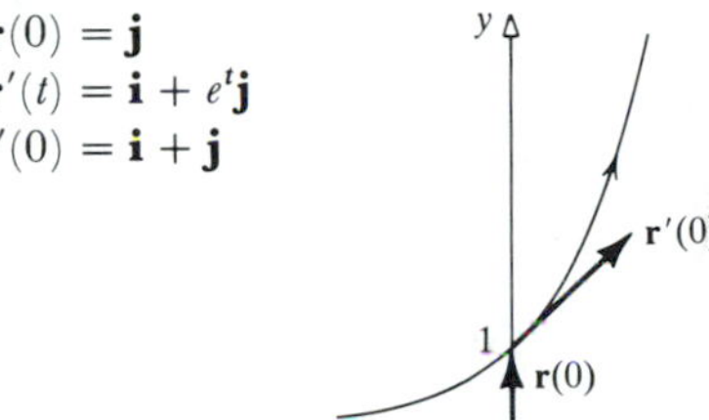

39. $\mathbf{r}(0) = -3\mathbf{j}$
$\mathbf{r}'(t) = 3\cos t\mathbf{i} + 3\sin t\mathbf{j}$
$\mathbf{r}'(0) = 3\mathbf{i}$

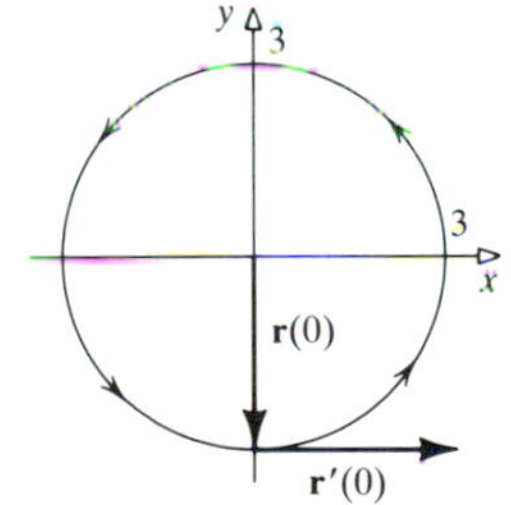

41. $\mathbf{r}(0) = 2\mathbf{i}$
$\mathbf{r}'(t) = -2\sin t\mathbf{i} - 3\cos t\mathbf{j}$
$\mathbf{r}'(0) = -3\mathbf{j}$

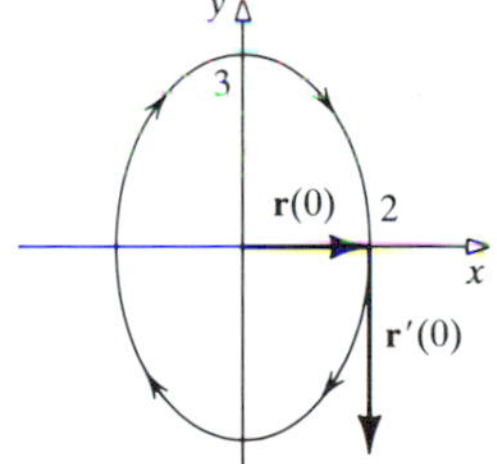

43. 0 **45.** $te^t + e^t - 2te^{-t} + t^2e^{-t}$ **47.** $\omega\cos\omega t - \omega\sin\omega t$ **49.** $12t^2$; $(2t + 10)\mathbf{i} - (2 + 10t)\mathbf{j} + 4(2t - t^3)\mathbf{k}$

51. $\cos 2t\sin t - \sin 2t\cos t$; $(2\cos 2t - \cos t)\mathbf{i} - (\sin t - 2\sin 2t)\mathbf{j} + (-\sin 2t\sin t - \cos 2t\cos t)\mathbf{k}$

53. $e^t - 4e^{-4t} - 2t$; $-2(e^{-2t} - 2te^{-2t})\mathbf{i} + (e^{2t} + e^{-t} + 2te^{2t} - te^{-t})\mathbf{j} + 3e^{-3t}\mathbf{k}$ **55.** $f''(t) = -f(t)$

57. (a) $\omega(-\sin\omega t\mathbf{i} + \cos\omega t\mathbf{j})$; $|\omega|$ (b) $2\mathbf{j} + 6t\mathbf{k}$; $2\sqrt{1 + 9t^2}$

59. Follow the pattern of the proof of theorem (15.1.7(b)). **61.** $\mathbf{u} \times \mathbf{v} = -(\mathbf{v} \times \mathbf{u})$

63. $\|\mathbf{r}(t)\|^2 = \mathbf{r}(t) \cdot \mathbf{r}(t)$; use theorem (15.1.7(c)). **65.** Differentiate $\|\mathbf{r}(t)\| = (\mathbf{r}(t) \cdot \mathbf{r}(t))^{1/2}$.

67. $y = (1 - x)^{2/3}$

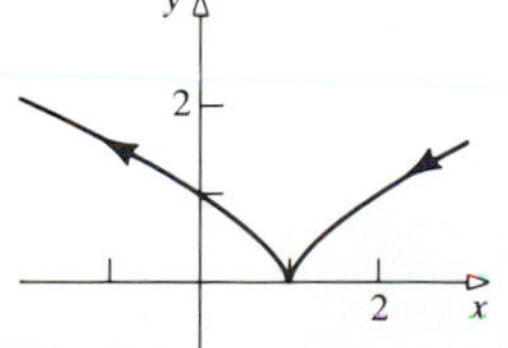

69. $y^2 = x/(1 - x)$, $0 \le x < 1$

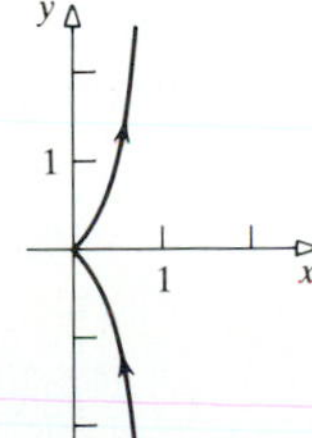

Exercise 15.2

1. $\mathbf{i} + 2\mathbf{j}$; $\mathbf{R}(u) = (1 + u)\mathbf{i} + (1 + 2u)\mathbf{j}$ **3.** $2\mathbf{i} - \mathbf{j}$; $\mathbf{R}(u) = 2(1 + u)\mathbf{i} - u\mathbf{j}$

5. $\frac{1}{2}\mathbf{i} - \frac{3}{4}\mathbf{j}$; $\mathbf{R}(u) = (1 + \frac{u}{2})\mathbf{i} - (\frac{1}{2} + \frac{3}{4}u)\mathbf{j}$ **7.** $\mathbf{i} - \mathbf{j}$; $\mathbf{R}(u) = (1 + u)\mathbf{i} + (1 - u)\mathbf{j}$

9. $-3\mathbf{i} + 2\mathbf{j} - \mathbf{k}$; $\mathbf{R}(u) = (1 - 3u)\mathbf{i} + 2u\mathbf{j} + (-5 - u)\mathbf{k}$ **11.** $-2\mathbf{i}$; $\mathbf{R}(u) = -2u\mathbf{i} + \mathbf{j} - 5\mathbf{k}$

13. $-\mathbf{i} + \sqrt{3}\mathbf{k}$; $\mathbf{R}(u) = (\sqrt{3} - u)\mathbf{i} + \mathbf{j} + (1 + \sqrt{3}u)\mathbf{k}$ **15.** $\mathbf{i} + \mathbf{j} + \mathbf{k}$; $\mathbf{R}(u) = (1 + u)\mathbf{i} + u\mathbf{j} + (1 + u)\mathbf{k}$

17. $\mathbf{T} = \frac{1}{\sqrt{5}}\mathbf{i} + \frac{2}{\sqrt{5}}\mathbf{j}$; $\mathbf{N} = \frac{-2}{\sqrt{5}}\mathbf{i} + \frac{1}{\sqrt{5}}\mathbf{j}$ **19.** $\mathbf{T} = \frac{2\mathbf{i} - \mathbf{j}}{\sqrt{5}}$; $\mathbf{N} = \frac{\mathbf{i} + 2\mathbf{j}}{\sqrt{5}}$

21. $\mathbf{T} = \frac{2}{\sqrt{13}}\mathbf{i} - \frac{3}{\sqrt{13}}\mathbf{j}$; $\mathbf{N} = \frac{-3}{\sqrt{13}}\mathbf{i} - \frac{2}{\sqrt{13}}\mathbf{j}$ **23.** $\mathbf{T} = \frac{1}{\sqrt{2}}\mathbf{i} - \frac{1}{\sqrt{2}}\mathbf{j}$; $\mathbf{N} = \frac{1}{\sqrt{2}}\mathbf{i} + \frac{1}{\sqrt{2}}\mathbf{j}$

25. $\mathbf{T} = \frac{-3}{\sqrt{14}}\mathbf{i} + \frac{2}{\sqrt{14}}\mathbf{j} - \frac{1}{\sqrt{14}}\mathbf{k}$; $\mathbf{N} = \mathbf{0}$ **27.** $\mathbf{T} = -\mathbf{i}$; $\mathbf{N} = -\mathbf{j}$ **29.** $\mathbf{T} = -\frac{1}{2}\mathbf{i} + \frac{\sqrt{3}}{2}\mathbf{k}$; $\mathbf{N} = -\frac{\sqrt{3}}{2}\mathbf{i} - \frac{1}{2}\mathbf{k}$

31. $\mathbf{T} = \frac{1}{\sqrt{3}}\mathbf{i} + \frac{1}{\sqrt{3}}\mathbf{j} + \frac{1}{\sqrt{3}}\mathbf{k}$; $\mathbf{N} = -\frac{1}{\sqrt{2}}\mathbf{i} + \frac{1}{\sqrt{2}}\mathbf{j}$ **33.** $(40\sqrt{40} - 13\sqrt{13})/27$ **35.** $\frac{33}{2}$ **37.** $\sqrt{6}$ **39.** $\sqrt{5}\pi$

41. $e - e^{-1}$ **43.** $(0, -1, 0)$; $(\frac{1}{4}, -\frac{3}{4}, \frac{1}{2})$; $(\frac{1}{4}, -\frac{3}{4}, -\frac{1}{2})$ **45.** $\cos^{-1}[-\pi/(\sqrt{4 + \pi^2}\sqrt{2})]$ **47.** $\sqrt{3}e^{2\pi} - \sqrt{3}$

49. 15

Exercise 15.3

1. $\mathbf{v}(t) = \mathbf{i} + 2t\mathbf{j}$
$\mathbf{v}(0) = \mathbf{i}$
$\mathbf{a}(t) = 2\mathbf{j}$
$\mathbf{a}(0) = 2\mathbf{j}$

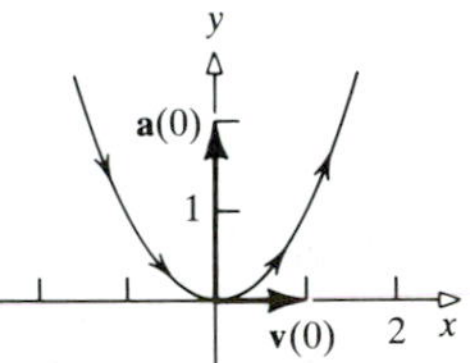

3. $\mathbf{v}(t) = 2t\mathbf{i} - \mathbf{j}$
$\mathbf{v}(0) = -\mathbf{j}$
$\mathbf{a}(0) = \mathbf{a}(t) = 2\mathbf{i}$

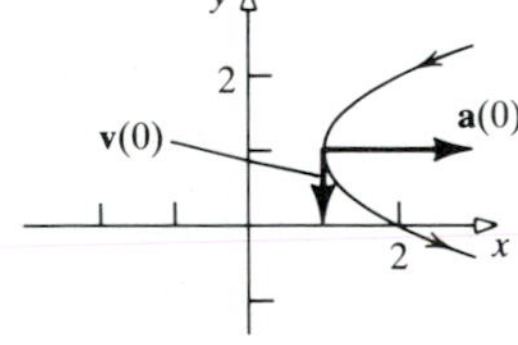

5.

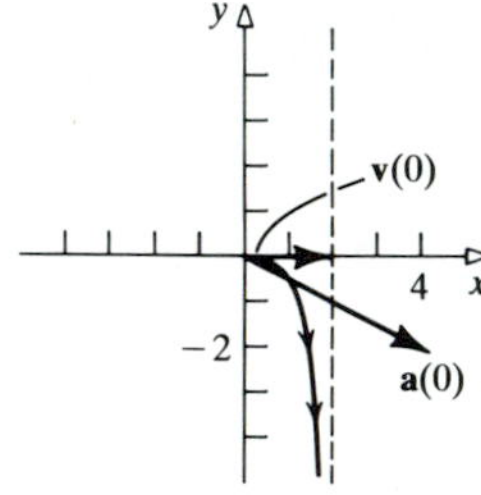

$\mathbf{v}(t) = \frac{2\mathbf{i} - (t^2 + 2t)\mathbf{j}}{(t + 1)^2}$

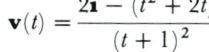

$\mathbf{v}(0) = 2\mathbf{i}$, $\mathbf{a}(t) = \frac{-4\mathbf{i} - 2\mathbf{j}}{(t + 1)^3}$, $\mathbf{a}(0) = -4\mathbf{i} - 2\mathbf{j}$

7. $\mathbf{v}(t) = e^t\mathbf{i} - e^{-t}\mathbf{j}$
$\mathbf{v}(0) = \mathbf{i} - \mathbf{j}$
$\mathbf{a}(t) = e^t\mathbf{i} + e^{-t}\mathbf{j}$
$\mathbf{a}(0) = \mathbf{i} + \mathbf{j}$

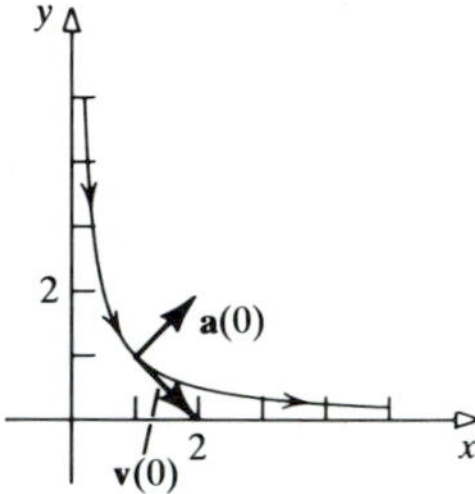

9. $\mathbf{v}(t) = \mathbf{i} + e^t\mathbf{j}$
$\mathbf{v}(0) = \mathbf{i} + \mathbf{j}$
$\mathbf{a}(t) = e^t\mathbf{j}$
$\mathbf{a}(0) = \mathbf{j}$

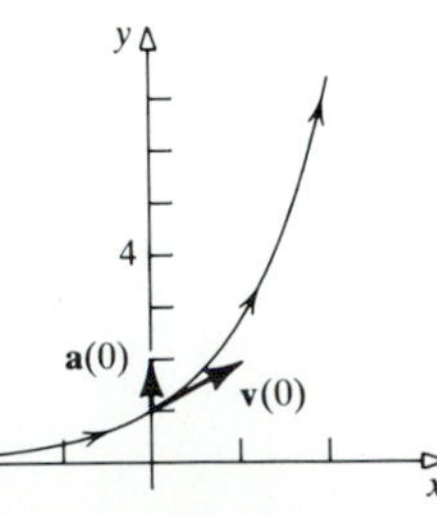

11. $\mathbf{v}(t) = 3 \cos t\mathbf{i} + 3 \sin t\mathbf{j}$
$\mathbf{v}(0) = 3\mathbf{i}$
$\mathbf{a}(t) = -3 \sin t\mathbf{i} + 3 \cos t\mathbf{j}$
$\mathbf{a}(0) = 3\mathbf{j}$

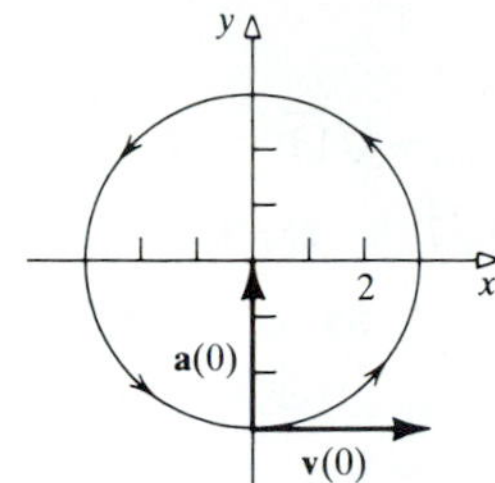

13. $\mathbf{v}(t) = -2 \sin t\mathbf{i} - 3 \cos t\mathbf{j}$
$\mathbf{v}(0) = -3\mathbf{j}$
$\mathbf{a}(t) = -2 \cos t\mathbf{i} + 3 \sin t\mathbf{j}$
$\mathbf{a}(0) = -2\mathbf{i}$

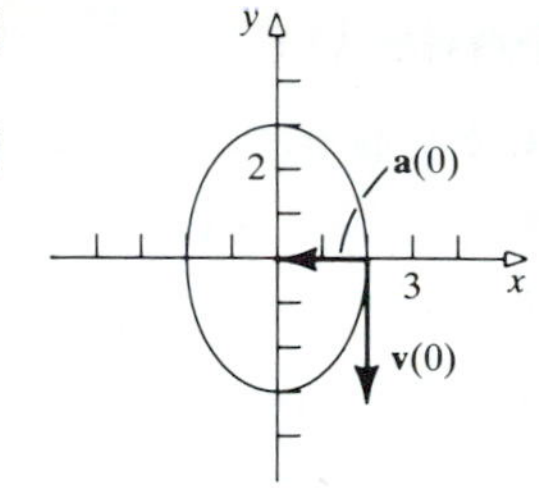

15. $2\mathbf{i} + \mathbf{j}$; $\mathbf{0}$; $\sqrt{5}$; $a_\mathbf{T} = a_\mathbf{N} = 0$

17. $-2 \sin t\mathbf{i} + 3 \cos t\mathbf{j}$; $-2 \cos t\mathbf{i} - 3 \sin t\mathbf{j}$; $\sqrt{4 \sin^2 t + 9 \cos^2 t}$; $a_\mathbf{T} = \dfrac{-5 \sin t \cos t}{\sqrt{4 \sin^2 t + 9 \cos^2 t}}$, $a_\mathbf{N} = \dfrac{6}{\sqrt{4 \sin^2 t + 9 \cos^2 t}}$

19. $-3\mathbf{i} + 2\mathbf{j} - \mathbf{k}$; $\mathbf{0}$; $\sqrt{14}$; $a_\mathbf{T} = a_\mathbf{N} = 0$ **21.** $-2 \sin 2t\mathbf{i} + 2 \cos 2t\mathbf{j}$; $-4 \cos 2t\mathbf{i} - 4 \sin 2t\mathbf{j}$; 2; $a_\mathbf{T} = 0$, $a_\mathbf{N} = 4$

23. $\dfrac{t^2 + 2t}{(t+1)^2}\mathbf{i} - \dfrac{1}{(t+1)^2}\mathbf{j} + \dfrac{1}{(t+1)^2}\mathbf{k}$; $\dfrac{2}{(t+1)^3}\mathbf{i} + \dfrac{2}{(t+1)^3}\mathbf{j} - \dfrac{2}{(t+1)^3}\mathbf{k}$; $\dfrac{\sqrt{t^4 + 4t^3 + 4t^2 + 2}}{(t+1)^2}$;
$a_\mathbf{T} = \dfrac{2t^2 + 4t - 4}{(t+1)^3\sqrt{t^4 + 4t^3 + 4t^2 + 2}}$; $a_\mathbf{N} = \dfrac{2\sqrt{2}}{|t+1|\sqrt{t^4 + 4t^3 + 4t^2 + 2}}$

25. $\mathbf{i} + 2t\mathbf{j} + 3t^2\mathbf{k}$; $2\mathbf{j} + 6t\mathbf{k}$; $\sqrt{1 + 4t^2 + 9t^4}$; $a_\mathbf{T} = \dfrac{4t + 18t^3}{\sqrt{1 + 4t^2 + 9t^4}}$, $a_\mathbf{N} = \dfrac{2\sqrt{1 + 9t^2 + 9t^4}}{\sqrt{1 + 4t^2 + 9t^4}}$

27. $(e^t\cos t - e^t\sin t)\mathbf{i} + (e^t\sin t + e^t\cos t)\mathbf{j} + e^t\mathbf{k}$; $(-2e^t\sin t)\mathbf{i} + (2e^t\cos t)\mathbf{j} + e^t\mathbf{k}$; $\sqrt{3}e^t$; $a_\mathbf{T} = \sqrt{3}e^t$, $a_\mathbf{N} = \sqrt{2}e^t$

29. $\dfrac{1}{t}\mathbf{i} + \dfrac{1}{2\sqrt{t}}\mathbf{j} + \dfrac{3t^{1/2}}{2}\mathbf{k}$; $-\dfrac{1}{t^2}\mathbf{i} - \dfrac{1}{4t^{3/2}}\mathbf{j} + \dfrac{3}{4t^{1/2}}\mathbf{k}$; $\dfrac{\sqrt{9t^3 + t + 4}}{2t}$; $a_\mathbf{T} = \dfrac{9t^3 - t - 8}{4t^2\sqrt{9t^3 + t + 4}}$, $a_\mathbf{N} = \dfrac{\sqrt{9t^4 + 81t^3 + t}}{2t^2\sqrt{9t^3 + t + 4}}$

31. Derivative of a constant is 0.

33. $\|\mathbf{v}\|$ constant implies $\|\mathbf{v}\|^2$ constant, and $(d/dt)(\mathbf{v} \cdot \mathbf{v}) = 0 = 2\mathbf{v} \cdot \mathbf{a}$, so $\mathbf{v}$ and $\mathbf{a}$ are orthogonal.

35. (a) 9 (b) Less; half as much **37.** 18,001 mph **39.** 504.5 miles **41.** 350 newtons **43.** 41,152 newtons

45. $\dfrac{d}{dt}\dfrac{1}{2}m\|\mathbf{v}(t)\|^2 = \dfrac{d}{dt}\dfrac{1}{2}m\mathbf{v} \cdot \mathbf{v} = m\mathbf{v} \cdot \mathbf{v}' = \mathbf{v} \cdot m\mathbf{a} = \mathbf{F} \cdot \mathbf{v}$ **47.** $\dfrac{d}{dt}[\mathbf{r}(t) \times m\mathbf{v}(t)] = \mathbf{r}' \times m\mathbf{v} + \mathbf{r} \times m\mathbf{a} = \mathbf{r} \times \mathbf{F} = \tau$

49. (a) $\mathbf{v} = \mathbf{r}'(t) = \mathbf{R}(t) + t\mathbf{R}'(t) = \cos \omega t\mathbf{i} + \sin \omega t\mathbf{j} + t\mathbf{v}_d$ (b) $\mathbf{a} = \mathbf{R}'(t) + \mathbf{R}'(t) + t\mathbf{R}''(t) = 2\mathbf{v}_d + t\mathbf{a}_d$

51. Let $\theta = t$ and use $x = r \cos \theta$, $y = r \sin \theta$.

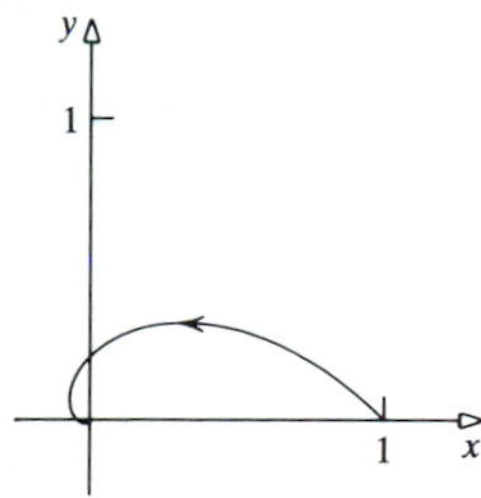

53. (a) $x^2 = -4(y - 2)$, $-4 \le x \le 4$ (b)

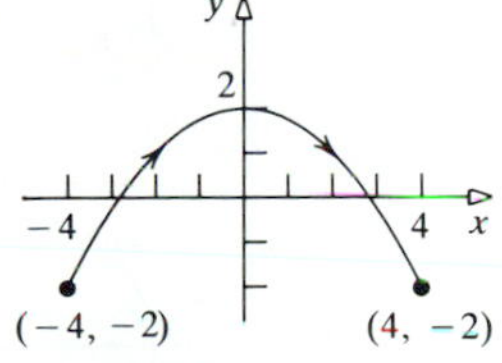

(c) $\mathbf{v} = \mathbf{0}$ when $t = k\pi$
$\mathbf{a} = 4\mathbf{i} + 8\mathbf{j}$ if k is odd
$\mathbf{a} = -4\mathbf{i} + 8\mathbf{j}$ if k is even

55. $\mathbf{a} = \mathbf{b} \times \mathbf{r}'$; $\mathbf{a} \cdot \mathbf{b} = 0$; $\dfrac{d}{dt}\|\mathbf{r}'(t)\|^2 = \dfrac{d}{dt}\mathbf{r}' \cdot \mathbf{r}' = 2\mathbf{r}' \cdot \mathbf{a} = 0$

Exercise 15.4

1. No, $\|\mathbf{r}'(t)\| = 4$ **3.** No, $\|\mathbf{r}'(t)\| = \sqrt{4t^2+1}$ **5.** No, $\|\mathbf{r}'(t)\| = \sqrt{13}$ **7.** Yes **9.** No, $\|\mathbf{r}'(t)\| = \sqrt{2}$
11. $2/5\sqrt{5}$ **13.** $\sqrt{2}$ **15.** $\frac{2}{27}$ **17.** $30/13\sqrt{13}$ **19.** $1/7a\sqrt{7}$ **21.** 1 **23.** $1/2\sqrt{2}$ **25.** $(3t^4)/[2(t^6+1)^{3/2}]$
27. 0 **29.** $\frac{4}{5}$ **31.** $\sqrt{2}/(e^{2t}+e^{-2t}+2)$ **33.** $1/[3a(1+t^2)^2]$ **35.** 6 **37.** $7\sqrt{14}/\sqrt{19}$ **39.** $\frac{5}{4}$ **41.** $2\sqrt{2}$
43. $12a$ **45.** $5\sqrt{10}/3$ **47.** 1 **49.** $3\sqrt{2}$ **51.** $\rho = [1+(2ax+b)^2]^{3/2}/2a$ is a minimum when $x = -b/2a$.
53. $(\sqrt{2}/2, -\ln\sqrt{2})$ **55.** $\pm(1/\sqrt[4]{5}, 1/3\sqrt[4]{125})$ **57.** $23/7\sqrt{7}$ **59.** $3/2\sqrt{2a}$ **61.** $8/(7\sqrt{7})$
63. $\kappa = \dfrac{(1/a)\cosh(x/a)}{[1+\sinh^2(x/a)]^{3/2}} = \dfrac{1}{a\cosh^2(x/a)}$ **65.** $3/(5\sqrt{5})$ **67.** $6/[t^2(4+9t^2)^3]^{1/2}$; $t=0$; no tangent line when $t=0$
69. $[a(a^2+b^2+b^2t^2)^{1/2}]/[2(a^2+b^2t^2)^{3/2}]$ **71.** $a_{\mathbf{N}} = 1$ **77.** $a=1$; $\kappa = a/(a^2+1)$
79. $\rho\mathbf{N}(t)$ has length ρ in direction $\mathbf{N}(t)$. Adding this to $\mathbf{r}(t)$ must give the center.
81. $\kappa = \dfrac{|y''|}{[1+(y')^2]^{3/2}}$, but $y''=0$ at a point of inflection.
83. From Section 15.2 we know that **T** and **N** are orthogonal; the definition of the cross product implies **B** is orthogonal to both **T** and **N**.
85. Differentiate $\mathbf{B} = \mathbf{T}\times\mathbf{N}$ and use $\dfrac{d\mathbf{T}}{ds} = \kappa(s)\mathbf{N}(s)$
87. $\kappa = \dfrac{1}{\sqrt{2}}$; $\mathbf{T}(s) = \dfrac{\cos s\mathbf{i} - \sin s\mathbf{j} + \mathbf{k}}{\sqrt{2}}$; $\mathbf{N}(s) = -\sin s\mathbf{i} - \cos s\mathbf{j}$; $\mathbf{B}(s) = \dfrac{\cos s\mathbf{i} - \sin s\mathbf{j} - \mathbf{k}}{\sqrt{2}}$

Exercise 15.5

1. $-\cos t\mathbf{i} - \sin t\mathbf{j} + \frac{1}{2}t^2\mathbf{k} + \mathbf{c}$ **3.** $\sin t\mathbf{i} - \cos t\mathbf{j} - t\mathbf{k} + \mathbf{c}$ **5.** $-32t\mathbf{k}$; $32t$; $-16t^2\mathbf{k}$
7. $(\sin t+1)\mathbf{i} + (-\cos t+1)\mathbf{j}$; $\sqrt{3+2(\sin t - \cos t)}$; $(-\cos t + t + 1)\mathbf{i} + (-\sin t + t + 1)\mathbf{j}$
9. $(e^t - e)\mathbf{i} - (t\ln t - t)\mathbf{j} + t^2\mathbf{k}$ **11.** 23,895 m; 53.1 sec; 3449 m **13.** $\mathbf{r}(t) = \frac{1200}{13}t\mathbf{i} + \frac{500}{13}t\mathbf{j} - \frac{1}{2}(9.8)t^2\mathbf{j}$; 724.55 m
15. $\sin 2\theta$ is maximum when $\theta = \pi/4$. **17.** 25,540 m **19.** $(-t^2/2\sqrt{2} + 3t + 1)\mathbf{i} + (-t^2/2\sqrt{2} + 4t + 2)\mathbf{j}$
21. 54.52 ft/sec **23.** 318 ft; 4.5 sec **25.** $\tan\alpha \approx 1.26$; $\alpha \approx 51.6°$ **27.** $\mathbf{r}(s) = \cos\dfrac{s}{\sqrt{2}}\mathbf{i} + \sin\dfrac{s}{\sqrt{2}}\mathbf{j} + \dfrac{s}{\sqrt{2}}\mathbf{k}$
29. Write **c** and **r** in terms of their components and integrate. **31.** Follow the same procedure as in Problem 29.

Exercise 15.6

1. 4.65×10^8 miles **3.** 248.25 earth years

Review Exercises

1.
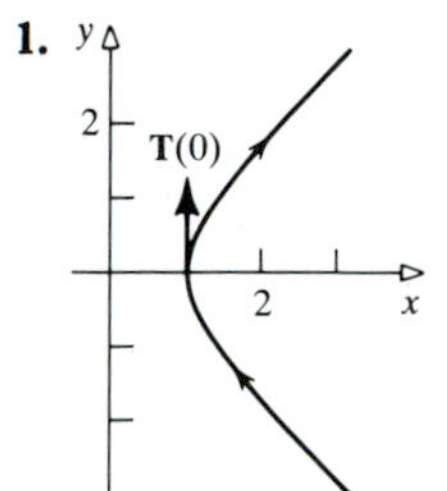

3.
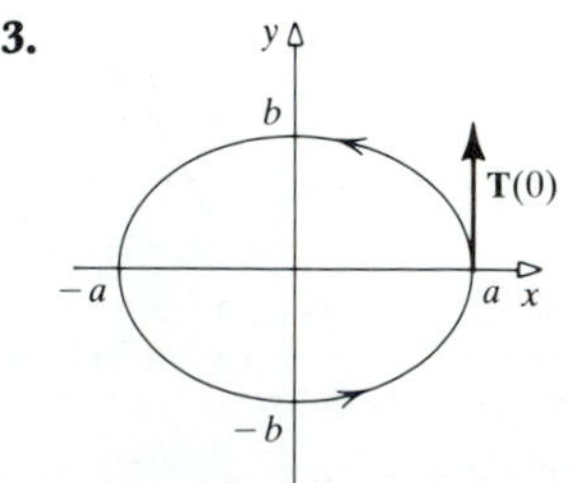

5. 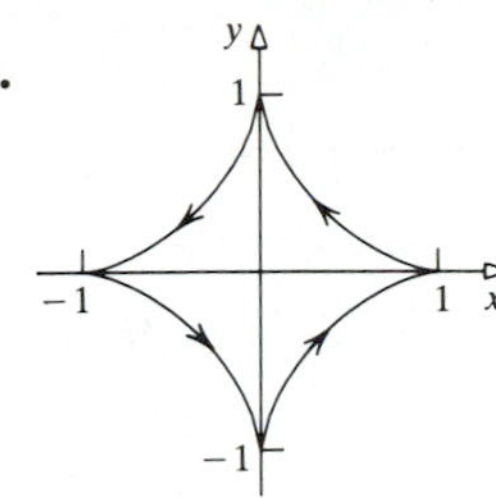

$\mathbf{T}(0)$ does not exist.

7. $\mathbf{T}(0) = \dfrac{1}{\sqrt{5}}\mathbf{i} + \dfrac{2}{\sqrt{5}}\mathbf{j}$; $\mathbf{N}(0) = -\mathbf{k}$

9. Problem 6: in the plane $z = 1$
Problem 7: on the sphere $x^2 + y^2 + z^2 = 1$
Problem 8: on the cone $x^2 + y^2 = z^2$

11. The curve spirals around the cone, making a complete revolution every 2π units.

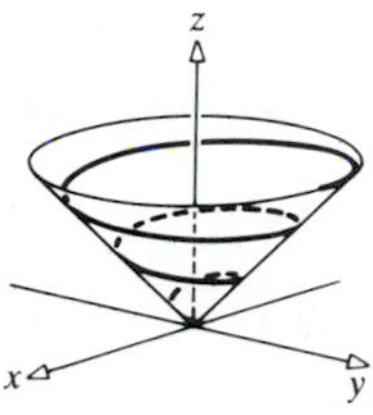

13. Choose $T \times N$ and $T \times N + N$ (Other answers are possible.) **15.** $\pi/2$

17. $(\sinh t)\mathbf{i} + (\cosh t)\mathbf{j}$; $(\cosh t)\mathbf{i} + (\sinh t)\mathbf{j}$; $\sqrt{\sinh^2 t + \cosh^2 t}$; $a_\mathbf{T} = \sqrt{\tanh 2t \sinh 2t}$, $a_\mathbf{N} = \sqrt{\operatorname{sech} 2t}$

19. $(-a \sin t)\mathbf{i} + (b \cos t)\mathbf{j}$; $(-a \cos t)\mathbf{i} - (b \sin t)\mathbf{j}$; $\sqrt{a^2\sin^2 t + b^2\cos^2 t}$; $a_\mathbf{T} = \dfrac{(a^2 - b^2)\sin t \cos t}{\sqrt{a^2\sin^2 t + b^2\cos^2 t}}$, $a_\mathbf{N} = \dfrac{ab}{\sqrt{a^2\sin^2 t + b^2\cos^2 t}}$

21. $-3\cos^2 t \sin t\mathbf{i} + 3\sin^2 t \cos t\mathbf{j}$; $(6\cos t \sin^2 t - 3\cos^3 t)\mathbf{i} + (6 \sin t \cos^2 t - 3\sin^3 t)\mathbf{j}$; $3|\cos t \sin t|$;
$a_\mathbf{T} = -3\sin^2 t + 3\cos^2 t$ (quadrants I and III), or $3\sin^2 t - 3\cos^2 t$ (quadrants II and IV), $a_\mathbf{N} = 3|\cos t \sin t|$

23. $\mathbf{i} + 2\mathbf{j} - \dfrac{5t}{\sqrt{1 - 5t^2}}\mathbf{k}$; $\dfrac{-5}{(1 - 5t^2)^{3/2}}\mathbf{k}$; $\dfrac{\sqrt{5}}{\sqrt{1 - 5t^2}}$; $a_\mathbf{T} = \dfrac{5\sqrt{5}t}{(1 - 5t^2)^{3/2}}$; $a_\mathbf{N} = \dfrac{5}{1 - 5t^2}$ **25.** $\tan \phi = y'$

27. $(\pi/2, 0)$ **29.** $(\frac{7}{4}, \frac{7}{4})$ **31.** $x^2 + (y - 2)^2 = 1$ **33.** $\left(x - \dfrac{\pi}{2}\right)^2 + (y + 1)^2 = 1$

35. $\left(x - \dfrac{5}{4}\right)^2 + \left(y - \dfrac{3\sqrt{3}}{4}\right)^2 = \dfrac{27}{16}$

37. Circle with radius $3/\sqrt{2}$ and center $(\frac{3}{2}, -\frac{1}{2}, 1)$ lying in the plane with equation $x + y - 2z + 1 = 0$

39. $\mathbf{v} = (\sec t - 1)\mathbf{i} + \sin t\mathbf{j}$; $\|\mathbf{v}\| = \sqrt{2 + \tan^2 t - 2\sec t + \sin^2 t}$; $\mathbf{r} = (\ln|\sec t + \tan t| - t)\mathbf{i} + (1 - \cos t)\mathbf{j}$

41. $\cos \alpha = \dfrac{x'(t_0)}{\|\mathbf{r}'(t_0)\|}$, $\cos \beta = \dfrac{y'(t_0)}{\|\mathbf{r}'(t_0)\|}$, $\cos \gamma = \dfrac{z'(t_0)}{\|\mathbf{r}'(t_0)\|}$, where $\|\mathbf{r}'(t_0)\| = \sqrt{[x'(t_0)]^2 + y'(t_0)]^2 + [z'(t_0)]^2}$

Challenge Exercises

1. Use Problem 25, Review Exercises.

3. $\mathbf{r}'(t) \cdot \mathbf{r}(t) = 0 \Rightarrow \dfrac{d}{dt}\|\mathbf{r}(t)\| = 0 \Rightarrow \|\mathbf{r}(t)\| =$ constant for all $t \Rightarrow$ a sphere.

$\mathbf{r}''(t) \cdot \mathbf{r}'(t) = 0 \Rightarrow \dfrac{dv}{dt} = 0 \Rightarrow$ speed is constant.

$\mathbf{r}'(t) \times \mathbf{r}(t) = \mathbf{0} \Rightarrow \mathbf{T}$ and $\mathbf{r}$ are parallel $\Rightarrow$ straight line.

5. $\mathbf{v} = -\sin t\mathbf{u}_r + 2(2 + \cos t)\mathbf{u}_\theta$; $\mathbf{a} = (-5\cos t - 8)\mathbf{u}_r - 4\sin t\mathbf{u}_\theta$ **7.** (a) The $\mathbf{u}_\theta$ component is 0.

9. 0.50310 **11.** The ellipsoid $\dfrac{x^2}{a^2}+\dfrac{y^2}{b^2}+\dfrac{z^2}{c^2}=1$

13. In each case, motion is along an ellipse, which is the intersection of the surface and a plane.

CHAPTER 16

Exercise 16.1

1. (a) 3 (b) 2 (c) 10 (d) $3(x+\Delta x)+2y+(x+\Delta x)y$ (e) $3x+2(y+\Delta y)+x(y+\Delta y)$

3. (a) 0 (b) 0 (c) $at+a^2$ (d) $\sqrt{(x+\Delta x)y}+x+\Delta x$ (e) $\sqrt{x(y+\Delta y)}+x$

5. (a) 4 (b) -1 (c) 0 (d) $3\sin t\cos t$ **7.** (a) -5 (b) 3 (c) $2a^2$

9. (a) 0 (b) -1 (c) 20 **11.** $e^{x^4+(2y-1)^2}$

13. $x \geq 0, y > 0$

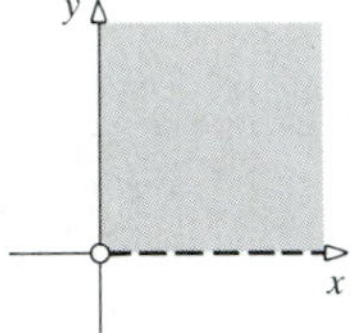

15. $x \geq 0, y \geq 0$ or $x \leq 0, y \leq 0$

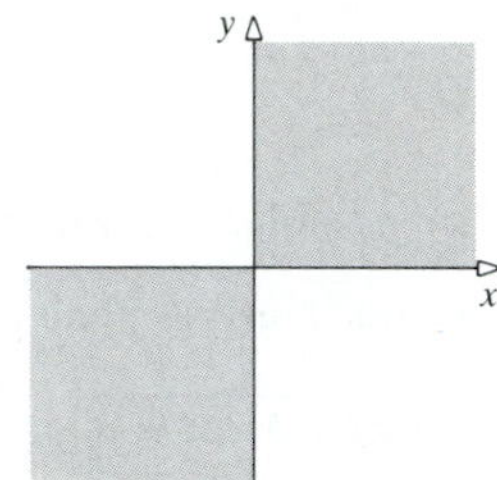

17. All x and y

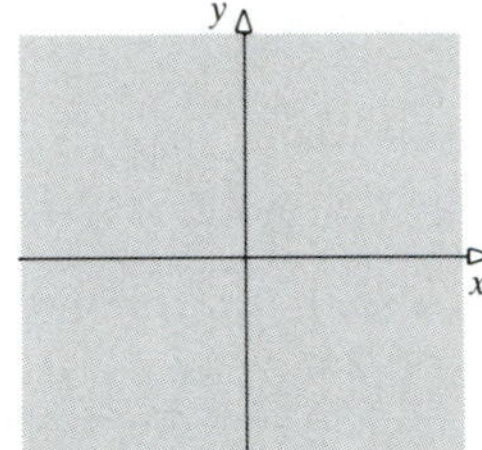

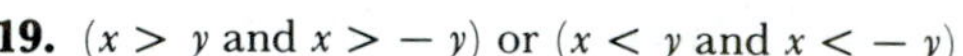

19. $(x > y$ and $x > -y)$ or $(x < y$ and $x < -y)$

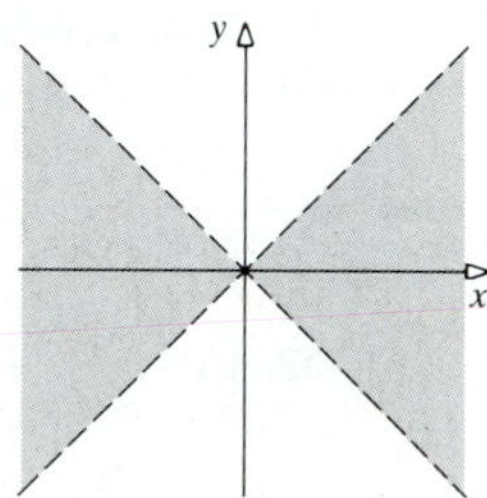

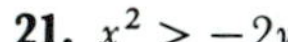

21. $x^2 > -2y$

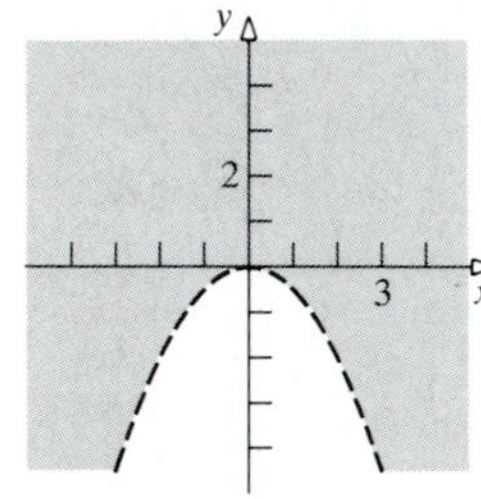

23. $x^2+y^2 > 4$

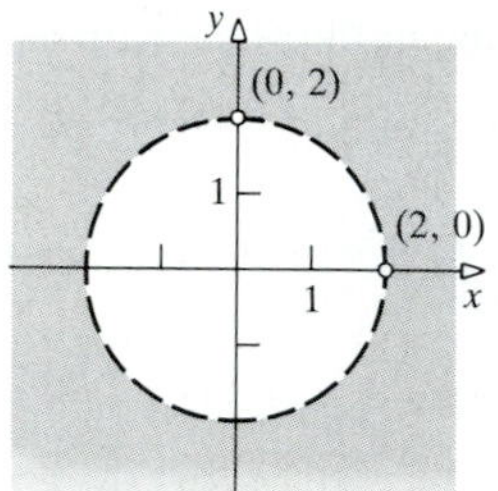

25. All x and y

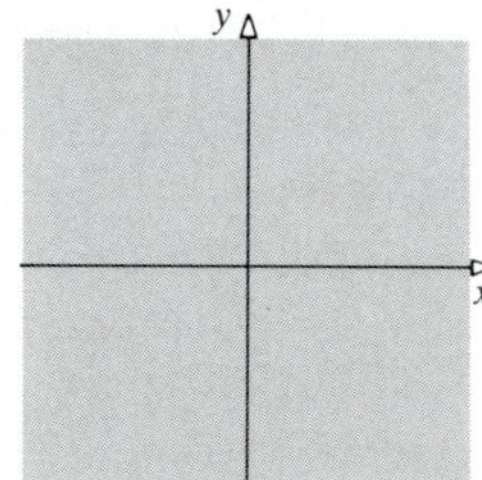

27. All x, y, z with $z \neq 0$

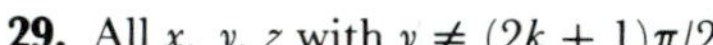

29. All x, y, z with $y \neq (2k+1)\pi/2$

31.

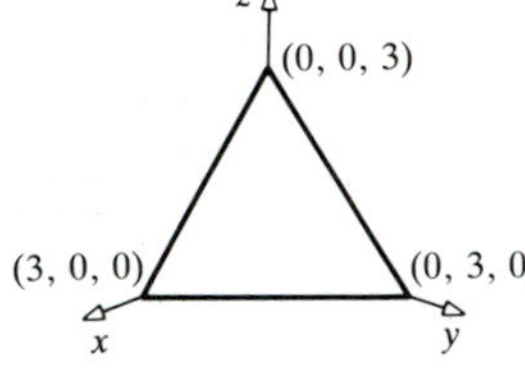

33.

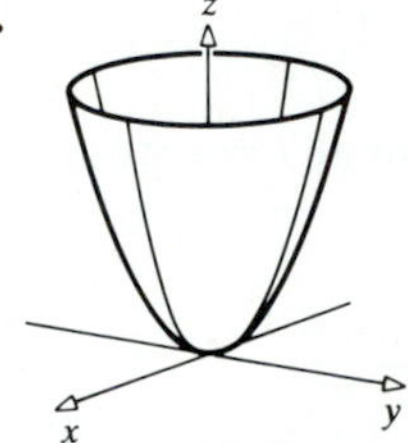

35.

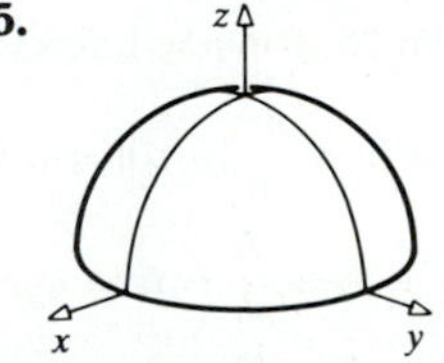

37.

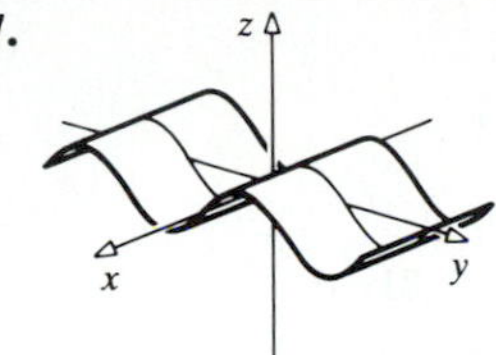

39.

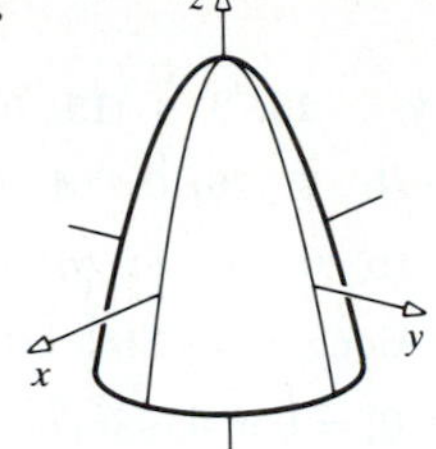

41.

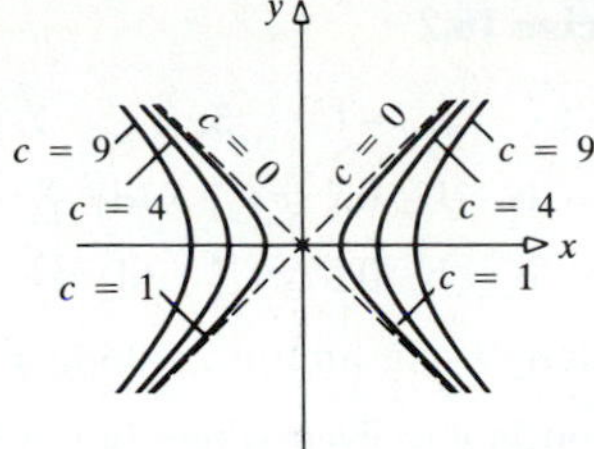

43.

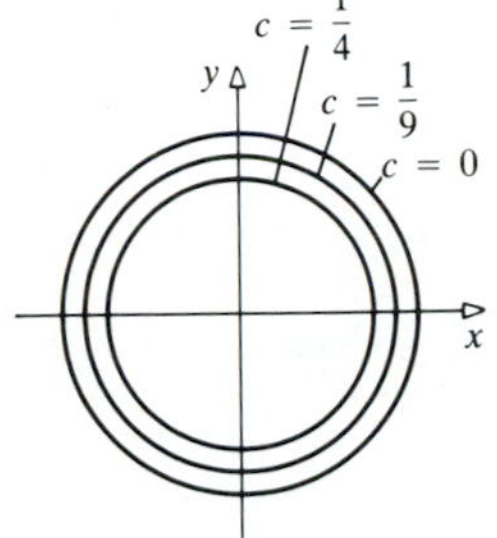

45.

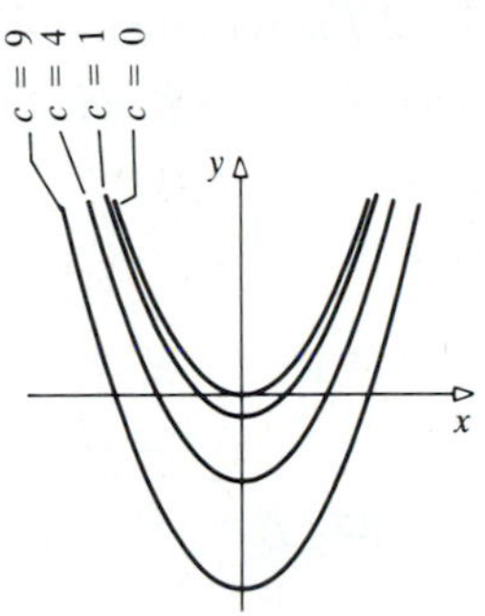

47.

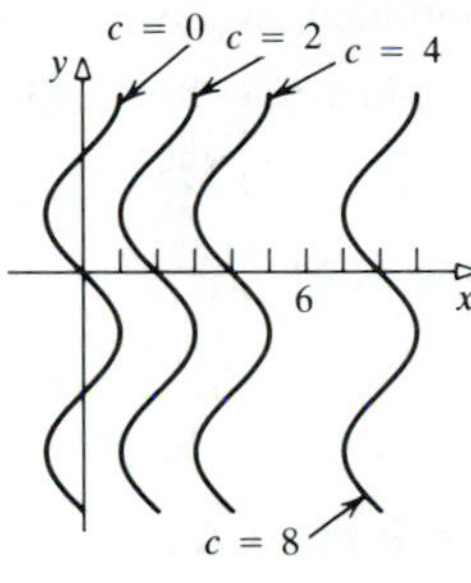

49. Spheres centered at the origin **51.** Parallel planes **53.** Circular cylinders centered on the z-axis

55.

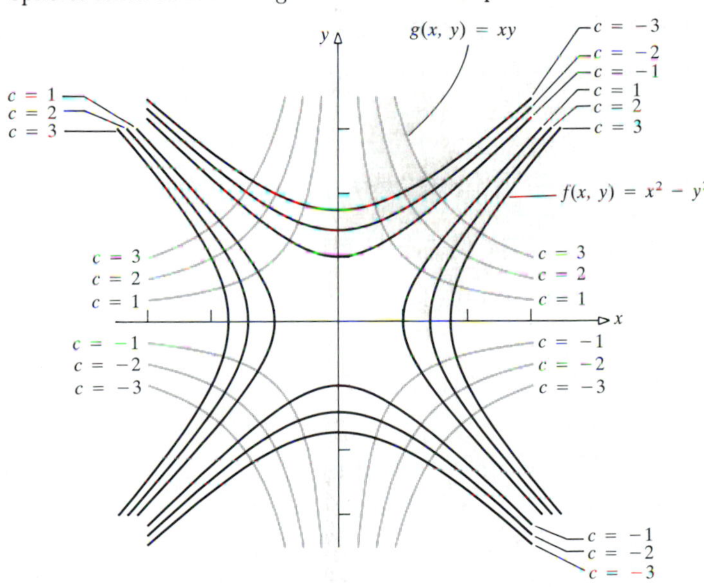

57. $S = 2xz + 2yz + xy$ **59.** $C = 4xy + 6yz + 4xz$

61. Circles centered at the origin; the surface is like a bowl with its low point (bottom) above the origin

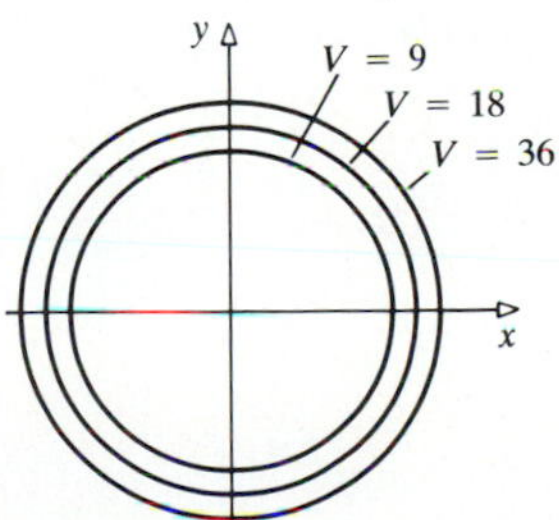

63. Circular cylinders centered on the x-axis

Exercise 16.2

1. -1 **3.** -1 **5.** 1 **7.** -1 **9.** 5 **11.** $\frac{1}{5}$ **13.** 3 **15.** 0 **17.** 0 **19.** 1

21. (a) 0 (b) 0 (c) 1 (d) $\frac{9}{11}$ (e) 0 **23.** (a) 0 (b) 0 (c) 0 (d) 0 (e) 0

25. (a) 0 (b) 0 (c) $\frac{3}{2}$ (d) $\frac{27}{82}$ (e) 0 **27.** (a) 1 (b) 0 (c) 1 (d) $\frac{2}{5}$ (e) 1 **29.** 0 **31.** 0

33. Along x-axis limit is 2; along y-axis limit is 1 **35.** Along x-axis limit is 0; along y-axis limit is -1

37. Continuous everywhere but at the origin. Setting $f(0, 0) = 0$ makes it continuous everywhere.

39. Continuous everywhere but at the origin. No value of f at the origin can make it continuous.

41. Use $(\sin u)/u \to 1$. **43.** $f(x, y) \to 0$ along the axes, $f(x, y) \to \frac{3}{2}$ along $y = x$

Exercise 16.3

1. $f_x = 2xy$; $f_y = x^2 + 12y$ **3.** $f_x = 3x^2/y^3$; $f_y = -3x^3/y^4$ **5.** $f_x = -e^y \sin x + e^x \sin y$; $f_y = e^y \cos x + e^x \cos y$

7. $f_x = 2e^{2x+3y}$; $f_y = 3e^{2x+3y}$ **9.** $f_x = x/(x^2 + y^2)$; $f_y = y/(x^2 + y^2)$

11. $f_x = 4y \sin(2xy)\cos(2xy)$; $f_y = 4x \sin(2xy)\cos(2xy)$ **13.** $f_x = yx^{y-1}$; $f_y = (\ln x)x^y$

15. $f_{xx} = 12$; $f_{yy} = 18$; $f_{xy} = f_{yx} = -8$

17. $f_{xx} = 3x(2y^2 - x^3)/(x^3 + y^2)^2$; $f_{yy} = 2(x^3 - y^2)/(x^3 + y^2)^2$; $f_{xy} = f_{yx} = -6x^2y/(x^3 + y^2)^2$

19. $f_{xx} = 4e^{2x+3y}$; $f_{yy} = 9e^{2x+3y}$; $f_{xy} = f_{yx} = 6e^{2x+3y}$

21. $f_{xx} = -2y^3 \sin(x^2y^3) - 4x^2y^6 \cos(x^2y^3)$; $f_{yy} = -6x^2y \sin(x^2y^3) - 9x^4y^4 \cos(x^2y^3)$;
$f_{xy} = f_{yx} = -6xy^2 \sin(x^2y^3) - 6x^3y^5 \cos(x^2y^3)$

23. $f_x = y + z$; $f_y = x + z$; $f_z = y + x$ **25.** $f_x = y \sin z - yz \cos x$; $f_y = x \sin z - z \sin x$; $f_z = xy \cos z - y \sin x$

27. $f_x = -yz/(x^2 + y^2)$; $f_y = xz/(x^2 + y^2)$; $f_z = \tan^{-1}(y/x)$ **29.** $f_x = f_y = z(x + y)^{z-1}$; $f_z = [\ln(x + y)](x + y)^z$

31. $f_x = yzx^{yz-1}$; $f_y = z(\ln x)x^{yz}$; $f_z = y(\ln x)x^{yz}$ **33.** $f_x = \dfrac{yz}{z^2 + x^2y^2}$; $f_y = \dfrac{xz}{z^2 + x^2y^2}$; $f_z = \dfrac{-xy}{z^2 + x^2y^2}$

35. $f_x(0, 0) = f_y(0, 0) = 1$ **37.** $z - 5 = 2(x - 1), y = 2$ **39.** $z - \sqrt{3}/2 = (-1/\sqrt{3})(y - 1/2), x = 0$

41. $z - \sqrt{15} = (-1/\sqrt{15})(y - 1), x = 4$ **43.** (a) $\frac{3}{25}$ (b) $\frac{4}{25}$

45. (b) $T_x(1, 0) = 200/(\ln 2)$; $T_x(0, 1) = 0$ (c) $T_y(2, 0) = 0$; $T_y(0, 2) = 100/(\ln 2)$

47. $\partial x/\partial r = \cos\theta$; $\partial x/\partial\theta = -r \sin\theta$; $\partial y/\partial r = \sin\theta$; $\partial y/\partial\theta = r \cos\theta$

49. $\partial u/\partial x = e^x \cos y = \partial v/\partial y$; $\partial u/\partial y = -e^x \sin y = -\partial v/\partial x$ **51.** $\partial u/\partial x = 2x$; $\partial u/\partial y = 8y$

53. $\partial w/\partial x = 2x$; $\partial w/\partial y = 2y - 3z$; $\partial w/\partial z = -3y$

55. $\partial^2 z/\partial x^2 = -\cos(x + y) - \cos(x - y)$; $\partial^2 z/\partial y^2 = -\cos(x + y) - \cos(x - y)$

57. $\partial^2 z/\partial x^2 = (y^2 - x^2)/(x^2 + y^2)^2$; $\partial^2 z/\partial y^2 = (x^2 - y^2)/(x^2 + y^2)^2$ **59.** $\dfrac{\partial^2 z}{\partial x^2} = \dfrac{2xy}{(x^2 + y^2)^2}$; $\dfrac{\partial^2 z}{\partial y^2} = \dfrac{-2xy}{(x^2 + y^2)^2}$

61. $\partial z/\partial x = [2xe^{x^2+y^2}/(x^2 + y^2)^2](x^2 + y^2 - 1)$; $\partial z/\partial y = [2ye^{x^2+y^2}/(x^2 + y^2)^2](x^2 + y^2 - 1)$

63. $\partial V/\partial T = nr/P$; $\partial T/\partial P = V/nr$; $\partial P/\partial V = -nrT/V^2$ **65.** $\partial v/\partial p = k/2\sqrt{pd}$; $\partial v/\partial d = (-k/2)\sqrt{p/d^3}$

67. $f_{xy} = f_{yx} = \dfrac{6x^2y - 2y^3}{(x^2 + y^2)^3}$ **69.** $f_{xy} = f_{yx} = \dfrac{x}{y}(\ln y)e^{x \ln y} + \dfrac{1}{y}e^{x \ln y}$

71. (a) $f_x(1, 6) = 6$; $f_y(1, 6) = 432$ (b) $\bar{f}_{x'}(a, b) = 219\sqrt{2}$; $\bar{f}_{y'}(a, b) = 213\sqrt{2}$

73. (a) $f_x(0, y) = \lim\limits_{\Delta x \to 0} \dfrac{y((\Delta x)^2 - y^2)}{(\Delta x)^2 + y^2} = -y$ (b) $f_y(x, 0) = \lim\limits_{\Delta y \to 0} \dfrac{x(x^2 - (\Delta y)^2)}{x^2 + (\Delta y)^2} = x$

(c) $f_{xy}(0, 0) = \lim\limits_{\Delta y \to 0} \dfrac{-\Delta y}{\Delta y} = -1$ (d) $f_{yx}(0, 0) = \lim\limits_{\Delta x \to 0} \dfrac{\Delta x}{\Delta x} = 1$

75. The difference quotients are $|h|/h$, so the limit as $h \to 0$ does not exist. The graph near the origin looks like the "point" of an ice cream cone. This is analogous to the absolute value function at $(0, 0)$.

77. $z - 1 = 4(y + 2), x = 1; z - 1 = 8(x - 1), y = -2$

79. $\partial^2 u/\partial x^2 = -2xyz/(x^2 + y^2)^2; \partial^2 u/\partial y^2 = 2xyz/(x^2 + y^2)^2; \partial^2 u/\partial z^2 = 0$

81. $f_x = \dfrac{1}{y} e^{x/y} - \dfrac{z}{x^2} e^{z/x}; f_y = -\dfrac{x}{y^2} e^{x/y} + \dfrac{1}{z} e^{y/z}; f_z = -\dfrac{y}{z^2} e^{y/z} + \dfrac{1}{x} e^{z/x}$

83. $u_{xx} = v_{yx}, u_{yy} = -v_{xy}$ and $v_{xy} = v_{yx}$ **85.** $f_{tt} = -c^2\cos(x + ct); f_{xx} = -\cos(x + ct)$

87. $f_x = -\ln \cos\sqrt{x}; f_y = \ln \cos\sqrt{y}$

Exercise 16.4

1. $2x\,dx + 2y\,dy$ **3.** $(\sin y + y \cos x)\,dx + (x \cos y + \sin x)\,dy$ **5.** $(e^x\cos y - e^{-x}\sin y)\,dx + (e^{-x}\cos y - e^x\sin y)\,dy$

7. $\dfrac{2x}{x^2 + y^2}\,dx + \dfrac{2y}{x^2 + y^2}\,dy$ **9.** $(2xy + z^2)\,dx + (x^2 + 2yz)\,dy + (y^2 + 2xz)\,dz$

11. $(e^{yz} + yze^{xz} + yze^{xy})\,dx + (xze^{yz} + e^{xz} + xze^{xy})\,dy + (xye^{yz} + xye^{xz} + e^{xy})\,dz$

13. $2e^t\left(\dfrac{dx}{x} + \dfrac{dy}{y} + \dfrac{dz}{z}\right) + e^t(\ln xy + \ln xz + \ln yz)\,dt$

15. (a) $\Delta z = 2xy\Delta y + x(\Delta y)^2 + y^2\Delta x + 2y\Delta x\Delta y + \Delta x(\Delta y)^2 - 2y\Delta x - 2x\Delta y - 2\Delta x\Delta y$
(b) Take $\eta_1 = 2y\Delta y + (\Delta y)^2 - 2\Delta y$; $\eta_2 = x\Delta y$

17. (a) $\Delta z = \dfrac{2xy\Delta y + x(\Delta y)^2 - y^2\Delta x}{x(x + \Delta x)}$ (b) $\eta_1 = \dfrac{y^2\Delta x - xy\Delta y}{x^2(x + \Delta x)}$; $\eta_2 = \dfrac{x^2\Delta y - xy\Delta x}{x^2(x + \Delta x)}$ (choices not unique)

19. 1.4 **21.** -0.3243 **23.** 4.003125 **25.** 0.1414 cm^3 **27.** 0.225 cm^2

29. $dx = \cos\theta\,dr - r\sin\theta\,d\theta$; $dy = \sin\theta\,dr + r\cos\theta\,d\theta$ **31.** 1.12% **33.** 3%

35. (a) $dA = \frac{1}{2}c\sin\alpha\,db + \frac{1}{2}b\sin\alpha\,dc + \frac{1}{2}bc\cos\alpha\,d\alpha$ (b) 6.356%

37. $f_x(0, 0) = f_y(0, 0) = 0$; f is not continuous at $(0, 0)$ (use $y = x$ and $y = 2x$), so f is not differentiable at $(0, 0)$.

39. $f_x(0, 0) = f_y(0, 0) = 0$; f is not continuous at $(0, 0)$ (use $y = x$ and $y = 2x$), so f is not differentiable at $(0, 0)$.

Exercise 16.5

1. $2\sin t\cos t - 4\sin 2t\cos 2t$ **3.** $2t^2e^{2t} + 2te^{2t} + 2te^{-2t} - 2t^2e^{-2t}$ **5.** $e^{\sqrt{t}}\left(\dfrac{\sin \pi t}{2\sqrt{t}}\right) + \pi e^{\sqrt{t}}\cos \pi t$

7. $e^{te^{t-t^2}}(1 + t - 2t^2)e^{t-t^2}$ **9.** $2e^{\sin^2 2t + \cos^2 t}(\sin 4t - \sin t\cos t)$ **11.** $\partial z/\partial u = 2ue^{2v} + 2v^2e^{2u}$; $\partial z/\partial v = 2u^2e^{2v} + 2ve^{2u}$

13. $\partial z/\partial u = 2uve^{u^2v}\sin[\ln(uv)] + (e^{u^2v}/u)\cos[\ln(uv)]$; $\partial z/\partial v = u^2e^{u^2v}\sin[\ln(uv)] + (e^{u^2v}/v)\cos[\ln(uv)]$

15. $\partial z/\partial u = (-v/u^2)e^{u^2+v^2} + 2ve^{u^2+v^2}$; $\partial z/\partial v = (e^{u^2+v^2}/u)(1 + 2v^2)$

17. $\partial z/\partial u = 2(5u - 3v)^2(2u + 3v)^3 + 10(2u + v)(5u - 3v)(2u + 3v)^3 + 6(2u + v)(5u - 3v)^2(2u + 3v)^2$;
$\partial z/\partial v = (5u - 3v)^2(2u + 3v)^3 - 6(2u + v)(5u - 3v)(2u + 3v)^3 + 9(2u + v)(5u - 3v)^2(2u + 3v)^2$

19. $\partial z/\partial u = 2(u^4 - v^8)/u(u^4 + v^8)$; $\partial z/\partial v = 4(v^8 - u^4)/v(v^8 + u^4)$

21. $\partial z/\partial u = \sin 2(u - v) - \sin 2(u + v)$; $\partial z/\partial v = -\sin 2(u - v) - \sin 2(u + v)$ **23.** $2te^{2t}(1 + t) + 2te^{-2t}(1 - t) - 4e^{4t}$

25. $e^{\sqrt{t}}\left[\dfrac{1}{2\sqrt{t}}\sin(\pi t)\cos(t/2) + \pi\cos(\pi t)\cos(t/2) - \dfrac{1}{2}\sin(\pi t)\sin(t/2)\right]$ **27.** $e^{2t}(1 + 2\ln t + 1/t - 2t^2$

29. $e^t(\sin t)(2\cos^2 t - \sin^2 t + \sin t\cos t)$ **31.** $\cos^2 t - \sin^2 t$ **33.** $y(-2x + y - 1)/x(x - 2y + 1)$

35. $-(\sin y + y\cos x)/(x\cos y + \sin x)$ **37.** $-y^{2/3}/x^{2/3}$

39. $\partial z/\partial x = -(z + 2xy^3)/(x + 6yz - 5)$; $\partial z/\partial y = -(3z^2 + 3x^2y^2)/(x + 6yz - 5)$

41. $\partial z/\partial x = -yz/(\cos z - y\sin z + xy)$; $\partial z/\partial y = -(\cos z + xz)/(\cos z - y\sin z + xy)$

43. $\partial z/\partial r = (\partial f/\partial x)\cos\theta + (\partial f/\partial y)\sin\theta$; $\partial z/\partial\theta = (\partial f/\partial x)(-r\sin\theta) + (\partial f/\partial y)(r\cos\theta)$

45. $\partial z/\partial u = (\partial f/\partial x)(1) + (\partial f/\partial y)(-1)$; $\partial z/\partial v = (\partial f/\partial x)(-1) + (\partial f/\partial y)(1)$

47. $\partial u/\partial x = x/u$; $\partial u/\partial y = y/u$; $\partial u/\partial z = z/u$; $\partial w/\partial x = (dw/du)(\partial u/\partial x)$; etc.

49. $\dfrac{\partial z}{\partial u} = f_x\left(\dfrac{u}{v}, \dfrac{v}{w}\right)\dfrac{1}{v}$; $\dfrac{\partial z}{\partial v} = f_x\left(\dfrac{u}{v}, \dfrac{v}{w}\right)\dfrac{-u}{v^2} + f_y\left(\dfrac{u}{v}, \dfrac{v}{w}\right)\dfrac{1}{w}$; $\dfrac{\partial z}{\partial w} = f_y\left(\dfrac{u}{v}, \dfrac{v}{w}\right)\dfrac{-v}{w^2}$

51. $\partial^2 z/\partial u^2 = (\partial^2 f/\partial x^2)(\partial g/\partial u)^2 + 2(\partial^2 f/\partial x\,\partial y)(\partial h/\partial u)(\partial g/\partial u) + (\partial^2 f/\partial y^2)(\partial h/\partial u)^2 + (\partial f/\partial x)(\partial^2 g/\partial u^2) + (\partial f/\partial y)(\partial^2 h/\partial u^2)$

53. $\partial z/\partial x = -F_x/F_z$; $\partial x/\partial y = -F_y/F_x$; $\partial y/\partial z = -F_z/F_y$

55. $\partial w/\partial x = 2(3y + 4z)(2x)^{3y+4z-1}$; $\partial w/\partial y = 3\ln(2x)(2x)^{3y+4z}$; $\partial w/\partial z = 4\ln(2x)(2x)^{3y+4z}$

57. Continue, using the result of Problem 43.

59. $14xy/(x - y^2)^3$

Review Exercises

1. $y = \sin x$

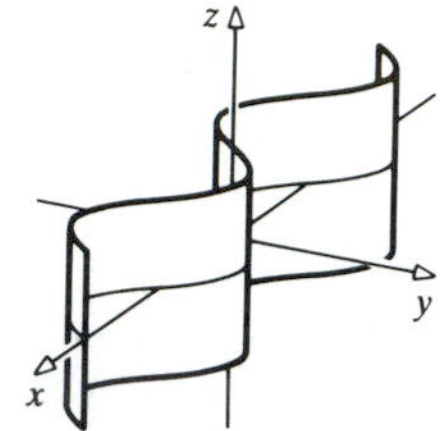

3. $z = e^x$ or $x = \ln z$

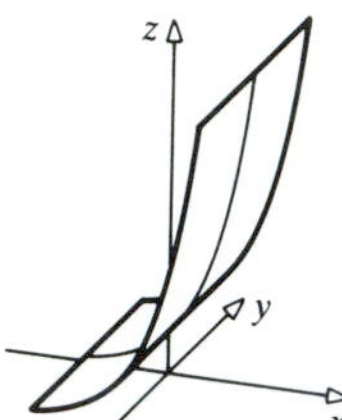

5. $\dfrac{\partial z}{\partial x} = 3x^2y^2 - 2y^4 + 6xy^3$; $\dfrac{\partial z}{\partial y} = 2x^3y - 8xy^3 + 9x^2y^2$ **7.** $e^{2x} + e^{2y} + 2e^{x+y}\sin(x + y)$ **9.** k/p; $-kT/p^2$

11. For all x and y

13. $f(x, y) \to 0$ along the x-axis, $f(x, y) \to \frac{1}{2}$ along the line $y = x$, so it cannot be made continuous at $(0, 0)$.

15. $\dfrac{\partial^2 w}{\partial x^2} = y^2ze^{xy}$; $\dfrac{\partial^2 w}{\partial y^2} = x^2ze^{xy}$; $\dfrac{\partial^2 w}{\partial z^2} = 0$; $\dfrac{\partial^2 w}{\partial y\,\partial x} = \dfrac{\partial^2 w}{\partial x\,\partial y} = (xyz + z)e^{xy}$; $\dfrac{\partial^2 w}{\partial x\,\partial z} = \dfrac{\partial^2 w}{\partial z\,\partial x} = ye^{xy}$; $\dfrac{\partial^2 w}{\partial y\,\partial z} = \dfrac{\partial^2 w}{\partial z\,\partial y} = xe^{xy}$

17. $w_{xx} = \dfrac{2xyz}{(x^2 + y^2)^2}$; $w_{yy} = \dfrac{-2xyz}{(x^2 + y^2)^2}$; $w_{zz} = 0$; $w_{xy} = w_{yx} = \dfrac{z(y^2 - x^2)}{(x^2 + y^2)^2}$; $w_{xz} = w_{zx} = \dfrac{-y}{x^2 + y^2}$; $w_{yz} = w_{zy} = \dfrac{x}{x^2 + y^2}$

19. $\dfrac{1}{2}$; $\dfrac{\sqrt{3}}{2}$ **21.** Yes. Let $r^2 = x^2 + y^2$; then $(\sin r^2)/r^2 \to 1$. **23.** $f_x = \dfrac{-y}{x\sqrt{x^2 - y^2}}$; $f_y = \dfrac{1}{\sqrt{x^2 - y^2}}$

25. $f(x) = -\dfrac{7}{2}\dfrac{y}{\sqrt{x + 2y}(3x - y)^{3/2}}$; $f(y) = \dfrac{7}{2}\dfrac{x}{\sqrt{x + 2y}(3x - y)^{3/2}}$

27. $\dfrac{\partial a}{\partial b} = \dfrac{b - c\cos A}{a}$; $\dfrac{\partial a}{\partial c} = \dfrac{c - b\cos A}{a}$; $\dfrac{\partial a}{\partial A} = \dfrac{bc}{a}\sin A$; $\dfrac{\partial c}{\partial A} = \dfrac{-bc\sin A}{c - b\cos A}$ **29.** $dz = \dfrac{y\,dx - x\,dy}{y\sqrt{y^2 - x^2}}$

31. $dw = \dfrac{dx}{x} + \dfrac{dy}{y} + \dfrac{dz}{z}$ **33.** 5% **35.** $\dfrac{\partial z}{\partial x} = \dfrac{x - yz}{xy}$; $\dfrac{\partial z}{\partial y} = \dfrac{y - xz}{xy}$

37. $\dfrac{dz}{dt} = e^{2t}((2t + 1)\cos(te^{2t}) - (t + 1)\cos(te^t)) - e^t\sin(te^t)$ **39.** $\dfrac{\partial u}{\partial r} = 2rs + 2s^2 - s$; $\dfrac{\partial u}{\partial s} = 4rs - 2s + r^2 - r$

Challenge Exercises

1. All points inside both the upper half of the unit sphere and the cone $x^2 + y^2 = z^2$

3. $f_x = f_y = 0$ **5.** $f_{rr} = n(n - 1)r^{n-2}\sin n\theta$; $f_{\theta\theta} = -n^2r^n\sin n\theta$. **7.** See Problem 5.

9. $z - f(x_0, y_0) = f_x(x_0, y_0)(x - x_0),\ y = y_0;\ z - f(x_0, y_0) = f_y(x_0, y_0)(y - y_0),\ x = x_0;$
$f_x(x_0, y_0)(x - x_0) + f_y(x_0, y_0)(y - y_0) - [z - f(x_0, y_0)] = 0$

CHAPTER 17

Exercise 17.1

1. $1 - 2\sqrt{3}$ **3.** $-3 - 4\sqrt{3}$ **5.** $\frac{1}{2}(1 + \sqrt{3})$ **7.** $-\frac{7}{10}$ **9.** $-\pi\sqrt{3}/12$ **11.** $3/\sqrt{5}$
13. $6\mathbf{i} + 4\mathbf{j};\ 14/\sqrt{5}$ **15.** $6\mathbf{i};\ 12/\sqrt{5}$ **17.** $\mathbf{j};\ \pi/\sqrt{\pi^2 + 9}$ **19.** $4\mathbf{i} + 4\mathbf{j};\ 4$ **21.** $\frac{3}{25}\mathbf{i} - \frac{4}{25}\mathbf{j};\ -1/(25\sqrt{2})$
23. $-4\mathbf{i};\ -4/\sqrt{11}$ **25.** $\frac{1}{3}\mathbf{i} + \frac{2}{3}\mathbf{j} + \frac{2}{3}\mathbf{k};\ -1/\sqrt{3}$ **27.** $\mathbf{i} + \mathbf{j};\ 3/\sqrt{5}$ **29.** $\frac{1}{\sqrt{5}}\mathbf{i} - \frac{2}{\sqrt{5}}\mathbf{j};\ 2\sqrt{5}$
31. $\frac{1}{\sqrt{2}}\mathbf{i} + \frac{1}{\sqrt{2}}\mathbf{j};\ \sqrt{2}$ **33.** $\frac{-6\mathbf{i} + 6\mathbf{j} + \pi\mathbf{k}}{\sqrt{72 + \pi^2}};\ \frac{\sqrt{72 + \pi^2}}{4}$ **35.** $6\mathbf{i} + 8\mathbf{j}$
37. $6\mathbf{i} - 4\sqrt{5}\mathbf{j}$ **39.** $\frac{2}{3}\mathbf{i} + 9\mathbf{j}$ **41.** $-\frac{1}{5};\ \mathbf{i} + \mathbf{j};\ -\mathbf{i} - \mathbf{j};\ \mathbf{i} - \mathbf{j}$ or $-\mathbf{i} + \mathbf{j}$
43. (a) $-(x^2 + y^2)^{-1/2}$ (b) 0 (c) $\frac{x}{x^2 + y^2}\mathbf{i} + \frac{y}{x^2 + y^2}\mathbf{j}$ (d) $\frac{-x}{x^2 + y^2}\mathbf{i} + \frac{-y}{x^2 + y^2}\mathbf{j}$
45. (a) $\sqrt{2}/2$ (b) $3\mathbf{i} - 4\mathbf{j};\ 5$ (c) $\pm(2\mathbf{i} + \mathbf{j})$ **47.** $-\mathbf{j}$ or $\frac{4}{5}\mathbf{i} - \frac{3}{5}\mathbf{j}$ **49.** $(2x\mathbf{i} - 2y\mathbf{j}) \cdot (y\mathbf{i} + x\mathbf{j}) = 0$
51. $\frac{2}{\sqrt{5}}\mathbf{i} + \frac{1}{\sqrt{5}}\mathbf{j}$ **53.** Use (17.1.4). **55.** $2\mathbf{i} - 2\mathbf{j} + \mathbf{k};\ 3/e^2$
57. $T_x = -x/\sqrt{4 - x^2}$; the temperature is 2(maximum) when $x = 0$ and decreases very rapidly to 0 as x gets closer to 2
59. $\nabla F \cdot (\mathbf{r} - \mathbf{r}_0) = 0$; $\mathbf{r}$ defines the tangent line. **61.** $m = -a/b$ **63.** $\pm 5/\sqrt{3}$
65. (a) Use $r^n = (x^2 + y^2)^{n/2}$. (b) Use the chain rule. (c) Apply the result from part (b).

Exercise 17.2

1. Tangent plane: $x - 2y + 3z - 14 = 0$; normal line: $\frac{x - 1}{2} = \frac{y + 2}{-4} = \frac{z - 3}{6}$

3. Tangent plane: $4x + 3y - z - 12 = 0$; normal line: $\frac{x - 2}{8} = \frac{y - 1}{6} = \frac{z + 1}{-2}$

5. Tangent plane: $x - 3y + 2z = 0$; normal line: $\frac{x - 1}{2} = \frac{y + 1}{-6} = \frac{z + 2}{4}$

7. Tangent plane: $z - 5 = -4(x + 2) + 2(y - 1)$; normal line: $\frac{x + 2}{-4} = \frac{y - 1}{2} = \frac{z - 5}{-1}$

9. Tangent plane: $z - 2 = 4(x - 1)$; normal line: $\frac{x - 1}{4} = \frac{z - 2}{-1},\ y = 0$

11. Tangent plane: $y + z - (\pi/2) = 0$; normal line: $x = 0,\ z = y - (\pi/2)$

13. Tangent plane: $2x + y - z - 18 = 0$; normal line: $\frac{x - 1}{2} = y - 8 = \frac{z + 8}{-1}$

15. $(3, -1, 11)$ **17.** $(x, -x, 0)$, x any real number **19.** $(\pm\frac{1}{2}, -1, \frac{7}{8}),\ (0, -1, 1)$
21. Each has the tangent plane $y + z - 1 = 0$.
23. If (x_0, y_0, z_0) is on the sphere, the normal line obeys $x = x_0 + 2x_0t,\ y = y_0 + 2y_0t,\ z = z_0 + 2z_0t$, which for $t = -\frac{1}{2}$ passes through $(0, 0, 0)$, the center of the sphere.

25. Tangent plane: $x + 4y - 5z - 4 = 0$; normal line: $\dfrac{x-1}{2} = \dfrac{y-2}{8} = \dfrac{z-1}{-10}$ **27.** $z = -\pi y + \dfrac{3\pi}{2}$, $x = 1$

29. $-2\mathbf{i} - 2\mathbf{j} - \mathbf{k}$

31. $F_x = 2x_0/a^2$; $F_y = 2y_0/b^2$; $F_z = -2z_0/c^2$; the cone has no tangent plane at the origin $[\nabla F(0, 0, 0) = \mathbf{0}]$.

33. $y + z = \pi$; $x = 0$, $y = z$

Exercise 17.3

1. $(0, 0)$, $(1, 0)$, $(-1, 0)$ **3.** $(0, 0)$, $(1, 1)$, $(-1, -1)$ **5.** $(0, 0)$ **7.** Local minimum at $(1, -2)$

9. Local minimum at $(2, -1)$ **11.** Local minimum at $(2, 2)$; saddle point at $(0, 0, 0)$ **13.** Saddle point at $(0, 2, 4)$

15. Local maximum at $(-1, -1)$; saddle point at $(0, 0, 0)$ **17.** Saddle point at $(3, -\frac{9}{2}, \frac{27}{4})$; no information about $(0, 0)$

19. None **21.** Local maximum at $(2k\pi, 2l\pi)$; local minimum at $((2k+1)\pi, (2l+1)\pi)$

23. Local minimum at $(k\pi, (-1)^k 3)$; saddle point at $\left((2k+1)\dfrac{\pi}{2}, 0, 6\right)$

25. Saddle point at $(0, 0, 1)$ **27.** Absolute maximum 4; absolute minimum 0

29. Absolute maximum $4 + 4\sqrt{2}$; absolute minimum -2 **31.** Absolute maximum 13, absolute minimum -7

33. (a) Complete the square to get $(3, -2, 1)$. (b) Local maximum at $(3, -2)$, for which $z = 1$

35. $S =$ Surface area; Base $= \sqrt{S/3} \times \sqrt{S/3}$, Height $= \frac{1}{2}\sqrt{S/3}$

37. A cube, each side of length $\sqrt{S/6}$ **39.** (a) $h = 14$ in., $w = 14$ in., $l = 28$ in. (b) $R = 28/\pi$ in., $l = 28$ in.

41. $h = L/2\sqrt{3}$, $x = L/3$; no, semicircular area is larger. **43.** $D = (\frac{1}{3}, \frac{1}{2})$

45. Maximum profit when $x = 15{,}250$ tons and $y = 4100$ tons

47. $g_{xx} = 2A$, $g_{xy} = 2B$, $g_{yy} = 2C$. Use theorem (17.3.6).

Exercise 17.4

1. Maximum at $(\sqrt{2}, 1)$, $(-\sqrt{2}, -1)$, $3\sqrt{2}$; minimum at $(-\sqrt{2}, 1)$, $(\sqrt{2}, -1)$, $-3\sqrt{2}$

3. Maximum at $(0, 1)$, 4; minimum at $(0, -1)$, -4

5. Maximum at $(\sqrt{2}, \frac{3}{2}\sqrt{2})$, $(-\sqrt{2}, -\frac{3}{2}\sqrt{2})$, 3; minimum at $(-\sqrt{2}, \frac{3}{2}\sqrt{2})$, $(\sqrt{2}, -\frac{3}{2}\sqrt{2})$, -3

7. Minimum at $(-6, \frac{9}{2}, \frac{225}{24})$, $-\frac{225}{12}$ **9.** $(\frac{3}{5}, -\frac{9}{5})$ **11.** $(1, \sqrt{2}/2)$ and $(-1, -\sqrt{2}/2)$

13. Base $= 2\sqrt[3]{3} \times 2\sqrt[3]{3}$ m; Height $= \sqrt[3]{3}$ m **15.** (a) Maximum 50 m; minimum 20 m (b) 3400 m^2 $(x = 30)$

17. Maximum $2(3 + 2\sqrt{2})$; minimum $2(3 - 2\sqrt{2})$

19. Use $\nabla V = \lambda \nabla g$, where $V = xyz$ and $g = 2xy + 2xz + 2zx - S = 0$; then $x = y = z = \sqrt{S/6}$.

Review Exercises

1. $\sqrt{\pi}$ **3.** $\sqrt{2}$ **5.** $-35/\sqrt{3}$ **7.** $\frac{1}{3}$ **9.** $3/\sqrt{10}$

11. Not necessarily; $\mathbf{v}_1 = \mathbf{i} + \mathbf{k}$ and $\mathbf{v}_2 = \mathbf{j} + \mathbf{k}$ satisfy the requirements at $(\frac{1}{2}, \frac{1}{2})$ on $f(x, y) = x^2 + y^2$, but $\mathbf{v}_1 \cdot \mathbf{v}_2 = 1 \neq 0$.

13. $\mathbf{r}(t) = (\sqrt{\pi/8}, \sqrt{\pi/8}, \ln\sqrt{2}) + t(\sqrt{\pi/2}, \sqrt{\pi/2}, -1)$; $\sqrt{\pi/2}\,x + \sqrt{\pi/2}\,y - z = \pi/2$

15. $\mathbf{r}(t) = (1, 1, 0) + t(1, -1, -1)$; $z = x - y$ **17.** $\mathbf{r}(t) = (1, 1, 1) + t(17, 9, 9)$; $17x + 9y + 9z = 35$

19. $\mathbf{r}(t) = t(0, 1, 0)$; $y = 0$ **21.** $\mathbf{r}(t) = (0, 1, 0) + t(1, 0, -1)$; $x - z = 0$

23. The required intersection is the circle $x^2 + y^2 = 1$, $z = 0$. The gradients are $2x\mathbf{i} + 2y\mathbf{j} + 2z\mathbf{k}$ and $2x\mathbf{i} + 2y\mathbf{j} - 2z\mathbf{k}$, which are equal if $z = 0$, so the normals are equal.

25. Relative minimum at $(-4, 2)$

27. $(k_1\pi + \frac{\pi}{2}, k_2\pi + \frac{\pi}{2})$ is a local maximum if k_1 and k_2 are both even or both odd and a local minimum otherwise; $(k_3\pi, k_4\pi)$ is a saddle point.

29. $\dfrac{8\sqrt{3}}{9}\,abc$ **31.** $\dfrac{abc}{27}$ **33.** Maximum at $(0, \pm 1)$, $(\pm 1, 0)$; minimum at $(\pm 1/\sqrt{2}, \pm 1/\sqrt{2})$ **35.** $\sqrt{3}/9, -\sqrt{3}/9$

37. $6 + 4\sqrt{2}, 6 - 4\sqrt{2}$ **39.** Equilateral triangle with sides of length $R\sqrt{3}$

Challenge Exercises

1. $3\sqrt{2}$ **3.** $(\frac{1}{2}, \frac{1}{2}, \frac{3}{2})$ **5.** $m = \frac{7}{10}, b = \frac{6}{5}$

7. $\dfrac{df}{d\theta} = \dfrac{\partial f}{\partial x}\dfrac{dx}{d\theta} + \dfrac{\partial f}{\partial y}\dfrac{dy}{d\theta} = cx(-k \sin \theta) + cy(k \cos \theta) = 0$; thus, $f(x, y)$ is constant on the circle.

9. $(-2, 4, 2)$, $(-2, 4, -2)$; Distance $= 2\sqrt{6}$

11. (b) See Problem 47, Exercise 17.3.
(c) $f_1(x, y) = x^3 + y^3$; $f_2(x, y) = x^4 + y^4$; $f_3(x, y) = -x^4 - y^4$; each has $AC - B^2 = 0$ at $(0, 0)$ and a saddle point, local minimum, and local maximum, respectively.

CHAPTER 18

Exercise 18.1

1. 36 **3.** 4 **5.** $\frac{47}{180}$ **7.** $\frac{2}{3}$ **9.** $\frac{2}{3}$ **11.** $\sqrt{2}$ **13.** $e - 1$ **15.** 0 **17.** 5 **19.** 1

21. $\frac{13}{2}\ln 2 - \frac{5}{2}\ln 5$ **23.** $(e^2 - 1)(e - 1)$ **25.** -2

Exercise 18.2

1. $\frac{1}{3}$ **3.** $\frac{9}{2}$ **5.** $\frac{1}{3}$ **7.** 6 **9.** $\frac{2}{27}$ **11.** $a^3/6$ **13.** $\pi/2$ **15.** $\frac{5}{6}$ **17.** $\frac{1}{2}(\ln 3)^2 - \frac{1}{2}(\ln 2)^2$ **19.** $\frac{36}{5}$

21. $\frac{423}{28}$ **23.** $\frac{1}{12}$ **25.** 72 **27.** $\frac{9}{2}$ **29.** 2 **31.** $\frac{3}{2} - 2\ln 2$ **33.** $\frac{2}{3}$ **35.** $\frac{8}{3}$ **37.** 4π **39.** 8 **41.** $\frac{56}{3}$

43. $\displaystyle\int_0^1 \left[\int_y^1 f(x, y)\,dx\right] dy$ **45.** $\displaystyle\int_0^a \left[\int_0^{\sqrt{a^2 - x^2}} f(x, y)\,dy\right] dx$

47. $\displaystyle\int_{-1}^0 \left[\int_{1-\sqrt{x+1}}^{1+\sqrt{x+1}} f(x, y)\,dy\right] dx + \int_0^8 \left[\int_{x/2}^{1+\sqrt{x+1}} f(x, y)\,dy\right] dx$ **49.** $\displaystyle\int_0^1 \left[\int_{y/3}^{y} xy\,dx\right] dy + \int_1^3 \left[\int_{y/3}^{1} xy\,dx\right] dy = 1$

51. $\displaystyle\int_0^1 \left[\int_{1-y}^1 (x + y)\,dx\right] dy + \int_1^2 \left[\int_0^{2-y} (x + y)\,dx\right] dy = \frac{3}{2}$ **53.** $\displaystyle\int_0^1 \left[\int_0^{1/2y} e^{y^2}\,dx\right] dy = \frac{1}{4}(e - 1)$

55. $\displaystyle\int_0^1 \left[\int_0^{\sqrt{1-y^2}} \frac{1}{\sqrt{1 - y^2}}\,dx\right] dy$; 1 **57.** $\displaystyle\int_0^1 \left[\int_0^x \sqrt{2 + x^2}\,dy\right] dx$; $\frac{1}{3}(3\sqrt{3} - 2\sqrt{2})$ **59.** $\displaystyle\int_0^1 \left[\int_0^{x^2} e^{y/x}\,dy\right] dx$; $\frac{1}{2}$ **61.** 4

63. 2 **65.** $\frac{1}{2}(e - 2)$ **67.** Inside the cylinder, $z^2 = 1 - x^2$, but above the triangular region, $0 \le x \le 1, 0 \le y \le x$

69. A solid of height 3, with base between $y = \sin x$, $0 \le x \le \pi$ **71.** (a) 0 (b) $\frac{4}{3}$

Exercise 18.3

1. $\pi/16 + 1/8$ **3.** $4\sqrt{2}/3$ **5.** Integral = 6 **7.** Integral = $\frac{2}{3}$

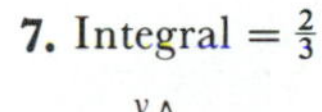

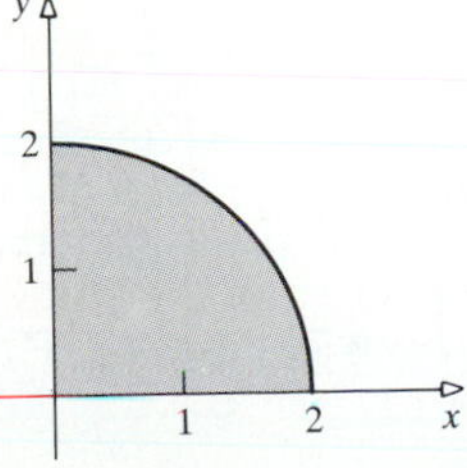

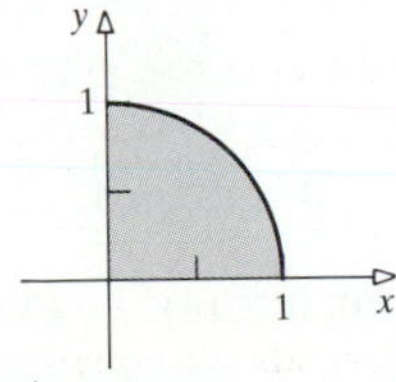

9. Integral = 0

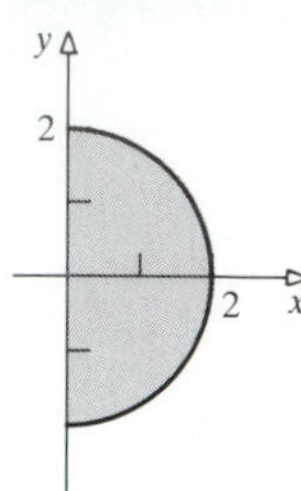

11. $\pi/4$ **13.** 12π **15.** $(\pi/4)\sin 1$ **17.** $(\pi/4)(e^{-1} - e^{-4})$ **19.** $\pi/8$

21. 21π **23.** $64\pi/3$ **25.** $-\frac{1}{2}$ **27.** $1 - \ln 2$ **29.** $(2a^3/9)(3\pi - 4)$ **31.** $16\pi/3$ **33.** $\pi/12$ **35.** $15\pi/4$

37. $(5\pi/6) + (7\sqrt{3}/8)$ **39.** $\int_0^{1/\sqrt{2}} \left[\int_{(1/2)-\sqrt{(1/4)-x^2}}^{x} dy\right] dx$ **41.** $\int_0^{\pi/4} \left[\int_0^{\sin\theta} f(r, \theta)\, dr\right] d\theta + \int_{\pi/4}^{\pi/2} \left[\int_0^{\cos\theta} f(r, \theta)\, dr\right] d\theta$

Exercise 18.4

1. $64; (\frac{3}{2}, \frac{8}{3})$ **3.** $\frac{31}{20}; (\frac{150}{217}, \frac{44}{93})$ **5.** $k/15; (\frac{15}{28}, \frac{5}{8})$ **7.** $5k; (\frac{6}{5}, \frac{7}{10})$ **9.** $5\pi k/3; (0, \frac{21}{20})$

11. $(2ak/3)(3\sqrt{3} - \pi); (0, 3a\sqrt{3}/2(3\sqrt{3} - \pi))$ **13.** 2ρ **15.** $\rho ab^3/3$ **17.** $8\rho/15$ **19.** $\rho ab^3\pi/4$

21. $2k\pi \ln(b/a)$ **23.** $ak(\pi - 2); (a\pi/2(\pi - 2), 0)$

25. If the rectangle is enclosed by $x = 0$, $x = a$, $y = 0$, $y = b$, then $\bar{x} = \dfrac{\rho ba^2/2}{\rho ba} = \dfrac{a}{2}$; $\bar{y} = \dfrac{\rho b^2 a/2}{\rho ba} = \dfrac{b}{2}$.

27. $(3.58, 0)$ **29.** If the triangle has vertices at $(0, 0)$, $(b, 0)$, $(0, h)$, then $m = \frac{1}{2}\rho bh$; $I_x = \rho h^3 b/12 = \frac{1}{6}mh^2$.

31. $(\bar{x}, \bar{y}) = (-\frac{574}{235}, -\frac{469}{470}); \frac{423}{28}$

Exercise 18.5

1. $(260 - 108\sqrt{3})/15$ **3.** $(\pi/6)(17\sqrt{17} - 1)$ **5.** $\pi a^2/3$ **7.** $(\pi/6)[(a^2 + 1)^{3/2} - 1]$ **9.** 16 **11.** $4\pi a^2$

13. $\frac{1}{12}(17\sqrt{17} - 1)$ **15.** $(\pi/6)(37\sqrt{37} - 5\sqrt{5})$ **17.** $\cos\gamma = c/\sqrt{A^2 + B^2 + C^2}$ (γ is a direction angle.)

Exercise 18.6

1. $\frac{2}{3}$ **3.** 1 **5.** $e - 1$ **7.** $-\frac{2}{3}$ **9.** 72 **11.** $\frac{5}{36}$ **13.** $\frac{1}{2240}$ **15.** 0

17. (*1*) $\int_0^6 \left\{\int_0^{3-(x/2)} \left[\int_0^{2-(x/3)-(2/3)y} f(x, y, z)\, dz\right] dy\right\} dx$; (*2*) $\int_0^3 \left\{\int_0^{6-2y} \left[\int_0^{2-(x/3)-(2/3)y} f(x, y, z)\, dz\right] dx\right\} dy$;

(*3*) $\int_0^6 \left\{\int_0^{2-(1/3)x} \left[\int_0^{3-(x/2)-(3/2)z} f(x, y, z)\, dy\right] dz\right\} dx$; (*4*) $\int_0^2 \left\{\int_0^{6-3z} \left[\int_0^{3-(x/2)-(3/2)z} f(x, y, z)\, dy\right] dx\right\} dz$;

(*5*) $\int_0^3 \left\{\int_0^{2-(2/3)y} \left[\int_0^{6-3z-2y} f(x, y, z)\, dx\right] dz\right\} dy$; (*6*) $\int_0^2 \left\{\int_0^{3-(3/2)z} \left[\int_0^{6-3z-2y} f(x, y, z)\, dx\right] dy\right\} dz$

19. $\frac{1}{8}$ **21.** $\frac{648}{5}$ **23.** $\frac{1184}{15}$ **25.** $\frac{1}{60}$ **27.** $81\pi/32$ **29.** $k\pi a^4 h/2$ **31.** $k/720$

33. $\int_{-\sqrt{3}}^{\sqrt{3}} \left\{\int_{-\sqrt{3-x^2}}^{\sqrt{3-x^2}} \left[\int_{\sqrt{3x^2+3y^2}}^{3} x^2yz\, dz\right] dy\right\} dx$

Exercise 18.7

1. $(-2, \pi/6, -5)$ or $(2, 7\pi/6, -5)$ **3.** $(\sqrt{2}, \pi/4, \sqrt{2})$ **5.** $(\sqrt{3}, 1, -5)$ **7.** $(2\sqrt{3}, 2, 2)$ **9.** $9\sqrt{3}$

11. $4\pi\rho a^3/3$ **13.** $(0, 0, \frac{4}{3})$ **15.** $(0, 0, 2.16)$ **17.** $\int_{-4}^{4} \left\{\int_{-\sqrt{16-x^2}}^{+\sqrt{16-x^2}} \left[\int_{-\sqrt{x^2+y^2}}^{\sqrt{16-x^2-y^2}} z^2x^4\, dz\right] dy\right\} dx$

19. $(2\pi\rho/3)(2\sqrt{2} - 1)$

Exercise 18.8

1. $(4, 5\pi/4, \pi/6)$ **3.** $(2, \pi/4, \pi/4)$ **5.** $(3.74, 1.11, 0.64)$ **7.** $(6, \pi/2, \pi/3)$

9. $(-1, -\sqrt{3}, -2\sqrt{3}/3)$; $(2, 4\pi/3, -2\sqrt{3}/3)$; $(4\sqrt{3}/3, 4\pi/3, 2\pi/3)$ **11.** $\pi^2/64$ **13.** $\sqrt{2}\pi/3 - (\pi/3)\ln(\sqrt{2} + 1)$

15. $4\pi\delta a^3/3$ **17.** $(0, 0, 2a/5)$ **19.** $8\pi/3$ **21.** $k\delta\pi a$

23. $$\int_{-a}^{a}\left\{\int_{-\sqrt{a^2-x^2}}^{\sqrt{a^2-x^2}}\left[\int_{-\sqrt{a^2-x^2-y^2}}^{\sqrt{a^2-x^2-y^2}} f(x, y, z)\,dz\right]dy\right\}dx$$
$$= \int_0^{2\pi}\left\{\int_0^a\left[\int_{-\sqrt{a^2-r^2}}^{\sqrt{a^2-r^2}} f(r\cos\theta, r\sin\theta, z)r\,dz\right]dr\right\}d\theta$$
$$= \int_0^{2\pi}\left\{\int_0^{\pi}\left[\int_0^a f(\rho\sin\phi\cos\theta, \rho\sin\phi\sin\theta, \rho\cos\phi)\rho^2\sin\phi\,d\rho\right]d\phi\right\}d\theta$$

25. $\displaystyle\int_{-\sqrt{8}}^{\sqrt{8}}\left\{\int_0^{\sqrt{8-x^2}}\left[\int_{-\sqrt{16-x^2-y^2}}^{-\sqrt{x^2+y^2}} x(x^2+y^2+z^2)\,dz\right]dy\right\}dx$ **27.** $\displaystyle\int_0^{\pi}\left\{\int_0^{\pi/2}\left[\int_0^3 \rho^3\sin\theta\sin^2\phi\,d\rho\right]d\phi\right\}d\theta$ **29.** 8π

31. $\bar{z} = 9/[8(2-\sqrt{2})]$ or $\frac{9}{16}(2+\sqrt{2})$ **33.** $k\pi(a^2 - b^2)/4$

Exercise 18.9

1. -2 **3.** 1 **5.** $(u+v)/v^2$ **7.** $(-uv-1)e^{u-v}$ **9.** -4 **11.** $\displaystyle\frac{1}{9}\int_0^4\int_0^3 (u+v)\,dv\,du = \frac{14}{3}$

13. $\displaystyle -\frac{1}{2}\int_0^2\int_{u-2}^0 \frac{u^2-v^2}{4}\,dv\,du = 0$ **15.** $\displaystyle\frac{1}{2}\int_0^2\int_{u-2}^0 u\sin v\,dv\,du = \left(\frac{1}{2} + \frac{1}{2}\cos 2\right)$ **17.** $\displaystyle 12\int_{-1}^1\int_{-\sqrt{1-u^2}}^{\sqrt{1-u^2}} 12uv\,dv\,du = 0$

19. $\displaystyle 12\int_{-1}^1\int_{-\sqrt{1-u^2}}^{\sqrt{1-u^2}}\int_{-\sqrt{1-u^2-v^2}}^{\sqrt{1-u^2-v^2}} 12uvw\,dw\,dv\,du = 0$ **21.** $\frac{423}{2401}$ **23.** $2(\sin 3 - \sin 1)$

Review Exercises

1. $(e-1)/6e$ **3.** $\sqrt{2} - 1$ **5.** (a) 0 (b) 32, -32, 0 (c) 0 **7.** $\displaystyle\int_0^1\left[\int_{e^y}^e \frac{dx}{x}\right]dy = \frac{1}{2}$ **9.** $abc/6$ **11.** $\frac{8}{15}abc$

13. $\displaystyle A = 4\int_0^r\int_0^{\sqrt{r^2-x^2}} dy\,dx = \pi r^2$ **15.** $\displaystyle A = 2\int_0^{\pi/2}\int_0^{2\cos\theta} r\,dr\,d\theta = \pi$

17. (a) $2\pi k\ln(b/a)$ (b) $\pi k(b^2 - a^2)$ approaches πkb^2 **19.** $(a/2, 2b/5)$

21. $\displaystyle S = \int_0^{2\pi}\int_0^a \sqrt{1 + (h/a)^2}\,dA = \pi a\sqrt{a^2 + h^2}$ **23.** $(\pi/6)(5\sqrt{5} - 1)$

25. A wedge-shaped figure with base, one vertical face and the slanted top all planar, and the remaining vertical face parabolic; $\displaystyle\int_0^1\int_0^{\sqrt{x}}\int_0^{1-x} dz\,dy\,dx = \int_0^1\int_0^{1-z}\int_0^{\sqrt{x}} dy\,dx\,dz = \int_0^1\int_0^{1-y^2}\int_{y^2}^1 dx\,dz\,dy =$ $\displaystyle\int_0^1\int_0^{1-x}\int_0^{\sqrt{x}} dy\,dz\,dx = \int_0^1\int_0^{\sqrt{1-z}}\int_{y^2}^1 dx\,dy\,dz$

27. $\displaystyle\int_a^b\left[\int_a^y f(x)\,dx\right]dy = \int_a^b\left[\int_x^b f(x)\,dy\right]dx = \int_a^b (b-x)f(x)\,dx$ **29.** $(\pi/6)abc$

31. $\displaystyle V = 8\int_0^{\pi/2}\int_0^{\pi/2}\int_1^2 \rho^2\sin\phi\,d\rho\,d\phi\,d\theta = \frac{28}{3}\pi$ **33.** $\displaystyle V = \int_0^{\pi/2}\int_0^2\int_0^{\sqrt{4-r^2}} \frac{z}{r}\,r\,dz\,dr\,d\theta = \frac{4\pi}{3}$

35. $\displaystyle V = \int_0^{\pi/2}\int_0^1\int_0^{1-(r\cos\theta)/2-(r\sin\theta)/3} r\,dz\,dr\,d\theta = \frac{\pi}{4} - \frac{5}{18}$ **37.** $\displaystyle\int_0^{\pi/2}\int_0^{\pi/2}\int_0^a \rho^2\sin\phi\,d\rho\,d\theta\,d\phi = \frac{\pi}{6}a^3$

39. $\iint_R (x+y)^3\,dx\,dy = \frac{1}{3}\int_a^b \int_c^d u^3\,du\,dv = \frac{1}{12}(d-c)(b^4-a^4)$.

41. $(8\pi/3)(16-6\sqrt{3})$ **43.** 16π **45.** $\dfrac{\pi h a^2}{12}$

Challenge Exercises

1. (a) $M(P) = (1/n)\Sigma f(u_i, v_i) = (1/n\Delta A)\Sigma f(u_i, v_i)\Delta A = (1/A)\Sigma f(u_i, v_i)\Delta A$
(b) Use the definition for double integral and divide by A.

3. Average value $= (1/\text{Area})\iint_R x\,dA = (1/m)(\bar{x}m) = \bar{x}$. Average value of $g(x, y) = y$ is $\bar{y}$.

5. (a) $I_0 = \iint_R (x^2+y^2)\rho(x, y)\,dA = \iint_R x^2\rho(x, y)\,dA + \iint_R y^2\rho(x, y)\,dA = I_x + I_y$

7. $S = 8\int_0^r \int_0^{\sqrt{r^2-x^2}} \sqrt{1+\frac{x^2}{z^2}+\frac{y^2}{z^2}}\,dy\,dx = 4\pi r^2$ **9.** $(2\pi/3)(5\sqrt{5}-4)$

CHAPTER 19

Exercise 19.1

1. All vectors are outwardly normal to circles centered at the origin, and their magnitudes increase directly with the radii of the circles.

3. All vectors on a given vertical line are parallel; as they move to the right, they become steeper and steeper with positive slope, and to the left they become steeper with negative slope.

5. At each point the vector is of magnitude 1 and is parallel to the x-axis.

7. At each point the vector is normal to the xy-plane and has magnitude equal to the distance of the point from the xy-plane.

9. All vectors are unit vectors outwardly normal to spheres centered at the origin.

Exercise 19.2

1. $5\sqrt{2}/6$ **3.** $\sqrt{2}/4$ **5.** (a) $\frac{1}{3}-(\pi/16)$ (b) $\frac{1}{3}-(\pi/16)$ **7.** (a) 3 (b) 3 (c) 0 (d) 0 **9.** (a) 5 (b) $\frac{123}{20}$
11. $\frac{1}{2}(1-\sin 1-\cos 1)$ **13.** C_1 is the same as C_2. **15.** $19+e^3-e^2+\cos 4-\cos 9$ **17.** 0 **19.** $-\frac{3}{2}$
21. $\frac{1}{12}(3e^4+6e^{-2}-12e+8e^3-5)$ **23.** 0 if n is even; $2\pi ab\left[\left(\frac{1}{2}\right)\left(\frac{3}{4}\right)\cdot\cdots\cdot\left(\frac{n}{n+1}\right)\right](b^{n-1}-a^{n-1})$ if n is odd
25. $\frac{a^2}{3}[(4\pi^2+1)^{3/2}-1]$ **27.** 3 **29.** $\frac{3}{2}$ **31.** π **33.** 1 **35.** $\frac{1}{3}(17\sqrt{17}-5\sqrt{5})$
37. $\int_C \rho(x, y, z)\sqrt{(dx/dt)^2+(dy/dt)^2+(dz/dt)^2}\,dt$ **39.** $20\pi+40\pi^2$ **41.** $\int_0^{2\pi} h\sqrt{(-R\sin t)^2+(R\cos t)^2}\,dt = 2\pi Rh$

Exercise 19.3

1. Yes **3.** Yes **5.** (a) $\partial P/\partial y = \partial Q/\partial x = 0$ (b) $f(x, y) = (x^2/2)+(y^2/2)$ (c) $\frac{19}{2}$
7. (a) $\partial P/\partial y = \partial Q/\partial x = 3$ (b) $f(x, y) = (x^3/3)+3xy$ (c) $-\frac{181}{3}$

9. (a) $\partial P/\partial y = \partial Q/\partial x = 60xy^2 - 12y^3$ (b) $f(x, y) = x^4 + 10x^2y^3 - 3xy^4 + y^5$ (c) 9 **11.** x^3y^2
13. $(x^2/2) + 3xy$ **15.** $x^2y - xy^2$ **17.** $(x^3/3) - (x^2/2) + xy^2 - e^y(y-1)$ **19.** $y \sin x - 2x \sin y$
21. $x^2 + y \sin x$ **23.** $f(x, y) = \tan^{-1}(y/x)$ and $(0, 0)$ is not in R. **25.** $k/\sqrt{x^2+y^2} + c$

Exercise 19.4

1. 1 **3.** $3\pi - \frac{1}{2}$ **5.** $-\frac{19}{15}$ **7.** $\frac{2}{3}$ **9.** $(3\pi ab/4)(a^2 - b^2)$ **11.** One gives 0; the other gives -1.
13. $\dfrac{-5(2^{15})}{3}k = -54{,}613.3k$ **15.** $1 - \dfrac{1}{\sqrt{5}}$ **17.** $\mathbf{F} \cdot \mathbf{r}' = 0$

Exercise 19.5

1. -12 **3.** -12 **5.** 0 **7.** $\pi/2$ **9.** $3\pi/4$ **11.** 0 **13.** $\frac{6}{35}$ **15.** $\frac{32}{9}$ **17.** $\frac{9}{2}$ **19.** 3π
21. Integrand $= \frac{1}{2}\nabla e^{x^2+y^2}$ or $\partial P/\partial y = \partial Q/\partial x$ **23.** $\dfrac{\partial}{\partial y}(e^x \sin y) = \dfrac{\partial}{\partial x}(e^x \cos y)$ **25.** -2π **27.** -2π

29. $$\iint_R \frac{\partial Q}{\partial x}\,dx\,dy = \int_c^d \left[\int_{g_1(y)}^{g_2(y)} \frac{\partial Q}{\partial x}\,dx\right] dy = \int_c^d Q(g_2(y), y)\,dy - \int_c^d Q(g_1(y), y)\,dy = \oint_C Q(x, y)\,dy$$

Exercise 19.6

1. $\frac{2}{3}\sqrt{14}$ **3.** $(51\sqrt{102} - \sqrt{2})/75$ **5.** $8\sqrt{3}/3$ **7.** $\sqrt{6}/4$ **9.** $\pi/2$ **11.** $\pi/4$ **13.** ρ **15.** 0 **17.** 3ρ
19. 3ρ **21.** 2π **23.** $\frac{160}{3}$ **25.** $(176\pi/3)\sqrt{2}$ **27.** $\frac{25}{2}\rho\pi$ **29.** Apply (2). **31.** $2k\pi a^3$

Exercise 19.7

1. $2x + 2y + 2z$ **3.** 4 **5.** $\frac{128}{3}$ **7.** 3 **9.** 4π **11.** 80π **13.** $\frac{2}{3}$
15. $\iint_\Sigma \mathbf{F} \cdot \mathbf{n}\,dS = 2\iiint_T (x + y + z)\,dV$; use cylindrical coordinates. **17.** $\iint_\Sigma \mathbf{F} \cdot \mathbf{n}\,dS = 3\iiint_T dV = 3V$
19. Decompose into top and sides, then use sec. 19.6, #29.
21. div $\mathbf{F}$ = div $\mathbf{G}$ = 14; use the divergence theorem. **23.** $\iiint_T \nabla \cdot \mathbf{F}\,dV = 4000\pi = \iint_\Sigma \mathbf{F} \cdot \mathbf{n}\,dS$
25. div $\mathbf{F} = 0$ **27.** div $\mathbf{F} = \dfrac{1}{\rho} - \dfrac{x^2}{\rho^3} + \dfrac{1}{\rho} - \dfrac{y^2}{\rho^3} + \dfrac{1}{\rho} - \dfrac{z^2}{\rho^3} = \dfrac{2}{\rho}$

Exercise 19.8

1. $\mathbf{0}$ **3.** $-x\mathbf{i} + xy\mathbf{j} + (1 - x)z\mathbf{k}$ **5.** $(-y^2 \cos z)\mathbf{i} + (6xyz - e^{2z})\mathbf{j} - 3xz^2\mathbf{k}$ **7.** $\mathbf{0}$ **9.** $-ze^{xz}\mathbf{j}$
11. $-(\mathbf{i} + \mathbf{j} + \mathbf{k})$ **13.** div(curl $\mathbf{F}$) $= R_{yx} - Q_{zx} - R_{xy} + P_{zy} + Q_{xz} - P_{yz} = 0$ **15.** (a), (b), (e)
17. Both equal -2π. **19.** Both equal 2π. **21.** 0 **23.** 4π **25.** $\nabla \times \mathbf{F} = \mathbf{0}$
27. Curl $\mathbf{F} = \mathbf{0}$; $(a^2/2) + ab - bc + (c^2/2)$ **29.** π; Hint: $\mathbf{n} = (\mathbf{i} + \mathbf{j} + \mathbf{k})\sqrt{3}$

Review Exercises

1. $\frac{5}{2}$ **3.** sin 1 **5.** -2 **7.** $e \sin 1$ **9.** $\frac{5}{63}$
11. $\nabla(fg) = (\nabla f)g + f(\nabla g)$, so $\int_c \nabla(fg)\,d\mathbf{r} = f(\mathbf{B})g(\mathbf{B}) - f(\mathbf{A})g(\mathbf{A}) = \int_c (\nabla f)g\,d\mathbf{r} + \int_c f(\nabla g)\,d\mathbf{r}$
13. π **15.** -2 **17.** 12π **19.** Any closed curve not containing (0, 0) in its interior
21. $\pi/2$ **23.** π **25.** 2π **27.** Area of Σ **29.** 4π **31.** $\nabla(\nabla f) = f_{xx} + f_{yy} + f_{zz} = 0$
33. Suppose $\mathbf{F} = P\mathbf{i} + Q\mathbf{j} + R\mathbf{k}$ is such a vector function; $R_y - Q_z = x$, $R_x - P_z = -y$, $Q_x - P_y = z \Rightarrow R_{xy} = Q_{xz} + 1$, $R_{yx} = P_{yz} - 1$, $Q_{xz} = P_{yz} + 1$; this requires that $R_{xy} = Q_{xz} + 1$ and $R_{xy} = Q_{xz} - 2$.
35. Show that curl $\mathbf{F} = \mathbf{0}$. **37.** $\frac{3}{2}$ **41.** Show that curl $\mathbf{F} = \nabla f \times \nabla g$.

Challenge Exercises

1. $\bar{x} = \dfrac{\iint_R x\,dA}{\iint_R dA} = \dfrac{1}{A}\iint_R \left(\dfrac{\partial Q}{\partial x} - \dfrac{\partial P}{\partial y}\right) dA$ (where $Q = x^2/2$, $P = 0$)

$= \dfrac{1}{2A}\oint_C x^2\,dy$

3. $\bar{x} = \bar{y} = 256a/315\pi$ **5.** $(0, 0, \frac{2}{3}a)$ **7.** Integral $= \int_a^b f(x(s), y(s))\,ds$

9. $\iint_R dx\,dy = \frac{1}{2}\iint_{R'}[(xy_t - yx_t)_s - (xy_s - yx_s)_t]\,ds\,dt$

$= \iint_{R'}(x_s y_t - x_t y_s)\,ds\,dt$

11. Note that $\dfrac{\partial Q}{\partial x} - \dfrac{\partial P}{\partial y} = 2\phi(x, y)$, so $\iint_R \phi(x, y)\,dx\,dy = \frac{1}{2}\int_c P\,dx + Q\,dy = \frac{1}{2}\int_{c'}(Px_s + Qy_s)\,ds + (Px_t + Qy_t)\,dt =$ $\frac{1}{2}\iint_{R'}\left[\dfrac{\partial}{\partial s}(Px_t + Qy_t) - \dfrac{\partial}{\partial t}(Px_s + Qy_s)\right] ds\,dt$. Expand the partials and use $\dfrac{\partial Q}{\partial x} = \dfrac{-\partial P}{\partial y} = \phi(x, y)$ to simplify.

CHAPTER 20

Exercise 20.1

1. First order; degree 1; linear; nonhomogeneous **3.** Fourth order; degree 1; linear

5. Second order; degree 1; linear; homogeneous **7.** Second order; degree 1; nonlinear

9. Second order; degree 3; nonlinear **11.** $\dfrac{d^2y}{dx^2} = e^x + 3e^{-x}$; $(e^x + 3e^{-x}) - (e^x + 3e^{-x}) = 0$

13. $\dfrac{d^2y}{dx^2} = -4\sin 2x$; $(-4\sin 2x) + 4(\sin 2x) = 0$ **15.** $\dfrac{dy}{dx} = \dfrac{9x^2}{(1 - x^3)^2}$; $\dfrac{9x^2}{(1 - x^3)^2} = x^2 \cdot \dfrac{9}{(1 - x^3)^2}$

17. $\dfrac{d^2y}{dx^2} = a^2C_1e^{ax} + a^2C_2e^{-ax}$; $(a^2C_1e^{ax} + a^2C_2e^{-ax}) - a^2(C_1e^{ax} + C_2e^{-ax}) = 0$

19. $\dfrac{dy}{dt} = \dfrac{a^2k}{(1 + akt)^2}$; $\dfrac{a^2k}{(1 + akt)^2} = k\left(a - \dfrac{a^2kt}{1 + akt}\right)^2 = k\left(\dfrac{a}{1 + akt}\right)^2$ **21.** $2yy' + 2xy' + 2y - 2x = 0$, so $y' = \dfrac{x - y}{x + y}$

23. $C = \dfrac{x^2 - y^2}{x}$ implies $\dfrac{x^2 - 2xyy' + y^2}{x^2} = 0$ and $y' = \dfrac{x^2 + y^2}{2xy}$ **25.** $\dfrac{dy}{dx} = 3 = \dfrac{3x}{x}$; $3 = 3 \cdot 1$

27. $\dfrac{dy}{dx} = \dfrac{1}{(1 - x)^2} = \left[\dfrac{1}{1 - x}\right]^2 = y^2$; $\dfrac{1}{1 - 0} = 1$

29. (a) $A = C_1\cos C_2$; $B = C_1\sin C_2$ (b) $A = -C_1\sin C_2$; $B = C_1\cos C_2$ **31.** $n = -3$ or 2

Exercise 20.2

1. $y = (\frac{1}{4}x^2 + C)^2$ **3.** $2\cot y = C - \ln\left|\dfrac{1 + x}{1 - x}\right|$ **5.** $y = -\ln|e^{-x} + C|$ **7.** $y = Cx^{-2} + \frac{5}{2}$

9. $y = C/(\cos x - 1)$ **11.** $y = 1 + Ce^{(-1/2)x^2}$ **13.** $y = Ce^{(1/2)(\ln x)^2}$ **15.** $y = \frac{1}{16}x^4$ or $y = (\frac{1}{4}x^2 - 2)^2$

17. $2\cot y = 2 + \ln 3 - \ln\left|\dfrac{1 + x}{1 - x}\right|$ **19.** $y = x$ **21.** $y = \frac{-3}{2}x^{-2} + \frac{5}{2}$ **23.** $y = 1/(1 - \cos x)$

25. Yes, degree 2 **27.** No **29.** Yes, degree 1 **31.** Yes, degree 0 **33.** Yes, degree 0 **35.** $y = Cx - x\ln|x|$

37. $x\ln x + Cx - y + 2x\ln\left|\dfrac{y}{x} + 1\right| = 0$ **39.** $\dfrac{1}{2}\ln(x^2 + y^2) + \tan^{-1}\left(\dfrac{y}{x}\right) = C$ **41.** $C = \ln(x^2y^2) + \dfrac{x^2}{y^2}$

43. $y^2 - x^2 = Cy$ **45.** $\dfrac{(y + \frac{3}{2})^2}{9/4} - \dfrac{x^2}{9/4} = 1$ **47.** $\ln 4 + \dfrac{1}{4} = \ln(x^2y^2) + \dfrac{x^2}{y^2}$

49. (a) $x = \dfrac{30}{14e^{-t\ln(14/9)} + 1}$ (b) $t = \dfrac{-\ln(\frac{1}{14})}{\ln(\frac{14}{9})} \approx 6$ days

Exercise 20.3

1. $(\partial/\partial y)(4x - 2y + 5) = -2 = (\partial/\partial x)(2y - 2x)$; $2x^2 - 2xy + 5x + y^2 = C$

3. $(\partial/\partial y)(a^2 - 2xy - y^2) = -2x - 2y = (\partial/\partial x)(-(x + y)^2)$; $a^2x - x^2y - xy^2 - \frac{1}{3}y^3 = C$

5. $\dfrac{\partial}{\partial y}\left(\dfrac{1}{y}\right) = \dfrac{-1}{y^2} = \dfrac{\partial}{\partial x}\left(\dfrac{-x}{y^2}\right)$; $y = Cx$

7. $\dfrac{\partial}{\partial y}[(x - 1)^{-1}y] = (x - 1)^{-1} = \dfrac{\partial}{\partial x}[\ln(2x - 2) + y^{-1}]$; $y\ln(2x - 2) + \ln|y| = C$

9. $\dfrac{\partial}{\partial y}[(x + 3)^{-1}\cos y] = \dfrac{-\sin y}{x + 3} = \dfrac{\partial}{\partial x}[-\sin y \ln(5x + 15) + y + 1]$; $\cos y \ln(5x + 15) + \frac{1}{2}y^2 + y = C$

11. $(\partial/\partial y)\cos(x + y^2) = -2y\sin(x + y^2) = (\partial/\partial x)[2y\cos(x + y^2)]$; $\sin(x + y^2) = C$

13. $(\partial/\partial x)e^{2x} = 2e^{2x} = (\partial/\partial y)(2ye^{2x} - x^2)$; $ye^{2x} - \frac{1}{3}x^3 = C$

15. $\dfrac{\partial}{\partial y}\left[\dfrac{1}{x + y} + y^2\right] = \dfrac{-1}{(x + y)^2} + 2y = \dfrac{\partial}{\partial x}\left[\dfrac{1}{x + y} + 2xy\right]$; $\ln|x + y| + xy^2 = C$

17. $(\partial/\partial y)(2y^3\sin 2x) = 6y^2\sin 2x = (\partial/\partial x)(-3y^2\cos 2x)$; $-y^3\cos 2x = C$ **19.** $x + xy^2 + \frac{1}{2}x^2y^2 + \frac{1}{2}y^2 = 3$

21. $x^2y + \cos x - y^2 = 0$

23. An equation with variables separated is $f(x)\,dx = g(y)\,dy$ or $f(x)\,dx - g(y)\,dy = 0$. But $(\partial/\partial y)f(x) = (\partial/\partial x)(-g(x)) = 0$.

25. $\dfrac{\partial}{\partial y}M \cdot \exp\left[\displaystyle\int \dfrac{M_y - N_x}{N}\,dx\right] = M_y \cdot \exp\left[\displaystyle\int \dfrac{M_y - N_x}{N}\,dx\right]$; $\dfrac{\partial}{\partial x}N \cdot \exp\left[\displaystyle\int \dfrac{M_y - N_x}{N}\,dx\right] = N_x \cdot \exp\left[\displaystyle\int \dfrac{M_y - N_x}{N}\,dx\right] +$
$N \cdot \exp\left[\displaystyle\int \dfrac{M_y - N_x}{N}\,dx\right] \cdot \dfrac{M_y - N_x}{N} = (N_x + M_y - N_x)\exp\left[\displaystyle\int \dfrac{M_y - N_x}{N}\,dx\right] = M_y \cdot \exp\left[\displaystyle\int \dfrac{M_y - N_x}{N}\,dx\right]$

27. (a) $y^3\,dx + 3x^2y\,dy = 0$ (b) $x = Cy^{-3}$ **29.** (a) $4x^2y^{3/4}\,dx + (x^3y^{-1/4} + y^{3/4})\,dy = 0$ (b) $\frac{4}{3}x^3y^{3/4} + \frac{4}{7}y^{7/4} = C$

31. (a) $(x^{-2}\cos y + x^{-1})\,dx + x^{-1}\sin y\,dy = 0$ (b) $-x^{-1}\cos y + \ln|x| = C$

Exercise 20.4

1. $y = 2x\ln x + Cx$ **3.** $y = xe^x + Cx$ **5.** $y = x^2 + (C/x)$ **7.** $y = -\frac{1}{3}e^{-x} + Ce^{2x}$ **9.** $r = \frac{1}{6}\theta^2 + C\theta^{-4}$

11. $y = (3x + C)(x + 1)^2$ **13.** $y = \sin x\cos x + C\cos x$ **15.** $y = (x^2/3) - (2/3) + (C/\sqrt{1 + x^2})$

17. $x^4(x^2 + 3y^2) = C$ **19.** $x = Cy - y^2$ **21.** $y = Ce^{x - x^2}$ **23.** $y = xe^{-x} + 5e^{-x}$ **25.** $y = e^x - (e^x/x) + (1/ex)$

27. $y = 2/(Cx - 3x^3)$ **29.** $x = 1/(Cy^2 - 2y^3)$ **31.** $x^2 = (e^{2y^2})/(C - 4y^3)$

Exercise 20.5

1. $\dfrac{x^2 + 2y^2}{y} = C$ **3.** $y = Ce^x$ **5.** $(x - 1)^2/2 + (y - 2)^2 = C$ (a family of ellipses)

7. After differentiating implicitly and eliminating C we get $y = 2xy' + y(y')^2$. Replace y' by $-1/y'$ and multiply by $(y')^2$. We get $y(y')^2 = -2xy' + y$.

9. (a) $s(t) = 20(1 - e^{(1/2\ln 17/20)t})$ (b) \$9.56 million (c) 11.3 years **11.** (a) 64 (b) 49 hours

13. (a) 86.53° (b) 175.2 seconds **15.** (a) $dy/dt = r - ky$ (b) $y = (r/k) - (r/k)e^{-kt}$

17. (a) 45 students (b) 26.55 days

19. $dy/dt = kmy - ky^2$, so $(d/dy)(dy/dt) = km - 2ky$, which is positive if $y < m/2$ and negative if $y > m/2$.

21. (a) $v(t) = -39.2 + 39.2e^{-t/4}$; $s(t) = 2156.8 - 39.2t - 156.8e^{-t/4}$ (b) -39.2 m/sec (c) -39.2 m/sec

23. (a) -276.7 ft/sec (b) 119 sec (c) -32 ft/sec **25.** $q = EC - ECe^{-t/RC}$ **27.** $I = \dfrac{E}{R} - \dfrac{E}{R}e^{(-R/L)t}$

29. $y(5) = 30 - (30/e) \approx 19$ lb; 30 lb

Exercise 20.6

1. $y = C_1e^{3x} + C_2e^x$ **3.** $y = C_1e^{-x} + C_2xe^{-x}$ **5.** $y = C_1e^{-2x} + C_2e^{1/2x}$ **7.** $y = e^{2x}[C_1 \cos 2x + C_2 \sin 2x]$

9. $y = e^{-x/2}[C_1 \cos(\sqrt{3}/2)x + C_2 \sin(\sqrt{3}/2)x]$ **11.** $y = e^{3x}$ **13.** $y = -2e^{3x-2} + 2xe^{3x-2}$

15. $y = C_1e^x + C_2e^{-x} + C_3e^{-2x}$ **17.** $y = C_1e^x + C_2 \sin 3x + C_3 \cos 3x$

19. $y = C_1e^{2x} + C_2xe^{2x} + C_3 \sin x + C_4 \cos x$ **21.** $y'' - 2y' - 8y = 0$ **23.** $y'' - y' = 0$ **25.** $y'' - 4y' + 13y = 0$

27. $y'' = 0$ **29.** $y'' - 2y' + 2y = 0$ **31.** $y = C_1x^{-2} + C_2x^{-1}$ **33.** $y = C_1x^2 + C_2x^3$

35. $y = x^{-2}(C_1 \sin(3 \ln x) + C_2 \cos(3 \ln x))$

37. (a) $y_1'(x) = ky_2'(x)$ and $W(x) = ky_2(x)y_2'(x) - y_2(x)ky_2'(x) = 0$
(b) Show that W satisfies the first order DE $dW/dx = -p \cdot W$ and has the form $W = Ce^{-\int p_1 dx}$ which either never vanishes on I or is identically 0.

39. (a) $\dfrac{r^2ke^{rx}}{P(r)} + \dfrac{brke^{rx}}{P(r)} + \dfrac{cke^{rx}}{P(r)} = \dfrac{ke^{rx}}{P(r)}(r^2 + br + c) = \dfrac{ke^{rx}}{P(r)}P(r) = ke^{rx}$ (b) $\dfrac{kxe^{rx}}{P'(r)}P(r) + \dfrac{ke^{rx}}{P'(r)}P'(r) = 0 + ke^{rx}$

(c) $\dfrac{kx^2e^{rx}}{P''(r)}P(r) + \dfrac{kxe^{rx}}{P''(r)}2P'(r) + \dfrac{2ke^{rx}}{P''(r)} = 0 + 0 + \dfrac{2ke^{rx}}{2} = ke^{rx}$

41.

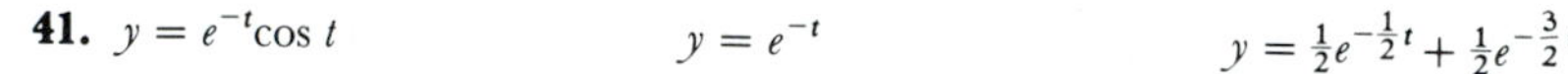

$y = e^{-t}\cos t$
$y = e^{-t}$
$y = \frac{1}{2}e^{-\frac{1}{2}t} + \frac{1}{2}e^{-\frac{3}{2}t}$

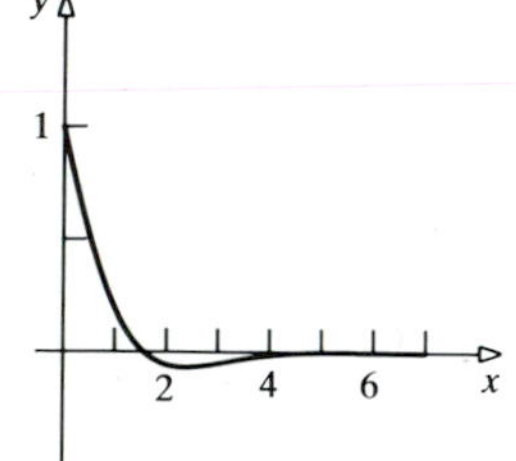

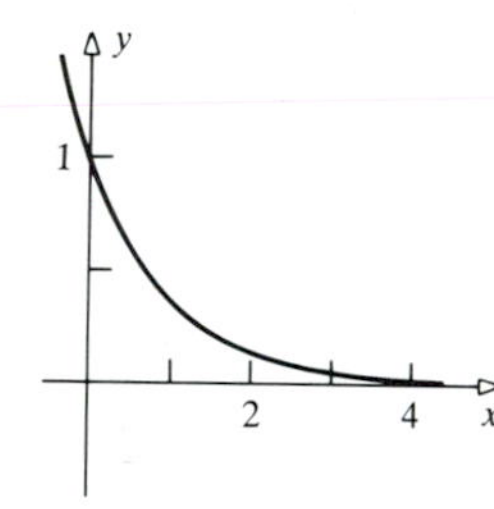

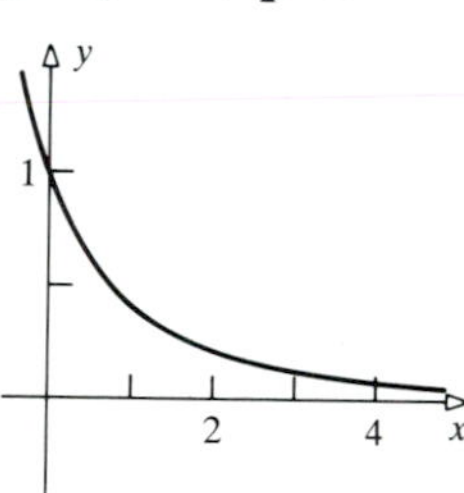

43. Overdamped: $y = 0$ means $e^{2rt} = -C_2/C$, which has at most one solution (since e^u is a one-to-one function). (One solution if $-C_2/C_1 > 0$; none otherwise.) Critically damped: $(C_1 + tC_2) = 0$; again there is at most one solution. (One solution if $C_2 \neq 0$; none if $C_2 = 0$.)

Exercise 20.7

1. $y = \frac{1}{2} \sin x \tan x + C_1 \cos x + C_2 \sin x$ **3.** $y = -(1 + \ln x)e^{-2x} + C_1e^{-2x} + C_2xe^{-2x}$

5. $y = \frac{1}{6}x^3e^x + C_1e^x + C_2xe^x$ **7.** $y = \frac{1}{10}e^x\cos x - \frac{3}{10}e^x\sin x + C_1e^{2x} + C_2e^{3x}$

9. $y = \frac{1}{10} \sin x + \frac{3}{10} \cos x + C_1e^x + C_2e^{2x}$ **11.** $y = \frac{1}{9}e^{-3x} + C_1e^{3x} + C_2xe^{3x}$ **13.** $y = \frac{1}{2} + C_1 \sin 2x + C_2 \cos 2x$

15. $y = \frac{1}{2}e^{4x} - \frac{1}{2}e^{2x} + C_1e^{3x} + C_2e^{-2x}$ **17.** $y = \frac{-3}{4} e^{-2x} + C_1e^{2x} + C_2e^{-x}$

19. $y = \frac{1}{4}x \sin 2x + C_1 \sin 2x + C_2 \cos 2x$ **21.** $y = xe^x + e^x(C_1 \cos x + C_2 \sin x)$

23. $y = \frac{-1}{2} x^2e^{2x} - xe^{2x} + C_1e^{2x} + C_2e^{3x}$ **25.** $y = \frac{1}{20}x \cos 2x - \frac{3}{20}x \sin 2x - \frac{19}{200} \cos 2x - \frac{1}{25} \sin 2x + C_1e^{2x} + C_2e^{-x}$

27. $y = \frac{1}{4}x^2\sin x + \frac{1}{4}x \cos x + C_1 \sin x + C_2 \cos x$ **29.** $y = (x^2 - 2) - \cos x \ln|\sec x + \tan x| + C_1 \cos x + C_2 \sin x$

31. $y = \frac{1}{2}e^x \ln(1 + e^{-x}) - \frac{1}{2}e^{3x}\ln(1 + e^{-x}) + \frac{1}{2}e^{2x} + C_1e^{3x} + C_2e^x$ **33.** $y = \frac{1}{6}x^3e^{2x} - (\ln x)e^{2x} + C_1e^{2x} + C_2xe^{2x}$

35. Using $y(t) = C_1 \cos kt + C_2 \sin kt + [r/(k^2 - h^2)]\cos ht$, show that $C_1 = -r/(k^2 - h^2)$ and $C_2 = 0$, then use the sum-to-product formulas from trigonometry.

Exercise 20.8

1. $y = a_0\left(1 - \frac{x^2}{2!} + \frac{x^4}{4!} - \frac{x^6}{6!} + \cdots + \frac{(-1)^n x^{2n}}{(2n)!} + \cdots\right) + a_1\left(x - \frac{x^3}{3!} + \frac{x^5}{5!} - \frac{x^7}{7!} + \cdots + \frac{(-1)^n x^{2n+1}}{(2n+1)!} + \cdots\right)$
or $y = a_0 \cos x + a_1 \sin x$

3. $$y = a_0\left[1 - \frac{1}{2 \cdot 3}x^3 + \frac{4}{2 \cdot 3 \cdot 5 \cdot 6}x^6 - \frac{4 \cdot 7}{2 \cdot 3 \cdot 5 \cdot 6 \cdot 8 \cdot 9}x^9 + \cdots\right] + a_1\left[x - \frac{2}{3 \cdot 4}x^4 + \frac{2 \cdot 5}{3 \cdot 4 \cdot 6 \cdot 7}x^7 - \frac{2 \cdot 5 \cdot 8}{3 \cdot 4 \cdot 6 \cdot 7 \cdot 9 \cdot 10}x^{10} + \cdots\right]$$

5. $y = a_0[1 - x^2] + a_1\left[x - \frac{1}{2 \cdot 3!}x^3 + \frac{1}{2^2 \cdot 5!}x^5 - \frac{3^2}{2^3 \cdot 7!}x^7 + \frac{5^2 \cdot 3^2}{2^4 \cdot 9!}x^9 - \cdots\right]$

7. $y = a_0\left[1 + \sum_{n=1}^{+\infty} \frac{(-3)^n x^{2n}}{2^n n!}\right] + a_1\left[x + \sum_{n=1}^{+\infty} \frac{(-3)^n x^{2n+1}}{3 \cdot 5 \cdot 7 \cdots (2n+1)}\right]$

9. $$y = a_0\left[1 + \frac{2}{4 \cdot 3 \cdot 2}x^4 + \frac{2^2}{8 \cdot 7 \cdot 6 \cdot 4 \cdot 3 \cdot 2}x^8 + \cdots\right] + a_1\left[x + \frac{2}{5 \cdot 4 \cdot 3}x^5 + \frac{2^2}{9 \cdot 8 \cdot 7 \cdot 5 \cdot 4 \cdot 3}x^9 + \cdots\right] + a_2\left[x^2 + \frac{2}{6 \cdot 5 \cdot 4}x^6 + \frac{2^2}{10 \cdot 9 \cdot 8 \cdot 6 \cdot 5 \cdot 4}x^{10} + \cdots\right]$$

11. $y = 1 - \frac{1}{2}x^2 + \frac{1}{8}x^4 - \frac{x^6}{48} + \cdots$ **13.** $y = 2 + x - \frac{3}{2}x^2 - \frac{1}{24}x^4 + \frac{1}{24}x^5 + \cdots$

15. $y = 1 - \frac{3}{2}x^2 - \frac{1}{6}x^3 + \frac{3}{8}x^4 + \frac{1}{10}x^5 + \cdots$ **17.** $y = 2 + x - x^2 + \frac{1}{6}x^3 - \frac{1}{4}x^4 + \frac{1}{24}x^5 + \cdots$

19. $y = 1 + x - \frac{1}{3}x^3 + \frac{1}{3}x^4 - \frac{7}{30}x^5 + \cdots$

Review Exercises

1. (a) separable (b) $(x - 1)(y^2 + 1)^{1/2} = C$ **3.** (a) linear, exact (b) $y = \frac{x^6}{7} + \frac{C}{x}$

5. (a) exact (b) $x^2y^3 + y^2\sin x - x^2 = C$ **7.** (a) separable, linear, exact (b) $y = C \sec x$

9. (a) linear (b) $y = -x - \frac{1}{2} + Cx^2$ **11.** $y = 3 + \frac{20}{x^2}$ **13.** $y = x - \frac{1}{x}$ **15.** $y = -\frac{1}{2}x - \frac{1}{2}$

17. $y = x^2\sin x + \left(\frac{12}{\pi^2} - 1\right)x^2$ **19.** $\sin x - x \cos x = \ln|y| + \frac{1}{y} + \pi - 1$ **21.** $y = C_1 \cos 3x + C_2 \sin 3x$

23. $y = C_1e^{-\frac{3}{2}x} \cos \frac{\sqrt{3}}{2}x + C_2e^{-\frac{3}{2}x} \sin \frac{\sqrt{3}}{2}x$ **25.** $y = C_1e^{3x} + C_2e^{-3x}$ **27.** $y = C_1 + C_2e^{(7/4)x}$

29. $y = C_1 \cos \frac{2}{\sqrt{3}}x + C_2 \sin \frac{2}{\sqrt{3}}x$ **31.** $y = 2e^{2x}\cos 3x - 3e^{2x}\sin 3x$ **33.** $y = 3 - e^{-x}$ **35.** $y = (2 - x)e^{2x-2}$

37. $y = \frac{-7}{50} e^x\sin x - \frac{1}{50}e^x\cos x + C_1e^{3x} + C_2e^{-2x}$ **39.** $y = x + \frac{1}{2}e^x + C_1 \sin x + C_2 \cos x$

41. $y = 3 - 2x \sin x + C_1 \sin x + C_2 \cos x$ **43.** $y = (e^x + e^{2x})\ln(1 + e^{-x}) + C_1e^{2x} + C_2e^x$

45. $y = \frac{1}{2} \sin x \tan x + C_1 \sin x + C_2 \cos x$ **47.** $y = a_0\left[1 + \frac{1}{2}x + \frac{1}{4}x^2 + \frac{1}{8}x^3 + \cdots\right] = a_0\left(\frac{2}{2 - x}\right)$

49. $r = r_0 - kt$ (k a positive constant) **51.** $y(t) = M + (y_0 - M)e^{-kt}$

Challenge Exercises

1. $\dfrac{du}{dt} = -ku$ means $u = u_0 e^{-kt}$, $l = u_0 - u = u_0(1 - e^{-kt})$. $r = \dfrac{l}{u} = \dfrac{u_0(1 - e^{-kt})}{u_0 e^{-kt}} = e^{kt} - 1$, so $t = \dfrac{1}{k}\ln(1 + r)$. Since $\frac{1}{2}u_0 = u_0 e^{-k(4.5 \times 10^9)}$, $k \approx 1.54 \times 10^{-10}$ and $\dfrac{r}{k} = \dfrac{0.054}{1.54 \times 10^{-10}} = 351$ million

3. $a = \dfrac{-gR^2}{x^2}$, since $a = \dfrac{k}{x^2}$ and $a = -g$ when $x = R$; $a = \dfrac{d^2x}{dt^2} = \dfrac{dv}{dt} = \dfrac{dv}{dx} \cdot \dfrac{dx}{dt} = \dfrac{dv}{dx} v$; $a = -\dfrac{gR^2}{x^2}$ and $v = v_0$ when $x = R$ yields $v^2 = v_0^2 - 2gR + \dfrac{2gR^2}{v}$.

5. Oscillatory motion occurs only with complex roots: $0 < B < 6$.

7. $h = (\sqrt{h_0} - kt)^2$; Time for three-fourths to run out $= \sqrt{h_0}/(2k)$; Time for rest $= (\sqrt{h_0}/k) - (\sqrt{h_0}/2k) = \sqrt{h_0}/2k$.

Index

59. $\int \frac{dx}{x^2\sqrt{x^2 \pm a^2}} = \mp \frac{\sqrt{x^2 \pm a^2}}{a^2 x} + C$

60. $\int (x^2 \pm a^2)^{3/2}\, dx = \frac{x}{8}(2x^2 \pm 5a^2)\sqrt{x^2 \pm a^2} + \frac{3a^4}{8} \ln\left|x + \sqrt{x^2 \pm a^2}\right| + C$

61. $\int \frac{dx}{(x^2 \pm a^2)^{3/2}} = \pm \frac{x}{a^2\sqrt{x^2 \pm a^2}} + C$

Forms Containing $\sqrt{a^2 - x^2}$

62. $\int \sqrt{a^2 - x^2}\, dx = \frac{x}{2}\sqrt{a^2 - x^2} + \frac{a^2}{2} \sin^{-1}\frac{x}{a} + C$

63. $\int x^2\sqrt{a^2 - x^2}\, dx = \frac{x}{8}(2x^2 - a^2)\sqrt{a^2 - x^2} + \frac{a^4}{8} \sin^{-1}\frac{x}{a} + C$

64. $\int \frac{\sqrt{a^2 - x^2}}{x}\, dx = \sqrt{a^2 - x^2} - a \ln\left|\frac{a + \sqrt{a^2 - x^2}}{x}\right| + C$

65. $\int \frac{\sqrt{a^2 - x^2}}{x^2}\, dx = -\frac{\sqrt{a^2 - x^2}}{x} - \sin^{-1}\frac{x}{a} + C$

66. $\int \frac{x^2}{\sqrt{a^2 - x^2}}\, dx = -\frac{x}{2}\sqrt{a^2 - x^2} + \frac{a^2}{2} \sin^{-1}\frac{x}{a} + C$

67. $\int \frac{dx}{x\sqrt{a^2 - x^2}} = -\frac{1}{a} \ln\left|\frac{a + \sqrt{a^2 - x^2}}{x}\right| + C$

68. $\int \frac{dx}{x^2\sqrt{a^2 - x^2}} = -\frac{\sqrt{a^2 - x^2}}{a^2 x} + C$

69. $\int (a^2 - x^2)^{3/2}\, dx = \frac{x}{4}(a^2 - x^2)^{3/2} + \frac{3a^2 x}{8}\sqrt{a^2 - x^2} + \frac{3a^4}{8} \sin^{-1}\frac{x}{a} + C$

70. $\int \frac{dx}{(a^2 - x^2)^{3/2}} = \frac{x}{a^2\sqrt{a^2 - x^2}} + C$

Forms Involving $\sqrt{2ax - x^2}$

71. $\int \sqrt{2ax - x^2}\, dx = \frac{x - a}{2}\sqrt{2ax - x^2} + \frac{a^2}{2} \cos^{-1}\left(\frac{a - x}{a}\right) + C$

72. $\int x\sqrt{2ax - x^2}\, dx = \frac{2x^2 - ax - 3a^2}{6}\sqrt{2ax - x^2} + \frac{a^3}{2} \cos^{-1}\left(\frac{a - x}{a}\right) + C$

73. $\int \frac{\sqrt{2ax - x^2}}{x}\, dx = \sqrt{2ax - x^2} + a \cos^{-1}\left(\frac{a - x}{a}\right) + C$

74. $\int \frac{\sqrt{2ax - x^2}}{x^2}\, dx = -\frac{2\sqrt{2ax - x^2}}{x} - \cos^{-1}\left(\frac{a - x}{a}\right) + C$

75. $\int \frac{dx}{\sqrt{2ax - x^2}} = \cos^{-1}\left(\frac{a - x}{a}\right) + C$

76. $\int \frac{x\, dx}{\sqrt{2ax - x^2}} = -\sqrt{2ax - x^2} + a \cos^{-1}\left(\frac{a - x}{a}\right) + C$

77. $\int \frac{x^2\, dx}{\sqrt{2ax - x^2}} = -\frac{(x + 3a)}{2}\sqrt{2ax - x^2} + \frac{3a^2}{2} \cos^{-1}\left(\frac{a - x}{a}\right) + C$

78. $\displaystyle\int \frac{dx}{x\sqrt{2ax-x^2}} = -\frac{\sqrt{2ax-x^2}}{ax} + C$

79. $\displaystyle\int \frac{\sqrt{2ax-x^2}}{x^n}\,dx = \frac{(2ax-x^2)^{3/2}}{(3-2n)ax^n} + \frac{n-3}{(2n-3)a}\int \frac{\sqrt{2ax-x^2}}{x^{n-1}}\,dx, \quad n \neq \frac{3}{2}$

80. $\displaystyle\int \frac{x^n\,dx}{\sqrt{2ax-x^2}} = -\frac{x^{n-1}\sqrt{2ax-x^2}}{n} + \frac{a(2n-1)}{n}\int \frac{x^{n-1}}{\sqrt{2ax-x^2}}\,dx$

81. $\displaystyle\int \frac{dx}{x^n\sqrt{2ax-x^2}} = \frac{\sqrt{2ax-x^2}}{a(1-2n)x^n} + \frac{n-1}{(2n-1)a}\int \frac{dx}{x^{n-1}\sqrt{2ax-x^2}}$

82. $\displaystyle\int \frac{dx}{(2ax-x^2)^{3/2}} = \frac{x-a}{a^2\sqrt{2ax-x^2}} + C$

83. $\displaystyle\int \frac{x\,dx}{(2ax-x^2)^{3/2}} = \frac{x}{a\sqrt{2ax-x^2}} + C$

Forms Containing Trigonometric Functions

84. $\displaystyle\int \sin^2 x\,dx = \frac{x}{2} - \frac{\sin 2x}{4} + C$

85. $\displaystyle\int \cos^2 x\,dx = \frac{x}{2} + \frac{\sin 2x}{4} + C$

86. $\displaystyle\int \tan^2 x\,dx = \tan x - x + C$

87. $\displaystyle\int \cot^2 x\,dx = -\cot x - x + C$

88. $\displaystyle\int \sec^3 x\,dx = \tfrac{1}{2}\sec x\tan x + \tfrac{1}{2}\ln|\sec x + \tan x| + C$

89. $\displaystyle\int \csc^3 x\,dx = -\tfrac{1}{2}\csc x\cot x + \tfrac{1}{2}\ln|\csc x - \cot x| + C$

90. $\displaystyle\int \sin^n x\,dx = -\frac{1}{n}\sin^{n-1}x\cos x + \frac{n-1}{n}\int \sin^{n-2}x\,dx$

91. $\displaystyle\int \cos^n x\,dx = \frac{1}{n}\cos^{n-1}x\sin x + \frac{n-1}{n}\int \cos^{n-2}x\,dx$

92. $\displaystyle\int \tan^n x\,dx = \frac{1}{n-1}\tan^{n-1}x - \int \tan^{n-2}x\,dx$

93. $\displaystyle\int \cot^n x\,dx = \frac{-1}{n-1}\cot^{n-1}x - \int \cot^{n-2}x\,dx$

94. $\displaystyle\int \sec^n x\,dx = \frac{1}{n-1}\tan x\sec^{n-2}x + \frac{n-2}{n-1}\int \sec^{n-2}x\,dx$

95. $\displaystyle\int \csc^n x\,dx = \frac{-1}{n-1}\cot x\csc^{n-2}x + \frac{n-2}{n-1}\int \csc^{n-2}x\,dx$

96. $\displaystyle\int \sin mx\sin nx\,dx = -\frac{\sin(m+n)x}{2(m+n)} + \frac{\sin(m-n)x}{2(m-n)} + C, \quad m^2 \neq n^2$

97. $\displaystyle\int \cos mx\cos nx\,dx = \frac{\sin(m+n)x}{2(m+n)} + \frac{\sin(m-n)x}{2(m-n)} + C, \quad m^2 \neq n^2$

98. $\displaystyle\int \sin mx\cos nx\,dx = -\frac{\cos(m+n)x}{2(m+n)} - \frac{\cos(m-n)x}{2(m-n)} + C, \quad m^2 \neq n^2$

99. $\displaystyle\int x\sin x\,dx = \sin x - x\cos x + C$

100. $\displaystyle\int x\cos x\,dx = \cos x + x\sin x + C$

101. $\displaystyle\int x^2\sin x\,dx = 2x\sin x + (2-x^2)\cos x + C$

102. $\displaystyle\int x^2\cos x\,dx = 2x\cos x + (x^2-2)\sin x + C$

103. $\displaystyle\int x^n\sin x\,dx = -x^n\cos x + n\int x^{n-1}\cos x\,dx$

104. $\displaystyle\int x^n\cos x\,dx = x^n\sin x - n\int x^{n-1}\sin x\,dx$

105. $\displaystyle\int \sin^m x\cos^n x\,dx = -\frac{\sin^{m-1}x\cos^{n+1}x}{m+n} + \frac{m-1}{m+n}\int \sin^{m-2}x\cos^n x\,dx$

$\displaystyle\qquad = \frac{\sin^{m+1}x\cos^{n-1}x}{m+n} + \frac{n-1}{m+n}\int \sin^m x\cos^{n-2}x\,dx$

(If $m = -n$ use formula 92 or 93.)